Higher Engineering Mathematics

About the Author

Dr. Bandaru Venkata Ramana obtained his Ph.D from the Indian Institute of Technology (IIT), Bombay in the year 1974. He has been a Post-Doctoral fellow of CSIR for one year. He has more than 30 years of experience in teaching the subject of Engineering Mathematics at IIT, Bombay (1970–1974), Regional Engineering College, Warangal (1975–1981), Jawaharlal Nehru Technological University (1981 onwards, more than 20 years), and Federal University of Technology, Nigeria (1983–1985 on overseas assignment). He has taught to all branches of B. Tech, M. Tech, and also to M. Sc. (Tech) Maths, M. Sc. (Tech) Physics, and MCA students.

Since 1987, he is working as the Professor and Head of the Department of Mathematics, JNTU, College of Engineering, Kakinada. Presently, he is Professor of Mathematics, Eritrea Institute of Technology, Eritrea.

He is the Chairman, Board of Studies in Mathematics for J.N. Technological University, Hyderabad.

He is a life member of the Indian Mathematical Society, Indian Society for Technical Education, and Indian Society for Technical and Applied Mechanics. He has to his credit several research publications in international journals.

Dr. Ramana can be contacted at ramanabv46@yahoo.com.

Higher Engineering Mathematics

B V Ramana

Professor of Mathematics
JNTU College of Engineering (Autonomous)
Kakinada, Andhra Pradesh

Presently, Professor of Mathematics
Eritrea Institute of Technology
Eritrea

McGraw Hill Education (India) Private Limited

Published by McGraw Hill Education (India) Private Limited,
Registered Office: Anjana Complex No: 5/90 A, Butt Road, St. Thomas Mount, Chennai – 600016, Tamil Nadu, India.

Higher Engineering Mathematics

40th reprint 2024

ISBN-13: 978-0-07-063419-0
ISBN-10: 0-07-063419-X

41 42 43 44 45 MAG 28 27 26 25 24

Typeset at Script Makers, 19, A-1B, DDA Market, Paschim Vihar, New Delhi 110063, and,
Printed and bound in India at Magic International Pvt., Plot No. 26E, Sector 31, Site-IV, Greater Noida 201306.

Visit us at: www.mheducation.co.in
Toll free in India: 1800 103 5875
Write to us at: support.india@mheducation.com
CIN: U80302TN2010PTC111532

To

my beloved parents
Sri Bandaru Subba Rao
and
Smt Bandaru Nagaratnam
who made me what I am today

Contents

PART—III
ORDINARY DIFFERENTIAL EQUATIONS

PART—V
FOURIER ANALYSIS AND PARTIAL DIFFERENTIAL EQUATIONS

PART—VI
COMPLEX ANALYSIS

Foreword

George Boole, an English mathematician, in 1854 has written a monumental work entitled "An investigation of the Laws of Thought". In 1938, Claude E. Shannon (MIT, USA), in his classical paper entitled" A symbolic analysis of relay and switching circuits" in the Truncations of AIEE, has developed the algebra of switching functions and showed how its structure is related to the ideas established by Boole. This is a classical example of how an abstract mathematics in the 19th century became an applied mathematical discipline in the 20th century. Such is the power of mathematics , the queen of sciences. In my view the aim of Engineering Mathematics is to make the student think **mathematically** and develop "*mathematical maturity*".

The need for a good text book on "Engineering Mathematics" for students of engineering and technology in India can be easily understood. Although several books are available, almost none of them have the right combination , simplicity , rig our, pedagogy and syllabus compatibility , dealing with all aspects of the course. I am confident that this present book will be able fill this void.

It gives me great pleasure to introduce "**HIGHER ENGINEERING MATHEMATICS**" by **B.V.RAMANA** the publication of which heralds the completion of book that caters completely and effectively from a modern point view of the students of Engineering mathematics and physics and computer science.

This book has been organized and executed with lot of care, dedication and passion for lucidity. The author has been an outstanding teacher and has vast and varied experience in India and abroad in the field of mathematics. A conscious attempt has been made to simplify the concepts to facilitate better understanding of the subject.

This book is self -contained, presentation is detailed, examples are simple, notations are modern and standard and finally the chapters are largely independent. The contents of the book are exhaustive containing Differential & integral calculus, Ordinary differential equations, Linear algebra, vector calculus, Fourier analysis, partial differential equations, complex function theory, probability & statistics, Numerical analysis and finally special topics Linear programming and calculus of variations.

Dr. Ramana, a senior most professor of Jawaharlal Nehru Technological University, Hyderabad deserves our praise and thanks for accomplishing this trying task. Tata Mcgraw-Hill, a prestigious publishing house, also deserves a pat for doing an excellent job.

I wish Dr. Ramana all success in his future endeavors.

Dr. K. Rajagopal

Preface

Mathematics is a necessary avenue to scientific knowledge which opens new vistas of mental activity. A sound knowledge of Engineering Mathematics is a 'sine qua non' for the modern engineer to attain new heights in all aspects of engineering practice.

This book is a self-contained, comprehensive volume covering the entire gamut of the course of Engineering Mathematics for 4 years' B.Tech program of I.I.Ts, N.I.Ts, and all other universities in India.

The contents of this book are divided into 8 parts as follows:

Part I: Preliminaries:

Ch. 1: Vector Algebra, Theory of Equations, and Complex Numbers, Matrices and Determinants, Sequences and Series, Analytical Solid Geometry, Calculus of Variations, Linear Programming, on website.

Part II: Differential and Integral Calculus

Ch. 2: Differential Calculus

Ch. 3: Partial Differentiation

Ch. 4: Maxima and Minima

Ch. 5: Curve Tracing

Ch. 6: Integral Calculus

Ch. 7: Multiple Integrals

Part III: Ordinary Differential Equations

Ch. 8: Ordinary Differential Equations: First Order and First Degree

Ch. 9: Linear Differential Equations of Second Order and Higher Order

Ch. 10: Series Solutions

Ch. 11: Special Functions—Gamma, Beta, Bessel and Legendre

Ch. 12: Laplace Transform

Part IV: Linear Algebra and Vector Calculus

Ch. 13: Matrices

Ch. 14: Eigen Values and Eigen Vectors

Ch. 15: Vector Differential Calculus: Gradient, Divergence and Curl

Ch. 16: Vector Integral Calculus

Part V: Fourier Analysis and Partial Differential Equations

Ch. 17: Fourier Series

Ch. 18: Partial Differential Equations

Ch. 19: Applications of Partial Differential Equations

Ch. 20: Fourier Integral, Fourier Transforms and Integral Transforms

Ch. 21: Linear Difference Equations and Z-Transforms

Part VI: Complex Analysis

Ch. 22: Complex Function Theory

Ch. 23: Complex Integration

Ch. 24: Theory of Residues

Ch. 25: Conformal Mapping.

Part VII: Probability and Statistics

Ch. 26: Probability

Ch. 27: Probability Distributions

Ch. 28: Sampling Distribution

Ch. 29: Estimation and Tests of Hypothesis

Ch. 30: Curve Fitting, Regression and Correlation Analysis

Ch. 31: Joint Probability Distribution and Markov Chains

Part VIII: Numerical Analysis

Ch. 32: Numerical Analysis

Ch. 33. Numerical Solutions of ODE and PDE

Web Supplement Besides the above, the following additional chapters are available at http://www.mhhe.com/ramanahem

1. Matrices and Determinants
2. Sequence and Series
3. Analytical Solid Geometry
4. Calculus of Variations
5. Linear Programming

The site also contains chapter-wise summary of all the chapters in the book.

This book is written in a lucid, easy to understand language. Each topic has been thoroughly covered in scope, content and also from the examination point of view. For each topic, several worked out examples, carefully selected to cover all aspects of the topic, are presented. This is followed by practice exercise with answers to all the problems and hints to the difficult ones. There are more than 1500 worked examples and 3500 exercise problems.

This textbook is the outcome of my more than 30 years of teaching experience of engineering mathematics at Indian Institute of Technology, Bombay (1970-74), National Institute of Technology,

Warangal (1975-81), J.N. Technological University, Hyderabad (since 1981), Federal University of Technology, Nigeria (1983-85), Eritrea Institute of Technology, Eritrea (since 2005).

I am hopeful that this 'new' exhaustive book will be useful to both students as well as teachers. If you have any queries, please feel free to write to me at: ramanabv48@rediffmail.com.

In spite of our best efforts, some errors might have crept in to the book. Report of any such error and all suggestions for improving the future editions of the book are welcome and will be gratefully acknowledged.

B V Ramana

Acknowledgements

Thanks giving is a sacred Indian culture. Although I owe a lot to many, due to space constraint, I am explicitly expressing my thanks to only few.

I express my thanks to

1. Professor K. Rajagopal, Vice-Chancellor, J.N. Technological University, Hyderabad
2. Professor B.C. Jinaga, Rector, JNTU
3. Professor K. Lalkishore, Registrar, JNTU
4. Professor K. Satya Prasad, Principal, JNTU College of Engineering, Kakinada
5. Professor M.P. Ranga Rao, IIT Bombay
6. Professor Gebrebrhan Ogbazgh, Vice-president (academic), Eritrea Institute of Technology
7. Dr C. Siva Ram, Ph.D. AT&T, USA
8. Professor S. Satya Sri, Rangaraya Medical College, Kakinada
9. Ms. Vibha Mahajan, Ms. Shukti Mukherjee, Ms. Mini Narayanan, and Mr. Sohan Gaur of Tata McGraw-Hill Publishing Co. Ltd., New Delhi.

My wife Smt B. Mahalakshmi deserves special thanks for her continued patience, care, support and encouragement all through this project.

Finally my sincere thanks to my senior brother Sri Bandaru Ramachandra Rao of Mandapeta for his moral support.

B V Ramana

Tata McGraw-Hill Publishing Co. Ltd would like to join the author in acknowledging the useful comments and suggestions of the following reviewers during the preparation of the book:

1. Prof. K.P. Patil, Head, Department of Mathematics, Veermata Jijabai Technology Institute (VJTI), Mumbai.
2. Dr. Meena Sankaranarayanan, Senior Research Fellow, Institute for Infocomm Research, Singapore.
3. Dr. V. K. Singh, Professor, Department of Mathematics, Indraprastha Engineering College, Ghaziabad.
4. Prof. A. Vincent Antony Kumar, Department of Mathematics, PSNA College of Engineering and Technology, Dindigul, Tamil Nadu
5. Dr. P.S. Wain, Head, Department of Applied Sciences, Sinhgad Institute of Technology, Lonavala, Pune.

6. Dr. M.N. Mehta, Department of Mathematics, SVNIT, Surat.
7. Dr. P.C. Agarwal, Head, Department of Mathematics, ISS Academy of Technical Education, Noida, UP.
8. Dr. S.P. Mohanty, Professor and Head, Department of Mathematics, College of Engineering and Technology, Biju Patnaik University of Technology, Bhubaneshwar.
9. Dr. D.V. Singh, Professor & Chairman, Department of Mathematics, NIT, Kurukshetra.
10. Prof. G. Prema, Department of Mathematics, Amrita Vishwa Vidyapeetham, Coimbatore.
11. Dr. B.B. Mishra, Associate Professor and Head, Department of Mathematics, KIIT (DU), Bhubaneshwar.

PART–I

PRELIMINARIES

- *Chapter 1* *Vector Algebra, Theory of Equations, and Complex Numbers*

Chapter 1

Preliminaries

Vector Algebra, Theory of Equations, and Complex Numbers

INTRODUCTION

The preliminary Chapter 1 contains an elementary treatment of Vector Algebra, Theory of Equations, and Complex Numbers. Vector Algebra and Analytical Solid Geometry are prerequisites for Vector Differential Calculus of Chapter 15 and 16. The theory of equations deals with the analytical solutions of cubic and quartic equations while several numerical methods for solutions of algebraic and transcendental equations are considered in the Chapter 32 on Numerical Analysis. Complex numbers is the introductory part for the complex function theory dealt in Chapters 22, 23, 24 and 25.

1.1 VECTOR ALGEBRA

Vectors are very useful in engineering mathematics since many rules of vector calculation are as simple as that of real numbers and is a shorthand simplifying several calculations.
Vector analysis was introduced by Gibbs.*

Scalars are physical quantities which possess only magnitude and are completely defined by a single real number.

Examples: Mass, temperature, volume, kinetic energy, salinity, length, voltage, time, work, electric charge.

Vectors are physical quantities which possess both magnitude and direction. Thus geometrically vectors are *directed line segment* (or an *arrow*) which are determined completely by their magnitude (length) and direction.

Examples: Force, velocity, acceleration, momentum, displacement, weight.

Vectors are denoted by lower case bold face type letters **a** or by an arrow overhead the letter as $\vec{a}$ or $\overline{a}$.

* Josiah Willard Gibbs (1839-1903), American mathematician.

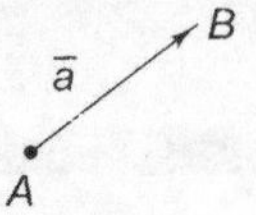

Fig. 1.1

A vector $\overline{a}$ depicted as an arrow (directed line segment) has a tail A known as the *initial point* (or origin) and a tip B known as the *terminal point* (or terminus).

The magnitude (or absolute value) of a vector $\overline{a}$ (length of the directed line segment) is denoted by $|\overline{a}|$ or a (non bold face). It is also known as *norm* (or Euclidean norm) of $\overline{a}$.

Unit vector is a vector of unit magnitude (i.e., of length 1). If $\overline{a}$ is any vector of non zero magnitude ($a \neq 0$), then $\frac{\overline{a}}{a}$ is a unit vector in the direction of $\overline{a}$ i.e., a unit vector is obtained by dividing it by its magnitude.

Zero (or null) vector is a vector of zero magnitude and has no specified direction.

Equality: Two vectors $\overline{a}$ and $\overline{b}$ are said to be equal, denoted as $\overline{a} = \overline{b}$, if both $\overline{a}$ and $\overline{b}$ have the same magnitude and same direction without regard to their initial points (also known as *free** vectors). Thus a vector can be moved parallel to itself without any change.

Negative of a vector $\overline{a}$, denoted by $-\overline{a}$, is a vector having the same magnitude as $\overline{a}$ but having opposite direction.

Note: Length of $-\overline{a}$ is *not* negative.

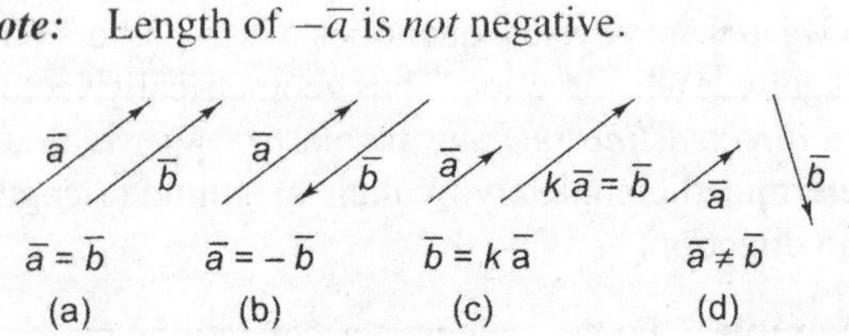

Fig. 1.2

Scalar multiplication $\alpha\overline{a}$ is a vector of magnitude $|\alpha|a$ and having the same direction of $\overline{a}$, if $\alpha > 0$ and of opposite direction if $\alpha < 0$.

* In contrast, *bound* vectors are fixed vectors and are restricted to their initial points (such as application of force *at a point*).

Here $|\alpha|$ is the absolute value of the scalar α.

Unequal: $\overline{a}$, $\overline{b}$, are said to be unequal, written as $\overline{a} \neq \overline{b}$ when magnitudes of $\overline{a}$ and $\overline{b}$ are different or direction of $\overline{a}$ and $\overline{b}$ or both magnitudes and direction of $\overline{a}$ and $\overline{b}$ are different.

In vector algebra, addition, subtraction and multiplication of vectors are introduced.

Addition

To obtain the sum or resultant of two vectors $\overline{a}$ and $\overline{b}$, move $\overline{b}$ so that initial point of $\overline{b}$ coincides with the terminal point of $\overline{a}$. Then the sum of $\overline{a}$ and $\overline{b}$, written as $\overline{a} + \overline{b}$, is defined by the vector (arrow) from the initial point of $\overline{a}$ to the terminal point $\overline{b}$. Similarly placing the tail of $\overline{a}$ with the tip of $\overline{b}$, the resultant $\overline{b} + \overline{a}$ is obtained. Thus the vector addition follows the parallelogram law of addition, i.e., the resultant of a parallelogram with $\overline{a}$ and $\overline{b}$ as adjacent sides is given by the diagonal of the parallelogram. Vector addition is commutative $\overline{a} + \overline{b} = \overline{b} + \overline{a}$ and is associative $a + (\overline{b} + \overline{c}) = (\overline{a} + \overline{b}) + \overline{c}$.

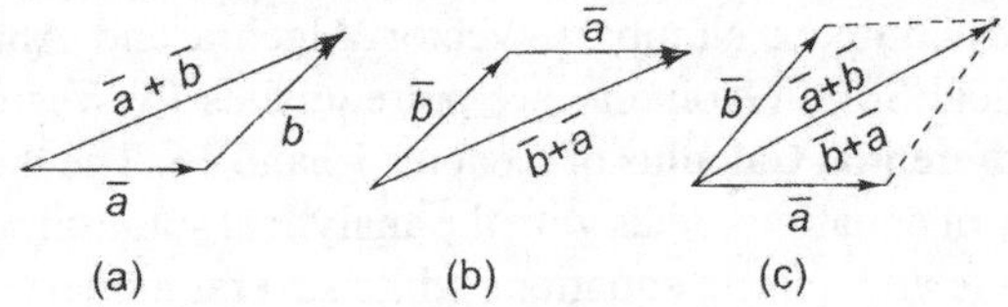

Fig. 1.3

Difference

$\overline{a} - \overline{b} = \overline{a} + (-\overline{b})$

Rectangular unit vectors:

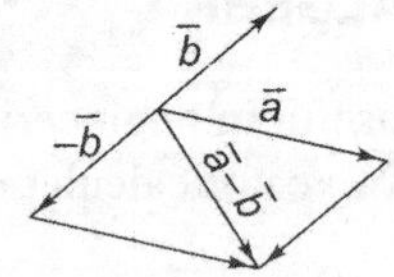

Fig. 1.4

Consider a right handed rectangular coordinate system. Let $\overline{i}, \overline{j}, \overline{k}$ be unit vectors along the positive X, Y and Z axes. Let $p(x, y, z)$ be any point and $0(0, 0, 0)$ be the origin.

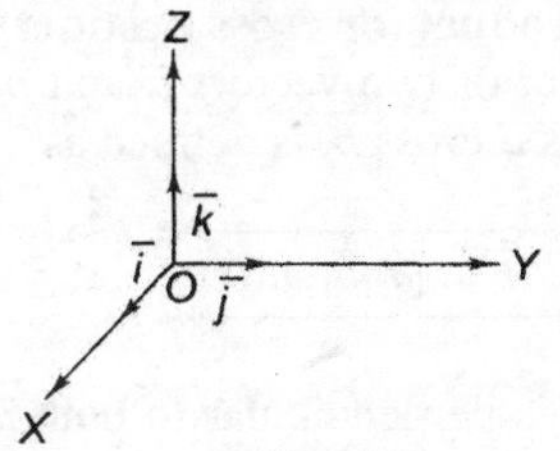

Fig. 1.5

Positive vector

$\overline{r}$ of a point $P(x, y, z)$ is $\overline{OP}$ with origin O as the initial point and P as the terminal point. Now $\overline{r}$ can be expressed as a linear combination of the unit vectors $\overline{i}, \overline{j}, \overline{k}$. Here $\overline{OA} = x\overline{i}$, $\overline{AB} = yj$ and $\overline{BP} = z\overline{k}$ since x, y, z are the lengths of OA, AB and BP respectively and $\overline{i}, \overline{j}, \overline{k}$ are unit vectors in the positive X, Y, Z axes. By vector addition

$$\boxed{\overline{r} = \overline{OP} = \overline{OA} + \overline{AB} + \overline{BP} = x\overline{i} + y\overline{j} + z\overline{k}}$$

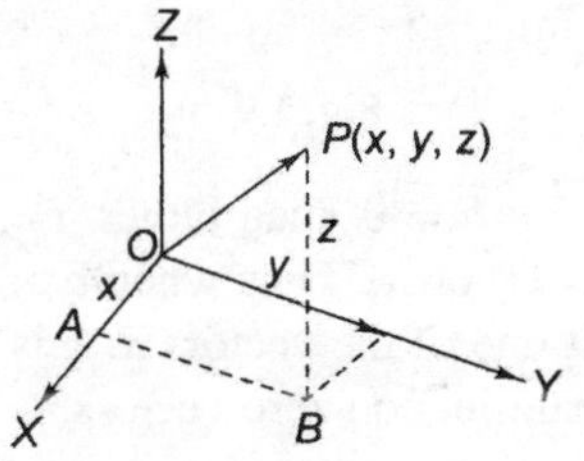

Fig. 1.6

Generalizing this, if $P(x_1, y_1, z_1)$ and $Q(x_2, y_2, z_2)$ be any two points, then the vector $\overline{PQ}$ can be represented as

$$\boxed{\overline{PQ} = (x_2 - x_1)\overline{i} + (y_2 - y_1)\overline{j} + (z_2 - z_1)\overline{k}}$$

Here $x_2 - x_1$, $y_2 - y_1$, $z_2 - z_1$ are known as *components* of $\overline{PQ}$ in the X, Y, Z directions. Now

$$|\overline{PQ}| = \sqrt{(x_2 - x_1)^2 + (y_2 - y_1)^2 + (z_2 - z_1)^2}$$

which is the distance between the points P and Q (and thus the length or modulus of the vector $\overline{PQ}$). With this the earlier definitions can be expressed in terms of the components. Suppose $\overline{a} = a_1\overline{i} + a_2\overline{j} + a_3\overline{k}$ and $\overline{b} = b_1\overline{i} + b_2\overline{j} + b_3\overline{k}$. Then

1. $\overline{a} = \overline{b}$ if $a_i = b_i$ for $i = 1, 2, 3$ i.e., their corresponding (respective) components are equal
2. $\overline{a} = 0$ if $a_1 = a_2 = a_3 = 0$
3. $\overline{a} \pm \overline{b} = (a_1 \pm b_1)i + (a_2 \pm b_2)j + (a_3 \pm b_3)k$ i.e., addition (subtraction) by adding (subtracting) the corresponding components.
4. $\alpha\overline{a} = \alpha a_1\overline{i} + \alpha a_2\overline{j} + \alpha a_3\overline{k}$ i.e., scalar multiplication amounts to multiplication of each component.

Multiplication

(a) **Inner product or scalar product or dot product** of two vectors $\overline{a}$ and $\overline{b}$, is denoted by $\overline{a} \cdot \overline{b}$, read as $\overline{a}$ dot $\overline{b}$, is defined as

$$\boxed{\overline{a} \cdot \overline{b} = |\overline{a}||\overline{b}| \cos\theta}$$

where θ is the angle between $\overline{a}$ and $\overline{b}$ and lies in the interval $0 \le \theta \le \pi$. Note that dot product is a scalar quantity. It is positive, zero or negative depending on whether θ is acute angle, right angle or obtuse angle. Thus for non-zero vectors.

$$\boxed{\overline{a} \cdot \overline{b} = 0 \text{ implies that } \theta = \tfrac{\pi}{2}}$$

i.e., $\overline{a}$ and $\overline{b}$ are perpendicular or orthogonal to each other. Now

$$\overline{a} \cdot \overline{a} = |\overline{a}||\overline{a}| \cos 0 = |\overline{a}|^2$$

or

$$\boxed{|\overline{a}| = \sqrt{\overline{a} \cdot \overline{a}} = \sqrt{a_1^2 + a_2^2 + a_3^2}}$$

For the unit vectors we have

$$\boxed{i \cdot j = j \cdot k = k \cdot i = 0} \text{ and}$$

$$\boxed{i \cdot i = j \cdot j = k \cdot k = 1}.$$

The dot product in the component form is $\overline{a} \cdot \overline{b} = (a_1\overline{i} + a_2\overline{j} + a_3\overline{k}) \cdot (b_1\overline{i} + b_2\overline{j} + b_3\overline{k})$

$$\boxed{\overline{a} \cdot \overline{b} = a_1b_1 + a_2b_2 + a_3b_3}$$

i.e., sum of the products of the corresponding components.

Properties

1. $\overline{a} \cdot \overline{b} = \overline{b} \cdot \overline{a}$ (Commutative)
2. $\overline{a} \cdot (\overline{b} + \overline{c}) = (\overline{a} \cdot \overline{b}) + (\overline{a} \cdot c)$ (Distributive)
3. $\overline{a} \cdot (\alpha\overline{b} + \beta\overline{c}) = \alpha\overline{a} \cdot b + \beta\overline{a} \cdot \overline{c}$ (Linearity)
4. $\overline{a} \cdot \overline{a} \ge 0$
   $\overline{a} \cdot \overline{a} = 0$ iff $\overline{a} = \overline{0}$ (Positive definiteness)

1. Angle The angle θ between two vectors $\bar{a}$ and $\bar{b}$ is defined as

$$\boxed{\cos\theta = \frac{\bar{a}\cdot\bar{b}}{|\bar{a}||\bar{b}|} = \frac{\bar{a}\cdot\bar{b}}{\sqrt{\bar{a}\cdot\bar{a}}\sqrt{\bar{b}\cdot\bar{b}}}}$$

Note:

Direction cosines of a vector $\bar{a}$ are the cosine of the angles which $\bar{a}$ makes with x, y, z axes and are given by $\bar{a}\cdot\frac{\bar{i}}{|\bar{a}|}, \bar{a}\cdot\frac{\bar{j}}{|\bar{a}|}, \bar{a}\cdot\frac{\bar{k}}{|\bar{a}|}$, respectively.

2. Projection Since $|\bar{b}|\cos\theta$ is the projection of $\bar{b}$ on $\bar{A}$, $\bar{a}\cdot\bar{b} = (|\bar{a}|)(|\bar{b}|\cos\theta)$ is the product of $|\bar{a}|$ and $|\bar{b}|\cos\theta$, ie. product of length of $\bar{a}$ and projection of $\bar{b}$ on $\bar{a}$. Similarly $\bar{a}\cdot\bar{b}$ can be interpreted as the product of length of $\bar{b}$ with the projection of $\bar{a}$ on $\bar{b}$. ie. dot product is length of either multiplied by projection of the other upon it. Thus the scalar projection of $\bar{a}$ in the direction of $\bar{b}$ is

$$\boxed{a\cos\theta = \bar{a}\cdot\frac{\bar{b}}{b}}$$

and vector projection of $\bar{a}$ in the direction of $\bar{b}$ is

$$(a\cos\theta)\frac{\bar{b}}{b} = \left(\bar{a}\cdot\frac{\bar{b}}{b}\right)\frac{\bar{b}}{b} = \left(\frac{\bar{a}\cdot\bar{b}}{b^2}\right)\bar{b}$$

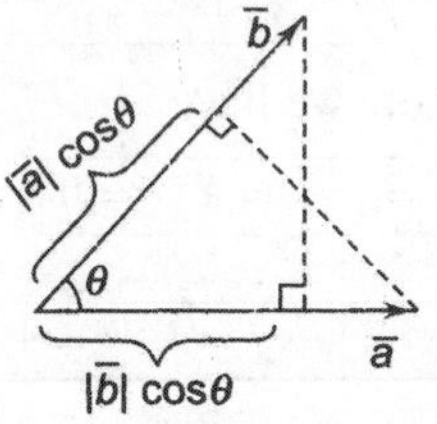

Fig. 1.7

3. Work done by a force The work done by a constant force $\bar{F}$ in moving an object through a distance d = (magnitude of force in the direction of motion) (distance moved)

$$= (F\cos\theta)(d) = \bar{F}\cdot\bar{d}$$

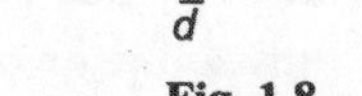

Fig. 1.8

(b) Vector product (or cross product)

Vector product of two vectors $\bar{a}$ and $\bar{b}$, denoted by $\bar{a}\times\bar{b}$, read as $\bar{a}$ cross $\bar{b}$, is defined as

$$\boxed{\bar{c} = \bar{a}\times\bar{b} = |\bar{a}||\bar{b}|\sin\theta\,\bar{u}}, \quad 0 \le \theta \le \pi$$

The vector $\bar{c}$ is perpendicular to both $\bar{a}$ and $\bar{b}$ such that $\bar{a}, \bar{b}, \bar{c}$ in this order form a *right handed system of vectors*. Here $\bar{u}$ is a unit vector in the direction of $\bar{a}\times\bar{b}$ as shown in Fig. 1.9.

Thus the modulus of $\bar{a}\times\bar{b}$ is the area of a parallelogram with $\bar{a}$ and $\bar{b}$ as adjacent sides as shown in Fig. 1.9(b).

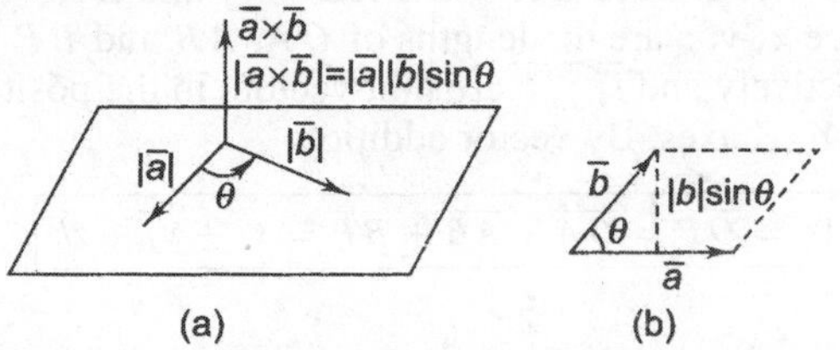

Fig. 1.9

Parallel If $\bar{a}\times\bar{b} = 0$, then either $\bar{a} = 0$ or $\bar{b} = 0$ or both or $\theta = 0$ or π. Thus when $\bar{a}\times\bar{b} = 0$, then either at least one of the vectors $\bar{a}, \bar{b}$ is zero or else $\bar{a}$ and $\bar{b}$ are parallel non zero vectors.

Corollary $\bar{a}\times\bar{a} = 0$ for any vector $\bar{a}$. Since the direction of $\bar{b}\times\bar{a}$ is opposite to that of $\bar{a}\times\bar{b}$, the cross product is anticommutative (i.e., *not* commutative) so

$$\boxed{\bar{a}\times\bar{b} = -\bar{b}\times\bar{a}}$$

However it is distributive

$$\bar{a}\times(\bar{b}+\bar{c}) = \bar{a}\times\bar{b} + \bar{a}\times\bar{c}$$

Now for the unit vectors $\bar{i}, \bar{j}, \bar{k}$ we have

$$\bar{i}\times\bar{i} = \bar{j}\times\bar{j} = \bar{k}\times\bar{k} = 0 \qquad \text{(parallel)}$$

$$\bar{i}\times\bar{j} = \bar{k},\ \bar{j}\times\bar{k} = \bar{i},\ \bar{k}\times\bar{i} = \bar{j} \qquad \text{(right handed)}$$

while

$$\bar{j}\times\bar{i} = -\bar{k},\ \bar{k}\times\bar{j} = -\bar{i},\ \bar{i}\times\bar{k} = -\bar{j} \qquad \text{(left handed)}$$

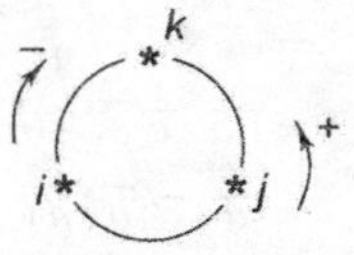

Fig. 1.10

In the component form

$$\bar{a} \times \bar{b} = (a_1\bar{i} + a_2\bar{j} + a_3\bar{k}) \times (b_1\bar{i} + b_2\bar{j} + b_3\bar{k})$$

$$= (a_2b_3 - a_3b_2)\bar{i} + (a_3b_1 - a_1b_3)\bar{j} + (a_1b_2 - a_2b_1)\bar{k}$$

$$= \begin{vmatrix} i & j & k \\ a_1 & a_2 & a_3 \\ b_1 & b_2 & b_3 \end{vmatrix}$$

Moment of a force $\bar{F}$ about a point P

The magnitude of the moment M (known as *torque*) of a force $\bar{F}$ about a point P is the product of F and the perpendicular distance of P from the line of action of force $\bar{F}$. Thus $M = (F)(PQ) = (F)(r \sin\theta)$ since $PQ = r\sin\theta$. Here $\overline{PR} = \bar{r}$ is the vector from P to the initial point R of $\bar{F}$. The direction of the moment is perpendicular to both $\bar{r}$ and $\bar{F}$. Thus the moment vector $\bar{M}$ is given by

$$\bar{M} = \bar{r} \times \bar{F}$$

where $|\bar{M}| = M = rF\sin\theta$

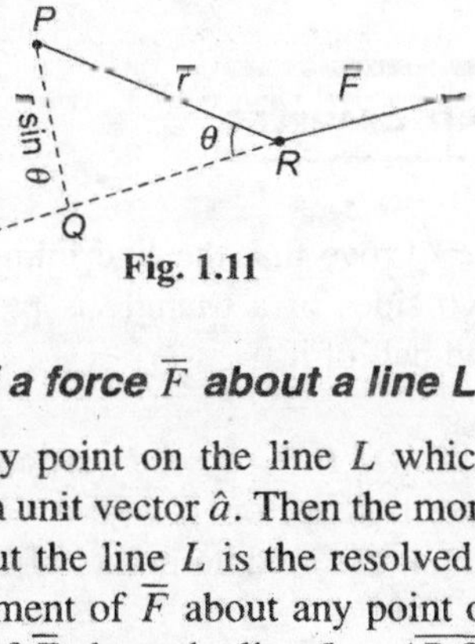

Fig. 1.11

Moment of a force $\bar{F}$ about a line L

Let P be any point on the line L which is in the direction of a unit vector $\hat{a}$. Then the moment of the force $\bar{F}$ about the line L is the resolved part along L of the moment of $\bar{F}$ about any point on L. Thus moment $\bar{M}$ of $\bar{F}$ about the line $L = (\overline{PQ} \times \bar{F}) \cdot \hat{a}$.

Fig. 1.12

Velocity of a rotating body

Suppose a rigid body rotates about an axis L through the point with angular speed ω. Let $\bar{r}$ be the position vector of any point P on the rigid body. Let θ be the angle between $\bar{r}$ and the axis of revolution L. Then the distance d of the point P from the axis is given by $r\sin\theta$

i.e., $$d = |\bar{r}|\sin\theta$$

since P travels in a circle of radius d. The magnitude of linear velocity $\bar{v}$ is

$$|\bar{v}| = \omega d = \omega|\bar{r}|\sin\theta = |\bar{\omega}||\bar{r}|\sin\theta$$

Also $\bar{v}$ must be perpendicular to both $\bar{\omega}$ and $\bar{r}$ such that $\bar{r}$, $\bar{\omega}$, $\bar{v}$ form a right handed system. Therefore the linear velocity $\bar{v}$ of the body is

$$\boxed{\bar{v} = \bar{\omega} \times \bar{r}}$$

Here $\bar{\omega}$ is known as the *angular velocity*.

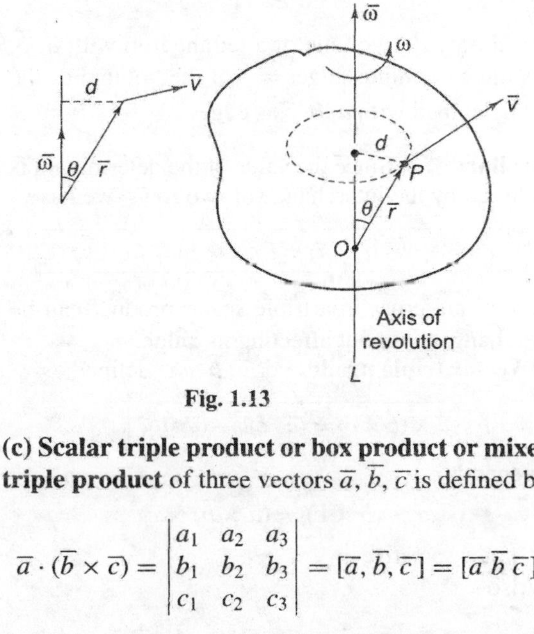

Fig. 1.13

(c) Scalar triple product or box product or mixed triple product of three vectors $\bar{a}$, $\bar{b}$, $\bar{c}$ is defined by

$$\bar{a} \cdot (\bar{b} \times \bar{c}) = \begin{vmatrix} a_1 & a_2 & a_3 \\ b_1 & b_2 & b_3 \\ c_1 & c_2 & c_3 \end{vmatrix} = [\bar{a}, \bar{b}, \bar{c}] = [\bar{a}\,\bar{b}\,\bar{c}]$$

Geometrically, the absolute value of the scalar triple product is the volume of a parallelepiped with $\bar{a}$, $\bar{b}$, $\bar{c}$ as the coterminus (or concurrent) edges. If $\bar{n}$ is a unit normal to the base parallelogram then area of the base parallelogram is $|\bar{b} \times \bar{c}|$ and height h of the terminal point of $\bar{a}$ above the parallelogram is $\bar{a} \cdot \bar{n}$. Then volume of parallelepiped = (height h) (area of base parallelogram)

$$= (\overline{a}\cdot\overline{n})(|\overline{b}\times\overline{c}|) = \overline{a}\cdot(|(\overline{b}\times\overline{c})|\overline{n}) = \overline{a}\cdot(\overline{b}\times\overline{c})$$

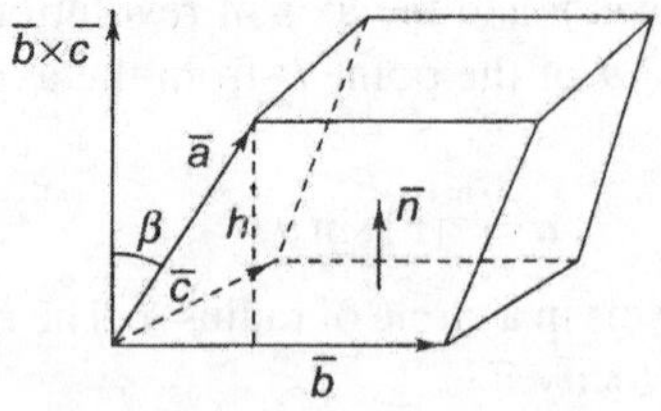

Fig. 1.14

Corollary 1: If $\overline{a}, \overline{b}, \overline{c}$ are coplanar (i.e., they lie in the same plane) then the volume of the parallelepiped is zero. Thus for coplanar vectors the scalar triple product is zero.
i.e., $\overline{a}, \overline{b}, \overline{c}$ are coplanar if $\overline{a}\cdot(\overline{b}\times\overline{c}) = 0$. Then $\overline{a}, \overline{b}, \overline{c}$ are said to be linearly dependent.

Corollary 2: Volume of a tetrahedron with $\overline{a}, \overline{b}, \overline{c}$ as the coterminus edges $= \frac{1}{6}$ of the volume of the parallelepiped with $\overline{a}, \overline{b}, \overline{c}$ as edges.

Corollary 3: Since the value of the determinant is unaltered by the interchange of two rows, we have

$$\boxed{\overline{a}\cdot(\overline{b}\times\overline{c}) = \overline{b}\cdot(\overline{c}\times\overline{a}) = \overline{c}\cdot(\overline{a}\times\overline{b}) = (\overline{a}\times\overline{b})\cdot\overline{c}}$$

i.e., dot and cross in a triple scalar product can be interchanged without affecting its value.
(d) Vector triple product of $\overline{a}, \overline{b}, \overline{c}$ is defined as

$$\boxed{\overline{a}\times(\overline{b}\times\overline{c}) = (\overline{a}\cdot c)\overline{b} - (\overline{a}\cdot\overline{b})\overline{c}}$$

In general

$$\overline{a}\times(\overline{b}\times\overline{c}) \neq (\overline{a}\times\overline{b})\times\overline{c}$$

i.e., not associative.
Also

$$(\overline{a}\times\overline{b})\times\overline{c} = (\overline{a}\cdot\overline{c})\overline{b} - (\overline{b}\cdot\overline{c})\overline{a}$$

Results

1. $\boxed{(\overline{a}\times\overline{b})\cdot(\overline{c}\times\overline{d}) = (\overline{a}\cdot\overline{c})(\overline{b}\cdot\overline{d}) - (\overline{a}\cdot\overline{d})(\overline{b}\cdot\overline{c})}$

Proof: Put $\overline{a}\times b = \overline{x}$ then

$$(\overline{a}\times\overline{b})\cdot(\overline{c}\times\overline{d}) = \overline{x}\cdot(\overline{c}\times\overline{d}) = (\overline{x}\times\overline{c})\cdot\overline{d}$$

$$= (\overline{a}\times\overline{b})\times\overline{c}\cdot\overline{d}$$

$$= [(\overline{a}\cdot\overline{c})\overline{b} - (\overline{b}\cdot\overline{c})\overline{a}]\cdot\overline{d}$$

$$= (\overline{a}\cdot\overline{c})(\overline{b}\cdot\overline{d}) - (\overline{b}\cdot\overline{c})(\overline{a}\cdot\overline{d})$$

2. $(\overline{a}\times\overline{b})\times(\overline{c}\times\overline{d}) = \overline{b}(\overline{a}\cdot\overline{c}\times\overline{d}) - \overline{a}(\overline{b}\cdot\overline{c}\times d)$

$$= [\overline{a}\overline{c}\overline{d}]\overline{b} - [\overline{b}\overline{c}\overline{d}]\overline{a}$$

$$= \overline{c}(\overline{a}\cdot\overline{b}\times\overline{d}) - \overline{d}(\overline{a}\cdot\overline{b}\times\overline{c})$$

$$= [\overline{a}\,\overline{b}\,\overline{d}]\,\overline{c} - [\overline{a}\,\overline{b}\,\overline{c}]\,\overline{d}$$

Proof: Put $\overline{a}\times\overline{b} = \overline{x}$. Then

$$(\overline{a}\times\overline{b})\times(\overline{c}\times\overline{d}) = \overline{x}\times(\overline{c}\times\overline{d}) = (\overline{x}\cdot\overline{d})\overline{c} - (\overline{x}\cdot\overline{c})\overline{d}$$

$$= (\overline{a}\times\overline{b}\cdot\overline{d})\overline{c} - (\overline{a}\times\overline{b}\cdot\overline{c})\overline{d}$$

$$= [\overline{a}\overline{b}\overline{d}]\overline{c} - [\overline{a}\overline{b}\overline{c}]\overline{d}$$

Linear independence: Set of vectors $\overline{a}, \overline{b}, \overline{c}$ are said to be linearly independent if $\alpha\overline{a} + \beta\overline{b} + \gamma\overline{c} = 0$ implies that $\alpha = \beta = \gamma = 0$. Otherwise they are said to be linearly dependent when not all α, β, γ are zero.

Vector Algebra

WORKED OUT EXAMPLES

Example 1: Prove that the line joining the mid-points of two sides of a triangle is parallel to the third side and half of it.

Solution: Let A, B, C be the vertices of the triangle with position vectors $\overline{a}, \overline{b}, \overline{c}$ wrt an origin O. Let D and E be the middle points of AC and BC respectively. Then

$$\overline{DE} = \overline{DA} + \overline{AB} + \overline{BE} = \frac{1}{2}\overline{CA} + \overline{AB} + \frac{1}{2}\overline{BC}$$

$$= \frac{1}{2}(\overline{a} - \overline{c}) + (\overline{b} - \overline{a}) + \frac{1}{2}(\overline{c} - \overline{b}) = \frac{\overline{b} - \overline{a}}{2} = \frac{1}{2}\overline{AB}$$

Thus DE is parallel to the third side AB and half of it.

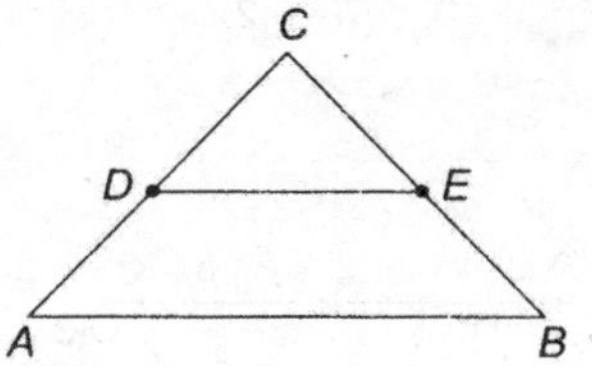

Fig. 1.15

Example 2: Prove that the vectors $\overline{a} = 3i + j - 2k$, $\overline{b} = -i + 3j + 4k$, $\overline{c} = 4i - 2j - 6k$ can form the sides of a triangle. Find the lengths of the medians of the triangle.

Solution: Consider $\alpha\overline{a} + \beta\overline{b} + \gamma\overline{c} = 0$ or

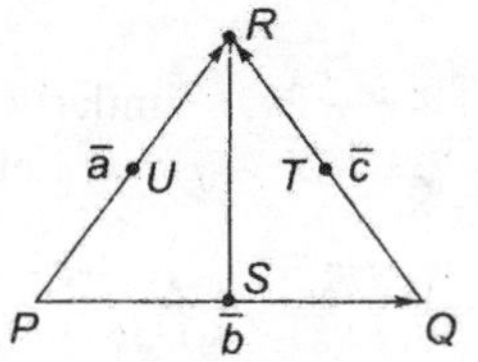

Fig. 1.16

$$\alpha(3i + j - 2k) + \beta(-i + 3j + 4k)$$
$$+ \gamma(4i - 2j - 6k) = 0$$

Equating each of the components of $\overline{i}, \overline{j}, \overline{k}$ to zero, we get

$$3\alpha - \beta + 4\gamma = 0$$
$$\alpha + 3\beta - 2\gamma = 0$$
$$-2\alpha + 4\beta - 6\gamma = 0$$

Solving $\beta = \gamma = -\alpha$. Choose $\alpha = 1$. Thus $-\overline{a} + \overline{b} + \overline{c} = 0$ i.e., three vectors are linearly dependent i.e., non-collinear. Therefore they form a plane triangle. Let S be the mid point of PQ. So that RS is a median. Now

$$\overline{RS} = \overline{RP} + \overline{PS} = -\overline{a} + \frac{1}{2}\overline{b}$$
$$= -(3i + j - 2k) + \frac{1}{2}(-i + 3j + 4k)$$
$$= -\frac{7}{2}\overline{i} + \frac{1}{2}\overline{j} + 4\overline{k}$$

$\therefore$ $|\overline{RS}|$ = length of the median

$$= \sqrt{\frac{49}{4} + \frac{1}{4} + 16} = \sqrt{\frac{114}{4}} = \frac{1}{2}\sqrt{114}$$

Similarly PT be another median. Then

$$\overline{PT} = \overline{PQ} + \frac{1}{2}\overline{QT} = \overline{b} + \frac{1}{2}\overline{c}$$
$$= (i + 3j + 4k) + \frac{1}{2}(4i - 2j - 6k)$$
$$= \overline{i} + 2\overline{j} + \overline{k}$$

So $|\overline{PT}|$ = length of median = $\sqrt{1^2 + 2^2 + 1^2} = \sqrt{6}$

Finally $\overline{QU} = \overline{QR} + \frac{1}{2}\overline{RU} = \overline{c} + \frac{1}{2}(-\overline{a})$

$$= (4i - 2j - 6k) - \frac{1}{2}(3i + j - 2k)$$
$$= \frac{5}{2}i - \frac{5}{2}j - 5k$$
$$|\overline{QU}| = \sqrt{\frac{25}{4} + \frac{25}{4} + 25} = \sqrt{\frac{150}{4}}$$

Example 3: If $\overline{a} = \overline{i} + \overline{j} + \overline{k}$, $\overline{b} = 2i - \overline{j} + 4\overline{k}$, $\overline{c} = 3i + 2j - \overline{k}$ then find a unit vector parallel to $2\overline{a} - 3\overline{b} + 4\overline{c}$.

Solution: $2\overline{a} - 3\overline{b} + 4\overline{c} = 2(i + j + k) - 3(2i - j + 4k) + 4(3i + 2j - k)$. So

$$2\overline{a} - 3\overline{b} + 4\overline{c} = 8i + 13j - 14\overline{k}$$
$$|2\overline{a} - 3\overline{b} + 4\overline{c}| = \sqrt{8^2 + 13^2 + 14^2}$$
$$= \sqrt{64 + 169 + 196} = \sqrt{429}$$

Unit vector parallel to $2\overline{a} - 3\overline{b} + 4\overline{c}$ is $\dfrac{2\overline{a} - 3\overline{b} + 4\overline{c}}{|2\overline{a} - 3\overline{b} + 4\overline{c}|}$.

$$= \frac{1}{\sqrt{429}}[8i + 13j - 14\overline{k}]$$

Example 4: Prove that $\overline{a} = \overline{i} - 3\overline{j} + 2\overline{k}$, $\overline{b} = 2i - 4j - k$ and $\overline{c} = 3i + 2j - k$ are linearly independent.

Solution: Consider $\alpha\overline{a} + \beta\overline{b} + \gamma\overline{c} = 0$. Then

$$\alpha(i - 3j + 2k) + \beta(2i - 4j - k)$$
$$+ \gamma(3i + 2j - k) = 0$$

Equating each component to zero, we have $\alpha + 2\beta + 3\gamma = 0$, $-3\alpha - 4\beta + 2\gamma = 0$, $2\alpha - \beta - \gamma = 0$. Solving $\alpha = \beta = \gamma = 0$ $\therefore$ $\overline{a}, \overline{b}, \overline{c}$ are linearly independent.

Example 5: Find the unit vector perpendicular to each of the vectors $2i + j + k$ and $i - 2j + k$.

Solution: Let $\overline{a} = a_1 i + a_2 j + a_3 k$ be the required unit vector so that $|\overline{a}| = \sqrt{a_1^2 + a_2^2 + a_3^2} = 1$. Since $\overline{a}$ is perpendicular to $2i + j + k$, we have $(a_1 i + a_2 j + a_3 k) \cdot (2i + j + k) = 0$

or $\quad 2a_1 + a_2 + a_3 = 0$

Similarly since $\overline{a}$ is perpendicular to $i - 2j + k$,

$$a_1 - 2a_2 + a_3 = 0.$$

Solving $5a_2 - a_3 = 0 \therefore a_3 = 5a_2$
So $a_1 = 2a_2 - a_3 = 2a_2 - 5a_2 = -3a_2$
Since $a_1^2 + a_2^2 + a_3^2 = 1$, we have

$$9a_2^2 + a_2^2 + 25a_2^2 = 1$$

$$a_2^2 = \frac{1}{35} \quad \text{or} \quad a_2 = \pm\frac{1}{\sqrt{35}}$$

Thus the unit vector are

$$\pm\frac{1}{\sqrt{35}}[-3i + j + 5k]$$

Example 6: Determine the projection of $\overline{a}$ on $\overline{b}$ and $\overline{b}$ on $\overline{a}$ given $a = \overline{i} + \overline{j} + \overline{k}$ and $\overline{b} = 2i - j + 5k$.

Solution: Projection of $\overline{a}$ on $\overline{b}$ is $\overline{a} \cdot \frac{\overline{b}}{|\overline{b}|}$

$$= (\overline{i} + \overline{j} + \overline{k}) \cdot \frac{(2\overline{i} - \overline{j} + 5\overline{k})}{\sqrt{4 + 1 + 25}} = \frac{2 - 1 + 5}{\sqrt{30}} = \frac{6}{\sqrt{30}}$$

projection of $\overline{b}$ on $\overline{a}$ is $\overline{b} \cdot \frac{\overline{a}}{|\overline{a}|}$

$$= (2\overline{i} - \overline{j} + 5\overline{k}) \cdot \frac{(\overline{i} + \overline{j} + \overline{k})}{\sqrt{3}} = \frac{6}{\sqrt{3}}$$

Example 7: Determine the three sides and angles of a triangle with vertices $(1, -1, 1)$, $(2, 3, -1)$, $(3, 0, 2)$.

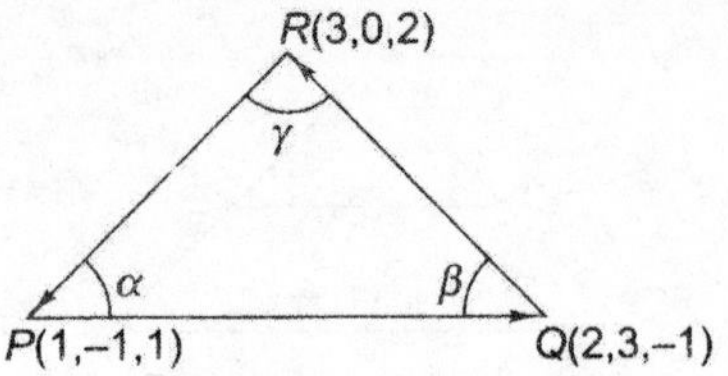

Fig. 1.17

Solution: Let $P(1, -1, 1)$, $Q(2, 3, -1)$ and $R(3, 0, 2)$ be the three vertices of the triangle. Then the side

$$\overline{PQ} = (2 - 1)i + (3 - (-1))j + (-1 - 1)k$$

i.e., $\overline{PQ} = i + 4j - 2k$, similarly $\overline{QR} = +i - 3j + 3k$ and $\overline{RP} = -2\overline{i} - \overline{j} - \overline{k}$. Let α, β, γ be the angles.

Then $\cos\alpha = \frac{\overline{PQ} \cdot \overline{PR}}{|\overline{PQ}||\overline{PR}|} = \frac{(i+4j-2k)\cdot(2i+j+k)}{\sqrt{1+16+4}\sqrt{4+1+1}}$

$\cos\alpha = \frac{2+4-2}{\sqrt{21}\sqrt{6}} = \frac{4}{\sqrt{21}\sqrt{6}}$ so $\alpha = \cos^{-1}\frac{4}{\sqrt{126}}$

Similarly $\cos\beta = \frac{\overline{QP} \cdot \overline{QR}}{|\overline{QP}||\overline{QR}|} = \frac{(-i-4j+2k)\cdot(-i-3j+3k)}{\sqrt{21}\sqrt{19}}$

$= \frac{-1+12+6}{\sqrt{21}\sqrt{19}} = \frac{+17}{\sqrt{399}}$

Finally $\cos\gamma = \frac{\overline{RP} \cdot \overline{RQ}}{|\overline{RP}||\overline{RQ}|} = \frac{(-2\overline{i}-\overline{j}-\overline{k})\cdot(-\overline{i}+3\overline{j}-3\overline{k})}{\sqrt{6}\sqrt{19}}$

$= \frac{+2-3+3}{\sqrt{6}\sqrt{19}} = \frac{2}{\sqrt{6}\sqrt{19}}$

Example 8: Forces $\overline{F}_1 = 3i + 5j + 6k$, $\overline{F}_2 = +\overline{i} + 2\overline{j} + \overline{k}$ and $\overline{F}_3 = 3\overline{i} + 8\overline{j}$ act on a particle P whose position vector is $3\overline{i} - 4\overline{j} + 2\overline{k}$. Determine the work done by the forces in a displacement of the particle to the point $Q(5, 2, 1)$.

Also find the vector moment of the resultant of the three forces acting at P about the point Q.

Solution: The resultant force $\overline{F} = \overline{F}_1 + \overline{F}_2 + \overline{F}_3$ so $\overline{F} = (3\overline{i} + 5\overline{j} + 6\overline{k}) + (\overline{i} + 2\overline{j} + \overline{k}) + (3i + 8j)$

$$\overline{F} = 7\overline{i} + 15\overline{j} + 7\overline{k}$$

displacement vector

$$= \overline{r} = \overline{PQ} = 2i + 6j - \overline{k}$$

since point P is $(3, -4, 2)$.

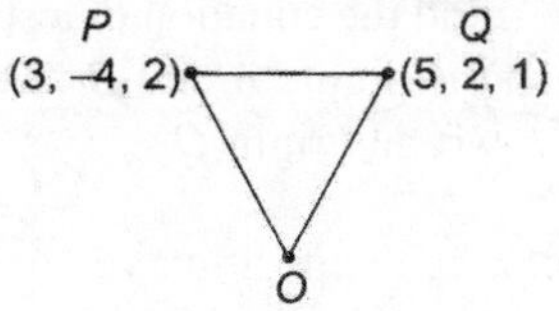

Fig. 1.18

Work done $= \overline{F} \cdot \overline{r}$

$$= (7\overline{i} + 15\overline{j} + 7\overline{k}) \cdot (2\overline{i} + 6\overline{j} - \overline{k})$$
$$= 14 + 90 - 7 = 97$$

Vector moment of the resultant force $\overline{F} = 7i + 15j + 7\overline{k}$ acting at $P(3, -4, 2)$ about the point $Q(5, 2, 1)$ is $\overline{QP} \times \overline{F}$

$$= (-2\overline{i} - 6\overline{j} + \overline{k}) \times (7\overline{i} + 15\overline{j} + 7\overline{k})$$

$$= \begin{vmatrix} i & j & k \\ -2 & -6 & 1 \\ 7 & 15 & 7 \end{vmatrix} = i(-42 - 15) - j(-14 - 7)$$
$$+ k(-30 + 42)$$
$$= -57\overline{i} + 21j + 12\overline{k}$$

Example 9: A rigid body is rotating at 5 radians per second about an axis OM where M is the point $3i - 4j + 2k$ relative to O. Find the magnitude of the linear velocity of the particle of the body at the point $5i + 2j + 3k$.

Solution:
Unit vector in the direction of the axis is $\frac{3i-4j+2k}{\sqrt{9+16+4}} = \frac{\overline{3i}-\overline{4j}+\overline{2k}}{\sqrt{29}}$

$$\therefore \text{ Angular velocity } \overline{\omega} = 5\left(\frac{3i - 4j + 2k}{\sqrt{29}}\right)$$

The point on the axis is M given by $3\overline{i} - 4j + 2k$. The point P on the right body is $5i + 2j + 3k$. So $\overline{MP} = (5i + 2j + 3k) - (3i - 4j + 2k)$

$$= 2\overline{i} + 6\overline{j} - \overline{k}$$

The linear velocity $\overline{v}$ at P is $\overline{\omega} \times \overline{MP}$

$$= \frac{5}{\sqrt{29}}(3i - 4j + 2k) \times (2i + 6j - k)$$

$$= \frac{5}{\sqrt{29}} \begin{vmatrix} i & j & k \\ 3 & -4 & 2 \\ 2 & 6 & -1 \end{vmatrix} = \frac{5}{\sqrt{29}}(-8i + 7j + 26k)$$

Magnitude of velocity $= \sqrt{\frac{25}{29}(64 + 49 + 676)}$

$$= \sqrt{\frac{25 \times 789}{29}} = \sqrt{\frac{19725}{29}} = 26.08$$

Example 10: Find the area of triangle with vertices at the points $A(3, -1, 2)$, $B(1, -1, -3)$ and $C(4, -3, 1)$.

Solution: We know that the area of a triangle is $\frac{1}{2}$ of the parallelogram. Thus area of triangle with $\overline{AB}$ and $\overline{AC}$ as the sides = $\frac{1}{2}$ area of the parallelogram with $\overline{AB}$ and $\overline{AC}$ as the adjacent sides $= \frac{1}{2}|\overline{AB} \times \overline{AC}|$.

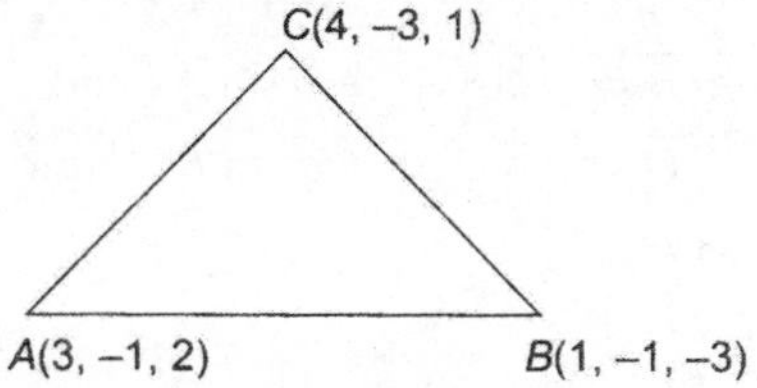

Fig. 1.19

Now $\overline{AB} = -2i + 0 - 5\overline{k}$ and $\overline{AC} = i - 2j - k$

$$\text{Area of } \Delta' = \frac{1}{2}\begin{vmatrix} i & j & k \\ -2 & 0 & -5 \\ 1 & -2 & -1 \end{vmatrix} = \frac{1}{2}| - 10i - 7j + 4k|$$

$$= \frac{1}{2}\sqrt{100 + 49 + 16} = \frac{1}{2}\sqrt{165}$$

Example 11: Find the area of a parallelogram with $\overline{A}$ and $\overline{B}$ as diagonals where $\overline{A} = 3\overline{i} + \overline{j} - 2\overline{k}$ and $\overline{B} = \overline{i} - 3\overline{j} + 4\overline{k}$.

Solution: Let $\overline{a}$ and $\overline{b}$ be the two adjacent sides of the parallelogram. Then

$$\overline{A} = \overline{a} + \overline{b} = 3i + j - 2\overline{k}$$

and

$$\overline{B} = \overline{a} - \overline{b} = i - 3j + 4k$$

Solving $\overline{a} = 2\overline{i} - \overline{j} + \overline{k}$ and $\overline{b} = \overline{i} + 2\overline{j} - 3\overline{k}$

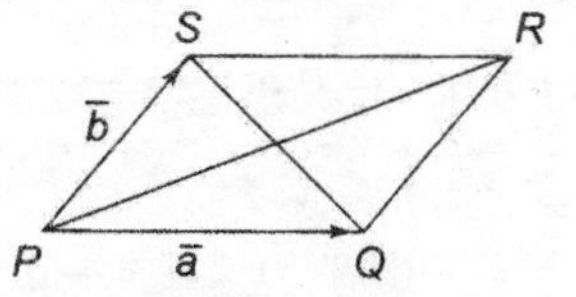

Fig. 1.20

Area of the required parallelogram

$$= |\bar{a} \times \bar{b}| = \begin{vmatrix} i & j & k \\ 2 & -1 & 1 \\ 1 & 2 & -3 \end{vmatrix} = |i + 7j + 5k| = \sqrt{75}$$

Example 12: Prove by vector methods that

(i) $\cos(A - B) = \cos A \cos B + \sin A \sin B$ and

(ii) $\sin(A - B) = \sin A \cdot \cos B - \sin B \cdot \cos A$

Solution: Let P and Q be any two points such that the position vector OP of length p makes an angle A with x-axis and position vector of length q makes an angle $B < A$. Then

$$\overline{OP} = p\cos A\bar{i} + p\sin A j \text{ and}$$

$$\overline{OQ} = q\cos B\bar{i} + q\sin Bj. \text{ Now}$$

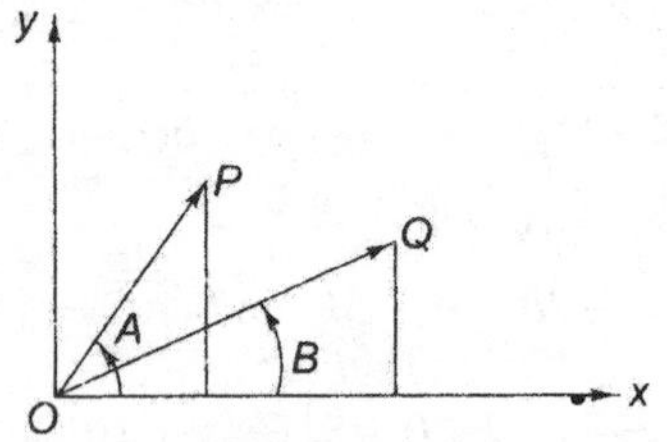

Fig. 1.21

a) $\overline{OP}\cdot\overline{OQ} = |\overline{OP}||\overline{OQ}|\cos\theta = pq\cos(A - B)$
$= (p\cos Ai + p\sin Aj)\cdot(q\cos Bi + q\sin Bj)$
$= pq(\cos A\cos B + \sin A\sin B)$
$\therefore\ \cos(A - B) = \cos A\cos B + \sin A\cdot\sin B$

b) Since $\overline{OQ}$, $\overline{OP}$ and $\overline{OQ}\times\overline{OP}$ form a right handed system, consider

$$\overline{OQ}\times\overline{OP} = |\overline{OQ}||\overline{OP}|\sin\theta\cdot\bar{k} = pq\sin(A - B)\bar{k}$$

$$= (q\cos Bi + q\sin Bj)\times(p\cos Ai + p\sin Aj)$$

$$= pq(\sin A\cos B - \sin B\cos A)\bar{k}$$

since $\bar{i}\times\bar{j} = \bar{k}$ while $\bar{j}\times\bar{i} = -\bar{k}$. Hence the result.

Example 13: If $\bar{a} + \bar{b} + \bar{c} = 0$ show that $\bar{a}\times\bar{b} = \bar{b}\times\bar{c} = \bar{c}\times\bar{a}$.

Solution: $\bar{a} + \bar{b} + \bar{c} = 0$ or $\bar{c} = -(\bar{a} + \bar{b})$.

Now $\bar{b}\times\bar{c} = \bar{b}\times(-\bar{a} - \bar{b}) = -\bar{b}\times\bar{a} - \bar{b}\times\bar{b}$
$= -\bar{b}\times\bar{a} = \bar{a}\times\bar{b}$

Similarly $\bar{c}\times\bar{a} = -(\bar{a} + \bar{b})\times\bar{a} = -\bar{a}\times\bar{a} - \bar{b}\times\bar{a}$
$= \bar{a}\times\bar{b}$

Example 14: Find the equation of a straight line L passing through the points A and B having position vectors $\bar{a}$ and $\bar{b}$ wrt an origin O.

Solution: Let $\bar{r}$ be the position vector of any point P on the straight line L, passing through A and B.

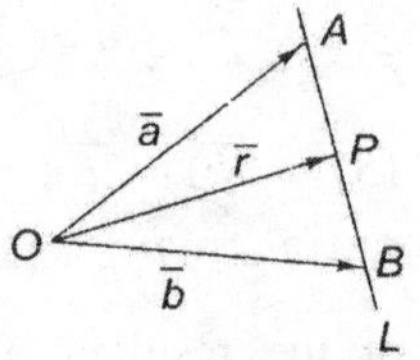

Fig. 1.22

Then $\overline{OP} = \bar{r} = \overline{OA} + \overline{AP}$ $\therefore$ $\overline{AP} = \bar{r} - \bar{a}$.
Similarly $\overline{OB} = \bar{b} = \overline{OA} + \overline{AB}$ $\therefore$ $\overline{AB} = \bar{b} - \bar{a}$
Since $\overline{AP}$ and $\overline{AB}$ are collinear vectors, there exists a scalar t such that $\overline{AP} = t\overline{AB}$ or

$$\bar{r} - \bar{a} = t(\bar{b} - \bar{a})$$

or $$\boxed{\bar{r} = \bar{a} + t(\bar{b} - \bar{a})}$$

Example 15: (a) Find the volume of a parallelepiped whose edges are $\bar{a} = 2i - 3j + 4k$, $\bar{b} = i + 2j - k$, $\bar{c} = 3i - j + 2\bar{k}$

(b) Find the volume of the tetrahedron having the following vertices (2, 1, 8), (3, 2, 9), (2, 1, 4), (3, 3, 10).

Solution: (a) Volume of parallelepiped

$$= \bar{a}\cdot(\bar{b}\times\bar{c}) = \begin{vmatrix} 2 & -3 & 4 \\ 1 & 2 & -1 \\ 3 & -1 & 2 \end{vmatrix} = |-7| = 7$$

(b) Let the four points be $A(2, 1, 8)$, $B(3, 2, 9)$, $C(2, 1, 4)$, $D(3, 3, 10)$. Then the three edges of the parallelepiped are $\overline{AB} = \bar{i} + \bar{j} + \bar{k}$, $\overline{BC} = -i - j - 5\bar{k}$, $\overline{CD} = i + 2j + 6\bar{k}$. Then volume of tetrahedron = $\frac{1}{6}$ of volume of parallelopiped with $\overline{AB}$, $\overline{BC}$, $\overline{CD}$ as the edges

$$= \frac{1}{6}\begin{vmatrix} 1 & 1 & 1 \\ -1 & -1 & -5 \\ 1 & 2 & 6 \end{vmatrix} = \frac{4}{6} = \frac{2}{3}.$$

Example 16: Find the constant b such that the three vectors $\bar{a} = 2\bar{i} - \bar{j} + \bar{k}$, $\bar{b} = i + 2j - 3k$ and $\bar{c} = 3i + bj + 5k$ are coplanar.

Solution: If the three vectors are coplanar (i.e., lie in the same plane) then the volume of the parallelepiped is zero i.e., $\bar{a}\cdot(\bar{b}\times\bar{c})=0$. So $\begin{vmatrix}2 & -1 & 1\\ 1 & 2 & -3\\ 3 & b & 5\end{vmatrix}=0$ or

$7b=-28$ or $b=-4$

Example 17: (a) Show that $(\bar{a}\times\bar{c})\times\bar{b}=0$ if

$$\bar{a}\times(\bar{b}\times\bar{c})=(\bar{a}\times\bar{b})\times\bar{c}$$

(b) Show that $(\bar{b}\times\bar{c})\times(\bar{a}\times\bar{d})+(\bar{c}\times\bar{a})\times(\bar{b}\times\bar{d})+(\bar{a}\times\bar{b})\times(\bar{c}\times\bar{d})=-2[abc]\bar{d}$

Solution: (a) $\bar{a}\times(\bar{b}\times\bar{c})=(\bar{a}\cdot\bar{c})\bar{b}-(\bar{a}\cdot\bar{b})\bar{c}=(\bar{a}\times\bar{b})\times\bar{c}$
$=(a\cdot c)\bar{b}-(\bar{b}-\bar{c})\bar{a}$
$\therefore\ (\bar{a}\cdot\bar{b})\bar{c}=(\bar{b}\cdot\bar{c})\bar{a}$
Now $(\bar{a}\times\bar{c})\times\bar{b}=(\bar{a}\cdot\bar{b})\cdot\bar{c}-(\bar{c}\cdot\bar{b})\bar{a}=0$ from the above result.
Note that $(\bar{a}\times\bar{b})\times(\bar{c}\times\bar{d})=\bar{b}[\bar{a}\bar{c}\bar{d}]-\bar{a}[\bar{b}\bar{c}\bar{d}]$ or $=c[abd]-d[\bar{a}\bar{b}\bar{c}]$.
Using the first formula for the first and third terms and using the second formula for the second term
(b) $(\bar{b}\times\bar{c})\times(\bar{a}\times\bar{d})+(\bar{c}\times\bar{a})\times(\bar{b}\times\bar{d})+(\bar{a}\times\bar{b})\times(\bar{c}\times\bar{d})$
$=[\bar{b}\bar{c}\bar{d}]\bar{a}-[\bar{b}\bar{c}\bar{a}]\bar{d}+[\bar{c}\bar{b}\bar{d}]\bar{a}-[\bar{a}\bar{b}\bar{d}]\bar{c}$
$+[\bar{a}\bar{b}\bar{d}]\bar{c}-[\bar{a}\bar{b}\bar{c}]\bar{d}=-2[\bar{a}\bar{b}\bar{c}]\bar{d}$.

Example 18: If $\bar{a},\bar{b},\bar{c}$ are any vectors prove that
(a) $\bar{a}\times(\bar{b}\times\bar{c})+\bar{b}\times(\bar{c}\times\bar{a})+\bar{c}\times(\bar{a}\times\bar{b})=0$
(b) $(\bar{a}\times\bar{b})\cdot(\bar{c}\times\bar{d})+(\bar{b}\times\bar{c})\cdot(\bar{a}\times\bar{d})+(\bar{c}\times\bar{a})\cdot(\bar{b}\times\bar{d})=0$

Solution: (a) LHS $=(a\cdot c)b-(a\cdot\bar{b})\bar{c}+(b\cdot\bar{a})\bar{c}-(\bar{b}\cdot\bar{c})\bar{a}+(\bar{c}\cdot\bar{b})\bar{a}-(\bar{c}\cdot\bar{a})\bar{b}=0$ since dot product is commutative so $\bar{a}\cdot\bar{c}=\bar{c}\cdot\bar{a},\bar{a}\cdot\bar{b}=\bar{b}\cdot\bar{a},\bar{b}\cdot\bar{c}=\bar{c}\cdot\bar{b}$
(b) Since $(\bar{a}\times\bar{b})\cdot(\bar{c}\times\bar{d})=(\bar{a}\cdot\bar{c})(\bar{b}\cdot\bar{d})-(\bar{a}\cdot\bar{d})(\bar{b}\cdot\bar{c})$
LHS $=(\bar{a}\cdot\bar{c})(\bar{b}\cdot\bar{d})-(\bar{a}\cdot\bar{d})(\bar{b}\cdot\bar{c})+(\bar{b}\cdot\bar{a})(\bar{c}\cdot\bar{d})-(\bar{b}\cdot\bar{d})(\bar{c}\cdot\bar{a})+(\bar{c}\cdot\bar{b})(\bar{a}\cdot\bar{d})-(\bar{c}\cdot\bar{d})(\bar{a}\cdot\bar{b})=0$, since dot product is commutative.

Example 19: Compute $(\bar{a}\times\bar{b})\times(\bar{c}\times\bar{d})$ (a) directly (b) by using formula when $\bar{a}=2i-2j+k$, $\bar{b}=i+8j-4k$, $\bar{c}=12i-4j-3k$, $\bar{d}=i+2j-k$

Solution: Direct $\bar{a}\times\bar{b}=\begin{vmatrix}i & j & k\\ 2 & -2 & 1\\ 1 & 8 & -4\end{vmatrix}=9j+18k$

$$\bar{c}\times\bar{d}=\begin{vmatrix}i & j & k\\ 12 & -4 & -3\\ 1 & 2 & -1\end{vmatrix}=10i+9j+28k$$

$$(\bar{a}\times\bar{b})\times(\bar{c}\times\bar{d})=\begin{vmatrix}i & j & k\\ 0 & 9 & 18\\ 10 & 9 & 28\end{vmatrix}=90(i+2j-k)$$

(b) Using formula $(\bar{a}\times\bar{b})\times(\bar{c}\times\bar{d})=[\bar{a}\bar{b}\bar{d}]\bar{c}-[\bar{a}\bar{b}\bar{c}]\bar{d}$

$$\text{Now } [\bar{a}\bar{d}\bar{b}]=\begin{vmatrix}2 & -2 & 1\\ 1 & 8 & -4\\ 1 & 2 & -1\end{vmatrix}=2(0)+6-6=0$$

$$[\bar{a}\,\bar{b}\bar{c}]=\begin{vmatrix}2 & -2 & 1\\ 1 & 8 & -4\\ 12 & -4 & -3\end{vmatrix}=-80+90-100=-90$$

$(\bar{a}\times\bar{b})\times(\bar{c}\times\bar{d})=0\cdot\bar{c}-(-90)(i+2j-k)=90(i+2j-k)$

EXERCISE

1. Show that the diagonals of a parallelogram bisect each other.

   Hint: P is point of intersection of diagonals of the parallelogram $ABCD$; with $\overline{AB}=a$, $\overline{AD}=b$. Then $\overline{AB}=AP-BP$, $\bar{a}=y(\bar{a}+\bar{b})-x(\bar{b}-\bar{a})$, $\bar{a},\bar{b}$, non collinear $x=y=\frac{1}{2}$.

2. If the mid-points of the consecutive sides of any quadrilateral are connected by straight lines show that the resulting quadrilateral is a parallelogram.

   Hint: $ABCD$ is a quadrilateral with $\overline{AB}=\bar{a}$, $\overline{BC}=\bar{b}$, $\overline{CD}=\bar{c}$, $\overline{DA}=\bar{d}$; P, Q, R, S are mid-points of AB, BC, CD, DA; $\bar{a}+\bar{b}+\bar{c}+\bar{d}=0$,

   $$\overline{PQ}=\tfrac{1}{2}(\bar{a}+\bar{b})=-\tfrac{1}{2}(\bar{c}+\bar{d})=\overline{SR},$$
   $$\overline{QR}=\tfrac{1}{2}(\bar{b}+\bar{c})=-\tfrac{1}{2}(\bar{d}+\bar{a})=\overline{PS}$$

3. Find $|2\bar{a}-3\bar{b}-5\bar{c}|$ if $\bar{a}=3i-2j+k$, $\bar{b}=2i-4j-3k$, $\bar{c}=-i+2j+2k$

Ans. $2\bar{a}-3\bar{b}-5\bar{c}=5i-2j+k$; $|2\bar{a}-3\bar{b}-5\bar{i}|=\sqrt{30}$

4. Determine a unit vector parallel to the resultant of the vectors $\bar{a} = 3i + 4j + 5k$, $\bar{b} = 2i + 3j + 3k$.

Ans. $(5\bar{i} + 7\bar{j} + 8\bar{k})/\sqrt{138}$

5. If *ABCDEFA* is a regular hexagon, determine, CD, DE, EF, FA given that $\overline{AB} = \bar{a}, \overline{BC} = \bar{b}$.

Ans. $\overline{CD} = \bar{b} - \bar{a}$, $\overline{DE} = -\bar{a}$, $\overline{EF} = -\bar{b}$, $\overline{FA} = \bar{a} - \bar{b}$

Hint: $\overline{AC} = \bar{a} + \bar{b}$, $\overline{AD} = \bar{b} - \bar{a}$, $\overline{DE} = -\bar{a}$, $\overline{EF} = -\bar{b}$

6. Prove that the points $A(0, 4, 1)$, $B(2, 3, -1)$, $C(4, 5, 0)$, $D(2, 6, 2)$ form a square.

Hint: $\overline{AB} = \overline{DC} = 2i - j - 2k$, $\overline{BC} = \overline{AD} = 2i + 2j + k$

7. Let E be the mid point of the side $\overline{CD}$ of a parallelogram *ABCD*. Prove that the diagonal *BD* and *AE* trisect each other.

Hint: $\overline{AB} = \bar{a}$, $\overline{AD} = \bar{b}$, $\overline{AE} = \bar{b} + \frac{\bar{a}}{2}$, $\overline{AF} = \alpha\overline{AE} = \alpha\left(\bar{b} + \frac{\bar{a}}{2}\right)$.

Also $\overline{AF} = \overline{AB} + \overline{BF} = \bar{a} + \beta(\bar{b} - \bar{a}) - \alpha\left(\bar{b} + \frac{\bar{a}}{2}\right)$. Since $\bar{a}, \bar{b}$ are non collinear, $\alpha - \beta = 0$, $\frac{\alpha}{2} - 1 + \beta = 0 \therefore \beta = \frac{2}{3}$.

8. Show that $\bar{a} = \frac{2i-2j+k}{3}$, $\bar{b} = \frac{i+2j+2k}{3}$, $\bar{c} = \frac{2i+j-2k}{3}$ are mutually orthogonal unit vectors.

Hint: $\bar{a} \cdot \bar{b} = \bar{b} \cdot \bar{c} = \bar{c} \cdot \bar{a} = 0$

9. Determine the projection of $\bar{a} = 2\bar{i} - 3\bar{j} + 6\bar{k}$ on $\bar{b} = i + 2\bar{j} + 2\bar{k}$. Also find the projection of $\bar{b}$ on $\bar{a}$.

Ans. $\frac{8}{3}, \frac{8}{7}$

Hint: $\bar{a} \cdot \frac{\bar{b}}{|b|}, \bar{b} \cdot \frac{\bar{a}}{|a|}$

10. Find the angle between $\bar{a} = 3i + 4j + 2k$ and $\bar{b} = 2i - 2j + 3k$

Ans. $\cos\theta = \frac{4}{\sqrt{29}\sqrt{17}}$

11. Determine the unit vector perpendicular to both $\bar{a} = \bar{i} - \bar{j} + \bar{k}$ and $\bar{b} = \bar{i} + 2\bar{j} - \bar{k}$

Ans. $\pm\frac{1}{\sqrt{14}}[\bar{i} - 2\bar{j} - 3\bar{k}]$

12. Determine unit vectors which make an angle $60°$ with $\bar{a} = \bar{i} - \bar{j}$ and angle $60°$ with $\bar{b} = \bar{i} + \bar{k}$

Ans. $\frac{-\bar{j} + \bar{k}}{\sqrt{2}}$, $\frac{\sqrt{2}}{6}(4\bar{i} - \bar{j} + \bar{k})$.

Hint: $a_1 i + a_2 j + a_3 k$ be the unit vector, $a_1^2 + a_2^2 + a_3^2 = 1$, $\cos 60 = \frac{1}{2} = \frac{a_1 - a_2}{\sqrt{2}}$, $\cos 60 = \frac{1}{2} = \frac{a_1 + a_3}{\sqrt{2}}$.

13. Prove that the points $A(5, -1, 1)$, $B(7, -4, 7)$, $C(1, -6, 10)$, $D(-1, -3, 4)$ form the vertices of a rhombus.

Hint: $\overline{AD} = \overline{BC}$, $\overline{AB} = \overline{DC}$, $|\overline{AD}| = |\overline{AB}| = 7$, $|\overline{AC}| \cdot \overline{BD} = 0$ i.e., diagonals at right angles. Also $\overline{AB}.\overline{AD} \neq 0$, $\overline{AB} \cdot \overline{BC} \neq 0$ corner angles not right angles.

14. Determine the sides and angles of triangle with vertices $i - 2j + 2k$, $2i - j - k$, $3i - j + 2k$

Ans. Sides are $i - 2j + 3k$, $-2i - j$, $i + 3j - 3k$. Angles are $\cos^{-1}\sqrt{\frac{5}{19}}$, $\cos^{-1}\sqrt{\frac{14}{19}}$, $90°$

15. Show that $\bar{a} = 2i + j - 3k$, $\bar{b} = \bar{i} - 4\bar{k}$, $\bar{c} = 4i + 3j - k$ are linearly dependent.

16. Show that $\bar{a} = (\bar{a} \cdot i)\bar{i} + (\bar{a} \cdot \bar{j})\bar{j} + (\bar{a} \cdot \bar{k})\bar{k}$ for any vector $\bar{a} = a_1\bar{i} + a_2\bar{j} + a_3\bar{k}$

17. Find the work done in moving an object along a vector $2\bar{i} - 5\bar{j} + 6\bar{k}$ when the applied force $\bar{F}$ is $3i - j + \bar{k}$

Ans. 17

18. Find the direction cosines of $\bar{a} = 2\bar{i} - 3\bar{j} + 4\bar{k}$ or find the angles which $\bar{a}$ makes with the coordinate axes.

Ans. $\cos\alpha = \frac{2}{\sqrt{29}}$, $\cos\beta = -\frac{3}{\sqrt{29}}$, $\cos\gamma = \frac{4}{\sqrt{29}}$

19. If $\bar{a} = 4i + 3j + k$, $\bar{b} = 2i - j + 2k$, find a unit vector in the direction of $\bar{a} \times \bar{b}$. Determine the angle between $\bar{a}$ and $\bar{b}$.

Ans. $\frac{7i-6j-10k}{\sqrt{185}}$, $\theta = \sin^{-1}\frac{\sqrt{185}}{3\sqrt{26}} = 62° \cdot 40'$

20. Find the work done in displacing a particle

from P to Q having position vectors $4i - 3j - 2k$, and $6i + j - 3\overline{k}$ when constant forces $\overline{F}_1 = 2i - 5j + 6k$, $F_2 = -i + 2j - k$, $F_3 = \overline{i} + \overline{j} + \overline{k}$ act on the particle.

Ans. -10

Hint: resultant force $\overline{F} = 2i - 2j + 6k$, displacement vector $\overline{d} = 2i + 4\overline{j} - \overline{k}$, $\overline{F} \cdot \overline{d} =$ work done.

21. Determine the torque about the point P, $2\overline{i} + \overline{j} - \overline{k}$ of a force $\overline{F} = 4\overline{i} + \overline{k}$ acting through the point A with position vector $i - j + 2k$.

Ans. 15.4

Hint: $|\overline{PA} \times \overline{F}| = |(-i - 2j + 3k) \times (4\overline{i} + \overline{k})| = |-2i + 13j + 8k|$ where $\overline{PA} = (\overline{i} - \overline{j} + 2\overline{k}) - (2i + j - k)$

22. Find the moment about a line L through the origin having direction of $2i + 2j + k$ due to a 30 kg force acting at a point $A(-4, 2, 5)$ in the direction of $12i - 4j - 3k$.

Ans. 89.23

Hint: $\overline{F} = 30(12i - 4j - 3k)/13$, moment of $\overline{F}$ about $O = \overline{OA} \times \overline{F} = (-4i + 2j + 5k) \times \overline{F} = \frac{60}{13}(7i + 24j - 4\overline{k})$.

Moment of $\overline{F}$ about the line $L = \frac{60}{13}(7i + 24j - 4\overline{k}) \cdot \frac{2i+2j+k}{3}$

23. A rigid body is spinning with angular velocity 27 radians/second about an axis parallel to $2i + j - 2k$ passing through the point $A, i + 3j - k$. Find the (linear) velocity of the point P on the body with position vector $4i + 8j + k$.

Ans. $9\sqrt{293}$

Hint: $\overline{v} = \overline{\omega} \times \overline{AP} = 27\left(\frac{2i+j-2k}{3}\right) \times (3i + 5j + 2k) = 9(12i - 10j + 7k)$

where $\overline{AP} = (4i + 8j + k) - (i + 3j - k)$

24. Find the volume of a parallelepiped whose coterminus edges are $\overline{a} = 3i - 2\overline{j} + 5k$, $\overline{b} = \overline{i} + \overline{j} + \overline{k}$, $\overline{c} = 2i + 4j - \overline{k}$

Ans. $|-11| = 11$

25. Determine the volume of the tetrahedron having vertices at (0, 1, 1), (1, 0, 0), (2, 2, 3), $(-1, 0, 4)$

Ans. $\frac{1}{6}(14) = \frac{7}{3}$

26. Determine the constant b such that the vectors $4\overline{i} + 2\overline{j} - \overline{k}$, $b\overline{i} + \overline{j} + \overline{k}$, $3i - j - 5k$ are coplanar

Ans. $b = -1$

27. If $\overline{a} = 3i - j + 2k$, $\overline{b} = 2i + j - k$, $\overline{c} = i - 2j + 2\overline{k}$ find $\overline{a} \times (\overline{b} \times \overline{c})$ (a) directly (b) using dot formula

Ans. $15i + 15j - 15k$

Hint: (a) $\overline{b} \times \overline{c} = -5j - 5k$ (b) $\overline{a} \cdot \overline{c} = 9$, $\overline{a} \cdot \overline{b} = 3$ so $\overline{a} \times (\overline{b} \times \overline{c}) = (a \cdot \overline{c})\overline{b} - (\overline{a} \cdot \overline{b})\overline{c} = 9(2i + j - k) - 3(i - 2j + 2k)$

28. Compute $(\overline{a} \times \overline{b}) \times (\overline{c} \times \overline{d})$ (a) directly (b) using the formula when $\overline{a} = 10i + 10j + 5k$, $\overline{b} = 5i - 2j - 14k$, $\overline{c} = 4i + 7j - 4k$, $\overline{d} = 2i - \overline{j} + k$

Ans. $-3810\overline{i} - 2250j + 1065\overline{k}$

29. If $\overline{a} \cdot (\overline{b} \times \overline{c}) = 0$ prove that either (a) $\overline{a}, \overline{b}, \overline{c}$ are coplanar but no two of them are collinear or (b) two of vectors $\overline{a}, \overline{b}, \overline{c}$ are collinear or (c) all of the vectors $\overline{a}, \overline{b}, \overline{c}$ are collinear.

30. Show that if four vectors $\overline{a}, \overline{b}, \overline{c}, \overline{d}$ are coplanar then $(\overline{a} \times \overline{b}) \times (\overline{c} \times \overline{d}) = 0$

1.2 THEORY OF EQUATIONS

Theory of equations includes solution of equations which are needed in the study of characteristic equations, zeros of Bessel functions, integration etc.

A *polynomial* or a *rational integral algebraic function* is a function

$$y = a_0x^n + a_1x^{n-1} + a_2x^{n-2} + \cdots + a_n$$

where $a_0, a_1, a_2, \ldots, a_n$ are constants called *coefficients* and n is a non negative integer called the *degree* of the polynomial.

For $n = 1, 2, 3, 4$, the functions are known as linear, quadratic, cubic and quartic functions. A constant may be regarded as a polynomial of degree zero.

An *algebraic function* is any function $y = f(x)$ which satisfies an equation of the form

$$P_0(x)y^n + P_1(x)y^{n-1} + \cdots + P_n(x) = 0$$

where $P_0(x)$, $P_1(x)$, ... $P_n(x)$ are polynomials in x. A *transcendental* function is one which is not an algebraic equation.

Trigonometric functions cos, sin, tan, logarithmic, exponential, hyperbolic functions are examples of transcendental functions.

An equation

$$f(x) = 0$$

is said to be an algebraic equation or transcendent equation depending on whether $f(x)$ is an algebraic or transcendental function.

Any value of x^*, for which the equation (1) is satisfied, is known as the *solution* of the equation i.e., $f(x^*) = 0$. In case of algebraic equations, solutions are also known as *roots* (or zeros) of the equation. Geometrically the curve (graph of) $y = f(x)$ crosses (meets) the x-axis at the point x^*. Theory of equations consists of methods of obtaining solutions of equations. For the linear equation $ax + b = 0$, the solution is $x = -\frac{b}{a}$, $a \neq 0$. For the quadratic equation $ax^2 + bx + c = 0$, the solutions (or roots) are given by $x = \frac{-b \pm \sqrt{b^2 - 4ac}}{2a}$, $a \neq 0$. The roots of the cubic equation can be obtained by Cardano's method, while the roots of quartic are obtained by Ferrari's method. However no literal equations (formulae) exist for algebraic equations of degree $n \geq 5$ or for any transcendental equations. In these cases, one has to resort to numerical methods to find an approximate solution (or root).

For an algebraic equation

$$f(x) = a_0x^n + a_1x^{n-1} + a_2x^{n-2} + \cdots + a_n = 0$$

(a) There are exactly n real or complex roots (where n is degree of the equation). Roots need not be distinct.

(b) has at least one real root whose sign is opposite to that of the last term a_n, if n is odd

(c) $f(a)$ is the remainder when $f(x)$ is divided by $(x - a)$

(d) Complex roots occur in conjugate pairs, when the coefficients $a_0, a_1, \cdots, a_n$ are real.

(e) The following relations hold good between the roots and coefficients of equation
 (i) $\frac{-a_1}{a_0}$ = sum of the roots
 (ii) $\frac{-a_2}{a_0}$ = sum of the product of the roots taken two at a time
 (iii) $\frac{-a_3}{a_0}$ = sum of the product of the roots taken three at a time.
 (iv) $(-1)^n \frac{a_n}{a_0}$ = product of the roots.

(f) At least one root lies between a and b if $f(a)$ and $f(b)$ are of different (opposite) signs.

1.3 CARDANO'S* METHOD

Cardano's method obtains algebraic solution of a cubic equation. Cardano published this result in 1545 in Ars Magna (also credited to Tartaglia and Ferraro and Vieta (1591)).

Consider the general type of cubic equation

$$a^*x^3 + b^*x^2 + c^*x + d^* = 0 \qquad (1)$$

Dividing by a^*, we get the equation

$$x^3 + bx^2 + cx + d = 0 \qquad (2)$$

This equation can be reduced to a simpler form by removing the x^2 term by the substitution

$$x = u - \frac{b}{3} \qquad (3)$$

so

$$\left(u - \frac{b}{3}\right)^3 + b\left(u - \frac{b}{3}\right)^2 + c\left(u - \frac{b}{3}\right) + d = 0$$

$$\left(u^3 - \frac{3b}{3}u^2 + 3u\frac{b^2}{9} - \frac{b^3}{27}\right) + b\left(u^2 - \frac{2b}{3}u + \frac{b^2}{9}\right)$$

$$+cu - \frac{cb}{3} + d = 0$$

or

$$u^3 + pu + q = 0 \qquad (4)$$

where $p = c - \frac{b^2}{3}$, $q = d - \frac{bc}{3} + \frac{2b^3}{27}$ (5)

To solve the standard cubic equation put

$$u = y + z \qquad (6)$$

* Girolamo Cardano (1501-1576), Italian mathematician.

so $u^3 = (y+z)^3 = y^3 + z^3 + 3yz(y+z)$

$$= y^3 + z^3 + 3yzu$$

or $\quad u^3 - 3yzu - (y^3 + z^3) = 0 \qquad (7)$

comparing (4) and (7), we get $p = -3yz$ and $q = -(y^3 + z^3)$

or $yz = -\frac{p}{3}$ i.e., $y^3z^3 = -\frac{p^3}{27}$

or $yz = \frac{-p}{3}$ i.e., $y^3z^3 = \frac{-p^3}{27}$

and $y^3 + z^3 = -q$

Then y^3 and z^3 are the solutions (roots) of the quadratic equation

$$t^2 + qt - \frac{p^3}{27} = 0 \qquad (8)$$

Since sum of the roots of (8) is $y^3 + z^3 = -q$ and product of the roots of (8) is $y^3z^3 = -\frac{p^3}{27}$. Solving the quadratic equation (8), we get

$$y^3 = \frac{-q + \sqrt{q^2 + 4\frac{p^3}{27}}}{2} = \frac{-q}{2} + \sqrt{\frac{q^2}{4} + \frac{p^3}{27}} \qquad (9)$$

and

$$z^3 = \frac{-q}{2} - \sqrt{\frac{q^2}{4} + \frac{p^3}{27}} \qquad (10)$$

Let the discriminant of (8) be

$$R = \frac{q^2}{4} + \frac{p^3}{27} \qquad (11)$$

Case (i) Suppose $R > 0$, then y^3 and z^3 are both real and the roots cubic equation (4) are $y+z, \omega y + \omega^2 z, \omega^2 y + \omega z$. (12)

Here $\omega = \frac{-1+\sqrt{3}i}{2}$, $\omega^2 = \frac{-1-\sqrt{3}i}{2}$ and $\omega \cdot \omega^2 = \omega^3 = 1$. ($1, \omega, \omega^2$ are cube roots of unity such that $1 + \omega + \omega^2 = 0$).

Hence the required roots of the given cubic equation (2) are

$$\left.\begin{aligned} x_1 &= u_1 - \tfrac{b}{3} = y + z - \tfrac{b}{3}, \\ x_2 &= u_2 - \tfrac{b}{3} = \omega y + \omega^2 z - \tfrac{b}{3} \\ x_3 &= u_3 - \tfrac{b}{3} = \omega^2 y + \omega z - \tfrac{b}{3} \end{aligned}\right\} \qquad (13)$$

Case (ii) If $R = 0$, then equal roots $y = z$. Then the roots of (4) are $y+z$, $y(\omega + \omega^2)$, $y(\omega + \omega^2)$

or $\quad 2y, -y, -y \qquad (14)$

(since $\omega + \omega^2 = -1$ and $y = z$)

Hence roots of (2) are

$$2y - \tfrac{b}{3}, -y - \tfrac{b}{3}, -y - \tfrac{b}{3} \qquad (15)$$

Case (iii) If $R < 0$ then y^3 and z^3 are complex. Suppose $y^3 = a + ib$ and $z^3 = a - ib$.

If the cube roots of these quantities are $m + in$ and $m - in$, then the roots of the cubic equation (4) are

$$\left.\begin{aligned} y + z &= (m+in) + (m-in) = 2m \\ \omega y + \omega^2 z &= \omega(m+in) + \omega^2(m-in) = -m - n\sqrt{3} \\ \omega^2 y + \omega z &= \omega^2(m+in) + \omega(m-in) = -m + n\sqrt{3} \end{aligned}\right\} \qquad (16)$$

Using De Moivre's result, this irreducible case Cardano's solution can be expressed in terms of trigonometric functions as follows:

Let the solution of (4) be

$$u = y + z = (a+ib)^{1/3} + (a-ib)^{1/3} \qquad (17)$$

Put $a = r\cos\theta$, $b = r\sin\theta$, so $r^2 = a^2 + b^2$, $\tan\theta = \frac{b}{a}$. Then

$$(a+ib)^{1/3} = \{r(\cos\theta + i\sin\theta)\}^{1/3}$$

$= r^{1/3}\left(\cos\frac{\theta+2k\pi}{3} + i\sin\frac{\theta+2k\pi}{3}\right)$, $k = 0, 1, 2$ (using De Moivre's Theorem) (18)

In a similar way $(a-ib)^{1/3} = r^{1/3}\left(\cos\frac{\theta+2k\pi}{3} - i\sin\frac{\theta+2k\pi}{3}\right)$, $k = 0, 1, 2$ (19)

Here $r^{1/3}$ is the arithmetical cube root of r. Thus substituting (18), (19) in (17) we get $u = 2r^{1/3}\cos\frac{\theta}{3}$, $2r^{1/3}\cos\frac{\theta+2\pi}{3}$, $2r^{1/3}\cos\frac{\theta+4\pi}{3}$. Hence the required roots of the given cubic equation (2) are

$$\left.\begin{aligned} &2r^{1/3}\cos\left(\tfrac{\theta}{3}\right) - \tfrac{b}{3},\ 2r^{1/3}\cos\left(\tfrac{\theta+2\pi}{3}\right) - \tfrac{b}{3} \\ &2r^{1/3}\cos\left(\tfrac{\theta+4\pi}{3}\right) - \tfrac{b}{3} \end{aligned}\right\} \qquad (20)$$

The numbers (13) (in case (i)), (15) (in case (ii)) and (20) (in case (iii)) are known as *Cardano's formulas* for the roots of a reduced cubic equation (4).

WORKED OUT EXAMPLES

Example 1: Solve $28x^3 - 9x^2 + 1 = 0$.

Solution: Rewriting $x^3 - \frac{9}{28}x^2 + \frac{1}{28} = 0$. To remove the x^2 term, put

$$x = u - \frac{b}{3} = u - \frac{1}{3}\left(-\frac{9}{28}\right) = u + \frac{3}{28}$$

Substituting

$$\left(u+\frac{3}{28}\right)^3-\frac{9}{28}\left(u+\frac{3}{28}\right)^2+\frac{1}{28}=0$$

or

$$u^3-\frac{27}{(28)^2}u+\frac{730}{(28)^3}=0$$

so $p=-27/(28)^2$ and $q=730/(28)^3$.
Let $u=y+z$. Then $y^3+z^3=-q=\ -730/\ (28)^3$ and $y^3z^3=\frac{-p^3}{27}=\frac{-(27)^3}{(28)^6}\cdot\frac{1}{27}$. Now y^3 and z^3 are the roots of the quadratic equation in t

$$t^2+\frac{730}{(28)^3}t-\frac{(27)^2}{(28)^6}=0$$

Here the discriminant

$$R=\left[\frac{730}{(28)^3}\right]^2+4\frac{(27)^2}{(28)^6}>0$$

so the values of both y and z are real. Solving for t,

$$\left(\frac{-730}{(28)^3}\pm\sqrt{\left(\frac{730}{(28)^3}\right)^2-4\frac{(27)^2}{(28)^6}}\right)\Bigg/2$$

$$=\frac{-730\pm 728}{2(28)^3}$$

so $y^3=-\frac{2}{2(28)^3}$ or $y=-\frac{1}{28}$

and $z^3=\frac{-729}{(28)^3}=-\left(\frac{9}{28}\right)^3$ or $z=\frac{-9}{28}$

Thus $u=y+z=-\frac{1}{28}-\frac{9}{28}=\frac{-10}{28}$

Hence $x=u+\frac{3}{28}=\frac{-10}{28}+\frac{3}{28}=\frac{-7}{28}=\frac{-1}{4}$ is a root of the given cubic equation. Similarly the other two roots are:

$$\omega y+\omega^2 z=\left(\frac{-1+\sqrt{3}i}{2}\right)\left(-\frac{1}{28}\right)$$

$$+\left(\frac{-1-\sqrt{3}i}{2}\right)\left(\frac{-9}{28}\right)$$

$$=-\frac{1}{2.28}[-10-8\sqrt{3}i]=\frac{5+4\sqrt{3}i}{28}$$

so $x=u+\frac{3}{28}=\frac{5+4\sqrt{3}i}{28}+\frac{3}{28}=$

$=\frac{8+4\sqrt{3}i}{28}=\frac{2+\sqrt{3}i}{7}$

Also

$$\omega^2 y+\omega z=\left(\frac{-1-\sqrt{3}i}{2}\right)\left(-\frac{1}{28}\right)+\left(\frac{-1+\sqrt{3}i}{2}\right)\left(\frac{-9}{28}\right)$$

$$=\frac{5-4\sqrt{3}i}{28}$$

so

$$x=u+\frac{3}{28}=\frac{5-4\sqrt{3}i}{28}+\frac{3}{28}$$

$$=\frac{8-4\sqrt{3}i}{28}=\frac{2-\sqrt{3}i}{7}$$

Thus the three roots of the given cubic equation are $-\frac{1}{4},\frac{2+\sqrt{3}i}{7},\frac{2-\sqrt{3}i}{7}$

Example 2: Solve $x^3-27x+54=0$.

Solution: Here $p=-27, q=54$. Since x^2 term is absent, no translation is needed.
So put $x=y+z$. Then $y^3+z^3=-q=-54$ and $y^3z^3=-\frac{p^3}{27}=-\frac{(-27)^3}{27}=(27)^2$. Thus y^3, z^3 are the roots of the quadratic in t:

$$t^2+54t+(27)^2=0$$

Its discriminant $R=(54)^2-4(27)^2=2916-2916=0$. Therefore t has equal roots i.e., $y^3=z^3$ or $y=z$.
So $y^3=\frac{-54}{2}\pm 0=-27$ or $y=-3=z$.
The required three roots of the given cubic equation are: $2y,-y,-y$ i.e., $-6, 3, 3$.

Example 3: Solve $x^3-3x^2-12x+16=0$

Solution: To get rid of the x^2 term, put $x=u-\frac{b}{3}=u-\frac{(-3)}{3}=u+1$. Then $(u+1)^3-3(u+1)^2-12(u+1)+16=0$.
$u^3+3u^2+3u+1-3(u^2+2u+1)-$
$-12u-12+16=0$ or

or $\quad u^3-15u+2=0$

Here $p=-15$, $q=2$. Now put $u=y+z$. Then $y^3+z^3=-q=-2$ and

$$y^3z^3=\frac{-p^3}{27}=-\frac{(-15)^3}{27}=\frac{(15)^3}{27}=5^3$$

Thus y^3 and z^3 are the roots of the quadratic in t given by

$$t^2+2t+5^3=0$$

It discriminant is

$$R = 4 - 4(125) = -496 < 0$$

So the roots y^3 and z^3 are complex conjugate.

$$y^3 = \frac{-2 \pm \sqrt{4 - 4(125)}}{2} = -1 + \sqrt{124}i$$

put $x = -1$, $y = \sqrt{124}$, $r = \sqrt{x^2 + y^2} = \sqrt{1 + 124} = \sqrt{125}$, $\tan\theta = \frac{y}{x} = \frac{\sqrt{124}}{-1} = -\sqrt{124}$ and $r^{1/3} = (125)^{1/6} = \sqrt{5}$

So $y = (-1 + \sqrt{124}i)^{1/3} = r^{1/3}\text{cis}\ \left(\frac{\theta+2k\pi}{2}\right)$, $k = 0, 1, 2$

$y = \sqrt{5}\text{cis}\left(\frac{\theta+2k\pi}{2}\right)$, $k = 0, 1, 2$

Here cis $=$ cos $+i$ sin

In a similar way

$z^3 = -1 - \sqrt{124}i$ or $z = (-1 - \sqrt{124} \cdot i)^{1/3}$

$z = \sqrt{5}\left(\cos\frac{\theta+2k\pi}{2} - i\sin\left(\frac{\theta+2k\pi}{2}\right)\right)$, $k = 0, 1, 2$

Since $u = y + z$, and $x = u + 1$, the required three roots of the given cubic equation are

$1 + 2\sqrt{5}\cos\frac{\theta}{3}, 1 + 2\sqrt{5}\cos\frac{\theta+2\pi}{3},$

$1 + 2\sqrt{5}\cos\frac{\theta+4\pi}{3}$ where $\theta = \tan^{-1}(-\sqrt{124})$

EXERCISE

Solve the following cubic equations by Cardano's method.

1. $x^3 - 15x + 126 = 0$

   Hint: $t^2 - 126t + 125 = 0$, $y^3 = 125$, $y = 5$, $z^3 = 1$, $z = 1$

Ans. $y + z = 5 + 1 = 6, \omega y + \omega^2 z = -3 + 2\sqrt{3}i$, $\omega^2 y + \omega z = -3 - 2\sqrt{3}i$

2. $x^3 - 15x^2 - 33x + 847 = 0$

   Hint: Put $x = u + 5, u = y + z, u^3 - 108y + 432 = 0$, $t^2 + 432t + 46656 = 0$. R = discriminant = 0. Equal roots $y^3 = z^3 = -216$, or $y = z = -6$

Ans. 11, 11, −7

3. $x^3 + 72x - 1720 = 0$

   Hint: $y^3 = 1728$, $y = 12$, $z^3 = -8$, $z = -2$

Ans. $y + z = 12 - 2 = 10, -5 \pm 7\sqrt{3}i$

4. $x^3 + 3x^2 - 144x + 540 = 0$

   Hint: Put $x = u - 1$, $u^3 - 147y + 686 = 0$, $t^2 + 686t + (343)^2 = 0$. Equal roots, $y^3 = z^3 = -343$ or $y = z = -7$

Ans. −15, 6, 6

5. $2x^3 + 3x^2 + 3x + 1 = 0$

Ans. $-\frac{1}{2}, \frac{-1\pm\sqrt{3}i}{2}$

6. $x^3 - 18x - 35 = 0$

Ans. $5, \frac{-5\pm\sqrt{3}i}{2}$

   Hint: $x = y + z$, $t^2 - 35t + 216 = 0$, $y^3 = 27$, $y = 3$, $v^3 = 8$, $v = 2$, $x = 3 + 2 = 5$, $3\omega + 2\omega^2 = \frac{-5+\sqrt{3}i}{2}$ and $3\omega^2 + 2\omega = \frac{-5-\sqrt{3}i}{2}$

7. $x^3 - 3x^2 + 3 = 0$

   Hint: Put $x = u + 1$, $u^3 - 3u + 1 = 0$, put $u = y + z$, $t^2 + t + 1 = 0$, $u^3 = \frac{-1+\sqrt{3}i}{2}$, $v^3 = \frac{-1-\sqrt{3}i}{2}$, $u = y + z = 2\cos\frac{2\pi}{9}, 2\cos\frac{8\pi}{9}$, $2\cos\frac{14\pi}{9}$

Ans. $1 + 2\cos\frac{2\pi}{9}, 1 + 2\cos\frac{8\pi}{9}, 1 + 2\cos\frac{14\pi}{9}$.

8. $x^3 + 6x + 2 = 0$

Ans. $A - B, \omega A - \omega^2 B, \omega^2 A - \omega B$ where $A = 2^{1/3}$, $B = 4^{1/3}$.

9. $x^3 - 9x^2 - 9x - 15 = 0$

   Hint: Put $x = u + 3$

Ans. $3 + A + B, 3 + \omega A + \omega^2 B, 3 + \omega^2 A + \omega B$ where $A = (24)^{1/3}$, $B = (72)^{1/3}$

10 $x^3 + 21x + 342 = 0$

Ans. $-6, 3 \pm 4\sqrt{3}i$

11. $x^3 - 6x^2 + 6x - 5 = 0$

   Hint: Put $x = u + 2$, $u^3 - 6u - 9 = 0$, $u = y + z$, $t^2 - 9t + 8 = 0$, $y^3 = 8$, $y = 2$, $z^3 = 1$, $z = 1$

Ans. $x = (y + z) + 2 = (2 + 1) + 2 = 5; \frac{-3\pm\sqrt{3}i}{2}$

1.4 FERRARI'S* METHOD

Ferrari's method obtains the general solution of a biquadratic (or quartic) equation (Fourth degree polynomial).

* Scipio Ferrari, Italian mathematician, pupil of Cardano.

Consider

$$x^4 + 2px^3 + qx^2 + 2rx + s = 0 \quad (1)$$

Add $(ax + b)^2$ on both sides of (1). Then

$$x^4 + 2px^3 + (q + a^2)x^2 + 2(r + ab)x + (s + b^2) = (ax + b)^2 \quad (2)$$

Here the two unknowns a and b are determined such that the LHS of (2) is a perfect square say $(x^2 + px + k)^2$.

Comparing the coefficients on both sides
$p^2 + 2k = q + a^2$, $pk = r + ab$, $k^2 = s + b^2$ (3)
Eliminating a and b from these equations we get

$$(pk - r)^2 = a^2b^2 = (p^2 + 2k - q)(k^2 - s)$$

or

$$2k^3 - qk^2 + 2k(pr - s) + (qs - p^2s - r^2) = 0 \quad (4)$$

Choose any root k of this resolvent cubic equation (4) which has always a real root (since odd degree, sign is that of its last term). Substituting k in (3), we obtain values of a and b. Now

$$(x^2 + px + k)^2 = (ax + b)^2$$

or $(x^2 + px + k) \pm (ax + b) = 0$

Thus the four roots of the biquadratric equation (1) are obtained from the two quadratic equations

$$x^2 + (p - a)x + (k - b) = 0 \quad (5)$$

and $x^2 + (p + a)x + (k + b) = 0$ (6)

Note: Abel has demonstrated the impossibility of obtaining algebraical solution of equations of degree higher than four. In such cases Horner's method of approximation is used to find numerical solution to any required degree of accuracy.

Worked Out Examples

Example 1: Solve $x^4 + 2x^3 - 7x^2 - 8x + 12 = 0$

Solution: Let the quartic equation be

$$x^4 + 2px^3 + qx^2 + 2rx + s = 0$$

so here $2p = 2, q = -7, 2r = -8, s = 12$ or $p = 1$, $q = -7$, $r = -4$, $s = 12$. The equation for cubic k is

$$2k^3 - qk^2 + 2k(pr - s) + (qs - p^2s - r^2) = 0$$
$$2k^3 + 7k^2 + 2k(-4 - 12) + (-84 - 12 - 16) = 0$$
$$2k^3 + 7k^2 - 32k - 112 = 0$$

By trial, $k = 4$ is a root of this cubic

$$(2 \cdot 4^3 + 7 \cdot 4^2 - 32 \cdot 4 - 112 = 128 + 112 - 128 - 112 = 0)$$

Substituting $k = 4$ in

$$p^2 + 2k = q + a^2, \quad pk = r + ab, \quad k^2 = s + b^2$$

we have

$$1 + 8 = -7 + a^2 \quad \therefore\ a = 4$$
$$1 \cdot 4 = -4 + ab \quad \therefore 8 = ab = 4b \quad \therefore\ b = 2$$
$$4^2 = 12 + b^2 = 12 + 4$$

Thus the four roots of the given quartic equation are obtained from the solutions of the two quadratic equations

$$x^2 + (p \mp a)x + (k \mp b) = 0$$

i.e., $x^2 + (1 - 4)x + (4 - 2) = x^2 - 3x + 2 = 0$

$$x^2 + (1 + 4)x + (4 + 2) = x^2 + 5x + 6 = 0$$

Solving the two quadratics

$$x^2 - 3x + 2 = (x - 1)(x - 2) = 0$$

and $x^2 + 5x + 6 = (x + 2)(x + 3) = 0$

The four roots of the biquadratic equation are 1, 2, -2, -3.

Aliter: The above problem can be done as follows: $x^4 + 2x^3 - 7x^2 - 8x + 12 = 0$.

Introducing an unknown λ, combine the terms x^4 and $2x^3$ to form a perfect square

$$(x^2 + x + \lambda)^2 + \{-x^2 - \lambda^2 - 2\lambda x - 2\lambda x^2\} - 7x^2 - 8x + 12 = 0$$

$$(x^2 + x + \lambda)^2 - \{(8 + 2\lambda)x^2 + (8 + 2\lambda)x + (\lambda^2 - 12)\} = 0$$

The second expression in { } brackets will be a perfect square if $b^2 = 4ac$ i.e.,

$$(8 + 2\lambda)^2 = 4(8 + 2\lambda)(\lambda^2 - 12)$$

or $2\lambda^2 - \lambda - 28 = 0$

Solving $\lambda = 4, -\frac{7}{2}$. With $\lambda = 4$, we have

$$(x^2 + x + 4)^2 - \{16x^2 + 16x + 4\} = 0$$

$$(x^2 + x + 4)^2 - \{2(2x + 1)\}^2 = 0$$

Thus the four roots are given by the roots of the two quadratic equations

$$(x^2 + x + 4) + 2(2x + 1) = 0 \text{ i.e., } x^2 + 5x + 6 = 0$$

and

$$(x^2 + x + 4) - 2(2x + 1) = 0 \text{ i.e., } x^2 - 3x + 2 = 0$$

The four roots are $-2, -3, 1, 2$

Example 2: Solve $x^4 - 8x^2 + 24x + 7 = 0$

Solution: Here $2p = 0$, i.e., $p = 0, q = -8, 2r = 24$ i.e., $r = 12$ and $s = 7$. The cubic for k is

$$2k^3 + 8k^2 + 2k(0 - 7) + (0 - 56 - 144) = 0$$

$$k^3 + 4k^2 - 7k - 100 = 0$$

which has $k = 4$ as a root $(64 + 64 - 28 - 100 = 0)$ solving for a, b

$$0 + 8 = -8 + a^2 \quad \therefore a = 4,$$

$$0 = 12 + 4b \;\therefore b = -3$$

$$16 = 7 + b^2 = 7 + 9$$

Thus the four roots are given by the roots of two quadratic equations $x^2 + (0 - 4)x + (4 - (-3)) = x^2 - 4x + 7 = 0$ and $x^2 + (0 - 4)x + (4 - 3) = x^2 + 4x + 1 = 0$

The roots are $2 \pm \sqrt{3}i$ and $-2 \pm \sqrt{3}$.

EXERCISE

Solve the following biquadratic equations by Ferrari's method.

1. $x^4 - 2x^3 - 5x^2 + 10x - 3 = 0$

   Hint: $2k^3 + 5k^2 - 4k - 7 = 0, k = -1, a = 2, b = 2, x^2 - 3x + 1 = 0, x^2 + x - 3 = 0$

Ans. $\frac{3\pm\sqrt{5}}{2}, \frac{-1\pm\sqrt{13}}{2}$

2. $x^4 - 2x^2 + 8x - 3 = 0$

   Hint: $x^2 + 2x - 1 = 0, x^2 - 2x + 3 = 0$

Ans. $-1 \pm \sqrt{2}, 1 \pm \sqrt{2}i$

3. $x^4 - 12x^3 + 41x^2 - 18x - 72 = 0$

   Hint: $x^2 - 5x - 6 = 0, x^2 - 7x + 12 = 0$

Ans. $-1, 3, 4, 6$

4. $x^4 - 10x^3 + 35x^2 - 50x + 24 = 0$

   Hint: $2k^3 - 35k^2 + 202k - 385 = 0,\ k = 5,\ a = 0,\ b = 1,\ x^2 - 5x + 4 = 0,\ x^2 - 5x + 6 = 0$

Ans. $1, 4, 2, 3$

5. $x^4 + 8x^3 + 9x^2 - 8x - 10 = 0$

   Hint: $2k^3 - 9k^2 - 12k + 54 = 0, k = \frac{9}{2}, a = 4, b = \frac{11}{2}, x^2 - 1 = 0, x^2 + 8x + 10 = 0$

Ans. $\pm 1, -4 \pm \sqrt{6}$

6. $x^4 - 3x^2 - 42x - 40 = 0$

Ans. $4, -1, -\frac{1}{2}(3 \pm \sqrt{31}\, i)$

7. $x^4 - 2x^3 - 12x^2 + 10x + 3 = 0$

Ans. $1, -3, 2 \pm \sqrt{5}$

8. $4x^4 - 20x^3 + 33x^2 - 20x + 4 = 0$

Ans. $2, 2, \frac{1}{2}, \frac{1}{2}$

9. $x^4 - 3x^2 - 6x - 2 = 0$

Ans. $1 \pm \sqrt{2}, -1 \pm i$

10. $x^4 - 10x^2 - 20x - 16 = 0$

Ans. $4, -2, -1 \pm i$

1.5 COMPLEX NUMBERS

Gauss first introduced the term "complex number" Cardano* first used complex numbers in solving cubic equations. Complex numbers find applications in electric circuits, mechanical vibrating systems. Argand wrote a short book on the geometric representation of complex numbers in 1806. Kuhn of Denzig was the first mathematician who proposed geometric representation of imaginary number i.

* Girolamo Cardano (1501-1576) Italian mathematician.

There are *no* real solutions to equations such as $x^2+2=0$ or $x^2+3x+4=0$. This led to the introduction of complex numbers. A *complex number* denoted by z is defined as

$$z = x + iy$$

where x and y are real, while $i = \sqrt{-1}$ is known as the *imaginary unit*. Here x is known as the real part of z and y as the imaginary part of z and are denoted as

x = Real part of z = Re(z)

y = Imaginary part of z = Im(z).

For $y = 0$, real numbers form a subset of the complex numbers. When $x = 0$, the complex number is known purely imaginary complex number z. It can be represented as an ordered pair (x, y) and thus as a point in a plane known as *complex plane* or *Argand* diagram*. Here the x-axis is called as real axis, while the y-axis as imaginary axis.

Equality:

Two complex numbers $z_1 = x_1 + iy_1$ and $z_2 = x_2 + iy_2$ are said to be equal if $x_1 = x_2$ and $y_1 = y_2$. Thus $z_1 = z_2$ when their corresponding real parts are equal and corresponding imaginary parts are also equal. Otherwise they are said to be *not* equal i.e., $z_1 \neq z_2$.

Note: However inequalities between complex numbers such as $z_1 > z_2$ or $z_1 \le z_2$ has no meaning since the field of complex numbers can *not* be ordered. Thus $2+3i > 3+7i$ or $-6-2i < 0$ have no meaning.

Zero

A complex number z is zero if both the real and imaginary parts x and y are zero.

Conjugate complex numbers or complex conjugate number of z denoted by $\overline{z}$ is defined as

$$\overline{z} = x - iy$$

i.e., z and $\overline{z}$ differ only in the sign of the imaginary part.

Trigonometric form of a Complex Number

Since every complex number is represented as a point in the complex plane and vice versa, the complex number $z = x + iy$ is geometrically represented by the position vector $\overline{OP}$ where O is the origin and P is the point (x, y). Let (r, θ) denote the polar coordinates of P; with origin treated as the pole and positive x-axis as the polar axis. Then

$$x = r\cos\theta,\ y = r\sin\theta$$

So $z = x + iy = r\cos\theta + i\sin\theta$

$$z = r(\cos\theta + i\sin\theta) = r \text{ cis } \theta$$

This expression is known as the *trigonometric* form or polar form. Here r is termed as *modulus* or *absolute value* or *norm* of the complex number and is denoted by $|z|$.

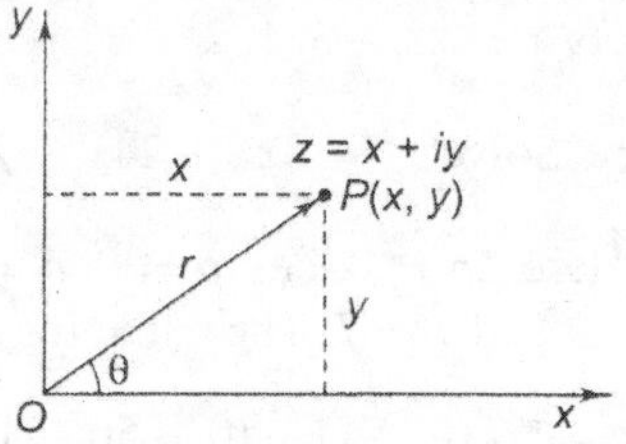

Fig. 1.23

Thus $\boxed{r = |z| = \sqrt{x^2+y^2}}$ represents the distance of z from origin. Here θ is the *argument* (or *amplitude* or *phase*) of z, denoted as $\boxed{\theta = \arg z = \tan^{-1}\frac{y}{x}}$. The polar form is also known as modulus-amplitude form.

Thus absolute value of z is

$$|z| = \sqrt{x^2+y^2}.$$

Amplitude, θ, the directed angle from positive x-axis is positive in the counterclockwise direction and is reckoned negative in the clockwise (opposite direction). θ is measured in radians. It is not unique, but multivalued and is determined up to $2\pi k$, for any integer k.

Principal value

Principal value of argument z is denoted by Arg z is the value of θ which lies in the interval $-\pi < \theta \le \pi$.

Note: Although z and $\overline{z}$ have the same moduli. Their arguments are equal in magnitude but differ in sign.

That is $\boxed{\arg z = -\arg \overline{z}}$

*Jean Robert Argand (1768-1822), French mathematician.

while $|z| = \sqrt{x^2 + y^2} = |\overline{z}|$

For any z,

1. $|z| \geq |Re(z)| \geq Re(z)$
2. $|z| \geq |Im(z)| \geq Im(z)$.

Complex Algebra

Addition

Sum of two complex numbers

$$z_1 + z_2 = (x_1 + iy_1) + (x_2 + iy_2) =$$
$$= (x_1 + x_2) + i(y_1 + y_2)$$

obtained by adding the corresponding real and imaginary parts.

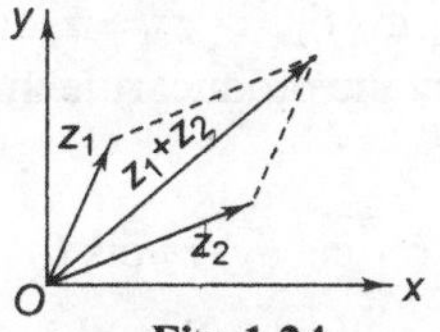

Fig. 1.24

So sum of two complex numbers $z_1 + z_2$ is given by the diagonal of the parallelogram with z_1 and z_2 as adjacent sides (Fig. 1.24).

Difference

Subtraction $z_1 - z_2$ is defined

$$z_1 - z_2 = (x_1 - y_1) + i(y_1 - y_2)$$

So $\boxed{|z_1 - z_2| = \sqrt{(x_1 - x_2)^2 + (y_1 - y_2)^2}}$

Thus the modulus of the difference between two numbers z_1 and z_2 equals to the distance between z_1 and z_2 [in the complex plane (see Fig. 1.25)].

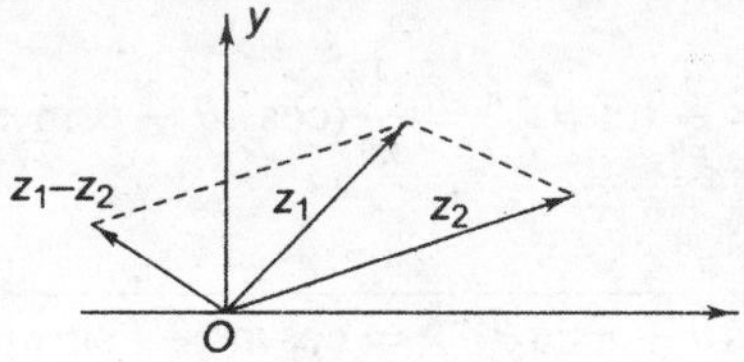

Fig. 1.25

Multiplication

$$z_1 \cdot z_2 = (x_1 + iy_1)(x_2 + iy_2) = (x_1x_2 - y_1y_2)$$
$$+ i(x_1y_2 + x_2y_1)$$

Since $i^2 = -1$, ($i^3 = -i$, $i^4 = 1$ etc. and in general $i^{4k} = 1$, $i^{4k+1} = i$, $i^{4k+2} = -1$, $i^{4k+3} = -i$).

In the polar form

Suppose $z_1 = r_1 \text{ cis } \theta_1$, $z_2 = r_2 \text{cis} \theta_2$ then

$$z_1z_2 = (r_1 \text{ cis } \theta_1)(r_2 \text{ cis } \theta_2)$$
$$= r_1r_2[\cos\theta_1 + i\sin\theta_1][\cos\theta_2 + i\sin\theta_2]$$
$$= r_1r_2[(\cos\theta_1\cos\theta_2 - \sin\theta_1\sin\theta_2)+$$
$$+i(\sin\theta_1 \cdot \cos\theta_2 + \cos\theta_1 \cdot \sin\theta_2)]$$

Thus $z = re^{i\theta} = z_1z_2$

$$\boxed{z_1z_2 = r_1r_2[\cos(\theta_1 + \theta_2) + i\sin(\theta_i + \theta_2)]}$$

Thus the modulus of the product is the product of the moduli i.e., $r = |z| = r_1r_2 = |z_1||z_2|$ and the argument of the product is the sum of the arguments, i.e., $\theta = \theta_1 + \theta_2$. or $\arg(z_1z_2) = \arg z_1 + \arg z_2$.

Division

$\frac{z_1}{z_2}$ is defined as the inverse operation of multiplication. Thus the quotient $z = \frac{z_1}{z_2}$ is defined as $z \cdot z_2 = z_1$.

In practice $\frac{z_1}{z_2}$ is obtained by multiplying the numerator or denominator by $\overline{z}_2$, the conjugate of z_2.

Thus $z = x + iy = \frac{z_1}{z_2} = \dfrac{z_1 \cdot \overline{z}_2}{z_2\ \overline{z}_2}$

$$= \frac{(x_1 + iy_1)(x_2 - iy_2)}{(x_2 + iy_2)(x_2 - iy_2)}$$

$$\boxed{\frac{z_1}{z_2} = \frac{(x_1x_2 + y_1y_2) + i(y_1x_2 - x_1y_2)}{x_2^2 + y_2^2}}$$

Thus $Re(z) = Re\left(\frac{z_1}{z_2}\right) = \frac{x_1x_2 + y_1y_2}{x_2^2 + y_2^2}$ and

$$Im(z) = Im\left(\frac{z_1}{z_2}\right) = \frac{y_1x_2 - x_1y_2}{x_2^2 + y_2^2}$$

Note that $z\overline{z} = x^2 + y^2 = |z|^2 = |\overline{z}|^2$.

Thus for any $z = x + iy$, ($\overline{z} = x - iy$),

1. $Re\ z = x = \frac{1}{2}(z + \overline{z})$
2. $Im\ z = y = \frac{1}{2i}(z - \overline{z})$
3. $z = \overline{z}$ then z must be real.
4. $\overline{z_1 \pm z_2} = \overline{z}_1 \pm \overline{z}_2$
5. $(\overline{z_1z_2}) = \overline{z}_1\overline{z}_2$

6. $\overline{\left(\frac{z_1}{z_2}\right)} = \frac{\overline{z_1}}{\overline{z_2}}$

In polar form

$$\frac{z_1}{z_2} = \frac{r_1(\cos\theta_1 + i\sin\theta_1)}{r_2(\cos\theta_2 + i\sin\theta_2)}$$

$$= \frac{r_1}{r_2}[\cos(\theta_1 - \theta_2) + i\sin(\theta_1 - \theta_2)]$$

Thus the modulus of the quotient is quotient of the moduli i.e., $r = \left|\frac{z_1}{z_2}\right| = \frac{r_1}{r_2} = \frac{|z_1|}{|z_2|}$ and the argument of the quotient is the difference between the arguments i.e., $\theta = \theta_1 - \theta_2$ or $\arg\left(\frac{z_1}{z_2}\right) = \arg z_1 - \arg z_2$.

Triangle inequality

Book work: Prove that

$$\boxed{|z_1 + z_2| \le |z_1| + |z_2|}$$

Proof Let $z_1 = x_1 + iy_2$, $z_2 = x_2 + iy_2$. Then consider $|z_1 + z_2| \le |z_1| + |z_2|$

$$\sqrt{(x_1 + x_2)^2 + (y_1 + y_2^2)} \le \sqrt{x_1^2 + y_1^2} + \sqrt{x_2^2 + y_2^2}.$$

Squaring on both sides

$$(x_1 + x_2)^2 + (y_1 + y_2)^2 \le (x_1^2 + y_1^2) + (x_2^2 + y_2^2) + 2\sqrt{x_1^2 + y_1^2} \cdot \sqrt{x_2^2 + y_2^2}$$

or

$$2(x_1 + x_2)(y_1 + y_2) \le 2\sqrt{x_1^2 + y_1^2} \cdot \sqrt{x_2^2 + y_2^2}$$

Squaring again on both sides

$$(x_1 + x_2)^2(y_1 + y_2)^2 \le (x_1^2 + y_1^2)(x_2^2 + y_2^2)$$

$$\text{or} \quad 2x_1x_2y_1y_2 \le x_1^2y_2^2 + x_2^2y_1^2$$

i.e., $(x_1y_2 - x_2y_1)^2 \ge 0$ which is always true.

Thus $|z_1 + z_2| \le |z_1| + |z_2|$

Note: Geometrically the triangle inequality states that the sum of the two sides of a triangle $|z_1| + |z_2|$ is greater than the third side of the triangle $|z_1 + z_2|$. The equality sign holds good i.e., $|z_1 + z_2| = |z_1| + |z_2|$ when the triangle degenerates into a straight line.

Result 1: Generalization

$$|z_1 + z_2 + z_3| \le |z_1| + |z_2| + |z_3|$$

or in general

$$\left|\sum_{i=1}^{n} z_i\right| \le \sum_{i=1}^{n} |z_i|$$

Result 2. $|z_1 - z_2| \ge ||z_1| - |z_2||$

Since $|z_1| = |z_1 - z_2 + z_2| \le |z_1 - z_2| + |z_2|$

Similarly

Result 3. $|z_1 + z_2| \ge ||z_1| - |z_2||$

Result 4. $|z_1 - z_2| \ge |z_1| - |z_2|$

Result 5. $|z_1 + z_2|^2 + |z_1 - z_2|^2 = 2(|z_1|^2 + |z_2|^2)$ (known as parallelogram equality)

(**Hint:** Use $|z_1 + z_2|^2 = r_1^2 + r_2^2 + 2r_1r_2\cos(\theta_2 - \theta_1)$). and $|z_1 - z_2|^2 = r_1^2 + r_2^2 - 2r_1r_2\cos(\theta_1 - \theta_2)$

De Moivre's Theorem: Power of complex numbers:

Let $z_1 = r_1 \text{ cis } \theta_1$, $z_2 = r_2\text{cis}\theta_2;, \cdots, z_n = r_n \text{ cis } \theta_n$. Then by the product rule in polar form, we get

$$z_1 \cdot z_2 \cdots z_n = (r_1 \text{ cis } \theta_1)(r_2 \text{ cis } \theta_2)\dots(r_n \text{ cis } \theta_n)$$

$$= (r_1r_2\dots r_n)(\text{cis } \theta_1)(\text{cis } \theta_2)\dots(\text{cis } \theta_n)$$

$$= (r_1r_2\dots r_n) \text{ cis } (\theta_1 + \theta_2 + \cdots + \theta_n)$$

Thus

$$\boxed{\text{cis } \theta_1 \cdot \text{cis } \theta_2 \dots \text{cis } \theta_n = \text{cis } (\theta_1 + \theta_2 + \cdots + \theta_n)}$$

If we choose $z_1 = z_2 = \cdots = z_n = z = r \text{ cis } \theta$ then $z^n = r^n \text{ cis } n\theta$ or for choice of $r = 1$, we get

$$\boxed{(\cos\theta + i\sin\theta)^n = \cos n\theta + i\sin n\theta}$$

which is known as De Moivre's theorem. In a similar way

$$\frac{1}{z} = \frac{1}{r\text{cis}\theta} = \frac{1}{r}(\text{cis}\theta)^{-1} = \frac{1}{r}(\cos\theta - i\sin\theta)$$

$$\frac{1}{z^n} = \frac{1}{r^n}(\text{cis}\theta)^{-n} = \frac{1}{r^n}(\cos n\theta - i\sin n\theta)$$

Thus

$$\boxed{(\cos\theta + i\sin\theta)^{-n} = \cos n\theta - i\sin n\theta}$$

De Moivre's theorem is valid for all rational values of n (including positive, negative integral values and zero).

$$\boxed{(\text{cis } m\theta)^n = \text{cis } mn\theta = (\text{cis } n\theta)^m}$$

1.6 ROOTS OF COMPLEX NUMBERS

Let $z = \omega^n$ where n is an integer 1, 2, 3, ... This is a single-valued function which associates a unique value z for each ω. Now consider the inverse function

$$\omega = z^{\frac{1}{n}} = \sqrt[n]{z}$$

which is multivalued, namely n valued. For any given $z \neq 0$ and the given integer n, there corresponds precisely n distinct values of ω which are known as the n roots of z. Thus the nth root of a complex number is another complex number whose nth power is equal to the radicand. To determine these n roots, let $\omega = R(\cos\phi + i\sin\phi)$ and $z = r(\cos\theta + i\sin\theta)$. Then

$$\begin{aligned}\omega^n &= [R(\cos\phi + i\sin\phi)]^n\\ &= R^n(\cos n\phi + i\sin n\phi)\\ &= z = r(\cos\theta + i\sin\theta)\end{aligned}$$

by using De Moivre's theorem. Since the moduli of equal complex numbers must be equal, while their amplitudes may circle with centre at origin and of radius $r^{1/n}$. They constitute the n vertices of a regular polygon of n sides inscribed in the circle, spaced at equal angular intervals of $\frac{2\pi}{n}$, beginning with the radius whose angle is $\frac{\theta}{n}$.

Principal value of $\omega = 2^{1/n}$ is obtained for $k = 0$ and by taking principal value of argument of z.

In particular, the nth roots of a real non-zero number A also has n values since the real number A can be expressed in trigonometric form as

$$A = |A|(\cos\theta + i\sin\theta) = |A|\text{cis } 0 \text{ for } A > 0$$

and

$$A = |A|(\cos\pi + i\sin\pi) = |A|\text{cis } \pi \text{ for } A < 0$$

Solution of Binomial Equation

The n roots of the binomial equation

$$x^n = A$$

differ by a multiple of 2π, we have

$R^n = r$ and $n\phi = \theta + 2k\pi$

or $R = \sqrt[n]{r}$ where the root is real positive and

$$\phi = \frac{\theta + 2k\pi}{n}$$

where k is an integer. For $k = 0, 1, 2, \cdots (n-1)$ we get n different roots since for these values of k, ϕ defines n distinct angles which identify n different complex numbers. But as k takes $n, n+1, \cdots$ or $-1, -2, \ldots$, the same angles ϕ are repeated again and again, thus giving the root values that coincide with those (already) obtained. Thus the nth *root of a complex number has n distinct values* given by

$$\omega = z^{\frac{1}{n}} = \sqrt[n]{z} = r^{\frac{1}{n}}\left[\cos\left(\frac{\theta + 2k\pi}{n}\right) + i\sin\left(\frac{\theta + 2k\pi}{n}\right)\right] \quad (1)$$

where $k = 0, 1, 2, \ldots, n-1$

Geometrically, these n values are given by

$$x = \sqrt[n]{A}\left(\cos\frac{2k\pi}{n} + i\sin\frac{2k\pi}{n}\right) \quad (2)$$

for $k = 0, 1, 2, \ldots, n-1$ when A is a real positive number. In particular the nth roots of unity (for $A = 1$) are given by

$$1^{1/n} = \left(\cos\frac{2k\pi}{n} + i\sin\frac{2k\pi}{n}\right), k = 0, 1, \ldots, n-1$$

similarly when A is a real negative number, then

$$x = \sqrt[n]{A}\left[\cos\left(\frac{\pi + 2k\pi}{n}\right) + i\sin\left(\frac{\pi + 2k\pi}{n}\right)\right] \quad (3)$$

for $k = 0, 1, 2, \ldots, (n-1)$. In particular the nth roots of -1 (for $A = -1$) are given by

$$(-1)^{\frac{1}{n}} = \cos\left(\frac{\pi + 2k\pi}{n}\right) + i\sin\left(\frac{\pi + 2k\pi}{n}\right)$$

for $k = 0, 1, 2, \ldots (n-1)$.

When A is a complex number, these n values of x are obtained from (1). Now the *general rational power* of a complex number is defined as

$$\begin{aligned}z^{p/q} = (x^{1/q})^p &= \left[r^{1/q}\text{cis}\frac{\theta + 2k\pi}{q}\right]^p\\ &= r^{p/q}\text{cis}\left(\frac{p}{q}(\theta + 2k\pi)\right)\end{aligned}$$

for $k = 0, 1, 2, \ldots, q-1$.

WORKED OUT EXAMPLES

Example 1: Express in the form $a + ib$ and find its modulus.

(a) $\dfrac{3+i}{5+5i}$ (b) $\dfrac{(1+i)(2+3i)}{4-i}$

Solution: (a) $\frac{3+i}{5+5i} = \frac{(3+i)(5-5i)}{(5+5i)(5-5i)} = \frac{15+5-10i}{25+25}$

$$= \frac{2-i}{5} = \frac{2}{5} - \frac{i}{5}$$

modulus of $\frac{3+i}{5+5i} = \left|\frac{2-i}{5}\right| = \sqrt{\left(\frac{2}{5}\right)^2 + \left(\frac{1}{5}\right)^2}$

$$= \sqrt{\frac{4}{25} + \frac{1}{25}} = \sqrt{\frac{1}{5}}.$$

(b) $\frac{(1+i)(2+3i)}{(4-i)} = \frac{(2-3+5i)(4+i)}{(4-i)(4+i)}$

$$= \frac{-4-5+19i}{17} = \frac{-9+19i}{17} = \frac{-9}{17} + \frac{19 \cdot i}{17}$$

modulus $= \left|\frac{-9+19i}{17}\right| = \sqrt{\left(\frac{9}{17}\right)^2 + \left(\frac{19}{17}\right)^2} = \sqrt{\frac{442}{289}}$

Example 2: Express $\sqrt{3} + i$ in the modulus amplitude form and find the principal argument of $-\sqrt{3} + i$.

Solution: $x + iy = -\sqrt{3} + i$, so $x = -\sqrt{3}$, $y = 1$, then $r = \sqrt{x^2 + y^2} = \sqrt{3+1} = \sqrt{4} = 2$, $x = -\sqrt{3} = r\cos\theta = 2\cos\theta$. So $\cos\theta = \frac{-\sqrt{3}}{2}$ and $y = 1 = r\sin\theta = 2\sin\theta$, so $\sin\theta = \frac{1}{2}$ · cos is negative and sine positive. θ in the 2nd quadrant $\theta = \pi - \frac{\pi}{6} \pm 2n\pi = \frac{5\pi}{6} \pm 2n\pi, n = 0, 1, 2, \ldots$

moulus amplitude form of $-\sqrt{3} + i$ is $2e^{i\left(\frac{5\pi}{6} \pm 2n\pi\right)}$ where 2 is the modulus and $\theta = \frac{5\pi}{6} \pm 2n\pi$ is the amplitude (or argument). The principal argument lies $-\pi \le \theta < \pi$. Thus $\frac{5\pi}{6}$ is the principal argument.

Example 3: Solve the equation $\frac{iy}{ix+1} - \frac{3y+4i}{3x+y} = 0$ given that x and y are real.

Solution:

$$\frac{iy}{ix+1} - \frac{3y+4i}{3x+y} = \frac{iy(-ix+1)}{(ix+1)(-ix+1)} - \frac{(3y+4i)}{(3x+y)}$$

$$= \frac{xy+iy}{1+x^2} - \frac{(3y+4i)}{(3x+y)} = 0$$

$$(xy+iy)(3x+y) - (1+x^2)(3y+4i) = 0$$

or

$$y(xy-3) + i(3xy + y^2 - 4x^2 - 4) = 0$$

$\Rightarrow y(xy-3) = 0$ and $3xy + y^2 - 4x^2 - 4 = 0$. If $y = 0$, then $x^2 + 1 = 0$. Since x is real, $x^2 + 1 =$ is not possible. Thus $y \neq 0$.

Assume $y \neq 0$ then $xy = 3$ or $y = \frac{3}{x}$. Eliminating
$3x \cdot \frac{3}{x} + \frac{9}{x^2} - 4x^2 - 4 = 0$
or $4x^4 - 5x^2 - 9 = 0$ or $(4x^2 - 9)(x^2 + 1) =$
Since x is real, $x^2 + 1 \neq 0$. Then $4x^2 - 9 = 0$
$x = \pm\frac{3}{2}$
Thus $x = 1.5$, $y = \pm 2$ are the solutions.

Example 4: Determine the curve represented b $z\bar{z} + (1+i)z + (1-i)\bar{z} = 0$.

Solution: $(x^2 + y^2) + (1+i)(x+iy) + (1-i)(x-iy) = 0$ or $x^2 + 2x + y^2 - 2y = 0$
Rewriting $(x+1)^2 + (y-1)^2 = (\sqrt{2})^2$; which the equation of a circle with centre at $(-1, 1)$ an of radius $\sqrt{2}$.

Example 5: Locate the points $z_1 = 9 + i$, $z_2 =$ $4 + 13i$, $z_3 = -8 + 8i$ and $z_4 = -3 - 4i$ in the A gand diagram and show that these four points for a square.

Solution: Distance between z_1 and z_2 = length the side AB

$$= |z_1 - z_2| = |9 - i - (4 - 13i)| = |5 + 12i|$$

$$= \sqrt{25 + 144} = \sqrt{169} = 13$$

Similarly

$$|z_1 - z_3| = |17 - 7i| = \sqrt{338}$$

$$|z_1 - z_4| = |12 - 5i| = \sqrt{144 + 25} = 13$$

$$|z_2 - z_3| = |12 - 5i| = 13$$

$$|z_2 - z_4| = |7 + 17i| = \sqrt{338}$$

$$|z_3 - z_4| = |-5 + 12i| = 13$$

Since $AB = AD = BC = CD$ (these four sides ar equal) the four points z_1, z_2, z_3, z_4 form a square i the Argand diagram (complex plane).

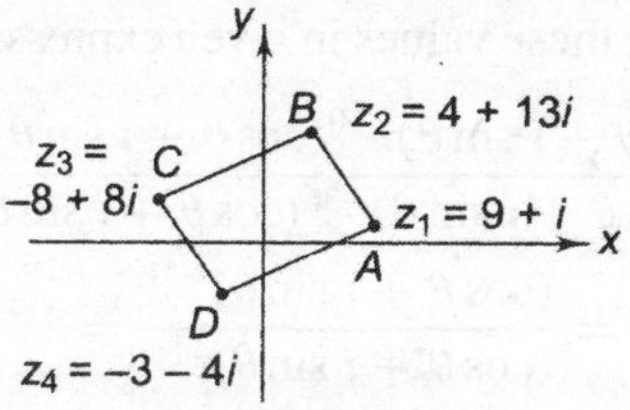

Fig. 1.26

Example 6: Show that multiplication of a complex number by 'i' corresponds to a counterclockwise rotation of the corresponding vector through the angle $\frac{\pi}{2}$.

Solution: Let the complex number be $z = x + iy = r(\cos\theta + i\sin\theta) = re^{i\theta}$. Then

$$iz = ire^{i\theta} = e^{i\frac{\pi}{2}} \cdot r \cdot e^{i\theta} = re^{i(\theta+\pi/2)}$$

Thus the argument of iz is $\theta + \frac{\pi}{2}$ which is $\frac{\pi}{2}$ more than the argument of z. Hence OZ is rotated through an angle $\frac{\pi}{2}$ in the counterclockwise direction.

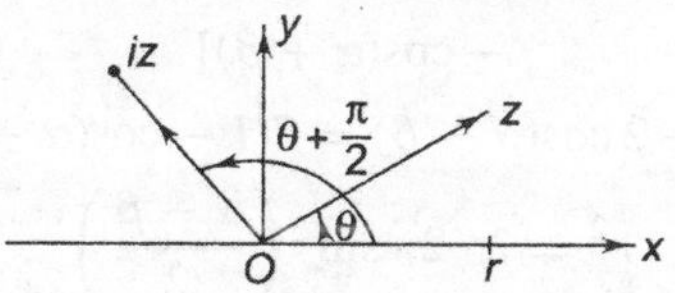

Fig. 1.27

Example 7: Determine the domain in the z-plane represented by (a) $3 < |z - 4| \le 5$ (b) $Im(z) < 6$ (c) $\frac{\pi}{4} < amp(z) < \frac{\pi}{2}$.

Solution: (a) $3 < |z - 4| \le 5$

$$|z - 4| = |x + iy - 4| = |(x - 4) + iy|$$

$$= \sqrt{(x-4)^2 + y^2}$$

From $3 < \sqrt{(x-4)^2 + y^2}$
we get $9 < (x-4)^2 + y^2$
From $\sqrt{(x-4)^2 + y^2} \le 5$, we get

$$(x-4)^2 + y^2 \le 25.$$

Thus the inequality represents the annulus region between two concentric circles with both centered at (4, 0) and of radii 3 and 5. The boundary of the outer circle with radius 5 is also included in the region.

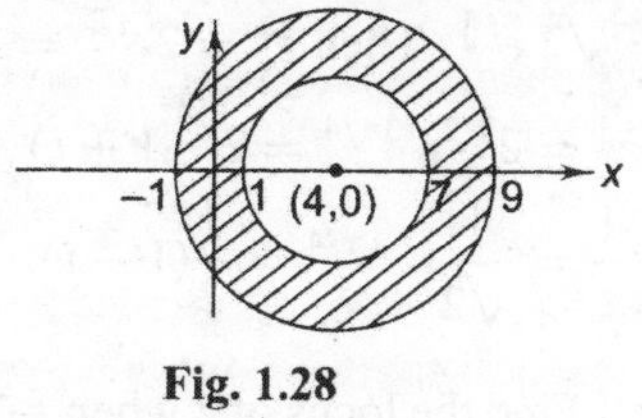

Fig. 1.28

(b) $Im(z) = y < 6$

The open half region below the line $y = 6$.

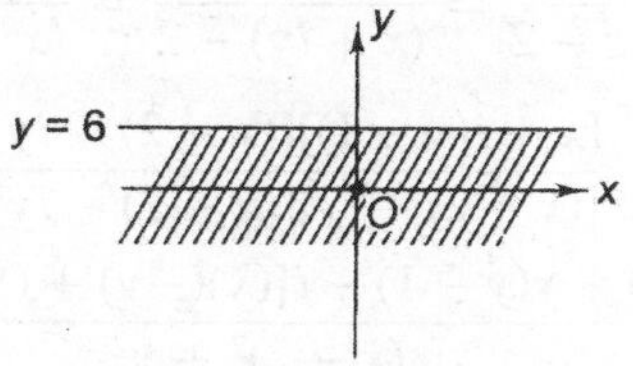

Fig. 1.29(a)

(c) The wedge region in the first quadrant bounded by the lines $\theta = \frac{\pi}{4}$, and $\theta = \frac{\pi}{2}$.

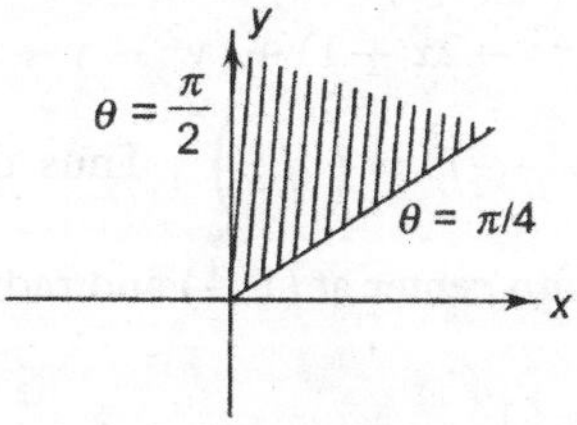

Fig . 1.29(b)

Example 8: If $z = 1 + i$, plot z^2, z^3, $\frac{1}{z}$ in the complex plane.

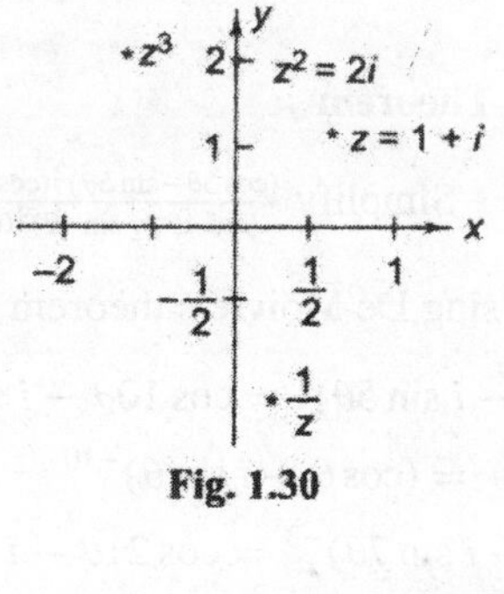

Fig. 1.30

Solution: $z = x + iy = 1 + i$, so $x = 1$, $y = 1$, then $r = \sqrt{x^2 + y^2} = \sqrt{2}$, $1 = x = \sqrt{2}\cos\theta$. Then $\cos\theta = \sin\theta = \frac{1}{\sqrt{2}}$. Thus $\theta = \frac{\pi}{4}$. Hence

$z = 1 + i = \sqrt{2}\, e^{i\frac{\pi}{4}}$. Then $z^2 = 2e^{i\frac{\pi}{2}} = 2i$

$$z^3 = 2^{3/2} e^{i3\pi/4} = 2(-1+i)$$

$$\frac{1}{z} = \frac{1}{\sqrt{2}} e^{-i\pi/4} = \frac{1}{2}(1-i)$$

Example 9: Find the locus of z when $\frac{z-i}{z-2}$ is purely imaginary.

Solution: $\frac{z-i}{z-2} = \frac{(x+iy)-i}{(x+iy)-2} = \frac{x+i(y-1)}{(x-2)+iy}$

$$= \frac{[x+i(y-1)][(x-2)-iy]}{[(x-2)+iy][(x-2)-iy]}$$

$$= \frac{x(x-2)+y(y-1)+i[(x)(-y)+(y-1)(x-2)]}{(x-2)^2+y^2}$$

Since $\frac{z-i}{z-2}$ is purely imaginary, its real part $\frac{x(x-2)+y(y-1)}{(x-2)^2+y^2}$ must be zero. Then $x(x-2)+y(y-1)=0$ or $x^2-2x+y^2-y=0$.
Rewriting $(x^2-2x+1)+(y^2-y+\frac{1}{4}) = \frac{5}{4}$ or $(x-1)^2+\left(y-\frac{1}{2}\right)^2 = \left(\sqrt{\frac{5}{4}}\right)^2$. Thus the locus of z is a circle with center at $(1, \frac{1}{2})$ and radius $\sqrt{\frac{5}{4}}$.

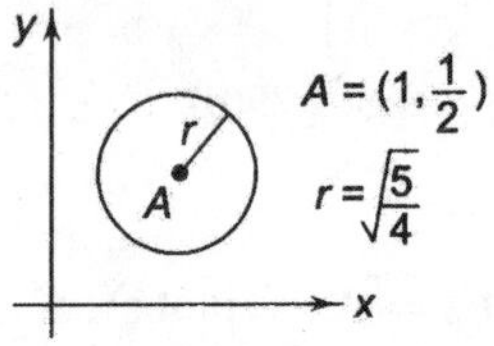

Fig. 1.31

De Moivre's Theorem

Example 10: Simplify $\frac{(\cos 5\theta - \sin 5\theta)^2 (\cos 7\theta + i \sin 7\theta)^{-3}}{(\cos 4\theta - i \sin 4\theta)^9 (\cos\theta + i\sin\theta)^5}$

Solution: Using De Moivre's theorem

$$(\cos 5\theta - i\sin 5\theta)^2 = \cos 10\theta - i\sin 10\theta$$

$$= (\cos\theta + i\sin\theta)^{-10}$$

$$(\cos 7\theta + i\sin 7\theta)^{-3} = \cos 21\theta - i\sin 21\theta$$

$$= (\cos\theta + i\sin\theta)^{-21}$$

$$(\cos 4\theta + i\sin 4\theta)^9 = \cos 36\theta - i\sin 36\,\theta$$

$$= (\cos\theta + i\sin\theta)^{-36}$$

Substituting these values in given expression, we get

$$\frac{(\cos\theta + i\sin\theta)^{-10}(\cos\theta + i\sin\theta)^{-21}}{(\cos\theta + i\sin\theta)^{-36}(\cos\theta + i\sin\theta)^{5}}$$

$$= \frac{(\cos\theta + i\sin\theta)^{-31}}{(\cos\theta + i\sin\theta)^{-31}} = 1$$

Example 11: Show that $[(\cos\alpha - \cos\beta) + i(\sin\alpha - \sin\beta)]^n + [(\cos\alpha - \cos\beta) - i(\sin\alpha - \sin\beta)]^n$

$$= 2^{n+1} \sin n\left(\alpha - \frac{\beta}{2}\right) \cdot \cos n\left(\frac{\pi + \alpha + \beta}{2}\right)$$

Solution: Put $\cos\alpha - \cos\beta = r\cos\theta$,
$\sin\alpha - \sin\beta = r\sin\theta$ (1)
Then $r^2 = (\cos\alpha - \cos\beta)^2 + (\sin\alpha - \sin\beta)^2$

$$= \cos^2\alpha + \cos^2\beta - 2\cos\alpha\cdot\cos\beta + \sin^2\alpha + \sin^2\beta - 2\sin\alpha\sin\beta$$

$$= 2 - 2\cos\alpha\cos\beta - 2\sin\alpha\sin\beta$$

$$= 2 - [\cos(\alpha+\beta) + \cos(\alpha-\beta)] - [\cos(\alpha-\beta) - \cos(\alpha+\beta)]$$

$$= 2 - 2\cos(\alpha-\beta) = 2(1-\cos(\alpha-\beta))$$

$$r^2 = 2\cdot 2\cdot \sin^2\left(\frac{\alpha-\beta}{2}\right)$$

So $r = 2\sin\left(\frac{\alpha-\beta}{2}\right)$ or $r^n = 2^n \sin n\left(\frac{\alpha-\beta}{2}\right)$ (2)
Now using (1) eliminating α, β in terms of r and θ, $[(\cos\alpha - \cos\beta) + i(\sin\alpha - \sin\beta)]^n + [(\cos\alpha - \cos\beta) - i(\sin\alpha - \sin\beta)]^n$

$$= (r\cos\theta + ir\sin\theta)^n + (r\cos\theta - ir\sin\theta)^n$$

$$= r^n[\cos n\theta + i\sin n\theta + \cos n\theta - i\sin n\theta]$$

$$= r^n \cdot 2\cdot\cos n\theta \quad (3)$$

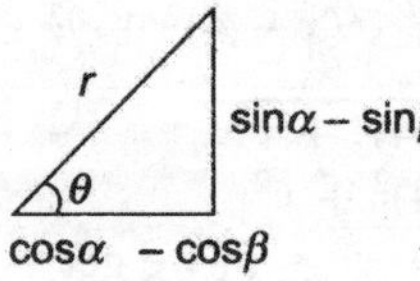

Fig. 1.32

Now from (1) $\tan\theta = \frac{r\sin\theta}{r\cos\theta} = \frac{\sin\alpha - \sin\beta}{\cos\alpha - \cos\beta}$, so

$$\cos\theta = \frac{\cos\alpha - \cos\beta}{r} = \frac{\cos\alpha - \cos\beta}{2\sin\left(\frac{\alpha-\beta}{2}\right)}$$

$$= \frac{-2\sin\frac{(\alpha+\beta)}{2}.\sin\frac{(\alpha-\beta)}{2}}{2\sin\frac{(\alpha-\beta)}{2}} = -\sin\frac{(\alpha+\beta)}{2}$$

$$\cos\theta = \cos\left(\frac{\pi}{2}+\frac{\alpha+\beta}{2}\right)$$

Thus $\theta = \frac{\pi}{2}+\frac{\alpha+\beta}{2}$

Then $\cos n\theta = \cos n\left(\frac{\pi}{2}+\frac{\alpha+\beta}{2}\right)$ (4)

- Substituting (3) and (4) in (2), we get the required result as

$$= 2^n \cdot \sin n\left(\frac{\alpha-\beta}{2}\right)\cdot 2\cdot \cos n\left(\frac{\pi}{2}+\frac{\alpha+\beta}{2}\right)$$

Example 12: If $z = \text{cis}\,\theta$ and $\omega = \text{cis}\,\phi$ show that

$$z^m\omega^n + z^{-m}\omega^{-n} = 2\cos(m\theta+n\phi).$$

Solution: Consider $z^m\omega^n + z^{-m}\omega^{-n}$
$= (\text{cis}\,\theta)^m(\text{cis}\,\phi)^n + (\text{cis}\,\theta)^{-m}(\text{cis}\,\phi)^{-n}$
(using De Movire's theorem)
$= (\cos m\theta + i\sin m\theta)(\cos n\phi + i\sin n\phi)$
$+ (\cos m\theta - i\sin m\theta)(\cos n\phi - i\sin n\phi)$
$= (\cos m\theta\cdot\cos n\phi - \sin m\theta\cdot\sin n\phi)$
$= \cos(m\theta+n\phi)$

Example 13: If $\sin\alpha + 2\sin\beta + 3\sin\gamma = 0$ and $\cos\alpha + 2\cos\beta + 3\cos\gamma = 0$ then prove that $\sin 3\alpha + 8\sin 3\beta + 27\sin 3\gamma = 18\sin(\alpha+\beta+\gamma)$ and $\cos 3\alpha + 8\cos 3\beta + 27\cos\gamma = 18\cos(\alpha+\beta+\gamma)$

Solution: Put $a = \text{cis}\,\alpha, b = \text{cis}\,\beta, c = \text{cis}\,\gamma$. Then
$a + 2b + 3c = \text{cis}\,\alpha + 2\text{cis}\,\beta + 3\text{cis}\,\gamma$
$= (\cos\alpha + i\sin\alpha) + 2(\cos\beta + i\sin\beta)$
$+ 3(\cos\gamma + i\sin\gamma)$
$= (\cos\alpha + 2\cos\beta + 3\cos\gamma) + i(\sin\alpha + 2\sin\beta + 3\sin\gamma)$
$= 0 + i\cdot 0 = 0$ using the given hypothesis.
Thus $(a+2b) = -3c$
Cubing on both sides

$$(a+2b)^3 = -27c^3$$

$$a^3 + 8b^3 + 3a^2\cdot 2b + 3a\cdot 4b^2 = -27c^3$$

$$a^3 + 8b^3 + 6ab(a+2b) = -27c^3$$

$$a^3 + 8b^3 + 6ab(-3c) = -27c^3$$

Then

$$a^3 + 8b^3 + 27c^3 = 18abc$$

$$(\text{cis}\alpha)^3 + 8(\text{cis}\beta)^3 + 27(\text{cis}\,\gamma)^3 = 18\cdot\text{cis}\alpha\,\text{cis}\beta\,\text{cis}\gamma$$

$$(\cos 3\alpha + i\sin 3\alpha) + 8(\cos 3\beta + i\sin 3\beta)$$
$$+ 27(\text{cis}\,3\gamma + i\sin 3\gamma) = 18\,\text{cis}(\alpha+\beta+\gamma)$$

or

$$(\cos 3\alpha + 8\cos 3\beta + 27\cos 3\gamma)$$
$$+ i(\sin 3\alpha + 8\sin 3\beta + 27\sin 3\gamma)$$
$$= 18[\cos(\alpha+\beta+\gamma) + i\sin(\alpha+\beta+\gamma)]$$

On comparing the real and imaginary parts on both sides, the two required results are obtained.

Worked Out Examples

Extraction of Roots

Example 1: Solve $z^4 + 1 = 0$ and locate the roots in the argand diagram.

Solution: $z^4 = -1$ so $z = (-1)^{\frac{1}{4}}$.
Consider $-1 = x + iy$ so $x = -1$, $y = 0$; then
$r = \sqrt{x^2+y^2} = \sqrt{1} = 1$
$-1 = x = r\cos\theta = \cos\theta, 0 = y = r\sin\theta$
$\therefore\ \theta = \pi$.
Thus $-1 = re^{i\theta} = 1\cdot e^{i\pi}$
Now $z = (-1)^{\frac{1}{4}} = r^{\frac{1}{n}}\text{cis}\left(\frac{\theta+2k\pi}{n}\right) = 1^{\frac{1}{4}}\text{cis}\left(\frac{(\pi+2k\pi)}{4}\right)$
i.e., $z = \cos\left(\frac{(\pi+2k\pi)}{4}\right) + i\sin\left(\frac{(\pi+2k\pi)}{4}\right)$;
with $k = 0, 1, 2, 3$

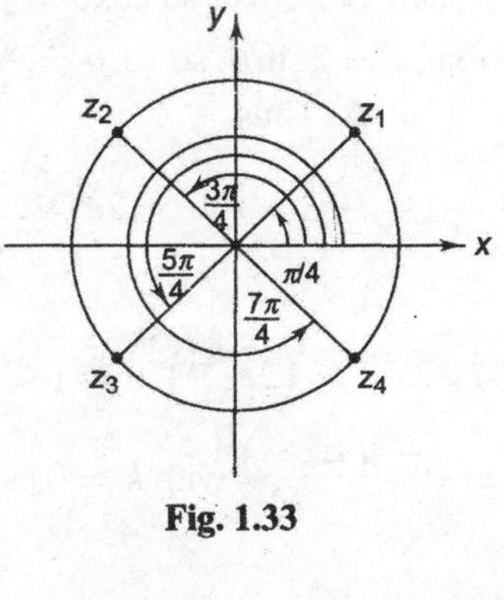

Fig. 1.33

For $k = 0$, $z_1 = \cos\frac{\pi}{4} + i\sin\frac{\pi}{4} = \frac{1}{\sqrt{2}} + \frac{i}{\sqrt{2}}$
$= \frac{1}{\sqrt{2}}(1+i)$
For $k = 1$, $z_2 = \cos\frac{3\pi}{4} + i\sin\frac{3\pi}{4} = -\frac{1}{\sqrt{2}} + \frac{i}{\sqrt{2}}$
$= \frac{1}{\sqrt{2}}(-1+i)$
For $k = 2$, $z_3 = \cos\frac{5\pi}{4} + i\sin\frac{5\pi}{4} = -\frac{1}{\sqrt{2}} - \frac{i}{\sqrt{2}}$
$= \frac{1}{\sqrt{2}}(-1-i)$
For $k = 3$, $z_4 = \cos\frac{7\pi}{4} + i\sin\frac{7\pi}{4} = \frac{1}{2} - \frac{i}{\sqrt{2}}$
$= \frac{1}{\sqrt{2}}(1-i)$

Thus the four roots of $(-1)^{\frac{1}{4}}$ are $\frac{1}{\sqrt{2}}(\pm 1 \pm i)$, i.e., z_1, z_2, z_3, z_4 are the solutions of the given equation. The four roots lie on a circle of radius 'one' making angles $\frac{\pi}{4}, \frac{3\pi}{4}, \frac{5\pi}{4}$ and $\frac{7\pi}{4}$ respectively with the positive x-axis.

Example 2: Find all solutions of the equation $z^4 - (1+4i)z^2 + 4i = 0$

Solution: $z^4 - z^2 - 4iz^2 + 4i =$

$$= z^2(z^2 - 1) - 4i(z^2 - 1) = 0$$

$$(z^2 - 1)(z^2 - 4i) = 0$$

Then $z^2 - 1 = 0$ and $z^2 = 4i$ or $z = \pm 1$ and $z = \pm\, 2\sqrt{i}$
But $\sqrt{i} = (e^{i\pi/2})^{1/2} = e^{i\pi/4}$ so $z = \pm 1$, $z = \pm 2e^{i\pi/4} = \pm 2\left(\frac{1}{\sqrt{2}} + i\frac{1}{\sqrt{2}}\right)$. The solutions are $\pm 1, \pm\sqrt{2}(1+i)$.

Example 3: Find all values of $(-1 + i\sqrt{3})^{3/2}$

Solution: Let $-1 + i\sqrt{3} = x + iy$, so $x = -1$, $y = \sqrt{3}$. Then $r = \sqrt{x^2+y^2} = \sqrt{1+3} = \sqrt{4} = 2$
$-1 = x = r\cos\theta = 2\cos\theta$, so $\cos\theta = -\frac{1}{2}$
$\sqrt{3} = y = r\sin\theta = 2\sin\theta$, so $\sin\theta = \frac{\sqrt{3}}{2}$
$\therefore\ \theta = \pi - \frac{\pi}{3} = \frac{2\pi}{3}$. Thus

$$-1 + i\sqrt{3} = re^{i\theta} = 2e^{i\frac{2\pi}{3}}$$

Now

$$(-1+i\sqrt{3})^{3/2} = \left(2e^{i\frac{2\pi}{3}}\right)^{3/2} = (8e^{i2\pi})^{1/2}$$

$$= \sqrt{8}e^{i\left(\frac{2\pi+2k\pi}{2}\right)} \text{ with } k = 0, 1$$

$$= \sqrt{8}e^{i\pi} = -\sqrt{8} = -2\sqrt{2} \text{ for } k = 0$$

$$= \sqrt{8}e^{i2\pi} = +\sqrt{8} = 2\sqrt{2} \text{ for } k = 1$$

Thus the solutions are $\pm 2\sqrt{2}$.

Example 4: Find the nth roots of unity or solve $z^n - 1 = 0$.

Solution: $z^n = 1$, or $z = 1^{1/n} = (1 \cdot e^{io})^{\frac{1}{n}}$ since $1 = x + iy$, $x = 1$, $y = 0$, $r = 1$, $\theta = 0$ so $1 = 1 \cdot e^{io}$. Then $z = 1^{1/n}e^{i\left(\frac{\theta + 2k\pi}{n}\right)} = e^{\frac{i2k\pi}{n}}$ with $k = 0, 1, \ldots, (n-1)$
For $k = 0$, $z = 1$
For $k = 1$, $\omega_1 = e^{i2\pi/n} = \cos\frac{2\pi}{n} + i\sin\frac{2\pi}{n}$
For $k = 2$, $\omega_2 = e^{i4\pi/n} = \left(\cos\frac{4\pi}{n} + i\sin\frac{4\pi}{n}\right)$
$= \left(\cos\frac{2\pi}{n} + i\sin\frac{2\pi}{n}\right)^2 = \omega_1^2$
For $k = 3$, $\omega_3 = e^{i6\pi/n} = \text{cis}\frac{6\pi}{n} = \left(\text{cis}\frac{2\pi}{n}\right)^3 = \omega_1^3$
Thus for $k = (n-1)$, $\omega_{n-1} = e^{i2(n-1)\frac{\pi}{n}} = \left(\text{cis}\frac{2\pi}{n}\right)^{n-1} = \omega_1^{n-1}$
Therefore, the n distinct roots of unity are $1, \omega, \omega^2, \omega^3, \ldots, \omega^{n-1}$. Here

$$\omega = e^{\frac{i2\pi}{n}} = \cos\frac{2\pi}{n} + i\sin\frac{2\pi}{n}$$

obtained for $k = 1$ is known as the primitive n^{th} root of 1 (i.e., root with the smallest non-zero angle). In the Argand diagram, these n roots of unity are represented by n distinct vertices of a regular polygon of n sides inscribed in a unit circle spaced at angular intervals of $\frac{2\pi}{n}$, beginning with one vertex at the point 1.

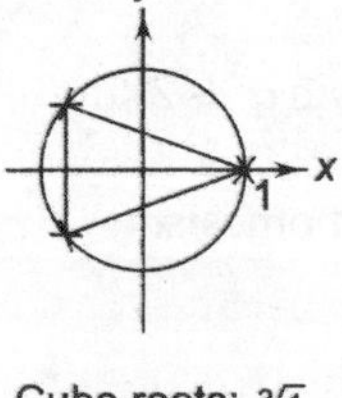

Cube roots: $\sqrt[3]{1}$
$1, -\frac{1}{2} \pm \frac{\sqrt{3}}{2}i$

Fig. 1.34

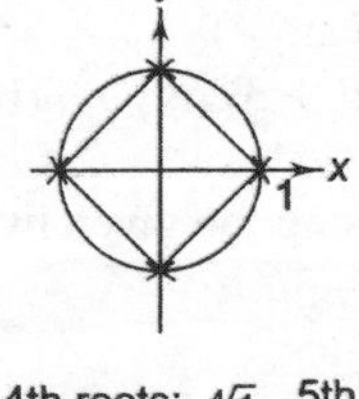

4th roots: $\sqrt[4]{1}$
$\pm 1, \pm i$

Fig. 1.35

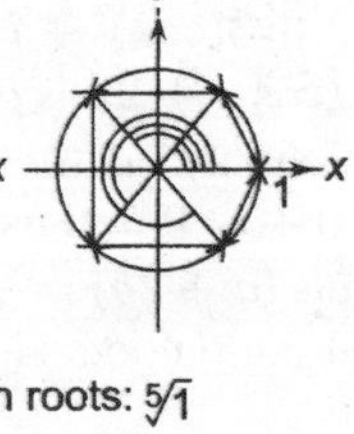

5th roots: $\sqrt[5]{1}$
$\theta = 0, \frac{2\pi}{5}, \frac{4\pi}{5}, \frac{6\pi}{5}, \frac{8\pi}{5}$

Fig. 1.36

Example 5: If ω is a complex 4th root of unity, prove that $1 + \omega + \omega^2 + \omega^3 = 0$

$k = 0, 1, 2, 3, 4$. Then ω corresponds root with $k = 1$. Thus for $k = 1$, $\omega = e^{i\pi/2} = i$. Now $1 + \omega + \omega^2 + \omega^3 = 1 + i + i^2 + i^3 = 1 + i - 1 - i = 0$

Example 6: Solve the equation $z^7 + z^4 + z^3 + 1 = 0$.

Solution: Observe that $z = -1$ is a solution of the given equation since $(-1)^7 + (-1)^4 + (-1)^3 + 1 = 0$.
By synthetic division,

1	1	0	0	1	1	0	0	1
		−1	1	−1	0	−1	1	−1
	1	−1	1	0	1	−1	1	0

$$z^7 + z^4 + z^3 + 1 = (z+1)(z^6 - z^5 + z^4 + z^2 - z + 1) = 0$$

Now consider

$$z^6 - z^5 + z^4 + z^2 - z + 1 = 0$$

Dividing by z^3,

$$z^3 - z^2 + z + \frac{1}{z} - \frac{1}{z^2} + \frac{1}{z^3} = 0$$

or

$$\left(z^3 + \frac{1}{z^3}\right) - \left(z^2 + \frac{1}{z^2}\right) + \left(z + \frac{1}{z}\right) = 0$$

Rewriting

$$\left[\left(z + \frac{1}{z}\right)^3 - 3\left(z + \frac{1}{z}\right)\right] - \left[\left(z + \frac{1}{z}\right)^2 - 2\right] + \left[z + \frac{1}{z}\right] = 0$$

or

$$\left(z + \frac{1}{z}\right)^3 - \left(z + \frac{1}{z}\right)^2 - 2\left(z + \frac{1}{z}\right) + 2 = 0$$

Put $\omega = z + \frac{1}{z}$, then

$$\omega^3 - \omega^2 - 2\omega + 2 = 0$$

for which $\omega = 1$ is a root (solution). By synthetic division.

1	1	−1	−2	+2
		1	0	−2
	1	0	−2	0

$$\omega^3 - \omega^2 - 2\omega + 2 = (\omega - 1)(\omega^2 - 2) = 0$$

Thus $\omega = 1$ and
$\omega^2 = 2$ or $z + \frac{1}{z} = 1$ and $\omega = z + \frac{1}{z} = \pm\sqrt{2}$.
Solving $z^2 - z + 1 = 0$ and $z^2 \pm \sqrt{2}z - 1 = 0$, we get $z = \frac{1 \pm \sqrt{1-4}}{2} = \frac{1 \pm \sqrt{3}i}{2}$ and

$$z = \frac{\pm\sqrt{2} \pm \sqrt{2-4}}{2} = \frac{\pm 1 \pm i}{\sqrt{2}}$$

Hence the seven solutions of the equation are

$z = -1, \dfrac{1 \pm \sqrt{3}i}{2}, \frac{\pm 1 \pm i}{\sqrt{2}}$.

Example 7: Determine the roots common to the equations $z^4 + 1 = 0$ and $z^6 - i = 0$.

Solution: The roots of $z^4 + 1 = 0$ are $z = (-1)^{1/4} = 1 \cdot \text{cis}\left(\frac{2k\pi + \pi}{4}\right)$ with $k = 0, 1, 2, 3$
i.e., $\text{cis}\frac{\pi}{4}, \text{cis}\frac{3\pi}{4}, \text{cis}\frac{5\pi}{4}, \text{cis}\frac{7\pi}{4}$.
Similarly the roots of $z^6 - i$ are $z = i^{1/6} = 1 \cdot \text{cis}\left(\frac{2k\pi + \frac{\pi}{2}}{6}\right)$ for $k = 0, 1, 2, 3, 4, 5$
i.e., $\text{cis}\frac{\pi}{12}, \text{cis}\frac{5\pi}{12}, \text{cis}\frac{9\pi}{12}, \text{cis}\frac{13\pi}{12}, \text{cis}\frac{17\pi}{12}, \text{cis}\frac{21\pi}{12}$.
Note that

$$\text{cis}\frac{3\pi}{4} = -\frac{1}{\sqrt{2}} + \frac{i}{\sqrt{2}} = \text{cis}\frac{9\pi}{12} = \text{cis}\frac{3\pi}{4}$$

and $\text{cis}\frac{7\pi}{4} = \frac{1}{\sqrt{2}} - \frac{i}{\sqrt{2}} = \text{cis}\frac{21\pi}{12}$ are the two common roots. Thus the $\frac{(-1+i)}{\sqrt{2}}$ and $\frac{1-i}{\sqrt{2}}$ are the common solutions.

Example 8: Find the equation whose roots are $2\cos\frac{2\pi}{7}, 2\cos\frac{4\pi}{7}, 2\cos\frac{6\pi}{7}$.

Solution: Let $z = \cos\theta + i\sin\theta = \text{cis}\theta$. Then

$$z^7 = (\text{cis}\theta)^7 = \cos 7\theta + i\sin 7\theta$$

For $\theta = 0, \frac{2\pi}{7}, \frac{4\pi}{7}, \frac{6\pi}{7}, \frac{8\pi}{7}, \frac{10\pi}{7}, \frac{12\pi}{7}$
$z^7 = \text{cis}7\theta = 1$ or $z^7 - 1 = 0$. Note that $\theta = 0$ corresponds to $z = 1$. Now by synthetic division

1	1	0	0	0	0	0	0	−1
		1	1	1	1	1	1	1
	1	1	1	1	1	1	1	0

Fig. 1.37

Rewrite

$$z^7 - 1 = (z-1)(z^6 + z^5 + z^4 + z^3 + z^2 + z + 1) = 0$$

Thus $\theta = \frac{2\pi}{7}, \frac{4\pi}{7}, \frac{6\pi}{7}, \frac{8\pi}{7}, \frac{10\pi}{7}, \frac{12\pi}{7}$ corresponds to the equation

$$z^6 + z^5 + z^4 + z^3 + z^2 + z + 1 = 0.$$

Dividing throughout by z^3,

$$z^3 + z^2 + z + 1 + \frac{1}{z} + \frac{1}{z^2} + \frac{1}{z^3} = 0$$

or $$\left(z^3 + \frac{1}{z^3}\right) + \left(z^2 + \frac{1}{z^2}\right) + \left(z + \frac{1}{z}\right) + 1 = 0$$

or $$\left(z + \frac{1}{z}\right)^3 - 3\left(z + \frac{1}{z}\right) + \left(z + \frac{1}{z}\right)^2 -$$

$$-2 + \left(z + \frac{1}{z}\right) + 1 = 0$$

i.e., $\left(z + \frac{1}{z}\right)^3 + \left(z + \frac{1}{z}\right)^2 - 2 \cdot \left(z + \frac{1}{z}\right) + 1 = 0.$

put $z + \frac{1}{z} = (\cos\theta + i\sin\theta) + (\cos\theta - i\sin\theta)$

$$= 2\cos\theta = \omega$$

Then the above solution reduces to

$$\omega^3 + \omega^2 - 2\omega + 1 = 0$$

observe that

$$\cos\frac{8\pi}{7} = \cos\left(2\pi - \frac{6\pi}{7}\right) = \cos\frac{6\pi}{7},$$

$$\cos\left(\frac{10\pi}{7}\right) = \cos\left(2\pi - \frac{4\pi}{7}\right) = \cos\frac{4\pi}{7} \text{ and}$$

$$\cos\left(\frac{12\pi}{7}\right) = \cos\left(2\pi - \frac{2\pi}{7}\right) = \cos\frac{2\pi}{7}$$

Hence $2\cos\frac{2\pi}{7}, 2\cos\frac{4\pi}{7}, 2\cos\frac{6\pi}{7}$ are the roots of the equation $\omega^3 + \omega^2 - 2\omega + 1 = 0$.

Example 9: Express $\sin^7\theta$ in sines of multiples of θ.

Solution: Let $z = \text{cis}\,\theta$, then $z - \frac{1}{z} = 2i\sin\theta$, $z^p = \text{cis}\,p\theta$, $\frac{1}{z^p} = \cos p\theta - i\sin p\theta$, so $z^p - \frac{1}{z^p} = 2i\sin p\theta$

Consider $(2i\sin\theta)^7 = 2^7 \cdot i^7 \cdot \sin^7\theta = \left(z - \frac{1}{z}\right)^7$
$= z^7 - 7 \cdot z^6 \cdot \frac{1}{z} + 21 \cdot z^5 \frac{1}{z^2} - 35z^4 \cdot \frac{1}{z^3} +$
$+35 \cdot z^3 \frac{1}{z^4} - 21 \cdot z^2 \frac{1}{z^5} + 7 \cdot z \cdot \frac{1}{z^6} - \frac{1}{z^7}$
$= \left(z^7 - \frac{1}{z^7}\right) - 7\left(z^5 - \frac{1}{z^5}\right) + 21\left(z^3 + \frac{1}{z^3}\right) -$
$35\left(z - \frac{1}{z}\right)$
$= 2i\sin 7\theta - 7 \cdot 2i\sin 5\theta + 21 \cdot 2i \cdot \sin 3\theta -$
$- 35 \cdot 2i\sin\theta$

Simplifying
$2^6\sin^7\theta = 35\sin\theta - 21\sin 3\theta + 7\sin 5\theta - \sin 7\theta$
Here the RHS is expressed in sines of multiples of θ.

Example 10: Show that
$32\sin^4\theta \cdot \cos^2\theta = \cos 6\theta - 2\cos 4\theta - \cos 2\theta + 2$

Solution: When $z = \text{cis}\,\theta$, then $z + \frac{1}{z} = 2\cos\theta$ and $z - \frac{1}{z} = 2i\sin\theta$; consider $z^p + \frac{1}{z^p} = 2\cos p\theta$.

Consider $(2i\sin\theta)^4(2\cos\theta)^2 = \left(z - \frac{1}{z}\right)^4\left(z + \frac{1}{z}\right)^2$
$= \left(z^4 - 4z^3 \cdot \frac{1}{z} + 6z^2 \cdot \frac{1}{z^2} - 4z \cdot \frac{1}{z^3} + \frac{1}{z^4}\right) \times$

$$\times\left(z^2 + 2 + \frac{1}{z^2}\right)$$

$$= \left(z^4 - 4z^2 + 6 - 4\frac{1}{z^2} + \frac{1}{z^4}\right)\left(z^2 + 2 + \frac{1}{z^2}\right)$$

$$= \left(z^6 + \frac{1}{z^6}\right) - 2\left(z^4 + \frac{1}{z^4}\right) - \left(z^2 + \frac{1}{z^2}\right) + 4.$$

Then

$$= 2\cos 6\theta - 2 \cdot 2\cos 4\theta - 2 \cdot \cos 2\theta + 4$$

$2^4 \cdot i^4 \sin^4\theta \cdot 2^2 \cdot \cos^2\theta = 2[\cos 6\theta - 2\cos 4\theta -$
$- \cos 2\theta + 4]$
$32 \cdot \sin^4\theta \cdot \cos^2\theta = \cos 6\theta - 2\cos 4\theta - \cos 2\theta +$
$+4$

EXERCISE

1. Express in $a + ib$ form and find modulus of
(a) $\frac{2+6i}{1-i}$ (b) $\frac{1+4i}{4+i}$ (c) $\frac{2-\sqrt{3}i}{1+i}$ (d) $\frac{(2+3i)}{(3-4i)}$

Ans. (a) $-2+4i, \sqrt{20}$ (b) $\frac{8+15i}{17}, 1$
(c) $\frac{1}{2}[2-\sqrt{3}-i(2+\sqrt{3})], \frac{7}{2}$ (d) $\frac{-6+17i}{2i}, \frac{\sqrt{13}}{5}$

2. Find the modulus and amplitude of
(a) $\frac{(3-\sqrt{2}i)^2}{1+2i}$ (b) $1+\sin\alpha+i\cos\alpha$
(c) $1-\cos\alpha+i\sin\alpha$ (d) $2+2\sqrt{3}i$

Ans. (a) $\frac{11\sqrt{5}}{5}, \frac{6\sqrt{2}+14}{12\sqrt{2}-7}$ (b) $\sqrt{2}\sqrt{1+\sin\alpha}, \frac{\pi}{4}-\frac{\alpha}{2}$
(c) $2\sin\frac{\alpha}{2}, \frac{\pi-\alpha}{2}$ (d) $4, \frac{\pi}{3}$

3. Show that $1+4i$, $2+7i$ and $3+10i$ are collinear

Hint: $2+7i-(1+4i)=1+3i, 3+10i-$
$-(1+4i)=2+6i$ so $1+3i=2(1+3i)=2+6i$

4. Determine the region in complex plane represented by (a) $1<|z+2i|\le 3$ (b) $Re(z)>3$
(c) $\frac{\pi}{6}\le \text{amp}(z)\le\frac{\pi}{3}$

Ans. (a) Annulus region between the concentric circles with centre at (0, 2) and radii 1 and 3.
(b) right open plane to the right of the line $x=3$.
(c) wedge region in the first quadrant bounded by the rays $\theta=\frac{\pi}{6}$ and $\theta=\frac{\pi}{3}$.

5. Determine the locus given by

$$|z-1|+|z+1|=3$$

Ans. Ellipse with foci at $z=\pm 1$ and major axis 3

6. If $z_1=2+4i$ and $z_2=3-5i$. Verify that
(a) $\overline{(z_1\pm z_2)}=\bar{z}_1\pm\bar{z}_2$ (b) $\overline{(z_1z_2)}=\bar{z}_1\bar{z}_2$
(c) $\overline{\left(\frac{z_1}{z_2}\right)}=\frac{\bar{z}_1}{\bar{z}_2}$

7. If $z_1=4+3i$ and $z_2=2-5i$, find
(a) z_1z_2 (b) $\frac{1}{z_1}$ (c) $Re(z_1^3)$ (d) $(Rez_1)^3$
(e) $z_1\bar{z}_2$ (f) $\bar{z}_1z_2$ (h) $\bar{z}_1/\bar{z}_2$

Ans. (a) $23-14i$ (b) $0.16-0.12i$ (c) -44
(d) 64 (e) $-7+26i$ (f) $-7-26i$
(h) $(-7-26i)/29$

8. If $z_1=-2+2i$ and $z_2=3i$ then find
(a) $|z_1z_2|$ (b) $\left|\frac{z_1}{z_2}\right|$
(c) $\arg z_1z_2$ (d) $\arg\frac{z_1}{z_2}$

Ans. (a) $6\sqrt{2}$ (b) $\frac{2\sqrt{2}}{3}$ (c) $\frac{-3\pi}{4}$ (d) $\frac{\pi}{4}$

Hint: $z_1z_2=-6-6i$, $\frac{z_1}{z_2}=\frac{2}{3}(1+i)$,
$\arg z_1=\frac{3\pi}{4}, \arg z_2=\frac{\pi}{2}$

9. Verify the triangle inequality for
(a) $z_1=2+3i$, $z_2=4-i$ (b) $z_1=1+i$, $z_2=7i$

Hint: (a) $|z_1+z_2|=|6+2i|=\sqrt{40}=6.32$,
$|z_1|+|z_2|=\sqrt{13}+\sqrt{17}=7.73$
(b) $|z_1+z_2|=|1+8i|=\sqrt{65}=8.063$,
$|z_1|=\sqrt{2}, |z_2|=7$,
$|z_1|+|z_2|=\sqrt{2}+7=1.414+7=8.414$

De Moivre's Theorem

10. Show that $(\cos 4\theta-i\sin 4\theta)^5\times(\cos 4\theta+i\sin 4\theta)^{-3}\times(\cos 3\theta+i\sin 3\theta)^4\times(\cos 5\theta+i\sin 5\theta)^4=1$

11. Simplify $(1+\cos\theta+i\sin\theta)^n(1+\cos\theta-i\sin\theta)^n$

Ans. $2^{n+1}\cdot\cos^n\frac{\theta}{2}\cdot\cos\frac{n\theta}{2}$

Hint: Put $1+\cos\theta=r\cos\alpha$, $\sin\theta=\sin\alpha$, $r=2\cos\frac{\theta}{2}, \alpha=\frac{\theta}{2}$.

12. If $z=\text{cis}\theta$, show that (a) $z^p+\frac{1}{z^p}=2\cos p\theta$
(b) $z^p-\frac{1}{z^p}=2i\sin p\theta$ (c) $\frac{z^{2n}+1}{z^{2n-1}+z}=\frac{\cos n\theta}{\cos(n-1)\theta}$.

13. Show that $\cos 3\alpha+\cos 3\beta+\cos 3\gamma=3\cos(\alpha+\beta+\gamma)$ and $\sin 3\alpha+\sin 3\beta+\sin 3\gamma=3\sin(\alpha+\beta+\gamma)$ if $\sin\alpha+\sin\beta+\sin\gamma=0$ and $\cos\alpha+\cos\beta+\cos\gamma=0$

Hint: Put $a=\text{cis}\,\alpha, b=\text{cis}\beta, c=\text{cis}\gamma, a^3+b^3+c^3=3abc$

14. Prove that
(a) $1+\cos\theta+\cos 2\theta+\cdots+\cos n\theta=\frac{1}{2}+\frac{\sin\left(n+\frac{1}{2}\right)\theta}{2\sin\frac{\theta}{2}}$
(b) $\sin\theta+\sin 2\theta+\cdots+\sin n\theta=\frac{1}{2}\cot\frac{\theta}{2}-\frac{\cos\left(n+\frac{1}{2}\right)\theta}{2\sin\frac{\theta}{2}}$

Hint: Use $1+z+z^2+\cdots+z^n=\frac{1-z^{n+1}}{1-z}$ when $z\neq 1$.

15. Show that
(a) $\cos 4\theta=\cos^4\theta-6\cos^2\theta\cdot\sin^2\theta+\sin^4\theta$
(b) $\sin 4\theta=4(\cos^3\theta\cdot\sin\theta-\cos\cdot\sin^3\theta)$

Hint: Expand $(\text{cis}\theta)^4$ by De Moivre's theorem and binomial expansion.

16. Expand
(a) $128 \cdot \cos^8 \theta$ in a series of cosines of multiples of θ
(b) $512 \cdot \sin^7 \theta \cdot \cos^3 \theta$ in a series of sines of multiples of 8.

Ans. (a) $\cos 8\theta + 8\cos 6\theta + 28\cos 4\theta + 56\cos 2\theta + 35$
(b) $14\sin 2\theta - 8\sin 4\theta - 3\sin 6\theta + 4\sin 8\theta - \sin 10\theta$

Roots

17. If ω is a complex cube root of unity, show that $1 + \omega + \omega^2 = 0$

Hint: $\omega = -\frac{1}{2} + i\frac{\sqrt{3}}{2}$, $\omega^2 = \frac{-1}{2} - \frac{\sqrt{3}}{2}i$

18. Solve the binomial equation $z^4 = 1$

Ans. $\pm 1, \pm i$

Hint: $1^{1/4} = \text{cis}\frac{0+2k\pi}{4}$, $k = 0, 1, 2, 3$

19. Find all the (a) square (b) cube roots of i

Ans. (a) $\pm\frac{1}{\sqrt{2}}(1+i)$ (b) $-i$, $\frac{\pm\sqrt{3}+i}{2}$

Hint: (a) $i^{1/2} = \text{cis}\left(\left(\frac{\pi}{2} + 2k\pi\right)/2\right)$, $k = 0, 1$
(b) $i^{1/3} = \text{cis}\left(\left(\frac{\pi}{2} + 2k\pi\right)/3\right)$, $k = 0, 1, 2$

20. Solve the binomial equation $z^5 = -32$

Ans. $2\text{cis}((\pi + 2k\pi)/5)$ for $k = 0, 1, 2, 3, 4$.

Hint: -32 in polar form is $32 \cdot \text{cis}\pi$, so $r = 32$, $\theta = \pi$

21. Find all roots of (a) $(-i)^{1/3}$ (b) $8^{1/6}$

Ans. (a) $i, \pm\frac{\sqrt{3}}{2} - \frac{i}{2}$ (b) $\pm\sqrt{2}, (\pm 1 \pm i\sqrt{3})/\sqrt{2}$

Hint: (a) $-i = e^{-i\pi/2}$, $r = 1$, $\theta = -\frac{\pi}{2}$, $(-i)^{1/3} = \text{cis}\left(\left(\frac{-\pi}{2} + 2k\pi\right)/3\right)$, $k = 0, 1, 2, 3$
(b) $8 = 8e^{io}, r = 8, \theta = 0, 8^{1/6} = \text{cis}\left(\frac{0+2k\pi}{6}\right)$, $k = 0, 1, 2, 3, 4, 5$

22. Find the fourth roots of $-8i$

Ans. $8^{1/4}\text{cis}\left(\left(\frac{3\pi}{2} + 2k\pi\right)/4\right)$ for $k = 0, 1, 2, 3$

Hint: $-8i = 8\text{cis}\frac{3\pi}{2}$.

23. Find all distinct values of (a) $(-1-i)^{4/5}$ (b) $((1+\sqrt{3}i)/2)^{3/4}$

Ans. (a) $2^{2/5}\text{cis}\left(\frac{5\pi+2k\pi}{5}\right)$, $k = 0, 1, 2, 3, 4$

Hint: (a) $-1 - i = \sqrt{2}e^{i5\pi/4}$, $(-1-i)^4 = 4^{1/5}\text{cis}5\pi$

Ans. (b) $\text{cis}(2k+1)\frac{\pi}{4}$ for $k = 0, 1, 2, 3$

Hint: (b) $\left(\frac{1}{2} + \frac{\sqrt{3}i}{2}\right)^3 = \left(\text{cis}\frac{\pi}{3}\right)^3 = \text{cis}\pi$

24. Solve $z^4 - z^3 + z^2 - z + 1 = 0$

Ans. $\text{cis}\frac{\pi}{5}, \text{cis}\frac{3\pi}{5}, \text{cis}\left(\frac{7\pi}{5}\right), \text{cis}\left(\frac{9\pi}{5}\right)$

Hint: $(z+1)(z^4 - z^3 + z^2 - z + 1) = z^5 + 1 = 0$ whose roots are $\text{cis}\left(\frac{(2k+1)\pi}{5}\right)$ for $k = 0, 1, 2, 3, 4$. Out of these for $k = 2$, $\text{cis}\pi = -1$ is deleted.

25. Find the equation whose roots are $2\cos\frac{\pi}{7}$, $2\cos\frac{3\pi}{7}$, $2\cos\frac{5\pi}{7}$.

Ans. $\omega^3 - \omega^2 - 2\omega + 1 = 0$

Hint: $z = \text{cis}\theta$, $z^7 + 1 = (z+1)(z^6 - z^5 + z^4 - z^3 + z^2 - z + 1) = 0$
$(-1)^{1/7} = \text{cis}\left(\frac{\pi+2k\pi}{7}\right)$ with $k = 0, 1, 2, 3, 4, 5, 6$, deleting $z = -1$ corresponding to $k = 3$, the six roots of $z^6 - z^5 + z^4 - z^3 + z^2 - z + 1 = 0$ are $z = \text{cis}\theta$, with $\theta = \frac{\pi}{7}, \frac{3\pi}{7}, \frac{5\pi}{7}, \frac{9\pi}{7}, \frac{11\pi}{7}, \frac{13\pi}{7}$. Dividing by z^3, introducing $\omega = z + \frac{1}{z}$, we have $[\omega^3 - 3\omega] - [\omega^2 - 2] + \omega = 0$.

PART—II

DIFFERENTIAL AND INTEGRAL CALCULUS

Chapter 2

Differential Calculus

INTRODUCTION

Calculus is one of the most beautiful intellectual achievements of human being. The mathematical study of change, motion, growth or decay is calculus. One of the most important ideas of differential calculus is derivative which measures the rate of change of a given function. Concept of derivative is very useful in engineering, science, economics, medicine and computer science. In this chapter we study the 300-year old Mitchel Rolle's theorem, Lagrange's mean value theorem which connects the average rate of change of a function over an interval with the instantaneous rate of change of the function at a point within that interval, generalized mean value theorem (Taylor's theorem) which enables to express any differentiable function in power series, namely Taylor's and Maclaurin's series. We also consider the problem of finding curvature, evolutes and envelope of a curve.

Suppose a function $y = f(x) = x^4 + e^{2x} + 3\sin 4x$ is differentiated w.r.t. x, then we get the first order derivative of y denoted by $y' = f' = 4x^3 + 2e^{2x} + 12\cos 4x$ which is itself a function of x. Differentiating y' again w.r.t. x, we get the second order derivative, denoted as $y'' = f'' = 12x^2 + 4e^{2x} - 48\sin 4x$, which is again a function of x. So by differentiating a second order derivative, we get the third order derivative and so on. Thus by differentiating a function $y = f(x)$, n times successively, we get the nth order derivative of y or simply nth derivative of y denoted by $y^{(n)}(x)$, $f^{(n)}(x)$, $D^n y$, $\frac{d^n y}{dx^n}$ or $y_n(x)$.

Note: The order of the derivative is taken in parentheses so as to avoid confusion with the exponent of a power. The order of the derivative is also denoted sometimes by Roman numerals for example 4th, 5th, 6th order derivatives are denoted by y^{iv}, y^{v}, y^{vi} etc.

2.1 DERIVATION OF *n*th DERIVATIVE OF SOME ELEMENTARY FUNCTIONS

Power Function

Consider $y = (ax+b)^m$, where m is any real number. Differentiating y w.r.t. x, successively, we get

$$y_1 = m(ax + b)^{m-1} \cdot a$$

$$y_2 = m(m-1)(ax+b)^{m-2} \cdot a^2$$

$$y_3 = m(m-1)(m-2)(ax+b)^{m-3} \cdot a^3$$

After n differentiations,

$$y_n = m(m-1)(m-2)\cdots(m-(n-1))a^n(ax+b)^{m-n}.$$

Case a: When m is a positive integer, then

$$y_n = \frac{m(m-1)(m-2)\cdots(m-(n-1)(m-n))\cdots 3\cdot 2\cdot 1}{(m-n)\cdots 3\cdot 2\cdot 1} \times$$

$$\times\, a^n(ax+b)^{m-n}$$

$$y_n = \frac{d^n}{dx^n}\left\{(ax+b)^m\right\} = \frac{m!}{(m-n)!}a^n(ax+b)^{m-n}.$$

Case b: When $m = n =$ a positive integer

$$y_n = \frac{d^n}{dx^n}\left\{(ax+b)^n\right\} = \frac{n!}{0!}a^n(ax+b)^0$$

$= n!a^n =$ a constant.

Case c: When $n > m$, then

$$y_n = 0$$

i.e., all the derivatives of order $m+1, m+2, m+3$ etc. are zero.

Case d: When $m = -1$, then

$$y = (ax+b)^{-1} = \frac{1}{ax+b}$$

and $y_n = (-1)(-2)(-3)\cdots(-n)a^n \cdot (ax+b)^{-1-n}$

$$y_n = \frac{d^n}{dx^n}\left\{\frac{1}{ax+b}\right\} = \frac{(-1)^n \cdot n!a^n}{(ax+b)^{n+1}}$$

Case e: If $y = \ln(ax+b)$, then

$$y_1 = \frac{a}{ax+b}$$

using result of Case (d) and differentiating $(n-1)$ times

$$y_n = \frac{(-1)^{n-1}(n-1)!\,a^n}{(ax+b)^n}$$

Thus $\dfrac{d^n}{dx^n}\left\{\ln(ax+b)\right\} = \dfrac{(-1)^{n-1}(n-1)!\,a^n}{(ax+b)^n}$.

Exponential Function

Consider $y = a^{mx}$.

Differentiating y w.r.t. x, successively, we get

$$y_1 = ma^{mx} \cdot \log_e a$$

$$y_2 = m^2 a^{mx} \;(\log_e a)^2$$

In general, $y_n = m^n a^{mx} (\log_e a)^n$.

In particular, when $a = e$

$$y_n = \frac{d^n}{dx^n}\{e^{mx}\} = m^n e^{mx} (\log_e e)^n = m^n e^{mx}.$$

Trigonometric Functions

Case a: Suppose $y = \cos(ax+b)$. Then

$$y_1 = -a \cdot \sin(ax+b) = a\cos\left(ax+b+\frac{\pi}{2}\right)$$

$$y_2 = -a^2 \cdot \cos(ax+b) = a^2\cos\left(ax+b+2\frac{\pi}{2}\right)$$

$$y_3 = +a^3 \cdot \sin(ax+b) = a^3\cos\left(ax+b+3\frac{\pi}{2}\right).$$

Generalizing,

$$y_n = \frac{d^n}{dx^n}\left\{\cos(ax+b)\right\} = a^n\cos\left(ax+b+n\frac{\pi}{2}\right).$$

In a similar way, we can get

Case b:

$$y_n = \frac{d^n}{dx^n}\left\{\sin(ax+b)\right\} = a^n\sin\left(ax+b+n\frac{\pi}{2}\right).$$

Product Functions

Case a: Consider $y = e^{ax} \cdot \cos(bx+c)$.
Differentiating,

$$y_1 = ae^{ax}\cos(bx+c) - e^{ax} \cdot b \cdot \sin(bx+c)$$

$$y_1 = e^{ax}\Big[a \cdot \cos(bx+c) - b\sin(bx+c)\Big].$$

To rewrite this in the form of cos, put

$$a = r\cos\phi, \qquad b = r\sin\phi.$$

Then

$$y_1 = e^{ax}\Big[r\cos\phi \cdot \cos(bx+c) - r\sin\phi \cdot \sin(bx+c)\Big]$$

$$y_1 = re^{ax}\Big[\cos(bx+c+\phi)\Big].$$

Here $r = \sqrt{a^2+b^2}$ and $\phi = \tan^{-1}\left(\frac{b}{a}\right)$.
Differentiating y, again w.r.t. x, we get

$$y_2 = rae^{ax}\cos(bx+c+\phi) - re^{ax} \cdot b \cdot \sin(bx+c+\phi).$$

Substituting for a and b,

$$y_2 = re^{ax} r\cos\phi \cdot \cos(bx+c+\phi) - r \cdot e^{ax} \cdot r\sin\phi \cdot \sin(bx+c+\phi)$$

$$= r^2 e^{ax}\Big[\cos\phi \cdot \cos(bx+c+\phi) - \sin\phi \cdot \sin(bx+c+\phi)\Big]$$

$$\therefore\; y_2 = r^2 e^{ax}\Big[\cos(bx+c+2\phi)\Big].$$

Observe that differentiation increases, the power of r and angle ϕ. Thus

$$y_3 = r^3 e^{ax}\Big[\cos(bx+c+3\phi)\Big].$$

In general,

$$y_n = \frac{d^n}{dx^n}\left\{e^{ax}\cos(bx+c)\right\} = r^n e^{ax}\cdot\cos(bx+c+n\phi).$$

Case b: In a similar way, we obtain

$$y_n = \frac{d^n}{dx^n}\left\{e^{ax}\sin(bx+c)\right\} = r^n e^{ax}\cdot\sin(bx+c+n\phi).$$

Case c: When the function $f(x)$ is the product of the powers of sine and cosine functions, then express $f(x)$ as the sum of the sine and cosines of multiples of the independent variable (angles) and use results in *Trigonometric Functions* (mentioned above).

Example:

$$\sin^2 x\cdot\cos 3x = \left(\frac{1-\cos 2x}{2}\right)\cos 3x$$

$$= \frac{1}{2}\cos 3x - \frac{1}{2}\cos 2x\cdot\cos 3x$$

$$= \frac{1}{2}\cos 3x - \frac{1}{2}\cdot\frac{1}{2}[\cos 5x + \cos x]$$

Case d: When the function $f(x)$ is an algebraic rational function then using partial fractions $f(x)$ can be decomposed into real linear factors and apply result *Power Function.* In case, $f(x)$ gets decomposed into complex linear factors, apply *Case (d)* and use DeMoivre's theorem $(\cos x \pm i\sin x)^n = \cos nx \pm i\sin nx$.

Example: $\frac{1}{x^2-5x+6} = \frac{1}{(x-2)(x-3)} = \frac{1}{x-3} - \frac{1}{x-2}$.

Example: $\frac{x}{x^2+a^2} = \frac{1}{2}\left[\frac{1}{x-ai} + \frac{1}{x+ai}\right]$.

Higher Derivatives of Sum

We have the obvious formulas

$$\frac{d^n}{dx^n}\left\{u(x)+v(x)\right\} = (u+v)^{(n)} = \frac{d^n u}{dx^n} + \frac{d^n v}{dx^n}$$

$$= u^{(n)} + v^{(n)}$$

and $$\frac{d^n}{dx^n}\left\{cu(x)\right\} = (cu)^{(n)} = c\frac{d^n u}{dx^n} = cu^{(n)}$$

where c is any arbitrary constant.

Worked Out Examples

Example 1: Find the fifth derivative of $x^3 \ln x$.

Solution: Let $y=x^3\ln x$ so $y_1=3x^2\ln x + x^3\cdot\frac{1}{x}$; $y_2=6x\ln x+3x^2\frac{1}{x}+2x$; $y_3=6\ln x+6\cdot x\frac{1}{x}+3+2$; $y_4=\frac{6}{x}+0$; $y_5=-\frac{6}{x^2}$.

Example 2: If $y=\frac{ax+b}{cx+d}$ prove that $2y_1y_3=3y_2^2$.

Solution:

$$y_1 = \frac{(cx+d)a-(ax+b)c}{(cx+d)^2} = \frac{ad-bc}{(cx+d)^2}$$

$$y_2 = \frac{-2c(ad-bc)}{(cx+d)^3}$$

$$y_3 = \frac{6c^2(ad-bc)}{(cx+d)^4}.$$

Now

$$2y_1y_3 = 2\cdot\frac{(ad-bc)}{(cx+d)^2}\cdot\frac{6c^2(ad-bc)}{(cx+d)^4}$$

$$= \frac{12c^2(ad-bc)^2}{(cx+d)^6} = 3\cdot\left[\frac{-2c(ad-bc)}{(cx+d)^3}\right]^2 = 3y_2^2.$$

Example 3: If $y=\sinh\left[m\log\left\{x+\sqrt{(x^2+1)}\right\}\right]$, prove that $(x^2+1)y_2+xy_1=m^3y$.

Solution:

$$y_1 = \cosh\left[m\ln\left\{x+\sqrt{(x^2+1)}\right\}\right]\cdot m\times$$

$$\times\frac{1}{\left(x+\sqrt{x^2+1}\right)}\left[1+\frac{1}{2}\frac{2x}{\sqrt{x^2+1}}\right]$$

$$y_1 = \frac{m}{\sqrt{x^2+1}}\cdot\cosh\left[m\ln\left\{x+\sqrt{x^2+1}\right\}\right].$$

Squaring on both sides

$$(1+x^2)y_1^2 = m^2\cosh^2\left[m\ln\left\{x+\sqrt{x^2+1}\right\}\right].$$

Differentiating

$$(1+x^2)2y_1y_2+2xy_1^2 = m^2\cdot 2\cdot\cosh\left[m\ln\left\{x+\sqrt{x^2+1}\right\}\right]\times$$

$$\times \sinh\left[m \ln\left\{x+\sqrt{x^2+1}\right\}\right] \cdot m \times$$

$$\times \frac{1}{x+\sqrt{x^2+1}} \cdot \left[1+\frac{1}{2}\frac{2x}{\sqrt{x^2+1}}\right]$$

$$\therefore \quad (1+x^2)2y_1y_2 + 2xy_1^2 = 2m^3 y_1 \cdot y$$

or $\quad (1+x^2)y_2 + xy_1 = m^3 y.$

Example 4: If $x = a\left\{\cos t + \frac{1}{2}\ln \tan^2\left(\frac{t}{2}\right)\right\}$ and $y = a \sin t$, then find $\frac{d^2y}{dx^2}$.

Solution: $\frac{dy}{dt} = a \cos t$

$$\frac{dx}{dt} = -a \sin t + \frac{1}{2}a\frac{1}{\tan^2 \frac{t}{2}} \cdot 2\tan\left(\frac{t}{2}\right) \cdot \frac{1}{2}\sec^2\left(\frac{t}{2}\right)$$

$$= a\left[-\sin t + \frac{1}{2\sin\left(\frac{t}{2}\right)\cdot\cos\left(\frac{t}{2}\right)}\right]$$

$$= a\left[-\sin t + \frac{1}{\sin t}\right]$$

$$\frac{dx}{dt} = \frac{a(1-\sin^2 t)}{\sin t} = \frac{a\cos^2 t}{\sin t}.$$

So $$\frac{dy}{dx} = \frac{\left(\frac{dy}{dt}\right)}{\left(\frac{dx}{dt}\right)} = a\cos t \cdot \frac{\sin t}{a\cos^2 t} = \tan t$$

$$\frac{d^2y}{dx^2} = \frac{d}{dx}\left(\frac{dy}{dx}\right) = \frac{d}{dx}(\tan t) = \frac{d}{dt}(\tan t)\frac{dt}{dx}$$

$$\frac{d^2y}{dx^2} = \sec^2 t \cdot \frac{\sin t}{a\cos^2 t} = \frac{\sin t}{a\cos^4 t}.$$

Example 5: Find $\frac{d^2y}{dx^2}$ if $x^3 + y^3 = 3axy$.

Solution: Differentiating the implicit equation

$$3x^2 + 3y^2 y_1 = 3ay + 3axy_1$$

$$(y^2 - ax)y_1 = (ay - x^2).$$

Differentiating

$$(2yy_1 - a)y_1 + (y^2 - ax)y_2 = ay_1 - 2x.$$

Substituting y_1

$$2y\left(\frac{ay-x^2}{y^2-ax}\right)^2 - 2a\left(\frac{ay-x^2}{y^2-ax}\right) + (y^2-ax)y_2 = -2x$$

$$y(ay-x^2)^2 - a(ay-x^2)(y^2-ax) + \frac{y_2}{2}(y^2-ax)^3$$

$$= -x(y^2-ax)^2$$

$$(a^2y^3 + x^4y - 2ax^2y^2) - a(ay^3 - a^2xy - x^2y^2 + ax^3)$$

$$+\frac{y_2}{2}(y^2-ax)^3 = -xy^4 - a^2x^3 + 2ax^2y^2$$

or

$$-3a\,x^2y^2 + xy(x^3+y^3) + a^3xy + \frac{y_2}{2}(y^2-ax)^3 = 0.$$

Since $x^3 + y^3 = 3axy$, we have

$$y_2 = \frac{-2a^3xy}{(y^2-ax)^3}$$

Example 6: Show that $D^{2n}(x^2-1)^n = (2n)!$

Solution: In the binomial series expansion of $(x^2 - 1)^n$, the highest power of x is x^{2n}. All the remaining terms will be degree less than $2n$. So when $(x^2 - 1)^n$ is differentiated $2n$ times, derivatives of all the terms except x^{2n}, become zero. Now

$$\frac{d^{2n}}{dx^{2n}}\left\{x^{2n}\right\} = 2n\cdot(2n-1)(2n-2)\cdots(2n-(2n-1))$$

$$= 2n(2n-1)(2n-2)\cdots 3\cdot 2\cdot 1 = (2n)!$$

Example 7: If $y = \sin^3 x$, find y_n.

Solution:

$$y = \sin^3 x = \sin x \cdot \sin^2 x$$

$$= \sin x\left(\frac{1-\cos 2x}{2}\right)$$

$$= \frac{1}{2}\sin x - \frac{1}{2}\cdot\sin x\cdot\cos 2x$$

$$= \frac{1}{2}\sin x - \frac{1}{4}\sin 3x + \frac{1}{4}\sin x$$

$$y = \frac{3}{4}\sin x - \frac{1}{4}\sin 3x$$

$$y_n = \frac{3}{4}\sin\left(x + n\frac{\pi}{2}\right) - \frac{1}{4}\cdot 3^n \cdot \sin\left(3x + \frac{n\pi}{2}\right).$$

Example 8: Find the nth derivative of

$$y = e^{2x}\cdot\cos x\cdot\sin^2 2x.$$

Solution: Rewriting

$$\cos x \cdot \sin^2 2x = \cos x \cdot \left(\frac{1-\cos 4x}{2}\right)$$

$$= \frac{1}{2}\cos x - \frac{1}{2}\cos x \cdot \cos 4x$$

$$= \frac{1}{2}\cos x - \frac{1}{4}\cos 5x - \frac{1}{4}\cos 3x.$$

So $\quad y = e^{2x}\left[\frac{1}{2}\cos x - \frac{1}{4}\cos 5x - \frac{1}{4}\cos 3x\right]$

$\therefore \quad y_n = \frac{1}{2}\frac{d^n}{dx^n}e^{2x}\cos x - \frac{1}{4}\frac{d^n}{dx^n}e^{2x}\cos x - \frac{1}{4}\frac{d^n}{dx^n}e^{2x}\cos 3x$

when $\quad y = e^{ax}\cos(bx+c)$, $\quad$ then

$$y_n = r^n e^{ax}\cos(bx + c + n\phi)$$

where $\quad r = \sqrt{a^2+b^2}, \qquad \phi = \tan^{-1}\left(\frac{b}{a}\right).$

So $\quad y_n = \frac{1}{2}\left(\sqrt{5}\right)^n e^{2x}\cdot\cos\left(x + n\tan^{-1}\frac{1}{2}\right)$

$$-\frac{1}{4}\left(\sqrt{29}\right)^n e^{2x}\cdot\cos\left(5x + n\tan^{-1}\frac{5}{2}\right)$$

$$-\frac{1}{4}\left(\sqrt{13}\right)^3 e^{2x}\cdot\cos\left(3x + n\tan^{-1}\frac{3}{2}\right).$$

Example 9: If $y = \frac{x+1}{x^2-4}$, find y_n.

Solution:

$$y = \frac{x+1}{x^2-4} = \frac{x+1}{(x-2)(x+2)} = \frac{3}{4}\cdot\frac{1}{x-2} + \frac{1}{4}\cdot\frac{1}{x+2}$$

$$y_n = \frac{3}{4}\cdot\frac{(-1)^n n!}{(x-2)^{n+1}} + \frac{1}{4}\cdot\frac{(-1)^n n!}{(x+2)^{n+1}}$$

Example 10: Determine $y_n(0)$ if $y = \frac{x^3}{x^2-1}$.

Solution:

$$y = \frac{x^3}{x^2-1} = \frac{x^3-1+1}{x^2-1} = \frac{(x-1)(x^2+x+1)}{x^2-1} + \frac{1}{x^2-1}$$

$$= \frac{x^2+x+1}{x+1} + \frac{1}{(x-1)(x+1)}$$

$$= \frac{x^2-1+1}{x+1} + 1 + \frac{1}{(x-1)(x+1)}$$

$$y = x + \frac{1}{x+1} + \frac{1}{2}\left[\frac{1}{x-1} - \frac{1}{x+1}\right]$$

$$y_n = \frac{d^n}{dx^n}\left\{\frac{x^3}{x^2-1}\right\} = 0 + \frac{1}{2}\cdot\frac{(-1)^n n!}{(x+1)^{n+1}} + \frac{1}{2}\frac{(-1)^n\cdot n!}{(x-1)^{n+1}}.$$

At $\ x = 0$, $y_n(0) = \frac{(-1)^n\cdot n!}{2}\left[\frac{1}{1^{n+1}} + \frac{1}{(-1)^{n+1}}\right]$

when n is even, $\quad y_n(0) = \frac{(-1)^n n!}{2}[1-1] = 0$

when n is odd, $\quad y_n(0) = \frac{(-1)^n n!}{2}\cdot 2 = -n!.$

Example 11: If $y = \tan^{-1}\frac{2x}{1-x^2}$, find y_n.
(UPTU 2002)

Solution: Differentiating y w.r.t. x,

$$y_1 = \frac{d}{dx}\tan^{-1}\frac{2x}{1-x^2} = \frac{1}{1+\left(\frac{2x}{1-x^2}\right)^2}\cdot\frac{d}{dx}\left(\frac{2x}{1-x^2}\right)$$

$$y_1 = \frac{(1-x^2)^2}{(1+x^4-2x^2+4x^2)}\cdot\frac{(1-x^2)2-2x(-2x)}{(1-x^2)^2}$$

$$= \frac{2(1+x^2)}{(1+x^2)^2} = \frac{2}{(1+x^2)}.$$

We know that

$$y_n = \frac{d^n}{dx^n}\left\{\frac{1}{x^2+a^2}\right\} = \frac{(-1)^n n!}{a^{n+2}}\sin(n+1)\theta\cdot\sin^{n+1}\theta$$

where $\theta = \cot^{-1}\left(\frac{x}{a}\right) = \tan^{-1}\left(\frac{a}{x}\right)$.
Now differentiating y_1, $(n-1)$ times

$$y_n = \frac{d^{n-1}}{dx^{n-1}}y_1 = \frac{d^{n-1}}{dx^{n-1}}\left\{\frac{2}{1+x^2}\right\}$$

$$= 2(-1)^{n-1}(n-1)!\sin n\theta\cdot\sin^n\theta$$

where $\quad \theta = \tan^{-1}\frac{1}{x} = \cot^{-1}x.$

Example 12: Find the nth derivative of

$$y = \frac{x}{x^2+x+1}.$$

Solution: The roots of $x^2+x+1=0$ are $z_{1,2} = \frac{-1\pm\sqrt{1-4}}{2}$

i.e., $\quad z_1 = \frac{-1+\sqrt{3}i}{2}, \quad z_2 = \frac{-1-\sqrt{3}i}{2}.$

In terms of linear complex factors,

$$y = \frac{x}{(x-z_1)(x-z_2)} = \frac{A}{x-z_1} + \frac{B}{x-z_2}$$

$$x = A(x - z_2) + B(x - z_1)$$

when $x = z_1$, $\quad z_1 = A(z_1 - z_2)$ or $A = \frac{z_1}{z_1 - z_2}$
when $x = z_2$, $\quad z_2 = B(z_2 - z_1)$ or $B = \frac{z_2}{z_2 - z_1}$.
Here $z_1 - z_2 = \sqrt{3}i = ai$ where $a = \sqrt{3}$. So

$$y = \frac{1}{z_1 - z_2}\left[\frac{z_1}{x - z_1} - \frac{z_2}{x - z_2}\right].$$

To find y_n, apply *Power Function Case (d)*, then

$$y_n = \frac{z_1}{z_1 - z_2} \cdot \frac{(-1)^n \cdot n! \cdot 1^n}{(x - z_1)^{n+1}} - \frac{z_2}{z_1 - z_2} \cdot \frac{(-1)^n n! 1^n}{(x - z_2)^{n+1}} \quad (1)$$

Now $x - z_1 = x - \left(\frac{-1+\sqrt{3}i}{2}\right) = \frac{2x+1-ai}{2}$.
Put $\frac{2x+1}{2} = r\cos\theta$, $\quad \frac{a}{2} = r\sin\theta$, then

$$r^2 = \left(\frac{2x+1}{2}\right)^2 + \frac{a^2}{4} = \frac{4x^2+1+4x}{4} + \frac{3}{4} = x^2 + x + 1.$$

So $r = \sqrt{x^2+x+1}$ and $\theta = \tan^{-1}\left(\frac{\sqrt{3}}{2x+1}\right)$
By De Moivre's theorem

$$(x - z_1)^{n+1} = (r\cos\theta - i\, r\sin\theta)^{n+1}$$
$$= (r\, e^{-i\theta})^{n+1} = r^{n+1}e^{-i(n+1)\theta} \quad (2)$$

Similarly,

$$(x + z_2)^{n+1} = (r\cos\theta + i\, r\sin\theta)^{n+1}$$
$$= (r\, e^{i\theta})^{n+1} = r^{n+1}e^{i(n+1)\theta} \quad (3)$$

Also

$$\frac{z_1}{z_1 - z_2} = \frac{-1+\sqrt{3}i}{2} \cdot \frac{1}{\sqrt{3}i} = \frac{3+\sqrt{3}i}{6} \quad (4)$$

and

$$\frac{z_2}{z_1 - z_2} = \frac{-1-\sqrt{3}i}{2} \cdot \frac{1}{\sqrt{3}i} = \frac{-3+\sqrt{3}i}{6} \quad (5)$$

Substituting (2), (3), (4), (5) in (1), we get

$$y_n = \frac{(-1)^n \cdot n!}{r^{n+1}} \cdot \frac{1}{6}\left[\left(3+\sqrt{3}i\right)e^{+i(n+1)\theta} - \right.$$
$$\left. -(-3+\sqrt{3}i)e^{-i(n+1)\theta}\right]$$
$$= \frac{(-1)^n n!}{r^{n+1}} \cdot \frac{1}{6}\left[(3+\sqrt{3}i)\{\cos(n+1)\theta + i\sin(n+1)\theta\}\right.$$
$$\left. +(3-\sqrt{3}i)\{\cos(n+1)\theta - i\sin(n+1)\theta\}\right]$$
$$= \frac{(-1)^n n!}{r^{n+1}} \cdot \frac{1}{6}\left[6\cos(n+1)\theta - 2\sqrt{3}\sin(n+1)\theta\right]$$

$$y_n = \frac{(-1)^n \cdot n!}{r^{n+1}} \cdot \left[\cos(n+1)\theta - \frac{1}{\sqrt{3}}\sin(n+1)\theta\right].$$

where $r = \sqrt{x^2+x+1}$ and $\theta = \tan^{-1}\left(\frac{\sqrt{3}}{2x+1}\right)$.

EXERCISE

Find the derivatives of y of indicated order (1 to 5).

1. $y = \left(\frac{1}{x}\right)^x$, $\quad y_{2(1)}$. $\quad$ *Ans.* 0
2. $y = 3x^8$, $\quad y^{(8)}$. $\quad$ *Ans.* $y^{(8)} = 3 \cdot 8!$
3. $y = 2\sqrt{x}$, $\quad y^{(4)}$. $\quad$ *Ans.* $y^{(4)} = -\frac{15}{8\sqrt{x^7}}$
4. $y = ax^5 + bx^4 + cx^3 + dx^2 + ex + f$, $\quad y^{(6)}$.

Ans. $y^{(6)} = 0$

5. $y = \tan x$, $\quad y'''$.

Ans. $y''' = 6\sec^4 x - 4\sec^2 x$

6. If $y = e^x \sin x$, prove that $y'' - 2y' + 2y = 0$.
7. If $p^2 = a^2\cos^2\theta + b^2\sin^2\theta$, prove that
$$p + \frac{d^2p}{d\theta^2} = \frac{a^2b^2}{p^3}.$$
8. Show that $y'' + \tan y \cdot (y')^2 = 0$ when $y = \tan^{-1}(\sinh x)$.
9. If $b^2x^2 + a^2y^2 = a^2b^2$, then find y_2 and y_3.

Ans. $y_2 = -\frac{b^4}{a^2y^3}$, $\quad y_3 = -\frac{3b^6x}{a^4y^5}$

10. If $\rho = \tan(\phi + \rho)$, find $\frac{d^3\rho}{d\phi^3}$.

Ans. $-\frac{2(5+8\rho^2+3\rho^4)}{\rho^8}$

11. Find y_2 when $x = a(t - \sin t)$, $y = a(1 - \cos t)$.

Ans. $\frac{d^2y}{dx^2} = -\frac{1}{4a\sin^4\left(\frac{t}{2}\right)}$

12. If $x = a\cos t$, $y = a\sin t$, find $\frac{d^3y}{dx^3}$.

Ans. $-\frac{3\cos t}{a^2\sin^5 t}$

13. Find y_{2n} and y_{2n+1} where $y = \sinh x$.

Ans. $\frac{d^{2n}}{dx^{2n}}(\sinh x) = \sinh x$, $\frac{d^{2n+1}}{dx^{2n+1}}(\sinh x) = \cosh x$

14. Find y_n where (a) $y = \frac{1-x}{1+x}$; (b) $y = e^x x$; (c) $y = x\sin x$.

Ans. (a) $2(-1)^n \cdot \frac{n!}{(1+x)^{n+1}}$; (b) $e^x(x+n)$; (c) $x \sin\left(x+n\frac{\pi}{2}\right) - n\cos\left(x+n\frac{\pi}{2}\right)$.

15. If $y = \cos^4 x$, find y_n.

Hint: $y = \cos^4 x = \left(\frac{1+\cos 2x}{2}\right)^2 = \frac{3}{8} + \frac{1}{2}\cos 2x + \frac{1}{8}\cos 4x$.

Ans. $\frac{1}{2}2^n \cos\left(2x+n\frac{\pi}{2}\right) + \frac{1}{8}4^n \cos\left(4x+n\frac{\pi}{2}\right)$

16. Find nth derivative of $y = e^{ax}\cos^2 x \cdot \sin x$.

Hint: $\cos^2 x \cdot \sin x = \frac{1+\cos 2x}{2}\sin x = \frac{1}{4}\sin x + \frac{1}{4}\sin 3x$.

Ans. $\frac{1}{4}(a^2+1)^{\frac{n}{2}} e^{ax} \cdot \sin\left(x + n\tan^{-1}\frac{1}{a}\right) + \frac{1}{4}(a^2+9)^{\frac{n}{2}} e^{ax} \cdot \sin\left(3x + n\tan^{-1}\frac{3}{a}\right)$

17. If $y = \cos x \cdot \cos 2x \cdot \cos 3x$ find y_n.

Hint: $\cos x \cdot \cos 2x \cdot \cos 3x = \frac{1}{4}[\cos 6x + \cos 4x + \cos 2x + 1]$.

Ans. $\frac{1}{4}\left[6^n \cos\left(6x+n\frac{\pi}{2}\right) + 4^n\cos\left(4x+n\frac{\pi}{2}\right) + 2^n\cos\left(2x+n\frac{\pi}{2}\right)\right]$.

18. Determine y_n if $y = \frac{x^2}{(x-1)^2(x+2)}$.

Hint: $\frac{x^2}{(x-1)^2(x+2)} = \frac{5}{9}\frac{1}{x-1} + \frac{1}{3}\frac{1}{(x-1)^2} + \frac{4}{9}\frac{1}{x+2}$.

Ans. $\frac{(-1)^n n!}{9}\left\{\frac{5}{(x-1)^{n+1}} + \frac{3}{(x-1)^{n+2}} + \frac{4}{(x+2)^{n+1}}\right\}$

19. Calculate y_n where $y = \frac{x}{x^2+a^2}$.

Hint: $\frac{x}{x^2+a^2} = \frac{1}{2}\left[\frac{1}{(x-ai)} + \frac{1}{x+ai}\right]$, use De Moivre's theorem.

Ans. $\frac{(-1)^n \cdot n!}{r^{n+1}} \cos(n+1)\theta$, $r = \sqrt{x^2+a^2}$, $\theta = \tan^{-1}\left(\frac{a}{x}\right)$.

20. If $y = \frac{1}{x^2+x+1}$, then find y_n.

Ans. $\frac{2(-1)^n \cdot n!}{\sqrt{3}r^{n+1}} \sin(n+1)\theta$ where $r = \sqrt{x^2+x+1}$, and $\theta = \cot^{-1}\left(\frac{2x+1}{\sqrt{3}}\right)$.

21. Find the nth derivative of (a) $y = \tan^{-1}\left(\frac{1+x}{1-x}\right)$; (b) $y = \sin^{-1}\left(\frac{2x}{1+x^2}\right)$.

Ans. (a) $(-1)^{n-1} \cdot (n-1)! \sin^n\theta \cdot \sin n\theta$ where $\theta = \cot^{-1} x$; (b) same as (a).

2.2 LEIBNITZ'S THEOREM (RULE or FORMULA)

Let $u(x)$ and $v(x)$ be two functions of x having derivatives of nth order. Then the nth derivative of the product of these two functions is

$$\frac{d^n}{dx^n}\{u(x)v(x)\} = (uv)_n = u_n v_0 + n_{c_1} u_{n-1} v_1 + n_{c_2} u_{n-2} v_2 + \cdots + n_{c_r} u_{n-r} v_r + \cdots + n_{c_n} u_0 v_n.$$

Proof by mathematical induction:
By direct differentiation, we have

$$\frac{d}{dx}(uv) = (uv)_1 = \frac{du}{dx}v + u\frac{dv}{dx} = u_1 v_0 + u_0 v_1$$

$$\frac{d^2}{dx^2}(uv) = (uv)_2 = u_2 v_0 + u_1 v_1 + u_1 v_1 + u_0 v_2$$
$$= u_2 v_0 + 2u_1 v_1 + u_0 v_2.$$

Here the subscripts indicate the orders of derivative and zero indices in the end terms indicate the functions themselves (i.e., derivatives of zero order). Thus

$$(uv)_3 = u_3 v_0 + u_2 v_1 + 2u_2 v_1 + 2u_1 v_2 + u_1 v_2 + u_0 v_3$$
$$= u_3 v_0 + 3u_2 v_1 + 3u_1 v_2 + u_0 v_3.$$

Similarly,

$$(uv)_4 = u_4 v_0 + 4u_3 v_1 + 6u_2 v_2 + 4u_1 v_3 + u_0 v_4.$$

Assume that the Leibnitz's theorem is valid for k. Then

$$(uv)_k = u_k v_0 + k_{c_1} u_{k-1} v_1 + k_{c_2} u_{k-2} v_2 + \cdots + kc_{r-1} u_{k-r+1} v_{r-1} + k_{c_r} u_{k-r} v_r + \cdots + kc_k u_0 v_k.$$

Differentiating the above both sides

$$(uv)_{k+1} = u_{k+1} v_0 + (u_k v_1 + k_{c_1} u_k \cdot v_1) + (k_{c_1} u_{k-1} v_2 + k_{c_2} u_{k-1} v_2) + k_{c_2} u_{k-2} v_3 + \cdots + kc_{r-1} u_{k-r+2} v_{r-1} + (k_{c_{r-1}} u_{k-r+1} v_r + k_{c_r} u_{k-r+1} v_r) + kc_r u_{k-r} v_{r+1} + \cdots + k_{c_k} u_1 v_k + kc_k u_0 v_{k+1}.$$

We know that

$$k_{c_{r-1}} + k_{c_r} = (k+1)\,c_r,$$
$$kc_k = (k+1)c_{(k+1)} = 1$$
$$1 + kc_1 = 1 + k = (k+1)c_1.$$

Therefore

$$(uv)_{k+1} = u_{k+1}v_0 + (k+1)c_1 u_k v_1 + (k+1)c_2 u_{k-1} v_2$$
$$+ \cdots + (k+1)c_r u_{k-r+1} v_r$$
$$+ \cdots + (k+1)c(k+1)u_0 v_{k+1}$$

i.e., theorem is valid for $k+1$ also.

Hence by mathematical induction the result is valid for any n.

Note: The Leibnitz's rule (or formula) is obtained by expanding $(u+v)^n$ by the binomial theorem and in the expansion obtained the exponents of the powers of u and v are replaced by (subscripts) indices that are the orders of the derivatives.

WORKED OUT EXAMPLES

Example 1: Find the nth derivative of $y = x^2 \sin x$ at $x = 0$.

Solution: $y = x^2 \sin x = \sin x \cdot x^2 = u(x)v(x)$. Applying Leibnitz's rule with $u = \sin x$, $v = x^2$

$$y_n = \frac{d^n}{dx^n}\sin x \cdot x^2 + n_{c_1}\frac{d^{n-1}}{dx^{n-1}}(\sin x)\cdot 2x$$
$$+ n_{c_2}\cdot\frac{d^{n-2}}{dx^{n-2}}(\sin x)\cdot 2 + 0.$$

We know that

$$\frac{d^n}{dx^n}\{\sin(ax+b)\} = a^n \sin\left(ax+b+\frac{n\pi}{2}\right)$$

$$y_n = x^2\cdot\sin\left(x+\frac{n\pi}{2}\right) + 2nx\cdot\sin\left(x+\frac{n\pi}{2}-\frac{\pi}{2}\right) +$$
$$+ n(n-1)\sin\left(x+\frac{n\pi}{2}-\pi\right)$$

$$y_n = (x^2 - n^2 + n)\sin\left(x+\frac{n\pi}{2}\right) - 2nx\cdot\cos\left(x+\frac{n\pi}{2}\right)$$

$$y_n(\text{at } x = 0) = (n - n^2)\sin\left(\frac{n\pi}{2}\right).$$

Example 2: Determine the nth derivative of $y = e^x \ln x$.

Solution: Applying Leibnitz's rule

$$y_n(e^x)_n\cdot\ln x + n_{c_1}(e^x)_{n-1}(\ln x)_1$$
$$+ n_{c_2}(e^x)_{n-2}(\ln x)_2 + \cdots + nc_n(e^x)(\ln x)_n.$$

Since $$\frac{d^n}{dx^n}ln(ax+b) = \frac{(-1)^{n-1}(n-1)!a^n}{(ax+b)^n}$$

and $$\frac{d^n}{dx^n}e^{mx} = m^n e^{mx}$$

$$y_n = e^x\left[\ln x + nc_1\frac{1}{x} - nc_2 x^{-2} + 2!n_{c_3}x^{-3} + \cdots\right.$$
$$\left. + (-1)^{n-1}(n-1)!nc_n x^{-n}\right].$$

Example 3: If $y = a\cos(\ln x) + b\sin(\ln x)$, prove that $x^2 y_{n+2} + (2n+1)xy_{n+1} + (n^2+1)y_n = 0$.

Solution: Differentiating y w.r.t. x,

$$y_1 = -a\cdot\sin(\ln x)\cdot\left(\frac{1}{x}\right) + b\cdot\cos(\ln x)\cdot\frac{1}{x}$$

$$xy_1 = -a\sin(\ln x) + b\cos(\ln x).$$

Differentiating

$$xy_2 + y_1 = -a\cdot\cos(\ln x)\cdot\frac{1}{x} - b\sin(\ln x)\cdot\frac{1}{x}$$

$$x^2y_2 + xy_1 = -[a\cos(\ln x) + b\sin(\ln x)] = -y$$

or $$x^2y_2 + xy_1 + y = 0.$$

Differentiating n times using Leibnitz's rule, we get

$$\left[x^2y_{n+2} + nc_1\cdot 2x\cdot y_{n+1} + nc_2\cdot 2\cdot y_n\right] +$$
$$+\left[xy_{n+1} + nc_1\cdot 1\cdot y_n\right] + y_n = 0.$$

Rewriting

$$x^2y_{n+2} + [2nx + x]y_{n+1} + [n(n-1) + n + 1]y_n = 0.$$

Example 4: Find the nth derivative of $y = x^{n-1}\cdot\ln x$ at $x = \frac{1}{2}$.

Solution: Differentiating

$$y_1 = (n-1)x^{n-2}\cdot\ln x + x^{n-1}\cdot\frac{1}{x}$$

$$xy_1 = (n-1)x^{n-1}\ln x + x^{n-1}$$

$$xy_1 = (n-1)y + x^{n-1}.$$

Differentiating $(n-1)$ times using Leibnitz's rule

$$[xy_n + (n-1)_{c_1}\cdot 1\cdot y_{n-1}] = (n-1)y_{n-1} + (n-1)!$$

since $$\frac{d^n(ax+b)^m}{dx^n} = \frac{m!}{(m-n)!}a^n(ax+b)^{m-n}$$

or $\quad xy_n = (n-1)! \quad$ i.e., $\quad y_n = \frac{(n-1)!}{x}$

at $\quad x = \frac{1}{2}, \quad y_n\left(\frac{1}{2}\right) = 2(n-1)!.$

Example 5: Determine $y_n(0)$ where $y = e^{m\cdot\cos^{-1}x}$.

Solution: Differentiating y w.r.t. x,

$$y_1 = e^{m\cdot\cos^{-1}x}\cdot m\cdot\left(\frac{-1}{\sqrt{1-x^2}}\right) = -\frac{m}{\sqrt{1-x^2}}\cdot y.$$

Squaring on both sides

$$(1-x^2)y_1^2 = m^2y^2.$$

Differentiating again

$$(1-x^2)2y_1y_2 - 2xy_1^2 = m^2\cdot 2yy_1$$

or
$$(1-x^2)y_2 - xy_1 = m^2y.$$

Using Leibnitz's rule, differentiate both sides n times

$$\left[(1-x^2)y_{n+2} + n_{c_1}(-2x)y_{n+1} + n_{c_2}(-2)y_n\right] -$$
$$-\left[xy_{n+1} + nc_1\cdot 1\cdot y_n\right] = m^2y_n$$

or

$$(1-x^2)y_{n+2} - (2n+1)xy_{n+1} - (n^2+m^2)y_n = 0.$$

Put $x = 0$, then

$$y_{n+2}(0) = (n^2+m^2)y_n(0).$$

At $\quad x = 0, \quad y(0) = e^{\frac{m\pi}{2}}$

$$y_1(0) = -me^{\frac{m\pi}{2}}$$

For $n = 0$, $\quad y_2(0) = (0^2+m^2)y_0(0) = m^2e^{\frac{m\pi}{2}}$

For $n = 1$, $\quad y_3(0) = (1^2+m^2)y_1(0)$

$$= (1^2+m^2)(-me^{\frac{m\pi}{2}})$$

For $n = 2$, $\quad y_4(0) = (2^2+m^2)y_2(0)$

$$= m^2(2^2+m^2)e^{\frac{m\pi}{2}}$$

For $n = 3$, $\quad y_5(0) = (3^2+m^2)y_3(0)$

$$= -m(1^2+m^2)(3^2+m^2)e^{\frac{m\pi}{2}}$$

In general, $\quad y_{2n}(0) = m^2(2^2+m^2)(4^2+m^2)\cdots\times$

$$\times\left((2n-2)^2+m^2\right)\cdot e^{\frac{m\pi}{2}}$$

$$y_{2n+1}(0) = -m(1^2+m^2)(3^2+m^2)\cdots\times$$
$$\times\left((2n-1)^2+m^2\right)e^{\frac{m\pi}{2}}.$$

Example 6: If $y = \sin(m\sin^{-1}x)$, prove that

$$(1-x^2)y_{n+2} - (2n+1)xy_{n+1} - (n^2-m^2)y_n = 0.$$

(UPTU 2002)

Solution: Differentiating y w.r.t. x,

$$y_1 = \cos(m\sin^{-1}x)\cdot m\cdot\frac{1}{\sqrt{1-x^2}}$$

or $\quad \sqrt{1-x^2}y_1 = m\cdot\cos(m\sin^{-1}x).$

Differentiating again w.r.t. x,

$$\sqrt{1-x^2}y_2 + \frac{1}{2}\frac{1}{\sqrt{1-x^2}}\cdot(-2x)\cdot y_1$$
$$= -m\sin(m\sin^{-1}x)\cdot m\cdot\frac{1}{\sqrt{1-x^2}}$$

or $\quad (1-x^2)y_2 - xy_1 + m^2y = 0.$

Differentiating n times by Leibnitz's rule

$$\left[(1-x^2)y_{n+2} + n\cdot(-2x)y_{n+1} + \frac{n(n-1)}{2}(-2)y_n\right] -$$
$$-\left[xy_{n+1} + n\cdot 1\cdot y_n\right] + m^2y_n = 0.$$

Collecting the terms

$$(1-x^2)y_{n+2} - (2n+1)y_{n+1} - (n^2-m^2)y_n = 0.$$

Exercise

1. Find the nth derivative of (a) $y = e^{ax}x^2$; (b) $y = \ln(1+x)$; (c) $y = \frac{1-x}{1+x}$; (d) $x\sin x$; (e) $e^x(2x+3)^3$; (f) $x^2e^x\cos x$.

Ans. **a.** $y^{(n)} = e^{ax}\left[a^nx^2 + 2na^{n-1}x + n(n-1)a^{n-2}\right]$

b. $(-1)^{n-1}(n-1)!/(1+x)^n$

c. $2(-1)^nn!/(1+x)^{n+1}$

d. $x\sin\left(x+n\frac{\pi}{2}\right) - n\cos\left(x+n\frac{\pi}{2}\right)$

e. $e^x\left\{(2x+3)^2 + 6n(2x+3)^2 + 12(n-1)(2x+3) + 8n(n-1)(n-2)\right\}$

f. $2^{\frac{(n-2)}{2}}e^x\left[2x^2\cos\left(x+n\frac{\pi}{4}\right) + 2^{\frac{3}{2}}nx\cdot\cos\left(x+\overline{n-1}\frac{\pi}{4}\right) + n(n-1)\cos\left(x+\overline{n-2}\frac{\pi}{4}\right)\right].$

2. If $y=\tan^{-1}\left(\frac{1+x}{1-x}\right)$, show that the nth derivative $y_n=(-1)^{n-1}(n-1)!\sin^n\theta\cdot\sin n\theta$ where $\theta=\cot^{-1}x$.

   Hint: $y_n=\frac{1}{2i}\left[\frac{1}{x-i}-\frac{1}{x+i}\right]_{n-1}=\frac{(-1)^{n-1}(n-1)!}{2i}\times$ $\times\left[\frac{1}{(x-i)^n}-\frac{1}{(x+i)^n}\right]$, put $x=r\cos\theta$, $1=r\sin\theta$, $x=\cot\theta$.

3. If $y=e^x\sin x$, prove that $y''-2y'+2y=0$.

4. Using Leibnitz's rule, differentiate n times, the Chebyshev D.E.

$$(1-x^2)y''-xy'+a^2y=0.$$

Ans. $(1-x^2)y_{n+2}-(2n+1)xy_{n+1}+(a^2-n^2)y_n=0$

5. If $\cos^{-1}\left(\frac{y}{b}\right)=\log\left(\frac{x}{n}\right)^n$, prove that

$$x^2y_{n+2}+(2n+1)xy_{n+1}+2n^2y_n=0$$

6. Find $y_n(0)$ if $y=(\sin^{-1}x)^2$.

   Hint: Put $x=0$ in $(1-x^2)y_{n+2}-$ $-(2n+1)xy_{n+1}-n^2y_n=0$.

Ans. $y_n(0)=0$ when n is odd and $y_n(0)=2\cdot 2^2\cdot 4^2\cdot 6^2\cdots(n-2)^2$ when n is even and $n\neq 2$.

7. Determine $y_n(0)$ if $y=e^{m\sin^{-1}x}$.

   Hint: Put $x=0$ in

$$(1-x^2)y_{n+2}-(2n+1)xy_{n+1}-(n^2+m^2)y_n=0.$$

Ans.

$$y_n(0)=\begin{cases} m^2(2^2+m^2)(4^2+m^2)\cdots((n-2)^2+m^2), & n \text{ even}\\ m(1^2+m^2)(3^2+m^2)\cdots((n-2)^2+m^2), & n \text{ odd}\end{cases}$$

8. If $y^{\frac{1}{m}}+y^{-\frac{1}{m}}=2x$, prove that

$$(x^2-1)y_{n+2}+(2n+1)xy_{n+1}+(n^2-m^2)y_n=0.$$

9. Find $y_n(0)$ when $y=\cos(m\sin^{-1}x)$.

   Hint: Put $x=0$ in

$$(1-x^2)y_{n+2}-(2n+1)xy_{n+1}+(m^2-n^2)y_n=0.$$

Ans. $y_n(0)=0$ if n is odd

$y_n(0)=m^2(2^2-m^2)(4^2-m^2)\cdots((n-2)^2-m^2)$ if n is even.

10. If $\sin^{-1}y=2\log(x+1)$, show that

$$(x+1)^2y_{n+2}+(2n+1)(x+1)y_{n+1}+(n^2+4)y_n=0.$$

2.3 ANGLE BETWEEN THE RADIUS VECTOR AND THE TANGENT

Let $r=f(\theta)$ be the equation of the curve in polar coordinates (r,θ). The formulas for changing from polar coordinates to the rectangular cartesian coordinates are

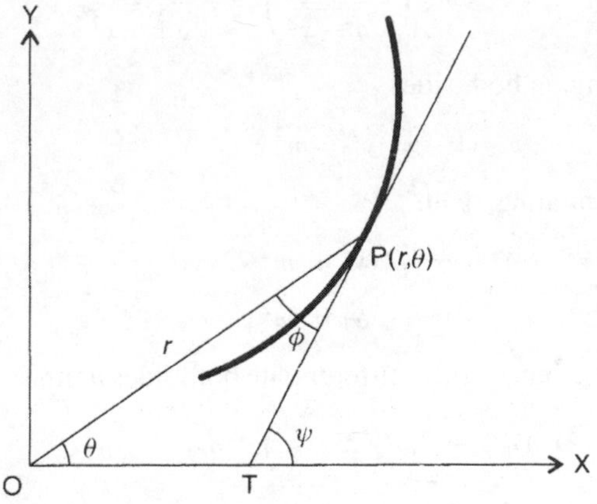

Fig. 2.1

$$x=r\cos\theta=f(\theta)\cos\theta \tag{1}$$

$$y=r\sin\theta=f(\theta)\sin\theta \tag{2}$$

These are the parametric equations of the given curve in terms of the parameter θ which is the polar angle. Let PT be the tangent to the curve at a point $P(r,\theta)$. Let ψ be the angle between the tangent PT and the positive x-axis. Let ϕ be the angle between the radius vector OP and the tangent PT i.e., $\angle TPO=\phi$. We know that

$$\text{slope of the tangent } PT=\tan\psi=\frac{dy}{dx}=\frac{\frac{dy}{d\theta}}{\frac{dx}{d\theta}} \tag{3}$$

Differentiating (1) and (2) w.r.t. the parameter θ

$$\frac{dx}{d\theta}=\frac{df}{d\theta}\cdot\cos\theta-f\sin\theta=f'\cos\theta-f\sin\theta \tag{4}$$

$$\frac{dy}{d\theta}=\frac{df}{d\theta}\cdot\sin\theta+f\cos\theta=f'\sin\theta+f\cos\theta \tag{5}$$

where $'$ dash denotes differentiation w.r.t. θ. Substituting (4) and (5) in (3), we have

$$\boxed{\tan\psi=\frac{f'\sin\theta+f\cos\theta}{f'\cos\theta-f\sin\theta}} \tag{6}$$

From the triangle OPT,

$$\theta + \phi + \angle OTP = \pi$$

$$\theta + \phi + (\pi - \psi) = \pi$$

$$\therefore \quad \phi = \psi - \theta \qquad (7)$$

From (6)

$$\tan\phi = \tan(\psi - \theta) = \frac{\tan\psi - \tan\theta}{1 + \tan\psi \cdot \tan\theta} \qquad (8)$$

Substitute values of $\tan\psi$ from (6) into (8)

$$\tan\phi = \frac{\frac{f'\sin\theta + f\cos\theta}{f'\cos\theta - f\sin\theta} - \tan\theta}{1 + \frac{f'\sin\theta + f\cos\theta}{f'\cos\theta - f\sin\theta}\cdot\tan\theta}$$

$$\tan\phi = \frac{(f'\sin\theta + f\cos\theta)\cos\theta - (f'\cos\theta - f\sin\theta)\sin\theta}{(f'\cos\theta - f\sin\theta)\cos\theta + (f'\sin\theta + f\cos\theta)\cdot\sin\theta}$$

$$= \frac{f(\cos^2\theta + \sin^2\theta)}{f'(\cos^2\theta + \sin^2\theta)} = \frac{f}{f'} = \frac{r}{\frac{dr}{d\theta}}$$

$$\therefore \quad \boxed{\tan\phi = r\frac{d\theta}{dr}} \qquad (9)$$

or

$$\frac{dr}{d\theta} = r\cot\phi.$$

Geometric meaning: Thus the derivative of the radius vector r w.r.t. the polar angle θ is equal to the length of the radius vector multiplied by the cotangent of the angle between the radius vector OP and the tangent PT to the curve at the point $P(r, \theta)$.

Corollary 1: Slope of the tangent PT:
For a given curve $r = f(\theta)$, calculate $\frac{dr}{d\theta} = f'(\theta)$ and substituting $r, \theta, f'(\theta)$ in (6), we get the slope of the tangent $\tan\psi$.

Corollary 2: Angle of intersection of two curves c_1 and c_2. Let PT_1 and PT_2 be the tangents to the two curves c_1 and c_2 respectively at the common point of intersection. Let ϕ_1 and ϕ_2 be the angles between the radius vector OP and tangents PT_1 and OP and PT_2 respectively.
Angle of intersection of two curves $= \angle T_1PT_2 = \alpha$

$$\alpha = \angle OPT_2 - \angle OPT_1 = \phi_2 - \phi_1 \qquad (10)$$

Corollary 3: The curves are said to intersect at right angle or intersect orthogonally if the angle between them $\alpha = 90^\circ$.

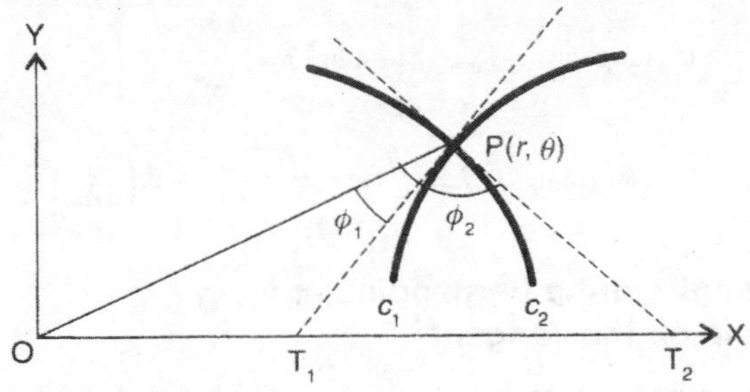

Fig. 2.2

Polar Subtangent and Tangent and Polar Subnormal and Normal

Let NOT be a straight line through the pole O and perpendicular to the radius vector OP. The tangent at P and the normal at P meets this line in T and N respectively. Then OT and ON are known as the polar subtangent and polar subnormal respectively,

$$\text{polar subtangent} = OT = r\cdot\tan\phi = r\cdot\frac{r\,d\theta}{dr} = r^2\frac{d\theta}{dr} \qquad (11)$$

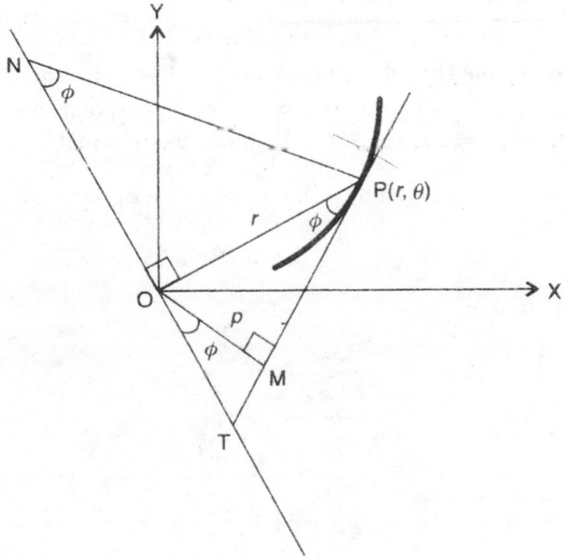

Fig. 2.3

where (9) is used to replace $\tan\phi$.

$$\text{Polar subnormal} = ON = r\cdot\cot\phi = r\cdot\frac{1}{r}\frac{dr}{d\theta} = \frac{dr}{d\theta} \qquad (12)$$

$$\text{Polar tangent}=PT=r\sqrt{1+\tan^2\phi}=r\sqrt{1+r^2\left(\frac{d\theta}{dr}\right)^2}$$

$$\text{Polar normal}=PN=r\sqrt{1+\cot^2\phi}=\sqrt{r^2+\left(\frac{dr}{d\theta}\right)^2}.$$

Length of the Perpendicular from Pole on the Tangent

Let p be the length of the perpendicular OM drawn from the pole on to the tangent PT. Then from the right angle triangle OPM

$$p = r\sin\phi.$$

Rewriting

$$\frac{1}{p^2} = \frac{1}{r^2\sin^2\phi} = \frac{1}{r^2}\text{cosec}^2\phi = \frac{1}{r^2}(1+\cot^2\phi)$$

$$\frac{1}{p^2} = \frac{1}{r^2}\left[1+\frac{1}{r^2}\left(\frac{dr}{d\theta}\right)^2\right] = \frac{1}{r^2}+\frac{1}{r^4}\left(\frac{dr}{d\theta}\right)^2$$

where (9) is used to replace $\cot\phi$.

Worked Out Examples

Example 1: For the parabola $\frac{2a}{r} = 1-\cos\theta$, show that (i) $\phi = \pi - \frac{\theta}{2}$; (ii) $p = a\text{cosec}\frac{\theta}{2}$; (iii) polar subtangent $= 2a\text{cosec}\theta$; (iv) Find polar subnormal.

Solution: Differentiating the equation of parabola

$$\frac{2a}{r} = 1-\cos\theta$$

w.r.t. θ, we get

$$-\frac{2a}{r^2} = \sin\theta\frac{d\theta}{dr}$$

or

$$\frac{d\theta}{dr} = -\frac{2a}{r^2\sin\theta}$$

i. Now $\tan\phi = r\frac{d\theta}{dr} = r\left(-\frac{2a}{r^2\sin\theta}\right) = -\frac{2a}{r}\cdot\frac{1}{\sin\theta}$

$$= -\frac{(1-\cos\theta)}{\sin\theta} = -\frac{2\sin^2\frac{\theta}{2}}{2\sin\frac{\theta}{2}\cdot\cos\frac{\theta}{2}} = -\tan\frac{\theta}{2}$$

$$\tan\phi = \tan\left(\pi-\frac{\theta}{2}\right)$$

$$\therefore \quad \phi = \pi - \frac{\theta}{2}$$

ii. $p = r\sin\phi = \frac{2a}{1-\cos\theta}\cdot\sin\left(\pi-\frac{\theta}{2}\right) = \frac{2a\sin\frac{\theta}{2}}{1-\cos\theta}$

$$= \frac{2a\sin\frac{\theta}{2}}{2\sin^2\frac{\theta}{2}} = a\text{ cosec }\frac{\theta}{2}$$

iii. Polar subtangent $= r^2\frac{d\theta}{dr} = -\frac{2a}{\sin\theta} =$

$$= \left|-2a\text{ cosec }\theta\right| = 2a\text{ cosec }\theta$$

iv. Polar subnormal $= \frac{dr}{d\theta} = \frac{r^2\sin\theta}{-2a} =$

$$= \frac{4a^2}{(1-\cos\theta)^2}\cdot\frac{\sin\theta}{-2a} = -\frac{2a\cdot 2\sin\frac{\theta}{2}\cdot\cos\frac{\theta}{2}}{4\sin^2\frac{\theta}{2}} = -a\cot\frac{\theta}{2}.$$

Example 2: Show that the curves $r = ae^\theta$ and $re^\theta = b$ intersect at right angles.

Solution: Point of intersection of the two curves

$$ae^\theta = r = be^{-\theta} \quad \text{or} \quad e^{2\theta} = \frac{b}{a}.$$

So
$$\theta = \frac{1}{2}\ln\left(\frac{b}{a}\right).$$

For the curve $r = ae^\theta$, $\quad \frac{dr}{d\theta} = ae^\theta \quad$ so

$$\tan\phi_1 = r\frac{d\theta}{dr} = (ae^\theta)\cdot a^{-1}e^{-\theta} = 1 \quad \therefore \quad \phi_1 = \frac{\pi}{4}.$$

For the curve $re^\theta = b$, $\quad \frac{dr}{d\theta} = -be^{-\theta} \quad$ so

$$\tan\phi_2 = r\frac{d\theta}{dr} = be^{-\theta}\cdot\frac{1}{-b}e^\theta = -1 \quad \therefore \quad \phi_2 = \pi - \frac{\pi}{4}.$$

Angle of intersection of the two curves is

$$\phi_2 - \phi_1 = \left(\pi - \frac{\pi}{4}\right) - \left(\frac{\pi}{4}\right) = \pi - \frac{\pi}{2} = \frac{\pi}{2}$$

i.e., they cut orthogonally.

Example 3: Find the angle of intersection of the curves $r = a$ and $r = 2a\cos\theta$.

Solution: For the curve $r = a$, $\quad \frac{dr}{d\theta} = 0 \quad$ so

$$\tan\phi_1 = r\frac{d\theta}{dr} = \infty \quad \therefore \quad \phi_1 = \frac{\pi}{2}$$

for the curve $r = 2a\cos\theta$, $\quad \frac{dr}{d\theta} = -2a\sin\theta$ so

$$\tan\phi_2 = r\frac{d\theta}{dr} = r\cdot\frac{1}{-2a\sin\theta} = \frac{2a\cos\theta}{-2a\sin\theta} = -\cot\theta$$

Points of intersection of the two curves are

$$a = r = 2a\cos\theta \quad \therefore \quad \cos\theta = \frac{1}{2}.$$

So, $\theta = \frac{\pi}{3}, \quad \frac{5\pi}{3}.$

Hence the points of intersection are $\left(a, \frac{\pi}{3}\right)$ and $\left(a, \frac{5\pi}{3}\right)$.

At $\theta = \frac{\pi}{3}$, $\tan\phi_1 = \infty$, so $\phi_1 = \frac{\pi}{2}$

$$\tan\phi_2 = -\cot\frac{\pi}{3} = -\frac{1}{\sqrt{3}}$$

Then $\phi_2 = \frac{\pi}{2} + \theta = \frac{\pi}{2} + \frac{\pi}{3}$

Angle of intersection is $\frac{\pi}{2} + \frac{\pi}{3} - \frac{\pi}{2} = \frac{\pi}{3}$.

Example 4: Prove that (i) for the curve $r = a\theta$, the polar subnormal is constant; (ii) for the curve $r\theta = a$ the polar subtangent is constant.

Solution:

i. For $r = a\theta$, $\frac{dr}{d\theta} = a =$ constant

Polar subnormal $= \frac{dr}{d\theta} = a =$ constant.

ii. For $r\theta = a$, $\frac{dr}{d\theta} = -\frac{a}{\theta^2}$

Polar subtangent $= r^2\frac{d\theta}{dr} = \frac{r^2\theta^2}{-a} = \frac{a^2}{-a} = -a =$ constant.

Example 5: For the curve $r^3 = a^3\cos 3\theta$, show that the normal at any point (r, θ) to the curve makes an angle 4θ with the initial line.

Solution: Differentiating w.r.t. r,

$$3r^2 = -3a^3\sin 3\theta \cdot \frac{d\theta}{dr}$$

$$\frac{d\theta}{dr} = \frac{-r^2}{a^3\sin 3\theta}.$$

Now $\tan\phi = r\frac{d\theta}{dr} = r\left(-\frac{r^2}{a^3\sin 3\theta}\right) = -\frac{a^3\cos 3\theta}{a^3\sin 3\theta}$

$$\tan\phi = -\cot 3\theta = \tan\left(\frac{\pi}{2} + 3\theta\right)$$

$$\therefore \quad \phi = \frac{\pi}{2} + 3\theta.$$

Slope of the tangent $\psi = \phi + \theta = \frac{\pi}{2} + 3\theta + \theta$

$\therefore$ Slope of the normal at any point (r, θ) is 4θ.

Example 6: Find the polar tangent and polar normal to the curve $\theta = \cos^{-1}\frac{r}{k} - \sqrt{\frac{k^2-r^2}{r^2}}$.

Solution: Differentiating w.r.t. r, we get

$$\frac{d\theta}{dr} = -\frac{1}{\sqrt{1-\left(\frac{r}{k}\right)^2}} \cdot \frac{1}{k} - \frac{1}{2}\sqrt{\frac{r^2}{k^2-r^2}}\,\frac{r^2(-2r)-(k^2-r^2)2r}{r^4}$$

$$= \frac{\sqrt{k^2-r^2}}{r^2}.$$

$$\text{Polar tangent} = r\sqrt{1 + r^2\left(\frac{d\theta}{dr}\right)^2}$$

$$= r\sqrt{1 + r^2\frac{(k^2-r^2)}{r^4}} = r\frac{k}{r} = k$$

$$\text{Polar normal} = \sqrt{r^2 + \left(\frac{dr}{d\theta}\right)^2} = \sqrt{r^2 + \frac{r^4}{k^2-r^2}}$$

$$= \sqrt{\frac{k^2r^2}{k^2-r^2}} = \frac{kr}{\sqrt{k^2-r^2}}.$$

EXERCISE

1. Show that, for the cardioid $r = a(1 - \cos\theta)$, (i) $\phi = \frac{\theta}{2}$; (ii) $p = 2a\sin^3\frac{\theta}{2}$; (iii) polar subtangent $= 2a\sin^2\frac{\theta}{2} \cdot \tan\frac{\theta}{2}$; (iv) polar subnormal $= a\sin\theta$.

2. Prove that the curves $r^m = a^m\cos m\theta$ and $r^m = a^m\sin m\theta$ intersects orthogonally.

   Hint: $r\frac{d\theta}{dr} = -\cot m\theta$, $\phi_1 = \frac{\pi}{2} + m\theta$, $r\frac{d\theta}{dr} = \tan m\theta$, $\phi_2 = m\theta$, $\phi_1 - \phi_2 = \frac{\pi}{2}$

3. Determine the angle of intersection of the curves $r = 3\cos\theta$, and $r = 1 + \cos\theta$.

   Hint: Points of intersection $\left(r = \frac{3}{2}, \theta = \frac{\pi}{3}\right)$, $\left(\frac{3}{2}, \frac{5\pi}{3}\right)$, $\tan\phi_1 = -\cot\frac{\pi}{3} = -\frac{1}{\sqrt{3}}$, $\tan\phi_2 = -\cot\frac{\theta}{2} = -\cot\frac{\pi}{6} = -\sqrt{3}$, $\tan(\phi_1 - \phi_2) = \frac{1}{\sqrt{3}}$, $\therefore \phi_1 - \phi_2 = \frac{\pi}{6}$

   Ans. $\frac{\pi}{6}$

4. Show that the angle of intersection of the curves $r = \sin\theta + \cos\theta$, $r = 2\sin\theta$ is $\frac{\pi}{4}$.

   Hint: $\theta = \frac{\pi}{4}$ is point of intersection, $\tan\phi_1 = \infty \therefore \phi_1 = \frac{\pi}{2}$, $\tan\phi_2 = 1$, $\therefore \phi_2 = \frac{\pi}{4}$, $\phi_1 - \phi_2 = \frac{\pi}{2} - \frac{\pi}{4} = \frac{\pi}{4}$.

5. Prove that the tangent to the logarithmic spiral $r = e^{a\theta}$ intersects the radius vector at a constant angle.

   Hint: $\frac{dr}{d\theta} = ae^{a\theta}$, $\cot\phi = \frac{1}{r}\frac{dr}{d\theta} = a \therefore \phi = \cot^{-1} a =$ const.

6. Prove that for the curve $r = ae^{m\theta^2}$, the ratio of polar subnormal to polar subtangent is proportional to θ^2.

7. Find ϕ for the following curves
   (a) $\frac{L}{r} = 1 + e\cos\theta$; (b) $r = a(1+\cos\theta)$; (c) $\frac{2a}{r} = 1 + \cos\theta$

Ans. (a) $\cot^{-1}\left(\frac{e\sin\theta}{1+e\cos\theta}\right)$; (b) $\frac{\pi}{2} + \frac{\theta}{2}$; (c) $\frac{\pi}{2} - \frac{\theta}{2}$.

8. Show that the following pair of curves cut orthogonally:
   (a) $\frac{2a}{r} = i + \cos\theta$, $\frac{2b}{r} = 1 - \cos\theta$; (b) $r = a\cos\theta$, $r = a\sin\theta$; (c) $r^2\sin 2\theta = a^2$, $r^2\cos 2\theta = b^2$; (d) $r = a(1+\cos\theta)$, $r = b(1-\cos\theta)$.

2.4 ROLLE'S THEOREM

If $f(x)$ is (i) *continuous in $[a, b]$* (ii) *derivable in (a, b) and* (iii) *$f(a) = f(b)$, then there exists at least one value $c \in (a, b)$ such that $f'(c) = 0$.*

Proof:

I. If $f(x) = 0$ for all x, then $f'(x) = 0$ for all x.

II. Since f is continuous in $[a, b]$, it is bounded and attains its maximum M and minimum m say at two numbers c and d lying in between a and b such that

$$f(c) = M \text{ and } f(d) = m.$$

Case a: If $M = m$, then f is a constant function so that f' is zero for all x in (a, b).

Case b: If $M \neq m$, then $M = f(c) \geq f(c+h)$ for values of h both positive and negative. Then

$$\frac{f(c+h) - f(c)}{h} \leq 0 \quad \text{for } h > 0 \tag{13}$$

and

$$\frac{f(c+h) - f(c)}{h} \geq 0 \quad \text{for } h < 0 \tag{14}$$

Since f is differentiable in (a, b) from (1) and (2) as $h \to 0$, we have

$$f'(c) \leq 0 \text{ and } f'(c) \geq 0$$

Hence $f'(c) = 0$ for some value c in (a, b). Similarly if $m = f(d) \leq f(d+h)$ for values of h both positive and negative it follows that

$$\frac{f(d+h) - f(d)}{h} \geq 0 \quad \text{for } h > 0$$

and

$$\frac{f(d+h) - f(d)}{h} \leq 0 \quad \text{for } h < 0$$

As $h \to 0$, $f'(d) \geq$ and $f'(d) \leq 0$. Hence $f'(d) = 0$.

Note 1: Geometrically, Rolle's theorem states that the tangents at c_1, c_2, c_3 are parallel to x-axis (see Fig. 2.4).

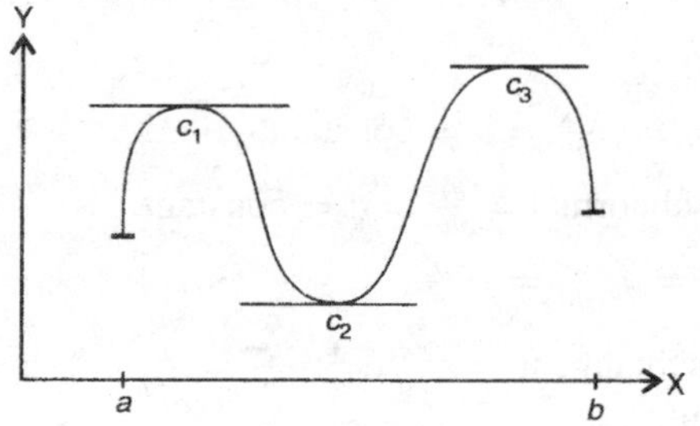

Fig. 2.4

Note 2: The function $y = f(x) = 1 - x^{\frac{2}{3}}$ in $[-1, 1]$ is not differentiable at origin O, it does not satisfy the condition of Rolle's theorem, so $f'(c) \neq 0$ for any $c \in (-1, 1)$ (refer Fig. 2.5).

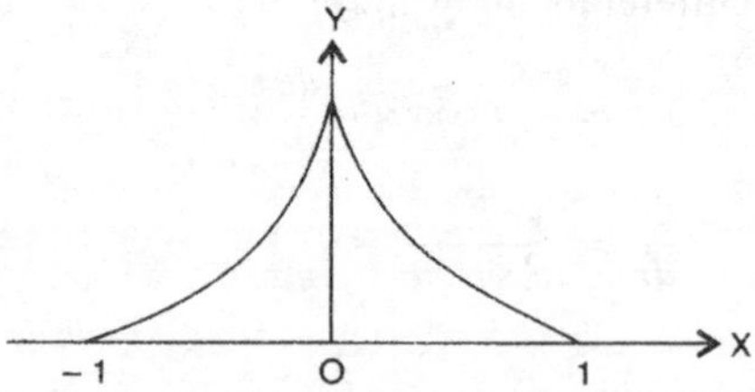

Fig. 2.5

Note 3: Alternate form: Let $b = a + h$ then $c = a + \theta h$ lies between a and b provided $0 < \theta < 1$. Thus the result of Rolle's theorem may be stated as

$$f'(c) = f(a + \theta h) = 0 \text{ with } 0 < \theta < 1.$$

Note 4: For $f(a) = f(b) = 0$, it follows from Rolle's theorem that there exists at least on $c \in (a, b)$ where $f'(c) = 0$. Thus the real root c of the equation $f'(x) = 0$ lies between a and b which are two adjacent real roots of the equation $f(x) = 0$, i.e., the real roots of equation $f'(x) = 0$ separate the real roots of the equation $f(x) = 0$.

Worked Out Examples

Verify Rolle's theorem for the following functions:

Example 1: $f(x) = x(x-2)e^{\frac{3x}{4}}$ in $(0, 2)$

Solution: $f(0) = 0$, $f(2) = 0$, f is continuous and differentiable, so by Rolle's theorem, $0 = f'(c)$. Here

$$f'(x) = [(x-2) + x + \frac{3}{4}(x)(x-2)]e^{\frac{3}{4}x}$$

$$= 0 \quad \text{or} \quad 3x^2 + 2x - 8 = 0$$

$\therefore c = -2$ or $\frac{8}{6}$ but $c = -2$ does not lie in $(0, 2)$ thus $c = \frac{8}{6} \in (0, 2)$.

Example 2: $f(x) = x^{2m-1}(a-x)^{2n}$ in $(0, a)$

Solution: $f(0) = f(a) = 0$. f is continuous and differentiable in $[0, a]$. By Rolle's theorem, $f'(c) = 0$ for some c in $(0, a)$.

$$\begin{aligned} f'(x) &= (2m-1)x^{2m-2}(a-x)^{2n} \\ &\quad + x^{2m-1} \cdot 2n(a-x)^{2n-1} \cdot (-1) \\ &= x^{2m-1} \cdot (a-x)^{2n}\left[(2m-1) \cdot \frac{1}{x} - \frac{2n}{a-x}\right] \end{aligned}$$

$$\therefore f'(c) = 0 \quad \text{if} \quad (2m-1)\frac{1}{c} - \frac{2n}{a-c} = 0$$

or
$$c = \frac{a(2m-1)}{(2m+2n-1)}.$$

Example 3: Find a root (solution) of the equation $x \ln x - 2 + x = 0$ lying in $(1, 2)$.

Solution: Choose $f(x) = (x-2) \ln x$ which is continuous, differentiable in $(1, 2)$ and further $f(1) = f(2) = 0$. Thus f satisfies all the 3 conditions of Rolle's theorem. So there exists a 'c' $\in (1, 2)$ such that $f'(c) = 0$. Here

$$f'(x) = \ln x + (x-2) \cdot \frac{1}{x}$$

$$f'(x) = \frac{x \ln x + x - 2}{x}$$

Thus $f'(c) = \frac{c \ln c + c - 2}{c} = 0$
or c is the root (solution) of the equation

$$x \ln x - 2 + x = 0.$$

Example 4: Show that the equation $f''(x) = 0$ has at least one real root between a and b if f, f', f'' are continuous in $a \le x \le b$ and the curve $y = f(x)$ crosses the x-axis at least at 3 distinct points between a and b inclusive.

Solution: Let c, d, e be the three points where $y = f(x)$ crosses the x-axis. Then $f(c) = f(d) = f(e) = 0$. Assume $a < c < d < e < b$. The function f satisfies Rolle's theorem in the two intervals (c, d) and (d, e), since f and f' are continuous and $f(c) = f(d) = f(e) = 0$. Therefore there exists at least one point in each interval (c, d) and (d, e) such that derivative is zero.
Let $c_1 \in (c, d)$ such that $f'(c_1) = 0$ and $c_2 \in (d, e)$ such that $f'(c_2) = 0$. Now the function f' satisfies Rolle's theorem because f', f'' are continuous and $f'(c_1) = f'(c_2) = 0$. Therefore by Rolle's theorem there exists a number c_3 in between c_1 and c_2 such that $f''(c_3) = 0$. Thus at least one root c_3 of the equation $f''(x) = 0$ lies in the interval (a, b).

Example 5: Without solving, show that the equation $x^4 + 2x^3 - 2 = 0$ has one and only one real root between 0 and 1.

Solution: Let $f(x) = x^4 + 2x^3 - 2$ which is continuous and differentiable in $(0, 1)$. $f'(x) = 4x^3 + 6x^2 = x^2(4x + 6) = 0$ only when $x = 0$ or $x = -\frac{6}{4}$ both of which do not lie in between 0 and 1. Thus $f'(x) \neq 0$ for any $x \in (0, 1)$. By Rolle's theorem, there do not exist a and b such that $f(a) = 0$ and $f(b) = 0$ i.e., the equation $f(x) = 0$ can not have two real roots. Further $f(0) = -2$ and $f(1) = 1$ are of opposite signs and $f'(x) > 0$ for every x in $(0, 1)$. Therefore $f(x)$ crosses the x-axis exactly once. In other words, $f(x) = 0$ has one and only one root in between 0 and 1.

EXERCISE

Verify Rolle's theorem for the following functions $f(x)$ in the indicated interval:

1. $\sin x/e^x$ in $[0, \pi]$

Ans. $c = \pi/4$

2. $x^3 - 12x$ in $[0, 2\sqrt{3}]$

Ans. $c = 2$

3. $\sin x$ in $[0, 2\pi]$

Ans. $\frac{\pi}{2}$ and $\frac{3\pi}{2}$

4. $|x|$ in $[-1, 1]$

Hint: Function is not differentiable at 0.

Ans. Rolle's theorem is not valid

5. $f(x) = \begin{cases} 1, \text{ when } x = 0 \\ x, \text{ when } 0 < x \le 1 \end{cases}$

Hint: Function is discontinuous at 0.

Ans. Rolle's theorem not applicable.

6. x^2 in $[1, 2]$

Hint: $f(1) = 1 \neq f(2) = 4$.

Ans. Rolle's theorem not applicable

7. $\ln [(x^2 + ab)/((a + b)x)]$ in $[a, b]$.

Ans. $c = \sqrt{ab}$

8. $2 + (x - 1)^{\frac{2}{3}}$ in $[0, 2]$

Hint: Not differentiable at $x = 1 \in (0, 2)$.

Ans. Rolle's theorem not applicable

9. $e^x(\sin x - \cos x)$ in $[\frac{\pi}{4}, \frac{5\pi}{4}]$

Ans. $c = \pi$

10. $2x^3 + x^2 - 4x - 2$ in $[-\sqrt{2}, \sqrt{2}]$

Ans. $\frac{2}{3}$ and -1

11. $(x - a)^m(x - b)^n$ in $[a, b]$ with $b > a$, $n > 1, m > 1$

Ans. $c = \frac{mb+na}{m+n}$

12. $x^3 - 4x$ in $[-2, 2]$

Ans. $\pm\sqrt{2/3}$

13. $x^3 - 3x^2 - x + 3$ in $[1, 3]$

Ans. $1 + 2/\sqrt{3}$

14. $\ln \{(x^2 + 6)/5x\}$ in $[2, 3]$

Ans. $c = \sqrt{6}$

15. $x(x + 3)e^{-x/2}$ in $[-3, 0]$

Ans. -2

16. $(x + 2)^3(x - 3)^4$ in $[-2, 3]$

Ans. $1/7$

17. $\tan x$ in $[0, \pi]$

Hint: $\tan x$ is discontinuous at $x = \pi/2$.

Ans. Rolle's theorem not applicable

18. $1 - (x - 3)^{\frac{2}{3}}$

Hint: $f(2) = f(4) = 0$, but f' does not exist at $x = 3 \in (2, 4)$.

Ans. Rolle's theorem not applicable

19. $\frac{x^2-4x}{x+2}$

Hint: The point of discontinuity $x = -2$ does not belong to $(0, 4)$ where $f(0) = f(4) = 0$.

Ans. $c = 2(\sqrt{3} - 1)$

20. Show that between any two roots of $e^x \cos x - 1 = 0$, there exists at least one root of $e^x \sin x - 1 = 0$.

Hint: Take $f(x) = e^x \cos x - 1$ in $[a, b]$ so $f' = e^x \cos x - e^x \sin x$. Since $e^x \cos x = 1$ so $f' = 1 - e^x \sin x$. Let a, b be two roots of $f(x)$. Then by Rolle's theorem $c \in (a, b) \exists$ $f'(c) = 0$ i.e., c is a root of $1 - e^x \sin x = 0$.

21. Show that polynomial equation

$$a_0x^n + a_1x^{n-1} + a_2x^{n-2} + \cdots + a_n = 0$$

has at least one real root in $(0, 1)$ if

$$\frac{a_0}{n+1} + \frac{a_1}{n} + \cdots + \frac{a_{n-1}}{2} + a_n = 0$$

and $a_0, a_1, \ldots, a_n$, are real numbers.

Hint: Take $f(x) = \frac{a_o}{n+1} x^{n+1} + \frac{a_1}{n} x^n + \cdots + \frac{a_{n-1}}{2} x^2 + a_nx$ in $[0, 1]$. Apply Rolle's theorem.

22. Prove that if a rational integral function $f(x)$ has n zeros between a and b then $f'(x)$ has

$(n-1)$ zeros in (a, b).

Hint: Suppose $x_1, x_2, \ldots x_n$ are the n zeros of f in (a, b) so $f(x_i) = 0$, for $i = 1, 2, \ldots n$. Then by Rolle's theorem applied to the $(n-1)$ intervals $(x_1, x_2), (x_2, x_3), \ldots (x_{n-1}, x_n) \exists c_i \ni f'(c_i) = 0$, for $i = 1, 2, \ldots n-1$.

Without solving exactly, show that the following equations have one and only one real root in the given interval.

23. $x^4 + 3x + 1 = 0$ in $(-2, -1)$.
24. $2x^3 - 3x^2 - 12x - 6 = 0$ in $(-1, 0)$.
25. If f and F are continuous in $[a, b]$ and derivable in (a, b) with $F'(x) \neq 0$ for every x in (a, b) then prove that there exists $c \in (a, b)$ such that

$$\frac{f'(c)}{F'(c)} = \frac{f(c) - f(a)}{F(b) - F(c)}.$$

Hint: Choose $\phi(x) = f(x)[F(b) - F(a)] - [f(x) - f(a)][F(x) - F(a)]$.
$\phi(b) = f(a)[F(b) - F(a)] = \phi(a)$,
ϕ satisfies Rolle's theorem. But
$\phi'(x) = f'(x)[F(b) - F(a)] - f'(x)[F(x) - F(a)] - F'(x)[f(x) - f(a)]$, $\phi'(c) = 0$.

2.5 LAGRANGE'S MEAN VALUE THEOREM

Let f be (i) continuous in $[a, b]$ and (ii) derivable in (a, b). Then there exists at least one value $c \in (a, b)$ such that

$$\frac{f(b) - f(a)}{b - a} = f'(c).$$

Proof: Choose $\phi(x) = f(x) + x \cdot A$, $x \in [a, b]$. Since f and x are continuous in $[a, b]$ and derivable in (a, b), therefore ϕ is continuous in $[a, b]$ and derivable in (a, b). Now choose the unknown constant A such that $f(b) + b \cdot A = \phi(b) = \phi(a) = f(a) + a \cdot A$ so that

$$A = \frac{f(b) - f(a)}{a - b}$$

Thus ϕ satisfies all the three conditions of the Rolle's theorem. Therefore by Rolle's theorem there exists a number $c \in (a, b)$ such that

$$0 = \phi'(c) = f'(c) + A$$

Thus $f'(c) = -A = \frac{f(b)-f(a)}{b-a}$.

Alternate Form of Lagrange's Mean Value Theorem

If $b = a + h$ then

$$\frac{f(a+h) - f(a)}{a + h - a} = f'(a + \theta h)$$

$c = a + \theta h$ lies between a and $b = (a + h)$ when $0 < \theta < 1$.

Thus $f(a+h) = f(a) + hf'(a + \theta h), 0 < \theta < 1$.

Note 1: Lagrange's Mean Value (LMV) theorem is generalization of Rolle's theorem. In the special case when $f(b) = f(a)$ then LMV theorem reduces to Rolle's theorem.

Note 2: Geometrically LMV theorem states that the tangent to the curve at D is parallel to the chord AB (see Fig. 2.6).

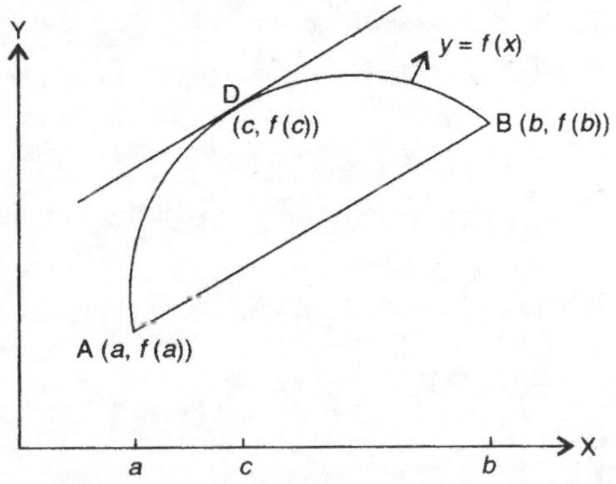

Fig. 2.6

Note 3: The average rate of change of f over the interval (a, b) given by $\frac{f(b)-f(a)}{b-a}$ is equal to $f'(c)$ which is the actual rate of change of f at some point of the interval (a, b).

Note 4: Using LMV theorem, approximate solution of equation $f(x) = 0$ can be obtained by Newton's method as follows:

Suppose $a + h$ is exact (actual) root so that

$$0 = f(a + h) = f(a) + h\, f'(a + \theta h), 0 < \theta < 1$$

Therefore $h \simeq -\frac{f(a)}{f'(a)}$.

Thus starting at a guess value 'a', h (correction) can be calculated approximately. By iteration a better root can be obtained.

Note 5: If M^* and m^* are the maximum and minimum of $f'(x)$ in $a < x < b$ then

$$m^*(b-a) < f(b) - f(a) < (b-a)M^*.$$

Note 6: Application of LMV theorem for sign of derivative.

Let $f(x)$ satisfy conditions of LMV theorem in $[a, b]$ and x_1, x_2 be any two points of $[a, b]$ such that $x_1 < x_2$. Applying LMV theorem in $[x_1, x_2]$, there exists a number $c \in (x_1, x_2)$ such that

$$f(x_2) - f(x_1) = (x_2 - x_1) f'(c) \quad (1)$$

I. If $f'(x) = 0$ for every $x \in (x_1, x_2)$ then

$$f(x_2) - f(x_1) = 0$$

or $f(x_2) = f(x_1) =$ constant
So f is a constant function in $[x, x_2]$.

II. If $f'(x) > 0$ then it follows from (1) that

$$f(x_2) - f(x_1) > 0 \text{ or } f(x_2) > f(x_1)$$

since $(x_2 - x_1) > 0$. So f is strictly increasing function.

III. If $f'(x) < 0$ then it follows from (1) that

$$f(x_2) - f(x_1) < 0 \text{ or } f(x_2) < f(x_1)$$

i.e., f is strictly decreasing function.

WORKED OUT EXAMPLES

Verify the Lagrange's Mean Value theorem:

Example 1: $f(x) = x^2$ in $(1, 5)$

Solution: f is continuous and differentiable in (2, 3) so by LMV theorem,

$$\frac{f(b) - f(a)}{b - a} = f'(c)$$

for some c in (a, b).
Here $f(5) = 25$, $f(1) = 1$, $f'(x) = 2x$. Therefore

$$\frac{25 - 1}{5 - 1} = 2c \quad \therefore c = 3 \in (1, 5).$$

Example 2: $f(x) = x^{\frac{2}{3}}, (-1, 1)$

Solution: LMV theorem is not applicable because f is not differentiable at $x = 0 \in (-1, 1)$.

Example 3: $f(x) = \cot \pi x, (-\frac{1}{2}, \frac{1}{2})$

Solution: LMV theorem is not valid because f is discontinuous at $0 \in (-\frac{1}{2}, \frac{1}{2})$.

Example 4: Deduce Lagrange's Mean Value theorem from Rolle's theorem.

Solution: Choose $g(x) = f(x) - f(a) - A(x - a)$ where A is a constant and f is continuous in $[a, b]$ and derivable in (a, b). Then $g(x)$ is also continuous in $[a, b]$ and derivable in (a, b). Further $g(a) = 0$ and determine A such that

$$g(b) = f(b) - f(a) - A(b - a) = 0$$

Solving $A = \dfrac{f(b) - f(a)}{b - a}$

Thus g satisfies all the conditions of Rolle's theorem. Therefore there exists a number $c \in (a, b)$ such that

$$g'(c) = 0$$

But $g'(x) = f'(x) - A$

so $0 = g'(c) = f'(c) - A$

$\therefore \quad f'(c) = A = \dfrac{f(b) - f(a)}{(b - a)}$

which is the result of Lagrange's Mean Value theorem.

Example 5: Show that $\frac{h}{1+h^2} < \tan^{-1} h < h$ when $h \neq 0$ and $h > 0$.

Solution: Take $f(x) = \tan^{-1} x$ in $0 \le x \le h$. By LMV theorem

$$\frac{\tan^{-1} h - \tan^{-1} 0}{h - 0} = \frac{1}{1 + c^2}$$

where $0 < c < h$.

or $\quad \tan^{-1} h = \frac{h}{1+c^2}$.

Since $0 < c < h \Rightarrow 0^2 < c^2 < h^2$

$$1 < 1 + c^2 < 1 + h^2$$

or $$h > \frac{h}{1+c^2} > \frac{h}{1+h^2}$$

or $$h > \frac{h}{1+c^2} = \tan^{-1} h > \frac{h}{1+h^2}.$$

Example 6: Separate the intervals in which the polynomial $f(x) = (4 - x^2)^2$ is increasing or decreasing.

Solution: $f'(x) = 2(4 - x^2)(-2x) = 4x(x^2 - 4)$. Since $f(2) = 0$ and $f(-2) = 0$, so $y = f(x)$ crosses x-axis at -2 and 2.

Note that

$f' > 0$ when $x > 0$ and $x > 2$ so f is increasing in $(2, \infty)$

$f' < 0$ when $x > 0$ and $x < 2$ so f is decreasing in $(0, 2)$

$f' > 0$ when $x < 0$ and $x > -2$ so f is increasing in $(-2, 0)$

$f' < 0$ when $x < 0$ and $x < -2$ so f is decreasing in $(-\infty, -2)$.

Example 7: Show that, for any $x \geq 0$

$$1 + x < e^x < 1 + xe^x.$$

Solution: Take $f(x) = e^x - (1 + x)$. Then $f'(x) = e^x - 1 \geq 0$ for any $x > 0$, so f is an increasing function and $f(0) = 0$. Therefore $f > 0$ for any $x \geq 0$

i.e., $$e^x - (1 + x) > 0 \text{ or } 1 + x < e^x \quad (1)$$

Similarly choose $g(x) = 1 + xe^x - e^x$.

$g(0) = 0$, and $g'(x) = e^x + xe^x - e^x = xe^x \geq 0$ for any $x \geq 0$. Thus g is an increasing function and therefore $g > 0$

i.e., $$1 + xe^x - e^x > 0$$

or $$e^x < 1 + xe^x \quad (2)$$

Results (1) and (2) form the required inequality.

Example 8: Show that for $x \in (0, 1)$

$$x < -\ln(1 - x) < x/(1 - x)$$

Solution: Rewrite the inequality as

$$-x > \ln(1 - x) > \frac{x}{x - 1} \quad (1)$$

or $$-1 > \frac{\ln(1 - x)}{x} > \frac{1}{x - 1}$$

Now choose $f(x) = \ln(1 - x)$ which is continu and differentiable in (0, 1). So applying LMV tl rem in $(0, x)$ for any x between 0 and 1.

$$\frac{\ln(1 - x) - \ln(1 - 0)}{x - 0} = f'(c) = -\frac{1}{1 - c}$$

$$= \frac{1}{x\theta - 1}$$

because c lies between 0 and x.

Here $0 < \theta < 1$.

Since $\theta x < x \Rightarrow \theta x - 1 < x - 1$

or $$\frac{1}{x\theta - 1} > \frac{1}{x - 1}$$

Also for $\theta > 0, x > 0, x\theta > 0$ or $-x\theta < 0$

$$1 - x\theta < 1$$

so $$\frac{1}{1 - x\theta} > 1$$

or $$\frac{1}{x\theta - 1} < -1$$

From (2), (3) and (4), we get the inequality (1)

i.e., $$-1 > \frac{1}{x\theta - 1} = \frac{\ln(1 - x)}{x} > \frac{1}{x - 1}.$$

Example 9: Calculate approximately the roo the equation $x^4 - 12x + 7 = 0$ near 2.

Solution: Choose $f(x) = x^4 - 12x + 7$

$f'(x) = 4x^3 - 12$

By LMV theorem, $f(a + h) = f(a) + hf'(a +$

so $$h \simeq -\frac{f(a)}{f'(a)}$$

$$f(2) = 16 - 24 + 7 = -1, f'(2) = 32 - 12 =$$

$$h \simeq -\frac{(-1)}{20} = +0.05$$

An approximate root $x = a + h = 2 + 0.05 = 2$

Applying LMV theorem again

$$h \simeq -\frac{f(2.05)}{f'(2.05)} = -\frac{0.061}{22.46}$$

$$= -0.0027$$

A better approximated root

$$= a + h = 2.05 - 0.0027$$

$$= 2.0473.$$

Example 10: Calculate approximately $\sqrt[5]{245}$ by using LMV theorem.

Solution: Choose $f(x) = x^{\frac{1}{5}}$, $a = 243$, $b = 245$ then $f'(x) = \frac{1}{5} \cdot x^{-\frac{4}{5}}$. By LMV theorem,

$$f(a+h) = f(a) + h \cdot f'(c)$$

Take $c = 243$ approximately

$$f(245) = f(243) + (245 - 243) \cdot f'(243)$$

$$(245)^{\frac{1}{5}} = (243)^{\frac{1}{5}} + 2\frac{1}{5}(243)^{-\frac{4}{5}}$$

$$= 3 + \frac{2}{5} \cdot \frac{1}{34} = 3.0049.$$

EXERCISE

Verify Lagrange's Mean Value theorem for the following functions $f(x)$ in the indicated interval:

1. $x(x-1)(x-2)$ in $(0, \frac{1}{2})$

Ans. $c = 0.236$

2. $\ln x$ in (e^2, e^3)

Ans. $c = (e-1)e^2$

3. $x^{\frac{1}{3}}$ in $(-1, 1)$

Hint: Derivative at $x = 0$ does not exist.

Ans. LMV theorem not applicable

4. $1 - 3x$ in $(1, 4)$

Hint: $\frac{f(4)-f(1)}{4-1} = -3 = f'(c) = -3$, true for any c.

Ans. any number in $(1, 4)$

5. e^{-x} in $(-1, 1)$

Ans. $c = \ln\left(\frac{2e}{e^2-1}\right) = -0.161$

6. $\sin^{-1} x$ in $(0, 1)$

Ans. $c = \frac{\sqrt{\pi^2-4}}{\pi} = 0.7712$

7. $\cos x$ in $(0, \frac{\pi}{2})$

Ans. $c = \sin^{-1}(2/\pi)$

8. If $f(x)$ is a quadratic expression, then the value of c of the Lagrange's Mean Value theorem in any interval $[a, a+h]$ is the mid-point of that interval.

Hint: Let $f(x) = Ax^2 + Bx + D$, then $c = a + \frac{h}{2} = a + \theta h$ so $\theta = \frac{1}{2}$. Use LMV theorem of the form

$$f(a+h) = f(a) = hf'(a+\theta h).$$

Separate the intervals in which the following function $f(x)$ is increasing or decreasing.

9. $2x^3 - 15x^2 + 36x + 1$

Ans. increasing in $(-\infty, 2)$ and $(3, \infty)$ decreasing in $(2, 3)$

10. $x^3 + 8x^2 + 5x - 2$

Ans. increasing in $(-\infty, -5)$ and $(-\frac{1}{3}, \infty)$ decreasing in $(-5, -\frac{1}{3})$

11. Show that $x - \frac{x^2}{2} < \ln(1+x) < x - \frac{x^2}{2(1+x)}$, for any $x > 0$.

Hint: Choose $f(x) = \ln(1+x) - (x - \frac{x^2}{2})$ and

$$g(x) = x - \frac{x^2}{2(1+x)} - \ln(1+x)$$

prove that f and g are increasing (i.e., $f' > 0$, $g' > 0$).

Note: $f(0) = g(0) = 0$, so $f > 0$ and $g > 0$.

12. Show that $\frac{\tan x}{x} > \frac{x}{\sin x}$ for $0 < x < \frac{\pi}{2}$.

Hint: Take $f(x) = \tan x \cdot \sin x - x^2$

$$f'' > 0,\ f'(0),\ \text{so } f' > 0$$

Also $f(0) = 0$, so $f > 0$.

13. Use LMV theorem to prove that if $0 < u < v < 1$

$$\frac{v-u}{1+v^2} < \tan^{-1} v - \tan^{-1} u < \frac{v-u}{1+u^2}$$

Deduce that $\frac{\pi}{4} + \frac{3}{25} < \tan^{-1}\frac{4}{3} < \frac{\pi}{4} + \frac{1}{6}$.

Hint: Take $f(x) = \tan^{-1} x$, $u < x < v$.

14. Show that for any x

$$x - \frac{x^3}{6} < \sin x < x - \frac{x^3}{6} + \frac{x^5}{120}.$$

15. For any x show that

$$1 - x < e^{-x} < 1 - x + \frac{x^2}{2}.$$

16. Using LMV theorem prove that

$$\frac{\pi}{3} - \frac{1}{5\sqrt{3}} > \cos^{-1}\frac{3}{5} > \frac{\pi}{3} - \frac{1}{8}.$$

17. Determine the root of the equation $x^3 + 5x - 10 = 0$ which lies in (1, 2) correct to two decimal places.

Ans. 1.43

18. Using Mean Value theorem calculate approximately $\sqrt[6]{65}$.

Hint: Take $f(x) = x^{\frac{1}{6}}$ in (64, 65).

Use $f(a+h) = f(a) + hf'(a)$.

Ans. 2.0052

19. A circular hole 4 inches in diameter and 1 foot deep in a metal block is drilled out to increase the diameter to 4.12 inches. Estimate the amount of metal removed.

Hint: $V = f(x) = \pi(12)x^2$inches.
By LMV theorem

$$f(2.06) - f(2) = \text{by MVT} = 0.06\ f'(x_1)$$
$$= 0.06(24\pi x_1)$$

where $2 < x_1 < 2.06$.

Ans. 2.88π in^3

20. Find an approximate value of the root of the equation $x^3 \quad 2x - 5 = 0$ in (2, 3).

Hint:

$$h = -\frac{f(2)}{f'(2)} = +0.1, \text{root} = 2 + 0.1$$

$$h = -\frac{f(2.1)}{f'(2.1)} = -0.00543$$

$$\text{root} = 2.1 - 0.00543 = 2.0946.$$

Ans. 2.0946

2.6 CAUCHY'S MEAN VALUE THEOREM

Let $f(x)$ and $g(x)$ be two functions which are both derivable in $[a, b]$ and $g'(x) \neq 0$ for any value of x in $[a, b]$. Then there exists at least one value c in between a and b such that

$$\frac{f(b) - f(a)}{g(b) - g(a)} = \frac{f'(c)}{g'(c)}$$

Proof: Define $\phi(x) = f(x) + Ag(x)$ where A is an unknown constant. ϕ is derivable in $[a, b]$ because f and g are derivable in $[a, b]$ by hypothesis. Choose A such that $\phi(b) = \phi(a)$ i.e.,

$$f(b) + Ag(b) = \phi(b) = \phi(a) = f(a) + Ag(a)$$

or $$A = \frac{f(b) - f(a)}{g(a) - g(b)}$$

with $g(a) - g(b) \neq 0$. If $g(a) - g(b) = 0$ then $g(a) = g(b)$ and g satisfies conditions of Rolle's theorem and then $g'(c) = 0$ for some c. This contradicts the hypothesis that $g'(x) \neq 0$ for any x. Thus $g(a) - g(b) \neq 0$. Now the new function ϕ satisfies the conditions of Rolle's theorem. Therefore there exists at least one $c \in (a, b)$ such that

$$0 = \phi'(c) = f'(c) + Ag'(c)$$

Thus

$$\frac{f'(c)}{g'(c)} = -A = \frac{f(b) - f(a)}{g(b) - g(a)}$$

with $g(b) - g(a) \neq 0$ as $g'(c) \neq 0$ for any c.

Alternate Form of Cauchy's Mean Value Theorem

Replacing b by $a + h$,

$$\frac{f'(a+\theta h)}{g'(a+\theta h)} = \frac{f(a+h) - f(a)}{g(a+h) - g(a)}, 0 < \theta < 1$$

Note: Cauchy's mean value theorem is a generalization of Lagrange's mean value theorem. When $g(x) = x$, CMV theorem reduces to LMV theorem.

WORKED OUT EXAMPLES

Example 1: Verify Cauchy's mean value (CMV) theorem for the functions

i. $f(x) = x^4, g(x) = x^2$ in the interval $[a, b]$
ii. $f(x) = \ln x, g(x) = \frac{1}{x}$ in $[1, e]$.

Solution: CMV theorem states that $\frac{f(b)-f(a)}{g(b)-g(a)} = \frac{f'(c)}{g'(c)}$

i. $f(x) = x^4, g(x) = x^2, [a, b]$

$f' = 4x^3, g' = 2x$

By CMV theorem,

$$\frac{b^4-a^4}{b^2-a^2}=\frac{f'(c)}{g'(c)}=\frac{4c^3}{2c}$$

$$\therefore c^2=\frac{1}{2}(a^2+b^2) \quad \therefore c=\frac{1}{\sqrt{2}}\sqrt{a^2+b^2}\in(a,b)$$

ii. $f'=\ln x,\ g(x)=\frac{1}{x},\ [1,e]$

$$f'=\frac{1}{x},\quad g'=-\frac{1}{x^2}$$

By CMV theorem,

$$\frac{\ln e-\ln 1}{\frac{1}{e}-1}=\frac{1}{c}\cdot(-c^2)=-c$$

$$\therefore c=\frac{e}{e-1}\in(1,e)$$

Example 2: If $0\le a<b<\pi/2$ show that

$$0<\cos a-\cos b<b-a.$$

Solution: Apply CMV theorem to functions $f(x)=\cos x$ and $g(x)=x$ in $[a,b]$. Then by CMV theorem

$$\frac{\cos b-\cos a}{b-a}=-\sin c$$

$$\cos a-\cos b=(b-a)\sin c$$

Since cos is decreasing in $(0,\pi/2)$, so $\cos a>\cos b$ for $a<b$. Thus $\cos a-\cos b>0$. Also in $(0,\pi/2)$ maximum value of $\sin x$ is 1. So since $c\in(a,b)$, and $\sin c<1$

so $\qquad (b-a)\sin c<(b-a)$

Thus

$$0<\cos a-\cos b<(b-a).$$

Example 3: Show that there exists a number $c\in(a,b)$ such that

$$2c\,[f(a)-f(b)]=f'(c)\cdot[a^2-b^2]$$

when f is continuous in $[a,b]$ and derivable in (a,b).

Solution: By applying CMV theorem to the two functions $f(x)$ and $g(x)=x^2$ in $[a,b]$,

$$\frac{f(b)-f(a)}{b^2-a^2}=\frac{f'(c)}{2c}$$

for some c in (a,b). Hence the result.

EXERCISE

Verify Cauchy's mean value theorem for the following pair of functions in the indicated interval:

1. $e^x,\ e^{-x}$, in the interval (a,b)

Ans. $c=\frac{a+b}{2}$

2. x^2, x in (a,b)

Ans. $c=\frac{a+b}{2}$

3. $\sin x,\ \cos x$ in (a,b)

Hint:

$$-\cot c=\frac{\cos c}{\sin c}=\frac{\sin b-\sin a}{\cos b-\cos a}$$

$$=\frac{2\sin\left(\frac{b-a}{2}\right)\cdot\cos\left(\frac{b+a}{2}\right)}{-2\sin\left(\frac{b-a}{2}\right)\cdot\sin\left(\frac{b+a}{2}\right)}$$

$$=-\cot\left(\frac{b+a}{2}\right)$$

Ans. $c=\frac{a+b}{2}$

4. $\sqrt{x},\ \frac{1}{\sqrt{x}}$ in (a,b) *Ans.* $c=\sqrt{ab}$

5. $\frac{1}{x^2},\ \frac{1}{x}$ in (a,b) *Ans.* $c=\frac{2ab}{a+b}$

6. $x^3,\ x^2$ in (a,b) *Ans.* $c=\frac{2(b^2+ab+a^2)}{3(a+b)}$

7. $2x+1,\ 3x-4$ in $(1,3)$

Ans. CMV theorem is satisfied for every point in the interval $(1,3)$

8. $x^3-3x^2+2x,\ x^3-5x^2+6x$ in $(0,\frac{1}{2})$

Ans. $c=\frac{5-\sqrt{13}}{6}$

9. $\frac{x^3}{4}-4x,\ x^2$ in $(0,3)$ *Ans.* $c=\frac{-1+\sqrt{37}}{3}$

10. $x^2,\ \sqrt{x}$ in $(1,4)$ *Ans.* $c=\left(\frac{15}{4}\right)^{\frac{2}{3}}$

11. Show that $\sin b-\sin a<b-a$ if $0<a<b<\frac{\pi}{2}$.

Hint: Take $f(x)=\sin x,\ g(x)=x$ in CMV theorem.

12. Show that $\sqrt{\frac{1-x}{1+x}}<\frac{\ln(1+x)}{\sin^{-1}x}<1$ if $0\le x<1$.

Hint: $f(x)=\ln(1+x),\ g(x)=\sin^{-1}x$.

2.7 GENERALIZED MEAN VALUE THEOREM

Taylor's Theorem

Let $f^{(n-1)}$ the $(n-1)$th derivative of f is continuous in $[a, a+h]$, $f^{(n)}$ the nth derivative exists in $(a, a+h)$ and p be a given positive integer. Then there exists at least one number θ lying between 0 and 1 such that

$$f(a+h) = f(a) + h\frac{f'(a)}{1!} + \frac{h^2}{2!}f''(a) + \cdots$$

$$+\frac{h^{n-1}}{(n-1)!}f^{(n-1)}(a) + R_n \quad (1)$$

where $$R_n = \frac{h^n(1-\theta)^{n-p}}{(n-1)!p}f^{(n)}(a+\theta h) \quad (2)$$

and $0 < \theta < 1$

Proof: $f, f', f'', \ldots, f^{(n-2)}$ are continuous in $[a, a+h]$ by the hypothesis that $f^{(n-1)}$ is continuous in $[a, a+h]$.

Define

$$\phi(x) = f(x) + (a+h-x)f'(x) + \frac{(a+h-x)^2}{2!}f''(x)$$

$$+\frac{(a+h-x)^3}{3!}f'''(x) + \cdots$$

$$+\frac{(a+h-x)^{n-2}}{(n-2)!}f^{(n-2)}(x) + \frac{(a+h-x)^{n-1}}{(n-1)!} \times$$

$$f^{(n-1)}(x) + (a+h-x)^p A \quad (3)$$

where A is unknown constant. Put $x = a+h$ in (3), we get

$$\phi(a+h) = f(a+h) + 0 + 0 \cdots + 0 \quad (4)$$

Put $x = a$ in (3), we get

$$\phi(a) = f(a) + \frac{h}{1!}f'(a) + \frac{h^2}{2!}f''(a)$$

$$+\frac{h^3}{3!}f'''(a) + \cdots + \frac{h^{n-2}}{(n-2)!}f^{(n-2)}(a)$$

$$+\frac{h^{n-1}}{(n-1)!}f^{(n-1)}(a) + h^p A \quad (5)$$

Choose the constant A such that $\phi(a) = \phi(a+h)$

Equating (3) and (4),

$$f(a+h) = \phi(a+h) = \phi(a) = f(a) + hf'(a)$$

$$+\frac{h^2}{2!}f''(a) + \cdots + \frac{h^{n-2}}{(n-2)!}f^{(n-2)}(a)$$

$$+\frac{h^{n-1}}{(n-1)!}f^{(n-1)}(a) + h^p A \quad (6)$$

The new function ϕ is continuous in $[a, a+h]$, derivable in $(a, a+h)$ and $\phi(a) = \phi(a+h)$ thus satisfying all the conditions of Rolle's theorem. Therefore it follows from Rolle's theorem that there exists at least one c between a and $a+h$ where

$$\phi'(c) = \phi'(a+\theta h) = 0.$$

Differentiating (3) w.r.t. x, we get

$$\phi' = f' - f' + (a+h-x)f'' + \frac{2(a+h-x)}{2!}(-1)f''$$

$$+\frac{(a+h-x)^2}{2!}f''' + \frac{3(a+h-x)^2}{3!}(-1)f''' + \cdots$$

$$+\frac{(n-2)(a+h-x)^{n-3}}{(n-2)!}f^{(n-2)}$$

$$+\frac{(a+h-x)^{n-2}}{(n-2)!}\cdot f^{(n-1)}$$

$$+\frac{(n-1)\cdot(a+h-x)^{n-2}}{(n-1)!}(-1)f^{(n-1)}$$

$$+\frac{(a+h-x)^{n-1}}{(n-1)!}f^{n} + p(a+h-x)^{p-1}A\cdot(-1).$$

or

$$\phi'(x) = \frac{(a+h-x)^{n-1}}{(n-1)!}f^{(n)}(x) - pA\cdot(a+h-x)^{p-1}$$

Thus

$$0 = \phi'(a+\theta h) = \frac{h^{n-1}}{(n-1)!}(1-\theta)^{n-1}f^{(n)}(a+\theta h)$$

$$-pA(1-\theta)^{p-1}h^{p-1}$$

Solving. $$A = \frac{h^{n-p}(1-\theta)^{n-p}}{(n-1)!p}f^{(n)}(a+\theta h) \quad (7)$$

with $h \neq 0$ and $1-\theta \neq 0$. Substituting A from (7) in (6), we get

$$f(a+h) = f(a) + hf'(a) + \frac{h^2}{2!}f''(a) + \frac{h^3}{3!}f'''(a) + \cdots + \frac{h^{n-1}}{(n-1)!}f^{(n-1)}(a) + \frac{h^n(1-\theta)^{n-p}}{(n-1)!p}f^{(n)}(a+\theta h) \quad (8)$$

with $0 < \theta < 1$.

Taylor's remainder after n terms R_n due to

1. Schlomilch & Roche: $R_n = \frac{h^n(1-\theta)^{n-p}f^{(n)}(a+\theta h)}{(n-1)!p}$
2. Cauchy: $p = 1 : R_n = \frac{h^n(1-\theta)^{n-1}}{(n-1)!}f^{(n)}(a+\theta h)$
3. Lagrange's: $p = n : R_n = \frac{h^n}{n!}f^{(n)}(a+\theta h)$

Putting $a + h = x$ or $h = x - a$ in (8) we get another convenient form of Taylor's theorem as

$$f(x) = f(a) + (x-a)\frac{f'(a)}{1!} + \frac{(x-a)^2}{2!}f''(a) + \frac{(x-a)^3}{3!}f'''(a) + \cdots + \frac{(x-a)^{n-1}}{(n-1)!}f^{(n-1)}(a) + \frac{(x-a)^n(1-\theta)^{n-p}}{(n-1)!p}f^{(n)}(a+\theta(x-a)) \quad (9)$$

with $0 < \theta < 1$.

Maclaurin's theorem is a special case of Taylor's theorem when $a = 0$. Thus putting $a = 0$ in (9)

$$f(x) = f(0) + xf'(0) + \frac{x^2}{2!}f''(0) + \frac{x^3}{3!}f'''(0) + \cdots + \frac{x^{n-1}}{(n-1)!}f^{(n-1)}(0) + \frac{x^n(1-\theta)^{n-p}}{(n-1)!p}f^{(n)}(\theta x) \quad (10)$$

Schlomilch remainder is given by $\frac{x^n(1-\theta)^{n-p}}{(n-1)!p} \cdot f^{(n)}(\theta x)$

Lagrange's Remainder for Maclaurin's Theorem: (Put p=n in Schlomilch Remainder)

$$\frac{x^n}{n!}f^{(n)}(\theta x)$$

Cauchy's Remainder for Maclaurin's Theorem: (Put p=1 in Schlomilch Remainder)

$$\frac{x^n(1-\theta)^{n-1}}{(n-1)!}f^{(n)}(\theta x)$$

WORKED OUT EXAMPLES

Example 1: State Maclaurin's theorem with Lagrange's form of remainder for $f(x) = \cos x$.

Solution: Maclaurin's theorem with Lagrange's form of remainder is given by

$$f(x) = f(0) + xf'(0) + \frac{x^2}{2!}f''(0) + \frac{x^3}{3!}f'''(0) + \cdots + \frac{x^{n-1}}{(n-1)!}f^{(n-1)}(0) + \frac{x^n}{n!}f^{(n)}(\theta x).$$

We know that $\frac{d^n}{dx^n}\{\cos(ax+b)\} = a^n \cdot \cos(ax + b + n\frac{\pi}{2})$.

Here $f(x) = \cos x$

$$\text{so} \quad f^{(n)}(x) = \frac{d^n}{dx^n}(\cos x) = \cos(x + \frac{n\pi}{2})$$

At $a = 0$, $f^{(n)}(0) = \cos\left(\frac{n\pi}{2}\right)$

Thus $\quad f(0) = \cos 0 = 1.$

$$f^{(2n)}(0) = \cos\left(2\frac{n\pi}{2}\right) = \cos(n\pi) = (-1)^n$$

$$f^{(2n+1)}(0) = \cos\left((2n+1)\frac{\pi}{2}\right) = 0$$

So coefficients of even powers of x will be $(-1)^n$ while coefficients of odd powers of x are all zero. Substituting these values of the $f^{(n)}(0)$ we have

$$\cos x = f(x) = 1 + 0 + \frac{x^2}{2!}(-1) + 0 + \frac{x^4}{4!}(+1) + \cdots + \frac{x^{2n}}{(2n)!}(-1)^n + \frac{x^{2n+1}}{(2n+1)!}(-1)^n(-1)\cos(\theta x)$$

so

$$\cos x = 1 - \frac{x^2}{2!} + \frac{x^4}{4!} - \frac{x^6}{6!} + \cdots + (-1)^n\frac{x^{2n}}{(2n)!} + \frac{(-1)^{n+1}x^{2n+1}}{(2n+1)!}\cos(\theta x).$$

Example 2: Verify Taylor's theorem for $f(x) = (1-x)^{\frac{5}{2}}$ with Lagrange's form of remainder up to 2 terms in the interval [0, 1].

Solution: $f(x) = (1-x)^{\frac{5}{2}}$ and $f'(x)$ are continuous in [0, 1] while $f''(x)$ is differentiable in (0, 1). Thus $f(x)$ satisfies the conditions of Taylor's theorem. In Taylor's theorem with Lagrange's form of remainder up to 2 terms, n = number of terms in the remainder $= 2 = p$ (since Lagrange's) and $a = 0$ and $x = 1$ (since interval is [0, 1]).

Thus

$$f(x) = f(0) + xf'(0) + \frac{x^2}{2!}f''(\theta x) \quad \text{with} \quad 0 < \theta < 1$$

$$f'(x) = -\frac{5}{2}(1-x)^{\frac{3}{2}},\ f''(x) = \frac{5}{2}\cdot\frac{3}{2}(1-x)^{\frac{1}{2}}$$

so $f(0)=1,\ f'(0)=-\frac{5}{2},\ f''(\theta h)=f''(\theta)=\frac{15}{4}(1-\theta)^{\frac{1}{2}}$, $f(1) = 0$

Substituting these values,

$$0 = 1 + 1.\left(\frac{-5}{2}\right) + \frac{1^2}{2!}\frac{15}{4}(1-\theta)^{\frac{1}{2}}$$

Solving for θ, we get $\theta = \frac{9}{25} = 0.36$ which lies between 0 and 1, thus verifying the Taylor's theorem.

Example 3: For every $x \geq 0$, show that

$$1 + x + \frac{x^2}{2} \leq e^x \leq 1 + x + \frac{x^2}{2}e^x.$$

Solution: The Maclaurin's theorem with Lagrange's form of remainder for the function $f(x) = e^x$ gives

$$e^x = 1 + x + \frac{x^2}{2!} + \frac{x^3}{3!} + \cdots + \frac{x^{n-1}}{(n-1)!} + \frac{x^n}{n!}e^{\theta x}$$

where $0 < \theta < 1$. Note that e^x and all its derivatives are continuous for any x. Taking remainder up to 2 terms i.e., $n = 2$, we have

$$e^x = 1 + x + \frac{x^2}{2!}e^{\theta x}$$

For $x \geq 0$ and $0 < \theta < 1$, $e^{\theta x} \leq e^x$ so that

$$1 + x + \frac{x^2}{2!}e^{\theta x} \leq 1 + x + \frac{x^2}{2!}e^x$$

Also for $x \geq 0$ and $0 < \theta < 1$, $e^{\theta x} > 1$ so that

$$1 + x + \frac{x^2}{2!} \leq 1 + x + \frac{x^2}{2!}e^{\theta x}$$

Thus

$$1 + x + \frac{x^2}{2!} \leq 1 + x + \frac{x^2}{2!}e^{\theta x} = e^x \leq 1 + x + \frac{x^2}{2!}e^x.$$

EXERCISE

1. Show that for any x

$$\sin x = x - \frac{x^3}{3!} + \frac{x^5}{5!} + \cdots + (-1)^{n-1}\cdot\frac{x^{2n-1}}{(2n-1)!}$$

$$+(-1)^n\frac{x^{2n}}{(2n)!}\sin\theta x$$

where $0 < \theta < 1$

Hint: Apply Maclaurin's theorem for $f(x) = \sin x$ with Lagrange's remainder.
Note: $f^{(n)}(x) = \sin\left(x + \frac{n\pi}{2}\right)$ and $f^{2n}(\theta x) = \sin(\theta x + n\pi) = (-1)^n \sin\theta x$.

2. Calculate the first four terms and the remainder after n terms of the Maclaurin's expression of $e^{ax}\cos bx$.

Hint:

$$\frac{d^n}{dx^n}\{e^{ax}\cos bx\}$$

$$= (a^2+b^2)^{n/2}e^{ax}\cdot\cos\left(bx + c + n\tan^{-1}\frac{b}{a}\right)$$

$$\tan\theta = \frac{b}{a},\ \cos\theta = a/\sqrt{a^2+b^2}.$$

Ans. $$e^{ax}\cos bx = 1 + ax + \frac{a^2-b^2}{2!}x^2 + \frac{a(a^2-3b^2)}{3!}x^3$$

$$+\cdots+\frac{x^n}{n!}(a^2+b^2)^{n/2}e^{a\theta x}\times$$

$$\times\cos\left(b\theta x + n\tan^{-1}\frac{b}{a}\right).$$

3. Verify Maclaurin's theorem for $f(x) = (1-x)^{5/2}$ with Lagrange's remainder up to 3 terms when $x = 1$

Hint: $f(x) = f(0) + xf'(0) + \frac{x^2}{2!}f''(0) + \frac{x^3}{3!}f'''(\theta h)$.

Ans. $\theta = \frac{11}{36} = 0.25 \in (0, 1)$

4. Verify Taylor's theorem for $f(x) = x^3 - 3x^2 + 2x$ in $[0, \frac{1}{2}]$ with Lagrange's remainder up to 2 terms.

Hint: $f\left(\frac{1}{2}\right) = f(0) + \frac{1}{2}f'(0) + \frac{(1/2)^2}{2!}f''(\theta)$.

Ans. $\theta = \frac{1}{6} \in \left(0, \frac{1}{2}\right)$

Using Taylor's theorem prove the following:

5. $\ln(1+x) < x - \frac{x^2}{2} + \frac{x^3}{3}$ for $x > 0$

Hint: For Lagrange's with 3 terms

$$\ln(1+x) = x - \frac{x^2}{2} + \frac{x^3}{3(1+\theta x)^3}; 0 < \theta < 1$$

and $(1+\theta x)^3 > 1$ for $x > 0$

Note: $\frac{d^n}{dx^n}\ln(1+x) = \frac{(-1)^{n-1}(n-1)!}{(1+x)^n}$.

6. $x - \frac{x^3}{6} < \sin x < x - \frac{x^3}{6} + \frac{x^5}{120}$, for $x > 0$

Hint: Maclaurin's with 2 and 5 terms

$$\sin x = x - \frac{x^3}{3!}\cos\theta_1 x, 0 < \theta_1 < 1$$

and

$$\sin x = x - \frac{x^3}{3!} + \frac{x^5}{5!}\cos\theta_2 x, 0 < \theta_2 < 1$$

and $\cos\theta_1 x \le 1$ and $\cos\theta_2 x \le 1$.

7. $1 - \frac{x^2}{2} \le \cos x \le 1 - \frac{x^2}{2} + \frac{x^2}{24}$, for any x.

8. $x - \frac{x^2}{2} \le \sin x \le x$, for $x > 0$

9. Show that for any $x > 0$ and $a > 0$

$$a^x = 1 + x\cdot\ln a + \frac{x^2}{2!}(\ln a)^2 + \cdots$$

$$+\frac{x^{n-1}}{(n-1)!}(\ln a)^{n-1} + \frac{x^n}{n!}a^{\theta x}(\ln a)^n$$

9. Show that for any $x > 0$ and $a > 0$

$$a^x = 1 + x\cdot\ln a + \frac{x^2}{2!}(\ln a)^2 + \cdots$$

$$+\frac{x^{n-1}}{(n-1)!}(\ln a)^{n-1} + \frac{x^n}{n!}a^{\theta x}(\ln a)^n$$

10. Find the first 3 terms and the Lagrange's remainder after n terms of the function $f(x) = e^{ax}\sin bx$.

Hint:

$$\frac{d^n}{dx^n}\{e^{ax}\sin bx\} = (a^2+b^2)^{n/2}e^{ax}\sin(bx+n\alpha)$$

so $f^{(n)}(0) = (a^2+b^2)^{n/2}\cdot\sin n\alpha.$

Ans.

$$bx + abx^2 + \frac{b(3a^2-b^2)}{3!}x^3 + \cdots$$

$$+\frac{x^n}{n!}(a^2+b^2)^{n/2}\cdot e^{a\theta x}\sin(b\theta x + n\alpha)$$

where $\tan\alpha = \frac{b}{a}$, $\sin\alpha = \frac{b}{\sqrt{a^2+b^2}}$,

$$\cos\alpha = \frac{a}{\sqrt{a^2+b^2}}$$

2.8 TAYLOR'S SERIES AND MACLAURIN'S SERIES EXPANSIONS

From the Taylor's theorem with Lagrange's form of remainder, we get the Taylor's formula

$$f(x) = f(a) + (x-a)\frac{f'(a)}{1!} + \frac{(x-a)^2}{2!}f''(a)$$
$$+\frac{(x-a)^3}{3!}f'''(a) + \cdots$$
$$+\frac{(x-a)^{n-1}}{(n-1)!}f^{n-1}(a)$$
$$+\frac{(x-a)^n}{n!}f^{(n)}(a+\theta(x-a)) \quad (1)$$

where $0 < \theta < 1$.

Note: Taylor's formula for $n = 1$, reduces to the Lagrange's mean value theorem (the law of means). Denote the first n terms of R.H.S. of (1) by $P_n(x)$ which is a polynomial of $(n-1)$th degree in the variable $(x-a)$ and the last term on R.H.S. of (1) by $R_n(x)$ which is the Lagrange's form of remainder after n terms i.e.,

$$P_n(x) = f(a) + (x-a)\frac{f'(a)}{1!} + \frac{(x-a)^2}{2!}f''(a) + \cdots$$
$$+\frac{(x-a)^{n-1}}{(n-1)!}f^{n-1}(a) \quad (2)$$

and

$$R_n(x) = \frac{(x-a)^n}{n!}f^n(a+\theta(x-a)) \quad (3)$$

Then $\quad f(x) = P_n(x) + R_n(x)$

or $\quad R_n(x) = f(x) - P_n(x)$

i.e., $\quad BC = AC - AB$

Thus for those values of x, for which $R_n(x)$ is small, the polynomial $P_n(x)$ yields an approximate

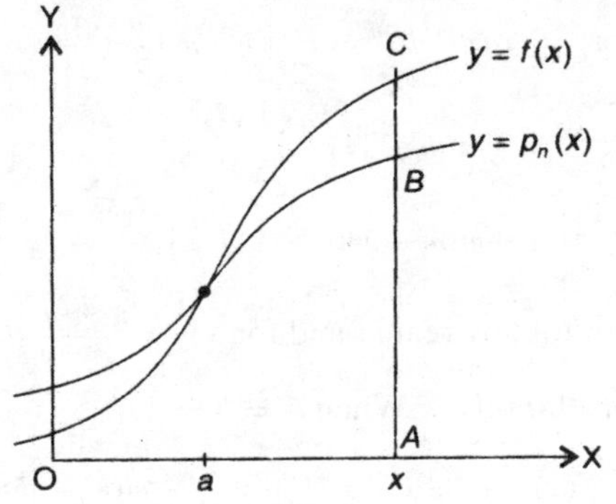

Fig. 2.7

representation of $f(x)$. When $a = 0$, the Taylor's formula (1) at origin (0, 0) reduces to

$$f(x) = f(0) + x\frac{f'(0)}{1!} + \frac{x^2}{2!}f''(0) + \frac{x^3}{3!}f'''(0) + \cdots + \frac{x^{n-1}}{(n-1)!}f^{(n-1)}(0) + \frac{x^n}{n!}f^{(n)}(\theta x) \qquad (4)$$

(4) is known as Maclaurin's formula.

In (1) as $n \to \infty$ if $R_n(x) \to 0$, then $f(x)$ is represented by infinite series in powers of $(x - a)$ and is given by

$$f(x) = f(a) + \frac{(x-a)}{1!}f'(a) + \frac{(x-a)^2}{2!}f''(a) + \cdots + \frac{(x-a)^n}{n!}f^{(n)}(a) + \cdots \qquad (5)$$

The power series (5) is known as Taylor's series. Rewriting in the summation form

$$f(x) = f(a) + \sum_{n=1}^{\infty} \frac{f^{(n)}(a)}{n!}(x-a)^n \qquad (6)$$

Thus (5) or (6) is the Taylor's series expansion of $f(x)$ about the point 'a' (or in powers of $x - a$). When $a = 0$, (6) reduces to

$$f(x) = f(0) + \sum_{n=1}^{\infty} \frac{f^{(n)}(0)}{n!}x^n \qquad (7)$$

The infinite series in powers of x given by (7) is called the Maclaurin's series expansion of $f(x)$ or simply Maclaurin's series of $f(x)$ (about origin).

Maclaurin Series Expansion of Some Elementary Functions

1. *Exponential function* $f(x) = e^x$.

   Differentiating n times, $f^{(n)}(x) = e^x$. So at $x = 0$, $f^{(n)}(0) = 1$. Substituting in the Maclaurin's formula (4), we obtain

$$f(x) = e^x = 1 + \frac{x}{1!} + \frac{x^2}{2!} + \frac{x^2}{3!} + \cdots + \frac{x^{n-1}}{(n-1)!} + \frac{x^n}{n!}e^{\theta x}$$

   where $0 < \theta < 1$.

   For $0 < \theta < 1$, $e^{\theta x}$ is bounded for any x($< e^x$ for $x > 0$ and < 1 for $x < 0$). Thus the Lagrange's remainder

$$R_n = \frac{x^n}{n!}e^{\theta x} \to 0 \text{ as } n \to \infty \text{ for any } x.$$

   Hence the Maclaurin series expansion of e^x is

$$e^x = 1 + \frac{x}{1!} + \frac{x^2}{2!} + \frac{x^3}{3!} + \cdots + \frac{x^n}{n!} + \cdots; -\infty < x < \infty$$

   Therefore e^x can be evaluated to any degree of accuracy by taking sufficient number of terms.

2. *Expansion of* $f(x) = \cos x$

   Differentiating n times, $f^{(n)}(x) = \cos\left(x + \frac{n\pi}{2}\right)$. so $f^{(n)}(0) = \cos n\frac{\pi}{2}$. The remainder

$$R_n(x) = \frac{x^n}{n!}f^{(n)}(\theta x) = \frac{x^n}{n!}\cos\left(\theta x + \frac{n\pi}{2}\right) \to 0 \text{ as } n \to \infty$$

   since $\left|\cos\left(\theta x + \frac{n\pi}{2}\right)\right| \le 1$, for all values of x.

   At $x = 0$, $f(0) = 1$, $f^{(1)}(0) = \cos\frac{\pi}{2} = 0$, $f^{(2)}(0) = \cos\pi = -1$, $f^{(3)}(0) = \cos\frac{3\pi}{2} = 0$, $f^{(4)}(0) = \cos 2\pi = 1$ and so on. Substituting these values in the Maclaurin series given by (7), we get

$$f(x) = \cos x = 1 + 0 - \frac{x^2}{2!} + 0 + \frac{x^4}{4!} + 0 - \frac{x^6}{6!} + \cdots$$

$$= 1 - \frac{x^2}{2!} + \frac{x^4}{4!} - \frac{x^6}{6!} + \cdots; \quad -\infty < x < \infty.$$

3. *Expansion of* $f(x) = \sin x$.

   We know that $f^n(x) = \sin\left(x + n\frac{\pi}{2}\right)$. So $f^{(n)}(0) = \sin n\frac{\pi}{2}$ then $f(0) = 0$, $f'(0) = \sin\frac{\pi}{2} = 1$, $f^{(2)}(0) = \sin\pi = 0$, $f^{(3)}(0) = \sin\frac{3\pi}{2} = -1$. $f^{(4)}(0) = \sin 2\pi = 0$ etc.

   Substituting these values in (7), we get the Maclaurin series expansion of $f(x)$ valid for any

x as

$$f(x) = \sin x = x - \frac{x^3}{3!} + \frac{x^5}{5!} - \frac{x^7}{7!} + \cdots, \quad -\infty < x < \infty$$

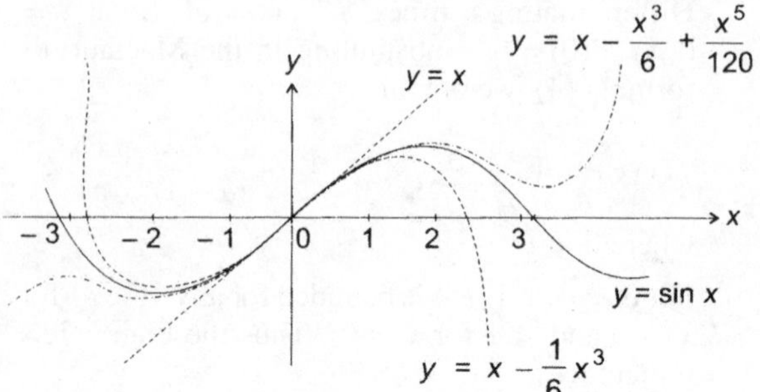

Fig. 2.8

Figure 2.8 shows the graphs of the function $f(x) = \sin x$ and the first three approximations

$$y = x,\ y = x - \frac{x^3}{3!},\ y = x - \frac{x^3}{3!} + \frac{x^5}{5!}$$

4. *Expansion of* $f(x) = \sinh x$.

   We know that $f^{(2n+1)}(x) = \cosh x$ and $f^{(2n)}(x) = \sinh x$ so $f^{(2n+1)}(0) = 1$, $f^{(2n)}(0) = 0$, $f(0) = 0$. Substituting in (7),

$$f(x) = \sinh x = x + \frac{x^3}{3!} + \frac{x^5}{5!} + \frac{x^7}{7!} + \cdots; \quad -\infty < x < \infty$$

5. *Expansion of* $f(x) = \cosh x$.

   Since $f^{(2n)}(x) = \cosh x$ and $f^{(2n+1)}(x) = \sinh x$, we have $f^{(2n)}(0) = 1$, $f^{(2n+1)}(0) = 0$, $f(0) = 1$. From (7),

$$f(x) = \cosh x = 1 + \frac{x^2}{2!} + \frac{x^4}{4!} + \frac{x^6}{6!} + \cdots; \quad -\infty < x < \infty$$

6. *Expansion of* $f(x) = (ax+b)^m$ where m is any real number and $x > -\frac{b}{a}$ (i.e., $ax + b > 0$) and $a \neq 0$. The nth derivative of f is

$$f^{(n)}(x) = m(m-1)(m-2)\cdots(m-n+1)a^n(ax+b)^{m-n}$$

   At $x = 0$, $f^{(n)}(0) = m(m-1)(m-2)\cdots(m-n+1)a^n \cdot b^{m-n}$ and $f(0) = b^m$.

   Substituting in (7), we obtain

$$f(x) = (ax+b)^m = b^m + m \cdot a \cdot b^{m-1} x + m(m-1)a^2 \cdot b^{m-2}\frac{x^2}{2!} + m(m-1)(m-2) \cdot a^3 b^{m-3} \cdot \frac{x^3}{3!} + \cdots$$

or
$$(ax+b)^m = b^m\left[1 + m\left(\frac{a}{b}\right)\frac{x}{1!} + m(m-1)\left(\frac{a}{b}\right)^2\frac{x^2}{2!} + m(m-1)(m-2)\left(\frac{a}{b}\right)^3\frac{x^3}{3!} + \cdots\right]$$

valid for any real m and for any $x > -\frac{b}{a}$, $a \neq 0$.

Corollary 1: When $a = b = 1$,

$$f(x) = (1+x)^m = 1 + m\frac{x}{1!} + m(m-1)\frac{x^2}{2!} + m(m-1)(m-2)\frac{x^3}{3!} + \cdots$$

valid for any real m and for any $x > -1$.

7. *Expansion of* $f(x) = \ln(1+x)$. We know that $f^{(n)}(x) = \frac{(-1)^{n-1}(n-1)!}{(1+x)^n}$ so

   $f^{(n)}(0) = (-1)^{n-1} \cdot (n-1)!$ and $f(0) = 0$. From (7),

$$f(x) = \ln(1+x) = 0 + x - \frac{x^2}{2!} + 2!\frac{x^3}{3!} - 3!\frac{x^4}{4!} + \cdots + (-1)^{n-1}(n-1)!\frac{x^n}{n!} + \cdots$$

or
$$\ln(1+x) = x - \frac{x^2}{2} + \frac{x^3}{3} - \frac{x^4}{4} + \frac{x^5}{5} \cdots + (-1)^{n-1}\frac{x^n}{n} + \cdots$$

   valid for $1 + x > 0$ i.e., $x > -1$.

8. *Expansion of* $f(x) = \ln(1-x)$.

   Here $f^{(n)}(x) = \frac{(-1)^{n-1} \cdot (n-1)!(-1)^n}{(1-x)^n}$, so $f^{(n)}(0) = -(n-1)!$

   From (7),

$$f(x) = \ln(1-x) = -x - \frac{x^2}{2} - \frac{x^3}{3} - \frac{x^4}{4} - \frac{x^5}{5} - \cdots - \frac{x^n}{n} - \cdots$$
$$= -\left(x + \frac{x^2}{2} + \frac{x^3}{3} + \frac{x^4}{4} + \cdots + \frac{x^n}{n} + \cdots\right)$$

   valid for $x < 1$.

Note: For any given function $f(x)$, in general, the nth derivative $f^{(n)}(x)$ can not be determined and therefore the nature of R_n not known. Therefore by assuming that $R_n \to 0$ as $n \to \infty$, formal Maclaurin's expansions of any given function is obtained from (7) by substituting f and its derivatives evaluated at $x = 0$. If the nth derivative is not obtainable, first few (5 to 6) terms are determined. Some times $f(x)$ can be expressed as sum or product of two or more infinite series.

WORKED OUT EXAMPLES

Example 1: Expand the polynomial

$$f(x) = x^5 + 2x^4 - x^2 + x + 1$$

in powers of $x + 1$.

Or

Obtain the Taylor's series expansion of $f(x)$ about the point $x = -1$.

Solution: $f(-1) = -1 + 2 - 1 - 1 + 1 = 0$

$$f'(x) = 5x^4 + 8x^3 - 2x + 1,\ f'(-1) = 5 - 8 + 2 + 1 = 0$$

$$f''(x) = 20x^3 + 24x^2 - 2,\ f''(-1) = -20 + 24 - 2 = 2$$

$$f'''(x) = 60x^2 + 48x,\ f'''(-1) = 60 - 48 = 12$$

$$f''''(x) = 120x + 48,\ f''''(-1) = -120 + 48 = -72$$

$$f'''''(x) = 120,\ f'''''(-1) = 120$$

$$f^n(x) = 0 \quad \text{for} \quad n > 5.$$

The Taylor series expansion of $f(x)$ about $x = a$.

$$f(x) = f(a) + (x-a)f'(a) + (x-a)^2 \frac{f''}{2!}(a) + \frac{(x-3)^3}{3!} f'''(a) + \cdots$$

Here $a = -1$. Substituting the above values

$$f(x) = x^5 + 2x^4 - x^2 + x + 1$$

$$= 0 + 0 + (x+1)^2 \cdot \frac{2}{2!} + 12 \cdot \frac{(x+1)^3}{3!} - 72 \cdot \frac{(x+1)^4}{4!} + 120 \frac{(x+1)^5}{5!}$$

$$f(x) = (x+1)^2 + 2(x+1)^3 - 3(x+1)^4 + (x+1)^5.$$

Example 2: Write Taylor's formula for the function $y = \sqrt{x}$ with Lagrange's remainder with $a = 1,\ n = 3$.

Solution:

$$y = f(x) = \sqrt{x},\ f(a) = f(1) = 1$$

$$f'(x) = \frac{1}{2\sqrt{x}},\ f'(1) = \frac{1}{2}$$

$$f''(x) = -\frac{1}{4}\frac{1}{x^{\frac{3}{2}}} = -\frac{1}{4}$$

$$f'''(x) = \frac{3}{8}\frac{1}{x^{\frac{5}{2}}} = \frac{1}{8}$$

$$f''''(x) = -\frac{15}{16}\frac{1}{x^{\frac{7}{2}}}$$

Taylor's formula with Lagrange's remainder up to 4 terms (i.e., $n = 3$)

$$f(x) = f(a) + (x-a)\frac{f'(a)}{1!} + (x-a)^2 \frac{f''(a)}{2!} + (x-a)^3 \frac{f'''(a)}{3!} + \frac{(x-a)^4}{4!} f''''(a + \theta(x-a))$$

Substituting f and its derivatives at $a = 1$,

$$f(x) = 1 + (x-1)\cdot\frac{1}{2} + (x-1)^2\left(-\frac{1}{4}\right)\frac{1}{2!} + (x-1)^3 \cdot \frac{3}{8} \cdot \frac{1}{3!} + (x-1)^4 \cdot \left(-\frac{15}{16}\right) \times \frac{1}{[1 + \theta(x-1)]^{\frac{7}{2}}} \cdot \frac{1}{4!}$$

$$= 1 + \frac{1}{2}(x-1) - \frac{1}{8}(x-1)^2 + \frac{1}{16}(x-1)^3 - \frac{5}{128}(x-1)^4[1 + \theta(x-1)]^{-\frac{7}{2}}.$$

Example 3: Expand $\sin x$ in powers of $\left(x - \frac{\pi}{2}\right)$.

Solution:

$$\sin(x) = \sin\left(x - \frac{\pi}{2} + \frac{\pi}{2}\right) = \sin\left(\left(x - \frac{\pi}{2}\right) + \frac{\pi}{2}\right)$$

$$= \sin\left(x - \frac{\pi}{2}\right) \cdot \cos\frac{\pi}{2} + \sin\frac{\pi}{2} \cdot \cos\left(x - \frac{\pi}{2}\right)$$

$$= 0 + 1 \cdot \cos\left(x - \frac{\pi}{2}\right).$$

We know that $\cos x = \sum_{n=0}^{\infty}(-1)^n \frac{x^{2n}}{(2n)!}$ so

$$\sin x = +\cos\left(x - \frac{\pi}{2}\right) = +\sum_{n=0}^{\infty}(-1)^n \frac{\left(x - \frac{\pi}{2}\right)^{2n}}{(2n)!}$$

The Taylor series expansion of $\sin x$ about $x = \frac{\pi}{2}$ is

$$\sin x = 1 - \left(x - \frac{\pi}{2}\right)^2 \frac{1}{2!} + \left(x - \frac{\pi}{2}\right)^4 \cdot \frac{1}{4!} - \left(x - \frac{\pi}{2}\right)^6 \frac{1}{6!} + \cdots$$

Example 4: Obtain the first 4 terms of the Taylor's series of $\cos x$ about $x = \frac{\pi}{4}$.

Solution: $\cos x = \cos\left(x - \frac{\pi}{4} + \frac{\pi}{4}\right) = \cos\left(t + \frac{\pi}{4}\right)$ where $t = x - \frac{\pi}{4}$

$$\cos x = \cos t \cdot \cos\frac{\pi}{4} - \sin t \cdot \sin\frac{\pi}{4} = \frac{1}{\sqrt{2}}(\cos t - \sin t)$$

$$= \frac{1}{\sqrt{2}}\left[\left(1 - \frac{t^2}{2!} + \frac{t^4}{4!} - \frac{t^6}{6!}\cdots\right) - \left(t - \frac{t^3}{3!} + \frac{t^5}{5!} - \frac{t^7}{7!} + \cdots\right)\right]$$

$$= \frac{1}{\sqrt{2}}\left[1 - t - \frac{t^2}{2!} + \frac{t^3}{3!} + \frac{t^4}{4!} - \frac{t^5}{5!} - \frac{t^6}{6!} + \frac{t^7}{7!} + \cdots\right]$$

$$= \frac{1}{\sqrt{2}}\left[1 - \left(x - \frac{\pi}{4}\right) - \left(x - \frac{\pi}{4}\right)^2 \cdot \frac{1}{2!} + \left(x - \frac{\pi}{4}\right)^3 \cdot \frac{1}{3!} + \left(x - \frac{\pi}{4}\right)^4 \cdot \frac{1}{4!} - \left(x - \frac{\pi}{4}\right)^5 \cdot \frac{1}{5!} + \cdots\right].$$

Example 5: Using Taylor's series expansion calculate an approximate value of $\sqrt{10}$.

Solution: Let $\sqrt{10} = \sqrt{(9+1)} = 3\left(1 + \frac{1}{9}\right)^{\frac{1}{2}}$ using,

$$(1 + x)^m = 1 + mx + \frac{m(m-1)}{2!}x^2 + \frac{m(m-1)(m-2)}{3!}x^3 + \cdots$$

with $x = \frac{1}{9}$, $m = \frac{1}{2}$, we get

$$3\left(1 + \frac{1}{9}\right)^{\frac{1}{2}} = 3\left[1 + \frac{1}{2} \cdot \frac{1}{9} + \frac{\frac{1}{2}\left(\frac{1}{2} - 1\right)}{2!}\left(\frac{1}{9}\right)^2 + \frac{\frac{1}{2}\left(\frac{1}{2} - 1\right)\left(\frac{1}{2} - 2\right)}{3!}\left(\frac{1}{9}\right)^3 - \cdots\right]$$

considering the first four terms,

$$\approx 3\left[1 + \frac{1}{18} - \frac{1}{8} \cdot \frac{1}{81} + \frac{1}{16} \cdot \frac{1}{729}\right] \approx 3.1629.$$

Example 6: Find the first seven terms of the Maclaurin series of $f(x) = \ln \sec x$.

Solution: $f(0) = \ln \sec 0 = \ln 1 = 0$
Differentiating w.r.t. x,

$$f'(x) = \frac{1}{\sec x} \cdot \sec x \cdot \tan x = \tan x,\ f'(0) = \tan 0 = 0$$

$$f''(x) = \sec^2 x,\ f''(0) = \sec^2 0 = 1$$

$$f''' = 2\sec x \cdot \sec x \cdot \tan x = 2\sec^2 x \cdot \tan x,\ f'''(0) = 0$$

$$f'''' = 4\sec x \cdot \sec x \cdot \tan x \cdot x + 2\sec^2 x \cdot \sec^2 x$$

$$f'''' = 4\sec^2 x \cdot \tan^2 x + 2\sec^4 x,\ f''''(0) = 0 + 2 = 2$$

$$f''''' = 8\sec^2 x \cdot \tan^3 x + 16\sec 4x \cdot \tan x,\ f'''''(0) = 0$$

$$f'''''' = 16\sec^2 x \cdot \tan^4 x + 88\sec^4 x \cdot \tan^2 x + 16\sec^6 x,\ f''''''(0) = 16$$

The Maclaurin series is

$$f(x) = f(0) + f'(0) \cdot \frac{x}{11} + f''(0) \cdot \frac{x^2}{2!} + f'''(0) \cdot \frac{x^3}{3!} + \cdots$$

Substituting the above values

$$f(x) = \ln \sec x = 0 + 1 \cdot \frac{x^2}{2!} + 0 + 2 \cdot \frac{x^4}{4!} + 0 + \frac{16x^6}{6!} + \cdots$$

$$f(x) = \frac{x^2}{2} + \frac{x^4}{12} + \frac{x^6}{45} + \cdots.$$

Example 7: If $y = \sin \ln(x^2 + 2x + 1)$, expand y in ascending powers of x up to x^6.

Solution: Differentiating y w.r.t. x,

$$y_1 = +\cos \ln(x^2 + 2x + 1) \cdot \left(\frac{1}{x^2 + 2x + 1}\right) \cdot (2x + 2)$$

$$y_1 = \frac{2}{x + 1} \cdot \cos \ln(x^2 + 2x + 1)$$

$$(x+1)y_1 = 2\cos\ln(x^2+2x+1)$$

Differentiating again w.r.t. x,

$$(x+1)y_2+y_1 = -2\cdot\sin\ln(x^2+2x+1)\times$$

$$\times\left(\frac{1}{x^2+2x+1}\right)\cdot(2x+2)$$

$$= -\frac{4}{x+1}\cdot\sin\ln(x^2+2x+1) = -\frac{4y}{(x+1)}$$

or $\quad (x+1)^2y_2+(x+1)y_1+4y=0.$

Differentiate 'n' times using Leibnitz's rule,

$$\left[(x+1)^2y_{n+2}+n\cdot 2(x+1)y_{n+1}+\frac{n(n-1)}{2}\cdot 2\cdot 1\cdot y_n\right]+$$

$$+\left[(x+1)y_{n+1}+n\cdot 1\cdot y_n\right]+4y_n=0$$

or

$$(x+1)^2y_{n+2}+(x+1)(2n+1)y_{n+1}+(4+n^2)y_n=0.$$

Recurrence relation: At $x=0$

$$y_{n+2}(0) = -\left[(2n+1)y_{n+1}(0)+(4+n^2)y_n(0)\right]$$

Now $y(0)=0,\ y_1(0)=2,$
Using recurrence relation

$$y_2(0) = -[y_1(0)+4y_0(0)] = -[2+4\cdot 0] = -2$$

$$y_3(0) = -[3y_2(0)+5y_1(0)] = -[3(-2)+5\cdot 2] = -4$$

$$y_4(0) = -[5y_3(0)+8y_2(0)] = -[5(-4)+8(-2)] = 36$$

$$y_5(0) = -[7y_4(0)+13y_3(0)] = [7(36)+13(-4)] = -200$$

$$y_6(0) = -[9y_5(0)+20y_4(0)] = -[9(-200)+20(36)] = 1080$$

Substituting these values in the Maclaurin series,

$$y=\sin\ln(x^2+2x+1) = 0+2\cdot\frac{x}{11}-2\frac{x^2}{2!}-4\cdot\frac{x^3}{3!}$$

$$+36\cdot\frac{x^4}{4!}-200\cdot\frac{x^5}{5!}+1080\cdot\frac{x^6}{6!}+\cdots$$

$$=2x-x^2-\frac{2}{3}x^3+\frac{3}{2}x^4-\frac{5}{3}x^5+\frac{3}{2}x^6+\cdots.$$

Example 8: Prove that

$$e^x\cos x = 1+x-\frac{2x^3}{3!}-2^2\frac{x^4}{4!}-2^2\frac{x^5}{5!}+2^3\frac{x^7}{7!}+\cdots.$$

Solution: R.H.S. in a Maclaurin series expansion of $y=f(x)=e^x\cos x\cdot$ We know that

$$y_n = f^{(n)}(x) = \frac{d^n}{dx^n}\{e^x\cos x\} = 2^{\frac{n}{2}}e^x\cdot\cos\left(x+n\frac{\pi}{4}\right).$$

For $n=4$, $y_4(x)=2^2\cdot e^x\cdot\cos(x+\pi) = -4e^x\cos x$ $= -4y$ i.e., $y_4=-4y$
Differentiating n times by Leibnitz's rule, we get the recurrence relation

$$y_{n+4}(x) = -4y_n(x)$$

At $x=0$, $\quad y_{n+4}(0) = -4y_n(0).$

Also $y(0)=f(0)=e^0\cdot\cos 0=1;\ y_1(0)=1,$ $y_2(0)=0;\ y_3(0)=-2,\ y_4(0)=-4,\ y_5(0)=-4,$ $y_6(0)=0,\ y_7(0)=8$ etc.
With these values, the Maclaurin series expansion of $e^x\cos x = 1+x-\frac{2}{3!}x^3-\frac{4}{4!}x^4-\frac{4}{5!}x^5+\frac{8}{7!}x^7+\cdots.$

Example 9: Expand $e^x\sin^2 x$ in ascending powers of x upto x^5.

Solution: Let $f(x)=e^x\cdot\sin^2 x$.
We know that

$$e^x=\sum_{n=0}^{\infty}\frac{x^n}{n!}\quad\text{and}\quad \sin x=\sum_{n=1}^{\infty}\frac{(-1)^{n+1}x^{2n-1}}{(2n-1)!}$$

$$f(x)=e^x\cdot\sin^2 x = \left(\sum_{n=0}^{\infty}\frac{x^n}{n!}\right)\left(\sum_{n=1}^{\infty}\frac{(-1)^{n+1}\cdot x^{2n-1}}{(2n-1)!}\right)$$

$$=\left(1+x+\frac{x^2}{2!}+\frac{x^3}{3!}+\frac{x^4}{4!}+\cdots\right)\times$$

$$\times\left(x-\frac{x^3}{3!}+\frac{x^5}{5!}-\frac{x^7}{7!}+\cdots\right)^2$$

$$=\left(1+x+\frac{x^2}{2!}+\frac{x^3}{3!}+\cdots\right)\times$$

$$\times\left(x^2+\cdots-2\frac{x^4}{3!}+\cdots\right)$$

Both the series are truncated, because we are considering terms upto x^5. Multiplying the series term by term, we get

$$f(x)=e^x\sin^2 x = x^2-\frac{2x^4}{3!}+x^3-2\frac{x^5}{3!}+\frac{x^4}{2!}+\frac{x^5}{3!}+\cdots$$

$$=x^2+x^3+\frac{1}{6}x^4+\frac{1}{6}x^5+\cdots.$$

Example 10: Using Maclaurin series show that

$$e^{x\cos x} = 1+x+\frac{x^2}{2}-\frac{x^3}{3}+\cdots.$$

Solution: We know that $e^x = \sum_{n=0}^{\infty} \frac{x^n}{n!}$

$$e^{x\cos x} = \sum_{n=0}^{\infty} \frac{(x\cos x)^n}{n!}$$

$$= 1 + x\cos x + \frac{(x\cos x)^2}{2!} + \frac{(x\cos x)^3}{3!} + \cdots$$

But $x\cos x = \sum_{n=0}^{\infty} (-1)^n \frac{x^{2n+1}}{(2n)!}$ so

$$e^{x\cos x} = 1 + \left(\sum_{n=0}^{\infty} (-1)^n \frac{x^{2n+1}}{(2n)!}\right) + \frac{1}{2!}\left(\sum_{n=0}^{\infty} (-1)^n \frac{x^{2n+1}}{(2n)!}\right)^2 + \frac{1}{3!}\left(\sum_{n=0}^{\infty} (-1)^n \frac{x^{2n+1}}{(2n)!}\right)^3 + \cdots$$

$$= 1 + \left(x - \frac{x^3}{2!} + \frac{x^5}{4!} - \frac{x^7}{6!} + \cdots\right) + \frac{1}{2!}\left(x - \frac{x^3}{2!} + \frac{x^5}{4!} - \frac{x^7}{6!} + \cdots\right)^2 + \frac{1}{3!}\left(x - \frac{x^3}{2!} + \frac{x^5}{4!} - \frac{x^7}{6!} + \cdots\right)^3 + \cdots$$

$$= 1 + x + \frac{1}{2!}x^2 + \left(-\frac{1}{2!} + \frac{1}{3!}\right)x^3 + \cdots$$

$$= 1 + x + \frac{x^2}{2} - \frac{x^3}{3} + \cdots.$$

EXERCISE

1. Expand the polynomial $x^4 - 5x^3 + 5x^2 + x + 2$ in powers of $x - 2$.

Ans. $-7(x-2) - (x-2)^2 + 3(x-2)^3 + (x-2)^4$

2. Expand $2x^3 + 7x^2 + x - 1$ about $x = 2$.

Ans. $45 + 53(x-2) + 19(x-2)^2 + 2(x-2)^3$

3. Find Taylor series expansion of $f(x)$ about 'a' where

   i. $f(x) = \ln x, a = 1$

   ii. $f(x) = \tan x, a = \frac{\pi}{4}$

   iii. $f(x) = \ln\cos x, a = \frac{\pi}{3}$

   iv. $f(x) = e^x, a = 3$

Ans. **i.** $(x-1) - \frac{(x-1)^2}{2!} + \frac{2!}{3!}(x-1)^3 - \frac{3!}{4!}(x-1)^4 + \cdots$

ii. $1 + 2(x - \frac{\pi}{4}) + 2(x - \frac{\pi}{4})^2 + \cdots$

iii. $-\ln 2 - \sqrt{3}\left(x - \frac{\pi}{3}\right) - 2\left(x - \frac{\pi}{3}\right)^2 + \cdots$

iv. $e^{x-3+3} = e^3\left[\sum_{n=0}^{\infty} \frac{(x-3)^n}{n!}\right]$

4. Prove that

$$f(x) = f(a) + 2\frac{(x-a)}{2} f'\left(\frac{x+a}{2}\right) + \frac{(x-a)^3}{8(3!)} f'''\left(\frac{(x+a)}{2}\right) + \frac{(x-a)^5}{32(5!)} f'''''\left(\frac{(x+a)}{2}\right) + \cdots$$

Hint: Expand $f(x) = f\left(\frac{x+a}{2} + \frac{x-a}{2}\right)$ and $f(a) = f\left(\frac{x+a}{2} + \frac{a-x}{2}\right)$ and subtract.

5. Show that

$$\tan^{-1}(x+h) = \tan^{-1}x = h\sin z \cdot \frac{\sin z}{1} - (h\sin z)^2 \frac{\sin 2z}{2} + (h\sin z)^2 \frac{\sin 3z}{3} + \cdots$$

where $z = \cot^{-1} x$.

Hint: Use Taylor's series, $x = \cot z$, $\frac{dz}{dx} = -\sin^2 z$, take $f(x) = \tan^{-1} x$.

6. Calculate the approximate value of cos 32° using Toylor series.

Hint: Expand $f(30+2) = \cos(30+2)$

Ans. 0.8482

7. Obtain the Maclaurin series of a^x

Hint: $f^{(n)}(x) = a^x(\log a)^n$, $f^{(n)}(0) = (\log a)^n$

Ans. $a^x = 1 + x\log a + \frac{1}{2!}(x\log a)^2 + \frac{1}{3!}(x\log a)^3 + \cdots + \frac{1}{n!}(x\log a)^n + \cdots$

8. If $f(x) = x^3 + 3x^2 + 15x - 10$, calculate the approximate value of $f(\frac{11}{10})$.

Hint: Use Taylor's series and expand $f(\frac{11}{10}) = f(x+h) = f(1 + \frac{1}{10})$; $f(1) = 9$, $f'(1) = 24$,

$f''(1) = 12$, $f'''(1) = 6$.

Ans. 11.461

9. Determine the approximate value of π using the Maclaurin series expansion of $\sin^{-1} x$.

Hint: $y = \sin^{-1} x = x + \frac{1^2}{3!}x^2 + \frac{1^2 \cdot 3^2}{5!}x^5 + \frac{1^2 \cdot 3^2 \cdot 5^2}{7!} \cdot x^7 + \cdots$, $(1 - x^2)y_{n+2} - (2n + 1)xy_{n+1} - n^2 y_n = 0$, $y_{n+2}(0) = n^2 y_n(0)$.

Ans. $\pi = 3.1386$

10. Expand $f(x) = e^{ax} \cdot \sin bx$ in ascending powers of x.

Hint:

$$f^{(n)}(0) = (a^2 + b^2)^{\frac{n}{2}} \sin n\theta;$$

$$\sin\theta = \frac{b}{\sqrt{a^2 + b^2}}; \quad \cos\theta = \frac{a}{\sqrt{a^2 + b^2}}$$

Ans. $e^{ax} \sin bx = bx + abx^2 + \frac{b(3a^2 - b^2)}{3!}x^3 + \cdots$

11. Obtain the Maclaurin series of $\tan x$.

Hint: $f(x) = \tan x$, $f(0) = 0$, $f'(x) = 1 + \tan^2 x$, $f'(0) = 1$, $f'' = 2(\tan x + \tan^3 x)$; $f''(0) = 0$, $f''' = 2 + 8\tan^2 x + 6\tan^4 x$, $f'''(0) = 2$, $f'''' = 16\tan x + 40\tan^3 x + 24\tan^5 x$, $f''''(0) = 0$ $f''''' = 16\sec^2 x + 120\tan^2 x \sec^2 x + 120\tan^4 x \sec^2 x$, $f'''''(0) = 16$

Ans. $\tan x = x + \frac{1}{3}x^3 + \frac{2}{15}x^5 + \cdots$

12. Find the Maclaurin series of $y = e^x \ln(1 + x)$.

Hint: $(1 + x)y_{n+2} + (n - x)y_{n+1} - (n + 1)y_n - e^x = 0$, $y(0) = 0$, $y_1(0) = 1$, $y_2(0) = 1$, $y_{n+2}(0) = -ny_{n+1}(0) + (n + 1)y_n(0) + 1$

Ans. $x + \frac{x^2}{2!} + \frac{2x^3}{3!} + \frac{9x^5}{5!} - \frac{35}{6!}x^6 + \cdots$

13. Write the Maclaurin formula for $y = \sqrt{1 + x}$ when $n = 2$ and with Lagrange's remainder

Ans. $\sqrt{1 + x} = 1 + \frac{1}{2}x - \frac{1}{8}x^2 + \frac{x^3}{16} \cdot (1 + \theta x)^{\frac{-5}{2}}$, $0 < \theta < 1$.

14. Show that

$$e^{a \sin^{-1} x} = 1 + ax + \frac{a^2}{2!}x^2 + \frac{a(1^2 + a^2)}{3!}x^3 + \frac{a^2(2^2 + a^2)}{4!}x^4 + \cdots.$$

Hint: $y = e^{a \sin^{-1} x}$, $y(0) = 1$, $y_1(0) = a$, $y_2(0) = a^2$, $(1 - x^2)y_{n+2} - (2n + 1)xy_{n+1} - (n^2 + a^2)y_n = 0$ and $y_{n+2}(0) = (n^2 + a^2)y_n(0)$

15. Show that $\tan^{-1} x = x - \frac{1}{3}x^3 + \frac{1}{5}x^5 + \cdots$ for $|x| < 1$. (This series is known as **Gregory's series**).

Hint: $(1 + x^2)y_1 = 1$, $(1 + x^2)y_{n+1} + 2nxy_n + n(n - 1)y_{n-1} = 0$, $y_{n+1}(0) = -n(n - 1)y_{n-1}(0)$.

2.9 INDETERMINATE FORMS

If two functions $f(x)$ and $g(x)$ are *both* zero at $x = a$, the fraction $\frac{f(a)}{g(a)}$ is said to assume the indeterminate form $\frac{0}{0}$. Although the function $F(x) = \frac{f(x)}{g(x)}$ is undefined (indeterminate) at $x = a$, it may however approach a limit as x approaches a. The process of determining such a limit, if it exists, is known as the evaluation of indeterminate forms. The L'Hospital's rule (theorem) allows the evaluation of indeterminate forms.

L'Hospital's Rule

Let $f(x)$ and $g(x)$ be continuous in an interval (c, d) containing $x = a$. Suppose $f(a) = 0$ and $g(a) = 0$. Also the derivatives of f and g exist and $g^{\mathrm{i}}(x) \neq 0$ in (c, d) except possibly at $x = a$. Then

$$\lim_{x \to a} \frac{f(x)}{g(x)} = \lim_{x \to a} \frac{f'(x)}{g'(x)} \qquad (1)$$

provided the limit on R.H.S. of (1) exists.

Proof: Since $f(x)$ and $g(x)$ satisfy conditions of Cauchy's mean value theorem in (c, d), we have

$$\frac{f(x) - f(a)}{g(x) - g(a)} = \frac{f'(e)}{g'(e)}$$

where $e \epsilon (c, d)$ such that $x < e < a$ or $a < e < x$. From hypothesis $f(a) = 0$, $g(a) = 0$. So

$$\frac{f(x)}{g(x)} = \frac{f'(e)}{g'(e)}$$

As $x \to a$, $e \to a$, we have

$$\lim_{x \to a} \frac{f(x)}{g(x)} = \lim_{x \to a} \frac{f'(x)}{g'(x)}$$

Note: In R.H.S. of (1) the functions $f(x)$ and $g(x)$ are differentiated w.r.t. x, separately, *not by using the quotient rule.*

Corollary 1: Suppose $f'(a) = 0$, and $g'(a) = 0$ in addition to $f(a) = 0$ and $g(a) = 0$. Then applying Cauchy's mean value theorem to $f'(x)$ and $g'(x)$, we get the L'Hospital's rule as

$$\lim_{x\to a} \frac{f(x)}{g(x)} = \lim_{x\to a} \frac{f''(a)}{g''(a)}$$

Corollary 2: If $f(a) = 0$, $g(a) = 0$, $f'(a) = 0$, $g'(a) = 0, \ldots, f^{(n-1)}(a) = 0$, $g^{(n-1)}(a) = 0$, and $g^{(n)}(a) \neq 0$ then

$$\lim_{x\to a} \frac{f(x)}{g(x)} = \lim_{x\to a} \frac{f^{(n)}(a)}{g^{(n)}(a)}$$

Corollary 3: As $x \to \infty$ if $\frac{f(x)}{g(x)}$ assumes indeterminate form $\frac{0}{0}$ then the L'Hospital's rule is

$$\lim_{x\to\infty} \frac{f(x)}{g(x)} = \lim_{x\to\infty} \frac{f'(x)}{g'(x)}$$

Results similar to Corollary 1, and Corollary 2 are valid as $x \to \infty$.

Indeterminate Forms of the Type $\frac{\infty}{\infty}$

Suppose $f(x) = \infty$ and $g(x) = \infty$ as $x \to a$ (or as $x \to \pm\infty$). Then $\frac{f(x)}{g(x)}$ assumes the indeterminate form $\frac{\infty}{\infty}$ which can also be evaluated by L'Hospital's rule.

$$\lim_{\substack{x\to a \\ (\text{or } x\to\pm\infty)}} \frac{f(x)}{g(x)} = \lim_{\substack{x\to a \\ (\text{or } x\to\pm\infty)}} \frac{f'(x)}{g'(x)}$$

Results similar to Corollary 1 and Corollary 2 are valid in this case also.

Important Note: While evaluating the indeterminate form $\frac{\infty}{\infty}$, in many cases, it is advised to convert the indeterminate from $\frac{\infty}{\infty}$ to the form $\frac{0}{0}$ as early as possible (say by rewriting as $\frac{\frac{1}{g(x)}}{\frac{1}{f(x)}}$). Otherwise the process of differentiation of the numerator $f(x)$ and denominator $g(x)$ may *never* terminate, thus complicating the problem.

Indeterminate Form $0 \cdot \infty$

As $x \to a$ (or $x \to \pm\infty$) if $f(x) \to 0$ and $g(x) \to \infty$ then the product $f(a) \cdot g(a)$ is undefined and is said to assume the indeterminate form $0 \cdot \infty$. This can be converted to the indeterminate form $\frac{0}{0}$ or $\frac{\infty}{\infty}$ by rewriting

$$f(x) \cdot g(x) = \frac{f(x)}{\left(\frac{1}{g(x)}\right)} \quad \text{or} \quad \frac{g(x)}{\left(\frac{1}{f(x)}\right)}$$

and apply L'Hospital's rule.

Indeterminate Form $\infty - \infty$

As $x \to a$(or $x \to \pm\infty$) if $f(x) \to \infty$ and $g(x) \to \infty$ then the difference $f(a) - g(a)$ is indeterminate of the form $\infty - \infty$. Rewriting the difference into a fraction of algebraic means, we get

$$f(x) - g(x) = \frac{\frac{1}{g(x)} - \frac{1}{f(x)}}{\frac{1}{f(x)g(x)}}$$

which $\frac{0}{0}$ form. Applying L'Hospital's rule, the indeterminate form $\infty - \infty$ is evaluated.

Indeterminate Forms of the Type 0^0, ∞^0, 1^∞

As $x \to a$ (or $x \to\neq \infty$), the expression $f(a)^{g(a)}$ is said to be (i) indeterminate form of type 0^0 if $f(x) \to 0$ and $g(x) \to 0$ (ii) indeterminate form of type ∞^0 if $f(x) \to \infty$ and $g(x) \to 0$ (iii) indeterminate form of type 1^∞ if $f(x) \to 1$ and $g(x) \to \infty$. To evaluate these forms consider.

$$y(x) = f(x)^{g(x)}$$

Taking logarithm

$$\ln y = g(x) \ln f(x)$$

which is of the form $0 \cdot \infty$ and can be evaluated as in *Indeterminate Form* $0 \cdot \infty$. Taking the limit as $x \to a$ (or $x \to \pm\infty$)

$$\lim_{x\to a} \ln y = k \text{ (say)}.$$

Then

$$k = \lim_{x\to a} (\ln y) = \ln\left(\lim_{x\to a} y\right) = \ln\left(\lim_{x\to a} f(x)^{g(x)}\right)$$

$$\therefore \lim_{x\to a} f(x)^{g(x)} = e^k.$$

Notes:

i. Maclaurin series expansion may be used to simplify the expressions.

ii. Standard limits such as $\lim_{x\to 0}\frac{\sin x}{x}=1$,

$\lim_{x\to 0}\frac{\tan x}{x}=1$, $\lim_{x\to 0}(1+x)^{\frac{1}{x}}=e$ may also be used.

WORKED OUT EXAMPLES

Type $\frac{0}{0}$ form

Example 1: Evaluate

(a) $\lim_{x\to 1}\frac{1+\ln x-x}{1-2x+x^2}$ (b) $\lim_{x\to 0}\frac{e^{ax}-e^{-ax}}{\ln(1+bx)}$

(c) $\lim_{x\to\frac{\pi}{4}}\frac{(\sec^2 x-2\tan x)}{1+\cos 4x}$

Solution:

a. At $x=1$, $\frac{1+\ln x-x}{1-2x+x^2}=\frac{1+\ln 1-1}{1-2\cdot 1+1^2}=\frac{0}{0}$, indeterminate. Applying L'Hospital's rule

$$\lim_{x\to 1}\frac{1+\ln x-x}{1-2x+x^2}=\lim_{x\to 1}\frac{(1+\ln x-x)^{\text{i}}}{(1-2x+x^2)^{\text{i}}}$$

$$=\lim_{x\to 1}\frac{0+\frac{1}{x}-1}{-2+2x}$$

which is again of the $\frac{0}{0}$ form. Applying L'Hospital's rule again, we get

$$\lim_{x\to 1}\frac{1+\ln x-x}{1-2x+x^2}=\lim_{x\to 1}\frac{(\frac{1}{x}-1)^{\text{i}}}{(-2+2x)^{\text{i}}}$$

$$=\lim_{x\to 1}\frac{-\frac{1}{x^2}}{2}=-\frac{1}{2}.$$

b. At $x=0$, $\frac{e^{ax}-e^{-ax}}{\ln(1+bx)}=\frac{1-1}{\ln 1}=\frac{0}{0}$ form.

Applying L'Hospital's rule

$$\lim_{x\to 0}\frac{e^{ax}-e^{-ax}}{\ln(1+bx)}=\lim_{x\to 0}\frac{\frac{d}{dx}(e^{ax}-e^{-ax})}{\frac{d}{dx}(\ln(1+bx))}$$

$$=\lim_{x\to 0}\frac{ae^{ax}+ae^{-ax}}{\frac{b}{1+bx}}=\frac{a+a}{\frac{b}{1}}=\frac{2a}{b}$$

c. $\lim_{x\to\frac{\pi}{4}}\frac{\sec^2 x-2\tan x}{1+\cos 4x}$: $\frac{0}{0}$ form. Applying the rule

$$=\lim_{x\to\frac{\pi}{4}}\frac{\frac{d}{dx}(\sec^2 x-2\tan x)}{\frac{d}{dx}(1+\cos 4x)}$$

$$=\lim_{x\to\frac{\pi}{4}}\frac{2\sec x\cdot\sec x\cdot\tan x-2\sec^2 x}{-4\sin 4x}$$

$$=\lim_{x\to\frac{\pi}{4}}\frac{\sec^2 x(\tan x-1)}{-2\sin 4x}:\frac{0}{0}$$

$$=\lim_{x\to\frac{\pi}{4}}\frac{2\sec x\cdot\sec x\cdot\tan x(\tan x-1)+\sec^2 x\cdot\sec^2 x}{-8\cos 4x}$$

$$=\frac{2\cdot 2\cdot 1\cdot(1-1)+2\cdot 2}{-8\cdot(-1)}=\frac{1}{2}$$

(since $\cos 4\frac{\pi}{4}=-1$, $\sec^2\frac{\pi}{4}=2$, $\tan\frac{\pi}{4}=1$).

Example 2: Evaluate

i. $\lim_{x\to 0}\frac{\cos x-\ln(1+x)-1+x}{\sin^2 x}$

ii. $\lim_{x\to 0}\frac{x^2+2\cos x-2}{x\sin^3 x}$

iii. $\lim_{x\to 0}\frac{\sin 2x+a\sin x}{x^3}$ is finite. Find a and the limit.

Solution:

i. $\frac{0}{0}$ form use Maclaurin series expansions for $\cos x$, $\ln(1+x)$ and $\sin x$. Then

$$\lim_{x\to 0}\frac{\cos x-\ln(1+x)-1+x}{\sin^2 x}$$

$$=\lim_{x\to 0}\frac{\left(1-\frac{x^2}{2!}+\frac{x^4}{4!}-\frac{x^6}{6!}+\cdots\right)\left(x-\frac{x^2}{2}+\frac{x^3}{3}-\frac{x^4}{4}+\cdots\right)-1+x}{\left(x-\frac{x^3}{3!}+\frac{x^5}{5!}-\frac{x^7}{7!}+\cdots\right)^2}$$

$$=\lim_{x\to 0}\frac{-\frac{x^3}{3}+\frac{7}{24}x^4+\cdots}{x^2-\frac{x^6}{(3!)^2}+\frac{x^{10}}{(5!)^2}+\cdots-\frac{2x^4}{3!}+\frac{2x^6}{5!}+\cdots}$$

$$=\lim_{x\to 0}\frac{-\frac{1}{3}x+\frac{7}{24}x^3+\cdots}{1-\frac{1}{3}x^2+\frac{1}{36}x^4+\cdots}=\frac{0}{1}=0.$$

ii. $\lim_{x\to 0}\frac{x^2+2\cos x-2}{x\cdot\sin^3 x}$

$$=\lim_{x\to 0}\frac{x^2+2\cos x-2}{x\cdot x^3}\cdot\frac{x^3}{\sin^3 x}$$

$$=\lim_{x\to 0}\frac{x^2+2\cos x-2}{x^4}\cdot\lim_{x\to 0}\left(\frac{x}{\sin x}\right)^3$$

$$=\lim_{x\to 0}\frac{x^2+2\cos x-2}{x^4}\cdot 1=\frac{0}{0}$$

since $\lim_{x\to 0}\frac{\sin x}{x}=1$. Now applying L'Hospital's rule

$$=\lim_{x\to 0}\frac{2x-2\sin x}{4x^3}=\lim_{x\to 0}\frac{1-\cos x}{6x^2}=\lim_{x\to 0}\frac{\sin x}{12x}$$

$$= \frac{1}{12}\lim_{x\to 0}\frac{\sin x}{x} = \frac{1}{12}\cdot 1 = \frac{1}{12}.$$

iii. At $x = 0$, $\frac{\sin 2x + a\sin x}{x^3}$ is $\frac{0}{0}$ for any a.

Applying L'Hospital's rule

$$\lim_{x\to 0}\frac{\sin 2x + a\sin x}{x^3}$$

$$= \lim_{x\to 0}\frac{2\cdot\cos 2x + a\cos x}{3x^2} = \infty.$$

In order that the limit is finite, choose $a = -2$ in which case it reduces to $\frac{0}{0}$ form. Applying L'Hospital's rule

$$= \lim_{x\to 0}\frac{-4\sin 2x + 2\sin x}{6x} = \frac{0}{0}$$

$$= \lim_{x\to 0}\frac{-4\cos 2x + \cos x}{3} = \frac{-4+1}{3} = -1.$$

Type $\frac{\infty}{\infty}$ form

Example 3: Evaluate (i) $\lim_{x\to\infty}\frac{x^n}{e^x}$

(ii) $\lim_{x\to\infty}\frac{x}{\sqrt{1+x^2}}$ (iii) $\lim_{x\to 0}\log_{\tan x}\tan 2x$.

Solution:

i. $\frac{\infty}{\infty}$ form. Applying L'Hospital's rule repeatedly n times

$$\lim_{x\to\infty}\frac{x^n}{e^x} = \lim_{x\to 0}\frac{\frac{d^n}{dx^n}x^n}{\frac{d^n}{dx^n}e^x} = \lim_{x\to\infty}\frac{n!}{e^x} = 0.$$

ii. $\frac{\infty}{\infty}$ form. Applying L'Hospital's rule

$$\lim_{x\to\infty}\frac{x}{\sqrt{1+x^2}} = \frac{1}{\frac{1}{2}(1+x^2)^{-\frac{1}{2}}\cdot 2x}$$

$$= \frac{\sqrt{1+x^2}}{x} : \frac{\infty}{\infty} \text{ form.}$$

Again applying L'Hospital's rule

$$= \lim_{x\to\infty}\frac{\frac{1}{2}(1+x^2)^{-\frac{1}{2}}\cdot 2x}{1}$$

$$= \lim_{x\to\infty}\frac{x}{\sqrt{1+x^2}} : \frac{\infty}{\infty} : \text{ which is the original function.}$$

Instead introduce $z = \frac{1}{x^2}$ then

$$\lim_{x\to\infty}\frac{x}{\sqrt{1+x^2}} = \lim_{z\to 0}\frac{x}{x\sqrt{\frac{1}{x^2}+1}}$$

$$= \lim_{z\to 0}\frac{1}{\sqrt{z+1}} = 1.$$

iii. $\frac{\infty}{\infty}$ form.

$$\lim_{x\to 0}\log_{\tan x}\tan 2x = \lim_{x\to 0}\frac{\log\tan 2x}{\log\tan x} : \frac{\infty}{\infty}$$

Applying L'Hospital's rule

$$= \lim_{x\to 0}\frac{1}{\tan 2x}\cdot 2\cdot\sec^2 2x\cdot\frac{\tan x}{\sec^2 x} : \frac{0}{0} \text{ form}$$

$$= \lim_{x\to 0}\frac{\sin x\cdot\cos x\cdot 2}{\sin 2x\cdot\cos 2x} = \lim_{x\to 0}\frac{2\cdot\frac{1}{2}\sin 2x}{\frac{1}{2}\sin 4x}$$

$$= \lim_{x\to 0}\cdot\frac{\sin 2x}{2x}\cdot\frac{4x}{\sin 4x} = 1\cdot 1 = 1.$$

Type $0\cdot\infty$ form

Example 4: Evaluate

i. $\lim_{x\to 1}\log(1-x)\cdot\cot\frac{\pi x}{2}$

ii. $\lim_{x\to a}(a-x)\cdot\tan\frac{\pi x}{2a}$.

Solution:

i. $\infty\cdot 0$ forms. Rewriting in $\frac{\infty}{\infty}$ form

$$\lim_{x\to 1}\log(1-x)\cdot\cot\frac{\pi x}{2} = \lim_{x\to 1}\frac{\ln(1-x)}{\tan\frac{\pi x}{2}}.$$

Applying L'Hospital's rule

$$= \lim_{x\to 1}\frac{\frac{1}{1-x}\cdot(-1)}{\frac{\pi}{2}\cdot\sec^2\frac{\pi x}{2}} : \frac{\infty}{\infty} \text{ form.}$$

Rewriting in $\frac{0}{0}$ form

$$= \lim_{x\to 1}\frac{-2}{\pi}\frac{\cos^2\frac{\pi x}{2}}{1-x} : \frac{0}{0} \text{ form.}$$

Applying L'Hospital's rule

$$= \lim_{x\to 1} -\frac{2}{\pi}\cdot\frac{2\cdot\cos\frac{\pi x}{2}\cdot\left(-\sin\frac{\pi x}{2}\right)\cdot\frac{\pi}{2}}{-1} = 0$$

ii. $0 \cdot \infty$ form. Rewriting

$$\lim_{x\to a}(a-x)\cdot\tan\frac{\pi x}{2a} = \lim_{x\to a}\frac{\tan\frac{\pi x}{2a}}{\left(\frac{1}{a-x}\right)} : \frac{\infty}{\infty} \text{ form.}$$

Applying L'Hospital's rule

$$= \lim_{x\to a}\frac{\frac{2a}{\pi}\cdot\sec^2\frac{\pi x}{2a}}{-\frac{1}{(a-x)^2}(-1)} : \frac{\infty}{\infty} \text{ form.}$$

Rewriting in $\frac{0}{0}$ form

$$= \lim_{x\to a}\frac{2a}{\pi}\cdot\frac{(a-x)^2}{\cos^2\left(\frac{\pi x}{2a}\right)} =$$

$$= \lim_{x\to a}\frac{2a}{\pi}\cdot\frac{(-2)(a-x)}{2\cdot\cos\left(\frac{\pi x}{2a}\right)\cdot\left(-\sin\frac{\pi x}{2a}\right)\cdot\frac{\pi}{2a}} : \frac{0}{0}$$

$$= \lim_{x\to a}\frac{-1}{\frac{\pi}{2a}\cdot\cos^2\frac{\pi x}{2a} - \frac{\pi}{2a}\sin^2\frac{\pi x}{2a}}$$

$$= \frac{-1}{0-\frac{\pi}{2a}} = \frac{2a}{\pi}.$$

Type $\infty - \infty$ form

Example 5: Evaluate

i. $\lim_{x\to\frac{\pi}{2}}(\sec x - \tan x)$

ii. $\lim_{x\to 1}\left[\frac{x}{x-1} - \frac{1}{\ln x}\right]$

iii. $\lim_{x\to 0}\left(\frac{1}{x^2} - \cot^2 x\right)$

Solution:

i. $\infty - \infty$ form. Rewriting

$$\lim_{x\to\frac{\pi}{2}}(\sec x - \tan x) = \lim_{x\to\frac{\pi}{2}}\left(\frac{1}{\cos x} - \frac{\sin x}{\cos x}\right)$$

$$= \lim_{x\to\frac{\pi}{2}}\left(\frac{1-\sin x}{\cos x}\right) : \frac{0}{0}$$

Applying L'Hospital's rule

$$\lim_{x\to\frac{\pi}{2}}\frac{\cos x}{\sin x} = \frac{0}{1} = 0.$$

ii. $\infty - \infty$ form. Rewriting

$$\lim_{x\to 1}\left[\frac{x}{x-1} - \frac{1}{\ln x}\right] = \lim_{x\to 1}\frac{x\ln x - (x-1)}{(x-1)\ln x} : \frac{0}{0} \text{ form.}$$

$$= \lim_{x\to 1}\frac{x\cdot\frac{1}{x} + \ln x - 1}{1\cdot\ln x + (x-1)\cdot\frac{1}{x}} : \frac{0}{0} \text{ form}$$

$$= \lim_{x\to 1}\frac{\frac{1}{x}}{\frac{1}{x}+\frac{1}{x^2}} = \frac{1}{1+1} = \frac{1}{2}.$$

iii. $\infty - \infty$ form. Rewriting

$$\lim_{x\to 0}\left(\frac{1}{x^2} - \cot^2 x\right) = \lim_{x\to 0}\left(\frac{1}{x^2} - \frac{\cos^2 x}{\sin^2 x}\right)$$

$$= \lim_{x\to 0}\left(\frac{\sin^2 x - x^2\cos^2 x}{x^2\sin^2 x}\right) : \frac{0}{0}$$

$$= \lim_{x\to 0}\frac{\sin^2 x - x^2\cos^2 x}{x^4}\cdot\lim_{x\to 0}\frac{x^2}{\sin^2 x}.$$

Since $\lim_{x\to 0}\frac{\sin x}{x} = 1$,

$$= \lim_{x\to 0}\left[\frac{1}{x^2}\cdot\frac{\sin^2 x}{x^2} - \frac{\cos^2 x}{x^2}\right]\cdot 1$$

$$= \lim_{x\to 0}\left(\frac{1}{x^2}\cdot 1 - \frac{\cos^2 x}{x^2}\right) : \infty - \infty$$

$$= \lim_{x\to 0}\left(\frac{1-\cos^2 x}{x^2}\right) : \frac{0}{0} \text{ form. Applying rule}$$

$$= \lim_{x\to 0}\frac{-2\cos x\cdot(-\sin x)}{2x}$$

$$= \lim_{x\to 0}\cos x\cdot\lim_{x\to 0}\frac{\sin x}{x} = 1.1 = 1.$$

Types 0^0, ∞^0, 1^∞

Example 6: Evaluate (0^0 form)

(i) $\lim_{x\to 0}(x)^x$ (ii) $\lim_{x\to\frac{\pi}{2}}(\cos x)^{\left(\frac{\pi}{2}-x\right)}$

(iii) $\lim_{x\to 0}(x+\sin x)^{\tan x}$.

Solution:

i. 0^0 form. Put $y = x^x$. Take logarithm, then $\lim_{x\to 0}\ln y = \lim_{x\to 0}x\ln x = 0\cdot\infty$ form. Rewriting

$$= \lim_{x\to 0}\frac{\ln x}{\frac{1}{x}} : \frac{\infty}{\infty} \text{ form} = \lim_{x\to 0}\frac{\frac{1}{x}}{-\frac{1}{x^2}} = \lim_{x\to 0} -x = 0$$

$$\therefore \lim_{x\to 0}x^x = \lim_{x\to 0}y = e^0 = 1.$$

ii. 0^0 form. Put $y=(\cos x)^{(\frac{\pi}{2}-x)}$. Take logarithm, then

$$\lim_{x\to\frac{\pi}{2}} \ln y = \lim_{x\to\frac{\pi}{2}} \left(\frac{\pi}{2}-x\right)\ln\cos x : 0.\infty \text{ form}$$

$$= \lim_{x\to\frac{\pi}{2}} \frac{\ln\cos x}{\frac{1}{(\frac{\pi}{2}-x)}} : \frac{\infty}{\infty} \text{ form. Applying rule}$$

$$= \lim_{x\to\frac{\pi}{2}} \frac{-\sin x}{\cos x}\cdot\frac{1}{\frac{-1}{(\frac{\pi}{2}-x)^2}(-1)}$$

$$= \lim_{x\to\frac{\pi}{2}} \frac{(\frac{\pi}{2}-x)^2}{\cot x} : \frac{0}{0} \text{ form}$$

$$= \lim_{x\to\frac{\pi}{2}} \frac{2(\frac{\pi}{2}-x)\cdot(-1)}{-\text{cosec}^2 x} = 0$$

$$\therefore \lim_{x\to\frac{\pi}{2}} (\cos x)^{(\frac{\pi}{2}-x)} = e^0 = 1$$

iii. 0^0 form. Put $y=(x+\sin x)^{\tan x}$. Take log

$$\lim_{x\to 0} \ln y = \lim_{x\to 0} \tan x\cdot\ln(x+\sin x) : 0\cdot\infty \text{ form.}$$

$$= \lim_{x\to 0} \frac{\ln(x+\sin x)}{\cot x} : \frac{\infty}{\infty} \text{ form}$$

$$= \lim_{x\to 0} \frac{1+\cos x}{x+\sin x}\cdot\frac{1}{(-\text{cosec}^2 x)}$$

$$= \lim_{x\to 0} -\frac{\sin^2 x\cdot(1+\cos x)}{(x+\sin x)} : \frac{0}{0}$$

$$= \lim_{x\to 0} -\frac{[2\sin x\cdot\cos x(1+\cos x)+\sin^2 x\cdot(-\sin x)]}{1+\cos x}$$

$$= \frac{0}{2} = 0.$$

$$\lim_{x\to 0} (x+\sin x)^{\tan x} = e^0 = 1.$$

Example 7: Evaluate (∞^0 form)

(i) $\lim_{x\to\infty} (e^x+x)^{\frac{1}{x}}$ (ii) $\lim_{x\to\frac{\pi}{2}} (\tan x)^{\tan 2x}$

(iii) $\lim_{x\to\infty} (\cosh x)^{\frac{1}{x}}$

Solution:

i. Put $y=(e^x+x)^{\frac{1}{x}}$. Take log

$$\lim_{x\to\infty} \ln y = \lim_{x\to\infty} \frac{1}{x}\cdot\ln(e^x+x) : 0\cdot\infty \text{ form}$$

$$= \lim_{x\to\infty} \frac{\ln(e^x+x)}{x} : \frac{\infty}{\infty} \text{ form}$$

$$= \lim_{x\to\infty} \frac{\frac{e^x+1}{e^x+x}}{1} : \frac{\infty}{\infty} \text{ form}$$

$$= \lim_{x\to\infty} \frac{e^x}{e^x+1} = \lim_{x\to\infty} \frac{e^x}{e^x} = 1.$$

$$\therefore \lim_{x\to\infty} (e^x+x)^{\frac{1}{x}} = e^1 = e.$$

ii. Put $y=(\tan x)^{\tan 2x}$. Take log

$$\lim_{x\to\frac{\pi}{2}} \ln y = \lim_{x\to\frac{\pi}{2}} \tan 2x\cdot\ln(\tan x) : 0\cdot\infty$$

$$= \lim_{x\to\frac{\pi}{2}} \frac{\ln(\tan x)}{\cot 2x} = \lim_{x\to\frac{\pi}{2}} \frac{\sec^2 x}{\tan x}\cdot\frac{1}{-2\cdot\text{cosec}^2 2x} : \frac{\infty}{\infty}$$

$$= \lim_{x\to\frac{\pi}{2}} \frac{\sin^2 2x\cdot}{(-2)\tan x\cos^2 x} = \lim_{x\to\frac{\pi}{2}} \frac{\sin^2 2x}{-\sin 2x}$$

$$= \lim_{x\to\frac{\pi}{2}} -\sin 2x = 0.$$

$$\therefore \lim_{x\to\frac{\pi}{2}} (\tan x)^{\tan 2x} = e^0 = 1.$$

iii. Put $y=(\cosh x)^{\frac{1}{x}}$, Take log

$$\lim_{x\to\infty} \ln y = \lim_{x\to\infty} \frac{\ln\cosh x}{x} = \lim_{x\to\infty} \frac{\sinh x}{\cosh x}\cdot\frac{1}{1} : \frac{\infty}{\infty}$$

$$= \lim_{x\to\infty} \frac{e^x-e^{-x}}{e^x+e^{-x}} = \lim_{x\to\infty} \frac{e^{2x}-1}{e^{2x}+1} = \lim_{x\to\infty} \frac{2e^{2x}}{2e^{2x}} = 1 : \frac{\infty}{\infty}$$

$$\therefore \lim_{x\to\infty} (\cosh x)^{\frac{1}{x}} = e^1 = e.$$

Example 8: Evaluate (1^∞ form)

(i) $\lim_{x\to 1} \left(\tan\frac{\pi x}{4}\right)^{\tan\frac{\pi x}{2}}$ (ii) $\lim_{x\to 0} \left(\frac{2x+1}{x+1}\right)^{\frac{1}{x}}$

(iii) $\lim_{x\to 0} \left(\frac{\sinh x}{x}\right)^{\frac{1}{x^2}}$ (iv) $\lim_{x\to 0} \left(\frac{a^x+b^x+c^x}{3}\right)^{\frac{1}{x}}$

Solution:

i. Put $y=(\tan\frac{\pi x}{4})^{\tan\frac{\pi x}{2}}$. Take log

$$\lim_{x\to 1} \ln y = \lim_{x\to 1} \tan\left(\frac{\pi x}{2}\right)\cdot\ln\tan\left(\frac{\pi x}{4}\right)$$

$$\cdot : \infty\cdot 0\cdot \text{ form}$$

$$= \lim_{x\to 1} \frac{\ln\tan(\frac{\pi x}{4})}{\cot(\frac{\pi x}{2})} =$$

$$= \lim_{x\to 1} \frac{\frac{\pi}{4}\sec^2(\frac{\pi x}{4})}{\tan\frac{\pi x}{4}}\cdot\frac{1}{(-\frac{\pi}{2})\text{cosec}^2\frac{\pi x}{2}} : \frac{0}{0} \text{ form}$$

$$=\frac{\frac{\pi}{4}\cdot(\sqrt{2})^2}{1}\cdot\frac{1}{\frac{-\pi}{2}\cdot 1}=-1$$

$$\therefore \lim_{x\to 1}\left(\tan\frac{\pi x}{4}\right)^{\tan\frac{\pi x}{2}}=e^{-1}=\frac{1}{e}.$$

ii. Put $y=\left(\frac{2x+1}{x+1}\right)^{\frac{1}{x}}$. Take log

$$\lim_{x\to 0}\ln y=\lim_{x\to 0}\frac{1}{x}\cdot\ln\left(\frac{2x+1}{x+1}\right): 0\cdot\infty \text{ form}$$

$$=\lim_{x\to 0}\frac{\ln\left(\frac{2x+1}{x+1}\right)}{x}=\frac{x+1}{2x+1}\cdot\frac{(x+1)2-(2x+1)\cdot 1}{(x+1)^2}\cdot\frac{1}{1}$$

$$=2-1=1$$

$$\therefore \lim_{x\to 0}\left(\frac{2x+1}{x+1}\right)^{\frac{1}{x}}=e^1.$$

iii. Expanding sinh$x$ in Maclaurin series

$$\lim_{x\to 0}\left(\frac{\sinh x}{x}\right)^{\frac{1}{x^2}}$$

$$=\lim_{x\to 0}\left(\frac{x+\frac{x^3}{3!}+\frac{x^4}{4!}+\frac{x^5}{5!}+\cdots}{x}\right)^{\frac{1}{x^2}}$$

$$=\lim_{x\to 0}\left[1+x^2\left(\frac{1}{3!}+\frac{x}{4!}+\frac{x^2}{5!}+\cdots\right)\right]^{\frac{1}{x^2}}$$

Put $t=\frac{1}{3!}+\frac{x}{4!}+\frac{x^2}{5!}+\cdots$, so $\lim_{x\to 0}t=\frac{1}{6}$.

Now

$$\lim_{x\to 0}\left(\frac{\sin hx}{x}\right)^{\frac{1}{x^2}}=\lim_{x\to 0}\left\{[1+tx^2]^{\frac{1}{tx^2}}\right\}^t=\lim_{x\to 0}e^t$$

$$=e^{\lim_{x\to 0}t}=e^{\frac{1}{6}}$$

iv. Put $y=\left(\frac{a^x+b^x+c^x}{3}\right)^{\frac{1}{x}}$. Take log

$$\lim_{x\to 0}\ln y=\lim_{x\to 0}\frac{1}{x}\cdot\ln\left(\frac{a^x+b^x+c^x}{3}\right):\ \frac{0}{0}\text{ form}$$

Apply rule

$$=\lim_{x\to 0}\frac{3}{a^x+b^x+c^x}\times$$

$$\times\frac{(a^x\log_e a+b^x\log_e b+c^x\log_e c)}{3}\cdot\frac{1}{1}$$

$$=\frac{3}{1+1+1}\,\frac{(1\cdot\log_e a+1\cdot\log_e b+1\cdot\log_e c)}{3}$$

$$=\frac{1}{3}\log_e abc$$

$$\therefore \lim_{x\to 0}\left(\frac{a^x+b^x+c^x}{3}\right)^{\frac{1}{x}}=e^{\frac{1}{3}\log_e abc}=(abc)^{\frac{1}{3}}.$$

EXERCISE

Evaluate

Type $\frac{0}{0}$

i. $\lim_{x\to 0}\frac{\sinh x-x}{\sin x-x\cos x}$

ii. $\lim_{x\to 4}\frac{x^2-16}{x^2+x-20}$

iii. $\lim_{x\to 1}\frac{\sqrt{x}-1+\sqrt{x-1}}{\sqrt{x^2-1}}$

iv. $\lim_{x\to 0}\frac{2\tan x\cdot\sec x}{xe^x}$

v. $\lim_{x\to 0}\frac{e^x-e^{\sin x}}{x-x\cos x}$

vi. $\lim_{x\to 0}\frac{xe^x-\ln(1+x)}{x^2}$

vii. $\lim_{x\to 1}\frac{x^2-x}{x-1-\ln x}$

viii. $\lim_{x\to 0}\frac{e^x-e^{-x}-2\ln(1+x)}{x\sin x}$

ix. $\lim_{x\to\infty}\frac{ax^2+b}{cx^2-d}$

x. Determine a and b such that

$$\lim_{x\to 0}\frac{x(1+a\cos x)-b\sin x}{x^3}=1$$

xi. $\lim_{x\to 0}\frac{(1+x)^{\frac{1}{x}}-e}{x}$

Hint: Expand $\ln(1+x)$ in (viii)

Ans. (1) $\frac{1}{2}$ (ii) $\frac{8}{9}$ (iii) $\frac{1}{\sqrt{2}}$ (iv) 2 (v) $\frac{1}{3}$ (vi) $\frac{1}{2}$ (vii) 2 (viii) 1 (ix) $\frac{a}{c}$ (x) $a=-\frac{5}{2}, b=-\frac{3}{2}$ (xi) $-\frac{e}{2}$.

Type $\frac{\infty}{\infty}$

(i) $\lim_{x\to\frac{\pi}{2}} \frac{\tan x}{\tan 3x}$ (ii) $\lim_{x\to a} \frac{\ln(x-a)}{\ln(e^x - e^a)}$ (iii) $\lim_{x\to\infty} \frac{x^2}{e^{x2}}$

(iv) $\lim_{x\to 0} \frac{\ln x}{\text{cosec} x}$ (v) $\lim_{x\to 0} \frac{\ln x}{\cot x}$ (vi) $\lim_{x\to\infty} \frac{e^x}{x^n}$.

Ans. (i)3 (ii) e^{2a} (iii) 0 (iv) 0 (v) 0 (vi) ∞.

Type $0\cdot\infty$

(i) $\lim_{x\to 0} x^n \ln x$ witn $n > 0$ (ii) $\lim_{x\to 0} x \ln \tan x$

(iii) $\lim_{\theta\to\pi} \ln(\theta - \pi).\tan\theta$ (iv) $\lim_{x\to\infty} xe^{-x}$

(v) $\lim_{x\to 0} xe^{\frac{1}{x}}$ (vi) $\lim_{x\to\frac{\pi}{2}} \tan x \cdot \tan 2x$.

Ans. (i) 0 (ii) 1 (iii) 0 (iv) 0 (v) ∞ (vi) -2.

Type $\infty - \infty$

(i) $\lim_{x\to 0}(\text{cosec} x - \cot)$ (ii) $\lim_{x\to 1}\left(\frac{x}{\ln x} - \frac{1}{\ln x}\right)$

(iii) $\lim_{x\to 0}\left(\frac{1}{\sin^2 x} - \frac{1}{x^2}\right)$ (iv) $\lim_{x\to\infty}(e^x - x)$

(v) $\lim_{x\to 0} \frac{1}{e^x - 1} - \frac{1}{x}$ (vi) $\lim_{x\to 2}\left\{\frac{1}{x-2} - \frac{1}{\ln(x-1)}\right\}$

Ans. (i) 0 (ii) -1 (iii) $\frac{1}{3}$ (iv) 0 (v) $-\frac{1}{2}$ (vi) $-\frac{1}{2}$.

Type $0^0 \infty^0 1^\infty$

(i) $\lim_{x\to a}(x-a)^{x-a}$ (ii) $\lim_{x\to 0}(\sin x)^x$

(iii) $\lim_{x\to\frac{\pi}{2}}(\sin x)^{\tan x}$ (iv) $\lim_{x\to 0}\left(\frac{\tan x}{x}\right)^{\frac{1}{x^2}}$

(v) $\lim_{x\to 0}(\cos x)^{\frac{1}{x^2}}$ (vi) $\lim_{x\to a}\left(2 - \frac{x}{a}\right)^{\tan\frac{\pi x}{2a}}$

(vii) $\lim_{x\to 0}(\cot x)^{\frac{1}{\ln x}}$ (viii) $\lim_{x\to\frac{\pi}{2}}\left(\frac{\pi}{2} - x\right)^{\tan x}$

(ix) $\lim_{x\to\infty}(1 + x^2)e^{-x}$ (x) $\lim_{x\to\frac{\pi}{4}}(\tan x)^{\tan 2x}$

(xi) $\lim_{x\to 1}(1 - x^2)^{\frac{1}{\ln(1-x)}}$ (xii) $\lim_{x\to 0}\left(\frac{a^x + b^x}{2}\right)^{\frac{1}{x}}$

(xiii) $\lim_{x\to 0}(\cos x)^{\frac{1}{x^2}}$ (xiv) $\lim_{x\to 0}\left(\frac{\sin x}{x}\right)^{\frac{1}{x^2}}$.

Ans. (i)1 (ii) 1 (iii) 1 (iv) $e^{\frac{1}{3}}$ (v) $e^{\frac{-1}{2}}$
(vi) $e^{\frac{2}{\pi}}$ (vii) e^{-1} (viii) 1 (ix)1 (x) $\frac{1}{e}$ (xi) e
(xii) $\sqrt{ab}$ (xiii) $e^{\frac{-1}{2}}$ (xiv) $e^{\frac{-1}{6}}$.

2.10 DERIVATIVES OF ARCS

Arc Length

The graph AB of a function $y = f(x)$ defined in the interval (a, b) be the arc of the curve AB. Join the $(n+1)$ points $a_0, a_1, a_2, \ldots, a_n$ on AB by broken lines $a_0a_1, a_1a_2, a_2a_3, \ldots, a_{n-1}a_n$.

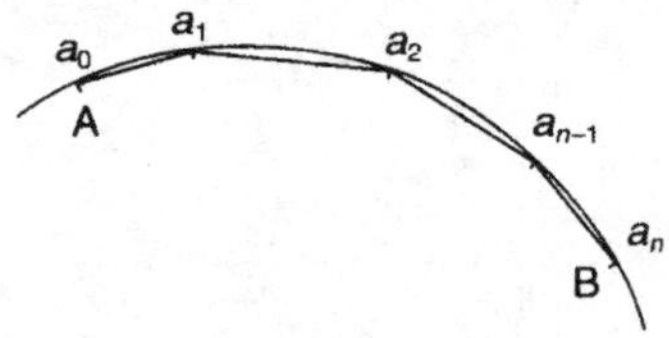

Fig. 2.9

Length of arc AB is defined to be the limit of the sum of the lengths of the n broken lines as the largest broken line $a_{i-1}a_i \to 0$.

Result:

$$\lim_{AB\to 0} \frac{\text{Length of arc } AB}{\text{Length of chord } AB} = \lim_{AB\to 0} \frac{\widehat{AB}}{\overline{AB}} = 1.$$

Derivative of Arc

Equation of the curve in the cartesian form

Let $A(x_0, y_0)$ be a fixed point of the curve $y = f(x)$. Let $P(x, y)$ and $Q(x + \Delta x, y + \Delta y)$ be any two neighbouring points on curve AB. Let s be the arc length AP measured from the fixed point A along the curve AB and denoted by $\widehat{AP}$. As P moves to Q, x changes to $x + \Delta x$, y changes to $y + \Delta y$, so the arc length s changes $s + \Delta s$ where $\Delta s = \widehat{PQ}$. Thus s is a function of x.

From the right angled triangle PQR

$$\overline{PQ}^2 = (\Delta x)^2 + (\Delta y)^2$$

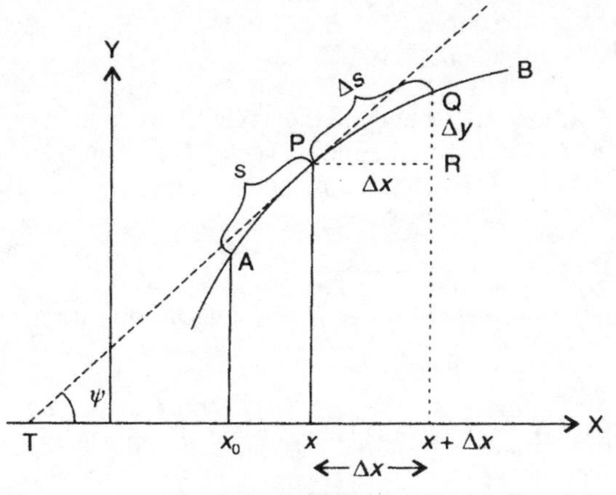

Fig. 2.10

where $\overline{PQ}$ = Length of the chord PQ, subtended by the arc PQ. Rewriting and dividing by $(\Delta x)^2$

$$\left(\frac{\overline{PQ}}{\Delta s}\right)^2\left(\frac{\Delta s}{\Delta x}\right)^2 = 1+\left(\frac{\Delta y}{\Delta x}\right)^2$$

As $Q \to P$, $\Delta x \to 0$ then chord $PQ \to 0$ and $\lim_{Q\to P} \frac{\widehat{PQ}}{\overline{PQ}} = 1$.

$$\lim_{\Delta x\to 0}\left(\frac{\overline{PQ}}{\widehat{PQ}}\right)^2\left(\frac{\Delta s}{\Delta x}\right)^2 = 1+\lim_{\Delta x\to 0}\left(\frac{\Delta y}{\Delta x}\right)^2$$

$$\left(\frac{ds}{dx}\right)^2 = 1+\left(\frac{dy}{dx}\right)^2$$

or
$$\frac{ds}{dx} = \sqrt{1+\left(\frac{dy}{dx}\right)^2} \qquad (2)$$

Here positive sign is taken before the radical, assuming that s increases as x increases, so $\frac{ds}{dx} > 0$. Thus the *differential of arc* is

$$ds = \sqrt{1+\left(\frac{dy}{dx}\right)^2}\,dx \qquad (3)$$

or
$$ds = \sqrt{(dx)^2+(dy)^2} \qquad (2^*)$$

Corollary 1: Let the equation of the curve be $x = f(y)$. Then

$$\frac{ds}{dy} = \frac{ds}{dx}\cdot\frac{dx}{dy} = \sqrt{1+\left(\frac{dy}{dx}\right)^2}\cdot\left(\frac{dx}{dy}\right)$$

$$= \sqrt{\left(\frac{dx}{dy}\right)^2+\left(\frac{dy}{dx}\right)^2\left(\frac{dx}{dy}\right)^2}$$

$$\frac{ds}{dy} = \sqrt{1+\left(\frac{dx}{dy}\right)^2} \qquad (3)$$

Corollary 2: Let ψ be the angle subtended by the tangent PT at P to the curve with the x-axis. Then $\frac{dy}{dx} = \tan\psi$. Then

$$\frac{ds}{dx} = \sqrt{1+\left(\frac{dy}{dx}\right)^2} = \sqrt{1+\tan^2\psi} = \sec\psi \quad \text{so}$$

$$\cos\psi = \frac{dx}{ds} \text{ and } \sin\psi = \tan\psi\cdot\cos\psi = \frac{dy}{dx}\cdot\frac{dx}{ds} = \frac{dy}{ds}.$$

Equation of the curve in the parametric form

$$x = x(t), \qquad y = y(t), \text{ then}$$

$$\frac{ds}{dt} = \frac{ds}{dx}\cdot\frac{dx}{dt} = \sqrt{1+\left(\frac{dy}{dx}\right)^2}\cdot\frac{dx}{dt}$$

$$= \sqrt{\left(\frac{dx}{dt}\right)^2+\left(\frac{dy}{dx}\right)^2\left(\frac{dx}{dt}\right)^2}$$

$$\frac{ds}{dt} = \sqrt{\left(\frac{dx}{dt}\right)^2+\left(\frac{dy}{dt}\right)^2}. \qquad (4)$$

Equation of the curve in polar coordinates: $r = r(\theta)$

On the curve AB, consider $P(r, \theta)$ and $Q(r+\Delta r, \theta+\Delta\theta)$ two neighbouring points. Let A be a fixed point. Assume that $\widehat{AP} = s$, $\widehat{PQ} = \Delta s$.

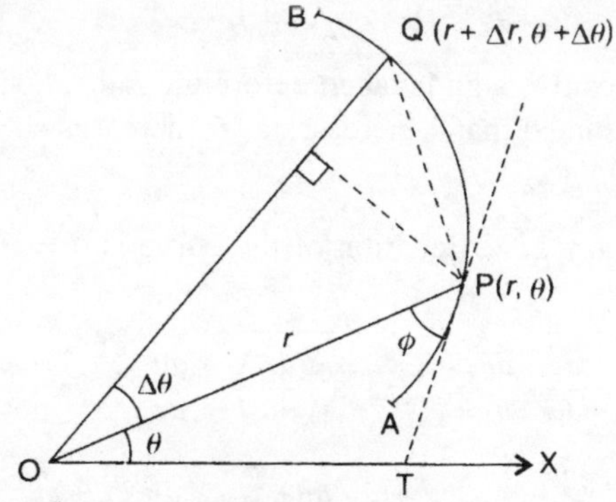

Fig. 2.11

Draw $\perp^r PN$ from P on to OQ. From the right angled triangle OPQ,

$$PN = r\sin\Delta\theta, \qquad ON = r\cos\Delta\theta$$

$$NQ = OQ - ON = (r+\Delta r) - r\cos\Delta\theta$$

$$= \Delta r + 2r\sin^2\left(\frac{\Delta\theta}{2}\right).$$

From the right angled triangle PNQ,

$$PQ^2 = PN^2 + NQ^2$$

$$PQ^2 = (r\sin\Delta\theta)^2 + \left(\Delta r + 2r\sin^2\frac{\Delta\theta}{2}\right)^2$$

Rewriting and dividing by $(\Delta\theta)^2$, we get

$$\left(\frac{PQ}{\widehat{PQ}}\cdot\frac{\widehat{PQ}}{\Delta\theta}\right)^2 = r^2\left(\frac{\sin\Delta\theta}{\Delta\theta}\right)^2$$

$$+\left(\frac{\Delta r}{\Delta\theta} + r\cdot\frac{\sin\frac{\Delta\theta}{2}}{\left(\frac{\Delta\theta}{2}\right)}\cdot\sin\left(\frac{\Delta\theta}{2}\right)\right)^2$$

As $Q\to P$, $\overline{PQ}\to 0$ so $\frac{PQ}{\widehat{PQ}}\to 1$. Also as $Q\to P$, $\Delta\theta\to 0$ so $\lim_{\Delta\theta\to 0}\frac{\sin\Delta\theta}{\Delta\theta} = 1$. With this

$$\lim_{Q\to P}\left(\frac{\widehat{PQ}}{\Delta\theta}\right)^2 = \lim_{\Delta\theta\to 0}\left(\frac{\Delta s}{\Delta\theta}\right)^2$$

$$= r^2\cdot 1 + \left(\frac{dr}{d\theta} + r\cdot 1\cdot 0\right)^2$$

Thus $$\left(\frac{ds}{d\theta}\right)^2 = r^2 + \left(\frac{dr}{d\theta}\right)^2$$

or $$\frac{ds}{d\theta} = \sqrt{r^2 + \left(\frac{dr}{d\theta}\right)^2} \qquad (5)$$

Here positive sign is taken before the radical, since it is assumed that s increases as θ increases so that $\frac{ds}{d\theta} > 0$.

Corollary 1: Equation of the curve is $\theta = \theta(r)$. Then,

$$\frac{ds}{dr} = \frac{ds}{d\theta}\cdot\frac{d\theta}{dr} = \sqrt{r^2 + \left(\frac{dr}{d\theta}\right)^2}\cdot\frac{d\theta}{dr}$$

$$= \sqrt{r^2\left(\frac{d\theta}{dr}\right)^2 + \left(\frac{dr}{d\theta}\cdot\frac{d\theta}{dr}\right)^2}$$

Thus $$\frac{ds}{dr} = \sqrt{1 + r^2\left(\frac{d\theta}{dr}\right)^2} \qquad (6)$$

Corollary 2: Let ϕ be the angle between the tangent PT at P and the radius vector OP. Then

$$r\frac{d\theta}{dr} = \tan\phi.$$

So $$\frac{ds}{dr} = \sqrt{1 + \left(\frac{rd\theta}{dr}\right)^2} = \sqrt{1+\tan^2\phi} = \sec\phi.$$

Then

$$\cos\phi = \frac{dr}{ds},\ \sin\phi = \tan\phi\cdot\cos\phi = r\frac{d\theta}{dr}\cdot\frac{dr}{ds} = r\frac{d\theta}{ds}.$$

WORKED OUT EXAMPLES

Example 1: Find $\frac{ds}{dx}$ when $y = c\cosh\frac{x}{c}$.

Solution: $\frac{dy}{dx} = c\cdot\frac{1}{c}\cdot\sinh\frac{x}{c}$

$$\frac{ds}{dx} = \sqrt{1+\left(\frac{dy}{dx}\right)^2} = \sqrt{1+\left(\sinh\frac{x}{c}\right)^2} = \cosh\frac{x}{c}.$$

Example 2: Find $\frac{ds}{dy}$ when $ax^2 = y^3$.

Solution: $2ax = 3y^2\frac{dy}{dx}$. So $\frac{dx}{dy} = \frac{3y^2}{2ax}$

$$\frac{ds}{dy} = \sqrt{1+\left(\frac{dx}{dy}\right)^2} = \sqrt{1+\frac{9}{4}\frac{y^4}{a^2x^2}} = \sqrt{1+\frac{9}{4}\frac{y}{a}}.$$

Example 3: Find $\frac{ds}{dt}$ for the curve $x = e^t\sin t$, $y = e^t\cos t$.

Solution: $\frac{dx}{dt} = e^t\sin t + e^t\cos t$, $\frac{dy}{dt} = e^t\cos t - e^t\sin t$

$$\left(\frac{ds}{dt}\right)^2 = \left(\frac{dx}{dt}\right)^2 + \left(\frac{dy}{dt}\right)^2$$

$$= e^{2t}(\sin t+\cos t)^2 + e^{2t}(\cos t - \sin t)^2$$

$$= e^{2t}\left[1 + 2\sin t\cos t + 1 - 2\sin t\cos t\right] = 2e^{2t}$$

$$\therefore \frac{ds}{dt} = \sqrt{2}\,e^t.$$

Example 4: Find $\frac{ds}{d\theta}$ for $r = a(1-\cos\theta)$.

Solution: $\frac{dr}{d\theta} = +a\sin\theta$

$$\frac{ds}{d\theta} = \sqrt{r^2 + \left(\frac{dr}{d\theta}\right)^2} = \sqrt{a^2(1-\cos\theta)^2 + a^2\sin^2\theta}$$

$$= a\sqrt{1+\cos^2\theta - 2\cos\theta + \sin^2\theta}$$

$$= a\cdot\sqrt{2}\sqrt{1-\cos\theta}$$

$$= 2a\sin\frac{\theta}{2}.$$

Example 5: Find $\frac{ds}{dr}$ for the curve $r\theta = a$.

Solution: $\theta = \frac{a}{r}$, so $\frac{d\theta}{dr} = -\frac{a}{r^2}$

$$\frac{ds}{dr} = \sqrt{1+r^2\left(\frac{d\theta}{dr}\right)^2} = \sqrt{1+r^2\cdot\frac{a^2}{r^4}}$$

$$= \sqrt{1+\frac{a^2}{r^2}} = \sqrt{1+\theta^2}.$$

Example 6: For the curve

$$\theta = \cos^{-1}\frac{r}{k} - \sqrt{\frac{k^2-r^2}{r^2}}$$

Show that $r\frac{ds}{dr}$ is a constant.

Solution: Differentiating

$$\theta = \cos^{-1}\frac{r}{k} - \sqrt{\frac{k^2-r^2}{r^2}}$$

w.r.t. r, we get

$$\frac{d\theta}{dr} = \frac{-1}{\sqrt{1-\left(\frac{r}{k}\right)^2}}\cdot\frac{1}{k} - \frac{1}{2}\sqrt{\frac{r^2}{k^2-r^2}}\times$$

$$\times\frac{r^2(-2r)-(k^2-r^2)2r}{r^4}$$

$$= -\frac{1}{\sqrt{k^2-r^2}} + \frac{1}{\sqrt{k^2-r^2}}\cdot\frac{k^2}{r^2}$$

$$= \frac{(k^2-r^2)}{r^2\sqrt{k^2-r^2}} = \frac{\sqrt{k^2-r^2}}{r^2}$$

Now

$$\frac{ds}{dr} = \sqrt{1+\left(\frac{rd\theta}{dr}\right)^2} = \sqrt{1+r^2\cdot\frac{k^2-r^2}{r^4}} = \frac{k}{r}$$

Then $r\frac{ds}{dr} = k =$ constant.

EXERCISE

1. Find $\frac{ds}{dx}$ for the curve

   (i) $y^2 = 4ax$ (ii) $3ay^2 = x^2(a-x)$
   (iii) $y = \ln\frac{e^x-1}{e^x+1}$ (iv) $y = a\ln\left\{\frac{a^2}{a^2-x^2}\right\}$

   Ans. (i) $\sqrt{1+\frac{a}{x}}$ (ii) $\sqrt{1+\frac{(2a-3x)^2}{12a(x-a)}}$ (iii) $\frac{e^{2x}+1}{e^{2x}-1}$
   (iv) $\frac{a^2+x^2}{a^2-x^2}$

2. Find $\frac{ds}{dy}$ for the curve $a^2y^2 = a^3 - x^3$ at $(a, 0)$.

   Hint: $\frac{dx}{dy} = \frac{2a^2y}{3x^2}$ at $(a, 0)$ is zero.

   Ans. 1

3. Find $\frac{ds}{dx}$ and $\frac{ds}{dy}$ for the curve $x^3 + xy^2 - 6y^2 = 0$ at $(3, 3)$.

   Hint: $3x^2 + (x-6)2yy_1 + y^2 = 0$. At $(3, 3)$, $y_1 = \frac{dy}{dx} = 2$

   Ans. $\frac{ds}{dx} = \sqrt{5}$, $\frac{ds}{dy} = \sqrt{\frac{5}{2}}$

4. Find $\frac{ds}{dx}$, $\frac{ds}{dy}$ and $\frac{ds}{d\theta}$ for the cycloid $x = a(\theta - \sin\theta)$, $y = a(1-\cos\theta)$.

   Hint: $\frac{dy}{dx} = \frac{\dot y}{\dot x} = \frac{a\sin\theta}{a(1-\cos\theta)} = \cot\frac{\theta}{2}$.
   $\frac{dx}{dy} = \tan\frac{\theta}{2}$

   $$\frac{ds}{dx} = \sqrt{\frac{2}{1-\cos\theta}} = \sqrt{\frac{2a}{y}},\ \frac{ds}{dy} = \sec\frac{\theta}{2}.$$

   $$\frac{ds}{dt} = \sqrt{2a(1-\cos\theta)} = \sqrt{2ay}$$

   Ans. $\sqrt{\frac{2a}{y}}$, $\sqrt{\frac{2a}{2a-y}}$, $\sqrt{2ay}$

5. Find $\frac{ds}{dt}$ for the astroid $x = a\cos^3 t$, $y = a\sin^3 t$.

   Hint: $\frac{dx}{dt} = -3a\cos^2 t\sin t$,
   $\frac{dy}{dt} = 3a\sin^2 t\cdot\cos t$

   Ans. $3a\sin t\cdot\cos t$

6. Determine $\frac{ds}{dt}$ for the ellipse $x = a\cos t$, $y = b\sin t$.

   Hint: $\dot x = -a\sin t$, $\dot y = b\cos t$

   Ans. $a\sqrt{1-e^2\cos^2 t}$ where $e =$ eccentricity.

$e = \frac{a^2-b^2}{a^2}$

7. Find $\frac{ds}{d\theta}$ for (i) $r = a(1+\cos\theta)$ (ii) $r = ae^{\theta \cot\alpha}$ (iii) $r^2 \sin 2\theta = 2a^2$.

Hint: (iii) $\frac{dr}{d\theta} = -r \cot 2\theta$

Ans. (i) $2a \cos\frac{\theta}{2}$ (ii) r cosec α (iii) $\frac{r^3}{2a^2}$

8. Find $\frac{ds}{dr}$ for (i) $r = a\theta$ (ii) $r^2 = a^2 \cos 2\theta$

Ans. (i) $\sqrt{1+\theta^2}$ (ii) $\sqrt{1+\cot^2 2\theta} = \operatorname{cosec} 2\theta$.

2.11 CURVATURE

The shape of a plane curve c is characterized by the degree of bentness or curvedness. A straight line has no bending while a circle has constant bending. Curvature of a curve is a measure of rate of change of bentness.

Angle of contingence of the arc AB of a curve c is the angle between the tangents at A and B to the curve c (see Fig. 2.12).

Given two arcs of the same length, the arc with greater angle of contingence is said to be more curved.

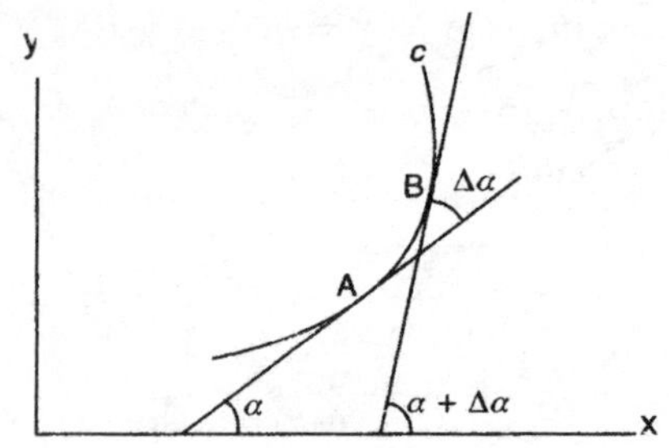

Fig. 2.12

Average curvature of an arc AB is

$$k_{av} = \frac{\Delta\alpha}{\widehat{AB}} = \frac{\text{angle of contingence}}{\text{length of the arc}}$$

Curvature of a curve c at a point A is denoted by k and is given by

$$k = \lim_{B\to A} k_{av} = \lim_{AB\to 0} \frac{\Delta\alpha}{\widehat{AB}} = \lim_{\Delta S\to 0} \frac{\Delta\alpha}{\Delta S} = \left|\frac{d\alpha}{ds}\right|.$$

Calculation of Curvature

Cartesian form Let the equation of the curve be given in cartesian form $y = f(x)$:

$$y = f(x) \text{ so } \tan\alpha = \frac{dy}{dx} = y_1$$

Then $\alpha = \tan^{-1}\frac{dy}{dx} = \tan^{-1} y_1$

$$\frac{d\alpha}{dx} = \frac{1}{1+y_1^2}\cdot y_2 \quad \text{where} \quad y_2 = \frac{d^2y}{dx^2}$$

we know that $\frac{ds}{dx} = \sqrt{1+y_1^2}$

Thus

$$k = \frac{d\alpha}{ds} = \frac{(d\alpha/dx)}{(ds/dx)} = \frac{y_2}{(1+y_1^2)}\cdot\frac{1}{\sqrt{1+y_1^2}}$$

$$\therefore k = \left|\frac{y_2}{(1+y_1^2)^{3/2}}\right| > 0$$

Cartesian form when $x = f(y)$:

$$k = \frac{\frac{d^2x}{dy^2}}{\left[1+\left(\frac{dx}{dy}\right)^2\right]^{3/2}}$$

Note: Useful when tangent is $\perp r$ to x-axis i.e., $\frac{dy}{dx} = \infty$ or $\frac{dx}{dy} = 0$.

Parametric form When equation of the curve is $x = x(t)$, $y = y(t)$ where t is the parameter:

$$\dot{x} = \frac{dx}{dt}, \dot{y} = \frac{dy}{dt} \text{ so } \frac{dy}{dx} = \frac{\frac{dy}{dt}}{\frac{dx}{dt}} = \frac{\dot{y}}{\dot{x}}$$

$$\frac{d^2y}{dx^2} = \frac{d}{dx}\left(\frac{\dot{y}}{\dot{x}}\right) = \frac{d}{dt}\left(\frac{\dot{y}}{\dot{x}}\right)\frac{dt}{dx} = \frac{1}{\dot{x}}\left[\frac{\dot{x}\ddot{y}-\ddot{x}\dot{y}}{\dot{x}^2}\right]$$

Substituting y_1 and y_2

$$k = \frac{|\dot{x}\ddot{y}-\ddot{x}\dot{y}|}{(\dot{x}^2+\dot{y}^2)^{3/2}}$$

Polar form When equation of the curve is $r = f(\theta)$, where r, θ are the polar coordinates given by $x = r\cos\theta$, $y = r\sin\theta$. Then

$$k = \frac{|r^2+2r_1^2-rr_2|}{(r^2+r_1^2)^{3/2}}$$

where $r_1 = \frac{dr}{d\theta}$ and $r_2 = \frac{d^2r}{d\theta^2}$.

Example: Curvature of a straight line $y = mx + c$.

Solution: $y_1 = \frac{dy}{dx} = m$, $y_2 = \frac{d^2y}{dx^2} = 0$, so $k = 0$ i.e., straight line has zero curvature.

Example: Curvature of a circle of given radius r (see Fig. 2.13).

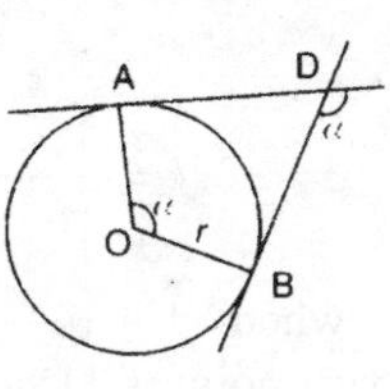

Fig. 2.13

Solution:

$$k_{av} = \frac{\text{angle of contingence}}{\text{length of arc}} = \frac{\alpha}{\alpha r}$$

k = curvature at any point A.

$$k = \lim_{B \to A} k_{av} = \lim_{B \to A} \frac{1}{r} = \lim_{\widehat{AB} \to 0} \frac{1}{r} = \frac{1}{r} = \text{constant}$$

The curvature of a circle is constant and is the reciprocal of the radius of the circle.

Implicit equation

Equation of the curve is $f(x, y) = 0$:
we know that $\frac{dy}{dx} = -\frac{f_x}{f_y}$ with $f_y \neq 0$.

$$\text{and} \qquad \frac{d^2y}{dx^2} = -\frac{|f_{xx} f_y^2 - 2 f_{xy} f_x f_y + f_{yy} f_x^2|}{f_y^3}$$

$$\text{with} \qquad f_y \neq 0$$

Here f_x, f_y, f_{xx} etc. are partial derivatives of f. Substituting y_1 and y_2 we get

$$k = \frac{f_{xx} f_y^2 - 2 f_{xy} f_x f_y + f_{yy} f_x^2}{(f_x^2 + f_y^2)^{\frac{3}{2}}}$$

Definition

Radius of curvature to a curve at a point is denoted by R and is the reciprocal of the curvature at that point. Thus

$$R = \frac{1}{k}$$

Cartesian form $y = f(x)$

$$R = \frac{(1 + y_1^2)^{\frac{3}{2}}}{y_2}$$

Cartesian form $x = f(y)$

$$R = \frac{[1 + (dx/dy)^2]^{\frac{3}{2}}}{\left(\frac{d^2x}{dy^2}\right)}$$

Parametric form $x = x(t), y = y(t)$

$$R = \frac{(\dot{x}^2 + \dot{y}^2)^{\frac{3}{2}}}{|(\dot{x}\ddot{y} - \ddot{x}\dot{y})|}$$

Polar form $r = f(\theta)$

$$R = \frac{(r^2 + r_1^2)^{\frac{3}{2}}}{|r^2 + 2r_1^2 - r r_2|}$$

Implicit equation $f(x, y) = 0$

$$R = \frac{(f_x^2 + f_y^2)^{\frac{3}{2}}}{|f_{xx} f_y^2 - 2 f_{xy} f_x f_y + f_{yy} f_x^2|}.$$

Worked Out Examples

Radius of curvature: Cartesian form

Example 1: Find the radius of curvature $y^2 = 2x(3 - x^2)$ at the points where the tangents are parallel to x-axis.

Solution: Differentiating the equation w.r.t. x.

$$2yy_1 = 6 - 6x^2 = 6(1 - x^2) \qquad (1)$$

At the points where the tangents are parallel to

$$x\text{-axis}, \; y_1 = 0 \quad \text{i.e.,} \quad x = \pm 1 \qquad (2)$$

Differentiating (1), $y_1^2 + yy_2 = -6x$

$$\text{At} \qquad x = \pm 1, \; y_2 = -\frac{6x}{y}$$

Take $x = 1$, so that $y^2 = 4$ or $y = \pm 2$, then $y_2 = -\frac{6}{\pm 2}$

Thus

$$R = \frac{(1 + y_1^2)^{\frac{3}{2}}}{y_2} = \frac{(1 + 0)^{\frac{3}{2}}}{\pm 3} = \left|\pm \frac{1}{3}\right| = \frac{1}{3}.$$

Example 2: Calculate R at $(a, 0)$ for $a^2y^2 = a^3 - x^3$.

Solution:

$$2a^2yy_1 = -3x^2 \quad \text{so} \quad y_1 = -\frac{3x^2}{2a^2y}$$

$$\text{At} \quad (a, 0),\ y_1 = \infty. \quad \text{or} \quad \frac{dx}{dy} = 0$$

So consider x as the dependent variable and y as the independent variable. i.e., $x = f(y)$.
Differentiating $a^2y^2 = a^3 - x^3$ w.r.t. y, we get

$$2a^2y = 0 - 3x^2\frac{dx}{dy} \quad (1)$$

$$\frac{dx}{dy} = \frac{2a^2y}{3x^2}$$

$$\text{At } (a, 0): \quad \frac{dx}{dy} = 0 \quad (2)$$

Differentiating (1) again w.r.t. y, we get

$$2a^2 = -6x\frac{dx}{dy} - 3x^2\frac{d^2x}{dy^2}$$

Using (2),

$$\frac{d^2x}{dy^2} = -\frac{2}{3}\frac{a^2}{x^2}$$

$$\text{At} \quad (a, 0), \quad \frac{d^2x}{dy^2} = -\frac{2}{3}$$

Then

$$R = \frac{\left[1 + \left(\frac{dx}{dy}\right)^2\right]^{\frac{3}{2}}}{\left|\frac{d^2x}{dy^2}\right|} = \frac{(1+0)^{\frac{3}{2}}}{\left|-\frac{2}{3}\right|} = \frac{3}{2}.$$

Example 3: Find R at (3, 3) for the implicit equation $x^3 + xy^2 - 6y^2 = 0$.

Solution: Differentiating implicitly w.r.t. x,

$$3x^2 + (x-6)\cdot 2yy_1 + 1\cdot y^2 = 0$$

$$\text{At } (3, 3). \quad 27 - 6\cdot 3y_1 + 9 = 0. \quad \therefore y_1 = 2$$

Differentiating again w.r.t. x,

$$6x + 2yy_1 + (x-6)[2y_1^2 + 2yy_2] + 2yy_1 = 0$$

$$\text{At } (3, 3). \ 18 + 12 + (-3)[8 + 6y^2] + 12 = 0$$

$$\therefore y_2 = 1$$

$$\text{Then} \quad R = \frac{(1+2^2)^{\frac{3}{2}}}{1} = 5^{\frac{3}{2}}.$$

Example 4: Determine the point on $y = 4x - x^2$ where the curvature is maximum.

Solution:

$$y_1 = 4 - 2x,\ y_2 = -2,\ k = \frac{|-2|}{[1 + (4-2x)^2]^{\frac{3}{2}}}$$

k is maximum when $1 + (4-2x)^2$ is minimum. The stationary point $\frac{d}{dx}[1 + (4-2x)^2] = 0$, $2(4-2x)(-2) = 0$ i.e., $x = 2$. Thus curvature is maximum at (2, 4).

EXERCISE

Find the radius of curvature of the following curves at the indicated point:

1. $x^3 + y^3 = 3axy$ at $\left(\frac{3a}{2}, \frac{3a}{2}\right)$

Ans. $3a/(8\sqrt{2})$

2. $y^2 = 4ax$ at (x, y)

Ans. $\frac{2}{\sqrt{a}}(x+a)^{\frac{3}{2}}$

3. $y = c \ln \sec(\frac{x}{c})$ at (x, y)

Ans. $c \sec(x/c)$

4. $x^{\frac{2}{3}} + y^{\frac{2}{3}} = a^{\frac{2}{3}}$ at (x, y)

   Hint: Use parametric equations $x = a\cos^3 t$, $y = a\sin^3 t$.

Ans. $3a^{\frac{2}{3}}x^{\frac{1}{3}}y^{\frac{1}{3}}$

5. $\frac{x^2}{a^2} + \frac{y^2}{b^2} = 1$ at $(a, 0)$ and $(0, b)$

Ans. $\frac{b^2}{a}$ and $\frac{a^2}{b}$

6. $x^2y = a(x^2 + y^2)$ at $(-2a, 2a)$

   Hint: $\frac{dy}{dx}$ at $(-2a, 2a)$ is ∞. Treat $x = f(y)$.

Ans. $|-2a|$

7. $xy^2 = a^3 - x^3$ at $(a, 0)$

   Hint: Since $\frac{dy}{dx}$ is ∞, take $x = f(y)$.

Ans. $|\frac{-3a}{2}|$

8. $y = c\cosh(x/c)$ at (x, y)

Ans. $\frac{y^2}{c}$

9. $xy = c^2$ at (x, y)

Ans. $(x^2 + y^2)^{\frac{3}{2}}/2xy$

10. $\sqrt{x} + \sqrt{y} = \sqrt{a}$ at $\left(\frac{a}{4}, \frac{a}{4}\right)$

Ans. $a/\sqrt{2}$

11. $x^{\frac{1}{3}} + y^{\frac{1}{3}} = 1$ at $(1/8, 1/8)$

Ans. $32/3$

12. $y^2(2 - x) = x^3$ at $(1, 1)$

Ans. $25/(3\sqrt{5})$

13. $x^3 + y^3 = 2a^3$ at (a, a)

Ans. $a/\sqrt{2}$

14. Find the point of greatest curvature on the curve $y = \ln x$
Hint: Stationary points: $\frac{dk}{dx} = \frac{1-2x^2}{(1+x^2)^{\frac{5}{2}}} = 0$.

Ans. $\left(\frac{1}{\sqrt{2}}, -\frac{1}{2}\ln 2\right)$

15. Prove that the ratio of the radii of curvature of the curves $xy = a^2$, $x^3 = 3a^2 y$ at points which have the same abscissa varies as the square root of the ratio of the ordinates.

Hint: $R_1 = \frac{(x_1^4+a^4)^{\frac{3}{2}}}{2a^2x_1^3}$, Radius of curvature to the first curve $xy = a^2$ with (x_1, y_1) point.
R_2 at (x_2, y_2) on second curve $x^3 = 3a^2 y$ is $\frac{(x_1^4+a^4)^{\frac{3}{2}}}{2x_1a^4}$

$$\frac{R_1}{R_2} = \frac{a^2}{x_1^2} = \frac{1}{\sqrt{3}}\sqrt{\frac{y_1}{y_2}}.$$

16. Find the point where the radius of curvature of the curve $x^2 y = a(x^2 + a^2/\sqrt{5})$ is minimum.

Ans. $x = a$, minimum curvature is $9a/10$

17. Show that $R_1^{-\frac{2}{3}} + R_2^{-\frac{2}{3}} = L^{-\frac{2}{3}}$ where R_1 and R_2 are the radii of curvature at the extremities of a focal chord of a parabola with latus rectum $2L$.

Hint: $y^2 = 4ax$ or $x = at^2$, $y = 2at$,

Latus rectum $L = 2a$

Extremities P(at t_1) and Q(at t_2) of focal chord

$$R = 2a(1 + t_1^2)^{\frac{3}{2}}.$$

18. Prove that $\left(\frac{2R}{a}\right)^{\frac{2}{3}} = \left(\frac{y}{x}\right)^2 + \left(\frac{x}{y}\right)^2$ where R is the radius of curvature of the curve $y = ax/(a + x)$

Hint: $R = -[1 + (y/x)^4]^{\frac{3}{2}} / \left[\left(\frac{2}{a}\right)\left(\frac{y}{x}\right)^3\right]$.

19. Show that $(R_1^{\frac{2}{3}} + R_2^{\frac{2}{3}})(ab)^{\frac{2}{3}} = a^2 + b^2$ where R_1 and R_2 are the radii of curvature at the extremities of the conjugate diameters of an ellipse $(x^2/a^2) + (y^2/b^2) = 1$.

Hint: $R = (b^4x^2 + a^4y^2)^{\frac{3}{2}}/(a^4b^4)$.
Coordinates of the extremities of the conjugate diameters are $(a\cos\theta, b\sin\theta)$ and $(-a\sin\theta, b\cos\theta)$.

20. The radius of curvature $R = \frac{ds}{d\psi}$ for a curve $s = s(\psi)$ where ψ is the angle the tangent to the curve makes with the x-axis. Find R for

i. catenary $S = c\tan\psi$
Ans. $c\sec^2\psi$
ii. cycloid $S = 4a\sin\psi$
Ans. $4a\cos\psi$
iii. cardioid $S = 4a\sin\frac{\psi}{3}$
Ans. $\frac{4}{3}a\cos\frac{\psi}{3}$
iv. parabola $S = a\ln(\tan\psi + \sec\psi) + a\tan\psi\sec\psi$
Ans. $2a\sec^3\psi$
v. $S = c\ln\sec\psi$
Ans. $\frac{1}{c}\cot\psi$.

Worked Out Examples

Radius of curvature for parametric curve

Find the radius of curvature for the following curves in the parametric form.

Example 1: $x = 6t^2 - 3t^4$, $y = 8t^3$.

Solution: Differentiating w.r.t. t

$$\dot{x} = 12t - 12t^3, \ddot{x} = 12 - 36t^2, \dot{y} = 24t^2, \ddot{y} = 48t$$

$$\dot{x}^2 + \dot{y}^2 = 12^2t^2(1 - t^2)^2 + 24^2t^4 = 12^2t^2(1 + t^2)^2$$

$$\dot{x}\ddot{y} - \ddot{x}\dot{y} = 12t(1 - t^2)(48t) - 12(1 - 3t^2)(24t^2)$$

$$= 12 \cdot 24 \cdot t^2(1 + t^2)$$

$$R = \frac{(\dot{x}^2 + \dot{y}^2)^{\frac{3}{2}}}{(\dot{x}\ddot{y} - \ddot{x}\dot{y})} = \frac{12^3t^3(1 + t^2)^3}{12 \cdot 24 \cdot t^2(1 + t^2)} = 6t(1 + t^2)^2.$$

Example 2: $x = e^t + e^{-t}$, $y = e^t - e^{-t}$ at $t = 0$.

Solution:

$$\dot{x} = e^t - e^{-t}, \ddot{x} = e^t + e^{-t}$$

$$\dot{y} = e^t + e^{-t},\ \ddot{y} = e^t - e^{-t}$$

At $t = 0, \dot{x} = 0, \ddot{x} = 2, \dot{y} = 2, \ddot{y} = 0$

$$R = \frac{(0 + 2^2)^{\frac{3}{2}}}{|-2 \cdot 2|} = 2.$$

Example 3: $x = \dfrac{a\cos t}{t},\ y = \dfrac{a\sin t}{t}.$

Solution:

$$\dot{x} = \frac{a[t(-\sin t) - \cos t \cdot 1]}{t^2}$$

$$\dot{y} = \frac{a[t\cos t - \sin t \cdot 1]}{t^2}$$

$$\dot{x}^2 + \dot{y}^2 = \frac{(1+t^2)a^2}{t^4}$$

$$\ddot{x} = a(-t^2\cos t + 2t\sin t + 2\cos t)/t^3$$

$$\ddot{y} = a(-t^2\sin t - 2t\cos t + 2\sin t)/t^3$$

$$\dot{x}\ddot{y} - \ddot{x}\dot{y} = a^2/t^2$$

$$R = \frac{a^3(1+t^2)^{\frac{3}{2}}}{t^6} \cdot \frac{t^2}{a^2} = \frac{a(1+t^2)^{\frac{3}{2}}}{t^4}.$$

Example 4: $x = a\ln(\sec t + \tan t),\ y = a\sec t.$

Solution:

$$\dot{x} = \frac{a}{\sec t + \tan t} \cdot (\sec t\tan t + \sec^2 t)$$

$$\dot{x} = a\sec t$$

$$\dot{y} = a\sec t\tan t$$

$$\ddot{x} = a\sec t \cdot \tan t,\ \ddot{y} = a[\sec t \cdot \tan^2 t + \sec t \cdot \sec^2 t]$$

$$\ddot{y} = a\sec t[\tan^2 t + \sec^2 t]$$

$$\dot{x}^2 + \dot{y}^2 = a^2\sec^2 t + a^2\sec^2 t\tan^2 t$$

$$= a^2\sec^2 t(1 + \tan^2 t) = a^2\sec^4 t$$

$$\dot{x}\ddot{y} - \ddot{x}\dot{y} = (a\sec t)(a\sec t)[\tan^2 t + \sec^2 t]$$

$$-(a^2\sec^2 t\tan^2 t)$$

$$= a^2\sec^4 t$$

$$R = \frac{(\dot{x}^2 + \dot{y}^2)^{\frac{3}{2}}}{\dot{x}\ddot{y} - \ddot{x}\dot{y}} = \frac{(a^2\sec^4 t)^{\frac{3}{2}}}{a^2\sec^4 t} = a\sec^2 t.$$

EXERCISE

Find the radius of curvature of the following parametric curves:

1. $x = 1 - t^2,\ y = t - t^3$ at $t = \pm 1$

Ans. $2\sqrt{2}$

2. $x = 2t^2 - t^4,\ y = 4t^3$ at $t = 1$

Ans. 18

3. $x = a(t - t^3/3),\ y = at^2$

Ans. $\frac{a}{2}(1+t^2)^2$

4. $x = \ln t,\ y = \frac{1}{2}(t + t^{-1})$

Ans. $\frac{t}{4}(1+t^2)^2$

5. Cycloid $x = a(t + \sin t),\ y = a(1 - \cos t)$

   Hint: $y_1 = \tan\frac{t}{2},\ y_2 = \dfrac{\sec^4 t/2}{4a}.$

Ans. $4a\cos\frac{t}{2}$

6. Cycloid $x = t - \sin t,\ y = 1 - \cos t$ at the highest point of the arch

   Hint: At $t = \pi$, highest point of arch.

Ans. 4

7. (a) Ellipse $x = a\cos t,\ y = b\sin t$, (b) circle

Ans. (a) $(a^2\sin^2\theta + b^2\cos^2\theta)^{\frac{3}{2}}/ab$,

(b) $a = b,\ R = \frac{a^3}{a^2} = a$

8. Tractrix $x = a[\cos t + \ln\tan(t/2)],\ y = a\sin t$

   Hint: $y_1 = \tan t,\ y_2 = (\sec^4 t \cdot \sin t)/a.$

Ans. $a\cot t$

9. $x = 3a\cos t - a\cos 3t,\ y = 3a\sin t - a\sin 3t$

   Hint:

$$y_1 = \tan 2t$$

$$y_2 = 1/(3a\cos^2 2t \cdot \sin t).$$

Ans. $3a\sin t$

10. $x = 3\sin t + \sin 3t,\ y = 3\cos t + \cos 3t$

Ans. $3\cot t$

11. $x = a(\cos t + t\sin t),\ y = a(\sin t - t\cos t)$

    Hint: $\dot{x}^2 + \dot{y}^2 = a^2t^2,\ \dot{x}\ddot{y} - \ddot{x}\dot{y} = a^2t^2.$

Ans. at

12. $x = a(\cos t + \ln\tan\frac{t}{2})$, $y = a\sin t$

Hint: $y_1 = a\cos t$, $y_2 = \sin t/(a\cos^4 t)$.

Ans. $a\cot t$

13. The tangents at two points A, B on the cycloid $x = a(t - \sin t)$, $y = a(1 - \cos t)$ are perpendicular. Show that $R_1^2 + R_2^2 = 16a^2$ where R_1 and R_2 are the radii of curvature at these points A and B.

Hint: $\dot{x}^2 + \dot{y}^2 = 4a^2\sin^2\frac{t}{2}$,
$\dot{x}\ddot{y} - \ddot{x}\dot{y} = -2a^2\sin^2\frac{t}{2}$.

$$R = -4a\sin\frac{t}{2}$$

Slope of tangent at t is $\cot t/2$

use perpendicularity: $\cot\frac{t_1}{2}\cdot\cot\frac{t_2}{2} = -1$.

14. $x = a\cos t(1 + \sin t)$, $y = a\sin t(1 + \cos t)$ at $t = -\pi/4$

Hint: $\dot{x} = \frac{a}{\sqrt{2}}$, $\dot{y} = \frac{a}{\sqrt{2}}$, $\ddot{x} = \frac{a(2\sqrt{2}-1)}{\sqrt{2}}$,
$\ddot{y} = \frac{a(2\sqrt{2}+1)}{\sqrt{2}}$, $\dot{x}^2 + \dot{y}^2 = a^2$, $\dot{x}\ddot{y} - \ddot{x}\dot{y} = a^2$.

Ans. a

15. $x = a\ln\tan\left(\frac{\pi}{4} + \frac{\theta}{2}\right)$, $y = a\sec\theta$

Ans. $a\sec^2\theta$

16. Astroid $x = a\cos^3 t$, $y = a\sin^3 t$

Hint: $\dot{x}^2 + \dot{y}^2 = 9a^2\sin^2 t\cos^2 t$, $\dot{x}\ddot{y} - \ddot{x}\dot{y} = -9a^2\sin^2 t\cos^2 t$.

Ans. $|-3a\sin t\cos t|$.

WORKED OUT EXAMPLES

Radius of curvature for polar curve

Example 1: Find the radius of curvature $r = e^{2\theta}$ at $\theta = \ln 2$.

Solution: $r_1 = \frac{dr}{d\theta} = 2e^{2\theta}$, $r_2 = \frac{d^2r}{d\theta^2} = 4e^{2\theta}$

$$R = \frac{(r^2 + r_1^2)^{\frac{3}{2}}}{|r^2 + 2r_1^2 - rr_2|} = \frac{(e^{4\theta} + 4e^{4\theta})^{\frac{3}{2}}}{(e^{4\theta} + 8e^{4\theta} - 4e^{4\theta})} = \frac{e^{6\theta}5^{\frac{3}{2}}}{e^{4\theta}5}$$

$$R = \sqrt{5}e^{2\theta} = \sqrt{5}\cdot 4 = \sqrt{80}$$

or at $\theta = \ln 2$, $r_1 = 2e^{2\ln 2} = 8$, $r_2 = 4e^{2\ln 2} = 16$,

$r = 4$

$$R = \frac{(16 + 64)^{\frac{3}{2}}}{16 + 128 - 64} = \sqrt{80}.$$

Example 2: Show that the radius of curvature at any point (r, θ) on the curve $r^2 = a^2\sec 2\theta$, is proportional to r^3.

Solution: Taking log: $2\ln r = \ln a^2 + \ln\sec 2\theta$. Differentiating w.r.t. θ,

$$2\cdot\frac{1}{r}\cdot\frac{dr}{d\theta} = 0 + \frac{1}{\sec 2\theta}\cdot\sec 2\theta\cdot\tan 2\theta\cdot 2$$

$$r_1 = \frac{dr}{d\theta} = r\tan 2\theta$$

so $$r_2 = \frac{d^2r}{d\theta^2} = \tan 2\theta\cdot\frac{dr}{d\theta} + r\cdot 2\cdot\sec^2 2\theta$$

$$r_2 = r\cdot\tan^2 2\theta + 2r\sec^2 2\theta$$

Then $r^2 + r_1^2 = r^2 + r^2\tan^2 2\theta = r^2\sec^2 2\theta$

$$r^2 + 2r_1^2 - rr_2 = r^2 + 2\cdot r^2\tan^2 2\theta - r(r\tan^2 2\theta + 2r\sec^2 2\theta) = -r^2\sec^2 2\theta$$

$$R = \frac{(r^2\sec^2 2\theta)^{\frac{3}{2}}}{|-r^2\sec^2 2\theta|} = r\sec 2\theta = r\cdot\frac{r^2}{a^2} = \frac{r^3}{a^2}.$$

Example 3: Determine the radius of curvature R at any point (r, θ) on the curve $r^n = a^n\sin n\theta$. Deduce R for the curves (i) $r = a\sin\theta$ and (ii) $r^2 = a^2\sin 2\theta$.

Solution: Taking log

$$n\ln r = \ln a^n + \ln\sin n\theta$$

Differentiating w.r.t. θ,

$$n\cdot\frac{1}{r}\frac{dr}{d\theta} = 0 + \frac{1}{\sin n\theta}\cdot n\cdot\cos n\theta$$

$$r_1 = \frac{dr}{d\theta} = r\cot n\theta$$

Differentiating again w.r.t. θ,

$$\frac{d^2r}{d\theta^2} = \frac{dr}{d\theta}\cdot\cot n\theta + r\cdot(-\mathrm{cosec}^2 n\theta)\cdot n$$

$$r_2 = \frac{d^2r}{d\theta^2} = r\cot^2 n\theta - nr\,\mathrm{cosec}^2 n\theta$$

So

$$r^2+2r_1^2-rr_2=r^2+2\cdot(r^2\cot^2 n\theta)$$
$$-r(r\cot^2 n\theta-nr\,\text{cosec}^2 n\theta)$$
$$=r^2[(1+\cot^2 n\theta)+n\,\text{cosec}^2 n\theta]$$
$$=r^2(1+n)\cdot\text{cosec}^2 n\theta$$

Also

$$(r^2+r_1^2)^{\frac{3}{2}}=[r^2+r^2\cot^2 n\theta]^{\frac{3}{2}}=r^3\text{cosec}^3 n\theta$$

Thus

$$R=\frac{(r^2+r_1^2)^{\frac{3}{2}}}{|r^2+2r_1^2-rr_2|}=\frac{r^3\text{cosec}^3 n\theta}{r^2(1+n)\text{cosec}^2 n\theta}$$
$$=\frac{r}{(1+n)}\frac{1}{\sin n\theta}$$
$$R=\frac{r}{(1+n)}\cdot\frac{a^n}{r^n}=\frac{a^n}{(1+n)r^{n-1}}$$

For $n=1$: Radius of curvature of $r=a\sin\theta$ is

$$r=\frac{a}{2}=\text{constant}$$

For $n=2$: Radius of curvature of $r^2=a^2\sin 2\theta$ is

$$r=\frac{a^2}{2r}.$$

Example 4: Prove that at the points in which the Archimedean spiral $r=a\theta$ intersects the hyperbolical spiral $r\theta=a$, their curvatures are in the ratio 3 : 1.

Solution: For the curve $r=a\theta$,

$$r_1=a,\qquad r_2=0$$

Let curvature of the curve $r=a\theta$ be k_1

$$k_1=\frac{r^2+2r_1^2-rr_2}{(r^2+r_1^2)^{\frac{3}{2}}}=\frac{a^2\theta^2+2a^2-0}{(a^2\theta^2+a^2)^{\frac{3}{2}}}$$

$$k_1=\frac{\theta^2+2}{a(\theta^2+1)^{\frac{3}{2}}}\tag{1}$$

Let curvature of the curve $r\theta=a$ be k_2.
Then $r=\frac{a}{\theta}$, $r_1=-\frac{a}{\theta^2}$, $r_2=\frac{2a}{\theta^3}$ so

$$k_2=\frac{\left(\frac{a}{\theta}\right)^2+2\left(-\frac{a}{\theta^2}\right)^2-\left(\frac{a}{\theta}\right)\left(\frac{2a}{\theta^3}\right)}{\left[\left(\frac{a}{\theta}\right)^2+\left(-\frac{a}{\theta^2}\right)^2\right]^{\frac{3}{2}}}=\frac{\theta^4}{a(\theta^2+1)^{\frac{3}{2}}}\tag{2}$$

The points of intersection of the two curves $r=a\theta$ and $r=\frac{a}{\theta}$ are given by

$$a\theta=r=\frac{a}{\theta}$$

or $\theta^2=1$ or $\theta=\pm 1$
Now from (1) and (2)

$$k_1|_{\theta=\pm 1}=\frac{3}{a2^{\frac{3}{2}}}\tag{3}$$

$$k_2|_{\theta=\pm 1}=\frac{1}{a2^{\frac{3}{2}}}\tag{4}$$

Thus k_1 and k_2 are in the ratio 3 : 1.

Example 5: For the cardioid $r=a(1+\cos\theta)$ show that the square of the radius of curvature at any point (r,θ) is proportional to r. Also find the radius of curvature when $\theta=0,\frac{\pi}{4},\frac{\pi}{2}$.

Solution: The pedal equation or $p-r$ equation (see Pages 2.51-2.52)

$$\frac{1}{p^2}=\frac{1}{r^2}+\frac{1}{r^4}\left(\frac{dr}{d\theta}\right)^2$$

For the cardioid $r=a(1+\cos\theta)$,

$$\frac{dr}{d\theta}=-a\sin\theta$$

So the pedal equation for cardioid is

$$\frac{1}{p^2}=\frac{1}{r^2}+\frac{1}{r^4}(a^2\sin^2\theta)=\frac{1}{r^4}[r^2+a^2\sin^2\theta]$$
$$=\frac{1}{r^4}[a^2(1+\cos\theta)^2+a^2\sin^2\theta]=\frac{2a^2}{r^4}(1+\cos\theta)$$
$$\frac{1}{p^2}=\frac{2a^2}{r^4}\cdot\frac{r}{a}=\frac{2a}{r^3}$$

or $$2ap^2=r^3$$

Differentiating w.r.t., 'r',

$$4ap\frac{dp}{dr}=3r^2$$

Radius of curvature in pedal form is (see Pages 2.51-2.52)

$$R=r\frac{dr}{dp}=r\cdot\left(\frac{4ap}{3r^2}\right)=\frac{4}{3}\frac{a}{r}p=\frac{4}{3}\frac{a}{r}\cdot\left(\frac{r^3}{2a}\right)^{\frac{1}{2}}$$
$$\therefore R=\frac{2}{3}\sqrt{(2ar)}$$

Thus $R^2 \propto r$

At $\theta = 0, r = 2a,$ so $R = \frac{2}{3}\sqrt{2a} \cdot \sqrt{2a} = \frac{4}{3}a$

At $\theta = \frac{\pi}{4}, r = a(1 + \frac{1}{\sqrt{2}}).$

so $R = \frac{2}{3}\sqrt{2a}\left[a\left(1 + \frac{1}{\sqrt{2}}\right)\right]^{\frac{1}{2}}$

At $\theta = \frac{\pi}{2}, r = a,$ so $R = \frac{2}{3}\sqrt{2a} \cdot \sqrt{a} = \frac{2\sqrt{2}}{3}a.$

Exercise

Find the radius of curvature R for the following curves at the indicated points

1. $r = \tan\theta$ at $\theta = \frac{3\pi}{4}$

Ans. $\sqrt{5}$

2. $r = 2\sin 3\theta$ at $\theta = \frac{\pi}{6}$

Ans. 1/5

3. $r = 2\cos 2\theta$ at $\theta = \frac{\pi}{6}$

Ans. $13\sqrt{13}/58$

4. $r^2 = a^2\cos 2\theta$

Hint: Convert to pedal equation $\frac{dp}{dr} = 3r^2/a^2$.

$$R = r\frac{dr}{dp} = r\left(\frac{a^2}{3r^2}\right) = \frac{a^2}{3r}.$$

Ans. $a^2/3r$

5. $r^2\cos 2\theta = a^2$

Hint: Pedal equation $p = a^2/r$, $R = \frac{rdr}{dp} = r \cdot (\frac{-r^2}{a^2})$.

Ans. r^3/a^2

6. $\sqrt{r}\frac{\cos\theta}{2} = \sqrt{a}$

Hint: Rewrite $r = a\sec^2\frac{\theta}{2}$.

Ans. $2r\sqrt{r/a}$

7. $r = ae^{\theta\cot\alpha}$

Hint: Pedal equation: $p = r\sin\alpha$.

Ans. $r\operatorname{cosec}\alpha$

8. $r(1 + \cos\theta) = 2a$

Hint: Pedal equation $p^2 = ar$.

Ans. $2r\sqrt{r/a}$

9. $\theta = \frac{\sqrt{r^2-a^2}}{a} - \cos^{-1}\frac{a}{r}$

Hint: Pedal equation $p^2 = r^2 - a^2$.

Ans. $\sqrt{r^2 - a^2}$

10. $r = a(1 - \cos\theta)$

Aliter: Pedal equation $2ap^2 = r^3$.

Ans. $\frac{2\sqrt{2a}}{3}\sqrt{(r/a)}$

11. $r^m = a^m\cos m\theta$

Ans. $a^m/[(m + 1)r^{m-1}]$

Deduce cases when,

i. $m = -1$, straight line $a = r\cos\theta$

ii. $m = 2, r^2 = a^2\cos 2\theta$ (Lemniscate of Bernoulli)

iii. $m = -2, r^2\cos 2\theta = a^2$ (Rectangular hyperbola)

iv. $m = 1, r = a\cos\theta$ (Circle of diameter a)

v. $m = \frac{1}{2}, r = \frac{a}{2}(1 + \cos\theta)$ (Cardioid)

vi. $m = -\frac{1}{2}, r(1 + \cos\theta) = 2a$ (Parabola)

vii. $m = \frac{1}{3}$, Cardioid.

12. Show that $9(R_1^2 + R_2^2) = 16a^2$, where R_1 and R_2 are the radii of curvature at the extremities of the chord through the pole of the cardioid $r = a(1 + \cos\theta)$.

Hint:

$$R = \frac{2\sqrt{2a}}{3}r^{\frac{1}{2}} = \frac{2\sqrt{2a}}{3}(1 + \cos\theta)^{\frac{1}{2}} = \frac{4a}{3}\cos\frac{\theta}{2}$$

$$R_2 = R|_{\theta+\pi} = \frac{4a}{3}\cos\left(\frac{\theta + \pi}{2}\right) = -\frac{4a}{3}\sin\frac{\theta}{2}$$

$$R_1 = R|_\theta = \frac{4a}{3}\cos\frac{\theta}{2}.$$

Pedal Equation:

Radius of Curvature for Pedal Curve

Let O be the pole and OX be the initial line (refer Fig. 2.14). Let $P(r, \theta)$ be any point on the curve c whose polar equation is $r = f(\theta)$. Let PT be the tangent at P to the curve. Draw ON perpendicular to PT and let ON = length of the perpendicular from pole O to the tangent at $P = p$. Let ϕ be the angle between the tangent PT and the radius vector OP. Then from the right angled triangle ONP,

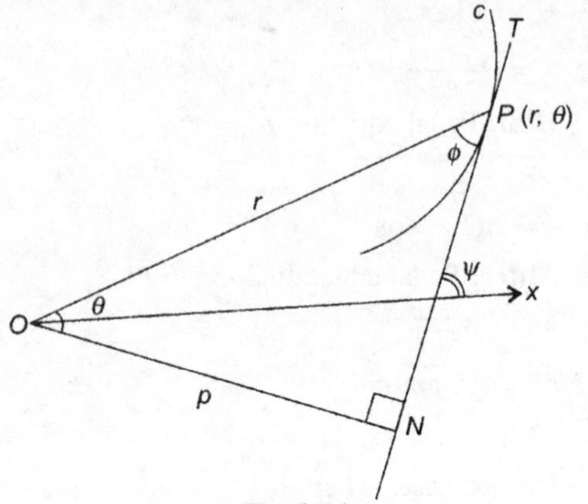

Fig. 2.14

$$\sin\phi = \frac{ON}{OP} = \frac{p}{r}$$

$$\therefore\ p = r\sin\phi \qquad (1)$$

Then

$$\frac{1}{p^2} = \frac{1}{r^2\sin^2\theta} = \frac{\operatorname{cosec}^2\phi}{r^2} = \frac{1+\cot^2\theta}{r^2} = \frac{1}{r^2} + \frac{1}{r^2\tan^2\phi}$$

we know that $\tan\phi = r\frac{d\theta}{dr} = \frac{r}{(dr/d\theta)}$

Thus

$$\frac{1}{p^2} = \frac{1}{r^2} + \frac{1}{r^2\left(\frac{r}{dr/d\theta}\right)^2} = \frac{1}{r^2} + \frac{1}{r^4}\left(\frac{dr}{d\theta}\right)^2 \qquad (2)$$

Pedal equation of curve c is the relation between p and r given by (1) or (2). It is also known as $p - r$ equation to the curve.

Note: If the polar equation of the curve is known, the pedal equation can be obtained by eliminating θ between $r = f(\theta)$ and (2). Sometimes it is more convenient to eliminate θ and ϕ between $r = f(\theta)$, $\tan\phi = r\frac{d\theta}{dr}$ and $p = r\sin\phi$.

Radius of curvature R, for pedal curve

Let $\qquad p = f(r) = r\sin\theta \qquad (1)$

be the pedal equation and

$$\psi = \theta + \phi \qquad (2)$$

Differentiating (1) w.r.t. 'r', we get

$$\frac{dp}{dr} = r\cos\phi\frac{d\phi}{dr} + \sin\phi$$

$$= r\frac{dr}{ds}\frac{d\phi}{dr} + r\frac{d\theta}{ds}$$

since $\cos\phi = \frac{dr}{ds}$ and $\sin\phi = r\frac{d\theta}{ds}$

Thus

$$\frac{dp}{dr} = r\frac{d}{ds}(\phi+\theta) = r\frac{d\psi}{ds} \quad \text{by (2)}$$

$$= r\frac{1}{R}$$

$$\therefore\ R = r\frac{dr}{dp}.$$

Worked Out Examples

Example 1: Find the radius of curvature at any point (r, θ) of the conic section $\frac{L}{r} = 1 + e\cos\theta$.

Solution: Differentiating w.r.t. θ,

$$-\frac{L}{r^2}\frac{dr}{d\theta} = 0 - e\sin\theta$$

$$\text{so} \quad \frac{dr}{d\theta} = \frac{r^2 e\sin\theta}{L}$$

The pedal equation is

$$\frac{1}{p^2} = \frac{1}{r^2} + \frac{1}{r^4}\left(\frac{dr}{d\theta}\right)^2 = \frac{1}{r^2} + \frac{1}{r^4}\frac{r^4e^2}{L^2}\sin^2\theta$$

From the given conic equation, $\cos\theta = \frac{L-r}{er}$, so

$$\sin^2\theta = 1 - \cos^2\theta = 1 - \left(\frac{L-r}{er}\right)^2 = \frac{e^2r^2 - (L-r)^2}{e^2r^2}$$

Thus

$$\frac{1}{p^2} = \frac{1}{r^2} + \frac{1}{L^2}\left[e^2 - \frac{L^2}{r^2} - 1 + \frac{2L}{r}\right] = \frac{e^2-1}{L^2} + \frac{2}{Lr}$$

Differentiating p w.r.t. r,

$$-\frac{2}{p^3}\cdot\frac{dp}{dr} = 0 - \frac{2}{L}\frac{1}{r^2}$$

$$R = \text{Radius of curvature} = r\frac{dr}{dp} = r\cdot\frac{Lr^2}{p^3} = L\left(\frac{r}{p}\right)^3.$$

Example 2: Determine R for $r\cos 2\theta = a$.

Solution: Differentiating w.r.t. θ,

$$\frac{dr}{d\theta}\cdot\cos 2\theta - r\cdot 2\cdot\sin 2\theta = 0$$

$$\frac{dr}{d\theta} = 2r\tan 2\theta$$

The pedal equation is

$$\frac{1}{p^2} = \frac{1}{r^2} + \frac{1}{r^4}\left(\frac{dr}{d\theta}\right)^2 = \frac{1}{r^2} + \frac{1}{r^4}(4r^2\tan^2 2\theta)$$

$$= \frac{1}{r^2}[1 + 4(\sec^2 2\theta - 1)] = \frac{1}{r^2}\left[-3 + 4\left(\frac{r}{a}\right)^2\right]$$

since $\sec 2\theta = \frac{r}{a}$ from the given equation.
Thus

$$\frac{1}{p^2} = \frac{4}{a^2} - \frac{3}{r^2}$$

Differentiating w.r.t. 'r',

$$-\frac{2}{p^3}\frac{dp}{dr} = 0 - \frac{3\cdot(-2)}{r^3}$$

$$R = r\frac{dr}{dp} = r\cdot\left(\frac{r^3}{-3p^3}\right) = \left|-\frac{r^4}{3p^3}\right|.$$

Example 3: Find the pedal equation of the curve $r = a\sec t, \theta = \tan t - t$ and find the radius of curvature at any point (r, θ).

Solution: $\frac{r}{a} = \sec t$, so $\cos t = \frac{a}{r}$ or $t = \cos^{-1}\frac{a}{r}$.

Then $\tan^2 t = \sec^2 t - 1 = \frac{r^2}{a^2} - 1 = \frac{r^2 - a^2}{a^2}$

or $\tan t = \frac{\sqrt{r^2 - a^2}}{a}$

Thus eleminating t, the polar equation is

$$\theta = \tan t - t = \frac{\sqrt{r^2 - a^2}}{a} - \cos^{-1}\frac{a}{r}.$$

Differentiating θ w.r.t. r,

$$\frac{d\theta}{dr} = \frac{1}{a}\cdot\frac{1}{2}\cdot\frac{1}{\sqrt{r^2 - a^2}}\cdot 2r - (-1)\cdot\frac{1}{\sqrt{1 - (\frac{a}{r})^2}}\cdot\left(-\frac{a}{r^2}\right)\times$$

$$r\frac{d\theta}{dr} = \frac{\sqrt{r^2 - a^2}}{a} = \tan\phi$$

So $\sin\phi = \frac{\sqrt{r^2 - a^2}}{\sqrt{a^2 + (r^2 - a^2)}} = \frac{\sqrt{r^2 - a^2}}{r}$

The pedal equation of the curve is

$$p = r\sin\phi = \sqrt{r^2 - a^2}$$

or $p^2 = r^2 - a^2$

Differentiating w.r.t. r, $2p\frac{dp}{dr} = 2r$ or $\frac{dr}{dp} = \frac{p}{r}$.
The radius of curvature R is

$$R = r\cdot\frac{dr}{dp} = r\cdot\frac{p}{r} = p.$$

Example 4: Write the $p - r$ equation of the polar curve $r^n = a^n\sin n\theta$ and find the radius of curvature to the curve.

Solution: Taking log

$$n\ln r = n\ln a + \ln\sin n\theta$$

Differentiating w.r.t. θ,

$$n\cdot\frac{1}{r}\frac{dr}{d\theta} = 0 + \frac{1}{\sin n\theta}\cdot n\cdot\cos n\theta$$

$$\frac{dr}{d\theta} = r\cot n\theta$$

we know that

$$\tan\phi = r\frac{d\theta}{dr} = r\cdot\frac{1}{r\cdot\cot n\theta} = \tan n\theta$$

Thus

$$\phi = n\theta$$

The required $p - r$ equation to the curve is

$$p = r\sin\phi = r\sin n\theta = r\cdot\left(\frac{r^n}{a^n}\right) = \frac{r^{n+1}}{a^n}$$

since $\sin n\theta = \frac{r^n}{a^n}$ from the equation of curve.
Differentiating w.r.t. r,

$$a^n\cdot\frac{dp}{dr} = (n+1)r^n$$

$$R = r\frac{dr}{dp} = r\cdot\frac{a^n}{(n+1)r^n} = \frac{a^n}{(n+1)r^{n-1}}.$$

Example 5: Find the pedal equation of $r = a\theta$ and find R.

Solution:

$$\frac{dr}{d\theta} = a$$

so pedal equation is

$$\frac{1}{p^2} = \frac{1}{r^2} + \frac{1}{r^4}\left(\frac{dr}{d\theta}\right)^2 = \frac{1}{r^2} + \frac{1}{r^4}(a^2) = \frac{r^2+a^2}{r^4}$$

Thus $p^2 = \frac{r^4}{r^2+a^2}$

Differentiating w.r.t. r,

$$2p\frac{dp}{dr} = \frac{(r^2+a^2)4r^3 - r^4(2r)}{(r^2+a^2)}$$

$$\frac{dp}{dr} = \frac{r(r^2+2a^2)}{(r^2+a^2)^{\frac{3}{2}}}$$

so $R = r\frac{dr}{dp} = \frac{(r^2+a^2)^{\frac{3}{2}}}{(r^2+2a^2)}$.

EXERCISE

Find the pedal curve of the polar curve $r = f(\theta)$ and find the radius of curvature R at any point (r, θ):

1. Ellipse $\frac{x^2}{a^2} + \frac{y^2}{b^2} = 1$ $(x = a\cos\theta,\ y = b\sin\theta)$

Ans. Pedal equation $\frac{1}{p^2} = \frac{1}{a^2} + \frac{1}{b^2} - \frac{r^2}{a^2b^2}$

$$R = a^2b^2/p^3$$

2. Parabola: $y^2 = 4a(x+a)$ or $\frac{2a}{r} = 1 - \cos\theta$

Ans. Pedal equation: $p^2 = ar$

$$R = 2r^{\frac{3}{2}}/\sqrt{a}$$

3. Cardioid: $r = a(1+\cos\theta)$

Ans. Pedal equation: $r^3 = 2ap^2$

$$R = \frac{2}{3}\sqrt{2ar}$$

4. Hyperbola: $\frac{x^2}{a^2} - \frac{y^2}{b^2} = 1$

Ans. Pedal equation: $r^2 = \frac{a^2b^2}{p^2} + a^2 - b^2$

$$R = a^2b^2/p^3$$

5. Rectangular hyperbola: $r^2\cos 2\theta = a^2$

Ans. Pedal equation: $p = a^2/r$

$$R = r^3/a^2$$

6. Equiangular spiral: $r = ae^{\theta\cot\alpha}$

Ans. Pedal equation: $p = r\sin\alpha$

$$R = r\operatorname{cosec}\alpha$$

7. Lemniscate $r^2 = a^2\cos 2\theta$

Ans. Pedal equation: $r^3 = a^2p$

$$R = a^2/3r$$

8. $x^2 + y^2 = 2ay$

Ans. Pedal equation: $r^2 = 2ap$

$$R = a$$

9. $r^m = a^m\cos m\theta$

Ans. Pedal equation: $r^{m+1} = a^m p$

$$R = a^m/\left[(m+1)r^{m-1}\right]$$

10. Astroid: $x = a\cos^3 t,\ y = a\sin^3 t$

or $x^{\frac{2}{3}} + y^{\frac{2}{3}} = a^{\frac{2}{3}}$

with pedal equation: $r^2 = a^2 - 3p^2$

Ans. $R = |-3p|$.

Newton's Method: To Find Radius of Curvature at the Origin

Let the given curve pass through the origin O (0, 0).

Case 1: If the x-axis is tangent to the curve at O then radius of curvature at origin is given by

$$R|_{\text{at}(0,0)} = \lim_{\substack{x\to0\\y\to0}}\left\{\frac{x^2}{2y}\right\}.$$

Proof: Since x-axis is tangent ot the curve at 0,

$$y_1(0) = \frac{dy}{dx}|_{\text{at}(0,0)} = 0 \tag{1}$$

Now

$$\lim_{\substack{x\to0\\y\to0}}\left\{\frac{x^2}{2y}\right\} = \lim\left\{\frac{2x}{2y_1}\right\} = \lim\left\{\frac{1}{y_2}\right\} = \frac{1}{y_2(0)} \tag{2}$$

where we have applied L'Hospital's rule as the limit is 0/0 indeterminate form.

Substituting $y_1(0)$ and $y_2(0)$ from (1) and (2) in

$$R\Big|_{(0,0)} = \frac{[1+y_1^2(0)]^{\frac{3}{2}}}{y_2(0)} = \frac{(1+0)^{\frac{3}{2}}}{y_2(0)}$$

$$= \frac{1}{y_2(0)} = \lim_{\substack{x\to 0\\ y\to 0}} \left\{\frac{x^2}{2y}\right\}$$

Case 2: If the y-axis is a tangent to the curve at the origin O then

$$R\Big|_{(0,0)} = \lim_{\substack{x\to 0\\ y\to 0}} \left\{\frac{y^2}{2x}\right\}$$

Case 3: Both x-axis and y-axis are not tangents to the curve $y = f(x)$ at origin O(0, 0).

By Maclaurin's theorem

$$y = f(0) + \frac{x}{1!} f_1(0) + \frac{x^2}{2!} f_2(0) + \cdots$$

$f(0) = 0$ since curve passes through the origin.

Then

$$R = \frac{(1+y_1^2)^{\frac{3}{2}}}{y_2} = \frac{(1+p^2)^{\frac{3}{2}}}{q}$$

where $p = f_1(0) = y_1(0)$ and $q = f_2(0) = y_2(0)$.

Note: Tangents at the origin to a curve are obtained by equating the lowest degree terms in the equation of the curve to zero.

Case 4: Radius of curvature at the pole (origin) when equation is in polar form: If the initial line is tangent at pole then

$$R_{\text{at pole}} = \lim_{\substack{x\to 0\\ y\to 0}} \left[\frac{x^2}{2y}\right] = \lim_{\substack{x\to 0\\ \theta\to 0}} \left[\frac{r\cos^2\theta}{2r\sin\theta}\right]$$

$$= \lim_{\substack{r\to 0\\ \theta\to 0}} \left[\frac{r}{2\theta}\right] = \lim \left[\frac{dr/d\theta}{2}\right]$$

$$R_{\text{at pole}} = \frac{1}{2} \lim_{\substack{r\to 0\\ \theta\to 0}} \left(\frac{dr}{d\theta}\right).$$

WORKED OUT EXAMPLES

Example 1: Find the radius of curvature at the origin for $x^4 - y^4 + x^3 - y^3 + x^2 - y^2 + y = 0$.

Solution: Equating to zero, the lowest term in the equation we get $y = 0$ (x-axis) is a tangent to the curve at the origin.

Divide the equation throughout by y, we get

$$x^2 \cdot \frac{x^2}{y} - y^3 + x \cdot \frac{x^2}{y} - y^2 + \frac{x^2}{y} - y + 1 = 0$$

We know that $\lim \frac{x^2}{y}$ as $x \to 0$ is $2R$

Taking the limit as $x \to 0$, $y \to 0$

$$0 \cdot 2R - 0 + 0 \cdot 2R - 0 + 2R - 0 + 1 = 0$$

$$\therefore \quad R = \left|-\frac{1}{2}\right|.$$

Example 2: Determine R at (0, 0) to the curve $x^3y - xy^3 + 2x^2y - 2xy^2 + 2y^2 - 3x^2 + 3xy - 4x = 0$.

Solution: Observe that $x = 0$ (y-axis) is a tangent to the curve at origin. Divide the equation throughout by x, we get

$$x^2y - y^3 + 2xy - 2y^2 + \frac{2y^2}{x} - 3x + 3y - 4 = 0$$

As $x \to 0$ and $y \to 0$

$$0 - 0 + 0 - 0 + 2 \cdot 2R - 0 + 0 - 4 = 0$$

$$\therefore R = 1.$$

Example 3: Calculate the radius of curvature at origin to the curve $a(y^2 - x^2) = x^3$.

Solution: Equating the lowest terms $y^2 - x^2$ to zero, note that $y = \pm x$ are the tangents to the curve at origin.

Substituting $y = px + q\frac{x^2}{2} + \cdots$ in the given equation

$$a\left[\left(px + q\frac{x^2}{2} + \cdots\right)^2 - x^2\right] = x^3$$

To find the unknowns p and q, equate to zero coefficients of x^2 on both sides

$$ap^2 - a = 0 \;\therefore\; p = \pm 1$$

Similarly equating to zero coefficients of x^3, we get

$$apq = 1, \quad q = \pm\frac{1}{a}$$

Now

$$R = \frac{(1+y_1^2)^{\frac{3}{2}}}{y_2} = \frac{(1+p^2)^{\frac{3}{2}}}{q} = \frac{(1+1)^{\frac{3}{2}}}{\pm\frac{1}{a}}$$

$R = \pm 2\sqrt{2}a.$

Example 4: Find the radius of curvature to the curve $x = at^2$, $y = 2at$ at the origin.

Solution: The curve is a parabola $y^2 = 4ax$ with y-axis as tangent at origin (0, 0).

$$R \text{ at origin} = \lim_{\substack{x\to 0\\ y\to 0}} \left(\frac{y^2}{2x}\right) = \lim_{t\to 0}\left(\frac{4a^2t^2}{2at^2}\right)$$

$$= \lim_{t\to 0} 2a = 2a.$$

EXERCISE

Find the radius of curvature at origin to the following curves:

1. $y^4 + x^3 + a(x^2 + y^2) - a^2y = 0$

   Hint: $y = 0$ (x-axis) is tangent to curve at origin.

Ans. $a/2$

2. $x^3 + y^3 - 2x^2 + 6y = 0$

   Hint: $y = 0$ is tangent to curve at origin.

Ans. 3/2

3. $2x^4 + 3y^4 + 4x^2y + xy - y^2 + 2x = 0$

   Hint: $x = 0$ (y-axis) is tangent to curve at origin.

Ans. 1

4. $y^2 = x^2(a + x)/(a - x)$

   Hint: $y = \pm x$ are tangents. $p = \pm 1, q = \pm \frac{2}{a}$.

Ans. $\sqrt{2}a$

5. $x^3 - 2x^2y + 3xy^2 - 4y^3 + 5x^2 - 6xy + 7y^2 - 8y = 0$

   Hint: x-axis is tangent.

Ans. 4/5

6. $y = x^4 - 4x^3 - 18x^2$

   Hint: x-axis is tangent.

Ans. 1/36

7. $x^3 + 3x^2y - 4y^3 + y^2 - 6x = 0$

   Hint: y-axis is tangent.

Ans. 3

8. $3x^2y - 3xy^2 + 2y^3 + 3x^2 - 3y^2 - 9y = 0$

   Hint: x-axis is tangent.

Ans. 3/2

9. $y^2 - 3xy - 4x^2 + x^3 + x^4y + y^5 = 0$

Ans. $\frac{85\sqrt{17}}{2}, 5\sqrt{2}$

10. $xy^2 = 4a^2(2a - x)$ at vertex $(2a, 0)$

    Hint: Put $x = 2a + X$, $y = Y$.
    New equation $(2a + X)Y^2 = -4a^2X$ has y-axis as tangent at origin.

Ans. $|-a|$

11. $5x^3 + 7y^3 + 4x^2y + xy^2 + 2x^2 + 3xy + y^2 + 4x = 0$

Ans. $|-2|$

12. $x^3 + y^3 = 3axy$

    Hint: Both x-axis and y-axis are tangents.

Ans. $3a/2$

13. $x = a(t + \sin t)$, $y = a(1 - \cos t)$

    Hint: Curve passes through origin. x-axis is tangent to the curve at origin.

$$R = \lim_{\substack{x\to 0\\ y\to 0}} \frac{x^2}{2y} = \lim_{\theta\to 0} \frac{a^2(0 + \sin\theta)^2}{2a(1 - \cos\theta)} = 4a.$$

Ans. $4a$

14. $r = a \sin n\theta$ at pole

    Hint: Curve passes through origin (pole) Initial line is tangent to the curve at pole

$$R = \lim_{\substack{x\to 0\\ \theta\to 0}} \left(\frac{1}{2}\frac{dr}{d\theta}\right) = \lim \frac{1}{2}(na\cos n\theta) = \frac{na}{2}.$$

Ans. $\frac{na}{2}$

15. $2x^4 + 4x^3y + xy^2 + 6y^3 - 3x^2 - 2xy + y^2 - 4x = 0$

    Hint: y-axis is tangent.

Ans. 2

16. $y - x = x^2 + 2xy + y^2$

    Hint: $y = x$ is tangent, $p = 1, q = 8$.

Ans. $a/2$.

Radius of Curvature, Centre of Curvature and Circle of Curvature

Let P(x, y) be any point on a curve $y = f(x)$. Let the tangent L at P makes an angle α with the x-axis (see Fig. 2.15). Construct a circle such that circle is tangent to L at P, lies on the same side of L as the curve and has the same curvature k at P. Then this circle is known as circle of curvature and its radius R is known as radius of curvature and c is known as the centre of circle of curvature with coordinates x and y. Thus

Radius of curvature to the curve at the point P is $R = \frac{1}{k}$

In cartesian form

$$R = \frac{1}{k} = \frac{(1+y_1^2)^{\frac{3}{2}}}{y_2}.$$

Centre of circle of curvature

$$X = x - R\sin\alpha = x - R\cdot\frac{y_1}{\sqrt{1+y_1^2}}$$

$$X = x - y_1\frac{(1+y_1^2)}{y_2}$$

Similarly

$$Y = y + R\cos\alpha = y + R\cdot\frac{1}{\sqrt{1+y_1^2}}$$

$$Y = y + \frac{(1+y_1^2)}{y_2}.$$

The equation of the **circle of curvature** to the given curve at the point P is the circle of radius R and with centre $C(x, y)$ given by

$$(x-X)^2 + (y-Y)^2 = R^2.$$

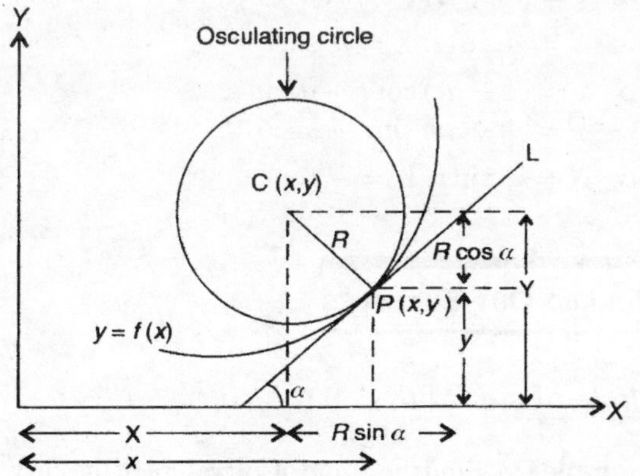

Fig. 2.15

Parametric Form

Radius of curvature in parametric form is

$$R = \frac{1}{k} = \frac{(\dot{x}^2+\dot{y}^2)^{\frac{3}{2}}}{(\dot{x}\ddot{y} - \ddot{x}\dot{y})}$$

Centre of circle of curvature is

$$X = x(t) - \frac{\dot{y}(\dot{x}^2+\dot{y}^2)}{(\dot{x}\ddot{y} - \ddot{x}\dot{y})}$$

$$Y = y(t) + \frac{\dot{x}(\dot{x}^2+\dot{y}^2)}{\dot{x}\ddot{y} - \ddot{x}\dot{y}}.$$

Worked Out Examples

Example 1: Find the centre of circle of curvature for $xy(x+y) = 2$ at $(1, 1)$.

Solution: The given curve is $x^2y + xy^2 = 2$
Differentiating w.r.t. x,

$$2xy + x^2y_1 + y^2 + 2xyy_1 = 0$$

$$y_1 = -\frac{(2xy+y^2)}{(x^2+2xy)},\quad y_1\big|_{(1,1)} = \frac{-(2+1)}{(1+2)} = -1$$

Differentiating again w.r.t. x,

$$2y + 2xy_1 + 2xy_1 + x^2y_2 + 2yy_1 + 2yy_1$$
$$+ 2xy_1^2 + 2xyy_2 = 0$$

$$y_2 = \frac{2(y + 2xy_1 + 2yy_1 + xy_1^2)}{-(2xy+x^2)},\quad y_2\big|_{(1,1)} = \frac{4}{3}$$

Centre of curvature

$$X = x - \frac{y_1(1+y_1^2)}{y_2} = 1 - \frac{(-1)(1+1)}{4/3} = \frac{5}{2}$$

$$Y = y + \frac{(1+y_1^2)}{y_2} = 1 + \frac{(1+1)}{4/3} = \frac{5}{2}$$

Coordinates of centre of curvature to the given curve at $(1, 1)$ are $\left(\frac{5}{2}, \frac{5}{2}\right)$.

Example 2: Determine the centre of curvature to the curve in parametric form $x = 3t^2$, $y = 3t - t^3$.

Solution: Differentiating x and y w.r.t. t,

$$\dot{x} = 6t, \ddot{x} = 6, \dot{y} = 3(1-t^2), \ddot{y} = -6t$$

$$\dot{x}^2 + \dot{y}^2 = 36t^2 + 9(1-t^2)^2 = 9(t^2+1)^2$$

$$\dot{x}\ddot{y} - \ddot{x}\dot{y} = 6t(-6t) - 6(3)(1-t^2) = -18(t^2+1)$$

Coordinates of the centre of curvature

$$X = x(t) - \frac{\dot{y}(\dot{x}^2+\dot{y}^2)}{\dot{x}\ddot{y} - \ddot{x}\dot{y}}$$

$$= 3t^2 - 3(1-t^2)\frac{9(t^2+1)^2}{-18(t^2+1)}$$

$$X(t) = \frac{3}{2}[1 + 2t^2 - t^4]$$

$$Y = y + \frac{\dot{x}(\dot{x}^2+\dot{y}^2)}{\dot{x}\ddot{y} - \ddot{x}\dot{y}}$$

$$= (3t - t^3) + 6t \cdot \frac{9(t^2+1)^2}{-18(t^2+1)}$$

$$Y = 3t - t^3 - 3t(t^2+1) = -4t^3$$

Centre of curvature $\left(\frac{3}{2}(1+2t^2-t^4), -4t^3\right)$.

EXERCISE

Find the centre of curvature of the following curve at the indicated point:

I. Cartesian form

1. $y^2 = 4ax$ at (x, y)

Ans. $(3x + 2a, -2x^{\frac{3}{2}}/\sqrt{a})$

2. $y = e^x$ at (0,1)

Ans. $(-2, 3)$

3. $y^2 = x^3$ at (1, 1)

Ans. $\left(-\frac{11}{2}, \frac{16}{3}\right)$

4. $y = \ln \sec x$ at $(\pi/3, \ln 2)$

Ans. $\left(\frac{\pi}{3} - \sqrt{3}, 1 + \ln 2\right)$

5. $y = (x^2+9)/x$ at (3, 6)

Ans. $(3, 15/2)$

6. $y = x^3 - 6x^2 + 3x + 1$ at $(1, -1)$

Ans. $(-36, -43/6)$

7. $x^3 + xy^2 - 6y^2 = 0$ at (3, 3)

Ans. $(-7, 8)$

8. $y = c\cosh(x/c)$ at (x, y)

Ans. $\left(x - y\frac{\sqrt{y^2-c^2}}{c}, 2y\right)$

9. $x^3 + y^3 = 2$ at (1, 1)

Ans. $\left(\frac{1}{2}, \frac{1}{2}\right)$

10. $xy = c^2$ at (c, c)

Ans. $(2c, 2c)$

11. $y^3 = a^2x$ at (x, y)

Ans. $\left(\frac{a^4+15y^4}{6a^2y}, \frac{a^4y+9y^5}{2a^4}\right)$

12. $y = x^4 - x^2$ at (0, 0)

Ans. $\left(0, -\frac{1}{2}\right)$

13. $y = x/(x+1)$ at (0, 0)

Ans. $(1, -1)$

14. $y = e^{-x^2}$ at (0, 1)

Ans. $\left(0, \frac{1}{2}\right)$

15. $y = x \ln x$ at the point where tangent is parallel to x-axis

Ans. $(1/e, (e^2-1)/e)$

II. Parametric form

1. $x = a(\theta - \sin\theta), y = a(1-\cos\theta)$

Ans. $X = a(\theta + \sin\theta), Y = -a(1-\cos\theta)$

2. $x = t^2, y = t^3$

Ans. $X = -t^2 - 9t^{\frac{4}{2}}, Y = 4t^3 + 4t/3$

3. $x = 3t, y = t^2 - 6$ at (a, b)

Ans. $X = -4a(20 + a^2), Y = b + (81 + 4a^2)/18$

4. $x = a(\cos t + t\sin t), y = a(\sin t - t\cos t)$

Ans. $X = a\cos t, Y = a\sin t$

5. $x = (1-at)\cos t + a\sin t$, $y = (1-at)\sin t - a\cos t$

Ans. $X = a\sin t, Y = -a\cos t$.

WORKED OUT EXAMPLES

Circle of curvature

Example 1: Find the circle of curvature of the curve $x + y = ax^2 + by^2 + cx^3$ at the origin.

Solution: Differentiating the given curve

$$x + y = ax^2 + by^2 + cx^3$$

w.r.t. 'x' implicitly, we get

$$1 + \frac{dy}{dx} = 2ax + 2by\frac{dy}{dx} + 3cx^2 \quad (1)$$

$$\frac{dy}{dx} = y_1 = \frac{1 - 2ax - 3cx^2}{2by - 1} \quad (2)$$

Differentiating (1) w.r.t. x,

$$0 + \frac{d^2y}{dx^2} = 2a + 2b\left(\frac{dy}{dx}\right)^2 + 2by \cdot \frac{d^2y}{dx^2} + 6cx$$

$$\frac{d^2y}{dx^2} = y_2 = +\frac{2a + 2by_1^2 + 6cx}{1 - 2by} \quad (3)$$

At origin (0, 0), $y_1 = -1$, $y_2 = 2(a+b)$

$$R = \text{Radius of curvature} = \frac{[1 + (-1)^2]^{\frac{3}{2}}}{y_2}$$

$$= \frac{2\sqrt{2}}{2(a+b)} = \frac{\sqrt{2}}{a+b} \quad (4)$$

The centre of circle of curvature is given by

$$X = x - \frac{y_1(1+y_1^2)}{y_2} = 0 - \frac{(-1)(1+(-1)^2)}{2(a+b)} = \frac{1}{a+b} \quad (5)$$

$$Y = y + \frac{(1+y_1^2)}{y_2} = 0 + \frac{(1+(-1)^2)}{2(a+b)} = \frac{1}{a+b} \quad (6)$$

The equation of the circle of curvature with centre given by (5) and (6) and radius by (4) is

$$\left(x - \frac{1}{a+b}\right)^2 + \left(y - \frac{1}{a+b}\right)^2 = \left(\frac{\sqrt{2}}{a+b}\right)^2$$

or $(a+b)(x^2+y^2) = 2(x+y)$.

Example 2: Determine the circle of curvature of the folium $x^3 + y^3 = 3axy$ at the point P$(3a/2, 3a/2)$.

Solution: By implicit differentiation

$$3x^2 + 3y^2y_1 = 3ay + 3axy_1$$

$$y_1 = \frac{ay - x^2}{y^2 - ax}$$

Differentiating w.r.t. x again

$$2x + 2yy_1^2 + y^2y_2 = ay_1 + ay_1 + axy_2$$

$$y_2 = \frac{2ay_1 - 2x - 2yy_1^2}{y^2 - ax}$$

At the point $P(3a/2, 3a/2)$

$$y_1 = \frac{a \cdot \frac{3a}{2} - \left(\frac{3a}{2}\right)^2}{\left(\frac{3a}{2}\right)^2 - a\frac{3a}{2}} = -1$$

$$y_2 = -\frac{32}{3a}$$

$$R = \text{Radius of curvature} = \left|\frac{[1+(-1)^2]^{\frac{3}{2}}}{(-32/3a)}\right|$$

$$R = \frac{3\sqrt{2}a}{16} \quad (1)$$

The centre of curvature

$$X = x - \frac{y_1(1+y_1^2)}{y_2} = \frac{3a}{2} - \frac{(-1)(2)}{(-32)} \cdot 3a = \frac{21a}{16} \quad (2)$$

$$Y = y + \frac{(1+y_1^2)}{y_2} = \frac{3a}{2} + \frac{2}{(-32)} \cdot 3a = \frac{21a}{16} \quad (3)$$

The required circle of curvature with centre given by (1) and (2) and radius (3) is

$$\left(x - \frac{21a}{16}\right)^2 + \left(y - \frac{21a}{16}\right)^2 = \left(\frac{3\sqrt{2}a}{16}\right)^2$$

or $(x^2+y^2) - \frac{21}{8}a(x+y) + \frac{432}{128}a^2 = 0.$

Example 3: Show that the parabolas $y = -x^2 + x + 1$, $x = -y^2 + y + 1$ have the same circle of curvature at the point (1, 1).

Solution: For the parabola

$$y = -x^2 + x + 1 \quad (1)$$

$\frac{dy}{dx} = -2x + 1$, $\frac{d^2y}{dx^2} = -2$. At (1, 1), $\frac{dy}{dx} = -1$, $\frac{d^2y}{dx^2} = -2$

R_1 = Radius of curvature of parabola (1) at (1,1)

$$= \left|\frac{[1+(-1)^2]^{\frac{3}{2}}}{-2}\right|$$

$$R_1 = \frac{\sqrt{8}}{2} = \sqrt{2}$$

Centre of circle for parabola (1) at (1, 1) is

$$X_1 = 1 - (-1)\frac{(1+1)}{-2} = 1 - 1 = 0.$$

$$Y_1 = 1 + \frac{(1+1)}{-2} = 1 - 1 = 0$$

Thus the equation of circle of curvature for the porabola (1) at the point (1,1) is

$$(x-0)^2 + (y-0)^2 = (\sqrt{2})^2 \text{ or } x^2 + y^2 = 2 \quad (2)$$

Similarly for the parabola

$$x = -y^2 + y + 1 \quad (3)$$

$\frac{dx}{dy} = -2y + 1,\ \frac{d^2x}{dy^2} = -2$ so at $P(1, 1)\frac{dx}{dy} = -1,$

$\frac{d^2x}{dy^2} = -2$

$R_2 =$ Radius of curvature for parabola (2) at (1, 1)

$$= \left|\frac{[1+(-1)^2]^{\frac{3}{2}}}{-2}\right| = \frac{\sqrt{8}}{2} = \sqrt{2}$$

Centre of curvature for parabola (2) at (1, 1) is

$$X = 1 - \frac{(-1)(1+1)}{-2} = 1 - 1 = 0,\ Y = 1 + \frac{(1+1)}{-2} = 0$$

So the equation of circle of curvature to the second parabola (2) at point (1, 1) is same as (2) given by

$$(x-0)^2 + (y-0)^2 = (\sqrt{2})^2 \text{ or } x^2 + y^2 = 2.$$

Example 4: Determine the circles of curvature at the vertices of an ellipse with semi-axes a, b.

Solution: Equation of ellipse $\frac{x^2}{a^2} + \frac{y^2}{b^2} = 1$

In the parametric form $x = a\cos t,\ y = b\sin t$

Differentiating w.r.t. the parameter 't'

$$\dot{x} = -a\sin t,\ \ddot{x} = -a\cos t,\ \dot{y} = b\cos t,\ \ddot{y} = -b\sin t$$

$$R = \frac{(\dot{x}^2 + \dot{y}^2)^{\frac{3}{2}}}{|\dot{x}\ddot{y} - \ddot{x}\dot{y}|} = \frac{(a^2\sin^2 t + b^2\cos^2 t)^{\frac{3}{2}}}{ab\sin^2 t + ab\cos^2 t}$$

$$R = \frac{(a^2\sin^2 t + b^2\cos^2 t)^{\frac{3}{2}}}{ab} \quad (1)$$

Centre of curvature

$$X = x - \frac{\dot{x}^2 + \dot{y}^2}{\dot{x}\ddot{y} - \ddot{x}\dot{y}} \cdot \dot{y}$$

$$= a\cos t - \frac{b \cdot \cos t(a^2\sin^2 t + b^2\cos^2 t)}{ab}$$

$$X = \frac{a^2 b\cos t + a^2 b\cos^3 t - a^2 b\cos t - b^3\cos^3 t}{ab}$$

$$X = \frac{(a^2 - b^2)\cos^3 t}{a} \quad (2)$$

Similarly

$$Y = y + \frac{\dot{x}(\dot{x}^2 + \dot{y}^2)}{\dot{x}\ddot{y} - \ddot{x}\dot{y}} = \frac{-(a^2 - b^2)\sin^3 t}{b}$$

The vertex $A(a, 0)$ corresponds to $t = 0$.

So at A

$$R|_{t=0} = \frac{b^2}{a},\ X|_{t=0} = \frac{a^2 - b^2}{a},\ Y|_{t=0} = 0$$

The circle of curvature at the vertex $A(a, 0)$ is

$$\left(x - \frac{a^2 - b^2}{a}\right)^2 + (y-0)^2 = \left(\frac{b^2}{a}\right)^2$$

$$\text{or} \quad x^2 + y^2 + 2x\left(\frac{b^2 - a^2}{a}\right) = (2b^2 - a^2)$$

Similarly the vertex $B(0, b)$ corresponds to $t = \frac{\pi}{2}$

$$R = \frac{a^2}{b},\ X = 0,\ Y = \frac{-(a^2 - b^2)}{b}$$

The circle of curvature at B is

$$(x-0)^2 + \left(y + \frac{(a^2 - b^2)}{b}\right)^2 = \left(\frac{a^2}{b}\right)^2$$

$$\text{or} \quad x^2 + y^2 + \frac{2(a^2 - b^2)y}{b} = (2a^2 - b^2)$$

EXERCISE

Circle of curvature

Determine the circle of curvature of the curve at the indicated point:

1. $2xy + x + y = 4$ at (1, 1)

Ans. $\left(x - \frac{5}{2}\right)^2 + \left(y - \frac{5}{2}\right)^2 = \frac{9}{2}$

2. $y^2 = 12x$ at (3, 6) and (0, 0)

Ans. At (3, 6) : $(x - 15)^2 + (y - (-6))^2 = (12\sqrt{2})^2$

At (0, 0) : $(x - 6)^2 + (y - 0)^2 = 6^2$

3. $\sqrt{x}+\sqrt{y}=\sqrt{a}$ at $(a/4, a/4)$

Ans. $\left(x-\frac{3a}{4}\right)^2+\left(y-\frac{3a}{4}\right)^2=\left(\frac{a}{\sqrt{2}}\right)^2$

4. $y^2=4ax$ at $(at^2, 2at)$

Ans. $x^2+y^2-6at^2x-4ax+4at^3y=3a^2t^4$

5. $y=mx+\frac{x^2}{a}$ at $(0, 0)$

Ans. $x^2+y^2=a(1+m^2)(y-mx)$

6. $x^3+y^3=2xy$ at $(1, 1)$

Ans. $\left(x-\frac{7}{8}\right)^2+\left(y-\frac{7}{8}\right)^2=\left(\frac{\sqrt{2}}{8}\right)^2$

7. $xy(x+y)=2$ at $(1, 1)$

Ans. $x^2+y^2+5x-5y+8=0$

8. $y=x^3+2x^2+x+1$ at $(0, 1)$

Ans. $x^2+y^2+x-3y+2=0$

9. $y^2=4ax$ at $(a, 2a)$

Ans. $x^2+y^2-10ax+4ay=3a^2$

10. Determine the values of a, b, c if $y=a+bx+cx^2$ and $xy=12$ have same circle of curvature at the point $(3, 4)$

Ans. $a=-12, b=-4, c=4/9$.

2.12 EVOLUTE

As a point P moves along a given curve c_1, the centre of curvature corresponding to P describes another curve c_2. The curve c_2 is known as the **Evolute** of the given curve c_1 and c_1 is known as the **Involute** of c_2.

Thus *Evolute* of a curve is the locus of the centres of curvature corresponding to points on the curve c.

Determination of Evolute

I. *Cartesian form* Let $y=f(x)$ be the equation of the given curve c in cartesian form. Then the coordinates of the centre of curvature given by

$$X=x-y_1(1+y_1^2)/y_2 \qquad (1)$$

$$Y=y(x)+(1+y_1^2)/y_2 \qquad (2)$$

form the parametric equations of the evolute of C expressed in terms of the parameter x.

In many cases, the parameter x can be eliminated between (1) and (2). This results in a relation between X and Y of the form $f(X, Y)=0$ which is the equation of the required evolute.

II. *Parametric form* Let the equation of the curve be in parametric form $x=x(t)$, $y=y(t)$ where t is the parameter. Then the parametric equations of the evolute are

$$X(t)=x(t)-\frac{\dot{y}(\dot{x}^2+\dot{y}^2)}{\dot{x}\ddot{y}-\ddot{x}\dot{y}}$$

$$Y(t)=y(t)+\frac{\dot{x}(\dot{x}+\dot{y}^2)}{\dot{x}\ddot{y}-\ddot{x}\dot{y}}$$

where $\cdot$ denotes differentiation w.r.t. 't'.

Result 1: The length of the arc of the evolute E between two points P_1 and P_2 equals to R_1-R_2 where R_1 and R_2 are the radius of curvatures of the given curve c (involute) at the two points on c corresponding to P_1 and P_2.

Result 2: Normal to a curve c (involute) is the tangent to its evolute.

WORKED OUT EXAMPLES

Example 1: Determine the parametric equations for the evolute of the curve $x=\frac{t^4}{4}$, $y=\frac{t^5}{5}$.

Solution: $\dot{x}=\frac{dx}{dt}=t^3$, $\ddot{x}=3t^2$, $\dot{y}=t^4$, $\ddot{y}=4t^3$. The coordinates of the centre of curvature in parametric form is

$$X=x(t)-\frac{\dot{y}(\dot{x}^2+\dot{y}^2)}{\dot{x}\ddot{y}-\ddot{x}\dot{y}}$$

$$=\frac{t^4}{4}-\frac{t^4(t^6+t^8)}{4t^6-3t^6}=-\frac{3}{4}t^4-t^6 \qquad (1)$$

$$Y=y(t)+\frac{\dot{x}(\dot{x}^2+\dot{y}^2)}{\dot{x}\ddot{y}-\ddot{x}\dot{y}}$$

$$=\frac{t^5}{5}+\frac{t^3(t^6+t^8)}{4t^6-3t^6}=\frac{6}{5}t^5+t^3 \qquad (2)$$

Note that the parameter t can not be eliminated between (1) and (2). Therefore the equation of the required evolute in the parametric equations $X = x(t)$ and $Y = y(t)$ are given by (1) and (2).

Example 2: Find the evolute of the hyperbola $\frac{x^2}{a^2} - \frac{y^2}{b^2} = 1$. Deduce the evolute of a rectangular hyperbola.

Solution: Differentiating the equation of hyperbola

$$\frac{x^2}{a^2} - \frac{y^2}{b^2} = 1 \tag{1}$$

w.r.t. x, we get

$$\frac{2x}{a^2} - \frac{2y\,y_1}{b_2} = 0$$

or
$$y_1 = \frac{dy}{dx} = \frac{b^2 x}{a^2 y} \tag{2}$$

so

$$y_2 = y'' = \frac{d^2y}{dx^2} = \frac{b^2}{a^4y^3}(a^2y^2 - b^2x^2) = -\frac{b^4}{a^2}\frac{1}{y^3} \tag{3}$$

since from (1)

$$\frac{y^2}{b^2} = \frac{x^2 - a^2}{a^2} \text{ or } a^2y^2 - b^2x^2 = -b^2a^2$$

Now the centre of curvature is

$$X = x - \left(\frac{b^2}{a^2}\frac{x}{y}\right)\cdot\left(1 + \frac{b^4x^2}{a^4y^2}\right)\cdot\left(\frac{a^2y^3}{-b^4}\right)$$

$$X = x\left[\frac{b^2a^4 + a^2b^2x^2 - a^4b^2 + b^4x^2}{b^2a^4}\right]$$

$$= \frac{x^3(b^2 + a^2)}{a^4} \tag{4}$$

Similarly,

$$Y = y + \left(1 + \frac{b^4x^2}{a^4y^2}\right)\cdot\left(\frac{-a^2y^3}{b^4}\right)$$

$$Y = y\left[\frac{-b^4a^2 + a^4y^2 + b^4x^2}{-b^4a^2}\right] = \frac{-y^3(a^2 + b^2)}{b^4} \tag{5}$$

From (4) :
$$x^2 = \left(\frac{a^4X}{a^2 + b^2}\right)^{\frac{2}{3}} \tag{6}$$

From (5) :
$$y^2 = \left(\frac{b^4Y}{a^2 + b^2}\right)^{\frac{2}{3}} \tag{7}$$

Then

$$1 = \frac{x^2}{a^2} - \frac{y^2}{b^2} = \frac{1}{a^2}\left(\frac{a^4X}{a^2 + b^2}\right)^{\frac{2}{3}} - \frac{1}{b^2}\left(\frac{b^4Y}{a^2 + b^2}\right)^{\frac{2}{3}}$$

Thus the required envelope is

$$(aX)^{\frac{2}{3}} - (bY)^{\frac{2}{3}} = (a^2 + b^2)^{\frac{2}{3}}$$

For a rectangular hyperbola with $a = b$ the envelope reduces to

$$X^{\frac{2}{3}} - Y^{\frac{2}{3}} = (2a)^{\frac{2}{3}}$$

Example 3: Show that the evolute of the deltoid $x = 2\cos t + \cos 2t$, $y = 2\sin t - \sin 2t$ is another deltoid three times the size of the given deltoid and has the equations

$$x = 3(2\cos t - \cos 2t), \quad y = 3(2\sin t + \sin 2t)$$

Solution: Differentiating x and y w.r.t. 't',

$$\dot{x} = -2\sin t - 2\sin 2t,\ \ddot{x} = -2\cos t - 4\cos 2t$$

$$\dot{y} = -2\cos t - 2\cos 2t,\ \ddot{y} = -2\sin t + 4\sin 2t$$

so $\dot{x}^2 + \dot{y}^2 = 4(\sin t + \sin 2t)^2 + 4(\cos t - \cos 2t)^2$

$$= 4(\sin^2 t + \sin^2 2t + 2\sin t\cdot\sin 2t)$$
$$+4(\cos^2 t + \cos^2 2t - 2\cos t\cdot\cos 2t)$$

$$\dot{x}^2 + \dot{y}^2 = 8(1 + \sin t\cdot\sin 2t - \cos t\cdot\cos 2t)$$

Also

$$\dot{x}\ddot{y} - \dot{y}\ddot{x} = -2(\sin t + \sin 2t)(-2\sin t + 4\sin 2t)$$
$$-2(\cos t - \cos 2t)(-2\cos t - 4\cos 2t)$$
$$= 4[\sin^2 t - 2\sin^2 2t - \sin t\cdot\sin 2t]$$
$$+4[\cos^2 t - 2\cos^2 2t + \cos t\cdot\cos 2t]$$
$$\dot{x}\ddot{y} - \dot{y}\ddot{x} = -4[1 + \sin t\cdot\sin 2t - \cos t\cdot\cos 2t]$$

so that $(\dot{x}^2 + \dot{y}^2)/(\dot{x}\ddot{y} - \dot{y}\ddot{x}) = -2$

Now the coordinates of the centre of curvature are

$$X = x(t) - \frac{\dot{y}(\dot{x}^2 + \dot{y}^2)}{\dot{x}\ddot{y} - \ddot{x}\dot{y}}$$

$$X = (2\cos t + \cos 2t) - (2\cos t - 2\cos 2t)(-2)$$

$$X = 3(2\cos t - \cos 2t)$$

Similarly

$$Y = y(t) + \frac{\dot{x}(\dot{x}^2 + \dot{y}^2)}{\dot{x}\ddot{y} - \ddot{x}\dot{y}}$$

$$= (2\sin t - \sin 2t) + (-2\sin t - 2\sin 2t)(-2)$$

$$Y = 3(2\sin t + 2\sin 2t)$$

Thus the evolute given by $X(t)$ and $Y(t)$ is the required deltoid.

Example 4: Calculate the total length of the evolute of the astroid $x^{\frac{2}{3}} + y^{\frac{2}{3}} = a^{\frac{2}{3}}$.

Solution: $x = a\cos^3 t$, $y = a\sin^3 t$ are the parametric equations of the given astroid (also known as four-cusped hypo-cycloid (refer Fig. 2.16).

$$y_1 = \frac{dy}{dx} = \frac{dy/dt}{dx/dt} = \frac{3a\sin^2\cos t}{-3a\cos^2 t \cdot \sin t} = -\tan t$$

$$y_2 = \frac{d^2y}{dx^2} = \frac{d}{dt}(-\tan t)\frac{dt}{dx}$$

$$= -\sec^2 t \times \frac{1}{-3a\cos^2 t \cdot \sin t}$$

so $$y_2 = \frac{1}{3a\cos^4 t \cdot \sin t}$$

Now the centre of curvature (X, Y) are

$$X = x - y_1\frac{(1+y_1^2)}{y_2}$$

$$= a\cos^3 t + \tan t(1+\tan^2 t)3a\cos^4 t\sin t$$

$$X = a\cos^3 t + 3a\sin^2 t\cos t$$

Similarly

$$Y = y + \frac{1+y_1^2}{y_2} = a\sin^3 t + 3a\cos^2 t \cdot \sin t$$

To eliminate the parameter t, consider

$$(X+Y)^{\frac{2}{3}} + (X-Y)^{\frac{2}{3}}$$

$$= a^{\frac{2}{3}}\left[(\cos t + \sin t)^2 + (\cos t - \sin t)^2\right]$$

Thus the required evolute for the astroid is given by

$$(X+Y)^{\frac{2}{3}} + (X-Y)^{\frac{2}{3}} = 2a^{\frac{2}{3}}$$

The radius of curvature R at any point of the given astroid is

$$R = \frac{(1+y_1^2)^{\frac{3}{2}}}{y_2}$$

$$= (1+\tan^2 t)^{\frac{3}{2}} \cdot 3a\cos^4 t \cdot \sin t = \frac{3a}{2}\sin 2t$$

Total length of the evolute $= L = 8$ times the length of arc of the evolute AB in the 1st quadrant. Radius of curvature of the astroid at the point A with $t = 0$ is $R_1 = \frac{3a}{2} \cdot \sin 2.0 = 0$.

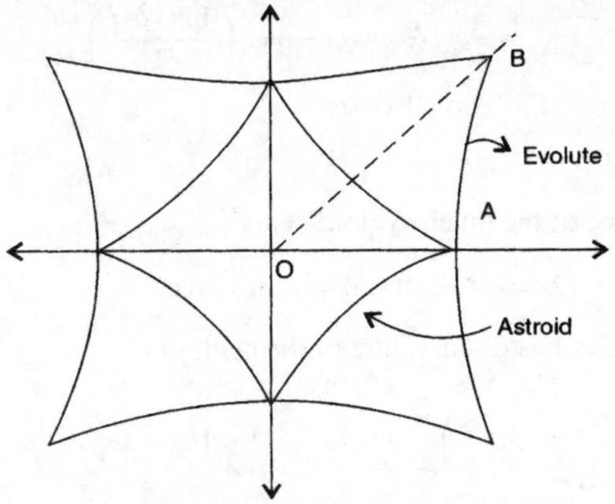

Fig. 2.16

Radius of curvature at B (with $t = \frac{\pi}{4}$) is $R_2 = \frac{3a}{2}$. Since the length of the arc of the evolute between two points is the difference between the radii of curvatures of the curve at the corresponding points, the length of the arc AB of the evolute $= R_2 - R_1 = \frac{3a}{2} - 0 = \frac{3a}{2}$.

Thus the total length of the evolute $= 8 \cdot \frac{3a}{2} = 12a$.

Example 5: Show that $R_1^2 + R_2^2 =$ constant if R_1 and R_2 are the radii of curvatures at the corresponding points of a cycloid and its evolute.

Solution: Consider a cycloid

$$x = a(t + \sin t),\ y = -a(1 - \cos t) \qquad (1)$$

$$\dot{x} = a(1 + \cos t),\ \dot{y} = -a\sin t$$

$$y_1 = \frac{\dot{y}}{\dot{x}} = -\frac{a\sin t}{a(1+\cos t)} = -\frac{-2\sin\frac{t}{2}\cdot\cos\frac{t}{2}}{2\cdot\cos^2\frac{t}{2}} = -\tan\frac{t}{2}$$

$$y_2 = -\sec^2\frac{t}{2}\cdot\frac{1}{2}\cdot\frac{1}{a(1+\cos t)} = -\frac{1}{2}\frac{\sec^2\frac{t}{2}}{2\cdot\cos^2\frac{t}{2}}.$$

$$y_2 = \frac{-1}{4a\cos^4\frac{t}{2}}$$

The centre of curvature of the given cycloid is

$$X = x - \left(-\tan\frac{t}{2}\right)\cdot\left(1+\tan^2\frac{t}{2}\right)\cdot\left(-4a\cos^4\frac{t}{2}\right)$$

$$X = a(t+\sin t) - 2a\cdot\sin 2\frac{t}{2} = a(t - \sin t)$$

Similarly

$$Y = y + \left(1+\tan^2\frac{t}{2}\right)\left(-4a\cos^4\frac{t}{2}\right)$$

$$= -a(1-\cos t) - 4a\cdot\left(\frac{1+\cos t}{2}\right)$$

$$Y = -a(3+\cos t)$$

$$Y + 2a = -a(1+\cos t)$$

Evolute of the given cycloid (1) is

$$X = a(t-\sin t),\ Y = -a(3+\cos t) \qquad (2)$$

R_1 = Radius of curvature of the cycloid (1)

$$R_1 = \left(1+\tan^2\frac{\theta}{2}\right)^{\frac{3}{2}}\cdot\left(-4a\cdot\cos^4\frac{t}{2}\right) = -4a\cos\frac{t}{2} \qquad (3)$$

R_2 = Radius of curvature of the evolute (2)

$$\dot{X} = a(1-\cos t),\ \dot{Y} = a\sin t$$

$$Y_1 = \frac{dY}{dX} = \frac{\dot{Y}}{\dot{X}} = \frac{a\sin t}{a(1-\cos t)}$$

$$= \frac{2\sin\frac{t}{2}\cdot\cos\frac{t}{2}}{2\cdot\sin^2\frac{t}{2}} = \cot\frac{t}{2}$$

$$Y_2 = \frac{d^2Y}{dX^2} = \frac{d}{dt}\left(\cot\frac{t}{2}\right)\cdot\frac{dt}{dX}$$

$$= -\mathrm{cosec}^2\frac{t}{2}\cdot\frac{1}{2}\cdot\frac{1}{a(1-\cos t)}$$

$$Y_2 = -\frac{1}{2a}\frac{\mathrm{cosec}^2\frac{t}{2}}{2\cdot\sin^2\frac{t}{2}} = \frac{-1}{4a\sin^4\frac{t}{2}}$$

$$\therefore R_2 = \left(1+\cot^2\frac{t}{2}\right)^{\frac{3}{2}}\cdot\left(-4a\sin^4\frac{t}{2}\right) = -4a\sin\frac{t}{2}. \qquad (4)$$

From (3) and (4)

$$R_1^2 + R_2^2 = 16a\cos^2\frac{t}{2} + 16a\sin^2\frac{t}{2} = 16a = \text{constant}.$$

EXERCISE

Find the evolute of the following curves:

1. $y = x^3$

Ans. $X(x) = \frac{x}{2} - \frac{9}{2}x^5,\ Y(x) = \frac{5}{2}x^3 - \frac{1}{6x}$

2. $y = e^x$

Ans. $X(x) = x - 1 - e^{2x},\ Y(x) = 2e^{2x} + e^{-x}$

3. Cycloid $x = a(t-\sin t),\ y = a(1-\cos t)$

Hint: $y_1 = \cot(t/2),\ y_2 = 1/(4a\sin^4(t/2))$.

Ans. $X(t) = a(t+\sin t),\ Y(t) = -a(1-\cos t)$

4. Cycloid $x = a(t+\sin t),\ y = a(1+\cos t)$

Ans. $X(t) = a(t-\sin t),\ Y(t) = -a(1+\cos t)$

5. Cardioid $x = 2\cos t \pm \cos 2t,\ y = 2\sin t \pm \sin 2t$

Hint: $\frac{\dot{x}^2+\dot{y}^2}{\dot{x}\ddot{y}-\ddot{x}\dot{y}} = \frac{8(1-\cos t)}{12(1-\cos t)}$ (for − sign).

Ans. Cardioid $X = \frac{1}{3}(2\cos t \mp \cos 2t)$,
$Y = \frac{1}{3}(2\sin t \mp \sin 2t)$

6. Nephroid $x = 3\cos t + \cos 3t$,
$y = 3\sin t - \sin 3t$

Hint: $y_1 = \frac{+12\sin^2 t\cdot\cos t}{-12\cos^2 t\cdot\sin t} = -\tan t$,
$y_2 = -1/(12\cos^4 t\cdot\sin t)$

Ans. Nephroid $X = 2(3\cos t - \cos 3t)$,
$Y = 2(3\sin t + \sin 3t)$

Note: In the above problems from 1 to 6 it is not possible to eliminate the parameter t between X and Y. Therefore the equation of the evolute is given in the parametric form $X = X(t),\ Y = Y(t)$.

7. Show that the evolute of the curve
$x = a(\cos t + t\sin t),\ y = a(\sin t - t\cos t)$
is a circle $x^2 + y^2 = a^2$.

Hint: $y_1 = \tan t,\ y_2 = 1/(at\cos^3 t)$.

Find the evolute of the following curves:

8. $y^2 = 4ax$ or in the parametric form
$x = a\cot^2 t,\ y = 2a\cot t$

Ans. $27aY^2 = 4(X-2a)^3$

9. $\frac{x^2}{a^2} + \frac{y^2}{b^2} = 1$

Hint: $y_1 = -b\cot t,\ y_2 = -b/(a^2\sin^3 t)$

Ans. $(aX)^{\frac{2}{3}} + (bY)^{\frac{2}{3}} = (a^2-b^2)^{\frac{2}{3}}$

10. $xy = c^2/2$

Ans. $(X+Y)^{\frac{2}{3}} - (X-Y)^{\frac{2}{3}} = 2C^{\frac{2}{3}}$

11. Calculate the length of the arc of the evolute of the parabola $y^2 = 4ax$ which is intercepted by the parabola.

Hint: Evolute is $27ay^2 = 4(x - 2a^3)$. Centre of curvature at any point (x, y) on parabola is $(3x + 2a, -y^3/4a^2)$. Points of intersection of parabola and evolute are $B(8a, -4\sqrt{2}a)$ and $C(8a, 4\sqrt{2}a)$. Length of evolute $=$ $2(\text{arc } AC) = 2(R_2 - R_1) = 2(6\sqrt{3}a - 2a)$

Ans. $4a(3\sqrt{3} - 1)$

12. Prove that the whole length of the evolute of the ellipse $\frac{x^2}{a^2} + \frac{y^2}{b^2} = 1$ is $4\left(\frac{a^2}{b} - \frac{b^2}{a}\right)$

Hint: Radius of curvature $R = \frac{(a^2\sin^2 t + b^2\cos^2 t)^{\frac{3}{2}}}{ab}$. Length of evolute of ellipse $= 4(R_2 - R_1)$ where R_2, R_1 are the radii of curvature at $t = 0$, and $t = \pi/2$.

2.13 ENVELOPES

One-parameter Family of Curves

Let $f(x, y, \alpha)$ be a function of the three independent variables x, y and α. For a particular given value of α, the equation

$$f(x, y, \alpha) = 0 \tag{1}$$

represents a plane curve in the xy-plane. Thus the one-parameter family of curves is the totality of all the curves having one common property obtained from (1) by assigning different values to α. The variable α is known as the parameter and the equation (1) $f(x, y, \alpha) = 0$ represents a one-parameter family (or system or group) of curves.

Example: $y = mx + 6$ represents a family of straight lines all passing through the point (0, 6) but having different slopes assigned by the parameter m.

Example: $(x - a)^2 + y^2 = 4$ represents a family of circles all with radius 2 but with different centers lying on x-axis for different values of the parameter a.

In a similar way, a two-parameter family of curves is represented by the equation $f(x, y, \alpha, \beta) = 0$ where α, β are the two parameters.

Example: Family of concentric and coaxial ellipses given by $\frac{x^2}{a^2} + \frac{y^2}{b^2} = 1$ with a, b as the parameters.

Note: If the two parameters α, β are connected by a relation, then the two-parameter family reduces to a one-parameter family of curves.

Envelope E of a given family of curves c is a curve which touches every member of the family of curves c and at each point of the envelope E is touched by some member of the family of curves c.

Example: The x-axis $y = 0$ is the envelope of the family of semicubical parabolas $y^2 - (x + b)^3 = 0$ with b as the parameter (see Fig. 2.17).

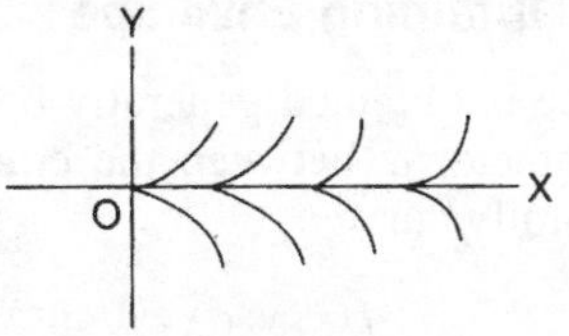

Fig. 2.17

Example: The two lines $x = a$ and $x = -a$ are the two envelopes to the family of circles

$$x^2 + (y - b)^2 = a^2$$

with centres on y-axis and of given radius a. Here b is the parameter (see Fig. 2.18).

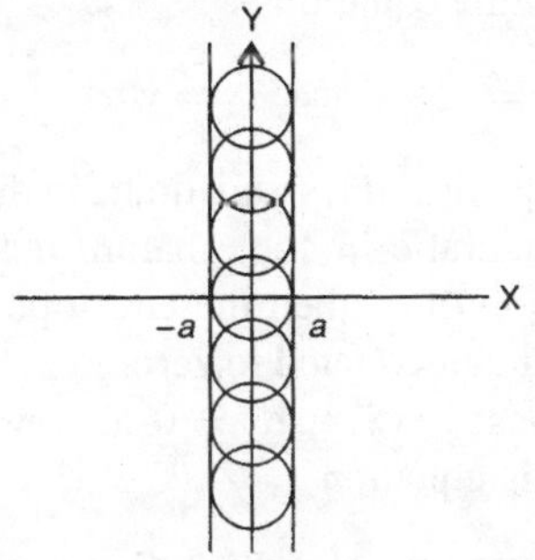

Fig. 2.18

Example: The family of circles $x^2 + (y - b)^2 = b^2$ with centres lying on y-axis, have no envelope (refer Fig. 2.19).

Thus a family of curves may have no envelope or unique envelope or several envelopes.

Envelope may be defined in a rigorous way as the locus of the limiting position of point of intersection of one member of the family with a neighbouring member as the later tends to the former.

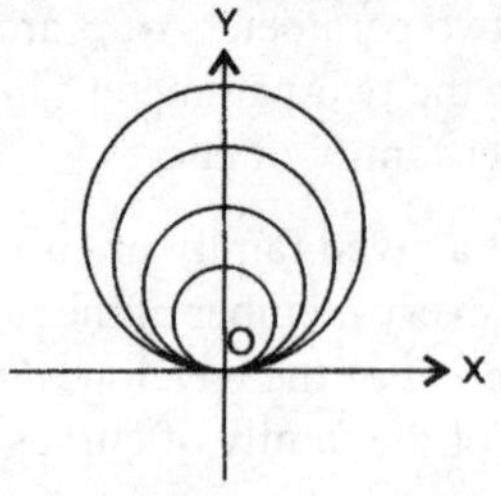

Fig. 2.19

Method of Obtaining Envelope

I. Envelope is obtained generally by eliminating the parameter α between the equation of the given family curves

$$f(x, y, \alpha) = 0 \quad (1)$$

and
$$\frac{\partial f}{\partial \alpha} = f_\alpha(x, y, \alpha) = 0 \quad (2)$$

where $\frac{\partial f}{\partial \alpha}$ is the partial derivative of f w.r.t. α.

II. In case α can not be eliminated between (1) and (2) then solve (1) and (2) for x and y in terms of α. Then the envelope is given in the parametric form by the equations

$$x = x(\alpha) \quad \text{and} \quad y = y(\alpha)$$

III. If the equation (1) is a quadratic in the parameter α or quadratic in some parameter λ which is a function of α, then the envelope is given by discriminant equated to zero.

Suppose $f(x, y, \alpha) = 0$ is rewritten as a quadratic equation

$$A\lambda^2 + B\lambda + C = 0 \quad (3)$$

where A, B, C are functions of x, y while λ is either α or function of α. Differentiating (3) w.r.t. λ,

$$2A\lambda + B = 0 \quad \text{or} \quad \lambda = -\frac{B}{2A} \quad (4)$$

Eliminating λ from (3) by using (4) we get the equation of the required envelope as

$$A\left(-\frac{B}{2A}\right)^2 + B\left(-\frac{B}{2A}\right) + C = 0$$

i.e., $\quad B^2 - 4AC = \text{discriminant} = 0$

IV. Envelope of the family of normals to a given curve C is the evolute of the curve C.

V. For a given two parameter family of curves

$$f(x, y, \alpha, \beta) = 0 \quad (5)$$

with a given relation $g(\alpha, \beta) = 0$ between the parameters α, β, (5) can be reduced to a one parameter family by elimination of one of the parameters say β in terms of α by using the given relation $g(\alpha, \beta)$. Then proceed as in I.

WORKED OUT EXAMPLES

Example 1: Find the envelope of the one parameter family of curves $y = mx + am^p$ where m is the parameter and a, p are constants.

Solution: Differentiate the given curves

$$y = mx + am^p \quad (1)$$

with respect to the parameter 'm'

$$O = x + apm^{p-1}$$

solving
$$m = \left(-\frac{x}{pa}\right)^{\frac{1}{(p-1)}} \quad (2)$$

using (2) eliminate m from (1)

$$y = \left(\frac{-x}{pa}\right)^{\frac{1}{(p-1)}} \cdot x + a\left(\frac{-x}{pa}\right)^{\frac{p}{(p-1)}}$$

$$y^{(p-1)} = \left(\frac{-x}{pa}\right) \cdot x^{(p-1)} + a^{(p-1)} \cdot \left(\frac{-x}{pa}\right)^p$$

or $\quad ap^p y^{p-1} = -x^p \cdot p^{p-1} + (-x)^p \quad (3)$

(3) is the equation of the required envelope of (1).

Example 2: Show that the family of straight lines $2y - 4x + \alpha = 0$ has no envelope, where α being the parameter.

Solution: Differentiating the given equation w.r.t. 'α' we get $0 + 0 + 1 = 0$ which is a contradiction. Note that the given family of straight lines $y = 2x - \frac{\alpha}{2}$ are all parallel with common slope

$m = 2$. Therefore no curve (envelope) exists which touches each member of this parallel straight lines.

Example 3: Determine the envelope of

$$x \sin t - y \cos t = at \quad (1)$$

where t is the parameter.

Solution: Differentiating (1) w.r.t. 't', we get

$$x \cos t + y \sin t = a \quad (2)$$

As 't' the parameter can not be eliminated between (1) and (2). Solve (1) and (2), for x and y in terms of t. For this, multiply (1) by $\sin t$ and (2) by $\cos t$

$$x \sin^2 t - y \sin t \cos t = at \sin t$$

$$x \cos^2 t + y \sin t \cos t = a \cos t$$

Adding, $x(t) = a(t \sin t + \cos t)$. Similarly multiplying (1) by $\cos t$ and (2) by $\sin t$ and subtracting, we get

$$y(t) = a(\sin t - t \cos t).$$

Example 4: (Leibnitz's problem) calculate the envelope of family of circles whose centres lie on the x-axis and radii are proportional to the abscissa of the centre.

Solution: Let $(b, 0)$ be the centre of any one of the family of circles with b as the parameter. Then the equation of the family of circles with centres on x-axis and radius proportional to the abscissa of the centre is

$$(x - b)^2 + y^2 = ab$$

where a is the proportionality constant. Differentiating the above equation w.r.t. 'b',

$$-2(x - b) + 0 = a \quad \therefore b = x + \frac{a}{2}$$

Eliminating the parameter b

$$\left(-\frac{a}{2}\right)^2 + y^2 = a\left(x + \frac{a}{2}\right) \quad \text{or} \quad y^2 = a\left(x + \frac{a}{4}\right)$$

The required envelope is the parabola.

Example 5: Prove that every member of the family of parabolas given by $y^2 = 4b(x - b)$ with b as the parameter is touched by one or the other of the straight lines $y = \pm x$.

Solution: Differentiating

$$y^2 = 4b(x - b)$$

w.r.t. 'b', we get

$$0 = 4x - 8b$$

$$b = \frac{x}{2}$$

Substituting b in the given equation

$$y^2 = 4\left(\frac{x}{2}\right)\left(x - \frac{x}{2}\right) = x^2$$

The required envelopes are the straight lines $y = \pm x$ which touch every parabola of the given family.

Example 6: Determine the envelope of the two parameter family of parabolas

$$\sqrt{\frac{x}{a}} + \sqrt{\frac{y}{b}} = 1$$

where the two parameters a and b are connected by the relation $a + b = c$ where c is a given constant.

Solution: Using the given relation

$$a + b = c$$

eliminate $b = c - a$ from the given family

$$\sqrt{\frac{x}{a}} + \sqrt{\frac{y}{c - a}} = 1 \quad (1)$$

which is now a one-parameter family of parabolas with a as the parameter.

Differentiating (1) w.r.t. 'a',

$$\sqrt{x} \cdot \left(-\frac{1}{2}\right)\frac{1}{a^{\frac{3}{2}}} + \sqrt{y} \cdot \left(-\frac{1}{2}\right) \times \frac{1}{(c - a)^{\frac{3}{2}}}(-1) = 0$$

$$\left(\frac{c - a}{a}\right)^{\frac{3}{2}} = \left(\frac{y}{x}\right)^{\frac{1}{2}}$$

$$\frac{c}{a} = \frac{x^{\frac{1}{3}} + y^{\frac{1}{3}}}{x^{\frac{1}{3}}}$$

$$\text{or} \quad a = \frac{cx^{\frac{1}{3}}}{x^{\frac{1}{3}} + y^{\frac{1}{3}}} \quad (2)$$

Substituting (2) in (1), we get the required envelope as

$$\left[x \cdot \frac{x^{\frac{1}{3}} + y^{\frac{1}{3}}}{cx^{\frac{1}{3}}}\right]^{\frac{1}{2}} + \left[y \cdot \frac{1}{\left\{c - \frac{cx^{\frac{1}{3}}}{x^{\frac{1}{3}} + y^{\frac{1}{3}}}\right\}}\right]^{\frac{1}{2}} = 1$$

$$\left[x^{\frac{2}{3}}\left(x^{\frac{1}{3}}+y^{\frac{1}{3}}\right)\right]^{\frac{1}{2}}+\left[y^{\frac{2}{3}}\left(x^{\frac{1}{3}}+y^{\frac{1}{3}}\right)\right]^{\frac{1}{2}}=c^{\frac{1}{2}}$$

or $$\left(x^{\frac{1}{3}}+y^{\frac{1}{3}}\right)\left[\left(x^{\frac{2}{3}}\right)^{\frac{1}{2}}+\left(y^{\frac{2}{3}}\right)^{\frac{1}{2}}\right]=c^{\frac{1}{2}}$$

$$\left(x^{\frac{1}{3}}+y^{\frac{1}{3}}\right)^{\frac{3}{2}}=c^{\frac{1}{2}}$$

Thus the envelope is the astroid given by

$$x^{\frac{1}{3}}+y^{\frac{1}{3}}=c^{\frac{1}{3}}.$$

Example 7: If $a^2+b^2=c$ show that the envelopes of the family of ellipses

$$\frac{x^2}{a^2}+\frac{y^2}{b^2}=1 \quad (1)$$

with a, b as parameters are the straight lines $\pm x \pm y=\sqrt{c}$.

Solution: Unlike the previous problem 6, where one parameter b is eliminated in terms of the other parameter a, in this problem treat b as a function of a. Thus differentiating equation of ellipses (1) w.r.t. 'a', we get

$$-\frac{2x^2}{a^3}-\frac{2y^2}{b^3}\cdot\frac{db}{da}=0 \quad (2)$$

Differentiating the given relation

$$a^2+b^2=c \quad (3)$$

w.r.t. 'a', we have

$$2a+2b\frac{db}{da}=0 \quad (4)$$

or $$\frac{db}{da}=-\frac{a}{b} \quad (5)$$

Substituting (5) in (2)

$$\frac{x^2}{a^4}=\frac{y^2}{b^4}$$

Rewriting

$$\frac{(x^2/a^2)}{a^2}=\frac{(y^2/b^2)}{b^2}=\frac{(x^2/a^2)+(y^2/b^2)}{(a^2+b^2)}=\frac{1}{c}$$

where equations (1) and (3) are used.

$$\therefore \frac{x^2}{a^4}=\frac{1}{c} \quad \text{and} \quad \frac{y^2}{b^4}=\frac{1}{c}$$

or $$a^2=\pm\sqrt{cx}, \quad b^2=\pm\sqrt{cy}$$

Using (3) again

$$c=a^2+b^2=\pm\sqrt{cx}\pm\sqrt{cy}$$

so the required envelopes are

$$\pm x\pm y=\sqrt{c}.$$

Example 8: Find the envelope of the family of circles passing through the origin and with their centres lying on the ellipse (in Fig. 2.20)

$$\frac{x^2}{a^2}+\frac{y^2}{b^2}=1$$

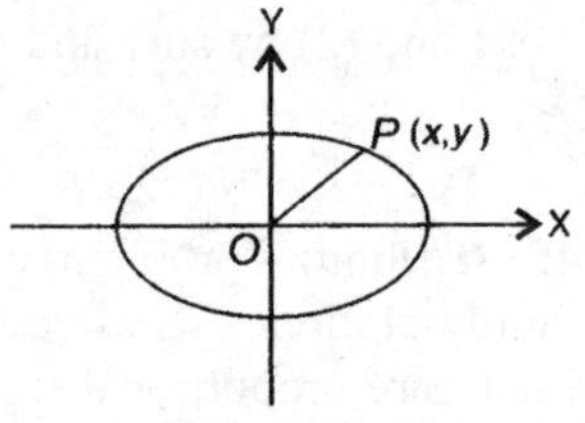

Fig. 2.20

Solution: Any point $P(x, y)$ lying on the given ellipse in the parametric form is $P(a\cos t, b\sin t)$ with t as the parameter. Since the family of circles pass through the origin $O(0,0)$ and have their centres as a point $P(x, y)$ on the ellipse, the equation of such family of circles is given by (with x, y as general point)

$$(x-a\cos t)^2+(y-b\sin t)^2=(\text{radius})^2=(OP)^2$$
$$=(a\cos t-0)^2+(b\sin t-0)^2 \quad (1)$$

Differentiating (1) w.r.t. the parameter 't'

$$2(x-a\cos t)(+a\sin t)+2(y-b\sin t)(-b\cos t)$$
$$=2a^2\cdot\cos t(-\sin t)+2b^2\sin t\cdot\cos t$$

Dividing throughout by $\sin t\cdot\cos t$

$$\frac{ax}{\cos t}-a^2-\frac{by}{\sin t}+b^2=b^2-a^2$$

so $$\tan t=\frac{by}{ax} \quad \text{(refer Fig. 2.21)}$$

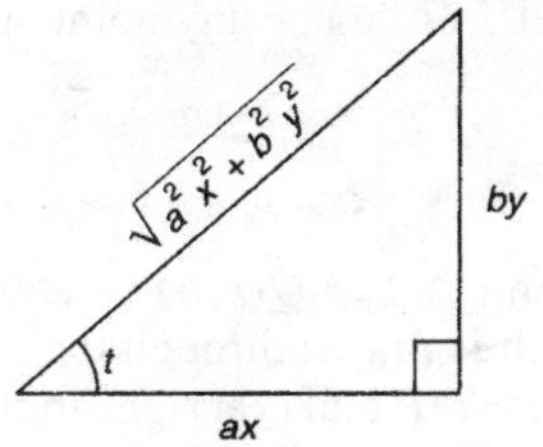

Fig. 2.21

Then

$$\sin t = \frac{by}{\sqrt{a^2x^2+b^2y^2}} = \frac{by}{h} \qquad (2)$$

and

$$\cos t = \frac{ax}{\sqrt{a^2x^2+b^2y^2}} = \frac{ax}{h} \qquad (3)$$

where $\quad h = \sqrt{a^2x^2+b^2y^2}$

using (2) and (3) eliminate the parameter 't' from (1)

$$\left(x - a\cdot\frac{ax}{h}\right)^2 + \left(y - b\cdot\frac{by}{h}\right)^2$$

$$= a^2\cdot\left(\frac{ax}{h}\right)^2 + b^2\left(\frac{by}{h}\right)^2$$

$$[x^2h^2 + a^4x^2 - 2a^2x^2h] + [y^2h^2 + b^4y^2 - 2b^2y^2h]$$

$$= a^4x^2 + b^4y^2$$

$$(x^2+y^2)(h^2) = 2h(a^2x^2+b^2y^2) = 2h\cdot h^2$$

so $\quad x^2 + y^2 = 2h = 2(a^2x^2+b^2y^2)^{\frac{1}{2}}$

Thus the required envelope is

$$(x^2+y^2)^2 = 4(a^2x^2+b^2y^2).$$

Example 9: Determine the evolute of the hyperbola $\frac{x^2}{a^2} - \frac{y^2}{b^2} = 1$.

Solution: We know that the evolute of a curve c is the envelope of the family of normals to the given curve c.

Let $P(a \cosh t, b \sinh t)$ be any point on the given hyperbola. Then

$$\frac{dx}{dt} = \dot{x} = \frac{d}{dt}(a \cosh t) = a \sinh t,\ \dot{y} = b \cosh t \quad \text{so}$$

$$\frac{dy}{dx} = \frac{\dot{y}}{\dot{x}} = \frac{a \cosh t}{a \sinh t} = \frac{b}{a}\coth t$$

Thus the equation to (any) normal to hyperbola at a point P is

$$(y - b \sinh t) = \frac{-a}{b \coth t}(x - a \cosh t) \qquad (1)$$

or

$$\frac{by}{\sinh t} - b^2 = -\frac{ax}{\cosh t} + a^2 \qquad (2)$$

Differentiating (2) w.r.t. the parameter 't', we get

$$-\frac{by \cosh t}{\sinh t^2} - O = -ax\cdot\frac{\sinh t}{\cosh^2 t} + O$$

$$\tanh t = -\left(\frac{by}{ax}\right)^{\frac{1}{3}}$$

Then

$$\sinh t = \mp\frac{(by)^{\frac{1}{3}}}{h} \quad \text{and} \quad \cosh t = \pm\frac{(ax)^{\frac{1}{3}}}{h} \qquad (3)$$

where $\quad h = \sqrt{(ax)^{\frac{2}{3}} - (by)^{\frac{2}{3}}}$

Using (3) eliminate 't' from (2). Thus

$$\frac{by}{-(by)^{\frac{1}{3}}}\cdot h + \frac{ax}{(ax)^{\frac{1}{3}}}\cdot h = a^2 + b^2$$

$$\left[(ax)^{\frac{2}{3}} - (by)^{\frac{2}{3}}\right]\left[(ax)^{\frac{2}{3}} - (by)^{\frac{2}{3}}\right]^{\frac{1}{2}} = a^2 + b^2$$

The required envelope of the normals to the given hyperbola

$$(ax)^{\frac{2}{3}} - (by)^{\frac{2}{3}} = (a^2+b^2)^{\frac{2}{3}}$$

is the evolute of the hyperbola.

Example 10: Find the envelope of

$$x \sec^2\theta + y\,\text{cosec}^2\theta = a$$

where θ is the parameter.

Solution: The given equation is rewritten as

$$x(1+\tan^2\theta) + y(1+\cot^2\theta) = a$$

or $\quad x\cdot\tan^4\theta + (x+y-a)\tan^2\theta + y = 0$

which is a quadratic equation in the parameter $\lambda = \tan^2\theta$. Therefore the required envelope is given by discriminant $= B^2 - 4AC = 0$

i.e., $\quad (x+y-a)^2 - 4(x)(y) = 0$

The envelope is

$$(x-y)^2 - 2ax - 2ay + a^2 = 0$$

Example 11: Show that the envelope of the family of circles described on the double ordinates of the ellipse $\frac{x^2}{a^2}+\frac{y^2}{b^2}=1$ as diameters, is the ellipse $\frac{x^2}{a^2+b^2}+\frac{y^2}{b^2}=1$.

Solution: The double ordinates on the ellipse

$$\frac{x^2}{a^2}+\frac{y^2}{b^2}=1$$

are $(a\cos t, b\sin t)$ and $(a\cos t, -b\sin t)$ with 't' as the parameter. The equation of the family of circles described on the double ordinates as diameter is $(x-a\cos t)(x-a\cos t)+(y-b\sin t)(y+b\sin t)=0$.

[*Note:* $(x-x_1)(x-x_2)+(y-y_1)(y-y_2)=0$ is equation of circle]

Rewriting the equation

$$x^2+y^2-2ax\cos t+a^2\cos^2 t-b^2\sin^2 t=0$$

or

$$(a^2+b^2)\cos^2 t-2ax\cdot\cos t+(x^2+y^2-b^2)=0$$

which is a quadratic in the parameter $\lambda=\cos t$. Therefore the envelope to this family of circle is $B^2-4AC=0$

i.e., $(-2ax)^2-4(a^2+b^2)(x^2+y^2-b^2)=0$

$$a^2x^2=(a^2x^2+b^2x^2)+(a^2+b^2)y^2-b^2(a^2+b^2)$$

Thus the envelope is the ellipse

$$\frac{x^2}{a^2+b^2}+\frac{y^2}{b^2}=1.$$

obtained by dividing throughout by $b^2(a^2+b^2)$.

Example 12: Find the envelope of the family of straight lines drawn through the extremities of and at right angles to the radii vectors of the limacon $r=a+b\cos\theta$ (see Fig. 2.22)

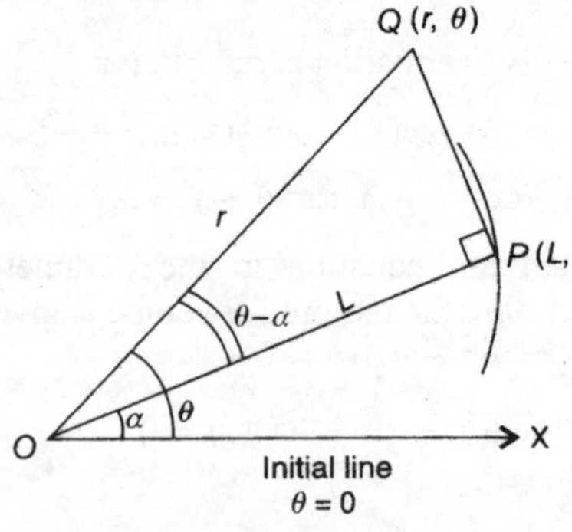

Fig. 2.22

Solution: Let $P(L,\alpha)$ be any point on the limacon

$$r=a+b\cos\theta \tag{1}$$

so $$L=a+b\cos\alpha \tag{2}$$

since P lies on (1). Let $Q(r,\theta)$ be any point on the line PQ, which is drawn through the extremity P of the radius vector OP and is at right angles to OP i.e., $\angle OPQ=90°$. The problem is to find the envelope of the family of straight lines PQ. From the right angled triangle OPQ,

$$\cos(\theta-\alpha)=\frac{L}{r}$$

or $$L=r\cos(\theta-\alpha) \tag{3}$$

Thus the equation of line PQ is (3), with L and α as parameters. Eliminate L, using (2) and (3).

$$a+b\cos\alpha=L=r\cos(\theta-\alpha). \tag{4}$$

So (4) is the equation of the family of straight lines passing through extremity of and perpendicular to the radius vectors. In (4) α is the only parameter. Differentiating (4) w.r.t. 'α', we get

$$0+b(-1)\sin\alpha=-r\sin(\theta-\alpha)\cdot(-1)$$

Solving for α

$$\tan\alpha=\frac{r\sin\theta}{r\cos\theta-b} \tag{5}$$

Let $$h=\sqrt{(r\sin\theta)^2+(r\cos\theta-b)^2} \tag{6}$$

$$h=\sqrt{r^2\sin^2\theta+r^2\cos^2\theta+b^2-2rb\cos\theta}$$

$$=\sqrt{r^2-2rb\cos\theta+b^2}$$

Then $$\cos\alpha=\frac{r\cos\theta-b}{h} \tag{7}$$

Rewrite (4) as

$$\frac{a}{\cos\alpha}+b=r\cos\theta+r\sin\theta\cdot\tan\alpha. \tag{8}$$

Eliminate the parameter 'α' from (8) using (5) and (7)

$$a\cdot\left(\frac{h}{r\cos\theta-b}\right)+b=r\cos\theta+r\sin\theta\left(\frac{r\sin\theta}{r\cos\theta-b}\right)$$

$$ah=(r\cos\theta-b)^2+r^2\sin^2\theta$$

$$ah=r^2-2rb\cos\theta+b^2=h^2$$

$$a=h=\sqrt{r^2-2rb\cos\theta+b^2}$$

Thus the required envelope is obtained by squaring

$$r^2 - 2b\cos\theta + (b^2 - a^2) = 0.$$

Example 13: Find the envelope of the system of straight lines $2y - 3tx + at^3 = 0$ where t is the parameter. Find the coordinates of the point of contact of the envelope and the particular member of the system when $t = 2$.

Solution: Differentiating the given system

$$2y - 3tx + at^3 = 0 \qquad (1)$$

w.r.t. 't', we get

$$0 - 3x + 3at^2 = 0 \qquad (2)$$

Multiplying (2) by t and subtracting from (1) we get $t = \frac{y}{x}$. Substituting this value of t in (1) we get the required envelope as

$$2y - 3x\left(\frac{y}{x}\right) + a\frac{y^3}{x^3} = 0$$

or $$ay^2 = x^3 \qquad (3)$$

From (2) $$x = at^2$$

For $$t = 2, \quad x = a(2)^2 = 4a \qquad (4)$$

Substituting x in (3)

$$ay^2 = 64a^3 \quad \therefore\ y = 8a$$

Thus the point of contact of the envelope with the given family when $t = 2$ is $(x = 4a, y = 8a)$.

EXERCISE

Find the envelope of the following family of curves with m as the parameter:

1. $y = mx + \frac{a}{m}$

Ans. parabola $y^2 = 4ax$

2. $y = mx + \sqrt{1 + m^2}$

Ans. circle $x^2 + y^2 = 1$

3. $y = mx + \sqrt{a^2m^2 + b^2}$

Ans. ellipse $\frac{x^2}{a^2} + \frac{y^2}{b^2} = 1$

4. $y = mx + am^3$

Ans. $8x^3 + 27ay^2 = 0$

5. $y = mx - 2am - am^3$

Ans. $27ay^2 = 4(x - 2a)^3$

Determine the envelope of following family of curves with t as the parameter:

6. $x\cos t - y\cot t = c$

Ans. $x^2 - y^2 = c^2$

7. $x\cos^n t + y \cdot \sin^n t = c$

Ans. $x^{\frac{2}{(2-n)}} + y^{\frac{2}{(2-n)}} = c^{\frac{2}{(2-n)}}$

8. $x\cos t + y\sin t = c\sin t\cos t$

Ans. $x^{\frac{2}{3}} + y^{\frac{2}{3}} = a^{\frac{2}{3}}$

9. $\frac{x}{a}\cos t + \frac{y}{b}\sin t = 1$

Ans. $\frac{x^2}{a^2} + \frac{y^2}{b^2} = 1$

10. $\frac{ax}{\cos t} + \frac{by}{\sin t} = a^2 - b^2$

Ans. $(ax)^{\frac{2}{3}} + (by)^{\frac{2}{3}} = (a^2 - b^2)^{\frac{2}{3}}$

11. $x\tan t + y\sec t = c$

Ans. $y^2 = a^2 + x^2$

12. $\frac{a^2\cos t}{x} - \frac{b^2\sin t}{y} = c$

Ans. $\frac{a^4}{x^2} + \frac{b^4}{y^2} = c^2$

13. Concentric circles of radius α

Hint: Equation $x^2 + y^2 = \alpha^2$, Differentiating $2\alpha = 0$, $x^2 + y^2 = 0 \Rightarrow x = 0, y = 0$ an isolated point.

Ans. No envelope

14. Show that the envelope of the family of circles which always pass through the vertex of the parabola $y^2 = 4ax$ and with centres of circles lying on the parabola is a cissoid $x^3 + y^2(x + 2a) = 0$.

Hint: Vertex of parabola is origin $(0, 0)$ centre of circle $(-f, -g)$ lies on parabola, so $f^2 = -4ag$ or $g = -f^2/4a$.
Equation of circle is $x^2 + y^2 + 2gx + 2fy = 0$ or $x^2 + y^2 - \frac{f^2}{2a}x + 2fy = 0$ which is a quadratic in the parameter $f : xf^2 - 4ayf - 2a(x^2 + y^2) = 0$.

15. Find the envelope of the family of circles whose diameter are double ordinates of a parabola $y^2 = 4ax$.

Hint: The Equation of required circle with centre at $(\alpha, 0)$, end points of diameter $(\alpha, y_1), (\alpha, -y_1)$, diameter $2y$, radius y_1 is $(x-\alpha)^2 + y^2 = y_1^2$. Since $\theta(\alpha, y_1)$ lies on parabola $y_1^2 = 4a\alpha$. So equation of family of circles is

$$(x-\alpha)^2 + y^2 = 4a\alpha$$

which is a quadratic in α

$$\alpha^2 - 2(x+2a)\alpha + (x^2+y^2) = 0.$$

Ans. Envelope is another parabola $y^2 = 4a(x+a)$

16. Find the envelope of the family of straight lines obtained by joining the feet of the two perpendiculars drawn from any point on the ellipse $\frac{x^2}{a^2} + \frac{y^2}{b^2} = 1$ to the two coordinate axes.

Hint: Any point P on ellipse is $(a\cos\theta; b\sin\theta)$. Coordinates of M and N the feet of perpendiculars PM and PN on to the x and y axis are $(a\cos\theta, 0)$ and $(0, b\sin\theta)$ respectively. So equation of family of straight lines MN is $y - 0 = \left(\frac{b\sin\theta - 0}{0 - a\cos\theta}\right)(x - a\cos\theta)$.

Ans. $\left(\frac{x}{a}\right)^{\frac{2}{3}} + \left(\frac{y}{b}\right)^{\frac{2}{3}} = 1$

17. A straight line of given length 2 slides with its extremities on two fixed straight lines which are at right angles. Determine the envelope of the family of circles drawn on the sliding line as diameter.

Hint: The sliding line AB of length L has extremities $A(L\cos\alpha, 0)$ and $B(0, L\sin\alpha)$. Equation of circle with AB as diameter is

$$(x - L\cos t)(x-0) + (y-0)(y - L\sin t) = 0.$$

Ans. $x^2 + y^2 = L^2$

18. Find the envelope of the family of straight lines of given length L, whose extremities slide on two fixed straight lines at right angles.

Hint: Equation of line is:
$x\sec\alpha + y\operatorname{cosec}\alpha = L$

Ans. $x^{\frac{2}{3}} + y^{\frac{2}{3}} = L^{\frac{2}{3}}$.

Quadratic equation in the parameter α or $\lambda = \lambda(\alpha)$

$A\lambda^2 + B\lambda + c = 0$, envelope is $B^2 - 4AC = 0$. Find the envelope of:

19. $(y - mx)^2 = a^2m^2 + b^2$, with m parameter

Ans. $\frac{x^2}{a^2} + \frac{y^2}{b^2} = 1$

20. $x\cos t + y\sin t = a\sec t$, t is parameter

Ans. $y^2 = 4a(a-x)$

21. $y = x\tan t - \frac{gx^2}{2u^2\cos t}$, t is the parameter

Hint: Quadratic in $\lambda = \tan t$ is

$$\left(\frac{g^2x^2}{2u^2}\right)\tan^2 t + (-x)\tan t + \left(y + \frac{gx^2}{2u^2}\right) = 0$$

Envelope is

$$B^2 - 4AC = (-x)^2 - 4\left(\frac{g^2x^2}{2u^2}\right)\left(y + \frac{gx^2}{2u^2}\right) = 0.$$

Ans. $u^4 = 2u^2gy + g^2x^2$

22. Straight lines which move such that the product of the intercepts on the two axes is constant.

Hint: $\frac{x}{a} + \frac{y}{b} = 1$ with $ab = c^2$ = constant. Quadratic in parameter 'a' is

$$\frac{y}{c^2}a^2 + (-1)\cdot a + x = 0$$

Envelope is:

$$B^2 - 4AC = (-1)^2 - 4\left(\frac{y}{c^2}\right)(x) = 0.$$

Ans. $xy = c^2/4$, rectangular hyperbola

23. $\frac{x^2}{\alpha^2} + \frac{y^2}{k^2 - \alpha^2} = 1$, α is the parameter

Hint: Quadratic in α^2 is

$$\alpha^4 - (x^2 - y^2 + k^2)\alpha^2 + k^2x^2 = 0$$

Envelope:

$$B^2 - 4AC = (x^2 - y^2 + k)^2 - 4\cdot 1\cdot k^2x^2 = 0.$$

Ans. Square with four lines

$$x + y + k = 0, \quad x + y - k = 0,$$
$$x - y + k = 0, \quad x - y - k = 0$$

24. $tx^3 + t^2y = a$, where t is the parameter

Ans. $x^6 + 4ay = 0$

25. Prove that all circles having for their diameters the radii vectors of a parabola touch a straight line or the curve $r \cos\theta + a \sin^2\theta = 0$ according as the radii vectors are drawn from the focus or the vertex.

Hint: Parabola $y^2 = 4ax$, vertex $A(0, 0)$, focus $S(a, 0)$, general point $P(at^2, 2at)$. Equation of circle on SP as diameter is (refer Fig. 2.23)

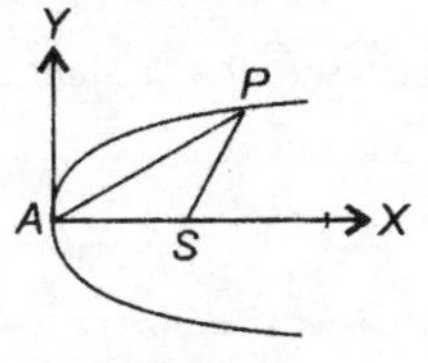

Fig. 2.23

$$(x - a)(x - at^2) + (y - 0)(y - at) = 0$$

This is a quadratic in t

$$a(a - x)t^2 - 2ay \cdot t + (x^2 + y^2 - ax) = 0$$

Envelope:

$$x[(x - a)^2 + y^2] = 0 \quad \therefore\ x = 0$$

Equation circle on AP as diameter

$$(x - 0)(x - at^2) + (y - 0)(y - 2at) = 0$$

Quadratic in t :

$$axt^2 + 2ayt - (x^2 + y^2) = 0$$

Envelope:

$$ay^2 + x(x^2 + y^2) = 0$$

In polar coordinate:

$$r \cos\theta + a \sin^2\theta = 0.$$

Evolute as envelope of normals

Evolute of a curve c is the envelope of the normals to that curve c. Using this, find the evolute of the following curves:

26. Parabola $y^2 = 4ax$

Hint: Equation of any normal to the parabola is

$$y = mx - 2am - am^3$$

where m is the parameter.

Ans. $27ay^2 = 4(x - 2a)^3$

27. Ellipse $\frac{x^2}{a^2} + \frac{y^2}{b^2} = 1$

Hint: Equation of normal at any point $(a \cos t, b \sin t)$ is

$$\frac{ax}{\cos t} - \frac{by}{\sin t} = a^2 - b^2.$$

Ans. $(ax)^{\frac{2}{3}} + (by)^{\frac{2}{3}} = (a^2 - b^2)^{\frac{2}{3}}$

28. $x = a(3 \cos t - 2 \cos^3 t)$, $y = a(3 \sin t - 2 \sin^3 t)$.

Hint: Equation of normal at 't' is $\frac{x}{\cos t} + \frac{y}{\sin t} = 4a$.

Ans. $x^{\frac{2}{3}} + y^{\frac{2}{3}} = (4a)^{\frac{2}{3}}$

29. Cycloid: $x = a(\cos t + t \sin t)$; $y = a(\sin t - a \cos t)$

Ans. $x^2 + y^2 = a^2$

30. Hyperbola $xy = c^2$

Ans. $(x + y)^{\frac{2}{3}} + (x - y)^{\frac{2}{3}} = (2a)^{\frac{2}{3}}$

31. Tractrix $x = a\left(\cos t + \ln \tan \frac{t}{2}\right)$, $y = a \sin t$

Ans. $y = a \cosh \frac{x}{a}$

Two-parameter family of curves

32. If $ab = 1$, find the envelope of $ax + by = 1$.

Ans. $4xy = 1$

33. Find the envelope of a system of concentric and coaxial ellipses of constant area.

Hint: Equation of ellipses $\frac{x^2}{a^2} + \frac{y^2}{b^2} = 1$ with the two parameters a, b connected by $ab = c^2$ (since area πab = constant).

Ans. $2xy = c^2$

34. If $a + b = c$, determine the envelope of the family of ellipses $x^2/a^2 + y^2/b^2 = 1$

Ans. $x^{\frac{2}{3}} + y^{\frac{2}{3}} = c^{\frac{2}{3}}$

35. Determine the envelop of the two parameter family of straight lines $\frac{x}{a} + \frac{y}{b} = 1$ when a and b are related by

i. $a^n + b^n = c^n$

ii. $a^m b^n = c^{m+n}$

iii. $\frac{a}{c} + \frac{b}{d} = 1$

Ans. **i.** $x^{\frac{n}{(n+1)}} + y^{\frac{n}{(n+1)}} = c^{\frac{n}{(n+1)}}$
ii. $(m+n)^{m+n} \cdot x^m y^n = m^n n^n c^{m+n}$
iii. $(x/c)^{\frac{1}{2}} + (y/d)^{\frac{1}{2}} = 1$

36. Find the envelope of the family of parabolas $\sqrt{\frac{x}{a}} + \sqrt{\frac{y}{b}} = 1$ when $ab = c^2$.

Ans. $16xy = c^2$, rectangular hyperbola

37. If $a^n + b^n = c^n$ determine the envelop of the family of curves $\frac{x^m}{a^m} + \frac{y^m}{b^m} = 1$.

Ans. $x^p + y^p = c^p$ where $p = mn/(m+n)$

Polar form

38. Prove that the envelopes of circles described on the central radii of a rectangular hyperbola is a lemniscate $r^2 = a^2 \cos 2\theta$.

Hint: General point on $x^2 - y^2 = a^2$ is P$(a \cosh t, a \sinh t)$. Equation of circle with (0, 0) and P as ends of a diameter is

$$(x-0)(x - a \cosh t) + (y-0)(y - a \sinh t) = 0$$

Envelope:

$$a^2(x^2 - y^2) = (x^2 + y^2)^2$$

In polar coordinates:

$$r^2 = a^2 \cos 2\theta.$$

39. Determine the envelope of the straight lines drawn through the extremities of and at right angles to the radii vectors of the equi-angular spiral $r = a\, e^{\theta \cot \alpha}$

Hint: If (R, ϕ) is any point on the straight line through (r, θ) and at right angles to radius vector is

$$R \cos(\phi - \theta) = r.$$

Ans. $r \sin \alpha = a\, e^{(\alpha - \pi/2)\cot \alpha} \cdot e^{(\theta \cot \alpha)}$

40. Show that the envelope of circles described on the radii vectors of $r = 2a \cos \theta$ as diameter is the cardioid $r = a(1 + \cos \theta)$.

Hint: Equation circle drawn on radius vector corresponding to the point (r, θ) on the curve $r = 2a \cos \theta$ as diameter is

$$R = r \cos(\phi - \theta) = 2a \cos \theta \cdot \cos(\phi - \theta)$$

41. Find the envelope of the family of circles whose centres lie on the rectangular hyperbola $x^2 - y^2 = a^2$ and which pass through the origin.

Hint: Any point on hyperbola is $P(a \sec \theta, a \tan \theta)$. Equation of circle with P as centre and OP as radius is

$$(x - a \sec \theta)^2 + (y - a \tan \theta)^2 = OP^2$$

$$= a^2 \sec^2 \theta + a^2 \tan^2 \theta$$

Envelope:

$$(x^2 + y^2)^2 = 4a^2(x^2 - y^2)$$

In polar coordinates is:

$$r^2 = 4a^2 \cos 2\theta$$

Ans. Envelope is lemniscate of Bernoulli $r^2 = 4a^2 \cos 2\theta$

42. Prove that the envelope of the circles whose centres lie on the rectangular hyperbola $xy = c^2$ and which pass through the origin is the lemniscate $r^2 = 8c^2 \sin 2\theta$.

43. Determine the envelope of the straight lines drawn through extremities of and at right angles to the radii vectors of the curve $r^n = a^n \cos n\theta$.

Hint: $P(L, \alpha)$ be any point on $r^n = a^n \cos n\theta$ so

$$L^n = a^n \cos n\alpha \quad \text{or} \quad L = a(\cos n\alpha)^{\frac{1}{n}}$$

$Q(r, \theta)$ be any point on the line PQ drawn at right angles to the radius vector OP, so

$$\frac{L}{r} = \cos(\theta - \alpha) \quad \text{or} \quad L = r \cos(\theta - \alpha).$$

Eliminating L, Equation of required line PQ is

$$r \cos(\theta - \alpha) = a(\cos n\alpha)^{\frac{1}{n}}.$$

Ans. $r^p = a^p \cdot \cos p\theta$ where $p = \frac{n}{1-n}$

Determine the envelope of the circles described on the radii vectors as diameter of the curve:

44. $r^n = a^n \cos n\theta$

Ans. $r^p = a^p \cos p\theta$ where $p = \frac{n}{1+n}$

45. $r = L/(L + e \cos\theta)$

Ans. $r^2(e^2 - 1) - 2Le\, r, \cos\theta + 2L^2 = 0.$

Chapter 3

Partial Differentiation

INTRODUCTION

Real world can be described in mathematical terms using parametric equations and functions such as trigonometric functions which describe cyclic, repetitive activity; exponential, logarithmic and logistic functions which describe growth and decay and polynomial functions which approximate these and most other functions. The problems in operations research, computer science, probability, statistics, fluid dynamics, economics, electricity etc. deal with functions of two or more independent variables. In this chapter we study the limit, continuity, partial derivative of such functions, Euler's theorem, Jacobians which determine the functional dependence and determination of errors and approximations of calculations.

3.1 FUNCTIONS OF SEVERAL VARIABLES: LIMIT AND CONTINUITY

The area of an ellipse is πab. It depends on two variables a and b; The total surface area of a rectangular parallelopiped is $2(xy + yz + zx)$ and it depends on 3 variables x, y, z; The velocity u of a fluid particle moving in space depends on 4 variables x, y, z, t. In transportation problem in operations research the cost function to be minimized is a function of several ($m \cdot n$: running into hundreds) variables (where m is the number of origins and n is the number of destinations). Thus functions of several variables plays a vital role in advanced Mathematics.

If $u = f(x, y, z, t)$ then x, y, z, t are known as the independent variables or arguments and u is known as the 'dependent variable' or 'value' of the function. In this section, we restrict to functions of two and three variables, although the analysis can easily be extended to several variables.

Function of Two Variables

If for every x and y a unique value $f(x, y)$ is associated, then f is said to be a function of the two independent variables x and y and is denoted by

$$z = f(x, y) \tag{1}$$

Geometrically, in three dimensional xyz-coordinate space (1) represents a surface. The values of x and y for which the function is defined is known as the domain of definition of the function:

Example: $z = \sqrt{a^2 - x^2 - y^2}$

$$\text{domain}: \quad x^2 + y^2 \le a^2$$

Function not defined when $x^2 + y^2 > a^2$.

Example: $z = x^y + y^x$

$$\text{domain}: \quad x > 0 \quad \text{and} \quad y > 0$$

δ-neighbourhood of a point (a, b) in the xy-plane is a square with centre at (a, b) bounded by the four lines $x = a - \delta$, $x = a + \delta$, $y = b - \delta$, $y = b + \delta$ i.e.,

$$a - \delta \le x \le a + \delta$$

$$b - \delta \le x \le b + \delta.$$

Note: Neighbourhood of a point (a, b) may also be defined as a circular disk with centre at (a, b) and of radius δ given by

$$(x-a)^2+(y-b)^2<\delta^2$$

concept of limit is paramount in defining continuity and differentiability of a function. Note that all the three concepts of limit, continuity and differentiability are point concepts i.e., defined at a point.

Limit: A function $f(x, y)$ is said to have a limit L as the point (x, y) approaches (a, b) and is denoted as

$$\lim_{(x,y)\to(a,b)} f(x, y) = L$$

if the value of $f(x, y)$ can be made as close (as we please) to the given finite number L for all those (x, y) in an appropriately chosen δ-neighbourhood of (a, b), i.e., for a given $\epsilon > 0$ we can find a δ such that

$$|f(x, y) - L| < \epsilon$$

for all (x, y) in the δ-neighbourhood

$$|x-a| < \delta \quad \text{and} \quad |y-b| < \delta$$

(or alternatively when $(x-a)^2+(y-b)^2 < \delta^2$).

Important Note: The limit of a function $f(x, y)$ of two variables is said to exist *only when* the *same* value is obtained for the limit along any path in the xy-plane from (x, y) to (a, b) say along $x \to a$ and $y \to b$ or along $y \to b$ and $x \to a$, etc.

Limit may or may not exist. If it exists limit must be unique.

Method of Obtaining Limit

Step I: Evaluate $\lim f(x, y)$ along path I: $x \to a$ and $x \to b$

Step II: Evaluate $\lim f(x, y)$ along path II: $y \to b$ and $x \to a$

If the limit values along path I and II are same, the limit exist. Otherwise not.

Step III: If $a = 0, b = 0$, evaluate limit along say path $y = mx$ or $y = mx^n$ also.

Results: If $\lim_{(x,y)\to(a,b)} f(x, y) = L$ and $\lim_{(x,y)\to(a,b)} g(x, y) = M$. Then

1. $\lim_{(x,y)\to(a,b)} (f \pm g) = L \pm M$
2. $\lim_{(x,y)\to(a,b)} (f \cdot g) = L \cdot M$
3. $\lim_{(x,y)\to(a,b)} (f/g) = L/M,$ provided $M \neq 0$.

Continuity A function $f(x, y)$ is said to be continuous at a point (a, b) if

$$\lim_{(x,y)\to(a,b)} f(x, y) = f(a, b) \tag{2}$$

i.e., the limit of f as (x, y) tends to $(a, b) =$ the value of f at (a, b).

A function is said to be continuous in a domain if it is continuous at every point of the domain.

(2) can also be written as

$$\lim_{(h,k)\to(0,0)} f(a+h, b+k) = f(a, b)$$

If f is not continuous at (a, b), it is said to be discontinuous at (a, b).

Result: If $f(x, y)$ and $g(x, y)$ are continuous at (a, b) then $f \pm g$, $f \cdot g$ and f/g are continuous at (a, b).

Test for Continuity at a Point *(a, b)*

Step I: $f(a, b)$ should be well defined

Step II: $\lim f(x, y)$ as $(x, y) \to (a, b)$ should exists (must be unique and same along any path)

Step III: The limit of $f =$ value of f at (a, b).

WORKED OUT EXAMPLES

Limits

Example 1: Evaluate $\lim_{(x,y)\to(0,0)} (x^2 + y^2)$

Solution:

$$\lim_{\substack{x\to0\\y\to0}} (x^2+y^2) = \lim_{y\to0} (y^2) = 0$$

$$\lim_{\substack{y\to0\\x\to0}} (x^2+y^2) = \lim_{x\to0} (x^2) = 0$$

Along $y = mx$

$$\lim_{\substack{y\to mx\\ x\to 0}} (x^2+y^2) = \lim_{x\to 0}(x^2+m^2x^2)$$

$$= \lim_{x\to 0}(1+m^2)x^2 = (1+m^2)\lim_{x\to 0}(x^2) = 0$$

Along $y = mx^2$

$$\lim_{\substack{y\to mx^2\\ x\to 0}} (x^2+y^2) = \lim_{x\to 0} x^2(1+m^2x^2) = 0.$$

Since the value of the limit along any path is same, the limit exists and the limit value is zero.

Example 2: If $f(x, y) = \frac{y^2-x^2}{x^2+y^2}$, show that

$$\lim_{x\to 0}\left\{\lim_{y\to 0} f(x,y)\right\} \neq \lim_{y\to 0}\left\{\lim_{x\to 0} f(x,y)\right\}$$

Solution: L.H.S. of the inequality:

$$\lim_{x\to 0}\left\{\lim_{y\to 0} f(x,y)\right\} = \lim_{x\to 0}\left\{\lim_{y\to 0}\frac{y^2-x^2}{x^2+y^2}\right\}$$

$$= \lim_{x\to 0}\frac{-x^2}{x^2} = \lim_{x\to 0} -1 = -1$$

R.H.S. of the inequality:

$$\lim_{y\to 0}\left\{\lim_{x\to 0} f(x,y)\right\} = \lim_{y\to 0}\left\{\lim_{x\to 0}\frac{y^2-x^2}{x^2+y^2}\right\}$$

$$= \lim_{y\to 0}\left\{\frac{y^2}{y^2}\right\} = \lim_{y\to 0} 1 = 1$$

Thus L.H.S. $= -1 \neq 1 =$ R.H.S.

[***Note:*** Since the limits along two different paths are not same, the limit does not exist.]

Example 3: Evaluate $\lim\limits_{(x,y)\to(0,0)}\frac{x^2y}{x^4+y^2}$

Solution:

I. $$\lim_{\substack{x\to 0\\ y\to 0}}\frac{x^2y}{x^4+y^2} = \lim_{y\to 0} 0 = 0$$

II. $$\lim_{\substack{y\to 0\\ x\to 0}}\frac{x^2y}{x^4+y^2} = \lim_{x\to 0} 0 = 0$$

III. $$\lim_{\substack{y=mx\\ x\to 0}}\frac{x^2y}{x^4+y^2} = \lim_{x\to 0}\frac{mx^3}{x^4+m^2x^2}$$

$$= \lim_{x\to 0}\frac{mx}{x^2+m^2} = 0$$

IV. $$\lim_{\substack{y=mx^2\\ x\to 0}}\frac{x^2y}{x^4+y^2} = \lim_{x\to 0}\frac{mx^4}{x^4+m^2x^4}$$

$$= \lim_{x\to 0}\frac{m}{1+m^2} = \frac{m}{1+m^2}$$

Since the limit along the last path $y = mx^2$ depends on m, ∴ limit does not exist.

Continuity

Example 4: If $f(x, y) = \frac{x^3-y^3}{x^2+y^2}$ when $x \neq 0, y \neq 0$ and $f(x, y) = 0$ when $x = 0, y = 0$, find out whether the function $f(x, y)$ is continuous at origin.

Solution: First calculate the limit of the function:

I. $$\lim_{\substack{x\to 0\\ y\to 0}}\frac{x^3-y^3}{x^2+y^2} = \lim_{y\to 0}\left(\frac{-y^3}{y^2}\right) = \lim_{y\to 0}(-y) = 0$$

II. $$\lim_{\substack{y\to 0\\ x\to 0}}\frac{x^3-y^3}{x^2+y^2} = \lim_{x\to 0}\left(\frac{x^3}{x^2}\right) = \lim_{x\to 0}(x) = 0$$

III. $$\lim_{\substack{y=mx\\ x\to 0}}\frac{x^3-y^3}{x^2+y^2} = \lim_{x\to 0}\frac{x^3-m^3x^3}{x^2+m^2x^2}$$

$$= \lim_{x\to 0}\frac{(1-m^3)}{(1+m^2)}\cdot x = 0$$

IV. $$\lim_{\substack{y=mx^2\\ x\to 0}}\frac{x^3-y^3}{x^2+y^2} = \lim_{x\to 0}\frac{x^3(1-m^3x^3)}{x^2(1+m^2x^2)}$$

$$= \lim_{x\to 0}\frac{(1-m^3x^3)}{(1+m^2x^2)}\cdot x = 0.$$

Since the limit along any path is same, the limit exists and equal to zero which is the value of the function $f(x, y)$ at the origin. Hence the function f is continuous at the origin.

Example 5: Discuss the continuity of the function $f(x, y) = \frac{x}{\sqrt{x^2+y^2}}$ when $(x, y) \neq (0, 0)$ and $f(x, y) = 2$ when $(x, y) = (0, 0)$.

Solution: At first, evaluate the limit

I. $$\lim_{\substack{x\to 0\\ y\to 0}}\frac{x}{\sqrt{x^2+y^2}} = 0$$

II. $\lim_{\substack{y\to 0\\x\to 0}} \frac{x}{\sqrt{x^2+y^2}} = \lim_{x\to 0} \frac{x}{\sqrt{x^2}} = \lim_{x\to 0} 1 = 1$

Since the limits along paths I and II are different, the limit itself does not exist. Therefore the function is discontinuous at the origin.

Example 6: Examine for continuity at origin of the function defined by

$$f(x, y) = \frac{x^2}{\sqrt{x^2+y^2}}, \quad \text{for } (x \neq 0, y \neq 0)$$
$$= 3, \quad \text{for } (x = 0, y = 0)$$

Redefine the function to make it continuous.

Solution: Initially, find the limit

I. $\lim_{\substack{x\to 0\\y\to 0}} \frac{x^2}{\sqrt{x^2+y^2}} = \lim_{y\to 0} 0 = 0$

II. $\lim_{\substack{y\to 0\\x\to 0}} \frac{x^2}{\sqrt{x^2+y^2}} = \lim_{x\to 0} \frac{x^2}{\sqrt{x^2}} = \lim_{x\to 0} x = 0$

III. $\lim_{\substack{y=mx\\x\to 0}} \frac{x^2}{\sqrt{x^2+y^2}} = \lim_{x\to 0} \frac{x^2}{x\sqrt{(1+m^2)}}$

$$= \frac{1}{\sqrt{1+m^2}} \lim_{x\to 0} x = 0$$

IV. $\lim_{\substack{y=mx^n\\x\to 0}} \frac{x^2}{\sqrt{x^2+y^2}} = \lim_{x\to 0} \frac{x^2}{x\sqrt{(1+m^2x^{2n-2})}}$

$$= \frac{0}{\sqrt{1+0}} = 0$$

Thus the limit along any path exists and is the same and the common value equals to zero i.e.,

$$\lim_{(x,y)\to(0,0)} \frac{x^2}{\sqrt{x^2+y^2}} = 0$$

However the value of the functions at origin is 3 i.e.,

$$f(0, 0) = 3$$

Therefore f is discontinuous at origin because

$$\lim_{(x,y)\to(0,0)} \frac{x^2}{\sqrt{x^2+y^2}} = 0 \neq 3 = f(0, 0)$$

The function can be 'made' continuous at origin by redefining the function as $f(0, 0) = 0$, since in this case

$$\lim_{(x,y)\to(0,0)} \frac{x^2}{\sqrt{x^2+y^2}} = 0 = f(0, 0).$$

EXERCISE

Limits

Evaluate the following limits:

1. $\lim_{\substack{x\to 2\\y\to 3}} \frac{x^2+y^3}{2x^2y}$ *Ans.* $\frac{31}{24}$

2. $\lim_{\substack{x\to 1\\y\to 2}} \frac{x^2+y}{3x+y^2}$ *Ans.* $\frac{3}{7}$

3. $\lim_{(x,y)\to(0,0)} \frac{xy}{y^2-x^2}$

   Hint: Along path $y = mx$, limit $= \frac{m}{m^2-1}$ which is different for different values of m.

   Ans. does not exist

4. $\lim_{\substack{x\to 1\\y\to 2}} \frac{2x^2y}{x^2+y^2+1}$ *Ans.* $\frac{2}{3}$

5. $\lim_{\substack{x\to 1\\y\to 1}} \frac{x(y-1)}{y(x-1)}$ *Ans.* does not exist

6. $\lim_{\substack{x\to 0\\y\to 0}} \frac{xy}{x^2+y^2}$ *Ans.* does not exist

7. $\lim_{\substack{x\to 0\\y\to 0}} \frac{x-y}{x^2+y^2}$ *Ans.* does not exist

8. $\lim_{\substack{x\to\infty\\y\to 3}} \frac{2xy-3}{x^3+4y^3}$ *Ans.* 0

Show that $\lim_{x\to 0}\left[\lim_{y\to 0} f(x, y)\right] \neq \lim_{y\to 0}\left[\lim_{x\to 0} f(x, y)\right]$ if

9. $f(x, y) = \frac{x-y}{2x+y}$ *Ans.* $\frac{1}{2} \neq -1$

10. $f(x, y) = \frac{x-y}{x+y}$ *Ans.* $1 \neq -1$.

Continuity

11. If $f(x, y) = \frac{(x^2-y^2)}{(x^2+y^2)}$ when $x \neq 0, y \neq 0$

    $= 0$ when $(x = 0, y = 0)$

    show that f is discontinuous at origin.

12. (a) Is the function $f(x, y) = \frac{xy(x^2-y^2)}{x^2+y^2}$ when $(x, y) \neq (0, 0)$ and $f(0, 0) = 4$ continuous at origin. (b) Redefine if necessary to make it continuous at $(0, 0)$.

Ans. **a.** not continuous: $\lim = 0 \neq 4 = f(0, 0)$

b. continuous if $f(0, 0) = 0$

13. If $f(x, y) = x^3 + y^2$ determine where the function is continuous.

Ans. continuous for every x and y i.e., everywhere

14. If $f(x, y) = \frac{x^3-y^3}{x^3+y^3}$, for $(x, y) \neq (0, 0)$

$= 15$, at $(0, 0)$

show that f is discontinuous at origin.

15. Find whether $f(x, y) = \frac{x^3y^3}{x^3+y^3}$ is continuous at $(0, 0)$ when (a) $f(0, 0)=-15$, (b) $f(0, 0)=0$.

Ans. **a.** not continuous: $\lim = 0 \neq -15 = f(0, 0)$

b. continuous since $\lim = 0 = 0 = f(0, 0)$

16. Given $f(x, y) = x^3 + 3y^2 + 2x + y$ for every (x, y) except at $(2, 3)$ where $f(2, 3) = 10$. Examine whether f is continuous at (a) point $(2, 3)$ (b) at any other points (c) can the function be made continuous at $(2, 3)$ by redefining f at $(2, 3)$.

Ans. **a.** discontinuous at $(2, 3)$

b. continuous for every x and y i.e., everywhere except at $(2, 3)$

c. f becomes continuous by redefining f at $(2, 3)$ as $f(2, 3) = 42$

17. **a.** Show that $f(x, y) = \frac{xy}{x^2+y^2}$, $x = y \neq 0$ is discontinuous at origin when $f(0, 0) = 0$.

b. Can it be made continuous by defining f?

Ans. **a.** since limit along $y = mx$ is $\frac{m}{1+m^2}$, not unique, the limit does not exist, so discontinuous

b. can not make f continuous by redefining f at $(0, 0)$ i.e., for any choice of $f(0, 0)$, since the limit at $(0, 0)$ does not exist.

18. Prove that $f(x, y)=\frac{x^2-y^3}{x^2-y^2}$ when $(x, y) \neq (0, 0)$ and $f(0, 0) = 0$ is discontinuous at origin.

Hint: $\lim f$ along $y = mx$ is not unique, depends on m. So limit does not exist.

Find out (and give reason) whether $f(x, y)$ is continuous at $(0, 0)$ if $f(0, 0) = 0$ and for $(x, y) \neq (0, 0)$ the function f is equal to:

19. $y/\sqrt{x^2 + y^2}$

Ans. discontinuous

20. $x/\left(1 + \sqrt{x^2 + y^2}\right)$

Ans. continuous

21. $xy/(x^2 + y^2)^{\frac{1}{2}}$

Ans. continuous

22. $(x^2 - y^2)/\sqrt{x^2 + y^2}$

Hint: continuous.

3.2 PARTIAL DIFFERENTIATION

A partial derivative of a function of several variables is the ordinary derivative with respect to one of the variables when all the remaining variables are held constant. Partial differentiation is the process of finding partial derivatives. All the rules of differentiation applicable to function of a single independent variable are also applicable in partial differentiation with the only difference that while differentiating (partially) with respect to one variable, all the *other* variables are treated (temporarily) as constants.

Consider a function u of three independent variables x, y, z, (refer Fig. 3.1)

$$u = f(x, y, z) \tag{1}$$

Keeping y, z constant and varying only x, the partial derivative of u with respect to x is denoted by $\frac{\partial u}{\partial x}$ and is defined as the limit

$$\frac{\partial u(x, y, z)}{\partial x} = \lim_{\Delta x \to 0} \frac{f(x + \Delta x, y, z) - f(x, y, z)}{\Delta x}.$$

Partial derivatives of u w.r.t. y and z can be defined similarly and are denoted by $\frac{\partial u}{\partial y}$ and $\frac{\partial u}{\partial z}$.

Notation: The partial derivative $\frac{\partial u}{\partial x}$ is also denoted by $\frac{\partial f}{\partial x}$ or $f_x(x, y, z)$ or f_x or $D_x f$ or $f_1(x, y, z)$ where the subscripts x and 1 denote the variable w.r.t. which the partial differentiation is carried out. Thus

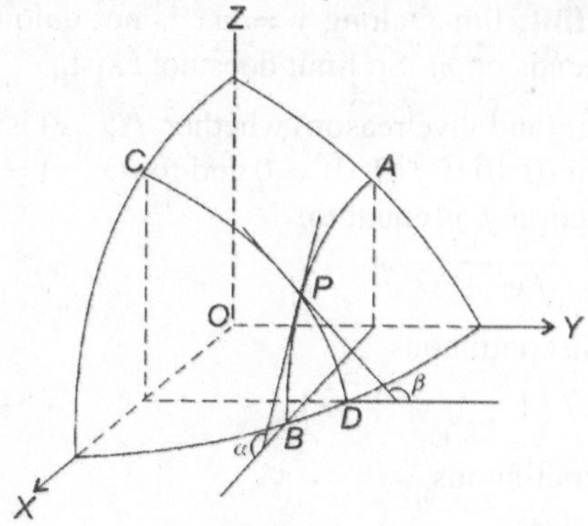

Fig. 3.1

we can have $\frac{\partial u}{\partial y} = \frac{\partial f}{\partial y} = f_y(x, y, z) = f_y = D_y f = f_2(x, y, z)$ etc. The value of a partial derivative at a point (a, b, c) is denoted by

$$\left.\frac{\partial u}{\partial x}\right|_{x=a, y=b, z=c} = \left.\frac{\partial u}{\partial x}\right|_{(a,b,c)} = f_x(a, b, c)$$

Geometrical interpretation of a partial derivative of a function of two variables: $z = f(x, y)$ represents the equation of a surface in xyz- coordinate system. Let APB the curve, which a plane through any point P on the surface parallel to the xz-plane, cuts. As point P moves along this curve APB, its coordinates z and x vary while y remains constant. The slope of the tangent line at P to APB represents the rate at which z changes w.r.t. x.

Thus

$$\frac{\partial z}{\partial x} = \tan\alpha = \text{slope of the curve } APB \text{ at the point } P$$

Similarly,

$$\frac{\partial z}{\partial y} = \tan\beta = \text{slope of the curve } CPD \text{ at } P.$$

Higher Order Partial Derivatives

Partial derivatives of higher order, of a function $f(x, y, z)$ are calculated by successive differentiation. Thus if $u = f(x, y, z)$ then

$$\frac{\partial^2 u}{\partial x^2} = \frac{\partial^2 f}{\partial x^2} = \frac{\partial}{\partial x}\left(\frac{\partial f}{\partial x}\right) = f_{xx} = f_{11}$$

$$\frac{\partial^2 u}{\partial x \partial y} = \frac{\partial^2 f}{\partial x \partial y} = \frac{\partial}{\partial x}\left(\frac{\partial f}{\partial y}\right) = f_{yx} = f_{21}$$

$$\frac{\partial^2 u}{\partial y \partial x} = \frac{\partial^2 f}{\partial y \partial x} = \frac{\partial}{\partial y}\left(\frac{\partial f}{\partial x}\right) = f_{xy} = f_{12}$$

$$\frac{\partial^2 u}{\partial y^2} = \frac{\partial^2 f}{\partial y^2} = \frac{\partial}{\partial y}\left(\frac{\partial f}{\partial y}\right) = f_{yy} = f_{22}$$

$$\frac{\partial^3 u}{\partial z^2 \partial y} = \frac{\partial}{\partial z}\left(\frac{\partial^2 f}{\partial z \partial y}\right) = \frac{\partial}{\partial z}\left(\frac{\partial}{\partial z}\left(\frac{\partial f}{\partial y}\right)\right)$$

$$= f_{yzz} = f_{233}$$

$$\frac{\partial^4 u}{\partial x \partial y \partial z^2} = \frac{\partial}{\partial x}\left(\frac{\partial^3 f}{\partial y \partial z^2}\right) = \frac{\partial}{\partial x}\left(\frac{\partial}{\partial y}\left(\frac{\partial^2 f}{\partial z^2}\right)\right)$$

$$= f_{zzyx} = f_{3321}.$$

The partial derivative $\frac{\partial f}{\partial x}$ obtained by differentiating once is known as first order partial derivative, while $\frac{\partial^2 f}{\partial x^2}, \frac{\partial^2 f}{\partial y^2}, \frac{\partial^2 f}{\partial x \partial y}, \frac{\partial^2 f}{\partial y \partial x}$ which are obtained by differentiating twice are known as second-order derivatives. 3rd order, 4th order derivatives involve 3, 4 times differentiation respectively.

Note 1: The crossed or mixed partial derivatives $\frac{\partial^2 f}{\partial y \partial x}$ and $\frac{\partial^2 f}{\partial x \partial y}$ are in general, equal

$$\frac{\partial^2 f}{\partial y \partial x} = \frac{\partial^2 f}{\partial x \partial y}$$

i.e., the order of differentiation is immaterial if the derivatives involved are continuous.

Note 2: In the subscript notation, the subscripts are written in the *same* order in which differentiation is carried out, while in the '∂' notation the order is opposite, for example,

$$\frac{\partial^2 u}{\partial y \partial x} = \frac{\partial}{\partial y}\left(\frac{\partial u}{\partial x}\right) = f_{xy}$$

Note 3: A function of 2 variables has two first order derivatives, four second order derivatives and 2^n of nth order derivatives. A function of m independent variables will have m^n derivatives of order n.

WORKED OUT EXAMPLES

Example 1: Find the first order partial derivatives $\frac{\partial w}{\partial x}$ and $\frac{\partial w}{\partial y}$ when:

(a) $w = e^x \cos y$ (b) $w = \tan^{-1}\frac{y}{x}$ (c) $w = \ln\sqrt{x^2 + y^2}$

Solution:

a. $$\frac{\partial w}{\partial x} = \cos y \frac{\partial}{\partial x}(e^x) = e^x \cos y$$

$$\frac{\partial w}{\partial y} = e^x \frac{\partial}{\partial y}(\cos y) = -e^x \sin y$$

b. $$\frac{\partial w}{\partial x} = \frac{1}{1+\left(\frac{y}{x}\right)^2} \cdot \frac{\partial}{\partial x}\left(\frac{y}{x}\right) = \frac{x^2}{x^2+y^2} \cdot \left(\frac{-y}{x^2}\right)$$

$$= \frac{-y}{x^2+y^2}$$

$$\frac{\partial w}{\partial y} = \frac{1}{1+\left(\frac{y}{x}\right)^2} \cdot \frac{\partial}{\partial y}\left(\frac{y}{x}\right) = \frac{x^2}{x^2+y^2} \cdot \frac{1}{x}$$

$$= \frac{x}{x^2+y^2}$$

c. $$\frac{\partial w}{\partial x} = \frac{1}{\sqrt{x^2+y^2}} \cdot \frac{1}{2}(x^2+y^2)^{-\frac{1}{2}} \cdot \frac{\partial}{\partial x}(x^2) = \frac{x}{x^2+y^2}$$

$$\frac{\partial w}{\partial y} = \frac{1}{\sqrt{x^2+y^2}} \cdot \frac{1}{2}(x^2+y^2)^{-\frac{1}{2}} \cdot \frac{\partial}{\partial y}(y^2) = \frac{y}{x^2+y^2}$$

Example 2: Find partial derivative of f with respect to each of the independent variable:

a. $f(x, y, z, w) = x^2 e^{2y+3z} \cos(4w)$

b. $f(r, \theta, z) = r(2 - \cos 2\theta)/(r^2 + z^2)$.

Solution:

a. $$f_x = \frac{\partial f}{\partial x} = e^{2y+3z} \cdot \cos(4w) \cdot \frac{\partial}{\partial x}(x^2)$$

$$f_x = 2x\, e^{2y+3z} \cos(4w) = 2f/x$$

$$f_y = x^2 e^{3z} \cos(4w) \frac{\partial}{\partial y} e^{2y} = x^2 e^{3z} \cos(4w) \cdot 2e^{2y}$$

$$f_y = 2f$$

$$f_z = x^2 e^{2y} \cos(4w) \frac{\partial}{\partial z} e^{3z} = x^2 e^{2y} \cos(4w) 3e^{3z}$$

$$f_z = = 3f$$

$$f_w = x^2 e^{2y+3z} \frac{\partial}{\partial w}(\cos(4w))$$

$$= x^2 e^{2y+3z} \cdot (-\sin 4w) \cdot 4$$

$$f_w = -4x^2 e^{2y+3z} \sin 4w$$

b. $$f = \frac{r(2 - \cos 2\theta)}{r^2 + z^2}$$

$$\frac{\partial f}{\partial r} = f_r = \frac{(r^2+z^2)(2 - \cos 2\theta) - r(2 - \cos 2\theta) \cdot 2r}{(r^2+z^2)^2}$$

$$f_r = \frac{(z^2 - r^2)(2 - \cos 2\theta)}{(r^2+z^2)^2}$$

$$\frac{\partial f}{\partial \theta} = f_\theta = \frac{r}{r^2+z^2} \cdot 2 \sin 2\theta$$

$$\frac{\partial f}{\partial z} = f_z = \frac{r(2 - \cos 2\theta)}{(r^2+z^2)^2}(-2z).$$

Example 3: Find $\frac{\partial^3 u}{\partial x \partial y \partial z}$ if $u = e^{x^2+y^2+z^2}$

Solution: $\frac{\partial u}{\partial z} = 2z\, e^{x^2+y^2+z^2}$

$$\frac{\partial^2 u}{\partial y \partial z} = \frac{\partial}{\partial y}\left(\frac{\partial u}{\partial z}\right) = \frac{\partial}{\partial y}(2z\, e^{x^2+y^2+z^2})$$

$$\frac{\partial^2 u}{\partial y \partial z} = 2z \cdot e^{x^2+y^2+z^2} \cdot 2y = 4yz\, e^{x^2+y^2+z^2}$$

$$\frac{\partial^3 u}{\partial x \partial y \partial z} = \frac{\partial}{\partial x}\left(\frac{\partial^2 u}{\partial y \partial z}\right) = \frac{\partial}{\partial x}(4yz\, e^{x^2+y^2+z^2})$$

$$= 4yz\, e^{x^2+y^2+z^2} \cdot 2x = 8xyz\, e^{x^2+y^2+z^2}$$

Thus $$\frac{\partial^3 u}{\partial x \partial y \partial z} = 8xyzu.$$

Example 4: Show that

$$V(x, y, z) = \cos 3x \cos 4y \sinh 5z$$

satisfies Laplace's equation

$$\frac{\partial^2 V}{\partial x^2} + \frac{\partial^2 V}{\partial y^2} + \frac{\partial^2 V}{\partial z^2} = 0.$$

Solution:

$$V_x = -3 \sin 3x \cdot \cos 4y \sinh 5z$$

$$V_{xx} = -9 \cos 3x \cos 4y \sinh 5z = -9V \quad (1)$$

$$V_y = -4 \cos 3x \sin 4y \cdot \sinh 5z$$

$$V_{yy} = -16 \cos 3x \cdot \cos 4y \sinh 5z = -16V \quad (2)$$

$$V_z = 5 \cos 3x \cos 4y \cosh 5z$$

$$V_{zz} = 25 \cos 3x \cos 4y \sinh 5z = 25V \quad (3)$$

Adding (1), (2) and (3), we get

$$V_{xx} + V_{yy} + V_{zz} = -9V - 16V + 25V = 0$$

Example 5: If $u = e^{a\theta} \cos(a \ln r)$ show that

$$\frac{\partial^2 u}{\partial r^2} + \frac{1}{r}\frac{\partial u}{\partial r} + \frac{1}{r^2}\frac{\partial^2 u}{\partial \theta^2} = 0$$

Solution:

$$u_r = e^{a\theta}\cdot(-\sin(a\ln r)\frac{a}{r} \qquad (1)$$

$$u_{rr} = -ae^{a\theta}\left[-\frac{1}{r^2}\sin(a\ln r) + \frac{1}{r}\cos(a\ln r)\cdot\frac{a}{r}\right]$$

$$u_{rr} = \frac{-ae^{a\theta}}{r^2}[-\sin(a\ln r) + a\cos(a\ln r)] \qquad (2)$$

$$u_\theta = a\,e^{a\theta}\cos(a\ln r).$$

$$u_{\theta\theta} = a^2 e^{a\theta}\cos(a\ln r) = a^2 u \qquad (3)$$

using (1), (2) and (3)

$$u_{rr} + \frac{1}{r}u_r + \frac{1}{r^2}u_{\theta\theta} = \left[\frac{ae^{a\theta}}{r^2}\sin(a\ln r) - \frac{a^2}{r^2}u\right]$$

$$+\left(\frac{-a}{r^2}e^{a\theta}\sin(a\ln r)\right)$$

$$+\left(\frac{a^2}{r^2}u\right)$$

$$= 0$$

Example 6: If $u = \ln(x^3 + y^3 - x^2y - xy^2)$ then show that $u_{xx} + 2u_{xy} + u_{yy} = -\frac{4}{(x+y)^2}$

Solution:

$$u_x = \frac{3x^2 - 2xy - y^2}{(x^3 + y^3 - x^2y - xy^2)}, \qquad (1)$$

$$u_y = \frac{3y^2 - x^2 - 2xy}{(x^3 + y^3 - x^2y - xy^2)} \qquad (2)$$

Adding (1) and (2), we get

$$u_x + u_y = \frac{(3x^2 - 2xy - y^2) + (3y^2 - x^2 - 2xy)}{(x^3 + y^3 - x^2y - xy^2)}$$

$$= \frac{2(x-y)^2}{(x^3 + y^3 - x^2y - xy^2)}$$

$$u_x + u_y = \frac{2(x-y)^2}{(x+y)(x^2+y^2-2xy)}$$

$$= \frac{2(x-y)^2}{(x+y)(x-y)^2} = \frac{2}{x+y} \qquad (3)$$

Now

$$u_{xx} + 2u_{xy} + u_{yy} = \frac{\partial^2 u}{\partial x^2} + \frac{2\partial^2 u}{\partial x\partial y} + \frac{\partial^2 u}{\partial y^2}$$

$$= \left(\frac{\partial}{\partial x} + \frac{\partial}{\partial y}\right)\left(\frac{\partial}{\partial x} + \frac{\partial}{\partial y}\right)u$$

$$= \left(\frac{\partial}{\partial x} + \frac{\partial}{\partial y}\right)\left(\frac{2}{x+y}\right) \quad \text{using (3)}$$

$$= \frac{\partial}{\partial x}\left(\frac{2}{x+y}\right) + \frac{\partial}{\partial y}\left(\frac{2}{x+y}\right)$$

$$= \frac{-2}{(x+y)^2} - \frac{2}{(x+y)^2} = \frac{-4}{(x+y)^2}.$$

EXERCISE

1. Find the first order partial derivatives of:

   a. $u = \frac{x-y}{x+y}$ c. $e^x \sin y$

   b. $u = \ln(x + \sqrt{x^2 - y^2})$ d. $u = x^{xy}$

   Ans. a. $u_x = \frac{2y}{(x+y)^2}$, $u_y = \frac{-2x}{(x+y)^2}$

   b. $u_x = (1 + x(x^2 - y^2)^{-\frac{1}{2}})/(x + \sqrt{x^2 - y^2})$

   $u_y = y(x^2 - y^2)^{-\frac{1}{2}}/(x + \sqrt{x^2 - y^2})$

   c. $u_x = e^x \sin y$, $u_y = e^x \cos y$

   d. $u_x = x^{xy}(y \log x + y)$; $u_y = x^{xy+1}\log x$

2. Find the partial derivative of the given function w.r.t. each variable:

   a. $f(x, y, z) = z\sin^{-1}(y/x)$

   b. $f(u, v, w) = (u^2 - v^2)/(v^2 + w^2)$

   c. $f(x, y, r, s) = \sin 2x \cosh 3r + \sinh 3y \cos 4s$

   Ans. a. $f_x = \frac{-yz}{\sqrt{x^4 - x^2y^2}}$, $f_y = \frac{|x|z}{x\sqrt{x^2-y^2}}$,

   $f_z = \sin^{-1}(y/x)$

   b. $f_u = \frac{2u}{v^2+w^2}$, $f_v = \frac{-2v(u^2+w^2)}{(v^2+w^2)^2}$,

   $f_w = \frac{-2w(u^2-v^2)}{(v^2+w^2)^2}$

   c. $f_x = 2\cos 2x \cosh 3r$,

   $f_y = 3\cosh 3y \cos 4s$

   $f_r = 3\sin 2x \sinh 3r$,

   $f_s = -4\sinh 3y \sin 4s$

3. If $w = \ln(2x + 2y) + \tan(2x - 2y)$ prove that

$$\frac{\partial^2 w}{\partial x^2} = \frac{\partial^2 w}{\partial y^2}$$

4. Verify that $w_{xy} = w_{yx}$ when
a. $w = \ln(2x + 3y)$
b. $w = xy^2 + x^2y^3 + x^3y^4$
c. $w = \tan^{-1} y/x$
d. $w = \ln(y \sin x + x \sin y)$.

5. Show that $u_x + u_y = u$ if $u = e^{x+y}/(e^x + e^y)$.

6. Prove that $w = f(x + ct) + g(x - ct)$ satisfies the wave equation $\frac{\partial^2 w}{\partial t^2} = c^2\frac{\partial^2 w}{\partial x^2}$ where c is a constant. Verify this when

$$w = 7\sin(3x + 3ct) + 9\cosh(5x - 5ct).$$

7. Show that $w_x + w_y + w_z = 0$ if $w = (y - z)(z - x)(x - y)$.

8. If $w = x^2y + y^2z + z^2x$, prove that

$$w_x + w_y + w_z = (x + y + z)^2.$$

9. Verify that V satisfies Laplace's equation

$$\frac{\partial^2 V}{\partial x^2} + \frac{\partial^2 V}{\partial y^2} + \frac{\partial^2 V}{\partial z^2} = 0$$

if (a) $V = x^2 + y^2 - 2z^2$, (b) $V = e^{3x+4y}\cos 5z$, (c) $V = (x^2 + y^2 + z^2)^{-\frac{1}{2}}$.

10. Find $\frac{\partial^3 w}{\partial x \partial y \partial z}$ if $w = e^{xyz}$.

Ans. $e^{xyz}(x^2y^2z^2 + 3xyz + 1)$

11. Show that $xw_x + yw_y + zw_z = 0$ when

$$w = \frac{y}{z} + \frac{z}{x}$$

12. If $w = r^m$ prove that

$$w_{xx} + w_{yy} + w_{zz} = m(m + 1)r^{m-2}$$

where $r^2 = x^2 + y^2 + z^2$.

13. For $n = 2$ or -3 show that $u = r^n(3\cos^2\theta - 1)$ satisfies the differential equation

$$\frac{\partial}{\partial r}\left(r^2\frac{\partial u}{\partial r}\right) + \frac{1}{\sin\theta}\frac{\partial}{\partial\theta}\left(\sin\theta\frac{\partial u}{\partial\theta}\right) = 0$$

14. Prove that $\left(\frac{\partial}{\partial x} + \frac{\partial}{\partial y} + \frac{\partial}{\partial z}\right)^2 u = \frac{-9}{(x+y+z)^2}$ if

$$u = \ln(x^3 + y^3 + z^3 - 3xyz)$$

Hint: Prove and use the result $u_x + u_y + u_z = \frac{3}{(x+y+z)}$

15. Prove (the Cauchy-Riemann equations in polar coordinates) $ru_r = v_\theta$, $rv_r + u_\theta = 0$ when

$$u = e^{r\cos\theta}\cos(r\sin\theta),\ v = e^{r\cos\theta}\sin(r\sin\theta).$$

16. Show that $yz_x + xz_y = x^2 - y^2$ if

$$e^{-z/(x^2-y^2)} = (x - y)$$

Hint: Solve for $z = (y^2 - x^2)\ln(x - y)$.

17. If $u = (x^2 + y^2 + z^2)^{-\frac{1}{2}}$ show that

$$(u_x)^2 + (u_y)^2 + (u_z)^2 = u^4.$$

18. Prove that $f(x, t) = a\sin bx \cdot \cos bt$ satisfies

$$\frac{\partial^2 f}{\partial x^2} = b^2\frac{\partial^2 f}{\partial t^2}.$$

19. Show that $\frac{\partial^3 u}{\partial x^2 \partial y} = \frac{\partial^3 u}{\partial x \partial y \partial x}$ if $u = x^y$.

3.3 VARIABLE TREATED AS CONSTANT

Consider $z = x^2 - y^2$. Introducing polar coordinates $x = r\cos\theta$, $y = r\sin\theta$, we have $r^2 = x^2 + y^2$ and $\theta = \tan^{-1}\frac{y}{x}$. To find $\frac{\partial z}{\partial\theta}$ with different variable treated as constant i.e., to find $\left(\frac{\partial z}{\partial\theta}\right)_r$, $\left(\frac{\partial z}{\partial\theta}\right)_x$, $\left(\frac{\partial z}{\partial\theta}\right)_y$.
Variable treated as constant: $\left(\frac{\partial z}{\partial\theta}\right)_r$ usually read as "partial derivative of z w.r.t. θ, with r held constant". However the important point is that z has been written as a function of the variables θ and r only and then differentiated w.r.t. θ, keeping r constant. Thus

$$z = x^2 - y^2 = r^2\cos^2\theta - r^2\sin^2\theta$$

$$\left(\frac{\partial z}{\partial\theta}\right)_r = 2r^2(\cos\theta(-\sin\theta) - \sin\theta\cos\theta)$$

$$= -4r^2\sin\theta\cos\theta \quad (1)$$

To find $\left(\frac{\partial z}{\partial\theta}\right)_x$ express z in terms of θ and x as $z^2 = x^2 - y^2 = x^2 - x^2\tan^2\theta$.

Now $$\left(\frac{\partial z}{\partial\theta}\right)_x = x^2(0 - 2\tan\theta\cdot\sec^2\theta) \quad (2)$$

To find $\left(\frac{\partial z}{\partial\theta}\right)_y$ express z in terms of θ and y as $z = x^2 - y^2 = y^2\cot^2\theta - y^2$

Now $$\left(\frac{\partial t}{\partial\theta}\right)_y = y^2(2\cdot\cot\theta\cdot(-\text{cosec}^2\theta)) \quad (3)$$

These three expressions (1), (2), (3) for $\frac{\partial z}{\partial \theta}$ have different values and are derivatives of three different functions.

Worked Out Examples

Example 1: If $z = x^2 + 2y^2$, $x = r\cos\theta$, $y = r\sin\theta$, find the partial derivatives:

a. $\left(\frac{\partial z}{\partial x}\right)_y$ **b.** $\left(\frac{\partial z}{\partial x}\right)_\theta$ **c.** $\left(\frac{\partial z}{\partial \theta}\right)_r$ **d.** $\frac{\partial^2 z}{\partial r \partial y}$

Solution: Here $z = x^2 + 2y^2$, $r^2 = x^2 + y^2$, $\tan\theta = \frac{y}{x}$

a. To get $\left(\frac{\partial z}{\partial x}\right)_y$, z should be expressed as function of x and y

i.e., $$z = x^2 + 2y^2$$

Differentiating partially w.r.t. x, with y held constant

$$\left(\frac{\partial z}{\partial x}\right)_y = 2x$$

b. Express z as function of x and θ

$$z = x^2 + 2y^2 = x^2 + 2x^2 + \tan^2\theta$$

Differentiating partially w.r.t. x, keeping θ constant

$$\left(\frac{\partial z}{\partial x}\right)_\theta = 2x + 4x\tan^2\theta = 2x(1 + 2\tan^2\theta)$$

c. Express z as function of r and θ

$$z = r^2\cos^2\theta + 2r^2\sin^2\theta$$

$$\left(\frac{\partial z}{\partial \theta}\right)_r = 2r^2\cos\theta(-\sin\theta) + 2r^2 2\cdot\sin\theta\cdot\cos\theta$$

$$= r^2\cdot\sin 2\theta$$

d. $z = (r^2 - y^2) + 2y^2 = r^2 + y^2$, $\frac{\partial z}{\partial y} = 2y$, $\frac{\partial^2 z}{\partial r \partial y} = 0$.

Example 2: If $x^2 = au + bv$; $y^2 = au - bv$ prove that

$$\left(\frac{\partial u}{\partial x}\right)_y\left(\frac{\partial x}{\partial u}\right)_v = \frac{1}{2} = \left(\frac{\partial v}{\partial y}\right)_x\left(\frac{\partial y}{\partial v}\right)_u$$

Solution: Solving

$$x^2 = au + bv \quad (1)$$

$$y^2 = au - bv \quad (2)$$

we get $$u = \frac{x^2 + y^2}{2a} \quad (3)$$

$$v = \frac{x^2 - y^2}{2b} \quad (4)$$

Differentiating (3) partially w.r.t. x, keeping y constant, we get

$$\left(\frac{\partial u}{\partial x}\right)_y = \frac{2x}{2a} = \frac{x}{a} \quad (5)$$

Differentiating (1) partially w.r.t. u, keeping v constant, we get

$$2x\left(\frac{\partial x}{\partial u}\right)_v = a \quad \therefore \left(\frac{\partial x}{\partial u}\right)_v = \frac{a}{2x} \quad (6)$$

From (5) and (6)

$$\left(\frac{\partial u}{\partial x}\right)_y\left(\frac{\partial x}{\partial u}\right)_v = \frac{x}{a}\cdot\frac{a}{2x} = \frac{1}{2}$$

Similarly differentiating (4) and (2) partially w.r.t. y and v respectively, we get

$$\left(\frac{\partial v}{\partial y}\right)_x = \frac{-2y}{2b} = \frac{-y}{b} \quad (7)$$

$$2y\left(\frac{\partial y}{\partial v}\right)_u = -b \quad \therefore \left(\frac{\partial y}{\partial v}\right)_u = \frac{-b}{2y} \quad (8)$$

From (7) and (8)

$$\left(\frac{\partial v}{\partial y}\right)_x\cdot\left(\frac{\partial y}{\partial v}\right)_u = \left(\frac{-y}{b}\right)\left(\frac{-b}{2y}\right) = \frac{1}{2}$$

Exercise

1. Let $z = x^2 - y^2$ and $x = r\cos\theta$, $y = r\sin\theta$.

Find **a.** $\left(\frac{\partial z}{\partial r}\right)_\theta$ **b.** $\left(\frac{\partial z}{\partial r}\right)_x$ **c.** $\left(\frac{\partial z}{\partial r}\right)_y$

a. Hint: $z = r^2 \quad (\cos^2\theta - \sin^2\theta)$
b. Hint: $z = 2x^2 - r^2$
c. Hint: $z = r^2 - 2y^2$

Ans. **a.** $2r(\cos^2\theta - \sin^2\theta)$
b. $-2r$

c. $2r$

2. Let $z = x^2 + 2y^2$ and $x = r\cos\theta$, $y = r\sin\theta$.

Find **a.** $\left(\frac{\partial z}{\partial y}\right)_r$ **b.** $\left(\frac{\partial z}{\partial \theta}\right)_x$ **c.** $\left(\frac{\partial z}{\partial r}\right)_x$

d. $\frac{\partial^2 z}{\partial y \partial \theta}$ **e.** $\frac{\partial^2 z}{\partial r \partial \theta}$.

Ans. **a.** $2y$ **b.** $4r^2\tan\theta$ **c.** $2x$ **d.** $-4x\operatorname{cosec}^2\theta$ **e.** $2r\sin 2\theta$

3. If $u = lx + my$, $v = mx - ly$ show that

$$\left(\frac{\partial u}{\partial x}\right)_y \left(\frac{\partial x}{\partial u}\right)_v = \frac{l^2}{l^2+m^2},$$

$$\left(\frac{\partial y}{\partial v}\right)_x \left(\frac{\partial v}{\partial y}\right)_u = \frac{l^2+m^2}{l^2}$$

4. If $f(p, v, t) = 0$ show that $\left(\frac{\partial p}{\partial T}\right)_v = -\left(\frac{\partial v}{\partial T}\right)_p \left(\frac{\partial p}{\partial v}\right)_T$

Hint: $\left(\frac{\partial p}{\partial T}\right)_v = -f_T/f_p$, $\left(\frac{\partial v}{\partial T}\right)_p = -f_T/f_v$, $\left(\frac{\partial p}{\partial v}\right)_T = -f_v/f_p$.

5. If $E = f(p, T)$ and $T = g(p, v)$ show that

$$\left(\frac{\partial E}{\partial p}\right)_v = f_p + f_T g_p$$

$$= \left(\frac{\partial E}{\partial p}\right)_T + \left(\frac{\partial E}{\partial T}\right)_p \left(\frac{\partial T}{\partial p}\right)_v$$

6. If $x = r\cos\theta$, $y = r\sin\theta$ find

a. $\left(\frac{\partial x}{\partial r}\right)_\theta$ **b.** $\left(\frac{\partial x}{\partial \theta}\right)_r$ **c.** $\left(\frac{\partial \theta}{\partial x}\right)_y$ **d.** $\left(\frac{\partial \theta}{\partial y}\right)_x$

e. $\left(\frac{\partial y}{\partial x}\right)_r$ **f.** $\left(\frac{\partial x}{\partial y}\right)_\theta$ **g.** $\left(\frac{\partial r}{\partial \theta}\right)_x$ **h.** $\left(\frac{\partial \theta}{\partial r}\right)_y$

Ans. **a.** $\cos\theta$ **b.** $-r\sin\theta$ **c.** $-r^{-1}\sin\theta$ **d.** $r^{-1}\cos\theta$ **e.** $-\cot\theta$ **f.** $\cot\theta$ **g.** $r\tan\theta$ **h.** $-r^{-1}\tan\theta$.

3.4 TOTAL DERIVATIVE

Total differential of a function f of three variables x, y, z is denoted by df and is defined as

$$df = \frac{\partial f}{\partial x}dx + \frac{\partial f}{\partial y}dy + \frac{\partial f}{\partial z}dz \tag{1}$$

whether or not x, y and z are independent of each other. Several types of dependence among x, y and z are considered now.

Chain Rules for Partial Differentiation

Total derivative Let $u = f(x, y)$ and x and y are themselves functions of a single independent variable t. Then the dependent variable f may be considered as truly a function of the one independent variable t via the intermediate variables x, y. Now the derivative of f w.r.t. 't' is known as the total derivative of f and is given by

$$\frac{df}{dt} = \frac{\partial f}{\partial x}\frac{dx}{dt} + \frac{\partial f}{\partial y}\frac{dy}{dt} \tag{2}$$

Generalizing this, if $u = f(x, y, z)$ and $x = x(t)$, $y = y(t)$, $z = z(t)$ then the total derivative of f is

$$\frac{df}{dt} = \frac{\partial f}{\partial x}\frac{dx}{dt} + \frac{\partial f}{\partial y}\frac{dy}{dt} + \frac{\partial f}{\partial z}\frac{dz}{dt} \tag{3}$$

This can easily be extended to function of several variables.

If $u = f(x, y, z)$ and suppose y and z are function of x, then f is a function of the one independent variable x. Here y and z are intermediate variables. Identifying t with x in (3), we obtain

$$\frac{df}{dx} = \frac{\partial f}{\partial x}\frac{dx}{dx} + \frac{\partial f}{\partial y}\frac{dy}{dx} + \frac{\partial f}{\partial z}\frac{dz}{dx}$$

$$\frac{df}{dx} = \frac{\partial f}{\partial x} + \frac{\partial f}{\partial y}\frac{dy}{dx} + \frac{\partial f}{\partial z}\frac{dz}{dx} \tag{4}$$

WORKED OUT EXAMPLES

Example 1: Find $\frac{du}{dt}$ as a total derivative and verify the result by direct substitution if $u = x^2 + y^2 + z^2$ and $x = e^{2t}$, $y = e^{2t}\cos 3t$, $z = e^{2t}\sin 3t$.

Solution: Here u is a function of x, y, z and x, y, z are in turn functions of t. Thus u is a function 't' via the intermediate variables x, y, z. Then

$$\frac{du}{dt} = \frac{\partial u}{\partial x}\frac{dx}{dt} + \frac{\partial u}{\partial y}\frac{dy}{dt} + \frac{\partial u}{\partial z}\frac{dz}{dt}$$

$$= 2x \cdot 2e^{2t} + 2y \cdot (2e^{2t}\cos 3t - 3e^{2t}\sin 3t)$$

$$+ 2z(2e^{2t}\sin 3t + 3e^{2t}\cos 3t)$$

Rewriting in terms of x, y, z

$$= 2x \cdot 2 \cdot x + 2 \cdot y(2y - 3 \cdot z) + 2z(2z + 3y)$$

$$= 4(x^2 + y^2 + z^2)$$

or in terms of t

$$\frac{du}{dt} = 4(e^{4t} + e^{4t}(\cos^2 3t + \sin^2 3t)) = 8e^{4t}$$

verification by direct substitution:

$$u = x^2 + y^2 + z^2 = e^{4t} + e^{4t}\cos^2 3t + e^{4t}\sin^2 3t = 2e^{4t}$$

$$\frac{du}{dt} = 8e^{4t}.$$

Example 2: Find the total differential coefficient of $x^2 y$ w.r.t. x when x, y are connected by

$$x^2 + xy + y^2 = 1.$$

Solution: Let $u = x^2 y$, then the total differential is

$$du = \frac{\partial u}{\partial x}dx + \frac{\partial u}{\partial y}dy$$

Thus the total differential coefficient of u w.r.t. x is

$$\frac{du}{dx} = \frac{\partial u}{\partial x} + \frac{\partial u}{\partial y}\frac{dy}{dx}$$

$$\frac{du}{dx} = 2xy + x^2\frac{dy}{dx}$$

From the implicit relation $f = x^2 + xy + y = 1$, we calculate

$$\frac{dy}{dx} = -\frac{f_x}{f_y} = -\frac{2x + y}{x + 2y}$$

so

$$\frac{du}{dx} = 2xy + x^2 \cdot \frac{dy}{dx} = 2xy + x^2\left(-\frac{(2x + y)}{(x + 2y)}\right)$$

$$\frac{du}{dx} = 2xy - \frac{x^2(2x + y)}{(x + 2y)}.$$

Example 3: The altitude of a right circular cone is 15 cm and is increasing at 0.2 cm/sec. The radius of the base is 10 cm and is decreasing at 0.3 cm/sec. How fast is the volume changing?

Solution: Let x be the radius and y be the altitude of the cone. So volume V of the right circular cone is

$$V = \frac{1}{3}\pi x^2 y.$$

Since x and y are changing w.r.t. time t, differentiate V w.r.t. t.

$$\frac{dV}{dt} = \frac{\partial V}{\partial x}\frac{dx}{dt} + \frac{\partial V}{\partial y}\frac{dy}{dt}$$

$$= \frac{1}{3}\pi\left(2xy\frac{dx}{dt} + x^2\frac{dy}{dt}\right)$$

It is given that at $x = 10$, $y = 15$, $\frac{dx}{dt} = -0.3$ and $\frac{dy}{dt} = 0.2$, substituting these values,

$$\frac{dV}{dt} = \frac{1}{3}\pi[2 \cdot 10 \cdot 15(-0.3) + 10^2(0.2)]$$

$$= \frac{-70}{3}\pi \text{ cm}^3/\text{sec}$$

i.e., volume is decreasing at the rate of $\frac{70\pi}{3}$.

EXERCISE

1. Find $\frac{du}{dt}$ when $u = \sin(x/y)$ and $x = e^t$, $y = t^2$. Verify the result by direct substitution.

   Ans. $\frac{t-2}{t^3}e^t \cos\left(\frac{e^t}{t^2}\right)$

2. Find $\frac{du}{dt}$ given $u = \sin^{-1}(x - y); x = 3t$, $y=4t^3$. Verify the result by direct substitution.

   Ans. $3(1 - t^2)^{-\frac{1}{2}}$

3. If $u = x^3 y e^z$ where $x = t, y = t^2$ and $z = \ln t$ find $\frac{du}{dt}$ at $t = 2$.

   Ans. $6t^5$; 192

4. Find $\frac{du}{dt}$ if $u = \tan^{-1}(y/x)$ and $x = e^t - e^{-t}$ and $y = e^t + e^{-t}$.

   Ans. $\frac{-2}{e^{2t}+e^{-2t}}$

5. If x, y are related by $x^2 - y^2 = 2$ and $u = \tan(x^2 + y^2)$ find $\frac{du}{dx}$

   Ans. $4x\sec^2(2x^2 - 2)$

6. If $u = \tan^{-1}(y/x)$ and $y = x^4$ find $\frac{du}{dx}$ at $x = 1$

   Ans. $\frac{3x^2}{1+x^6}$; $\frac{3}{2}$ at $x = 1$

7. In order that the function $u = 2xy - 3x^2 y$ remains constant, what should be the rate of change of y (w.r.t. t) given that x increases at the rate of 2 cm/sec at the instant when $x = 3$ cm and $y = 1$ cm.

   Ans. $\frac{dy}{dt} = -\frac{32}{21}$ cm/sec; y must decrease at the rate of $\frac{32}{21}$ cm/sec.

8. Find the rate at which the area of a rectangle is increasing at a given instant when the

sides of the rectangle are 4 ft and 3 ft and are increasing at the rate of 1.5 ft/sec and 0.5 ft/sec respectively.

Ans. 6.5 sq. ft/sec

9. Find **a.** $\frac{dz}{dx}$ and **b.** $\frac{dz}{dy}$ given $z = xy^2 + x^2y$, $y = \ln x$

Ans. **a.** Here x is the independent variable

$$\frac{dz}{dx} = \frac{\partial z}{\partial x} + \frac{\partial z}{\partial y}\frac{dy}{dx} = y^2 + 2xy + 2y + x$$

b. Here y is the independent variable

$$\frac{dz}{dy} = \frac{\partial z}{\partial y} + \frac{\partial z}{\partial x}\frac{dx}{dy} = xy^2 + 2x^2y + 2xy + x^2$$

Find the differential of the following functions:

10. $f(x, y) = x\cos y - y\cos x$

Ans. $df = (\cos y + y\sin x)dx - (x\sin y + \cos x)dy$

11. $u(x, y, z) = e^{xyz}$

Ans. $du = e^{xyz}(yz\,dx + zx\,dy + xy\,dz)$

Find $\frac{du}{dt}$ for the following functions:

12. $u = x^2 - y^2$, $x = e^t\cos t$, $y = e^t\sin t$ at $t = 0$.

Ans. $2e^{2t}(\cos 2t - \sin 2t)$; At $t = 0$, $\frac{du}{dt} = 2$

13. $u = \ln(x + y + z)$; $x = e^{-t}$, $y = \sin t$, $z = \cos t$

Ans. $\frac{\cos t - \sin t - e^{-t}}{\cos t + \sin t + e^{-t}}$

14. $u = \sin(e^x + y)$, $x = f(t)$, $y = g(t)$

Ans. $\frac{du}{dt} = [\cos(e^x + y)]e^x f'(t) + [\cos(e^x + y)]g'(t)$

15. $u = x^y$ when $y = \tan^{-1} t$, $x = \sin t$

Ans. $y \cdot x^{y-1}\cos t + x^y \ln x \cdot \frac{1}{1+t^2}$.

3.5 PARTIAL DIFFERENTIATION OF COMPOSITE FUNCTIONS: CHANGE OF VARIABLES

In the study of heat equation, wave equation and Laplace's equation, it is very often required to transform the representing partial differential equations in cartesian coordinate system to cylindrical, spherical or orthogonal curvilinear systems by changing the variables through partial differentiation of composite functions (function of a function).

Let $u = f(x, y, z)$ and x, y, z are functions of two (or more) independent variables say s and t as $x = x(s, t)$, $y = y(s, t)$, $z = z(s, t)$. Then f is considered as function of s and t via the intermediate variables x, y, z. Now the derivative of f w.r.t. t is partial but not total. Keeping s constant, Equation (3) is modified as

$$\left(\frac{\partial f}{\partial t}\right)_s = \frac{\partial f}{\partial x}\left(\frac{\partial x}{\partial t}\right)_s + \frac{\partial f}{\partial y}\left(\frac{\partial y}{\partial t}\right)_s + \frac{\partial f}{\partial z}\left(\frac{\partial z}{\partial t}\right)_s \quad (5)$$

The subscript s in (5) indicates the variable which is held constant. With this convention $\frac{\partial f}{\partial x}$ may be written as $\left(\frac{\partial f}{\partial x}\right)_{y,z}$ and so on. However following the convention that $\frac{\partial f}{\partial x}$, without subscripts, indicates the result of differentiating f w.r.t. the explicitely appearing variable x, while holding all other explicitly appearing variables (here y and z) constant. With this convention, the subscript s in (5) may be omitted and rewritten as

$$\frac{\partial f}{\partial t} = \frac{\partial f}{\partial x}\frac{\partial x}{\partial t} + \frac{\partial f}{\partial y}\frac{\partial y}{\partial t} + \frac{\partial f}{\partial z}\frac{\partial z}{\partial t} \quad (6)$$

In a similar way, we get

$$\frac{\partial f}{\partial s} = \frac{\partial f}{\partial x}\frac{\partial x}{\partial s} + \frac{\partial f}{\partial y}\frac{\partial y}{\partial s} + \frac{\partial f}{\partial z}\frac{\partial z}{\partial s} \quad (7)$$

Equations (6) and (7) are known as ***chain rules for partial differentiation.***

WORKED OUT EXAMPLES

Example 1: If $u = x^2 - y^2$, $x = 2r - 3s + 4$, $y = -r + 8s - 5$ find $\frac{\partial u}{\partial r}$ and $\frac{\partial u}{\partial s}$.

Solution: Here u is a function of x, y which are functions of s, t. So by chain rule

$$\frac{\partial u}{\partial r} = \frac{\partial u}{\partial x}\frac{\partial x}{\partial r} + \frac{\partial u}{\partial y}\frac{\partial y}{\partial r}$$

$$= 2x \cdot 2 + (-2y)(-1) = 2(2x + y)$$

$$\frac{\partial u}{\partial s} = \frac{\partial u}{\partial x}\frac{\partial x}{\partial s} + \frac{\partial u}{\partial y}\frac{\partial y}{\partial s}$$

$$= 2x \cdot (-3) + (-2y)8 = -6x - 16y.$$

Example 2: Find $\frac{\partial w}{\partial \theta}$ and $\frac{\partial w}{\partial \phi}$ given that $w(x, y, z)=f(x^2+y^2+z^2)$ where $x=r\cos\theta\cdot\cos\phi$, $y=r\cos\theta\cdot\sin\phi$, $z=r\sin\theta$.

Solution: Put $u=x^2+y^2+z^2$. Then $w(x,y,z)=f(u)$ where $u=u(x,y,z)$ and x, y, z are functions of r, θ, ϕ. So

$$\frac{\partial w}{\partial \theta}=\frac{\partial w}{\partial u}\cdot\frac{\partial u}{\partial \theta}=\frac{\partial w}{\partial u}\left(\frac{\partial u}{\partial x}\frac{\partial x}{\partial \theta}+\frac{\partial u}{\partial y}\frac{\partial y}{\partial \theta}+\frac{\partial u}{\partial z}\frac{\partial z}{\partial \theta}\right)$$

$$=\frac{\partial w}{\partial u}(2x\cdot(-r\sin\theta\cos\phi)+2y(-r\sin\theta\sin\phi)$$
$$+2z(r\cos\theta))$$

Substituting x, y, z in terms of r, θ, ϕ, we get

$$=2\frac{\partial w}{\partial u}[-r^2\sin\theta\cos\theta\cos^2\phi-r^2\sin\theta\cos\theta\sin^2\phi$$
$$+r^2\sin\theta\cos\theta]$$

$$=2r^2\frac{\partial w}{\partial u}[-\sin\theta\cos\theta+\sin\theta\cos\theta]=0$$

Now

$$\frac{\partial w}{\partial \phi}=\frac{\partial w}{\partial u}\left(\frac{\partial u}{\partial x}\frac{\partial x}{\partial \phi}+\frac{\partial u}{\partial y}\frac{\partial y}{\partial \phi}+\frac{\partial u}{\partial z}\frac{\partial z}{\partial \phi}\right)$$

$$=\frac{\partial w}{\partial u}(2x\cdot(-r\cos\theta\cdot\sin\phi)$$
$$+2y(r\cos\theta\cos\phi)+2z\cdot 0)$$

$$=2\frac{\partial w}{\partial u}(-r^2\cos^2\theta\cdot\cos\phi\sin\phi$$
$$+r^2\cos^2\theta\sin\phi\cos\phi)$$

$$\frac{\partial w}{\partial \phi}=2r^2\frac{\partial w}{\partial u}(0)=0.$$

Example 3: If $V=f(2x-3y, 3y-4z, 4z-2x)$ prove that $6V_x+4V_y+3V_z=0$.

Solution: Put $u=2x-3y$, $v=3y-4z$, $w=4z-2x$. Then $V=f(u,v,w)$ and u, v, w are functions of x, y, z. So

$$V_x=\frac{\partial V}{\partial x}=\frac{\partial V}{\partial u}\frac{\partial u}{\partial x}+\frac{\partial V}{\partial v}\frac{\partial v}{\partial x}+\frac{\partial V}{\partial w}\frac{\partial w}{\partial x}$$

$$V_x=V_u\cdot 2+V_v\cdot 0+V_w(-2)=2(V_u-V_w) \qquad (1)$$

$$V_y=\frac{\partial V}{\partial y}=\frac{\partial V}{\partial u}\frac{\partial u}{\partial y}+\frac{\partial V}{\partial v}\frac{\partial v}{\partial y}+\frac{\partial V}{\partial w}\frac{\partial w}{\partial y}$$

$$V_y=V_u(-3)+\frac{\partial V}{\partial v}(3)+\frac{\partial V}{\partial w}\cdot 0=3(V_v-V_u) \qquad (2)$$

Similarly

$$V_w=V_u\cdot 0+V_v(-4)+V_w\cdot 4=4(V_w-V_v) \qquad (3)$$

Multiplying (1), (2) and (3) by 6, 4, 3 respectively and adding, we get

$$6V_x+4V_y+3V_z=6(2(V_u-V_w))+4(3(V_v-V_u)$$
$$+3(4(V_w-V_v))$$
$$=12(V_u-V_w+V_v-V_u+V_w-V_v)$$
$$=0$$

Example 4: If V is a function of u, v where $u=x-y$ and $v=xy$ prove that

$$x\frac{\partial^2 V}{\partial x^2}+y\frac{\partial^2 V}{\partial y^2}=(x+y)\left(\frac{\partial^2 V}{\partial u^2}+xy\frac{\partial^2 V}{\partial v^2}\right)$$

Solution: Since $V=f(u,v)$ and u, v are functions of x, y use chain rule, to differentiate V w.r.t. x. Note that $u_x=1, v_x=y, u_y=-1, v_y=x$

$$\frac{\partial V}{\partial x}=\frac{\partial V}{\partial u}\frac{\partial u}{\partial x}+\frac{\partial V}{\partial v}\frac{\partial v}{\partial x}=\frac{\partial V}{\partial u}\cdot 1+\frac{\partial V}{\partial v}\cdot y$$

$$\frac{\partial^2 V}{\partial x^2}=\frac{\partial}{\partial x}\left(\frac{\partial V}{\partial x}\right)=\frac{\partial}{\partial x}\left(\frac{\partial V}{\partial u}+y\frac{\partial V}{\partial v}\right)$$

$$=\frac{\partial}{\partial u}\left(\frac{\partial V}{\partial u}+y\frac{\partial V}{\partial v}\right)\frac{\partial u}{\partial x}+\frac{\partial}{\partial v}\left(\frac{\partial V}{\partial u}+y\frac{\partial V}{\partial v}\right)\frac{\partial v}{\partial x}$$

$$\frac{\partial^2 V}{\partial x^2}=\left(\frac{\partial^2 V}{\partial u^2}+y\frac{\partial^2 V}{\partial u\partial v}\right)\cdot 1+\left(\frac{\partial^2 V}{\partial v\partial u}+y\frac{\partial^2 V}{\partial v^2}\right)y \qquad (1)$$

Differentiating V w.r.t. y by chain rule

$$\frac{\partial V}{\partial y}=\frac{\partial V}{\partial u}\frac{\partial u}{\partial y}+\frac{\partial V}{\partial v}\frac{\partial v}{\partial y}=\frac{\partial V}{\partial u}(-1)+\frac{\partial V}{\partial v}x$$

$$\frac{\partial^2 V}{\partial y^2}=\frac{\partial}{\partial y}\left(\frac{\partial V}{\partial y}\right)=\frac{\partial}{\partial y}\left(-\frac{\partial V}{\partial u}+x\frac{\partial V}{\partial v}\right)$$

$$=\frac{\partial}{\partial u}\left(-\frac{\partial V}{\partial u}+x\frac{\partial V}{\partial v}\right)\frac{\partial u}{\partial y}$$
$$+\frac{\partial}{\partial v}\left(-\frac{\partial V}{\partial u}+x\frac{\partial V}{\partial v}\right)\frac{\partial v}{\partial y}$$

$$\frac{\partial^2 V}{\partial y^2}=\left(-\frac{\partial^2 V}{\partial u^2}+x\frac{\partial^2 V}{\partial u\partial v}\right)(-1)$$
$$+\left(-\frac{\partial^2 V}{\partial v\partial u}+x\frac{\partial^2 V}{\partial v^2}\right)x \qquad (2)$$

Multiply (1) by x and (2) by y and adding, we get

$$x\frac{\partial^2 V}{\partial x^2}+y\frac{\partial^2 V}{\partial y^2}$$

$$=\left(x\frac{\partial^2 V}{\partial u^2}+xy\frac{\partial^2 V}{\partial u\partial v}+xy\frac{\partial^2 V}{\partial v\partial u}+xy^2\frac{\partial^2 V}{\partial v^2}\right)$$

$$+\left(+y\frac{\partial^2 V}{\partial u^2}-xy\frac{\partial^2 V}{\partial u\partial v}-xy\frac{\partial^2 V}{\partial v\partial u}+x^2y\frac{\partial^2 V}{\partial v^2}\right)$$

Since $\dfrac{\partial^2 V}{\partial u\partial v}=\dfrac{\partial^2 V}{\partial v\partial u}$

The R.H.S. gets simplified to

$$x\frac{\partial^2 V}{\partial x^2}+y\frac{\partial^2 V}{\partial y^2}=(x+y)\frac{\partial^2 V}{\partial u^2}+xy(x+y)\frac{\partial^2 V}{\partial v^2}$$

$$=(x+y)\left(\frac{\partial^2 V}{\partial u^2}+xy\frac{\partial^2 V}{\partial v^2}\right).$$

EXERCISE

1. If $V=u^2v$ and $u=e^{x^2-y^2}$, $v=\sin(xy^2)$ find $\frac{\partial V}{\partial x}, \frac{\partial V}{\partial y}$.

Ans. $V_x=e^{2(x^2-y^2)}[4x\sin(xy^2)+y^2\cos(xy^2)]$

$V_y=2e^{2(x^2-y^2)}y[x\cos(xy^2)-2\sin(xy^2)]$

2. Find $\frac{\partial V}{\partial s}$ if $V=(x+y)/(1-xy)$ and $x=\tan(2r-s^2)$, $y=\cot(r^2s)$.

Ans. $-(1+x^2)(1+y^2)(2s+r^2)/(1-xy)^2$

3. If $z=\frac{\sin u}{\cos v}$ where $u=\frac{\cos y}{\sin x}$ and $v=\frac{\cos x}{\sin y}$ find $\frac{\partial z}{\partial x}$.

Ans. $-(u\cot x\cos u\sin y+z\sin v\cdot\sin x)/(\cos v\sin y)$

4. Prove that $\frac{\partial z}{\partial u}-\frac{\partial z}{\partial v}=x\frac{\partial z}{\partial x}-y\frac{\partial z}{\partial y}$ if z is a function of x and y and $x=e^u+e^{-v}$, $y=e^{-u}-e^v$.

5. If $\cos z=u+v$ and $u=\sin x$, $v=\cos x$ find the total derivative of z w.r.t. x i.e., $\frac{dz}{dx}$.

Ans. $\frac{dz}{dx}=\frac{\partial z}{\partial u}\frac{du}{dx}+\frac{\partial z}{\partial v}\frac{dv}{dx}=\frac{\sin x-\cos x}{\sqrt{1-(u+v)^2}}$

6. Prove that $y\frac{\partial V}{\partial y}-x\frac{\partial V}{\partial x}=x^2V^3$ if

$$V=(1-2xy+y^2)^{-\frac{1}{2}}.$$

7. Show that $\left(\frac{\partial u}{\partial x}\right)^2+\left(\frac{\partial u}{\partial y}\right)^2=\left(\frac{\partial u}{\partial r}\right)^2+\frac{1}{r^2}\left(\frac{\partial u}{\partial \theta}\right)^2$ when u is a function of x, y where $x=r\cos\theta$, $y=r\sin\theta$.

8. If $V=f(r,s,t)$ and $r=x/y$, $s=y/z$, $t=z/x$ show that $x\frac{dV}{dx}+y\frac{dV}{dy}+z\frac{dV}{dz}=0$.

9. Change the Laplacian equation in cartesian coordinates $\frac{\partial^2 u}{\partial x^2}+\frac{\partial^2 u}{\partial y^2}=0$ into polar coordinates.

Hint: Use the change of variables $x=r\cos\theta$, $y=r\sin\theta$ or $r=\sqrt{x^2+y^2}$, $\theta=\tan^{-1}(y/x)$, calculate $u_x=u_rr_x+u_\theta\theta_x$ etc.

Ans. $\frac{\partial^2 u}{\partial r^2}+\frac{1}{r}\frac{\partial u}{\partial r}+\frac{1}{r^2}\frac{\partial^2 u}{\partial\theta^2}=0$

10. If $V=f(u,v)$ where $u=a\cosh x\cos y$, $v=a\sinh x\sin y$ prove that $\frac{\partial^2 V}{\partial x^2}+\frac{\partial^2 V}{\partial y^2}=\frac{a^2}{2}(\cosh 2x-\cos 2y)\left(\frac{\partial^2 V}{\partial u^2}+\frac{\partial^2 V}{\partial v^2}\right)$.

11. Transform the Laplacian equation $\frac{\partial^2 u}{\partial x^2}+\frac{\partial^2 u}{\partial y^2}=0$ by change of variables from x, y to r, θ when $x=e^r\cos\theta$, $y=e^r\sin\theta$.

Hint: Use $r=\ln\sqrt{x^2+y^2}$, $\theta=\tan^{-1}(y/x)$.

Ans. $e^{-2r}\left(\frac{\partial^2 u}{\partial r^2}+\frac{\partial^2 u}{\partial\theta^2}\right)=0$

12. Transform the Laplacian $\frac{\partial^2 V}{\partial u^2}+\frac{\partial^2 V}{\partial v^2}$ by changing the variables from u, v to x, y when $x=u\cos\theta-v\sin\theta$, $y=u\sin\theta+v\cos\theta$.

Hint: Use chain rule $\frac{\partial V}{\partial u}=\frac{\partial V}{\partial x}\frac{\partial x}{\partial u}+\frac{\partial V}{\partial y}\frac{\partial y}{\partial u}$.

Ans. $\frac{\partial^2 V}{\partial x^2}+\frac{\partial^2 V}{\partial y^2}$

13. Express $\nabla f=\frac{\partial f}{\partial x}\bar{i}+\frac{\partial f}{\partial y}\bar{j}$ in plane polar coordinates.

Hint: Use $x=r\cos\theta$, $y=r\sin\theta$, use chain rule.

Ans. $\nabla f=\left(f_r\cos\theta-f_\theta\frac{\sin\theta}{r}\right)\bar{i}+\left(\sin\theta f_r+\frac{\cos\theta}{r}f_\theta\right)\bar{j}$

14. If $x=r\cos\theta$, $y=r\sin\theta$ prove that

$$\left(\frac{\partial^2 r}{\partial x^2}\right)\left(\frac{\partial^2 r}{\partial y^2}\right)=\left(\frac{\partial^2 r}{\partial x\partial y}\right)^2$$

Hint: $r_{xx} = \frac{1}{r} - \frac{x^2}{r^3}, r_{yy} = \frac{1}{r} - \frac{y^2}{r^3}, r_{xy} = \frac{-xy}{r^3}$.

15. If $u = x^2 + 2xy - y \ln z$ where

$$x = s + t^2, y = s - t^2, z = 2t$$

find $\frac{\partial u}{\partial s}, \frac{\partial u}{\partial t}$ at (1, 2, 1)

Ans. $\frac{\partial u}{\partial s} = 4x + 2y - \ln z$, at (1, 2, 1) : 8

$\frac{\partial u}{\partial t} = 4yt + 2t \ln z - \frac{2y}{z}$, at (1, 2, 1) : $8t - 4$.

3.6 DIFFERENTIATION OF AN IMPLICIT FUNCTION

An implicit function of x and y is an equation of the form

$$f(x, y) = 0$$

which can not necessarily be solved for one of the variables say x in terms of the other variable say y.

For example

$$x^2 + y^2 + a^2 = 0 \qquad (1)$$

is an implicit function which can not be solved for say x in terms of y explicitly. If (1) defines y as a function of x, the derivative of y w.r.t. x can be calculated in terms of f, without solving (1) explicitly for x in the form $y = y(x)$, by differentiating (1) partially w.r.t. to x as

$$\frac{\partial f}{\partial x}\frac{dx}{dx} + \frac{\partial f}{\partial y}\frac{dy}{dx} = 0 \qquad (2)$$

solving (2) we get

$$\frac{dy}{dx} = -\frac{\frac{\partial f}{\partial x}}{\frac{\partial f}{\partial y}} = -\frac{f_x}{f_y} \quad \text{provided } f_y \neq 0 \qquad (3)$$

Higher derivative $\frac{d^2y}{dx^2}$ of an implicit function (1) can be obtained by differentiating (3) on both sides, keeping in mind that the arguments on the right of (3) are x and y and that y itself is the function of x defined by (1).

Differentiating (3) w.r.t. x, noting that f_x and f_y as composite functions of x, we have

$$\frac{d^2y}{dx^2} = \frac{\left[\frac{\partial^2 f}{\partial x^2} + \frac{\partial^2 f}{\partial y \partial x}\frac{dy}{dx}\right]\frac{\partial f}{\partial y} - \left[\frac{\partial^2 f}{\partial x \partial y} + \frac{\partial^2 f}{\partial y^2}\frac{dy}{dx}\right]\frac{\partial f}{\partial x}}{\left(\frac{\partial f}{\partial y}\right)^3} \qquad (4)$$

Substituting from (3) $\frac{dy}{dx} = -\frac{f_x}{f_y}$ in (4) and rearranging, we get

$$\frac{d^2y}{dx^2} = -\frac{\left[f_{xx}(f_y)^2 - 2f_{xy}f_x f_y + f_{yy}(f_x)^2\right]}{(f_y)^3}, \qquad (5)$$

$$\text{if } f_y \neq 0$$

Thus $\frac{dy}{dx}$ and $\frac{d^2y}{dx^2}$, given by (3) and (5) respectively, are expressed in terms of the partial derivatives $f_x, f_y, f_{xx}, f_{xy}, f_{yy}$.

Implicit Function of Three Variables

Let $f(x, y, z) = 0$ be the equation of an implicit function of three variables x, y, z. Suppose y and z are functions of x, then f is a function of one independent variable x and y, z are intermediate variables.

Keeping z constant, differentiating w.r.t. x, we get

$$\frac{\partial f}{\partial x} + \frac{\partial f}{\partial y}\frac{\partial y}{\partial x} = 0$$

solving $\quad \frac{\partial y}{\partial x} = -\frac{f_x}{f_y}, \quad$ provided $\quad f_y \neq 0$

Similarly differentiating w.r.t. x, holding y constant

$$\frac{\partial f}{\partial x} + \frac{\partial f}{\partial z}\frac{\partial z}{\partial x} = 0$$

solving $\quad \frac{\partial z}{\partial x} = -\frac{f_x}{f_y}, \quad$ if $\quad f_z \neq 0$.

WORKED OUT EXAMPLES

Implicit function of two variables

Example 1: Find $\frac{dy}{dx}$ from the given implicit function f connecting x and y:

a. $f(x, y) = x \sin(x - y) - (x + y) = 0$

$$\frac{dy}{dx} = -\frac{\frac{\partial f}{\partial x}}{\frac{\partial f}{\partial y}} = -\frac{[\sin(x - y) + x \cdot \sin(x - y) \cdot 1 - 1]}{x \cos(x - y) \cdot (-1) - 1}.$$

b. $x^y = y^x$

Taking log, $f = y \log x - x \log y = 0$

$$\frac{dy}{dx} = -\frac{f_x}{f_y} = -\frac{\frac{y}{x} - \log y}{\log x - \frac{x}{y}}.$$

Example 2: Find $\frac{dy}{dx}$ when $y^{x^y} = \sin x$.

Solution:

$$\text{Taking log, } f(x, y) = x^y \ln y - \ln \sin x = 0 \quad (1)$$

$$\text{Put } z = x^y, \text{ so } \ln z = y \ln x \quad (2)$$

Differentiating (2) w.r.t. x and y, we get

$$\frac{1}{z} z_x = y \cdot \frac{1}{x} \quad \text{so } z_x = \frac{y}{x} \cdot z = \frac{y}{x} \cdot x^y = yx^{y-1} \quad (3)$$

$$\frac{1}{z} z_y = \ln x \quad \text{so } z_y = z \ln x = x^y \ln x \quad (4)$$

Differentiating (1) partially w.r.t. x and y, we get

$$\frac{\partial f}{\partial x} = f_x = \frac{\partial}{\partial x}(x^y) \ln y - \frac{1}{\sin x} \cdot \cos x$$

using result (3)

$$f_x = yx^{y-1} \cdot \ln y - \cot x \quad (5)$$

$$\frac{\partial f}{\partial y} = f_y = \frac{\partial}{\partial y}(x^y) \ln y + x^y \cdot \frac{1}{y} - 0$$

using result (4)

$$f_y = x^y \ln x \cdot \ln y + \frac{x^y}{y} \quad (6)$$

Substituting (5) and (6) in

$$\frac{dy}{dx} = \frac{-f_x}{f_y} = \frac{-(yx^{y-1} \ln y - \cot x)}{x^y \ln x \cdot \ln y + x^y y^{-1}}$$

Example 3: Find $\frac{d^2y}{dx^2}$ if $x^5 + y^5 = 5a^3x^2$.

Solution: Let $f(x, y) = x^5 + y^5 - 5a^3x^2 = 0$
Differentiating f w.r.t. x, y, we get

$$f_x = 5x^4 - 10a^3x, \; f_{xy} = 0$$

$$f_{xx} = 20x^3 - 10a^3,$$

$$f_y = 5y^4; \; f_{yy} = 20y^3$$

Substituting these values in

$$\frac{d^2y}{dx^2} = -\frac{f_{xx}(f_y)^2 - 2f_{xy} \cdot f_x \cdot f_y + f_{yy}(f_x)^2}{(f_y)^3}$$

$$= -\frac{(20x^3 - 10a^3)(25y^8) - 0 + 20y^3(5x^4 - 10a^3x)^2}{125y^{12}}$$

$$= -\frac{y^5(20x^3 - 10a^3) + 20(x^8 - 4a^3x^5 + 4a^6x^2)}{5y^9}$$

Replace $y^5 = 5a^3x^2 - x^5$

$$= -\frac{(5a^3x^2 - x^5)(4x^3 - 2a^3) + 4(x^8 - 4a^3x^5 + 4a^6x^2)}{y^9}$$

$$= \frac{6a^3x^2(a^3 + x^3)}{y^9}.$$

Example 4: Compute $\frac{\partial z}{\partial x}$ and $\frac{\partial y}{\partial x}$ at $(1, -1, 2)$ if $x^2 + y^2 + z^2 = a^2$.

Solution: $f(x, y, z) = x^2 + y^2 + z^2 - a^2 = 0$ is the equation of implicit function. Differentiating f partially w.r.t. x,

$$\frac{\partial f}{\partial x} = 2x, \; \left.\frac{\partial f}{\partial x}\right|_{1,-1,2} = 2$$

Differentiating w.r.t. z

$$\frac{\partial f}{\partial z} = 2z, \; \left.\frac{\partial f}{\partial z}\right|_{1,-1,2} = 4$$

$$\text{so} \quad \left.\frac{\partial z}{\partial x}\right|_{(1,-1,2)} = -\left.\frac{f_x}{f_z}\right|_{(1,-1,2)} = \frac{-2}{4} = -\frac{1}{2}$$

Differentiating w.r.t. y

$$\frac{\partial f}{\partial y} = 2y, \; \left.\frac{\partial f}{\partial y}\right|_{1,-1,2} = -2$$

$$\text{so} \quad \frac{\partial y}{\partial x} = -\frac{f_x}{f_y} = \frac{-2}{-2} = 1.$$

Example 5: If $xy^3 - yx^3 = 6$ is the equation of a curve, find the slope and the equation of the tangent line at the point $(1, 2)$.

Solution: Differentiating $xy^3 - yx^3 = 6$ implicitly w.r.t. x, we get

$$y^3 + 3xy^2\frac{dy}{dx} - \frac{dy}{dx}x^3 - 3yx^2 = 0$$

$$\text{At } (x = 1, y = 2), \; 8 + 12y' - y' - 6 = 0$$

$$\text{slope at } (1, 2) \text{ is } \frac{dy}{dx} = -\frac{2}{11}.$$

Equation of the tangent line at $(1, 2)$ is

$$\frac{y - 2}{x - 1} = -\frac{2}{11}$$

$$\text{or} \quad 2x + 11y - 24 = 0.$$

EXERCISE

Find the derivative $\frac{dy}{dx}$ from the given implicit function $f(x, y) = c$. **Hint** : use $\frac{dy}{dx} = \frac{-f_x}{f_y}$:

1. $x^3 + y^3 = 3ax^2$

Ans. $\frac{dy}{dx} = (2ax - x^2)/y^2$

2. $(\cos x)^y = (\sin y)^x$

Ans. $\frac{\sin y(y \sin x + \cos x \ln \sin y)}{\cos x(\sin y \ln \cos x - x \cos y)}$

3. $\sin(xy) = e^{xy} + x^2y$

Ans. $\frac{-y(\cos xy - e^{xy} - 2x)}{x(\cos xy - e^{xy} - x)}$

4. $ax^2 + 2hxy + by^2 = c$

Ans. $-(ax + hy)/(hx + by)$

5. $x^y + y^x = c$

Ans. $-\frac{yx^{y-1} + y^x \ln y}{xy^{x-1} + x^y \ln x}$

Find $\frac{d^2y}{dx^2}$ from the implicit function $f(x, y) = c$

6. $x^4 + y^4 = 4a^2xy$

Ans. $\frac{2a^2xy(3a^4 + x^2y^2)}{(a^2x - y^3)^3}$

7. $ax^2 + 2hxy + by^2 + 2gx + 2fy + c = 0$

Ans. $\frac{abc + 2fgh - af^2 - bg^2 - ch^2}{(hx + by + f)^3}$

8. $x\sqrt{1 - y^2} + y\sqrt{1 - x^2} = a$

Ans. $\frac{a}{(1 - x^2)^{\frac{3}{2}}}$

9. Find $\frac{\partial y}{\partial u}$ and $\frac{\partial y}{\partial x}$ if $\ln uy + y \ln u = x$

Ans. $\frac{\partial y}{\partial u} = -(y^2 + y)/(u + yu \ln u)$

$\frac{\partial y}{\partial x} = y/(1 + y \ln u)$

10. Find $\frac{du}{dx}$ given $u = x \ln xy$ and $x^3 + y^3 + 3xy - 1 = 0$

Ans. $1 + \ln a^2 - x(x^2 + y)/(y(y^2 + x))$ at $x = y = a$ is $\ln a^2$

11. If $z(z^2 + 3x) + 3y = 0$, prove that

$$\frac{\partial^2 z}{\partial x^2} + \frac{\partial^2 z}{\partial y^2} = \frac{-2z(x - 1)}{(z^2 + x)^4}$$

12. Find $\frac{\partial u}{\partial x}$ and $\frac{\partial u}{\partial y}$ if $u = f(x + u, yu)$

Ans. $\frac{\partial u}{\partial x} = f_1/(1 - f_1 - yf_2)$

$\frac{\partial u}{\partial y} = uf_2/(1 - f_1 - yf_2)$

where f_1 is differentiation w.r.t. to the first variable $x + u$; f_2 is w.r.t. yu

13. If $ye^{xy} = \sin x$, find $\frac{dy}{dx}$ and $\frac{d^2y}{dx^2}$ at (0, 0)

Ans. $\frac{dy}{dx} = 1$, $\frac{d^2y}{dx^2} = 0$

14. If $x^y = y^x$ find $\frac{dy}{dx}$ at (2, 4)

Ans. $y' = 4(\ln 2 - 1)/(2 \ln 2 - 1)$

15. Find $\frac{\partial y}{\partial x}$ and $\frac{\partial z}{\partial x}$ at (0, 1, 2) if $z^3 + xy - y^2z = 6$

Ans. $\frac{\partial y}{\partial x} = -\left.\frac{y}{x - 2yz}\right|_{(0,1,2)} = \frac{1}{4}$

$\frac{\partial z}{\partial x} = -\left.\frac{y}{3z^2 - y^2}\right|_{(0,1,2)} = -\frac{1}{11}$

16. For the curve $xe^y + ye^x = 0$, find the equation of the tangent line at the origin.

Ans. $x + y = 0$.

3.7 EULER'S THEOREM

Homogeneous Function

A polynomial in x and y is said to be homogeneous if all its terms are of the same degree. Generalizing this property to include non-polynomials, a function $f(x, y)$ in two variables x and y is said to be a homogeneous function of degree n if for any positive number λ,

$$f(\lambda x, \lambda y) = \lambda^n f(x, y)$$

This definition can be further enlarged to include transcendental functions also as follows. A function $f(x, y)$ is said to be homogeneous of degree n if it can be expressed as

$$x^n \phi\left(\frac{y}{x}\right) \quad \text{or} \quad y^n \psi\left(\frac{x}{y}\right)$$

Here n need not be an integer, n could be positive, negative or zero.

Example:

1. $3x^2 - 2xy + \frac{15}{2}y^2$ is homogeneous of degree 2
2. $\frac{\sqrt{y}+\sqrt{x}}{y+x}$ is homogeneous of degree $-\frac{1}{2}$
3. $\sin\left(\frac{y}{x}\right) + \tan^{-1}\left(\frac{x}{y}\right)$ is homogeneous of degree zero
4. $\left(\frac{x+y}{xy} + x^{\frac{2}{3}}e^{\frac{x}{y}}\right)y^{-\frac{5}{3}}$ is not homogeneous
5. $x^{\frac{1}{3}}y^{-\frac{2}{3}} + x^{\frac{2}{3}}y^{-\frac{1}{3}}$ is not homogeneous

Homogeneous function f of three variables x, y, z of degree n can be expressed as

$$f = x^n\phi\left(\frac{y}{x}, \frac{z}{x}\right) \quad \text{or} \quad y^n\psi\left(\frac{x}{y}, \frac{z}{y}\right) \quad \text{or} \quad z^n x\left(\frac{x}{z}, \frac{y}{z}\right)$$

Euler's Theorem on Homogeneous Functions

Theorem: *If f is a homogeneous function of x, y of degree n then*

$$x\frac{\partial f}{\partial x} + y\frac{\partial f}{\partial y} = nf \tag{1}$$

Proof: Since f is a homogeneous function of degree n, f can be written in the form

$$f(x, y) = x^n\phi\left(\frac{y}{x}\right) \tag{2}$$

Differentiating partially w.r.t. x and y, we get

$$\frac{\partial f}{\partial x} = nx^{n-1}\phi + x^n \cdot \phi' \cdot \left(\frac{-y}{x^2}\right) \tag{3}$$

$$\frac{\partial f}{\partial y} = x^n\phi' \cdot \frac{1}{x} \tag{4}$$

Multiplying (3) by x and (4) by y and adding we have

$$x\frac{\partial f}{\partial x} + y\frac{\partial f}{\partial y} = nx^n\phi - x^{n-1}y\phi' + yx^{n-1}\phi'$$

$$= \eta x^n\phi = n.f$$

Thus differential operator $x\frac{\partial}{\partial x} + y\frac{\partial}{\partial y}$ operating on a homogeneous function f of degree n amounts to multiplication of f by n.

Corollary 1: If f is a homogeneous function of degreen n, then

$$x^2\frac{\partial^2 f}{\partial x^2} + 2xy\frac{\partial^2 f}{\partial x\partial y} + y^2\frac{\partial^2 f}{\partial y^2} = n(n-1)f \tag{5}$$

Proof: Differentiating Euler's result (1) w.r.t. x and y respectively, we get

$$\frac{\partial f}{\partial x} + x\frac{\partial^2 f}{\partial x^2} + y\frac{\partial^2 f}{\partial y\partial x} = n\frac{\partial f}{\partial x} \tag{6}$$

$$x\frac{\partial^2 f}{\partial x\partial y} + \frac{\partial f}{\partial y} + y\frac{\partial^2 f}{\partial y^2} = n\frac{\partial f}{\partial y} \tag{7}$$

Multiplying (6) by x and (7) by y and adding, we have

$$x^2 f_{xx} + y^2 f_{yy} + 2xy f_{xy} = (n-1)(xf_x + yf_y)$$

$$= n(n-1)f$$

where we have used Euler's theorem (1) and assumed that $f_{xy} = f_{yx}$.

Euler's Theorem for Three Variables

Theorem: *If f is a homogeneous function of three independent variables x, y, z of order n, then*

$$x f_x + y f_y + z f_z = nf \tag{8}$$

Proof: Express f as

$$f = x^n\phi\left(\frac{y}{x}, \frac{z}{x}\right) = x^n\phi(u, v) \tag{9}$$

where $u = y/x$, $v = z/x$

Differentiating (9) partially w.r.t. x, y, z, respectively

$$f_x = \eta x^{n-1}\phi + x^n\left(\frac{\partial\phi}{\partial u}\frac{\partial u}{\partial x} + \frac{\partial\phi}{\partial v}\frac{\partial v}{\partial x}\right)$$

$$f_x = \eta x^{n-1}\phi + x^n\left[\frac{\partial\phi}{\partial u}\cdot\left(\frac{-y}{x^2}\right) + \frac{\partial\phi}{\partial v}\left(\frac{-z}{x^2}\right)\right] \tag{10}$$

$$f_y = x^n\left[\frac{\partial\phi}{\partial u}\cdot\frac{\partial u}{\partial y} + \frac{\partial\phi}{\partial v}\cdot\frac{\partial v}{\partial y}\right] = x^n\cdot\frac{\partial\phi}{\partial u}\cdot\frac{1}{x} + 0 \tag{11}$$

$$f_z = x^n\left[\frac{\partial\phi}{\partial u}\cdot\frac{\partial u}{\partial z} + \frac{\partial\phi}{\partial v}\cdot\frac{\partial v}{\partial z}\right] = x^n\left[0 + \frac{\partial\phi}{\partial v}\left(\frac{1}{x}\right)\right] \tag{12}$$

Multiplying (10), (11), (12) by x, y, z respectively and adding the resultant equations, we get

$$x f_x + y f_y + z f_z = \eta x^n\phi - x^{n-1}\left(y\frac{\partial\phi}{\partial u} + z\frac{\partial\phi}{\partial v}\right)$$

$$+ yx^{n-1}\frac{\partial\phi}{\partial u} + zx^{n-1}\frac{\partial\phi}{\partial v}$$

$$= \eta x^n\phi = nf.$$

WORKED OUT EXAMPLES

Example 1: Find the degree of the following homogeneous functions:

a. $x^2 - 2xy + y^2$ **d.** $x^{\frac{1}{3}} y^{-\frac{4}{3}} \tan^{-1}(y/x)$

b. $\log y - \log x$ **e.** $3x^2yz + 5xy^2z + 4x^4$

c. $(\sqrt{x^2+y^2})^3$ **f.** $[z^2/(x^4+y^4)]^{\frac{1}{3}}$

Ans:

a. 2

b. $\log y - \log x = \ln\left(\frac{y}{x}\right) = x^0 \ln\left(\frac{y}{x}\right)$ degree zero

c. $(\sqrt{x^2+y^2})^3 = x^3\sqrt{1+\left(\frac{y}{x}\right)^2}$ degree 3

d. $x^{\frac{1}{3}} y^{-\frac{4}{3}} \tan^{-1}(y/x) = x^{-1} \cdot x^{-\frac{4}{3}} y^{-\frac{4}{3}} \tan^{-1}\frac{y}{x} = x^{-1}\left(\frac{x}{y}\right)^{\frac{4}{3}} \tan^{-1}\frac{y}{x}$. degree: -1

e. degree 4

f. $\left[\frac{z^2}{x^4+y^4}\right]^{\frac{1}{3}} = \left[\frac{1}{z^2}\frac{z^4}{x^4+y^4}\right]^{\frac{1}{3}} = z^{-\frac{2}{3}}\left[\frac{1}{\left(\frac{x}{z}\right)^4+\left(\frac{y}{z}\right)^4}\right]$

degree $= -2/3$.

Example 2: Verify Euler's theorem for the following functions by computing both sides of Euler's Equation (1) directly:

i. $(ax+by)^{\frac{1}{3}}$ **ii.** $x^{+\frac{1}{3}} y^{-\frac{4}{3}} \tan^{-1}(y/x)$

Solution: **i.** $f = (ax+by)^{\frac{1}{3}}$ is homogeneous function of degree $\frac{1}{3}$

Differentiating f partially w.r.t. x and y, we get

$$f_x = \frac{\partial f}{\partial x} = \frac{1}{3}(ax+by)^{-\frac{2}{3}} \cdot a$$

$$f_y = \frac{\partial f}{\partial y} = \frac{1}{3}(ax+by)^{-\frac{2}{3}} \cdot b$$

Multiplying by x and y and adding, we get the L.H.S. of (1)

$$x\,f_x + y\,f_y = \frac{1}{3}(ax+by)^{-\frac{2}{3}}ax + \frac{1}{3}(ax+by)^{-\frac{2}{3}}\,by$$

$$= \frac{1}{3}(ax+by)^{-\frac{2}{3}}(ax+by)$$

$$= \frac{1}{3}(ax+by)^{\frac{1}{3}} = \frac{1}{3}f.$$

Since f is homogeneous function of degree $\frac{1}{3}$ the R.H.S. of (1) is $nf = \frac{1}{3}f$.

Thus

$$x\,f_x + y\,f_y = \text{L.H.S.} = \frac{1}{3}f = \text{R.H.S.}$$

ii. $f = x^{\frac{1}{3}} y^{-\frac{4}{3}} \tan^{-1}(y/x)$ is homogeneous function of degree -1

$$f_x = \frac{1}{3}x^{-\frac{2}{3}}y^{-\frac{4}{3}}\tan^{-1}\left(\frac{y}{x}\right) + x^{\frac{1}{3}}y^{-\frac{4}{3}} \cdot \frac{1}{1+\left(\frac{y}{x}\right)^2} \cdot \left(\frac{-y}{x^2}\right)$$

$$f_y = x^{\frac{1}{3}}\left(-\frac{4}{3}\right)y^{-\frac{7}{3}}\tan^{-1}\left(\frac{y}{x}\right) + x^{\frac{1}{3}}y^{-\frac{4}{3}} \cdot \frac{1}{1+\left(\frac{y}{x}\right)^2}\frac{1}{x}$$

so

$$xf_x + yf_y$$

$$= \frac{1}{3} \cdot x^{\frac{1}{3}}y^{-\frac{4}{3}}\tan^{-1}(y/x) + x^{\frac{4}{3}}y^{-\frac{4}{3}}\left(\frac{-y}{x^2+y^2}\right)$$

$$-\frac{4}{3}x^{\frac{1}{3}}y^{-\frac{4}{3}}\tan^{-1}(y/x) + x^{\frac{4}{3}}y^{-\frac{1}{3}} \cdot \frac{1}{x^2+y^2}$$

$$= -x^{\frac{1}{3}}y^{-\frac{4}{3}}\tan^{-1}(y/x) = -f.$$

Example 3: If $u = \log\frac{x^2+y^2}{x+y}$, prove that

$$x\,u_x + y\,u_y = 1$$

Solution: Let

$$f = e^u = \frac{x^2+y^2}{x+y} = \frac{x^2\left(1+\left(\frac{y}{x}\right)^2\right)}{x\left(1+\frac{y}{x}\right)} = x\phi\left(\frac{y}{x}\right)$$

f is a homogeneous function of degree 1.

Applying Euler's theorem for the function f, we get

$$xf_x + yf_y = n \cdot f = f.$$

Since $f = e^u$, $f_x = e^u \cdot u_x$, $f_y = e^u u_y$

$$\text{so} \quad x \cdot e^u u_x + ye^u u_y = f = e^u$$

since $e^u \neq 0$, $x\,u_x + y\,u_y = 1$.

Example 4: Show that $x\,u_x + y\,u_y + z\,u_z = -2\cot u$ when

$$u = \cos^{-1}\left(\frac{x^3+y^3+z^3}{ax+by+cz}\right)$$

Solution: Let

$$f = \cos u = \frac{x^3+y^3+z^3}{ax+by+cz}$$

Here f is a homogeneous function of degree 2 in the three variables x, y, z. By Euler's theorem

$x f_x + y f_y + z f_z = 2f$. Now differentiating f w.r.t. x, y, z respectively, we get

$$f_x = -\sin u \cdot u_x,\ f_y = -\sin u\, u_y,\ f_z = -\sin u\, u_z.$$

Substituting

$$x f_x + y f_y + z f_z = -\sin u(x\, u_x + y\, u_y + z\, u_z) = 2f$$
$$= 2\cos u$$

or

$$x\, u_x + y\, u_y + z\, u_z = \frac{-2\cos u}{\sin u} = -2\cot u$$

Example 5: Prove that $x\, u_x + y\, u_y = \frac{5}{2}\tan u$ if $u = \sin^{-1}\left(\frac{x^3+y^3}{\sqrt{x}+\sqrt{y}}\right)$

Solution: Let $f = \sin u = \frac{x^3+y^3}{\sqrt{x}+\sqrt{y}}$ then f is a homogeneous function of degree $\frac{5}{2}$ since

$$f = \frac{x^3}{\sqrt{x}}\left(\frac{1+(y/x)^3}{1+(y/x)^{\frac{1}{2}}}\right) = x^{\frac{5}{2}}\phi(y/x).$$

Applying Euler's theorem for f, we have

$$x f_x + y f_y = nf = \frac{5}{2}f.$$

Since $f = \sin u$, $f_x = \cos u \cdot u_x$, $f_y = \cos u\, u_y$ so that

$$x \cdot \cos u\, u_x + y \cos u\, u_y = \frac{5}{2}f = \frac{5}{2} \cdot \sin u$$

$$x\, u_x + y\, u_y = \frac{5}{2}\frac{\sin u}{\cos u} = \frac{5}{2}\tan u$$

Example 6: If $u = x^3y^2\sin^{-1}(y/x)$ show that

$$x\, u_x + y\, u_y = 5u$$

and $$x^2 u_{xx} + 2xy\, u_{xy} + y^2 u_{yy} = 20u$$

Solution: Rewriting $u = \frac{x^2}{x^2} \cdot x^3y^2 \sin^{-1}(y/x) = x^5\left(\frac{y}{x}\right)^2 \sin^{-1}\frac{y}{x}$

$u = x^5\phi\left(\frac{y}{x}\right)$ so u is a homogeneous function of degree 5.

Applying Euler's theorem to u, we get

$$x\, u_x + y\, u_y = nu = 5u$$

Differentiating the above equation partially w.r.t. x and y, we have

$$x\, u_{xx} + u_x + y\, u_{yx} = 5u_x$$
$$x\, u_{xy} + u_y + y\, u_{yy} = 5u_y.$$

Multiplying by x and y and adding, we get

$$(x^2u_{xx} + xu_x + xy\, u_{yx}) + (xy\, u_{xy} + yu_y + y^2u_{yy})$$
$$= 5(xu_x + yu_y)$$

or

$$x^2u_{xx} + 2xy\, u_{xy} + y^2u_{yy} = (5-1)(x\, u_x + y\, u_y)$$
$$= (5-1)\cdot 5u = 20u$$

Here we replaced $x\, u_x + y\, u_y$ by $5u$ (from Euler's theorem) and assumed $u_{yx} = u_{xy}$.

EXERCISE

1. Determine the degree of the following homogeneous functions:
   a. $\sqrt{x^2 - xy}$
   b. $\sin^{-1}\left(\frac{y}{x}\right)$
   c. $\frac{x^3-y^3}{x+y}$
   d. $\frac{ax+by+cz}{Ax^6+By^6+Cz^6}$
   e. $x^2(x^2-y^2)^{\frac{1}{3}}/(x^2+y^2)^{\frac{2}{3}}$
   f. $2x^3y^2 + 3x^2y^3 + 6xy^4 - 8y^5$

Ans. **a.** 1 **b.** 0 **c.** 2 **d.** 5 **e.** 4/3 **f.** 5

2. Verify Euler's theorem
   a. $\sqrt{x^2+y^2}$
   b. $\cos^{-1}\left(\frac{x}{y}\right)$
   c. $(ax+by)^{\frac{3}{2}}$
   d. $x^2(x^2-y^2)^3/(x^2+y^2)^3$
   e. $x^2y^2/(x+y)$
   f. $\cos^{-1}\left(\frac{x}{y}\right) + \cot^{-1}\left(\frac{y}{x}\right)$
   g. $(x^{\frac{1}{3}} + y^{\frac{1}{3}})/(x^{\frac{1}{4}} - y^{\frac{1}{4}})$
   h. $xy/(x+y)$
   i. $(x^2+xy+y^2)^{-1}$
   j. $\frac{1}{x^2} + \frac{1}{xy} + \frac{\log x - \log y}{x^2+y^2}$
   k. $\log\frac{x^4+y^4}{x+y}$

3. If $f(x, y) = \sec^{-1}\left(\frac{x^3+y^3}{x-y}\right)$ show that $x f_x + y f_y = 2 \cot f$.

4. If $u = \sin^{-1}\left(\frac{x}{y}\right) + \tan^{-1}\left(\frac{y}{x}\right)$ show that $x u_x + y u_y = 0$.

5. If $f = \ln\left(\frac{x^4+y^4}{x+y}\right)$ then $x f_x + y f_y = 3$.

6. If $u = \sin^{-1}\left(\frac{x^2+y^2}{x+y}\right)$ show that $x u_x + y u_y =$ $\tan u$ and $x^2 u_{xx} + 2xy\, u_{xy} + y^2 u_{yy} = \tan^3 u$.

7. If $u = \sin^{-1}\left(\frac{x^2 y^2}{x+y}\right)$ then $x u_x + y u_y =$ $3 \tan u$.

8. If $f = x\, g\left(\frac{y}{x}\right) + h\left(\frac{y}{x}\right)$ show that $x^2 f_{xy} +$ $2xy\, f_{yy} + y^2 f_{yy} = 0$.

9. If $u = y^2 e^{\frac{y}{x}} + x^2 \tan^{-1}\left(\frac{x}{y}\right)$ show that

$$x u_x + y u_y = 2u$$

and $$x^2 u_{xx} + 2xy\, u_{xy} + y^2 u_{yy} = 2u.$$

10. If $\tan u = \frac{x^3+y^3}{x-y}$ show that $x u_x + y u_y =$ $\sin 2u$ and $x^2 u_{xx} + 2xy\, u_{xy} + y^2 u_{yy} =$ $2 \cos 3u \cdot \sin u$.

11. If $\cos u = \frac{x^2+y^2}{\sqrt{x}+\sqrt{y}}$ then $x u_x + y u_y =$ $-\frac{3}{2} \cdot \cot u$.

12. If $\ln u = x^2 y^2/(x + y)$ then show that

$$x u_x + y u_y = 3u \ln u.$$

13. If $u = 3x^4 \cot^{-1}\left(\frac{y}{x}\right) + 16y^4 \cos^{-1}\left(\frac{x}{y}\right)$ then prove that $x u_{xx} + 2yx\, u_{xy} + y^2 u_{yy} = 12u$.

14. Show that $x^2 u_{xx} + 2xy\, u_{xy} + y^2 u_{yy} =$ $\frac{\tan u}{12}\left(\frac{13}{12} + \frac{\tan^2 u}{12}\right)$ when $(\sqrt{x} + \sqrt{y}) \sin^2 u =$ $x^{\frac{1}{3}} + y^{\frac{1}{3}}$.

15. Verify Euler's theorem for

$$f = \frac{z}{x+y} + \frac{y}{z+x} + \frac{x}{y+z}.$$

16. If $u = \cos^{-1}\left(\frac{x^5 - 2y^5 + 6z^5}{\sqrt{ax^3+by^3+cz^3}}\right)$ then show that $x u_x + y u_y + z u_z = -\frac{7}{2} \cot u$.

17. Prove that

$$\left(x\frac{\partial}{\partial x} + y\frac{\partial}{\partial y}\right)^3 f = x^3 f_{xxx} + 3x^2 y\, f_{xxy} + 3xy^2 f_{xyy} + y^3 f_{yyy}$$
$$= n(n-1)(n-2) f$$

if $f(x, y)$ is a homogeneous function of degree n.

18. Prove that $4x\, f_x + 4y\, f_y + \sin 2f = 0$ if $(\sqrt{x} + \sqrt{y}) \cot f - x - y = 0$.

19. Show that $x u_x + y u_y + z u_z = 0$ when $u = \frac{y}{z} + \frac{z}{x}$.

Hint: u is homogeneous function of degree zero.

3.8 JACOBIAN

Jacobian* is a functional determinant (whose elements are functions) which is very useful in transformation of variables from cartesian to polar, cylindrical and spherical coordinates in multiple integrals (Chapter 7). Let $u(x, y)$ and $v(x, y)$ be two given functions of two independent variables x and y.

The **Jacobian** of u, v with respect to x, y denoted by $J\left(\frac{u, v}{x, y}\right)$ or $\frac{\partial(u, v)}{\partial(x, y)}$ is a second order functional determinant defined as

$$J\left(\frac{u, v}{x, y}\right) = \frac{\partial(u, v)}{\partial(x, y)} = \begin{vmatrix} \frac{\partial u}{\partial x} & \frac{\partial u}{\partial y} \\ \frac{\partial v}{\partial x} & \frac{\partial v}{\partial y} \end{vmatrix}$$

Note: Obviously $J\left(\frac{u, v}{u, v}\right) = \frac{\partial(u, v)}{\partial(u, v)} = \begin{vmatrix} 1 & 0 \\ 0 & 1 \end{vmatrix} = 1$

Similarly the Jacobian of three functions u, v, w of three independent variables x, y, z is defined as

$$J\left(\frac{u, v, w}{x, y, z}\right) = \frac{\partial(u, v, w)}{\partial(x, y, z)} = \begin{vmatrix} \frac{\partial u}{\partial x} & \frac{\partial u}{\partial y} & \frac{\partial u}{\partial z} \\ \frac{\partial v}{\partial x} & \frac{\partial v}{\partial y} & \frac{\partial v}{\partial z} \\ \frac{\partial w}{\partial x} & \frac{\partial w}{\partial y} & \frac{\partial w}{\partial z} \end{vmatrix}$$

*Carl Gustav Jacob Jacobi (1804–1851), German mathematician.

In a similar way, Jacobian of n functions in n variables can be defined.

Two Important Properties of Jacobians

BookWork I. If $J = \dfrac{\partial(u, v)}{\partial(x, y)}$ and $J^* = \dfrac{\partial(x, y)}{\partial(u, v)}$ then $JJ^* = 1$

i.e., $J = \dfrac{\partial(u, v)}{\partial(x, y)} = \dfrac{1}{J^*} = \dfrac{1}{\frac{\partial(x,y)}{\partial(u,v)}}$

Proof: Let

$$u = f(x, y) \text{ and } v = g(x, y) \quad (1)$$

be two given functions in terms of x and y, which are transformations from u, v to x, y. Solving (1) for x and y, we get x and y as functions in terms of the two independent variables u and v as

$$x = \phi(u, v) \text{ and } y = \psi(u, v) \quad (2)$$

known as inverse transformation from x, y to u, v.

Differentiating partially w.r.t. u and v, we get

$$1 = \frac{\partial u}{\partial u} = \frac{\partial u}{\partial x}\frac{\partial x}{\partial u} + \frac{\partial u}{\partial y}\frac{\partial y}{\partial u} = u_x x_u + u_y y_u \quad (3)$$

$$0 = \frac{\partial u}{\partial v} = \frac{\partial u}{\partial x}\frac{\partial x}{\partial v} + \frac{\partial u}{\partial y}\frac{\partial y}{\partial v} = u_x x_v + u_y y_v \quad (4)$$

$$0 = \frac{\partial v}{\partial u} = \frac{\partial v}{\partial x}\frac{\partial x}{\partial u} + \frac{\partial v}{\partial y}\frac{\partial y}{\partial u} = v_x x_u + v_y y_u \quad (5)$$

$$1 = \frac{\partial v}{\partial v} = \frac{\partial v}{\partial x}\frac{\partial x}{\partial v} + \frac{\partial v}{\partial y}\frac{\partial y}{\partial v} = v_x x_v + v_y y_v \quad (6)$$

Consider

$$J \cdot J^* = \begin{vmatrix} u_x & u_y \\ v_x & v_y \end{vmatrix} \begin{vmatrix} x_u & x_v \\ y_u & y_v \end{vmatrix}$$

$$= \begin{vmatrix} u_x & u_y \\ v_x & v_y \end{vmatrix} \begin{vmatrix} x_u & y_u \\ x_v & y_v \end{vmatrix}$$ By interchanging rows and columns in the second determinant.

Multiplying the determinant row-wise,

$$J \cdot J^* = \begin{vmatrix} u_x x_u + u_y y_u & u_x x_v + u_y y_v \\ v_x x_u + v_y y_v & v_x x_v + v_y y_v \end{vmatrix}$$

Substituting (3), (4), (5), (6) above, we get

$$J \cdot J^* = \begin{vmatrix} 1 & 0 \\ 0 & 1 \end{vmatrix} = 1$$

Chain Rule for Jacobians

Book Work II. If u, v are functions of r, s and r, s are themselves functions of x, y then

$$\frac{\partial(u, v)}{\partial(x, y)} = \frac{\partial(u, v)}{\partial(r, s)}\frac{\partial(r, s)}{\partial(x, y)}$$

i.e., $$J\left(\frac{u, v}{x, y}\right) = J\left(\frac{u, v}{r, s}\right) J\left(\frac{r, s}{x, y}\right)$$

Proof: Differentiating u, v partially w.r.t. x, y

$$\frac{\partial u}{\partial x} = \frac{\partial u}{\partial r}\frac{\partial r}{\partial x} + \frac{\partial u}{\partial s}\frac{\partial s}{\partial x} \quad (1)$$

$$\frac{\partial u}{\partial y} = \frac{\partial u}{\partial r}\frac{\partial r}{\partial y} + \frac{\partial u}{\partial s}\frac{\partial s}{\partial y} \quad (2)$$

$$\frac{\partial v}{\partial x} = \frac{\partial v}{\partial r}\frac{\partial r}{\partial x} + \frac{\partial v}{\partial s}\frac{\partial s}{\partial x} \quad (3)$$

$$\frac{\partial v}{\partial y} = \frac{\partial v}{\partial r}\frac{\partial r}{\partial y} + \frac{\partial v}{\partial s}\frac{\partial s}{\partial y} \quad (4)$$

From definition of Jacobian

$$\frac{\partial(u, v)}{\partial(r, s)} \cdot \frac{\partial(r, s)}{\partial(x, y)} = \begin{vmatrix} u_r & u_s \\ v_r & v_s \end{vmatrix} \begin{vmatrix} r_x & r_y \\ s_x & s_y \end{vmatrix}$$

By interchanging the rows and columns in second determinant

$$= \begin{vmatrix} u_r & u_s \\ v_r & v_s \end{vmatrix} \begin{vmatrix} r_x & s_x \\ r_y & s_y \end{vmatrix}$$

Multiplying the determinant row-wise

$$= \begin{vmatrix} u_r r_x + u_s s_x & u_r r_y + u_s s_y \\ v_r r_x + v_s s_x & v_r r_y + v_s s_y \end{vmatrix}$$

Using (1), (2), (3), (4)

$$\frac{\partial(u, v)}{\partial(r, s)} \cdot \frac{\partial(r, s)}{\partial(x, y)} = \begin{vmatrix} u_x & u_y \\ v_x & v_y \end{vmatrix} = \frac{\partial(u, v)}{\partial(x, y)}$$

Thus Jacobians behave in a certain way like derivatives.

Note: All the above results of Jacobian of two variables can be extended similarly to n (any number of) variables.

Standard Jacobians

Jacobians for change of variables from cartesian coordinates to

i. polar coordinate $x = r\cos\theta$, $y = r\sin\theta$

ii. cylindrical coordinates $x = r\cos\theta$, $y = r\sin\theta, z = z$

iii. spherical coordinates $x = r\sin\theta\cos\phi$, $y = r\sin\theta\sin\phi$ and $z = r\cos\theta$

i. $x = r\cos\theta,\ y = r\sin\theta$, so $x_r = \cos\theta$, $x_\theta = -r\sin\theta,\ y_r = \sin\theta,\ y_\theta = r\cos\theta$

$$\frac{\partial(x,y)}{\partial(r,\theta)} = \begin{vmatrix} x_r & x_\theta \\ y_r & y_\theta \end{vmatrix} = \begin{vmatrix} \cos\theta & -r\sin\theta \\ \sin\theta & r\cos\theta \end{vmatrix}$$

$$= r(\cos^2\theta + \sin^2\theta) = r$$

Solving for r, θ we have

$$r = \sqrt{x^2+y^2},\ \theta = \tan^{-1}\left(\frac{y}{x}\right)$$

so $r_x = \dfrac{x}{r}, r_y = \dfrac{y}{r}, \theta_x = \dfrac{-y}{r^2}, \theta_y = \dfrac{x}{r^2}$

$$\frac{\partial(r,\theta)}{\partial(x,y)} = \begin{vmatrix} r_x & r_y \\ \theta_x & \theta_y \end{vmatrix} = \begin{vmatrix} \frac{x}{r} & \frac{y}{r} \\ \frac{-y}{r^2} & \frac{x}{r^2} \end{vmatrix} = \frac{x^2}{r^3} + \frac{y^2}{r^3} = \frac{r^2}{r^3} = \frac{1}{r}$$

ii. $x = r\cos\theta,\ y = r\sin\theta,\ z = z$ so $x_r = \cos\theta$, $x_\theta = -r\sin\theta, x_z = 0, y_r = \sin\theta, y_\theta = r\cos\theta$, $y_z = 0, z_r = 0, z_\theta = 0, z_z = 1$

$$J = \frac{\partial(x,y,z)}{\partial(r,\theta,z)} = \begin{vmatrix} x_r & x_\theta & x_z \\ y_r & y_\theta & y_z \\ z_r & z_\theta & z_z \end{vmatrix} = \begin{vmatrix} \cos\theta & -r\sin\theta & 0 \\ \sin\theta & r\cos\theta & 0 \\ 0 & 0 & 1 \end{vmatrix}$$

$$= r(\cos^2 + \sin^2\theta) = r$$

$$J^* = \frac{\partial(r,\theta,z)}{\partial(x,y,z)} = \frac{1}{J} = \frac{1}{r}$$

iii. $x = r\sin\theta\cos\phi,\ y = r\sin\theta\sin\phi$, $z = r\cos\theta$

$$\frac{\partial(x,y,z)}{\partial(r,\theta,\phi)}$$

$$= \begin{vmatrix} \sin\theta\cos\phi & r\cos\theta\cos\phi & -r\sin\theta\sin\phi \\ \sin\theta\sin\phi & r\cos\theta\sin\phi & r\sin\theta\cos\phi \\ \cos\theta & -r\sin\theta & 0 \end{vmatrix}$$

$$= r^2\sin\theta\cos^2\theta + r^2\sin\theta\sin^2\theta = r^2\sin\theta$$

Using property (ii)

$$\frac{\partial(r,\theta,\phi)}{\partial(x,y,z)} = \frac{1}{\frac{\partial(x,y,z)}{\partial(r,\theta,\phi)}} = \frac{1}{r^2\sin\theta}.$$

Worked Out Examples

Example 1: Find the Jacobian $\frac{\partial(u,v)}{\partial(x,y)}$ in each of the following:

i. $u = x\sin y,\ v = y\sin x$

ii. $u = e^x\sin y,\ v = x + \log\sin y$

Solution: By definition

$$\frac{\partial(u,v)}{\partial(x,y)} = \begin{vmatrix} \frac{\partial u}{\partial x} & \frac{\partial u}{\partial y} \\ \frac{\partial v}{\partial x} & \frac{\partial v}{\partial y} \end{vmatrix}$$

i. $u = x\ \sin y, v = y\ \sin x$ so $u_x = \sin y, u_y = x\cos y, v_x = y\cos x, v_y = \sin x$

$$\text{Jacobian} = \begin{vmatrix} \sin y & x\cos y \\ y\cos x & \sin x \end{vmatrix}$$

$$= \sin x\sin y - xy\cos x\cos y$$

ii. $u = e^x\sin y, v = x + \log\sin y$ so $u_x = e^x\sin y$, $u_y = e^x\cos y,\ v_x = 1,\ v_y = \frac{\cos y}{\sin y}$

$$\frac{\partial(u,v)}{\partial(x,y)} = \begin{vmatrix} e^x\sin y & e^x\cos y \\ 1 & \frac{\cos y}{\sin y} \end{vmatrix}$$

$$= e^x\cos y - e^x\cos y = 0.$$

Example 2: Calculate $J = \frac{\partial(u,v)}{\partial(x,y)}$ and $J^* = \frac{\partial(x,y)}{\partial(u,v)}$. verify that $J \cdot J^* = 1$ given

i. $u = x + \frac{y^2}{x}, v = \frac{y^2}{x}$

ii. $x = e^u\cos v,\ y = e^u\sin v$

i. $u = x + \frac{y^2}{x}, v = \frac{y^2}{x}$

$$J = \frac{\partial(u,v)}{\partial(x,y)} = \begin{vmatrix} 1 - \frac{y^2}{x^2} & \frac{2y}{x} \\ \frac{-y^2}{x^2} & \frac{2y}{x} \end{vmatrix}$$

$$= \frac{2y}{x} - \frac{2y^3}{x^3} + \frac{2y^3}{x^3} = \frac{2y}{x}$$

Solving for x, y in terms of u and v, we have $u = x + \frac{y^2}{x} = x + v \quad \therefore \quad x = u - v, v = \frac{y^2}{x}$,

$y^2 = vx,\ y^2 = v(u-v) \quad \therefore \quad y = \sqrt{v(u-v)}$

$$J^* = \frac{\partial(x,y)}{\partial(u,v)} = \begin{vmatrix} x_u & x_v \\ y_u & y_v \end{vmatrix} = \begin{vmatrix} 1 & -1 \\ \frac{1}{2}\frac{v}{\sqrt{v(u-v)}} & \frac{1}{2}\frac{(u-2v)}{\sqrt{v(u-v)}} \end{vmatrix}$$

$$J^* = \frac{1}{2}\frac{1}{\sqrt{v(u-v)}}[u-2v+v] = \frac{1}{2}\frac{u-v}{\sqrt{v(u-v)}} = \frac{1}{2}\frac{x}{y}$$

Verification: $J \cdot J^* = \frac{2y}{x} \cdot \frac{1}{2}\frac{x}{y} = 1$

ii. $x = e^u \cos v,\ y = e^u \sin v$

$$J = \frac{\partial(x,y)}{\partial(u,v)} = \begin{vmatrix} e^u \cos v & -e^u \sin v \\ e^u \sin v & e^u \cos v \end{vmatrix}$$

$$= e^{2u}(\cos^2 v + \sin^2 v) = e^{2u}$$

Solving for u, v in terms of x, y

$$\tan v = \frac{e^u \sin v}{e^u \cos v} = \frac{y}{x} \quad \therefore \quad v = \tan^{-1}\frac{y}{x}$$

$$x^2 + y^2 = e^{2u}(\cos^2 v + \sin^2 v) = e^{2u}$$

$$\therefore \quad u = \frac{1}{2}\ln(x^2+y^2).$$

$$J^* = \frac{\partial(u,v)}{\partial(x,y)} = \begin{vmatrix} \frac{x}{x^2+y^2} & \frac{y}{x^2+y^2} \\ \frac{-y}{x^2+y^2} & \frac{x}{x^2+y^2} \end{vmatrix}$$

$$J^* - \frac{x^2+y^2}{(x^2+y^2)^2} = \frac{1}{x^2+y^2} = \frac{1}{e^{2u}}$$

Verification: $J \cdot J^* = e^{2u}\frac{1}{e^{2u}} = 1.$

Example 3: Calculate $\frac{\partial(x,y,z)}{\partial(u,v,w)}$ if $u = \frac{2yz}{x},\ v = \frac{3zx}{y},\ w = \frac{4xy}{z}$.

Solution:

$$\frac{\partial(x,y,z)}{\partial(u,v,w)} = \begin{vmatrix} x_u & x_v & x_w \\ y_u & y_v & y_w \\ z_u & z_v & z_w \end{vmatrix}$$

$$= \begin{vmatrix} \frac{-2yz}{x^2} & \frac{2z}{x} & \frac{2y}{x} \\ \frac{3z}{y} & \frac{-3zx}{y^2} & \frac{3x}{y} \\ \frac{4y}{z} & \frac{4x}{z} & \frac{-4xy}{z^2} \end{vmatrix}$$

Expanding the determinant

$$= \frac{-2yz}{x^2}\left[\frac{12x^2yz}{y^2z^2} - \frac{12x^2}{yz}\right] - \frac{2z}{x}\left[\frac{-12xyz}{yz^2} - \frac{12xy}{yz}\right]$$

$$+ \frac{2y}{x}\left[\frac{12xz}{yz} + \frac{12xyz}{zy^2}\right]$$

$$= 0 + 48 + 48 = 96.$$

Example 4: Calculate $\frac{\partial(u,v)}{\partial(r,\theta)}$ if $u = 2axy$, $v = a(x^2 - y^2)$ where $x = r\cos\theta,\ y = r\sin\theta$.

Solution: Since u, v are functions of x, y which are themselves functions of r, θ, use chain rule for Jacobians. Thus

$$\frac{\partial(u,v)}{\partial(r,\theta)} = \frac{\partial(u,v)}{\partial(x,y)}\frac{\partial(x,y)}{\partial(r,\theta)}$$

Given $u = 2axy,\ v = a(x^2-y^2),\ u_x = 2ay,\ u_y = 2ax,\ v_x = 2ax,\ v_y = -2ay$

$$\frac{\partial(u,v)}{\partial(x,y)} = \begin{vmatrix} 2ay & 2ax \\ 2ax & -2ay \end{vmatrix}$$

$$= -4a^2(y^2+x^2) = -4a^2r^2$$

Since $x^2 + y^2 = r^2\cos^2\theta + r^2\sin^2\theta = r^2$

Also $\frac{\partial(x,y)}{\partial(r,\theta)} = \begin{vmatrix} \cos\theta & -r\sin\theta \\ \sin\theta & r\cos\theta \end{vmatrix} = r$

Hence $\frac{\partial(u,v)}{\partial(r,\theta)} = \frac{\partial(u,v)}{\partial(x,y)}\frac{\partial(x,y)}{\partial(r,\theta)} = (-4a^2r^2)r = -4a^2r^3.$

Example 5: If $x = \sqrt{vw},\ y = \sqrt{wu},\ z = \sqrt{uv}$ and $u = r\sin\theta\cdot\cos\phi,\ v = r\sin\theta\sin\phi,\ w = r\cos\theta$, calculate $\frac{\partial(x,y,z)}{\partial(r,\theta,\phi)}$.

Solution: Since x, y, z are functions of u, v, w which are in turn functions of r, θ, ϕ, so use chain rule for Jacobians. Thus

$$\frac{\partial(x,y,z)}{\partial(r,\theta,\phi)} = \frac{\partial(x,y,z)}{\partial(u,v,w)}\frac{\partial(u,v,w)}{\partial(r,\theta,\phi)}$$

Consider

$$\frac{\partial(x,y,z)}{\partial(u,v,w)} = \begin{vmatrix} x_u & x_v & x_w \\ y_u & y_v & y_w \\ z_u & z_v & z_w \end{vmatrix}$$

$$= \begin{vmatrix} 0 & \frac{1}{2}\sqrt{\frac{w}{v}} & \frac{1}{2}\sqrt{\frac{v}{w}} \\ \frac{1}{2}\sqrt{\frac{w}{u}} & 0 & \frac{1}{2}\sqrt{\frac{u}{w}} \\ \frac{1}{2}\sqrt{\frac{v}{u}} & \frac{1}{2}\sqrt{\frac{u}{v}} & 0 \end{vmatrix}$$

$$= \frac{1}{8}\left[\sqrt{\frac{w}{v}\frac{v}{u}\frac{u}{v}} + \sqrt{\frac{v}{w}\frac{w}{u}\frac{u}{v}}\right] = \frac{2}{8} = \frac{1}{4}$$

We know already that Jacobian for spherical coordinates is

$$\frac{\partial(u, v, w)}{\partial(r, \theta, \phi)} = r^2 \sin\theta$$

thus $\dfrac{\partial(x, y, z)}{\partial(r, \theta, \phi)} = \dfrac{\partial(x, y, z)}{\partial(u, v, w)}\dfrac{\partial(u, v, w)}{\partial(r, \theta, \phi)} = \dfrac{1}{4}r^2 \sin\theta$

Example 6: Calculate $\frac{\partial(u,v,w)}{\partial(x,y,z)}$ if

$$u = x/\sqrt{(1-r^2)},\ v = y/\sqrt{(1-r^2)},\ w = z/\sqrt{(1-r^2)}$$

where $r^2 = x^2 + y^2 + z^2$.

Solution: Given $r^2 = x^2+y^2+z^2$, $r = \sqrt{x^2+y^2+z^2}$, $r_x = \frac{1}{2}\frac{2x}{\sqrt{x^2+y^2+z^2}} = \frac{x}{r}$. Similarly $r_y = \frac{y}{r}$, $r_z = \frac{z}{r}$.

Differentiating $u = x/\sqrt{(1-r^2)}$ w.r.t. x, we get

$$u_x = \frac{1}{\sqrt{1-r^2}} + x\cdot\left(-\frac{1}{2}\right)\cdot\frac{-2r}{(1-r^2)^{\frac{3}{2}}}\cdot r_x$$

$$u_x = \frac{1}{\sqrt{1-r^2}} + \frac{rx}{(1-r^2)^{\frac{3}{2}}}\cdot\frac{x}{r} = \frac{(1-r^2)+x^2}{(1-r^2)^{\frac{3}{2}}}$$

Put $e = (1-r^2)^{3/2}$, so $u_x = \frac{1-r^2+x^2}{e}$

By symmetry

$$v_y = \frac{(1-r^2)+y^2}{e}, \quad w_z = \frac{(1-r^2)+z^2}{e}$$

Differentiating u w.r.t. y, we get

$$u_y = x\cdot\left(-\frac{1}{2}\right)\cdot\frac{1}{(1-r^2)^{\frac{3}{2}}}(-2r)\cdot r_y = \frac{xr}{(1-r^2)^{\frac{3}{2}}}\cdot\frac{y}{r}$$

$$u_y = \frac{xy}{(1-r^2)^{\frac{3}{2}}} = \frac{xy}{e}$$

In a similar way, we have

$$u_z = \frac{xz}{e}, \quad v_x = \frac{yx}{e}, \quad v_z = \frac{yz}{e}$$

$$w_x = \frac{xz}{e}, \quad w_y = \frac{yz}{e}$$

Thus

$$\frac{\partial(u, v, w)}{\partial(x, y, z)} = \begin{vmatrix} \frac{(1-r^2)+x^2}{e} & \frac{xy}{e} & \frac{xz}{e} \\ \frac{xy}{e} & \frac{(1-r^2)+y^2}{e} & \frac{yz}{e} \\ \frac{xz}{e} & \frac{yz}{e} & \frac{(1-r^2)+z^2}{e} \end{vmatrix}$$

$$= \frac{1}{(1-r^2)^{\frac{9}{2}}}\begin{vmatrix} 1-r^2+x^2 & xy & xz \\ xy & 1-r^2+y^2 & yz \\ xz & yz & 1-r^2+z^2 \end{vmatrix}$$

$$= (1-r^2)^{\frac{-9}{2}}[(1-r^2+x^2)\{(1-r^2+y^2)(1-r^2+z^2) - y^2z^2\} - (1-r^2)x^2y^2 - (1-r^2)x^2z^2]$$

$$= (1-r^2)^{\frac{-9}{2}}[(1-r^2+x^2)(1-r^2+y^2)(1-r^2+z^2) - (1-r^2)(y^2z^2+x^2y^2+x^2z^2) - x^2y^2z^2]$$

$$= (1-r^2)^{\frac{-9}{2}}[(1-r^2)^3 + (1-r^2)^2(x^2+y^2+z^2)]$$

$$= (1-r^2)^{\frac{-9}{2}}(1-r^2)^2 = (1-r^2)^{-\frac{5}{2}}.$$

Example 7: Verify the chain rule for Jacobians if $x = u$, $y = u\tan v$, $z = w$.

Solution:

$$J = \frac{\partial(x, y, z)}{\partial(u, v, w)} = \begin{vmatrix} 1 & 0 & 0 \\ \tan v & u\sec^2 v & 0 \\ 0 & 0 & 1 \end{vmatrix} = u\sec^2 v$$

Solving for u, v, w in terms of x, y, z we have $u=x$, $\tan v = \frac{y}{u} = \frac{y}{x}$, $v = \tan^{-1}\frac{y}{x}$, $w = z$

$$J^* = \frac{\partial(u, v, w)}{\partial(x, y, z)} = \begin{vmatrix} 1 & 0 & 0 \\ \frac{-y}{x^2+y^2} & \frac{x}{x^2+y^2} & 0 \\ 0 & 0 & 1 \end{vmatrix}$$

$$J^* = \frac{x}{x^2+y^2} = \frac{1}{x\left[1+\left(\frac{y}{x}\right)^2\right]} = \frac{1}{u\sec^2 v}$$

Thus $J\cdot J^* = u\sec^2 v\cdot\dfrac{1}{u\sec^2 v} = 1.$

EXERCISE

Find the Jacobian $\frac{\partial(u,v)}{\partial(x,y)}$ when

1. $u = 3x + 5y, v = 4x - 3y$

Ans. -29

2. $x + y = u, y = uv$

Ans. $(x + y)^{-1}$

Hint: Solving $u = x + y, v = y/(x + y)$

3. $u = (x + y)/(1 - xy), v = \tan^{-1} x + \tan^{-1} y$

Ans. 0

Verify the chain rule for Jacobians

i.e., $J \cdot J^* = \frac{\partial(u, v)}{\partial(x, y)} \frac{\partial(x, y)}{\partial(u, v)} = 1$ for the following

4. $x = u(1 - v), y = uv$

Ans. $J = u, J^* = (x + y)^{-1} = u^{-1}$

5. $u = x + y, v = xy$

Ans. $J = x - y$

6. Show that $\frac{\partial(u,v)}{\partial(r,\theta)}=6r^3 \sin 2\theta$ given $u=x^2-2y^2$, $v = 2x^2 - y^2$ and $x = r \cos\theta, y = r \sin\theta$.

Hint: use $\frac{\partial(u,v)}{\partial(r,\theta)} = \frac{\partial(u,v)}{\partial(x,y)} \frac{\partial(x,y)}{\partial(r,\theta)}$, Also $\frac{\partial(x,y)}{\partial(r,\theta)} = r$.

7. Calculate Jacobian of u, v, w w.r.t. x, y, z when $u = yz/x, v = zx/y, w = xy/z$

Ans. 4

8. Find $\partial(u, v)/\partial(r, \theta)$ if $u = 2xy, v = x^2 - y^2$ and $x = r \cos\theta, y = r \sin\theta$.

Ans. $4r^3$

9. $u = x^2 + y^2, v = y, x = r \cos\theta$, $y = r \sin\theta$.

Ans. $2xr$

10. If $X = u^2v, Y = uv^2$ and $u = x^2 - y^2$, $v = xy$ find $\frac{\partial(X,Y)}{\partial(x,y)}$.

Hint: Use $\frac{\partial(X,Y)}{\partial(x,y)} = \frac{\partial(X,Y)}{\partial(u,v)} \frac{\partial(u,v)}{\partial(x,y)}$ (chain rule).

Ans. $6x^2y^2(x^2 + y^2)(x^2 - y^2)^2$

11. Find $\frac{\partial(u,v,w)}{\partial(x,y,z)}$ if $u = x^2, v = \sin y, w = e^{-3z}$

Ans. $-6e^{-3z} x \cos y$

12. If $u = x + y + z, uv = y + z, uvw = z$ find $\frac{\partial(x,y,z)}{\partial(u,v,w)}$

Ans. u^2v

13. $u = xyz, v = xy + yz + zx, w = x + y + z$

Ans. $(x - y)(y - z)(z - x)$

14. $x = \frac{1}{2}(u^2 - v^2), y = uv, z = w$

Ans. $(u^2 + v^2)^{-1}$

15. $u = 3x + 2y - z, v = x - y + z$, $w = x + 2y - z$

Ans. -2.

3.9 FUNCTIONAL DEPENDENCE

Let $u = f(x, y), v = \phi(x, y)$ be two given differentiable functions of the two independent variables x and y. Suppose these functions u and v are connected by a relation $F(u, v) = 0$, where F is differentiable. Then these functions u and v are said to be functionally dependent on one another (i.e., one function say u is a function of the second function v) if the partial derivatives u_x, u_y, v_x and v_y are not all zero simultaneously.

Necessary and sufficient condition for functional dependence can be expressed in terms of a determinant as follows: Differentiating $F(u, v) = 0$ partially w.r.t. x and y, we get

$$\frac{\partial F}{\partial u}\frac{\partial u}{\partial x} + \frac{\partial F}{\partial v}\frac{\partial v}{\partial x} = 0$$

$$\frac{\partial F}{\partial u}\frac{\partial u}{\partial y} + \frac{\partial F}{\partial v}\frac{\partial v}{\partial y} = 0$$

A non-trivial solution $F_u \neq 0, F_v \neq 0$ to this system exists if the coefficient determinant is zero.

$$\begin{vmatrix} u_x & v_x \\ u_y & v_y \end{vmatrix} = \begin{vmatrix} u_x & u_y \\ v_x & v_y \end{vmatrix} = 0$$

Result: Two functions u and v are **functionally dependent** if their Jacobian

$$J\left(\frac{u, v}{x, y}\right) = \frac{\partial(u, v)}{\partial(x, y)} = \begin{vmatrix} u_x & u_y \\ v_x & v_y \end{vmatrix} = 0$$

If Jacobian is not equal to zero, then u and v are said to be **functionally independent**.

Extending this concept, three given functions $u(x, y, z)$, $v(x, y, z)$, $w(x, y, z)$ of three independent variables x, y, z, connected by the relation $F(u, v, w)$ are functionally dependent if first order derivatives of u, v, w w.r.t. x, y, z are **not** all zero simultaneously or if

$$J\left(\frac{u, v, w}{x, y, z}\right) = \frac{\partial(u, v, w)}{\partial(x, y, z)} = \begin{vmatrix} u_x & u_y & u_z \\ v_x & v_y & v_z \\ w_x & w_y & w_z \end{vmatrix} = 0$$

If Jacobian is non-zero, the functions u, v, w are said to be functionally independent.

Note: m functions of n variables are **always** functionally dependent when $m > n$.

Worked Out Examples

Determine which of the following functions are functionally dependent. Find a functional relation between them in case they are functionally dependent.

Example 1: $u = e^x \sin y,\ v = e^x \cos y$.

Solution:

$$\text{Jacobian} = \frac{\partial(u, v)}{\partial(x, y)} = \begin{vmatrix} u_x & u_y \\ v_x & v_y \end{vmatrix}$$

$$= \begin{vmatrix} e^x \sin y & e^x \cos y \\ e^x \cos y & -e^x \sin y \end{vmatrix}$$

$$= e^x(-\sin^2 y - \cos^2 y) = -e^x \neq 0$$

$\therefore\ u, v$ are functionally independent.

Example 2: $u = \frac{x}{y},\ v = \frac{x+y}{x-y}$.

Solution:

$$J = \frac{\partial(u, v)}{\partial(x, y)} = \begin{vmatrix} \frac{1}{y} & \frac{-x}{y^2} \\ \frac{-2y}{(x-y)^2} & \frac{2x}{(x-y)^2} \end{vmatrix}$$

$$= \frac{2xy}{(x-y)^2} - \frac{2xy}{(x-y)^2} = 0$$

$\therefore\ u, v$ are functionally dependent

$$v = \frac{x+y}{x-y} = \frac{y\left(\frac{x}{y}+1\right)}{y\left(\frac{x}{y}-1\right)} = \frac{u+1}{u-1}$$

$\therefore\ v = \frac{u+1}{u-1}$ is the functional relation between u and v.

Example 3: $u = x^2e^{-y} \cosh z,\ v = x^2e^{-y} \sinh z$, $w = 3x^4e^{-2y}$.

Solution:

$$J\left(\frac{u, v, w}{x, y, z}\right) = \frac{\partial(u, v, w)}{\partial(x, y, z)} = \begin{vmatrix} u_x & u_y & u_z \\ v_x & v_y & v_z \\ w_x & w_y & w_z \end{vmatrix}$$

$$= \begin{vmatrix} 2xe^{-y} \cosh z & -x^2e^{-y} \cosh z & x^2e^{-y} \sinh z \\ 2xe^{-y} \sinh z & -x^2e^{-y} \sinh z & x^2e^{-y} \cosh z \\ 12x^3e^{-2y} & -6x^4e^{-2y} & 0 \end{vmatrix}$$

$$= 12x^7e^{-4y}(\cosh^2 z - \sinh^2 z)$$

$$-12x^7e^{-4y}(\cosh^2 z - \sinh^2 z) = 0$$

$\therefore\ u, v, w$ are functionally dependent

$$3u^2 - 3v^2 = 3(x^4e^{-2y} \cosh^2 z - x^4e^{-2y} \sinh^2 z)$$

$$= 3x^4e^{-2y} = w.$$

Exercise

Determine whether the following functions are functionally dependent or not. If functionally dependent, find the functional relation between them:

1. $u = \frac{x^2-y^2}{x^2+y^2},\ v = \frac{2xy}{x^2+y^2}$

Ans. dependent, $u^2 + v^2 = 1$

2. $u = \sin x + \sin y;\ v = \sin(x + y)$

Ans. independent

3. $u = \frac{x+y}{x-y},\ v = \frac{xy}{(x-y)^2}$

Ans. dependent, $u^2 = 1 + 4v$

4. $u = \frac{x-y}{(x+a)},\ v = \frac{(x+a)}{(y+a)},\ a = \text{const}$

Ans. dependent, $v = \frac{1}{1-u}$

5. $u = x^2 + y^2 + 2xy + 2x + 2y$, $v = e^x \cdot e^y$

Ans. dependent, $u = (\log v)^2 + 2 \log v$

6. $u = x + y + z$, $v = x^2 + y^2 + z^2$, $w = x^3 + y^3 + z^3 - 3xyz$

Ans. dependent, $2w = u(3v - v^2)$

7. $u = xe^y \sin z$, $v = xe^y \cos z$, $w = x^2e^{2y}$

Ans. dependent, $u^2 + v^2 = w$

8. $u = x + y + z$, $v = x^3 + y^3 + z^3 - 3xyz$, $w = x^2 + y^2 + z^2 - xy - yz - zx$

Ans. dependent, $uw = v$

9. $u = 4x^2 + 9y^2 + 16z^2$, $v = 2x + 3y + 4z$, $w = 12xy + 16xz + 24yz$

Ans. dependent, $v^2 - u - w = 0$

10. $u = \frac{3x^2}{2(y+z)}$, $v = \frac{2(y+z)}{3(x-y)^2}$, $w = \frac{x-y}{x}$

Ans. dependent, $uvw^2 = 1$

11. $u = \frac{x}{y-z}$, $v = \frac{y}{z-x}$, $w = \frac{z}{x-y}$

Ans. dependent, $vw + wu + uv + 1 = 0$

12. $u = \sin^{-1} x + \sin^{-1} y$, $v = x\sqrt{1-y^2} + y\sqrt{1-x^2}$

Ans. dependent, $v = \sin u$.

3.10 ERRORS AND APPROXIMATIONS

If $z = f(x, y)$, then the total differential of z, denoted by dz, is given by

$$dz = \frac{\partial f}{\partial x} dx + \frac{\partial f}{\partial y} dy \qquad (1)$$

If x increases by an increment Δx and y increases by an increment Δy, then the total increment in z, denoted by Δz, is

$$\Delta z = f(x + \Delta x, y + \Delta y) - f(x, y)$$

or

$$f(x + \Delta x, y + \Delta y) = f(x, y) + \Delta z$$

But $\Delta z \approx dz$.

Replacing dz by (1), we have the approximate formula

$$f(x + \Delta x, y + \Delta y) \approx f(x, y) + \frac{\partial f(x, y)}{\partial x} dx + \frac{\partial f(x, y)}{\partial y} dy \qquad (2)$$

Thus the value of a function at a point can be obtained approximately if the value of the function and its derivatives at a neighbouring point are known.

Errors in measured data will result in error in the estimated value. For example, small error in the measurement of radius of a sphere will introduce a corresponding error in the volume of the sphere $V = \frac{4\pi}{3} r^3$. *Absolute error*, denoted by Δx, is the error in x. Error may be positive or negative. *Relative or proportional error* in x is $\frac{\Delta x}{x}$ or $\frac{dx}{x}$ since $\Delta x = dx$. *Percentage error* in x is given by $100\frac{dx}{x}$.

Example: If the error is 0.05 cm in measuring a dimension of length of 2 cm, then the absolute error is 0.05 cm, the relative error is $\frac{0.05}{2} = 0.025$ cm and the percentage error is $100 \times 0.025 = 2.5$.

For a function $z = f(x, y)$, the actual error Δz in z can be calculated approximately by using the differential dz, for given errors Δx, Δy in x and y respectively.

WORKED OUT EXAMPLES

Example 1: Using differentials, calculate approximately the value of $f(0.999)$ where $f(x) = 2x^4 + 7x^3 - 8x^2 + 3x + 1$.

Solution: Choose $x = 1$ and $\Delta x = -0.001$ so that $x + \Delta x = 1 + (-0.001) = 0.999$. Thus to calculate

$$f(0.999) = f(x + \Delta x) = f(x) + \Delta f$$
$$\approx f(x) + f'(x)\Delta x \approx f(1) + f'(1)(-0.001).$$

Here $f(1) = 2.1 + 7.1 - 8.1 + 3.1 + 1 = 5$ and

$$f'(x) = 8x^3 + 21x^2 + 16x + 3 \quad \text{so}$$
$$f'(1) = 8.1 + 21.1 + 16.1 + 3 = 16.$$

Thus the approximate value of

$$f(0.999) \approx 5 + 16(-0.001) = 5 - 0.016 = 4.984.$$

Example 2: Considering the volume of a spherical shell as an increment of volume of a sphere, calculate approximately the volume of a spherical shell whose inner diameter is 8 inches and whose thickness is $\frac{1}{16}$ inch (Fig. 3.2).

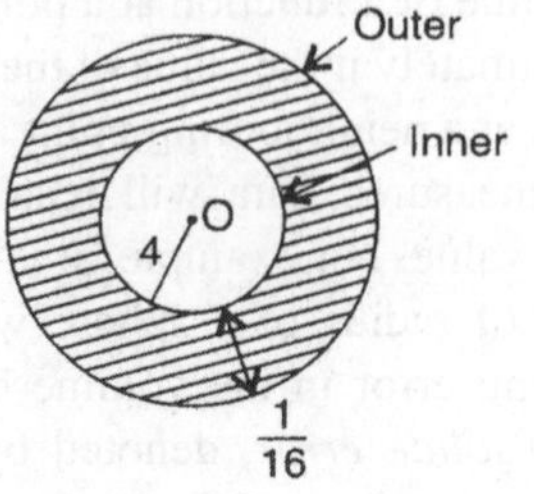

Fig. 3.2

Solution: Volume of a sphere of radius r is $V = \frac{4}{3}\pi r^3$. Now volume of the spherical shell ΔV is the difference between volume V_o of outer sphere and volume V_i of inner sphere. Thus volume of the spherical shell $= \Delta V = V_o - V_i$. Here radius of the outer sphere $r = 4$ in (diameter is 8 inches) choose $r = 4$ and $dr = \frac{1}{16}$. Then

$$V\left(4+\frac{1}{16}\right) - V(4) = V(r+\Delta r) - V(r) =$$

$$= \Delta V \approx dV = \frac{4}{3}\cdot 3\pi r^2 dr \text{ at } r = 4 \text{ and } dr = \frac{1}{16}$$

$$= 4\pi(4)^2\left(\frac{1}{16}\right) = 4\pi \text{ cubic inches.}$$

Example 3: The time T of a complete oscillation of a simple pendulum of length L is governed by the equation $T = 2\pi\sqrt{\frac{L}{g}}$ where g is a constant.

i. Find the approximate error in the calculated value of T corresponding to an error of 2% in the value of L.

ii. By what percentage should the length be changed in order to correct a loss of 2 minutes per day?

Solution:

i. Taking log: $\ln T = \ln 2\pi + \frac{1}{2}\ln\frac{L}{g}$.

Differentiating $\frac{dT}{T} = 0 + \frac{1}{2}\frac{g}{L}\cdot\frac{dL}{g} = \frac{1}{2}\frac{dL}{L}$

Error relation: $\frac{dT}{T} = \frac{1}{2}\frac{dL}{L}$

Error in value of L is 2% i.e., $\frac{dL}{L}\times 100 = 2$. Then $100\times\frac{dT}{T} = \frac{1}{2}\frac{dL}{L}\times 100 = \frac{1}{2}2 = 1$. The percentage error in calculate value of T is 1.

ii. Loss of error in value of T is 2 minutes per day i.e., $\frac{dT}{T} = (-2)\times\frac{1}{60}\times\frac{1}{24}$

From the error relation

$$\frac{dT}{T}\times 100 = -2\times\frac{1}{60}\times\frac{1}{24}\times 100 = \frac{1}{2}\frac{dL}{L}\times 100$$

$$\text{or}\quad 100\times\frac{dL}{L} = -\frac{1}{360}\times 100 = -0.2777 \approx -0.278\%$$

Example 4: The diameter and height of a right circular cylinder are measured to be 5 and 8 inches respectively. If each of these dimensions may be in error by ± 0.1 inch, find the relative percentage error in volume of the cylinder.

Solution: Let x be the diameter and y be the height of the cylinder then

$$V = \text{volume of cylinder} = \pi\left(\frac{x}{2}\right)^2 y$$

$$= \frac{1}{4}\pi x^2 y.\ \left(\because \text{radius} = \frac{x}{2}\right)$$

$$\text{Differential}: \quad dV = \frac{1}{4}\pi\cdot 2xy\cdot dx + \frac{1}{4}\pi x^2\,dy$$

$$100\times\frac{dV}{V} = 2\frac{dx}{x}\times 100 + \frac{dy}{y}\times 100$$

Given x = diameter=5 inches, height=y=8 inches, error $dx = dy = \pm 0.1$. So

$$100\times\frac{dV}{V} = \pm\left(2\cdot\frac{0.1}{5} + \frac{0.1}{8}\right) = \pm 0.0525$$

Thus the percentage error in volume is ± 0.0525.

EXERCISE

1. Using differential calculate approximately
   (a) $(2.98)^3$ (b) $\sqrt{4.05}$ (c) $\frac{1}{2.1}$ (d) $(83.7)^{\frac{1}{4}}$
   (e) $y(1.997)$ where $y(x)=x^4 - 2x^3 + 9x + 7$

   Hint: Choose $y = f(x)$, x and Δx as follows

   a. x^3, $x = 3$, $\Delta x = -0.02$
   b $y = \sqrt{x}$, $x = 4$, $\Delta x = 0.5$
   c. $y = \frac{1}{x}$, $x = 2$, $\Delta x = 0.1$
   d. $y = x^{\frac{1}{4}}$, $x = 81$, $\Delta x = 2.7$

Ans. (a) 26.46 (b) 2.13 (c) 0.475 (d) 3.025 (e) 24.949

2. If the radius of a sphere is measured as 5 inches with a possible error of 0.02 inches, find approximately the greatest possible error and percentage error in the computed value of the volume.

   Hint: $V = \frac{4}{3}\pi r^3$, $dV = 4\pi r^2 dr$, $r = 5$, $dr = 0.02$, $dV = \pm 2\pi$, $V = \frac{500\pi}{3}$.

Ans. ± 0.012, $\pm 1.2\%$

3. The quantity Q of water flowing over a V-notch is given by the formula $Q = cH^{\frac{5}{2}}$ where H is the head of water and c is a constant. Find the error in Q if the error in H is 1.5%.

   Hint: $\frac{dQ}{Q} = \frac{5}{2}\frac{dH}{H} = \frac{5}{2}(1.5)$

Ans. 3.75%

4. Calculate the percentage increase in the pressure p corresponding to a reduction of $\frac{1}{2}\%$ in the volume V, if the p and V ae related by $pV^{1.4} = c$ where c is a constant:

   Hint: $100 \times \frac{dp}{p} = -1.4\frac{dV}{V} \times 100 =$
   $= -(1.4) \times \left(-\frac{1}{2}\right)$

Ans. 0.7

5. Find the possible error in (a) surface area (b) volume of a sphere of radius r if r is measured as 18.5 inches with a possible error of 0.1 inch.

Ans. (a) 14.8π sq. in (b) 136.9π cubic inch

6. Show that the relative error in c due to a given error in θ is minimum when $\theta = 45°$ if $c = k\tan\theta$.

   Hint: $\frac{dc}{c} = \frac{2d\theta}{\sin 2\theta}$ minimum when $\sin 2\theta$ is greatest i.e., $2\theta = 90$.

7. Considering the area of a circular ring as an increment of area of a circle, find approximately the area of a ring whose inner and outer radii are 3 in and 3.02 in respectively.

   Hint: $A = \pi r^2$, Area of circular ring $= A(3.02) - A(3) = 2\pi r\, dr$ with $r = 3$, $dr = 0.02$

Ans. 0.12π

8. Find the percentage error in calculated value of volume of a right circular cone whose attitude is same as the base radius and is measured as 5 inches with a possible error of 0.02 inches.

   Hint: $V = \pi r^2 h = \pi r^3$ ($\because r = h$). $\frac{dV}{V} = 3\frac{dr}{r}$, $r = 5$, $dr = 0.02$

Ans. 1.2%

9. Calculate the error in R if $RI = E$ and possible errors in E and I are 20% and 10% respectively

   Hint: $\frac{dR}{R} \times 100 = \left(\frac{dE}{E} - \frac{dI}{I}\right) \times 100 = 20 - 10 = 10$

Ans. 10%

10. The diameter and the height of a right circular cylinder are measured as 4 cm and 6 cm respectively, with a possible error of 0.1 cm. Find approximately the maximum possible error in the computed value of the volume and surface area.

    Hint: $V = \pi r^2 h = \frac{\pi}{4}D^2 H$, $dV = \frac{\pi}{4}[2DHdD + D^2 dH]$
    $S = 2\pi rh = \pi DH$, $ds = \pi[HdD + DdH]$.

Ans. 1.6π cu. cm; π sq. cm

3.11 DIFFERENTIATION UNDER INTEGRAL SIGN: LEIBNITZ'S RULE

We know from the fundamental theorem on integral calculus that if $f(x)$ is a continuous function and $\phi(x) = \int_a^x f(t)dt$, then

$$\frac{d\phi}{dx} = \phi'(x) = \frac{d}{dx}\int_a^x f(t)dt = f(x) \qquad (1)$$

i.e., the derivative of a definite integral w.r.t. the upper limit is equal to the integrand in which the variable of integration t' is replaced by the upper limit 'x'

Example 1: (i) $\frac{d}{dx}\int_0^x e^{-t^2}dt = e^{-x^2}$

(ii) $\frac{d}{dx}\int_1^x \frac{1}{t}dt = \frac{1}{x}$, $x > 0$

Now if $F(x)$ is some antiderivative of a continuous function $f(x)$, then the Newton-Leibnitz formula $\int_a^b f(x)dx = F(b) - F(a)$ yields a practical and convenient method of computing definite integrals in cases where the anti-derivative of the integrand is known. This general method has extended the range of applications of definite integral to technology, mechanics, astronomy etc.

Differentiating integrals depend on a parameter:

Consider the definite integral

$$I(\alpha) = \int_a^b f(x, \alpha)dx \tag{2}$$

in which the integrand $f(x, \alpha)$ is dependent on a parameter α. The value of the definite integral (2) changes as the parameter α varies. It is difficult to integrate (2) in many cases.

Now to differentiate the integral (2) w.r.t. the parameter α, without having to first carry out integration and then differentiation, we use the Leibnitz's rule.

Book work 1. Leibnitz's formula (rule)
If $f(x, \alpha)$ and $\frac{\partial f}{\partial \alpha}$ are continuous functions when $c \le \alpha \le d$ and $a \le x \le b$, then

$$\boxed{\frac{d}{d\alpha}\int_a^b f(x,\alpha)dx = \int_a^b \frac{\partial f(x,\alpha)}{\partial \alpha}dx} \tag{3}$$

i.e., the order of differentiation and integration can be interchanged.

Proof: Consider

$$\frac{I(\alpha+\Delta\alpha)-I(\alpha)}{\Delta\alpha} = \frac{1}{\Delta\alpha}\left[\int_a^b f(x,\alpha+\Delta\alpha)dx - \int_a^b f(x,\alpha)dx\right]$$

$$= \int_a^b \frac{f(x,\alpha+\Delta\alpha)-f(x,\alpha)}{\Delta\alpha}dx$$

By Lagrange's mean value theorem

$$\frac{f(x,\alpha+\Delta\alpha)-f(x,\alpha)}{\Delta\alpha} = \frac{\partial f}{\partial\alpha}(x,\alpha+\theta\Delta\alpha)$$

$$= \frac{\partial f}{\partial\alpha}(x,\alpha)+\varepsilon$$

Here $0<\theta<1$ and ε which depends on x, α, $\Delta\alpha$ tends to zero as $\Delta\alpha \to 0$.

As $\Delta\alpha \to 0$

$$\lim_{\Delta\alpha\to 0}\frac{I(\alpha+\Delta\alpha)-I(\alpha)}{\Delta\alpha} = \lim_{\Delta\alpha\to 0}\left[\int_a^b \frac{\partial f}{\partial\alpha}(x,\alpha)dx + \int_a^b \varepsilon\, dx\right]$$

Thus

$$\frac{dI}{d\alpha} = \frac{d}{d\alpha}\int_a^b f(x,\alpha)dx = \int_a^b \frac{\partial f(x,\alpha)}{\partial\alpha}dx$$

General Leibnitz's rule

Book work 2. If $f(x,\alpha)$ and $f_\alpha(x,\alpha)$ are continuous and if the limits of integration a and b are functions of α then

$$\boxed{\frac{d}{d\alpha}\int_{a(\alpha)}^{b(\alpha)} f(a,\alpha)dx = \int_{a(\alpha)}^{b(\alpha)}\frac{\partial f(x,\alpha)}{\partial\alpha}dx + f[b(\alpha),\alpha]\frac{db}{d\alpha} - f[a(\alpha),\alpha]\frac{da}{d\alpha}} \tag{4}$$

Proof: Consider the definite integral

$$I(\alpha) = \int_{a(\alpha)}^{b(\alpha)} f(x,\alpha)dx = \phi(\alpha, a(\alpha), b(\alpha))$$

which depends on the parameter α through the integrand and intermediate arguments $a(\alpha)$, $b(\alpha)$, the limits of integration. From chain rule,

$$\frac{dI}{d\alpha} = \frac{\partial\phi}{\partial\alpha} + \frac{d\phi}{\partial a}\frac{\partial a}{d\alpha} + \frac{\partial\phi}{\partial b}\frac{db}{d\alpha} \tag{5}$$

Using (1), the fundamental theorem on integral calculus, we have

$$\frac{\partial\phi}{\partial b} = \frac{\partial}{\partial b}\int_a^b f(x,\alpha)dx = f(b(\alpha),\alpha) \tag{6}$$

$$\frac{\partial\phi}{\partial a} = \frac{\partial}{\partial a}\int_a^b f(x,\alpha)dx = -\frac{\partial}{\partial a}\int_b^a f(x,\alpha)dx$$

$$= -f(a(\alpha),\alpha) \tag{7}$$

Now we know from (3) (in the above book work.

$$\frac{\partial\phi}{\partial\alpha} = \frac{\partial}{\partial\alpha}\int_a^b f(x,\alpha)dx = \int_a^b \frac{\partial f(x,\alpha)}{\partial\alpha}dx \tag{8}$$

Here the limits of integration a and b are treated as constants since the dependence of ϕ on α here is through the integrand $f(x,\alpha)$ substituting (6), (7), (8) in (5), we get the result.

Note: Leibnitz's rule is applicable even when one of the limits of integration is infinite.

WORKED OUT EXAMPLES

Example 1: Apply Leibnitz's rule $\frac{d}{d\alpha}\int_{-2\alpha^2}^{-\alpha} e^{\alpha x^3}\,dx$.

Solution: Here $I(\alpha)=\int_{-2\alpha^2}^{-\alpha} e^{\alpha x^3}dx$

$=\int_{a(\alpha)}^{b(\alpha)} f(x,\alpha)dx$, so $b(\alpha)=-\alpha$, $a(\alpha)$ $=-2\alpha^2$. $f(x,\alpha)=e^{\alpha x^3}$. Applying Leibnitz's rule

$$\frac{dI}{d\alpha}=\frac{d}{d\alpha}\int_{-2\alpha^2}^{-\alpha} e^{\alpha x^3}dx=\int_{-2\alpha^2}^{-\alpha} x^3\cdot e^{\alpha x^3}dx+$$

$$+(-1)\cdot e^{\alpha\cdot(-\alpha)^3}-(-4\alpha)e^{\alpha(-2\alpha^2)^3}$$

$$=\int_{-2\alpha^2}^{-\alpha} x^3 e^{\alpha x^3}dx-e^{-\alpha^4}+4\alpha e^{-8\alpha^7}$$

Example 2: Using Leibnitz's rule, show that $\frac{d}{d\alpha}\int_0^{\alpha^2}\tan^{-1}\frac{x}{\alpha}dx=2\alpha\tan^{-1}\alpha-\frac{1}{2}\log(\alpha^2+1)$. Verify this result by direct integration followed by differentiation.

Solution: Here $b(\alpha)=\alpha^2$, $a(\alpha)=0$, $f(x,\alpha)=\tan^{-1}\frac{x}{\alpha}$. Using Leibnitz's rule

$$\frac{d}{d\alpha}\int_0^{\alpha^2}\tan^{-1}\frac{x}{\alpha}dx=\int_0^{\alpha^2}\frac{1}{1+\left(\frac{x}{\alpha}\right)^2}\cdot\left(\frac{-x}{\alpha^2}\right)dx$$

$$+2\alpha\cdot\tan^{-1}\frac{\alpha^2}{\alpha}-0$$

$$=-\int_0^{\alpha^2}\frac{1}{2}\frac{d(x^2+\alpha^2)}{(x^2+\alpha^2)}+2\alpha\tan^{-1}\alpha$$

$$=-\frac{1}{2}\ln(x^2+\alpha^2)\Big|_{x=0}^{\alpha^2}+2\alpha\tan^{-1}\alpha$$

$$=-\frac{1}{2}\ln(\alpha^4+\alpha^2)+\frac{1}{2}\ln\alpha^2+2\alpha\tan^{-1}\alpha$$

$$=2\alpha\tan^{-1}\alpha-\frac{1}{2}\ln(\alpha^2+1)$$

Direct verification:

Integrating

$$\int_0^{\alpha^2}\tan^{-1}\frac{x}{\alpha}dx=\alpha\left[\frac{x}{\alpha}\tan^{-1}\frac{x}{\alpha}-\log\sqrt{\frac{x^2}{\alpha^2}+1}\right]\Bigg|_{x=0}^{\alpha^2}$$

$$=\alpha^2\tan^{-1}\frac{\alpha^2}{\alpha}-\alpha\log\sqrt{\alpha^2+1}$$

Now differentiating w.r.t. 'α'

$$\frac{d}{d\alpha}\int_0^{\alpha^2}\tan^{-1}\frac{x}{\alpha}dx=2\alpha\cdot\tan^{-1}\alpha+\alpha^2\cdot\frac{1}{1+\alpha^2}$$

$$-\alpha\cdot\frac{1}{\sqrt{\alpha^2+1}}\cdot\frac{1}{2}\frac{1}{\sqrt{\alpha^2+1}}-\log\sqrt{\alpha^2+1}$$

$$=2\alpha\cdot\tan^{-1}\alpha+\frac{\alpha^2}{1+\alpha^2}-\frac{\alpha^2}{\alpha^2+1}-\log\sqrt{\alpha^2+1}$$

$$=2\alpha\tan^{-1}\alpha-\frac{1}{2}\ln(\alpha^2+1)$$

Example 3: Using Leibnitz rule evaluate the definite integral $\int_0^1 x^m(\log x)^n dx$.

Solution: Consider $I(m)=\int_0^1 x^m dx$. Using Leibnitz rule

$$\frac{dI}{dm}=\frac{d}{dm}\int_0^1 x^m dx=\int_0^1\frac{\partial}{\partial m}(x^m)dx$$

$$=\int_0^1 x^m\cdot\log x dx$$

Differentiating w.r.t. 'm' again

$$\frac{d^2I}{dm^2}=\frac{d}{dm}\int_0^1 x^m\log x\,dx=\int_0^1\frac{\partial}{\partial m}(x^m\log x)dx$$

$$=\int_0^1 x^m\cdot\log x\cdot\log x\,dx=\int_0^1 x^m(\log x)^2dx$$

Thus differentiating 'n' times w.r.t. m

$$\frac{d^nI}{dm^n}=\int_0^1 x^m(\log x)^n dx$$

From reduction formula we know that

$$\int_0^1 x^m(\log x)^n dx=\frac{x^{m+1}}{m+1}\cdot(\log x)^n\Big|_{x=0}^{n}$$

$$-\frac{n}{m+1}\cdot\int_0^1 x^m(\log x)^{n-1}dx$$

$$= -\frac{n}{(m+1)} \cdot \int_0^1 x^m (\log x)^{n-1} dx$$

Applying reduction formula again

$$= \frac{-n}{m+1}\left[\frac{x^{m+1}}{m+1} \cdot (\log x)^{n-1} - \frac{n-1}{m+1}\int_0^1 x^m (\log x)^{n-2} dx\right]$$

$$= (-1)^2 \frac{n(n-1)}{(m+1)^2}\left[\int_0^1 x^m (\log x)^{n-2} dx\right]$$

By repeated application

$$= \frac{(-1)^{(n-1)} \cdot n \cdot (n-1)(n-2) \cdots (n-(n-1))}{(m+1)^n} \times \left[\int_0^1 x^m dx\right]$$

$$= \frac{(-1)^{(n-1)} \cdot n!}{(m+1)^n} \cdot \left.\frac{x^{m+1}}{m+1}\right|_{x=0}^{1} = \frac{(-1)^n \cdot n!}{(m+1)^{n+1}}$$

Thus

$$\int_0^1 x^m (\log x)^n dx = \frac{(-1)^n n!}{(m+1)^{n+1}}$$

Example 4: Evaluate $\int_0^\infty x^n e^{-x} dx$

Solution: Consider $I(\alpha) = \int_0^\infty e^{-\alpha x} dx$. Differentiating w.r.t. 'α' by Leibnitz's rule we get

$$\frac{dI}{d\alpha} = \frac{d}{d\alpha}\int_0^\infty e^{-\alpha x} dx = \int_0^\infty \frac{\partial}{\partial x}(e^{-\alpha x}) dx$$

$$= -\int_0^\infty x e^{-\alpha x} dx$$

Differentiating successively $(n-1)$ times,

$$\frac{d^n I}{d\alpha^n} = (-1)^n \int_0^\infty x^n \cdot e^{-\alpha x} dx \quad (1)$$

But we know that $I(\alpha) = \int_0^\infty e^{-\alpha x} dx$

$$I(\alpha) = \left.\frac{e^{-\alpha x}}{-\alpha}\right|_0^\infty = \frac{1}{\alpha} \quad (2)$$

So differentiating (2) n times and equating with (1) we get

$$(-1)^n \int_0^\infty x^n e^{-\alpha x} dx = \frac{d^n I}{d\alpha^n} = \frac{d^n}{d\alpha^n}(\alpha^{-1})$$

$$= (-1)^n \cdot \frac{n!}{\alpha^{n+1}}$$

Thus $\int_0^\infty x^n e^{-\alpha x} dx = \frac{n!}{\alpha^{n+1}}$ for $n = 0, 1, 2, \ldots$
when $\alpha = 1$, $\int_0^\infty x^n e^{-x} dx = n!$; for $n = 0, 1, 2, \ldots$

Example 5: If $\int_0^\infty \frac{x^\alpha}{1+x^3} dx = \frac{\pi}{3}\operatorname{cosec}\left(\frac{\alpha+1}{3}\pi\right)$, evaluate $\int_0^\infty \frac{(\ln x)^2}{(1+x^3)} dx$. Hence or otherwise evaluate $\int_0^\infty \frac{x \ln x}{(1+x^3)} dx$.

Solution: Consider $I(\alpha) = \int_0^\infty \frac{x^\alpha}{1+x^3} dx$. Differentiating w.r.t. 'α' by Leibnitz rule

$$\frac{dI}{d\alpha} = \frac{d}{d\alpha}\int_0^\infty \frac{x^\alpha}{1+x^3} dx = \int_0^\infty \frac{\partial}{\partial x}\left(\frac{x^\alpha}{1+x^3}\right) dx$$

$$\frac{dI}{d\alpha} = \int_0^\infty \frac{x^\alpha}{(1+x^3)}(\log x) dx \quad (1) \quad \left(\because \frac{dx^\alpha}{d\alpha} = x^\alpha \log x\right)$$

Differentiating once more with respect to 'α'

$$\frac{d^2 I}{d\alpha^2} = \frac{d}{d\alpha}\int_0^\infty \frac{x^\alpha (\log x)}{(1+x^3)} dx = \int_0^\infty \frac{\partial}{\partial\alpha}\left(\frac{x^\alpha \log x}{1+x^3}\right) dx$$

$$\frac{d^2 I}{d\alpha^2} = \int_0^\infty \frac{x^\alpha \cdot \log x}{1+x^3} \cdot \log x\, dx = \int_0^\infty \frac{x^\alpha (\log x)^2}{1+x^3} dx \quad (2)$$

since

$$I(\alpha) = \int_0^\infty \frac{x^\alpha}{1+x^3} dx = \frac{\pi}{3}\operatorname{cosec}\left(\frac{\alpha+1}{3}\pi\right) \quad (3)$$

differentiating the RHS of (3) twice wrt α we have

$$\frac{dI}{d\alpha} = \frac{\pi}{3}[-\operatorname{cosec} y\pi \cdot \cot y\pi] \cdot \frac{\pi}{3} \quad (4)$$

where $y = \left(\frac{\alpha+1}{3}\right)\pi$

and

$$\frac{d^2 I}{d\alpha^2} = -\frac{\pi^2}{9}[\cot y(-\operatorname{cosec} y) \cdot \cot y + (-\operatorname{cosec}^2 y) \cdot \operatorname{cosec} y]\frac{\pi}{3}$$

$$\frac{d^2 I}{d\alpha^2} = \frac{\pi^3}{27}[\cot^2 y + \operatorname{cosec}^2 y] \cdot \operatorname{cosec} y \quad (5)$$

Equating (2) and (5)

$$\int_0^\infty \frac{x^\alpha (\log x)^2}{1+x^3} dx = \frac{d^2 I}{d\alpha^2} =$$

$$\frac{\pi^3}{27}[\cot^2 y + \text{cosec}^2 y]\text{cosec } y \qquad (6)$$

Put $\alpha = 0$ in (6) then $y = \frac{\pi}{3}$, so

$$\int_0^\infty \frac{(\log x)^2}{1+x^3}dx = \frac{\pi^3}{27}\left[\cot^2\frac{\pi}{3} + \text{cosec}^2\frac{\pi}{3}\right]\cdot \text{cosec}\frac{\pi}{3}$$

$$= \frac{\pi^3}{27}\left[\frac{1}{3}+\frac{4}{3}\right]\frac{2}{\sqrt{3}} = \frac{10\pi^3}{8\sqrt{3}}$$

Put $\alpha = 1$ in (1) and (4)

$$\int_0^\infty \frac{x\log x}{1+x^3}dx = \left.\frac{dI}{d\alpha}\right|_{\alpha=1} = \frac{-\pi^2}{9}\text{cosec}\frac{2\pi}{3}\cdot\cot\frac{2\pi}{3}$$

$$= \frac{-\pi^2}{9}\cdot\frac{2}{\sqrt{3}}\left(-\frac{1}{\sqrt{3}}\right) = \frac{2\pi^2}{27}$$

Example 6: Show that $y(x)$ satisfies the initial value problem (IVP) where

(a) $y(x) = \frac{1}{6}\int_a^x (x-t)^3 f(t)dt$,
IVP : $y''''(x) = f(x)$, $y(a) = y'(a) = y''(a) = y'''(a) = 0$

(b) $y(x) = e^x + \int_0^x t^2\cosh(x-t)dt$,
IVP: $y'' - y = 2x$; $y(0) = y'(0) = 1$

Solution: (a) Differentiating w.r.t. (the parameter) 'x' four times, we get

$$\frac{dy}{dx} = \frac{1}{6}\frac{d}{dx}\int_a^x (x-t)^3 f(t)dt$$

$$= \frac{1}{6}\int_a^x \frac{\partial}{\partial x}\left[(x-t)^3 f(t)\right]dt$$

$$\frac{dy}{dx} = \frac{1}{6}\left[\int_a^x 3(x-t)^2 f(t)dt + 1\cdot(x-x)^3 f(x) - 0\right]$$

$$\frac{d^2y}{dx^2} = \frac{1}{6}\left[\int_a^x 3\cdot 2\cdot(x-t)f(t)dt + 3(x-x)^2 f(x) - 0\right]$$

$$\frac{d^3y}{dx^3} = \frac{d}{dx}\int_a^x (x-t)f(t)dt$$

$$\frac{d^3y}{dx^3} = \int_a^x 1\cdot f(t)dt + 1\cdot(x-x)f(x) - 0$$

$$\frac{d^4y}{dx^3} = y'''' = \frac{d}{dx}\int_a^x f(t)dt = f(x)$$

Thus y satisfies the D.E. $y'''' = f(x)$.
At $x = a$, from the above results, $y(a) = y'(a) = y''(a) = y'''(a) = 0$

(b) Differentiating w.r.t. x twice

$$\frac{dy}{dx} = e^x + \int_0^x t^2\cdot\sinh(x-t)dt$$

$$+ 1\cdot x^2\cosh(x-x) - 0$$

$$\frac{d^2y}{dx^2} = e^x + \int_0^x t^2\cosh(x-t)dt$$

$$+ 1\cdot x^2\cdot\sinh(x-x) - 0 + 2x = y(x) + 2x$$

$\therefore y'' - y = 2x$, the DE is satisfied.
Note that at $x = 0$, $y(0) = y'(0) = 1$ from the above.

Exercise

Evaluate the following integrals $I(\alpha)$ using Leibnitz's rule

1. $\int_0^\infty x^2 e^{-x^2}dx$

Ans. $\frac{\sqrt{\pi}}{4}$

Hint: Differentiate w.r.t. 'α' the well-known result

$$\int_0^\infty e^{-\alpha x^2}dx = \frac{1}{2}\sqrt{\frac{\pi}{\alpha}}$$

2. $I(\alpha) - \int_0^\infty e^{-x}\frac{\sin\alpha x}{x}dx$

Ans. $\tan^{-1}\alpha$

Hint: $I'(\alpha) = \int_0^\infty e^{-x}\cos\alpha x\,dx = \frac{1}{1+\alpha^2}$, integrating $I(\alpha) = \tan^{-1}\alpha + c$, use $I(0) = 0$

3. $I(\alpha) = \int_0^1 \frac{x^\alpha - 1}{\log x}dx$, $\alpha \ge 0$

Ans. $\log(1+\alpha)$

Hint: $I'(\alpha) = \int_0^1 x^\alpha dx = \frac{1}{1+\alpha}$, integrate and use $I(0) = 0$

4. $I(\alpha) = \int_0^\alpha \frac{\log(1+\alpha x)}{1+x^2}dx$ and deduce $\int_0^1 \frac{\log(1+x)}{1+x^2}dx$

Ans. $\frac{1}{2}\log(1+\alpha^2)\tan^{-1}\alpha$; with $\alpha = 1$, $\frac{\pi}{8}\log_e 2$

Hint: $I'(\alpha) = \int_0^\alpha \frac{x}{(1+\alpha x)(1+x^2)}dx + \frac{\log(1+\alpha^2)}{1+\alpha^2}$, use partial fraction $\frac{x}{(1+\alpha x)(1+x^2)} = -\frac{\alpha}{1+\alpha^2}\frac{1}{1+\alpha x} + \frac{1}{2(1+\alpha^2)}\cdot\frac{2x}{1+x^2} + \frac{\alpha}{1+\alpha^2}\frac{1}{1+x^2}$.

Integrating (R.H.S.), $I'(\alpha) = \frac{\log(1+\alpha^2)}{2(1+\alpha^2)} + \frac{\alpha \cdot \tan^{-1}\alpha}{1+\alpha^2}$.

Integrating both sides and use $I(0) = 0$.

5. If $I(\alpha) = \int_{-\alpha}^{\alpha^2} \cos(\alpha x^2)dx$. Find $I'(\alpha)$.

Ans. $I'(\alpha) = -\int_{-\alpha}^{\alpha^2} x^2 \sin(\alpha x^2)dx + 2\alpha\cos(\alpha^5) + \cos(\alpha^3)$

6. $\int_0^\infty \frac{\tan^{-1}\alpha x}{x(1+x^2)}dx$ where $\alpha \geq 0$.

Ans. $\frac{\pi}{2}\log(1+\alpha)$

Hint: $I'(\alpha) = \int_0^\infty \frac{1}{(1+x^2)(1+\alpha^2x^2)}dx$

$= \frac{1}{1-\alpha^2}\int_0^\infty \left(\frac{1}{1+x^2} - \frac{\alpha^2}{1+\alpha^2x^2}\right)dx$

$= \frac{\pi}{2(1+\alpha)}$, Integrate and use $I(0) = 0$.

7. $\int_0^\infty e^{-x^2}\cos\alpha x\, dx$

Ans. $\frac{1}{2}\sqrt{\pi}e^{-\frac{1}{4}\alpha^2}$

Hint: $I'(\alpha) = -\frac{\alpha}{2}$, Integrating $I(\alpha) = ce^{-\frac{1}{4}\alpha^2}$, use $I(0) = \frac{\sqrt{\pi}}{2}$.

8. $\int_0^\pi \frac{\log(1+\sin\alpha\cdot\cos x)}{\cos x)}dx$

Ans. $\pi\alpha$

9. $\int_0^\infty xe^{-x^2}\sin\alpha x\, dx$

Ans. $\frac{\sqrt{\pi}}{4}\cdot\alpha e^{-\alpha^2/4}$

10. $\int_0^1 \frac{x^3-1}{\ln x}dx$

Ans. ln 4

Hint: $I(\alpha) = \int_0^1 \frac{x^\alpha - 1}{\ln x}dx$, $I'(\alpha) = \frac{1}{\alpha+1}$, Integrate and use $I(0) = 0$

11. Show that $y(x) = \frac{1}{x}\int_a^x (x-t)f(t)dt$ satisfies the initial value problem $(xy)'' = f(x)$, $y(a) = y'(a) = 0$.

12. Show that $x(t) = \frac{1}{m\omega}\int_0^t \sin\omega(t-\tau)f(\tau)d\tau$ satisfies forced harmonic oscillator: $mx'' + kx = f(t)$, $x(0) = x'(0) = 0$; $\omega = \sqrt{\frac{k}{m}}$

Chapter 4

Maxima and Minima

INTRODUCTION

To optimize something means to maximize or minimize some aspects of it. An important application of multivariable differential calculus is finding the maximum and minimum values of functions of several variables and determining where they occur. In the study of stability of the equilibrium states of mechanical and physical systems, determination of extrema is of greatest importance. Lagrange multipliers method developed by Lagrange in 1755 is a powerful method for finding extreme values of constrained functions in economics, in designing multistage rockets in engineering, in geometry etc.

4.1 TAYLOR'S THEOREM FOR FUNCTION OF TWO VARIABLES

Functions of two or more variables often can be expanded in power series which generalize the familiar one-dimensional expansion. Let $f(x, y)$ be a function of two independent variables x and y. Let $P(x, y)$ and $Q(x+h, y+k)$ be two neighbouring points. Then, $f(x+h, y+k)$, the value of f at Q can be expressed in terms of f and its derivatives at P.

Keeping y temporarily constant, $f(x+h, y+k)$ is treated as a function of the single variable x and expanded as follows using Taylor's theorem.

$$f(x+h, y+k) = f(x, y+k) + \frac{h\,\partial f(x, y+k)}{\partial x} + \frac{h^2}{2!}\frac{\partial^2 f(x, y+k)}{\partial x^2} + \frac{h^3}{3!}\frac{\partial^3 f(x, y+k)}{\partial x^3} + \cdots \quad (1)$$

Now the first term on the R.H.S. of (1) is expanded as function of y, treating x temporarily constant

$$f(x, y+k) = f(x, y) + \frac{k\,\partial f(x, y)}{\partial y} + \frac{k^2}{2!}\frac{\partial^2 f(x, y)}{\partial y^2} + \frac{k^3}{3!}\frac{\partial^3 f(x, y)}{\partial y^3} + \cdots \quad (2)$$

To get the 2nd, 3rd, 4th terms on the R.H.S. of (1), differentiate (2) partially w.r.t. x, once, twice, thrice, etc., yielding

$$\frac{\partial f(x, y+k)}{\partial x} = \frac{\partial f(x, y)}{\partial x} + \frac{k\partial^2 f(x, y)}{\partial x \partial y} + \frac{k^2}{2!}\frac{\partial^3 f(x, y)}{\partial x \partial y^2} + \frac{k^3}{3!}\frac{\partial^4 f(x, y)}{\partial x \partial y^3} + \cdots \quad (3)$$

$$\frac{\partial^2 f(x, y+k)}{\partial x^2} = \frac{\partial^2 f(x, y)}{\partial x^2} + \frac{k\partial^3 f(x, y)}{\partial x^2 \partial y} + \frac{k^2}{2!}\frac{\partial^4 f(x, y)}{\partial x^2 \partial y^2} + \frac{k^3}{3!}\frac{\partial^5 f(x, y)}{\partial x^2 \partial y^3} + \cdots \quad (4)$$

$$\frac{\partial^3 f(x, y+k)}{\partial x^3} = \frac{\partial^3 f(x, y)}{\partial x^3} + \frac{k\partial^4 f(x, y)}{\partial x^3 \partial y} + \frac{k^2}{2!}\frac{\partial^5 f(x, y)}{\partial x^3 \partial y^2} + \frac{k^3}{3!}\frac{\partial^6 f(x, y)}{\partial x^3 \partial y^3} + \cdots \quad (5)$$

Using (2), (3), (4), (5), Equation (1) becomes after rearrangement,

$$f(x+h, y+k) = f(x, y) + \left(\frac{h\,\partial f(x, y)}{\partial x} + \frac{k\,\partial f(x, y)}{\partial y}\right)$$
$$+\frac{1}{2!}\left[\frac{h^2\partial^2 f(x, y)}{\partial x^2} + \frac{2hk\,\partial^2 f(x, y)}{\partial x\partial y}\right.$$
$$\left.+\frac{k^2\,\partial^2 f(x, y)}{\partial y^2}\right] + \frac{1}{3!}\left[\frac{h^3\,\partial^3 f(x, y)}{\partial x^3}\right.$$
$$+\frac{3h^2k\,\partial^3 f(x, y)}{\partial x^2\partial y} + \frac{3hk^2\,\partial^3 f(x, y)}{\partial x\partial y^2}$$
$$\left.+\frac{k^3\,\partial^3 f(x, y)}{\partial y^3}\right] + \text{higher order terms.} \quad (6)$$

In the suffix notation,

$$f(x+h, y+k) = f(x, y) + [hf_x(x, y) + kf_y(x, y)]$$
$$+\frac{1}{2!}\left[h^2 f_{xx}(x, y) + 2hkf_{xy}(x, y)\right.$$
$$\left.+ k^2 f_{yy}(x, y)\right] + \frac{1}{3!}\left[h^3 f_{xxx}(x, y)\right.$$
$$+3h^2kf_{xxy}(x, y) + 3hk^2 f_{xyy}(x, y)$$
$$\left.+ k^3 f_{yyy}(x, y)\right] + \cdots \quad (7)$$

Equation (7) represents $f(x+h, y+k)$ in a power series of ascending powers of h and k.

For any specific point (a, b) replace (x, y) by (a, b) in (7) resulting in

$$f(a+h, b+k) = f(a, b) + [hf_x(a, b) + kf_y(a, b)]$$
$$+\frac{1}{2!}\left[h^2 f_{xx}(a, b) + 2hkf_{xy}(a, b)\right.$$
$$\left.+ k^2 f_{yy}(a, b)\right] + \frac{1}{3!}[h^3 f_{xxx}(a, b)$$
$$+3h^2kf_{xxy}(a, b) + 3hk^2 f_{xyy}(a, b)$$
$$+k^3 f_{yyy}(a, b)] + \cdots \quad (8)$$

Alternative Form

Equation (8) can be expressed in ascending powers of $(x-a)$ and $(y-b)$ by replacing h by $x-a$ and k by $y-b$ which gives

$$f(x, y) = f(a, b) + [(x-a)f_x(a, b)$$
$$+(y-b)\cdot f_y(a, b)] + \frac{1}{2!}\left[(x-a)^2 f_{xx}(a, b)\right.$$
$$\left.+ 2(x-a)(y-b)f_{xy}(a, b) + (y-b)^2 f_{yy}(a, b)\right]$$
$$+\frac{1}{3!}\left[(x-a)^3 f_{xxx}(a, b) + 3(x-a)^2(y-b)\,\times\right.$$
$$\times f_{xxy}(a, b) + 3(x-a)(y-b)^2 f_{xyy}(a, b)$$
$$\left.+ (y-b)^3 f_{yyy}(a, b)\right] + \cdots \quad (9)$$

Equation (9) which expands $f(x, y)$ in infinite power series in powers (terms) of $(x-a)$ and $(y-b)$ is known as ***Taylor's series*** or ***Taylor's expansion*** or ***Taylor's series expansion*** of $f(x, y)$ about the point (a, b).

Introducing the operator notation,

$$\Delta = h\frac{\partial}{\partial x} + k\frac{\partial}{\partial y}$$

so that

$$\Delta^2 = \left(h\frac{\partial}{\partial x} + k\frac{\partial}{\partial y}\right)^2 = h^2\frac{\partial^2}{\partial x^2} + 2hk\frac{\partial^2}{\partial x\partial y} + k^2\frac{\partial^2}{\partial y^2}$$
$$\Delta^3 = \left(h\frac{\partial}{\partial x} + k\frac{\partial}{\partial y}\right)^3 = h^3\frac{\partial^3}{\partial x^3} + 3h^2k\frac{\partial^3}{\partial x^2\partial y} +$$
$$+ 3hk^2\frac{\partial^3}{\partial x\partial y^2} + k^3\frac{\partial^3}{\partial y^3}$$

Equation (6) can be rewritten now as

$$f(x+h, y+k) = f(x, y) + \Delta f(x, y) + \frac{1}{2!}\Delta^2 f(x, y)$$
$$+\frac{1}{3!}\Delta^3 f(x, y) + \cdots$$

$$\text{Alternatively } f(x, y) = \sum_{n=0}^{\infty}\frac{1}{n!}\left(h\frac{\partial}{\partial x} + k\frac{\partial}{\partial y}\right)^n f(a, b)$$
$$= \sum_{n=0}^{\infty}\frac{\Delta^n}{n!}f(a, b).$$

Maclaurin's series expansion is a special case of Taylor's series when the expansion is about the origin (0, 0). Thus putting $a = 0, b = 0$ in (9), we get

$$f(x, y) = f(0, 0) + [x\cdot f_x(0, 0) + yf_y(0, 0)]$$
$$+\frac{1}{2!}\left[x^2 f_{xx}(0, 0) + 2xy\, f_{xy}(0, 0)\right.$$
$$\left.+ y^2 f_{yy}(0, 0)\right] + \frac{1}{3!}\left[x^3 f_{xxx}(0, 0)\right.$$
$$+3x^2yf_{xxy}(0, 0) + 3xy^2 f_{xyy}(0, 0)$$

$$+ y^3 f_{yyy}(0,0)\Big] + \cdots \qquad (10)$$

Thus the Maclaurin's series expansion of $f(x, y)$ given by (10) is a series of powers of x and y.

WORKED OUT EXAMPLES

Example 1: Use Taylor's theorem to expand $f(x, y) = x^2 + xy + y^2$ in powers of $(x - 1)$ and $(y - 2)$.

Solution: Differentiating $f(x, y) = x^2 + xy + y^2$ partially w.r.t. x and y, we get

$$f_x = 2x + y,\ f_y = x + 2y,\ f_{xy} = 1,\ f_{xx} = 2,\ f_{yy} = 2,$$

$$f_{xxx} = 0,\ f_{xxy} = 0,\ f_{yyx} = 0,\ f_{yyy} = 0$$

The Taylor's series expansion of $f(x, y)$ in powers of $(x - 1)$ and $(y - 2)$ is

$$\begin{aligned}
f(x, y) = {} & f(1,2) + [(x-1)f_x(1,2) \\
& +(y-2)f_y(1,2)] + \frac{1}{2!}\Big[(x-1)^2 f_{xx}(1,2) \\
& + 2(x-1)(y-2)f_{xy}(1,2) + (y-2)^2 f_{yy}(1,2)\Big] \\
& + \frac{1}{3!}\Big[(x-1)^3 f_{xxx}(1,2) \\
& +3(x-1)^2(y-2)f_{xxy}(1,2) \\
& +3(x-1)(y-2)^2 f_{yyx}(1,2) \\
& + (y-2)^3 f_{yyy}(1,2)\Big] + \cdots
\end{aligned}$$

Here, $f(1,2) = 7$, $f_x(1,2) = 4$, $f_y(1,2) = 5$, $f_{xy}(1,2) = 1$, $f_{xx} = f_{yy} = 2$, etc.
Substituting these values

$$\begin{aligned}
f(x, y) = {} & 7 + 4(x-1) + 5(y-2) + \frac{1}{2!}\Big[2(x-1)^2 \\
& + 2(x-1)(y-2) + 2(y-2)^2\Big] + 0 + \cdots.
\end{aligned}$$

Example 2: Expand $f(x, y) = e^{x+y}$ in Taylor's series up to terms of second degree in the form $a_0 + b_1x + b_2y + c_1x^2 + c_2xy + c_3y^2 + \cdots$

a. by direct use of Taylor's theorem

b. by expanding e^{x+y} in a series of powers of $x + y$

c. by multiplying together the separate expansions of e^x and e^y.

Solution:

a. $f = e^{x+y}$, $f_x = e^{x+y}$, $f_y = e^{x+y}$, $f_{xx} = e^{x+y}$, $f_{yy} = e^{x+y}$, $f_{xy} = e^{x+y}$ etc. Since the series in powers of x and y, the expansion is about $(0, 0)$ (i.e., Maclaurin's series) so $f = f_x = f_y = f_{xx} = f_{yy} = f_{xy}$ at $(0,0) = 1$.
By Taylor's theorem

$$e^{x+y} = 1 + x + y + \frac{x^2 + 2xy + y^2}{2!} + \cdots$$

b. Expanding in powers of $x + y$

$$\begin{aligned}
e^{x+y} &= \sum_{n=0}^{\infty} \frac{(x+y)^n}{n!} = 1 + (x+y) \\
&\quad + \frac{(x+y)^2}{2!} + \frac{(x+y)^3}{3!} + \cdots
\end{aligned}$$

c. Termwise series multiplication

$$\begin{aligned}
e^{x+y} = e^x \cdot e^y &= \left(\sum_{n=0}^{\infty} \frac{x^n}{n!}\right)\left(\sum_{n=0}^{\infty} \frac{y^n}{n!}\right) \\
&= \left(1 + x + \frac{x^2}{2!} + \frac{x^3}{3!} + \cdots\right) \\
&\quad \left(1 + y + \frac{y^2}{2!} + \frac{y^3}{3!} + \cdots\right) \\
&= 1 + (x+y) + \frac{x^2}{2!} + xy + \frac{y^2}{2!} + \cdots
\end{aligned}$$

Example 3: Expand $f(x, y) = e^y \ln(1 + x)$ in powers of x and y and verify the result by direct expansion.

Solution: $f = e^y \ln(1 + x)$

$$f_x = e^y \frac{1}{1+x},\quad f_y = e^y \ln(1+x),\quad f_{xy} = \frac{e^y}{1+x}$$

$$f_{xx} = \frac{-e^y}{(1+x)^2},\quad f_{yy} = e^y \ln(1+x),$$

$$f_{xxx} = \frac{2e^y}{(1+x)^3},\quad f_{yyy} = e^y \ln(1+x),$$

$$f_{xxy} = \frac{-e^y}{(1+x)^2},\quad f_{yyx} = \frac{e^y}{1+x}$$

Evaluating these derivations at $x = 0$, $y = 0$,

$$f(0,0) = 0,\quad f_x(0,0) = 1,\quad f_y(0,0) = 0,\ f_{xy} = 1$$

$$f_{xx}(0,0) = -1,\quad f_{yy}(0,0) = 0,\quad f_{xxx}(0,0) = 2$$

$f_{yyy}(0,0)=0,\quad f_{xxy}(0,0)=-1,\quad f_{yyx}(0,0)=1$

The Taylor's series expansion up to 3rd degree terms is

$$e^y\ln(1+x)=f(0,0)+xf_x(0,0)+yf_y(0,0)$$
$$+\frac{1}{2!}\left[x^2f_{xx}(0,0)+2xy\,f_{xy}(0,0)\right.$$
$$\left.+y^2f_{yy}(0,0)\right]+\frac{1}{3!}\left[x^3f_{xxx}(0,0)\right.$$
$$+3x^2yf_{xxy}(0,0)+3xy^2f_{yyx}(0,0)$$
$$\left.+y^3f_{yyy}(0,0)\right]+\cdots$$
$$=0+x\cdot 1+0+\frac{1}{2!}\left[-x^2+2.1.xy+0\right]$$
$$+\frac{1}{3!}\left[2x^3-3x^2y+3xy^2+0\right]$$
$$=x-\frac{x^2}{2}+xy+\frac{x^3}{3}-\frac{x^2y}{2}+\frac{xy^2}{2}+\cdots$$

Verification by series multiplication:

we know that $e^x=\sum_{n=0}^{\infty}\frac{x^n}{n!}$ and

$$\ln(1+x)=x-\frac{x^2}{2}+\frac{x^3}{3}-\cdots$$

so $$e^y\ln(1+x)=\left(1+y+\frac{y^2}{2!}+\frac{y^3}{3!}+\cdots\right)$$
$$\times\left(x-\frac{x^2}{2}+\frac{x^3}{3}\cdots\right)$$

Multiplying term by term up to 3rd degree

$$e^y\ln(1+x)=x-\frac{x^2}{2}+\frac{x^3}{3}+xy-\frac{x^2y}{2}+\frac{xy^2}{2}+\cdots$$

Example 4: Find Taylor's expansion of $f(x,y)=\cot^{-1}xy$ in powers of $(x+0.5)$ and $(y-2)$ up to second degree terms. Hence compute $f(-0.4,2.2)$ approximately.

Solution: Here $f(x,y)=\cot^{-1}xy$

$$f_x=\frac{-y}{1+x^2y^2},\quad f_y=\frac{-x}{1+x^2y^2},$$
$$f_{xx}=\frac{2xy^3}{(1+x^2y^2)^2},\quad f_{yy}=\frac{2x^3y}{(1+x^2y^2)^2},$$
$$f_{xy}=\frac{(x^2y^2-1)}{(1+x^2y^2)^2}$$

Evaluating these derivatives at the point $x=-\frac{1}{2}$, $y=2$

$$f(x,y)=\cot^{-1}(x,y)\text{ at }x=-\frac{1}{2},\ y=2,$$
$$f\left(-\frac{1}{2},2\right)=\cot^{-1}\left(-\frac{1}{2}\cdot 2\right)=\cot^{-1}(-1)=\frac{3\pi}{4}$$
$$f_x=-1,\ f_y=\frac{1}{4},\ f_{xy}=0$$
$$f_{xx}=-2,\ f_{yy}=-\frac{1}{8}$$

Expanding $\cot^{-1}xy$ in Taylor's series in powers of $(x+0.5)$ and $(y-2)$, we get

$$f(x,y)=\cot^{-1}xy=f(-0.5,2)+(x+0.5)f_x(-0.5,2)$$
$$+(y-2)f_y(-0.5,2)+\frac{1}{2!}[(x+0.5)^2\times$$
$$f_{xx}(-0.5,2)+2(x+0.5)(y-2)\times$$
$$f_{xy}(-0.5,2)+(y-2)^2f_{yy}(-0.5,2)]+\cdots$$
$$=\frac{3\pi}{4}-(x+0.5)+\frac{y-2}{4}+\frac{1}{2}[-2(x+0.5)^2$$
$$-\frac{1}{8}(y-2)^2]+\cdots$$

Put $x=-0.4$ and $y=2.2$ to compute

$$\cot^{-1}((-0.4),(2.2))=f(-0.4,2.2)$$
$$=\frac{3\pi}{4}-(0.1)+\frac{.2}{4}$$
$$-(0.1)^2-\frac{1}{16}(.2)^2$$
$$=2.29369.$$

Exercise

1. Expand $f(x,y)=x^3+y^3+xy^2$ in powers of $(x-1)$ and $(y-2)$ using Taylor's series.

Ans. $13+7(x-1)+16(y-2)+3(x-1)^2+4(x-1)(y-2)+7(y-2)^2+(x-1)^3+(x-1)\cdot(y-2)^2+(y-2)^3$

2. Obtain Taylor's expansion of $(1+x-y)^{-1}$ in powers of $(x-1)$ and $(y-1)$.

Ans. $1-x+y+x^2-2xy+y^2+\cdots$

3. Expand $\cos x\cos y$ in powers of x and y up to fourth degree terms

Ans. $1-\frac{1}{2}(x^2+y^2)+\frac{1}{24}(x^4+6x^2y^2+y^4)+\cdots$

4. Obtain the expansion of e^{xy} in powers of $(x-1)$ and $(y-1)$.

Ans. $e\left[1+(x-1)+(y-1)+\frac{(x-1)^2}{2!}\right.$ $\left.+(x-1)(y-1)+\frac{(y-1)^2}{2!}+\cdots\right]$

5. Find the Taylor's expansion of $e^x \cos y$ about the point $x=1$, $y=\frac{\pi}{4}$.

Ans. $\frac{e}{\sqrt{2}}\left[1+(x-1)-\left(y-\frac{\pi}{4}\right)+\frac{(x-1)^2}{2!}\right.$ $\left.-(x-1)\left(y-\frac{\pi}{4}\right)-\frac{\left(y-\frac{\pi}{4}\right)^2}{2!}+\cdots\right]$

6. Find the Maclaurin's expansion of $e^x \ln(1+y)$ up to terms of 3rd degree.

Ans. $y+xy-\frac{y^2}{2}+\frac{(x^2y-xy^2)}{2}+\frac{y^3}{3}+\cdots$

7. Expand $e^{ax}\sin by$ about origin up to 3rd degree terms

Ans. $(by+abxy)+\frac{1}{6}(3a^2bx^2y-b^3y^3)+\cdots$

8. Find Taylor's expansion of x^y about $(1,1)$.

Ans. $1+(x-1)+(x-1)(y-1)+\frac{1}{2}(x-1)^2+\cdots$

9. Expand $(xy+hk+hy+xk)/(x+y+h+k)$ in powers of h and k up to second degree terms.

Hint: Take $f(x,y)=\frac{xy}{x+y}$.

Ans. $\frac{xy}{x+y}+\frac{y^2}{(x+y)^3}h+\frac{x^2}{(x+y)^2}k-\frac{y^2}{(x+y)^3}h^2+$ $\frac{2xy}{(x+y)^3}hk-\frac{x^2}{(x+y)^3}k^2+\ldots$

10. Calculate $\ln\left[(1.03)^{\frac{1}{3}}+(0.98)^{\frac{1}{4}}-1\right]$ approximately by using Taylor's expansion up to first order terms.

Hint: Take $f(x,y)=\ln\left[x^{\frac{1}{3}}+y^{\frac{1}{4}}-1\right]$ and expand $f(x+h,y+k)$ in powers of h and k. Choose $x=1$, $y=1$, $h=0.03$, $k=-0.02$.

Ans. 0.005

11. Compute $\tan^{-1}(0.9/1.1)$ approximately.

Hint: Take $f(x,y)=\tan^{-1}(y/x)$, expand in Taylor's series about (1,1) $\tan^{-1}(y/x)=\frac{\pi}{4}-\frac{1}{2}(x-1)+\frac{1}{2}(y-1)+\frac{1}{4}(x-1)^2-\frac{1}{4}(y-1)^2$. Now put $x=1.1$ and $y=0.9$.

Ans. 0.6904

12. Find Taylor's expansion of $\sqrt{1+x+y^2}$ in powers of $(x-1)$ and $(y-0)$.

Ans. $\sqrt{2}\left[1+\frac{x-1}{4}-\frac{(x-1)^2}{32}+\frac{y^2}{4}+\cdots\right]$.

4.2 MAXIMA AND MINIMA OF FUNCTIONS OF TWO VARIABLES: WITH AND WITHOUT CONSTRAINTS

Let $z=f(x,y)$ be a function of two independent variables x and y.

Relative maximum: $f(x,y)$ is said to have a relative maximum at a point (a,b) if

$$f(a,b)>f(a+h,b+k)$$

for small positive or negative values of h and k i.e., $f(a,b)$ the value of the function f at (a,b) is greater than the value of the function f at all points in some small neighbourhood of (a,b).

Relative minimum is similarly defined. $f(x,y)$ has a relative minimum at (a,b) if

$$f(a,b)<f(a+h,b+k).$$

Denote $[f(a+h,b+k)-f(a,b)]$ by $\Delta f(a,b)$ or simply by Δ

i.e., $$\Delta=f(a+h,b+k)-f(a,b)$$

then f has a maximum at (a,b) if Δ has the *same negative* sign for all small values of h,k; i.e., $\Delta<0$ a minimum at (a,b) if Δ has the same positive sign

i.e., $$\Delta>0.$$

Extremum is a point which is either a maximum or minimum. The value of the function f at an extremum (maximum or minimum) point is known as the extremum (maximum or minimum) value of the function f.

Geometrically, $z=f(x,y)$ represents a surface. The maximum is a point on the surface (hill top) from which the surface descends (comes down) in every direction towards the xy-plane (see Fig. 4.1). The minimum is the bottom of depression from which the surface ascends (climbs up) in every direction (refer Fig. 4.1). In either case, the tangent planes to the surface at a maximum or minimum point is horizontal (parallel to xy-plane) and perpendicular to z-axis.

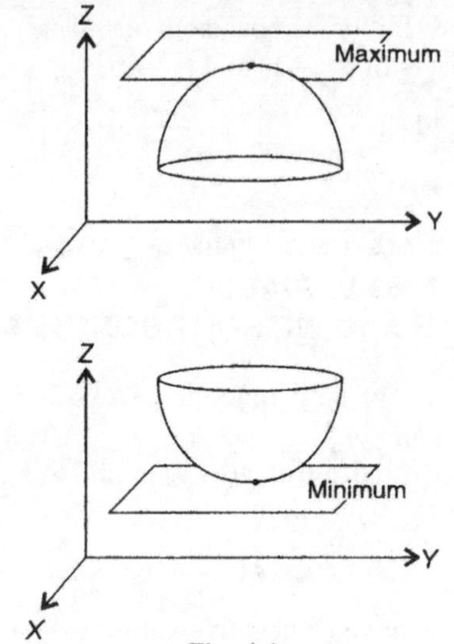

Fig. 4.1

Saddle point or minimax is a point where function is neither maximum nor minimum. At such point f is maximum in one direction while minimum in another direction.

Geometrically such a surface (looks like the leather seat on back of a horse) (Fig. 4.2) forms a ridge rising in one direction and falling in another direction.

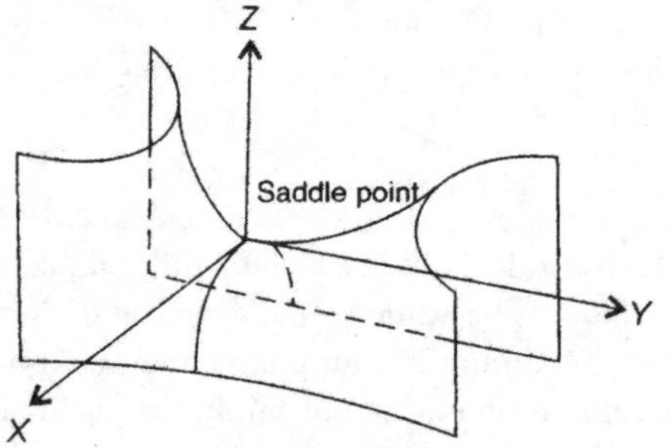

Fig. 4.2

Example: $z = xy$, hyperbolic paraboloid has a saddle point at the origin.

Necessary and Sufficient Conditions for Extrema of a Function *f* of Two Variables

By Taylor's theorem,

$$f(a+h, b+k) = f(a, b) + \left[hf_x(a, b) + kf_y(a, b)\right]$$

$$+\frac{1}{2!}\left[h^2 f_{xx}(a, b) + 2hkf_{xy}(a, b) + k^2 f_{yy}(a, b)\right] + \cdots \quad (1)$$

Neglecting higher order terms of h^2, hk, k^2, etc. since h, k are small, the above expansion reduces to

$$\Delta = [f(a+h, b+k) - f(a, b)]$$
$$= hf_x(a, b) + kf_y(a, b) \quad (2)$$

The necessary condition that Δ has the same positive or same negative sign is when $f_x(a, b) = 0$ and $f_y(a, b) = 0$ (even though h and k can take both positive and negative values).

With $f_x(a, b) = 0$ and $f_y(a, b) = 0$, expansion (1), neglecting higher order terms h^3, k^3, h^2k etc reduces to

$$\Delta = \frac{1}{2!}\left[h^2 f_{xx}(a, b) + 2hkf_{xy}(a, b) + k^2 f_{yy}(a, b)\right] \quad (3)$$

Denote $f_{xx}(a, b) = r$, $f_{yy}(a, b) = t$, $f_{xy}(a, b) = s$.

From (3) we observe that the nature of sign of Δ depends on the nature of sign of $h^2r + 2hks + k^2t$.

Rewriting

$$\text{sign of } \Delta = \text{sign of } (h^2r + 2hks + k^2t)$$
$$= \text{sign of } \left(\frac{h^2r^2 + 2hkrs + k^2tr}{r}\right)$$
$$= \text{sign of } \left\{\frac{(hr+ks)^2 + k^2(rt - s^2)}{r}\right\} \quad (4)$$

If $rt - s^2 > 0$ then the numerator in R.H.S. of (4) is positive. In that case sign of Δ = sign of r.

Thus $\Delta < 0$ if $rt - s^2 > 0$ and $r < 0$

and $\Delta > 0$ if $rt - s^2 > 0$ and $r > 0$.

Therefore the sufficient (Lagrange's) conditions for extrema are:

I. f attains (has) a **maximum** at (a, b) if $rt - s^2 > 0$ and $r < 0$.

II. f attains a **minimum** at (a, b) if $rt - s^2 > 0$ and $r > 0$.

III. **Saddle point:** If $rt - s^2 < 0$ then $\Delta > 0$ or < 0 depending on h and k. Therefore f has a saddle point (minimax) at (a, b) if

$$rt - s^2 < 0$$

IV. Failure case: If $rt - s^2 = 0$, further investigation is needed to determine the nature of function f.

Method of Finding Extrema of $f(x, y)$

1. Solving $f_x = 0$ and $f_y = 0$ yields **critical or stationary point** P of f.
2. Calculate $r = f_{xx}$, $s = f_{xy}$, $t = f_{yy}$ at the critical point P.
3. **a.** Maximum: If $rt - s^2 > 0$ and $r < 0$ then f has a maximum at P.
   b. Minimum: If $rt - s^2 > 0$ and $r > 0$ then f has a minimum at P.
   c. Saddle point: If $rt - s^2 < 0$ then f has neither maximum nor minimum.
   d. Failure case: If $rt - s^2 = 0$, further investigation needed.

Note: Extrema occur only at stationary points. However stationary points need not be extrema.

Examples

a. $f = 1 - x^2 - y^2$, $f_x = -2x$, $f_y = -2y$
$\therefore$ Stationary point $(0, 0)$, $r = r_{xx} = -2$, $t = f_{yy} = -2$, $s = 0$ so $rt - s^2 = (-2)(-2) - 0 = 4 > 0$ and $r = -2 < 0$. $(0, 0)$ is maximum point of f and the maximum value of f is 1.

b. $f = x^2 + y^2$, $f_x = 2x$, $f_y = 2y$, $f_{xx} = 2$, $f_{yy} = 2$, $f_{xy=0}$ $\therefore$ stationary point $(0, 0)$. $rt - s^2 = 2.2 - 0 = 4 > 0$ and $r = 2 > 0$. $(0, 0)$ is a minimum point of f and the minimum value of f is 0.

c. $f = xy$; $f_x = y$, $f_y = x$, $f_{xx} = 0$, $f_{yy} = 0$, $f_{xy} = 1$ $\therefore$ stationary point $(0, 0)$.

$$rt - s^2 = 0.0 - 1 = -1 < 0$$

So $(0, 0)$ is a saddle point of f

d. i. $f = 1 - x^2y^2$,
$f_x = -2xy^2$, $f_y = -2x^2y$
$\therefore$ stationary point $(0, 0)$
$f_{xx} = -2y^2$, $f_{yy} = -2x^2$, $f_{xy} = -4xy$
At stationary point $(0, 0)$, $f_{xx} = f_{yy} = f_{xy} = 0$
By inspection f has a maximum at $(0, 0)$.

ii. $f = x^2y^2$, $f_x = 2xy^2$, $f_y = 2x^2y$,
$\therefore$ stationary point $(0, 0)$, $f_{xx} = 2y^2$, $f_{yy} = 2x^2$, $f_{xy} = 4xy$
At stationary point $(0, 0)$, $f_{xx} = f_{yy} = f_{xy} = 0$
However observe that f has a minimum at $(0, 0)$.

iii. $f = x^3y^2$, $f_x = 3x^2y^2$, $f_y = 2x^3y$
$\therefore$ stationary point $(0, 0)$
$f_{xx} = 6xy^2$, $f_{yy} = 2x^3$, $f_{xy} = 6x^2y$
At stationary point $(0, 0)$, $f_{xx} = f_{yy} = f_{xy} = 0$.
Note that $(0, 0)$ is a saddle point.

Worked Out Examples

Example 1: Find the maximum and minimum values of $f(x, y) = x^3 + 3xy^2 - 15x^2 - 15y^2 + 72x$.

Solution: Differentiating f partially w.r.t. x and y,

$$f_x = \frac{\partial f}{\partial x} = 3x^2 + 3y^2 - 30x + 72$$

$$f_y = \frac{\partial f}{\partial y} = 6xy - 30y$$

The stationary (critical) points are given by $f_x = 0$ and $f_y = 0$
From

$$f_y = 6xy - 30y = 0$$

so $$6y(x - 5) = 0$$

Thus either $y = 0$ or $x = 5$
Since $f_x = 3x^2 + 3y^2 - 30x + 72 = 0$,

for $y = 0$, $3x^2 - 30x + 72 = 0$

so $x = 6$ or 4

for $x = 5$, $75 + 3y^2 - 150 + 72 = 0$

so $y = \pm 1$

Thus the four stationary points are given by

$$(6, 0), (4, 0), (5, 1), (5, -1)$$

To determine the nature of these points, calculate

f_{xx}, f_{yy} and f_{xy}

$$f_{xx} = A = 6x - 30,\ f_{xy} = B = 6y,\ C = f_{yy} = 6x - 30$$

so $AC - B^2 = (6x - 30)^2 - 36y^2 = 36\{(x-5)^2 - y^2\}$

i. At the stationary point (6, 0), we have $A = 36 - 30 = 6 > 0$ and $AC - B^2 = 36 > 0$. So (6, 0) is a minimum point of the given function f and the minimum value of f at (6, 0) is $6^3 + 0 - 15.36 + 72.6 = 108$.

ii. At $(4, 0): A = 24 - 30 = -6 < 0$ and $AC - B^2 = 36 > 0$. So a maximum occurs at the point (4, 0) and the maximum value of f at (4,0) is 112.

iii. At (5, 1), $A = 0$, $AC - B^2 = -36 < 0$. (5, 1) is a saddle point (it is neither maximum nor minimum).

iv. At (5, −1), $A = 0$, $AC = B^2 = -36 < 0$. So (5, −1) is neither a maximum nor a minimum (it is a saddle point).

Example 2: Find the shortest distance from origin to the surface $xyz^2 = 2$.

Solution: Let d be the distance from origin (0, 0, 0) to any point (x, y, z) of the given surface then

$$d = \sqrt{(x-0)^2 + (y-0)^2 + (z-0)^2}$$

or
$$d^2 = x^2 + y^2 + z^2$$

Eliminate z^2 using the equation of the surface $xyz^2 = 2$. So replace z^2 by $\frac{2}{xy}$

$$\therefore\ d^2 = x^2 + y^2 + \frac{2}{xy} = f(x, y)$$

$$f_x = 2x - \frac{2}{x^2y},\ f_y = 2y - \frac{2}{xy^2}.$$

Solving $f_x = 0$ and $f_y = 0$, we get

$$\frac{x^3y - 1}{x^2y} = 0 \text{ and } \frac{xy^3 - 1}{xy^2} = 0$$

$$x^3y = 1 = xy^3 \text{ or } xy(x^2 - y^2) = 0$$

since $x \neq 0$, $y \neq 0$, so $x = \pm y = 1$

Thus the two stationary points are (1, 1) and (−1, −1)

$$f_{xx} = 2 + \frac{4}{x^3y},\ f_{yy} = 2 + \frac{4}{xy^3},\ f_{xy} = \frac{2}{x^2y^2}$$

At $(1, 1): f_{xx} = 6 > 0$, $f_{xx} \cdot f_{yy} - f_{xy}^2 = 6 \cdot 6 - 4 = 32 > 0$

At $(-1, -1)$, $f_{xx} = +6$, $f_{xx}f_{yy} - f_{xy}^2 = 32$

So minimum occurs at $(1, 1, \sqrt{2})$ and $(-1, -1, \sqrt{2})$. The shortest distance is $\sqrt{1^2 + 1^2 + (\sqrt{2})^2} = \sqrt{4} = 2$.

Example 3: The temperature T at any point (x, y, z) in space is $T(x, y, z) = kxyz^2$ where k is a constant. Find the highest temperature on the surface of the sphere

$$x^2 + y^2 + z^2 = a^2.$$

Solution: Eliminating the variable z, using $z^2 = a^2 - x^2 - y^2$, we get

$$T(x, y, z) = kxyz^2 = kxy(a^2 - x^2 - y^2) = F(x, y)$$

$$F_x = ky(a^2 - 3x^2 - y^2),\ F_y = kx(a^2 - x^2 - 3y^2)$$

The stationary points are given by ($x = 0$, $y = 0$) or solution of

$$3x^2 + y^2 = a^2$$

$$x^2 + 3y^2 = a^2$$

Solving $x = y = \pm\frac{a}{2}$

$$F_{xx} = -6kxy,\ F_{yy} = -6kxy,\ F_{xy} = k(a^2 - 3x^2 - 3y^2)$$

At (0, 0), $F_{xx} = 0 = F_{yy}$, $F_{xy} = ka^2$

$\therefore 0.0 - ka^2 < 0$ so (0, 0) is a saddle point.

At both the points $\left(\frac{a}{2}, \frac{a}{2}\right)$ and $\left(\frac{-a}{2}, \frac{-a}{2}\right)$, $F_{xx} = -6\frac{ka^2}{4} < 0$ and

$$F_{xx} \cdot F_{yy} - F_{xy}^2 = \frac{9}{4}k^2a^4 - \frac{a^4k^2}{4} = 2k^2a^4 > 0$$

$\therefore T$ attains a maximum value at both these points $\left(\frac{a}{2}, \frac{a}{2}\right)$ and $\left(-\frac{a}{2}, -\frac{a}{2}\right)$. The maximum value of T is $k \cdot \frac{a^2}{4}\left(\frac{a^2}{2}\right) = \frac{ka^4}{8}$.

Example 4: Find the shortest distance between the lines

$$\frac{x-3}{1} = \frac{y-5}{-2} = \frac{z-7}{1} \quad (1)$$

and
$$\frac{x+1}{7} = \frac{y+1}{-6} = \frac{z+1}{1} \quad (2)$$

Solution: Equating each of the fractions of (1) to λ, we get $x = 3 + \lambda$, $y = 5 - 2\lambda$, $z = 7 + \lambda$. Thus any point P on the first line (1) is given by

$$(3+\lambda, 5-2\lambda, 7+\lambda).$$

Similarly any point Q on the second line (2) is $(-1+7\mu, -1-6\mu, -1+\mu)$.

The distance between the given two lines is

$$PQ=\sqrt{(3+\lambda+1-7\mu)^2+(5-2\lambda+1+6\mu)^2+(7+\lambda+1-\mu)^2}$$

Consider $f(\lambda, \mu)=(PQ)^2=6\lambda^2+86\mu^2-40\lambda\mu+105$. The problem is to find minimum value of f as a function of the two variables λ, μ.

$$f_\lambda = 12\lambda - 40\mu,\ f_\mu = 172\mu - 40\lambda.$$

Solving $12\lambda - 40\mu = 0$ and $172\mu - 40\lambda = 0$, we get, $\lambda = 0, \mu = 0$ as the only stationary point

$$f_{\lambda\lambda} = 12,\ f_{\mu\mu} = 172,\ f_{\lambda\mu} = -40$$

Now $\quad f_{\lambda\lambda}\cdot f_{\mu\mu} - f_{\lambda\mu}^2 = (12)\cdot(172) - (-40)^2 > 0$

Since $f_{\lambda\lambda} = 12 > 0$ and $f_{\lambda\lambda} f_{\mu\mu} - f_{\lambda\mu}^2 > 0$, a minimum occurs at $\lambda = 0, \mu = 0$. The minimum, (shortest) distance is given by

$$PQ = \sqrt{4^2+6^2+8^2} = \sqrt{116} = 2\sqrt{29}$$

EXERCISE

1. Test the functions for maxima, minima and saddle points:
   a. $x^4+y^4-x^2-y^2+1$
   b. $x^2+2y^2+3z^2-2xy-2yz-2$

Ans. **a.** maximum at (0,0)
maximum value is 1.
minima at four points $(\pm 1/\sqrt{2}, \pm 1/\sqrt{2})$
minimum value at these 4 points is $\frac{1}{2}$.
Saddle points at four points $(0, \pm 1/\sqrt{2})$, $(\pm 1/\sqrt{2}, 0)$
b. maximum at (1,1), minimum at $(-1, -1)$

2. Find the extrema of $f(x, y)$:

$$(x^2+y^2)e^{6x+2x^2}$$

Ans. minima at (0,0) (minimum value 0) and at $(-1, 0)$ (minimum value e^{-4}).
Saddle point at $\left(-\frac{1}{2}, 0\right)$

3. Examine the following function $f(x, y)$ for extrema:

$$\sin x + \sin y + \sin(x+y)$$

Ans. Maximum at $(\pi/3, \pi/3)$, maximum value $\frac{3\sqrt{3}}{2}$.

4. Find the shortest distance from the origin to the plane $x-2y-2z=3$.

Ans. Shortest distance is 1 (from (0,0,0)) to the point $\left(\frac{1}{3}, -\frac{2}{3}, -\frac{2}{3}\right)$ on the plane.

5. Given $ax+by+cz=p$ find the minimum value of $x^2+y^2+z^2$.

Ans. $p^2/(a^2+b^2+c^2)$

6. Find the shortest distance between the lines

$$\frac{x-2}{3} = \frac{y-6}{-2} = \frac{z-5}{-2}$$

and $$\frac{x-5}{2} = \frac{y-3}{1} = \frac{z-8}{6}.$$

Ans. Shortest distance is 3 between the points (5, 4, 3) and (3, 2, 2).

7. If the perimeter of a triangle is constant, prove that the area of this triangle is maximum when the triangle is equilateral.

Hint: $2s = a+b+c$ where a, b, c are sides
Area of $\Delta = \sqrt{s(s-a)(s-b)(s-c)}$
Maximum when $a = b = c = \frac{2s}{3}$.

8. Find the volume of the largest rectangular parallelopiped with edges parallel to the axes, that can be inscribed in the:
   a. ellipsoid $\frac{x^2}{a^2}+\frac{y^2}{b^2}+\frac{z^2}{c^2}=1$
   b. sphere
   c. equation of ellipsoid is $4x^2+4y^2+9z^2 = 36$.

Ans. **a.** volume $\frac{8abc}{3\sqrt{3}}$, parallelopiped has dimensions

$$x = \frac{a}{\sqrt{3}},\ y = \frac{b}{\sqrt{3}},\ z = \frac{c}{\sqrt{3}}$$

b. special case $a = b = c$

$$\text{volume}: \frac{8a^3}{3\sqrt{3}},\ x = y = z = \frac{a}{\sqrt{3}}$$

c. volume: $16\sqrt{3}$ (with $a=3, b=3, c=2$)

9. Show that the rectangular solid of maximum volume that can be inscribed in a sphere is a cube.

Hint: V be the volume of the rectangular solid with length, breadth and height x, y, z

$$V = xyz$$

diagonal of solid $= \sqrt{x^2+y^2+z^2} = d =$ diameter of sphere. Eliminate $z = \sqrt{d^2-x^2-y^2}$ so

$$v = xy\sqrt{d^2-x^2-y^2} = f(x, y).$$

10. Find the dimensions of a rectangular box, with open top, so that the total surface area of the box is a minimum, given that the volume of the box is constant say V.

    Hint: $S = xy + 2xz + 2yz$, eliminate $z = \frac{V}{xy}$ where V is a given constant so that

$$S = xy + \frac{2V}{y} + \frac{2V}{x} = f(x, y).$$

Ans. $x = y = 2z = (2V)^{1/3}$

11. Find the dimensions of the rectangular box, with open top, of maximum capacity whose surface area is 432 sq. cm.

Ans. 12, 12, 6

12. If the total surface area of a closed rectangular box is 108 sq. cm, find the dimensions of the box having maximum capacity.

Ans. $\sqrt{18}, \sqrt{18}, \sqrt{18}$

13. An aquarium with rectangular sides and bottom (and no top) is to hold 32 litres. Find its dimensions so that it will use the least amount of material.

    Hint: Work as Example 10 with $V = 32$.

Ans. 4, 4, 2.

4.3 LAGRANGE'S* METHOD OF UNDETERMINED MULTIPLIERS

In many practical and theoretical problems, it is required to find the maximum or minimum of a function of several variables, where the variables are connected by some given relation or condition known as a constraint. Thus if $f(x, y, z)$ is a function of 3 independent variables, where x, y, z are related by a known constraint $g(x, y, z) = 0$, then the problem of constrained extrema consists of finding the

$$\text{Extrema of} \quad u = f(x, y, z) \tag{1}$$

$$\text{subject to} \quad g(x, y, z) = 0 \tag{2}$$

This problem can be solved by (a) elimination method (b) Implicit differentiation method (c) Lagrange's multiplier's method.

a. In elimination method, the constraint (2) is solved for say one variable z in terms of the other variables x and y. Then z is eliminated from $f(x, y, z)$ resulting in a function of two variables x and y only. The disadvantage of this method is that many times, (2) may not be solvable and in case of solution also the amount of algebra will be generally enormous.

b. In implicit differentiation method, no elimination of variables is done but derivatives are eliminated by calculating them through implicit differentiation. This method also suffers due to more labour involved.

c. The very useful Lagrange's method of undetermined multiplier's introduces an additional unknown constant λ known as Lagrange multiplier. Since the stationary values occur when $f_x = f_y = f_z = 0$, so the total differential

$$df = f_x dx + f_y dy + f_z dz = 0 \tag{3}$$

Differential of the constraint (2) is

$$dg = g_x dx + g_y dy + g_z dz = 0 \tag{4}$$

Multiplying (4) by λ and adding to (3), we get

$$(f_x + \lambda g_x)dx + (f_y + \lambda g_y)dy + (f_z + \lambda g_z)dz = 0 \tag{5}$$

Since x, y, z are independent variables (5) implies that

$$f_x + \lambda g_x = 0 \tag{6}$$

$$f_y + \lambda g_y = 0 \tag{7}$$

$$f_z + \lambda g_z = 0 \tag{8}$$

Solving the four Equations (2), (6), (7), (8) for the four unknowns x, y, z, λ, we get the required stationary points of $f(x, y, z)$ subject to the

*Joseph Louis Lagrange (1736–1813).

constraint (2). Thus the method of Lagrange's multipliers consists of:

Step I. From the auxiliary equation

$$F(x, y, z) = f(x, y, z) + \lambda\phi(x, y, z) \qquad (9)$$

Step II. Partially differentiate F in (9) w.r.t. x, y, z respectively

Step III. Solve the four equations

$$F_x = 0, F_y = 0, F_z = 0$$

and the constraint (2) for the Lagrange multiplier λ and stationary values x, y, z.

Advantages

1. The stationary value $f(x, y, z)$ can be determined from (2), (6), (7), (8) even without determining x, y, z explicitly.
2. This method can be extended to a function of several 'n' variables $x_1, x_2, x_3, \ldots x_n$ and subject to many (more than one) 'm' constraints by forming the auxiliary equation

$$F(x_1, x_2, \ldots, x_n) = f(x_1, x_2, \ldots, x_n) + \sum_{i=1}^{m} \lambda_i \phi_i(x_1, x_2, \ldots, x_n).$$

The stationary values are obtained by solving the $n + m$ equations consisting of n equations $\frac{\partial F}{\partial x_i} = 0$, for $i = 1, 2, 3, \ldots n$ and the m constraint, $\phi_i = 0$ for $i = 1, 2, 3, \ldots m$.

Disadvantages

1. Nature of the stationary points can not be determined. Further investigation needed.
2. Equations (6), (7), (8) are only necessary conditions but not sufficient.

Worked Out Examples

Example 1: Find the maximum value of $x^m y^n z^p$ when $x + y + z = a$.

Solution: This is a constrained maximum problem where the function $f(x, y, z) = x^m y^n z^p$, subjected to the constraint $x + y + z = a$. So consider the auxiliary function

$$F(x, y, z) = x^m y^n z^p + \lambda(x + y + z - a) \qquad (1)$$

Differentiating (1) partially w.r.t. x, y, z and equating to zero we get

$$F_x = \frac{\partial F}{\partial x} = m x^{m-1} y^n z^p + \lambda = 0 \qquad (2)$$

$$F_y = \frac{\partial F}{\partial y} = n x^m y^{n-1} z^p + \lambda = 0 \qquad (3)$$

$$F_z = \frac{\partial F}{\partial z} = p x^m y^n z^{p-1} + \lambda = 0 \qquad (4)$$

Solving for x, y, z, $\frac{m}{x} f + \lambda = 0$ so $x = -\frac{mf}{\lambda}$. Similarly $y = -\frac{nf}{\lambda}$ and $z = -\frac{pf}{\lambda}$.

Substituting these values of x, y, z in the given constraint, we have

$$x + y + z = -\left(\frac{m}{\lambda} + \frac{n}{\lambda} + \frac{p}{\lambda}\right) f = a$$

Solving, we get the value of λ as

$$\lambda = -\frac{(m + n + p)}{a} f.$$

Using this λ, we get

$$x = -\frac{mf}{\lambda} = -\frac{mf \cdot (-a)}{f(m + n + p)} = \frac{am}{m + n + p}$$

Similarly $$y = -\frac{nf}{\lambda} = \frac{an}{m + n + p}$$

and $$z = -\frac{pf}{\lambda} = \frac{ap}{m + n + p}$$

Thus the maximum value is

$$x^m y^n z^p = \left(\frac{am}{m + n + p}\right)^m \left(\frac{an}{m + n + p}\right)^n \left(\frac{ap}{m + n + p}\right)^p$$

$$= \frac{a^{m+n+p} \cdot m^m \cdot n^n \cdot p^p}{(m + n + p)^{m+n+p}}.$$

Example 2: Find the maximum and minimum distances from the origin to the curve

$$3x^2 + 4xy + 6y^2 = 140.$$

Solution: The distance d from the origin (0, 0) to any point (x, y) is given by

$$d^2 = x^2 + y^2 = f(x, y)$$

To find extrema of $f(x, y)$ subject to the condition that the point (x, y) lies on the curve $3x^2 + 4xy + 6y^2 = 140$.

So the auxiliary function is

$$F(x, y) = (x^2 + y^2) + \lambda(3x^2 + 4xy + 6y^2 - 140)$$

$$F_x = 2x + \lambda(6x + 4y) = 0,$$

$$F_y = 2y + \lambda(12y + 4x) = 0$$

Solving for $\lambda = -\frac{x}{(3x+2y)} = -\frac{y}{6y+2x}$

$$-\lambda = \frac{x^2}{3x^2 + 2xy} = \frac{y^2}{6y^2 + 2xy} = \frac{x^2 + y^2}{3x^2 + 4xy + 6y^2}$$

$$\therefore -\lambda = \frac{f}{140}$$

Substituting λ in $F_x = 0$ and $F_y = 0$, we get

$$(140 - 3f)x - 2fy = 0$$

$$-2fx + (140 - 6f)y = 0$$

This system has non-trivial solution if

$$\begin{vmatrix} 140 - 3f & -2f \\ -2f & (140 - 6f) \end{vmatrix} = 0$$

$$\text{i.e., } (140 - 3f)(140 - 6f) - 4f^2 = 0$$

$$14f^2 - 1260f + 140^2 = 0$$

$$f^2 - 90f - 1400 = 0$$

$$(f - 70)(f - 20) = 0$$

$$\therefore f = 70, 20$$

Thus the maximum and minimum distances are $\sqrt{70}$, $\sqrt{20}$.

Example 3: A wire of length b is cut into two parts which are bent in the form of a square and circle respectively. Find the least value of the sum of the areas so found.

Solution: Let x and y be two parts into which the given wire is cut so that $x + y = b$. Suppose the piece of wire of length x is bent into a square so that each side is $\frac{x}{4}$ and thus the area of the square is $\frac{x}{4} \cdot \frac{x}{4} = \frac{x^2}{16}$.

Suppose the wire of length y is bent into a circle with the perimeter y. So the area of this circle so formed is

$$\pi(\text{radius})^2 = \pi\left(\frac{y}{2\pi}\right)^2 = \frac{\pi y^2}{4\pi^2} = \frac{y^2}{4\pi}$$

Thus to find the minimum of the sum of the two areas subject to the constraint that sum is b. So the auxiliary equation is

$$F(x, y) = \left(\frac{x^2}{16} + \frac{y^2}{4\pi}\right) + \lambda(x + y - b)$$

$$F_x = \frac{x}{8} + \lambda = 0, \; F_y = \frac{y}{2\pi} + \lambda = 0$$

Solving $x = -8\lambda$, $y = -2\pi\lambda$.

Substituting these values in the constraint $x + y = -8\lambda - 2\pi\lambda = b$

$$\therefore +\lambda = -\frac{b}{8 + 2\pi}$$

Thus $x^* = -8\lambda = \frac{8b}{8+2\pi}$, $y^* = -2\pi\lambda = \frac{2\pi b}{8+2\pi}$

The least value of the sum of the areas of the square and circle is

$$f(x, y) = \left.\frac{x^2}{16} + \frac{y^2}{4\pi}\right|_{x^*, y^*} = \frac{64b^2}{16(8 + 2\pi)^2} + \frac{4\pi^2 b^2}{(8 + 2\pi)^2}$$

$$= \frac{b^2(\pi + 4)}{4(\pi + 4)^2} = \frac{b^2}{4(\pi + 4)}.$$

Example 4: Find the dimensions of a rectangular box of maximum capacity whose surface area is given when (a) box is open at the top (b) box is closed.

Solution: Let x, y, z be the dimensions of the rectangular box so that its volume V is

$$V = xyz \tag{1}$$

The total surface area of the box is

$$nxy + 2yz + 2zx = S = \text{given constant} \tag{2}$$

Here $n = 1$, the box is open at the top
$n = 2$, the box is closed (on all sides)

The constrained maximum problem is to maximize V subject to constraint (2).
So the auxiliary function is

$$F(x, y, z) = xyz + \lambda(nxy + 2yz + 2zx - S) \tag{3}$$

$$F_x = yz + \lambda(ny + 2z) = 0 \tag{4}$$

$$F_y = xz + \lambda(nx + 2z) = 0 \tag{5}$$

$$F_z = xy + \lambda(2y + 2x) = 0 \tag{6}$$

Multiplying (4), (5), (6) by x, y, z respectively and adding, we get

$$3xyz + \lambda[2(nxy + 2yz + 2zx)] = 0$$

or $3 \cdot V + 2\lambda \cdot S = 0$ using (1) and (2)

solving, $$\lambda = -\frac{3V}{2S} \quad (7)$$

Substituting value of λ from (7) in (4), (5), (6)

$$yz - \frac{3V}{2S}(ny + 2z) = 0 \text{ or } yz - \frac{3xyz}{2S}(ny + 2z) = 0$$

$$nxy + 2xz = \frac{2S}{3} \quad (8)$$

Similarly $$nxy + 2yz = \frac{2S}{3} \quad (9)$$

$$2yz + 2zx = \frac{2S}{3} \quad (10)$$

From (8) – (9), $x = y$ (11)

From (9) – (10), $ny = 2z$ (12)

Substituting (11) and (12) in the given constraint (2)

$$n \cdot x \cdot x + 4x\frac{nx}{2} = S \quad \therefore \quad 3nx^2 = S$$

$$\therefore \quad x^2 = \frac{S}{3n}$$

a. When box is open at the topes $n = 1$

$$\therefore \; x^2 = \frac{S}{3} \quad \text{or } x = \sqrt{\frac{S}{3}}$$

The dimensions of the open top box are

$$x = y = \sqrt{\frac{S}{3}}, \quad z = \frac{1}{2}\sqrt{\frac{S}{3}}$$

b. When the box is closed: $n = 2$

$$x^2 = \frac{S}{6} \text{ or } x = \sqrt{\frac{S}{6}}, x = y = z$$

The dimensions are

$$x = y = z = \sqrt{\frac{S}{6}}.$$

Example 5: Suppose a closed rectangular box has length twice its breadth and has constant volume V. Determine the dimensions of the box requiring least surface area (sheet metal).

Solution: Let x be the breadth so that the length is $2x$ and y be the height of the closed rectangular box. Its volume is $x \cdot 2x \cdot y = 2x^2y = V$ (given). The surface area (6 faces) S is given by

$$S = 2(2x.x) + 2(2x.y) + 2(x.y) = 4x^2 + 6xy$$

Thus the problem is to minimize

$$f(x, y) = 4x^2 + 6xy \quad (1)$$

Subject to the condition that

$$x^2y = \frac{V}{2} \text{ (known)} \quad (2)$$

The auxiliary function is

$$F(x, y) = 4x^2 + 6xy + \lambda\left(x^2y - \frac{V}{2}\right) \quad (3)$$

Differentiating (3) w.r.t. x and y and equating to zero, we get

$$F_x = 8x + 6y + 2\lambda xy = 0 \quad (4)$$

$$F_y = 6x + \lambda x^2 = 0 \quad (5)$$

Solving (5), $$\lambda = -\frac{6}{x} \quad \text{or} \quad x = -\frac{6}{\lambda} \quad (6)$$

Substituting x from (6) in (4), we get

$$\frac{-48}{\lambda} + 6y - 12y = 0 \quad \text{or} \quad y = -\frac{8}{\lambda} \quad (7)$$

Substituting (6) and (7) in (2), we have

$$\lambda^3 = -\frac{576}{V} \quad \text{or} \quad \lambda = -\left(\frac{576}{V}\right)^{\frac{1}{3}} \quad (8)$$

using (8); from (6) and (7), we get

$$x = \frac{-6}{\lambda} = 6 \cdot \left(\frac{V}{576}\right)^{\frac{1}{3}} = \left(\frac{3V}{8}\right)^{\frac{1}{3}} \quad (9)$$

$$y = \frac{-8}{\lambda} = 8\left(\frac{V}{576}\right)^{\frac{1}{3}} = \left(\frac{8V}{9}\right)^{\frac{1}{3}} \quad (10)$$

The least surface area with these dimensions (9) (10) is

$$S = 4x^2 + 6xy = 4 \cdot \left(\frac{3V}{8}\right)^{\frac{2}{3}} + 6 \cdot \left(\frac{3V}{8}\right)^{\frac{1}{3}}\left(\frac{8V}{9}\right)^{\frac{1}{3}}$$

On simplification

$$S = (3^5V^2)^{\frac{1}{3}} = (243V^2)^{\frac{1}{3}}.$$

EXERCISE

1. Find the extremum values of $\sqrt{x^2 + y^2}$ when $13x^2 - 10xy + 13y^2 = 72$.

Ans. maximum 3, minimum 2

2. Find the minimum value of $x^2 + y^2 + z^2$ subject to the condition $\frac{1}{x} + \frac{1}{y} + \frac{1}{z}$.

Ans. minimum is 27

3. Divide 24 into three parts such that the continued product of the first, square of the second and the cube of the third may be maximum.

Ans. 4, 8, 12, maximum value: $4 \cdot 8^2 \cdot 12^3$

4. Determine the perpendicular distance of the point (a, b, c) from the plane $lx+my+nz=0$ by the Lagrange's method.

Ans. minimum distance $= \frac{la+mb+nc}{\sqrt{l^2+m^2+n^2}}$

5. Determine the point in the plane $3x - 4y + 5z = 50$ nearest to the origin.

Ans. $(3, -4, 5)$

6. Find the maximum and minimum distances of the point $(3, 4, 12)$ from the unit sphere with centre at origin.

Hint:

$$F = (x-3)^2 + (y-4)^2 + (z-12)^2 + \lambda(x^2 + y^2 + z^2 - 1),$$

$$x = \frac{3}{1 \pm \sqrt{f}}, \ y = \frac{4}{1 \pm \sqrt{f}}, \ z = \frac{12}{1 \pm \sqrt{f}}$$

where $f = (x-3)^2 + (y-4)^2 + (z-12)^2$.

Ans. maximum 14 and minimum 12

7. Determine the point on the paraboloid $z = x^2 + y^2$ which is closest to the point $(3, -6, 4)$.

Ans. $(1, -2, 6)$

8. **a.** Find the dimensions of the rectangular box, without top, of maximum capacity whose surface is 108 sq. cm.

b. What are the dimensions when the box is closed (on all sides).

Ans. **a.** 6, 6, 3 maximum capacity (volume) $= 6 \cdot 6 \cdot 3 = 108$

b. $\sqrt{18}, \sqrt{18}, \sqrt{18}$

9. **a.** If the surface of a rectangular box, with open top, is 432 sq. cm, find the dimensions of the box having maximum capacity (volume).

b. If the box is closed (on all sides) what are the dimensions.

Ans. **a.** 12, 12, 6 for open top box

b. $\sqrt{72}, \sqrt{72}, \sqrt{72}$ for closed box

10. Find the length and breadth of a rectangle of maximum area that can be inscribed in the ellipse $4x^2 + 9y^2 = 36$.

Ans. length $= 3\sqrt{2}/2$, breadth: $\sqrt{2}$, maximum area of the rectangle is 12

11. Find the volume of the largest rectangular parallelopiped that can be inscribed in the ellipsoid of revolution $4x^2 + 4y^2 + 9z^2 = 36$.

Ans. maximum volume of rectangular parallelopiped $16\sqrt{3}$

12. Find the dimensions of a rectangular box, with open top, of given capacity (volume) such that the sheet metal (surface area) required is least.

Hint: Auxiliary function.

$$F(x, y, z) = xy + 2xz + 2yz + \lambda(xyz - V).$$

Ans. $x = y = 2z = (2V)^{\frac{1}{3}}$ where $V =$ Volume of the box (given).

Chapter 5

Curve Tracing

INTRODUCTION

Curve tracing is an analytical method of drawing an approximate shape of a curve by the study of some of its important characteristics such as symmetry, intercepts, asymptotes, tangents, multiple points, region of existence, sign of the first and second derivatives. Knowledge of curve tracing is useful in application of integration in finding length, area, volume etc. In this chapter, we study tracing of standard and other curves in the (a) cartesian (b) polar and (c) parametric form.

5.1 CURVE TRACING : CURVES IN CARTESIAN FORM

Plane algebraic curve of nth degree is represented by

$$f(x, y) = ay^n + (bx + c)y^{n-1} + (dx^2 + ex + f)y^{n-2} + \ldots + u_n(x) = 0 \quad (1)$$

where $a, b, c, d, e, f \ldots$ are all constants and $u_n(x)$ is a polynomial in x of degree n.

General procedure for tracing the algebraic curve (1) consists of the study of the following characteristics of the curve.

Symmetry

A plane algebraic curve (1) is

a. **Symmetric about x-axis:** if only even powers of y occur in (1) i.e., if y is replaced by $-y$ in (1), the Equation (1) remains unaltered or in other words

$$f(x, -y) = f(x, y).$$

Example: $y^2 = 4ax$ (see Fig. 5.1).

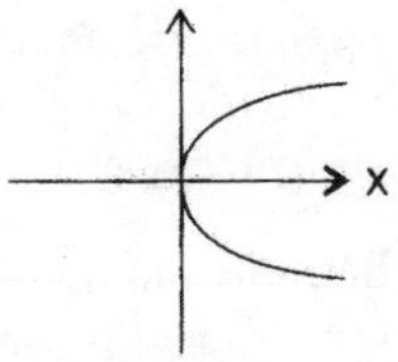

Fig. 5.1

b. **Symmetric about y-axis:** if only even powers of x appear in (1) i.e., $f(-x, y) = f(x, y)$.

Example: $x^2 = 4ay$ (refer Fig. 5.2).

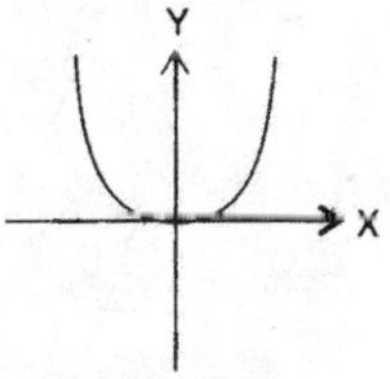

Fig. 5.2

c. **Symmetric about both x- and y-axes:** if only even powers of x and y appear in (1) i.e., $f(-x, -y) = f(x, y)$.

Example: $x^2 + y^2 = a^2$ (see Fig. 5.3).

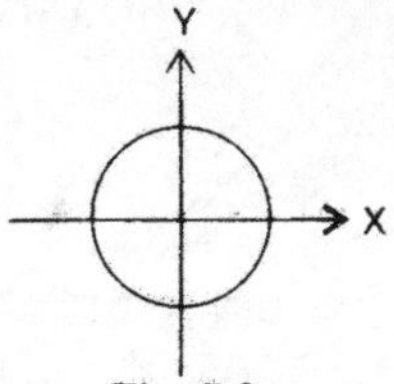

Fig. 5.3

d. Symmetric about origin: if equation remains unaltered when x and y are replaced by $-x$ and $-y$ i.e., $f(-x, -y) = f(x, y)$

Example: $x^5 + y^5 = 5a^2x^2y$ (Fig. 5.4).

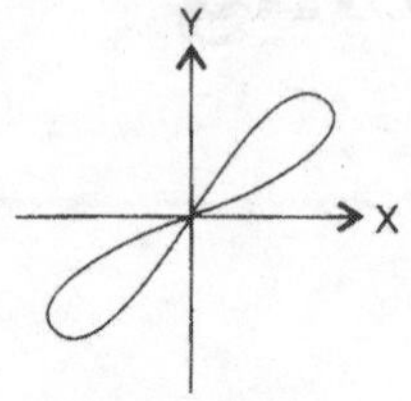

Fig. 5.4

Note: Curve symmetric about both x- and y-axes is also symmetric about origin but not the converse (because of the presense of odd powers)

e. Symmetric about the line $y = x$: if the equation remains unaltered if x and y are replaced by y and x i.e., x, y are interchanged or

$$f(x, y) = f(y, x)$$

Example: $x^3 + y^3 = 3axy$ (Fig. 5.5).

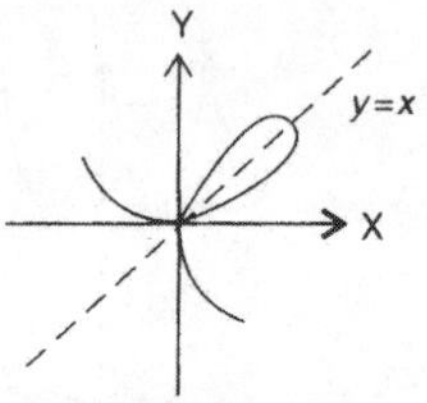

Fig. 5.5

f. Symmetric about the line $y = -x$: if

$$f(x, y) = f(-y, -x)$$

i.e., x is replaced by $-y$ and y is replaced by $-x$.

Example: $x^3 - y^3 = 3axy$ (Fig. 5.6).

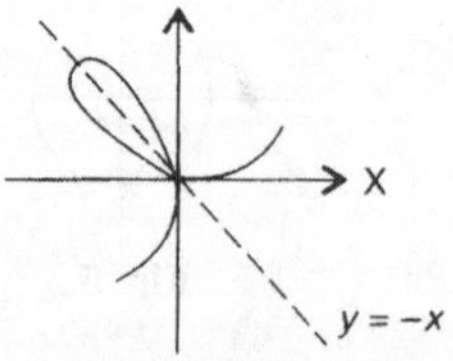

Fig. 5.6

Region or Extent

is obtained by solving y in terms of x or vice versa. Real horizontal extent is defined by values of x for which y is defined. Real vertical extent is defined by values of y for which x is defined.

Imaginary region: is the region in which the curve does not exist. In such region y becomes imaginary (undefined) for values of x or vice versa.

Asymptotes

a. Parallel to x-axis: are obtained by equating to zero the coefficients of the highest powers of x in (1).

b. Parallel to y-axis: are obtained by equating to zero the coefficients of the highest powers of y in (1).

Example: $x^2y - y - x = 0$ is of 3rd degree has maximum number of 3 asymptotes.

y is the coefficient of highest power of x i.e., of x^2. Thus asymptote parallel to x-axis is $y = 0$. $(x^2 - 1)$ is the coefficient of highest power of y i.e., of y. Thus asymptotes parallel to y-axis are $x^2 - 1 = 0$ or $x = \pm$ are two asymptotes parallel to y-axis.

c. No vertical or horizontal asymptotes: In cases (a) and (b) if the coefficients are constants or has imaginary (no real) factors, then curve has no asymptotes.

Example: $x^3 + y^3 = 3axy$, no asymptotes parallel to x- and y-axes because coefficients of x^3 and y^3 are constant one.

d. Oblique asymptotes (not parallel to x-axes and y-axes): are given by $y = mx + c$ where $m = \lim_{x\to\infty}\left(\frac{x}{y}\right)$ and $c = \lim_{x\to\infty}(y - mx)$

Example:

$$y = \frac{x^2 + 2x - 1}{x}, \quad m = \lim_{x\to\infty}\left(\frac{y}{x}\right)$$

$$= \lim\left(\frac{x^2 + 2x - 1}{x^2}\right) = 1$$

$$c = \lim_{x\to\infty}(y - mx) = \lim\left(\frac{x^2 + 2x + 1}{x} - x\right) = 2$$

So $y = x + 2$ is an oblique (inclined) asymptote.

e. Oblique asymptotes (when curve is represented by implicit equation $f(x, y) = 0$): are given by $y = mx + c$ where

m is solution of $\phi_n(m) = 0$ and

c is solution of $c\phi'_n(m) + \phi_{(n-1)}(m) = 0$ or

$c = \frac{-\phi_{n-1}(m)}{\phi'_n(m)}$

Here $\phi_n(m)$ and $\phi_{n-1}(m)$ are obtained by putting $x = 1$ and $y = m$ in the collection of highest degree terms of degree n and in the collection of the next highest degree terms of degree $(n - 1)$.

Example: $x^3 + y^3 = 3axy$, highest degree $n = 3$.

$\phi_3(m) = 1 + m^3$ (obtained by putting $x = 1$ and $y = m$ in $x^3 + y^3$). Real solution $m = -1$. $\phi_2(m) = -3am$ (obtained by putting $x = 1$ and $y = m$ in the next highest degree 2 term $-3axy$)

$$c = -\frac{(-3am)}{(3m^2)} = \frac{a}{m}. \quad \text{At } m = -1, c = -a$$

Thus,

$$y = mx + c = -x - a \quad \text{or} \quad y + x + a = 0$$

is the oblique asymptote.

Origin

If origin (0, 0) lies on the curve then there will be no constant term in equation (1).

Tangents to the Curve at the Origin

When the curve passes through origin, then the tangents to the curve at this origin are obtained by equating to zero the group (or collection) of the lowest degree terms in (1).

Example: $y^2 = 4ax$, lowest degree term $4ax$ equated to zero gives $x = 0$ (y-axis) as tangent to curve at origin.

Example: $x^3 + y^3 = 3axy$, lowest degree term $3axy$ equated to zero gives $xy = 0$ or $x = 0$ and $y = 0$ are the two tangents to the curve at origin.

Example: $a^2y^2 = a^2x^2 - x^4$, lowest degree term $(y^2 - x^2)$ equating to zero gives $y = \pm x$ as the two tangents at origin.

Intercepts

a. The x-intercept (i.e., the point where the curve meets the x-axis) is obtained by putting $y = 0$ in (1) and solving for x. Similarly y-intercept (where curve crosses y-axis) is obtained by putting $x = 0$ in (1) and solving for y.

b. Points of intersection: when curve is symmetric about the line $y = \pm x$, the points of intersection are obtained by putting $y = \pm x$ in Equation (1).

c. Tangents to the curve at other points (h, k) can be obtained by shifting the origin to these points (h, k) by the substitution $x = X + h$, $y = Y + k$ and calculating the tangents at origin in the new XY-plane.

Note: At point where $\frac{dy}{dx} = 0$, the tangent is parallel to x-axis i.e., horizontal. At point where $\frac{dy}{dx} = \infty$, the tangent is vertical i.e., parallel to y-axis.

Sign of First Derivative $\frac{dy}{dx}$

In an interval $a \le x \le b$ if

a. $\frac{dy}{dx} > 0$ then curve is increasing in $[a, b]$

b. $\frac{dy}{dx} < 0$ then curve is decreasing in $[a, b]$

c. If at $x = x_0$, $\frac{dy}{dx} = 0$ then (x_0, y_0) is a stationary point where maxima and minima can occur.

Sign of Second Derivative $\frac{d^2y}{dx^2}$

In an interval $a \le x \le b$ if

a. $\frac{d^2y}{dx^2} > 0$ then curve is convex or concave upward (holds water)

b. $\frac{d^2y}{dx^2} < 0$ then curve is concave downward (spills water)

c. A point at which $\frac{d^2y}{dx^2} = 0$ is known as an inflection point where the curve changes the direction of concavity from downward to upward or vice versa.

Multiple Point (or Singular Point)

A point through which r branches of a curve pass is called a multiple point of the rth order and has r

tangents. Thus at a double point two branches of the curve pass.

Double point is classified as a node, a cusp or an isolated (or conjugate) point according as the two tangents are real distinct, coincident or imaginary.

Multiple points are obtained by solving for (x, y) the three equations

$$f_x(x, y) = 0,\ f_y(x, y) = 0,\ f(x, y) = 0$$

The slopes of the tangents at a double point are the roots of

$$f_{yy}\left(\frac{dy}{dx}\right)^2 + 2f_{xy}\frac{dy}{dx} + f_{xx} = 0$$

Thus the point (x, y) will be a double point and will be a node, cusp or conjugate according as values of $\left(\frac{dy}{dx}\right)$ are real/distinct, equal or imaginary i.e., according as $(f_{xy})^2 - f_{xx}f_{yy} > 0, = 0, < 0$.

Example: $y^2 = x^2(x - 1)$. So $f(x, y) = y^2 - x^3 + x^2 = 0$, $f_x = -3x^2 + 2x = 0$, $f_y = 2y = 0$.

$\therefore$ $(0, 0)$ is a double point. Equating to zero lowest degree terms $x^2 + y^2$, we get imaginary tangents. So $(0, 0)$ is a conjugate point.

Example: $y^2(a - x) = x^3$. So $f(x, y) = x^3 - ay^2 + xy^2 = 0$, $f_x = 3x^2 + y^2 = 0$, $f_y = 2y(x - a) = 0$, $(0, 0)$ is a double point. Tangents at origin are given by equating to zero the lowest degree terms i.e., $y^2 = 0$.

$\therefore$ x-axis is coincident tangent: $(0, 0)$ is a cusp.

Example: $f(x, y) = x^3 - y^2 - 7x^2 + 4y + 15x - 13 = 0$, $f_x = 3x^2 - 14x + 15 = 0$, $f_y = -2y + 4 = 0$. out of $(3, 2)$, $(5/3, 2)$ only $(3, 2)$ lies on curve. $f_{xy} = 0$, $f_{xx} = 6x$, $f_{yy} = -2$. At $(3, 2)$, $f_{xx} = 18$, $f_{xy} = 0$, $f_{yy} = -2$ so $(f_{xy})^2 - f_{xx}f_{yy} > 0$.

$\therefore$ $(3, 2)$ is a node.

WORKED OUT EXAMPLES

Example 1: Trace the curve $y = (x^2 - x - 6)(x - 7)$.

Solution: Equation is $y = (x + 2)(x - 3)(x - 7)$

1. No symmetry.
2. Origin does not lie on curve.
3. No asymptotes.
4. *Intercepts: x-intercepts:* $x = -2, 3, 7$. Curve crosses x-axis at A$(-2, 0)$, B$(3, 0)$ and C$(7, 0)$ *y-intercepts:* $y = 42$.

   Curve crosses y-axis at D$(0, 42)$.
5. *Maxima and Minima:*

   $$\frac{dy}{dx} = 3x^2 - 16x + 1 = 3(x - x_1)(x - x_2)$$

   where

   $$x_1 = \frac{8 - \sqrt{61}}{3} = 0.063, \quad x_2 = \frac{8 + \sqrt{61}}{3} = 5.27$$

   Thus the stationary points where $y' = 0$ are x_1 and x_2.

   $$\frac{d^2y}{dx^2} = 6x - 16 = 2(3x - 8)$$

   At $x_1 = .063$, $y'' < 0$, so y attains maximum value 42.0314 at $x_1 = 0.063$ i.e., at B(0.063, 42.0314). At $x_2 = 5.27$, $y'' > 0$, so y attains minimum value -28.152 at $x_2 = 5.27$ i.e., $E(5.27, -28.152)$.
6. *Inflection point:* is at $x = \frac{8}{3} = 2.66$ since $y'' = 0$ at $x = \frac{8}{3}$.
7. *Extent:* $-\infty < x < \infty$ since y is defined for all x.
8. *Sign of first derivative:*

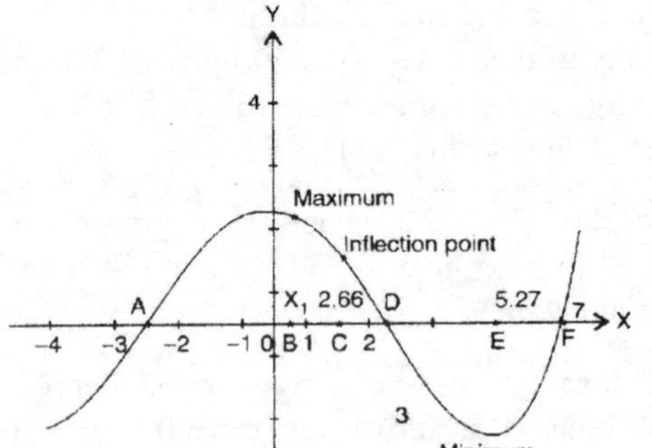

Fig. 5.7

Interval	Sign of y $y = (x+2)(x-3)(x-7)$	Quad-rant	Sign of y' $y' = 3(x-x_1)(x-x_2)$	Nature of curve
$-\infty < x < -2$	$y < 0$	III	$y^1 > 0$	increasing
$-2 < x < 0$	$y > 0$	II	$y' > 0$	increasing
$0 < x < .063$	$y > 0$	I	$y' > 0$	increasing
$.063 < x < 3$	$y > 0$	I	$y' < 0$	decreasing
$3 < x < 5.27$	$y < 0$	IV	$y' < 0$	decreasing
$5.27 < x < 7$	$y < 0$	IV	$y' > 0$	increasing
$7 < x < \infty$	$y > 0$	I	$y' > 0$	increasing

$A(-2, 0)$, $B(0.063, 42.0314)$, $C(2.66, 6.74)$, $D(3, 0)$, $E(5.27, -28.152)$, $F(7, 0)$, $x_1 = 0.063$, $x_2 = 5.27$

Using the above knowledge the graph of the given curve is as shown above (in Fig. 5.7).

Example 2: Trace the curve $y = \frac{x^2-3x}{(x-1)}$.

Solution:

1. Curve is not symmetric
2. Origin lies on the curve
3. Tangent at origin is $y = 3x$ obtained by equating the lowest degree term $y - 3x$ to zero.
4. Intercepts: x intercept: put $y = 0$, then

$$x(x-3) = 0.$$

$\therefore$ Curve crosses x-axis at $x = 0$ and $x = 3$

y-intercepts: put $x = 0$, then $y = 0$.

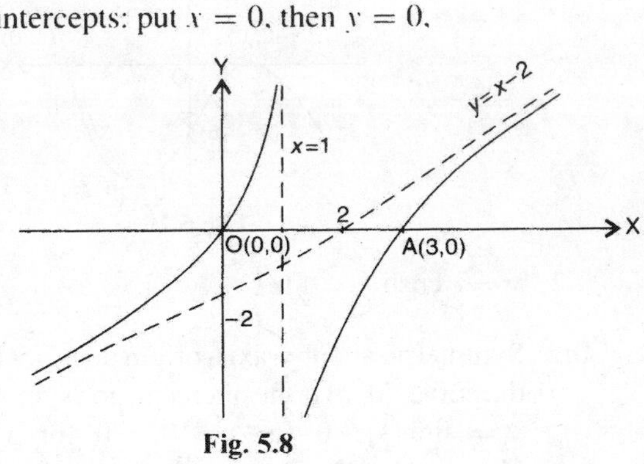

Fig. 5.8

Thus curve meets the x-axis at $O(0, 0)$, $A(3, 0)$ (see Fig. 5.8)

5. *Extent or Region:* y is defined for all x, except at $x = 1$ where it is discontinuous. Thus the region of definition is $\{-\infty < x < \infty\} - \{1\}$ or $R - \{1\}$.
6. *Asymptotes*

 a. No horizontal asymptote because co-efficient of x^2 is constant.

 b. $x = 1$ is the vertical asymptote which is obtained by equating to zero the coefficient of the highest power in y i.e., $(x-1) = 0$.

 c. *Oblique asymptote:* $m = \lim_{x\to\infty} \frac{y}{x} = \lim \frac{x(x-3)}{(x-1)} \cdot \frac{1}{x} = 1$

$$c = \lim_{x\to\infty} (y - mx) = \lim_{x\to\infty} \left(\frac{-2x}{x-1}\right) = -2.$$

 Oblique asymptote is $y = mx + c = x - 2$.

7. *Maxima and Minima:* $\frac{dy}{dx} = \frac{x^2-2x+3}{(x-1)^2}$, $y' > 0$ for any x except $x = 1$. $y' = 0$ when $x^2 - 2x + 3 = 0$ whose roots are imaginary. So no stationary points and therefore no maximum and no minimum.
8. *Inflection point*:

$$\frac{d^2y}{dx^2} = -\frac{4}{(x-1)^3}$$

 No inflection points since $y'' \neq 0$.

9. *Sign of derivative:*

Interval	Sign of y	Quad-rant	Sign of y'	Nature of curve
$-\infty < x < 0$	$y < 0$	III	$y' > 0$	increasing
$0 < x < 1$	$y > 0$	I	$y' > 0$	increasing
$1 < x < 3$	$y < 0$	IV	$y' > 0$	increasing
$3 < x < \infty$	$y > 0$	I	$y' > 0$	increasing

Example 3: Trace the curve

$$a^2y^2 = x^2(2a - x)(x - a)$$

Solution: Equation is $a^2y^2 = -x^4 + 3ax^3 - 2a^2x^2$.

1. Curve is symmetric about x-axis only.
2. Origin lies on the curve.
3. *Intercepts*: y-intercept: $(0, 0)$.
 x-intercepts: $(0, 0)$, $(a, 0)$, $(2a, 0)$.

4. *Region:* Solving for y, we get

$$y = \pm\frac{1}{a}\sqrt{x^2(2a-x)(x-a)}$$

y is not imaginary if both $x - a > 0$ and

$$(2a - x) > 0 \text{ i.e., } a < x < 2a$$

Thus curve exists only between $x = a$ and $x = 2a$. Curve has two branches one above and other below x-axis.

5. *Multiple point:*

Tangents at origin are obtained by equating to zero terms of lowest degree i.e., $a^2y^2 + 2a^2x^2 = 0$ which has imaginary values. Thus tangents at origin are imaginary. Therefore origin O is an isolated point.

6. *Asymptotes:* No horizontal and no vertical asymptotes because coefficients of x^4 and y^2 are constants. No oblique asymptotes because $m = \lim\limits_{x\to\infty}\frac{y}{x} = \lim\limits_{x\to\infty}\sqrt{(2a-x)(x-a)} = \infty$ (Fig. 5.9).

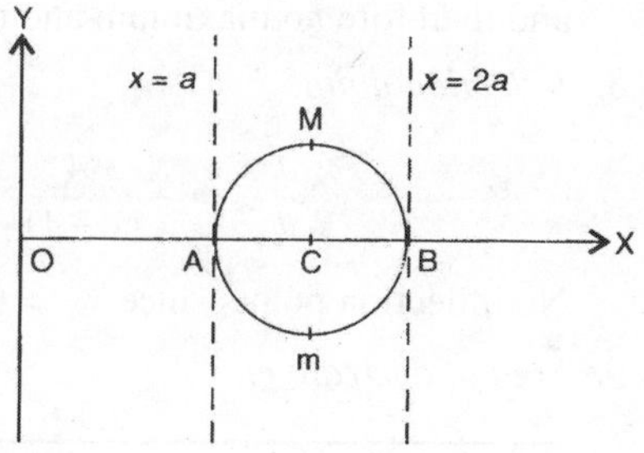

Fig. 5.9

7. *Loop:* curve crosses at $A(a, 0)$ and $B(2a, 0)$ and is symmetric about x-axis. Therefore curve has a loop between A and B.

8. *Maxima and Minima:* Derivative $\frac{dy}{dx} = \pm\frac{2x^2-\frac{9}{2}ax+2a^2}{a(x-a)^{1/2}(2a-x)^{1/2}}$ stationary points where $y' = 0$ are given by $2x^2 - \frac{9}{2}ax + 2a^2 = 0$ or $x = \frac{a(9\pm\sqrt{17})}{8} = \frac{a}{8}(9 \pm 4.1231)$. Out of these $x_1 = \frac{a}{8}(9 - 4.1231) = 0.6096a$ does not lie in the region of interest $a < x < 2a$. The other value $x_2 = \frac{a}{8}(9 + 4.1231) = 1.64a$ is the required value where a maximum value $M = 0.7872a$ occurs on the upper branch of the curve. Similarly the minimum value $m = -0.7872a$ occurs on the lower branch at $x_2 = 1.64a$ (because of symmetry). $C(1.64a, \pm.7872a)$.

9. Tangents at A and B are vertical because $\frac{dy}{dx} = \infty$ at $x = a$ and $x = 2a$. Thus $x = a$ and $x = 2a$ are the tangents to the curve at the points A and B.

10. *Sign of derivative:*

For $y = +\frac{1}{a}\sqrt{x^2(2a-x)(x-a)}$

the derivative is $y' = \frac{-2(x-x_1)(x-x_2)}{a(x-a)^{\frac{1}{2}}(2a-x)^{\frac{1}{2}}}$

For $x_1 = .6a < a < x < x_2 = 1.64a$, then $y > 0$, $y' > 0$, increasing.

For $x_2 = 1.64a < x < 2a$, then $y > 0$, $y' < 0$, decreasing.

Similar arguments for the other branch of the curve below the x-axis i.e., $y < 0$.

EXERCISE

Trace the following curves giving the salient points:

1. $y = x^3 - 12x - 16$.

Ans. Not symmetric, does not pass through origin, $(-2, 0)$, $(0, -16)$, $(4, 0)$ are intercepts, maximum at $x = -2$ and minimum at $x = 2$, inflection point at $(0, -16)$, y is increasing in $(-\infty, -2]$, and $[2, \infty)$ and decreasing in $(-2, 2)$, no asymptotes (Fig. 5.10).

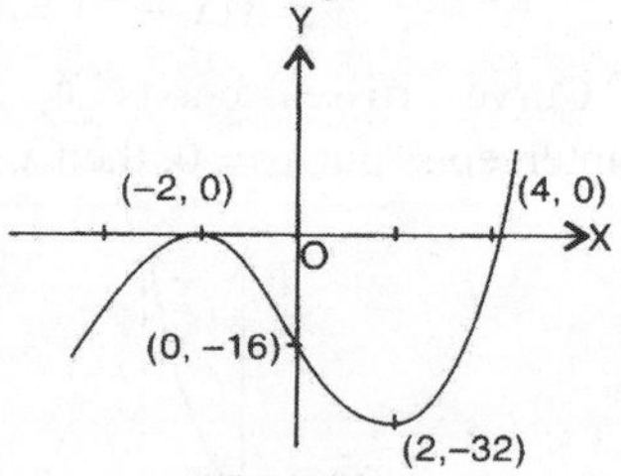

Fig. 5.10

2. $y = c\cosh\frac{x}{c} = \frac{c}{2}[e^{\frac{x}{c}} + e^{-\frac{x}{c}}]$.

Ans. Symmetric about y-axis, origin does not lie on the curve, $(0, c)$ is the intercept, no asymptotes, $y' = \sinh\frac{x}{c} > 0$ for $x > 0$, < 0 for $x < 0$. Curve increasing in $(0, \infty)$ and decreasing in $(-\infty, 0)$, tangent at origin $y = c$ is horizontal (Fig. 5.11).

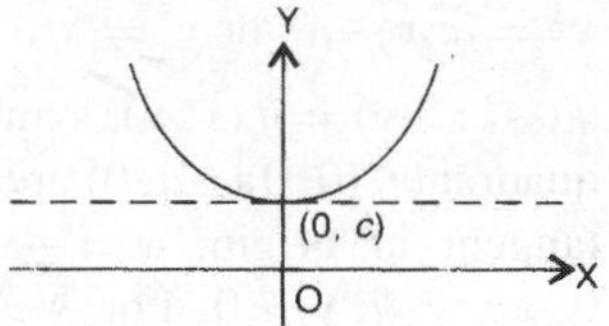

Fig. 5.11

3. $y = (x^2 + 1)/(x^2 - 1)$.

Ans. Symmetric about y-axis, origin not on curve, $(0, -1)$ intercept point, $x = \pm 1$ vertical asymptotes, $y=1$ horizontal asymptote, no inclined asymptote, maximum value -1 at $x = 0$, curve does not exist between the lines $y = -1$ and $y = 1$, no inflection points (Fig. 5.12).

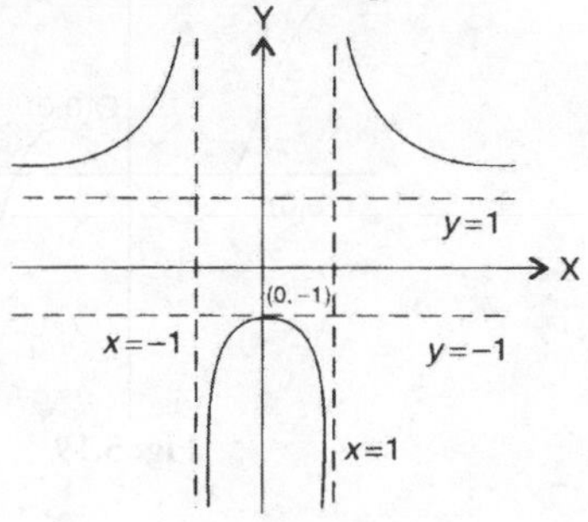

Fig. 5.12

4. $y^2x^2 = x^2 - a^2$.

Ans. Symmetric about both axes, origin not on curve, $(-a, 0)$, $(a, 0)$ are intercepts and $x = a$, $x = -a$ are tangents at these points, $y = \pm 1$ are asymptotes, curve is defined only when $x \geq a$ or $x \leq -a$, no curve between $x = -a$ and $x = a$, $y' = \pm a^2/(x^2\sqrt{x^2 - a^2}) \neq 0$, for $y' > 0$ in $(-\infty, -a)$ and (a, ∞) curve is increasing in these intervals (Fig. 5.13).

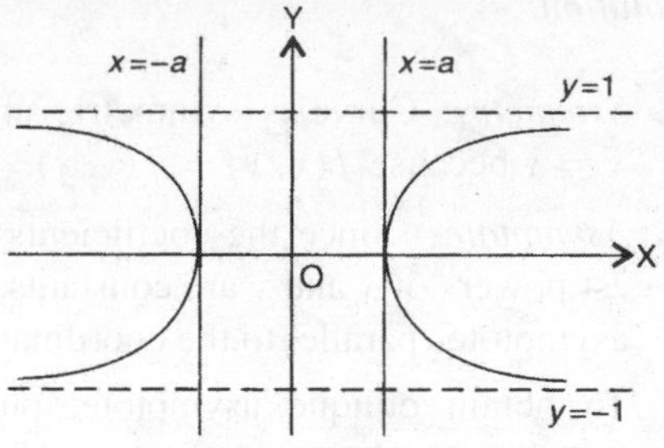

Fig. 5.13

5. $y^2(x^2 - 1) = 2x - 1$.

Ans. Symmetric about x-axis, not through origin, $(0, 1)$, $(0, -1)$, $(\frac{1}{2}, 0)$ are intercepts, $x = \frac{1}{2}$ is tangent at $(\frac{1}{2}, 0)$, $x = \pm 1$, $y = 0$ are asymptotes; curve exists in the region $-1 < x < \frac{1}{2}$ and $x > 1$, $y = \pm\sqrt{\frac{2x-1}{x^2-1}}$, $y' = \pm(-x^2 + x + 1)/[(2x - 1)^{1/2}(x^2 - 1)^{3/2}]$ (Fig. 5.14).

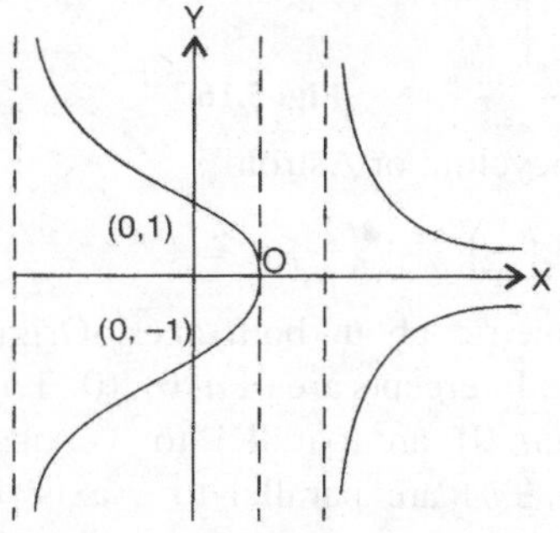

Fig. 5.14

6. $x^3 + y^3 = 3ax^2$, $a > 0$.

Ans. Not symmetric, origin is cusp; $x = 0$ is tangent thereat, $(0, 0)$, $(3a, 0)$ are intercepts; $x = 3a$ is tangent at $(3a, 0)$; $y + x = a$ is asymptote, meets curve at $(\frac{a}{3}, \frac{2a}{3})$; curve does not exist in 3rd quadrant because both x and y cannot be negative, $y' = x(2a - x)/y^2$ (Fig. 5.15).

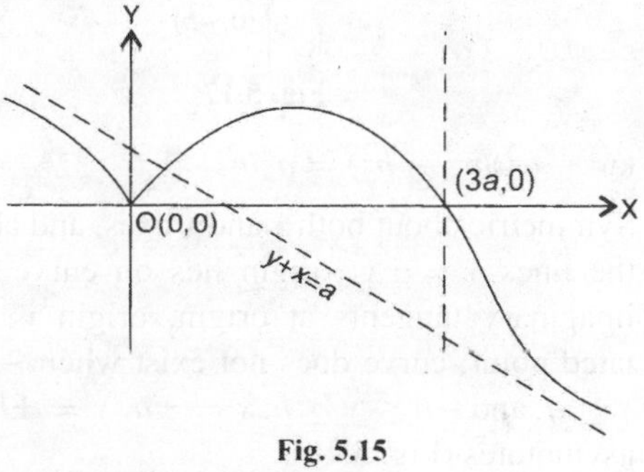

Fig. 5.15

7. $9ay^2 = x(x - 3a)^2$.

Ans. Symmetric about x-axis, origin on curve; $x = 0$ is tangent at origin 0: $(3a, 0)$ is intercept; tangents at $(3a, 0)$ are $y = \pm(x - 3a)/\sqrt{3}$; curve exists only for $x > 0$ (in Ist and IV quadrants), no asymptotes; $y = \sqrt{x}(x - 3a)/(3\sqrt{a})$; $y' = (x - a)/(2\sqrt{ax})$ (Fig. 5.16).

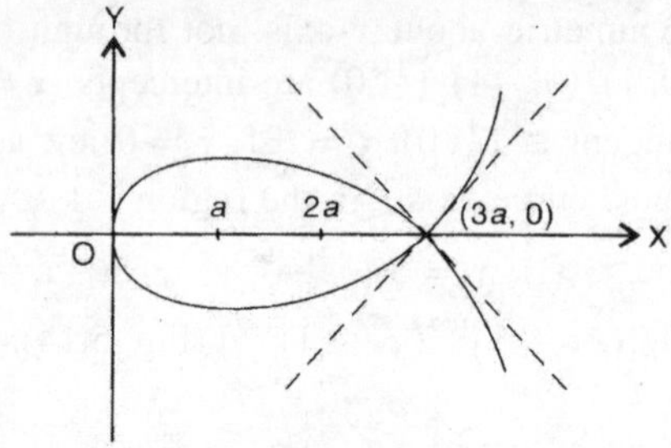

Fig. 5.16

8. Hypocycloid or Astroid

$$\left(\frac{x}{a}\right)^{2/3}+\left(\frac{y}{b}\right)^{2/3}=1.$$

Ans. Symmetric about both axes. Origin not on curve; Intercepts are $(\pm a, 0)$, $(0, \pm b)$ tangents at $(\pm a, 0)$ are parallel to x-axis, tangents at $(0, \pm b)$ are parallel to y-axis; curve exists when $-a < x < a$ and $-b < y < b$; no asymptotes (Fig. 5.17).

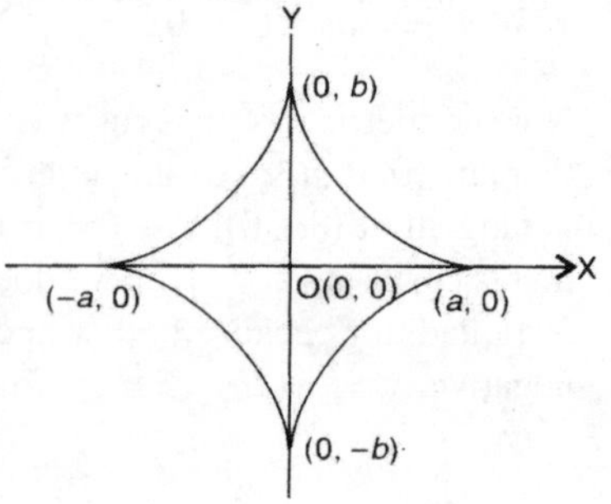

Fig. 5.17

9. $(x^2 - a^2)(y^2 - b^2) = a^2b^2$.

Ans. Symmetric about both x and y axes, and about the lines $y = \pm x$; origin lies on curve, has imaginary tangents at origin, origin is isolated point; curve does not exist when $-a \le x \le a$, and $-b \le y \le b$; $x = \pm a$, $y = \pm b$ are asymptotes (Fig. 5.18).

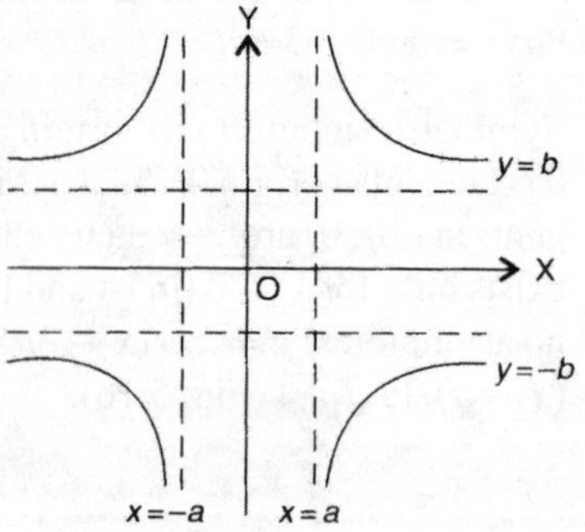

Fig. 5.18

10. $y^3 = a^2x - x^3$ or $y^3 = x(a^2 - x^2)$

Ans. $f(-x, -y) = f(x, y)$, symmetry in opposite quadrants, $(0, 0)(\pm a, 0)$ are intercepts, y-axis tangent at origin, $y = -x$ asymptote. For $0 < x < a, y > 0$. For $x > a, y < 0$. When $-a < x < 0$ then $y < 0$, also $-a > x$ then $y > 0$; $y' = \frac{a^2-3x^2}{3y^2}$ Max at $x = \frac{a}{\sqrt{3}}$, min at $x = \frac{-a}{\sqrt{3}}$. Also $y' < 0$ when $(-\infty, \frac{-a}{\sqrt{3}}]$ and $(\frac{a}{\sqrt{3}}, \infty)$ so curve is decreasing. Finally $y' > 0$ in $\left(\frac{-a}{\sqrt{3}} \text{to} \frac{a}{\sqrt{3}}\right)$ so curve is increasing (Fig. 5.19).

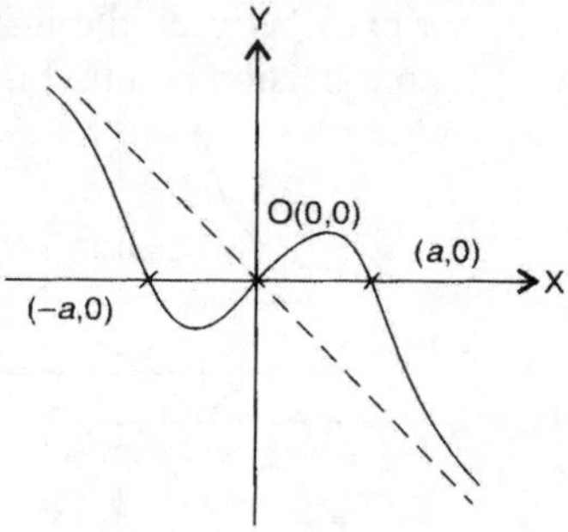

Fig. 5.19

5.2 CURVE TRACING: STANDARD CURVES IN CARTESIAN FORM

WORKED OUT EXAMPLES

Example 1: Folium of Descartes: $x^3+y^3=3axy$.

Solution:

1. *Symmetry:* Curve is symmetric about the line $y = x$ because $f(x, y) = f(y, x)$.
2. *Asymptotes:* Since the coefficients of the highest powers of x and y are constants, there are no asymptotes parallel to the coordinate axis.

   To obtain oblique asymptote, put $y = mx + c$ in the equation $x^3 + (mx + c)^3 = 3ax(mx + c)$ or $(1 + m^3)x^3 + 3x^2(m^2c - am) + 3x(cm - ca) + c^3 = 0$ This equation will give

$$1 + m^3 = 0 \text{ and } m^2c - am = 0.$$

Solving $m = -1$ and $c = -a$. Oblique asymptote is $y = -x - a$ or $y + x + a = 0$.

3. *Origin:* O(0, 0) is a point on the curve. Tangents at origin: Equating to zero the lowest degree term xy i.e., $xy = 0$, observe that $x = 0$ and $y = 0$ are tangents to the curve at the origin.

4. *Intercepts:* there are no x-intercepts and no y intercepts except the origin (0, 0) (Fig. 5.20) because by putting $x = 0$ in equation $0 + y^3 = 0$ so $y = 0$ (similarly by putting $y = 0$, $x^3 + 0 = 0$ so $x = 0$).

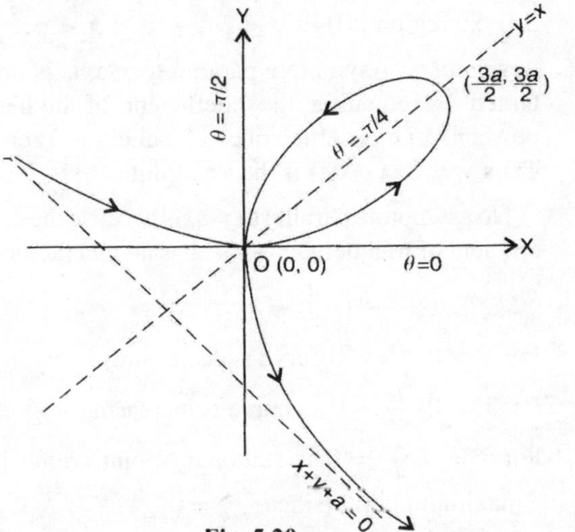

Fig. 5.20

5. *Intersection of curve with the line $y = x$:* put $y = x$ in equation: $x^3 + x^3 = 3a \cdot x \cdot x$ or $x^2(2x - 3a) = 0$ so $x = 0$ or $x = \frac{3a}{2} = y$. Thus line $y = x$ meets the curve in two points (0, 0) and $\left(\frac{3a}{2}, \frac{3a}{2}\right)$.

$$\text{Derivative}: \frac{dy}{dx} = \frac{ay - x^2}{y^2 - ax}$$

At $\left(\frac{3a}{2}, \frac{3a}{2}\right)$, $\frac{dy}{dx} = -1$.

Equation of tangent to the curve at the point $\left(\frac{3a}{2}, \frac{3a}{2}\right)$ is

$$\left(y - \frac{3a}{2}\right) = -\left(x - \frac{3a}{2}\right) \quad \text{or } x + y - 3a = 0$$

Thus this tangent is parallel to the asymptote $x + y + a = 0$

6. *Region:*
   a. When both x and y are negative simultaneously equation of the curve is not satisfied

$$(-x)^3 + (-y)^3 = 3a(-x)(-y)$$
$$-(x^3 + y^3) = 3axy$$

   (R.H.S. is positive while L.H.S. is negative). Thus no part of the curve exists in the 3rd quadrant.

   b. To study the variation of y with x, put $x = r\cos\theta$, $y = r\sin\theta$ in the equation. $r^3(\cos^3\theta + \sin^3\theta) = 3ar^2 \sin\theta\cos\theta$

$$r = \frac{3a\sin\theta\cos\theta}{\cos^3\theta + \sin^3\theta}$$

$\theta = 0$	$\pi/4$	$\pi/3$	$\pi/2$	$2\pi/3$	$3\pi/4$	π
$r = 0$	$\frac{3\sqrt{2}a}{2} = 2.12a$	$\frac{6\sqrt{3}a}{1+3\sqrt{3}} = 1.51a$	0	$\frac{6\sqrt{3}a}{3\sqrt{3}-1} = -2.46a$	∞	0

$$\frac{dr}{d\theta} = \frac{3a(\cos\theta - \sin\theta)[1 + \sin\theta\cos\theta + \sin^2\theta\cos^2\theta]}{(\cos^3\theta + \sin^3\theta)}$$

$$\frac{dr}{d\theta} = 0 \text{ when } \cos\theta = \sin\theta \text{ i.e., at } \theta = \frac{\pi}{4}.$$

$$\frac{dr}{d\theta} > 0 \text{ in } (0, \pi/4) \text{ and } < 0 \text{ in } (\pi/4, \pi/2).$$

Thus the curve has a loop between $\theta = 0$ to $\frac{\pi}{2}$ since r increases first as θ increases from 0 to $\pi/2$ from 0 to $3\sqrt{2}a/2$ as θ increases from 0 to $\pi/4$, then r decreases from $3\sqrt{2}a/2$ to 0 as θ varies from $\pi/4$ to $\pi/2$.

Example 2: Cissoid: $y^2(a - x) = x^3$, $a > 0$.

Solution:

1. Curve is symmetric about x-axis since y^2 is present in the equation.
2. Origin lies on the curve.
3. Tangents at origin: Equating to zero the lowest degree terms i.e., $y^2 = 0$. Thus x-axis is a coincidental tangent to the curve at the origin. Hence origin is a cusp.
4. *Intercepts:* only intercept point is $x = 0$, $y = 0$. i.e., origin is the only point where curve meets the coordinate axes.

5. *Asymptotes:* Asymptotes parallel to y-axis are given by equating to zero, the coefficient of the highest degree terms in y i.e., $a - x = 0$. Thus $x = a$ is the asymptote. For $a > 0$ the asymptote $x = a$ is parallel to y-axis and lies to the right $(y > 0)$.

   Since coefficient of x^3 is constant, no asymptotes parallel to x-axis.

6. *Extent:* solving $y = \pm x\left(\frac{x}{a-x}\right)^{\frac{1}{2}}$.
   y is imaginary when $x < 0$.
   y is imaginary when $x > a$.
   Thus curve does not exist for $x < 0$ and $x > a$.
   Hence curve exists only for $0 \le x < a$.

7. *Sign of derivative*

$$\frac{dy}{dx} = \pm\frac{\sqrt{x}(\frac{3}{2}a - x)}{(a-x)(a-x)^{1/2}}$$

In the first qauadrant for $0 < x < a$, $\frac{dy}{dx} > 0$ curve increases in the first quadrant (Fig. 5.21).

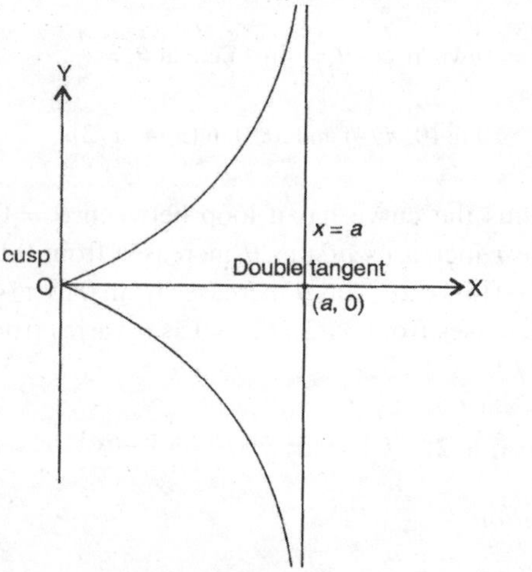

Fig. 5.21

Example 3: $y = 8a^3/(x^2 + 4a^2)$

Solution:

1. Curve is symmetric about y-axis.
2. Curve does not pass through origin.
3. *y-intercept*: curve meets y-axis at $(0, 2a)$.
   x-intercept: curve does not cross x-axis.
4. *Tangent at* $(0, 2a)$: $\frac{dy}{dx} = \frac{-16a^3x}{(x^2+4a^2)^2}$

   So $\frac{dy}{dx} = 0$ when $x = 0$. (then $y = 2a$).

   Thus at $(0; 2a)$, the tangent is parallel to x-axis i.e., the horizontal line $y = 2a$ is tangent at $(0, 2a)$.

5. *Region:* y is always positive for any value of x (because of the presence of x^2 term). Thus curve lies above the x-axis $(y > 0)$. Also solving the equation $x = 2a\left(\frac{2a-y}{y}\right)^{\frac{1}{2}}$.
   x is imaginary when $y < 0$ or $y > 2a$.

   Curve does not exist below the x-axis (i.e., $y < 0$) and above the horizontal line $y = 2a$ (i.e., $y > 2a$). So region is $0 < y < 2a$, $-\infty < x < \infty$.

6. *Asymptotes:* Asymptote parallel to x-axis is obtained by equating the coefficient of highest power of x i.e., coefficient of x^2 namely y to zero. Thus $y = 0$ (x-axis) is the asymptote.

   No asymptote parallel to y-axis because the coefficient of y namely $x^2 + 4a^2$ has no real factors.

7. *Sign of derivative:* $\frac{dy}{dx} = -\frac{16a^3x}{(x^2+4a^2)^2}$.

   For $x > 0$, $\frac{dy}{dx} < 0$ curve is decreasing.

   For $x < 0$, $\frac{dy}{dx} > 0$ so curve is increasing.

   For $x = 0$, $\frac{dy}{dx} = 0$ is stationary point which is a maximum point (since $\frac{d^2y}{dx^2} = \frac{16a^3(3x^2-4a^2)}{(x^2+4a^2)^3} < 0$ at $x = 0$).

8. Inflection points occur at $3x^2 - 4a^2 = 0$ where

$$\frac{d^2y}{dx^2} = 0 \text{ i.e., at } x_{1,2} = \pm 2a/\sqrt{3}$$

9. Variation of y w.r.t. x (Fig. 5.22)

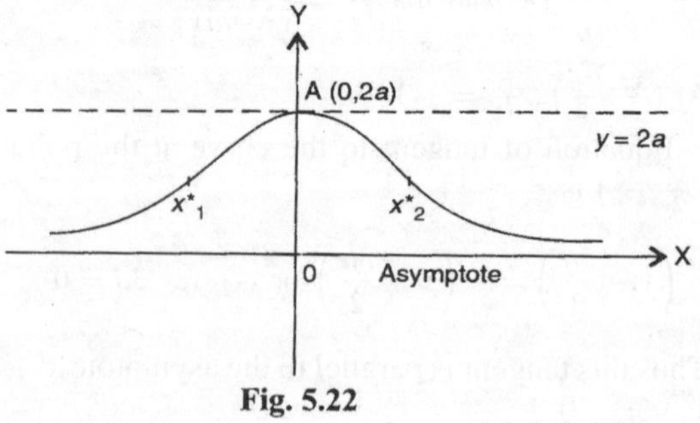

Fig. 5.22

$x = -\infty$	$-2a$	$-a$	0	a	$2a$	∞
$y = 0$	a	$1.6a$	$2a$	$1.6a$	a	0

Example 4: Lemniscate of Bernoulli

$$y^2(a^2 + x^2) = x^2(a^2 - x^2)$$

or $$x^2(x^2 + y^2) = a^2(x^2 - y^2).$$

Solution:

1. Curve is symmetric about both x-axis and y-axis since it contains even powers of x and y.
2. Origin lies on the curve.
3. Tangents at origin are obtained by equating the lowest degree term $(y^2 - x^2)a^2$ to zero. Thus $y = \pm x$ are the two tangents to the curve at origin. Hence origin is a node because there two real distinct tangents.
4. *Asymptotes:* no asymptotes parallel to x-axis because the coefficient of x^4 is constant. No asymptotes parallel to y-axis because the coefficient of y^2 has no real factors.
5. *Intercepts:* x-intercept is $(0, 0)$. y-intercept is $x^2(a^2 - x^2) = 0$ or $x = 0$ or $x \pm a$.

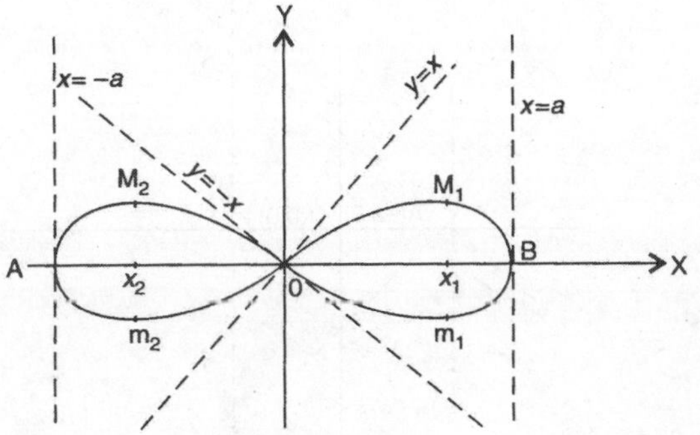

Fig. 5.23

Thus curve meets the x-axis at $A(-a, 0)$ and $B(a, 0)$. To find the tangents at A and B, shift the origin to $(\pm a, 0)$. So put $X = x + a$, $Y = y + 0$ in the equation of the curve.

Then $$Y^2[a^2 + (X + a)^2] = (X + a)^2[a^2 - (X \pm a)^2]$$

or $$Y^2[2a^2 + 2aX + X^2] = (X + a^2)^2[-2aX - X^2]$$

Equating to zero the lowest degree terms the tangent at the next origin i.e., $-2aX^3 = 0$ i.e., new y-axis. Thus $X = x \pm a = 0$ or $x = \pm a$ are the tangents to the curve at B and A.

6. *Region:* Solving the equation for y, we get

$$y = \pm x\sqrt{\frac{a^2 - x^2}{a^2 + x^2}}$$

y is defined only when $a^2 - x^2 \geq 0$ or $-a \leq x \leq a$. Thus curve exists only when $-a \leq x \leq a$ i.e., curve lies between the lines $x = -a$ and $x = a$.

Sign of derivative: $$\frac{dy}{dx} = \frac{\pm[a^4 - 2a^2x^2 - x^4]}{(a^2 + x^2)^{3/2}(a^2 - x^2)^{1/2}}.$$

Stationary points: $\frac{dy}{dx} = 0$ when $x^4 + 2a^2x^2 - a^4 = 0$

or at $$x = \pm a\sqrt{2} - 1, x_1 = \sqrt{-1 + \sqrt{2}a},$$

and $$x_2 = -\sqrt{-1 + \sqrt{2}a}$$

or $$x_1 = \cdot 64a \quad \text{and} \quad x_2 = -\cdot 64a$$

Also $\frac{dy}{dx} \to \infty$ when $x \to +a$ and $x \to -a$.

Thus tangents are parallel to x-axis at x_1 and x_2 while parallel to y-axis at $x = a$ and $x = -a$.

Suppose $y > 0$ then y is considered with + sign. Then for $-a < x < x_2$ and $0 < x < x_1$, $y' > 0$ curve is increasing in these two intervals.

For $x_2 < x < 0$ and $x_1 < x < a$, $y' < 0$ curve is decreasing in these two intervals. Similarly for $y < 0$ take y' with $-$ve sign. Then for $-a < x < x_2$ and $0 < x < x_1$, $y' < 0$, curve decreasing.

For $x_2 < x < 0$ and $x_1 < x < a$, $y' > 0$, curve increasing. M_1, M_2, m_1, m_2 are the extrema at x_1, x_2 (Fig. 5.23).

Example 5: $y = x + \frac{1}{x}$

Solution:

1. Curve is symmetric about the origin since $f(-x, -y) = f(x, y)$ (but is not symmetric about both x- and y-axis because of the odd powers of x and y).
2. Origin does not lie on the curve.
3. *No intercepts:* Rewriting the equation we get $x^2 - xy + 1 = 0$. This has no solutions when $x = 0$ $(1 = 0)$ and $y = 0$, $(x^2 + 1 = 0)$.

4. *Region:* y is defined for any value of x except $x = 0$ where it is discontinous. Thus Region is $\{-\infty < x < \infty\} - \{0\}$ i.e., $R - \{0\}$.

5. *Asymptotes:* As $x \to 0$, $y \to \pm\infty$ so y-axis is an asymptote. No horizontal asymptote because coefficient of x^2 is constant 1.
Oblique asymptote: $y = mx + c$

$$m = \lim_{x\to\infty} \frac{y}{x}$$

$$m = \lim_{x\to\infty} 1 + \frac{1}{x^2} = 1$$

$$c = \lim_{x\to\infty} (y - mx) = \lim_{x\to\infty} \frac{1}{x} = 0$$

Thus the oblique asymptote is the line $y = x$.

6. *No multiple points:* $f(x, y) = x^2 - xy + 1 = 0$. $f_x(x, y) = 2x - y = 0$, $f_y(x, y) = -x = 0$ has no solution.

7. *Sign of derivative:* $\frac{dy}{dx} = \frac{x^2-1}{x^2} = \frac{(x-1)(x+1)}{x^2}$ (Fig. 5.24)

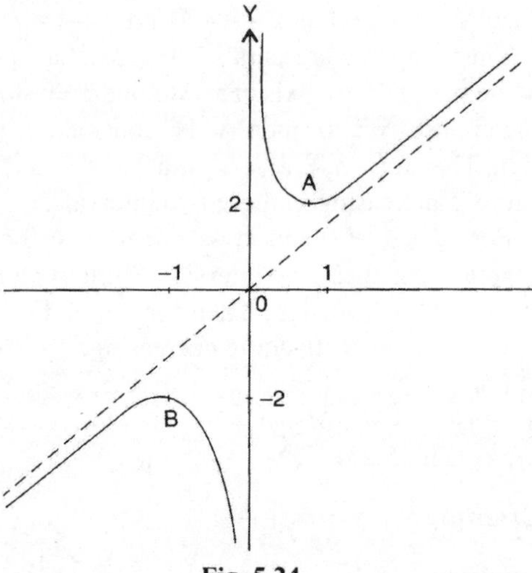

Fig. 5.24

a. *Stationary points:* $x^2 - 1 = 0$ or $x = \pm 1$
$\frac{d^2y}{dx^2} = \frac{2}{x^3}$, so at $x = 1$, $y'' > 0$, attains minimum. At $x = -1$, $y'' < 0$, attains maximum. Thus $A(1, 2)$ is minimum, while $B(-1, -2)$ is maximum.

b. Tangents to curve at A, B are parallel to x-axis.

c. For $0 < x < 1$, $y' < 0$, curve is decreasing.
For $1 < x < \infty$, $y' > 0$, curve is increasing.

Similarly (also by symmetry about origin).
For $-\infty < x < -1$, $y' > 0$, curve is increasing.
$-1 < x < 0$, $y' < 0$, curve is decreasing.

8. *No inflection points:* since $y'' \neq 0$.

Example 6: $y^2(x - a) = x^2(x + a)$, $a > 0$

Solution:

1. Curve is symmetric about x-axis only.
2. Curve passes through origin.
3. *Tangents at origin:* Equating to zero group of terms with lowest degree i.e., $a(x^2 + y^2) = 0$. The two tangents $y = \pm ix$ are imaginary.
4. Therefore origin is an isolated point.
5. *Intercepts:* y-intercept: $x = 0$ then $y = 0$ so origin (0, 0) is the y-intercept point.
x-intercept: $y = 0$ then $x = 0$ or $x = -a$. Thus $A(-a, 0)$ is x-intercept.
The tangent at A is $x = a$ which is obtained by putting $x = X - a$ and $Y = y = 0$ in the equation resulting in $y^2X = (X + a)^2(X + 2a)$ and equating to zero the lowest term i.e., $X = 0$ or $x + a = 0$ i.e., $x = -a$ (Fig. 5.25).

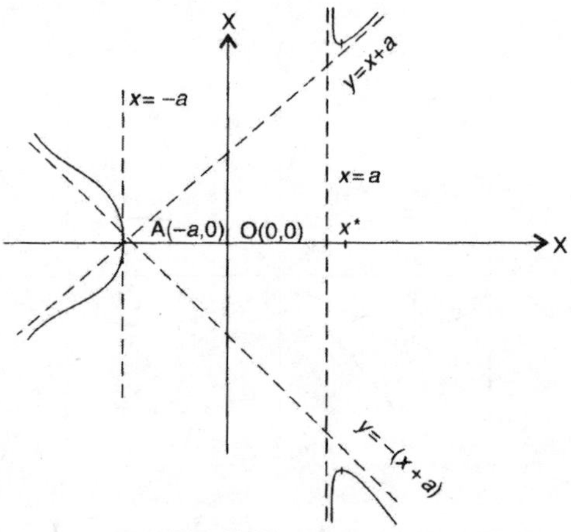

Fig. 5.25

6. Equation of curve is of third degree. The maximum number of asymptotes is three. Asymptote parallel to y-axis: equating to zero coefficient of y^2 i.e., $x - a = 0$ or $x = a$ is the vertical asymptote.

Asymptotes parallel to x-axis: no horizontal asymptote since coefficient of x^3 is constant. Oblique asymptotes: $\phi_3(m) = m^2 - 1$ (obtained by putting) $x = 1$ and $y = m$ in 3rd degree terms $xy^2 - x^3$. So $m = \pm 1$ are solution of $\phi_3(m) = 0$. $\phi_2(m) = -a(x^2 + y^2) = -a(1 + m^2)$, (put $x = 1, y = m$).
$c = -\frac{(-a)(1+m^2)}{2m}$. At $m = \pm 1, c = \pm a$.
Thus the two oblique asymptotes are

$$y = x + a \quad \text{and} \quad y = -x - a.$$

7. *Region:* Solving for y, we get two branches of the curve as

$$y = \pm x\sqrt{\frac{x + a}{x - a}}$$

y is imaginary when $-a < x < a$. Curve does not exist between the lines $x = -a$ and $x = a$ (except the isolated point 0).

8. *Sign of derivative:* $\frac{dy}{dx} = \pm \frac{(x^2 - ax - a^2)}{(x-a)^{\frac{3}{2}}(x+a)^{\frac{1}{2}}}$

   a. *Stationary points:* $\frac{dy}{dx} = 0$ when $x^2 - ax - a^2 = 0$. Out of the two stationary points $x = \frac{1}{2}(1 \pm \sqrt{5})a$ only $x^* = \frac{1}{2}(1 + \sqrt{5})a = 1.62a$ is considered, because for $x = \frac{1}{2}(1 - \sqrt{5})a$, y becomes imaginary. Maximum and minimum occur at x^*.

   b. For the branch of curve $y = +x\sqrt{\frac{x+a}{x-a}}$, the derivative $\frac{dy}{dx} = +\frac{x^2 - ax - x^2}{(x-a)^{3/2}(x+a)^{\frac{1}{2}}}$.
   Then for $-\infty < x < -a$, $\frac{dy}{dx} < 0$, curve is decreasing, for $a < x < \infty$, $\frac{dy}{dx} > 0$, curve is increasing.

   c. $x = a$ and $x = -a$ are vertical tangents to the curve because $\frac{dy}{dx} = \infty$ when $x = a$ or $x = -a$.

Example 7: Strophoid: $y^2(a + x) = x^2(b - x)$.

Solution:

1. Curve is symmetric about x-axis.
2. It passes through origin $O(0, 0)$.
3. *Tangents at origin*: are $y = \pm\sqrt{\frac{b}{a}}x$ which are obtained by equating the lowest degree terms $bx^2 - ay^2$ to zero.
4. Origin is a node since there are two real distinct tangents.
5. *Intercepts*: x-intercept: Put $y = 0$

$$x^2(b - x) = 0 \quad \text{or} \quad x = 0 \quad \text{or} \quad b.$$

The curve meets x-axis at $(0, 0)$ and $A(b, 0)$. y-intercept: put $x = 0$ then $y = 0$, so $O(0, 0)$ is the y-intercept (Fig. 5.26).

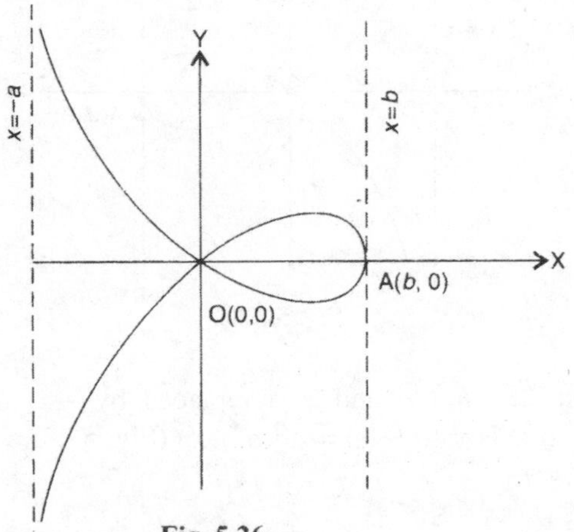

Fig. 5.26

6. *Asymptotes:* No horizontal asymptote since coefficient of x^3 is constant.

   $x = -a$ is the vertical asymptote obtained by equating the coefficient of y^2 namely $a + x$ to zero.

7. *Loop:* since curve crosses the x-axis (x-intercepts) at $O(0, 0)$ and $A(b, 0)$ and is symmetric about x-axis, a loop exists between O and A.
8. *Region:* solving for y, we get

$$y = \pm x\left(\frac{b - x}{a + x}\right)^{\frac{1}{2}}$$

y becomes imaginary when $x > b$ and when $x < -a$. Thus the curve exists only between $x = -a$ and $x = b$ i.e., for $-a < x < b$.

9. *Derivative:*

$$\frac{dy}{dx} = \frac{(-2x^2 - 3ax + bx + 2ab)}{2(a + x)^{\frac{3}{2}}(b - x)^{\frac{1}{2}}}$$

$\frac{dy}{dx} = \infty$ at $x = b$. So the tangent to the curve at $(b, 0)$ is vartical (parallel to y-axis). Therefore curve cuts the x-axis at right angle at $(b, 0)$.

Case 1: $b = a$ (Fig. 5.27) equation is $y^2(a + x) = x^2(a - x)$.

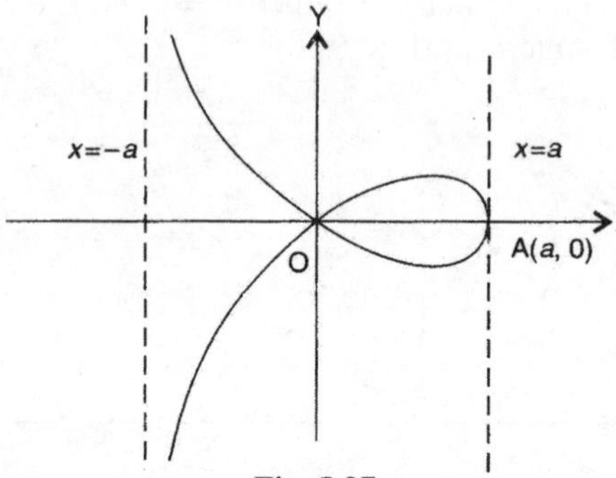

Fig. 5.27

Case 2: $b = a$ and x is replaced by $-x$, then equation is $y^2(a - x) = x^2(a + x)$ (Fig. 5.28).

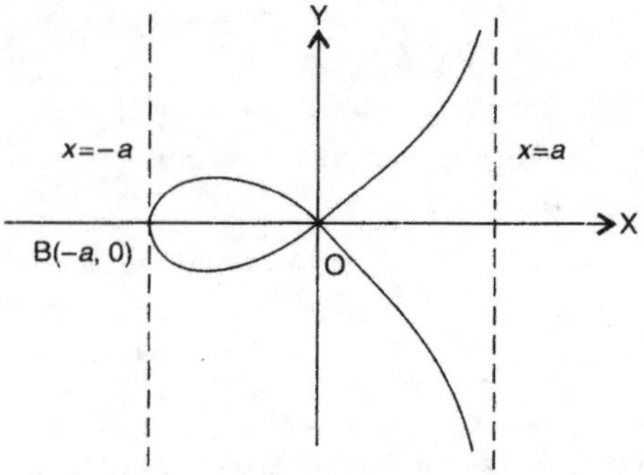

Fig. 5.28

Example 8: $y^2 = (x - a)(x - b)(x - c)$ with a, b, c all positive.

Solution: **Case 1:** Let $a < b < c$.

1. Curve is symmetric about x-axis only.
2. Does not pass through origin.
3. *Intercepts*: x-intercept: put $y = 0$. Then $(x - a)(x - b)(x - c) = 0$.
   Curve crosses the x-axis at $A(a, 0)$, $B(b, 0)$ and $D(c, 0)$.

   y-intercept: put $x = 0$.

   $$y^2 = -abc$$

   no solution, no y-intercepts. Curve does not cut the y-axis (Fig. 5.29).

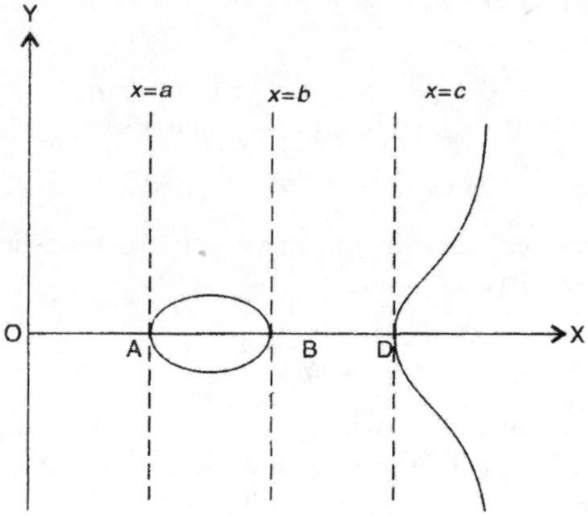

Fig. 5.29

4. Tangents at points A, B, D: are obtained by shifting the origin to these points. For example put $x = X + a$, equation becomes $Y^2{=}X(X{+}(a{-}b))(X{+}(a{-}c))$. Terms of least degree in x are $(a{-}b)(a{-}c)X$. The tangent at A is $X{=}0$ or $x{-}a{=}0$ or $x{=}a$. Similarly $x{=}b$ and $x{=}c$ are the tangents to the curve at B and D.
5. *Region:* Solving for y, we get

   $$y = \pm\sqrt{(x - a)(x - b)(x - c)}$$

   when $x < a$, y is imaginary when $b < x < c$, y is imaginary. Thus curve exists when

   i. $a < x < b$

   ii. $x > c$.
6. No horizontal or vertical asymptotes since coefficients of x^3 and y^2 are constants.
7. *Loop:* Curve crosses the x-axis between A and B and is symmetric about x-axis. Therefore curve has a loop between A and B.
8. y^2 increases and tends to infinity as x takes values bigger than c and tends to infinity.
9. *Derivative:* $\frac{dy}{dx} = \frac{(x-b)(x-c)+(x-c)(x-a)+(x-a)(x-b)}{2\sqrt{(x-a)(x-b)(x-c)}}$

   $$\frac{dy}{dx} \to \infty \text{ at } x = a, b, c$$

Thus the tangents at A, B, D are vertical as proved already in 4.

Case 2: Let $b = c$ and $a < b$. Then the equation reduces to $y^2 = (x-a)(x-b)^2$.

1. Curve is symmetric about x-axis.
2. Origin does not lie on the curve.
3. *Intercepts*: no y-intercepts x-intercepts: put $y = 0$, $(x-a)(x-b)^2 = 0$. Curve meets x-axis at $x = a$ and $x = b$. i.e., at $A(a, 0)$ and $B(b, 0)$.
4. *Region:* solving for y,

$$y = \pm(x-a)^{\frac{1}{2}}(x-b)$$

y is imaginary when $x < a$.
Curve exists only when $x \geq a$.

5. *Loop:* curve has a loop between A and B as curve crosses x-axis at A and B and is symmetric about x-axis (see Fig. 5.30).

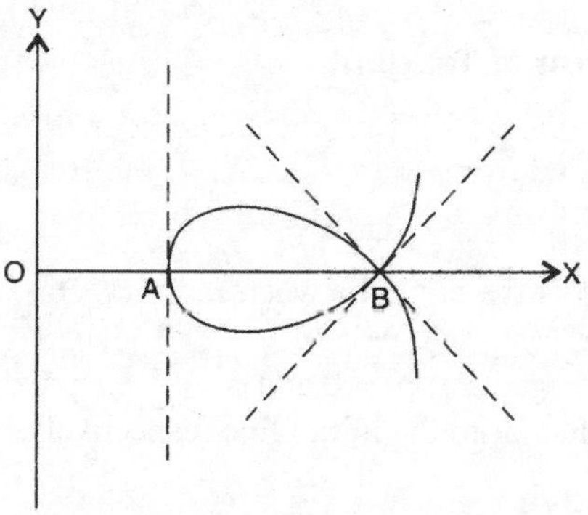

Fig. 5.30

6. For $x > b$ and x tends to ∞, y^2 also tends to ∞.
7. *Derivative:* $\frac{dy}{dx} = \pm\frac{[3x-2a-b]}{2(x-a)^{\frac{1}{2}}}$
8. Tangents at A and B. At $A(a, 0)$, $\frac{dy}{dx} = \infty$, so tangent at A, $x = a$ is vertical. At $B(b, 0)$, $\frac{dy}{dx} = \pm(b-a)^{\frac{1}{2}}$. So there are two distinct tangents at B given by

$$y = \pm m(x-b) = \pm(b-a)^{\frac{1}{2}}(x-b)$$

9. B is a node because there are two real distinct tangents to the curve.

Case 3: Let $a = b = c$. Then the equation is

$$y^2 = (x-a)^3$$

1. Curve is symmetric about x-axis.
2. Origin does not lie on the curve.
3. No y-intercepts. $x = a$ is the x-intercept.
4. *Region*: For $x < a$, y is imaginary. So curve exists only when $x \geq a$.
5. No asymptotes because coefficients of y^2 and x^3 are constants.
6. *Derivative*: $\frac{dy}{dx} = \pm\frac{3}{2}(x-a)^{\frac{1}{2}}$.
   At $x = a$, $\frac{dy}{dx} = 0$ twice.
   So x-axis is common tangent to the two branches of the curve at the point $A(a, 0)$. Thus A is cusp (Fig. 5.31).

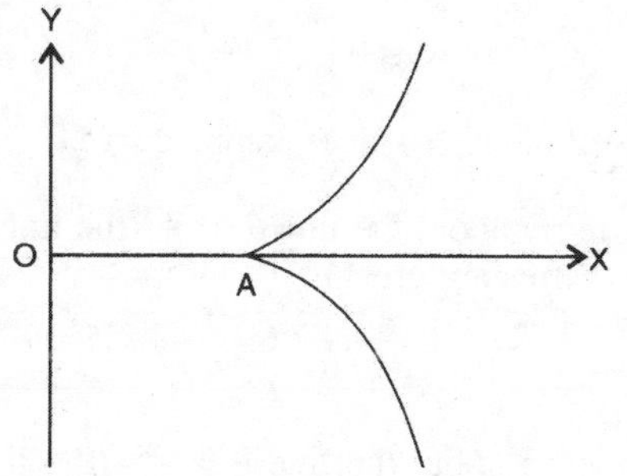

Fig. 5.31

7. As $x > b$ and x tends to infinity, y also tends to infinity.

5.3 CURVE TRACING: POLAR CURVES

The general equation of a curve in polar coordinates (r, θ) in the explicit form is $r = f(\theta)$ or $\theta = f(r)$ and in the implicit form is $F(r, \theta) = 0$.

Salient points to trace polar curves.

Symmetry

a. Curve is symmetric about the initial line $\theta = 0$ (usually the positive x-axis in the cartesian form) if the equation remains unaltered when θ is replaced by $-\theta$ i.e., $f(r, -\theta) = f(r, \theta)$.

Example: $r = a(1 \pm \cos\theta)$.

b. Curve is symmetric about the line $\theta = \frac{\pi}{2}$ (passing through the pole and perpendicular to the initial line) which is usually the positive y-axis in the cartesian form if the equation does not change when θ is replaced by $\pi - \theta$ i.e., $f(r, \theta) = f(r, \pi - \theta)$.

Example: $r = a(1 \pm \sin\theta)$.

c. Curve is symmetric about the pole (usually the origin) if equation remains unchanged when r is replaced by $-r$ i.e., $f(r, \theta) = f(-r, \theta)$. Equation contains only even powers of r.

Example: $r^2 = a\cos 2\theta$.

d. Curve is symmetric about pole if

$$f(r, \theta) = f(r, \theta + \pi)$$

Example: $r = 4\tan\theta$, kappa curve.

e. Symmetric about the line $\theta = \frac{\pi}{4}$ (the line $y = x$ in the cartesian form) if

$$f(r, \theta) = f(r, \frac{\pi}{2} - \theta)$$

f. Symmetric about the line $\theta = \frac{3\pi}{4}$ (the line $y = -x$ in cartesian form) if

$$f(r, \theta) = f(r, \frac{3\pi}{2} - \theta)$$

Pole

If $r = f(\theta_1) = 0$ for some $\theta = \theta_1 =$ constant then curve passes through the pole and the tangent at the pole is $\theta = \theta_1$

Example: At $\theta = \pi, r = a(1 + \cos\theta) = 0$

Asymptote

If $\lim_{\theta \to \theta_1} r = \infty$ then an asymptote to the curve exists and is given by the equation

$$r \sin(\vartheta - \theta_1) = f^1(\theta_1)$$

where θ_1 is the solution of $\frac{1}{f(\theta)} = 0$.

Points of Intersection

Points of intersection of the curve with the initial line and line $\theta = \pi/2$ are obtained by putting $\theta = 0$ and $\theta = \pi/2$ respectively in the polar equation.

Region or Extent

a. If a and b are the least and greatest values of r such that $a < r < b$ then curve lies in the annulus region between the two circles of radii a and b.

Example: $r = a\sin 2\theta$; since max $\sin 2\theta = 1$, curve lies in the circle $r = a$.

b. Curve does not exist for values of θ for which r is imaginary.

Example: $r^2 = a^2\cos 2\theta$; for $\frac{\pi}{4} < \theta < \frac{3\pi}{2}$, $\cos 2\theta$ is $-$ve so r is imaginary, hence no curve exists in this region.

c. For equations involving periodic functions generally θ varies from 0 to 2π.

Direction of Tangent

At any point (r, θ): Tangent at a point on the curve is determined from

$$\tan\phi = r\frac{d\theta}{dr}$$

where ϕ is the angle between radius vector and the tangent.

The tangents at $\phi = 0$ and $\phi = \frac{\pi}{2}$ are parallel and perpendicular to the initial line respectively.

Derivative

If $\frac{dr}{d\theta} > 0$, then r increases while,
if $\frac{dr}{d\theta} < 0$, then r decreases.

Loop

If curve meets the initial line at points A and B and the curve is symmetric about the initial line, then a loop of the curve exists between A and B.

Curves of the type $r = a\sin n\theta$ and $r = a\cos n\theta$ (are called roses) consist of either n or $2n$ equal (similar) loops (leaves) according as n is odd or even. Divide each quadrant into n equal parts and plot r for θ.

$n = 1$ corresponds to a circle.

Some times it may be advantageous to convert polar curves to cartesian curves by the transromation $x = r\cos\theta$, $y = r\sin\theta$.

Result:

$\theta =$	0	$\pi/6$	$\pi/4$	$\pi/3$	$\pi/2$	$2\pi/3$	$3\pi/4$	$5\pi/6$	π	$7\pi/6$	$4\pi/3$	$3\pi/2$	2π
	0	30	45	60	90	120	135	150	180	210	240	270	360
$\sin\theta =$	$\sqrt{\frac{0}{4}}$	$\sqrt{\frac{1}{4}}$	$\sqrt{\frac{2}{4}}$	$\sqrt{\frac{3}{4}}$	$\sqrt{\frac{4}{4}}$	$\sqrt{\frac{3}{4}}$	$\sqrt{\frac{2}{4}}$	$\sqrt{\frac{1}{4}}$	$\sqrt{\frac{0}{4}}$	$-\frac{1}{2}$	$-\frac{\sqrt{3}}{2}$	-1	0
$\cos\theta =$	$\sqrt{\frac{4}{4}}$	$\sqrt{\frac{3}{4}}$	$\sqrt{\frac{2}{4}}$	$\sqrt{\frac{1}{4}}$	$\sqrt{\frac{0}{4}}$	$-\sqrt{\frac{1}{4}}$	$-\sqrt{\frac{2}{4}}$	$-\sqrt{\frac{3}{4}}$	$-\sqrt{\frac{4}{4}}$	$-\sqrt{\frac{3}{4}}$	$-\sqrt{\frac{2}{4}}$	0	1
	Max				Min				Max			Min	Max

WORKED OUT EXAMPLES

Example 1: **Lemniscate** $r^2 = a^2\sin 2\theta$ (Fig. 5.32).

Solution:

1. *Symmetry:* **a.** Curve is symmetric about the pole since equation contains even power of r i.e., $f(-r, \theta) = f(r, \theta)$.

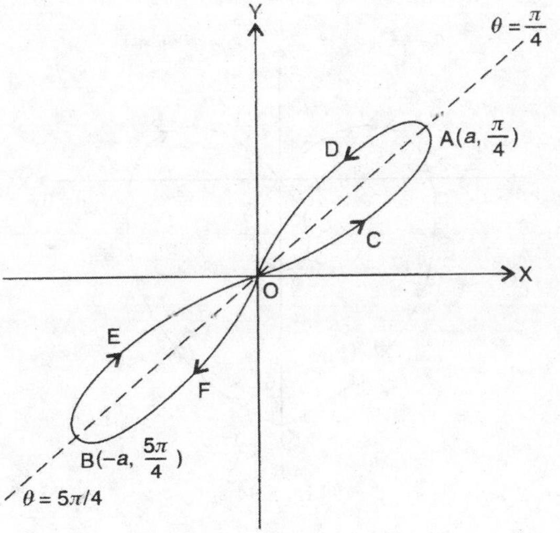

Fig. 5.32

b. Curve is symmetric about the line $\theta = \frac{\pi}{4}$ since

$$f(r, \frac{\pi}{2} - \theta) = r^2 - a^2\sin 2\left(\frac{\pi}{2} - \theta\right)$$

$$= r^2 - a^2\sin(\pi - 2\theta)$$

$$= r^2 - a^2\sin 2\theta = f(r, \theta)$$

2. *Pole:* Curve passes through the pole (origin 0) since at $\theta = 0$ and $\theta = \frac{\pi}{2}$, the value of $r^2 = a^2\sin 2\cdot 0 = 0$. Thus the two tangents to the curve at the pole are given by $\theta = 0$ and $\theta = \frac{\pi}{2}$ (i.e., x and y-axis). Thus pole is a node.

3. No asymptote since r is finite for any value of θ.

4. *Intersection:* Curve meets the lines $\theta = \frac{\pi}{4}$ and $\theta = \frac{5\pi}{4}$ at $A(a, \frac{\pi}{4})$ and $B(-a, \frac{5\pi}{4})$ respectively since $r^2 = a^2\sin 2(\frac{\pi}{4}) = a^2\sin\frac{\pi}{2} = a^2$ i.e., $r = \pm a$

5. *Region*:
   a. Curve lies completely within a circle of radius 'a' since the maximum value of $\sin 2\theta$ is 1.
   b. *Imaginary*: $\sin\theta$ curve is negative when θ is between π and 2π, then $\sin 2\theta$ is negative between $\frac{\pi}{2}$ to π. Thus r^2 is negative and therefore r is imaginary for $\frac{\pi}{2} < \theta < \pi$. Similarly r is imaginary for $\frac{3\pi}{2} < \theta < 2\pi$. Hence curve does not exist in the 3rd and 4th quadrants.

6. *Loop:*

θ :	0	$\frac{\pi}{8}$	$\frac{\pi}{4}$	$\frac{3\pi}{8}$	$\frac{\pi}{2}$
2θ :	0	$\frac{\pi}{4}$	$\frac{\pi}{2}$	$\frac{3\pi}{4}$	π
r :	0	·7a	a	·7a	0
Point:	0	C	A	D	0

Since curve is symmetric about the line $\theta = \frac{\pi}{4}$ ($y = x$ in the cartesian form), the arc OCA gets reflected about the line $\theta = \frac{\pi}{4}$ and thus forms the loop $OCADO$ in the first quadrant.

Due to symmetry about the pole (origin) this loop $OCADO$ in the first quadrant gets reflected to a loop $OFBEO$ in the third quadrant.

7. Tangent: $\tan\phi = r\frac{d\theta}{dr} = \frac{a\sqrt{\sin 2\theta}\cdot\sqrt{\sin 2\theta}}{a\cos 2\theta} = \tan 2\theta$.

Then $\phi = 2\theta$. So when $\theta = \frac{\pi}{4}, \phi = \frac{\pi}{2}$, $\tan\phi = \infty$. Thus the tangent to curve at $A(a, \frac{\pi}{4})$ is perpendicular to the initial line. On a similar argument, tangent to curve at $B(-a, \frac{3\pi}{4})$ is perpendicular to initial line.

Example 2: Three leaved rose $r = a\cos 3\theta$.

Solution:

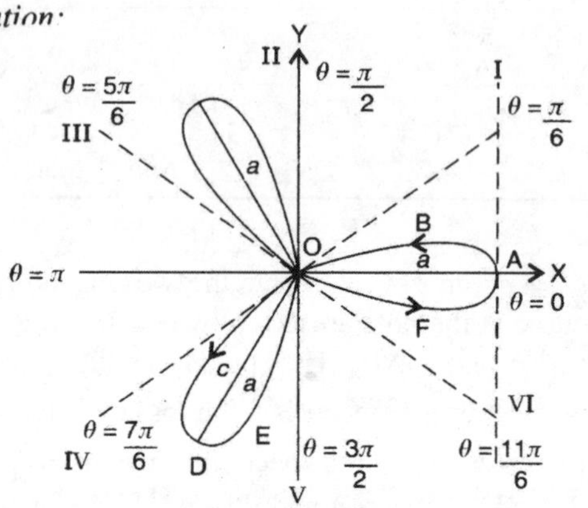

Fig. 5.33

1. *Symmetry:* Curve is symmetric about the initial line, since $r(-\theta) = a\cos 3(-\theta) = a\cos 3\theta = r(\theta)$
2. *Pole:* Curve passes through the pole O when $r = a\cos 3\theta = 0$ i.e., when $\cos 3\theta = 0$ or for
$$3\theta = \frac{\pi}{2}, \frac{3\pi}{2}, \frac{5\pi}{2}, \frac{7\pi}{2}, \frac{9\pi}{2}, \frac{11\pi}{2}$$
Thus the curve passes through the pole when
$$\theta = \frac{\pi}{6}, \frac{3\pi}{6} = \frac{\pi}{2}, \frac{5\pi}{6}, \frac{7\pi}{6}, \frac{9\pi}{6} = \frac{3\pi}{2}, \frac{11\pi}{6}$$
The tangents to the curve at pole O are given by the 6 lines I: $\theta = \frac{\pi}{6}$; II: $\theta = \frac{\pi}{2}$; III: $\theta = \frac{5\pi}{6}$; IV: $\theta = \frac{7\pi}{6}$; V: $\theta = \frac{3\pi}{2}$; VI: $\theta = \frac{11\pi}{6}$.
Hence pole is a node.
3. *Asymptote*: No asymptote since r is finite for any θ
4. Curve intersects the initial line $\theta = 0$ at $A(a, 0)$ only (Fig. 5.33).
5. *Region:* curve lies completely within a circle $r = a$ since maximum value of $\cos 3\theta$ is one.
6. *Tangent*: $\tan\phi = r\frac{d\theta}{dr} = \frac{\cos 3\theta}{-3\sin 3\theta}$. At $\theta = 0$, $\tan\phi = \infty$. So the tangent at the point $A(a, 0)$ is perpendicular to the initial line.
7. For $n = 3$, the curve has 3 loops.
8. *Variation of r:*

θ :	0	$\frac{\pi}{12}$	$\frac{\pi}{6}$	$\frac{3\pi}{12}$	$\frac{4\pi}{12}$	$\frac{5\pi}{12}$	$\frac{6\pi}{12}$
3θ :	0	$\frac{\pi}{4}$	$\frac{\pi}{2}$	$\frac{3\pi}{4}$	π	$\frac{5\pi}{4}$	$\frac{3\pi}{2}$
r:	a	·7a	0	−·7a	−a	−·7a	0
point:	A	B	O	C	D	E	O.

9. *Loop:* As θ varies from $\frac{\pi}{6}$ to $\frac{\pi}{2}$, r varies from 0 to $-a$ to 0 thus forming a loop *OCDEO*. Due to symmetry about initial line, this loop *OCDEO* gets reflected to the loop in the second quadrant. Similarly the arc *ABO* gets reflected as *OFA* thus forming the loop *ABOFA*. Hence the three loops (or leaves).

Example 3: Limacon $r = 2(1 - 2\sin\theta)$.

Solution:

1. Curve is symmetric about the line $\theta = \frac{\pi}{2}$ through pole and perpendicular to the initial line because $\sin(\pi - \theta) = \sin\theta$ i.e., $f(r, \pi - \theta) = f(r, \theta)$ (Fig. 5.34).

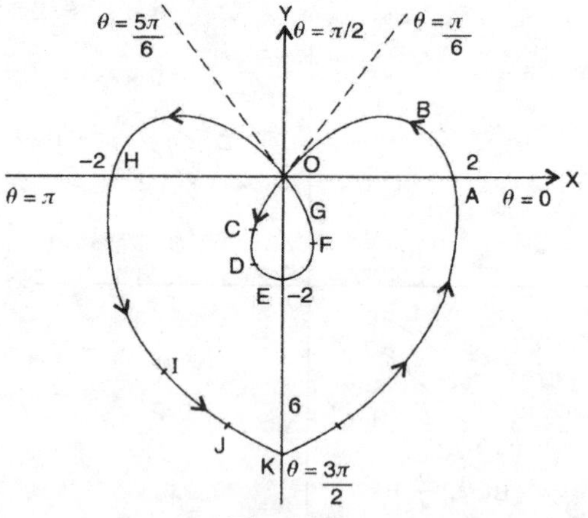

Fig. 5.34

2. *Pole:* Curve passes through the pole (origin 0) since $r = 2(1 - 2\sin\theta) = 0$ when $\theta = \frac{\pi}{6}$ and $\frac{5\pi}{6}$ ($\because$ $\sin\frac{\pi}{6} = \sin\frac{5\pi}{6} = \frac{1}{2}$). The tangents to the curve at pole are given by two lines $\theta = \frac{\pi}{6}$ and $\theta = \frac{5\pi}{6}$. Thus pole is a node.

3. Curve meets the initial line $\theta = 0$ at the two points $A(2, 0)$ and $H(2, \pi)$. Curve meets the line $\theta = \frac{\pi}{2}$ at the two points $E(-2, \frac{\pi}{2})$ and $K(6, \frac{3\pi}{2})$.
4. *Asymptote:* No asymptote since r is finite for any θ.
5. *Region:* Since $|\sin\theta| \le 1$, the entire curve lies within a circle $r = 6$ (of radius 6).
6. *Variation of r:*

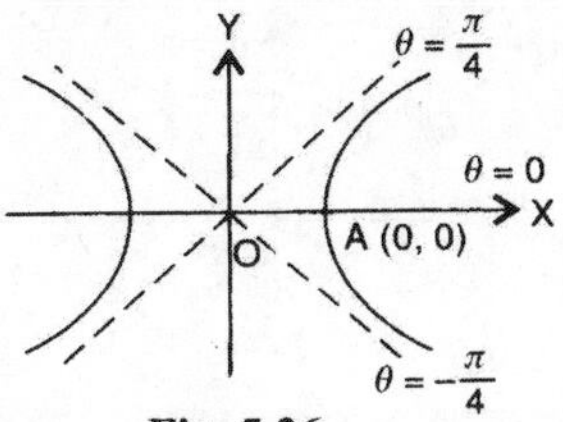

Fig. 5.36

$\theta =$	0	$\frac{\pi}{12}$	$\frac{\pi}{6}$	$\frac{\pi}{4}$	$\frac{\pi}{3}$	$\frac{\pi}{2}$	$\frac{2\pi}{3}$	$\frac{3\pi}{4}$	$\frac{5\pi}{6}$	π	$\frac{7\pi}{6}$	$\frac{4\pi}{3}$	$\frac{3\pi}{2}$	2π
$\sin\theta =$	0	·26	0·5	$\frac{1}{\sqrt{2}}$	$\frac{\sqrt{3}}{2}$	1	$\frac{\sqrt{3}}{2}$	$\frac{1}{\sqrt{2}}$	$\frac{1}{2}$	0	$-\frac{1}{2}$	$-\frac{\sqrt{3}}{2}$	-1	0
$r =$	2	·965	0	$-$·83	$-$1·46	$-$2	$-$1·46	$-$·83	0	2	4	5·464	6	2
Point:	A	B	O	C	D	E	F	G	O	H	I	J	K	A

As θ varies from $\frac{3\pi}{2}$ to $\frac{\pi}{2}$, r varies from 6 to -2, traversing the curve *KABOCDE* which due to symmetry about $\theta = \frac{\pi}{2}$ gives a limacon with an inner loop.

Exercise

Trace the following curves stating the salient points:

1. $r = a\sin 2\theta$.

Ans. Symmetric about the lines $\theta = \frac{\pi}{4}$, and $\theta = \frac{3\pi}{4}$ and about the line $\theta = \frac{\pi}{2}$; pole lies on curve, pole is node, four tangents at pole are $\theta = 0, \frac{\pi}{2}, \pi, \frac{3\pi}{2}$; $r \le a$; $\theta = \frac{\pi}{4}$ meets curve at $A(a, \frac{\pi}{4})$; for $0 \le \theta \le \frac{\pi}{2}, 0 \le r \le a$, first loop in first quadrant; other 3 loops by symmetry: for $n = 2$ there are 4 similar loops (refer Fig. 5.35).

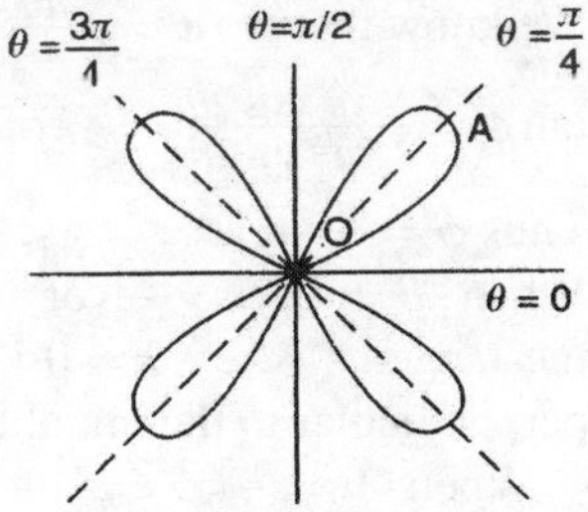

Fig. 5.35

2. $r^2 = a^2\sec 2\theta$ or $r^2\cos 2\theta = a^2$.

Ans. Symmetric about $\theta = 0, \theta = \frac{\pi}{2}$, and $\theta = \frac{\pi}{4}$. Curve exists for $r > a$; $\theta = \pm\frac{\pi}{4}$ are asymptotes ($r \to \infty$ as $\theta \to \pm\frac{\pi}{4}$); meets $\theta = 0$ line at $A(a, 0)$ (Fig. 5.36).

3. $r = a(1 + \sin\theta)$.

Ans. Symmetric about $\theta = \frac{\pi}{2}$, pole O lies on curve, $\theta = \frac{-\pi}{2}$ is tangent at 0; curve meets A$(a, 0)$, and B$(0, 2a)$; $\phi = \frac{\pi}{4} + \frac{\theta}{2}$; tangent $\theta = \frac{\pi}{2}$ is $\perp^r$ to initial line; $r \le 2a$ (Fig. 5.37).

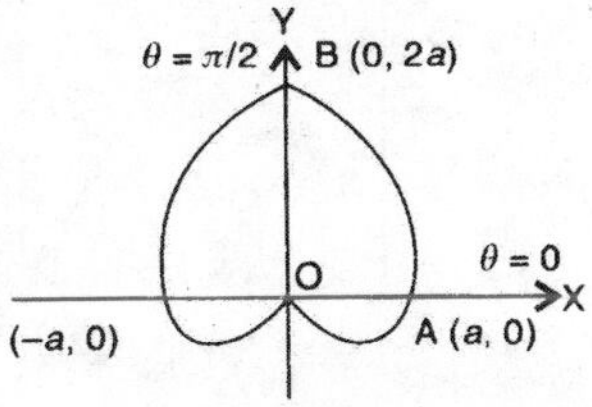

Fig. 5.37

4. $r = a(1 - \cos\theta)$.

Ans. Symmetric about $\theta = 0$, pole lies on curve, $\theta = 0$ is tangent at pole, intersection points $(a, \frac{\pi}{2}), (a, \frac{3\pi}{2}), (2a, \pi)$; tangent at $\theta = \pi$, vertical, curve lies within $r = 2a$ (Fig. 5.38).

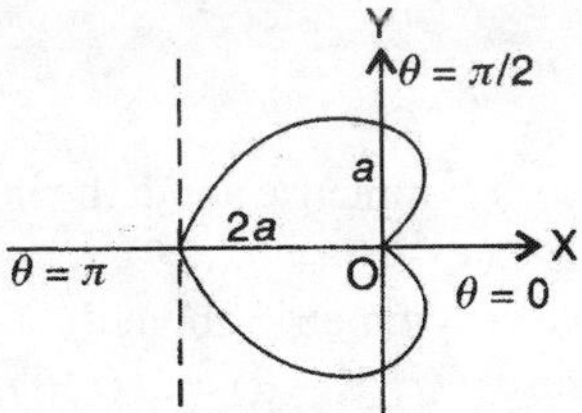

Fig. 5.38

5. $r = 1 + \sqrt{2}\cos\theta$.

Ans. Symmetric about $\theta = 0$, pole lies on the curve when $\theta = \frac{2\pi}{3}, \frac{7\pi}{6}$; lies within $r = 2{\cdot}414$ (Fig. 5.39).

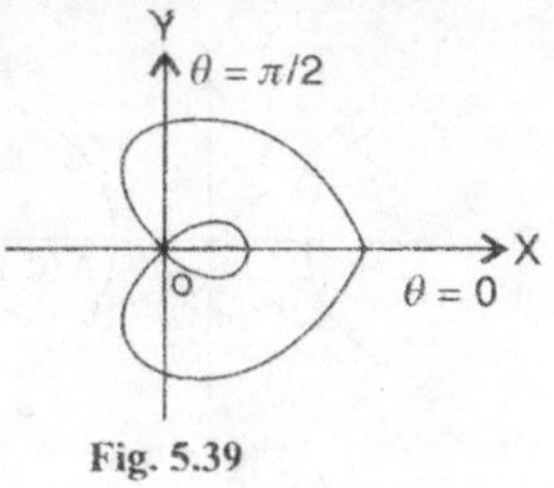

Fig. 5.39

6. Cissoid: $r = a\sin^2\theta/\cos\theta$

Ans. Cartesian form $r = a\frac{(y/r)^2}{(x/r)}$ or $(x^2+y^2)x = ay^2$ or $y^2(a-x) = x^3$. Symmetric about x-axis, passes through origin, origin is a cusp, $x = a$ is asymptote. Curve exists for $0 < x < a$ (refer Fig. 5.40).

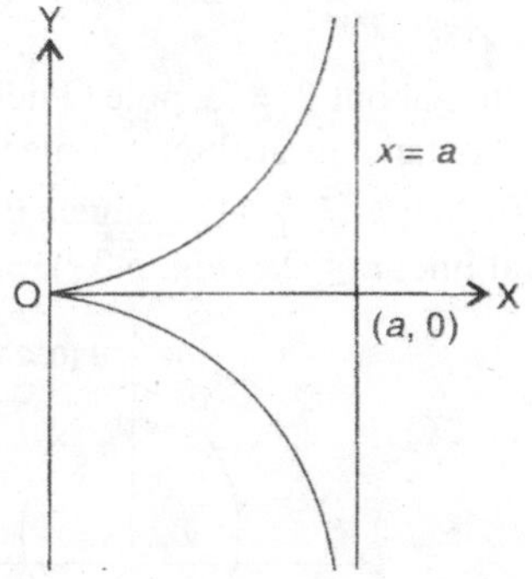

Fig. 5.40

5.4 CURVE TRACING: STANDARD POLAR CURVES

Example: Lemniscate of Bernoulli,

$$r^2 = a^2\cos 2\theta.$$

Solution:

I. *Symmetry:*

a. Curve is symmetric about the initial line because $f(r,-\theta) = f(r,\theta)$.

b. Curve is symmetric about the line $\theta = \pi/2$ because $f(r,\pi-\theta) = f(r,\theta)$.

c. Curve is symmetric about the pole because $f(-r,\theta) = f(r,\theta)$.

II. *Pole:* Pole lies on the curve because $r = 0$ when $\cos 2\theta = 0$ or $2\theta = \pm\frac{\pi}{2}$ or $\theta = \pm\frac{\pi}{4}$. Thus there two distinct real tangents $\theta = \pm\frac{\pi}{4}$ to the curve at the pole. Hence pole is a node.

III. *Asymptote:* No asymptote because r is finite for any value of θ.

IV. Points of intersection of the curve with the initial line are $A(a,0)$, $B(-a,0)$ which are obtained by putting $\theta = 0$ in the equation. (i.e., $r^2 = a^2.\cos 2\cdot 0 = a^2$ $\therefore$ $r = \pm a$ and $\theta = 0$).

V. *Region:* Cosine curve is positive between $\frac{-\pi}{2}$ and $\frac{\pi}{2}$. Thus $\cos 2\theta \geq 0$ when $\frac{-\pi}{2} \leq 2\theta \leq \frac{\pi}{2}$ or $\frac{-\pi}{4} \leq \theta \leq \frac{\pi}{4}$.

Similarly, cosine curve is negative between $\frac{\pi}{2}$ to $\frac{3\pi}{2}$. Thus $\cos 2\theta < 0$ when $\frac{\pi}{2} < 2\theta < \frac{3\pi}{2}$ or $\frac{\pi}{4} < \theta < \frac{3\pi}{4}$. Thus $r^2 = a^2\cos 2\theta < 0$ when $\frac{\pi}{4} < \theta < \frac{3\pi}{4}$. Hence curve does not exist between the lines $\theta = \frac{\pi}{4}$ and $\theta = \frac{3\pi}{4}$ (Fig. 5.41).

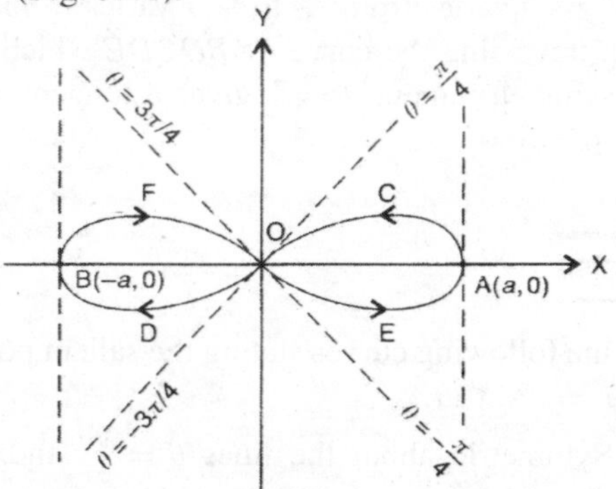

Fig. 5.41

VI. *Direction of tangent:* Differentiating the given equation

$$r^2 = a^2\cos 2\theta$$

w.r.t. θ, we get $\frac{dr}{d\theta} = \frac{-2a^2\sin 2\theta}{r} = \frac{-a\sin 2\theta}{\sqrt{\cos 2\theta}}$.

We know that $\tan\phi = r\frac{d\theta}{dr} = r\cdot\frac{(-r)}{a^2\sin 2\theta}$.

$$\tan\phi = -\frac{a^2\cos 2\theta}{a^2\sin 2\theta} = -\cot 2\theta = \tan\left(2\theta + \frac{\pi}{2}\right).$$

Thus $\phi = \frac{\pi}{2} + 2\theta$.

When $\theta = 0, \phi = \frac{\pi}{2}$ or $\tan\theta = \infty$. Thus the tangents when $\theta = 0$ i.e., at $r = \pm a$ are perpendicular to the initial line.

When $\theta = \frac{\pi}{4}, \phi = \frac{\pi}{2} + 2.\frac{\pi}{4} = \pi$ so the radius vector itself is the tangent to curve at pole (corresponding to $\theta = \pi/4$).

VII. *Derivative:* $\frac{dr}{d\theta} = -\frac{a\sin 2\theta}{\sqrt{\cos 2\theta}}$.

Variation of θ	Variation of r	Sign of derivative	Nature of curve	Curve traced
$0 \le \theta \le \frac{\pi}{4}$	$a, 0{\cdot}70a, 0$	Negative	Decreasing	ACO
$\frac{\pi}{4} < \theta < \frac{3\pi}{4}$	r is imaginary		Does not exist	
$\frac{3\pi}{4} < \theta < \pi$	$0, -0{\cdot}70a, -a$	Positive	Increasing	ODB

VIII. *Loop:* Since curve crosses the initial line at A and O and is symmetric about the initial line, a loop of the curve $ACOEA$ exists between O and A. Since curve is symmetric about the line $\theta = \frac{\pi}{2}$, this loop $ACOEA$ is reflected as another loop $ODBFO$.

Example: Four leaved Rose: $r = a\cos 2\theta;\ a > 0$.

Solution: Since $n = 2$ the curve consists of $2n = 2{\cdot}2 = 4$ equal (similar) loops. Divide each quadrant into $(n =)2$ equal parts (Fig. 5.42).

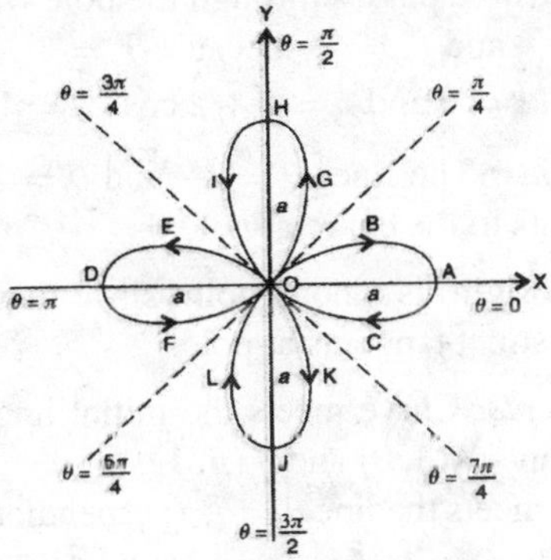

Fig. 5.42

1. *Symmetry:*
   a. Curve is symmetric about the initial line since
   $$r(-\theta) = a\cos\,(2(-\theta)) = a\cos 2\theta = r(\theta)$$
   b. Curve is symmetric about the line $\theta = \frac{\pi}{2}$ which is perpendicular to the initial line since $r(\pi - \theta) = a\cos 2(\pi - \theta) = a\cos 2\pi \cdot \cos 2\theta = a\cos 2\theta = r(\theta)$.
2. *Pole:* Curve passes through the pole when $r = a\cos 2\theta = 0$ i.e., $\cos 2\theta = 0$ so when $2\theta = \frac{\pi}{2}, \frac{3\pi}{2}, \frac{5\pi}{2}, \frac{7\pi}{2}$.

   Thus curve passes through the pole when $\theta = \frac{\pi}{4}, \frac{3\pi}{4}, \frac{5\pi}{4}, \frac{7\pi}{4}$.

   The tangents to the curve at pole are given by the lines $\theta = \frac{\pi}{4}, \theta = \frac{3\pi}{4}, \theta = \frac{5\pi}{4}$ and $\theta = \frac{7\pi}{4}$.
3. *Asymptote:* No asymptote since r is finite for any θ. In fact $r \le a$ since maximum value of $\cos 2\theta = 1$. Thus the entire θ curve lies within the circle $r = a$.
4. *Value of r:* The following table gives values of $r = a\cos 2\theta$ for different values of θ. (Note: $\frac{1}{\sqrt{2}} = 0 \cdot 70$).

$\theta =$	0	$\frac{\pi}{8}$	$\frac{\pi}{4}$	$\frac{3\pi}{8}$	$\frac{\pi}{2}$	$\frac{5\pi}{8}$	$\frac{3\pi}{4}$	$\frac{7\pi}{8}$	π
$2\theta =$	0	$\frac{\pi}{4}$	$\frac{\pi}{2}$	$\frac{3\pi}{4}$	π	$\frac{5\pi}{4}$	$\frac{3\pi}{2}$	$\frac{7\pi}{4}$	2π
$r =$	a	$\cdot 7a$	0	$-\cdot 7a$	$-a$	$-\cdot 7a$	0	$\cdot 7a$	a
$\theta =$	$\frac{9\pi}{8}$	$\frac{5\pi}{4}$	$\frac{11\pi}{8}$	$\frac{3\pi}{2}$	$\frac{13\pi}{8}$	$\frac{7\pi}{4}$	$\frac{15\pi}{8}$	2π	
$2\theta =$	$\frac{9\pi}{4}$	$\frac{5\pi}{2}$	$\frac{11\pi}{4}$	3π	$\frac{13\pi}{4}$	$\frac{7\pi}{2}$	$\frac{15\pi}{4}$	4π	
$r =$	$\cdot 7a$	0	$-\cdot 7a$	$-a$	$-\cdot 7a$	0	$\cdot 7a$	a	

As θ varies from 0 to $\frac{\pi}{4}$, r varies from a to 0 thus traversing the curve ABO. Due to symmetry about initial line ABO gets reflected as OCA. The loop $ABOCA$ gets reflected as $OEDFO$ due to symmetry about line $\theta = \frac{\pi}{2}$. Similarly OGH forms loop $OGHIO$ due to symmetry about the line $\theta = \frac{\pi}{2}$. This loop gets reflected as another loop $OLJKO$ due to symmetry about the initial line.

Example: Three leaved Rose; $r = a\sin 3\theta$.

Solution: For $n = 3$ odd, the curve consists of three loops (Fig. 5.43).

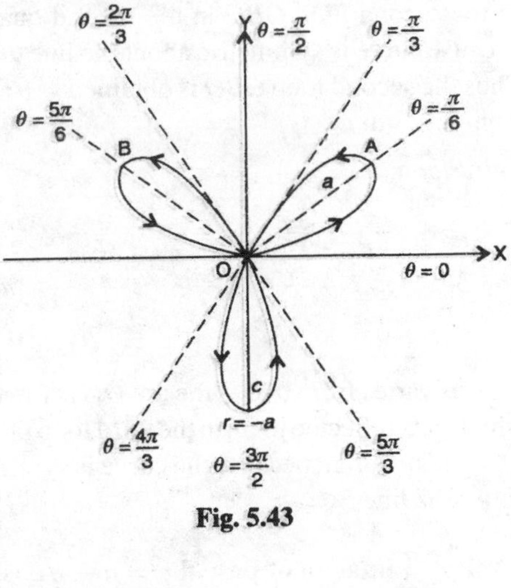

Fig. 5.43

1. *Symmetry:* Curve is symmetric about the line $\theta=\frac{\pi}{2}$ passing through the pole 0 and perpendicular to the initial line, since

$$r(\pi-\theta)=a\sin(3(\pi-\theta))$$
$$=a\sin 3\theta=r(\theta)$$

2. *Pole:* Curve passes through the pole 0 when $r=a\sin 3\theta=0$ i.e., when $3\theta=0,\pi,2\pi,3\pi,4\pi,5\pi$. Thus for $\theta=0,\frac{\pi}{3},\frac{2\pi}{3},\pi,\frac{4\pi}{3},\frac{5\pi}{3}$, curve passes through the pole. The tangents to the curve at the pole are given by the lines $\theta=0,\theta=\frac{\pi}{3},\theta=\frac{2\pi}{3}$, $\theta=\pi,\theta=\frac{4\pi}{3}$ and $\theta=\frac{5\pi}{3}$. Pole is a node since $\theta=\frac{\pi}{3}$ and $\theta=\frac{2\pi}{3}$ are two real distinct tangents at 0.
3. *Asymptote*: No asymptote since r is always finite for any θ. In fact maximum value of r is a since maximum value of $\sin 3\theta$ is 1. Thus the entire curve lies within the region of the circle $r=a$.
4. *Intersection*: Curve meets the line $\theta=\frac{\pi}{2}$ at $r=a\sin 3(\frac{\pi}{2})=-a$.
5. *Variation of r*: Variation of r as θ varies from 0 to $\frac{\pi}{3}$.

θ:	0	$\frac{\pi}{12}$	$\frac{\pi}{6}$	$\frac{\pi}{4}$	$\frac{\pi}{3}$
3θ:	0	$\frac{\pi}{4}$	$\frac{\pi}{2}$	$\frac{3\pi}{4}$	π
r:	a	$\cdot 7a$	a	$\cdot 7a$	0

6. *Loops*: As θ varies from 0 to $\frac{\pi}{3}$, r varies from 0 to a to 0. So the curve traverses the loop *OAO* in the first quadrant. This loop *OAO* is reflected as the second loop *OBO* in the second quadrant because curve is symmetric about the line $\theta=\frac{\pi}{2}$. Thus the second loop *OBO* is obtained as θ varies from $2\pi/3$ to π.
7. *Third loop*:

θ:	$\frac{\pi}{3}$	$\frac{5\pi}{12}$	$\frac{\pi}{2}$	$\frac{7\pi}{12}$	$\frac{2\pi}{3}$
3θ:	π	$\frac{5\pi}{4}$	$\frac{3\pi}{2}$	$\frac{7\pi}{4}$	2π
r:	0	$-\cdot 7a$	$-a$	$-\cdot 7a$	0

As θ varies from 0 to $\frac{\pi}{2}$ the arc *OC* is traversed which gets reflected to form the third loop *OCO* in the third quadrant because the curve is symmetric about the line $\theta=\frac{\pi}{2}$.

Example: Limacon of pascal $r=a+b\cos\theta$.

Solution: Limacons are polar curves whose equations are of the form

$$r=a+b\sin\theta, r=a-b\sin\theta,$$
$$r=a+b\cos\theta, r=a-b\cos\theta$$

with $a>0$ and $b>0$. We get a

Limacon with inner loop when $a/b<1$ or $a<b$; a cardioid when $a=b$; dimpled Limacon when $1<a/b<2$ and convex Limacon when $\frac{a}{b}\geq 2$.

For the curve $r=a+b\cos\theta$ consider three cases $a<b$, $a=b$ and $a>b$

Case 1: $a<b$: Limacon with inner loop: $r=1+2\cos\theta$; $a=1<2=b$

Solution:

1. *Symmetry:* Curve is symmetric about the initial line since $r(-\theta)=1+2\cos(-\theta)=1+2\cos\theta=r(\theta)$.
2. *Pole:* Curve passes through the pole when $\theta=\frac{2\pi}{3}$ and $\theta=\frac{4\pi}{3}$ because $r=1+2\cos\frac{2\pi}{3}=1-1=0$ and $r=1+2\cos\frac{4\pi}{3}=1-1=0$.
3. *Tangents:* The lines $\theta=\frac{2\pi}{3}$ and $\theta=\frac{4\pi}{3}$ are the tangents to the curve at pole 0.
4. Pole (origin) is a node point since there are two real distinct tangents at pole.
5. *Intercepts:* Curve meets the initial line $\theta=0$ at the points $A(3,0)$ and $f(\pi,1)$.
Curve meets the line $\theta=\frac{\pi}{2}$ (perpendicular to the initial line) at $D(\frac{\pi}{2},1)$ and $H(\frac{3\pi}{2},1)$ (Fig. 5.44).

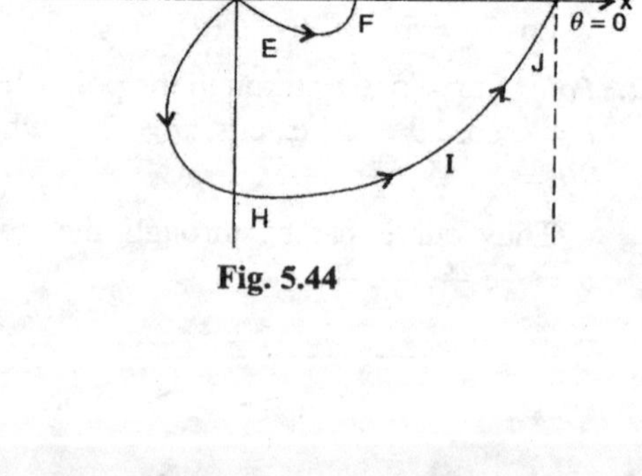

Fig. 5.44

6. *Asymptote:* No asymptote since r is finite for any θ.

7. *Region:* Since maximum value of $\cos\theta$ is 1, $r \le 3$. Thus the entire curve lies inside the circle $r = 3$.

8. *Derivative*: $\tan\phi = r\frac{d\theta}{dr} = \frac{1+2\cos\theta}{-2\sin\theta}$. For $\theta = 0$ and $\frac{\pi}{2}$, $\tan\phi = \infty$ i.e., $\phi = \frac{\pi}{2}$. Thus the tangents to the curve at $\theta = 0$ and $\theta = \frac{\pi}{2}$ are perpendicular to the initial line.

9. Value of r:

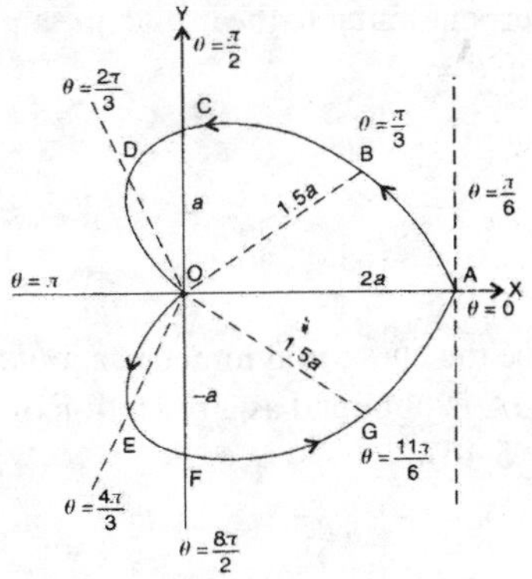

Fig. 5.45

$\theta =$	0	$\frac{\pi}{6}$	$\frac{\pi}{3}$	$\frac{\pi}{2}$	$\frac{2\pi}{3}$	$\frac{5\pi}{6}$	π	$\frac{7\pi}{6}$	$\frac{4\pi}{3}$	$\frac{3\pi}{2}$	$\frac{5\pi}{3}$	$\frac{11\pi}{6}$	2π
$r =$	3	$1+\sqrt{3}$	2	1	0	$1-\sqrt{3}$	-1	$1-\sqrt{3}$	0	1	2	$1+\sqrt{3}$	3
Point:	A	B	C	D	O	E	F	G	O	H	I	J	A

Note: $\sqrt{3} = 1.732$.

10. *Loops: Inner loop:* As the curve crosses the initial line at O and F and is symmetric about the initial line, a loop exists between O and F. Similarly another loop exists between O and A. The curve is traced as shown in the figure.

Case 2: $a = b$: **Cardioid:** $r = a(1 + \cos\theta)$.

Solution:

1. *Symmetric:* Curve is symmetric about the initial linc.

2. *Pole:* Curve passes through pole O since $r = a(1 + \cos(\pi)) = a(1 - 1) = 0$ when $\theta = \pi$. The tangent to the curve at the pole is $\theta = \pi$.

3. *Asymptote:* No asymptote since r is finite for any θ.

4. *Region:* Curve lies within the circle $r = 2a$ since maximum value of $\cos\theta$ is 1.

5. *Intersection:* Curve meets the initial line at $A(2a, 0)$ and $0(0, \pi)$ curve intersects the line $\theta = \frac{\pi}{2}$ at $C(a, \frac{\pi}{2})$ and $F(a, \frac{3\pi}{2})$.

6. Value of r:

θ:	0	$\frac{\pi}{3}$	$\frac{\pi}{2}$	$\frac{2\pi}{3}$	π	$\frac{4\pi}{3}$	$\frac{3\pi}{2}$	$\frac{11\pi}{6}$	2π
r:	$2a$	$\frac{3a}{2}$	a	$\frac{a}{2}$	0	$\frac{a}{2}$	a	$\frac{3a}{2}$	$2a$
Point:	A	B	C	D	O	E	F	G	A

(Fig. 5.45)

7. *Tangent:*

$$\tan\phi = r\frac{d\theta}{dr} = \frac{a(1+\cos\theta)}{-a\sin\theta}$$

$$= -\cot\frac{\theta}{2}$$

$$= \tan(\frac{\theta}{2} + \frac{\pi}{2}).$$

So $\phi = \frac{\theta}{2} + \frac{\pi}{2}$. When $\theta = 0, \phi = \frac{\pi}{2}, \tan\phi = \infty$. Then tangent at A$(2a, 0)$ is perpendicular to the initial line.

Case 3: $a > b$: **Dimpled limacon** $r = 3 + 2\cos\theta$

1. Curve is symmetric about the initial line.

2. Curve does not pass through pole because for any $\theta, r = 3 + 2\cos\theta \ne 0$ (otherwise $|\cos\theta| = |\frac{-3}{2}| > 1$ which is not true for any θ).

3. Curve meets the initial line at $A(5, 0)$ and at $E(1, \pi)$.

   Curve meets the line $\theta = \frac{\pi}{2}$ at $C(3, \frac{\pi}{2})$ and $F\left(3, \frac{3\pi}{2}\right)$.

4. No asymptote since r is finite.

5. *Region:* Curve lies within a circle of radius 5 since $\cos\theta$ maximum value is 1.

6. *Tangent:* $\tan\phi = r\frac{d\theta}{dr} = \frac{3+2\cos\theta}{-2\sin\theta}$.

   No tangent to the curve is parallel to the initial line since $\tan\phi \ne 0$ for any θ. The tangent at $A(5, 0)$

is perpendicular to the initial line since $\tan\theta = \infty$ when $\theta = 0$

7. *Value of r:*

$\theta =$	0	$\frac{\pi}{3}$	$\frac{\pi}{2}$	$\frac{2\pi}{3}$	π
$r =$	5	4	3	2	1
Point:	A	B	C	D	E

Since the curve is symmetric about the initial line EFGA is obtained as a reflection of ABCDE (see Fig. 5.46).

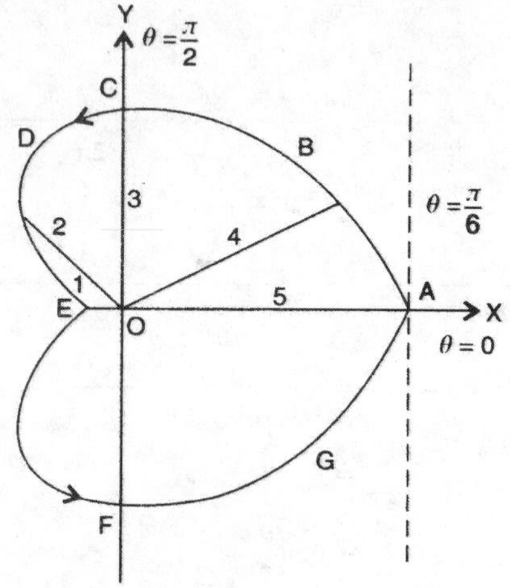

Fig. 5.46

5.5 CURVE TRACING: PARAMETRIC CURVES

Let $x = f(t)$ and $y = g(t)$ be the parametric equations of a curve with t as parameter.

Case 1: If t can be eliminated between $x = f(t)$ and $y = g(t)$ we may obtain a cartesian equation in x and y.

Example: $x = a\cos t$, $y = b\sin t$, Eliminating t, $\frac{x^2}{a^2} + \frac{y^2}{b^2} = 1$ is the cartesian equation.

Example: $x = a\sin^2 t$, $y = a\frac{\sin^2 t}{\cos t}$, Eliminating t, $y^2(a - x) = x^3$ is cartesian equation, of cissoid.

Case 2: Suppose t can not be eliminated.

1. *Symmetry:* Symmetric about x-axis if $x = f(t)$ is even and $y = g(t)$ is odd function of t. Symmetric about y-axis if $x = f(t)$ is odd and $y = g(t)$ is an even function of t.
2. *Origin:* Determine t for which $x = 0$ and $y = 0$.
3. *Intercept:* y-intercept obtained for values of t for which $x = 0$, x-intercept for values of t for which $y = 0$.
4. Find the least and greatest values of x and y.
5. *Asymptotes:* $\lim_{t\to t_1} x = \infty$, $\lim_{t\to t_1} y = \infty$. Then $t = t_1$ is an asymptote.
6. *Tangents:* Vertical and horizontal tangents exist at points where $\frac{dy}{dx} = \frac{(dy/dt)}{(dx/dt)}$ is ∞ or 0.
7. Curve does not exist when x or y is imaginary.
8. If $x = f(t)$ and $y = g(t)$ are periodic functions of t having a common period, it is enough to trace the curve for one period.

WORKED OUT EXAMPLES

Example 1: Cycloid

$$x = a(t - \sin t);\ y = a(1 - \cos t)$$

Solution: $y = 0$ when $\cos t = 1$ i.e., when $t = 0, 2\pi, 4\pi$, etc. Curve meets x-axis when $t = 0, 2\pi, 4\pi, 6\pi$, etc.

Since $|\cos t| \le 1, 0 \le y \le 2a$, curve lies above the x-axis and below the horizontal line $y = 2a$. At $t = 0, x = 0, y = 0$. Thus origin O lies on the curve. Curve is symmetric about y-axis since x is an odd function and y is an even function of t. For any t, x and y are finite, no asymptote.

$$\frac{dx}{dt} = a(1 - \cos t)$$

$$\frac{dy}{dt} = a\sin t$$

$$\frac{dy}{dx} = \frac{(dx/dt)}{(dy/dt)} = \frac{a\sin t}{a(1 - \cos t)}$$

$$= \frac{2\cdot\sin\frac{t}{2}\cdot\cos\frac{t}{2}}{2\cdot\sin^2\frac{t}{2}} = \frac{\cos\frac{t}{2}}{\sin\frac{t}{2}} = \cot\frac{t}{2}$$

t:	0	$\frac{\pi}{2}$	π	$\frac{3\pi}{2}$	2π
x:	0	$a(\frac{\pi}{2}-1)$ (0.57a)	$a\pi$	$a(\frac{3\pi}{2}+1)$ (5.71a)	$2\pi a$ (6.28a)
y:	0	a	$2a$	a	0
$\frac{dy}{dx}$:	∞	1	0	-1	∞
point:	O	A	B	C	D

x and y are defined for all t.

Tangents at the points O and D are vertical while tangent at B is horizontal ($y = 2a$) and tangents at A and C are of slope 1 and -1. B is known as vertex and x-axis as base. Curve is periodic of period 2π in the interval [0 to 2π] with t as the parameter. Curve repeats over intervals of $[0, 2a\pi]$ (refer Fig. 5.47).

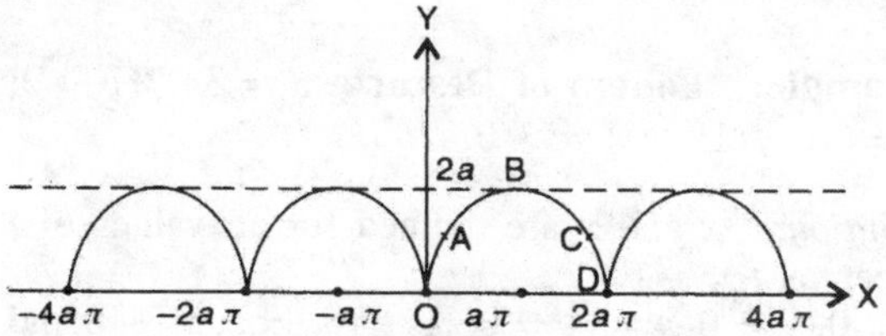

Fig. 5.47

EXERCISE

Trace the following curves:

1. Cycloid $x = a(t + \sin t)$, $y = a(1 - \cos t)$.

Ans. $t = 0$, then $x = 0$, $y = 0$, origin lies. Symmetric about y-axis since x is odd and y is even function of t. Curve lies between $y = 0$ and $y = 2a$. Periodic of period 2π. Tangent at O horizontal, tangents at $t = \pm\pi$ are vertical. No asymptote, $dy/dx = \tan(t/2)$ (Fig. 5.48).

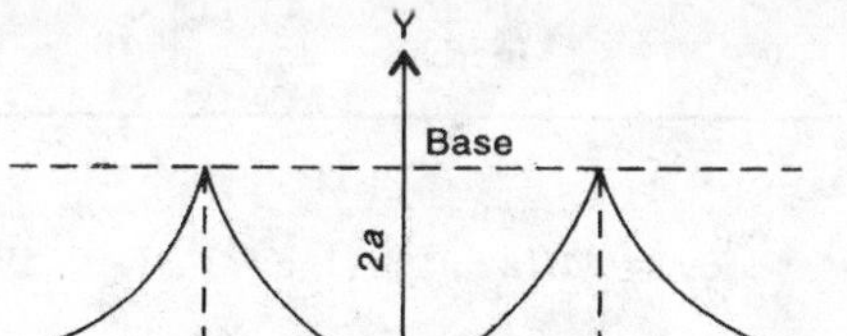

Fig. 5.48

2. *Tractrix*: $x = a[\cos t + \frac{1}{2}\ln\tan^2(\frac{t}{2})]$; $y = a\sin t$.

Ans. Symmetric about x-axis since x is even and y odd function of t (Fig. 5.49). Symmetric about y-axis, origin does not on curve, y lies between $\pm a$, x-axis is asymptote $(0, \pm a)$ are intersection with y-axis at $t = \pm\frac{\pi}{2}$, at which tangents are parallel to y-axis since $\frac{dy}{dx} = \tan t$.

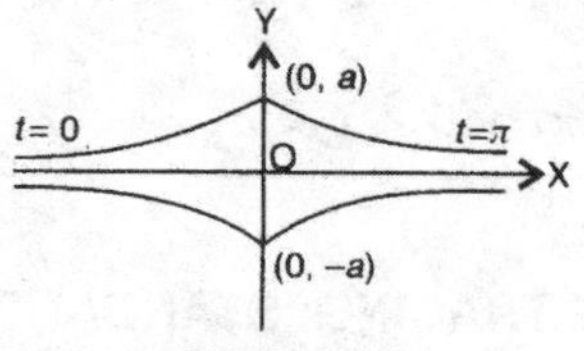

Fig. 5.49

3. $x = a(t + \sin t)$, $y = a(1 + \cos t)$.

Ans. Symmetric about y-axis, since x is odd and y is even function of t. Origin does not lie on curve. $(0, 2a)$ is y-intercept.
Curve lies between $y = 0$ and $y = 2a$. No asymptote, $\frac{dy}{dx} = -\tan\frac{\theta}{2}$. Tangent at $(0, 2a)$ is horizontal. Tangent, at $\theta = \pm\pi$ i.e., $(\pm a\pi, 0)$ are vertical (refer Fig. 5.50).

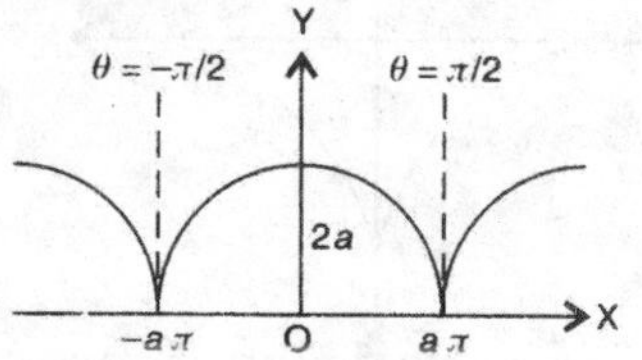

Fig. 5.50

4. $x = a\sin 2t(1 + \cos 2t)$, $y = a\cos 2t(1 - \cos 2t)$ (see Fig. 5.51).

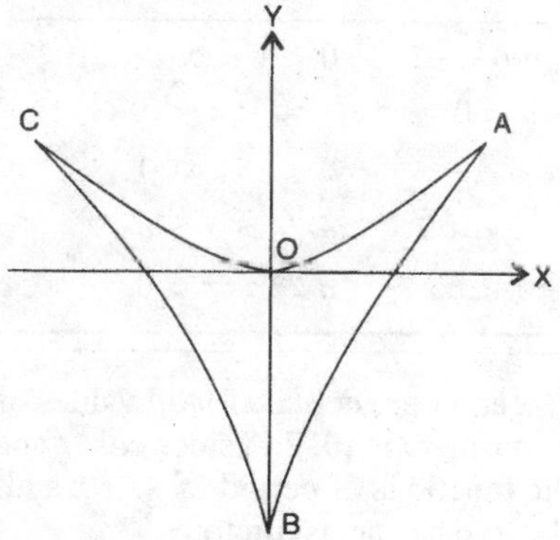

Fig. 5.51

Ans. x and y are periodic of period π. Origin lies on curve. y-intercept $B(0, 2a)$; $\frac{dx}{dt} = 4a\cos 3t\cos t$, $\frac{dy}{dt} = 4a\cos 3t\sin t$, $\frac{dy}{dx} = \tan t$, $A(\frac{3\sqrt{3}a}{4}, \frac{a}{4})$, $C(\frac{-3\sqrt{3}a}{4}, \frac{a}{4})$.

Range of t	Range of x	Range of y	Quadrant	Sign of y'	Nature of curve $y = h(x)$	Sign of y''	Type of curve
$0 < t < \frac{\pi}{2}$	$0 < x < a$	$0 < y < a$	I	$-$ve	Decreases	> 0	Concave
$\frac{\pi}{2} < t < \pi$	$-a < x < 0$	$0 < y < a$	II	$+$ve	Increases	> 0	Concave
$\pi < t < \frac{3\pi}{2}$	$-a < x < 0$	$-a < y < 0$	III	$-$ve	Decreases	< 0	Convex
$\frac{3\pi}{2} < t < 2\pi$	$0 < x < a$	$-a < y < 0$	IV	$+$ve	Increases	< 0	Convex

5.6 CURVE TRACING: STANDARD PARAMETRIC CURVES

Example: Four Cusped Hypocycloid or Astroid. $x = a\cos^3 t$, $y = a\sin^3 t$, with $a > 0$ (Fig. 5.52).

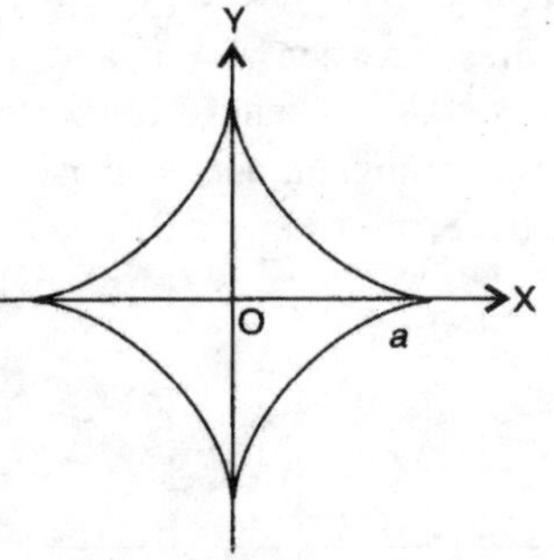

Fig. 5.52

Solution: x and y are defined for all values of t. It is enough to consider t in $[0, 2\pi]$ since $\cos^3 t$ and $\sin^3 t$ are periodic functions of period 2π. As x and y vary in $[-a, a]$ curve has no asymptotes.

$$\frac{dx}{dt} = -3a\cos^2 t \sin t,$$

$$\frac{dy}{dt} = 3a\sin^2 t \cos t$$

$$\frac{dy}{dx} = \frac{3a\sin^2 t\cos t}{-3a\cos^2 t\sin t} = -\tan t.$$

$\frac{dy}{dx}$ at $t = \pi/2, 3\pi/2$ is ∞. Therefore the tangents at $t = \pi/2$ and $t = 3\pi/2$ are vertical. Similarly horizontal tangents occur at $t = 0, \pi, 2\pi$ where $\frac{dy}{dx} = 0$.

$\frac{d^2y}{dx^2} = 1/(3a\cos^4 t \sin t)$.

Example: Folium of Descartes: $x = 3at/(1+t^3)$; $y = 3at^2/(1+t^3)$, $a > 0$.

Solution: x and y are defined for all values of t except at $t = -1$.
$x = 0$, $y = 0$ when $t = 0$. As $t \to \pm\infty$, $x \to 0$ and $y \to 0$

$$\frac{dx}{dt} = 6a\left(\frac{1}{2} - t^3\right)/(1+t^3)^2$$

Range of t	Range of x	Range of y	Sign of y'	Nature of curve	Quadrant
$-\infty < t < -1$	$0 < x < \infty$	$-\infty < y < 0$	$-$ve	Decreases	IV
$-1 < t < 0$	$-\infty < x < 0$	$0 < y < \infty$	$-$ve	Decreases	II
$0 < t < \frac{1}{\sqrt[3]{2}}$	$0 < x < a\sqrt[3]{4}$	$0 < y < a\sqrt[3]{2}$	$+$ve	Increases	I
$\frac{1}{\sqrt[3]{2}} < t < \sqrt[3]{2}$	$a\sqrt[3]{4} > x > a\sqrt[3]{2}$	$a\sqrt[3]{2} < y < a\sqrt[3]{4}$	$-$ve	Decreases	I
$\sqrt[3]{2} < t < \infty$	$a\sqrt[3]{2} > x > 0$	$a\sqrt[3]{4} < y < 0$	$+$ve	Increases	I

$$\frac{dy}{dt} = 3at(2 - t^3)/(1+t^3)^2$$

$$\frac{dy}{dx} = t(2 - t^3)/[2(\frac{1}{2} - t^3)].$$

Tangents at $(x = a\sqrt[3]{2}, y = a\sqrt[3]{4})$ is horizontal since $\frac{dy}{dx} = 0$ at this point $t = \sqrt[3]{2}$. Tangent at $(x = a\sqrt[3]{4}, y = a\sqrt[3]{2})$ is vertical since $\frac{dy}{dx} = \infty$ at this point $t = \sqrt[3]{2}$.

Asymptote: $m = \lim_{x\to\infty} \frac{y}{x} = \lim_{t\to -1} \frac{3at^2(1+t^3)}{3at(1+t^3)} = -1$

$$c = \lim_{x\to\infty} (y - mx) = \lim_{t\to -1} \left[\frac{3at^2}{1+t^3} - (-1) \cdot \frac{3at}{1+t^3} \right] = -a.$$

So $y = -x - a$ or $y + x + a = 0$ is the asymptote (Fig. 5.53).

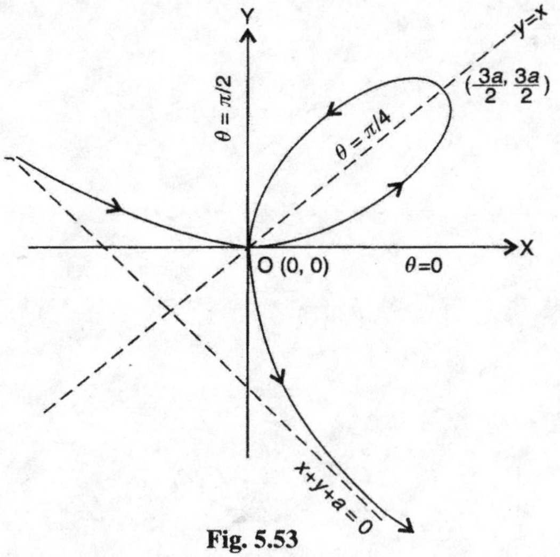

Fig. 5.53

Chapter 6

Integral Calculus

INTRODUCTION

Integral calculus is the study of finding a function from information about its rate of change. Application of integral calculus includes finding areas of irregular plane regions, length of curves, volume, surface area of solid of revolution, mass, moment of inertia, centre of gravity. Successive application of a reduction formula, which connects an integral with another integral of lower order, enables to evaluate the given integral. Sine-integral function (useful in optics), gamma, beta, functions are some important improper integrals.

6.1 REDUCTION FORMULAE

Reduction formulae reduces a given integral to a known integration form by the repeated application of integration by parts. Here reduction formulae of only trigonometric functions $\sin^n x$, $\cos^n x$, $\tan^n x$, $\cot^n x$, $\sec^n x$, $\operatorname{cosec}^n x$, $\sin^n x \cdot \cos^n x$ are considered.

Reduction Formula for $\int \sin^n x\, dx$; $n > 0$

Consider

$$I = \int \sin^n x\, dx = \int \sin^{n-1} x \cdot \sin x\, dx$$

$$= -\int \sin^{n-1} x \cdot d(\cos x)$$

Integrating by parts. $\int u dv = uv - \int v du$

$$I = \sin^{n-1} x(-\cos x) -$$

$$-\int (-\cos x)\cdot(n-1)\sin^{n-2} x \cdot \cos x\, dx$$

$$= -\cos x \cdot \sin^{n-1} x + (n-1)\int \sin^{n-2} x \cdot \cos^2 x\, dx$$

$$= -\cos x \cdot \sin^{n-1} x + (n-1)\int \sin^{n-2} x(1-\sin^2 x)dx$$

$$I = -\cos x \cdot \sin^{n-1} x - (n-1)\int \sin^n x\, dx +$$

$$+(n-1)\cdot\int \sin^{n-2} x\, dx \quad \text{or}$$

$$(1+n-1)I = -\cos x \cdot \sin^{n-1} x + (n-1)\int \sin^{n-2} x\, dx$$

$$\int \sin^n x\, dx = -\frac{1}{n}\cos x \cdot \sin^{n-1} x + \frac{n-1}{n}\int \sin^{n-2} x\, dx \quad (1)$$

Reduction Formula for $\int \cos^n x\, dx$; $n > 0$

Put $x = \frac{\pi}{2} - y$, $dx = -dy$ in (1) then

$$\int \cos^n y\, dy = \frac{1}{n}\sin y \cos^{n-1} y + \frac{n-1}{n}\int \cos^{n-2} y\, dy \quad (2)$$

Note 1: In the reduction formulae (1) and (2), the powers of $\sin x$ and $\cos x$ are reduced by 2.

Note 2: Instead of reduction formulae, alternatively the following substitution methods can also be used, to evaluate (1) and (2).

a. Put $\cos x = t$ in $\int \sin^n x\, dx$ when the power (index) n is odd and integrate the integral in t. Then replace t by $\cos x$.

b. Put $\sin x = t$ in $\int \cos^n x\, dx$ when the power (index) n is odd.

c. When the index n is an even positive integer and n is small, then express powers of $\sin x$ and $\cos x$ in terms of multiple angles using $\sin^2 x = \frac{1-\cos 2x}{2}$, $\cos^2 x = \frac{1+\cos 2x}{2}$, $\sin x \cdot \cos x = \frac{\sin 2x}{2}$

d. When n is large positive integer, use DeMoivre's theorem to express $\sin^n x$ and $\cos^n x$ in terms of multiple angles as follows: put $z = \cos x + i\sin x$, then $\frac{1}{z} = \cos x - i\sin x$, $2\cos x = z + \frac{1}{z}$, and $2i\sin x = z - \frac{1}{z}$ so that $2\cos nx = z^n + \frac{1}{z^n}$ and $2i\sin nx = z^n - \frac{1}{z^n}$. Now $(2\cos x)^n = 2^n\cos^n \cdot x = \left(z + \frac{1}{z}\right)^n =$ $= z^n + nc_1 z^{n-2} + \cdots + nc_{n-1}\frac{1}{z^{n-2}} + \frac{1}{z^n}$.

Using $nc_1 = nc_{n-1}$, $nc_2 = nc_{n-2}$ etc.

$$2^n\cos^n x = (z^n + \frac{1}{z^n}) + nc_1(z^{n-2} + \frac{1}{z^{n-2}}) + \cdots$$

$$= 2\cos nx + nc_1 \cdot 2 \cdot \cos(n-2)x + \cdots$$

$$2^n\int \cos^n x\,dx = 2\left[\frac{\sin nx}{n} + \frac{nc_1}{(n-2)}\cdot\sin(n-2)x + \cdots\right]$$

Similarly, $(2i\sin x)^n = \left(y - \frac{1}{y}\right)^n$.

Wallis' Formula*

Prove that

$$I_n = \int_0^{\pi/2}\sin^n x\,dx = \int_0^{\pi/2}\cos^n x\,dx$$

$$= \begin{cases} \frac{n-1}{n}\cdot\frac{n-3}{n-2}\cdot\frac{n-5}{n-4}\cdots\frac{2}{3}\cdot I_1 & \text{when } n \text{ is odd} \\ \frac{n-1}{n}\cdot\frac{n-3}{n-2}\cdot\frac{n-5}{n-4}\cdots\frac{1}{2}\cdot I_0 & \text{when } n \text{ is even.} \end{cases} \quad (3)$$

Here

$$I_1 = \int_0^{\pi/2}\sin x\,dx = \int_0^{\pi/2}\cos x\,dx = 1$$

and $$I_0 = \int_0^{\pi/2} dx = \frac{\pi}{2}.$$

Solution: Taking the limits 0 to $\frac{\pi}{2}$ in the reduction formula (1), we have

$$I_n = \int_0^{\pi/2}\sin^n x\,dx = -\frac{\cos x\cdot\sin^{n-1}x}{n}\bigg|_0^{\pi/2} +$$

$$+ \frac{n-1}{n}\int_0^{\pi/2}\sin^{n-2}x\,dx$$

$$I_n = 0 + \frac{n-1}{n}\cdot I_{n-2}$$

Using this recurrence relation

$$I_{n-2} = \frac{n-3}{n-2}I_{n-4}$$

$$I_{n-4} = \frac{n-5}{n-4}I_{n-6}$$

– – – – – – – – – –

– – – – – – – – – –

$$I_3 = \frac{2}{3}I_1 \qquad \text{when } n \text{ is odd}$$

$$I_2 = \frac{1}{2}I_0 \qquad \text{when } n \text{ is even}$$

Substituting these values and using $I_1 = 1$, $I_0 = \frac{1}{2}$

$$I_n = \frac{n-1}{n}\cdot\frac{n-3}{n-2}\cdot\frac{n-5}{n-4}\cdots\frac{2}{3}\cdot \quad \text{when } n \text{ is odd}$$

$$I_n = \frac{n-1}{n}\cdot\frac{n-3}{n-2}\cdot\frac{n-5}{n-4}\cdots\frac{1}{2}\cdot\frac{\pi}{2} \quad \text{when } n \text{ is even.} \quad (3)$$

By the substitution $x = \frac{\pi}{2} - y$, we have

$$I_n = \int_0^{\frac{\pi}{2}}\sin^n x\,dx = \int_0^{\frac{\pi}{2}}\cos^n x\,dx$$

Note: The numerator of I_n consists of products of numbers starting from $(n-1)$ and decreasing by 2. The denominator of I_n consists of products of numbers starting from n and decreasing by 2. In either case the process terminates with the last quotient as $\frac{2}{3}$ or $\frac{1}{2}$ according as n is odd or even. In the event when n is even the last factor $\frac{1}{2}$ is multiplied by $\frac{\pi}{2}$.

Corollary 1: Certain definite integrals can be reduced to Wallis's formula (3) by simple trigonometric substitution. Let n be a positive integer. Then

a. Putting $x = a\sin\theta$,

$$\int_0^a \frac{x^n dx}{\sqrt{a^2 - x^2}} = \int_0^{\pi/2}\frac{a^n\cdot\sin^n\theta\cdot a\cos\theta d\theta}{\cos\theta}$$

$$= a^n\int_0^{\pi/2}\sin^n\theta d\theta$$

b. Putting $x = a\tan\theta$,

$$\int_0^\infty \frac{dx}{(a^2+x^2)^n} = \int_0^{\pi/2}\frac{a\sec^2\theta d\theta}{a^{2n}(\sec^2\theta)^n}$$

$$= \frac{1}{a^{2n-1}}\int_0^{\pi/2}\cos^{2n-2}\theta d\theta$$

c. Putting $x = a\tan\theta$

$$\int_0^\infty \frac{dx}{(a^2+x^2)^{n+\frac{1}{2}}} = \frac{1}{a^{2n+1}}\int_0^{\pi/2}\frac{a\sec^2\theta d\theta}{\sec^{2n+1}\theta}$$

$$= \frac{1}{a^{2n}}\int_0^{\pi/2}\cos^{2n-1}\theta d\theta$$

---

*John Wallis (1616–1703), English mathematician.

Reduction Formula for $\int \tan^n x\, dx$

$$\int \tan^n x\, dx = \int \tan^{n-2} x \cdot \tan^2 x\, dx$$

$$= \int \tan^{n-2} x \cdot (\sec^2 x - 1)dx$$

$$= \int \tan^{n-2} x \cdot \sec^2 x\, dx - \int \tan^{n-2} x\, dx$$

$$= \int \tan^{n-2} x \cdot d(\tan x) - \int \tan^{n-2} x\, dx$$

$$\int \tan^n x\, dx = \frac{\tan^{n-1} x}{n-1} - \int \tan^{n-2} x\, dx \quad (4)$$

This formula reduces power of $\tan x$ by 2. After repeated application, the integral on R.H.S. reduces to $I_1 = \int \tan x\, dx = \log \sec x$ if n is odd or reduces to $I_2 = \int \tan^2 x\, dx = \tan x - x$ if n is even. Similarly,

Reduction Formula for $\int \cot^n x\, dx$

$$\int \cot^n x\, dx = \int \cot^{n-2} x(\operatorname{cosec}^2 x - 1)dx$$

$$= -\frac{\cot^{n-1} x}{n-1} - \int \cot^{n-2} x\, dx \quad (5)$$

If n is odd, $I_1 = \int \cot x\, dx = \ln \sin x$ or
If n is even, $I_2 - \int \cot^2 x\, dx = -\cot x - x$

Reduction Formula for $\int \sec^n x\, dx$

$$\int \sec^n x\, dx = \int \sec^{n-2} x \cdot \sec^2 x\, dx = \int \sec^{n-2} x \cdot d(\tan x).$$

Integrating by parts

$$= \tan x \cdot \sec^{n-2} x - \int \tan x \cdot (n-2) \cdot \sec^{n-3} x \cdot \sec x \cdot \tan x\, dx$$

$$= \tan x \cdot \sec^{n-2} x - (n-2) \int \tan^2 x \cdot \sec^{n-2} x\, dx$$

$$= \tan x \cdot \sec^{n-2} x - (n-2) \int \sec^{n-2} x \cdot (\sec^2 x - 1)dx$$

$$= \tan x \cdot \sec^{n-2} x - (n-2)\left[\int \sec^n x\, dx - \int \sec^{n-2} x\, dx\right]$$

$$[1+(n-2)] \int \sec^n x\, dx = \tan x \cdot \sec^{n-2} x + (n-2) \int \sec^{n-2} x\, dx$$

$$\int \sec^n x\, dx = \frac{\tan x \cdot \sec^{n-2} x}{n-1} + \frac{n-2}{n-1} \int \sec^{n-2} x\, dx \quad (6)$$

The powers of $\sec x$ are reduced by 2.
If n is odd, $I_1 = \int \sec x\, dx = \ln(\sec x + \tan x)$ or
If n is even, $I_2 = \int \sec^2 x\, dx = \tan x$

Reduction Formula for $\int \operatorname{cosec}^n \cdot x\, dx$

Similarly, we get

$$\int \operatorname{cosec}^n x\, dx = -\frac{\cot x \cdot \operatorname{cosec}^{n-2} x}{n-1} + \frac{n-2}{n-1} \int \operatorname{cosec}^{n-2} x\, dx \quad (7)$$

If n is odd, $I_1 = \int \operatorname{cosec} x\, dx = \ln \tan \frac{x}{2}$ or
If n is even, $I_2 = \int \operatorname{cosec}^2 x\, dx = -\cot x$.

Reduction Formula for $\int \sin^m x \cdot \cos^n x\, dx$; $m, n > 0$

Consider

$$\int \sin^m x \cdot \cos^n x\, dx = \int \sin^{m-1} x \cdot \sin x \cdot \cos^n x\, dx$$

$$= \int \sin^{m-1} x \cdot d\left(-\frac{\cos^{n+1} x}{n+1}\right)$$

Integrating by parts

$$= -\frac{\sin^{m-1} x \cdot \cos^{n+1} x}{(n+1)} + \int \frac{\cos^{n+1} x}{n+1} \cdot (m-1) \times \sin^{m-2} x \cdot \cos x\, dx$$

$$= -\frac{\sin^{m-1} x \cdot \cos^{n+1} x}{(n+1)} + \frac{m-1}{n+1} \int \cos^n x \cdot \sin^{m-2} x \times (1 - \sin^2 x)\, dx$$

$$= -\frac{\sin^{m-1} x \cdot \cos^{n+1} x}{(n+1)} + \frac{m-1}{n+1}\left[\int \cos^n x \cdot \sin^{m-2} x\, dx - \int \cos^n x \cdot \sin^m x\, dx\right]$$

$$\left(1 + \frac{m-1}{n+1}\right) \int \sin^m x \cdot \cos^n x\, dx$$

$$= -\frac{\sin^{m-1} x \cdot \cos^{n+1} x}{(n+1)} + \frac{m-1}{n+1} \int \cos^n x \cdot \sin^{m-2} x\, dx$$

Therefore the reduction formula is

$$\int \sin^m x \cdot \cos^n x\,dx = -\frac{\sin^{m-1} x \cdot \cos^{n+1} x}{m+n} + \frac{m-1}{m+n}\int \sin^{m-2} x \cdot \cos^n x\,dx \tag{8}$$

Similary considering

$$\int \sin^m x \cos^n x\,dx = \int \cos^{n-1} x \cdot \cos x \cdot \sin^m x\,dx = \int \cos^{n-1} x \cdot d\left(\frac{\sin^{m+1} x}{m+1}\right)$$

and integrating by parts as above we get another form of the reduction formula as

$$\int \sin^m x \cdot \cos^n x\,dx = \frac{\sin^{m+1} x \cdot \cos^{n-1} x}{m+n} + \frac{n-1}{m+n}\int \sin^n x \cdot \cos^{m-2} x\,dx \tag{9}$$

In addition to (8) and (9), we can obtain the following other four reduction formulae.

$$\int \sin^m x \cdot \cos^n x\,dx = \frac{\sin^{m+1} x \cos^{n-1} x}{m+1} + \frac{n-1}{m+1}\int \sin^{m+2} x \cdot \cos^{n-2} x\,dx \tag{10}$$

Note 1: (10) is useful when m is negative integer and n is positive integer

$$\int \sin^m x \cdot \cos^n x\,dx = -\frac{\sin^{m+1} x \cdot \cos^{n+1} x}{n+1} + \frac{m+n+2}{n+1}\int \sin^m x \cdot \cos^{n+2} x\,dx \tag{11}$$

Note 2: (11) is useful when n is a negative integer. See WE 16 on Page 6.11.

$$\int \sin^m x \cdot \cos^n x\,dx = \frac{\sin^{m+1} x \cdot \cos^{n+1} x}{m+1} + \frac{m+n+2}{m+1}\int \sin^{m+2} x \cdot \cos^n x\,dx \tag{12}$$

Note 3: (12) is useful when n is a negative integer

$$\int \sin^m x \cdot \cos^n x\,dx = -\frac{\sin^{m-1} x \cdot \cos^{n+1} x}{n+1} + \frac{m-1}{n+1}\int \sin^{m-2} x \cdot \cos^{n+2} x\,dx \tag{13}$$

Note 4: (13) is useful when m is positive and n is negative integer.

Evaluation of Definite Integral $\int_0^{\frac{\pi}{2}} \sin^m x \cdot \cos^n x\,dx$ when both m and n are Positive Integers

Introduce the notation

$$I_{m,n} = \int_0^{\frac{\pi}{2}} \sin^m x \cdot \cos^n x\,dx \tag{14}$$

and using the reduction formula (8), we obtain

$$I_{m,n} = \int_0^{\frac{\pi}{2}} \sin^m x \cdot \cos^n x\,dx = -\left.\frac{\sin^{m-1} x \cdot \cos^{n+1} x}{m+n}\right|_0^{\frac{\pi}{2}} + \frac{m-1}{m+n}\int_0^{\frac{\pi}{2}} \sin^{m-2} x \cdot \cos^n x\,dx$$

Since the first term on R.H.S. is zero at both the limits, we get

$$I_{m,n} = \frac{m-1}{m+n} \cdot I_{m-2,n} \tag{15}$$

Using this recurrence relation, we obtain

$$I_{m-2,n} = \frac{m-3}{m+n-2} I_{m-4,n}$$

$$I_{m-4,n} = \frac{m-5}{m+n-4} I_{m-6,n}$$

- - - - - - - - - - - - - -

$$I_{3,n} = \frac{2}{3+n} I_{1,n} \quad \text{when } m \text{ is odd.}$$

$$I_{2,n} = \frac{1}{2+n} I_{0,n} \quad \text{when } m \text{ is even.}$$

Observe that

$$I_{1,n} = \int_0^{\frac{\pi}{2}} \sin x \cdot \cos^n x\,dx = -\left.\frac{\cos^{n+1} x}{n+1}\right|_0^{\frac{\pi}{2}} = \frac{1}{n+1} \tag{16}$$

and

$$I_{0,n} = \int_0^{\frac{\pi}{2}} \cos^n x \, dx \qquad (17)$$

Substituting these values, we have

(a) When m is odd and n may be odd or even

$$I_{m,n} = \frac{(m-1)}{(m+n)} \cdot \frac{(m-3)}{(m+n-2)} \times$$

$$\times \frac{(m-5)}{(m+n-4)} \cdots \frac{2}{3+n} \cdot \frac{1}{1+n}$$

Rewriting

$$I_{m,n} = \frac{(m-1)(m-3)(m-5)\cdots 2 \cdot 1}{(m+n)(m+n-2)(m+n-4)\cdots(n+3)(n+1)} \times$$

$$\times \left[\frac{(n-1)(n-3)(n-5)\cdots}{(n-1)(n-3)(n-5)\cdots}\right]$$

(b) When m is even and n is odd

$$I_{m,n} = \frac{(m-1)(m-3)(m-5)\cdots}{(m+n)(m+n-2)\cdots} \times$$

$$\times \frac{1}{2+n}\left(\frac{n-1}{n} \cdot \frac{n-3}{n-2} \cdots \cdot \frac{2}{3}\right)$$

(c) When m is even and n is even

$$I_{m,n} = \frac{(m-1)(m-3)\cdots 1}{(m+n)(m+n-2)\cdots(n+2)} \times$$

$$\times \left(\frac{n-1}{n} \cdot \frac{n-3}{n-2} \cdots \frac{1}{2} \cdot \frac{\pi}{2}\right)$$

General result

$$I_{m,n} = \int_0^{\frac{\pi}{2}} \sin^m x \cdot \cos^n x \, dx = \frac{NR}{DR} \cdot e$$

where NR = product of two sets of factors each commencing from $(m-1)$ and $(n-1)$ and decreasing by 2 at a time, while DR = product of factors starting from $(m+n)$ and decreasing by 2 at a time, descending ultimately to 1 or 2 according as the first factor of the set is odd or even. Here $e = \frac{\pi}{2}$ only when **both** m and n are even. In all other cases $e = 1$.

Corollary 1: Certain definite (proper and improper) integrals reduce to the $I_{m,n}$ form by simple substitutions. Let n and m be positive integers.

a. Putting $x = 2a\sin^2\theta$

$$\int_0^{2a} x^m \sqrt{2ax - x^2}\,dx$$

$$= \int_0^{2a} x^{m+\frac{1}{2}} \sqrt{2a - x}\,dx = \int_0^{\frac{\pi}{2}} (2a)^{m+\frac{1}{2}} \sin^{2m+1}\theta \times$$

$$\times \sqrt{2a} \cdot \cos\theta \cdot 4a \cdot \sin\theta \cdot \cos\theta\, d\theta$$

$$= (2a)^{m+2} \cdot 2 \int_0^{\frac{\pi}{2}} \sin^{2m+2}\theta \cdot \cos^2\theta\, d\theta$$

b. Putting $x = a\tan\theta$

$$\int_0^\infty \frac{x^n dx}{(a^2+x^2)^m} = \frac{a^{2n}}{a^{2m}} \int_0^{\pi/2} \frac{\tan^2\theta \cdot a \cdot \sec^2\theta}{\sec^{2m}\theta} d\theta$$

$$= a^{2n-2m+1} \int_0^{\pi/2} \sin^n\theta \cdot \cos^{2m-n-2}\theta\, d\theta$$

c. Putting $x = a\tan\theta$

$$\int_0^\infty \frac{x^n dx}{(a^2+x^2)^{m+\frac{1}{2}}}$$

$$= \int_0^{\pi/2} \frac{a^n \sin^n\theta}{\cos^n\theta} \cdot \frac{1}{a^{2m+1} . (\sec^2\theta)^{\frac{2m+1}{2}}} \cdot a \cdot \sec^2\theta\, d\theta$$

$$= a^{n-2m} \int_0^{\pi/2} \sin^n\theta \cdot \cos^{2m-n-1}\theta\, d\theta$$

Note: Instead of the six reduction formulae (8), (9), (10), (11), (12), (13), the integral $\int \sin^m x \cdot \cos^n x\,dx$ can be evaluated using certain substitutions as follows.

a. If the power (index) m of $\sin x$ is odd, then put $\cos x = t$.

b. If the power n of $\cos x$ is odd, then put $\sin x = t$.

c. If $m+n$ is negative even integer, then put $\tan x = t$.

In all the three above cases, integrand becomes a function of t and is easily integrated.

d. If both m and n are even integers and

i. m, n are small, then convert $\cos x$ and $\sin x$ into cosine and sines of multiple angles using $\cos^2 x = \frac{1+\cos 2x}{2}$, $\sin^2 x = \frac{1-\cos 2x}{2}$, $\sin x \cdot \cos x = \frac{\sin 2x}{2}$.

ii. m, n are large, use DeMoivre's theorem and use $z = \cos x + i\sin x$, $(2\cos x) = (z + \frac{1}{z})$ and $(2i\sin x) = z - \frac{1}{z}$ etc.

WORKED OUT EXAMPLES

Reduction formulae for $\int \sin^n x\, dx$, $\int \cos^n x\, dx$

Example 1: Evaluate $\int \sin^5 x\, dx$

a. by reduction formula

b. by substitution

c. Hence find $\int_0^{\pi/2} \sin^5 x\, dx$.

Solution:

a. From reduction formula (1) on page 6.1

$$\int \sin^n x\, dx = -\frac{\sin^{n-1} x}{n} \cdot \cos x + \frac{n-1}{n} \int \sin^{n-2} x\, dx$$

Put $n = 5$

$$\int \sin^5 x\, dx = -\frac{1}{5} \sin^4 x \cdot \cos x + \frac{4}{5} \int \sin^3 x\, dx$$

Put $n = 3$

$$\int \sin^3 x\, dx = -\frac{\sin^2 x}{3} \cdot \cos x + \frac{2}{3} \int \sin x\, dx$$

$$\int \sin^5 x\, dx = -\frac{1}{5} \cdot \sin^4 x \cdot \cos x + \frac{4}{5}\left[\frac{-\sin^2 x \cdot \cos x}{3} + \frac{2}{3}(-\cos x)\right]$$

$$= -\frac{1}{5} \sin^4 x \cdot \cos x - \frac{4}{15} \sin^2 x \cdot \cos x - \frac{8}{15} \cos x$$

b. Since the index $n = 5$ is odd, put $\cos x = t$

$$\int \sin^5 x\, dx = \int \sin^5 x \cdot \frac{dt}{-\sin x} = -\int \sin^4 x dt$$

$$= -\int (1 - \cos^2 x)^2 dt = -\int (1 - t^2)^2 dt$$

$$= -\int (1 + t^4 - 2t^2) dt = -\left[t + \frac{t^5}{5} - \frac{2t^3}{3}\right]$$

$$= -\left[\cos x + \frac{1}{5} \cos^5 x - \frac{2}{3} \cos^3 x\right]$$

This result can be rewritten as in (a).

$$= -\left[\cos x + \frac{1}{5} \cdot \cos x (1 - \sin^2 x)^2 \cdot - \frac{2}{3} \cdot \cos x (1 - \sin^2 x)\right]$$

$$= -\frac{8}{15} \cos x - \frac{4}{15} \sin^2 x \cdot \cos x - \frac{1}{5} \sin^4 x \cdot \cos x.$$

c.
$$\int_0^{\frac{\pi}{2}} \sin^5 x\, dx = \left[-\frac{8}{15} \cos x - \frac{4}{15} \sin^2 x \cdot \cos x - \frac{1}{5} \cdot \sin^4 x \cdot \cos x\right]_0^{\frac{\pi}{2}}$$

$$= 0 + 0 + 0 - \left(-\frac{8}{15}\right) = \frac{8}{15}$$

Example 2: Without using reduction formula, evaluate $\int \sin^6 x\, dx$ and hence find $\int_0^{\pi/2} \sin^6 x\, dx$. Verify by Wallis's formula.

Solution: Since power $n = 6$ is even, use $\sin^2 x = \frac{1 - \cos 2x}{2}$ to rewrite

$$\int \sin^6 x\, dx = \int (\sin^2 x)^3 dx$$

$$= \int \left(\frac{1 - \cos 2x}{2}\right)^3 dx$$

$$= \frac{1}{8} \int (1 - 3\cos 2x + 3\cos^2 2x - \cos^3 2x) dx$$

Using $\cos^2 x = \frac{1 + \cos 2x}{2}$, we have

$$= \frac{1}{8} \int \left[1 - 3\cos 2x + 3\left(\frac{1 + \cos 4x}{2}\right) - \cos 2x \left(\frac{1 + \cos 4x}{2}\right)\right] dx$$

$$= \frac{1}{8} \int \left(\frac{5}{2} - \frac{7}{2} \cos 2x + \frac{3}{2} \cos 4x - \frac{1}{2} \cdot \cos 2x \cdot \cos 4x\right) dx$$

$$= \frac{1}{8} \int \left(\frac{5}{2} - \frac{7}{2} \cos 2x + \frac{3}{2} \cos 4x - \frac{1}{2} \cdot \frac{1}{2} [\cos 6x + \cos 2x]\right) dx$$

$$= \frac{1}{8} \int \left(\frac{5}{2} - \frac{15}{4} \cos 2x + \frac{3}{2} \cos 4x - \frac{1}{4} \cos 6x\right) dx$$

$$= \frac{1}{8} \left[\frac{5}{2} x - \frac{15}{8} \sin 2x + \frac{3}{8} \sin 4x - \frac{1}{24} \sin 6x\right]$$

So

$$\int_0^{\pi/2} \sin^6 x\, dx$$

$$= \frac{1}{8}\left[\frac{5}{2}x - \frac{15}{8}\sin 2x + \frac{3}{8}\sin 4x - \frac{1}{24}\sin 6x\right]_0^{\pi/2}$$

$$= \frac{1}{8}\cdot\frac{5}{2}\cdot\frac{\pi}{2} = \frac{5\pi}{32}$$

Verification:

$$\int_0^{\pi/2} \sin^6 x\, dx = \frac{(6-1)(6-3)(6-5)}{6\cdot(6-2)(6-4)}\cdot\frac{\pi}{2}$$

$$= \frac{5\cdot 3\cdot 1}{6\cdot 4\cdot 2}\cdot\frac{\pi}{2} = \frac{5\pi}{32}$$

Example 3: Evaluate $\int \cos^3 x\, dx$ (i) by reduction formula (ii) by substitution (iii) hence find $\int_0^{\pi/2} \cos^3 x\, dx$.

Solution:

i. Use the reduction formula (2) on page 6.1

$$\int \cos^n x\, dx = \frac{1}{n}\sin x\cdot\cos^{n-1} x + \frac{n-1}{n}\int \cos^{n-2} x\, dx$$

Put $n = 3$

$$\int \cos^3 x\, dx = \frac{1}{3}\sin x\cdot\cos^2 x + \frac{2}{3}\int \cos x\, dx$$

$$= \frac{1}{3}\sin x\cdot\cos^2 x + \frac{2}{3}\cdot\sin x$$

ii. Since the power of cos is $n = 3$ is odd, use the substitution $\sin x = t$, $\cos x\, dx = dt$

$$\int \cos^3 x\, dx = \int \cos^2 x\cdot\cos x\cdot dx$$

$$= \int (1 - \sin^2 x)\cos x\cdot dx$$

$$= \int (1-t^2)dt = t - \frac{t^3}{3} = \sin x - \frac{1}{3}\sin^3 x$$

Rewriting

$$= \sin x - \frac{1}{3}\cdot\sin x\cdot(1-\cos^2 x)$$

$$= \frac{2}{3}\sin x + \frac{1}{3}\sin x\cdot\cos^2 x$$

iii. $\int_0^{\pi/2} \cos^3 x\, dx = \frac{(3-1)}{3} = \frac{2}{3}$ from Wallis's formula

From above integral (ii) also

$$\int_0^{\pi/2} \cos^3 x\, dx = \frac{2}{3}\sin x + \frac{1}{3}\sin x\cdot\cos^2 x\Big|_0^{\pi/2}$$

$$= \frac{2}{3} + 0 - 0 - 0 = \frac{2}{3}$$

Example 4: Evaluate $\int_0^5 \frac{x^6 dx}{\sqrt{25-x^2}}$.

Solution: Put $x = 5\sin\theta$, $x = 0$ then $\theta = 0$. $x = 5$, then $\theta = \frac{\pi}{2}$.

$$\int_0^5 \frac{x^6 dx}{\sqrt{25-x^2}} = \int_0^{\pi/2} \frac{5^6 \sin^6\theta}{5\cos\theta}\cdot 5\cdot\cos\theta d\theta$$

$$= 5^6\cdot\frac{5}{6}\cdot\frac{3}{4}\cdot\frac{1}{2}\cdot\frac{\pi}{2} = \frac{5^7}{32}\pi = \frac{78125}{32}\pi$$

(by Wallis's formula, with $n = 6$, even).

Example 5:

$$\int_0^\infty \frac{dx}{(a^2+x^2)^7}$$

Solution: Put $x = a\tan\theta$, $dx = a\sec^2\theta d\theta$, when $x = 0$, then $\theta = 0$; when $x = \infty$, then $\theta = \frac{\pi}{2}$.

$$\int_0^\infty \frac{dx}{(a^2+x^2)^7}$$

$$= \int_0^{\pi/2} \frac{a\cdot\sec^2\theta\cdot d\theta}{a^{14}\sec^{14}\theta d\theta} = \frac{1}{a^{13}}\int_0^{\pi/2} \cos^{12}\theta d\theta$$

$$= \frac{1}{a^{13}}\cdot\frac{11}{12}\cdot\frac{9}{10}\cdot\frac{7}{8}\cdot\frac{5}{6}\cdot\frac{3}{4}\cdot\frac{1}{2}\cdot\frac{\pi}{2} = \frac{1}{a^{13}}\cdot\frac{231}{2048}.$$

Example 6: Evaluate $\int_0^\infty \frac{dx}{(16+x^2)^{\frac{9}{2}}}$

Solution: Put $x = 4\tan\theta$, limits: $\theta : 0$ to $\frac{\pi}{2}$. $dx = 4\cdot\sec^2\theta d\theta$

$$\int_0^\infty \frac{dx}{(16+x^2)^{\frac{9}{2}}} = \int_0^{\pi/2} \frac{4\cdot\sec^2\theta\cdot d\theta}{4^9\sec^9\theta}$$

$$= \frac{1}{4^8}\int_0^{\pi/2} \cos^7\theta d\theta = \frac{1}{4^8}\cdot\frac{6}{7}\cdot\frac{4}{5}\cdot\frac{2}{3}$$

$$= \frac{1}{4^6\cdot 7.5} = \frac{1}{143360}$$

Reduction formula for $\int \tan^n x\,dx$, $\int \cot^n x\,dx$, $\int \sec^n x\,dx$, $\int \operatorname{cosec}^n x\,dx$

Example 7: Evaluate $\int_0^{\pi/4} \tan^6 x\,dx$.

Solution: Use the reduction formula (4) on page 6.3

$$\int \tan^n x\,dx = \frac{\tan^{n-1} x}{n-1} - \int \tan^{n-2} x\,dx$$

Put $n = 6$, $\int \tan^6 x\,dx = \frac{\tan^5 x}{5} - \int \tan^4 x\,dx$

Put $n = 4$, $\int \tan^4 x\,dx = \frac{\tan^3 x}{3} - \int \tan^2 x$

Put $n = 2$, $\int \tan^2 x\,dx = \frac{\tan x}{1} - \int dx$

Substituting, $\int \tan^6 x\,dx = \frac{\tan^5 x}{5} - \frac{\tan^3 x}{3} + \tan x - x$

Now,

$$\int_0^{\pi/4} \tan^6 x\,dx = \frac{\tan^5 x}{5} - \frac{\tan^3 x}{3} + \tan x - x \Big|_0^{\frac{\pi}{4}}$$

$$= \left(\frac{1}{5} - \frac{1}{3} + 1 - \frac{\pi}{4}\right) - 0 = \frac{13}{15} - \frac{\pi}{4}$$

Example 8: Find $\int_{\pi/4}^{\pi/2} \cot^4 x\,dx$.

Solution: Use the reduction formula (5) on page 6.3

$$\int \cot^n x\,dx = -\frac{\cot^{n-1} x}{(n-1)} - \int \cot^{n-2} x\,dx$$

Put $n = 4$, $\int \cot^4 x\,dx = -\frac{\cot^3 x}{3} - \int \cot^2 x\,dx$

Put $n = 2$, $\int \cot^2 x\,dx = -\frac{\cot x}{1} - \int dx$

Substituting,

$$\int \cot^4 x\,dx = -\frac{\cot^3 x}{3} + \cot x + x$$

$$\int_{\frac{\pi}{4}}^{\pi} \cot^4 x\,dx = \left(-\frac{\cot^3 x}{3} + \cot x + x\right)\Big|_{\frac{\pi}{4}}^{\pi}$$

$$= \left(0 + 0 + \frac{\pi}{2}\right) - \left(-\frac{1}{3} + 1 + \frac{\pi}{4}\right)$$

$$= \frac{\pi}{4} - \frac{2}{3} = \frac{3\pi - 8}{12}$$

Example 9: If $I_n = \int_0^{\pi/4} \tan^n x\,dx$, prove that

i. $(n-1)(I_n + I_{n-2}) = 1$

ii. $(n-1)(I_{n+1} + I_{n-1}) = 1$

iii. Evaluate I_5

iv. Hence find $\int_0^a x^5(2a^2 - x^2)^{-3}\,dx$.

Solution: From the reduction formula (4) on page 6.3

i. $I_n = \frac{\tan^{n-1} x}{n-1}\Big|_0^{\pi/4} - I_{n-2} = \frac{1}{n-1} - I_{n-2}$

So $(n-1)(I_n + I_{n-2}) = 1$

ii. Put $n = n+1$

$$n(I_{n+1} + I_{n-1}) = 1$$

iii. Put $n = 5$ in (i): $4(I_5 + I_3) = 1$

Put $n = 3$ in (i): $2(I_3 + I_1) = 1$

$I_5 = +\frac{1}{4} \cdot -I_3 = \frac{1}{4} - (\frac{1}{2} - I_1) = -\frac{1}{4} + I_1$

But $I_1 = \int_0^{\pi/4} \tan x\,dx = \ln \sec x\big|_0^{\pi/4} = \ln\sqrt{2}$

Thus $I_5 = -\frac{1}{4} + \ln\sqrt{2}$

iv. Put $x = \sqrt{2}a\sin\theta$, $dx = \sqrt{2}a \cdot \cos\theta\,d\theta$

Limits: $x = 0$ then $\theta = 0$, $x = a$ then $\theta = \frac{\pi}{4}$

$$\int_0^a x^5(2a^2 - x^2)^{-3}dx$$

$$= \int_0^{\pi/4} (\sqrt{2}a)^5 \sin^5\theta(2a^2)^{-3}\cos^{-6}\theta \cdot \sqrt{2}a \cdot \cos\theta\,d\theta$$

$$= (\sqrt{2}a)^6(\sqrt{2}a)^{-6}\int_0^{\pi/4} \frac{\sin^5\theta}{\cos^5\theta}d\theta = I_5$$

$$= \ln\sqrt{2} - \frac{\pi}{4} \quad \text{[from (iii)]}$$

Example 10: Determine (i) $\int_0^{\pi/4} \sec^5 x\,dx$ (ii) $\int_{\pi/4}^{\pi/2} \operatorname{cosec}^3 x\,dx$

Solution: **i.** Use reduction formula (6) on page 6.3

$$\int \sec^n x\,dx = \frac{\tan x \cdot \sec^{n-2} x}{n-1} + \frac{n-2}{n-1}\int \sec^{n-2} x\,dx$$

Putting $n = 5, 3, 1$ and substituting

$$\int_0^{\pi/4} \sec^5 x\,dx = \frac{1}{4}\left[\sec^3 x \cdot \tan x + \frac{3}{2}\left\{\sec x \cdot \tan x + \ln(\tan x + \sec x)\right\}\right]_0^{\pi/4}$$

$$= \frac{1}{4}\left[(\sqrt{2})^3 \cdot 1 + \frac{3}{2}\left\{\sqrt{2} \cdot 1 + \ln(1 + \sqrt{2})\right\}\right]$$

ii. Use reduction formula (7) on page 6.3

$$(n-1)\int \operatorname{cosec}^n x\,dx = -\cot x\cdot\operatorname{cosec}^{n-2}x\cdot$$
$$+(n-2)\int \operatorname{cosec}^{n-2}x\,dx$$

Putting $n=3, 1$ and substituting

$$\int_{\pi/4}^{\pi/2}\operatorname{cosec}^3 x\,dx$$
$$=\frac{-1}{2}[\operatorname{cosec}x\cdot\cot x+\ln(\operatorname{cosec}x+\cot x)]_{\pi/4}^{\pi/2}$$
$$=-\frac{1}{2}\left[(0-\sqrt{2})+\ln(1)-\ln(\sqrt{2+1})\right]$$

iii. Put $x=a\tan\theta$, $dx=a\sec^2\theta d\theta$, limits: θ: 0 to $\frac{\pi}{2}$

$$\int_0^\infty \frac{x^4}{(a^2+x^2)^5}dx=\int_0^{\frac{\pi}{2}}\cdot\frac{\tan^4\theta}{a^{10}}\cdot\frac{\sec^2\theta}{\sec^{10}\theta}d\theta$$
$$=\frac{1}{a^{10}}\int_0^{\frac{\pi}{2}}\sin^4\theta\cdot\cos^4\theta d\theta$$
$$=\frac{1}{a^{10}}\cdot\frac{3\cdot 3}{8\cdot 6\cdot 4\cdot 2}\cdot\frac{\pi}{2}=\frac{3\pi}{256}$$

Reduction formulae for $\int \sin^m x\cdot\cos^n x\,dx$

Example 11: Evaluate

i. $\int_0^{\pi/2}\sin^4 x\cdot\cos^5 x\,dx$

ii. $\int_0^{\pi/2}\sin^8 x\cdot\cos^{12} x\,dx$

iii. $\int_0^\infty \frac{x^4dx}{(a^2+x^2)^5}$

Solution:

i. Here $m=4=$ even, $n=5=$ odd

$$\int_0^{\pi/2}\sin^4 x\cdot\cos^5 x\,dx=\frac{3\cdot 1\cdot 4\cdot 2\cdot}{9\cdot 7\cdot 5\cdot 3\cdot 1}=\frac{8}{315}$$

ii. Here $m=8$, $n=12$ are both even.

$$\int_0^{\frac{\pi}{2}}\sin^8 x\cdot\cos^{12}x\,dx$$
$$=\frac{7\cdot 5\cdot 3\cdot 11\cdot 9\cdot 7\cdot 5\cdot 3}{20\cdot 18\cdot 16\cdot 14\cdot 12\cdot 10\cdot 8\cdot 6\cdot 4\cdot 2}\cdot\frac{\pi}{2}$$
$$=\frac{539\pi}{3670016}$$

Example 12: Find $\int \sin^5 x\cdot\cos^{\frac{3}{4}} x\,dx$.

Solution: Since $m=5$ is odd, put $\cos x=t$, so $-\sin x\,dx=dt$.

$$\int \sin^5 x\cdot\cos^{\frac{3}{4}}x\,dx$$
$$=\int \sin^4 x\cdot\sin x\cdot\cos^{\frac{3}{4}}x\,dx$$
$$=\int(1-\cos^2 x)^2\sin x\cdot\cos^{\frac{3}{4}}x\,dx$$
$$=\int(1-t^2)^2\cdot t^{\frac{3}{4}}(-dt)$$
$$=-\int t^{\frac{3}{4}}(1+t^4-2t^2)dt$$
$$=-\left[\frac{4}{7}t^{\frac{7}{4}}+\frac{4}{23}t^{\frac{23}{4}}-\frac{8}{15}t^{\frac{15}{4}}\right]$$
$$=-\frac{4}{7}\cos^{\frac{7}{4}}x-\frac{4}{23}\cos^{\frac{23}{4}}x+\frac{8}{15}\cos^{\frac{15}{4}}x$$

Example 13: Evaluate $\int \operatorname{cosec}x\cdot\cot^5 x\,dx$.

Solution:

$$\int\operatorname{cosec}x\cdot\cot^5 x\,dx=\int\frac{1}{\sin x}\cdot\frac{\cos^5 x}{\sin^5 x}dx$$
$$=\int\sin^{-6}x\cdot\cos^5 x\,dx$$

Here $n=5$ is odd, so put $\sin x=t$, $\cos x\,dx=dt$

$$=\int\sin^{-6}x\cdot(1-\sin^2 x)^2\cdot\cos x\,dx=\int\frac{1}{t^6}(1-t^2)^2dt$$
$$=\int\frac{1+t^4-2t^2}{t^6}dt=-\frac{1}{5}\frac{1}{t^5}-\frac{1}{t}-\frac{2}{3}\frac{1}{t^3}$$
$$=-\operatorname{cosec}x-\frac{2}{3}\operatorname{cosec}^3 x-\frac{1}{5}\operatorname{cosec}^5 x$$

Example 14: Calculate $\int\frac{dx}{\sin^3 x\cdot\cos^5 x}$.

Solution: Here $m=-3$, $n=-5$, so $m+n=-3-5=-8$ is a negative even integer. Put $\tan x=t$, $\sec^2 x\,dx=dt$, then $\sin x=\frac{t}{\sqrt{1+t^2}}$, and $\cos x=\frac{1}{\sqrt{1+t^2}}$ (Fig. 6.1).

$$\int\frac{dx}{\sin^3 x\cdot\cos^5 x}$$

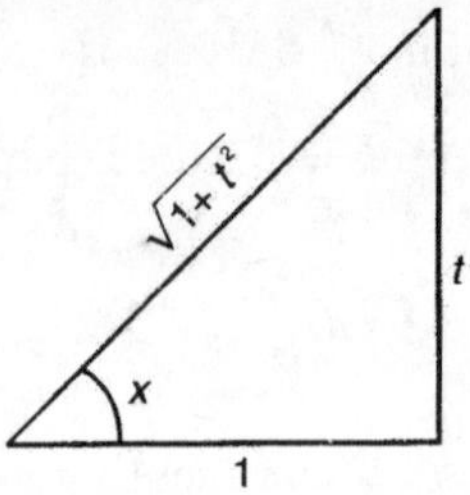

Fig. 6.1

$$= \int \left(\frac{t}{\sqrt{1+t^2}}\right)^{-3} \left(\frac{1}{\sqrt{1+t^2}}\right)^{-5} \frac{1}{(1+t^2)} dt$$

$$= \int \frac{(1+t^2)^3}{t^3} dt = \int \frac{1+3t^2+3t^4+t^6}{t^3} dt$$

$$= -\frac{1}{2}\frac{1}{t^2} + 3\ln t + 3\frac{t^2}{2} + \frac{t^4}{4}$$

$$= 3\ln\tan x + \frac{3}{2}\tan^2 x - \frac{1}{2}\cot^2 x + \frac{1}{4}\tan^4 x$$

Example 15: Evaluate $\int_0^{\frac{\pi}{4}} \sin^2 x \cdot \cos^4 x \, dx$ by (i) substitution method (ii) using DeMoivre's theorem.

Solution: Since $m = 2$, $n = 4$ are both even integers

i. Replace $\sin^2 x$ by $\frac{1-\cos 2x}{2}$ and $\cos^2 x$ by $\frac{1+\cos 2x}{2}$

$$\int_0^{\pi/4} \sin^2 x \cdot \cos^4 x \, dx$$

$$= \int_0^{\pi/4} \frac{(1-\cos 2x)}{2}\left(\frac{1+\cos 2x}{2}\right)^2 dx$$

$$= \frac{1}{8}\int_0^{\pi/4} (1+\cos 2x - \cos^2 2x - \cos^3 2x) dx$$

Put $2x = t$, $dx = \frac{1}{2}dt$, limits: t: 0 to $\frac{\pi}{2}$

$$= \frac{1}{16}\int_0^{\frac{\pi}{2}} (1+\cos t - \cos^2 t - \cos^3 t) dt$$

$$= \frac{1}{16}\left[\frac{\pi}{2} + 1 - \frac{1}{2}\cdot\frac{\pi}{2} - \frac{2}{3}\right] = \frac{1}{8}\left(\frac{\pi}{8} + \frac{1}{6}\right)$$

ii. Put $z = \cos x + i\sin x$, then $\frac{1}{z} = \cos x - i\sin x$, $z + \frac{1}{z} = 2\cos x$, $z - \frac{1}{z} = 2i\sin x$.

Consider

$$(2i\sin x)^2(2\cos x)^4 = \left(z - \frac{1}{z}\right)^2\left(z + \frac{1}{z}\right)^4$$

$$= \left(z^2 - \frac{1}{z^2}\right)^2\left(z + \frac{1}{z}\right)^2$$

$$= \left(z^4 + \frac{1}{z^4} - 2\right)\left(z^2 + \frac{1}{z^2} + 2\right)$$

$$= \left(z^4 + \frac{1}{z^4}\right)\left(z^2 + \frac{1}{z^2}\right) + 2\left(z^4 + \frac{1}{z^4}\right) - 2\left(z^2 + \frac{1}{z^2}\right) - 4 \quad (*)$$

Now,

$$z^n = (\cos x + i\sin x)^n = \cos nx + i\sin nx$$

$$\frac{1}{z^n} = (\cos x + i\sin x)^{-n} = \cos nx - i\sin nx$$

$$z^n + \frac{1}{z^n} = 2\cos nx \quad \text{and} \quad z^n - \frac{1}{z^n} = 2i\sin nx$$

$$(2i\sin x)^2(2\cos x)^4$$
$$= -2^2\sin^2 x \cdot 2^4 \cdot \cos^4 x = -2^6\sin^2 x \cdot \cos^4 x$$

using (*)

$$= (2\cos 4x)(2\cos 2x) + 2.2\cdot\cos 4x - 2.2\cos 2x - 4$$

$$\sin^2 x \cdot \cos^4 x = -\frac{1}{2^6}\left[4\cdot\frac{1}{2}(\cos 6x + \cos 2x) + 4\cos 4x - 4\cos 2x - 4\right]$$

$$= -\frac{1}{2^4}\left[\frac{1}{2}\cos 6x + \cos 4x - \frac{1}{2}\cos 2x - 1\right]$$

Integrating both sides w.r.t. x,

$$\int_0^{\pi/4} \sin^2 x \cdot \cos^4 x \, dx$$

$$= -\frac{1}{16}\left[\frac{1}{2}\cdot\frac{\sin 6x}{6} + \frac{\sin 4x}{4} - \frac{1}{2}\cdot\frac{\sin 2x}{2} - x\right]_0^{\pi/4}$$

$$= -\frac{1}{16}\left[\left(\frac{1}{12}\sin\frac{3\pi}{2} + \frac{1}{4}\sin\pi - \frac{1}{4}\sin\frac{\pi}{2} - \frac{\pi}{4}\right) - 0\right]$$

$$= -\frac{1}{16}\left[-\frac{1}{12} + 0 - \frac{1}{4} - \frac{\pi}{4}\right] = \frac{1}{8}\left[\frac{\pi}{8} + \frac{1}{6}\right]$$

Example 16: Find $\int_0^{\pi/4} \frac{\sin^2 x}{\cos^3 x}\, dx$.

Solution: Use the reduction formula (11) on page 6.4

$$\int \sin^m x \cdot \cos^n x\, dx = -\frac{\sin^{m+1} x \cdot \cos^{n+1} x}{n+1} \cdot$$

$$+\frac{m+n+2}{n+1}\int \sin^m x \cdot \cos^{n+2} x\, dx$$

Here $m = 2, n = -3$

$$\int_0^{\pi/4} \sin^2 x \cos^{-3} x\, dx$$

$$= -\frac{\sin^3 x \cdot \cos^{-2} x}{-2} + \left(\frac{1}{-2}\right)\int \sin^2 x \cos^{-1} x\, dx$$

$$= \frac{1}{2}\frac{\sin^3 x}{\cos^2 x} - \frac{1}{2}\int \frac{1-\cos^2 x}{\cos x} dx$$

$$= \frac{1}{2}\frac{\sin^3 x}{\cos^2 x} - \frac{1}{2}\int (\sec x - \cos x) dx$$

$$= \frac{1}{2}\frac{\sin^3 x}{\cos^2 x} - \frac{1}{2}\ln\,(\sec x + \tan x) + \frac{1}{2}\sin x$$

$$= \frac{1}{2}\sin x\left(\frac{\sin^2 x}{\cos^2 x} + 1\right) - \frac{1}{2}\ln\,(\sec x + \tan x)$$

$$= \frac{1}{2}\sec x \cdot \tan x - \frac{1}{2}\ln\,(\sec x + \tan x)\Big|_0^{\pi/4}$$

$$= \frac{1}{2}\left[\sqrt{2} - \frac{1}{2}\ln\,(\sqrt{2}+1) - \ln\,\sqrt{2}\right]$$

EXERCISE

1. Evaluate

   (a) $\int \sin^4 x\, dx$; (b) $\int_0^{\pi/2} \sin^7 x\, dx$;
   (c) $\int_0^{\pi/4} \cos^6 2t dt$; (d) $\int_0^{\pi/2} \cos^9 x\, dx$

   Hint: a. Put $n = 4, 2$

   c. Put $2t = x$, limits: 0 to $\frac{\pi}{2}$

   Ans. **a.** $-\frac{\cos x \sin^2 x}{4} - \frac{3}{8}\cos x \cdot \sin x + \frac{3}{8}x$

   b. $\frac{6}{7} \cdot \frac{4}{5} \cdot \frac{2}{3} = \frac{16}{35}$

   c. $\frac{1}{2}\left[\frac{5}{6} \cdot \frac{3}{4} \cdot \frac{1}{2} \cdot \frac{\pi}{2}\right] = \frac{5\pi}{64}$

   d. $\frac{8}{9} \cdot \frac{6}{7} \cdot \frac{4}{5} \cdot \frac{2}{3} \cdot 1 = \frac{128}{315}$, *Note*: $n = 9$ is odd

2. Solve Example 1 without using reduction formulae

   Hint:

   a. Express $\sin^4 x = (\sin^2 x)^2 = \left(\frac{1-\cos 2x}{2}\right)^2 = \frac{1}{8}(3 - 4\cos 2x + \cos 4x)$

   b. Since $n = 7$ is odd, put $\cos x = t$, $-\sin x\, dx = dt$, $\sin^7 x\, dx = \sin^6 x \cdot \sin x\, dx = = (1-\cos^2 x)^6 \sin x\, dx = -(1-t^2)^6 dt$

   c. Express $\cos^6 x = (\cos^2 x)^3 = \left(\frac{1+\cos 2x}{2}\right)^3$

   d. Since $n{=}9{=}$odd, put $\sin x{=}t$, $\cos x\, dx{=}dt$ $\cos^9 x\, dx = (1 - \sin^2 x)^8 \cdot \cos x\, dx = = (1-t^2)dt$

   Ans. Same as in Example 1

3. Use Wallis's formula to evaluate

   (i) $\int_0^{\pi/2} \sin^8 x \cdot \cos^4 x\, dx$; (ii) $\int_0^{\pi/2} \sin^2 x \times \times \cos^6 x\, dx$; (iii) $\int_0^{\pi/4} 8\cos^4 x \cdot \sin^4 x\, dx$

   Hint:

   i. $\sin^8 x{\cdot}\cos^4 x{=}\sin^8 x(1{-}\sin^2 x)^2{=}\sin^8 x{-}2\sin^{10} x + \sin^{12} x$, $I_8 - 2I_{10} + I_{12}$

   ii. $\sin^2 x \cos^6 x{=}(1{-}\cos^2 x)(\cos^6 x)$, $I_6{-}I_8$

   iii. $(\cos x \cdot \sin x)^4 = \left(\frac{\sin 2x}{2}\right)^4$, put $2x = t$, $\frac{1}{4}\int_0^{\pi/2} \sin^4 t dt$

   Ans. **i.** $\frac{7\pi}{2048}$ **ii.** $\frac{5\pi}{256}$ **iii.** $\frac{3\pi}{64}$

4. Evaluate the integrals

   (i) $\int_0^a \frac{x^7 dx}{\sqrt{a^2-x^2}}$; (ii) $\int_0^\infty \frac{dx}{(a^2+x^2)^{\frac{7}{2}}}$;

   (iii) $\int_0^{2a} \frac{x^3 dx}{\sqrt{2ax-x^2}}$; (iv) $\int_0^1 x^{\frac{3}{2}}(1-x)^{\frac{1}{2}} dx$

   Hint:

   i. Put $x = a\sin\theta, \theta; 0$ to $\frac{\pi}{2}$

   ii. Put $x = a\tan\theta, \theta; 0$ to $\frac{\pi}{2}$

   iii. $\frac{x^3}{\sqrt{2ax-x^2}} = \frac{x^3 . x^{-1/2}}{\sqrt{2a-x}}$, put $x = 2a\sin^2\theta$,

   limits: θ: 0 to $\frac{\pi}{2}$, $\int_0^{\pi/2} 16a^3 \sin^6\theta d\theta$

   iv. Put $x = \sin^2\theta$, $dx = 2\sin\theta \cdot \cos\theta d\theta$

Limits: θ: 0 to $\frac{\pi}{2}$, $2[\int_0^{\pi/2}\sin^4\theta d\theta - \int_0^{\pi/2}\sin^6\theta d\theta]$

Ans. (i) $\frac{16}{35}a^7$ (ii) $\frac{8}{15a^6}$ (iii) $\frac{5\pi a^3}{2}$ (iv) $\frac{\pi}{16}$

5. Evaluate

(i) $\int_0^{\pi/4}\tan^4 x\,dx$; (ii) $\int_0^{\pi/4}\tan^5 x\,dx$

(iii) $\int_{\pi/4}^{\pi/2}\text{cosec}^5 x\,dx$; (iv) $\int\sec^4 x\,dx$

(v) $\int_{\pi/6}^{\pi/2}\text{cosec}^5 x\,dx$; (vi) $\int_0^a (a^2+x^2)^{5/2}dx$

Hint: Put $x = a\tan\theta$,

Limits: θ: 0 to $\frac{\pi}{4}$, $a^6\int_0^{\pi/4}\sec^7\theta d\theta$ for (vi)

Ans. **i** $-\frac{2}{3}+\frac{\pi}{4}$

ii. $-\frac{1}{4}+\frac{1}{2}\ln 2$

iii. $\frac{7}{4\sqrt{2}}+\frac{3}{8}\ln(\sqrt{2}+1)$

iv. $\frac{1}{3}\sec^2 x\tan x+\frac{2}{3}\tan x$

v. $\frac{11\sqrt{3}}{4}+\frac{3}{8}\ln(2+\sqrt{3})$

vi. $a^6\left\{\frac{67\sqrt{2}}{48}+\frac{5}{16}\ln(1+\sqrt{2})\right\}$

6. Evaluate

(i) $\int\sec x\cdot\tan^5 x\,dx$; (ii) $\int\sec^{\frac{4}{3}}\text{cosec}^{\frac{8}{3}}\,dx$

Hint:

i. Put $\cos x = t$, $\int\frac{\sin^4 x\cdot\sin x\,dx}{\cos^6 x} = \int\frac{(1-t^2)(-dt)}{t^6}$

ii. $\int\frac{dx}{\cos^{\frac{4}{3}}x\cdot\sin^{\frac{8}{3}}x} = \int\frac{dx}{\cos^{\frac{4}{3}}x\cdot\cos^{\frac{8}{3}}x\cdot\tan^{\frac{8}{3}}x} =$

$\int\frac{\sec^4 x\,dx}{\tan^{\frac{8}{3}}x}$, put $\tan x = t$, $\int\frac{(1+t^2)}{t^{\frac{8}{3}}}dt$

Ans. **i.** $\frac{1}{5}\sec^5 x-\frac{2}{3}\sec^3 x+\sec x$

ii. $-\frac{3}{5}\cot^{\frac{5}{3}}x+3\tan^{\frac{1}{3}}x$

7. Evaluate

i. $\int_0^{\pi/2}\sin^9 x\cdot\cos^3 x\,dx$

ii. $\int_0^{\pi/2}\cdot\sin^8 x\cdot\cos^4 x\,dx$

iii. $\int_0^{\pi/2}\sin^5 x\cdot\cos^6 x\,dx$

iv. $\int_0^{\pi/2}\sin^6 x\cdot\cos^8 x\,dx$

v. $\int_0^{\pi/6}\sin^3 6\theta\cdot\cos^4 3\theta d\theta$

Hint: $\cos^4 3\theta(2\sin 3\theta\cos 3\theta)^3$, put $3\theta=t$ for (v)

Ans. **i** $\frac{8\cdot6\cdot4\cdot2\cdot2\cdot}{12\cdot10\cdot8\cdot6\cdot4\cdot2} = \frac{1}{60}$

ii. $\frac{7\cdot5\cdot3\cdot3\cdot}{12\cdot10\cdot8\cdot6\cdot4\cdot2\cdot}\cdot\frac{\pi}{2} = \frac{7\pi}{2048}$

iii. $\frac{4\cdot2\cdot5\cdot3\cdot}{11\cdot9\cdot7\cdot5\cdot3\cdot} = \frac{8}{693}$

iv. $\frac{5\cdot3\cdot7\cdot5\cdot3}{14\cdot12\cdot10\cdot8\cdot6\cdot4\cdot2}\cdot\frac{\pi}{2} = \frac{5\pi}{4096}$

v. $\frac{1}{15}$

8. Evaluate the integrals in Example 7 without using reduction formulae.

Hint:

i. Put $\cos x = t\sin\theta m = 9$ is odd or put $\sin x = t$ since $n = 3$ is odd

ii. Since $m = 8, n = 4$ are both even integers,

$$\sin^8 x\cdot\cos^4 x = \left(\frac{1-\cos 2x}{2}\right)^4\cdot\left(\frac{1+\cos 2x}{2}\right)^2$$

iii. Since $m = 5$ is odd put $\cos x = t$

iv. Since $m = 6, n = 8$ are both even integers

$$\sin^6 x\cdot\cos^8 x = \left(\frac{1-\cos 2x}{2}\right)^3\left(\frac{1+\cos 2x}{2}\right)^4$$

Ans. Same as in Example 7

Note: (ii) and (iv) can also be solved using demoveis theorem: see WOE 15 (ii) on Page 6.10.

9. Evaluate

(i) $\int\sin^3 x\cdot\sec^7 x\,dx$; (ii) $\int\frac{\cos^3 x}{\text{cosec}^{3/4}x}dx$

Hint:

i. Put $\tan x = t$ since $m+n = 3-7 = -4$

ii. put $\sin x = t$ since $n = 3$ is odd

Ans. **i.** $\frac{1}{4}\tan^4 x+\frac{1}{6}\tan^6 x$

ii. $\frac{4}{7}\sin^{7/4}x-\frac{4}{15}\sin^{15/4}x$

10. Evaluate the following integral

(i) $\int_0^{2a}x\sqrt{2ax-x^2}dx$; (ii) $\int_0^\infty\frac{x^2}{(1+x^2)^4}dx$;

(iii) $\int_0^1 x^4(1-x^2)^{3/2}dx$

Hint:

i. Put $x = 2a\sin^2\theta$

ii. Put $x = \tan\theta$

iii. Put $x = \sin t$

Ans. **i.** $\frac{\pi a^3}{2}$

ii. $\frac{\pi}{32}$

iii. $\frac{3\pi}{256}$

11. Evaluate

(i) $\int \cos^3 x \cdot \operatorname{cosec}^4 x\, dx$
(ii) $\int \sin^2 x \cdot \sec^6 x\, dx$

Hint:

i. Put $\sin x = t$

ii. Put $\tan x = t$

Ans. **i.** $-\frac{1}{3}\operatorname{cosec}^3 x + \operatorname{cosec} x$

ii. $\frac{\tan^3 x}{3} + \frac{\tan^5 x}{5}$.

6.2 AREA OF A PLANE REGION: QUADRATURE

Quadrature is the process of determining the plane area bounded by a given set of plane curves.

Area of a Curvilinear Trapezoid

Suppose $y = f(x)$ be a function defined on the interval $[a, b]$ and assume that $f(x) \geq 0$.

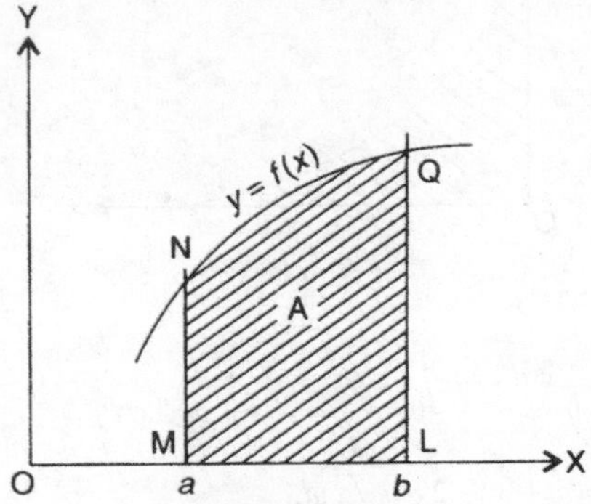

Fig. 6.2

Then the area A of the curvilinear trapezoid $MLQN$ bounded by the given curve (graph of) $y = f(x)$, the x-axis and the two ordinates (straight lines) $x = a$ and $x = b$ is numerically equal to the definite integral:

$$A = \int_a^b f(x)dx \tag{1}$$

Similarly, if the curve is defined by the function $x = f(y)$ on the interval $[c, d]$, then

$$\text{Area} = \int_c^d f(y)dy$$

If $f(x) \leq 0$ on $[a, b]$, then $\int_a^b f(x)dx \leq 0$. Then

$$A = \left|\int_a^b f(x)dx\right|$$

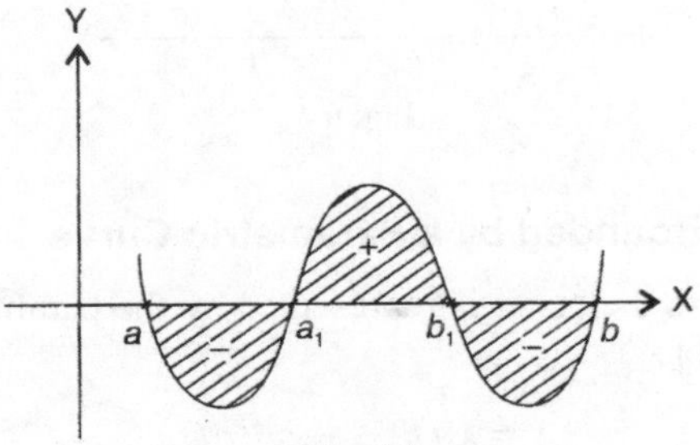

Fig. 6.3

Thus if $f(x)$ changes sign on the interval $[a, b]$ a finite number of times i.e., if the graph (curve) of $y = f(x)$ crosses the x-axis several times, then the total area bounded by the curve is the sum of the areas above and below the x-axis, with absolute value taken for the areas when $f(x) \leq 0$ i.e., below the x-axis or equivalently

$$A = \int_a^b |f(x)|dx$$

Example: For the curve shown in the figure

$$A = \left|\int_a^{a_1} f(x)dx\right| + \int_{a_1}^{b_1} f(x)dx + \left|\int_{b_1}^{b} f(x)dx\right|$$

$$= \int_a^b |f(x)|dx$$

Area between Two Curves

The area bounded by the curves $y = f_2(x)$, $y = f_1(x)$, and the ordinates $x = a$, $x = b$ is given by

$$A = \int_a^b f_2(x)dx - \int_a^b f_1(x)dx$$

$$= \int_a^b \left[f_2(x) - f_1(x)\right]dx \quad (2)$$

Provided $f_2(x) \geq f_1(x)$.

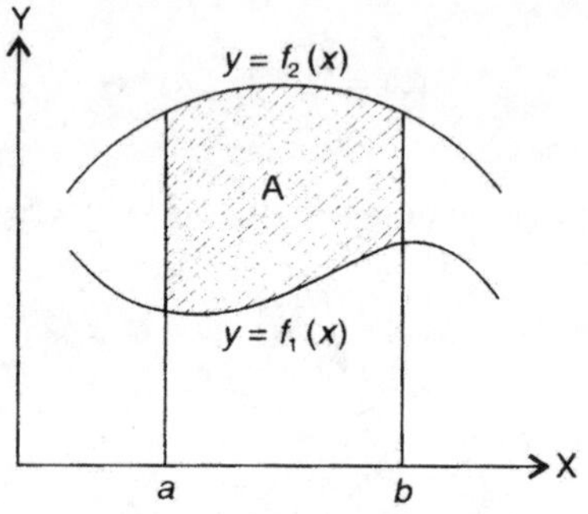

Fig. 6.4

Area Bounded by a Parametric Curve

Suppose the curve is represented by the parametric equations

$$x = x(t), \qquad y = y(t)$$

with $\alpha \leq t \leq \beta$ and $x(\alpha) = a$, $x(\beta) = b$. Then

$$A = \int_a^b f(x)dx = \int_a^b y\,dx = \int_\alpha^\beta y(t)\frac{dx}{dt}dt \quad (3)$$

Area Bounded by a Polar Curve

Suppose the equation of the given curve in polar coordinates (r, θ) be

$$r = f(\theta)$$

where $\alpha \leq \theta = \beta$. Let r_i be the length of the radius vector corresponding to some angle θ_i between θ_{i-1} and θ_i. Divide the given area OAB into n parts by radius vectors $\theta_0 = \alpha, \theta = \theta_1, \ldots, \theta_n = \beta$. Then the area of the sector OAB bounded by the polar curve $r = f(\theta)$ and the radius vectors $\theta = \alpha$ and $\theta = \beta$ is

$$A = \lim_{n\to\infty} \sum_{i=1}^{n} \frac{1}{2}r_i^2\Delta\theta_i = \frac{1}{2}\int_\alpha^\beta r^2 d\theta = \frac{1}{2}\int_\alpha^\beta [f(\theta)]^2\,d\theta$$

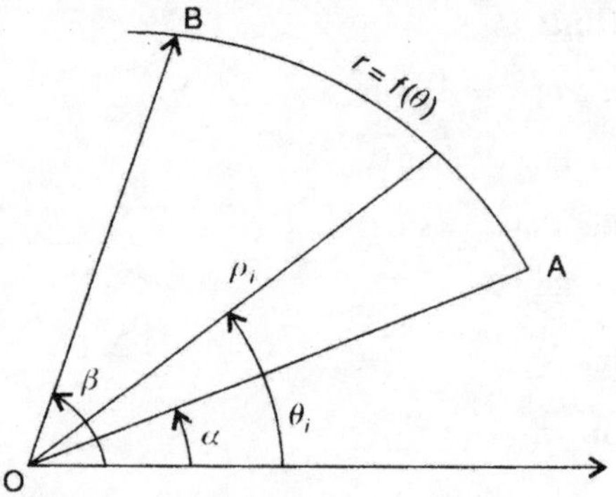

Fig. 6.5

since the area of any typical circular sector with radius r_i and central angle $\Delta\theta_i$ is given by

$$\Delta A_i = \frac{1}{2}r_i^2\Delta\theta_i$$

Thus area of sector OAB is

$$A = \frac{1}{2}\int_\alpha^\beta [f(\theta)]^2 d\theta \quad (4)$$

WORKED OUT EXAMPLES

Example 1: Find the area bounded by the parabolic arc $\sqrt{x} + \sqrt{y} = \sqrt{a}$ and the coordinate axes.

Solution: Curve intersects x-axis at $(a, 0)$ and y-axis at $(0, a)$ (Fig. 6.6). So area is

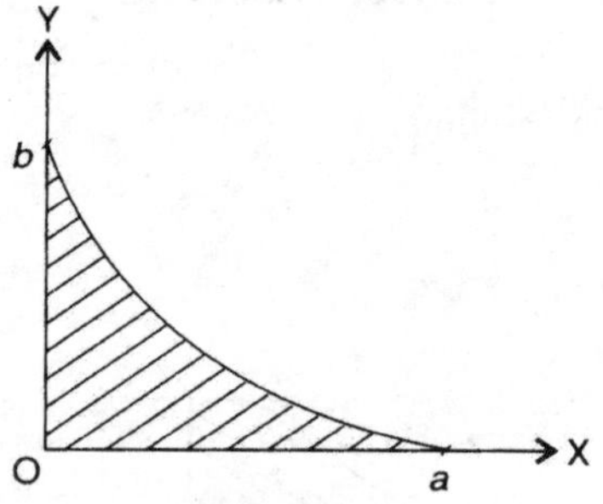

Fig. 6.6

$$A = \int_{x=0}^a y\,dx = \int_0^a (\sqrt{a} - \sqrt{x})^2 dx$$

$$= \int_0^a (a + x - 2\sqrt{a}\sqrt{x})dx$$

$$= ax + \frac{x^2}{2} - 2\sqrt{a}\frac{2}{3}x^{\frac{3}{2}} \cdot \Big|_0^a = a^2 + \frac{a^2}{2} - \frac{4}{3}a^2 = \frac{1}{6}a^2$$

Example 2: Find the total area between the cubic $y = 2x^3 - 3x^2 - 12x$, x-axis and its maximum and minimum ordinates.

Solution: $y' = 6x^2 - 6x - 12$, $y'' = 12x - 6$
Stationary points: $y' = 6x^2 - 6x - 12 = 0$
$\therefore x = -1, 2$

$$y''\Big|_{x=-1} = -18 < 0 \quad \text{and} \quad y''\Big|_{x=2} = 18 > 0$$

Thus y attains maximum at $x = -1$ and minimum at $x = 2$ (Fig. 6.7).

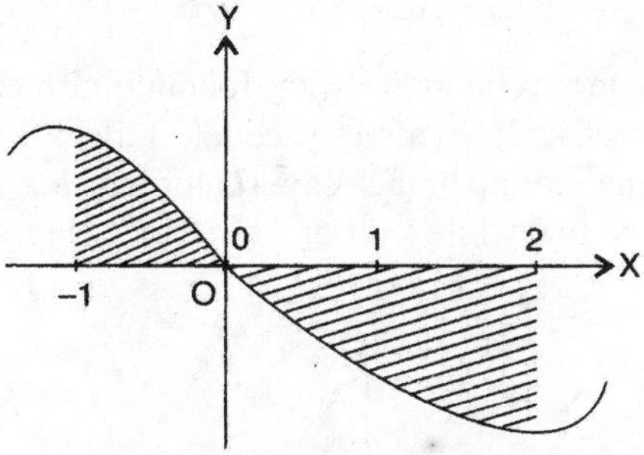

Fig. 6.7

The cubic crosses the x-axis when

$0 = y = 2x^3 - 3x^2 - 12x$ or when $x = 0, \frac{3 \pm \sqrt{104}}{4}$.

Here only $x = 0$ lies between -1 and 2.

Further $\quad y\Big|_{x=-\frac{1}{2}} = 5 > 0 \quad$ and

$$y\Big|_{x=1} = -13 < 0$$

Thus $f(x) \geq 0$ in $[-1, 0]$ and $f(x) \leq 0$ in $[0, 2]$

$$A = \int_{-1}^{2} f(x)dx = \int_{-1}^{0} f(x)dx + \left|\int_0^2 f(x)dx\right|$$

$$= -\left[\frac{1}{2} + 1 - 6\right] + |(8 - 8 - 24)| = \frac{9}{2} + 24 = \frac{57}{2}$$

Example 3: Determine the area between the cubic $y = x^3$ and the parabola $y = 4x^2$.

Solution: The points of intersection of the two curves are

$$4x^2 = y = x^3$$

$$x^2(x - 4) = 0$$

i.e., $x = 0$ and $x = 4$.

Since at $\quad x = 1,\ y\Big|_{x=1} = 4x^2\Big|_{x=1} = 4,$

while $\quad y\Big|_{x=1} = x^3\Big|_{x=1} = 1$

so $y = 4x^2$ is the upper curve and $y = x^3$ is the lower curve (Fig. 6.8). The area bounded between the two curves is

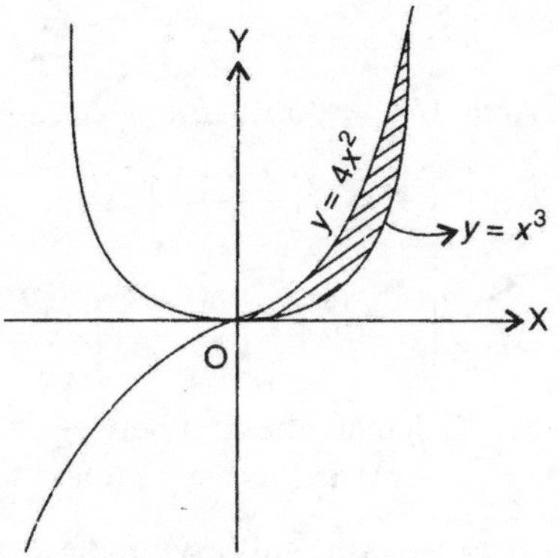

Fig. 6.8

$$A = \int_{x=0}^{4} (4x^2 - x^3)dx = \frac{4x^3}{3} - \frac{x^4}{4}\Big|_{x=0}^{4} = \frac{64}{3}$$

Example 4: Show that the area bounded by the three straight lines $x + 2y = 2$, $y - x = 1$ and $2x + y = 7$ is 6.

Solution: The points of intersection of the two lines

$$x + 2y = 2 \quad \text{(I)} \quad \text{and} \quad -x + y = 1 \quad \text{(II)}$$

is $B(0, 1)$ and of (I) and $2x + y = 7$ (III) is $C(4, -1)$

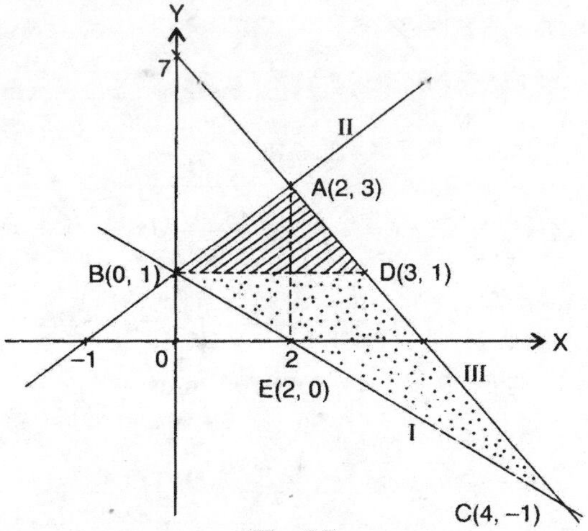

Fig. 6.9

and of II and III is $A(2,3)$ (refer Fig. 6.9).

Area ABC = Area ABD + Area BDC

$$= \int_{y=-1}^{1}\left[\left(\frac{7-y}{2}\right)-(2-2y)\right]dy + \int_{y=1}^{3}\left[\left(\frac{7-y}{2}\right)-(y-1)\right]dy$$

$$= 3 + 3 = 6$$

Aliter: Area ABC = Area ABE + Area AEC

$$= \int_{x=0}^{2}\left[(x+1)-\left(\frac{2-x}{2}\right)\right]dx + \int_{x=2}^{4}\left[(7-2x)-\left(\frac{2-x}{2}\right)\right]dx = 6$$

Example 5: Calculate the area between the curve $y^2(a+x)=(a-x)^3$ and its asymptotes.

Solution: $x=-a$ is the only asymptote to the given curve. Since the curve is symmetric about the x-axis, area bounded by the given curve and the asymptote MN is twice the area BAN (Fig. 6.10).

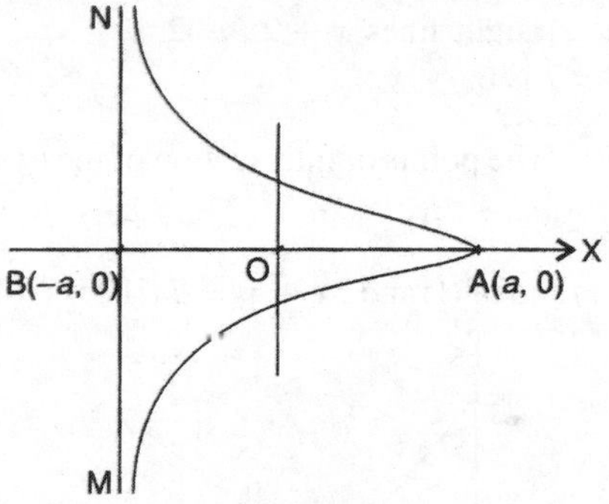

Fig. 6.10

$$A = 2\int_{x=-a}^{a} y\,dx = 2\int_{-a}^{a}\sqrt{\frac{(a-x)^3}{(a+x)}}dx$$

$$= 2\int_{-a}^{a}(a-x)\sqrt{\frac{a-x}{a+x}}dx = 2\int_{-a}^{a}\frac{(a-x)^2}{a^2-x^2}dx.$$

Put $x=a\sin\theta$, $dx=a\cos\theta d\theta$, limits: $-\frac{\pi}{2}$ to $\frac{\pi}{2}$.

$$A = 2\int_{-\frac{\pi}{2}}^{\frac{\pi}{2}}\frac{a^2(1-\sin\theta)^2}{a\cdot\cos\theta}\cdot a\cos\theta d\theta$$

$$= 2a^2\int_{-\frac{\pi}{2}}^{\frac{\pi}{2}}(1+\sin^2\theta)d\theta - 2a^2\int_{\frac{\pi}{2}}^{\frac{\pi}{2}}2\sin\theta d\theta$$

$$= 4a^2\int_0^{\frac{\pi}{2}}(1+\sin^2\theta)d\theta = 4a^2\left[\frac{\pi}{2}+\frac{1}{2}\cdot\frac{\pi}{2}\right] = 3\pi a^2$$

Example 6: Find the whole area bounded by the four infinite branches of the tractrix.

$$x = a\cos t + \frac{1}{2}a\ln\tan^2\frac{t}{2}, \qquad y = a\sin t.$$

Solution: Variation of x, y w.r.t. t,

t:	0	$\frac{\pi}{2}$	π	$-\frac{\pi}{2}$
x:	$-\infty$	0	∞	0
y:	0	a	0	$-a$

The whole area bounded by the four infinite branches of the tractrix, by symmetry, equals 4 times the area under one infinite branch say AD for which x varies from $-\infty$ to 0 while t varies from 0 to $\frac{\pi}{2}$ (Fig. 6.11).

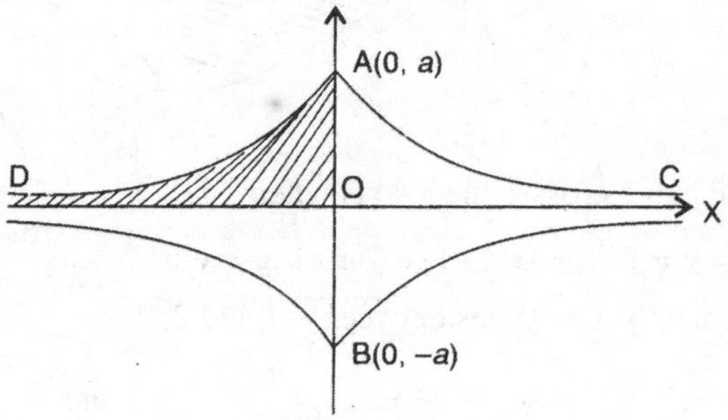

Fig. 6.11

$$\text{Required area } A = 4\int_0^{\frac{\pi}{2}} y(t)\frac{dx}{dt}dt \qquad (1)$$

Now
$$\frac{dx}{dt} = -a\sin t + \frac{1}{2}a\cdot\frac{2}{\tan\frac{t}{2}}\cdot\frac{1}{2}\cdot\sec^2\frac{t}{2}$$

$$= -a\sin t + \frac{a}{2\sin\frac{t}{2}\cdot\cos\frac{t}{2}}$$

$$= -a\sin t + \frac{a}{\sin t} = \frac{a}{\sin t}(1-\sin^2 t)$$

$$= \frac{a}{\sin t}\cdot\cos^2 t \qquad (2)$$

Substituting (2) in (1), we have

$$A = 4\int_0^{\frac{\pi}{2}}(a\sin t)\left(\frac{a}{\sin t}\cdot\cos^2 t\right)dt$$

$$= 4a^2 \int_0^{\frac{\pi}{2}} \cos^2 t dt = 4a^2 \cdot \frac{1}{2} \cdot \frac{\pi}{2} = \pi a^2.$$

Example 7: Find the common area included between the parabolas $y^2 = 4a(x+a)$, and $y^2 = 4b(b-x)$.

Solution: The points of intersection of the two parabolas are $4a(x+a) = y^2 = 4b(b-x)$ or $x = \frac{b^2-a^2}{b+a}$ i.e., $x = (b-a)$ or in B and D (Fig. 6.12).

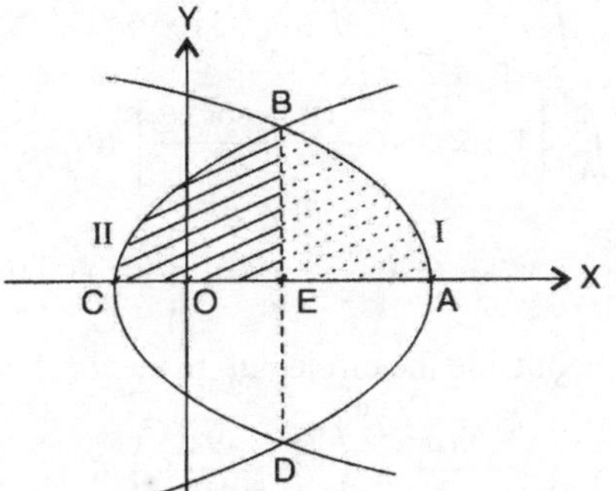

Fig. 6.12

The parabola I: $y^2 = 4b(b-x)$ meets the x-axis in $A(b,0)$, meets the parabola II: $y^2 = 4a(x+a)$ in $B(b-a, 2\sqrt{ab})$ and in $D(b-a, -2\sqrt{ab})$. Here $E(b-a, 0)$. Also parabola II meets x-axis in $C(-a, 0)$. Since both the parabolas I and II are symmetric about the x-axis, the required common area A between the two parabolas I and II is twice the area $ABCEA$.

$A = 2$ Area of $ABCEA = 2$[Area of ABE + Area of BCE]

$$= 2\left[\int_E^A y_{\text{I}} dx + \int_C^E y_{\text{II}} dx\right]$$

where suffix I, II indicate the respective parabolas.

$$A = 2\int_{b-a}^{b} 2\sqrt{b}\sqrt{(b-x)}dx + 2\int_{-a}^{b-a} 2\sqrt{a}\sqrt{(x+a)}dx$$

$$= 4\sqrt{b}\cdot(b-x)^{\frac{3}{2}}\cdot\frac{2}{3}(-1)\cdot\Big|_{x=b-a}^{b} +$$

$$+4\sqrt{a}\cdot(x+a)^{\frac{3}{2}}\cdot\frac{2}{3}\cdot\Big|_{-a}^{b-a}$$

$$= \frac{8}{3}\sqrt{b}\left[0+(b-(b-a))^{3/2}\right] + \frac{8}{3}\sqrt{a}\left[(b-a+a)^{\frac{3}{2}}+0\right]$$

$$= \frac{8}{3}\sqrt{b}a^{3/2} + \frac{8}{3}\sqrt{a}b^{3/2} = \frac{8}{3}\sqrt{ab}\cdot(a+b).$$

Example 8: Find the whole area of the curve
(i) $r = a\cos n\theta$ (ii) $r = a\sin n\theta$ (iii) $r = a\sin 3\theta$
(iv) $r = a\cos 4\theta$ (v) $r = a\cos 3\theta + b\sin 3\theta$

Solution: The curves of the type $r = a\sin n\theta$ or $r = a\cos n\theta$ consists of either n or $2n$ equal loops respectively based on whether n is odd or even (Fig. 6.13).

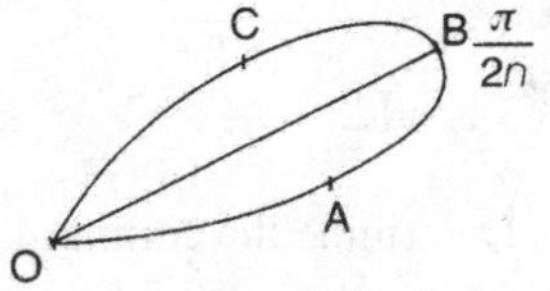

Fig. 6.13

i. Consider the curve $r = a\cos n\theta$

Area of one loop = twice the area of OBC

$= 2\int_0^{\frac{\pi}{2n}} \frac{1}{2}r^2 d\theta = a^2\int_0^{\frac{\pi}{2n}} \cos^2 n\theta d\theta$

Put $n\theta = t$, so $d\theta = \frac{1}{n}dt$ and limits are $\theta = 0$, then $t = 0$, $\theta = \frac{\pi}{2n}$ then $t = \frac{\pi}{2}$. So area of one loop $= \frac{a^2}{n}\int_0^{\frac{\pi}{2}} \cos^2 t \cdot dt = \frac{a^2}{n}\cdot\frac{1}{2}\cdot\frac{\pi}{2} = \frac{\pi a^2}{4n}$.

Then the whole area of the curve = (number of loops) × (area of one loop)

$$= n\cdot\frac{\pi a^2}{4n} = \frac{\pi a^2}{4} \quad \text{when } n \text{ is odd}$$

$$= 2n\cdot\frac{\pi a^2}{4n} = \frac{\pi a^2}{2} \quad \text{when } n \text{ is even}$$

ii. When $r = a\sin n\theta$,

Area of one loop $= \frac{a^2}{n}\int_0^{\frac{\pi}{2}} \sin^2 t\, dt = \frac{a^2}{n}\cdot\frac{1}{2}\frac{\pi}{2} = \frac{\pi a^2}{4n}$.

So whole area $= \frac{\pi a^2}{4}$ or $\frac{\pi a^2}{2}$ according as n is odd or even.

iii. When $r = a\sin 3\theta$, take $n = 3$ (odd). Then the entire area $= \frac{\pi a^2}{4}$.

iv. When $r = a\cos 4\theta$, take $n = 4$ (even). Then the entire area $= \frac{\pi a^2}{2}$.

v. When $r = a\cos 3\theta + b\sin 3\theta$, introduce

$$\sin 3\alpha = \frac{a}{\sqrt{a^2+b^2}}, \cos 3\alpha = \frac{b}{\sqrt{a^2+b^2}}. \quad \text{Then}$$

$$r = \sqrt{a^2+b^2}\,[\sin 3\alpha \cdot \cos 3\theta + \cos 3\alpha \cdot \sin 3\theta]$$

$$r = \sqrt{a^2+b^2} \cdot \sin\{3(\theta+\alpha)\}$$

This curve is obtained by the rotation of the curve $r = \sqrt{a^2+b^2}\sin 3\theta$ through an angle α. Then entire area of the loops of the two curves is same. Thus entire area of the curve $r = a\cos 3\theta + b\sin 3\theta$ is the same as the entire area of the curve $r = \sqrt{a^2+b^2}\sin 3\theta$.

$$\text{Required area} = \frac{\pi\left(\sqrt{a^2+b^2}\right)^2}{4} = \frac{\pi(a^2+b^2)}{4}$$

since $n = 3$ is odd.

Example 9: Determine the common area between the circle $r = \frac{3}{2}a$ and the cardioid $r = a(1+\cos\theta)$. Deduce the area which is external to the circle but inside the cardioid.

Solution: The points of intersection of the two curves are $\frac{3}{2}a = a(1+\cos\theta)$ i.e., $\theta = \pm\frac{\pi}{3}$ or $B\left(\frac{3}{2}a, \frac{\pi}{3}\right)$, $C\left(\frac{3}{2}a, -\frac{\pi}{3}\right)$. Here $D\left(\frac{3}{2}a, 0\right)$ and $A(2a, 0)$, $E\left(a, \frac{\pi}{2}\right)$. Common area included between the circle and the cardioid is twice the area of the shaded region $ODBEO$ because both the curves circle and cardioid are symmetric about the initial line (x-axis) (Fig. 6.14). Again area of $ODBEO$ = Area of ODB + Area of $OBEO$ common area

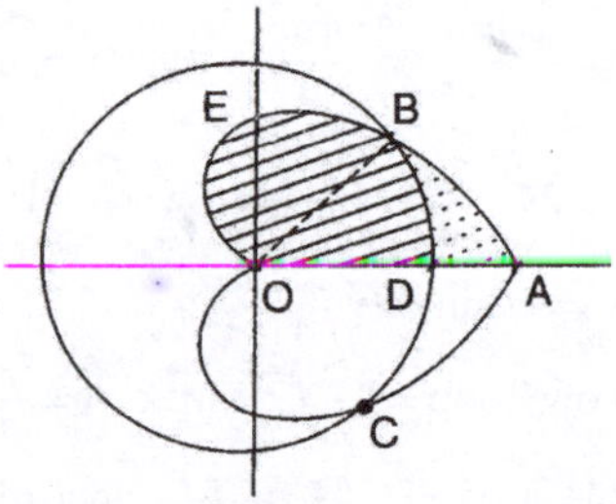

Fig. 6.14

$$= 2\left[\int_0^{\frac{\pi}{3}} \frac{1}{2}r^2 d\theta + \int_{\pi/3}^{\pi} \frac{1}{2}r^2 d\theta\right]$$

$$= \int_0^{\frac{\pi}{3}} \frac{9}{4}a^2 d\theta + \int_{\frac{\pi}{3}}^{\pi} a^2(1+\cos\theta)^2 d\theta$$

$$= \frac{9}{4}a^2\theta\Big|_0^{\frac{\pi}{3}} + a^2\int_{\frac{\pi}{3}}^{\pi}\left[1 + 2\cos\theta + \frac{1+\cos 2\theta}{2}\right]d\theta$$

$$= \frac{3\pi a^2}{4} + a^2\left[\frac{3}{2}\theta + 2\sin\theta + \frac{\sin 2\theta}{4}\right]_{\frac{\pi}{3}}^{\pi}$$

$$= \frac{3\pi a^2}{4} + a^2\left[\frac{3}{2}\cdot\frac{2\pi}{3} + 2\left(0 - \frac{\sqrt{3}}{2}\right) + \frac{1}{4}\left(0 - \frac{\sqrt{3}}{2}\right)\right]$$

$$A = \frac{7\pi a^2}{4} - \frac{9}{8}\sqrt{3}a^2 = \left(\frac{7\pi}{4} - \frac{9\sqrt{3}}{8}\right)a^2$$

Now area of the cardioid = twice of area of $OABEO$

$$= 2\cdot\frac{1}{2}\int_0^{\pi} r^2 d\theta = \frac{1}{2}\int_0^{\pi} a^2(1+\cos\theta)^2 d\theta$$

$$= a^2\int_0^{\pi}\left[1 + 2\cos\theta + \frac{1+\cos 2\theta}{2}\right]d\theta$$

$$= a^2\left[\frac{3}{2}\theta + 2\sin\theta + \frac{\sin 2\theta}{4}\right]_0^{\pi} = a^2\left[\frac{3}{2}\pi + 0 + 0\right] = \frac{3\pi a^2}{2}$$

Now area outside the circle but inside the cardioid is

$$= \frac{3\pi a^2}{2} - \left(\frac{7\pi}{4} - \frac{9\sqrt{3}}{8}\right)a^2$$

Exercise

Find the area bounded by the following curve (1 to 7).

1. $y = 2x + x^2 - x^3$, x-axis, $x = -1$, $x = 1$

   Hint: Curve crosses x-axis at $-1, 0, 2$. So

   $\text{Area} = \left|\int_{-1}^{0}\right| + \int_0^1 = \left|-\frac{5}{12}\right| + \frac{13}{12} = \frac{3}{2}$

   Ans. $\frac{3}{2}$

2. $y = 2x^4 - x^2$, x-axis, its two minimum ordinates.

   Hint: $x = \pm\frac{1}{2}$ minimum ordinates

   Ans. $\frac{7}{120}$

3. Parabola $y = x^2$ and the line $y = x$.

   Hint: $= \int_0^1 (x - x^2)dx$

   Ans. $\frac{1}{6}$

4. $x^2 = 2ay$, $y = 2a$.

Hint: $\int_0^{2a}\left[\sqrt{2ay} - (-\sqrt{2ay})\right]\cdot dy$

Ans. $\frac{16}{3}a^2$

5. $y = \sin x$, x-axis, $0 \le x \le 2\pi$.

Hint: $\int_0^{\pi} \sin x\, dx + \left|\int_{\pi}^{2\pi} \sin x\, dx\right|$

Ans. 4

6. $y^2 = x$, $y = x^2$.

Hint: $\int_0^1(\sqrt{x} - x^2)dx$

Ans. $\frac{1}{3}$

7. Ellipse $\frac{x^2}{a^2} + \frac{y^2}{b^2} = 1$ or $x = a\cos t$, $y = b\sin t$.

Hint: $\int_0^a \frac{b}{a}\sqrt{a^2 - x^2}dx$ or $2\int_{\pi}^{0}(b\sin t)(-a\sin t)dt$

Ans. πab

8. Find the area of the loop of the curve

$a^4y^2 = x^5(2a - x)$

Hint: Loop between $x = 0$, $x = 2a$, curve is symmetric about x-axis,

$A = 2\int_0^{2a} \frac{1}{a^2}x^{\frac{5}{2}}\sqrt{2a - x}dx$

Ans. $\frac{5\pi a^2}{4}$

9. Show that the area of the loop of the curve $ay^2 = x^2(a - x)$ is $\frac{8a^2}{15}$.

Hint: $A = 2\int_0^a x\sqrt{\frac{a-x}{a}}\, dx$

10. Find the area included between the curve

$y^2(2a - x) = x^3$ and its asymptote.

Hint: $x{=}2a$ is asymptote. $A{=}2\int_0^{2a}\sqrt{\frac{x^3}{2a-x}}dx$

Ans. $3\pi a^2$

11. Determine the area enclosed by the curve

$a^2x^2 = y^3(2a - y)$

Hint: Loop between $y = 0$ to $y = 2a$. Curve is symmetric about y-axis

$A = 2\int_0^{2a} \frac{1}{a}y\sqrt{y(2a - y)}\,dy$

Ans. πa^2

12. Obtain the area enclosed between one arch of the cycloid $x = a(\theta \mp \sin\theta)$, $y = a(1 - \cos\theta)$ and its base.

Hint: $\int_{x=0}^{2\pi a} y\,dx{=}\int_0^{\frac{\pi}{2}} y(t)\frac{dx}{d\theta}\cdot d\theta{=}\int_0^{\pi/2} a(1-\cos\theta)\cdot a(1-\cos\theta)d\theta$ or $2a^2\int_0^{\pi}(\theta + \sin\theta)(\sin\theta)d\theta$

Ans. $3\pi a^2$

13. Find the area of the loop and area between the curve $x(x^2 + y^2) = a(x^2 - y^2)$ and its asymptote.

Ans. $\frac{4-\pi}{2}a^2$, $\frac{4+\pi}{2}a^2$; $x = -a$ is the asymptote. Loop is between $x = 0$ to $x = a$, symmetry about x-axis.

14. Find the area of the hypocycloid. Deduce the area of the astroid.

Hint: Eq. $\frac{x^{\frac{2}{3}}}{a^{\frac{2}{3}}} + \frac{y^{\frac{2}{3}}}{b^{\frac{2}{3}}} = 1$, parametric

$x = a\cos^3 t$, $y = b\sin^3 t$

$$A = 4\int_{\pi/2}^{0}(b\sin^3 t)(-3a\cos^2 t\cdot\sin t)\,dt$$

Ans. $\frac{3}{8}\pi ab$, $\frac{3}{8}\pi a^2$ (Put $a = b$)

15. Find the area common to the parabola $y^2 = ax$ and the circle $x^2 + y^2 = 4ax$.

Hint: $A = 2\int_0^{3a}\sqrt{ax}dx + 2\int_{3a}^{4a}\sqrt{4ax - x^2}\,dx$

Ans. $\left(3\sqrt{3} + \frac{4}{3}\pi\right)a^2$

16. Compute the area bounded by the lemniscate

$$r^2 = a^2\cos 2\theta.$$

Hint: $A = 4\cdot\int_0^{\frac{\pi}{4}}\frac{1}{2}\cdot a^2\cdot\cos 2\theta\, d\theta$

Ans. a^2

17. Find the area of the limacon $r = a + b\cos\theta$ (with $a > b$). Deduce the area of the cardioids $r = a(1 + \cos\theta)$, and $r = a(1 - \cos\theta)$. When $a < b$ find the sum of the areas of the loops of the curve $r = a + b\cos\theta$.

Ans. $\pi\left(a^2 + \frac{b^2}{2}\right)$, $\frac{3\pi a^2}{2}$ (with $b = a$), $\frac{3\pi a^2}{2}$ (with $b = -a$), $\pi\left(a^2 + \frac{b^2}{2}\right)$

18. Find the area common to the circles $r = a\sqrt{2}$ and $r = 2a\cos\theta$.

Hint:

$$A = 2\left[\frac{1}{2}\int_0^{\frac{\pi}{4}}(a\sqrt{2})^2 d\theta + \int_{\frac{\pi}{4}}^{\frac{\pi}{2}}(2a\cos\theta)^2 d\theta\right]$$

Ans. $a^2(\pi - 1)$

19. Find the common area between the cardioids

$r = a(1 + \cos\theta)$ and $r = a(1 - \cos\theta)$

Hint: $A = 4\int_0^{\frac{\pi}{2}} \frac{1}{2}a^2(1 - \cos\theta)^2\, d\theta$

Ans. $\frac{a^2}{2}\cdot(3\pi - 8)$

20. Let PQ be the common tangent to the two loops of the lemniscate $r^2 = a^2\cos 2\theta$ with pole at 0. Then find the area bounded by the line PQ and the arcs OP and OQ of the curve (Fig. 6.15).

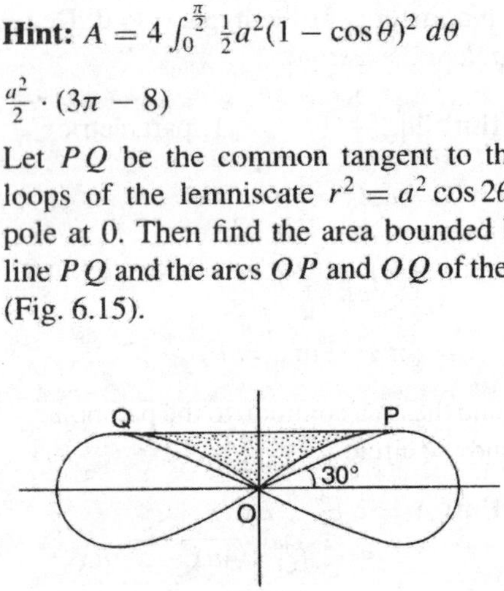

Fig. 6.15

Hint: $\phi = \pi = \theta + \phi$, $\cot\phi = -\tan 2\theta$, $\pi = \phi = \theta + \frac{\pi}{2} + 2\theta$ $\therefore$ $\theta = 30°$, $OP = \frac{a}{\sqrt{2}} = OQ$. $A = \Delta OPQ - 2\int_{\frac{\pi}{6}}^{\frac{\pi}{4}} \frac{1}{2}a^2\cos 2\theta\, d\theta$.

Ans. $\frac{a^2}{8}(3\sqrt{3} - 4)$

6.3 LENGTH OF PLANE CURVE: RECTIFICATION

Rectification is the process of determining the length of arc of a plane curve whose equation may be given in cartesian, parametric cartesian or polar form.

Cartesian Form

Let S be the length of the arc of the plane curve c included between two points A and B whose abscissa are a and b. Let $y = f(x)$ be the equation of the curve c in the cartesian form, with $a \le x \le b$. Divide the curve into n segments by n points $P_k(x_k, y_k)$. Then the length S = limit of the sum of the lengths of the n line segments $= AP_1 + P_1P_2 + \cdots + P_{n-1}B$. By mean value theorem the length of the kth line segment is (refer Fig. 6.16)

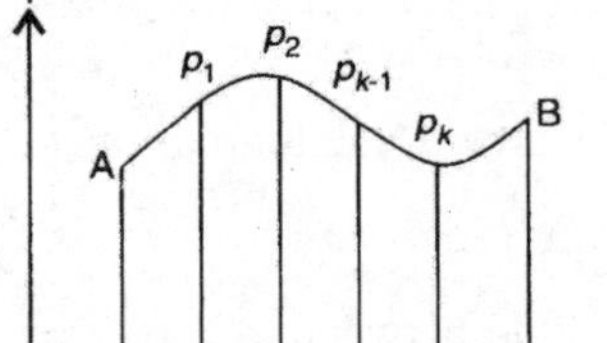

Fig. 6.16

$$P_{k-1}P_k = \sqrt{(x_k - x_{k-1})^2 + (y_k - y_{k-1})^2}$$

$$= \sqrt{(x_k - x_{k-1})^2 + \{f'(x_k^*)(x_k - x_{k-1})\}^2}$$

$$= \sqrt{1 + (f'(x_k^*))^2}\cdot\Delta x_k \quad \text{where } \Delta x_k = x_k - x_{k-1}$$

Therefore,

$$S = \lim_{n\to\infty}\sum_{k=1}^{n}\left\{1 + [f'(x_k^*)]^2\right\}^{\frac{1}{2}}\Delta x_k$$

$$S = \int_a^b \sqrt{1 + [f'(x)]^2}dx$$

$$= \int_a^b \sqrt{1 + \left(\frac{dy}{dx}\right)^2}\, dx.$$

Corollary 1: If the equation of the curve is $x = f(y)$ then the length of the arc of the plane curve included between two points whose ordinates are c, d is

$$S = \int_c^d \sqrt{1 + \left(\frac{dx}{dy}\right)^2}\, dy$$

Parametric Form

When the equation of the curve is in the parametric form $x = x(t)$, $y = y(t)$ with the parameter t varying from t_1 to t_2, then

$$S = \int_{t_1}^{t_2} \sqrt{\left(\frac{dx}{dt}\right)^2 + \left(\frac{dy}{dt}\right)^2}\, dt$$

$$= \int_{t_1}^{t_2} \sqrt{\dot{x}^2 + \dot{y}^2}\, dt$$

since $\dfrac{dy}{dx} = \dfrac{dy/dt}{dx/dt} = \dfrac{\dot{y}}{\dot{x}}$ Here dot $(\cdot)$ denotes differentiation w.r.t. the parameter t.

Polar Form

The length of the arc of the curve $r = f(\theta)$ included between two points whose vectorial angles are $\theta = \theta_1$, and $\theta = \theta_2$ is

$$S - \int_{\theta_1}^{\theta_2} \sqrt{r^2 + \left(\frac{dr}{d\theta}\right)^2}\, d\theta$$

Corollary 1: When the equation of the curve is $\theta = g(r)$ then the length included between two points whose radii vectors are r_1 and r_2 is

$$S = \int_{r_1}^{r_2} \sqrt{1 + r^2 \left(\frac{d\theta}{dr}\right)^2}\, dr$$

WORKED OUT EXAMPLES

Example 1: Find the length of the curve

$$y = \ln\,(e^x + 1) - \ln\,(e^x - 1) \quad \text{from} \quad x = 1 \text{ to } x = 2$$

Solution: Differentiating y w.r.t. x

$$\frac{dy}{dx} = \frac{e^x}{e^x + 1} - \frac{e^x}{e^x - 1} = -\frac{2e^x}{e^{2x} - 1}$$

$$1 + \left(\frac{dy}{dx}\right)^2 = 1 + \frac{4e^{2x}}{(e^{2x} - 1)^2} = \frac{e^{4x} + 1 - 2e^{2x} + 4e^{2x}}{(e^{2x} - 1)^2}$$

$$= \left(\frac{e^{2x} + 1}{e^{2x} - 1}\right)^2$$

Length of the curve:

$$S = \int_a^b \sqrt{1 + \left(\frac{dy}{dx}\right)^2}\, dx = \int_1^2 \frac{e^{2x} + 1}{e^{2x} - 1} dx$$

$$S = \int_1^2 \frac{(e^{2x} - 1) + 1 + 1}{e^{2x} - 1} dx$$

$$= \int_1^2 dx + 2\int_1^2 \frac{1}{e^{2x} - 1} dx$$

$$= 1 + 2\int_1^2 \frac{dx}{(e^x)^2 - 1}$$

Put $e^x = \sec t$, $dx = \tan t\, dt$, $\cos t = \frac{1}{e^x}$

Consider

$$\int_1^2 \frac{dx}{(e^x)^2 - 1} = \int \frac{\tan t\, dt}{\tan^2 t} = \int \cot t\, dt = \ln\, \sin t$$

$$= \ln\left(\frac{\sqrt{e^{2x} - 1}}{e^x}\right)\Bigg|_1^2 = \ln\, \frac{e^4 - 1}{e^2} - \frac{\ln\, e^2 - 1}{e}$$

$$= \ln\, \frac{e^2 + 1}{e}$$

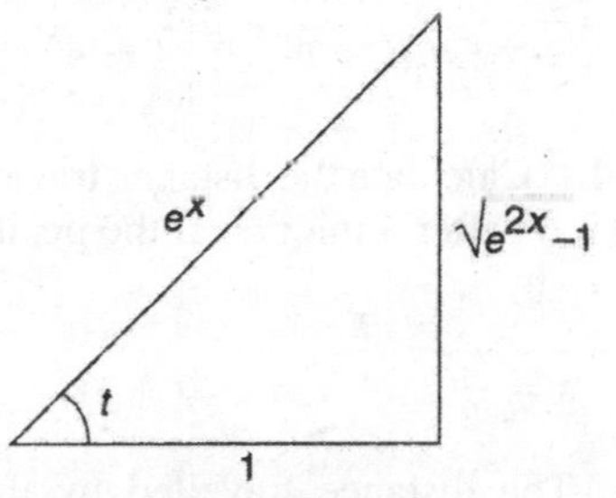

Fig. 6.17

Thus the length of the curve $S = 1 + 2\,\ln\,\frac{e^2+1}{e}$.

Example 2: Find the length of the curve

$$8x = y^4 + 2y^{-2} \quad \text{from} \quad y = 1 \text{ to } y = 2$$

Solution: Here y is the independent variable. Differentiating w.r.t. y, we get

$$\frac{dx}{dy} = \frac{4y^3 - 4y^{-3}}{8} = \frac{1}{2}\left(y^3 - \frac{1}{y^3}\right)$$

Length of the curve:

$$S = \int_{y_1}^{y_2} \sqrt{1 + \left(\frac{dx}{dy}\right)^2}\, dy$$

$$= \int_1^2 \sqrt{1 + \frac{1}{4}\left(y^3 - \frac{1}{y^3}\right)^2}\, dy = \int_1^2 \frac{y^6 + 1}{2y^3} dy$$

$$S = \frac{1}{2}\left[\frac{y^4}{4} - \frac{y^{-2}}{2}\right]_1^2 = \frac{33}{16}$$

Example 3: Determine the length of the curve whose parametric equation is

$$x = e^{-t}\cos t,\ y = e^{-t}\sin t,\ 0 \le t \le \frac{\pi}{2}$$

Solution: Differentiating x and y w.r.t. 't'

$$\frac{dx}{dt} = -e^{-t}\cos t - e^{-t}\sin t,$$

$$\frac{dy}{dt} = -e^{-t}\sin t + e^{-t}\cos t$$

$$\left(\frac{dx}{dt}\right)^2 + \left(\frac{dy}{dt}\right)^2 = e^{-2t}\cos^2 t + e^{-2t}\sin^2 t$$
$$+2e^{-2t}\sin t\cos t + e^{-2t}\sin^2 t$$
$$+e^{-2t}\cos^2 t - 2e^{-2t}\sin t\cos t$$
$$= 2e^{-2t}$$

Length of the curve

$$S = \int_{t_1}^{t_2} \sqrt{\left(\frac{dx}{dt}\right)^2 + \left(\frac{dy}{dt}\right)^2}\, dt$$

$$= \int_0^{\frac{\pi}{2}} \sqrt{2e^{-2t}}\, dt = \sqrt{2}\cdot\left.\frac{e^{-t}}{-1}\right|_0^{\frac{\pi}{2}} = \sqrt{2}(1 - e^{-\frac{\pi}{2}})$$

Example 4: Calculate the distance travelled by the particle P(x, y) after 4 minutes if the position at any time is given by

$$x = \frac{t^2}{2}, \qquad y = \frac{1}{3}(2t + 1)^{\frac{3}{2}}.$$

Solution: The distance travelled by the particle $P(x, y)$ is the length of the curve whose parametric equation is given above when t varies from 0 to 4. So differentiating x, y w.r.t. t

$$\frac{dx}{dt} = t; \qquad \frac{dy}{dt} = (2t + 1)^{\frac{1}{2}}$$

$$\left(\frac{dx}{dt}\right)^2 + \left(\frac{dy}{dt}\right)^2 = t^2 + 2t + 1 = (t + 1)^2$$

distance travelled by P = Length of the curve

$$= \int_0^4 \sqrt{(t + 1)^2}\, dt \quad = \left.\frac{t^2}{2} + t\right|_0^4 = 12$$

Example 5: Determine the length (perimeter) of one loop of the curve $6ay^2 = x(x - 2a)^2$.

Solution: The curve is symmetric about the x-axis and crosses the x-axis at the points $x = 0$ and $x = 2a$. The length of the loop of the curve is twice the length of OBA (see Fig. 6.18).

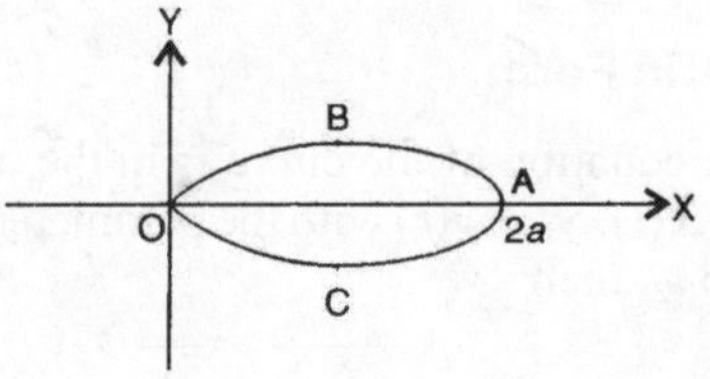

Fig. 6.18

$$\text{Length of curve } OBA = \int_0^{2a} \sqrt{1 + y'^2}\, dx$$

Differentiating y w.r.t. x

$$\frac{dy}{dx} = \frac{3x - 2a}{\sqrt{2 \cdot 4ax}}, \quad \text{so} \quad 1 + y'^2 = \frac{(3x + 2a)^2}{24ax}$$

Length of the loop

$$= 2\int_0^{2a} \frac{3x + 2a}{\sqrt{24ax}} dx$$

$$= \frac{2}{\sqrt{24a}}\left[3\frac{x^{\frac{3}{2}}}{\frac{3}{2}} + 2a\frac{x^{\frac{1}{2}}}{\frac{1}{2}}\right]_0^{2a}$$

$$= \frac{4}{\sqrt{24a}}\left[(2a)^{\frac{3}{2}} + 2a \cdot (2a)^{\frac{1}{2}}\right] = \frac{8a}{\sqrt{3}}$$

Example 6: Find the perimeter of the curve

$$r = a(\cos\theta + \sin\theta); \qquad 0 \le \theta \le \pi$$

Solution: Differentiating r w.r.t. θ

$$\frac{dr}{d\theta} = a(-\sin\theta + \cos\theta)$$

$$r^2 + \left(\frac{dr}{d\theta}\right)^2 = a^2(\cos\theta + \sin\theta)^2$$
$$+a^2(-\sin\theta + \cos\theta)^2$$
$$= 2a^2$$

$$\text{Length of the curve} = \int_0^{\pi} \sqrt{r^2 + \left(\frac{dr}{d\theta}\right)^2}\, d\theta$$

$$= \sqrt{2}a \cdot \theta|_0^{\pi} = \sqrt{2}a\pi$$

Example 7: Find the length of the spiral $r = e^{\alpha\theta}$ from the pole to the point (r, θ).

Solution: $\frac{dr}{d\theta} = \alpha e^{\alpha\theta}$

$$\sqrt{r^2 + \left(\frac{dr}{d\theta}\right)^2} = \sqrt{e^{2\alpha\theta} + \alpha^2 e^{2\alpha\theta}}$$

$$\text{Length of the spiral} = \int_0^\theta \sqrt{[e^{2\alpha\theta} + \alpha^2 e^{2\alpha\theta}]}d\theta$$

$$= \sqrt{1+\alpha^2}\int_0^\theta e^{\alpha\theta}d\theta$$

$$= \sqrt{1+\alpha^2}\left.\frac{e^{\alpha\theta}}{\alpha}\right|_0^\theta$$

$$= \frac{\sqrt{1+\alpha^2}}{\alpha}(e^{\alpha\theta} - 1) = \frac{\sqrt{1+\alpha^2}}{\alpha}(r-1)$$

Example 8: What is the length of the loop of the curve $x = t^2$, $y = t - \frac{t^3}{3}$.

Solution: $t = \sqrt{x}$, $y = \sqrt{x} - \frac{x^{\frac{3}{2}}}{3}$ so the equation in cartesian form is

$$3y = 3\sqrt{x} - x^{\frac{3}{2}} = \sqrt{x}(3-x)$$

So loop of the curve lies between $x = 0$ and $x = 3$. Differentiating x and y w.r.t. t

$$\frac{dx}{dt} = 2t, \quad \frac{dy}{dt} = 1 - t^2$$

$$\sqrt{\left(\frac{dx}{dt}\right)^2 + \left(\frac{dy}{dt}\right)^2} = \sqrt{(2t)^2 + (1-t^2)^2}$$

$$= \sqrt{(1+t^2)^2} = 1 + t^2$$

$$\text{Length of the loop} = \int_{-\sqrt{3}}^{\sqrt{3}}(1+t^2)dt$$

$$= 2\left.\left(\frac{t^3}{3} + t\right)\right|_0^{\sqrt{3}} = 4\sqrt{3}$$

Example 9: Determine the length of the loop of the curve $r = a(\theta^2 - 1)$.

Solution: As θ varies from $-1, 0, 1$, r takes the values $0, -a, 0$. Thus travelling the loop of the given curve. Differentiating r w.r.t. θ, we get

$$\frac{dr}{d\theta} = 2a\theta, \text{ so } r^2 + \left(\frac{dr}{dt}\right)^2 = a^2(\theta^2-1)^2 + (2a\theta)^2$$

$$\therefore \sqrt{r^2 + \left(\frac{dr}{d\theta}\right)^2} = \sqrt{a^2(\theta^2+1)^2} = a(\theta^2+1)$$

Length of the loop of the curve is

$$= \int_{-1}^{1} a(\theta^2+1)d\theta = a\left.\left(\frac{\theta^3}{3} + \theta\right)\right|_{-1}^{1} = \frac{8a}{3}$$

Example 10: Show that the curve

$$x = a(\theta - \sin\theta), \; y = a(1 - \cos\theta)$$

is divided in the ratio 1:3 at $\theta = 2\pi/3$.

Solution: $\frac{dx}{d\theta} = a(1-\cos\theta)$, $\frac{dy}{d\theta} = a\sin\theta$

$$\left(\frac{dx}{d\theta}\right)^2 + \left(\frac{dy}{d\theta}\right)^2 = a^2(1-\cos\theta)^2 + a^2\sin^2\theta$$

$$= 2a^2(1-\cos\theta) = 4a^2\sin^2\frac{\theta}{2}$$

Length of the curve as θ varies from 0 to 2π is

$$L = \int_0^{2\pi}\sqrt{4a^2\sin^2\frac{\theta}{2}}d\theta = 2a\int_0^{2\pi}\sin\frac{\theta}{2}d\theta$$

$$= -2a \cdot 2\cos\frac{\theta}{2} = -4a\left.\cos\frac{\theta}{2}\right|_0^{2\pi}$$

$$= -4a[-1-1] = 8a$$

Length of the curve as θ varies from 0 to $\frac{2\pi}{3}$ is M

$$M = \int_0^{\frac{2\pi}{3}} 2a\sin\frac{\theta}{2}d\theta = -4a\left.\cos\frac{\theta}{2}\right|_0^{\frac{2\pi}{3}}$$

$$= -4a\left[\frac{1}{2} - 1\right] = 2a$$

Thus L is divided in the ratio 1:3 when $\theta = \frac{2\pi}{3}$.

EXERCISE

Cartesian form

Find the length of the curve

1. $y = x\sqrt{x}$ from $x = 0$ to $x = 4/3$

Ans. $\frac{56}{27}$

2. $3y = 2(x^2+1)^{\frac{3}{2}}$ from $x = 0$ to $x = 3$

Ans. 21

3. $x = \ln y$, between the points whose ordinates are $\frac{3}{4}$ and $\frac{4}{3}$

Hint: Treat y as the independent variable.

Ans. $\ln \frac{3}{2} + \frac{5}{12}$

4. Find the perimeter of a circle of radius b.

Hint: Take equation of circle as $x^2 + y^2 = a^2$, $S = 2\int_{-a}^{a} \sqrt{1+y^{12}}dx$

Ans. $2\pi b$

5. Calculate the length of the arc of parabola $y^2 = 4ax$ cut-off by the latus rectum.

Ans. $2a\left[\sqrt{2} + \ln\left(1+\sqrt{2}\right)\right]$

6. What is the length of the catenary (hanging chain; hyperbolic cosine curve) $y = \frac{1}{2}a(e^{\frac{x}{a}} + e^{-\frac{x}{a}})$ from $x = -a$ to $x = a$.

Hint: $y = a \cosh\left(\frac{x}{a}\right)$

Ans. $a(e - e^{-1})$

7. Show that the length of the arc of the curve

$$y = \ln \tanh (x/2)$$

from $x = 1$ to $x = 2$ is $\ln\,[(e^2+1)/e]$.

8. Determine the distance a particle which moves along a curve $20x = 3(4t^2 - 20t + 9)$ is 4 minutes starting at $t = \frac{1}{2}$.

Ans. $\frac{5}{6}\left[\frac{156}{25} + \ln 5\right]$

9. Find the length of the curve $y^2 = x(1-\frac{x}{3})^2$ from origin to the ordinate when $x = a$.

Ans. $\frac{(3+a)}{3}\sqrt{a}$

10. Find the perimeter of the loop of the curve

$$3ay^2 = x(x-a)^2$$

Hint: Perimeter $= 2\int_0^a \sqrt{1+y'^2}dx$ since curve is symmetric about x-axis and lies between 0 to 9.

Ans. $\frac{4a}{\sqrt{3}}$

11. Calculate the perimeter of the loop of the curve

$$9ay^2 = (x-2a)(x-5a)^2.$$

Hint: Curve lies between $x = 2a$ and $x = 5a$ and is symmetric about y-axis. Perimeter is $2\int_{2a}^{5a} \sqrt{1+y'^2}dx$.

Ans. $4\sqrt{3}a$

12. Determine the total length of the curve

$$x^2(a^2 - x^2) = 8a^2y^2$$

Hint: Curve is symmetric about both x-axis and y-axis and lies between $x = -a$ to $x = a$.

$$\text{Total length} = 4 \times \int_0^a \sqrt{1+y'^2}dx$$

Ans. $\pi a\sqrt{2}$

13. What is the length of the arc of the parabola $y^2 = 12x$ cut-off by its latus rectum.

Hint: Length is twice that form (0, 0) to (3, 6).

Ans. $6\left[\sqrt{2} + \ln\left(1+\sqrt{2}\right)\right]$

14. Find the length of an arc of the curve

$$x^2 = a^2(1 - e^{\frac{y}{a}})$$

measured from (0, 0) to any point (x, y).

Ans. $a \ln\left(\frac{a+x}{a-x}\right) - x$

15. Determine the length of the parabola $y^2 = 4ax$ cut-off by the line $3y = 8x$.

Ans. $\left(\frac{15}{16} + \ln 2\right)a$

16. Find the length of the loop of the curve $9ay^2 = x(x-3a)^2$.

Hint: Curve is symmetric about x-axis and meets x-axis at $x = 0$, and $x = 3a$

$$S = \frac{1}{2\sqrt{a}}\int_0^{3a} \frac{x+a}{\sqrt{x}}dx$$

Ans. $2a\sqrt{3}$

Parametric form

Find the length of the curve whose equation is given in the parametric form:

1. $x = t^3 - 3t$; $y = 3t^2$; t: from 0 to 1

Ans. 4

2. $x = a(t + \sin t); y = a(1 - \cos t), t : -\pi$ to π

Ans. 8a

3. $x = e^t \sin t, y = e^t \cos t, t : 0$ to $\frac{\pi}{2}$

Ans. $\sqrt{2}(e^{\frac{\pi}{2}} - 1)$

4. $x = a(\cos t + t \sin t), y = a(\sin t - t \cos t)$ from $t = 0$ to $t = \pi/2$

Ans. $\frac{a\pi^2}{8}$

5. Find the distance travelled by a particle $P(x, y)$ whose position at any time t is given by

$$x = \frac{1}{3}(2t+3)^{\frac{3}{2}}, y = \frac{t^2}{2} + t$$

in 3 minutes commencing from $t = 0$.

Ans. $\frac{21}{2}$

6. (a) Determine the total length of the four-cusped hypocycloid given by

$$x = a\cos^3\theta, y = b\sin^3\theta$$

or

$$\left(\frac{x}{a}\right)^{\frac{2}{3}} + \left(\frac{y}{b}\right)^{\frac{2}{3}} = a^{\frac{2}{3}}.$$

Hint: (a) Curve is symmetric about both x-axis and y-axis. For one cusp: θ varies from 0 to $\pi/2$

$$\text{Total length} = 4\int_0^{\frac{\pi}{2}} \sqrt{\left(\frac{dx}{d\theta}\right)^2 + \left(\frac{dy}{d\theta}\right)^2}\, d\theta$$

Ans. $4(a^2 + ab + b^2)/(a + b)$.

6. (b) Deduce whole length of the astroid given by

$$x = a\cos^3\theta, = y = a\sin^3\theta$$

Hint: take $a = b$

Ans. 6a

7. Calculate the length of the arc of the cycloid $x = a(t - \sin t), y = a(1 - \cos t)$ between two cusps.

Hint: Between two cusps t varies from 0 to 2π.

Ans. 8a

8. Determine the length of the curve given by $x = a\sin 2t(1 + \cos 2t); y = a\cos 2t(1 - \cos 2t)$

measured from origin to any point (x, y).

Ans. $\frac{4a}{3}\sin 3t$

Polar coordinates

1. Find the perimeters of the curves:
   a. $r = a\cos\theta$ (circle with centre at origin and radius a)
   b. $r = a\sin\theta$
   c. $r = 2a\cos\theta$ (circle with centre at $r = a$ and radius a).

Ans. (a) πa (b) πa (c) $2a\pi$

2. Show that the arc of the upper half of the curve

$$r = a(1 - \cos\theta)$$

is bisected at $\theta = 2\pi/3$.

3. Calculate the length of the curve

$$r = a\sin\theta + b\cos\theta, 0 \le \theta \le 2\pi$$

Ans. $\pi\sqrt{a^2 + b^2}$

4. **a.** Find the perimeter of the cardioid

$$r = (1 + \cos\theta).$$

   b. Show that the arc of the upper half is bisected at $\theta = \pi/3$.
   c. Also show that the length of the part of the curve which lies on the side of the line $4r = 3a\sec\theta$ remote from the pole is equal to $4a$.

Hint: (b) Curve is symmetrical about initial line $\theta = 0$ and for the upper half of the curve, θ varies from 0 to π. The length of the curve as θ varies from 0 to $\frac{\pi}{3}$ is $2a$ i.e., half of $4a$, therefore, bisected at $\theta = \pi/3$.

Hint: (c) Line meets cardioid at $\theta = \frac{\pi}{3}$; length = twice $\int_0^{\frac{\pi}{3}} \sqrt{r^2 + \left(\frac{dr}{d\theta}\right)^2}\, d\theta$.

Ans. (a) $8a$

5. Determine the perimeter of the curve

$$r = a\sin^3\left(\frac{\theta}{3}\right)$$

Hint: As θ varies from $0, \frac{3\pi}{2}, 3\pi, r$ takes c, a, c, perimeter = 2 length from $\theta = 0$ to $\frac{3\pi}{2}$

Ans. $\frac{3a\pi}{2}$

6. Find the length of the arc of the parabola

$$r = 2a/(1 + \cos\theta)$$

cut-off but its latus rectum.

Ans. $\left[\sqrt{2} + \ln(1 + \sqrt{2})\right] \cdot 2a$

7. Prove that the length of the curve

$$r = a\cos^3\left(\frac{\theta}{3}\right) \quad \text{is } 8a\pi/2$$

8. Determine the whole length of the lemniscate

$$r^2 = a^2\cos 2\theta$$

Hint: 4 times the length as θ varies from 0 to $\frac{\pi}{4}$ since curve is symmetric about both x-axis and y-axis (i.e., $\theta = 0$ line and $\theta = \frac{\pi}{2}$ line).

Ans. $\sqrt{2}\pi a\left\{1 + \left(\frac{1}{2}\right)^2 + \left(\frac{1\cdot 3}{2\cdot 4}\right)^2 \cdot \left(\frac{1}{2}\right)^2 + \left(\frac{1\cdot 3\cdot 5}{2\cdot 4\cdot 6}\right)^2\left(\frac{1}{2}\right)^2 + \cdots\right\}$

9. Find the length of the arc of the equiangular spiral $r = a\,e^{\theta\cot\alpha}$ between the points for which the radii vectors are r_1 and r_2.

Hint: $\frac{dr}{d\theta} = r\cot\alpha$, $\frac{d\theta}{dr} = \frac{1}{r}\tan\alpha$

$$\text{length} = \int_{r_1}^{r_2}\sqrt{1 + r^2\left(\frac{d\theta}{dr}\right)^2}\,dr$$

$$= \int_{r_1}^{r_2}\sqrt{1 + \tan^2\alpha}\,dr = \int_{r_1}^{r_2}\sec\alpha\,dr$$

Ans. $(r_2 - r_1)\sec\alpha$

10. Calculate the length of the arc of the hyperbolic spiral $r = a/\theta$ between $r = a$ and $r = 2a$.

Ans. $a\left[\sqrt{5} - \sqrt{2} + \ln\left(\frac{2+\sqrt{8}}{1+\sqrt{5}}\right)\right]$

11. Determine the whole length of a loop of the curve

$$r = a(1 + \cos 2\theta)$$

Ans. $\frac{2}{3}\sqrt{3}\left[2\sqrt{3} + \ln\left(\sqrt{3} + 2\right)\right]a$

6.4 VOLUME OF SOLID OF REVOLUTION

A solid of revolution is generated by revolving a plane area R about a line L in the plane. Line L is known as the axis of revolution. Line L does not intersect the plane area R but may touch the boundary of R.

Examples:

1. Sphere is a solid of revolution generated by revolving the semicircular region R about its diameter L (Fig. 6.19).

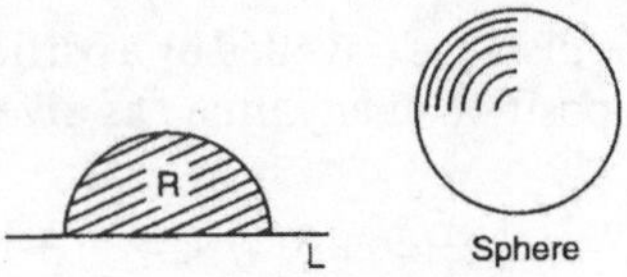

Fig. 6.19

2. Right circular cylinder is a solid of revolution obtained by revolving a rectangle R about its edge L (Fig. 6.20).

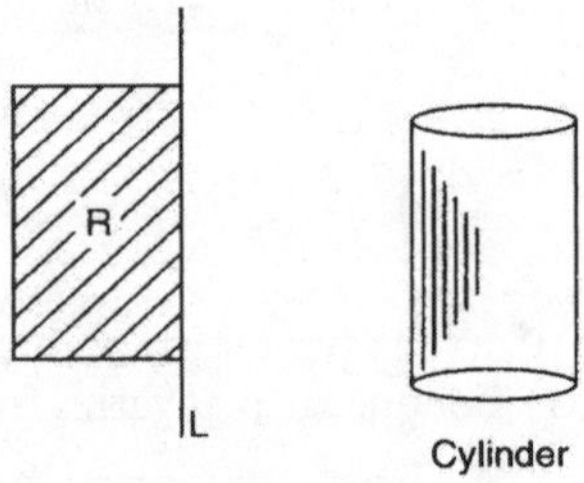

Fig. 6.20

3. Parabola about x-axis generates paraboloid (Fig. 6.21).

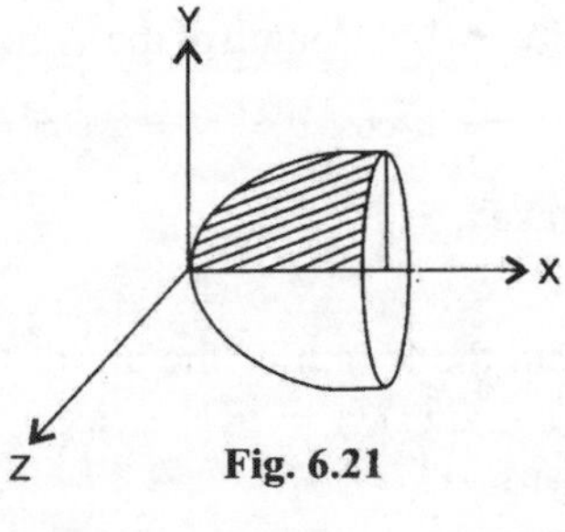

Fig. 6.21

4. Circular region R about L not touching it, produces a torus (see Fig. 6.22).

The volume of solid of revolution may be obtained by (a) *cylindrical disc (C.D.) method* (b) *cylindrical shell (C.S.) method.*

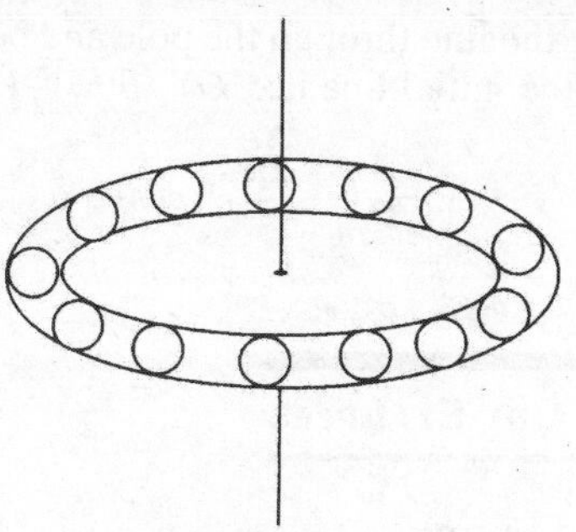

Fig. 6.22

Cartesian Form

Cylindrical disc method

I. Axis of revolution L is a part of the boundary of the plane area. Consider the plane area $ABCD$ bounded by the curve $y = f(x)$, x-axis, ordinates $x = a$ and $x = b$ as shown in Fig. 6.23. When the plane area $ABCD$ is revolved about x-axis, a solid of revolution is obtained, one quarter of which is shown in Fig. 6.23. The volume of an element circular disk of radius y

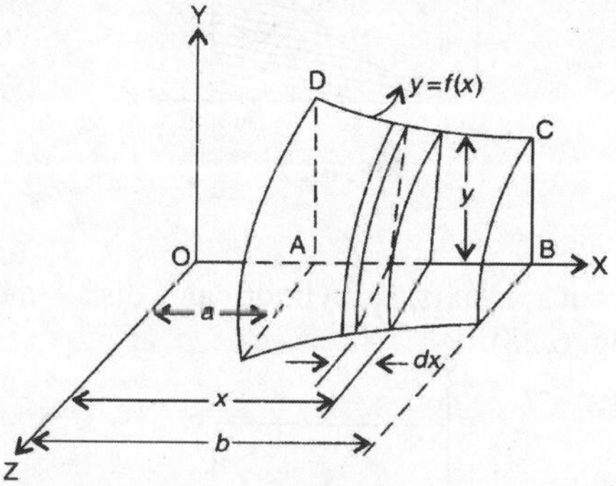

Fig. 6.23

and thickness dx is $\pi y^2 dx$. Integrating these elements, the volume V of solid of revolution obtained by revolving about the x-axis the plane area bounded by $y = f(x)$, $x = a$, $x = b$, x-axis is

$$V = \int_a^b \pi y^2\, dx$$

Similarly, when plane area bounded by the curve $x = g(y)$, $y = c$, $y = d$, y-axis, is revolved about y-axis,

$$V = \int_c^d \pi x^2\, dy$$

II. *Any axis of revolution*:

$$V = \int_a^b \pi r^2\, dh$$

where r=perpendicular distance from the curve to the axis of revolution AB (Fig. 6.24).

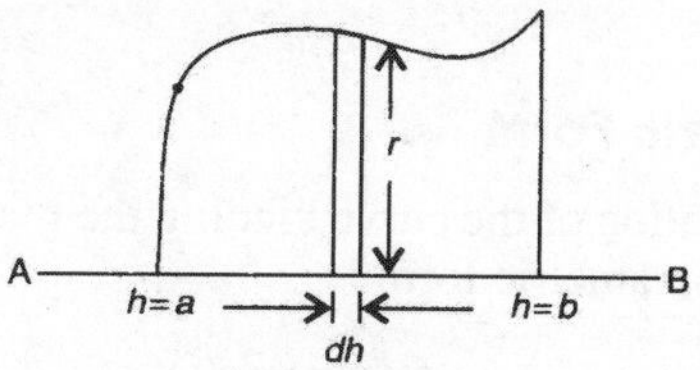

Fig. 6.24

III. *The plane area is bounded by two curves*: Let the plane area bounded by two curves $y = y_1(x)$ lower curve, $y = y_2(x)$ upper curve, the ordinates $x = a$, $x = b$ is revolved about x-axis, then volume of solid of revolution generated is the difference between the volume generated by the upper curve and lower curve. Thus

$$V = \int_a^b \pi y_2^2\, dx - \int_a^b \pi y_1^2\, dx = \int_a^b \pi (y_2^2 - y_1^2) dx$$

where y_2 and y_1 are the ordinates of the upper and lower curves.

Cylindrical shell method

Axis of rotation AB is not part of the boundary of the plane area $DEFG$, volume element generated by revolving a rectangular strip about an axis AB (Fig. 6.25).

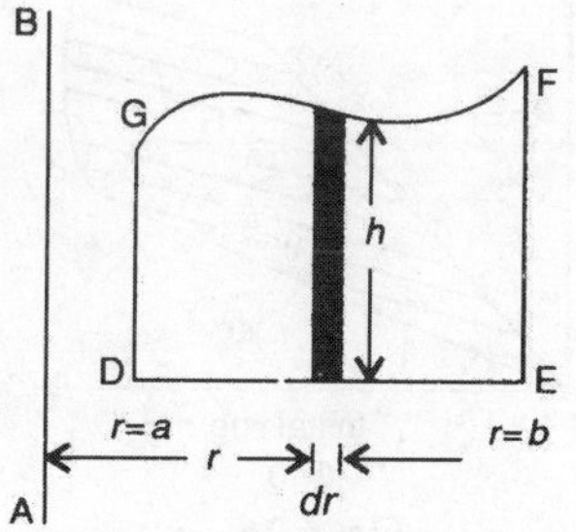

Fig. 6.25

$$dV = (\text{mean circumference}) \times (\text{height}) \times (\text{thickness})$$

$$dV = (2\pi r)(h)(dr)$$

So $$V = \int_{r=a}^{r=b} 2\pi r h \quad dr$$

If area $DEFG$ is rotated about y-axis (AB) thus

$$V = \int_{x=a}^{x=b} (2\pi x)(y)dx$$

Similarly, about x-axis,

$$V = \int_{y=c}^{y=d} (2\pi y)(x)dy$$

Parametric Form

If the equation of the curve binding the plane area is given in parametric form

$$x = f(t),\ y = g(t)$$

with the parameter t varying between t_1 and t_2, then the volume about x-axis

$$V = \int_{t_1}^{t_2} \pi y^2 \frac{dx}{dt} dt = \int_{t_1}^{t_2} \pi g^2(t) \frac{df(t)}{dt} dt$$

Similarly, about y-axis

$$V = \int_{t_1}^{t_2} \pi x^2 \frac{dy}{dt} dt = \int_{t_1}^{t_2} \pi f^2(t) \frac{dg(t)}{dt} dt$$

Polar Form

The volume of the solid of revolution generated by revolving the plane area R, bounded by the curve c whose equation is given in polar form $r = f(\theta)$, and radii vectors $\theta = \theta_1, \theta = \theta_2$ (Fig. 6.26).

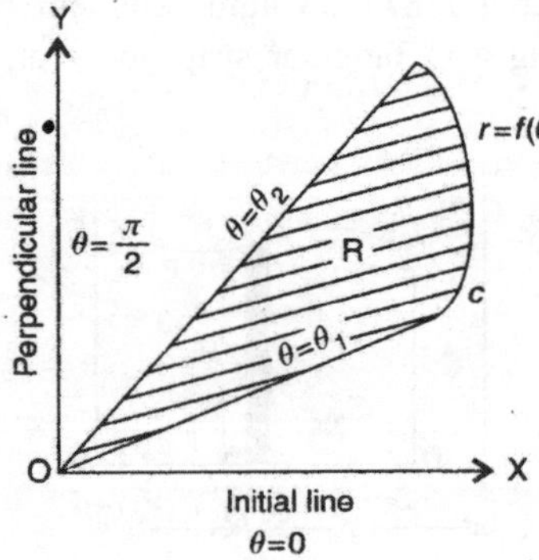

Fig. 6.26

I. About the initial line $OX(\theta = 0)$

$$V = \int_{\theta_1}^{\theta_2} \frac{2\pi}{3} r^3 \sin\theta \quad d\theta$$

II. About the line through the pole and perpendicular to the initial line i.e., OY $\left(\theta = \frac{\pi}{2}\right)$

$$V = \int_{\theta_1}^{\theta_2} \frac{2\pi}{3} r^3 \cos\theta \, d\theta$$

Worked Out Examples

Cartesian Form

Example 1: The equation of the curve OP is $y = x^2$. Find the volume of solid of revolution generated when the area OAP, bounded by $y = x^2$, $x = 3$ and x-axis, is revolved about (a) x-axis (b) y-axis (c) line AP (d) Line BP (Fig. 6.27).

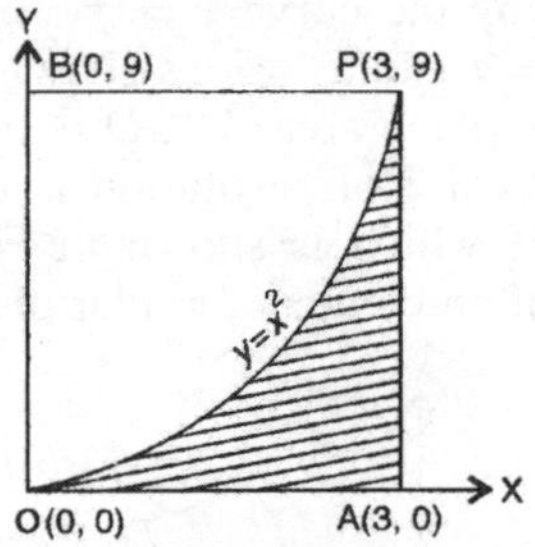

Fig. 6.27

Solution:

a. About x-axis: (cylindrical disk method) (Fig. 6.28).

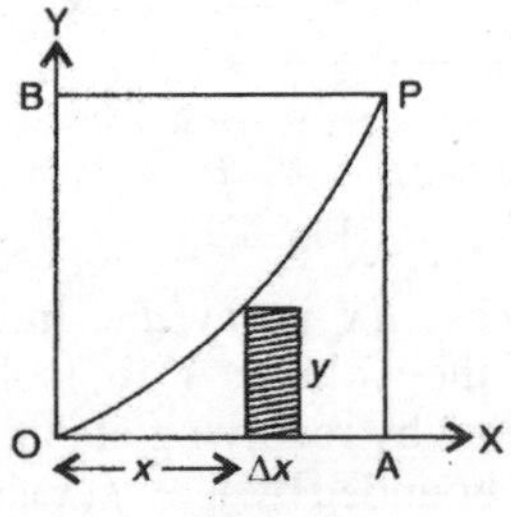

Fig. 6.28

A rectangle of height y and thickness Δx revolved about x-axis (Line OA), generates a solid of revolution, a circular disk, of volume $\pi y^2 dx$. Since x varies from 0 to 3, the volume V of solid of revolution obtained by revolving the area OAP about x-axis is

$$V = \int_{x=0}^{3} \pi y^2 \, dx = \pi \int_0^3 (x^2)^2 dx = \frac{243\pi}{5}$$

b. About y-axis: (cylindrical shell method) (Fig. 6.29) when a rectangle of height y and thickness Δx is revolved about y-axis (line

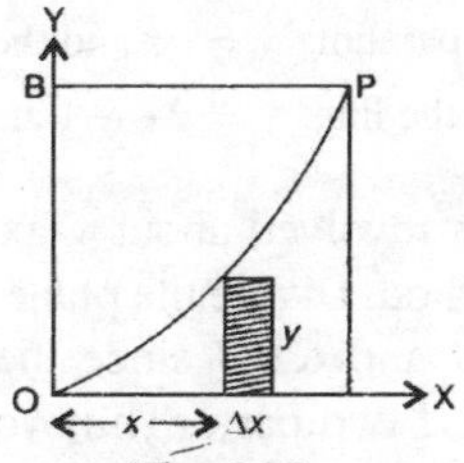

Fig. 6.29

OB), it generates a cylindrical shell whose circumference is $2\pi x$, height y and thickness Δx so that its volume is

$$2\pi x \cdot y \cdot \Delta x$$

Thus the volume of solid of revolution generated by revolving the area OAP about y-axis is

$$V = \int_0^3 (2\pi x)(y)dx = \int_0^3 2\pi x \cdot x^2 dx$$

$$= 2\pi \left.\frac{x^4}{4}\right|_0^3 = \frac{81\pi}{2}$$

c. About the line AP (Fig. 6.30).

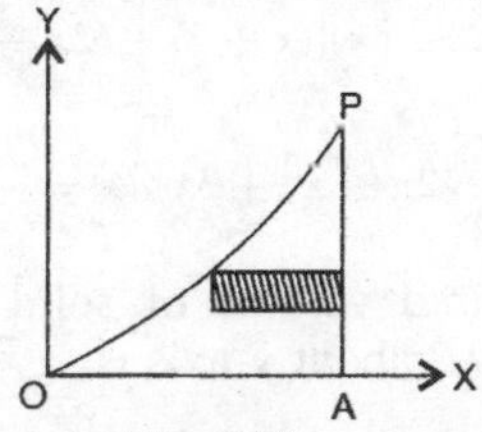

Fig. 6.30

i. By cylindrical disk method: When a horizontal rectangle of length $(3 - x)$ and thickness Δy is revolved about the line AP, it generates a circular disk of volume $\pi(3 - x)^2$, so the volume generated by revolving OAP about AP is

$$V = \int_0^9 \pi(3 - x)^2 dy = \int_0^9 \pi(3 - \sqrt{y})^2 dy$$

$$= \pi\left.\left(9y + \frac{y^2}{2} - 12y^{\frac{3}{2}}\right)\right|_0^9 = \frac{27\pi}{2}$$

ii. Aliter: By cylindrical shell method: (Fig. 6.31): By revolving a vertical rectangle of height y and thickness Δx about the

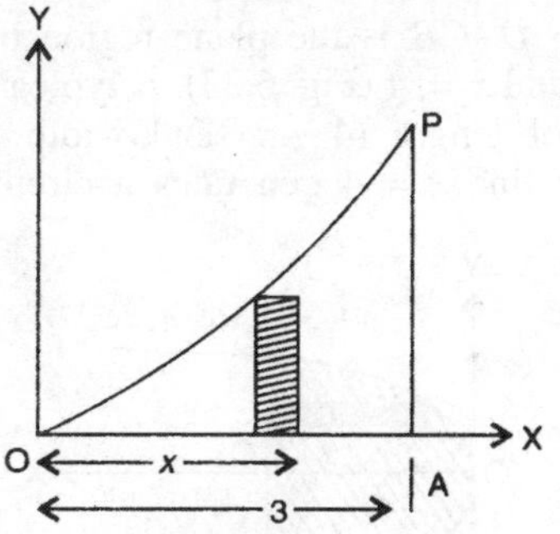

Fig. 6.31

line AP, a cylindrical shell of circumference $2\pi(3 - x)$, height y and thickness Δx is generated. Its volume is

$$2\pi(3 - x) \cdot y \cdot \Delta x$$

So the volume required is

$$V = \int_0^3 2\pi(3 - x) \cdot y\, dx = \int_0^3 2\pi(3 - x)(x^2)dx$$

$$= 2\pi\left[x^3 - \frac{x^4}{4}\right]_0^3 = \frac{27\pi}{2}$$

d. About the line BP (Fig. 6.32). A horizontal strip of length $(3 - x)$ and width Δy, revolved about the line BP generates a cylindrical shell of circumference $2\pi(9 - y)$, height $(3 - x)$ and width Δy. Thus the required volume

$$V = \int_0^9 2\pi(9 - y) \cdot (3 - x)dy$$

$$= 2\pi \int_0^9 (9 - y)(3 - \sqrt{y})dy = \frac{567\pi}{5}$$

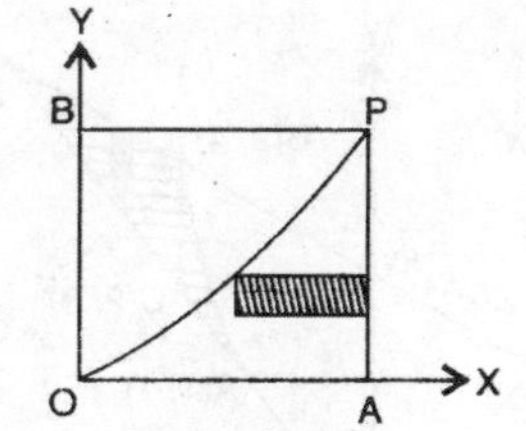

Fig. 6.32

Example 2: Determine the volume of solid generated by revolving the plane area bounded by $y^2 = 4x$ and $x = 4$ about the line $x = 4$.

Solution: $OACB$ is the plane region bounded by $y^2 = 4x$ and $x = 4$ (Fig. 6.33). A typical horizontal element of length $(4 - x)$ and width dy rotated about the line $x = 4$ generates a circular disk of

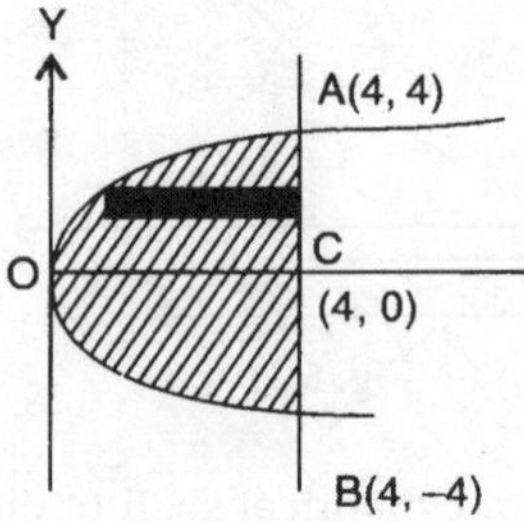

Fig. 6.33

volume $\pi(4 - x)^2 dy$. Here y varies from -4 to 4. Thus the required volume generated by rotating $OACB$ about $x = 4$ is

$$V = \int_{-4}^{4} \pi(4 - x)^2 \cdot dy = \int_{-4}^{4} \pi \left(4 - \frac{y^2}{4}\right)^2 dy$$

$$= 2\pi \int_0^4 \left(16 + \frac{y^4}{16} - 2y^2\right) dy$$

$$= 2\pi \left[16y + \frac{y^5}{80} - \frac{2}{3}y^3\right]_0^4 = \frac{1024}{15}\pi$$

Example 3: Calculate the volume of the solid of revolution generated by revolving about y-axis the plane area bounded by the straight lines $y = x + 2$, $y = 2x - 1$ and outside of the parabola $y = x^2$.

Solution: The plane area which is outside of the parabola and included between the two straight lines is $ACDB$ shown shaded in Fig. 6.34.

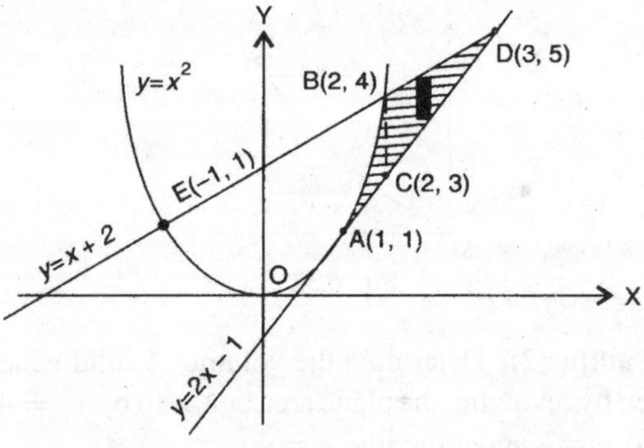

Fig. 6.34

[The points of intersection are:

$A(1, 1)$ between parabola $y = x^2$ and the st line $2x - 1 = y$

$B(2, 4)$ between parabola $y = x^2$ and the line $y = x + 2$

$D(3, 5)$ between the lines $y = 2x - 1$ and $y = x + 2$]

Since $ACBD$ is revolved about y-axis, apply cylindrical shell method. Divide the plane arc $ACBD$ into two parts ACB and CDB since the variation y is different in these two parts. Thus volume V generated by revolving $ACDB$ about y-axis is sum of the volumes V_1 and V_2 generated by revolving ACB and CDB about y-axis respectively. These volumes V_1 and V_2 are obtained by cylindrical shell method.

For the plane area ACB, the height of a vertical element is $(x^2) - (2x - 1)$. So the volume V_1 of the solid of revolution generated by revolving ACB about y-axis is

$$V_1 = \int_1^2 (2\pi x)[(x^2) - (2x + 1)]dx$$

$$= 2\pi \left[\frac{x^4}{4} - \frac{2x^3}{3} + \frac{x^2}{2}\right]_1^2 = \frac{7\pi}{6}$$

Similarly, for CDB, we have

$$V_2 = \int_2^3 (2\pi x)[(x + 2) - (2x - 1)]dx$$

$$= \int_2^3 2\pi(-x^2 + 3x)dx = \frac{7\pi}{3}$$

Thus the required volume of solid generated by revolving $ACDB$ about y-axis is

$$V = V_1 + V_2 = \frac{7\pi}{6} + \frac{7\pi}{3} = \frac{7\pi}{2}$$

Example 4: Find the volume generated when the plane area bounded by the parabola $y + x^2 + 3x - 6 = 0$ and the line $x + y = 3$ is revolved about the line.

Solution: For the representative rectangle of the Fig. 6.35, the height is $\frac{x+y-3}{\sqrt{2}}$ and the width is $\sqrt{(\Delta x)^2 + (\Delta y)^2} = \sqrt{2}\Delta x$. By circular disc method, the volume of the solid of revolution generated by revolving the plane area $ACBD$ about the line BDA is

$$V = \int_{-3}^{1} \pi \left(\frac{x + y - 3}{\sqrt{2}}\right)^2 \cdot (\sqrt{2}dx)$$

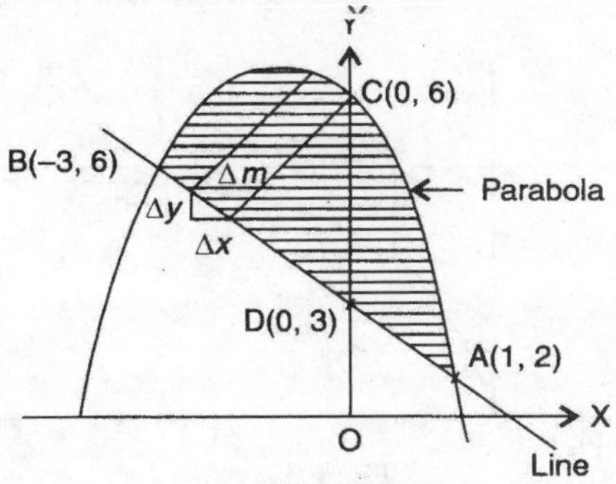

Fig. 6.35

$$= \frac{\sqrt{2}\pi}{2} \int_{-3}^{1} [x + (-x^2 - 3x + 6) - 3]^2 dx$$

$$= \frac{\pi}{\sqrt{2}} \int_{-3}^{1} (-x^2 - 2x + 3)^2 dx = \frac{256}{15}\sqrt{2}\pi$$

Example 5: Find the volume of the solid generated by revolving about the x- axis, the smaller area bounded by the circle $x^2 + y^2 = 2$ and the semicubical parabola $y^3 = x^2$.

Solution: The smaller area *OACB* bounded by the circle and parabola is shaded in the Fig. 6.36. Consider a typical vertical element whose length is $y_2 - y_1$ where y_2 and y_1 are the ordinates of the outer curve circle and the inner curve parabola respectively.

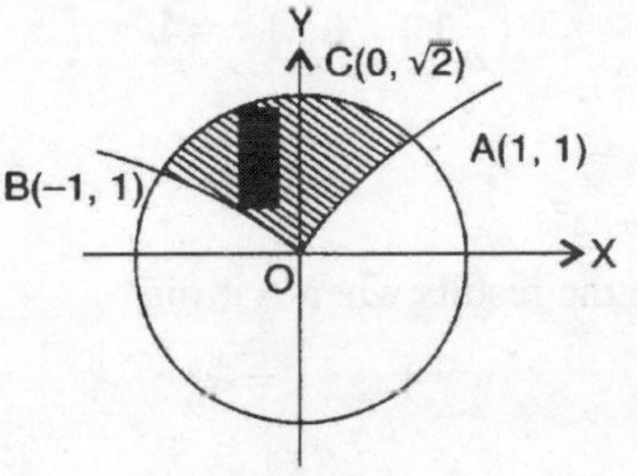

Fig. 6.36

This vertical element revolved about x-axis generates a circular ring whose outer and inner radii are the respective ordinates of the outer and inner curves. Thus the volume generated by $OACB$ by revolving about x-axis is

$$V = \left(\int \pi y_2^2 dy\right) - \left(\int \pi y_1^2 dy\right)$$

$$= \pi \int (y_2^2 - y_1^2) dy$$

$$V = \pi \int_{-1}^{1} [(2 - x^2) - (x^{\frac{4}{3}})] dx$$

$$= \pi \left[2x - \frac{x^3}{3} - \frac{3}{7}x^{\frac{7}{3}}\right]_{-1}^{1} = \frac{52}{21}\pi$$

Example 6: What is the volume generated by revolving the area enclosed by the loop of the curve $y^4 = x(4 - x)$ about x-axis.

Solution: Curve meets the x axis at $x = 0$ and $x = 4$ and is symmetric about the x-axis. The area OAO enclosed by the loop of the curve is shaded in Fig. 6.37. By circular disc method

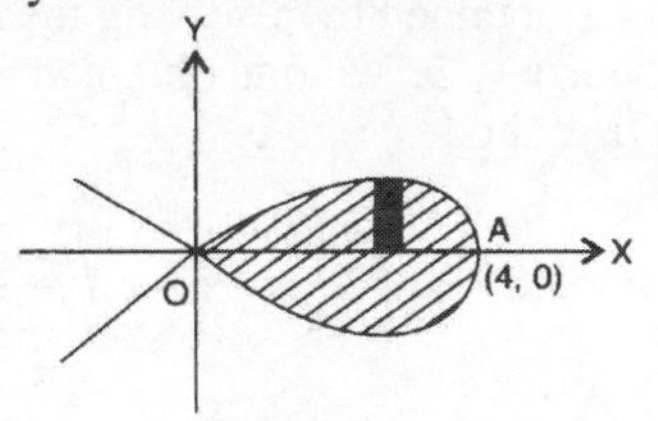

Fig. 6.37

$$V = \int \pi y^2 dx = \pi \int_0^4 \sqrt{x(4 - x)} dx$$

Put $\sqrt{x} = 2\cos t$

$$V = \pi \int_{-\frac{\pi}{2}}^{0} 2\cos t \cdot 2\sin t \cdot (-4) 2\cos t \cdot \sin t \, dt$$

$$V = 32\pi \int_0^{\frac{\pi}{2}} \cos^2 t \sin^2 t \, dt = 32\pi \cdot \frac{1 \cdot 1}{4 \cdot 2} \cdot \frac{\pi}{2} = 2\pi^2$$

Example 7: Determine the volume of the solid of revolution generated by revolving the loop of the curve $ay^2 = x^2(a - x)$ about the straight line $y = b$.

Solution: The loop lies between the points $O(0, 0)$ and $A(a, 0)$. Solving the given equation for y,

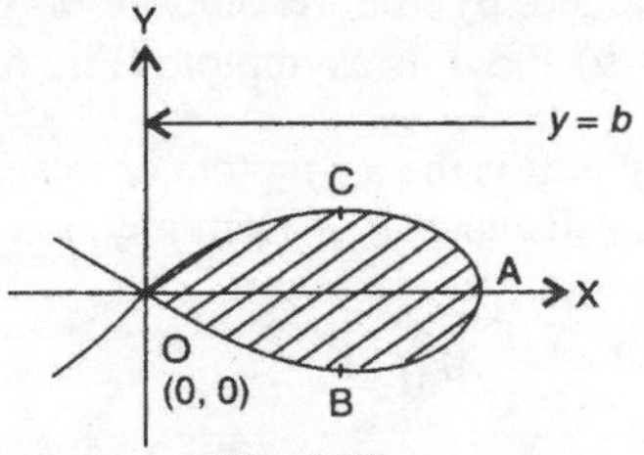

Fig. 6.38

$$y = \pm x\sqrt{\frac{a-x}{a}}$$

So the lower part of the loop below (x-axis) (see Fig. 6.38)

OBA is $\qquad y = -x\sqrt{\frac{a-x}{a}}$ and

the upper part of the loop above (x-axis)

OCA is $\qquad y = x\sqrt{\frac{a-x}{a}}$

The volume V obtained by revolving the loop about the line $y = b$ is the difference of the two volumes V_{OBA} and V_{OCA} obtained by revolving the arcs OBA and OCA about $y = b$. Volume obtained by revolving the arc OBA about $y = b$ is

$$V_{OBA} = \int_0^a \pi y^2 dx = \pi \int_0^a \left[b - \left(-x\sqrt{\frac{a-x}{a}}\right)\right]^2 dx$$

Similarly,

$$V_{OCA} = \int_0^a \pi y^2 dx = \pi \int_0^a \left[b - \frac{x\sqrt{a-x}}{\sqrt{a}}\right]^2 dx$$

$$V = V_{OBA} - V_{OCA} = \pi \int_0^a \frac{4bx(a-x)^{\frac{1}{2}}}{\sqrt{a}} dx$$

Put $x = a\cos^2 t$. As x varies from 0 to a, t varies from $\frac{\pi}{2}$ to 0

$$V = \frac{4\pi b}{\sqrt{a}} \int_{\frac{\pi}{2}}^0 a\cos^2 t\sqrt{a}\sin t \cdot 2a \cdot \cos t(-\sin t)dt$$

$$= 8\pi a^2 b \int_0^{\frac{\pi}{2}} \sin^2 t \cdot \cos^3 t\, dt = 8\pi a^2 b \cdot \frac{2}{5} \cdot \frac{1}{3}$$

$$V = \frac{16\pi a^2 b}{15}$$

Example 8: Find the volume of the solid of revolution obtained by the revolution of the curve $yx = a(x - b)$ about its asymptote (Fig. 6.39).

Solution: $y = a$ is the asymptote of the curve. The perpendicular distance PM from any point $P(x, y)$ on the curve to the asymptote is

$$PM = a - y$$

x varies from b to ∞.

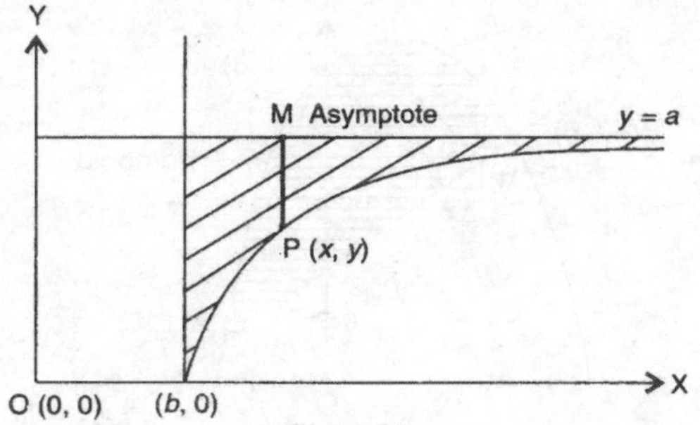

Fig. 6.39

The volume of solid of revolution obtained

$$V = \int \pi(PM)^2 dx = \int_b^\infty \pi(a-y)^2 dx$$

$$= \int_b^\infty \pi\left(a - \frac{a(x-b)}{x}\right)^2 dx = \pi a^2 \int_b^\infty \frac{b^2}{x^2} dx$$

$$= \pi a^2 b^2 \left(-\frac{1}{x}\right)_b^\infty = \pi a^2 b$$

Parametric Form

Example 1: Calculate the volume of the solid of revolution generated by revolving the hypocycloid

$$\left(\frac{x}{a}\right)^{\frac{2}{3}} + \left(\frac{y}{b}\right)^{\frac{2}{3}} = 1$$

a. about x-axis

b. about y-axis

Deduce the results when Astroid

$$x^{\frac{2}{3}} + y^{\frac{2}{3}} = a^{\frac{2}{3}}$$

is revolved.

c. about x-axis

d. about y-axis

Solution: The parametric equations of hypocycloid are

$$x = a\cos^3 t,\; y = b\sin^3 t \qquad (1)$$

The curve is symmetric about both x-axis and y-axis and meets them at the points $ABCD$ (Fig. 6.40).

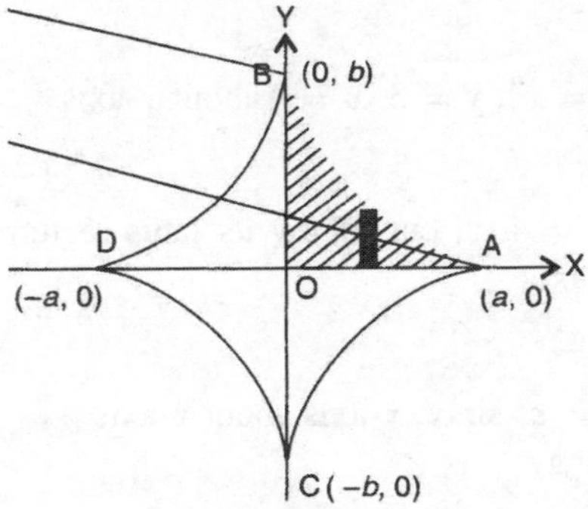

Fig. 6.40

a. *Revolution about x-axis*: The solid of revolution obtained by revolving the plane area $BACOB$ enclosed by the arc BAC is the same as the solid obtained by revolving the plane area $ABOA$ contained by the arc BA in the first quadrant. Also due to symmetry about y-axis, the required volume V is twice the volume obtained by revolving arc BA in the first quadrant. Thus

$$V = 2\int_0^a \pi y^2 dx = 2\int_0^a \pi y^2 \frac{dx}{dt} dt \qquad (2)$$

Using the parametric Equation (1)

$$\frac{dx}{dt} = 3a \cdot \cos^2 t(-\sin t)$$

and t varies from $\frac{\pi}{2}$ to 0 as x varies from 0 to a. With this (2) becomes

$$V = 2\int_{\frac{\pi}{2}}^0 \pi(b^2 \sin^6 t)\cdot(-3a\cos^2 t \cdot \sin t)dt$$

$$= 6\pi ab^2 \int_0^{\frac{\pi}{2}} \sin^7 t \cdot \cos^2 t\, dt$$

$$= 6\pi ab^2 \frac{6\cdot 4\cdot 2\cdot 1}{9\cdot 7\cdot 5\cdot 3} = \frac{32}{105}\pi ab^2$$

b. *Revolution about y-axis*: By similar arguments, volume generated by revolving hypocycloid about y-axis is

$$V = 2\cdot\int_0^b \pi x^2 dy = 2\pi\int_0^b x^2\frac{dy}{dt}dt$$

$$= 2\pi\int_0^{\frac{\pi}{2}} a^2\cos^6 t\cdot 3b\cdot\sin^2 t\cdot\cos t\, dt$$

$$= 6\pi a^2 b\int_0^{\frac{\pi}{2}} \sin^2 t\cos^7 t\, dt = 6\pi a^2 b\frac{1\cdot 6\cdot 4\cdot 2}{9\cdot 7\cdot 5\cdot 3}$$

$$= \frac{32\pi a^2 b}{105}$$

c. *Astroid* is special case of hypocycloid for $b = a$ volume generated by revolving astroid about

x-axis is $\qquad V = \frac{32}{105}a^3$

d. Due to *symmetry about both x and y-axis* volume generated by revolving astroid about y-axis is also $V = \frac{32}{105}a^3$.

Example 2: Determine the volume of solid of revolution generated by revolving the curve whose parametric equation are

$$x = 2t + 3,\ y = 4t^2 - 9$$

about x-axis for $t_1 = -\frac{3}{2}$ to $t_2 = \frac{3}{2}$.

Solution: Required volume V is

$$V = \int \pi y^2 dx = \int \pi y^2 \frac{dx}{dt} dt$$

using the parametric equations

$$= \int_{-\frac{3}{2}}^{\frac{3}{2}} \pi(4t^2 - 9)^2(2)dt = 4\pi\int_0^{\frac{3}{2}}(16t^4 + 81 - 72t^2)$$

$V = 1296\pi$

Aliter: Eliminating t, the equation of the curve in the cartesian form is $y = x^2 - 6x$ which meet x-axis at $x = 0$ and $x = 6$ and is shown in the Fig. 6.41. Revolving the arc OBA about x-axis

$$V = \pi\int_0^6 y^2 dx = \pi\int_0^6 (x^2 - 6x)^2 dx$$

$= 1296\pi$.

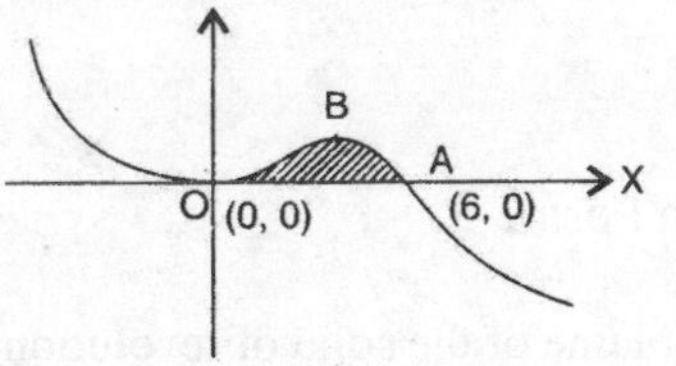

Fig. 6.41

Polar Form

Example 1: Find the volume generated by the revolution of the curve $r = 2a\cos\theta$ about the initial line.

Solution: Volume generated by revolution about initial line is given by

$$V=\int_{\theta_1}^{\theta_2}\frac{2}{3}\pi r^3\sin\theta\,d\theta$$

$$=\frac{2}{3}\pi\int_0^{\frac{\pi}{2}}(2a\cos\theta)^3\sin\theta\,d\theta$$

$$=\frac{16\pi a^3}{3}\int_0^{\frac{\pi}{2}}\cos^3\theta\cdot\sin\theta\,d\theta$$

$$=\frac{16\pi a^3}{3}\cdot\frac{2}{4}\cdot\frac{1}{2}=\frac{4\pi a^3}{3}$$

Example 2: The arc of the cardioid: $r=a(1+\cos\theta)$ included between $\theta=-\pi/2$ and $\theta=\pi/2$ is rotated about the line $\theta=\pi/2$. Find the volume of the solid of revolution.

Solution: Volume of solid generated by revolution about the line $\theta=\pi/2$ is

$$V=\int_{\theta_1}^{\theta_2}\frac{2}{3}\pi r^3\cos\theta\,d\theta$$

$$=\frac{2}{3}\pi\int_{-\frac{\pi}{2}}^{\frac{\pi}{2}}a^3(1+\cos\theta)^3\cdot\cos\theta\,d\theta$$

$$=\frac{2\pi a^3}{3}\cdot 2\int_0^{\frac{\pi}{2}}(\cos^4\theta+3\cos^3\theta+3\cos^2\theta+\cos\theta)d\theta$$

$$=\frac{4\pi a^3}{3}\left[1+\frac{3}{4}\cdot\frac{1}{2}\cdot\frac{\pi}{2}+3\cdot\frac{2}{3}+3\cdot\frac{1}{2}\cdot\frac{\pi}{2}\right]$$

$$=\frac{\pi a^3}{4}[16+5\pi]$$

EXERCISE

Cartesian Form

Find the volume of the solid of revolution generated by revolving the plane area bounded by the given curves about the indicated line.

1. $y=x^3$, $y=0$, $x=2$ about x-axis

Ans. $\frac{128\pi}{7}$

2. $y=2x+1$, $y=0$, $x=1$, $x=2$ about x-axis

Ans. $\frac{49\pi}{3}$

3. $y=x^3$, $y=8$, $x=0$ about y-axis

Ans. $\frac{96\pi}{5}$

4. $y^2=4ax$, cut off by its latus rectum, about y-axis

Ans. $\frac{4\pi a^3}{5}$

5. $y=e^x\sin x$, x-axis about x-axis

Ans. $\frac{\pi}{8}(e^{2\pi}-1)$

6. Parabola $y^2=8x$, latus rectum $x=2$
   a. About x-axis
   Hint: $V=\int_0^2\pi y^2\,dx=\int_0^2 8x\,dx$
   Ans. 16π
   b. About its latus rectum $x=2$
   Hint 1: $V=\int_{-4}^{4}\pi(2-x)^2dy$ by cylindrical disc method.
   Ans. $\frac{256}{15}\pi$
   Hint 2: By cylindrical shell method
   $$V=\int_0^2(2\pi(2-x))\cdot\left(4\sqrt{2x}\right)\cdot dx$$
   Ans. $\frac{256}{15}\pi$

7. $y=1-x^2$, $y=0$ about $x=1$

Ans. $\frac{8\pi}{3}$

8. Circle $x^2+y^2=a^2$ about a diameter

Ans. $\frac{4}{3}\pi a^3$

9. $2x+y=2$, $x=0$, $y=0$ about y-axis

Ans. $\frac{2\pi}{3}$

10. Parabola $y^2=4ax$, $x=0$, $x=ah$ about x-axis

Ans. $2\pi ah^2$

11. **a.** Ellipse $\frac{x^2}{a^2}+\frac{y^2}{b^2}=1$ about x-axis

Ans. $\frac{4}{3}\pi ab^2$

b. About y-axis

Ans. $\frac{4}{3}\pi a^2b$

12. $y=x\sin mx$, $x=0$, $x=\frac{2\pi}{m}$ about x-axis

Ans. $\frac{\pi^2}{6}\frac{(8\pi^2-3)}{m^3}$

13. $\sqrt{x}+\sqrt{y}=\sqrt{a}$, $x=0$, $y=0$ about x-axis

Ans. $\frac{\pi a^3}{12}$

14. $y^2 = x(2x-1)^2$ about x-axis

Ans. $\frac{\pi}{48}$

15. $y = 2x, y = x, x + y = 6$ about x-axis

Hint: The plane region $OBAC$ where $O(0,0), B(3,3), A(2,4), C\left(\frac{3}{2}, 3\right)$; is to be divided into two parts OBC and CBA and each volume to be calculated by shell method.

$$V_{OBC} = \int_0^3 (2\pi y)\cdot\left(y - \frac{y}{2}\right)dy$$

$$V_{ABC} = \int_3^4 (2\pi y)\cdot\left[(6-y) - \frac{y}{2}\right]\cdot dy$$

Ans. 14π

Bounded by two curves: $\int_a^b \pi(y_2^2 - y_1^2)dx$:

16. $y = x^2, y = 2x$ about x-axis

Hint: $V = \int_0^2 \pi\left[(2x)^2 - (x^2)^2\right]dx$

Element of volume generated by a vertical element of area is a circular ring with outer radius $y_2 = 2x$ and inner radius $y_1 = x^2$.

Ans. $\frac{64\pi}{15}$

17. $y = x^2, y = x$ about y-axis

Hint: $V = \int_0^1 \pi\left[(\sqrt{y})^2 - (y)^2\right]dy$

As above outer radius $x_2 = \sqrt{y}$ and inner radius $x_1 = y$.

Ans. $\frac{\pi}{6}$

18. $y^2 = ax^3, x^2 = ay^3$ about x-axis

Hint: $V = \int_0^{1/a} \pi\left[\left(\frac{x^{\frac{2}{3}}}{a^{\frac{1}{3}}}\right)^2 - \left(\sqrt{a}x^{\frac{3}{2}}\right)^2\right]dx$

since points of intersection of the two curves are $(0,0), (1/a, 1/a)$.

Ans. $\frac{5\pi}{28a^3}$

19. $y^2 = 4ax, x = 2$, about y-axis

Hint: $V = \int_{-4}^{4}\pi(2)^2dy - \int_{-4}^{4}\pi(x)^2dy$

y varies from -4 to 4 since $y^2 = 4ax$ and $x = 2$ intersects at $(2,4)$ and $(2,-4)$.

Ans. $\frac{128}{5}\pi$

20. $y = 4x - x^2$, x-axis, about the line $y = 6$

Hint: $V = \pi\int_0^4\left[(6)^2 - (6-y)^2\right]dx =$

$\pi\int_0^4(48x - 28x^2 + 8x^3 - x^4)dx.$

Ans. $1408\pi/15$

21. $y^2 = 4ax, 27ay^2 = 4(x-2a)^3$, about x-axis

Hint: Points of intersection of the two curves are $B(8a, 4a\sqrt{2}), C(8a, -4a\sqrt{2})$

second curve meets the x-axis when $x = 2a$ (i.e., $A(2a, 0)$ is point on the 2nd curve).

volume = volume generated by the area under arc OB

−volume generatd by the area under arc AB

$$= \int_0^{8a}\pi(4ax)dx - \int_{2a}^{8a}\pi\frac{4(x-2a)^3}{27a}dx$$

Ans. $80\pi a^3$

Loop Volumes

Find the volume of the solid of revolution generated by revolving the plane area enclosed by the loop of the following curve about the indicated line

22. $y^2(a-x) = x^2(a+x)$, about the x-axis

Hint: $V = \int_{-a}^{0}\pi y^2dx = \pi\int_{-a}^{0}\frac{x^2(a+x)}{(a-x)}dx$

$= \pi\int_{-a}^{0}\left[-x^2 - 2ax - 2a^2 + \frac{2a^3}{(a-x)}\right]dx$

Ans. $2\pi a^3\left[\ln 2 - \frac{2}{3}\right]$

23. $y^2(a+x) = x^2(3a-x), 0 \le x \le 3a,$ about x-axis

Hint: $V = \int_0^{3a}\pi y^2dx = \pi\int_0^{3a}\frac{x^2(3a-x)}{(a+x)}dx$

$= \pi\int_0^{3a}\left(-x^2+4ax-4a^2+\frac{4a^3}{x+a}\right)dx$

Ans. $\pi a^3(4\ln 4 - 3)$

24. $2ay^2 = x(x-a)^2$, about x-axis

Ans. $\frac{\pi a^3}{24}$

25. $y^2 = x^2(1-x^2)$, about x-axis

Ans. $\frac{2\pi}{15}$

26. $a^2x^2 = y^3(2a-y)$, about y-axis

Ans. $\frac{8}{5}\pi a^3$

27. $a^2y^2 = x^2(2a-x)(x-a)$, about x-axis

Ans. $\frac{23\pi a^3}{60}$

28. $2ay^2 = x(x-a)^2$, about the line $y = a$
Hint: Loop lies between $O(0, 0)$ and $A(a, 0)$

Since, $y = \pm\frac{\sqrt{x}(x-a)}{\sqrt{2a}}$

lower part of the loop (below x-axis)

be OBA : $y = \frac{-\sqrt{x}(a-x)}{\sqrt{2a}}$

upper part OCA : $y = \frac{\sqrt{x}(a-x)}{\sqrt{2a}}$

volume = volume$_{OBA}$ − volume$_{OCA}$

$$= \pi\int_0^a\left[a - \left(-\frac{\sqrt{x}(a-x)}{\sqrt{2a}}\right)\right]^2 dx$$

$$-\pi\int_0^a\left[a - \frac{\sqrt{x}(a-x)}{\sqrt{2a}}\right]^2 dx$$

Ans. $\frac{8\sqrt{2}\pi a^3}{15}$

29. $ay^2 = x^2(a-x)$, about x-axis

Ans. $\pi a^3/12$

30. $y^2 = x^2(1-x^2)$ about y-axis

Hint: Use shell method: $V = \int_0^1 (2\pi x)\cdot(2y)\cdot dx$

Ans. $\pi^2/4$.

Any Axis of Revolution: Asymptote

Find the volume of the solid obtained by the revolution of the area enclosed by the curve about its asymptote (or any axis)

31. Cissoid $y^2(2a-x) = x^3$

Hint: $x = 2a$ is the asymptote to the given curve.

MP = perpendicular distance between curve and asymptote = $(2a - x)$

$$V = 2\pi\int_0^\infty (2a-x)^2 dy$$

$$= 2\pi\int_0^{2a}(3a-x)\sqrt{(2a-x)}\sqrt{x}dx$$

Ans. $2\pi^2 a^3$

32. $xy^2 = a^2(a-x)$

Hint: y-axis is the asymptote.
Curve is symmetric above x-axis

$$V = \text{twice}\int_0^\infty \pi x^2 dy = 2\pi a^6\int_0^\infty (y^2+a^2)^{-2}dy$$

Ans. $\pi^2 a^3/2$

33. $(a^2 + x^2)y = a^3$

Hint: x-axis is the asymptote, curve symmetric about y-axis

$$V = \text{twice}\int_0^\infty \pi y^2 dx = 2\pi a^6\int_0^\infty (a^2+x^2)^{-2}dx$$

Ans. $\pi^2 a^3/2$

34. $y^2(2a-x) = x^3$, $a > 0$

Hint: $x = 2a$ is the asymptote.
Curve is symmetric about x-axis

$$V = \text{twice}\int_0^\infty \pi(2a-x)^2 dy. \text{ Put } x = 2a\cos^2 t$$

Ans. $2\pi^2 a^3$

35. Witch of Agnesi: $xy^2 = 4a^2(2a-x)$, $a > 0$

Hint: y-axis is the asymptote.
Curve is symmetric about x-axis

$$V = \text{twice}\int_0^\infty \pi x^2 dy = 128\pi a^6\int_0^\infty \frac{dy}{(y^2+4a^2)^2}$$

Ans. $4\pi^2 a^3$

36. Cissoid $y^2(a-x) = a^2 x$

Hint: $x = a$ is the asymptote. Curve is symmetric about x-axis

$$V = \text{twice}\int_0^\infty \pi(a-x)^2 dy = 2\pi a^2\int_0^\infty \frac{dy}{(y^2+a^2)^2}$$

Ans. $\pi^2 a^3/2$

37. Parabola $y^2 = 4ax$, cut-off by its latus rectum, revolved about its directrix.

Hint: Curve is symmetric about x-axis. Latus rectum meets parabola at $x = 2a$ and $x = -2a$

PM = perpendicular distance from parabola to directrix = $a + x$

$$V = \text{twice}\int_0^{2a}\pi(ax)^2 dy$$

$$= 2\pi\int_0^{2a}\left(a + \frac{y^2}{4a}\right)^2 dy$$

Ans. $128\pi a^3/15$.

Parametric Form

38. Find the volume of the solid obtained by revolving about x-axis, the plane area enclosed by one arch of the cycloid

$$x = a(t + \sin t)$$
$$y = a(1 + \cos t)$$

and the x-axis.

Hint: t varies from 0 to π.

Ans. $5\pi a^3$

39. Determine the volume the (reel shaped) solid formed by the revolution of the cycloid

$$x = a(t + \sin t)$$
$$y = a(1 - \cos t)$$

about x-axis.

Hint: t varies from 0 to π.

Ans. $\pi^2 a^3$

40. The cycloid $x = a(t - \sin t)$, $y = a(1 - \cos t)$ rotates about its base. Calculate the volume of the solid generated.

Hint: t varies from 0 to 2π.

Ans. $5\pi^2 a^3$

41. Find the volume of solid generated by revolving the loop of the curve

$$x = t^2, \; y = t - \frac{t^3}{3}$$

Hint: t varies from 0 to $\sqrt{3}$ (since $y = \frac{t^3 - 3t}{3} = 0$)

Ans. $3\pi/4$

42. The tractrix $x = a \cos t + a \ln \tan \frac{t}{2}$, $y = a \sin t$ is revolved about its asymptote. Find the volume of the solid so generated.

Hint: x-axis is the asymptote. Curve is symmetric about both x-axis and y-axis. t varies from 0 to $\pi/2$ for one arc

$$\text{Volume} = \text{twice} \int_0^{\frac{\pi}{2}} \pi y^2 \frac{dx}{dt} dt$$
$$= 2\int_0^{\frac{\pi}{2}} \pi a^3 \cos^2 t \sin t\, dt$$

Ans. $2\pi a^3/3$

43. Prove that the volume of solid generated by revolving the cissoid

$$x = 2a \sin^2 t, \qquad y = 2a \frac{\sin^3 t}{\cos t}$$

(with $-\pi/2 < t < \pi/2$) about its asymptote, is $2\pi a^3$.

44. If the ellipse $x = a \cos t$, $y = b \sin t$ is revolved about the line $x = 2a$, show that the volume of the solid generated is $4\pi^2 a^2 b$.

45. If one arch of the cycloid

$$x = a(t + \sin t), \qquad y = a(1 - \cos t)$$

(with $-\pi < t < \pi$) is revolved about y-axis, prove that the volume of the solid generated is $\frac{\pi^3}{6}(9a^3 - 16)$.

Polar Form

Find the volume of the solid generated by revolving the given curve in polar form about the indicated line:

46. Cardioid $r = a(1 + \cos\theta)$ about the initial line

Hint: θ varies from 0 to π.

Ans. $8\pi a^3/3$

47. Cardioid $r = a(1 - \cos\theta)$ about the initial line

Hint: θ varies from 0 to π.

Ans. $8\pi a^3/3$

48. $r^2 = a^2 \cos 2\theta$ about the initial line

Hint: Curve is symmetric about both x-axis and y-axis. θ varies from 0 to $\frac{\pi}{4}$.

volume = twice volume generated by one arch in the 1st quadrant

$$= 2\int_0^{\frac{\pi}{4}} \frac{2}{3}\pi r^3 \sin\theta\, d\theta$$
$$V = 2\pi a^3 \int_0^{\pi/4} \sin^2\theta (3\sin\theta - 4\sin^3\theta) \times \sqrt{\cos 2\theta}\, d\theta$$

Ans. $\frac{\pi a^3}{2}\left[\frac{1}{\sqrt{2}} \ln(\sqrt{2}+1) - \frac{1}{3}\right]$

49. $r^2 = a^2 \cos 2\theta$, about the line $\theta = \frac{\pi}{2}$

Hint:

$$V = \text{twice} \int_0^{\frac{\pi}{4}} \frac{2}{3}\pi r^3 \cos\theta \, d\theta, \text{put} \sqrt{2}\sin\theta = \sin\phi$$

$$= \frac{4\pi a^3}{3\sqrt{2}} \int_0^{\frac{\pi}{2}} \cos^4\phi \, d\phi$$

$$= \frac{4\pi a^3}{3\sqrt{2}} \cdot \frac{3\cdot 1}{4\cdot 2} \cdot \frac{\pi}{2}$$

Ans. $\frac{\pi^2 a^3}{4\sqrt{2}}$

50. $r^2 = a^2 \cos 2\theta$ about a tangent at the pole

Ans. $\frac{\pi a^3}{12}\left[\frac{3}{\sqrt{2}} \ln\,(1+\sqrt{2}) - 1\right]$

51. $r^2 = a^2 \cos\theta$ about initial line

Ans. $\frac{8\pi a^3}{15}$

52. Loop of the curve $r = a \cos 3\theta$ lying between $\theta = -\pi/6$ and $\theta = \pi/6$ about the initial line.

Ans. $19\pi a^3/960$

53. Area lying within the cardioid

$$r = 2a(1+\cos\theta)$$

and outside the parabola.

$$r = \frac{2a}{1+\cos\theta}$$

about the initial line (Fig. 6.42).

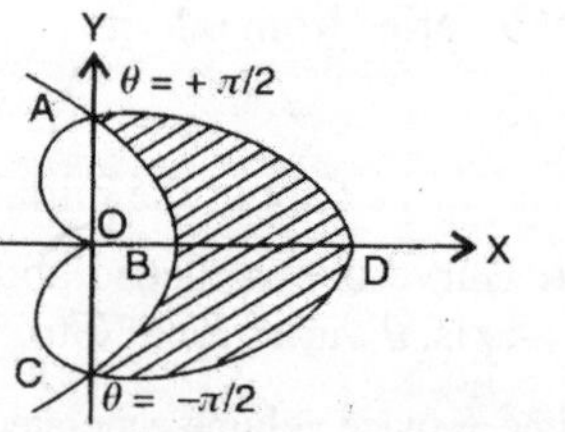

Fig. 6.42

Hint: Points of intersection are $\theta = \frac{-\pi}{2}$ and $\frac{\pi}{2}$. Both curves are symmetric about x-axis.

volume = volume generated by area under the cardioid $ODAO$

−volume generated by area under the parabola $OBAO$

$$= \int_0^{\frac{\pi}{2}} \frac{2}{3}\pi r_2^3 \sin\theta \, d\theta - \int_0^{\frac{\pi}{2}} \frac{2}{3}\pi r_1^3 \sin\theta \, d\theta$$

where r_2 is r of cardioid, r_1 is r of parabola.

Ans. $18\pi a^3$

54. $r = a + b\cos\theta$ with $a < b$ about the initial line.

Hint: θ varies from 0 to π, curve symmetric about the initial line

$$V = \frac{2\pi}{3} \int_0^{\pi} r^3 \sin\theta \, d\theta$$

$$= \frac{2\pi}{3} \int_0^{\pi} (a + b\cos\theta)^3 \sin\theta \, d\theta$$

$$V = \frac{2\pi}{3} \int_{a+b}^{a-b} t^3 \frac{dt}{b} \quad \text{with} \quad a + b\cos\theta = t$$

Ans. $4\pi a(a^2 + b^2)/3$

6.5 AREA OF THE SURFACE OF A SOLID OF REVOLUTION

Let $y = f(x)$ be a plane curve c in the xy-plane included between the ordinates $x = a$ and $x = b$. Let P(x, y) be any point on c (Fig. 6.43). When the chord $PQ = \Delta S$ is revolved about the x-axis, a solid of revolution is generated which is the frutum of cone of slant height $PQ = \Delta S$ and radii y and $y + \Delta y$. Hence the area (of the surface of solid of revolution of) of this elementary belt is $2\pi y \Delta S$. Dividing the curve c into n parts and summing up the areas, of these elementary belts from $x = a$ to $x = b$, we get the surface area of solid of revolution.

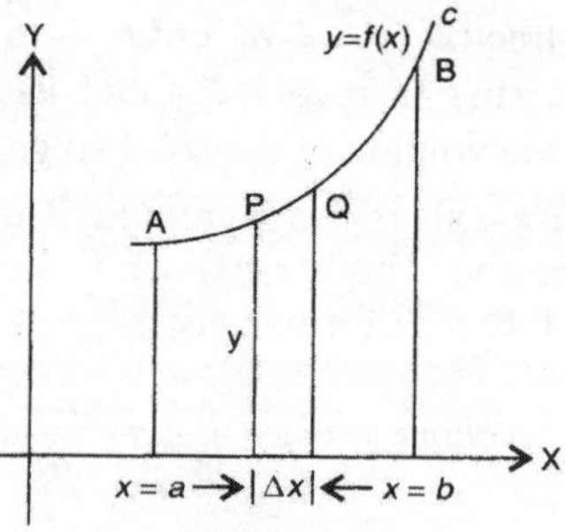

Fig. 6.43

Cartesian Form

a. Area of the surface of the solid of revolution generated by revolving the arc AB of the curve $y = f(x)$ about the x-axis is given by

$$S = \int_{AB} 2\pi y\, ds = \int_{x=a}^{x=b} 2\pi y \sqrt{1 + \left(\frac{dy}{dx}\right)^2}\, dx$$

$$\left[\text{Note that } \frac{ds}{dx} = \sqrt{1 + \left(\frac{dy}{dx}\right)^2}\right]$$

b. Area of the surface generated by revolving an arc AB of the curve $x = g(y)$ about y-axis is

$$S = \int_{AB} 2\pi x\, ds = \int_{y=c}^{y=d} 2\pi x \sqrt{1 + \left(\frac{dx}{dy}\right)^2}\, dy$$

$$\left[\text{Note that } \frac{ds}{dy} = \sqrt{1 + \left(\frac{dx}{dy}\right)^2}\right]$$

Parametric Form: $x = x(t)$, $y = y(t)$

a. About x-axis:

$$S = \int_{t=t_1}^{t_2} 2\pi y(t) \cdot \sqrt{\left(\frac{dx}{dt}\right)^2 + \left(\frac{dy}{dt}\right)^2}\, dt$$

b. About y-axis

$$S = \int_{t=t_1}^{t_2} 2\pi x(t) \sqrt{\left(\frac{dx}{dt}\right)^2 + \left(\frac{dy}{dt}\right)^2}\, dt$$

Polar Form: $r = f(\theta)$

a. About the x-axis: initial line $\theta = 0$

$$S = \int_{\theta=\theta_1}^{\theta_2} 2\pi y \frac{ds}{d\theta} d\theta$$

$$= \int_{\theta=\theta_1}^{\theta_2} 2\pi (r \sin\theta) \sqrt{r^2 + \left(\frac{dr}{d\theta}\right)^2}\, d\theta$$

Here replace r by $f(\theta)$.

b. About the y-axis perpendicular line through the pole $\theta = \pi/2$

$$S = \int_{\theta=\theta_1}^{\theta_2} 2\pi x \frac{ds}{d\theta} d\theta$$

$$= \int_{\theta=\theta_1}^{\theta_2} 2\pi (r \cos\theta) \sqrt{r^2 + \left(\frac{dr}{d\theta}\right)^2}\, d\theta$$

Here replace r by $f(\theta)$.

About Any (axis) Line L

$$S = \int 2\pi (PM) ds$$

where PM is the perpendicular distance from a point P of the given curve to the axis of revolution (Line L) (Fig. 6.44)

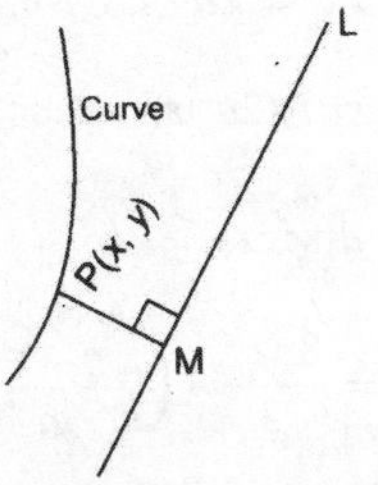

Fig. 6.44

a. *Limits for* x: If limits for x are given as $x = a$ to $x = b$ then

$$S = \int_{x=a}^{x=b} 2\pi (PM) \sqrt{1 + \left(\frac{dy}{dx}\right)^2}\, dx$$

Here PM is expressed in terms of x.

b. *Limits for* y: If limits for y are given as $y = c$ to $y = d$ then

$$S = \int_{y=c}^{y=d} 2\pi (PM) \sqrt{1 + \left(\frac{dx}{dy}\right)^2}\, dy$$

Here PM is expressed in terms of y.

WORKED OUT EXAMPLES

Example 1: Find the area of the surface of the solid of revolution generated by revolving the parabola $y^2 = 4ax, 0 \le x \le 3a$ about the x-axis.

Solution: $2yy' = 4a$, $y' = \frac{2a}{y}$, $1 + y'^2 = 1 + \left(\frac{2a}{y}\right)^2$ $= \frac{y^2+4a^2}{y^2}$.

$\sqrt{1 + y'^2} = \frac{4ax+4a^2}{y^2} = \frac{2\sqrt{a}}{y}\sqrt{a+x}$

The area S of the curved surface of the solid generated by revolving the arc of the parabola $y^2 = 4ax$

included between the ordinates $x = 0$ and $x = 3a$, about the x-axis is given by

$$S = \int 2\pi y \sqrt{1 + \left(\frac{dy}{dx}\right)^2}\, dx$$

$$= 2\pi \int_0^{3a} y\left(\frac{2\sqrt{a}}{y} \cdot \sqrt{a+x}\right) dx$$

$$= 4\pi\sqrt{a} \int_0^{3a} \sqrt{a+x}\, dx$$

put $x = a\tan^2 t$, $dx = 2a \cdot \tan t \cdot \sec^2 t\, dt$

$$S = 4\pi\sqrt{a} \int (\sqrt{a}\sec t)(2a \tan t \sec^2 t\, dt)$$

$$= 8\pi a^2 \int \sec^2 t\, d(\sec t)$$

$$S = 8\pi a^2 \cdot \frac{\sec^3 t}{3} = \frac{8\pi a^2}{3} \cdot \left(1 + \frac{x}{a}\right)^{\frac{3}{2}}\Bigg|_{x=0}^{3a} = \frac{56\pi a^2}{3}.$$

Example 2: Determine the surface area of the paraboloid generated by revolving the curve $y = x^2$ included between $x = 0$ and $x = \frac{6}{5}$ about y- axis.

Solution: $y' = 2x$, $\sqrt{1 + y'^2} = \sqrt{1 + 4x^2}$

$$\text{Surface area} = S = \int_0^{\frac{6}{5}} 2\pi x \sqrt{1 + \left(\frac{dy}{dx}\right)^2} \cdot dx$$

$$S = \int_0^{\frac{6}{5}} 2\pi x\sqrt{1 + 4x^2}\, dx$$

put $1 + 4x^2 = t$

$$= \frac{2\pi}{8} \int t^{\frac{1}{2}} dt = \frac{\pi}{6} t^{\frac{3}{2}}$$

$$= \frac{\pi}{6}(1 + 4x^2)^{\frac{3}{2}}\Big|_0^{\frac{6}{5}}$$

$$S = \frac{1036\pi}{375}$$

Example 3: Show that the area of the surface generated when the loop of the curve

$$9ay^2 = x(3a - x)^2$$

revolves about the x-axis is $3\pi a^2$.

Solution: Curve is symmetric about x-axis. The loop of the curve lies between $x = 0$ and $x = 3a$ (Fig. 6.45).

Differentiating the equation of the curve

$$18ay\, y' = (3a - x)^2 - 2x(3a - x)$$

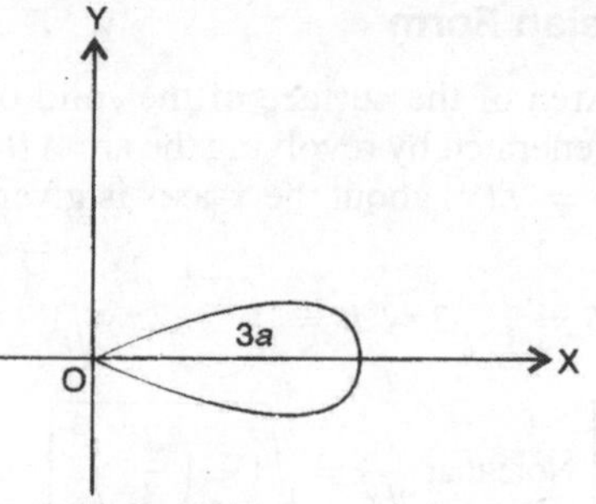

Fig. 6.45

$$1 + y'^2 = \frac{1}{18^2 a^2 y^2}\Big[(3a - x)^2 - 2x(3a - x) + 18^2 a^2 y^2\Big]$$

$$\sqrt{1 + y'^2} = \frac{1}{18ay}\Big[(3a - x)^2 9(a - x)^2 + 36ax(3a - x)^2\Big]^{\frac{1}{2}}$$

$$= \frac{3}{18ay}(3a - x)(a + x)$$

Surface area obtained by revolution about x-axis is

$$S = \int 2\pi y \sqrt{1 + \left(\frac{dy}{dx}\right)^2}\, dx$$

$$= \int_0^{3a} 2\pi y \cdot \frac{3}{18ay}(3a - x)(a + x) dx$$

$$S = \frac{\pi}{3a}\left[3a^2 x + 2a\frac{x^2}{2} - \frac{x^3}{3}\right]_0^{3a} = 3\pi a^2$$

Example 4: Calculate the area of the surface of revolution generated by revolving the cardioid $x = 2\cos\theta - \cos 2\theta$; $y = 2\sin\theta - \sin 2\theta$ about the x-axis.

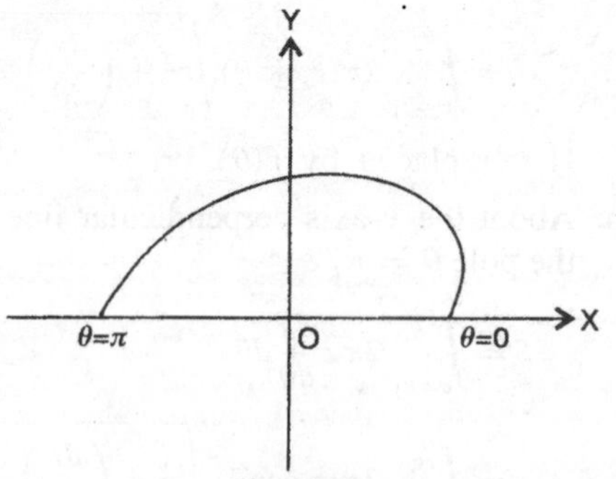

Fig. 6.46

Solution: The required surface is generated by revolving the arc of the curve from $\theta = 0$ to $\theta = \pi$ (Fig. 6.46).

$$\frac{dx}{d\theta} = -2\sin\theta + 2\sin 2\theta,$$

$$\frac{dy}{d\theta} = 2\cos\theta - 2\cos 2\theta$$

$$\left(\frac{dx}{d\theta}\right)^2 + \left(\frac{dy}{d\theta}\right)^2 = 8(1 - \sin\theta\cdot\sin 2\theta - \cos\theta\cdot\cos 2\theta)$$

$$= 8(1 - \cos\theta)$$

The area of surface by revolving about x-axis is

$$S = \int_{\theta_1}^{\theta_2} 2\pi\, y(\theta)\cdot\sqrt{\left(\frac{dx}{d\theta}\right)^2 + \left(\frac{dy}{d\theta}\right)^2}\, d\theta$$

$$= 2\pi\int_0^{\pi}(2\sin\theta - \sin 2\theta)2\sqrt{2}\sqrt{(1-\cos\theta)}d\theta$$

$$= 8\sqrt{2}\pi\int_0^{\pi}\sin\theta(1-\cos\theta)^{\frac{3}{2}}d\theta$$

$$S = \frac{16\sqrt{2}}{5}\pi(1-\cos\theta)^{\frac{5}{2}}\Big|_0^{\pi} = \frac{128\pi}{5}$$

Example 5: Find the area of the surface generated by revolving the curve with parametric equations

$$x(t) = 3t(t-2),\ y(t) = 8t^{\frac{3}{2}}$$

with $0 \le t \le 1$, about the y-axis.

Solution: $\frac{dx}{dt} = 6t - 6$, $\frac{dy}{dt} = 12t^{\frac{1}{2}}$

$$\sqrt{\left(\frac{dx}{dt}\right)^2 + \left(\frac{dy}{dt}\right)^2} = \sqrt{36(t-1)^2 + 144t}$$

$$= \sqrt{36(t+1)^2} = 6(t+1)$$

Surface area obtained by revolving about y-axis is

$$S = \int 2\pi\, x(t)\sqrt{\left(\frac{dx}{dt}\right)^2 + \left(\frac{dy}{dt}\right)^2}\, dt$$

$$= 2\pi\int_0^1 [3t(t-2)][6(t+1)]dt$$

$$S = 12\pi\left[\frac{3t^4}{4} - \frac{3t^3}{3} - \frac{6t^2}{2}\right]_0^1 = 39\pi$$

Example 6: The arc of the cardioid

$$r = a(1+\cos\theta)$$

included between $-\frac{\pi}{2} \le \theta \le \frac{\pi}{2}$ is rotated about the line $\theta = \frac{\pi}{2}$. Show that the area of the surface thus generated is $48\sqrt{2}\pi a^2/5$.

Solution: Cardioid is symmetric about the initial line OB. As θ varies from $-\frac{\pi}{2}$ to $\frac{\pi}{2}$, the part of arc of the cardioid is ABC (Fig. 6.47). Since this arc ABC is revolved about the line AOC $\left(\theta = \frac{\pi}{2}\right)$, the area of the surface of the solid of revolution generated is twice the area of the surface generated by revolving the

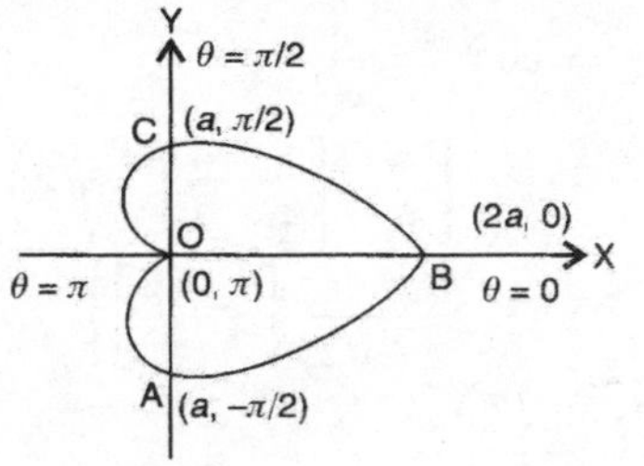

Fig. 6.47

arc BC about the line $\theta = \frac{\pi}{2}(OC)$. Thus the area of the surface of the solid of revolution generated by rotating the arc ABC (for which θ varies from $-\frac{\pi}{2}$ to $\frac{\pi}{2}$) about the line $\theta = \frac{\pi}{2}$ (OC) is

$$S = \text{twice}\int_0^{\frac{\pi}{2}} 2\pi\, x\sqrt{r^2 + \left(\frac{dr}{d\theta}\right)^2}\, d\theta$$

$$S = 2\int_0^{\frac{\pi}{2}} 2\pi(r\cos\theta)\sqrt{r^2 + \left(\frac{dr}{d\theta}\right)^2}\, d\theta$$

Differentiating $\frac{dr}{d\theta} = -a\sin\theta$

$$r^2 + \left(\frac{dr}{d\theta}\right)^2 = a^2(1+\cos\theta)^2 + a^2\sin^2\theta$$

$$\sqrt{r^2 + \left(\frac{dr}{d\theta}\right)^2} = \sqrt{a^2 2(1+\cos\theta)} = a\sqrt{2(1+\cos\theta)}$$

$$= a\sqrt{2\cdot 2\cdot\cos^2\frac{\theta}{2}} = 2a\cos\frac{\theta}{2}$$

Substituting this value and replacing r by $a(1+\cos\theta)$

$$S = 4\pi\int_0^{\frac{\pi}{2}}[a(1+\cos\theta)](\cos\theta)\left(2a\cos\frac{\theta}{2}\right)d\theta$$

$$= 8\pi a^2\int_0^{\frac{\pi}{2}} 2\cos^2\frac{\theta}{2}\left(\cos^2\frac{\theta}{2} - \sin^2\frac{\theta}{2}\right)\cos\frac{\theta}{2}d\theta$$

$$= 32\pi a^2 \left[\int_0^{\frac{\pi}{4}} \cos^5 t \, dt - \int_0^{\frac{\pi}{4}} \cos^3 t \cdot \sin^2 t \, dt \right]$$

where $\frac{\theta}{2} = t$, $d\theta = 2\,dt$, t : varies from 0 to $\frac{\pi}{4}$

$$= 32\pi a^2 \left[\frac{\cos^4 t \cdot \sin t}{5} + \frac{4}{5} \left\{ \frac{\sin t \cdot \cos^2 t}{3} + \frac{2}{3} \sin t \right\} \right]_0^{\frac{\pi}{4}}$$

$$-32\pi a^2 \left[\frac{\sin^3 t \cdot \cos^2 t}{5} + \frac{2}{5} \frac{\sin^3 t}{3} \right]_0^{\frac{\pi}{4}}$$

Since $\sin \frac{\pi}{4} = \cos \frac{\pi}{4} = \frac{1}{\sqrt{2}}$

$$S = 32\pi a^2 \left\{ \frac{1}{5\sqrt{2}} \left[\frac{1}{4} + \frac{4}{3} \left(\frac{1}{2} + 2 \right) \right] - \frac{1}{2\sqrt{25}} \left[\frac{1}{2} + \frac{2}{3} \right] \right\}$$

$$S = \frac{48\sqrt{2}\pi a^2}{5}$$

Exercise

Cartesian Form

1. Find the surface area of a sphere of radius a.

Hint:

$$S = 2\pi \int_{-a}^{a} y \sqrt{1 + \left(\frac{dy}{dx}\right)^2} \, dx$$

$$= 2\pi \int_{-a}^{a} y \sqrt{1 + \frac{x^2}{y^2}} dx = 2\pi a \int_{-a}^{a} dx$$

Ans. $4\pi a^2$

2. Determine the area of the curved surface of the reel generated by revolving the part of the parabola $y^2 = 4ax$, cut-off by the latus rectum about the tangent at the vertex.

Hint: y-axis is the tangent to the parabola at the origin. Required surface area = twice surface area generated by the upper half of the parabola

$$= 2 \int_0^a 2\pi x \sqrt{1 + \left(\frac{dy}{dx}\right)^2} \, dx = 4\pi \int_0^a x \sqrt{1 + \frac{a}{x}} dx$$

$$= 4\pi \int_0^a \sqrt{\left(x + \frac{a}{2}\right)^2 - \left(\frac{a}{2}\right)^2} dx.$$

Ans. $\pi a^2 \left[3\sqrt{2} - \ln\left(\sqrt{2} + 1\right)\right]$

3. Calculate the area of the surface of revolution generated by revolving the arc of the parabola $y^2 = 12x$ from $x = 0$ to $x = 3$ about x-axis.

Ans. $24(2\sqrt{2} - 1)\pi$

4. Find the cost of plating of the front portion of the parabolic reflector (Fig. 6.48) of an automobile head light of 12 cm diameter and 4 cm deep if the cost of plating is Rs. 5/cm^2.

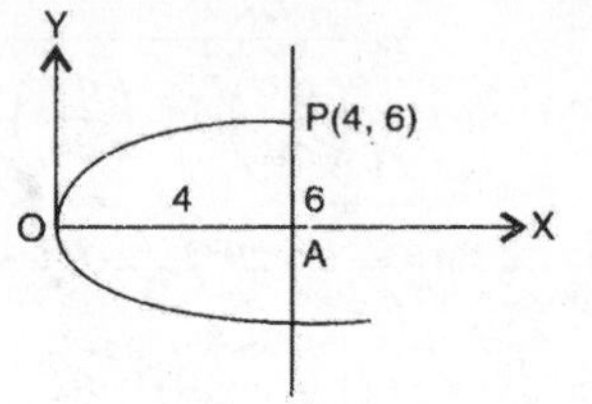

Fig. 6.48

Hint: $6^2 = 4a \cdot 4 \quad \therefore \; a = \frac{9}{4}$

Equation of parabola: $y^2 = 9x$

$$\text{Surface area} = \int_0^4 2\pi y \sqrt{1 + \left(\frac{dy}{dx}\right)^2} \, dx$$

$$= 6\pi \int_0^4 \sqrt{x} \sqrt{1 + \frac{9}{4x}} dx = 154 \text{ cm}^2$$

Cost of plating = Rs. (154 × 5) = Rs. 770.

Ans. Rs. 770

5. Show that the area of the surface of revolution generated by revolving about the x-axis the arc of $y^2 + 4x = 2 \ln y$ from $y = 1$ to $y = 3$ is $32\pi/3$.

6. Determine the surface area of the solid "catenoid" obtained by revolving the catenary $y = c \cosh \frac{x}{c}$ from the vertex to any point (x, y), about the x-axis.

Ans. $\pi c \left\{x + \frac{c}{2} \sinh\left(\frac{2x}{c}\right)\right\}$

7. Prove that the area of the surface generated by revolving the curve $x = y^3$ included between the ordinates $y = 0$ and $y = 1$, about the y-axis, is $\frac{\pi}{27}[10^{\frac{3}{2}} - 1]$.

8. Find the area of the surface of revolution generated by revolving one arch of the curve $y = \sin x$ about the x-axis.

Hint: x varies from 0 to $\pi/2$.

Ans. $\pi\left[\sqrt{2} + \ln\,(1 + \sqrt{2})\right]$

9. Show that the area of the surface formed by rotating the curve $y^2 = x^3$ from $x = 0$ to $x = 4$ about the y-axis is $\frac{128\pi}{1215}(1 + 125\sqrt{10})$.

10. Calculate the area of the surface generated by revolving the arc of the curve $9y^2 = (2x - 1)^3$, with $\frac{1}{2} \le x \le 2$ about the y-axis.

11. A sphere of radius 'b' is cut by two parallel planes which are at a distance 'h' apart. Find the surface area of the sphere included between the planes. Deduce the surface area of a hemisphere.

Hint: Area $= \int_{x=c}^{x=c+h} 2\pi y\sqrt{1 + \left(\frac{dy}{dx}\right)^2}\,dx = \int_c^{c+h} 2\pi y \cdot \left(\frac{b}{y}\right) dx = 2\pi bh$ since $x^2 + y^2 = b^2$.

Special case: hemisphere: $h = b$,
Area $= 2\pi b \cdot b = 2\pi b^2$.

Ans. $2\pi bh$; hemisphere: $2\pi b^2$

12. Determine the surface area of the prolate spheroid generated by revolving an ellipse of eccentricity e about the major axis.

Hint: Given $e < 1$, $b^2 = a^2(1 - e^2)$, equation of ellipse $\frac{x^2}{a^2} + \frac{y^2}{b^2} = 1$. Required area = twice the surface area of revolution generated by revolving the arc of the ellipse in the first quadrant about x-axis

$$= 2\int_0^a 2\pi y\sqrt{1 + \left(\frac{dy}{dx}\right)}dx$$

$$= \frac{4\pi b}{a}\int_0^a \sqrt{a^2 - x^2}\sqrt{1 + (1 - e^2)^2\frac{x^2}{y^2}}dx$$

$$= \frac{4\pi b}{a}\int_0^a \sqrt{a^2 - e^2x^2}dx$$

$$= \frac{4\pi be}{a}\int_0^a \sqrt{\left(\frac{a}{e}\right)^2 - x^2\,dx}$$

Ans. $2 \cdot \pi ab \cdot \left[\sqrt{(1 - e^2)} + \frac{\sin^{-1} e}{e}\right]$

Find the area of the surface of the solid generated by revolution of the loop of the curve about x-axis:

13. $3ay^2 = x(x - a)^2$

Hint: Loop lies between $x = 0$ and $x = a$.

Ans. $\pi a^2/3$

14. $8a^2y^2 = a^2x^2 - x^4$ about x-axis

Ans. $\pi a^2/4$

15. If the closed portion of the curve $9ay^2 = (a - x)(x + 2a)^2$ with $a > 0$ is revolved through two right angles about the x-axis, find the area of the surface of revolution so generated

Ans. $3\pi a^2$.

Parametric Form

Find the area of the surface of revolution of the solid generated by the revolving, the curve whose parametric equations are given, about the indicated line:

1. $x = t^3 - 3t$, $y = 3t^2$, $0 \le t \le 1$, about x-axis

Ans. $48\pi/5$

2. $x = t^2$, $y = t - \frac{t^3}{3}$, $0 \le t \le \sqrt{3}$, about x-axis

Ans. 3π

3. Tractrix $x = a\,\cos t + \frac{1}{2}a\,\ln\tan^2\frac{t}{2}$, $y = a\,\sin t$ about the x-axis

Hint: Curve is symmetric about both X and Y axis (Fig. 6.49).

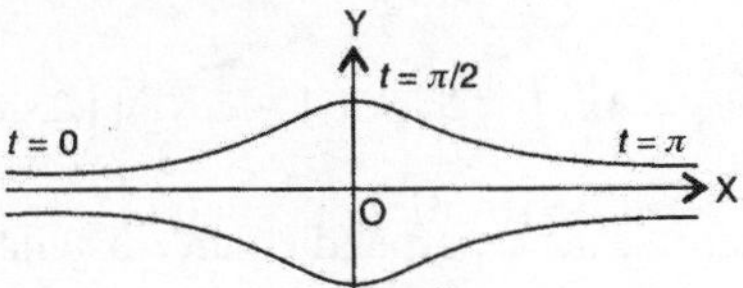

Fig. 6.49

Area = twice $\int_{t=0}^{\frac{\pi}{2}} 2\pi y\sqrt{\left(\frac{dx}{dt}\right)^2 + \left(\frac{dy}{dt}\right)^2}\,dt$

$$= 4\pi \int_0^{\frac{\pi}{2}} a \sin t \cdot \frac{a \cos t}{\sin t} \cdot dt$$

Ans. $4\pi a^2$

4. A quadrant of a circle of radius b about its chord.

Hint: $x = b \cos t$, $y = b \sin t$

$$\text{Area} = \text{twice} \int_0^{\frac{\pi}{4}} 2\pi(PL)\sqrt{\left(\frac{dx}{dt}\right)^2 + \left(\frac{dy}{dt}\right)^2}\, dt$$

$$= 4\pi \int_0^{\frac{\pi}{4}} \left(x - b \cos\frac{\pi}{4}\right)(b)dt$$

where PL is the perpendicular distance from circle to chord.

Ans. $2\sqrt{2}\pi b^2 \left(1 - \frac{\pi}{4}\right)$

5. $x = a \cos^2 t$, $y = a \sin^2 t$, about x-axis

Ans. $12\pi a^2/5$

6. $x = a \cos^3 t$, $y = a \sin^3 t$ about x-axis

Ans. $12\pi a^2/5$

7. Cycloid $x = a(t - \sin t)$, $y = a(1 - \cos t)$, about the base

Ans. $64\pi a^2/3$

8. Arch of cycloid $x = a(t - \sin t)$, $y = a(1 - \cos t)$ about the line $y = 2a$.

Hint:

$$\text{Area} = \text{twice} \int_0^{\pi} 2\pi(PL)\sqrt{\left(\frac{dx}{dt}\right)^2 \left(\frac{dy}{dt}\right)^2}\, dt$$

$$= 4\pi \int_0^{\pi} (2a - y)\left(\sqrt{2(1 - \cos t)}\right) dt$$

$$= 4\pi \int_0^{\pi} [2a - a(1 - \cos t)]\left(2 \sin\frac{t}{2}\right) dt$$

where PL = perpendicular distance from cycloid to the line $y = 2a$.

Ans. $32\pi a^2/3$

9. Cycloid $x = a(t + \sin t)$, $y = a(1 - \cos t)$ about the tangent at the vertex.

Hint: Curve is symmetric about y-axis.

$$\text{Area} = \text{twice} \int_0^{\pi} 2\pi y\sqrt{\left(\frac{dx}{dt}\right)^2 + \left(\frac{dy}{dt}\right)^2}\, dt$$

$$= 16\pi \int_0^{\pi} a^2(1 - \cos\theta) \cdot \cos\frac{\theta}{2} d\theta$$

Ans. $32\pi a^2/3$

10. Show that the ratio of the areas of the surface formed by revolving the arch of the cycloid $x = a(t + \sin t)$, $y = a(1 + \cos t)$ between two consecutive cusps about the x-axis to the area enclosed by the cycloid and x-axis is $64/9$.

Polar Form

Find the area of the surface of the solid generated by revolving the curve in polar form about the indicated line:

1. $r = a(1 \pm \cos\theta)$, about the initial line

Hint: For $r = a(1 + \cos\theta)$

$$\text{Area} = \int_0^{\pi} 2\pi y\sqrt{r^2 + \left(\frac{dr}{d\theta}\right)^2}\, d\theta$$

$$= \int_0^{\pi} 2\pi[a \sin\theta(1 + \cos\theta)]\, 2a\frac{\cos\theta}{2} d\theta$$

where $y = r \sin\theta = a(1 + \cos\theta)\sin\theta$. Similar result for $r = a(1 - \cos\theta)$.

Ans. $32\pi a^2/5$

2. $r = 2a \cos\theta$, about the initial line

Hint:

$$\text{Area} = \int_0^{\frac{\pi}{2}} 2\pi(r \sin\theta)[4a^2 \cos^2\theta + 4a^2 \sin^2\theta]^{\frac{1}{2}} d\theta$$

Ans. $4\pi a^2$

3. $r^2 = a^2 \cos 2\theta$, about the initial line

Hint:

$$\text{Area} = \text{twice} \int_0^{\frac{\pi}{4}} 2\pi\left(a\sqrt{\cos 2\theta} \cdot \sin\theta\right) d\theta \times$$

$$\times \frac{a}{\sqrt{\cos 2\theta}} d\theta$$

Ans. $4\pi a^2\left(1-\frac{1}{\sqrt{2}}\right)$

4. $r^2 = a^2\cos 2\theta$, about the line $\theta = \frac{\pi}{2}$

Ans. $\pi^2 a^3/(4\sqrt{2})$

5. $r^2 = a^2\cos 2\theta$, about a tangent at the pole

Hint: $\theta = \frac{\pi}{4}$ and $\theta = \frac{3\pi}{4}$ are the tangents to curve at the pole.

Ans. $4\pi a^2$

6. $r = 4 + 2\cos\theta$ about the intial line

Ans. $375\pi/5$

6.6 IMPROPER INTEGRALS

By definition of a *regular* (or *proper*) definite integral

$$\int_a^b f(x)dx \tag{1}$$

it is assumed that the limits of integration are finite and that the integrand $f(x)$ is continuous for every value of x in the interval $a \le x \le b$. If at least one of these conditions is violated, then the integral is known as an *improper integral* (or *singular* or *generalized* or *infinite integral*). More generally improper integrals are classified into three kinds.

Improper Integral

Improper integral of the *first kind*. It is a definite integral in which one or both limits of integration are infinite ie. the interval of integration is *not* finite. These are defined by the following relations.

(a) $$\int_a^\infty f(x)dx = \lim_{b\to\infty}\int_a^b f(x)dx \tag{2}$$

It is said to have a singularity at the upper limit

(b) $$\int_{-\infty}^b f(x)dx = \lim_{a\to-\infty}\int_a^b f(x)dx \tag{3}$$

(c) $$\int_{-\infty}^\infty f(x)dx = \lim_{\substack{b\to\infty\\ a\to-\infty}}\int_a^b f(x)dx \tag{4}$$

or

$$\int_{-\infty}^\infty f(x)dx = \lim_{a\to-\infty}\int_a^0 f(x)dx + \lim_{b\to\infty}\int_0^b f(x)dx \tag{4*}$$

If both the limits exist, we can write

$$\int_{-\infty}^\infty f(x)dx = \lim_{r\to\infty}\int_{-r}^r f(x)dx \tag{4**}$$

The improper integral is said to *converge* (or exist) when the limits in RHS of (2), (3), (4) or (4*) or (4**) exists (and are finite). Otherwise it is said to *diverge* (when either of the limits fail to exist). The limit in the RHS of (4**) is known as the *Cauchy's principal value* of the improper integral.

Geometrically for $f(x) \ge 0$, the improper integral $\int_a^\infty f(x)dx$ denotes the area of an unbounded (infinite region) lying between the curve $y = f(x)$, the ordinate $x = a$ and axis of abscissa, i.e. x-axis.

Example 1: Consider the improper integral $\int_0^\infty \frac{dx}{1+x^2}$. Here $y = f(x) = \frac{1}{1+x^2}$.

Solution: By definition of improper integral $\int_0^\infty \frac{dx}{1+x^2} = \lim_{b\to\infty}\int_0^b \frac{dx}{1+x^2} = \lim_{b\to\infty} \tan^{-1} x\Big|_0^b$
$= \lim_{b\to\infty}\tan^{-1} b = \frac{\pi}{2}$

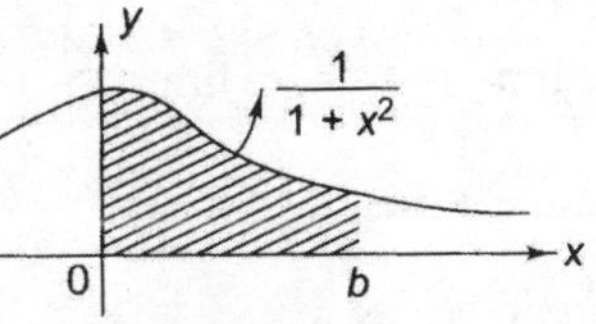

Fig. 6.50

Thus while the shaded area in Fig. 6.50 represents the definite integral $\int_0^b \frac{dx}{1+x^2}$, the shaded area in Fig. 6.51 represents the improper integral $\int_0^\infty \frac{dx}{1+x^2}$. Thus this integral represents the area of an infinite curvilinear trapezoid cross-hatched in Fig. 6.51.

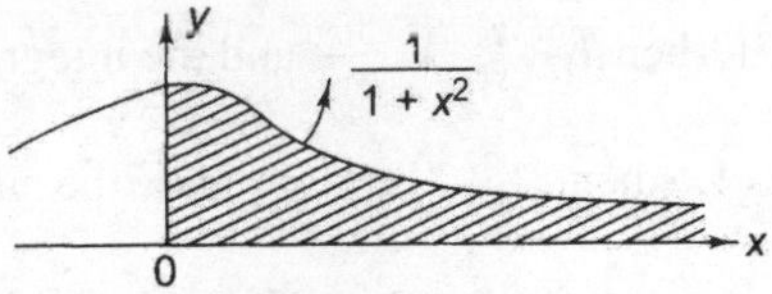

Fig. 6.51

For a convergent integral

$$\int_0^\infty f(x)dx = \int_a^b f(x)dx + \int_b^\infty f(x)dx$$

the basic contribution is made by its finite (proper) part (the first integral in the RHS) while the contri-

bution of (the 2nd integral in RHS), the singularity is arbitrarily small for large values of b.
Whereas for a divergent integral, the value of the integral tends to infinity or oscillates (without having a definite limit) since the value of the integral essentially depends on b.

Comparison Test for Convergence or Divergence of an Improper Integral by p-Integral

Book work: Prove that the p-integral

$$\int_1^\infty \frac{dx}{x^p}$$

converges when $p > 1$ and diverges when $p \le 1$.

Proof: For $p \neq 1$, $\int_1^b \frac{dx}{x^p} = \frac{1}{1-p} x^{1-p}\Big|_1^b$

$$= \frac{1}{1-p}(b^{1-p} - 1)$$

Now

$$\int_1^\infty \frac{dx}{x^p} = \lim_{b\to\infty} \int_1^b \frac{dx}{x^p} = \lim_{b\to\infty} \frac{1}{1-p}(b^{1-p} - 1).$$

Consequently.

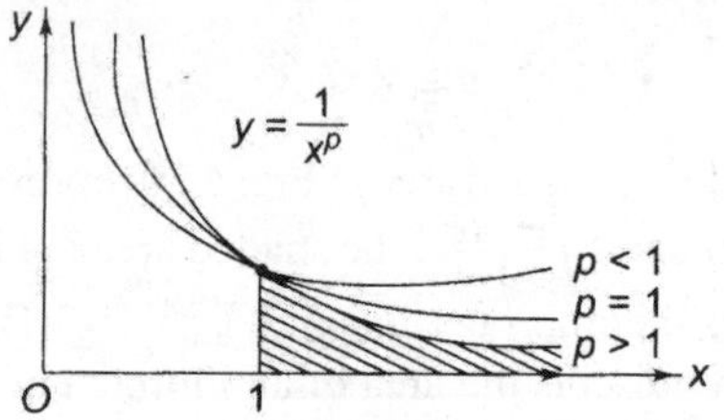

Fig. 6.52

If $p > 1$, then $\int_1^\infty \frac{dx}{x^p} = \frac{1}{p-1}$ and the integral converges.

If $p < 1$, then $\int_1^\infty \frac{dx}{x^p} = \infty$ and the integral diverges.

When $p = 1$, $\int_1^\infty \frac{dx}{x} = \ln x\Big|_1^\infty = \infty$ and the integral diverges

Results

Theorem 1. *Let $f(x)$ and $g(x)$ be non-negative functions. Suppose*
1. $0 \le f(x) \le g(x)$ for $x \ge a$.

If $\int_a^\infty g(x)dx$ converges, then $\int_a^\infty f(x)dx$ also converges and $\int_a^\infty f(x)dx \le \int_a^\infty g(x)dx$.

Theorem 2. *Suppose $0 \le g(x) \le f(x)$. If $\int_a^\infty g(x)dx$ diverges, then $\int_a^\infty f(x)dx$ also diverges.*

Thus the convergence or divergence of an improper integral is determined by comparing it with a simple integral, the p-integral (of book work I)

Theorem 3. *For a function $f(x)$ which changes its sign, if $\int_a^\infty |f(x)|dx$ converges, then $\int_a^\infty f(x)dx$ also converges, ie. $\int_a^\infty f(x)dx$ is said to converge absolutely.*

Improper Integral of the Second Kind

It is one in which both the limits of integration are finite, but the integrand $f(x)$ is not defined (infinite) or discontinuous at a point between a and b inclusive. Suppose $f(x) \to \infty$ as $x \to a$, then the integral has a singularity at the lower limit a. Then this singularity is 'cut-off' by considering

$$\int_{a+\varepsilon}^b f(x)dx$$

where ε is a small positive number. Thus for a convergent improper integral of the second kind

$$\int_a^b f(x)dx = \lim_{\varepsilon\to 0} \int_{a+\varepsilon}^b f(x)dx \tag{6}$$

which ignores the contribution of the singularity. Similarly when $f(x)$ is discontinuous at the upper limit b then

$$\int_a^b f(x)dx = \lim_{\varepsilon\to 0} \int_a^{b-\varepsilon} f(x)dx \tag{7}$$

Finally when $f(x)$ has a singularity at an intermediate point c, i.e. $a < c < b$, then

$$\int_a^b f(x)dx = \lim_{\varepsilon\to 0} \int_a^{c-\varepsilon} f(x)dx + \lim_{\varepsilon\to 0} \int_{c+\varepsilon}^b f(x)dx \tag{8}$$

The RHS limit above is known as the *Cauchy's principal value of the integral.*

When the limits in the RHS of (6), (7) or (8) fails to exist (or infinite) then the improper integral is said to diverge.

WORKED OUT EXAMPLES

Example 1: Consider the improper integral of the second kind $\int_0^3 \frac{dx}{(x-2)^{2/3}}$ whose integrand $f(x) = (x-2)^{-2/3}$ has a singularity at $x = 2$ which lies in the interval (0, 3). By (8)

$$\int_0^3 \frac{dx}{(x-2)^{2/3}} = \lim_{c\to 2-\varepsilon} \int_0^c \frac{dx}{(x-2)^{2/3}} + \lim_{c\to 2+\varepsilon} \int_c^3 \frac{dx}{(x-2)^{2/3}}$$

The first integral in RHS is

$$\lim_{c\to 2-\varepsilon} [3(6-1)^{1/3} - 3(0-1)^{1/3}] = 6$$

The second integral in RHS is

$$\lim_{c\to 2+\varepsilon} [3(3-1)^{1/3} - 3(0-1)^{1/3}] = 3(2^{1/3}) - 3$$

Thus the value of the improper integral is $6 + 3\sqrt[3]{2} - 3 = 3 + 3\sqrt[3]{2}$

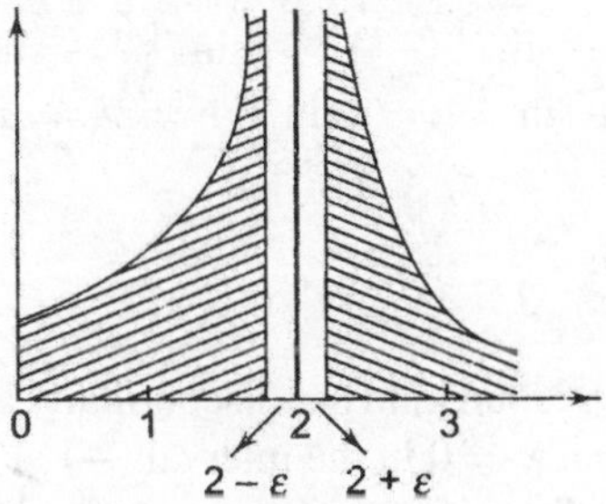

Fig. 6.53

Example 2: Consider $\int_{-2}^{2} \frac{dx}{x^2}$ which has a singularity at $x = 0$. The point of discontinuity $x = 0$ of the integrand $f(x) = x^{-2}$ lies in the interval of integration (−2, 2). Now by (8)

$$\int_{-2}^{2} \frac{dx}{x^2} = \lim_{c\to 0-} \int_{-2}^{c} \frac{dx}{x^2} + \lim_{c\to 0+} \int_{c}^{+2} \frac{dx}{x^2}$$

Now $\lim_{c\to 0-} \int_{-2}^{2} \frac{dx}{x^2} = -\lim_{c\to 0-} \frac{1}{x}\Big|_{-2}^{c}$

$$= -\lim_{c\to 0-} \left[\frac{1}{c} - \frac{1}{-2}\right] = \infty$$

The integral diverges in [−2, 0].

Similarly

$$\lim_{c\to 0+} \int_c^{+2} \frac{dx}{x^2} = -\lim_{c\to 0+} \left(2 - \frac{1}{c}\right) = \infty$$

This integral also diverges in [0, 2].

Hence the given integral $\int_{-2}^{2} \frac{dx}{x^2}$ diverges on the interval [−2, 2].

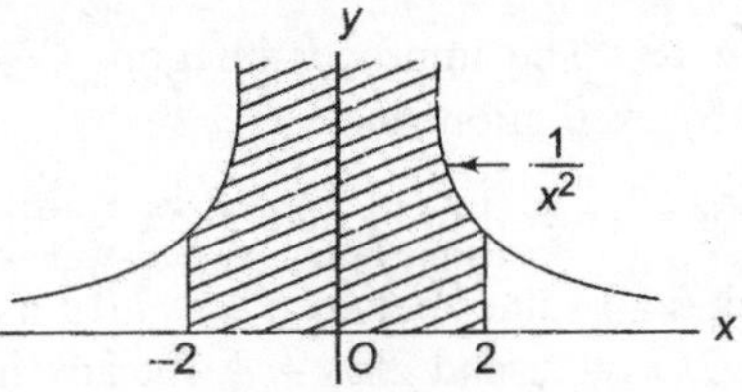

Fig. 6.54

Suppose ignoring the presence of the discontinuity at $x = 0$, we evaluate the integral obtaining

$$\int_{-2}^{2} \frac{dx}{x^2} = -\frac{1}{x}\Big|_{-2}^{2} = -\left(\frac{1}{2} - \frac{1}{-2}\right) = -1$$

which is impossible as is evident from the figure.

Comparison Test for Convergence or Divergence of Improper Integral of the Second Kind

Book work: Prove that $\int_a^c \frac{k\,dx}{(x-c)^p}$ converges for $p < 1$ and diverges for $p \geq 1$.

Here k is constant.

Proof: For $p = 1$,

$$\int_a^c \frac{k\,dx}{(x-c)^p} = \int_a^c k\frac{dx}{x-c} = k\ln(x-c)\Big|_a^c$$

$$\int_a^c \frac{k\,dx}{(x-c)} = \lim_{c\to c+} k\ln(x-c)|_a^c$$

$$= k[\ln(a-c) - \ln 0] = \infty$$

For $p \neq 1$.

$$\int_a^c \frac{k\,dx}{(x-c)^p} = \lim_{c\to c+} \left[k\frac{(x-c)^{-p+1}}{-p+1}\right]\Bigg|_a^c$$

= infinity when $p > 1$

= finite when $p < 1$.

Hence the result.

Note 1: Suppose $f(x)$ has several points of discontinuity at $a_1, a_2, \ldots a_n$ in the interval $[a, b]$. Then the improper integral is defined by

$$\int_a^b f(x)dx = \int_a^{a_1} f(x)dx + \int_{a_1}^{a_2} f(x)dx + \ldots + \int_{a_n}^{b} f(x)dx,$$

converges only when each of the improper integrals on the RHS above converges. Even if one of these integrals diverge, then $\int_a^b f(x)dx$ is said to diverge.

Note 2: Theorems similar to theorems 1, 2, 3 are applicable to improper integrals of second kind also.

Example 1: The improper integral $\int_0^\infty \cos x\, dx$ diverges by oscillation since

$$\int_0^\infty \cos x\, dx = \lim_{b\to\infty} \int_0^b \cos x\, dx = \lim_{b\to\infty} \sin b$$

which takes all values between -1 and 1 as b varies between $2n\pi - \frac{\pi}{2}$ and $2n\pi + \frac{\pi}{2}$ for any integer n. Thus integral diverges without becoming infinite.

In the improper integral of *third kind*, either the limits of integration may be infinite or the integrand is discontinuous or both.

Example 2: $\int_0^\infty e^{-x}x^{n-1}\, dx$ has an infinite upper limit and the integrand $e^{-x}x^{n-1}$ is discontinuous at the lower limit 0.

Worked Out Examples

Test for convergence or divergence of the following improper integrals and hence evaluate them.

Example 1: $\displaystyle\int_{-\infty}^{\infty} \frac{dx}{a^2+x^2}$

Solution: Improper integral of first kind, with both limits infinite. Now

$$\int_{-\infty}^{\infty} \frac{dx}{a^2+x^2} = \lim_{r\to\infty} \int_{-r}^{r} \frac{dx}{a^2+x^2} = \lim_{r\to\infty} \frac{1}{a}\tan^{-1}\frac{x}{a}\Big|_{-r}^{r}$$
$$= \lim_{r\to\infty} \left[\frac{1}{a}\right]\left[\tan^{-1}\frac{r}{a} - \tan^{-1}\frac{(-r)}{a}\right]$$
$$= \frac{2}{a}\lim_{r\to\infty} \tan^{-1}\frac{r}{a} = \frac{2}{a}\cdot\frac{\pi}{2} = \frac{\pi}{a}$$

Integral convergent.

Example 2: $\displaystyle\int_0^1 \frac{dx}{x^{0.9999}}$

Solution: Second kind with singularity (discontinuity) at the lower limit 0. So

$$\int_0^1 \frac{dx}{x^{0.9999}} = \lim_{a\to 0+\varepsilon} \int_a^1 \frac{dx}{x^{0.9999}} = \lim_{a\to 0+} \frac{x^{-0.9999+1}}{-0.9999+1}\Big|_a^1$$
$$= \lim_{a\to 0+} \frac{1}{0.0001}\left[1^{0.0001} - a^{0.0001}\right]$$
$$= \frac{1}{0.0001}[1-0] = 10000$$

Integral convergent.

Example 3: $\displaystyle\int_0^{\pi/2} \tan x\, dx$

Solution: Second kind with discontinuity at upper limit $\frac{\pi}{2}$. Now

$$\int_0^{\pi/2} \tan x\, dx = \lim_{a\to\frac{\pi}{2}-\varepsilon} \int_0^a \tan x\, dx$$
$$= \lim_{a\to\frac{\pi}{2}-} \ln\sec x\big|_0^a = \lim_{a\to\frac{\pi}{2}-} [\ln\sec a - \ln\sec 0]$$
$$= \left[\ln\sec\frac{\pi}{2} - \ln 1\right] = \ln\infty - 0 = \infty$$

Integral divergent.

Example 4: $\displaystyle\int_{-a}^{a} \frac{dx}{\sqrt{a^2-x^2}}$

Solution: Second kind with discontinuites at both the limits. Now

$$\int_{-a}^{a} \frac{dx}{\sqrt{a^2-x^2}} = \lim_{c\to -a+} \int_c^0 \frac{dx}{\sqrt{a^2+x^2}} + \lim_{c\to a-} \int_0^c \frac{dx}{\sqrt{a^2-x^2}}$$
$$= \lim_{c\to -a+} \sin^{-1}\frac{x}{a}\Big|_c^0 + \lim_{c\to a-} \sin^{-1}\frac{x}{a}\Big|_0^c$$
$$= \left[\sin^{-1}0 - \sin^{-1}\left(\frac{-a}{a}\right)\right] + \left[\sin^{-1}\frac{a}{a} - \sin^{-1}0\right]$$
$$= \sin^{-1}1 + \sin^{-1}1 = 2\sin^{-1}1 = 2\frac{\pi}{2} = \pi$$

Example 5: $\displaystyle\int_{-1}^{1} \frac{dx}{x^{2/3}}$

Solution: Second kind has discontinuity at an intermediate point $x = 0$ in the interval $(-1, 1)$

$$\int_{-1}^{1} \frac{dx}{x^{2/3}} = \lim_{a\to 0-} \int_{-1}^{a} \frac{dx}{x^{2/3}} + \lim_{a\to 0+} \int_a^1 \frac{dx}{x^{2/3}}$$
$$= \lim_{a\to 0-} \left[3x^{1/3}\big|_{-1}^{a}\right] + \lim_{a\to 0+} \left[3x^{1/3}\big|_a^1\right]$$
$$= [3\cdot 0 - 3(-1)^{1/3}] + [3\cdot 1^{1/3} - 3\cdot 0] = 3+3$$
$$= 6$$

Example 6: $\displaystyle\int_0^\infty \frac{dx}{x^3}$

Solution: This is third kind involving infinite upper limit and unboundedness at the lower limit 0. Now

$$\int_0^\infty \frac{dx}{x^3} = \lim_{\substack{b\to\infty\\ a\to 0+}} \int_a^b \frac{dx}{x^3} = \lim_{\substack{b\to\infty\\ a\to 0+}} \left[-\frac{1}{2x^2}\right]\Big|_{x=a}^{b}$$
$$= \lim_{\substack{b\to\infty\\ a\to 0+}} -\frac{1}{2}\left[\frac{1}{b^2} - \frac{1}{a^2}\right] = -\frac{1}{2}[0-\infty] = \infty$$

Integral diverges.

Example 7: $\int_a^\infty x \sin x\, dx$ with $a > 0$

Solution:

$$\int_a^\infty x \sin x\, dx = \lim_{b\to\infty} \int_a^b x \sin x\, dx$$

Integrating by parts

$$= \lim_{b\to\infty} [-x \cdot \cos x + \sin x]_a^b = \pm\infty$$

since $\cos x$ and $\sin x$ oscillate between ± 1. Thus the limit of the integral oscillates infinitely.

Example 8: Test for convergence of the integration $\int_1^\infty \frac{dx}{x^4(1+e^{2x})}$

Solution: For $x \geq 1$, $\frac{1}{x^4(1+e^{2x})} \leq \frac{1}{x^4}$

Now

$$\int_1^\infty \frac{dx}{x^4} = \lim_{b\to\infty} \int_1^b \frac{dx}{x^4} = \lim_{b\to\infty} -\frac{1}{3x^3}\Big|_1^b$$

$$= \frac{1}{3} \text{(convergent by } p\text{-integral for } p = 3 > 1)$$

By comparison using Theorem 1, the given integral converges and its value is less than $\frac{1}{3}$.

Example 9: Test for convergence $\int_2^\infty \frac{x^2+1}{\sqrt{x^5}} dx$

Solution: For $x \geq 2$,

$$\frac{x^2+1}{\sqrt{x^5}} > \frac{x^2}{\sqrt{x^5}} = \frac{1}{\sqrt{x}}$$

Now

$$\int_1^\infty \frac{dx}{\sqrt{x}} = \lim_{b\to\infty} 2\sqrt{x}\Big|_1^b = \infty$$

(diverges by p-integral for $p = \frac{1}{2} < 1$)
By comparison, using theorem 2, the given integral diverges.

Example 10: Test for convergence

$$\int_1^\infty \frac{\sin 2x}{x^5} dx$$

Solution: Since $\sin x$ takes all values between -1 to 1, the integrand $\frac{\sin 2x}{x^5}$ also oscillates between -1 and 1. ie., takes both positive and negative values. Now $\left|\frac{\sin 2x}{x^5}\right| \leq \frac{1}{x^5}$ and

$$\int_1^\infty \frac{dx}{x^5} = \lim_{b\to\infty} \int_1^b \frac{dx}{x^5} = \lim_{b\to\infty} \left[-\frac{1}{4x^4}\Big|_1^b\right]$$

$$= \lim_{b\to\infty} \left[\frac{1}{4} - \frac{1}{4b^4}\right] = \frac{1}{4}$$

Therefore the integral $\int_1^\infty \left|\frac{\sin 2x}{x^5}\right| dx$ converge. By theorem 3, the integral $\int_1^\infty \frac{\sin 2x}{x^5} dx$ also converges (absoloutely).

EXERCISE

Evaluate the following improper integrals

1. $\int_0^\infty e^{-x} dx$

Ans. 1, converges

2. $\int_0^t \ln x\, dx$

Ans. -1, converges

3. $\int_0^1 \frac{x\, dx}{\sqrt{1-x^2}}$

Ans. 1, converges

4. $\int_1^\infty \frac{dx}{x\sqrt{x^2-1}}$

Ans. $\frac{\pi}{2}$, converges

5. $\int_{-1}^1 \frac{dx}{x^5}$

Ans. Diverges

6. $\int_1^\infty \ln\left(\frac{1}{x}\right) dx$

Ans. Diverges

7. $\int_{-\infty}^0 \frac{dx}{(1-3x)^2}$

Ans. $\frac{1}{3}$, converges

8. $\int_{-\infty}^0 \cos hx\, dx$

Ans. Diverges.

9. $\int_0^\infty \frac{dx}{(1+x)\sqrt{x}}$

Ans. π, converges

10. $\int_{-1}^{1} \sqrt{\frac{1+x}{1-x}}dx$

Ans. π, converges

Hint: $\sin^{-1} b - \sqrt{1-b^2}$

11. $\int_0^\infty \frac{4z^2dz}{(z^2+1)^2}$

Ans. π, converges

(Integrals 10 and 11 are same. Use substitution $z = \sqrt{\frac{1+x}{1-x}}$)

12. $\int_5^\infty \frac{dx}{x\log x}$

Ans. Diverges

13. $\int_0^3 \frac{dx}{(x-1)^{2/3}}$

Ans. $3+3\sqrt[3]{2}$, converges

14. $\int_1^\infty \frac{dx}{x^{1.00001}}$

Ans. 100000, converges

15. $\int_{-1}^{1} \frac{dx}{x^{2/5}}$

Ans. converges

16. $\int_{-\infty}^{1} e^x\, dx$

Ans. e, converges

17. $\int_{-\infty}^{\infty} \frac{dx}{a+2bx+cx^2}$

Ans. $\frac{\pi}{\sqrt{ac-b^2}}$, converges

Hint: $\frac{1}{c}\left[1\Big/\left[\left(x+\frac{b}{c}\right)^2+\left(\sqrt{\frac{ac-b}{c^2}}\right)^2\right]\right]$

Test for convergence

18. $\int_0^\infty \frac{dx}{\sqrt[3]{x^2+1}}$

Ans. Diverges

Hint: Compare with $x^{-2/3}$ which is divergent, $p=\frac{2}{3}<1$

19. $\int_0^\infty \frac{dx}{\sqrt{x^3+1}}$

Ans. Converges

Hint: Compare with $x^{-3/2}$ which is convergent, $p=\frac{3}{2}>1$

20. $\int_2^\infty \frac{3x+5}{x^5+7}dx$

Ans. Converges

Hint: Compare with $\frac{3}{x^4}$ which is convergent, $p=4>1$

21. $\int_0^5 \frac{\sin 3x}{x^{5/2}}dx$

Ans. Diverges

Hint: As $x\to 0$ $\frac{\sin 3x}{x^{5/2}} \sim \frac{3x}{x^{5/2}} = \frac{3}{x^{3/2}}$

22. $\int_0^\infty \frac{dx}{x^4+2}$

Ans. Converges

Hint: $\int_0^\infty = \int_0^1 + \int_1^\infty$, first integral converges, second integral converges by comparison with integrand $\frac{1}{x^4}$ which is p-integral with $p=4>1$.

23. $\int_0^\infty \frac{x^{3.2}dx}{x^4+100}$

Ans. Diverges

Hint: $\int_0^\infty = \int_0^1 + \int_1^\infty$, first integral converges, $\int_1^\infty$ diverges by comparison with integrand $x^{-0.8}$ which is p-integral with $p<1$.

24. $\int_0^4 \frac{\sin^2 x\, dx}{\sqrt{x}(x-1)}$

Ans. Converges

Hint: Compare with $x^{-3/2}$ which is p integral with $p=\frac{3}{2}>1$ and is convergent.

25. $\int_0^1 \frac{dx}{x^2\cos x}$

Ans. Diverges

Hint: As $x\to 0$, $\frac{1}{x^2\cos x} \sim \frac{1}{x^2}$ which is divergent for $p>1$.

Chapter 7

Multiple Integrals

INTRODUCTION

Double integrals are sometimes easier to evaluate when we change the order of integration or when we change to polar coordinates, over regions whose boundaries are given by polar equations. General change of variables involves evaluation of multiple integrals by substitution. The aim of substitution is to replace complicated integrals by one that are easier to evaluate. Change of variables simplifies the integrand, the limits of integration or both. Evaluation of triple integral is simplified by the use of cylindrical or spherical coordinates in problems of physics, engineering or geometry involving a cylinder, cone, or sphere. Dirichlet's integral is useful in the evaluation of certain double and triple integrals by expressing them in terms of beta and gamma functions.

Multiple integral is a natural extension of an (ordinary) definite integral to a function of two variables (double integral) or three variables (triple integral) or more variables. Double and triple integrals are useful in evaluating area, volume, mass, centroid and moments of inertia of plane and solid regions.

7.1 DOUBLE INTEGRAL

Let D be a closed* and bounded** domain in the XY-plane bounded by a simple closed curve c. Let $f(x, y)$ be a given continuous function in D. Divide D into n parts and form the sum

$$\sum_{i=1}^{n} f(P_i)\Delta S_i$$

where $f(P_i)$ is the value of f at an arbitrary point P_i of the subdomain whose area is ΔS_i. Double integral of $f(x, y)$ over the domain D is the limit of the above sum as $n \to \infty$ and is denoted by I_D as

$$I_D = \iint_D f(P)\,dS = \iint_D f(x, y)\,dx\,dy$$

$$= \lim_{\Delta S_i \to 0} \sum_{i=1}^{n} f(P_i)\,\Delta S_i \qquad (1)$$

Here D is known as the domain (region) of integration.

The following properties of double integrals follow from the definition (1).

Properties of Double Integral

1. $\iint_D a\, f(x, y) dS = a \iint_D f(x, y) dS$, $a =$ constant.

2. $\iint_D [f(x, y) + g(x, y)] dS = \iint_D f(x, y) dS. +$
$+ \iint_D g(x, y) dS$

3. $\iint_D f(x, y) dS = \iint_{D_1} f(x, y) dS + \iint_{D_2} f(x, y) dS$

where D is the union of disjoint domains D_1 and D_2.

*Boundary included i.e., part of the region.
**Can be enclosed in a circle of finite radius.

Regular Domain

A domain D in XY-plane bounded by a curve c is said to be regular in the Y-direction, if straight lines passing through an interior point (a point which does not lie on the boundary c) and parallel to Y-axis meets c in two points A and B (Fig. 7.1).

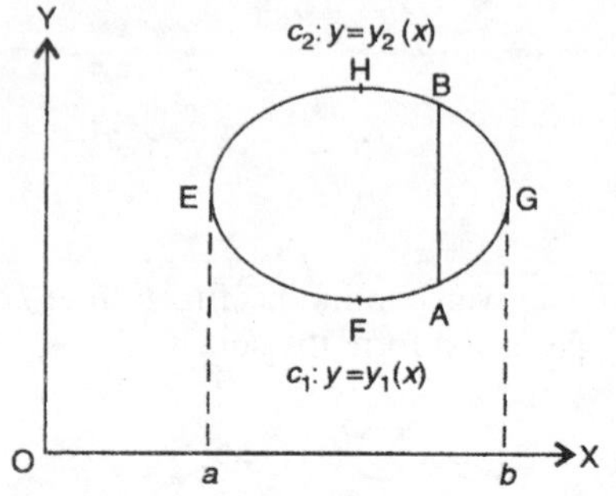

Fig. 7.1

In such a case, the domain D is bounded by the two curves $c_2 : GHE : y = y_2(x)$ and $c_1 : EFG : y = y_1(x)$ and the ordinates $x = a$ and $x = b$ such that

$$y_2(x) \le y_1(x)$$

and $$a < b$$

Here c is the sum of the two curves c_1 and c_2.

In other words, "A domain D is said to be "regular" domain if it is regular in both X and Y directions".

Examples of Regular Domain: Regions bounded by rectangle, triangle, circle, ellipses.

Example of Non-regular Domain: Annulus region.

Evaluation of a Double Integral

A double integral can be evaluated by successive single integrations i.e., as a two-fold iterated (repeated) integral as follows (if D is regular in y-direction):

$$I_D = \int_{x=a}^{b} \left\{ \int_{y=y_1(x)}^{y_2(x)} f(x, y)dy \right\} dx \qquad (2)$$

where integration is performed *first* with respect to y (within the braces). With the substitution of the limits $y_1(x)$ and $y_2(x)$, the integrand becomes a function of x alone, which is *then* integrated with respect to x from a to b.

In a similar way, for a domain D (regular in x-direction) which is bounded above by EHF $x = x_2(y)$ and bounded below by EGF: $x = x_1(y)$ and the abscissa $y = d$ and $y = e$ (Fig. 7.2).

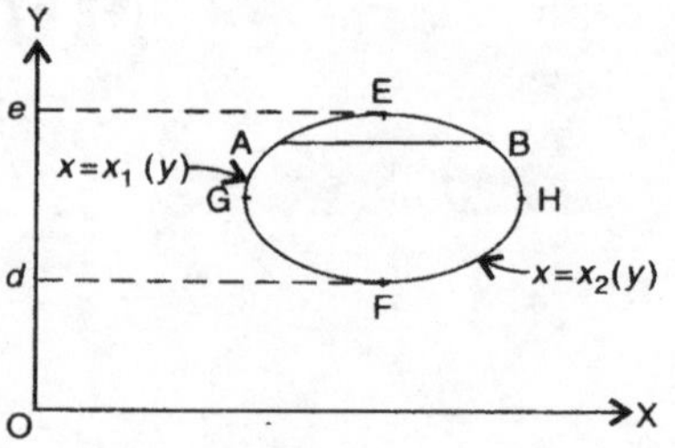

Fig. 7.2

The double integral is evaluated as

$$I_D = \int_{y=d}^{e} \left\{ \int_{x=x_1(y)}^{x_2(y)} f(x, y)dx \right\} dy \qquad (3)$$

In this case the integration is *first* performed with respect to x and *then* later with respect to y.

Note 1: The braces in (2) and (3) can be omitted, since it is conventional to integrate first with respect to a variable whose differential appears first (for example in (2) the order of integration is first y and later x).

Note 2: If all the four limits are constants then the order of integration can be done in either way i.e., integration first with respect to x and later w.r.t. y or first w.r.t. y and later w.r.t. x, yielding the same result, provided the limits are taken (for x and y) accordingly.

Note 3: When D is regular (in both x and y directions), draw a rough sketch of D the domain of integration to fix the limits of integration. Choose (or change) the order of integration (see Section 7.5) and use (2) or (3) whichever is easier for integration.

Note 4: Suppose D is such that the lower curve $AFE : y = y_1(x)$ consists of two different curves

$$AF : y_1 = \phi(x), \quad a \le x \le e$$

$$FE : y_1 = \psi(x), \quad e \le x \le b$$

while the upper curve is $AGB : y = y_2(x)$ (Fig. 7.3). Then $D{=}AFEBGA{=}AFGA{+}FEBG{=} D_1 + D_2$, so that

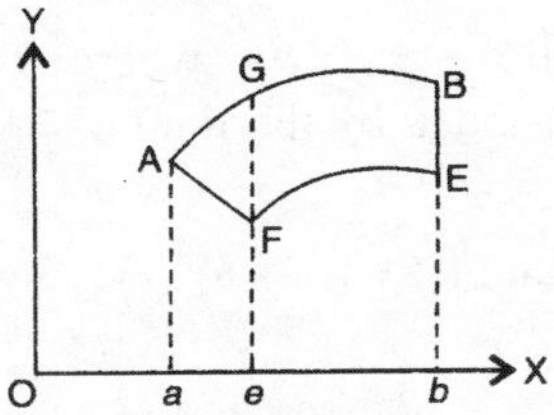

Fig. 7.3

$$\iint_D f(x, y)dy\,dx = \int_a^b \int_{y_1(x)}^{y_2(x)} f(x, y)dy\,dx$$
$$= \iint_{D_1} f(x, y)dy\,dx + \iint_{D_2} f(x, y)dy\,dx$$
$$= \int_a^e \int_{\phi(x)}^{y_2(x)} f(x, y)dy\,dx + \int_e^b \int_{\psi(x)}^{y_2(x)} f(x, y)dy\,dx.$$

(see Worked Out Example 4).

WORKED OUT EXAMPLES

Evaluate the following double integrals as two-fold iterated integral:

Example 1: $I_D = \int_1^{\ln 8} \int_0^{\ln y} e^{x+y} dx\,dy$

Solution:

$$I_D = \int_1^{\ln 8} e^{x+y}\Big|_0^{\ln y} dy = \int_1^{\ln 8} (e^{y+\ln y} - e^y)dy$$

$$I_D = \int_1^{\ln 8} e^y(y-1)dy = \left(ye^y - e^y - e^y\right)\Big|_1^{\ln 8}$$
$$= 8\ln 8 - 16 + e$$

Example 2: $I_D = \iint_D (x^2 + y^2)dx\,dy$ where D is bounded by $y = x$ and $y^2 = 4x$.

Solution:

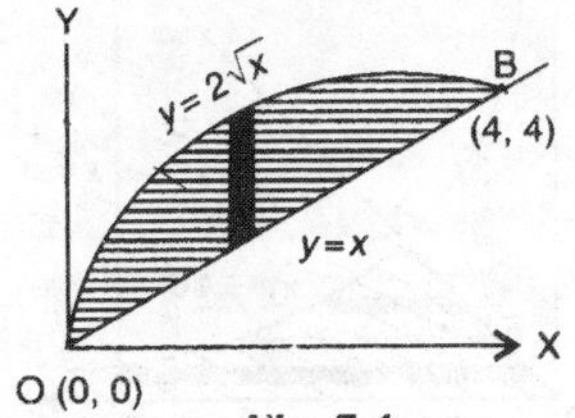

Fig. 7.4

$$I_D = \int_0^4 \int_x^{2\sqrt{x}} (x^2 + y^2)dy\,dx$$
$$= \int_0^4 x^2y + \frac{y^3}{3}\Big|_x^{2\sqrt{x}} dx$$
$$= \int_0^4 \left(2x^{\frac{5}{2}} + \frac{8}{3}x^{\frac{3}{2}} - \frac{4}{3}x^3\right) dx$$
$$= \frac{768}{35}$$

Example 3: $I_D = \iint_D x^3y\,dx\,dy$ where D is the region enclosed by the ellipse $\frac{x^2}{a^2} + \frac{y^2}{b^2} = 1$ in the first quadrant.

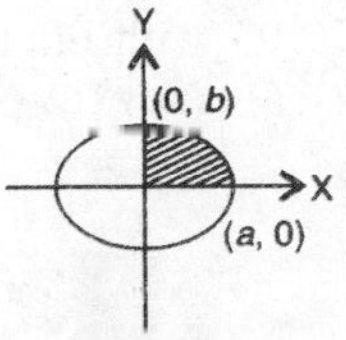

Fig. 7.5

Solution:

$$I_D = \int_0^a \int_{y=0}^{\frac{b}{a}\sqrt{a^2-x^2}} x^3y\,dy\,dx$$

$$I_D = \int_0^a \frac{x^3y^2}{2}\Big|_0^{\frac{b}{a}\sqrt{a^2-x^2}} dx = \frac{b^2}{2a^2}\int_0^a (a^2x^3 - x^5)dx$$

$$I_D = \frac{b^2}{2a^2}\left[\frac{a^2x^4}{4} - \frac{x^6}{6}\right]_0^a = \frac{b^2a^4}{24}$$

Example 4: $I_D = \iint_D x^2 dx\,dy$ where D is the region in the first quadrant bounded by the hyperbola $xy = 16$ and the lines $y = x$, $y = 0$ and $x = 8$.

Solution: If horizontal strips are considered, the upper curve ABE consists of two curves AB and BE. It is necessary to consider the region D as the union of two disjoint regions D_1 and D_2 as

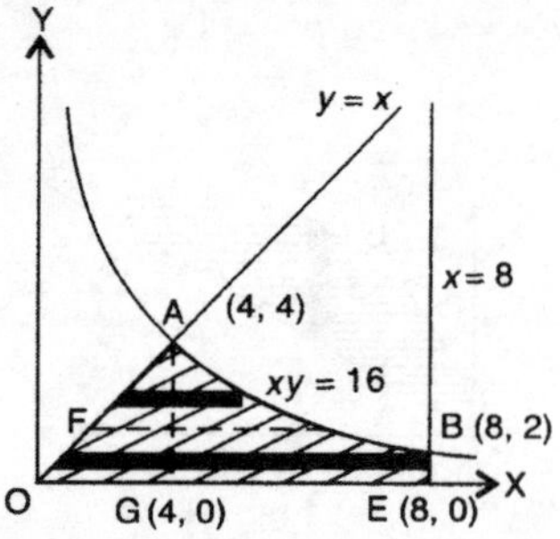

Fig. 7.6

$D = OGE\ BAFO = D_1 + D_2 = FBAF + OGEBF$

$$I_D = \iint_D x^2 dx\,dy = \iint_{D_1} + \iint_{D_2}$$

$$= \int_2^4 \int_{x=y}^{16/y} x^2 dx\,dy + \int_0^2 \int_{x=y}^{8} x^2 dx\,dy$$

$$= \frac{1}{3}\int_0^4 \left(\frac{16^3}{y^3} - y^3\right) dy + \frac{1}{3}\int_0^2 (8^3 - y^3) dy = 448$$

Note: This example can also be considered with vertical strip as follows: $D = R_1 + R_2 = OGAFO + GEBAG$ with limits for R_1: $y : 0$ to x; $x : 0$ to 4, for R_2: $y : 0$ to $16/x$; $x :$ 4 to 8.

Exercise

Evaluate the following double integrals as two-fold iterated integrals:

1. $\int_3^4 \int_1^2 \frac{dy\,dx}{(x+y)^2}$

Ans. ln (25/24)

2. $\int_0^a \int_o^{y_2/a} e^{x/y} dx\,dy$

Ans. $a^2/2$

3. $\int_{-\frac{1}{2}}^1 \int_{-x}^{1+x} (x^2 + y) dy\,dx$

Ans. $\frac{63}{32}$

4. $\int_0^\pi \int_o^x x \sin y\, dy\, dx$

Ans. $2 + \pi^{2/2}$

5. $\iint_D (4xy - y^2) dx\,dy$ where D is the rectangle bounded by $x = 1, x = 2, y = 0, y = 3$.

Ans. 18

6. $\iint_D (x^2 + y^2) dx\,dy$ where D is the region bounded by $y = x$, $y = 2x$ and $x = 1$ in the first quadrant.

Ans. $\frac{5}{6}$

7. $\iint_D (1 + x + y) dx\,dy$ where D is the region bounded by the lines $y = -x, x = \sqrt{y}$, $y = 2, y = 0$.

Hint: Limits: $x : -y$ to $\sqrt{y}$; $y : 0$ to 2.

Ans. $\frac{44}{15}\sqrt{2} + \frac{13}{3}$

8. $\iint_D xy\, dx\, dy$ where D is the domain bounded by the parabola $x^2 = 4ay$, the ordinates $x = a$ and x-axis.

Ans. $a^4/3$

9. $\int_0^1 \int_0^{\sqrt{1+x^2}} (1 + x^2 + y^2)^{-1} dx\,dy$

Ans. $\frac{\pi}{4} \log(1 + \sqrt{2})$

10. $\iint_D (x + y)^2 dx\,dy$ where D is the region bounded by the ellipse $\frac{x^2}{a^2} + \frac{y^2}{b^2} = 1$.

Ans. $\pi ab(a^2 + b^2)/4$

11. $\int_0^a \int_{x/a}^x \frac{x\,dy\,dx}{x^2+y^2}$

Ans. $a\left[\frac{\pi}{4} - \tan^{-1}\left(\frac{1}{a}\right)\right]$

12. $\iint_R \sqrt{xy(a - x - y)}$ where R is the region bounded by $x = 0, y = 0, x + y = a$.

Ans. $2\pi a^{7/2}/105$

7.2 APPLICATION OF DOUBLE INTEGRAL

Area of a Plane Region

The area A of a plane regular region (domain) D is given by a two-fold iterated integral

$$A = \iint_D ds = \iint_D dx\,dy$$

$$= \int_{x=a}^{b} \int_{y=y_1(x)}^{y_2(x)} dy\,dx = \int_{y=d}^{e} \int_{x=x_1(y)}^{x_2(y)} dx\,dy$$

Mass Contained in a Plane Region

Let $f(x, y) > 0$ be the surface density (mass/unit area) of a given plane region D. Then the amount (quantity) of mass M contained in the plane region

D is given by

$$M = \iint_D f(x, y)dx\,dy$$

Centre of Gravity (Centroid) of a Plane Region D

The coordinates (x_c, y_c) of the centre of gravity (centroid) of a plane region D with surface density $f(x, y)$ and containing mass M are

$$x_c = \frac{\iint_D x \cdot f(x, y)dx\,dy}{M}, \quad y_c = \frac{\iint_D y \cdot f(x, y)dx\,dy}{M}$$

Moment of Inertia of a Plane Region

Moments of inertia of a plane region D (with surface density $f(x, y)$) relative to x-axis, y-axis and origin O are respectively given by

$$I_{xx} = \iint_D y^2 f(x, y)dx\,dy$$

$$I_{yy} = \iint_D x^2 f(x, y)dx\,dy$$

$$I_o = I_{xx} + I_{yy} - \iint_D (x^2 + y^2) f(x, y)dx\,dy$$

I_o is also known as polar moment of inertia.

Volume under a Surface of a Solid as a Double Integral

Let $z = f(x, y) > 0$ be the equation of a surface. Let the curve c be the boundary of the plane domain D in the XY-plane. Further let V be the volume of the solid Q, under the surface $z = f(x, y)$ (i.e., bounded above by the surface) and above the XY-plane (i.e., bounded below by $z = 0$) and a cylindrical surface whose generator are parallel to the z-axis, while the directrix Q is c (see Fig. 7.7).

Then the double integral of $f(x, y)$ taken over D gives the volume V under the surface $z = f(x, y)$

$$V = \iint_D f(x, y)dx\,dy.$$

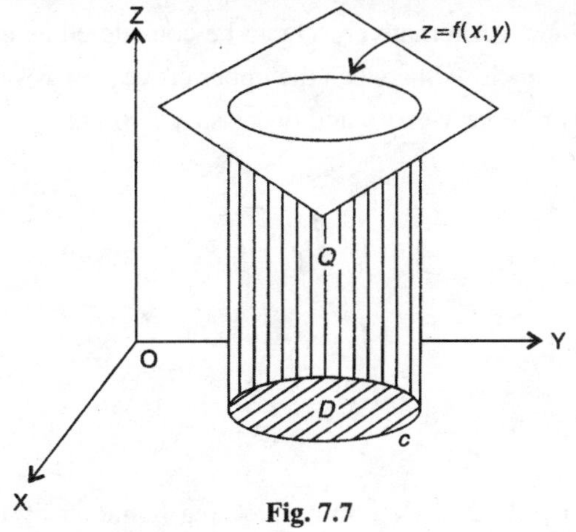

Fig. 7.7

WORKED OUT EXAMPLES

Area of Plane Region: Cartesian Coordinates

Example 1: Find the area bounded by the curves $y^2 = x^3$ and $x^2 = y^3$.

Solution: Area $= A = \iint_D dx\,dy = \iint_D dy\,dx$ the plane region D can be covered by varying y from the upper curve $y = y_2(x) = x^{\frac{2}{3}}$ to the lower curve $y = y_1(x) = x^{\frac{3}{2}}$, while x varies from 0 to 1 (refer Fig. 7.8). Thus

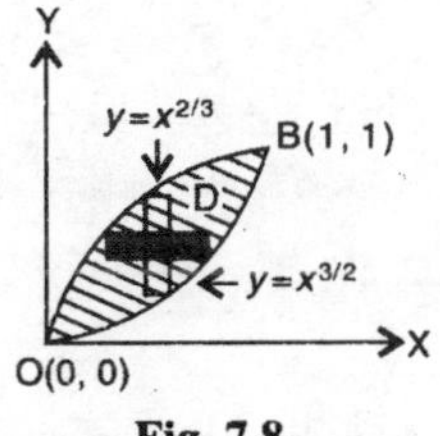

Fig. 7.8

$$A = \int_0^1 \int_{x^{\frac{3}{2}}}^{x^{\frac{2}{3}}} dy\,dx$$

$$= \int_0^1 (x^{\frac{2}{3}} - x^{\frac{3}{2}})dx = \frac{3}{5}x^{\frac{5}{3}} - \frac{2}{5}x^{\frac{5}{2}}\Big|_0^1$$

$$A = \frac{3}{5} - \frac{2}{5} = \frac{1}{5}$$

Aliter: Alternatively D can be considered as Domain enclosed between the upper curve $x = y^{\frac{2}{3}}$ and lower curve $x = y^{\frac{3}{2}}$ and obscissa $y = 0$ and $y = 1$. Thus

$$A = \int_0^1 \int_{x=y^{\frac{3}{2}}}^{y^{\frac{2}{3}}} dx\, dy$$

$$= \int_0^1 (y^{\frac{2}{3}} - y^{\frac{3}{2}})dy$$

$$= \frac{3}{5}y^{\frac{5}{3}} - \frac{2}{5}y^{\frac{5}{2}}\Big|_0^1 = \frac{1}{5}$$

Example 2: Determine the area bounded by the curves $xy = 2, 4y = x^2$ and $y = 4$.

Solution: Here $A\left(\frac{1}{2}, 4\right), B(4, 4), E(2, 1), F(2, 4)$ observe that one of (lower) curve consists of more than one curve (equation) (Fig. 7.9), so the given domain $D = D_1 + D_2$ where $D_1 : AEFA$ bounded by $y = \frac{2}{x}, y = 4, x = \frac{1}{2}, x = 1$ and $D_2 : FEBF$ bounded by $y = \frac{x^2}{4}, y = 4, x = 1$ and $x = 4$, so that

$$A = \iint_D dx\, dy = \iint_{D_1} + \iint_{D_2}$$

$$= \int_{\frac{1}{2}}^2 \int_{y=\frac{2}{x}}^4 dy\, dx + \int_2^4 \int_{y=\frac{x^2}{4}}^4 dy\, dx$$

$$= \int_{\frac{1}{2}}^2 \left(4 - \frac{2}{x}\right) dx + \int_2^4 \left(4 - \frac{x^2}{4}\right) dx$$

$$= 4x - 2\ln x|_{\frac{1}{2}}^2 + 4x - \frac{x^3}{12}\Big|_2^4$$

$$A = \frac{28}{3} - 2\ln 4$$

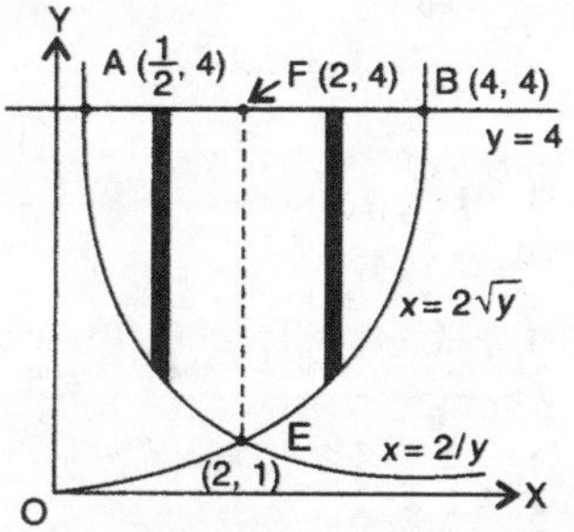

Fig. 7.9

Note: Alternatively by conveniently taking $y = \frac{x^2}{4}$ as the upper curve and $y = 2/x$ as the lower curve (refer Fig. 7.10), this problem can be solved easily as follows:

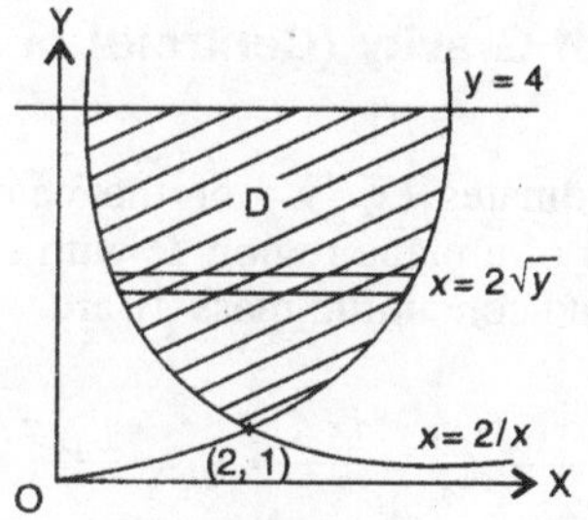

Fig. 7.10

Aliter : $A = \int_{y=1}^4 \int_{x=\frac{2}{y}}^{2\sqrt{y}} dx\, dy = \int_1^4 \left(2\sqrt{y} - \frac{2}{y}\right) dy$

$$= 2\left[\frac{y^{\frac{3}{2}}}{(3/2)} - 2\ln y\right]_1^4 = \frac{28}{3} - 2\ln 4$$

EXERCISE

Area of plane region: Cartesian coordinates

Find the area of the following domains in XY-plane bounded by the indicated curves

1. $y = x, y = x^2$

Ans. $\frac{1}{6}$

2. $y = 3x - x^2, y = x$

Ans. $\frac{4}{3}$

3. $3x = 4 - y^2, x = y^2$

Hint: Easy to integrate with,

limits: $x : y^2$ to $(4 - y^2)/3$, $y : -1$ to 1.

Ans. $\frac{16}{9}$

4. $y = x^2$ and $y = x + 2$

Hint: Easy to integrate with

limits: $y : x^2$ to $x + 2$; $x : -1$ to 2.

Ans. $\frac{9}{2}$

5. Ellipse $\frac{x^2}{a^2} + \frac{y^2}{b^2} = 1$.

   Hint: Limits: $y : \pm b\sqrt{1 - \frac{x^2}{a^2}}$; $x = \pm a$ or

   $x : \pm a\sqrt{1 - \frac{y^2}{b^2}}$, $y = \pm b$.

Ans. πab

6. $\sqrt{x} + \sqrt{y} = \sqrt{a}$ and $x + y = a$

   Hint: Limits: $y : a + x - 2\sqrt{ax}$ to $(a-x)$, $x : 0$ to a.

Ans. $\frac{a^2}{3}$

7. Parabola $y^2 = 4ax$ and straight line $x + y = 3a$

   Hint: Easy to integrate with limits: $x : 3a - y$ to $y^2/4a$; $y : 0$ to a.

Ans. $\frac{10a^2}{3}$

8. Common to the two parabolas

   $y^2 = 4a(x + a)$, $y^2 = 4b(b - x)$

Ans. $\frac{8(a + b)\sqrt{ab}}{3}$

9. $y = \sin x$, $y = \cos x$, $x = 0$

Ans. $\sqrt{2} - 1$

10. $3y^2 = 25x$ and $5x^2 = 9y$

    Hint: Limits: $y : \frac{5}{9}x^2$ to $5\sqrt{x}/\sqrt{3}$, $x : 0$ to 3

Ans. 5

11. $x = y - y^2$ and $x + y = 0$

Ans. $\frac{4}{3}$

WORKED OUT EXAMPLES

Mass, centroid and moments of inertia

Example 1: Find the mass, coordinates of the centre of gravity and moments of inertia relative to x-axis, y-axis and origin of a rectangle $0 \le x \le 4$, $0 \le y \le 2$ having mass density xy (see Fig. 7.11).

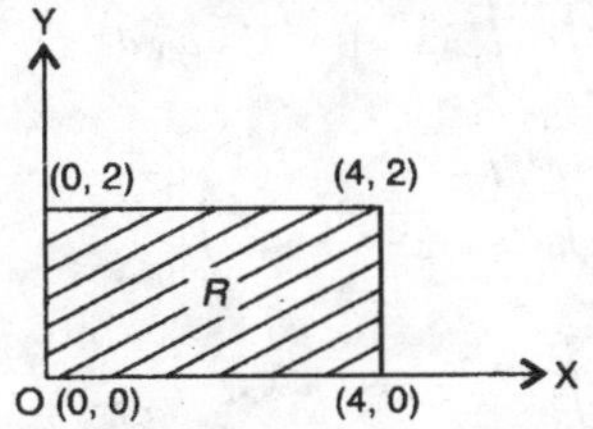

Fig. 7.11

Solution: Here density $f(x, y) = xy$

$$\text{Mass } M = \iint_R f(x, y)dx\,dy$$

$$M = \int_0^4 \int_0^2 (xy)dy\,dx = \int_0^4 \left.\frac{xy^2}{2}\right|_0^2 dx$$

$$= \int_0^4 2x\,dx = 16$$

Let x_c, y_c be the coordinates of the centre of gravity of R then

$$x_c = \frac{1}{M}\iint_R x\,f(x, y)dx\,dy = \frac{1}{16}\int_0^4\int_0^2 x(xy)dy\,dx$$

$$x_c = \frac{1}{16}\int_0^4 x^2\left.\frac{y^2}{2}\right|_0^2 dx = \frac{1}{8}\int_0^4 x^2 dx = \frac{8}{3}$$

$$y_c = \frac{1}{M}\iint_R yf(x, y)dx\,dy = \frac{1}{16}\int_0^4\int_0^2 y(xy)dy\,dx$$

$$= \frac{1}{16}\int_0^4 x\left.\frac{y^3}{3}\right|_0^2 dx = \frac{1}{6}\int_0^4 x\,dx = \frac{4}{3}$$

Moment of inertia relative to x-axis

$$I_x = \iint_R y^2 f(x, y)dx\,dy = \int_0^4\int_0^2 y^2(xy)dy\,dx$$

$$= \int_0^4 x\left.\frac{y^4}{4}\right|_0^2 dx = 4\int_0^4 x\,dx = 4\left.\frac{x^2}{2}\right|_0^4 = 32$$

Similarly

$$I_y = \iint_R x^2 f(x, y)dx\,dy = \int_0^4\int_0^2 x^2(xy)dy\,dx$$

$$I_y = \int_0^4 x^3\left.\frac{y^2}{2}\right|_0^2 dx = 2\int_0^4 x^3 dx = 2\left.\frac{x^4}{4}\right|_0^4 = 128$$

$$I_0 = \iint_R (x^2+y^2) f(x,y)\,dx\,dy$$

$$= \int_0^4 \int_0^2 (x^2+y^2)xy\,dy\,dx$$

$$= \int_0^4 \left(\frac{x^3y^2}{2} + \frac{xy^4}{4} \right)\Bigg|_0^2 dx$$

$$= \int_0^4 (2x^3+4x)dx$$

$$I_0 = \left(\frac{2x^4}{4} + \frac{4x^2}{2} \right)\Bigg|_0^4 = 128+32=160$$

Thus $I_0 = I_x + I_y$.

Example 2: Find the mass, centroid and moments of inertia relative to x-axis, y-axis and origin of the plane region R having mass density $x+y$ and bounded by the parabola $x = y - y^2$ and the straight line

$$x + y = 0$$

Solution: The plane region R shown in the (Fig. 7.12) is OAB with $O(0,0)$, $A(0,1)$, $B(-2,2)$ since the points of intersection of the parabola $x = y - y^2$ and $x+y=0$ are $(-2,2)$ and $(0,0)$.

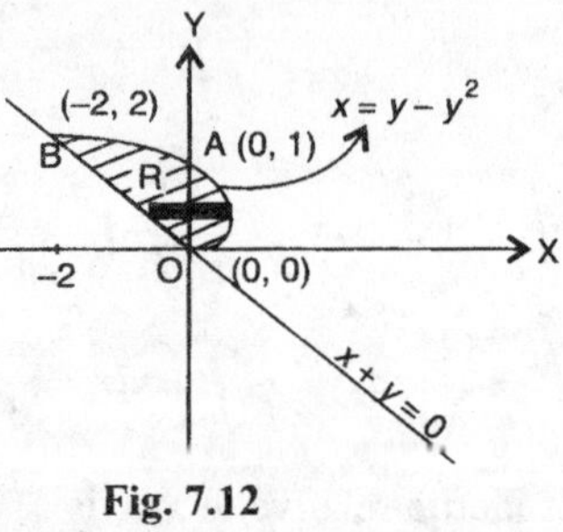

Fig. 7.12

This region R can be covered with x varying from $x = y - y^2$ to $x = -y$ and y from 0 to 2. Here the density function is $f(x,y) = x+y$. Thus the mass M contained in R is

$$M = \iint_R f(x,y)\,dx\,dy = \int_0^2 \int_{x=-y}^{y-y^2} (x+y)\,dx\,dy$$

$$= \int_0^2 \left(\frac{x^2}{2} + xy \right)\Bigg|_{-y}^{y-y^2} dy$$

$$= \int_0^2 \left(\frac{1}{2}y^4 - 2y^3 + 2y^2 \right) dy$$

$$= \left(\frac{1}{2}\frac{y^5}{5} - 2y^4 + 2\frac{y^3}{3} \right)\Bigg|_0^2$$

$$M = \frac{8}{15}$$

$$x_c = \frac{1}{M} \iint_R x f(x,y)\,dx\,dy$$

$$= \frac{15}{8} \int_0^2 \int_{-y}^{y-y^2} x(x+y)\,dx\,dy$$

$$= \int_0^2 \left(\frac{x^3}{3} + \frac{x^2}{2}y \right)\Bigg|_{-y}^{y-y^2} dy$$

$$= \int_0^2 \frac{y^3}{6}[4 - 12y + 9y^2 - 2y^3]dy$$

$$= \frac{1}{6}\left[\frac{4y^4}{4} - \frac{12y^5}{5} + \frac{9y^6}{6} - \frac{2y^7}{7} \right]_0^2 = \frac{-8}{35}$$

$$y_c = \frac{1}{M} \iint y\,f(x,y)\,dx\,dy$$

$$= \frac{15}{8} \int_0^2 \int_{-y}^{y-y^2} y(x+y)\,dx\,dy$$

$$= \frac{15}{8} \int_0^2 \left(\frac{yx^2}{2} + y^2x \right)\Bigg|_{-y}^{y-y^2} dy$$

$$= \frac{15}{8} \int_0^2 \frac{1}{2}\left(y^5 - 3y^4 + \frac{5}{2}y^3 \right) dy$$

$$y_c = \frac{15}{16}\left[\frac{y^6}{6} - \frac{3y^5}{5} + \frac{5}{2}\frac{y^4}{4} \right]_0^2 = \frac{15}{16}\frac{22}{15} = \frac{11}{8}$$

$$I_x = \int_0^2 \int_{-y}^{y-y^2} y^2(x+y)\,dx\,dy$$

$$= \int_0^2 \left(\frac{y^2x^2}{2} + y^3x \right)\Bigg|_{-y}^{y-y^2} dy$$

$$I_x = \int_0^2 \left(\frac{y^6}{2} - 2y^5 + 2y^4 \right) dy$$

$$= \frac{y^7}{14} - \frac{2y^6}{6} + \frac{2y^5}{5}\Bigg|_0^2 = \frac{64}{105}$$

$$I_y = \int_0^2 \int_{-y}^{y-y^2} x^2(x+y)\,dx\,dy$$

$$= \int_0^2 \left(\frac{x^4}{4} + \frac{x^3}{3}y\right)\Bigg|_{-y}^{y-y^2} dy$$

$$= \frac{1}{12}\int_0^2 (3y^8 - 16y^7 + 30y^6 - 24y^5 + 8y^4)dy$$

$$I_y = \frac{1}{12}\left[\frac{3y^9}{9} - \frac{16y^8}{8} + \frac{30y^7}{7} - \frac{24y^6}{6} + \frac{8y^5}{5}\right]_0^2$$

$$= \frac{2^8}{7 \times 3 \times 5} = \frac{256}{105}$$

$$I_0 = I_x + I_y = \frac{64}{105} + \frac{256}{105} = \frac{320}{105} = \frac{64}{21}$$

EXERCISE

Mass, Centroid, and Moments of Inertia

1. Determine the mass of a circular lamina of radius b if the surface density of the material at any point $P(x, y)$ is proportional to the distance of the point (x, y) from the centre of the circle.

   Hint: Density $= f(x, y) = k\sqrt{x^2 + y^2}$, k constant.

Ans. $\frac{2}{3}k\pi b^3$

2. Find the mass contained in the ellipse $\frac{x^2}{a^2} + \frac{y^2}{b^2} = 1$ with surface density $(x + y)^2$.

   Hint: Limits: $y = \pm\frac{b}{a}\sqrt{a^2 - x^2}$, $x : \pm a$.

Ans. $\frac{\pi ab}{4}(a^2 + b^2)$

3. Find the centroid of the region in the first quadrant bounded by $\left(\frac{x}{a}\right)^{\frac{2}{3}} + \left(\frac{y}{b}\right)^{\frac{2}{3}} = 1$, $x = 0$, $y = 0$ and having surface density kxy with k constant.

   Hint: Put $x = a\cos^3\theta$, $y = b\sin^3\theta$ with θ varying from 0 to $\frac{\pi}{2}$.

Ans. $\left(\frac{128}{429}a, \frac{128}{429}b\right)$

4. Find the centroid of the upper half of the circle $x^2 + y^2 = a^2$.

Ans. $\left(0, \frac{\pi a^2}{2}\right)$

5. Find the centroid of the plane region bounded by $y^2 + x = 0$ and $y = x + 2$.

   Hint: Points of intersection: $(-1, 1)$, $(-4, -2)$.

   Limits: $x = y - 2$ to $-y^2$, $y : -2$ to 1.

Ans. $\left(-\frac{8}{5}, -\frac{1}{2}\right)$

6. Determine the coordinates of the centre of gravity of the plane region bounded by $y = \ln x$, $y = 1$, $y = 0$, $x = 0$

   Hint: Points of intersection $(1, 0)$, $(e, 1)$.

   Limits: $x : 0$ to e^y; $y : 0$ to 1.

Ans. $\left(\frac{e+1}{4}, \frac{1}{e-1}\right)$

7. Calculate the centroid of the area bounded by the parabola $x^2 + 4y - 16 = 0$ and x-axis.

   Hint: $x_c = 0$ due to symmetry about y-axis.

   Limits: $y : 0$ to $(16 - x^2)/4$; $x : \pm 4$.

Ans. $\left(0, \frac{8}{5}\right)$

8. Find the centroid of the area bounded by the parabolas $y^2 = 20x$, $x^2 = 20y$.

   Hint: Limits: $y : \sqrt{20x}$ to $\frac{x^2}{20}$; $x : 0$ to 20.

Ans. (9, 9)

9. Find moment of inertia of circular region of radius b relative to its centre O.

Ans. $\frac{\pi b^4}{2}$

10. Compute the moment of inertia relative to the origin of the area of a rectangle bounded by the straight lines $x = 0$, $x = a$, $y = 0$, $y = b$.

Ans. $\frac{ab(a^2 + b^2)}{3}$

11. Compute the moment of inertia of the area of the circle $(x - a)^2 + (y - b)^2 = 2a^2$ relative to the y-axis.

Ans. $3\pi a^4$

12. Find the moment of inertia of the region bounded by the parabola $y^2 = ax$ and the straight line $x = a$ relative to the straight line

$y = -a$.

Ans. $\frac{8}{5}a^4$

Calculate I_x, I_y for the regions bounded by

13. $4y = x^3$, $y = |x|$

Ans. $I_x = \frac{4}{5}$, $I_y = \frac{4}{3}$

14. $y = e^x$, $y = e$, $x = 0$

Ans. $I_x = (1 + 2e^3)/9$
$I_y = 2(3 - e)/3$

15. Find the centre of gravity and the moments of inertia I_x, I_y, I_0 of the region, $0 \le y \le \sqrt{1 - x^2}, 0 \le x \le 1$ and $f(x, y) = 1$ be the density of mass.

Ans. Mass $M = \frac{\pi}{4}$, $\bar{x} = \bar{y} = \frac{4}{3\pi}$ (by symmetry),
$I_x = \frac{\pi}{16}$, $I_y = \frac{\pi}{16}$, $I_{0} = I_x + I_y = \frac{\pi}{8}$.

16. A thin plate of uniform thickness and constant density ρ covers the region of xy-plane and is bounded by $y = x^2$ and $y = x + 2$. Find the mass M. Find its moment of inertia I_y about the y-axis.

Hint: Limits: $y : x^2$ to $x + 2$, $x : -1$ to 2.

Ans. $I_y = \frac{63}{20}\rho$, Mass $M = \frac{9}{2}\rho$.

WORKED OUT EXAMPLES

Volume of solid by double integral

Example 1: Find the volume of the tetrahedron in space cut from the first octant by the plane $6x + 3y + 2z = 6$.

Solution: The tetrahedron is bounded below by $z = 0$ (xy-plane) and bounded above by the surface $z = \frac{6-6x-3y}{2} = f(x, y)$.

The projection R of this surface $z = f(x, y)$ onto the xy-plane is the triangular region OAB bounded by the lines $x = 0$, $y = 0$ and $6x + 3y = 6$. This injected region R can be covered by varying y from 0 to $2 - 2x$, and x from 0 to 1. Thus the required volume V of the tetrahedron is given by (Fig. 7.13)

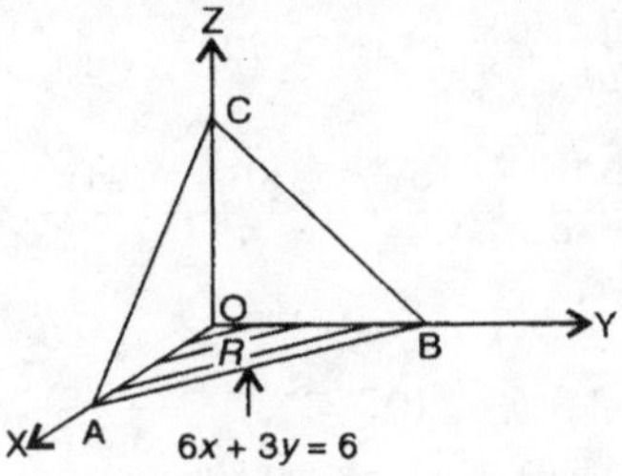

Fig. 7.13

$$V = \iint_R f(x, y)dx\,dy$$

$$= \int_0^1 \int_{y=0}^{2-2x} \frac{6 - 6x - 3y}{2} dy\,dx$$

$$= \int_0^1 (3 - 3x)y - \frac{3}{2}\frac{y^2}{2}\Big|_0^{2-2x} dx$$

$$= 3\int_0^1 (1 - x)^2 dx = 1$$

Example 2: Calculate the volume of a solid whose base is in a xy-plane and is bounded by the parabola $y = 4 - x^2$ and the straight line $y = 3x$, while the top of the solid is in the plane $z = x + 4$ (see Fig. 7.14).

Solution: The points of intersection of the two curves $y = 4 - x^2$ and $y = 3x$ are $(-4, -12)$, $(1, 3)$. The projection R of $z = x + 4$ in the xy-plane is bounded by these curves. So R can be covered by varying y from $3x$ (lower curve) to $4 - x^2$ (top curve) and x from -4 to 1. Thus the required volume V is

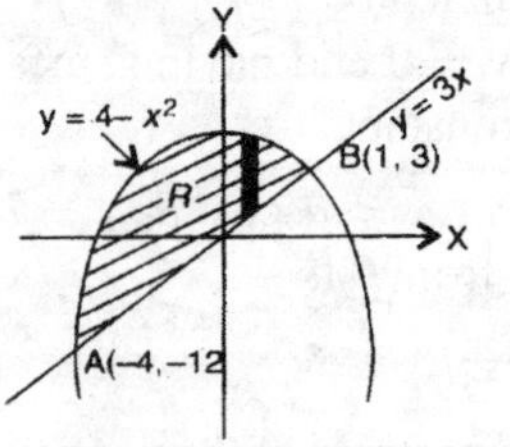

Fig. 7.14

$$V = \int_{-4}^{1} \int_{3x}^{4-x^2} (x + 4)dy\,dx$$

$$= \int_{-4}^{1} (x+4)(4-x^2-3x)dx$$

$$= \int_{-4}^{1} (-x^3-7x^2-8x+16)dx$$

$$= \frac{-x^4}{4} - \frac{7x^3}{3} - \frac{8x^2}{2} + 16x\Big|_{-4}^{1} = \frac{625}{12}$$

Example 3: Determine the volume of the space below the paraboloid $x^2+y^2+z-4=0$ and above the square in the xy-plane with vertices at (0, 0), (0, 1), (1, 0), (1, 1).

Solution: As shown in Fig. 7.15 the top surface is $z = 4 - x^2 - y^2 = f(x, y)$. The projection R of this surface on the xy plane into the square is the square itself with vertices $O(0,0)$, $A(1,0)$, $B(1,1)$, $C(0,1)$. Thus,

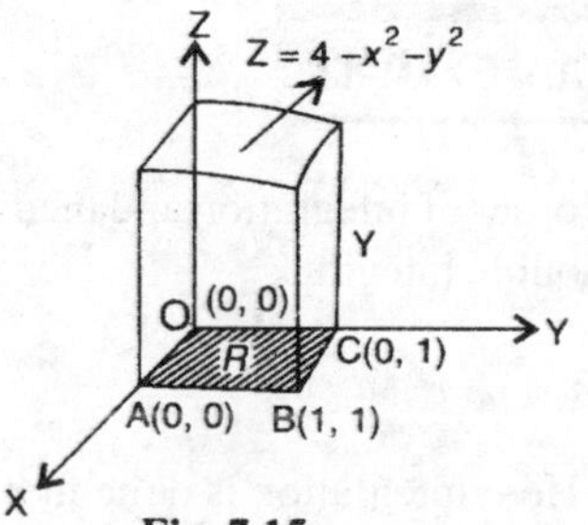

Fig. 7.15

$$V = \iint_R f(x,y)dx\,dy = \int_0^1 \int_0^1 (4-x^2-y^2)dx\,dy$$

$$V = \int_0^1 \left(4y - x^2y - \frac{y^3}{3}\right)\Big|_0^1 dx$$

$$= \int_0^1 \left(\frac{11}{3} - x^2\right)dx = \frac{11}{3}x - \frac{x^3}{3}\Big|_0^1 = \frac{10}{3}$$

Example 4: Find the volume of the solid under the surface $az = x^2 + y^2$ and whose base R is the circle $x^2+y^2=a^2$ (Fig. 7.16).

Solution: Here $f(x, y) = (x^2+y^2)/a$.

$$V = \iint_R f(x,y)dx\,dy$$

$$V = \int_{-a}^{a} \int_{-\sqrt{a^2-x^2}}^{\sqrt{a^2-x^2}} \frac{x^2+y^2}{a} dy\,dx$$

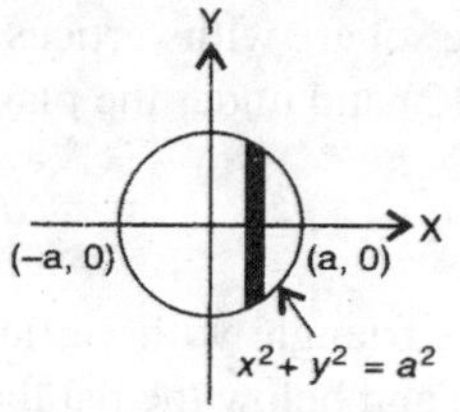

Fig. 7.16

Integration in cartesian coordinates x, y will be cumbersome. Instead by introducing polar coordinates, $x = r\cos\theta$, $y = r\sin\theta$, then $x^2+y^2 = r^2\cos^2\theta + r^2\sin^2\theta = r^2$. The base R is covered by varying r from 0 to a and θ from 0 to 2π. Thus, (see Fig. 7.16)

$$V = \int_0^{2\pi} \int_0^a \left(\frac{r^2}{a}\right) r\,dr\,d\theta$$

$$= \int_0^{2\pi} \frac{r^4}{4a}\Big|_0^a d\theta = \frac{a^3}{4}2\pi = \frac{\pi a^3}{2}$$

EXERCISE

Volume of Solids by Double Integral

By means of double integral, find the volume V of the following solids bounded by

1. $x = 0,\ y = 0,\ z = 0,\ \frac{x}{a} + \frac{y}{b} + \frac{z}{c} = 1$

Ans. $\frac{abc}{6}$

2. The cylinder: $x^2+y^2=4$ and the planes $y + z = 4$ and $z = 0$.

   Hint: $z = f(x, y) = 4 - y$.

   Limits: x : 0 to $\sqrt{4-y^2}$, y : −2 to 2.

Ans. 16π

3. $y = x^2,\ x = y^2,\ z = 0;\ z = 12 + y - x^2$.

Ans. 549/144

4. $z = 0,\ x^2+y^2=1,\ x+y+z=3$.

Ans. 3π

5. Vertical plane $zx + y = z$, coordinate planes, top plane $z = 1 + y$.

   Hint: $z = f(x, y) = 1 + y$.

   Limits: y : 0 to $2 - 2x$, x : 0 to 1.

Ans. $\frac{5}{3}$

6. Above the square with vertices at (0, 0), (2, 0), (0, 2), (2, 2) and under the plane $z = 8 - x + y$.

Ans. 32

7. Above the triangle with vertices (0, 0), (2, 0) and (2, 1) and below the paraboloid $z = 24 - x^2 - y^2$

Ans. $\frac{131}{6}$

8. Under the surface $z = y(x+2)$ and over the area bounded by $x + y = 0$, $y = 1$, $y = \sqrt{x}$.

Ans. 9/8

9. Paraboloid $z = 4 - x^2 - y^2$ and the xy-plane.

Hint: $z = f(x, y) = 4 - x^2 - y^2$;
$R : x^2 + y^2 = 4$.

Limits: $y = \pm\sqrt{4 - x^2}$ and $x : \pm 2$.

Ans. 8π

10. $z = 0$, $z = x + y + 2$ and inside the cylinder $x^2 + y^2 = 16$.

Hint: $z = f(x, y) = x + y + z$.

Limits: $y : 0$ to $\sqrt{16 - x^2}$, $x : 0$ to 4.

Ans. $\left(\frac{128}{3} + 8\pi\right)$

11. Above by paraboloid $x^2 + 4y^2 = z$, below by plane $z = 0$ and laterally by the cylinder $y^2 = x$ and $x^2 = y$.

Hint: $z = f(x, y) = x^2 + 4y^2$.

Limits: $y : x^2$ to $\sqrt{x}$, $x : 0$ to 1.

Ans. 3/7

12. Find the volume of a sphere of radius b.

Hint: $f(x, y) = \sqrt{b^2 - x^2 - y^2}$.

Limits: $y = \pm\sqrt{a^2 - x^2}$, $x = \pm b$.

Ans. $\frac{4}{3}\pi b^3$.

7.3 CHANGE OF ORDER OF INTEGRATION: DOUBLE INTEGRAL

As already seen, for the double integral with variable limits

$$I_D = \iint_D f(x, y)ds \tag{1}$$

The limits of integration can be fixed from a rough sketch of the domain of integration. Then (1) can be evaluated as a two-fold iterated integral using either

$$I_D = \int_{x=a}^{b} \int_{y=y_1(x)}^{y_2(x)} f(x, y)dy\,dx \tag{2}$$

or

$$I_D = \int_{y=d}^{e} \int_{x=x_1(y)}^{x_2(y)} f(x, y)dx\,dy \tag{3}$$

In each specific problem, depending upon the type of the domain D and/or the nature of integrand, choose either of the form (2) or (3) whichever is easier to evaluate. Thus in several problems, the evaluation of double integral becomes easier with the change of order of integration, which of course, changes the limits of integration also.

WORKED OUT EXAMPLES

Change the order of integration and then evaluate the following double integrals:

Example 1: $\int_0^2 \int_1^{e^x} dy\,dx$

Solution: Here integration is done first w.r.t. y and later w.r.t. x. So y varies from 1 to e^x while x varies from 0 to 2. Thus the domain of integration ABE is shaded as shown in Fig. 7.17, and is bounded by the curve $y = e^x$ and the straight line $y = 1$, $x = 0$ and $x = 2$. Here A(0, 1), B(2, 1), E(2, e^2).

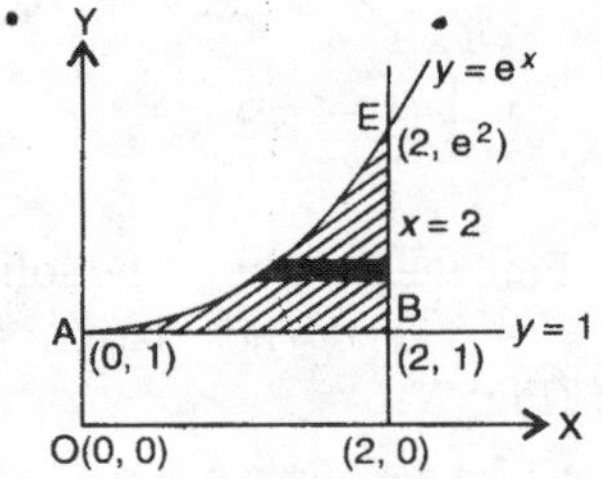

Fig. 7.17

When the order of integration is changed, integrate first w.r.t. x and later w.r.t. y. For this purpose, consider a horizontal strip whose length x varies from $x = \ln y$ to $x = 2$ and width y varies from $y = 1$ to

$y = e^2$. Thus the given integral can be written with change of order of integration as follows:

$$I_D = \int_0^2 \int_1^{e^x} dy\, dx = \int_1^{e^2} \int_{x=lny}^{2} dx\, dy$$

$$= \int_1^{e^2} (2 - lny) dy$$

$$= (2y - ylny + y)|_1^{e^2} = e^2 - 3$$

Example 2: $\int_{-2}^{1} \int_{x^2+4x}^{3x+2} dy\, dx$

Solution: Since the integration is done first w.r.t. y and later w.r.t. x, the domain D of integration is bounded by the following curves: $y = x^2 + 4x$, and the straight lines $y = 3x + 2$, $x = -2$ and $x = 1$ as shown shaded in Fig. 7.18. Here $A(-2, -4)$, $B(0, 0)$, $E(1, 5)$, $F(0, 2)$, $G(-\frac{2}{3}, 0)$. Considering horizontal strip, x varies from the upper curve $x = \sqrt{4 + y} - 2$ to the lower curve $x = (y - 2)/3$ and then y varies from -4 to 5. Changing the order of integration to first x and later to y, the double integral becomes

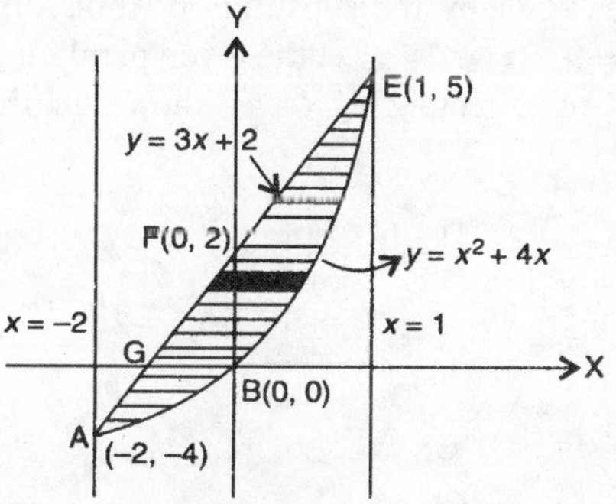

Fig. 7.18

$$\int_{-2}^{1} \int_{y=x^2+4x}^{3x+2} dy\, dx$$

$$= \int_{-4}^{5} \int_{x=\frac{y-2}{3}}^{\sqrt{y+4}-2} dx\, dy$$

$$= \int_{-4}^{5} \left[\sqrt{y+4} - 2 - \left(\frac{y-2}{3}\right)\right] dy$$

$$= \left[\frac{2}{3}(y+4)^{3/2} - \frac{y^2}{6} - \frac{4}{3}y\right]_{-4}^{5} = \frac{9}{2}$$

EXERCISE

Change (reverse) the order of integration and then evaluate the following double integrals:

1. $\int_1^2 \int_3^4 (x + y) dx\, dy$.

   Hint: $\int_3^4 \int_1^2 (x + y) dx\, dy$.

Ans. 5

2. $\int_0^1 \int_x^{\sqrt{x}} xy\, dy\, dx$

   Hint: Limits: $x : y^2$ to y; $y : 0$ to 1.

Ans. $\frac{1}{24}$

3. $\int_0^1 \int_{x^2}^{2-x} xy\, dy\, dx$.

   Hint: D = Domain = sum of two domains = $D_1 + D_2$.

   Limits: $D_1 : x : 0$ to $\sqrt{y}$, $y : 0$ to 1; $D_2 : x : 0$ to $2 - y$; $y : 1$ to 2.

Ans. $\frac{3}{8}$

4. $\int_0^2 \int_{y^3}^{4\sqrt{2y}} y^2 dx\, dy$.

   Hint:

   Limits: $y : x^2/32$ to $x^{\frac{1}{3}}$; $y : 0$ to 8.

Ans. $\frac{160}{21}$

5. $\int_0^a \int_{\frac{y^2}{a}}^{2a-y} xy\, dx\, dy$.

   Hint: Domain $D = D_1 + D_2$.

   Limits: $D_1 : y : 0$ to $\sqrt{ax}$, $x : 0$ to a; $D_2 : y : 0$ to $2a - x$, $x : 0$ to 2a.

Ans. $\frac{3a^4}{8}$

6. $\int_0^1 \int_{y^2}^{y^{1/3}} f(x, y) dx\, dy$.

Ans. $\int_0^1 \int_{x^3}^{\sqrt{x}} f(x, y) dy\, dx$

7. $\int_{-1}^2 \int_{-x}^{2-x^2} f(x, y) dy\, dx$.

Ans. $\int_{-2}^1 \int_{-y}^{\sqrt{2-y}} f(x, y) dx\, dy$
$+ \int_1^2 \int_{-\sqrt{2-y}}^{\sqrt{2-y}} f(x, y) dx\, dy$

8. $\int_0^a \int_0^{\sqrt{2ay-y^2}} f(x, y) dx\, dy$.

Ans. $\int_0^a \int_{a-\sqrt{a^2-x^2}}^{a} f(x, y) dy\, dx$

9. $\int_0^{\sqrt{2}}\int_{-\sqrt{4-2y^2}}^{\sqrt{4-2y^2}} y\,dx\,dy.$

Hint:

Limits: $y : 0$ to $\sqrt{(4-x^2)/2}$, $x : -2$ to 2.

Ans. $\frac{8}{3}$

10. $\int_0^a \int_{\frac{x}{a}}^{\sqrt{\frac{x}{a}}} (x^2+y^2)dy\,dx.$

Hint:

Limits: $x : ay^2$ to ay; $y : 0$ to 1.

Ans. $\frac{a^3}{28}+\frac{a}{20}$

11. $\int_0^a \int_0^{\sqrt{a^2-x^2}} \sqrt{a^2-x^2-y^2}\,dy\,dx.$

Hint:

Limits: $x : 0$ to $\sqrt{a^2-y^2}$, $y : 0$ to a.

Ans. $\frac{\pi a^3}{6}$

12. $\int_0^{2a}\int_0^{\sqrt{2ay-y^2}} dx\,dy.$

Hint:

Limits: $y : a \pm \sqrt{a^2-x^2}$, $x : 0$ to a.

Ans. $\frac{\pi a^2}{2}$

7.4 GENERAL CHANGE OF VARIABLES IN DOUBLE INTEGRAL

In several cases, the evaluation of double integrals becomes easy when there is a change of variables.

Let D be domain in xy-plane and let x, y be the rectangular cartesian coordinates of any point P in D. Let u, v be new variables in domain D^* such that x, y and u, v are connected through the continuous functions (transformations).

$$x = g(u, v),\ y = h(u, v) \tag{1}$$

Then u, v are said to be curvilinear coordinates of point P^* in D^* which uniquely corresponds to P in D. Solving (1) for u and v, we get

$$u = g^*(x, y),\ v = h^*(x, y) \tag{2}$$

Then a given double integral in the given (old) variables x, y can be transformed to a double integral in terms of new variables u, v as follows:

$$\iint_D f(x, y)dx\,dy = \iint_{D^*} F(u, v)|J|du\,dv \tag{3}$$

Here $f(x, y) = f(x(u, v),\ y(u, v)) = F(u, v)$ and J is the Jacobian (functional determinant) defined as

$$J = J\left(\frac{x, y}{u, v}\right) = \frac{\partial(x, y)}{\partial(u, v)} = \begin{vmatrix} \frac{\partial x}{\partial u} & \frac{\partial x}{\partial v} \\ \frac{\partial y}{\partial u} & \frac{\partial y}{\partial v} \end{vmatrix}$$

(3) is knows as formula for transformation of coordinates in double integral.

WORKED OUT EXAMPLES

Example 1: Evaluate $\iint_R (x+y)^2 dx\,dy$ where R is region bounded by the parallelogram $x+y=0$, $x+y=2$, $3x-2y=0$, $3x-2y=3$.

Solution: By changing the variables x, y to the new variables u, v, by the substitution (transformation) $x+y=u$, $3x-2y=v$, the given parallelogram R reduces to a rectangle R^* as shown in the (Fig. 7.19):

$$\frac{\partial(u, v)}{\partial(x, y)} = \begin{vmatrix} \frac{\partial u}{\partial x} & \frac{\partial u}{\partial y} \\ \frac{\partial v}{\partial x} & \frac{\partial v}{\partial y} \end{vmatrix} = \begin{vmatrix} 1 & 1 \\ 3 & -2 \end{vmatrix} = -5$$

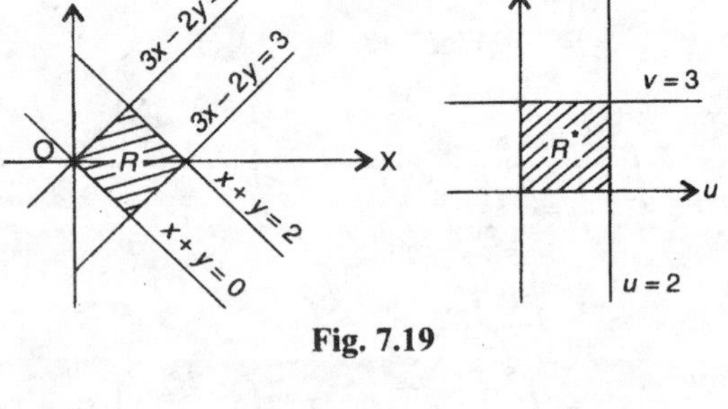

Fig. 7.19

So required Jacobian $J = \frac{\partial(x, y)}{\partial(u, v)} = -\frac{1}{5}$.

Since, $u = x + y = 0$ and $u = x + y = 2$, u varies from 0 to 2, while v varies from 0 to 3 since $3x-2y=v=0$, $3x-2y=v=3$. Thus the given integral in terms of the new variables u, v is

$$\iint_R (x+y)^2 dx\,dy = \iint_{R^*} u^2 \left|\frac{1}{-5}\right| du\,dv$$

$$= \frac{1}{5}\int_0^3 \int_0^2 u^2 du\,dv$$

$$= \frac{1}{5}\int_0^3 \left.\frac{u^3}{3}\right|_0^2 dv = \frac{8}{15}\cdot v\Big|_0^3 = \frac{8}{5}$$

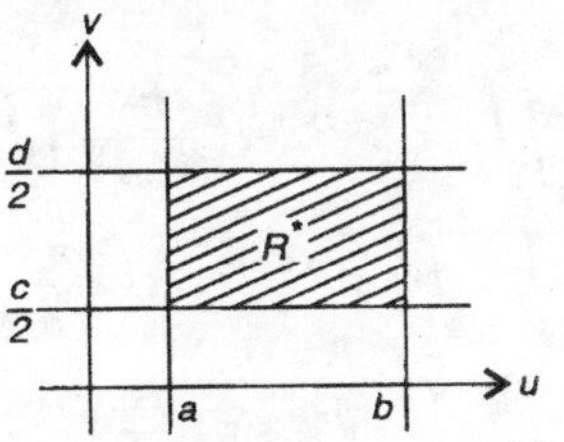

Fig. 7.21

Example 2: Evaluate $\iint_R (x^2+y^2)dx\,dy$ where R is the region in the first quadrant bounded by $x^2 - y^2 = a, x^2 - y^2 = b, 2xy = c, 2xy = d,$ $0 < a < b, 0 < c < d$.

Solution: Introducing $x^2 - y^2 = u, xy = v$, the given region R gets transformed to a rectangle R^* in the uv plane given by

$$a < u < b; \quad \frac{c}{2} < v < \frac{d}{2}$$

Since

$$\frac{\partial(u,v)}{\partial(x,y)} = \begin{vmatrix} \frac{\partial u}{\partial x} & \frac{\partial u}{\partial y} \\ \frac{\partial v}{\partial x} & \frac{\partial v}{\partial y} \end{vmatrix} = \begin{vmatrix} 2x & -2y \\ y & x \end{vmatrix} = 2(x^2+y^2)$$

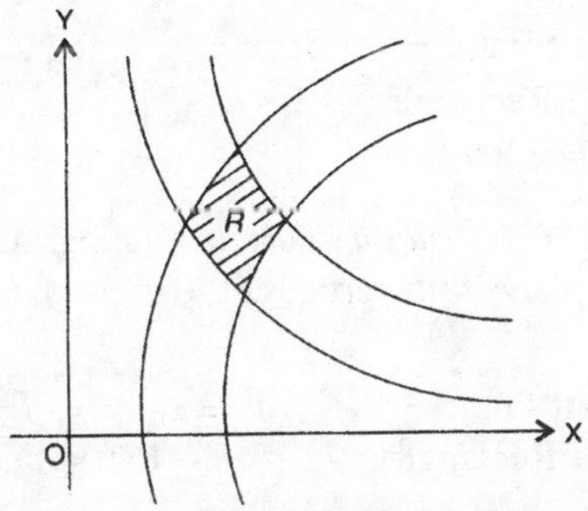

Fig. 7.20

so that Jacobian $= J = \frac{\partial(x,y)}{\partial(u,v)} = \frac{1}{2(x^2+y^2)}$. Thus,

$$\iint_R (x^2+y^2)dx\,dy$$

$$= \iint_{R^*} (x^2+y^2)\left|\frac{1}{2(x^2+y^2)}\right| du\,dv$$

$$= \int_{\frac{c}{2}}^{\frac{d}{2}} \int_a^b du\,dv = \frac{1}{2}(b-a)\cdot\left(\frac{d}{2}-\frac{c}{2}\right)$$

$$= \frac{(b-a)(d-c)}{4}$$

Example 3: Evaluate $\iint_R e^{-(x+y)} \sin\left(\frac{\pi y}{x+y}\right) dx\,dy$ where R is the entire first quadrant in the xy-plane.

Solution: In the first quadrant both x and y vary from 0 to ∞. Put $x + y = u, y = v$ so that $x = u - v, y = v$ (Fig. 7.22).

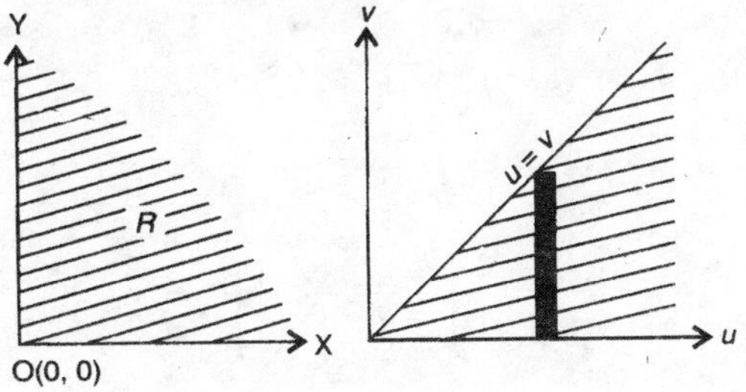

Fig. 7.22

i. *Origin* O(0, 0) *in xy-plane:* $0 = u - v$, $0 = v, u = 0$. Thus $(0, 0)$ in xy-plane corresponds to $(0, 0)$ in u, v-plane.

ii. *x-axis:* $y = 0$ i.e., $v = 0$, so $x = u - 0 = u$ since $x \geq 0, u = x \geq 0$.

Thus x-axis gets transformed to u-axis ($v = 0, u \geq 0$)

iii. *y-axis:* $x = 0$ so $0 = x = u - v$, so that $u = v$ since $y \geq 0, u = v = y \geq 0$.

Thus y-axis gets mapped to the line $u = v$ in the uv-plane.

iv. *Any point* $x > 0, y > 0$.

Since $v = y > 0$ and $u - v = x > 0$ so that $u > v$.

Thus any point $(x > 0, y > 0)$ corresponds to (u, v) where $u > v > 0$. Thus the entire first quadrant in xy-plane gets transformed to the region between u-axis and the line $u = v$.

Thus

$$I = \iint_R e^{-(x+y)} \sin\left(\frac{\pi y}{x+y}\right) dx\, dy$$

$$= \int_0^{\infty}\int_0^{\infty} e^{-(x+y)} \sin\left(\frac{\pi y}{x+y}\right) dy\, dx$$

$$= \int_0^{\infty}\int_{v=0}^{u} e^{-u} \sin\left(\frac{\pi v}{u}\right)\cdot 1 \cdot dv\, du$$

since

$$\text{the Jacobian} = J = \frac{\partial(x,y)}{\partial(u,v)} = \begin{vmatrix} \frac{\partial x}{\partial u} & \frac{\partial x}{\partial u} \\ \frac{\partial y}{\partial u} & \frac{\partial y}{\partial v} \end{vmatrix} = \begin{vmatrix} 1 & -1 \\ 0 & 1 \end{vmatrix} = 1$$

$$= \int_0^{\infty} e^{-u}\cdot\left(-\cos\frac{\pi v}{u}\right)\cdot\frac{u}{\pi}\Bigg|_0^u du$$

$$= \int_0^{\infty} -\frac{u}{\pi} e^{-u}\cdot(-1 - 1) du$$

$$= \frac{2}{\pi}\int_0^{\infty} u e^{-u} du$$

$$= \frac{2}{\pi}\left[\frac{ue^{-4}}{-1} - e^{-u}\right]_0^{\infty} = \frac{2}{\pi}$$

Example 4: $\int_0^1 \int_0^x \sqrt{x^2+y^2}\,dx\,dy$

Solution: Region of integration R is the triangle bounded by $y = 0$, $x = 1$ and $y = x$ (Fig. 7.23).

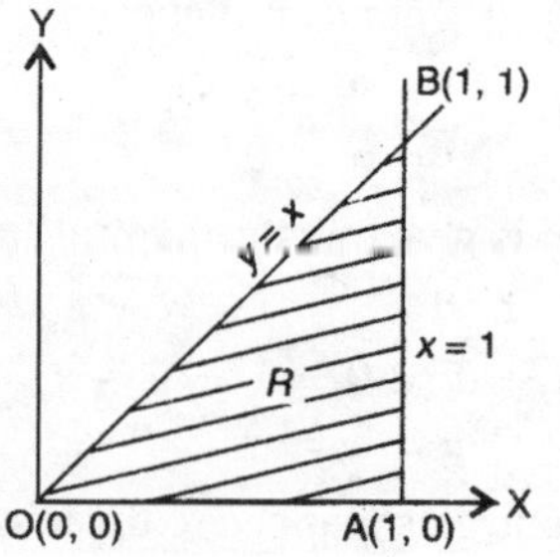

Fig. 7.23

Put $x = u$, $y = uv$

$$J = \text{Jacobian} = \frac{\partial(x,y)}{\partial(u,v)} = \begin{vmatrix} \frac{\partial x}{\partial u} & \frac{\partial x}{\partial v} \\ \frac{\partial y}{\partial u} & \frac{\partial y}{\partial v} \end{vmatrix}$$

$$= \begin{vmatrix} 1 & 0 \\ v & u \end{vmatrix} = u$$

In the given region R, x: varies from 0 to 1 while y varies from 0 to x. Since $u = x$, so u varies from 0 to 1. Similarly since $0 \le y = uv \le x = u$ so v varies from 0 to 1. Thus,

$$\int_0^1\int_0^x \sqrt{x^2+y^2}\,dx\,dy$$

$$= \int_0^1\int_0^1 u\sqrt{1+v^2}\,u\,du\,dv = \frac{1}{3}\int_0^1 \sqrt{1+v^2}\,dv$$

$$= \frac{1}{3}\left[\frac{v\sqrt{(1+v^2)}}{2} + \frac{1}{2}\sinh^{-1} v\right]_0^1$$

$$= \frac{1}{3}\left[\frac{\sqrt{2}}{2} + \frac{1}{2}\sinh^{-1} 1\right]$$

EXERCISE

Evaluate the following integrals by changing the variables:

1. $\iint_D (y-x)dx\,dy$; D: region in xy-plane bounded by the straight lines $y = x + 1$, $y = x - 3$, $y = -\frac{1}{3}x + \frac{7}{3}$, $y = -\frac{1}{3}x + 5$.

   Hint: Put $u = y - x$, $v = y + \frac{1}{3}x$; $J = -\frac{3}{4}$

   D^* : Rectangle: $-3 \le u \le 1$, $\frac{7}{3} \le v \le 5$.

Ans. -8

2. $\iint_R (x+y)^2 dx\,dy$; R: parallelogram in the xy-plane with vertices (1, 0), (3, 1), (2, 2), (0, 1).

   Hint: Put $u = x + y$, $v = x - 2y$, $J = -\frac{1}{3}$

   D^* : Rectangle: $-2 \le u \le 1$, $1 \le v \le 4$.

Ans. 21

3. $\int_0^1 \int_0^{1-x} e^{\frac{y}{(x+y)}} dy\,dx$.

   Hint: Put $u = x + y$, $uv = y$.

Ans. $\frac{(e-1)}{2}$

4. $\iint_D [xy(1-x-y)]^{\frac{1}{2}} dx\,dy$; D: is region bounded by the triangle with the sides $x = 0$, $y = 0$, $x + y = 1$.

   Hint: Put $u = x + y$, $uv = y$, $J = u$,

   D^* : square $0 \le u \le 1$, $0 \le v \le 1$.

Ans. $\frac{2\pi}{105}$

5. $\iint_R (x-y)^4 e^{x+y} dx\, dy$; R : square with vertices at (1, 0), (2, 1), (1, 2), (0, 1).

Hint: Since the square is bounded by the lines $x+y=1, x+y=3; x-y=1, x-y=-1,$ put $x+y=u, \quad x-y=v;$ $J=-\frac{1}{2},$
D^* : rectangle :$1 \le u \le 3, -1 \le v \le 1$.

Ans. $\frac{(e^3-e)}{5}$

6. $\iint_R xy\, dx\, dy$; R: Region in the first quadrant bounded by the hyperbolas $x^2-y^2=a^2$ and $x^2-y^2=b^2$ and the circles $x^2+y^2=c^2$, $x^2+y^2=d^2$ with $0<a<b<c<d$.

Hint: Put $x^2-y^2=u, x^2+y^2=v;$
$J=8xy$, R^*: rectangle $a^2 \le u \le b^2$, $c^2 \le v \le d^2$.

Ans. $\frac{1}{8}(b^2-a^2)(d^2-c^2)$

7. Find the area of the curvilinear quadrilateral bounded by the four parabolas $y^2=ax$, $y^2=bx, x^2=cy, x^2=dy$.

Hint: Put $y^2=u^3x, x^2=b^3y; J=-3u^2v^2$.

Limits: $u : a^{\frac{1}{3}}$ to $b^{\frac{1}{3}}; v : c^{\frac{1}{3}}$ to $d^{\frac{1}{3}}$.

Ans. $\frac{(b-a)(d-c)}{3}$

8. $\iint_D e^{(x-y)/(x+y)} dx\, dy$; D: triangle bounded by $y=0, x=1$ and $y=x$. Use $x=u-uv$, $y-uv$ to transform the double integrals.

Ans. $\frac{(e^2-1)}{4e}$

9. $\int_0^e \int_{\alpha x}^{\beta x} f(x, y) dy\, dx$.

Ans. $\int_{\frac{\alpha}{1+\alpha}}^{\frac{\beta}{1+\beta}} \int_0^{\frac{e}{1-v}} f(u-uv, uv) u\, du\, dv$

10. $\int_0^c \int_0^b f(x, y) dy\, dx$.

Ans. $\int_0^{\frac{b}{b+c}} \int_0^{\frac{c}{1-v}} f(u-uv, uv) u\, du\, dv$
$+ \int_{\frac{b}{b+c}}^{b} \int_0^{\frac{b}{v}} f(u-uv, uv) u\, du\, dv$

11. $\int_0^\infty \int_0^\infty \frac{x^2+y^2}{1+(x^2-y^2)^2} e^{-2xy} dx\, dy$.

Hint: Use the substitution: $u=x^2-y^2$, $v=2xy$.

Ans. $\frac{\pi}{4}$

12. Evaluate $\iint_R (x^2+y^2) dx\, dy$ where R is the region shown in Fig. 7.24.

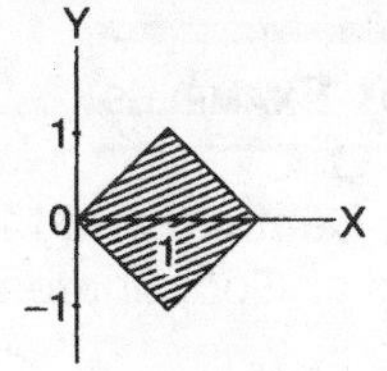

Fig. 7.24

Hint: Use $x+y=u, x-y=v$, so that $x=\frac{1}{2}(u+v), y=\frac{1}{2}(u-v), J=-\frac{1}{2},$ $0 \le u \le 2, 0 \le v \le 2$.

Ans. $\frac{8}{3}$

Change of Variables: Cartesian to Polar Coordinates

Double Integrals in Polar Coordinates

For a double integral in cartesian coordinates x, y, the change of variables to polar coordinates r, θ can be done through the transformation

$$x = r\cos\theta,\ y = r\sin\theta$$

(i.e., $u=r, v=\theta$ in general change of variables.) The Jacobian in this case is

$$J = J\left(\frac{x, y}{r, \theta}\right) = \frac{\partial(x, y)}{\partial(r, \theta)} = \begin{vmatrix} \frac{\partial x}{\partial r} & \frac{\partial x}{\partial \theta} \\ \frac{\partial y}{\partial r} & \frac{\partial y}{\partial \theta} \end{vmatrix}$$

$$= \begin{vmatrix} \cos\theta & -r\sin\theta \\ \sin\theta & r\cos\theta \end{vmatrix}$$

$$J = r(\cos^2\theta + \sin^2\theta) = r$$

Thus the double integral in cartesian coordinates x, y gets transformed to double integral in polar coordinates as follows:

$$\iint_D f(x, y) dx\, dy = \iint_{D^*} f(r\cos\theta, r\sin\theta)|r| dr\, d\theta$$

$$= \int_{\theta=\alpha}^{\beta} \int_{r=\phi_1(\theta)}^{\phi_2(\theta)} F(r, \theta) r\, dr\, d\theta$$

where $F(r, \theta) = f(r\cos\theta, r\sin\theta) = f(x, y)$ and D^* is the corresponding domain in polar coordinates. Area in polar coordinates

$$A = \iint_D ds = \iint_D dx\, dy = \iint_{D^*} r\, dr\, d\theta$$

WORKED OUT EXAMPLES

Area, Mass, Centroid and Moments of Inertia: In Polar Coordinates

Example 1: Evaluate, $I = \int_0^{2\pi} \int_{a \sin\theta}^{a} r\, dr\, d\theta$.

Solution:

$$I = \int_{\theta=0}^{2\pi} \int_{r=a\sin\theta}^{a} r\, dr\, d\theta$$

$$I = \int_0^{2\pi} \frac{r^2}{2}\Big|_{a\sin\theta}^{a} d\theta = \frac{1}{2}\int_0^{2\pi}(a^2 - a^2\sin^2\theta)d\theta$$

$$= \frac{a^2}{2}\int_0^{2\pi}\cos^2\theta\, d\theta = \frac{a^2}{2}\int_0^{2\pi}\left(\frac{1+\cos 2\theta}{2}\right)d\theta$$

$$I = \frac{a^2}{4}[\theta - \sin 2\theta]_0^{2\pi} = \frac{\pi a^2}{2}$$

Example 2: $I = \int_0^\infty \int_0^\infty e^{-(x^2+y^2)} dx\, dy$.

Solution: Using polar coordinates

$$I = \int_0^{\frac{\pi}{2}} \int_0^\infty e^{-r^2} r\, dr\, d\theta$$

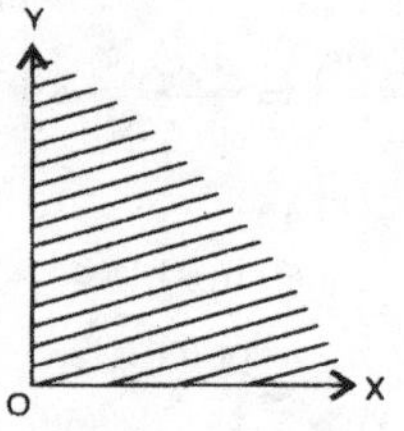

Fig. 7.25

since the first quadrant in xy-plane ($x : 0$ to ∞ and $y : 0$ to ∞) is covered when $r : 0$ to ∞ and $\theta = 0$ to $\pi/2$ (refer Fig. 7.25)

$$I = \frac{\pi}{2}\frac{1}{2}\int_0^\infty e^{-r^2} d(r^2) = \frac{\pi}{4}\frac{e^{-r^2}}{-1}\Big|_0^\infty = \frac{\pi}{4}$$

Example 3: Calculate the area which is inside the cardioid $r = 2(1+\cos\theta)$ and outside the circle $r = 2$.

Solution: $r = 2$ is a circle centred at origin and of radius 2. The shaded area in Fig. 7.26 is the region R which is outside the given circle and inside the cardioid. So r varies from the circle 2 to the cardioid $2(1+\cos\theta)$, while θ varies from $-\pi/2$ to $\pi/2$. Since R is symmetric about x-axis, the required area A of the region R is given by

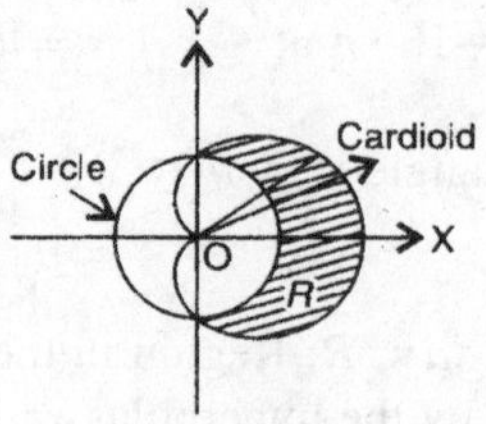

Fig. 7.26

$$A = \iint_R dA = \int_{-\frac{\pi}{2}}^{\frac{\pi}{2}} \int_{r=2}^{2(1+\cos\theta)} r\, dr\, d\theta$$

$$= 2\int_0^{\frac{\pi}{2}} \int_2^{2(1+\cos\theta)} r\, dr\, d\theta = 2\int_0^{\frac{\pi}{2}} \frac{r^2}{2}\Big|_2^{2(1+\cos\theta)} d\theta$$

$$= 4\int_0^{\frac{\pi}{2}} (2\cos\theta + \cos^2\theta)d\theta$$

$$= 4\left[2\sin\theta + \frac{1}{2}\theta + \frac{1}{4}\sin 2\theta\right]_0^{\frac{\pi}{2}} = \pi + 8.$$

Example 4: Find the centroid of the area inside $\rho = \sin\theta$ and outside $\rho = 1 - \cos\theta$.

Solution: In the required region R, shaded in Fig. 7.27, r varies from the top curve the circle $r = \sin\theta$ to the lower curve the cardioid $r = 1 - \cos\theta$; while θ varies from 0 to $\frac{\pi}{2}$. Thus,

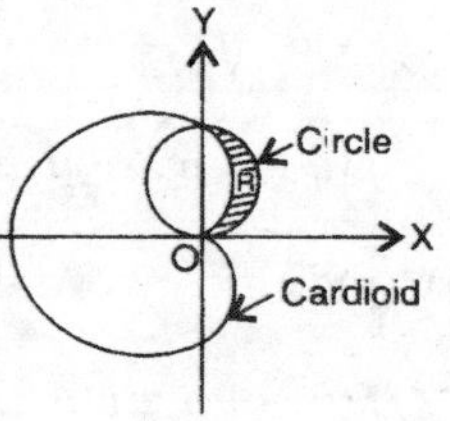

Fig. 7.27

$$\text{Area of region } R = A = \iint_R dA =$$

$$A = \int_0^{\frac{\pi}{2}} \int_{r=1-\cos\theta}^{\sin\theta} r\, dr\, d\theta = \int_0^{\frac{\pi}{2}} \frac{r^2}{2}\Big|_{1-\cos\theta}^{\sin\theta} d\theta$$

$$A = \frac{1}{2}\int_0^{\frac{\pi}{2}} (2\cos\theta - 1 - \cos 2\theta)d\theta = \frac{4-\pi}{4}.$$

Let the coordinates of the centroid be (x_c, y_c). Then

$$x_c = \frac{\iint_R x\,dA}{A} = \frac{4}{(4-\pi)}\int_0^{\frac{\pi}{2}}\int_{1-\cos\theta}^{\sin\theta} r\cos\theta\, r\,dr\,d\theta$$

$$= \frac{4}{4-\pi}\int_0^{\frac{\pi}{2}} \left.\frac{r^3}{3}\right|_{1-\cos\theta}^{\sin\theta} d\theta$$

$$= \frac{4}{3(4-\pi)}\cdot\int_0^{\frac{\pi}{2}} (\sin^3\theta - 1 + 3\cos\theta - 3\cos^2\theta + \cos^3\theta)\cos\theta\,d\theta$$

$$x_c = (15\pi - 44)/48$$

$$y_c = \frac{\iint_R y\,dA}{A} = \frac{4}{4-\pi}\int_0^{\frac{\pi}{2}}\int_{1-\cos\theta}^{\sin\theta} (r\sin\theta) r\,dr\,d\theta$$

$$= \frac{4}{(4-\pi)}\cdot\frac{1}{3}\int_0^{\frac{\pi}{2}} (\sin^3\theta - 1 + 3\cos\theta - 3\cos^2\theta + \cos^3\theta)\sin\theta\,d\theta$$

$$y_c = \frac{3\pi - 4}{12(4-\pi)}.$$

Example 5: Determine the moments of Inertia I_x, I_y and I_0 of the plane region in the first quadrant which is inside the circle $r = 4a\cos\theta$ and outside the circle $r = 2a$.

Solution: For the shaded region R shown in Fig. 7.28, r varies from (inside) curve $r = 2a$ to the (outside) curve $r = 4a\cos\theta$, while θ varies from 0 to $\frac{\pi}{3}$ (since $\theta = \pi/3$ is the point of intersection of the two circles in the first quadrant i.e., $2a = \rho = 4a\cos\theta$; $\cos\theta = \frac{1}{2}$ so $\theta = \frac{\pi}{3}$).

Thus area A is

$$A = \iint_R dA = \int_0^{\frac{\pi}{3}}\int_{2a}^{4a\cos\theta} r\,dr\,d\theta$$

$$= \frac{1}{2}\int_0^{\frac{\pi}{3}}\left[(4a\cos\theta)^2 - (2a)^2\right]d\theta$$

$$= 2a^2\int_0^{\frac{\pi}{3}}\left[\frac{4(1+\cos 2\theta)}{2} - 1\right]d\theta$$

$$A = 2a^2\frac{\pi}{3} + 4a^2\cdot\sin\frac{2\pi}{3} = \frac{2\pi + 3\sqrt{3}}{3}a^2.$$

Now

$$I_x = \iint_R y^2\,dA = \int_0^{\frac{\pi}{3}}\int_{2a}^{4a\cos\theta} (r\sin\theta)^2 r\,dr\,d\theta$$

$$= \frac{1}{4}\int_0^{\frac{\pi}{3}}\left[(4a\cos\theta)^4 - (2a)^4\right]\sin^2\theta\,d\theta$$

$$I_x = 4a^4\int_0^{\frac{\pi}{3}}(16\cos^4\theta - 1)\sin^2\theta\,d\theta = \frac{4\pi + 9\sqrt{3}}{6}a^4$$

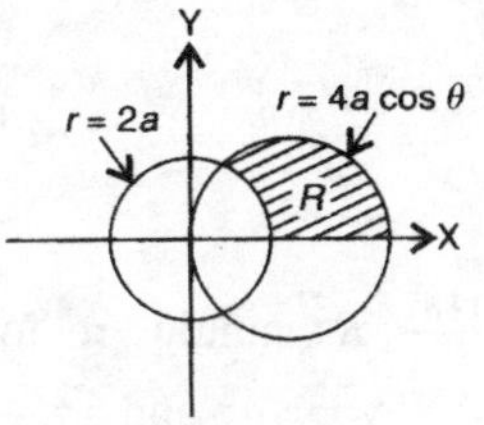

Fig. 7.28

Similarly,

$$I_y = \iint_R x^2\,dA = \int_0^{\frac{\pi}{3}}\int_{2a}^{4a\cos\theta} (r\cos\theta)^2 r\,dr\,d\theta$$

$$= \frac{1}{4}\int_0^{\frac{\pi}{3}}\left[(4a\cos\theta)^4 - (2a)^4\right]\cos^2\theta\,d\theta$$

$$I_y = 4a^4\int_0^{\frac{\pi}{3}}(16\cos^6\theta - \cos^2\theta)d\theta = \frac{12\pi + 11\sqrt{3}}{2}a^4$$

Then,

$$I_0 = I_x + I_y = \frac{20\pi + 21\sqrt{3}}{3}a^4$$

EXERCISE

Area, mass centroid and moments of inertia: In polar coordinates

Using polar coordinates, evaluate the following double integrals:

1. $\int_{\frac{b}{2}}^{b}\int_0^{\frac{\pi}{2}} \rho\,d\theta\,d\rho$.

Ans. $\frac{3\pi b^2}{16}$

2. $\int_0^{\pi}\int_{2\sin\theta}^{4\sin\theta} r^3\,dr\,d\theta$.

Ans. $\frac{45\pi}{2}$

3. $\int_0^{\pi}\int_0^{\cos\theta} \rho\sin\theta\,d\rho\,d\theta$.

Ans. $\frac{1}{3}$

4. $\int_0^{2a}\int_0^{\sqrt{2ax-x^2}} dy\,dx.$

Hint: Limits: $r: 0$ to $2a\cos\theta$, $\theta: 0$ to $\frac{\pi}{2}$.

Ans. $\frac{\pi a^2}{2}$

5. $\int_0^a\int_0^{\sqrt{a^2-y^2}}(x^2+y^2)dx\,dy.$

Hint: Limits: $r: 0$ to a, $\theta: 0$ to $\frac{\pi}{2}$.

Ans. $\frac{\pi a^4}{8}$

6. $\int_0^a\int_{\sqrt{ax-x^2}}^{\sqrt{a^2-x^2}}\frac{dy\,dx}{\sqrt{a^2-x^2-y^2}}.$

Ans. a

7. $\iint_R\frac{x^2y^2dx\,dy}{(x^2+y^2)}$; R: annulus region between the circles $x^2+y^2=a^2$ and $x^2+y^2=b^2$, with $b>a$.

Ans. $\frac{\pi(b^4-a^4)}{16}$

8. $\iint_D xy(x^2+y^2)^{\frac{n}{2}}dx, dy$; D is the region in the first quadrant bounded by the circle $x^2+y^2=4$. Assume that $n+3>0$.

Ans. $\frac{2^{n+3}}{n+4}$

9. Evaluate $\int_0^{\pi}\int_0^{a}r^3\sin\theta\cos\theta\,dr\,d\theta$ by transforming it into cartesian coordinates.

Hint: $\int_{-a}^{a}\int_0^{\sqrt{a^2-x^2}}xy\,dy\,dx=0.$

Ans. 0

10. Find the area inside the circle $\rho=4\sin\theta$ and outside the Lemniscate $\rho^2=8\cos 2\theta$.

Hint: $A=A_1+A_2$

Limit:

$A_1:\rho: 2\sqrt{2\cos 2\theta}$ to $4\sin\theta$; $\theta: \frac{\pi}{6}$ to $\frac{\pi}{4}$

$A_2:\rho: 0$ to $4\sin\theta$; $\theta: \frac{\pi}{4}$ to $\frac{\pi}{2}$

Ans. $\left(\frac{8\pi}{3}+4\sqrt{3}-4\right)$

11. Compute the area of a loop of the curve $r=a\sin 2\theta$.

Ans. $\frac{\pi a^2}{8}$

12. Find the entire area bounded by the lemniscate $\rho^2=a^2\cos 2\theta$.

Ans. a^2

13. Find the centre of gravity of the area of the cardioid $r=a(1\pm\cos\theta)$.

Hint: Solve $r=a(1+\cos\theta)$, results for $r=a(1-\cos\theta)$ is exactly similar. $\overline{y}=0$ due to symmetry.

Limit: $r=0$ to $a(1-\cos\theta)$; $\theta: -\pi$ to π.

Ans. $(\pm\frac{5a}{6}, 0)$

14. Determine I_x for the region: $0\le y\le\sqrt{1-x^2}$, $0\le x\le 1$ and $f(x,y)=1$ density mass.

Hint: Use polar coordinates, $0\le r\le 1$, $0\le\theta\le\frac{\pi}{2}$.

Ans. $I_x=\frac{\pi}{16}$

15. Find the moment of inertia, about the y-axis of the area enclosed by the cardioid

$$r=a(1-\cos\theta)$$

Hint: Put $x=r\cos\theta$, Jacobian : r.

Limit: $r: 0$ to $a(1-\cos\theta)$, $\theta: 0$ to 2π. Use the reduction formula

$$\int_0^{2\pi}\cos^n\theta\,d\theta=\left.\frac{\cos^{n-1}\theta\sin\theta}{n}\right|_0^{2\pi}+\frac{n-1}{n}\int_0^{2\pi}\cos_\theta^{n-2}d\theta$$

for $n=2,3,4,5,6$.

Ans. $I_y=\frac{49\pi a^4}{32}$

16. Find the area of the cardioid $r=a(1+\cos\theta)$.

Hint: Limits: $r: 0$ to $a(1+\cos\theta)$, $\theta: -\pi$ to π

Ans. $\frac{3\pi a^2}{2}$

17. Find the area which is inside the circle $r=3a\cos\theta$ and outside the cardioid $r=a(1+\cos\theta)$

Hint: Limits: $r: a(1+\cos\theta)$ to $3a\cos\theta$, $\theta: -\frac{\pi}{3}$ to $\frac{\pi}{3}$.

Ans. πa^2

18. Find the centroid of a loop of the lemniscate $r^2=a^2\cos 2\theta$.

Hint: Limits: $r : 0$ to $a\sqrt{\cos 2\theta}, \theta : -\frac{\pi}{4}$ to $\frac{\pi}{4}$ (one loop) use polar coordinates (put $x = r\cos\theta$; $J = r$).

Ans. $x = \frac{\pi a\sqrt{2}}{8}$, centroid lies on the initial line.

7.5 TRIPLE INTEGRALS

Triple integral is a generalization of a double integral. Let V be a given three-dimensional domain in space, bounded by a closed surface S. Let $f(x, y, z)$ be a continuous function in V of the rectangular coordinates x, y, z.

Divide V into subdomains Δv_i. Let $f(P_i)$ be the value of f at an arbitrary point P_i of Δv_i. Then a triple integral of f over the domain V, denoted by $\iiint_V f(P)dV$, is defined as

$$\lim_{\Delta v_i \to 0} \Sigma f(P_i)\Delta v_i = \iiint_V f(P)dV$$

$$= \iiint_V f(x, y, z)dx\,dy\,dz \quad (1)$$

However, the triple integral is seldom evaluated directly from its definition (1) as a limit of a sum.

Evaluation of a Triple Integral

Regular three-dimensional domain

V is said to be a regular three-dimensional domain if (i) every straight line parallel to z-axis and drawn through an interior (i.e., not lying on the boundary S) point of V cuts the surface S at two points (ii) entire V is projected on the xy-plane into a regular two-dimensional domain D.

Examples: Parallelopiped, ellipsoid, tetrahedron.

Let the equations of the surfaces bounding a regular domain V below and above be $z = z_1(x, y)$ and $z = z_2(x, y)$ respectively (see Fig. 7.29).

Let the projection D of V onto xy-plane be bounded by the curves $y = y_1(x)$ and $y = y_2(x)$ and $x = a, x = b$.

Then the three fold integral I_V of a continuous function $f(x, y, z)$ over a regular domain V is

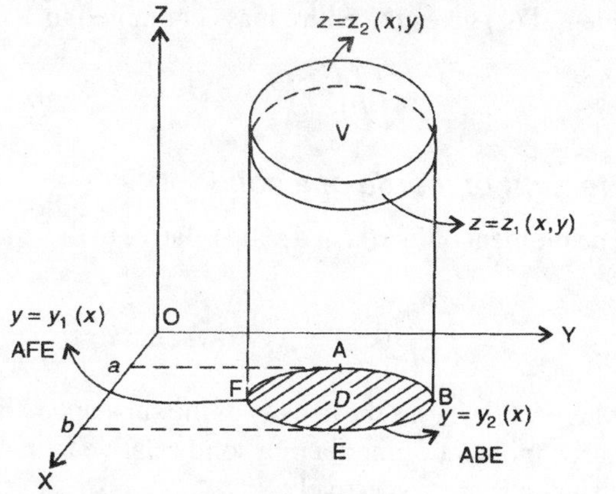

Fig. 7.29

defined as

$$I_V = \int_{x=a}^{x=b} \left[\int_{y=y_1(x)}^{y=y_2(x)} \left\{ \int_{z=z_1(x,y)}^{z=z_2(x,y)} f(x, y, z)\,dz \right\} dy \right] dx$$

Here the limits of integration are chosen to cover the domain V by varying z from lower surface $z = z_1(x, y)$ to the upper surface $z = z_2(x, y)$ and covering the projection D by varying y from $y_1(x)$ to $y_2(x)$ and x from a to b.

In the three-fold iterated integral, integration is done first with respect to z (i.e., within the braces) with the substitution of limits for z, the next integration is carried with respect to y (i.e., within the square brackets). This results in an integrand which is a function of x alone which is then integrated w.r.t. x between a and b.

Note: When V is projected on to xz-plane or yz-plane instead of xy-plane, then the order of integration and the limits are to be rewritten appropriately.

Applications of Triple Integrals

Volume

Volume of a solid contained in the domain V is given by the triple integral (1) with $f(x, y, z) = 1$ i.e., volume $= \iiint_V dx\,dy\,dz$.

Mass

If $\gamma(x, y, z) > 0$ is the volume density (mass/unit volume) of distribution of mass over V then the triple integral (1) gives the entire mass contained in V

$$\text{Mass} = \iiint_V \gamma(x, y, z)dx\,dy\,dz$$

Moment of inertia of a solid

The moment of inertia of a solid relative to the z-axis is

$$I_{zz} = \iiint_V (x^2 + y^2)\gamma(x, y, z)\,dx\,dy\,dz$$

where $\gamma(x, y, z)$ is the density of the substance. Similarly, moment of inertia of a solid relative to x-axis and y-axis are respectively

$$I_{xx} = \iiint_V (y^2 + z^2)\gamma \cdot dx\,dy\,dz$$

$$I_{yy} = \iiint_V (x^2 + z^2)\gamma\,dx\,dy\,dz$$

Centre of gravity of a solid: (x_c, y_c, z_c)

$$x_c = \frac{\iiint_V x\gamma dV}{\iiint_V \gamma dV},\ y_c = \frac{\iiint_V y\gamma dV}{\iiint \gamma dV}$$

$$z_c = \frac{\iiint z\gamma dV}{\iiint \gamma dV}.$$

WORKED OUT EXAMPLES

Example 1: Evaluate

$$I_V = \int_0^1 \int_0^x \int_0^{x+y} (x + y + z)dz\,dy\,dx$$

Solution: First integrating with respect to z, we get

$$I_V = \int_0^1 \int_0^x \left. (x + y)z + \frac{z^2}{2} \right|_0^{x+y} dy\,dx$$

$$= \frac{3}{2}\int_0^1 \int_0^x (x + y)^2 dy\,dx$$

$$= \frac{3}{2}\int_0^1 \int_0^x (x^2 + y^2 + 2xy)dy\,dx$$

Integrating now with respect to y, we get

$$I_V = \frac{3}{2}\int_0^1 \left. x^2y + \frac{y^3}{3} + xy^2 \right|_0^x dx = \frac{7}{2}\int_0^1 x^3 dx$$

Finally integrating with respect to x

$$I_V = \left. \frac{7}{2}\frac{x^4}{4} \right|_0^1 = \frac{7}{8}$$

Example 2: Find the volume bounded by the elliptic paraboloids $z = x^2 + 9y^2$ and $z = 18 - x^2 - 9y^2$

Solution: The two surfaces intersect on the elliptic cylinder $x^2 + 9y^2 = z = 18 - x^2 - 9y^2$ i.e., $x^2 + 9y^2 = 9$. The projection of this volume onto xy-plane is the plane region D enclosed by ellipse having the same equation $\frac{x^2}{3^2} + \frac{y^2}{1^2} = 1^2$ as shown in (Fig. 7.30).

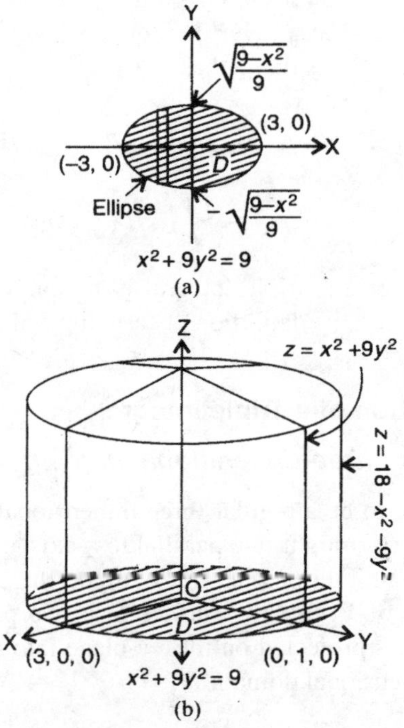

Fig. 7.30

This volume can be covered as follows:

z: from $z_1(x, y) = x^2 + 9y^2$ to $z_2(x, y) = 18 - x^2 - 9y^2$

y: from $y_1(x) = -\sqrt{\frac{9-x^2}{9}}$ to $y_2(x) = \sqrt{\frac{9-x^2}{9}}$

x: from -3 to 3

Thus the volume V bounded by the elliptic paraboloids

$$V = \int_{-3}^{3}\int_{y_1(x)}^{y_2(x)}\int_{z_1(x,y)}^{z_2(x,y)} dz\,dy\,dx$$

$$= \int_{-3}^{3}\int_{y_1(x)}^{y_2(x)} \left[(18 - x^2 - 9y^2) - (x^2 + 9y^2)\right] dy\,dx$$

$$= 2\int_{-3}^{3}\int_{y_1(x)}^{y_2(x)} (9 - x^2 - 9y^2) dy\,dx$$

$$= 2\int_{-3}^{3} (9y - x^2y - 3y^3)\Big|_{-\sqrt{\frac{9-x^2}{9}}}^{\sqrt{\frac{9-x^2}{9}}} dx$$

$$= \frac{8}{9}\int_{-3}^{3} (9 - x^2)^{\frac{3}{2}} dx$$

$$= 72\int_0^{\pi} \sin^4\theta\, d\theta \qquad \text{where } x = 3\cos\theta$$

$$= 72\int_0^{\pi} \left(\frac{1 - \cos 2\theta}{2}\right)^2 d\theta = 27\pi$$

Example 3: Find the total mass of the region in the cube $0 \le x \le 1, 0 \le y \le 1, 0 \le z \le 1$ with density at any point given by xyz.

Solution:

$$Mass = \int_0^1\int_0^1\int_0^1 xyz\,dx\,dy\,dz$$

$$= \int_0^1\int_0^1 \frac{x^2}{2}yz\Big|_0^1 dy\,dz$$

$$= \frac{1}{2}\int_0^1 z\frac{y^2}{2}\Big|_0^1 dz = \frac{1}{4}\int_0^1 z\,dz$$

$$= \frac{1}{4}\frac{z^2}{2}\Big|_0^1 = \frac{1}{8}$$

Example 4: Find the mass, centroid of the tetrahedron bounded by the coordinate planes and the plane $\frac{x}{a} + \frac{y}{b} + \frac{z}{c} = 1$.

Solution: Yet ρ be the constant density of the substance (mass/unit volume). Mass M in the tetrahedron $= \iiint_V \rho dx\,dy\,dz$ (refer Fig. 7.31).

$$M = \rho\int_{x=0}^{a}\int_{y=0}^{b\left(1-\frac{x}{a}\right)}\int_{z=0}^{c\left(1-\frac{x}{a}-\frac{y}{b}\right)} dz\,dy\,dx$$

$$= \rho\int_0^a\int_0^{b\left(1-\frac{x}{a}\right)} c\left(1 - \frac{x}{a} - \frac{y}{b}\right) dy\,dz$$

$$= \rho c\int_0^a \left[\left(1 - \frac{x}{a}\right)y - \frac{y^2}{2b}\right]\Big|_0^{b\left(1-\frac{x}{a}\right)} dx$$

$$= \frac{cb\rho}{2}\int_0^a \left(1 - \frac{x}{a}\right)^2 dx = \frac{\rho bc}{2}\cdot\frac{a}{3} = \frac{\rho abc}{6}$$

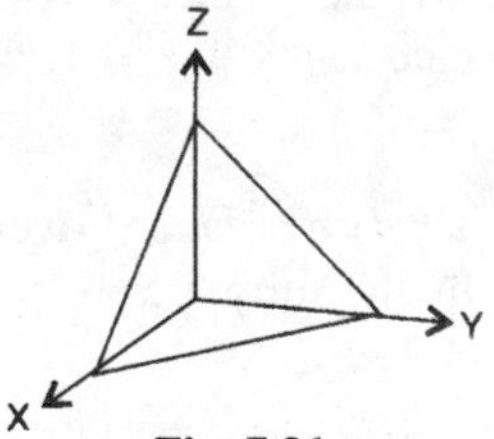

Fig. 7.31

Let (x_c, y_c, z_c) be the coordinates of the centroid. Then,

$$M\cdot x_c = \rho\int_0^a\int_0^{b\left(1-\frac{x}{a}\right)}\int_0^{c\left(1-\frac{x}{a}-\frac{y}{b}\right)} x\,dz\,dy\,dx$$

$$= \rho\int_0^a\int_0^{b\left(1-\frac{x}{a}\right)} cx\left(1 - \frac{x}{a} - \frac{y}{b}\right) dy\,dx$$

$$= c\rho\int_0^a \left[x\left(1 - \frac{x}{a}\right)y - \frac{xy^2}{2b}\right]\Big|_0^{b\left(1-\frac{x}{a}\right)} dx$$

$$= c\rho b\int_0^a x\left(1 - \frac{x}{a}\right)^2 dx = \rho bc\cdot\frac{a^2}{12}$$

$$x_c = \frac{\rho a^2 bc}{12}\cdot\frac{6}{\rho abc} = \frac{a}{4}$$

Similarly, $y_c = \frac{b}{4}, \quad z_c = \frac{c}{4}$.

EXERCISE

Evaluate the following triple integrals:

1. $\int_0^2\int_1^z\int_0^{yz} xyz\,dx\,dy\,dz$.

Ans. $\frac{7}{2}$

2. $\int_1^2 y\,dy \int_y^{y^2} dx \int_0^{\ln x} e^z dz =$
$\int_1^2 \int_y^{y^2} \int_0^{\ln x} e^z dz\,dx\,dy.$

Ans. $\frac{47}{24}$

3. $\int_0^a \int_0^x \int_0^{y+x} e^{x+y+z} dz\,dy\,dx.$

Ans. $\frac{e^{4a}}{8} - \frac{3}{4}e^{2a} + e^a - \frac{3}{8}.$

4. $\int_0^{\frac{\pi}{2}} \int_x^{\frac{\pi}{2}} \int_0^{xy} \cos\frac{z}{x} dz\,dy\,dx.$

Ans. $\frac{\pi}{2} - 1$

5. $\iiint_V \frac{dx\,dy\,dz}{(x+y+z+1)^3}$ taken over the volume bounded by the planes $x = 0, y = 0, z = 0$ and the plane $x + y + z = 1$.

Ans. $\frac{\ln 2}{2} - \frac{5}{16}$

6. $\iiint x^2yz\,dx\,dy\,dz$ taken over the volume bounded by the surface $x^2 + y^2 = 9$, $z = 0, z = 2$.

Ans. $\frac{648}{5}$

7. $\int_0^a \int_0^x \int_0^y xyz\,dz\,dy\,dx.$

Ans. $\frac{a^6}{48}$

8. Compute the three-fold iterated integral of the function xyz over the domain bounded by the planes $x = 0, y = 0, z = 0, x + y + z = 1$.

Hint: $I = \int_0^1 \int_0^{1-x} \int_0^{1-x-y} xyz\,dz\,dy\,dx.$

Ans. $\frac{1}{720}$

9. Find the volume of the tetrahedron bounded by the plane $\frac{x}{a} + \frac{y}{b} + \frac{z}{c} = 1$ and the coordinate planes.

Hint: Limits: $z : 0$ to $c\left(1 - \frac{x}{a} - \frac{y}{b}\right)$, $y : 0$ to $b\left(1 - \frac{x}{a}\right)$, $x : 0$ to a.

Ans. $\frac{|abc|}{6}$

10. Compute the volume of the ellipsoid of semi-axes a, b, c. Hence derive the volume of a sphere.

Hint: Limits: $z : \pm c\sqrt{1 - \frac{x^2}{a^2} - \frac{y^2}{b^2}}$,
$y = \pm b\sqrt{\frac{1-x^2}{a^2}}, x = \pm a.$
For sphere: put $a = b = c$.

Ans. $V = \frac{4\pi abc}{3}$
V of sphere $= \frac{4\pi a^3}{3}$

11. Compute the volume of the solid enclosed between the two surfaces elliptic paraboloids $z = 8 - x^2 - y^2$ and $z = x^2 + 3y^2$.

Hint: Projection of the volume on to xy-plane is the ellipse $x^2 + 2y^2 = 4$, so limits are z: $x^2 + 3y^2$ to $8 - x^2 - y^2$; $g: \pm\sqrt{\frac{4-x^2}{2}}$, $x: \pm 2$.

Ans. $8\pi\sqrt{2}$

12. Compute the volume of the solid bounded by the plane $2x + 3y + 4z = 12$, xy-plane and the cylinder $x^2 + y^2 = 1$.

Hint: Limits: $z : 0$ to $\frac{1}{4}(12 - 2x - 3y)$, $x, y :$ $x^2 + y^2 \le 1$.

Ans. 3π

13. Find the volume of the solid common to the two cylinders $x^2 + y^2 = a^2, x^2 + z^2 = a^2$.

Hint: Limits: $z: \pm\sqrt{a^2 - x^2}$, $y: \pm\sqrt{a^2 - x^2}$, $x : \pm a$.

Ans. $\frac{16a^3}{3}$

14. Compute the volume in the first octant bounded by the cylinder $x = 4 - y^2$ and the planes $z = y, x = 0, z = 0$.

Ans. 4

15. Find the volume of the solid bounded by $y = x^2, x = y^2, z = 0, z = 12 + y - x^2$

Hint: Limits: $z : 0$ to $12 + y - x^2$, $y : \sqrt{x}$ to x^2, $x : 0$ to 1.

Ans. $\frac{549}{144}$

16. Compute the volume bounded by $xy = z, z = 0$ and $(x - 1)^2 + (y - 1)^2 = 1$

Ans. π.

7.6 GENERAL CHANGE OF VARIABLES IN A TRIPLE INTEGRAL

Let the functions $x = f(u, v, w)$, $y = g(u, v, w)$ and $z = h(u, v, w)$ be the transformations from cartesian coordinates x, y, z to the curvilinear coordinates u, v, w. Then the Jacobian $J\left(\frac{x,y,z}{u,v,w}\right)$ is given by the 3rd order determinant.

$$J = J\left(\frac{x, y, z}{u, v, w}\right) = \begin{vmatrix} \frac{\partial x}{\partial u} & \frac{\partial x}{\partial v} & \frac{\partial x}{\partial w} \\ \frac{\partial y}{\partial u} & \frac{\partial y}{\partial v} & \frac{\partial y}{\partial w} \\ \frac{\partial z}{\partial u} & \frac{\partial z}{\partial v} & \frac{\partial z}{\partial w} \end{vmatrix}.$$

Let $F(x, y, z)$ be a continuous function defined in a domain V in the xyz coordinate system. Then a triple integral in cartesian coordinates x, y, z can be transformed to a triple integral in the curvilinear coordinates u, v, w as follows:

$$\iiint_V F(x, y, z) dx\, dy\, dz = \iiint_{V^*} F^*(u, v, w) \times |J| du\, dv\, dw$$

where $F^*(u, v, w) = F(f(u, v, w), \quad g(u, v, w), h(u, v, w))$ and V^* is the corresponding domain in the curvilinear coordinates u, v, w.

Triple Integral in Cylindrical Coordinates

Cylindrical coordinates ρ, θ, z are particularly useful in problems of solids having axis of symmetry. The transformation of cartesian coordinates x, y, z in terms of cylindrical coordinates ρ, θ, z are given by $x = \rho \cos\theta$, $y = \rho \sin\theta$, $z = z(\rho = u, \theta = v, z = w)$ so that the Jacobian is given by

$$J = J\left(\frac{x, y, z}{\rho, \theta, z}\right) = \begin{vmatrix} \cos\theta & -\rho\sin\theta & 0 \\ \sin\theta & \rho\cos\theta & 0 \\ 0 & 0 & 0 \end{vmatrix} = \rho$$

Thus,

$$\iiint_V F(x, y, z) dx\, dy\, dz = \iiint_{V^*} F^*(\rho, \theta, z) |\rho| d\rho\, d\theta\, dz$$

where $F(x, y, z) = F(\rho\cos\theta, \rho\sin\theta, z) = F^*(\rho, \theta, z)$

Triple Integral in Spherical Coordinates

In problems having symmetry with respect to a point 0 (generally the origin), it would be convenient to use spherical coordinates with this point chosen as origin.

Coordinate transformations from x, y, z to the spherical coordinates ρ, θ, ϕ are given by $x = \rho\sin\theta\cos\phi$, $y = \rho\sin\theta\sin\phi$, $z = \rho\cos\theta$ (i.e., $u = \rho$, $v = \theta$, $w = \phi$) so that the Jacobian

$$J = J\left(\frac{x, y, z}{\rho, \theta, \phi}\right) = \begin{vmatrix} \sin\theta\cos\phi & \rho\cos\theta\cos\phi & -\rho\sin\theta\sin\phi \\ \sin\theta\sin\phi & \rho\cos\theta\sin\phi & \rho\sin\theta\cos\phi \\ \cos\theta & -\rho\sin\theta & 0 \end{vmatrix} = \rho^2\sin\theta$$

Thus

$$\iiint_V F(x, y, z) dx\, dy\, dz = \iiint_{V^*} F^*(\rho, \theta, \phi) |\rho^2 \sin\theta| d\rho\, d\theta\, d\phi$$

where $F(x, y, z) = F(\rho\sin\theta\cos\phi, \rho\sin\theta\sin\phi, \rho\cos\theta) = F^*(\rho, \theta, \phi)$.

Worked Out Examples

Example 1: By transforming into cylindrical coordinates evaluate the integral $\iiint (x^2 + y^2 + z^2) dx\, dy\, dz$ taken over the region $0 \le z \le x^2 + y^2 \le 1$.

Solution: Introducing cylindrical polar coordinates $x = r\cos\theta$, $y = r\sin\theta$, $z = z$, the given integral becomes

$$\int_0^1 \iint_R (x^2 + y^2 + z^2) dx\, dy\, dz = \int_0^1 \int_0^{2\pi} \int_0^1 (r^2\cos^2\theta + r^2\sin^2\theta + z^2) r\, dr\, d\theta\, dz$$

where R: circular region bounded by the circle of radius one and centre at origin: $x^2 + y^2 = 1$, so that r varies from 0 to 1 and θ from 0 to 2π.

$$= 2\pi \int_0^1 \left. \frac{r^4}{4} + \frac{r^2}{2} z^2 \right|_0^1 dz = 2\pi \int_0^1 \left(\frac{1}{4} + \frac{z^2}{2}\right) dz$$

$$= 2\pi \left[\frac{z}{4} + \frac{z^3}{6}\right]_0^1 = 2\pi\left[\frac{1}{4} + \frac{1}{6}\right] = \frac{5\pi}{6}$$

Example 2: Use spherical coordinates to evaluate the integral $\iiint \frac{dx\, dy\, dz}{\sqrt{x^2+y^2+z^2}}$ taken over the region V

in the first octant bounded by the cones $\theta = \frac{\pi}{4}$ and $\theta = \arctan z$ and the sphere $x^2 + y^2 + z^2 = 6$ (Fig. 7.32).

Solution: Introducing spherical coordinate

$$x = \rho \sin\theta \cos\phi,$$

$$y = \rho \sin\theta \sin\phi$$

$$z = \rho \cos\theta$$

we have

$$\sqrt{x^2 + y^2 + z^2}$$

$$= \sqrt{\rho^2 \sin^2\theta \cos^2\rho + \rho^2 \sin^2\theta \sin^2\phi + \rho^2 \cos^2\theta}$$

$$= \rho$$

Jacobian $= \rho^2 \sin\theta$.
Radius of the sphere is $\sqrt{6}$.
Thus the given integral in spherical coordinates is

$$I = \iiint \frac{dx\,dy\,dz}{\sqrt{x^2 + y^2 + z^2}}$$

$$= \int_0^{\frac{\pi}{2}} \int_{\frac{\pi}{4}}^{\arctan 2} \int_0^{\sqrt{6}} \frac{1}{\rho} \cdot \rho^2 \sin\theta \cdot d\rho\, d\theta\, d\phi$$

since restricted to first octant, ρ varies from 0 to $\sqrt{6}$, θ : cone angle varies from $\theta_1 = \frac{\pi}{4}$ to $\theta_2 = \arctan 2$ and ϕ from 0 to $\frac{\pi}{2}$.

$$I = 3 \int_0^{\frac{\pi}{2}} \int_{\frac{\pi}{4}}^{\arctan 2} \sin\theta\, d\theta\, d\phi$$

$$= -3 \int_0^{\frac{\pi}{2}} \left(\frac{1}{\sqrt{5}} - \frac{1}{\sqrt{2}} \right) d\phi$$

$$= \frac{3\pi}{2} \left(\frac{1}{\sqrt{2}} - \frac{1}{\sqrt{5}} \right).$$

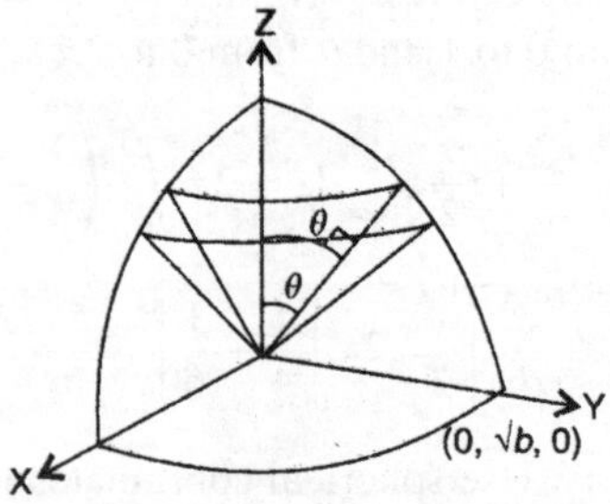

Fig. 7.32

Example 3: Evaluate

$$\iiint_V xyz(x^2 + y^2 + z^2)^{\frac{n}{2}}\, dx\, dy\, dz$$

taken through the positive octant of the sphere $x^2 + y^2 + z^2 = b^2$ provided $n + 5 > 0$.

Solution: Since the geometry involves sphere, introduce spherical coordinates $x = \rho \sin\theta \cos\phi$, $y = \rho \sin\theta \sin\phi$, $z = \rho \cos\theta$ and the Jacobian $J = \rho^2 \sin\theta$. Substituting these transformations.

$$\iiint_V xyz(x^2 + y^2 + z^2)^{\frac{n}{2}}\, dx\, dy\, dz$$

$$= \int_{\phi=0}^{\frac{\pi}{2}} \int_{\theta=0}^{\frac{\pi}{2}} \int_0^b (\rho \sin\theta \cos\phi)(\rho \sin\theta \sin\phi) \times$$

$$\times (\rho \cos\theta) \cdot (\rho^2)^{\frac{n}{2}} \rho^2 \sin\theta d\rho\, d\theta\, d\phi$$

$$= \int_0^b \rho^{n+5} d\rho \int_0^{\frac{\pi}{2}} \sin^3\theta \cos\theta\, d\theta \int_0^{\frac{\pi}{2}} \cos\phi \sin\phi\, d\rho$$

$$= \left.\frac{\rho^{n+6}}{n+6}\right|_0^b \frac{2}{4 \cdot 2} \cdot \frac{1 \cdot 1}{2} = \frac{b^{n+6}}{8(n+6)} \text{ provided } n + 5 > 0.$$

Example 4: Find the volume of the solid surrounded by the surface $\left(\frac{x}{a}\right)^{\frac{2}{3}} + \left(\frac{y}{b}\right)^{\frac{2}{3}} + \left(\frac{z}{c}\right)^{\frac{2}{3}} = 1$.

Solution: The desired volume V of the solid is

$$V = \iiint dx\, dy\, dz$$

Here the limits of integration is not easy. Instead, by a transformation the given surface becomes a sphere as follows.

Put $\frac{x}{a} = u^3$, $\frac{y}{b} = v^3$, $\frac{z}{c} = w^3$ so that

$$\left(\frac{x}{a}\right)^{\frac{2}{3}} + \left(\frac{y}{b}\right)^{\frac{2}{3}} + \left(\frac{z}{c}\right)^{\frac{2}{3}} = u^2 + v^2 + w^2 = 1$$

which is the equation of a sphere of radius 1 and with centre at origin in the new variables u, v, w.

The Jacobian J

$$J = \frac{\partial(x, y, z)}{\partial(u, v, w)} = \begin{vmatrix} \frac{\partial x}{\partial u} & \frac{\partial x}{\partial v} & \frac{\partial x}{\partial w} \\ \frac{\partial y}{\partial u} & \frac{\partial y}{\partial v} & \frac{\partial y}{\partial w} \\ \frac{\partial z}{\partial u} & \frac{\partial z}{\partial v} & \frac{\partial z}{\partial w} \end{vmatrix}$$

$$= \begin{vmatrix} 3au^2 & 0 & 0 \\ 0 & 3bv^2 & 0 \\ 0 & 0 & 3cw^2 \end{vmatrix}$$

$$J = 27abc\, u^2 v^2 w^2.$$

Thus the required volume in terms of the new variables u, v, w is

$$V = \iiint_V dx\,dy\,dz$$

$$= \iiint_{V^*} 27abc\,u^2 v^2 w^2\,du\,dv\,dw$$

where V^* is sphere $u^2 + v^2 + w^2 = 1$.

To evaluate this integral, introduce again spherical coordinates

$$u = r\sin\theta\cos\phi,\ v = r\sin\theta\sin\phi,\ w = r\cos\theta$$

$$V = 27abc\,8\int_0^{\frac{\pi}{2}}\int_0^{\frac{\pi}{2}}\int_0^1 (r^2\sin^2\theta\cos^2\phi)\times$$

$$\times\,(r^2\sin^2\theta\sin^2\phi)(r^2\cos^2\theta)(r^2\sin\theta)dr\,d\theta\,d\phi$$

since the Jacobian $J^* = \frac{\partial(u,v,w)}{\partial(r,\theta,\phi)} = r^2\sin\theta$ and since to describe the positive octant of the sphere, $0\le r\le 1$, $0\le\phi\le\frac{\pi}{2}, 0\le\theta\le\frac{\pi}{2}$.

Since all the limits are constant, V can be rewritten as

$$V = 216abc\int_0^1 r^8\,dr\int_0^{\frac{\pi}{2}}\sin^5\theta\cos^2\theta\,d\theta$$

$$\int_0^{\frac{\pi}{2}}\sin^2\phi\cos^2\phi\,d\phi$$

$$V - 216abc\left.\frac{r9}{9}\right|_0^1\left(\frac{\pi}{16}\right)\cdot\left(\frac{8}{35.3}\right)$$

$$= 216\cdot\frac{abc}{9}\cdot\frac{\pi}{16}\cdot\frac{8}{35.3} = \frac{4abc}{35}.$$

Example 5: Evaluate $\iiint\left[\frac{(1-x-y-z)}{xyz}\right]^{\frac{1}{2}} dx\,dy\,dz$ taken over the volume bounded by the planes $x=0, y=0, z=0$ and $x+y+z=1$ (Fig. 7.33).

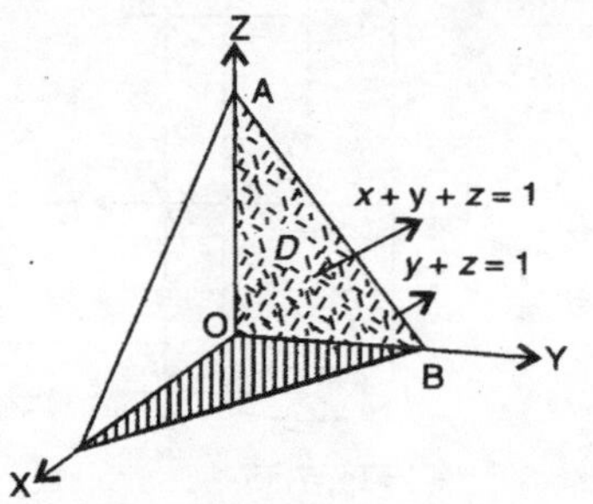

Fig. 7.33

Solution: Put

$$x + y + z = u,\ y + z = uv \text{ and } z = uvw \tag{1}$$

Solving (1)

$$x = u(1-v),\ y = uv(1-w) \text{ and } z = uvw \tag{2}$$

The Jacobian J is

$$J = \frac{\partial(x,y,z)}{\partial(u,v,w)} = \begin{vmatrix} 1-v & -u & 0 \\ v(1-w) & u(1-w) & -uv \\ vw & uw & uv \end{vmatrix} = u^2 v \tag{3}$$

Let D, the triangular region OAB be the projection of the surface on the zy-plane. Then the tetrahedral volume is covered with the following limits:

$$0 \le z \le 1 \tag{4}$$

$$0 \le y \le 1 - z \tag{5}$$

$$0 \le x \le 1 - z - y \tag{6}$$

Since $z = uvw$, from (4),

$$0 \le uvw \le 1 \tag{7}$$

Since $y = uv(1-w)$, from (5),

$$0 \le uv(1-w) \le 1 - uvw$$

or $\quad 0 \le uv \le 1 \quad (8)$

Since $x = u(1-v)$, from (6)

$$0 \le u(1-v) \le 1 - uvw - uv(1-w)$$

or $\quad 0 \le u \le 1 \quad (9)$

Using (9) in (7) and (8) the limits of integration for the new variables u, v, w are

$$0 \le u \le 1,\quad 0 \le v \le 1,\quad 0 \le w \le 1 \tag{10}$$

Using (1), (2) and (10) the given integral gets transformed to

$$I = \int_0^1\int_0^{1-z}\int_0^{1-z-y}\left(\frac{1-x-y-z}{xyz}\right)^{\frac{1}{2}} dx\,dy\,dz$$

$$= \int_0^1\int_0^1\int_0^1\left\{\frac{(1-u)}{[u(1-v)][uv(1-w)][uvw]}\right\}^{\frac{1}{2}}\times$$

$$\times\,u^2 v\,du\,dv\,dw$$

Since all the limits are constants, this integral can be

rewritten as

$$= \int_0^1 \sqrt{u(1-u)}du \int_0^1 \frac{dv}{\sqrt{1-v}} \int_0^1 \frac{dw}{\sqrt{w(1-w)}}$$

put $u = \sin^2\theta$, $v = \sin^2\phi$, $w = \sin^2 t$

$$= \int_0^{\frac{\pi}{2}} \sin\theta \cdot \cos\theta \cdot 2\sin\theta\cos\theta\, d\theta$$

$$\int_0^{\frac{\pi}{2}} \frac{2\sin\phi\cos\phi}{\cos\phi} d\phi \int_0^{\frac{\pi}{2}} \frac{2\sin t\cos t}{\sin t\cos t} dt$$

$$= 2\cdot\frac{1\cdot 1}{4\cdot 2}\cdot\frac{\pi}{2}\cdot 2\cdot 1\cdot 2\cdot\frac{\pi}{2} = \frac{\pi^2}{4}.$$

Example 6: Determine the mass M of a hemisphere of radius b with centre at the origin (Fig. 7.34), if the density F of its substance at each point (x, y, z) is proportional to the distance of this point from the base i.e., $F = kz$.

Solution: Equation of sphere $x^2 + y^2 + z^2 = b^2$. Equation of the upper part of the hemisphere

$$z = \sqrt{b^2 - x^2 - y^2}.$$

In cylindrical coordinates $x = r\cos\theta$, $y = r\sin\theta$ the equation of the hemisphere becomes

$$z = \sqrt{b^2 - r^2\cos^2\theta - r^2\sin^2\theta} = \sqrt{b^2 - r^2}.$$

Hence mass M is

$$M = \iiint_V (kz) r\, dr\, d\theta\, dz$$

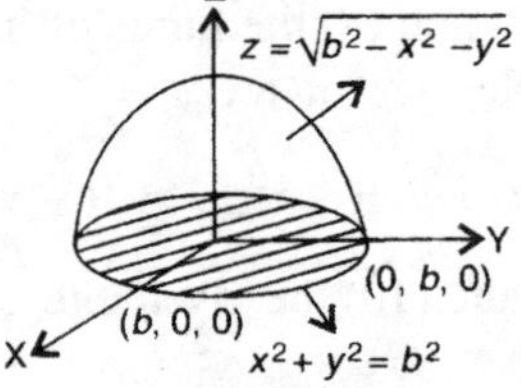

Fig. 7.34

The hemisphere can be covered as follows:

z: varying from $z = 0$ to $z = \sqrt{b - x^2 - y^2}$ (upper surface)

r: varying from 0 to b

θ: varying from 0 to 2π.

Thus

$$M = \int_0^{2\pi} \left\{ \int_0^b \left(\int_0^{\sqrt{b^2-r^2}} kz\, dz \right) r\, dr \right\} d\theta$$

$$= 2\pi \int_0^b k\frac{z^2}{2}\Big|_0^{\sqrt{b^2-r^2}} r\, dr$$

$$= \pi k \int_0^b (b^2 - r^2) r\, dr = \pi k \left[b^2\frac{r^2}{2} - \frac{r^4}{4} \right]_0^b$$

$$M = \frac{\pi k b^4}{4}.$$

Example 7: Compute the mass of a sphere of radius b if the density varies inversely as the square of the distance from the centre.

Solution: Density $F(x, y, z) = \frac{k}{x^2+y^2+z^2}$.

$$M = \iiint_V F(x, y, z) dv$$

In spherical coordinates

$$M = \iiint \frac{k}{x^2 + y^2 + z^2} dx\, dy\, dz$$

$$= 8\int_0^{\frac{\pi}{2}} \int_0^{\frac{\pi}{2}} \int_0^b \frac{k}{r^2}(r^2 \sin\theta) dr\, d\theta\, d\phi$$

$$= 8kb \cdot 1 \cdot \frac{\pi}{2} = 4k\pi b.$$

Example 8: Compute the moment of inertia of a right circular cylinder of altitude $2h$ and radius b, relative to the diameter of its median section with density equals to k, a constant.

Solution: Choose z-axis along the axis of the cylinder and the origin at its centre of symmetry as shown

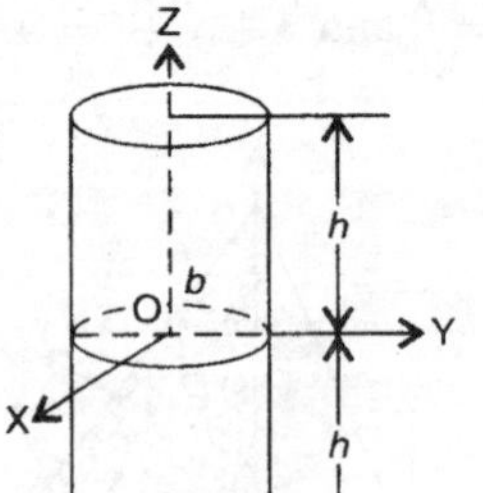

Fig. 7.35

in Fig. 7.35. Moment of inertia of the cylinder relative to the x-axis:

$$I_{xx} = \iiint_V (y^2 + z^2)k \, dx\, dy\, dz$$

Introducing cylindrical coordinates $x = r\cos\theta$, $y = r\sin\theta$, we have

$$I_{xx} = k\int_0^{2\pi}\int_0^b\int_{-h}^h (z^2 + r^2\sin^2\theta)dz\, r\, dr\, d\theta$$

$$= k\int_0^{2\pi}\int_0^b \left.\frac{z^3}{3} + zr^2\sin^2\theta\right|_{-h}^{h} r\, dr\, d\theta$$

$$= k\int_0^{2\pi}\int_0^b \left(\frac{2h^3}{3} + 2h\, r^2\sin^2\theta\right) r\, dr\, d\theta$$

$$= k\int_0^{2\pi} \left. 2\frac{h^3}{3}\frac{r^2}{2} + 2h\frac{r^4}{4}\sin^2\theta\right|_0^b d\theta$$

$$= k\int_0^{2\pi}\left(\frac{h^3b^2}{3} + \frac{hb^4}{2}\sin^2\theta\right)d\theta$$

$$= k\left[\frac{2\pi h^3b^2}{3} + \frac{hb^4}{2}\right]$$

EXERCISE

1. Find the volume of the portion of the sphere $x^2 + y^2 + z^2 = b^2$ lying inside the cylinder $x^2 + y^2 = bx$.

   Hint: Use cylindrical coordinate $x = r\cos\theta$, $y = r\sin\theta$, $z = z$ then equation of sphere is $r^2 + z^2 = b^2$ and equation of cylinder is $r = b\cos\theta$.

   $$\text{Volume } V = 2\int_0^{\pi}\int_0^{b\cos\theta}\int_0^{\sqrt{b^2-r^2}} r\, dz\, dr\, d\theta.$$

   Ans. $\frac{2b^3(3\pi-4)}{9}$

2. Evaluate $\iiint xyz\, dx\, dy\, dz$ over the positive octant of the sphere $x^2 + y^2 + z^2 = b^2$ by transforming to spherical polar coordinates.

   Hint: Limits: $r : 0$ to a; $\theta : 0$ to $\frac{\pi}{2}$, $\phi : 0$ to $\frac{\pi}{2}$

   Ans. $\frac{b^6}{48}$

3. Evaluate $\iiint \frac{dx\, dy\, dz}{\sqrt{(x^2+y^2+z^2)}}$ taken over the common region between the cone $z = x^2 + y^2$ and the cylinder $x^2 + y^2 = 1$ bounded by the plane $z = 1$ in the positive octant.

   Hint: Limits:

   a. In cartesian coordinates

   $z : 1$ to $\sqrt{x^2 + y^2}$, $y : 0$ to $\sqrt{1 - x^2}$; $x : 0$ to 1

   b, In spherical coordinates

   $r : 0$ to $\sec\theta$; $\theta : 0$ to $\frac{\pi}{4}$; $\phi : 0$ to $\frac{\pi}{2}$.

   Ans. $\frac{(\sqrt{2}-1)\pi}{4}$

4. Evaluate $\iiint z^2\, dx\, dy\, dz$ taken over the volume bounded by the surfaces $x^2 + y^2 = a^2$, $x^2 + y^2 = z$ and $z = 0$.

   Hint: Transform into spherical coordinates.

   Ans. $\frac{\pi a^8}{12}$

5. Evaluate $\iiint (x^2 + y^2 + z^2)^{-1}\, dx\, dy\, dz$ taken throughout the volume of the sphere $x^2 + y^2 + z^2 = b^2$.

   Hint: Limits: $r : 0$ to b, $\theta : 0$ to π, $\phi : 0$ to 2π.

   Ans. $4\pi b$

6. Use cylindrical coordinates, to evaluate

   $$\iiint_V (x^2 + y^2)dx\, dy\, dz$$

   taken over the region V bounded by the paraboloid $z = 9 - x^2 - y^2$ and the plane $z = 0$.

   Hint: Equation of paraboloid is $z = 9 - \rho^2$ so that z varies from 0 to $9 - \rho^2$, $\rho : 0$ to 3, $\theta : 0$ to 2π Jacobian: r, and integrand $x^2 + y^2 = \rho^2$. Thus,

   $$\int_0^{2\pi}\int_0^3\int_0^{9-\rho^2} (\rho^2)(\rho\, dz\, d\rho\, d\theta).$$

   Ans. $\frac{243\pi}{2}$

7. Use spherical coordinates to evaluate

   $$\int_0^1\int_0^{\sqrt{1-x^2}}\int_0^{\sqrt{1-x^2-y^2}} \frac{dx\, dy\, dz}{\sqrt{1 - x^2 - y^2 - z^2}}$$

   Ans. $\frac{\pi^2}{8}$

8. Evaluate the triple integrals taken over the tetrahedral volume enclosed by the planes $x = 0$, $y = 0$, $z = 0$ and $x + y + z = 1$.

   a. $\iiint [xyz(1 - x - y - z)]^{\frac{1}{2}} dx\,dy\,dz$.

   b. $\iiint (x + y + z)^2 xyz\,dx\,dy\,dz$.

   c. $\iiint \cdot e^{(x+y+z)^3} dx\,dy\,dz$.

   Hint: Use the transformation $u = x + y + z$, $uv = y + z$, $uvw = z$, so that $x = u(1 - v)$, $y = uv(1 - w)$ and $z = uvw$ with limits for $u, v, w : 0$ to 1 and Jacobian $u^2 v$.

   See Worked Out Example 5 on Page 192.

Ans. a. $\frac{\pi^2}{1920}$
b. $\frac{1}{960}$
c. $\frac{(e-1)}{6}$

9. Find the volume cut from the sphere of radius b by the cone $\phi = \alpha$. Hence deduce the volumes of the hemisphere and sphere.

   Hint: Use spherical coordinates.

   Limit: $r : 0$ to a; $\theta : 0$ to 2π, $\phi : 0$ to α.

Ans. $\frac{2\pi b^3(1-\cos\alpha)}{3}$

hemisphere volume: $(\alpha = \frac{\pi}{2}) : \frac{2\pi b^3}{3}$

sphere volume: $(\alpha = \pi) : \frac{4\pi b^3}{3}$

10. Find the mass M of the tetrahedron bounded by the coordinate planes and the plane $\frac{x}{a} + \frac{y}{b} + \frac{z}{c} = 1$ given that the density F at any point (x, y, z) is $kxyz$.

    Hint: Limits: $z : 0$ to $1 - \frac{x}{a} - \frac{y}{b}$, $y : 0$ to $b\left(1 - \frac{x}{a}\right)$, $x : 0$ to a.

Ans. $\frac{ka^2b^2c^2}{720}$

11. Find the mass of a solid in the form of the positive octant of the sphere $x^2 + y^2 + z^2 = 9$ given that the density at any point is $2xyz$.

Ans. 30.375

12. Find the centroid of the tetrahedron bounded by the coordinate planes and the plane $x + y + z = 1$. The density at any point is varying as its distance from the face $z = 0$.

Ans. $\left(\frac{1}{5}, \frac{1}{5}, \frac{2}{5}\right)$

13. Find the centroid of the solid octant of the ellipsoid $\left(\frac{x}{a}\right)^2 + \left(\frac{y}{b}\right)^2 + \left(\frac{z}{c}\right)^2 = 1$ if the density at any point of the solid varies as xyz.

Ans. $\left(\frac{16a}{35}, \frac{16b}{35}, \frac{16c}{35}\right)$

14. Find the centre of gravity of a homogeneous solid sphere of radius b.

    Hint: With cylindrical coordinates, the equation of hemisphere is $z = \sqrt{b^2 - x^2 - y^2} = \sqrt{b^2 - r^2}$ so that limits are: $z : 0$ to $\sqrt{b^2 - r^2}$; $r : 0$ to a, $\theta : 0$ to 2π.

Ans. $\bar{x} = \bar{y} = 0; \bar{z} = \frac{3b}{8}$

15. Compute the moment of inertia of a circular cone relative to the diameter of the base.

Ans. $\frac{\pi hr^2(2h^2+3r^2)}{60}$

Note: h is the altitude and r is the radius of the base of the cone.

16. Find the moment of inertia of a right circular cone relative to its axis.

Ans. $\frac{\pi hr^4}{10}$

17. Find the volume of the cone $0 \le z \le h(a - r)/a$.

    Hint: Limits: $r : 0$ to a, $\theta : 0$ to 2π.

Ans. $\frac{\pi a^2 h}{3}$

18. Find the volume and mass contained in a solid below the plane $z = 1 + y$, bounded by the coordinate planes and the vertical plane $2x + y = 2$ and having density $f(x, y, z) = x + z$.

    Hint: Limits: $z : 0$ to $1 + y$, $y : 0$ to $2 - 2x$, $x : 0$ to 1.

Ans. Volume: $\frac{5}{3}$, Mass: 2

19. Find the z-coordinate of the centroid of a uniform solid cone of height h equal to the radius of the base r (i.e., $h = r$). Also find the moment of inertia of the solid about its axis. Assume ρ to be the constant density of the solid.

Ans. Mass $M = \frac{\rho\pi h^3}{3}$

z coordinate of centroid $= \bar{z} = \frac{3h}{4}$

M.I. about z-axis: $I = \frac{\rho\pi h^5}{10} = \frac{3}{10} Mh^2$

20. Find the moment of inertia about the z-axis of the solid ellipsoid inside

$$\frac{x^2}{a^2}+\frac{y^2}{b^2}+\frac{z^2}{c^2}=1$$

Hint: Make the change of variables $x = ax^*, y = by^*, z = cz^*$ than equation is $x^{*2}+y^{*2}+z^{*2}=1$ i.e., sphere of radius 1. By symmetry $\iiint x^{*2}dV^* = \iiint y^{*2}dV^* = \iiint z^{*2}dV^* = \frac{1}{3}\iiint r^{*2}dV^*$ where $r^{*2} = x^{*2}+y^{*2}+z^{*2}$.

Ans. $I = \rho abc(a^2+b^2)\frac{4\pi}{15} = \frac{1}{5}M(a^2+b^2)$

where $M = \text{mass} = \frac{4}{3}\pi\rho abc$.

7.7 DIRICHLETS* INTEGRAL

Dirichlet's integral is useful in the evaluation of certain double and triple integrals by expressing them in terms of beta and gamma functions which can be evaluated numerically.

Using Dirichlet integral, plane area, volume of a solid region, mass, centroid can be calculated in a simpler way.

Book work I Let D be the triangular region in the xy-plane bounded by $x \geq 0, y \geq 0, x+y \leq b$. Then prove that the double integral over D

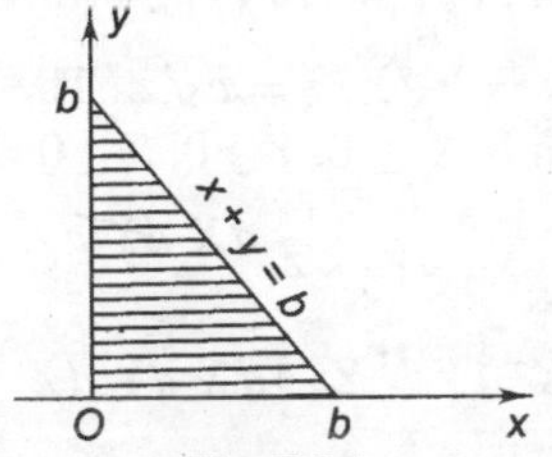

Fig. 7.36

$$\iint_D x^p y^q dxdy = b^{p+q+2}\frac{\Gamma(p+1)\Gamma(q+1)}{\Gamma(p+q+3)}$$

Proof: Introducing $X = \frac{x}{b}, Y = \frac{y}{b}$ the given integral in X, Y becomes

$$\iint_D x^p y^q dx\,dy = \iint_{D^*} (bX)^p (bY)^q b^2 dX\,dY$$

where the transformed region D^* is the triangular region bounded by $X \geq 0, Y \geq 0, X+Y \leq 1$.

$$= b^{p+q+2}\int_{X=0}^{1}\int_{Y=0}^{1-X} X^pY^q dY\,dX$$

$$= b^{p+q+2}\int_{X=0}^{1} X^p \left.\frac{Y^{q+1}}{q+1}\right|_{Y=0}^{1-X} dX$$

$$= \frac{b^{p+q+2}}{q+1}\int_0^1 X^p(1-X)^{q+1}dX$$

$= \frac{b^{p+q+2}}{q+1}\cdot\beta(p+1,\ q+2)$ from the definition of β function

$$= \frac{b^{p+q+2}}{(q+1)}\cdot\frac{\Gamma(p+1)\Gamma(q+2)}{\Gamma(p+q+3)} = \frac{b^{p+q+z}}{(q+1)}\frac{\Gamma(p+1)(q+1)\Gamma(q+1)}{\Gamma(p+q+3)}$$

since $\Gamma(q+2) = (q+1)\Gamma(q+1)$. Hence the result.

Book work II Dirichlet's integral

Let V be the solid region, tetrahedron in the first octant, bounded by the coordinate plane and the plane $x+y+z=b$. Then the triple integral taken over V.

$$\iiint_V x^py^qz^r dx\,dy\,dz = b^{p+q+r+3}\cdot\frac{\Gamma(p+1)\Gamma(q+1)\Gamma(r+1)}{\Gamma(p+q+r+4)}$$

Proof: Introducing $X = \frac{x}{b}, Y = \frac{y}{b}, Z = \frac{z}{b}$, the region V^* in xyz-system transforms to V^* in XYZ coordinate system given by $X \geq 0, Y \geq 0, Z \geq 0$ and $X+Y+Z \leq 1$, which is a tetrahedron in the first octant. Then the triple integral T.I.

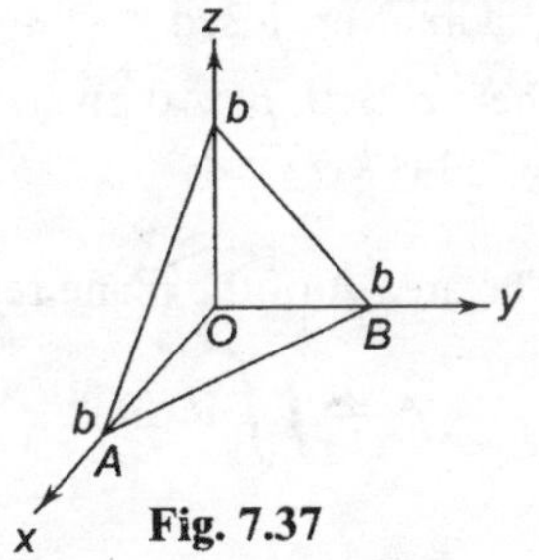

Fig. 7.37

$$\text{T.I.} = \iiint_V x^py^qz^r dx\,dy\,dz$$

$$= \iiint_{V^*} (bX)^p(bY)^q(bZ)^r b^3 dX\,dY\,dZ$$

$$= b^{p+q+r+3}\int_{X=0}^{1}\int_{Y=0}^{1-X}\int_{Z=0}^{1-X-Y} X^pY^qZ^r\,dZ\,dY\,dX$$

$$= \frac{b^{p+q+r+3}}{r+1}\int_{X=0}^{1}\int_{Y=0}^{1-X} X^pY^q(1-X-Y)^{r+1}dY\,dX$$

Introducing the variables $u = X+Y$ and $uv = Y$,

* Peter Gustav Lejeune Dirichlet (1805-1859), German mathematician

the triangular region AOX in the XY-plane (given by $X=0, Y=0, X+Y=1$) gets transformed to a square in the uv-plane (given by $u=0, v=0, u=1, v=1$). The jacobian $\frac{\partial(X,Y)}{\partial(u,v)} = \begin{vmatrix} 1-v & v \\ -u & u \end{vmatrix} = u$ and $X = u(1-v), Y = uv$ with this,

$$\text{T.I.} = \frac{b^{p+q+r+3}}{r+1}\int_{v=0}^{1}\int_{u=0}^{1} u^p(1-v)^p(uv)^q \times$$

$$\times(1-u)^{r+1}\cdot u\,du\,dv$$

$$= \frac{b^{p+q+r+3}}{r+1}\left[\int_{u=0}^{1} u^{p+q+1}(1-u)^{r+1}du\right]\times$$

$$\times\left[\int_{v=0}^{1} v^q(1-v)^p dv\right]$$

$$= \frac{b^{p+q+r+3}}{r+1}\beta(p+q+2, r+2)\cdot\beta(q+1, p+1)$$

$$= \frac{b^{p+q+r+3}}{(r+1)}\frac{\Gamma(p+q+2)(\Gamma(r+2)}{\Gamma(p+q+r+4)}\cdot\frac{\Gamma(q+1)\Gamma(p+1)}{\Gamma(p+q+2)}$$

$$= b^{p+q+r+3}\cdot\frac{\Gamma(p+1)\Gamma(q+1)\Gamma(r+1)}{\Gamma(p+q+r+4)}$$

since $\Gamma(r+2) = (r+1)\Gamma(r+1)$. Hence the result.

Note: This triple integral can also be evaluated directly by putting $x+y+z=u, y+z=uv, z=uvw$.

WORKED OUT EXAMPLES

Example 1: Find the area and the mass contained in the first quadrant enclosed by the curve $\left(\frac{x}{a}\right)^\alpha + \left(\frac{y}{b}\right)^\beta = 1$, where $\alpha > 0, \beta > 0$ given that density at any point $p(x, y)$ is $k\sqrt{xy}$.

Solution: The area A of the plane region is

$$A = \iint_D dx\,dy$$

put $\left(\frac{x}{a}\right)^\alpha = X$, $\left(\frac{y}{b}\right)^\beta = Y$ then $x = aX^{1/\alpha}$, $y = bY^{1/\beta}$. Then

$$A = \iint_{D^*} a\frac{1}{\alpha}X^{\frac{1}{\alpha}-1}\cdot dX\cdot b\cdot\frac{1}{\beta}Y^{\frac{1}{\beta}-1}dY$$

where $X \geq 0, Y \geq 0, X+Y \leq 1$. So

$$A = \frac{ab}{\alpha\beta}\iint_{D^*} X^{\frac{1}{\alpha}-1}Y^{\frac{1}{\beta}-1}dX\,dY$$

Using book work I

$$A = \frac{ab}{\alpha\beta}\cdot\beta\left(\frac{1}{\alpha}, \frac{1}{\beta}\right) = \frac{ab}{\alpha\beta}\frac{\Gamma\left(\frac{1}{\alpha}\right)\Gamma\left(\frac{1}{\beta}\right)}{\Gamma\left(\frac{1}{\alpha}+\frac{1}{\beta}\right)}$$

Now the total mass M contained in the plane region A is

$$M = \iint_D P(x, y)dx\,dy = \iint_D k\sqrt{xy}dx\,dy$$

$$= k\iint_{D^*}\sqrt{a}X^{\frac{1}{2\alpha}}\cdot\sqrt{b}Y^{\frac{1}{2\beta}}\cdot\frac{a}{\alpha}X^{\frac{1}{\alpha}-1}\frac{b}{\beta}Y^{\frac{1}{\beta}-1}dX\,dY$$

$$= k\frac{(ab)^{\frac{3}{2}}}{\alpha\beta}\iint_{D^*} X^{\frac{3}{2\alpha}-1}Y^{\frac{3}{2\beta}-1}dX\,dY$$

$$= k\frac{(ab)^{\frac{3}{2}}}{\alpha\beta}\cdot\frac{\Gamma\left(\frac{3}{2\alpha}\right)\Gamma\left(\frac{3}{2\beta}\right)}{\Gamma\left(\frac{3}{2\alpha}+\frac{3}{2\beta}\right)}$$

Example 2: Find the volume and the mass contained in the solid region in the first octant of the ellipsoid $\frac{x^2}{a^2}+\frac{y^2}{b^2}+\frac{z^2}{c^2} = 1$ if the density at any point $\rho(x, y, z)$ is $kxyz$. Also find the coordinates of the centroid (UPTU 2002).

Solution: Volume V of the solid region is $V = \iiint_V dx\,dy\,dz$. Here V is the region $x \geq 0, y \geq 0, z \geq 0, \frac{x^2}{a^2}+\frac{y^2}{b^2}+\frac{z^2}{c^2} \leq 1$.

Put $\left(\frac{x}{a}\right)^2 = X, \left(\frac{y}{b}\right)^2 = Y, \left(\frac{z}{c}\right)^2 = Z$. Then $x = a\sqrt{X}, y = b\sqrt{Y}, z = c\sqrt{Z}$. The new region in XYZ-system is $X \geq 0, Y \geq 0, Z \geq 0, X+Y+Z \leq 1$

$$V = \iiint_{V^*}\frac{a}{2}\frac{1}{\sqrt{X}}\frac{b}{2}\frac{1}{\sqrt{Y}}\frac{c}{2}\frac{1}{\sqrt{Z}}\,dX\,dY\,dZ$$

$$= \frac{abc}{8}\iiint_{V^*} X^{-\frac{1}{2}}Y^{-\frac{1}{2}}Z^{-\frac{1}{2}}dX\,dY\,dZ$$

$$= \frac{abc}{8}\frac{\Gamma\left(\frac{1}{2}\right)\Gamma\left(\frac{1}{2}\right)\Gamma\left(\frac{1}{2}\right)}{\Gamma\left(-\frac{1}{2}-\frac{1}{2}-\frac{1}{2}+4\right)} = \frac{abc}{8}\frac{\pi^{3/2}}{\Gamma\left(\frac{5}{2}\right)}$$

$$V = \frac{abc}{8}\frac{\pi\cdot\sqrt{\pi}}{\frac{3}{2}\cdot\frac{1}{2}\sqrt{\pi}} = \frac{\pi abc}{6}$$

Now the mass M contained is

$$M = \iiint \rho(x, y, z)dx\,dy\,dz = k\iiint_V xyz\,dx\,dy\,dz.$$

With the change of variables.

$$M = k\iiint_{V^*} a\sqrt{X}b\sqrt{Y}c\sqrt{Z}\times$$

$$\times\frac{a}{2\sqrt{X}}\cdot\frac{b}{2\sqrt{Y}}\frac{c}{2\sqrt{Z}}dX\,dY\,dZ$$

$= k\frac{a^2b^2c^2}{8}\iiint\limits_{V^*} dX\,dY\,dZ$. Using book work II

$= k\frac{a^2b^2c^2}{8}\frac{\Gamma(1)\Gamma(1)\Gamma(1)}{\Gamma(0+0+0+4)}$ since $p = q = r = 0$

$M = \frac{ka^2b^2c^2}{8}\cdot\frac{1}{3!} = \frac{ka^2b^2c^2}{48}$

Let $(\overline{X}, \overline{Y}, \overline{Z})$ be the coordinates of the centroid of the solid region V. Then

$$\overline{X} = \frac{\iiint\limits_V x\rho(x, y, z)dx\,dy\,dz}{\iiint\limits_V \rho(x, y, z)dx\,dy\,dz}$$

The denominator integral is the mass M calculated as $ka^2b^2c^2/48$. Now consider

$\iiint\limits_V x\rho(x, y, z)dx\,dy\,dz$

$= k\iiint\limits_{V^*} a\cdot\sqrt{X}\cdot a\cdot\sqrt{X}\cdot b\cdot\sqrt{Y}\cdot c\cdot\sqrt{Z}\times$

$\times\frac{a}{2\cdot\sqrt{X}}\cdot\frac{b}{2\cdot\sqrt{Y}}\cdot\frac{c}{2\cdot\sqrt{Z}}\cdot dX\,dY\,dZ$

$= \frac{ka^3b^2c^2}{8}\iiint\limits_{V^*} X^{\frac{1}{2}}dX\,dY\,dZ$. Using book work II with $p = \frac{1}{2}, q = r = 0$

$= \frac{ka^3b^2c^2}{8}\frac{\Gamma\left(\frac{3}{2}\right)\cdot\Gamma(1)\Gamma(1)}{\Gamma\left(\frac{1}{2}+0+0+4\right)} = \frac{ka^3b^2c^2}{8}\frac{\frac{1}{2}\Gamma\left(\frac{1}{2}\right)}{\Gamma\left(\frac{9}{2}\right)}$

$= \frac{ka^3b^2c^2}{8}\frac{\frac{1}{2}\cdot\Gamma\left(\frac{1}{2}\right)}{\frac{7}{2}\cdot\frac{5}{2}\cdot\frac{3}{2}\cdot\frac{1}{2}\cdot\Gamma\left(\frac{1}{2}\right)} = \frac{ka^3b^2c^2}{105}$

$\overline{X} = \frac{ka^3b^2c^2}{105}\frac{48}{ka^2b^2c^2} = \frac{16}{35}a$. Similarly $\overline{Y} = \frac{16}{35}b$, $\overline{Z} = \frac{16}{35}c$

Exercise

1. Evaluate $\iint\limits_D x^p y^q\,dx\,dy$ where D is the region bounded by $x = 0, y = 0$ and $\left(\frac{x}{a}\right)^m + \left(\frac{y}{b}\right)^n = 1$.

Ans. $\frac{a^{p+1}b^{q+1}}{mn}\frac{\Gamma\left(\frac{p+1}{m}\right)\Gamma\left(\frac{q+1}{n}\right)}{\Gamma\left(\frac{p+1}{m}+\frac{q+1}{n}+1\right)}$

2. Find the mass of the region in the xy-plane bounded by $x = 0$, $y = 0$, $x + y = 1$ with density $k\sqrt{xy}$.

Ans. $k\frac{\pi}{24}$

3. Find the area enclosed by the curve $\left(\frac{x}{a}\right)^{2m} + \left(\frac{y}{b}\right)^{2n} = 1$, m, n being positive integers

Ans. $\frac{ab}{4mn}\frac{\Gamma\left(\frac{1}{2m}\right)\Gamma\left(\frac{1}{2n}\right)}{\Gamma\left(\frac{1}{2m}+\frac{1}{2n}+1\right)}$

Hint: Put $\left(\frac{x}{a}\right)^{2m} = X$, $\left(\frac{y}{b}\right)^{2n} = Y$, area $= \iint a\frac{1}{2m}X^{\frac{1}{2m}-1}\cdot b\frac{1}{2n}Y^{\frac{1}{2n}-1}dX\,dY$.
Hence find the area bounded by the astroid $\left(\frac{x}{a}\right)^{2/3} + \left(\frac{y}{b}\right)^{2/3} = 1$

Ans. $\frac{3ab\pi}{32}$. Hint put $m = n = \frac{1}{3}$.

4. Determine the area enclosed by the curve $\left(\frac{x}{a}\right)^4 + \left(\frac{y}{b}\right)^{10} = 1$

Ans. $\frac{ab}{40}\frac{\Gamma\left(\frac{1}{4}\right)\Gamma\left(\frac{1}{10}\right)}{\Gamma\left(\frac{27}{20}\right)}$

Hint: Take $m = 2, n = 5$ in above example 3.

5. Express $\iint\limits_D x^p y^q\,dx\,dy$ in terms of gamma functions where D is the region of the ellipse $\frac{x^2}{a^2} + \frac{y^2}{b^2} = 1$ in the first octant.

Ans. $\frac{1}{4}a^{p+1}b^{q+1}\frac{\Gamma\left(\frac{p+1}{2}\right)\Gamma\left(\frac{q+1}{2}\right)}{\Gamma\left(\frac{p+1+q+1}{2}+1\right)}$

Hint: Take $m = 2, n = 2$ in above example 1.

6. Evaluate $\iiint\limits_V x^{\alpha-1}y^{\beta-1}z^{\gamma-1}dx\,dy\,dz$ where V is the closed region in the first octant bounded by the surface $\left(\frac{x}{a}\right)^p + \left(\frac{y}{b}\right)^q + \left(\frac{z}{c}\right)^r = 1$ and coordinate planes $x = 0, y = 0, z = 0$.

Ans. $\frac{a^\alpha b^\beta c^\gamma}{pqr}\frac{\Gamma\left(\frac{\alpha}{p}\right)\Gamma\left(\frac{\beta}{q}\right)\Gamma\left(\frac{\gamma}{r}\right)}{\Gamma\left(\frac{\alpha}{p}+\frac{\beta}{q}+\frac{\gamma}{r}+1\right)}$

Hint: Put $\left(\frac{x}{a}\right)^p = X$, $\left(\frac{y}{b}\right)^q = Y$, $\left(\frac{z}{c}\right)^r = Z$, then $X + Y + Z = 1$. Integrand of the triple integral is

$a^{\alpha-1}X^{(\alpha-1)/p}.b^{\beta-1}Y^{(\beta-1)/q}c^{\gamma-1}Z^{(\gamma-1)/r}\times$

$\times\frac{a}{p}X^{\frac{1}{p}-1}\cdot\frac{b}{q}Y^{\frac{1}{q}-}\cdot\frac{c}{y}Z^{\frac{1}{r}-1}$

$= \frac{a^\alpha b^\beta c^\gamma}{pqr}\iiint X^{\frac{\alpha}{p}-1}Y^{\frac{\beta}{q}-1}Z^{\frac{\gamma}{r}-1}dX\,dY\,dZ$

Apply book work II.

7. Find the volume of the solid bounded by the coordinate planes and the surface $\left(\frac{x}{a}\right)^{\frac{1}{2}} + \left(\frac{y}{b}\right)^{\frac{1}{2}} + \left(\frac{z}{c}\right)^{\frac{1}{2}} = 1$

Ans. $\frac{abc}{90}$

Hint: Put $p = q = r = \frac{1}{2}, \alpha = \beta = \gamma = 1$ in above example

$$\frac{a^1\cdot b^1\cdot c^1}{\frac{1}{2}\cdot\frac{1}{2}\cdot\frac{1}{2}}\cdot\frac{\Gamma(2)\Gamma(2)\Gamma(2)}{\Gamma(2+2+2+1)}=\frac{8abc}{6!}=\frac{abc}{90}$$

8. If V is the solid region in the first octant bounded by the unit sphere $x^2+y^2+z^2=1$. Find the mass of the ellipsoid $\left(\frac{x}{a}\right)^2+\left(\frac{y}{b}\right)^2+\left(\frac{z}{c}\right)^2=1$ with density $\rho(x,y,z)=(x+y+z)^4$

Ans. $4\pi(a^2+b^2+c^2)abc/35$

Hint: Put $x+y+z=u$, $y+z=uv$, $z=uvw$.

Jacobian $=\frac{\partial(x,y,z)}{\partial(u,v,w)}=u^2v$, $x=u(1-v)$, $y=uv$, $z=uvw$

9. If D is the positive quadrant of the ellipse $\left(\frac{x}{a}\right)^2+\left(\frac{y}{b}\right)^2=1$, evaluate $\iint\limits_D x^py^q dx\,dy$

Ans. $\frac{a^{p+1}\cdot b^{q+1}}{4}\beta\left(\frac{p+1}{2},\frac{q+1}{2}+1\right)$

Hint: $\left(\frac{x}{a}\right)^2=X$, $\left(\frac{y}{b}\right)^2=Y$, $x=a\sqrt{X}$, $y=b\sqrt{Y}$, $dx=\frac{a}{2\sqrt{X}}dX$, $dy=\frac{b}{2\sqrt{Y}}dY$

$$DI=\iint a^p\cdot X^{\frac{p}{2}}\cdot b^qY^{\frac{q}{2}}\cdot\frac{a}{2\sqrt{X}}\cdot\frac{b}{2\sqrt{Y}}dXdY$$

$$=\frac{a^{p+1}b^{q+1}}{4}\iint X^{\frac{p}{2}-\frac{1}{2}}Y^{\frac{q}{2}-\frac{1}{2}}dXdY$$

$$=\frac{a^{p+1}\cdot b^{q+1}}{4}\cdot\frac{\Gamma\left(\frac{p+1}{2}\right)\Gamma\left(\frac{q+1}{2}\right)}{\Gamma\left(\frac{p+1+q+1}{2}+1\right)}$$

$$=\frac{a^{p+1}b^{q+1}}{4}\beta\left(\frac{p+1}{2},\frac{q+1}{2}\right)$$

10. Determine the area of a circle, centroid of a quadrant of a circle by evaluating $\iint\limits_D x^py^q dx\,dy$ over the region D which is the positive quadrant of the circle $x^2+y^2=a^2$.

Ans. area $=\pi a^2$, centroid $\left(\frac{4a}{3\pi},\frac{4a}{3\pi}\right)$

Hint: Put $\frac{x}{q}=\sqrt{X}$, $\frac{y}{a}=\sqrt{Y}$, $dx=\frac{a}{2\sqrt{X}}dX$, $dy=\frac{a}{2\sqrt{Y}}dY$, $X+Y=1$

$$\iint x^py^q dx\,dy=\iint a^pX^{\frac{p}{2}}a^qY^{\frac{q}{2}}\cdot\frac{a}{2\sqrt{X}}\frac{a}{2\sqrt{Y}}dX\,dY$$

$$=\frac{a^{p+q+2}}{4}\iint X^{\frac{p}{2}-\frac{1}{2}}Y^{\frac{q}{2}-\frac{1}{2}}dX\,dY$$

$$=\frac{a^{p+q+2}}{4}\cdot\frac{\Gamma\left(\frac{p+1}{2}\right)\Gamma\left(\frac{q+1}{2}\right)}{\Gamma\left(\frac{p+q}{2}+2\right)}$$

Area : $p=0$, $q=0$ area = 4 times $\iint dxdy$

$\overline{x}=\frac{\iint x\,dy\,dx}{\iint dy\,dx}$. Numerator: $p=1$, $q=0$,

$$=\frac{a^3}{4}\frac{\Gamma(1)\Gamma\left(\frac{1}{2}\right)}{\Gamma\left(\frac{5}{2}\right)}=\frac{a^3}{3}$$

$$\overline{x}=\frac{a^3/3}{(\pi a^2/4)}=\frac{4a}{3\pi}$$

11. Determine the volume and mass contained in the tetrahedron in the first octant bounded by the coordinate planes $x=0$, $y=0$, $z=0$ and the plane $\frac{x}{a}+\frac{y}{b}+\frac{z}{c}=1$; if the density at any point $\rho(x,y,z)$ is $kx^2y^3z^4$

Ans. $\frac{abc}{6}$, $\frac{ka^3b^4c^5}{332640}$

Hint: Put $\frac{x}{a}=X$, $\frac{y}{b}=Y$, $\frac{z}{c}=Z$, so $X+Y+Z=1$.

$$V=\iiint dx\,dy\,dz=\iiint abc\,dX\,dY\,dZ$$

$$=abc\cdot\frac{\Gamma(1)\Gamma(1)\Gamma(1)}{\Gamma(0+0+0+4)}$$

where BW II with $p=q=r=0$ is applied.

$$M=\iiint kxyx\,dx\,dy\,dz$$

$$=\iiint ka^2X^2b^3Y^3c^4Z^4a\,dXb\,dY\,c\,dZ$$

$=ka^3b^4c^5\frac{\Gamma(3)\Gamma(4)\Gamma(5)}{\Gamma(2+3+4+4)}$ where BW II with $p=2$, $q=3$, $r=4$ is used; $M=ka^3b^4c^5\cdot\frac{2!3!4!}{12!}$ $M=\frac{ka^3b^4c^5}{332640}$

12. If V is tetrahedron in the first octant bounded by $x=0$, $y=0$, $z=0$, $x+y+z=1$, show that

$$\iiint\limits_V x^py^qz^rf(x+y+z)dx\,dy\,dz$$

$$=\frac{\Gamma(p+1)\Gamma(q+1)\Gamma(r+1)}{\Gamma(p+q+r+4)}\int_0^1 u^{p+q+r+2}f(u)du$$

and the coordinate planes, evaluate $\iiint\limits_V x^py^qz^rdx\,dy\,dz$

Ans. $\frac{1}{8}\frac{\Gamma\left(\frac{p+1}{2}\right)\Gamma\left(\frac{q+1}{2}\right)\Gamma\left(\frac{r+1}{2}\right)}{\Gamma\left(\frac{p+1+q+1+r+1}{2}+1\right)}$

Hint: Put $x^2=X$, $y^2=Y$, $z^2=Z$ then $dx=\frac{1}{2\sqrt{X}}dX$, $x=X^{\frac{1}{2}}$ with $X+Y+Z=1$

$$\frac{1}{8}\iiint\limits_{V*}X^{\frac{p}{2}-\frac{1}{2}}Y^{\frac{q}{2}-\frac{1}{2}}Z^{\frac{r}{2}-\frac{1}{2}}dX\,dY\,dZ$$

$$=\frac{1}{8}\beta\left(\frac{p+1}{2},\frac{q+1}{2},\frac{r+1}{2}\right)$$

HIGHER ENGINEERING MATHEMATICS

PART—III
ORDINARY DIFFERENTIAL EQUATIONS

Chapter 8

Ordinary Differential Equations: First Order and First Degree

INTRODUCTION

To describe, understand and predict the behaviour of a physical process or system, a "mathematical model" is constructed by relating the variables by means of one or more equations. Usually these equations describing the system in motion are "differential equations" involving derivatives which measures the rates of change. The behaviour and interaction of components of the system at later times is described by the "solutions" of these differential equations.

In this chapter we consider the simplest of these differential equations which is of first order. We study the solutions of differential equations which are: variables separable, homogeneous, non-homogeneous, exact, non-exact using integrating factors, linear, Bernoulli, higher degree, Clairaut's, Lagrange's. We consider geometrical applications and physical problems of law of natural growth, natural decay, Newton's law of cooling, velocity of escape from earth and simple electrical circuits.

8.1 INTRODUCTION TO MATHEMATICAL MODELING

Scientific model is an abstract and simplified description of a given phenomenon and is most often based on mathematical structures.

Historically following the invention of calculus by Newton (1642-1727) and Leibnitz (1646-1716), there is a burst of activity in mathematical sciences. Early mathematical modeling problems include boundary value problems in vibration of strings, elastic bars and columns of air due to Taylor (1685-1731), Daniel Bernoulli (1700-1782), Euler (1707-1783) and d'Alembert (1717-1783).

Modeling is a technique of transforming a physical problem to a "mathematical model". Thus a mathematical model describes a natural process or a physical system in mathematical terms, representing an idealization by simplifying the reality by ignoring negligible details of the natural process and emphasizing on only its essential manifestations. Such model yielding reproducible results, can be used for prediction. Thus a mathematical model essentially expresses a physical system in terms of a functional relationship of the kind:

Dependent variable = function of independent variables, parameters and forcing functions

A model should be general enough to explain the phenomenon but not too complicated precluding analysis. Mathematical formulation of problems involving continuously or discretely changing quantities leads to ordinary or partial differential equations, linear or non-linear equations, integral equations or a combination of these.

Example of a Mathematical Model of a Mechanical System

Consider a mechanical oscillator consisting of a block of a mass m lying on a table and restrained laterally by an ordinary spring. The displacement $x(t)$ of the spring as a function of time t is governed by

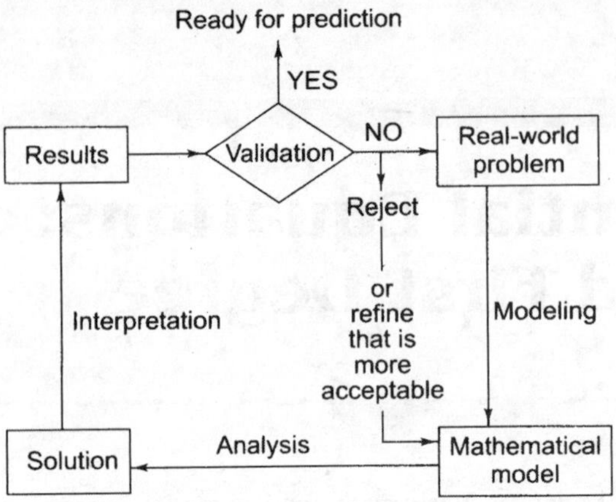

Fig. 8.1 Scheme of mathematical modeling

a second order differential equation which arises out of the Newton's second law as follows:

(mass)(acceleration) = sum of forces

or
$$m\frac{d^2x}{dt^2} = F - F_s - F_f - F_a$$

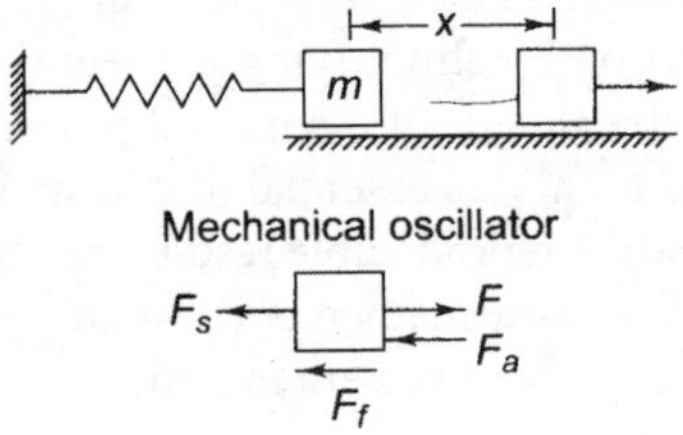

Fig. 8.2

Here $F(t)$, the applied force, is the forcing function. The force F_s exerted by the spring on the mass is generally assumed to be of the form $F_s(x) = kx$; where k the spring stiffness is the parameter of the material. The sliding friction force F_f exerted on the bottom of the mass may be assumed to be proportional to the velocity i.e. $F_f = c\frac{dx}{dt}$, where c is known as the damping coefficient, another parameter. Neglecting the aerodynamic drag F_a we get an approximate model of the system described by

$$m\frac{d^2x}{dt^2} = F - kx - c\frac{dx}{dt} - 0$$

or
$$\boxed{m\frac{d^2x}{dt^2} + c\frac{dx}{dt} + kx = f(t)}$$

Thus the mathematical model of the physical mechanical oscillator is described by a differential equation which expresses the relationship between the dependent variable, x the displacement as a function of the independent variable t, the parameters k and c and the forcing $F(x)$. Models representing the electrical, chemical, mechanical, and civil systems are abundant.

A mechanical system describing the forced oscillations of a mass-spring system and an electrical system describing an RLC-circuit are both represented by a linear second order non-homogeneous ordinary differential equation. This demonstrates the unifying power of "mathematical modeling" through which entirely different physical systems can be studied by the same mathematical model. The analogy between mechanical and electrical systems simplifies the study of mechanical systems since the electrical systems are easy to construct and easy to measure accurately.

The power of modeling is seen from the differential equation

$$\frac{d^2x}{dt^2} + \omega^2 x = 0$$

which represents simple harmonic motion, also the motion of a particle with constant angular velocity along a circle and also the free undamped vertical motion of a mass-spring system and finally the motion of the bob of a simple pendulum in a vertical plane.

8.2 BASIC DEFINITIONS

A "**differential equation**" (D.E.) is an equation involving (connecting) an unknown (or sought-for) function y of one or more independent variables $x, t, \ldots$ and its derivatives.

Differential equations are classified into two categories "ordinary and partial" depending on the number of independent variables appearing in the equation.

Ordinary Differential Equation (O.D.E.)

An ordinary differential equation is a D.E. in which the dependent variable y depends only on one independent variable say x (so that the derivatives of y are ordinary derivatives).

Example: $F(x, y, y', y'', \ldots, y^{(n)}) = 0$.

Notation: The first derivative $\frac{dy}{dx}$ is denoted by y', second derivative $\frac{d^2y}{dx^2}$ by y'', etc.

Partial Differential Equation (P.D.E.)

A partial differential equation is one in which y depends on two or more independent variables say $x, t, \ldots$ (so that the derivatives of y are partial derivatives.)

Example: $F\left(x, t, y, \frac{\partial y}{\partial x}, \frac{\partial y}{\partial t}, \frac{\partial^2 y}{\partial x^2}, \ldots, \frac{\partial^m y}{\partial x^k \partial t^l}\right) = 0.$

Order

of a D.E. is the order of the highest derivative appearing in the equation.

Degree

of a D.E. is the degree of the highest ordered derivative (when the derivatives are cleared of radicals and fractions).

Linear

An nth order O.D.E. in the dependent variable y is said to be linear in y if

i. y and all its derivatives are of degree one.

ii. No product terms of y and/or any of it derivatives are present.

iii. No tanscendental functions of y and/or its derivatives occur.

Non-linear

O.D.E. is an O.D.E. that is *not* linear. The general form of an nth order linear O.D.E. in y with variables coefficient is

$$a_0(x)\frac{d^n y}{dx^n} + a_1(x)\frac{d^{n-1}y}{dx^{n-1}} + \ldots + a_{n-1}(x)\frac{dy}{dx} + a_n(x)y = b(x)$$

where R.H.S. $b(x)$ and all the coefficients $a_0(x), a_1(x), \ldots, a_n(x)$ are given functions of x and $a_0(x) \neq 0$.

If all the coefficients $a_0, a_1, \ldots, a_n$ are constants then the above equation is known as nth order linear O.D.E. with constant coefficients.

Note: A linear D.E. is of first degree but a first degree D.E. need not be linear since it may contain nonlinear terms such as y^2, $y^{\frac{-1}{2}}$, e^y, $\sin y$, etc.

[see Examples 3 and 6 in Exercise]

Solution or integral or primitive

Solution or integral or primitive of a D.E. is any function which satisfies the equation i.e., reduces it to an identity.

Note 1: A D.E. may have a unique solution or several solution, or no solution.

In Explicit solutions the dependent variable can be expressed explicitly in terms of the independent variable, like $y = f(x)$. Otherwise the solution is said to be an *implicit* solution where $F(x, y) = 0$ where $F(x, y)$ is an implicit function.

Note 2: *General (or complete) solution* of an nth order D.E. will have n arbitrary constants.

Particular solution

Particular solution is a solution obtained from the general solution by choosing particular values of the arbitrary constants. *Integral curve* of D.E. is the graph of the general/particular solution of D.E.

Initial (boundary) value problem

Initial (boundary) value problem is one in which a solution to a D.E. is obtained subject to conditions on the unknown function and its derivative specified at one (two or more) value(s) of the independent variable. Such conditions are called *initial (boundary) conditions.*

General (or *complete) integral* of a D.E. is an implicit function $\phi(x, y, c) = 0$.

Particular integral

Particular integral is one obtained from general integral for a particular value of constant C.

Singular solutions of a D.E. are (unusual or odd) solutions of D.E. which can not be obtained from the general solution.

In Chapter 8, first order first degree (both linear and non-linear) O.D.E's are studied.

EXERCISE

Classify each of the following D.Es by its kind, order, degree and linearity:

S. No.	Differential Equation	Ans: Kind	Order	Degree	Linearity
1.	$\frac{dy}{dx} = kx^2$	ordinary	1	1	yes
2.	$\frac{dy}{dx} + P(x)y = y^n Q(x)$	ordinary	1	1	no (yes for $n = 0, 1$)
3.	$e^x dx + e^y dy = 0$	ordinary	1	1	nonlinear (in x and y)
4.	$\left(\frac{d^3y}{dx^2}\right)^4 - 6x^2\left(\frac{dy}{dx}\right)^8 + e^y = \sin xy$	ordinary	3	4	no
5.	$y\frac{d^2y}{dx^2} + \sin x = 0$	ordinary	2	1	no
6.	$x^2 dy + y^2 dx = 0$	ordinary	1	1	no
7.	$\frac{d^4y}{dx^4} + 3\left(\frac{d^2y}{dx^2}\right)^5 + 5y = 0$	ordinary	4	1	no
8.	$y^2 dx + (3xy - 1)dy = 0$	ordinary	1	1	nonlinear in y linear in x
9.	$k(y'')^2 = [1 + (y'')^2]^3$	ordinary	2	2	no
10.	$\frac{\partial u}{\partial t} = k\left(\frac{\partial^2 u}{\partial x^2} + \frac{\partial^2 u}{\partial y^2}\right)$	partial	2	1	yes
11.	$\frac{\partial^2 Y}{\partial t^2} = a^2 \frac{\partial^2 Y}{\partial x^2}$	partial	2	1	yes
12.	$\left(\frac{dr}{ds}\right)^3 = \sqrt{\frac{d^2r}{ds^2} + 1}$	ordinary	2	1	no

8.3 FIRST ORDER FIRST DEGREE DIFFERENTIAL EQUATIONS

A first order first degree ordinary differential equation contains only y', known functions of x and perhaps terms of y itself. Thus the general form is

$$F(x, y, y') = 0$$

which may be solved for y' and rewritten in the explicit form

$$\frac{dy}{dx} = f(x, y)$$

The Cauchy or initial value problem (IVP) is the problem of finding the solution (function) $y = y(x)$ of the above D.E. satisfying the initial condition $y(x_0) = y_0$ (i.e., value of y is y_0 when $x = x_0$ where x_0 and y_0 are given number). Thus for the first order D.E., the general solution (G.S.) $y = \phi(x, c)$ contains one arbitrary constant C. Geometrically, infinitely many (integral) curves are obtained by varying the one "parameter" C, forming a "one-parameter family of (integral) curves." So a particular solution is a particular (specific) integral curve of this family.

Normally it is easy to verify that a given function is a solution of a D.E. while it is very difficult to find the solution of even the first order first degree O.D.E.

$$\frac{dy}{dx} = f(x, y)$$

But on the other hand, there are certain standard types of first order first degree D.E's for which solutions can be readily obtained by standard methods such as

A. Variables separable (Section 8.4)
B. Homogeneous equation (Section 8.5)
C. Non-homogeneous equation reducible to homogeneous equation (Section 8.6)
D. Exact differential equation (Section 8.7)
E. Non-exact differential equations that can be made exact with the help of integrating factors (Section 8.8)
F. Linear first order equation (Section 8.9)
G. Bernoulli's equation (Section 8.10).

8.4 VARIABLES SEPARABLE OR SEPARABLE EQUATION

An equation of the form

$$F(x)\,G(y)\,dx + f(x)\,g(y)\,dy = 0 \tag{1}$$

is called an equation with variables separable or simply a separable equation, because the variables x and y can be *separated*.

Rewriting

$$\frac{F(x)}{f(x)}dx + \frac{g(y)}{G(y)}dy = 0 \tag{2}$$

or $$M(x)dx + N(y)dy = 0 \quad (3)$$

where $M(x) = F(x)/f(x)$ is a function of x only and $N(y) = g(y)/G(y)$ is a function of y only.

Integrating, we get the one-parameter family of solutions as

$$\int M(x)dx + \int N(y)dy = C \quad (4)$$

where C is the arbitrary constant. Since division by $f(x)$ and $G(y)$ are involved in (2), meaningful valid solutions can be obtained provided $f(x) \neq 0$ and $G(y) \neq 0$.

Note: In an equation of the form

$$\frac{dy}{dx} = f(ax + by + c)$$

(i.e., R.H.S. is a function of the variable $ax + by + c$) can be changed into a separable equation by the substitution $z = ax + by + c$.

WORKED OUT EXAMPLES

Variables separable

Solve the following:

Example 1: $\tan x \cdot \sin^2 y\, dx + \cos^2 x \cdot \cot y\, dy = 0$

Solution: Rewriting

$$\tan x \cdot \sec^2 x \cdot dx + \cot y \cdot \operatorname{cosec}^2 y dy = 0$$

Integrating

$$\tan x \cdot d(\tan x) - \cot y d(\cot y) = 0$$

$$\therefore \quad \frac{\tan^2 x}{2} - \frac{\cot^2 y}{2} = C$$

Example 2: $\frac{dy}{dx} = e^{2x-y} + x^3 e^{-y}$

Solution: Rewriting $\frac{dy}{dx} = (e^{2x} + x^3)e^{-y}$

Separating the variables, we get

$$e^y dy = (e^{2x} + x^3)dx$$

Integrating

$$e^y = \frac{e^{2x}}{2} + \frac{x^4}{4} + C$$

Example 3: $y' = \sin^2(x - y + 1)$

Solution: Put $z = x - y + 1$, so $\frac{dz}{dx} = 1 - \frac{dy}{dx} + 0$.

Substituting z and $\frac{dy}{dx}$, we get a separable equation as

$$1 - \frac{dz}{dx} = \sin^2 z$$

or $\frac{dz}{dx} = 1 - \sin^2 z = \cos^2 z$

Separating the variables

$$\sec^2 z dz = dx$$

Integrating

$$\tan z = x + c$$

or $$\tan (x - y + 1) = x + c$$

Example 4: Show that the particular solution of

$$(x^2 + 1)\frac{dy}{dx} + (y^2 + 1) = 0,\ y(0) = 1, \text{ is } y = \frac{1 - x}{1 + x}$$

Solution: Separating the variables

$$\frac{dy}{y^2 + 1} + \frac{dx}{x^2 + 1} = 0$$

Integrating $\tan^{-1} y + \tan^{-1} x = c_0$.

$\tan(\tan^{-1} y + \tan^{-1} x) = \tan c$

Using $$\tan (a + b) = \frac{\tan a + \tan b}{1 - \tan a \cdot \tan b}$$

$$\frac{y + x}{1 - xy} = \tan c$$

when $x = 0$, $y = 1$ then $\frac{1+0}{1-0} = \tan c$

$$\therefore \quad \frac{y + x}{1 - xy} = 1$$

Solving, $y = \frac{1-x}{1+x}$ is the required particular solution.

EXERCISE

Variables separable

Solve the following:

1. $4xydx + (x^2 + 1)dy = 0$

Ans. $y(x + 1)^2 = c$

2. $(x + 4)(y^2 + 1)dx + y(x^2 + 3x + 2)dy = 0$

Ans. $3(x^2 + x)y = x^3 - 3x + c$

3. $(xy + x)dx = (x^2y^2 + x^2 + y^2 + 1)dy$

Ans. $\ln (x^2 + 1) = y^2 - 2y + 4 \ln |c(y + 1)|$

4. $y' = x \tan (y - x) + 1$

Ans. $\log \sin (y - x) = \frac{x^2}{2} + c$

5. $(x - y)^2 y' = a^2$

Ans. $a\log\left(\frac{x-y-a}{x-y+a}\right)=2y+c$

6. Obtain particular solution $2xyy'=1+y^2$; $y(2)=3$.

Ans. $y^2=5x-1$

7. $y'=(x+y)^2$

Hint: Put $x+y=z$.

Ans. $(x+y)=\tan\ (x+c)$

8. $(2x-4y+5)y'+x-2y+3=0$

Hint: Put $z=x-2y$.

Ans. $4x-8y+\ln|4x-8y+11|=c$

9. Solve $\frac{dy}{dx}=(4x+y+1)^2$, $y(0)=1$.

Hint: Put $4x+y+1=z$.

Ans. $4x+y+1=2\tan\ (2x+\frac{\pi}{4})$

10. $xyy'=\left(\frac{1+y^2}{1+x^2}\right)(1+x+x^2)$

Ans. $\frac{1}{2}\ln\ (1+y^2)=\ln x+\tan^{-1}x+c$

11. $x^3e^{2x^2+3y^2}dx-y^3e^{-x^2-2y^2}dy=0$

Ans. $25(3x^2-1)e^{3x^2}+9(5y^2+1)e^{-5y^2}=c$

12. $y'=\frac{(y-1)(x-2)(y+3)}{(x-1)(y-2)(x+3)}$

Ans. $(x+1)(y+3)^5=c(y-1)(x+3)^5$.

8.5 HOMOGENEOUS EQUATION–REDUCTION TO SEPARABLE FORM

Differential Equation of the Form $y'=g(\frac{y}{x})$

Homogeneous function

A function $f(x,y)$ is said to be homogeneous function of degree n in the variables x and y if for any t,

$$f(tx,ty)=t^n f(x,y)$$

Examples:

1. $f(x,y)=4x^2-3xy+y^2$, homogeneous of degree 2

since $f(tx,ty)=4t^2x^2-3tx.ty+t^2y^2$
$=t^2(4x^2-3xy+y^2)=t^2f(x,y)$

2. $f(x,y)=(x^3+y^3)e^{\frac{2x}{y}}+4xy^2$, homogeneous of degree 3.

3. $f(x,y)=x^3+\sin x\cdot e^y$ is *not* homogeneous.

A first order equation $y'=f(x,y)$ is said to be homogeneous* if $f(x,y)$ is a homogeneous function of degree zero. i.e., $f(x,y)$ will depend only on $\frac{y}{x}$ i.e., $f(x,y)$ is of the form $g(\frac{y}{x})$.

Alternatively $M(x,y)dx+N(x,y)dy=0$ is said to be homogeneous if $M(x,y)$ and $N(x,y)$ are both homogeneous of the same degree.

Method to Solve Homogeneous Equation $y'=f(x,y)=g(\frac{y}{x})$

I. Put

$$u=\frac{y}{x}\quad\text{or}\quad y=ux$$

and

$$\frac{dy}{dx}=u+x\frac{du}{dx}$$

in the given D.E., which reduces to a separable equation.

$$u+x\frac{du}{dx}=g(u)$$

II. Separating the variables and integrating, the solution is obtained as

$$\int\frac{du}{g(u)-u}=\int\frac{dx}{x}+c$$

III. Replace u by y/x, in the solution obtained in II.

WORKED OUT EXAMPLES

Solve the following:

Example 1: $(x+2y)dx+(2x+y)dy=0$

Solution: Rewriting $y'=-\frac{(x+2y)}{(2x+y)}=f(x,y)$

This D.E. is a homogeneous, as the R.H.S. function $f(x,y)$ is homogeneous of degree 0.

Put $u=\frac{y}{x}$, $\frac{xdu}{dx}+u=\frac{dy}{dx}$ in given D.E., we get

* The term "homogeneous" is used here loosely in the sense that $f(x,y)$ or $M(x,y)$, $N(x,y)$ are homogeneous functions. But strictly speaking homogeneous D.E. is defined later in a more appropriate way.

$$u+\frac{xdu}{dx}=\frac{-(1+2u)}{(2+u)}$$

Rearranging

$$\frac{xdu}{dx}=\frac{-(1+2u)}{(2+u)}-u=\frac{-(u^2+4u+1)}{u+2}$$

Separating the variables, we get

$$-\frac{dx}{x}=\frac{(u+2)du}{(u^2+4u+1)}=\frac{1}{2}\frac{d(u^2+4u+1)}{(u^2+4u+1)}$$

Integrating

$$-2\ln x=\ln\,(u^2+4u+1)+c_0$$

$$x^2(u^2+4u+1)=c$$

$$x^2\left(\frac{y^2}{x^2}+4\frac{y}{x}+1=c\right)$$

Thus $\quad y^2+4xy+x^2=c$

Example 2: $(1+2e^{\frac{x}{y}})+2e^{\frac{x}{y}}\cdot(1-\frac{x}{y})y'=0$

Solution: Rewriting

$$(1+2e^{\frac{x}{y}})dx+2e^{\frac{x}{y}}(1-\frac{x}{y})dy=0$$

Put $\frac{x}{y}=u$, so that $dx=ydu+udy$
D.E. becomes

$$(1+2e^u)(udy+ydu)+2e^u(1-u)dy=0$$

$$(u+2e^u)dy+y(1+2e^u)du=0$$

Separating the variables

$$\frac{dy}{y}+\frac{1+2e^u}{u+2e^u}du=0$$

Integrating

$$\ln\,y+\ln\,(u+2e^u)=\ln c$$

$$y(u+2e^u)=c$$

Replacing u, we get

$$y\left(\frac{x}{y}+2e^{\frac{x}{y}}\right)=c$$

Example 3: $(y+\sqrt{x^2+y^2})dx-xdy=0$,
$y(1)=0$.

Solution: Rewriting

$$\frac{dy}{dx}=\frac{y+\sqrt{x^2+y^2}}{x}=\frac{y}{x}+\sqrt{1+\left(\frac{y}{x}\right)^2}$$

is homogeneous. Put $y=ux$ and $\frac{dy}{dx}=u+x\frac{du}{dx}$ in D.E., we get

$$u+x\frac{du}{dx}=u+\sqrt{1+u^2}$$

or

$$x\frac{du}{dx}=\sqrt{1+u^2}$$

Separating the variables

$$\frac{dx}{x}=\frac{du}{\sqrt{u^2+1}}$$

Integrating

$$\ln|x|+\ln|c|=\ln|u+\sqrt{u^2+1}|$$

or $u+\sqrt{u^2+1}=cx$
Replacing u, we have

$$\frac{y}{x}+\sqrt{\frac{y^2}{x^2}+1}=cx$$

or

$$y+\sqrt{x^2+y^2}=cx^2$$

Put $y=0$, when $x=1$, then $c=1$
So the required solution is

$$y+\sqrt{x^2+y^2}=x^2.$$

EXERCISE

Solve the following:

1. $(2xy+3y^2)dx-(2xy+x^2)dy=0$

Ans. $y^2+xy=cx^3$

2. $(x^3+y^2\sqrt{x^2+y^2})dx-xy\sqrt{x^2+y^2}dy=0$

Ans. $(x^2+y^2)^{\frac{3}{2}}=x^3\ln cx^3$

3. $(2x-5y)dx+(4x-y)dy=0,\ y(1)=4$

Ans. $(2x+y)^2=12(y-x)$

4. $x\cdot\sin\frac{y}{x}\frac{dy}{dx}=y\sin\frac{y}{x}+x$

Ans. $\cos\frac{y}{x}+\log cx=0$

5. $x^2y^1=3(x^2+y^2)\tan^{-1}\frac{y}{x}+xy$

Ans. $y=x\tan cx^3$

6. $(x^2 + xy)dy = (x^2 + y^2)dx$

Ans. $(x - y)^2 = cx \cdot e^{\frac{-y}{x}}$

7. $x^2ydy + (x^3 + x^2y - 2xy^2 - y^3)dx = 0$

Ans. $\log\left\{\frac{c(y-x)}{x^4(y+x)}\right\} = \frac{2x}{x+y}$

8. $xy^1 = y + x \cdot \cos^2(\frac{y}{x}),\ y(1) = \frac{\pi}{4}$

Ans. $1 + \ln x = \tan\frac{y}{x}$

9. $y^1 = \frac{6x^2-5xy-2y^2}{6x^2-8xy+y^2}$

Ans. $(y - x)(y - 3x)^9 = c(y - 2x)^{12}$

10. $[2x \cdot \sin\frac{y}{x} + 2x \cdot \tan\frac{y}{x} - y\cos\frac{y}{x} - y\sec^2\frac{y}{x}]dx + [x \cdot \cos\frac{y}{x} + x\sec^2\frac{y}{x}]dy = 0.$

Ans. $x^2(\sin\frac{y}{x} + \tan\frac{y}{x} = c)$.

8.6 NON-HOMOGENEOUS EQUATIONS REDUCIBLE TO HOMOGENEOUS FORM

Consider the non-homogeneous equation

$$\frac{dy}{dx} = \frac{a_1x + b_1y + c_1}{a_2x + b_2y + c_2} \tag{1}$$

where $a_1, b_1, c_1, a_2, b_2, c_2$ are all constants.

Case 1: If $\frac{a_1}{a_2} \neq \frac{b_1}{b_2}$, i.e., $\begin{vmatrix} a_1 & b_1 \\ a_2 & b_2 \end{vmatrix} \neq 0$ then the transformation, (shift of origin)

$$x = x_1 + h,\ y = y_1 + k \tag{2}$$

reduces the non-homogeneous Equation (1) to the homogeneous equation of the form

$$\frac{dy_1}{dx_1} = \frac{a_1x_1 + b_1y_1}{a_2x_1 + b_2y_2} \tag{3}$$

Here the unknown constants h, k in (2) are determined by solving the pair of equations

$$a_1h + b_1k + c_1 = 0$$

$$a_2h + b_2k + c_2 = 0$$

Now that Equation (3) is homogeneous in the new variables x_1 and y_1, it can be solved as in 8.5.

Case 2: If $\frac{a_1}{a_2} = \frac{b_1}{b_2}$, i.e., $\begin{vmatrix} a_1 & b_1 \\ a_2 & b_2 \end{vmatrix} = 0$, then the transformation

$$z = a_1x + b_1y$$

reduces (1) to a separable equation in the variables x and z, which can be solved as in 8.4.

WORKED OUT EXAMPLES

Solve the following:

Example 1: $(2x^2 + 3y^2 - 7)xdx = (3x^2 + 2y^2 - 8)y\,dy$

Solution: Put $x^2 = X$, $y^2 = Y$, so that $2x\,dx = dX$, $2y\,dy = dY$

$$(2X + 3Y - 7)dX = (3X + 2Y - 8)dY$$

$\frac{dY}{dX} = \frac{2X+3Y-7}{3X+2Y-8}$, not homogeneous

Put $X = X_1 + h$, $Y = Y_1 + k$

$$\frac{dY_1}{dX_1} = \frac{2X_1 + 3Y_1 + (2h + 3k - 7)}{3X_1 + 2Y_1 + (3h + 2k - 8)}$$

To convert this into a homogeneous equation, Put $2h + 3k - 7 = 0$, $3h + 2k - 8 = 0$.

Solving $h = 2$, $k = 1$, with these values

$$\frac{dY_1}{dX_1} = \frac{2X_1 + 3Y_1}{3X_1 + 2Y_1}$$

To solve this put $u = \frac{Y_1}{X_1}$

$$u + X_1\frac{du}{dX_1} = \frac{2X_1 + 3uX_1}{3X_1 + 2uX_1} = \frac{2 + 3u}{3 + 2u}$$

$$X_1\frac{du}{dX_1} = \frac{2 + 3u}{3 + 2u} - u = \frac{2(1 - u^2)}{3 + 2u}$$

Separating

$$\frac{2dX_1}{X_1} = \left(\frac{3 + 2u}{1 - u^2}\right)du = \frac{3du}{1 - u^2} + \frac{2udu}{1 - u^2}$$

But $$\frac{1}{1 - u^2} = \frac{1}{(1 - u)(1 + u)} = \frac{1}{2}\left[\frac{1}{(1 - u)} + \frac{1}{1 + u}\right]$$

Integrating

$$2\int\frac{dX_1}{X_1} = 3 \cdot \frac{1}{2}\left[\int\frac{du}{1 - u} + \frac{du}{1 + u}\right] - 2\int\frac{udu}{u^2 - 1}$$

$$4\ln X_1 + 2\ln c = 3\ln\left(\frac{u + 1}{u - 1}\right) - 2\ln\,(u^2 - 1)$$

$$c^2X_1^4 = \frac{(u+1)^3}{(u-1)^3} \cdot \frac{1}{(u^2-1)^2}$$

$$c^2X_1^4 = \frac{u+1}{(u-1)^5}$$

Replacing $u = \frac{Y_1}{X_1}$, we get

$$c^2 = \frac{(Y_1 + X_1)}{(Y_1 - X_1)^5}$$

Replacing $X_1 = X - 2$, $Y_1 = Y - 1$, we have

$$c^2 = \frac{(Y - 1 + X - 2)}{[Y - 1 - (X-2)]^5}$$

$$c^2(x^2 - y^2 + 1)^5 = (x^2 + y^2 - 3)$$

Example 2: $\frac{dy}{dx} = \frac{y-x}{y-x+2}$

Solution: Since this is non-homogeneous, if $x = x_1 + h$, $y = y_1 + k$ then

$$\frac{dy}{dx} = \frac{y_1 + k - (x_1 + h)}{y_1 + k - (x_1 + h) + 2}$$

$$= \frac{(y_1 - x_1) + (k - h)}{(y_1 - x_1) + (k - h) + 2}$$

To convert this to homogeneous, put

$$k - h = 0$$

$$k - h = 2$$

which has *no* solution. So this substitute method *fails* in this case. But on the other hand observe that by introducing

$$z = y - x$$

The given equation becomes separable.

$$\frac{dz}{dx} = \frac{dy}{dx} - 1$$

or

$$\frac{dy}{dx} = \frac{dz}{dx} + 1 = \frac{z}{z+2}$$

$$\frac{dz}{dx} = \frac{z}{z+2} - 1 = \frac{-2}{z+2}$$

Separating the variables and integrating

$$\int (z+2)dz + 2\int dx = c$$

$$\frac{z^2}{2} + 2z + 2x = c$$

Replacing z by $y - x$, we get

$$(y-x)^2 + 4(y-x) + 4x = c$$

or

$$(y-x)^2 + 4y = c$$

Example 3: $(x-2y+1)dx + (4x-3y-6)dy = 0$

Solution: This equation is nonhomogeneous.

Since $\frac{a_1}{a_2} = \frac{1}{4} \neq \frac{b_1}{b_2} = \frac{-2}{-3} = \frac{2}{3}$

i.e., $\begin{vmatrix} 1 & -2 \\ 4 & -3 \end{vmatrix} = 5 \neq 0$

we can make the substitution

$$x = x_1 + h, \; y = y_1 + k$$

with this, the given equation becomes

$$[x_1 + h - 2(y_1 + k) + 1]dx_1$$
$$+ [4(x_1 + h) - 3(y_1 + k) - 6]dy_1 = 0$$

or $[(x_1 - 2y_1) + (h - 2k + 1)]dx_1 + [(4x_1 - 3y_1) + (4h - 3k - 6)]dy_1 = 0$

This reduces to homogeneous if

$$h - 2k + 1 = 0 \quad \text{and} \quad 4h - 3k - 6 = 0$$

Solving $h = 3$, $k = 2$.

Thus the homogeneous equation in the new variables x_1 and y_1 is

$$\frac{dy_1}{dx_1} = \frac{x_1 - 2y_1}{3y_1 - 4x_1}$$

Introduce $u = \frac{y_1}{x_1}$, $x_1 \frac{du}{dx_1} + u = \frac{dy_1}{dx_1}$

$$x_1 \frac{du}{dx_1} + u = \frac{1-2u}{3u-4}$$

$$x_1 \frac{du}{dx_1} = \frac{1-2u}{3u-4} - u = \frac{1 - 2u - 3u^2 + 4u}{3u-4}$$

$$x_1 \frac{du}{dx_1} = \frac{1 + 2u - 3u^2}{3u - 4}$$

Separating the variables

$$\left(\frac{3u-4}{3u^2 - 2u - 1}\right) du = \frac{-dx_1}{x_1}$$

Integrating

$$\frac{1}{2}\ln|3u^2 - 2u - 1| - \frac{3}{4}\ln\left|\frac{3u-3}{3u+1}\right| = -\ln|x_1| + \ln|c_1|$$

$$\ln(3u^2-2u-1)^2 - \ln\left|\frac{3u-3}{3u+1}\right|^3 = \ln\left(\frac{c_1^4}{x_1^4}\right)$$

$$\ln\left|\frac{(3u+1)^5}{3(u-1)}\right| = \ln\left(\frac{c_1^4}{x_1^4}\right)$$

$x_1^4(3u+1)^5 = c|u-1|$ where $c = 3c_1^4$

Replacing u, $|3y_1+x_1|^5 = c|y_1-x_1|$.

Replacing $y_1 = y-2$, $x_1 = x-3$,

we get $|x+3y-9|^5 = c|y-x+1|$.

Exercise

Solve the following:

1. $(3x-y-9)y' = (10-2x+2y)$

Ans. $y-2x+7 = c(x+y+1)^4$

2. $y' = \frac{ax+by-a}{bx+ay-b}$

Ans. $(y-x+1)^{(a+b)/a}(y+x-1)^{(a-b)/a}$

3. $(2x-5y+3)dx-(2x+4y-6)dy = 0$

Ans. $(4y-x-3)(y+2x-3)^2 = c$

4. $(3y-7x+7)dx+(7y-3x+3)dy = 0$

Ans. $(y-x+1)^2(y+x-1)^5 = c$

5. $y' = \frac{2x+3y+1}{3x-2y-5}$

Ans. $\ln\ [(x-1)^2+(y+1)^2] - 3\tan^{-1}\left(\frac{y+1}{x-1}\right) = c$

6. $y' = \frac{2x+y-1}{4x+2y+5}$

Hint: Put $2x+y = z$

Ans. $10y-5x+7\ln(10x+5y+9) = c$

7. $(2x+2y+1)dx+(x+y-1)dy = 0$

Hint: Put $x+y = z$.

Ans. $3\ln\ (x+y+2)-2x-y = c$

8. $y' = (x-2y+3)/(2x-4y+5)$

Hint: Put $x-2y = z$.

Ans. $x^2-4xy+4y^2+6x-10y = c$

9. $y' + \frac{ax+hy+g}{hx+by+f} = 0$

Ans. $ax^2+2hxy+by^2+2gx+2fy+c = 0$.

8.7 EXACT DIFFERENTIAL EQUATIONS

The (total) differential of a function $f(x, y)$ is denoted by df and is given by

$$df = \frac{\partial f}{\partial x}dx + \frac{\partial f}{\partial y}dy \qquad (1)$$

Consider the differential equation

$$M(x, y)dx + N(x, y)dy = 0 \qquad (2)$$

Suppose there exists a function $f(x, y)$ such that

$$\frac{\partial f}{\partial x} = M(x, y) \qquad (3)$$

and

$$\frac{\partial f}{\partial y} = N(x, y) \qquad (4)$$

Using (3) and (4) then the given D.E. (2) becomes

$$0 = Mdx + Ndy = \frac{\partial f}{\partial x}dx + \frac{\partial f}{\partial y}dy = df$$

i.e., $df = 0$

Integrating $f(x, y) = c =$ arbitrary constant.

In this case, the L.H.S. expression of (2) $Mdx + Ndy$ is said to be an *exact differential* and the differential Equation (2) is called an *exact differential equation*.

Necessary Condition for Exactness

Differentiating (3) and (4) partially w.r.t. y and x respectively, we get

$$\frac{\partial M}{\partial y} = \frac{\partial}{\partial y}\left(\frac{\partial f}{\partial x}\right) = \frac{\partial^2 f}{\partial y \partial x} = \frac{\partial^2 f}{\partial x \partial y}$$

$$= \frac{\partial}{\partial x}\left(\frac{\partial f}{\partial y}\right) = \frac{\partial N}{\partial x}$$

Thus the necessary condition for D.E. (1) to be an exact D.E. is

$$\frac{\partial M}{\partial y} = \frac{\partial N}{\partial x}$$

Method of Finding *f*

Step I. Integrating (3) partially w.r.t. x, we get

$$f(x, y) = \int M(x, y)dx + g(y) \qquad (5)$$

where $g(y)$ is the constant of integration which depends only on y.

Step II. To find $f(y)$.

Differentiate (5) partially w.r.t. y and equating it to (4), we get

$$\frac{\partial}{\partial y}\int M dx + \frac{dg}{dy} = \frac{\partial f}{\partial y} = N$$

so that

$$\frac{dg}{dy} = N - \frac{\partial}{\partial y}\int M dx$$

Integrating w.r.t. y, we have

$$g(y) = \int\left[N - \frac{\partial}{\partial y}\int M dx\right] dy + c_1 \quad (6)$$

Note: Similar result can be obtained from the above procedure starting with (4) also.

Step III. Substituting $g(y)$ from (6) in (5)

The required general solution of the exact D.E. (2) is

$$f(x, y) = c_2$$

To sum up:

Test for exactness of D.E. (1): $\frac{\partial M}{\partial y} = \frac{\partial N}{\partial x}$

Method to find f : f is determined from (5) and (6). Then the general solution of (1) is the equation

$$f(x, y) = c$$

(and *not* simply the function $f(x, y)$).

Worked Out Examples

Determine which of the following equations are exact and solve the ones that are exact:

Example 1: $e^y dx + (xe^y + 2y)dy = 0$

Solution: $M = e^y$, $N = xe^y + 2y$

so that $$\frac{\partial M}{\partial y} = e^y = \frac{\partial N}{\partial x}$$

Hence the given equation is exact.
Integrating w.r.t. x, $\frac{\partial f}{\partial x} = e^y = M$ yields

$$f = \int e^y dx + g(y) = xe^y + g(y)$$

Differentiating w.r.t. y

$$xe^y + \frac{dg}{dy} = \frac{\partial f}{\partial y} = N = xe^y + 2y$$

i.e., $$\frac{dg}{dy} = 2y$$

Integrating $g(y) = y^2 + c_1$.
Substituting $g(y)$ in f

$$f = xe^y + y^2 + c_1$$

Thus the desired solution is

$$f = xe^y + y^2 + c_1 = c_2$$

i.e., $xe^y + y^2 = c_3$ where $c_3 = c_2 - c_1$.

Example 2: $(3x^2y + \frac{y}{x})dx + (x^3 + \ln x)dy = 0.$

Solution: $M = 3x^2y + \frac{y}{x}$, $N = x^3 + \ln x$

$$\frac{\partial M}{\partial y} = 3x^2 + \frac{1}{x} = \frac{\partial N}{\partial x}$$

So equation is exact.
Integrating w.r.t. x, $\frac{\partial f}{\partial x} = M = 3x^2y + \frac{y}{x}$ yields

$$f = x^3y + y\ln x + g(y)$$

Differentiating w.r.t. y and equating the result to N, we have

$$x^3 + \ln x + \frac{dg}{dy} = \frac{\partial f}{\partial y} = N = x^3 + \ln x$$

so that $$\frac{dg}{dy} = 0$$

i.e., $g(y) = \text{constant} = c_1$.
Required solution is

$$x^3y + y\ln x + c_1 = c_2$$

i.e., $x^3y + y\ln x = c_3$, where $c_3 = c_2 - c_1$.

Example 3: $(\cos x - x\cos y)dy - (\sin y + y\sin x)dx = 0$

Solution: $M = -\sin y - y\sin x$
$N = \cos x - x\cos y$

$$\frac{\partial M}{\partial y} = -\cos y - \sin x = \frac{\partial N}{\partial x}, \text{ exact.}$$

Integrating $\frac{\partial f}{\partial x} = M = -\sin y - y\sin x$ w.r.t. x we get

$$f(x, y) = -x\sin y + y\cos x + h(y)$$

Differentiating w.r.t. y and equating it to N

$$-x\cdot\cos y+\cos x+\frac{\partial h}{\partial y}=\frac{\partial f}{\partial y}=N$$

$$=\cos x-x\cos y$$

$\therefore\ \frac{dh}{dy}=0$ i.e., $h=$ constant.

The required solution is

$$y\cos x-x\sin y=c$$

Example 4: $xdy+2y^2dx=0$

Solution: $M=2y^2, N=x$, Differentiating

$$\frac{\partial M}{\partial y}=4y,\ \frac{\partial N}{\partial x}=0$$

So that $\frac{\partial M}{\partial y}\neq\frac{\partial N}{\partial x}$

The D.E. is not exact.

EXERCISE

Determine which of the following equations are exact and solve the ones that are exact:

1. $(2x^3-xy^2-2y+3)dx-(x^2y+2x)dy=0$

Ans. $x^4-x^2y^2-4xy+6x=c$

2. $(\cos x.\cos y-\cot x)dx-(\sin x\sin y)dy=0$

Ans. $\sin x\cdot\cos y=\ln(c\sin x)$

3. $(y-x^3)dx+(x+y^3)dy=0$

Ans. $4xy-x^4+y^4=c$

4. $(y+xy^2+x^2y^3)dx+(x-x^2y+x^3y^2)dy=0$

Ans. not exact

5. $(y^2e^{xy^2}+4x^3)dx+(2xye^{xy^2}-3y^2)dy=0$

Ans. $x^4-y^3+e^{xy^2}=c$

6. $(\sin x\cdot\tan y+1)dx+\cos x\cdot\sec^2 y dy=0$

Ans. not exact

7. $(\sin x\cdot\sin y-xe^y)dy=(e^y+\cos x\cdot\cos y)\,dx$

Ans. $xe^y+\sin x\cdot\cos y=c$

8. $(x^2+y^2-a^2)xdx+(x^2-y^2-b^2)ydy=0$

Ans. $x^4+2x^2y^2-y^4-2a^2x^2-2b^2y^2=c$

9. $(\sin x\cdot\cosh y)dx-(\cos x\sinh y)dy=0$, $y(0)=3$

Ans. $\cos x\cdot\cosh y=10.07$

10. $\left[\frac{y}{(x+y)^2}-1\right]dx+\left[1-\frac{x}{(x+y)^2}\right]dy=0$

Ans. $x+y^2-x^2=c(x+y)$

11. $y'=\frac{y-2x}{2y-x}$; $y(1)=2$

Ans. $x^2-xy+y^2=3.$

8.8 REDUCTION OF NON-EXACT DIFFERENTIAL EQUATIONS: USING INTEGRATING FACTORS

Consider a D.E.

$$M(x,y)dx+N(x,y)dy=0 \tag{1}$$

which is *not* exact. Suppose there exists a function $F(x,y)$ such that

$$F(x,y)[Mdx+Ndy]=0 \tag{2}$$

is exact, then $F(x,y)$ is called an **integrating factor** (I.F.) of D.E. (1). There may exist several integrating factors or may not, since exact D.E. are relatively rare.

Some methods to find an I.F. to a nonexact D.E. $Mdx+Ndy=0$ are

Case 1: Method of inspection (or "Grouping" of terms)

Case 2: $(My-Nx)/N=g(x)$

Case 3: $(Nx-My)/M=h(y)$

Case 4: Homogeneous D.E. with $xM+yN\neq 0$

Case 5: D.E. of the form $yg(xy)dx+xh(xy)dy=0$

Case 6: D.E. of the form $x^ay^b(mydx+nxdy)+x^cy^d(pydx+qxdy)=0$

Case 1: Integrating Factors by Inspection (Grouping of Terms)

Not very often, but sometimes I.F. can be obtained by inspection largely upon experience and recognition of "regrouping" the terms of the given equation appropriately such that they form the part of certain common exact differentials listed below for ready reference:

S. No.	Group of terms	Integrating factor	Exact differential
1	$xdy+ydx$	1	$d(x, y)$
2	$xdy+ydx$	$\frac{1}{xy}$	$\frac{xdy+ydx}{xy} = d\{\ln(xy)\}$
3	$xdy+ydx$	$\frac{1}{(xy)^n}$, $n \neq 1$	$\frac{xdy+ydx}{(xy)^n} = d\left\{\frac{(xy)^{1-n}}{(1-n)}\right\}$
4	$xdx+ydy$	$\frac{1}{x^2+y^2}$	$\frac{xdx+ydy}{x^2+y^2} = \frac{1}{2}d\{\ln(x^2+y^2)\}$
5	$xdx+ydy$	$\frac{1}{(x^2+y^2)^n}$, $n \neq 1$	$\frac{xdx=ydy}{(x^2+y^2)^n} = d\left\{\frac{(x^2+y^2)^{1-n}}{2(1-n)}\right\}$
6	$xdx+ydy$	2	$2(xdx+ydy)=d(x^2+y^2)$
7	$xdy-ydx$	$\frac{1}{x^2}$	$\frac{xdy-ydx}{x^2} = d\left(\frac{y}{x}\right)$
8	$xdy-ydx$	$\frac{1}{y^2}$	$\frac{-(ydx-xdy)}{y^2} = -d\left(\frac{x}{y}\right)$
9	$xdy-ydx$	$\frac{1}{xy}$	$\frac{dy}{y} - \frac{dx}{x} = d\left(\ln\frac{y}{x}\right)$
10	$xdy-ydx$	$\frac{1}{x^2+y^2}$	$\frac{xdy-ydx}{x^2+y^2} = \frac{\frac{xdy-ydx}{x^2}}{1+\left(\frac{y}{x}\right)^2} = d\left(\tan^{-1}\frac{y}{x}\right)$

Note: A very simple D.E. $ydx - xdy = 0$ has several I.F. like x^{-2}, y^{-2}, $(xy)^{-1}$, $(x^2+y^2)^{-1}$. Use the appropriate D.E. depending on the given D.E.

WORKED OUT EXAMPLES

Case 1: Solve the following:

Example 1: $y(y^3 - x)dx + x(y^3 + x)dy = 0$

Solution: $y^4dx - xydx + xy^3dy + x^2dy = 0$
Regrouping the terms,

$$y^3(ydx + xdy) + x(xdy - ydx) = 0$$

$$y^3d(xy) + x \cdot x^2 d\left(\frac{y}{x}\right) = 0$$

$$d(xy) + \left(\frac{x}{y}\right)^3 d\left(\frac{y}{x}\right) = 0$$

or

$$d(xy) + \left(\frac{y}{x}\right)^{-3} d\left(\frac{x}{y}\right) = 0$$

Integrating

$$xy - \frac{1}{2}\left(\frac{y}{x}\right)^{-2} = c_1$$

Rearranging $2xy^3 - x^2 = cy^2$ where $c = 2c_1$

Example 2: $(x^3y^3 + 1)dx + x^4y^2dy = 0$

Solution: $x^3y^3dx + dx + x^4y^2dy = 0$
Dividing by x throughout

$$x^2y^3dx + \frac{dx}{x} + x^3y^2dy = 0$$

Regrouping

$$x^2y^2(ydx + xdy) + \frac{dx}{x} = 0$$

$$(xy)^2d(xy) + \frac{dx}{x} = 0$$

$$\frac{(xy)^3}{3} + \ln x = c.$$

Example 3: $y(x^3e^{xy} - y)dx + x(xy + x^3e^{xy})dy = 0.$

Solution: $yx^3e^{xy}dx - y^2dx + xydy + x^4e^{xy}dy = 0.$

Regrouping, $x^3e^{xy}(ydx + xdy) + y(xdy - ydx) = 0$

$$x^3d(e^{xy}) + y \cdot x^2 \cdot \left(\frac{xdy - ydx}{x^2}\right) = 0.$$

Dividing throughout by x^3,

$$d(e^{xy}) + \frac{y}{x} \cdot d\left(\frac{y}{x}\right) = 0$$

Integrating $e^{xy} + \left(\frac{y}{x}\right)^2 \frac{1}{2} = c.$

Example 4:

$$(x^{n+1} \cdot y^n + ay)dx + (x^ny^{n+1} + ax)dy = 0$$

Solution: Regrouping the terms

$$x^n y^n(xdx + ydy) + a(ydx + xdy) = 0$$

Dividing throughout by $x^n y^n$

$$(xdx + ydy) + \frac{a\,d(xy)}{(xy)^n} = 0$$

If $n \neq 1$, Integrating

$$\frac{x^2 + y^2}{2} + a \cdot \frac{1}{(-n+1)(xy)^{n-1}} = c_0$$

$(n-1)(x^2 + y^2 - c)(xy)^{n-1} = 2a$, where $c = 2c_0$

If $n = 1$, $\frac{x^2+y^2}{2} + a \ln xy = c$.

Example 5: $y(x^2y^2 - 1)dx + x(x^2y^2 + 1)dy = 0.$

Solution: $x^2y^3dx - ydx + x^3y^2dy + xdy = 0$

Regrouping $x^2y^2(ydx + xdy) + (xdy - ydx) = 0$

$$x^2y^2d(xy) + x^2\frac{(xdy - ydx)}{x^2} = 0$$

$$x^2y^2d(xy) + x^2d\left(\frac{y}{x}\right) = 0$$

The second term in L.H.S. is *not* an exact differential, Dividing by x^2; $y^2d(xy) + d\left(\frac{y}{x}\right) = 0$.

Now multiply throughout by $x^k y^n$, we get

$$x^k y^{n+2}d(xy) + x^k y^n d\left(\frac{y}{x}\right) = 0$$

The first term in the L.H.S. becomes an exact differential if $k = n + 2$, while the second term in L.H.S. becomes an exact differential if $n = -k$. Solving these two equations

$$k = n + 2$$

$$n = -k$$

We get $n = -1$, $k = 1$.

Substituting these values, the D.E. reduces to

$$x^1 \cdot y^{-1+2}d(xy) + x^1y^{-1}d\left(\frac{y}{x}\right) = 0$$

or

$$xyd(xy) + \frac{x}{y}d\left(\frac{y}{x}\right) = 0$$

$$xyd(xy) + \frac{d\left(\frac{y}{x}\right)}{\left(\frac{y}{x}\right)} = 0$$

Integrating

$$\frac{(xy)^2}{2} + \ln\left(\frac{y}{x}\right) = c$$

Example 6: $(y + x)dy = (y - x)dx$

Solution: $ydy + xdy - ydx + xdx = 0$

Regrouping $(ydy + xdx) + (xdy - ydx) = 0$

$$\frac{1}{2}d(y^2 + x^2) + (xdy - ydx) = 0$$

Dividing throughout by $(x^2 + y^2)$, we get

$$\frac{1}{2}\frac{1}{(x^2 + y^2)}d(x^2 + y^2) + \frac{xdy - ydx}{x^2 + y^2} = 0$$

Rewriting $\frac{1}{2}d\,(\ln\,(x^2 + y^2)) + \left(\frac{\frac{xdy - ydx}{x^2}}{1 + \left(\frac{y}{x}\right)^2}\right) = 0$

$$\frac{1}{2}d\,(\ln\,(x^2 + y^2)) + d\left(\tan^{-1}\left(\frac{y}{x}\right)\right) = 0$$

Integrating

$$\ln\sqrt{x^2 + y^2} + \tan^{-1}\frac{y}{x} = c$$

EXERCISE

Solve the following D.E. (by regrouping the terms):

1. $(4x^3y^3 - 2xy)dx + (3x^4y^2 - x^2)dy = 0$

Ans. $x^4y^3 - x^2y = c$

2. $3x^2ydx + (y^4 - x^3)dy = 0$

Ans. $3x^3 + y^4 = cy$

3. $(x^3 + xy^2 + y)dx + (y^3 + x^2y + x)dy = 0$

Ans. $(x^2 + y^2) = c - 4xy$

4. $ydx + (x + x^3y^2)dy = 0$

Ans. $2x^2y^2 \cdot \ln\,(cy) = 1$

5. $y(x^3 - y)dx - x(x^3 + y)dy = 0$

Hint: After regrouping multiply by $x^k y^n$ and determine k and n to make the D.E. exact.

Ans. $x^2 + 2y = cxy^2$

6. $ydx - xdy = xy^3dy$

Ans. $\ln\frac{x}{y} = \frac{1}{3}y^3 + c$

7. $xdy = (x^5 + x^3y^2 + y)dx$

Ans. $\tan^{-1}\frac{x}{y} = -\frac{1}{4}x^4 + c$

8. $xdy = (y + x^2 + 9y^2)dx$

Ans. $\tan^{-1}\frac{3y}{x} = 3x + c$

9. $y(2xy + e^x)dx = e^x dy$

Ans. $\frac{e^x}{y} + x^2 = c$

10. $(y^2 e^{xy^2} + 4x^3)dx + (2xye^{xy^2} - 3y^2)dy = 0$

Ans. $e^{xy^2} + x^4 - y^3 = c.$

Case 2: When I.F. is a Function of *x* alone: $(My - Nx)/N = g(x)$

When $F(x, y)$ is an I.F. then

$$F(Mdx + Ndy) = 0$$

is exact therefore

$$\frac{\partial}{\partial y}(FM) = \frac{\partial}{\partial x}(FN)$$

or $$F\frac{\partial M}{\partial y} + M\frac{\partial F}{\partial y} = F\frac{\partial N}{\partial x} + N\frac{\partial N}{\partial x}$$

Solving we get the formula for I.F. $f(x, y)$ as

$$\frac{1}{F}\left(N\frac{\partial F}{\partial x} - M\frac{\partial F}{\partial y}\right) = \frac{\partial M}{\partial y} - \frac{\partial N}{\partial x} \quad (1)$$

(which is a very difficult partial D.E.)

Suppose I.F. 'F' is a function of x alone. In this case (1) reduces to

$$\frac{1}{F}\frac{dF}{dx} = \frac{1}{N}\left(\frac{\partial M}{\partial y} - \frac{\partial N}{\partial x}\right) \quad (2)$$

Since L.H.S. of (2) is a function of x alone say $g(x)$

i.e., $$\frac{1}{F}\frac{dF}{dx} = g(x) = \frac{1}{N}\left(\frac{\partial M}{\partial y} - \frac{\partial N}{\partial x}\right) \quad (3)$$

Then integrating $\frac{1}{F}\frac{dF}{dx} = g(x)$, we get

$$F(x) = e^{\int g(x)dx}$$

where $g(x) = (M_y - N_x)/N$.

Case 3: When I.F. is a Function of *y* alone: $(Nx - My)/M = h(y)$

Suppose I.F. 'F' is a function of y alone, in which case (1) reduces to

$$\frac{1}{F}\frac{dF}{dy} = -\frac{\left(\frac{\partial M}{\partial y} - \frac{\partial N}{\partial x}\right)}{M} = h(y)$$

then the I.F. 'F' is obtained by integration as

$$F(y) = e^{\int h(y)dy}$$

where $h(y) = (Nx - My)/M$.

Case 4

If $M(x, y)$ and $N(x, y)$ are homogeneous functions of the same degree then $(xM + yN)^{-1}$ is an I.F. of $Mdx + Ndy = 0$, provided $xM + yN \neq 0$.

In case $xM + yN = 0$ then $\frac{1}{x^2}$ or $\frac{1}{y^2}$ or $\frac{1}{xy}$ are I.F's.

Worked Out Examples

Case 2:

Solve the following:

Example 1: $y(2x^2 - xy + 1)dx + (x - y)dy = 0$

Solution: Here $M = 2yx^2 - xy^2 + y$, $N = x - y$

$$\frac{\partial M}{\partial y} = 2x^2 - 2xy + 1 \neq 1 = \frac{\partial N}{\partial x}, \quad \textit{not exact}$$

Since

$$\frac{1}{N}\left(\frac{\partial M}{\partial y} - \frac{\partial N}{\partial x}\right) = \frac{2x^2 - 2xy + 1 - 1}{x - y} = 2x = f(x),$$

is a function of x only, we get an I.F. as

$$IF = e^{\int 2x dx} = e^{x^2}$$

Multiplying the given D.E. by I.F. e^{x^2}, we get

$$ye^{x^2}(2x^2 - xy + 1)dx + e^{x^2}(x - y)dy = 0$$

which is of the form

$$M^*dx + N^*dy = 0$$

[This is exact since $\frac{\partial M^*}{\partial y} = (2x^2 - 2xy + 1)e^{x^2} = \frac{\partial N^*}{\partial x}$]

Integrate

$$\frac{\partial f}{\partial y} = e^{x^2}(x - y) = N^*$$

partially w.r.t. y, we have

$$f(x, y) = e^{x^2}\cdot\left(xy - \frac{y^2}{2}\right) + h(x)$$

Differentiating partially w.r.t. x and equating it to M^*, we get

$$e^{x^2}\left[xy - \frac{y^2}{2}\right]\cdot 2x + e^{x^2}[y] + \frac{dh}{dx}$$

$$= M^* = ye^{x^2}(2x^2 - xy + 1)$$

Simplifying, $\frac{dh}{dx} = 0$ so $h =$ constant.
Thus the general solution is

$$e^{x^2}(2xy - y^2) = c$$

Example 2: $(x - y)dx - dy = 0,\ y(0) = 2.$

Solution: $M = x - y,\ N = -1,\ M_y = -1,$
$N_x = 0$, not exact.
$\frac{1}{N}(M_y - N_x) = \frac{1}{-1}[-1 - 0] = 1$ is a function of x.
I.F.$= e^{\int 1dx} = e^x$.
Multiplying D.E. by I.F.
$(x - y)e^x dx - e^x dy = 0$

Rewriting
$$xe^x dx - ye^x dx - e^x dy = 0$$
$$xe^x dx - \left(d(ye^x)\right) = 0$$
$$xe^x dx + e^x dx - e^x dx - d(ye^x) = 0$$
$$(x - 1)e^x dx + e^x dx - d(ye^x) = 0$$
$$d((x - 1)e^x) - d(ye^x) = 0$$

Solution
$(x - 1)e^x - ye^{-x} = c$
Put $x = 0,\ y = 2$, so that $c = -3$.
$\therefore\ (x - 1)e^x - ye^{-x} = -3$ is the particular solution.

Case 3:

Solve the following:

Example 3: $(3x^2y^4 + 2xy)dx + (2x^3y^3 - x^2)dy = 0$

Solution: Here $M = 3x^2y^4 + 2xy,\ N = 2x^3y^3 - x^2$.
$M_y = 12x^2y^3 + 2x \neq 6x^2y^3 - 2x = N_x$, not exact.
Since

$$\frac{1}{M}(N_x - M_y) = \frac{6x^2y^3 - 2x - (12x^2y^3 + 2x)}{3x^2y^4 + 2xy}$$
$$= -\frac{2}{y} = g(y) = \text{function of } y \text{ alone},$$

we get an I.F.$= e^{\int g(y)dy}$

$$= e^{-\int \frac{2}{y}dy} = e^{-2\ln y}e = \frac{1}{y^2}$$

Multiplying the given D.E. throughout by $\frac{1}{y^2}$, we have

$$\left(3x^2y^2 + \frac{2x}{y}\right)dx + \left(2x^3y - \frac{x^2}{y^2}\right)dy = 0$$

[Since $M_y^* = 6x^2y - \frac{2x}{y^2} = N_x^*$, this D.E. is exact]
Rearranging the terms

$$(3x^2y^2dx + 2x^3ydy) + \left(\frac{2x}{y}dx - \frac{x^2}{y^2}dy\right) = 0$$
$$y^2d(x^3) + x^3d(y^2) + \frac{1}{y}d(x^2) + x^2d\left(\frac{1}{y}\right) = 0$$

Regrouping

$$d(x^3y^2) + d\left(\frac{x^2}{y}\right) = 0$$

Integrating

$$x^3y^2 + \frac{x^2}{y} = c$$

is the general solution.

Case 4:

Example 4: $y(y^2 - 2x^2)dx + x(2y^2 - x^2)dy = 0$

Solution: Here $M = y(y^2 - 2x^2)$ and $N = x(2y^2 - x^2)$ are both homogeneous functions of degree 3.
Since $xM + yN = xy(y^2 - 2x^2) + yx(2y^2 - x^2)$
$= 3(xy)(y^2 - x^2) \neq 0$ unless $y = x$, D.E. has an

$$\text{I.F} = \frac{1}{xM + yN} = \frac{1}{3xy(y^2 - x^2)}$$

Multiplying the D.E. by I.F., we get

$$\frac{y(y^2 - 2x^2)}{3xy(y^2 - x^2)}dx + \frac{x(2y^2 - x^2)}{3xy(y^2 - x^2)}dy = 0$$

Rewriting

$$\frac{(y^2 - x^2) - x^2}{x(y^2 - x^2)}dx + \frac{y^2 + (y^2 - x^2)}{y(y^2 - x^2)}dy = 0$$

or
$$\frac{dx}{x} - \frac{xdx}{y^2 - x^2} + \frac{ydy}{y^2 - x^2} + \frac{dy}{y} = 0$$

Regrouping

$$d(\ln xy) + \frac{1}{2}\frac{d(y^2 - x^2)}{(y^2 - x^2)} = 0$$

Integrating

$$d\left(\ln\left\{x^2y^2(y^2 - x^2)\right\}\right) = 0$$

we get

$$\ln x^2y^2(y^2-x^2)=c$$

or $$x^2y^2(y^2-x^2)=c_1, \quad \text{where} \quad c_1=e^c.$$

Case 5: D.E. of the Form— $yg(xy)dx+xh(xy)dy=0$

If the D.E. $Mdx+Ndy=0$ is in the form

$$yg(xy)dx+xh(xy)dy=0$$

where $g(xy)$ and $h(xy)$ are functions of the argument (product) xy and $g(xy)\neq h(xy)$ then

$$\frac{1}{xM-yN}=\frac{1}{xy\{g(xy)-h(xy)\}}$$

is an integrating factor provided $xM-yN\neq 0$.

Note: If $xM-yN=0$ then $\frac{M}{N}=\frac{y}{x}$ and the given D.E. reduces to $xdy+ydx=0$ with $xy=c$ as its solution.

WORKED OUT EXAMPLES

Example 1:

$$(x^2y^2+xy+1)ydx+(x^2y^2-xy+1)xdy=0$$

Solution: Here $M=(x^2y^2+xy+1)y$,
$N=(x^2y^2-xy+1)/x$ so

$$M_y=3x^2y^2+2xy+1\neq 3x^2y^2-2xy+1=N_x$$

D.E. is not exact. But $M=yg(xy)$ and $N=xh(xy)$ so the given D.E. is of the form

$$yg(xy)dx+xh(xy)dy=0$$

which has an integrating factor given by

$$\frac{1}{xM-yN}=\frac{1}{xy(x^2y^2+xy+1)-xy(x^2y^2-xy+1)}=\frac{1}{2x^2y^2}$$

Multiplying D.E. with I.F. $\frac{1}{x^2y^2}$, we get

$$\frac{(x^2y^2+xy+1)ydx}{x^2y^2}+\frac{(x^2y^2-xy+1)x}{x^2y^2}dy=0$$

Rearranging

$$\left(ydx+\frac{dx}{x}+\frac{dx}{x^2y}\right)+\left(xdy-\frac{dy}{y}+\frac{dy}{xy^2}\right)=0$$

Regrouping the terms

$$(ydx+xdy)+\left(\frac{ydx}{x^2y^2}+\frac{xdy}{x^2y^2}\right)+\left(\frac{dx}{x}-\frac{dy}{y}\right)=0$$

or $$d(xy)+\frac{d(xy)}{(xy)^2}+d(\ln\frac{x}{y})=0$$

Integrating

$$xy-\frac{1}{xy}+\ln\frac{x}{y}=c$$

Case 6

The D.E. is of the form

$$x^ay^b(mydx+nxdy)+x^cy^d(pydx+qxdy)=0$$

where a,b,m,n,c,d,p,q are all constants and $mp-nq\neq 0$ has an integrating factor of the form

$$x^hy^k$$

where the unknown constants h,k are determined from the two equations

$$\frac{a+h+1}{m}=\frac{b+k+1}{n}$$

and

$$\frac{c+h+1}{p}=\frac{d+k+1}{q}$$

WORKED OUT EXAMPLES

Example 1: $(2y^2+4x^2y)dx+(4xy+3x^3)dy=0$

Solution: Here $M=2y^2+4x^2y$, $N=4xy+3x^3$, $M_y=4y+4x^2\neq 4y+9x^2=N_x$, not exact. It is also not homogeneous. It is *not* $M=yh(xy)$ and $N=xg(xy)$ form. So let us try to find an integrating factor of the form

$$x^hy^k$$

Consider the D.E.

$$2y^2dx+4x^2ydx+4xydy+3x^3dy=0$$

Rearranging the terms

$$x^2(4ydx+3xdy)+y(2ydx+4xdy)=0$$

Comparing this with

$$x^ay^b(mydx+nxdy)+x^cy^d(pydx+qxdy)=c$$

here $a=2, b=0, m=4, n=3.$

$$c = 0, d = 1, p = 2, q = 4$$

Also $mp - nq = 8 - 12 = -4 \neq 0$

The unknown constants in the integrating factor are determined from the following:

$$\frac{a+h+1}{m} = \frac{b+k+1}{n} \text{ i.e., } \frac{2+h+1}{4} = \frac{0+k+1}{3}$$

i.e., $4k - 3h = 5$

$$\frac{c+h+1}{p} = \frac{d+k+1}{q} \text{ i.e., } \frac{0+h+1}{2} = \frac{1+k+1}{4}$$

i.e., $k - 2 = 0$

Solving for h, k, we get $h = 1, k = 2$.

Thus the required integrating factor is

$$x^1 \cdot y^2.$$

Multiplying the given D.E. by this integrating factor xy^2, we get

$$xy^2(2y^2 + 4x^2y)dx + xy^2(4xy + 3x^3)dy = 0$$

$$2xy^4 + 4x^3y^3dx + 4x^2y^3dy + 3x^4y^2dy = 0$$

Regrouping the terms (1st and 3rd) and (2nd and 4th)

$$d(x^2y^4) + d(x^4y^3) = 0$$

Integrating $x^2y^4 + x^4y^3 = c$ is the solution.

EXERCISE

Case 2: Solve the following:

1. $(4xy + 3y^2 - x)dx + x(x + 2y)dy = 0$

Ans. $x^3(4xy + 4y^2 - x) = c$

2. $y(x + y)dx + (x + 2y - 1)dy = 0$

Ans. $y(x - 1 + y) = ce^{-x}$

3. $2xydy - (x^2 + y^2 + 1)dx = 0$

Ans. $y^2 - x^2 + 1 = cx$

4. $(3xy - 2ay^2)dx + (x^2 - 2axy)dy = 0$

Ans. $x^3y - ax^2y^2 = c$

5. $2\sin\ (y^2)dx + xy\cos\ (y^2)dy = 0, y(2) = \sqrt{\frac{\pi}{2}}$

Ans. $x^4 \sin\ (y^2) = 16$

6. $(2x^3y^2 + 4x^2y + 2xy^2 + xy^4 + 2y)dx + 2(y^3 + x^2y + x)dy = 0$

Ans. $(2x^2y^2 + 4xy + y^4)e^{x^2} = c.$

Case 3: Solve the following:

1. $(y + xy^2)dx - xdy = 0$

Ans. $\frac{x}{y} + \frac{x^2}{2} = c$

2. $y(x + y + 1)dx + x(x + 3y + 2)dy = 0$

Ans. $xy^2(x + 2y + 2) = c$

3. $3(x^2 + y^2)dx + x(x^2 + 3y^2 + 6y)dy = 0$

Ans. $xe^y(x^2 + 3y^2) = c$

4. $(xy^3 + y)dx + 2(x^2y^2 + x + y^4)dy = 0$

Ans. $3x^2y^4 + 6xy^2 + 2y^6 = c$

5. $(y^4 + 2y)dx + (xy^3 + 2y^4 - 4x)dy = 0$

Ans. $(y^3 + 2)x + y^4 = cy^2$

6. $2xydx + (y^2 - x^2)dy = 0, y(2) = 1$

Ans. $x^2 + y^2 = 5y.$

Case 4: Solve the following:

1. $x^2ydx - (x^3 + y^3)dy = 0$

Ans. $y = ce^{-x^3/(3y^3)}$

2. $(x^2y - 2xy^2)dx - (x^3 - 3x^2y)dy = 0$

Ans. $\frac{x}{y} - 2\log x + 3\log y = c$

3. $(x^4 + y^4)dx - xy^3dy = 0$

Ans. $y^4 = 4x^4 \ln x + cx^4$

4. $y^2dx + (x^2 - xy - y^2)dy = 0$

Ans. $(x - y)y^2 = c(x + y)$

5. $(y - x)dx + (y + x)dy = 0$

Ans. $\ln\ (x^2 + y^2)^{\frac{1}{2}} - \tan^{-1}\left(\frac{x}{y}\right) = c.$

Case 5: Solve the following:

1. $y(x^2y^2 + 2)dx + x(2 - 2x^2y^2)dy = 0$

Ans. $x = cy^2e^{\frac{1}{(x^2y^2)}}$

2. $y(xy + 2x^2y^2)dx + x(xy - x^2y^2)dy = 0$

Ans. $\frac{2}{3}\ln x - \frac{1}{3}\ln y - \frac{1}{3xy} = c$

3. $(xy\sin xy + \cos xy)ydx + (xy\sin xy - \cos xy)xdy = 0$

Ans. $x\sec xy = cy$

4. $y(1+xy)dx + (1-xy)xdy = 0$

Ans. $\ln\frac{x}{y} - \frac{1}{xy} = c$

5. $(x^3y^3 + x^2y^2 + xy + 1)ydx + (x^3y^3 - x^2y^2 + xy + 1)xdy = 0$

Ans. $x^2y^2 - 2xy\log cy = 1$

6. $(2xy^2 + y)dx = (x + 2x^2y - x^4y^3)dy = 0$

Ans. $y = ce^{\frac{(-3xy+1)}{(3x^3y^3)}}$

Case 6: Solve the following:

1. $x(4ydx + 2xdy) + y^3(3ydx + 5xdy) = 0$

Ans. $x^4y^2 + x^3y^3 = c$

2. $(8ydx + 8xdy) + x^2y^3(4ydx + 5xdy) = 0$

Ans. $4x^2y^2 + x^4y^5 = c$

3. $x^3y^3(2ydx + xdy) - (5ydx + 7xdy) = 0$

Ans. $x^3y^3 + 2 = cx^{\frac{5}{3}}y^{\frac{7}{3}}$

4. $y(xy + 2x^2y^2)dx + x(xy - x^2y^2)dy = 0$

Ans. $2\ln x - \ln y - \frac{1}{xy} = c$

8.9 LINEAR DIFFERENTIAL EQUATION: FIRST ORDER

"Leibnitz's Linear Equation"

The general form of a first order linear differential equation in the dependent variable y is

$$A(x)\frac{dy}{dx} + B(x)y = c(x)$$

By dividing throughout by $A(x)$ we get the standard form of the linear equation of first order which is also known as Leibnitz's* linear equation.

$$\frac{dy}{dx} + P(x)y = Q(x) \qquad (1)$$

The important feature of this equation is that it is linear (i.e., of first degree) in y and its derivative y', and does not contain product terms of y and y'.

Here $P(x)$ and $Q(x)$ may be *any* given functions of x.

Because of this, note that a linear equation is of first degree but first degree equation need not be linear (in y i.e., it can contain terms like $y^{\frac{23}{5}}$ $y\sin y$, e^y etc.)

Homogeneous: If the R.H.S. $Q(x)$ is zero for all x then Equation (1) is said to be homogeneous; otherwise (i.e., $Q(x) \neq 0$) it is said to be non-homogeneous.

The non-homogeneous Equation (1) has an integrating factor $v(x)$ depending only on x. Multiplying (1) by $v(x)$ and rewriting it, we get

$$[v(x)P(x)y - v(x)Q(x)]dx + v(x)dy = 0 \qquad (2)$$

This Equation (2) which is in the form $Mdx + Ndy = 0$ with $M = vPy - vQ$ and $N = v$ is exact and should satisfy the condition.

$$\frac{\partial M}{\partial y} = \frac{\partial N}{\partial x} \quad \text{i.e.,} \quad vP = \frac{dv}{dx}$$

which is a separable equation.

To get v, we integrate

$$\int\frac{dv}{v} = \int Pdx \quad \text{i.e.,} \quad \ln v = \int Pdx$$

$$\text{or} \quad v(x) = e^{\int P(x)dx} \qquad (3)$$

Thus $v(x)$ given by (3) is the required integrating factor of the first order non-homogeneous linear differential Equation (1).

To find the solution of (1), multiply (1) throughout by the integrating factor (3).

$$\frac{dy}{dx}\cdot e^{\int Pdx} + Py\cdot e^{\int Pdx} = Q\cdot e^{\int Pdx}$$

Rewriting

$$d(ye^{\int Pdx}) = Qe^{\int Pdx}$$

Integrating, we get

$$ye^{\int Pdx} = \int Qe^{\int Pdx}dx \qquad (4)$$

* Gottfried Wilhelm Leibnitz (1646–1716), German mathematician.

Method of solving linear equation:

I. Put the given equation in the standard form (1).

II. Determine the integrating factor $v(x) = e^{\int P dx}$.

III. Multiply equation throughout by the integrating factor $v(x)$.

IV. Solve the resultant exact equation.

Note 1: The L.H.S. of (4) is *always* the product of y and the integrating factor $v(x)$.

Note 2: In some cases, when given D.E. in non-linear in y, it would be much convenient to treat x as the dependent variable instead of y and solve the equation $\frac{dx}{dy} + P^*(y)x = Q^*(y)$ which is linear in x (see Worked Out Examples 6, 7 and also Exercise Examples 11–15).

Worked Out Examples

Example 1: Solve $y' = 4y + 2x - 4x^2$.

Solution: Standard form $y' - 4y = 2x - 4x^2$ with $P(x) = -4$, $Q(x) = 2x - 4x^2$

I.F: $e^{\int P dx} = e^{\int -4 dx} = e^{-4x}$

Multiplying by I.F.

$$e^{-4x} \cdot y' - 4ye^{-4x} = (2x - 4x^2)e^{-4x}$$

Rewriting

$$\frac{d}{dx}(ye^{-4x}) = (2x - 4x^2)e^{-4x}$$

Integrating

$$ye^{-4x} = \int (2x - 4x^2)e^{-4x} dx$$

$$= 2\int xe^{-4x} dx + \int x^2 d(e^{-4x})$$

Integrating by parts

$$ye^{-4x} = \int 2xe^{-4x} dx + x^2 e^{-4x} - \int e^{-4x} \cdot 2x dx + c$$

$$ye^{-4x} = x^2 e^{-4x} + c$$

Solution is $y = x^2 + ce^{4x}$.

Example 2: Solve $xy' + 2y - x \sin x = 0$.

Solution: Standard form $y' + \frac{2y}{x} = \sin x$

with $\quad Q = \sin x, \; P = 2/x$

I.F. $= e^{\int P dx} = e^{\int \frac{2}{x} dx} = e^{\ln x^2} = x^2$

Multiplying by I.F. x^2

$$x^2 y' + 2xy = x^2 \sin x$$

$\frac{d}{dx}(yx^2) = x^2 \sin x$

Integrating

$$yx^2 = \int x^2 \sin x \cdot dx = \int x^2 d(\cos x)$$

Integrating by parts

$$yx^2 = -x^2 \cos x + \int 2x \cdot \cos x dx$$

$$yx^2 = -x^2 \cos x + 2x \sin x + 2\cos x + c$$

Example 3:

$$\{y(1 - x\tan x) + x^2 \cos x\}dx - xdy = 0.$$

Solution: Standard form

$$y' + \frac{(x\tan x - 1)}{x} y = x\cos x$$

with $\quad P(x) = \frac{x\tan x - 1}{x}, \; Q(x) = x\cos x$

I.F. $\exp\left\{\int \left(\frac{x\tan x - 1}{x}\right) dx\right\} = \exp\{-\ln\cos x - \ln x\}$

$$\text{I.F.} = \frac{1}{x\cos x}$$

$$\frac{1}{x\cos x} \cdot y' + \frac{(x\tan x - 1)}{x} \cdot \frac{1}{x\cos x} \cdot y = 1$$

Rewriting

$$\frac{d}{dx}\left(\frac{y}{x\cos x}\right) = 1$$

Integrating

$$\frac{y}{x\cos x} = \int dx + c = x + c$$

Solution is $y = x^2 \cos x + cx\cos x$.

Example 4: Solve $x(1 - 4y)dx - (x^2 + 1)dy = 0$ with $y(2) = 1$.

Solution: Standard form

$$y' + \frac{4x}{x^2 + 1} \cdot y = \frac{x}{x^2 + 1}$$

with $P(x)=\frac{4x}{x^2+1}$ and $Q(x)=\frac{x}{x^2+1}$

I.F.$= e^{\int P dx} = e^{\int \frac{4x}{x^2+1}dx} = e^{2\ln(x^2+1)} = (x^2+1)^2$

Multiplying by I.F. $(x^2+1)^2$, we get an exact equation

$$(x^2+1)^2 y' + 4x(x^2+1)y = x(x^2+1)$$

Rewriting

$$\frac{d}{dx}\{y(x^2+1)^2\} = x^3 + x$$

Integrating

$$y\cdot(x^2+1)^2 = \int (x^3+x)dx + c = \frac{x^4}{4} + \frac{x^2}{2} + c$$

$$4y(x^2+1)^2 = x^4 + 2x^2 + 4c$$

Since $y(2)=1$, put $x=2$ and $y=1$

$$4.25 = 16 + 8 + 4c \quad \therefore\ c = 19$$

The particular solution is

$$4y(x^2+1)^2 = x^4 + 2x^2 + 76$$

Example 5: Solve $\frac{dy}{dx} = \frac{1}{(1+x^2)}(e^{\tan^{-1}x} - y)$.

Solution: Standard form

$$\frac{dy}{dx} + \frac{1}{1+x^2}y = \frac{e^{\tan^{-1}x}}{1+x^2}$$

with $P(x)=\frac{1}{1+x^2}$, $Q(x)=\frac{e^{\tan^{-1}x}}{1+x^2}$.

I.F. $= e^{\int P(x)dx} = e^{\int \frac{dx}{1+x^2}} = e^{\tan^{-1}x}$

Multiplying by I.F. $e^{\tan^{-1}x}$, we get

$$e^{\tan^{-1}x}\cdot\frac{dy}{dx} + \frac{1}{1+x^2}\cdot y\cdot e^{\tan^{-1}x} = \frac{(e^{\tan^{-1}x})^2}{(1+x^2)}$$

Rewriting

$$\frac{d}{dx}\{y\cdot e^{\tan^{-1}x}\} = \frac{(e^{\tan^{-1}x})^2}{(1+x^2)}$$

Integrating

$$y\cdot e^{\tan^{-1}x} = \int \frac{(e^{\tan^{-1}x})^2}{(1+x^2)}dx + c$$

$$= \int (e^{\tan^{-1}x})^2\cdot d\cdot(\tan^{-1}x) + c$$

$$y\cdot e^{\tan^{-1}x} = \frac{(e^{\tan^{-1}x})^2}{2} + c$$

Example 6: Solve $y^2dx + (3xy-1)dy = 0$.

Solution: Rewriting in standard form

$$\frac{dy}{dx} = \frac{y^2}{1-3xy}$$

This is *not* linear in y (because of the presence of the term y^2). This is also *not* exact, not homogeneous, nor separable. Instead if we swap the roles of x and y by treating x as dependent variable and y as the independent variable, the given equation can be rearranged as

$$\frac{dx}{dy} + \frac{3}{y}x = \frac{1}{y^2}$$

with $P(y)=\frac{3}{y}$ and $Q(y)=\frac{1}{y^2}$

Now we get.

I.F. $= e^{\int P(y)dy} = e^{\int \frac{3}{y}dy} = e^{3\ln y} = e^{\ln y^3} = y^3$

Multiplying by I.F. y^3, we have

$$y^3\frac{dx}{dy} + 3xy^2 = y$$

Rewriting

$$\frac{d}{dy}(xy^3) = y$$

Integrating w.r.t. y

$$xy^3 = \int y dy + c = \frac{y^2}{2} + c$$

$$2xy^3 - y^2 = 2c$$

Example 7: Solve $\frac{dy}{dx} + \frac{y\ln y}{x-\ln y} = 0$.

Solution: This equation is *not* linear in y because of the presence of the term $\ln y$. It is neither separable, nor homogeneous nor exact. But with x taken as dependent variable the equation can be rewritten as

$$\frac{dx}{dy} + \frac{1}{y\ln y}\cdot x = \frac{1}{y}$$

which is the standard form of first order linear differential equation with $P(y)=\frac{1}{y\ln y}$, $Q(y)=\frac{1}{y}$ so that

I.F. is $e^{\int P(y)dy} = \exp\left\{\int \frac{dy}{y\ln y}\right\} = e^{\ln(\ln y)} = \ln y$

Multiplying with the I.F. $\ln y$, we get the exact equation

$$\ln y\cdot\frac{dx}{dy} + \frac{1}{y}x = \frac{\ln y}{y}$$

or
$$\frac{d}{dy}(x \cdot \ln y) = \frac{\ln y}{y}$$

Integrating w.r.t. y

$$x \cdot \ln y = \int \frac{\ln y}{y} dy = \int \ln y d(\ln y) = \frac{(\ln y)^2}{2} + c$$

So the solution is

$$2x \ln y = (\ln y)^2 + 2c$$

EXERCISE

Solve the following:

1. $2(y - 4x^2)dx + xdy = 0$

Ans. $x^2y = 2x^4 + c$

2. $y' + y \cot x = 2x \operatorname{cosec} x$

Ans. $y = (x^2 + c) \operatorname{cosec} x$

3. $y' + y = \frac{1}{1+e^{2x}}$

Ans. $y = e^{-x} \cdot \tan^{-1} e^x + ce^{-x}$

4. $(1 + x^2)dy + 2xydx = \cot x dx$

Ans. $y = \frac{\log (\sin x)+c}{(1+x^2)}$

5. $y' + 2y = e^x(3 \sin 2x + 2 \cos 2x)$

Ans. $y = ce^{-2x} + e^x \cdot \sin 2x$

6. $y' + y \tan x + \sin 2x,\ y(0) = 1$

Ans. $y = 3 \cos x - 2 \cos^2 x$

7. $y' + y = e^{e^x}$

Ans. $ye^x = e^{e^x} + c$

8. $[y + (x + 1)^2 e^{3x}]dx - (x + 1)dy = 0$

Ans. $y = (\frac{1}{3}e^{3x} + c)(x + 1)$

9. $dx + (3y - x)dy = 0$

Ans. $x - 3y - 3 = ce^y$

10. $\frac{dI}{dt} + 2I = 10e^{-2t},\ I = 0$ when $t = 0$

Ans. $I = 10te^{-2t}$

11. $ydx + (3x - xy + 2)dy = 0$

Hint: Examples 11 to 15 are linear when x is considered as the dependent variable.

Ans. $xy^3 = 2y^2 + 4y + 4 + ce^y$

12. $(1 + y^2)dx + (x - \tan^{-1} y)dy = 0$

Ans. $x = \tan^{-1} y - 1 + ce^{-\tan^{-1} y}$

13. $y^2dx + (xy - 2y^2 - 1)dy = 0$

Ans. $xy = y^2 + \ln y + c$

14. $dx - (x + y + 1)dy = 0$

Ans. $x = -(y + 2) + ce^y$

15. $ydx - (x + 2y^3)dy = 0$

Ans. $x = y^3 + cy$

8.10 BERNOULLI* EQUATION

Nonlinear Equation Reducible to Linear Form

A first order first degree differential equation of the form

$$\frac{dy}{dx} + P(x)y = Q(x) \cdot y^a \tag{1}$$

is known as Bernoulli equation, which is nonlinear for any value of the real number a (except for $a = 0$ and 1).

For $a = 0$, (1) reduces to Linear first order D.E. discussed in Section 8.9.

For $a = 1$, (1) reduces to linear separable D.E., considered in Section 8.4.

For any a, other than 0 and 1, the nonlinear first order D.E. (1) can be reduced to linear D.E. by the substitution

$$z = y^{1-a} \tag{2}$$

Substituting $\frac{dz}{dx} = (1 - a)y^{-a}\frac{dy}{dx}$ and (2) in (1), we get

$$\frac{1}{1-a}y^a\frac{dz}{dx} + P(x) \cdot y^a z = Q(x)y^a$$

Simplifying, we get

$$\frac{dz}{dx} + (1 - a)P(x)z = (1 - a)Q(x)$$

which is a linear first order D.E. in the standard form, discussed in Section 8.9.

*Jakob Bernoulli (1654–1705), Swiss mathematician.

Method of finding solution to Bernoulli equation.

Step I. Rewrite the given equation in standard Bernoulli equation.

Step II. Identify a, $P(x)$ and $Q(x)$.

Step III. Introduce a new variable

$$z = y^{1-a}$$

and obtain the resultant first order linear equation in z.

Step IV. Solve linear equation in z by the method discussed in Section 8.9.

Note: A nonlinear equation of the form

$$f'(y)\frac{dy}{dx} + P(x)f(y) = Q(x) \quad (3)$$

can be reduced to linear equation in z by the substitution

$$z = f(y)$$

(see Worked Out Example 5 and Exercise Examples 7, 8, 9, 10 on pages 8.24 and 8.25)

Now consider the Bernoulli equation

$$\frac{dy}{dx} + P(x)y = Q(x)y^a \quad (1)$$

Multiplying throughout by $(1-a)y^{-a}$, we get

$$(1-a)y^{-a}\frac{dy}{dx} + (1-a)P(x)y^{1-a} = (1-a)Q(x) \quad (4)$$

Let $f(y) = y^{1-a}$, so that $f'(y) = (1-a)y^{-a}$

The Equation (4) becomes

$$f'(y)\frac{dy}{dx} + (1-a)P(x)f(y) = (1-a)Q(x)$$

Thus we observe that Bernoulli Equation (1) is a special case of Equation (3).

Worked Out Examples

Example 1: Solve $3y' + xy = xy^{-2}$.

Solution: Rewriting $y' + \frac{x}{3}y = \frac{x}{3}y^{-2}$

This is a Bernoulli equation with $a = -2$

Introducing $z = y^{1-a} = y^{1-(-2)} = y^3$, so that

$$\frac{dz}{dx} = 3y^2\frac{dy}{dx}$$

Substituting in D.E.

$$\frac{1}{3}\frac{dz}{dx} + \frac{1}{3}xz = \frac{1}{3}x$$

This is a linear equation but is also a separable equation

$$\frac{dz}{dx} = x(1-z)$$

or

$$\frac{dz}{z-1} = -x\,dx$$

Integrating $\ln(z-1) = -\frac{x^2}{2} + c_0$

$$(z-1) = e^{-\frac{x^2}{2}} \cdot c$$

where $c = e^{c_0}$

Replacing z, we get the solution as

$$y^3 = 1 + ce^{-\frac{x^2}{2}}$$

Example 2: Solve $\cos x\,dy = y(\sin x - y)dx$.

Solution: $y' - y \cdot \tan x = -\sec x \cdot y^2$

Bernoulli with $a = 2$

Put $\quad z = y^{1-a} = y^{1-2} = y^{-1}$,

$$\frac{dz}{dx} = -\frac{1}{y^2}\frac{dy}{dx}$$

Substituting

$$\frac{dz}{dx} + z \cdot \tan x = \sec x$$

which is a linear equation with I.F.$= e^{\int \tan x d} = \sec x$.

Multiplying throughout by I.F. $\sec x$, we get

$$\sec x \cdot \frac{dz}{dx} + z \cdot \sec x \cdot \tan x = \sec^2 x$$

or

$$\frac{d}{dx}(z \sec x) = \sec^2 x$$

Integrating $z \sec x = \int \sec^2 x\,dx + c$

$$z \sec x = \tan x + c$$

Replacing z,

$$\frac{1}{y}\sec x = \tan x + c$$

or $$\sec x = y(\tan x + c).$$

Example 3: Solve $2xyy' = y^2 - 2x^3$, $y(1) = 2$.

Solution: Rewriting $y' - \frac{1}{2x}y = -x^2y^{-1}$ which is a Bernoulli equation with $a = -1$ so that the substitution is $z = y^{1-a} = y^{1-(-1)} = y^2$, and $\frac{dz}{dx} = 2y\frac{dy}{dx}$. With these the D.E. takes the linear form as

$$\frac{dz}{dx} - \frac{1}{x}z = -2x^2$$

The I.F. is $e^{\int -\frac{1}{x}dx} = e^{-\ln x} = e^{\ln\frac{1}{x}} = \frac{1}{x}$

By multiplying D.E. with I.F.: $\frac{1}{x}$, we get

$$\frac{1}{x}\frac{dz}{dx} - \frac{1}{x^2}z = -2x$$

or $$\frac{d}{dx}(z \cdot \frac{1}{x}) = -2x$$

Integrating $z \cdot \frac{1}{x} = \int -2x dx + c = -x^2 + c$

$$z = -x^3 + cx$$

Replacing z, $y^2 = cx - x^3$

Since $y(1) = 2$, $2^2 = c \cdot 1 - 1^3$ $\therefore$ $c = 5$. Thus the required solution is $y^2 = x(5 - x^2)$.

Example 4: Solve $(xy^5 + y)dx - dy = 0$.

Solution: Rewriting $y' - y = xy^5$. This is Bernoulli equation with $a = 5$ so that put $z = y^{1-a} = y^{1-5} = y^{-4}$. Also $\frac{dz}{dx} = -4 \cdot y^{-5}\frac{dy}{dx}$.

Substituting in the D.E., we get

$$\frac{dz}{dx} + 4z = -4x$$

I.F. for this linear equation is $e^{\int 4dx} = e^{4x}$

Multiplying D.E. with I.F. e^{4x}, we get

$$e^{4x} \cdot \frac{dz}{dx} + 4z \cdot e^{4x} = -4xe^{4x}$$

or $$\frac{d}{dx}(z \cdot e^{4x}) = -4xe^{4x}$$

Integrating $z \cdot e^{4x} = -4\int xe^{4x}dx + c$

Integrating by parts

$$ze^{4x} = -xe^{4x} + \frac{1}{4}e^{4x} + c$$

Replacing z by y^{-4}, we have

$$y^{-4}e^{4x} = -xe^{4x} + \frac{1}{4}e^{4x} + c$$

Example 5: Solve

$$y' - 2\cos x \cdot \cot y + \sin^2 x \cdot \operatorname{cosec} y \cdot \cos x = 0$$

Solution: Rewriting $\sin y \cdot y' = 2\cos x \cdot \cos y - \cos x . \sin^2 x$ or

$-\sin y\frac{dy}{dx} + (\cos y)(2\cos x) = \sin^2 x \cdot \cos x$

Put $v = \cos y$, so that $\frac{dv}{dy} = -\sin y$

with this substitution the given equation becomes

$$\frac{dv}{dx} + 2v\cos x = \sin^2 x \cdot \cos x$$

This is a linear equation for which the I.F. is $e^{\int 2\cos x dx} = e^{2\sin x}$

Then the solution is

$$ve^{2\sin x} = \int e^{2\sin x} . \sin^2 x \cdot \cos x dx$$

$$= \frac{1}{2}e^{2\sin x} . \sin^2 x - \frac{1}{2}e^{2\sin x} . \sin x + \frac{1}{4}e^{2\sin x} + c$$

Replacing v by $\cos y$, we get the required solution as

$$\cos y = \frac{1}{2}\sin^2 x - \frac{1}{2}\sin x + \frac{1}{4} + ce^{-2\sin x}$$

EXERCISE

1. $(y - y^2x^2\sin x)dx + xdy = 0$

Ans. $yx(c + \cos x) = 1$

2. $y^1 + y + y^2(\sin x - \cos x) = 0$

Ans. $y(ce^x - \sin x) = 1$

3. $\frac{dz}{dx} + \left(\frac{z}{x}\right)\log z = \frac{z}{x}(\log z)^2$

Hint: Put $v = \log z$

Ans. $(1 + cx)\log z = 1$

4. $y^1 + \frac{y}{2x} = \frac{x}{y^3}$, $y(1) = 2$

Ans. $x^2y^4 = x^4 + 15$

Hint: Take x as dependent variable in examples 5, 6.

5. $dx - (x^2y^3 + xy)dy = 0$

Ans. $x(2-y^2)+cxe^{\frac{-y^2}{2}}=1$

6. $y^2dx+(xy-x^3)dy=0$

Ans. $2x^2-3y=cx^2y^3$

7. $xy''-3y'=4x^2$

Hint: Put $y'=v$

Ans. $y=c_1x^4-\frac{4}{3}x^3+c_2$

8. $y'=Ay-By^n$ with A, B, n; constants and $n \neq 0, 1$

Ans. $y^{1-n}=\left[\frac{B}{A}+ce^{(1-n)Ax}\right]$

9. $xy'+y=x^3y^6$

Ans. $(cx^2+2.5)x^3y^5=1$

10. $y^2 \cdot y'-y^3\tan x-\sin x\cdot\cos^2 x=0$

Ans. $2y^3+\cos^3 x=2c\sec^3 x$

Hint: Put $f(y)=v$ in Examples 11, 12, 13, 14, to reduce the given equation to linear form.

11. $y'+x\sin 2y=x^3\cos^2 y$

Ans. $2y=(x^2-1)+2ce^{-x^2}$

12. $[(x+1)^4+2\sin y^2]dx-2y(x+1)\times\cos y^2dy=0$

Ans. $2\sin y^2=(x+1)^4+2c(x+1)^2$

13. $(4e^{-y}\sin x-1)dx-dy=0$

Ans. $e^y=2(\sin x-\cos x)+ce^{-x}$

14. $y'-\cot y+x\cot y=0$

Ans. $\sec y=x+1+ce^x$

15. $x(4y-8y^{-3})dx+dy=0$

Ans. $y^4=(2+ce^{-8x^2})$.

8.11 FIRST ORDER NONLINEAR DIFFERENTIAL EQUATIONS

A differential equation of first order and higher (more than first) degree is of the form

$$f(x, y, y')=0 \tag{1}$$

or

$$f(x, y, p)=0 \tag{1*}$$

where $\frac{dy}{dx}\equiv y'=p$. The degree of D.E. (of p) in (1) is more than one. So (1*) is a nonlinear first order D.E.

Example: $p^3-p^2(\cos x+e^x)+5py+2y^2=0$, first order, 3rd degree, nonlinear

A first order D.E. of nth degree has the general form

$$p^n+a_1(x, y)p^{n-1}+a_2(x, y)p^{n-2}+\cdots+a_{n-1}(x, y)p+a_n(x, y)=0 \tag{2}$$

Here the coefficients $a_1(x, y), a_2(x, y), \ldots a_n(x, y)$ are functions of x and y.

In several cases, (2) can be solved by reducing (2) to first order and first degree (n) equation (s), by (a) solving for p (b) solving for y (c) solving for x.

Equations Solvable for *p*

If the LHS of (2), which is an nth degree polynomial in p, can be resolved into n linear real factor then (2) takes the form

$$(p-b_1)(p-b_2)\cdots(p-b_n)=0 \tag{3}$$

where $b_1, b_2, \ldots, b_n$ are all functions of x and y. Equating the n factors in the RHS of (3) to zero, the solution of the nth degree equation (2) reduces to the problem of solving n first order and first degree differential equations given by

$$\frac{dy}{dx}=b_1(x, y), \quad \frac{dy}{dx}=b_2(x, y), \cdots, \quad \frac{dy}{dx}=b_n(x, y)$$

Solving these n equations, we obtain

$$f_1(x, y, c)=0, \; f_2(x, y, c)=0, \cdots, f_n(x, y, c)=0 \tag{4}$$

Thus the solution of (2) is given by (4) or by the product of the functions in (4)

i.e., $\boxed{f_1(x, y, c)\cdot f_2(x, y, c)\cdots f_n(x, y, c)=0}$

Equations Solvable for *y*

Solving the given equation $f(x, y, p)=0$ for y, we get

$$y=F(x, p) \tag{1}$$

Differentiating (1) w.r.t. 'x' we get

$$\frac{dy}{dx}=\frac{\partial F}{\partial x}+\frac{\partial F}{\partial p}\frac{dp}{dx}.$$

Since $p = \frac{dy}{dx}$, this equation takes the form

$$p = \phi\left(x, p, \frac{dp}{dx}\right) \tag{2}$$

which is a first order and first degree differential equation in the variable p. Solving (2), suppose we obtain its solution as

$$\psi(x, p, c) = 0 \tag{3}$$

Then eliminating 'p' between (1) and (3), we get the required solution of (1). When it is not possible to eliminate p, equations expressing x and y interms of p as $x = x(p, c)$, $y = y(p, c)$ will give the parametric representation of the solution of (1) in the parameter p.

Equations Solvable for x

Solving $f(x, y, p) = 0$ for x, we obtain

$$x = F(y, p) \tag{4}$$

which on differentiation w.r.t. y gives

$$\frac{dx}{dy} = \frac{1}{p} = \frac{\partial F}{\partial y} + \frac{\partial F}{\partial p}\frac{dp}{dy} \tag{5}$$

which is first order and first degree differential equation in p,

$$\frac{1}{p} = \phi\left(x, p, \frac{dp}{dy}\right). \tag{6}$$

Suppose $\psi(x, p, c) = 0$ be the solution of (6). Then the required solution is obtained by eliminating p between the equations $x = F(y, p)$ and $\psi(x, p, c) = 0$. When elimination of p is not possible we can express $x = x(p, c)$ and $y = y(p, c)$ which will be the parametric representation of the solution.

Worked Out Examples

Equations Solvable for p

Example 1: Solve $x^2\left(\frac{dy}{dx}\right)^2 + xy\frac{dy}{dx} - 6y^2 = 0$

Solution: This is a first order, second degree, non-linear homogeneous DE (all the three terms are non-linear). Introducing $\frac{dy}{dx} = p$, the given equation takes the form

$$x^2p^2 + xyp - 6y^2 = 0$$

Factorizing, $(px + 3y)(px - 2y) = 0$
or $(px + 3y) = 0$ and $(px - 2y) = 0$
Solving

$$x\frac{dy}{dx} + 3y = 0, \quad \frac{dy}{y} + 3\frac{dx}{x} = 0, \quad yx^3 = c$$

$$x\frac{dy}{dx} - 2y = 0, \quad \frac{dy}{y} - \frac{2dx}{x} = 0, \quad \frac{y}{x^2} = c$$

The primitive of the given DE is

$$(yx^3 - c)(y - cx^2) = 0$$

Example 2: Solve $p^3 + 2xp^2 - y^2p^2 - 2xy^2p = 0$

Solution: This is a first order, 3rd degree, nonlinear homogeneous DE. Rewriting

$$\begin{cases} p^2(p + 2x) - y^2p(p + 2x) = \\ = (p + 2x)p(p - y^2) = 0 \end{cases}$$

Thus $p + 2x = 0$, $p = 0$, $p - y^2 = 0$.

Solving $y + x^2 = c$, $y = c$, $-\frac{1}{y} = x + c$.

So the general solution is

$$(y + x^2 - c)(y - c)(xy + cy + 1) = 0$$

Example 3: Solve $2p^3 - (2x + 4\sin x - \cos x)\cdot p^2 - (x\cos x - 4x\sin x + \sin 2x)p + x\cdot\sin 2x = 0$

Solution: Observe that $p = x$ satisfies the given DE. i.e., $2x^3 - (2x + 4\sin x - \cos x)x^2 - (x\cos x - 4x\sin x + \sin 2x)x + x\sin 2x = 0$. Thus $(p - x)$ is factor. Rewriting the given DE $(p - x)[2p^2 - (4\sin x - \cos x)p - \sin 2x] = 0$ or $(p - x)[2p(p - 2\sin x) + \cos x(p - 2\sin x)] = 0$
Thus $(p - x)(p - 2\sin x)(2p + \cos x) = 0$.
Equating the three factors to zero

$$p - x = 0 \text{ or } \frac{dy}{dx} = x \quad \therefore \; y = \frac{x^2}{2} + c$$

$p - 2\sin x = 0$ or $dy - 2\sin x dx = 0$

$$\therefore\ y + 2\cos x = c$$

$2p + \cos x = 0$ or $2dy + \cos x dx = 0$

$$\therefore\ 2y + \sin x = c$$

The general solution is

$$(2y - x^2 - c)(y + 2\cos x - c)(2y + \sin x - c) = 0$$

Equations Solvable for p

EXERCISE

Solve

1. $p^2 - 5p + 6 = 0$

Ans. $(y - 3x - c)(y - 2x - c) = 0$

Hint: $(p - 3)(p - 2) = 0$

2. $4y^2p^2 + 2pxy(3x + 1) + 3x^3 = 0$

Ans. $(x^2 + 2y^2 - c)(x^3 + y^2 - c) = 0$

Hint: $(2py + x)(2py + 3x^2) = 0$

3. $p^2 + 2py\cot x - y^2 = 0$

Ans. $[y(1 + \cos x) - c][y(1 - \cos x) - c] = 0$

Hint: $(p + y\cot x - y\,\text{cosec}\, x)(p + y\cot x + y\,\text{cosec}\, x) = 0$

4. $p - \frac{1}{p} - \frac{x}{y} + \frac{y}{x} = 0$

Ans. $(xy - c)(x^2 - y^2 - c) = 0$

Hint: $\left(p + \frac{y}{x}\right)\left(p - \frac{x}{y}\right) = 0$

5. $xyp^2 + p(3x^2 - 2y^2) - 6xy = 0$

Ans. $(y - cx^2)(y^2 + 3x^2 + c) = 0$

Hint: $(py + 3x)(px - 2y) = 0$

6. $p^3 - (y + 2x - e^{x-y})p^2 + (2xy - 2xe^{x-y} - ye^{x-y})p + 2xye^{x-y} = 0$

Ans. $(y - ce^x)(y - x^2 - c)(e^y + e^x - c) = 0$

Hint: $(p - y)(p - 2x)(p + e^{x-y}) = 0$

7. $yp^2 + (x - y)p - x = 0$

Ans. $(y - x - c)(y^2 + x^2 - c) = 0$

Hint: $(p - 1)\left(p + \frac{x}{y}\right) = 0$

8. $p^4 - (x + 2y + 1)p^3 + (x + 2y + 2xy)p^2 - 2xyp = 0$

Ans. $(y - c)(y - x - c)(2y - x^2 - c)\times(y - ce^{2x}) = 0$

Hint: $p(p - 1)(p - x)(p - 2y) = 0$

9. $xyp^2 + (x^2 + xy + y^2)p + x^2 + xy = 0$

Ans. $(2xy + x^2 - c)(x^2 + y^2 - c) = 0$

Hint: $(xp + x + y)(yp + x) = 0$

10. $(x^2 + x)p^2 + (x^2 + x - 2xy - y)p + y^2 - xy = 0$

Ans. $[y - c(x + 1)][y + x\ln cx] = 0$

Hint: $[(x + 1)p - y][xp + x - y] = 0$

WORKED OUT EXAMPLES

Equations Solvable for y

Example 1: Find the general solution of $3x^4p^2 - xp - y = 0$

Solution: This is a first order, second degree, non-linear homogeneous DE which can be solved for y. Thus

$$y = 3x^4p^2 - xp$$

Differentiating w.r.t. x both sides, we get

$$\frac{dy}{dx} = 12x^3p^2 + 6x^4p\frac{dp}{dx} - p - x\frac{dp}{dx}$$

or

$$(2p - 12x^3p^2) + (x - 6x^4p)\frac{dp}{dx} = (1 - 6x^3p)\left(2p + x\frac{dp}{dx}\right) = 0$$

Equating the second order to zero we have

$$2p + x\frac{dp}{dx} = 0 \text{ or } 2\frac{dx}{x} + \frac{dp}{p} \text{ so } px^2 = c$$

Eliminating p from the given DE by using $p = \frac{c}{x^2}$, we get

$$y = 3x^4\left(\frac{c}{x^2}\right)^2 - x\cdot\frac{c}{x^2} = 3c^2 - \frac{c}{x}$$

The required general solution is

$$xy = c(3cx - 1)$$

Example 2: Solve $p \tan p - y + \log \cos p = 0$

Solution: Solving $y = p \tan p + \log \cos p$. Differentiating w.r.t. x both sides, we get

$$\frac{dy}{dx} = \tan p \cdot \frac{dp}{dx} + p \cdot \sec^2 p \cdot \frac{dp}{dx} + \frac{1}{\cos p} \cdot (-\sin p) \cdot \frac{dp}{dx}$$

or $$p\left(1 - \sec^2 p \cdot \frac{dp}{dx}\right) = 0$$

Considering the second factor,

$$1 - \sec^2 p \frac{dp}{dx} = 0$$

solving $dx = \sec^2 p dp$ or $x = \tan p + c$.
Since p cannot be eliminated, the general solution in the parametric form (with parametric p) is $x = \tan p + c$, $y = p \tan p + \log \cos p$

EXERCISE

Solve

1. $p^2x^4 - px - y = 0$

Ans. $y = c^2x - c$

Hint: $\left(2p + x\frac{dp}{dx}\right)(1 - 2x^3p) = 0$

2. $2p^2y - p^3x + 16x^2 = 0$

Ans. $2 + c^2y - c^3x^2 = 0$

Hint: $(p^3 + 32x)\left(p - x\frac{dp}{dx}\right) = 0$

3. $yp + p^2 - x = 0$

Ans. $x = -\frac{p}{p_1} \ln (p + p_1) + \frac{cp}{p_1}$, $y = -p - \frac{1}{p_1} \ln (p + p_1) + \frac{c}{p_1}$ where $p_1 = \sqrt{p^2 - 1}$

Hint: $\frac{dx}{dp} + x \cdot \frac{1}{p^3 - p} = -\frac{p}{p^2 - 1}$

4. $2x + p^2 - y + px = 0$

Ans. $x = 2(2 - p) + ce^{-p/2}$, $y = 8 - p^2 + (2 + p)ce^{-p/2}$

Hint: $\frac{dx}{dp} + \frac{1}{2}x = -p$

5. $p^3 + mp^2 - ay - amx = 0$

Ans. $ax = c + \frac{3p^2}{2} - mp + m^2 \log\ (p + m)$ and $ay = -m[c + \frac{3}{2}p^2 - mp + m^2 \log\ (p + m)] + mp^2 + p^3$

Hint: $a dx = \frac{p(3p + 2m)}{(p + m)} dp = \left(3p - m + \frac{m^2}{p + m}\right) dp$

6. $2px + \tan^{-1}(xp^2) - y = 0$

Ans. $y = 2\sqrt{cx} + \tan^{-1} c$

Hint: $\left(p + 2x\frac{dp}{dx}\right)\left(1 + \frac{p}{1 + x^2p^4}\right) = 0$

WORKED OUT EXAMPLES

Equations Solvable for x

Example 1: Find the primitive of $p^2 - xp + y = 0$

Solution: Solving for x, we obtain

$$x = \frac{p^2 + y}{p}$$

Differentiating w.r.t. 'y', we get

$$\frac{dx}{dy} = \frac{1}{p} = \frac{dp}{dy} + \frac{\left(p - y\frac{dp}{dy}\right)}{p^2}$$

$$(p^2 - y)\frac{dp}{dy} = 0$$

so $\frac{dp}{dy} = 0$ with solution $p = c =$ constant. Eliminating p from the given equation

$$c^2 - xc + y = 0 \text{ or } y = cx - c^2$$

which is the required primitive.

Example 2: Solve $xp^2 - yp - y = 0$

Solution: Solving for x we obtain

$$x = \frac{y(1 + p)}{p^2}$$

Differentiating w.r.t. 'y', we get

$$\frac{dx}{dy}=\frac{1}{p}=\frac{1+p}{p^2}+y\left(\frac{p^2-(1+p)2p}{p^4}\right)\frac{dp}{dy}$$

or $\frac{dp}{dy}=\frac{p}{2+p}\cdot\frac{1}{y}$

Separating the variables and integrating

$$\frac{2+p}{p}dp=\frac{dy}{y}$$

$$\ln p^2+p=\ln y+c_1 \text{ or}$$

$$\ln (p^2e^p)=\ln y+c_1 \text{ so}$$

$y=cp^2e^p$ and $x=\frac{1+p}{p^2}\cdot y=c(1+p)e^p$ which is required primitive in the parametric form.

EXERCISE

Equation Solvable for x

Obtain the primitive for the following equatons:

1. $6p^2y^2-y+3px=0$

Ans. $y^3=3cx+6c^2$

Hint: $2p+y\frac{dp}{dy}=0,\ py^2=c$

2. $4y^2+p^3=2xyp$

Ans. $2y=c(c-x)^2$

Hint: $p-2y\frac{dp}{dy}=0,\ p^2=cy$

3. $3py+4x=p^3y$

Ans. $y=c\cdot P,\ x=\frac{1}{4}cp(p^2-3)\cdot P$ where $P^{-1}=-(p^2-4)^{9/10}\cdot(p^2+1)^{3/5}$

Hint: $\frac{dy}{y}+\frac{3p(p^2-1)dp}{(p^2-4)(p^2+1)}=0;\ \frac{3p(p^2-1)}{(p^2-4)(p^2+1)}=\frac{9}{10}\left(\frac{1}{p+2}+\frac{1}{p-2}\right)+\frac{3}{5}\frac{2p}{p^2+1}$

4. $y^2p^3-y+2px=0$

Ans. $y^2=2cx+c^3$

Hint: $p+y\frac{dp}{dy}=\frac{d}{dy}(py)=0,\ py=c$

5. $3px-y+6p^2y^2=0$

Ans. $y^3=3c(x+2c)$

Hint: $2p+y\frac{dp}{dy}=0,\ py^2=c$

6. $p^2+y-x=0$

Ans. $x=c-2[p+\log(p-1)]$,
$y=c-2\left[\frac{1}{2}p^2+p+\log(p-1)\right]$

Hint: $\frac{dp}{dy}=\frac{1-p}{2p^2}$

8.12 CLAIRAUT'S* EQUATION

Clairaut's equation is a first order and higher degree DE of the form

$$y=x\frac{dy}{dx}+\psi\left(\frac{dy}{dx}\right) \tag{1}$$

This equation (1) is linear in y and x. Here ψ is a known function of $\frac{dy}{dx}$.

Introducing $\frac{dy}{dx}=p$, equation (1) takes the form

$$y=xp+\psi(p) \tag{2}$$

Differentiating both sides of (2) w.r.t. x, we have

$$\frac{dy}{dx}=p+x\frac{dp}{dx}+\frac{d\psi}{dp}\cdot\frac{dp}{dx}$$

since p itself is a function of x. On simplification

$$\left[x+\frac{d\psi}{dp}\right]\frac{dp}{dx}=0$$

Thus
$$\frac{dp}{dx}=0 \tag{3}$$

and
$$x+\frac{d\psi}{dp}=0 \tag{4}$$

Integrating (3), we get $p=\text{constant}=c$. Substituting $p=c$ in (2), the *complete integral* of the Clairaut's equation (1) is given by

$$y=x\cdot c+\psi(c) \tag{5}$$

which is a one parameter family of straight lines; with c as the parameter. Thus the complete solution of Clairaut's equation is obtained simply by replacing $\frac{dy}{dx}$ (i.e., p) by constant c. Besides the complete integral (5), one may obtain a "*singular solution*"

* Alexis Claude Clairaut (1713-1765), French mathematician.

of Clairaut's equation which satisfies the Clairaut's equation (1) but *not* obtained from the complete integral (5) for any value of c.

Now solving (4), we get p as a function of x. Thus (2) becomes

$$y = x \cdot p(x) + \psi(p(x)) \qquad (6)$$

Differentiating (6) w.r.t. x, we get

$$\frac{dy}{dx} = p + x\frac{dp}{dx} + \frac{d\psi}{dp} \cdot \frac{dp}{dx}$$

$$= p + \left[x + \frac{d\psi}{dp}\right]\frac{dp}{dx}.$$

Using (4), $\frac{dy}{dx} = p + 0 \cdot \frac{dp}{dx} = p$

Substituting (6), Clairaut's DE (1), is satisfied, i.e., $x \cdot p + \psi(p) = y = x \cdot \frac{dy}{dx} + \psi\left(\frac{dy}{dx}\right) = x \cdot p + \psi(p)$ since $\frac{dy}{dx} = p$.

Thus the singular solution (6) known as *p-discriminant* equation is readily obtained by eliminating the parameter p from the two equations (2) and (4):

$$y = xp + \psi(p)$$

$$x + \frac{d\psi}{dp} = 0$$

Note 1: Singular solution (6) is the envelope of the family of straight lines represented by the complete integral (5).

Note 2: Singular solution can also be obtained by eliminating the constant c form the two equations (5) and (4):

$$y = xc + \psi(c)$$

$$x + \frac{d\psi(c)}{dc} = 0$$

Note 3: Equations which are not in the Clairaut's form can be reduced to Clairaut's form by suitable substitutions (transformations) (see W.E. 3, 4, 5, 6).

WORKED OUT EXAMPLES

Example 1: Solve $y = xy' - (y')^2$

Solution: Clairaut's equation is $y = xp - p^2$. Its general solution is obtained by replacing p by a constant c. Thus $y = xc - c^2$ is the required complete integral.

To obtain the singular solution, differentiate the general solution w.r.t. 'c'. Then

$$0 = x - 2c \quad \therefore \quad c = \frac{x}{2}$$

Eliminating c from the G, S, we get

$$y = cx - c^2 = \frac{x}{2} \cdot x - \frac{x^2}{4} = \frac{x^2}{4}$$

$$\text{or} \quad x^2 = 4y$$

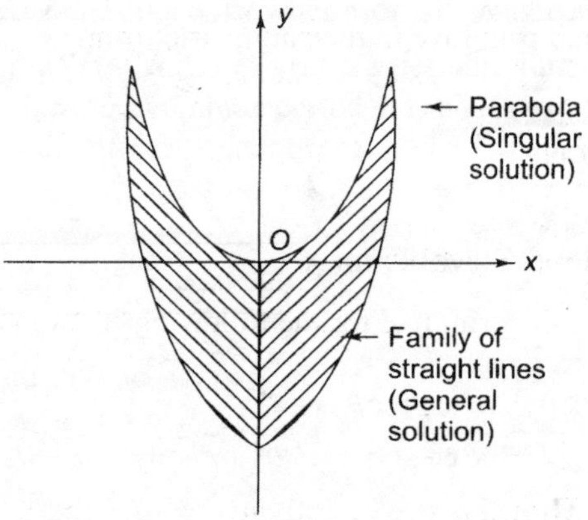

Fig. 8.3

Thus the singular solution $x^2 = 4y$ (which is a parabola) is the envelope of the one parameter family of straight lines $y = cx - c^2$ (representing the general solution).

Example 2: Solve $p = \ln(px - y)$.

Solution: Rewriting $e^p = px - y$ or $y = xp - e^p$ which is Clairaut's equation. The complete solution is obtained by replacing p by c. Thus $y = cx - e^c$ is the complete integral or $c = \ln\,(cx - y)$.

To obtain the singular solution, differentiate

$$c = \ln\,(cx - y)$$

w.r.t. c then $1 = \frac{x}{cx-y}$. Solving $c = \frac{x+y}{x}$. Eliminating c from the complete integeral.

$$y = \frac{x+y}{x} \cdot x - e^{\frac{x+y}{x}} \text{ or } \ln x = \frac{x+y}{x}$$

The singular solution is $y = x(\ln x - 1)$ (which cannot be obtained from the complete solution for any value of c).

Example 3: Find the primitive $y = 4xp - 16y^3p^2$

Solution: This equation is *not* in the Clairaut's form. Multiplying the given equation by y^3, we have

$$y^4 = 4xy^3p - 16y^6p^2$$

put $y^4 = v$ so $4y^3\frac{dy}{dx} = \frac{dv}{dx}$. Then

$$v = x\frac{dv}{dx} - \left(\frac{dv}{dx}\right)^2$$

which is a Clairaut's equation in v. Its general solution is obtained by replacing $\frac{dv}{dx}$ by a constant c. Thus the general solution is

$$v = xc - c^2$$

or

$$y^4 = cx - c^2$$

Example 4: Solve $\sin y\cos^2 x = \cos^2 yp^2 + \sin x\cdot\cos x\cos yp$.

Solution: To reduce to Clairaut's form, put $\sin y = u$ and $\sin x = v$. Then $\frac{\cos y}{\cos x}\cdot\frac{dy}{dx} = \frac{du}{dv}$. Dividing the given equation throughtout by $\cos^2 x$, we get

$$\sin y = \frac{\cos^2 y}{\cos^2 x}\cdot p^2 + \sin x\cdot\frac{\cos y}{\cos x}\cdot p$$

Substituting for $p \equiv \frac{dy}{dx} = \frac{\cos x}{\cos y}\frac{du}{dv}$ the DE transforms to

$$u = \left(\frac{du}{dv}\right)^2 + v\left(\frac{du}{dv}\right)$$

which is in the Clairaut's form. The general solution is

$$u = cv + c^2$$

or

$$\sin y = c\cdot\sin x + c^2$$

Example 5: Solve $(p-1)e^{3x} + p^3e^{2y} = 0$.

Solution: To reduce to Clairaut's form, put $e^x = X$, $e^y = Y$; then $P = \frac{dY}{dX} = \frac{e^y}{e^x}\frac{dy}{dx} = \frac{Y}{X}p$. Substituting for $p = \frac{X}{Y}\frac{dY}{dX} = \frac{X}{Y}P$ the given equation reduces to

$$\left(\frac{X}{Y}P - 1\right)X^3 + \frac{X^3}{Y^3}P^3Y^2 = 0$$

or

$$Y = YP + P^3$$

which is Clairaut's DE. Its general solution is

$$Y = Xc + c^3$$

or

$$e^y = ce^x + c^3$$

Example 6: Solve $(x^2+y^2)(1+p)^2 = 2(x+y)(1+p)(x+yp) - (x+yp)^2$

Solution: Rewriting $x^2 + y^2 = \frac{2(x+y)(x+yp)}{(1+p)} - \left(\frac{x+yp}{1+p}\right)^2$. Put $x^2+y^2 = v$ so $2x + 2yp = \frac{dv}{dx}$ and put $x + y = u$ so $1 + p = \frac{du}{dx}$. Then

$$\frac{dv}{du} = \frac{2(x+yp)}{(1+p)}$$

Substituting the given equation reduces to i.e., $v = u\frac{dv}{du} - \frac{1}{4}\left(\frac{dv}{du}\right)^2$ which is a Clairaut's DE. Its general solution is $v = uc - \frac{1}{4}c^2$ or $x^2 + y^2 = c(x+y) - \frac{1}{4}c^2$

Exercise

Clairaut's Equations (CE)

1. Find the general solution (GS) and singular solution (SS) the Clairaut's equation $y = xp + \frac{ap}{\sqrt{1+p^2}}$.

Ans. GS: $y = cx + \frac{ac}{\sqrt{1+c^2}}$; SS: $x^{2/3} + y^{2/3} = a^{2/3}$

Hint: Parametric form of SS: $x = -a/(1+c^2)^{3/2}$, $y = ac^3/(1+c^2)^{3/2}$

2. Determine GS and SS of $p = \sin(y - xp)$.

Ans. GS: $y = cx + \sin^{-1} c$, SS: $y = (x^2-1) + \sin^{-1}\left(\frac{x^2-1}{x}\right)$

Hint: $y = px + \sin^{-1} p$ is CE.

3. Find GS and SS of $y = px + p^3$

Ans. GS: $y = cx + c^3$, SS: $27y^2 = -4x^3$

4. Solve $(x^2-1)p^2 - 2xyp + y^2 - 1 = 0$

Ans. GS: $(x^2-1)c^2 - 2xyc + y^2 - 1 = 0$, SS: $x^2 + y^2 = 1$.

Hint: $(y-xp)^2-1-p^2=0$, $y=xp\pm\sqrt{1-p^2}$, two DE's each of Clairaut's form.

5. Solve $y=px+\log p$

Ans. GS: $y=cx+\log c$

6. Solve $p^2y+px^3-x^2y=0$

Ans. $y^2=cx^2+c^2$

Hint: use $u=x^2$, $v=y^2$

7. Solve $y=xy'+\frac{a}{2y'}$

Ans. GS: $y=cx+\frac{a}{2c}$, one parameter family of straight lines. SS: $y^2=2ax$ parabola, which is the envelope of the family of straight lines.

8. Solve $y=px+2p^2$

Ans. GS: $y=cx+c^2$, SS: $x^2=-8y$ (parabola)

9. Solve $y=3px+6p^2y^2$

Ans. GS: $y^3=cx+\frac{2}{3}c^2$, SS: $8y^3+3x^2=0$ (semicubical parabola)

10. Solve $a^2p=(py+x)(px-y)$

Ans. $y^2=cx^2-a^2c/(c+1)$

Hint: Use $u=x^2$, $v=y^2$, $v=uP-a^2P/(P+1)$

11. $y+y^2+p(2y-2xy-x+2)+xp^2(x-2)=0$

Ans. GS: $(y-cx+2c)(y-cx+1)=0$

Hint: $(y-px+2p)(y-px+1)=0$, each of Clairaut's form

12. $(px^2+y^2)(px+y)=(p+1)^2$

Ans. $c^2(x+y)-cxy-1=0$

Hint: Use $x+y=u$, $xy=v$, $P=\frac{dv}{du}$, $v=Pu-\frac{1}{P}$.

8.13 LAGRANGE'S EQUATION

Lagrange's equation is a first order and higher degree DE of the form

$$y=x\phi\left(\frac{dy}{dx}\right)+\psi\left(\frac{dy}{dx}\right) \quad (1)$$

Here ϕ and ψ are known functions of $\frac{dy}{dx}$. Equation (1) is linear in x and y. Note that Clairaut's equation is a particular case of Lagrange's equation when $\phi\left(\frac{dy}{dx}\right)\equiv\frac{dy}{dx}$.

Putting $y'\equiv\frac{dy}{dx}=p$, equation (1) takes the form

$$y=x\phi(p)+\psi(p) \quad (2)$$

Differentiating both sides of (2) w.r.t. x, we get

$$\frac{dy}{dx}=\phi(p)+x\cdot\frac{d\phi}{dp}\cdot\frac{dp}{dx}+\frac{d\psi}{dp}\cdot\frac{dp}{dx}$$

or $$p-\phi(p)=[x\phi'(p)+\psi'(p)]\frac{dp}{dx} \quad (3)$$

Rewriting (3) with x as a function of p, we get

$$\frac{dx}{dp}-x\cdot\frac{\phi'(p)}{p-\phi(p)}=\frac{\psi'(p)}{p-\phi(p)} \quad (4)$$

which is a linear DE in x. Integrating (4), we obtain

$$x=\chi(p,c) \quad (5)$$

Now the *p-discriminant* equation obtained by eliminating the parameter p from the two equations (2) and (5) is of the form $\Phi(x,y,c)=0$ which is the required *complete integral* of the Lagrange's equation.

The general solution of (1) is also written in the parametric form

$$x=\chi(p,c)$$

and $y=\chi(p,c)\cdot\phi(p)+\psi(p)$
Here p is the parameter.
When p = constant = c, equation (3) is satisfied provided LHS of (3) is zero i.e., c is the root of the equation $c-\phi(c)=0$. Thus the singular solution of (1) is

$$y=x\phi(c)+\psi(c)$$

where c satisfies the condition $\phi(c)=c$.

Lagranges' Equation

WORKED OUT EXAMPLES

Example 1: Solve $y=xy'^2-\frac{1}{y'}$.

Solution: This is a Lagrange's equation. Put $y' = p$ then $y = xp^2 - \frac{1}{p}$. Differentiating

$$dy = pdx = x \cdot 2p \cdot dp + p^2 dx + \frac{1}{p^2} dp$$

since $\frac{dy}{dx} = p$ i.e., $dy = p\,dx$. Rewriting

$$(p - p^2)\frac{dx}{dp} + x(-2p) = \frac{1}{p^2}$$

or $\frac{dx}{dp} + x\frac{2}{(p-1)} = \frac{-1}{p^3(p-1)}$

This is a linear DE in x with I.F. $e^{\int\left(\frac{2\,dp}{p-1}\right)} = (p-1)^2$.

Integrating

$$x \cdot (p-1)^2 = \int \frac{-1}{p^3(p-1)} \cdot (p-1)^2 \cdot dp$$

$$= \int \frac{1-p}{p^3} dp = -\frac{1}{2p^2} + \frac{1}{p} + c^*$$

$$\therefore \quad x = \frac{cp^2 + 2p - 1}{2p^2(p-1)^2} \tag{1}$$

Substituting x in the expression for y, we get

$$y = xp^2 - \frac{1}{p} = \frac{cp^2 + 2p - 1}{2p^2(p-1)^2} \cdot p^2 - \frac{1}{p}$$

i.e., $$y = \frac{cp^2 + 2p - 1}{2(p-1)^2} - \frac{1}{p} \tag{2}$$

The general solution in the parametric form is given by (1) and (2) with parameter p.

EXERCISE

Integrate the following equations

1. $y = 2xy' + \ln y'$

Ans. $x = \frac{c}{p^2} - \frac{1}{p}$, $y = \ln p + \frac{2c}{p} - 2$

Hint: $\frac{dx}{dp} + \frac{2}{p}x = -\frac{1}{p^2}$

2. $y = 2xy' + \sin y'$

Ans. $x = \frac{c}{p^2} - \frac{\cos p}{p^2} - \frac{\sin p}{p}$, $y = \frac{2c}{p} - \frac{2\cos p}{p}$ $- \sin p$

Note: $y = 0$ is SS satisfying DE.

Hint: $\frac{dx}{dp} + \frac{2}{p}x = -\frac{\cos p}{p}$

3. $y = xy'^2 + y'^2$

Ans. GS: $y = (c + \sqrt{x+1})^2$

Hint: $\frac{dx}{dp} + x\frac{2}{p-1} = \frac{2}{1-p}$

Note: $y = 0$ is SS

4. $y = x(1 + y') + y'^2$

Ans. $x = ce^{-p} - 2p + 2$,
$y = c(p+1)e^{-p} - p^2 + 2$

5. $y = y\left(\frac{dy}{dx}\right)^2 + 2xy'$

Ans. $4cx = 4c^2 - y^2$

8.14 FORMATION OF ORDINARY DIFFERENTIAL EQUATIONS BY ELIMINATION OF ARBITRARY CONSTANTS

Consider the equation

$$y = x^2 + c \tag{1}$$

where c is an "arbitrary constant".

For each particular value of c, equation (1) represents a different parabola. Since c can vary, c is often called a "parameter" to distinguish it from the main variables x and y. Thus equation (1) represents a "one-parameter family of curves" in the xy-plane, namely a family of parabolas.

Thus the general equation

$$f(x, y, c) = 0 \tag{2}$$

represents a one-parameter family of curves with c as the parameter.

Now differentiating (1) w.r.t. x, we get

$$\frac{dy}{dx} = 2x \tag{3}$$

Thus, *each* member of the one-parameter family of parabolas (1) has the "property" that the slope at any point is $2x$.

Hence differential Equation (3), satisfied by *all* members of the family (1), is often called the "differential equation of the family".

Now differentiating (2) w.r.t. x and eliminating the arbitrary constant c we obtain a D.E.

$$F(x, y, y') = 0 \tag{4}$$

Thus (4) is the "differential equation" of the "one-parameter family of curves (2)", expressing the "property" common to *all* the curves of the family (2).

Similarly all circles of unit radius represented by

$$(x - c_1)^2 + (y - c_2)^2 = 1$$

with c_1 and c_2 as arbitrary constants forms a "two-parameter family" of curves (in this case circles). Method of forming D.E. by elimination of arbitrary constants.

Step I. Differentiate (2) w.r.t. x.

Step II. Eliminate the arbitrary constant c between (2) and equation obtained in Step I.

Note 1: If there are n arbitrary constants present in the equation of an "n-parameter family of curves" then in step I, differentiate successively n times and eliminate the n constants from the $n + 1$ equations.

Note 2: Sometimes, during differentiation itself, the arbitrary constants may get eliminated.

Worked Out Examples

Example 1: Obtain (form) the differential equation from each of the following functions by the elimination of arbitrary constants:

a. $y = c_1 e^x + c_2 e^{-2x}$ (1)

Solution: Differentiating w.r.t. x

$$y' = c_1 e^x - 2c_2 e^{-2x} \tag{2}$$

Since elimination of c_1 and c_2 is not possible at this stage, (see note 1) differentiate once more w.r.t. x

$$y'' = c_1 e^x + 4c_2 e^{-2x} \tag{3}$$

From (1) $c_1 e^x = y - c_2 e^{-2x}$ (4)

Eliminate c_1 by substituting (4) in (2) and (3)

$$y' = (y - c_2 e^{-2x}) - 2c_2 e^{-2x} = y - 3c_2 e^{-2x} \tag{5}$$

$$y'' = (y - c_2 e^{-2x}) + 4c_2 e^{-2x} = y + 3c_2 e^{-2x} \tag{6}$$

Eliminate c_2 by adding (5) and (6)

$$y' + y'' = 2y$$

Thus the D.E. is $y'' + y' - 2y = 0$.

b. $y = ax^2 + bx + c$

Solution: Differentiating w.r.t. x

$$y' = 2ax + b$$

Differentiating again

$$y'' = 2a$$

Differentiating once more, we get the required D.E.

$$y''' = 0$$

which is free of the given 3 arbitrary constants.

c. $y = a\cos(\ln x) + b\sin(\ln x)$

Solution: Differentiating w.r.t. x

$$y' = \frac{-a}{x}\sin(\ln x) + \frac{b}{x}\cdot\cos(\ln x)$$

$$xy' = -a\sin(\ln x) + b\cos(\ln x)$$

Differentiating once again w.r.t. x

$$y' + xy'' = \frac{-a}{x}\cdot\cos(\ln x) - \frac{b}{x}\sin(\ln x)$$

$$xy' + x^2y'' = -[a\cos(\ln x) + b\sin(\ln x)]$$

$$= -y$$

Thus the required D.E. is

$$x^2y'' + xy' + y = 0.$$

Example 2: Obtain the differential equation of the family of plane curves which are:

a. *all circles of unit radius:*

Solution: The general evaluation of all circles with unit radius and centre at any point (a, b) is

$$(x - a^2) + (y - b)^2 = 1^2 \tag{1}$$

Here a and b are arbitrary constants. Differentiating (1) w.r.t. x, we get

$$2(x - a) + 2(y - b)y' = 0 \tag{2}$$

Differentiating this once more w.r.t. x

$$1 + (y - b)y'' + y'^2 = 0 \tag{3}$$

Solving (3),

$$(y - b) = \frac{-(1 + y'^2)}{y''} \tag{4}$$

Substituting (4) in (2), we get

$$x - a = -(y - b)y' = \frac{(1 + y'^2)}{y''} y' \qquad (5)$$

Substituting (4) and (5) in (1), we have

$$(x - a)^2 + (y - b)^2 = \left[\frac{1 + y'^2}{y''} \cdot y'\right]^2 + \left[\frac{1 + y'^2}{y''}\right] = 1$$

$$(1 + y'^2)^2[y^2 + 1] = y''^2$$

Thus the required D.E. is

$$(1 + y'^2)^3 = y''^2$$

b. *family of parabolas with foci at the origin and axes along the x-axis:*

Solution: The equation is

$$y^2 = 4ax + 4a^2 \qquad (1)$$

Differentiating (1) w.r.t. x

$$2yy' = 4a \qquad (2)$$

Substituting 'a' from (2) in (1), we get

$$y^2 = \left(2yy'x + 4 \cdot \left(\frac{yy'}{2}\right)\right)^2$$

So the required D.E. is

$$2xy' + y(y')^2 - y = 0$$

c. *family of confocal central conics:*

$$\frac{x^2}{a^2 + c} + \frac{y^2}{b^2 + c} = 1 \qquad (1)$$

with a, b fixed, c arbitrary constant.

Solution: Differentiating (1) w.r.t. x

$$\frac{2x}{a^2 + c} + \frac{2yy'}{b^2 + c} = 0 \qquad (2)$$

Solving $yy'(a^2 + c) = -x(b^2 + c)$

$$c = \frac{-(xb^2 + yy'a^2)}{yy' + x} \qquad (3)$$

Substituting c from (3) in (1)

$$\frac{x^2}{\left(\frac{x(a^2 - b^2)}{yy' + x}\right)} + \frac{y^2}{\left(\frac{yy'(b^2 - a^2)}{yy' + x}\right)} = 1$$

Simplifying, we get

$$(yy' + x)(xy' - y) = (a^2 - b^2)y'$$

EXERCISE

1. Form (obtain) the differential equation by eliminating the arbitrary constants from each of the following equations:

   a. $x^3 - 3x^2y = c$

   Ans. $xy' + 2y - x = 0$

   b. $y \sin x - xy^2 = c$

   Ans. $y' = \frac{y(y - \cos x)}{(\sin x - 2xy)}$

   c. $y = c(1 - e^{-\frac{x}{c}})$

   Ans. $y \ln y' + x(1 - y') = 0$

   d. $y = A \sin (Bx + c)$

   where B is a parameter, *not* to be eliminated.

   Ans. $y'' + B^2y = 0$

   e. $y = a\, e^{3x} + be^{-x}$

   Ans. $y'' + 2y' - 3y = 0$

   f. $y = c_1 e^{ax} \cdot \cos bx + c_2 e^{ax} \sin bx$

   Here a, b are parameters *not* to be eliminated.

   Ans. $y'' - 2a\, y' + (a^2 + b^2)y = 0$

   g. $y = cx + \frac{b}{c}$, parameter 'b' not to be eliminated.

   Ans. $x(y')^2 - yy' + b = 0$

   h. $y = c_1 e^x + c_2e^{2x} + c_3 e^{3x}$

   Ans. $y''' - 3y'' + 11y' - 6y = 0$

   i. $y = c_1 e^x + c\, e^{-x} + c_3 \cos x + c_4 \sin x$

   Ans. $y'^2 - y = 0$

2. Obtain (form) the differential equation of family of plane curves which are:

   a. family of circles having their centres on the y-axis

Hint: $x^2 + (y - b)^2 = r^2$

Ans. $xy'' - (y')^3 - y' = 0$

b. parabolas having their vertices at the origin and their foci on the y-axis

Hint: $x^2 = 4ay$

Ans. $y' = \frac{2y}{x}$

c. all tangents to the parabola $y^2 = 2x$

Hint: $cy = x + \frac{c^2}{2}$

Ans. $2x(y')^2 - 2yy' + 1 = 0$

d. all straight lines at a unit distance from the origin

Ans. $(xy' - y)^2 = 1 + (y')^2$

e. all circles $r = 2a\,(\sin\theta - \cos\theta)$

Ans. $(\cos\theta - \sin\theta)dr + r(\cos\theta + \sin\theta)d\theta = 0$

f. all cissoids $r = a\sin\theta\,\tan\theta$

Ans. $\sin\theta \cdot \cos\theta\, dr - r(1 + \cos^2\theta)d\theta = 0$

g. all strophoids $r = a\,(\sec\theta + \tan\theta)$

Ans. $dr - r\sec\theta\, d\theta = 0.$

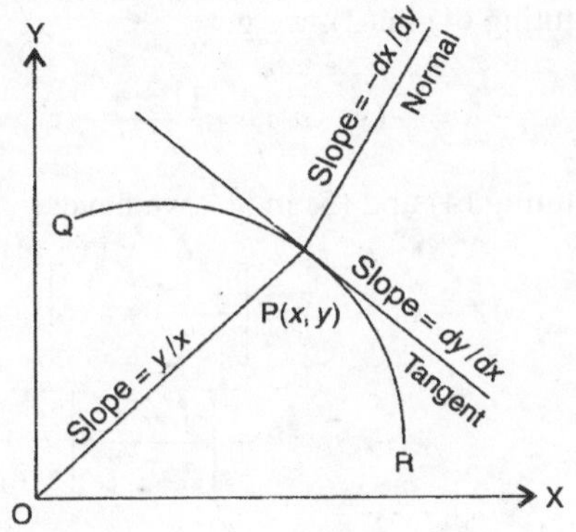

Fig. 8.4

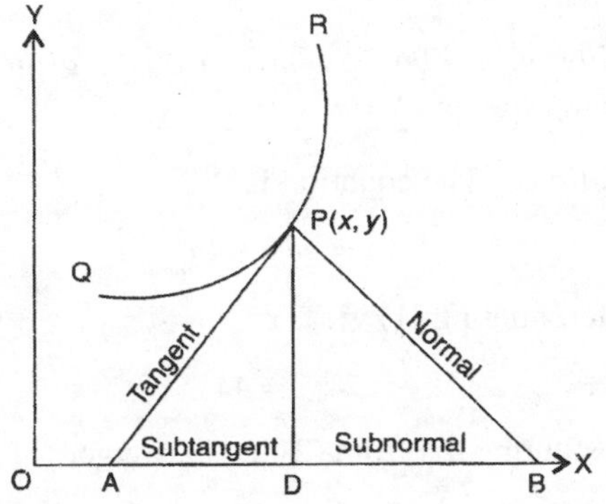

Fig. 8.5

8.15 GEOMETRICAL APPLICATIONS

Tangent, Normal, Subtangent and Subnormal

If the analytical representation of a "property" of a curve involves the derivative, then the solution of the resulting differential equation represents a one-parameter family of curves, such that each member of the family having this "property".

The following "properties" of curves involving derivatives are listed below:

Rectangular coordinates Let $P(x, y)$ be any (general) point of a curve QPR given by $f(x, y) = 0$ (Fig. 8.4).

Let (X, Y) be any (general) point on the tangent AP (refer Fig. 8.5). Then the equation of the tangent AP at the point $P(x, y)$ is

a. $$Y - y = \frac{dy}{dx}(X - x) \quad (1)$$

Similarly if (X, Y) is any point on the normal PB then the equation of the normal line PB at $P(x, y)$ is

b. $$Y - y = -\frac{dx}{dy}(X - x) \quad (2)$$

c. *Intercept of tangent line:*

on x-axis: $\quad x - \frac{y}{y'} \quad (3)$

on y-axis: $\quad y - xy' \quad (4)$

(3) and (4) are obtained from (1) by putting $Y = 0$ and $X = 0$ respectively.

d. *Intercept of normal line:*

on x-axis: $\quad x + yy' \quad (5)$

on y-axis: $\quad y + \frac{x}{y'} \quad (6)$

(5) and (6) obtained from (2) by putting $Y = 0$ and $X = 0$ respectively.

e *Length of tangent line from P*:

to x-axis: $$\frac{y\sqrt{1+y'^2}}{y'} \quad (7)$$

to y-axis: $$x\sqrt{1+y'^2} \quad (8)$$

f. *Length of normal line from* P:

to x-axis: $$y\sqrt{1+y'^2} \quad (9)$$

to y-axis: $$\frac{x\sqrt{1+y'^2}}{y'} \quad (10)$$

g. *Length of subtangent AD*: $\frac{y}{y'}$ (11)

h. *Length of subnormal DB*: yy' (12)

Worked Out Examples

Example 1: The slope at any point (x, y) of a curve is $1 + y/x$. If the curve passes through $(1, 1)$ find its equation.

Solution:

$$\frac{dy}{dx} = 1 + \frac{y}{x}$$

This is a linear equation

$$\frac{dy}{dx} - \frac{1}{x}y = 1$$

with $$\text{I.F.} = e^{\int -\frac{dx}{x}} = e^{-\ln x} = \frac{1}{x}$$

The general solution is

$$y \cdot \frac{1}{x} = \int \frac{1}{x} \cdot 1 \cdot dx + c = \ln x + c$$

$$y = cx + x \ln x$$

Since it passes through $(1, 1)$, we put $x = 1, y = 1$ then

$$1 = c \cdot 1 + 1 \cdot 0 \qquad \therefore\ c = 1$$

Thus the required equation of the curve is

$$y = x(1 + \ln x)$$

Example 2: The tangent line to a curve at any point (x, y) on it has its intercept on the x-axis always equal to $\frac{x}{2}$. Find the equation of the curve if it passes through $(1, 2)$.

Solution: Let (X, Y) be any point on the normal line to the curve at a point (x, y) on the curve. Then the equation of the tangent line, having slope y' is

$$Y - y = y'(X - x)$$

The x-intercept: X is obtained by putting $Y = 0$.

$$X = x - \frac{y}{y'}$$

It is given that this x-intercept $= \frac{x}{2}$

Thus $$x - \frac{y}{y'} = \frac{x}{2}$$

or $$\frac{y}{y'} = \frac{x}{2}$$

i.e., $$y' = \frac{2y}{x}$$

Integrating $y = cx^2$.
Since it passes through $(1, 2)$

$$2 = c \cdot 1 \qquad \therefore c = 2$$

The required curve is

$$y = 2x^2$$

Example 3: Determine the equation of the tangent, equation of the normal, lengths of the tangent, normal, subtangent and subnormal to the curve $y = 3x^2$ at the point $P(-1, 3)$.

Solution:

$$y' = 6x, \quad y'|_P = -6, \; y|_P = 3$$

Equation of tangent:

$$Y - 3 = -6(X + 1) \quad \text{i.e.,} \quad Y + 6X + 3 = 0$$

Equation of normal:

$$Y - 3 = \frac{1}{6}(X + 1) \quad \text{i.e.,} \quad 6Y - X - 4 = 0$$

Length of tangent:

$$\left| y\sqrt{1 + \left(\frac{dx}{dy}\right)^2} \right| = 3\sqrt{1 + \frac{1}{36}} = \sqrt{\frac{37}{4}}$$

Length of normal:

$$\left| y\sqrt{1+\left(\frac{dy}{dx}\right)^2} \right| = 3\sqrt{1+36} = 3\sqrt{37}$$

Length of subtangent:

$$\left| y\frac{dx}{dy} \right| = \left| 3 \cdot \frac{1}{-6} \right| = -\frac{1}{2}$$

Length of subnormal:

$$\left| y\frac{dy}{dx} \right| = \left| 3 \cdot (-6) \right| = 18$$

Example 4: For the curve $y = a \ln (x^2 - a^2)$, show that the sum of the (lengths of) tangent and subtangent varies as the product of the coordinates of the point.

Solution:

$$y' = \frac{2ax}{x^2 - a^2}, \quad 1 + y'^2 = \left(\frac{x^2+a^2}{x^2-a^2}\right)^2$$

Length of tangent:

$$T = \frac{y\sqrt{1+y'^2}}{y'} = a \cdot \ln (x^2 - a^2) \cdot \frac{x^2+a^2}{x^2-a^2} \cdot \frac{x^2-a^2}{2ax}$$

$$= \frac{(x^2+a^2)}{2x} \ln (x^2 - a^2)$$

Length of subtangent:

$$S.T. = \frac{y}{y'} = a \ln (x^2 - a^2) \cdot \frac{(x^2 - a^2)}{2ax}$$

Thus

$$T + S.T. = \frac{(x^2+a^2)}{2x} \ln (x^2 - a^2) + \frac{x^2 - a^2}{2x} \ln (x^2 - a^2)$$

$$= x \cdot \ln (x^2 - a^2) = x \cdot \frac{y}{a} = \frac{xy}{a}$$

$=$ product of the coordinates.

Example 5: Determine the curve in which the length of the subnormal is proportional to the square of the ordinate.

Solution: Length of subnormal $= y\frac{dy}{dx} \propto y^2$

$$y\frac{dy}{dx} = a\,y^2$$

where a is proportionality constant.

Solving
$$\frac{dy}{y} = a\,dx$$
$$\ln y = ax + c_1$$
$$y = c\,e^{ax}$$

Example 6: Find the equation of the curve in which the portion of the tangent included between the co-ordinate axes is divided in the ratio $m : n$ at the point of contact. Hence or otherwise show that when the point of contact is the mid point, then the curve is a rectangular hyperbola (see Fig. 8.6).

Solution: Let $V(X, Y)$ be any point on the tangent QPR. Then the equation of the tangent QPR to the curve SPT at the point of contact $P(x, y)$ is

$$Y - y = \frac{dy}{dx}(X - x)$$

x-intercept of the tangent (obtained by putting $Y=0$) is $x - \frac{y}{y'}$

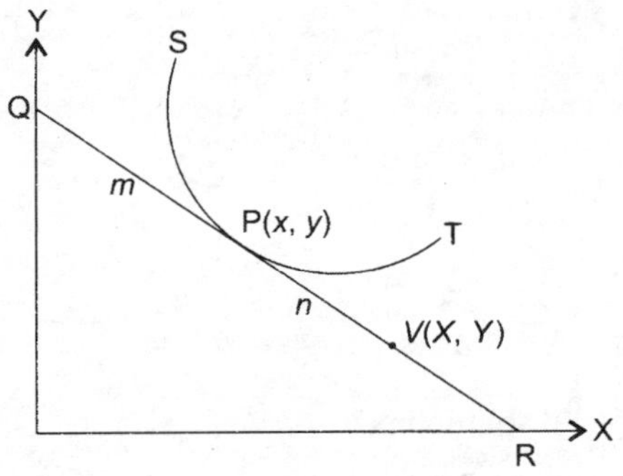

Fig. 8.6

Similarly y-intercept is $y + xy'$
Thus $Q(0, y + xy')$ and $R(x - \frac{y}{y'}, 0)$.
Since P divide QR in the ratio $m : n$, the abscissa x of the point of contact P is

$$x = \frac{m\left(x - \frac{y}{y'}\right) + n \cdot 0}{m + n}$$

Rewriting $\quad y' = -\dfrac{my}{nx}$

which is a separable equation.

Integrating

$$n\frac{dy}{y} = -m\frac{dx}{x}$$

we have

$$x^m y^n = \text{constant}.$$

Special Case: When p is the midpoint $m = n$ so that the curve is

$$xy = c$$

which is a rectangular hyperbola.

EXERCISE

1. Find the curve passing through (2, 1) and having the property that at any point (x, y) of the curve, the intercept of the tangent on the y-axis is equal to $2xy^2$.

Ans. $x - x^2y + 2y = 0$

2. Find the equation of a curve through (2, 1) and having y intercept of its tangent line at any point is always equal to the slope at that point.

Ans. $3y = x + 1$

3. Find the equation of the tangent, normal and lengths of tangent, normal, subtangent and subnormal to the curve $y = \cos 2x$ at the point $P(0, 1)$.

Ans. Equation of tangent: $y = 1$
Equation of normal: $x = 0$
Length of tangent: ∞
Length of normal: 1
Length of subtangent: ∞
Length of subnormal: 0

4. Show that the subnormal at any point of the curve

$$y^2x^2 = a^2(x^2 - a^2)$$

varies inversely as the cube of its abscissa.

5. Prove that the subtangent at any point of the curve

$$x^m y^n = a^{m+n}$$

varies as the abscissa.

6. For the exponential curve

$$y = ae^{\frac{x}{b}}$$

show that the subtangent at any point is constant b.

7. Find the equation of the curve for which the subnormal at any point has a constant value b.

Ans. $y^2 = 2bx + c$

8. Find the shape of a reflector such that the light coming from a fixed source is reflected in parallel rays.

Ans. parabolic reflector: $y^2 = 2cx + c^2$

9. Determine the curve which passes through $(1, e)$ and is such that at each point of the curve the subtangent is proportional to the square of the abscissa.

Ans. $k \ln y = -\frac{1}{x} + k + 1$
(k is the proportionality constant)

10. Prove that a curve, for which the slope of the tangent at any point of the curve is proportional to the abscissa of the point of tangency, is a parabola.

8.16 ORTHOGONAL* TRAJECTORIES** OF CURVES

In Cartesian and Polar Coordinates

Given one (first) family of curves, if there exists another (second) family of curves such that each curve of the first family cuts each curve of the second family at right angles,*** then the first family is said to be orthogonal trajectories of the second family and vice versa. In other words each curve in either family is orthogonal (i.e., perpendicular) to every curve in the other family. In such a case the two families are said to be mutually orthogonal and each family is said to be the orthogonal trajectories of the other family.

*Greek: orthogonal (right angle, 90° or perpendicular).
**Latin: trajectories (cut across).
***Two curves cut at right angles if the angle between their corresponding tangents at the point of intersection is 90°.

Classical examples of orthogonal trajectories are:

i. Meridians and parallels on world globe.
ii. Curves of steepest descent and contour lines on a map.
iii. Curves of electric force and equipotential lines (constant voltage).
iv. Stream lines and equipotential lines (of constant velocity potential).
v. Lines of heat flow and isothermal curves.

Case I: Rectangular coordinates: method of finding orthogonal trajectories of a family of curves

$$F(x, y, c) = 0 \tag{1}$$

Step I. Obtain the differential equation of the family of curves (1) by eliminating the arbitrary constant c, resulting in

$$\frac{dy}{dx} = f(x, y) \tag{2}$$

Step II. Consider the differential equation

$$\frac{dy}{dx} = -\frac{1}{f(x, y)} \tag{3}$$

which will be the differential equation of the family of orthogonal trajectories (since the slope curves of first family is $f(x, y)$, while the slope of curves second family is $-\frac{1}{f(x,y)}$, so that their product $f \cdot \left(-\frac{1}{f}\right) = -1$ i.e., the curves of first family are at right angles to the curves of the second family).

Step III. Solving Equation (3) yields

$$G(x, y, d) = 0 \tag{4}$$

The family of curves (4) is the required orthogonal trajectories (O.T.) of the family of curves (1), with d as the parameter.

Case II: Polar coordinates Consider a curve c whose equation is expressed in polar coordinates

$$\tan\psi = r\frac{d\theta}{dr} \tag{5}$$

where ψ is the angle measured positive in counter clockwise direction from radius vector to the tangent line at any point P (see Fig. 8.7).

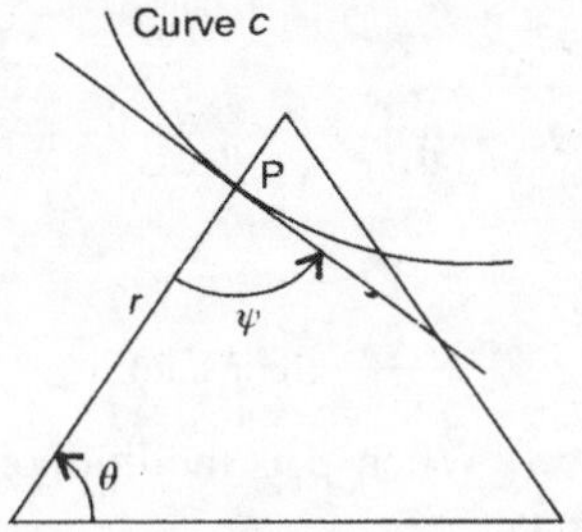

Fig. 8.7

Orthogonality The two curves c_1 and c_2 are said to intersect orthogonally at the point P (Fig. 8.8) if

$$\psi_2 = \psi_1 + \frac{\pi}{2} \tag{6}$$

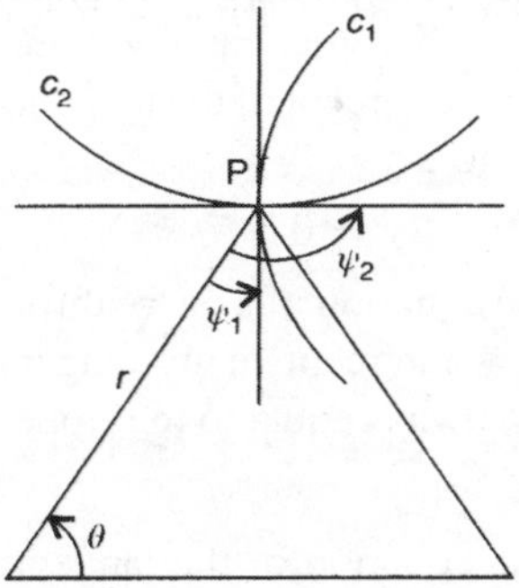

Fig. 8.8

so that

$$\tan\psi_2 = \tan\left(\psi_1 + \frac{\pi}{2}\right) = -\cot\psi_1 = -\frac{1}{\tan\psi_1} \tag{7}$$

i.e., $\tan\psi_2 \cdot \tan\psi_1 = -1$

From (5) we observe that at the point of intersection P, the value of the product $r\frac{d\theta}{dr}$ for one curve c_1 must be the negative reciprocal (see (7)) of value of that product for the other curve i.e., if (5) is the differential equation corresponding to c_1 then the differential equation corresponding to its orthogonal trajectory c_2 is given by $\tan\psi = -\frac{1}{r}\frac{dr}{d\theta}$.

Method of obtaining orthogonal trajectories in polar coordinates

Step I. Suppose the first family of curves has differential equation

$$P(r, \theta)dr + Q(r, \theta)d\theta = 0 \tag{8}$$

Then

$$\frac{d\theta}{dr} = -\frac{P}{Q}$$

$$r\frac{d\theta}{dr} = -r\frac{P}{Q} \quad (9)$$

Step II. The differential equation corresponding to the orthogonal trajectories (by (7)) is

$$r\frac{d\theta}{dr} = \frac{Q}{rP}$$

i.e., $$Q\,dr - r^2 P\,d\theta = 0 \quad (10)$$

Step III. Solve (10). The solution of (10) is the required orthogonal trajectories.

Self-orthogonal A given family of curves is said to be "self-orthogonal" if its family of orthogonal trajectories is the same as the given family (see Worked Out Example 6 and Exercise Examples 18, 19, 20).

Worked Out Examples

O.T.: In rectangular coordinates

Find the orthogonal trajectories (O.T.) of each of the following one-parameter family of curves:

Example 1: $xy = c$

Solution:

D.E. $y + x\frac{dy}{dx} = 0$

$\therefore \quad \frac{dy}{dx} = \frac{-y}{x}$

D.E. corresponding to O.T. is

$$\frac{dy}{dx} = \frac{x}{y}$$

Solving, $y^2 - x^2 = 2c$ is the required orthogonal trajectories.

Example 2: $e^x + e^{-y} = c$.

Solution:

D.E. $e^x - e^{-y} \cdot y' = 0$ or $y' = e^{x+y}$.

D.E. corresponding to O.T. is

$$y' = -e^{-(x+y)}$$

Solving

$$\frac{dy}{dx} = \frac{-e^{-x}}{e^y}$$

$$e^y dy + e^{-x} dx = 0$$

$$e^y - e^{-x} = k$$

Example 3: $y^2 = cx$.

Solution:

D.E. $2yy' = c$

Eliminating c, we get

$$2yy' = c = \frac{y^2}{x}$$

$$y' = \frac{y}{2x}$$

D.E. corresponding to O.T. is

$$y' = -\frac{2x}{y}$$

Solving $\frac{y^2}{2} + x^2 = c$.

Example 4: Show that family of curves $x^2 + 4y^2 = c_1$ and $y = c_2 x^4$ are (mutually) orthogonal (to each other).

Solution: Slope of tangent of any curve of the first family of curves is obtained by differentiating it w.r.t. x.

i.e., $$2x + 8yy' = 0$$

or $$y_1' = -\frac{2x}{8y} = -\frac{x}{4y} \quad (*)$$

Similarly for the second family by eliminating c_2

$$\frac{dy}{dx} = 4x^3 c_2 = 4x^3 \cdot \frac{y}{x^4} = \frac{4y}{x}$$

i.e., $$y_2' = \frac{4y}{x} \quad (**)$$

The given two families are orthogonal if the product of their slopes is -1. From (*) and (**)

$$y_1' \cdot y_2' = \left(\frac{-x}{4y}\right)\left(\frac{4y}{x}\right) = -1$$

Hence the result.

Example 5: Find particular member of orthogonal trajectories of $x^2 + cy^2 = 1$ passing through the point (2, 1).

Solution:

D.E. $2x + 2cyy' = 0$

Eliminating c

$$x + yy'c = 0, x + yy' \cdot \left(\frac{1 - x^2}{y^2}\right) = 0$$

i.e.. $$y' = \frac{xy}{x^2 - 1}$$

D.E. corresponding to O.T.

$$y' = \frac{1 - x^2}{xy}$$

Solving

$$y\,dy = \frac{1 - x^2}{x}dx = \frac{dx}{x} - x\,dx$$

$$\frac{y^2}{2} + \frac{x^2}{2} = \ln x + c_1$$

$$x^2 = c_2 e^{x^2 + y^2} \qquad (*)$$

(*) is the required O.T.

To find the particular member of this O.T.

Put $x = 2$ when $y = 1$ in (*)

$$4 = c_2 e^5 \qquad \therefore \quad c_2 = 4e^{-5}$$

Thus the particular curve of O.T. passing through the point (2, 1) is

$$x^2 = 4e^{-5}e^{x^2 + y^2}$$

Example 6: Show that the family of parabolas $y^2 = 4cx + 4c^2$ is "self-orthogonal".

Solution:

D.E. $2yy' = 4c + 0$

Substituting $c = \frac{yy'}{2}$ in given equation, we get

$$y^2 = 4x \cdot \left(\frac{yy'}{2}\right) + 4\left(\frac{yy'}{2}\right)^2$$

$$y^2 = 2xyy' + y^2y'^2 \qquad (*)$$

Put $p = y'$ so that

$$y^2 = 2xyp + y^2p^2 \qquad (**)$$

This is the D.E. of the given family of parabolas.

In order to get D.E. corresponding to the O.T. replace y' by $-\frac{1}{p}$ in (*). Then

$$y^2 = 2xy\left(-\frac{1}{p}\right) + y^2\left(-\frac{1}{p}\right)^2$$

$$p^2y^2 = 2xy(-p) + y^2$$

Rewriting

$$y^2 = 2xyp + p^2y^2$$

which is same as equation (**). Thus D.E. (*) is D.E. for the given family and its orthogonal trajectories. Hence the given family is "self-orthogonal".

EXERCISE

Rectangular coordinates

Find the orthogonal trajectories (O.T.) of each of the following family of curves. [Here c, a, b, k are all constants.]

1. $x - 4y = c$

Ans. $4x + y = k$

2. $x^2 + y^2 = c^2$

Ans. $y = kx$

3. $x^2 - y^2 = c$

Ans. $xy = k$

4. $y^2 = cx^3$

Ans. $(x + 1)^2 + y^2 = a^2$

5. $y = c\,(\sec x + \tan x)$

Ans. $y^2 = 2(k - \sin x)$

6. $x^2 - y^2 = cx$

Ans. $y(y^2 + 3x^2) = k$

7. $y^2 = \frac{x^3}{a - x}$

Ans. $(x^2 + y^2)^2 = b(2x^2 + y^2)$

8. $(a + x)y^2 = x^2(3a - x)$

Ans. $(x^2 + y^2)^5 = cy^3(5x^2 + y^2)$

9. $y = cx^2$

Ans. $\frac{x^2}{2} + y^2 = c^*$

10. Circles through origin with centres on the x-axis.

Ans. Circles through origin with centres on y-axis.

11. Family of parabolas through origin and focii on y-axis.

Ans. Ellipses with centres at origin and foci on x-axis.

12. The family of ellipses having centre at the origin, a focus at the point $(c, 0)$ and semi-major axis of length $2c$.

Ans. $y = cx^{\frac{4}{3}}$

13. Given $x^2 + 3y^2 = cy$, find that member of the orthogonal trajectories which passes through the point $(1, 2)$

Ans. $y^2 = x^2(3x + 1)$

14. Given $y = ce^{-2x} + 3x$, find that member of the O.T. which passes through point $(0, 3)$

Ans. $9x - 3y + 5 = -4e^{6(3-y)}$

15. Find constant 'e' such that $y^3 = c_1 x$ and $x^2 + ey^2 = c_2$ are orthogonal to each other.

Ans. $e = \frac{1}{3}$

16. Find the value of constant d such that the parabolas $y = c_1 x^2 + d$ are the orthogonal trajectories of the family of ellipses $x^2 + 2y^2 - y = c_2$.

Ans. $d = \frac{1}{4}$

17. Show that the family of parabolas $y^2 = 2cx + c^2$ is "self-orthogonal".

18. Find orthogonal trajectories of

$$\frac{x^2}{a} + \frac{y^2}{a-b} = 1$$

Here a is an arbitrary constant.

Ans. $a \ln x = \frac{x^2}{2} + \frac{y^2}{2} + c$

19. Find the orthogonal trajectories of a system of confocal and coaxial parabolas.

Ans. The family of confocal and coaxial parabolas by having x-axis as their axis is given by $y^2 = 4a(x + a)$ is "self-orthogonal".

20. Show that the family of confocal conics

$$\frac{x^2}{a^2 + c} + \frac{y^2}{b^2 + c} = 1$$

is "self-orthogonal". Here a and b are given constants.

21. Find the O.T. of $x^p + cy^p = 1, p = \text{constant}$.

Ans. $y^2 = \frac{2x^{2-p}}{2-p} - x^2 + c$, if $p \neq 2$

$c_1 x^2 = e^{x^2+y^2}$ if $p = 2$

22. Show that the two families of one parameter family of curves $u(x, y) = c_1$ and $v(x, y) = c_2$ are mutually orthogonal provided they satisfy the (Cauchy-Riemann) equations $u_x = v_y$ and $u_y = -v_x$.

WORKED OUT EXAMPLES

O.T.: In polar coordinates

Find the orthogonal trajectories (O.T.) of each of the following family of curves:

Example 1: $r = a\cos^2\theta$

Solution:

Differentiating $r = a\cos^2\theta$ w.r.t. r

$$dr = 2a\cos\theta(-\sin\theta)d\theta$$

Eliminating $a = \frac{r}{\cos^2\theta}$, we get

$$dr = -2 \cdot \frac{r}{\cos^2\theta} \cdot \cos\theta \cdot \sin\theta \cdot d\theta$$

$$r\frac{d\theta}{dr} = -\frac{1}{2\tan\theta}$$

D.E. corresponding to O.T. is

$$r\frac{d\theta}{dr} = 2\tan\theta$$

Solving

$$\frac{d\theta}{\tan\theta} = 2\frac{dr}{r}$$

$$\ln\ \sin\theta = 2\ln\ r + c$$

So the O.T. is given by

$$r^2 = b\sin\theta.$$

Example 2: $r^2 = a \sin 2\theta$

Solution:

Differentiating w.r.t. r

$$2r\,dr = 2a \cos 2\theta\, d\theta$$

Eliminating $a = \frac{r^2}{\sin 2\theta}$, we get

$$r\,dr = \frac{r^2}{\sin 2\theta} \cdot \cos 2\theta\, d\theta$$

$$r\frac{d\theta}{dr} = \tan 2\theta$$

D.E. of O.T. is

$$r\frac{d\theta}{dr} = -\cot 2\theta$$

Solving

$$-\frac{\sin 2\theta}{\cos 2\theta}d\theta = \frac{dr}{r}$$

Integrating $\ln \cos 2\theta = 2 \ln r + c$

O.T. $\therefore$ $r^2 = b \cos 2\theta$

EXERCISE

Polar coordinates

Find the orthogonal trajectories (O.T.) of each of the following family of curves:

1. Cardioids: $r = a(1 + \cos\theta)$

Ans. $r = b(1 - \cos\theta)$

Note: Family of cardioids is self-orthogonal.

2. Confocal and coaxial parabolas (self-orthogonal)

$$r = \frac{2a}{(1+\cos\theta)}$$

Ans. $r = \frac{2b}{1-\cos\theta}$

3. $\left(r^2 + \frac{k^2}{r}\right)\cos\theta = d$

d being a parameter

Ans. $(r^2 - k^2)\sin\theta = cr$

4. $r = 2a(\sin\theta + \cos\theta)$

Ans. $r = 2b(\sin\theta - \cos\theta)$

5. $r = 4a \sec\theta \cdot \tan\theta$

Ans. $r^2(1 + \sin^2\theta) = b^2$

6. $r = a(1 + \sin^2\theta)$

Ans. $r^2 = b\cos\theta \cdot \cot\theta$

7. Cissoids $r = a \sin\theta \tan\theta$

Ans. $r^2 = b(1 + \cos^2\theta)$

8. $r = \frac{k}{1+2\cos\theta}$

Ans. $r^2 \sin^3\theta = b(1 + \cos\theta)$

9. Stophoids $r = a(\sec\theta + \tan\theta)$

Ans. $r = b\,e^{-\sin\theta}$

8.17 LAW OF NATURAL GROWTH

Let $x(t)$ be the population at any time t. Assume that population grows at a rate directly proportional to the amount of population present at that time. Then the D.E. governing this phenomena is the first order, first degree linear equation

$$\frac{dx}{dt} = kx$$

where k is the proportionality constant. Here $k > 0$ since this is a growth phenomena. The solution of the D.E. is

$$x(t) = c\,e^{kt}$$

where c is constant of integration. Here, c, k are determined from the two given (initial) conditions.

WORKED OUT EXAMPLES

Example: A bacterial population B is known to have a rate of growth $\propto$ to B itself. If between noon and 2 PM the population triples, at what time, no controls being exerted, should B become 100 times what it was at noon.

Solution: Equation is

$$\frac{dB}{dt} = kB$$

whose solution is $B(t) = ce^{kt}$. Let B_0 be the initial population at $t = 0$, using this condition

$$B_0 = ce^0 \quad \therefore \quad c = B_0$$

Thus
$$B = B_0 e^{kt}$$

Since population triples i.e., becomes $3B_0$ between noon and 2 PM i.e., in two hours, we use this condition to find k

$$3B_0 = B_0 e^{k \cdot 2}$$

Thus
$$k = \frac{1}{2} \ln 3 = 0.54930$$

Hence the population rule is

$$B(t) = B_0 e^{0.54930 t}$$

To find the time at which population becomes 100 times the original i.e., 100 B_0 we put $B = 100\,B_0$ in the above equation and solve for t.

$$100\,B_0 = B_0 e^{0.54930\,t}$$

Solving

$$t = \frac{\ln 100}{0 \cdot 54930} = 8.3837015$$

i.e., at 8.383 PM population becomes 100 times the original population.

EXERCISE

1. In a certain culture of bacteria the rate of increase is proportional to the number present. (a) If it is found that the number doubles in 4 hours, how many may be expected at the end of 12 hours. (b) If there are 10^4 at the end of 3 hours and $4 \cdot 10^4$ at the end of 5 hours, how many were in the beginning.

Ans. **a.** 8 times the original number

b. $\frac{10^4}{8}$ bacteria at the beginning

2. If the population of a country doubles in 50 years, in how many years will it treble under the assumption that the rate of increase is proportional to the number of inhabitants.

Ans. 79 years

3. In a certain bacteria culture the rate of increase in the number of bacteria is $\propto$ to the number present. (a) If the number triples in 5 hrs how many will be present in 10 hrs, (b) when will the number present be 10 times the number initially present.

Ans. **a.** 9 times the original number

b. 10.48 hours

4. The number N of bacteria in a culture grew at a rate proportional to N. The value of N was initially 100 and increased to 332 in one hour, what would be the value of N after $1\frac{1}{2}$ hours.

Ans. $604.9 \approx 605$

5. In a culture of yeast, the active ferment doubles itself in 3 hours. Assuming that the quantity increases at a rate proportional to itself, determine the number of times it multiplies itself in 15 hours.

Ans. It multiplies itself 32 times.

6. Find the time required for a sum of money to double itself at 5% per annum compounded continuously.

Ans. 13.9 years.

8.18 LAW OF NATURAL DECAY

The DE

$$\frac{dm}{dt} = -km, \quad k > 0$$

describes the decay phenomena where it is assumed that the material $m(t)$ at any time t decays at a rate which is proportional to the amount present. The solution is

$$m(t) = c\,e^{-kt}$$

If initially at, $t = 0$, m_0 is the amount present then

$$m(t) = m_0 e^{-kt}$$

WORKED OUT EXAMPLES

Example: Radium decomposes at a rate $\propto$ the quantity of radium present. Suppose that it is found that in 25 years approximately 1.1% of a certain quantity of radium has decomposed. Determine approximately how long will it take for one-half of the original amount of radium to decompose.

Solution: Let m be the amount of radioactive material radium present at time t. Let m_0 be initial (original) amount of radium at $t = 0$. By the decay rule

$$\frac{dm}{dt} = -km$$

whose solution is

$$m(t) = ce^{-kt} \quad (1)$$

using the initial condition $m(0) = m_0$ in (1), we get

$$m_0 = m(0) = ce^0$$

$$\therefore \quad c = m_0$$

Thus

$$m = m_0 e^{-kt}$$

Since 1.1% of original mass of radium decays in 25 years, the amount of mass of radium present when $t = 25$ years is $\left(1 - \frac{1.1}{100}\right) m_0$. Using this in (1)

$$\left(1 - \frac{1.1}{100}\right) m_0 = m(25) = m_0 e^{-k(25)}$$

Thus

$$k = -\frac{1}{25} \ln (1 - 0.011) = 0.000443$$

In order to find the time taken for half the radium to disintegrate (to decay) put $m(t = t^*) = \frac{1}{2} m_0$ in (1).

$$\frac{1}{2} m_0 = m_0 e^{-kt^*}$$

Solving for $t = \frac{\ln 2}{k} = 1564.66 \approx 1565$ years.

EXERCISE

1. Radium decomposes at a rate $\propto$ to the amount present. If a fraction p of the original amount disappears in 1 year, how much will remain at the end of 21 years ?

Ans. $\left(1 - \frac{1}{p}\right)^{21}$ times the original amount.

2. Find the half-life of uranium, which disintegrates at a rate $\propto$ to the amount present at any instant given that m_1 and m_2 grams of uranium are present at t_1 and t_2 respectively.

Ans. $T = \frac{(t_2 - t_1) \log 2}{\log (m_1/m_2)}$

3. If half-life of uranium is 1500 years. (a) Find percentage of original amount that will remain after 4500 years, (b) Find in how many years will only $\frac{1}{10}$ of the original amount remain?

Ans. $m = m_0 e^{-0.000461t}$; (a) $\frac{1}{8}$th or 12.5% of the original amount will remain after 4500 years; (b) $t = 4985$ years.

4. Radium disintegrates at a rate $\propto$ to its mass. When mass is 10 mgm, the rate of disintegration is 0.051 mgm per day. How long will it take for the mass to be reduced to 10 to 5 mgm?

Ans. 135.9 days.

8.19 NEWTON'S* LAW OF COOLING

Physical experiments show that the time (t) rate of change $\frac{dT}{dt}$ of the temperature T of a body is proportional to the difference between T and the temperature T_A of the (ambient) surrounding medium. This is known as Newton's law of cooling. Taking the unknown proportionality constant as k, the equation governing the Newton's law of cooling is a first order first degree linear separable differential equation.

$$\frac{dT}{dt} = -k(T - T_A)$$

Here proportionality constant is taken as $-k$ so that $k > 0$. Separating the variables

$$\frac{dT}{(T - T_A)} = -k\,dt$$

Integrating

$$\ln (T - T_A) = -kt + c_0$$

Solution is

*Sir Isaac Newton (1642–1727), English physicist and mathematician.

$$T - T_A = e^{-kt+c_0} = c\,e^{-kt}$$

where $c = e^{c_0}$. Thus the solution to Newton's law of cooling is

$$T(t) = T_A + c\,e^{-kt} \qquad (1)$$

where the arbitrary constant c will be found by using the initial condition $T(t = 0) = T_0$.

Method of Solving the Problem of Newton's Law of Cooling

I. Identify T_A, the temperature of the surrounding medium, so that the general solution is given by (1).

II. Use two conditions given to determine the constant of integration c and unknown proportionality constant k.

III. Substituting c and k, obtained from step II, in (1) (a) the value of T for a given time t or (b) the value of time t for a given temperature T can be determined from (1).

WORKED OUT EXAMPLES

Example 1: A body of temperature 80°F is placed in a room of constant temperature 50°F at time $t = 0$. At the end of 5 minutes the body has cooled to a temperature of 70°F. (a) Find the temperature of the body at the end of 10 minutes. (b) When will the temperature of the body be 60°F? (c) After how many minutes will the temperature of the body be within 1°F of the constant 50°F temperature of the room?

Solution: Let T be the temperature of the body. Then the $T(t) = 50 + ce^{-kt}$ since $T_A = 50$. Apply the condition $T(0) = 80$, to find c:

$$80 = 50 + ce^0 \qquad \therefore\ c = 30$$

Thus $\quad T(t) = 50 + 30\,e^{-kt}$

use the condition $T(5) = 70$ to determine k:

$$70 = 50 + 30\,e^{-k5}$$

so that

$$k = \frac{1}{5}\ln\left(\frac{3}{2}\right)$$

i.e., $\quad k = 0.08109$

Thus the required solution which gives the temperature of the body at any time t is

$$T(t) = 50 + 30\,e^{-0.08109t}$$

a. $T(10) = 50 + 30\,e^{-0.08109(10)} \simeq 63.33$°F

b. $60 = 50 + 30\,e^{-0.08109t}$

Solving

$$t = 5\left(\frac{\ln\frac{1}{3}}{\ln\frac{2}{3}}\right) \approx 13.55 \text{ mts}$$

c. $51 = 50 + 30\,e^{-0.08109t}$

Solving

$$t = 5\left(\frac{\ln\frac{1}{30}}{\ln\frac{2}{3}}\right) \approx 41.91 \text{ mts}$$

Example 2: If a substance cools from 370 k to 330 k in 10 mts, when the temperature of the surrounding air is 290 k, find the temperature of the substance after 40 mts.

Solution: Here $T_A = 290$ so that the solution is

$$T(t) = 290 + c\,e^{-kt}$$

Use condition $T(0) = 370$ to find c

$$370 = 290 + c \cdot e^0$$

$$\therefore \quad c = 80$$

Thus

$$T(t) = 290 + 80\,e^{-kt}$$

Use condition $T(10) = 330$ to find k

Thus

$$330 = 290 + 80\,e^{-k\cdot 10}$$

so that

$$k = \frac{\ln 2}{10} = 0.069314718$$

The required solution is

$$T(t) = 290 + 80\,e^{-0.0693t}$$

Putting $t = 40$ mts in the above solution

$$T(40) = 290 + 80 \cdot e^{-0.0693[40]}$$
$$= 295$$

EXERCISE

1. Water at temperature 100 °C cools in 10 min to 80°C in a room of temperature 25°C. (a) Find the temperature of water after 20 min. when is the temperature (b) 40°C (c) 26°C ?

Ans. (a) 65.3°C; (b) 52 min, (c) 139 min.

2. Water at temperature 10°C takes 5 min to warm up to 20°C in a room at temperature 40°C. (a) Find the temperature after 20 min; after $\frac{1}{2}$ hr, (b) When will the temperature be 25°C ?

Ans. (a) 34.1°C, 37.4°C; (b) 8.5 min.

3. A copper ball is heated to a temperature of 100°C. Then at time $t = 0$ it is placed in water which is maintained at a temperature of 30°C. At the end of 3 mts the temperature of the ball is reduced to 70°C. Find the time at which the temperature of the ball drops to 31°C.

Ans. $22.78 \approx 23$ mts.

4. A thermometer reading 18°F is brought into a room the temperature of which is 70°F. One minute later the thermometer reading is 31°F. Find the temperature reading 5 minutes after the thermometer is first brought into the room.

Ans. $57.80235 \approx 58$°F.

5. A body is heated to 110°C and placed in air at 10°C. After 1 hour its temperature is 60°. How much additional time is required for it to cool to 30°C ?

Ans. $\frac{\log 5}{\log 2} - 1 = 2.3223 - 1 = 1.3223$ hours.

8.20 VELOCITY OF ESCAPE FROM EARTH

Consider the problem of determining the minimum velocity with which a body (projectile) be projected vertically upwards in the radial direction from the earth so that the body will escape from earth and will not return to earth. Here it is assumed that retardation effect due to air resistance and gravitational pull of other celestial bodies is neglected.

Let m be the mass of the body and M be the mass of earth. Let r be the distance between the (centre of the) earth and (the centre of gravity of) the body. Let R be the radius of earth.

According to Newton's second law and the universal law of gravitation and ignoring all forces (frictional, magnetic or other) except for gravity, the vertical motion of the body is described by the second order differential equation

$$m\frac{d^2r}{dt^2} = -k\frac{M \cdot m}{r^2} \tag{1}$$

where k is the gravitational constant. The minus sign indicates that the acceleration is negative.

Expressing the acceleration in terms of velocity and distance $\frac{d^2r}{dt^2} = \frac{dv}{dt} = \frac{dv}{dr}\frac{dr}{dt} = v\frac{dv}{dr}$, the equation (1) can be written as

$$v\frac{dv}{dr} = -\frac{kM}{r^2}$$

Here v is the velocity of the body. Separating the variables and integrating, we get

$$\frac{v^2}{2} = \frac{kM}{r} + c \tag{2}$$

Suppose v_0 is the launching velocity with which the body leavies the earth's surface then the initial conditions are:

$$\text{for } t = 0,\ r = R,\ \frac{dr}{dt} = v_0 \tag{3}$$

Using (3) in (2), we have

$\frac{v_0^2}{2} = \frac{kM}{k} + c$ or $c = \frac{v_0^2}{2} - \frac{kM}{R}$

Substituting c in (2), we get

$$\frac{v^2}{2} = \frac{kM}{r} + \frac{v_0^2}{2} - \frac{kM}{R}$$

$$\text{or} \quad v^2 = \frac{2kM}{r} + \left(v_0^2 - \frac{2kM}{R}\right) \tag{4}$$

As $r \to \infty$, $\frac{kM}{r} \to 0$, the velocity is always positive if $v^2 > 0$. Thus for any r and m, $v^2 > 0$ provided

$$v_0^2 - \frac{2kM}{R} \geq 0$$

if $v^2 > 0$. Thus for any r and m, $v^2 > 0$ provided

$$v_0^2 - \frac{2kM}{R} \geq 0$$

or $$v_0 \geq \sqrt{\frac{2kM}{R}}$$

Thus the minimum velocity for the body to escape from earth is

$$v_0 = \sqrt{\frac{2kM}{R}}.$$

Since the acceleration due to gravity g is given by

$$g = \frac{kM}{R^2}$$

We have $$M = \frac{gR^2}{k} \qquad (5)$$

Hence the velocity of escape v_0 is rewritten as

$$\boxed{v_0 = \sqrt{\frac{2k}{R} \cdot \frac{gR^2}{k}} = \sqrt{2gR}} \qquad (6)$$

Note that velocity of escape v_0 is independent of m, the mass of the body. It depends only on M and R.

For radius of earth $R = 63.10^7$ cm and $g = 981$ cm/sec^2, the velocity of escape from earth is

$$v_0 = \sqrt{2(981)(63.10^7)} \approx 11.2.10^5 \text{ cm}$$
$$= 11.2 \text{ km/sec}$$

Using (5) in (4) we get

$$\boxed{v^2 = \frac{2gR^2}{r} + v_0^2 - 2gR} \qquad (7)$$

Thus the body projected with an initial (launching) velocity v_0 from the earth's surface in a radial direction will travel with velocity v given by (7). If $v_0^2 < 2gR$ then for some critical value r, the velocity v given by (7) becomes zero, in which case the body will stop and return to earth since the velocity will change from positive to negative.

Cor: If the body is carried by a rocket and is separated from the rocket at a distance of r^* miles from the earth's surface then the velocity of escape

$$v_0 = \sqrt{\frac{2gr^2}{R+r^*}}$$

In this case the differential equation is

$$v\frac{dv}{dr} = -\frac{Mg}{(R+r^*)^2}$$

which on integrating gives

$$v^2 = \frac{2R^2g}{r} + \left(v_0^2 - 2\frac{R^2g}{R+r^*}\right)$$

WORKED OUT EXAMPLES

Example 1: Assuming the radius of moon as $R = 1080$ miles and acceleration of gravity at the surface of the moon as 0.165 g where g is the acceleration of gravity at the surface of the earth, determine the velocity of escape for the moon.

Solution: Velocity of escape $v_0 = \sqrt{2g^*R^*}$. Here $R^* = 1080$ miles, $g^* = .165$ g $= (.165)(6.09) \times 10^{-3}$ so $v_0^2 = 2(.165)(6.09) \times 10^{-3} \times 1080 = 2.170476$ or $v_0 = 1.47325 \approx 1.5$ miles/sec.

Example 2: Determine the velocity of escape for a body which is carried by a rocket and is separated from it at a distance of 1000 miles from the earth's surface.

Solution: The velocity of escape in this case is $v_0 = \sqrt{\frac{2gR^2}{R+r^*}}$. Here R = radius of earth = 3960 miles, $r^* = 1000$ miles, $g = 6.1 \times 10^{-3}$ miles/sec^2. So

$$v_0 = \sqrt{\frac{2(6.1 \times 10^{-3}) \times (3960)^2}{3960 + 1000}}$$
$$= \sqrt{\frac{(1.218)(156816)}{4960}}$$

$v_0 = \sqrt{38.5084} = 6.2055173$ miles/sec^2 (or 9.985 km/sec^2) which is smaller than $v_0 = 6.95$ miles/sec when launched from the surface of earth ($r^* = 0$).

EXERCISE

1. Suppose a body is carried by a rocked and is separated from the rocket at a distance of 200 miles from the earth's surface, determine the minimum velocity at this point, sufficient for escape from the earth.

Ans. $v_0 = 6.7923$ miles/sec (10.913 km/sec)

the acceleration of gravity at the surface of the body and R^* the radius in miles.

(a) Venus : $g^* = 0.85$ g, $R^* = 3800$
(b) Mars: $g^* = 0.38$ g, $R^* = 2100$
(c) Jupiter: $g^* = 2.6$ g, $R^* = 43,000$
(d) Sun: $g^* = 28$ g, $R^* = 432000$

Ans. (a) 6.3 (b) 3.1 (c) 37 (d) 380 miles/sec

Hint: $g = 6.1 \times 10^{-3}$ miles/sec

8.21 SIMPLE ELECTRIC CIRCUITS

Electric current is a flow of charges, measured in *amperes (A)*. Electric current flows due to a difference in the electric potential or *voltage* measured in *volts* (v), just as heat flows from one point to the other due to a temperature difference. An *electric circuit* consists of a source of electric energy (*electromotive force*) and elements such as *resistors*, *inductors* or *voltage* and *capacitors*. A mathematical model of an electric circuit is represented by linear (first or second order) differential equations. To form such an equation, the following relationships are needed:

The voltage drop E_R across a resistor is proportional to the instantaneous current $I(t)$ through it:

$$\boxed{E_R = RI} \quad \text{(Ohm's law)} \qquad (1)$$

Here t is the time and the constant of proportionality R is known as the *resistance* of the resistor, measured in ohms (Ω). Resistor uses (consumes) energy. Resistor is represented by —/\/\/\/—. The voltage drop E_L across an inductor is proportional to the instantaneous time rate of change of the current:

$$E_L = L\frac{dI}{dt} \qquad (2)$$

Here the constant of proportionality L is known as *inductance* of the inductor and is measured in *henrys* (H). An inductor opposes a change in current and is represented by —ooooo—. Inductor causes inertia effect in electricity just as mass in mechanics. The voltage drop E_c across a capacitor is proportional to the instantaneous electric charge Q on the capacitor.

$$E_c = \frac{1}{c}Q \qquad (3)$$

Here c is called the *capacitance* and is measured in farads (F) and is represented by —| |—. A capacitor stores energy. The charge Q is measured in *coulombs*. Since current is the time rate of change of charge,

$$I(t) = \frac{dQ}{dt} \qquad (4)$$

Equation (3) may be written as

$$E_c = \frac{1}{c}\int_{t_0}^{t} I(u)du \qquad (5)$$

To determine the current $I(t)$ in an electric circuit, a differential equation is formed using the Kirchhoff's* voltage law (KVL) which states that *the algebraic sum of all the instantaneous voltage drops around any closed loop is zero or the voltage impressed on a closed loop is equal to the sum of the voltage drops in the rest of the loop.* Consider two simple cases of *series or one-loop* electric circuits.

I. *RL*-circuit

By (1), the voltage drop across the resistor is RI. By (2), the voltage drop across the inductor is $L\frac{dI}{dt}$. Now applying Kirchhoff's law to the RL-circuit, the sum of the two voltage drops must be equal to electromotive force $E(t)$. Thus the current $I(t)$ in the RL-circuit is determined by the first order linear differential equation

$$\boxed{L\frac{dI}{dt} + RI = E(t)}. \qquad (6)$$

Rewriting (6), we have

$$\frac{dI}{dt} + \frac{R}{L}I = \frac{E(t)}{L}$$

which has an integrating factor $e^{\int \frac{R}{L}dt} = e^{\frac{R}{L}t} = e^{\alpha t}$ where $\alpha = \frac{R}{L}$.

*Gustav Robert Kirchhoff (1824-1887), German physicist.

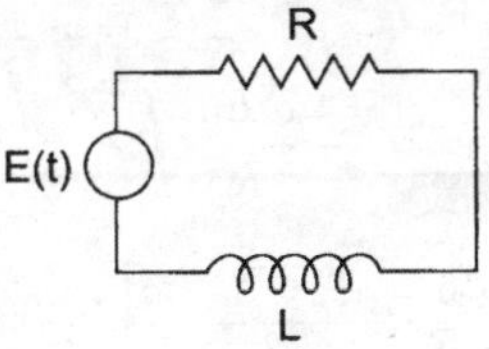

RL-circuit

Fig. 8.9

Then the general solution of (6) is

$$I(t)\cdot e^{\alpha t} = \int \frac{E(t)}{L}\cdot e^{\alpha t}\,dt + c$$

or $$\boxed{I(t) = e^{-\alpha t}\left[\int \frac{E(t)}{L}\cdot e^{\alpha t}\,dt + c\right]} \qquad (7)$$

Case (a): Suppose $E = E_0 =$ constant. Then (7) simplifies to

$$\boxed{I(t) = e^{-\alpha t}\left[\frac{E_0}{L}\cdot\frac{e^{\alpha t}}{\alpha} + c\right] = \frac{E_0}{R} + ce^{-\alpha t}} \qquad (8)$$

As $t \to \infty$, $I(t) \to \frac{E_0}{R}$ = constant. Here $\frac{L}{R} = \frac{1}{\alpha}$ is known as *inductive time constant*.

Case (b): Suppose $E = E_0 \sin \omega t$. Then (7) reduces

$$I(t) = e^{-\alpha t}\left[\frac{E_0}{L}\int e^{\alpha t}\sin\omega t\,dt + c\right]$$

since $\int e^{at}\sin bt\,dt = \frac{e^{at}}{(a^2+b^2)}(a \sin bt - b\cos bt)$,

$$I(t) = ce^{-\frac{R}{L}t} + \frac{E_0}{L\left(\frac{R^2}{L^2}+\omega^2\right)}\left(\frac{R}{L}\sin\omega t - \omega\cdot\cos\omega t\right)$$

or

$$I(t) = ce^{-\frac{R}{L}t} + \frac{E_0}{R^2+\omega^2L^2}(R\sin\omega t - \omega L\cdot\cos\omega t)$$

Using $a\cos x + b\sin x = \sqrt{a^2+b^2}\sin\ (x\pm\theta)$, where $\tan\theta = \frac{\sin\theta}{\cos\theta} = \pm\frac{a}{b}$, the trigonometric terms can be expressed in "phase-angle" form as

$$\boxed{I(t) = ce^{-\frac{R}{L}t} + \frac{E_0}{\sqrt{R^2+\omega^2L^2}}\sin\ (\omega t - \theta)} \qquad (9)$$

where $\theta = \tan^{-1}\frac{\omega L}{R}$. The current I in (9) is expressed as the sum of an exponential and sinusoidal terms. As $t \to \infty$, the first term tends to zero. It is known as the *transient* term. The second sinusoidal term corresponds to the *steady-state* which is free of e^{-t}. The period is $2\pi/\omega$ and phase is θ. The steady-state solution is permanent, periodic and has the same period as that of the applied external force.

Case (c): Suppose $E = E_0\cos\omega t$. Then (7) reduces to

$$I(t) = e^{-\alpha t}\left[\frac{E_0}{L}\int e^{\alpha t}\cos\omega t\,dt + c\right]$$

Since $\int e^{at}\cos bt\,dt = \frac{e^{at}}{(a^2+b^2)}(a\cos bt + b\sin bt)$,

$$I(t) = ce^{-\frac{R}{L}t} + \frac{E_0}{R^2+\omega^2L^2}(R\cos\omega t + \omega\cdot L\cdot\sin\omega t$$

Using $a\cos x + b\sin x = \sqrt{a^2+b^2}\cos\quad (x\pm\theta)$ where

$$\tan\theta = \frac{\sin\theta}{\cos\theta} = \mp\frac{b}{a}$$

we have

$$\boxed{I(t) = ce^{-\frac{R}{L}t} + \frac{E_0}{\sqrt{R^2+\omega^2L^2}}\cos\ (\omega t - \theta)} \qquad (10)$$

where $\theta = \tan^{-1}\frac{\omega L}{R}$.

II. RC-circuit

Using (1), (5) and Kirchhoff's law, we get the integro-differential equation.

$$RI + \frac{1}{c}\int I(t)dt = E(t)$$

which on differentiation reduces to

$$R\frac{dI}{dt} + \frac{1}{c}I = \frac{dE}{dt} \qquad (11)$$

or $$\frac{dI}{dt} + \frac{1}{Rc}I = \frac{1}{R}\frac{dE}{dt}$$

This is a first order linear differential equation with integrating factor $e^{\int \frac{1}{cR} dt} = e^{\frac{t}{cR}}$.

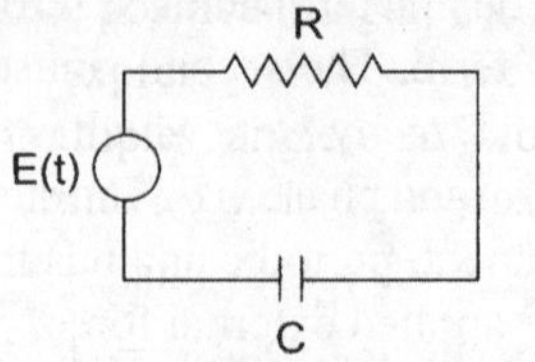

RC-circuit

Fig. 8.10

The general solution of (11) is

$$I(t) \cdot e^{\frac{t}{cR}} = \int \frac{1}{R} \frac{dE}{dt} \cdot e^{\frac{t}{cR}} \cdot dt + c$$

or $$I(t) = e^{-\frac{t}{cR}} \left[\int \frac{1}{R} \frac{dE}{dt} e^{\frac{t}{cR}} dt + c \right] \qquad (12)$$

Case (a): Suppose $E = E_0 = \text{constant}$. Then $\frac{dE}{dt} = 0$. The solution (12) reduces to

$$\boxed{I(t) = ce^{-\frac{t}{cR}}} \qquad (13)$$

Here RC is known as capacitive time constant of the circuit.

Case (b): Suppose $E = E_0 \sin \omega t$. Then

$$\frac{dE}{dt} = \omega E_0 \cos \omega t$$

From (12), we have

$$I(t) = ce^{-\frac{t}{cR}} + \frac{1}{R} \int e^{\frac{t}{cR}} \cdot \omega E_0 \cdot \cos \omega t \, dt$$

$$= ce^{-\frac{t}{cR}} + \frac{\omega E_0}{R} \frac{e^{\frac{t}{cR}}}{\left(\frac{1}{cR}\right)^2 + \omega^2} \left[\frac{1}{cR} \cos \omega t + \omega \sin \omega t \right]$$

$$= ce^{-\frac{t}{cR}} + \frac{\omega E_0 c}{1 + (\omega c R)^2} (\cos \omega t + \omega Rc \cdot \sin \omega t)$$

$$\boxed{I(t) = ce^{-\frac{t}{cR}} + \frac{\omega E_0 c}{1 + (\omega c R)^2} \sin(\omega t - \theta)} \qquad (14)$$

where $\tan \theta = -\frac{1}{\omega Rc}$. The first term which is transient tends to zero as $t \to \infty$ while the second sinusoidal term corresponds to steady-state.

Case (b) Suppose $E = E_0 \cos \omega t$. Then

$$\frac{dE}{dt} = -\omega E_0 \sin \omega t$$

From (12), we have

$$I(t) = ce^{-\frac{t}{cR}} - \frac{\omega E_0}{R} \int e^{\frac{t}{cR}} \cdot \sin \omega t \, dt$$

$$= ce^{-\frac{t}{cR}} - \frac{\omega E_0}{R} \cdot \frac{e^{\frac{t}{cR}}}{\left(\frac{1}{cR}\right)^2 + \omega^2} \left[\frac{1}{cR} \sin \omega t - \omega \cos \omega t \right]$$

$$= ce^{-\frac{t}{cR}} + \frac{\omega E_0 c}{1 + (\omega Rc)^2} (\omega Rc \cos \omega t - \sin \omega t)$$

$$\boxed{I(t) = ce^{-\frac{t}{cR}} + \frac{\omega E_0 c}{\sqrt{1 + (\omega Rc)^2}} \cos(\omega t + \theta)} \qquad (15)$$

where $\tan \theta = -\frac{1}{\omega Rc}$

WORKED OUT EXAMPLES

RL-circuit

Example 1: Find the current at any time $t > 0$ in a circuit having in series a constant electromotive force 40 V, a resistor 10 Ω, and an inductor 0.2 H given that initial current is zero. Find the current when $E(t) = 150 \cos 200\, t$.

Solution: Equation for RL circuit is

$$L \frac{dI}{dt} + RI = E(t)$$

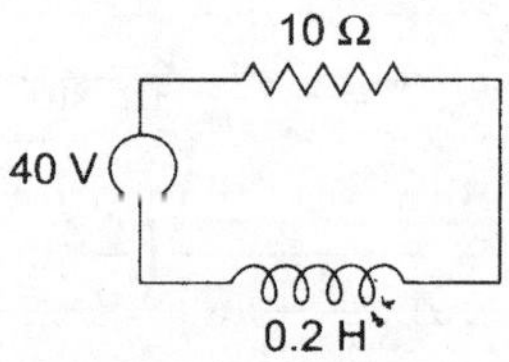

Fig. 8.11

(a) $L = 0.2,\ R = 10,\ E = 40$

$$0.2 \frac{dI}{dt} + 10I = 40$$

$$\frac{dI}{dt} + 50I = 200$$

Its general solution is

$$I(t) \cdot e^{50t} = \int 200 \cdot e^{50t} + c = 200 \cdot \frac{e^{50t}}{50} + c$$

$$I(t) = e^{-50t}[4e^{50t} + c]$$

At $t = 0$, $I = 0$, so $0 = [4 + c]$ or $c = -4$. The current $I(t)$ is given by

$$I(t) = 4(1 - e^{-50t})$$

(b) Here $E(t) = 150 \cos 200\, t v$. So equation is

$$\frac{dI}{dt} + 50I = 750 \cos 200\, t$$

The general solution is

$$I(t) \cdot e^{50t} = 750 \int e^{50t} \cdot \cos 200t dt + c$$

$$= c + 750 \cdot \frac{e^{50t}}{(2500 + 40000)}(50 \cos 200t + 200 \sin 200t)$$

At $t = 0$, $I = 0$ so

$$0 = \frac{3}{170} 50 + c \quad \therefore \; c = -\frac{15}{17}$$

The current $I(t)$ is given by

$$I(t) = \frac{3}{170}(50 \cos 200t + 200 \sin 200t) - \frac{15}{17} e^{-50t}$$

RC-circuit

Example 1: A capacitor $c = 0.01$ F in series with a resistor $R = 20$ ohms is charged from a battery $E_0 = 10$ V. Assuming that initially the capacitor is completely uncharged, determine the charge $Q(t)$, voltage $v(t)$ on the capacitor and the current $I(t)$ in the circuit.

Solution: Equation is

$$RI + \frac{Q}{c} = E$$

$$20I + \frac{Q}{0.01} = 10$$

or

$$\frac{dQ}{dt} + 5Q = 0.5$$

The general solution is

$$Q \cdot e^{5t} = 0.5 \int e^{5t} dt = 0.5 \cdot \frac{e^{5t}}{5} + c$$

$$Q = 0.1 + ce^{-5t}$$

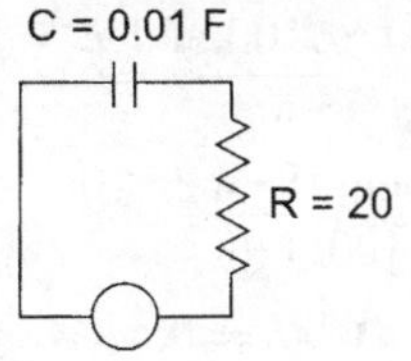

Fig. 8.12

At $t = 0$, $Q = 0$, so $0 = 0.1 + c \quad \therefore \; c = -0.1$
Thus the charge $Q(t) = 0.1(1 - e^{-5t})$
Voltage $v(t) = \frac{Q(t)}{c} = \frac{0.1(1-e^{-5t})}{0.01} = 10(1 - e^{-5t})$.
The current $I(t) = \frac{dQ}{dt} = 0.5e^{-5t}$.

EXERCISE

1. A generator having emf 100 volts is connected in series with a 10 ohm resistor and an inductor of 2 henries. If the switch is closed at a time $t = 0$, determine the current at time $t > 0$.

Ans. $I = 10(1 - e^{-5t})$

Hint: $\dot{I} + 5I = 50$, $I(0) = 0$.

2. Solve the above example when the generator is replaced by one having an emf of $20 \cos 5t$ volts.

Ans. $I = \cos 5t + \sin 5t - e^{-5t}$

Hint: $\dot{I} + 5I = 10 \cos 5t$, $I(0) = 0$, $c = -1$

3. A decaying emf $E = 200e^{-5t}$ is connected in series with a 20 ohm resistor and 0.01 farad capacitor. Find the charge and current at any time assuming $Q = 0$ at $t = 0$. Show that the charge reaches a maximum, calculate it and find the time when it is reached.

Ans. $Q(t) = 10te^{-5t}$, $\quad I(t) = 10e^{-5t} - 50\, t\, e^{-5t}$. Maximum of $Q = 10 \cdot \frac{1}{5} \cdot e^{-1} = \frac{2}{e} \sim 0.74$ coulombs when $t = \frac{1}{5}$ sec.

Hint: $\dot{Q} + 5Q = 10e^{-5t}$, to find maximum $\frac{dQ}{dt} = 0$.

4. Find the current $I(t)$ in the RL-circuit with $R = 10$ ohms, $L = 100$ henries, $E = 40$ volts

when $0 \le t \le 100$ and $E = 0$ when $t > 100$ and $I(0) = 4$

Ans. $I = 4$ when $0 \le t \le 100$, $I = 4 \cdot e^{10} e^{-t/10}$ when $t > 100$

Hint: $\dot{I} + 0.1I = 0.4$, $0 \le t \le 100$, $\dot{I} + 0.1I = 0$ when $t > 100$, use $I = 4$ when $t = 100$.

5. Find $I(t)$ in an RL-circuit with $E = 10$ V, $R = 5$ ohms $L = (10 - t)$ henry, when $0 \le t \le 10$ sec, and $L = 0$ when $t > 10$ sec and $I(0) = 0$.

Ans. $I = 2 - 2(1 - 0.1t)^5$ when $0 \le t \le 10$; when $t > 10$, $I = 2$

Hint: $(10 - t)\dot{I} + 5I = 10$ when $0 \le t \le 10$.

6. Solve the (RL-circuit) equation $L\frac{dI}{dt} + RI = E(t)$ when (a) $E(t) = E_0$ and the initial current is I_0.
(b) Solve the problem when $L = 3$ henries, $R = 15$ ohms, emf is the 60 cycle sine wave of amplitude 110 volts and $I(t = 0) = 0$.

Ans. (a) $I(t) = \frac{E_0}{R}(1 - e^{-Rt/L}) + I_0 e^{-Rt/L}$

(b) $I(t) = \frac{22}{3} \cdot \frac{\sin 120\pi t - 24\pi \cos 120\pi t + 24\pi e^{-5t}}{1 + 576\pi^2}$

Hint: $3\dot{I} + 15I = 100 \sin 120\pi t$.

7. Solve the (RC-circuit) equation $R\frac{dQ}{dt} + \frac{Q}{c} = E$ with $R = 10$ ohms, $c = 10^{-3}$ farad and $E(t) = 100 \sin 120\pi t$ assuming $Q(t = 0) = 0$. Find $I(t)$ given $I(t = 0) = 5$.

Ans. $Q(t) = \frac{\sin(120\pi t - \phi)}{2\sqrt{(25 + 36\pi^2)}} + \frac{3\pi e^{-100t}}{25 + 36\pi^2}$

$$I(t) = \frac{60\pi}{\sqrt{25 + 36\pi^2}} \cos(120\pi t - \phi)$$

$$- \left(\frac{300\pi}{25 + 36\pi^2} - 5\right) e^{-100t}$$

Hint: $\dot{Q} + 100Q = 10 \sin 120\pi t$, $\sin\phi = \frac{12\pi}{\sqrt{100 + 144\pi^2}}$, $\cos\phi \frac{10}{\sqrt{100 + 144\pi^2}}$

8. Determine the current at time $t > 0$ in a series RL-circuit having an emf given by $E(t) = 100 \sin 40t\, V$, a resistor of 10 Ω and an inductor of 0.5 H given that initial current is zero. Find the period and the phase angle.

Ans. $I(t) = 2(\sin 40t - 2\cos 40t) + 4e^{-20t}$, period $\frac{\pi}{20}$, phase angle $\phi \approx -1.11$

Hint: $\dot{I} + 20I = 200 \sin 40t$, $I(t) = 4.47 \sin(40t - 1.11) + 4e^{-20t}$.

9. Find the current in RC-circuit with $R = 10$, $c = 0.1$, $E(t) = 110 \sin 314t$, $I(0) = 0$

Ans. $I(t) = 0.035(\cos 314t + 314 \sin 314t - e^{-t})$

10. Determine the charge and current at time $t > 0$ in a RC-circuit with $R = 10$, $c = 2 \times 10^{-4}$, $E = 100$ V given that $Q(t = 0) = 0$.

Ans. $Q(t) = (1 - e^{-500t})/50$, $I(t) = 10e^{-500t}$.

Chapter 9

Linear Differential Equations of Second Order and Higher Order

INTRODUCTION

The simple harmonic motion, oscillations of mass-spring system, RLC-circuit, oscillations of a simple pendulum are all described by nonhomogeneous (or homogeneous) linear second order differential equations. Higher order differential equations appear in problems involving deflections of loaded beams (4th order), mechanical spring systems having several springs connected in tandem or series of electrical circuits containing several loops. In this chapter, we consider methods of obtaining solutions of homogeneous and nonhomogeneous differential equations of second and higher order with constant coefficients. We also consider the Cauchy's and Legendre differential equations with variable coefficients. We present two powerful general methods: method of variation of parameters and method of undetermined coefficients. Method of obtaining solution of system of simultaneous linear differential equations is presented. Finally we consider four important engineering applications: the simple harmonic motion, mass-spring system, RLC-circuit and simple pendulum.

9.1 LINEAR INDEPENDENCE AND DEPENDENCE

Two functions $y_1(x)$ and $y_2(x)$ are said to be linearly independent if

$$k_1 y_1(x) + k_2 y_2(x) = 0 \tag{1}$$

implies k_1 and k_2 are both zeros i.e., when y_1 or y_2 can not be expressed as proportional to the other.

Otherwise, y_1 and y_2 are linearly dependent if (1) holds for some constants k_1 and k_2 not both zero.

Criterion for linear dependence or independence of two functions y_1 and y_2

Define the Wronski* determinant or Wronskian

$$w(y_1, y_2) = \begin{vmatrix} y_1 & y_2 \\ y_1' & y_2' \end{vmatrix} = y_1 y_2' - y_2 y_1' \tag{2}$$

Results:

i. y_1, y_2 are linearly independent if $w \neq 0$

ii. otherwise linearly dependent when $w = 0$.

WORKED OUT EXAMPLES

Determine whether the following functions y_1 and y_2 are linearly dependent or independent:

Example 1: $y_1 = \cos ax$, $y_2 = \sin ax$ with $a \neq 0$.

Solution:

$$w = \begin{vmatrix} \cos ax & \sin ax \\ -a \sin ax & a \cos ax \end{vmatrix} = a(\cos^2 ax + \sin^2 ax) = a \neq 0$$

So y_1 and y_2 are linearly independent.

Example 2: $y_1 = \ln x$, $y_2 = \ln x^n$ with n non negative integer.

Solution:

$$w = \begin{vmatrix} \ln x & \ln x^n \\ \frac{1}{x} & \frac{nx^{n-1}}{x^n} \end{vmatrix} = \frac{n}{x}\ln x - \frac{1}{x}\ln x^n = 0$$

So y_1 and y_2 are linearly dependent.

* I.M. Hone (1778–1853) Polish mathematician, who changed his name to Wronski.

EXERCISE

Show that the following pair of functions are linearly independent (with $w \neq 0$):

1. e^x, xe^x
2. e^x, x^2
3. $1, x$
4. $x^2, x^2 \ln x$
5. $x^2, x^{\frac{1}{2}}$
6. $\cos 2\pi x, \sin 2\pi x$
7. $x^{\frac{3}{2}}, x^{-\frac{3}{2}}$
8. $x^4, x^4 \ln x$
9. $x^a \cos(2 \ln x), x^a \sin(2 \ln x)$
10. $e^{ax} \sin bx, e^{ax} \cos bx$
11. e^{ax}, e^{-ax}

9.2 LINEAR DIFFERENTIAL EQUATIONS OF SECOND ORDER WITH VARIABLE COEFFICIENTS

Standard form of linear D.E. of 2nd order with variable coefficients is

$$y'' + P(x)y' + Q(x)y = F(x) \tag{1}$$

Here $P(x)$, $Q(x)$ and $F(x)$, are known functions of x. P and Q are known as coefficients of D.E. (1).

Non-homogeneous

D.E. (1) is said to be non-homogeneous if the R.H.S. of (1) $F(x) \neq 0$. Otherwise

Homogeneous

when $F(x) = 0$. Thus

$$y'' + P(x)y' + Q(x)y = 0 \tag{2}$$

is known as the reduced or corresponding or *complimentary* homogeneous linear Equation of (1).

Superposition or Linearity Principle or Fundamental Theorem for Homogeneous D.E.

Theorem: *If $y_1(x)$ and $y_2(x)$ are any two linearly independent solutions of the homogeneous D.E. (2), then the general solution of the homogeneous D.E. (2) is given by*

$$y(x) = c_1 y_1(x) + c_2 y_2(x) \tag{3}$$

where c_1 and c_2 are two arbitrary constants.

Proof: Differentiating (3) w.r.t. x twice, we get

$$y' = c_1 y_1' + c_2 y_2' \tag{4}$$

$$y'' = c_1 y_1'' + c_2 y_2'' \tag{5}$$

Substituting (3), (4), (5) in (2), we have

$$(c_1 y_1'' + c_2 y_2'') + P(c_1 y_1' + c_2 y_2') + Q(c_1 y_1 + c_2 y_2) = 0$$

Rewriting

$$c_1(y_1'' + Py_1' + Qy_1) + c_2(y_2'' + Py_2' + Qy_2) = 0 \tag{6}$$

Since y_1 and y_2 are solutions of (2) i.e.,

$$y_1'' + P y_1' + Qy_1 = 0 \quad \text{and} \quad y_2'' + P y_2' + Qy_2 = 0$$

Equation (2) is identically satisfied by (3). Thus (3) is the general solution of (2) (since it contains two arbitrary constants c_1 and c_2).

Note: Above superposition principle is *not* applicable to non-homogeneous or nonlinear equations.

9.3 SECOND ORDER DIFFERENTIAL EQUATIONS WITH CONSTANT COEFFICIENTS; HOMOGENEOUS

Standard form

$$y'' + ay' + by = 0 \tag{1}$$

where the coefficients a and b are constants. Consider the function

$$y = e^{mx} \tag{2}$$

then $y' = me^{mx}$, $y'' = m^2 e^{mx}$

Substituting these values of y, y', y'' in (1), we get

$$(m^2 + am + b)e^{mx} = 0$$

Since $e^{mx} \neq 0$, $y = e^{mx}$ is a solution of (1) if m satisfies the quadratic equation

$$m^2 + am + b = 0 \tag{3}$$

Equation (3) is known as auxiliary equation (A.E.) or characteristic equation of (1).

Observation A.E. (3) is obtained from D.E. (1) by replacing y by 1, y' by m and y'' by m^2.

The general solution (G.S.) of the homogeneous D.E. (1) is obtained depending on the nature of the two roots of the auxiliary Equation (3) as follows:

Case 1: Two distinct real roots

When the two roots m_1, m_2 of A.E. (3) are real distinct, then $e^{m_1 x}$ and $e^{m_2 x}$ form a linearly independent set of solutions to D.E. (1). By superposition principle, the general solution of (1) is

$$y = c_1 e^{m_1 x} + c_2 e^{m_2 x}$$

where c_1 and c_2 are any two arbitrary constants.

Case 2: Real double (repeated) root

In case of repeated double root, $e^{m_1 x}$ and $xe^{m_1 x}$ are linearly independent so that the general solution of (1) by linearity principle is

$$y = c_1 e^{mx} + xc_2 e^{mx} = (c_1 + c_2 x)e^{mx}$$

Case 3: Complex conjugate roots.

For complex conjugate roots e^{p+iq} and e^{p-iq} are linearly independent solutions of (1) therefore the general solution of (1) is

$$y = c_1^* e^{(p+iq)x} + c_2^* e^{(p-iq)x}$$

$$= c_1^* e^{px}(e^{iqx}) + c_2^* e^{px} e^{-iqx}$$

$$= e^{px}\left[c_1^*(\cos qx + i\sin qx) + c_2^*(\cos qx - i\sin qx)\right]$$

$$= e^{px}\left[(c_1^* + c_2^*)\cos qx + (c_1^* - c_2^*)i \cdot \sin qx\right]$$

$$y = e^{px}\left[c_1 \cos qx + c_2 \sin qx\right]$$

These results are listed in the following table.

The general solution of (1) in these three cases is obtained as follows:

Case	Nature of the two roots of A.E. (3)	Set of linearly independent solutions of (1)	General solution of (1)
1.	Distinct real roots m_1, m_2	$e^{m_1 x}, e^{m_2 x}$	$y = c_1 e^{m_1 x} + c_2 e^{m_2 x}$
2.	Real double root (two equal roots) (repeated roots) $m = m_1 = m_2$	e^{mx}, xe^{mx}	$y = (c_1 + c_2 x)e^{mx}$
3.	Complex conjugate $m_1 = p + iq$ $m_2 = p - iq$	$e^{px}\cos qx$, $e^{px}\sin qx$	$y = e^{px} \times (c_1 \cos qx + c_2 \sin qx)$

WORKED OUT EXAMPLES

Distinct real roots: second order

Solve the following:

Example 1: $y'' - y' - 12y = 0$

Solution: Rewriting the given D.E. in operator form

$$(D^2 - D - 12)y = 0$$

so that the corresponding auxiliary equation or characteristic equation is obtained by replacing D by m in the given D.E. as $m^2 - m - 12 = 0$.

$m^2 - m - 12 = (m+3)(m-4) = 0$. Thus there are two real distinct roots $-3, 4$.

General solution: $y(x) = c_1 e^{-3x} + c_2 e^{4x}$ where c_1 and c_2 are arbitrary constants.

Example 2: $y'' - 3y' + 2y = 0$, $y(0) = -1$, $y'(0) = 0$

Solution:

A.E.: $m^2 - 3m + 2 = 0$

$(m-2)(m-1) = 0$.

$m = 1, 2$ are two real distinct roots

G.S.: $y = c_1 e^x + c_2 e^{2x}$

Particular solution using condition $y(0) = -1$ in the G.S., we get

$$-1 = y(0) = c_1 + c_2 \tag{1}$$

using the condition $y'(0) = 0$ is derivative of G.S. we have

$$y' = c_1 e^x + 2c_2 e^{2x}$$

$$0 = y'(0) = c_1 + 2c_2 \tag{2}$$

Solving (1) and (2), we get

$$c_1 = -2, c_2 = 1$$

Thus the particular solution is

$$y = -2e^x + e^{2x}$$

EXERCISE

Solve the following:

1. $y'' - 3y' + 2y = 0$

Ans. $y = c_1 e^x + c_2 e^{2x}$

2. $y'' - 5y' + 6y = 0$

Ans. $y = c_1 e^{2x} + c_2 e^{3x}$

3. $4y'' - 12y' + 5y = 0$

Ans. $y = c_1 e^{\frac{x}{2}} + c_2 e^{\frac{5x}{2}}$

4. $2y'' + y' - 6y = 0$

Ans. $y = c_1 e^{\frac{3x}{2}} + c_2 e^{-2x}$

5. $y'' + y' - 2y = 0$

Ans. $y = c_1e^{-2x} + c_2e^{x}$

6. $y'' - 6y' + 8y = 0,\ y(0) = 1,\ y'(0) = 6$

Ans. $y = -e^{2x} + 2e^{4x}$

7. $y'' - y' - 12y = 0,\ y(0) = 3,\ y'(0) = 5$

Ans. $y = 2e^{4x} + e^{-3x}$

8. $y'' + y' - 2y = 0,\ y(0) = 4,\ y'(0) = -5$

Ans. $y = e^{x} + 3e^{-2x}$

9. $y'' - 4y' + y = 0$

Ans. $y = c_1e^{(2+\sqrt{3})x} + c_2e^{(2-\sqrt{3})x}$

10. $y'' + 2y' - 8 = 0$

Ans. $y = c_1e^{2x} + c_2e^{-4x}$.

WORKED OUT EXAMPLES

Equal (or repeatd or double) roots: second order

Solve the following:

Example 1: $y'' + 3y' + 2.25y = 0$

Solution:

A.E.: $m^2 + 3m + 2.25 = 0$.

$\left(m + \frac{3}{2}\right)^2 = 0$ i.e., $m = -\frac{3}{2}$ is a double (repeated) root.

G.S: $(y = c_1 + c_2x)e^{-\frac{3x}{2}}$.

Example 2: $y''-6y'+9y = 0,\ y(0) = 2,\ y'(0) = 8$

Solution:

A.E.: $m^2 - 6m + 9 = 0$

$(m - 3)^2 = 0$ i.e., $m = 3$ is a double root

G.S: $y = (c_1 + c_2x)e^{3x}$

using $y(0) = 2$ in G.S, we get $2 = y(0) = c_1$

using $y'(0) = 8$ in $y' = 3e^{3x}(c_1 + c_2x) + e^{3x}\cdot c_2$ we get

$$8 = 3c_1 + c_2 \quad \therefore c_2 = 2$$

Thus the particular solution is $y = 2(1 + x)e^{3x}$

EXERCISE

Equal (or repeated or double) root: second order

Solve the following:

1. $y'' + 8y' + 16y = 0$

Ans. $y = (c_1 + c_2x)e^{-4x}$

2. $y'' - 6y' + 9y = 0$

Ans. $y = (c_1 + c_2x)e^{3x}$

3. $4y'' - 4y' + y = 0$

Ans. $y = (c_1 + c_2x)e^{\frac{x}{2}}$

4. $y'' - 8y' + 16y = 0$

Ans. $y = (c_1 + c_2x)e^{4x}$

5. $16y'' - 8y' + y = 0$

Ans. $y = (c_1 + c_2x)e^{\frac{x}{4}}$

6. $y'' + 6y' + 9y = 0,\ y(0) = 2,\ y'(0) = -3$

Ans. $y = (3x + 2)e^{-3x}$

7. $y'' - 4y' + 4y = 0,\ y(0) = 3,\ y'(0) = 1$

Ans. $y = (3 - 5x)e^{2x}$

8. $y'' + 4y' + 4y = 0,\ y(0) = 3,\ y'(0) = 7$

Ans. $y = (13x + 3)e^{-2x}$

9. $9y'' - 30y' + 25y = 0$

Ans. $y = (c_1 + c_2x)e^{\frac{5x}{3}}$

10. $y'' + 2ky' + k^2y = 0$

Ans. $y = (c_1 + c_2x)e^{-kx}$

WORKED OUT EXAMPLES

Complex conjugate roots: Second order

Solve the following:

Example 1: $y'' + 5y' + 12.5y = 0$.

Solution:

A.E.: $m^2 + 5m + 12.5 = 0$.

So $m = \frac{-5\pm 5i}{2}$ are the complex conjugate roots.

$p = -\frac{5}{2}, q = \frac{5}{2}$,

G.S.: $y = e^{-\frac{5x}{2}}\left[c_1 \cos \frac{5}{2}x + c_2 \sin \frac{5x}{2}\right]$

Example 2: $y''-2y'+5y=0, y(0) = -3, y'(0) = 1$

Solution:

A.E.: $m^2 - 2m + 5 = 0.$

So $m = 1 \pm 2i$ are the complex conjugate roots.

$p = 1, q = 2,$

G.S.: $y = e^x\left[c_1 \cos 2x + c_2 \sin 2x\right]$

using $y(0) = -3$ in G.S., we get

$$-3 = e^0[c_1 + 0] \quad \therefore \quad c_1 = -3$$

Differentiating G.S. w.r.t., x

$$y' = e^x[c_1 \cos 2x + c_2 \sin 2x] + e^x[-2c_1 \sin 2x + 2c_2 \cos 2x]$$

using $y'(0) = 1$ in the above equation, we have

$$1 = y'(0) = e^0[c_1 \cdot 1 + 0] + e^0[0 + 2c_2 \cdot 1]$$

$$1 = c_1 + 2c_2. \qquad \text{Thus } c_2 = 2$$

The required particular solution is

$$y = e^x[-3 \cos 2x + 2 \sin 2x]$$

EXERCISE

Complex conjugate roots: second order

Solve the following:

1. $y'' + 9y = 0$

Ans. $y = c_1 \sin 3x + c_2 \cos 3x$

2. $y'' - 6y' + 25y = 0$

Ans. $y = e^{3x}(c_1 \cos 4x + c_2 \sin 4x)$

3. $y'' + 6y' + 11y = 0$

Ans. $y = e^{-3x}(c_1 \sin \sqrt{2}x + c_2 \cos \sqrt{2}x)$

4. $y'' - 4y' + 13y = 0$

Ans. $y = e^{2x}(c_1 \sin 3x + c_2 \cos 3x)$

5. $y'' - 4y' + 29y = 0, y(0) = 0, y'(0) = 5$

Ans. $y = e^{2x} \sin 5x$

6. $y'' + 0.2y' + 4.01y = 0, y(0) = 0, y'(0) = 2$

Ans. $y = e^{-0.1x} \sin 2x$

7. $y'' + 6y' + 13y = 0, y(0) = 3, y'(0) = -1$

Ans. $y = e^{-3x}(4 \sin 2x + 3 \cos 2x)$

8. $9y'' + 6y' + 5y = 0, y(0) = 6, y'(0) = 0$

Ans. $y = 3e^{-\frac{x}{3}}\left(\sin \frac{2x}{3} + 2 \cos \frac{2x}{3}\right)$

9. $y'' + 2a \cos \alpha y' + a^2 y = 0$

Ans. $y = e^{-ax \cos \alpha}[c_1 \cos (ax \sin \alpha) + c_2 \sin (ax \sin \alpha)]$

10. $y'' + y' + y = 0$

Ans. $y = e^{-\frac{x}{2}}\left[c_1 \cos \frac{\sqrt{3}}{2}x + c_2 \sin \frac{\sqrt{3}}{2}x\right]$

9.4 HIGHER ORDER LINEAR HOMOGENEOUS DIFFERENTIAL EQUATIONS

Analysis considered in 9.3 for second order homogeneous equations with constant coefficients can be extended in a similar way to equations of higher order three or more. Consider the nth order homogeneous equation

$$a_0 \frac{d^n y}{dx^n} + a_1 \frac{d^{n-1} y}{dx^{n-1}} + a_2 \frac{d^{n-2} y}{dx^{n-2}} + \cdots + a_{n-1}\frac{dy}{dx} + a_n y = 0 \tag{1}$$

Here $a_0, a_1, a_2, \ldots, a_n$ are all constants.

Introducing the notation of differential operator $D \equiv \frac{d}{dx}$ and higher order operators as $D^2 = \frac{d^2}{dx^2}, \ldots, D^n = \frac{d^n}{dx^n}$, etc, the given equation can be rewritten with this notation as

$$a_0 D^n y + a_1 D^{n-1} y + a_2 D^{n-2} y + \cdots + a_{n-1} Dy + a_n y = 0$$

or

$$(a_0 D^n + a_1 D^{n-1} + a_2 D^{n-2} + \cdots + a_{n-1} D + a_n)y = 0 \tag{2}$$

or

$$f(D)y = 0 \tag{3}$$

where $f(D) = a_0 D^n + a_1 D^{n-1} + \cdots + a_{n-1} D + a_n$ is function of D.

Observation

The auxiliary equation of (1) is obtained by replacing y by 1, y' by m, y'' by $m^2, \ldots, y^{(n)}$ by m^n in D.E.

(1). Thus the A.E.

$$a_0m^n + a_1m^{n-1} + a_2m^{n-2} + \cdots + a_{n-1}m + a_n = 0 \quad (4)$$

is an nth degree polynomial in m i.e.,

$$f(m) = 0 \quad (5)$$

The general solution of (1) containing n arbitrary constants is obtained according as the nature of the roots of A.E. (4), as follows:

Case	Nature of the n roots of A.E. (4)	Linearly independent solution of (1)	General solution of (1)
I.	n distinct and real roots $m_1, m_2, m_3 \cdots$	$e^{m_1x}, e^{m_2x}, e^{m_3x} \ldots$	$y = c_1e^{m_1x} + c_2e^{m_2x} + c_3e^{m_3x} + \cdots$
II.	Two equal and real roots, $n-2$ distinct roots $m_1, m_1, m_3, m_4 \cdots$	$e^{m_1x}, xe^{m_1x}, e^{m_3x}, e^{m_4x}$	$y = (c_1 + c_2x)e^{m_1x} + c_3e^{m_3x} + c_4e^{m_4x} + \cdots$
III.	Three equal real roots, $n-3$ distinct roots $m_1, m_1, m_1, m_4, m_5, \ldots$	$e^{m_1x}, xe^{m_1x}, x^2e^{m_1x}, e^{m_4x}, \ldots$	$y = (c_1 + c_2x + c_3x^2)e^{m_1x} + c_4e^{m_4x} + c_5e^{m_5x} + \cdots$
IV.	Two complex conjugate roots, $n-2$ distinct real roots $m_1 = p + iq$ $m_2 = p - iq$ $m_3, m_4, \ldots$	$e^{px}\cos qx, e^{px}\sin qx, e^{m_3x}, e^{m_4x}, \ldots$	$y = e^{px}(c_1\cos qx + c_2\sin qx) + c_3e^{m_3x} + c_4e^{m_4x} + \cdots$
V.	Two equal complex conjugate roots, $n-4$ distinct real roots $m_1 = m_2 = p + iq,$ $m_3 = m_4 = p - iq,$ $m_5, m_6, \ldots$	$e^{px}\cos qx, xe^{px}\cos qx, e^{px}\sin qx, xe^{px}\sin qx, e^{m_5x}, e^{m_6x}, \ldots$	$y = e^{px}[(c_1 + c_2x)\cos qx + (c_3 + c_4x)\sin qx] + c_5e^{m_5x} + \cdots$

WORKED OUT EXAMPLES

Distinct real roots: higher order

Solve the following:

Example 1: $y''' - 4y'' + y' + 6y = 0$

Solution:

A.E.: $m^3 - 4m^2 + m + 6 = 0.$

Observe that $m = -1$ is a root of this equation.

By synthetic division

$$\begin{array}{r|rrrr} -1 & 1 & -4 & 1 & 6 \\ & & -1 & 5 & -6 \\ \hline & 1 & -5 & 6 & 0 \end{array}$$

We obtain the factorization as

$$(m - (-1))(m^2 - 5m + 6) = 0$$

or

$$(m + 1)(m - 2)(m - 3) = 0$$

Thus there 3 real distinct roots $-1, 2, 3$.

G.S.: $y = c_1e^{-x} + c_2e^{2x} + c_3e^{3x}$.

Example 2: $y''' - 3y'' - y' + 3y = 0.$

Solution:

A.E.: $m^3 - 3m^2 - m + 3 = 0.$

Observe that $m = 1$ is a root $(1 - 3 - 1 + 3 = 0)$.

By synthetic division

$$\begin{array}{r|rrrr} 1 & 1 & -3 & -1 & 3 \\ & & 1 & -2 & -3 \\ \hline & 1 & -2 & -3 & 0 \end{array}$$

We obtain the factorization as

$$(m-1)(m^2-2m-3)$$
$$=(m-1)(m+1)(m-3)=0$$

3 real distinct roots $-1, 1, 3$

G.S.: $y=c_1e^{-x}+c_2e^{x}+c_3e^{3x}$.

Example 3: $4y^{(4)}-8y'''-7y''+11y'+6y=0$

Solution:

A.E: $4m^4-8m^3-7m^2+11m+6=0$.

Observe that -1 is a root
$(4+8-7-11+6=0)$

Note also that 2 is a root
$(64-64-28+22+6=0)$

By synthetic division

−1	4	−8	−7	11	6
		−4	12	−5	−6
2	4	−12	5	6	0
		8	−8	−6	
	4	−4	−3	0	

We obtain the factorization as

$$(m+1)(m-2)(4m^2-4m-3)=0$$
$$(m+1)(m-2)\left(m+\frac{1}{2}\right)\left(m-\frac{3}{2}\right)=0$$

4 distinct real roots, $-\frac{1}{2}, -1, \frac{3}{2}, 2$

G.S.: $y=c_1e^{-\frac{x}{2}}+c_2e^{-x}+c_3e^{\frac{3x}{2}}+c_4e^{2x}$.

EXERCISE

Higher order: Homogeneous: Distinct roots

Solve the following:

1. $y'''-4y''+y'+6y=0$

Ans. $y=c_1e^{-x}+c_2e^{2x}+c_3e^{3x}$

2. $y'''-6y''+11y'-6y=0$

Ans. $y=c_1e^{x}+c_2e^{2x}+c_3e^{3x}$

3. $y'''-3y''-y'+3y=0$

Ans. $y=c_1e^{x}+c_2e^{-x}+c_3e^{3x}$

4. $y'''-6y''+11y'-6y=0$, $y(0)=0$, $y'(0)=0$, $y''(0)=2$

Ans. $y=e^{x}-2e^{2x}+e^{3x}$

5. $y'''-9y''+23y'-15y=0$

Ans. $y=c_1e^{x}+c_2e^{3x}+c_3e^{5x}$

6. $y'''-2y''-3y'=0$

Ans. $y=c_1+c_2e^{3x}+c_3e^{-x}$

WORKED OUT EXAMPLES

Double (repeated) root: higher order

Solve the following:

Example 1: $4y'''+4y''+y'=0$.

Solution:

A.E.: $4m^3+4m^2+m=0$.

$m(4m^2+4m+1)=m\left(m+\frac{1}{2}\right)^2=0$

so $m=0, -\frac{1}{2}, -\frac{1}{2}$ are the roots of which $m=-\frac{1}{2}$ is a repeated (double) root.

G.S.: $y=c_1\cdot e^{o\cdot x}+(c_2+xc_3)e^{-\frac{x}{2}}$

Example 2: $y''''+6y'''+9y''=0$

Solution:

A.E.: $m^4+6m^3+9m^2=0$

$m^2(m^2+6m+9)=m^2(m+3)^2=0$

so $m=0, 0, -3, -3$ are the roots

i.e., $m=0$ and $m=-3$ each is a double root.

G.S.: $y=(c_1+c_2x)e^{o\cdot x}+(c_3+c_4x)e^{-3x}$.

Example 3: $y'''''-y'''=0$

Solution:

A.E.: $m^5-m^3=0$

$m^3(m^2-1)=0$

so $m=0$ is a triple (repeated 3 times) root and $m=-1, 1$ are distinct real roots.

G.S.: $y=c_1e^{-x}+c_2e^{x}+(c_3+xc_4+x^2c_5)e^{o\cdot x}$

$y=c_1e^{-x}+c_2e^{x}+(c_3+xc_4+x^2c_5)$

Exercise

Higher order: Homogeneous: repeated roots

1. $y''' - 4y'' - 3y' + 18y = 0$

Ans. $y = (c_1 + c_2x)e^{3x} + c_3e^{-2x}$

2. $y''' - 5y'' + 7y' - 3y = 0$

Ans. $y = (c_1 + c_2x)e^{x} + c_3e^{3x}$

3. $y'''' - 5y''' + 6y'' + 4y' - 8y = 0$

Ans. $y = (c_1 + c_2x + c_3x^2)e^{2x} + c_4e^{-x}$

4. $y''' - 3y'' + 4y = 0,\ \ y(0) = 1,\ \ y'(0) = -8,$ $y''(0) = -4$

Ans. $y = \frac{32}{9}e^{-x} - \frac{23}{9}e^{2x} + \frac{2}{3}xe^{2x}$

5. $y''' - 6y'' + 12y' - 8y = 0$

Ans. $y = (c_1 + c_2x + c_3x^2)e^{2x}$

6. $y'''' = 0$

Ans. $y = c_1 + c_2x + c_3x^2 + c_4x^3$

7. $y''''' - 2y'''' + y''' = 0$

Ans. $y = c_1 + c_2x + c_3x^2 + (c_4 + c_5x)e^{x}$

8. $y'''' - 3y''' + 2y'' = 0,\ \ y(0) = 2,\ \ y'(0) = 0,$ $y''(0) = 2,\ y'''(0) = 2$

Ans. $y = 2(e^x - x).$

Worked Out Examples

Complex (conjugate) roots: higher order

Solve the following:

Example: $y^{(4)} + 18y'' + 81y = 0.$

Solution:

A.E.: $m^4 + 18m^2 + 81 = 0.$

$(m^2 + 9)^2 = 0$ so $m^2 = -9$ is a repeated root

$m = \pm 3i$ is repeated (double) root

i.e., $m = +3i, +3i, -3i, -3i$ are the 4 roots.

Solutions are $y_1 = e^{3ix}$, $y_2 = xe^{3ix}$ and $y_3 = e^{-3ix}$ and $y_4 = xe^{-3ix}$.

Thus the G.S. is

$$y = (c_1 + c_2x)e^{3ix} + (c_3 + c_4x)e^{-3ix}$$
$$= (c_1 + c_2x)(\cos 3x + i\sin 3x)$$
$$+ (c_3 + c_4 \cdot x)(\cos 3x - i\sin 3x)$$

Exercise

Higher order: Homogeneous: complex roots

1. $y''' + y'' + 4y'' + 4 = 0$

Ans. $y = c_1e^{-x} + c_2\cos 2x + c_3\sin 2x$

2. $y''' - y'' + y' - y = 0$

Ans. $y = c_1e^{x} + c_2\sin x + c_3\cos x$

3. $y'''' - 3y''' - 2y'' + 2y' + 12y = 0$

Ans. $y = c_1e^{2x} + c_2e^{3x} + e^{-x}(c_3\sin x + c_4\cos x)$

4. $y'''' - 4y''' + 14y'' - 20y' + 25y = 0$

Ans. roots are $1 + 2i, 1 - 2i, 1 + 2i, 1 - 2i$ (each pair of conjugate complex roots is double)

$$y = e^x[(c_1 + c_2x)\sin 2x + (c_3 + c_4x)\cos 2x]$$

or $y = c_1e^x\sin 2x + c_2xe^x\sin 2x + c_3e^x\cos 2x + c_4xe^x\cos 2x$

5. $y'''' + 4y = 0$

Ans. roots are $-1 \pm i$ and $1 \pm i$

$$y = e^{-x}(c_1\cos x + c_2\sin x) + e^{x}(c_3\cos x + c_4\sin x)$$

6. $y'''' + 8y'' + 16y = 0$

Ans. $y = (c_1 + c_2x)\sin 2x + (c_3 + c_4x)\cos 2x$

7. $y'''''' + 3y'''' + 3y'' + y = 0$

Ans. $y = (c_1 + c_2x + c_3x^2)\sin x + (c_4 + c_5x + c_6x^2)\cos x$

8. $y'''''' - 2y''' + y = 0$

Ans. $y = (c_1 + c_2x)e^x + e^{-\frac{x}{2}}\left[(c_3 + c_4x)\cos\frac{\sqrt{3}}{2}x + (c_5 + c_6x)\sin\frac{\sqrt{3}}{2}x\right]$

9.5 NON-HOMOGENEOUS EQUATIONS

Consider the non-homogeneous linear differential equation of nth order with constant coefficients

$$(a_0D^n + a_1D^{n-1} + \cdots + a_{n-1}D + a_n)y = F(x) \quad (1)$$

Then

$$(a_0D^n + a_1D^{n-1} + \cdots + a_{n-1}D + a_n)y = 0 \quad (2)$$

is the corresponding (or complementary or reduced) homogeneous equation of (1) (obtained by putting the R.H.S. $F(x) = 0$ in D.E. (1)).

A general solution

A general solution of the non-homogeneous equation (1) is a solution of the form

$$y(x) = y_c(x) + y_p(x) \quad (3)$$

where $y_c(x) = c_1y_1(x) + c_2y_2(x) + \cdots + c_ny_n(x)$ (4)

is the general solution of the corresponding homogeneous Equation (2), containing n arbitrary constants and $y_p(x)$ is any (particular) solution of the non-homogeneous Equation (1) containing no arbitrary constants.

Complementary Function (C.F.): $y_c(x)$

$y_c(x)$ which is the general solution of the reduced homogeneous Equation (2) is more often known as complementary function (C.F.) although it is also called complementary solution or complementary integral.

Particular Integral (P.I.): $y_p(x)$

$y_p(x)$ is more often called as particular integral (also particular solution of the non-homogeneous Equation (1)).

Thus the

General solution = complimentary function
+ particular integral

generally abbreviated as

$$\boxed{\text{G.S.} = \text{C.F.} + \text{P.I.}} \quad (5)$$

Method of Obtaining Particular Integral (P.I.):

Inverse operator: $D^{-1}F(x) = \frac{1}{D}F(x) = \int F(x)dx$. If $Du(x) = v(x)$ then $u(x) = D^{-1}v(x)$. Here D^{-1} is known as inverse operator of D such that

$$DD^{-1}(F(x)) = F(x) \quad \text{i.e.,} \quad DD^{-1} = 1$$

when D is differential operator then D^{-1} also denoted by $\frac{1}{D}$ represents integral operator $\int$. Thus in a similar way $\frac{1}{D^2}, \frac{1}{D^3}, \ldots$ etc. denotes integration twice, thrice etc. w.r.t., x.

Equation (1) can now be rewritten as

$$f(D)y = F(x) \quad (6)$$

where

$$f(D) = a_0D^n + a_1D^{n-1} + \cdots + a_{n-1}D + a_n \quad (7)$$

is a function of the differential operator D. Then the particular integral (P.I.) of

$$f(D)y = F(x) \quad (6)$$

is
$$y_p = \frac{1}{f(D)}F(x) \quad (8)$$

where $\frac{1}{f(D)}$ is the inverse operator of $f(D)$, such that

$$f(D)\left\{\frac{1}{f(D)}F(x)\right\} = F(x) = \frac{1}{f(D)}\{f(D)F(x)\}.$$

Book Work: Prove that if

$$(D - a)y = F(x) \quad (9)$$

Then
$$y = \frac{1}{(D-a)}F(x) = e^{ax}\int F(x)e^{-ax}dx \quad (10)$$

Proof: The D.E. $(D - a)y = F(x)$ is a first order linear equation $\frac{dy}{dx} - ay = F(x)$ with I.F. as $e^{\int -adx} = e^{-ax}$. Then the solution of (9) is

$$ye^{-ax} = \int e^{-ax} \cdot F(x)dx + c$$

Thus a particular solution of (9) (with $C = 0$) is

$$\boxed{y = e^{ax}\int F(x)e^{-ax}dx} \quad (10)$$

Special Case i. When $F(x) = b$ = constant, then

$$(D - a)y = b$$

Integrating $ye^{-ax} = \int be^{-ax}dx = \frac{be^{-ax}}{-a}, a \neq 0$

$$\boxed{y = \frac{1}{D-a}b = \frac{b}{-a} \quad \text{with} \quad a \neq 0} \tag{11}$$

Special Case ii. If R.H.S. of (8) is $F(x) = e^{ax}$, then (10) reduces to

$$y = e^{ax}\int F(x)e^{-ax}dx = e^{ax}\int e^{ax}e^{-ax}dx$$

$$y = xe^{ax}$$

Thus

$$\boxed{y_p = \left(\frac{1}{D-a}\right)e^{ax} = xe^{ax}} \tag{12}$$

Special Case iii. $(D-a)^r y = F(x) = e^{ax}$ then repeated application of result (ii) above by integrating r times successively, we get

$$\boxed{y_p = \frac{1}{(D-a)^r}e^{ax} = \frac{x^r}{r!}e^{ax}} \tag{13}$$

Using the inverse operator (short-cut) methods, a particular integral y_p of the non-homogeneous D.E.

$$f(D)y = F(x)$$

can be obtained when $F(x)$, the R.H.S. function has the following forms:

a. $F(x) = e^{ax+b}$

b. $F(x) = \sin(ax+b)$ or $\cos(ax+b)$

c. $F(x) = x^m$ or polynomial in x

d. $F(x) = e^{ax}v(x)$, exponential shift

e. $F(x) = xv(x)$

These and several other useful results of obtaining particular integral are presented in a tabular form at the end of Section 9.5.

Method of Obtaining General Solution of a Non-homogeneous Differential Equations

Step I. Obtain C.F. which is the general solution of the corresponding homogeneous equation.

Step II. P.I. is obtained depending on the nature of the R.H.S. function $F(x)$ using appropriate results listed in the table.

Step III. The G.S.: $y =$ C.F. + P.I..

[If initial conditions are specified, the constants in C.F. are evaluated using IC's].

P.I. when $F(x) = e^{ax+b}$

Consider the nth order linear non-homogeneous D.E. with constant coefficients and with R.H.S. function $F(x) = e^{ax+b}$ given by

$$(a_0D^n + a_1D^{n-1} + \cdots + a_{n-1}D + a_n)y$$
$$= F(x) = e^{ax+b} \tag{1}$$

or
$$f(D)y = e^{ax+b} \tag{2}$$

where

$$f(D) = a_0D^n + a_1D^{n-1} + \cdots + a_{n-1}D + a_n \tag{3}$$

We know that $\frac{d}{dx}e^{ax+b} = De^{ax+b} = ae^{ax+b}$, $D^2e^{ax+b} = a^2e^{ax+b}$, $D^3e^{ax+b} = a^3e^{ax+b}$ ..., $D^ne^{ax+b} = a^ne^{ax+b}$.

Substituting these values in (3)

$$f(D)e^{ax+b} = (a_0D^n + a_1D^{n-1} + \cdots$$
$$+ a_{n-1}D + a_n)e^{ax+b}$$
$$= (a_0a^n + a_1a^{n-1} + \cdots$$
$$+ a_{n-1}a + a_n)e^{ax+b}$$

$$f(D)e^{ax+b} = f(a)e^{ax+b} = e^{ax+b}f(a) \tag{4}$$

where $f(a)$, a constant is obtained, from $f(D)$ by replacing D by 'a'.

Operating on both sides of (4) by $\frac{1}{f(D)}$, we get

$$\frac{1}{f(D)}f(D)e^{ax+b} = \frac{1}{f(D)}f(a)e^{ax+b}$$
$$= f(a)\frac{1}{f(D)}e^{ax+b}$$
$$e^{ax+b} = f(a)\frac{1}{f(D)}e^{ax+b}$$

Therefore
$$\frac{1}{f(D)}e^{ax+b} = \frac{1}{f(a)}e^{ax+b}$$

provided $f(a) \neq 0$.

Result 1: When $f(a) \neq 0$, then the

$$\text{P.I.} = y_p = \frac{1}{f(D)}e^{ax+b} = \frac{1}{f(a)} \cdot e^{ax+b}.$$

Result 2: Suppose $f(a) = 0$

If $f(a)=0$ then $(m-a)$ is a factor of $f(m)$ so that $(D-a)$ is a factor of $f(D)$ i.e.,

$$f(D)=(D-a)\phi(D)$$

such that $\phi(a)\neq 0$. Then

$$\frac{1}{f(D)}e^{ax+b}=\frac{1}{(D-a)\phi(D)}e^{ax+b}$$

$$=\frac{1}{\phi(a)}\frac{1}{D-a}e^{ax} \text{ (from Result I)}$$

$$=\frac{1}{\phi(a)}\cdot xe^{ax+b}$$

Result 3: Suppose $f(D)=(D-a)^r\psi(D)$ with $\psi(a)\neq 0$

Then

$$\frac{1}{f(D)}e^{ax+b}=\frac{1}{(D-a)^r\psi(D)}e^{ax+b}$$

$$=\frac{1}{\psi(a)}\frac{1}{(D-a)^r}e^{ax+b}$$

$$=\frac{1}{\psi(a)}\cdot\frac{x^r}{r!}e^{ax+b}$$

Worked Out Examples

P.I. when $F(x)=e^{ax+b}$

Cases: I_a, I_b, I_c, I_d and II_a, II_b, II_c

Solve the following:

Example 1: $(D^2+2D+1)y=2e^{3x}$

Solution:

C.F.: Here A.E. $m^2+2m+1=0$ i.e., $(m+1)^2=0$ so $m=-1$ is a double root. Thus the C.F. is

$$y_c=(c_1+c_2x)e^{-x}$$

P.I.: $$y_p=\frac{1}{D^2+2D+1}2e^{3x}$$

$$=2\cdot\frac{1}{3^2+2\cdot 3+1}e^{3x}=\frac{e^{3x}}{8}$$

Hence

G.S.: $y=y_c+y_p=(c_1+c_2x)e^{-x}+\dfrac{e^{3x}}{8}$.

Example 2: $(D^3-2D^2-5D+6)y=2e^x+4e^{3x}+7e^{-2x}+8e^{2x}+15$.

Solution:

C.F.: Here A.E. is $m^3-2m^2-5m+6=0$ having roots $m=1,3,-2$. Thus the C.F. is

$$y_c=c_1e^x+c_2e^{3x}+c_3e^{-2x}$$

P.I.: $$y_p=\frac{1}{D^3-2D^2-5D+6}\times$$
$$\times[(2e^x+4e^{3x}+7e^{-2x}+8e^{2x}+15)]$$

$$=\frac{2}{(D+2)(D-3)(D-1)}e^x$$
$$+\frac{4}{(D-1)(D+2)(D-3)}e^{3x}$$
$$+\frac{7}{(D-1)(D-3)(D+2)}e^{-2x}$$
$$+\frac{8}{D^3-2D^2-5D+6}e^{2x}$$
$$+\frac{15}{D^3-2D^2-5D+6}\cdot 1$$

Since $m=1,3,-2$ are roots of A.E., the 1st, 2nd and 3rd terms in the R.H.S. are evaluated using result I_c whereas the 4th and 5th terms by applying result II_a because $m=2$ and $m=0$ are not roots of A.E.

Thus

$$y_p=\frac{2}{-6}\frac{1}{D-1}e^x+\frac{4}{10}\frac{1}{D-3}e^{3x}$$
$$+\frac{7}{15}\frac{1}{D+2}e^{-2x}$$
$$+\frac{8}{2^3-2\cdot 2^2-5\cdot 2+6}\cdot e^{2x}$$
$$+\frac{15}{0-0-0+6}$$

$$\therefore\ y_p=-\frac{1}{3}xe^x+\frac{2}{5}xe^{3x}+\frac{7}{15}xe^{-2x}-2e^{2x}+\frac{15}{6}$$

Hence G.S.: $y=y_c+y_p$ i.e.,

$$y=c_1e^x+c_2e^{3x}+c_3e^{-2x}-\frac{1}{3}xe^x$$
$$+\frac{2}{5}xe^{3x}+\frac{7}{15}xe^{-2x}-2e^{2x}+\frac{15}{6}$$

Example 3: $(D^2-p^2)y = \sinh px$

Solution:

C.F.: Here A.E. is $m^2-p^2=0$ having $m=\pm p$ as the roots. Thus the C.F. is

$$y_c = c_1e^{px}+c_2e^{-px}$$

P.I.: Since p and $-p$ are roots of $m^2-p^2=0$.

Rewriting

$$y_p = \frac{1}{(D^2-p^2)}\sinh px$$

$$= \frac{1}{(D+p)(D-p)}\cdot \sinh px$$

$$= \frac{1}{(D+p)(D-p)}\left(\frac{e^{px}-e^{-px}}{2}\right)$$

$$= \frac{1}{2}\frac{1}{(D+p)(D-p)}e^{px}$$

$$-\frac{1}{2}\frac{1}{(D-p)(D+p)}e^{-px}$$

$$= \frac{1}{2}\frac{1}{p+p}\cdot\frac{1}{D-p}e^{px}-\frac{1}{2}\frac{1}{-p-p}\frac{1}{D+p}e^{-px}$$

Applying result I_c

$$= \frac{1}{2}\frac{1}{2p}x\cdot e^{px}+\frac{1}{2}\frac{1}{2p}x\cdot e^{-px}$$

$$= \frac{x}{2p}\left\{\frac{1}{2}(e^{px}+e^{-px})\right\} = \frac{x}{2p}\cosh px$$

Hence G.S.: y

$$y = y_c+y_p = c_1e^{px}+c_2e^{-px}+\frac{x}{2p}\cosh px.$$

Example 4: $(D-2)^3y = e^{2x}$

Solution:

C.F.: A.E. is $(m-2)^3=0$. So $m=2$ is root of order 3 i.e., repeated three times. Thus C.F. is

$$y_c = (c_1+c_2x+c_3x^2)e^{2x}$$

P.I.: $y_p = \frac{1}{(D-2)^3}e^{2x}$

Case II_a cannot be applied since $f(2)=(D-2)^3$ at $D=2$ is zero. Using result I_d with $n=3$

$$y_p = \frac{1}{(D-2)^3}e^{2x} = \frac{x^3}{3!}e^{2x}$$

Hence G.S.:

$$y = y_c+y_p = (c_1+c_2x+c_3x^2)\,e^{2x}+\frac{x^3}{6}e^{2x}.$$

EXERCISE

Solve the following:

1. $(D^2-5D+6)y = e^{4x}$

   Ans. $y = c_1e^{2x}+c_2e^{3x}+\frac{1}{2}e^{4x}$

2. $(D^2-a^2)y = e^{2x}$

   Ans. $y = c_1e^{ax}+c_2e^{-ax}+\frac{e^{2x}}{3}$ for $a\neq 2$

   $y = c_1e^{2x}+c_2e^{-2x}+\frac{1}{4}xe^{2x}$ for $a=2$

3. $(D^2+4)y = e^{3x}$

   Ans. $y = (c_1e^{2ix}+c_2e^{-2ix})+\frac{1}{13}e^{3x}$

4. $(D^2+2D-8)y = e^{-3x}+e^{-4x}$

   Ans. $y = c_1e^{2x}+c_2e^{-4x}-\frac{e^{-3x}}{5}-\frac{xe^{-4x}}{6}$

5. $(D^2-D+1)y = \sinh x$

   Ans. $y = e^{\frac{x}{2}}\left[c_1\cos\frac{\sqrt{3}x}{2}+c_2\sin\frac{\sqrt{3}x}{2}\right]$

   $+\frac{1}{6}(3e^x-e^{-x})$

6. $(D^2+4D+5)y = -2\cosh x$, with $y(0)=0$, $y'(0)=1$

   Ans. $y = \frac{3}{5}e^{-2x}(\cos x+3\sin x)-\frac{e^x}{10}-\frac{e^{-x}}{2}$

7. $(D^2-2aD+a^2)y = e^{ax}$

   Ans. $y = (c_1+c_2x)e^{ax}+\frac{x^2}{2}e^{ax}$

8. $(D^3+6D^2+9D)y = e^{-3x}$

   Ans. $y = c_1+(c_2+xc_3)e^{-3x}-\frac{x^2e^{-3x}}{6}$

9. $(D^3-5D^2+8D-4)y = e^{2x}+2e^x+3e^{-x}$

   Ans. $y = c_1e^x+c_2e^{2x}+c_3xe^{2x}+\frac{1}{2}x^2e^{2x}$

   $+2xe^x-\frac{e^{-x}}{6}$

10. $(D+2)(D-1)^2y = e^{-2x}+2\sinh x$

    Ans. $y = c_1e^{-2x}+(c_2+c_3x)e^x+xe^{-2x}+\frac{x^2e^x}{6}$

    $+\frac{e^{-x}}{4}$

11. $(D^3-12D+16)y = (e^x+e^{-2x})^2$

    Ans. $y = (c_1+c_2x)e^{2x}+c_3e^{-4x}+\frac{x^2e^{2x}}{12}+\frac{2e^{-x}}{27}$

    $+\frac{xe^{-4x}}{36}$

12. $D(D+1)^2y = 12e^{-x}$

Ans. $y = c_1 + (c_2 + xc_3)e^{-x} - 6x^2e^{-x}$

13. $D^2(D-1)^3(D+1)y = e^x$

Ans. $y = (c_1 + c_2x) + (c_3 + c_4x + c_5x^2)e^x + c_6e^{-x} + \frac{x^3e^x}{12}$

14. $y'' + 4y' + 13y = 18e^{-2x}$, $y(0) = 0$, $y'(0) = 4$

Ans. $y = e^{-2x}\left(\frac{4}{3}\sin 3x - 2\cos 3x\right) + 2e^{-2x}$

P.I. When $F(x) = \sin(ax+b)$ or $\cos(ax+b)$

(Case: $\text{III}_a, \text{III}_b, \text{III}_c, \text{III}_d, \text{III}_e, \text{III}_f$)

Let $\phi(D^2)$ be a rational function of D^2. Since

$$D\sin(ax+b) = a\cos(ax+b)$$

$$D^2\sin(ax+b) = -a^2\sin(ax+b)$$

$$D^3\sin(ax+b) = -a^3\cos(ax+b)$$

$$D^4\sin(ax+b) = a^4\sin(ax+b) = (-a^2)^2\sin(ax+b)$$

It follows that

$\phi(D^2)\sin(ax+b) = \phi(-a^2)\sin(ax+b)$ where $\phi(-a^2)$ is obtained by replacing D^2 in $\phi(D^2)$ by "$-a^2$". Operating with $\frac{1}{\phi(D^2)}$ on both sides

$$\frac{1}{\phi(D^2)}\cdot\phi(D^2)\sin(ax+b) = \frac{1}{\phi(D^2)}\phi(-a^2)\sin(ax+b)$$

$$\sin(ax+b) = \phi(-a^2)\cdot\frac{1}{\phi(D^2)}\sin(ax+b)$$

since $\phi(-a^2)$ is a constant. Thus

$$\frac{1}{\phi(D^2)}\sin(ax+b) = \frac{1}{\phi(-a^2)}\sin(ax+b)$$

provided $\phi(-a^2) \neq 0$. A similar result follows

$$\frac{1}{\phi(D^2)}\cos(ax+b) = \frac{1}{\phi(-a^2)}\cos(ax+b)$$

Note: When $f(D)$ contains terms of odd powers $D, D^3, \ldots$ etc. even after application of the above result, multiply by D so as to get even powers of $D^2, D^4, \ldots$ etc. interpreting D as differentiation and $\frac{1}{D}$ as integration.

Corollary: If $\phi(-a^2) = 0$ then

$$\phi(D^2) = (D^2 + a^2)\psi(D^2)$$

with $\psi(-a^2) \neq 0$. Thus

$$\frac{1}{\phi(D^2)}\sin(ax+b) = \frac{1}{(D^2+a^2)\psi(D^2)}\sin(ax+b)$$

$$= \frac{1}{\psi(-a^2)}\cdot\frac{1}{D^2+a^2}\sin(ax+b)$$

$$= \frac{1}{\psi(-a^2)}\cdot\left(\frac{-x\cos(ax+b)}{2a}\right)$$

Similarly,

$$\frac{1}{\phi(D^2)}\cos(ax+b) = \frac{1}{(D^2+a^2)\psi(D^2)}\cos(ax+b)$$

$$= \frac{1}{\psi(-a^2)}\frac{1}{D^2+a^2}\cos(ax+b)$$

$$= \frac{1}{\psi(-a^2)}\left(\frac{x\sin(ax+b)}{2a}\right)$$

WORKED OUT EXAMPLES

Example 1: $(D^2 - 4D - 5)y = e^{2x} + 3\cos(4x+3)$

Solution: C.F.: The A.E. is $m^2 - 4m - 5 = 0$. The roots are $m = -1, 5$ so that the C.F. is

$$y_c = c_1e^{-x} + c_2e^{5x}.$$

P.I. $$y_p = \frac{1}{D^2 - 4D - 5}[e^{2x} + 3\cos(4x+3)]$$

$$= I_1 + I_2$$

$$I_1 = \frac{1}{2^2 - 4\cdot 2 - 5}e^{2x} = -\frac{1}{9}e^{2x}$$

$$I_2 = \frac{3}{-4^2 - 4D - 5}\cos(4x+3)$$

$$= -\frac{3}{4D+21}\cos(4x+3)$$

Since D^2 terms are not present, we rewrite

$$I_2 = -\frac{3(4D-21)}{(4D+21)(4D-21)}\cdot\cos(4x+3)$$

$$I_2 = \frac{-12D+63}{16D^2 - 21^2}\cos(4x+3)$$

Replace D^2 by -4^2. Then

$$I_2 = \frac{-12D+63}{16(-4^2) - 21^2}\cdot\cos(4x+3)$$

$$= \frac{1}{697}[-12D(\cos(4x+3)) + 63\cos(4x+3)]$$

$$= \frac{-1}{697}[48\sin(4x+3) + 63\cos(4x+3)].$$

Example 2: $(D^2+4)y = \sin 3x + \cos 2x$.

Solution:

C.F.: Here A.E. is $m^2+4=0$ having complex roots $m=\pm 2i$. Thus C.F. is

$$y_c = c_1\cos 2x + c_2 \sin 2x$$

P.I. $$y_p = \frac{1}{D^2+4}(\sin 3x + \cos 2x)$$

$$= \frac{1}{D^2+4}\sin 3x + \frac{1}{D^2+4}\cos 2x$$

$$= I_1 + I_2$$

consider $$I_1 = \frac{1}{D^2+4}\sin 3x$$

Replace D^2 by -3^2 since $f(-a^2) = f(-3^2) \neq 0$

$$I_1 = \frac{1}{-3^2+4}\sin 3x = -\frac{1}{5}\sin 3x$$

consider

$$I_2 = \frac{1}{D^2+4}\cdot\cos 2x$$

Since $f(-a^2) = f(-2^2) = 0$ we apply result III_e

$$I_2 = \frac{x\sin 2x}{4}$$

$$y_p = I_1 + I_2 = -\frac{1}{5}\sin 3x + \frac{x}{4}\sin 2x$$

G.S.: $$y = y_c + y_p = c_1\cos 2x + c_2\sin 2x$$

$$-\frac{1}{5}\sin 3x + \frac{x}{4}\sin 2x$$

Example 3: $(D^2+5D-6)y = \sin 4x \cdot \sin x$.

Solution:

C.F.: Here A.F. is $m^2+5m-6=0$. It has roots $m_1 = 1, -6$. So C.F. is

$$y_c = c_1e^x + c_2e^{-6x}$$

P.I.: $$y_p = \frac{1}{D^2+5D-6}\sin 4x\cdot\sin x$$

$$= \frac{1}{D^2+5D-6}\cdot\frac{1}{2}[\cos 3x - \cos 5x]$$

$$= \frac{1}{2}[I_1 - I_2]$$

Consider

$$I_1 = \frac{1}{D^2+5D-6}\cdot\cos 3x$$

Replacing D^2 by -3^2, we get

$$I_1 = \frac{1}{-3^2+5D-6}\cos 3x = \frac{1}{5(D-3)}\cos 3x$$

To get D^2 terms in denominar, rewrite as

$$I_1 = \frac{1}{5}\frac{(D+3)}{(D-3)(D+3)}\cdot\cos 3x = \frac{1}{5}\frac{(D+3)}{D^2-3^2}\cos 3x$$

Replacing $D^2 = -3^2$, we get

$$I_1 = \frac{1}{5}\frac{(D+3)}{-3^2-3^2}\cos 3x = -\frac{1}{90}(D+3)(\cos 3x)$$

$$= -\frac{1}{90}[-3\sin 3x + 3\cos 3x] = \frac{1}{30}[\sin 3x - \cos 3x]$$

consider

$$I_2 = \frac{1}{D^2+5D-6}\cdot\cos 5x$$

Replacing D^2 by -5^2, we get

$$I_2 = \frac{1}{-5^2+5D-6}\cos 5x = \frac{1}{5D-31}\cos 5x$$

In order to get D^2 terms, we rewrite

$$I_2 = \frac{5D+31}{(5D-31)(5D+31)}\cdot\cos 5x$$

$$= \frac{(5D+31)}{25D^2-31^2}\cos 5x$$

Replacing D^2 by -5^2, we get

$$= \frac{(5D+31)}{25(-5)^2-31^2}\cos 5x$$

$$= -\frac{1}{1586}(5D+31)(\cos 5x)$$

$$= -\frac{1}{1586}[-25\sin 5x + 31\cos 5x]$$

Thus

$$y_p = \frac{1}{2}[I_1 - I_2] = \frac{1}{2}\left[\frac{\sin 3x - \cos 3x}{30} + \frac{31\cos 5x - 25\sin 5x}{1586}\right]$$

Hence G.S.: $y = y_c + y_p$

Example 4:

$$(D^2-4D+1)y = \cos x\cdot\cos 2x + \sin^2 x$$

Solution:

C.F.: Here A.E. is $m^2 - 4m + 1 = 0$ with real distinct roots $m_1 = 2 + \sqrt{3}$ and $m_2 = 2 - \sqrt{3}$ so that the C.F. is

$$y_c = c_1 e^{(2+\sqrt{3})x} + c_2 e^{(2-\sqrt{3})x}$$

P.I.: Note that

$$\cos x \cdot \cos 2x = \frac{1}{2}(\cos 3x + \cos x)$$

and $$\sin^2 x = \frac{1}{2}[1 - \cos 2x]$$

so that $\cos x \cdot \cos 4x + \sin^2 x = \frac{1}{2}[\cos 3x + \cos x + 1 - \cos 2x]$

$$y_p = \frac{1}{D^2 - 4D + 1}[\cos x \cdot \cos 4x + \sin^2 x]$$

$$= \frac{1}{2}\frac{1}{D^2 - 4D + 1}[\cos 3x + \cos x + 1 - \cos 2x]$$

$$= \frac{1}{2}[I_1 + I_2 + I_3 - I_4]$$

Here

$$I_1 = \frac{1}{D^2 - 4D + 1}\cos 3x$$

Replacing D^2 by -3^2

$$I_1 = \frac{1}{-3^2 - 4D + 1}\cos 3x = \frac{-1}{4}\frac{1}{D + 2}\cos 3x$$

To introduce D^2 terms we rewrite the above as

$$I_1 = -\frac{1}{4}\frac{D - 2}{(D + 2)(D - 2)} \cdot \cos 3x$$

$$= -\frac{1}{4}\frac{(D - 2)}{(D^2 - 2^2)}\cos 3x$$

Replacing D^2 by -3^2, we get

$$I_1 = -\frac{1}{4}\frac{(D - 2)}{-3^2 - 2^2}\cos 3x = \frac{1}{52}(D - 2)\cos 3x$$

$$= \frac{1}{52}[-3\sin 3x - 2\cos 3x]$$

Similarly,

$$I_2 = \frac{1}{D^2 - 4D + 1} \cdot \cos x$$

$$= \frac{1}{-1^2 - 4D + 1}\cos x, \quad \because D^2 = -1^2$$

$$= -\frac{1}{4}\frac{1}{D}\cos x = -\frac{1}{4}\int \cos x\,dx = -\frac{1}{4}\sin x$$

Also

$$I_3 = \frac{1}{D^2 - 4D + 1} \cdot 1 = \frac{1}{D^2 - 4D + 1}e^{o \cdot x}$$

$$= \frac{1}{0 - 4 \cdot 0 + 1} = \frac{1}{1} = 1$$

where D is replaced by $a = 0$

Finally

$$I_4 = \frac{1}{D^2 - 4D + 1} \cdot \cos 2x = \frac{1}{-2^2 - 4D + 1}\cos 2x$$

where D^2 is replaced by -2^2

$$I_4 = -\frac{1}{(4D + 3)}\cos 2x$$

Rewriting this to get D^2 terms we have

$$I_4 = -\frac{(4D - 3)}{(4D + 3)(4D - 3)} \cdot \cos 2x$$

$$= -\frac{(4D - 3)}{16D^2 - 3^2}\cos 2x$$

Now replacing D^2 by -2^2, we get

$$I_4 = -\frac{(4D - 3)}{16(-2^2) - 3^2}\cos 2x$$

$$= \frac{(4D - 3)}{73}\cos 2x$$

$$= \frac{1}{73}(4(-2)\sin 2x - 3\cos 2x)$$

$$I_4 = -\frac{1}{73}(8\sin 2x + 3\cos 2x)$$

Thus the P.I.

$$y_p = \frac{1}{2}[I_1 + I_2 + I_3 - I_4]$$

$$y_p = \frac{1}{2}\left[-\frac{1}{52}(3\sin 3x + 2\cos 3x) - \frac{1}{4}\sin x + 1 - \frac{1}{73}(8\sin 2x + 3\cos 2x)\right]$$

G.S.: $y = y_c + y_p$

$$y = c_1 e^{(2+\sqrt{3})x} + c_2 e^{(2-\sqrt{3})x} + 1 - \frac{1}{4}\sin x - \frac{1}{73}(3\cos 2x + 8\sin 2x) - \frac{1}{52}(3\sin 3x + 2\cos 3x)$$

EXERCISE

Solve the following:

1. $(D^4+10D^2+9)y = \cos\,(2x+3)$

Ans. $y = c_1\cos x + c_2\sin x + c_3\cos 3x + c_4\cdot\sin 3x - \frac{1}{15}\cos\,(2x+3)$

2. $(D^2+2D+5)y = 6\sin 2x + 7\cos 2x$

Ans. $y = e^{-x}(c_1\sin 2x + c_2\cos 2x) + 2\sin 2x - \cos 2x$

3. $(D^3+D^2+D+1)y = \sin 2x + \cos 3x$

Ans. $y = c_1e^{-x} + c_2\cos x + c_3\sin x + \frac{1}{15}(2\cos 2x - \sin 2x) - \frac{1}{80}(3\sin 3x + \cos 3x)$

4. $(D^2+4)y = \sin x + \sin 2x$

Ans. $y = c_1\sin 2x + c_2\cos 2x + \frac{\sin x}{3} - \frac{x\cos 2x}{4}$

5. $(D^2-8D+9)y = 8\sin 5x$

Ans. $y = c_1e^{(4+\sqrt{7})x} + c_2e^{(4-\sqrt{7})x} + \frac{1}{29}(5\cos 5x - 2\sin 5x)$

6. $(D^2+16)y = e^{-3x} + \cos 4x$

Ans. $y = c_1\cos 4x + c_2\sin 4x + \frac{1}{25}e^{-3x} + \frac{x}{8}\sin 4x$

7. $(D^2-2D+2)y = e^x + \cos x$

Ans. $y = e^x(c_1\cos x + c_2\sin x) + \left(\frac{\cos x - 2\sin x}{5}\right)$.

8. $(D^2+9)y = \cos^2 x$

Ans. $y = c_1\cos 3x + c_2\sin 3x + \frac{1}{18} + \frac{1}{10}\cos 2x$

9. $(D^2+2D+1)y = e^{2x} - \cos^2 x$

Ans. $y = (c_1+c_2x)e^{-x} + \frac{1}{2} + \frac{1}{5}(2\sin 2x + \cos 2x)$

10. $(D^2+1)y = \cos x$

Ans. $y = c_1\cos x + c_2\sin x + \sin x\ln\sin x - x\cos x$

11. $(D^2-4D+13)y = 8\sin 3x$, $y(0)=1$, $y'(0)=2$

Ans. $y = \frac{1}{5}\left[e^{2x}(\sin 3x + 2\cos 3x) + \sin 3x + 3\cos 3x\right]$

12. $(D^4+2D^2n^2+n^4)y = \cos mx$

Ans. $y = (c_1\cos\eta x + c_2\sin\eta x)(c_3+c_4x) + \frac{1}{\eta^2-m^2}\cos mx$, with $m \neq \eta$

13. $(D^2+4)y = \cos x\cos 3x$

Ans. $y = (c_1\cos 2x + c_2\sin 2x) - \frac{1}{24}\cos 4x + \frac{x}{8}\sin 2x$

14. $(2D^2-2D+1)y = \sin 3x\cdot\cos 2x$

Ans. $y = e^{\frac{x}{2}}\left[c_1\cos\frac{x}{2} + c_2\sin\frac{x}{2}\right] + \frac{10\cos 5x - 49\sin 5x}{5002} + \frac{2\cos x - \sin x}{10}$

15. $(D^3+4D)y = \sin 2x$

Ans. $y = c_1 + c_2\cos 2x + c_3\sin 2x - \frac{x}{8}\sin 2x$.

P.I. When $F(x) = x^m$, m being a Positive Integer

Case IV: Consider $f(D)y = x^m$ so that

$$\text{P.I.} = y_p = \frac{1}{f(D)}x^m$$

Expanding $\frac{1}{f(D)}$ in ascending power of D, we get

$$y_p = [a_0 + a_1D + a_2D^2 + \cdots + a_mD^m]x^m$$

since all the terms beyond D^m are omitted as $D^nx^m = 0$ when $n > m$.

This result can be extended when $F(x) = P_m(x)$ a polynomial in x of degree m so that

$$y_p = [a_0 + a_1D + a_2D^2 + \cdots + a_mD^m][P_m(x)]$$

In particular for

$$(D+a)y = P_m(x)$$

we get

$$\text{P.I.} = y_p = \frac{1}{D+a}[P_m(x)] = \frac{1}{a\left[1+\frac{D}{a}\right]}P_m(x)$$

$$= \frac{1}{a}\left[1+\frac{D}{a}\right]^{-1}P_m(x)$$

$$= \frac{1}{a}\left[1 - \frac{D}{a} + \frac{D^2}{a^2} + \cdots + (-1)^m\frac{D^m}{a^m}\right]P_m(x)$$

wherein terms of order higher than m are omitted.

WORKED OUT EXAMPLES

Solve the following:

Example 1: $(D^6 - D^4)y = x^2$.

Solution:

C.F.: Here A.E. is $m^6 - m^4 = m^4(m^2 - 1) = 0$. Thus $m = 0$ is root repeated 4 times and $m \pm 1$ are the other roots. So the C.F. is

$$y_c = (c_1 + c_2x + c_3x^2 + c_4x^3) + c_5e^x + c_6e^{-x}$$

P.I.: Rewriting $(D^6 - D^4)y = D^4(D^2 - 1)y = x^2$ so that

$$y_p = \frac{1}{D^4(D^2-1)}x^2 = \frac{1}{D^4}\left[\frac{1}{(D^2-1)}x^2\right]$$

Consider $y_p = \frac{1}{(D^2-1)}x^2 = -\frac{1}{(1-D^2)}x^2$
Expanding in Binomial series

$$= -(1-D^2)^{-1}x^2$$
$$= -[1 + D^2 + D^4 + D^6 + \cdots]x^2$$
$$= -x^2 - D^2x^2 + 0 + 0\cdots = -x^2 - 2$$

Thus

$$y_p - \frac{1}{D_4}\left[\frac{1}{(D^2-1)}x^2\right] - \frac{-1}{D^4}[x^2 + 2]$$

Here $\frac{1}{D^4}$ means successive integration 4 times (i.e., integrate w.r.t. x sequentially one after another 4 times).

So

$$\frac{1}{D^4}x^2 = \frac{1}{D^3}\int x^2dx = \frac{1}{D^3}\frac{x^3}{3} = \frac{1}{D^2}\frac{1}{D}\frac{x^3}{3}$$
$$= \frac{1}{D^2}\int\frac{x^3}{3}dx$$
$$= \frac{1}{D^2}\frac{x^4}{12} = \frac{1}{D}\frac{1}{D}\frac{x^4}{12} = \frac{1}{D}\int\frac{x^4}{12}dx = \frac{1}{D}\frac{x^5}{60}$$
$$= \int\frac{x^5}{60}dx = \frac{x^6}{360}$$

Similarly,

$$\frac{1}{D^4}[2] = \frac{1}{D^3}2x = \frac{1}{D^2}x^2 = \frac{1}{D}\frac{x^3}{3} = \frac{x^4}{12}$$

Thus

$$y_p = -\left[\frac{x^6}{360} + \frac{x^4}{12}\right].$$

Hence the G.S. $y = y_e + y_p$

$$y = (c_1 + c_2x + c_3x^2 + c_4x^3 + c_5e^x + c_6e^{-x}) - \left[\frac{x^6}{360} + \frac{x^4}{12}\right]$$

Example 2: $(D^2 - 1)y = 2x^4 - 3x + 1$.

Solution:

C.F.: Here A.E. is $m^2 - 1 = 0$ so that $m = \pm 1$ are the two roots. Thus the C.F. is

$$y_c = c_1e^x + c_2e^{-x}$$

P.I.: $$y_p = \frac{1}{(D^2-1)}(2x^4 - 3x + 1)$$
$$= -\frac{1}{1-D^2}(2x^4 - 3x + 1)$$
$$= -[1 + D^2 + D^4 + D^6 + \cdots][2x^4 - 3x + 1]$$
$$= -[1 + D^2 + D^4][2x^4 - 3x + 1]$$
$$= -\left[(2x^4 - 3x + 1) + (24x^2) + 48\right]$$

G.S.:
$$y = y_c + y_p = c_1e^x + e_2e^{-x} - [2x^4 + 24x^2 - 3x + 49]$$

Example 3: $(D^3 - 1)y = x^5 + 3x^4 - 2x^3$

Solution:

C.F.: Here A.E. is $m^3 - 1 = 0$ having 3 roots $m = 1, \frac{-1\pm\sqrt{3}i}{2}$. Thus C.F. y_c is

$$y_c = c_1e^x + e^{-\frac{x}{2}}\left[c_2\cos\frac{\sqrt{3}}{2}x + c_3\sin\frac{\sqrt{3}}{2}x\right]$$

P.I.: $$y_p = \frac{1}{(D^3-1)}(x^5 + 3x^4 - 2x^3)$$
$$= \frac{-1}{(1-D^3)}[x^5 + 3x^4 - 2x^3]$$
$$= -[1 + D^3 + D^6 + D^9 + \cdots][x^5 + 3x^4 - 2x^3]$$
$$= -\Big[(x^5 + 3x^4 - 2x^3) +$$
$$D^3(x^5 + 3x^4 - 2x^3) + 0 + 0 + \cdots\Big]$$
$$= -\left[(x^5 + 3x^4 - 2x^3) + (60x^2 + 72x^- + 12)\right]$$

Hence G.S.:

$$y = y_c + y_p$$

$$y = c_1 e^x + e^{-\frac{x}{2}}\left[c_2 \cos\frac{\sqrt{3}}{2}x + c_3 \sin\frac{\sqrt{3}}{2}x\right]$$

$$-[x^5 + 3x^4 - 2x^3 + 60x^2 + 72x + 12]$$

Example 4: $(D^2+2)y = x^3+x^2+e^{-2x}+\cos 3x.$

Solution:

C.F.: Here A.E. is $m^2 + 2 = 0$ with complex roots $m = \pm\sqrt{2}i$ so that the C.F. is

$$y_c = (c_1 \cos\sqrt{2}x + c_2 \sin\sqrt{2}x)$$

P.I.:

$$y_p = \frac{1}{(D^2+2)}(x^3 + x^2 + e^{-2x} + \cos 3x)$$

$$= \frac{1}{(D^2+2)}(x^3 + x^2) + \frac{1}{D^2+2}e^{-2x}$$

$$+\frac{1}{D^2+2}\cos 3x$$

$$= I_1 + I_2 + I_3$$

Consider

$$I_1 = \frac{1}{D^2+2}(x^3+x^2) = \frac{1}{2\left(1+\frac{D^2}{2}\right)}(x^3+x^2)$$

$$= \frac{1}{2}\left[1 - \frac{D^2}{2} + \frac{D^4}{4} + \cdots\right][x^3 + x^2]$$

Since x^3 is the higher degree, omit terms containing D^4 and higher orders. Thus

$$I_1 = \frac{1}{2}\left[1 - \frac{D^2}{2}\right][x^3 + x^2]$$

$$= \frac{1}{2}\left[x^3 + x^2 - \frac{1}{2}(6x+2)\right]$$

$$I_1 = \frac{1}{2}[x^3 + x^2 - 3x - 1]$$

Consider $I_2 = \frac{1}{D^2+2}e^{-2x}$.

Replacing D by -2, we get

$$I_2 = \frac{1}{(-2)^2+2}e^{-2x} = \frac{1}{6}e^{-2x}$$

Consider

$$I_3 = \frac{1}{D^2+2}\cos 3x$$

Replacing D^2 by -3^2, we get

$$I_3 = \frac{1}{-3^2+2}\cos 3x = -\frac{1}{7}\cos 3x$$

Thus

$$y_p = \frac{1}{2}[x^3 + x^2 - 3x - 1] + \frac{1}{6}e^{-2x} - \frac{1}{7}\cos 3x$$

G.S.:

$$y = y_c + y_p$$

$$y = \left(c_1 \cos\sqrt{2}x + c_2 \sin\sqrt{2}x\right)$$

$$+\frac{1}{2}(x^3 + x^2 - 3x - 1) + \frac{1}{6}e^{-2x} - \frac{1}{7}\cos 3x$$

EXERCISE

Solve the following:

1. $(D^2 + 3D + 2)y = x^3 + x^2$

Ans. $y = c_1 e^{-x} + c_2 e^{-2x} + \frac{x^3}{2} - \frac{7x^2}{4} + \frac{15x}{4} - \frac{31}{8}$

2. $(D^3 - D)y = 1 + x^5$

Ans. $y = c_1 + c_2 e^x + c_3 e^{-x}$
$- \left(x + 60x^2 + 5x^4 + \frac{x^6}{6}\right)$

3. $(D^4 + D^3 + D^2)y = 5x^2$

Ans. $y = (c_1 + c_2 x)$
$+e^{\frac{-x}{2}}\left[c_3 \cos\frac{\sqrt{3}}{2}x + c_4 \sin\frac{\sqrt{3}}{2}x\right]$
$+\frac{5}{12}x^4 - \frac{5}{3}x^3 + 10x - 10$

4. $(2D^2 + 2D + 3)y = x^2 + 2x - 1$

Ans. $y = e^{-\frac{x}{2}}\left[c_1 \cos\frac{\sqrt{5}}{2}x + c_2 \sin\frac{\sqrt{5}}{2}x\right] + \frac{x^2}{3}$
$+ \frac{2x}{9} - \frac{25}{27}$

5. $(D^2 + D + 1)y = x^3$.

Ans. $y = e^{\frac{-x}{2}}\left[c_1 \cos\frac{\sqrt{3}}{2}x + c_2 \sin\frac{\sqrt{3}}{2}x\right] + x^3$
$-3x^2 + 6.$

6. $(D^2 + 4D + 4)y = x^2 + 2x$ with $y(0) = 0$, $y'(0) = 0$.

Ans. $y = -\frac{3}{8}(1+2x)e^{-2x} + \frac{1}{8}(2x^2+3)$

7. $(D^3 - 2D + 4)y = x^4 + 3x^2 - 5x + 2$

Ans. $y = c_1e^{-2x} + e^x(c_2\cos x + c_3\sin x) + \frac{x^4}{4} + \frac{x^3}{2} + \frac{3x^2}{2} - \frac{5x}{4} - \frac{7}{8}$

8. $(D^4 + 2D^3 - 3D^2)y = x^2 + 3e^{2x} + 4\sin x$

Ans. $y = c_1 + c_2x + c_3e^x + c_4e^{-3x} - \frac{x^2}{108}(3x^2 + 8x + 28) + \frac{3}{20}e^{2x} + \frac{2}{5}(\cos x + 2\sin x)$

9. $(D^2 + 3D + 2)y = e^{-x} + x^2 + \cos x$

Ans. $y = c_1e^{-x} + c_2e^{-2x} - e^{-x} + \frac{2x^2-6x+7}{4}$

$+\frac{1}{10}(\cos x + 3\sin x)$

10. $(D+1)^2y = e^{-x} + x^2$

Ans. $y = (c_1 + c_2x)e^{-x} + \frac{1}{2}\frac{x^2}{2}e^{-x} + x^2 - 4x + 6$

11. $(D^3 - D^2 - 6D)y = 1 + x^2$

Ans. $y = c_1 + c_2e^{-2x} + c_3e^{3x}$

$-\frac{1}{6}\left(\frac{23x}{18} - \frac{x^2}{6} + \frac{x^3}{3}\right)$

12. $(D^3 - D)y = 2x + 1 + 4\cos x + 2e^x$

Ans. $y = c_1 + c_2e^x + c_3e^{-x} + x\,e^x - (x^2 + x) - 2\sin x$.

Exponential Shift

(Case: V)

Book Work: Prove that

$$\frac{1}{f(D)}e^{ax}V(x) = e^{ax}\frac{1}{f(D+a)}V(x)$$

where $V(x)$ is any function of x and $f(D+a)$ is obtained by replacing D by $D+a$ in $f(D)$.

Proof: Let u be a function of x, then

$$\frac{d}{dx}\left(e^{ax}u(x)\right) = D\left(e^{ax}u(x)\right) = ae^{ax}u + e^{ax}Du$$

$$D(e^{ax}u) = e^{ax}(au + Du) = e^{ax}(D+a)u$$

Differentiating once more w.r.t. x

$$D^2(e^{ax}u) = D\left[e^{ax}(D+a)u\right]$$

$$= ae^{ax}(D+a)u + e^{ax}[D^2 + aD]u$$

$$D^2(e^{ax}u) = e^{ax}[D^2u + 2aDu + a^2u] = e^{ax}[D+a]^2u$$

By mathematical induction

$$D^r(e^{ax}u) = e^{ax}(D+a)^r u$$

Thus substituting these values

$$f(D)(e^{ax}u) = e^{ax}f(D+a)(u) \qquad (1)$$

Put

$$f(D+a)(u(x)) = V(x) \qquad (2)$$

Then

$$u(x) = \frac{1}{f(D+a)}V(x) \qquad (3)$$

Substituting (3) in (1)

$$f(D)\left(e^{ax}\frac{1}{f(D+a)}V(x)\right) = e^{ax}V(x) \qquad (4)$$

operating with $\frac{1}{f(D)}$ on both sides of (4), we have

$$\frac{1}{f(D)}\cdot f(D)\cdot\left(e^{ax}\frac{1}{f(D+a)}V(x)\right) = \frac{1}{f(D)}\{e^{ax}V(x)\} \qquad (5)$$

Thus

$$\frac{1}{f(D)}\{e^{ax}V(x)\} = e^{ax}\frac{1}{f(D+a)}V(x)$$

Worked Out Examples

Solve the following:

Example 1: $(D^2 - 4D + 3)y = e^x\cos 2x + \cos 3x$.

Solution:

C.F. Here A.E. is $m^2 - 4m + 3 = 0$ with roots $m = 1, 3$ so that the C.F. is $y_c = c_1e^x + c_2e^{3x}$

P.I.:

$$y_p = \frac{1}{D^2 - 4D + 3}(e^x\cos 2x + \cos 3x)$$

$$= \frac{1}{D^2 - 4D + 3}(e^x\cos 2x)$$

$$+\frac{1}{D^2 - 4D + 3}(\cos 3x) = I_1 + I_2$$

Applying shift result to I_1 replace D by $D + a = D + 1$, we get

$$I_1 = \frac{1}{D^2 - 4D + 3}(e^x\cos 2x)$$

$$= \frac{e^x}{(D+1)^2 - 4(D+1) + 3}\cdot\cos 2x$$

$$I_1 = \frac{e^x}{D^2 - 2D}\cos 2x$$

Replacing D^2 by -2^2 we have

$$I_1 = \frac{e^x}{-2^2 - 2D}\cos 2x = \frac{-e^x}{2}\frac{1}{D+2}\cos 2x$$

In order to get D^2 terms we rewrite the above as

$$= -\frac{e^x}{2}\cdot\frac{D-2}{(D+2)(D-2)}\cos 2x$$

$$= -\frac{e^x}{2}\frac{D-2}{D^2-2^2}\cdot\cos 2x$$

Replace D^2 by -2^2 then

$$I_1 = -\frac{e^x}{2}\frac{(D-2)}{-2^2-2^2}\cos 2x = \frac{e^x}{16}(D-2)(\cos 2x)$$

$$= \frac{e^x}{16}[-2\sin 2x - 2\cos 2x]$$

$$= -\frac{e^x}{8}[\sin 2x + \cos 2x]$$

Consider

$$I_2 = \frac{1}{D^2 - 4D + 3}\cdot\cos 3x$$

Replacing D^2 by -3^2, we get

$$I_2 = \frac{1}{-3^2 - 4D + 3}\cos 3x = -\frac{1}{2(2D+3)}\cos 3x$$

Rewrite this to get D^2 terms as

$$I_2 = -\frac{1}{2}\frac{2D-3}{(2D+3)(2D-3)}\cdot\cos 3x$$

$$= -\frac{1}{2}\frac{(2D-3)}{(4D^2-3^2)}\cos 3x$$

Replace D^2 by -3^2 then

$$I_2 = -\frac{1}{2}\frac{2D-3}{4(-3^2)-3^2}\cos 3x = \frac{1}{90}(2D-3)(\cos 3x)$$

$$I_2 = \frac{1}{90}[-6\sin 3x - 3\cos 3x]$$

$$= -\frac{1}{30}[2\sin 3x + \cos 3x]$$

Thus

$$y_p = I_1 + I_2$$

$$= -\frac{e^x}{8}[\sin 2x + \cos 2x] - \frac{1}{30}[2\sin 3x + \cos 3x]$$

Hence G.S.:

$$y = y_c + y_p u$$

i.e.,

$$y = c_1 e^x + c_2 e^{3x} - \frac{e^x}{8}[\sin 2x + \cos 2x]$$
$$- \frac{1}{30}[2\sin 3x + \cos 3x].$$

Example 2: $(D^2 + 2)y = x^2 e^{3x} + e^x \cos 2x$

Solution:

C.F.: Here the A.E. is $m^2 + 2 = 0$ with roots $m = \pm\sqrt{2}i$ so that the C.F. is

$$y_c = (c_1 \cos\sqrt{2}x + c_2 \sin\sqrt{2}x)$$

P.I.: $$y_p = \frac{1}{D^2+2}[x^2 e^{3x} + e^x\cdot\cos 2x]$$

$$= \frac{1}{D^2+2}[e^{3x}\cdot x^2] + \frac{1}{D^2+2}[e^x\cdot\cos 2x]$$

$$= I_1 + I_2$$

Applying shift result with $a = 3$ replace D by $D+3$ in I_1.

Then

$$I_1 = \frac{1}{D^2+2}[e^{3x}\cdot x^2] = \frac{e^{3x}}{(D+3)^2+2}[x^2]$$

$$= \frac{e^{3x}}{D^2+6D+11}x^2$$

$$= \frac{e^{3x}}{11\left[1+\frac{D^2+6D}{11}\right]}\cdot x^2$$

Expanding in powers of $\frac{D^2+6D}{11}$, we get

$$I_1 = \frac{e^{3x}}{11}\left[1 - \left(\frac{D^2+6D}{11}\right) + \left(\frac{D^2+6D}{11}\right)^2 - \left(\frac{D^2+6D}{11}\right)^3 + \cdots\right]x^2$$

Since x^2 is the highest power, discord terms of D^3 and higher orders. Then

$$I_1 = \frac{e^{3x}}{11}\left[1 + \left(-\frac{D^2}{11} - \frac{6D}{11}\right) + \left(\frac{36}{11^2}D^2\right)\right]x^2$$

$$= \frac{e^{3x}}{11}\left[x^2 - \frac{12x}{11} + \frac{50}{121}\right]$$

Consider

$$I_2 = \frac{1}{D^2+2} e^x \cos 2x$$

Applying shift result, replace D by $D+1$ in the above then

$$I_2 = \frac{e^x}{(D+1)^2+2} \cdot \cos 2x = \frac{e^x}{D^2+2D+3} \cos 2x$$

Replace D^2 by -2^2, we get

$$I_2 = \frac{e^x}{-2^2+2D+3} \cos 2x = \frac{e^x}{2} \frac{1}{2D-1} \cos 2x$$

To have D^2 terms, rewrite the above as

$$I_2 = e^x \cdot \frac{2D+1}{(2D-1)(2D+1)} \cos x$$

$$= \frac{e^x(2D+1)}{4D^2-1^2} \cos 2x$$

Replace D^2 by -2^2, we get

$$I_2 = \frac{e^x(2D+1)}{4\cdot(-2^2)-1^2} \cos 2x = \frac{e^x(2D+1)}{-17} \cos 2x$$

$$I_2 = -\frac{e^x}{17}[-4 \sin 2x + \cos 2x]$$

Thus $y_p = I_1 + I_2 = \frac{e^{3x}}{11}\left[x^2 - \frac{12x}{11} + \frac{50}{121}\right] + \frac{e^x}{17}$ $[4 \sin 2x - \cos 2x]$

Hence G.S.: $y = y_c + y_p$

$$y = [c_1 \cos \sqrt{2}x + c_2 \sin \sqrt{2}x] + \frac{e^{3x}}{11}$$

$$\times \left[x^2 - \frac{12x}{11} + \frac{50}{121}\right] + \frac{e^x}{17}[4 \sin 2x - \cos 2x]$$

EXERCISE

Solve the following:

1. $(D^2 - 4D + 3)y = 2xe^{3x} + 3e^x \cos 2x$

Ans. $y = c_1e^x + c_2e^{3x} + \frac{e^{3x}}{2}(x^2 - x)$ $-\frac{3e^x}{8}(\cos 2x + \sin 2x)$

2. $(D^2 + 5D + 6)y = e^{-2x}(\sec^2 x)(1 + 2\tan x)$

Ans. $y = c_1e^{-2x} + c_2e^{-3x} + e^{-2x} \tan x$

3. $(D^2 + 4)y = e^x \sin^2 x$

Ans. $y = c_1 \cos 2x + c_2 \sin 2x + \frac{e^x}{2}\left[\frac{1}{5} - \frac{1}{17}(4 \sin 2x + \cos 2x)\right]$

4. $(D^2 + 4D + 3)y = e^{-x} \sin x + x$

Ans. $y = c_1e^x + c_2e^{-3x} + \frac{1}{16}e^x(2x^2 - x) - \frac{1}{27}(9x^3 + 18x^2 + 42x + 40)$

5. $(D^4 - 1)y = \cos x \cdot \cosh x$

Ans. $y = c_1e^x + c_2e^{-x} + c_3 \cos x + c_4 \cos x - \frac{1}{5} \cos x \cdot \cosh x$

6. $(D^2 - 4D + 4)y = e^{2x} \cos^2 x,$

Ans. $y = (c_1 + c_2x)e^{2x} + e^{2x}\left(\frac{x^2}{4} - \frac{1}{8}\cos 2x\right)$

7. $(D^2 + 4D + 5)y = e^{-2x}(1 + \cos x)$

Ans. $y = e^{-2x}\left(c_1 \cos x + c_2 \sin x + 1 + \frac{x}{2} \sin x\right)$

8. $(D^2 - 6D + 13)y = 8e^{3x} \sin 2x$

Ans. $y = e^{3x}(c_1 \cos 2x + c_2 \sin 2x - 2x \cos 2x)$

9. $(D^2 - 4)y = x \sinh x$

Ans. $y = c_1e^{2x} + c_2e^{-2x} - \frac{x}{2} \sinh x - \frac{2}{9} \cosh x$

10. $(D^3 - 2D + 4)y = e^x \sin x$

Ans. $y = c_1e^{-2x} + (c_2 \cos x + c_3 \sin x)e^x + (3 \sin x + \cos x)\frac{xe^x}{20}$

11. $(D^2 - 2D + 1)y = x^2e^{3x}$

Ans. $y = (c_1 + c_2x)e^x + (2x^2 - 4x + 3)\frac{e^{3x}}{8}$

P.I. When $F(x) = xV(x)$

(Case: VI)

Book Work: Prove that

$$\frac{1}{f(D)}\{xV(x)\} = x\frac{1}{f(D)}V(x) - \frac{f'(D)}{\{f(D)\}^2}V(x)$$

where $V(x)$ is any function of x and $f'(D)$ is the derivative of $f(D)$ w.r.t. D.

Proof: Let $z = xU(x)$ where U is any function of x.
Then $Dz = xDU + U$
$D^2z = D(xDU + U) = xD^2U + DU + DU$
$D^2z = xD^2U + 2DU$
$D^3z = D(xD^2U + 2DU) = xD^3U + 3DU$
By mathematical induction

$$D^r z = xD^rU + rD^{r-1}U = xD^rU + \left(\frac{d}{dD}D^r\right)U$$

Substituting these values in $f(D)$, we get

$$f(D)(xU) = xf(D)U + \left(\frac{d}{dD}f(D)\right)U \quad (1)$$

Put $\quad f(D)U = V(x) \quad (2)$

Then $\quad U = \frac{1}{f(D)}V \quad (3)$

Substituting U from (3) in (1)

$$f(D)\left[x\cdot\frac{1}{f(D)}V\right] = xf(D)\cdot\frac{1}{f(D)}V + \left[\frac{d}{dD}f(D)\right]\left[\frac{1}{f(D)}V\right]$$

Rearranging the terms

$$xV = f(D)\left[x\frac{1}{f(D)}V\right] - f'(D)\cdot\frac{1}{f(D)}V \quad (4)$$

Operating with $\frac{1}{f(D)}$ on both sides of (4), we get

$$\frac{1}{f(D)}\{xV\} = \frac{1}{f(D)}f(D)\left[x\frac{1}{f(D)}V\right] - \frac{1}{f(D)}f'(D)\frac{1}{f(D)}V$$

Rewriting

$$\frac{1}{f(D)}\{xV\} = x\frac{1}{f(D)}V - \frac{f'(D)}{\{f(D)\}^2}V$$

$$\frac{1}{f(D)}\{xV\} = \left[x\frac{1}{f(D)} - \frac{f'(D)}{\{f(D)\}^2}\right]V$$

WORKED OUT EXAMPLES

Solve the following:

Example 1: $(D^2 - 2D + 1)y = xe^x \sin x$

Solution:

C.F.: Here $m^2 - 2m + 1 = 0$ is the A.E. with $m = 1$ as a double root so that the C.F. y_c is

$$y_c = (c_1 + c_2x)e^x$$

P.I.: $y_p = \frac{1}{(D^2 - 2D + 1)}x(e^x \sin x)$

$$= \frac{1}{(D-1)^2}e^x(x \sin x)$$

using shift result with $a = 1$ so that D is replaced by $D + 1$, we get

$$y_p = \frac{e^x}{[(D+1)-1]^2}(x\sin x) = \frac{e^x}{D^2}x \sin x$$

Applying result VI

$$\frac{1}{D^2}x(\sin x) = x\cdot\frac{1}{D^2}\sin x - \frac{2D}{D^4}\sin x \quad (5)$$

$$= x(-\sin x) - 2\cos x \quad (6)$$

Thus $\quad y_p = e^x[-x\sin x - 2\cos x] \quad (7)$

Hence G.S.: $y = y_c + y_p$

$$y = (c_1 + c_2x)e^x - e^x(x \sin x + 2\cos x)$$

Example 2: $(D^2 - 1)y = x^2 \cos x$

Solution:

C.F.: Here A.E. is $m^2 - 1 = 0$ with $m = \pm 1$ as the roots so that the C.F. is

$$y_c = c_1e^x + c_2e^{-x}$$

P.I.:

$$y_p = \frac{1}{(D^2-1)}\cdot x^2\cos x = \frac{1}{D^2-1}x\cdot(x\cos x)$$

Apply result VI

$$y_p = \frac{x}{D^2-1}(x\cos x) - \frac{2D}{(D^2-1)^2}\cdot x\cos x = I_1 - 2I_2$$

Consider

$$I_1 = \frac{x}{D^2-1}(x\cos x)$$

Applying result VI once again

$$I_1 = x\cdot\left[x\cdot\frac{1}{D^2-1}\cos x - \frac{2D}{(D^2-1)^2}\cos x\right]$$

Replacing D^2 by -1^2 in the denominator

$$I_1 = x^2\cdot\frac{1}{-1^2-1}\cos x - \frac{x2D}{(-1^2-1)^2}\cos x$$

$$= -\frac{-x^2\cos x}{2} + \frac{x\sin x}{2}$$

Consider

$$I_2 = \frac{D}{(D^2-1)^2}x\cos x, \text{ applying VI}$$

$$= D\left[\frac{x}{(D^2-1)^2}\cdot \cos x - \frac{2(D^2-1)\cdot 2D}{(D^2-1)^4}\cos x\right]$$

Replacing D^2 by -1^2 in the denominator

$$I_2 = D\left[\frac{x}{(-1^2-1)^2}\cos x - \frac{4D}{(-1^2-1)^3}\cos x\right]$$

$$= \frac{1}{4}D[x\cos x] + \frac{1}{2}D^2(\cos x)$$

$$= \frac{1}{4}[\cos x - x\sin x] - \frac{1}{2}\cos x$$

Thus

$$y_p = I_1 - 2I_2 = \frac{1}{2}[-x^2\cos x + x\sin x]$$

$$-\frac{1}{2}[\cos x - x\sin x] + \cos x$$

$$= \left(\frac{1-x^2}{2}\right)\cos x + x\sin x$$

Ans. $y = c_1e^x + c_2e^{-x} + x\sin x + \frac{(1-x^2)\cos x}{2}$

Example 3: $(D^2-1)y = x\sin x + x^2e^x$

Solution:
C.F.: Here A.E. is $m^2 - 1 = 0$ with $m = \pm 1$ as roots so that the C.F. $y_c = c_1e^x + c_2e^{-x}$.
P.I.: $y_p = \frac{1}{D^2-1}x\sin x + \frac{1}{D^2-1}x^2e^x = I_1 + I_2$
Applying result VI for I_1, we get

$$I_1 = \frac{1}{D^2-1}x\sin x$$

$$= \frac{x}{D^2-1}\sin x - \frac{2D}{(D^2-1)^2}\sin x$$

Replacing D^2 by -1^2 in the denominator

$$I_1 = \frac{x}{-1^2-1}\sin x - \frac{2}{(-1^2-1)^2}\cdot D\sin x$$

$$= -\frac{x\sin x}{2} - \frac{\cos x}{2}$$

$$I_1 = -\frac{1}{2}(x\sin x + \cos x)$$

Applying result VI to I_2, we have

$$I_2 = \frac{1}{D^2-1}x^2e^x$$

$$= x\cdot\frac{1}{D^2-1}(xe^x) - \frac{2D}{(D^2-1)^2}(xe^x)$$

Applying result VI once again to each of the two terms in the R.H.S., we get

$$I_2 = x\left[x\frac{1}{D^2-1}e^x - \frac{2D}{(D^2-1)}e^x\right]$$

$$-2D\left[x\frac{1}{(D^2-1)^2}e^x - \frac{2(D^2-1)2D}{(D^2-1)^4}e^x\right]$$

$$= \left[x^2\frac{1}{(D+1)(D-1)}e^x - \frac{2xD}{(D+1)^2(D-1)^2}e^x\right]$$

$$-2D\left[x\frac{1}{(D+1)^2(D-1)^2}e^x\right.$$

$$\left.-\frac{4D}{(D+1)^3(D-1)^3}e^x\right]$$

Replacing D by 1, we get

$$= \left[x^2\frac{1}{2(D-1)}e^x - \frac{2xD}{4(D-1)^2}e^x\right]$$

$$-2D\left[\frac{x}{4}\frac{1}{(D-1)^2}e^x - \frac{4}{8}\frac{D}{(D-1)^3}e^x\right].$$

Applying result I_c, I_d

$$I_2 = \frac{x^2}{2}xe^x - \frac{x}{e}D\cdot\left(\frac{x^2e^x}{2!}\right)$$

$$-2D\left[\frac{x}{4}\cdot\frac{x^2e^x}{2!} - \frac{1}{2}D\frac{x^3e^x}{3!}\right]$$

$$= \frac{x^3e^x}{2} - \frac{x}{4}[x^2e^x + 2xe^x] - \frac{1}{4}[x^3e^x + 3x^2e^x]$$

$$+\frac{1}{6}[x^3e^x + 6x^2e^x + 6xe^x]$$

$$= xe^x\left[\frac{x^2}{2} - \frac{x^2}{4} - \frac{x}{2} - \frac{x^2}{4} - \frac{3}{4}x + \frac{1}{6}x^2 + x + 1\right]$$

$$I_2 = xe^x\left[\frac{x^2}{6} - \frac{x}{4} + 1\right] = \frac{xe^x}{12}[2x^2 - 3x + 12]$$

$$y_p = I_1 + I_2 = -\frac{1}{2}(x\sin x + \cos x)$$

$$+\frac{xe^x}{12}(2x^2 - 3x + 12)$$

Thus G.S.: $y = y_c + y_p$

$$y = c_1e^x + c_2e^{-x} - \frac{1}{2}(x\sin x + \cos x)$$

$$+\frac{xe^x}{12}(2x^2 - 3x + 12)$$

EXERCISE

Solve the following:

1. $(D^2+3D+2)y = x\sin 2x$

Ans. $y = c_1e^{-x}+c_2e^{-2x} - \frac{(30x-7)}{200}\cos 2x - \left(\frac{5x-12}{100}\right)\sin 2x$

2. $(D^2+2D+1)y = x\cos x$

Ans. $y = (c_1+c_2x)e^{-x} + \frac{x}{2}\sin x + \frac{1}{2}(\cos x - \sin x)$

3. $(D^2-1)y = x^2\sin 3x$

Ans. $y = c_1e^x + c_2e^{-x}\frac{-x^2}{10}\cdot\sin 3x - \frac{3x}{25}\cos 3x + \frac{13}{250}\sin 3x$

4. $(D^3-3D^2-6D+8)y = xe^{-3x}$

Ans. $y = c_1e^x + c_2e^{4x} + c_3e^{-2x} - \frac{e^{-3x}}{784}(28x+39)$

5. $(D^2-1)y = x\sin x + e^x(1+x^2)$

Ans. $y_1 = c_1e^x + c_2e^{-x} - \frac{1}{2}(x\sin x + \cos x) + \frac{e^x}{2}\left(\frac{x^3}{3} - \frac{x^2}{2} + \frac{3x}{2} - \frac{3}{4}\right)$

6. $(D^2+4)y = x\sin^2 x$

Ans. $y = c_1\cos 2x + c_2\sin 2x + \frac{x}{8} - \frac{x\cos 2x + 2x^2\sin 2x}{32}$

7. $(D^2+1)y = x^2e^{2x} + x\cos x$

Ans. $y = (c_1\cos x + c_2\sin x) + \frac{e^{2x}}{5}\left[x^2 - \frac{8}{5}x + \frac{22}{25}\right] + \left(\frac{x^2}{4} - \frac{1}{8}\right)\sin x + \frac{x}{4}\cos x$

8. $(D^4-2D^3-3D^2+4D+4)y = x^2e^x$

Ans. $y_1 = (c_1+c_2x)e^{-x} + (c_3+c_4x)e^{2x} + \frac{e^x}{4}\left(x^2-2x-\frac{3}{2}\right)$

9. $(D^2-5D+6)y = xe^{4x}$

Ans. $y = c_1e^{2x} + c_2e^{3x} + e^{4x}\frac{(2x-3)}{4}$

10. $(D^2+a^2)y = \sec ax$

Ans. $y = c_1\cos ax + c_2\sin ax + \frac{x}{a}\sin ax + \frac{1}{a^2}\cos ax\cdot\log(\cos ax)$

Table of Particular Integrals

Inverse operator (short) methods to find particular integral of (6) when R.H.S. $F(x)$ is of the form

Case	Form of $F(x)$	Particular Integral (P.I.): y_p	Subject to the condition that	Remarks/ observation
I_a	Any function of x	$y_p = \frac{1}{D-a}F(x) = e^{ax}\int F(x)e^{-ax}dx$	$c = 0$	Substitute F(x) and integrate w.r.t., x
I_b	$F(x) = b =$ constant	$y_p = \frac{1}{D-a}b = -\frac{b}{a}$	$a \neq 0$	
I_c	$F(x) = e^{ax}$	$y_p = \frac{1}{D-a}e^{ax} = xe^{ax}$		
I_d	$F(x) = e^{ax}$	$y_p = \frac{1}{(D-a)^n}e^{ax} = \frac{x^n}{n!}e^{ax}$		Repeated factor n times
I_e	Any function of x	If $f(D) = (D-a_1)(D-a_2)\cdots(D-a_n)$ Resolving into partial fractions $y_p = \frac{1}{f(D)}F(x) = \left(\frac{A_1}{D-a_1} + \frac{A_2}{D-a_2} + \cdots + \frac{A_n}{D-a_n}\right)F(x)$ Applying result I_a $y_p = A_1e^{a_1x}\int F(x)e^{-a_1x}dx + \cdots + A_ne^{a_nx}\int F(x)e^{-a_nx}dx$		Most general case Resolve $F(D)$ into factors Resolve $\frac{1}{f(D)}$ into partial fractions Apply result I_a to each term
II_a	e^{ax+b}	$y_p = \frac{1}{f(D)}e^{ax+b} = \frac{1}{f(a)}e^{ax+b}$	$f(a) \neq 0$	Replace D by 'a' in $f(D)$
II_b	Constant $= b = be^{0x}$ with $a = 0$	$y_p = \frac{1}{f(D)}b = \frac{b}{f(0)}$	$f(0) \neq 0$	Replace D by zero in $f(D)$
II_c	e^{ax+b}	$y_p = \frac{1}{f(D)}e^{ax+b} = \frac{1}{(D-a)^n g(D)}e^{ax+b}$ $= \frac{1}{g(a)}\frac{1}{(D-a)^n}e^{ax+b}$ $y_p = \frac{1}{g(a)}\frac{x^n}{n!}e^{ax+b}$	$g(a) \neq 0$	Replace D by 'a' in $g(D)$ and apply result of I_d

(Contd.)

(Continued)

Case	Form of $F(x)$	Particular Integral (P.I.): y_p	Subject to the condition that	Remarks/ observation
III_a	$\sin(ax+b)$	$y_p = \frac{1}{f(D^2)}\sin(ax+b)$ $= \frac{1}{f(-a^2)}\sin(ax+b)$	$f(-a^2) \neq 0$	Replace D^2 by $-a^2$ in $f(D)$ when terms of $D, D^3, \ldots$ are present,
III_b	$\cos(ax+b)$	$y_p = \frac{1}{f(D^2)}\cos(ax+b)$ $= \frac{1}{f(-a^2)}\cos(ax+b)$	$f(-a^2) \neq 0$	Multiply by D in order to get $D^2, D^4, \ldots$, etc. terms. Then replace D^2 by $-a^2$, etc.
III_c	$\sin(ax+b)$	$y_p = \frac{1}{D^2+a^2}\sin(ax+b)$ $= -\frac{x\cdot\cos(ax+b)}{(2a)}$	$a \neq 0$	sin is replaced by cos and multiplied by $-\frac{x}{2a}$
III_d	$\sin(ax+b)$	$y_p = \frac{1}{(D^2+a^2)^n}\sin(ax+b)$ $= \frac{x^n}{(2a)^n n!}\sin(ax - n\frac{\pi}{2})$	$a \neq 0$	
III_e	$\cos(ax+b)$	$y_p = \frac{1}{D^2+a^2}\cos(ax+b)$ $= \frac{x\cdot\sin(ax+b)}{2a}$	$a \neq 0$	cos is replaced by sin and multiplied by $\frac{x}{2a}$
III_f	$\cos(ax+b)$	$y_p = \frac{1}{(D^2+a^2)^n}\cos(ax+b)$ $= \frac{x^n}{(2a)^n}\frac{1}{n!}\cos(ax - \frac{n\pi}{2})$	$a \neq 0$	
IV	x^m	$y_p = \frac{1}{f(D)}x^m = [f(D)]^{-1}x^m$ $y_p = [a_0 + a_1 D + a_2 D^2 + \cdots$ $+ a_m D^m]x^m$	$a_0 \neq 0$	Expand $[f(D)]^{-1}$ is ascending powers of D and delete all terms beyond D^m (since $D^n x^m = 0$ when $n > m$)
V	$e^{ax}V(x)$	$y_p = \frac{1}{f(D)}e^{ax}V(x) = e^{ax}\frac{1}{f(D+a)}V(x)$ (Exponential shift)	$f(D+a) \neq 0$	Replace D by $D+a$ in $f(D)$. Then evaluate $\frac{1}{f(D+a)}V(x)$ by the above methods
VI	$xV(x)$	$y_p = \frac{1}{f(D)}xV(x)$ $= x\frac{1}{f(D)}V(x) - \frac{f^{1}(D)}{\{f(D)\}^2}V(x)$		

9.6 DIFFERENTIAL EQUATIONS WITH VARIABLE COEFFICIENTS: REDUCIBLE TO EQUATIONS WITH CONSTANT COEFFICIENTS

Special Case A: Euler or Cauchy or Euler-Cauchy or Cauchy-Euler Differential Equation or Equi-dimensional equation

The equation of nth order is of the form

$$(a_0 x^n D^n + a_1 x^{n-1} D^{n-1} + \cdots + a_{n-1} x D + a_n)y = F(x) \qquad (1)$$

where $a_0, a_1, a_2, \ldots, a_n$ are all constants

and $F(x)$ is a function of x

By the substitution

$$x = e^t \qquad (2)$$

the Euler Equation (1) can be transformed into a linear D.E. with constant coefficients with t as the independent variable, which can be solved by the methods described in the earlier Sections 9.4 and 9.5.

From (2), $t = \ln x$ and $\frac{dx}{dt} = e^t$ so that

$$\frac{dy}{dx} = \frac{dy}{dt}\frac{dt}{dx} = e^{-t}\frac{dy}{dt} = \frac{1}{x}\frac{dy}{dt} \qquad (3)$$

Introducing $\mathcal{D} \equiv \frac{d}{dt}$, we have

$$\frac{dy}{dx} = Dy = \frac{1}{x}\frac{dy}{dt} = \frac{1}{x}\mathcal{D}y \qquad (4)$$

or $\quad xDy = \mathcal{D}y$

Similarly, differentiating (3) once more w.r.t. x, we

get

$$\frac{d^2y}{dx^2} = \frac{d}{dx}\left[e^{-t}\frac{dy}{dt}\right] = \frac{d}{dt}\left[e^{-t}\frac{dy}{dt}\right]\frac{dt}{dx}$$

$$= e^{-t}\left[e^{-t}\frac{d^2y}{dt^2} - e^{-t}\frac{dy}{dt}\right] = e^{-2t}\left[\frac{d^2y}{dt^2} - \frac{dy}{dt}\right]$$

Thus

$$\frac{d^2y}{dx^2} = D^2y = e^{-2t}\left[\mathcal{D}^2y - \mathcal{D}y\right] = \frac{1}{x^2}[\mathcal{D}^2y - \mathcal{D}y]$$

Thus

$$x^2D^2y = (\mathcal{D}^2 - \mathcal{D})y = \mathcal{D}(\mathcal{D}-1)y \qquad (5)$$

Similarly, differentiating again, we get

$$x^3D^3 = \mathcal{D}^3 - 3\mathcal{D}^2 + 2\mathcal{D} = \mathcal{D}(\mathcal{D}-1)(\mathcal{D}-2) \qquad (6)$$

$$x^4D^4 = \mathcal{D}^4 - 6\mathcal{D}^3 + 11\mathcal{D}^2 - 6\mathcal{D}$$

$$= \mathcal{D}(\mathcal{D}-1)(\mathcal{D}-2)(\mathcal{D}-3) \qquad (7)$$

and by mathematical induction

$$x^rD^r = \mathcal{D}(\mathcal{D}-1)(\mathcal{D}-2)\ldots(\mathcal{D}-(r-1)) \qquad (8)$$

Method of solving Cauchy-Euler equation

I. Put $x = e^t$ and use (4), (5), (6), etc. to transform Euler equation to D.E. with constant coefficients.

II. Solve this D.E. with t as the independent variable by methods described in Sections 9.3 to 9.5.

III. Replace t by $\ln x$ in the solution obtained in step II.

The second order equation

$$ax^2y'' + bxy' + cy = Q(x)$$

reduces to

$$a\left(\frac{d^2y}{dt^2} - \frac{dy}{dt}\right) + b\left(\frac{dy}{dt}\right) + cy = Q^*(t)$$

$$a\mathcal{D}^2y + (b-a)\mathcal{D}y + cy = Q^*(t)$$

where $Q^*(t)$ is $Q(x)$ with x replaced by e^t.

Worked Out Examples

Solve the following:

Example 1: $x^2y'' - 3xy' + 3y = 0$ with $y(1) = 0$, $y'(1) = -2$

Solution: Substituting (2), (3), (4), (5) in the given D.E., we get

$$\left(\frac{d^2y}{dt^2} - \frac{dy}{dt}\right) - 3\frac{dy}{dt} + 3y = 0$$

or $\frac{d^2y}{dt^2} - 4\frac{dy}{dt} + 3y = 0$

This is 2nd order D.E. whose A.E. is

$$m^2 - 4m + 3 = 0$$

The roots are $m = 1, 3$

G.S. $\quad y(t) = c_1e^t + c_2e^{3t}$

$\quad y(x) = c_1x + c_2x^3$

Since $\quad 0 = y(1) = c_1 + c_2$

$\quad -2 = y^1(1) = c_1 + 3c_2$

$c_1 = -c_2$ so that $c_2 = -1$ thus $c_1 = 1$

Hence the particular solution

$$y = x - x^3$$

Example 2: $x^3\frac{d^3y}{dx^3} + 3x^2\frac{d^2y}{dx^2} + x\frac{dy}{dx} + y = x + \ln x$

Solution: Substituting (2), (4), (5), (6) in the given D.E., we get

$$\mathcal{D}(\mathcal{D}-1)(\mathcal{D}-2)y + 3\mathcal{D}(\mathcal{D}-1)y + \mathcal{D}y + y = e^t + t$$

or

$$\mathcal{D}^3y + y = e^t + t$$

C.F.: The A.E. is $m^3 + 1 = 0$ having roots $m = -1, \frac{1\pm\sqrt{3}i}{2}$ so that the C.F. y_c is

$$y_c = c_1e^{-t} + e^{\frac{t}{2}}\left(c_2\cos\frac{\sqrt{3}}{2}t + c_3\sin\frac{\sqrt{3}}{2}t\right)$$

P.I. : $$y_p = \frac{1}{\mathcal{D}^3+1}\{e^t + t\} = \frac{1}{\mathcal{D}^3+1}e^t + \frac{1}{\mathcal{D}^3+1}t$$

$$= \frac{1}{1^3+1}e^t + \{1 - \mathcal{D}^3 + \mathcal{D}^6 + \ldots\}t$$

$$= \frac{e^t}{2} + t - 0 + 0 + \ldots$$

G.S: $y = y_c + y_p$

$$y = c_1 e^{-t} + e^{\frac{t}{2}}\left(c_2 \cos\frac{\sqrt{3}}{2}t + c_3 \sin\frac{\sqrt{3}}{2}t\right) + \frac{e^t}{2} + t$$

Replacing t by $\ln x$

$$y(x) = \frac{c_1}{x} + \sqrt{x}\left(c_2 \cos\frac{\sqrt{3}}{2}\ln x + c_3 \sin\frac{\sqrt{3}}{2}\ln x\right)$$

$$+\frac{x}{2} + \ln x.$$

Example 3: $x^2\frac{d^2y}{dx^2} - 3x\frac{dy}{dx} + 5y = x^2 \sin(\log x)$

Solution: Using (2), (3), (4), (5) the D.E. reduces to

$$\left(\frac{d^2y}{dt^2} - \frac{dy}{dt}\right) - 3\left(\frac{dy}{dt}\right) + 5y + e^{2t}\sin t$$

$$\frac{d^2y}{dt^2} - 4\frac{dy}{dt} + 5y = e^{2t}\sin t$$

C.F.: Here A.E. is $m^2 - 4m + 5 = 0$ having roots $m = 2 \pm i$ so that the C.F. y_c is

$$y_c = e^{2t}(c_1 \cos t + c_2 \sin t)$$

P.I. $$y_p = \frac{1}{\mathcal{D}^2 - 4\mathcal{D} + 5}e^{2t}\sin t$$

Using shift result replace $\mathcal{D}$ by $\mathcal{D}+2$

$$y_p = \frac{e^{2t}}{(\mathcal{D}+2)^2 - 4(\mathcal{D}+2) + 5}\sin t$$

$$y_p = \frac{e^{2t}}{\mathcal{D}^2 + 1}\sin t$$

Using result III$_c$

$$= e^{2t}\left(\frac{-t\cdot\cos t}{2}\right)$$

G.S.: $y = y_c + y_p$

$$y(t) = e^{2t}(c_1 \cos t + c_2 \sin t)\frac{-te^{2t}}{2}\cos t$$

Replacing t by $\ln x$

$$y(x) = x^2(c_1 \cos(\log x) + c_2 \sin(\log x)$$

$$-\log x \cdot \frac{x^2}{2}\cdot\cos(\log x)$$

EXERCISE

Solve the following:

1. $x^2y'' + 3\cdot 5xy' + y = 0$

Ans. $y = c_1x^{-\frac{1}{2}} + c_2x^{-2}$

2. $x^2y'' - 3xy' + 3y = 0$

Ans. $c_1x^3 + c_2x^2$

3. $4x^2y'' - 4xy' + 3y = 0$

Ans. $y = c_1x^{\frac{1}{2}} + c_2x^{\frac{3}{2}}$

4. $9x^2y'' + 3xy' + y = 0$

Ans. $y = (c_1 + c_2\ln x)x^{\frac{1}{3}}$

5. $x^3y''' - 3x^2y'' + 6xy' - 6y = 0$

Ans. $y = c_1x + c_2x^2 + c_3x^3$

6. $x^3y''' - x^2y'' - 6xy' + 18y = 0$

Ans. $y = (c_1 + c_2\ln x)x^3 + c_3x^{-2}$

7. $x^2y'' - 2xy' + 2y = 4x^3$

Ans. $y = c_1x^2 + c_2x + 2x^3$

8. $x^2y'' - xy' + 4y = \cos\ln x + x\sin\ln x$

Ans. $y = c_1\cdot x\cdot\cos(\sqrt{3}\ln x) + c_2x\sin(\sqrt{3}\ln x) + \frac{1}{13}(3\cos\ln x - 2\sin\ln x) + \frac{1}{2}x\cdot\sin\ln x$

9. $x^2y'' + 4xy' + 2y = e^x$

Ans. $y = c_1x^{-1} + c_2x^{-2} + x^{-2}e^x$

10. $x^2y'' - 4xy' + 6y = 4x - 6$

Ans. $y = c_1x^2 + c_2x^3 + 2x - 1$

11. $x^2y'' - 2xy' - 10y = 0$ with $y(1) = 5$, $y'(1) = 4$

Ans. $y = \frac{3}{x^2} + 2x^5$

12. $x^2y'' - 4xy' + 4y = 4x^2 - 6x^3$, $y(2) = 4$, $y'(2) = -1$

Ans. $y = \frac{5x}{3} - 2x^2 + 3x^3 - \frac{23x^4}{24}$

13. $x^2y'' - 2xy' + 2y = \ln^2 x - \ln x^2$

Ans. $y = c_1x + c_2x^2 + \frac{1}{2}(\ln^2 x + \ln x) + \frac{1}{4}$

14. $x^3y''' - 8x^2y'' + 28xy' - 40y = -\frac{9}{x}$

Ans. $y = c_1x^2 + c_2x^4 + c_3x^5 + \frac{x^{-1}}{10}$

15. $x^3y''' + 2x^2y'' = x + \sin(\ln x)$

Ans. $y = c_1 + c_2x + c_3 \ln x + x \ln x + \frac{1}{2}(\cos \ln x + \sin \ln x)$

16. $x^2y'' + 5xy' + 4y = x^2 + 16(\ln x)^2$

Ans. $y = c_1x^{-2} + c_2x^{-2}\ln x + \frac{x^2}{16} + 4(\ln x)^2 - 8\ln x + 6$

17. $x^2y'' + 3xy' + y = \frac{1}{(1-x)^2}$

Ans. $y = \left(\frac{1}{x}\right)(c_1 + c_2 \ln x) + \frac{1}{x}\ln\frac{x}{1-x}$.

Special Case B:

Legendre Linear Equation

The Equation of nth order is of the form

$$[a_0(ax+b)^n D^n + a_1(ax+b)^{n-1}D^{n-1} + \cdots + a_{n-1}(ax+b)D + a_n]y = F(x) \quad (1)$$

where $a, b, a_0, a_1, a_2, \ldots, a_{n-1}, a_n$ are all constants. This Equation (1) can be reduced to (Cauchy-Euler equation) and then to a linear differential equation with constant coefficients by the transformation

$$ax + b = e^t \quad (2)$$

Solving for $$x = \frac{e^t - b}{a} \quad (3)$$

Differentiating (2) w.r.t. 't', we have

$$a\frac{dx}{dt} = e^t \quad (4)$$

Now

$$\frac{dy}{dx} = \frac{dy}{dt}\cdot\frac{dt}{dx} = ae^{-t}\frac{dy}{dt} = \frac{a}{e^t}\frac{dy}{dt}$$

$$(ax+b)Dy = a\mathcal{D}y \quad (5)$$

Differentiating again w.r.t., x

$$\frac{d^2y}{dx^2} = \frac{d}{dx}\left(\frac{dy}{dx}\right) = \frac{d}{dt}\left(\frac{dy}{dx}\right)\frac{dt}{dx}$$

$$= ae^{-t}\cdot\frac{d}{dt}\left(ae^{-t}\frac{dy}{dt}\right)$$

$$\frac{d^2y}{dx^2} = a^2e^{-2t}\left[\frac{d^2y}{dt^2} - \frac{dy}{dt}\right]$$

$$(ax+b)^2D^2y = a^2[\mathcal{D}^2 - \mathcal{D}]y \quad (6)$$

Similarly $$(ax+b)^3D^3y = a^3[\mathcal{D}^3 - 3\mathcal{D}^2 + 2\mathcal{D}]y \quad (7)$$

Method of Solving Legendre Equation

I. Identify a and b. Using the substitution $ax + b = e^t$, and (3), (4), (5), (6), (7), etc., the given Legendre equation reduces to D.E. with constant coefficients.

II. Solve D.E. obtained in I with t as independent variable, by standard methods described in Section. 9.3 to 9.5

III. Replace t by $\ln(ax+b)$ in the solution obtained in II.

Note: The second order equation

$$a_0(ax+b)^2y'' + b_0(ax+b)y' + c_0 y = Q(x)$$

reduces to

$$a_0 a^2(\mathcal{D}^2 - \mathcal{D})y + b_0 a\mathcal{D}y + C_0 y = Q^*(t)$$

or

$$a_0 a^2\mathcal{D}^2y + (b_0 a - a_0 a^2)\mathcal{D}y + C_0 y = Q^*(t)$$

Worked Out Examples

Solve the following:

Example: $(2x+5)^2y'' - 6(2x+5)y' + 8y = 6x$

Solution:

Here $a_0 = 1, b_0 = -6, c_0 = 8, Q(x) = 6x, a = 2, b = 5$. Replace $2x + 5 = e^t$ and use (3), (4), (5) (6) in the given Legendre D.E. which then reduces to a linear equation

$$4\left[\frac{d^2y}{dt^2} - \frac{dy}{dt}\right] - 12\frac{dy}{dt} + 8y = 3(e^t - 5)$$

$$\frac{d^2y}{dt^2} - 4\frac{dy}{dt} + 2y = \frac{1}{4}(3e^t - 15)$$

C.F.: The A.E. is $m^2 - 4m + 2 = 0$ with roots $m = 2 \pm \sqrt{2}$ so that the C.F. y_c is

$$y_c = c_1e^{(2+\sqrt{2})t} + c_2e^{(2-\sqrt{2})t}$$

P.I.: $y_p = \frac{1}{D^2-4D+2}\cdot\frac{1}{4}(3e^t-15)$

$= \frac{3}{4}\frac{1}{D^2-4D+2}e^t - \frac{15}{4}\frac{1}{D^2-4D+2}\cdot 1$

$= \frac{3}{4}\frac{1}{1-4+2}e^t - \frac{15}{4}\frac{1}{0-0+2}$

$= -\frac{3}{4}e^t - \frac{15}{8}$

Then G.S.: $y = y_c + y_p$

$$y = c_1 e^{(2+\sqrt{2})t} + c_2 e^{(2-\sqrt{2})t} - \frac{3}{4}e^t - \frac{15}{8}.$$

Replacing t by $\ln\,(2x+5)$, we get

$$y(x) = c_1(2x+5)^{2+\sqrt{2}} + c_2(2x+5)^{2-\sqrt{2}} - \frac{3}{4}(2x+5) - \frac{15}{8}.$$

EXERCISE

Solve the following:

1. $(x+2)^2y'' + 3(x+2)y' - 3y = 0$

Ans. $y = c_1(x+2) + c_2(x+2)^{-3}$

2. $(2x+1)^2y'' - 2(2x+1)y' + 4y = 0$

Ans. $y = c_1(2x+1) + c_2(2x+1) + \ln\,(2x+1)$

3. $(x+1)^2y'' - 3(x+1)y' + 4y = x^2 + x + 1$

Ans. $y = (c_1 + c_2\ln x)x^2 + \frac{(\ln x)^2}{2}x^2 - x + \frac{1}{4}$

4. $(3x+2)^2y'' + 3(3x+2)y' - 36y = 3x^2 + 4x + 1$

Ans. $y = c_1(3x+2)^2 + c_2(3x+2)^{-2} + \frac{1}{108}[(3x+2)^2\ln\,(3x+2) + 1]$

5. $(x+1)^2y'' + (x+1)y' - y = \ln\,(x+1)^2 + x - 1$

Ans. $y = c_1(x+1) + c_2(x+1)^{-1} - \ln\,(x+1)^2 + \frac{1}{2}(x+1)\ln\,(x+1) + 2$

6. $(1+x)^2y'' + (1+x)y' + y = 2\sin[\log(1+x)]$

Ans. $y = c_1\cos[\log(1+x)] + c_2\sin[\log(1+x)] - \log(1+x)\cdot\cos x$

7. $(1+4x)^2y'' + (1+4x)y' + 4y = 8(1+4x)^2$

Ans. $y = c_1(1+4x)^a + c_2(1+4x)^b + \frac{8}{41}(1+4x)^2$ with $a = 6+2\sqrt{5}$, $b = 6-2\sqrt{5}$

8. $(2x+3)^2y'' + (2x+3)y' - 2y = 24x^2$

Ans. $y = c_1u^{-\frac{1}{2}} + c_2u + \frac{3}{5}u^2 - 6u\ln\,(u) - 27$ with $u = 2x+3$

9. $(x+1)^3y'' + 3(x+1)^2y' + (x+1)y = 6\ln\,(x+1)$

Ans. $y(x+1) = c_1 + c_2\ln(x+1) + \ln 3(x+1)$

10. $(x-2)^2y'' - 3(x-2)y' + 4y = x$

Ans. $y = (x-2)^2\cdot[c_1 + c_2\ln\,(x-2)] + x - \frac{3}{2}$.

9.7 METHOD OF VARIATION OF PARAMETERS

Method of variation of parameters enables to find the general solution of any linear non-homogeneous D.E. of second order even (with variable coefficients also) provided its complimentary function is given (known). The particular integral of the non-homogeneous equation is obtained by varying the parameters i.e., by replacing the arbitrary constants in the C.F. by variable functions.

Consider a linear non-homogeneous second order D.E. with variable coefficients

$$y'' + P(x)y' + Q(x)y = R(x) \qquad (1)$$

Suppose the complimentary functions y_c of (1) is *given* as

$$y_c = c_1y_1(x) + c_2y_2(x) \qquad (2)$$

In method of variation of parameters the arbitrary constants c_1 and c_2 in (2) are replaced by two unknown functions $u_1(x)$ and $u_2(x)$

Thus the particular integral y_p of (1) is

$$y_p = u_1(x)y_1(x) + u_2(x)y_2(x) \qquad (3)$$

where

$$u_1(x) = -\int \frac{R(x)y_2(x)}{W}dx \quad (4)$$

$$u_2(x) = \int \frac{R(x)y_1(x)}{W}dx \quad (5)$$

Here $W = $ Wronskian of $y_1, y_2 = \begin{vmatrix} y_1 & y_2 \\ y_1' & y_2' \end{vmatrix}$

$$= y_1y_2' - y_1'y_2 \neq 0 \quad (6)$$

Hence the required general solution of (1) is

$$y = y_c + y_p$$

where y_c and y_p are given by (2) and (3).

Method of Obtaining a Particular Integral of (1) by Variation of Parameters

I. Obtain the complimentary function y_c of (1) as

$$y_c = c_1y_1(x) + c_2y_2(x)$$

II. Calculate Wronskian w by (6)

Identify $R(x), y_1(x), y_2(x)$

Determine $u_1(x)$ and $u_2(x)$ by (4) and (5).

Particular integral

$y_p = u_1(x)y_1(x) + u_2(x)y_2(x)$.

Note: There is another powerful "method of undetermined coefficients" which is not discussed here.

WORKED OUT EXAMPLES

Solve the following D.E. by variation of parameters method:

Example 1: $(D^2 - 1)y = 2(1 - e^{-2x})^{-\frac{1}{2}}$

Solution: C.F.: Here A.E. is $m^2 - 1 = 0$ with, $m = \pm 1$ as the roots so that the C.F. y_c is

$$y_c = c_1e^x + c_2e^{-x}$$

So $y_1 = c^x$ and $y_2 = e^{-x}$ form two independent solutions to the corresponding homogeneous equation.

Wronskian $w = y_1y_2' - y_2y_1' = e^x(-e^{-x}) - e^{-x}e^x = -2$

Assume the particular solution y_p as

$$y_p = u_1(x)y_1(x) + u_2(x)y_2(x)$$

where

$$u_1(x) = -\int \frac{fy_2}{w}dx = -\int \frac{2}{(1-e^{-2x})^{\frac{1}{2}}} \cdot \frac{e^{-x}}{-2}dx$$

Here $f(x) = $ R.H.S. of the given equation$= \frac{2}{(1-e^{-2x})^{\frac{1}{2}}}$

So $u_1 = \int \frac{e^{-x}dx}{(1-e^{-2x})^{\frac{1}{2}}}$

Put $e^{-x} = t, -e^{-x}dx = dt$

Thus $u_1 = \int \frac{dt}{\sqrt{1-t^2}} = -\sin^{-1}(t) = -\sin^{-1}(e^{-x})$

Similarly

$$u_2 = \int \frac{fy_1dx}{w} = \int \frac{2}{\sqrt{1-e^{-2x}}} \cdot \frac{e^x}{-2}dx$$

$$= -\int \frac{e^x dx}{\sqrt{1-e^{-2x}}} = -\int \frac{e^{2x}dx}{\sqrt{e^{2x}-1}}$$

Put $e^{2x} = t, 2e^{2x}dx = dt$

So $$u_2 = -\frac{1}{2}\int \frac{dt}{\sqrt{t-1}} = -\frac{1}{2}\frac{(t-1)^{\frac{1}{2}}}{\frac{1}{2}}$$

$$= -(t-1)^{\frac{1}{2}} = -(e^{2x}-1)^2$$

$$y_p = -\sin^{-1}(e^{-x})e^x - (e^{2x}-1)^{\frac{1}{2}} \cdot e^x$$

Hence G.S.: $y = y_c + y_p$ so that

$$y = c_1e^x + c_2e^{-x} - e^x \cdot \sin^{-1}(e^{-x}) - (e^{2x}-1)^{\frac{1}{2}}e^{-x}$$

Example 2: $(D^2 + 1)y = \operatorname{cosec} x \cdot \cot x$.

Solution:

C.F.: Here A.E. is $m^2 + 1 = 0$ with complex roots $m = \pm i$ so that the C.F. y_c is

$$y_c = c_1 \cos x + c_2 \sin x$$

Take $y_1 = \cos x$ and $y_2 = \sin x$. They form the system of independent solutions. The Wronskian

$$w = y_1y_2' - y_2y_1' = \cos x \cdot \cos x - \sin x \cdot (-\sin x) = 1$$

Assume that the particular integral y_p of the form

$$y_p = u_1(x)y_1(x) + u_2(x)y_2(x)$$

where $u_1 = -\int \frac{fy_2}{w}dx$.
Here $f = \text{cosec}x \cdot \cot x$

$$\text{Thus } u_1 = -\int \frac{\text{cosec} \cdot \cot x \cdot \sin x}{1}dx = -\ln|\sin x|$$

Similarly,

$$u_2 = \int \frac{fy_1}{w}dx = \int \frac{\text{cosec}x \cdot \cot x \cdot \cos x}{1}dx$$

$$= \int \cot^2 x dx = -\cot x - x$$

Hence

$$y_p = -\ln|\sin x| \cdot \cos x - (x + \cot x)\sin x$$

Thus G.S.: $y = y_c + y_p$ is

$$y = c_1 \cos x + c_2 \sin x - \cos x \cdot \ln|\sin x|$$
$$-x \sin x - \sin x \cdot \cot x$$

Example 3: $(D^2 + 2D + 1)y = e^{-x}\ln x$

Solution:

C.F.: Here A.E. is $m^2 + 2m + 1 = 0$ with $m = -1$ as the double root so that the C.F. y_c is

$$y_c = (c_1 + c_2 x)e^{-x}.$$

Take $y_1 = e^{-x}$ and $y^2 = xe^{-x}$ as the fundamental system. Now the Wronskian w is

$$w = y_1y_2' - y_2y_1' = e^{-x}(e^{-x} - xe^{-x})$$
$$-(xe^{-x})(-e^{-x}) = e^{-2x}$$

Assume the P.I. y_p as

$$y_p = u_1(x)y_1(x) + u_2(x)y_2(x)$$

where $u_1 = -\int \frac{fy_2}{w}dx$ with $f = e^{-x} \cdot \ln x$

$$u_1 = -\int \frac{e^{-x} \cdot \ln x \cdot xe^{-x}}{e^{-2x}}dx$$

$$= -\int x \cdot \ln x \cdot dx$$

$$u_1 = \frac{x^2}{2}\left[\frac{1}{2} - \ln x\right]$$

Also

$$u_2 = \int \frac{fy_2}{w}dx = \int \frac{e^{-x}\ln x \cdot e^{-x}}{e^{-2x}}dx$$

$$= \int \ln x dx = x\ln x - x$$

Thus $y_p = \frac{x^2}{2}\left[\frac{1}{2} - \ln x\right]e^{-x} + e^{-2x}[x\ln x - x]$.
Hence G.S: $y = y_c + y_p$

$$y = c_1e^{-x} + c_2xe^{-x} + \frac{x^2}{2}\left(\frac{1}{2} - \ln x\right)e^{-x}$$
$$+e^{-2x}(x \cdot \ln x - x)$$

EXERCISE

Solve the following D.E. by the method of variation of parameters:

1. $(D^2 + 1)y = cscx$

Ans. $y = c_1 \cos x + c_2 \sin x - x\cos x + \sin x \ln(\sin x)$

2. $(D^2 - 2D + 1)y = x^{\frac{3}{2}}e^x$

Ans. $y = [c_1 + c_2x + \frac{4}{35}x^{\frac{7}{2}}]e^x$

3. $(D^2 - 1)y = e^{-2x} \cdot \sin(e^{-x})$

Ans. $y = c_1e^x + c_2e^{-x} - e^x\cos(e^{-x}) - \sin(e^{-x})$

4. $(D^2 - 3D + 2)y = \frac{1}{1+e^{-x}}$

Ans. $y = c_1e^x + c_2e^{2x} + e^x \cdot \ln(1 + e^{-x}) + e^{2x}[\ln(1 + e^{-x}) - e^{-x}]$

5. $(D^2 - 6D + 9)y = \frac{e^{3x}}{x^2}$

Ans. $y = c_1e^{3x} + c_2xe^{3x} - e^{3x}\ln x$

6. $(D^2 - 2D)y = e^x \sin x$

Ans. $y = c_1 + c_2e^{2x} - \frac{e^x}{2}\sin x$

7. $(D^3 + D)y = cscx$

Ans. $y = c_1 + c_2\cos x + c_3\sin x - \ln(cscx + \cot x) - \cos x\ln \sin x - x\sin x$

8. $(D^2 - 2D + 2)y = e^x \tan x$

Ans. $y = e^x(c_1\cos x + c_2\sin x) - e^x \cdot \cos x \log(\sec x + \tan x)$

9. $(D^2 + 3D + 2)y = e^x + x^2$

Ans. $y = c_1e^{-x} + c_2e^{-2x} + \frac{e^x}{6} + \left(\frac{x^2}{2} - \frac{3x}{2} + \frac{7}{4}\right)$

10. $(D^2 + a^2)y = x\cos ax$

Ans. $$c_1\cos ax + c_2\sin ax + \frac{\cos ax}{2a} \times \left[\frac{\sin 2ax}{4a^2} - \frac{x\cos 2ax}{2a}\right] + \frac{\sin ax}{2a} \times \left[\frac{x^2}{2} + \frac{x\sin 2ax}{2a} - \frac{\cos 2ax}{4a^2}\right]$$

11. $(D^2 + 1)y = \log\cos x$

Ans. $y = c_1\cos x + c_2\sin x + (\log\cos x - 1) + \sin x \cdot \log(\sec x + \tan x)$

12. $(D^2 + 4)y = 4\sec^2 2x$

Ans. $y = c_1\cos 2x + c_2\sin 2x - 1 + \sin 2x \cdot \ln(\sec 2x + \tan 2x)$

13. $(D^2 + 3D + 2)y = \frac{1}{1+e^x}$

Ans. $y = c_1e^{-x} + c_2e^{-2x} + (e^{-x} + e^{-2x})\log(1 + e^x)$

14. $(D^2 - 1)y = e^{-x}\sin e^{-x} + \cos e^{-x}$

Ans. $y = c_1e^x + c_2e^{-x} - e^x\sin e^{-x}$

15. $(D^2 + 1)y = x\cos 2x$

Ans. $y = c_1\cos x + c_2\sin x - \frac{x}{2}\cos 2x + \frac{4}{9}\sin 2x$

16. $(D^2 - 3D + 2)y = xe^x + 2x$

Ans. $c_1e^x + c_2e^{2x} - \frac{x^2}{2}e^x - xe^{-x} + x + \frac{3}{2}$.

9.8 THE METHOD OF UNDETERMINED COEFFICIENTS

The particular integral of an nth order linear non-homogeneous D.E. with constant coefficients

$$(a_0D^n + a_1D^{n-1} + \cdots + a_{n-1}D + a_n)y = F(x) \quad (1)$$

can be determined by the method of undetermined coefficients, provided the R.H.S. function $F(x)$ in (1) is an exponential function, polynomial in x, cosine, sine or sums or products of such functions. The derivatives of such functions $F(x)$ have a form similar to $F(x)$ itself and are finite in number. Although the class of such functions $F(x)$ is quite restricted, they include functions of frequent occurrence. This method is relatively simple. The particular integral y_p of (1) is assumed in a form similar to the R.H.S. function $F(x)$ and involving undetermined (unknown) coefficients which are then determined by substitution of y_p in D.E. (1). The advantage of this method is that for a wrong choice of y_p, or with few terms leads to contradiction, while choice of too many terms makes superfluous coefficients zero. Thus this method is self correcting. However, this method fails when $F(x) = \sec x, \tan x$ etc. since the number of new terms obtained by differentiation is infinite. Table 9.4 gives the form (choice) of P.I. y_p for a specific $F(x)$.

Procedure:

Let $m_1, m_2, \ldots, m_n$ be the roots of auxiliary equation $f(m) = 0$ where $f(D)y = F(x)$ is the given D.E. Here $f(D) = a_0D^n + a_1D^{n-1} + a_2D^{n-2} + \ldots + a_{n-1}D + a_n$. Let $S = \{y_1(x), y_2(x), \ldots, y_n(x)\}$ be the set of n linearly independent solutions of the homogeneous equation $f(D)y = 0$ corresponding to the n roots $m_1, m_2, \ldots, m_n$. Then the complementary function y_c is given by $y_c = c_1y_1 + c_2y_2 + \ldots + c_ny_n$.

To find the particular integral by the method of undetermined coefficients

I. Straight case

If the R.H.S. function $F(x)$ is ***not*** a member of the solution set, then choose P.I. y_p from the above table depending on the nature of $F(x)$.

Example: $(D^2 - 3D + 2)y = 4e^{3x}$ has A.E.: $m^2 - 3m + 2 = 0$ with roots $m_1 = 1, m_2 = 2$ and set of L.I. solutions, $S = \{e^x, e^{2x}\}$. C.F. is $y_c = c_1e^x + c_2e^{2x}$.

Since the R.H.S. e^{3x} is *not* a member of the solution set, choose P.I. y_p as ce^{3x}. Substituting y_p in the given D.E., we determine the unknown coefficient c as: $9ce^{3x} - 3 \cdot 3ce^{3x} + 2ce^{3x} = 4e^{3x}$ $\therefore 2c = 4$ or $c = 2$, So $y_p = 2e^{3x}$. (Check : $y_p = \frac{4}{D^2-3D+2}e^{3x} = \frac{4}{9-3\cdot 3+2}e^{3x} = 2e^{3x}$).

Table 9.4

S. No.	R.H.S. function $F(x)$	Choice of particular integral y_p
1.	ke^{ax}	ce^{ax}
2.	$k\sin(ax+b)$ or $k\cos(ax+b)$	$c_1\sin(ax+b)+c_2\cos(ax+b)$
3.	$ke^{ax}\sin(bx+d)$ or $ke^{ax}\cos(bx+d)$	$c_1e^{ax}\sin(bx+d)+c_2e^{ax}\cos(bx+d)$
4.	kx^n where $n=0,1,2,\ldots$	$c_0+c_1x+c_2x^2+\cdots+c_{n-1}x^{n-1}+c_nx^n$
5.	kx^ne^{ax} for $n=0,1,2,\ldots$	$e^{ax}\{c_0+c_1x+c_2x^2+\cdots+c_{n-1}x^{n-1}+c_nx^n\}$
6.	$kx^n\sin(ax+b)$ or $kx^n\cos(ax+b)$	$a_0\sin(ax+b)+b_0\cos(ax+b)+$ $+a_1\cdot x\cdot\sin(ax+b)+b_1\cdot x\cdot\cos(ax+b)+$ $+a_2x^2.\sin(ax+b)+b_2x^2\cos(ax+b)+$ $-----------------$ $+a_nx^n\sin(ax+b)+b_nx^n\cos(ax+b)$
7.	$kx^ne^{dx}\sin(ax+b)$ or $kx^ne^{dx}\cos(ax+b)$	$e^{dx}\{a_0\sin(ax+b)+b_0\cos(ax+b)+$ $+a_1x\sin(ax+b)+b_1x\cdot\cos(ax+b)+$ $-----------------$ $+a_nx^n\sin(ax+b)+b_nx^n\cos(ax+b)\}$

Here $k, a, b, d, c, c_1, c_2, \ldots, c_n, a_0, b_0, a_1, b_1, \ldots, a_n, b_n$ are all constants and n is a positive integer i.e., $n=0,1,2,3,\ldots$.

II. Sum case

When the R.H.S. $F(x)$ is a combination (sum) of the functions in column 1 of the table, then P.I. is chosen as a combination of the corresponding functions in second column and proced as in straight case I.

Note: Here also as in case I, the terms of R.H.S. $F(x)$ are *not* members of the solution set S.

Example: $(D^3-2D^2+D-2)y=5\cos 2x-6x^2$

A.E.: $m^3-2m^2+m-2=0$, $(m-2)(m^2+1)=0$

i.e., $m=2, \pm i$, $S=\{e^{2x}, \cos x, \sin x\}$;

C.F.: $y_c=c_1e^{2x}+c_2\cos x+c_3\sin x$.

Choose $y_p=A\cos 2x+B\sin 2x+cx^2+Dx+E$.

Substitute y_p in D.E. and solve for A, B, C, D, E.

$(8A\sin 2x-8B\cos 2x)-2(-4A\cos 2x-4B\sin 2x+2C)+$

$+(-2A\sin 2x+2B\cos 2x+2Cx+D)-$

$-2(A\cos 2x+B\sin 2x+Cx^2+Dx+E)=5\cos 2x-6x^2$.

Equating coefficients of:

x^2: $-2C=-6 \therefore C=3$

x : $2C-2D=0 \therefore D=C=3$

const: $-4C+D-2E=0 \therefore E=-\frac{3C}{2}=-\frac{9}{2}$

$\left.\begin{array}{l}\cos 2x\text{: } 6A-6B=5\\ \sin 2x\text{: } A+B=0\end{array}\right\} A=-B=\frac{5}{12}$

Required P.I. y_p is

$y_p=\frac{5}{12}\cos 2x-\frac{5}{12}\sin 2x+3x^2+3x-\frac{9}{2}$.

III. Modified case

When any term of $F(x)$ is a *member* of the solution set S, then the method fails if we choose y_p from the table. In such cases, the choice from the table should be modified as follows:

A. If a term u of $F(x)$ is also a term of the complementary function (i.e., $u \epsilon S=$ solution set) then the choice from the table corresponding to u should be multiplied by

a. x if u corresponds to a simple root of C.F.

b. x^2 if u corresponds to a double root C.F.

c. x^s if u corresponds to a s-fold root of C.F.

Example: $(D-2)^3y = 6e^{2x}$; A.E.: $m = 2, 2, 2$, $S = \{e^{2x}, xe^{2x}, x^2e^{2x}\}$ C.F.:$y_c = c_1e^{2x} + c_2xe^{2x} + c_3x^2e^{2x}$. But e^{2x} on R.H.S. ϵS i.e., e^{2x} is a term of the C.F. So the choice of Ae^{2x} should be multiplied by x^3. Thus the modified choice of P.I. is x^3Ae^{2x}.

B. Suppose x^ru is a term of $F(x)$ and u is a term of C.F. corresponding to an s-fold root then the choice from the table corresponding to x^ru should be multiplied by x^s.

Example: $(D-3)^4(D+4)y = x^3e^{3x} + 6x^2$
Normal choice from the table would have been

$$y_p^* = (Ax^3e^{3x} + Bx^2e^{3x} + Cxe^{3x} + De^{3x}) + +(Ex^2 + Fx + G).$$

But e^{3x} is a term of the C.F. and $m = 3$ is 4th order root. Here

$$y_c = c_1e^{3x} + c_2xe^{3x} + c_3x^2e^{3x} + c_4x^3e^{3x}.$$

So the modified (corrected) choice of P.I. in this case is

$$y_p = x^4(Ax^3e^{3x} + Bx^2e^{3x} + Cxe^{3x} + De^{3x}) + +(Ex^2 + Fx + G).$$

WORKED OUT EXAMPLES

Straight case

Solve the following examples by the method of undetermined coefficients.

Example 1: $y'' - 3y' + 2y = 4x^2$.

Solution: A.E.: $m^2 - 3m + 2 = 0$, $m = 1, 2$.
C.F.: $y_c = c_1e^x + c_2e^{2x}$.
P.I.: R.H.S. $F(x) = kx^n (n = 2)$ so choose from the table P.I. as $y_p = Ax^2 + Bx + C$
Differentiating $y_p' = 2Ax + B$, $y_p'' = 2A$.

Substituting in the given D.E.

$$2A - 3(2Ax + B) + 2(Ax^2 + Bx + C) = 4x^2$$

Equating the coefficients of like powers of x:

$x^2 : 2A = 4$ $\quad\therefore A = 2$
$x : -6A + 2B = 0$ $\quad\therefore B = 6$
$x^0 : 2A - 3B + 2C = 0$ $\quad\therefore C = 7$

Required P.I. by the method of undetermined coefficients is

$$y_p = 2x^2 + 6x + 7$$

G.S.: $y = y_c + y_p = c_1e^x + c_2e^{2x} + 2x^2 + 6x + 7$.

Example 2: $y'' + 6y' + 5y = 2e^x + 10e^{5x}$.

Solution: A.E.: $m^2 + 6m + 5 = 0$, $m = -1, -5$
C.F.:$y_c = c_1e^{-x} + c_2e^{-5x}$
P.I.: Choose $y_p = Ae^x + Be^{5x}$
$y_p' = Ae^x + 5Be^{5x}$, $y'' = Ae^x + 25Be^{5x}$

$$(Ae^x + 25Be^{5x}) + 6(Ae^x + 5Be^{5x}) + +5(Ae^x + Be^{5x}) = 2e^x + 10e^{5x}.$$

Equating the coefficients of e^x and e^{5x} :
$e^x : A + 6A + 5A = 2$ $\quad\therefore A = \frac{1}{6}$
e^{5x}: $25B + 30B + 5B = 10$, $B = \frac{1}{6}$
$y_p = \frac{1}{6}e^x + \frac{1}{6}e^{5x}$
$y = y_c + y_p = c_1\,e^{-x} + c_2\,e^{-5x} + \frac{1}{6}(e^x + e^{5x})$.

Example 3: $y'' + 2y' + 4y = 13\cos(4x-2)$.

Solution: A.E.: $m^2 + 2m + 4=0$, $m= -1 \pm \sqrt{3}i$
$y_c = e^{-x}[c_1\cos\sqrt{3}x + c_2\sin\sqrt{3}x]$
P.I.: Choose $y_p = A\cos(4x-2) + B\sin(4x-2)$

$$y_p' = -4A\sin(4x-2) + 4B\cos(4x-2)$$

$$y_p'' = -16A\cos(4x-2) - 16B\sin(4x-2)$$

$$[-16A\cos(4x-2) - 16B\sin(4x-2)] + +2[-4A\sin(4x-2) + 4B\cos(4x-2)] + +4[A\cos(4x-2) + B\sin(4x-2)] = 13\cos(4x-2)$$

Equating coefficients of $\cos(4x-2)$ and $\sin(4x-2)$,
$-16A + 8B + 4A = 13$, $-16B - 8A + 4B = 0$.

Solving $A = -\frac{3}{4}$, $B = \frac{1}{2}$, so

$$y_p = -\frac{3}{4}\cos(4x-2) + \frac{1}{2}\sin(4x-2)$$

$$y = y_c + y_p = e^{-x}[c_1 \cos\sqrt{3}x + c_2 \sin\sqrt{3}x] - \frac{3}{4}\cos(4x-2) + \frac{1}{2}\sin(4x-2).$$

Example 4: $y'' - y = e^x \sin 2x$.

Solution: A.E.: $m^2 - 1 = 0$, $m = \pm 1$
C.F.: $y_c = c_1 e^x + c_2 e^{-x}$
P.I.: Choose $y_p = Ae^x \sin 2x + Be^x \cos 2x$

$$y_p' = 2Ae^x \cos 2x + Ae^x \sin 2x + Be^x \cos 2x - 2Be^x \sin 2x$$

$$y'' = (A-2B)(e^x \sin 2x + 2e^x \cos 2x) + +(2A+B)(e^x \cos 2x - 2e^x \sin 2x)$$

Substituting in D.E.

$$[2(A-2B) + (2A+B)]e^x \cos 2x + +[A - 2B - 2(2A+B)]e^x \sin 2x - -[Ae^x \sin 2x + Be^x \cos 2x] = e^x \sin 2x.$$

Equating the coefficients of

$$e^x \sin 2x : A - 2B - 4A - 2B - A = 1 \quad \text{or} \quad A + B = -\frac{1}{4}$$

$$e^x \cos 2x : 2A - 4B + 2A + B - B = 0 \quad \text{or} \quad A - B = 0$$

Solving $A = B = -\frac{1}{8}$

$$y_p = -\frac{1}{8}(e^x)(\sin 2x + \cos 2x).$$

$$y = y_c + y_p = c_1 e^x + c_2 e^{-x} - \frac{1}{8}e^x(\sin 2x + \cos 2x).$$

Sum case

Example 5: $y'' - 9y = x^3 + e^{2x} - \sin 3x$.

Solution: $m^2 - 9 = 0$, $m = \pm 3$,
C.F.: $y_c = c_1 e^{-3x} + c_2 e^{3x}$
P.I.: Choose $y_p = Ax^3 + Bx^2 + Cx + D + Ee^{2x} + F\sin 3x + G\cos 3x$

$$y_p' = 3Ax^2 + 2Bx + C + 2Ee^{2x} + 3F\cos 3x - 3G\sin 3x$$

$$y_p'' = 6Ax + 2B + 4Ee^{2x} - 9F\sin 3x - 9G\cos 3x.$$

Substituting

$$(6Ax + 2B + 4Ee^{2x} - 9F\sin 3x - 9G\cos 3x) - -9(Ax^3 + Bx^2 + Cx + D + Ee^{2x} + F\sin 3x + G\cos 3x) = x^3 + e^{2x} - \sin 3x.$$

Equating the coefficients of:

$$x^3 : -9A = 1, \qquad A = -\frac{1}{9}$$

$$x^2 : -9B = 0 \qquad \therefore B = 0$$

$$x : 6A - 9C = 0 \qquad \therefore C = -\frac{2}{27}$$

$$x^0 : 2B - 9D = 0 \qquad \therefore D = 0$$

$$e^{2x} : 4E - 9E = 1 \qquad \therefore E = -\frac{1}{5}$$

$$\sin 3x : -9F - 9F = -1 \quad \therefore F = \frac{1}{18}$$

$$\cos 3x : -9G - 9G = 0 \qquad \therefore G = 0$$

$$y_p = -\frac{1}{9}x^3 - \frac{2}{27}x - \frac{1}{5}e^{2x} + \frac{1}{18}\sin 3x.$$

$$y = y_c + y_p = c_1 e^{-3x} + c_2 e^{3x} - \frac{1}{9}x^3 - \frac{2}{27}x - \frac{1}{5}e^{2x} + +\frac{1}{18}\sin 3x.$$

Modified case

Example 6: $y''' + y' = 2x^2 + e^{2x} + 4\sin x$.

Solution: A.E. is $m^3 + m = 0$ or $m = 0, \pm i$
C.F.: $y_c = c_1 + c_2 \sin x + c_3 \cos x$.
Normally as per the table one would have assumed the P.I. as,

$$y_p = Ax^2 + Bx + C + De^{2x} + E\sin x + F\cos x.$$

But note that $\sin x$, a term in R.H.S. function, is a term in the C.F. i.e., member of solution set $S = \{1, \sin x, \cos x\}$. Therefore the corresponding terms in y_p should be multiplied by x. Also note that the term x^2 in R.H.S. also corresponds to a term in C.F. Thus the modified (correct) choice of P.I. y_p is

$$y_p = x(Ax^2 + Bx + C) + De^{2x} + x(E\sin x + F\cos x)$$

$$y_p = (Ax^3 + Bx^2 + Cx) + De^{2x} + Ex\sin x + Fx\cos x$$

$$y_p' = 3Ax^2 + 2Bx + C + 2De^{2x} + E\sin x + +Ex\cos x + F\cos x - Fx\sin x$$

$$y_p'' = 6Ax + 2B + 4De^{2x} + 2E\cos x - Ex\sin x - -2F\sin x - Fx\cos x$$

$$y_p''' = 6A + 8De^{2x} - 3E\sin x - Ex\cos x - 3F\cos x + +Fx\sin x.$$

Substituting in D.E.

$$(6A+8De^{2x}-3E\sin x-Ex\cos x-3F\cos x+Fx\sin x)+$$
$$+(3Ax^2+2Bx+C+2De^{2x}+E\sin x+Ex\cos x$$
$$+F\cos x-Fx\sin x)=2x^2+e^{2x}+4\sin x.$$

Equating the coefficients of

$$x^2: 3A=2 \qquad \therefore A=\frac{2}{3}$$
$$x: 2B=0 \qquad \therefore B=0$$
$$x^0: 6A+C=0 \qquad \therefore C=-6A=-6\cdot\frac{2}{3}=-4$$
$$e^{2x}: 8D+2D=1 \qquad \therefore D=\frac{1}{10}$$
$$\sin x: -3E+E=4 \qquad \therefore E=-2$$
$$\cos x: -3F+F=0 \qquad \therefore F=0$$
$$x\cos x: -E+E=0 \qquad \therefore E=0$$
$$x\sin x: F+F=0 \qquad \therefore F=0$$

$$y_p=\frac{2}{3}x^3-4x+\frac{1}{10}e^{2x}-2x\sin x.$$

Example 7: $(D^2+9)y=x^2\cos 3x.$

Solution: A.E.: $m^2+9=0$, $m=\pm 3i$
C.F.: $y_c=c_1\cos 3x+c_2\sin 3x$.
In the normal case, the choice of P.I. from table is

$$y_p=Ax^2\cos 3x+Bx^2\sin 3x+Cx\cos 3x+$$
$$+Dx\sin 3x+E\cos 3x+F\sin 3x.$$

But in the R.H.S. expression $F(x)=x^2\cos 3x$, the term $\cos 3x$ is also a term (part) of the C.F. corresponding to a root of multiplicity one i.e., $\cos 3x$ is a member of the solution set. So the modified (corrected) choice of P.I. is

$$y_p=x\Big[Ax^2\cos 3x+Bx^2\sin 3x+Cx\cos 3x+$$
$$+Dx\sin 3x+E\cos 3x+F\sin 3x\Big].$$

Note that $H\cos 3x+K\sin 3x$ is *not* included in the P.I. y_p, because these terms are already present in the C.F. (and therefore are superfluous i.e., even if we take these two terms in y_p, finally coefficients H and K will become zero). So

$$y_p=Ax^3\cos 3x+Bx^3\sin 3x+Cx^2\cos 3x+$$
$$+Dx^2\sin 3x+Ex\cos 3x+Fx\sin 3x.$$

Differentiating twice

$$y_p'=3Ax^2\cos 3x-3Ax^3\sin 3x+3Bx^2\sin 3x-$$
$$-3Bx^3\cos 3x+2Cx\cos 3x-3Cx^2\sin 3x+$$
$$+2Dx\sin 3x-3Dx^2\cos 3x++E\cos 3x-$$
$$-3Ex\sin 3x+F\sin 3x-3Fx\cos 3x. \quad \text{or}$$

$$y_p'=(3Ax^2-3Bx^3+2Cx-3Dx^2+E-3Fx)\cos 3x+$$
$$+(-3Ax^3+3Bx^2-3Cx^2+2Dx-3Ex+F)\sin 3x$$
$$y_p''=(6Ax-9Bx^2+2C-6Dx+0-3F)\cos 3x-$$
$$-3(3Ax^2-3Bx^3+2Cx-3Dx^2+E-3Fx)\sin 3x+$$
$$+3(-3Ax^3+3Bx^2-3Cx^2+2Dx-3Ex+F)\cos 3x+$$
$$+(-9Ax^2+6Bx-6Cx+2D-3E)\sin 3x.$$

Substituting in D.E. and equating the coefficients on both sides of

$$x^2\cos 3x: 9B+9B-9C+9C=1 \qquad \therefore B=\frac{1}{18}$$
$$x^2\sin 3x: -9A-9A-9D+9D=0 \qquad \therefore A=0$$
$$x^3\cos 3x: -9A+9A=0$$
$$x^3\sin 3x: -9B+9B=0$$
$$x\cos 3x: 6A+6D+6D-9E+9E=0 \quad \therefore D=0$$
$$x\sin 3x: 6B-6C-6C-9F+9F=0,$$
$$B-2C=0,\ C=\frac{1}{36}$$
$$\cos 3x: 2C+3F+3F=0,\ F=\frac{1}{3}C=\frac{1}{108}$$
$$\sin 3x: 2D-3E-3E=0 \qquad \therefore E=0$$
$$y_p=\frac{1}{18}x^3\sin 3x+\frac{1}{36}x^2\cos 3x+\frac{1}{108}x\sin 3x.$$

Example 8: $(D-2)^3y=17e^{2x}$.

Solution: C.F.: A.E. is $m-2=0$, $m=2,2,2$. So $y_c=(c_1+c_2x+c_3x^2)e^{2x}$.
The choice of P.I. y_p as Ae^{2x} is not sufficient, since the term e^{2x} in the R.H.S. of D.E. is a term of the C.F. corresponding to a root of multiplicity $m=3$. So the modified choice of P.I. y_p is

$$y_p=x^3(Ae^{2x})=Ax^3e^{2x}$$

Differentiating,

$$y'_p = (3x^2 + 2x^3)Ae^{2x}$$

$$y''_p = (x + 2x^2 + \frac{4}{6}x^3)6Ae^{2x}$$

$$y'''_p = (1 + 6x + 6x^2 + \frac{8}{6}x^3)6Ae^{2x}.$$

Substituting in $(D^3 - 6D^2 + 12D - 8)y = 17e^{2x}$ we get

$$6Ae^{2x}(1+6x+6x^2+\frac{8}{6}x^3) - 36(x+2x^2+\frac{4}{6}x^3)Ae^{2x} +$$

$$+12Ae^{2x}(3x^2 + 2x^3) - 8Ae^{2x}(x^3) = 17e^{2x}.$$

Equating the coefficients on both sides of

$$e^{2x} : 6A = 17, \quad \text{so} \quad A = \frac{17}{6};$$

$$x : 36A - 36A = 0, \quad x^2 : 36A - 72A + 36A = 0,$$

$$x^3 : 8A - 24A + 24A - 8A = 0. \qquad \text{Thus}$$

$$y_p = \frac{17}{6}x^3 e^{2x}$$

Thus

$$y = y_c + y_p = (c_1 + c_2x + c_3x^2)e^{2x} + \frac{17}{6}x^3e^{2x}.$$

[Note that the P.I. can be obtained in a simpler way by operator method, where $y_p = \frac{1}{(D-2)^3}(17e^{2x}) = 17 \cdot \frac{x^3}{3!}e^{2x} = \frac{17}{6}x^3e^{2x}$].

EXERCISE

Method of undetermined coefficients

Solve the following D.E. by the method of undetermined coefficients. (Here G.S.= $y = y_c + y_p$, where y_c is the complementary function and y_p is the particular integral.)

Straight and sum cases:

1. $y'' - 2y' - 3y = 2e^{4x}$

Ans. $y_c = c_1e^{3x} + c_2e^{-x}$, $y_p = \frac{2}{5}e^{4x}$

2. $y'' + 2y' + 5y = 6\sin 2x + 7\cos 2x$

Ans. $y_c = e^{-x}[c_1\cos 2x + c_2\sin 2x]$, $y_p = 2\sin 2x - \cos 2x$

3. $y'' - 2y' = e^x \sin x$

Ans. $y_c = c_1 + c_2e^{2x}$, $y_p = -\frac{1}{2}e^x \sin x$

4. $y'' + y' - 2y = 2x - 40\cos 2x$

Ans. $y_c = c_1e^x + c_2e^{-2x}$, $y_p = -\frac{1}{2} - x + 6\cos 2x - 2\sin 2x$

5. $y'' - 2y' + 3y = x^3 + \sin x$

Ans. $y_c = e^x(c_1\cos\sqrt{2}x + c_2\sin\sqrt{2}x)$

$$y_p = \frac{1}{3}x^3 + \frac{2}{3}x^2 + \frac{2}{9}x - \frac{8}{29} + \frac{1}{4}(\sin x + \cos x)$$

6. $y'' + 4y = 8x^2$

Ans. $y_c = c_1\cos 2x + c_2\sin 2x$, $y_p = 2x^2 - 1$

7. $y'' + 2y' + 5y = 1.25e^{0.5x} + 40\cos 4x - 55\sin 4x$, with $y(0) = 0.2$ and $y'(0) = 60.1$

Ans. $y_c = e^{-x}(c_1\cos 2x + c_2\sin 2x)$

$y_p = 0.2e^{0.5x} + 5\sin 4x$

particular solution satisfying initial conditions

$$y = 20e^{-x}\sin 2x + 0.2e^{0.5x} + 5\sin 4x.$$

8. $y'' + 2y' + 4y = 2x^2 + 3e^{-x}$

Ans. $y_c = e^{-x}(c_1\cos\sqrt{3}x + c_2\sin\sqrt{3}x)$

$y_p = \frac{x^2}{2} - \frac{x}{2} + e^{-x}$

9. $y'' - y = e^{3x}\cos 2x - e^{2x}\sin 3x$

Ans. $y_c = c_1e^x + c_2e^{-x}$

$$y_p = \frac{1}{30}e^{2x}(2\cos 3x + \sin 3x) + \frac{1}{40}e^{3x}(\cos 2x + 3\sin 2x)$$

10. $y'' - 9y = x + e^{2x} - \sin 2x$

Ans. $y_c = c_1e^{3x} + c_2e^{-3x}$

$$y_p = -\frac{x}{9} - \frac{e^{2x}}{5} + \frac{1}{13}\sin 2x$$

Modified case

11. $y'' + y' - 6y = 10e^{2x} - 18e^{3x} - 6x - 11$

Ans. $y_c = c_1e^{-3x} + c_2e^{2x}$

$y_p = 2xe^{2x} - 3e^{3x} + x + 2$

Hint: Choose $y_p = x(Ae^{2x}) + Be^{3x} + Cx + D$.

12. $y''' + 2y'' - y' - 2y = e^x + x^2$

Ans. $y_c = c_1e^x + c_2e^{-x} + c_3e^{-2x}$

$$y_p = -\frac{x^2}{2} + \frac{x}{2} - \frac{5}{4} + \frac{1}{6}xe^x$$

Hint: Choose $y_p = x(Ae^x) + Bx^2 + Cx + D$.

13. $y'' - 3y' + 2y = e^x$

Ans. $y_c = c_1e^x + c_2e^{2x}$

$y_p = -xe^x$

Hint: Choose $y_p = x(Ae^x)$.

14. $y'' + y = \sin x$

Ans. $y_c = c_1 \cos x + c_2 \sin x$

$$y_p = -\frac{1}{2}x \sin x$$

Hint: Choose $y_p = x(A\cos x + B\sin x)$.

15. $y'' - 4y' + 4y = x^3e^{2x} + xe^{2x}$

Ans. $y_c = c_1e^{2x} + c_2xe^{2x}$

$$y_p = \frac{1}{20}x^5e^{2x} + \frac{1}{6}x^3e^{2x}$$

Hint: Choose $y_p = x^2(Ax^3e^{2x} + Bx^2e^{2x} + Cxe^{2x} + De^{2x})$.

16. $y'' - 3y' + 2y = 2x^2 + e^x + 2xe^x + 4e^{3x}$

Ans. $y_c = c_1e^x + c_2e^{2x}$

$$y_p = x^2 + 3x + \frac{7}{2} + 2e^{3x} - x^2e^x - 3xe^x$$

Hint: Choose $y_p = (Ax^2 + Bx + C) + x(Dxe^x + Ee^x) + Fe^{3x}$.

17. $y'''' + y'' = 3x^2 + 4\sin x - 2\cos x$

Ans. $y_c = c_1 + c_2x + c_3 \sin x + c_4 \cos x$

$$y_p = \frac{x^4}{4} - 3x^2 + x\sin x + 2x\cos x$$

Hint: Choose $y_p = x^2(Ax^2 + Bx + C) + x(D\sin x + E\cos x)$.

18. $y'' + 2y = x^2 \sin 2x$

Ans. $y_c = c_1 \cos 2x + c_2 \sin 2x$

$$y_p = -\frac{1}{12}x^3\cos 2x + \frac{1}{16}x^2\sin 2x + \frac{1}{32}x\cos 2x$$

Hint: Choose $y_p = x(Ax^2\cos 2x + Bx^2\sin 2x + Cx\cos 2x + Dx\sin 2x + E\cos 2x + F\sin 2x)$.

19. $y''' - y' = 4e^{-x} + 3e^{2x}$ with initial conditions $y(0) = 0$, $y'(0) = -1$, $y''(0) = 2$

Ans. $y_c = c_1 + c_2e^x + c_3e^{-x}$

$$y_p = 2xe^{-x} + \frac{1}{2}e^{2x}$$

particular solution obtained by using initial conditions $y = -\frac{9}{2} + 4e^{-x} + 2xe^{-x} + \frac{1}{2}e^{2x}$

Hint: Choose $y_p = x(Ae^{-x}) + Be^{2x}$.

20. $y''' - y'' - 4y' + 4y = 2x^2 - 4x - 1 + 2x^2e^{2x} + 5xe^{2x} + e^{2x}$

Ans. $y_c = c_1e^x + c_2e^{2x} + c_3e^{-2x}$

$$y_p = \frac{x^2}{2} + \frac{1}{6}x^3e^{2x}$$

Hint: Choose $y_p = (Ax^2 + Bx + C) + x(Dx^2e^{2x} + Exe^{2x} + Fe^{2x})$.

9.9 SYSTEM OF SIMULTANEOUS LINEAR D.E. WITH CONSTANT COEFFICIENTS

In several applied mathematics problems, there are more than one dependent variables, each of which is a function of one independent variable, usually say time t. The formulation of such problems leads to a system (or a family) of simultaneous linear D.E. with constant coefficients. Such a system can be solved by the method of elimination, Laplace transform method, method using matrices and short-cut operator methods. Here only the method of elimination is considered.

Note: The number of simultaneous equations in the system = number of dependent variables.

Method of Elimination

Consider a system of 2 O.D.E. in 2 dependent variables x and y and one independent variable t given by

$$f_1(D)x + g_1(D)y = h_1(t) \quad (1)$$
$$f_2(D)x + g_2(D)y = h_2(t)$$

where f_1, f_2, g_1, g_2, are all functions of the differential operator $D = \frac{d}{dt}$.

I. Eliminate y from the given system, resulting in a D.E. exclusively in x alone.

II. Solve this D.E. for x.

III. Substituting x (obtained in step II) in a similar manner obtain a D.E. only in y.

IV. Solve the D.E. obtained in III for y.

Note 1: When the system consists of n O.D.E. in n dependent variables, repeat the above procedure for each of the n dependent variables.

Note 2: In the elimination process in step I, D may be treated as if it is a (ordinary) variable.

Note 3: The number of independent arbitrary constants appearing in the general solution of the system of D.E. (1) is equal to the degree of D in the coefficient determinant

$$\Delta = \begin{vmatrix} f_1(D) & g_1(D) \\ f_2(D) & g_2(D) \end{vmatrix}$$

provided $\Delta \neq 0$.

Note 4: If $\Delta = 0$, the system (1) is dependent.

WORKED OUT EXAMPLES

Solve the following:

Example 1: Solve

$$\frac{dx}{dt} = 3x + 8y \quad (1)$$

$$\frac{dy}{dt} = -x - 3y \quad (2)$$

$$\text{with} \quad x(0) = 6,\ y(0) = -2 \quad (3)$$

Solution: Solving (2) for x, we get

$$x = -3y - \frac{dy}{dt} \quad (4)$$

Differentiating (4) w.r.t. 't', we have

$$\frac{dx}{dt} = -3\frac{dy}{dt} - \frac{d^2y}{dt^2} \quad (5)$$

Substituting (4) and (5) in (1), we obtain

$$\frac{d^2y}{dt^2} - y = 0$$

whose general solution is

$$y(t) = c_1e^t - c_2e^{-t} \quad (6)$$

Substituting (6) in (4) we get

$$x(t) = -4c_1e^t - 2c_2e^{-t} \quad (7)$$

Thus the general solution of the given system of Equations (1) and (2) is given by (6) and (7).

We use the initial conditions (3) to determine c_1 and c_2

$$6 = -4c_1 - 2c_2$$

$$-2 = c_1 + c_2$$

solving we find $c_1 = -1, c_2 = -1$.

Substituting these values in (6) and (7), we get the particular solution

$$x = 4e^t + 2e^{-t},\ y = -e^t - e^{-t}$$

Example 2:

$$(D^2 + D + 1)x + (D^2 + 1)y = e^t \quad (1)$$

$$(D^2 + D)x + D^2y = e^{-t} \quad (2)$$

Solution: To find x, eliminate y from (1) and (2), by multiplying (1) by D^2 and (2) by $(D^2 + 1)$ and subtracting

$$(D^2 + D + 1)D^2x - (D^2 + D)(D^2 + 1)x$$

$$= D^2(e^t) - (D^2 + 1)e^{-t}$$

$$\left[D^4 + D^3 + D^2 - (D^4 + D^3 + D^2 + D)\right]x$$

$$= e^t - e^{-t} - e^{-t} - Dx = e^t - 2e^{-t}$$

Integrating w.r.t. 't', we get

$$x(t) = -e^t - 2e^{-t} + c_1 \quad (3)$$

To find y, substitute (3) in (2)

$$(D^2 + D)(-e^t - 2e^{-t} + c_1) + D^2y = e^{-t}$$

$$-e^{+t} - 2e^{-t} + 0 - e^t + 2e^{-t} + 0 + D^2y = e^{-t}$$

$$D^2y = 2e^t + e^t$$

Integrating w.r.t. 't', we get

$$Dy = 2e^t - e^{-t} + c_2$$

Integrating once more w.r.t. 't', we get

$$y(t) = 2e^t + e^t + c_2 t + c_3. \quad (4)$$

Since $\begin{vmatrix} D^2+D+1 & D^2+1 \\ D^2+D & D^2 \end{vmatrix} = -D$ of degree 1 in D, there should be only one arbitrary constant.

Substituting (3) and (4) in (2), we get

$$(D^2 + D)(-e^t - 2e^{-t} + c_1)$$
$$+D^2(2e^t + e^{-t} + c_2 t + c_3) = e^{-t}$$
$$(-e^t - 2e^{-t} - e^t + 2e^{-t})$$
$$+(2e^t + e^{-t} + 0 + 0) = -e^{-t}$$

Equation (2) satisfied.

So substitute x and y from (3) and (4) in (1)

$$(D^2 + D + 1)(-e^t - 2e^{-t} + c_1)$$
$$+(D^2 + 1)(2e^t + e^{-t} + c_2 t + c_3) = e^t$$
$$\left[-e^t - 2e^{-t} - e^t + 2e^{-t} + (-e^t - 2e^{-t} + c_1)\right]$$
$$+[2e^t + e^{-t} + 0 + 0 + (2e^t + e^{-t} + c_2 t + c_3)] = e^t$$

i.e., $c_1 + c_2 t + c_3 = 0 \;\therefore\; c_2 t + c_3 = -c_1$. Thus the G.S. of the system with one arbitrary constant c_1

$$x(t) = -e^t - 2e^{-t} + c_1$$
$$y(t) = 2e^t + e^{-t} - c_1$$

EXERCISE

Solve the following:

1. $\frac{dx}{dt} = y + 1,\ \frac{dy}{dt} = x + 1$

Ans. $x = c_1 e^t + c_2 e^{-t} - 1$

$y = c_1 e^t - c_2 e^{-t} - 1$

2. $\frac{dx}{dt} = 2y - 1,\ \frac{dy}{dt} = 1 + 2x$

Ans. $x(t) = c_1 e^{2t} + c_2 e^{-2t} - \frac{1}{2}$

$y(t) = c_1 e^{2t} - c_2 e^{-2t} + \frac{1}{2}$

3. $\frac{dx}{dt} - 3x - 6y = t^2,\ \frac{dy}{dt} + \frac{dx}{dt} - 3y = e^t$

Ans. $x(t) = c_3 \cos 3t + c_4 \sin 3t + \frac{3}{5}e^t - \frac{t^2}{3} + \frac{2t}{9} + \frac{2}{27}$

$y(t) = \left(\frac{1}{2}c_4 - \frac{1}{2}c_3\right)\cos 3t + \left(\frac{-c_3}{2} - \frac{c_4}{2}\right)\sin 3t - \frac{e^t}{5} - \frac{2t}{9}$

4. $(D + 6)y - Dx = 0$

$(3 - D)x - 2Dy = 0$

with $x = 2,\ y = 3$ at $t = 0$

Ans. $x = 4e^{2t} - 2e^{-3t}$

$y = e^{2t} + 2e^{-3t}$

5. $(D + 5)x + (D + 3)y = e^{-t}$

$(2D + 1)x + (D + 1)y = 3$

Ans. $x(t) = -\frac{2}{3}c_1 e^t - \frac{c_2}{3}e^{-2t} - 9$

$y(t) = c_1 e^t + c_2 e^{-2t} + \frac{15}{2} + \frac{e^{-t}}{2}$

6. $(D + 1)x + 2y = 1$

$2x + (D - 2)y = t$

Ans. $x = c_1 e^{3t} + 2c_2 e^{-2t} + \frac{t}{3} + \frac{4}{9}$

$y = -2c_1 e^{3t} + c_2 e^{-2t} - \frac{2t}{3} + \frac{1}{9}$

7. $(D - 3)x + 2(D + 2)y = 2\sin t$

$2(D + 1)x + (D - 1)y = \cos t$

Ans. $x = c_1 e^{-5t} + c_2 e^{-\frac{t}{3}} + \frac{8\sin t + \cos t}{65}$

$y = -\frac{4}{3}c_1 e^{-5t} + c_2 e^{-\frac{t}{3}} + \frac{61\sin t - 33\cos t}{130}$

8. $(D + 2)x + (D - 1)y = -\sin t$

$(D - 3)x + (D + 2)y = 4\cos t$

Ans. $x = \frac{3}{5}c_1 e^{\frac{-t}{8}} + \frac{2}{5}\sin t - \frac{1}{5}\cos t$

$y = c_1 e^{-\frac{t}{8}} + \sin t + \cos t$

9. $Dx - (D+1)y = -e^t$

$x + (D-1)y = e^{2t}$

Ans. $x = (c_1 - c_2)\cos t + (c_1 + c_2)\sin t + \frac{3e^{2t}}{5}$

$y = c_1 \cos t + c_2 \sin t + 2\frac{e^{2t}}{5} + \frac{e^t}{2}$

10. $(D^2+4)x - 3Dy = 0$

$3Dx + (D^2+4)y = 0$

Ans. $x = c_1 \cos 4t + c_2 \sin 4t + c_3 \cos t + c_4 \sin t$

$y = c_2 \cos 4t - c_1 \sin 4t - c_4 \cos t + c_3 \sin t$

11. $D^2 y = x - 2$

$D^2 x = y + 2$

Ans. $x = c_1 \sin t + c_2 \cos t + c_3 e^t + c_4 e^{-t} + 2$

$y = -c_1 \sin t - c_2 \cos t + c_3 e^t + c_4 e^{-t} - 2$

12. $(D+1)x + (D-1)y = e^t$

$(D^2 + D + 1)x + (D^2 - D + 1)y = t^2$

Ans. $x = \frac{t^2}{2} - t + \frac{e^t}{2}$

$y = \frac{t^2}{2} + t - \frac{3}{2}e^t$

13. $D^2 x + 2Dy + 8x = 32t$

$D^2 y + 3Dx - 2y = 60e^{-t}$

with $x(0) = 6$, $x^1(0) = 8$,
$y(0) = -24$, $y^1(0) = 0$

Ans. $x = 12\cos 2t + 2e^{-2t} - 8e^{-t} + 4t$

$y = -12\sin 2t + 6e^{-2t} - 36e^{-t} + 6$

14. $2D^2 x + 3Dy - 4 = 0$

$2D^2 y - 3Dx = 0$

with $x = y = Dx = Dy = 0$ at $t = 0$

Ans. $x = \frac{8}{9}(1 - \cos \frac{3}{2}t)$

$y = \frac{4}{3}t - \frac{8}{9}\sin \frac{3}{2}t$

9.10 METHOD OF REDUCTION OF ORDER*

Suppose $y_1(x)$ is known non-trivial solution of the second order homogeneous linear equation

$$a_0(x)\frac{d^2y}{dx^2} + a_1(x)\frac{dy}{dx} + a_2(x)y = 0 \qquad (1)$$

Then the second linearly independent non-trivial solution $y_2(x)$ of (1) can be obtained by the *method of reduction of order*. This method makes use of a transformation of the form

$$y_2(x) = y_1(x)v(x) \qquad (2)$$

which reduces the second order equation (1) to a first order differential equation which is then integrate for the unknown function $v(x)$.

Rewrite (1) in the standard form

$$\frac{d^2y}{dx^2} + P(x)\frac{dy}{dx} + Q(x)y = 0 \qquad (3)$$

where $P(x) = \frac{a_1(x)}{a_0(x)}$, $Q(x) = \frac{a_2(x)}{a_0(x)}$ with $a_0(x) \neq 0$. Differentiating (2) w.r.t. x, we get

$$\frac{dy_2}{dx} = \frac{dy_1}{dx}\cdot v(x) + y_1(x)\frac{dv}{dx} \qquad (4)$$

$$\frac{d^2y_2}{dx^2} = \frac{d^2y_1}{dx^2}v(x) + 2\frac{dy_1}{dx}\cdot\frac{dv}{dx} + y_1\frac{d^2v}{dx^2} \qquad (5)$$

Substituting (2), (4), (5) in (3), we obtain

$$\left[y_1\frac{d^2v}{dx^2} + 2\frac{dy_1}{dx}\cdot\frac{dv}{dx} + v\frac{d^2y_1}{dx^2}\right] + P(x)\left[y_1\frac{dv}{dx} + \frac{dy_1}{dx}v\right] + Q(x)y_1(x)v(x) = 0$$

Rewriting

$$y_1\frac{d^2v}{dx^2} + \left[P(x)y_1 + 2\frac{dy_1}{dx}\right]\frac{dv}{dx} + v(x)\left[\frac{d^2y_1}{dx^2} + P(x)\frac{dy_1}{dx} + Q(x)y_1\right] = 0$$

*Credited to Joseph Louis Lagrange (1736-1813).

Since $y_1(x)$ is a (known) solution of (1), the last term in the above equation becomes zero. Now introducing $\omega = \frac{dv}{dx}$, the above equation reduces to a first order linear differential equation in the new dependent variable ω, given by

$$\frac{d\omega}{dx} + \left[P(x) + 2\frac{y_1'}{y_1}\right]\omega = 0 \qquad (6)$$

Separating the variables and integrating, we get

$$\frac{d\omega}{\omega} + 2\frac{dy_1}{y_1} + P(x)dx = 0$$

or

$$\ln|\omega| + \ln y_1^2 = -\int P(x)dx + \ln|c|$$

Then

$$\boxed{\omega(x) = \frac{c}{y_1^2}e^{-\int P(x)dx}} \qquad (7)$$

Choosing, $c = 1$

$$\omega(x) = \frac{dv}{dx} = \frac{1}{y_1^2}e^{-\int P\,dx}$$

On integration

$$v(x) = \int \frac{1}{y_1^2}e^{-\int P\,dx}dx$$

Thus we obtain the required second solution $y_2(x)$ as

$$\boxed{y_2(x) = y_1(x)v(x) = y_1(x)\int \frac{1}{y_1^2}e^{-\int P(x)dx}\cdot dx} \qquad (8)$$

It can easily be verified that $y_1(x)$ and $y_2(x)$ are linearly independent since

$$\omega(y_1, y_2) = \begin{vmatrix} y_1 & y_2 \\ y_1' & y_2' \end{vmatrix} = \begin{vmatrix} y_1 & y_1 v \\ y_1' & y_1'v + y_1 v' \end{vmatrix} = y_1^2 v'$$

$$= e^{\int -P\,dx} \neq 0$$

Hence y_1, y_2 form the basis for D.E. (1) and the general solution of (1) is given by

$$y(x) = c_1 y_1(x) + c_2 y_2(x)$$

Note: This method of reduction of order is useful only when *one* solution is *known*.

WORKED OUT EXAMPLES

Example 1: Verify that $y = e^{2x}$ is a solution of $(2x+1)y'' - 4(x+1)y' + 4y = 0$. Hence find the general solution.

Solution: Differentiating $y' = 2e^{2x}$, $y'' = 4e^{2x}$. Substituting y, y', y'' in the given D.E., we get

$$(2x+1)4e^{2x} - 4(x+1)2e^{2x} + 4e^{2x} = 0$$

or $[8x + 4 - 8x - 8 + 4]e^{2x} = 0$ or $0 = 0$

Thus $y_1 = e^{2x}$ is a solution. To obtain a second linearly independent solution, by the method of reduction of order, assume

$$y_2(x) = y_1(x)v(x)$$

The given D.E. in the standard form is

$$y'' - \frac{4(x+1)}{(2x+1)}y' + \frac{4}{(2x+1)}y = 0$$

So $P(x) = -\frac{4(x+1)}{(2x+1)}$. Then

$$\omega(x) = \frac{1}{y_1^2}e^{-\int P\,dx}$$

Now

$$\int -P\,dx = -\int -\frac{4(x+1)}{(2x+1)}dx$$

$$= \int \left(\frac{4x+2}{2x+1} + \frac{2}{2x+1}\right)dx$$

$$= 2x + \ln\,(2x+1)$$

Then

$$\omega = \frac{1}{(e^{2x})^2}e^{2x+\ln\,(2x+1)} = \frac{e^{2x}}{(e^{2x})^2}\cdot(2x+1)$$

$$\omega(x) = \frac{2x+1}{e^{2x}}$$

Now $v(x) = \int \omega(x)dx = \int \frac{2x+1}{e^{2x}}dx$

Integrating by parts

$$v(x) = (2x+1)\frac{e^{-2x}}{-2} - 2\cdot\frac{e^{-2x}}{4}$$

The required second solution

$$y_2(x) = y_1(x)v(x) = e^{2x}\left[-\frac{2x+1}{2}\cdot\frac{1}{e^{2x}} - \frac{1}{2}\frac{1}{e^{2x}}\right]$$

$$= -x - 1 = -(x+1)$$

Then the general solution is

$$y(x) = c_1 y_1(x) + c_2 y_2(x) = c_1 e^{2x} + c_2(x+1)$$

Example 2: Obtain the basis for the equation $\sin^2 x \cdot \frac{d^2y}{dx^2} - 2y = 0$ if $y_1(x) = \cot x$ is a solution. Also write the general solution.

Solution: In the standard form the equation is

$$\frac{d^2y}{dx^2} - 2(\operatorname{cosec}^2 x)y = 0$$

So $P(x) = 0$. The $e^{-\int P\,dx} = e^0 = 1$.
Now $\omega(x) = \frac{1}{y_1^2}e^{-\int P\,dx} = \frac{1}{\cot^2 x}\cdot 1 = \tan^2 x$
So $v(x) = \int \omega(x)dx = \int \tan^2 x\,dx$

$$= \int (\sec^2 x - 1)dx = [\tan x - x]$$

Then $y_2(x) = y_1(x)v(x) = \cot x[\tan x - x]$
$= 1 - x\cot x$.

The basis consisting of the two linearly independent solutions is $\{y_1, y_2\} = \{\cot x, 1 - x\cot x\}$. The general solution is

$$y(x) = c_1 y_1(x) + c_2 y_2(x) = c_1 \cot x + c_2[1 - x\cot x]$$

EXERCISE

Determine a linearly independent solution by reduction of order given y_1 as a solution, for the following differential equations.

1. $(x^2+1)y'' - 2xy' + 2y = 0,\ y_1(x) = x$

Ans. $y_2(x) = x^2 - 1$

Hint: $\omega(x) = \left(\frac{x^2+1}{x^2}\right)$, $v(x) = x - \frac{1}{x}$

2. $x^2y'' - 4xy' + 4y = 0,\ y_1 = x$

Ans. $y_2 = x^4$

Hint: $P(x) = \frac{-4}{x}$, $\omega(x) = \frac{1}{x^2}\cdot x^4 = x^2$, $v(x) = \frac{x^3}{3}$

3. $x^2y'' - xy + y = 0,\ y_1 = x$

Ans. $y_2 = x\cdot \ln x$

Hint: $P(x) = -\frac{1}{x}$, $\omega(x) = \frac{1}{x^2}\cdot x$, $v(x) = x\int \frac{dx}{x}$

4. $y'' - y' = 0,\ y_1 = c$

Ans. $y_2 = e^x$

5. $x^2y'' - 5xy' + 9y = 0,\ y_1 = x^3$

Ans. $y_2 = x^3 \ln x$

Hint: $\omega(x) = \frac{1}{x^6}\cdot e^{\int \frac{5dx}{x}} = \frac{x^5}{x^6} = \frac{1}{x}$, $v(x) = \int \frac{dx}{x} = \ln x$.

6. $x^2y'' + xy' + \left(x^2 - \frac{1}{4}\right)y = 0,\ y_1 = \frac{\cos x}{\sqrt{x}}$

Ans. $y_2(x) = \frac{\sin x}{\sqrt{x}}$

Hint: $\omega(x) = \sec^2 x$, $v(x) = \int \sec^2 x \cdot dx = \tan x$

7. $xy'' - (2x-1)y' + (x-1)y = 0,\ y_1 = e^x$

Ans. $y_2 = e^x \ln x$

8. $y'' - 4xy' + (4x^2 - 2)y = 0,\ y_1 = e^{x^2}$

Ans. $y_2 = xe^{x^2}$

9.11 HIGHER ORDER LINEAR EQUATIONS WITH VARIABLE COEFFICIENTS

No general procedure exists for solving second and higher order linear equations with variable coefficients. However by transformations, some types of second - (or higher) - order differentia equations can be reduced to first (or lower) - order equations which can be solved by earlier methods. We consider the following types.

I. Equation of the Form $\frac{d^n y}{dx^n} = f(x)$

Integrating this exact equation directly, we get an equation of lower $(n-1)$ degree as

$$\frac{d^{n-1}}{dx^n} = \int f(x)dx + c_1 = g(x) + c_1$$

which is also exact.

By integrating directly again we have

$$\frac{d^{n-2}y}{dx^{n-2}} = \int (g(x) + c_1)dx + c_2$$

$$= \int g(x)dx + c_1(x) + c_2$$

The solution is obtained by repeated direct integration.

II. Equation of the form $\frac{d^2y}{dx^2} = f(y)$

This equation becomes exact by multiplying both the sides with $2\frac{dy}{dx}$. Then

$$2\frac{dy}{dx} \cdot \frac{d^2y}{dx^2} = f(y)\frac{dy}{dx}$$

or

$$\frac{d}{dx}\left(\frac{dy}{dx}\right)^2 = 2f(y)\frac{dy}{dx}$$

Integrating w.r.t. x, we get

$$\left(\frac{dy}{dx}\right)^2 = 2\int f(y)dy + c_1 = g(y) + c_1$$

Since the variables are separable

$$\frac{dy}{\sqrt{g(y) + c_1}} = dx$$

which on integration yields the required solution as

$$\int \frac{dy}{\sqrt{g(y) + c_1}} = x + c_2$$

III. Equation not Explicitly Containing the Unknown Dependent Variable *y*

When the dependent variable y is absent, the equation takes the form

$$F\left(\frac{d^ny}{dx^n}, \frac{d^{n-1}y}{dx^{n-1}}, \ldots, \frac{dy}{dx}, x\right) = 0$$

The order of this equation can be reduced by one with the subsitution $\frac{dy}{dx} = p$, $\frac{d^2y}{dx^2} = \frac{dp}{dx} \cdots \frac{d^ny}{dx^n} = \frac{d^{n-1}p}{dx^{n-1}}$. Then the reduced equation with p as the dependent variable is solved. In general a substitution of the form $\frac{d^my}{dx^m} = q$ will reduce the order of the equation by m. Thus the substitution $\frac{dy}{dx} = p$ reduces a second order equation to a first order equation in p as a function of x. The equation

$$\frac{d^2y}{dx^2} = f\left(x, \frac{dy}{dx}\right)$$

reduces to a first order equation

$$\frac{dp}{dx} = f(x, p)$$

which on integration gives

$$p = p(x, c_1)$$

Since $\frac{dy}{dx} = p = p(x, c_1)$, the general solution is obtained as

$$y = \int p(x, c_1)dx + c_2$$

IV. Equation not Explicitly Containing the Independent Variable *x*

In the equation of the form

$$F\left(\frac{d^ny}{dx^n}, \frac{d^{n-1}}{dx^{n-1}}, \ldots, \frac{dy}{dx}, y\right) = 0$$

The substitution $\frac{dy}{dx} = p$ reduces it by order one. Considering p as a function of y (and not of x as before), we have

$$\frac{d^2y}{dx^2} = \frac{dp}{dx} = \frac{dp}{dy}\frac{dy}{dx} = p\frac{dp}{dy}$$

Similarly

$$\frac{d^3y}{dx^3} = \frac{d}{dy}\left(\frac{d^2y}{dx^2}\right)\frac{dy}{dx} = \frac{d}{dy}\left(p\frac{dp}{dy}\right)p$$

$$= p^2\frac{d^2p}{dy^2} + p\left(\frac{dp}{dy}\right)^2$$

The equation

$$\frac{d^2y}{dx^2} = f\left(y, \frac{dy}{dx}\right)$$

takes the form

$$p\frac{dp}{dy} = f(y, p)$$

which on integration w.r.t. y yields

$$p = p(y, c_1)$$

or $$\frac{dy}{dx} = p(y, c_1)$$

Separating the variables

$$\frac{dy}{p(y, c_1)} = dx$$

which on integration gives the general solution

$$\phi(x, y, c_1, c_2) = 0$$

V. Solution by Change of Independent Variable

Assume a relation $z = z(x)$ then $\frac{dy}{dx} = \frac{dy}{dz} \cdot \frac{dz}{dx}$. Now

$$\frac{d^2y}{dx^2} = \frac{d}{dx}\left(\frac{dy}{dz}\frac{dz}{dx}\right) = \frac{d}{dx}\left(\frac{dy}{dz}\right)\frac{dz}{dx} + \frac{dy}{dz} \cdot \frac{d^2z}{dx^2}$$

$$= \frac{d}{dz}\left(\frac{dy}{dz}\right)\frac{dz}{dx} \cdot \frac{dz}{dx} + \frac{dy}{dz} \cdot \frac{d^2z}{dx^2}$$

$$\frac{d^2y}{dx^2} = \frac{d^2y}{dz^2}\left(\frac{dz}{dx}\right)^2 + \frac{dy}{dz} \cdot \frac{d^2z}{dx^2}$$

with the above relation the D.E.

$$\frac{d^2y}{dx^2} + P(x)\frac{dy}{dx} + Q(x)y = R(x)$$

gets transformed to a D.E. in the independent variable z as

$$\left(\frac{d^2y}{dz^2} \cdot \left(\frac{dz}{dx}\right)^2 + \frac{dy}{dz} \cdot \frac{d^2z}{dx^2}\right) + P(x) \cdot \frac{dy}{dz} \cdot \frac{dz}{dx}$$
$$+Q(x)y = R(x)$$

or $$\frac{d^2y}{dz^2} + P_1(x)\frac{dy}{dz} + Q_1(x)y = R_1(x)$$

where $P_1(x) = \frac{\frac{d^2z}{dx^2} + p\frac{dz}{dx}}{S^*}$ and $Q_1(x) = \frac{Q(x)}{S^*}$, $R_1(x) = \frac{R(x)}{S^*}$, $S^* = \left(\frac{dz}{dx}\right)^2$

Here P_1, Q_1, R_1 will be converted to functions of z by eliminating x using the relation $z = z(x)$. Now choose

$$\left(\frac{dz}{dx}\right)^2 = Q(x)$$

To get real functions, the sign before $Q(x)$ is chosen as positive always. Integrating

$$z = \int Q(x)dx = z(x)$$

omitting the constant of integration. Now calculate P_1, Q_1, R_1 with this $z = z(x)$. When P, Q, R turns out to be constants (or zero), the new equation in z can be solved for $y(z)$. By replacing z in terms of x, the required solution $y(x)$ is obtained.

WORKED OUT EXAMPLES

I. $\frac{d^ny}{dx^n} = f(x)$

Example 1: Solve $\frac{d^3y}{dx^3} = \log x$

Solution: Integrating w.r.t. x, we get

$$\frac{d^2y}{dx^2} = \int \log x\, dx + c_1 = x \ln x - x + c_1$$

Integrating again wrt x, we have

$$\frac{dy}{dx} = \int x \ln x\, dx - \frac{x^2}{2} + c_1x + c_2$$

$$= \frac{x^2}{2}\ln x - \frac{x^2}{4} - \frac{x^2}{2} + c_1x + c_2$$

Integrating again,

$$y(x) = \frac{1}{2}\int x^2 \ln x\, dx - \frac{3}{4}\frac{x^3}{3} + c_1\frac{x^2}{2} + c_2x + c_3$$

$$y = \frac{1}{2}\left[\frac{x^3}{3}\ln x - \frac{x^3}{9}\right] - \frac{1}{4}x^3 + \frac{c_1}{2}x^2 + c_2x + c_3$$

$$36y = 6x^3 \ln x - 11x^3 + c_1^*x^2 + c_2^*x + c_3^*$$

II. $\frac{d^ny}{dx^n} = f(y)$

Example 2: Solve $y^3y'' = a$ with $y = 1$, $y' = 0$ at $x = 0$

Solution: Multiplying both sides by $2y'$, we have

$$2y'y'' = 2y' \cdot a \cdot y^{-3}$$

$$\frac{d}{dx}(y')^2 = 2ay^{-3}\frac{dy}{dx}$$

Integrating w.r.t. x

$$(y')^2 = 2a\frac{y^{-2}}{-2} + c_1$$

At $x = 0$, $y = 1$, $y' = 0$ so $0 = -a + c_1 \therefore c_1 = a$

Then

$$(y')^2 = \left(a - \frac{a}{y^2}\right) \quad \text{or} \quad y' = \sqrt{\frac{a(y^2-1)}{y^2}}$$

Separating the variables

$$\frac{y\,dy}{\sqrt{a}\sqrt{y^2-1}} = dx$$

Integrating

$$\frac{1}{2\sqrt{a}}\frac{\sqrt{(y^2-1)}}{\frac{1}{2}} = x + c_2$$

when $x = 0$, $y = 1$ so $0 = 0 + c_2 \therefore c_2 = 0$

Thus $(y^2 - 1) = ax^2$

III. Equation not explicitly containing y

Example 3: Solve $x^2\frac{d^3y}{dx^3} - 4x\frac{d^2y}{dx^2} + 6\frac{dy}{dx} = 4$

Solution: Put $\frac{dy}{dx} = p$, then $\frac{d^2y}{dx^2} = \frac{dp}{dx}$, $\frac{d^3y}{dx^3} = \frac{d^2p}{dx^2}$. With these substitution, the given D.E. reduces to a second order D.E. in the dependent variable p with variable coefficients given by

$$x^2\frac{d^2p}{dx^2} - 4x\frac{dp}{dx} + 6p = 4$$

This is Euler-Cauchy equation. So put $x = e^t$ then $\mathcal{D}(\mathcal{D}-1)p - 4\mathcal{D}p + 6p = 4$ where $\mathcal{D} = \frac{d}{dt}$.

or $$\frac{d^2p}{dt^2} - 5\frac{dp}{dt} + 6p = 4$$

The auxilary equation is $m^2 - 5m + 6 = 0$ with real distinct roots $m = 2, 3$. So the complementary function is

$$p_c(t) = c_1e^{2t} + c_2e^{3t}$$

particular integral

$$p_p(t) = \frac{1}{\mathcal{D}^2 - 5\mathcal{D} + 6}4 = \frac{4}{6} = \frac{2}{3}$$

The general solution is

$$p(t) = c_1e^{2t} + c_2e^{3t} + \frac{2}{3}$$

or $$\frac{dy}{dx} = c_1x^2 + c_2x^3 + \frac{2}{3}$$

Integrating w.r.t. x

$$y(x) = c_1\frac{x^3}{3} + c_2\frac{x^4}{4} + \frac{2}{3}x + c_3$$

IV. Equations not explicitly containing x

Example 4: Solve. $y(1 - \ln y)\frac{d^2y}{dx^2} + (1 + \ln y)\left(\frac{dy}{dx}\right)^2 = 0$

Solution: Put $\frac{dy}{dx} = p$. Then $\frac{d^2y}{dx^2} = \frac{d}{dx}(p) = \frac{dp}{dy}\cdot\frac{dy}{dx}$ ie, $\frac{d^2y}{dx^2} = p\frac{dp}{dy}$. Here we consider p as a function of y with these substitutions, the given equation reduces to

$$y(1 - \ln y)p\frac{dp}{dy} + (1 + \ln y)p^2 = 0$$

Separating the variables

$$\frac{dp}{p} + \frac{(1 + \ln y)}{y(1 - \ln y)}dy = 0$$

Introducing $t = \ln y$, we have $dt = \frac{dy}{y}$, so

$$\frac{dp}{p} + \frac{(1+t)}{(1-t)}\cdot dt = 0$$

or $\frac{dp}{p} - \left[1 + \frac{2}{t-1}\right]dt = 0$

Integrating $\ln p - t - 2\ln(t-1) = c_1$

or $p = c_2(t-1)^2e^t$

i.e., $\frac{dy}{dx} = c_2(\ln y - 1)^2e^{\ln y}$

Separating the variables

$$\frac{dy}{y(\ln y - 1)^2} = c_2\,dx$$

or $c_2dx = \frac{dt}{(t-1)^2}$ where $t = \ln y$

Integrating

$$c_2x + c_3 = -\frac{1}{t-1}$$

The general solution is

$$(1 - \ln y)(c_2 x + c_3) = 1$$

V. Change of independent variable

Example 5: Solve $\frac{d^2y}{dx^2} + \tan x \frac{dy}{dx} + y \cos^2 x = 0$

Solution: $P(x) = \tan x$, $Q(x) = \cos^2 x$. The transformation $z = f(x)$ is so chosen that $P_1 = \frac{d^2z}{dx^2} + P\frac{dz}{dx} = 0$ or with $u = \frac{dz}{dx}$,

$$\frac{du}{dx} + u \cdot \tan x = 0$$

Integrating $\ln u = \ln \cos x + c_1$ or $u = c_2 \cos x$.
Since $u = \frac{dz}{dx} = c_2 \cos x$ we have $z(x) = c_2 \sin x$
Now $Q_1 = \frac{Q}{u^2} = \frac{\cos^2 x}{(c_2 \cos x)^2} = \frac{1}{c_2^2}$
Then the transformed equation with z as the new independent variable is

$$\frac{d^2y}{dz^2} + P_1\frac{dy}{dz} + Q_1 y = 0$$

or

$$\frac{d^2y}{dz^2} + 0 + \frac{1}{c_2^2}y = 0$$

Integrating $y(z) = A\cos\left(\frac{z}{c_2}\right) + B\sin\left(\frac{z}{c_2}\right)$
or $y(x) = A\cos(\sin x) + B\sin(\sin x)$

Example 6: Solve $y'' - y'\cot x - y\sin^2 x = \cos x - \cos^3 x$

Solution: The equation in the standard form has $P(x) = -\cot x$, $Q(x) = -\sin^2 x$, $R(x) = \cos x - \cos^3 x$. Choose the new independent variable z such that
$S^* = \left(\frac{dz}{dx}\right)^2 = \sin^2 x$ or $\frac{dz}{dx} = \sin x$. On integration, $z(x) = -\cos x$. With this z, the given equation reduces to $\frac{d^2y}{dz^2} + p_1\frac{dy}{dz} + Q_1 y = R_1$ where

$$P_1 = \frac{\frac{d^2z}{dx^2} + P\frac{dz}{dx}}{S^*} = \frac{\cos x - \cot x \cdot \sin x}{\sin^2 x} = 0$$

$$Q_1 = \frac{Q}{S^*} = \frac{-\sin^2 x}{\sin^2 x} = -1$$

and

$$R_1 = \frac{R}{S^*} = \frac{\cos x - \cos^3 x}{\sin^2 x} = \frac{\cos x(1-\cos^2 x)}{\sin^2 x}$$
$$= \cos x$$

Thus the D.E. with z as the independent variable is

$$\frac{d^2y}{dz^2} + 0 - 1 \cdot y = \cos x = -z$$

The complementary function is $y = c_1 e^z + c_2 e^{-z}$ and particular integral $y = \frac{1}{D^2-1}(-z) = +\frac{1}{(1-D^2)}z = +[1 + D^2 - D^4 + \cdots]z = +z$. Thus the general solution is

$$y(z) = c_1 e^z + c_2 e^{-z} + z \text{ or}$$
$$y(x) = c_1 e^{-\cos x} + c_2 e^{\cos x} - \cos x$$

EXERCISE

I. $\frac{d^n y}{dx^n} = f(x)$

Solve

1. $y''' = xe^x$

Ans. $y = xe^x - 3e^x + \frac{1}{2}c_1x^2 + c_2x + c_3$

2. $y'' = x^2 \sin^2 x$

Ans. $y = -x^2 \sin x - 4x\cos x + 6\sin x + c_1x + c_2$

3. $y''' = x + \ln x$

Ans. $y = \frac{x^4}{24} + \frac{x^3}{6}\ln x - \frac{11}{35}x^3 + c_1x^2 + c_2x + c_3$

4. $y''' = \sin^2 x$

Ans. $y = \frac{x^3}{12} + \frac{\sin 2x}{16} + \frac{c_1x^2}{2} + c_2x + c_3$

5. $y^{(n)} = x^m$

Ans. $y = \frac{m!}{(m+n)!}x^{m+n} + c_1x^{n-1} + c_2x^{n-2} + \cdots + c_n$

6. $x^2y'' = \ln x$

Ans. $y = -\frac{1}{2}(\ln x)^2 + \ln x - c_1x + c_2$

II. $\frac{d^2y}{dx^2} = f(y)$

Solve

7. $y'' = \sec^2 y \cdot \tan y$ with $y = 0$ and $y' = 1$ when $x = 0$

Ans. $x + c_2 = \frac{1}{\sqrt{c_1 - 1}} \sin^{-1}\left\{\sin y \Big/ \sqrt{\left(\frac{c_1 - 1}{c_1}\right)}\right\}$,

$c_1 = 1, c_2 = 0$, so $y = \sin^{-1} x$.

8. $y'' = 2(y^3 + y)$ with $y = 0$, $y' = 1$ when $x = 0$

Ans. $y = \tan x$

9. $y'' = e^{-2y}$

Ans. $c_1 e^y = \cosh\,(c_1 x + c_2)$

10. $\sin^3 y \cdot y'' = \cos y$

Ans. $\sqrt{1 + c_1} \cdot \sin\,(x + c_2) + \sqrt{\frac{1 + c_1}{c_1}} \cos y = 0$

11. $y'' = 3\sqrt{y}$ with $y = 1$, $y' = 2$ when $x = 0$

Ans. $x = 2(y^{\frac{1}{4}} - 1)$

12. $y'' + a^2 y^{-2} = 0$

Ans. $\sqrt{c_1 y^2 + y} - \sqrt{2} a c_1 x + c_2 = \frac{1}{\sqrt{c_1}} \ln\left\{\sqrt{(c_1 y + 1)} + \sqrt{c_1 y}\right\}$

III. $\frac{d^2y}{dx^2} = f\left(x, \frac{dy}{dx}\right)$

Solve

13. Catenary: $ay'' = \sqrt{1 + (y')^2}$ with $y = a$ and $y' = 0$ at $x = 0$

Ans. $y = a \cosh\left(\frac{x}{a}\right)$,

Hint: $a\frac{dp}{dx} = \sqrt{1 + p^2}$, $p = \sinh\left(\frac{x}{a} + c_1\right)$, $c_1 = c_2 = 0$

14. $(1 - x^2)y'' - xy' - 2 = 0$

Ans. $y = (\sin^{-1} x)^2 + c_1 \sin^{-1} x + c_2$

15. $y'' + y' + (y')^3 = 0$

Ans. $y = -\sin^{-1}(c_1 e^{-x}) + c_2$

16. $xy'' - \sqrt{1 + (y')^2} = 0$

Ans. $2y = c_1 \frac{x^2}{2} - \frac{1}{c_1} \ln x + c_2$

17. $y'''' \cdot y''' = 1$

Ans. $y = \frac{8\sqrt{2}}{105}(x + c_1)^{7/2} + \frac{1}{2} c_2 x^2 + c_3 x + c_4$

18. $y'''' - \cot x \cdot y''' = 0$

Ans. $y = c_1 \cos x + c_2 x^2 + c_3 x + c_4$

IV. $\frac{d^2y}{dx^2} = f\left(y, \frac{dy}{dx}\right)$

19. $3y'' = y^{-5/3}$

Ans. $x + c_2 = \pm\frac{1}{c_1^2}\sqrt{c_1 y^{2/3} - 1}(c_1 y^{2/3} + 2)$

20. $yy'' - (y')^2 = y^2 \ln y$

Ans. $\ln y = \sqrt{c_1} \cdot \sinh\,(x + c_2)$

21. $yy'' + y'(y' - 2y) = 0$

Ans. $y^2 + c_1 = c_2 e^{2x}$

22. $2yy'' - (y')^2 - 1 = 0$

Ans. $a^2(x + b)^2 = 4(ay - 1)$

23. $yy'' - 2(y')^2 - y^2 = 0$

Ans. $\sqrt{c_1} \cdot y \cdot \cos\,(x + c_2) = 1$

V. Change of Independent Variable

24. $(1 + x^2)y'' + 2x(1 + x^2)y' + 4y = 0$

Ans. $y(1 + x^2) = c_1(1 - x^2) + 2c_2 x$

25. $\sin^2 x y'' + \sin x \cdot \cos x y' + 4y = 0$

Ans. $y = c_1 \cos\,(2 \log \tan\frac{x}{2}) + c_2 \sin\,(2 \log \tan\frac{x}{2})$

26. $\cos x y'' + y' \sin x - 2y \cos^3 x = 2\cos^5 x$

Ans. $y = c_1 e^z + c_2 e^{-z}$ where $z = \sqrt{2} \sin x$

27. $xy'' - y' + 4x^3 y = x^5$

Ans. $y = c_1 \cos x^2 + c_2 \sin x^2 + \frac{x^2}{4}$

28. $xy'' - y' - 4x^3 y = 8x^3 \sin x^2$

Ans. $y = c_1 x^2 + c_2 e^{-x^2} - \sin x^2$

29. $xy'' + (4x^2 - 1)y' + 4x^3 y = 2x^3$

Ans. $y = e^{-x^2}(c_1 \cos x^2 + c_2 \sin x^2) + \frac{1}{2}$

30. $y'' + (3 \sin x - \cot x)y' + 2y \sin^2 x = \sin^2 x \cdot e^{-\cos x}$

Ans. $y = c_1 e^{\cos x} + c_2 e^{2\cos x} + \frac{1}{6} e^{-\cos x}$

9.12 SIMPLE HARMONIC MOTION

A particle is said to be in simple harmonic motion if the acceleration of the particle is proportional to its displacement i.e.,

$$\frac{d^2x}{dt^2} = -\omega^2 x$$

$$\text{or} \quad \ddot{x} + \omega^2 x = 0 \qquad (1)$$

where $x(t)$ is the displacement of the particle at any time t, from a fixed reference point O. The general solution of the second order linear homogeneous D.E. (1), known as harmonic D.E., is

$$x(t) = c_1 \cos \omega t + c_2 \sin \omega t \tag{2}$$

Introducing $c_1 = A \sin \phi_0$, $c_2 = A \cos \phi_0$, so that $A = \sqrt{c_1^2 + c_2^2}$ and $\tan \phi_0 = \frac{c_1}{c_2}$, (2) can be written as $x(t) = A \sin (\omega t + \phi_0)$. (3)

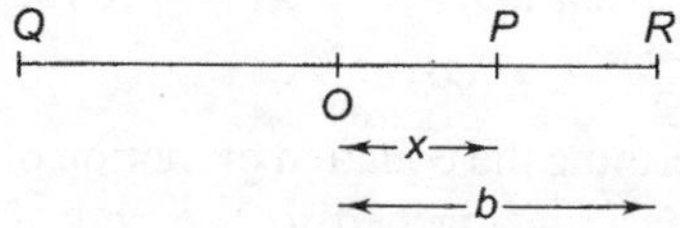

Fig. 9.1

Suppose the particle starts from rest from the point R which is at a distance 'b' from 0, then the initial conditions are $x = b$, $\dot{x} = 0$ at $t = 0$.

From (2)

$$b = x(0) = c_1 \cdot 1 + c_2 \cdot 0 \qquad \therefore\ c_1 = b$$

Differentiating (2), the velocity of the particle is

$$v = \frac{dx}{dt} = \omega(-b \sin \omega t + c_2 \cos \omega t)$$

Using the second initial condition

$$0 = \dot{x}(0) = \omega(-b \cdot 0 + c_2 \cdot 1)\ \therefore\ c_2 = 0$$

Thus the displacement x and velocity v are given respectively by

$$\boxed{x(t) = b \cos \omega t} \tag{4}$$

and

$$v = \frac{dx}{dt} = -b\omega \sin \omega t = -\omega\sqrt{b^2 - x^2} \tag{5}$$

Thus the particle oscillates between the two extremities R and Q where the velocity is zero. The velocity of the particle attains maximum at origin O. The oscillations (motion) are called *harmonic*. The integral curves are sine curves. *Amplitude* of the simple harmonic motion is the constant A, which is the maximum displacement of particle from origin (equilibrium position).

Periodic time or *period of oscillation* is the time for a complete oscillation and is given by $\frac{2\pi}{\omega}$. Note that x and $\dot{x}$ given by (4) and (5), involving cos and sin, are periodic functions of period $\frac{2\pi}{\omega}$ i.e., $x\left(t + \frac{2\pi}{\omega}\right) = x(t)$. The part of the curve between P and Q represents one period of the motion.

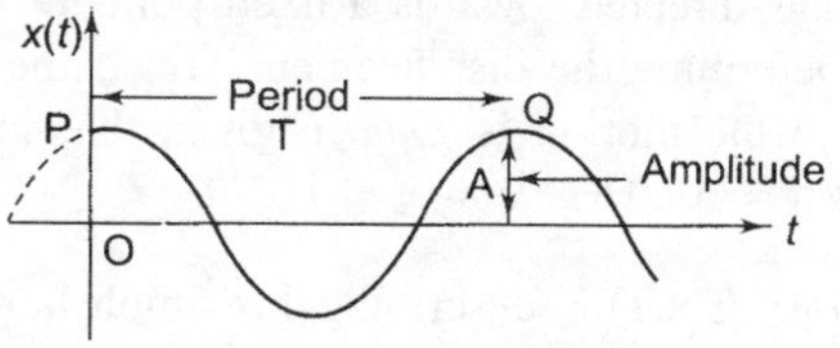

Fig. 9.2

Frequency is the number of oscillations per second during the time 2π and is given by $\frac{1}{\text{periodic time}} = \frac{\omega}{2\pi}$.

Here ϕ_0 is the initial phase.

The power of modeling is seen by the equation (1) which also represents the motion of a particle with constant angular velocity along a circle, also the free undamped vertical motion of a mass-spring system and also the motion of the bob of a simple pendulum in a vertical plane.

WORKED OUT EXAMPLES

Example 1: Compute the time required for a particle, in simple harmonic motion with amplitude 20 cm and periodic time 4 seconds, in passing between two points which are at distances 15 cm and 5 cm from the origin O.

Solution: Here $b = 20$, $4 = \frac{2\pi}{\omega}$ or $\omega = \frac{2\pi}{\omega} = \frac{\pi}{2}$

Then $x(t) = b \cos \omega t = 20 \cos \frac{\pi t}{2}$

When $t = t_1$, given $x = x_1 = 5$

When $t = t_2$, given $x = x_2 = 15$

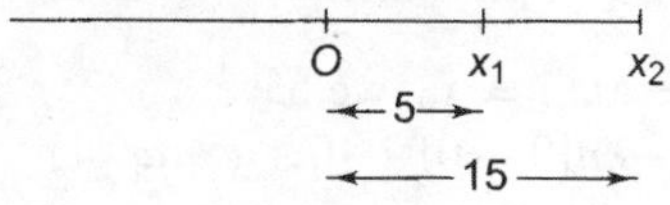

Fig. 9.3

Thus $5 = x_1 = 20 \cos \frac{\pi t_1}{2}$ and $15 = x_2 = 20 \cos \frac{\pi t^2}{2}$.

Then $\frac{\pi t_2}{2} - \frac{\pi t_1}{2} = \cos^{-1} \frac{15}{20} - \cos^{-1} \frac{5}{20}$ or the time

taken by particle in passing between x_1, and x_1 is given by
$t_2 - t_1 = \frac{2}{\pi}\left[\cos^{-1}\frac{3}{4} - \cos^{-1}\frac{1}{4}\right]$

Example 2: A particle of mass m moves in a straight line under the action of force mn^2x which is always directed towards a fixed point O on the line. Determine the displacement $x(t)$ if the resistance to the motion is $2\lambda mnv$ given that initially $x = 0$, $\dot{x} = x_0$. Here $0 < \lambda < 1$.

Solution: The D.E. describing this simple harmonic motion is

$$m\ddot{x} = -2\lambda mn\dot{x} - mn^2x$$

$$\text{or} \quad \frac{d^2x}{dt^2} + 2\lambda n\frac{dx}{dt} + n^2x = 0$$

The auxiliary equation is

$$r^2 + 2\lambda nr + n^2 = 0$$

$$\text{with roots } r = \frac{-2\lambda n \pm \sqrt{4\lambda^2n^2 - 4n^2}}{2}$$

$$r = -\lambda n \pm n\sqrt{1-\lambda^2}i$$

The general solution is

$$x(t) = e^{-\lambda nt}[c_1 \cos n\omega t + c_2 \sin n\omega t]$$

where $\omega = \sqrt{1-\lambda^2}$. Using the initial condition

$$0 = x(0) = c_1 \cdot 1 + c_2 \cdot 0 \quad \therefore c_1 = 0$$

Differentiating $x(t)$ w.r.t. 't'

$$\dot{x} = -\lambda ne^{-\lambda nt}[c_1 \cos n\omega t + c_2 \sin n\omega t]$$

$$+e^{-\lambda nt}[-c_1 n\omega \sin n\omega t + c_2 n\omega \cos n\omega t]$$

Since at $t = 0$, $\dot{x} = x_0$, we have
$x_0 = \dot{x} = -\lambda n[0+0] + [0 + c_2 n\omega \cdot 1]$
$\therefore c_2 = \frac{x_0}{n\omega}$
Then the displacement $x(t)$ is given by

$$x(t) = e^{-\lambda nt} \cdot \frac{x_0}{n\omega} \cdot \sin n\omega t$$

with $\omega = \sqrt{1-\lambda^2}$

EXERCISE

1. Find the time of a complete oscillation in a simple harmonic motion if $x = x_1$, $x = x_2$, and $x = x_3$ when $t = 1$, $t = 2$, $t = 3$ seconds respectively.

Ans. $2\pi/\theta$ where $\cos\theta = \frac{x_1+x_3}{2x_2}$

Hint: $x(t) = b\cos\omega t$, $x_1 = b\cos\omega$, $x_2 = b\cos 2\omega$, $x_3 = b\cos 3\omega$, $\frac{x_1+x_3}{2x_2} = \frac{\cos\omega+\cos 3\omega}{2\cos 2\omega} = \cos\omega$

2. Determine the equation of motion of a particle of mass m attached to one end of stretched elastic horizontal string whose other end is fixed. Find the displacement x of the particle if $x = x_0$, $\dot{x} = 0$ when $t = 0$.

Ans. $\ddot{x} + \frac{g}{e}x = \frac{gL}{e}$ where L is the natural length of the string, e is the elongation due to weight mg. $x(t) = (x_0 - L)\cos\omega t + L$ where $\omega = \sqrt{\frac{g}{e}}$.

Hint: G.S.: $x(t) = c_1\cos\omega t + c_2\sin\omega t + L$.

3. Find the period of a particle of mass m, in simple harmonic motion, attached to the middle point of an elastic string (of natural length $2a$) stretched between two points Q and R which are $4a$ apart.

Ans. Period $2\pi/\omega$ where $\omega^2 = 2\lambda/am$ where λ is the modulus.

9.13 MASS-SPRING MECHANICAL SYSTEM

Airplanes, bridges, ships, machines, cars etc. are vibrating mechanical systems. The simplest mechanical system is the mass-spring system which consists of a coil spring of natural length L, suspended vertically from a fixed point support (such as a ceiling or beam). A constant mass 'm' attached to the lower end of the spring, stretches the spring to a length $(L+e)$ and comes to rest which is known as the static equilibrium position. Here $e > 0$ is the static deflection due to the hanging the mass on the spring. Now the mass is set in motion either by pushing or pulling the mass from equilibrium position and/or by imparting a non-zero velocity (downward or upward) to the mass (in the equilibrium position). Since the motion takes place in the vertical direction, we consider the downward direction as positive. In order to determine the displacement $x(t)$ of the mass from the static equilibrium position, we use Newton's second law and Hooke's law. The mass m is subjected to the following forces:

(a) a gravitational force mg acting downwards.

(b) a spring restoring force $-k(x(t)+e)$ due to displacement of the spring from rest (equilibrium) position.

(c) a frictional (or damping or resistance) force of the medium, opposing the motion and of magnitude- $-c\dot{x}(t)$.

(d) an external force $F(t)$.

The differential equation (D.E.) describing the motion of the mass is obtained by Newton's second law as

$$m\ddot{x}(t) = mg - k(x(t)+e) - c\dot{x}(t) + F(t)$$

Here $k > 0$ is known spring constant or stiffness of the spring, $c \geq 0$ is known as damping constant, g is gravitational constant. Since the force on the mass exerted by the spring must be equal and opposite to

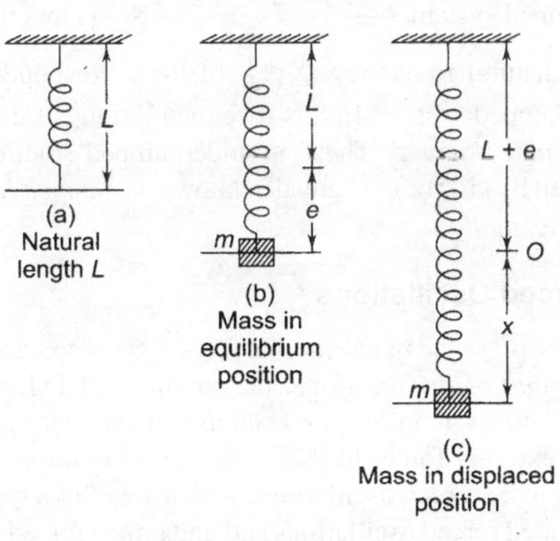

Fig. 9.4 Forced damped mass-spring system

the gravitational force on the mass, we have $ke = mg$. Thus the D.E. modeling the motion of mass is

$$m\ddot{x}(t) + c\dot{x}(t) + kx(t) = F(t)$$

which is a second order linear non-homogenous equation with constant coefficients. The displacement (or motion) of the mass at any time t is $x(t)$ which is the solution of D.E. (1). Let us consider three important cases of D.E. (1) referred to as free motion, damped motion and forced motion.

Free, Undamped Oscillations of a Spring

In the absence of external force ($F(t) = 0$) and neglecting the damping force ($c = 0$), D.E. (1) reduce to

$$m\ddot{x} + kx = 0 \quad (2)$$

which is the *harmonic oscillator equation*. Putting $\omega^2 = \frac{k}{m}$, the equation (2) takes the form

$$\ddot{x} + \omega^2 x = 0$$

whose general solution is sinusoidal given by

$$x(t) = c_1 \cos \omega t + c_2 \sin \omega t \quad (3)$$

Introducing $c_1 = A\cos\phi$, $c_2 = -A\sin\phi$, equation (3) can be rewritten as

$$x = A\cos\phi\cos\omega t - A\sin\phi\sin\omega t = A\cos(\omega t + \phi)$$

i.e. $$\boxed{x(t) = A\cos(\omega t + \phi)} \quad (4)$$

where $A = \sqrt{c_1^2 + c_2^2}$, $\tan\phi = -\frac{c_2}{c_1}$. The constant A is called the *amplitude* of the motion and gives the maximum (positive) displacement of the mass from its equilibrium position. Thus the free, undamped motion of the mass is a simple harmonic motion, which is periodic. The *period* of motion is the time interval between two successive maxima and is given by

$$T = \frac{2\pi}{\omega} = 2\pi\sqrt{\frac{m}{k}}$$

The *natural frequency* (or simply frequency) of the motion (or harmonic oscillator) is the reciprocal of the period, which gives the number of oscillations/second. Thus the natural frequency is the undemped frequency (i.e. frequency of the system with out damping)

Free, Damped Motion of a Mass

Every system has some damping, otherwise the system continues to move forever. Damping force opposes oscillations. Damping not only decreases the amplitude but also alters the natural frequency of the system. With external force absent ($F(t) = 0$) and damping present ($c \neq 0$), the D.E. (1) takes the form

$$m\ddot{x} + c\dot{x} + kx = 0 \qquad (5)$$

whose auxiliary equation is

$$r^2 + 2br + \omega^2 = 0 \qquad (6)$$

Here $2b = \frac{c}{m}$ and $\omega^2 = \frac{k}{m}$. The roots of (6) are

$$r = \frac{-2b \pm \sqrt{4b^2 - 4\omega^2}}{2} = -b \pm \sqrt{b^2 - \omega^2} \qquad (7)$$

The motion of mass depends on the damping through the nature of the discriminant $b^2 - \omega^2$.

Case (i) $b^2 - \omega^2 > 0$ i.e., $c^2 > 4\,mk$

Since $b > \omega$, the roots $r_1 = -b + \sqrt{b^2 - \omega^2}$ and $r_2 = -b - \sqrt{b^2 - \omega^2}$ are distinct, real negative numbers. The general solution is

$$x(t) = c_1 e^{r_1 t} + c_2 e^{r_2 t} \qquad (8)$$

which tends to zero as $t \to \infty$. Thus the damping is so great that no oscillations can occur. The motion (or system) is said to be *over critically damped* (or simply *overdamped*).

Case (ii) $b^2 - \omega^2 = 0$ or $b = \omega$. Here both the roots are equal, real negative number $-b$. The general solution is

$$x(t) = (c_1 + c_2 t)e^{-bt} \qquad (9)$$

which tends to zero as $t \to \infty$. Thus the damping is just enough to prevent oscillations so that the motion is no longer oscillatory. In this case the motion is said to be *critically damped.*

In both cases (i) and (ii) the displacement $x(t)$ is asymptotic and approaches zero (the equilibrium position) as $t \to \infty$.

Case (iii) when $b < \omega$ then $b^2 - \omega^2 < 0$ so the roots of the auxiliary equation (7) are complex conjugate given by $-b \pm \sqrt{\omega^2 - b^2}\, i$. The general solution is

$$x(t) = e^{-bt}[c_1 \cos \sqrt{\omega^2 - b^2}\, t + c_2 \sin \sqrt{\omega^2 - b^2}\, t \qquad (10)$$

which can be written in the alternative form

$$x(t) = Ae^{-bt}[\cos\left(\left(\sqrt{\omega^2 - b^2}\right) \cdot t + \phi\right)] \qquad (11)$$

Here $A = \sqrt{c_1^2 + c_2^2}$ and $\phi = \tan^{-1}\left(-\frac{c_1}{c_2}\right)$.

Since $A > 0$ and $b > 0$ the first factor is in (11). The *time-varying amplitude* or *the damping factor* $Ae^{-bt} \to 0$ monotonically as $t \to \infty$. The other factor $\cos\left(\left(\sqrt{\omega^2 - b^2}\right) \cdot t + \phi\right)$ in (11) is periodic, oscillatory representing simple harmonic motion.

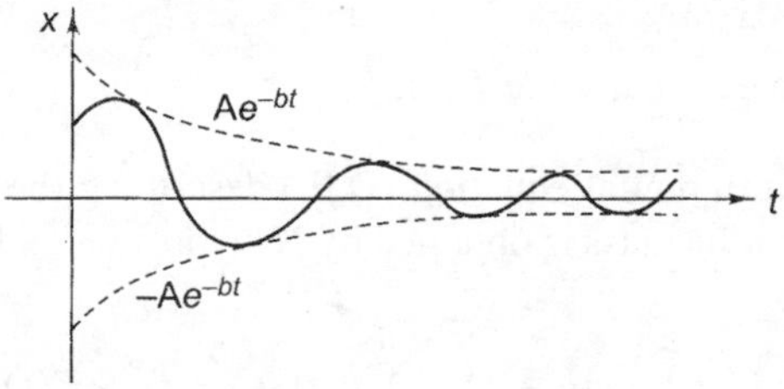

Fig. 9.5

The solution (11) which is the product of these two factors is a damped oscillatory motion in which the oscillations die (damped) out. Resonance never occurs in this case since the frequency of the free damped system $\frac{\sqrt{\omega^2 - b^2}}{2\pi} = \frac{1}{2\pi}\sqrt{\frac{k}{m} - \frac{c^2}{4m^2}}$ is less than the natural frequency $\frac{1}{2\pi}\sqrt{\frac{k}{m}}$ of the corresponding undamped system. In this case the motion is said to be *underdamped.* The free underdamped solutions given by (10) or (11) are also known as *damped harmonic motion.*

Forced Oscillations

In the presence of an external force $F(t)$, also known as *input* or *driving force*, the solutions of D.E. (1) are known as *output* or *response* of the system to the external force. In this case, the oscillations are said to be forced oscillations, which are of two types damped forced oscillations and undamped forced oscillations.

Case (i): Damped forced oscillations

The D.E. in this case is

$$m\ddot{x} + c\dot{x} + kx = F(t) \tag{1}$$

Suppose the external impressed forces $F(t)$ is a periodic sinusoidal force with amplitude F_1 and circular frequency β of the form

$$F(t) = F_1 \cos \beta t, \quad (F_1 > 0; \beta > 0)$$

Then D.E. (1) takes the form

$$m\ddot{x} + c\dot{x} + kx = F_1 \cos \beta t$$

or $$\ddot{x} + 2b\dot{x} + \omega^2 x = E_1 \cos \beta t \tag{12}$$

where $2b = \frac{c}{m}$, $\omega^2 = \frac{k}{m}$, $E_1 = \frac{F_1}{m}$. Assume that $b < \omega$. Then the complementary function of (12) is

$$x_c = Ae^{-bt} \cos \left(\left(\sqrt{\omega^2 - b^2} \right) \cdot t + \phi \right) \tag{13}$$

which approaches zero as $t \to \infty$. The particular integral of (12) is

$$x_p = \frac{1}{D^2 + 2bD + \omega^2} E_1 \cos \beta t$$

$$= \frac{E_1}{\omega^2 - \beta^2 + 2bD} \cos \beta t$$

$$= \frac{E_1[\omega^2 - \beta^2 - 2bD]}{(\omega^2 - \beta^2)^2 - 4b^2D^2} \cos \beta t$$

$$= \frac{E_1[\omega^2 - \beta^2 - 2bD] \cos \beta t}{(\omega^2 - \beta^2)^2 + 4b^2\beta^2}$$

$$x_p = \frac{E_1}{(\omega^2 - \beta^2)^2 + 4b^2\beta^2} [(\omega^2 - \beta^2)^2 \cos \beta t + \, + 2b\beta \sin \beta t]$$

Put $\cos \theta = \frac{\omega^2 - \beta^2}{(\omega^2 - \beta^2)^2 + 4b^2\beta^2}$, $\sin \theta = \frac{2b\beta}{(\omega^2 - \beta^2)^2 + 4b^2\beta^2}$

Then x_p can be written in the phase angle form as

$$x_p = E_1 \cdot \cos\ (\beta t - \theta) \tag{14}$$

The general solution is

$$x(t) = x_c + x_p \tag{15}$$

where x_c and x_p are respectively given by (13) and (14). The first terms x_c in (15) given by $Ae^{-bt} \cos \left(\left(\sqrt{\omega^2 - b^2} \right) t + \phi \right)$ known as *transient term*, tends to zero as $t \to \infty$ and represents the damped oscillations that would be the entire motion of the corresponding free motion (i.e. when external force $F_1 \cos \omega t$ was absent). The second term x_p in (15) given by (14), known as *steady-state* term represents a simple harmonic motions of period, $\frac{2\pi}{\beta}$ which results from the presence of the external force $F(t)$ whose period is also $\frac{2\pi}{\beta}$. *Hence after a sufficiently long time, the output corresponding to a purely sinusoidal input will be practically a harmonic oscillation whose frequency is that of the input.*

Case (ii): Undamped forced oscillations: Resonance

In the undamped case $c = 0$ and D.E. (1) takes the form

$$m\ddot{x} + kx = F(t) = F_1 \cos \beta t$$

or $$\ddot{x} + \omega^2 x = E_1 \cos \beta t \tag{16}$$

where $\omega^2 = \frac{k}{m}$, $E_1 = \frac{F_1}{m}$ and ω, E_1, β are positive constants. The complementary function of (16) is

$$x_c = c_1 \cos \omega t + c_2 \sin \omega t \tag{17}$$

Thus every free solution (17) of D.E. (16) is periodic with frequency ω. The frequency of the driving force in (16) is of frequency β. Now we study the nature of solution of D.E. (16).

When $\omega = \beta$: In this case the particular integral of D.E. (16), which is the forced solution, is

$$x_p = \frac{1}{D^2 + \omega^2} E_1 \cos \beta t = \frac{E_1}{2\omega} t \cdot \sin \beta t \tag{18}$$

Thus when $\omega = \beta$, the general solution of D.E. (16) is $x = x_c + x_p$ or

$$x = c_1 \cos \omega t + c_2 \sin \omega t + \frac{E_1}{2\omega} \cdot t \cdot \sin \beta t \tag{19}$$

The forced solution (the particular integral (18)) grows with time and becomes larger and larger (because of the presence of 't'). Thus in an undamped system if $\omega = \beta$ i.e., the frequency β of external force matches (equals) with the natural frequency

ω. The phenomenon of unbounded oscillations occurs, which is known as *resonance*. In resonance, for a bounded input the system responds with an unbounded output. Thus resonance, the phenomenon of excitation of large oscillation, is undesirable because the system may get destroyed due to these unwanted large vibrations.

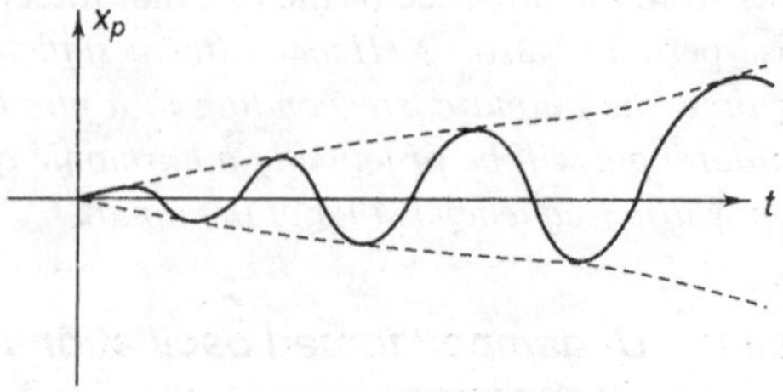

Forced solution x_p (18) in the case of resonance

Fig. 9.6

Worked Out Examples

Free, undamped oscillations:

Example 1: An 8 lb weight is placed at one end of a spring suspended from the ceiling. The weight is raised to 5 inches above the equilibrium position and left free. Assuming the spring constant 12 lb/ft, find the equation of motion, displacement function $x(t)$, amplitude, period, frequency and maximum velocity.

Solution: $\frac{\omega}{g}\ddot{x} = -kx$. Here $\omega = 8$, $g = 32$, $k = 12$. The equation of motion is $\frac{8}{32}\ddot{x} + 12x = 0$ or

$$\frac{d^2x}{dt^2} + 48x = 0$$

The auxiliary equation $r^2 + 48 = 0$ has two complex conjugate roots given by $r = \pm 4\sqrt{3}i$. The displacement function $x(t)$ is

$$x(t) = C_1 \cos \omega t + C_2 \sin \omega t$$

where $\omega = 4\sqrt{3}$. The given initial conditions are $x = -5$ inches $= \frac{-5}{12}$ ft and $\dot{x} = 0$ at $t = 0$. Using $x = \frac{-5}{12}$ at $t = 0$, we get $\frac{-5}{12} = x(0) = c_1 \cdot 1 + c_2 \cdot 0$ $\therefore$ $c_1 = \frac{-5}{12}$.

Differentiating $x(t)$ w.r.t. 't' we get

$$\dot{x} = -c_1\omega \sin \omega t + c_2\omega \cos \omega t$$

Using $\dot{x} = 0$ at $t = 0$, we get

$$0 = \dot{x} = -c_1\omega \cdot 0 + c_2\omega \cdot 1 \quad \therefore \quad c_2 = 0$$

$$x(t) = -\frac{5}{12}\cos 4\sqrt{3}t = \frac{5}{12}\sin\left(4\sqrt{3}t - \frac{\pi}{2}\right)$$

Amplitude is $\frac{5}{12}$ feet

Period $T = \frac{2\pi}{\omega} = \frac{2\pi}{4\sqrt{3}} = \frac{\pi\sqrt{3}}{6}$ sec

Frequency : $f = \frac{6}{\pi\sqrt{3}}$ cycles/sec

Maximum velocity: $\omega \cdot (\text{Amplitude}) = (4\sqrt{3})\left(\frac{5}{12}\right) = \frac{5}{\sqrt{3}}$ ft/sec^2.

Free, Damped Oscillation

Example 2: A 2 lb weight suspended from one end of a spring stretches it to 6 inches. A velocity of 5 ft/sec^2 upwards is imparted to the weight at its equilibrium position. Suppose a damping force βv acts on the weight. Here $0 < \beta < 1$ and $v = \dot{x} =$ velocity. (a) Determine the position and velocity of the weight at any time (b) Express the displacement function $x(t)$ in the amplitude-phase form. (c) Find the amplitude, period, frequency, maximum velocity (d) Determine the values of β for which the system is critically damped, over damped or oscillatory. (e) Discuss the case for $\beta = 0.6$.

Solution: Since a 2 lb weight stretches to $\frac{1}{2}$ feet, from Hookes law, we get
$2 = k \cdot \frac{1}{2}$ or $k = 4$ is spring constant. Then the D.E. of motion is

$$\frac{\omega}{g}\ddot{x} + \beta\dot{x} + kx = 0$$

or
$$\frac{2}{32}\ddot{x} + \beta\dot{x} + 4x = 0$$

$$\frac{d^2x}{dt^2} + 16\beta\frac{dx}{dt} + 64x = 0$$

The auxiliary equation is

$$r^2 + 16\beta + 64 = 0$$

with roots $r = \dfrac{-16\beta \pm \sqrt{256\beta^2 - 256}}{2}$

$= 8(-\beta \pm \sqrt{(\beta^2 - 1)})$

$r = 8(-\beta \pm \sqrt{1-\beta^2}i) = 8(-\beta \pm \alpha i)$ where $\alpha = \sqrt{1-\beta^2}$

The general solution is

$$x(t) = e^{-8\beta t}[c_1 \cos 8\alpha t + c_2 \sin 8\alpha t]$$

Using the initial condition $x = 0$ at $t = 0$,

$$0 = 1 \cdot [c_1 \cdot 1 \cdot + c_2 \cdot 0] \quad \therefore\ c_1 = 0$$

Differentiating $x(t)$ w.r.t. 't', we get

$$\dot{x} = -8\beta e^{-8\beta t}[c_1 \cos 8\alpha t + c_2 \sin 8\alpha t]$$

$$+\ e^{-8\beta t}[-c_1 8\alpha \sin 8\alpha t + c_2 8\alpha \cdot \cos 8\alpha t]$$

Initially $\dot{x} = -5$ so at $t = 0$,

$$-5 = \dot{x} = -8\beta \cdot 1[c_1 \cdot 1 + c_2 \cdot 0] + 1 \cdot [-c_1 \cdot 8\alpha \cdot 0 +$$

$$+\ c_2 \cdot 8\alpha \cdot 1]$$

$$\therefore\ c_2 = \frac{-5}{8\alpha}$$

(a) Thus the displacement function is

$$x(t) = e^{-8\beta t}\left(\frac{-5}{8\alpha}\right) \sin 8\alpha t$$

and velocity downward positive is

$$v = \dot{x} = \frac{-5}{8\alpha} e^{-8\beta t}[-8\beta \sin 8\alpha t + 8\alpha \cos 8\alpha t]$$

$$\dot{x} = \frac{5}{\alpha} e^{-\beta t}[-\alpha \cos 8\alpha t + 8\beta \sin 8\alpha t]$$

(b) Rewriting $x(t)$ as

$$x(t) = \frac{5}{8\alpha} e^{-8\beta t} \sin\ (8\alpha t + \pi)$$

which is in the amplitude-phase form

(c) Amplitude: $A(t) = \frac{5}{8\alpha} e^{-8\beta t}$

Period: $T = \frac{2\pi}{\omega} = \frac{2\pi}{8\alpha} = \frac{\pi}{4\alpha}$

Frequency: $f = \frac{1}{T} = \frac{4\alpha}{\pi}$

Maximum velocity: $\omega \cdot A = 8\alpha\left(\frac{5}{8\alpha} e^{-8\beta t}\right) = 5e^{-8\beta t}$

(d) If the discrimenent of the auxiliary equations

$$\beta^2 \gtreqless 0$$

accordingly the system is said to be over or critically or under damped. Thus the given (mass-spring) system is

(i) over damped if $\beta^2 > 1$

(ii) critically damped if $\beta^2 = 1$

(iii) under damped if $\beta^2 < 1$ (i.e., oscillatory)

(e) If $\beta = 0.6$ then

$$x(t) = \frac{-5}{6.4} e^{-4.8t} \sin 6.4t$$

$$v(t) = e^{-4.8t}[3.75 \sin 6.4t - 5 \cos 6.4t]$$

Amplitude: $\frac{5}{6.4} e^{-4.8t}$

Period: $\frac{\pi}{3.2}$

Frequency: $\frac{3.2}{\pi}$

Since $\beta^2 = 0.36 < 1$, the system is damped i.e., it is oscillatory

Forced Damped Oscillations

Example 3: A 16 lb weight is suspended from a spring having a constant 5 lb/ft. Assume that an external force given by 24 sin $10t$ and a damping force $4v$ are acting on the spring. Initially the weight is at rest at its equilibrium position.

(a) Find the position of the weight at any time.

(b) Indicate the transient and steady-state solutions

(c) Find the amplitude, period and frequency of the steady-state solution

(d) Determine the velocity of the weight at any time.

Solution: The D.E. is

$$\frac{\omega}{g}\ddot{x} = -c\dot{x} - kx + F(t)$$

or $\frac{16}{32}\ddot{x} + 4\dot{x} + 5x = 24 \sin 10t$

or $\frac{d^2x}{dt^2} + 8\frac{dx}{dt} + 10x = 48 \sin 10t$

which is a second order linear non-homogeneous D.E. with constant coefficients. Its auxiliary equation is

$r^2 + 8r + 10 = 0$ or $r = \frac{-8 \pm \sqrt{64-40}}{2}$

so two real distinct roots $r_1 = -4 - \sqrt{6}$, $r_2 = -4 + \sqrt{6}$. Then the complementary function is

$$x_c(t) = c_1 e^{r_1 t} + c_2 e^{r_2 t}$$

Now the particular integral is

$$x_p = \frac{1}{D^2+8D+10} 48 \sin 10t = \frac{48}{-100+8D+10} \sin 10t$$

$$= \frac{48(8D+90)}{(8D+90)(8D-90)} \sin 10t = \frac{48(8D+90)}{64D^2-8100} \sin 10t$$

$$= \frac{48[80 \cos 10t + 90 \sin 10t]}{64(-100)-8100}$$

$$x_p = \frac{-48}{1450}(8 \cos 10t + 9 \sin 10t)$$

The position of weight at any time is

$$x(t) = c_1 e^{r_1 t} + c_2 e^{r_2 t} - \frac{48}{1450}(8 \cos 10t + 9 \sin 10t).$$

Use the initial condition: $x = 0$ at $t = 0$

$$0 = x(0) = c_1 + c_2 - \frac{48}{1450}(8) \quad \therefore\ c_1 + c_2 = \frac{384}{1450}$$

Differentiating $x(t)$ w.r.t. 't', we get

$$\dot{x}(t) = r_1 c_1 e^{r_1 t} + r_2 c_2 e^{r_2 t} - \frac{48}{1450}[-80 \sin 10t + 90 \cos 10t]$$

Use the initial condition: $\dot{x} = 0$ at $t = 0$.

$$0 = \dot{x}(0) = r_1 c_1 + r_2 c_2 - \frac{48 \times 9}{145}$$

$$(-4-\sqrt{6})c_1 + (-4+\sqrt{6})c_2 = \frac{-432}{145}$$

Solving for c_1 and c_2, we get the required solution as

(a) $x(t) = 0.96e^{-1.56t} - 0.695e^{-6.45t}$
$- 0.298 \sin 10t \quad - 0.265 \cos 10t$

(b) Transient solution is

$$0.960\, e^{-1.56t} - 0.695\, e^{-6.45t}$$

Steady-state solution is

$$-0.298 \sin 10t - 0.265 \cos 10t$$

$$= 0.397 \sin(10t + 3.87)$$

(c) Steady-state part:
Amplitude = 0.397 feet
Period = $\frac{2\pi}{\omega} = \frac{2\pi}{10} = \frac{\pi}{5}$ sec
Frequency = $\frac{5}{\pi}$ cycles/sec.

(d) The velocity of the weight at any time is

$$\dot{x}(t) = -1.5e^{-1.56t} + 4.483\, e^{-6.45t} - 2.98 \cos 10t +$$
$$+ 2.65 \sin 10t$$

Resonance : Forces Undamped Oscillations

Example 4: A 32 lb weight is suspended from a spring having constant 4 lb/ft. Prove that the motion is one of resonance if a force 16 sin 2t is applied and damping force is negligible. Assume that initially the weight is at rest in the equilibrium position.

Solution: The D.E. describing this phenomena is

$$\frac{32}{32}\ddot{x} + 4x = 16 \sin 2t$$

The roots of the auxiliary equation $r^2 + 4 = 0$ are complex conjugate $\pm 2i$. The complementary function is

$$x_c(t) = c_1 \cos 2t + c_2 \sin 2t$$

The particular integral is

$$x_p = \frac{1}{D^2+4} 16 \sin 2t = \frac{-16 \cdot t \cdot \cos 2t}{2 \cdot 2} = -4t \cos 2t$$

. Then the general solution is

$$x(t) = c_1 \cos 2t + c_2 \sin 2t - 4t \cos 2t$$

Using $x = 0$ at $t = 0$

$$0 = x(0) = c_1 \cdot 1 + c_2 \cdot 0 - 0 \quad \therefore\ c_1 = 0$$

Differentiating $x(t)$ w.r.t. 't', we get

$$\dot{x} = 2c_2 \cos 2t - 4 \cos 2t + 8t \sin 2t$$

Using $\dot{x} = 0$ at $t = 0$

$$0 = \dot{x}(0) = 2c_2 \cdot 1 - 4.1 + 0 \quad \therefore\ c_2 = 2$$

The position of weight at any time is

$$x(t) = 2 \sin 2t - 4t \cos 2t = \sqrt{2^2 + 4t^2} \sin(2t - \phi)$$

and its velocity is

$$\dot{x} = 4 \cos 2t - 4 \cos 2t + 8t \sin 2t = 8t \sin 2t$$

Frequency of the external force is $\frac{2}{2\pi} = \frac{1}{\pi}$ cycles/sec. Natural frequency (of the free undamped) system is $\frac{2}{2\pi} = \frac{1}{\pi}$ cycles/sec. Therefore *resonance* occurs in the system because the frequency of the external force equals to the natural frequency of the system.

EXERCISE

Free Undamped Oscillations

1. An 8 lb weight is placed at the lower end of a coil spring suspended from the ceiling. The weight comes to rest in its equilibrium position, thereby stretching the spring 6 inches. The weight is then pulled down 3 inches below its equilibrium position and released at $t = 0$ with an initial velocity of 1 ft/sec directed downward. Neglecting air resistance of the medium and assuming that no external forces are present, determine the amplitude, period and frequency of the resulting motion.

Ans. Amplitude $\frac{\sqrt{5}}{8} \approx 0.280$ft, period $= \frac{2\pi}{8} = \frac{\pi}{4}$ sec.
Frequency: $\frac{4}{\pi}$ oscillations/sec.

Hint: $\frac{8}{32}\ddot{x} + 16x = 0,\ x(0) = \frac{1}{4},\quad \dot{x} = 1$ roots $\pm 8i,\quad x(t) = c_1 \cos 8t + c_2 \sin 8t$, $c_1 = \frac{1}{8}, c_2 = \frac{1}{4}, x = \frac{\sqrt{5}}{8} \cos\,(8t + \phi)$

2. Determine the displacement of a body of weight 10 kg attached to a spring given that 20 kg weight will stretch the spring to 10 cm. Find the maximum velocity and period of oscillations

Ans. $x(t) = 0.2 \cos 14t$, Max. velocity $= 14(0.2) =$ 2.8 m/sec.
period: $\frac{2\pi}{14} = 0.45$ sec

Hint Spring constant $k = 200$ kg/m; Tension $= k(0.05 + x)$, $\frac{\omega}{g}\ddot{x} = \omega - T$, $\frac{10}{9.8}\ddot{x} =$ $10 - (10 + 200x)$ or $\ddot{x} + 14^2 x = 0$.

3. A spring is such that it would be stretched to 6 inches by 12 lb weight. If the weight is pulled down 4 inches below the equilibrium point and given an upward velocity of 2 feet/sec, determine the motion of the weight assuming no damping. Find also the amplitude and period .

Ans: $x(t) = \frac{1}{3}\cos 8t - \frac{1}{4}\sin 8t$

Hint: $k = 24$ lb/ft, $\frac{12}{32}\ddot{x} + 24x = 0$, amplitude $= \frac{5}{12}$, Period: $\frac{\pi}{4}$

4. A 6 lb weight stretches a spring 6 inches. If the weight is pulled 4 inches below the equilibrium position, find the motion of weight. Find the amplitude, period, frequency and position, velocity and acceleration of the weight $\frac{1}{2}$ sec after it has been released.

Ans: $x = \frac{1}{3}\cos 8t$, Amplitude $\frac{1}{3}$ ft, period: $\frac{2\pi}{8} = \frac{\pi}{4}$ sec, frequency: $\frac{8}{2\pi} = \frac{4}{\pi}$ cycles/sec; At $t = \frac{1}{2}$ sec, $x = -0.219$, $\dot{x} = 2.01$, $\ddot{x} = 14.0$

Hint: $k = 12$, $\ddot{x} + 64x = 0$

5. An 8 lb weight is placed on a spring with spring constant $k = 12$ lb/ft. Find the motion amplitude, period and frequency if the weight is raised 5 inches and then thrust upward with velocity 5 ft/sec.

Ans: $x(t) = \frac{5}{12}\cos 4\sqrt{3}t + \frac{5\sqrt{3}}{12}\sin 4\sqrt{3}t$, Amplitude: 5/6, Period $\pi/2\sqrt{3}$, Frequency $= 2\sqrt{3}/\pi$

Hint: $\ddot{x} + 48x = 0$, $x(0) = -\frac{5}{12}$, $\dot{x}(0) = -5$.

Free Damped Oscillations

6. A 32 lb weight is suspended from a coil spring stretches the spring to 2 ft. The weight is then pulled down 6 inches from the equilibrium position and released at $t = 0$. Find the motion of the weight and determine the nature of motion if the resistance of the medium is
(a) $4\dot{x}$, (b) $8\dot{x}$, (c) $10\dot{x}$.

Ans: $x(t) = e^{-2t}(c_1 \sin 2\sqrt{3}t + c_2 \cos 2\sqrt{3}t)$, $c_1 = \sqrt{3}/6$, $c_2 = \frac{1}{2}$, $x = \frac{\sqrt{3}}{3}e^{-2t}\cdot \cos\left(2\sqrt{3}t - \frac{\pi}{6}\right)$. Motion is damped oscillatory with period $2\pi/\left(2\sqrt{3}\right)$; with damping factor $e^{-2t}/\sqrt{3}$.

Hint: $\ddot{x} + 4\dot{x} + 16x = 0$, $x(0) = \frac{1}{2}$, $\dot{x}(0) = 0$, roots $-2 \pm 2\sqrt{3}$

(b) $x(t) = (c_1 + c_2 t)e^{-4t}$, $c_1 = \frac{1}{2}$, $c_2 = 2$, Motion is critically damped, and $x \to 0$ as $t \to \infty$

Hint: $\ddot{x} + 8\dot{x} + 16x = 0$, roots: $-4, -4$

(c) $x(t) = \frac{2}{3}e^{-2t} - \frac{1}{6}e^{-8t}$, overdamped case

Hint: $\ddot{x} + 10\dot{x} + 16x = 0$, roots: $-2, -8$

7. A 3 lb weight on a spring stretches it to 6 inches. Suppose a damping force βv is present ($\beta > 0$). Show that the motion is (a) critically damped if $\beta = 1.5$ (b) overdamped if $\beta > 1.5$ (c) oscillatory if $\beta < 1.5$

Hint: $\ddot{x} + \frac{32}{3}\beta\dot{x} + 64x = 0$

8. A weight of 980 gm is attached to a spring with spring constant $k = 20$ gm/cm. The resistance is $\frac{1}{4}v$. Find the motion of weight if it pulled down $\frac{1}{4}$ cm below its equilibrium position and then released. Also find the time it takes the damping factor to drop to $\frac{1}{10}$ of its initial value.

Ans: $x(t) = e^{-0.05t}$ [0.25 cos(4.5)t+ 0.003 sin(4.5)t], time $t = 46$ sec.

Hint: $10\ddot{x} + \dot{x} + 200x = 0$, roots $-0.05 \pm 4.5i$ $x(0) = \frac{1}{4}$cm, $\dot{x}(0) = 0$, damping factor $= re^{-0.05t}$. Here r is constant of proportionality. At time t, damping factor $= \frac{1}{10}r$ or $\frac{1}{10}r = re^{-0.05t}$ or $e^{t/20} = 10$.

9. A 2 lb weight is pulled 6 inches below its equilibrium position and then released. Assuming a spring constant $k = 16$, damping force $2\dot{x}$. Determine whether the motion is overdamped or critically damped.

Ans: Critically damped

Hint: $\ddot{x} + 32\dot{x} + 256x = 0$, roots: $-16, -16$. $x(t) = e^{-16t}(c_1 + c_2t)$, $x = 6$ inches and $\dot{x} = 0$ at $t = 0$, $c_1 = \frac{1}{2}$, $c_2 = 8$.

Forced Oscillations

10. If weight $\omega = 16$ lb, spring constant $k = 10$ lb/ft, damping force $2\dot{x}$, external force $F(t)$ is $5\cos 2t$, find the motion of the weight given $x(0) = \dot{x}(0) = 0$. Write the transient and steady-state solutions. Describe the nature of these solutions.

Ans: $x(t) = e^{-2t}\left[-\frac{3}{8}\sin 4t - \frac{1}{2}\cos 4t\right] + +\frac{1}{2}\cos 2t + \frac{1}{4}\sin 2t$.

Transient solution: $\frac{5e^{-2t}}{8}\cos(4t - 0.64)$, representing a damped oscillatory motion, steady-state solution: $\frac{\sqrt{5}}{4}\cos(2t - 0.46)$, representing a simple harmonic motion with amplitude $\frac{\sqrt{5}}{4}$ and period π.

Hint: $\ddot{x} + 4\dot{x} + 20x = 10\cos 2t$, $x(0) = \dot{x}(0) = 0$. $x(t) = \frac{5e^{-2t}}{8}\cos(4t - \phi) + \frac{\sqrt{5}}{4}\cos(2t - \theta)$ with $\tan\theta = \frac{1}{2}$, $\tan\phi = \frac{3}{4}$.

11. Determine the transient and steady-state solutions of mechanical system with weight $w = 6$ lb, stiffness constant $k = 12$, damping force $1.5\dot{x}$, external force 24 cos 8t and initial conditions $x = \frac{1}{3}$ ft, $\dot{x} = 0$.

Ans: Transient: $\frac{e^{-4t}}{9}(3\cos 4\sqrt{3}t - 11\sqrt{3}\sin 4\sqrt{3}t)$. Steady-state $2\sin 8t$.

Hint: $\ddot{x} + 8\dot{x} + 64x = 128\cos 8t$, root: $-4 \pm 4\sqrt{3}i$.

Find the steady-state and transient oscillations of the mechanical system corresponding to the following D.E.

12. $\ddot{x} + 3\dot{x} + 2x = 10\sin t$

Ans: $\sin t - 3\cos t$, $c_1e^{-2t} + c_2e^{-t}$

13. $\ddot{x} + 2\dot{x} + 2x = \sin 2t - 2\cos 2t$, $x(0) = \dot{x}(0) = 0$

Ans: $-0.5\sin 2t$; $e^{-t}\sin t$

Resonance: Forced Undamped Oscillations

14. If a weight 6 lbs hangs from a spring with constant $k = 12$ and no damping force exists, find the motion of weight when an external force 3 cos $8t$ acts. Initially $x = 0$ and $\dot{x} = 0$. Determine whether resonance occurs.

Ans: $x(t) = t\sin 8t$, resonance occurs.

Hint: $\ddot{x} + 64x = 16\cos 8t$, solution is $x(t) = c_1\cos 8t + c_2\sin 8t + t\sin 8t$, use I.C., $c_1 = c_2 = 0$.

15. A 64 lb weight is attached to the lower end of a coil spring with spring constant 18 lb/ft. An external force 3 cos ωt is applied to the system. If the weight is pulled down to 6 inches below its

equilibrium position and released from rest at $t = 0$, (a) Assuming a damping force $4\dot{x}$ determine the resonance frequency of the resulting motion (b) If there is no damping, determine the value of ω that gives rise to undamped resonance

Ans: (a) resonance frequency $\frac{1}{2\pi}\sqrt{\frac{18}{2}-\frac{1}{2}\left(\frac{16}{4}\right)} \approx 0.42$ cycles/sec. Thus resonance occurs when $\omega = \sqrt{7} \approx 2.65$. (b) C = damping constant = 0, thus undamped resonance occurs when $\omega = 3$.

Hint: (a) $2\ddot{x} + 4\dot{x} + 18x = 3\cos\omega t$ (b) $\ddot{x} + 9x = \frac{3}{2}\cos\omega t$, C.F. : $x_c = c_1 \sin 3t + c_2 \cos 3t$ with natural frequency $\frac{3}{2\pi}$. Initial conditions $x(0) = \frac{1}{2}, \dot{x}(0) = 0$, then $x(t) = \frac{1}{2}\cos 3t + \frac{1}{4}t \sin 3t$.

16. Determine forced solutions if resonance occurs in a mechanical system consisting of a weight ω attached to a spring with stiffness constant $k = 24$ lb/ft and an applied external force $2\cos 2t$. What should be the value of the weight w in this case?

    Ans: Forced solution: $\frac{1}{12}t\sin 2t$; weight w = 6 g.

    Hint: $\omega\ddot{x} + 24gx = 2g\cos 2t$, since resonance occurs, period of free oscillations $\frac{2\pi}{\sqrt{\left(\frac{24g}{\omega}\right)}}$ = period of forced oscillations π so $w = 6$ g.

17. Determine whether resonance occurs in a system consisting of a weight 32 lb attached to a spring with constant $k = 4$ lb/ft and external force $16\sin 2t$ and no damping force present. Initially $x = \frac{1}{2}$ and $\dot{x} = -4$.

    Ans: Resonance occurs

    Hint: $\ddot{x} + 4x = 16\sin 2t$, $x(t) = \frac{1}{2}\cos 2t - 4t\cos 2t$. Natural frequency of system $= \frac{2}{2\pi}$ and frequency of the forcing function $= \frac{2}{2\pi}$.

9.14 RLC-CIRCUIT

An RLC-series circuit consists of a resistor, a conductor, a capacitor and an emf as shown in the figure.

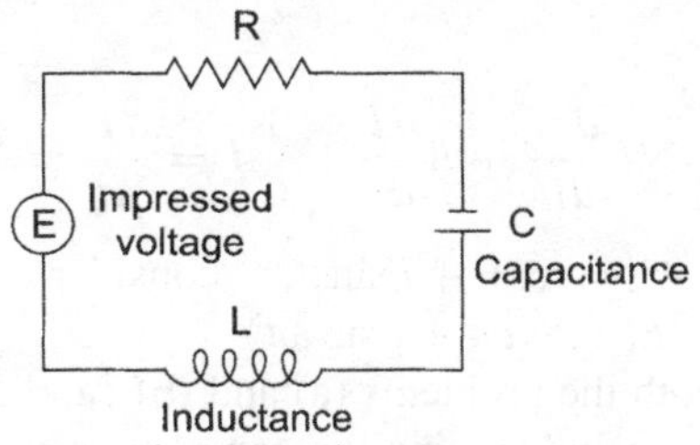

RLC-circuit

Fig. 9.7

Using the Kirchhoff's law, the sum of the voltage drops across the three elements inductor, resistor and capacitance equal to the external source E. Thus the RLC-circuit is modeled by

$$\boxed{L\frac{dI}{dt} + RI + \frac{1}{C}Q = E(t)} \tag{1}$$

which contains two dependent variable Q and I. Since $I = \frac{dQ}{dt}$, the above equation can be written as

$$\boxed{L\frac{d^2Q}{dt^2} + R\frac{dQ}{dt} + \frac{1}{C}Q = E(t)} \tag{2}$$

which contains only one dependent variable Q. Differentiating (1) w.r.t. 't', we obtain

$$\boxed{L\frac{d^2I}{dt^2} + R\frac{dI}{dt} + \frac{1}{C}I = \frac{dE}{dt}} \tag{3}$$

which contains only one dependent variable I. Thus the charge Q and current I at any time in the RLC circuit are obtained as solutions of (2) and (3) which are both linear 2nd order non homogeneous ordinary differential equations. The equation (3) is used more often, since current $I(t)$ is more important than charge $Q(t)$, in most of the practical problems. The RLC-circuit reduces to an RL-circuit in absence of capacitor and to RC-circuit when no inductor is present.

Example 1: Determine the current $I(t)$ in an RLC-circuit with (a) emf $E(t) = E_0\cos\omega t$ (b) emf $E(t) = E_0 \sin\omega t$.

Solution: The differential equation representing the RLC-circuit with I as the dependent variable is given by

$$L\frac{d^2I}{dt^2}+R\frac{dI}{dt}+\frac{1}{C}I=\frac{dE}{dt} \quad (3)$$

Since $e^{i\theta}=\cos\theta+i\sin\theta$, consider $E(t)=E_0e^{i\omega t}=E_0(\cos\omega t+i\sin\omega t)$
so that both the problems (a) and (b) can be solved simultaneously by using the complex form for emf $E(t)$. Thus

$$L\frac{d^2I}{dt^2}+R\frac{dI}{dt}+\frac{1}{c}I=E_0\cdot\omega i e^{i\omega t} \quad (4)$$

The general solution of (4) = Complimentary function (C.F.) + particular integral (P.I.). Complementary Function : I_C
The auxiliary equation is

$$Lm^2+Rm+\frac{1}{C}=0$$

$$m^2+\frac{R}{L}m+\frac{1}{LC}=0$$

With two distinct roots

$$m_{1,2}=\frac{-\frac{R}{L}\pm\sqrt{\left(\frac{R}{L}\right)^2-4\cdot\frac{1}{CL}}}{2}$$

$$m_{1,2}=-\frac{R}{2L}\pm\frac{1}{2L}\sqrt{R^2-\frac{4L}{c}} \quad (5)$$

$$m_1=-a+b$$

and $m_2=-a-b$ where $b=\frac{1}{2L}\sqrt{R^2-\frac{4L}{C}}, a=\frac{R}{2L}$
The complimentary function I_C is given by

$$I_C=c_1e^{m_1t}+c_2e^{m_2t} \quad (6)$$

Particular Integral: I_p

Assume the particular solution of (3) as

$$I_p=A\,e^{i\omega t} \quad (7)$$

where A is an undetermined constant obtained by substituting (7) in (3). Then

$$L(i\omega)^2Ae^{i\omega t}+Ri\omega Ae^{i\omega t}+\frac{1}{C}Ae^{i\omega t}=i\,E_0\omega e^{i\omega t}$$

or $\left(-\omega^2L+i\omega R+\frac{1}{C}\right)Ae^{i\omega t}=E_0\omega i e^{i\omega t}$. Then

$$A=\frac{E_0\,\omega\,i}{\left(-\omega^2L+\frac{1}{C}\right)+i\,\omega R}$$

$$=\frac{E_0\,i}{-\left(\omega L-\frac{1}{\omega C}\right)+i\,R}$$

The particular solution i.e., Re($I p$). Thus when $E=E_0\cos\omega t$, the steady-state solution is

$$I_p=\frac{E_0}{\sqrt{R^2+S^2}}\cdot\cos(\omega t-\delta) \quad (10)$$

Similarly when $E=E_0\sin\omega t$, the steady-state solution is the imaginary part of I_p, namely,

$$I_p=\frac{E_0}{\sqrt{R^2+S^2}}\sin(\omega t-\delta) \quad (11)$$

Complex number $z=R+is$ is known as the *complex impedance*. Its magnitude $\sqrt{R^2+S^2}$ is known as simply *impedance* R. The real part of z is *resistance* while S, the imaginary part of z is called the *reactance* and $\frac{1}{z}$ is called the *admittance*.
Put $S=\omega L-\frac{1}{\omega c}$ Then

$$A=\frac{E_0\,i}{-S+i\,R}=\frac{E_0}{R+iS}$$

Then $I_p=Ae^{i\omega t}=\frac{E_0}{R+iS}e^{i\omega t}$ (8)
Here S is known as *reactance* of the circuit. When $S=0$, the amplitude of I is greatest and the circuit is in resonance.
Using the polar form of a complex number

$$a+ib=re^{i\theta}=\sqrt{a^2+b^2}e^{i\theta}$$

where $r=\sqrt{a^2+b^2}$ and $\theta=\tan^{-1}\frac{b}{a}$, we can write I_p as

$$I_p=\frac{E_0}{\sqrt{R^2+S^2}\cdot e^{i\delta}}e^{i\omega t}=\frac{E_0}{\sqrt{R^2+S^2}}e^{i}(\omega t-\delta)$$

$$I_p=\frac{E_0\cdot[\cos(\omega t-\delta)+i\sin(\omega t-\delta)]}{\sqrt{R^2+S^2}} \quad (9)$$

where $\tan\delta=\frac{S}{R}$. The quantity $\sqrt{R^2+S^2}$ is known as impedance of the circuit.
For an impressed voltage $E(t)=E_0\cos\omega t$ = Real part of $(E_0\,e^{i\omega t})$ = Re $(E_0e^{i\omega t})$, the steady-state current in the RLC-circuit is given by the real part. Since $R>0$, the transient solution

$$I_C = c_1 e^{m_1 t} + c_2 e^{m_2 t} \to 0$$

as $t \to \infty$. Thus after long time, the out put will be a harmonic oscillation given by either (10) or (11) as the case may be.

RLC-circuit

WORKED OUT EXAMPLES

Example 1: A circuit consists of an inductance of 0.05 henrys, a resistance of 5 ohms and a condensor of capacitance 4×10^{-4} farad. If $Q = I = 0$ when $t = 0$, find $Q(t)$ and $I(t)$ when (a) there is a constant emf of 110 volts (b) there is an alternating emf 200 cos $100t$. (c) Find the steady-state solution in (b).

Solution:

(a) Here $L = 0.05$,

$$R = 5, C = 4 \times 10^{-4}, E = 110$$

The differential equation of the RLC-circuit

$$L\frac{d^2Q}{dt^2} + R\frac{dQ}{dt} + \frac{Q}{C} = E(t)$$

now takes the form

$$0.05\frac{d^2Q}{dt^2} + 5\frac{dQ}{dt} + \frac{Q}{4 \times 10^{-4}} = 110$$

or $\frac{d^2Q}{dt^2} + 100\frac{dQ}{dt} + 50000\,Q = 2200$

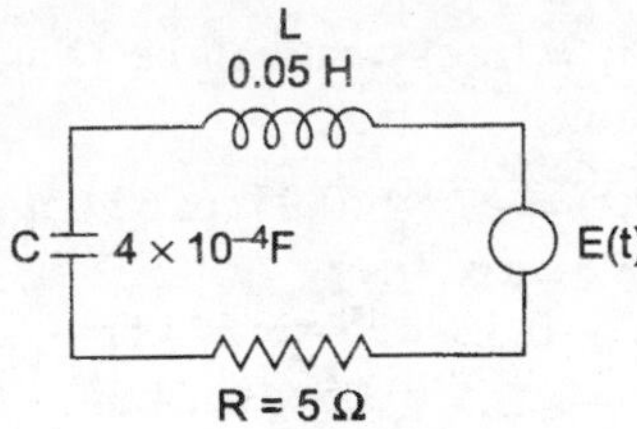

Fig. 9.8

C.F.: A.E.: $m^2 + 100m + 50{,}000 = 0$ with roots

$$m = -50 \pm 50\sqrt{19}\,i$$

Complementary function

$$Q_F = e^{-50t}(A\cos 50\sqrt{19}t + B\sin 50\sqrt{19}t)$$

Particular integral:

$$Q_P = \frac{1}{D^2 + 100D + 50{,}000}2200 = \frac{2200}{50{,}000} = \frac{11}{250}$$

The general solution is $Q = Q_C + Q_F$,

$$Q = e^{-50t}(A\cos\omega t + B\sin\omega t) + \frac{11}{250}$$

where $\omega = 50\sqrt{19}$. Since $Q = 0$ at $t = 0$, we have

$$0 = A + \frac{11}{250} \quad \text{or} \quad A = -\frac{11}{250}$$

Now differentiating Q w.r.t. 't', get

$$I(t) = \frac{dQ}{dt} = e^{-50t}[-\omega A\sin\omega t + B\omega\cos\omega t]$$

$$-\,50\,e^{-50t}(A\cos\omega t + B\sin\omega t)$$

Since $I = 0$ at $t = 0$, we have

$$0 = B\omega - 50\,A \quad \text{or} \quad B = \frac{50A}{\omega}$$

i.e., $B = \frac{50}{50\sqrt{19}} \cdot \left(-\frac{11}{250}\right) = -\frac{11\sqrt{19}}{4750}$. Thus

$$Q(t) = e^{-50t}\left(-\frac{11}{250}\cos 50\sqrt{19}t - \frac{11\sqrt{19}}{4750}\sin 50\sqrt{19}t\right) + \frac{11}{250}$$

and

$$I(t) = e^{-50t}[(B\omega - 50A)\cos\omega t + (-\omega A - 50B)\sin\omega t]$$

$$I(t) = \frac{44}{\sqrt{19}}e^{-50t}\sin 50\sqrt{19}t$$

(b) When $E(t) = 200\cos 100t$; the equation is

$$\frac{d^2Q}{dt^2} + 100\frac{dQ}{dt} + 50{,}000\,Q = 40{,}000\cos 100t$$

for which the complementary is same as in the above case (a).

The particular integral Q_p is

$$Q_P = \frac{1}{D^2 + 100D + 50{,}000}40{,}000\cos 100t$$

$$= \frac{40{,}000}{-10{,}000 + 100D + 50{,}000} \cdot \cos 100t$$

$$= \frac{40(D+400)}{(D-400)(D+400)} \cos 100t$$

$$= \frac{40(D+400)}{-10{,}000 - 1{,}60{,}000} \cos 100t$$

$$= \frac{-4}{17000}[-100 \sin 100t - 400 \cos 100t]$$

$$Q_p = \frac{4 \sin 100t + 16 \cos 100t}{170}$$

Thus

$$Q(t) = e^{-50t}\left(-\frac{16}{170}\cos \omega t - \frac{12\sqrt{19}}{1615}\sin \omega t\right)$$

$$+ \tfrac{4}{170}(4 \cos 100t + \sin 100t)$$

Differentiating Q w.r.t. 't', we get

$$I(t) = e^{-50t}\left(-\frac{40}{17}\cos \omega t + \frac{1640}{323}\sqrt{19} \cdot \sin \omega t\right)$$

$$+ \frac{40}{17}(\cos 100t - 4 \sin 100t)$$

where $\omega = 50\sqrt{19}$

(c) The steady-state solutions are obtained in the limiting case as $t \to \infty$. Taking $t \to \infty$, we get the steady-state solutions as

$$Q(t) = \frac{4}{170}(4 \cos 100t + \sin 100t)$$

and

$$I(t) = \frac{40}{17}(\cos 100t - 4 \sin 100t)$$

RLC-circuits

EXERCISE

1. Determine the charge on the capacitor at any time $t > 0$ in a circuit in series having an emf given by $E(t) = 100 \sin 60t$ V, a resistor of 2 Ω, an inductor of 0.1 H and a capacitor of $\frac{1}{260}$ farads if the initial current and initial charge on the capacitor are both zero. Find the steady-state solution.

Ans. $Q(t) = \frac{6e^{-10t}}{61}(6 \sin 50t + 5 \cos 50t) - \frac{5}{\sqrt{61}}(5 \sin 60t + 6 \cos 60t)$

or $Q(t) = \frac{6\sqrt{61}}{61}e^{-10t}\cos(50t - \phi)$ $\frac{-5\sqrt{61}}{61}\cos(60t - \theta)$ where $\cos\phi = 5/\sqrt{61}$, $\sin\phi = 6/\sqrt{61}, \cos\theta = 6/\sqrt{61}, \sin\theta = 5/\sqrt{61}$.

Steady-state solution: $Q(t) = -\frac{5}{61}$ $(5 \sin 60t + 6 \cos 60t)$

Hint: $\frac{d^2Q}{dt^2} + 20\frac{dQ}{dt} + 2600Q = 1000 \sin 60t$, roots of A.E.: $r^2 + 20r + 2600$ are $-10 \pm 50i$; use $Q(0) = I(0) = 0$

2. Assuming $Q = I = 0$ at $t = 0$, in an RLC-circuit having a source of voltage $E(t) = 155 \sin 377t$, $R = 100\ \Omega$, $L = 0.1$ henry, $C = 10^{-3}$ farad, determine the current at any instant of time.

Ans. $I(t) = -0.042\,e^{-10t} + 0.526\,e^{-990t} - 0.484 \cos 377t + 1.380 \sin 377t$.

Hint: $0.1\frac{d^2I}{dt^2} + 100\frac{dI}{dt} + 1000I = (155)(377 \cos 377t)$ A.E.: $0.1r^2 + 100r + 1000 = 0$ has roots $r_1 = -10$, $r_2 = -990$

3. An electric circuit consists of an inductance of 0.1 henery, a resistance of 20 ohms and a condenser of capacitance 25 micro farads (one micro = 10^{-6}). Find the charge Q and the current I at any time t, with the following initial conditions (a) $Q = 0.05$ coulomb, $I = \frac{dQ}{dI} = 0$ when $t = 0$ (b) $Q = 0.05, I = -0.2$ ampere when $t = 0$ (c) what will be Q and I after a long time?

Ans. (a) $Q(t) = e^{-100t}(0.05 \cos 624.5t + 0.008 \sin 624.5t)$

$I(t) = -0.32\,e^{-100t} \sin 624.5t$

(b) $Q = e^{-100t}(0.05 \cos 624.5t + 0.0077 \sin 624.5t)$

$I = e^{-100t}(-0.2 \cos 624.5t - 32.0 \sin 624.5t)$

(c) Q and $I \to 0$ as $t \to \infty$ since both solution are transient containing e^{-t}.

Hint: $\ddot{Q} + 200\,\dot{Q} + 400{,}000\,Q = 0$.

4. Determine Q and I in an RLC-circuit with $L = 0.05$ H, $R = 20\ \Omega$, $C = 100$ micro F, emf $E = 100$ V. With $Q = 0$, $I = 0$ at $t = 0$.

Ans. $Q = e^{-200t}(-0.01 \cos 400t - 0.005 \sin 400t) + 0.01$, $I = 5e^{-200t} \sin 400t$

Hint: $\ddot{Q} + 400\,\dot{Q} + 200{,}000\,Q = 2000$

5. (a) Solve the above problem 4 when emf $E(t) = 100 \cos 200t$ (b) Find the steady-state solutions.

Ans. $Q = e^{-200t}[-0.01 \cos 400t - 0.0075 \sin 400t] + 0.01 \cos 200t + 0.005 \sin 200t$

$$I = e^{-200t}[-\cos 400t + 5.5 \sin 400t] - 2 \sin 200t + \cos 200t$$

(b) Steady-state solutions are obtained by taking limit as $t \to \infty$. They are

$$Q = 0.01 \cos 200t + 0.005 \sin 200t$$

$$I = \cos 200t - 2 \sin 200t$$

Hint: $\ddot{Q} + 400\,\dot{Q} + 200{,}000\,Q = 2000 \cos 200t$

6. (a) Determine Q and I in the RLC-circuit with $L = 0.5$ H, $R = 6\ \Omega$, $C = 0.02$ F, $E(t) = 24 \sin 10t$ and initial conditions $Q = I = 0$ at $t = 0$ (b) state the steady-state and transient solutions.

Ans. (a) $Q = \frac{e^{-6t}}{10}[4 \cos 8t + 3 \sin 8t] - \frac{2}{5} \cdot \cos 10t$

$I = \frac{e^{-6t}}{10}[-32 \sin 8t + 24 \cos 8t] - \frac{6}{10}e^{-6t}[4 \cos 8t + 3 \sin 8t] + 4 \sin 10t$

(b) Transient solutions of Q and I are

$$Q_{t_r} = \frac{e^{-6t}}{10}[4 \cos 8t + 3 \sin 8t]$$

$$I_{t_r} = \frac{e^{-6t}}{10}[-50 \sin 8t]$$

Steady-state solutions of Q_s and I_s are $Q_s = -\frac{2}{5} \cos 10t$, $I_s = 4 \sin 10t$

7. (a) Find Q and I in the RLC-circuit with $L = 2$, $R = 4$, $C = 0.05$, $E = 100$ and $Q(0) = \dot{Q}(0) = 0$. (b) Find the steady-state solutions.

Ans. (a) $Q(t) = 5 - \frac{5}{3}e^{-t}(3 \cos 3t + \sin 3t)$

$$I(t) = \frac{5}{3}e^{-t}(3 \cos 3t + \sin 3t) - \frac{5}{3}e^{-t}(3 \cos 3t - 9 \sin 3t)$$

(b) Steady-state solutions: as $t \to \infty$, $Q = 5$, $I = 0$.

8. Find $I(t)$ in the RLC-circuit with $E(t) = 100 \sin 200t$ V, $R = 40\ \Omega$, $L = 0.25$ H, $C = 4 \times 10^{-4}$ F and $I(0) = 0$ and $Q(0) = 0.01$.

Ans. $I(t) = e^{-80t}(-4.588 \sin 60t + 1.247 \cos 60t) - 1.247 \cos 200t + 1.331 \sin 200t$

9.15 SIMPLE PENDULUM

A simple pendulum consists of a mass m (the bob) at the end of a straight wire of variable length $L(t)$ where t is the time. It is assumed that the mass of the wire is negligible compared to the mass of the bob.

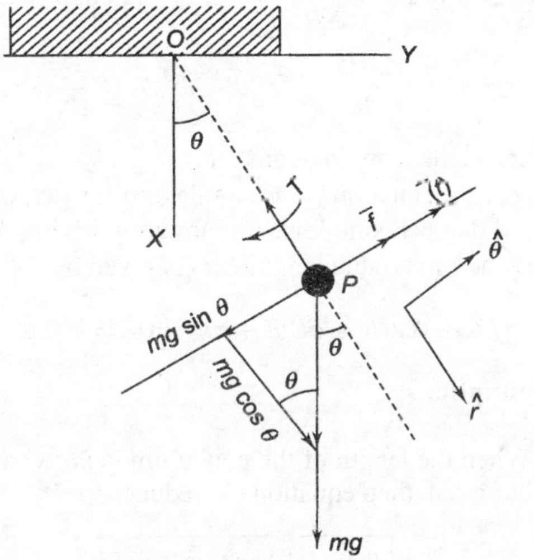

Fig. 9.9

The pendulum is suspended from a fixed point of support O. The bob is free to move in a vertical plane due to force of gravity and given external force $F(t)$. Let θ be the angle which the wire OP makes with the vertical OX at time t, positive when measured counterclockwise. The equation of angular motion $\theta(t)$ will be obtained by modeling using Newton's second law.

Assume that the moving pendulum experiences a viscous damping force $\bar{f}$ proportional to its linear velocity and opposing the motion. The other forces acting on the bob are the tension of the wire $\bar{T}$, gravity and a given external force $F(t)$. The tangential and normal components of the force of gravity mg are $-mg\sin\theta$ and $mg\cos\theta$. Let $\hat{r}$ and $\hat{\theta}$ be unit vectors, where $\hat{\theta}$ points in the direction of increasing θ. Then

$$\bar{v} = \frac{d}{dt}(L\hat{r}) = L\dot{\theta}\hat{\theta} + \frac{dL}{dt}\hat{r}$$

$$\bar{a} = \frac{d}{dt}\left[L\dot{\theta}\,\hat{\theta} + \frac{dL}{dt}\hat{r}\right]$$

$$\bar{a} = L\ddot{\theta}\,\hat{\theta} + \frac{dL}{dt}\dot{\theta}\hat{\theta} + \frac{d^2L}{dt^2}\hat{r} + \frac{dL}{dt}\frac{d\hat{r}}{dt} - L\dot{\theta}^2\hat{r}$$

Using Newton's second law we get two equations (along $\bar{r}$ and along $\hat{\theta}$)

$$-mL\dot{\theta}^2 + \frac{d^2L}{dt^2} = -T + mg\cos\theta$$

$$mL\ddot{\theta} = -\frac{2dL}{dt}\dot{\theta} - cL\dot{\theta} - mg\sin\theta + F(t)$$

Here c is the damping constant.

Thus the equation of the angular motion $\theta(t)$ of the forced damped simple pendulum with driving force $F(t)$ and of variable length $L(t)$ is given by

$$mL\ddot{\theta} + (2m\dot{L} + cL)\dot{\theta} + mg\sin\theta = F(t) \qquad (1)$$

Assumptions:

1. When the length of the pendulum is *not* variable but fixed, then equation (1) reduces to

$$\boxed{mL\ddot{\theta} + cL\dot{\theta} + mg\sin\theta = F(t)} \qquad (2)$$

2. Free undamped nonlinear equation of pendulum (with $c = 0$, $F(t) = 0$) is

$$mL\ddot{\theta} = -mg\sin\theta$$

$$L\frac{d^2\theta}{dt^2} = -g\sin\theta$$

Multiplying both sides by $2\frac{d\theta}{dt}$ on both sides

$$2\frac{d\theta}{dt}\cdot L\cdot\frac{d^2\theta}{dt^2} = -2g\cdot\sin\theta\frac{d\theta}{dt}$$

$$L\frac{d}{dt}\left(\frac{d\theta}{dt}\right)^2 = 2g\frac{d}{dt}(\cos\theta)$$

Integrating

$$L\left(\frac{d\theta}{dt}\right)^2 = 2g\cos\theta + c_1 \qquad (3)$$

$$\frac{d\theta}{\sqrt{2g\cos\theta + c_1}} = \pm\frac{dt}{\sqrt{L}} \qquad (4)$$

This integral (4) cannot be expressed in terms of elementary functions.

When the maximum displacement is α, then the angular velocity $\dot{\theta} = 0$. Put $\theta = \alpha$ and $\dot{\theta} = 0$ in (3).

Then $LO = 2g\cos\alpha + c_1 \therefore c_1 = -2g\cos\alpha$

Substituting c_1 in (3), we get

$$L\dot{\theta}^2 = 2g\cos\theta - 2g\cos\alpha$$

or $$\dot{\theta}^2 = \frac{2g}{L}(\cos\theta - \cos\alpha)$$

Since $\cos A = 1 - 2\sin^2\frac{A}{2}$, we can rewrite

$$\dot{\theta}^2 = \frac{4g}{L}\left(\sin^2\frac{\alpha}{2} - \sin^2\frac{\theta}{2}\right)$$

Change the variable of integration from θ to ϕ by the substitution

$$\sin\frac{\theta}{2} = \sin\frac{\alpha}{2}\cdot\sin\phi \qquad (6)$$

and integrating w.r.t. 't', we get

$t = \sqrt{\frac{L}{g}}\int_0^{\phi}\frac{d\phi}{\sqrt{1-k^2\sin^2\phi}}$. where $k^2 = \sin^2\frac{\alpha}{2}$.

This integral (6) is known as an elliptic integral of the first kind.

3. Equation (2) is non-linear second order non-homogeneous differential equation (with constant coefficients) with non-linearity arising through the term '$\sin\theta$'. Since

$$\sin\theta = \theta - \frac{\theta^3}{3!} + \frac{\theta^5}{5!}\cdots$$

for sufficiently small θ, the term $\sin\theta$ can be replaced by θ. Thus we get an approximate *linearized pendulum equation* with damping as

$$\boxed{mL\ddot{\theta} + cL\dot{\theta} + mg\theta = F(t)} \quad (7)$$

4. Free damped motion of the pendulum is obtained from (7) with $F(t) = 0$ as

$$mL\ddot{\theta} + cL\dot{\theta} + mg\theta = 0 \quad (8)$$

or
$$\frac{d^2\theta}{dt^2} + \frac{c}{m}\frac{d\theta}{dt} + \frac{g}{L}\theta = 0$$

This is second order linear homogeneous equation with auxiliary equation

$$\lambda^2 + \frac{c}{m}\lambda + \frac{g}{L} = 0$$

Its roots are $\lambda = -\frac{c}{m} \pm \sqrt{\frac{c^2}{m^2} - \frac{4g^2}{L^2}}$

Case (i) When $\frac{c}{m} > \frac{2g}{L}$ then the roots λ_1 and λ_2 are real and district. In this case the general solution is

$$\theta(t) = c_1 e^{\lambda_1 t} + c_2 e^{\lambda_2 t}$$

Case (ii) when $\frac{c}{m} = \frac{2g}{L}$, the roots are real and equal then the solution is

$$\theta(t) = (c_1 + tc_2)\cdot e^{\lambda t}$$

Case (iii) When $\frac{c}{m} < \frac{2g}{L}$ the roots are complex conjugate then the complete solution is

$$\theta(t) = e^{-kt}\cdot\left[c_1 \cos\sqrt{\omega^2 - k^2}\, t + c_2 \sin\sqrt{\omega^2 - k^2} t\right] \quad (9)$$

Here $k = \frac{c}{m}, \omega = \frac{2g}{L}$. Thus the pendulum motion is oscillatory with period $\frac{2\pi}{\sqrt{\omega^2 - k^2}}$ when $k < \omega$

5. Free undamped linear equation is obtained from (7) with $F(t) = 0$ and $c = 0$

$$mL\ddot{\theta} + mg\theta = 0$$

or
$$\frac{d^2\theta}{dt^2} + \frac{g}{L}\theta = 0 \quad (10)$$

The roots of the auxiliary equation are

$$\lambda^2 + \frac{g}{L} = 0 \text{ or } \lambda = \pm\sqrt{\frac{g}{L}}i$$

The general solution of (10) is

$$\theta(t) = c_1 \cos\sqrt{\frac{g}{L}}t + c_2 \sin\sqrt{\frac{g}{L}}t \quad (11)$$

Thus the motion of the bob is simple harmonic and the time of an oscillation is $2\pi\sqrt{\frac{L}{g}}$. Observe that the period $2\pi\sqrt{\frac{L}{g}}$ is independent of the initial displacement. A beat or a swing of a pendulum is the movement of the bob from one end to the other constituting half an oscillation. Thus the time for one beat is $\pi\sqrt{\frac{L}{g}}$. Suppose a pendulum of length L makes n beats in time T, then

T = time of n beats = $n\cdot$ time for one beat

$$T = n\cdot\pi\sqrt{\frac{L}{g}}$$

or

$$n = \frac{T}{\pi}\sqrt{\frac{g}{L}}$$

Now taking log and differentiating both sides

$$\boxed{\frac{dn}{n} = \frac{1}{2}\left(\frac{dg}{g} - \frac{dL}{L}\right)} \quad (12)$$

Thus the gain or loss of oscillations of a pendulum due to change in g or L can be determined from (12).

Case (i) If L = constant and g is a variable, then

$$\frac{dn}{n} = \frac{1}{2}\frac{dg}{g}$$

Case (ii) : If g = constant and L is a variable, then

$$\frac{dn}{n} = -\frac{dL}{2L}$$

WORKED OUT EXAMPLES

Example 1: If a clock, loses 5 seconds/day, determine the alteration required in the length of the pendulum in order that the clock keeps correct time.

Solution: Since 24 hours = 24 × 60 × 60 = 86400 seconds, a seconds pendulum beats 86400 times a day. As the clock loses 5 seconds/day

$$\frac{n+dn}{n}=\frac{86400-5}{86400}$$

or $\frac{dn}{n}=\frac{86395}{86400}-1$

Assuming g as constant we have

$$\frac{86395}{86400}-1=\frac{dn}{n}=-\frac{dL}{2L}$$

or $\frac{dL}{L}=+2\left(1-\frac{86395}{86400}\right)=\frac{10}{86400}=\frac{1}{8640}$

i.e., $dL=\frac{1}{8640}L$. So the length of the pendulum be shortened by $\frac{1}{8640}$ of its original length.

Example 2: Find the angular motion $\theta(t)$ of a forced undamped pendulum whose equation is given by $\ddot{\theta}+\omega_0^2 t=F_0 \sin kt$ where ω_0 and F_0 are constants; if $x=\dot{x}=0$ at $t=0$.

Solution: The roots of the auxiliary equation are $\lambda^2+\omega_0^2=0$ or $\lambda=\pm\omega_0 i$. The complementary function $\theta_c(t)=c_1\cos\omega_0 t+c_2\sin\omega_0 t$. The particular integral

$\theta_p=\frac{F_0}{D^2+\omega_0^2}\sin kt=F_0\cdot\left(\frac{-t}{2\omega_0}\right)\cdot\cos\omega_0 t$ when $k=\omega_0$.

Then the general solution is

$$\theta(t)=\theta_c(t)+\theta_p(t)$$

$$\theta(t)=c_1\cos\omega_0 t+c_2\sin\omega_0 t-\frac{F_0 t}{2\omega_0}\cos\omega_0 t$$

At $t=0, 0=\theta(0)=c_1+0-0 \therefore c_1=0$

Differentiating θ w.r.t. t

$$\dot{\theta}=-\omega_0 c_1\sin\omega_0 t+c_2\omega_0\cos\omega_0 t-\frac{F_0}{2\omega_0}\cos\omega_0 t$$

$$+\frac{F_0 t}{2}\sin\omega_0 t$$

At $t=0, 0=\dot{\theta}(0)=0+c_2\omega_0-\frac{F_0}{2\omega_0}+0$

$\therefore c_2=\frac{F_0}{2\omega_0 2}$

The required solution is $\theta(t)=\frac{F_0}{2\omega_0^2}\sin\omega_0 t-\frac{F_0 t}{2\omega_0}\cos\omega_0 t$.

$$\theta(t)=\frac{F_0}{2\omega_0^2}(\sin\omega_0 t-t\omega_0\cdot\cos\omega_0 t)$$

EXERCISE

1. Determine the change in gravity g in order to correct a clock with a seconds pendulum which is losing 10 seconds/day at a place where $g=32$ ft/sec^2.

Ans. g must be increased by 0.0074 ft/sec^2

Hint: $\frac{dn}{n}=\left(\frac{86400-10}{86400}\right)=\frac{dg}{2g}$ or $dg=\left(1-\frac{1}{4320}\right)32=31.9925$

2. Compare the acceleration due to gravity at two places if a seconds pendulum gains 10 seconds/day at one place and looses 10 seconds/day at another place.

Ans. 4321/4319

3. If the length of a pendulum is increased in the ratio 900:901, determine how many seconds a clock would lose/day.

Ans. Clock loses 48 seconds/day

Hint: $\frac{L+dL}{dL}=\frac{901}{900}$, $\frac{dn}{n}=\frac{-dl}{2l}=\frac{-1}{1800}$, $dn=\frac{-86400}{1800}=-48$

4. If a simple undamped linearized undriven pendulum is 8 feet long and swings with an amplitude of 1 rad, compute (a) The angular velocity of the pendulum at its lowest point (b) its acceleration at the ends of its paths.

Ans. ± 2 rad/sec (b) ± 4 rad sec^2

Chapter 10

Series Solutions

INTRODUCTION

In the earlier Chapter 9, methods of obtaining solutions to second and higher order linear non-homogeneous differential equations with constant coefficients were studied.

Euler-Cauchy and Legendre equations, two special cases of differential equations with variable coefficients were also considered earlier, in 9.5.

In general, the solutions to differential equations with variable coefficients such as Bessel's equation, Legendre's equation and hypergeometric equation can *not* be expressed as finite linear combination of known elementary functions. However in such cases solutions can be obtained in the form of infinite power series. In this chapter, two methods *power series* method and an extension of the power series method the *Frobenius* method* (*generalized power series* method) are considered for solving differential equations with variable coefficients.

We consider the important concept of orthoganality and the process of Gram-Schmidt orthogonalization. Finally we also study the Sturm-Liouville problems.

10.1 CLASSIFICATION OF SINGULARITIES

Consider a homogeneous linear second order differential equation with variable coefficients:

$$a_0(x)\frac{d^2y}{dx^2} + a_1(x)\frac{dy}{dx} + a_2(x)y = 0 \tag{1}$$

Assuming $a_0(x) \neq 0$, the above equation is written in the standard (normalized) form as

$$\frac{d^2y}{dx^2} + P(x)\frac{dy}{dx} + Q(x)y = 0 \tag{2}$$

where $P(x) = \frac{a_1(x)}{a_0(x)}$ and $Q(x) = \frac{a_2(x)}{a_0(x)}$

Analytic

A function $f(x)$ is said to be analytic at x_0 if $f(x)$ has Taylor's series expansion about x_0 given by

$$\sum_{n=0}^{\infty} \frac{f^{(n)}(x_0)}{n!}(x - x_0)^n$$

exists and converges to $f(x)$ for all x in some open interval including x_0.

If a function $f(x)$ is not analytic at x_0 then it is said to be **singular** at x_0.

Regular or Ordinary Point (O.P.)

A point x_0 is said to be a regular or ordinary point of the differential equation (2), equivalently (1), if *both* $P(x)$ and $Q(x)$ are analytic at x_0.

Singular Point

A point x_0 is said to be a singular point of (2) if either of $P(x)$ or $Q(x)$ or both are *not* analytic at x_0.

Singular points are classified as regular singular point and irregular point as follows:

*Ferdinand Georgy Frobenius (1849-1917) German mathematician.

Regular Singular Point (RSP)

A singular point x_0 of differential equation (2) is called a *regular singular point* if *both* $(x - x_0)P(x)$ and $(x - x_0)^2\, Q(x)$ are analytic at x_0. On the other hand, if either $(x - x_0)P(x)$ or $(x - x_0)^2 Q(x)$ or both are *not* analytic at x_0, then x_0 is called an *irregular singular point* (ISP).

WORKED OUT EXAMPLES

Example 1: $y'' + (x^2 + 1)y' + (x^3 + 2x^2 + 3x)y = 0$. Since $P(x) = x^2 + 1$ and $Q(x) = x^3 + 2x^2 + 3x$ are polynomials, are *both* analytic everywhere, any point x is an ordinary or regular point of the given differential equation.

Example 2: $(1 - x^2)y'' + 2y' - 3y = 0$. Standard form: $y'' + \frac{2}{1-x^2}y' - \frac{3}{1-x^2}y = 0$. Since both $P(x) = \frac{2}{1-x^2}$. and $Q = -\frac{3}{1-x^2}$ are not analytic at $x = \pm 1$, the given DE has two singular points at $x = +1$ and $x = -1$. Further since $(x - 1)P(x) = (x - 1)\cdot\frac{2}{(1-x^2)} = \frac{2}{1+x}$ is analytic at $x = 1$ and $(x - 1)^2 Q(x) = (x - 1)^2 \cdot \frac{(-3)}{(1-x^2)} = +\frac{3(x-1)}{(x+1)}$ is analytic at $x = 1$.

Therefore $x = 1$ is a regular singular point. Similarly since both $(x + 1)P(x)$ and $(x + 1)^2 Q(x)$ are analytic at $x = -1$, the point $x = -1$ is also a regular singular point

Example 3: $x^3(x - 1)y'' + 2(x - 1)y' + 5xy = 0$ Here $P(x) = \frac{2(x-1)}{x^3(x-1)} = \frac{2}{x^3}$ and $Q(x) = \frac{5}{x^2(x-1)}$. $x = 0$ and $x = 1$ are singular points. Further $x = 0$ is an irregular singular point since $x \cdot P(x) = \frac{2\cdot(x-1)}{x^2(x-1)}$ is *not* analytic (although $x^2 Q(x) = \frac{5}{x-1}$ is analytic at $x = 0$). (Note *both* must be analytic). However $x = 1$ is a regular singular point since *both* $(x - 1)P(x) = \frac{2(x-1)}{x^3}$ and $(x - 1)^2 Q(x) = \frac{5(x-1)}{x^2}$ are analytic at $x = 1$.

EXERCISE

Locate and classify the ordinary points (OP), regular singular point (RSP) and irregular singular point (ISP) of the following differential equations

1. $x^2(x^2 - 4)y'' + 2x^3y' + 3y = 0$

Ans. $x = 0, \pm 2$ are regular singular points (R.S.P.) No I.S.P. All points, except $x = 0, \pm 2$ are O.P.'s.

2. $(x^2 + 1)(x - 4)^3y'' + (x - 4)^2y' + y' = 0$

Ans. $x = \pm i$ are RSP, $x = 4$ is ISP. All other points are O.P.'s

3. $e^x y'' + 2y' - xy = 0$

Ans. No singular points. Any point is a OP

4. $x^2y'' + y' + y = 0$

Ans. $x = 0$ is I.S.P.

5. $x(x - 1)^3y'' + 2(x - 1)^3 + 3y = 0$

Ans. Except $x = 0$, $x = 1$, all other points are OP $x = 0$ is RSP, $x = 1$ is ISP

6. $y'' + (x^3 + x^2 + 1)y' - 3(x^2 - 4x - 2)y = 0$

Ans. No singular points. Any point is OP

7. $(x - 1)xy'' + x^2y' + y = 0$

Ans. $x = 0, 1$ are S.P. all other points are O.P.s $x = 0$ is RSP, $x = 1$ is also RSP.

8. $x^2(x - 2)y'' + 2(x - 2)y' + (x + 1)y = 0$

Ans. $x = 0, 2$ are S.P., $x = 0$ is ISP, $x = 2$ is RSP, all other points O.P.

9. $x(x - 1)^2(x + 2)y'' + x^2y' - (x^3 + 2x - 1)y = 0$

Ans. $x = 0$, RSP, $x = 1$ is ISP, $x = -2$ is RSP. All other points O.P.

10. $x^4(x^2 + 1)(x - 1)^2y'' + 4x^3(x - 1)y' + (x + 1)y = 0$

Ans. $x = \pm i, 1$ are RSP; $x = 0$ is ISP.

11. $(x^4 - 2x^3 + x^2)y'' + 2(x - 1)y' + x^2y = 0$

Ans. $x = 1$ is RSP, $x = 0$ is ISP

10.2 POWER SERIES SOLUTION

An expression of the form $c_0 + c_1(x - x_0) + c_2(x - x_0)^2 + \cdots + c_n(x - x_0)^n + \cdots$
or in the summation form

$$\sum_{n=0}^{\infty} c_n(x - x_0)^n \qquad (3)$$

is known as a *power series* of the variable x in powers of $(x - x_0)$ (or about the point x_0). The constants c_0, $c_1, c_2, \cdots c_n \cdots$ are known as the *coefficients* and x_0 is known as the *centre* (of expansion) of the power series (1). Since n takes only positive integral values, the power series (1) does *not* contain negative or fractional powers. So power series (1) contains only positive powers.

Theorem: *If x_0 is a regular (or ordinary) point of differential equation (2), then a general solution of (2) is obtained as a linear combination of two linearly independent power series solutions of the form*

$$\sum_{n=0}^{\infty} c_n(x - x_0)^n \tag{3}$$

and these power series both converge in some interval $|x - x_0| < R$ (with $R > 0$).

Power Series Method:

Step I: Assume that $y = \sum_{n=0}^{\infty} c_n(x - x_0)^n$ (4)

be the solution (2).

Step II: Substitute, y, y', y'' obtained by differentiating (4) termwise, in (2). Collect the coefficients of like powers of $(x - x_0)$. This converts the differential equation (2) into the form

$$k_0 + k_1(x - x_0) + k_2(x - x_0)^2 + \cdots = 0 \tag{5}$$

Here $k_i (i = 0, 1, 2, 3, \ldots)$ are functions of certain coefficients c_n.

Step III: If (4) is solution of (2), *all* $k_i's$ must be zero. Solve $k_0 = 0$, $k_1 = 0$, $k_2 = 0 \ldots$ for the unknown coefficients $c_n's$.

Generally this leads to a *recurrence relation* between $c_n's$, which helps to determine unknown coefficients in terms of the other known coefficients.

Thus $c_n's$ are determined by equating to zero each power of $(x - x_0)$ in (5).

Step IV: Substitution of these $c_n's$ in (4) gives the required power series solution of (2).

WORKED OUT EXAMPLES

Example 1: Find a power series solution in powers of x of the differential equation

$$x(x+1)y' - (2x+1)y = 0$$

Solution: Assume that $y = \sum_{n=0}^{\infty} c_n x^n$ is a power series solution in power of x (i.e., about $x_0 = 0$). Differentiating y w.r.t. x we get.

$$y' = \sum_{n=0}^{\infty} nc_n x^{n-1}$$

Substituting y and y' in the given equation, we get

$$x(x+1)\sum_{n=0}^{\infty} nc_n x^{n-1} - (2x+1)\sum_{n=0}^{\infty} c_n x^n = 0.$$

$$\sum_{n=0}^{\infty} nc_n x^{n+1} + \sum_{n=0}^{\infty} nc_n x^n - 2\cdot\sum_{n=0}^{\infty} c_n x^{n+1} - \sum_{n=0}^{\infty} c_n x^n = 0.$$

or $$\sum_{n=0}^{\infty}(n-2)c_n x^{n+1} + \sum_{n=0}^{\infty}(n-1)c_n x^n = 0$$

Rewrite the first summation in the LHS so that x in each of the summations in LHS will have the common exponent n. Put $n + 1 = m$, then $\sum_{m=1}^{\infty}(m-3)c_{m-1}x^m$, since $n = m - 1$. But m is a dummy variable. So the first summation may be written as $\sum_{n=1}^{\infty}(n-3)c_{n-1}x^n$. Thus the given D.E. takes the form

$$\sum_{n=1}^{\infty}(n-3)c_{n-1}x^n + \sum_{n=0}^{\infty}(n-1)c_n x^n = 0$$

Although x has the same exponent in the two summations, the range of the summations are different. The common range is from 1 to ∞. Expanding the second summation for $n = 0$ (that *do* not belong to the common range 1 to ∞) we can rewrite the above equation as

$$\sum_{n=1}^{\infty}\left[(n-3)c_{n-1} + (n-1)c_n\right]x^n - c_0 \cdot x^0 = 0$$

Equating to zero, the coefficients of x^n: $n = 0 : x^0$: coefficient of x^0 is c_0 equated to zero. Thus $c_0 = 0$.
$n \geq 1 : x^n : (n-3)c_{n-1} + (n-1)c_n = 0$

or $(n-1)c_n = -(n-3)c_{n-1}$ for $n \geq 1$
For $n = 1, 0 \cdot c_1 = 2c_0 = 2 \cdot 0 = 0$.
Thus c_1 is arbitrary (can take any value).
For $n = 2$, $c_2 = c_1$
For $n = 3$, $2c_3 = 0 \quad \therefore c_3 = 0$
For $n > 3$, $c_n = 0$ because of the presence of the factor $(n-3)$.
For example $n = 4$, $3c_4 = -4c_3 = 0$ since $c_3 = 0$.
For $n = 5$, $4c_5 = -2c_4 = 0$ since $c_4 = 0$ and so on.
Thus $c_0 = 0$, $c_1 = c_2$ and $c_n = 0$ for $n \geq 3$. Then the power series solution reduces to the form $y = c_1x + c_2x^2 = c_1(x + x^2)$
since $c_1 = c_2$.

Example 2: Using power series method solve $(1 - x^2)y'' - 2xy' + 2y = 0$ (which is a particular case of Legendre's equation with $n = 1$).

Solution: Assume $y = \sum_{n=0}^{\infty} c_nx^n$. Differentiating, $y' = \sum_{n=0}^{\infty} nc_nx^{n-1}$, $y'' = \sum_{n=0}^{\infty} n(n-1)c_nx^{n-2}$. Substituting in the given equation

$$(1 - x^2)\sum_{n=0}^{\infty} n(n-1)c_nx^{n-2} - 2x\sum_{n=0}^{\infty} nc_nx^{n-1} + $$
$$+ 2\sum c_nx^n = 0$$

or
$$\sum_{n=2}^{\infty} n(n-1)c_nx^{n-2} - \sum_{n=0}^{\infty} [n(n-1) + 2n - 2]c_nx^n = 0$$

since for $n = 0$ and $n = 1$ the first two terms in the first summation vanish.
Put $m = n - 2$ so that the first summation can be written as

$$\sum_{n=0}^{\infty} (n+2)(n+1)c_{n+2}x^n - \sum_{n=0}^{\infty} (n^2 + n - 2)c_nx^n = 0$$

Equating to zero the coefficient of x^n,

$$(n+2)(n+1)c_{n+2} = (n+2)(n-1)c_n$$

or $\quad (n+1)c_{n+2} = (n-1)c_n$.
For $n = 0$, $c_2 = -c_0$

For $n = 1$, $2c_3 = 0 \cdot c_1$ (because of the presence of the factor $(n-1)$. So c_1 is arbitrary and *all* the coefficients with odd suffix will be zero i.e. $c_3 = c_5 = c_7 = c_9 = \cdots = 0$ (because $4c_5 = 2c_3 = 0$ etc.).
For even suffix coefficients

$$c_{2m+2} = \frac{(2m-1)}{(2m+1)}c_{2m} \quad \text{for } m = 0, 1, 2, \ldots$$

For $m = 0$, $c_2 = -c_0$, for $m = 1$, $c_4 = \frac{1}{3}c_2 = -\frac{1}{3}c_0$, for $m = 2$, $c_6 = \frac{3}{5}c_4 = \frac{3}{5} \cdot \left(-\frac{1}{3}\right)c_0 = -\frac{1}{5}c_0$ etc.
Thus the required power series solution involving two arbitrary constants is

$$y = c_0 + c_1x + c_2x^2 + 0 + c_4x^4 + 0 + c_6x^6 + 0 + c_8x^8 + \cdots$$

$$y = c_1x + c_0\left[1 - x^2 - \frac{1}{3}x^4 - \frac{1}{5}x^6 - \frac{1}{7}x^8 \cdots\right]$$

Note: Except $x \pm 1$, all other points are regular points of the given DE.

Example 3: Solve the initial value problem $xy'' + y' + 2y = 0$ with $y(1) = 2$, $y'(1) = 4$.

Solution: Since the initial conditions are prescribed at $x = 1$, we obtain a power series solution in powers of $(x-1)$ of the form

$$y = \sum_{n=0}^{\infty} c_n(x-1)^n$$

The procedure is simplified by introducing $t = x - 1$. In this case the initial value problem transforms to solution of the differential equation

$$(t+1)\frac{d^2y}{dt^2} + \frac{dy}{dt} + 2y = 0 \qquad (1)$$

with $y(t=0) = 2$ and $y'(t=0) = 4$. So assume the power series solution of the form

$$y(t) = \sum_{n=0}^{\infty} c_nt^n \qquad (2)$$

Substituting (2) in (1), we get

$$(t+1)\sum_{n=2}^{\infty} n(n-1)c_nt^{n-2} + \sum_{n=1}^{\infty} nc_nt^{n-1}$$

$$+2\cdot\sum_{n=0}^{\infty} c_n t^n = 0.$$

$$\sum_{n=2}^{\infty} n(n-1)c_n t^{n-1} + \sum_{2}^{\infty} n(n-1)c_n t^{n-2}$$

$$+\sum_{1}^{\infty} nc_n t^{n-1} + 2\sum_{0}^{\infty} c_n t^n = 0$$

To have a common exponent n for t put $n-1=m$ in first summation, put $n-2=m$ in second and put $n-1=m$ in third summation. Then the above equation takes the form

$$\sum_{n=1}^{\infty}(n+1)nc_{n+1}t^n + \sum_{n=0}^{\infty}(n+2)(n+1)c_{n+2}t^n$$

$$+\sum_{n=0}^{\infty}(n+1)c_{n+1}t^n + 2\sum_{n=0}^{\infty}c_n t^n = 0$$

Thus the recurrence relation is obtained by equating to zero the coefficient of t^n as

$$(n+1)^2 c_{n+1} + (n+1)(n+2)c_{n+2} + 2c_n = 0$$

For $n \geq 0$. Solving

$$c_{n+2} = \frac{-[2c_n + (n+1)^2 c_{n+1}]}{(n+1)(n+2)},\ \text{for } n \geq 0$$

For $n=0$, $c_2 = \dfrac{-[2c_0 + c_1]}{2}$

For $n=1$, $c_3 = \dfrac{-[2c_1 + 4c_2]}{6} = \dfrac{2}{3}c_0$

For $n=2$, $c_4 = \dfrac{-[2c_2 + 9c_3]}{12}$

For $n=3$, $c_5 = \dfrac{-[2c_3 + 16c_4]}{20}$ etc.

Substituting these coefficients c_n's in (2), we get

$$y(t) = c_0 + c_1 t + c_2 t^2 + c_3 t^3 + c_4 t^4 + c_5 t^5 + \cdots \tag{3}$$

Using the initial condition $y(t=0)=2$, we have

$$2 = y(0) = c_0. \quad \therefore \quad c_0 = 2.$$

Differentiating (3) w.r.t. 't', we get

$$y'(t) = c_1 + 2c_2 t + 3c_3 t^2 + \cdots$$

Using the second initial condition $y'(t=0)=4$, we get

$$4 = y'(0) = c_1 \quad \therefore c_1 = 4$$

with $c_0 = 2$, $c_1 = 4$, we get $c_2 = -4$, $c_3 = \frac{4}{3}$, $c_4 = -\frac{1}{3}$, $c_5 = \frac{-2}{15}$ etc. Substituting these values

$$y(t) = 2 + 4t - 4t^2 + \frac{4}{3}t^3 - \frac{1}{3}t^4 + \frac{2}{15}t^5 + \cdots$$

Replacing t by $x-1$, we get the required power series solution in powers of $(x-1)$ as

$$y(x) = 2 + 4(x-1) - 4(x-1)^2 + \frac{4}{3}(x-1)^3 - \frac{1}{3}(x-1)^4 + \frac{2}{15}(x-1)^5 + \cdots$$

EXERCISE

Obtain power series solution in powers of x for the following differential equations (1 to 9)

1. $y' = ky$

Ans. $y = c_0 e^{kx}$

Hint: Recurrence relation (RR): $c_n = \frac{k^n}{n!}c_0$

2. $(1-x^2)y' - y = 0$

Ans. $y = c_0\left[1 + x + \frac{x^2}{2} + \frac{x^3}{2} + \frac{3}{8}x^4 + \frac{11}{40}x^5 + \cdots\right]$

Hint: RR: $c_{n+1} = \dfrac{(n-1)c_{n-1} + c_n}{n+1}$ for $n \geq 1$, $c_1 = c_0$

3. $xy' - (x+2)y - 2x^2 - 2x = 0$

Ans. $y = 2x + c_2 x^2 e^x$

Hint: RR: $c_n = \dfrac{c_{n-1}}{n-2}$ for $n > 2$, $c_0 = 0$, $c_1 = 2$, c_2 arbitrary constant.

4. $y' - 2xy = 0$

Ans. $y = c_0 e^{x^2}$

Hint: $c_1 = c_3 = c_5 = \cdots = 0$, $c_2 = c_0$, $c_4 = \dfrac{c_2}{2}$, $c_6 = \dfrac{c_0}{3!}$ etc.

5. $y'' - 3y' + 2y = 0$

Ans. $y = c_0 + c_1 x + \left(\frac{3}{2}c_1 - c_0\right) x^2 + \left(\frac{7}{6}c_1 - c_0\right) x^3 + \cdots$ put $c_0 = A + B$, $c_1 = A + 2B$, then $y = Ae^x + Be^{2x}$

Hint: RR: $c_{m+2} = \frac{3(m+1)c_{m+1} - 2c_m}{(m+1)(m+2)}$, $m \geq 0$

6. $y'' - xy' + y = 0$

Ans. $y = c_1 x + c_0 \left[1 - \frac{1}{2!}x^2 - \frac{1}{4!}x^4 - \frac{3}{6!}x^6 - \frac{3 \cdot 5}{8!}x^8 + \cdots\right]$

Hint: c_1 arbitrary, $c_3 = c_5 = c_7 = \cdots = 0$.
RR: $c_{2m+2} = \frac{(2m-1)}{2(m+1)(2m+1)} c_{2m}$, for $m = 0, 1, 2, \cdots$

7. $y'' + y = 0$

Ans. $y = c_0 \cos x + c_1 \sin x$

8. $y'' + xy' + (x^2 + 2)y = 0$

Ans. $y = c_0 \left(1 - x^2 + \frac{1}{4}x^4 + \cdots\right) + c_1 \left(x - \frac{1}{2}x^3 + \frac{3}{40}x^5 + \cdots\right)$

Hint: RR: $c_{n+2} = -\frac{(n+2)c_n + c_{n-2}}{(n+1)(n+2)}$ for $n \geq 2$,
$c_2 = -c_0$, $c_3 = -\frac{1}{2}c_1$, $c_0 c_1$ arbitrary

9. $(x^2 - 1)y'' + 3xy' + xy = 0$, $y(0) = 4$, $y'(0) = 6$

Ans. $y = 4 + 6x + \frac{11}{3}x^3 + \frac{1}{2}x^4 + \frac{11}{4}x^5 + \cdots$

Hint: $y = c_0\left(1 + \frac{1}{6}x^3 + \frac{1}{8}x^5 + \cdots\right) + c_1\left(x + \frac{1}{2}x^3 + \frac{1}{12}x^4 + \frac{3}{8}x^5 + \cdots\right)$
RR: $c_{n+2} = \frac{n(n+2)c_n + c_{n-1}}{(n+1)(n+2)}$ for $n \geq 2$,
$c_2 = 0$, c_1, c_0 arbitrary.

10. Solve $y'' = y$ in power series in power of $(x - 1)$.

Ans. $y = c_0 \cosh(x-1) + c_1 \sinh(x-1)$.

Hint: $c_{n+2} = \frac{c_n}{(n+1)(n+2)}$, $c_0 c_1$ arbitrary.
Put $x - 1 = t$, $y = c_0\left(1 + \frac{1}{2!}t^2 + \frac{1}{4!}t^4 + \cdots\right) + c_1\left(t + \frac{t^3}{3!} + \frac{t^5}{5!} + \cdots\right)$

11. Obtain power series solution about $x = 1$ for equation $y'' + (x-1)^2 y' - 4(x-1)y = 0$.

Ans. $y = c_0 \sum_{n=0}^{\infty} \frac{4(-1)^n (x-1)^{3n}}{3^n (3n-1)(3n-4)n!} + c_1(x-1) + \frac{1}{4}(x-1)^4 c_1$.

Hint: c_0, c_1 arbitrary,
$c_{3k} = (-1)^k \frac{[(-4)(-1)(2)\cdots(3k-1)}{(3 \cdot 6 \cdot 9 \cdots 3k)(2 \cdot 5 \cdot 8 \cdots (3k-1))} c_0$

12. Determine power series solution with centre at $x = -3$ for $y'' - 2(x+3)y' - 3y = 0$

Ans. $y = c_0\left[1 + \sum_{n=1}^{\infty} \frac{3 \cdot 7 \cdot 11 \cdots (4n-1)}{(2n)!}(x+3)^{2n}\right] + c_1\left[(x+3) + \sum_{1}^{\infty} \frac{5 \cdot 9 \cdot 13 \cdots (4n+1)}{(2n+1)!}(x+3)^{2n+1}\right]$

13. Solve $y'' + xy = 0$ in power series about $x = 0$.

Ans. $y = c_0\left(1 - \frac{x^3}{3!} + \frac{4x^6}{6!} - \frac{2 \cdot 8}{7!}x^9 \cdots\right) + c_1\left(x - \frac{2x^4}{4!} + \frac{10x^7}{7!} + \cdots\right)$

Hint: RR: $c_{n+2} = \frac{-c_{n-1}}{(n+2)(n+1)}$, $c_2 = 0$

14. Solve the initial value problem $(x-1)y'' + xy' + y = 0$, $y(0) = 2$, $y'(0) = -1$

Ans. $y = 2 - x + x^2 + \frac{x^4}{4} + \frac{3x^5}{20}$

Hint: $y = c_0\left(1 + \frac{x^2}{2} + \frac{x^3}{6} + \frac{5x^4}{24} + \frac{19x^5}{120} + \cdots\right) + c_1\left(x + \frac{x^3}{3} + \frac{x^4}{6} + \frac{x^5}{6} + \cdots\right)$
$c_2 = \frac{c_0}{2}$, RR: $c_n = \frac{(n-2)c_{n-1} + c_{n-2}}{n}$ for $n \geq 3$, c_0 arbitrary.

15. Solve the initial value problem $(x-1)y'' + y' + 2(x-1)y = 0$, $y(4) = 5$, $y'(4) = 0$.

Ans. $y(x) = 5\left[1 - (x-4)^2 + \frac{1}{9}(x-4)^3 + \frac{5}{36}(x-4)^4 + \cdots\right]$

Hint: $y(x) = c_0\left[1 - (x-4)^2 + \frac{1}{9}(x-4)^3 + \frac{5}{36}(x-4)^3 + \cdots\right] + c_1\left[(x-4) - \frac{1}{6}(x-4)^2 - \frac{8}{27}(x-4)^3 + \frac{5}{108}(x-4)^4 + \cdots\right.$

RR: $c_{n+2} = -\frac{n+1}{3(n+2)}c_{n+1} - \frac{2}{(n+1)(n+2)}c_n - \frac{2c_{n-1}}{3(n+2)(n+1)}$ for $n = 0, 1, 2 \ldots$. Here c_0, c_1 are arbitrary.

10.3 FROBENIUS METHOD

In 10.2 a power series solution $\sum_{n=0}^{\infty} c_n(x - x_0)^n$ was obtained for differential equation *only* when x_0 is an ordinary (or regular) point. Solution near regular singular point x_0 can be obtained by an extension of the power series method known as *Frobenius method* (or *generalized power series method*). In this method we consider a series of the form

$$|x - x_0|^r \sum_{n=0}^{\infty} c_n(x - x_0)^n \tag{1}$$

known as *Frobenius* series. Here r is an unknown (real or complex) constant to be determined.

Theorem: *Let x_0 be a regular singular point of the differential equation*

$$\frac{d^2y}{dx^2} + P(x)\frac{dy}{dx} + Q(x)y = 0 \tag{2}$$

Then (2) has at least one nontrivial solution of the form

$$y = |x - x_0|^r \sum_{n=0}^{\infty} c_n(x - x_0)^n \tag{3}$$

which is convergent (valid) in some deleted interval about x_0, $0 < |x - x_0| < R$ (with $R > 0$).

Note: In the following analysis, for simplicity, we consider the interval as $0 < (x - x_0) < R$. Solutions valid in the interval $-R < (x - x_0) < 0$, can be obtained simply by replacing $(x - x_0)$ by $-(x - x_0)$.

Frobenius Method

Step I. Assume a solution of the form

$$y = (x - x_0)^r \sum_{n=0}^{\infty} c_n(x - x_0)^n \tag{4}$$

for the differential equation (2) with x_0 as a regular singular point. The series (4) is valid in $0 < x - x_0 < R$. Here the exponent r is chosen so that $c_0 \neq O$, which simply means that the highest possible power of x is factored out.

Step II. Differentiate (4) obtaining

$$\frac{dy}{dx} = \sum_{n=0}^{\infty} (n + r)c_n(x - x_0)^{n+r-1} \tag{5}$$

and

$$\frac{d^2y}{dx^2} = \sum_{n=0}^{\infty} (n + r)(n + r - 1)c_n(x - x_0)^{n+r-2} \tag{6}$$

Substitute (4), (5), (6) is the differential equation (2). Collecting the coefficients of like powers of $(x - x_0)$, equation (2) takes the form

$$K_0(x - x_0)^{r+k} + K_1(x - x_0)^{r+k+1} + K_2(x - x_0)^{r+k+2} + \cdots = 0$$

Here the coefficients K_i, for $i = 0, 1, 2, \ldots$, are functions of the exponent r and the constant coefficients c_n. Also k is an integer.

Step III. Since (4) is assumed to be the solution of (2), we must have

$$K_0 = K_1 = K_2 = \cdots = 0$$

Here K_0 is the coefficient of the *lowest* power $r + k$ of $(x - x_0)$. The equation $K_0 = 0$ is a quadratic equation in r, known as the *indicial* equation of the differential equation (2). The roots r_1 and r_2 of the indicial equation are known as the *exponents* of (2) and are the only possible values for the constant r in the assumed solution (4). Generally r_1 is taken as the *larger* root so that $r_1 > r_2$ (or $Re(r_1) > Re(R_2)$ in the case of complex conjugate roots).

Step IV. Now that r is known, solving the equations $K_1 = 0$, $K_2 = 0$, $K_3 = 0 \ldots$, we determine completely the unknown constant coefficients c_n's.

Step V. Let $y_1(x)$ and $y_2(x)$ be the two nontrivial linearly independent solutions of the differential equation (2). Then the general solution of (2) is $y(x) = Ay_1(x) + By_2(x)$, where A and B are arbitrary constants. Then with the known exponent r and the known coefficients c_n's, $y_1(x)$, one of the two solution of (2) is of the form (4). The form of the second (other) solution $y_2(x)$ may be similar to (4) (with different r and different coefficients) or may contain a logarithmic term. The form of $y_2(x)$ will be indicated by the indicial equation. There are three cases:

Case 1. Distinct roots not differing by an integer 1, 2, 3, . . .

Suppose $r_1 - r_2 \neq N$, where N is a non-negative integer (i.e. $r_1 - r_2 \neq 0, 1, 2, 3, \ldots$). Then

$$y_1(x) = |x - x_0|^{r_1} \sum_{n=0}^{\infty} c_n (x - x_0)^n \tag{7}$$

$$\text{and } y_2(x) = |x - x_0|^{r_2} \sum_{n=0}^{\infty} b_n (x - x_0)^n \tag{8}$$

Here $c_0 \neq 0$, $b_0 \neq 0$.

Case 2. Roots differing by an integer 1, 2, 3, . . .

Suppose $r_1 - r_2 = N$ where N is a positive integer. Then

$$y_1(x) = |x - x_0|^{r_1} \sum_{n=0}^{\infty} c_n (x - x_0)^n \tag{7}$$

and

$$y_2(x) = |x - x_0|^{r_2} \sum_{n=0}^{\infty} b_n (x - x_0)^n + A^* y_1(x) \ln |x - x_0| \tag{9}$$

Here $c_0 \neq 0$, $b_0 \neq 0$ and the constant A^* may or may not be zero.

Case 3. Double root: $r_1 = r_2$

Suppose $r_1 - r_2 = 0$. Then

$$y_1(x) = |x - x_0|^{r_1} \sum_{n=0}^{\infty} c_n (x - x_0)^n \tag{7}$$

with $c_0 \neq 0$ and

$$y_2(x) = |x - x_0|^{r_1+1} \sum_{n=0}^{\infty} b_n (x - x_0)^n + y_1(x) \ln |x - x_0| \tag{10}$$

Note 1: Series solutions in (7), (8), (9), (10) are convergent in some deleted interval $0 < |x - x_0| < R$.

Note 2: If $r_1 - r_2 = N$, where $N = 1, 2, 3, \cdots$ sometimes it is possible to obtain the general solution using the *smaller root alone*, without bothering to find explicitly the solution corresponding to the larger root.

Note 3: In case 2, $r_1 - r_2 = N$ where $N = 1, 2, 3, \ldots$, the second solution $y_2(x)$ may or may not contain the logarithmic term $A^* y_1(x) \ln |x - x_0|$. In some cases A^* is zero, so y_2 is same as (8).

Note 4: In case 3, $r_1 - r_2 = 0$ (double root), $y_2(x)$ *always* contain the logarithmic term $y_1(x) \ln |x - x_0|$ and is *never* of the simple form (8).

Note 5: In case 2: roots differing by an integer in the logarithmic case, take the first solution $y_1(x, r)$, then the second solution is $\frac{\partial y_1(x,r)}{\partial r}$ i.e., derivative of $y_1(x, r)$ partially w.r.t. the exponent r. Then taking r as the *smaller* root r_2, we get the two linearly independent solutions $y_1(x, r_2)$ and $\frac{\partial y_1}{\partial r}(x, r_2)$. Here the second solution y_2 contains the logarithmic term.

Note 6: In both cases 2 and 3, the second solution containing logarithmic term can be obtained by reduction of order.

WORKED OUT EXAMPLES

Case 1. Roots not differening by an integer

Example 1: Find the general solution of

$$8x^2 y'' + 10xy' - (1 + x)y = 0$$

Solution: The given equation in the standard form

$$y'' + \frac{10}{8x} y' - \frac{(1 + x)}{8x^2} y = 0$$

So $x = 0$ is a regular singular point since $x P(x) = x \frac{10}{8x} = \frac{10}{8}$ and $x^2 Q(x) = x^2 \frac{(1+x)}{8x^2} = \frac{(1+x)}{8}$ are both analytic at the point $x = 0$. To obtain the solutions about a regular singular point $x_0 = 0$ we use the Frobenius method. Assume the solution of the given DE as

$$y = (x - 0)^r \sum_{n=0}^{\infty} c_n x^n = \sum_{n=0}^{\infty} c_n x^{n+r}$$

Differentiating y w.r.t. x twice, we get

$$y' = \sum_{n=0}^{\infty} (n+r)c_n x^{n+r-1}$$

and

$$y'' = \sum_{n=0}^{\infty} (n+r)(n+r-1)c_n x^{n+r-2}$$

Substituting y, y', y'' in the given DE:

$$8x^2 \sum_0^{\infty} (n+r)(n+r-1)c_n x^{n+r-2}$$

$$+ 10x \sum_0^{\infty} (n+r)c_n x^{n+r-1}$$

$$-(1+x)\sum_0^{\infty} c_n x^{n+r} = 0$$

or

$$\sum_{n=0}^{\infty} [8(n+r)(n+r-1) + 10(n+r) - 1]c_n x^{n+r}$$

$$-\sum_{n=0}^{\infty} c_n x^{n+r+1} = 0$$

Equating to zero the coefficient of lowest power $n+r$ of x i.e. $0+r$ of x or x^r, we get $[8(0+r)(0+r-1) + 10(0+r) - 1]c_0 = 0$
Since it is assumed that $c_0 \neq 0$

$$8r^2 + 2r - 1 = 0$$

which is the required indicial equation with roots $r_{12} = \frac{-2\pm 6}{16}$ or $r_1 = \frac{1}{4}$ and $r_2 = -\frac{1}{2}$. Here r_1, r_2 are real, distinct and $r_1 - r_2 = \frac{1}{4} - \left(-\frac{1}{2}\right) = \frac{3}{4} \neq N$ where N is an integer $0, 1, 2, 3, \ldots$. The present problem is case 1. To obtain the recurrence relation the last summation in above equation is rewritten. Then

$$\sum \{8(n+r)(n+r-1) + 10(n+r) - 1]c_n$$

$$-c_{n-1}\}x^{n+r} = 0$$

Thus the recurrence relation is

$$[8(n+r)^2 + 2(n+r) - 1]c_n = c_{n-1}$$

To obtain $y_1(x)$, put $r = \frac{1}{4}$. Then the recurrence relation reduces to

$$\left[8\left(n+\frac{1}{4}\right)^2 + 2\left(n+\frac{1}{4}\right) - 1\right]c_n = c_{n-1}$$

or

$$c_n = \frac{1}{2n(4n+3)}c_{n-1}$$

For $n = 1$, $c_1 = \frac{1}{2\cdot 1\cdot(7)} \cdot c_0$
$n = 2$, $c_2 = \frac{1}{2\cdot 2\cdot(11)}c_1$
$n = 3$, $c_3 = \frac{1}{2\cdot 3\cdot(15)}c_2$
$n = n-1$, $c_{n-1} = \frac{1}{2(n-1)(4n-1)}c_{n-2}$
$n = n$, $c_n = \frac{1}{2n(4n+3)}c_{n-1}$

Then multiplying these, we can express c_n in terms of c_0:
$c_1 \cdot c_2 \cdot c_2 \ldots c_{n-1} \cdot c_n = \frac{c_0}{2\cdot 1\cdot 7} \cdot \frac{c_1}{2\cdot 2\cdot 11} \cdot \frac{c_2}{2\cdot 3\cdot 15} \ldots \times$
$\times \ldots \frac{c_{n-2}}{2(n-1)(4n-1)} \cdot \frac{c_{n-1}}{2n(4n+3)}$

Thus $c_n = \frac{c_0}{2^n \cdot n! 7\cdot 11\cdot 15 \ldots (4n-1)(4n+3)}$
Hence the first solution

$$y_1(x) = \sum_{n=0}^{\infty} c_n x^{n+\frac{1}{4}} = c_0 x^{\frac{1}{4}} + \sum_{n=1}^{\infty} c_n x^{n+\frac{1}{4}}$$

$$y_1(x) = c_0 x^{1/4} + c_0 \sum_{n=1}^{\infty} \frac{x^{n+\frac{1}{4}}}{2^n \cdot n! 7\cdot 11 \ldots (4n-1)(4n+3)}$$

Generally, c_0 is assumed to be equal to 1. In a similar way, to obtain the second solution $y_2(x)$, put $r = -\frac{1}{2}$ in the recurrence relation which then reduces to

$$c_n = \frac{c_{n-1}}{2n(4n-3)}$$

Then $c_1 \cdot c_2 \cdot c_3 \ldots c_{n-1} \cdot c_n = \frac{c_0}{2\cdot 1} \cdot \frac{c_1}{4\cdot 5} \times$
$\times \frac{c_2}{6\cdot 9} \cdots \frac{c_{n-2}}{(2n-2)(4n-7)} \frac{c_{n-1}}{2n(4n-3)}$

$$c_n = \left[\frac{1}{2}\cdot\frac{1}{4}\cdot\frac{1}{6}\cdots\frac{1}{2(n-1)}\cdot\frac{1}{2n}\right]\left[\frac{1}{1}\frac{1}{5}\frac{1}{9}\cdots\frac{1}{4n-7}\frac{1}{4n-3}\right]c_0$$

$$c_n = \frac{1}{2^n \cdot n!} \frac{c_0}{1\cdot 5\cdot 9 \ldots (4n-3)}$$

Choose $c_0 = 1$. Then

$$y_2(x) = \sum_{n=0}^{\infty} c_n x^{n-\frac{1}{2}} = c_0 x^{-\frac{1}{2}} + \sum_{n=1}^{\infty} c_n x^{n-\frac{1}{2}}$$

$$y_2(x) = x^{-\frac{1}{2}} + \sum_{n=1}^{\infty} \frac{x^{n-\frac{1}{2}}}{2^n \cdot n! 1\cdot 5\cdot 9 \ldots (4n-3)}$$

The required general solution is

$$y(x) = Ay_1(x) + By_2(x)$$

Case 2: Roots differing by an integer: Non-logarithmic case

Example 2: Using method of Frobenius, obtain series solution in power of x for $x(1+x)y'' + (x+5)y' - 4y = 0$

Solution: Here $x = 0$ is a regular singular point of the given differential equation (DE). Assume the solution in the form

$$y = \sum_{r=0}^{\infty} c_n x^{n+r}$$

Substituting y, y' and y'' in DE, we get

$$x(1+x)\sum_0^{\infty}(n+r)(n+r-1)c_n x^{n+r-2}$$

$$+(x+5)\sum_0^{\infty}(n+r)c_n x^{n+r-1} - 4\sum c_n x^{n+r} = 0$$

or $$\sum_0^{\infty}(n+r)(n+r-1)c_n x^{n+r-1}+$$

$$+\sum_0^{\infty}(n+r)(n+r-1)c_n x^{n+r} + \sum_0^{\infty}(n+r)c_n x^{n+r}+$$

$$+5\sum(n+r)c_n x^{n+r-1} - 4\sum_0^{\infty} c_n x^{n+r} = 0$$

The indicial equation is obtained by equating to zero the coefficient of lowest power $r-1$ of x namely (from first and fourth summations.

$r(r-1)+5r = 0$ or $r(r+4) = 0$ i.e., $r = 0, -4$

The roots $r_1 = 0$ and $r_2 = -4$ are real, distinct and differ by an integer ie, $r_1 - r_2 = 0 - (-4) = 4$. To obtain the recurrence relation, the indices in the second, third and fifth summations are shifted to $n-1$ so we get

$$\sum_0^{\infty}[(n+r)(n+r-1)+5(n+r)]c_n x^{n+r-1}+$$

$$+\sum_{n=1}^{\infty}[(n-1+r)(n+r-2)+$$

$$(n+r-1)-4]c_{n-1}\cdot x^{n+r-1} = 0$$

The required recurrence relation is $[(n+r)(n+r-1)+5(n+r)]c_n$ $+[(n+r-1)(n+r-2)+(n+r-1)-4]c_{n-1} = 0$ valid for $n \geq 1$.

Start with the smaller root $r_2 = -4$ (with a hope that (!) both the two linearly independent solutions $y_1(x)$ and $y_2(x)$ are obtained in this case itself). With $r = -4$, the recurrence relation reduces to $[(n-4)(n-5)+5(n-4)]c_n$ $+[(n-5)(n-6)+(n-5)-4]c_{n-1} = 0$

or $n(n-4)c_n = -(n-7)(n-3)c_{n-1}$ for $n \geq 1$.

Now, for $n = 1$, $c_1 = -\frac{(-6)(-2)}{1\cdot(-3)}c_0 = 4c_0$

For $n = 2$, $c_2 = -\frac{(-5)(-1)}{2(-2)}c_1 = \frac{5}{4}c_1 = 5c_0$

For $n = 3$, $c_3 = 0$ because of the presence of $(n-3)$ factor in the recurrence relation.

For $n = 4$,

$$4\cdot 0\cdot c_4 = -(-3)(4-3)c_3 = 0$$

Therefore c_4 is arbitrary since for any c_4 the above equation is satisfied.

For $n = 5$, $c_5 = \frac{4}{5}c_4$

For $n = 6$, $c_6 = \frac{1}{5}c_4$

For $n = 7$, $c_7 = 0\cdot c_6 = 0$ (because of $(n-7)$)

Therefore $c_n = 0$ for $n \geq 7$ (since $c_7 = 0$) i.e., $c_8, c_9, c_{10}, c_{11}\ldots$ are all zero.

Thus we had two arbitrary constant's c_0 and c_4. The solution is

$$y_2(x) = \sum_{n=0}^{\infty} c_0 x^{n+r} = \sum_{n=0}^{\infty} c_0 x^{n-4}$$

Substituting the coefficients $c_1 = 4c_0$, $c_2 = 5c_0$, $c_3 = 0, c_5 = \frac{4}{5}c_4, c_6 = \frac{1}{5}c_4, c_7 = c_8 = c_9 = \ldots = 0$ we get

$$y_2(x) = c_0x^{-4} + c_1x^{-3} + c_2x^{-2} + c_3x^{-1} + c_4 + c_5x + c_6x^2 + c_7x^3 + c_8x^4 + c_9x^5 + \ldots$$

$$y_2(x) = c_0(x^{-4} + 4x^{-3} + 5x^{-2}) + c_4\left(1 + \frac{4}{5}x + \frac{1}{5}x^2\right)$$

or $y_2(x) = c_0x^{-4}(1+4x+5x^2)+$ $c_4x^0\left(1+\frac{4}{5}x+\frac{1}{5}x^2\right)$

Indeed $y_2(x)$ itself is the required general solution containing two arbitrary constants c_0 and c_4.

Verification: Suppose we investigate the solution for the larger root $r_1 = 0$. Then the recurrence relation becomes

$$c_n = \frac{-(n-3)(n+1)}{n(n+4)}c_{n-1} \quad \text{for } n \geq 1$$

Now for $n = 1$, $c_1 = \frac{4}{5}c_0$
for $n = 2$, $c_2 = \frac{1}{4}c_1 = \frac{1}{5}c_0$
for $n = 3$, $c_3 = 0$
for $n > 3$, $c_n = 0$ i.e., $c_4 = c_5 = c_6 = \ldots = 0$
Then the solution corresponding to the larger indicial root $r_1 = 0$ is

$$y_1(x) = \sum_{n=0}^{\infty} c_n x^{n+r} = \sum_{n=0}^{\infty} c_n x^n$$

$$= c_0 + c_1 x + c_2 x^2 + c_3 x^3 + c_4 x^4 + \ldots$$

$$= c_0 + \frac{4}{5}c_0 x + \frac{1}{5}c_0 x^2 + 0 + 0 + 0 + \ldots$$

$$y_1(x) = c_0\left(1 + \frac{4}{5}x + \frac{1}{5}x^2\right)$$

This solution $y_1(x)$ is *already* contained in (a part of) the solution $y_2(x)$ (see note 2 on page 10.8).

Case 3: Roots differing by an integer: Logarithmic case

Example 3: Determine two linearly independent solutions about $x = 0$ for

$$4x^2y'' + 2x(2-x)y' - (1+3x)y = 0$$

Solution: $x = 0$ is a regular singular point. Assume $y = \sum_{n=0}^{\infty} c_n x^{n+r}$ to be the series solution. Substituting y, y', y'' the equation becomes

$$4\sum_0^{\infty}(n+r)(n+r-1)c_n x^{n+r} + 4\sum_0^{\infty}(n+r)c_n x^{n+r}$$

$$-2\sum_0^{\infty}(n+r)c_n x^{n+r+1} - \sum_0^{\infty} c_n x^{n+r} -$$

$$-3\sum_0^{\infty} c_n x^{n+r+1} = 0$$

Equating to zero the coefficient of the lowest power x^r (from 1st, 2nd, 4th summations), we get

$$4(r)(r-1) + 4(r) - 1 = 0$$

$$4r^2 - \tfrac{1}{2} = 0 \text{ or } r = +\tfrac{1}{2}, -\tfrac{1}{2}$$

Here $r_1 = \frac{1}{2}$, $r_2 = -\frac{1}{2}$ so $r_1 - r_2 = \frac{1}{2} - (-\frac{1}{2}) = 1$.

Thus the difference is a non-zero positive integer. From the larger root $r_1 = \frac{1}{2}$, only one solution can be obtained. From the smaller root, $r_2 = -\frac{1}{2}$, two solutions can be obtained. Rewriting

$$\sum_0^{\infty}[4(n+r)(n+r-1) + 4(n+r) - 1]c_n x^{n+r} -$$

$$-\sum_1^{\infty}[2(n+r-1)+3]c_{n-1}x^{n+r} = 0$$

The indicial equation is

$$[4(n+r)^2 - 1]c_n = [2(n+r)+1]c_{n-1}$$

or $$c_n = \frac{c_{n-1}}{2r+2n-1} \quad \text{for } n \geq 1$$

Now for $n = 1$, $c_1 = \frac{c_0}{2r+1}$
$n = 2$, $c_2 = \frac{c_1}{2r+3}$
$n = 3$, $c_3 = \frac{c_2}{2r+5}$

$n = n-1$, $c_{n-1} = \frac{c_{n-2}}{2r+2n-3}$
$n = n$, $c_n = \frac{c_{n-1}}{2r+2n-1}$
$c_1 \cdot c_2 \cdot c_3 \ldots c_{n-1} \cdot c_n = \frac{c_0}{2r+1} \cdot \frac{c_1}{2r+3} \times$
$\times \cdot \frac{c_2}{2r+5} \cdots \frac{c_{n-2}}{2r+2n-3} \cdot \frac{c_{n-1}}{2r+2n-1}$

$$\therefore \; c_n = \frac{c_0}{(2r+1)(2r+3)(2r+5)\ldots(2r+2n-3)(2r+2n-1)}$$

for $n \geq 1$. With this c_n, the solution takes the form

$$y = \sum_{n=0}^{\infty} c_n x^{n+r}$$

$$y(x,r) = c_0 \sum_{n=0}^{\infty} \frac{x^{n+r}}{(2r+1)(2r+3)(2r+5)\ldots(2r+2n-3)(2r+2n-1)}.$$

Since the factor $(2r+1)$ in the denominator vanishes for the value of the smaller root $r_2 = -\frac{1}{2}$, expand the series for $n = 0, 1$ and choose $c_0 = (2r+1)$ so that $c_1 = 1$.

$$y(x,r) = (2r+1)x^r + 1 \cdot x^{r+1} +$$

$$+ \sum_{n=2}^{\infty} \frac{x^{n+r}}{(2r+3)(2r+5)\cdots(2r+2n-1)}$$

Here $c_n = \frac{c_1}{(2r+3)(2r+5)\cdots(2r+2n-1)}$ for $n \geq 2$

The two linearly independent solutions are given by $y(x,r)$ at $r_2 = -\frac{1}{2}$ and $\frac{\partial y(x,r)}{\partial r}$ at $r_2 = -\frac{1}{2}$. Differentiating y partially w.r.t. r we get

$$\frac{\partial y(x,r)}{\partial r} = 2x^r + (2r+1)x^r \ln x + x^{r+1}\ln x +$$

$$+\sum_{n=2}^{\infty} \frac{x^{n+r}\cdot \ln x}{(2r+3)(2r+5)\cdots(2r+2n-1)}+$$

$$+\sum_{n=2}^{\infty} x^{n+r}\frac{\partial}{\partial r}\left[\frac{1}{(2r+3)(2r+5)\cdots(2r+2n-1)}\right]$$

For $u(x) = u_1(x)u_2(x)\ldots u_n(x)$, then

$$\frac{du}{dx} = u(x)\left[\frac{\frac{du_1}{dx}}{u_1} + \frac{\frac{du_2}{dx}}{u_2} + \cdots + \frac{\frac{du_n}{dx}}{u_n}\right]$$

Using this result

$$\frac{d}{dr}[(2r+3)^{-1}(2r+5)^{-1}\cdots(2r+2n-1)^{-1}]$$

$$= \frac{1}{(2r+3)(2r+5)\ldots(2r+2n-1)}\left[\frac{-(2r+3)^{-2}\cdot 2}{(2r+3)^{-1}} - \frac{(2r+5)^{-2}\cdot 2}{(2r+5)^{-1}}\right.$$
$$\left.\cdots - \frac{(2r+2n-1)^{-2}\cdot 2}{(2r+2n-1)^{-1}}\right]$$

$$= \frac{-2}{(2r+3)(2r+5)\ldots(2r+2n-1)}\times$$
$$\times\left[\frac{1}{2r+3}+\frac{1}{2r+5}+\ldots+\frac{1}{2r+2n-1}\right]$$

Substituting this value

$$\frac{\partial y(x,r)}{\partial r} = y(x,r)\cdot \ln x + 2x^r -$$

$$-\sum_{n=2}^{\infty}\frac{2\cdot x^{n+r}}{(2r+3)\ldots(2r+2n-1)}\left[\frac{1}{2r+3}+\cdots+\frac{1}{2r+2n-1}\right]$$

Now the required two linearly independent solutions are given by $y_1(x) = y(x,r)$ at $r = -\frac{1}{2}$

$$= y\left(x, r=-\frac{1}{2}\right)$$

$$= 0\cdot x^{-\frac{1}{2}} + x^{-\frac{1}{2}+1} + \sum_{n=2}^{\infty}\frac{x^{n-\frac{1}{2}}}{2\cdot 4\cdot 6\ldots 2(n-1)}$$

$$= x^{\frac{1}{2}} + \sum_{n=2}^{\infty}\frac{x^{n-\frac{1}{2}}}{2^{n-1}(n-1)!}$$

$$y_1(x) = \sum_{n=1}^{\infty}\frac{x^{n-\frac{1}{2}}}{2^{n-1}(n-1)!}$$

The second solution $y_2(x)$ is given by $y_2(x) = \frac{\partial y(x,r)}{\partial r}$ at $r = -\frac{1}{2}$

$$= y_1(x)\cdot \ln x + 2x^{-\frac{1}{2}} -$$

$$-\sum_{n=2}^{\infty}\frac{2\cdot x^{n-\frac{1}{2}}}{2\cdot 4\cdot 6\ldots 2(n-1)}\left[\frac{1}{2}+\frac{1}{4}+\ldots+\frac{1}{2(n-1)}\right]$$

$$= y_1(x)\ln x + 2x^{-\frac{1}{2}} - \sum_{n=2}^{\infty}\frac{2\cdot x^{n-\frac{1}{2}}}{2^{n-1}\cdot(n-1)!}\frac{1}{2}H_{n-1}$$

where $H_{n-1} = 1 + \frac{1}{2} + \frac{1}{3} + \ldots + \frac{1}{n-1}$ is the partial sum of the harmonic series. Hence the second solution is

$$y_2(x) = y_1(x)\ln x + 2x^{-\frac{1}{2}} - \sum_{n=2}^{\infty}\frac{H_{n-1}}{2^{n-1}}\frac{x^{n-\frac{1}{2}}}{(n-1)!}$$

Case 4. Double root:

Example 4: Using method of Frobenius obtain two linearly independent solutions about $x = 0$ for $x^2y'' + x(x-1)y' + (1-x)y = 0$

Solution: Here $x_0 = 0$ is a regular singular point of the given DE. Assume the solution in the form

$$y = \sum_{n=0}^{\infty} c_n x^{n+r}$$

Substituting y, y', y'' in DE, we get

$$\sum_0^{\infty}(n+r)(n+r-1)c_nx^{n+r} + \sum(n+r)c_nx^{n+r+1} -$$

$$-\sum(n+r)c_nx^{n+r} + \sum c_nx^{n+r} - \sum c_nx^{n+r+1} = 0$$

The indicial equation is (with $n = 0$ from 1st, 3rd and 4th summations)

$$r(r-1) - r + 1 = 0 \text{ or } (r-1)^2 = 0$$

Thus the indicial roots are equal $r_1 = r_2 = 1$. The recurrence relation for $n \geq 1$ is $[(n+r)(n+r-1) - (n+r) + 1]c_n + [(n+r-1) - 1]c_{n-1} = 0$. Put $r = 1$. Then the recurrence relation takes the form

$$c_n = -\frac{(n-1)}{n^2}c_{n-1} \quad \text{for} \quad n \geq 1$$

Now $c_n = 0$ for $n \geq 1$ because of the factor $(n-1)$. Thus $c_1 = c_2 = c_3 = c_4 = \cdots = 0$. So the first solution is

$$y_1(x) = \sum_{n=0}^{\infty} c_nx^{n+1} = c_0x$$

Choosing $c_0 = 1$, $y_1(x) = x$.
The second solution $y_2(x)$ is obtained by reduction of order. The standard form of the given DE is

$$y'' + \frac{x(x-1)}{x^2}y' + \frac{(1-x)}{x^2}y = 0$$

So $p(x) = \frac{x-1}{x}$. Assume the linearly independent solution $y_2 = uy_1$ where $u = \int \frac{1}{y_1^2}e^{-\int p\,dx}dx$.
Here $e^{-\int p\,dx} = e^{-\int\left(1-\frac{1}{x}\right)dx} = e^{\ln x - x} = xe^{-x} = \frac{x}{e^x}$. Then

$$u = \int \frac{1}{x^2}\cdot\frac{x}{e^x}dx = \int \frac{dx}{xe^x}$$

$$= \int \frac{1}{x}e^{-x}dx = \int \frac{1}{x}\cdot\sum_{n=0}^{\infty}\frac{(-x)^n}{n!}dx$$

$$= \sum_{n=0}^{\infty}\frac{(-1)^n}{n!}\int x^{n-1}dx$$

$$u = \int\frac{dx}{x} + \sum_{n=1}^{\infty}\frac{(-1)^n}{n!}\int x^{n-1}dx$$

$$u = \ln x + \sum_{n=1}^{\infty}\frac{(-1)^n}{n!}\cdot\frac{x^n}{n}$$

Now the second solution is

$$y_2 = uy_1 = ux = \left[\ln x + \sum\frac{(-1)^n}{n!}\frac{x^n}{n}\right]x$$

$$y_2 = x\ln x + \sum_{n=1}^{\infty}\frac{(-1)^n\cdot x^{n+1}}{n!\cdot n}$$

Case 5: Double root: solution in powers of $(x - x_0)$

Example 5: Using Frobenius method, obtain series solution about $x = 2$ for $x(x-2)y'' + 2(x-1)y' - 2y = 0$.

Solution: $x = 2$ is a regular singular point of the given DE. Put $x - 2 = t$. Then $x = t + 2$, $dx = dt$. So the given DE in t takes the form

$$(t+2)t\frac{d^2y}{dt^2} + 2(t+1)\frac{dy}{dt} - 2y = 0 \quad (1)$$

Now assume a solution of the form

$$y(t) = \sum_{n=0}^{\infty} c_n t^{n+r} \quad (2)$$

Substituting y, $\frac{dy}{dt}$, $\frac{d^2y}{dt^2}$ in (1), we get

$$(t+2)t\sum(n+r)(n+r-1)c_nt^{n+r-2} +$$
$$+2(t+1)\sum(n+r)c_nt^{n+r-1} - 2\sum c_nt^{n+r} = 0$$

or

$$\sum_0^{\infty}(n+r)(n+r-1)c_nt^{n+r} +$$
$$+2\sum_0^{\infty}(n+r)(n+r-1)c_nt^{n+r-1} +$$
$$+2\sum_0^{\infty}(n+r)c_nt^{n+r} + 2\sum_0^{\infty}(n+r)c_nt^{n+r-1} -$$
$$-2\sum_0^{\infty}c_nt^{n+r} = 0$$

Indicial equation is obtained by equating to zero the coefficient of lowest power of t namely t^{r-1} at $(n = 0)$
i.e., $2r(r-1) + 2r = 0$ or $r^2 = 0$ or $r = 0, 0$
roots are equal. To obtain the recurrence relation shift the index of first, third and fourth summations. Then

$$[(n+r-1)(n+r-2) + 2(n+r-1) - 2]\,c_{n-1}$$
$$+ 2\,[(n+r)(n+r-1) + 2(n+r)]\,c_n = 0$$

or

$$c_n = -\frac{[(n+r)^2 - (n+r) - 2]}{2(n+r)^2}c_{n-1}, \text{ for } n \geq 1$$

Take $r = 0$. Then

$$c_n = -\frac{(n^2 - n - 2)}{2n^2}c_{n-1}$$

$$c_n = -\frac{(n-2)(n+1)}{2n^2}c_{n-1} \text{ for } n \geq 1$$

For $n = 1$, $c_1 = c_0$,
For $n = 2$, $c_2 = 0$ so $c_3 = c_4 = c_5 = c_6 \cdots = 0$.
Thus the first solution is

$$y_1(t) = \sum_{n=0}^{\infty}c_nt^{n+0} = c_0 + c_1t = c_0 + c_0t$$

or $\quad y_1(x) = c_0(1 + (x - 2))$

The second solution y_2 is obtained by reduction of order. Rewriting the given DE in standard form

$$\frac{d^2y}{dt^2} + \frac{2(t+1)}{t(t+2)}\frac{dy}{dt} - \frac{2}{t(t+2)}y = 0$$

Here $P(t) = \frac{2(t+1)}{t(t+2)}$. So

$$e^{-\int P(t)dt} = e^{-\int \frac{dt}{d} + \frac{dt}{t+2}}$$

$$= e^{-\ln(t)(t+2)} = \frac{1}{t(t+2)}$$

since $\frac{2(t+1)}{t(t+2)} = \frac{A}{t} + \frac{B}{t+2} = \frac{1}{t} + \frac{1}{t+2}$.
The second solution is

$$y_2(t) = y_1 U$$

where $U = \int \frac{1}{y_1^2} e^{-\int P(t)dt} dt$.

Now

$$U = \int \frac{1}{(1+t)^2} \cdot \frac{1}{t(t+2)} dt$$

By partial fractions

$$\frac{1}{t(t+2)(t+1)^2} = \frac{A}{t} + \frac{B}{t+2} + \frac{C}{t+1} + \frac{D}{(t+1)^2}$$

$$= \frac{1}{2}t - \frac{1}{2}\frac{1}{t+2} + 0 - \frac{1}{(t+1)^2}$$

So

$$U = \int \frac{1}{2}\frac{dt}{t} - \frac{1}{2}\int \frac{dt}{t+2} - \int \frac{dt}{(t+1)^2}$$

$$= \frac{1}{2}\ln\left(\frac{t}{t+2}\right) + \frac{1}{(t+1)}$$

Then

$$y_2 = y_1 U = (1+t)\left[\frac{1}{2}\ln\left(\frac{t}{t+2}\right) + \frac{1}{t+1}\right]$$

$$y_2(t) = 1 + \frac{1}{2}(t+1)\cdot\ln\left(\frac{t}{t+2}\right)$$

or

$$y_2(x) = 1 + \frac{1}{2}(x-1)\ln\left(\frac{x-2}{x}\right).$$

Exercise

Using Frobenius method, obtain two linearly independent solutions about $x_0 = 0$, for the following differential equations.

Distinct roots not differing by an integer

1. $2x^2y'' - xy' + (x-5)y = 0$

Ans. $y = c_1x^{5/2}\left(1 - \frac{1}{9}x + \frac{1}{198}x^2 - \frac{1}{7722}x^3 + \cdots\right)$
$+ c_2x^{-1}\left(1 + \frac{1}{5}x + \frac{1}{30}x^2 + \frac{1}{90}x^3 + \cdots\right)$
$= c_1y_1 + c_2y_2$

Hint: Indicial equation: $2r^2 - 3r - 5 = 0$, $r_1 = \frac{5}{2}, r_2 = -1$
RR: $[2(n+r)(n+r-1) - (n+r) - 5]c_n + c_{n-1} = 0$ for $n \geq 1$.
For $r_1 = \frac{5}{2}$, $c_n = \frac{c_{n-1}}{n(2n+7)}$, for $r = -1$, $c_n = \frac{c_{n-1}}{n(2n-7)}$.

2. $2xy'' + (1+x)y' - 2y = 0$

Ans. $y_1 = x^{\frac{1}{2}} + \sum_{n=1}^{\infty} \frac{(-1)^n 3x^{n+\frac{1}{2}}}{2^n n!(2n-3)(2n-1)(2n+1)}$ and

$$y_2 = 1 + 2x + \frac{1}{3}x^2$$

Hint: Indicial equation: $r(2r-1) = 0$, $r_1 = \frac{1}{2}, r_2 = 0$
RR: $(n+r)(2n+2r-1)c_n + (n+r-3)c_{n-1} = 0$
For $r_1 = \frac{1}{2}$, $c_n = \frac{(-1)^n 3c_0}{2^n n!(2n-3)(2n-1)(2n+1)}$
For $r_1 = 0$, $b_n = \frac{-(n-3)b_n - 1}{n(2n-1)}$

3. $2x^2y'' + xy' + (x^2 - 3)y = 0$

Ans. $y_1 = c_0x^{3/2}\left(1 - \frac{1}{18}x^2 + \frac{1}{936}x^4 \cdots\right)$, $y_2(x) = c_0x^{-1}\left(1 + \frac{1}{2}x^2 - \frac{1}{24}x^4 + \cdots\right)$

Hint: Indicial equation: $2r^2 - r - 3 = 0$, $r_1 = \frac{3}{2}, r_2 = -1$
RR: $[2(n+r)(n+r-1) + (n+r) - 3]c_n + c_{n-2} = 0$.
For $r_1 = \frac{3}{2}$, $c_n = \frac{c_{n-2}}{n(2n+5)}$,
For $r = -1$, $c_n = \frac{c_{n-2}}{n(2n-5)}$

4. $2x(x-1)y'' + 3(x-1)y' - y = 0$

Ans. $y_1 = c_0 - c_0\sum_{n=1}^{\infty}\frac{1}{4n^2-1}x^n$, $y_2 = c_0(1-x)x^{-\frac{1}{2}}$

Hint: $r_1 = 0$, $r_2 = -\frac{1}{2}$, RR: $(2n + 2r + 1)c_n = (2n + 2r - 3)c_{n-1}$ for $r_1 = 0$, $c_n = \frac{2n-3}{2n+1}c_{n-1}$, for $r_2 = 0$, $c_n = \frac{(n-2)}{n}c_{n-1}$, and $c_n = 0$ for $n \geq 2$.

5. $2xy'' + 5(1 + 2x)y' + 5y = 0$

Ans. $y_1 = 1 + \sum_{1}^{\infty} \frac{3(-5)^n x^n}{n!(2n+1)(2n+3)}$,

$y_2 = x^{-3/2} - 10x^{-1/2}$

Hint: $r_1 = 0$, $r_2 = -\frac{3}{2}$, RR: $(n + r)(2(n + r) + 3)c_n = -5(2(n + r) - 1)c_{n-1}$, for $r_1 = 0$, $c_n = \frac{-5(2n-1)}{n(2n+3)}c_{n-1}$

For $r = -\frac{3}{2}$, $c_n = \frac{-10(n-2)}{2n\left(n-\frac{3}{2}\right)}c_{n-1}$, $c_n = 0$, for $n \geq 2$

6. $4xy'' + 2y' + y = 0$

Ans. $y_1 = c_0\left(-\frac{x}{2!} + \frac{x^2}{4!} - \frac{x^3}{6!} + \cdots\right) = c_0 \cos\sqrt{x}$

$y_2 = c_0(x^{1/2}) \cdot \left(1 - \frac{x}{3!} + \frac{x^2}{5!} + \cdots\right) = c_0 \cdot \sin\sqrt{x}$

Hint: $r_1 = \frac{1}{2}$, $r_2 = 0$

RR: $2(r + n)(2r + 2n - 1)c_n + c_{n-1} = 0$ for $n > 1$.

For $r_2 = 0$, $c_n = \frac{-1}{2n(2n-1)} \cdot c_{n-1}$

For $r_1 = \frac{1}{2}$, $c_n = -\frac{1}{2n(2n+1)}c_{n-1}$

Difference of roots a positive integer: Non-Logarithmic Case

7. $x^2y'' - xy' - \left(x^2 + \frac{5}{4}\right)y = 0$

Ans. $y_1 = x^{5/2}\left[1 + \sum_{n=1}^{\infty} \frac{x^{2n}}{[2\cdot4\cdot6\cdots2n][5\cdot7\cdot9\cdots(2n+3)]}\right]$

$y_2 = x^{-1/2}\left[1 - \frac{x^2}{2} - \frac{x^4}{2\cdot4} - \sum_{n=3}^{\infty} \frac{x^{2n}}{[2\cdot4\cdot6\cdots2n][3\cdot5\cdot7\cdots(2n-3)]}\right]$

Hint: $r_1 = \frac{5}{2}$, $r_2 = -\frac{1}{2}$,

RR: $\left[(n + r)(n + r - 1) - (n + r) - \frac{5}{4}\right]c_n - c_{n-2} = 0$, for $n \geq 2$

For $r_1 = \frac{5}{2}$, $c_n = \frac{c_{n-2}}{n(n+3)}$, for $r_2 = -\frac{1}{2}$, $c_n = \frac{c_{n-2}}{n(n-3)}$

8. $xy'' - (4 + x)y' + 2y = 0$

Ans. $y_1 = c_0\left(1 + \frac{1}{2}x + \frac{1}{12}x^2\right)$

$y_2 = c_5\left(x^5 + \sum_{n=6}^{\infty} \frac{60x^n}{(n-5)!n(n-1)(n-2)}\right)$

Hint: $r_1 = 5$, $r_2 = 0$

RR: $(n + r)(n + r - 5)c_n - (n + r - 3)c_{n-1} = 0$

For $r_2 = 0$, $n(n - 5)c_n - (n - 3)c_{n-1} = 0$, c_0 and c_5 are arbitrary, for $n > 5$, $c_n = \frac{(n-3)(c_{n-1})}{n(n-5)}$.

9. $x^2y'' + 2x(x - 2)y' + 2(2 - 3x)y = 0$

Ans. $y_1 = c_0(x - 2x^2 + 2x^3)$

$y_2 = c_3\left[x^4 + \sum_{n=4}^{\infty} \frac{(-2)^{n-3}}{n!} 6 \cdot x^{n+1}\right]$

Hint: $r_1 = 4$, $r_2 = 1$. For $r_2 = 1$, $c_n = \frac{-2}{n}c_{n-1}$. For $r_1 = 4$, $c_n = \frac{-2}{n+3}c_{n-1}$.

10. $x(1 + x)y'' + (x + 5)y' - 4y = 0$ about $x = -1$

Ans. $y = c_0[1 + (x + 1) + \frac{1}{2}(x + 1)^2]$

$+\frac{c_5}{12}\left[\sum_{n=5}^{\infty} (n - 4)(n - 3)(n + 1)(x + 1)^n\right]$

11. $(1 - x^2)y'' + 2xy' + y = 0$

Ans. $y_1 = c_0\left(1 - \frac{x^2}{2} + \frac{x^4}{8} + \cdots\right)$

$y_2 = c_1\left(x - \frac{x^3}{5} + \frac{x^5}{40} \cdots\right)$

Hint: $r_1 = 1$, $r_2 = 0$

RR: $(r + n + 1)(r + n)c_{n+1} = [(r + n)^2 - 5(r + n) + 3]c_{n-1}$

For $r_2 = 0$, $c_{n+1} = \frac{n^2-5n+3}{n(n+1)}c_{n-1}$, c_0 and c_1 are arbitrary.

Difference of roots a positive integer: Logarithmic case

12. $x^2y'' + (x^2 - 3x)y' + 3y = 0$

Ans. $y_1 = c_2x^3e^{-x}$

$y_2 = \left(-\frac{x}{2} - \frac{x^2}{2} + \frac{3}{4}x^3 - \frac{1}{4}x^4 + \ldots\right) + \frac{1}{2}x^3e^{-x}\ln x$

Hint: $r_1 = 3$, $r_2 = 1$

RR: $[(n + r)(n + r - 1) - 3(n + r)]c_n + (n + r - 1)c_{n-1} = 0$, $n \geq 1$

For $r_1 = 3$, $c_n = \frac{c_{n-1}}{n}$, $n \geq 1$. For $r_1 = 1$, solution involves logarithm, obtained by reduction of order.

13. $x^2y'' + x(1 - x)y' - (1 + 3x)y = 0$

Ans. $y_1 = -3x - \sum_{n=3}^{\infty} \frac{(n+1)x^{n-1}}{(n-2)!}$

$y_2 = y_1 \ln x + x^{-1} - 2 - \sum_{n=0}^{\infty}\left[\frac{1-(n+3)H_n}{n!}\right]x^{n+1}$

Here H_n = partial sum of Harmonic series

$H_n = \left[1 + \frac{1}{2} + \frac{1}{3} + \frac{1}{4} + \cdots + \frac{1}{n}\right]$

Hint: $r_1 = 1, r_2 = -1$,

RR: $(n+r+1)(n+r-1)c_n = (n+r+2)c_{n-1}$

For $n \geq 1$, $c_n = \frac{(n+r+2)c_0}{(r+2)[r(r+1)\cdots(r+n-1)]}$

First solution

$y_1 = y(x,r) = (r+1)x^r + \frac{(r+1)(r+3)x^{1+r}}{r(r+2)} + \frac{(r+4)x^{2+r}}{r(r+2)} + \sum_{n=3}^{\infty}\frac{(n+r+2)x^{n+r}}{(r+2)r[(r+2)(r+3)\cdots(r+n-1)]}$

Second solution $y_2 = \left.\frac{\partial y(x,r)}{\partial r}\right|$ at $r = -1$.

14. $(x^2 - x)y'' - xy' + y = 0$

Ans. $y_1 = x$, $y_2 = x \ln x + 1$

Hint: $r_1 = 1, r_2 = 0$,

RR: $(n+r-1)c_n - (n+r+1)(n+r)c_{n+1} = 0$

For $r_1 = 1$, $c_{n+1} = \frac{n^2}{(n+2)(n+1)}c_n$

Second solution by reduction of order.

15. $xy'' - 3y' + xy = 0$

Ans. $y_1 = -\frac{1}{2^2 \cdot 4}x^4 + \frac{x^6}{2 \cdot 2^2 \cdot 4 \cdot 6} - \frac{x^8}{2 \cdot 2^2 \cdot 4^2 \cdot 6 \cdot 8} + \cdots$

$y_2 = y_1 \ln x + 1 + \frac{1}{2^2}x^2 + \frac{1}{2^5 2!}x^4 - \frac{1}{2^6 3!1!}\left(1 + \frac{1}{2} + \frac{1}{3}\right)x^6 + \cdots$

Hint: $r_1 = 4, r_2 = 0$

RR: $(n+r)(n+r-4)c_n = -c_{n-2}$, for $n \geq 2$

Second solution by $\frac{\partial y}{dr}$ with $r = 0$ where

$\overline{y}(x,r) = x^r\left[r - \frac{rx^2}{(2+r)(r-2)} + \frac{x^4}{(r-2)(r+2)(r+4)} - \frac{x^6}{(r-2)(r+2)^2(r+4)(r+6)} + \cdots\right]$

16. $(x - x^2)y'' - (1+3x)y' - y = 0$

Ans. $y_1 = c_0[-2x^2 - 6x^3 + \cdots]$

$y_2 = y_1 \log x + c_0[1 - x - 5x^2 - 11x^3 + \cdots]$

Hint: $r_1 = 2, r_2 = 0$, First solution $y(x,r)$ at $r = 0$

$y(x,r) = c_0 x^r\left[r + \frac{r(r+1)}{r-1}x + \frac{(r+1)(r+2)}{r-1}x^2 + \frac{(r+1)(r+2)(r+3)}{(r-1)(r+1)}x^3 + \cdots\right]$

Second solution is $\frac{\partial y(x,r)}{\partial r}$ at $r = 0$.

Equal roots (double root)

17. $x(x-1)y'' + (3x-1)y' + y = 0$

(special case of Hypegeometric equation)

Ans. $y_1 = \frac{1}{1-x}$, $y_2 = \frac{\ln x}{1-x}$

Hint: $r_1 = r_2 = 0$, RR: $c_{n+1} = c_n$. So $c_0 = c_1 = c_2 = \cdots = 1$, second solution by reduction of order

18. $x^2y'' + 3xy' + (1-2x)y = 0$

Ans. $y_1 = x^{-1} + \sum_{n=1}^{\infty}\frac{2^n x^{n-1}}{(n!)^2}$

$y_2 = y_1 \cdot \ln x - \sum_{n=1}^{\infty}\frac{2^{n+1} \cdot H_n \cdot x^{n-1}}{(n!)^2}$

Hint: $r_1 = r_2 = -1$

RR: $(n+r+1)^2c_n - 2c_{n-1} = 0$ for $n \geq 1$. Also for $n \geq 1$, $c_n = \frac{2^n c_0}{[(r+2)(r+3)\cdots(r+n+1)]^2} = c_n(r)$

First solution $y(x,r) = x^r + \sum_{n=1}^{\infty} c_n(r) \cdot x^{n+r}$ at $r = -1$

Second solution $\frac{\partial y(x,r)}{\partial r}$ at $r = -1$

19. $x^2y'' - x(1+x)y' + y = 0$

Ans. $y_1 = c_0xe^x$, $y_2 = c_0xe^x \ln x - c_0x\left[x - \frac{3}{4}x^2 + \frac{11}{6^2}x^3 + \cdots\right]$

Hint: $r_1 = r_2 = 1$

RR: $c_n = \frac{c_{n-1}}{(n+r-1)}$ for $n \geq 1$. Also $c_n = \frac{c_0}{r(r+1)(r+2)\cdots(r+n-1)} = c_n(r)$

First solution

$y_1 = c_0x^r\left[1 + \frac{x}{r} + \frac{x^2}{r(r+1)} + \frac{x^3}{r(r+1)(r+2)} + \cdots\right]$ at $r = 1$.

Second solution is $y_2(x) = \frac{\partial y_1(x,r)}{\partial r}$ at $r = 1$ or second solution by reduction of order.

20. $(x - x^2)y'' + (1-x)y' - y = 0$

Ans. $y_1 = c_0\left(1 + x + \frac{1}{2}x^2 + \frac{1}{2}\,\frac{5}{9}x^3 + \cdots\right)$

$y_2 = c_0 \log x \cdot \left(1 + x + \frac{1}{2}x^2 + \frac{1}{2}\,\frac{5}{9}x^3 + \cdots\right) + c_0\left(-2x - x^2 - \frac{14}{27}x^3 \cdots\right)$

Hint: $r_1 = r_2 = 0$

RR: $(r+n)^2c_n = [(r+n-1)^2 + 1]c_{n-1}$

First solution $y(x,r)$ at $r = 0$

$y_1 = c_0x^r\left[1 + \frac{(r^2+1)}{(r+1)^2}x + \frac{[(r+1)^2+1][r^2+1]}{(r+2)^2(r+1)^2}x^2 + \cdots\right]$

Second solution $\frac{\partial y(x,r)}{\partial r}$ at $r = 0$

21. $xy'' + y' + xy = 0$

Ans. $y_1 = c_0\left[1 - \frac{1}{2}x^2 + \frac{1}{2^24^2}x^4 - \frac{1}{2^24^26^2}x^6 + \cdots\right]$
$= J_0(x) =$ Bessel's function of first kind

$y_2 = y_1 \ln x + c_0\left[\frac{1}{2^2}x^2 - \frac{1}{2^24^2}\left(1+\frac{1}{2}\right)x^4 + \frac{1}{2^24^26^2}\left(1+\frac{1}{2}+\frac{1}{3}\right)x^6 + \ldots\right]$
$= y_0(x)$
= Bessel's function of second kind

Hint: $r_1 = r_2 = 0, \quad a_1 = a_3 = a_5 = a_7 = \cdots = 0$

First solution $y(x, r)$ at $r = 0$

$y(x,r) = c_0x^r\left[1 - \frac{x^2}{(r+2)^2} + \frac{x^4}{(r+2)^2(r+4)^2} - \frac{x^6}{(r+2)^2(r+4)^2(r+6)^2} + \cdots\right]$

Second solution is $\frac{\partial y(x,r)}{\partial r}$ at $r = 0$.

10.4 ORTHOGONALITY OF FUNCTIONS

Orthogonality is one of the most useful concept ever introduced in applied mathematics. Let $f(x)$ and $g(x)$ be any two functions which are piecewise continuous on the interval $a < x < b$. The *inner product* of f and g, denoted by $\langle f, g\rangle$ is defined as the number

$$\langle f, g\rangle = \int_a^b f(x)g(x)dx$$

The interval $a < x < b$ is known as the *fundamental interval*. The *norm* of $f(x)$, denoted by $||f||$, is defined by the non-negative number

$$\boxed{||f|| = \sqrt{\int_a^b [f(x)]^2dx} = \langle f, f\rangle^{\frac{1}{2}}}$$

Orthogonal: Two functions f and g are said to be orthogonal on the interval $a < x < b$ if their inner product is zero i.e.,

$$\boxed{\int_a^b f(x)g(x)dx = 0}.$$

An infinite sequence (system or set) of functions $\phi_1(x), \phi_2(x), \ldots, \phi_n(x), \ldots$ is said to be orthogonal on the interval $[a, b]$ if these functions are mutually Orthogonal i.e.,

$$\boxed{\int_a^b \phi_m(x)\phi_n(x)dx = 0 \quad \text{for } m \neq n}$$

Any function $f(x)$, with non-zero norm, is said to be *normalized* by dividing it by its norm i.e., $\frac{f(x)}{||f(x)||}$

Suppose the norm of any of these functions is not zero, then the new sequence of functions

$$\frac{\phi_1(x)}{||\phi_1(x)||}, \frac{\phi_2(x)}{||\phi_2(x)||}, \ldots, \frac{\phi_n(x)}{||\phi_n(x)||}, \ldots$$

are said to be *orthonormal* because they are orthogonal and each have norm unity. Thus new set of functions

$$\psi_n(x) = \frac{\phi_n(x)}{||\phi_n(x)||} \quad \text{for } n = 1, 2, 3, \ldots$$

are orthonormal if

$$\langle \psi_m, \psi_n\rangle = \int_a^b \psi_m(x)\psi_n(x)dx = \delta_{mn} = \begin{cases} 0 & \text{for } m \neq n \\ 1 & \text{for } m = n \end{cases}$$

Here δ_{mm} is the Kronecker* delta.

More generally these definitions are modified w.r.t. a *weight function* $P(x)$ as follows:
Inner product of f and g w.r.t. the weight function $P(x) > 0$ is

$$\langle f, g\rangle = \int_a^b P(x)f(x)g(x)dx$$

Two functions f and g are said to be orthogonal w.r.t. the weight function $P(x)$ on the interval $a < x < b$ if

$$\int_a^b P(x)f(x)g(x)dx = 0$$

Finally a set of functions $\phi_n(x)$ on (a, b) are said to be orthogonal w.r.t. the weight function $P(x) > 0$ if all pairs of distinct functions in the set are orthogonal i.e.,

$$\int_a^b P(x)\phi_m(x)\phi_n(x)dx = 0 \quad \text{for } m \neq n$$

Norm is similarly defined as

*Leopold Kronecker (1823-1891), German mathematician.

$$||f|| = \langle f, f \rangle^{\frac{1}{2}} = \int_a^b P(x)[f(x)]^2 dx \geq 0$$

Note 1: Orthogonality w.r.t. $P(x) = 1$ is simply referred to as "orthogonal".

Note 2: Weight functions other than unity will occur in Bessel functions, Harmite polynomials, Lagurre polynomials and several Sturm-Liouville problems.

Example 1: Hermite polynomials $H_n(x)$ are orthogonal w.r.t. the weight function e^{-x^2} on the internal $-\infty < x < \infty$. i.e.,

$$\int_{-\infty}^{\infty} e^{-x^2} H_m(x)H_n(x)dx = \begin{cases} 0 \text{ if } m \neq n \\ 2^n n! \sqrt{\pi} \text{ if } m = n \end{cases}$$

Generalized Fourier Series

A vector in three-dimensional space, in analytic geometry is represented in terms of an orthonormal set of vectors $\bar{i}, \bar{j}, \bar{k}$ as

$$\bar{A} = A_1\bar{i} + A_2\bar{j} + A_3\bar{k}$$

Extending this concept, a vector in an n-dimensional Euclidean space E_n can be represented as

$$\bar{A} = A_1\bar{e}_1 + A_2\bar{e}_2 + \ldots + A_n\bar{e}_n = \sum_{i=1}^{n} A_i\bar{e}_i$$

where the system of vectors $\{\bar{e}_1, \bar{e}_2, \ldots, \bar{e}_n\}$ form an orthonormal set in E_n. In a similar way any given function $f(x)$ belonging to $c_p(a_1 b)$, the space of piecewise continuous functions on the interval $a < x < b$, can be represented in terms of an orthonormal set of functions $\phi_n(x)$, $(n = 1, 2, 3, \ldots)$ in the space $c_p(a_1 b)$ w.r.t. a weight function $P(x)$, as an infinite series of the form

$$f(x) = a_0\phi_0(x) + a_1\phi_1(x) + a_2\phi_2(x) + \cdots + a_n\phi_n(x) + \cdots \quad (1)$$

The infinite series (1) is known as the *generalized Fourier series* of $f(x)$ w.r.t. the orthonormal set $\{\phi_n(x)\}$ on the interval $a < x < b$ with weight function $P(x)$. Series (1) is also known as *orthogonal expansion* or *eigenfunction expansion* (in case $\phi(x)$ are eigenfuntions of a Sturm-Liouville problem). The coefficients $c_n's$ are known as *Fourier* constants, which can be determined using the orthogonality properties as follows. Multiplying (1) on both sides by $P(x)\phi_n(x)$ and integrating w.r.t. x from a to b, we get

$$\int_a^b P(x)f(x)\phi_n(x)dx = \int_a^b P(x)\phi_n(x)\sum_{i=0}^{\infty} a_i\phi_i(x)dx$$

$$= \int_a^b P(x)\sum_{i=0}^{\infty} a_i\phi_n(x)\phi_i(x)dx$$

$$= \sum_{i=0}^{\infty} a_i \int_a^b P(x)\phi_n(x)\phi_i(x)dx$$

$$= a_n$$

since $\int_a^b P(x)\phi_n(x)\phi_m(x)dx = 0$ for $m \neq n$.

Thus $\boxed{a_n = \int_a^b P(x)f(x)\phi_n(x)dx}$ (2)

for $n = 0, 1, 2, 3, \ldots$.

Completeness

Given a set of orthonormal functions $\{\phi_n(x)\}$, in general, *no* nontrivial function $f(x)$ exists which is orthogonal to *all* $\phi_n(x)$. Thus an orthonormal set of functions $\{\phi_n(x)\}$ is said to be *complete* if any function $f(x)$ which is orthogonal to *all* $\phi_n(x)$ is a *null function* (which is identically zero with zero norm). In other words, the orthogonal set $\{\phi_n(x)\}$ is complete, if *each* function f can have formal representation (1).

Convergence

If the orthonormal set $\{\phi_n(x)\}$ is complete, then the formal representative of any function $f(x)$ by the infinite series (1) is valid and converges to $f(x)$ at all points of continuity and converges to the mean value $\frac{1}{2}[f(x+) + f(x-)]$ at points of discontinuity.

The classical example of orthogonal expansion (1) is the Fourier series, which is the daily bread of the physicist and engineer. Fourier series is discussed in detail in Chapter 17.

WORKED OUT EXAMPLES

Example 1: Show that 1, $\cos 4nx$, $\sin 4nx$, $n = 1, 2, 3 \ldots$ on $0 \le x \le \frac{\pi}{2}$ are orthogonal. Find the corresponding orthonormal set of functions.

Solution: Consider for any $n = 1, 2, 3 \cdots$ $\int_0^{\frac{\pi}{2}} 1 \cdot \cos 4nx = \frac{1}{4}\int_0^{2\pi} \cos nt\, dt$ where $4x = t$

$$= \frac{1}{4} \cdot \left.\frac{\sin nt}{-n}\right|_0^{2\pi} = 0$$

So, the functions 1 and $\cos 4nx$ for $n = 1, 2, 3 \ldots$ are orthogonal. Similarly $\int_0^{\frac{\pi}{2}} 1 \cdot \sin 4nx dx = \frac{1}{4}\int_0^{2\pi} \sin nt dt = \frac{1}{4} \cdot \left.\frac{\cos nt}{n}\right|_0^{2\pi} = 0$

Also

$\int_0^{\frac{\pi}{2}} \cos 4mx \cdot \sin 4n\, x\, dx$

$= \frac{1}{4}\int_0^{2\pi} \cos mt \cdot \sin nt\, dt = 0$

Thus the set of functions 1, $\cos 4nx$, $\sin 4nx$ for $n = 1, 2, 3 \cdots$ are mutually orthogonal. Now since

$$\int_0^{\frac{\pi}{2}} \cdot 1 \cdot 1 \cdot dx = x\Big|_0^{\frac{\pi}{2}} = \frac{\pi}{2},$$

therefore norm of the function 1 is $\sqrt{\frac{\pi}{2}}$. Similarly since $\int_0^{\frac{\pi}{2}} \cos 4nx \cdot \cos 4nx\, dx = \frac{1}{4}\int_0^{2\pi} \cos^2 nt\, dt$

$= \frac{1}{4} \cdot \pi$

So the norm of the function $\cos 4nx$ is $\sqrt{\frac{\pi}{4}}$.

Finally norm of $\sin 4nx$ is $\sqrt{\frac{\pi}{4}}$. Then the orthonormal set of functions are obtained by dividing the functions by their norms. Thus the corresponding orthonormal set of functions are $\sqrt{\frac{2}{\pi}}$; $\frac{2}{\sqrt{\pi}} \cdot \cos 4nx$; $\frac{2}{\sqrt{\pi}} \sin 4nx$ for $n = 1, 2, \ldots$

Example 2: Prove that the functions $f_1(x) = b$, and $f_2(x) = x^3$ are orthogonal on the interval $(-a, a)$ where a and b are real constants. Determine constants A and B such that the function $f_3(x) = 1 + Ax + Bx^2$ is orthogonal to both $f_1(x)$ and $f_2(x)$ on $(-a, a)$.

Solution: f_1, f_2 are orthogonal if $\int_{-a}^{a} f_1(x)\, f_2(x)\, dx = 0$. So consider $\int_{-a}^{a} b \cdot x^3 dx = b\int_{-a}^{a} x^3 dx = b \cdot 0 = 0$. Thus f_1 and f_2 are orthogonal. Since f_3 is orthogonal to f, we have

$$\int_{-a}^{a} (1 + Ax + Bx^2) \cdot 1\, dx = 0$$

or $$\int_{-a}^{a} dx + A\int_{-a}^{a} x dx + B\int_{-a}^{a} x^2 dx = 0$$

$2a + A \cdot 0 + \frac{B}{3} 2a^3 = 0 \quad \therefore \; B = \frac{-3}{a^2}$

Also since f_3 is orthogonal to f_2, we have

$$\int_{-a}^{a} (1 + Ax + Bx^2) x^3 dx = 0$$

or $\int_{-a}^{a} x^3 dx + A\int_{-a}^{a} x^4 dx + B\int_{-a}^{a} x^5 dx = 0$

$0 + A \cdot \frac{2a^5}{5} + B \cdot 0 = 0 \quad \therefore A = 0$

Example 3: Prove that the Laguerre* polynomials $f_1(x) = 1 - x$, $f_2(x) = 1 - 2x + \frac{1}{2}x^2$, $f_3(x) = 1 - 3x + \frac{3}{2}x^2 - \frac{1}{6}x^3$ are orthogonal w.r.t. the weight function e^{-x} on $0 \le x \le \infty$. Determine the corresponding orthonormal functions.

Solution: Consider

$\int_0^{\infty} P(x) f_1(x) f_2(x) dx =$

$\int_0^{\infty} e^{-x}(1 - x)\left(1 - 2x + \frac{x^2}{2}\right) dx$

$= \int_0^{\infty} e^{-x}\left(1 - 3x + \frac{5}{2}x^2 - \frac{1}{2}x^3\right) dx$

We know that $\int_0^{\infty} x^n e^{-x} dx = n!$ for n positive integer. So

$$\int_0^{\infty} e^{-x} f_1 \cdot f_2 dx = 1 \cdot 1 - 3 \cdot 1 + \frac{5}{2} \cdot 2! - \frac{1}{2} 3!$$

$$= 1 - 3 + 5 - 3 = 0$$

Thus f_1 and f_2 are orthogonal w.r.t. e^{-x} on the interval $(0, \infty)$.

Now consider

$$\int_0^{\infty} e^{-x} f_1(x) f_3(x) dx$$

$$= \int_0^{\infty} e^{-x}(1 - x)\left(1 - 3x + \tfrac{3}{2}x^2 - \tfrac{1}{6}x^3\right) dx$$

$$= \int_0^{\infty} e^{-x}\left[1 - 4x + \frac{9}{2}x^2 - \frac{5}{3}x^3 + \frac{1}{6}x^4\right] dx$$

*Edmond Laguerre (1834-1886), French mathematician.

$= 1 \cdot 1 - 4 \cdot 1 + \frac{9}{2} \cdot 2! - \frac{5}{3} \cdot 3! + \frac{1}{6} \cdot 4!$
$= 1 - 4 + 9 - 10 + 4 = 0$

So f_1 and f_3 are orthogonal.
Finally consider
$\int_0^\infty e^{-x} f_2(x) f_3(x) dx$
$= \int_0^\infty e^{-x}\left(1 - 2x + \frac{x^2}{2}\right)\left(1 - 3x + \frac{3}{2}x^2 - \frac{1}{6}x^3\right) dx$
$= \int_0^\infty e^{-x}\left[1 - 5x + 8x^2 - \frac{28}{6}x^3 + \frac{13}{12}x^4 - \frac{1}{12}x^5\right] dx$
$= 1 \cdot 1 - 5 \cdot 1 + 8 \cdot 2! - \frac{28}{6} \cdot 3! + \frac{13}{12} \cdot 4! - \frac{1}{12} \cdot 6!$
$= 1 - 5 + 16 - 28 + 26 - 10 = 43 - 43 = 0$
Thus the functions f_1, f_2, f_3 are orthogonal w.r.t. e^{-x} on $(0, \infty)$.
Norm of the function $f_1(x)$ w.r.t. e^{-x} on $(0, \infty)$ is $\sqrt{\int_0^\infty e^{-x} \cdot f_1(x) f_1(x) dx}$.
Since $\int_0^\infty e^{-x}(1 - x)^2 dx$
$= \int_0^\infty e^{-x}(1 + x^2 - 2x) dx$
$= 1 \cdot 1 + 1 \cdot 2 - 2 \cdot 1 = 1$
So the norm of f_1 is $||f_1|| = 1$.
Similarly consider
$\int_0^\infty e^{-x}\left(1 - 2x + \frac{x^2}{2}\right)^2 dx$
$= \int_0^\infty e^{-x}\left[1 + 4x^2 + \frac{x^4}{4} - 4x - 2x^3 + x^2\right] dx$
$= 1 \cdot 1 - 4 \cdot 1 + 5 \cdot 2! - 2 \cdot 3! + \frac{1}{4} \cdot 4! = 1$
Then norm of f_2 is 1.
Also

$$\int_0^\infty e^{-x}\left(1 - 3x + \frac{3}{2}x^2 - \frac{1}{6}x^3\right)^2 dx$$

$$= \int_0^\infty e^{-x}\left[1 + 9x^2 + \frac{9}{4}x^4 + \frac{1}{36}x^6 - 6x + 3x^2 - \frac{1}{3}x^3 - 9x^3 + x^4 - \frac{1}{2}x^5\right] dx$$

$$= \int_0^\infty e^{-x}\left[1 - 6x + 12x^2 - \frac{28}{3}x^3 + \frac{13}{4}x^4 - \frac{1}{2}x^5 + \frac{1}{36}x^6\right] dx$$

$= 1 - 6 \cdot 1 + 12 \cdot 2! - \frac{28}{3} \cdot 3! + \frac{13}{4} \cdot 4! - \frac{1}{2} \cdot 5! + \frac{1}{36} \cdot 6! = 1$
So norm of f_3 is one. Thus the orthonormal set is itself i.e., $f_1 = (1 - x)$, $f_2 = \left(1 - 2x + \frac{x^2}{2}\right)$ and $f_3(x) = 1 - 3x + \frac{3}{2}x^2 - \frac{1}{6}x^3$

Orthogonality of Laguerre polynomials

Example 4: Prove that the Laguerre polynomials $L_n(x)$ are orthonormal w.r.t. the weight function e^{-x} on the interval $0 < x < \infty$.

Solution: The Laguerre polynomial $L_m(x)$ is defined by the generating function

$$\sum_{m=0}^{\infty} s^m L_m(x) = \frac{1}{(1 - s)} e^{-\left(\frac{xs}{1-s}\right)} \quad (1)$$

Also

$$\sum_{n=0}^{\infty} t^n L_n(x) = \frac{1}{(1 - t)} e^{-\left(\frac{xt}{1-t}\right)} \quad (2)$$

Multiplying (1) and (2), we get
$\left(\sum_{m=0}^{\infty} s^m \cdot L_m(x)\right)\left(\sum_{n=0}^{\infty} t^n \cdot L_n(x)\right)$
$= \frac{1}{(1-s)} \cdot \frac{1}{(1-t)} \cdot e^{-\left(\frac{xs}{1-s} + \frac{xt}{1-t}\right)}$
$\sum_{m,n=0}^{\infty} s^m \cdot t^n \cdot L_m(x) \cdot L_n(x) = \frac{1}{(1-s)(1-t)} e^{-\left(\frac{xs}{1-s} + \frac{xt}{1-t}\right)}$
Multiplying both sides by the weight function $P(x) = e^{-x}$ and integrating w.r.t. x from 0 to ∞, we have

$$\sum_{m,n=0}^{\infty}\left[\int_0^\infty e^{-x} L_m(x) \cdot L_n(x) dx\right] s^m \cdot t^n$$

$$= \frac{1}{(1-s)(1-t)} \int_0^\infty \exp\left\{-x\left(1 + \frac{s}{1-s} + \frac{t}{1-t}\right)\right\} dx$$

Integrating the RHS we get

$$= \frac{1}{(1 - s)(1 - t)} \cdot \left(1 + \frac{s}{1 - s} + \frac{t}{1 - t}\right)^{-1}$$

$$= \frac{1}{(1-s)(1-t)} \frac{(1-s)(1-t)}{(1-s)(1-t) + s(1-t) + t(1-s)} = \frac{1}{1-st}$$

$$= \sum_{n=0}^{\infty} (st)^n$$

Now compare the coefficients of like power of (st) on both sides. For $m \neq n$, the coefficients of terms containing $s^m t^n$ in the LHS should be zero (because no such terms exists in the RHS). Thus

$$\int_0^\infty e^{-x} L_m(x) L_n(x) dx = 0 \text{ for } m \neq n$$

For $m = n$, the coefficient of $(st)^n$ in the LHS is one. So

$$\int_0^\infty e^{-x} L_n^2(x)dx = 1 \text{ for } m = n$$

Thus the Laguerre polynomials $L_n(x)$ are orthonormal w.r.t. the weight function e^{-x} on the interval $(0, \infty)$.

EXERCISE

Orthogonality

Show that the set of functions $\{\phi_n(x)\}$ are orthogonal on the interval $a < x < b$. Normalize and obtain the orthonormal set $\{\psi_n(x)\}$, where

1. $\phi_n(x) = \sin\frac{n\pi x}{c}$ in $0 < x < c; n = 1, 2, 3, \cdots$

Ans. $\psi_n(x) = \sqrt{\frac{2}{c}}\sin\left(\frac{n\pi x}{c}\right), n = 1, 2, 3, \cdots$

Hint: $\int_0^c \sin\frac{n\pi x}{c}\cdot\sin\frac{m\pi x}{c}dx = \begin{cases} 0 \text{ for } m \neq n \\ \frac{c}{2} \text{ for } m = n \end{cases}$

2. $\phi_n(x) = \cos\frac{n\pi x}{c}$ in $0 < x < c, n = 0, 1, 2, 3, \ldots$

Ans. $\psi_0(x) = \frac{1}{\sqrt{c}}, \psi_n(x) = \sqrt{\frac{2}{c}}\cos\frac{n\pi x}{c}, n = 0, 1, 2, 3, \cdots$

Hint: $\int_0^c \cos\frac{n\pi x}{c}, \cos\frac{m\pi x}{c}dx = \begin{cases} 0 \text{ for } m \neq n \\ \frac{c}{2} \text{ for } m = n \end{cases}$

3. $1, \cos nx, \sin nx, n = 1, 2, 3, \cdots, [-\pi, \pi]$

Ans. $\frac{1}{\sqrt{2\pi}}, \frac{1}{\sqrt{\pi}}\cos n\pi, \frac{1}{\sqrt{\pi}}\sin nx$

4. $1, \cos\frac{n\pi x}{L}, \sin\frac{n\pi x}{L}, n = 1, 2, 3, \cdots, [-L, L]$

Ans. $\frac{1}{\sqrt{2L}}, \frac{1}{\sqrt{L}}\cos\left(\frac{n\pi x}{L}\right), \frac{1}{\sqrt{L}}\sin\frac{n\pi x}{L}$

5. $1, \cos nx, n = 1, 2, 3, \cdots, [0, \pi]$

Ans. $\frac{1}{\sqrt{\pi}}, \sqrt{2/\pi}\cos nx$

6. $\sin nx, n = 1, 2, 3, \cdots [0, \pi]$.

Ans. $\sqrt{\frac{2}{\pi}}\sin nx$

7. $\sin n\pi x, n = 1, 2, 3, \cdots \left(-\frac{\pi}{\omega} \le x \le \frac{\pi}{\omega}\right)$

Ans. $\sqrt{\frac{\omega}{\pi}}\cdot\sin n\pi x$

8. $1, \cos 2nx, n = 1, 2, 3, \cdots, 0 \le x \le \pi$

Ans. $\frac{1}{\sqrt{\pi}}, \sqrt{\frac{2}{\pi}}\cos 2nx$

9. $1, \cos\left(\frac{2n\pi x}{L}\right), \sin\left(\frac{2n\pi x}{L}\right), n = 1, 2, 3, \cdots, -\frac{L}{2} \le x \le \frac{L}{2}$

Ans. $\frac{1}{\sqrt{L}}, \sqrt{\frac{2}{L}}\cos\left(\frac{2n\pi x}{L}\right), \sqrt{\frac{2}{L}}\sin\left(\frac{2n\pi x}{L}\right)$

10. Find the function $f_3(x) = 1 + Ax + Bx^2$ such that $f_1 = 1$, $f_2(x) = x$, $f_3(x)$ and are orthogonal to each other on $(-1, 1)$.

Ans. $A = 0, B = 3$

10.5 STURM*- LIOUVILLE** PROBLEMS

A Sturm-Liouville problem is a boundary value problem consisting of a linear homogeneous second order ordinary differential equation, together with a pair of homogeneous boundary conditions.

Consider any equation of the form

$$a_0(x)\frac{d^2y}{dx^2} + a_1(x)\frac{dy}{dx} + [a_2(x) + \lambda a_3(x)]y = 0$$

defined on a finite interval $a < x < b$. This can be written, assuming $a_0(x) \neq 0$, $\frac{d^2y}{dx^2} + \frac{a_1(x)}{a_0(x)}\frac{dy}{dx} + \left[\frac{a_2(x)}{a_0(x)} + \lambda\frac{a_3(x)}{a_0(x)}\right]y = 0$. Multiply the above equation throughout by $r(x) = e^{\int\frac{a_1(x)}{a_0(x)}dx}$. Then the equation takes the form

$r(x)\frac{d^2y}{dx^2} + r(x)\frac{a_1(x)}{a_0(x)}\frac{dy}{dx} + [q(x) + \lambda P(x)]\,y = 0$

where $q(x) = \frac{a_2(x)}{a_0(x)}r(x)$, $P(x) = \frac{a_3(x)}{a_0(x)}r(x)$.

Combining the first two terms, this equation can be rewritten as

$$\frac{d}{dx}\left[r(x)\frac{dy}{dx}\right] + [q(x) + \lambda P(x)]y = 0 \quad (1)$$

Here the functions $P(x)$, $q(x)$, $r(x)$, $r'(x)$ are continuous real valued functions on $a \le x \le b$ and are free of the parameter λ. Also $P(x) > 0$ and $r(x) > 0$ when $a \le x \le b$, while $q(x)$ is non-positive. Equation (1) is known as *Sturm-Liouville* equation. Suppose $y(x)$ is required to satisfy the homogeneous *separated* boundary conditions

$$k_1 y(a) + k_2 y'(a) = 0 \quad (2)$$

*J.C.F. Sturm (1803-1855), Swiss mathematician.

** J. Liouville (1809-1882), French mathematician.

and

$$l_1 y(b) + l_2 y'(b) = 0 \tag{3}$$

where k_1, k_2, l_1, l_2 are real constants independent of λ. The conditions (2), (3) are called separated because (2) is satisfied at $x = a$ and (3) is satisfied at $x = b$. Further assume that *both* k_1 and k_2 are not zero in (2) and also *both* l_1 and l_2 are not zero in (3).

A *regular Sturm-Liouville problem* consists of the Sturm-Liouville equation (1) and the homogeneous boundary conditions (2) and (3).

The obvious solution $y(x) = 0$ is known as a trivial solution and is of no interest. Only non-trivial solutions $y \neq 0$ are of practical use. Although $p(x)$, $q(x)$, $r(x)$, a, b, k_1, k_2, m_1, m_2 are all specified, λ is a free parameter. The value of λ for which a non-trivial solution $y(x)$ exists is known as *eigen value or characteristic value* or *eignwerte* of the Sturm-Liouville problem and $y(x)$ is known as the *eigen function* or *characteristic function*. Thus the Sturm-Liouville problem in an eignvalue problem. For any $c \neq 0, cy(x)$, is also an eigen function of the problem since equation (1) and the boundary conditions are linear. *Spectrum* of the problem is the set of eigenvalues.

In *singular* Sturm-Liouville problem, at least one of the regularity conditions is violated as follows:

(a) $r(a) = 0$ and $r(b) \neq 0$ (4)
Then the first boundary condition (2) at 'a' is dropped while (3) still holds, and y is bounded at 'a'. Omitting of (2) amounts to $k_1 = k_2 = 0$.

(b) $r(b) = 0$ and $r(a) \neq 0$. (5)
Then the second boundary condition (3) at 'b' is dropped while (2) still holds and y is bounded at b. Omitting of (3) amounts to $l_1 = l_2 = 0$.

(c) $r(a) = r(b) = 0$ (6)
Then both the boundary conditions (2) and (3) are dropped and y is bounded at a and b.

In *periodic* Sturm-Liouville problem $r(a) = r(b) \neq 0$ and the boundary conditions (2) and (3) are replaced by *non-separated periodic boundary conditions* given by

$$y(a) = y(b) \tag{7}$$

$$y'(a) = y'(b) \tag{8}$$

Orthogonality of Eigen Functions

Sturm-Liouville Theorem

The eigen functions $y_m(x)$ and $y_n(x)$ corresponding to *distinct* eigenvalues λ_m and λ_n of a regular (or singular or periodic) Sturm-Liouville problem are orthogonal on $[a, b]$ with respect to the weight function $P(x)$ i.e.,

$$\int_a^b P(x) y_m(x) y_n(x) dx = 0 \text{ when } m \neq n \tag{9}$$

Proof: I consider a regular Sturm-Liouville problem defined by (1), (2), (3). Since $y_m(x)$ and $y_n(x)$ are solutions of (1), we have

$$(r\, y_n')' + (q + \lambda_n p) y_n = 0 \tag{10}$$

and

$$(r\, y_m')' + (q + \lambda_m p) y_m = 0 \tag{11}$$

Multiplying (10) by y_m and (11) by y_n and subtracting, we get

$$(\lambda_m - \lambda_n) p y_m y_n = y_m (r\, y_n')' - y_n (r y_m')'$$
$$= \left[y_m r y_n' - y_n r y_m'\right]'$$
$$= [r(y_m y_n' - y_n y_m')]'$$

Integrating both sides w.r.t. x on $[a, b]$,

$$(\lambda_m - \lambda_n) \int_a^b P y_m y_n dx$$
$$= \int_a^b \frac{d}{dx}[r(y_m y_n' - y_n y_m']dx$$
$$= (r)[y_m y_n' - y_n y_m']\Big|_{x=a}^{b}$$

$= r(b)[y_m(b) y_n'(b) - y_n(b) y_m'(b)]$
$- r(a)[y_m(a)\, y_n'(a) - y_n(a) y_m'(a)]$

Introducing the Wronskian of y_n, y_m we have

$$\Delta(x) = \begin{vmatrix} y_n(x) & y_n'(x) \\ y_m(x) & y_m'(x) \end{vmatrix}$$

Thus

$$(\lambda_m - \lambda_n) \int_a^b P(x) y_m(x) y_n(x) dx$$
$$= r(b)\Delta(b) - r(a)\Delta(a) \tag{12}$$

Since $y_n(x)$ and $y_m(x)$ satisfy the boundary condition (2) we get

$$k_1 y_n(a) + k_2 y_n'(a) = 0$$

$$k_1 y_m(a) + k_2 y'_m(a) = 0$$

For a regular problem, both k_1 and k_2 are not zero. So necessary condition for the existance of a non-trivial solution (where *both* k_1 and k_2 are *not* zero) for the above homogeneous equations in k_1, k_2 is that the coefficient determinant should be zero i.e.

$$\begin{vmatrix} y_n(a) & y'_n(a) \\ y_m(a) & y'_m(a) \end{vmatrix} = \Delta(a) = 0 \qquad (13)$$

Similarly from the regular boundary condition (3) for which *both* l_1 and l_2 are not zero, we get

$$\begin{vmatrix} y_n(b) & y'_n(b) \\ y_m(b) & y'_m(b) \end{vmatrix} = \Delta(b) = 0 \qquad (14)$$

Note that $(\lambda_n - \lambda_m) \neq 0$ since λ_m, λ_n are distinct. Now using (13) and (14), $\Delta(a) = \Delta(b) = 0$, the RHS of (12) is zero, so we have

$$\int_a^b P(x) y_m(x) y_n(x) dx = 0 \qquad (9)$$

which is the condition for orthogonality of y_m and y_n w.r.t. $P(x)$ over $[a, b]$.

II. Consider the three cases in the singular problem

(a) $r(a) = 0$. Since (3) is valid we have $\Delta(b) = 0$ from (14). So RHS of (12) is zero and hence we get (9).

(b) $r(b) = 0$. Since (2) is valid, from (13) we have $\Delta(a) = 0$. Again RHS of (12) is zero, hence we get (9).

(c) When $r(a) = r(b) = 0$ from (12), condition (9) follows.

III Periodic Problem: Since $r(a) = r(b)$, the RHS of (12) reduces to $r(b)[\Delta(b) - \Delta(a)]$. Now

$$\begin{aligned} \Delta(b) - \Delta(a) &= [y_m(b) \cdot y'_n(b) - y_n(b) y'_m(b)] \\ &\quad - [y_m(a) y'_n(a) - y_n(a) \cdot y'_m(a)] \\ &= [y_m(b) y'_n(b) - y_m(a) \cdot y'_n(a)] + [y_n(a) y'_m(a) \\ &\quad - y_n(b) y'_m(b)] \end{aligned}$$

Since $y(a) = y(b)$ and $y'(a) = y'(b)$ we have $\Delta(b) - \Delta(a) = 0$. Thus RHS of (12) is zero and hence the result.

Corollary: Eigenvalues and eigen functions of the Sturm-Liouville Theorem are real.

Proof: Suppose the eigenvalue $\lambda = \alpha + i\beta$ is complex; with y as the corresponding eigen function. Then $\bar{\lambda}$ is the eigen value with $\bar{y}$ as the corresponding eigen function. Taking conjugate of (1), (2), (3), we get

$$\overline{\frac{d}{dx}\left[r(x)\frac{dy}{dx}\right] + [q(x) + \lambda P(x)]y} = 0$$

$$\frac{d}{dx}\left[r(x)\frac{d\bar{y}}{dx}\right] + [q(x) + \bar{\lambda} P(x)]\bar{y} = 0$$

since $r(x)$, $q(x)$, $P(x)$ are all real and

$$k_1 \bar{y}(a) + k_2 \bar{y}'(a) = 0$$

$$l_1 \bar{y}(b) + l_2 \bar{y}'(b) = 0$$

Thus the nontrivial solution $\bar{y}$ is the eigen function corresponding to the eigenvalue $\bar{\lambda}$. If $\beta \neq 0$ then λ and $\bar{\lambda}$ are distinct. So by the Sturm-Liouville theorem, y and $\bar{y}$ should be orthogonal on $[a, b]$ w.r.t. the weight function $P(x)$. Thus

$$\int_a^b P(x) y(x) \bar{y}(x) dx = 0$$

$$\text{or} \quad \int_a^b P(x)(u^2 + v^2) dx = 0$$

But since $P(x) > 0$ and $y\bar{y} = u^2 + v^2 > 0$ so the above integral should have positive value. This contradiction is due to the wrong assumption that $\beta \neq 0$. So $\beta = 0$ or λ is real. Thus the eigen function $y(x)$ is also real except for the possible imaginary *constant* multiplicative factor.

WORKED OUT EXAMPLES

Laguerre Equation

Example 1: Express the Laguerre differential equation $xy'' + (1 - x)y' + ny = 0$, $x \neq 0$ in the form of a Sturm-Liouville equation.

Solution: Here $a_0(x) = x$, $a_1(x) = 1 - x$, $a_2(x) = 0$, $a_3(x) = n$. Dividing by x the given equation takes the form

$$y'' + \left(\frac{1-x}{x}\right) y' + \frac{ny}{x} = 0$$

Then

$$r(x) = e^{\int \frac{1-x}{x} dx} = e^{\ln x - x} = xe^{-x}$$

Multiplying throughout by the integrating factor $r(x) = xe^{-x}$, we have

$$xe^{-x}y'' + \frac{(1-x)}{x} \cdot xe^{-x} \cdot y' + \frac{ny}{x} \cdot xe^{-x} = 0$$

$$\frac{d}{dx}[xe^{-x}y'] + (ne^{-x})y = 0$$

which is the Sturm-Liouville equation with $r(x) = xe^{-x}$, $q(x) = 0$, $P(x) = e^{-x}$, $\lambda = n$.

Self-Adjoint

Example 2: Express the Sturm-Liouville equation using the linear differential operator L defined by $L = \frac{d}{dx}\left[r(x)\frac{d}{dx}\right] + q(x)$. Show that L is a self-adjoint operator.

Solution: The Sturm-Liouville equation

$$\frac{d}{dx}\left[r(x)\frac{dy}{dx}\right] + [q(x) + \lambda P(x)]y = 0$$

in terms of the operator L is

$$L[y(x)] + \lambda P(x)y(x) = 0$$

For a general second-order linear differential operator L defined by

$$L[y(x)] = A(x)y'' + B(x)y' + C(x)y$$

its adjoint operator L^* is defined as

$$L^*[y(x)] = (Ay)'' - (By)' + Cy$$

$$= (A'y + Ay')' - B'y - By' + Cy$$

$$= A''y + 2A'y' + Ay'' - B'y - By' + Cy$$

$$= Ay'' + (2A' - B)y' + [A'' - B' + C]y$$

Now L is said to be self-adjoint operator if $L = L^*$. So the conditions for self-adjoint operator is $B = 2A' - B$ and $C = A'' - B' + C$ or $A' = B$. Now for the given operator L, we have

$$L(y) = \frac{d}{dx}\left[r(x)\frac{dy}{dx}\right] + q(x)$$

$$= r(x)y'' + r'(x)y' + q(x)$$

We have $A(x) = r(x)$, $B(x) = r'(x)$, $C = q(x)$. Since $A'(x) = r'(x) = B(x)$, the operator L is a self adjoint operator.

Regular

Example 3: (a) Find the characteristic values and normalized characteristic functions of the Sturm-Liouville problem $(xy')' + \frac{\lambda}{x}y = 0$, with boundary $y(1) = 0$, $y(b) = 0$ on the interval $1 < x < b$ (b) Expand a piecewise differentable function $f(x)$ defined in the interval $1 < x < b$ in terms of the characteristic functions obtained in (a). Determine the generalized Fourier series when (c) $f(x) = 1$ (d) $f(x) = x$.

Solution: The given equation $x^2y'' + xy' + \lambda y = 0$ is an Euler-Cauchy equation which can be converted to DE with constant coefficients by the substitution $x = e^t$. The $xDy = \mathcal{D}y$, $x^2D^2y = \mathcal{D}(\mathcal{D}-1)y$ where $D = \frac{d}{dx}$ and $\mathcal{D} = \frac{d}{dt}$. Then the transformed equation with t as the independent variable is

$\mathcal{D}(\mathcal{D}-1)y + \mathcal{D}y + \lambda y = 0$ or $\mathcal{D}^2y + \lambda y = 0$ i.e., $\frac{d^2y}{dt^2} + \lambda y = 0$ in the interval $\ln 1 < \ln x < \ln b$, $0 < t < \ln b$ with the boundary conditions $y(t = 0) = 0$, $y(t = \ln b) = 0$. The general solution of DE is

$$y(t) = c_1 \cos\sqrt{\lambda}t + c_2 \sin\sqrt{\lambda}t$$

Since $0 = y(t = 0) = c_1 \cdot 1 + c_2 \cdot 0$, we get $c_1 = 0$. Since

$$0 = y(t = \ln b) = c_2 \cdot \sin\sqrt{\lambda}\ln b,$$

We get

$$\sin\sqrt{\lambda}\ln b = 0$$

because $c_2 \neq 0$ (otherwise we get $c_1 = 0$, $c_2 = 0$, leading to trivial solution).

$$\therefore \quad \sqrt{\lambda}\ln b = \pi n$$

Thus the characteristic values are given by

$$\lambda_n = \left(\frac{\pi n}{\ln b}\right)^2 \quad \text{for } n = 1, 2, 3, \ldots$$

The corresponding characteristic functions are

$$y_n(t) = c_2 \sin\sqrt{\lambda}t = \sin\left(\frac{\pi n}{\ln b}\right)t$$

or

$$y_n(x) = \sin\left(\frac{\pi n \ln x}{\ln b}\right)$$

where c_2 is taken as 1 for convenience. Here the weight function is $P(x) = \frac{1}{x}$.
The square of norm of $y_n(x)$ is

$$||y_n(x)||^2 = \int_1^b \frac{1}{x}\cdot\sin^2\left(\frac{\pi n \ln x}{\ln b}\right)dx$$

$$= \int_0^{\ln b} \sin^2\left(\frac{\pi n t}{\ln b}\right)dt$$

$$= \ln b \int_0^1 \sin^2(\pi n y)dy = \frac{\ln b}{2}$$

Recall that

$$\int_0^c \sin\frac{m\pi x}{c}\cdot\sin\frac{n\pi x}{c}dx = \begin{cases} 0 \text{ for } m \neq n \\ \frac{c}{2}\text{for } m = n \end{cases}$$

Here $t = \ln x$, $y = \frac{t}{\ln b}$. Thus norm is $\sqrt{\frac{\ln b}{2}}$. So the normalized characteristic functions are

$$\phi_n(x) = \frac{y_n(x)}{||y_n(x)||} = \sqrt{\frac{2}{\ln b}}\sin\left(\frac{\pi n \ln x}{\ln b}\right)$$

for $n = 1, 2, 3, \ldots$. Now the generalized Fourier series expansion of $f(x)$ in terms of the normalized characteristic functions $\phi_n(x)$ of the given Sturm-Liouville problem is

$$f(x) = \sum_{n=1}^{\infty} A_n\phi_n(x) = \sum_{n=1}^{\infty} A_n\sqrt{\frac{2}{\ln b}}\cdot\sin\left(\frac{\pi n \ln x}{\ln b}\right)$$

valid in the interval $1 < x < b$. To determine the unknown Fourier constants A_n's multiple both sides by $\frac{1}{x}\sin\left(\frac{\pi m \ln x}{\ln b}\right)$ and integrate from w.r.t. x from 1 to b. Then

$$\int_1^b \frac{1}{x} f_1(x)\cdot\sin\left(\frac{\pi m \ln x}{\ln b}\right)dx$$

$$= \sum_{n=1}^{\infty} A_n\sqrt{\frac{2}{\ln b}}\int_1^b \frac{1}{x}\cdot\sin\left(\frac{\pi n \ln x}{\ln b}\right)\cdot\sin\frac{(\pi m \ln x)}{\ln b}dx$$

where we have assumed termwise integration. Note that $\phi_n(x)$ are orthogonal functions in the interval $1 < x < b$ w.r.t. the weight function $\frac{1}{x}$ i.e.,

$$\int_1^b \frac{1}{x}\phi_n(x)\phi_m(x)dx$$

$$= \int_1^b \frac{1}{x}\cdot\sin\left(\frac{\pi n \ln x}{\ln b}\right)\sin\left(\frac{\pi m \ln x}{\ln b}\right)dx$$

$$= \begin{cases} 0 & \text{for } m \neq n \\ \frac{\ln b}{2} & \text{for } m = n \end{cases}$$

Thus for $m = n$, in the RHS all the coefficients of A_n's will vanish except when $m = n$. So the Fourier coefficients are

$$\sqrt{\frac{2}{\ln b}}\left(\frac{\ln b}{2}\right)A_n = \int_1^b \frac{1}{x}\cdot f(x)\cdot\sin\left(\frac{\pi n \ln x}{\ln b}\right)dx$$

$$\text{or } A_n = \sqrt{\frac{2}{\ln b}}\int_0^b \frac{1}{x}f(x)\cdot\sin\left(\frac{\pi n \ln x}{\ln b}\right)dx$$

for $m = 1, 2, 3, \ldots$

(c) Expansion when $f(x) = 1$. In this case

$$A_n = \sqrt{\frac{2}{\ln b}}\times\int_0^b \frac{1}{x}\cdot 1\sin\left(\frac{n\pi \ln x}{\ln b}\right)dx$$

$$A_n = \sqrt{\frac{2}{\ln b}}\times\int_0^{\ln b}\sin\left(\frac{n\pi t}{\ln b}\right)dt$$

where $t = \ln x$.

$$A_n = \sqrt{\frac{2}{\ln b}}\times\left[-\left(\frac{\ln b}{n\pi}\right)\cos\left(\frac{n\pi t}{\ln b}\right)\right]\Bigg|_{t=0}^{\ln b}$$

$$A_n = \sqrt{\frac{2}{\ln b}}\left(\frac{\ln b}{n\pi}\right)(1-\cos n\pi)$$

$$= \frac{\sqrt{2\ln b}}{n\pi}(1-\cos n\pi)$$

$$f(x) = 1 = \sqrt{2\ln b}\cdot\sum_{n=1}^{\infty}\frac{1}{n\pi}(1-\cos n\pi)\sin\left(\frac{\pi n \ln x}{\ln b}\right)$$

(d) When $f(x) = x$. In this case

$$A_n = \sqrt{\frac{2}{\ln b}}\int_1^b \frac{1}{x}\cdot x\cdot\sin\left(\frac{n\pi \ln x}{\ln b}\right)dx$$

$$= \sqrt{\frac{2}{\ln b}}\cdot\int_0^{\ln b} e^t\cdot\sin\left(\frac{\pi n t}{\ln b}\right)dt \text{ where } t = \ln x$$

$$= \sqrt{\frac{2}{\ln b}}\cdot b\ln b\int_0^1 e^y\cdot\sin(\pi n y)dy \text{ where } y = \frac{t}{\ln b}$$

$$= b\sqrt{2\ln b}\left[\frac{e^y}{1+\pi^2 n^2}(1\cdot\sin\pi ny - \pi n\cos\pi ny)\right]\Bigg|_{y=0}^{1}$$

$$= \frac{b\sqrt{2\ln b}}{1+\pi^2 n^2}[e(-\pi n(-1)^n - \pi n(-1)]$$

$$A_n = \frac{b\pi n\sqrt{2\ln b}}{1+\pi^2 n^2}[e(-1)^n + 1]$$

$$\therefore f(x) = x = \sum_{n=1}^{\infty}\frac{2b}{1+\pi^2 n^2}[1+e(-1)^n]\sin\left(\frac{\pi n\ln x}{\ln b}\right)$$

Singular

Example 4: Expand $f(x) = 1 - x$ on $0 < x < 1$ in terms of the eigen functions of the Sturm-Liouville problem $(1-x^2)y'' - 2xy' + \lambda y = 0$ with $y'(0) = 0$ and $y(1)$ is bounded. Write the first three terms of the expansion.

Solution: The given DE is Legendre's equation on $0 < x < 1$ which is singular at the end point $x = 1$. By the substitution $y(x) = \sum_{k=0}^{\infty} a_k x^k$, we get the general solution as

$$y(x) = a_0\left[1 - \frac{\lambda}{2}x^2 - \frac{(6-\lambda)\lambda}{24}x^4 - \frac{(20-\lambda)(6-\lambda)}{720}\lambda x^6 \dots\right] +$$

$$+ a_1\left[x + \frac{(2-\lambda)}{6}x^3 + \frac{(12-\lambda)(2-\lambda)}{120}x^5 + \dots\right]$$

$$y(x) = a_0 y_1(x) + a_1 y_2(x)$$

Differentiating w.r.t. x, we have

$$y'(x) = a_0 y_1'(x) + a_1 y_2'(x)$$

Now
$y_1'(x) = 0 - \lambda x - \frac{\lambda(6-\lambda)}{6}x^3 - \frac{\lambda(20-\lambda)(6-\lambda)}{120}x^5 + \cdots$
So $y_1'(0) = 0$
Also $y_2'(x) = 1 + \frac{2-\lambda}{2}x^2 + \frac{(12-\lambda)(2-\lambda)}{24}x^4 + \cdots$
So $y_2'(0) = 1$
From the boundary condition

$$0 = y'(0) = a_0 y_1'(0) + a_1 y_2'(0) = a_0\cdot 0 + a_1\cdot 1$$

$$\therefore \qquad a_1 = 0$$

Thus

$$y(x) = a_0 y_1(x) = P_{2n}(x) \quad \text{for } n = 0, 1, 2, \dots$$

where $P_{2n}(x)$ is Legendre polynomial with even powers. So the eigen functions of the Sturm-Liouville problem are $P_{2n}(x)$ for $n = 0, 1, 2, \dots$ which are orthogonal w.r.t. the weight function 1. Norm of $P_{2n}(x)$ is $\sqrt{\int_0^1 P_{2n}(x)P_{2n}(x)dx} = \sqrt{\frac{1}{2(2n)+1}}\sqrt{\frac{1}{4n+1}}$. Normalized eigen functions are $\sqrt{4n+1}P_{2n}(x)$. The eigen-function expansion of the given function $f(x) = 1 - x$ on $0 < x < 1$ is given by

$$f(x) = 1 - x = \sum_{n=0}^{\infty} a_n P_{2n}(x)$$

where $a_n = (4n+1)\int_0^1 (1-x)P_{2n}(x)dx$
Now for $n = 0$, $P_0(x) = 1$, so

$$a_0 = 1\cdot\int_0^1 (1-x)\cdot 1\cdot dx = \left(x - \frac{x^2}{2}\right)\Bigg|_{x=0}^{1} = \frac{1}{2}$$

For $n = 1$, $P_2(x) = \frac{1}{2}(3x^2 - 1)$, so

$$a_1 = 5\int_0^1 (1-x)\frac{1}{2}(3x^2-1)dx$$

$$= \frac{5}{2}\left[\frac{3x^3}{3} - x - \frac{3x^4}{4} + \frac{x^2}{2}\right]\Bigg|_{x=0}^{1} = -\frac{5}{8}$$

For $n = 2$, $P_4(x) = \frac{1}{8}(35x^4 - 30x^2 + 3)$ so

$$a_2 = 9\int_0^1 (1-x)\cdot\frac{1}{8}(35x^4 - 30x^2 + 3)dx$$

$$= \frac{9}{8}\left[35\frac{x^5}{5} - 30\frac{x^3}{3} + 3x - 35\frac{x^6}{6} + 30\frac{x^4}{4} - \frac{3x^2}{2}\right]\Bigg|_{x=0}^{1}$$

$$= \frac{9}{48}.$$

Thus the expansion of $f(x) = 1 - x$ in terms of the eigen functions is

$$f(x) = 1 - x = \frac{1}{2}P_0(x) - \frac{5}{8}P_2(x) + \frac{9}{48}P_4(x)\dots,$$

Periodic

Example 5: Find the eigenvalues and eigen functions of the periodic Sturm-Liouville problem $y'' + \lambda^2 y = 0$, $y(0) = y(2L)$ and $y'(0) = y'(2L)$. Verify orthogonality by direct calculations.

Solution: The general solution of $y'' + \lambda^2 y = 0$ is $y(x) = c_1 \cos \lambda x + c_2 \sin \lambda x$. Using the periodic boundary condition $y(0) = y(2L)$ we get $c_1 \cdot 1 + c_2 \cdot 0 = y(0) = y(2L) = c_1 \cos 2L\lambda + c_2 \sin 2L\lambda$. This is valid if $L\lambda = \pi n$ so $\lambda = \frac{\pi n}{L}$. Now $y'(x) = -c_1\lambda \sin \lambda x + c_2 \lambda \cos \lambda x$. Using the second condition $y'(0) = y'(2L)$, we have $-c_1\lambda \cdot 0 + c_2\lambda \cdot 1 = y'(0)$ $= y'(2L) = -c_1\lambda \sin 2L\lambda + c_2\lambda \cos 2L\lambda$ which is valid for $L\lambda = \pi n$ or $\lambda = \frac{\pi n}{L}$. Thus the eigenvalues are $\lambda = \frac{\pi n}{L}$ for $n = 0, 1, 2, \ldots$ and the corresponding eigen functions are $\cos \frac{\pi n x}{L}$ and $\sin \frac{\pi n x}{L}$ for $n = 0, 1, 2, \ldots$. These eigen functions are orthogonal w.r.t. the weight function $P(x) = 1$ since it is known

$$\int_0^{2\pi} \cos \frac{n\pi x}{L} \cdot \cos \frac{m\pi x}{L} dx = 0 \text{ for } m \neq n$$

and

$$\int_0^{2\pi} \sin \frac{n\pi x}{L} \cdot \sin \frac{m\pi x}{L} dx = 0 \text{ when } m \neq n$$

Example 6: Determine the eigen values and normalized eigen functions of the regular Sturm-Liouville problem

$$(x^2 y')' + \lambda y = 0,\ y(1) = 0,\ y(b) = 0,\ 1 < x < b$$

Solution: The given equation is

$$x^2 y'' + 2xy' + \lambda y = 0$$

Put $y = Y/\sqrt{x}$ then $y' = \frac{1}{\sqrt{x}} Y' - \frac{1}{2x^{3/2}} Y$ and $y'' = \frac{1}{\sqrt{x}} y'' - \frac{1}{2x^{3/2}} Y' - \frac{1}{2x^{3/2}} Y' - \frac{1}{2}\left(-\frac{3}{2}\right) \frac{1}{x^{5/2}} Y$. Then the given equation transforms to $x^2 \left[\frac{1}{\sqrt{x}} Y'' - \frac{1}{x^{3/2}} Y' + \frac{3}{4} \frac{1}{x^{5/2}} Y\right]$ $+2x \left[\frac{1}{\sqrt{x}} Y' - \frac{1}{2x^{3/2}} y\right] + \lambda \cdot \frac{Y}{\sqrt{x}} = 0$

or $x^{3/2} Y'' + x^{1/2} Y' + \left(\lambda - \frac{1}{4}\right) \frac{Y}{\sqrt{x}} = 0$

or $[xY']' + \mu \frac{Y}{x} = 0$ where $\mu = \lambda - \frac{1}{4}$.

From Example 3 above, the eigen values and normalized eigen functions of the present problem are

$$\lambda_n = \mu_n + \frac{1}{4} = \frac{1}{4} + \left(\frac{n\pi}{\log b}\right)^2 \text{ and}$$

$$\phi_n(x) = \sqrt{\frac{2}{x \log b}} \sin\left(\frac{n\pi \log x}{\log b}\right)$$

for $n = 1, 2, 3$

EXERCISE

Reduce each of the following differential equations (1 to 6) to the Sturm-Liouville equation form indicating the weight function $P(x)$.

1. $(1 - x^2)y'' - xy' + n^2 y = 0$ (Chebyshev equation)

Ans. $[(1 - x^2)^{1/2} y']' + \left[\frac{n^2}{(1-x^2)^{1/2}}\right] y = 0$, $P(x) = (1 - x^2)^{-1/2}$

Hint: $r(x) = (1 - x^2)^{1/2}$, $q(x) = 0$, $P(x) = (1 - x^2)^{-1/2}$, $\lambda = n^2$

2. $y'' - 2xy' + 2ny = 0$ (Hermite equation)

Ans. $[e^{-x^2} y']' + [2ne^{-x^2}]y = 0$, $P(x) = e^{-x^2}$

Hint: $r(x) = e^{-x^2}$, $q(x) = 0$, $P(x) = e^{-x^2}$, $\lambda = 2n$

3. $xy'' + 2y' + (x + \lambda)y = 0$

Ans. $(x^2 y')' + (x^2 + \lambda x)y = 0$, $P(x) = x$

4. $y'' + y' \cot x + \lambda y = 0$

Ans. $(y' \sin x)' + \lambda y \sin x = 0$, $P(x) = \sin x$

5. $y'' + ay' + (b + \lambda)y = 0$

Ans. $(e^{ax} y')' + e^{ax}(b + \lambda)y = 0$, $P(x) = e^{ax}$

6. $xy'' + (c - x)y' - ay + \lambda y = 0$

Ans. $(x^c e^{-x} y')' + x^{c-1} e^{-x}(-a + \lambda)y = 0$, $P(x) = x^{c-1} e^{-x}$

Obtain the eigen values λ_n and eigen functions $y_n(x)$ of the given Sturm-Liouville problem. Find the norm and determine $\phi_n(x)$, normalized eigen functions. Verify that the eigen functions are orthogonal w.r.t. the weight function $P(x)$. Finally expand $f(x)$ in generalized Fourier series in terms of these eigen functions in the interval $a < x < b$.

7. $y'' + \lambda y = 0, \quad y(0) = y(L) = 0, \quad f(x) = x, \quad 0 < x < L$

Ans. $\lambda_n = \frac{n^2\pi^2}{L^2}$, $y_n(x) = \sin\left(\frac{n\pi x}{L}\right)$, $P(x) = 1$ norm $= \sqrt{\frac{L}{2}}$, $\phi_n(x) = \sqrt{\frac{2}{L}} \sin \frac{n\pi x}{L}$, $n = 1, 2, 3 \ldots$, $f(x) = x = \frac{2L}{\pi} \sum_{n=1}^{\infty} \frac{(-1)^{n+1}}{n} \sin \frac{n\pi x}{L}$

8. $y'' + \lambda y = 0, 0 < x < L, y'(0) = 0, hy(L) + y'(L) = 0, h > 0$

Ans. $\lambda_n = \alpha_n^2$ when $\tan \alpha_n L = \frac{h}{\alpha_n}, \alpha_n > 0, y_n(x) = \cos \alpha_n x, n = 1, 2, 3, \ldots,$ norm $= \sqrt{\frac{2h}{hL + \sin^2 \alpha_n L}}$,

$\phi_n(x) = \sqrt{\frac{2h}{hL + \sin^2 \alpha_n^2}} \cos \alpha_n x$

9. $(xy')' + \frac{\lambda}{x} y = 0, \quad y'(1) = 0, \quad hy(b) + y'(b) = 0, 1 < x < b; h > 0$

Ans. $\lambda_n = \alpha_n^2, \quad \tan(\alpha_n \log(b)) = \frac{hb}{\alpha_n}, \quad \alpha_n > 0, y_n(x) = \cos(\alpha_n \log x), \quad n = 1, 2, \ldots,$ norm $= \sqrt{\frac{2hb}{hb \log b + \sin^2(\alpha_n \log b)}}$

10. $y'' + \lambda y = 0, y'(0) = 0, y(L) = 0$

Ans. $\lambda_n = \alpha_n^2, \phi_n(x) = \sqrt{\frac{2}{c}}, \cos \alpha_n x, n = 1, 2, \ldots,$ $\alpha_n = \frac{(2n-1)\pi}{2L}$

11. $y'' + \lambda y = 0, y(0) = 0, y'(L) = 0$

Ans. $\lambda_n = \left[(2n+1)\frac{\pi}{2L}\right]^2, n = 0, 1, 2, \ldots, y_n(x) = \sin(2n+1)\frac{\pi x}{2L}$

12. $(xy')' + \frac{\lambda}{x} y = 0, y(1) = 0, y'(e) = 0$

Ans. $\lambda_n = \left[(2n+1)\frac{\pi}{2}\right]^2, \quad n = 0, 1, 2, \ldots, y_n(x) = \sin\left(n + \frac{1}{2}\right)\pi \ln |x|$

13. $y'' + \lambda y = 0, \quad y(1) = 0, \quad y(0) - 2y'(0) = 0,$ $0 < x < 1$ (mixed boundary conditions).

Ans. λ_n are solutions of $\tan\sqrt{\lambda} = -2\sqrt{\lambda}, \sqrt{\lambda_n} \sim (2n-1)\frac{\pi}{2}, \quad \lambda_1 = 3.3731, \quad \lambda_2 = 23.19, \quad \lambda_3 = 62.67, \phi_n(x) = 2\sqrt{\lambda_n} \cos\sqrt{\lambda_n} x + \sin\sqrt{\lambda_n} x$

14. $y'' - 2y' + \lambda y = 0, y(0) = 0, y(\pi) = 0,$ $0 < x < \pi$

Ans. $\lambda_n = 1 + n^2$, for $n = 1, 2, 3, \ldots,$ $y_n(x) = e^x \sin nx$

15. Periodic: $\quad y'' + \lambda y = 0, \quad y(-L) = y(L), y'(-L) = y'(L), -L < x < L$

Ans. $\lambda_0 = 0, \quad y_0(x) = 1, \quad \lambda_n = \left(\frac{n\pi}{L}\right)^2, \quad y_n(x) = \cos\frac{n\pi x}{L}$ and $\sin\frac{n\pi x}{L}$, $n = 1, 2, 3, \ldots$

16. $y'' + \lambda^2 y = 0, y(0) = 0, hy(L) + y'(L) = 0$

Ans. $\lambda_n = \frac{z_n}{L}$ where $\sin z_n = -\alpha z_n \cos z_n, \alpha = \frac{1}{Lh}$, $y_n(x) = \sin \lambda_n x$.

10.6 GRAM*-SCHMIDT** ORTHOGONALIZATION PROCESS

Let $X = [x_1, x_2, \ldots, x_n]^T$ and $Y = [y_1, y_2, \ldots, y_n]^T$ be two vectors of $V_n(R)$, the n-dimensional Euclidean space of all real n-component vectors. *Inner product* of the two vectors X and Y denoted by $X \cdot Y$, is a scalar defined as

$$X \cdot Y = \sum_{i=1}^{\infty} x_i y_i = x_1 y_1 + x_2 y_2 + \ldots + x_n y_n$$

Orthogonal vectors

Two vectors X and Y are said to be orthogonal if $X \cdot Y = 0$ i.e., their inner product is zero.

Norm or length

Magnitude of a vector X, denoted by $||X||$, is defined by a non negative real number

$$||X|| = \sqrt{X \cdot X} = \sqrt{\sum_{i=1}^{n} x_i^2} = \sqrt{x_1^2 + x_2^2 + \ldots + x_n^2}$$

Normalized Vector

It is a unit vector whose norm or magnitude is unity. Thus given any non-zero vector X, we can define $Y = \frac{X}{||X||}$ which is a normalized vector with norm 1.

---

* Jorgen P. Gram (1850-1916).
** Erhardt Schmidt (1876-1959).

Orthonormal Set of Vectors

A set of n non-zero vectors $X_1, X_2, X_3, \ldots, X_n$, *are said to be (mutually) orthogonal* if $X_i X_j = 0$ for $i \neq j$.

This orthogonal set $\{X_i, X_2, \ldots, X_n\}$ can be normalized by dividing these vectors by their corresponding norms $||X_i||$. Now the set of vectors $\left\{\frac{X_1}{||X_1||}, \frac{X_2}{||X_2||}, \ldots, \frac{X_n}{||X_n||}\right\}$ is an orthogonal set of unit vectors. Introducing $Y_i = \frac{X_i}{||X_i||}$, the set of vectors $\{Y_1, Y_2, \ldots, Y_n\}$, is said to be an orthonormal set of vectors. Thus for an orthonormal set

$$Y_i \cdot Y_j = \delta_{ij} = \begin{cases} 0 \text{ for } i \neq j \\ 1 \text{ for } i = j \end{cases}$$

Gram-Schmidt orthogonalization Process

Let $X_1, X_2, \ldots, X_n$ be a given non-orthogonal basis of $V_n(R)$. Then an orthonormal basis of vectors $\left\{z_i = \frac{Y_i}{||Y_i||}\right\}$ can be constructed using the Gram-Schmidth orthogonalization process as follows. Choose $Y_1 = X_1$

Put $Y_2 = X_2 + a \cdot Y_1$

Since Y_1 and Y_2 are orthonormal, we have

$Y_1 \cdot Y_2 = Y_1 \cdot (X_2 + ay_1) = Y_1 \cdot X_2 + ay_1 \cdot Y_1 = 0$. So

$$a = -\frac{Y_1 \cdot X_2}{Y_1 \cdot Y_1}$$

Thus $Y_2 = X_2 - \frac{Y_1 \cdot X_2}{Y_1 \cdot Y_1} Y_1$.

Now put $Y_3 = X_3 + bY_2 + cY_1$

Since Y_1, Y_2, Y_3 are mutually orthogonal we have

$$Y_1 \cdot Y_3 = Y_1 \cdot (X_3 + by_2 + XY_1) = 0 \text{ and}$$

or $Y_1 \cdot X_3 + bY_1 \cdot Y_2 + cY_1 \cdot Y_1 = 0$

or $\quad c = \frac{-Y_1 \cdot X_3}{Y_1 \cdot Y_1}$

and $Y_2 \cdot Y_3 = Y_2 \cdot (X_3 + bY_2 + cY_1) = 0$

$$= Y_2 \cdot X_3 + bY_2 \cdot Y_2 + cY_2 \cdot Y_1 = 0$$

$$\text{or} \qquad b = \frac{-Y_2 \cdot X_3}{Y_2 \cdot Y_2}$$

Thus substituting values of b and c, we get

$$Y_3 = X_3 - \frac{Y_2 \cdot X_3}{Y_2 \cdot Y_2} Y_2 - \frac{Y_1 \cdot X_3}{Y_1 \cdot Y_1} Y_1$$

In a similar way, the remaining vectors can be obtained $Y_j = X_j - \sum_{i=1}^{j-1} \frac{(X_j \cdot Y_i)}{(Y_i \cdot Y_i)} Y_i$ for $j = 1, 2, 3, \ldots N$.

Define $z_i = \frac{Y_i}{||Y_i||}$ then $\{z_i\}$ form an orthonormal basis.

Expansion in Terms of Orthogonal Basis

Theorem: *If $\{e_1, e_2, \ldots e_k\}$ is an orthogonal basis, then any given vector $\overline{u}$ can be expanded in terms of the vectors $e_1, e_2, \ldots e_k$ as*

$$\overline{u} = \frac{u \cdot e_1}{e_1 \cdot e_1} e_1 + \ldots + \frac{u \cdot e_k}{e_k \cdot e_k} e_k = \sum_{j=1}^{k} \frac{(u \cdot e_j)}{(e_j \cdot e_j)} e_j$$

Further if $\{\hat{e}_1, \hat{e}_2, \ldots, \hat{e}_k\}$ is an orthonormal basis then

$$\overline{u} = (u \cdot \hat{e}_1)\hat{e}_1 + \ldots + (u \cdot \hat{e}_k)\hat{e}_k = \sum_{j=1}^{k} (u \cdot \hat{e}_j)\hat{e}_j$$

where $\hat{e}_1, \hat{e}_2 \ldots \hat{e}_k$ are unit vectors.

Proof: Suppose u is expanded in terms of $e_1, e_2, \ldots e_k$ as

$$u = \alpha_1 e_1 + \alpha_2 e_2 + \ldots + \alpha_k e_k$$

where $\alpha_1, \alpha_2, \ldots, \alpha_k$ are unknown coefficients. Taking inner product with e_1 we get

$u \cdot e_1 = \alpha_1 e_1 \cdot e_1 + \alpha_2 e_2 \cdot e_1 + \ldots + \alpha_k e_k \cdot e_1$

Since $\{e_1, e_2, \ldots, e_k\}$ is an orthogonal set, we have $e_2 \cdot e_1 = 0, e_3 \cdot e_1 = 0, \ldots, e_k \cdot e_1 = 0$. So

$$\alpha_1 = \frac{u \cdot e_1}{e_1 \cdot e_1}$$

Thus taking inner product with e_j, we get

$$\alpha_j = \frac{u \cdot e_j}{e_j \cdot e_j}$$

Substituting these values we have

$$u = \sum_{j=1}^{k} \alpha_j u_j = \sum_{j=1}^{k} \frac{u \cdot e_j}{e_j \cdot e_j} u_j$$

If e_j's are unit vectors then $e_j \cdot e_j = 1$ and the second result in terms of unit vectors follows.

Gram-Schmidt orthogonalization process for functions

Suppose $f_1(x), f_2(x), \ldots, f_n(x), \ldots$ be a sequence of piecewise continuous non-orthogonal functions defined in the interval $a \le x \le b$. Assume that each of these functions have non-zero norm. Then using Gram-Schmidt orthogonalization process we can construct an orthonormal set of functions $\{\phi_n(x)\}$, $n = 1, 2, 3, \ldots$ w.r.t. a weight function $P(x)$ over (a, b).

I. Choose $\phi_1(x) = \frac{f_1(x)}{||f_1(x)||}$ (1)

where $||f_1(x)||^2 = \int_a^b P(x) f_1^2(x)\, dx$

II Take $g_2(x) = f_2(x) + c\phi_1(x)$ (2)

Determine c using orthogonality between $g_2(x)$ and $\phi_1(x)$. Then

$$\int_a^b P(x) g_2(x) \phi_1(x) dx = 0$$

or

$$\int_a^b P(x)[f_2 + c_1\phi_1]\phi_1 dx = 0$$

$$\text{so}\quad c_1 = -\frac{\int_a^b P f_2 \phi_1 dx}{\int_a^b P\phi_1^2 dx} \tag{3}$$

Substituting c from (3) in (2) we get $g_2(x)$ and therefore

$$\phi_2(x) = \frac{g_2(x)}{||g_2(x)||} \tag{4}$$

and $\phi_1(x)$ form an orthonormal set of functions.

III. Continuing this process we choose $g_3(x) = f_3(x) + c_1\phi_1(x) + c_2\phi_2(x)$ (5)

Since ϕ_1 and g_3 are orthogonal, we have

$$\int_a^b P g_3 \phi_1 dx = \int_a^b P(f_3 + c_1\phi_1 + c_2\phi_2)\phi_1 dx = 0$$

$$\text{so } c_1 = -\frac{\int_a^b P f_3 \phi_1 dx}{\int_a^b P\phi_1^2 dx} \tag{6}$$

Since ϕ_1, ϕ_2 are orthogonal. Similarly since ϕ_2 and g_3 are orthogonal we have

$$\int_a^b P g_3 \phi_2 dx = \int_a^b P(f_3 + c_1\phi_1 + c_2\phi_2)\phi_2 dx = 0$$

$$\text{so}\quad c_2 = -\frac{\int_a^b p f_3 \phi_2 dx}{\int_a^b p \phi_2 \phi_2 dx} \tag{7}$$

Substituting (6) and (7) in (5), $g_3(x)$ is completely determined. Then

$$\phi_3(x) = \frac{g_3(x)}{||g_3(x)||}$$

and ϕ_1, ϕ_2 form an orthonormal set. Continuing this process we get an orthonormal set of functions $\phi_1(x), \phi_2(x), \ldots, \phi_n(x), \ldots$ where $\phi_n(x) = \frac{g_n(x)}{||g_n(x)||}$ and

$$g_n(x) = f_n(x) - \phi(x)\int_a^b P(x) f_n(x)\phi_1(x) dx$$

$$-\phi_2(x)\int_a^b P(x) f_n(x)\phi_2(x) dx - \cdots -$$

$$-\phi_{n-1}(x)\int_a^b P(x) f_n(x)\phi_{n-1}(x) dx$$

or

$$g_n(x) = f_n(x) - \sum_{i=1}^{n-1} \phi_i(x) \int_a^b P(x)\, f_n(x)\, \phi_i(x) dx$$

Here $||g_n(x)||^2 = \int_a^b P(x) g_n^2(x) dx$

WORKED OUT EXAMPLES

Example 1: If $X_1 = [1, 2, 1]^T$, $X_2 = [2, 1, 2]^T$, $X_3 = [2, 1, -4]^T$ find (a) the inner product of each pair (b) the length of each vector (c) a vector orthogonal to both X_1 and X_2 (d) a vector orthogonal to both X_1 and X_3.

Solution: (a) Inner product of $X_1 X_2$ is $X_1^T X_2 = [1, 2, 1]\begin{bmatrix}2\\1\\2\end{bmatrix}$

$$= 1\cdot 2 + 2\cdot 1 + 1\cdot 2 = 2 + 2 + 2 = 6$$

$$X_1^T X_3 = [1\ 2\ 1]\begin{bmatrix}2\\1\\-4\end{bmatrix} = 1\cdot 2 + 2\cdot 1 + 1\cdot(-4) = 0$$

$$X_2^T X_3 = [2\ 1\ 2]\begin{bmatrix}2\\1\\-4\end{bmatrix} = 2\cdot 2 + 1\cdot 1 + 2\cdot(-4) = -3$$

(b) Length of the vector $X_1 = \|X_1\| = \sqrt{X \cdot X}$

$$X \cdot X = [1\ 2\ 1]\begin{bmatrix}1\\2\\1\end{bmatrix}, \text{ so } \|X_1\| = \sqrt{6}$$

Length of $X_2 = \|X_2\| = \sqrt{X_2 \cdot X_2}$

$$= \sqrt{[2\ 1\ 2]\begin{bmatrix}2\\1\\2\end{bmatrix}} = \sqrt{4+1+4} = \sqrt{9} = 3$$

Length of $X_3 = \|X_3\| = \sqrt{X_3 \cdot X_3}$

$$= \sqrt{[2\ 1\ -4]\begin{bmatrix}2\\1\\-4\end{bmatrix}} = \sqrt{4+1+16} = \sqrt{21}$$

(c) Let z_1 be the vector which is orthogonal to both X_1 and X_2. Then

$$[X_1\ \ X_2\ \ 0] = \begin{bmatrix}1 & 2 & 0\\2 & 1 & 0\\1 & 2 & 0\end{bmatrix}$$

The cofactors of the elements of the column of zeros

are $+\begin{vmatrix}2 & 1\\1 & 2\end{vmatrix}, -\begin{vmatrix}1 & 2\\1 & 2\end{vmatrix}, +\begin{vmatrix}1 & 2\\2 & 1\end{vmatrix}$

So $z_1 = [3\ 0\ -3]^T = [1\ 0\ -1]^T$

(d) If z_2 is the vector which is orthogonal to both X_1 and X_3 then the cofactors of the elements of the column of zeros in

$$[X_1 X_3 0] - \begin{bmatrix}1 & 2 & 0\\2 & 1 & 0\\1 & -4 & 0\end{bmatrix} \text{ are}$$

$$+\begin{vmatrix}2 & 1\\1 & -4\end{vmatrix}, -\begin{vmatrix}1 & 2\\1 & -4\end{vmatrix}, +\begin{vmatrix}1 & 2\\2 & 1\end{vmatrix} \text{ i.e., } [-9\ 6\ -3]$$

$$= [3\ \ -2\ \ 1]$$

Example 2: Using Gram-Schmidt process construct an orthonormal set of basis vectors of $V_3(R)$ for the given vectors $X_1 = [1, -1, 0]^T$

$$X_2 = [2, -1, -2]^T, X_3 = [1, -1, -2]^T$$

Solution: Choose $Y_1 = X_1 = [1, -1, 0]^T$. Then

$$Y_2 = X_2 - \frac{Y_1 \cdot X_2}{Y_1 \cdot Y_1} Y_1$$

Here $Y_1 \cdot Y_2 = X_1 \cdot X_2 = [1\ \ -1\ \ 0]\begin{bmatrix}2\\-1\\-2\end{bmatrix}$

$= 2 + 1 + 0 = 3$

and $Y_1 \cdot Y_1 = X_1 \cdot X_1 = [1\ \ -1\ \ 0]\begin{bmatrix}1\\-1\\0\end{bmatrix} =$
$1 + 1 + 0 = 2$

$$\text{so } Y_2 = \begin{bmatrix}2\\-1\\-2\end{bmatrix} - \frac{3}{2}\begin{bmatrix}1\\-1\\-0\end{bmatrix} = \begin{bmatrix}\frac{1}{2}\\\frac{1}{2}\\-2\end{bmatrix}$$

Now $Y_3 = X_3 - \frac{Y_2 \cdot X_3}{Y_2 \cdot Y_2} Y_2 - \frac{Y_1 \cdot X_3}{Y_1 \cdot Y_1} Y_1$

$$\text{Here } Y_2 \cdot X_3 = [\tfrac{1}{2}\tfrac{1}{2} - 2]\begin{bmatrix}1\\-1\\-2\end{bmatrix} = \tfrac{1}{2} - \tfrac{1}{2} + 4 = 4$$

$$Y_2 \cdot Y_2 = \begin{bmatrix}\frac{1}{2}\frac{1}{2} - 2\end{bmatrix}\begin{bmatrix}\frac{1}{2}\\\frac{1}{2}\\-2\end{bmatrix} = \frac{1}{4} + \frac{1}{4} + 4 = \frac{9}{2}$$

$$Y_1 \cdot X_3 = [1\ \ -1\ \ 0]\begin{bmatrix}1\\-1\\-2\end{bmatrix} = 1 + 1 + 0 = 2$$

$$\text{So}\quad Y_3 = \begin{bmatrix}1\\-1\\-2\end{bmatrix} - \frac{4}{\left(\frac{9}{2}\right)}\begin{bmatrix}\frac{1}{2}\\\frac{1}{2}\\-2\end{bmatrix} - \frac{2}{2}\begin{bmatrix}1\\-1\\0\end{bmatrix} = \begin{bmatrix}-\frac{4}{9}\\-\frac{4}{9}\\-\frac{2}{9}\end{bmatrix}$$

$$\text{Now } Y_3 \cdot Y_3 = [-\tfrac{4}{9}\tfrac{-4}{9}\tfrac{-2}{9}]\begin{bmatrix}-\frac{4}{9}\\-\frac{4}{9}\\-\frac{2}{9}\end{bmatrix} = \frac{16}{81} + \frac{16}{81} + \frac{4}{81}$$

$$= \frac{36}{81} = \frac{4}{9}$$

Normalizing the vectors we get the orthonormal set of vectors

$$z_1 = \frac{Y_1}{\|Y_1\|} = \frac{1}{\sqrt{2}}\begin{bmatrix}1\\-1\\0\end{bmatrix}, z_2 = \frac{Y_2}{\|Y_2\|} = \frac{\sqrt{2}}{3}\begin{bmatrix}\frac{1}{2}\\\frac{1}{2}\\-2\end{bmatrix}$$

$$\text{and } z_3 = \frac{Y_3}{\|Y_3\|} = \frac{3}{2}\begin{bmatrix}-\frac{4}{9}\\-\frac{4}{9}\\-\frac{2}{9}\end{bmatrix} = -\begin{bmatrix}\text{46pt]}\frac{2}{3}\\\frac{2}{3}\\\frac{1}{3}\end{bmatrix} = \frac{-1}{3}\begin{bmatrix}2\\2\\1\end{bmatrix}$$

Example 3: Given $X_1 = [1, 1, -1]^T$ and $X_2 = [2, 1, 0]^T$ obtain an orthonormal basis of $V_3(R)$.

Solution: Take $Y_1 = X_1 = [1, 1, -1]^T$. Let us construct Y_2 by Gram-Schmidt orthogonal process. So $Y_2 = X_2 - \frac{Y_1 \cdot X_2}{Y_1 \cdot Y_1} Y_1$. Here $Y_1 \cdot Y_1 = X_1 \cdot X_1 = [1, 1, -1]\begin{bmatrix}1\\1\\-1\end{bmatrix} = 1 + 1 + 1 = 3$ and $Y_1 \cdot X_2 =$

$$X_1 \cdot X_2 = [1 \ \ 1 \ \ -1]\begin{bmatrix}2\\1\\0\end{bmatrix} = 2 + 1 + 0 = 3. \text{ Thus}$$

$$Y_2 = \begin{bmatrix}2\\1\\0\end{bmatrix} - \frac{3}{3}\begin{bmatrix}1\\1\\-1\end{bmatrix} = \begin{bmatrix}1\\0\\1\end{bmatrix}, Y_2 \cdot Y_2 = [1 \ 0 \ 1]\begin{bmatrix}1\\0\\1\end{bmatrix} = 2$$

Normalizing the vectors

$$z_1 = \frac{Y_1}{\|Y_1\|} = \frac{1}{\sqrt{3}}\begin{bmatrix}1\\1\\-1\end{bmatrix}, z_2 = \frac{Y_2}{\|Y_2\|} = \frac{1}{\sqrt{2}}\begin{bmatrix}1\\0\\1\end{bmatrix}$$

The third orthonormal vector z_3 is the vectors of the cofactors of the column of zeros in the matrix

$$[z_1 \quad z_2 \quad 0] = \begin{bmatrix}\frac{1}{\sqrt{3}} & \frac{1}{\sqrt{2}} & 0\\ \frac{1}{\sqrt{3}} & 0 & 0\\ -\frac{1}{\sqrt{3}} & \frac{1}{\sqrt{2}} & 0\end{bmatrix}$$

$$\text{i.e.} \begin{vmatrix}\frac{1}{\sqrt{3}} & 0\\ -\frac{1}{\sqrt{3}} & \frac{1}{\sqrt{2}}\end{vmatrix}, -\begin{vmatrix}\frac{1}{\sqrt{3}} & \frac{1}{\sqrt{2}}\\ -\frac{1}{\sqrt{3}} & \frac{1}{\sqrt{2}}\end{vmatrix}, \begin{vmatrix}\frac{1}{\sqrt{3}} & \frac{1}{\sqrt{2}}\\ \frac{1}{\sqrt{3}} & 0\end{vmatrix}$$

$$\text{or } z_3 = \left[\frac{1}{\sqrt{6}}, \frac{2}{\sqrt{6}}, -\frac{1}{\sqrt{6}}\right]^T$$

Example 4: (a) Expand the vector $u = (9, -2, 4)$ in terms of the orthogonal basis $\{e_1, e_2, e_3\}$ of R^3 where $e_1 = (2, 1, 3)$, $e_2 = (1, -2, 0)$, $e_3 = (6, 3, -5)$ (b) what is the expansion in terms orthonormal vectors $\hat{e}_1, \hat{e}_2, \hat{e}_3$.

Solution: (a) Here $\overline{u} \cdot e_1 = (9, -2, 4) \cdot (2, 1, 3) = 18 - 2 + 12 = 28$,
$\overline{u} \cdot e_2 = (9, -2, 4) \cdot (1, -2, 0) = 9 + 4 = 13$,
$\overline{u} \cdot e_3 = (9, -2, 4) \cdot (6, 3, -5) = 54 - 6 - 20 = +\ 28$. Also $e_1 \cdot e_1 = 4 + 1 + 9 = 14$, $e_2 \cdot e_2 = 1 + 4 + 0 = 5$, $e_3 \cdot e_3 = 36 + 9 + 25 = 70$. Then

$$(9, -2, 4) = \overline{u} = \frac{u \cdot e_1}{e_1 \cdot e_1}e_1 + \frac{u \cdot e_2}{e_2 \cdot e_2}e_2 + \frac{u \cdot e_3}{e_3 \cdot e_3}e_3$$
$$= \frac{28}{14}e_1 + \frac{13}{5}e_2 + \frac{28}{70}e_3$$
$$(9, -2, 4) = \overline{u} = 2e_1 + \frac{13}{5}e_2 + \frac{2}{5}e_3$$

(b) Normal of base vectors are $\|e_1\| = \sqrt{e_1 \cdot e_1} = \sqrt{14}$, $\|e_2\| = \sqrt{5}$, $\|e_3\| = \sqrt{70}$. Then $\hat{e}_1 = \frac{e_1}{\|e_1\|} = \frac{e_1}{\sqrt{14}}$, $\hat{e}_2 = \frac{e_2}{\sqrt{5}}$, $\hat{e}_3 = \frac{e_3}{\sqrt{70}}$
Then $\overline{u} = (u \cdot \hat{e}_1)\hat{e}_1 + (u \cdot \hat{e}_2)\hat{e}_2 + (u \cdot \hat{e}_3)\hat{e}_3$

$$= \frac{28}{\sqrt{14}}\hat{e}_1 + \frac{13}{\sqrt{5}}\hat{e}_2 + \frac{28}{\sqrt{70}}\hat{e}_3$$
$$= 2\sqrt{14}\,\hat{e}_1 + \frac{13}{\sqrt{5}}\hat{e}_2 + \frac{2}{5}\sqrt{70}\,\hat{e}_3$$

Example 5: Using Gram-Schmidt orthogonalization process, construct an orthonormal set for the given functions $f_1(x) = 1$, $f_2(x) = x$, $f_3(x) = x^2$ w.r.t. the weight function e^{-x} over the interval $0 < x < \infty$.

Solution: Take $\phi_1(x) = \frac{f_1(x)}{\|f_1(x)\|}$. Here $f_1(x) = 1$ norm of $f_1(x)$ w.r.t. e^{-x} over $(0, \infty)$ is $\|f_1(x)\| = \int_0^\infty e^{-x} \cdot 1 \cdot dx = \left.\frac{e^{-x}}{-1}\right|_0^\infty = 1$. Thus $\phi_1(x) = \frac{1}{1} = 1$. Now take

$$g_2(x) = f_2(x) + c\phi_1(x)$$

and determine c using the fact that $g_2(x)$ and $\phi_1(x)$ are orthogonal. Then

$$\int_0^\infty e^{-x} \cdot g_2(x) \cdot \phi_1(x)dx = 0$$

$$\int_0^\infty e^{-x}(f_2(x) + c\phi_1(x)\phi_1(x))dx = 0$$

$$\text{So } c = -\frac{\int_0^\infty e^{-x} \cdot x dx}{\int_0^\infty e^{-x} \cdot 1 \cdot 1 \cdot dx} = -\int_0^\infty xe^{-x}dx$$

$$= -\left[x \cdot \frac{e^{-x}}{-1} - 1 \cdot e^{-x}\right]_0^\infty = -1$$

Thus $g_2(x) = f_2(x) + c\phi_1(x) = x - 1 \cdot 1 = x - 1$. Norm of $g_2(x)$ is $\|g_2(x)\|$ where

$$\|g_2(x)\|^2 = \int_0^\infty e^{-x}(x-1)^2dx = \int_0^\infty e^{-x}(x^2 + 1 - 2x)dx$$
$$= 1 \cdot 2! + 1 \cdot 1 + 2 \cdot 1 = 1$$

Then $\phi_2(x) = \frac{g_2(x)}{\|g_2(x)\|} = \frac{x-1}{1} = x - 1.$

Now take

$$g_3(x) = f_3(x) + c_1\phi_1(x) + c_2\phi_2(x)$$

We determine c_1 and c_2 using orthogonality property between g_3, ϕ_1 and g_3, ϕ_2. Then since g_3 and ϕ_1 are orthogonal, we have

$$\int_0^\infty e^{-x} g_3(x)\phi_1(x)dx = 0$$

or $\int_0^\infty e^{-x}(f_3 + c_1\phi_1 + c_2\phi_2)\phi_1 dx = 0$

so $c_1 = -\frac{\int_0^\infty e^{-x} f_3\phi_1 dx}{+\int_0^\infty e^{-x}\phi_1\phi_1 dx} = -\int_0^\infty e^{-x} f_3\phi_1 dx$

$$c_1 = -\int_0^\infty e^{-x}x^2 \cdot 1 \cdot dx = -1 \cdot 2! = -2$$

Also since g_3 and ϕ_2 are orthogonal, we have $\int_0^\infty e^{-x} g_3(x)\phi_2(x)dx = 0$

or $\int_0^\infty e^{-x}(f_3 + c_1\phi_1 + c_2\phi_2)\phi_2 dx = 0$

$$c_2 = -\frac{\int_0^\infty e^{-x} f_3\phi_2 dx}{\int_0^\infty e^{-x}\phi_2\phi_2 dx} = -\int_0^\infty e^{-x} f_3\phi_2 dx$$

$$c_2 = -\int_0^\infty e^{-x}x^2(x-1)dx = -[3! - 2!] = -4$$

Therefore

$$g_3(x) = f_3 + c_1\phi_1 + c_2\phi_2$$
$$= x^2 - 2 \cdot 1 - 4(x-1)$$
$$g_3(x) = x^2 - 4x + 2$$

Now

$$\|g_3(x)\|^2 = \int_0^\infty e^{-x} g_3^2(x)d$$
$$= \int_0^\infty e^{-x}(x^2 - 4x + 2)^2 dx$$
$$= \int_0^\infty e^{-x}(x^4 - 16x^2 + 4 - 8x^3 - 16x + 4x^2)dx$$
$$= 4! - 8 \cdot 3! + 20 \cdot 2! - 16 \cdot 1 + 4 = 4$$
$$\therefore \phi_3(x) = \frac{g_3(x)}{\|g_3(x)\|} = \frac{x^2 - 4x + 2}{4}$$

Thus the three functions

$$\phi_1(x) = 1, \phi_2(x) = x - 1, \phi_3(x) = \frac{x^2 - 4x + 2}{4}$$

which are constructed by Gram-Schmidh orthogonalization using the given function $f_1(x) = 1$, $f_2(x) = x$, $f_3(x) = x^2$ form an othonormal set of functions wrt e^{-x} over $(0, \infty)$.

EXERCISE

1. If $X_1 = [1, 2, 3]^T$ and $X_2 = [2, -3, 4]^T$ find (a) their inner product (b) length of each

Ans. (a) 8 (b) $\sqrt{14}, \sqrt{29}$

2. (a) Prove that $X_1 = \left[\frac{1}{3}, \frac{-2}{3}, \frac{-2}{3}\right]^T$ and $X_2 = \left[\frac{2}{3}, \frac{-1}{3}, \frac{-2}{3}\right]^T$ are orthogonal (b) Find a vector X_3 orthogonal to both X and Y.

Ans. (a) $X_1 \cdot X_2 = 0$

(b) $X_3 = \left[\frac{-2}{3}, \frac{-2}{3}, \frac{1}{3}\right]^T$

Hint: Elements of X_3 are the cofactors of the elements of the column of zeros in the matrix

$$\begin{bmatrix} \frac{1}{3} & \frac{2}{3} & 0 \\ -\frac{2}{3} & -\frac{1}{3} & 0 \\ -\frac{2}{3} & \frac{2}{3} & 0 \end{bmatrix}$$

3. Using Gram-Schmidt process, construct an orthogo-normal basis of $V_3(R)$ given a basis of vectors X_1, X_2, X_3 where
(a) $X_1 = [1, 1, 1]^T, X_2 = [1, -2, 1]^T, X_3 = [1, 2, 3]^T$
(b) $X_1 = [2, 1, 3]^T, X_2 = [1, 2, 3]^T, X_3 = [1, 1, 1]^T$
(c) $X_1 = [1, 0, 1]^T, X_2 = [1, 3, 1]^T, X_3 = [3, 2, 1]$
(d) $X_1 = [2, -1, 0], X_2 = [4, -1, 0], X_3 = [4, 0, -1]$
(e) $X_1 = [1, 1, 0], X_2 = [2, -1, 1], X_3 = [1, 0, 3]$

Ans. (a) $\frac{1}{\sqrt{3}}[1, 1, 1], \frac{1}{\sqrt{6}}[1, -2, 1], \frac{1}{\sqrt{2}}[-1, 0, 1]$
(b) $\frac{1}{\sqrt{14}}[2, 1, 3], \frac{1}{\sqrt{42}}[-4, 5, 1], \frac{1}{\sqrt{3}}[1, 1, -1]$
(c) $\frac{1}{\sqrt{2}}[1, 0, 1], [0, 1, 0], \frac{1}{\sqrt{2}}[1, 0, -1]$
(d) $\frac{1}{\sqrt{5}}[2, -1, 0], \frac{1}{\sqrt{5}}[1, 2, 0], [0, 0, -1]$
(e) $\frac{1}{\sqrt{2}}[1, 1, 0], \frac{1}{\sqrt{22}}[3, -3, 2], \frac{1}{\sqrt{11}}[-1, 1, 3]$

4. Show that $X_1 = (1, 0, 0, 0)$, $X_2 = \left(0, \frac{1}{\sqrt{2}}, 0, \frac{1}{\sqrt{2}}\right)$, $X_3 = \left(0, \frac{1}{\sqrt{2}}, 0 - \frac{1}{\sqrt{2}}\right)$ form an orthonormal set.

5. Find scalars α, β, γ and the vectors u_1, u_2, u_3 such that $u_1 = u, u_2 = u + \alpha v, u_3 = u + \beta v + \gamma w$ is a non-zero orthogonal set. Normalize the set, given

$$u = (1, 3, 0),\ v = (2, 3, 0),\ w = (2, 1, -3)$$

Ans. $u_1 = u = (1, 3, 0)$, $u_2 = u - \frac{10}{11}v = [-\frac{9}{11}, \frac{3}{11}, 0]$,
$u_3 = u - \frac{5}{4}v + \frac{3}{4}w = [0, 0, \frac{-9}{4}]$, normalized vector

$$X_1 = \frac{u_1}{\|u_1\|} = \left(\frac{1}{\sqrt{10}}, \frac{3}{\sqrt{10}}, 0\right),$$

$$X_2 = \frac{u_2}{\|u_2\|} = \left(-\frac{9}{\sqrt{90}}, \frac{3}{\sqrt{90}}, 0\right)$$

$$X_3 = \frac{u_3}{\|u_3\|} = [0, 0, -1]$$

6. Find all non-zero vectors (if any) orthogonal to the following vectors (a) (3, 0, −1) (b) (1, 3, 4, 0) and (2, −1, 0, 5)

Ans. (a) $u = [\frac{\alpha}{3}, \beta, \alpha]$ where α, β are arbitrary
(b) $u = [-15\alpha - 4\beta, 5\alpha - 8\beta, 7\beta, 7\alpha]$ where α, β are arbitrary.

Hint. Assume $u = [\alpha, \beta, \gamma]$ solve for α, β, γ using orthogonality condition $u \cdot u_1 = 0$ etc.

7. Expand $u = (1, 0, 0, 0)$ in terms of orthogonal basis vectors $e_1 = (2, 0, -1, -5)$, $e_2 = (2, 0, -1, 1)$, $e_3 = (0, 1, 0, 0)$, $e_4 = (1, 0, 2, 0)$

Ans. $\frac{1}{15}e_1 + \frac{1}{3} \cdot e_2 + 0 \cdot e_3 + \frac{1}{5}e_4$

8. (a) Expand (4, 3, −3, 6) in terms of orthogonal basics vectors $e_1 = (1, 0, 2, 0)$, $e_2 = (0, 1, 0, 0)$, $e_3 = (-2, 0, 1, 5)$, $e_4 = (-2, 0, 1, -1)$ (b) What is the expansion in terms of unit vectors

Ans. (a) $-\frac{2}{5}e_1 + 3e_2 + \frac{19}{30}e_3 - \frac{17}{6}e_4$
(b) $-\frac{2}{\sqrt{5}}\hat{e}_1 + 3\hat{e}_2 + \frac{19}{\sqrt{30}}\hat{e}_3 - \frac{17}{\sqrt{6}}\hat{e}_4$.

Using Gram-Schmidt orthogonalization process construct an orthonormal set of vectors from the given functions $f_1(x)$, $f_2(x)$ and $f_3(x)$ over the interval (a, b).

9. $f_1(x) = 1, f_2(x) = x, f_3(x) = x^2, 0 < x < 1$.

Ans. $\phi_1(x) = \frac{1}{\sqrt{2}}, \phi_2(x) = \sqrt{\frac{3}{2}}x, \phi_3(x) = \frac{1}{2}\sqrt{\frac{5}{2}}(2x^2 - 1)$

10. $f_1(x) = 1, f_2(x) = x, f_3(x) = x^2, -1 \le x \le 1$

Ans. $\phi_0 = \frac{1}{\sqrt{2}}, \phi_1 = \frac{x}{\sqrt{2/3}}, \phi_2 = \frac{x^2 - \frac{2}{3}}{\sqrt{2/5}}$.

Chapter **11**

Special Functions—Gamma, Beta, Bessel and Legendre

INTRODUCTION

Algebraic function $f(x)$ is obtained by the algebraic operations of addition, subtraction, multiplication, division and square rooting of x polynomial and rational functions are such functions. Transcendental functions include trigonometric functions (sine, cosine, tan) exponential, logarithmic and hyperbolic functions.

Algebraic and transcendental functions together constitute the elementary functions. Special functions (or higher functions) are functions other than the elementary functions such as Gamma, Beta functions (expressed as integrals) Bessel's functions, Legendre polynomials (as solutions of ordinary differential equations). Special functions also include Laguerre, Hermite, Chebyshev polynomials, error function, sine integral, exponential integral, Fresnel integrals, etc.

Many integrals which can not be expressed in terms of elementary functions can be evaluated in terms of beta and gamma functions.

Heat equation, wave equation and Laplace's equation with cylindrical symmetry can be solved in terms of Bessel's functions, with spherical symmetry by Legendre polynomials. We consider Fourier-Legendre series and Fourier-Bessel series. Chebyshev-polynomials which are useful in approximation theory are also presented.

11.1 GAMMA FUNCTION

Gamma function denoted by $\Gamma(p)$ is defined by the improper integral which is dependent on the parameter p,

$$\Gamma(p) = \int_0^\infty e^{-t} t^{p-1}\, dt, \quad (p > 0) \tag{1}$$

Gamma function is also known as Euler's integral of the second kind.

Integrating by parts

$$\Gamma(p+1) = \int_0^\infty e^{-t} t^{p}\, dt$$

$$= -e^{-t} t^p \Big|_0^\infty + p \int_0^\infty e^{-t} t^{p-1}\, dt$$

$$= 0 + p\Gamma(p)$$

Thus $\quad \Gamma(p+1) = p\Gamma(p) \quad (2)$

(2) is known as the functional relation or reduction or recurrence formula for gamma function.

Result:
$\Gamma(n+a) = (n+a-1)(n+a-2)(n+a-3)\cdots a \cdot \Gamma(a)$, n is integer.

By definition

$$\Gamma(1) = \int_0^\infty e^{-t}\, dt = \frac{e^{-t}}{-1}\Big|_0^\infty = 1 \tag{3}$$

By the reduction formula (2),

$$\Gamma(2) = 1 \cdot \Gamma(1) = 1$$

and $\quad \Gamma(3) = 2 \cdot \Gamma(2) = 2 \cdot 1 = 2!$

and in general when p is a positive integer n

$$\Gamma(n+1) = n\Gamma(n) = n(n-1)\,\Gamma(n-1)$$

$$= n(n-1)(n-2)\,\Gamma(n-2)$$

$$= n \cdot (n-1)(n-2)\cdots 3 \cdot 2 \cdot 1 = n!$$

Thus for $p = n$, positive integer

$$\Gamma(n+1) = n! \tag{4}$$

For this reason, Gamma function is regarded as the generalization of the elementary factorial function.

Gamma function for negative values of p i.e., $p < 0$: Rewrite (2) as left-marching recurrence formula,

$$\Gamma(p) = \frac{\Gamma(p+1)}{p} \tag{5}$$

As $p \to 0$, $\quad \Gamma(0) = \lim_{p\to 0} \frac{\Gamma(1)}{p} = \lim_{p\to 0} \frac{1}{p} \to \infty$

Thus $\Gamma(0)$ is undefined and it follows from (5) that $\Gamma(-1), \Gamma(-2), \Gamma(-3)$, etc. are all undefined.

Repeated application of (5) results in

$$\Gamma(p) = \frac{\Gamma(p+1)}{p} = \frac{\Gamma(p+2)}{p(p+1)} = \cdots$$

$$= \frac{\Gamma(p+k+1)}{p(p+1)\cdots(p+k)} \tag{6}$$

Relation (6) is used to find gamma function for $p < 0$ (except at $p = 0, -1, -2, -3, \ldots$).

Hence gamma function is continuous for any $p > 0$ and is discontinuous at $p = 0, -1, -2, -3, \ldots$. Thus $\Gamma(p)$ is defined for all p, except for zero and negative integers.

Standard Results

1. $\Gamma\left(\frac{1}{2}\right) = \sqrt{\pi}$

By definition $\Gamma\left(\frac{1}{2}\right) = \int_0^\infty t^{-\frac{1}{2}} e^{-t}\, dt$, put $t = u^2$ then $\Gamma\left(\frac{1}{2}\right) = 2\int_0^\infty e^{-u^2} du$

$$\Gamma\left(\frac{1}{2}\right) \cdot \Gamma\left(\frac{1}{2}\right) = \left[2\int_0^\infty e^{-u^2} du\right]\left[2\int_0^\infty e^{-v^2} dv\right]$$

$$= 4\int_0^\infty \int_0^\infty e^{-(u^2+v^2)} du\, dv$$

This double integral in the first quadrant is evaluated by changing to polar coordinates $u = r\cos\theta$, $v = r\sin\theta$, $J = r$

$$\left[\Gamma\left(\frac{1}{2}\right)\right]^2 = 4\int_{\theta=0}^{\frac{\pi}{2}} \int_{r=0}^\infty e^{-r^2} r\, dr\, d\theta$$

$$= 4\int_0^{\frac{\pi}{2}} -\frac{1}{2} e^{-r^2}\Big|_{r=0}^\infty d\theta$$

$$= 2\int_0^{\frac{\pi}{2}} d\theta = 2\cdot\theta\Big|_0^{\frac{\pi}{2}} = 2\cdot\frac{\pi}{2} = \pi.$$

Hence $\Gamma\left(\frac{1}{2}\right) = \sqrt{\pi}$

2. $\Gamma\left(\frac{p+1}{q}\right) = qa^{(p+1)/q} \int_0^\infty x^p e^{-ax^q} dx$; p, q, a are positive constants

Put $y = ax^q$ then $dy = aqx^{q-1}dx$

$$\int_0^\infty x^p e^{-ax^q} dx = \int_0^\infty \left[\left(\frac{y}{a}\right)^{\frac{1}{q}}\right]^p e^{-y} \cdot \frac{1}{aqx^{q-1}} dy$$

$$= \left[qa^{(p+1)/q}\right]^{-1} \int_0^\infty y^{(p+1-q)/q} e^{-y} dy$$

$$= \frac{\Gamma\left(\frac{p+1}{q}\right)}{qa^{\frac{(p+1)}{q}}}.$$

3. $\Gamma(n+1) = (m+1)^{n+1}(-1)^n \int_0^1 x^m (\ln x)^n dx$ where n is a positive integer and $m > -1$.

Put $x = e^{-y}$ then $dx = -e^{-y}dy = -x\, dy$

$$\int_0^1 x^m (\ln x)^n dx = \int_0^\infty e^{-my} \cdot (-y)^n e^{-y} dy$$

$$= (-1)^n \int_0^\infty y^n \cdot e^{-(m+1)y} dy,$$

Put $(m+1)y = u$

$$= (-1)^n \int_0^\infty \frac{u^n}{(m+1)^n} \cdot e^{-u} \cdot \frac{du}{m+1}$$

$$= \frac{(-1)^n}{(m+1)^{n+1}} \int_0^\infty e^{-u} \cdot u^n du = \frac{(-1)^n}{(m+1)^{n+1}} \cdot \Gamma(n+1)$$

11.2 BETA FUNCTION

Beta function $\beta(p, q)$ defined by

$$\beta(p, q) = \int_0^1 x^{p-1}(1-x)^{q-1} dx, \; (p > 0, q > 0) \tag{1}$$

is convergent for $p > 0, q > 0$. This function is also known as Euler's integral of the first kind.

Standard Results

1. Symmetry: $\beta(p, q) = \beta(q, p)$ (2)

$$\beta(p, q) = \int_0^1 x^{p-1}(1-x)^{q-1} dx. \quad \text{Put } x = 1-y$$

$$= \int_1^0 (1-y)^{p-1} \cdot y^{q-1}(-dy)$$

$$= \int_1^0 y^{q-1}(1-y)^{p-1}dy = \beta(q, p).$$

2. Beta function in terms of trigonometric functions

$$\beta(p, q) = 2\int_0^{\frac{\pi}{2}} \sin^{2p-1} x \cdot \cos^{2q-1} x\, dx \quad (3)$$

Putting $x = \sin^2\theta$, $dx = 2\sin\theta \cdot \cos\theta\, d\theta$ and $1 - x = \cos^2\theta$,

$$\beta(p, q) = \int_0^1 x^{p-1}(1-x)^{q-1}dx$$

$$= \int_0^{\frac{\pi}{2}} \sin^{2p-2}\theta \cdot \cos^{2q-2}\theta \cdot 2 \times \sin\theta \cdot \cos\theta\, d\theta$$

$$\beta(p, q) = 2\int_0^{\frac{\pi}{2}} \sin^{2p-1}\theta \cdot \cos^{2q-1}\theta\, d\theta$$

$$= 2 \cdot I_{2p-1,2q-1}$$

or $$I_{p,q} = \int_0^{\frac{\pi}{2}} \sin^p\theta \cos^q\theta\, d\theta$$

$$= \frac{1}{2}\beta\left(\frac{p+1}{2}, \frac{q+1}{2}\right), p > -1, q > -1 \quad (4)$$

3. Beta function expressed as an improper integral

$$\beta(p, q) = \int_0^\infty \frac{y^{p-1}}{(1+y)^{p+q}}dy$$

$$= \int_0^\infty \frac{y^{q-1}dy}{(1+y)^{p+q}} \quad (5)$$

Putting $x = \frac{y}{1+y}$ or $y = \frac{x}{1-x}$, limits for y are 0 to ∞.

$$\beta(p, q) = \int_0^1 x^{p-1}(1-x)^{q-1}dx$$

$$= \int_0^\infty \frac{y^{p-1}}{(1+y)^{p-1}} \cdot \left(\frac{1}{1+y}\right)^{q-1} \cdot \frac{dy}{(1+y)^2}$$

$$= \int_0^\infty \frac{y^{p-1}}{(1+y)^{p+q}}dy$$

$$= \int_0^\infty \frac{y^{q-1}}{(1+y)^{q+p}}dy.$$

The last integral follows from symmetry.

4. Relation between β and Γ functions

$$\beta(p, q) = \frac{\Gamma(p)\Gamma(q)}{\Gamma(p+q)} \quad (6)$$

By definition

$$\Gamma(p) = \int_0^\infty x^{p-1}e^{-x}\, dx,$$

Put $x = t^2$, $dx = 2t\, dt$

$$\Gamma(p) = \int_0^\infty t^{2p-2} \cdot e^{-t^2} \cdot 2t\, dt$$

$$= 2\int_0^\infty t^{2p-1} \cdot e^{-t^2}\, dt$$

Then $$\Gamma(p)\,\Gamma(q) = \left[2\int_0^\infty x^{2p-1}e^{-x^2}dx\right] \times \left[2\int_0^\infty y^{2q-1}e^{-y^2}dy\right]$$

Here t, x, y are dummy variables.

$$= 4\int_0^\infty\int_0^\infty x^{2p-1} \cdot y^{2q-1} \cdot e^{-(x^2+y^2)}dxdy.$$

Introduce polar coordinates

$$x = r\cos\theta, \quad y = r\sin\theta.$$

As x, y vary in the first quadrant (i.e., $0 < x < \infty, 0 < y < \infty$), r varies from 0 to ∞ and θ from 0 to $\frac{\pi}{2}$. Jacobian: r

$$\Gamma(p)\,\Gamma(q) = 4\int_0^{\frac{\pi}{2}}\int_0^\infty (r\cos\theta)^{2p-1} \cdot (r\sin\theta)^{2q-1} \times e^{-r^2} \cdot r\, dr\, d\theta$$

$$= 4\left[\int_0^{\frac{\pi}{2}} \sin^{2q-1}\theta \cdot \cos^{2p-1}\theta\, d\theta\right] \times \left[\int_0^\infty e^{-r^2} \cdot r^{2p+2q-1} \cdot dr\right]$$

Using result 2 above in this page and putting $r^2 = t$

$$\Gamma(p)\,\Gamma(q) = 4 \cdot \left[\frac{1}{2}\beta(p, q)\right]\left[\frac{1}{2}\int_0^\infty e^{-t} \cdot t^{p+q-1}dt\right]$$

$\Gamma(p)\,\Gamma(q) = \beta(p, q) \cdot \Gamma(p+q)$, hence the result.

5. $\Gamma(p)\,\Gamma(1-p) = \frac{\pi}{\sin p\pi}, 0 < p < 1$

Put $q = 1 - p$ in (6) and use (5)

$$\frac{\Gamma(p)\Gamma(1-p)}{\Gamma(p+1-p)} = \beta(p, 1-p)$$

$$= \int_0^\infty \frac{x^{p-1}}{1+x} dx = \frac{\pi}{\sin p\pi}$$

(which follows from residue theorem) and since $\Gamma(1) = 1$

6. $\Gamma\left(\frac{1}{2}\right) = \sqrt{\pi}$

Put $p = q = \frac{1}{2}$ in (6) and use (3)

$$\frac{\Gamma\left(\frac{1}{2}\right)\Gamma\left(\frac{1}{2}\right)}{\Gamma\left(\frac{1}{2}+\frac{1}{2}\right)} = \beta\left(\frac{1}{2}, \frac{1}{2}\right) = 2\int_0^{\frac{\pi}{2}} \sin^\circ\theta \cos^\circ\theta \, d\theta$$

$$= 2 \cdot \frac{\pi}{2} = \pi$$

Since $\Gamma(1) = 1$, $\Gamma\left(\frac{1}{2}\right) \cdot \Gamma\left(\frac{1}{2}\right) = \pi$

7. $\int_0^{\frac{\pi}{2}} \sin^n x \, dx = \frac{1}{2}\beta\left(\frac{n+1}{2}, \frac{1}{2}\right) = \frac{\Gamma\left(\frac{n+1}{2}\right)}{\Gamma\left(\frac{n+2}{2}\right)} \frac{\sqrt{\pi}}{2}$

which follows from (4) with $p = n$ and $q = 0$

Similarly with $p = 0$, $q = n$ from (4), we get

8. $\int_0^{\frac{\pi}{2}} \cos^n x \, dx = \frac{1}{2}\beta\left(\frac{1}{2}, \frac{n+1}{2}\right) = \frac{\Gamma\left(\frac{n+1}{2}\right)}{\Gamma\left(\frac{n+2}{2}\right)} \frac{\sqrt{\pi}}{2}$

9. Legendre's duplication formula for Γ function:

$$\sqrt{\pi}\Gamma(2p) = 2^{2p-1} \cdot \Gamma(p)\Gamma\left(p + \frac{1}{2}\right)$$

Putting $p = q$ in (3), we get

$$\beta(p, p) = 2\int_0^{\frac{\pi}{2}} \sin^{2p-1}\theta \cdot \cos^{2p-1}\theta \, d\theta$$

$$= 2\int_0^{\frac{\pi}{2}} (\sin\theta \cdot \cos\theta)^{2p-1} d\theta$$

$$= 2\int_0^{\frac{\pi}{2}} \left(\frac{\sin 2\theta}{2}\right)^{2p-1} d\theta$$

$$= \frac{2}{2^{2p-1}} \int_0^{\frac{\pi}{2}} (\sin 2\theta)^{2p-1} d\theta$$

Put $2\theta = t$,

$$\beta(p, p) = \frac{1}{2^{2p-1}} \cdot \int_0^\pi \sin^{2p-1} t \, dt$$

$$= \frac{1}{2^{2p-1}} \left[2\int_0^{\frac{\pi}{2}} \sin^{2p-1} t \, dt\right]$$

$$\beta(p, p) = \frac{1}{2^{2p-1}} \cdot \beta\left(p, \frac{1}{2}\right)$$

since from (3) with $q = \frac{1}{2}$,

$$\beta(p, \frac{1}{2}) = 2\int_0^{\frac{\pi}{2}} \sin^{2p-1} x \, dx.$$

Using (6), express β in terms of Γ functions

$$2^{2p-1} \cdot \frac{\Gamma(p)\Gamma(p)}{\Gamma(p+p)} = \frac{\Gamma(p)\Gamma\left(\frac{1}{2}\right)}{\Gamma\left(p+\frac{1}{2}\right)}$$

or $$2^{2p-1} \cdot \Gamma(p)\Gamma\left(p + \frac{1}{2}\right) = \Gamma\left(\frac{1}{2}\right)\Gamma(2p)$$

$$= \sqrt{\pi}\Gamma(2p)$$

10. $I = \int_0^{\frac{\pi}{2}} \sin^p\theta \, d\theta = \int_0^{\frac{\pi}{2}} \cos^p\theta \, d\theta =$

a. $\frac{1\cdot3\cdot5\cdots(p-1)}{2\cdot4\cdot6\cdots p} \cdot \frac{\pi}{2}$ if p is an even positive integer and

b. I $\frac{2\cdot4\cdot6\cdots(p-1)}{1\cdot3\cdot5\cdots p}$ if p is an odd positive integer.

a. $\int_0^{\frac{\pi}{2}} \sin^p\theta \, d\theta = \frac{\Gamma\left(\frac{1}{2}(p+1)\right)\Gamma\left(\frac{1}{2}\right)}{2\Gamma\left(\frac{1}{2}(p+2)\right)}$ from result 7.

If $p = 2r$,

$$I = \frac{\Gamma\left(r+\frac{1}{2}\right)\sqrt{\pi}}{2\Gamma(r+1)}$$

$$= \frac{\left(r-\frac{1}{2}\right)\left(r-\frac{3}{2}\right)\cdots\frac{3}{2}\cdot\frac{1}{2}\cdot\Gamma\left(\frac{1}{2}\right)\sqrt{\pi}}{2\cdot r!}$$

$$I = \frac{(2r-1)(2r-3)\cdots3\cdot1}{2r\cdot(2r-2)\cdot(2r-4)\cdots2} \cdot \frac{\pi}{2}$$

b. If $p = 2r+1$,

$$I = \frac{\Gamma(r+1)\,\Gamma\left(\frac{1}{2}\right)}{2\Gamma\left(r+\frac{3}{2}\right)}$$

$$= \frac{r!\sqrt{\pi}}{2\left(r+\frac{1}{2}\right)\left(r-\frac{1}{2}\right)\left(r-\frac{3}{2}\right)\cdots\frac{3}{2}\cdot\frac{1}{2}\Gamma\left(\frac{1}{2}\right)}$$

$$= \frac{2^r\cdot r!}{(2r+1)(2r-1)\cdots5\cdot3\cdot1}$$

$$= \frac{2\cdot4\cdot6\cdots(2r-2)\cdot2r}{1\cdot3\cdot5\cdots(2r-1)(2r+1)}$$

11. $\beta(p,q)=\beta(p+1,q)+\beta(p,q+1)$

By definition

$$\beta(p+1,q)+\beta(p,q+1)$$

$$=\int_0^1 x^p(1-x)^{q-1}dx+\int_0^1 x^{p-1}(1-x)^q\,dx$$

$$=\int_0^1 x^{p-1}(1-x)^{q-1}\left[x+(1-x)\right]dx$$

$$=\int_0^1 x^{p-1}(1-x)^{q-1}dx=\beta(p,q).$$

12. $\beta(m,n)=\frac{(m-1)!\,(n-1)!}{(m+n-1)!}$ for m, n positive integers

From the above result 11

$$\beta(m,n)=\beta(m+1,n)+\beta(m,n+1)$$

Expressing in Γ functions using (6)

$$\beta(m,n)=\frac{\Gamma(m+1)\,\Gamma(n)}{\Gamma(m+n+1)}+\frac{\Gamma(m)\,\Gamma(n+1)}{\Gamma(m+n+1)}$$

Since m and n are positive integers

$$=\frac{m!(n-1)!+(m-1)!n!}{(m+n)!}$$

$$=\frac{(m-1)!(n-1)![m+n]}{(m+n)!}=\frac{(m-1)!(n-1)!}{(m+n-1)!}$$

13. $\int_0^1 x^p(1-x^q)^r dx=\frac{1}{q}\beta\left(\frac{p+1}{q},r+1\right)$

Put $x^q=y$, $qx^{q-1}dx=dy$ or $qy^{\left(\frac{q-1}{q}\right)}dx=dy$

$$\int_0^1 x^p(1-x^q)^r dx=\int_0^1 y^{\left(\frac{p}{q}\right)}(1-y)^r\frac{1}{qy^{(q-1)/q}}\cdot dy$$

$$=\frac{1}{q}\int_0^1 y^{(p-q+1)/q}(1-y)^r dy$$

$$=\frac{1}{q}\beta\left(\frac{p-q+1}{q}+1,r+1\right)$$

$$=\frac{1}{q}\beta\left(\frac{p+1}{q},r+1\right)$$

Note: When $q=1$, $\int_0^1 x^p(1-x)^r\,dx=\beta\,(p+1,r+1)$.

14. $\int_0^1 \frac{x^{m-1}(1-x)^{n-1}}{(b+cx)^{m+n}}dx=\frac{1}{b^n(b+c)^m}\cdot\beta(m,n)$

Put $y=\frac{x(1+a)}{b+cx}$ then $x=\frac{yb}{1+a-cy}$,

$$1-x=\frac{(1+a)-y(c+b)}{(1+a-cy)},\quad b+cx=\frac{b(1+a)}{(1+a-cy)}$$

and $dx=\dfrac{(b+cx)dy}{(1+a-cy)}=\dfrac{b(1+a)dy}{(1+a-cy)^2}$.

Now y varies from 0 to $\frac{1+a}{b+c}=e$ say

$$\int_0^1\frac{x^{m-1}(1-x)^{n-1}}{(b+cx)^{m+n}}dx$$

$$=\int_0^e\left(\frac{yb}{1+a-cy}\right)^{m-1}\left[\frac{(1+a)-y(c+b)}{1+a-cy}\right]^{n-1}$$

$$\times\frac{(1+a-cy)^{m+n}}{\left[b(1+a)\right]^{m+n}}\frac{b(1+a)\,dy}{(1+a-cy)^2}$$

$$=\int_0^e\frac{1}{b^n(1+a)^{m+n-1}}y^{m-1}$$

$$\times\left[(1+a)-y(c+b)\right]^{n-1}dy$$

$$=\frac{1}{b^n}\frac{1}{(1+a)^m}\int_0^e y^{m-1}\left(1-\frac{1}{e}y\right)^{n-1}dy$$

Put $\frac{1}{e}y=t$

$$=\frac{1}{b^n}\frac{1}{(1+a)^m}\int_0^1 (et)^{m-1}(1-t)^{n-1}\cdot e\,dt$$

$$=\frac{1}{b^n}\frac{1}{(1+a)^m}\cdot\left(\frac{1+a}{b+c}\right)^m\cdot\int_0^1 t^{m-1}(1-t)^{n-1}dt$$

$$=\frac{1}{b^n}\frac{1}{(b+c)^m}\cdot\beta(m,n).$$

15. When $c=1$, $b=a$, from above result 14,

$$\int_0^1\frac{x^{m-1}(1-x)^{n-1}}{(a+x)^{m+n}}dx=\frac{1}{a^n(1+a)^m}\beta(m,n).$$

Worked Out Examples

Gamma and Beta functions

Example 1: Compute (a) $\Gamma(4.5)$ (b) $\Gamma(-3.5)$ (c) $\Gamma\left(\frac{1}{4}\right)\Gamma\left(\frac{3}{4}\right)$ (d) $\beta\left(\frac{5}{2},\frac{3}{2}\right)$ (e) $\frac{\Gamma\left(n+\frac{1}{2}\right)}{\Gamma(n+1)}$.

Solution:

a. Using $\Gamma(p+1)=p\Gamma(p)$

$$\Gamma(4.5)=\Gamma(3.5+1)=3.5\Gamma(3.5)=(3.5)(2.5)\Gamma(2.5)$$
$$=(3.5)(2.5)(1.5)(.5)\Gamma(.5)$$
$$=6.5625\sqrt{\pi}=11.62875$$

b. Using $\Gamma(p)=\frac{\Gamma(p+1)}{p}$

$$\Gamma(-3.5)=\frac{\Gamma(-3.5+1)}{-3.5}=\frac{\Gamma(-2.5)}{-3.5}=\frac{\Gamma(-2.5+1)}{(-3.5)(-2.5)}$$
$$=\frac{\Gamma(-1.5)}{(3.5)(2.5)}=\frac{\Gamma(-1.5+1)}{-(3.5)(2.5)(1.5)}$$
$$=\frac{\Gamma(.5)}{(3.5)(2.5)(1.5)(.5)}=\frac{\sqrt{\pi}}{(3.5)(2.5)(1.5)(.5)}$$
$$=.270019$$

c. Using $\Gamma(n)\Gamma(1-n)=\frac{\pi}{\sin n\pi}$

$$\Gamma\left(\frac{1}{4}\right)\Gamma\left(1-\frac{1}{4}\right)=\Gamma\left(\frac{1}{4}\right)\Gamma\left(\frac{3}{4}\right)$$
$$=\frac{\pi}{\sin\frac{\pi}{4}}=\sqrt{2}\pi=4.444$$

d.
$$\beta\left(\frac{5}{2},\frac{3}{2}\right)=\frac{\Gamma\left(\frac{5}{2}\right)\Gamma\left(\frac{3}{2}\right)}{\Gamma\left(\frac{5}{2}+\frac{3}{2}\right)}=\frac{\frac{3}{2}\Gamma\left(\frac{3}{2}\right)\cdot\Gamma\left(\frac{3}{2}\right)}{3!}$$
$$=\frac{1}{4}\left[\frac{1}{2}\Gamma\left(\frac{1}{2}\right)\right]^2=\frac{1}{4}\cdot\frac{1}{4}\cdot\pi=\cdot 1964$$

e.
$$\Gamma\left(n+\frac{1}{2}\right)=\left(n+\frac{1}{2}-1\right)\left(n+\frac{1}{2}-2\right)$$
$$\times\left(n+\frac{1}{2}-3\right)\cdots\frac{1}{2}\Gamma\left(\frac{1}{2}\right)$$
$$=\frac{(2n-1)}{2}\left(\frac{2n-3}{2}\right)\left(\frac{2n-5}{2}\right)$$
$$\times\cdots\frac{3}{2}\cdot\frac{1}{2}\cdot\sqrt{\pi}$$
$$=\frac{[(2n-1)(2n-3)(2n-5)\cdots 1\cdot\sqrt{\pi}\,]}{2^n}$$

Since $\Gamma(n+1)=n!$ thus

$$\frac{\Gamma\left(n+\frac{1}{2}\right)}{\Gamma(n+1)}=\frac{1\cdot 3\cdot 5\cdots(2n-3)(2n-1)}{2^n\cdot n!}\sqrt{\pi}.$$

Example 2: Evaluate $I=\int_0^\infty x^4 e^{-x^4}dx$.

Solution: Put $x^4=t$, $4x^3dx=dt$, $dx=\frac{1}{4}t^{-\frac{3}{4}}dt$

$$I=\int_0^\infty t\cdot e^{-t}\cdot\frac{t^{-\frac{3}{4}}}{4}\cdot dt=\frac{1}{4}\int_0^\infty e^{-t}\cdot t^{\frac{1}{4}}dt$$
$$=\frac{1}{4}\Gamma\left(1+\frac{1}{4}\right)=\frac{1}{4}\Gamma\left(\frac{5}{4}\right).$$

Example 3: Evaluate $I=\int_0^1\sqrt[3]{x\ln\left(\frac{1}{x}\right)}\,dx$.

Solution: Put $\ln\left(\frac{1}{x}\right)=t$, $x=e^{-t}$, $dx=-e^{-t}\,dt$

$$I=\int_\infty^0(e^{-t}\cdot t)^{\frac{1}{3}}(-e^{-t})dt$$
$$=\int_0^\infty t^{\frac{1}{3}}e^{-\frac{4t}{3}}dt,\quad\text{Put }\frac{4t}{3}=y$$
$$=\int_0^\infty\left(\frac{3}{4}\right)^{\frac{4}{3}}e^{-y}y^{\frac{1}{3}}dy=\left(\frac{3}{4}\right)^{\frac{4}{3}}\Gamma\left(\frac{1}{3}+1\right)$$
$$=\left(\frac{3}{4}\right)^{\frac{4}{3}}\Gamma\left(\frac{4}{3}\right).$$

Example 4: Evaluate

a. $\int_0^{\frac{\pi}{2}}\sin^{10}\theta\,d\theta$

b. $\int_0^{\frac{\pi}{2}}\cos^9\theta\,d\theta$

c. $\int_0^{\frac{\pi}{2}}\sin^6\theta\cdot\cos^7\theta\,d\theta$

d. $\int_0^{\frac{\pi}{2}}\left(\frac{\sqrt[3]{\sin 8x}}{\sqrt{\cos x}}\right)dx$

Solution:

a. $\int_0^{\frac{\pi}{2}}\sin^{10}\theta\,d\theta=\frac{1\cdot3\cdot5\cdot7\cdot9}{2\cdot4\cdot6\cdot8\cdot10}\frac{\pi}{2}=\frac{63}{256}\pi$, since $p=8$ is even

b. $\int_0^{\frac{\pi}{2}}\cos^9\theta\,d\theta=\frac{2\cdot4\cdot6\cdot8}{1\cdot3\cdot5\cdot7\cdot9}$, $\quad p=9$ is odd

c.
$$\int_0^{\frac{\pi}{2}}\sin^6\theta\cdot\cos^7\theta\,d\theta$$
$$=\frac{1}{2}\beta\left(\frac{6+1}{2},\frac{7+1}{2}\right)=\frac{1}{2}\frac{\Gamma\left(\frac{7}{2}\right)\Gamma(4)}{\Gamma\left(\frac{15}{2}\right)}$$
$$=\frac{1}{2}\frac{\left(\frac{7}{2}-1\right)\left(\frac{7}{2}-2\right)\left(\frac{7}{2}-3\right)\cdot 3!}{\left(\frac{15}{2}-1\right)\left(\frac{15}{2}-2\right)\left(\frac{15}{2}-3\right)\left(\frac{15}{2}-4\right)\left(\frac{15}{2}-5\right)\left(\frac{15}{2}-6\right)\left(\frac{15}{2}-7\right)}$$
$$=\frac{2^4}{3\cdot7\cdot11\cdot13}$$

d. $\int_0^{\frac{\pi}{2}} \frac{\sqrt[3]{\sin 8x}}{\sqrt{\cos x}} dx$

$$= \int_0^{\frac{\pi}{2}} \sin^{\frac{8}{3}} x \cdot \cos^{-\frac{1}{2}} x \, dx = \frac{1}{2}\beta\left(\frac{\frac{8}{3}+1}{2}, \frac{-\frac{1}{2}+1}{2}\right)$$

$$= \frac{1}{2}\beta\left(\frac{11}{6}, \frac{1}{4}\right) = \frac{1}{2}\frac{\Gamma\left(\frac{11}{6}\right)\Gamma\left(\frac{1}{4}\right)}{\Gamma\left(\frac{11}{6}+\frac{1}{4}\right)}$$

$$= \frac{1}{2}\cdot\frac{\left(\frac{5}{6}\right)\Gamma\left(\frac{5}{6}\right)\Gamma\left(\frac{1}{4}\right)}{\frac{13}{12}\cdot\frac{1}{12}\Gamma\left(\frac{1}{12}\right)} = \frac{60}{13}\frac{\Gamma\left(\frac{5}{6}\right)\Gamma\left(\frac{1}{4}\right)}{\Gamma\left(\frac{1}{12}\right)}.$$

Example 5: Evaluate $I = \int_0^1 \left(\frac{x}{1-x^3}\right)^{\frac{1}{2}} dx.$

Solution: $I = \int_0^1 x^{\frac{3}{2}}(1-x^3)^{-\frac{1}{2}} dx.$ put $x^3 = t$

$$I = \int_0^1 t^{\frac{1}{2}} \cdot (1-t)^{-\frac{1}{2}} \cdot \frac{1}{3} t^{-\frac{2}{3}} dt$$

$$= \frac{1}{3}\int_0^1 t^{-\frac{1}{6}} \cdot (1-t)^{-\frac{1}{2}} dt = \frac{1}{2}\beta\left(1-\frac{1}{6}, 1-\frac{1}{2}\right)$$

$$= \frac{1}{3}\beta\left(\frac{5}{6}, \frac{1}{2}\right) = \frac{1}{3}\frac{\Gamma\left(\frac{5}{6}\right)\cdot\Gamma\left(\frac{1}{2}\right)}{\Gamma\left(\frac{5}{6}+\frac{1}{2}\right)}$$

$$= \frac{1}{3}\frac{\Gamma\left(\frac{5}{6}\right)\sqrt{\pi}}{\frac{1}{3}\Gamma\left(\frac{1}{3}\right)} = \frac{\sqrt{\pi}}{\Gamma\left(\frac{1}{3}\right)}\cdot\Gamma\left(\frac{1}{3}+\frac{1}{2}\right).$$

using duplication form $\Gamma\left(\frac{1}{3}+\frac{1}{2}\right)\Gamma\left(\frac{1}{3}\right) = \frac{\sqrt{\pi}\Gamma\left(\frac{2}{3}\right)}{2^{\frac{2}{3}-1}}$

Also $\Gamma\left(\frac{1}{3}\right)\Gamma\left(1-\frac{1}{3}\right) = \Gamma\left(\frac{1}{3}\right)\Gamma\left(\frac{2}{3}\right) = \frac{\pi}{\sin\frac{\pi}{3}}.$

Substituting

$$= \frac{\sqrt{\pi}}{\Gamma\left(\frac{1}{3}\right)}\frac{\sqrt{\pi}\cdot\Gamma\left(\frac{2}{3}\right)}{\Gamma\left(\frac{1}{3}\right)\cdot 2^{\frac{1}{3}}}$$

$$= \frac{\pi}{\left[\Gamma\left(\frac{1}{3}\right)\right]^2\cdot 2^{\frac{1}{3}}}\cdot\frac{\pi}{\Gamma\left(\frac{1}{3}\right)\cdot\sin\left(\frac{\pi}{3}\right)}$$

$$I = \frac{\pi^2 2^{-\frac{1}{3}}}{\left[\Gamma\left(\frac{1}{3}\right)\right]^3\cdot\sin\frac{\pi}{3}}.$$

Example 6: Evaluate $I = \int_0^\infty a^{-bx^2} dx.$

Solution: Put $a^{-bx^2} = e^{-t}$, $-bx^2 \ln a = -t$
$2bx \ln a \, dx = dt$, also $x = \left(\frac{t}{b\ln a}\right)^{\frac{1}{2}}$

So $$dx = \frac{t^{-\frac{1}{2}}dt}{(2b \ln a)^{\frac{1}{2}}}$$

$$I = \int_0^\infty \frac{e^{-t}\cdot t^{-\frac{1}{2}}dt}{(2b \ln a)^{\frac{1}{2}}} = \frac{1}{(2b \ln a)^{\frac{1}{2}}}\cdot\Gamma\left(1-\frac{1}{2}\right)$$

$$= \frac{\sqrt{\pi}}{(2b \ln a)^{\frac{1}{2}}}.$$

Example 7: Evaluate

$$I = \left[\int_0^\infty x\, e^{-x^8} dx\right]\times\left[\int_0^\infty x^2\, e^{-x^4} dx\right].$$

Solution: $I = I_1 \times I_2.$
Put $x^8 = t$ in I_1, $x = t^{\frac{1}{8}}$, $dx = \frac{1}{8}t^{-\frac{7}{8}}dt$

$$I_1 = \int_0^\infty x\, e^{-x^8}\, dx = \int_0^\infty t^{\frac{1}{8}}\cdot e^{-t}\cdot\frac{1}{8}t^{-\frac{7}{8}}dt$$

$$I_1 = \frac{1}{8}\int_0^\infty t^{-\frac{3}{4}}e^{-t}dt = \frac{1}{8}\Gamma\left(1-\frac{3}{4}\right) = \frac{1}{8}\Gamma\left(\frac{1}{4}\right)$$

Put $x^4 = t$ in I_2, so $x = t^{\frac{1}{4}}$, $dx = \frac{1}{4}t^{-\frac{3}{4}}dt$

$$I_2 = \int_0^\infty x^2 e^{-x^4} dx = \int_0^\infty t^{\frac{1}{2}}\cdot e^{-t}\cdot\frac{1}{4}t^{-\frac{3}{4}}dt$$

$$= \frac{1}{4}\int_0^\infty t^{-\frac{1}{4}}e^{-t}dt = \frac{1}{4}\Gamma\left(1-\frac{1}{4}\right) = \frac{1}{4}\Gamma\left(\frac{3}{4}\right)$$

Thus

$$I = I_1\cdot I_2 = \frac{1}{8}\Gamma\left(\frac{1}{4}\right)\cdot\frac{1}{4}\Gamma\left(\frac{3}{4}\right)$$

$$= \frac{1}{32}\Gamma\left(\frac{1}{4}\right)\Gamma\left(1-\frac{1}{4}\right) = \frac{1}{32}\cdot\frac{\pi}{\sin\frac{\pi}{4}} = \frac{\sqrt{2}\pi}{32}$$

Since $\Gamma(n)\,\Gamma(1-n) = \frac{\pi}{\sin n\pi}.$

Example 8: Show that

$$\int_0^\infty \cos\left(bz^{\frac{1}{n}}\right) dz = \frac{\Gamma(n+1)\cdot\cos\frac{n\pi}{2}}{b^n}.$$

Solution: Put $z = x^n$, $x = z^{\frac{1}{n}}$, $dz = nx^{n-1}dx$

$$\int_0^\infty \cos\left(bz^{\frac{1}{n}}\right)dz = \int_0^\infty nx^{n-1}\cdot\cos(bx)dx$$

$$= \text{Real part of}\left\{\int_0^\infty n\,x^{n-1}\cdot e^{-ibx}dx\right\}$$

But $\int_0^\infty x^{n-1}e^{-ibx}dx$

$$= \int_0^\infty \left(\frac{t}{ib}\right)^{n-1} \cdot e^{-t} \cdot \frac{dt}{ib} = \frac{1}{(ib)^n}\Gamma(n)$$

where $t = ibx$.

$$\int_0^\infty \cos(bz^{\frac{1}{n}})dz = \text{Re}\left\{\frac{n \cdot \Gamma(n)}{b^n}(i)^{-n}\right\}$$

$$= \text{Re}\left\{\frac{\Gamma(n+1)}{b^n}\left(\cos\frac{\pi}{2} + i\sin\frac{\pi}{2}\right)^{-n}\right\}$$

$$= \text{Re}\left\{\frac{\Gamma(n+1)}{b^n}\left(\cos\frac{n\pi}{2} - i\sin\frac{n\pi}{2}\right)\right\}$$

$$= \frac{\Gamma(n+1)}{b^n} \cdot \cos\frac{n\pi}{2}.$$

Example 9: Prove that $\int_{-1}^{1}(1-t^2)^n dt = \frac{2^{n+1} \cdot n!}{1\cdot 3\cdot 5\cdots(2n+1)}$ for $n = 0, 1, 2, \ldots$

Solution: Put $t = \sin\theta$, $1 - t^2 = 1 - \sin^2\theta = \cos^2\theta$, $dt = \cos\theta d\theta$

Limits for θ are $-\frac{\pi}{2}$ to $\frac{\pi}{2}$. Thus

$$\int_{-1}^{1}(1-t^2)^n dt$$

$$= \int_{-\frac{\pi}{2}}^{\frac{\pi}{2}} \cos^{2n}\theta \cdot \cos\theta\, d\theta = 2\int_0^{\frac{\pi}{2}} \cos^{2n+1}\theta\, d\theta$$

$$= 2 \cdot \frac{2\cdot 4\cdot 6\cdot 8\cdots(2n)}{1\cdot 3\cdot 5\cdots(2n+1)} = \frac{2\cdot 2^n \cdot 1\cdot 2\cdot 3\cdots n}{1\cdot 3\cdot 5\cdots(2n+1)}$$

$$= \frac{2^{n+1}\cdot n!}{1\cdot 3\cdot 5\cdots(2n+1)}$$

Example 10: Show that $\int_0^\infty \frac{x^2dx}{(1+x^4)^3} = \frac{5\pi\sqrt{2}}{128}$.

Solution: Put $x = \sqrt{\tan\theta}$, $dx = \frac{1}{2}\frac{1}{\sqrt{\tan\theta}} \cdot \sec^2\theta\, d\theta$

$$\int_0^\infty \frac{x^2dx}{(1+x^4)^3}$$

$$= \int_0^{\frac{\pi}{2}} \frac{\tan\theta \cdot \frac{1}{2}(\tan\theta)^{-\frac{1}{2}} \cdot \sec^2\theta\, d\theta}{(1+\tan^2\theta)^3}$$

$$= \frac{1}{2}\int_0^{\frac{\pi}{2}} (\tan\theta)^{\frac{1}{2}} \cdot \sec^{-4}\theta\, d\theta$$

$$= \frac{1}{2}\int_0^{\frac{\pi}{2}} \sin^{\frac{1}{2}}\theta \cdot \cos^{\frac{7}{2}}\theta\, d\theta$$

$$= \frac{1}{2}\cdot\frac{1}{2}\cdot\beta\left(\frac{1+\frac{1}{2}}{2}, \frac{1+\frac{7}{2}}{2}\right) = \frac{1}{4}\beta\left(\frac{3}{4}, \frac{9}{4}\right)$$

$$= \frac{1}{4}\cdot\frac{\Gamma\left(\frac{3}{4}\right)\Gamma\left(\frac{9}{4}\right)}{\Gamma\left(\frac{3}{4}+\frac{9}{4}\right)} = \frac{1}{4}\frac{\Gamma\left(\frac{3}{4}\right)\cdot\frac{5}{4}\cdot\frac{1}{4}\Gamma\left(\frac{1}{4}\right)}{\Gamma(3)}$$

$$= \frac{1}{4}\cdot\frac{5}{16}\cdot\frac{1}{2!}\Gamma\left(\frac{1}{4}\right)\Gamma\left(\frac{3}{4}\right) = \frac{5}{128}\cdot\pi\sqrt{2}$$

since $\Gamma\left(\frac{1}{4}\right)\Gamma\left(\frac{3}{4}\right) = \sqrt{2}\pi$.

Example 11: Prove that

$$\int_0^{\frac{\pi}{2}} \frac{\cos^{2m-1}\theta \cdot \sin^{2n-1}\theta \cdot d\theta}{(a\cos^2\theta + b\sin^2\theta)^{m+n}} = \frac{\beta(m,n)}{2a^m b^n}$$

Solution: Put $\tan\theta = t$, $d\theta = \cos^2\theta\, dt$, $\sin\theta = t\cos\theta$

$$\int_0^\infty \frac{\cos^{2m-1}\theta \cdot \sin^{2n-1}\theta\, d\theta}{(a\cos^2\theta + b\sin^2\theta)^{m+n}}$$

$$= \int_0^\infty \frac{\cos^{2m-1}\theta \cdot t^{2n-1} \cdot \cos^{2n-1}\theta \cdot \cos^2\theta\, dt}{(a\cos^2\theta + bt^2\cos^2\theta)^{m+n}}$$

$$= \int_0^\infty \frac{\cos^{2m+2n}\theta \cdot t^{2n-1}dt}{\cos^{2m+2n}\theta(a+bt^2)^{m+n}}, \quad \text{Put } \sqrt{b}t = \sqrt{ay}$$

$$= \int_0^\infty \frac{\left(\frac{ay}{b}\right)^{\left(\frac{2n-1}{2}\right)} \cdot \frac{a}{b}\frac{1}{2\sqrt{y}}dy}{(a+ay)^{m+n}}$$

$$= \frac{a^n}{2b^n a^{m+n}}\int_0^\infty \frac{y^{n-1}dy}{(1+y)^{m+n}} = \frac{1}{2a^m b^n}\cdot\beta(m,n).$$

Example 12: Evaluate

$$I = \int_0^1 x^{\frac{3}{2}}(1-x^2)^{\frac{5}{2}}dx.$$

Solution: From result 13 Page 5 with $p = \frac{3}{2}, q = 2, r = \frac{5}{2}$

$$I = \frac{1}{q}\beta\left(\frac{p+1}{q}, r+1\right) = \frac{1}{2}\beta\left(\frac{\frac{3}{2}+1}{2}, \frac{5}{2}+1\right)$$

$$= \frac{1}{2}\beta\left(\frac{5}{4}, \frac{7}{2}\right) = \frac{1}{2}\frac{\Gamma\left(\frac{5}{4}\right)\Gamma\left(\frac{7}{2}\right)}{\Gamma\left(\frac{5}{4}+\frac{7}{2}\right)}$$

$$= \frac{1}{2}\frac{1}{4}\Gamma\left(\frac{1}{4}\right)\cdot\frac{5}{2}\cdot\frac{3}{2}\cdot\frac{1}{2}\Gamma\left(\frac{1}{2}\right)\frac{1}{\Gamma\left(\frac{19}{4}\right)}$$

Since $\Gamma\left(\frac{19}{4}\right) = \frac{15}{4}\cdot\frac{11}{4}\cdot\frac{7}{4}\cdot\frac{3}{4}\cdot\Gamma\left(\frac{3}{4}\right)$.

$$I = \frac{4\Gamma\left(\frac{1}{4}\right)\sqrt{\pi}}{\left[121\Gamma\left(\frac{3}{4}\right)\right]}.$$

EXERCISE

Gamma and Beta functions

1. Compute (a) $\frac{\Gamma(6)}{2\Gamma(3)}$ (b) $\frac{\Gamma\left(\frac{5}{2}\right)}{\Gamma\left(\frac{1}{2}\right)}$ (c) $\frac{\Gamma(3)\Gamma(2.5)}{\Gamma(5.5)}$ (d) $\Gamma\left(-\frac{5}{2}\right)$.

Ans. (a) 30 (b) $\frac{3}{4}$ (c) $\frac{16}{315}$ (d) $-\frac{8\sqrt{\pi}}{15}$

2. Evaluate (a) $\int_0^\infty \sqrt{y}e^{-y^2}dy$ (b) $\int_0^\infty 3^{-4z^2}dz$ (c) $\int_0^1 \frac{dx}{\sqrt{-\ln x}}$.

Ans. (a) $\frac{\sqrt{\pi}}{3}$ (b) $\frac{\sqrt{\pi}}{(4\sqrt{\ln 3})}$ (c) $\sqrt{\pi}$

3. Evaluate (a) $\int_0^1 x^4(1-x)^3dx$ (b) $\int_0^2 \frac{x^2dx}{\sqrt{2-x}}$ (c) $\int_0^a y^4\sqrt{a^2-y^2}\,dy$.

Ans. (a) $\frac{1}{280}$ (b) $\frac{64\sqrt{2}}{15}$ (c) $\frac{\pi a^6}{32}$.

4. Evaluate (a) $\int_0^{2\pi} \sin^8\theta\, d\theta$ (b) $\int_0^{\frac{\pi}{2}} \cos^6\theta\, d\theta$ (c) $\int_0^{\frac{\pi}{2}} \sin^4\theta\cdot\cos^5\theta\, d\theta$.

Ans. (a) $I = 4\int_0^{\frac{\pi}{2}} \sin^8\theta d\theta = \frac{4\cdot 1\cdot 3\cdot 5\cdot 7}{(2\cdot 4\cdot 6\cdot 8)}\frac{\pi}{2} = \frac{35\pi}{64}$
(b) $\frac{5\pi}{32}$ (c) $\frac{8}{315}$

5. Show that $\int_0^2 x\sqrt[3]{8-x^3}dx = \frac{16\pi}{(9\sqrt{3})}$

6. Find $\int_0^{\frac{\pi}{2}} \sqrt{\cot\theta}\, d\theta$.

Ans. $\frac{1}{2}\Gamma\left(\frac{1}{4}\right)\Gamma\left(\frac{3}{4}\right) = \frac{1}{2}\pi\sqrt{2}$

7. Show that $\int_0^{\frac{\pi}{2}} (\sqrt{\tan\theta} + \sqrt{\sec\theta})\, d\theta = \frac{1}{2}\Gamma\left(\frac{1}{4}\right)\left\{\Gamma\left(\frac{3}{4}\right) + \frac{\sqrt{(\pi)}}{\Gamma\left(\frac{3}{4}\right)}\right\}$.

8. Prove that $\int_0^1 x^4\left[\ln\left(\frac{1}{x}\right)\right]^3 dx = \frac{6}{625}$.

9. Show that $\left[\int_0^1 x^2(1-x^4)^{-\frac{1}{2}}dx\right] \times \left[\int_0^1 (1+x^4)^{-\frac{1}{2}}dx\right] = \frac{\pi}{4\sqrt{2}}$.

10. Prove that $\left[\int_0^{\frac{\pi}{2}} \sqrt{\sin\theta}d\theta\right]\left[\int_0^{\frac{\pi}{2}} (\sin\theta)^{-\frac{1}{2}}d\theta\right] = \pi$.

11. Show that $\int_0^{\frac{\pi}{2}} \sin^7\theta\cdot\cos^7\theta\, d\theta = \frac{1}{280}$.

12. Evaluate $\int_0^a x^3(a^3-x^3)^5dx$.

Ans. $\frac{a^{19}\cdot 3^5}{19\cdot 16\cdot 13\cdot 7}$.

13. Prove that $\int_0^1 x^m(\ln x)^n dx = \frac{(-1)^n n!}{(m+1)^{n+1}}$ where n is a positive integer and $m > -1$.

14. Prove that $\int_0^\infty \frac{t^2dt}{1+t^4} = \frac{\pi}{\sqrt{2}}$.

Hint: Put $t = \sqrt{\tan\theta}$.

15. Show that the area under the normal curve $y = \frac{1}{\sigma\sqrt{2\pi}}\cdot e^{-\frac{x^2}{2\sigma^2}}$ and x- axis is unity.

16. Show that $\frac{\beta(p,q)}{p+q} = \frac{\beta(p,q+1)}{q} = \frac{\beta(p+1,q)}{p}$.

17. Prove that $\beta(m,n) = \frac{1}{2}\int_0^\infty \frac{x^{m-1}+x^{n-1}}{(1+x)^{m+n}}$.
Hint: Use symmetry property of β function.

18. $\int_0^\infty x^{-\frac{3}{2}}(1-e^{-x})dx$.

Ans. $2\sqrt{\pi}$.

19. Show that $\int_b^a (x-b)^{m-1}(a-x)^{n-1}dx = (a-b)^{m+n-1}\cdot\beta(m,n)$
Hint: Put $x = \frac{(t-b)}{(a-b)}$.

20. Prove that $\int_0^\infty e^{-x^4}dx = \frac{1}{4}\Gamma\left(\frac{1}{4}\right)$.

21. Evaluate $\int_0^\infty \frac{x^a}{a^x}dx$.

Ans. $\frac{\Gamma(a+1)}{(\ln a)^{a+1}}$

22. Show that $\int_0^1 \frac{x^{p-1}+x^{q-1}}{(1+x)^{p+q}}dx = \beta(p,q)$
Hint: From (5) $\beta(p,q) = \int_0^\infty \frac{y^{q-1}}{(1+y)^{p+q}}dx = \int_0^1 + \int_0^\infty$. Put $y = \frac{1}{z}$ in 2nd integral.

23. Show that $\int_{-1}^1 \sqrt{\frac{1+t}{1-t}}dt = \pi$.

24. Evaluate $\int_0^\infty \frac{x^8(1-x^6)}{(1+x)^{24}}dx$.

Ans. 0

25. Prove that $\int_0^\infty e^{-ax}\cdot x^{n-1}dx = \frac{\Gamma(n)}{a^n}$ where a and n are positive.

11.3 BESSEL'S FUNCTIONS

The boundary value problems (such as the one-dimensional heat equation) with cylindrical symmetry (independent of θ) reduces to two ordinary differential equations by the separation of variables technique. One of them is the most important differential equation known as **Bessel's* differential equation**

$$x^2\frac{d^2y}{dx^2} + x\frac{dy}{dx} + (x^2-p^2)y = 0$$

$$x^2y'' + xy' + (x^2-p^2)y = 0 \quad (1)$$

Here p, which is a given constant (not necessarily an integer) is known as the order of the Bessel's equation.

* Friedrich Wilhelm Bessel (1784–1846) German mathematician.

Bessel's Functions (Cylindrical functions)

Bessel's functions (Cylindrical functions) are series solution of the Bessel's differential Equation (1) obtained by Frobenius method.

Assume that p is real and non-negative. Assume the series solution of (1) as

$$y(x) = \sum_{m=0}^{\infty} a_m x^{m+r} \qquad (a_0 \neq 0) \tag{2}$$

To determine the unknown coefficients a_m and power (exponent) r, substitute (2) in (1), we get

$$\sum_{m=0}^{\infty}(m+r)(m+r-1)a_m x^{m+r} + \sum_{m=0}^{\infty}(m+r)a_m x^{m+r}$$
$$+\sum_{m=0}^{\infty} a_m x^{m+r+2} - p^2 \sum_{m=0}^{\infty} a_m x^{m+r} = 0$$

Now equate the sum of the coefficients of x^{s+r} to zero. For $s=0$ and $s=1$, the contribution comes from first, second and fourth series (not from third series because it starts with x^{r+2}). For $s \geq 2$, all the four terms contribute. Thus sum of the coefficients of powers of r, $r+1$ and $s+r$ are respectively given by

$$r(r-1)a_0 + ra_0 - p^2 a_0 = 0 \qquad (s=0) \tag{4}$$

$$(r+1)ra_1 + (r+1)a_1 - p^2 a_1 = 0 \qquad (s=1) \tag{5}$$

$$(s+r)(s+r-1)a_s + (s+r)a_s + a_{s-2} - p^2 a_s = 0 \quad (s = 2, 3\cdots) \tag{6}$$

Solving (4), we get the indicial equation

$$(r+p)(r-p) = 0 \tag{7}$$

Solutions of (7) are $r_1 = p(\geq 0)$ and $r_2 = -p$.

Case 1: $r_1 = p$

With $r_1 = p$, Equation (5) becomes $(2p+1)a_1 = 0$ so $a_1 = 0$

Rewrite (6) as

$$(s+r+p)(s+r-p)a_s + a_{s-2} = 0$$

Substituting $r = p$, this becomes

$$s(s+2p)a_s + a_{s-2} = 0 \tag{8}$$

or
$$a_s = -\frac{a_{s-2}}{s(s+2p)}$$

For
$$s = 3, \quad a_3 = -\frac{a_1}{3(3+2p)}$$

Since $a_1 = 0$ and $p \geq 0$, then $a_3 = 0$. Thus from (8) it follows that

$$a_3 = 0, \quad a_5 = 0, \quad a_7 = 0 \text{ etc.}$$

i.e., all coefficients with odd subscripts are zero. Rewriting (8) with $s = 2m$, we have

$$2m(2m+2p)a_{2m} + a_{2n-2} = 0$$

Solving

$$a_{2m} = -\frac{1}{2^2 m(m+p)} \cdot a_{2m-2}, \quad m = 1, 2, \ldots$$

Thus
$$a_2 = -\frac{a_0}{2^2(1+p)}$$

$$a_4 = -\frac{a_2}{2^2 \cdot 2(2+p)} = \frac{a_0}{2^4 2!(p+1)(p+2)}$$

In general

$$a_{2m} = \frac{(-1)^m a_0}{2^{2m} \cdot m!(p+1)(p+2)\cdots(p+m)},$$
$$m = 1, 2, \ldots \tag{9}$$

a_0 which is arbitrary may be taken as

$$a_0 = \frac{1}{2^p \Gamma(p+1)}$$

Then
$$a_2 = -\frac{a_0}{2^2(p+1)} = -\frac{1}{2^2 \cdot 2^p (p+1)\Gamma(p+1)}$$
$$= \frac{-1}{2^{2+p}\Gamma(p+2)}$$

since $\Gamma(\alpha+1) = \alpha\Gamma(\alpha)$.
Similarly,

$$a_4 = \frac{-a_2}{2^2 \cdot 2 \cdot (p+2)} = \frac{1}{2^2 \cdot 2 \cdot 2^{2+p} \cdot (p+2)\Gamma(p+2)}$$
$$= \frac{1}{2^{4+p} \cdot 2!\Gamma(p+3)}$$

In general

$$a_{2m} = \frac{(-1)^m}{2^{2m+p} \cdot m!\Gamma(p+m+1)} \quad \text{for } m = 1, 2, \ldots \tag{10}$$

By substituting these coefficients from (10) in (2) and observing that $a_1 = a_3 = a_5 = \cdots = 0$, a particular solution of the Bessel's Equation (1) is obtained as

$$J_p(x) = x^p \sum_{m=0}^{\infty} \frac{(-1)^m x^{2m}}{2^{2m+p} \cdot m!\Gamma(p+m+1)} \tag{11}$$

(11) is known as the Bessel's function of the first kind of order p, which converges for all x (by ratio test).

Case 2: For $r_2 = -p$
By replacing p by $-p$ in (11), we get a second linearly independent solution of (1) as

$$J_{-p}(x) = x^{-p} \sum_{m=0}^{\infty} \frac{(-1)^m x^{2m}}{2^{2m-p} m! \Gamma(m-p+1)} \tag{12}$$

Hence the general solution of Bessel's Equation (1) for all $x \neq 0$ is

$$y(x) = c_1 J_p(x) + c_2 J_{-p}(x) \tag{13}$$

provided p is not an integer.

Linear Dependence of Bessel's Functions: J_n and J_{-n}

Assume that $p = n$ where n is an integer.
Then from (11), we get

$$J_n(x) = x^n \sum_{m=0}^{\infty} \frac{(-1)^m x^{2m}}{2^{2m+n} \cdot m! \Gamma(n+m+1)}$$

Since $\Gamma(n+1) = n!$, we have $\Gamma(n+m+1) = (n+m)!$

$$J_n(x) = x^n \sum_{m=0}^{\infty} \frac{(-1)^m x^{2m}}{2^{2m+n} \cdot m!(m+n)!} \tag{14}$$

Book Work: Prove that $J_n(x)$ and $J_{-n}(x)$ are linearly dependent because

$$J_{-n}(x) = (-1)^n J_n(x) \qquad \text{for } n = 1, 2, 3, \ldots$$

Proof: Replacing p by $-n$ in (11), we get

$$J_{-n}(x) = \sum_{m=0}^{\infty} \frac{(-1)^m x^{2m-n}}{2^{2m-n} \cdot m! \Gamma(m-n+1)} \tag{15}$$

When $m - n + 1 \leq 0$ or $m \leq (n-1)$, the gamma function of zero or negative integers is infinite. Therefore for $m = 0$ to $n-1$, the coefficients in (15) become zero. So m starts at n. Thus

$$J_{-n}(x) = \sum_{m=n}^{\infty} \frac{(-1)^m x^{2m-n}}{2^{2m-n} \cdot m!(m-n)!}$$

since $\Gamma(m-n+1) = (m-n)!$

Put $m - n = s$ then s varies from 0 to ∞.

$$J_{-n}(x) = \sum_{s=0}^{\infty} \frac{(-1)^{s+n} x^{2(s+n)-n}}{2^{2(s+n)-n}(s+n)!s!}$$

$$= (-1)^n \sum_{s=0}^{\infty} \frac{(-1)^s x^{2s+n}}{2^{2s+n} \cdot s!(s+n)!}$$

$$J_{-n}(x) = (-1)^n J_n(x). \tag{16}$$

Generating Function

Generating function of a sequence of functions $f_n(x)$ is

$$G(u, x) = \sum_{n=-\infty}^{\infty} f_n(x) \cdot u^n$$

which generates $f_n(x)$ i.e., $f_n(x)$ appear as coefficients of powers of u.

Theorem: *Prove that the generating function for Bessel's functions of integral order is*

$$e^{\frac{1}{2}x\left(t-\frac{1}{t}\right)} \tag{17}$$

Proof: If $e^{\frac{1}{2}x\left(t-\frac{1}{t}\right)}$ is the generating function of Bessel function then the coefficients of different powers of t in the expansion of (17) are the Bessel's functions of different integral orders.

Consider

$$e^{\frac{1}{2}x\left(t-\frac{1}{t}\right)} = e^{\frac{xt}{2}} \cdot e^{-\frac{x}{2t}}$$

Expanding in series, we get

$$= \left[1 + \frac{xt}{2} + \frac{1}{2!}\left(\frac{xt}{2}\right)^2 + \frac{1}{3!}\left(\frac{xt}{2}\right)^3 + \cdots\right] \times$$

$$\times \left[1 - \frac{x}{2t} + \frac{1}{2!}\left(\frac{x}{2t}\right)^2 - \frac{1}{3!}\left(\frac{x}{2t}\right)^3 \cdots\right] \tag{18}$$

Case 1: $n = 0$.
The coefficient of $t^0 = 1$ in the expansion (18) is

$$1 - \left(\frac{x}{2}\right)^2 + \left(\frac{1}{2!}\right)^2 \left(\frac{x}{2}\right)^4$$

$$- \left(\frac{1}{3!}\right)^2 \left(\frac{x}{2}\right)^6 + \left(\frac{1}{4!}\right)^2 \left(\frac{x}{2}\right)^8 - \cdots$$

$$= \sum_{m=0}^{\infty} \frac{(-1)^m}{(m!)^2} \left(\frac{x}{2}\right)^{2m} = J_0(x). \tag{19}$$

Case 2: Positive powers of $t : t^n$

The coefficient of t^n in the above expansion (18) is

$$\frac{1}{n!}\left(\frac{x}{2}\right)^n - \frac{1}{(n+1)!}\left(\frac{x}{2}\right)^{n+2} + \frac{1}{2!}\frac{1}{(n+2)!}\left(\frac{x}{2}\right)^{n+4} + \cdots$$

$$= \sum_{m=0}^{\infty} \frac{(-1)^m}{m!(n+m)!}\left(\frac{x}{2}\right)^{n+2m}$$

$$= J_n(x). \qquad (20)$$

Case 3: Negative powers of $t : t^{-n}$

The coefficient of t^{-n} in the expansion (18) is

$$\frac{(-1)^n}{n!}\left(\frac{x}{2}\right)^n + \left(\frac{x}{2}\right)\frac{(-1)^{n+1}}{(n+1)!}\left(\frac{x}{2}\right)^{n+1} + \frac{1}{2!}\left(\frac{x}{2}\right)^2\frac{(-1)^{n+2}}{(n+2)!}\left(\frac{x}{2}\right)^{n+2} + \cdots$$

$$= (-1)^n \sum_{m=0}^{\infty} \frac{(-1)^m}{m!(n+m)!}\left(\frac{x}{2}\right)^{n+2m}$$

$$= (-1)^n J_n(x) = J_{-n}(x) \qquad (21)$$

Thus from (19), (20) and (21), we have

$$e^{\frac{x}{2}\left(t-\frac{1}{t}\right)} = \sum_{n=-\infty}^{\infty} J_n(x)t^n.$$

Equation Reducible to Bessel's Equation

The differential equation

$$x^2\frac{d^2y}{dx^2} + x\frac{dy}{dx} + (\lambda^2x^2 - p^2)y = 0 \qquad (22)$$

where λ is a parameter, can be reduced Bessel's differential equation of order p in t,

$$t^2\frac{d^2y}{dt^2} + t\frac{dy}{dt} + (t^2 - p^2)y = 0 \qquad (23)$$

where $t = \lambda x$ $\left(\text{so } \frac{dy}{dx} = \lambda\frac{dy}{dt}, \frac{d^2y}{dx^2} = \lambda^2\frac{d^2y}{dt^2}\right)$.

For p non-integral, the general solution of Equation (23) is

$$y = c_1 J_n(t) + c_2 J_{-n}(t).$$

Thus the general solution of Equation (22) is

$$y(x) = c_1 J_n(\lambda x) + c_2 J_{-n}(\lambda x)$$

when p is non-integral.

Orthogonality of Bessel's Functions

Prove that

$$\int_0^a x J_n(\alpha x)J_n(\beta x)dx = \begin{cases} 0, & \text{if } \alpha \neq \beta \\ \frac{a^2}{2}J_{n+1}^2(a\alpha), & \text{if } \alpha = \beta \end{cases}$$

where α and β are roots of $J_n(ax) = 0$.

Proof: Let $u = J_n(\alpha x)$ and $v = J_n(\beta x)$ respectively be the solutions of the equations

$$x^2u'' + xu' + (\alpha^2x^2 - n^2)u = 0 \qquad (1)$$

and

$$x^2v'' + xv' + (\beta^2x^2 - n^2)v = 0 \qquad (2)$$

Multiplying (1) by $\frac{v}{x}$ and (2) by $\frac{u}{x}$ and subtracting

$$x(u''v - uv'') + (u'v - uv') + (\alpha^2 - \beta^2)xuv = 0$$

or

$$\frac{d}{dx}\left\{x(u'v - uv')\right\} = (\beta^2 - \alpha^2)xuv \qquad (3)$$

Integrating both sides of (3) from $x = 0$ to a

$$(\beta^2 - \alpha^2)\int_0^a xuv\, dx = \left[x(u'v - uv')\right]_0^a$$

$$= a\left[u'(a)v(a) - u(a)v'(a)\right] \qquad (4)$$

where $'$ denotes differentiation w.r.t., x.

Now $$u' = \frac{d}{dx}u = \frac{d}{dx}J_n(\alpha x) = \alpha J_n'(\alpha x) \qquad (5)$$

Similarly, $$v' = \frac{dv}{dx} = \frac{d}{dx}J_n(\beta x) = \beta J_n'(\beta x) \qquad (6)$$

Substituting u' and v' from (5) and (6) in (4), we get

$$\int_0^a x J_n(\alpha x)J_n(\beta x)dx = \frac{a}{\beta^2 - \alpha^2}\left[\alpha J_n'(\alpha a)J_n(\beta a) - \beta J_n(\alpha a)J_n'(\beta a)\right] \qquad (7)$$

Case 1: Suppose α and β are two distinct roots of $J_n(ax) = 0$ then $J_n(a\alpha) = J_n(a\beta) = 0$.

Thus for $\alpha \neq \beta$

$$\int_0^a x J_n(\alpha x)J_n(\beta x)dx = 0 \qquad (8)$$

(8) is known as the orthogonality relation for Bessel's functions.

Case 2: Suppose $\beta = \alpha$; then R.H.S. of (4) is $\frac{0}{0}$ form. Assuming α as a root of $J_n(ax) = 0$, evaluate

R.H.S. of (4) as $\beta \to \alpha$

$$\lim_{\beta\to\alpha} \int_0^a x\, J_n(\alpha x) J_n(\beta x) dx$$

$$= \lim_{\beta\to\alpha} \left(\frac{a}{\beta^2-\alpha^2}\right)\left[\alpha J_n'(a\alpha) J_n(a\beta) - 0\right]$$

Since $J_n(a\alpha) = 0$.

Now applying L'Hospital's rule (differentiating w.r.t., β), we get

$$= \lim_{\beta\to\alpha} \frac{a}{2\beta}\left[\alpha J_n'(a\alpha)\cdot a\, J_n'(a\beta)\right]$$

$$= \frac{a^2}{2}\left[J_n'(a\alpha)\right]^2$$

In the recurrence relation IV on Page 11.14

$$J_{n+1}(x) = \frac{n}{x} J_n(x) - J_n'(x)$$

Put $x = a\alpha$, then $J_{n+1}(a\alpha) = \frac{n}{a\alpha} J_n(a\alpha) - J_n'(a\alpha)$. Since α is a root, $J_n(a\alpha) = 0$. Then

$$J_n'(a\alpha) = -J_{n+1}(a\alpha)$$

Thus for $\alpha \neq \beta$,

$$\int_0^a x\, J_n(\alpha x) J_n(\beta x) dx = \frac{a^2}{2}\left[J_n'(a\alpha)\right]^2$$

$$= \frac{a^2}{2}\left[J_{n+1}(a\alpha)\right]^2$$

Note: Put $x = a\alpha$ in the recurrence relation VI on Page 11.14

$$J_{n-1}(a\alpha) + J_{n+1}(a\alpha) = \frac{2n}{a\alpha} J_n(a\alpha).$$

Since $J_n(a\alpha) = 0$, $J_{n-1}(a\alpha) = -J_{n+1}(a\alpha)$. Thus

$$\int_0^a x\, J_n(\alpha x) J_n(\beta x) dx = \frac{a^2}{2}\left[J_{n-1}(a\alpha)\right]^2$$

Recurrence Relations (or identities) for Bessel's Functions

Valid for any p.

Prove that

I. $\frac{d}{dx}\left\{x^p J_p(x)\right\} = x^p J_{p-1}(x)$

Proof: From (11)

$$J_p(x) = \sum_{m=0}^{\infty} \frac{(-1)^m x^{2m+p}}{2^{2m+p}\cdot m!\Gamma(m+p+1)}$$

So

$$\frac{d}{dx}\left\{x^p J_p(x)\right\} = \frac{d}{dx}\left\{\sum_{m=0}^{\infty} \frac{(-1)^m x^{2m+2p}}{2^{2m+p}\cdot m!\Gamma(m+p+1)}\right\}$$

$$= \sum_{m=0}^{\infty} \frac{(-1)^m\cdot(2m+2p)x^{2m+2p-1}}{2^{2m+p}\cdot m!(m+p)\Gamma(m+p)}$$

$$= x^p \sum_{m=0}^{\infty} \frac{(-1)^m x^{2m+(p-1)}}{2^{2m+(p-1)}.m!\Gamma(m+(p-1)+1)}$$

$$= x^p J_{p-1}(x)$$

II. $\frac{d}{dx}\left\{x^{-p} J_p(x)\right\} = -x^{-p} J_{p+1}(x).$

Proof: Multiplying (11) by x^{-p} and differentiating

$$\frac{d}{dx}\left\{x^{-p} J_p(x)\right\} = \frac{d}{dx}\left\{\sum_{m=0}^{\infty} \frac{(-1)^m\cdot x^{2m}}{2^{2m+p}\cdot m!\Gamma(m+p+1)}\right\}$$

$$= \sum_{m=1}^{\infty} \frac{(-1)^m\cdot 2m\cdot x^{2m-1}}{2^{2m+p}\cdot m!\Gamma(m+p+1)}$$

since for $m = 0$, the first term in R.H.S. is zero.

$$= \sum_{m=1}^{\infty} \frac{(-1)^m\cdot x^{2m-1}}{2^{2m+p-1}\cdot(m-1)!\Gamma(m+p+1)}$$

Put $s = m - 1$ or $m = s + 1$ then

$$= \sum_{s=0}^{\infty} \frac{(-1)^{s+1}\cdot x^{2(s+1)-1}}{2^{2(s+1)+p-1}\cdot s!\Gamma(s+1+p+1)}$$

$$= -x^{-p} \sum_{s=0}^{\infty} \frac{(-1)^s x^{2s+(p+1)}}{2^{2s+(p+1)} s!\Gamma((s+1)+p+1)}$$

$$= -x^{-p}\cdot J_{p+1}(x).$$

III. $\frac{d}{dx}\left\{J_p(x)\right\} = J_{p-1}(x) - \frac{p}{x} J_p(x)$

or $x J_p'(x) = x J_{p-1}(x) - p J_p(x)$

Proof: From recurrence relation (I)

$$\frac{d}{dx}\left\{x^p J_p(x)\right\} = x^p J_{p-1}(x)$$

Performing the differentiation in the L.H.S.,

$$x^p \cdot \frac{d}{dx}\left\{J_p(x)\right\} + px^{p-1} \cdot J_p(x) = x^p J_{p-1}(x)$$

or $$J_p'(x) + \frac{p}{x} J_p(x) = J_{p-1}(x)$$

or $$J_p'(x) = J_{p-1}(x) - \frac{p}{x} J_p(x)$$

IV. $J_p'(x) = \frac{p}{x} J_p(x) - J_{p+1}(x)$

Proof: From recurrence relation (II)

$$\frac{d}{dx}\left\{x^{-p} J_p(x)\right\} = -x^{-p} J_{p+1}(x)$$

Performing the differentiation in the L.H.S.,

$$x^{-p} \cdot \frac{d}{dx} J_p(x) - px^{-p-1} J_p(x) = -x^{-p} J_{p+1}(x)$$

or $$J_p'(x) - \frac{p}{x} J_p(x) = -J_{p+1}(x)$$

or $$J_p'(x) = \frac{p}{x} J_p(x) - J_{p+1}(x)$$

V. $J_p'(x) = \frac{1}{2}\left\{J_{p-1}(x) - J_{p+1}(x)\right\}$ is obtained by adding recurrence relations (III) and (IV)

VI. $J_{p-1}(x) + J_{p+1}(x) = \frac{2p}{x} J_p(x)$ is obtained by subtracting (IV) from (III).

Elementary Bessel's Functions

Bessel's functions J_p of orders $p = \pm\frac{1}{2}, \pm\frac{3}{2}, \pm\frac{5}{2}, \ldots$ are elementary and can be expressed in terms of sine and cosines and powers of x.

Result 1: $J_{\frac{1}{2}}(x) = \sqrt{\frac{2}{\pi x}} \cdot \sin x.$

Proof: With $p = \frac{1}{2}$, (11) reduces to

$$J_{\frac{1}{2}}(x) = \sqrt{x} \sum_{m=0}^{\infty} \frac{(-1)^m x^{2m}}{2^{2m+\frac{1}{2}} \cdot m!\Gamma\left(m + \frac{3}{2}\right)}$$

Now

$$\Gamma\left(m + \frac{3}{2}\right) = \left(m + \frac{1}{2}\right)\left(m - \frac{1}{2}\right)\left(m - \frac{3}{2}\right)\cdots \times \frac{3}{2} \cdot \frac{1}{2} \Gamma\left(\frac{1}{2}\right)$$

$$= \frac{(2m+1)(2m-1)(2m-3)\cdots 3 \cdot 1 \cdot \sqrt{\pi}}{2^{m+1}}$$

Also

$$2^{2m+1} \cdot m! = 2^{m+1} \cdot 2^m \cdot m!$$
$$= 2^{m+1} \cdot 2^m (m)(m-1)\cdots 2 \cdot 1$$
$$= 2^{m+1} \cdot (2m)(2m-2)\cdots 4 \cdot 2.$$

Thus

$$2^{2m+1} \cdot m! \cdot \Gamma\left(m + \frac{3}{2}\right)$$

$$= \left[2^{m+1} \cdot 2m \cdot (2m-2)\cdots 4 \cdot 2\right] \times \left[(2m+1)(2m-1)\cdots 3 \cdot 1\right] \cdot 2^{-(m+1)} \cdot \sqrt{\pi}$$

$$= (2m+1)!\sqrt{\pi}$$

Then

$$J_{\frac{1}{2}}(x) = \sqrt{\frac{2}{x}} \sum_{m=0}^{\infty} \frac{(-1)^m x^{2m-1}}{2^{2m+1} \cdot m!\Gamma\left(m + \frac{3}{2}\right)}$$

$$= \sqrt{\frac{2}{x}} \sum_{m=0}^{\infty} \frac{(-1)^m x^{2m-1}}{(2m+1)!\sqrt{\pi}}$$

$$= \sqrt{\frac{2}{\pi x}} \cdot \sum_{m=0}^{\infty} \frac{(-1)^m x^{2m-1}}{(2m+1)!}$$

$$= \sqrt{\frac{2}{\pi x}} \cdot \sin x.$$

Result 2: In the recurrence relation I, put $p = \frac{1}{2}$ then

$$\frac{d}{dx}\left\{\sqrt{x} J_{\frac{1}{2}}(x)\right\} = \sqrt{x} J_{-\frac{1}{2}}(x)$$

$$\frac{d}{dx}\left\{\sqrt{x}\sqrt{\frac{2}{\pi x}} \cdot \sin x\right\} = \sqrt{x} J_{-\frac{1}{2}}(x)$$

$$\sqrt{\frac{2}{\pi}} \cos x = \sqrt{x} J_{-\frac{1}{2}}(x)$$

or $$J_{-\frac{1}{2}}(x) = \sqrt{\frac{2}{\pi x}} \cos x.$$

Similarly with $p = \frac{1}{2}$, we get from recurrence relation VI.

Result 3:

$$J_{-\frac{1}{2}}(x) + J_{\frac{3}{2}}(x) = \frac{1}{x} J_{\frac{1}{2}}(x)$$

or $$J_{\frac{3}{2}}(x) = \frac{1}{x} J_{\frac{1}{2}}(x) - J_{-\frac{1}{2}}(x)$$

Using result (1) and (2) for $J_{\frac{1}{2}}$ and $J_{-\frac{1}{2}}$, we get

$$J_{\frac{3}{2}}(x) = \sqrt{\frac{2}{\pi x}}\left(\frac{\sin x}{x} - \cos x\right).$$

Similarly with $p = -\frac{1}{2}$ in recurrence relation VI

Result 4: $$J_{-\frac{3}{2}}(x) = -\frac{1}{x}J_{-\frac{1}{2}}(x) - J_{\frac{1}{2}}(x)$$

$$= -\sqrt{\frac{2}{\pi x}}\left(\frac{\cos x}{x} + \sin x\right).$$

Integrals of Bessel's Functions

Integrating the recurrence relation

$$\frac{d}{dx}\left\{x^p J_p(x)\right\} = x^p J_{p-1}(x), \quad \text{we get}$$

$$\int x^p J_{p-1}(x)dx = x^p J_p(x) + c \tag{1}$$

For $p = 1$, $$\int x J_0(x)\,dx = x J_1(x) + c \tag{2}$$

Integrating the recurrence relation

$$\frac{d}{dx}\left\{x^{-p} J_p(x)\right\} = -x^{-p} J_{p+1}(x), \quad \text{we get}$$

$$\int x^{-p} J_{p+1}(x)dx = -x^{-p} J_p(x) + c \tag{3}$$

For $p = 0$, $$\int J_1(x)\,dx = -J_0(x) + c \tag{4}$$

In general $\int x^m J_n(x)dx$ for m and n integers with $m + n \geq 0$ can be integrated by parts completely if $m + n$ is odd. But when $m + n$ is even, the integral depends on the residual integral $\int J_0(x)dx$ which has been tabulated.

Integrating $$J'_p(x) = \frac{1}{2}\left[J_{p-1}(x) - J_{p+1}(x)\right]$$

$$2J_p(x) = \int J_{p-1}(x)\,dx - \int J_{p+1}(x)dx$$

or $$\int J_{p+1}(x)dx = \int J_{p-1}(x)dx - 2J_p(x).$$

Bessel's Function of Second Kind of Order *n* or Neumann Function

When n is integral, $J_n(x)$ and $J_{-n}(x)$ are linearly dependent and do not constitute the solution.

Let $y = u(x)\,J_n(x)$ be a solution of (1). Substituting in (1),

$$x^2(u''J_n + 2u'J'_n + uJ''_n) + x(u'J_n + uJ'_n)$$
$$+(x^2 - n^2)uJ_n = 0$$

or $$u\left\{x^2J''_n + xJ'_n + (x^2 - n^2)J_n\right\} + x^2u''J_n$$
$$+2x^2u'J'_n + xu'J_n = 0$$

Since J_n is a solution of (1), the first term is zero. Dividing throughout by $x^2u'J_n$, we get

$$\frac{u''}{u} + 2\frac{J'_n}{J_n} + \frac{1}{x} = 0$$

Integrating $\ln(u'J_n^2 \cdot x) = \ln B$ or $xu'J_n^2 = B$. Thus

$$u' = \frac{B}{xJ_n^2}$$

Integrating

$$u = B\int\frac{dx}{xJ_n^2} + c$$

Hence $y = AJ_n(x) + BY_n(x)$ is the complete solution of (1) where

$$Y_n(x) = J_n(x)\cdot\int\frac{dx}{x[J_n(x)]^2}$$

$Y_n(x)$ is known as Bessel's function of second kind of order n or Neumann function.

Worked Out Examples

Example 1: Find $J_0(x)$ and $J_1(x)$.

Solution: Put $n = 0$ in

$$J_n(x) = \sum_{m=0}^{\infty}\frac{(-1)^m x^{2m+n}}{2^{2m+n}\cdot m!(m+n)!}$$

Then $$J_0(x) = \sum_{m=0}^{\infty}\frac{(-1)^m x^{2m}}{2^{2m}(m!)^2} = 1 - \frac{1}{1!}\left(\frac{x}{2}\right)^2$$
$$+\frac{1}{(2!)^2}\left(\frac{x}{2}\right)^4 - \left(\frac{1}{3!}\right)^2\left(\frac{x}{2}\right)^6 + \cdots$$

For $n = 1$,

$$J_1(x) = \sum_{m=0}^{\infty} \frac{(-1)^m x^{2m+1}}{2^{2m+1} m!(m+n)!}$$

or
$$J_1(x) = \frac{x}{2}\left[1 - \frac{1}{1!2!}\left(\frac{x}{2}\right)^2 + \frac{1}{2!3!}\left(\frac{x}{2}\right)^4 - \frac{1}{3!4!}\left(\frac{x}{2}\right)^6 + \cdots\right].$$

Example 2: Show that $J_n(x)$ is an even function when n is even and odd function when n is odd.

Solution: Suppose n is even.

$$J_n(-x) = (-x)^n \sum_{m=0}^{\infty} \frac{(-1)^m(-x)^{2m}}{2^{2m+n} \cdot m!(m+n)!}$$

For n even $(-1)^n = 1$ and $(-1)^{2m} = 1$

$$J_n(-x) = x^n \sum_{m=0}^{\infty} \frac{(-1)^m x^{2m}}{2^{2m+n} \cdot m!(m+n)!} = J_n(x)$$

Thus $J_n(x)$ is an even function.

Suppose n is odd. Then $(-1)^n = -1$, so

$$J_n(-x) = -x^n \sum_{m=0}^{\infty} \frac{(-1)^m x^{2m}}{2^{2m+n} m!(m+n)!} = -J_n(x).$$

Thus $J_n(x)$ is an odd function.

Example 3: Express $J_6(x)$ in terms of $J_0(x)$ and $J_1(x)$.

Solution: Rewriting the recurrence relation (VI)

$$J_{n+1}(x) = \frac{2n}{x} J_n(x) - J_{n-1}(x) \quad (1)$$

Put $n = 1, 2, 3, 4, 5$ in (1), we get (suppressing the argument x)

$$J_2 = \frac{2}{x} J_1 - J_0 \quad (2)$$

$$J_3 = \frac{4}{x} J_2 - J_1 \quad (3)$$

$$J_4 = \frac{6}{x} J_3 - J_2 \quad (4)$$

$$J_5 = \frac{8}{x} J_4 - J_3 \quad (5)$$

$$J_6 = \frac{10}{x} J_5 - J_4 \quad (6)$$

Substituting (3) in (2),

$$J_3 = \left\{\frac{8}{x^2} - 1\right\} J_1 - \frac{4}{x} J_0 \quad (7)$$

Substituting (7) and (2) in (4),

$$J_4 = \left[\frac{48}{x^3} - \frac{8}{x}\right] J_1 + \left(1 - \frac{24}{x^2}\right) J_0 \quad (8)$$

Substituting (8), and (3) in (5)

$$J_5 = \left(\frac{384}{x^4} - \frac{72}{x^2} - 1\right) J_1 + \left(\frac{12}{x} - \frac{192}{x^3}\right) J_0 \quad (9)$$

Substituting (9) and (8) in (6), we get

$$J_6 = \frac{10}{x}\left\{\left(\frac{384}{x^4} - \frac{72}{x^2} - 1\right) J_1 + \left(\frac{12}{x} - \frac{192}{x^3}\right) J_0\right\} - \left(\frac{48}{x^3} - \frac{8}{x}\right) J_1 - \left(1 - \frac{24}{x^2}\right) J_0$$

$$J_6(x) = \left(\frac{3840}{x^4} - \frac{768}{x^3} - \frac{2}{x}\right) J_1(x) + \left(\frac{144}{x^2} - 1 - \frac{1920}{x^4}\right) J_0(x).$$

Example 4: Express $J_{\frac{7}{2}}(x)$ in terms of sine and cosine functions.

Solution: Put $n = \frac{5}{2}$ in the recurrence relation

$$J_{n+1}(x) = \frac{2n}{x} J_n(x) - J_{n-1}(x) \quad (1)$$

so
$$J_{\frac{7}{2}}(x) = \frac{5}{x} J_{\frac{5}{2}}(x) - J_{\frac{3}{2}}(x) \quad (2)$$

Put $n = \frac{3}{2}$ in (1) then

$$J_{\frac{5}{2}}(x) = \frac{3}{x} J_{\frac{3}{2}}(x) - J_{\frac{1}{2}}(x) \quad (3)$$

Put $n = \frac{1}{2}$ in (1) then

$$J_{\frac{3}{2}}(x) = \frac{1}{x} J_{\frac{1}{2}}(x) - J_{-\frac{1}{2}}(x)$$

Since
$$J_{\frac{1}{2}}(x) = \sqrt{\frac{2}{\pi x}} \sin x, \; J_{-\frac{1}{2}}(x) = \sqrt{\frac{2}{\pi x}} \cos x$$

$$J_{\frac{3}{2}}(x) = \sqrt{\frac{2}{\pi x}}\left(\frac{\sin x}{x} - \cos x\right) \quad (4)$$

Substituting (4) and $J_{\frac{1}{2}}(x)$ in (3)

$$J_{\frac{5}{2}}(x) = \frac{3}{x}\left[\sqrt{\frac{2}{\pi x}}\left(\frac{\sin x}{x} - \cos x\right)\right] - \sqrt{\frac{2}{\pi x}} \sin x$$

$$J_{\frac{5}{2}}(x) = \sqrt{\frac{2}{\pi x}}\left[\left(\frac{3 - x^2}{x^2}\right) \sin x - \frac{3}{x} \cos x\right] \quad (5)$$

Substituting (5) and (4) in (2), we get

$$J_{\frac{7}{2}}(x) = \frac{5}{x}\cdot\sqrt{\frac{2}{\pi x}}\left[\left(\frac{3-x^2}{x^2}\right)\sin x - \frac{3}{x}\cos x\right]$$

$$-\sqrt{\frac{2}{\pi x}}\left(\frac{\sin x}{x} - \cos x\right)$$

$$= \sqrt{\frac{2}{\pi x}}\left[\left(\frac{15-6x^2}{x^3}\right)\sin x + \left(\frac{15}{x^2}-1\right)\cos x\right].$$

Example 5: Show that $J_1''(x) = -J_1(x) + \frac{1}{x}J_2(x)$.

Solution: Put $n = 1$ in the recurrence relation III

$$J_n'(x) = J_{n-1}(x) - \frac{n}{x}J_n(x) \quad (1)$$

Then $$J_1'(x) = J_0(x) - \frac{J_1(x)}{x} \quad (2)$$

Differentiating (2) w.r.t., 'x'

$$J_1''(x) = J_0'(x) + \frac{1}{x^2}J_1(x) - \frac{1}{x}J_1'(x) \quad (3)$$

Put $n = 0$, in (1) then

$$J_0'(x) = J_{-1}(x) - 0 \quad (4)$$

But $$J_{-n}(x) = (-1)^n J_n(x)$$

So with $n = +1$, $J_{-1}(x) = -J_1(x)$ (5)

Substituting (5) in (4),

$$J_0'(x) = J_{-1}(x) = J_1(x) \quad (6)$$

Put (6) and (2) in (3), we get

$$J_1'' = -J_1(x) + \frac{1}{x^2}J_1(x) - \frac{1}{x}\left\{J_0(x) - \frac{J_1(x)}{x}\right\} \quad (7)$$

Put $n = 1$ in recurrence relation (VI)

$$J_{n+1}(x) = \frac{2n}{x}J_n(x) - J_{n-1}(x)$$

Then $$J_2(x) = \frac{2}{x}J_1(x) - J_0(x) \quad (8)$$

or $$J_0(x) = \frac{2}{x}J_1(x) - J_2(x) \quad (9)$$

Using (9) eliminate J_0 from (7) then

$$J_1'' = -J_1(x) + \frac{1}{x^2}J_1(x) - \frac{1}{x}\left\{\frac{2}{x}J_1(x) - J_2(x)\right\} + \frac{J_1(x)}{x^2}$$

$$J_1''(x) = -J_1(x) + \frac{1}{x}J_2(x).$$

Example 6: Prove that

$$\frac{d}{dx}\left\{J_n^2(x)\right\} = \frac{x}{2n}\left[J_{n-1}^2(x) - J_{n+1}^2(x)\right]$$

Solution: $\frac{d}{dx}\left\{J_n^2(x)\right\} = 2\cdot J_n(x)J_n'(x)$.

Using recurrence relation (V)

$$J_n'(x) = \frac{1}{2}\left\{J_{n-1} - J_{n+1}\right\}$$

$$\frac{d}{dx}\{J_n^2\} = 2\cdot J_n(x)\cdot\left[\frac{1}{2}\left(J_{n-1} - J_{n+1}\right)\right]$$

From recurrence relation VI

$$J_n(x) = \frac{x}{2n}\left[J_{n-1}(x) + J_{n+1}(x)\right]$$

$$\frac{d}{dx}\left\{J_n^2(x)\right\} = \frac{x}{2n}\left[J_{n-1}(x) + J_{n+1}(x)\right] \times \left[J_{n-1}(x) - J_{n+1}(x)\right]$$

$$= \frac{x}{2n}\left[J_{n-1}^2(x) - J_{n+1}^2(x)\right].$$

Example 7: Show that

$$J_2'(x) = \left(1 - \frac{4}{x^2}\right)J_1(x) + \frac{2}{x}J_0(x).$$

Solution: In recurrence relation

$$J_n'(x) = J_{n-1}(x) - \frac{n}{x}J_n(x)$$

Put $n = 2$, $$J_2'(x) = J_1(x) - \frac{2}{x}J_2(x) \quad (1)$$

Since $$J_{n+1}(x) + J_{n-1}(x) = \frac{2n}{x}J_n(x)$$

for $n = 1$, $$J_2(x) + J_0(x) = \frac{2}{x}J_1(x) \quad (2)$$

Substituting J_2 from (2) in (1)

$$J_2'(x) = J_1(x) - \frac{2}{x}\left[\frac{2}{x}J_1(x) - J_0(x)\right]$$

$$J_2'(x) = \left(1 - \frac{4}{x^2}\right)J_1(x) + \frac{2}{x}J_0(x).$$

Example 8: Evaluate $\int J_5(x)dx$.

Solution: Putting $n = 4$ in

$$\int J_{p+1}(x)dx = \int J_{p-1}(x)dx - 2J_p(x),$$

we get

$$\int J_5(x)dx = \int J_3(x)dx - 2J_4(x) \qquad (1)$$

Again with $p = 2$

$$\int J_3(x)dx = \int J_1(x)dx - 2J_2(x) \qquad (2)$$

Also we know that

$$\int J_1(x)dx = -J_0(x) + c \qquad (3)$$

Substituting (2) and (3) in (1)

$$\int J_5(x)dx = \left[-J_0(x) + c\right] - 2J_2(x) - 2J_4(x)$$

Example 9: Evaluate $\int x^2 J_1(x)dx$.

Solution: Put $p = 2$ in

$$\int x^p J_{p-1}(x)dx = x^p J_p(x) + c$$

Then $$\int x^2 J_1(x)dx = x^2 J_2(x) + c$$

But $$J_2(x) = \left[\frac{2}{x}J_1(x) - J_0(x)\right]$$

$$\int x^2 J_1(x)dx = x^2\left[\frac{2}{x}J_1(x) - J_0(x)\right] + c$$

$$= 2xJ_1(x) - x^2 J_0(x) + c.$$

Example 10: Evaluate $\int x^3 J_3(x)dx$.

Solution: Integrating by parts (suppressing argument x)

$$\int x^3 J_3 dx = \int x^5[x^{-2}J_3]dx = x^5[x^{-2}J_2]$$

$$- \int -x^{-2}J_2 \cdot 5x^4 dx$$

$$= -x^3 J_2 + 5\int x^2 J_2 dx$$

Now $$\int x^2 J_2 dx = \int x^3[x^{-1}J_2]dx = x^3[-x^{-1}J_1]$$

$$- \int -x^{-1}J_1 3x^2 dx$$

$$= -x^2 J_1 + 3\int xJ_1\, dx$$

But $$\int xJ_1(x)dx = -\int xJ_0'(x)dx$$

$$= -\left[xJ_0 - \int J_0(x)dx\right]$$

Substituting

$$\int x^3 J_3(x)dx = -x^3 J_2(x) + 5\Big\{-x^2 J_1(x)$$

$$+3\left[-xJ_0(x) + \int J_0 dx\right]\Big\}$$

$$= -x^3 J_2(x) - 5x^2 J_1(x) - 15xJ_0(x)$$

$$+15\int J_0(x)dx.$$

Note: $\int J_0(x)dx$ can not be integrated but its values are tabulated.

EXERCISE

1. Show that **a.** $J_0'(x) = -J_1(x)$

   b. $\frac{d}{dx}(xJ_1) = xJ_0$.

2. Express $J_{\frac{3}{2}}$, $J_{-\frac{3}{2}}$ in terms of sin and cos.

   Ans. $J_{\frac{3}{2}} = \sqrt{\frac{2}{\pi x}}\left(\frac{\sin x}{x} - \cos x\right)$;

   $J_{-\frac{3}{2}} = -\sqrt{\frac{2}{\pi x}}\left(\sin x + \frac{\cos x}{x}\right)$

3. Express $J_{\frac{5}{2}}$, $J_{-\frac{5}{2}}$ in terms of sin and cos.

   Ans. $J_{\frac{5}{2}} = \sqrt{\frac{2}{\pi x}}\left\{\frac{3-x^2}{x^2}\sin x - \frac{3}{x}\cos x\right\}$

   $J_{-\frac{5}{2}} = \sqrt{\frac{2}{\pi x}}\left\{\frac{3}{x}\sin x + \left(\frac{3}{x^2} - 1\right)\cos x\right\}$

4. Express $J_4(x)$ in terms of J_0 and J_1.

   Ans. $J_4 = \left(\frac{48}{x^3} - \frac{8}{x}\right)J_1(x) - \left(\frac{24}{x^2} - 1\right)J_0(x)$

5. Express $J_5(x)$ in terms of J_0 and J_1.

   Ans. $J_5 = \left(\frac{384}{x^4} - \frac{72}{x^2} - 1\right)J_1(x) + \left(\frac{12}{x} - \frac{192}{x^3}\right)J_0(x)$

6. Prove that $J_n'' = \frac{1}{4}[J_{n-2} - 2J_n + J_{n+2}]$.

7. Show that $\frac{d}{dx}\{xJ_n \cdot J_{n+1}\} = x[J_n^2 - J_{n+1}^2]$.

8. Prove that $J_0'' = \frac{1}{2}[J_2 - J_0]$.

9. Show that $\frac{d}{dx}\left[J_n^2+J_{n+1}^2\right]=2\left[\frac{n}{x}J_n^2-\frac{n+1}{x}J_{n+1}^2\right]$.

10. Prove that $J_n'''=\frac{1}{8}[J_{n-3}-3J_{n-1}+3J_{n+1}-J_{n+3}]$.

11. Show that $2J_0''(x) = J_2(x) - J_0(x)$.

12. Evaluate $\int J_3(x)dx$.

Ans. $-2J_1 - 2J_2 + c + \int J_0(x)dx$.

13. Evaluate: $\int x^3 J_0(x)dx$

Ans. $x^3 J_1(x) - 2x^2 J_2(x) + c$

14. Evaluate: $\int x^4 J_1(x)dx$

Ans. $(8x^2 - x^4)J_0(x) + (4x^3 - 16x)J_1(x)$

15. Establish the Jacobi series:

a. $\cos(x\cos\theta) = J_0 - 2J_2\cos 2\theta + 2J_4\cos 4\theta - \cdots$

b. $\sin(x\cos\theta) = 2[J_1\cos\theta - J_3\cos 3\theta + J_3\cos 5\theta - \cdots]$.

Hint: $e^{\frac{x}{2}\left(t-\frac{1}{t}\right)} = J_0 + \left(t-\frac{1}{t}\right)J_1 + \left(t^2-\frac{1}{t^2}\right)J_2 + \left(t^3-\frac{1}{t^3}\right)J_3 + \cdots$ obtained from generating function using $J_{-n}(x) = (-1)^n J_n(x)$. Put $t = \cos\theta + i\sin\theta$, $\frac{1}{t} = \cos\theta - i\sin\theta$, then $t^p + \frac{1}{t^p} = 2\cos p\theta$, $t^p - \frac{1}{t^p} = 2i\sin p\theta$, $t - \frac{1}{t} = 2i\sin\theta$. Equate real and imaginary parts. Replace θ by $\frac{\pi}{2} - \theta$.

11.4 DIFFERENTIAL EQUATIONS REDUCIBLE TO BESSEL'S EQUATION

Various differential equations which are not Bessel's equations can be reduced to Bessel's equation by changing the dependent or and independent variable.

The differential equation

$$x^2y''+x(a+2bx^p)y'+[c+dx^{2q}+b(a+p-1)x^p + b^2x^{2p}]y = 0 \quad (1)$$

can be transformed to Bessel's equation in the new variables X and Y where

$$y = x^{(1-a)/2}e^{-(b/p)x^p}\,Y \text{ and } x = \left(\frac{qX}{\sqrt{|d|}}\right)^{1/q}.$$

Then the general solution of (1) is given by

$$y(x) = x^{\alpha}e^{-\beta x^p}[c_1 J_\upsilon(\lambda x^q) + c_2 Y_\upsilon(\lambda x^q)] \quad (2)$$

Here $\alpha = \frac{1-a}{2}$, $\beta = \frac{b}{p}$, $\lambda = \frac{\sqrt{|d|}}{q}$, $\upsilon = \frac{\sqrt{(1-a)^2-4c}}{2q}$ (3)

We assume that $d \neq 0$, $p \neq 0$, $q \neq 0$ and $(1-a^2) \geq 4c$. Also if $d < 0$, replace J_ν and Y_ν by I_ν and k_ν respectively. When υ is not an integer, replace Y_ν and k_ν by $J_{-\nu}$ and $I_{-\nu}$ respectively.

Corollary: The general solution of

$$x^r y'' + rx^{r-1}y' + (ax^s + bx^{r-2})y = 0 \quad (4)$$

is $y = x^\alpha[c_1 J_\nu(\lambda x^\gamma) + c_2 Y_\nu(\lambda x^\gamma)]$ (5)

where $\alpha = \frac{1-r}{2}$, $\gamma = \frac{2-r+s}{2}$, $\lambda = \frac{2\sqrt{|a|}}{2-r+s}$,

$$\upsilon = \frac{\sqrt{(1-r)^2 - 4b}}{2-r+s} \quad (6)$$

we assume that $2 - r + s \neq 0$ and $(1-r)^2 \geq 4b$. If $a < 0$, replace J_ν and y_ν by I_ν and k_ν respectively.

WORKED OUT EXAMPLES

Reduce the given differential equation to the Bessel's equation and solve.

Example 1: $y'' + \left(\varepsilon^2 - \frac{4n^2-1}{4x^2}\right)y = 0$

Solution: Rewriting the given D.E.

$$x^2y'' + \left(\varepsilon^2x^2 - \frac{4n^2-1}{4}\right)y = 0.$$

Comparing this with D.E. (1), we have $a = 0$, $b = 0$, $c = -\frac{(4n^2-1)}{4}$, $d = \varepsilon^2$, $q = 1$. So from relations (3), $\alpha = \frac{1-0}{2} = \frac{1}{2}$, $\beta = 0$, $\lambda = \frac{\varepsilon}{1}$, $\upsilon = \frac{\sqrt{1+4n^2-1}}{1} = 2n$.

Then the general solution of the given D.E. is $y(x) = x^{1/2}[c_1 J_{2n}(\varepsilon x) + c_2 Y_{2n}(\varepsilon x)]$

Example 2: $y'' + 2y' + \left(x^2 + 1 - \frac{2}{x^2}\right)y = 0.$

Solution: Rewriting

$$x^2y'' + 2x^2y' + (x^4 + x^2 - 2)y = 0$$

Comparing with D.E. (1), we have $a = 0$, $b = 1$, $p = 1, c = -2, d = 1, q = 2$. Then from (3) $\alpha = \frac{1}{2}$, $\beta = \frac{1}{1} = 1, \lambda = \frac{1}{2}, v = \frac{\sqrt{(1-0)^2-4(-2)}}{4} = \frac{3}{4}$. Thus the general solution of given D.E. is

$$y = x^{1/2}e^{-x}\left[c_1 J_{3/4}\left(\frac{1}{2}x^2\right) + c_2 Y_{\frac{3}{4}}\left(\frac{1}{2}x^2\right)\right]$$

since $v = \frac{3}{4}$ is not an integer, we can replace $Y_{\frac{3}{4}}$ by $J_{-3/4}$, thus the general solution takes the form

$$y = \sqrt{x}e^{-x}\left[c_1 J_{3/4}\left(\frac{x^2}{2}\right) + c_2 J_{-3/4}\left(\frac{x^2}{2}\right)\right]$$

Example 3: $(x^5y')' = y$

Solution: Rewriting $x^5y'' + 5x^4y' - y = 0$. Comparing this equation with (4), we have $r = 5$, $a = -1, s = 0, b = 0$ then from relations (5), $\alpha = \frac{1-5}{2} = -2$, $\gamma = \frac{2-5+0}{2} = -\frac{3}{2}$, $\lambda = \frac{2}{2-5+0} = \frac{-2}{3}$ and $\nu = \frac{\sqrt{(1-5)^2-0}}{2-5+0} = \frac{4}{-3}$. Thus the general solution from (6) is $y = x^{-2}\left[c_1 I_{4/3}\left(-\frac{2}{3}x^{-3/2}\right) + c_2 I_{-4/3}\left(-\frac{2}{3}x^{-3/2}\right)\right]$ since $a = -1 < 0$

Example 4: Show that the general solution of $y'' + \frac{1}{x}y' + \left(1 - \frac{1}{4x^2}\right)y = 0$ is $\sqrt{x} \cdot y = c_1 \sin x + c_2 \cos x$.

Solution: Comparing with D.E. (1), we have $a = 1$, $b = 0$, $c = -\frac{1}{4}$, $d = 1$, $q = 1$. So from (3) $\alpha = 0$, $\beta = 0$, $\lambda = 1$, $v = \frac{1}{2}$. Thus the general solution is $y(x) = c_1\ J_{\frac{1}{2}}(x) + c_2 Y_{\frac{1}{2}}(x)$. Since $v = \frac{1}{2}$ is not an integer,

$$y(x) = c_1 J_{1/2}(x) + c_2 J_{-1/2}(x)$$

$$= c_1\sqrt{\frac{2}{\pi x}} \sin x + c_2 \frac{\sqrt{2}}{\pi x} \cos x$$

or $\quad \sqrt{x}\, y(x) = c_1^* \sin x + c_2^* \cos x$

EXERCISE

Reduce the given differential equations to the Bessel's equation and solve.

1. $y'' + \frac{1}{2x}y' + \frac{1}{16}\left(x^{-3/2} + \frac{15}{16}x^{-2}\right)y = 0$

Ans. $y = x^{1/4}(c_1 J_{1/4}(x^{1/4}) + c_2 J_{-1/4}(x^{1/4}))$

Hint: $a = \frac{1}{2}$, $b = 0$, $c = \frac{15}{256}$, $d = \frac{1}{16}$, $q = \frac{1}{4}$, $\alpha = \frac{1}{4}$, $\beta = 0$, $\lambda = 1$, $v = \frac{1}{4}$

2. $81x^2y'' + 27xy' + (9x^{2/3} + 8)y = 0$

Ans. $y(x) = x^{1/3}[c_1 J_{1/3}(x^{1/3}) + c_2 J_{-1/3}(x^{1/3})]$

Hint: $a = \frac{1}{3}$, $b = 0$, $c = \frac{8}{81}$, $d = \frac{1}{9}$, $q = \frac{1}{3}$, $\alpha = \frac{1}{3}$, $\beta = 0$, $\lambda = 1$, $v = \frac{1}{3}$

3. $y'' + 3\sqrt{x}y = 0$

Ans. $y(x) = \sqrt{x}\left[c_1 J_{2/5}\left(\frac{4}{5}\sqrt{3}x^{5/4}\right) + c_2 Y_{2/5}\left(\frac{4}{5}\sqrt{3}x^{5/4}\right)\right]$

Hint: $a = 0, b = 0, c = 0, d = 3, q = \frac{5}{4}, \alpha = \frac{1}{2}$, $\beta = 0$, $\lambda = \frac{\sqrt{3}}{(5/4)}$, $v = \frac{2}{5}$.

4. $xy'' + 3y' + y = 0$

Ans. $y(x) = x^{-1}\left[c_1 J_2(2\sqrt{x}) + c_2 Y_2\left(2\sqrt{x}\right)\right]$

Hint: $a = 3, b = 0, c = 0, d = 1, q = \frac{1}{2}, \alpha = -1$, $\beta = 0$, $\lambda = 2$, $v = 2$

5. $(xy')' - 5x^3y = 0$

Ans. $y(x) = c_1\ I_0\left(\frac{\sqrt{5}}{2}x^2\right) + c_2 k_0\left(\frac{\sqrt{5}}{2}x^2\right)$

Hint: $a = 1$, $b = 0$, $c = 0$, $d = -5$, $q = 2$, $\alpha = 0$, $\beta = 0$, $\lambda = \frac{\sqrt{5}}{2}$, $v = 0$

6. $y'' + \frac{y'}{x} + \left(1 - \frac{1}{9x^2}\right)y = 0$

Ans. $y(x) = c_1 J_{1/3}(x) + c_2 J_{-1/3}(x)$

Hint: $a = 1$, $b = 0$, $c = -\frac{1}{9}$, $d = 1$, $q = 1$, $\alpha = 0$, $\beta = 0$, $\lambda = 1$, $v = \frac{1}{3}$

7. $x^2y'' + xy' + \left(x^2 - \frac{1}{6.25}\right)y = 0$

Ans. $y(x) = c_1 J_{2/5}(x) + c_2 J_{-2/5}(x)$

Hint: $a = 1, b = 0, c = -\frac{1}{6.25}, d = 1, q = 1$, $\alpha = 0$, $\beta = 0$, $v = \frac{2}{5}$

8. $x^2y'' + x(4x^4 - 3)y' + (4x^8 - 5x^2 + 3)y = 0$

Ans. $y(x) = x^2 e^{-x^{4/2}}[c_1 I_1(\sqrt{5}x) + c_2 k_1\sqrt{5}x)]$

Hint: $a = -3, b = 2, p = 4, c = 3, d = -5, q = 1, \alpha = 2, \beta = \frac{1}{2}, \lambda = \sqrt{5}, v = 1$

9. $\left(\frac{1}{x}y'\right)' + \left(\frac{1}{x^2} + \frac{1}{x^3}\right)y = 0$

Ans. $y = x[c_1 J_0(2\sqrt{x}) + c_2 Y_0(2\sqrt{x})]$

Hint: $r = -1,\ s = -2,\ a = b = 1,\ \alpha = 1, v = \frac{1}{2}, \lambda = 2, v = 0$

10. $x^2y'' - 2xy' + (2 - x^3)y = 0$

Ans. $y = x^{3/2}\left[c_1 I_{1/3}\left(\frac{2}{3}x^{3/2}\right) + c_2 I_{-1/3}\left(\frac{2}{3}x^{3/2}\right)\right]$

Hint: $a = -2, b = 0, c = 2, d = -1, q = \frac{3}{2}, \alpha = \frac{3}{2}, \beta = 0, \lambda = 2/3, v = 1/3$

11. $x^2y'' + xy' + 8x^2 = y$

Ans. $y = c_1 J_1(2\sqrt{2}x) + c_2 Y_1(2\sqrt{2}x)$

Hint: $a = 1,\ b = 0,\ c = -1,\ d = 8,\ q = 1, \alpha = 0, \beta = 0, \lambda = 2\sqrt{2}, v = 1$

12. $4y'' + 9xy = 0$

Ans. $y = \sqrt{x}\left[c_1 J_{1/3}(x^{3/2}) + c_2 J_{-1/3}(x^{3/2})\right]$

Hint: $a = 0, b = 0, c = 0, d = \frac{9}{4}, q = \frac{3}{2}, \alpha = \frac{1}{2}, \beta = 0, \lambda = 1, v = \frac{1}{3}$

13. $9x^2y'' + 9xy' + (36x^4 - 16)y = 0$

Ans. $y(x) = c_1 J_{2/3}(x^2) + c_2 J_{-2/3}(x^2)$

Hint: $a = 1,\ b = 0,\ c = -\frac{16}{9},\ d = 4,\ q = 2, \alpha - 0, \beta - 0, \lambda - 1, v = \frac{2}{3}$

14. $y'' + k^2x^4y = 0$

Ans. $y = \sqrt{x}\left[c_1 J_{1/4}\left(\frac{1}{2}kx^2\right) + c_2 Y_{1/4}\left(\frac{1}{2}kx^2\right)\right]$

Hint: $a = 0,\ b = 0,\ c = 0,\ d = k^2,\ q = 2, \alpha = \frac{1}{2}, \beta = 0, \lambda = \frac{k}{2}, v = \frac{1}{4}$

15. $2xy'' + 4y' + xy = 0$

Ans. $y = [c_1 J_{1/2}(x/\sqrt{2}) + c_2 J_{-1/2}(x/\sqrt{2})]x^{-1/2}$

Hint: $a = 2,\ b = 0,\ c = 0,\ d = \frac{1}{2},\ q = +1, \alpha = -\frac{1}{2}\ \beta = 0, \lambda = \frac{1}{\sqrt{2}}, v = \frac{1}{2}.$

11.5 LEGENDRE* FUNCTIONS

The boundary value problems with spherical symmetry (independent of θ) by the application of separation of variables reduces two ordinary differential equations. One of them is the very important differential equation

$$(1 - x^2)\frac{d^2y}{dx^2} - 2x\frac{dy}{dx} + n(n+1)y = 0 \tag{1}$$

known as the Legendre's differential equation.

The parameter n is given integer, (although it could be a real number).

The solution of Legendre's Equation (1) is known as Legendre's function of order n.

Assume a power series solution of (1) as

$$y(x) = \sum_{m=0}^{\infty} a_m x^m \tag{2}$$

Substitute (2) and its derivatives in (1), then

$$(1 - x^2)\sum_{m=2}^{\infty} m(m-1)a_m x^{m-2} - 2x\sum_{m=1}^{\infty} m a_m x^{m-1} + k\sum_{m=0}^{\infty} a_m x^m = 0$$

where $k = n(n+1)$. Rewriting

$$\sum_{m=2}^{\infty} m(m-1)a_m x^{m-2} - \sum_{m=2}^{\infty} m(m-1)a_m x^m - 2\sum_{m=1}^{\infty} m a_m x^m + k\sum_{m=0}^{\infty} a_m x^m = 0 \tag{3}$$

(3) is an identity since (2) is a solution of (1). So equate the sum of the coefficients of each power of x to zero.

Coefficient of x^0 arise from 1st and fourth series in (3). Thus

$$2a_2 + n(n+1)a_0 = 0 \tag{4}$$

coefficient of x^1 arise from 1st, 3rd and 4th series in (3). So

$$6a_3 + [-2 + n(n+1)]a_1 = 0 \tag{5}$$

All the four series in (3) contribute coefficients of x^s for $s \geq 2$. Thus

$$(s+2)(s+1)a_{s+2} + [-s(s-1) - 2s + n(n+1)]a_s = 0 \tag{6}$$

Solving (6)

$$a_{s+2} = -\frac{(n-s)(n+s+1)}{(s+2)(s+1)}a_s \quad \text{for} \quad s = 0, 1, 2, \ldots \tag{7}$$

---

* Adrien Marie Legendre (1752–1833), French mathematician.

since $-s(s-1)-2s+n(n+1)$

$$=-s^2+s-2s+n^2+n$$
$$=(n^2-s^2+n-s)$$
$$=(n-s)(n+s+1).$$

(7) is known as a recurrence relation or recursion formula, which determines all coefficients in terms of a_1 or a_0. Here a_0 and a_1 are arbitrary constants, to be chosen appropriately. Thus

$$a_2=\frac{-(n)(n+1)}{2!}a_0, \quad a_3=\frac{-(n-1)(n+2)}{3!}a_1$$
$$a_4=\frac{-(n-2)(n+3)}{4\cdot 3}a_2, \quad a_5=\frac{-(n-3)(n+4)}{5\cdot 4}a_3$$
$$a_4=\frac{(n-2)n(n+1)(n+3)}{4!}a_0,$$
$$a_5=\frac{-(n-3)(n-1)(n+2)(n+4)}{5!}a_1$$

In general, the coefficients with even subscripts are

$$a_{2m}=\frac{-(n-2m+2)(n+2m-1)}{(2m)(2m-1)}a_{2m-2}$$

and the coefficients with odd subscripts are

$$a_{2m+1}=\frac{-(n-2m+1)(n+2m)}{(2m)(2m+1)}a_{2m-1}$$

Substituting these coefficients in (2), we get

$$y(x)=a_0y_1(x)+a_1y_2(x) \tag{8}$$

where $y_1(x)=1-\frac{n(n+1)}{2!}x^2+\frac{(n-2)n(n+1)(n+3)}{4!}x^4-\cdots$ (9)

and $y_2(x)=x-\frac{(n-1)(n+2)}{3!}x^3+\frac{(n-3)(n-1)(n+2)(n+4)}{5!}x^5-\cdots$ (10)

Both the series (9) and (10) converge for $|x|<1$. y_1 and y_2 are linearly independent (i.e., y_1/y_2 is not a constant) because y_1 contains only even powers of x while y_2 contains only odd powers of x. Thus $y(x)$ given by (8) is the general solution of Legendre's Equation (1) and is valid for $-1<x<1$.

Legendre Polynomials

Assume that the parameter n is non-negative integer i.e., $n \geq 0$, n integer. Since $(n-s)$ appears in the recurrence relation (7), with $s=n$, the coefficients $a_{n+2}, a_{n+4}, a_{n+6}$, etc., are all zero i.e., $a_{m+2}=0$ when $m \geq n$. If n is even, then $y_1(x)$ reduces to a polynomial of degree n (while $y_2(x)$ remains an infinite series).

Similarly if n is odd, then $y_2(x)$ becomes a polynomial of degree n (while $y_1(x)$ remains an infinite series).

In either of these cases, the series which reduces to a finite sum (a polynomial), multiplied by some constant, is known as the Legendre polynomial or zonal harmonic of order n denoted by $P_n(x)$. The series which remain infinite is known as the Legendre's function of the second kind denoted by $Q_n(x)$. Thus for a non-negative integer n, the general solution (2) of Legendre's Equation (1) is the sum a polynomial solution and an infinite series solution i.e.,

$$y(x)=AP_n(x)+BQ_n(x) \tag{11}$$

Note: $Q_n(x)$ is unbounded at $x=\pm 1$.

Derivation of Legendre Polynomial $P_n(x)$

Rewriting (7),

$$a_m=\frac{-(m+2)(m+1)}{(n-m)(n+m+1)}a_{m+2} \quad \text{for } m \leq n-2 \tag{12}$$

(12) expresses all non-vanishing coefficients in terms of a_n, which is coefficient of the highest power of x i.e., x^n in the polynomial. The arbitrary coefficient a_n may be chosen as

$$a_n=1 \quad \text{for } n=0$$

and

$$a_n=\frac{(2n)!}{2^n(n!)^2} \quad \text{for } n=1,2,\ldots \tag{13}$$

For this choice of a_n, $P_n(x=1)=1$.

The non-vanishing coefficients are obtained from (12) and (13)

$$a_{n-2}=\frac{-n(n-1)}{2(2n-1)}a_n \quad (\text{for } m=n-2)$$
$$=\frac{-n(n-1)}{2(2n-1)}\cdot\frac{(2n)!}{2^n(n!)^2}$$

$$= \frac{-n(n-1)\cdot 2n(2n-1)(2n-2)!}{2(2n-1)2^n\cdot n(n-1)!n(n-1)(n-2)!}$$

$$a_{n-2} = \frac{-(2n-2)!}{2^n(n-1)!(n-2)!}$$

Similarly, for $m = n-4$, we get

$$a_{n-4} = \frac{-(n-2)(n-3)}{4(2n-3)} a_{n-2} = \frac{(2n-4)!}{2^n 2!(n-2)!(n-4)!}$$

In general for $n - 2m \geq 0$

$$a_{n-2m} = \frac{(-1)^m(2n-2m)!}{2^n m!(n-m)!(n-2m)!} \tag{14}$$

By inserting these coefficients (14) in (9) or (10), we get a polynomial $P_n(x)$ known as the Legendre's polynomial of degree n and is given by (Fig. 11.1)

$$P_n(x) = \sum_{m=0}^{M} \frac{(-1)^m(2n-2m)!}{2^n m!(n-m)!(n-2m)!}\cdot x^{n-2m} \tag{15}$$

where $M = \frac{n}{2}$ or $\frac{n-1}{2}$ according as n is even or n is odd (whichever is an integer).

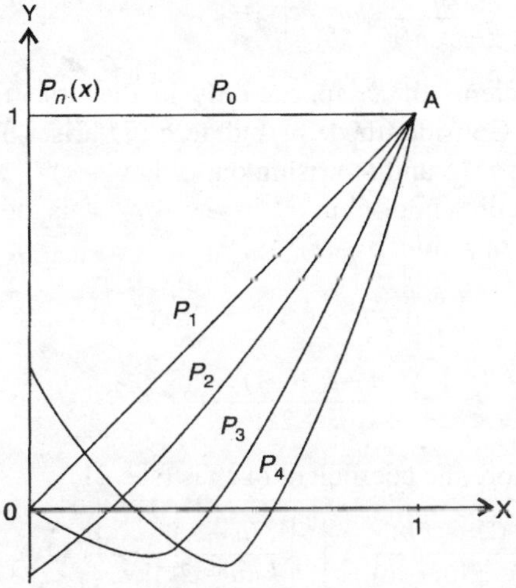

Fig. 11.1

In particular,

$P_0(x) = 1$ (for $n = 0, M = 0$)

$P_1(x) = x$ (for $n = 1, M = 0$)

$P_2(x) = \frac{1}{2}(3x^2 - 1)$ (for $n = 2, M = 1, m = 0$ to 1)

$P_3(x) = \frac{1}{2}(5x^3 - 3x)$ (for $n = 3, M = 1, m = 0$ to 1)

$P_4(x) = \frac{1}{8}(35x^4 - 30x^2 + 3)$

(for $n = 4, M = 2, m = 0$ to 2)

$P_5(x) = \frac{1}{8}(63x^5 - 70x^3 + 15x)$

(for $n = 5, M = 2, m = 0$ to 2)

$P_6(x) = \frac{1}{16}(231x^6 - 315x^4 + 105x^2 - 5)$

(for $n = 6, M = 3, m = 0$ to 3).

Note 1: At $x = 1$, $P_n(x = 1) = P_n(1) = 1$.

Note 2: Any polynomial $f(x)$ of degree n can be expressed in terms of $P_n(x)$ as

$$f(x) = \sum_{m=0}^{n} c_m P_m(x).$$

Rodrigue's* Formula

Show that

$$P_n(x) = \frac{1}{n!2^n}\frac{d^n}{dx^n}(x^2-1)^n \tag{16}$$

Proof: Let

$$v = (x^2-1)^n$$

Then $\frac{dv}{dx} = n(x^2-1)^{n-1}\cdot 2x$

or $-(1-x^2)\frac{dv}{dx} = 2nx\cdot(x^2-1)^n = 2nxv$

i.e., $$(1-x^2)\frac{dv}{dx} + 2nxv = 0 \tag{17}$$

Differentiating (17), $(n+1)$ times by Leibnitz's rule

$$\left[(1-x^2)\frac{d^{n+2}v}{dx^{n+2}} + (n+1)(-2x)\frac{d^{n+1}v}{dx^{n+1}} - (n+1)n\frac{d^n v}{dx^n}\right]$$

$$+2n\left[x\frac{d^{n+1}v}{dx^{n+1}} + (n+1)\cdot 1\cdot\frac{d^n v}{dx^n}\right] = 0$$

or $(1-x^2)\frac{d^{n+2}v}{dx^{n+2}} - 2x\frac{d^{n+1}v}{dx^{n+1}} + n(n+1)\frac{d^n v}{dx^n} = 0$

Put $U = \frac{d^n v}{dx^n}$ then

$$(1-x^2)\frac{d^2U}{dx^2} - 2x\frac{dU}{dx} + n(n+1)U = 0$$

which is a Legendre's equation of order n and has a finite series solution $P_n(x)$. Thus

$$U = C\,P_n(x) \tag{18}$$

* Olinde Rodrigues (1794–1851), French mathematician.

Here C is an arbitrary constant which is determined by equating the coefficient of x^n on both sides of (18) i.e.,

$$\frac{d^n}{dx^n}(x^2-1)^n = \frac{d^n}{dx^n}v = U = C \cdot P_n(x) \quad (18)$$

The coefficient of x^n in $P_n(x)$ is $\frac{(2n)!}{2^n(n!)^2}$ (obtained by putting $m = 0$ in (15)).

The coefficient of x^n in L.H.S. of (18) arises solely from the n-fold differentiation of the term of highest degree i.e., x^{2n}

$$2n(2n-1)(2n-2)\cdots(2n-(n-1)).$$

$$= (2n)(2n-1)(2n-2)\cdots(n+1)\cdot\frac{n!}{n!}$$

$$= \frac{(2n)!}{n!}$$

Thus $$\frac{(2n)!}{n!} = C \cdot \frac{(2n)!}{2^n(n!)^2}$$

or $$C = 2^n \cdot n! \quad (19)$$

Putting C from (19) in (18), we get the Rodrigue's formula

$$P_n(x) = \frac{1}{C}U = \frac{1}{2^n n!}\cdot\frac{d^n}{dx^n}\left\{(x^2-1)^n\right\}$$

Generating Function for Legendre Polynomials

Prove that

$$(1-2xt+t^2)^{-\frac{1}{2}} = \sum_{n=0}^{\infty} t^n \cdot P_n(x) \quad (20)$$

Proof:

$$(1-y)^{-n} = 1 + ny + \frac{n(n+1)}{1\cdot 2}y^2 + \frac{n(n+1)(n+2)}{1\cdot 2\cdot 3}y^3 + \cdots$$

$$(1-y)^{-\frac{1}{2}} = 1 + \frac{1}{2}y + \frac{\frac{1}{2}\cdot\frac{3}{2}}{2!}y^2 + \frac{\frac{1}{2}\cdot\frac{3}{2}\cdot\frac{5}{2}}{3!}y^3 + \cdots$$

$$(1-y)^{-\frac{1}{2}} = 1 + \frac{1}{2}y + \frac{1\cdot 3}{2\cdot 4}y^2 + \frac{1\cdot 3\cdot 5}{2\cdot 4\cdot 6}y^3 + \cdots + \frac{1\cdot 3\cdots(2n-1)}{2\cdot 4\cdots 2n}\cdot y^n + \cdots$$

Rewriting

$$\frac{1\cdot 3\cdot 5\cdots(2n-1)}{2\cdot 4\cdot 6\cdots 2n} = \frac{1\cdot 3\cdot 5\cdots(2n-1)}{2\cdot 4\cdot 6\cdots 2n}\cdot\frac{2\cdot 4\cdots 2n}{2\cdot 4\cdots 2n}$$

$$= \frac{1\cdot 2\cdot 3\cdot 4\cdot 5\cdots(2n-1)\cdot 2n}{2^n\cdot n!\,2^n\cdot n!} = \frac{(2n)!}{2^{2n}(n!)^2}$$

Now using this result, expand

$$(1-2xt+t^2)^{-\frac{1}{2}}$$

$$= \left[1-(2xt-t^2)\right]^{-\frac{1}{2}} = \left[1-\left\{t(2x-t)\right\}\right]^{-\frac{1}{2}}$$

$$= 1 + \frac{2!}{2^2(1!)^2}t\cdot(2x-t) + \frac{4!}{2^4(2!)^2}t^2(2x-t)^2 + \cdots + \frac{(2(n-k))!}{2^{2(n-k)}\left\{(n-k)!\right\}^2}t^{n-k}(2x-t)^{n-k} + \cdots + \frac{(2n)!}{2^{2n}\cdot(n!)^2}t^n(2x-t)^n + \cdots \quad (21)$$

Coefficients of t^n appear only in the first $(n+1)$ terms. Consider the $(n-k)$th term: t^n arises as product of t^{n-k} and t^k arising out of $(2x-t)^{n-k}$. Thus the coefficient of t^n in $t^{n-k}\cdot(2x-t)^{n-k}$ is the coefficient of t^k in $(2x-t)^{n-k}$

i.e., $$(n-k)_{C_k}(2x)^{(n-k)-k}\cdot(-1)^k$$

$$= \frac{(n-k)!(-1)^k}{k!(n-2k)!}\cdot(2x)^{n-2k} \quad (22)$$

Therefore the coefficient of t^n is (see 21)

$$\left[\frac{(2n-2k)!}{2^{2n-2k}\{(n-k)!\}^2}\right]\cdot\left[\frac{(n-k)!}{k!(n-2k)!}(-1)^k(2x)^{n-2k}\right]$$

$$= \frac{(-1)^k(2n-2k)!}{2^n k!(n-k)!(n-2k)!}\cdot x^{n-2k}$$

collecting and summing up for k all the coefficients of t^n from the first $(n+1)$ terms, we get

$$\sum_{k=0}^{M}\frac{(-1)^k(2n-2k)!}{2^n(n-k)!k!(n-2k)!}\cdot x^{n-2k} = P_n(x)$$

where $M = \frac{n}{2}$ or $\frac{n-1}{2}$ according as n is even or odd.

Thus the Legendre polynomials $P_0(x)$, $P_1(x)$, $P_2(x)\cdots P_n(x)\cdots$ appear as coefficients of t^0, t^1, $t^2,\ldots t^n\ldots$ etc. in the expansion of $(1-2xt+t^2)^{-\frac{1}{2}}$. Hence $(1-2xt+t^2)^{-\frac{1}{2}}$ is the generating function of the Legendre polynomials i.e.,

$$(1-2xt+t^2)^{-\frac{1}{2}}=\sum_{n=0}^{\infty}P_n(x)\cdot t^n \qquad (20)$$

Result 1: $P_n(1)=1$ for any n.

Put $x=1$ in (20). Then

$$\sum_{n=0}^{\infty}P_n(1)t^n=(1-2t+t^2)^{-\frac{1}{2}}$$

$$=\left((1-t)^2\right)^{-\frac{1}{2}}=(1-t)^{-1}$$

$$=1+t+t^2+\cdots+t^n+\cdots$$

Equating the coefficients of t^n on both sides

$$P_n(1)=1 \qquad \text{for any } n.$$

Result 2: $P_n(-1)=(-1)^n$ for any n.

Put $x=-1$ in (20). Then

$$\sum_{n=0}^{\infty}P_n(-1)t^n=(1+2t+t^2)^{-\frac{1}{2}}$$

$$=\left[(1+t)^2\right]^{-\frac{1}{2}}=(1+t)^{-1}$$

$$=1-t+t^2-\cdots+(-1)^n t^n+\cdots$$

Equating the coefficients of t^n, $P_n(-1)=(-1)^n$.

Result 3:

$$P_n(0)=\begin{cases}0, & \text{when } n \text{ is odd}\\ (-1)^{\frac{n}{2}}\cdot\frac{1\cdot3\cdot5\cdots(n-1)}{2\cdot4\cdot6\cdots n}, & \text{if } n \text{ is even}\end{cases}$$

Put $x=0$ in (20). Then

$$\sum_{n=0}^{\infty}P_n(0)t^n=(1+t^2)^{-\frac{1}{2}}$$

$$=1-\frac{1}{2}t^2+\frac{\left(-\frac{1}{2}\right)\left(-\frac{1}{2}+1\right)}{1\cdot2}t^4+\cdots$$

$$+\frac{\left(-\frac{1}{2}\right)\left(-\frac{1}{2}+1\right)\cdots\left(-\frac{1}{2}-(n-1)\right)}{1\cdot2\cdot3\cdots n}t^{2n}+\cdots$$

$$=1-\frac{1}{2}t^2+\frac{1\cdot3}{2\cdot4}t^4+\cdots$$

$$+\frac{(-1)^n1\cdot3\cdot5\cdots(2n-1)}{2\cdot4\cdot6\cdots2n}\cdot t^{2n}+\cdots$$

Equating the coefficients of t^{2m}

$$P_{2m}(0)=(-1)^m\frac{1\cdot3\cdot5\cdots(2m-1)}{2\cdot4\cdot6\cdots2m}$$

Since the R.H.S. contains only even powers of t, coefficients of odd powers of t are all zero i.e., $P_{2m+1}(0)=0$.

Recurrence Relations for $P_n(x)$

I. $P_n'(x)=x\,P_{n-1}'(x)+n\,P_{n-1}(x)$

Proof: From Rodrigue's formula

$$P_n(x)=\frac{1}{2^n n!}\frac{d^n}{dx^n}(x^2-1)^n$$

Differentiating w.r.t., x

$$\frac{d}{dx}\{P_n(x)\}=P_n'(x)=\frac{1}{2^n\cdot n!}\cdot\frac{d^n}{dx^n}\left\{n(x^2-1)^{n-1}\cdot2x\right\}$$

$$P_n'(x)=\frac{1}{2^{n-1}\cdot(n-1)!}\frac{d^n}{dx^n}\left\{x(x^2-1)^{n-1}\right\}$$

Differentiating by Leibnitz's rule

$$=\frac{1}{2^{n-1}(n-1)!}\left[x\cdot\frac{d^n}{dx^n}(x^2-1)^{n-1}+\right.$$

$$\left.+n\cdot1\cdot\frac{d^{n-1}(x^2-1)^{n-1}}{dx^{n-1}}\right]$$

$$=x\cdot\frac{d}{dx}\left\{\frac{1}{2^{n-1}(n-1)!}\frac{d^{n-1}}{dx^{n-1}}(x^2-1)^{n-1}\right\}$$

$$+n\frac{1}{2^{n-1}(n-1)!}\frac{d^{n-1}}{dx^{n-1}}(x^2-1)^{n-1}$$

$$P_m'x=x\cdot\frac{d}{dx}P_{n-1}+nP_{n-1}=xP_{n-1}'(x)+nP_{n-1}(x).$$

II. $P_{n+1}'(x)-P_{n-1}'(x)=(2n+1)P_n(x)$.

Proof:

$$P_{n+1}'(x)=\frac{d}{dx}P_{n+1}(x)$$

$$=\frac{d}{dx}\left\{\frac{1}{2^{n+1}(n+1)!}\frac{d^{n+1}}{dx^{n+1}}(x^2-1)^{n+1}\right\}$$

$$=\frac{1}{2^{n+1}(n+1)!}\cdot\frac{d^{n+1}}{dx^{n+1}}\left\{(n+1)x(x^2-1)^n\cdot2x\right\}$$

$$= \frac{1}{2^n}\frac{1}{n!}\frac{d^{n+1}}{dx^{n+1}}\left\{x(x^2-1)^n\right\}$$

$$= \frac{1}{2^n}\cdot\frac{1}{n!}\frac{d^n}{dx^n}\left\{x\cdot n(x^2-1)^{n-1}\cdot 2x+(x^2-1)^n\right\}$$

$$= \frac{1}{2^{n-1}\cdot(n-1)!}\frac{d^n}{dx^n}\left\{x^2(x^2-1)^{n-1}\right\}+P_n(x)$$

$$= \frac{1}{2^{n-1}\cdot(n-1)!}\frac{d^n}{dx^n}\left\{(x^2-1+1)\right.$$

$$\left.\times(x^2-1)^{n-1}\right\}+P_n(x)$$

$$= \frac{1}{2^{n-1}(n-1)!}\cdot\frac{d^n}{dx^n}\cdot(x^2-1)^n$$

$$+\frac{1}{2^{n-1}(n-1)!}\frac{d^n}{dx^n}(x^2-1)^{n-1}+P_n$$

$$= 2n\cdot P_n(x)+\frac{d}{dx}P_{n-1}(x)+P_n(x)$$

$$P_n'(x)-P_{n-1}'(x)=(2n+1)P_n(x).$$

III. $x\,P_n'(x)=n\,P_n(x)+P_{n-1}'(x).$

Proof: In the recurrence relation I

$$P_n'=x\,P_{n-1}'+n\,P_{n-1}$$

replace n by $n+1$ and rewrite

$$x\,P_n'=P_{n+1}'-(n+1)P_n$$

From recurrence relation II,

$$P_{n+1}'=P_{n-1}'+(2n+1)P_n$$

Substituting P_{n+1}', we get

$$x\,P_n'=P_{n-1}'+(2n+1)P_n-(n+1)P_n$$

$$x\,P_n'=P_{n-1}'+n\,P_n$$

IV. $(1-x^2)P_{n-1}'=n(x\,P_{n-1}-P_n)$

From R.R. I: $\quad P_n'=x\,P_{n-1}'+n\,P_{n-1}\qquad(1)$

From R.R. III: $\quad x\,P_n'=P_{n-1}'+n\,P_n\qquad(2)$

Multiply (1) by x and subtract (2) from (1)

$$0=x[x\,P_{n-1}'+n\,P_{n-1}]-[P_{n-1}'+n\,P_n]$$

$$0=(x^2-1)(P_{n-1}')+n(x\,P_{n-1}-P_n)$$

or $(1-x^2)P_{n-1}'=n(x\,P_{n-1}-P_n)$.

V. $(x^2-1)P_n'=n(x\,P_n-P_{n-1})$

From R.R. I: $\quad P_n'=x\,P_{n-1}'+n\,P_{n-1}\qquad(1)$

From R.R. III: $\quad x\,P_n'=P_{n-1}'+n\,P_n\qquad(2)$

Multiply (2) by x and subtract (1) from (2)

$$(x^2-1)P_n'=n\,x\,P_n-n\,P_{n-1}=n(x\,P_n-P_{n-1}).$$

VI. $(n+1)P_{n+1}=(2n+1)x\,P_n-n\,P_{n-1}$

From R.R. IV: $\quad (1-x^2)P_{n-1}'=n[x\,P_{n-1}-P_n]$

Replace n by $n+1$ and rewrite

$$(n+1)P_{n+1}=(n+1)x\,P_n-(1-x^2)P_n'$$

From R.R. V: $\quad (x^2-1)P_n'=n(x\,P_n-P_{n-1})$

Eliminating P_n' using R.R. V.

$$(n+1)P_{n+1}=(n+1)x\,P_n+n(x\,P_n-P_{n-1})$$

$$(n+1)P_{n+1}=(2n+1)x\,P_n-n\,P_{n-1}.$$

Orthogonality of Legendre Polynomials

Prove that

a. $\int_{-1}^{1}P_m(x)P_n(x)dx=0 \qquad \text{if } m\neq n$

This is known as the orthogonality property of Legendre polynomial

b. $\int_{-1}^{1}P_n^2(x)dx=\frac{2}{2n+1} \qquad \text{for } n=0,1,2,\ldots$

Proof: (a) Let $P_m(x)$ and $P_n(x)$ be Legendre polynomials of order m and n satisfying respectively the Legendre equations

$$(1-x^2)P_m''-2x\,P_m'+m(m+1)P_m=0$$

or rewritten as

$$\left[(1-x^2)P_m'\right]'=-m(m+1)P_m\qquad(1)$$

and $$\left[(1-x^2)P_n'\right]'=-n(n+1)P_n\qquad(2)$$

Multiply (1) by P_n and (2) by P_m and add

$$\left[n(n+1) - m(m+1)\right] P_n(x) \cdot P_m(x)$$

$$= \left[(1-x^2)P_m'\right]' P_n - \left[(1-x^2)P_n'\right]' P_m \qquad (3)$$

Put $c = n(n+1) - m(m+1) = n^2 - m^2 + (n - m) = (n-m)(n-m+1) \neq 0$ when $m \neq n$.

Integrate both sides of (3) w.r.t., x from -1 to 1

$$I = c \int_{-1}^{1} P_n(x) P_m(x) dx$$

$$= \int_{-1}^{1} \left[(1-x^2)P_m'\right]' P_n \, dx - \int_{-1}^{1} \left[(1-x^2)P_n'\right]' P_m \, dx$$

$$= I_1 - I_2$$

Consider

$$I_1 = \int_{-1}^{1} \left[(1-x^2)P_m'\right]' P_n \, dx$$

$$= \int_{-1}^{1} \frac{d}{dx}\left[(1-x^2)\frac{dP_m}{dx}\right] P_n \, dx. \text{ Integrating by parts}$$

$$= P_n \cdot (1-x^2)\frac{dP_m}{dx}\Big|_{-1}^{1} - \int_{-1}^{1} (1-x^2)P_m' P_n' \, dx$$

$$= 0 - \int_{-1}^{1} (1-x^2)P_m' P_n' \, dx$$

Similarly,

$$I_2 = \int_{-1}^{1} \left[(1-x^2)P_n'\right]' P_m dx$$

$$= -\int_{-1}^{1} (1-x^2)P_n' P_m' \, dx$$

Thus

$$c \int_{-1}^{1} P_n(x) P_m(x) dx = 0 \quad \text{since } I_1 = I_2$$

For $m \neq n, c \neq 0$, therefore $\int_{-1}^{1} P_n P_m \, dx = 0$

(b) From Rodrigue's formula

$$2^n \cdot n! P_n(x) = \frac{d^n}{dx^n}(x^2-1)^n \qquad (1)$$

Multiplying (1) with itself and integrating w.r.t., x from -1 to 1, and denoting $\frac{d}{dx}$ by D, we have

$$(2^n \cdot n!)^2 \int_{-1}^{1} P_n^2(x) dx$$

$$= \int_{-1}^{1} D^n(x^2-1)^n \left[D^n(x^2-1)\right] dx$$

Integrating by parts

$$= D^n(x^2-1)^n \cdot D^{n-1}(x^2-1)^n \Big|_{-1}^{1}$$

$$- \int_{-1}^{1} \left[D^{n+1}(x^2-1)^n\right]\left[D^{n-1}(x^2-1)^n\right] dx$$

The first term on the R.H.S. becomes zero at $x = \pm 1$ because after $(n-1)$ differentiations of $(x^2-1)^n$, the factor (x^2-1) will still be present.

So

$$(2^n n!)^2 \int_{-1}^{1} P_n^2 dx = -\int_{-1}^{1} \left[D^{n+1}(x^2-1)^n\right] \times \left[D^{n-1}(x^2-1)^n\right] dx$$

Integrating by parts $(n-1)$ times, we get

$$(2^n n!)^2 \int_{-1}^{1} P_n^2 dx = (-1)^n \int_{-1}^{1} (x^2-1)^n \times D^{n+1+n-1}(x^2-1)^n \, dx$$

But $D^{2n}(x^2-1)^n = (2n)!$

$$= (-1)^n \int_{-1}^{1} (x^2-1)^n \cdot (2n)! \, dx$$

$$= 2(-1)^n (2n)! \int_{0}^{1} (x^2-1)^n \, dx$$

$$= 2(2n)! \int_{0}^{1} (1-x^2)^n \, dx$$

Put $x = \sin\theta$, limits for θ are 0 to $\frac{\pi}{2}$

$$= 2(2n)! \int_{0}^{\frac{\pi}{2}} \cos^{2n}\theta \cos\theta \, d\theta$$

$$= 2(2n)! \int_{0}^{\frac{\pi}{2}} \cos^{2n+1}\theta \, d\theta$$

$$= 2(2n)! \left[\frac{2n \cdot (2n-2) \cdots 4 \cdot 2}{(2n+1)(2n-1) \cdots 2 \cdot 1}\right]$$

$$(2^n n!)^2 \int_{-1}^{1} P_n^2 dx = 2(2n)! \cdot \left[\frac{2^{2n}(n!)^2}{(2n+1)!}\right]$$

$$\therefore \int_{-1}^{1} P_n^2(x) dx = \frac{2(2n)!}{(2n+1)} = \frac{2}{(2n+1)}.$$

WORKED OUT EXAMPLES

Example 1: Find $P_6(x)$.

Solution:

$$P_n(x) = \sum_{m=0}^{M} \frac{(-1)^m (2n-2m)!}{2^n m!(n-m)!(n-2m)!} x^{n-2m}$$

Here $n = 6 =$ even, $M = \frac{n}{2} = \frac{6}{2} = 3$,
m: varies from 0 to 3

$$P_6(x) = \sum_{m=0}^{3} \frac{(-1)^m (12-2m)!}{2^6 m!(6-m)!(6-2m)!} x^{6-2m}$$

$$= \frac{12!x^6}{2^6 \cdot 6!6!} - \frac{10!x^4}{2^6 \cdot 1!5!4!} + \frac{8!x^2}{2^6 2!4!2!} - \frac{6!x^0}{2^6 3!3!}$$

$$P_6(x) = \frac{1}{16}[231x^6 - 315x^4 + 105x^2 - 5].$$

Example 2: Express x^5 in terms of Legendre polynomials.

Solution: We know that $P_5(x) = \frac{1}{8}(63x^5 - 70x^3 + 15x)$
Solving $x^5 = \frac{8}{63}P_5 + \frac{70}{63}x^3 - \frac{15}{63}x$
Since $P_3 = \frac{(5x^3-3x)}{2}$, we have

$$x^3 = \frac{2}{5}P_3 + \frac{3}{5}x = \frac{2}{5}P_3 + \frac{3}{5}P_1$$

Substituting x^3

$$x^5 = \frac{8}{63}P_5 + \frac{70}{63}\left[\frac{2}{5}P_3 + \frac{3}{5}P_1\right] - \frac{15}{63}P_1 \quad \text{since } P_1 = x$$

$$x^5 = \frac{8}{63}P_5 + \frac{28}{63}P_3 + \frac{27}{63}P_1.$$

Example 3: Express the polynomial $f(x) = 4x^3 - 2x^2 - 3x + 8$ in terms of Legendre polynomials.

Solution: Since the polynomial $f(x)$ is of degree 3, write

$$f(x) = C_0P_0 + C_1P_1 + C_2P_2 + C_3P_3$$

$$4x^3 - 2x^2 - 3x + 8 = C_0 \cdot 1 + C_1x + C_2\left(\frac{3x^2-1}{2}\right) + C_3\left(\frac{5x^3-3x}{2}\right)$$

Equating the coefficients of like powers of x on both sides, we get

$$x^3: \quad 4 = \frac{5}{2}C_3 \quad \therefore C_3 = \frac{8}{5}$$

$$x^2: \quad -2 = \frac{3}{2}C_2 \quad \therefore C_2 = \frac{-4}{3}$$

$$x: \quad -3 = C_1 - \frac{3C_3}{2} \quad \therefore C_1 = \frac{-3}{5}$$

$$x^0: \quad 8 = C_0 - \frac{C2}{2} \quad \therefore C_0 = \frac{22}{3}$$

Thus

$$4x^3 - 2x^2 - 3x + 8 = \frac{22}{3}P_0 - \frac{3}{5}P_1 - \frac{4}{3}P_2 + \frac{8}{5}P_3.$$

Example 4: Show that (a) $P_n(-x) = (-1)^n P_n(x)$ and (b) $P_n'(-x) = (-1)^{n+1}P_n'(x)$.

Solution:

$$P_n(x) = \sum_{m=0}^{M} \frac{(-1)^m (2n-2m)!}{2^n \cdot m!(n-m)!(n-2m)!} x^{n-2m}$$

$$P_n(-x) = \sum_{m=0}^{M} \frac{(-1)^m (2n-2m)!}{2^n m!(n-m)!(n-2m)!} (-x)^{n-2m}$$

$$P_n(-x) = (-1)^n(-1)^{-2m} \cdot P_n(x) = (-1)^n P_n(x)$$

Differentiating w.r.t., x

$$(-1)P_n(-x) = (-1)^n \cdot P_n'(x)$$

$$\therefore P_n'(-x) = (-1)^{n+1}P_n'(x).$$

Example 5: Prove that

$$(2n+1)(1-x^2)P_n'(x) = n(n+1)\Big[P_{n-1}(x) - P_{n+1}(x)\Big]$$

Solution: Consider R.R. V

$$(x^2-1)P_n' = n(x\,P_n - P_{n-1}) \tag{1}$$

Multiplying (1) by $-(2n+1)$, we get

$$(2n+1)(1-x^2)P_n' = n(2n+1)(P_{n-1} - x\,P_n) \tag{2}$$

Solving R.R. 6 for P_n

$$(n+1)P_{n+1} = (2n+1)x\,P_n - n\,P_{n-1},$$

We get

$$(2n+1)x\,P_n = (n+1)P_{n+1} + n\,P_{n-1} \tag{3}$$

Using (3) eliminate P_n from (2) then

$$(2n+1)(1-x^2)P_n' = n(2n+1)P_{n-1} - -n\left[(n+1)P_{n+1} + n\,P_{n-1}\right]$$
$$= (2n^2+n-n^2)P_{n-1} - n(n+1)P_{n+1}$$
$$(2n+1)(1-x^2)P_n' = n(n+1)[P_{n-1} - P_{n+1}].$$

Example 6: Show that

a. $\int_{-1}^{1} f(x)P_n(x)dx = \frac{(-1)^n}{2^n n!}\int_{-1}^{1} f^{(n)}(x)\times (x^2-1)^n dx$

Hence deduce that

b. $\int_{-1}^{1} x^m P_n(x)dx = \begin{cases} 0, & \text{if } m<n \\ \frac{2^{n+1}\cdot(n!)^2}{(2n+1)!}, & \text{if } m=n \end{cases}$

Solution: Using Rodrigue's formula

$$\int_{-1}^{1} f(x)P_n(x)dx = \int_{-1}^{1} f(x)\cdot\frac{1}{2^n\cdot n!}\frac{d^n}{dx^n}(x^2-1)^n dx$$

Integrating by parts

$$= \frac{1}{2^n\cdot n!}\left[\left\{f(x)\cdot\frac{d^{n-1}}{dx^{n-1}}(x^2-1)^n\right\}\Bigg|_{-1}^{1} - \int_{-1}^{1} f'(x)\cdot\frac{d^{n-1}}{dx^{n-1}}(x^2-1)^n dx\right]$$

The first term in the R.H.S. becomes zero at both the limits $x=\pm1$ because $\frac{d^{n-1}}{dx^{n-1}}(x^2-1)^n$ contains

a. Factor of x^2-1. Integrating by parts $(n-1)$ times

$$= \frac{1}{2^n\cdot n!}(-1)^n\int_{-1}^{1}(x^2-1)^n\cdot\frac{d^n f(x)}{dx^n}dx.$$

b. Take $f(x)=x^m$, then

$$\frac{d^n f}{dx^n} = \frac{d^n x^m}{dx^n} = 0$$

whenever $m<n$. Thus

$$\int_{-1}^{1} x^m P_n(x)dx = 0$$

c. Take $f(x)=x^n$. Then

$$\frac{d^n f}{dx^n} = \frac{d^n x^n}{dx^n} = n!$$

Thus

$$\int_{-1}^{1} x^n P_n(x)dx = \frac{1}{2^n\cdot n!}(-1)^n\int_{-1}^{1}(x^2-1)^n\cdot n!dx$$
$$= \frac{2}{2^n}\int_0^1 (1-x^2)^n dx. \quad \text{put } x=\sin\theta$$
$$= \frac{2}{2^n}\int_0^{\frac{\pi}{2}}\cos^{2n}\theta\, d\theta$$
$$= \frac{2}{2^n}\left[\frac{2n\cdot(2n-2)\cdots4\cdot2}{(2n+1)(2n-1)\cdots2\cdot1}\right],$$

using result 10 on page 11.4

$$= \frac{2}{2^n}\left[\frac{2^{2n}\cdot(n!)^2}{(2n+1)!}\right] = \frac{2^{n+1}\cdot(n!)^2}{(2n+1)!}.$$

EXERCISE

1. Find $P_5(x)$

Ans. $P_5(x) = \frac{1}{8}(63x^5 - 70x^3 + 15x)$

2. Express x^0, x, x^2, x^3 and x^4 in terms of Legendre polynomials

Ans. $1 = P_0(x)$, $x = P_1(x)$, $x^2 = \frac{2}{3}P_2(x) + \frac{1}{3}P_0(x)$, $x^3 = \frac{2}{5}P_3(x) + \frac{3}{5}P_1(x)$, $x^4 = \frac{8}{35}P_4(x) + \frac{4}{7}P_2(x) + \frac{7}{35}P_0(x)$

Express the polynomial $f(x)$ in terms of Legendre polynomials where

3. $f(x) = x^4 + 3x^3 - x^2 + 5x - 2$

Ans. $f(x) = \frac{8}{35}P_4 + \frac{6}{5}P_3 - \frac{2}{21}P_2 + \frac{34}{5}P_1 - \frac{224}{105}P_0$

4. $f(x) = 1 - 3x + 3x^2$

Ans. $2P_0 - 3P_1 + 2P_2$

5. $f(x) = 2x + 10x^3$

Ans. $8P_1 + 4P_3$

6. $f(x) = x^3 + 2x^2 - x - 3$

Ans. $\frac{2}{5}P_3 + \frac{4}{3}P_2 - \frac{2}{5}P_1 - \frac{7}{3}P_0$

7. Show that

$\int_{-1}^{1} x^2 P_{n-1} \cdot P_{n+1} dx = \frac{2n(n+1)}{(2n-1)(2n+1)(2n+3)}$

Hint: Use R.R. VI, $x P_{n-1} = \frac{(n P_n + (n-1) P_{n-2})}{(2n-1)}$

Replace n by $n+2$,

$x P_{n+1} = \frac{\left((n+2) P_{n+2} + (n+1) P_n\right)}{(2n+3)}$

Multiply, integrate −1 to 1, use orthogonality.

8. Prove that
   a. $P_{2n+1}(0) = 0$,
   b. $P_{2n}(0) = \frac{(-1)^n (2n)!}{2^{2n} \cdot (n!)^2}$

   Hint: Use generating function $\sum P_n(x) t^n = (1 - 2xt + t^2)^{-\frac{1}{2}}$. Put $x = 0$, equate coefficients of even powers and odd powers on t on both sides.

9. Show that
   a. $P_n'(1) = \frac{n(n+1)}{2}$
   b. $P_n'(-1) = (-1)^n \frac{n(n+1)}{2}$.

10. Prove that

$$\int_{-1}^{1} (1 - x^2) P_m' P_n' dx = \begin{cases} 0, & \text{if } m \neq n \\ \frac{2n(n+1)}{2n+1}, & \text{if } m = n \end{cases}$$

Hint: Multiply Legendre's equation for P_m by P_n, integrate −1 to 1, use orthogonality.

11. Show that $\int_{-1}^{1} P_n(x) dx = 0$, for $n \neq 0$

    Hint: Use Rodrigue's formula, differentiate $(n-1)$ times.

12. Prove that

$$\int_{-1}^{1} P_n(x) \cdot (1 - 2xt + t^2)^{-\frac{1}{2}} dx = \frac{2t^n}{2n+1}$$

when n is a positive integer.

Hint: Express $(1 - 2xt + t^2)^{-\frac{1}{2}}$ as $\sum t^n P_n(x) dx$, use orthogonality.

11.6 FOURIER-LEGENDRE AND FOURIER-BESSEL SERIES

Another two important generalized Fourier series are Fouries-Legendre series and Fouries-Bessel series.

Fourier-Legendre Series

The Fourier-Legendre series is an eigen function expansion of a given function $f(x)$ on the interval $-1 \leq x \leq 1$ w.r.t. the weight function $P(x) = 1$ and is given by

$$f(x) = \sum_{n=0}^{\infty} a_n P_n(x)$$

where $P_n(x)$ are Legendre polynomials. Here

$$a_n = \frac{\int_{-1}^{1} f(x) P_n(x) dx}{\int_{-1}^{1} P_n^2(x) dx} = \frac{(2n+1)}{2} \int_{-1}^{1} f(x) P_n(x) dx$$

$$\text{since} \int_{-1}^{1} P_n^2(x) dx = \frac{2}{2n+1}$$

Fourier-Bessel Series

The Fourier-Bessel series is an orthogonal expansion of a given function $f(x)$ defined on the interval $0 \leq x \leq R$ w.r.t. the weight function $P(x) = x$ and in terms of orthogonal Bessel functions $J_n(K_{1n} x)$, $J_n(K_{2n} x), \ldots$, where n is fixed and $R \cdot K_{mn}$ are the zeros of the Bessel functions, (i.e., $J_n(R K_{mn}) = 0$). It is given by

$$\boxed{f(x) = \sum_{m=1}^{\infty} a_m J_n(K_{mn} x)}$$

or

$$f(x) = a_1 J_n(K_{1n} x) + a_2 J_n(K_{2n} x) + a_3 J_n(K_{3n} x) + \cdots$$

Here

$$\boxed{a_m = \frac{2}{R^2 J_{n+1}^2 (K_{mn} R)} \int_0^R x \cdot f(x) \cdot J_n(K_{mn} x) dx}$$

for $m = 1, 2, 3, \ldots$

WORKED OUT EXAMPLES

Example 1: Compute the first three non-vanishing terms in the Fourier-Legendre series over the interval $(-1, 1)$ of the function

$$f(x) = \begin{cases} \frac{1}{2\in}, & |x| < \in \\ 0, & \in < |x| < 1 \end{cases}$$

Solution: Let the Fourier-Legendrge series be

$$f(x) = \sum_{n=0}^{\infty} a_n P_n(x)$$

where $a_n = \frac{2n+1}{2} \int_{-1}^{1} f(x) P_n(x) dx$

For $n = 0$

$$a_0 = \frac{1}{2} \int_{-1}^{1} f(x) P_0(x) dx = \frac{1}{2} \int_{-1}^{1} f(x) \cdot 1 \cdot dx$$

$$a_0 = \frac{1}{2} \int_{-\in}^{\in} \frac{1}{2\in} dx = \frac{1}{4\in} x|_{-\in}^{\in} = \frac{2\in}{4\in} = \frac{1}{2}$$

For $n = 1$,

$$a_1 = \frac{3}{2} \int_{-1}^{1} f(x) P_1(x) dx = \frac{3}{2} \int_{-\in}^{\in} \frac{1}{2\in} x\, dx = 0$$

When n is odd, $P_n(x)$ is an odd (polynomial) function. So $f(x)P_n(x)$ is odd since $f(x)$ is constant. Therefore $a_n = 0$ for all n odd.
Now for $n = 2$,

$$a_2 = \frac{5}{2} \int_{-\in}^{\in} \frac{1}{2\in} P_2(x) dx$$

$$a_2 = \frac{5}{4\in} \int_{-\in}^{\in} \frac{1}{2}(3x^2 - 1) dx = \frac{5}{8\in} \left[x^3 - \frac{x^2}{2} \right]_{-\in}^{\in}$$

$$= \frac{5}{4}(\in^2 - 1)$$

Now

$$a_4 = \frac{9}{2} \int_{-\in}^{\in} \frac{1}{2\in} P_4(x) dx$$

$$= \frac{9}{4\in} \int_{-\in}^{\in} \frac{1}{8}(35x^4 - 30x^2 + 3) dx$$

$$= \frac{9}{32\in} [7x^5 - 10x^3 + 3x] \Big|_{-\in}^{\in}$$

$$a_4 = \frac{9}{16} [7\in^4 - 10\in^2 + 3]$$

Thus the Fourier-Legendre series

$$f(x) = \frac{1}{2} P_0(x) - 0 + \frac{5}{4}(\in^2 - 1) P_2(x) + 0$$
$$+ \frac{9}{16} [7\in^4 - 10\in^2 + 3] P_4(x) + 0 \ldots$$

Example 2: Expand $f(x) = \cos \frac{\pi}{2} x$ in Fourier-Legendare series.

Solution: Here $a_n = \frac{2n+1}{2} \int_{-1}^{1} f(x) P_n(x) dx$

or $$a_n = \frac{2n+1}{2} \int_{-1}^{1} \cos \frac{\pi}{2} x \cdot P_n(x) dx.$$

For $n = 0$,

$$a_0 = \frac{1}{2} \int_{-1}^{1} \cos \frac{\pi}{2} x \cdot 1 dx = \frac{1}{2} \cdot \frac{2}{\pi} \cdot \sin \frac{\pi}{2} x \Big|_{-1}^{1}$$

$$a_0 = \frac{2}{\pi} = 0.6366.$$

When n is odd, $P_n(x)$ is odd. So the product $\cos \frac{\pi}{2} x \cdot P_n(x)$ is odd and therefore $a_n = 0$ for n odd. Thus, for $n = 1$,

$$a_1 = \frac{3}{2} \int_{-1}^{1} \left(\cos \frac{\pi}{2} x \right) P_1(x) dx =$$

$$= \frac{3}{2} \int_{-1}^{1} x \cdot \cos \frac{\pi}{2} x dx$$

$$a_1 = \frac{3}{2}\left[x \cdot \frac{2}{\pi}\sin\frac{\pi}{2}x + 1 \cdot \frac{4}{\pi^2}\cos\frac{\pi}{2}x\right]_{-1}^{1} = 0$$

For $n = 2$, $a_2 = \frac{5}{2}\int_{-1}^{1} \cos\frac{\pi}{2}x \cdot P_2(x)dx$

$$a_2 = \frac{5}{2}\int_{-1}^{1} \cos\frac{\pi}{2}x \cdot \left[\frac{1}{2}(3x^2-1)\right]dx =$$

$$= \frac{15}{4}\int_{-1}^{1} x^2\cos\frac{\pi}{2}x - \frac{5}{4}\int_{-1}^{1}\cos\frac{\pi}{2}x dx$$

$$a_2 = \frac{15}{4}\left[x^2\frac{2}{\pi}\cdot\sin\frac{\pi}{2}x + 2x\cdot\frac{4}{\pi^2}\cos\frac{\pi}{2}x + 2\cdot\frac{8}{\pi^3}\sin\frac{\pi}{2}x\right]_{-1}^{1} - \frac{5}{4}\left[\frac{2}{\pi}\cdot\sin\frac{\pi}{2}x\right]_{-1}^{1}$$

$$a_2 = \frac{15}{4}\left[\frac{4}{\pi}+0+\frac{32}{\pi^3}\right] - \frac{5}{4}\left[\frac{4}{\pi}\right]$$

Now

$$a_4 = \frac{9}{2}\int_{-1}^{1}\frac{1}{8}(35x^4-30x^2+3)\cos\frac{\pi}{2}xdx$$

Recall that

$$\int_{-1}^{1} x^n\cos\frac{\pi}{2}xdx = \int_{-\frac{\pi}{2}}^{\frac{\pi}{2}}\left(\frac{2}{\pi}\right)^{n+1}\cdot t^n\cos t\,dt$$

where $\frac{\pi}{2}x = t$. For n odd, this integral is zero since the integrand is odd.
For n even,

$$I_n = \int_{-1}^{1} x^n\cos\frac{\pi}{2}xdx = 2\left(\frac{2}{\pi}\right)^{n+1}\int_{0}^{\frac{\pi}{2}} t^n\cos t\,dt$$

Now by reduction formula

$$I_n = 2\left(\frac{2}{\pi}\right)^{n+1}\left[n\cdot\left(\frac{\pi}{2}\right)^{n-1} - n(n-1)I_{n-2}\right]$$

for $n > 1$.
So

$$I_2 = 2\left(\frac{2}{\pi}\right)^3\left[2\cdot\left(\frac{\pi}{2}\right) - 2I_0\right];$$

Here

$$I_0 = \frac{4}{\pi}$$

then,

$$I_2 = 4\left(\frac{2}{\pi}\right)^3\left[\frac{\pi}{2}-\frac{4}{\pi}\right],$$

$$I_4 = 2\left(\frac{2}{\pi}\right)^5\left[4\left(\frac{\pi}{2}\right)^3 - 12I_2\right] =$$

$$= 8\left(\frac{2}{\pi}\right)^5\left[\left(\frac{\pi}{2}\right)^3 - 3\cdot4\left(\frac{2}{\pi}\right)^3\left(\frac{\pi}{2}-\frac{4}{\pi}\right)\right]$$

$$a_4 = \frac{9}{16}\int_{-1}^{1}(35x^4-30x^2+3)\cos\frac{\pi}{2}x\,dx$$

$$a_4 = \frac{9}{16}[35I_4 - 30I_2 + 3I_0]$$

Similarly for $n = 6$,

$$a_6 = \frac{13}{2}\int_{-1}^{1} P_6(x)\cos\frac{\pi}{2}x\,dx$$

$$a_6 = \frac{13}{2}\int_{-1}^{1}\frac{1}{16}(231x^6-315x^4+105x^2-5)\cos\frac{\pi}{2}x\,dx$$

$$a_6 = \frac{13}{32}[231I_6 - 315I_4 + 105I_2 - I_0]$$

Here

$$I_6 = 2\left(\frac{2}{\pi}\right)^7\left[6\left(\frac{\pi}{2}\right)^5 - 30I_4\right]$$

Thus, the Legendre series is

$$\cos\frac{\pi}{2}x = 0.6366P_0 - 0.6871P_2 + 0.0518P_4 - 0.0013P_6 + \cdots$$

Fourier-Bessel Series

Example 1: Expand $f(x) = x^4$, in Fourier-Bessel series in terms of Bessel functions of order zero ($n = 0$) and w.r.t. the weight function $P(x) = x$ over the interval $0 < x < R$.

Solution: With $n = 0$, weight function x and interval $0 < x < R$, the required Fourier Bessel series is given by

$$f(x) = a_1 J_0(K_{10}x) + a_2 J_0(K_{20}x) + a_3 J_0(K_{30}x) + \cdots$$

where

$$a_m = \frac{2}{R^2 J_1^2(K_{m1}R)} \int_0^R x \cdot f(x) J_0(K_{m0}x)dx$$

Here $f(x) = x^4$, so

$$a_m = \frac{2}{R^2 J_1^2(K_{m1}R)} \int_0^R x^5 J_0(K_{m0}x)dx$$

Put $K_{mo}x = t$ then $K_{mo}dx = dt$, then

$$\int_0^R x^5 J_0(K_{mo}x)dx = \int_0^{RK_{mo}} \frac{1}{K_{mo}^5} t^5 \cdot J_0(t)\frac{dt}{K_{mo}}$$

$$= \frac{1}{(K_{mo})^6} \int_0^{RK_{mo}} t^5 J_0(t)dt$$

we know that

$$\int x^n J_{n-1}(x)dx = x^n J_n(x) + c$$

Integrating by parts

$$\int_0^{RK_{m0}} t^5 J_0(t)dt = \int_0^{RK_{m0}} t^4(tJ_0)dt$$

$$= t^4(tJ_1)\Big|_0^{RK_{m0}} - \int_0^{RK_{m0}} (tJ_1)4t^3dt$$

Similarly

$$\int_0^{RK_{m0}} t^4 J_1 dt = \int_0^{RK_{m0}} t^2(t^2 J_1)dt =$$

$$(t^2 J_2)t^2\Big|_0^{RK_{m0}} - \int (t^2 J_2)2t\, dt$$

Finally

$$\int_0^{RK_{m0}} t^3 J_2 dt = t^3 J_3\Big|_0^{RK_{m0}}$$

Thus substituting these values, we get

$$\int_0^{RK_{m0}} t^5 J_0 dt = t^5 J_1 - 4[t^4 J_2 - 2t^3 J_3]\Big|_{t=0}^{RK_{m0}}$$

$$= (RK_{mo})^5 J_1 - 4(RK_{mo})^4 J_2 + 8(RK_{mo})^3 J_3$$

Then the Fourier coefficients

$$a_m = \frac{1}{(K_{m0})^6} \frac{2}{R^2 J_1^2(RK_{m1})}[(RK_{m0})]^3 \times$$

$$\times[(RK_{m0})^2 J_1 - 4RK_{m0}J_2 + 8J_3]$$

or

$$a_m = 2R[R^2 K_{m0}^2 J_1(K_{m0}R) - 4RK_{m0}J_2(K_{m0}R) +$$

$$8J_3(K_{m0}R)]/[K_{m0}^3 J_1^2(K_{mo}R)]$$

Example 2: Find Fourier-Bessel series of $f(x) = x - x^3$ in terms of Bessel functions of order one ($n = 1$) over the interval $0 < x < 1$.

Solution: The required Fourier-Bessel series with $n = 1$ over the interval $0 < x < 1$ is $f(x) = a_1 J_1(K_{11}x) + a_2 J_l(K_{21}x) + a_3 J_1(K_{31}x) + \cdots$ where

$$a_m = \frac{2}{J_2^2(K_{m1}R)} \int_0^1 x \cdot f(x) J_1(K_{m1}x)dx$$

Here $f(x) = x - x^3$ so

$$\int_0^1 xf(x)J_1(K_{m1}x)dx = \int_0^1 x(x - x^3)J_1(K_{m1}x)dx$$

$$= \int_0^1 x^2 J_1(K_{m1}x)dx - \int_0^1 x^4 J_1(K_{m1}x)dx = I_1 - I_2$$

Put $K_{m1}x = t$ then

$$I_1 = \int_0^1 x^2 J_1(K_{m1}x)dx = \int_0^{K_{m1}} \left(\frac{t}{K_{m1}}\right)^2 J_1(t)\cdot \frac{dt}{K_{m1}}$$

$$= \frac{1}{(K_{m1})^3}\int_0^{K_{m1}} t^2 J_1(t)dt = \frac{1}{(K_{m1})^3}\cdot t^2 J_2(t)\Big|_{t=0}^{K_{m1}}$$

$$I_1 = \frac{1}{K_{m1}^3}\cdot K_{m1}^2 J_2(K_{m1}) = \frac{J_2(K_{m1})}{K_{m1}}$$

Now

$$I_2 = \int_0^1 x^4 J_1(K_{m1}x)dx = \int_0^{K_{m1}} \left(\frac{t}{K_{m1}}\right)^4 J_1(t)\cdot \frac{dt}{K_{m1}}$$

$$= \frac{1}{K_{m1}^5}\int_0^{K_{m1}} t^2(t^2 J_1)dt = \frac{1}{K_{m1}^5}[(t^2 J_2)t^2 - \int (t^2 J_2)2t\, dt$$

But $\int t^3 J_2 dt = t^3 J_3$. Thus

$$I_2 = \frac{1}{K_{m1}^5}[K_{m1}^4 J_2(K_{m1}) - 2K_{m1}^3 J_3(K_{m1})].$$

$$I_2 = \frac{1}{K_{m1}^2}[K_{m1} J_2(K_{m1}) - 2J_3(K_{m1})]$$

Therefore

$$a_m = \frac{2}{J_2^2(K_{m1}R)}\left[\frac{J_2(K_{m_1})}{K_{m1}} - \frac{1}{K_{m1}^2}(K_{m1} J_2(Km1) - 2J_3(K_{m_1}))\right]$$

or

$$a_m = \frac{4J_3(K_{m1})}{K_{m1}^2 J_2^2(K_{m1}R)}$$

EXERCISE

Generalited Fourier Series

Legende Series

1. Expand $f(x)$ in Fourier-Legendre series where
$$f(x) = \begin{cases} 0, & -1 < x < 0 \\ x, & 0 < x < 1 \end{cases}$$

Ans. $f(x) = \frac{1}{4}P_0(x) + \frac{1}{2}P_1(x) + \frac{5}{16}P_2(x)$
$- \frac{3}{32}P_4(x) + \frac{13}{256}P_6(x) + \ldots$

2. Develop the Legendre series for $f(x) = \sin \pi x$

Ans. $f(x) = 0.95493 P_1(x) - 1.15824 P_3(x) + 0.21429 P_5(x) - -0.01664 P_7(x) + 0.00069 P_9(x) - 0.00002 P_{11}(x) + \cdots$

3. Find the Fourier-Legendre series expansion for $f(x) = x^3 + x$ on $-1 \le x \le 1$.

Ans. $[8P_1(x) + 2P_3(x)]/5$

4. Obtain Legendre series for $f(x) = 1$, $0 < x < 1$.

Ans. $1 = \sum_{n=0}^{\infty} (-1)^n \frac{(2n+3)}{(2n+2)} \frac{(2n)!}{(2^{2n}n!)^2} P_{2n+1}(x).$

5. Expand $f(x) = \begin{cases} 0, & -1 < x < 0 \\ 1, & 0 < x < 1 \end{cases}$ is a series of Legendre polynomials.

Ans. $f(x) = \frac{1}{2}P_0(x) + \sum_{n=1}^{\infty} \frac{1}{2}\{P_{n-1}(0) - P_{n+1}(0)\}P_n(x)$

Hint: $A_0 = \frac{1}{2}$, $A_1 = \frac{3}{4}$, $A_2 = 0$, $A_3 = -\frac{7}{16}$, $A_4 = 0$, $A_5 = \frac{11}{32}$.

Bessel Series

6. Find the Fourier-Bessel series expansion of $f(x) = 4x - x^3$ in terms of Bessel functions

of order one ($n = 1$) in the interval [0, 2] satisfying the boundary condition $y(2) = 0$ and bounded at the origin

Ans. $4x - x^3 = -16\sum_{n=1}^{\infty}\frac{J_1(K_{n1}x)}{K_{n1}^3 J_0(2K_{n1})}$

Hint: K_{m1} are zeros of $J_1(2K) = 0$ and

$$a_n = \int_0^2 x(4x - x^3)J_1(K_{n1}x)dx/(2J_2^2(2K_{n1})),$$

$$\int_0^2 x^4 J_1(K_{n1}x)dx = \left(\frac{-16}{K_{n1}} + \frac{32}{K_{n1}^3}\right)J_0(2K_{n1}).$$

7. Show that $\frac{x}{2} = \sum_{J=1}^{\infty}\frac{J_1(K_{j1}x)}{K_{j1}J_2(K_{j1})} = 0 < x < 1$

Hint: Expand $\frac{x}{2}$ in Bessel functions of order one in $0 < x < 1$.

8. Obtain the Fourier-Bessel series expansion of $f(x) = R^2 - x^2$, $0 < x < R$ in terms of $J_0(K_{n0}x)$.

Ans. $\sum_{j=1}^{\infty} a_J J_0(K_{j0}x); a_j = \frac{4J_2(K_{j0}R)}{K_{j0}^2 J_1^2(K_{j0}R)}$

9. Prove that $\frac{c}{2} = \sum_{j=1}^{\infty}\frac{J_0(K_{j0}x)}{K_{j0}J_1(K_{j0}c)}$, $0 < x < c$

10. Develop $f(x) = x^2$ in terms of Bessel functions of order $2(n = 2)$ in the interval $0 < x < 2$.

Ans. $x^2 = 4\sum_{n=1}^{\infty}\frac{J_2(K_{n2}x)}{K_{n2}J_3(2K_{n2})}$

11. Show that $x^3 = \sum_{n=1}^{\infty}\frac{6J_1(K_{n1}x)}{K_{n1}^2 J_2^2(3K_{m1})}\times$

$\times[3K_{n1}J_2(3K_{n1}) - 2J_3(3K_{n1})]$

Hint: Expand x^3 in $0 < x < 3$ in J_1's. Also

$$\int_0^3 x^4 J_1(K_{n1}x)dx = \frac{81K_{n1}J_2(3K_{n1}) - 54J_3(3K_{n1})}{K_{n1}^2}$$

11.7 CHEBYSHEV* POLYNOMIAL

Chebyshev polynomials (of the first kind) are very useful in approximation work.

The second order homogeneous differential equation with variable coefficients

$$(1 - x^2)\frac{d^2y}{dx^2} - x\frac{dy}{dx} + \lambda y = 0 \qquad (1)$$

is known as *Chebyshev equation*. Here, $-1 < x < 1$ and $y(-1)$, $y^1(-1)$, $y(1)$ and $y^1(1)$ are to be bounded.

Changing the variable by

$$x = \cos\theta \qquad (2)$$

The Chebyshev D.E. (1) reduces to

$$\frac{d^2Y}{d\theta} + \lambda Y = 0, \qquad 0 < \theta < \pi \qquad (3)$$

where $Y(\theta) = y(x(\theta)) = y(\cos\theta)$ since $1 - x^2 = \sin^2\theta$, $\frac{dy}{dx} = \frac{dy}{d\theta}\frac{d\theta}{dx} = -\frac{1}{\sin\theta}\frac{dy}{d\theta}$ and

$$\frac{d^2y}{dx^2} = \frac{d}{d\theta}\left(-\frac{1}{\sin\theta}\frac{dy}{d\theta}\right)\frac{d\theta}{dx} = \frac{1}{\sin^2\theta}\frac{d^2y}{d\theta^2} - \frac{\cos\theta}{\sin^3\theta}\frac{dy}{d\theta}$$

The general solution of (3) in terms of θ is

$$\left.\begin{aligned} Y(\theta) &= A\cos\sqrt{\lambda}\theta + B\sin\sqrt{\lambda}\theta \quad \text{when } \lambda \neq 0 \\ Y(\theta) &= C + D\theta \quad \text{when } \lambda = 0 \end{aligned}\right\} \qquad (4)$$

At $\theta = 0 (x = 1)$ and $\theta = \pi (x = -1)$, the solutions $\cos\sqrt{\lambda}\theta$, $\sin\sqrt{\lambda}\theta$, 1 and θ are bounded.

However

$$y'(x) = -A\sin(\sqrt{\lambda}\cos^{-1}x)\sqrt{\lambda}\left(\frac{-1}{\sqrt{1-x^2}}\right) +$$

$$+B\cos\left(\sqrt{\lambda}\cos^{-1}x\right)\sqrt{\lambda}\left(-\frac{1}{\sqrt{1-x^2}}\right) \quad \text{for } \lambda \neq 0$$

and

$$y'(x) = 0 + D\left(\frac{-1}{\sqrt{1-x^2}}\right) \quad \text{for } \lambda = 0$$

and bounded at $x = \pm 1$ only if $B = D = 0$ and $\sqrt{\lambda} = n = 1, 2, 3, \ldots$. Thus the eigen functions of (1) in terms of θ are. $\cos n\theta$, $n = 1, 2, \ldots$ and 1 or

* Pafnuti Chebyshev (1821–1894), Russian mathematician, often transliterated as Tchebichef or Tschebysheff.

equivalently, $\cos n\theta$, for $n = 0, 1, 2, \ldots$. Thus the eigenvalues are

$$\boxed{\lambda_n = n^2} \quad (5)$$

and eigen functions of Chebyshev D.E. (1), are

$$\boxed{T_n(x) = \cos\sqrt{\lambda}\theta = \cos n\theta} \quad (6)$$

$$\boxed{T_n(x) = \cos(n\cos^{-1}x)}, \quad n = 0, 1, 2, \ldots$$

Here T is used in honour of Chebyshev.

Note: Since $T_n(1) = 1$ and $T_n(-1) = (-1)^n$ (See W.E.: 6) so for $-1 \le x \le 1$, $|T_n(x)| \le 1$.

Chebyshev Polynomial

Consider

$$T_n(x) = \cos(n\theta)$$

$$= \frac{1}{2}[e^{in\theta} + e^{-in\theta}]$$

$$= \frac{1}{2}[(\cos\theta + i\sin\theta)^n + (\cos\theta - i\sin\theta)^n]$$

$$= \frac{1}{2}[x + i\sqrt{1-x^2}]^n + \frac{1}{2}[x - i\sqrt{1-x^2}]^n$$

Expanding by binomial series

$$T_n(x) = \frac{1}{2}\sum_{m=0}^{n} {}^nc_m x^{n-m}(i\sqrt{1-x^2})^m +$$

$$+\frac{1}{2}\sum_{m=0}^{n} {}^nc_m x^{n-m}(-i\sqrt{1-x^2})^m$$

$$= \frac{1}{2}\sum_{m=0}^{n} {}^nc_m x^{n-m}\{1^n + (-1)^m\}i^m(1-x^2)^{m/2}$$

when m is odd, $1^m + (-1)^m = 0$. When m is even, $1^m + (-1)^m = 2$.

Take $m = 2r$ and $m \le n$ i.e., $r \le \frac{n}{2}$. Put $N = \left[\frac{n}{2}\right]$,

$$T_n(x) = \sum_{r=0}^{N} {}^nc_{2r} x^{n-2r} i^{2r}(1-x^2)^r$$

$$= \sum_{r=0}^{N} {}^nc_{2r}(-1)^r(1-x^2)^r x^{n-2r}$$

since $i^{2r} = (i^2)^r = (-1)^r$. Thus

$$\boxed{T_n(x) = \sum_{m=0}^{N}(-1)^m \frac{n!}{(2m)!(n-2m)!}(1-x^2)^m x^{n-2m}} \quad (7)$$

where $N = \left[\frac{n}{2}\right]$ is $\frac{n}{2}$ when n is even and $N = \frac{n+1}{2}$ when n is odd.

$T_n(x)$ given by (7) is an nth degree polynomial in x.

$T_n(x)$ which is a solution of D.E. (1) is known as Chebyshev polynomial (of the first kind).

In the expanded form, the Chebyshev polynomial is given by

$$T_n(x) = 2^{n-1}\left[x^n - \frac{n}{1!2^2}x^{n-2} + \frac{n(n-3)}{2!2^4}x^{n-4} - \frac{n(n-4)(n-5)}{3!2^6}x^{n-6} + \cdots\right]$$

Derivation of Chebyshev Polynomials

Since $T_n(x) = \cos n\theta$, for $n = 0$

$T_0(x) = \cos 0 = 1$

For $n = 1$, $T_1(x) = \cos\theta = x$

For $n = 2$, $T_2(x) = \cos 2\theta = 2\cos^2\theta - 1 = 2x^2 - 1$

For $n = 3$, $T_3(x) = \cos 3\theta = 4\cos^3\theta - 3\cos\theta = 4x^3 - 3x$

For $n = 4$, $T_4(x) = \cos 4\theta = 8\cos^4\theta - 8\cos^2\theta + 1 = 8x^4 - 8x^2 + 1$

For $n = 5$, $T_5(x) = \cos 5\theta = 16\cos^5\theta - 20\cos^3\theta + 5\cos\theta = 16x^5 - 20x^3 + 5x$

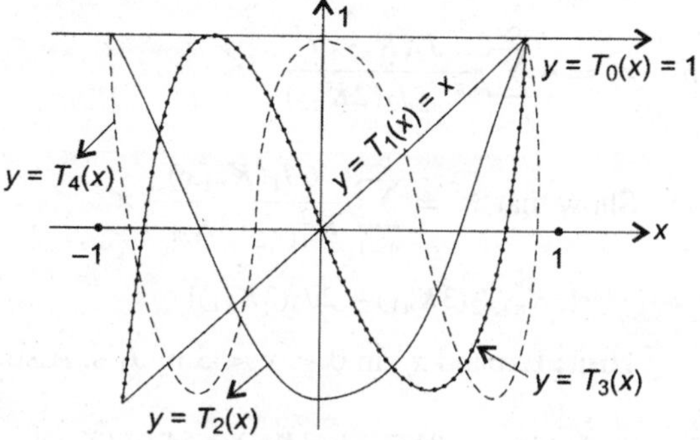

Fig. 11.2

Graphs of $T_0(x)$, $T_1(x)$, $T_2(x)$, $T_3(x)$, $T_4(x)$,

$T_6(x) = 32x^6 - 48x^4 + 18x^2 - 1.$

Thus $T_n(x)$ is even polynomial when n is even and odd polynomial when n is odd.

Powers of x in terms of $T_n(x)$

We can express x^n in terms of the Chybyshev polynomials $T_n(x)$ by solving the above results. Thus

$$1 = T_0$$

$$x = T_1$$

$$x^2 = \frac{1}{2}(T_2 + 1) = \frac{1}{2}(T_0 + T_2)$$

$$x^3 = \frac{1}{4}(T_3 + 3x) = \frac{1}{4}(3T_1 + T_3)$$

$$x^4(T_4 + 8x^2 - 1)/8$$

or

$$x^4 = \frac{1}{8}\left[T_4 + 8\left\{\frac{1}{2}(T_0 + T_2)\right\} - T_0\right]$$

$$= \frac{1}{8}[3T_0 + 4T_2 + T_4]$$

$$x^5 = (T_5 + 20x^3 - 5x)/16$$

$$x^5 = \frac{1}{16}\left[T_5 + 20\left\{\frac{1}{4}(3T_1 + T_3)\right\} - 5T_1\right]$$

$$= \frac{1}{16}[10T_1 + 5T_3 + T_5]$$

Now

$$x^6 = (T_6 + 48x^4 - 18x^2 + 1)/32$$

or

$$x^6 = \left[T_6 + \frac{48}{8}(3T_0 + 4T_2 + T_4) - \frac{18}{2}(T_0 + T_2) + T_0\right]\Big/32$$

Thus

$$x^6 = \frac{1}{32}[10T_0 + 15T_2 + 6T_4 + T_6]$$

Note that the coefficient of x^n in $T_n(x)$ is always 2^{n-1} (see exercise example 7 on page 11.44).

Zeros of $T_n(x)$

Equating to zero, we get

$$T_n(x) = \cos(n\cos^{-1} x) = 0$$

Then

$$n\cos^{-1} x = (2k+1)\frac{\pi}{2} \text{ for } k = 0, 1, 2, \cdots (n-1)$$

or $x = \cos\frac{(2k+1)\pi}{2n}$ are the n simple zeros of $T_n(x)$.

Extrema of $T_n(x)$

The stationary points of $T_n(x)$ are given by

$$T_n'(x) = \frac{n \cdot \sin(n\cos^{-1} x)}{\sqrt{1-x^2}} = 0.$$

or $n\cos^{-1} x = K\pi$. Thus $x_k = \cos\left(\frac{K\pi}{n}\right)$ for $K = 1, 2, \ldots, n-1$ are the $(n-1)$ stationary points. The extrema at these $(n-1)$ points are

$$T_n(x_k) = \cos(n\cos^{-1} x_k) = \cos(K\pi) = (-1)^k$$

Also we know that (worked example 6) at the end points of the interval $(-1, 1)$, $T_n(-1) = (-1)^n$ and $T_n(1) = 1$. Thus $T_n(x)$ attains extrema at $(n+1)$ points given by

$$x_K = \cos\left(\frac{K\pi}{n}\right)$$

where $K = 0, 1, 2, \ldots, n$.

Integrals of $T_n(x)$

With $x = \cos\theta$, $dx = -\sin\theta d\theta$,

$$\int T_n(x)dx = \int \cos(n\theta)\cdot(-\sin\theta)d\theta$$

$$= \frac{1}{2}\int(\sin(n-1)\theta - \sin(n+1)\theta)d\theta$$

$$= \frac{1}{2}\left[\frac{\cos(n+1)\theta}{n+1} - \frac{\cos(n-1)\theta}{n-1}\right]$$

$$= \frac{1}{2}\left[\frac{1}{n+1}T_{n+1}(x) - \frac{1}{n-1}\cdot T_{n-1}(x)\right] \text{ for } n \geq 2$$

For $n = 0$, $\int T_0(x)dx = x = T_1$ and for $n = 1$, $\int T_1(x)dx = \frac{x^2}{2} = \frac{1}{4}[T_0 + T_2]$.

Generating Function

Prove that

$$\frac{1-xt}{1-2xt+t^2} = \sum_{n=0}^{\infty} T_n(x)\cdot t^n, -1 < t < 1 \quad (8)$$

Note: LHS of (8) is known as the generating function of the Chebyshev polynomials.
Proof: Put $x = \cos\theta$ then

$$\text{L.H.S.} = \frac{1-xt}{1-2xt+t^2} = \frac{1-t\cdot\cos\theta}{1-2t\cos\theta+t^2}$$

Replacing $\cos\theta$ by $(e^{i\theta}+e^{-i\theta})/2$, we have

$$\text{L.H.S.} = \frac{1}{2}\left[\frac{2-t(e^{i\theta}+e^{-i\theta})}{1-t(e^{i\theta}+e^{-i\theta})+t^2}\right]$$

$$= \frac{1}{2}\left[\frac{2-t(e^{i\theta}+e^{-i\theta})}{(1-te^{i\theta})(1-te^{-i\theta})}\right]$$

$$= \frac{1}{2}\left[\frac{1}{1-te^{i\theta}}+\frac{1}{1-te^{-i\theta}}\right]$$

Expanding by binonial series

$$= \frac{1}{2}\left[\sum_{n=0}^{\infty}(te^{i\theta})^n+\sum_{n=0}^{\infty}(te^{-i\theta})^n\right]$$

$$= \frac{1}{2}\sum_{n=0}^{\infty} t^n(e^{in\theta}+e^{-in\theta}).$$

Thus

$$\frac{1-xt}{1-2xt+t^2} = \sum_{n=0}^{\infty} t^n\cdot\cos n\theta = \sum_{n=0}^{\infty} T_n(x)t^n$$

Orthogonality

Prove that

$$\int_{-1}^{1}\frac{1}{\sqrt{1-x^2}}\cdot T_m(x)\cdot T_n(x)dx = \begin{cases} 0 & \text{if } m\neq n \\ \frac{\pi}{2} & \text{if } m=n\neq 0 \\ \pi & \text{if } m=n=0\end{cases} \quad (9)$$

ie, the Chebyshev polynomials are orthogonal on $(-1, 1)$.w.r.t. the weight function

$$\frac{1}{\sqrt{1-x^2}}$$

Proof: We know that $x=\cos\theta$, $dx = -\sin\theta d\theta$, $\sqrt{1-x^2}=\sin\theta$, $T_m(x)=\cos m\theta$, $T_n(x)=\cos n\theta$, and limits: $\theta=0$ when $x=1$ and $\theta=\pi$ when $x=-1$.

Now

$$I = \int_{-1}^{1}\frac{1}{\sqrt{1-x^2}}\cdot T_m(x)T_n(x)dx$$

$$= \int_{\pi}^{0}\frac{1}{\sin\theta}\cdot\cos m\theta\cdot\cos n\theta\cdot(-\sin\theta)d\theta$$

$$= \int_{0}^{\pi}\cos m\theta\cdot\cos n\theta\, d\theta$$

Case (i) when $m\neq n$

$$I = \frac{1}{2}\int_{0}^{\pi}[\cos(m+n)\theta+\cos(m-n)\theta]d\theta$$

$$= \frac{1}{2}\left[\frac{\sin(m+n)\theta}{m+n}+\frac{\sin(m-n)\theta}{m-n}\right]\Bigg|_0^{\pi} = 0$$

Case (ii) when $m=n\neq 0$

$$I = \int_0^{\pi}\cos^2 n\theta\, d\theta = \frac{1}{2}\int_0^{\pi}(1+\cos 2n\theta)d\theta$$

$$= \frac{1}{2}\left[\theta+\frac{\sin 2n\theta}{2n}\right]\Bigg|_0^{\pi} = \frac{\pi}{2}$$

Case (iii) when $m=n=0$

$$I = \int_0^{\pi} 1\cdot 1\cdot d\theta = \pi$$

Recall that

$$\int_0^{\pi}\cos m\theta\cdot\cos n\theta\cdot d\theta = \begin{cases}0, & m\neq n\\ \frac{\pi}{2}, & m=n\neq 0\\ \pi, & m=n=0\end{cases}$$

Recurrence Relations

I. Prove that $T_{n+1}(x)+T_{n-1}(x) = 2xT_n(x)$.
Proof:

$$T_{n+1}(x)+T_{n-1}(x) = \cos(n+1)\theta+\cos(n-1)\theta$$

$$= 2\cos n\theta\cdot\cos\theta = 2T_n(x)\cdot x = 2xT_n(x)$$

II. Prove that

$$(1-x^2)T_n'(x) = n\{T_{n-1}(x)-xT_n(x)\}$$

Proof: Differentiating w.r.t. x

$$\frac{d}{dx}T_n(x)=\frac{d}{dx}\{\cos(n\theta)\}=\frac{d}{dx}\{\cos(n\cos^{-1}x)\}$$

$$=-\sin(n\cos^{-1}x)\cdot n\cdot\left(-\frac{1}{\sqrt{1-x^2}}\right)$$

$$\sqrt{1-x^2}T_n'(x)=n\cdot\sin(n\theta)$$

Multiplying by

$$\sqrt{1-x^2}=\sin\theta$$

we get

$$(1-x^2)T_n'(x)=n\cdot\sin\theta\cdot\sin(n\theta)$$

$$=\frac{1}{2}n\cdot[\cos(n-1)\theta-\cos(n+1)\theta]$$

$$=\frac{n}{2}[T_{n-1}(x)-T_{n+1}(x)]$$

Eliminate $T_{n+1}(x)$ using RRI i.e., $T_{n+1}(x)=2xT_n(x)-T_{n-1}(x)$ then

$$(1-x^2)T_n'(x)=\frac{n}{2}[T_{n-1}(x)-2xT_n(x)+T_{n-1}(x)]$$

$$=n[T_{n-1}(x)-xT_n(x)]$$

Chebyshev Series

The Chebyshev D.E. (1) can be written as

$$y''-\frac{x}{(1-x^2)}y'+\frac{\lambda}{(1-x^2)}y=0$$

with an integrating factor $e^{+\int\frac{-x}{(1-x^2)}dx}=e^{+\frac{1}{2}\ln(1-x^2)}=\sqrt{1-x^2}$. Multiplying throughout by this integrating factor, the D.E. takes the form

$$\sqrt{1-x^2}y''-\frac{x}{\sqrt{1-x^2}}y'+\frac{\lambda}{\sqrt{1-x^2}}y=0$$

or in the Sturm-Liouville equation form

$$\left[\sqrt{1-x^2}y'\right]'+\frac{1}{\sqrt{1-x^2}}\lambda y=0 \qquad (10)$$

with the weight function $P(x)=\frac{1}{\sqrt{1-x^2}}$. The eigen-values of (10) are

$$\lambda_n=n^2,\qquad n=0,1,2,\ldots \qquad (5)$$

and eigen functions are

$$T_n(x)=\cos(n\cos^{-1}x) \qquad (6)$$

Any given piecewise continuous function $f(x)$ on $-1<x<1$, can be expanded in terms of the eigen functions (6) as

$$f(x)=a_0T_0(x)+a_1T_1(x)+a_2T_2(x)+\cdots$$

or

$$\boxed{f(x)=\sum_{n=0}^{\infty}a_nT_n(x)} \qquad (11)$$

The infinite series (11) is known as Chebyshev series which expresses $f(x)$ in terms of Chebyshev polynomials $T_n(x)$. The unknown coefficients a_n's are determined using the orthogonality property (9). Multiplying (11) on both sides $T_n(x)\cdot\frac{1}{\sqrt{1-x^2}}$ and integrating w.r.t. x from -1 to 1, we get

$$\int_{-1}^{1}\frac{1}{\sqrt{1-x^2}}f(x)T_n(x)dx=$$

$$\int_{-1}^{1}\frac{1}{\sqrt{1-x^2}}T_n\left(\sum_{m=0}^{\infty}a_iT_m(x)\right)dx$$

$$=\sum_{m=0}^{\infty}a_i\int_{-1}^{1}\frac{1}{\sqrt{1-x^2}}T_n(x)\cdot T_m(x)dx$$

From (9) for $m=n=0$, the RHS reduces to $a_0\pi$ while all other coefficients are zero (because of orthogonality). Thus

$$\boxed{a_0=\frac{1}{\pi}\int_{-1}^{1}\frac{1}{\sqrt{1-x^2}}f(x)dx} \qquad (12)$$

since $T_0(x)=1$.

From (9) for $m=n\neq 0$, the RHS reduces to $\frac{\pi a_n}{2}$. Thus

$$\boxed{a_n=\frac{2}{\pi}\int_{-1}^{1}\frac{1}{\sqrt{1-x^2}}f(x)T_n(x)dx} \qquad (13)$$

for $n=1,2,3,\ldots$

Worked Out Examples

Example 1: Prove that $T_n(x)-2xT_{n-1}(x)+T_{n-2}(x)=0$

Solution: Consider $T_n(x) + T_{n-2}(x)$

$$= \cos n\theta + \cos(n-2)\theta$$

$$= \cos n\theta + \cos n\theta \cdot \cos 2\theta + \sin n\theta \cdot \sin 2\theta$$

$$= \cos n\theta(1 + \cos 2\theta) + \sin n\theta \cdot 2\sin\theta.\cos\theta$$

$$= \cos n\theta \cdot 2 \cdot \cos^2\theta + 2\sin\theta - \cos\theta \cdot \sin n\theta$$

$$= 2\cos\theta[\cos n\theta \cdot \cos\theta + \sin\theta \cdot \sin n\theta]$$

$$= 2\cos\theta[\cos(n-1)\theta]$$

$$= 2x \cdot T_{n-1}(x)$$

Example 2: Prove that

$$T_{m+n}(x) + T_{m-n}(x) = 2T_m(x) \cdot T_n(x)$$

Solution:

$$T_{m+n}(x) + T_{m-n}(x)$$

$$= \cos(m+n)\theta + \cos(m-n)\theta$$

$$= [\cos m\theta \cdot \cos n\theta - \sin m\theta \cdot \sin n\theta] +$$

$$[\cos m\theta \cdot \cos n\theta + \sin m\theta \cdot \sin n\theta]$$

$$= 2\cos m\theta \cdot \cos n\theta = 2T_m(x)T_n(x).$$

Example 3: Show that

$$2[T_n(x)]^2 = 1 + T_{2n}(x)$$

Solution: Consider

$$2[T_n(x)]^2 - T_{2n}(x) =$$

$$= 2\cos^2 n\theta - \cos 2n\theta = 2\cos^2 n\theta - (\cos^2 n\theta - \sin^2 n\theta)$$

$$= \cos^2 n\theta + \sin^2 n\theta = 1$$

Example 4: Prove that

$$\int_{-1}^{1} x^6(1-x^2)^{-\frac{1}{2}} T_8(x)dx = 0$$

Solution: Put $x = \cos\theta$, $1 - x^2 = \sin^2\theta$, when $x = 1$, $\theta = 0$ and $x = -1$, $\theta = \pi$. Substituting in the integral

$$I = \int_{-1}^{1} x^6(1-x^2)^{-\frac{1}{2}} T_8(x)dx$$

$$= \int_{\pi}^{0} \cos^6\theta \cdot \frac{1}{\sin\theta} \cdot T_8(\cos\theta) \cdot (-)\sin\theta d\theta$$

$$= \int_0^{\pi} \cos^6\theta \cdot \cos 8\theta d\theta$$

Now $\cos^6\theta = \cos^3\theta \cdot \cos^3\theta$

$$= \frac{1}{4}[\cos 3\theta + 3\cos\theta]\frac{1}{4}[\cos 3\theta + 3\cos\theta]$$

$$= \frac{1}{16}[\cos^2 3\theta + 9\cos^2\theta + 6\cos\theta\cos 3\theta]$$

$$= \frac{1}{16}\left[\left(\frac{1+\cos 6\theta}{2}\right) + 9\left(\frac{1+\cos 2\theta}{2}\right) + 3(\cos 4\theta + \cos 2\theta)\right]$$

$$I = \int_0^{\pi} \frac{1}{32}[\cos 6\theta + 6\cos 4\theta + 15\cos 2\theta + 10] \times$$

$$\times \cos 8\theta d\theta = 0$$

since

$$\int_0^{\pi} \cos m\theta \cdot \cos n\theta d\theta = 0$$

for $m \neq n$

Example 5: Derive $T_6(x)$ and $T_7(x)$.

Solution: Put $n = 5$ in the recurrence relation $T_{n+1}(x) = 2xT_n(x) - T_{n-1}(x)$, then

$$T_6(x) = 2xT_5(x) - T_4(x)$$

Substituting $T_4(x) = 8x^4 - 8x^2 + 1$ and

$$T_5(x) = 16x^5 - 20x^3 + 5x$$

We have

$$T_6(x) = 2x[16x^5 - 20x^3 + 5x] - [8x^4 - 8x^2 + 1]$$

$$T_6(x) = 32x^6 - 48x^4 + 18x^2 - 1$$

Put $n = 6$ in the recurrence relation. Then

$$T_7(x) = 2xT_6(x) - T_5(x)$$

$$= 2x[32x^6 - 48x^4 + 18x^2 - 1] - [16x^5 - 20x^3 + 5x]$$

$$T_7(x) = 64x^7 - 112x^5 + 56x^3 - 7x.$$

Example 6: Prove that

(a) $T_n(1) = 1$

(b) $T_n(-1) = (-1)^n$

(c) $T_{2n}(0) = (-1)^n$

(d) $T_{2n+1}(0) = 0$

Solution: By definition

$$T_n(x) = \cos(n\cos^{-1} x) \quad (1)$$

(a) Put $x = 1$ in (1). Then

$$T_n(1) = \cos(n\cos^{-1} 1) = \cos(n \cdot 0) = 1$$

(b) Put $x = -1$ in (1). Then

$$T_n(-1) = \cos(n\cos^{-1}(-1)) = \cos(n\pi) = (-1)^n$$

[(c, d)] Put $x = 0$ in (1). Then

$$T_n(0) = \cos(n \cdot \cos^{-1} 0) = \cos\left(\frac{n\pi}{2}\right)$$

For $n = 2m =$ even,

$$T_n(0) = T_{2m}(0) = \cos\left(\frac{2m\pi}{2}\right) = (-1)^m$$

For $n = 2m+1 =$ odd

$$T_n(0) = T_{2m+1}(0) = \cos\left\{(2m+1)\frac{\pi}{2}\right\}$$

$$= \cos\left(m\pi + \frac{\pi}{2}\right) = (-1)^m \cdot \cos\frac{\pi}{2} = 0$$

Example 7: Express the polynomial $16x^4 + 12x^3 + 6x^2 + 4x - 1$ in terms of $T_n(x)$.

Solution: Substitute the values of x^4, x^3, x^2, x in the given polynomial; then

$$16x^4 + 12x^3 + 6x^2 + 4x - 1$$

$$= 16\frac{1}{8}[3T_0 + 4T_2 + T_4] +$$

$$+12 \cdot \frac{1}{4}[3T_1 + T_3] + 6 \cdot \frac{1}{2}(T_0 + T_2) - T_0$$

$$= 2T_4 + 3T_3 + 11T_2 + 9T_1 + 8T_0$$

Example 8: Expand $f(x) = x^4 + 3x^3 + 2x^2 + 5x + 1$ as a Chebyshev series in terms of $T_n(x)$. Verify the result by direct substitution of x^n in $f(x)$.

Solution: The Chebyshev series in terms of Chebyshev polynomials is

$$f(x) = a_0T_0(x) + a_1T_1(x) + a_2T_2(x) + \cdots$$

Here

$$a_0 = \frac{1}{\pi}\int_{-1}^{1} \frac{f(x)}{\sqrt{1-x^2}}dx = \frac{1}{\pi}\int_{-1}^{1} \frac{x^4+3x^3+2x^2+5x+1}{\sqrt{1-x^2}}dx$$

Note that $\sqrt{1-x^2}$ is an even function and $\int_{-a}^{a} f(x)dx = 0$ when $f(x)$ is an odd function.

Thus $a_0 = \frac{2}{\pi}\int_0^1 \frac{x^4+2x^2+1}{\sqrt{1-x^2}}dx$. Put $x = \cos\theta$, so $dx = -\sin\theta d\theta$, $\sqrt{1-x^2} = \sqrt{1-\cos^2\theta} = \sin\theta$, limits for $\theta : x = 0, \theta = \frac{\pi}{2}$ and $x = 1, \theta = 0$. Then

$$a_0 = \frac{2}{\pi}\int_{+\frac{\pi}{2}}^{0} \frac{\cos^4\theta + 2\cos^2\theta + 1}{\sin\theta}(-\sin\theta)d\theta$$

$$a_0 = \frac{2}{\pi}\int_0^{\frac{\pi}{2}} (\cos^4\theta + 2\cos^2\theta + 1)d\theta$$

Recall that $\int_0^{\frac{\pi}{2}} \cos^n\theta d\theta = \frac{(n-1)(n-3)(n-5)\cdots}{n(n-2)(n-4)\cdots} \times (\frac{\pi}{2}$ only when n is even).

$$a_0 = \frac{2}{\pi}\left[\frac{3 \cdot 1}{4 \cdot 2} \cdot \frac{\pi}{2} + 2 \cdot \frac{1}{2} \cdot \frac{\pi}{2} + \frac{\pi}{2}\right]$$

$$= \frac{3}{8} + 1 + 1 = \frac{19}{8}$$

Now

$$a_n = \frac{2}{\pi}\int_{-1}^{1} \frac{f(x)T_n(x)}{\sqrt{1-x^2}}dx \quad \text{for} \quad n = 1, 2, 3, \ldots$$

So

$$a_1 = \frac{2}{\pi}\int_{-1}^{1} \frac{(x^4 + 3x^3 + 2x^2 + 5x + 1)x}{\sqrt{1-x^2}}dx$$

$$a_1 = \frac{4}{\pi}\int_0^1 \frac{3x^4 + 5x^2}{\sqrt{1-x^2}}dx$$

$$= \frac{4}{\pi}\int_0^{\frac{\pi}{2}} (3\cos^4\theta + 5\cos^2\theta)d\theta$$

$$a_1 = \frac{4}{\pi}\left[3\cdot\frac{3\cdot 1}{4\cdot 2} + 5\cdot\frac{1}{2}\right]\frac{\pi}{2} = \frac{29}{4}$$

Now

$$a_2 = \frac{2}{\pi}\int_{-1}^{1}\frac{(x^4+3x^3+2x^2+5x+1)(2x^2-1)}{\sqrt{1-x^2}}dx$$

$$= \frac{4}{\pi}\int_{0}^{1}\frac{(2x^6+3x^4-1)}{\sqrt{1-x^2}}dx$$

$$= \frac{4}{\pi}\int_{0}^{\pi/2}(2\cos^6\theta + 3\cos^4\theta - 1)d\theta$$

$$a_2 = \frac{4}{\pi}\left[2\cdot\frac{5\cdot 3\cdot 1}{6\cdot 4\cdot 2} + 3\cdot\frac{3\cdot 1}{4\cdot 2} - 1\right]\frac{\pi}{2} = \frac{3}{2}$$

Now

$$a_3 = \frac{2}{\pi}\int_{-1}^{1}\frac{(x^4+3x^3+2x^2+5x+1)}{\sqrt{1-x^2}}(4x^3-3x)dx$$

$$= \frac{4}{\pi}\int_{0}^{1}\frac{(12x^6+11x^4-15x^2)}{\sqrt{1-x^2}}dx$$

$$= \frac{4}{\pi}\left[\int_{0}^{\pi/2}(12\cos^6\theta + 11\cos^4\theta - 15\cdot\cos^2\theta)d\theta\right]$$

$$= \frac{4}{\pi}\left[12\cdot\frac{5\cdot 3\cdot 1}{6\cdot 4\cdot 2} + 11\cdot\frac{3\cdot 1}{4\cdot 2} - 15\cdot\frac{1}{2}\right]\frac{\pi}{2} = \frac{3}{4}$$

Finally

$$a_4 = \frac{2}{\pi}\int_{-1}^{1}\frac{(x^4+3x^3+2x^2+5x+1)}{\sqrt{1-x^2}}(8x^4-8x^2+1)dx$$

$$a_4 = \frac{4}{\pi}\int_{0}^{1}\frac{8x^8+8x^6-7x^4-6x^2+1)}{\sqrt{1-x^2}}dx$$

$$= \frac{4}{\pi}\int_{0}^{\frac{\pi}{2}}(8\cos^8\theta + 8\cos^6\theta - 7\cos^4\theta - 6\cos^2\theta + 1)d\theta$$

$$= \frac{4}{\pi}\left[8\cdot\frac{7\cdot5\cdot3\cdot1}{8\cdot6\cdot4\cdot2} + 8\cdot\frac{5\cdot3\cdot1}{6\cdot4\cdot2} - 7\cdot\frac{3\cdot1}{4\cdot2} - 6\cdot\frac{1}{2} + 1\right]\frac{\pi}{2}$$

$$a_4 = \frac{1}{8}$$

All the remaining coefficients are zero i.e., $a_n = 0$ for $n \geq 5$ because of the result

$$\int_{-1}^{1}\frac{x^m T_n(x)}{\sqrt{1-x^2}}dx = 0 \quad \text{when } m < n$$

(see W.E. 9 on page 11.43).

Here the given function $f(x) = x^4 + 3x^2 + 2x^2 + 5x + 1$ is a polynomial of degree $m = 4$ and $n \geq 5$.

Thus the required Chebyshev series is

$$f(x) = x^4 + 3x^3 + 2x^2 + 5x + 1$$

$$= \frac{19}{8}T_0 + \frac{29}{8}T_1 + \frac{3}{2}T_2 + \frac{3}{4}T_3 + \frac{1}{8}T_4$$

Direct Verification

Substitution x^4, x^3, x^2, x, 1 in terms of $T_n(x)$, we have

$$x^4 + 3x^3 + 2x^2 + 5x + 1 = \frac{1}{8}[3T_0 + 4T_2 + T_4]$$

$$+3\cdot\frac{1}{4}[3T_1 + T_3] + 2\cdot\frac{1}{2}(T_0 + T_2) + 5\cdot T_1 + T_0$$

$$= \frac{19}{8}T_0 + \frac{29}{4}T_1 + \frac{3}{2}T_2 + \frac{3}{4}T_3 + \frac{1}{8}T_4$$

Example 9: Show that $\int_{-1}^{1}\frac{x^m T_n(x)}{\sqrt{1-x^2}}dx = 0$ when $m < n$.

Solution: By Chebyshev series

$$x^m = \sum_{i=0}^{m} a_i T_i(x)$$

For m even, all the odd coefficients $a_1, a_3, \ldots$ will be zero, while for m odd, the even coefficients a_0, a_2, $a_4 \ldots$ will be zero.

Now

$$\int_{-1}^{1}\frac{x^m T_n(x)}{\sqrt{1-x^2}}dx = \int_{1}^{1}\frac{\left[\sum_{i=0}^{m} a_i T_i(x)\right]T_n(x)}{\sqrt{1-x^2}}dx\sqrt{1-x^2}dx$$

$$= \sum_{i=0}^{m} a_i \int_{-1}^{1}\frac{T_i(x)\cdot T_n(x)}{\sqrt{1-x^2}}dx$$

For $m < n$ since the index i takes the values $0, 1, 2, 3, \ldots, m$, all, these indices are less than n and therefore not equal to n. Using orthogonality of Chebyshev polynomials the integral on the RHS vanishes for $i = 0, 1, 2, \ldots, m$ each of which is not equal to n.

EXERCISE

1. Prove that $[T_n(x)]^2 - T_{n+1}(x) \cdot T_{n-1}(x) = 1 - x^2$

Hint: LHS $= \cos^2 n\theta - \cos(n+1)\theta \cdot \cos(n-1)\theta$ $= \cos^2\theta - (\cos^2 n\theta - \sin^2\theta) = \sin^2\theta = 1 - x^2$

2. Prove that

$$\frac{1-t^2}{1-2tx+t^2} = T_0(x) + 2\sum_{n=1}^{\infty} T_n(x) \cdot t^n$$

Hint: Put $x = \cos\theta = \frac{1}{2}(e^{i\theta} + e^{-i\theta})$ in LHS

$$\sum_{r,s=0}^{\infty} t^{r+s} e^{i(r-s)\theta} - \sum_{r,s=0}^{\infty} t^{r+s+2} \cdot e^{i(r-s)\theta}$$

coefficient of t^0, $(r = 0, s = 0) : T_0(x)$.
coefficient of t' : $(r = 0, s = 1)$: $2x = 2T_1(x)$
coefficient of $t^n : (s = n - r) : 2\cos n\theta = 2T_n(x)$.

3. Prove that $T_n(x)$ is a polynomial of nth degree in x.

4. Express the following polynomials in $T_n(x)$
(a) $x^3 + 3x^2 - 5x + 2$
(b) $5x^4 - x^3 + 3$
(c) $12x^3 + 6x^2 + 4x + 1$
(d) $16x^4 + 4x^3 + 2x^2 + 4x + 5$

Ans. (a) $\frac{1}{4}T_3 + \frac{3}{2}T_2 - \frac{17}{4}T_1 + \frac{7}{2}T_0$
(b) $\frac{5}{8}T_4 - \frac{1}{4}T_3 + \frac{5}{2}T_2 + \frac{9}{4}T_1 + \frac{15}{8}T_0$
(c) $3T_3 + 3T_2 + 13T_1 + 4T_0$
(d) $2T_4 + T_3 + 9T_2 + 7T_1 + 12T_0$

5. Express x^7 in terms of $T_n(x)$

Ans. $x^7 = (T_7 + 7T_5 + 21T_3 + 35T_1)/64$

Hint: $x^7 = (T_7 + 112x^5 - 56x^3 + 7x)/64$

6. Obtain the Chebyshev series of $f(x) = x^3 + x$.

Ans. $f(x) = x^3 + x = \frac{1}{4}(T_3 + 7_1)$.

Hint: $a_0 = 0, a_1 = \frac{7}{4}, a_2 = 0, a_3 = \frac{1}{4}, a_n = 0$ for $n \geq 4$.
Note that $f(x) = x^3 + x$ is an odd function.

7. Show that the leading coefficient of x^n in $T_n(x)$ is 2^{n-1}.

Hint: $T_n(x) = \cos n\theta = Re(e^{in\theta}) = Re(\cos\theta)^n$ $= Re[\cos^n\theta - n_{c_2}\cos^{n-1}\theta \cdot \sin^2\theta + n_{n_4}\cos^{n-4}\theta \cdot \sin^4\theta + \cdots] = x^n$ $-nc_2x^{n-2} \cdot (1-x^2) + n_{c_4}x^{n-4}(1-x^2) + \cdots]$, coefficient of x^n is $1 + n_{c_2} + n_{c_4} + \cdots = 2^{n-1}$.

8. Show that Chebyshev polynomials are solutions of Chebyshev differential equation.

Hint: $y = T_n(x) = \cos(n\cos^{-1}x)$, $y' = \frac{n\sin}{\sqrt{1-x^2}}(n\cos^{-1}x)$, $y'' = \frac{nx}{(1-x^2)^{3/2}} \times$ $\cdot \sin(n\cos^{-1}x)$ or $(1-x^2)y'' = xy' - n^2y$

9. Determine $T_7(x)T_8(x), T_9(x), T_{10}(x)$ in powers of x

Ans. $T_7(x) = 64x^7 - 112x^5 + 56x^3 - 7x$
$T_8(x) = 128x^8 - 256x^6 + 160x^4 - 32x^2 + 1$
$T_9(x) = 256x^9 - 576x^7 + 432x^5 - 120x^3 + 9x$
$T_{10}(x) = 512x^{10} - 1280x^8 + 1120x^6 - 400x^4 + 50x^2 - 1$

Chapter

12

Laplace Transform

INTRODUCTION

Laplace transform is useful since (i) particular solution is obtained without first determining the general solution, (ii) non homogeneous equations are solved without obtaining the complementary integral (iii) solutions of mechanical or electrical problems involving discontinuous force function (RHS function $F(x)$) or periodic functions other than cos and sine are obtained easily (iv) system of DE, PDE and integral equations.

Before the advent of calculators and computers, logarithms were extensively used to replace multiplication (or division) of two large numbers by addition (or subtraction) of two numbers. The crucial idea which made the Laplace* transform (L.T.) a very powerful technique is that it replaces operations of calculus by operations of algebra. For example, with the application of Laplace transform to an initial value problem, consisting of an ordinary (or partial) differential equation (O.D.E.) together with initial conditions (I.C.) is reduced to a problem of solving an algebraic equation (with any given initial conditions automatically taken care) as shown in Fig. 12.1.

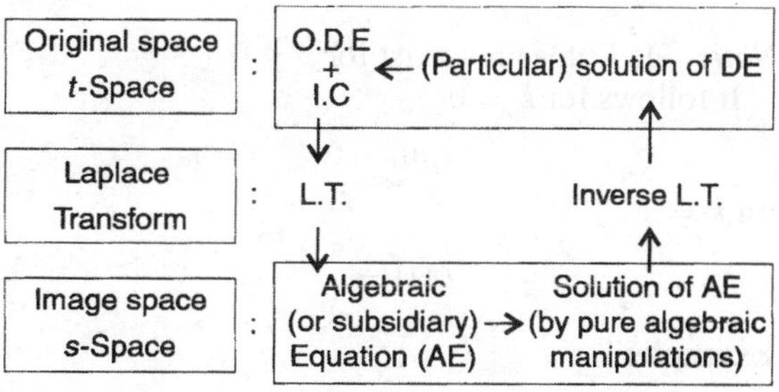

Fig. 12.1

12.1 LAPLACE TRANSFORM

Definition: Let $f(t)$ be a given function defined for all $t \geq 0$. Laplace transform of $f(t)$ denoted by $L\{f(t)\}$ or simply $L\{f\}$ is defined as

$$L\{f(t)\} = \int_0^\infty e^{-st} f(t) dt = F(s) \qquad (1)$$

L is known as Laplace transform operator. The original given function $f(t)$ known as determining function depends on t, while the new function to be determined $F(s)$, called as generating function, depends only on s (because the improper integral on the R.H.S. of (1) is integrated with respect to t).

$F(s)$ in (1) is known as the Laplace transform of $f(t)$.

Equation (1) is known as direct transform or simply transform, in which $f(t)$ is given and $F(s)$ is to be determined.

Thus Laplace transform transforms one class of complicated functions $f(t)$ to produce another class of simpler functions $F(s)$.

Notation: Original given functions are denoted by lower case letters say $f(t)$ and their Laplace transforms by the *same* letters in capitals say $F(s)$. (Sometimes Laplace transform of $f(t)$ is denoted by $\overline{f(s)}$.

* Pierre Simon Marquis De Laplace (1749–1827), French mathematician, known as the Newton of France and teacher to Napoleon Bonaparte.

Laplace Transform of Some Elementary Functions

Example: Find Laplace transform of

$$f(t) = k$$

where k is a constant and $t \geq 0$.

Solution:

$$L\{f(t)\} = L\{k\} = \int_0^\infty e^{-st} k\, dt = -\frac{k}{s} e^{-st}\Big|_0^\infty$$

$$L\{k\} = \frac{k}{s} \quad \text{when} \quad s > 0$$

Note: L.T. does not exist for $s \leq 0$
It follows for $k = 0$

$$L\{0\} = 0$$

for $k = 1$

$$L\{1\} = \frac{1}{s}$$

Example:

$$L\{e^{at}\} = \int_0^\infty e^{-st} \cdot e^{at} dt = \int_0^\infty e^{-(s-a)t} dt$$

$$= \frac{e^{-(s-a)t}}{-(s-a)}\Big|_{t=0}^\infty = \frac{1}{s-a}$$

Thus $L\{e^{at}\} = \frac{1}{s-a}$ when $s > a$
Note: L.T. does not exist when $s \leq a$.

Example: Let $f(t) = t^b$, where b is non-negative real number i.e., $b > 0$, then

$$L\{t^b\} = \int_0^\infty e^{-st} t^b dt, \quad \text{put } st = x$$

$$= \int_0^\infty e^{-x} \left(\frac{x}{s}\right)^b \frac{dx}{s} = \frac{1}{s^{b+1}} \int_0^\infty e^{-x} x^b dx.$$

$L\{t^b\} = \frac{1}{s^{b+1}} \Gamma(b+1)$ when $s > 0$ and $b + 1 > 0$
Here the Gamma function is defined as

$$\Gamma(\alpha) = \int_0^\infty e^{-t} t^{\alpha - 1} dt, \text{ with } \alpha > 0.$$

Example: When $b = n =$ non-negative integer, then

$$L\{t^n\} = \frac{1}{s^{n+1}} \Gamma(n+1) = \frac{n!}{s^{n+1}}$$

Since $\Gamma(n+1) = n!$

Thus

$$L\{t^n\} = \frac{n!}{s^{n+1}} \text{ for } n = 0, 1, 2, 3, \ldots.$$

Example: When $b = \frac{1}{2}$

$$L\{t^{\frac{1}{2}}\} = \frac{\Gamma\left(\frac{1}{2}+1\right)}{s^{\frac{1}{2}+1}} = \frac{\frac{1}{2}\Gamma\left(\frac{1}{2}\right)}{s^{\frac{3}{2}}} = \frac{\sqrt{\pi}}{2s^{\frac{3}{2}}}$$

since $\Gamma(\alpha + 1) = \alpha\Gamma(\alpha)$ and $\Gamma\left(\frac{1}{2}\right) = \sqrt{\pi}$.

Example: When $b = -\frac{1}{2}$

$$L\{t^{\frac{-1}{2}}\} = \frac{\Gamma\left(-\frac{1}{2}+1\right)}{s^{-\frac{1}{2}+1}} = \frac{\Gamma\left(\frac{1}{2}\right)}{s^{\frac{1}{2}}} = \sqrt{\frac{\pi}{s}}.$$

Worked Out Examples

Laplace transform of piecewise continuous functions

Example 1: (refer Fig. 12.2)

$$\begin{aligned} f(t) &= 0, \quad 0 < t < 1 \\ &= t, \quad 1 < t < 4 \\ &= 0, \quad t > 4 \end{aligned}$$

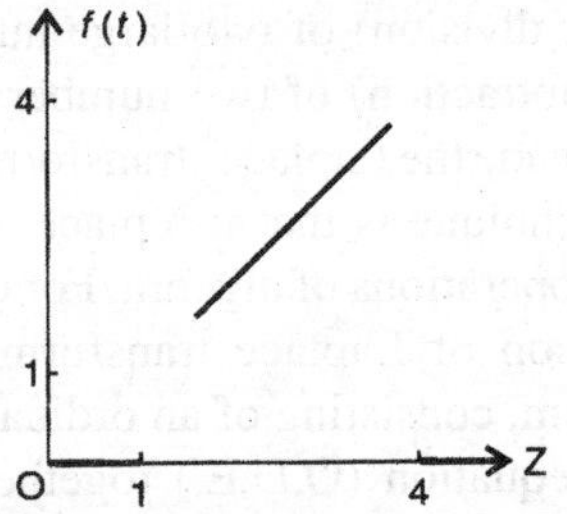

Fig. 12.2

Solution:

$$L\{f\} = \int_0^\infty e^{-st} f(t) dt$$

$$= \int_0^1 0 \cdot e^{-st} dt + \int_1^4 t \cdot e^{-st} dt + \int_4^\infty 0 \cdot e^{-st} dt$$

Integrating by parts

$$= -\frac{t}{s} e^{-st} - \frac{e^{-st}}{s^2}\Big]_1^4$$

$$= e^{-s}\left(\frac{1}{s} + \frac{1}{s^2}\right) - e^{-4s}\left(\frac{4}{s} - \frac{1}{s^2}\right).$$

EXERCISE

Piecewise continuous functions

Find the Laplace Transform (L.T.) of the following piecewise continuous functions:

1. $f(t) = \begin{cases} 4, & 0<t<1 \\ 3, & t>1 \end{cases}$

Ans. $\frac{1}{s}(4 - e^{-s})$

2. $f(t) = \begin{cases} \sin 2t, & 0<t\le\pi \\ 0, & t>\pi \end{cases}$

Ans. $\frac{2(1-e^{-\pi s})}{s^2+4}$

3. $f(t) = \begin{cases} e^t, & 0<t<1 \\ 0, & t>1 \end{cases}$

Ans. $\frac{e^{1-s}-1}{1-s}$

4. $f(t) = \begin{cases} t, & 0\le t<1 \\ 2-t, & 1\le t<2 \\ 0, & t\ge 2 \end{cases}$

Ans. $\frac{1}{s^2}\left(1 - 2e^{-s} + e^{-2s}\right)$.

12.2 APPLICATIONS, ADVANTAGES AND SUFFICIENT CONDITIONS FOR EXISTENCE OF LAPLACE TRANSFORM

Applications

Laplace transform is very useful in obtaining solution of linear differential equations, both ordinary and partial, solution of system of simultaneous differential equations, solution of integral equations, solution of linear difference equations and in the evaluation of definite integrals.

Advantages

1. With the application of Laplace transform, particular solution of differential equation (D.E.) is obtained directly without the necessity of first determining general solution and then obtaining the particular solution (by substitution of initial conditions).

2. L.T. solves non-homogeneous D.E. without the necessity of first solving the corresponding homogeneous D.E.

3. L.T. is applicable not only to continuous functions but also to piecewise continuous functions, complicated periodic functions, step functions and impulse functions.

4. Laplace transforms of various functions are readily available (in tabulated form). In Section 12.12 Laplace transforms of some most often used functions are tabulated.

Sufficient Conditions for the Existence of Laplace Transform of $f(t)$

The L.T. of $f(t)$ exists i.e., the improper integral in the R.H.S. of (1) converges (has a finite value) when the following sufficient conditions are satisfied:

a. $f(t)$ is piecewise (or sectionally) continuous i.e., $f(t)$ is continuous in every subinterval and has finite limits at end points of each of these subintervals and

b. $f(t)$ is of exponential order of γ i.e., there exists M, γ such that $|f(t)| < Me^{\gamma t}$. In other words functions of exponential order do *not* grow faster than $e^{\gamma t}$.

Example: Since $\lim\limits_{t\to\infty} \frac{t^2}{e^{3t}} =$ finite, $f(t) = t^2$ is of exponential order say 3.

Example: Since $\lim\limits_{t\to\infty} \frac{e^{t^2}}{e^{\gamma t}} =$ not finite, $f(t) = e^{t^2}$ is *not* of exponential order.

Note: Above conditions (a) and (b) are not necessary conditions.

12.3 GENERAL PROPERTIES OF LAPLACE TRANSFORM

Although theoretically $F(s)$, the Laplace transform of $f(t)$ is obtained from the definition (1) (in Section 12.2), in practice most of the time Laplace transforms are obtained by the judicial application of some of the following important properties. In a nutshell, they are:

1. Linearity property states that L.T. of a linear combination (sum) is the linear combination (sum) of Laplace transforms.

2. In change of scale, where the argument t of f is multiplied by a constant a, s is replaced by $\frac{s}{a}$ in $F(s)$ and then multiplied by $\frac{1}{a}$.

3. First shift theorem proves that multiplication of $f(t)$ by e^{at} amounts to replacement of s by $s-a$ in $F(s)$.
4. L.T. of a derivative f' amounts to multiplication of $F(s)$ by s (approximately but for the constant $-f(0)$).
5. L.T. of an integral of f amounts to division of $F(s)$ by s.
6. Multiplication of $f(t)$ by t^n amounts to differentiation of $F(s)$ n times w.r.t. s (with $(-1)^n$ as sign).
7. Division of $f(t)$ by t amounts to integration of $F(s)$ between the limits s to ∞.
8. Second shift theorem proves that the L.T. of shifted function $f(t-a)u(t-a)$ is obtained by multiplying $F(s)$ by e^{-at}. The above important properties are tabulated in Section 12.12.

Linearity Property (Principle)

Book Work: If $L\{f(t)\}=F(s)$ and $L\{g(t)\}=G(s)$

Then
$$L\{c_1 f(t)+c_2 g(t)\} = c_1 L\{f(t)\}+c_2 L\{g(t)\} = c_1 F(s)+c_2 G(s)$$

where c_1 and c_2 are any two constants.

Proof:

$$L\{c_1 f(t)+c_2 g(t)\}$$
$$= \int_0^\infty e^{-st}\{c_1 f(t)+c_2 g(t)\}dt$$
$$= c_1\int_0^\infty e^{-st} f(t)dt + c_2\int_0^\infty e^{-st} g(t)dt$$
$$= c_1 L\{f(t)\}+c_2 L\{g(t)\}$$
$$= c_1 F(s)+c_2 G(s).$$

Thus Laplace transform is a linear operator, additive and homogeneous (like the differential operator $D=\frac{d}{dx}$).

This result can easily be generalized to more than two functions.

Example: Find $L\{\cosh at\}$

Solution: $L\{\cosh at\} = L\left\{\frac{e^{at}+e^{-at}}{2}\right\}$

By Linearity property

$$L\{\cosh at\} = \frac{1}{2}L\{e^{at}\}+\frac{1}{2}L\{e^{-at}\}$$
$$L\{\cosh at\} = \frac{1}{2}\frac{1}{s-a}+\frac{1}{2}\frac{1}{s+a} = \frac{s}{s^2-a^2},\ s>a\ge 0.$$

Similarly,

Example: $L\{\sinh at\} = \frac{a}{s^2-a^2}$

Example:

$$L\{e^{iwt}\} = \frac{1}{s-iw} = \frac{s+iw}{(s-iw)(s+iw)} = \frac{s+iw}{s^2+w^2}$$

since $e^{i\theta}=\cos\theta+i\sin\theta$, by linearity property

$$L\{e^{iwt}\} = L\{\cos wt+i\sin wt\}$$
$$= L\{\cos wt\}+iL\{\sin wt\}$$
$$= \frac{s+iw}{s^2+w^2}$$

Comparing the real and imaginary parts on both sides, we have

$$L\{\cos wt\} = \frac{s}{s^2+w^2}$$

and

$$L\{\sin wt\} = \frac{w}{s^2+w^2}.$$

WORKED OUT EXAMPLES

Linearity property

Find the Laplace transform of the following:

Example 1: $e^{at}-e^{bt}$

Solution:

$$L\{e^{at}-e^{bt}\} = L\{e^{at}\}-L\{e^{bt}\}$$
$$= \frac{1}{s-a}-\frac{1}{s-b} = \frac{a-b}{(s-a)(s-b)}$$

Example 2: $\cos^2 kt$

Solution:

$$L\{\cos^2 kt\} = L\left\{\frac{1+\cos 2kt}{2}\right\}$$
$$= \frac{1}{2}L\{1\}+\frac{1}{2}L\{\cos 2kt\} = \frac{1}{2}\frac{1}{s}+\frac{1}{2}\frac{s}{s^2+4k^2}.$$

Example 3: $(5e^{2t}-3)^2$

Solution:

$$\begin{aligned}L\{(5e^{2t}-3)^2\} &= L\{25e^{4t}-30e^{2t}+9\}\\ &= 25L\{e^{4t}\}-30L\{e^{2t}\}+9L\{1\}\\ &= 25\cdot\frac{1}{s-4}-30\frac{1}{s-2}+9\cdot\frac{1}{s}.\end{aligned}$$

Example 4: $3t^4-2t^3+4e^{-3t}-2\sin 5t+3\cos 2t$

Solution:

$$L\{3t^4-2t^3+4e^{-3t}-2\sin 5t+3\cos 2t\}$$

$$=3L\{t^4\}-2L\{t^3\}+4L\{e^{-3t}\}-2L\{\sin 5t\}$$

$$+3L\{\cos 2t\}$$

$$=3\frac{4!}{t^5}-2\frac{3!}{t^4}+4\frac{1}{s+3}-2\cdot\frac{5}{s^2+5^2}+3\cdot\frac{s}{s^2+2^2}.$$

Example 5: $\cos\sqrt{t}$

Solution: Expanding in series

$$\cos\sqrt{t}=\sum_{n=0}^{\infty}\frac{(-1)^n(t^{\frac{1}{2}})2n}{(2n)!}=\sum\frac{(-1)^n}{(2n)!}t^n$$

$$L\{\cos\sqrt{t}\}=L\left\{\sum\frac{(-1)^n}{(2n)!}t^n\right\}$$

By linearity principle

$$=\sum L\left\{\frac{(-1)^n}{(2n)!}t^n\right\}$$

$$=\sum_{n=0}^{\infty}\frac{(-1)^n}{(2n)!}L\{t^n\}$$

$$=\sum_{n=0}^{\infty}\frac{(-1)^n}{(2n)!}\frac{n!}{s^{n+1}}.$$

EXERCISE

Linearity property

Find the Laplace transform of the following:

1. $\left(\sqrt{t}\pm\frac{1}{\sqrt{t}}\right)^3$

Ans. $\frac{\sqrt{\pi}}{4}\left(\frac{3}{s^{\frac{5}{2}}}\pm\frac{6}{s^{\frac{3}{2}}}+\frac{12}{s^{\frac{1}{2}}}\pm\frac{8}{s^{-\frac{1}{2}}}\right)$

2. $\sin^2 kt$

Ans. $\frac{2k^2}{s(s^2+4k^2)}$

3. $4e^{5t}+6t^3-3\sin 4t+2\cos 2t$

Ans. $\frac{4}{s-5}+\frac{36}{s^4}-\frac{12}{s^2+16}+\frac{2s}{s^2+4}$

4. $\cos^3 at$

Ans. $\frac{s(s^2+7a^2)}{(s^2+a^2)(s^2+9a^2)}$

5. $\cos 3t\cdot\cos 2t\cos t$

Ans. $\frac{1}{4}\left(\frac{s}{s^2+36}+\frac{s}{s^2+16}+\frac{s}{s^2+4}+\frac{1}{s}\right)$

6. $\sin\sqrt{t}$

Hint: Use $\Gamma\left(n+\frac{1}{2}\right)=\frac{1\cdot3\cdot5\cdots(2n-1)}{2^n}\sqrt{\pi}$ for n positive integer.

Ans. $\frac{\sqrt{\pi}}{2s\sqrt{s}}\sum_{n=1}^{\infty}\frac{1}{(n-1)!}\left(-\frac{1}{4s}\right)^{n-1}=\frac{\sqrt{\pi}}{2s\sqrt{s}}e^{-\left(\frac{1}{4s}\right)}.$

Change of Scale Property

Book Work: If $L\{f(t)\}=F(s)$ then

$$L\{f(at)\}=\frac{1}{a}F\left(\frac{s}{a}\right).$$

Proof:

$$L\{f(at)\}=\int_0^{\infty}e^{-st}f(at)dt$$

Put $at=u$, $a\,dt=du$

$$L\{f(at)\}=\int_0^{\infty}e^{-s\left(\frac{u}{a}\right)}f(u)\frac{du}{a}$$

$$=\frac{1}{a}\int_0^{\infty}e^{-\frac{s}{a}u}f(u)du$$

$$L\{f(at)\}=\frac{1}{a}F\left(\frac{s}{a}\right)$$

Thus when the argument t of f is multiplied by a then s is replaced by $\frac{s}{a}$ in transform $F(s)$ and is multiplied by $\frac{1}{a}$.

WORKED OUT EXAMPLES

Change of scale

Example 1: If $L\{f(t)\}=\frac{e^{-\frac{1}{s}}}{s}$ find $L\{e^{-t}f(3t)\}$.

Solution: From change of scale property with $a = 3$, replace s by $\frac{s}{3}$

$$L\{f(3t)\} = \frac{1}{3}\frac{e^{-\frac{3}{s}}}{\frac{s}{3}} = \frac{e^{-\frac{3}{s}}}{s}$$

Applying first shift theorem

$$L\{e^{-t}f(3t)\} = \frac{e^{-\frac{3}{(s+1)}}}{s+1}$$

Obtained by replacing s by $s - a = s - (-1) = s + 1$.

EXERCISE

Change of scale

1. If $L\left\{\frac{\sin t}{t}\right\} = \tan^{-1}\left(\frac{1}{s}\right)$ find $L\left\{\frac{\sin at}{t}\right\}$

Ans. $\tan^{-1}(a/s)$

2. If $L\{f(t)\} = \frac{20-4s}{s^2-4s+20}$ find $L\{f(3t)\}$

Ans. $4(15 - s)/(s^2 - 12s + 180)$

3. If $L\{\sin t\} = \frac{1}{s^2+1}$, find $L\{\sin 3t\}$

Ans. $\frac{3}{s^2+9}$.

First Shifting or First Translation or s-Shift Theorem: Replacement of s by s–a in transform

Book Work: If $L\{f(t)\} = F(s)$ then

$$L\{e^{at}f(t)\} = F(s-a)$$

Proof: Consider $F(s-a) = \int_0^\infty e^{-(s-a)t} f(t)dt$

$$F(s-a) = \int_0^\infty e^{-sa} \cdot \{e^{at}f(t)\}\, dt$$

$$= L\{e^{at}f(t)\}$$

Thus the Laplace transform of $f(t)$ multiplied by e^{at} is obtained by replacing s by $s - a$ in $F(s)$ which is the Laplace transform of $f(t)$.

Example: Find $L\{e^{at}\sin at\}$

Solution: We know that $L\{\sin at\} = \frac{a}{s^2+a^2}$.

By first shifting theorem, replace s by $s - a$ in $F(s)$, the Laplace transform of $\sin at$ i.e.,

$$L\{e^{at}\sin at\} = \left.\frac{a}{s^2+a^2}\right|_{s=s-a} = \frac{a}{(s-a)^2+a^2}$$

Similarly,

Example: $L\{e^{at}\cos at\} = \left.\frac{s}{s^2+a^2}\right|_{s=s-a} = \frac{s-a}{(s-a)^2+a^2}$

Example: $L\{e^{at}\sin wt\} = \frac{w}{(s-a)^2+w^2}$

Example: $L\{e^{at}\cos wt\} = \frac{s-a}{(s-a)^2+w^2}$

Example: $L\{e^{at}\cdot t^n\} = \left.\frac{n!}{s^{n+1}}\right|_{s=s-a} = \frac{n!}{(s-a)^{n+1}}$.

WORKED OUT EXAMPLES

Find the Laplace transform of $f(t)$:

Example 1: $f(t) = t^{\frac{7}{2}}e^{3t}$

Solution:

$$L\{t^{\frac{7}{2}}\} = \frac{\Gamma\left(\frac{7}{2}+1\right)}{s^{\frac{7}{2}+1}} = \frac{\frac{7}{2}\cdot\frac{5}{2}\cdot\frac{3}{2}\cdot\frac{1}{2}\Gamma\left(\frac{1}{2}\right)}{s^{\frac{9}{2}}} = \frac{105\sqrt{\pi}}{16s^{\frac{9}{2}}}$$

By first shift theorem,

$$L\{e^{3t}\cdot t^{\frac{7}{2}}\} = \left.\frac{105\sqrt{\pi}}{16s^{\frac{9}{2}}}\right|_{\text{at } s=s-3} = \frac{105\sqrt{\pi}}{16(s-3)^{\frac{9}{2}}}$$

Example 2: $\{3t^5 - 2t^4 + 4e^{-5t} - 3\sin 6t + 4\cos 4t\}e^{2t}$

Solution:

$$L\{3t^5 - 2t^4 + 4e^{-5t} - 3\sin 6t + 4\cos 4t\}$$

$$= 3L\{t^5\} - 2L\{t^4\} + 4L\{e^{-5t}\}$$

$$-3L\{\sin 6t\} + 4L\{\cos 4t\}$$

$$= 3\frac{5!}{s^6} - 2\frac{4!}{s^5} + 4\frac{1}{s+5} - 3\frac{6}{s^2+36} + 4\frac{s}{s^2+16}$$

Applying first shift theorem,

$$L\{(3t^5 - 2t^4 + 4e^{-5t} - 3\sin 6t + 4\cos 4t)e^{2t}\}$$

$$= \frac{360}{s^6} - \frac{48}{s^5} + \frac{4}{s+5} - \frac{18}{s^2+36} + \frac{4s}{s^2+16}\Bigg|$$

with s replaced by $s-2$

$$= \frac{360}{(s-2)^6} - \frac{48}{(s-2)^5} + \frac{4}{s+3}$$

$$- \frac{18}{(s-2)^2+36} + \frac{4(s-2)}{(s-2)^2+16}.$$

Example 3: $f(t) = \cosh at \cdot \cos bt$

Solution:

$$L\{\cosh at \cdot \cos bt\} = L\left\{\frac{1}{2}(e^{at}+e^{-at}) \cdot \cos bt\right\}$$

$$= \frac{1}{2}L\{e^{at}\cos bt\} + \frac{1}{2}L\{e^{-at}\cos bt\}$$

By first shift theorem,

$$= \frac{1}{2}\frac{s}{s^2+b^2}\bigg|_{s=s-a} + \frac{1}{2}\frac{s}{s^2+b^2}\bigg|_{s=s+a}$$

$$= \frac{1}{2}\frac{s-a}{(s-a)^2+b^2} + \frac{1}{2}\frac{s+a}{(s+a)^2+b^2}.$$

EXERCISE

Find the Laplace transform of $f(t)$ given by:

1. $(t+2)^2 e^t$

Ans. $(4s^2-4s+2)/(s-1)^3$

2. $e^{-4t}\cosh 2t$

Ans. $(s+4)/(s^2+8s+12)$

3. $\sinh at \sin at$

Ans. $2a^2 s/(s^4+4a^4)$

4. $e^{2t}(3\sin 4t - 4\cos 4t)$

Ans. $(20-4s)/(s^2-4s+20)$

5. $e^{-t}\sin^2 t$

Ans. $2/\left[(s+1)(s^2+2s+5)\right]$

6. $t \sin at$

Ans. $2as/(s^2+a^2)^2$

7. $t \cos at$

Hint: For problems 6 and 7

$$L\{t\,e^{iat}\} = \frac{1}{s^2}\bigg|_{s=s-ia} = \frac{1}{(s-ia)^2}\frac{(s+ia)^2}{(s+ia)^2}$$

$$= \frac{s^2-a^2+2ias}{(s^2+a^2)^2}$$

Comparing the real and imaginary parts on both sides, the results 6, 7 are obtained.

Ans. $(s^2-a^2)/(s^2+a^2)^2$

8. $\frac{t^{n-1}}{1-e^{-t}}$

Hint:

$$\frac{t^{n-1}}{1-e^{-t}} = t^{n-1}(1-e^{-t})^{-1}$$

$$= t^{n-1}(1+e^{-t}+e^{-2t}+e^{-3t}+\cdots)$$

$$= \sum_{m=0}^{\infty} t^{n-1}e^{-mt}.$$

Ans. $\sum_{m=0}^{\infty} \frac{\Gamma(n)}{(s+m)^n}$

9. If $L\{f(t)\} = F(s)$ then prove that

   a. $L\{\cosh at \cdot f(t)\} = \frac{1}{2}[F(s-a)+F(s+a)]$

   b. $L\{\sinh at \cdot f(t)\} = \frac{1}{2}[F(s-a)-F(s+a)]$

   Hint: Express cosh, sinh in e^{at} and use first shifting theorem.

10. $e^{-2t}\sin^3 t$

Ans. $\frac{3}{4}\frac{1}{s^2+4s+5} - \frac{3}{4}\frac{1}{s^2+4s+13}$

11. $\sin^4 t \cdot e^{2t}$

Ans. $\frac{1}{8}\left[\frac{3}{s-2} - \frac{4(s-2)}{(s-2)^2+4} + \frac{s-2}{(s-2)^2+16}\right]$

12. $e^{4t}\sin 2t \cdot \cos t$

Ans. $\frac{1}{2}\left[\frac{3}{(s-4)^2+9} + \frac{1}{(s-4)^2+1}\right]$.

Laplace Transform of Derivatives (Multiplication by s)

Differentiation of $f(t)$ is replaced by multiplication of the transform $F(s) = L(f)$ by s, roughly.

Theorem: *Laplace transform of the derivative of $f(t)$ (Multiplication by s)*

$$L\{f'\} = s\,L\{f\} - f(0) \qquad (s > 0)$$

Proof:

$$L\{f'\} = \int_0^{\infty} e^{-st} f'\,dt : \qquad \text{Integrating by parts}$$

$$= e^{-st} \cdot f\Big|_0^{\infty} + s\int_0^{\infty} e^{-st} f\,dt$$

since f is of exponential order the first term is R.H.S. becomes zero at the upper limit ∞. Thus

$$L\{f'\} = -f(0) + sL\{f\}.$$

Result 1: By applying the above theorem to f'', we have

$$L\{f''\} = s\,L\{f'\} - f'(0)$$
$$= s\,\{s\,L\{f\} - f(0)\} - f'(0)$$
$$\therefore \quad L\{f''\} = s^2L\{f\} - s\,f(0) - f'(0)$$

Result 2: Similarly, for L.T. of derivatives of order n:

$$L\{f^{(n)}\} = s^nL\{f\} - s^{n-1}f(0) - s^{n-2}f'(0)\cdots - f^{(n-1)}(0).$$

Note: L.T. of functions (listed in the L.T. table) can also be (re)derived by the application of transform of derivative theorem.

Worked Out Examples

Find Laplace transform of the following functions using the theorem of transform of derivative:

Example 1: $f(t) = t^2$

Solution: $f(0) = 0,\ f'(0) = 0,\ f''(t) = 2$
By theorem

$$L\{f''\} = s^2L\{f\} - sf(0) - f'(0)$$
$$L\{2\} = s^2L\{t^2\} - s\cdot 0 - 0$$
$$\therefore \quad L\{t^2\} = \frac{1}{s^2}L\{2\} = \frac{2}{s^2}L\{1\} = \frac{2}{s^2}\frac{1}{s} = \frac{2}{s^3}.$$

Example 2: $f(t) = t\cdot\cos at$

Solution:

$$f' = -at\cdot\sin at + \cos at$$
$$f'' = -a(at\cos at + \sin at) - a\sin at$$
$$f'' = -a^2t\cos at - 2a\sin at$$
$$f(0) = 0,\ f'(0) = 1$$

Using theorem for 2nd derivative

$$L\{-a^2t\cos at - 2a\sin at\} = s^2L\{t\cos at\} - s\cdot 0 - 1.$$

Rearranging

$$(s^2 + a^2)L\{t\cos at\} = 1 - 2a\,L\{\sin at\}$$
$$= 1 - 2a\cdot\frac{a}{s^2+a^2}$$
$$= \frac{s^2-a^2}{s^2+a^2}$$
$$\therefore \quad L\{t\cos at\} = \frac{s^2-a^2}{(s^2+a^2)^2}.$$

Example 3: Given $L\{\sin\sqrt{t}\} = \frac{\sqrt{\pi}}{2s^{\frac{3}{2}}}e^{-\frac{1}{4s}}$. Prove that $L\left\{\frac{\cos\sqrt{t}}{\sqrt{t}}\right\} = \left(\frac{\pi}{s}\right)^{\frac{1}{2}}e^{-\frac{1}{4s}}$.

Solution: Take $f(t) = \frac{\cos\sqrt{t}}{\sqrt{t}}$

Since $\frac{\cos\sqrt{t}}{2\sqrt{t}}$ is the derivative of $\sin\sqrt{t}$, choose

$$g(t) = \sin\sqrt{t}, \quad \text{then} \quad g'(t) = \frac{\cos\sqrt{t}}{2\sqrt{t}}$$
$$g(0) = 0$$

Using theorem $L\{g'\} = s\,L\{g\} - g(0)$ we have

$$L\left\{\frac{\cos\sqrt{t}}{2\sqrt{t}}\right\} = s\cdot L\{\sin\sqrt{t}\} - 0$$
$$\frac{1}{2}L\left\{\frac{\cos\sqrt{t}}{\sqrt{t}}\right\} = s\cdot\sqrt{\pi}\cdot\frac{e^{-\frac{1}{4s}}}{2s^{\frac{3}{2}}}$$
$$L = \left\{\frac{\cos\sqrt{t}}{\sqrt{t}}\right\} = \sqrt{\frac{\pi}{s}}\cdot e^{-\frac{1}{4s}}.$$

Exercise

Find the Laplace transform of the following functions using the theorem on transform of derivatives:

1. t^n — *Ans.* $\frac{n!}{s^{n+1}}$
2. $\sin^2 t$ — *Ans.* $\frac{2}{s(s^2+4)}$
3. cos at — *Ans.* $\frac{s}{s^2+a^2}$
4. t^2 sin at — *Ans.* $\frac{2a}{(s^2+a^2)^2} + \frac{4a(s^2-a^2)}{(s^2+a^2)^3}$
5. t cosh at — *Ans.* $\frac{s^2+a^2}{(s^2-a^2)^2}$

Laplace Transform of the Integral of a Function

Just as division is the inverse operation of multiplication, differentiation and integration are inverse operations. Thus, while differentiation of a function corresponds to multiplication of the transform by s, integration of a function corresponds to division of the transform by s.

Theorem: *Integration of $f(t)$ (Division by s). If $L\{f(t)\} = F(s)$, then*

$$L\left\{\int_0^t f(u)du\right\} = \frac{F(s)}{s}, \quad s > 0$$

Proof: Let $g(t) = \int_0^t f(u)du$. Then $g'(t) = f(t)$ and $g(0) = 0$

By derivative theorem

$$F(s) = L\{f(t)\} = L\{g'(t)\} = s\,L\{g(t)\} - g(0)$$

$$= s \cdot L\{g(t)\}$$

$$\therefore \quad L\{g(t)\} = \frac{F(s)}{s}$$

or

$$L\left\{\int_0^t f(u)du\right\} = \frac{F(s)}{s}.$$

Result: Similarly if $L\{f(t)\} = F(s)$, then

$$L\left\{\int_0^t dt_1 \int_0^{t_1} f(u)du\right\} = \frac{F(s)}{s^2}$$

The double integral can also be written briefly as

$$\int_0^t\int_0^t f(t)dt^2$$

Generalization for nth integral

$$L\left\{\int_0^t\int_0^t \underset{n \text{ times}}{\cdots} \int_0^t f(t)dt\right\} = \frac{F(s)}{s^n}.$$

WORKED OUT EXAMPLES

Example 1: $L\left\{\int_0^t \frac{1-e^{-u}}{u}du\right\}$.

Solution: Here the integrand is $f(t) = (1-e^{-t})/t$, we know that

$$L\{f(t)\} = L\left\{\frac{1-e^{-t}}{t}\right\} = F(s) = \ln\left(1+\frac{1}{s}\right)$$

Using the theorem of L.T. of integrals

$$L\left\{\int_0^t \frac{1-e^{-u}}{u}du\right\} = \frac{1}{s}\cdot\ln\left(1+\frac{1}{s}\right).$$

Example 2: $L\left\{\int_0^t\int_0^t\int_0^t \cos au\,du\,du\,du\right\}$

Solution: Here integrand is $f(t) = \cos at$

$$L\{f(t)\} = L\{\cos at\} = \frac{s}{s^2+a^2}$$

Using the theorem on L.T. of integrals

$$L\left\{\int_0^t \cos au\,du\right\} = \frac{1}{s}\cdot\frac{s}{s^2+a^2} = \frac{1}{s^2+a^2}$$

Applying repeatedly.

$$L\left\{\int_0^t\int_0^t \cos au\,du\,du\right\} = \frac{1}{s}\cdot\frac{1}{s^2+a^2}$$

$$\therefore \quad L\left\{\int_0^t\int_0^t\int_0^t \cos au\,du\,du\,du\right\} = \frac{1}{s^2}\frac{1}{s^2+a^2}.$$

Example 3: $L\left\{\int_0^t \frac{\sin u}{u}du\right\}$.

Solution: The integrand is $f(t) = \frac{\sin t}{t}$, we know that

$$L\{\sin t\} = \frac{1}{s^2+1}$$

Using division by t

$$L\left\{\frac{\sin t}{t}\right\} = \int_s^\infty \frac{du}{u^2+1} = \tan^{-1}\frac{1}{u}\Big|_s^\infty = \tan^{-1}\frac{1}{s}$$

$\therefore$ Using theorem on L.T. of integral

$$L\left\{\int_0^t \frac{\sin u}{u}du\right\} = \frac{1}{s}\tan^{-1}\frac{1}{s}.$$

Example 4: Show that $\int_{t=0}^\infty\int_{u=0}^t \frac{e^{-t}\sin u}{u}du\,dt = \frac{\pi}{4}$

Solution: The given double integral on the L.H.S. is the Laplace transform of $\int_0^t \frac{\sin u}{u}du$ with $s=1$. We know that

$$L\left\{\int_0^t \frac{\sin u}{u}du\right\} = \int_{t=0}^\infty\int_{u=0}^t \frac{e^{-st}\sin u}{u}du\,dt$$

$$= \frac{1}{s}\tan^{-1}\frac{1}{s}\Big|_{s=1} = \tan^{-1}1 = \frac{\pi}{4}.$$

Example 5: $L\left\{\int_0^t ue^{-u}\cdot \sin 4u\, du\right\}$

Solution:

$$L\{\sin 4t\} = \frac{4}{s^2+4^2} = \frac{4}{s^2+16}$$

$$L\{e^{-t}\sin 4t\} = \frac{4}{(s+1)^2+16} = \frac{4}{s^2+2s+17}$$

$$L\{t\,e^{-t}\sin 4t\} = -\frac{d}{ds}\left[\frac{4}{s^2+2s+17}\right]$$

$$= \frac{4(2s+2)}{(s^2+2s+17)^2}$$

$$\therefore\quad L\left\{\int_0^t u\,e^{-u}\sin 4u\,du\right\} = \frac{8}{s}\frac{(s+1)}{(s^2+2s+17)^2}.$$

Example 6: $L\left\{\sinh ct \int_0^t e^{au}\sinh bu\,du\right\}$

Solution:

$$L\{\sinh bt\} = \frac{b}{s^2-b^2}$$

$$L\{e^{at}\sinh bt\} = \frac{b}{(s-a)^2-b^2}$$

$$L\left\{\int_0^t e^{au}\sinh bu\,du\right\} = \frac{1}{s}\cdot\frac{b}{(s-a)^2-b^2} \quad (*)$$

Now consider

$$L\left\{\sinh ct\int_0^t e^{au}\sinh bu\,du\right\}$$

$$= L\left\{\left(\frac{e^{ct}-e^{-ct}}{2}\right)\int_0^t e^{au}\sinh bu\,du\right\}$$

$$= \frac{1}{2}L\left\{e^{ct}\int_0^t e^{au}\sinh bu\,du\right\}$$

$$+\frac{1}{2}L\left\{e^{-ct}\int_0^t e^{au}\sinh bu\,du\right\}$$

Using (*)

$$= \frac{1}{2}\frac{b}{s[(s-a)^2-b^2]}\bigg|_{s=s-c} + \frac{1}{2}\frac{b}{s[(s-a)^2-b^2]}\bigg|_{s=s+c}$$

$$= \frac{1}{2}\frac{b}{(s-c)[(s-c-a)^2-b^2]}$$

$$+\frac{1}{2}\frac{b}{(s+c)[(s+c+-a^2)-b^2]}.$$

Exercise

Find the Laplace Transform of:

1. $\int_0^t e^{-u}\cos u\,du$ *Ans.* $\frac{1}{s}\frac{(s+1)}{(s^2+2s+2)}$
2. $\int_0^t e^{u}\cdot\frac{\sin u}{u}du$ *Ans.* $\frac{1}{s}\cot^{-1}(s-1)$
3. $\int_0^t \frac{e^{-4u}\sin 3u}{u}du$ *Ans.* $\frac{1}{s}\cot^{-1}\left(\frac{s+4}{3}\right)$
4. $\cosh t\int_0^t e^{u}\cosh u\,du$

   Ans. $\frac{1}{2}\left[\frac{s-2}{(s-1)^2(s-3)}+\frac{s}{(s+1)^2(s-1)}\right]$
5. $e^{-4t}\int_0^t \frac{\sin 3u}{u}du$ *Ans.* $\frac{1}{s+4}\cot^{-1}\left(\frac{s+4}{3}\right)$

Differentiation of Transforms: Multiplication by *t*

Differentiation of the transform of a function corresponds to the multiplication of the function by $-t$.

Theorem: *If* $L\{f(t)\} = F(s)$ *then*

$$L\{-t\,f(t)\} = \frac{d}{ds}F(s)$$

or

$$L\{t\cdot f(t)\} = -\frac{d}{ds}F(s)$$

In general,

$$L\{t^n f(t)\} = (-1)^n\frac{d^n}{ds^n}F(s), \quad \text{for } n = 1, 2, 3, \ldots$$

Proof: We know that $F(s) = \int_0^\infty e^{-st}f(t)dt$.

By Leibnitz's rule for differentiating under the integral sign

$$\frac{dF(s)}{ds} = \frac{d}{ds}\int_0^\infty e^{-st}f(t)dt = \int_0^\infty f(t)\left(\frac{\partial}{\partial s}e^{-st}\right)dt$$

$$= \int_0^\infty f(t)\cdot(-t\,e^{-st})dt$$

$$= -\int_0^\infty e^{-st}\{t\,f(t)\}\,dt = -L\{t\,f(t)\}$$

$$\therefore\quad L\{t\,f(t)\} = -\frac{dF(s)}{ds}$$

By mathematical induction the result for *n*th derivative follows.

WORKED OUT EXAMPLES

Example 1: Show that $L\{t \sin at\} = \frac{1}{(s-a)^2}$.

Solution:

$$L\{\sin at\} = \frac{a}{s^2+a^2}$$

$$\therefore \quad L\{t \cdot \sin at\} = -\frac{d}{ds}\left\{\frac{a}{s^2+a^2}\right\} = \frac{2as}{(s^2+a^2)^2}.$$

Example 2: $L\{t \cdot e^{-2t} \sin t\}$.

Solution:

$$L\{\sin\} = \frac{1}{s^2+1}$$

$$L\{e^{-2t} \cdot \sin t\} = \frac{1}{(s+2)^2+1} = \frac{1}{s^2+4s+5}$$

by first shift theorem.

Using theorem on multiplication by t

$$L\left\{t \cdot (e^{-2t} \cdot \sin t)\right\} = -\frac{d}{ds}\left\{\frac{1}{s^2+4s+5}\right\}$$

$$= \frac{2s+4}{(s^2+4s+5)^2}.$$

Example 3: $L\left\{(t^2-3t+2)\sin 3t\right\}$.

Solution:

$$L\{\sin 3t\} = \frac{3}{s^2+3^2} = \frac{3}{s^2+9}$$

$$L\left\{(t^2-3t+2)\sin 3t\right\} = L\{t^2 \cdot \sin 3t\} - 3L\{t \cdot \sin 3t\} + 2L\{\sin 3t\}.$$

Using multiplication by t

$$= (-1)^2\frac{d^2}{ds^2}\left\{\frac{3}{s^2+9}\right\} - 3 \cdot (-1)\frac{d}{ds}\left\{\frac{3}{s^2+9}\right\} + 2 \cdot \frac{3}{s^2+9}$$

$$= \frac{(-6+24s^2)}{(s^2+9)^3} + 3 \cdot \frac{(-6s)}{(s^2+9)^2} + \frac{6}{s^2+9}$$

$$= \frac{6s^4-18s^3+126s^2-162s+432}{(s^2+9)^3}$$

Example 4: Show that $\int_0^\infty t^2 e^{-4t} \cdot \sin 2t\, dt = \frac{11}{500}$.

Solution: L.H.S. is rewritten as

$\int_0^\infty e^{-4t} \cdot (t^2 \sin 2t)dt$

i.e., it is L.T. of $t^2 \sin 2t$ with $s = 4$. We know that $L\{\sin 2t\} = \frac{2}{s^2+2^2} = \frac{2}{s^2+4}$.

Using multiplication by t

$$L\{t^2 \sin 2t\} = (-1)^2\frac{d^2}{ds^2}\left\{\frac{2}{s^2+4}\right\} = \frac{d}{ds}\left\{\frac{-4s}{(s^2+4)^2}\right\}$$

$$= -4 \cdot \left.\frac{(4-3s^2)}{(s^2+4)^3}\right|_{s=4} = \frac{11}{500}.$$

EXERCISE

Find Laplace theorem using the theorem of differentiation of L.T.:

1. $t \cos at$ *Ans.* $\frac{s^2-a^2}{(s^2+a^2)^2}$
2. $t^2 \sin t$ *Ans.* $\frac{6s^2-2}{(s^2+1)^3}$
3. $t \cdot \sinh 2t$ *Ans.* $\frac{4s}{(s^2-4)^2}$
4. $t(3\sin 2t - 2\cos 2t)$ *Ans.* $\frac{8+12s-2s^2}{(s^2+4)^2}$
5. $t^3 \cdot \cos t$ *Ans.* $\frac{6s^4-36s^2+6}{(s^2+1)^4}$
6. $t \cdot e^{at} \cdot \sin at$ *Ans.* $\frac{2a}{(s-a)(s^2-2as^2+2a^2)^2}$

Prove the following:

7. $\int_0^\infty t \cdot e^{-3t} \cdot \sin t\, dt = \frac{3}{50}$

   Hint: L.T. of $t \sin t$ with $s = 3$
8. $\int_0^\infty t^3 e^{-t} \cdot \sin t\, dt = 0$

   Hint: L.T. of $t^3 \sin t$ with $t = 1$
9. $\int_0^\infty e^{-3t} t \cos t\, dt = \frac{2}{25}$

   Hint: L.T. of $t \cos t$ with $s = 3$
10. $\int_0^\infty e^{-2t} \sin^3 t\, dt = \frac{6}{65}$

Ans. L.T. of $\sin^3 t$ with $s = 2$.

Integration of Transform: Division by t

Integration of the transform of a function $f(t)$ corresponds to the division of $f(t)$ by t.

Theorem: *If* $L\{f(t)\} = F(s)$ *then*

$$L\left\{\frac{f(t)}{t}\right\} = \int_s^\infty F(u)du$$

Proof: $F(s) = L\{f(t)\} = \int_0^\infty e^{-st} f(t)dt$
Integrating on both sides w.r.t. s from s to ∞, we have

$$\int_s^\infty F(u)du = \int_s^\infty \int_0^\infty e^{-st} f(t)dt.$$

Since s and t are independent variables, interchanging the order of integration, we get

$$\int_s^\infty F(u)du = \int_0^\infty f(t)\left\{\int_s^\infty e^{-st}ds\right\}dt$$

$$= \int_0^\infty f(t)\left(\frac{e^{-st}}{-t}\right)\Big|_{s=s}^{\infty} dt$$

$$= \int_0^\infty f(t)\cdot\left(\frac{-e^{-st}}{-t}\right)dt$$

$$= \int_0^\infty e^{-st}\left(\frac{f(t)}{t}\right)dt = L\left\{\frac{f(t)}{t}\right\}.$$

WORKED OUT EXAMPLES

Find Laplace theorem of the following functions $f(t)$:

Example 1: $\frac{e^{-at}-e^{-bt}}{t}$

Solution:

$$L\{e^{-at} - e^{-bt}\} = L\{e^{-at}\} - L\{e^{-bt}\}$$

$$= \frac{1}{s+a} - \frac{1}{s+b}.$$

Division by t amounts to integration

$$L\left\{\frac{1}{t}(e^{-at} - e^{-bt})\right\}$$

$$= \int_s^\infty \left(\frac{1}{u+a} - \frac{1}{u+b}\right)du = \ln\left(\frac{u+a}{u+b}\right)\Big|_s^\infty$$

$$= 0 - \ln\left(\frac{s+a}{s+b}\right) = \ln\left(\frac{s+b}{s+a}\right).$$

Example 2: $\frac{\sin^2 t}{t}$.

Solution:

$$L(\sin^2 t) = L\left(\frac{1-\cos 2t}{2}\right) = L\left(\frac{1}{2}\right) - \frac{1}{2}L(\cos 2t)$$

$$= \frac{1}{2}\frac{1}{s} - \frac{1}{2}\frac{s}{s^2+2^2}.$$

Using division by t

$$L\left\{\frac{1}{t}\cdot\sin^2 t\right\} = \frac{1}{2}\int_s^\infty \left(\frac{du}{u} - \frac{u\,du}{u^2+4}\right)$$

$$= \frac{1}{2}\ln s - \frac{1}{4}\ln(s^2+4)\Big|_s^\infty$$

$$= \ln\left(\frac{\sqrt{s}}{(s^2+4)^{\frac{1}{4}}}\right)\Big|_s^\infty = \frac{1}{4}\ln\left(\frac{s^2+4}{s^2}\right).$$

Example 3: $\frac{\cos at - \cos bt}{t}$.

Solution:

$$L(\cos at - \cos bt) = L(\cos at) - L(\cos bt)$$

$$= \frac{s}{s^2+a^2} - \frac{s}{s^2+b^2}$$

Using theorem on integration

$$L\left\{\frac{1}{t}(\cos at - \cos bt)\right\}$$

$$= \int_s^\infty \left(\frac{s}{s^2+a^2} - \frac{s}{s^2+b^2}\right)ds$$

$$= \frac{1}{2}\ln\left(\frac{s^2+a^2}{s^2+b^2}\right)\Big|_s^\infty = \frac{1}{2}\left[0 - \ln\left(\frac{s^2+a^2}{s^2+b^2}\right)\right]$$

$$= \frac{1}{2}\ln\left(\frac{s^2+b^2}{s^2+a^2}\right).$$

Example 4: $\int_0^\infty \frac{\cos 6t - \cos 4t}{t}dt$.

Solution: This integral can be looked upon as

$$\int_0^\infty e^{-0\cdot t}\cdot\left(\frac{\cos 6t - \cos 4t}{t}\right)dt$$

so that the given integral is L.T. of $\frac{\cos 6t - \cos 4t}{t}$ with $s = 0$. From above Worked Out Example 3.

$$L\left(\frac{\cos 6t - \cos 4t}{t}\right)_{\text{with } s=0}$$

$$= \frac{1}{2}\ln\left(\frac{s^2+b^2}{s^2+a^2}\right)\Bigg|_{\substack{s=0\\a=6\\b=4}} = \ln\frac{2}{3}.$$

Example 5: Evaluate $\int_0^\infty e^{-t}\frac{\sin^2 t}{t}dt$.

Solution: The given integral is the Laplace transform of $\frac{\sin^2 t}{t}$ with $s = 1$, i.e.,

$$L\left\{\frac{\sin^2 t}{t}\right\}\Bigg|_{s=1} = \int_0^\infty e^{-st} \cdot \frac{\sin^2 t}{t} dt\Bigg|_{s=1}$$

$$= \int_0^\infty e^{-t} \frac{\sin^2 t}{t} dt.$$

From Example 2 above, we know that

$$L\left\{\frac{\sin^2 t}{t}\right\} = \frac{1}{4} \ln\left(\frac{s^2+4}{s^2}\right)$$

Thus

$$\int_0^\infty e^{-t} \frac{\sin^2 t}{t} dt = L\left\{\frac{\sin^2 t}{t}\right\}_{\text{at } s=1}$$

$$= \frac{1}{4} \ln\left(\frac{s^2+4}{s^2}\right)\Bigg|_{s=1} = \frac{1}{4} \ln 5.$$

EXERCISE

1. $\frac{1-e^{2t}}{t}$ *Ans.* $\ln \frac{s-2}{s}$
2. $\frac{\sinh t}{t}$ *Ans.* $\frac{1}{2} \ln\left(\frac{s+1}{s-1}\right)$
3. $\frac{1-\cos at}{t}$ *Ans.* $\frac{1}{2} \ln \frac{s^2+a^2}{s}$
4. $\frac{\sin 3t \cdot \cos t}{t}$

   Ans. $\frac{1}{2}\left[\pi - \tan^{-1}\left(\frac{s}{4}\right) - \tan^{-1}\left(\frac{s}{2}\right)\right]$
5. Show that $\int_0^\infty \frac{e^{-3t}-e^{-6t}}{t} dt = \ln 2$

   Hint: Use Example 1 with $a = 3, b = 6$ and $s = 0$
6. Show that $\int_0^\infty \frac{\sin^2 t}{t^2} dt = \frac{\pi}{2}$
7. Find the L.T. of $\left\{\int_0^t e^t \cdot \frac{\sin t}{t} dt\right\}$

   Ans. $\frac{1}{s} \cot^{-1}(s-1)$
8. Show that $\int_0^\infty \cdot e^{-t} \cdot \frac{\sin^2 t}{t} dt = \frac{1}{2} \ln 5$
9. $\int_0^\infty e^{-2t} \frac{(2\sin t - 3\sinh t)}{t} dt = 2\cot^{-1} 2 + \frac{3}{2} \log\left(\frac{1}{3}\right)$

   Hint: L.T. of $\frac{2\sin t - 3\sinh t}{t}$ with $s = 2$.

Unit step function ("Heavisides" unit function)*

It is defined as (Fig. 12.3)

$$u(t-a) = 0 \quad \text{if } t < a$$
$$= 1 \quad \text{if } t > a$$

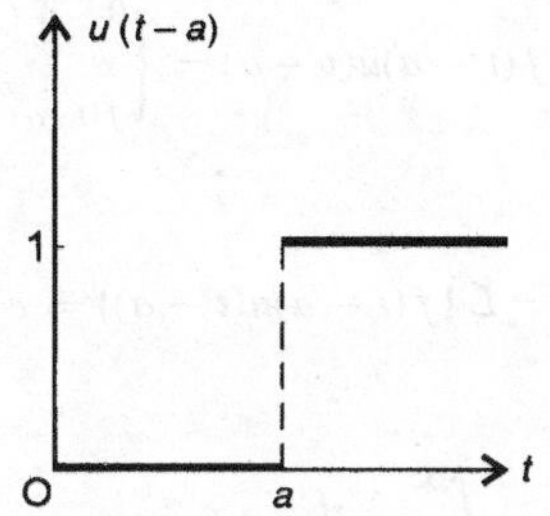

Fig. 12.3

Note: $u(t-a)$ is also denoted as $u_a(t)$. In particular when $a = 0$,

$$u(t) = 0 \quad \text{if } t < 0$$
$$= 1 \quad \text{if } t > 0$$

Multiplying a given function $f(t)$ with the "engineering function" the unit step function $u(t-a)$, several effects can be produced as shown in Fig. 12.4.

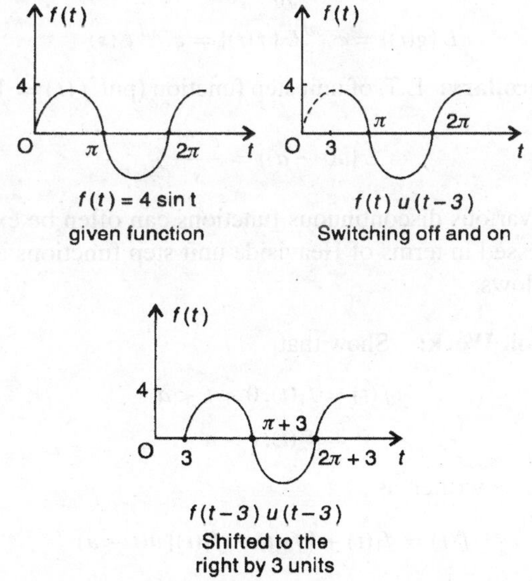

Fig. 12.4

*Oliver Heaviside (1850–1925), English electrical engineer.

Second Translation (Or Second Shifting Theorem)

t-Shifting (replacing t by (t–a) in f (t))

Theorem: *If* $L\{f(t)\} = F(s)$ *and the shifted function*

$$g(t) = f(t-a)u(t-a) = \begin{cases} 0 & \text{if } t<a \\ f(t-a) & \text{if } t>a \end{cases}$$

Then

$$L\{g(t)\} = L\{f(t-a)u(t-a)\} = e^{-as}F(s).$$

Proof:

$$L\{g(t)\} = \int_0^\infty e^{-st}g(t)dt$$

$$= \int_0^a e^{-st}g(t)dt + \int_a^\infty e^{-st}g(t)dt$$

$$= 0 + \int_a^\infty e^{-st}f(t-a)dt$$

Put $t-a=u$, $dt=du$ then

$$L\{g(t)\} = \int_0^\infty e^{-s(u+a)}f(u)du$$

$$= e^{-sa}\int_0^\infty e^{-su}f(u)du$$

$$L\{g(t)\} = e^{-sa}L\{f(t)\} = e^{-sa}F(s)$$

Corollary: L.T. of unit step function (put $f(t)=1$)

$$L\{u(t-a)\} = \frac{e^{-as}}{s}$$

Various discontinuous functions can often be expressed in terms of Heaviside unit step functions as follows:

Book Work: Show that

$$f(t) = f_1(t), 0<t<a$$
$$= f_2(t), t>a$$

can be written as

$$f(t) = f_1(t) + \{f_2(t) - f_1(t)\}u(t-a)$$

Proof: When $t<a$, $u(t-a)=0$ so that

$$f(t) = f_1(t) \quad \text{for } t<a$$

when $t>a$, $u(t-a)=1$ so that

$$f(t) = f_1(t) + \{f_2(t) - f_1(t)\}$$
$$= f_2(t) \quad \text{for } t>a.$$

In general, if

$$\begin{aligned} f(t) &= f_1(t) && \text{for } 0<t<a_1 \\ &= f_2(t) && \text{for } a_1<t<a_2 \\ &\equiv \\ &= f_{n-1}(t) && \text{for } a_{n-2}<t<a_{n-1} \\ &= f_n(t) && \text{for } t>a_{n-1} \end{aligned}$$

then

$$f(t) = f_1(t) + \{f_2(t) - f_1(t)\}u(t-a_1) + \cdots$$
$$+ \{f_n(t) - f_{n-1}(t)\}u(t-a_{n-1}).$$

WORKED OUT EXAMPLES

Find the Laplace transform of the following functions using second translation theorem:

Example 1: Express in terms of Heaviside's unit step function

$$\begin{aligned} f(t) &= \sin t, && 0<t<\pi \\ &= \sin 2t, && \pi<t<2\pi \\ &= \sin 3t, && t>2\pi. \end{aligned}$$

Solution: Let

$$f_1(t) = \sin t,\ f_2(t) = \sin 2t,\ f_3(t) = \sin 3t$$

so that

$$f(t) = f_1(t) + (f_2 - f_1)u(t-\pi) + (f_3 - f_2)u(t-2\pi)$$
$$= \sin t + (\sin 2t - \sin t)u(t-\pi)$$
$$+ (\sin 3t - \sin 2t)u(t-2\pi).$$

Example 2: Given

$$\begin{aligned} f(t) &= e^{-t}, && 0<t<3 \\ &= 0, && t>3 \end{aligned}$$

write $f(t)$ in terms of Heaviside's unit step function and hence find the Laplace transform of $f(t)$.

Solution: Let $f_1(t) = e^{-t}$ and $f_2(t) = 0$ then $f(t)$ can be written as

$$f(t) = f_1(t) + \{f_2(t) - f_1(t)\}\,u(t-3)$$
$$= e^{-t} + (0 - e^{-t})u(t-3)$$
$$= e^{-t}\,(1 - u(t-3))$$
$$L\{f(t)\} = L\left\{e^{-t} - e^{-t}u(t-3)\right\}$$
$$= L\{e^{-t}\} - L\left\{e^{-t}u(t-3)\right\}$$
$$= \frac{1}{s+1} - \frac{e^{-3(s+1)}}{s+1}$$

Since $L\{u(t-3)\} = e^{-3s}/s$ and $L\{e^{-t}\cdot u(t-3)\} = e^{-3(s+1)}/(s+1)$.

Example 3: $g(t) = 0, \quad 0 < t < 5$

$= t - 3, \quad t > 5$

Solution: To apply the t-shift theorem, express the functional values $t-3$ for $t > 5$ in terms of $t-5$ i.e., express $t-3$ as $(t-5)+2$ and rewrite

$$g(t) = 0, \quad 0 < t < 5$$
$$= (t-5) + 2, \quad t > 5$$

Thus $g(t) = u(t-5)f(t-5)$ where $f(t) = t+2$, $t > 0$.
Applying t-shift theorem

$$L\{u(t-5)f(t-5)\} = e^{-5s}F(s)$$

where

$$F(s) = L\{f(t)\} = L\{t+2\} = L\{t\} + L\{2\}$$
$$= \frac{1}{s^2} + \frac{2}{s}$$
$$\therefore \quad L\{g(t)\} = L\{u(t-5)f(t-5)\} = e^{-5s}\cdot F(s)$$
$$= e^{-5s}\left[\frac{1}{s^2} + \frac{2}{s}\right].$$

Example 4: $g(t) = 0, \quad 0 < t < \frac{\pi}{2}$

$= \sin t, \quad t > \frac{\pi}{2}$

Solution: Express $\sin t$ in terms of $t - \frac{\pi}{2}$ by observing that $\sin t = \cos(t - \frac{\pi}{2})$ hence

$$g(t) = u\left(t - \frac{\pi}{2}\right)\cdot f\left(t - \frac{\pi}{2}\right) = \begin{cases} 0, & 0<t<\frac{\pi}{2} \\ \cos\left(t-\frac{\pi}{2}\right), & t>\frac{\pi}{2} \end{cases}$$

where $f(t) = \cos t, \quad t > 0$

$$L\{g(t)\} = L\left\{u\left(t-\frac{\pi}{2}\right)\cdot f\left(t-\frac{\pi}{2}\right)\right\} = e^{-\frac{\pi}{2}s}\cdot F(s)$$

where $F(s) = L\{f(t)\} = L\{\cos t\} = \frac{s}{s^2+1}$

$$\therefore \quad L\{g(t)\} = \frac{se^{-\frac{\pi}{2}s}}{s^2+1}.$$

Example 5: $4\sin(t-3)u(t-3)$

Solution: We know that

$$L\{4\sin t\} = F(s) = \frac{4}{s^2+1}$$

Applying t-shift

$$L\{4\sin(t-3)u(t-3)\} = e^{-3s}F(s) = e^{-3s}\frac{4}{(s^2+1)}.$$

EXERCISE

Find the Laplace transform of the following:

1. $f(t) = \begin{cases} 1, & 0<t<2 \\ 2, & 2<t<4 \\ 3, & 4<t<6 \\ 0, & t>6 \end{cases}$

Ans. $\frac{1+e^{-2s}+e^{-4s}-3e^{-6s}}{s}$

2. $f(t) = \begin{cases} t, & 0<t<3 \\ 3, & t>3 \end{cases}$

Ans. $\frac{1}{s^2}(1 - e^{-3s})$

3. Express $f(t)$ in terms of heavisides unit step function

$$f(t) = \begin{cases} t^2, & 0<t<2 \\ 4t, & t>2 \end{cases}$$

Ans. $t^2 + (4t - t^2)u(t-2)$

4. $f(t) = \begin{cases} \sin t, & t>\pi \\ \cos t, & t<\pi \end{cases}$

Ans. $\cos t + (\sin t - \cos t)u(t-\pi)$.

Find L.T. by expressing $f(t)$ in unit step functions:

5. $f(t) = \begin{cases} t^2, & 0<t\le 2 \\ 0, & t>2. \end{cases}$

Ans. $t^2\,[u(t) - u(t-2)]\;\frac{2(1-e^{-2s})}{s^3} - \frac{4e^{-2s}(1+s)}{s^2}$.

6. $f(t) = \begin{cases} 2, & 0<t<\pi \\ 0, & \pi<t<2\pi \\ \sin t, & t>2\pi \end{cases}$

Ans. $f(t) = 2 - 2u(t-\pi) + u(t-2\pi)\sin t$.
$F(s) = \frac{2}{s} - \frac{2e^{-\pi s}}{s} + \frac{e^{-2\pi s}}{s^2+1}$

7. $4u(t-\pi)\cos t$

Ans. $-4e^{-\pi s}\cdot s/(s^2+1)$

8. $f(t) = t-1, \quad 1<t<2$
$\quad\;\; = 3-t, \quad 2<t<3$

Ans. $(e^{-s} - 2e^{-2s} + e^{-3s})/s^2$

9. Staircase function

$f(t) = 1, \quad 0<t<1$
$\quad\; = 2, \quad 1<t<2$
$\quad\; = 3, \quad 2<t<3$
$\quad\; \ldots \quad \ldots$
$\quad\; \ldots \quad \ldots$

Ans. $\frac{1}{s(1-e^{-s})}$

10. Saw tooth function (Fig. 12.5)

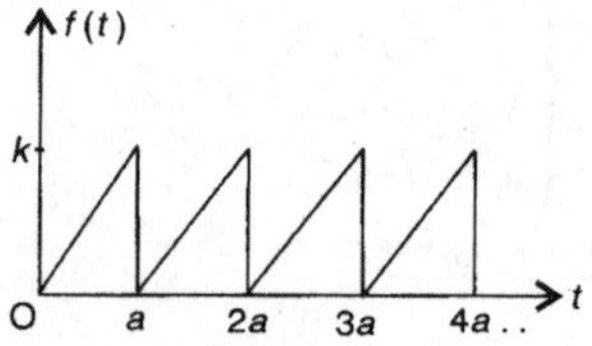

Fig. 12.5

Ans. $\frac{k}{as^2} - \frac{ke^{-as}}{s}\left(\frac{1}{1-e^{-as}}\right)$.

Dirac's* Delta Function (or Unit Impulse Function)

Forces (like earthquake) that produce large effects on a system when applied for a very short time interval can be represented by an impulse function which is a discontinuous function and is highly irregular from the mathematical point of view.

Impulse of a forces $f(t)$ in the interval $(a, a+\epsilon)$

$$= \int_a^{a+\epsilon} f(t)\,dt$$

Now define the function

$$f_\epsilon(t-a) = \begin{cases} 0 & \text{for } t<a \\ \frac{1}{\epsilon} & \text{for } a\le t\le a+\epsilon \\ 0 & \text{for } t>a \end{cases}$$

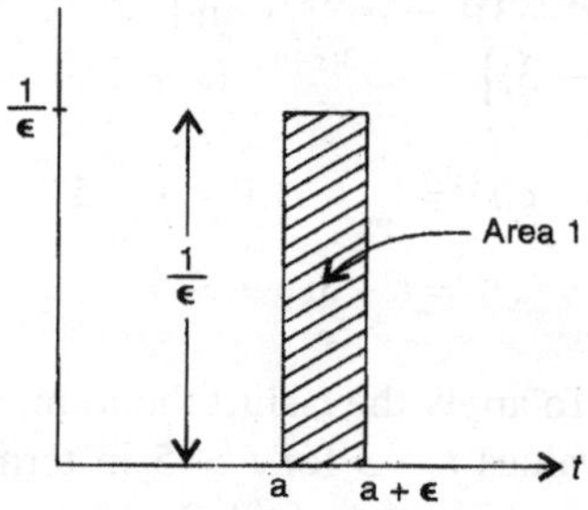

Fig. 12.6

This can also be represented in terms of two unit step functions as follows

$$f_\epsilon(t-a) = \frac{1}{\epsilon}[u(t-a) - u(t-(a+\epsilon))]$$

Note that

$$\int_0^\infty f_\epsilon(t-a)\,dt = \int_0^a 0 + \int_a^{a+\epsilon}\frac{1}{\epsilon}\,dt + \int_{a+\epsilon}^\infty 0 = 1$$

Thus the impulse I_ϵ is 1.

Taking Laplace transform

$$L\{f_\epsilon(t-a)\} = \frac{1}{\epsilon}L\{u(t-a) - u(t-(a+\epsilon))\}$$

$$= \frac{1}{\epsilon s}[e^{-as} - e^{-(a+\epsilon)s}]$$

$$= e^{-as}\,.\,\frac{(1-e^{-\epsilon s})}{\epsilon s}$$

Dirac delta function (or unit impulse function) denoted by $\delta(t-a)$ is defined as the limit of $f_\epsilon(t-a)$ as $\epsilon\to 0$ i.e.,

$$\delta(t-a) = \lim_{\epsilon\to 0} f_\epsilon(t-a).$$

* Paul Dirac (1902–1984), English physicist, Nobel prize winner.

Then the Laplace transform of Dirac delta function is obtained as

$$L\{\delta(t-a)\} = \lim_{\epsilon\to 0} L\{f_\epsilon(t-a)\}$$

$$= \lim_{\epsilon\to 0} e^{-as}\cdot\frac{(1-e^{-\epsilon s})}{\epsilon s}$$

$$L\{\delta(t-a)\} = e^{-as}.$$

Thus the Dirac delta function is a "generalized function" defined as

$$\delta(t-a) = \begin{cases}\infty & \text{when } t=a\\ 0 & \text{otherwise}\end{cases}$$

subject to $\int_0^\infty \delta(t-a)\,dt = 1$.

WORKED OUT EXAMPLES

Example: A beam has its ends clamped at $x=0$ and $x=L$. A concentrated load W acts vertically downwards at the point $x=\frac{L}{3}$. Find the resulting deflection.

Solution: The differential equation for deflection is

$$\frac{d^4y}{dx^4} = \frac{W}{EI}\delta\left(x-\frac{L}{3}\right) \tag{1}$$

Taking the Laplace transform on both sides of (1)

$$s^4Y - s^3y(0) - s^2y'(0) - sy''(0) - y'''(0) = \frac{W}{EI}e^{-\frac{Ls}{3}} \tag{2}$$

where $Y = L\{y(x)\}$.

Since the end $x=0$ is clamped, we have $y(0)=0$, $y'(0)=0$. Put $y''(0)=c_1$ and $y'''(0)=c_2$ where c_1 and c_2 are constants. Then (2) reduces to

$$s^4Y - sc_1 - c_2 = \frac{W}{EI}e^{-\frac{Ls}{3}}$$

Solving,

$$Y = \frac{c_1}{s^3} + \frac{c_2}{s^4} + \frac{W}{EI}\frac{e^{-\frac{Ls}{3}}}{s^4} \tag{3}$$

Taking the inverse Laplace transform of (3),

$$y(x) = c_1\frac{x^2}{2!} + c_2\frac{x^3}{3!} + \frac{W}{EI}\frac{\left(x-\frac{L}{3}\right)^3}{3!}\left[u\left(x-\frac{L}{3}\right)\right] \tag{4}$$

or

$$y(x) = c_1\frac{x^2}{2} + c_2\frac{x^3}{6} \quad \text{when } 0 < x < \frac{L}{3} \tag{5}$$

$$y(x) = c_1\frac{x^2}{2} + c_2\frac{x^3}{6} + \frac{W}{6EI}\left(x-\frac{L}{3}\right)^3 \quad \text{when } \frac{L}{3} < x < L \tag{6}$$

To determine c_1 and c_2 use the condition that the other end $x=L$ is clamped i.e., $y(L)=y'(L)=0$. From (6)

$$0 = y(L) = c_1\frac{L^2}{2} + c_2\frac{L^3}{6} + \frac{W}{EI}\frac{4}{81}L^3 \tag{7}$$

Differentiating (6) and putting $x=L$,

$$0 = y'(L) = c_1L + c_2\frac{L^2}{2} + \frac{W}{EI}\frac{2}{9}L^2 \tag{8}$$

Solving (7) and (8), we get

$$c_1 = \frac{12}{81}L\frac{W}{EI} \tag{9}$$

and

$$c_2 = -\frac{60}{81}\frac{W}{EI} \tag{10}$$

Thus the deflection of the beam is given by (5) and (6) where c_1 and c_2 are given by (9) and (10).

EXERCISE

1. An impulsive voltage $E\delta(t)$ is applied to a circuit consisting of L, R, C in series with zero initial conditions. Find the limit of I as $t\to 0$ where I is the current at any subsequent time t.

   Hint: Equation of circuit is

   $$L\frac{dI}{dt} + RI + \frac{1}{c}\int_0^t I\,dt = E\cdot\delta(t)$$

   where $I(0)=0$.

   Ans. $I = E/L$

2. A beam is simply supported at its end $x=0$ and is clamped at the other end $x=L$. It carries a load W at $x=\frac{L}{4}$. Find the resulting deflection at any point.

Hint: D.E. for deflection is $\frac{d^4y}{dx^4} = \frac{W}{EI}\delta\left(x - \frac{L}{4}\right)$ with boundary conditions $y(0) = y''(0) = 0$ and

$$y(L) = y'(L) = 0.$$

Ans. $y = c_1x + \frac{1}{6}c_2x^3, \quad 0 < x < \frac{L}{4}$

$$y = c_1x + \frac{1}{6}c_2x^3 + \frac{W}{6EI}\left(x - \frac{L}{4}\right)^3, \quad \frac{L}{4} < x < L$$

where $c_1 = \frac{9WL^2}{256EI}, \quad c_2 = -\frac{81W}{128EI}$

3. Obtain the deflection of a weighless beam of length L and freely supported at ends, when a concentrated load W acts at $x = a$.

Ans. $y(x) = \frac{W}{6EI}\left[\frac{ab(L+b)}{L}x - \frac{b}{L}x^3\right], \quad 0 < x < a$

$$y(x) = \frac{W}{6EI}\left[\frac{ab(L+b)}{L}x - \frac{b}{L}x^3 + (x-a)^3\right],$$

$$a < x < L$$

4. Determine the response of the damped mass–spring system governed by $y'' + 3y' + 2y = r(t)$, $y(0) = 0$, $y'(0) = 0$ where $r(t)$ is (a) square wave

$$r(t) = u(t-1) - u(t-2)$$

and (b) the unit impulse at time $t = 1$

$$r(t) = \delta(t-1)$$

Hint: Subsidiary equation is

$$s^2Y + 3sY + 2Y = e^{-s}$$

Ans. **a.**

$$y(t) = \begin{cases} 0, & 0 < t < 1 \\ \frac{1}{2} - e^{-(t-1)} + \frac{1}{2}e^{-2(t-1)}, & \ < t < 2 \\ -e^{-(t-1)} + e^{-(t-2)} + \frac{1}{2}e^{-2(t-1)} - \frac{1}{2}e^{-2(t-2)}, & \text{for } t < 2 \end{cases}$$

b. $y(t) = \begin{cases} 0, & 0 \le t < 1 \\ e^{-(t-1)} - e^{-2(t-1)}, & t > 1 \end{cases}$

12.4 LAPLACE TRANSFORM OF PERIODIC FUNCTION

A function $f(t)$ is said to be a periodic function of period $T > 0$ if

$$f(t) = f(T+t) = f(2T+t) = \cdots = f(nT+t).$$

Example: $\sin t$, $\cos t$ are periodic functions of period 2π.

Theorem: *The Laplace transform of a piecewise periodic function $f(t)$ with period p is*

$$L\{f(t)\} = \frac{1}{1 - e^{-ps}}\int_0^p e^{-st}\cdot f(t)dt; s > 0$$

Proof:

$$L\{f(t)\} = \int_0^\infty e^{-st} f(t)dt \quad \text{by definition}$$

$$= \int_0^p e^{-st} f(t)dt + \int_p^{2p} e^{-st} f(t)dt$$

$$+ \int_{2p}^{3p} e^{-st} f(t)dt + \cdots$$

Put $t = u + p$ in the 2nd integral,

$t = u + 2p$ in the 3rd integral

$\vdots$

$t = u + (n-1)p$ in the nth integral

Then the new limits for each integral are 0 to p and by periodicity

$$f(t+p) = f(t), \ f(t+2p) = f(t)$$

and so on.

Therefore

$$L\{f(t)\} = \int_0^p e^{-su} f(u)du + \int_0^p e^{-s(u+p)} f(u)du$$

$$+ \int_0^p e^{-s(u+2p)}\cdot f(u)du + \cdots$$

$$= [1 + e^{-sp} + e^{-2sp} + \cdots]\int_0^p e^{-su} f(u)du$$

$$L\{f\} = \frac{1}{1 - e^{-ps}}\int_0^p e^{-st} f(t)dt, \ s > 0$$

since the bracketed quantity in R.H.S. is geometric series

i.e., $\frac{1}{1-r} = 1 + r + r^2 + r^3 + \cdots, \ |r| < 1$ with $r = e^{-ps}$

Result: Thus the Laplace transform of a periodic function $f(t)$ of period p is obtained by integrating $e^{-st}f(t)$ in the interval $(0, p)$ with respect to t and multiplying the resultant by the factor $(1 - e^{-ps})^{-1}$.

Worked Out Examples

Find the Laplace transform of the following periodic functions:

Example 1: Half wave rectifier (Fig. 12.7)

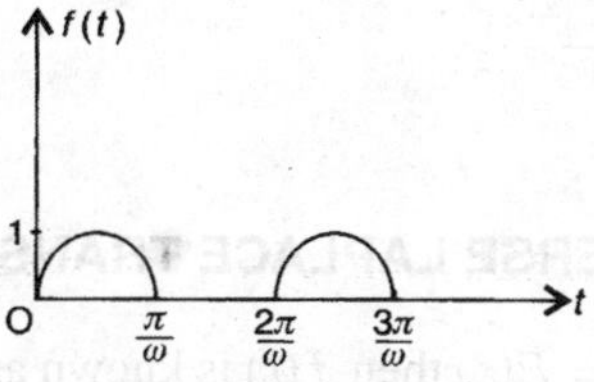

Fig. 12.7

$$f(t) = \sin wt, \quad 0 < t < \frac{\pi}{w}$$
$$= 0, \quad \frac{\pi}{w} < t < \frac{2\pi}{w}$$

and periodic of period $\frac{2\pi}{w}$.

Solution:

$$L(f(t)) = \frac{1}{1-e^{-ps}} \int_0^{p} e^{-st} f(t) dt$$
$$= \frac{1}{1-e^{-2\pi s/w}} \int_0^{\frac{2\pi}{w}} e^{-st} \cdot f(t) dt$$
$$= \frac{1}{1-e^{-2\pi s/w}} \left[\int_0^{\frac{\pi}{w}} e^{-st} \cdot \sin wt\, dt + \int_{\frac{\pi}{w}}^{\frac{2\pi}{w}} 0 \right]$$

Consider $\int_0^{\frac{\pi}{w}} e^{-st} \cdot \sin wt\, dt$

$$= \text{Im. p. of} \left\{ \int_0^{\frac{\pi}{w}} e^{(-s+iw)t} dt \right\}$$
$$= \text{Im. p. of} \left\{ \frac{1}{-s+iw} \right\} e^{(-s+iw)t} \Big|_0^{\frac{\pi}{w}}$$
$$= \text{Im. p. of} \left\{ \frac{(s+iw)}{(s^2+w^2)} \left(1 + e^{-s\pi/w}\right) \right\}$$
$$= \frac{w}{s^2+w^2} \left(e^{-s\pi/w} + 1\right)$$

$$\therefore \quad L\{f(t)\} = \frac{1}{1-e^{-2\pi s/w}} \cdot \frac{w}{s^2+w^2} \left(1 + e^{-s\pi/w}\right)$$
$$= \frac{w}{(s^2+w^2)} \cdot \frac{1}{(1-e^{-\pi s/w})}.$$

Example 2: Saw-tooth wave (Fig. 12.8)

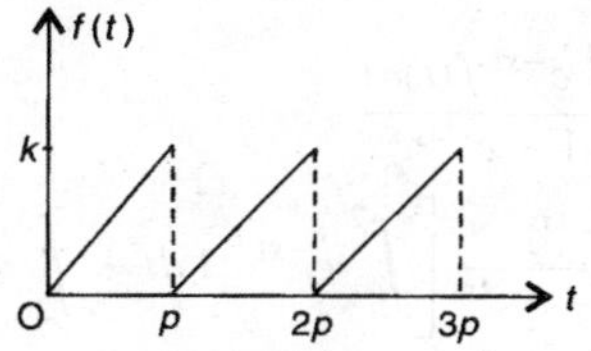

Fig. 12.8

$$f(t) = \frac{k}{p} t, \quad 0 < t < p$$

and

$$f(t+p) = f(t)$$

Solution:

$$L(f(t)) = \frac{1}{1-e^{-sp}} \int_0^p e^{-st} \cdot \left(\frac{k}{p} t\right) dt,$$

Integrating by parts

$$\int_0^p t \cdot e^{-st} dt = -\frac{t}{s} e^{-st} \Big|_0^p + \frac{1}{s} \int_0^p e^{-st} dt$$
$$= -\frac{t}{s} e^{-st} \Big|_0^p - \frac{1}{s^2} e^{-st} \Big|_0^p$$
$$= -\frac{p}{s} e^{-sp} - \frac{1}{s^2}(e^{-sp} - 1)$$

$$\therefore \quad L(f(t)) = \frac{1}{(1-e^{-sp})} \frac{k}{p} \times \left[-\frac{p}{s} e^{-sp} - \frac{1}{s^2}(e^{-sp} - 1) \right]$$
$$= \frac{k}{ps^2} - \frac{ke^{-sp}}{s(1-e^{-ps})}, \; s > 0$$

Example 3: (refer Fig. 12.9)

$$f(t) = 1, \quad 0 \le t < 2$$
$$= -1, \quad 2 \le t < 4$$
$$f(t+4) = f(t)$$

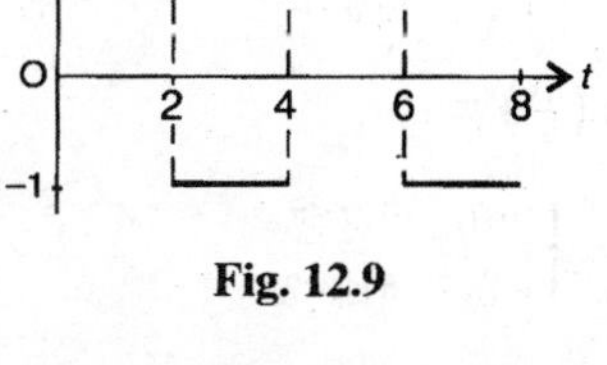

Fig. 12.9

Solution: Here $p = 4$. Applying theorem

$$L\{f(t)\} = \frac{\int_0^4 e^{-st} f(t)dt}{1-e^{-4s}}$$

$$= \frac{1}{1-e^{-4s}}\left[\int_0^2 e^{-st}\cdot 1\,dt + \int_2^4 e^{-st}(-1)dt\right]$$

$$= \frac{1}{1-e^{-4s}}\cdot\left[\frac{e^{-st}}{-s}\Big|_0^2 + \frac{e^{-st}}{s}\Big|_2^4\right]$$

$$= \frac{1}{1-e^{-4s}}\left(\frac{1}{s}\right)\left[-e^{-2s}+1+e^{-4s}-e^{-2s}\right]$$

$$= \frac{1-e^{-2s}}{s(1+e^{-2s})}.$$

EXERCISE

Find the Laplace transform of the following periodic functions:

1. $f(t) = t, \quad 0 < t < a$
   $= -t + 2a, \quad a < t < 2a$

Ans. $\frac{1}{s^2}\tanh\frac{as}{2}$

2. $f(t) = t^2, 0 < t < \alpha$ and $f(t+\alpha) = f(t)$

Ans. $\frac{1}{(1-e^{-\alpha s})s^3}\left[2 - e^{-\alpha s}(\alpha^2 s^2 + 2\alpha s + 2)\right]$

3. $f(t) = 1, \quad 0 < t < 1$
   $= 0, \quad 1 < t < 2 \quad$ Period 3
   $= -1, \quad 2 < t < 3$

Ans. $\frac{1}{s}\left[\frac{3}{(1-e^{-3s})} - \frac{1}{(1-e^{-s})} - 1\right]$

4. $f(t) = t, \quad 0 < t < \pi$
   $= 0, \quad \pi < t < 2\pi \quad$ Period 2π

Ans. $\frac{1}{s^2}\left[(1-e^{-\pi s}) - \frac{\pi}{s}e^{-\pi s}\right]/(1-e^{-2\pi s})$

5. $f(t) = \begin{cases} a, & \text{for } 0 \le t \le a \\ -a, & \text{for } a < t \le 2a \end{cases}$

Ans. $\frac{a}{s}\tanh\left(\frac{as}{2}\right)$

6. $f(t) = \begin{cases} \cos t, & 0 < t \le \pi \\ -1, & \pi \le t < 2\pi \end{cases}$

Ans. $\frac{s}{(1+s^2)(1-e^{-\pi s})} - \frac{e^{-s\pi}}{s(1+e^{-\pi s})}$

7. $f(t) = \sin t, 0 \le t \le \pi$

Ans. $\frac{\coth(\pi s/3)}{1+s^2}$

8. $f(t) = \begin{cases} 1+t, & 0 \le t < 1 \\ 3-t, & 1 \le t < 2, \quad f(t+2) = f(t) \end{cases}$

Ans. $\frac{1}{s} + \frac{1-e^{-s}}{1+e^{-s}}\frac{1}{s^2}.$

12.5 INVERSE LAPLACE TRANSFORM

If $L\{f(t)\} = F(s)$ then $f(t)$ is known as the inverse Laplace transform (I.L.T.) or inverse transform or simply inverse of $F(s)$ and is denoted by $L^{-1}\{F(s)\}$. Thus

$$f(t) = L^{-1}\{F(s)\} \tag{1}$$

L^{-1} is known as the inverse Laplace transform operator and is such that

$$LL^{-1} = L^{-1}L = 1$$

In the inverse problem (1), $F(s)$ is given (known) and $f(t)$ is to be determined.

Note: I.L.T. of $F(s)$ need not exist for all $F(s)$.

Inverse Laplace Transform of Some Elementary Functions

Example: Find $f(t)$ the inverse Laplace transform of $F(s) = \frac{k}{s}$ where k is a constant.

Solution: We know that $L\{k\} = \frac{k}{s}$. Taking inverse Laplace transform on both sides, we have

$$k = L^{-1}\left\{\frac{k}{s}\right\}$$

That $\quad f(t) = k = L^{-1}\left\{\frac{k}{s}\right\} = L^{-1}\{F(s)\}$

Similarly, since $L\{e^{at}\} = \frac{1}{s-a}$ it follows that

$$L^{-1}\left\{\frac{1}{s-a}\right\} = e^{at}$$

In a similar way, since $L\{\cos at\} = \frac{s}{s^2+a^2}$, we readily get

$$L^{-1}\left\{\frac{s}{s^2+a^2}\right\} = \cos at$$

and so on. Thus for each direct Laplace transform result (listed in table of L.T. in Section 12.12) a corresponding inverse Laplace transform result can be read accordingly.

The proficiency in solving O.D.E./P.D.E. using L.T. is practically synonymous with the proficiency in determining inverse Laplace transform (I.L.T.).

Evaluation of I.L.T, $L^{-1}\{F(s)\}$ essentially reduces to expressing $F(s)$ as some combination of functions, each of whose I.L.T. can be read from the L.T. tables or otherwise. Some methods for finding I.L.T. are:

a. Use of L.T. tables(12.12 on page 12.41)

b. Use of some important properties of I.L.T.(12.11 on page 12.41)

c. Convolution(12.8 on page 12.33)

d. Use of partial fractions.(12.7 on page 12.30)

12.6 GENERAL PROPERTIES OF INVERSE LAPLACE TRANSFORM

For each property (Theorem) on Laplace transform, there is a corresponding property (Theorem) on inverse Laplace transform which readily follow from the definitions.

Linearity Property

Book Work: Let $L\{f(t)\} = F(s)$ and $L\{g(t)\} = G(s)$ then

$$L^{-1}\{c_1F(s)+c_2G(s)\} = c_1L^{-1}\{F(s)\}+c_2L^{-1}\{G(s)\}$$

$$= c_1f(t)+c_2g(t)$$

where c_1 and c_2 are any two constants.

Proof: From the Linearity property for L.T., we know that

$$L\{c_1f(t)+c_2g(t)\} = c_1L\{f(t)\}+c_2L\{g(t)\}$$

$$= c_1F(s)+c_2G(s).$$

Taking inverse L.T. on either side, we get

$$c_1f(t)+c_2g(t) = L^{-1}\{c_1F(s)+c_2G(s)\}$$

Since $f(t) = L^{-1}\{F(s)\}$ and $g(t) = L^{-1}\{G(s)\}$, we have

$$c_1L^{-1}\{F(s)\}+c_2L^{-1}\{G(s)\} = L^{-1}\{c_1F(s)+c_2G(s)\}.$$

Result: Thus L^{-1} the Inverse Laplace Transform operator is a linear operator.

Note: This result can readily be extended to more than two functions.

First Shift or Translation Theorem

Book Work: If $L^{-1}\{F(s)\} = f(t)$ then

$$L^{-1}\{F(s-a)\} = e^{at}f(t) = e^{at}L^{-1}\{F(s)\}$$

Proof: From the first translation property on L.T., we have

$$L\{e^{at}f(t)\} = F(s-a)$$

then $$e^{at}f(t) = L^{-1}\{F(s-a)\}.$$

Result: Thus if s is replaced by $s-a$ in $F(s)$ then $f(t)$ is multiplied by e^{at}.

Change of Scale Property

Book Work: $L^{-1}\{F(ks)\} = \frac{1}{k}f\left(\frac{t}{k}\right)$

Proof: From change of scale property for L.T.

$$L\{f(at)\} = \frac{1}{a}F\left(\frac{s}{a}\right)$$

Take $a = \frac{1}{k}$ then

$$L\left\{f\left(\frac{t}{k}\right)\right\} = kF(ks)$$

or $$f\left(\frac{t}{k}\right) = L^{-1}\{kF(ks)\}$$

Thus $$L^{-1}\{F(ks)\} = \frac{1}{k}f\left(\frac{t}{k}\right).$$

WORKED OUT EXAMPLES

Linearity property

Find the inverse Laplace transform of the following:

Example 1: $\frac{2s+1}{s^2-4}$

Solution:

$$L^{-1}\left\{\frac{2s+1}{s^2-4}\right\}=L^{-1}\left\{\frac{2s}{s^2-4}\right\}+L^{-1}\left\{\frac{1}{s^2-4}\right\}$$

$$=2\cdot\cosh 2t+\frac{1}{2}\sinh 2t.$$

Example 2: $\frac{3(s^2-2)^2}{2s^5}$

Solution:

$$L^{-1}\left\{\frac{3(s^4-4s^2+4)}{2s^5}\right\}$$

$$=L^{-1}\left\{\frac{3}{2}\frac{1}{s}-6\frac{1}{s^3}+6\frac{1}{s^5}\right\}$$

$$=\frac{3}{2}L^{-1}\left\{\frac{1}{s}\right\}-6L^{-1}\left\{\frac{1}{s^3}\right\}+6L^{-1}\left\{\frac{1}{s^5}\right\}$$

$$=\frac{3}{2}-6\cdot\frac{t^2}{2!}+6\cdot\frac{t^4}{4!}=\frac{3}{2}-3t^2+\frac{1}{4}t^4.$$

Example 3: $\frac{1}{s}e^{-\frac{1}{\sqrt{s}}}$

Solution: We know that

$$e^{-\frac{1}{\sqrt{s}}}=\sum_{n=0}^{\infty}\left(-\frac{1}{\sqrt{s}}\right)^n\frac{1}{n!}=\sum_{n=0}^{\infty}\frac{(-1)^n}{n!}\frac{1}{s^{\frac{n}{2}}}$$

$$\frac{1}{s}e^{-\frac{1}{\sqrt{s}}}=\sum_{n=0}^{\infty}\frac{(-1)^n}{n!}\frac{1}{s^{\frac{n}{2}+1}}$$

$$L^{-1}\left\{\frac{1}{s}e^{-\frac{1}{\sqrt{s}}}\right\}=L^{-1}\left\{\sum_{n=0}^{\infty}\frac{(-1)^n}{n!}\frac{1}{s^{\frac{n}{2}+1}}\right\}$$

$$=\sum_{n=0}^{\infty}\frac{(-1)^n}{n!}L^{-1}\left\{\frac{1}{s^{\frac{n}{2}+1}}\right\}$$

$$=\sum_{n=0}^{\infty}\frac{(-1)^n}{n!}\frac{t^{\frac{n}{2}}}{\Gamma\left(\frac{n}{2}+1\right)}.$$

Exercise

Find the inverse Laplace transform $f(t)$ of each of the following functions $F(s)$:

1. $\frac{3}{s+4}$ *Ans.* $3e^{-4t}$
2. $\frac{8s}{s^2+16}$ *Ans.* $8\cdot\cos 4t$
3. $\frac{1}{2s-5}$ *Ans.* $\frac{1}{2}e^{\frac{5t}{2}}$
4. $\frac{6}{s^2+4}$ *Ans.* $3\sin 2t$
5. $\frac{3s-12}{s^2+8}$

   Ans. $3\cos 2\sqrt{2}t-3\sqrt{2}\sin 2\sqrt{2}t$
6. $(2s-5)/(s^2-9)$ *Ans.* $2\cosh 3t-\frac{5}{3}\sinh 3t$
7. $s^{-\frac{7}{2}}$ *Ans.* $8t^{\frac{5}{2}}/(15\sqrt{\pi})$
8. $\frac{s+1}{s^{\frac{4}{3}}}$ *Ans.* $(t^{-\frac{2}{3}}+3t^{\frac{1}{3}})/\Gamma\left(\frac{1}{3}\right)$
9. $\left(\frac{\sqrt{s}-1}{s}\right)^2$ *Ans.* $1+t-\frac{4t^{\frac{1}{2}}}{\sqrt{\pi}}$
10. $\frac{3s-8}{4s^2+25}$ *Ans.* $\frac{3}{4}\cos\frac{5t}{2}-\frac{4}{5}\sin\frac{5t}{2}$
11. $\frac{5s+10}{9s^2-16}$ *Ans.* $\frac{5}{9}\cosh\frac{4t}{3}+\frac{5}{6}\sinh\frac{4t}{3}$
12. $\frac{3(s^2-1)^2}{2s^5}+\frac{4s-18}{9-s^2}+\frac{(s+1)(2-s^{\frac{1}{2}})}{s^{\frac{5}{2}}}$

   Ans. $\frac{1}{2}-t-\frac{3}{2}t^2+\frac{1}{16}t^4+4\frac{t^{\frac{1}{2}}}{\sqrt{\pi}}+\frac{8t^{\frac{3}{2}}}{3\sqrt{\pi}}$

   $-4\cosh 3t+6\sinh 3t$
13. $\frac{1}{s}\sin\left(\frac{1}{s}\right)$ *Ans.* $\sum_{n=0}^{\infty}\frac{(-1)^{n-1}}{[(2n-1)!]^2}t^{2n-1}$
14. $\frac{1}{s}e^{-\frac{1}{s}}$ *Ans.* $\sum_{n=0}^{\infty}\frac{(-1)^n t^n}{(n!)^2}$

Hint: For Examples 13, 14, expand sin and exponential in series of s.

Worked Out Examples

Change of scale property

Example: Find

$$L^{-1}\left\{\frac{64}{81s^4-256}\right\}$$

Solution: We know that

$$L^{-1}\left\{\frac{a^3}{s^4-a^4}\right\}=\frac{1}{2}L^{-1}\left\{\frac{a}{s^2-a^2}-\frac{a}{s^2+a^2}\right\}$$

$$=\frac{1}{2}(\sinh at-\sin at)$$

Rewriting

$$L^{-1}\left\{\frac{64}{81s^4-256}\right\}=L^{-1}\left\{\frac{4^3}{(3s)^4-4^4}\right\}$$

$$=L^{-1}\{F(3s)\} \text{ with } a=4$$

where $F(s)=\frac{a^3}{s^4-a^4}$ and
$f(t)=\frac{1}{2}(\sinh at-\sin at)$.
Thus applying change of scale property (with $a=3$)

$$L^{-1}\{F(3s)\}=\frac{1}{3}f\left(\frac{t}{3}\right)=\frac{1}{3}\left[\frac{1}{2}\sinh 4\frac{t}{3}-\sin\frac{4t}{3}\right].$$

EXERCISE

Change of scale property

1. If $L^{-1}\left\{e^{-\frac{1}{s}}/\sqrt{s}\right\}=\frac{\cos 2\sqrt{t}}{\sqrt{\pi t}}$ show that

$$L^{-1}\left\{\frac{e^{-\frac{a}{s}}}{\sqrt{s}}\right\}=\frac{\cos 2\sqrt{at}}{\sqrt{\pi t}}, \quad \text{when} \quad a>0.$$

2. If $L^{-1}\{F(s)\}=f(t)$ prove that

$$L^{-1}\{F(as+b)\}=\frac{1}{a}e^{-(b/a)t}f\left(\frac{t}{a}\right) \quad \text{where } a>0.$$

Hint: Take s as $as+b$ in the definition of L.T.

3. If $L^{-1}\left\{\frac{s}{(s^2+1)^2}\right\}=\frac{t\sin t}{2}$ show that

$$L^{-1}\left\{\frac{8s}{(4s^2+1)^2}\right\}=\frac{t}{2}\sin\frac{t}{2}.$$

4. If $L^{-1}\left\{\frac{s^2-1}{(s^2+1)^2}\right\}=t\cos t$ show that

$$L^{-1}\left\{\frac{9s^2-1}{(9s^2+1)^2}\right\}=\frac{t}{9}\cos\frac{t}{3}.$$

WORKED OUT EXAMPLES

First shift theorem

Find the inverse transform:

Example 1: $\frac{1}{(s+a)^{n+1}}$, n: non-negative integer

Solution:

$$L^{-1}\left\{\frac{1}{(s+a)^{n+1}}\right\}=e^{-at}L^{-1}\left\{\frac{1}{s^{n+1}}\right\}$$

by shift theorem.

$$=e^{-at}\frac{t^n}{n!}.$$

Example 2: $\frac{s}{(s+a)^2+b^2}$.

Solution:

$$L^{-1}\left[\frac{s}{(s+a)^2+b^2}\right]=e^{-at}L^{-1}\left[\frac{s-a}{s^2+b^2}\right]$$

by shift theorem

$$=e^{-at}\left\{L^{-1}\left[\frac{s}{s^2+b^2}\right]\frac{-a}{b}L^{-1}\left[\frac{b}{s^2+b^2}\right]\right\}$$

$$=e^{-at}\left\{\cos bt-\frac{a}{b}\cdot\sin bt\right\}.$$

Example 3: $\frac{3s}{s^2-25}$.

Solution: $L^{-1}\left\{\frac{3s}{s^2-25}\right\}=3L^{-1}\left\{\frac{s}{s^2-5^2}\right\}=3\cosh 5t$.

Example 4: $\frac{1}{s^{\frac{3}{2}}}$.

Solution:

$$L^{-1}\left\{\frac{1}{s^{\frac{3}{2}}}\right\}=\frac{t^{\frac{1}{2}}}{\Gamma\left(\frac{3}{2}\right)}=\frac{t^{\frac{1}{2}}}{\frac{1}{2}\Gamma\left(\frac{1}{2}\right)}=2\sqrt{\frac{t}{\pi}}$$

Example 5: $\frac{3s+1}{(s+1)^4}$.

Solution:

$$L^{-1}\left(\frac{3s+1}{(s+1)^4}\right)=e^{-t}L^{-1}\left(\frac{3(s-1)+1}{s^4}\right)$$

$$=e^{-t}\left\{3L^{-1}\left(\frac{1}{s^3}\right)-2L^{-1}\left(\frac{1}{s^4}\right)\right\}$$

$$=e^{-t}\left\{3\cdot\frac{t^2}{2!}\cdot-2\cdot\frac{t^3}{3!}\right\}$$

$$=e^{-t}\left\{\frac{3}{2}t^2-\frac{1}{3}t^3\right\}.$$

Example 6: $\frac{s+1}{s^2-6s+25}$.

Solution: $L^{-1}\left(\frac{s+1}{s^2+6s+25}\right)=L^{-1}\left(\frac{s+1}{(s+3)^2+16}\right)$.

Applying shift theorem.

$$=e^{-3t}L^{-1}\left(\frac{(s-3)+1}{s^2+16}\right)=e^{-3t}\left\{L^{-1}\left\{\frac{s-2}{s^2+16}\right\}\right\}$$

$$=e^{-3t}\cdot\left\{\cos 4t-\frac{1}{2}\sin 4t\right\}.$$

Example 7: $F(as+b)$.

Solution: $F(as+b)=\int_0^\infty e^{-(as+b)t}f(t)dt$

$$=\int_0^\infty e^{-ast}\cdot e^{-bt}\cdot f(t)dt$$

Put $at=u,\ dt=\frac{du}{a}$

$$F(as+b)=\int_0^\infty e^{-su}\cdot e^{-\frac{b}{a}u}\cdot f\left(\frac{u}{a}\right)\cdot\frac{1}{a}\cdot du$$

$$=\frac{1}{a}\int_0^\infty e^{-su}\left\{e^{-\frac{b}{a}u}\cdot f\left(\frac{u}{a}\right)\right\}du$$

$$=\frac{1}{a}L\left\{e^{-\frac{bu}{a}}\cdot f\left(\frac{u}{a}\right)\right\}.$$

Result: $L^{-1}\{F(as+b)\}=\frac{1}{a}e^{-\frac{bt}{a}}\cdot f\left(\frac{t}{a}\right)$.

Example 8: $\frac{1}{\sqrt{2s+3}}$

Solution:

$$L^{-1}\left\{\frac{1}{\sqrt{2s+3}}\right\}=\frac{1}{\sqrt{2}}L^{-1}\left\{\frac{1}{\left(s+\frac{3}{2}\right)^{\frac{1}{2}}}\right\}.$$

By shift theorem

$$=\frac{1}{\sqrt{2}}e^{\frac{-3t}{2}}\cdot L^{-1}\left\{\frac{1}{s^{\frac{1}{2}}}\right\}$$

$$=\frac{1}{\sqrt{2}}e^{\frac{-3t}{2}}\cdot\frac{t^{\frac{-1}{2}}}{\Gamma\left(\frac{1}{2}\right)}=\frac{1}{\sqrt{2\pi}}t^{-1/2}\cdot e^{\frac{-3t}{2}}.$$

EXERCISE

First shift theorem

Find the inverse Laplace transform of the following:

1. $\frac{5}{(s+2)^5}$ *Ans.* $\frac{5}{24}t^4e^{-2t}$
2. $\frac{4s+12}{s^2+8s+16}$ *Ans.* $4e^{-4t}(1-t)$
3. $\frac{s}{(s+1)^5}$ *Ans.* $\frac{e^{-t}}{24}(4t^3-t^4)$
4. $\frac{s}{(s+1)^{\frac{5}{2}}}$ *Ans.* $\frac{2t^{\frac{1}{2}}(3-2t)}{3\sqrt{\pi}}$
5. $\frac{1}{\sqrt[3]{8s-27}}$ *Ans.* $\frac{t^{\frac{-2}{3}}e^{\frac{27t}{8}}}{2\Gamma\left(\frac{1}{3}\right)}$
6. $\frac{3s-14}{s^2-4s+8}$ *Ans.* $e^{2t}(3\cos 2t-4\sin 2t)$
7. $\frac{5s-2}{s^2+4s+8}$

   Ans. $\frac{e^{\frac{-2t}{3}}}{15}\{25\cos 2\sqrt{5}t/3-24\sqrt{5}\sin 2\sqrt{5}t/3\}$
8. $\frac{3s+2}{4s^2+12s+9}$ *Ans.* $\frac{3}{4}e^{\frac{-3t}{2}}-\frac{5}{8}te^{\frac{-3t}{2}}$
9. $\frac{8s+20}{s^2-12s+32}$

   Ans. $2e^{6t}(4\cosh 2t+17\sinh 2t)$
10. $\frac{1}{(s^2+2s+5)^2}$ *Ans.* $\frac{e^{-t}}{16}(\sin 2t-2t\cos 2t)$

Inverse L.T. of Derivatives

$L^{-1}\{F^{(n)}(s)\}=(-1)^n\cdot t^nf(t),\ n=1,2,3,\ldots$.

WORKED OUT EXAMPLES

Example 1: $\frac{s+1}{(s^2+2s+2)^2}$.

Solution:

$$L^{-1}\left\{\frac{s+1}{(s^2+2s+2)^2}\right\}=L^{-1}\left\{\frac{s+1}{((s+1)^2+1)^2}\right\}$$

$$=e^{-t}\cdot L^{-1}\left\{\frac{s}{(s^2+1)^2}\right\}$$

We know that $\frac{d}{ds}\left\{\frac{1}{s^2+1}\right\}=\frac{-2s}{(s^2+1)^2}$

so that $\frac{s}{(s^2+1)^2}=-\frac{1}{2}\frac{d}{ds}\left\{\frac{1}{s^2+1}\right\}$

But $L^{-1}\left\{\frac{1}{s^2+1}\right\}=\sin t$

Using I.L.T. of derivatives [(with $n = 1$ (one differentiation)]

$$L^{-1}\left\{\frac{s+1}{(s^2+2s+2)^2}\right\} = e^{-t}L^{-1}\left\{\frac{s}{(s^2+1)^2}\right\}$$

$$= e^{-t}\cdot\frac{-1}{2}L^{-1}\left\{\frac{d}{ds}\left(\frac{1}{(s^2+1)}\right)\right\}$$

$$= -\frac{1}{2}e^{-t}\cdot(-1)^1\cdot t^1\cdot L^{-1}\left\{\frac{1}{s^2+1}\right\}$$

$$= \frac{1}{2}e^{-t}\cdot t\cdot \sin t.$$

Example 2: $\frac{1}{(s-a)^3}$.

Solution: We know that $L^{-1}\left(\frac{1}{s-a}\right) = e^{at}$

Let $F(s) = \frac{1}{(s-a)}$ so that

$$f(t) = L^{-1}\{F(s)\} = e^{at}$$

Then

$$F'(s) = \frac{-1}{(s-a)^2},\ F''(s) = \frac{(-1)(-2)}{(s-a)^3} = \frac{2}{(s-a)^3}$$

By I.L.T. of derivatives with $n = 2$

$$L^{-1}\left\{\frac{1}{(s-a)^3}\right\} = \frac{1}{2}L^{-1}\{F''(s)\} = \frac{(-1)^2}{2}t^2 f(t)$$

$$= 1.\frac{t^2\cdot e^{at}}{2} = \frac{t^2e^{at}}{2}.$$

Example 3: $\frac{1}{2}\ln\frac{s^2+b^2}{s^2+a^2}$.

Solution: Here $F(s) = \frac{1}{2}\frac{\ln(s^2+b^2)}{(s^2+a^2)}$

$$F(s) = \frac{1}{2}\left[\ln(s^2+b^2) - \ln(s^2+a^2)\right]$$

$$F'(s) = \frac{1}{2}\frac{2s}{s^2+b^2} - \frac{1}{2}\frac{2s}{s^2+a^2}$$

$$= \frac{s}{s^2+b^2} - \frac{s}{s^2+a^2}$$

$$L^{-1}\{F'(s)\} = L^{-1}\left\{\frac{s}{s^2+b^2} - \frac{s}{s^2+a^2}\right\}$$

$$= L^{-1}\left\{\frac{s}{s^2+b^2}\right\} - L^{-1}\left\{\frac{s}{s^2+a^2}\right\}$$

$$= \cos bt - \cos at$$

Using I.L.T. of derivatives

$$-tf(t) = L^{-1}\{F'(s)\} = \cos bt - \cos at$$

Thus

$$f(t) = -\frac{(\cos bt - \cos at)}{t}.$$

Example 4: $\cot^{-1}\left(\frac{s+a}{b}\right)$.

Solution: $F(s) = \cot^{-1}\left(\frac{s+a}{b}\right)$, $F'(s) = -\frac{1}{1+\left(\frac{s+a}{b}\right)^2}\cdot\frac{1}{b}$

$F'(s) = \frac{1}{b}\frac{-b^2}{(s+a)^2+b^2}$, so

$$L^{-1}\{F'(s)\} = -L^{-1}\left\{\frac{b}{(s+a^2)+b^2}\right\} = -e^{-at}\cdot\sin bt$$

Since $\quad -tf(t) = L^{-1}\{F'(s)\} = -e^{-at}\sin bt$

Thus $\quad f(t) = \frac{e^{-at}}{t}\sin bt.$

EXERCISE

Find the inverse Laplace transform of the following:

1. $\frac{1}{(s-a)^n}$, $n = 1, 2, 3$ *Ans.* $\frac{t^{n-1}\cdot e^{at}}{(n-1)!}$
2. $\frac{s}{(s^2+a^2)^2}$ *Ans.* $\frac{t\sin at}{2a}$
3. $\frac{1}{s^2+4s+5}$ *Ans.* $\frac{te^{-2t}\sin t}{2}$
4. $\log\left(\frac{s+a}{s+b}\right)$ *Ans.* $\frac{e^{-at}-e^{-bt}}{-t}$
5. $\tan^{-1}\frac{2}{s^2}$ *Ans.* $\frac{2}{t}\cdot\sinh t\cdot\sin t$
6. $\log\frac{s(s+1)}{(s^2+4)}$ *Ans.* $\frac{2\cos 2t - e^{-t} - 1}{t}$
7. $\log\left(1-\frac{a^2}{s^2}\right)$ *Ans.* $\frac{2}{t}(1-\cosh at)$
8. $\log\frac{s+1}{s-1}$ *Ans.* $\frac{2\sinh t}{t}$
9. $\cot^{-1}\left(\frac{s+3}{2}\right)$ *Ans.* $\frac{2}{t}e^{-3t}\sin 2t$.

Inverse Laplace Transform of Integrals

If $L^{-1}\{F(s)\} = f(t)$ then $L^{-1}\{\int_s^\infty F(u)du\} = \frac{f(t)}{t}$

WORKED OUT EXAMPLES

Evaluate the following:

Example 1: $L^{-1}\left\{\frac{1}{2}\ln\left(\frac{s+1}{s-1}\right)\right\}$

Solution:

$$\frac{1}{2}\ln\left(\frac{s+1}{s-1}\right)=\frac{1}{2}[\ln(s+1)-\ln(s-1)]$$

$$=\frac{1}{2}\int_s^\infty\left(\frac{-du}{u+1}+\frac{du}{u-1}\right)$$

But we know that

$$\frac{1}{2}L^{-1}\left\{\frac{1}{s-1}-\frac{1}{s+1}\right\}=\frac{e^t-e^{-t}}{2}=\sinh t$$

Thus applying I.L.T. of integrals

$$L^{-1}\left\{\frac{1}{2}\ln\left(\frac{s+1}{s-1}\right)\right\}$$

$$=L^{-1}\left\{\int_s^\infty\frac{1}{2}\left(\frac{du}{u-1}-\frac{du}{u+1}\right)\right\}$$

$$=\frac{f(t)}{t}=\frac{\sinh t}{t}$$

Example 2: Evaluate

$$L^{-1}\left\{\int_s^\infty\left(\frac{u}{u^2+a^2}-\frac{u}{u^2+b^2}\right)du\right\}$$

Solution: Consider $F(s)=\frac{s}{s^2+a^2}-\frac{s}{s^2+b^2}$

$$f(t)=L^{-1}\{F(s)\}=L^{-1}\left\{\frac{s}{s^2+a^2}-\frac{s}{s^2+b^2}\right\}$$

$$=\cos at-\cos bt$$

By I.L.T. of Integral

$$L^{-1}\{\int_s^\infty F(u)du\}$$

$$=L^{-1}\left\{\int_s^\infty\left(\frac{u}{u^2+a^2}-\frac{u}{u^2+b^2}\right)du\right\}=\frac{f(t)}{t}$$

$$=\frac{\cos at-\cos bt}{t}.$$

Exercise

Evaluate the following:

1. $L^{-1}\left\{\int_s^\infty\left(\frac{1}{u}-\frac{1}{u+1}\right)du\right\}$

Hint: $L^{-1}\left\{\frac{1}{s}-\frac{1}{s+1}\right\}=1-e^{-t}$.

Ans. $\frac{1-e^{-t}}{t}$

2. $L^{-1}\left\{\int_s^\infty\ln\left(\frac{u+2}{u+1}\right)du\right\}$

Hint: $L^{-1}\left\{\ln\left(\frac{s+2}{s+1}\right)\right\}=\frac{e^{-t}-e^{-2t}}{t}$.

Ans. $\frac{e^{-t}-e^{-2t}}{t^2}$

3. $L^{-1}\left\{\int_s^\infty\tan^{-1}\left(\frac{2}{u^2}\right)du\right\}$

Hint: $L^{-1}\left\{\tan^{-1}\frac{2}{s^2}\right\}=\frac{2\sin t\sinh t}{t^2}$.

Ans. $\frac{2\sin t\sinh t}{t^3}$

Multiplication by *s*

Book Work: If $L^{-1}\{F(s)\}=f(t)$ and $f(0)=0$ then

$$L^{-1}\{sF(s)\}=\frac{df}{dt}.$$

Proof: We know that

$$L\left\{\frac{df}{dt}\right\}=sF(s)-f(0)=sF(s)$$

Then $$\frac{df}{dt}=L^{-1}\{sF(s)\}.$$

Thus Multiplication by s amounts to differentiating $f(t)$ w.r.t. t.

Worked Out Examples

Example: $L^{-1}\left\{\frac{s}{s^2-a^2}\right\}$.

Solution: We know that

$$\frac{1}{a}L^{-1}\left\{\frac{a}{s^2-a^2}\right\}=\frac{\sinh at}{a},\text{ and }\sinh 0=0$$

$$L^{-1}\left\{s\cdot\frac{1}{s^2-a^2}\right\}=\frac{d}{dt}\frac{\sinh at}{a}=\frac{a\cosh at}{a}$$

$$L^{-1}\left\{\frac{s}{s^2-a^2}\right\}=\cosh at.$$

Exercise

Find the inverse Laplace transform:

1. $\frac{s^2}{s^2+1}$ *Ans.* $\cos t$

2. $\frac{s}{(s^2+1)^2}$ **Hint:** $L^{-1}\left\{\frac{1}{(s^2+1)^2}\right\}=\frac{\sin t-t\cos t}{2}$.

Ans. $\frac{1}{2}t\sin t$

Division by Powers of s

Integration of a function $f(t)$ amounts to division of transform $F(s)$ by s

Theorem: $L^{-1}\left\{\frac{F(s)}{s}\right\} = \int_0^t f(u)du.$

Proof: Let $g(t) = \int_0^t f(u)du$. Then $g'(t) = f(t)$, $g(0) = 0$

Thus $\quad L\{g'(t)\} = sL\{g(t)\} - g(0) = sL\{g(t)\}$

But since $\quad g'(t) = f(t)$

$$F(s) = L\{f(t)\} = L\{g'(t)\} = sL\{g(t)\}$$

$$\therefore \quad L\{g(t)\} = \frac{F(s)}{s}$$

$$\therefore \quad L^{-1}\left\{\frac{F(s)}{s}\right\} = g(t) = \int_0^t f(u)du$$

Similarly, $L^{-1}\left\{\frac{F(s)}{s^2}\right\} = \int_0^t \int_0^v f(u)du\,dv.$

In general,

$$L^{-1}\left\{\frac{F(s)}{s^n}\right\} = \int_0^t \int_0^t \cdots \int_0^t f(t)dt^n.$$

Worked Out Examples

Find the inverse Laplace transform $f(t)$ of each of the following functions $F(s)$:

Example 1: $\frac{s+2}{s^2(s+3)}$

Solution:

$$L^{-1}\left\{\frac{s+2}{s^2(s+3)}\right\} = L^{-1}\left\{\frac{1}{s(s+3)}\right\}$$

$$+L^{-1}\left\{\frac{2}{s^2(s+3)}\right\} = R_1 + R_2$$

We know that $L^{-1}\left\{\frac{1}{s+3}\right\} = e^{-3t} \cdot L^{-1}\left\{\frac{1}{s}\right\} = e^{-3t}$

$$R_1 = L^{-1}\left\{\frac{1}{s} \cdot \frac{1}{s+3}\right\}$$

$$= \int_0^t e^{-3u}du \text{ by ILT of Integral}$$

$$= \frac{e^{-3u}}{-3}\Big|_0^t = \frac{1}{3}\left[1 - e^{-3t}\right]$$

$$R_2 = 2L^{-1}\left\{\frac{1}{s^2}\frac{1}{s+3}\right\} = 2L^{-1}\left\{\frac{1}{s} \cdot \frac{1}{s(s+3)}\right\}$$

$$= 2\int_0^t \frac{1}{3}(1 - e^{-3u})du = \frac{2}{3}\left[u - \frac{e^{-3u}}{-3}\right]_0^t$$

$$= \frac{2}{3}\left[t + \frac{e^{-3t}}{3} - \frac{1}{3}\right]$$

$$\therefore \quad L^{-1}\left\{\frac{s+2}{s^2(s+3)}\right\} = \frac{1}{3} - \frac{1}{3}e^{-3t} + \frac{2}{3}t + \frac{2}{9}e^{-3t} - \frac{2}{9}$$

$$= \frac{1}{9} + \frac{2}{3}t - \frac{1}{9}e^{-3t}.$$

Example 2: $\frac{1}{s^3(s+1)}$

Solution: We know that

$$L^{-1}\left\{\frac{1}{s+1}\right\} = e^{-t}L\left\{\frac{1}{s}\right\} = e^{-t}$$

Applying I.L.T. of integrals

$$L^{-1}\left\{\frac{1}{s}\frac{1}{s+1}\right\} = \int_0^t e^{-u}du = \frac{e^{-u}}{-1}\Big|_0^t = 1 - e^{-t}$$

Now consider

$$L^{-1}\left\{\frac{1}{s} \cdot \frac{1}{s(s+1)}\right\} = \int_0^t (1 - e^{-u})du = u - \frac{e^{-u}}{-1}\Big]_0^t$$

$$= t + e^{-t} - 1$$

Finally $L^{-1}\left\{\frac{1}{s} \cdot \frac{1}{s^2(s+1)}\right\} = \int_0^t (u + e^{-u} - 1)du$

$$= -u + \frac{u^2}{2} + \frac{e^{-u}}{-1}\Big]_0^t = -t + \frac{t^2}{2} - e^{-t} + 1$$

$$= 1 - t + \frac{1}{2}t^2 - e^{-t}$$

Example 3: $\frac{1}{(s^2+a^2)^2}$

Solution: Rewrite

$$\frac{1}{(s^2+a^2)^2} = \frac{1}{s} \cdot \frac{s}{(s^2+a^2)^2}$$

We know that

$$L^{-1}\left\{\frac{s}{(s^2+a^2)^2}\right\} = +t\frac{\sin at}{2a}$$

(Using I.L.T. for derivatives).

Now

$$L^{-1}\left\{\frac{1}{(s^2+a^2)^2}\right\} = L^{-1}\left\{\frac{1}{s}\cdot\frac{s}{(s^2+a^2)^2}\right\}$$

Using I.L.T. for integral $= \int_0^t -u\cdot\frac{\sin au}{2a}du.$

Integrating by parts

$$= -\frac{1}{2a^2}t\cdot\cos at\Big|_0^t + \frac{1}{2a^2}\cdot\frac{\sin at}{a}\Big|_0^t$$

$$= \frac{1}{2a^3}\{\sin at - at\cdot\cos at\}.$$

Example 4: $\frac{s^2}{(s^2+a^2)^2}$

Solution: Rewriting

$$\frac{s^2}{(s^2+a^2)^2} = \frac{s^2+a^2-a^2}{(s^2+a^2)^2} = \frac{1}{(s^2+a^2)} - \frac{a^2}{(s^2+a^2)^2}$$

$$L^{-1}\left\{\frac{s^2}{(s^2+a^2)^2}\right\}$$

$$= L^{-1}\left\{\frac{1}{s^2+a^2}\right\} - a^2L^{-1}\left\{\frac{1}{(s^2+a^2)^2}\right\}$$

(use the result in Example 3 for second term in R.H.S.)

$$= \frac{\sin at}{a} - a^2\cdot\frac{1}{2a^3}\{\sin at - at\cos at\}$$

$$= \frac{1}{2a}\{\sin at + at\cos at\}.$$

EXERCISE

Given $F(s)$ find the inverse Laplace transform $f(t)$:

1. $\frac{1}{s^3(s^2+1)}$ *Ans.* $\frac{t^2}{2}+\cos t - 1$
2. $\frac{s}{(s-2)^5(s+1)}$

Ans. $e^{2t}\left(\frac{t^4}{36}+\frac{t^3}{54}-\frac{t^2}{54}+\frac{t}{81}-\frac{1}{243}\right)+\frac{e^{-t}}{243}$

3. $\frac{s^2}{(s^2-4s+5)^2}$ *Ans.* $te^{2t}(\cos t - \sin t)$
4. $\frac{1}{s}\left(\frac{s-a}{s+a}\right)$ *Ans.* $2e^{-at}-1$
5. $\frac{1}{s^2}\left(\frac{s+1}{s^2+1}\right)$ *Ans.* $1+t-\cos t-\sin t$
6. $\frac{1}{s^4-2s^3}$ *Ans.* $(e^{26}-1-2t-2t^2)/8$
7. $\frac{1}{s(s^2+a^2)}$ *Ans.* $(1-\cos at)/a^2$
8. $\frac{1}{s(s+a)^3}$

Ans. $-\frac{t^2e^{-at}}{2a}-\frac{te^{-at}}{a^2}-\frac{1}{a^3}(e^{-at}-1)$

9. $\frac{1}{s^2(s^2+a^2)}$ *Ans.* $\frac{1}{a^2}\left(t-\frac{\sin at}{a}\right)$.

Second Shifting Theorem

If $L\{f(t)\} = F(s)$ then from second shifting theorem for L.T., we have

$$L\{f(t-a)u(t-a)\} = e^{-as}F(s).$$

Therefore

$$f(t-a)u(t-a) = L^{-1}\{e^{-as}F(s)\}.$$

WORKED OUT EXAMPLES

Find the inverse Laplace transform of the following:

Example 1: $\frac{e^{-s}}{\sqrt{s+1}}$

Solution: Let $F(s) = \frac{1}{\sqrt{s+1}}$ so that

$$f(t) = L^{-1}\{F(s)\} = L^{-1}\left\{\frac{1}{(s+1)^{\frac{1}{2}}}\right\}$$

$$= e^{-t}L^{-1}\left\{\frac{1}{s^{\frac{1}{2}}}\right\} = e^{-t}\cdot\frac{t^{\frac{-1}{2}}}{\sqrt{\pi}}$$

Using t-shift

$$L^{-1}\left\{\frac{e^{-s}}{\sqrt{s+1}}\right\} = L^{-1}\{e^{-s}\cdot F(s)\} = f(t-1)\cdot u(t-1)$$

$$= \left\{e^{-(t-1)}\cdot\frac{(t-1)^{\frac{-1}{2}}}{\sqrt{\pi}}\right\}u(t-1)$$

$$= \begin{cases} e^{-(t-1)}\cdot\frac{(t-1)^{\frac{-1}{2}}}{\sqrt{\pi}}, & t>1 \\ 0, & 0<t<1\end{cases}$$

Example 2: $(5-3e^{-3s}-2e^{-7s})/s$

Solution:

$$L^{-1}\left\{\frac{5-3e^{-3s}-2e^{-7s}}{s}\right\}$$

$$= 5L^{-1}\left\{\frac{1}{s}\right\} - 3L^{-1}\left\{e^{-3s}\frac{1}{s}\right\} - 2L^{-1}\left\{e^{-7s}\cdot\frac{1}{s}\right\}$$

Applying the second shifting theorem

$$= 5.1 - 3\cdot u(t-3) - 2u(t-7)$$

Since $L^{-1}\left\{\frac{1}{s}\right\} = 1$

Also $u(t-3) = 0, \quad 0 < t < 3$

$= 1, \quad t > 3$

and $u(t-7) = 0, \quad 0 < t < 7$

$= 1, \quad t > 7$

Therefore

$$L^{-1}\left\{\frac{5 - 3e^{-3s} - 2e^{-7s}}{s}\right\}$$

$$= \begin{cases} 5 - 3\cdot 0 - 2\cdot 0, & 0 < t < 3 \\ 5 - 3\cdot 1 - 2\cdot 0, & 3 < t < 7 \\ 5 - 3\cdot 1 - 2\cdot 1, & t > 7 \end{cases}$$

$$= \begin{cases} 5, & 0 < t < 3 \\ 2, & 3 < t < 7 \\ 0, & t > 7. \end{cases}$$

Example 3: $(2+5s)/(s^2 e^{4s})$.

Solution: Take $F(s) = \frac{2+5s}{s^2} = \frac{2}{s^2} + \frac{5}{s}$. Then

$$f(t) = L^{-1}\{F(s)\} = L^{-1}\left\{\frac{2}{s^2} + \frac{5}{s}\right\}$$

$$= 2L^{-1}\left\{\frac{1}{s^2}\right\} + 5L^{-1}\left\{\frac{1}{s}\right\}$$

$$= 2t + 5$$

Using result

$$L^{-1}\{e^{-as}F(s)\} = f(t-a)u(t-a)$$

We get (with $a = 4$)

$$L^{-1}\left\{e^{-4s}\left(\frac{2}{s^2} + \frac{5}{s}\right)\right\} = f(t-4)\cdot u(t-4)$$

$$= \begin{cases} 0, & 0 < t < 4 \\ f(t-4), & t > 4 \end{cases}$$

Since $f(t) = 2t + 5$,

so $f(t-4) = 2(t-4) + 5 = 2t - 3$

$$\therefore \quad L^{-1}\left\{e^{-4s}\left(\frac{2}{s^2} + \frac{5}{s}\right)\right\} = \begin{cases} 0, & 0 < t < 4 \\ 2t - 3, & t > 4 \end{cases}$$

Example 4: $\frac{1}{s^2 - e^{-as}}$

Solution:

$$\frac{1}{s^2 - e^{-as}} = \frac{1}{s^2}\left[1 - \frac{e^{-as}}{s^2}\right]^{-1} = \frac{1}{s^2}\sum_{n=0}^{\infty}\left(\frac{e^{-as}}{s^2}\right)^n$$

$$= \sum_{n=0}^{\infty}\frac{e^{-ans}}{s^{2n+2}} = \sum_{n=0}^{\infty}\frac{e^{-ans}}{s^{(2n+1)+1}}$$

Since $L^{-1}\left\{\frac{1}{s^{(2n+1)+1}}\right\} = \frac{t^{2n+1}}{(2n+1)!}$.

Therefore

$$\frac{1}{s^2 - e^{-as}} = L^{-1}\left\{\sum_{n=0}^{\infty}\frac{e^{-ans}}{s^{2n+2}}\right\} = \sum_{n=0}^{\infty}L^{-1}\left\{\frac{e^{-ans}}{s^{2n+2}}\right\}$$

$$= \sum_{n=0}^{\infty}\frac{(t-an)^{2n+1}}{(2n+1)!}u(t-an).$$

EXERCISE

Using t-shift find the inverse Laplace transform $f(t)$ for each of the following functions $F(s)$:

1. $\frac{(s+1)e^{-\pi s}}{s^2+s+1}$

Ans. $\frac{e^{-\frac{1}{2}(t-\pi)}}{\sqrt{3}}\left\{\sqrt{3}\cos\frac{\sqrt{3}}{2}(t-\pi) + \sin\frac{\sqrt{3}}{2}(t-\pi)\right\}\times$
$\times u(t-\pi)$

2. $(e^{-4s} - e^{-7s})/s^2$

Ans. $\begin{cases} 0, & 0 < t < 4 \\ t-4, & 4 < t < 7 \\ 3, & t > 7 \end{cases}$

3. $\frac{e^{4-3s}}{(s+4)^{\frac{5}{2}}}$

Ans. $\frac{4(t-3)^{\frac{3}{2}}e^{-4(t-4)}}{3\sqrt{\pi}}u(t-3)$

4. $\frac{e^{-3s}}{s^2-2s+5}$

Ans. $\frac{1}{2}e^{(t-3)}\sin 2(t-3)\cdot u(t-3)$

5. $\frac{e^{-s}}{(s+1)^3}$

Ans. $\frac{1}{2}\cdot e^{-(t-1)}\cdot(t-1)^2\cdot u(t-1)$

6. $\frac{se^{\frac{-s}{2}} + \pi e^{-s}}{s^2+\pi^2}$

Ans. $\sin\pi t\left\{u\left(t-\frac{1}{2}\right) - u(t-1)\right\}$

7. $\frac{s}{s^2-5s+6}e^{-2s}$

Ans. $\left\{-2e^{2(t-2)}+3e^{3(t-2)}\right\}u(t-2)$

8. $\frac{3}{s}-\frac{4e^{-s}}{s^2}+\frac{4e^{-3s}}{s^2}$

Ans. $3-4(t-1)u(t-1)+4(t-3)u(t-3)$

9. $\frac{e^{-3s}}{(s+1)^3}$

Evaluate $f(2), f(5), f(7)$

Ans. $f(2)=0,\ f(5)=2e^{-2},\ f(7)=8e^{-4}$

10. $\frac{e^{-3s}}{(s-4)^2}$

Ans. $(t-3)e^{4(t-3)}u(t-3)$

11. $\frac{se^{-\pi s}}{s^2+9}$

Ans. $-\cos 3t\cdot u(t-\pi)$

12. $\frac{1}{s+e^{-s}}$

Ans. $\sum_{n=0}^{\infty}\frac{(-1)^n}{n!}(t-n)^n u(t-n).$

12.7 USE OF PARTIAL FRACTIONS TO FIND INVERSE L.T.

Application of L.T. to a D.E. results in a subsidiary (algebraic) equation which usually comes out as a rational function $Y(s)=\frac{P(s)}{Q(s)}$ where $P(s)$ and $Q(s)$ are polynomials in s. When the degree of $P(s)\le$ degree of $Q(s)$, then the rational function $\frac{P(s)}{Q(s)}$ can be written as the *sum* of simpler rational functions, called partial fractions, depending on the nature of factors of the denominator $Q(s)$ as follows:

Factor in denominator	Corresponding partial fraction
a. Non-repeated linear factor $ax+b$ (occurring once)	$\frac{A}{ax+b}$ with $A\ne 0$
b. Repeated linear factor $(ax+b)^r$ (occurring r times)	$\frac{A_1}{ax+b}+\frac{A_2}{(ax+b)^2}+\cdots+\frac{Ar}{(ax+b)^r}$ with $A_r\ne 0$
c. Non-repeated quadratic factor ax^2+bx+c	$\frac{Ax+B}{ax^2+bx+c}$ with at least one of A, B non-zero
d. Repeated quadratic factor $(ax^2+bx+c)^r$ (occurring r times)	$\frac{A_1x_1+B_1}{ax^2+bx+c}+\frac{A_2x+B_2}{(ax^2+bx+c)^2}+\cdots+\frac{A_rx+B_r}{(ax^2+bx+c)^r}$ with at least one of Ar, B_r non-zero.

Here $A, B, A_1, B_1, \ldots, A_r, B_r$ are all constants.

By finding the inverse L.T. of each of these partial fractions, $L^{-1}\{Y(s)\}=L^{-1}\left\{\frac{P(s)}{Q(s)}\right\}$ can be determined.

WORKED OUT EXAMPLES

Find the inverse Laplace transform $f(t)$ of each of the following functions $F(s)$:

Distinct non-repeated linear factors

Example 1: $\frac{3s+7}{s^2-2s-3}$

Solution: Using partial fractions, we have

Method 1:

$$\frac{3s+7}{s^2-2s-3}=\frac{3s+7}{(s-3)(s+1)}=\frac{A}{s-3}+\frac{B}{s+1}$$

Multiplying both sides by $(s-3)(s+1)$,

$$3s+7=A(s+1)+B(s-3)=(A+B)s+A-3B$$

Equating the corresponding coefficients on both sides

$A+B=3,\ A-3B=7$ Then $A=4,\ B=-1$

Thus $\frac{3s+7}{(s-3)(s+1)}=\frac{4}{s-3}-\frac{1}{s+1}$

$$L^{-1}\left\{\frac{3s+7}{(s-3)(s+1)}\right\}=4L^{-1}\left\{\frac{1}{s-3}\right\}-L^{-1}\left\{\frac{1}{s+1}\right\}$$
$$=4e^{3t}-e^{-t}.$$

Method 2:

Put $s=3,\ 16=4A\quad\therefore\quad A=4,$

Put $s=-1,\ 4=-4B\quad\therefore\quad B=-1$

Method 3: Multiplying by $(s-3)$ and taking the limit as s tends to 3

$$\lim_{s\to 3}\frac{3s+7}{s+1}=A+\lim_{s\to 3}\frac{B(s-3)}{(s+1)},\quad \text{then } A=4.$$

Similarly, multiplying by $(s+1)$ and letting $s\to -1$, we get

$$\lim_{s\to -1}\frac{3s+7}{s-3}=\lim_{s\to -1}\frac{A(s+1)}{(s-3)}+B\quad \text{Then } B=-1$$

Applying inverse transform, we get the above result.

Linear repeated factors

Example 2: $\frac{s^3+6s^2+14s}{(s+2)^4}$

Solution:

$$\frac{s^3+6s^2+14s}{(s+2)^4} = \frac{A}{(s+2)^4}+\frac{B}{(s+2)^3}+\frac{C}{(s+2)^2}+\frac{D}{(s+2)}$$

$$s^3+6s^2+14s = A+B(s+2)+C(s+2)^2+D(s+2)^3$$
$$= Ds^3+(6D+C)s^2+(12D+4C+B)s +(8D+4C+2B+A)$$

Equating the coefficients of s
$A=-12,\ B=2,\ C=0,\ D=1$

$$L^{-1}\left\{\frac{s^3+6s^2+14s}{(s+2)^4}\right\} = -12L^{-1}\left\{\frac{1}{(s+2)^4}\right\}+2L^{-1}\left\{\frac{1}{(s+2)^3}\right\} +L^{-1}\left\{\frac{1}{s+2}\right\}$$
$$= -12e^{2t}L^{-1}\left\{\frac{1}{s^4}\right\}+2e^{-2t}L^{-1}\left\{\frac{1}{s^3}\right\}+e^{-2t}$$
$$= -2e^{-2t}t^3+e^{-2t}t^2+e^{-2t}$$
$$= e^{-2t}\{1+t^2-2t^3\}.$$

Non-repeated Quadratic Factors

Example 3: $\frac{s^2+2s+3}{(s^2+2s+2)(s^2+2s+5)}$

Solution:

Method 1:

$$\frac{s^2+2s+3}{(s^2+2s+2)(s^2+2s+5)} = \frac{As+B}{s^2+2s+2}+\frac{Cs+D}{s^2+2s+5}$$

$$s^2+2s+3 = (As+B)(s^2+2s+5)+(Cs+D)(s^2+2s+2)$$
$$= (A+C)s^3+(2A+B+2C+D)s^2 +(5A+2B+2C+2D)s+5B+2D$$

Comparing coefficients of s on either side

$$A+C=0,\ 2A+B+2C+D=1,$$
$$5A+2B+2C+2D=2,$$
$$5B+2D=3$$

By solving these equations, we get

$$A=0,\ B=\frac{1}{3},\ C=0,\ D=\frac{2}{3}$$

$$L^{-1}\left\{\frac{s^2+2s+3}{(s^2+2s+2)(s^2+2s+5)}\right\} = L^{-1}\left\{\frac{\frac{1}{3}}{s^2+2s+2}\right\}+L^{-1}\left\{\frac{\frac{2}{3}}{s^2+2s+5}\right\}$$
$$= \frac{1}{3}L^{-1}\left\{\frac{1}{(s+1)^2+1}\right\}+\frac{2}{3}L^{-1}\left\{\frac{1}{(s+1)^2+4}\right\}$$
$$= \frac{1}{3}e^{-t}\sin t+\frac{2}{3}\cdot\frac{1}{2}\cdot e^{-t}\cdot\sin 2t$$
$$= \frac{1}{3}e^{-t}(\sin t+\sin 2t).$$

Method 2: Reduction of non-repeated quadratic factors to non-repeated linear factors using complex numbers

$$\frac{s^2+2s+3}{[(s^2+2s+2)][s^2+2s+5]} = \frac{s^2+2s+3}{[(s+1-i)(s+1+i)][(s+1-2i)(s+1+2i)]}$$
$$= \frac{A}{s+1-i}+\frac{B}{s+1+i}+\frac{C}{s+1-2i}+\frac{D}{s+1+2i}.$$

Solving $A=\frac{1}{6i},\ B=-\frac{1}{6i},\ C=\frac{1}{6i},\ D=\frac{-1}{6i}$

$$L^{-1}\left\{\frac{s^2+2s+3}{(s^2+2s+2)(s^2+2s+5)}\right\} = \frac{e^{-(1-i)t}}{6i}-\frac{e^{-(1+i)t}}{6i}+\frac{e^{-(1-2i)t}}{6i}-\frac{e^{-(1+2i)t}}{6i}$$
$$= \frac{1}{3}e^{-t}\left(\frac{e^{it}-e^{-it}}{2i}\right)+\frac{1}{3}e^{-t}\left(\frac{e^{2it}-e^{-2it}}{2i}\right)$$
$$= \frac{1}{3}e^{-t}\cdot\sin t+\frac{1}{3}e^{-t}\sin 2t.$$

Example 4: $\frac{a(s^2-2a^2)}{s^4+4a^4}$

Solution: Rewriting

$$s^4 + 4a^4 = (s^2 + 2a^2)^2 - (2as)^2.$$
$$= (s^2 + 2as + 2a^2)(s^2 - 2as + 2a^2)$$

$$\frac{a(s^2 - 2a^2)}{s^4 + 4a^4} = \frac{a(s^2 - 2a^2)}{(s^2 + 2as + 2a^2)(s^2 - 2as + 2a^2)}$$
$$= \frac{1}{2}\frac{-(s+a)}{s^2 + 2as + 2a^2} + \frac{1}{2}\frac{s-a}{s^2 - 2as + 2a^2}$$
$$= \frac{1}{2}\left[\frac{-(s+a)}{(s+a)^2 + a^2} + \frac{s-a}{(s-a)^2 + a^2}\right]$$

$$\therefore\ L^{-1}\left\{\frac{a(s^2 - 2a^2)}{s^4 + 4a^4}\right\}$$
$$= \frac{1}{2}L^{-1}\left\{\frac{-(s+a)}{(s+a)^2 + a^2}\right\} + \frac{1}{2}L^{-1}\left\{\frac{s-a}{(s-a)^2 + a^2}\right\}$$
$$= \frac{-1}{2}e^{-at}\cos at + \frac{1}{2}e^{at}\cos at$$
$$= \cos at\left[\frac{-1}{2}e^{-at} + \frac{e^{at}}{2}\right] = \cos at \cdot \sinh at.$$

Repeated quadratic factors

Example 5: $\frac{s^3 - 3s^2 + 6s - 4}{(s^2 - 2s + 2)^2}$

Solution:

$$\frac{s^3 - 3s^2 + 6s - 4}{(s^2 - 2s + 2)^2} = \frac{As + B}{(s^2 - 2s + 2)^2} + \frac{Cs + D}{(s^2 - 2s + 2)}$$

$$s^3 - 3s^2 + 6s - 4 = As + B + (Cs + D)(s^2 - 2s + 2)$$
$$= Cs^3 + (D - 2c)s^2$$
$$+(A + 2c - 2D)s + B + 2D.$$

Equating and solving $A = 2, B = -2, C = 1, D = -1$, and rewriting $s^2 - 2s + 2 = (s-1)^2 + 1$

$$\text{I.T.} = L^{-1}\left\{\frac{2s - 2}{[(s-1)^2 + 1]^2}\right\} + L^{-1}\left\{\frac{s-1}{(s-1)^2 + 1}\right\}$$
$$= e^t L^{-1}\left\{\frac{2s}{(s^2 + 1)^2} + \frac{s}{(s^2 + 1)}\right\}$$
$$= 2e^{-t}L^{-1}\left\{\frac{s}{(s^2 + 1)^2}\right\} + e^{-t} \cdot \cos t$$
$$= 2e^t \cdot \frac{t}{2} \cdot \sin t + e^{-t} \cdot \cos t = e^t\{t \sin t + \cos t\}.$$

EXERCISE

Find the inverse Laplace transform $f(t)$ of each of the following functions $F(s)$:

1. $\frac{s-2}{s^2+5s+6}$ *Ans.* $-4e^{-2t} + 5e^{-3t}$
2. $\frac{2s^2-4}{(s+1)(s-2)(s-3)}$ *Ans.* $\frac{-1}{6}e^{-t} - \frac{4}{3}e^{2t} + \frac{7}{2}e^{3t}$
3. $\frac{s^2-7s+24}{s^3-7s^2+14s-8}$ *Ans.* $6e^t - 7e^{2t} + 2e^{4t}$
4. $\frac{s+17}{(s-1)(s+3)}$ *Ans.* $\frac{9}{2}e^t - \frac{7}{2}e^{-3t}$
5. $\frac{2s^2-6s+5}{s^3-6s^2+11s-6}$ *Ans.* $\frac{1}{2}e^t - e^{2t} + \frac{5}{2}e^{3t}$
6. $\frac{5s}{s^2+4s+4}$ *Ans.* $5e^{-2t}(1 - 2t)$
7. $\frac{5}{(s-2)^4}$ *Ans.* $\frac{5t^2e^{2t}}{6}$
8. $\frac{7}{(2s+1)^3}$ *Ans.* $\frac{7}{16}t^2e^{\frac{-t}{2}}$
9. $\frac{s+2}{s^2+4s+7}$ *Ans.* $e^{-2t} \cdot \cos\sqrt{3}t$
10. $\frac{2s+12}{s^2+6s+13}$

    Ans. $e^{-3t}(2\cos 2t + 3\sin 2t)$
11. $\frac{s^2+9s-9}{s^3-9s}$ *Ans.* $1 + 3\sinh 3t$
12. $\frac{s}{(s^2-2s+2)(s^2+2s+2)}$ *Ans.* $\frac{1}{2}\sin t \cdot \sinh t$
13. $\frac{3s^3-3s^2-40s+36}{(s^2-4)^2}$ *Ans.* $(5t + 3)e^{-2t} - 2te^{2t}$
14. $\frac{2s^3-s^2-1}{(s+1)^2(s^2+1)^2}$

    Ans. $\frac{1}{2}\sin t + \frac{1}{2}t\cos t - te^{-t}$
15. $\frac{5s^2-7s+17}{(s-1)(s^2+4)}$

    Ans. $3e^t + 2\cos 2t - \frac{5}{2}\sin 2t$
16. $\frac{2s^2+15s+7}{(s+1)^2(s-2)}$ *Ans.* $(2t - 3)e^{-t} + 5e^{2t}$
17. $\frac{s+1}{(s^2+1)(s^2+4)}$

    Ans. $\frac{1}{6}(2\cos t - 2\cos 2t + 2\sin t - \sin 2t)$
18. $\frac{10}{s(s^2-2s+5)}$

    Ans. $2 - e^t(2\cos 2t - \sin 2t)$
19. $\frac{1}{s(s+1)^2}$ *Ans.* $1 - e^{-t} - te^{-t}$

20. $\frac{s^2+8s+27}{(s+1)(s^2+4s+13)}$

Ans. $2e^{-t}+e^{-2t}(\sin 3t-\cos 3t)$

21. $\frac{s}{s^4+s^2+1}$

Hint: $s^4+s^2+1=(s^2+s+1)(s^2-s+1)$

Ans. $\frac{1}{\sqrt{3}}\left[e^{\frac{t}{2}}\sin\frac{\sqrt{3}}{2}t-e^{\frac{-t}{2}}\sin\frac{\sqrt{3}}{2}t\right]$

22. $\frac{s}{s^4+4a^4}$

Hint: $s^4+4a^4=(s^2+2a^2)^2-(2as)^2$
$=(s^2+2as+2a^2)(s^2-2as+2a^2)$.

Ans. $\frac{1}{2a^2}\sin at\cdot\sinh at$

23. $\frac{5s+3}{(s-1)(s^2+2s+5)}$

Ans. $e^t-e^{-t}\left(\cos 2t-\frac{3}{2}\sin 2t\right)$

24. $\frac{s^2}{s^4+4a^4}$

Hint: Use hint of above Example 22.

Ans. $\frac{1}{2a}[\sinh at\cos at+\cosh at\cdot\sin at]$.

12.8 CONVOLUTION

Convolution is used to find inverse Laplace transforms in solving differential equations and integral equations.

Suppose two Laplace transforms $F(s)$ and $G(s)$ are given. Let $f(t)$ and $g(t)$ be their inverse Laplace transforms respectively i.e., $f(t)=L^{-1}\{F(s)\}$ and $g(t)=L^{-1}\{G(s)\}$. Then the inverse $h(t)$ of the *product* of transforms $H(s)=F(s)G(s)$ can be calculated from the known inverse $f(t)$ and $g(t)$.

Convolution

$h(t)$ of $f(t)$ and $g(t)$, denoted by $(f*g)(t)$ is defined as

$$h(t)=(f*g)(t)=\int_0^t f(u)g(t-u)du$$

$f*g$ is called the convolution or faltung* of f and g and can be regarded as a "generalized product" of these functions.

Result: Thus to find the inverse transform of product of transforms $H(s)=F(s)G(s)$, calculate $h(t)=f*g$ which is the convolution of f and g. Therefore one should tactfully rewrite $H(s)$ as a product of $F(s)$ and $G(s)$ in such a way that the corresponding inverses $f(t)$ and $g(t)$ are readily known from transform tables or other means.

Convolution Theorem

Prove that

$$L\{h(t)\}=L\{f*g\}=H(s)=F(s)\cdot G(s)$$

or $L^{-1}\{F(s)\cdot G(s)\}=h(t)=f*g$.

Proof: From the definition of Laplace transform

$$L\{f*g\}=\int_0^\infty e^{-st}(f*g)dt$$
$$=\int_0^\infty e^{-st}\left[\int_0^t f(\tau)g(t-\tau)d\tau\right]dt$$
$$=\int_0^\infty\int_0^t e^{-st}f(\tau)g(t-\tau)d\tau\,dt,$$
$$=\iint_R e^{-st}f(\tau)g(t-\tau)d\tau\,dt$$

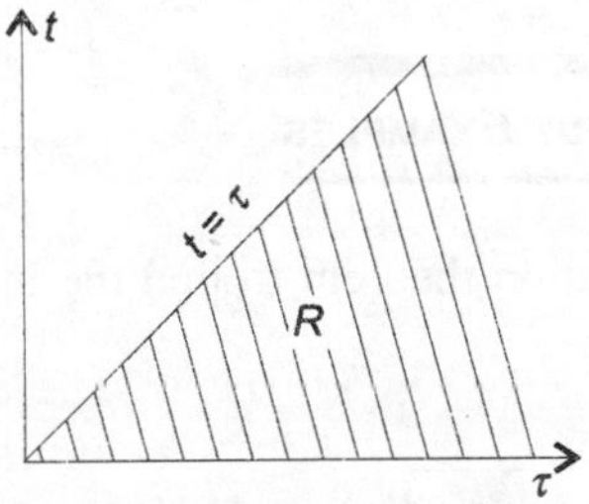

Fig. 12.10

where R is the 45° wedge bounded by the lines $\tau=0$ and $t=\tau$ (see Fig. 12.10). Change the variables τ, t to the new variables u, v by the transformation

$$u=t-\tau$$
$$v=\tau$$

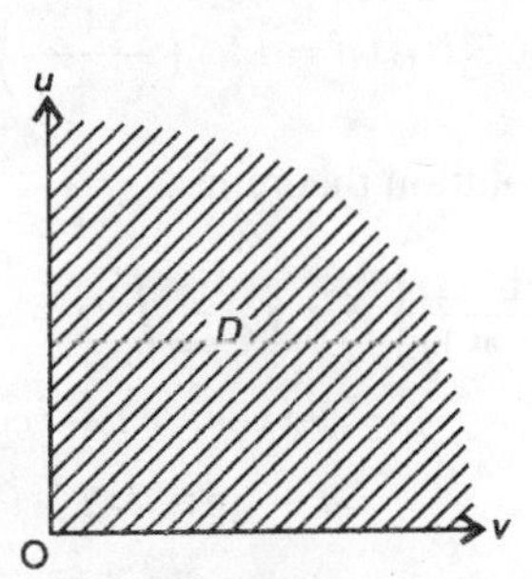

Fig. 12.11

---
*German for folding.

The Jacobian $J = 1$. Thus the double integral over R transforms to a double integral over D the first quadrant of the new uv-plane (i.e., $u > 0, v > 0$) (refer Fig. 12.11). Thus

$$L\{f * g\} = \iint_D e^{-s(u+v)} f(v)g(u)du\,dv$$

$$= \int_0^\infty \int_0^\infty e^{-s(u+v)} f(v)g(u)du\,dv$$

$$= \int_0^\infty e^{-sv} f(v)dv \int_0^\infty e^{-su} g(u)du$$

$$= L\{f\} \cdot L\{g\}$$

Valid Properties

1. $f * g = g * f$ Commutative
2. $(f * g) * v = f * (g * v)$ Associative
3. $f * (g_1 + g_2) = f * g_1 + f * g_2$ Distributive
4. $f * 0 = 0 * f = 0$.

Worked Out Examples

Use convolution theorem to find the inverse of the following:

Example 1: $\frac{1}{s^2(s^2+1)}$

Solution: Rewriting

$$\frac{1}{s^2(s^2+1)} = \frac{1}{s^2} \cdot \frac{1}{s^2+1} = F(s) \cdot G(s)$$

so that $f(t) = L^{-1}(F(s)) = L^{-1}\left(\frac{1}{s^2}\right) = t$

$$g(t) = L^{-1}(G(s)) = L^{-1}\left(\frac{1}{s^2+1}\right) = \sin t$$

$\therefore$ By convolution theorem

$$L^{-1}\left\{\frac{1}{s^2(s^2+1)}\right\} = f * g = \int_0^t (t-u) \cdot \sin u\,du$$

$$= -t\cos t + t + t\cos t - \sin t$$

$$= t - \sin t.$$

Example 2: $\frac{1}{(s^2+a^2)^2}$

Solution: Rewriting

$$\frac{1}{(s^2+a^2)^2} = \left(\frac{1}{s^2+a^2}\right)\left(\frac{1}{s^2+a^2}\right)$$

Here $f(t) = L^{-1}\left(\frac{1}{s^2+a^2}\right) = \frac{1}{a}\sin at$

Similarly, $g(t) = \frac{1}{a}\sin at$

$$\therefore L^{-1}\left\{\frac{1}{(s^2+a^2)^2}\right\} = f * g$$

$$= \int_0^t \frac{\sin au}{a} \cdot \frac{\sin a(t-u)}{a} du$$

$$= \frac{1}{2a^2}\int_0^t [\cos a(2u-t) - \cos at]du$$

$$\because \sin A \cdot \sin B = \frac{\cos(A-B) - \cos(A+B)}{2}$$

$$= \frac{1}{2a^2}\left[\frac{\sin a(2u-t)}{2a} - \cos at \cdot u\right]_{u=0}^{a}$$

$$= \frac{1}{2a^2}\left[\frac{\sin at}{2a} - t\cos at - \frac{\sin(-at)}{2a}\right]$$

$$= \frac{1}{2a^3}[\sin at - at\cos at]$$

Example 3: $\frac{16}{(s-2)(s+2)^2}$

Solution: Rewriting $\frac{1}{(s-2)} \cdot \frac{1}{(s+2)^2} = F(s) \cdot G(s)$

We know that $f(t) = L^{-1}\{F(s)\} = L^{-1}\left\{\frac{1}{s-2}\right\} = e^{2t}$

$$g(t) = L^{-1}\{G(s)\} = L^{-1}\left\{\frac{1}{(s+2)^2}\right\}$$

$$= e^{-2t}L^{1}\left\{\frac{1}{s^2}\right\} = te^{-2t}$$

Applying convolution theorem

$$L^{-1}\left\{\frac{16}{(s-2)(s+2)^2}\right\} = 16 \cdot g * s = 16\int_0^t ue^{-2u}e^{2(t-u)}du$$

$$= 16e^{2t}\int_0^t ue^{-4u}du$$

$$= 16e^{2t}\left[\frac{ue^{-4u}}{-4} - 1\frac{e^{-4u}}{16}\right]_{u=0}^{t}$$

$$= e^{2t} - e^{-2t} - 4te^{-2t}.$$

Example 4: $\frac{1}{s(s+1)(s+2)}$

Solution: Rewriting

$$\frac{1}{s(s+1)(s+2)} = \frac{1}{s(s+1)} \cdot \frac{1}{s+2}$$

Consider

$$\frac{1}{s(s+1)} = \frac{1}{s} \cdot \frac{1}{s+1}$$

so that $f(t) = 1, g(t) = e^{-t}$

$$L^{-1}\left\{\frac{1}{s}\frac{1}{s+1}\right\} = f * g = \int_0^t 1.e^{-u} du$$

$$= 1 - e^{-t} = h(t)$$

Also $L^{-1}\left\{\frac{1}{s+2}\right\} = J(t) = e^{-2t}$

$$\therefore L^{-1}\left\{\frac{1}{s(s+1)} \cdot \frac{1}{s+2}\right\} = h * J$$

$$= \int_0^t e^{-2(t-u)} \cdot (1 - e^u)\, du$$

$$= e^{-2t} \int_0^t (e^{2u} - e^u)\, du$$

$$= e^{-2t}\left[\frac{e^{2u}}{2}\Big|_0^t - e^u\Big|_0^t\right]$$

$$= e^{-2t}\left[\frac{e^{2t}}{2} - \frac{1}{2} - (e^t - 1)\right]$$

$$= \frac{1}{2} + \frac{1}{2}e^{-2t} - e^{-t}.$$

EXERCISE

Use convolution theorem to find inverse Laplace transform of the following:

1. $\frac{1}{(s+a)(s+b)}$ *Ans.* $\frac{e^{-at}-e^{-bt}}{b-a}$
2. $\frac{1}{s(s^2+9)}$ *Ans.* $(1 - \cos 3t)/9$
3. $\frac{1}{s^2(s+3)}$ *Ans.* $(-1 + 3t + e^{-3t})/9$
4. $\frac{s^2}{(s^2+a^2)^2}$ *Ans.* $\frac{1}{2a}(\sin at + at \cos at)$
5. $\frac{s}{(s^2+a^2)^3}$ *Ans.* $\frac{t}{8a^3}(\sin at - at \cos at)$
6. $\frac{s^2}{(s^2+a^2)(s^2+b^2)}$ *Ans.* $\frac{a \sin at - b \sin bt}{a^2-b^2}$
7. $\frac{1}{(s+1)(s^2+1)}$ *Ans.* $\frac{1}{2}(\sin t - \cos t + e^{-t})$
8. $\frac{1}{(s+1)(s+9)^2}$ *Ans.* $\frac{e^{-t}}{64}[1 - e^{-8t}(1 + 8t)]$
9. $\frac{1}{s^2(s-a)}$ *Ans.* $\frac{1}{a^2}(e^{at} - at - 1)$
10. $\frac{s}{(s^2+a^2)^2}$ *Ans.* $\frac{t \sin at}{2a}$
11. $\frac{a}{s^2(s^2+a^2)}$ *Ans.* $(at - \sin at)/a^2$
12. $\frac{1}{(s^2+4)(s+1)^2}$

    Ans. $\frac{e^{-t}}{50}[10e^{-t} - (3 \sin 2t + 4 \cos 2t)]$
13. $\frac{1}{s^3(s^2+1)}$ *Ans.* $\frac{t^2}{2} + \cos t - 1$
14. $\frac{1}{[s^2(s^2-a^2)}$ *Ans.* $\frac{1}{a^2}\left[-t + \frac{\sinh at}{a}\right]$
15. $\frac{1}{(s+2)^2(s-2)}$ *Ans.* $\frac{1}{16}[e^{2t} - (4t + 1)e^{-2t}]$.

12.9 APPLICATION OF LAPLACE TRANSFORM TO DIFFERENTIAL EQUATIONS WITH CONSTANT COEFFICIENTS

Laplace transform is especially suitable to obtain the solution of linear non-homogeneous ordinary differential equations with constant coefficients, when all the boundary conditions are specified for the unknown function and its derivatives at a single point.

Consider the initial value problem

$$\frac{d^2y}{dt^2} + a\frac{dy}{dt} + by = r(t) \tag{1}$$

$$y(t = 0) = k_0, \quad y'(t = 0) = k_1 \tag{2}$$

where a, b, k_0, k_1 are all constants and $r(t)$ is a function of t.

Method of solution to D.E. by L.T.

Step I. Apply Laplace transform on both sides of the given differential Equation (1), resulting in a subsidiary equation

$$[s^2Y - sy(0) - y'(0)] + a[sY - y(0)] + bY = R(s) \tag{3}$$

where $Y = L\{y(t)\}$ and $R(s) = L\{r(t)\}$.

Replace $y(0)$, $y'(0)$ using given initial conditions (2).

Step II. Solve (3) algebraically for $Y(s)$, usually to a sum of partial fractions.

Step III. Apply inverse Laplace tranform to $Y(s)$ obtained in step II. This yields the solution of O.D.E. (1) satisfying the initial conditions (2) as

$$y(t) = L^{-1}\{Y(s)\}.$$

WORKED OUT EXAMPLES

Homogeneous

Solve the following using Laplace transform:

Example 1: $y'' - 2y' - 8y = 0, y(0) = 3, y'(0) = 6$

Solution: Applying L.T.

$$(s^2Y - 3s - 6) - 2(sY - 3) - 8Y = 0$$

Solving,

$$Y(s) = \frac{3s}{s^2 - 2s - 8} = \frac{3s}{(s-4)(s+2)}.$$

Using partial fractions

$$Y(s) = \frac{2}{s-4} + \frac{1}{s+2}$$

Applying I.L.T.

$$y(t) = L^{-1}(Y(s)) = 2L^{-1}\left(\frac{1}{s-4}\right) + L^{-1}\left(\frac{1}{s+2}\right)$$

$$= 2e^{4t} + e^{-2t}.$$

Non-homogeneous

Example 2: $y'' + 2y' + 5y = e^{-t}\sin t, \ y(0) = 0, \ y'(0) = 1.$

Solution: Using L.T.

$$[s^2Y - 0 - 1] + 2[sY - 0] + 5Y = L(e^{-t}\sin t)$$

$$= \frac{1}{(s+1)^2 + 1}$$

Solving $Y = \frac{s^2+2s+3}{(s^2+2s+5)(s^2+2s+2)}$

By partial fractions,

$$\frac{s^2 + 2s + 3}{(s^2 + 2s + 5)(s^2 + 2s + 2)} = \frac{As + B}{s^2 + 2s + 5} + \frac{Cs + D}{s^2 + 2s + 2}$$

$$s^2 + 2s + 3 = (As + B)(s^2 + 2s + 2) + (Cs + D)(s^2 + 2s + 5)$$
$$= s^3(A + C) + s^2(2A + 2C + B + D) + s(2A + 5C + 2B + 2D) + 2B + 5D.$$

Equaring coefficients of s on either side

$$A + C = 0, \quad 2A + 2C + B + D = 1,$$
$$2A + 5C + 2B + 2D = 2, \quad 2B + 5D = 3$$

$$\therefore \ A = 0, \ B = \tfrac{1}{3}, \ C = 0, \ D = \tfrac{2}{3}$$

$$Y(s) = \frac{1}{3}\frac{1}{s^2 + 2s + 5} + \frac{2}{3}\frac{1}{s^2 + 2s + 2}$$

Rewriting

$$Y(s) = \frac{1}{3}\frac{1}{(s+1)^2 + 2^2} + \frac{2}{3}\frac{1}{(s+1)^2 + 1}.$$

Applying I.T.

$$y(t) = L^{-1}(Y) = \frac{1}{3}L^{-1}\left(\frac{1}{(s+1)^2 + 2^2}\right) + \frac{2}{3}L^{-1}\left(\frac{1}{(s+1)^2 + 1^2}\right)$$

Using first shift theorem

$$= \frac{1}{3}e^{-t}L^{-1}\left(\frac{1}{s^2+1}\right) + \frac{2}{3}e^{-t}L^{-1}\left(\frac{1}{s^2+2^2}\right)$$

$$y(t) = \frac{e^{-t}}{3}[\sin t + \sin 2t].$$

Example 3: $y'' + n^2y = a\sin(nt + 2), y(0) = 0, y'(0) = 0.$

Solution:

$y'' + n^2y = a[\sin nt \cdot \cos 2 + \cos nt \cdot \sin 2]$
Applying L.T., $L(y'') + n^2L(y) =$
$a \cdot \cos 2 \cdot L(\sin nt) + a\sin 2 \cdot L(\cos nt).$

$$[s^2Y - sy(0) - y'(0)] + n^2Y$$
$$= \frac{n}{s^2+n^2} \cdot a \cdot \cos 2 + \frac{s}{s^2+n^2} \cdot a\sin 2$$

Solving for Y

$$Y(s) = \frac{n}{(s^2+n^2)^2} \cdot a \cdot \cos 2 + \frac{s}{(s^2+n^2)^2} \cdot a \cdot \sin 2.$$

Applying I.T.

$$y(t) = n \cdot a \cdot \cos 2 \cdot L^{-1}\left\{\frac{1}{(s^2+n^2)^2}\right\}$$

$$+a \cdot \sin 2 \cdot L^{-1}\left\{\frac{s}{(s^2+n^2)^2}\right\} \quad (1)$$

From I.L.T. tables, we know that (2nd term in R.H.S.)

$$L^{-1}\left\{\frac{s}{(s^2+n^2)^2}\right\} = \frac{t \cdot \sin nt}{2n} \quad (2)$$

To find 1st term in R.H.S.

$$L^{-1}\left\{\frac{1}{(s^2+n^2)^2}\right\} = L^{-1}\left\{\frac{1}{s} \cdot \frac{s}{(s^2+n^2)^2}\right\}$$

$$= \frac{1}{2n}\int_0^t t \cdot \sin nt\, dt$$

because

$$L^{-1}\left\{\frac{F(s)}{s}\right\} = \int_0^t f(t)dt.$$

$$\therefore \quad L^{-1}\left\{\frac{1}{(s^2+n^2)^2}\right\} = \frac{1}{2n}\int_0^t t \cdot \sin t\, nt\, dt$$

$$= \frac{1}{2n^3}[-nt \cos nt + \sin nt] \quad (3)$$

Thus substituting (2) and (3) in (1), we get

$$y(t) = an \cdot \cos 2 \cdot \frac{1}{2n^3}[-nt \cos nt + \sin nt]$$

$$+a \sin 2 \frac{t}{2n} \sin nt$$

$$= \frac{a}{2n^2}[-nt \cdot \cos 2 \cdot \cos nt + \cos 2 \cdot \sin nt$$

$$+nt \cdot \sin 2 \cdot \sin nt]$$

$$= \frac{a}{2n^2}[\sin nt \cdot \cos 2 - nt(\cos nt \cdot \cos 2$$

$$- \sin nt \cdot \sin 2)]$$

$$= \frac{a}{2n^2}[\sin nt \cdot \cos 2 - nt \cos(nt+2)].$$

Example 4: Solve $y''' - 3y'' + 3y' - y = t^2 e^t$, $y(0) = 1$, $y'(0) = 0$, $y''(0) = -2$.

Solution: Applying L.T. to D.E.

$$L(y''') - 3L(y'') + 3L(y') - L(y) = L(t^2 e^t)$$

$$\left[s^3 Y - s^2 y(0) - sy'(0) - y''(0)\right]$$

$$-3[s^2 Y - sy(0) - y'(0)]$$

$$+3[sY - y(0)] - Y = \frac{2}{(s-1)^3}$$

Using the initial condition $y(0) = 1$, $y'(0) = 0$, $y''(0) = -2$, and solving for Y

$$(s^3 - 3s^2 + 3s - 1)Y - s^2 + 3s - 1 = \frac{2}{(s-1)^3}$$

$$Y = \frac{s^2 - 3s + 1}{(s-1)^3} + \frac{2}{(s-1)^6}$$

$$= \frac{s^2 - 2s + 1 - s}{(s-1)^3} + \frac{2}{(s-1)^6}$$

$$= \frac{(s-1)^2 - (s-1) - 1}{(s-1)^3} + \frac{2}{(s-1)^6}$$

$$Y = \frac{1}{s-1} - \frac{1}{(s-1)^2} - \frac{1}{(s-1)^3} + \frac{2}{(s-1)^6}.$$

Applying I.L.T.

$$y(t) = L^{-1}(Y) = L^{-1}\left(\frac{1}{s-1}\right) - L^{-1}\left(\frac{1}{(s-1)^2}\right)$$

$$-L^{-1}\left(\frac{1}{(s-1)^3}\right) + 2L^{-1}\left(\frac{1}{(s-1)^6}\right)$$

$$y(t) = e^t - te^t - \frac{t^2 e^t}{2} + \frac{t^5 e^t}{60}$$

where we have used the first shift theorem.

Example 5: Find the general solution of the D.E. in the above Example 4.

Solution: Since the initial conditions are arbitrary assume $y(0) = a$, $y'(0) = b$, $y''(0) = c$.
Then

$$(s^2 Y - as^2 - bs - c) - 3(s^2 Y - as - b)$$

$$+3(sY - a) - Y = \frac{2}{(s-1)^3}$$

$$Y = \frac{as^2 + (b - 3a)s + (3a - 3b + c)}{(s-1)^3} + \frac{2}{(s-1)^6}$$

By partial fractions

$$Y = \frac{c_1}{(s-1)^3} + \frac{c_2}{(s-1)^2} + \frac{c_3}{(s-1)^1} + \frac{2}{(s-1)^6}$$

where c_1, c_2, c_3 are constants depending on a, b, c.
Applying I.T., and using first shift theorem

$$y(t) = c_1 \frac{t^2}{2} e^t + c_2 t e^t + c_3 e^t + \frac{t^5}{60} e^t.$$

EXERCISE

Use L.T. to solve each of the following I.V.P. consisting of a D.E. with I.C:

1. $y' - y = 0$, general solution

   Hint: Assume $y(0) = A =$ constant

Ans. $y = Ae^t$

2. $y' - y = e^{3t}$, $y(0) = 2$

Ans. $y = (3e^t + e^{3t})/2$

3. $y'' + y' = 0$, general solution

   Hint: Assume $y(0) = A$, $y'(0) = B$

Ans. $y = C + De^{-t}, C = A + B, D = -B$

4. $y'' + y = 2e^t$, $y(0) = 0$, $y'(0) = 2$

Ans. $y = e^t + \cos t + \sin t$

5. $y'' - 6y' + 9y = 0$, $y(0) = 2$, $y'(0) = 9$

Ans. $y = (3t + 2)e^{3t}$

6. $y'' + 4y = 9t$, $y(0) = 0$, $y'(0) = 7$

Ans. $y = \frac{9}{4}t + \frac{19}{8}\sin 2t$

7. $y'' + 7y' + 10y = 4e^{-3t}$, $y(0) = 0$,

   $y'(0) = -1$

Ans. $y = e^{-2t} - 2e^{-3t} + e^{-5t}$

8. $y'' - 8y' + 15y = 9te^{2t}$, $y(0) = 5$,

   $y'(0) = 10$

Ans. $y = 4e^{2t} + 3te^{2t} + 3e^{3t} - 2e^{5t}$

9. $y'' + y = t\cos 2t$, $y(0) = 0$, $y'(0) = 0$

Ans. $y = \frac{4}{9}\sin 2t - \frac{5}{9}\sin t - \frac{1}{3}t\cos 2t$

10. $y'' + n^2 y = a\sin(nt + \theta)$, $y(0) = y'(0) = 0$

Ans. $y = \frac{a}{2n^2}[\sin nt\cos\theta - nt\cos(nt + \theta)]$

11. $y'' + y = \sin t \cdot \sin 2t$, $y(0) = 1$, $y'(0) = 0$

Ans. $y = \frac{15}{16}\cos t + \frac{t}{4}\sin t + \frac{1}{16}\cos 3t$

12. $y'' + y = e^{-2t}\sin t$, $y(0) = 0$, $y'(0) = 0$

Ans. $y = \frac{1}{8}(\sin t - \cos t) + \frac{e^{-2t}}{8}(\sin t + \cos t)$

13. $y''' + 4y'' + 5y' + 2y = 10\cos t$

    $y(0) = 0$, $y'(0) = 0$, $y''(0) = 3$

Ans. $y = -e^{-2t} + 2e^{-t} - 2te^{-t} - \cos t + 2\sin t$

14. $y''' - y = e^t$

    $y(0) = y'(0) = y''(0) = 0$

Ans. $y = \frac{t}{3}e^t + \frac{e^{\frac{-1}{2t}}}{18}\left\{9\cos\frac{\sqrt{3}}{2}t + \frac{5\sqrt{3}}{2}\sin\frac{\sqrt{3}}{2}t\right\} - \frac{e^t}{2}$

15. $y^{iv} - 16y = 30\sin t$

    $y''(0) = 0$, $y'''(0) = -18$,
    $y''(\pi) = 0$, $y'''(\pi) = -18$

Ans. $y(t) = -\frac{1}{8}e^{2t} + \frac{1}{8}e^{-2t} + 7\sin t - \frac{9}{2}\sin 2t.$

12.10 APPLICATION OF LAPLACE TRANSFORM TO SYSTEM OF SIMULTANEOUS DIFFERENTIAL EQUATIONS

Laplace Transform can also be used to solve a system (or family) of m simultaneous ordinary differential equations in m dependent variables which are functions of the independent variable t. Consider a family of two simultaneous D.E. in the 2 dependent variables x and y which are functions of t.

$$a_1\frac{d^2x}{dt^2} + a_2\frac{d^2y}{dt^2} + a_3\frac{dx}{dt} + a_4\frac{dy}{dt} + a_5x + a_6y = R_1(t) \qquad (1)$$

$$b_1\frac{d^2x}{dt^2} + b_2\frac{d^2y}{dt^2} + b_3\frac{dx}{dt} + b_4\frac{dy}{dt} + b_5x + b_6y = R_2(t) \qquad (2)$$

Initial conditions:

$$x(0) = c_1,\ y(0) = c_2,\ x'(0) = c_3,\ y'(0) = c_4 \qquad (3)$$

Here $a_1, a_2, \ldots a_6, b_1, b_2, \ldots b_6, c_1, c_2, c_3, c_4$ are all constants and $R_1(t)$ and $R_2(t)$ are functions of t.

Method of solution to system of D.E

Step I. Apply Laplace transform on both sides of each of the two D.E. (1) and (2). This reduces (1) and (2) to two algebraic equations in $X(s)$ and $Y(s)$ where $X(s) = L\{x(t)\}$ and $Y(s) = L\{y(t)\}$.

$$a_1\{s^2X - sx(0) - x'(0)\} + a_2\{s^2Y - sy(0) - y'(0)\}$$
$$+a_3\{sX - x(0)\} + a_4\{sY - y(0)\}$$
$$+a_5X + a_6Y = Q_1(s) \quad (4)$$

$$b_1\{s^2X - sx(0) - x'(0)\} + b_2\{s^2Y - sy(0) - y'(0)\}$$
$$+b_3\{sX - x(0)\} + b_4\{sY - y(0)\}$$
$$+b_5X + b_6Y = Q_2(s) \quad (5)$$

Use the initial conditions (3) and substitute for $x(0)$, $x'(0)$, $y(0)$, $y'(0)$.

Step II. Solve (4) and (5) for $X(s)$ and $Y(s)$.

Step III. The required solution is obtained by taking the inverse Laplace transform of $X(s)$ and $Y(s)$ as

$$x(t) = L^{-1}\{X(s)\} \quad \text{and}$$
$$y(t) = L^{-1}\{Y(s)\}.$$

WORKED OUT EXAMPLES

Example 1: $2\frac{dx}{dt} + \frac{dy}{dt} - x - y = e^{-t}$

$$\frac{dx}{dt} + \frac{dy}{dt} + 2x + y = e^t, \quad x(0) = 2,\ y(0) = 1$$

Solution: Taking L.T. of the given D.E., we get

$$2[sX(s) - x(0)] + sY(s) - y(0) - X(s) - Y(s) = \frac{1}{s+1}$$

$$sX(s) - x(0) + sY(s) - y(0) + 2X(s) + Y(s) = \frac{1}{s-1}$$

Using I.C. $x(0) = 2$, $y(0) = 1$,

$$(2s-1)X(s) + (s-1)Y(s) = \frac{5s+6}{s+1}$$

$$(s+2)X(s) + (s+1)Y(s) = \frac{3s-2}{s-1}$$

Solving

$$X(s) = \frac{2(s+4)}{s^2+1}$$

$$Y(s) = \frac{s^3 - 12s^2 - s + 14}{(s^2+1)(s^2-1^2)} = \frac{s-13}{s^2+1} + \frac{1}{s^2-1}$$

Taking inverse Laplace transform, we have

$$x(t) = L^{-1}\{X(s)\} = L^{-1}\left\{\frac{2s+8}{s^2+1}\right\} = 2\cos t + 8\sin t$$

$$y(t) = L^{-1}\left\{\frac{s-13}{s^2+1} + \frac{1}{s^2-1}\right\}$$
$$= \cos t - 13\sin t + \sinh t$$

Example 2: $\frac{d^2x}{dt^2} - 3\frac{dx}{dt} - \frac{dy}{dt} + 2y = 14t + 3$

$$\frac{dx}{dt} - 3x + \frac{dy}{dt} = 1$$
$$x(0) = 0,\ x'(0) = 0,\ y(0) = 6.5$$

Solution: Taking L.T. of the given D.E., we have

$$[s^2X - sx(0) - x'(0)] - 3[sX - x(0)]$$
$$-[sY - y(0)] + 2Y = 14\frac{1}{s^2} + 3\frac{1}{s}$$

$$[sX - x(0)] - 3X + [sY - y(0)] = \frac{1}{s}$$

Use I.C.: $x(0) = 0$, $x'(0) = 0$, $y(0) = 6.5 = \frac{13}{2}$

$$s(s-3)X + (2-s)Y = \frac{28 + 6s - 13s^2}{2s^2}$$

$$(s-3)X + sY = \frac{2+13s}{2s}$$

Solving

$$Y(s) = \frac{13s^3 + 15s^2 - 6s - 28}{2s^2(s^2+s-2)}$$

$$X(s) = \frac{4(7-6s)}{s(s-3)(s+2)(s-1)}$$

Taking inverse Laplace transform

$$y(t) = L^{-1}\{Y(s)\} = L^{-1}\left\{\frac{13s^3 + 15s^2 - 6s - 28}{2s^2(s^2 + s - 2)}\right\}$$

$$= L^{-1}\left\{\frac{A}{s^2} + \frac{B}{s} + \frac{C}{s-1} + \frac{D}{s+2}\right\}$$

$$= L^{-1}\left\{\frac{7}{s^2} + \frac{5}{s} - \frac{1}{s-1} + \frac{5}{2}\frac{1}{s+2}\right\}$$

$$y(t) = 7t + 5 - e^{+t} + \frac{5}{2}e^{-2t}$$

Similarly,

$$x(t) = L^{-1}\left\{\frac{4(7-6s)}{s(s-3)(s+2)(s-1)}\right\}$$

$$= L^{-1}\left\{\frac{2}{s} - \frac{1}{2}\frac{1}{s-1} - \frac{1}{2}\frac{1}{s-3} - \frac{1}{s+2}\right\}$$

$$x(t) = 2 - \frac{1}{2}e^{t} - \frac{1}{2}e^{3t} - e^{-2t}$$

EXERCISE

Solve the following system of equations:

1. $\frac{dx}{dt} = 2x - 3y;\ \frac{dy}{dt} = y - 2x,\ x(0) = 8,$
$y(0) = 3$

Ans. $x(t) = 5e^{-t} + 3e^{4t};\ y(t) = 5e^{-t} - 2e^{4t}$

2. $\frac{dx}{dt} + y \sin t;\ \frac{dy}{dt} + x = \cos t,\ x(0) = 2,$
$y(0) = 0$

Ans. $x(t) = e^t + e^{-t} = 2\cosh t,$

$y(t) = \sin t - 2\sinh t$

3. $\frac{dx}{dt} - 6x + 3y = 8e^t;\ \frac{dy}{dt} - 2x - y = 4e^t,$
$x(0) = -1,\ y(0) = 0$

Ans. $x(t) = -2e^t + e^{4t};\ y(t) = -\frac{2}{3}e^t + \frac{2}{3}e^{4t}$

4. $\frac{2dx}{dt} + \frac{4dy}{dt} + x - y = 3e^t;$

$\frac{dx}{dt} + \frac{dy}{dt} + 2x + 2y = e^t,\ x(0) = 1,\ y(0) = 0$

Ans. $x(t) = e^{-2t} - te^t;\ y(t) = \frac{e^t}{3} - \frac{e^{-2t}}{3} + te^t$

5. $\frac{dx}{dt} - \frac{dy}{dt} - 2x + 2y = 1 - 2t$

$\frac{d^2x}{dt^2} + 2\dot{x} + x = 2,\ x(0) = y(0) = x'(0) = 0$

Ans. $x = 2 - 2e^{-t}(1 + t);$

$y(t) = 2 - t - 2(1 + t)e^{-t}$

6. $\frac{dx}{dt} + 2\frac{d^2y}{dt^2} = e^{-t}$

$\frac{dx}{dt} + 2x - y = 1$

$x(0) = y(0) = y'(0) = 0$

Ans. $x(t) = 1 + e^{-t} - e^{-at} - e^{-bt}$

$y(t) = 1 + e^{-t} - be^{-at} - ae^{-bt}$

where $a = \frac{1}{2}(2 - \sqrt{2}),\ b = \frac{1}{2}(2 + \sqrt{2}).$

7. $\frac{3dx}{dt} + \frac{dy}{dt} + 2x = 1;\ \frac{dx}{dt} + 4\frac{dy}{dt} + 3y = 0$

$x(0) = 3,\ y(0) = 0$

Ans. $x = (5 - 2e^{-t} - 3e^{\frac{-6t}{11}})/10;$

$y(t) = (e^{-t} - e^{\frac{-6t}{11}})/5$

8. $\frac{dx}{dt} = y + e^t;\ \frac{dy}{dt} = \sin t - x,$

$x(0) = 1,\ y(0) = 0$

Ans. $x(t) = (e^t + \cos t + 2\sin t - t\cos t)/2$

$y(t) = (t\sin t - e^t + \cos t - \sin t)/2$

9. $\frac{d^2x}{dt^2} = 2x + 3y + e^{2t};\ \frac{d^2y}{dt^2} = -x - 2y$

$x(0) = y(0) = 1,\ x'(0) = y'(0) = 0$

Ans. $x(t) = \frac{1}{4}(3e^t + 7e^{-t}) - \frac{1}{10}(19\cos t -$
$2\sin t) + \frac{2}{5}e^{2t}$

$y(t) = \frac{-1}{12}(3e^t + 7e^{-t}) + \frac{1}{10}(19\cos t -$
$2\sin t) - \frac{1}{15}e^{2t}$

10. $\frac{d^2x}{dt^2} + \frac{dy}{dt} + 3x = 15e^{-t}$

$\frac{d^2y}{dt^2} - \frac{4dx}{dt} + 3y = 15\sin 2t$

$x(0) = 35,\ x'(0) = -48,\ y(0) = 27,$

$y'(0) = -55$

Ans. $x(t) = 30\cos t - 15\sin 3t + 3e^{-t} + 2\cos 2t$

$y(t) = 30\cos 3t - 60\sin t - 3e^{-t} + \sin 2t.$

12.11 TABLE OF GENERAL PROPERTIES OF LAPLACE TRANSFORM

$$F(s) = \int_0^{\infty} e^{-st} f(t)dt$$

S.No.	Name	Laplace transform	Inverse Laplace transform
1.	Definition	$L\{f(t)\} = F(s)$	$L^{-1}\{F(s)\} = f(t)$
2.	Linearity	$af_1(t) + bf_2(t)$	$aF_1(S) + bF_2(s)$
3.	Change of Scale	$f(at)$	$\frac{1}{a}F\left(\frac{s}{a}\right)$
4.	First Shifting (s-shift) theorem	$e^{at} f(t)$	$F(s-a)$
5.	Second Shifting (t-shift) Theorem	$u(t-a) = \begin{cases} f(t-a), & t>a \\ 0, & t<a \end{cases}$	$e^{-as} F(s)$
6.	Derivative (Multiplication by s)	$f'(t)$	$sF(s) - f(0)$
7.	Second Derivative (Multiplication by s^2)	$f''(t)$	$s^2F(s) - sf(0) - f'(0)$
8.	nth Derivative (Multiplication by s^n)	$f^{(n)}(t)$	$s^n F(s) - s^{(n-1)} f(0) - s^{(n-2)} f'(0) \ldots - f^{n-1}(0)$
9.	Integral: Division by s	$\int_0^t f(u)du$	$\frac{F(s)}{s}$
10.	Multiple Integral: Division by s^n	$\int_0^t \cdots \int_0^t f(u)du^n = \int_0^t \frac{(t-u)^{n-1}}{(n-1)!} f(u)du$	$\frac{F(s)}{s^n}$
11.	Multiplication by t	$-tf(t)$	$F'(s)$
12.	Multiplication by t^2	$t^2 f(t)$	$F''(s)$
13.	Multiplication by t^n	$(-1)^n t^n f(t)$	$F^{(n)}(s)$
14.	Division by t	$\frac{f(t)}{t}$	$\int_s^{\infty} F(u)du$
15.	Convolution	$f(t) * g(t) = \int_0^t f(u)g(t-u)du = \int_0^t f(t-u)g(u)du = L^{-1}\{F(s)G(s)\}$	$F(s)G(s) = L(f * g)$
16.	f-periodic with Period p	$f(t) = f(t+p)$	$\frac{1}{1-e^{-sp}} \cdot \int_0^p e^{-su} f(u)du$

12.12 TABLE OF SOME LAPLACE TRANSFORMS

S.No.	Laplace transform	Inverse Laplace transform
1.	$L\{f(t)\} = F(s)$	$L^{-1}\{F(s)\} = f(t)$
2.	1	$\frac{1}{s}$
3.	t	$\frac{1}{s^2}$
4.	t^2	$\frac{2!}{s^3}$
5.	$t^n, n = 0, 1, 2, \ldots\ldots$	$\frac{n!}{s^{n+1}}$
6.	t^a, (a positive)	$\frac{\Gamma(a+1)}{s^{a+1}}$
7.	e^{at}	$\frac{1}{s-a}$
8.	$\frac{t^{n-1}e^{at}}{(n-1)!}$	$\frac{1}{(s-a)}^{n}$, $n = 1, 2, 3 \ldots\ldots,$

(Contd.)

S.No.	Laplace transform	Inverse Laplace transform
9.	$\frac{t^{k-1}e^{at}}{\Gamma(k)}$	$\frac{1}{(s-a)^k}, k > 0$
10.	$\sin at$	$\frac{a}{s^2+a^2}$
11.	$\cos at$	$\frac{s}{s^2+a^2}$
12.	$e^{bt} \sin at$	$\frac{a}{(s-b)^2+a^2}$
13.	$e^{bt} \cos at$	$\frac{s-b}{(s-b)^2+a^2}$
14.	$\sinh at$	$\frac{a}{s^2-a^2}$
15.	$\cosh at$	$\frac{s}{s^2-a^2}$
16.	$e^{bt} \sinh at$	$\frac{a}{(s-b)^2-a^2}$
17.	$e^{bt} \cosh at$	$\frac{s-b}{(s-b)^2-a^2}$
18.	$u(t-a)$	$\frac{e^{-as}}{s}$
19.	$f(t-a) \cdot u(t-a)$	$e^{-as}F(s)$
20.	$\frac{e^{bt}-e^{at}}{b-a}$	$\frac{1}{(s-a)(s-b)}, a \neq b$
21.	$\frac{be^{bt}-ae^{at}}{b-a}$	$\frac{s}{(s-a)(s-b)}, a \neq b$
22.	$\frac{\sin at - at \cos at}{2a^3}$	$\frac{1}{[(s^2+a^2)]^2}$
23.	$\frac{t \sin at}{2a}$	$\frac{s}{[(s^2+a^2)]^2}$
24.	$\frac{\sin at + at \cos at}{2a}$	$\frac{s^2}{[(s^2+a^2)]^2}$
25.	$\cos at - \frac{1}{2} at \sin at$	$\frac{s^3}{[(s^2+a^2)]^2}$
26.	$t \cos at$	$\frac{s^2-a^2}{[(s^2+a^2)]^2}$
27.	$\frac{at \cosh at - \sinh at}{2a^3}$	$\frac{1}{[(s^2-a^2)]^2}$
28.	$\frac{t \sinh at}{2a}$	$\frac{s}{[(s^2-a^2)]^2}$
29.	$(\sinh at + at \cosh at)/2a$	$\frac{s^2}{[(s^2-a^2)]^2}$
30.	$\cosh at + \frac{1}{2} at \sinh at$	$\frac{s^3}{[(s^2-a^2)]^2}$
31.	$t \cosh at$	$\frac{s^2+a^2}{[(s^2-a^2)]^2}$
32.	$\frac{t^2 \sin at}{2a}$	$\frac{3s^2-a^2}{(s^2+a^2)^3}$
33.	$\frac{1}{2} t^2 \cos at$	$\frac{s^3-3a^2s}{(s^2+a^2)^3}$
34.	$\frac{1}{6} t^3 \cos at$	$\frac{s^4-6a^2s^2+a^4}{(s^2+a^2)^4}$
35.	$\frac{t^3 \sin at}{24a}$	$\frac{s^3-a^2s}{(s^2+a^2)^4}$
36.	$\frac{e^{bt}-e^{at}}{t}$	$\log \left\| \frac{s-a}{s-b} \right\|$
37.	$\frac{\sin t}{t}$	$\tan^{-1} \frac{1}{s}$
38.	$\Gamma^1(1) - \log t$	$\frac{\log s}{s}$

PART–IV
LINEAR ALGEBRA AND VECTOR CALCULUS

Chapter 13

Matrices

INTRODUCTION

The term matrix was apparently coined by Sylvester about 1850, but was introduced first by Cayley in 1860. By a 'matrix' we mean an "arrangement" or "rectangular array" of numbers. The elegant "shorthand" representation of an array of many numbers as a single object and perform calculations makes matrices very useful. Matrices (plural of matrix) find applications in solution of system of linear equations, probability, mathematical economics, quantum mechanics, electrical networks, curve fitting, transportation problems, frameworks in mechanics. Matrices are easily amenable for computers.

A brief revision of matrices, types, properties is presented.

A matrix is a rectangular array of $m \cdot n$ numbers (or functions) arranged in m rows (horizontal lines) and n columns l (vertical lines). These numbers known as elements or entries are enclosed in brackets [ ] or ( ) or ‖ ‖.

The order of such matrix is $m \times n$ and is said to be a rectangular matrix.

Notation

Elements of a matrix are located by the double subscript ij where i denotes the row and j the column.

Null or Zero matrix is a matrix with all elements zero.

Equality

Two matrices A and B are equal if they are of the same order and $a_{ij} = b_{ij}$, for every i, j.

Sum (difference)

$C = A \pm B$ where $c_{ij} = a_{ij} \pm b_{ij}$ (and A and B are conformable i.e., of the same order). Scalar multiplication: $C = kA$ where $c_{ij} = ka_{ij}$ i.e., every element of A is multiplied by constant k.

Matrix multiplication:

$C_{m\times n} = A_{m\times p} B_{p\times n}$ where $c_{ij} = \sum_{k=1}^{p} a_{ik}\, b_{kj}$

Transpose of a matrix $A_{m\times n}$ is denoted by $A^T_{n\times m}$ obtained by interchanging rows and columns.

Result: $(AB)^T = B^T A^T$.

Square matrix

$A: m = n$, when the number of rows equals to the number of columns, known as n-square matrix.

The elements a_{ii} are known as diagonal elements

Trace: $\sum_{i=1}^{n} a_{ii}$ = sum of the diagonal elements.

Singular matrix: if $|A| = 0$

Non-singular matrix: if $|A| \neq 0$

Upper triangular matrix A: $a_{ij} = 0$ for $i > j$

Lower triangular matrix A: $a_{ij} = 0$ for $i < j$

Diagonal matrix A: $a_{ij} = 0$ when $i \neq j$

Scalar matrix A: a diagonal matrix with $a_{ii} = k$ for every i and k is a constant.

Identity matrix: is a scalar matrix with $k = 1$

i.e., $$I_3 = \begin{bmatrix} 1 & 0 & 0 \\ 0 & 1 & 0 \\ 0 & 0 & 1 \end{bmatrix}$$

Note: All the above definitions are only for square matrices.

Row matrix (vector) is a matrix having only one row.

Column matrix (vector) is a matrix having only one column.

Matrix addition and multiplication is associative but not (necessarily) commutative.

i.e., $A+(B+C)=(A+B)+C$ and $A(BC)=(AB)C$.

Distributive: $A(B+C)=AB=AC$.

Power of a matrix: A^n is a matrix obtained by multiplying A by itself n times.

13.1 INVERSE OF A MATRIX

Consider only square matrices.

Inverse of a n-square matrix A is denoted by A^{-1} and is defined such that

$$AA^{-1}=A^{-1}A=I$$

where I is $n\times n$ unit matrix.

Result 1: Inverse of A exists only if $|A|\neq 0$ i.e., is A is non-singular.

Result 2: Inverse of a matrix is unique.
If B, C are two inverses of A then $(CA)B=C(AB)$, $IB=CI$ i.e., $B=C$, so inverse is unique.

Result 3: Inverse of a product is the product of inverses in the reverse order
i.e., $(AB)^{-1}=B^{-1}A^{-1}$
since $(AB)(B^{-1}A^{-1})=A(BB^{-1})A^{-1}=AIA^{-1}$ $=AA^{-1}=I$.

Result 4: For a diagonal matrix D with d_{ii} as diagonal elements, D^{-1} is a diagonal matrix with reciprocals $1/d_{ii}$ as the diagonal elements.

Result 5: Transposition and inverse are commutative i.e.,
$(A^{-1})^T=(A^T)^{-1}$.

Taking transpose of $AA^{-1}=A^{-1}A=I_n$
$(A^{-1})^T A^T=A^T(A^{-1})^T=I^T=I$ i.e.,
$(A^{-1})^T$ is the inverse of A^T or $(A^{-1})^T=(A^T)^{-1}$.

Result 6: $(A^{-1})^{-1}=A$.

Taking inverse of $(AA^{-1})=I$,
$(AA^{-1})^{-1}=(A^{-1})^{-1}A^{-1}=I^{-1}=I=A\,A^{-1}$.
Thus $A=(A^{-1})^{-1}$.

Inverse by Adjoint Matrix

Minor M_{ij} of an element a_{ij} of a $n\times n$ square matrix A is the determinant of the $(n-1)$ square matrix of A obtained by deleting the ith row and jth column from A.

Cofactor A_{ij} of a_{ij} of A is a signed minor

$$\text{i.e.,}\quad A_{ij}=(-1)^{i+j}M_{ij}$$

Adjoint of a Matrix A

Adjoint of a matrix is denoted by adj A is the transpose of a n- square matrix $[A_{ij}]$ where the elements A_{ij} are the cofactors of a_{ij} of A.

$$\text{i.e.,}\quad \text{adj } A=\begin{bmatrix} A_{11} & A_{12} & A_{13}\cdots A_{1n} \\ A_{21} & A_{22} & A_{23}\cdots A_{2n} \\ \cdots\cdots & \cdots\cdots & \cdots\cdots\cdots\cdots \\ A_{n1} & A_{n2} & A_{n3}\cdots A_{nn}\end{bmatrix}^T=$$

$$\text{adj } A=\begin{bmatrix} A_{11} & A_{21} & A_{31}\cdots A_{n1} \\ A_{12} & A_{22} & A_{32}\cdots A_{n2} \\ \cdots\cdots & \cdots\cdots & \cdots\cdots\cdots\cdots \\ A_{1n} & A_{2n} & A_{3n}\cdots A_{nn}\end{bmatrix}$$

Result: adj $(AB)=$ (adj A)(adj B)

Inverse of a matrix can be calculated by several methods.

Inverse from the adjoint:

$$A^{-1}=\frac{\text{adj } A}{|A|}.$$

WORKED OUT EXAMPLES

Inverse of a matrix

Example: Find the adjoint and inverse of

$$A=\begin{bmatrix} 2 & 3 & 4 \\ 4 & 3 & 1 \\ 1 & 2 & 4\end{bmatrix}.$$

Solution:

$$\text{Adjoint of } A=\begin{bmatrix} A_{11} & A_{12} & A_{13} \\ A_{21} & A_{22} & A_{23} \\ A_{31} & A_{32} & A_{33}\end{bmatrix}^T$$

where A_{ij} are the cofactors of the element a_{ij}. Thus minors of a_{ij} are

$$M_{11}=\begin{vmatrix} 3 & 1 \\ 2 & 4\end{vmatrix}=10,\quad M_{12}=\begin{vmatrix} 4 & 1 \\ 1 & 4\end{vmatrix}=15$$

Similarly,

$$M_{13}=\begin{vmatrix} 4 & 3 \\ 1 & 2\end{vmatrix}=5,\quad M_{21}=\begin{vmatrix} 3 & 4 \\ 2 & 4\end{vmatrix}=4,$$

$$M_{22} = \begin{vmatrix} 2 & 4 \\ 1 & 4 \end{vmatrix} = 4, \quad M_{23} = \begin{vmatrix} 2 & 3 \\ 1 & 2 \end{vmatrix} = 1,$$

$$M_{31} = \begin{vmatrix} 3 & 4 \\ 3 & 1 \end{vmatrix} = -9, \quad M_{32} = \begin{vmatrix} 2 & 4 \\ 4 & 1 \end{vmatrix} = 14$$

$$M_{33} = \begin{vmatrix} 2 & 3 \\ 4 & 3 \end{vmatrix} = -6.$$

Cofactors $A_{ij} = (-1)^{i+j} M_{ij}$

$$\text{Adjoint of } A = \begin{bmatrix} 10 & -15 & 5 \\ -4 & 4 & -1 \\ -9 & +14 & -6 \end{bmatrix}^T$$

$$= \begin{bmatrix} 10 & -4 & -9 \\ -15 & 4 & +14 \\ 5 & -1 & -6 \end{bmatrix}$$

$$|A| = 2(12-2) - 3(16-1) + 4(8-3)$$

$$= 20 - 45 + 20 = 40 - 45 = -5$$

$$A^{-1} = \frac{1}{|A|} \text{ adj } A = \frac{1}{-5} \begin{bmatrix} 10 & -4 & -9 \\ -15 & 4 & 14 \\ 5 & -1 & -6 \end{bmatrix}$$

or

$$A^{-1} = \frac{1}{5} \begin{bmatrix} -10 & 4 & 9 \\ 15 & -4 & -14 \\ -5 & 1 & 6 \end{bmatrix}.$$

EXERCISE

Inverse of a matrix

Find the inverse of the matrix A, by adjoint matrix:

1. $\begin{bmatrix} 1 & 1 & 3 \\ 1 & 3 & -3 \\ -2 & -4 & -4 \end{bmatrix}$

Ans. $|A| = -8$

$$\text{adj } A = \begin{bmatrix} -24 & -8 & -12 \\ 10 & 2 & 6 \\ 2 & 2 & 2 \end{bmatrix}$$

$$A^{-1} = \frac{1}{4} \begin{bmatrix} 12 & 4 & 6 \\ -5 & -1 & -3 \\ -1 & -1 & -1 \end{bmatrix}$$

2. $\begin{bmatrix} 1 & 2 & 3 \\ 1 & 3 & 4 \\ 1 & 4 & 3 \end{bmatrix}$

Ans. $|A| = -2$

$$\text{adj } A = \begin{bmatrix} -7 & 6 & -1 \\ 1 & 0 & -1 \\ 1 & -2 & 1 \end{bmatrix}$$

$$A^{-1} = \frac{1}{2} \begin{bmatrix} 7 & -6 & 1 \\ -1 & 0 & 1 \\ -1 & 2 & -1 \end{bmatrix}$$

3. $\begin{bmatrix} -1 & 1 & 2 \\ 3 & -1 & 1 \\ -1 & 3 & 4 \end{bmatrix}$

Ans. $|A| = 10$

$$\text{adj A} = \begin{bmatrix} -7 & 2 & 3 \\ -13 & -2 & 7 \\ 8 & 2 & -2 \end{bmatrix}, A^{-1} = \frac{1}{|A|} \text{ adj } A$$

$$A^{-1} = \begin{bmatrix} -0.7 & 0.2 & 0.3 \\ -1.3 & -0.2 & 0.7 \\ 0.8 & 0.2 & -0.2 \end{bmatrix}$$

4. $\begin{bmatrix} 7 & 6 & 2 \\ -1 & 2 & 4 \\ 3 & 3 & 8 \end{bmatrix}$

Ans. $|A| = 130,$

$$\text{adj } A = \begin{bmatrix} 4 & -42 & 20 \\ 20 & 50 & -30 \\ -9 & -3 & 20 \end{bmatrix}$$

13.2 RANK OF A MATRIX

Let A be a rectangular matrix of order $m \times n$.

Submatrix

Submatrix of a matrix A is any matrix obtained from A by omitting some rows and columns in A.

A is a submatrix of itself (obtained by deleting zero rows and columns).

Rank

Rank of a matrix A is the positive integer r such that there exists at least one r-rowed square matrix with non-vanishing determinant while every $(r+1)$ or more rowed matrices have vanishing determinants.

Thus rank of a matrix is the largest order of a non-zero minor of matrix.

Rank of A is denoted by $r(A)$.

Result: Rank of A and A^T is same.

Note 1: Rank of a null matrix is zero.

Note 2: For a rectangular matrix A of order $m \times n$, rank of $A \leq \min(m, n)$ i.e., rank can not exceed the smaller of m and n.

Note 3: For a n-square matrix, if rank $= n$ then $|A| \neq 0$ i.e., A is non-singular.

Note 4: For any square matrix, if rank $< n$, then $|A| = 0$ i.e., A is singular.

Elementary Row Transformations (Operation) on a Matrix

1. R_{ij}: Interchange of the ith and jth rows.
2. $R_{i(k)}$: Multiplication of every element of ith row by a non-zero scalar k.
3. $R_{ij(k)}$: Addition to the elements of ith row, of k times the corresponding elements of the jth row.

In a similar way, elementary column transformations (operations) are denoted by $C_{ij}, C_{i(k)}, C_{ij(k)}$ where the row in the above definitions is replaced by column.

WORKED OUT EXAMPLES

Inverse by Gauss-Jordan

Example: Find the inverse of A by Gauss-Jordan method where

$$A = \begin{bmatrix} 1 & 2 & 3 \\ 2 & 4 & 5 \\ 3 & 5 & 6 \end{bmatrix}.$$

Solution: Consider $A|I$ and apply elementary row operations on both A and I until A gets transformed to I.

$$\left[\begin{array}{ccc:ccc} 1 & 2 & 3 & 1 & 0 & 0 \\ 2 & 4 & 5 & 0 & 1 & 0 \\ 3 & 5 & 6 & 0 & 0 & 1 \end{array}\right] \underset{\substack{R_{21(-2)} \\ R_{31(-3)}}}{\sim} \left[\begin{array}{ccc:ccc} 1 & 2 & 3 & 1 & 0 & 0 \\ 0 & 0 & -1 & -2 & 1 & 0 \\ 0 & -1 & -3 & -3 & 0 & 1 \end{array}\right]$$

$$\underset{\substack{R_{23} \\ R_{2(-1)} \\ R_{3(-1)}}}{\sim} \left[\begin{array}{ccc:ccc} 1 & 2 & 3 & 1 & 0 & 0 \\ 0 & 1 & 3 & 3 & 0 & -1 \\ 0 & 0 & 1 & 2 & -1 & 0 \end{array}\right]$$

$$\underset{\substack{R_{23(-3)} \\ R_{13(-3)}}}{\sim} \left[\begin{array}{ccc:ccc} 1 & 2 & 0 & -5 & 3 & 0 \\ 0 & 1 & 0 & -3 & 3 & -1 \\ 0 & 0 & 1 & 2 & -1 & 0 \end{array}\right]$$

$$\underset{R_{12(-2)}}{\sim} \left[\begin{array}{ccc:ccc} 1 & 0 & 0 & 1 & -3 & 2 \\ 0 & 1 & 0 & -3 & 3 & -1 \\ 0 & 0 & 1 & 2 & -1 & 0 \end{array}\right] = [I|A^{-1}]$$

Thus $A^{-1} = \begin{bmatrix} 1 & -3 & 2 \\ -3 & 3 & -1 \\ 2 & -1 & 0 \end{bmatrix}$.

EXERCISE

By Gauss-Jordan elimination

1. $\begin{bmatrix} 1 & 3 & 3 \\ 1 & 4 & 3 \\ 1 & 3 & 4 \end{bmatrix}$

Ans. $A^{-1} = \begin{bmatrix} 7 & -3 & -3 \\ -1 & 1 & 0 \\ -1 & 0 & 1 \end{bmatrix}$

2. $\begin{bmatrix} 2 & 4 & 3 & 2 \\ 3 & 6 & 5 & 2 \\ 2 & 5 & 2 & -3 \\ 4 & 5 & 14 & 14 \end{bmatrix}$

Ans. $A^{-1} = \begin{bmatrix} -23 & 29 & \frac{-64}{5} & \frac{-18}{5} \\ 10 & -12 & \frac{26}{5} & \frac{7}{5} \\ 1 & -2 & \frac{6}{5} & \frac{2}{5} \\ 2 & -2 & \frac{3}{5} & \frac{1}{5} \end{bmatrix}$

3. $\begin{bmatrix} 0 & 1 & 3 \\ 1 & 2 & 3 \\ 3 & 1 & 1 \end{bmatrix}$

Ans. $A^{-1} = \frac{1}{2}\begin{bmatrix} 1 & -1 & 1 \\ -8 & 6 & -2 \\ 5 & -3 & 1 \end{bmatrix}$

4. $\begin{bmatrix} 9 & 7 & 3 \\ 5 & -1 & 4 \\ 3 & 4 & 1 \end{bmatrix}$

Ans. $A^{-1} = -\frac{1}{35}\begin{bmatrix} -17 & 5 & 31 \\ 7 & 0 & -21 \\ 23 & -15 & -44 \end{bmatrix}$

5. $\begin{bmatrix} -1 & -3 & 3 & -1 \\ 1 & 1 & -1 & 0 \\ 2 & -5 & 2 & -3 \\ -1 & 1 & 0 & 1 \end{bmatrix}$

Ans. $A^{-1} = \begin{bmatrix} 0 & 2 & 1 & 3 \\ 1 & 1 & -1 & -2 \\ 1 & 2 & 0 & 1 \\ -1 & 1 & 2 & 6 \end{bmatrix}$

6. $\begin{bmatrix} 1 & 2 & -1 \\ -1 & 1 & 2 \\ 2 & -1 & 1 \end{bmatrix}$

Ans. $A^{-1} = \frac{1}{14}\begin{bmatrix} 3 & -1 & 5 \\ 5 & 3 & -1 \\ -1 & 5 & 3 \end{bmatrix}$

7. $\begin{bmatrix} 4 & -1 & 1 \\ 2 & 0 & -1 \\ 1 & -1 & 3 \end{bmatrix}$

Ans. $A^{-1} = \begin{bmatrix} -1 & 2 & 1 \\ -7 & 11 & 6 \\ -2 & 3 & 2 \end{bmatrix}$

Equivalent matrices

Two matrices A and B are said to be equivalent, denoted by $A \sim B$, if one matrix say A can be obtained from B by a sequence of elementary transformations.

Row-equivalence

Two matrices A and B are said to be row-equivalent if A can be reduced to B by a sequence of elementary *row* transformations or vice versa.

Determination of Rank of a Matrix A

Let A be a rectangular matrix of order $m \times n$.

I. Enumeration: Evaluate all the minors such that a minor of r is non-zero and every minor of $(r+1)$ or more is zero:

Note: This is impracticable for matrices of higher order.

II. Apply only **elementary row operations** on A. Then the number of non-zero rows is the rank of A.

III. Normal form N of a matrix A of rank r is one of the froms

$$N = I_r, \begin{bmatrix} I_r & 0 \\ 0 & 0 \end{bmatrix}, \quad [I_r \quad 0], \begin{bmatrix} I_r \\ 0 \end{bmatrix}$$

where I_r is an identity matrix of order r. By the application of both elementary row and column operations, a matrix of rank r can be reduced to normal form. Then the rank of A is r.

IV. Echelon Form.* Row Reduced Echelon form: The number of non-zero rows in an Echelon form is the rank.

Result: Equivalent matrices have the same order and same rank because elementary transformations do not alter (effect) its order and rank.

Worked Out Examples

Rank of a matrix

Determine the rank of the following matrices:

Example 1: $A = \begin{bmatrix} 4 & 2 & 3 \\ 8 & 4 & 6 \\ -2 & -1 & -1.5 \end{bmatrix}$

Solution: Rank of $A \le 3$ since A is of 3rd order.
$|A| = 4(-6+6) - 2(-12+12) + 3(-8+8) = 0$

Since $|A| = 0$, rank of $A < 3$ i.e., $r(A) \le 2$

Consider the determinants of 2nd order submatrices

$$\begin{vmatrix} 4 & 2 \\ 8 & 4 \end{vmatrix} = 0, \quad \begin{vmatrix} 2 & 3 \\ 4 & 6 \end{vmatrix} = 0, \quad \begin{vmatrix} 4 & 3 \\ 8 & 6 \end{vmatrix} = 0,$$

$$\begin{vmatrix} 4 & 2 \\ -2 & -1 \end{vmatrix} = 0, \begin{vmatrix} 4 & 3 \\ -1 & -1.5 \end{vmatrix} = 0, \begin{vmatrix} 4 & 3 \\ -2 & -1.5 \end{vmatrix} = 0,$$

Since all 2nd order submatrices have zero determinants i.e., 2nd order minors are all zero. So $r(A) < 2$.
Since A is a non-zero matrix $r(A) > 0$.
Thus the rank of A is one.

Aliter: Apply elementary row operations on A

$$\begin{matrix} R_{21(-2)} \\ R_{31(\frac{1}{2})} \end{matrix} \sim \begin{bmatrix} 4 & 2 & 3 \\ 0 & 0 & 0 \\ 0 & 0 & 0 \end{bmatrix}$$

The number of non-zero rows is one. So the rank of A is one.

*A matrix $A = (a_{ij})$ is an echelon matrix or is said to be in echelon form, if the number of zeros preceding the first non-zero entry (known as distinguished elements) of a row increases row by row until only zero rows remain.
In row reduced echelon matrix, the distinguished elements are unity and are the only non-zero entry in their respective columns.

Example 2: $A = \begin{bmatrix} 1 & 5 & 4 \\ 0 & 3 & 2 \\ 2 & 3 & 10 \end{bmatrix}, B = \begin{bmatrix} 1 & 1 & 1 \\ 2 & 2 & 2 \\ 3 & 3 & 3 \end{bmatrix}$

Find rank of A, rank of B, rank of $A+B$, rank of AB and rank of BA.

Solution: $A = \begin{bmatrix} 1 & 5 & 4 \\ 0 & 3 & 2 \\ 2 & 13 & 10 \end{bmatrix} R_{31(-2)} \sim \begin{bmatrix} 1 & 5 & 4 \\ 0 & 3 & 2 \\ 0 & 3 & 2 \end{bmatrix}$

$$R_{32(-1)} \sim \begin{bmatrix} 1 & 5 & 4 \\ 0 & 3 & 2 \\ 0 & 0 & 0 \end{bmatrix}$$

Rank of A is 2 since the number of non-zero rows is 2.

$$B = \begin{bmatrix} 1 & 1 & 1 \\ 2 & 2 & 2 \\ 3 & 3 & 3 \end{bmatrix} \begin{matrix} R_{21(-1)} \\ R_{31(-3)} \end{matrix} \sim \begin{bmatrix} 1 & 1 & 1 \\ 0 & 0 & 0 \\ 0 & 0 & 0 \end{bmatrix},$$

$\therefore \quad r(B) = 1$

$$A+B = \begin{bmatrix} 2 & 6 & 5 \\ 2 & 5 & 4 \\ 5 & 16 & 13 \end{bmatrix} \begin{matrix} R_{21(-2)} \\ R_{31(-\frac{5}{2})} \end{matrix} \sim \begin{bmatrix} 2 & 6 & 5 \\ 0 & -7 & -6 \\ 0 & -1 & \frac{1}{2} \end{bmatrix},$$

$r(A+B) = 3$

$$AB = \begin{bmatrix} 23 & 23 & 23 \\ 12 & 12 & 12 \\ 58 & 58 & 58 \end{bmatrix} \begin{matrix} R_{1(\frac{1}{23})} \\ R_{21(-12)} \\ R_{31(-58)} \end{matrix} \sim \begin{bmatrix} 1 & 1 & 1 \\ 0 & 0 & 0 \\ 0 & 0 & 0 \end{bmatrix},$$

$r(AB) = 1$

$$BA = \begin{bmatrix} 3 & 21 & 16 \\ 6 & 42 & 32 \\ 9 & 63 & 48 \end{bmatrix} \begin{matrix} R_{21(-2)} \\ R_{31(-3)} \end{matrix} \sim \begin{bmatrix} 1 & 1 & 1 \\ 0 & 0 & 0 \\ 0 & 0 & 0 \end{bmatrix},$$

$r(BA) = 1.$

Note: Rank of product $\leq$ rank of either.

Example 3:

$$A = \begin{bmatrix} 1 & 2 & -2 & 3 \\ 2 & 5 & -4 & 6 \\ -1 & -3 & 2 & -2 \\ 2 & 4 & -1 & 6 \end{bmatrix}_{4\times 4}$$

$$\begin{matrix} R_{21(-2)} \\ R_{31(1)} \\ R_{41(-2)} \end{matrix} \sim \begin{bmatrix} 1 & 2 & -2 & 3 \\ 0 & 1 & 0 & 0 \\ 0 & -1 & 0 & 1 \\ 0 & 0 & 3 & 0 \end{bmatrix}$$

$$\begin{matrix} R_{32(1)} \\ R_{34} \\ R_{3(\frac{1}{3})} \end{matrix} \sim \begin{bmatrix} 1 & 2 & -2 & 3 \\ 0 & 1 & 0 & 0 \\ 0 & 0 & 1 & 0 \\ 0 & 0 & 0 & 1 \end{bmatrix}$$

$\therefore \quad r(A) = 4$

Example 4:

$$A = \begin{bmatrix} 3 & 4 & 5 & 6 & 7 \\ 4 & 5 & 6 & 7 & 8 \\ 5 & 6 & 7 & 8 & 9 \\ 10 & 11 & 12 & 13 & 14 \\ 15 & 16 & 17 & 18 & 19 \end{bmatrix}_{5\times 5}$$

$$\begin{matrix} R_{12(-1)} \\ R_{1(-1)} \end{matrix} \sim \begin{bmatrix} 1 & 1 & 1 & 1 & 1 \\ 4 & 5 & 6 & 7 & 8 \\ 5 & 6 & 7 & 8 & 9 \\ 10 & 11 & 12 & 13 & 14 \\ 15 & 16 & 17 & 18 & 19 \end{bmatrix}$$

$$\begin{matrix} R_{21(-4)} \\ R_{31(-5)} \\ R_{41(-10)} \\ R_{51(-15)} \end{matrix} \sim \begin{bmatrix} 1 & 1 & 1 & 1 & 1 \\ 0 & 1 & 2 & 3 & 4 \\ 0 & 1 & 2 & 3 & 4 \\ 0 & 1 & 2 & 3 & 4 \\ 0 & 1 & 2 & 3 & 4 \end{bmatrix} \begin{matrix} R_{32(-1)} \\ R_{42(-1)} \\ R_{52(-1)} \end{matrix} \sim \begin{bmatrix} 1 & 1 & 1 & 1 & 1 \\ 0 & 1 & 2 & 3 & 4 \\ 0 & 0 & 0 & 0 & 0 \\ 0 & 0 & 0 & 0 & 0 \\ 0 & 0 & 0 & 0 & 0 \end{bmatrix}$$

Rank of A is 2 since the number of non-zero rows is 2.

Example 5: Determine the values of b such that the rank of A is 3.

Solution:

$$A = \begin{bmatrix} 1 & 1 & -1 & 0 \\ 4 & 4 & -3 & 1 \\ b & 2 & 2 & 2 \\ 9 & 9 & b & 3 \end{bmatrix}$$

$$\begin{matrix} R_{21(-4)} \\ R_{31(-2)} \\ R_{41(-9)} \end{matrix} \sim \begin{bmatrix} 1 & 1 & -1 & 0 \\ 0 & 0 & 1 & 1 \\ b-2 & 0 & 4 & 2 \\ 0 & 0 & b+9 & 3 \end{bmatrix}$$

$$\begin{matrix} R_{32(-4)} \\ R_{42(-3)} \end{matrix} \sim \begin{bmatrix} 1 & 1 & -1 & 0 \\ 0 & 0 & 1 & 1 \\ b-2 & 0 & 0 & -2 \\ 0 & 0 & b+6 & 0 \end{bmatrix}$$

$$R_{43} \sim \begin{bmatrix} 1 & 1 & -1 & 0 \\ 0 & 0 & 1 & 1 \\ 0 & 0 & b+6 & 0 \\ b-2 & 0 & 0 & -2 \end{bmatrix}$$

Cases:

i. If $b=2$, $|A| = 1\cdot 0\cdot 8\cdot(-2)=0$, rank of $A=3$

ii. If $b=-6$, no. of non-zero rows is 3, rank of $A=3$.

Echelon form

Example 6: Reduce A to Echelon form and then to its row canonical form (or row reduced Echelon form) where

$$A=\begin{pmatrix}1 & 3 & -1 & 2\\ 0 & 11 & -5 & 3\\ 2 & -5 & 3 & 1\\ 4 & 1 & 1 & 5\end{pmatrix}$$

Hence find the rank of A.

Solution: Applying elementary row operations on A

$$\begin{matrix}R_{31(-2)}\\ R_{41(-4)}\end{matrix}\sim\begin{bmatrix}1 & 3 & -1 & 2\\ 0 & 11 & -5 & 3\\ 0 & -11 & 5 & -3\\ 0 & -11 & 5 & -3\end{bmatrix}$$

$$\begin{matrix}R_{32(1)}\\ R_{42(1)}\\ \sim\end{matrix}\begin{bmatrix}1 & 3 & -1 & 2\\ 0 & 11 & -5 & 3\\ 0 & 0 & 0 & 0\\ 0 & 0 & 0 & 0\end{bmatrix}\quad \text{This is Echelon form.}$$

$$\begin{matrix}R_{2(\frac{1}{11})}\\ \sim\end{matrix}\begin{bmatrix}1 & 3 & -1 & 2\\ 0 & 1 & -\frac{5}{11} & \frac{3}{11}\\ 0 & 0 & 0 & 0\\ 0 & 0 & 0 & 0\end{bmatrix}$$

$$R_{12(-3)}\begin{bmatrix}1 & 0 & \frac{4}{11} & \frac{13}{11}\\ 0 & 1 & -\frac{5}{11} & \frac{3}{11}\\ 0 & 0 & 0 & 0\\ 0 & 0 & 0 & 0\end{bmatrix}$$

This is the row canonical or row reduced Echelon form.

Rank of A is 2 since there are two non-zero rows.

EXERCISE

Rank of a matrix

Find the rank of the matrix:

1. $\begin{bmatrix}1 & 2 & 3\\ 2 & 3 & 1\\ 3 & 1 & 2\end{bmatrix}$ *Ans.* 3

2. $\begin{bmatrix}1 & 1 & -1 & 1\\ 1 & -1 & 2 & -1\\ 3 & 1 & 0 & 1\end{bmatrix}$ *Ans.* 2

3. $\begin{bmatrix}2 & 3 & -1 & -1\\ 1 & -1 & -2 & -4\\ 3 & 1 & 3 & -2\\ 6 & 3 & 0 & -7\end{bmatrix}$ *Ans.* 3

4. $\begin{bmatrix}3 & -2 & 0 & -1 & -7\\ 0 & 2 & 2 & 1 & -5\\ 1 & -2 & -3 & -2 & 1\\ 0 & 1 & 2 & 1 & -6\end{bmatrix}$ *Ans.* 4

5. $\begin{bmatrix}1 & 2 & 3 & 2\\ 2 & 3 & 5 & 1\\ 1 & 3 & 4 & 5\end{bmatrix}$ *Ans.* 2

6. $\begin{bmatrix}1 & 2 & 3 & 4 & 5\\ 2 & 3 & 4 & 5 & 6\\ 3 & 4 & 5 & 6 & 7\\ 4 & 5 & 6 & 7 & 8\end{bmatrix}$ *Ans.* 2

7. $\begin{bmatrix}1 & 2 & 1 & 2\\ 1 & 3 & 2 & 2\\ 2 & 4 & 3 & 4\\ 3 & 7 & 4 & 6\end{bmatrix}$ *Ans.* 3

8. $\begin{bmatrix}1 & 2 & -3 & 4 & 9\\ 1 & 0 & -1 & 1 & 1\\ 3 & -1 & 1 & 0 & -1\\ -1 & 1 & 0 & 2 & 9\\ 3 & 1 & 0 & 3 & 9\end{bmatrix}$ *Ans.* 4

9. $\begin{bmatrix}1 & 2 & 3\\ 1 & 4 & 2\\ 2 & 6 & 5\end{bmatrix}$ *Ans.* 2

10. $\begin{bmatrix}0 & 1 & -3 & -1\\ 0 & 0 & 1 & 1\\ 3 & 1 & 0 & 2\\ 1 & 1 & -2 & 0\end{bmatrix}$ *Ans.* 2

11. $\begin{bmatrix}3 & 0 & 2 & 2\\ -6 & 42 & 24 & 54\\ 21 & -21 & 0 & -15\end{bmatrix}$ *Ans.* 2

12. $\begin{bmatrix} 1 & 2 & -1 & 4 \\ 2 & 4 & 3 & 5 \\ -1 & -2 & 6 & -7 \end{bmatrix}$ *Ans.* 2

13. $\begin{bmatrix} 1 & 2 & 3 & 0 \\ 2 & 4 & 3 & 2 \\ 3 & 2 & 1 & 3 \\ 6 & 8 & 7 & 5 \end{bmatrix}$ *Ans.* 3

14. $\begin{bmatrix} 1 & 2 & 3 \\ 2 & 3 & 1 \\ -2 & -3 & -1 \end{bmatrix}$ *Ans.* 2

15. $\begin{bmatrix} 2 & -2 & 0 & 6 \\ 4 & 2 & 0 & 2 \\ 1 & -1 & 0 & 3 \\ 1 & -2 & 1 & 2 \end{bmatrix}$ *Ans.* 3

16. $\begin{bmatrix} 2 & 1 & 3 & 4 \\ 0 & 3 & 4 & 1 \\ 2 & 3 & 7 & 5 \\ 2 & 5 & 11 & 6 \end{bmatrix}$ *Ans.* 3

Echelon form

Find the Echelon form and row reduced echelon form of the matrix and hence find the rank:

17. $\begin{bmatrix} 1 & -2 & 3 & -1 \\ 2 & -1 & 2 & 2 \\ 3 & 1 & 2 & 3 \end{bmatrix}$

Ans. $\begin{pmatrix} 1 & -2 & 3 & -1 \\ 0 & 3 & -4 & 4 \\ 0 & 0 & 7 & -10 \end{pmatrix}$,

$\begin{pmatrix} 1 & 0 & 0 & \frac{15}{7} \\ 0 & 1 & 0 & -\frac{4}{7} \\ 0 & 0 & 1 & -\frac{10}{7} \end{pmatrix}$, rank = 3

18. $\begin{bmatrix} 1 & 2 & -5 \\ -4 & 1 & -6 \\ 6 & 3 & -4 \end{bmatrix}$

Ans. $\begin{pmatrix} 1 & 2 & -5 \\ 0 & 9 & -26 \\ 0 & 0 & 0 \end{pmatrix}$, $\begin{pmatrix} 1 & 0 & 7/9 \\ 0 & 1 & -26/9 \\ 0 & 0 & 0 \end{pmatrix}$

rank = 2

19. $\begin{bmatrix} 0 & 1 & 3 & -2 \\ 0 & 4 & -1 & 3 \\ 0 & 0 & 2 & 1 \\ 0 & 5 & -3 & 4 \end{bmatrix}$

Ans. $\begin{bmatrix} 0 & 1 & 3 & -2 \\ 0 & 0 & -13 & 11 \\ 0 & 0 & 0 & 35 \\ 0 & 0 & 0 & 0 \end{bmatrix}$, $\begin{bmatrix} 0 & 1 & 0 & 0 \\ 0 & 0 & 1 & 0 \\ 0 & 0 & 0 & 1 \\ 0 & 0 & 0 & 0 \end{bmatrix}$

rank = 3

20. $\begin{bmatrix} 2 & 3 & -2 & 5 & 1 \\ 3 & -1 & 2 & 0 & 4 \\ 4 & -5 & 6 & -5 & 7 \end{bmatrix}$

Ans. $\begin{pmatrix} 2 & 3 & -2 & 5 & 1 \\ 0 & -11 & 10 & -15 & 5 \\ 0 & 0 & 0 & 0 & 0 \end{pmatrix}$,

$\begin{pmatrix} 1 & 0 & \frac{4}{11} & \frac{5}{11} & \frac{13}{11} \\ 0 & 1 & -\frac{10}{11} & \frac{15}{11} & -\frac{5}{11} \\ 0 & 0 & 0 & 0 & 0 \end{pmatrix}$, rank = 2

13.3 NORMAL FORM

Procedure to Obtain Normal Form

Consider

$$A_{m\times n} = I_{m\times m}\, A_{m\times n}\, I_{n\times n}$$

Apply elementary row operations on A and on the prefactor $I_{m\times m}$ and apply elementary column operations on A and on the postfactor $I_{n\times n}$, such that A on the L.H.S. reduces to normal form. Then $I_{m\times m}$ reduces to $P_{m\times m}$ and $I_{n\times n}$ reduces to $Q_{n\times n}$; resulting in $N = PAQ$.

Here P and Q are non-singular matrices.

Thus for any matrix of rank r, there exist non-singular matrices P and Q such that

$$PAQ = N = \begin{bmatrix} I_r & 0 \\ 0 & 0 \end{bmatrix}.$$

WORKED OUT EXAMPLES

Example 1: Find the non-singular matrices P and Q such that the normal form of A is PAQ where

$$A = \begin{bmatrix} 1 & 3 & 6 & -1 \\ 1 & 4 & 5 & 1 \\ 1 & 5 & 4 & 3 \end{bmatrix}_{3\times 4}$$

Hence find its rank.

Solution: Consider $A_{3\times4} = I_{3\times3}\, A_{3\times4}\, I_{4\times4}$

$$\begin{bmatrix}1&3&6&-1\\1&4&5&1\\1&5&4&3\end{bmatrix} = \begin{bmatrix}1&0&0\\0&1&0\\0&0&1\end{bmatrix} A I_4$$

$$\begin{matrix}R_{21(-1)}\\R_{31(-1)}\\\text{pre}\end{matrix}\begin{bmatrix}1&3&6&-1\\0&1&-1&2\\0&2&-2&4\end{bmatrix} = \begin{bmatrix}1&0&0\\-1&1&0\\-1&0&1\end{bmatrix} A I_4$$

$$\begin{matrix}R_{32(-2)}\\\text{pre}\end{matrix}\begin{bmatrix}1&3&6&-1\\0&1&-1&2\\0&0&0&0\end{bmatrix} = \begin{bmatrix}1&0&0\\-1&1&0\\1&-2&1\end{bmatrix} A I_4$$

$$\begin{matrix}C_{21(-3)}\\C_{31(-6)}\\C_{41(1)}\\\text{post}\end{matrix}\begin{bmatrix}1&0&0&0\\0&1&-1&2\\0&0&0&0\end{bmatrix} = \begin{bmatrix}1&0&0\\-1&1&0\\1&-2&1\end{bmatrix} \times A \begin{bmatrix}1&-3&-6&1\\0&1&0&0\\0&0&1&0\\0&0&0&1\end{bmatrix}$$

$$\begin{matrix}C_{32(1)}\\C_{42(-2)}\end{matrix}\begin{bmatrix}1&0&0&0\\0&1&0&0\\0&0&0&0\end{bmatrix} = \begin{bmatrix}1&0&0\\-1&1&0\\1&-2&1\end{bmatrix} \times A \begin{bmatrix}1&-3&-9&7\\0&1&1&-2\\0&0&1&0\\0&0&0&1\end{bmatrix}$$

Thus $I_2 = PAQ$ where

$$P = \begin{bmatrix}1&0&0\\-1&1&0\\1&-2&1\end{bmatrix},\ Q = \begin{bmatrix}1&-3&-9&7\\0&1&1&-2\\0&0&1&0\\0&0&0&1\end{bmatrix}$$

Rank of A is 2.

Example 2: Find P and Q such that the normal form of

$$A = \begin{bmatrix}1&-1&-1\\1&1&1\\3&1&1\end{bmatrix}$$

is PAQ.

Hence find the rank of A.

Solution: Consider

$$A_{3\times3} = I_{3\times3}\, A_{3\times3}\, I_{3\times3}$$

$$\begin{bmatrix}1&-1&-1\\1&1&1\\3&1&1\end{bmatrix} = \begin{bmatrix}1&0&0\\0&1&0\\0&0&1\end{bmatrix} A \begin{bmatrix}1&0&0\\0&1&0\\0&0&1\end{bmatrix}$$

$$\begin{matrix}C_{21(1)}\\C_{31(1)}\\\text{post}\end{matrix}\begin{bmatrix}1&0&0\\1&2&2\\3&4&4\end{bmatrix} = \begin{bmatrix}1&0&0\\0&1&0\\0&0&1\end{bmatrix} A \begin{bmatrix}1&1&1\\0&1&0\\0&0&1\end{bmatrix}$$

$$\begin{matrix}R_{21(-1)}\\R_{31(-3)}\\\text{pre}\end{matrix}\begin{bmatrix}1&0&0\\0&2&2\\0&4&4\end{bmatrix} = \begin{bmatrix}1&0&0\\-1&1&0\\-3&0&1\end{bmatrix} A \begin{bmatrix}1&1&1\\0&1&0\\0&0&1\end{bmatrix}$$

$$\begin{matrix}R_{2(\frac{1}{2})}\\R_{3(\frac{1}{4})}\\\text{pre}\end{matrix}\begin{bmatrix}1&0&0\\0&1&1\\0&1&1\end{bmatrix} = \begin{bmatrix}1&0&0\\-\frac{1}{2}&\frac{1}{2}&0\\-\frac{3}{4}&0&\frac{1}{4}\end{bmatrix} A \begin{bmatrix}1&1&1\\0&1&0\\0&0&1\end{bmatrix}$$

$$\begin{matrix}R_{32(-1)}\\\text{pre}\end{matrix}\begin{bmatrix}1&0&0\\0&1&1\\0&0&0\end{bmatrix} = \begin{bmatrix}1&0&0\\-\frac{1}{2}&\frac{1}{2}&0\\-\frac{1}{4}&-\frac{1}{2}&\frac{1}{4}\end{bmatrix} A \begin{bmatrix}1&1&1\\0&1&0\\0&0&1\end{bmatrix}$$

$$\begin{matrix}C_{32(-1)}\\\text{post}\end{matrix}\begin{bmatrix}1&0&0\\0&1&0\\0&0&0\end{bmatrix} = \begin{bmatrix}1&0&0\\-\frac{1}{2}&\frac{1}{2}&0\\-\frac{1}{4}&-\frac{1}{2}&\frac{1}{4}\end{bmatrix} A \begin{bmatrix}1&1&0\\0&1&-1\\0&0&1\end{bmatrix}$$

Thus the L.H.S. is in the normal form $\begin{bmatrix}I_2&0\\0&0\end{bmatrix}$.

Hence

$$P_{3\times3} = \begin{bmatrix}1&0&0\\-\frac{1}{2}&\frac{1}{2}&0\\-\frac{1}{4}&-\frac{1}{2}&\frac{1}{4}\end{bmatrix} \text{ and } Q_{3\times3} = \begin{bmatrix}1&1&0\\0&1&-1\\0&0&1\end{bmatrix}$$

Rank of $A = 2$.

EXERCISE

Determine the non-singular matrices P and Q such that PAQ is in the normal form for A. Hence find the rank of A.

1. $A = \begin{bmatrix}1&1&2\\1&2&3\\0&-1&-1\end{bmatrix}$

Ans. $P = \begin{bmatrix}1&0&0\\-1&1&0\\-1&1&1\end{bmatrix},\ Q = \begin{bmatrix}1&-1&-1\\0&1&-1\\0&0&1\end{bmatrix}$

rank = 2

2. $A = \begin{bmatrix} 3 & 2 & -1 & 5 \\ 5 & 1 & 4 & -2 \\ 1 & -4 & 11 & -19 \end{bmatrix}$

Ans. $P = \begin{bmatrix} 0 & 0 & 1 \\ 0 & \frac{1}{3} & -\frac{5}{3} \\ \frac{1}{2} & -\frac{1}{3} & \frac{1}{6} \end{bmatrix}$,

$Q = \begin{bmatrix} 1 & \frac{4}{17} & \frac{9}{119} & \frac{9}{217} \\ 0 & \frac{1}{7} & -\frac{1}{7} & -\frac{1}{7} \\ 0 & 0 & -\frac{1}{17} & 0 \\ 0 & 0 & 0 & \frac{1}{31} \end{bmatrix}$, rank = 2

3. $A = \begin{bmatrix} 1 & 2 & 3 \\ 2 & -1 & 0 \\ 3 & 1 & 2 \end{bmatrix}$

Ans. $P = \begin{bmatrix} 1 & 0 & 0 \\ \frac{2}{5} & -\frac{1}{5} & 0 \\ 1 & 1 & -1 \end{bmatrix}$, $Q = \begin{bmatrix} 1 & -2 & -\frac{3}{5} \\ 0 & 1 & -\frac{6}{5} \\ 0 & 0 & 1 \end{bmatrix}$

rank = 3

4. $A = \begin{bmatrix} 1 & -1 & -1 & 2 \\ 4 & 2 & 2 & -1 \\ 2 & 2 & 0 & -2 \end{bmatrix}$

Ans. $P = \begin{bmatrix} 1 & 0 & 0 \\ -\frac{2}{3} & \frac{1}{6} & 0 \\ -\frac{1}{3} & \frac{1}{3} & -\frac{1}{2} \end{bmatrix}$,

$Q = \begin{bmatrix} 1 & 1 & 0 & -\frac{1}{2} \\ 0 & 1 & -1 & -\frac{3}{2} \\ 0 & 0 & 1 & 0 \\ 0 & 0 & 0 & 1 \end{bmatrix}$, rank = 3

5. $A = \begin{bmatrix} 1 & 2 & 3 & -2 \\ 2 & -2 & 1 & 3 \\ 3 & 0 & 4 & 1 \end{bmatrix}$

Ans. $P = \begin{bmatrix} 1 & 0 & 0 \\ -2 & 1 & 0 \\ -1 & -1 & 1 \end{bmatrix}$,

$Q = \begin{bmatrix} 1 & \frac{1}{3} & -\frac{4}{3} & \frac{-1}{3} \\ 0 & -\frac{1}{6} & -\frac{5}{6} & \frac{7}{6} \\ 0 & 0 & 1 & 0 \\ 0 & 0 & 0 & 1 \end{bmatrix}$, rank = 2

6. $A = \begin{bmatrix} 1 & -1 & 2 & -1 \\ 4 & 2 & -1 & 2 \\ 2 & 2 & -2 & 0 \end{bmatrix}$

Ans. $P = \begin{bmatrix} 1 & 0 & 0 \\ -\frac{2}{3} & \frac{1}{6} & -\frac{1}{2} \\ -\frac{1}{3} & \frac{1}{3} & -\frac{1}{2} \end{bmatrix}$,

$Q = \begin{bmatrix} 1 & 1 & 0 & -\frac{1}{2} \\ 0 & 1 & -1 & \frac{3}{2} \\ 0 & 0 & 0 & 1 \\ 0 & 0 & 1 & 0 \end{bmatrix}$, rank = 3.

13.4 SYSTEM OF LINEAR NON-HOMOGENEOUS EQUATIONS

A system (or family) of m linear algebraic equations in n unknowns $x_1, x_2, \cdots x_n$ is a set of equations of the form

$$\begin{aligned} a_{11}x_1 + a_{12}x_2 + \cdots + a_{1n}x_n &= b_1 \\ a_{21}x_1 + a_{22}x_2 + \cdots + a_{2n}x_n &= b_2 \\ \cdots\cdots\cdots\cdots\cdots\cdots\cdots\cdots \\ \cdots\cdots\cdots\cdots\cdots\cdots\cdots\cdots \\ a_{m1}x_1 + a_{m2}x_2 + \cdots + a_{mn}x_n &= b_m \end{aligned} \qquad (1)$$

The numbers a_{ij} are known as coefficients and b_i are known as (R.H.S.) constants of the system.

(1) can be represented as

$$\sum_{j=1}^{n} a_{ij}x_j = b_i \; ; \; i = 1 \text{ to } m$$

Non-homogeneous system: When all b_i are not zero, i.e., at least one b_i is non-zero.

Homogeneous system: If $b_i = 0, i = 1$ to m (all R.H.S. constants are zero).

Solution of system (1) is a set of numbers $x_1, x_2, \cdots, x_n$ which satisfy (simultaneously) all the equations of the system (1).

Trivial solution is a solution where all x_i are zero i.e., $x_1 = x_2 = \cdots = x_n = 0$.

Matrix Representation

Let $X = \begin{bmatrix} x_1 \\ x_2 \\ \vdots \\ x_n \end{bmatrix}_{n\times 1}$, $B = \begin{bmatrix} b_1 \\ b_2 \\ \vdots \\ b_m \end{bmatrix}_{m\times 1}$

be two column vectors.

$$\text{Let } A = \begin{bmatrix} a_{11} & a_{12} & a_{13} & \cdots a_{1n} \\ a_{21} & a_{22} & a_{23} & \cdots a_{2n} \\ \vdots & \vdots & \vdots & \vdots \\ a_{m1} & a_{m2} & a_{m3} & \cdots a_{mn} \end{bmatrix}_{m\times n}$$

Here

A = Coefficient matrix of the system (1)
B = (R.H.S.) constant vector
X = Solution (vector)

Then the system (1) can be represented as

$$A_{m\times n} X_{n\times 1} = B_{m\times 1}$$

Augmented matrix $[A|B]$ or $\tilde{A}$ of system (1) is obtained by augmenting A by the column B

$$\text{i.e., } \tilde{A} = [A|B] = \begin{bmatrix} a_{11} & a_{12} & \cdots a_{1n} & b_1 \\ a_{21} & a_{22} & \cdots a_{2n} & b_2 \\ \vdots & \vdots & \vdots & \vdots \\ a_{m1} & a_{m2} & \cdots a_{mn} & b_m \end{bmatrix}$$

Consistent: System is said to be consistent if (1) has at least one solution.
Inconsistent if system has no solution at all.

Solution of System of Linear Equations

We consider two methods of obtaining solution of system of n linear equations in n unknowns. They are

i. Cramer's rule **ii.** matrix inverse

Cramer's rule (Solution by determinants)

a. If A is non-singular i.e., D = determinant of A = $|A| \neq 0$. Then system (1) has a unique solution given by

$$x_i = \frac{Di}{D} \quad \text{for} \quad i = 1, 2, \cdots n$$

where D_i is the determinant obtained from D by replacing the ith column in D by constant column vector B.

b. For homogeneous system with $D \neq 0$, only trivial solution exists.

c. For homogeneous system with $D = 0$, non-trivial solutions exists.

Note: Cramer's rule is not suitable for computations.

Matrix inversion method

Consider the system of n equations in n unknowns represented by

$$AX = B$$

where A is n-square non-singular matrix. Premultiplying by A^{-1} on both sides, we get

$$A^{-1}AX = A^{-1}B$$

or $$X = A^{-1}B$$

which is the required solution.

Here A^{-1}, the inverse of A is obtained by Gauss-Jordan method: (see Page 13.4)

Consider $\quad A|I$

Apply *only* elementary row operations on both A and I such that A is reduced to an identity matrix I, then I gets transformed to A^{-1} i.e.,

$$A^{-1} \; A\big|A^{-1}I$$

$$I\big|A^{-1}$$

Consistency of System of Linear Equations

Consider m linear equations in n unknowns so that

$$A_{m\times n} X_{n\times 1} = B_{m\times 1}$$

Fundamental theorem

I. If rank of A and rank of the augmented matrix $\tilde{A}$ are equal, then the system is consistent.

a. If $r(A) = r(\tilde{A}) = n$

then unique solution exists.

b. If $r(A) = r(\tilde{A}) < n$

then infinitely many solutions exist in terms of $(n - r)$ arbitrary constants.

II. If rank of A is not equal to rank of $\tilde{A}$ then the system is inconsistent and has no solution at all.

Procedure

1. Determine $r(A)$ and $r(\tilde{A})$.
2. If $r(A) \neq r(\tilde{A})$, system inconsistent, no solutions.
3. If $r(A) = r(\tilde{A}) = n$

Then the unique solution may be obtained by Cramer's rule or matrix inversion method.

4. If $r(A) = r(\tilde{A}) < n$

Then rewrite $x_1, \cdots x_r$ variables (whose coefficient submatrix has rank r) in terms $(n-r)$ variables and solve by Gaussian elimination or Gauss-Jordan elimination method.

WORKED OUT EXAMPLES

Example 1: Solve by Cramer's rule

$$x + y + z = 11$$
$$2x - 6y - z = 0$$
$$3x + 4y + 2z = 0.$$

Solution:

$$D = \begin{vmatrix} 1 & 1 & 1 \\ 2 & -6 & -1 \\ 3 & 4 & 2 \end{vmatrix} = 11,\ D_1 = \begin{vmatrix} 11 & 1 & 1 \\ 0 & -6 & -1 \\ 0 & 4 & 2 \end{vmatrix} = -88,$$

$$D_2 = \begin{vmatrix} 1 & 11 & 1 \\ 2 & 0 & -1 \\ 3 & 0 & 2 \end{vmatrix} = -77,\ D_3 = \begin{vmatrix} 1 & 1 & 11 \\ 2 & -6 & 0 \\ 3 & 4 & 0 \end{vmatrix} = 286$$

The unique solution $x = \frac{D_1}{D} = \frac{-88}{11} = -8$

$$y = \frac{D_2}{D} = \frac{-77}{11} = -7,\quad z = \frac{D_3}{D} = \frac{286}{11} = 26$$

Thus $x = -8, \quad y = -7, \quad z = 26.$

Example 2: Solve by calculating the inverse by adjoint method

$$x_1 + x_2 + 2x_3 = 4$$
$$2x_1 + 5x_2 - 2x_3 = 3$$
$$x_1 + 7x_2 - 7x_3 = 5.$$

Solution: The given system is written as $AX = B$ where

$$A = \begin{bmatrix} 1 & 1 & 2 \\ 2 & 5 & -2 \\ 1 & 7 & -7 \end{bmatrix},\ X = \begin{bmatrix} x_1 \\ x_2 \\ x_3 \end{bmatrix},\ B = \begin{bmatrix} 4 \\ 3 \\ 5 \end{bmatrix}$$

Inverse by adjoint

$$|A| = \begin{vmatrix} 1 & 1 & 2 \\ 2 & 5 & -2 \\ 1 & 7 & -7 \end{vmatrix} = 9$$

$$A^{-1} = \frac{1}{|A|} \text{adj } A = \frac{1}{|A|} \begin{bmatrix} A_{11} & A_{12} & A_{13} \\ A_{21} & A_{22} & A_{23} \\ A_{31} & A_{32} & A_{33} \end{bmatrix}^T$$

where A_{ij} are cofactor of the element a_{ij}

$$A^{-1} = \frac{1}{9} \begin{bmatrix} -21 & 12 & 9 \\ 21 & -9 & -6 \\ -12 & 6 & 3 \end{bmatrix}^T = \frac{1}{3} \begin{bmatrix} -7 & 7 & -4 \\ 4 & -3 & 2 \\ 3 & -2 & 1 \end{bmatrix}$$

The solution to the given system is

$$X = \begin{bmatrix} x_1 \\ x_2 \\ x_3 \end{bmatrix} = A^{-1}B = \frac{1}{3} \begin{bmatrix} -7 & 7 & -4 \\ 4 & -3 & 2 \\ 3 & -2 & 1 \end{bmatrix} \begin{bmatrix} 4 \\ 3 \\ 5 \end{bmatrix} = \frac{1}{3} \begin{bmatrix} -27 \\ 17 \\ 11 \end{bmatrix}$$

i.e., $x_1 = -27/3, x_2 = 17/3, \quad x_3 = 11/3.$

Example 3: Solve by calculating the inverse by elementary row operations

$$x_1 + x_2 + x_3 + x_4 = 0$$
$$x_1 + x_2 + x_3 - x_4 = 4$$
$$x_1 + x_2 - x_3 + x_4 = -4$$
$$x_1 - x_2 + x_3 + x_4 = 2$$

Solution: The system is written as $AX = B$ where

$$A = \begin{bmatrix} 1 & 1 & 1 & 1 \\ 1 & 1 & 1 & -1 \\ 1 & 1 & -1 & 1 \\ 1 & -1 & 1 & 1 \end{bmatrix},\ X = \begin{bmatrix} x_1 \\ x_2 \\ x_3 \\ x_4 \end{bmatrix},\ B = \begin{bmatrix} 0 \\ 4 \\ -4 \\ 2 \end{bmatrix}$$

Inverse by elementary row operations

$$[A|I] = \left[\begin{array}{cccc|cccc} 1 & 1 & 1 & 1 & 1 & 0 & 0 & 0 \\ 1 & 1 & 1 & -1 & 0 & 1 & 0 & 0 \\ 1 & 1 & -1 & 1 & 0 & 0 & 1 & 0 \\ 1 & -1 & 1 & 1 & 0 & 0 & 0 & 1 \end{array}\right] \begin{array}{l} R_{21(-1)}, R_{31(-1)}, \\ R_{41(-1)} \text{ and} \\ R_{2(-1)}, R_{3(-1)}, \\ R_{4(-1)} \end{array}$$

$$\sim \left[\begin{array}{cccc|cccc} 1 & 1 & 1 & 1 & 1 & 0 & 0 & 0 \\ 0 & 0 & 0 & 2 & 1 & -1 & 0 & 0 \\ 0 & 0 & 2 & 0 & 1 & 0 & -1 & 0 \\ 0 & 2 & 0 & 0 & 1 & 0 & 0 & -1 \end{array}\right] \begin{array}{l} R_{24} \\ R_{2(\frac{1}{2})} \\ R_{3(\frac{1}{2})} \\ R_{4(\frac{1}{2})} \end{array}$$

$$\sim \left[\begin{array}{cccc|cccc} 1 & 1 & 1 & 1 & 1 & 0 & 0 & 0 \\ 0 & 1 & 0 & 0 & \frac{1}{2} & 0 & 0 & -\frac{1}{2} \\ 0 & 0 & 1 & 0 & \frac{1}{2} & 0 & -\frac{1}{2} & 0 \\ 0 & 0 & 0 & 1 & \frac{1}{2} & -\frac{1}{2} & 0 & 0 \end{array}\right] \begin{array}{l} R_{14(-1)} \\ R_{13(-1)} \\ R_{12(-1)} \end{array}$$

$$\sim \left[\begin{array}{cccc|cccc} 1 & 0 & 0 & 0 & -\frac{1}{2} & \frac{1}{2} & \frac{1}{2} & \frac{1}{2} \\ 0 & 1 & 0 & 0 & \frac{1}{2} & 0 & 0 & -\frac{1}{2} \\ 0 & 0 & 1 & 0 & \frac{1}{2} & 0 & -\frac{1}{2} & 0 \\ 0 & 0 & 0 & 1 & \frac{1}{2} & -\frac{1}{2} & 0 & 0 \end{array}\right]$$

Thus $A^{-1} = \frac{1}{2}\begin{bmatrix} -1 & 1 & 1 & 1 \\ 1 & 0 & 0 & -1 \\ 1 & 0 & -1 & 0 \\ 1 & -1 & 0 & 0 \end{bmatrix}$

The required solution is

$$X = A^{-1}B = \frac{1}{2}\begin{bmatrix} -1 & 1 & 1 & 1 \\ 1 & 0 & 0 & -1 \\ 1 & 0 & -1 & 0 \\ 1 & -1 & 0 & 0 \end{bmatrix}\begin{bmatrix} 0 \\ 4 \\ -4 \\ 2 \end{bmatrix} = \begin{bmatrix} 1 \\ -1 \\ 2 \\ -2 \end{bmatrix}$$

i.e., $x_1 = 1, x_2 = -1, x_3 = 2, x_4 = -2$.

Example 4: Solve

$$2x_1 - 2x_2 + 4x_3 + 3x_4 = 9$$
$$x_1 - x_2 + 2x_3 + 2x_4 = 6$$
$$2x_1 - 2x_2 + x_3 + 2x_4 = 3$$
$$x_1 - x_2 + x_4 = 2$$

Solution: Apply elementary row operation on $[A|B]$

$$[A|B] = \left[\begin{array}{cccc|c} 2 & -2 & 4 & 3 & 9 \\ 1 & -1 & 2 & 2 & 6 \\ 2 & -2 & 1 & 2 & 3 \\ 1 & -1 & 0 & 1 & 2 \end{array}\right] \begin{array}{l} R_{12} \\ R_{21(-2)} \\ R_{41(-1)} \\ R_{31(-2)} \\ R_{2(-1)} \\ R_{3(-1)} \\ R_{4(-1)} \end{array} \left[\begin{array}{cccc|c} 1 & -1 & 2 & 2 & 6 \\ 0 & 0 & 0 & 1 & 3 \\ 0 & 0 & 3 & 2 & 9 \\ 0 & 0 & 2 & 1 & 4 \end{array}\right]$$

$$\begin{array}{c} R_{34(-1)} \\ \sim \end{array} \left[\begin{array}{cccc|c} 1 & -1 & 2 & 2 & 6 \\ 0 & 0 & 0 & 1 & 3 \\ 0 & 0 & 1 & 1 & 5 \\ 0 & 0 & 2 & 1 & 4 \end{array}\right]$$

$$\begin{array}{c} R_{32} \\ R_{43} \\ \sim \end{array} \left[\begin{array}{cccc|c} 1 & -1 & 2 & 2 & 6 \\ 0 & 0 & 1 & 1 & 5 \\ 0 & 0 & 2 & 1 & 4 \\ 0 & 0 & 0 & 1 & 3 \end{array}\right]$$

$$\begin{array}{c} R_{32(-2)} \\ R_{3(-1)} \\ \sim \end{array} \left[\begin{array}{cccc|c} 1 & -1 & 2 & 2 & 6 \\ 0 & 0 & 1 & 1 & 5 \\ 0 & 0 & 0 & 1 & 6 \\ 0 & 0 & 0 & 1 & 3 \end{array}\right]$$

$$\begin{array}{c} R_{43(-1)} \\ R_{4(-1)} \\ \sim \end{array} \left[\begin{array}{cccc|c} 1 & -1 & 2 & 2 & 6 \\ 0 & 0 & 1 & 1 & 5 \\ 0 & 0 & 0 & 1 & 6 \\ 0 & 0 & 0 & 0 & 3 \end{array}\right]$$

rank of $(A) = 3 \neq 4 =$ rank of $[A|B]$

So the given system is inconsistent and therefore has no solution.

Example 5: Solve

$$3x + 3y + 2z = 1$$
$$x + 2y = 4$$
$$10y + 3z = -2$$
$$2x - 3y - z = 5.$$

Solution:

$$[A|B] = \left[\begin{array}{ccc|c} 3 & 3 & 2 & 1 \\ 1 & 2 & 0 & 4 \\ 0 & 10 & 3 & -2 \\ 2 & -3 & -1 & 5 \end{array}\right] R_{12} \sim \left[\begin{array}{ccc|c} 1 & 2 & 0 & 4 \\ 3 & 3 & 2 & 1 \\ 0 & 10 & 3 & -2 \\ 2 & -3 & -1 & 5 \end{array}\right]$$

$$\begin{array}{c} R_{21(-3)} \\ \sim \\ R_{41(-2)} \end{array} \left[\begin{array}{ccc|c} 1 & 2 & 0 & 4 \\ 0 & -3 & 2 & -11 \\ 0 & 10 & 3 & -2 \\ 0 & -7 & -1 & -3 \end{array}\right]$$

$$\begin{array}{c} R_{2(-\frac{1}{3})} \\ R_{32(-10)} \\ R_{42(7)} \end{array} \sim \left[\begin{array}{ccc|c} 1 & 2 & 0 & 4 \\ 0 & 1 & -\frac{2}{3} & \frac{11}{3} \\ 0 & 0 & \frac{29}{3} & \frac{-116}{3} \\ 0 & 0 & \frac{-17}{3} & \frac{68}{3} \end{array}\right]$$

$$\begin{array}{c} R_{3(\frac{3}{29})} \\ R_{43(\frac{17}{3})} \end{array} \sim \left[\begin{array}{ccc|c} 1 & 2 & 0 & 4 \\ 0 & 1 & \frac{-2}{3} & \frac{11}{3} \\ 0 & 0 & 1 & -\frac{116}{29} \\ 0 & 0 & 0 & 0 \end{array}\right]$$

$r(A) = 3 = r(A|B) = n =$ number of variables. The system is consistent and has unique solution. Solving

$$z = -\frac{116}{29} = -4$$

$$y - \frac{2}{3}z = \frac{11}{3} \quad \text{or} \quad y = \frac{11}{3} + \frac{2}{3}(-4) = 1$$

$$x + 2y + 0 = 4 \quad \text{or} \quad x = 4 - 2 = 2$$

i.e., $x = 2, y = 1, z = -4$.

Example 6: Solve

$$x_1 + x_2 - x_3 = 0$$
$$2x_1 - x_2 + x_3 = 3$$
$$4x_1 + 2x_2 - 2x_3 = 2.$$

Solution: By applying elementary row operations

$$[A|B] = \left[\begin{array}{ccc|c} 1 & 1 & -1 & 0 \\ 2 & -1 & 1 & 3 \\ 4 & 2 & -2 & 2 \end{array}\right] \underset{\sim}{\overset{R_{21(-2)}}{R_{31(-4)}}} \left[\begin{array}{ccc|c} 1 & 1 & -1 & 0 \\ 0 & -3 & 3 & 3 \\ 0 & -2 & 2 & 2 \end{array}\right]$$

$$\underset{R_{3(-\frac{1}{2})}}{\overset{R_{2(-\frac{1}{3})}}{\sim}} \left[\begin{array}{ccc|c} 1 & 1 & -1 & 0 \\ 0 & 1 & -1 & -1 \\ 0 & 1 & -1 & -1 \end{array}\right]$$

$$\underset{\sim}{R_{32(-1)}} \left[\begin{array}{ccc|c} 1 & 1 & -1 & 0 \\ 0 & 1 & -1 & -1 \\ 0 & 0 & 0 & 0 \end{array}\right]$$

$r(A) = 2 = r(A|B) < 3 = n =$ number of variables.

The system is consistent but has infinite number of solutions in terms of $n - r = 3 - 2 = 1$ variable. Choose $x_3 = k =$ arbitrary constant.

Solving $x_2 - x_3 = -1$ or $x_2 = x_3 - 1 = k - 1$.

$x_1 + x_2 - x_3 = 0$ or $x_1 = -x_2 + x_3 = -k + 1 + k = 1$

Thus the solutions are

$x_1 = 1, x_2 = k - 1, x_3 = k,$ where k is arbitrary.

Example 7: Determine the values of a and b for which the system

$$x + 2y + 3z = 6$$
$$x + 3y + 5z = 9$$
$$2x + 5y + az = b$$

has (i) no solution (ii) unique solution (iii) infinite number of solutions. Find the solutions in case (ii) and (iii).

Solution:

$$[A|B] = \left[\begin{array}{ccc|c} 1 & 2 & 3 & 6 \\ 1 & 3 & 5 & 9 \\ 2 & 5 & a & b \end{array}\right] \overset{R_{21(-1)}}{R_{31(-2)}} \left[\begin{array}{ccc|c} 1 & 2 & 3 & 6 \\ 0 & 1 & 2 & 3 \\ 0 & 1 & a-6 & b-12 \end{array}\right]$$

$$R_{32(-1)} \sim \left[\begin{array}{ccc|c} 1 & 2 & 3 & 6 \\ 0 & 1 & 2 & 3 \\ 0 & 0 & a-8 & b-15 \end{array}\right]$$

Case 1: $a = 8, b \neq 15, r(A) = 2 \neq 3 = r(A|B)$, inconsistent, no solution.

Case 2: $a \neq 8$, b any value, $r(A) = 3 = r(A|B) = n =$ number of variables, unique solution, $z = \frac{b-15}{a-8}$,

$y = (3a - 2b + 6)/(a - 8), x = z = (b - 15)/(a - 8)$.

Case 3: $a = 8, b = 15, r(A) = 2 = r(A|B) < 3 = n$, infinite solutions with $n - r = 3 - 2 = 1$ arbitrary variable. $x = k, y = 3 - 2k, z = k$, with k arbitrary.

EXERCISE

Solve the following:

1. $5x + 3y + 7z = 4,$
   $3x + 26y + 2z = 9,$
   $7x + 2y + 10z = 5.$

Ans. $x = (7 - 16k)/11, y = (3 + k)/11, z = k,$ k arbitrary

2. $x_1 + x_2 - 2x_3 + x_4 + 3x_5 = 1,$
   $2x_1 - x_2 + 2x_3 + 2x_4 + 6x_5 = 2,$
   $3x_1 + 2x_2 - 4x_3 - 3x_4 - 9x_5 = 3.$

Ans. $x_1 = 1, x_2 = 2a, x_3 = a, x_4 = -3b, x_5 = b$ where a and b are arbitrary constants

3. $x_1 + x_2 + 2x_3 + x_4 = 5,$
   $2x_1 + 3x_2 - x_3 - 2x_4 = 2,$
   $4x_1 + 5x_2 + 3x_3 = 7.$

Ans. No solution, system inconsistent

4. Using A^{-1} (inverse of the coefficient matrix)

$$2x_1 + x_2 + 5x_3 + x_4 = 5,$$
$$x_1 + x_2 - 3x_3 - 4x_4 = -1,$$
$$3x_1 + 6x_2 - 2x_3 + x_4 = 8,$$
$$2x_1 + 2x_2 + 2x_3 - 3x_4 = 2.$$

Ans. $x_1 = 2, x_2 = 1/5, x_3 = 0, x_4 = 4/5.$ unique solution

Hint.

$$A^{-1} = \frac{1}{120}\begin{bmatrix} 120 & 120 & 0 & -120 \\ -69 & -73 & 17 & 80 \\ -15 & -35 & -5 & 40 \\ 24 & 8 & 8 & -40 \end{bmatrix}$$

5. $2x_1 + 3x_2 - x_3 = 1,$

$3x_1 - 4x_2 + 3x_3 = -1,$

$2x_1 - x_2 + 2x_3 = -3,$

$3x_1 + x_2 - 2x_3 = 4.$

Ans. Inconsistent, no solution

6. $3x_1 + 2x_2 + x_3 = 3,$

$2x_1 + x_2 + x_3 = 0,$

$6x_1 + 2x_2 + 4x_3 = 6.$

Ans. Inconsistent, no solution.

7. $-x_1 + x_2 + 2x_3 = 2,$

$3x_1 - x_2 + x_3 = 6,$

$-x_1 + 3x_2 + 4x_3 = 4.$

Ans. $x_1 = 1, x_2 = -1, x_3 = 2$, unique solution

8. $7x + 16y - 7z = 4,$

$2x + 5y - 3z = -3,$

$x + y + 2z = 4.$

Ans. Inconsistent, no solution

9. $x + y + z = 4,$

$2x + 5y - 2z = 3,$

$x + 7y - 7z = 5.$

Ans. Inconsistent, no solution

10. $2x + y - z = 0,$

$2x + 5y + 7z = 52,$

$x + y + z = 9.$

Ans. unique solution $x = 1, y = 3, z = 5$

Find the values of a and b for which the system has (i) no solution (ii) unique solution (iii) infinitely many solutions for:

11. $2x + 3y + 5z = 9,$

$7x + 3y - 2z = 8,$

$2x + 3y + az = b.$

Ans. **i.** no solution of $a = 5, b \neq 9$;

ii. unique solution $a \neq 5$, b any value;

iii. infinitely many solutions $a = 5, b = 9$

12. $x + y + z = 6,$

$x + 2y + 3z = 10,$

$x + 2y + az = b.$

Ans. **i.** $a = 3, b \neq 10$ inconsistent

ii. $a \neq 3$, b any value, unique solution

iii. $a = 3, b = 10$ infinite solutions

13. Test for consistency

$-2x + y + z = a,$

$x - 2y + z = b,$

$x + y - 2z = c.$

where a, b, c are constants

Ans. i. If $a + b + c \neq 0$, inconsistent

ii. If $a + b + c = 0$, infinite solutions

14. Solve the system

$x + y + z = 6,$

$2x - 3y + 4z = 8,$

$x - y + 2z = 5$ by

i. Cramer's rule

ii. Matrix inversion

iii. Gauss-Jordan.

Ans. **i.** $x_1 = 1, \quad x_2 = 2, \quad x_3 = 3, \quad \Delta = -1,$

$\Delta_1 = -1, \Delta_2 = -2, \Delta_3 = -3$

ii. $A^{-1} = \begin{bmatrix} 2 & 3 & -7 \\ 0 & -1 & 2 \\ -1 & -2 & 5 \end{bmatrix}$

iii. $\left[\begin{array}{ccc|c} 1 & 1 & 1 & 6 \\ 2 & -3 & 4 & 8 \\ 1 & -1 & 2 & 5 \end{array}\right] \sim \left[\begin{array}{ccc|c} 1 & 0 & 0 & 1 \\ 0 & 1 & 0 & 2 \\ 0 & 0 & 1 & 3 \end{array}\right].$

13.5 SYSTEM OF HOMOGENEOUS EQUATIONS

Solution to a System of m Homogeneous Equations in n Unknowns

Result 1: If $r < m$, omit $m - r$ equations such that the coefficient matrix of the remaining equations still has rank r. Rewrite r unknowns in terms of $n - r$ arbitrary unknowns and solve.

Result 2: If $m < n$, system has non-trivial solutions.

Result 3: If $m = n$, system has non-trivial solutions if its coefficient determinant is zero.

Note: A homogeneous system always has a trivial solution since $r[A|B] = r[A|O] = r[A]$ for any A.

WORKED OUT EXAMPLES

Example 1: Determine b such that the system of homogeneous equations

$$2x + y + 2z = 0$$

$$x + y + 3z = 0$$

$$4x + 3y + bz = 0$$

has (i) Trivial solution (ii) non-trivial solution.
Find the non-trivial solution.

Solution: The coefficient matrix A is

$$A = \begin{bmatrix} 2 & 1 & 2 \\ 1 & 1 & 3 \\ 4 & 3 & b \end{bmatrix} \overset{R_{12}}{\sim} \begin{bmatrix} 1 & 1 & 3 \\ 2 & 1 & 2 \\ 4 & 3 & b \end{bmatrix}$$

$$\begin{matrix} R_{21(-2)} \\ \sim R_{31(-4)} \end{matrix} \begin{bmatrix} 1 & 1 & 3 \\ 0 & -1 & -4 \\ 0 & -1 & b-12 \end{bmatrix}$$

$$R_{32(-1)} \sim \begin{bmatrix} 1 & 1 & 3 \\ 0 & 1 & 4 \\ 0 & 0 & b-8 \end{bmatrix}$$

Case 1: If $b \neq 8$ then $r(A) = r(A|B) = 3$ = number of variables. i.e., $|A| \neq 0$. System has only trivial solution $x = 0$, $y = 0$; $z = 0$.

Case 2: If $b = 8$ then $r(A) = r(A|B) = 2 < 3 = n$. System has non-trivial solutions in terms of $n - r = 3 - 2 = 1$ arbitrary variable. Solving the system

$$x + y + 3z = 0$$

$$y + 4z = 0$$

Choose z as arbitrary say $z = k$ = arbitrary constant. Then $y = -4z = -4k$ and $x = -y - 3z = 4k - 3k = k$.

Thus the infinite number of non-trivial solutions are obtained for different values of k as

$$x = k,\ y = -4k,\ z = k.$$

Example 2: Solve

$$x + y - 2z + 3w = 0$$

$$x - 2y + z - w = 0$$

$$4x + y - 5z + 8w = 0$$

$$5x - 7y + 2z - w = 0.$$

Solution: The coefficient matrix A is

$$A = \begin{bmatrix} 1 & 1 & -2 & 3 \\ 1 & -2 & 1 & -1 \\ 4 & 1 & -5 & 8 \\ 5 & -7 & 2 & -1 \end{bmatrix}$$

$$\begin{matrix} R_{21(-1)} \\ R_{31(-4)} \\ R_{41(-5)} \\ \sim \end{matrix} \begin{bmatrix} 1 & 1 & -2 & 3 \\ 0 & -3 & 3 & -4 \\ 0 & -3 & 3 & -4 \\ 0 & -12 & 12 & -16 \end{bmatrix}$$

$$\begin{matrix} R_{2(-1)} \\ R_{32(1)} \\ R_{42(4)} \\ \sim \end{matrix} \begin{bmatrix} 1 & 1 & -2 & 3 \\ 0 & 3 & -3 & 4 \\ 0 & 0 & 0 & 0 \\ 0 & 0 & 0 & 0 \end{bmatrix}$$

$r(A) = r(A|B) = 2 < 4 = n$ = number of variables.

Non-trivial solutions exist in terms of $n - r = 4 - 2 = 2$ variables.

Choose $z = k_1$, and $w = k_2$. Then solving

$$x + y - 2z + 3w = 0$$

$$3y - 3z + 4w = 0$$

We get

$$y = \frac{1}{3}(3z - 4w) = z - \frac{4}{3}w = k_1 - \frac{4}{3}k_2$$

$$x = -y + 2z - 3w = -k_1 + \frac{4}{3}k_2 + 2k_1 - 3k_2$$

$$x = k_1 - \frac{5}{3}k_2$$

where k_1 and k_2 are arbitrary constants.

EXERCISE

Solve the system of homogeneous equations:

1. $x + 2y + 3z = 0,$
   $3x + 4y + 4z = 0,$

$$7x + 10y + 12z = 0.$$

Ans. Trivial solution $x = y = z = 0$ since $r(A) = 3 = n$

2. $$4x + 2y + z + 3w = 0,$$
$$6x + 3y + 4z + 7w = 0,$$
$$2x + y + w = 0.$$

Ans. $y = -2k_1 - k_2, z = -k_2, x = k_1, w = k_2$ where k_1 and k_2 are arbitrary constants, giving infinite number of solutions

3. $$x + y - 3z + 2w = 0,$$
$$2x - y + 2z - 3w = 0,$$
$$3x - 2y + z - 4w = 0,$$
$$-4x + y - 3z + w = 0.$$

Ans. Trivial solution $x = y = z = 0$, since $r(A) = 4 = n$

4. $$x_1 + x_2 + x_3 + x_4 = 0,$$
$$x_1 + 3x_2 + 2x_3 + 4x_4 = 0,$$
$$2x_1 + x_3 - x_4 = 0.$$

Ans. $x_1 = -\frac{1}{2}k_1 + \frac{1}{2}k_2, x_2 = -\frac{1}{2}k_1 \frac{-3}{2}k_2,$ $x_3 = k_1, x_4 = k_2$ where k_1 and k_2 are arbitrary constants giving infinite number of solutions: $r(A) = 2, n = 4$

5. $$3x + 2y + z = 0,$$
$$2x + 3z = 0,$$
$$y + 5z = 0,$$
$$x + 2y + 3z = 0.$$

Ans. $x = 0 = y = z$ is the only (trivial) solution since $r(A) = 3 = n$

6. $$2x + 3y - 4z + w = 0,$$
$$x - y + z + 2w = 0,$$
$$5x - z + 7w = 0,$$
$$7x + 8y - 11z + 5w = 0.$$

Ans. $z = k_1, w = k_2, x = (k_1 - 7k_2)/5,$ $y = (6k_1 + 3k_2)/5$ where k_1, k_2 are arbitrary constants

7. $$x + 3y - 2z = 0,$$
$$2x - y + 4z = 0,$$
$$x - 11y + 14z = 0.$$

Ans. $z = k, x = -10k/7, y = 8k/7, k$ arbitrary

8. $$x_1 + 3x_2 + 2x_3 = 0,$$
$$2x_1 - x_2 + 3x_3 = 0,$$
$$3x_1 - 5x_2 + 4x_3 = 0,$$
$$x_1 + 17x_2 + 4x_3 = 0.$$

Ans. $x_1 = 11k, x_2 = k, x_3 = -7k, k$ is arbitrary $r(A) = 2, n = 3$

9. Determine the values of b for which the system of equations has non-trivial solutions. Find them.

$$(b - 1)x + (4b - 2)y + (b + 3)z = 0,$$
$$(b - 1)x + (3b + 1)y + 2bz = 0,$$
$$2x + (3b + 1)y + 3(b - 1)z = 0.$$

Ans. **i.** $b = 0, x = y = z$
ii. $b = 3, x = -5k_1 - 3k_2, y = k_1, z = k_2$ where k_1 and k_2 are arbitrary

10. Find the values of b for which the system has non-trivial solutions. Find them

$$2x + 3by + (3b + 4)z = 0,$$
$$x + (b + 4)y + (4b + 2)z = 0,$$
$$x + 2(b + 1)y + (3b + 4)z = 0.$$

Ans. **i.** $b \neq \pm 2$, only trivial solution $x = y = z = 0$
ii. $b = 2, x = 0, z = k, y = -5k/3, k$ arbitrary
iii. $b = -2, x = 4k, y = z = k, k$ arbitrary.

13.6 GAUSSIAN ELIMINATION METHOD

Gaussian elimination method is an exact method which solves a given system of equations in n unknowns by transforming the coefficient matrix, into an upper triangular matrix and then solve for the unknowns by back substitution.

Consider a system of n equations in n unknowns

$$a_{11}x_1 + a_{12}x_2 + a_{13}x_3 + \cdots + a_{1n}x_n = a_{1,n+1} \quad (1)$$
$$a_{21}x_1 + a_{22}x_2 + a_{23}x_3 + \cdots + a_{2n}x_n = a_{2,n+1} \quad (2)$$
$$\cdots\cdots\cdots\cdots\cdots\cdots\cdots\cdots\cdots\cdots$$
$$a_{n1}x_1 + a_{n2}x_2 + a_{n3}x_3 + \cdots + a_{nn}x_n = a_{n,n+1} \quad (n)$$

Eliminate the unknown x_1 from the $(n-1)$ equations namely (2), (3), $\dots (n-1)$, (n) by subtracting the multiple $\frac{a_{i1}}{a_{11}}$ of the first equation from the ith equation, for $i = 2, 3, 4, \cdots, n$. Now eliminate x_2 from the $(n-2)$ equations of the resultant system. By this procedure, we arrive at a derived system as follows:

$$a_{11}x_1 + a_{12}x_2 + \cdots + a_{1n}x_n = a_{1,n+1} \quad (1)$$

$$a_{22}^{(1)} + x_2 + \cdots + a_{2n}^{(1)}x_n = a_{2,n+1}^{(2)} \quad (2^*)$$

$$a_{33}^{(2)}x_3 + \cdots + a_{3n}^{(2)}x_n = a_{3,n+1}^{(2)} \quad (3^*)$$

$$\cdots\cdots\cdots\cdots\cdots\cdots$$

$$a_{nn}^{(n-1)}x_n = a_{n,n+1}^{(n-1)} \quad (n^*)$$

In the forward elimination process, the coefficients are given by

$$\boxed{a_{ij}^{(k)} = a_{ij}^{(k-1)} - \frac{a_{ik}^{(k-1)}}{a_{kk}^{(k-1)}}a_{kj}^{(k-1)}}$$

where $k = 1, 2, \dots, n-1$

$j = k+1, \dots, n+1$

$i = k+1, \dots, n$

and $a_{ij}^{(0)} = a_{ij}$

Back substitution

Now the drived system (1), (2*), (3*) $\cdots (n^*)$ is solved by back substitution. Solve equation (n^*) for the unknown x_n. Substituting this x_n in (n^*-1) equation, solve for x_{n-1}. Continuing this process, x_1 is solved from the first equation. Thus

$$\boxed{x_i = \frac{1}{a_{ii}^{(i-1)}}\left[a_{i,n+1}^{(i-1)} - \sum_{j=i+1}^{n} a_{ij}^{(i-1)}x_j\right]}$$

for $i = n, n-1, n-2, \dots, 3, 2, 1.$

Check sum

Initially, for the given system, write row sum, the sum of the coefficients in each row, in the $(n+2)$nd column. Perform the same operations on the elements of this column also. Now in the absence of computational errors, at any stage, the row sum element in $(n+2)$nd row, will be equal to the sum of the elements of the corresponding transformed row.

Gauss-Jordan elimination method

Apply elementary row operations on both A and B such that A reduces to the normal form I_r. Then the solution is obtained (without the necessity of back substitution).

WORKED OUT EXAMPLES

Example 1: Solve by Gaussian elimination method, the following system of equations:

$$2x_1 + 2x_2 + x_3 + 2x_4 = 7$$
$$-x_1 + 2x_2 + x_4 = -2$$
$$-3x_1 + x_2 + 2x_3 + x_4 = -3$$
$$-x_1 + 2x_4 = 0.$$

Solution: Arranging in tabular form, we get

Table 13.1

Row No.	x_1	x_2	x_3	x_4	b	Check Sum	Explanation
[1]	2	2	1	2	7	14	Equation 1
[2]	−1	2	0	1	−2	0	Equation 2
[3]	−3	1	2	1	−3	−2	Equation 3
[4]	−1	0	0	2	0	1	Equation 4
[5]	1	1	$\frac{1}{2}$	1	$\frac{7}{2}$	7	$R_1\left(\frac{1}{2}\right)$
[6]	0	3	$\frac{1}{2}$	2	$\frac{3}{2}$	7	$R_{25(1)}$
[7]	0	4	$\frac{7}{2}$	4	$\frac{15}{2}$	19	$R_{35(3)}$
[8]	0	1	$\frac{1}{2}$	3	$\frac{7}{2}$	8	$R_{45(1)}$
[9]	0	1	$\frac{1}{2}$	3	$\frac{7}{2}$	8	R_{86}
[10]	0	3	$\frac{1}{2}$	2	$\frac{3}{2}$	7	R_{78}
[11]	0	4	$\frac{7}{2}$	4	$\frac{15}{2}$	19	
[12]	0	0	−1	−7	−9	−17	$R_{10,9(-3)}$
[13]	0	0	$\frac{3}{2}$	−8	$\frac{-13}{2}$	−13	$R_{11,9(-4)}$
[14]	0	0	0	$\frac{-37}{2}$	−20	$\frac{-77}{2}$	$R_{13,12\left(\frac{3}{2}\right)}$

Here $R_{ij(k)}$ denotes a row operation in which the kth multiples of jth row are added to the corresponding elements of the ith row. Also, R_{ij}: interchange of ith and jth rows.

Check sum: The sum of the elements of any row must be equal to check sum (otherwise errors in operations). The given system of equations has reduced to an upper triangular matrix. Now using back substitution, solve [14] (row) equation

$$x_4 = \frac{40}{37} = 1.08.$$

Solve [13] equation

$$x_3 + 7x_4 = 9 \quad \text{or} \quad x_3 = 1.4324$$

Solve [9]: $x_2 = -0.4562$

Solve [5]: $x_1 = 2.1600$

The solution is ($x_1 = 2.16, x_2 = -0.4562, x_3 = 1.4324, x_4 = 1.08$).

Example 2: Solve the system by (i) Gaussian elimination method (ii) Gauss-Jordan method

$$\begin{aligned} 2x_1 + 5x_2 + 2x_3 - 3x_4 &= 3 \\ 3x_1 + 6x_2 + 5x_3 + 2x_4 &= 2 \\ 4x_1 + 5x_2 + 14x_3 + 14x_4 &= 11 \\ 5x_1 + 10x_2 + 8x_3 + 4x_4 &= 4 \end{aligned}$$

Solution: Consider the augmented matrix $[A|B]$

$$[A|B] = \left[\begin{array}{cccc|c} 2 & 5 & 2 & -3 & 3 \\ 3 & 6 & 5 & 2 & 2 \\ 4 & 5 & 14 & 14 & 11 \\ 5 & 10 & 8 & 4 & 4 \end{array}\right] \begin{matrix} R_{21(-1)} \\ R_{32(-1)} \\ R_{43(-1)} \end{matrix}$$

$$\sim \left[\begin{array}{cccc|c} 2 & 5 & 2 & -3 & 3 \\ 1 & 1 & 3 & 5 & -1 \\ 1 & -1 & 9 & 12 & 9 \\ 1 & 5 & -6 & -10 & -7 \end{array}\right] \begin{matrix} R_{41}, R_{23} \\ R_{21(-1)} \\ R_{31(-1)} \\ R_{41(-2)} \end{matrix}$$

$$\sim \left[\begin{array}{cccc|c} 1 & 5 & -6 & -10 & -7 \\ 0 & -6 & 15 & 22 & 16 \\ 0 & -4 & 9 & 15 & 6 \\ 0 & -5 & 14 & 17 & 17 \end{array}\right] \begin{matrix} R_{24(-1)} \\ R_{2(-1)}, R_{32(-4)} \\ R_{3(-1)}\ R_{42(-5)} \\ R_{4(-1)} \end{matrix}$$

$$\sim \left[\begin{array}{cccc|c} 1 & 5 & -6 & -10 & -7 \\ 0 & 1 & -1 & -5 & 1 \\ 0 & 0 & -5 & 5 & -10 \\ 0 & 0 & 9 & -8 & 22 \end{array}\right] \begin{matrix} R_{3(-\frac{1}{5})} \\ R_{43(-9)} \end{matrix}$$

$$[A|B] \sim \left[\begin{array}{cccc|c} 1 & 5 & -6 & -10 & -7 \\ 0 & 1 & -1 & -5 & 1 \\ 0 & 0 & 1 & -1 & 2 \\ 0 & 0 & 0 & 1 & 4 \end{array}\right]$$

By back substitution: $x_4 = 4$

$$x_3 - x_4 = 2 \quad \text{or} \quad x_3 = 2 + x_4 = 2 + 4 = 6$$

$$x_2 - x_3 - 5x_4 = 1 \quad \text{or} \quad x_2 = 27$$

$$x_1 + 5x_2 - 6x_3 - 10x_4 = -7 \quad \text{or} \quad x_1 = -66$$

Thus $x_1 = -66, x_2 = 27, x_3 = 6, x_4 = 4$.

Gauss-Jordan method:

$$\left[\begin{array}{cccc|c} 1 & 5 & -6 & -10 & -7 \\ 0 & 1 & -1 & -5 & 1 \\ 0 & 0 & 1 & -1 & 2 \\ 0 & 0 & 0 & 1 & 4 \end{array}\right] \begin{matrix} R_{34(1)} \\ R_{24(5)} \\ R_{14(10)} \\ \sim \end{matrix} \left[\begin{array}{cccc|c} 1 & 5 & -6 & 0 & 33 \\ 0 & 1 & -1 & 0 & 21 \\ 0 & 0 & 1 & 0 & 6 \\ 0 & 0 & 0 & 1 & 4 \end{array}\right]$$

$$\begin{matrix} R_{23(1)} \\ \sim \\ R_{13(6)} \end{matrix} \left[\begin{array}{cccc|c} 1 & 5 & 0 & 0 & 69 \\ 0 & 1 & 0 & 0 & 27 \\ 0 & 0 & 1 & 0 & 6 \\ 0 & 0 & 0 & 1 & 4 \end{array}\right] \begin{matrix} R_{12(-5)} \\ \sim \end{matrix} \left[\begin{array}{cccc|c} 1 & 0 & 0 & 0 & -66 \\ 0 & 1 & 0 & 0 & 27 \\ 0 & 0 & 1 & 0 & 6 \\ 0 & 0 & 0 & 1 & 4 \end{array}\right]$$

$\therefore x_1 = -66, x_2 = 27, x_3 = 6, x_4 = 4.$

EXERCISE

Solve the following system of equations by Gaussian elimination method.

1. $x_1 + 2x_2 - x_3 = 3, \quad 3x_1 - x_2 + 2x_3 = 1, \quad 2x_1 - 2x_2 + 3x_3 = 2, \quad x_1 - x_2 + x_3 = -1$

   Ans. $x_1 = 1, x_2 = 4, x_3 = 4$

2. $2x_1 + x_2 + 3x_3 = 1, \quad 4x_1 + 4x_2 + 7x_3 = 1, \quad 2x_1 + 5x_2 + 9x_3 = 3$

   Ans. $x_1 = -\frac{1}{2}, x_2 = -1, x_3 = 1$

3. $2x_1 - 7x_2 + 4x_3 = 9, \quad x_1 + 9x_2 - 6x_3 = 1, \quad -3x_1 + 8x_2 + 5x_3 = 6$

   Ans. $x_1 = 4, x_2 = 1, x_3 = 2$

4. $2x_1 + 2x_2 + 4x_3 = 18, \quad x_1 + 3x_2 + 2x_3 = 13, \quad 3x_1 + x_2 + 3x_3 = 14$

   Ans. $x_1 = 1, x_2 = 2, x_3 = 3$

5. $2x_1 + x_2 + x_3 = 10, \quad 3x_1 + 2x_2 + 3x_3 = 18, \quad x_1 + 4x_2 + 9x_3 = 16$

   Ans. $x_1 = 7, x_2 = -9, x_3 = 5$

6. $2x_1 + x_2 + 4x_3 = 12, \quad 8x_1 - 3x_2 + 2x_3 = 20, \quad 4x_1 + 11x_2 - x_3 = 33$

   Ans. $x_1 = 3, x_2 = 2, x_3 = 1$

7. $x_1 + 4x_2 - x_3 = -5, \quad x_1 + x_2 - 6x_3 = -12,$

$3x_1 - x_2 - x_3 = 4$

Ans. $x_1 = \frac{117}{71}, x_2 = -\frac{81}{71}, x_3 = \frac{148}{71}$

8.
$$10x_1 - 7x_2 + 3x_3 + 5x_4 = 6,$$
$$-6x_1 + 8x_2 - x_3 - 4x_4 = 5,$$
$$3x_1 + x_2 + 4x_3 + 11x_4 = 2,$$
$$5x_1 - 9x_2 - 2x_3 + 4x_4 = 7$$

Ans. $x_1 = 5, x_2 = 4, x_3 = -7, x_4 = 1.$

13.7 LU-DECOMPOSITIONS

The Gaussian elimination with back substitution, Gauss-Jordan elimination, computing A^{-1} then $x = A^{-1}B$ and Cramer's rule are some of the direct (non-iterative) methods for solving system of linear equations. Gauss-Jordan elimination produces both solution for one or more R.H.S. vector B and also A^{-1}. Its principal weakness is (i) it requires all RHS $\overline{B}$ to be stored and manipulated and (ii) when A^{-1} is not required. The usefulness of Gaussian elimination with back substitution is primarily pedagogical. It stands between full elimination schemes such as Gauss-Jordan and triangular decomposition. *LU*-decomposition or triangular decomposition (triangular factorization) is a different approach in which the coefficient matrix A is factored into the product of a lower matrix L and an upper triangular matrix U i.e.,

$$A = LU.$$

Since a matrix that is either upper triangular or lower triangular is called "triangular", so *LU*-decomposition is also referred to as triangular factorization. *LU* method can be easily adopted to solve a system with new R.H.S. B with great economy of effort.

It is popular because storage of space can be economized and accumulates sums in double precision (Example: LINPAK (1979) computer program of Argonne National Labs).

Solution of Linear System by LU-Decomposition

A non singular matrix A is said to have a triangular factorization or *LU*-decomposition if A can be expressed as the product of a lower triangular matrix L with ones on its main diagonal and an upper triangular matrix U. i.e.,

$$A = LU$$

For $n = 4$, we have $A_{4\times4} = L_{4\times4}U_{4\times4}$

$$\begin{bmatrix} a_{11} & a_{12} & a_{13} & a_{14} \\ a_{21} & a_{22} & a_{23} & a_{24} \\ a_{31} & a_{32} & a_{33} & a_{34} \\ a_{41} & a_{42} & a_{43} & a_{44} \end{bmatrix}$$
$$= \begin{bmatrix} 1 & 0 & 0 & 0 \\ l_{21} & 1 & 0 & 0 \\ l_{31} & l_{32} & 1 & 0 \\ l_{41} & l_{42} & l_{43} & 1 \end{bmatrix} \begin{bmatrix} u_{11} & u_{12} & u_{13} & u_{14} \\ 0 & u_{22} & u_{23} & u_{24} \\ 0 & 0 & u_{33} & u_{34} \\ 0 & 0 & 0 & u_{44} \end{bmatrix}$$

The condition of non singularity of A implies that $u_{kk} \neq 0$ for all k. Now consider the system of equations

$$AX = B$$

or
$$LUX = B$$

Put $Y = UX$ then

$$LY = B$$

and
$$UX = Y$$

Solve first $LY = B$ for Y using forward substitution and then solve $UX = Y$ for X using backward substitution. Here X is the required solution vector.

LU-decomposition is also known as Doolittle's method. Another variation of *LU*-decomposition is crout's reduction or Cholosky's reduction in which the upper triangular matrix U has ones on its main diagonal (instead of L) in the triangular decomposition $A = LU$

Note that *LU* decompostion is not unique. Any matrix A with all non-zero diagonal elements (i.e., $a_{ii} \neq 0$ for $i = 1$ to n) can be factored in infinite number of ways.

Example 1:

$$\begin{bmatrix} 2 & -1 & -1 \\ 0 & -4 & 2 \\ 6 & -3 & 1 \end{bmatrix} = \begin{bmatrix} 2 & 0 & 0 \\ 0 & -4 & 0 \\ 6 & 0 & 4 \end{bmatrix} \begin{bmatrix} 1 & -\frac{1}{2} & -\frac{1}{2} \\ 0 & 1 & -\frac{1}{2} \\ 0 & 0 & 1 \end{bmatrix} = LU$$

or

$$= \begin{bmatrix} 1 & 0 & 0 \\ 0 & 1 & 0 \\ 3 & 0 & 1 \end{bmatrix} \begin{bmatrix} 2 & -1 & -1 \\ 0 & -4 & 2 \\ 0 & 0 & 4 \end{bmatrix} = LU$$

or

$$= \begin{bmatrix} 1 & 0 & 0 \\ 0 & 2 & 0 \\ 3 & 0 & 1 \end{bmatrix} \begin{bmatrix} 2 & -1 & -1 \\ 0 & -2 & 1 \\ 0 & 0 & 4 \end{bmatrix} = LU$$

and so on.

Example 2:

$$A = LU = \begin{bmatrix} l_{11} & 0 & 0 \\ l_{21} & l_{22} & 0 \\ l_{31} & l_{32} & l_{33} \end{bmatrix} \begin{bmatrix} 1 & u_{12} & u_{13} \\ 0 & 1 & u_{23} \\ 0 & 0 & 1 \end{bmatrix}$$

The non-zero diagonal entries in L can be shifted to U.

$$A = \begin{bmatrix} 1 & 0 & 0 \\ \frac{l_{21}}{l_{11}} & 1 & 0 \\ \frac{l_{31}}{l_{11}} & \frac{l_{32}}{l_{22}} & 1 \end{bmatrix} \begin{bmatrix} l_{11} & 0 & 0 \\ 0 & l_{22} & 0 \\ 0 & 0 & l_{33} \end{bmatrix} \begin{bmatrix} 1 & u_{12} & u_{13} \\ 0 & 1 & u_{23} \\ 0 & 0 & 1 \end{bmatrix}$$

$$= \begin{bmatrix} 1 & 0 & 0 \\ \frac{l_{21}}{l_{11}} & 1 & 0 \\ \frac{l_{31}}{l_{11}} & \frac{l_{32}}{l_{22}} & 1 \end{bmatrix} \begin{bmatrix} l_{11} & l_{11}u_{12} & l_{11}l_{13} \\ 0 & l_{22} & l_{22}u_{23} \\ 0 & 0 & l_{33} \end{bmatrix}$$

which is another LU decomposition of A. However of the entire set of LU decompositions, choose the pair in which L has ones on its diagonal.

13.8 LU-DECOMPOSITION FROM GAUSSIAN ELIMINATION

Theorem: If A is a square matrix which can be reduced to echlon from U without using any row interchanges, then A has a LU decomposition and can be factored as $A = LU$ where L is a lower triangular matrix with ones on its main diagonal.

Explanation: In solving a system $AX = B$ of n equations in n unknowns, use the Gaussian elimination method to reduce A to an echlon form (upper triangular matrix) U. We assume that no row interchanges are necessary in this process. Then the multipliers l_{ij} used in the Gaussian elimination process will form the subdiagonal enteries in the lower triangular matrix L.

Step I: Use Gaussian elimination to reduce A to echolon form U, without using any row interchanges. Keep track of the multipliers used to introduce zeros below the leading diagonal elements of A.

Step II: In each position below the main diaginal (consisting of ones) of L, place the *negative* of the multiplier used to introduce zeros in that position in U.

The LU-decomposition can also be obtained by solving the equations in l_{ij} and u_{ij} as follows. Suppose

$$A = \begin{pmatrix} a_{11} & a_{12} & a_{13} \\ a_{21} & a_{22} & a_{23} \\ a_{31} & a_{32} & a_{33} \end{pmatrix} = LU =$$

$$= \begin{pmatrix} 1 & 0 & 0 \\ l_{21} & 1 & 0 \\ l_{31} & l_{32} & 1 \end{pmatrix} \cdot \begin{pmatrix} u_{11} & u_{12} & u_{13} \\ 0 & u_{22} & u_{23} \\ 0 & 0 & u_{33} \end{pmatrix}$$

From the first row elements.

$u_{11} = a_{11}, u_{12} = a_{12}, u_{13} = a_{13}$

From 2nd row elements:

$l_{21}u_{11} = a_{21}, l_{21}u_{12} + u_{22} = a_{22},$

$$l_{21}u_{13} + u_{23} = a_{23}$$

From 3rd row elements

$$l_{31}u_{11} = a_{31}, l_{31}u_{12} + l_{32}u_{22} = a_{32},$$

$$l_{31}u_{13} = l_{32}u_{23} + u_{33} = a_{33}.$$

Solving we get u_{11}, u_{12}, u_{13}, then l_{21}, u_{22}, u_{23} followed by l_{31}, l_{32}, u_{33}.

LU-Decomposition by Gaussian Elimination

WORKED OUT EXAMPLES

Example 1: Solve $AX = B$ by LU-decomposition using Gaussian elimination where

$$A = \begin{pmatrix} 2 & 4 & -6 \\ 1 & 5 & 3 \\ 1 & 3 & 2 \end{pmatrix}$$

and

(a) $B^T = (-4, 10, 5)$, (b) $B^T = (20, 49, 32)$

Solution: Since A has all non-zero diagonal elements, we can factor A as LU. Use Gaussian elimination to reduce A to echlon form U, without using any row interchanges.

$$A = \begin{pmatrix} 2 & 4 & -6 \\ 1 & 5 & 3 \\ 1 & 3 & 2 \end{pmatrix}$$

Step I: The multiplier $\left(-\frac{1}{2}\right)$ is used to reduce the element $a_{21} = 1$ to zero. The operation is $R_{21}\left(-\frac{1}{2}\right)$. So $m_{21} = -\frac{1}{2}$. Similarly the multiplier $\left(-\frac{1}{2}\right)$ is used to reduce the element $a_{31} = 1$ to zero i.e., $R_{31}\left(-\frac{1}{2}\right)$. So $m_{31} = -\frac{1}{2}$. This results

$$A \sim \begin{pmatrix} 2 & 4 & -6 \\ 0 & 3 & 6 \\ 0 & 1 & 5 \end{pmatrix}$$

Use the multiplier $\left(-\frac{1}{3}\right)$ to reduce the element $a_{32} = 1$ to zero i.e., $R_{32}\left(-\frac{1}{3}\right)$. So the multiplier is $m_{32} = -\frac{1}{3}$. This yields the echelon form (or upper triangular matrix) U of A as

$$A \sim \begin{pmatrix} 2 & 4 & -6 \\ 0 & 3 & 6 \\ 0 & 0 & 3 \end{pmatrix}$$

Step II: The lower triangular matrix L is obtained by simply placing the *negative* of the multipliers used in introducing zeros in that position in U i.e., $l_{21} = -m_{21} = -\left(-\frac{1}{2}\right) = +\frac{1}{2}$, $l_{31} = -m_{31} = \frac{1}{2}$ and $l_{32} = -m_{32} = \frac{1}{3}$

$$L = \begin{pmatrix} 1 & 0 & 0 \\ l_{21} & 1 & 0 \\ l_{31} & l_{32} & 0 \end{pmatrix} = \begin{pmatrix} 1 & 0 & 0 \\ \frac{1}{2} & 1 & 0 \\ \frac{1}{2} & \frac{1}{3} & 0 \end{pmatrix}$$

Thus the LU (factorization) decomposition of A by Gaussian elimination is

$$A = \begin{pmatrix} 2 & 4 & -6 \\ 1 & 5 & 3 \\ 1 & 3 & 2 \end{pmatrix} = LU = \begin{pmatrix} 1 & 0 & 0 \\ \frac{1}{2} & 1 & 0 \\ \frac{1}{2} & \frac{1}{3} & 0 \end{pmatrix} \begin{pmatrix} 2 & 4 & -6 \\ 0 & 3 & 6 \\ 0 & 0 & 3 \end{pmatrix}$$

Now to solve $AX = B$, $LUX = B$ put $UX = Y$ so $LY = B$. First we solve $LY = B$ for Y by using forward substitution

(a) $B^T = (-4, 10, 5)$

$$\begin{pmatrix} 1 & 0 & 0 \\ \frac{1}{2} & 1 & 0 \\ \frac{1}{2} & \frac{1}{3} & 1 \end{pmatrix} \begin{pmatrix} y_1 \\ y_2 \\ y_3 \end{pmatrix} = \begin{pmatrix} -4 \\ 10 \\ 5 \end{pmatrix}$$

By forward substitution

$$y_1 = -4$$

$$\frac{1}{2}y_1 + y_2 = 10 \text{ so } y_2 = 10 - \frac{1}{2}y_1 = 12$$

$$\tfrac{1}{2}y_1 + \tfrac{1}{3}y_2 + y_3 = 5 \text{ so } y_3 = 5 - \tfrac{1}{2}y_1, -\tfrac{1}{3}y_2 = 3.$$

Thus $Y^T = (-4, 12, 3)$.

Now solve $UX = Y$ using backward substitution.

$$\begin{pmatrix} 2 & 4 & -6 \\ 0 & 3 & 6 \\ 0 & 0 & 3 \end{pmatrix} \begin{pmatrix} x_1 \\ x_2 \\ x_3 \end{pmatrix} = \begin{pmatrix} -4 \\ 12 \\ 3 \end{pmatrix}$$

So $3x_3 = 3$ or $x_3 = 1$

$$3x_2 + 6x_3 = 12 \text{ so } x_2 = \frac{12 - 6x_3}{3} = 2$$

$$2x_1 + 4x_2 - 6x_3 = -4 \text{ so } x_1 = \frac{-4 + 6x_3 - 4x_2}{2} = -3$$

Solution: $X^T = (-3, 2, 1)$

(b) $B^T = (20, 49, 32)$

$$LY = B$$

$$\begin{pmatrix} 1 & 0 & 0 \\ \frac{1}{2} & 1 & 0 \\ \frac{1}{2} & \frac{1}{3} & 1 \end{pmatrix} \begin{pmatrix} y_1 \\ y_2 \\ y_3 \end{pmatrix} = \begin{pmatrix} 20 \\ 49 \\ 32 \end{pmatrix}$$

Solving $y_1 = 20$

$\frac{1}{2}y_1 + y_2 = 49$ so $y_2 = 49 - \frac{1}{2}y_1 = 39$

$$\frac{1}{2}y_1 + \frac{1}{3}y_2 + y_3 = 32 \text{ so } y_3 = 32 - \frac{1}{2}y_1 - \frac{1}{3}y_2 = 9$$

Thus $Y^T = (20, 39, 9)$

Now solve $UX = Y$

$$\begin{pmatrix} 2 & 4 & -6 \\ 0 & 3 & 6 \\ 0 & 0 & 3 \end{pmatrix}\begin{pmatrix} x_1 \\ x_2 \\ x_3 \end{pmatrix} = \begin{pmatrix} 20 \\ 39 \\ 9 \end{pmatrix}$$

Solving $3x_3 = 9$ so $x_3 = 3$

$$3x_2 + 6x_3 = 39 \quad \text{so } x_2 = \frac{39 - 6x_3}{3} = 7$$

$$2x_1 + 4x_2 - 6x_3 = 20 \text{ so } x_1 = \frac{20 + 6x_3 - 4x_2}{2} = 5$$

Solution: $X^T = (5, 7, 3)$.

Example 2: Solve the system

$$3x_1 - 6x_2 - 3x_3 = -3$$
$$2x_1 \quad + 6x_3 = -22$$
$$-4x_1 + 7x_2 + 4x_3 = 3$$

Solution: $A = \begin{bmatrix} 3 & -6 & -3 \\ 2 & 0 & 6 \\ -4 & 7 & 4 \end{bmatrix}$, $B = \begin{bmatrix} -3 \\ -22 \\ 3 \end{bmatrix}$

consider

$$A = \begin{bmatrix} 3 & -6 & -3 \\ 2 & 0 & 6 \\ -4 & 7 & 4 \end{bmatrix} = LU =$$

$$= \begin{bmatrix} 1 & 0 & 0 \\ l_{21} & 1 & 0 \\ l_{31} & l_{32} & 1 \end{bmatrix}\begin{bmatrix} u_{11} & u_{12} & u_{13} \\ 0 & u_{22} & u_{23} \\ 0 & 0 & u_{33} \end{bmatrix}$$

From first row: $u_{11} = 3, u_{12} = -6, u_{13} = -3$

First 2nd row: $l_{21}u_{11} = 2, l_{21} \cdot 3 = 2, l_{21} = \frac{2}{3}$

$l_{21}u_{12} + u_{22} = 0, u_{22} = -l_{21}u_{12} = -\frac{2}{3} \cdot (-6) = 4$

$l_{21}u_{13} + u_{23} = 6, u_{23} = 6 - l_{21}u_{13}$

$= 6 - \frac{2}{3}(-3) = 8$

From 3rd row: $l_{31}u_{11} = -4, l_{31} = -\frac{4}{3}$

$l_{31}u_{12} + l_{32}u_{22} = 7, l_{32} = \frac{7 - l_{31}u_{12}}{u_{22}} = \frac{7 - \left(-\frac{4}{3}\right)(-6)}{4}$

so $l_{32} = -\frac{1}{4}$

$l_{31}u_{13} + l_{32}u_{23} + u_{33} = 4$

so $u_{33} = 4 - l_{31}u_{13} - l_{32}u_{23}$

$u_{33} = 4 - \left(-\frac{4}{3}\right) \cdot (-3) - \left(\frac{-1}{4}\right) \cdot 8 = 0 + 2 = +2$

$$A = \begin{bmatrix} 1 & 0 & 0 \\ \frac{2}{3} & 1 & 0 \\ \frac{-4}{3} & -\frac{1}{4} & 1 \end{bmatrix}\begin{bmatrix} 3 & -6 & 3 \\ 0 & 4 & 8 \\ 0 & 0 & +2 \end{bmatrix}$$

$$LY = B$$

$$\begin{bmatrix} 1 & 0 & 0 \\ \frac{2}{3} & 1 & 0 \\ -\frac{4}{3} & -\frac{1}{4} & 1 \end{bmatrix}\begin{bmatrix} y_1 \\ y_2 \\ y_3 \end{bmatrix} = \begin{bmatrix} -3 \\ -22 \\ 3 \end{bmatrix}$$

Solving $y_1 = -3$, $\frac{2}{3}y_1 + y_2 = -22$

$$y_2 = -22 - \frac{2}{3}y_1 = -20$$

So $$-\frac{4}{3}y_1 - \frac{1}{4}y_2 + y_3 = 3$$

$$y_3 = 3 + \frac{4}{3}y_1 + \frac{1}{4}y_2 = 3 - 4 - 5 = -6$$

$$UX = Y$$

$$\begin{bmatrix} 3 & -6 & -3 \\ 0 & 4 & 8 \\ 0 & 0 & 2 \end{bmatrix}\begin{bmatrix} x_1 \\ x_2 \\ x_3 \end{bmatrix} = \begin{bmatrix} -3 \\ -20 \\ -6 \end{bmatrix}$$

solving $2x_3 = -6$ or $x_3 = -3$

$4x_2 + 8x_3 = -20$ or $x_2 = \frac{-20-8x_3}{4} = 1$

$3x_1 - 6x_2 - 3x_3 = -3$ or $x_1 = \frac{-3+6x_2+3x_3}{3} = -2$

solution $X^T = [-2, 1, -3]$.

Exercise

1. Let $A = \begin{bmatrix} a & b \\ c & d \end{bmatrix}$. If $a \neq 0$ find LU decomposition.

Ans. $\begin{bmatrix} 1 & 0 \\ \frac{c}{a} & 1 \end{bmatrix}\begin{bmatrix} a & b \\ 0 & \frac{ad-bc}{a} \end{bmatrix}$

2. Let $A = \begin{bmatrix} 0 & 1 \\ 1 & 0 \end{bmatrix}$. Show that A has no LU-decomposition.

Hint: Since $a_{11} = 0$, $a_{21} = 1$ can not be made zero.

Solve the following system of equations by *LU*-decomposition.

3. $3x_1 - 6x_2 = 0$, $-2x_1 + 5x_2 = 1$

Ans. $x_1 = 2, x_2 = 1$

Hint: $L = \begin{pmatrix} 1 & 0 \\ -\frac{2}{3} & 1 \end{pmatrix}$, $U = \begin{pmatrix} 3 & -6 \\ 0 & 1 \end{pmatrix}$

4. $2x_1 + 8x_2 = -2$, $-x_1 - x_2 = -2$

Ans. $x_1 = 3$, $x_2 = -1$

Hint: $L = \begin{pmatrix} 1 & 0 \\ -\frac{1}{2} & 1 \end{pmatrix}$, $U = \begin{pmatrix} 2 & 8 \\ 0 & 3 \end{pmatrix}$

5. $-5x_1 - 10x_2 = -10$, $6x_1 + 5x_2 = 19$

Ans. $x_1 = 4, x_2 = -1$

Hint: $L = \begin{pmatrix} 1 & 0 \\ -\frac{6}{5} & 1 \end{pmatrix}$, $U = \begin{pmatrix} -5 & -10 \\ 0 & -7 \end{pmatrix}$

6. $2x_1 - 2x_2 - 2x_3 = -4$
$- 2x_2 + 2x_3 = -2$
$-x_1 + 5x_2 + 2x_3 = 6$

Ans. $x_1 = -1, x_2 = 1, x_3 = 0$

Hint: $L = \begin{pmatrix} 1 & 0 & 0 \\ 0 & 1 & 0 \\ -\frac{1}{2} & -2 & 1 \end{pmatrix}$,

$U = \begin{pmatrix} 2 & -2 & -2 \\ 0 & -2 & 2 \\ 0 & 0 & 5 \end{pmatrix}$

7. $\begin{bmatrix} -1 & 0 & 1 & 0 \\ 2 & 3 & -2 & 6 \\ 0 & -1 & 2 & 0 \\ 0 & 0 & 1 & 5 \end{bmatrix} \begin{bmatrix} x_1 \\ x_2 \\ x_3 \\ x_4 \end{bmatrix} = \begin{bmatrix} 5 \\ -1 \\ 3 \\ 7 \end{bmatrix}$

Ans. $x_1 = -3$, $x_2 = 1$, $x_3 = 2$, $x_4 = 1$

Hint: $L = \begin{bmatrix} 1 & 0 & 0 & 0 \\ -2 & 1 & 0 & 0 \\ 0 & -\frac{1}{3} & 1 & 0 \\ 0 & 0 & \frac{1}{2} & 1 \end{bmatrix}$,

$U = \begin{bmatrix} -1 & 0 & 1 & 0 \\ 0 & 3 & 0 & 6 \\ 0 & 0 & 2 & 2 \\ 0 & 0 & 0 & 4 \end{bmatrix}$

8. Solve $\begin{bmatrix} 4 & 8 & 4 & 0 \\ 1 & 5 & 4 & -3 \\ 1 & 4 & 7 & 2 \\ 1 & 3 & 0 & -2 \end{bmatrix} \begin{bmatrix} x_1 \\ x_2 \\ x_3 \\ x_4 \end{bmatrix} = \begin{bmatrix} b_1 \\ b_2 \\ b_3 \\ b_4 \end{bmatrix}$

(a) $B^T = [8, -4, 10, -4]$

(b) $B^T = [28, 13, 23, 4]$

Ans. (a) $Y^T = [8, -6, 12, 2]$, $X^T = [3, -1, 1, 2]$

(b) $Y^T = [28, 6, 12, 1]$, $X^T = [3, 1, 2, 1]$

Hint: $L = \begin{bmatrix} 1 & 0 & 0 & 0 \\ \frac{1}{4} & 1 & 0 & 0 \\ \frac{1}{4} & \frac{2}{3} & 1 & 0 \\ \frac{1}{4} & \frac{1}{3} & -\frac{1}{2} & 1 \end{bmatrix}$,

$U = \begin{bmatrix} 4 & 8 & 4 & 0 \\ 0 & 3 & 3 & -3 \\ 0 & 0 & 4 & 4 \\ 0 & 0 & 0 & 1 \end{bmatrix}$

Find *LU* decomposition (triangular factorization) $A = LU$

9. $\begin{bmatrix} 4 & 3 & -1 \\ -2 & -4 & 5 \\ 1 & 2 & 6 \end{bmatrix}$

Ans. $L = \begin{bmatrix} 1 & 0 & 0 \\ -0.5 & 1 & 0 \\ 0.25 & -0.5 & 1 \end{bmatrix}$

$U = \begin{bmatrix} 4 & 3 & 1 \\ 0 & -2.5 & 4.5 \\ 0 & 0 & 8.5 \end{bmatrix}$

10. $\begin{bmatrix} -5 & 2 & -1 \\ 1 & 0 & 3 \\ 3 & 1 & 6 \end{bmatrix}$

Ans. $L = \begin{bmatrix} 1 & 0 & 0 \\ -0.2 & 1 & 0 \\ -0.6 & 5.5 & 1 \end{bmatrix}$

$U = \begin{bmatrix} -5 & 2 & -1 \\ 0 & 0.4 & 2.8 \\ 0 & 0 & -10 \end{bmatrix}$

11. $\begin{bmatrix} 1 & 1 & 0 & 4 \\ 2 & -1 & 5 & 0 \\ 5 & 2 & 1 & 2 \\ -3 & 0 & 2 & 6 \end{bmatrix}$

$$Ans.\ L = \begin{bmatrix} 1 & 0 & 0 & 0 \\ 2 & 1 & 0 & 0 \\ 5 & 1 & 1 & 0 \\ -3 & -1 & -1.75 & 1 \end{bmatrix}$$

$$U = \begin{bmatrix} 1 & 1 & 0 & 4 \\ 0 & -3 & 5 & -8 \\ 0 & 0 & -4 & -10 \\ 0 & 0 & 0 & -7.5 \end{bmatrix}$$

13.9 SOLUTION TO TRIDIAGONAL SYSTEMS

Band matrix is a $n \times n$ square matrix A with the property that $a_{ij} = 0$ whenever $i + p \le j$ or $j + q \le i$ for integers p and q with $p > 1$ and $q < n$. The *band width* of such matrix is defined to be $w = p + q - 1$.

Example:

$$A = \begin{bmatrix} 8 & 3 & 0 \\ 2 & 6 & -1 \\ 0 & 6 & -9 \end{bmatrix}$$

A is band matrix with $p = 2, q = 2$ and band width 3.

In band matrices, all the non-zero entries are concentrated about the diagonal.

Tridiagonal matrix is a band matrix of width 3 with $p = q = 2$. Thus *tridiagonal* matrices are those that have non-zero elements *only* on the diagonal a_{ii} or super diagonal $a_{i,i+1}$ or subdiagonal $a_{i+1,i}$. So $a_{ij} = -0$ if $|i - j| > 1$.

Example:

$$B = \begin{bmatrix} -4 & 2 & 0 & 0 & 0 \\ 1 & -4 & 1 & 0 & 0 \\ 0 & 1 & -4 & 1 & 0 \\ 0 & 0 & 1 & -4 & 1 \\ 0 & 0 & 0 & 2 & -4 \end{bmatrix}$$

Note: Non-zero elements (entries) occur only on the diagonal and in the positions adjacent to the diagonal.

Most often tridiagonal matrices occur in cubic spline interpolation and numerical solution (crank-Nicolson method) of PDE involving heat equation.

13.10 CROUT REDUCTION FOR TRIDIAGONAL LINEAR SYSTEMS

Consider a tridiagonal linear system of n equations in n unknowns.

$$a_{11}x_1 + a_{12}x_2 \cdots = b_1$$

$$a_{21}x_1 + a_{22}x_2 + a_{23}x_3 \cdots = b_2$$

$$\cdots$$

$$a_{n-1,n-2}x_{n-2} + a_{n-1,n-1}x_{n-1} + a_{n-1,n}x^n = b_{n-1}$$

$$a_{n,n-1}x_{n-1} + a_{nn}x_n = b_n$$

with the tridiagonal coefficient matrix

$$A = \begin{bmatrix} a_{11} & a_{12} & 0 & \cdots & 0 \\ a_{21} & a_{22} & a_{23} & & \vdots \\ 0 & a_{32} & a_{33} & a_{34} & 0 \\ & \ddots & \ddots & \ddots & \\ 0 & & \ddots & \ddots & a_{n-1,n} \\ 0 & \cdots & 0 & a_{n,n-1} & a_{nn} \end{bmatrix}$$

In the case of tridiagonal matrix A having large number of zeros in regular patterns, the computational effort is reduced due to the structure of A. Using Crout or Doolittle factorization algorithm, A can be factored into L and U where L is lower triangular matrix and U is an upper triangular matrix with one's on its main diagonal.

Here

$$L = \begin{bmatrix} l_{11} & 0 & \cdots & & 0 \\ l_{21} & l_{22} & \ddots & & \vdots \\ 0 & \ddots & \ddots & \ddots & \vdots \\ \vdots & \ddots & \ddots & \ddots & 0 \\ 0 & \cdots & 0 & l_{n,n-1} & l_{nn} \end{bmatrix},$$

$$U = \begin{bmatrix} 1 & u_{12} & 0 & \cdots & 0 \\ 0 & 1 & \ddots & \ddots & \vdots \\ \vdots & \ddots & \ddots & \ddots & 0 \\ 0 & & \ddots & \ddots & u_{n-1,n} \\ 0 & \cdots & \cdots & 0 & 1 \end{bmatrix}$$

Since A has only $(3n - 2)$ non-zero entries, there are only $(3n - 2)$ conditions to be applied to determine the entries of L and U. There are $(2n - 1)$ undetermined entries in L and $(n-1)$ undetermined entries of U, which totals the number of conditions $(3n-2)$.

Carrying out the multiplication LU, we get

$$a_{11} = l_{11}$$

$$a_{i,i-1} = l_{i,i-1} \quad \text{for} \quad i = 2, 3, \ldots n \quad (1)$$

$$a_{ii} = l_{i,i-1}u_{i-1,i} + l_{ii} \quad \text{for} \quad i = 2, 3, \ldots n \quad (2)$$

$$a_{i,i+1} = l_{ii}u_{i,i+1} \quad \text{for} \quad i = 1, 2, \ldots n-1 \quad (3)$$

From (2), non-zero off-diagonal terms in L are calculated first. Then using (3) and (2) obtain alternately the remainder of the entries in U and L.

Thus the tridiagonal system can be solved by LU decomposition followed by forward and backward substitution. The LU decomposition can be obtained using Gaussian elimination as is done in the earlier section (without tedious calculations of l_{ii} and u_{ii} see W.E.2).

If A is a symmetric $(a_{ij} = a_{ji})$ and positive definite (i.e., $V^TAV > 0$ for all $V \neq 0$). Then by Cholesky decomposition we can factorize A as

$$A = LL^T$$

This factorization is sometimes referred to as taking the square root of the matrix A. Instead of seeking arbitrary lower triangular matrix L and upper triangular matrix U, Cholesky decomposition constructs a lower triangular matrix L whose transpose L^T can itself serves as upper triangular matrix U.

WORKED OUT EXAMPLES

Example 1: Solve the following tridiagonal system.

$$x_1 - x_2 = 0$$

$$-2x_1 + 4x_2 - 2x_3 = -1$$

$$-x_2 + 2x_3 = 1.5$$

Solution: The tridiagonal matrix is

$$A = \begin{bmatrix} 1 & -1 & 0 \\ -2 & 4 & -2 \\ 0 & -1 & 2 \end{bmatrix}$$

LU-decomposition:

$$\begin{bmatrix} 1 & -1 & 0 \\ -2 & 4 & -2 \\ 0 & -1 & 2 \end{bmatrix} = \begin{bmatrix} l_{11} & 0 & 0 \\ l_{21} & l_{22} & 0 \\ l_{31} & l_{32} & l_{33} \end{bmatrix} \begin{bmatrix} 1 & u_{12} & u_{13} \\ 0 & 1 & u_{23} \\ 0 & 0 & 1 \end{bmatrix}$$

From first column: $l_{11} = a_{11} = 1$, $l_{21} = a_{21} = -2$, $l_{31} = 0$

From 2nd column: $l_{11}u_{12} = -1$, $u_{12} = \frac{-1}{l_{11}} = \frac{-1}{1} = -1$

$l_{21}u_{12} + l_{22} = 4$, $l_{22} = 4 - l_{21}u_{12} = 4 - (-2)(-1) = 2$,
$l_{21}u_{13} + l_{22}u_{23} = -2$,
Also $l_{11}u_{13} = 0$ so $u_{13} = 0$

$$0 + 2 \cdot u_{23} = -2 \quad \text{so} \quad u_{23} = -1$$

From 3rd row: $l_{31} = 0$

$$l_{31}u_{12} + l_{32} = -1 \quad \therefore \quad l_{32} = -1$$

$$l_{31}u_{13} + l_{32}u_{23} + l_{33} = 2$$

$$0 - 1 \cdot (-1) + l_{33} = 2 \quad \therefore \quad l_{33} = 1$$

Thus

$$\begin{bmatrix} 1 & -1 & 0 \\ -2 & 4 & -2 \\ 0 & -1 & 2 \end{bmatrix} = A = LU =$$

$$= \begin{bmatrix} 1 & 0 & 0 \\ -2 & 2 & 0 \\ 0 & -1 & 1 \end{bmatrix} \begin{bmatrix} 1 & -1 & 0 \\ 0 & 1 & -1 \\ 0 & 0 & 1 \end{bmatrix}$$

Now $AX = LUX = B$
Put $UX = Y$ so $LY = B$

$$\begin{bmatrix} 1 & 0 & 0 \\ -2 & 2 & 0 \\ 0 & -1 & 1 \end{bmatrix} \begin{bmatrix} y_1 \\ y_2 \\ y_3 \end{bmatrix} = B = \begin{bmatrix} 0 \\ -1 \\ \frac{3}{2} \end{bmatrix}$$

solving $y_1 = 0$, $y_2 = -\frac{1}{2}$, $y_3 = 1$
From $UX = Y$, we have

$$\begin{bmatrix} 1 & -1 & 0 \\ 0 & 1 & -1 \\ 0 & 0 & 1 \end{bmatrix} \begin{bmatrix} x_1 \\ x_2 \\ x_3 \end{bmatrix} = \begin{bmatrix} 0 \\ -\frac{1}{2} \\ 1 \end{bmatrix}$$

solving $x_3 = 1$, $x_2 = x_1 = \frac{1}{2}$

Example 2: Solve

$$\begin{bmatrix} 2 & 1 & 0 & 0 & 0 \\ 1 & 2 & 1 & 0 & 0 \\ 0 & 1 & 2 & 1 & 0 \\ 0 & 0 & 1 & 2 & 1 \\ 0 & 0 & 0 & 1 & 2 \end{bmatrix} \begin{bmatrix} x_1 \\ x_2 \\ x_3 \\ x_4 \\ x_5 \end{bmatrix} = \begin{bmatrix} b_1 \\ b_2 \\ b_3 \\ b_4 \\ b_5 \end{bmatrix}$$

Also solve when $b_1 = 1, b_2 = 0, b_3 = 2, b_4 = 3, b_5 = -1$

Solution: Assume $A = LU$ where

$$L = \begin{bmatrix} 1 & 0 & 0 & 0 & 0 \\ l_{21} & 1 & 0 & 0 & 0 \\ l_{31} & l_{32} & 1 & 0 & 0 \\ l_{41} & l_{42} & l_{43} & 1 & 0 \\ l_{51} & l_{52} & l_{53} & l_{54} & 0 \end{bmatrix},$$

$$U = \begin{bmatrix} u_{11} & u_{12} & u_{13} & u_{14} & u_{15} \\ 0 & u_{22} & u_{23} & u_{24} & u_{25} \\ 0 & 0 & u_{33} & u_{34} & u_{35} \\ 0 & 0 & 0 & u_{44} & u_{45} \\ 0 & 0 & 0 & 0 & u_{55} \end{bmatrix}$$

Now we reduce the tridiagonal matrix A to echolon form using Gaussian elimination method without any row interchanges.

$$\begin{bmatrix} 2 & 1 & 0 & 0 & 0 \\ 1 & 2 & 1 & 0 & 0 \\ 0 & 1 & 2 & 1 & 0 \\ 0 & 0 & 1 & 2 & 1 \\ 0 & 0 & 0 & 1 & 2 \end{bmatrix} \underset{R_{21}\left(-\frac{1}{2}\right)}{\sim} \begin{bmatrix} 2 & 1 & 0 & 0 & 0 \\ 0 & \frac{3}{2} & 1 & 0 & 0 \\ 0 & 1 & 2 & 1 & 0 \\ 0 & 0 & 1 & 2 & 1 \\ 0 & 0 & 0 & 1 & 2 \end{bmatrix}$$

$$\underset{R_{32}\left(-\frac{2}{3}\right)}{\sim} \begin{bmatrix} 2 & 1 & 0 & 0 & 0 \\ 0 & \frac{3}{2} & 1 & 0 & 0 \\ 0 & 0 & \frac{4}{3} & 1 & 0 \\ 0 & 0 & 1 & 2 & 1 \\ 0 & 0 & 0 & 1 & 2 \end{bmatrix} \underset{R_{43}\left(-\frac{3}{4}\right)}{\sim} \begin{bmatrix} 2 & 1 & 0 & 0 & 0 \\ 0 & \frac{3}{2} & 1 & 0 & 0 \\ 0 & 0 & \frac{4}{3} & 1 & 0 \\ 0 & 0 & 0 & \frac{5}{4} & 1 \\ 0 & 0 & 0 & 1 & 2 \end{bmatrix}$$

$$\underset{R_{54}\left(-\frac{4}{5}\right)}{\sim} U = \begin{bmatrix} 2 & 1 & 0 & 0 & 0 \\ 0 & \frac{3}{2} & 1 & 0 & 0 \\ 0 & 0 & \frac{4}{3} & 1 & 0 \\ 0 & 0 & 0 & \frac{5}{4} & 1 \\ 0 & 0 & 0 & 0 & \frac{6}{5} \end{bmatrix}$$

Then multipliers are $-\frac{1}{2}, -\frac{2}{3}, -\frac{3}{4}, -\frac{4}{5}$ so

$$L = \begin{bmatrix} 1 & 0 & 0 & 0 & 0 \\ \frac{1}{2} & 1 & 0 & 0 & 0 \\ 0 & \frac{2}{3} & 1 & 0 & 0 \\ 0 & 0 & \frac{3}{4} & 1 & 0 \\ 0 & 0 & 0 & \frac{4}{5} & 1 \end{bmatrix}$$

Now solve by forward substitution

$$\begin{bmatrix} 1 & 0 & 0 & 0 & 0 \\ \frac{1}{2} & 1 & 0 & 0 & 0 \\ 0 & \frac{2}{3} & 1 & 0 & 0 \\ 0 & 0 & \frac{3}{4} & 1 & 0 \\ 0 & 0 & 0 & \frac{4}{5} & 1 \end{bmatrix} \begin{bmatrix} y_1 \\ y_2 \\ y_3 \\ y_4 \\ y_5 \end{bmatrix} = \begin{bmatrix} b_1 \\ b_2 \\ b_3 \\ b_4 \\ b_5 \end{bmatrix}$$

Solving $y_1 = b_1,\ y_2 = b_2 - \frac{1}{2}b_1$

$$y_3 = b_3 - \frac{2}{3}b_2 + \frac{1}{3}b_1$$

$$y_4 = b_4 - \frac{3}{4}b_3 + \frac{1}{2}b_2 - \frac{1}{4}b_1$$

$$y_5 = b_5 - \frac{4}{5}b_4 + \frac{3}{5}b_3 - \frac{2}{5}b_2 + \frac{1}{5}b_1$$

For $b_1 = 1,\ b_2 = 0,\ b_3 = 2,\ b_4 = 3,\ b_5 = -1$,

$$y_1 = 1,\ y_2 = -\frac{1}{2},\ y_3 = \frac{7}{3},\ y_4 = \frac{21}{4},\ y_5 = 0.$$

Now solve by backward substitution

$$\begin{bmatrix} 2 & 1 & 0 & 0 & 0 \\ 0 & \frac{3}{2} & 1 & 0 & 0 \\ 0 & 0 & \frac{4}{3} & 1 & 0 \\ 0 & 0 & 0 & \frac{5}{4} & 1 \\ 0 & 0 & 0 & 0 & \frac{6}{5} \end{bmatrix} \begin{bmatrix} x_1 \\ x_2 \\ x_3 \\ x_4 \\ x_5 \end{bmatrix} = B = \begin{bmatrix} 1 \\ -\frac{1}{2} \\ \frac{7}{3} \\ \frac{21}{4} \\ 0 \end{bmatrix}$$

solving $x_5 = 0,\ x_4 = \frac{21}{5},\ x_3 = -\frac{13}{10},$
$x_2 = \frac{8}{15},\ x_1 = \frac{7}{30}$

Thus the solution is

$$X^T = \left[\frac{7}{30}, \frac{8}{15}, -\frac{13}{10}, \frac{21}{5}, 0\right]$$

Note the amount of simplification in calculation of L and U.

EXERCISE

Solve the tridiagonal systems.

1. $3x_1 + x_2 = -1$
   $2x_1 + 4x_2 + x_3 = 7$
   $2x_2 + 5x_3 = 9$

Ans. $x_1 = -0.999995,\ x_2 = 1.999999,\ x_3 = 1$

2. $2x_1 - x_2 = 3$
$-x_1 + 2x_2 - x_3 = -3$
$-x_2 + 2x_3 = 1$

Ans. $x_1 = 1, x_2 = -1, x_3 = 0$

3. $0.5x_1 + 0.25x_2 = 0.35$
$0.35x_1 + 0.8x_2 + 0.4x_3 = 0.77$
$0.25x_2 + x_3 + 0.5x_4 = -0.5$
$x_3 - 2x_4 = -2.25$

Ans. $x_1 = -0.09357762, x_2 = 1.587155$
$x_3 = -1.16743, x_4 = 0.5412842$

4. $2x_1 - x_2 = 1$
$-x_1 + 2x_2 - x_3 = 0$
$-x_2 + 2x_3 - x_4 = 0$
$-x_3 + 2x_4 = 1$

Ans. $x_1 = x_2 = x_3 = x_4 = 1$

Hint:

$$L = \begin{bmatrix} 2 & 0 & 0 & 0 \\ -1 & \frac{3}{2} & 0 & 0 \\ 0 & -1 & \frac{4}{3} & 0 \\ 0 & 0 & -1 & \frac{5}{4} \end{bmatrix}, U = \begin{bmatrix} 1 & -\frac{1}{2} & 0 & 0 \\ 0 & 1 & -\frac{2}{3} & 0 \\ 0 & 0 & 1 & -\frac{3}{4} \\ 0 & 0 & 0 & 1 \end{bmatrix}$$

5. $2x_1 + x_2 = 1$
$x_{i-1} + 2x_i + x_{i+1} = 0, 2 \le i \le n-1$
$x_{n-1} + 2x_n = (-1)^{n+1}$
for $n = 10$.

Ans. $x_i = (-1)^{i-1}$ for $1 \le i \le 10$

Chapter 14

Eigen Values and Eigen Vectors

INTRODUCTION

Suppose A is a 2×2 matrix and X is a non-zero vector such that AX is a scalar multiple of X say $AX = \lambda X$. Then geometrically each vector on the line through the origin determined by X gets mapped back onto the same line under multiplication by A. The algebraic eigen value problem consists of determination of such vectors X, known as eigen vectors, such scalars λ, known as eigen values. Thus the finding of non zero vectors that get mapped into scalar multiples of themselves under a linear operator are most important in the study of vibrations of beams, probability (Markov process), Economics (Leontief model), genetics, quantum mechanics, population dynamics and geometry. For example in a mechanical system, they represent the normal modes of vibration. EISPACK is a package of programs in FORTRAN for solving eigen value problems [Smith et. al (1976) in "Matrix Eigen-system routines-EISPACK guide", Lecture notes in computer science, vol. 6, Springer-Verlag, New York].

14.1 LINEAR TRANSFORMATION

Consider a set of n linear equations

$$\left.\begin{aligned} y_1 &= a_{11}x_1 + a_{12}x_2 + \cdots + a_{1n}x_n \\ y_2 &= a_{21}x_1 + a_{22}x_2 + \cdots + a_{2n}x_n \\ &\cdots\cdots\cdots\cdots\cdots\cdots\cdots\cdots \\ y_n &= a_{n1}x_1 + a_{n2}x_2 + \cdots + a_{nn}x_n \end{aligned}\right\} \qquad (1)$$

Let $y = [y_1, y_2, y_3, \ldots, y_n]^T$, $X = [x_1, x_2, x_3, \ldots, x_n]^T$

$$A = \begin{bmatrix} a_{11} & a_{12} & a_{13} & \cdots & a_{1n} \\ a_{21} & a_{22} & a_{23} & \cdots & a_{2n} \\ \cdots & \cdots & \cdots & \cdots & \cdots \\ a_{n1} & a_{n2} & a_{n3} & \cdots & a_{nn} \end{bmatrix}_{n \times n}$$

Then set of n equations (1) can be represented as

$$Y = AX \qquad (2)$$

which transforms the set of n variables $(x_1, x_2, \ldots, x_n)$ to the set of n variables $(y, y_2, \ldots, y_n)$. Thus (2) is a transformation which transforms X into Y. Here A is known as the matrix of the transformation.

Linear transformation

(2) is said to be linear if it is additive and homogeneous i.e.,

$$\begin{aligned} A(c_1X_1 + c_2X_2) &= c_1AX_1 + c_2AX_2 \\ &= c_1Y_1 + c_2Y_2 \end{aligned}$$

where c_1 and c_2 are constants. Inverse transformation of (2) is $X = A^{-1}Y$.

14.2 EIGEN VALUES AND EIGEN VECTORS

Only square matrices are considered in this chapter.

Let A be a $n \times n$ matrix. Suppose the linear transformation $Y = AX$ transforms X into a scalar multiple of itself i.e.,

$$AX = Y = \lambda X$$

i.e., X is an invariant vector.

Then the unknown scalar λ is known as an eigen value of the matrix A and the corresponding non zero vector X as eigen vector. Thus eigen values or characteristic values or latent or proper values are scalars λ which satisfy the equation

$$AX = \lambda X \tag{3}$$

for an $X \neq 0$

or $$AX - \lambda IX = 0$$

$$(A - \lambda I)X = 0 \tag{4}$$

(4) represents a system of n homogeneous equations in the n variables $x_1, x_2, \ldots x_n$. (4) has non-trivial solutions if the coefficient matrix $(A - \lambda I)$ is singular i.e.,

$$|A - \lambda I| = 0 \tag{5}$$

or

$$\begin{vmatrix} a_{11} - \lambda & a_{12} & a_{13} \cdots a_{1n} \\ a_{21} & a_{22} - \lambda & a_{23} \cdots a_{2n} \\ \vdots & & \\ a_{n1} & a_{n2} & a_{n3} \cdots a_{nn} - \lambda \end{vmatrix} = 0 \tag{6}$$

Expansion of the determinant gives a nth degree polynomial $P_n(\lambda)$ known as the **characteristic polynomial** of A. (6) is known as the **characteristic equation** of A. Thus eigen values of n- square matrix A are the roots of the characteristic Equation (6). Hence A can have at least one and at most n eigen values.

Degree of characteristic polynomial = order of matrix A.

Spectrum of A is the set of all eigen values of A.

The eigen vector X corresponding to an eigen value λ is obtained by solving the homogeneous system (4) with this known eigen value λ. However eigen vector X is not unique, since kX is also an eigen vector where $k \neq 0$ constant ($AX = \lambda X$, multiplying by k, $kAX = k\lambda X$ or $A(kX) = \lambda(kX)$. Thus kX is an eigen vector of A).

Note 1: If all the n eigen values of A are distinct, then there correspond n distinct linearly independent eigen vectors.

Note 2: For an eigen value of A, repeated (twice or more), there may correspond one or several linearly independent eigen vectors. Thus the set of eigen vectors may or may not form a set of n linearly independent vectors.

Note 3: Algebraic multiplicity of an eigen value λ is the order of the eigen value as a root of the characteristic polynomial (i.e., if λ is a double root then algebraic multiplicity is 2).

Note 4: Geometric multiplicity of λ is the number of linearly independent eigen vectors corresponding to λ.

Procedure to Obtain Eigen Values and Eigen Vectors

1. Solve the characteristic equation $|A - \lambda I| = 0$ for eigen values λ_i. If A is of nth order, the number of eigen values are n or less than n (with repeated real roots or complex conjugate pairs).
2. For a specific eigen value λ_i, solve the homogeneous system of equations $(A - \lambda_i I)X = 0$.

Observation: See Note 1 and Note 2 above.

14.3 PROPERTIES OF EIGEN VALUES AND EIGEN VECTORS

1. **Real and complex eigen values:** If A is real, its eigen values are real or complex conjugates in pairs.

Proof: Expanding the characteristic equation

$$|A - \lambda I| = \begin{vmatrix} a_{11} - \lambda & a_{12} & \cdots & a_{1n} \\ a_{21} & a_{22} - \lambda & \cdots & a_{2n} \\ \vdots & & & \\ a_{n1} & a_{n2} & \cdots & a_{nn} - \lambda \end{vmatrix} = 0$$

we get an nth degree polynomial in λ

$$|A - \lambda I| = (-1)^n \Big[\lambda^n - \sigma_1 \lambda^{n-1} + \sigma_2 \lambda^{n-2} \ldots + (-1)^{n-1} \sigma_{n-1} \lambda + (-1)^n \sigma_n\Big] = 0$$

Since A is real, the roots of the characteristic polynomial will be real or complex conjugate pairs.

Let $\lambda_1, \lambda_2, \ldots, \lambda_n$ be the roots of the characteristic equation.

2. **Trace:** σ_1 = sum of the diagonal elements of A = trace of A = sum of the roots of the polynomial equation $= \lambda_1 + \lambda_2 + \cdots + \lambda_n$.

Thus, trace of A = sum of the eigen values of A.

3. **Determinant of** $A = \sigma_n = |A|$
$= \lambda_1 \cdot \lambda_2 \cdot \lambda_3 \cdots \lambda_n.$

Thus, the determinant of A = product of eigen values of A.

4. **Transpose:** A and A^T has same eigen values. Since the diagonal elements in the determinants of A and A^T are same, the determinant $|A - \lambda I|$ and $|A^T - \lambda I|$ are equal; hence have the same eigen values.

$$|A| = |A^T|, |A - \lambda I| = |(A - \lambda I)^T| = |A^T - (\lambda I)^T| = |A^T - \lambda I|$$

5. **Non-zero eigen values:** If all the eigen values are non-zero, then $|A| \neq 0$ since $|A|$ = product of the eigen values. i.e., A is non singular.
6. **Singular matrix:** $|A| = 0$. If at least one eigen value is zero then $|A|$ = product of eigen values $= \lambda_1, \lambda_2, \ldots, \lambda_n = 0$ i.e., A is singular.
7. **Inverse:** A^{-1} exists iff 0 is not an eigen value of A.
Eigen values of A^{-1} are $\frac{1}{\lambda_1}, \frac{1}{\lambda_2}, \ldots \frac{1}{\lambda_n}$,
i.e., the reciprocals of the eigen values of A:

$$AX = \lambda X, \text{ premultiply by } A^{-1},$$
$$A^{-1}AX = A^{-1}\lambda X = \lambda A^{-1} X$$
$$\therefore X = \lambda A^{-1} X \text{ or } A^{-1}X = \frac{1}{\lambda} X$$

8. **"Spectral shift":** $A \mp kI$ has eigen values $\lambda_i \pm k$ and has the same eigen vectors of A: characteristic polynomial of $A \mp kI$ is $|A \mp kI - \lambda I| = |A - (\lambda \mp k)I|$ which has $\lambda \mp k$ as the eigen values.
9. **Scalar multiples:** kA has eigen values $k\lambda_i$. Multiplying $(A - \lambda I)X = 0$ by k, $k(A - \lambda I)X = 0$, characteristic equation is $|k(A - \lambda I)| = |kA - k\lambda I| = 0$, thus kA has eigen values $k\lambda$.
10. **Powers:** A^m has eigen values λ^m.

By induction : $X = A^0X = \lambda^0 X = X, m = 0$ true
$AX = \lambda X, \quad m = 1$

Suppose $A^k X_j = \lambda_j^k X_j$ is true

premultiplying by A, $AA^k X_j = A\lambda_j^k X_j$

$$A^{k+1}X_j = \lambda_j^k AX_j = \lambda_j^k \lambda_j X = \lambda_j^{k+1} X$$

true for $k + 1$, hence by induction A^m has λ^m as eigen value.

11. **Spectral mapping threorem:** The polynomial matrix $P(A) = k_m A^m + k_{m-1}A^{m-1} + \cdots + k_1 A + k_0 I$ has the eigen values $P(\lambda_j) = k_m\lambda_j^m + k_{m-1}\lambda_j^{m-1} + \cdots + k_1\lambda_j + k_0$

From 9, 10 kA^m has eigen value $k\lambda^m$.

To prove additive:

$$\begin{aligned}(k_2A^2 + k_1A + k_0I)X &= k_2A^2X + k_1AX + k_0IX \\ &= k_2\lambda^2X + k_1\lambda X + k_0IX \\ &= (k_0\lambda^2 + k_1\lambda + k_0)X\end{aligned}$$

Thus $k_0\lambda^2 + k_1\lambda + k_0$ is the eigen value of $k_2A^2 + k_1A + k_0I$.

12. **Characteristic Vector cannot correspond to two distinct characteristic values:** Suppose X_1 corresponds to λ_1 and λ_2 where $\lambda_1 \neq \lambda_2$. Then $(A - \lambda_1 I)X_1 = 0$, $(A - \lambda_2 I)X_1 = 0$, subtracting $(\lambda_1 - \lambda_2)IX_1 = 0$, which implies $X_1 = 0$ since $\lambda_1 \neq \lambda_2$. But $X_1 \neq 0$ (eigen vectors are chosen to be non-zero). Hence a contradiction.
13. **Eigen values of diagonal,** upper triangular or lower triangular matrices are the principal diagonal elements since the characteristic polynomial becomes the product of diagonal elements namely

$$|D - \lambda I| = (a_{11} - \lambda)(a_{22} - \lambda)(a_{33} - \lambda)\cdots(a_{nn} - \lambda).$$

Thus characteristic values are the diagonal elements a_{ii}.

14. If λ is an eigen value of an orthogonal matrix A then $\frac{1}{\lambda}$ is also an eigen value of A. By definition of orthogonal matrix, $A^T = A^{-1}$. But A and A^T have same eigen values. Also eigen values of A^{-1} are reciprocals of eigen values of A. Thus if λ is an eigen value of A then $\frac{1}{\lambda}$ is eigen value of $A^{-1} = A^T$. Since A and A^T has same eigen values, λ and $\frac{1}{\lambda}$ are eigen values of A.
15. **Orthogonal:** Two vectors X and Y are said to be orthogonal if $X^TY = Y^TX = 0$.

WORKED OUT EXAMPLES

Find the eigen values and eigen vectors of:

Example 1: $A = \begin{pmatrix} 8 & -4 \\ 2 & 2 \end{pmatrix}$

Solution: The eigen values are the roots of the characteristic equation

$$\begin{vmatrix} 8-\lambda & -4 \\ 2 & 2-\lambda \end{vmatrix} = 0$$

i.e., $(8-\lambda)(2-\lambda)+8=0$

or $\lambda^2 - 10\lambda + 24 = 0,$

$$(\lambda-4)(\lambda-6)=0$$

The two distinct eigen values are $\lambda = 4, 6$.

Eigen vector corresponding to eigen value $\lambda = 4$:

$$(A-\lambda I)X = 0$$

$$\begin{pmatrix} 8-4 & -4 \\ 2 & 2-4 \end{pmatrix}\begin{pmatrix} x_1 \\ x_2 \end{pmatrix} = \begin{pmatrix} 4 & -4 \\ 2 & -2 \end{pmatrix}\begin{pmatrix} x_1 \\ x_2 \end{pmatrix} = 0$$

$$4x_1 - 4x_2 = 0$$

$$2x_1 - 2x_2 = 0 \quad \therefore x_1 = x_2$$

$$\overline{X}_1 = C_1\begin{pmatrix} 1 \\ 1 \end{pmatrix}$$

$\overline{X}_2$ corresponding to $\lambda = 6$

$$\begin{pmatrix} 8-6 & -4 \\ 2 & 2-6 \end{pmatrix}\begin{pmatrix} x_1 \\ x_2 \end{pmatrix} = \begin{pmatrix} 2 & -4 \\ 2 & -4 \end{pmatrix}\begin{pmatrix} x_1 \\ x_2 \end{pmatrix} = 0$$

$$2x_1 - 4x_2 = 0 \quad \therefore x_1 = 2x_2$$

$$\overline{X}_2 = C_2\begin{pmatrix} 1 \\ 2 \end{pmatrix}$$

Example 2: $A^T = \begin{pmatrix} 8 & 2 \\ -4 & 2 \end{pmatrix}$

Solution: Characteristic equation is $\begin{vmatrix} 8-\lambda & 2 \\ -4 & 2-\lambda \end{vmatrix} = 0$ characteristic equation is $\lambda^2 - 10\lambda + 24 = 0$ same as the ch. equation of A. Thus the eigen values of A and A^T are same. However the eigen vectors are not the same.

For $\lambda = 4$,

$$\begin{pmatrix} 8-4 & 2 \\ -4 & 2-4 \end{pmatrix}\begin{pmatrix} x_1 \\ x_2 \end{pmatrix} = 0$$

$$4x_1 + 2x_2 = 0 \quad \text{or} \quad x_2 = -2x_1$$

$$X_1 = C_1\begin{pmatrix} -2 \\ 1 \end{pmatrix}$$

For $\lambda = 6$,

$$\begin{pmatrix} 8-6 & 2 \\ -4 & 2-6 \end{pmatrix}\begin{pmatrix} x_1 \\ x_2 \end{pmatrix} = 0$$

$$2x_1 + 2x_2 = 0$$

$$X_2 = C_2\begin{pmatrix} 1 \\ -1 \end{pmatrix}$$

Example 3: $A^{-1} = \frac{1}{24}\begin{bmatrix} 2 & 4 \\ -2 & 8 \end{bmatrix}.$

Solution: Characteristic equation is $|A^{-1} - \lambda I| = 0$

$$\begin{vmatrix} \frac{1}{12}-\lambda & \frac{1}{6} \\ -\frac{1}{12} & \frac{1}{3}-\lambda \end{vmatrix} = \left(\frac{1}{12}-\lambda\right)\left(\frac{1}{3}-\lambda\right) + \frac{1}{12}\cdot\frac{1}{6} = 0$$

$$24\lambda^2 - 10\lambda + 1 = 0, \ \left(\lambda - \frac{1}{4}\right)\left(\lambda - \frac{1}{6}\right) = 0$$

The eigen values of A^{-1} are $\frac{1}{4}, \frac{1}{6}$ which are the reciprocals of the eigen values 4, 6 of A. Also the eigen vectors of A^{-1} and A are same.

For $\lambda = \frac{1}{4}$,

$$\begin{pmatrix} \frac{1}{12}-\frac{1}{4} & \frac{1}{6} \\ -\frac{1}{12} & \frac{1}{3}-\frac{1}{4} \end{pmatrix}\begin{pmatrix} x_1 \\ x_2 \end{pmatrix} = 0$$

$$-2x_1 + x_2 = 0 \quad \therefore x_1 = x_2$$

$$X_1 = C_1\begin{pmatrix} 1 \\ 1 \end{pmatrix}$$

For $\lambda = \frac{1}{6}$,

$$\begin{pmatrix} \frac{1}{12}-\frac{1}{6} & \frac{1}{6} \\ -\frac{1}{12} & \frac{1}{3}-\frac{1}{6} \end{pmatrix}\begin{pmatrix} x_1 \\ x_2 \end{pmatrix} = 0$$

$$-x_1 + 2x_2 = 0 \quad \therefore x_1 = 2x_2$$

$$X_2 = C_2\begin{pmatrix} 1 \\ 2 \end{pmatrix}.$$

Example 4: $B = kA$ where $k = -\frac{1}{2}$

Solution: $B = -\frac{1}{2}A = \begin{pmatrix} -4 & +2 \\ -1 & -1 \end{pmatrix}$

Characteristic equation of B is

$$|B - \lambda I| = \begin{vmatrix} -4-\lambda & 2 \\ -1 & -1-\lambda \end{vmatrix} = 0$$

$$(4+\lambda)(1+\lambda)+2=0 \quad \text{or} \quad \lambda^2 + 5\lambda + 6 = 0$$

So the eigen values of B are $-2, -3$ which are $-\frac{1}{2}$ times of eigen values 4, 6 of A. Also the eigen vectors of B and A are same.

For $\lambda = -2$,

$$\begin{bmatrix} -4+2 & 2 \\ -1 & -1+2 \end{bmatrix}\begin{bmatrix} x_1 \\ x_2 \end{bmatrix} = 0 \quad \therefore x_1 = x_2$$

$$X_1 = C_1\begin{bmatrix} 1 \\ 1 \end{bmatrix}$$

For $\lambda = -3$,

$$\begin{bmatrix} -4+3 & 2 \\ -1 & -1+3 \end{bmatrix}\begin{bmatrix} x_1 \\ x_2 \end{bmatrix} = 0,$$

$$-x_1 + 2x_2 = 0$$

$$X_2 = C_2\begin{bmatrix} 1 \\ 2 \end{bmatrix}.$$

Example 5: $A^2 = \begin{pmatrix} 56 & -40 \\ 20 & -4 \end{pmatrix}$

Solution: Characteristic equation of A^2 is

$$\begin{vmatrix} 56-\lambda & -40 \\ 20 & -4-\lambda \end{vmatrix} = 0$$

$$\lambda^2 - 52\lambda + 576 = (\lambda - 16)(\lambda - 36) = 0$$

So eigen values of A^2 are 16, 36 which are square of the eigen values 4, 6 of A. Also the eigen vectors of A and A^2 are same

For $\lambda = 16$,

$$\begin{bmatrix} 56-16 & -40 \\ 20 & -4-16 \end{bmatrix}\begin{bmatrix} x_1 \\ x_2 \end{bmatrix} = 0$$

$$x_1 \quad x_2 = 0 \quad \text{i.e., } x_1 = x_2$$

$$X_1 = C_1\begin{pmatrix} 1 \\ 1 \end{pmatrix}$$

For $\lambda = 36$,

$$\begin{bmatrix} 56-36 & -40 \\ 20 & -4-36 \end{bmatrix}\begin{bmatrix} x_1 \\ x_2 \end{bmatrix} = 0,$$

$$x_1 - 2x_2 = 0 \quad \text{i.e., } x_1 = 2x_2$$

$$X_2 = C\begin{pmatrix} 1 \\ 2 \end{pmatrix}.$$

Example 6:

$$B = A \pm kI = \begin{pmatrix} 8 & -4 \\ 2 & 2 \end{pmatrix} \pm k\begin{pmatrix} 1 & 0 \\ 0 & 1 \end{pmatrix}$$

$$= \begin{pmatrix} 8 \pm k & -4 \\ 2 & 2 \pm k \end{pmatrix}$$

Solution: Characteristic equation of B is

$$|B - \lambda I| = 0$$

$$\begin{vmatrix} 8 \pm k - \lambda & -4 \\ 2 & 2 \pm k - \lambda \end{vmatrix} = 0$$

i.e.,
$$(8 \pm k - \lambda)(2 \pm k - \lambda) + 8 = 0$$

$$\lambda^2 - (10 \pm 2k)\lambda + (k^2 \pm 10k + 24) = 0$$

roots are $\frac{10 \pm 2}{2} \pm k$ i.e., $4 \pm k$ and $6 \pm k$ which are 4, 6 of A with $\pm k$.

Eigen vectors of B and A are same

For $\lambda = 4 \pm k$,

$$\begin{bmatrix} 8 \pm k - (4 \pm k) & -4 \\ 2 & 2 \pm k - (4 \pm k) \end{bmatrix}\begin{bmatrix} x_1 \\ x_2 \end{bmatrix} = 0$$

$$4x_1 - 4x_2 = 0 \quad \text{or} \quad x_1 = x_2 \text{ etc.}$$

Example 7: $D = 2A^2 - \frac{1}{2}A + 3I$

Solution:

$$D = 2\begin{pmatrix} 56 & -40 \\ 20 & -4 \end{pmatrix} - \frac{1}{2}\begin{pmatrix} 8 & -4 \\ 2 & 2 \end{pmatrix} + 3\begin{pmatrix} 1 & 0 \\ 0 & 1 \end{pmatrix}$$

$$= \begin{pmatrix} 111 & -78 \\ 39 & -6 \end{pmatrix}$$

$$\text{ch. eq. of } D \text{ is } \begin{bmatrix} 111-\lambda & -78 \\ 39 & -6-\lambda \end{bmatrix} = 0$$

or
$$\lambda^2 - 105\lambda + 2376 = (\lambda - 33)(\lambda - 72) = 0$$

Thus the eigen values of D are 33 and 72.

Note that $33 = 2 \cdot 16 - \frac{1}{2} \cdot 4 + 3$ and $72 = 2 \cdot 36 - \frac{1}{2} \cdot 6 + 3$ i.e., eigen value of D is $2\lambda^2 - \frac{1}{2}\lambda + 3$ where λ is the eigen value of A. The eigen vectors of D and A are same.

For $\lambda = 33$,

$$\begin{bmatrix} 111-33 & -78 \\ 39 & -6-33 \end{bmatrix}\begin{bmatrix} x_1 \\ x_2 \end{bmatrix} = 0$$

i.e.,
$$78x_1 - 78x_2 = 0$$

i.e.,
$$x_1 = x_2 \text{ etc.}$$

Example 8: Find the sum and product of eigen values of A.

Solution: Sum of eigen values of $A = 4 + 6 = 10 = \text{trace of } A = a_{11} + a_{22} = 8 + 2 = 10$.

Product of eigen values of $A = 4.6 = 24 = |A| = 16 + 8 = 24$.

Example 9: Find the eigen values and eigen vectors of

$$A = \begin{pmatrix} 3 & 1 & 4 \\ 0 & 2 & 6 \\ 0 & 0 & 5 \end{pmatrix}$$

Solution: For upper triangular, lower triangular and diagonal matrices, the eigen values are given by the diagonal elements.
The characteristic eq.

$$|A-\lambda I| = \begin{vmatrix} 3-\lambda & 1 & 4 \\ 0 & 2-\lambda & 6 \\ 0 & 0 & 5-\lambda \end{vmatrix} = 0$$

i.e., $(3-\lambda)(2-\lambda)(5-\lambda)=0$

So eigen values of A are 3, 2, 5 which are the diagonal elements of A.

Eigen vector X_1 for $\lambda = 3$

$$\begin{bmatrix} 0 & 1 & 4 \\ 0 & -1 & 6 \\ 0 & 0 & 2 \end{bmatrix}\begin{bmatrix} x_1 \\ x_2 \\ x_3 \end{bmatrix} = 0$$

i.e., $x_2+4x_3=0, -x_2+6x_3=0, 2x_3=0$

$$x_2=0, x_3=0, x_1=\text{arbitrary},$$

$$X_1 = C_1\begin{pmatrix} 1 \\ 0 \\ 0 \end{pmatrix}.$$

Eigen vector X_2 for $\lambda = 2$,

$$x_1+x_2+4x_3=0, \quad 6x_3=0, 3x_3=0$$

$\therefore \quad x_3=0, \quad x_1=-x_2, \quad X_2 = C_1\begin{pmatrix} 1 \\ -1 \\ 0 \end{pmatrix}$

For $\lambda = 5$,

$$-2x_1+x_2+4x_3=0$$
$$-3x_2+6x_3=0$$

i.e., $x_1=3x_3, x_2=2x_3$

$$X_3 = C_2\begin{pmatrix} 3 \\ 2 \\ 1 \end{pmatrix}.$$

Find the eigen values and eigen vectors of:

Example 10:

$$A = \begin{bmatrix} 1 & 0 & -1 \\ 1 & 2 & 1 \\ 2 & 2 & 3 \end{bmatrix}$$

Determine whether the eigen vectors are orthogonal.

Solution:

Characteristic equation is $\begin{vmatrix} 1-\lambda & 0 & -1 \\ 1 & 2-\lambda & 1 \\ 2 & 2 & 3-\lambda \end{vmatrix}$

$$=\lambda^3-6\lambda^2+11\lambda-6=0$$
$$(\lambda-1)(\lambda-2)(\lambda-3)=0$$

So $\lambda = 1, 2, 3$ are three distinct eigen values of A

For $\lambda = 1$,

$$\begin{bmatrix} 0 & 0 & -1 \\ 1 & 1 & 1 \\ 2 & 2 & 2 \end{bmatrix}\begin{bmatrix} x_1 \\ x_2 \\ x_3 \end{bmatrix} = 0, \quad \begin{matrix} x_3=0 \\ x_1+x_2=0, \end{matrix}$$

$$X_1 = C_1\begin{bmatrix} 1 \\ -1 \\ 0 \end{bmatrix}$$

For $\lambda = 2$,

$$\begin{bmatrix} -1 & 0 & -1 \\ 1 & 0 & 1 \\ 2 & 2 & 1 \end{bmatrix}\begin{bmatrix} x_1 \\ x_2 \\ x_3 \end{bmatrix} = 0, \quad \begin{matrix} x_1+x_3=0 \\ 2x_1+2x_2+x_3=0 \end{matrix}$$

$$x_1=-x_3, \quad x_2=\frac{1}{2}x_3, \quad X_2 = C\begin{bmatrix} 2 \\ -1 \\ -2 \end{bmatrix}$$

For $\lambda = 3$,

$$\begin{bmatrix} -2 & 0 & -1 \\ 1 & -1 & 1 \\ 2 & 2 & 0 \end{bmatrix}\begin{bmatrix} x_1 \\ x_2 \\ x_3 \end{bmatrix} = 0, \quad \begin{matrix} x_1=-x_2 \\ x_1=-\frac{1}{2}x_3, \end{matrix}$$

$$X_3 = C\begin{bmatrix} 1 \\ 1 \\ -2 \end{bmatrix}$$

Thus there are three linearly independent eigen vectors X_1, X_2, X_3 corresponding to the three distinct eigen values. Since $X_1^T X_2 = 3 \neq 0, X_2^T X_3 = 5 \neq 0, X_3^T X_1 = 0$.
Therefore only X_1 and X_3 are orthogonal.

Example 11:

$$A = \begin{bmatrix} 1 & 2 & 2 \\ 0 & 2 & 1 \\ -1 & 2 & 2 \end{bmatrix}$$

Determine the algebraic and geometric multiplicity.

Solution:

Characteristic equation is $\begin{vmatrix} 1-\lambda & 2 & 2 \\ 0 & 2-\lambda & 1 \\ -1 & 2 & 2-\lambda \end{vmatrix} = 0$

$$\lambda^3-5\lambda^2+8\lambda-4=(\lambda-1)(\lambda-2)^2=0$$

So $\lambda = 1, 2, 2$ are eigen values with $\lambda = 2$ repeated twice (double root) of multiplicity 2.

The algebraic multiplicity of the eigen value $\lambda = 2$ is 2.

For $\lambda = 1$,

$$\begin{bmatrix} 0 & 2 & 2 \\ 0 & 1 & 1 \\ -1 & 2 & 1 \end{bmatrix}, \quad \begin{matrix} x_2 = -x_3 \\ x_1 = -x_3 \end{matrix}$$

$$X_1 = C\begin{bmatrix} 1 \\ 1 \\ -1 \end{bmatrix}$$

For $\lambda = 2$,

$$\begin{bmatrix} -1 & 2 & 2 \\ 0 & 0 & 1 \\ -1 & 2 & 0 \end{bmatrix}, \quad \begin{matrix} x_3 = 0 \\ x_1 = 2x_2 \end{matrix}$$

$$X_2 = C\begin{bmatrix} 2 \\ 1 \\ 0 \end{bmatrix}$$

Thus only one eigen vector X_2 corresponds to the repeated eigen value $\lambda = 2$.

The geometric multiplicity of eigen value $\lambda = 2$ is one.

Example 12: $A = \begin{bmatrix} 0 & 1 & 0 \\ 0 & 0 & 1 \\ 1 & -3 & 3 \end{bmatrix}$

Determine the algebraic and geometric multiplicity.

Solution: Characteristic equation is

$$\lambda^3 - 3\lambda^2 + 3\lambda - 1 = (\lambda - 1)^3 = 0$$

$\lambda = 1, 1, 1$ is an eigen value of algebraic multiplicity 3.

For $\lambda = 1$,

$$\begin{aligned} -x_1 + x_2 &= 0 \qquad \therefore x_1 = x_2 \\ -x_2 + x_3 &= 0 \qquad\quad x_2 = x_3 \\ x_1 - 3x_2 + 2x_3 &= 0 \end{aligned}$$

$$X = C\begin{bmatrix} 1 \\ 1 \\ 1 \end{bmatrix}$$

Thus only one eigen value X corresponds to the thrice repeated eigen value $\lambda = 1$, so geometric multiplicity is one.

Example 13: $A = \begin{bmatrix} 2 & 1 & 1 \\ 1 & 2 & 1 \\ 0 & 0 & 1 \end{bmatrix}$

Determine the algebraic and geometric multiplicity.

Solution: Characteristic equation is

$$\begin{vmatrix} 2-\lambda & 1 & 1 \\ 1 & 2-\lambda & 1 \\ 0 & 0 & 1-\lambda \end{vmatrix} = (1-\lambda)(\lambda - 1)(\lambda - 3) = 0$$

Thus $\lambda = 1, 1, 3$ are the eigen values of A. So the algebraic multiplicity of eigen value $\lambda = 1$ is two.
For $\lambda = 3$,

$$\begin{bmatrix} -1 & 1 & 1 \\ 1 & -1 & 1 \\ 0 & 0 & -2 \end{bmatrix} \sim \begin{matrix} x_3 = 0 \\ x_1 = x_2, \end{matrix} \quad X_1 = C\begin{bmatrix} 1 \\ 1 \\ 0 \end{bmatrix}$$

For $\lambda = 1$,

$$\begin{bmatrix} 1 & 1 & 1 \\ 1 & 1 & 1 \\ 0 & 0 & 0 \end{bmatrix} \sim \begin{bmatrix} 1 & 1 & 1 \\ 0 & 0 & 0 \\ 0 & 0 & 0 \end{bmatrix}, \begin{matrix} n = 3 \\ r = 1 \end{matrix}$$

$$n - r = 3 - 1 = 2 \quad \text{arbitrary}$$

$$x_1 + x_2 + x_3 = 0 \quad \text{or} \quad x_1 = -x_2 - x_3$$

where x_2 and x_3 are arbitrary

For a choice of $x_2 = 0$, $x_3 =$ arbitrary.

$$X_2 = C\begin{bmatrix} 1 \\ 0 \\ -1 \end{bmatrix}$$

For a choice of $x_2 \neq 0$, $x_3 = 0$

$$X_3 = C\begin{bmatrix} 1 \\ -1 \\ 0 \end{bmatrix}$$

Thus for the repeated eigen value $\lambda = 1$, there corresponds two linearly independent eigen vectors X_2 and X_3. So the geometric multiplicity of eigen value $\lambda = 1$ is 2.

Example 14: Find the eigen values of the orthogonal matrix

$$B = \frac{1}{3}\begin{bmatrix} 1 & 2 & 2 \\ 2 & 1 & -2 \\ 2 & -2 & 1 \end{bmatrix}$$

Solution: The characteristic equation of

$$A = \begin{bmatrix} 1 & 2 & 2 \\ 2 & 1 & -2 \\ 2 & -2 & 1 \end{bmatrix} \text{ is}$$

$$\begin{vmatrix} 1-\lambda & 2 & 2 \\ 2 & 1-\lambda & -2 \\ 2 & -2 & 1-\lambda \end{vmatrix} = \lambda^3 - 3\lambda^2 - 9\lambda + 27 = 0$$

or $(\lambda - 3)^2(\lambda + 3) = 0$

The eigen values of A are 3, 3 – 3, so the eigen values of $B = \frac{1}{3}A$ are 1, 1, –1.

Note that $\lambda = 1$ is an eigen value of B then its reciprocal $\frac{1}{\lambda} = \frac{1}{1} = 1$ is also an eigen value of B.

Example 15: Find the inverse transformation of

$$y_1 = x_1 + 2x_2 + 5x_3$$

$$y_2 = -x_2 + 2x_3$$

$$y_3 = 2x_1 + 4x_2 + 11x_3$$

Solution: With $Y = [y_1\ y_2\ y_3]^T$, $X = [x_1\ x_2\ x_3]^T$, the coefficient matrix

$$A = \begin{bmatrix} 1 & 2 & 5 \\ 0 & -1 & 2 \\ 2 & 4 & 11 \end{bmatrix}. \text{ Its } |A| = -1$$

$$\text{Adj } A = \begin{bmatrix} -19 & -2 & 9 \\ 4 & 1 & -2 \\ 2 & 0 & -1 \end{bmatrix}$$

Thus the inverse transformation is

$$X = \begin{bmatrix} x_1 \\ x_2 \\ x_3 \end{bmatrix} = A^{-1}Y = \begin{bmatrix} 19 & 2 & -9 \\ -4 & -1 & 2 \\ -2 & 0 & 1 \end{bmatrix} \begin{bmatrix} y_1 \\ y_2 \\ y_3 \end{bmatrix}$$

$$= \begin{bmatrix} 19y_1 & +2y_2 & -9y_3 \\ -4y_1 & -y_2 & +2y_3 \\ -2y_1 & & +y_3 \end{bmatrix}$$

EXERCISE

Find the eigen values and eigen vectors of:

1. $\begin{bmatrix} -5 & 2 \\ 2 & -2 \end{bmatrix}$

Ans. $\lambda^2 + 7\lambda + 6 = 0, \lambda = -1, -6, \begin{bmatrix} 1 \\ 2 \end{bmatrix}, \begin{bmatrix} 2 \\ -1 \end{bmatrix}$

2. $\begin{bmatrix} 6 & 8 \\ 8 & -6 \end{bmatrix}$

Ans. $10, -10, \begin{bmatrix} 2 \\ 1 \end{bmatrix}, \begin{bmatrix} 1 \\ -2 \end{bmatrix}$

3. $\begin{bmatrix} 1 & 2 \\ 1 & 0 \end{bmatrix}$

Ans. $2, -1, \begin{bmatrix} 2 \\ 1 \end{bmatrix}, \begin{bmatrix} 1 \\ -1 \end{bmatrix}$

4. $\begin{bmatrix} 1 & 2 \\ 3 & 2 \end{bmatrix}$

Ans. $4, -1, \begin{bmatrix} 2 \\ 3 \end{bmatrix}, \begin{bmatrix} 1 \\ -1 \end{bmatrix}$

5. $\begin{bmatrix} -2 & 2 & -3 \\ 2 & 1 & -6 \\ -1 & -2 & 0 \end{bmatrix}$

Ans. $5, -3, -3, \begin{bmatrix} 1 \\ 2 \\ -1 \end{bmatrix}, \begin{bmatrix} -2 \\ 1 \\ 0 \end{bmatrix}, \begin{bmatrix} 3 \\ 0 \\ 1 \end{bmatrix}$

6. $\begin{bmatrix} 1 & 1 & 3 \\ 1 & 5 & 1 \\ 3 & 1 & 1 \end{bmatrix}$

Ans. $\lambda^3 - 7\lambda^2 + 36 = 0$, $\lambda = -2, 3, 6$ $\begin{bmatrix} -1 \\ 0 \\ 1 \end{bmatrix}, \begin{bmatrix} +1 \\ -1 \\ 1 \end{bmatrix}, \begin{bmatrix} 1 \\ 2 \\ 1 \end{bmatrix}$

7. $\begin{bmatrix} -3 & -7 & -5 \\ 2 & 4 & 3 \\ 1 & 2 & 2 \end{bmatrix}$

Ans. $(\lambda - 1)^3 = 0$, $\lambda = 1, 1, 1$, $\begin{bmatrix} -3 \\ 1 \\ 1 \end{bmatrix}$

8. $\begin{bmatrix} 8 & -6 & 2 \\ -6 & 7 & -4 \\ 2 & -4 & 3 \end{bmatrix}$

Ans. $\lambda^3 - 18\lambda^2 + 45\lambda = 0$, $\lambda = 0, 3, 15$ $\begin{bmatrix} 1 \\ 2 \\ 2 \end{bmatrix}, \begin{bmatrix} 2 \\ 1 \\ -2 \end{bmatrix}, \begin{bmatrix} 2 \\ -2 \\ 1 \end{bmatrix}$

9. $\begin{bmatrix} 2 & 2 & 1 \\ 1 & 3 & 1 \\ 1 & 2 & 2 \end{bmatrix}$

Ans. $\lambda^3 - 7\lambda^2 + 11\lambda - 5 = 0$
$\lambda = 5, 1, 1,$ $\begin{bmatrix} 1 \\ 1 \\ 1 \end{bmatrix}, \begin{bmatrix} 1 \\ 0 \\ -1 \end{bmatrix}, \begin{bmatrix} 2 \\ -1 \\ 0 \end{bmatrix}$

10. $\begin{bmatrix} 3 & 10 & 5 \\ -2 & -3 & -4 \\ 3 & 5 & 7 \end{bmatrix}$

Ans. $\lambda^3 - 7\lambda^2 + 16\lambda - 12 = 0$
$\lambda = 2, 2, 3$, for $\lambda = 2$, $\begin{bmatrix} 5 \\ 2 \\ -3 \end{bmatrix}$,

For $\lambda = 3$, $\begin{bmatrix} 1 \\ 1 \\ -2 \end{bmatrix}$

11. $\begin{bmatrix} 6 & -2 & 2 \\ -2 & 3 & -1 \\ 2 & -1 & 3 \end{bmatrix}$

Ans. $\lambda^3 - 12\lambda^2 + 36\lambda - 32 = 0$
$\lambda = 2, 2, 8$, $\begin{bmatrix} -1 \\ 0 \\ 2 \end{bmatrix}, \begin{bmatrix} 1 \\ 2 \\ 0 \end{bmatrix}$,

For $\lambda = 8$, $\begin{bmatrix} 2 \\ -1 \\ 1 \end{bmatrix}$

12. $\begin{bmatrix} 2 & 1 & 0 \\ 0 & 2 & 1 \\ 0 & 0 & 2 \end{bmatrix}$

Ans. $(\lambda - 2)^3 = 0$
$\lambda = 2, 2, 2$, $\begin{bmatrix} 1 \\ 0 \\ 0 \end{bmatrix}$

13. $\begin{bmatrix} 1 & -4 & -1 & -4 \\ 2 & 0 & 5 & -4 \\ -1 & 1 & -2 & 3 \\ -1 & 4 & -1 & 6 \end{bmatrix}$

Ans. $\lambda^4 - 5\lambda^3 + 9\lambda^2 - 7\lambda + 2 = 0$
$\lambda = 2, 1, 1, 1,$

For $\lambda = 2$, $\begin{bmatrix} 2 \\ 3 \\ -2 \\ -3 \end{bmatrix}$, for $\lambda = 1$, $\begin{bmatrix} 3 \\ 6 \\ -4 \\ -5 \end{bmatrix}$

14. $\begin{bmatrix} 2 & -2 & 2 \\ 1 & 1 & 1 \\ 1 & 3 & -1 \end{bmatrix}$

Ans. $\lambda^3 - 2\lambda^2 - 4\lambda + 8 = 0$
$\lambda = 2, 2, -2$
For $\lambda = 2$, $[0 \quad 1 \quad 1]^T$
For $\lambda = -2$, $[-4 \quad -1 \quad 7]^T$

15. $\begin{bmatrix} 3 & -2 & -5 \\ 4 & -1 & -5 \\ -2 & -1 & -3 \end{bmatrix}$

Ans. $(\lambda + 5)(\lambda - 2)^2 = 0$, $\lambda = 5, 2, 2$
For $\lambda = 5$, $X_1 = [3 \quad 2 \quad 4]^T$
For $\lambda = 2$, $X_2 = [1 \quad 3 \quad -1]$

16. Find the sum and product of the eigen values of

$$A = \begin{bmatrix} 2 & 3 & -2 \\ -2 & 1 & 1 \\ 1 & 0 & 2 \end{bmatrix}$$

Ans. sum = trace = $2 + 1 + 2 = 5$
product = $|A| = 21$.

14.4 CAYLEY-HAMILTON THEOREM

Every square matrix satisfies its own characteristic equation.

Proof: Let A be an n-square matrix. Let $D(\lambda)$ be the characteristic polynomial of A, given by

$$D(\lambda) = |\lambda I - A| = \lambda^n + C_{n-1}\lambda^{n-1} + C_{n-2}\lambda^{n-2} + \cdots + C_1\lambda + C_0 \quad (1)$$

Let $B(\lambda)$ be the adjoint of $(\lambda I - A)$. The elements of $B(\lambda)$ are cofactors of the matrix $(\lambda I - A)$ and are polynomials in λ of degree not exceeding $n - 1$. Thus

$$B(\lambda) = B_{n-1}\lambda^{n-1} + B_{n-2}\lambda^{n-2} + \cdots + B_1\lambda + B_0 \quad (2)$$

where B_i are n-square matrices whose elements are functions of the elements of A and independent of λ. We know that

$$(\lambda I - A) \cdot \text{adj}\,(\lambda I - A) = |\lambda I - A| I$$

$$(\lambda I - A) \cdot B(\lambda) = |\lambda I - A| I$$

From (1) and (2), we have

$$(\lambda I - A)\left(B_{n-1}\lambda^{n-1} + B_{n-2}\lambda^{n-2} + \cdots + B_1\lambda + B_0\right)$$

$$= I(\lambda^n + C_{n-1}\lambda^{n-1} + \cdots + C_1\lambda + C_0) \quad (3)$$

Equating the like powers of λ on both sides of (3), we get

$$B_{n-1} = I$$
$$B_{n-2} - AB_{n-1} = C_{n-1}I$$
$$B_{n-3} - AB_{n-2} = C_{n-2}I$$
$$\cdots\cdots\cdots\cdots$$
$$B_0 - AB_1 = C_1 I$$
$$-AB_0 = C_0 I$$

Multiplying both sides of the above matrix equations by $A^n, A^{n-1}, A^{n-2}, \ldots, A, I$ respectively, we have

$$A^n B_{n-1} = A^n$$
$$A^{n-1}B_{n-2} - A^n B_{n-1} = C_{n-1}A^{n-1}$$
$$A^{n-2}B_{n-3} - A^{n-1}B_{n-2} = C_{n-2}A^{n-2}$$
$$\cdots\cdots\cdots\cdots$$
$$AB_0 - A^2B_1 = C_1 A$$
$$-AB_0 = C_0 I$$

By adding all the above equations, we get

$$0 = A^n + C_{n-1}A^{n-1} + C_{n-2}A^{n-2} + \cdots + C_1A + C_0I \quad (4)$$

since all the terms on the L.H.S. cancel each other. Thus A satisfies its own characteristic equation.

Inverse by Cayley-Hamilton Theorem

Multiplying (4) by A^{-1}

$$0 = A^{n-1} + C_{n-1}A^{n-2} + C_{n-2}A^{n-3} + \cdots + C_1 I + C_0 A^{-1}$$

Solving for A^{-1}, we get

$$A^{-1} = \frac{-1}{C_0}\left[A^{n-1} + C_{n-1}A^{n-2} + C_{n-2}A^{n-3} + \cdots + C_1 I\right].$$

Note: A^{-1} exists only if $C_0 =$ determinant of A is not equals to zero.

Worked Out Examples

Example 1: Verify Cayley-Hamilton theorem for the matrix $A = \begin{bmatrix} 1 & 2 \\ 2 & -1 \end{bmatrix}$. Find A^{-1}. Determine A^8.

Solution: The characteristic equation is

$$|A - \lambda I| = \begin{vmatrix} 1-\lambda & 2 \\ 2 & -1-\lambda \end{vmatrix} = 0$$

or $$(\lambda - 1)(1 + \lambda) - 4 = 0$$

so $$\lambda^2 - 5 = 0$$

$$A^2 = A \cdot A = \begin{bmatrix} 1 & 2 \\ 2 & -1 \end{bmatrix}\begin{bmatrix} 1 & 2 \\ 2 & -1 \end{bmatrix} = \begin{vmatrix} 5 & 0 \\ 0 & 5 \end{vmatrix} = 5I$$

or $$A^2 - 5I = 0$$

Thus A satisfies the characteristic equation. To find A^{-1}, multiply $A^2 - 5I = 0$ by A^{-1}.

$$A^{-1} \cdot A^2 - 5A^{-1}I = 0$$

or $$A - 5A^{-1} = 0$$

So $$A^{-1} = \frac{1}{5}A = \frac{1}{5}\begin{bmatrix} 1 & 2 \\ 2 & -1 \end{bmatrix}$$

To find A^8, multiply $A^2 - 5I = 0$ by A^6

$$A^6 \cdot A^2 - 5I \cdot A^6 = 0$$
$$A^8 = 5A^6 = 5 \cdot A^2 \cdot A^2 \cdot A^2 = 5 \cdot (5I)(5I)(5I)$$
$$A^8 = 625I$$

Example 2: Verify Cayley-Hamilton theorem for the matrix

$$A = \begin{bmatrix} 1 & 2 & 3 \\ 2 & 4 & 5 \\ 3 & 5 & 6 \end{bmatrix}$$

and hence find the inverse of A. Find A^4. Express $B = A^8 - 11A^7 - 4A^6 + A^5 + A^4 - 11A^3 - 3A^2 + 2A + I$ as a quadratic polynomial in A. Find B.

Solution: The characteristic equation of A is

$$|A - \lambda I| = \begin{vmatrix} 1-\lambda & 2 & 3 \\ 2 & 4-\lambda & 5 \\ 3 & 5 & 6-\lambda \end{vmatrix}$$

$$= (1-\lambda)[(4-\lambda)(6-\lambda) - 25]$$

$$-2[2(6-\lambda)-15]$$

$$+3[10-3(4-\lambda)]=0,$$

$$\lambda^3-11\lambda^2-4\lambda+1=0.$$

Cayley-Hamilton theorem is verified if A satisfies the above characteristic equation i.e.,

$$A^3-11A^2-4A+I=0$$

$$A^2=A\cdot A=\begin{bmatrix}1&2&3\\2&4&5\\3&5&6\end{bmatrix}\begin{bmatrix}1&2&3\\2&4&5\\3&5&6\end{bmatrix}$$

$$=\begin{bmatrix}14&25&31\\25&45&56\\31&56&70\end{bmatrix}$$

$$A^3=A\cdot A^2=\begin{bmatrix}1&2&3\\2&4&5\\3&5&6\end{bmatrix}\begin{bmatrix}14&25&31\\25&45&56\\31&56&70\end{bmatrix}$$

$$=\begin{bmatrix}157&283&353\\283&510&636\\353&636&793\end{bmatrix}$$

verification

$$A^3-11A^2-4A+I$$

$$=\begin{bmatrix}157&283&353\\283&510&636\\353&636&793\end{bmatrix}-11\begin{bmatrix}14&25&31\\25&45&56\\31&56&70\end{bmatrix}$$

$$-4\begin{bmatrix}1&2&3\\2&4&5\\3&5&6\end{bmatrix}+\begin{bmatrix}1&0&0\\0&1&0\\0&0&1\end{bmatrix}=\begin{bmatrix}0&0&0\\0&0&0\\0&0&0\end{bmatrix}$$

To find A^{-1}: From characteristic equation $A^{-1}=-A^2+11A+4I$. So.

$$A^{-1}=-\begin{bmatrix}14&25&31\\25&45&56\\31&56&70\end{bmatrix}+11\begin{bmatrix}1&2&3\\2&4&5\\3&5&6\end{bmatrix}+$$

$$+4\begin{bmatrix}1&0&0\\0&1&0\\0&0&1\end{bmatrix}=\begin{bmatrix}1&-3&2\\-3&3&-1\\2&-1&0\end{bmatrix}$$

To find A^4: From Cayley-Hamilton theorem

$$A^3-11A^2-4A+I=0$$

or $$A^3=11A^2+4A-I$$

Multiplying both sides by A

$$A^4=A\cdot A^3=A(11A^2+4A-I)=11A^3+4A^2-A$$

$$=11\begin{bmatrix}157&283&353\\283&510&636\\353&636&793\end{bmatrix}+4\begin{bmatrix}14&25&31\\25&45&56\\31&56&70\end{bmatrix}$$

$$-\begin{bmatrix}1&2&3\\2&4&5\\3&5&6\end{bmatrix}$$

$$=\begin{bmatrix}1782&3211&4004\\3211&5786&7215\\4004&7215&8997\end{bmatrix}$$

To find B: Rewrite

$$B=A^8-11A^7-4A^6+A^5+A^4-11A^3$$
$$-3A^2+2A+I$$
$$=A^5(A^3-11A^2-4A+I)$$
$$+A(A^3-11A^2-4A+I)+A^2+A+I$$
$$=A^5(0)+A(0)+A^2+A+I$$

since A satisfies the characteristic equation. Thus

$$B=A^2+A+I=\begin{bmatrix}14&25&31\\25&45&56\\31&56&70\end{bmatrix}$$

$$+\begin{bmatrix}1&2&3\\2&4&5\\3&5&6\end{bmatrix}+\begin{bmatrix}1&0&0\\0&1&0\\0&0&1\end{bmatrix}$$

$$B=\begin{bmatrix}16&27&34\\27&50&61\\34&61&77\end{bmatrix}$$

Example 3: Determine A^{-1}, A^{-2}, A^{-3} if

$$A=\begin{bmatrix}4&6&6\\1&3&2\\-1&-4&-3\end{bmatrix}$$

Solution: The characteristic equation of A is

$$|A-\lambda I|=\begin{bmatrix}4-\lambda&6&6\\1&3-\lambda&2\\-1&-4&-3-\lambda\end{bmatrix}$$

$$=\lambda^3-4\lambda^2-\lambda+4=0$$

It follows from Cayley-Hamilton theorem that

$$A^3 - 4A^2 - A + 4I = 0$$

Multiplying by A^{-1},

$$A^{-1}A^3 - 4A^{-1}A^2 - A^{-1}\cdot A + A^{-1}4I = 0$$

Solving $A^{-1} = \frac{1}{4}[I + 4A - A^2]$

$$A^2 = A\cdot A = \begin{bmatrix} 16 & 18 & 18 \\ 5 & 7 & 6 \\ -5 & -6 & -5 \end{bmatrix}$$

$$A^{-1} = \frac{1}{4}\begin{bmatrix} 1 & 0 & 0 \\ 0 & 1 & 0 \\ 0 & 0 & 1 \end{bmatrix} + 4\begin{bmatrix} 4 & 6 & 6 \\ 1 & 3 & 2 \\ -1 & -4 & -3 \end{bmatrix}$$

$$-\frac{1}{4}\begin{bmatrix} 16 & 18 & 18 \\ 5 & 7 & 6 \\ -5 & -6 & -5 \end{bmatrix}$$

$$A^{-1} = \frac{1}{4}\begin{bmatrix} 1 & 6 & 6 \\ -1 & 6 & 2 \\ 1 & -10 & -6 \end{bmatrix}$$

Multiplying A^{-1} by A^{-1}, we have

$$A^{-2} = A^{-1}A^{-1} = A^{-1}\frac{1}{4}[I + 4A - A^2] = \frac{1}{4}[A^{-1} + 4I - A]$$

$$A^{-2} = \frac{1}{4}\begin{bmatrix} \frac{1}{4} & -\frac{9}{2} & -\frac{9}{2} \\ -\frac{5}{4} & \frac{5}{2} & -\frac{3}{2} \\ \frac{5}{4} & \frac{3}{2} & \frac{11}{2} \end{bmatrix}$$

Similarly,

$$A^{-3} = A^{-1}A^{-2} = A^{-1}[A^{-1} + 4I - A]\frac{1}{4}$$

$$= \frac{1}{4}[A^{-2} + 4A^{-1} - I] = \frac{1}{64}\begin{bmatrix} 1 & 78 & 78 \\ -21 & 90 & 26 \\ 21 & -154 & -90 \end{bmatrix}$$

Exercise

Verify Cayley-Hamilton theorem for the matrix:

1. $\begin{bmatrix} 2 & 5 \\ 1 & -3 \end{bmatrix}$

Ans. Characteristic polynomial: $\lambda^2 + \lambda - 11$

2. $\begin{bmatrix} 2 & -3 \\ 7 & -4 \end{bmatrix}$

Ans. Characteristic polynomial: $\lambda^2 + 2\lambda + 13$

3. $\begin{bmatrix} 1 & 4 & -3 \\ 0 & 3 & 1 \\ 0 & 2 & -1 \end{bmatrix}$

Ans. Characteristic equation: $\lambda^3 - 3\lambda^2 - 3\lambda + 5 = 0$

4. Verify Cayley-Hamilton theorem for

$$A = \begin{bmatrix} 1 & 4 \\ 2 & 3 \end{bmatrix}. \text{ Find } A^{-1}.$$

Find $B = A^5 - 4A^4 - 7A^3 + 11A^2 - A - 10I$

Ans. Characteristic equation: $\lambda^2 - 4\lambda - 5 = 0$,

$$A^{-1} = \frac{1}{5}\begin{bmatrix} -3 & 4 \\ 2 & -1 \end{bmatrix},\ B = A + 5I = \begin{bmatrix} 6 & 4 \\ 2 & 8 \end{bmatrix}$$

5. Use Cayley-Hamilton theorem to find A^{-1} if

$$A = \begin{bmatrix} 1 & 1 & 3 \\ 1 & 3 & -3 \\ -2 & -4 & -4 \end{bmatrix}$$

Ans. Characteristic equation: $\lambda^3 - 20\lambda + 8 = 0$

$$A^{-1} = \frac{1}{4}\begin{bmatrix} 12 & 4 & 6 \\ -5 & -1 & -3 \\ -1 & -1 & -1 \end{bmatrix}$$

6. Find A^{-1} and A^4 if

$$A = \begin{bmatrix} 1 & 0 & 3 \\ 2 & 1 & -1 \\ 1 & -1 & 1 \end{bmatrix}$$

Ans. Characteristic equation: $\lambda^3 - 3\lambda^2 - \lambda + 9 = 0$,

$$A^{-1} = \frac{1}{9}\begin{bmatrix} 0 & 3 & 3 \\ 3 & 2 & -7 \\ 3 & -1 & -1 \end{bmatrix}$$

$$A^4 = \begin{bmatrix} 7 & -30 & 42 \\ 18 & -13 & 46 \\ -6 & -14 & 17 \end{bmatrix}$$

7. Find A^{-1} for

$$A = \begin{bmatrix} 2 & -1 & 1 \\ -1 & 2 & -1 \\ 1 & -1 & 2 \end{bmatrix}$$

Ans. Characteristic equation: $\lambda^3 - 6\lambda^2 + 9\lambda - 4 = 0$,

$$A^{-1} = \frac{1}{4}\begin{bmatrix} 3 & 1 & -1 \\ 1 & 3 & 1 \\ -1 & 1 & 3 \end{bmatrix}$$

8. Compute A^{-1}, A^{-2}, A^3 and A^4 if

$$A = \begin{bmatrix} 1 & 1 & 2 \\ 3 & 1 & 1 \\ 2 & 3 & 1 \end{bmatrix}$$

Ans. Characteristic equation: $\lambda^3 - 3\lambda^2 - 7\lambda - 11 = 0$

$$A^{-1} = \frac{1}{11}\begin{bmatrix} -2 & 5 & -1 \\ -1 & -3 & 5 \\ 7 & -1 & -2 \end{bmatrix},$$

$$A^{-2} = \frac{1}{121}\begin{bmatrix} -8 & -24 & 29 \\ 40 & -1 & -24 \\ -27 & 40 & -8 \end{bmatrix}$$

$$A^3 = \begin{bmatrix} 42 & 31 & 29 \\ 45 & 39 & 31 \\ 53 & 45 & 42 \end{bmatrix},$$

$$A^4 = \begin{bmatrix} 193 & 160 & 144 \\ 224 & 177 & 160 \\ 272 & 224 & 193 \end{bmatrix}$$

9. If $A = \begin{bmatrix} 2 & 1 & 1 \\ 0 & 1 & 0 \\ 1 & 1 & 2 \end{bmatrix}$ find A^{-1}. Find $B =$ $A^8 - 5A^7 + 7A^6 - 3A^5 - 5A^3 + 8A^2 - 2A + I$

Ans. Characteristic equation: $\lambda^3 - 5\lambda^2 + 7\lambda - 3 = 0$

$$A^{-1} = \frac{1}{3}\begin{bmatrix} 2 & -1 & -1 \\ 0 & 3 & 0 \\ -1 & -1 & 2 \end{bmatrix}, B = \begin{bmatrix} 8 & 5 & 5 \\ 0 & 3 & 0 \\ 5 & 5 & 8 \end{bmatrix}$$

10. Verify Cayley-Hamilton theorem and hence find A^{-1} for

$$A = \begin{bmatrix} 0 & 1 & 0 & 0 \\ 0 & 0 & 0 & 1 \\ 0 & 0 & 1 & 0 \\ 1 & 0 & 0 & 0 \end{bmatrix}$$

Ans. Characteristic equation: $\lambda^4 - \lambda^3 - \lambda + 1 = 0$

$$A^{-1} = \begin{bmatrix} 0 & 0 & 0 & 1 \\ 1 & 0 & 0 & 0 \\ 0 & 0 & 1 & 0 \\ 0 & 1 & 0 & 0 \end{bmatrix}$$

11. Find $B = A^6 - 4A^5 + 8A^4 - 12A^3 + 14A^2$ if $A = \begin{bmatrix} 1 & 2 \\ -1 & 3 \end{bmatrix}$

Ans. Characteristic equation: $\lambda^2 - 4\lambda + 5 = 0$,

$$B = 5I - 4A = \begin{bmatrix} +1 & -8 \\ 4 & -7 \end{bmatrix}$$

12. Find A^{-1} if $A = \begin{bmatrix} 7 & 2 & -2 \\ -6 & -1 & 2 \\ 6 & 2 & -1 \end{bmatrix}$

Ans. Characteristic equation: $\lambda^3 - 5\lambda^2 + 7\lambda - 3 = 0$,

$$A^{-1} = \frac{1}{3}\begin{bmatrix} -3 & -2 & 2 \\ 6 & 5 & -2 \\ -6 & -2 & 5 \end{bmatrix}.$$

14.5 DIAGONALIZATION POWERS OF A MATRIX

Consider only square matrices of order *n*.

Similar matrix

A is said to be similar to *B* if there exists a non singular matrix *P* such that

$$B = P^{-1}AP$$

This transformation of *A* to *B* is known as similarity transformation.

Invariant Eigen Values

1. Similar matrices *A* and *B* have same eigen values

2. Further if *X* is an eigen vector of *A* then $Y = P^{-1}X$ is an eigen vector of the matrix *B*.

Proof:

1. Suppose *B* is similar to *A* i.e., $B = P^{-1}AP$.

Consider the characteristic polynomial of B

$$|B-\lambda I| = |P^{-1}AP - \lambda I| = |P^{-1}AP - \lambda P^{-1}IP|$$
$$= |P^{-1}(A-\lambda I)P| = |P^{-1}||A-\lambda I||P|$$
$$= |A-\lambda I|$$

since $|P^{-1}||P| = |P^{-1}P| = |I| = 1$.

Thus A and B have the same characteristic polynomial and therefore has the same eigen values.

2. Let X be an eigen vector of A so that $AX = \lambda X$. Consider

$$B = P^{-1}AP$$

Post multiplying by P^{-1}

$$BP^{-1} = (P^{-1}AP)P^{-1} = (P^{-1}A)(PP^{-1}) = P^{-1}A$$

Post multiply by X

$$B(P^{-1}X) = P^{-1}AX = P^{-1}(AX)$$
$$= P^{-1}\lambda X = \lambda(P^{-1}X)$$

Thus $P^{-1}X$ is an eigen vector of B corresponding to the eigen value λ.

Diagonalization

A n-square matrix A with n linearly independent eigen vectors is similar to a diagonal matrix D whose diagonal elements are the eigen values of A.

Proof: Let $X_1, X_2, \ldots, X_n$ be the n linearly independent eigen vectors of A corresponding to n eigen values $\lambda_1, \lambda_2, \ldots, \lambda_n$. Construct P, known as **modal matrix,** having $X_1, X_2, \ldots, X_n$ as the n column vectors i.e.,

$$P_{n\times n} = [X_1 X_2 \cdots X_n].$$

Since $X_1, X_2, \ldots, X_n$ are linearly independent, P^{-1} exists.

Consider

$$AP = A[X_1 X_2 \cdots X_n] = [AX_1\ AX_2 \cdots AX_n]$$
$$= [\lambda_1 X_1\ \lambda_2 X_2 \cdots \lambda_n X_n]$$
$$= [X_1 X_2 \cdots X_n]\begin{bmatrix} \lambda_1 & 0 & 0\cdots 0 \\ 0 & \lambda_2 & 0\cdots 0 \\ \vdots & & \\ 0 & 0 & \cdots \lambda_n \end{bmatrix}$$

$$AP = PD$$

where D is the diagonal matrix with eigen values of A as the principal diagonal elements. D is known as **spectral matrix.**

Pre multiplying by P^{-1} on both sides

$$B = P^{-1}AP = P^{-1}PD = (P^{-1}P)D = ID = D.$$

Powers of a Matrix A

Consider $\quad D = P^{-1}AP$

Then $\quad D^2 = (P^{-1}AP)(P^{-1}AP)$

$$= P^{-1}A(PP^{-1})AP$$
$$= P^{-1}A \cdot IAP = P^{-1}AAP$$
$$= P^{-1}A^2P$$

Similarly, $\quad D^3 = P^{-1}A^3P$

Thus $\quad D^n = P^{-1}A^nP$

To obtain A^n, pre-multiply by P and post-multiply by P^{-1},

$$PD^nP^{-1} = P(P^{-1}A^nP)P^{-1} = (PP^{-1})A^n(PP^{-1})$$
$$= IA^nI = A^n$$
$$\therefore\ A^n = PD^nP^{-1}.$$

WORKED OUT EXAMPLES

Example 1: Find a matrix P which diagonalizes the matrix

$$A = \begin{bmatrix} 4 & 1 \\ 2 & 3 \end{bmatrix}. \text{ Verify that } P^{-1}AP = D$$

where D is the diagonal matrix. Hence find A^6.

Solution: A is diagonalizable by P whose columns are the linearly independent eigen vectors of A.

The characteristic equation of A is

$$|A-\lambda I| = \begin{bmatrix} 4-\lambda & 1 \\ 2 & 3-\lambda \end{bmatrix} = 0$$

$(4-\lambda)(3-\lambda) - 2 = \lambda^2 - 7\lambda + 10$
$= (\lambda-2)(\lambda-5) = 0,$

so $\lambda = 2, 5$ are two distinct eigen values of A.

For $\lambda = 2$, $2x_1 + x_2 = 0$, $x_2 = -2x_1$, $X_1 = \begin{bmatrix} 1 \\ -2 \end{bmatrix}$

For $\lambda = 5$, $-x_1 + x_2 = 0$, $x_2 = x_1$, $X_2 = [1, 1]^T$

Thus the matrix P which diagonalizes A is

$$P = [X_1, X_2] = \begin{bmatrix} 1 & 1 \\ -2 & 1 \end{bmatrix}$$

Verification:

$$P^{-1} = \frac{1}{3}\begin{bmatrix} 1 & -1 \\ 2 & 1 \end{bmatrix}$$

$$P^{-1}AP = \frac{1}{3}\begin{bmatrix} 1 & -1 \\ 2 & 1 \end{bmatrix}\begin{bmatrix} 4 & 1 \\ 2 & 3 \end{bmatrix}\begin{bmatrix} 1 & 1 \\ -2 & 1 \end{bmatrix}$$

$$= \frac{1}{3}\begin{bmatrix} 2 & -2 \\ 10 & 5 \end{bmatrix}\begin{bmatrix} 1 & 1 \\ -2 & 1 \end{bmatrix}$$

$$P^{-1}AP = \frac{1}{3}\begin{bmatrix} 6 & 0 \\ 0 & 15 \end{bmatrix} = \begin{bmatrix} 2 & 0 \\ 0 & 5 \end{bmatrix} = D = \text{diagonal matrix}$$

D contains eigen values 2, 5 as diagonal elements.
To find A^6:

$$A^6 = PD^6P^{-1} = \begin{bmatrix} 1 & 1 \\ -2 & 1 \end{bmatrix}\begin{bmatrix} 2^6 & 0 \\ 0 & 5^6 \end{bmatrix}\frac{1}{3}\begin{bmatrix} 1 & -1 \\ 2 & 1 \end{bmatrix}$$

$$A^6 = \frac{1}{3}\begin{bmatrix} 64 & 15625 \\ -128 & 15625 \end{bmatrix}\begin{bmatrix} 1 & -1 \\ 2 & 1 \end{bmatrix}$$

$$= \frac{1}{3}\begin{bmatrix} 31314 & 15561 \\ 31122 & 15753 \end{bmatrix}$$

$$\therefore \quad A^6 = \begin{bmatrix} 10438 & 5187 \\ 10374 & 5251 \end{bmatrix}$$

Example 2: Diagonalize

$$A = \begin{bmatrix} 1 & 6 & 1 \\ 1 & 2 & 0 \\ 0 & 0 & 3 \end{bmatrix}$$

and hence find A^8. Find the modal matrix.

Solution: The non-singular square matrix P containing eigen vectors of A as columns, diagonalizes A.

$$\text{The ch. eq. of } A \text{ is } \begin{bmatrix} 1-\lambda & 6 & 1 \\ 1 & 2-\lambda & 0 \\ 0 & 0 & 3-\lambda \end{bmatrix} = 0$$

$$\text{i.e., } (\lambda + 1)(\lambda - 3)(\lambda - 4) = 0$$

so eigen values of A are $\lambda = -1, 3, 4$

For $\lambda = -1$,

$$\begin{aligned} 2x_1 + 6x_2 + x_3 &= 0 \\ x_1 + 3x_2 + 0 &= 0 \\ 4x_3 &= 0 \end{aligned} \qquad \begin{aligned} \therefore x_3 &= 0 \\ x_1 &= -3x_2 \end{aligned}$$

$$X_1 = \begin{bmatrix} -3 \\ 1 \\ 0 \end{bmatrix}$$

For $\lambda = 3$,

$$\begin{aligned} -2x_1 + 6x_2 + x_3 &= 0 \\ x_1 - x_2 &= 0 \end{aligned} \qquad \begin{aligned} \therefore x_1 &= x_2 \\ x_3 &= -4x_2\text{'} \end{aligned}$$

$$X_2 = \begin{bmatrix} 1 \\ 1 \\ -4 \end{bmatrix}$$

For $\lambda = 4$,

$$\begin{aligned} -3x_1 + 6x_2 + x_3 &= 0 \\ x_1 - 2x_2 &= 0 \\ -x_3 &= 0 \end{aligned} \qquad \begin{aligned} \therefore x_3 &= 0 \\ x_2 &= 2x_2\text{'} \end{aligned}$$

$$X_3 = \begin{bmatrix} 2 \\ 1 \\ 0 \end{bmatrix}.$$

Thus

$$P = \begin{bmatrix} -3 & 1 & 2 \\ 1 & 1 & 1 \\ 0 & -4 & 0 \end{bmatrix}$$

is the modal matrix.
To find P^{-1}:

$$\left[\begin{array}{ccc|ccc} -3 & 1 & 2 & 1 & 0 & 0 \\ 1 & 1 & 1 & 0 & 1 & 0 \\ 0 & -4 & 0 & 0 & 0 & 1 \end{array}\right]$$

$$\begin{matrix} R_{12} \\ R_{21(3)} \end{matrix} \sim \left[\begin{array}{ccc|ccc} 1 & 1 & 1 & 0 & 1 & 0 \\ 0 & 4 & 5 & 1 & 3 & 0 \\ 0 & -4 & 0 & 0 & 0 & 1 \end{array}\right]$$

$$\begin{matrix} R_{32(1)} \\ \sim \end{matrix} \left[\begin{array}{ccc|ccc} 1 & 1 & 1 & 0 & 1 & 0 \\ 0 & 4 & 5 & 1 & 3 & 0 \\ 0 & 0 & 5 & 1 & 3 & 1 \end{array}\right]$$

$$\begin{matrix} R_{2}\left(\frac{1}{4}\right) \\ R_3\left(\frac{1}{5}\right) \\ \sim \end{matrix} \left[\begin{array}{ccc|ccc} 1 & 1 & 1 & 0 & 1 & 0 \\ 0 & 1 & \frac{5}{4} & \frac{1}{4} & \frac{3}{4} & 0 \\ 0 & 0 & 1 & \frac{1}{5} & \frac{3}{5} & \frac{1}{5} \end{array}\right]$$

$$\begin{matrix} R_{23}\left(-\frac{5}{4}\right) \\ R_{13(-1)} \\ \sim \end{matrix} \left[\begin{array}{ccc|ccc} 1 & 1 & 0 & -\frac{1}{5} & \frac{2}{5} & -\frac{1}{5} \\ 0 & 1 & 0 & 0 & 0 & -\frac{1}{4} \\ 0 & 0 & 1 & \frac{1}{5} & \frac{3}{5} & \frac{1}{5} \end{array}\right]$$

$$\underset{R_{12(-1)}}{\sim}\left[\begin{array}{ccc|ccc} 1 & 0 & 0 & -\frac{1}{5} & \frac{2}{5} & \frac{1}{20} \\ 0 & 1 & 0 & 0 & 0 & -\frac{1}{4} \\ 0 & 0 & 1 & \frac{1}{5} & \frac{3}{5} & \frac{1}{5} \end{array}\right]$$

Thus

$$P^{-1} = \frac{1}{20}\begin{bmatrix} -4 & 8 & 1 \\ 0 & 0 & -5 \\ 4 & 12 & 4 \end{bmatrix}$$

Diagonalization:

$D = P^{-1}AP$

$$= \frac{1}{20}\begin{bmatrix} -4 & 8 & 1 \\ 0 & 0 & -5 \\ 4 & 12 & 4 \end{bmatrix}\begin{bmatrix} 1 & 6 & 1 \\ 1 & 2 & 0 \\ 0 & 0 & 3 \end{bmatrix}\begin{bmatrix} -3 & 1 & 2 \\ 1 & 1 & 1 \\ 0 & -4 & 0 \end{bmatrix}$$

$$= \frac{1}{20}\begin{bmatrix} 4 & -8 & -1 \\ 0 & 0 & -15 \\ 16 & 48 & 16 \end{bmatrix}\begin{bmatrix} -3 & 1 & 2 \\ 1 & 1 & 1 \\ 0 & -4 & 0 \end{bmatrix}$$

$$= \frac{1}{20}\begin{bmatrix} -20 & 0 & 0 \\ 0 & 60 & 0 \\ 0 & 0 & 80 \end{bmatrix} = \begin{bmatrix} -1 & 0 & 0 \\ 0 & 3 & 0 \\ 0 & 0 & 4 \end{bmatrix}$$

To find A^8:

$A^8 = PD^8P^{-1}$

$$= \begin{bmatrix} -3 & 1 & 2 \\ 1 & 1 & 1 \\ 0 & -4 & 0 \end{bmatrix}\begin{bmatrix} (-1)^8 & 0 & 0 \\ 0 & 3^8 & 0 \\ 0 & 0 & 4^8 \end{bmatrix}\frac{1}{20}\begin{bmatrix} -4 & 8 & 1 \\ 0 & 0 & -5 \\ 4 & 12 & 4 \end{bmatrix}$$

$$= \begin{bmatrix} -3 & 1 & 2 \\ 1 & 1 & 1 \\ 0 & -4 & 0 \end{bmatrix}\begin{bmatrix} 1 & 0 & 0 \\ 0 & 6561 & 0 \\ 0 & 0 & 65536 \end{bmatrix}\begin{bmatrix} -4 & 8 & 1 \\ 0 & 0 & 5 \\ 4 & 12 & 4 \end{bmatrix}\frac{1}{20}$$

$$= \frac{1}{20}\begin{bmatrix} -3 & 6561 & 131072 \\ 1 & 6561 & 65536 \\ 0 & -26244 & 0 \end{bmatrix}\begin{bmatrix} -4 & 8 & 1 \\ 0 & 0 & 5 \\ 4 & 12 & 4 \end{bmatrix}$$

$$= \frac{1}{20}\begin{bmatrix} 524300 & 1572840 & 491480 \\ 262140 & 786440 & 229340 \\ 0 & 0 & 131220 \end{bmatrix}$$

$$\therefore A^8 = \begin{bmatrix} 26215 & 78642 & 24574 \\ 13107 & 39322 & 11467 \\ 0 & 0 & 6561 \end{bmatrix}$$

EXERCISE

Diagonalize the following matrices. Find the modal matrix P which diagonalizes (transforms) A

1. $\begin{bmatrix} 5 & 3 \\ 3 & 5 \end{bmatrix}$

Ans. $P = \begin{bmatrix} 1 & 1 \\ 1 & -1 \end{bmatrix}, D = \begin{bmatrix} 8 & 0 \\ 0 & 2 \end{bmatrix}$

2. $\begin{bmatrix} 0 & 1 \\ 0 & 0 \end{bmatrix}$

Ans. not diagonizable since only one eigen vector $\begin{bmatrix} k \\ 0 \end{bmatrix}$ exists.

3. $\begin{bmatrix} 5 & 4 \\ 1 & 2 \end{bmatrix}$

Ans. $P = \begin{bmatrix} 4 & 1 \\ 1 & -1 \end{bmatrix}, D = \begin{bmatrix} 6 & 0 \\ 0 & 1 \end{bmatrix}$

4. $\begin{bmatrix} 2 & -2 \\ -2 & 5 \end{bmatrix}$

Ans. $P = \begin{bmatrix} 1 & 2 \\ -2 & 1 \end{bmatrix}, D = \begin{bmatrix} 6 & 0 \\ 0 & 1 \end{bmatrix}$

5. $\begin{bmatrix} 1 & 2 \\ 2 & 1 \end{bmatrix}$

Ans. $P = \begin{bmatrix} 1 & 1 \\ 1 & -1 \end{bmatrix}, D = \begin{bmatrix} 3 & 0 \\ 0 & -1 \end{bmatrix}$

6. $\begin{bmatrix} 1 & 2 \\ 5 & 4 \end{bmatrix}$

Ans. $P = \begin{bmatrix} 1 & 2 \\ -1 & 5 \end{bmatrix}, D = \begin{bmatrix} -1 & 0 \\ 0 & 6 \end{bmatrix}$

7. $\begin{bmatrix} 4 & 1 \\ 3 & 2 \end{bmatrix}$ hence find A^5

Ans. $P = \begin{bmatrix} 1 & 1 \\ -3 & 1 \end{bmatrix}, D = \begin{bmatrix} 1 & 0 \\ 0 & 5 \end{bmatrix}$,

$$A^5 = \begin{bmatrix} 2344 & 781 \\ 2343 & 782 \end{bmatrix}$$

8. $\begin{bmatrix} 1 & 1 \\ -1 & 1 \end{bmatrix}$

Ans. no real eigen values, $\lambda = 1 \pm i$ so not diagonalizable over real.

Modal matrix over complex

$$\begin{pmatrix} 1 & 1 \\ i & -i \end{pmatrix}, D = \begin{pmatrix} 1+i & 0 \\ 0 & 1-i \end{pmatrix}.$$

9. $\begin{bmatrix} 7.3 & 0.2 & -3.7 \\ -11.5 & 1.0 & 5.5 \\ 17.7 & 1.8 & -9.3 \end{bmatrix}$

Ans. characteristic equation $\lambda^3 + \lambda^2 - 12\lambda = 0$ eigen values 3, −4, 0.

$$\text{Modal matrix} = \begin{bmatrix} -1 & 1 & 2 \\ 3 & -1 & 1 \\ -1 & 3 & 4 \end{bmatrix},$$

$$D = \begin{bmatrix} 3 & 0 & 0 \\ 0 & -4 & 0 \\ 0 & 0 & 0 \end{bmatrix}$$

10. $\begin{bmatrix} 1 & 1 & 3 \\ 1 & 5 & 1 \\ 3 & 1 & 1 \end{bmatrix}$ hence find A^4.

Ans. characteristic equation $(\lambda + 2)(\lambda - 3)(\lambda - 6) = 0, \lambda = -2, 3, 6$

$$\text{Modal matrix } P = \begin{bmatrix} -1 & 1 & 1 \\ 0 & -1 & 2 \\ 1 & 1 & 1 \end{bmatrix},$$

$$D = \begin{bmatrix} -2 & 0 & 0 \\ 0 & 3 & 0 \\ 0 & 0 & 6 \end{bmatrix},$$

$$A^4 = \begin{bmatrix} 251 & 485 & 235 \\ 485 & 1051 & 485 \\ 235 & 485 & 251 \end{bmatrix}.$$

11. $\begin{bmatrix} 8 & -6 & 2 \\ -6 & 7 & -4 \\ 2 & -4 & 3 \end{bmatrix}$

Ans. $\lambda^3 - 18\lambda^2 + 45\lambda = 0, \lambda = 0, 3, 15$

$$P = \begin{bmatrix} 1 & 2 & 2 \\ 2 & 1 & -2 \\ 2 & --2 & 1 \end{bmatrix}, \quad D = \begin{bmatrix} 0 & 0 & 0 \\ 0 & 3 & 0 \\ 0 & 0 & 15 \end{bmatrix}$$

12. Find A^8 for $A = \begin{bmatrix} 1 & 1 & 1 \\ 0 & 2 & 1 \\ -4 & 4 & 3 \end{bmatrix}$

Ans. $(1 - \lambda)(\lambda - 2)(\lambda - 3) = 0, \lambda = 1, 2, 3,$

$$P = \begin{bmatrix} 1 & 1 & 1 \\ 2 & 1 & 1 \\ -2 & 0 & 1 \end{bmatrix}$$

$$A^8 = \begin{bmatrix} -12099 & 12355 & 6305 \\ -12100 & 12356 & 6305 \\ -13120 & 13120 & 6561 \end{bmatrix}$$

13. Find A^5 for $A = \begin{bmatrix} 11 & -4 & -7 \\ 7 & -2 & -5 \\ 10 & -4 & -6 \end{bmatrix}$

Ans. $\lambda = 0, 1, 2$

$$P = \begin{bmatrix} 1 & 1 & -2 \\ 1 & -1 & -1 \\ 1 & 2 & -2 \end{bmatrix}, A^5 = \begin{bmatrix} 191 & -64 & -127 \\ 97 & -32 & -65 \\ 190 & -64 & -126 \end{bmatrix}$$

14. Find A^4 for $A = \begin{bmatrix} 3 & -1 & 1 \\ -1 & 5 & -1 \\ v1 & -1 & 3 \end{bmatrix}$

Ans. $\lambda = 2, 3, 6,$

$$P = \begin{bmatrix} 1 & 1 & 1 \\ 0 & 1 & -2 \\ -1 & 1 & 1 \end{bmatrix}, A^4 = \begin{bmatrix} 251 & -405 & 235 \\ -405 & 891 & -405 \\ 235 & -405 & 251 \end{bmatrix}$$

15. $\begin{bmatrix} 7 & -2 & 1 \\ -2 & 10 & -2 \\ 1 & -2 & 7 \end{bmatrix}$

Ans. $\lambda^3 - 24\lambda^2 + 180\lambda - 432 = 0, \lambda = 6, 6, 12$

$$P = \begin{bmatrix} \frac{1}{\sqrt{2}} & \frac{1}{\sqrt{3}} & \frac{1}{\sqrt{6}} \\ 0 & \frac{1}{\sqrt{3}} & -\frac{2}{\sqrt{6}} \\ -\frac{1}{\sqrt{2}} & \frac{1}{\sqrt{3}} & \frac{1}{\sqrt{6}} \end{bmatrix}$$

16. $\begin{bmatrix} +1 & -3 & 2 \\ -3 & 7 & -5 \\ 2 & -5 & 8 \end{bmatrix}$

Ans. $\lambda = 1, -2, 18,$

$$P = \begin{bmatrix} 1 & 3 & -1 \\ 0 & 1 & 1 \\ 0 & 0 & 2 \end{bmatrix}$$

14.6 REAL MATRICES: SYMMETRIC, SKEW-SYMMETRIC, ORTHOGONAL. QUADRATIC FORM

A matrix $A = (a_{ij})$ is said to be a real matrix if every element a_{ij} of A is real. A real **square** matrix $A = (a_{ij})$ is said to be

a. Symmetric if $A^T = A$ i.e., $a_{ji} = a_{ij}$

b. Skew-symmetric if $A^T = -A$, i.e., $a_{ji} = -a_{ij}$

c. Orthogonal if $A^T = A^{-1}$

The determinant of an orthogonal matrix is ± 1 since

$$1 = |I| = |AA^{-1}| = |AA^T| = |A||A^T|$$
$$= |A||A| = |A|^2 \quad \text{i.e., } |A| = \pm 1.$$

Orthogonal Transformation

which geometrically represents a rotation, is a transformation

$$Y = AX$$

where A is an orthogonal matrix.

Norm of a vector X denoted by $||X||$ is

$$||X|| = \sqrt{X^T X}$$

represents the length of the vector X.

Orthonormal System (Set) of Vectors

A set of vectors $X_1, X_2, \ldots, X_n$ are said to form an orthonormal system if

$$X_i^T X_j = \delta_{ij} = \begin{cases} 1, & \text{if } i = j \\ 0, & \text{if } i \neq j \end{cases}$$

i.e, vectors are mutually orthogonal and normalized

Note: Column (row) vectors of an orthogonal matrix form an orthonormal system of vectors (proof on Page 14.27).

Quadratic Form

Quadratic form in n variables $x_1, x_2, \ldots, x_n$ is an expression of the form

$$Q = X^T AX = \sum_{i=1}^{n} \sum_{j=1}^{n} a_{ij} x_i x_j$$

$$= a_{11}x_1^2 + a_{12}x_1x_2 + \cdots + a_{1n}x_1x_n$$
$$+ a_{21}x_2x_1 + a_{22}x_2^2 + \cdots + a_{2n}x_2x_n$$
$$+ \cdots\cdots\cdots\cdots\cdots$$
$$+ a_{n1}x_nx_1 + a_{n2}x_nx_2 + \cdots + a_{nn}x_n^2.$$

Here A is known as the coefficient matrix.

Rewriting

$$Q = X^T AX = a_{11}x_1^2 + (a_{12} + a_{21})x_1x_2 + \cdots$$
$$+ (a_{1n} + a_{n1})x_1x_n$$
$$+ a_{22}x_2^2 + (a_{23} + a_{32})x_2x_3 + \cdots$$
$$+ (a_{2n} + a_{n2}) \cdots x_2x_n$$
$$+ \cdots\cdots\cdots\cdots + a_{nn}x_n^2$$

Put $c_{ij} = \frac{1}{2}(a_{ij} + a_{ji})$ then $c_{ij} = c_{ji}$ and $c_{ij} + c_{ji} = a_{ij} + a_{ji}$. Thus the quadratic form can be rewritten as

$$Q = \sum_{i=1}^{n} \sum_{j=1}^{n} c_{ij} x_i x_j$$

where $c = \frac{1}{2}[A + A^T]$ is a real symmetric matrix. Thus the coefficient matrix in a quadratic form can always be assumed (can be constructed) as a real symmetric matrix.

WORKED OUT EXAMPLES

Example 1: Show that any square matrix A can be written as the sum of a symmetric matrix B and skew-symmetric matrix C.

Solution: Consider $B = \frac{1}{2}(A + A^T)$ since

$$B^T = \frac{1}{2}(A + A^T)^T = \frac{1}{2}(A^T + A^{T^T})$$
$$= \frac{1}{2}(A^T + A) = \frac{1}{2}(A + A^T) = B,$$

so B is symmetric.

Similarly, $C = \frac{1}{2}(A - A^T)$ is skew-symmetric because

$$C^T = \frac{1}{2}(A - A^T)^T = \frac{1}{2}(A^T - A^{T^T})$$
$$= \frac{1}{2}(A^T - A) = -(A - A^T) = -C$$

Now

$$A=(B+C)=\frac{1}{2}\left[(A+A^T)+(A-A^T)\right]=A.$$

Example 2: Show that

$$A=\frac{1}{2}\begin{bmatrix}-1&1&1&1\\1&-1&1&1\\1&1&-1&1\\1&1&1&-1\end{bmatrix}\text{ is orthogonal.}$$

Solution: Consider $A\cdot A^T$

$$AA^T=\frac{1}{2}\begin{bmatrix}-1&1&1&1\\1&-1&1&1\\1&1&-1&1\\1&1&1&-1\end{bmatrix}\frac{1}{2}\begin{bmatrix}-1&1&1&1\\1&-1&1&1\\1&1&-1&1\\1&1&1&-1\end{bmatrix}$$

$$=\frac{1}{4}\begin{bmatrix}4&0&0&0\\0&4&0&0\\0&0&4&0\\0&0&0&4\end{bmatrix}=I$$

i.e., $A^T=A^{-1}$ $\therefore$ A is orthogonal.

Example 3: Determine a, b, c so that A is orthogonal, where

$$A=\begin{bmatrix}0&2b&c\\a&b&-c\\a&-b&c\end{bmatrix}$$

Solution: For orthogonal matrix $AA^T=I$ so

$$AA^T=\begin{bmatrix}0&2b&c\\a&b&-c\\a&-b&c\end{bmatrix}\begin{bmatrix}0&a&a\\2b&b&-b\\c&-c&c\end{bmatrix}$$

$$=\begin{bmatrix}4b^2+c^2&2b^2-c^2&-2b^2+c^2\\2b^2-c^2&a^2+b^2+c^2&a^2-b^2-c^2\\-2b^2+c^2&a^2-b^2-c^2&a^2+b^2+c^2\end{bmatrix}=I$$

Solving $2b^2-c^2=0,\ a^2-b^2-c^2=0$ (non-diagonal elements of I)

$$c=\pm\sqrt{2}b,\ a^2=b^2+c^2=b^2+2b^2=3b^2,\ a=\pm\sqrt{3}b$$

From diagonal elements of I,

$$4b^2+c^2=1,\quad 4b^2+2b^2=1$$

$$\therefore b=\pm\frac{1}{\sqrt{6}},\quad c=\pm\frac{1}{\sqrt{3}},\ a=\pm\frac{1}{\sqrt{2}}$$

Example 4: If $X_1=\frac{1}{3}[2\ \ -1\ \ \ 2]^T$

and $X_2=k[3\ \ -4\ \ -5]^T$

where $k=\frac{1}{\sqrt{50}}$, construct an orthogonal matrix

$$A=[X_1\quad X_2\quad X_3]$$

Solution: Let $X_3=[a_1\ a_2\ a_3]^T$ be the undetermined vector. Since A is orthogonal, the columns vectors of A form an orthonormal system $X_i^TX_j=\delta_{ij}$

$$X_1^TX_2=\begin{bmatrix}\frac{2}{3}&-\frac{1}{3}&\frac{2}{3}\end{bmatrix}\begin{bmatrix}3k\\-4k\\-5k\end{bmatrix}$$

$$=2k+\frac{4}{3}-\frac{10}{3}k=0,\text{ true}$$

$\therefore$ X_1 and X_2 are orthogonal.

$$X_1^TX_3=\begin{bmatrix}\frac{2}{3}&-\frac{1}{3}&\frac{2}{3}\end{bmatrix}\begin{bmatrix}a_1\\a_2\\a_3\end{bmatrix}$$

$$=\frac{1}{3}[2a_1-a_2+2a_3]=0 \qquad (1)$$

$$X_2^TX_3=[3k\quad -4k\quad -5k]\begin{bmatrix}a_1\\a_2\\a_3\end{bmatrix}$$

$$=(3a_1-4a_2-5a_3)k=0 \qquad (2)$$

Since X_3 should be normalized

$$X_3^TX_3=[a_1\quad a_2\quad a_3]\begin{bmatrix}a_1\\a_2\\a_3\end{bmatrix}$$

$$=a_1^2+a_2^2+a_3^2$$

$$1=||X_3||=\sqrt{X_3^TX_3}=\sqrt{a_1^2+a_2^2+a_3^2} \qquad (3)$$

Solving (1) (2) (3), we get a_1, a_2, a_3

$$2a_1-a_2+2a_3=0$$

$$3a_1-4a_2+5a_3=0$$

$$a_1^2+a_2^2+a_3^2=1$$

So $a_1=-\frac{13}{5}a_3,\quad a_2=-\frac{16}{5}a_3,$

$$a_3^2=\frac{25}{550},\qquad a_3=\frac{1}{\sqrt{22}}$$

$\therefore \quad a_1 = -\frac{13}{5}k_1, \quad a_2 = -\frac{16}{5}k_1, a_3 = k_1$

where $k_1 = \frac{1}{\sqrt{22}}$

Thus the required orthogonal matrix A is

$$A = \begin{bmatrix} \frac{2}{3} & 3k & -\frac{13}{5}k_1 \\ -\frac{1}{3} & -4k & -\frac{16}{5}k_1 \\ \frac{2}{3} & -5k & k_1 \end{bmatrix}.$$

Example 5: Find a real symmetric matrix C of the quadratic form $Q = x_1^2 + 3x_2^2 + 2x_3^2 + 2x_1x_2 + 6x_2x_3$

Solution: The coefficient matrix of Q is

$$A = \begin{bmatrix} 1 & 2 & 0 \\ 0 & 3 & 6 \\ 0 & 0 & 2 \end{bmatrix}, \text{ so } C = \text{symmetric matrix} = \frac{1}{2}[A + A^T]$$

$$C = \frac{1}{2}\left\{\begin{bmatrix} 1 & 2 & 0 \\ 0 & 3 & 6 \\ 0 & 0 & 2 \end{bmatrix} + \begin{bmatrix} 1 & 0 & 0 \\ 2 & 3 & 0 \\ 0 & 6 & 2 \end{bmatrix}\right\} = \begin{bmatrix} 1 & 1 & 0 \\ 1 & 3 & 3 \\ 0 & 3 & 2 \end{bmatrix}$$

Notes : The simplest of way writing C is

1. Put coefficients of square terms as the diagonal elements.
2. Place $\frac{1}{2}$ of a_{ij}, the coefficient of $x_i\, x_j$ at c_{ij} and the remaining $\frac{1}{2}$ of a_{ij} at c_{ji} i.e., $c_{ij} = c_{ji} = \frac{1}{2}a_{ij}$ such that $c_{ij} + c_{ji} = \frac{1}{2}(a_{ij} + a_{ij}) = a_{ij}$. For example 6, coefficient of $x_2\, x_3$ is equidistributed as 3 and 3 to c_{23} and c_{32}.

EXERCISE

1. Express A as the sum of a symmetric and skew-symmetric matrix where

$$A = \begin{bmatrix} 4 & 2 & -3 \\ 1 & 3 & -6 \\ -5 & 0 & -7 \end{bmatrix}$$

Ans. $A + A^T = \frac{1}{2}\begin{bmatrix} 8 & 3 & -8 \\ 3 & 6 & -6 \\ -8 & -6 & -14 \end{bmatrix}$ symmetric,

$A - A^T = \frac{1}{2}\begin{bmatrix} 0 & 1 & 2 \\ -1 & 0 & -6 \\ -2 & 6 & 0 \end{bmatrix}$ skew–symmetric

2. If A and B are square matrices of the same order and A is symmetric, then prove that $B^T AB$ is also symmetric.

Hint: $(B^T AB)^T = B^T A^T B^{T^T}$ but $A^T = A$, $= B^T AB$.

3. Prove that the inverse of a non-singular symmetric matrix A is symmetric.

Hint: $(A^{-1})^T = (AT)^{-1} = A^{-1}$ (since $(AT)^{-1} = (A^{-1})^T$ and $A = A^T$ by symmetry).

4. Write $A = \begin{bmatrix} 3 & -4 & -1 \\ 6 & 0 & -1 \\ -3 & 13 & -4 \end{bmatrix}$ as the sum of a symmetric R and skew-symmetric S.

Ans. $R = \frac{1}{2}[A + A^T] = \begin{bmatrix} 3 & 1 & -2 \\ 1 & 0 & 6 \\ -2 & 6 & -4 \end{bmatrix}$,

$S = \frac{1}{2}[A - A^T] = \begin{bmatrix} 0 & -5 & 1 \\ 5 & 0 & -7 \\ -1 & 7 & 0 \end{bmatrix}$

5. Show that the eigen values of the skew-symmetric matrix

$$A = \begin{bmatrix} 0 & 9 & -12 \\ -9 & 0 & 20 \\ 12 & -20 & 0 \end{bmatrix}$$

are purely imaginary or zero.

Ans. eigen values are $0, -25i, 25i$ (see Page 14.27)

Find real symmetric matrix C such that $Q = X^T CX$ where

6. $Q = 6x_1^2 - 4x_1x_2 + 2x_2^2$

Ans. $\begin{pmatrix} 6 & -2 \\ -2 & 2 \end{pmatrix}$

7. $Q = 2(x_1 - x_2)^2$

Ans. $\begin{pmatrix} 2 & -1 \\ -1 & 2 \end{pmatrix}$

8. $Q = (x_1 + x_2 + x_3)^2$

Ans. $\begin{bmatrix} 1 & 1 & 1 \\ 1 & 1 & 1 \\ 1 & 1 & 1 \end{bmatrix}$

9. $Q = 4x_1x_3 + 2x_2x_3 + x_3^2$

Ans. $\begin{bmatrix} 0 & 0 & 2 \\ 0 & 0 & 1 \\ 2 & 1 & 1 \end{bmatrix}$

Verify that the following matrices are orthogonal:

10. $\frac{1}{3}\begin{bmatrix} 1 & 2 & 2 \\ 2 & 1 & -2 \\ 2 & -2 & 1 \end{bmatrix}$

Hint: Show that $AA^T = I$.

11. $\begin{bmatrix} 0 & 1 & 0 \\ 1 & 0 & 0 \\ 0 & 0 & 1 \end{bmatrix}$

12. Prove that the product AB of two symmetric matrices A and B is symmetric if $AB = BA$.

Hint: $(AB)^T = B^TA^T = BA$ since $A = A^T, B = B^T$, so $(AB)^T = BA = AB$ then AB is symmetric. Provided $AB = BA$.

14.7 CANONICAL FORM: or SUM OF THE SQUARES FORM

Of a real quadratic form $Q = X^TAX$ is Y^TDY or

$$\lambda_1 y_1^2 + \lambda_2 y_2^2 + \cdots + \lambda_n y_n^2 \qquad (1)$$

which is obtained by an orthogonal transformation $X = PY$. Here P is known as **modal** matrix. D is known as **spectral** matrix. D is a diagonal matrix with the eigen values of A as the diagonal elements.

Let r be the rank of A and n be the number of variables $x_1, x_2, \cdots, x_n$ in the quadratic form. Then:

Index S of a quadratic form is the number of positive square terms in the canonical form.

Signature of a quadratic form is the difference between the number of positive and negative square terms in the canonical form.

Definiteness A real nonsingular quadratic form $Q = X^TAX$ (with $|A| \neq 0$) is said to be

Positive definite: If rank and index are equal i.e., $r = n$, $s = n$ or if all the eigen values of A are positive

Negative definite: If index equals to zero i.e., $r = n, s = 0$ or if all the eigen values of A are negative.

Positive semi-definite: If rank and index are equal but less than n

i.e., $$s = r < n, (|A| = 0)$$

or all eigen values of A are non-negative (≥ 0) and at least one eigen value is zero.

Negative semi definite: If index zero

i.e., $$s = 0, r < n, (|A| = 0)$$

or all eigen values of A are non-positive (≤ 0) and at least one eigen value is zero.

Indefinite: Quadratic form is said to be indefinite in any other case or some eigen values are positive and some eigen values are negative.

Note: If Q is negative definite (semi-definite) then $-Q$ is positive definite (semi-definite).

WORKED OUT EXAMPLES

Example 1: Determine the nature, index and signature of the quadratic form $2x_1^2 + 2x_2^2 + 3x_3^2 + 2x_1x_2 - 4x_1x_3 - 4x_2x_3$.

Solution: The real symmetric matrix A associated with the Q.F. is

$$A = \begin{bmatrix} 2 & 1 & -2 \\ 1 & 2 & -2 \\ -2 & -2 & 3 \end{bmatrix}$$

Its characteristic equation is

$$\begin{vmatrix} 2-\lambda & 1 & -2 \\ 1 & 2-\lambda & -2 \\ -2 & -2 & 3-\lambda \end{vmatrix}$$

$$= \lambda^3 - 7\lambda^2 + 7\lambda - 1 = 0$$

$$= (\lambda - 1)(\lambda - (3 + \sqrt{8}))(\lambda - (3 - \sqrt{8})) = 0$$

The eigen values are $\lambda = 1, 0.1715, 3.1715$ which are all positive. So Q.F. is positive definite.

Index: 3, Signature : $3 - 0 = 3$.

Example 2: Find the nature, index and signature of Q.F.

$$2x_1x_2 + 2x_1x_3 + 2x_2x_3$$

Solution: $A = \begin{bmatrix} 0 & 1 & 1 \\ 1 & 0 & 1 \\ 1 & 1 & 0 \end{bmatrix}$

Characteristic equation is

$$\begin{vmatrix} -\lambda & 1 & 1 \\ 1 & -\lambda & 1 \\ 1 & 1 & -\lambda \end{vmatrix} = 0$$

$$\lambda^3 - 3\lambda - 2 = 0 \quad \text{or} \quad (\lambda + 1)^2(\lambda - 2) = 0$$

The eigen values are $2, -1, -1$, some are positive and some are negative. So the Q.F. is indefinite.

Index: 1, Signature: $1 - 2 = -1$.

Example 3: Identify the nature, index and signature of the Q.F.

$$x_1^2 + 4x_2^2 + x_3^2 - 4x_1x_2 + 2x_3x_1 - 4x_2x_3$$

Solution: $A = \begin{bmatrix} 1 & -2 & 1 \\ -2 & 4 & -2 \\ 1 & -2 & 1 \end{bmatrix}$

Characteristic equation is

$$|A - \lambda I| = \begin{vmatrix} 1-\lambda & -2 & 1 \\ -2 & 4-\lambda & -2 \\ 1 & -2 & 1-\lambda \end{vmatrix} = \lambda^2(\lambda - 6) = 0$$

Eigen values are $\lambda = 0, 0, 6$. So Q.F. is positive semi definite.

Index: 3, Signature: 3.

Example 4: Classify the Q.F. and find the index and signature of

$$-3x_1^2 - 3x_2^2 - 3x_3^2 - 2x_1x_2 - 2x_1x_3 + 2x_2x_3$$

Solution: $A = \begin{bmatrix} -3 & -1 & -1 \\ -1 & -3 & 1 \\ -1 & 1 & -3 \end{bmatrix}$

Characteristic equation is

$$\begin{vmatrix} -3-\lambda & -1 & -1 \\ -1 & -3-\lambda & 1 \\ -1 & 1 & -3-\lambda \end{vmatrix} = 0$$

$$\lambda^3 + 9\lambda^2 + 24\lambda + 16 = (\lambda + 1)(\lambda + 4)^2 = 0$$

All the eigen values $-1, -4, -4$, are negative. So Q.F. is negative definite.

Index: 0, Signature: $0 - 3 = -3$.

Note:

$$Q = 3x_1^2 + 3x_2^2 + 3x_3^2 + 2x_1x_2 + 2x_1x_3 - 2x_2x_3$$

is positive definite.

EXERCISE

Determine the nature, index and signature of the quadratic form:

1. $x_1^2 + 2x_2^2 + 3x_3^2 + 2x_2x_3 - 2x_3x_1 + 2x_1x_2$

Ans. indefinite (eigen value: $1, 1, -2$), index: 2, signature: 1

2. $5x_1^2 + 26x_2^2 + 10x_3^2 + 4x_2x_3 + 14x_3x_1 + 6x_1x_2$

Ans. positive semi definite (eigen value: $5, 0, 5$), index: 3, signature: 3

3. $x_1^2 + 5x_2^2 + x_3^2 + 2x_1x_2 + 2x_2x_3 + 6x_3x_1$

Ans. indefinite (eigen value: $-2, 3, 6$), index: 2, signature: 1

4. $3x_1^2 + 5x_2^2 + 3x_3^2 - 2x_2x_3 + 2x_3x_1 - 2x_1x_2$

Ans. positive definite (eigen value: $2, 3, 6$), index: 3, signature: 3

5. $8x_1^2 + 7x_2^2 + 3x_3^2 - 12x_1x_2 - 8x_2x_3 + 4x_3x_2$

Ans. positive semi definite (eigen value: $3, 0, 15$), index: 3, signature: 3

6. $6x_1^2 + 3x_2^2 + 3x_3^2 - 4x_1x_2 - 2x_2x_3 + 4x_1x_3$

Ans. positive definite (eigen value: $8, 2, 2$), index: 3, signature: 3

7. $-4x_1^2 - 2x_2^2 - 13x_3^2 - 4x_1x_2 - 8x_2x_3 - 4x_1x_3$

Ans. negative definite, index: 0, signature: -3

8. $-3x_1^2 - 3x_2^2 - 7x_3^2 - 6x_1x_2 - 6x_2x_3 - 6x_3x_1$

Ans. negative definite, index: 0, signature -3.

14.8 TRANSFORMATION (REDUCTION) OF QUADRATIC FORM TO CANONICAL FORM

Let Q be the quadratic form given by

$$Q = X^T AX = \sum_{i=1}^{n}\sum_{j=1}^{n} a_{ij}x_i x_j \quad (1)$$

The coefficient matrix A is real symmetric therefore has n linearly independent orthonormal set of eigen vectors corresponding n eigen values (which need not necessarily be distinct). Let $\hat{P}$ be normalized modal matrix of A. Then $\hat{P}$ is an orthogonal matrix.

Thus the transformation

$$X = \hat{P}Y \quad (2)$$

is an orthogonal transformation. This transfers the quadratic form Q to canonical form, as follows:

We know that P diagonalizes A. Thus

$$\hat{P}^{-1}A\hat{P} = D$$

$$A = \hat{P}D\hat{P}^{-1} = \hat{P}D\hat{P}^T \quad (3)$$

since $\hat{P}^{-1} = \hat{P}^T$ by virtue of orthogonality

Substituting (3) in (1)

$$Q = X^T AX = X^T \hat{P}D\hat{P}^T X = (X^T\hat{P})(D)(\hat{P}^T X) \quad (4)$$

Pre-multiplying (2) by $\hat{P}^{-1}$, we get

$$\hat{P}^{-1}X = \hat{P}^{-1}\hat{P}Y = Y$$

So $$Y = \hat{P}^{-1}X = \hat{P}^T X \quad (5)$$

since $\hat{P}^{-1} = \hat{P}^T$

Taking transpose of this equation

$$Y^T = (\hat{P}^T X)^T = X^T\hat{P} \quad (6)$$

Using (5) and (6) in (4), we have

$$Q = X^T AX = Y^T DY$$

$$= [y_1, y_2 \dots y_n]\begin{bmatrix} \lambda_1 & 0 & 0 & 0 \\ 0 & \lambda_2 & \cdots & 0 \\ \vdots & & & \\ 0 & \cdots & \cdots & \lambda_n \end{bmatrix}\begin{bmatrix} y_1 \\ y_z \\ \vdots \\ y_n \end{bmatrix}$$

$$Q = \lambda_1 y_1^2 + \lambda_2 y_2^2 + \cdots + \lambda_n y_n^2 \quad (7)$$

(7) is known as the *canonical form* or "*sum of the squares form*" or "*principal axes form*".

Procedure to Reduce Quadratic Form to Canonical Form

1. Identify the real symmetric matrix associated with the quadratic form Q.
2. Determine the eigen values of A.
3. The required canonical form is given by (7)
4. Form the modal matrix containing the n eigen vectors of A as n column vectors. Normalize. Then

$$X = \hat{P}Y$$

is the required orthogonal transformation which reduces Q.F. to C.F.

WORKED OUT EXAMPLES

Example 1: Find the orthogonal transformation which transforms the quadratic form

$$x_1^2 + 3x_2^2 + 3x_3^2 - 2x_2x_3$$

to canonical form (or "sum of squares form" or "principal axes form). Determine the index, signature and nature of the quadratic form.

Solution: Let $X = [x_1x_2x_3]^T$, $Y = [y_1y_2y_3]^T$. Let P be the non-singular orthogonal matrix, containing the (three) eigen vectors of the coefficient matrix A of the given quadratic form. Then $X = \hat{P}Y$ is the required non-singular linear transformation that transforms (reduces) the given quadratics form to canonical form. Here $\hat{P}$ is the normalized modal matrix P. The coefficient matrix A of the given quadratic form is

$$A = \begin{bmatrix} 1 & 0 & 0 \\ 0 & 3 & -1 \\ 0 & -1 & 3 \end{bmatrix}$$

The characteristic equation of A is

$$|A - \lambda I|$$

$$= \begin{vmatrix} 1-\lambda & 0 & 0 \\ 0 & 3-\lambda & -1 \\ 0 & -1 & 3-\lambda \end{vmatrix} \begin{matrix} =\lambda^3 - 7\lambda^2 + 14\lambda - 8 \\ =(\lambda-1)(\lambda-2)(\lambda-4)=0 \end{matrix}$$

So there are three distinct real eigen values $\lambda = 1, 2, 4$ of A.

For $\lambda = 1$,

$$\begin{matrix} 0 & 0 & 0 \\ 0 & 2 & -1 \\ 0 & -1 & 2 \end{matrix} \sim \begin{matrix} 2x_2 = x_3 \\ x_2 = 2x_3 \end{matrix}$$

$\therefore \quad x_2 = x_3 = 0, x_1 = \text{arbitrary},$

The eigen vector X_1 associated with $\lambda = 1$ is

$$X_1 = [1 \quad 0 \quad 0]^T$$

For $\lambda = 2$,

$$\begin{matrix} -x_1 + 0 + 0 = 0 \\ x_2 - x_3 = 0 \\ -x_2 + x_3 = 0 \end{matrix} \quad \therefore \quad \begin{matrix} x_1 = 0 \\ x_2 = x_3 \end{matrix}$$

$$X_2 = [0 \quad 1 \quad 1]^T$$

For $\lambda = 3$,

$$\begin{matrix} -3 & 0 & 0 \\ 0 & -1 & -1 \\ 0 & -1 & -1 \end{matrix} \sim \begin{matrix} x_1 = 0 \\ x_2 = -x_3 \end{matrix}$$

$$X_3 = [0 \quad 1 \quad -1]^T$$

Thus the modal matrix P is

$$P = \begin{bmatrix} 1 & 0 & 0 \\ 0 & 1 & 1 \\ 0 & 1 & -1 \end{bmatrix}$$

The norm of the eigen vector X_1 is

$$\|X_1\| = \sqrt{1^2 + 0 + 0} = 1,$$

$$\|X_2\| = \sqrt{0 + 1^2 + 1^2} = \sqrt{2},$$

$$\|X_3\| = \sqrt{1^2 + 1^2} = \sqrt{2}.$$

Then the normalized modal matrix $\hat{P}$ is

$$\hat{P} = \begin{bmatrix} \frac{1}{1} & 0 & 0 \\ 0 & \frac{1}{\sqrt{2}} & \frac{1}{\sqrt{2}} \\ 0 & \frac{1}{\sqrt{2}} & -\frac{1}{\sqrt{2}} \end{bmatrix} = \frac{1}{\sqrt{2}} \begin{bmatrix} \sqrt{2} & 0 & 0 \\ 0 & 1 & 1 \\ 0 & 1 & -1 \end{bmatrix}$$

To find inverse of P:

$$\left[\begin{array}{ccc|ccc} 1 & 0 & 0 & 1 & 0 & 0 \\ 0 & 1 & 1 & 0 & 1 & 0 \\ 0 & 1 & -1 & 0 & 0 & 1 \end{array}\right]$$

$$\sim \begin{matrix} R_{32(-1)} \\ R_{3(-\frac{1}{2})} \\ R_{23(-1)} \end{matrix} \left[\begin{array}{ccc|ccc} 1 & 0 & 0 & 1 & 0 & 0 \\ 0 & 1 & 0 & 0 & \frac{1}{2} & \frac{1}{2} \\ 0 & 0 & 1 & 0 & \frac{1}{2} & -\frac{1}{2} \end{array}\right]$$

Thus

$$P^{-1} = \begin{bmatrix} 1 & 0 & 0 \\ 0 & \frac{1}{2} & \frac{1}{2} \\ 0 & \frac{1}{2} & -\frac{1}{2} \end{bmatrix}$$

and the normalized P^{-1} is

$$\hat{P}^{-1} = \frac{1}{\sqrt{2}} \begin{bmatrix} \sqrt{2} & 0 & 0 \\ 0 & 1 & 1 \\ 0 & 1 & -1 \end{bmatrix}$$

Diagonalization:

$$\hat{P}^{-1} A \hat{P} = \frac{1}{\sqrt{2}} \begin{bmatrix} \sqrt{2} & 0 & 0 \\ 0 & 1 & 1 \\ 0 & 1 & -1 \end{bmatrix} \begin{bmatrix} 1 & 0 & 0 \\ 0 & 3 & -1 \\ 0 & -1 & 3 \end{bmatrix}$$

$$\times \frac{1}{\sqrt{2}} \begin{bmatrix} \sqrt{2} & 0 & 0 \\ 0 & 1 & 1 \\ 0 & 1 & -1 \end{bmatrix}$$

$$= \frac{1}{\sqrt{2}} \begin{bmatrix} \sqrt{2} & 0 & 0 \\ 0 & 2 & 2 \\ 0 & 4 & -4 \end{bmatrix} \begin{bmatrix} \sqrt{2} & 0 & 0 \\ 0 & 1 & 1 \\ 0 & 1 & -1 \end{bmatrix}$$

$$= \frac{1}{2} \begin{bmatrix} 2 & 0 & 0 \\ 0 & 4 & 0 \\ 0 & 0 & 8 \end{bmatrix}$$

Then

$$\hat{P}^{-1} A \hat{P} = \begin{bmatrix} 1 & 0 & 0 \\ 0 & 2 & 0 \\ 0 & 0 & 4 \end{bmatrix} = D = \text{diagonal matrix}$$

with the eigen values of A as the diagonal elements.

Transformation (reduction) to canonical form:

Quadratic form (Q.F.)

$$Q = x_1^2 + 3x_2^2 + 3x_3^2 - 2x_2x_3$$

$$= [x_1 \quad x_2 \quad x_3] \begin{bmatrix} 1 & 0 & 0 \\ 0 & 3 & -1 \\ 0 & -1 & 3 \end{bmatrix} \begin{bmatrix} x_1 \\ x_2 \\ x_3 \end{bmatrix} = X^T A X$$

Put $X = \hat{P}Y$ and $X^T = (\hat{P}Y)^T = Y^T \hat{P}^T$.

So $Q = X^T A X = Y^T \hat{P}^T A \hat{P} Y = Y^T (\hat{P}^T A \hat{P}) Y$

But we know that $\hat{P}$ is an orthogonal matrix because

$$\hat{P}\,\hat{P}^T = \frac{1}{\sqrt{2}} \begin{bmatrix} \sqrt{2} & 0 & 0 \\ 0 & 1 & 1 \\ 0 & 1 & -1 \end{bmatrix} \frac{1}{\sqrt{2}} \begin{bmatrix} \sqrt{2} & 0 & 0 \\ 0 & 1 & 1 \\ 0 & 1 & -1 \end{bmatrix}$$

$$= \frac{1}{2} \begin{bmatrix} 2 & 0 & 0 \\ 0 & 2 & 0 \\ 0 & 0 & 2 \end{bmatrix} = \begin{bmatrix} 1 & 0 & 0 \\ 0 & 1 & 0 \\ 0 & 0 & 1 \end{bmatrix} = I$$

Thus

$$\hat{P}^T = \hat{P}^{-1}$$

So $\quad$ Q.F. $= X^T A X = Y^T(\hat{P}^{-1} A\hat{P})Y$

But through diagonalization

$$\hat{P}^{-1} A \hat{P} = D$$

Therefore

$$Q = X^T A X = Y^T D Y$$

$$= [y_1 \quad y_2 \quad y_3]\begin{bmatrix}1 & 0 & 0\\ 0 & 2 & 0\\ 0 & 0 & 4\end{bmatrix}\begin{bmatrix}y_1\\ y_2\\ y_3\end{bmatrix}$$

$$= [y_1 \quad 2 \cdot y_2 \quad 4y_3]\begin{bmatrix}y_1\\ y_2\\ y_3\end{bmatrix}$$

$$= y_1^2 + 2y_2^2 + 4y_3^2$$

This is the required canonical form (or sum of squares form).

Orthogonal transformation:

$$X = \begin{bmatrix}x_1\\ x_2\\ x_3\end{bmatrix} = \hat{P}Y = \frac{1}{\sqrt{2}}\begin{bmatrix}\sqrt{2} & 0 & 0\\ 0 & 1 & 1\\ 0 & 1 & -1\end{bmatrix}\begin{bmatrix}y_1\\ y_2\\ y_3\end{bmatrix}$$

so $x_1 = y_1$; $x_2 = \frac{1}{\sqrt{2}}(y_2 + y_3)$, $x_3 = \frac{1}{\sqrt{2}}(y_2 - y_3)$ is the orthogonal transformation which reduces the Q.F. to canonical form.

Index is 3 for the Q.F. since the number of positive terms in the canonical form is 3. i.e., $S = 3$, rank r is 3. The number of variables is $n = 3$.

Signature of Q.F. is $2s - r = 6 - 3 = 3$ (difference between number of positive and negative terms in C.F.).

The given Q.F. is *positive definite* because $r = 3 = n$ and $s = 3 = n$.

Example 2: By **Lagrange's reduction** transform the quadratic form $X^T AX$ to sum of the squares form for

$$A = \begin{bmatrix}1 & 2 & 4\\ 2 & 6 & -2\\ 4 & -2 & 18\end{bmatrix}$$

Solution:

$$\text{Q.F.} = X^T AX = [x_1 \quad x_2 \quad x_3]\begin{bmatrix}1 & 2 & 4\\ 2 & 6 & -2\\ 4 & -2 & 18\end{bmatrix}\begin{bmatrix}x_1\\ x_2\\ x_3\end{bmatrix}$$

$$\text{Q.F.} = [x_1+2x_2+4x_3 \quad 2x_1+6x_2-2x_3 \quad 4x_1-2x_2+18x_3] \times \begin{bmatrix}x_1\\ x_2\\ x_3\end{bmatrix}$$

$$= x_1^2 + 6x_2^2 + 18x_3^2 + 4x_1x_2 + 8x_1x_3 - 4x_2x_3$$

$$= [x_1^2 + 4x_1(x_2 + 2x_3)] + 6x_2^2 + 18x_3^2 - 4x_2x_3$$

$$= [x_1^2 + 4x_1(x_2 + 2x_3) + 2^2(x_2 + 2x_3)^2]$$
$$-2^2(x_2 + 2x_3)^2 + 6x_2^2 + 18x_3^2 - 4x_2x_3$$

$$= [x_1 + 2(x_2 + 2x_3)]^2 + 2x_2^2 + 2x_3^2 - 20x_2x_3$$

$$= [x_1 + 2(x_2 + 2x_3)]^2 + 2[x_2^2 - 10x_2x_3] + 2x_3^2$$

$$= [x_1 + 2(x_2 + 2x_3)]^2$$
$$+2[x_2^2 - 10x_2x_3 + 5^2x_3^2] - 2 \cdot 5^2x_3^2 + 2x_3^2$$

$$= [x_1 + 2(x_2 + 2x_3)]^2 + 2[x_2 - 5x_3]^2 - 48x_3^2$$

$$\text{Q.F.} = y_1^2 + 2y_2^2 - 48y_3^2$$

where

$$y_1 = x_1 + 2(x_2 + 2x_3),$$
$$y_2 = x_2 - 5x_3,$$
$$y_3 = x_3.$$

Index: $S = 2$, $(n = 3, r = 3)$,
Signature: $2s - r = 2 \cdot 2 - 3 = 1$ (or $2 - 1 = 1$).

EXERCISE

Transform (reduce) the quadratic form to canonical form (or "sum of squares form" or "principal axes form") by orthogonal transformation.

State matrix for transformation (i.e., modal matrix).

1. $17x_1^2 - 30x_1x_2 + 17x_2^2$

Ans. $A = \begin{bmatrix}17 & -15\\ -15 & +17\end{bmatrix}$, $\lambda = 2, 32$,

$P = \begin{bmatrix}1 & -1\\ 1 & 1\end{bmatrix}\frac{1}{\sqrt{2}}$, C.F.: $2y_1^2 + 32y_2^2$

2. $3x_1^2+5x_2^2+3x_3^2-2x_2x_3+2x_1x_3-2x_1x_2$

Ans. $A=\begin{bmatrix}3&-1&1\\-1&5&-1\\1&-1&3\end{bmatrix}$, $\lambda=2,3,6$,

$P=\begin{bmatrix}\frac{1}{\sqrt{2}}&\frac{1}{\sqrt{3}}&\frac{1}{\sqrt{6}}\\0&\frac{1}{\sqrt{3}}&-\frac{2}{\sqrt{6}}\\-\frac{1}{\sqrt{2}}&\frac{1}{\sqrt{3}}&\frac{1}{\sqrt{6}}\end{bmatrix}$, C.F.: $2y_1^2+3y_2^2+6y_3^2$

3. $5x_1^2+26x_2^2+10x_3^2+4x_2x_3+6x_1x_2+14x_1x_3$

Ans. $A=\begin{bmatrix}5&3&7\\3&26&2\\7&2&10\end{bmatrix}$, $\lambda=5,\frac{121}{3},0$,

$P=\begin{bmatrix}1&-\frac{3}{5}&-\frac{16}{11}\\0&1&\frac{1}{11}\\0&0&1\end{bmatrix}$, C.F.: $5y_1^2+\frac{121}{3}y_2^2$

4. $2(x_1x_2+x_2x_3+x_3x_1)$; nature of Q.F.

Ans. $A=\begin{bmatrix}0&1&1\\1&0&+1\\1&+1&0\end{bmatrix}$, $\lambda=2,-1,-1$,

$P=\begin{bmatrix}\frac{1}{\sqrt{3}}&-\frac{1}{\sqrt{2}}&\frac{1}{\sqrt{6}}\\\frac{1}{\sqrt{3}}&\frac{1}{\sqrt{2}}&\frac{1}{\sqrt{6}}\\\frac{1}{\sqrt{3}}&0&-\frac{2}{\sqrt{6}}\end{bmatrix}$,

C.F.: $2y_1^2-y_2^2-y_3^2$

Nature: Indefinite

5. $2(x_1^2+x_1x_2+x_2^2)$

Ans. $A=\begin{bmatrix}2&1\\1&2\end{bmatrix}$, $\lambda=1,3$,

$P=\frac{1}{\sqrt{2}}\begin{bmatrix}1&1\\-1&1\end{bmatrix}$, C.F.: $y_1^2+3y_2^2$

6. $2x_1^2+x_2^2-3x_3^2-8x_2x_3-4x_3x_1+12x_1x_2$, find index

Ans. $A=\begin{bmatrix}2&6&-2\\6&1&-4\\-2&-4&-3\end{bmatrix}$, $\lambda=1,-1,-1$,

$P=\begin{bmatrix}a&-3b&\frac{11c}{17}\\0&b&\frac{2b}{17}\\0&0&c\end{bmatrix}$

where $a=1/\sqrt{2}, b=1/\sqrt{17}, c=\sqrt{(17/81)}$,

C.F.: $y_1^2-y_2^2-y_3^2$, Index = 1

7. $3x_1^2-2x_2^2-x_3^2-4x_1x_2+12x_2x_3+8x_1x_3$

Ans. $A=\begin{bmatrix}3&-2&4\\-2&-2&6\\4&6&-1\end{bmatrix}$, $\lambda=3,6,-9$,

$P=\frac{1}{3}\begin{bmatrix}2&2&1\\-2&1&2\\-1&2&2\end{bmatrix}$, C.F.: $3y_1^2+6y_2^2-9y_3^2$

8. $8x_1^2+7x_2^2+3x_3^2+12x_1x_2+4x_1x_3-8x_2x_3$, find the rank, index, signature and nature.

Ans. $A=\begin{bmatrix}8&-6&2\\-6&7&-4\\2&-4&3\end{bmatrix}$, $\lambda=3,0,15$,

$P=\frac{1}{3}\begin{bmatrix}2&1&2\\1&2&-2\\-2&2&1\end{bmatrix}$, C.F.: $3y_1^2+15y_3^2$

rank of Q.F.: 2, index: 2, signature 2, positive definite.

Lagrange's reduction

9. $x_1^2+2x_2^2-7x_3^2-4x_1x_2+8x_1x_3$

Ans. $(x_1-2x_2+4x_3)^2-2(x_2-4x_3)^2+9x_3^2$

10. $2x_1^2+5x_2^2+19x_3^2-24x_4^2+8x_1x_2+12x_1x_3+8x_1x_4+18x_2x_3-8x_2x_4-16x_3x_4$

Ans. $2(x_1+2x_2+3x_3+2x_4)^2-3(x_2+x_3+4x_4)^2+4(x_3-2x_4)^2$

11. $2x_1^2+7x_2^2+5x_3^2-8x_1x_2-10x_2x_3+4x_1x_3$

Ans. $2(x_1-2x_2-x_3)^2-(x_2+x_3)^2+4x_3^2$

12. Coefficient matrix $A=\begin{bmatrix}1&-1&0&2\\-1&4&6&4\\0&6&11&8\\2&4&8&8\end{bmatrix}$

Hint: QF: $x_1^2+4x_2^2+11x_3^2+8x_4^2-2x_1x_2+4x_1x_4+12x_2x_3+8x_2x_4+16x_3x_4$

Ans. $(x_1-x_2+2x_3)^2+3(x_2+2x_3+2x_4)^2-(x_3+4x_4)^2+8x_4^2$.

14.9 COMPLEX MATRICES: HERMITIAN, SKEW-HERMITIAN, UNITARY MATRICES

In a complex matrix A, the elements are complex or real.

$\overline{A} = (\overline{a_{ij}})$ is matrix obtained by replacing each a_{ij} of A by its complex conjugate $\overline{a_{ij}}$. A complex square matrix A is said to be

a. Hermitian if $\overline{A}^T = A$ i.e, $\overline{a}_{ji} = a_{ij}$
b. Skew-Hermitian if $\overline{A}^T = -A$ i.e, $\overline{a}_{ji} = -a_{ij}$
c. Unitary if $\overline{A}^T = A^{-1}$

Thus the Hermitian, Skew-Hermitian and Unitary matrices are respectively the natural generalization of the real symmetric, Skew-symmetric and orthogonal matrics to complex matrices.

It follows from the definition, that the diagonal elements of a Hermitian matrix are always real because $\overline{a}_{ji} = a_{ji}$ means a_{ij} must be real. Similarly, for a Skew-Hermitian matrix $\overline{a}_{ii} = -a_{ii}$ means that the diagonal elements are purely imaginary or zero.

Unitary system

of row and column vectors

$$\overline{X}_i^T X_j = \delta_{ij} = \begin{cases} 1 & \text{if } i = j \\ 0 & \text{if } i \neq j \end{cases}$$

which is a direct analog of orthonormal system. Hermitian form H (generalization of the real quadratic form) is given by

$$H = \overline{X}^T A X = \sum_{i=1}^{n} \sum_{j=1}^{n} a_{ij} \overline{x}_i x_j$$

where A is Hermitian matrix.

Similarly, Skew-Hermitian form S is given by $S = \overline{X}^T AX$ where A is Skew-Hermitian matrix.

Theorem: *For any X, Hermitian form is real.*

Proof: $\overline{H} = \overline{(\overline{X}^T A X)} = X^T \overline{A}\overline{X} = X^T A^T \overline{X}$ since A is Hermitian i.e., $\overline{A} = A^T$. As the R.H.S. is a scalar, transposition does not change its values. Thus $\overline{H} = X^T A^T \overline{X} = (X^T A^T \overline{X})^T = \overline{X}^T AX = H$, so H is real.

Similarly,

Theorem: *For any X, Skew-Hermitian is purely imaginary or zero.*

Proof: $\overline{S} = \overline{(\overline{X}^T AX)} = X^T \overline{A}\overline{X} = -X^T A^T \overline{X}$ since A is Skew-Hermitian i.e., $\overline{A} = -A^T$. Taking transpose of R.H.S., we have

$$\overline{S} = -(X^T A^T \overline{X})^T = -\overline{X}^T AX = -S$$

So S is purely imaginary or zero.

Theorem: *The column (and also row) vectors of a unitary matrix form a unitary system.*

Proof: Let $c_1, c_2, \ldots, c_n$ be the n column vectors of a n-squared unitary matrix A. Consider

$$I = A^{-1}A = \overline{A}^T A = \begin{bmatrix} \overline{c}_1^T \\ \overline{c}_2^T \\ \vdots \\ \overline{c}_n^T \end{bmatrix}_{n\times 1} [c_1, c_2, \ldots, c_n]_{1\times n}$$

since for a unitary matrix $A^{-1} = \overline{A}^T$. Then

$$I_{n\times n} = \begin{bmatrix} \overline{c}_1^T c_1 & \overline{c}_1^T c_2 & \ldots & \overline{c}_1^T c_n \\ \overline{c}_2^T c_1 & \overline{c}_2^T c_2 & \ldots & \overline{c}_2^T c_n \\ \ldots & \ldots & \ldots & \ldots \\ \overline{c}_n^T c_1 & \overline{c}_n^T c_2 & \ldots & \overline{c}_n^T c_n \end{bmatrix}_{n\times n}$$

Thus

$$\overline{c}_i^T c_j = \delta_{ij} = \begin{cases} 0 & \text{for } i \neq j \\ 1 & \text{for } i = j \end{cases}$$

Hence the column vectors (in a similar way row vectors) of a unitary matrix A form a unitary system.

Corollary 1: If A is real orthogonal matrix, it follows that

$$c_i^T c_j = \begin{cases} 0 & \text{for } i \neq j \\ 1 & \text{for } i = j \end{cases}$$

Thus the column (and also the row) vectors of an orthogonal matrix form an orthonormal system (of vectors).

Theorem: *Prove that the eigen values of*

a. *Hermitian matrix A are real*

b. *Skew-Hermitian S are purely imaginary or zero*

c. *Unitary matrix U have absolute value 1.*

Proof: Let λ be an eigen value and $X(\neq 0)$ be the corresponding eigen vector. Then

$$\overline{X}^T X = \overline{x}_1 x_1 + \overline{x}_2 x_2 + \cdots + \overline{x}_n x_n = |x_1|^2 + \cdots + |x_n|^2$$

is real and not equals to zero.

a. Let A be Hermitian

$$AX = \lambda X$$

Pre-multiplying by $\overline{X}^T$, we have

$$\overline{X}^T AX = \overline{X}^T \lambda X = \lambda \overline{X}^T X$$

$$\lambda = \frac{\overline{X}^T AX}{\overline{X}^T X}$$

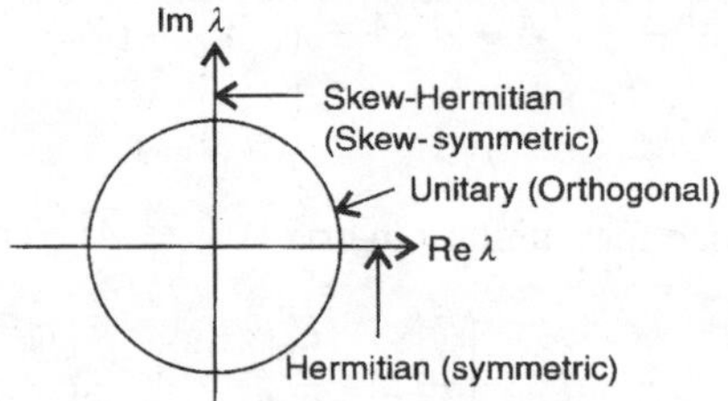

Fig. 14.1

Hence λ the eigen value of a Hermitian matrix is real because the numerator $\overline{X}^T AX$ is a Hermitian form which is always a real (and $\overline{X}^T X$ the denominator is also real).

b. Let S be Skew-Hermitian (refer Fig. 14.1)

$$SX = \lambda X$$

pre multiplying by $\overline{X}^T$, we have

$$\overline{X}^T SX = \overline{X}^T \lambda X = \lambda \overline{X}^T X$$

$$\lambda = \frac{\overline{X}^T SX}{\overline{X}^T X}$$

Since the numerator $\overline{X}^T SX$ is a Skew-Hermitian form which is purely imaginary or zero, therefore λ the eigen value of the Skew-Hermitian matrix is purely imaginary or zero.

c. Let U be a unitary matrix

$$UX = \lambda X \quad (1)$$

Taking conjugate transpose of (1)

$$(\overline{UX})^T = (\overline{U}\,\overline{X})^T = \overline{X}^T \overline{U}^T$$

$$= \overline{X}^T U^{-1} = (\overline{\lambda X})^T = \overline{\lambda}\overline{X}^T$$

i.e., $$\overline{X}^T U^{-1} = \overline{\lambda}\overline{X}^T \quad (2)$$

since $\overline{U}^T = U^{-1}$ for unitary matrix U and transposition on R.H.S. does not affect the scalar λ. Post multiplying, R.H.S. of (2) by R.H.S. of (1) and L.H.S. of (2) by L.H.S. of (1), we get

$$(\overline{X}^T U^{-1})(UX) = (\overline{\lambda}\overline{X}^T)(\lambda X)$$

$$\overline{X}^T (U^{-1}U)X = (\overline{\lambda}\lambda)(\overline{X}^T X)$$

$$\overline{X}^T X = (\lambda\overline{\lambda})\overline{X}^T X$$

Since $U^{-1}U = I$. Thus

$$\lambda\overline{\lambda} = |\lambda|^2 = 1$$

Since $\overline{X}^T X \neq 0$. Hence eigen values of unitary matrix are of absolute value 1.

Corollary 2: Eigen values of (a) real symmetric matrix are real (b) Skew-symmetric are purely imaginary or zero (c) orthogonal matrix real or complex conjugates in pairs and have absolute value 1.

Properties of Unitary Matrix

1. Product of two unitary matrices is unitary.

Proof: Let A and B be unitary matrices so that $A^T = \overline{A}^{-1}$, $B^T = \overline{B}^{-1}$.

Consider $(\overline{AB})^{-1} = (\overline{A}\,\overline{B})^{-1} = \overline{B}^{-1}\overline{A}^{-1}$

$= B^T A^T$ Since A, B are unitary

$= (AB)^T$ by transposition rule

Thus $(\overline{AB})^{-1} = (AB)^T$ $\therefore$ AB is unitary.

Corollary 1: Product of two orthogonal matrices is orthogonal.

2. Inverse of a unitary matrix is unitary.

Proof: For a unitary matrix A, $\overline{A}^T = A^{-1}$

or $$\overline{A^T} = \overline{A}^T = A^{-1}$$

$$A^T = \overline{A^{-1}}$$

$$\left((A^{-1})^{-1}\right)^T = \overline{A^{-1}}$$

Denote A^{-1} by B then

$$(B^{-1})^T = \overline{B}$$

or

$$B^{-1} = \overline{B}^T$$

So B is unitary i.e., A^{-1} is unitary.

Corollary 2: Inverse of an orthogonal matrix is orthogonal.

3. Transpose of a unitary matrix is unitary.

Proof: For unitary matrix A, $\overline{A^T} = A^{-1}$

or

$$(\overline{A^T}) = A^{-1}$$

Taking transpose on both sides

$$(\overline{A^T})^T = (A^{-1})^T = (A^T)^{-1}$$

since transposition and inverse taking are commutative. Taking $B = A^T$

$$\overline{B}^T = B^{-1}$$

Thus $B = A^T$ is also unitary.

Corollary 3: Transpose of an orthogonal matrix is orthogonal.

Properties of Hermitian and Skew-Hermitian Matrices

Book Work:

1. Show that any square matrix A can be written as the sum of a Hermitian and Skew-Hermitian matrices.

Proof: Choose $B = \frac{1}{2}(A + \overline{A}^T)$ and $C = \frac{1}{2}(A - \overline{A}^T)$. Then

$$B^T = \frac{1}{2}(A + \overline{A}^T)^T = \frac{1}{2}(A^T + \overline{A})$$

$$\overline{B} = \frac{1}{2}\overline{(A + \overline{A}^T)} = \frac{1}{2}(\overline{A} + \overline{\overline{A}^T}) = \frac{1}{2}(\overline{A} + A^T)$$

$$\therefore \quad B^T = \overline{B}$$

Thus B is Hermitian. Now

$$\overline{C} = \frac{1}{2}\overline{(A - \overline{A}^T)} = \frac{1}{2}(\overline{A} - \overline{\overline{A}^T}) = \frac{1}{2}(\overline{A} - A^T)$$

$$C^T = \frac{1}{2}(A - \overline{A}^T)^T = \frac{1}{2}(A^T - \overline{A}) = -\overline{C}$$

Therefore C is Skew-Hermitian. Thus $A = B + C = \frac{1}{2}(A + \overline{A}^T) + \frac{1}{2}(A - \overline{A}^T) = A$ is expressed as sum of Hermitian and Skew-Hermitian matrices.

2. If A, B are Hermitian, prove that $AB - BA$ Skew-Hermitian.

Proof:

$$\overline{(AB - BA)}^T = (\overline{AB} - \overline{BA})^T = (\overline{A}\,\overline{B} - \overline{B}\,\overline{A})^T$$

$$= (\overline{A}\,\overline{B})^T - (\overline{B}\,\overline{A})^T = \overline{B}^T\overline{A}^T - \overline{A}^T\overline{B}^T$$

$$= BA - AB = -(AB - BA),$$

since A, B are Hermitian, $\overline{A}^T = A$ and $\overline{B}^T = B$.
Thus $\overline{(AB - BA)}^T = -(AB - BA)$
Therefore $AB - BA$ is Skew-Hermitian.

3. If A is Hermitian (Skew-Hermitian) then (iA) is Skew-Hermitian (Hermitian).

Proof: Suppose A is Hermitian. Then

$$(\overline{iA})^T = (\overline{i}\,\overline{A})^T = (-i\overline{A})^T = -i\overline{A}^T$$

since transposition does not effect scalar i. Thus

$$(\overline{iA})^T = -i\overline{A}^T = -iA$$

since for Hermitian A, $\overline{A}^T = A$. Hence iA is Skew-Hermitian. [Similarly let A be Skew-Hermitian.

$$(\overline{iA})^T = (\overline{iA})^T = (-i\overline{A})^T = -i\overline{A}^T = iA$$

since $-\overline{A}^T = A$ for a Skew-Hermitian. Thus iA is Hermitian.]

WORKED OUT EXAMPLES

Example 1: If

$$A = \begin{bmatrix} 2 & 3+2i & -4 \\ 3-2i & 5 & 6i \\ -4 & -6i & 3 \end{bmatrix}$$

Then show that A is Hermitian and iA is Skew-Hermitian.

Solution:

$$\overline{A} = \begin{bmatrix} 2 & 3-2i & -4 \\ 3+2i & 5 & -6i \\ -4 & 6i & 3 \end{bmatrix}$$

$$\overline{A}^T = \begin{bmatrix} 2 & 3+2i & -4 \\ 3-2i & 5 & 6i \\ -4 & -6i & 3 \end{bmatrix} = A$$

Thus A is Hermitian.

Let

$$B = iA = i\begin{bmatrix} 2 & 3+2i & -4 \\ 3-2i & 5 & +6i \\ -4 & -6i & 3 \end{bmatrix}$$

$$= \begin{bmatrix} 2i & -2+3i & -4i \\ 2+3i & 5i & -6 \\ -4i & 6 & 3i \end{bmatrix}$$

$$\overline{B} = \begin{bmatrix} -2i & -2-3i & 4i \\ 2-3i & -5i & -6 \\ 4i & 6 & -3i \end{bmatrix},$$

$$B^T = \begin{bmatrix} 2i & 2+3i & -4i \\ -2+3i & 5i & 6 \\ -4i & -6 & 3i \end{bmatrix}$$

Thus $\overline{B} = -B^T$ or B is Skew-Hermitian.

Note: From Book Work 3, (Page 450) it follows that for A Hermitian, iA is Skew-Hermitian.

Example 2: Show that

$$A = \begin{bmatrix} 2 & 3+4i \\ 3-4i & 2 \end{bmatrix}$$

is Hermitian. Find its eigen values and eigen vectors.

Solution:

$$\overline{A} = \begin{bmatrix} 2 & 3-4i \\ 3+4i & 2 \end{bmatrix}, \overline{A}^T = \begin{bmatrix} 2 & 3+4i \\ 3-4i & 2 \end{bmatrix} = A$$

Thus A is Hermitian.

(Note that the diagonal elements of A are real.)

The characteristic equation for A is

$$|A - \lambda I| = \begin{vmatrix} 2-\lambda & 3+4i \\ 3-4i & 2-\lambda \end{vmatrix} = 0$$

or

$$(2-\lambda)^2 - (3+4i)(3-4i) = 4 + \lambda^2 - 4\lambda - [9+16] = 0$$

i.e.,

$$\lambda^4 - 4\lambda - 21 = (\lambda+3)(\lambda-7) = 0$$

Eigen values of A, Hermitian matrix are real $-3, 7$.

For $\lambda = -3$,

$$\begin{bmatrix} 5 & 3+4i \\ 3-4i & 5 \end{bmatrix}\begin{bmatrix} x_1 \\ x_2 \end{bmatrix} = 0$$

$$x_1 = -\left(\frac{3+4i}{5}\right)x_2$$

The eigen vector corresponding to $\lambda = -3$ is

$$X_1 = \begin{bmatrix} -3-4i \\ 5 \end{bmatrix}$$

For $\lambda = 7$,

$$\begin{bmatrix} -5 & 3+4i \\ 3-4i & -5 \end{bmatrix}\begin{bmatrix} x_1 \\ x_2 \end{bmatrix} = 0$$

$$x_1 = \frac{3+4i}{5}x_2$$

The eigen vector corresponding to $\lambda = 7$ is

$$X_2 = \begin{bmatrix} 3+4i \\ 5 \end{bmatrix}$$

Example 3: Show that

$$A = \begin{bmatrix} i & 0 & 0 \\ 0 & 0 & i \\ 0 & i & 0 \end{bmatrix}$$

is Skew-Hermitian and also unitary. Find the eigen values and eigen vectors.

Solution:

$$\overline{A} = \begin{bmatrix} -i & 0 & 0 \\ 0 & 0 & -i \\ 0 & -i & 0 \end{bmatrix}, \overline{A}^T = \begin{bmatrix} -i & 0 & 0 \\ 0 & 0 & -i \\ 0 & -i & 0 \end{bmatrix} = -A$$

Thus A is Skew-Hermitian.

Consider

$$A\overline{A}^T = \begin{bmatrix} i & 0 & 0 \\ 0 & 0 & i \\ 0 & i & 0 \end{bmatrix}\begin{bmatrix} -i & 0 & 0 \\ 0 & 0 & -i \\ 0 & -i & 0 \end{bmatrix} = \begin{bmatrix} 1 & 0 & 0 \\ 0 & 1 & 0 \\ 0 & 0 & 1 \end{bmatrix} = I$$

Thus $\overline{A}^T = A^{-1}$, i.e, A is unitary matrix also. The characteristic equation of A is

$$|A - \lambda I| = \begin{vmatrix} i-\lambda & 0 & 0 \\ 0 & 0-\lambda & i \\ 0 & i & 0-\lambda \end{vmatrix} \begin{aligned} &= (i-\lambda)(\lambda^2+1) \\ &= \lambda^3 - i\lambda^2 + \lambda - i = 0 \\ &= (\lambda+i)(\lambda-i)^2 = 0 \end{aligned}$$

The eigen values of A are $\lambda = -i, i, i$ which are purely imaginary (for Skew-Hermitian) and are of absolute value unity (i.e., $|-i| = |i| = 1$)

For $\lambda = -i$,

$$\begin{bmatrix} 2i & 0 & 0 \\ 0 & i & i \\ 0 & i & i \end{bmatrix} \begin{bmatrix} x_1 \\ x_2 \\ x_3 \end{bmatrix} = 0$$

Solving

$$x_1 = 0, \quad x_2 = -x_3,$$

eigen vector corresponding to $\lambda = -i$ is

$$X_1 = \begin{bmatrix} 0 \\ 1 \\ -1 \end{bmatrix}$$

For $\lambda = i$

$$\begin{bmatrix} 0 & 0 & 0 \\ 0 & -i & i \\ 0 & i & -i \end{bmatrix} \begin{bmatrix} x_1 \\ x_2 \\ x_3 \end{bmatrix} = 0$$

Solving

$$x_1 = \text{arbitrary}, \quad x_2 = x_3$$

Choose x_1, so that two linearly independent eigen vectors are obtained (with $x_1 = 0$, $x_2 = 1$ and $x_1 = 1, x_2 = 0$)

$$X_2 = \begin{bmatrix} 0 \\ 1 \\ 1 \end{bmatrix} \quad \text{and} \quad X_3 = \begin{bmatrix} 1 \\ 0 \\ 0 \end{bmatrix}$$

Example 4: Find the Hermitian form H for

$$A = \begin{bmatrix} 0 & i & 0 \\ -i & 1 & -2i \\ 0 & 2i & 2 \end{bmatrix} \quad \text{with} \quad X = \begin{bmatrix} i \\ 1 \\ -i \end{bmatrix}$$

Solution:

$$H = \overline{X}^T AX = [-i \quad 1 \quad i] \begin{bmatrix} 0 & i & 0 \\ -i & 1 & -2i \\ 0 & 2i & 2 \end{bmatrix} \begin{bmatrix} i \\ 1 \\ -i \end{bmatrix}$$

$$= [-i \quad 1+1-2 \quad 0] \begin{bmatrix} i \\ 1 \\ -i \end{bmatrix} = 1, \text{ real.}$$

Example 5: Determine the Skew-Hermitian form S for

$$A = \begin{pmatrix} 2i & 3i \\ 3i & 0 \end{pmatrix} \quad \text{with} \quad X = \begin{bmatrix} 4i \\ -5 \end{bmatrix}$$

Solution:

$$S = \overline{X}^T AX = [-4i \quad -5] \begin{bmatrix} 2i & 3i \\ 3i & 0 \end{bmatrix} \begin{bmatrix} 4i \\ -5 \end{bmatrix}$$

$$= (8 - 15i \quad 12) \begin{pmatrix} 4i \\ -5 \end{pmatrix} = 32i + 60 - 60$$

$$= 32i, \quad \text{purely imaginary.}$$

EXERCISE

1. Determine for what values of the numbers a and b, $c = aA + bB$ is Skew-Hermitian given that A and B are Skew-Hermitian.

Ans. Both a and b must be real.

2. Show that the eigen vectors X_i, X_j corresponding to two distinct eigen values λ_i, λ_j of a Hermitian matrix H are orthogonal i.e., $\overline{X}_i^T X_j = 0$.

   Hint: $HX_i = \lambda_i X_i$, $HX_j = \lambda_j X_j$ (1), $\overline{X}_i^T \overline{H}^T = \overline{\lambda}_i \overline{X}_i^T$ or $\overline{X}_i^T H = \lambda_i \overline{X}_i^T$ (2) since $\overline{H}^T = H$, $\overline{\lambda}_i = \lambda_i$, pre multiply (1) by $\overline{X}_i^T$ and post multiply (2) by X_j and subtract.

3. Prove that $A = \begin{bmatrix} 4 & 1-3i \\ 1+3i & 7 \end{bmatrix}$ is Hermitian matrix. Find its eigen values.

Ans. characteristic equation: $\lambda^2 - 11\lambda + 18 = 0$, eigen values 9, 2.

4. Show that $B = \begin{bmatrix} 3i & 2+i \\ -2+i & -i \end{bmatrix}$ is Skew-Hermitian. Find its eigen values.

Ans. characteristic equation: $\lambda^2 - 2i\lambda + 8 = 0$, eigen values $4i, -2i$.

5. Prove that $C = \begin{bmatrix} \frac{i}{2} & \frac{\sqrt{3}}{2} \\ \frac{\sqrt{3}}{2} & \frac{i}{2} \end{bmatrix}$ is unitary matrix. Find its eigen values.

Ans. $\lambda^2 - i\lambda - 1 = 0, \lambda = (\sqrt{3}+i)/2, (-\sqrt{3}+i)/2$

6. If $A = \begin{bmatrix} 0 & 1+2i \\ -1+2i & 0 \end{bmatrix}$ show that $(I - A)(I + A)^{-1}$ is a unitary matrix.

Hint: $I - A = \begin{bmatrix} 1 & -1-2i \\ 1-2i & 1 \end{bmatrix}$,

$(I + A)^{-1} = \begin{bmatrix} 1 & -1-2i \\ 1-2i & 1 \end{bmatrix}$

7. Find the eigen vectors of the Hermitian matrix

$$A = \begin{pmatrix} a & b+ic \\ b-ic & k \end{pmatrix}.$$

Ans. $\lambda_{1,2} = [(a+k) \pm (a-k)^2 + 4(b^2+c^2)]/2$

eigen vectors: $\begin{bmatrix} \dfrac{-(b^2+c^2)}{(a-\lambda)(b-ic)} & 1 \end{bmatrix}^T_{\text{at } \lambda=\lambda_1,\lambda_2}$

8. Find the eigen vectors of the Skew-Hermitian matrix $A = \begin{bmatrix} 2i & 3i \\ 3i & 0 \end{bmatrix}$

Ans. $\lambda_{1,2} = (1 \pm \sqrt{10})i$, eigen vectors: $\left(1 \pm \frac{\sqrt{10}-1}{3}\right)^T$

9. Show that $A = \begin{bmatrix} a+ic & -b+id \\ b+id & a-ic \end{bmatrix}$ is unitary matrix if $a^2+b^2+c^2+d^2 = 1$.

Hint: $|A| = (a^2+c^2) + (b^2+d^2)$

10. Find the Hermitian form of

$$A = \begin{bmatrix} 3 & 2-i \\ 2+i & 4 \end{bmatrix} \quad \text{with} \quad X = \begin{bmatrix} 1+i \\ 2i \end{bmatrix}$$

Ans. 34

11. Find the Skew-Hermitian form for

a. $A = \begin{pmatrix} i & 0 \\ 0 & -i \end{pmatrix}$ with $X = \begin{pmatrix} 1 \\ i \end{pmatrix}$,

Ans. 0

b. $A = \begin{pmatrix} 2i & 4 \\ -4 & 0 \end{pmatrix}$ with $X = \begin{pmatrix} x_1 \\ x_2 \end{pmatrix}$

Ans. $2i|x_1|^2 + 8i\,\mathrm{Im}(\overline{x}_1 x_2)$

12. Find the Hermitian form of $A = \begin{pmatrix} 0 & i \\ -i & 0 \end{pmatrix}$, $X = \begin{pmatrix} 1 \\ i \end{pmatrix}$

Ans. -2

13. Show that the column (and also row) vectors of the unitary matrix

$$A = \frac{1}{2}\begin{bmatrix} 1+i & -1+i \\ 1+i & 1-i \end{bmatrix}$$

form an orthonormal system.

14. Determine the eigen values and eigen vectors of the unitary matrix $\frac{1}{\sqrt{2}}\begin{bmatrix} 1 & i \\ -i & -1 \end{bmatrix}$.

Ans. eigen values $1, -1$, eigen vectors $[1\ i \pm i\sqrt{2}]^T$

15. Find the Skew-Hermitian form for

$$A = \begin{bmatrix} -i & 1 & 2+i \\ -1 & 0 & 3i \\ -2+i & 3i & i \end{bmatrix} \quad \text{with} \quad X = \begin{bmatrix} 0 \\ 1 \\ 2 \end{bmatrix}$$

Ans. $16i$.

14.10 SYLVESTER'S LAW OF INERTIA

Let $Q(x_1, x_2 \ldots x_n) = \sum_{i=1}^{n}\sum_{j=1}^{n} a_{ij}x_ix_j = X^TAX$ be any real quadratic form with the coefficient matrix A given by $A = \begin{bmatrix} a_{11} & a_{12} & \ldots & a_{1n} \\ a_{21} & a_{22} & \ldots & a_{2n} \\ \\ a_{n1} & a_{n2} & \ldots & a_{nn} \end{bmatrix}$

Let 'r' be the rank of A.

Sylvester's law of inertia states that any real quadratic form $Q(x_1, x_2, \ldots, x_n)$ can be transformed by a regular linear transformation (substitution) into the form

$$Q^*(y_1, y_2, \ldots y_n) = y_1^2 + y_2^2 + \ldots + y_{s_1}^2 - y_{s+1}^2 - y_{s+2}^2 - \cdots - y_{s_1+s_2}^2$$

where $s_1 + s_2 = r$, While the substitution which transforms $Q(x_1, x_2, \ldots x_n)$ to $Q^*(y_1, y_2, \ldots y_n)$ is not unique, however the number of positive terms (signs) s_1 as well as the number of negative terms (signs) $s_2 = r - s_1$ in the resulting term $Q^*(y_1, y_2 \ldots y_n)$ is always the same. Here s_1 is

[1] James Joseph Sylvester (1814–1897), British mathematician who in 1850 introduced for the first time the word "matrix" (in the sense of "The mother of determinants").

known as *index* and $s_1 - s_2$ is known as the 'signature' of the real quadratic form $Q(x_1, x_2, \dots x_n)$. Introducing $y = [y_1, y_2, \dots y_n]^T$ and

$$S = \begin{bmatrix} 1 & & & & & & & & & & \\ & 1 & & & & & & & & \bigcirc & \\ & & 1 & & & & & & & & \\ & & & \ddots & & & & & & & \\ & & & & 1 & & & & & & \\ & & & & & -1 & & & & & \\ & & & & & & -1 & & & & \\ & & & & & & & \ddots & & & \\ & & & & & & & & -1 & & \\ & & & & & & & & & 0 & \\ & & & & & & & & & & \ddots \\ \bigcirc & & & & & & & & & & & 0 \end{bmatrix}$$

we can rewrite

$$\begin{aligned} Q^*(y_1, y_2, \dots y_n) &= y_1^2 + y_2^2 + \dots y_{s_1}^2 \\ &\quad - y_{s_1+1}^2 - y_{s_1+2}^2 \dots - y_{s_1+s_2}^2 \\ &= Y^T S Y \end{aligned}$$

The diagonal matrix S with diagonal elements $+1$, -1 and 0's only, is known as "*Sylvester's canonical form*". The total number of non zero entries in S is equal to r, the rank of A.

WORKED OUT EXAMPLES

Example 1: Find the rank, index, and signature of the Sylvester's canonical form

$$S = \begin{bmatrix} 1 & 0 & 0 & 0 & 0 \\ 0 & 1 & 0 & 0 & 0 \\ 0 & 0 & 1 & 0 & 0 \\ 0 & 0 & 0 & -1 & 0 \\ 0 & 0 & 0 & 0 & 0 \end{bmatrix}$$

Solution: r = rank = the number of non zero rows = 4; index = s_1 = number of positive signs (ones) = 3, s_2 = number of negative signs (ones) = 1, signature = $s_1 - s_2 = 3 - 1 = 2$ (note $r = 4 = s_1 + s_2 = 3 + 1$).

EXERCISE

Find the rank, index, and signature of the Sylvester's canonical form.

1. $\begin{bmatrix} 1 & 0 & 0 \\ 0 & -1 & 0 \\ 0 & 0 & -1 \end{bmatrix}$

   Ans. $r = 3$, $s_2 = 2$, s_1 = index = 1, signature = $1 - 2 = -1$

2. $\begin{bmatrix} 1 & 0 & 0 \\ 0 & -1 & 0 \\ 0 & 0 & 0 \end{bmatrix}$

   Ans. $r = 2$, s_1 = index = 1, $s_2 = 1$, signature = $1 - 1 = 0$

3. $\begin{bmatrix} 1 & 0 & 0 & 0 \\ 0 & 1 & 0 & 0 \\ 0 & 0 & -1 & 0 \\ 0 & 0 & 0 & 0 \end{bmatrix}$

   Ans. $r = 3$, s_1 = index = 2, $s_2 = 1$, signature = $2 - 1 = 1$

4. $\begin{bmatrix} 1 & 0 & 0 & 0 \\ 0 & 1 & 0 & 0 \\ 0 & 0 & 1 & 0 \\ 0 & 0 & 0 & -1 \end{bmatrix}$

   Ans. $r = 4$, s_1 = index = 3, $s_2 = 1$, signature = $3 - 1 = 2$

Chapter 15

Vector Differential Calculus: Gradient, Divergence and Curl

INTRODUCTION

Principal application of vector functions is the analysis of motion in space. The gradient defines the normal to the tangent plane, the directional derivative gives the rate of change in any given direction. If $\overline{F}$ is the velocity field of a fluid flow, then divergence of $\overline{F}$ at a point $P(x, y, z)$ (flux density) is the rate at which fluid is (diverging) piped in or drained away at P, and the curl $\overline{F}$ (or circulation density) is the vector of greatest circulation in flow. We express grad, div and curl in general curvilinear coordinates and in cylindrical and spherical coordinates which are useful in engineering, physics or geometry involving a cylinder or cone or a sphere.

In this Chapter 15, vector differential calculus is considered, which extends the basic concepts of (ordinary) differential calculus to vector functions, by introducing derivative of a vector function and the new concepts of gradient, divergence and curl.

15.1 VECTOR DIFFERENTIATION

Definitions

Scalar function

Scalar function of a scalar variable t is a function $F = F(t)$ which uniquely associates a scalar $F(t)$ for every value of the scalar t in an interval $[a, b]$.

Scalar field

Scalar field is a region in space such that for every point P in this region, the scalar function f associates a scalar $f(P)$.

Scalar function of a vector variable $\overline{u}$ is a function $F = F(\overline{u})$ which uniquely associates a scalar $F(\overline{u})$ for every vector $\overline{u}$.

Vector function

Vector function of a scalar variable t is a function $\overline{F} = \overline{F}(t)$ which uniquely associates a vector $\overline{F}$ for each scalar t.

Vector field

Vector field is a region in space such that with every point P in that region, the vector function $\overline{V}$ associates a vector $\overline{V}(P)$.

Vector function

Vector function of a vector variable $\overline{u}$ is $\overline{F} = \overline{F}(\overline{u})$ if for every $\overline{u}$ a unique vector $\overline{F}(\overline{u})$ is associated.

Derivative

Derivative of a vector function $\overline{F}(u)$ with respect to a scalar variable u is denoted by and is defined as

$$\frac{d\overline{F}}{du} = \lim_{\Delta u \to 0} \frac{\overline{F}(u + \Delta u) - \overline{F}(u)}{\Delta u}.$$

Let $\overline{i}, \overline{j}, \overline{k}$ be three mutually orthogonal unit vectors in the direction of the x, y, z-coordinate axes such that $\overline{i}, \overline{j}, \overline{k}$ form a right handed triad (i.e., $i \cdot i = 1, i \cdot j = 0, \overline{i} \cdot k = 0, j \cdot j = 1, \ldots$ etc.).

Derivative in the Component Form

Let $\overline{F}(u) = F_1(u)\overline{i} + F_2(u)\overline{j} + F_3(u)\overline{k}$ in the component form with $F_1(u)$, $F_2(u)$ and $F_3(u)$ as components of $\bar{F}$ in the x, y, z-coordinate axes. Then

$$\frac{d\bar{F}}{du} = \frac{dF_1}{du}\bar{i} + \frac{dF_2}{du}\bar{j} + \frac{dF_3}{du}\bar{k}.$$

Thus the derivative of a vector function $\bar{F}$ with respect to a scalar variable u is the vector whose components are the derivatives of the components F_1, F_2, F_3 of $\bar{F}$ with respect to u.

Results: Most of the basic rules of differentiation that are true for a scalar function of a scalar variable hold good for vector function of a scalar variable, provided the order of factors in vector products is maintained.

1. $\frac{d\bar{C}}{du} = 0 \quad (\bar{C} = \text{constant vector})$
2. $\frac{d}{du}[\bar{F}(u) \pm \bar{G}(u)] = \frac{d\bar{F}}{du} \pm \frac{d\bar{G}}{du}$
3. $\frac{d}{du}[\alpha(u)\bar{F}(u)] = \alpha(u)\frac{d\bar{F}}{du} + \bar{F}\frac{d\alpha}{du}$
4. $\frac{d}{du}[\bar{F}(u)\cdot\bar{G}(u)] = \frac{d\bar{F}}{du}\cdot\bar{G} + \bar{F}\cdot\frac{d\bar{G}}{du}$
5. $\frac{d}{du}[\bar{F}(u) \times \bar{G}(u)] = \bar{F} \times \frac{d\bar{G}}{du} + \frac{d\bar{F}}{du} \times \bar{G}$
6. $\frac{d}{du}[\bar{A}(u)\cdot\bar{B}(u) \times \bar{C}(u)] = \bar{A}\cdot\bar{B} \times \frac{d\bar{C}}{du} + \bar{A}\cdot\frac{d\bar{B}}{du} \times \bar{C} + \frac{d\bar{A}}{du}\cdot\bar{B} \times \bar{C}$.
7. $\frac{d}{du}[\bar{A} \times (\bar{B} \times \bar{C})] = \bar{A} \times \left(\bar{B} \times \frac{d\bar{C}}{du}\right) + \bar{A} \times \left(\frac{d\bar{B}}{du} \times \bar{C}\right) + \frac{d\bar{A}}{du} \times (\bar{B} \times \bar{C})$.

Velocity and Acceleration

Let $\bar{r}$ be the position vector of a point P $(x(t), y(t), z(t))$ in space where t is the scalar time. Then $\bar{r}$ in the component form is

$$\bar{r} = \bar{r}(t) = x(t)\bar{i} + y(t)\bar{j} + z(t)\bar{k}$$

Derivative of $\bar{r}$ denoted by $\dot{\bar{r}}$ is defined as

$$\frac{d\bar{r}}{dt} = \dot{\bar{r}} = \lim_{\Delta t \to 0} \frac{\bar{r}(t + \Delta t) - \bar{r}(t)}{\Delta t} = \frac{dx}{dt}\bar{i} + \frac{dy}{dt}\bar{j} + \frac{dz}{dt}\bar{k}$$

$\dot{\bar{r}}$ and $\ddot{\bar{r}}$ denote the velocity and acceleration of a particle with position vector $\bar{r}$.

Unit Tangent Vector

Let s be the arc length reckoned (measured) from a fixed point M_0 of a space curve c whose equation is $\bar{r} = \bar{r}(s)$. Then the unit tangent vector of c is

$$\frac{d\bar{r}}{ds} = \frac{dx}{ds}\bar{i} + \frac{dy}{ds}\bar{j} + \frac{dz}{ds}\bar{k}$$

such that

$$\left|\frac{d\bar{r}}{ds}\right| = \sqrt{\left(\frac{dx}{ds}\right)^2 + \left(\frac{dy}{ds}\right)^2 + \left(\frac{dz}{ds}\right)^2} = 1.$$

Partial Derivatives of a Vector Function $\bar{F}$

which depends on more than one scalar variables u, v, w: The partial derivative of $\bar{F}$ with respect to u is

$$\frac{\partial\bar{F}}{\partial u} = \lim_{\Delta u \to 0} \frac{\bar{F}(u + \Delta u, v, w) - \bar{F}(u, v, w)}{\Delta u}$$

In the component form, if
$\bar{F}(u,v,w) = F_1(u,v,w)i + F_2(u,v,w)j + F_3(u,v,w)k$
then the partial derivative of $\bar{F}$ with respect to say u is obtained by taking the partial derivatives of the components F_1, F_2, F_3 of $\bar{F}$ with respect to u. i.e.,

$$\frac{\partial\bar{F}}{\partial u} = \frac{\partial F_1}{\partial u}\bar{i} + \frac{\partial F_2}{\partial u}\bar{j} + \frac{\partial F_3}{\partial u}\bar{k}$$

Higher order partial derivatives can be obtained similarly.

WORKED OUT EXAMPLES

Example 1: Find the magnitude of the velocity and acceleration of a particle which moves along the curve $x = 2\sin 3t$, $y = 2\cos 3t$, $z = 8t$ at any time $t > 0$. Find unit tangent vector to the curve.

Solution: The position vector $\bar{r}$ of the particle is

$$\bar{r}(t) = x(t)i + y(t)j + z(t)\bar{k}$$

$$\bar{r}(t) = 2\sin 3t\bar{i} + 2\cos 3tj + 8t\bar{k}$$

$$\text{Velocity} = \bar{V} = \frac{d\bar{r}}{dt} = 6\cos 3t\, i - 6\sin 3t\, j + 8\bar{k}$$

$$\text{Acceleration} = \bar{a} = \dot{\bar{V}} = \frac{d\dot{\bar{r}}}{dt}$$

$$= -18\sin 3t\, i - 18\cos 3t\, j + 0$$

$$|\bar{V}| = \sqrt{36\cos^2 3t + 36\sin^2 3t + 64}$$

$$= \sqrt{36 + 64} = 10$$

$$|\bar{a}| = \sqrt{18^2\sin^2 3t + 18^2\cos^2 3t} = 18.$$

$$\text{Unit tangent vector} = \frac{d\bar{r}}{dt} \Big/ \left|\frac{d\bar{r}}{dt}\right|$$

$$= \frac{1}{10}[6\cos 3t\bar{i} - 6\sin 3t\bar{j} + 8\bar{k}]$$

Example 2: If $\bar{A} = t^2 i - tj + (2t+1)\bar{k}$, $\bar{B} = (2t-3)i + j - tk$ find (a) $\frac{d}{dt}(\bar{A}\cdot\bar{B})$ (b) $\frac{d}{dt}(\bar{A}\times\bar{B})$ (c) $\frac{d}{dt}|A+B|$ (d) $\frac{d}{dt}\left(\bar{A}\times\frac{d\bar{B}}{dt}\right)$ at $t = 1$.

Solution:

a. $\bar{A}\cdot B = t^2(2t-3) - t + (2t+1)(-t)$

$$\frac{d}{dt}(\bar{A}\cdot\bar{B}) = 6t^2 - 6t - 1 - 4t - 1\Big|_{\text{at } t=1} = -6$$

b. $$\bar{A}\times\bar{B} = \begin{vmatrix} i & j & k \\ t^2 & -t & (2t+1) \\ 2t-3 & 1 & -t \end{vmatrix}$$

$$\bar{A}\times\bar{B} = i(t^2 - 2t - 1) + j(+t^3 + 4t^2 - 4t - 3) + \bar{k}(3t^2 - 3t)$$

$$\frac{d}{dt}(\bar{A}\times\bar{B}) = (2t-2)\bar{i} + (3t^2 + 8t - 4)\bar{j} + (6t-3)\bar{k}$$

$$\text{At } t = 1,\ \frac{d}{dt}(\bar{A}\times\bar{B}) = 7\bar{j} + 3\bar{k}$$

c. $$\bar{A}+\bar{B} = (t^2 + 2t - 3)\bar{i} + (1-t)j + (t+1)\bar{k}$$

$$|\bar{A}+\bar{B}| = \sqrt{(t^2+2t-3)^2 + (1-t)^2 + (t+1)^2}$$

$$= \sqrt{t^4 + 4t^3 - 12t + 11}$$

$$\frac{d}{dt}|\bar{A}+\bar{B}| = \frac{4t^3 + 12t^2 - 12}{2\sqrt{t^4 + 4t^3 - 12t + 11}} \text{ at } t = 1 \text{ is } 1.$$

d. $\frac{d\bar{B}}{dt} = 2\bar{i} + 0 - \bar{k}$

$$\bar{A}\times\frac{d\bar{B}}{dt} = \begin{vmatrix} i & j & k \\ t^2 & -t & (2t+1) \\ 2 & 0 & -1 \end{vmatrix}$$

$$= t\bar{i} + (t^2 + 4t - 2)j + 2t\bar{k}$$

$$\frac{d}{dt}\left(\bar{A}\times\frac{d\bar{B}}{dt}\right) = i + (2t+4)\bar{j} + 2\bar{k} \text{ at } t = 1$$

$$\text{is } \bar{i} + 6j + 2k$$

Aliter : $$\frac{d}{dt}\left(\bar{A}\times\frac{d\bar{B}}{dt}\right) = \frac{d\bar{A}}{dt}\times\frac{d\bar{B}}{dt} + \bar{A}\times\frac{d^2\bar{B}}{dt^2}$$

$$\frac{d\bar{A}}{dt} = 2ti - j + 2\bar{k}$$

$$\frac{d\bar{A}}{dt}\times\frac{d\bar{B}}{dt} = \begin{vmatrix} i & j & k \\ 2t & -1 & 2 \\ 2 & 0 & -1 \end{vmatrix}$$

$$= i + (2t+4)j + 2k$$

Also $$\frac{d^2\bar{B}}{dt^2} = 0 + 0 + 0$$

so that $$\bar{A}\times\frac{d^2\bar{B}}{dt^2} = 0$$

Thus $$\frac{d}{dt}\left(\bar{A}\times\frac{d\bar{B}}{dt}\right) = i + (2t+4)j + 2k \text{ at } t = 1$$

$$\text{is } i + 6j + 2k.$$

Example 3: If $\bar{A} = \cos xy\, i + (3xy - 2x^2)j - (3x + 2y)\bar{k}$ find

$$\frac{\partial\bar{A}}{\partial x}, \frac{\partial\bar{A}}{\partial y}, \frac{\partial^2 A}{\partial x^2}, \frac{\partial^2 A}{\partial y^2}, \frac{\partial^2 A}{\partial x\partial y}, \frac{\partial^2 A}{\partial y\partial x}.$$

Solution:

$$\frac{\partial\bar{A}}{\partial x} = -y\sin xy\, i + (3y - 4x)j - (3\bar{k})$$

$$\frac{\partial\bar{A}}{\partial y} = -x\sin xy\bar{i} + (3xj) - 2\bar{k}$$

$$\frac{\partial^2\bar{A}}{\partial x^2} = -y^2\cos xy\, i - 4j$$

$$\frac{\partial^2 A}{\partial y^2} = -x^2\cos xy\, i$$

$$\frac{\partial^2 A}{\partial x\partial y} = (-\sin xy - xy\cos xy)i + 3j.$$

Example 4: Prove that $\bar{A}\cdot\frac{d\bar{A}}{dt} = 0$ if $\bar{A}$ is a constant vector.

Solution: For any vector $\bar{A}$, $\bar{A}\cdot\bar{A} = A^2$. Differentiating w.r.t., t

$$\frac{d}{dt}(\bar{A}\cdot\bar{A}) = \bar{A}\cdot\frac{d\bar{A}}{dt} + \frac{d\bar{A}}{dt}\cdot\bar{A} = 2A\frac{dA}{dt}$$

$$2\bar{A}\cdot\frac{d\bar{A}}{dt} = 2A\frac{dA}{dt}$$

If A is a vector of constant magnitude $\frac{dA}{dt} = 0$ so that

$$\bar{A}\cdot\frac{d\bar{A}}{dt} = 0.$$

EXERCISE

1. If $\bar{A} = 5t^2 i + tj - t^3 k$ and $\bar{B} = \sin t i - \cos t j$, find (a) $\frac{d}{dt}(\bar{A}\cdot\bar{B})$, (b) $\frac{d}{dt}(\bar{A}\times\bar{B})$ (c) $\frac{d}{dt}(\bar{A}\cdot\bar{A})$.

Ans. **a.** $(5t^2-1)\cos t+11t\sin t$
b. $(t^3\sin t-3t^2\cos t)i-(t^3\cos t+3t^2\sin t)j+(5t^2\sin t-11t\cos t-\sin t)\overline{k}$
c. $100t^3+2t+6t^5$

2. Find (a) $\frac{d}{du}(\overline{A}\cdot\overline{B})$ (b) $\frac{d}{du}(\overline{A}\times\overline{B})$ if $\overline{A}(u)=2ui-3u^2j+u^3k$, $\overline{B}(u)=\sin ui-uk$.

Ans. $2\sin u+2u\cos u-4u^3; 9u^2i+(u^3\cos u+3u^2\sin u+4u)j+(3u^2\cos u+6u\sin u)\overline{k}$

3. If $\overline{A}=2ti-t^2j+t^3k$, $\overline{B}=-ti+t^2k$ and $\overline{C}=t^3j-2tk$ find $\frac{d}{dt}(\overline{A}\cdot\overline{B}\times\overline{C})$ at $t=1$.

Ans. $-12t^5+8t^3-7t^6, -11$

4. If $\overline{A}=\sin ui+\cos uj+uk$, $\overline{B}=\cos ui-\sin uj-3k$ and $\overline{C}=2i+3j-k$ find $\frac{d}{du}(\overline{A}\times(\overline{B}\times\overline{C}))$ at $u=0$.

Ans. $7i+6j-6k$

5. If $\overline{A}=(2x^2y-x^4)i+(e^{xy}-y\sin x)j+(x^2\cos y)\overline{k}$ find (a) $\frac{\partial A}{\partial x}$, (b) $\frac{\partial A}{\partial y}$, (c) $\frac{\partial^2 A}{\partial x^2}$, (d) $\frac{\partial^2 A}{\partial y^2}$, (e) $\frac{\partial^2 A}{\partial x\partial y}$, (f) $\frac{\partial^2 A}{\partial y\partial x}$.

Ans. **a.** $(4xy-4x^3)i+(ye^{xy}-y\cos x)j+2x\cos yk$
b. $2x^2i+(xe^{xy}-\sin x)j-x^2\sin y\overline{k}$
c. $(4y-12x^2)\overline{i}+(y^2e^{xy}+y\sin x)j+2\cos yk$
d. $\overline{0}+x^2e^{xy}j-x^2\cos y\overline{k}$
e, f. $4xi+(xye^{xy}+e^{xy}-\cos x)j-2x\sin yk$

6. Find $\frac{\partial^3}{\partial x^2\partial z}(f\overline{A})$ at the point $(2,-1,1)$ if
$f=xy^2z$, $\overline{A}=xzi-xy^2j+yz^2k$.

Ans. $4y^2zi-2y^4j; 4i-2j$

7. Find $\frac{\partial^2(\overline{A}\times\overline{B})}{\partial x\partial y}$ at $(1,0,-2)$ if $\overline{A}=x^2yzi-2xz^3j+xz^2\overline{k}$, $\overline{B}=2zi+yj-x^2k$.

Ans. $-4i-8j$

8. Prove that $\overline{A}\times\frac{d\overline{A}}{dt}=0$ if $\overline{A}(t)$ has constant (fixed) direction.

Hint: Take $\overline{A}=a(t)\overline{B}(t)$ where $a(t)=|\overline{A}|$ and $\overline{B}(t)$ is a unit vector in the direction of $\overline{A}$ so that $\frac{d\overline{B}}{dt}=0$.

9. Given the curve $x=t^2+2$, $y=4t-5$, $z=2t^2-6t$ find the unit tangent vector at the point $t=2$.

Ans. $ti+2j+(2t-3)k/\left(\sqrt{5t^2-12t+13}\right)$; $(2i+2j+k)/3$

10. Find the angle between the tangents to the curve $\overline{r}=t^2i+2tj-t^3k$ at the points $t=\pm1$.

Hint: $\overline{T}_1\cdot\overline{T}_2=T_1T_2\cos\theta$.

Ans. $\theta=\cos^{-1}(9/17)$

11. Determine the magnitude of velocity and acceleration at $t=0$ of a particle moving along a curve whose parametric equations are $x=e^{-t}$, $y=2\cos 3t$, $z=2\sin 3t$; where t is the time.

Ans.
$$\overline{V}=-e^{-t}i-6\sin 3t\,j+6\cos 3t\,k$$
$$\overline{a}=e^{-t}i-18\cos 3t\,j-18\sin 3t\,k$$
$|\overline{V}|$ at $t=0$ is $\sqrt{37}$; $|\overline{a}|$ at $t=0$ is $\sqrt{325}$

12. A particle moves along the curve $x=2t^2$, $y=t^2-4t$, $z=3t-5$ where t is the time. Find the components of its velocity and acceleration at time $t=1$ in the direction $\overline{i}-3\overline{j}+2\overline{k}$.

Hint: Component of $\overline{V}$= dot product of $\overline{V}$ with unit vector in the direction of $i-3j+2k$.

Ans. $8\sqrt{14}/7; -\sqrt{14}/7$

13. If a, b, w are constants show that the acceleration of a particle with displacement vector $\overline{r}=a\cos wt+b\sin wt$ is always directed towards the origin.

Hint: $\overline{a}=\ddot{\overline{r}}=-w^2\overline{r}$.

14. Find the angle between the directions of the velocity and acceleration vectors at time t of a particle with position vector $\overline{r}=(t^2+1)i-2tj+(t^2-1)k$.

Ans. $\text{arc}\cos t\sqrt{2}/\sqrt{2t^2+1}$

15. Prove that $\frac{d}{du}(\overline{A}\times\overline{B})=\overline{C}\times(\overline{A}\times\overline{B})$ if
$$\frac{d\overline{A}}{du}=\overline{C}\times\overline{A}\quad\text{and}\quad\frac{d\overline{B}}{du}=\overline{C}\times\overline{B}$$

Hint:
$$\frac{d}{du}(\overline{A}\times\overline{B})=\overline{A}\times\frac{d\overline{B}}{du}+\frac{d\overline{A}}{du}\times\overline{B}$$
$$=\overline{A}\times(\overline{C}\times\overline{B})+(\overline{C}\times\overline{A})\times\overline{B}$$

$$= (A \cdot B)C - (A \cdot C)B - (\bar{B} \cdot \bar{A})C$$

$$+ (B \cdot C)A$$

$$= \bar{C} \times (\bar{A} \times \bar{B}).$$

15.2 DIRECTIONAL DERIVATIVE, GRADIENT OF A SCALAR FUNCTION AND CONSERVATIVE FIELD

In vector differential calculus, it is very convenient to introduce the symbolic linear vector differential "Hamiltonian" operator **del** defined and denoted as

$$\nabla = \bar{i}\frac{\partial}{\partial x} + \bar{j}\frac{\partial}{\partial y} + \bar{k}\frac{\partial}{\partial z} \qquad (1)$$

This operator read as del (or nabla) is *not* a vector (neither has magnitude nor direction) but combines both differential and vectorial properties analogous to those of ordinary vectors.

Directional Derivative

If $f = f(x, y, z)$ then the partial derivatives $\frac{\partial f}{\partial x}, \frac{\partial f}{\partial y}, \frac{\partial f}{\partial z}$ are the derivatives (rates of change) of f in the "direction" of the coordinate axes OX, OY, OZ respectively. This concept can be extended to define a derivative of f in a "given" direction $\overline{PQ}$ (Fig. 15.1).

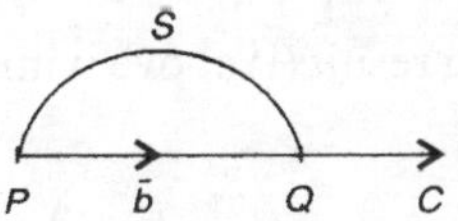

Fig. 15.1

Let P be a point in space and $\bar{b}$ be a unit vector from P in the given direction. Let s be the arc lengths measured from P to another point Q along the ray C in the direction of $\bar{b}$. Now consider

$$f(s) = f(x, y, z) = f((s), y(s), z(s))$$

Then
$$\frac{df}{ds} = \frac{\partial f}{\partial x}\frac{dx}{ds} + \frac{\partial f}{\partial y}\frac{dy}{ds} + \frac{\partial f}{\partial z}\frac{dz}{ds} \qquad (2)$$

The *directional derivative* of f at the point P in the given direction $\bar{b}$ is $\frac{df}{ds}$ given by (2). $\frac{df}{ds}$ gives the rate of change of f in the direction of $\bar{b}$.

Since
$$\frac{dx}{ds}\bar{i} + \frac{dy}{ds}\bar{j} + \frac{dz}{ds}\bar{k} = \bar{b} = \text{unit vector} \qquad (3)$$

Using the del operator defined by (1) $\frac{df}{ds}$ given by (2) can be rewritten as

$$\frac{df}{ds} = \left(\bar{i}\frac{\partial f}{\partial x} + \bar{j}\frac{\partial f}{\partial y} + \bar{k}\frac{\partial f}{\partial z}\right) \cdot \left(\frac{dx}{ds}\bar{i} + \frac{dy}{ds}\bar{j} + \frac{dz}{ds}\bar{k}\right)$$

$$\frac{df}{ds} = \left[\left(\bar{i}\frac{\partial}{\partial x} + \bar{j}\frac{\partial}{\partial y} + \bar{k}\frac{\partial}{\partial z}\right) f\right] \cdot \bar{b} = \nabla f \cdot \bar{b} \qquad (4)$$

Thus the directional derivative of f at P is the component (dot product) of ∇f in the direction of (with) unit vector $\bar{b}$.

The directional derivative in the direction of any (non unit) vector $\bar{a}$ is

$$\frac{df}{ds} = \nabla f \cdot \left(\frac{\bar{a}}{|\bar{a}|}\right) \qquad (5)$$

Equation (4) introduces the vector quantity, the ***gradient of a scalar function*** $f(x, y, z)$ or gradient f denoted by ∇f and defined as

$$\nabla f = \left(\bar{i}\frac{\partial}{\partial x} + \bar{j}\frac{\partial}{\partial y} + \bar{k}\frac{\partial}{\partial z}\right) f$$

$$= \bar{i}\frac{\partial f}{\partial x} + \bar{j}\frac{\partial f}{\partial y} + \bar{k}\frac{\partial f}{\partial z} = \text{grad } f = \text{vector}$$

$$\nabla f = \left[\frac{\partial f}{\partial x}, \frac{\partial f}{\partial y}, \frac{\partial f}{\partial z}\right].$$

Properties of gradient

1. Projection of ∇f in any direction is equal to the derivative of f in that direction.
2. The gradient of f is in the direction of the normal to the level surface $f(x, y, z) = C =$ constant. So, the angle between any two surfaces $f(x, y, z) = C_1$ and $g(x, y, z) = C_2$ is the angle between their corresponding normals given by ∇f and ∇g respectively.
3. The gradient at P is in the direction of maximum increase of f at P.
4. The modulus of the gradient is equal to the largest directional derivative at a given point P.

i.e., $\max \left.\frac{df}{ds}\right|_P = \left|\nabla f\right|_P$

$$= \sqrt{\left(\frac{\partial f}{\partial x}\right)^2 + \left(\frac{\partial f}{\partial y}\right)^2 + \left(\frac{\partial f}{\partial z}\right)^2}\,\Bigg|_{\text{at } P.}$$

These properties thus state that the vector gradient f indicates the direction and magnitude of maximum change of a scalar function f at a given point.

Normal derivative

$\frac{df}{dn} = \nabla f \cdot \bar{n}$ where $\bar{n}$ is the unit normal to the surface $f =$ constant.

Conservative

A vector function $\bar{A}$ is said to be a conservative vector field if $\bar{A} = \nabla f$ i.e., $\bar{A}$ is the gradient of a scalar function f. In this case f is known as the *potential function* of $\bar{A}$.

Instead of applying (operating) on a scalar function f, if del is applied to a vector function $\bar{A}$, we get divergence and curl: (see Sections 15.3, 15.4)

WORKED OUT EXAMPLES

Example 1: If $\bar{A} = 2x^2 i - 3yzj + xz^2 k$ and $f = 2z - x^3 y$ find (i) $A \cdot \nabla f$ and (ii) $A \times \nabla f$ at the point $(1, -1, 1)$.

Solution: Here $\frac{\partial f}{\partial x} = -3x^2 y$, $\frac{\partial f}{\partial y} = -x^3$, $\frac{\partial f}{\partial z} = 2$ so that

$$\nabla f = \frac{\partial f}{\partial x} i + \frac{\partial f}{\partial y} j + \frac{\partial f}{\partial z} k$$

$$\nabla f = -3x^2 yi - x^3 j + 2k$$

i.
$$\bar{A} \cdot \nabla f = (2x^2 i - 3yzj + xz^2 k) \cdot (-3x^2 yi - x^3 j + 2k) = -6x^4 y + 3x^3 yz + 2xz^2$$

$$\bar{A} \cdot \nabla f \Big|_{1,-1,1} = 6 - 3 + 2 = 5$$

ii.
$$\bar{A} \times \nabla f = \begin{vmatrix} i & j & k \\ 2x^2 & -3yz & xz^2 \\ -3x^2 y & -x^3 & 2 \end{vmatrix}$$

Expanding the determinant

$$\bar{A} \times \nabla f = (-6yz + x^4 z^2) i - j(4x^2 + 3x^3 yz^2) + (-2x^5 - 9x^2 y^2 z)\bar{k}$$

$$\bar{A} \times \nabla f \Big|_{1,-1,1} = 7i - j - 11\bar{k}.$$

Example 2: Evaluate

i. ∇r^n iv. ∇r

ii. $\nabla |\bar{r}|^3$ v. $\nabla (\ln r)$

iii. $\nabla (3r^2 - 4\sqrt{r} + 6r^{-\frac{1}{3}})$ vi. $\nabla (r^{-1})$

Solution:

i. $\nabla r^n = i\frac{\partial r^n}{\partial x} + j\frac{\partial r^n}{\partial y} + k\frac{\partial r^n}{\partial z}$

$$\frac{\partial r^n}{\partial x} = nr^{n-1}\frac{\partial r}{\partial x} = nr^{n-1}\frac{x}{r}$$

$$\frac{\partial r^n}{\partial x} = nr^{n-2} x$$

Similarly, $\frac{\partial r^n}{\partial y} = nr^{n-2} y$, and

$$\frac{\partial r^n}{\partial z} = nr^{n-2} z.$$

Then $\nabla r^n = nr^{n-2}(xi + yj + zk) = nr^{n-2}\bar{r}$

ii. Put $n = 3$ in the result (i) above

$$\nabla r^3 = 3r^{3-2}\bar{r} = 3r\bar{r}$$

iii. $\nabla (3r^2 - 4\sqrt{r} + 6r^{-\frac{1}{3}})$

$$= 3\nabla r^2 - 4\nabla r^{\frac{1}{2}} + 6\nabla r^{-\frac{1}{3}}$$

Applying result (i) above with $n = 2, \frac{1}{2}, -\frac{1}{3}$, we get

$$= 3(2r^{2-2}\bar{r}) - 4\left(\frac{1}{2} r^{\frac{1}{2}-2}\bar{r}\right) + 6\left(-\frac{1}{3} r^{-\frac{1}{3}-2}\bar{r}\right)$$

$$= \left(6 - 2r^{-\frac{3}{2}} - 2r^{-\frac{7}{3}}\right)\bar{r}$$

iv. $\nabla r = 1r^{1-2}\bar{r} = \frac{\bar{r}}{r}$

v. $\nabla f = \nabla \ln r = \frac{1}{r}\nabla r = \frac{1}{r}\frac{\bar{r}}{r} = \frac{\bar{r}}{r^2}$

vi. $\nabla (r^{-1}) = -1 \cdot r^{-1-2}\bar{r} = -\bar{r}/r^3.$

Example 3: Find the directional derivative of $f(x, y, z) = 4e^{2x-y+z}$ at the point $(1, 1, -1)$ in the direction toward the point $(-3, 5, 6)$.

Solution:

$$\nabla f = 4e^{2x-y+z}(2i - j + k)$$

$$\nabla f\Big|_{(1,1,-1)} = 4(2i - j + k)$$

A unit vector $\hat{a}$ from the point $(1, 1, -1)$ in the direction toward the point $(-3, 5, 6)$ is

$$\hat{a} = \frac{-4i + 4j + 7k}{\sqrt{16 + 16 + 49}}$$

$$= \frac{-4i + 4j + 7k}{9}$$

The required directional derivative is

$$\nabla f\Big|_{(1,1,-1)} \cdot \hat{a} = 4(2i - j + k)\cdot \frac{(-4i + 4j + 7k)}{9}$$

$$= -\frac{20}{9}.$$

Example 4: Find the values of the constants a, b, c so that the directional derivative of $f = axy^2 + byz + cz^2x^3$ at $(1, 2, -1)$ has a maximum of magnitude 64 in a direction parallel to the z-axis.

Solution: Since $\overline{k}$ is a unit vector parallel to the z-axis, the maximum of magnitude of the directional derivative of f at $(1, 2, -1)$ in the direction parallel to z-axis is given by

$$\nabla f_{\text{at }(1,2,-1)} \cdot \overline{k}$$

Here $$\nabla f = (ay^2 + 3x^2cz^2)i + (2axy + bz)j + (by + 2czx^3)k$$

So that

$$\nabla f_{\text{at }(1,2-1)} = (4a + 3c)i + (4a - b)\overline{j} + (2b - 2c)\overline{k}$$

$$\text{Maximum} = \nabla f_{\text{at }(1,2,-1)} \cdot k$$

$$= [(4a + 3c)i + (4a - b)j + (2b - 2c)k] \cdot k$$

$$= (2b - 2c)$$

It is given in the problem that this maximum is 64. Thus

$$2b - 2c = 64$$

or $$b - c = 32 \qquad (1)$$

Since ∇f is in the direction of z-axis, it is perpendicular to the x and y-axes
Thus

$$\nabla f\Big|_{(1,2,-1)} \cdot i = [(4a + 3c)i + (4a - b)j + (2b - 2c)k] \cdot i$$

$$= 4a + 3i = 0 \qquad (2)$$

Similarly,

$$\nabla f \cdot j = 4a - b = 0 \qquad (3)$$

Solving the Equations (1), (2), (3), we get

$$a = 6, b = 24, c = -8.$$

Example 5: Find the constants a and b so that the surface $ax^2 - byz = (a + 2)x$ will be orthogonal to the surface $4x^2y + z^3 = 4$ at the point $(1, -1, 2)$.

Solution: The given surfaces

$$f = ax^2 - byz - (a + 2)x = 0 \qquad (1)$$

and $$g = 4x^2y + z^3 = 4 \qquad (2)$$

are orthogonal at the point P $(1, -1, 2)$ provided the normals to (1) and (2) at P are at right angles. The normal to surface (1) is given by ∇f,

$$\nabla f\Big|_{1,-1,2} = \{2ax - (a + 2)\}\overline{i} - bz\overline{j} - by\overline{k}\ \Big|_{1,-1,2}$$

$$\nabla f = (a - 2)i - 2bj + bk \qquad (3)$$

and normal to surface (2) by ∇g,

$$\nabla g\Big|_{1,-1,2} = 8xyi + 4x^2j + 3z^2k\ \Big|_{1,-1,2}$$

$$\nabla g = -8i + 4j + 12k \qquad (4)$$

The orthogonality condition is

$$0 = \nabla f \cdot \nabla g = [(a - 2)i - 2bj + bk] \cdot [-8i + 4j + 12k]$$

$$0 = -2a + 4 + b \qquad (5)$$

Since $(1, -1, 2)$ lies on the surface (1), we have

$$a + 2b - (a + 2) = 0$$

i.e., $$b = 1.$$

So from (5), $a = \frac{5}{2}$.

Example 6: If $\nabla f = (y^2 - 2xyz^3)i + (3 + 2xy - x^2z^3)j + (6z^3 - 3x^2yz^2)\overline{k}$, find f if $f(1, 0, 1) = 8$.

Solution: Since

$$\frac{\partial f}{\partial x}i + \frac{\partial f}{\partial y}j + \frac{\partial f}{\partial z}k = \nabla f = (y^2 - 2xyz^3)i + (3 + 2xy - x^2z^3)j + (6z^3 - 3x^2yz^2)k$$

We have

$$\frac{\partial f}{\partial x} = y^2 - 2xyz^3 \tag{1}$$

$$\frac{\partial f}{\partial y} = 3 + 2xy - x^2z^3 \tag{2}$$

$$\frac{\partial f}{\partial z} = 6z^3 - 3x^2yz^2 \tag{3}$$

Integrating (1), (2), (3) partially w.r.t., x, y, z respectively, we get

$$f = xy^2 - x^2yz^3 + c_1(y, z) \tag{4}$$

$$f = 3y + xy^2 - x^2yz^3 + c_2(x, z) \tag{5}$$

$$f = \frac{6}{4}z^4 - x^2yz^3 + c_3(x, y) \tag{6}$$

where c_1, c_2, c_3 arbitrary functions of the variables indicated.

To find $c_1(y, z)$, differentiate (4) partially w.r.t. z and equate it with (3). Thus

$$0 - 3x^2yz^2 + \frac{\partial c_1}{\partial z} = \frac{\partial f}{\partial z} = 6z^3 - 3x^2yz^2$$

So

$$\frac{\partial c_1}{\partial z} = 6z^3 \tag{7}$$

Integrating (7) partially w.r.t. 'z', we get

$$c_1(y, z) = \frac{6}{4}z^4 + c_4(y) \tag{8}$$

where c_4 is a function of y alone.

Substituting (8) in (4), we have

$$f = xy^2 - x^2yz^3 + \frac{3}{2}z^4 + c_4(y) \tag{9}$$

To find c_4, differentiate (9) partially w.r.t. y and equate it with (2), we get

$$2xy - x^2z^3 + 0 + \frac{dc_4}{dy} = \frac{\partial f}{\partial y} = 3 + 2xy - x^2z^3$$

So

$$\frac{dc_4}{dy} = 3 \tag{10}$$

Integrating (10) w.r.t. y, we have

$$c_4(y) = 3y + c_5 \tag{11}$$

where c_5 is a pure arbitrary constant.

Substituting (11) in (9), we get the required

$$f(x, y, z) = xy^2 - x^2yz^3 + \frac{3}{2}z^4 + 3y + c_5$$

Since $8 = f(1, 0, 1) = 1 - 0 + 0 + 3 + c_5$

$$\therefore \quad c_5 = 4.$$

Hence $f = xy^2 - x^2yz^3 + \frac{3}{2}z^4 + 3y + 4.$

Similar result can be obtained by starting from (5) or (6).

Example 7: Find $f(r)$ such that $\nabla f = \frac{\bar{r}}{r^5}$ and $f(1) = 0$.

Solution: It is given that

$$\frac{\partial f}{\partial x}i + \frac{\partial f}{\partial y}j + \frac{\partial f}{\partial z}k = \nabla f = \frac{\bar{r}}{r^5} = \frac{xi + yj + zk}{r^5}$$

so $\frac{\partial f}{\partial x} = \frac{x}{r^5}, \ \frac{\partial f}{\partial y} = \frac{y}{r^5}, \text{ and } \frac{\partial f}{\partial z} = \frac{z}{r^5}$

We know that

$$df = \frac{\partial f}{\partial x}dx + \frac{\partial f}{\partial y}dy + \frac{\partial f}{\partial z}dz = \frac{x}{r^5}dx + \frac{y}{r^5}dy + \frac{z}{r^5}dz$$

$$df = \frac{xdx + ydy + zdz}{r^5} = \frac{rdr}{r^5} = r^{-4}dr$$

Integrating $f(r) = \frac{r^{-3}}{-3} + c$

Since $0 = f(1) = -\frac{1}{3} + c$

so $c = \frac{1}{3}$

Thus $f(r) = \frac{1}{3} - \frac{1}{3}\frac{1}{r^3}.$

EXERCISE

1. Find ∇f if $f = \ln(x^2 + y^2 + z^2)$.

Ans. $2(xi + yj + zk)/(x^2 + y^2 + z^2)$

2. If $f(x, y, z) = 3x^2y - y^3z^2$, find ∇f and $|\nabla f|$ at $(1, -2, -1)$.

Ans. $\nabla f = -12i - 9j - 16k, |\nabla f| = \sqrt{481}$

3. If $f = 2xz^4 - x^2y$, find ∇f and $|\nabla f|$ at $(2, -2, -1)$.

Ans. $10\,i - 4j - 16k, 2\sqrt{93}$

4. Find ∇f when $f = (x^2 + y^2 + z^2)\, e^{-\sqrt{x^2+y^2+z^2}}$.

Ans. $(2 - r)e^{-r}\bar{r}$

5. If $U = 3x^2y$, $V = xz^2 - 2y$ evaluate $\nabla[\nabla U . \nabla V]$.

Ans. $(6yz^2 - 12x)\bar{i} + 6xz^2\bar{j} + 12xyz\bar{k}$

6. Find a unit normal to the surface $x^2y + 2xz = 4$ at the point $(2, -2, 3)$.

Ans. $\pm\frac{1}{3}(i - 2j - 2k)$

7. Find the unit outward drawn normal to the surface $(x-1)^2 + y^2 + (z+2)^2 = 9$ at the point $(3, 1, -4)$.

Ans. $(2i + j - 2k)/3$

8. Determine a unit vector normal to the surface $xy^3z^2 = 4$ at the point $(-1, -1, 2)$.

Ans. $\pm(\bar{i} + 3\bar{j} - \bar{k})/\sqrt{11}$.

9. What is the directional derivative of $f = xy^2 + yz^3$ at the point $(2, -1, 1)$ in the direction of the normal to the surface $x \ln z - y^2 = -4$ at $(-1, 2, 1)$.

Ans. $\frac{15}{\sqrt{17}}$

10. Find the directional derivative of $f = x^2yz + 4xz^2$ at $(1, -2, -1)$ in the direction $2\bar{i} - \bar{j} - 2\bar{k}$.

Ans. $\frac{37}{3}$

11. Find the directional derivative of $f = xy + yz + zx$ in the direction of vector $i + 2j + 2k$ at the point $(1, 2, 0)$.

Ans. $\frac{10}{3}$

12. Determine the directional derivative of $f = xy^2 + yz^3$ at the point $(2, -1, 1)$ in the direction of the vector $i + 2j + 2k$.

Ans. $-\frac{11}{3}$

13. Find the maximal directional derivative of x^3y^2z at $(1, -2, 3)$.

Ans. $4\sqrt{91}$

14. **a.** In what direction from the point $(2, 1, -1)$ is the directional derivative of $f = x^2yz^3$ a maximum?

b. What is the magnitude of this maximum?

Ans. **a.** The directional derivative is a maximum in the direction of $\nabla f = -4i - 4j + 12k$.

b. The magnitude of this maximum is $4\sqrt{11}$.

15. Find the direction in which temperature changes most rapidly with distance from the point $(1, 1, 1)$ and determine the maximum rate of change if the temperature at any point is given by $f(x, y, z) = xy + yz + zx$.

Ans. Maximum direction is $2i + 2j + 2k$, maximum: $2\sqrt{3}$.

16. In what direction from $(3, 1, -2)$ is the directional derivative of $f = x^2y^2z^4$ maximum. Find also the magnitude of the maximum.

Ans. $96(\bar{i} + 3j - 3k)$; $96\sqrt{19}$

17. Find the angle between the surfaces $x^2 + y^2 + z^2 = 9$ and $z = x^2 + y^2 - 3$ at the point $(2, -1, 2)$.

Ans. The acute angle $= \cos^{-1}\left(\frac{8\sqrt{21}}{63}\right) = 54^\circ \cdot 25$.

18. Find the angle of intersection of the spheres $x^2 + y^2 + z^2 = 29$ and $x^2 + y^2 + z^2 + 4x - 6y - 8z = 47$ at $(4, -3, 2)$.

Ans. $\theta = \cos^{-1}(19/29)$

19. Determine the angle between the normals to the surface $xy = z^2$ at the points $(4, 1, 2)$ and $(3, 3, -3)$.

Ans. $\cos^{-1}(1/\sqrt{22})$

20. Calculate the angle between the normals to the surface $2x^2 + 3y^2 = 5z$ at the points $(2, -2, 4)$ and $(-1, -1, 1)$.

Ans. $\theta = \cos^{-1}(65/(\sqrt{233}\sqrt{77}))$

21. If $\nabla f = 2xyz^3 i + x^2z^3 j + 3x^2yz^2 k$, find $f(x, y, z)$ if $f(1, -2, 2) = 4$.

Ans. $f = x^2yz^3 + 20$

22. Find f given $\nabla f = 2xi + 4yj + 8z\bar{k}$.

Ans. $f = x^2 + 2y^2 + 4z^2$

23. **a.** Determine f when $\nabla f = (zy\,i + xz\,j - xy\,k)/z^2$.

b. If $\nabla f = xy\,i + 2xy\,j$ find f.

Ans. **a.** $f = xy/z$

b. f does not exist

24. Prove that $\nabla(f/g) = (g\nabla f - f\nabla g)/g^2$.

Hint: Use quotient formula for derivative of f/g.

15.3 DIVERGENCE

Divergence of a vector function $\overline{A}(x, y, z)$ is written as divergence of $\overline{A}$ or div of $\overline{A}$ and denoted by $\nabla\cdot\overline{A}$ is defined as

$$\nabla\cdot\overline{A}=\left(i\frac{\partial}{\partial x}+\overline{j}\frac{\partial}{\partial y}+k\frac{\partial}{\partial z}\right)\cdot(\overline{A})$$

If $\overline{A}=A_1 i+A_2 j+A_3 k$, then

$$\nabla\cdot\overline{A}=\left(i\frac{\partial}{\partial x}+\overline{j}\frac{\partial}{\partial y}+k\frac{\partial}{\partial z}\right)\cdot(A_1 i+A_2 j+A_3 k)$$

$$\nabla\cdot\overline{A}=\frac{\partial A_1}{\partial x}+\frac{\partial A_2}{\partial y}+\frac{\partial A_3}{\partial z}=\text{a scalar quantity}$$

Note that $\nabla\cdot\overline{A}\neq A\cdot\nabla$ because L.H.S. $\nabla\cdot\overline{A}$ is a scalar quantity, whereas the R.H.S. $\overline{A}\cdot\nabla=A_1\frac{\partial}{\partial x}+A_2\frac{\partial}{\partial y}+A_3\frac{\partial}{\partial z}$ is a scalar differential operator.

Physically the divergence of $\overline{A}$ at point P constitutes the volume density of the flux of $\overline{A}$ at P. i.e., divergence measures outflow minus inflow.

A point P in a vector field $\overline{A}$ is said to be a source (sink) if divergence $\overline{A} > (<)0$.

Solenoidal Function

$\overline{A}$ is said to be solenoidal if divergence $\overline{A}=0$ (at all points of function).

WORKED OUT EXAMPLES

Example 1: Evaluate divergence of $(2x^2z\overline{i}-xy^2zj+3yz^2k)$ at the point (1, 1, 1).

Solution: Divergence of $\overline{A}$

$$=\text{div }\overline{A}=\nabla\cdot\overline{A}$$

$$=\frac{\partial A_1}{\partial x}+\frac{\partial A_2}{\partial y}+\frac{\partial A_3}{\partial z}$$

Here $\quad \overline{A}=2x^2z\overline{i}-xy^2z\overline{j}+3yz^2k$

$$=A_1 i+A_2 j+A_3 k$$

so that $\quad \frac{\partial A_1}{\partial x}=\frac{\partial}{\partial x}(2x^2z)=4xz$

$$\frac{\partial A_2}{\partial y}=\frac{\partial}{\partial y}(-xy^2z)=-2xyz$$

$$\frac{\partial A_3}{\partial z}=\frac{\partial}{\partial z}(3yz^2)=6yz$$

Thus $\quad \nabla\cdot\overline{A}=4xz-2xyz+6yz,$

$$\nabla\cdot\overline{A}\Big|_{(1,1,1)}=8.$$

Example 2: Determine the constant b such that

$$\overline{A}=(bx+4y^2z)i+(x^3\sin z-3y)j-(e^x+4\cos x^2y)\overline{k}$$

is solenoidal.

Solution: $\nabla\cdot\overline{A}=b-3=0 \quad \therefore b=3.$

Example 3: Find the directional derivative of $\nabla\cdot\overline{U}$ at the point (4, 4, 2) in the direction of the corresponding outer normal of the sphere $x^2+y^2+z^2=36$ where $\overline{U}=x^4i+y^4j+z^4k$.

Solution: Let $f=\nabla\cdot\overline{U}=\nabla\cdot(x^4i+y^4j+z^4\overline{k})=4(x^3+y^3+z^3)$

$$\nabla f\Big|_{(4,4,2)}=12\ (x^2i+y^2j+z^2k)\Big|_{4,4,2}$$

$$=48(4i+4j+k)$$

Normal to the sphere $g=x^2+y^2+z^2=36$ is

$$\nabla g\Big|_{4,4,2}=2(xi+yj+zk)\ \Big|_{\text{at }4,4,2}=4(2i+2j+k)$$

$$\overline{a}=\text{unit normal}=\frac{\nabla g}{|\nabla g|}=\frac{4(2i+2j+k)}{\sqrt{64+64+16}}$$

$$=\frac{2i+2j+k}{3}$$

The required directional derivative is

$$\nabla f\cdot\overline{a}=48(4i+4j+k)\cdot\frac{(2i+2j+k)}{3}$$

$$=16(8+8+1)=272.$$

Example 4: $\nabla\cdot(r^3\overline{r})$.

Solution: Since $\nabla\cdot(f\overline{A})=f\nabla\cdot\overline{A}+\nabla f\cdot\overline{A}$

$$\nabla\cdot(r^3\overline{r})=r^3\nabla\cdot\overline{r}+\overline{r}\cdot\nabla r^3$$

$$=3r^3+\overline{r}\cdot[3r^{3-2}\overline{r}]$$

$$=3r^3+3r\overline{r}\cdot\overline{r}$$

$$=3r^3+3r\,r^2=6r^3.$$

Example 5: If f and g are solutions of the Laplace equation show that

$$\nabla\cdot(f\nabla g-g\nabla f)=0$$

Solution:

$$\nabla\cdot(f\nabla g-g\nabla f)=\nabla\cdot(f\nabla g)-\nabla\cdot(g\nabla f)$$

$$=f\nabla\cdot\nabla g+\nabla f\cdot\nabla g$$

$$-g\nabla\cdot\nabla f + \nabla g\cdot\nabla f$$
$$= f\nabla^2 g + \nabla f\cdot\nabla g$$
$$-g\nabla^2 f - \nabla g\cdot\nabla f = 0$$

Note that $\nabla\cdot\nabla f = \nabla^2 f$.
Since f and g satisfy Laplace's equation we have $\nabla^2 f = 0$ and $\nabla^2 g = 0$, and also $\nabla f\cdot\nabla g = \nabla g\cdot\nabla f$ by commutative property.

Example 6: Find $\nabla(\nabla\cdot\bar{A})$ where $\bar{A} = \bar{r}/r$.

Solution: Consider

$$\nabla\cdot\bar{A} = \nabla\cdot\left(\frac{\bar{r}}{r}\right) = r^{-1}\nabla\cdot\bar{r} + \bar{r}\cdot\nabla r^{-1}$$
$$= 3r^{-1} + \bar{r}\cdot(-r^{-1-2}\bar{r})$$
$$= 3r^{-1} - r^{-3}\bar{r}\cdot\bar{r}$$
$$= 3r^{-1} - r^{-3}r^2 = 2r^{-1}$$

So $$\nabla(\nabla\cdot A) = \nabla\left(\nabla\cdot\frac{\bar{r}}{r}\right) = \nabla(2r^{-1}) = 2\nabla r^{-1}$$
$$= 2(-1)r^{-1-2}\bar{r} = -2r^{-3}\bar{r}.$$

EXERCISE

1. Prove that $\nabla\cdot\bar{r} = 3$.
2. Find $\nabla\cdot\bar{A}$ when $\bar{A} = (x\bar{i} + y\bar{j} + z\bar{k})/r$ where $r = \sqrt{x^2+y^2+z^2}$.

Ans. $2/r$

3. Calculate $\nabla\cdot(3x^2i + 5xy^2j + xyz^3k)$ at the point $(1, 2, 3)$.

Ans. 80

4. If $\bar{A} = 3xyz^2\bar{i} + 2xy^3\bar{j} - x^2yz\bar{k}$ and $f = 3x^2 - yz$ find (i) $\nabla\cdot\bar{A}$ (ii) $\bar{A}\cdot\nabla f$ (iii) $\nabla\cdot(f\bar{A})$ (iv) $\nabla\cdot\nabla f$.

Ans. (i) 4 (ii) −15 (iii) 1 (iv) 6

5. Find $\nabla\cdot\left[(e^y\sin x\cos z)i + e^{-x}\sin y\cos z\,j + z^2e^z k\right]$.

Ans. $e^y\cos x\cdot\ \cos z + e^{-x}\cdot\ \cos y\cos z + (z^2 + 2z)e^z$

6. Show that $\bar{A} = 3y^4z^2\bar{i} + 4x^3z^2j - 3x^2y^2\bar{k}$ is solenoidal.
7. Prove that $\bar{A} = (2x^2 + 8xy^2z)i + (3x^3y - 3xy)\bar{j} - (4y^2z^2 + 2x^3z)\bar{k}$ is not solenoidal but $\bar{B} = xyz^2\bar{A}$ is solenoidal.
8. Determine the constant b such that $\bar{A} = (bx^2y + yz)\bar{i} + (xy^2 - xz^2)\bar{j} + (2xyz - 2x^2y^2)\bar{k}$ has zero divergence (i.e., $\nabla\cdot\bar{A} = 0$).

Ans. $b = -2$

9. Evaluate $\nabla\cdot[r\nabla(1/r^3)]$.

Ans. $3r^{-4}$

10. Find most general $f(r)$ such that $f(r)\bar{r}$ is solenoidal.

Ans. $f(r) = c/r^3$ where c is an arbitrary constant

11. Show that $\nabla f\times\nabla g$ is solenoidal.
Hint: $\nabla\cdot(\nabla f\times\nabla g) = \nabla g\cdot(\nabla\times\nabla f) - \nabla f\cdot(\nabla\times\nabla g) = 0$.
12. Prove that $\bar{A} = (y^2 - z^2 + 3yz - 2x)\bar{i} + (3xz + 2xy)\bar{j} + (3xy - 2xz + 2z)\bar{k}$ is both solenoidal and irrotational.
13. Show that $\nabla\cdot(f\bar{A}) = 5f$ where $f = x^2 + y^2 + z^2$ and $\bar{A} = x\bar{i} + y\bar{j} + z\bar{k}$.
14. Find the directional derivative of $\nabla\cdot\bar{U}$ at the point $(4, 4, 2)$ in the direction of the corresponding outer normal of the sphere $x^2 + y^2 + z^2 = 36$ where $\bar{U} = xz\bar{i} + yx\bar{j} + zy\bar{k}$.

Ans. 5/3

15. Show that the vector field $\bar{V} = \frac{a(xi+yj)}{x^2+y^2}$ is a "source" or "sink" field according as $a > 0$ or $a < 0$.
Hint: If $\nabla\cdot\bar{V} > 0$ then $\bar{V}$ is a source field and if $\nabla\cdot\bar{V} < 0$ it is a sink field.
16. If $f = x^2yz$ and $g = xy - 3z^2$, calculate $\nabla(\nabla f\cdot\nabla g)$.

Ans. $2(y^3 + 3x^2y - 6xy^2)z\bar{i} + 2(3xy^2 + x^3 - 6x^2y)z\bar{j} + 2(xy^2 + x^3 - 3x^2y)y\bar{k}$.

15.4 CURL

Curl of $\bar{A}$, denoted by $\nabla\times\bar{A}$ also known as rotation $\bar{V}$ or rot of $\bar{V}$ is defined as

$$\text{curl of } \bar{A} = \nabla\times\bar{A} = \left(i\frac{\partial}{\partial x} + j\frac{\partial}{\partial y} + k\frac{\partial}{\partial z}\right)\times(A_1i + A_2j + A_3k)$$

$$= \begin{vmatrix} i & j & k \\ \frac{\partial}{\partial x} & \frac{\partial}{\partial y} & \frac{\partial}{\partial z} \\ A_1 & A_2 & A_3 \end{vmatrix}$$

$$= i\left(\frac{\partial A_3}{\partial y} - \frac{\partial A_2}{\partial z}\right) + j\left(\frac{\partial A_1}{\partial z} - \frac{\partial A_3}{\partial x}\right) + k\left(\frac{\partial A_2}{\partial x} - \frac{\partial A_1}{\partial y}\right)$$

$\nabla \times \bar{A} =$ a vector quantity.

Irrotational Field

A vector point function $\bar{A}$ is said to be irrotational, if curl of $\bar{A}$ is zero at every point where $\bar{A}$ is defined. Otherwise it is said to be rotational. The curl of any vector point function, in general, gives the measure of the angular velocity at any point of the vector field.

WORKED OUT EXAMPLES

Example 1: Find the curl of $\bar{V} = e^{xyz}(\bar{i} + \bar{j} + \bar{k})$ at the point (1, 2, 3).

Solution: Curl of $\bar{V}$

$$= \nabla \times \bar{V} = \begin{vmatrix} i & j & k \\ \frac{\partial}{\partial x} & \frac{\partial}{\partial y} & \frac{\partial}{\partial z} \\ e^{xyz} & e^{xyz} & e^{xyz} \end{vmatrix}.$$

$$= e^{xyz}\left[(xz-xy)\bar{i}-(yz-xy)j+(yz-xz)\bar{k}\right]$$

$$\nabla \times \bar{V}\Big|_{1,2,3} = e^6\left[\bar{i} - 4j + 3\bar{k}\right].$$

Example 2: Prove that $\nabla \times \nabla f = 0$ for any $f(x, y, z)$.

Solution:

$$\nabla \times \nabla f = \begin{vmatrix} i & j & k \\ \frac{\partial}{\partial x} & \frac{\partial}{\partial y} & \frac{\partial}{\partial z} \\ \frac{\partial f}{\partial x} & \frac{\partial f}{\partial y} & \frac{\partial f}{\partial z} \end{vmatrix}$$

$$= i\left(\frac{\partial^2 f}{\partial y \partial z} - \frac{\partial^2 f}{\partial z \partial y}\right) - j\left(\frac{\partial^2 f}{\partial x \partial z} - \frac{\partial^2 f}{\partial z \partial x}\right) + k\left(\frac{\partial^2 f}{\partial x \partial y} - \frac{\partial^2 f}{\partial y \partial x}\right) = 0$$

since $f_{yz} = f_{zy}$, $f_{xz} = f_{zx}$ and $f_{xy} = f_{yx}$.

Note: Since $\bar{V}=\nabla\phi$ for a conservative field, $\nabla \times \bar{V} = \nabla \times \nabla\phi = 0$. Thus for a conservative field $\bar{V}$, we have $\nabla \times \bar{V} = 0$.

Example 3: If $f(r)$ is differentiable and $r = \sqrt{x^2 + y^2 + z^2}$ show that $f(r)\bar{r}$ is irrotational. Hence deduce that (i) $r^n\bar{r}$ is irrotational (ii) $\nabla \times \bar{r} = 0$.

Solution: Here $f(r)\bar{r} = f(r)[xi + yj + zk]$

$$\nabla \times (f(r)\bar{r}) = \begin{vmatrix} i & j & k \\ \frac{\partial}{\partial x} & \frac{\partial}{\partial y} & \frac{\partial}{\partial z} \\ xf(r) & yf(r) & zf(r) \end{vmatrix}$$

$$= i\left(\frac{z\partial f}{\partial y} - \frac{y\partial f}{\partial z}\right) - j\left(\frac{z\partial f}{\partial x} - \frac{x\partial f}{\partial z}\right) + k\left(\frac{y\partial f}{\partial x} - \frac{x\partial f}{\partial y}\right)$$

Here $\frac{\partial f(r)}{\partial y} = \frac{\partial}{\partial r} f(r) \cdot \frac{\partial r}{\partial y} = f'(r) \cdot \frac{y}{r}$

since $\frac{\partial r}{\partial y} = \frac{\partial}{\partial y}\sqrt{x^2 + y^2 + z^2}$

$$= \frac{1}{2}\frac{2y}{\sqrt{x^2 + y^2 + z^2}} = \frac{y}{r}$$

Similarly, $\frac{\partial f}{\partial x} = f'\frac{x}{r}$ and $\frac{\partial f}{\partial z} = f'\frac{z}{r}$

Substituting these values

$$\nabla \times (f\bar{r}) = i\left[zf'\frac{y}{r} - yf'\frac{z}{r}\right] - j\left[zf'\frac{x}{r} - xf'\frac{z}{r}\right] + k\left[yf'\frac{x}{r} - xf'\frac{y}{r}\right]$$

$$= 0$$

i. with $f(r) = r^n$.

$$\nabla \times (r^n\bar{r}) = \nabla(f(r)\bar{r}) = 0$$

follows from the above result.

ii. with $n = 0$, $\nabla \times \bar{r} = 0$ from above result (i).

Example 4: Prove that $\bar{A} = (6xy + z^3)i + (3x^2 - z)j + (3xz^2 - y)k$ is irrotational. Find a scalar function $f(x, y, z)$ such that $\bar{A} = \nabla f$.

Solution:

$$\nabla \times \bar{A} = \begin{vmatrix} i & j & k \\ \frac{\partial}{\partial x} & \frac{\partial}{\partial y} & \frac{\partial}{\partial z} \\ 6xy + z^3 & 3x^2 - z & 3xz^2 - y \end{vmatrix}$$

$$= i[-1 - (-1)] - j[3z^2 - 3z^2] + k[6x - 6x] = 0$$

Therefore $\bar{A}$ is irrotational.

To find f : $\bar{A} = \nabla f = \frac{\partial f}{\partial x} i + \frac{\partial f}{\partial y} j + \frac{\partial f}{\partial z} k$

comparing components of i, j, k on either side

$$\frac{\partial f}{\partial x} = 6xy + z^3 \qquad (1)$$

$$\frac{\partial f}{\partial y} = 3x^2 - z \qquad (2)$$

$$\frac{\partial f}{\partial z} = 3xz^2 - y \qquad (3)$$

Integrating (1) partially w.r.t. x, we get

$$f = 3x^2y + xz^3 + c_1(y, z) \qquad (4)$$

Differentiating (4) partially w.r.t. y and equating it with (2), we get

$$3x^2 + 0 + \frac{\partial c_1}{\partial y} = \frac{\partial f}{\partial y} = 3x^2 - z$$

i.e.,
$$\frac{\partial c_1}{\partial y} = -z \qquad (5)$$

Integrating (5) partially w.r.t. y, we have

$$c_1(y, z) = -zy + c_2(z) \qquad (6)$$

Substituting (6) in (4)

$$f = 3x^2y + xz^3 - zy + c_2(z) \qquad (7)$$

Differentiating (7) partially w.r.t. z and equating it with (3), we get

$$0 + 3xz^2 - y + \frac{dc_2}{dz} = 3xz^2 - y$$

So
$$\frac{dc_2}{dz} = 0$$

i.e., c_2 = a pure constant (independent of z)
Thus the required scalar function

$$f = 3x^2y + xz^3 - zy + c_2.$$

Example 5: Find curl curl of $\bar{A} = x^2y\bar{i} - 2xzj + 2yz\bar{k}$ at the point (1, 0, 2).

Solution:

$$\nabla \times \bar{A} = \begin{vmatrix} i & j & k \\ \frac{\partial}{\partial x} & \frac{\partial}{\partial y} & \frac{\partial}{\partial z} \\ x^2y & -2xz & 2yz \end{vmatrix}$$

$$\nabla \times \bar{A} = i[2z + 2x] - j[0 - 0] + \bar{k}[-2z - x^2]$$

Now
$$\nabla \times (\nabla \times \bar{A}) = \begin{vmatrix} i & j & k \\ \frac{\partial}{\partial x} & \frac{\partial}{\partial y} & \frac{\partial}{\partial z} \\ 2(z + x) & 0 & -(2z + x^2) \end{vmatrix}$$

$$= i[0 - 0] - j[-2x - 2] + k[0 - 0]$$

$$= 2(x + 1)\bar{j}$$

$$\nabla \times (\nabla \times A)\Big|_{\text{at } (1,0,2)} = 2(1 + i)j = 4\bar{j}.$$

EXERCISE

1. Prove that $\nabla \times \bar{r} = 0$.
2. Find the curl of $yzi + 3zxj + z\bar{k}$ at (2, 3, 4).

Ans. $-6i + 3j + 8k$

3. Find $\nabla \times \left[(yz - 2x^2y)i + x(y^2 - z^2)j + 2xy(z - xy)\bar{k}\right]$ at the point (1, 1, 1).

Ans. $4x(z - xy)\bar{i} + (y - 2yz + 4xy^2)j + (2x^2 + y^2 - z^2 - z)\bar{k}$; $3\bar{j} + \bar{k}$

4. If $f = x^2yz$, $g = xy - 3z^2$, calculate $\nabla \cdot (\nabla f \times \nabla g)$.

Ans. zero

5. Determine curl of $xyz^2i + yzx^2j + zxy^2\bar{k}$ at the point (1, 2, 3).

Ans. $xy(2z - x)\bar{i} + yz(2x - y)j + zx(2y - z)\bar{k}$; $10\bar{i} + 3k$

6. If $\bar{A}$ and $\bar{B}$ are irrotational show that $\nabla \cdot (\bar{A} \times \bar{B}) = 0$.
7. Determine the constants a and b such that curl of $(2xy + 3yz)\bar{i} + (x^2 + axz - 4z^2)j + (3xy + 2byz)k = 0$.

Ans. $a = 3, b = 4$.

8. Find the value of constant b such that

$$\bar{A} = (bxy - z^3)\bar{i} + (b - 2)x^2j + (1 - b)xz^2k$$

has its curl identically equal to zero.

Ans. $b = 4$

9. Evaluate $\nabla \times (\bar{r}r^{-2})$. Find f such that $\bar{r}r^{-2} = -\nabla f$ with $f(a) = 0$ where $a > 0$.

Ans. $f = \ln(a/r)$

10. Determine the constants a, b, c so that

$$\bar{A} = (x + 2y + az)i + (bx - 3y - z)j + (4x + cy + 2z)\bar{k}$$

is irrotational. Find a scalar function $f(x, y, z)$ such that $\bar{A} = \nabla f$.

Ans. i. $a = 4, b = 2, c = -1$
ii. $f = \frac{x^2}{2} - \frac{3y^2}{2} + z^2 + 2xy + 4xz - yz$

11. Prove that $\bar{A} = (x^2 - yz)i + (y^2 - zx)j + (z^2 - xy)k$ is irrotational and find the scalar potential f such that $\bar{A} = \nabla f$.

Ans. $f(x, y, z) = \frac{x^3+y^3+z^3}{3} - xyz$

12. Show that $\nabla \times (\nabla \times (\nabla \times (\nabla \times \bar{A}))) = \nabla^4 \bar{A}$ where $\bar{A}$ is a solenoidal vector.

13. Prove that $(y^2 - z^2 + 3yz - 2x)\bar{i} + (3xz + 2xy)j + (3xy - 2xz + 2z)$ is both solenoidal and irrotational.

14. Prove that $\nabla \cdot (\nabla \times \bar{A}) = 0$.

15.5 RELATED PROPERTIES OF GRADIENT, DIVERGENCE AND CURL OF SUMS

The gradient, divergence and curl are distributive with respect to the sum and difference of functions:

1. $\nabla(f \pm g) = \nabla f \pm \nabla g$

2. $\nabla \cdot (\bar{A} \pm \bar{B}) = (\nabla \cdot \bar{A}) \pm (\nabla \cdot \bar{B})$

3. $\nabla \times (\bar{A} \pm \bar{B}) = (\nabla \times \bar{A}) \pm (\nabla \times \bar{B})$.

The above results follow, since derivative of sum or difference of scalars or vectors is sum or difference of the derivatives of scalars or vectors. For example,

$$\nabla \cdot (\bar{A} \pm \bar{B}) = \nabla \cdot ((A_1 \pm B_1)i + (A_2 \pm B_2)j + (A_3 \pm B_3)k)$$

$$= \frac{\partial}{\partial x}(A_1 \pm B_1) + \frac{\partial}{\partial y}(A_2 \pm B_2) + \frac{\partial}{\partial z}(A_3 \pm B_3)$$

$$= \left(\frac{\partial A_1}{\partial x} + \frac{\partial A_2}{\partial y} + \frac{\partial A_3}{\partial z}\right) \pm \left(\frac{\partial B_1}{\partial x} + \frac{\partial B_2}{\partial y} + \frac{\partial B_3}{\partial z}\right)$$

$$= \nabla \cdot \bar{A} \pm \nabla \cdot \bar{B}.$$

Gradient, Divergence and Curl of Products

1. $\nabla(fg) = f\nabla g + g\nabla f$

2. $\nabla \cdot (f\bar{A}) = f\nabla \cdot \bar{A} + (\nabla f) \cdot \bar{A}$

3. $\nabla \times (f\bar{A}) = f\nabla \times \bar{A} + (\nabla f) \times \bar{A}$

4. $\nabla(\bar{A} \cdot \bar{B}) = (B \cdot \nabla)\bar{A} + (A \cdot \nabla)\bar{B} + \bar{B} \times (\nabla \times A) + A \times (\nabla \times B)$

5. $\nabla \cdot (\bar{A} \times \bar{B}) = \bar{B} \cdot (\nabla \times \bar{A}) - \bar{A} \cdot (\nabla \times \bar{B})$

6. $\nabla \times (\bar{A} \times \bar{B}) = (B \cdot \nabla)\bar{A} - B(\nabla \cdot A) - (A \cdot \nabla)B + A(\nabla \cdot B)$.

The results 1, 2, 3 follow from the fact that the derivative of a product of scalar functions is the product of the derivatives of the scalar functions.

For example,

$$\nabla \times (f\bar{A}) = \nabla \times (f(A_1 i + A_2 j + A_3 k))$$

$$= \nabla \times (fA_1 i + fA_2 j + fA_3 k)$$

$$= \begin{vmatrix} i & j & k \\ \frac{\partial}{\partial x} & \frac{\partial}{\partial y} & \frac{\partial}{\partial z} \\ fA_1 & fA_2 & fA_3 \end{vmatrix}$$

$$= i\left[\frac{\partial(fA_3)}{\partial y} - \frac{\partial}{\partial z}(fA_2)\right] - j\left[\frac{\partial}{\partial x}(fA_3) - \frac{\partial}{\partial z}(fA_1)\right] + k\left[\frac{\partial}{\partial x}(fA_2) - \frac{\partial}{\partial y}(fA_1)\right]$$

Expanding the product of the derivatives and rearranging the terms, we get

$$= f\left[i\left(\frac{\partial A_3}{\partial y} - \frac{\partial A_2}{\partial z}\right) - j\left(\frac{\partial A_3}{\partial x} - \frac{\partial A_1}{\partial z}\right) + k\left(\frac{\partial A_2}{\partial x} - \frac{\partial A_1}{\partial y}\right)\right] + i\left(A_3\frac{\partial f}{\partial y} - A_2\frac{\partial f}{\partial z}\right) - j\left(A_3\frac{\partial f}{\partial x} - A_1\frac{\partial f}{\partial z}\right) + k\left(A_2\frac{\partial f}{\partial x} - A_1\frac{\partial f}{\partial y}\right)$$

$$\nabla \times (f\bar{A}) = f\nabla \times \bar{A} + (\nabla f) \times \bar{A}.$$

Example 1: Prove that $\nabla \cdot (\bar{A} \times \bar{B}) = \bar{B} \cdot (\nabla \times \bar{A}) - \bar{A} \cdot (\nabla \times \bar{B})$.

Solution:

$$\bar{A} \times \bar{B} = \begin{vmatrix} i & j & k \\ A_1 & A_2 & A_3 \\ B_1 & B_2 & B_3 \end{vmatrix}$$

$$\bar{A} \times \bar{B} = i(A_2B_3 - A_3B_2) - j(A_1B_3 - A_3B_1) + k(A_1B_2 - A_2B_1)$$

$$\nabla \cdot (\bar{A} \times \bar{B}) = \frac{\partial}{\partial x}(A_2B_3 - A_3B_2) - \frac{\partial}{\partial y}(A_1B_3 - A_3B_1) + \frac{\partial}{\partial z}(A_1B_2 - A_2B_1).$$

Expanding the derivatives of the products and rearranging the 12 terms in to 2 groups of 6 terms each, we get

$$\nabla \cdot (\bar{A} \times \bar{B}) = \left[B_1\left(\frac{\partial A_3}{\partial y} - \frac{\partial A_2}{\partial z}\right) + B_2\left(\frac{\partial A_1}{\partial z} - \frac{\partial A_3}{\partial x}\right) + B_3\left(\frac{\partial A_2}{\partial x} - \frac{\partial A_1}{\partial y}\right)\right] - \left[A_1\left(\frac{\partial B_3}{\partial y} - \frac{\partial B_2}{\partial z}\right) + A_2\left(\frac{\partial B_1}{\partial z} - \frac{\partial B_3}{\partial x}\right) + A_3\left(\frac{\partial B_2}{\partial x} - \frac{\partial B_1}{\partial y}\right)\right]$$

$$\nabla \cdot (\bar{A} \times \bar{B}) = \bar{B} \cdot (\nabla \times \bar{A}) - \bar{A} \cdot (\nabla \times \bar{B}).$$

Example 2: Prove that

$$\nabla \times (\bar{A} \times \bar{B}) = (B \cdot \nabla)\bar{A} - \bar{B}(\nabla \cdot \bar{A}) - (\bar{A} \cdot \nabla)\bar{B} + \bar{A}(\nabla \cdot \bar{B}).$$

Solution:

$$\nabla \times (\bar{A} \times \bar{B}) = \left(\bar{i}\frac{\partial}{\partial x} + \bar{j}\frac{\partial}{\partial y} + \bar{k}\frac{\partial}{\partial z}\right) \times (\bar{A} \times \bar{B})$$
$$= i \times \frac{\partial}{\partial x}(\bar{A} \times \bar{B}) + j \times \frac{\partial}{\partial y}(\bar{A} \times \bar{B}) + \bar{k} \times \frac{\partial}{\partial z}(\bar{A} \times \bar{B})$$

Expanding the derivative of the products, we get 6 terms,

$$= \left[i \times \left(\frac{\partial \bar{A}}{\partial x} \times \bar{B}\right) + i \times \left(\bar{A} \times \frac{\partial \bar{B}}{\partial x}\right)\right] + \left[\bar{j} \times \left(\frac{\partial \bar{A}}{\partial y} \times \bar{B}\right) + \bar{j} \times \left(\bar{A} \times \frac{\partial \bar{B}}{\partial y}\right)\right] + \left[\bar{k} \times \left(\frac{\partial \bar{A}}{\partial z} \times \bar{B}\right) + \bar{k} \times \left(\bar{A} \times \frac{\partial \bar{B}}{\partial z}\right)\right] \quad (1)$$

Since $\bar{a} \times (\bar{b} \times \bar{c}) = (\bar{a} \cdot \bar{c})\bar{b} - (\bar{a} \cdot \bar{b})\bar{c}$,

$$\bar{i} \times \left(\frac{\partial \bar{A}}{\partial x} \times \bar{B}\right) = (\bar{i} \cdot \bar{B})\frac{\partial \bar{A}}{\partial x} - \left(\bar{i} \cdot \frac{\partial \bar{A}}{\partial x}\right)\bar{B}$$
$$= (\bar{B} \cdot i)\frac{\partial \bar{A}}{\partial x} - \left(\bar{i} \cdot \frac{\partial}{\partial x}\bar{A}\right)\bar{B} \quad (2)$$

We get similar results for the 3rd and 5th terms in the R.H.S. of (1). Collecting these 3 terms from the R.H.S. of (1), namely 1st, 3rd and 5th terms and using the summation notation with respect to i (and x), we get

$$i \times \left(\frac{\partial \bar{A}}{\partial x} \times \bar{B}\right) + j \times \left(\frac{\partial A}{\partial y} \times B\right) + \bar{k} \times \left(\frac{\partial A}{\partial z} \times \bar{B}\right)$$
$$= \sum \bar{i} \times \left(\frac{\partial \bar{A}}{\partial x} \times \bar{B}\right)$$
$$= \sum \left(B \cdot i\frac{\partial}{\partial x}\right)\bar{A} - \sum \left(i\frac{\partial}{\partial x} \cdot A\right)\bar{B}$$

Since the summation is with respect to i, we get

$$= \left(\bar{B} \cdot \sum i\frac{\partial}{\partial x}\right)\bar{A} - \left\{\left(\sum i\frac{\partial}{\partial x}\right) \cdot \bar{A}\right\}\bar{B}$$

$$\sum \bar{i} \times \left(\frac{\partial \bar{A}}{\partial x} \times \bar{B}\right) = (\bar{B} \cdot \nabla)\bar{A} - (\nabla \cdot \bar{A})\bar{B} \quad (3)$$

In a similar manner, interchanging the roles of $\bar{A}$ and $\bar{B}$, for the remaining 3 terms namely 2nd, 4th and 6th terms of (1), we get

$$\sum i \times \left(\bar{A} \times \frac{\partial \bar{B}}{\partial x}\right) = (\nabla \cdot \bar{B})\bar{A} - (\bar{A} \cdot \nabla)\bar{B} \quad (4)$$

Adding (3) and (4) the required result is obtained.

Example 3: Prove that

$$\nabla(\bar{A} \cdot \bar{B}) = \bar{A} \times (\nabla \times \bar{B}) + \bar{B} \times (\nabla \times \bar{A}) + (\bar{A} \cdot \nabla)\bar{B} + (\bar{B} \cdot \nabla)\bar{A}.$$

Solution: $\nabla(\bar{A}\cdot\bar{B}) = i\frac{\partial}{\partial x}(\bar{A}\cdot\bar{B}) + j\frac{\partial}{\partial y}(\bar{A}\cdot\bar{B}) + k\frac{\partial}{\partial z}(\bar{A}\cdot\bar{B})$.

Expanding the derivative of the product terms and rearranging the 6 terms, we get

$$\nabla(\bar{A}\cdot\bar{B}) = \sum i\frac{\partial}{\partial x}(\bar{A}\cdot\bar{B})$$

$$= \left(\sum i\frac{\partial}{\partial x}\bar{A}\right)\cdot\bar{B} + \bar{A}\cdot\sum i\frac{\partial \bar{B}}{\partial x} \quad (1)$$

Consider

$$\bar{A}\times(\nabla\times\bar{B}) = \bar{A}\times\left(\left(\sum i\frac{\partial}{\partial x}\right)\times\bar{B}\right)$$

$$= \bar{A}\times\left(\sum i\times\frac{\partial\bar{B}}{\partial x}\right)$$

Using triple cross product result

$$= \sum\left(\bar{A}\cdot\frac{\partial\bar{B}}{\partial x}\right)i - \left(\bar{A}\cdot\sum i\right)\frac{\partial\bar{B}}{\partial x}$$

$$= \sum i\left(\bar{A}\cdot\frac{\partial\bar{B}}{\partial x}\right) - \left(\bar{A}\cdot\sum i\frac{\partial}{\partial x}\right)\bar{B}$$

$$\bar{A}\times(\nabla\times\bar{B}) = \sum i\left(\frac{\partial\bar{B}}{\partial x}\cdot\bar{A}\right) - (\bar{A}\cdot\nabla)\bar{B}$$

Rewriting, we have

$$\bar{A}\times(\nabla\times\bar{B}) + (\bar{A}\cdot\nabla)\bar{B} = \left(\sum i\frac{\partial\bar{B}}{\partial x}\right)\cdot\bar{A}$$

$$= \bar{A}\cdot\left(\sum i\frac{\partial\bar{B}}{\partial x}\right) \quad (2)$$

Similarly (interchanging the roles of $\bar{A}$ and $\bar{B}$), we get

$$\bar{B}\times(\nabla\times\bar{A}) + (\bar{B}\cdot\nabla)\bar{A} = \bar{B}\cdot\left(\sum i\frac{\partial\bar{A}}{\partial x}\right) \quad (3)$$

Addition of (2) and (3) gives the desired result.

15.6 SECOND-ORDER DIFFERENTIAL OPERATOR

It is a two-fold application of the operator ∇ to function.

Laplacian Operator ∇^2

$$\text{div grad } f = \nabla\cdot(\nabla f)$$

$$= \nabla\cdot\left(i\frac{\partial f}{\partial x} + j\frac{\partial f}{\partial y} + k\frac{\partial f}{\partial z}\right)$$

$$= \frac{\partial}{\partial x}\left(\frac{\partial f}{\partial x}\right) + \frac{\partial}{\partial y}\left(\frac{\partial f}{\partial y}\right) + \frac{\partial}{\partial z}\left(\frac{\partial f}{\partial z}\right)$$

$$= \frac{\partial^2 f}{\partial x^2} + \frac{\partial^2 f}{\partial y^2} + \frac{\partial^2 f}{\partial z^2}$$

$$= \left(\frac{\partial^2}{\partial x^2} + \frac{\partial^2}{\partial y^2} + \frac{\partial^2}{\partial z^2}\right)f$$

$$= \nabla^2 f = \Delta f$$

Thus the scalar differential operator (read as "nabla squared" or "delta")

$$\nabla^2 = \Delta = \frac{\partial^2}{\partial x^2} + \frac{\partial^2}{\partial y^2} + \frac{\partial^2}{\partial z^2}$$

is known as the Laplacian operator.

Thus we have following second order differential operators:

1. $\nabla\cdot\nabla f$ = div grad $f = \nabla^2 f = \Delta f = \frac{\partial^2 f}{\partial x^2} + \frac{\partial^2 f}{\partial y^2} + \frac{\partial^2 f}{\partial z^2}$
2. $\nabla\times\nabla f$ = curl grad $f = 0$
3. $\nabla\cdot\nabla\times\bar{A}$ = div cur $\bar{A} = 0$
4. $\nabla\times(\nabla\times\bar{A})$ = curl curl $\bar{A} = \nabla(\nabla\cdot\bar{A}) - \nabla^2\bar{A}$
   (4) may be rewritten as
5. $\nabla(\nabla\cdot\bar{A})$ = grad div $\bar{A} = \nabla\times(\nabla\times\bar{A}) + \nabla^2\bar{A}$

The possible combinations of second order differential operators are tabulated below:

	Scalar field f	Vector field $\bar{A}$	
	grad	**div**	**curl**
grad	—	grad div$\bar{A}$	—
div	div grad f $= \Delta f$	—	div curl $\bar{A} = 0$
curl	curl grad $f = 0$	—	curl curl $\bar{A}$ = grad div $\bar{A}$ $-\Delta\bar{A}$

Example 1: Prove that $\nabla\times\nabla f = 0$ for any scalar function f.

Solution: $\nabla f = \frac{\partial f}{\partial x} i + \frac{\partial f}{\partial y} j + \frac{\partial f}{\partial z} \bar{k}$ so

$$\nabla \times \nabla f = \begin{vmatrix} i & j & k \\ \frac{\partial}{\partial x} & \frac{\partial}{\partial y} & \frac{\partial}{\partial z} \\ \frac{\partial f}{\partial x} & \frac{\partial f}{\partial y} & \frac{\partial f}{\partial z} \end{vmatrix}$$

$$= i\left(\frac{\partial^2 f}{\partial y \partial z} - \frac{\partial^2 f}{\partial z \partial y}\right) - j\left(\frac{\partial^2 f}{\partial x \partial z} - \frac{\partial^2 f}{\partial z \partial x}\right) + k\left(\frac{\partial^2 f}{\partial x \partial y} - \frac{\partial^2 f}{\partial y \partial z}\right)$$

$$= 0 + 0 + 0 = \bar{0}$$

Since $f_{yz} = f_{zy}, f_{xz} = f_{zx}, f_{xy} = f_{yx}$.

Note: Gradient field describing a motion, in this case, is known as "irrotational".

If gradient field is not a velocity field, then it is known as "conservative".

Example 2: Prove that $\nabla \cdot (\nabla \times \bar{A}) = 0.$ for any vector function $\bar{A}$.

Solution:

$$\nabla \times \bar{A} = i\left(\frac{\partial A_3}{\partial y} - \frac{\partial A_2}{\partial z}\right) - j\left(\frac{\partial A_3}{\partial x} - \frac{\partial A_1}{\partial z}\right) + k\left(\frac{\partial A_2}{\partial x} - \frac{\partial A_1}{\partial y}\right)$$

So that

$$\nabla \cdot (\nabla \times \bar{A}) = \frac{\partial}{\partial x}\left[\left(\frac{\partial A_3}{\partial y} - \frac{\partial A_2}{\partial z}\right)\right] + \frac{\partial}{\partial y}\left[\frac{\partial A_1}{\partial z} - \frac{\partial A_3}{\partial x}\right] + \frac{\partial}{\partial z}\left(\frac{\partial A_2}{\partial x} - \frac{\partial A_1}{\partial y}\right)$$

$$= \frac{\partial^2 A_3}{\partial x \partial y} - \frac{\partial^2 A_2}{\partial x \partial z} + \frac{\partial^2 A_1}{\partial y \partial z} - \frac{\partial^2 A_3}{\partial y \partial x} + \frac{\partial^2 A_2}{\partial z \partial x} - \frac{\partial^2 A_1}{\partial z \partial y} = 0.$$

Example 3: Show that $\nabla \times (\nabla \times \bar{A}) = \nabla(\nabla \cdot \bar{A}) - \nabla^2 \bar{A}$.

Solution:

$$\nabla \times (\nabla \times \bar{A}) = \begin{vmatrix} i & j & k \\ \frac{\partial}{\partial x} & \frac{\partial}{\partial y} & \frac{\partial}{\partial z} \\ \frac{\partial A_3}{\partial y} - \frac{\partial A_2}{\partial z} & \frac{\partial A_1}{\partial z} - \frac{\partial A_3}{\partial x} & \frac{\partial A_2}{\partial x} - \frac{\partial A_1}{\partial y} \end{vmatrix}$$

Expanding the determinant, we have

$$\nabla \times (\nabla \times \bar{A}) = \left(\frac{\partial^2 A_2}{\partial y \partial x} - \frac{\partial^2 A_1}{\partial y^2} - \frac{\partial^2 A_1}{\partial z^2} + \frac{\partial^2 A_3}{\partial z \partial x}\right)\bar{i} + \left(\frac{\partial^2 A_3}{\partial z \partial y} - \frac{\partial^2 A_2}{\partial z^2} - \frac{\partial^2 A_2}{\partial x^2} + \frac{\partial^2 A_1}{\partial x \partial y}\right)\bar{j} + \left(\frac{\partial^2 A_1}{\partial x \partial z} - \frac{\partial^2 A_3}{\partial x^2} - \frac{\partial^2 A_3}{\partial y^2} + \frac{\partial^2 A_2}{\partial y \partial z}\right)\bar{k}$$

Rearranging the 12 terms into 2 groups of 6 terms each, we get

$$\nabla \times (\nabla \times \bar{A}) = \left[i\frac{\partial}{\partial x}\left\{\frac{\partial A_1}{\partial x} + \frac{\partial A_2}{\partial y} + \frac{\partial A_3}{\partial z}\right\} - i\left(\frac{\partial^2 A_1}{\partial x^2} + \frac{\partial^2 A_1}{\partial y^2} + \frac{\partial^2 A_1}{\partial z^2}\right)\right]$$
$$+ \left[j\frac{\partial}{\partial y}\left\{\frac{\partial A_1}{\partial x} + \frac{\partial A_2}{\partial y} + \frac{\partial A_3}{\partial z}\right\} - j\left(\frac{\partial^2 A_2}{\partial x^2} + \frac{\partial^2 A_2}{\partial y^2} + \frac{\partial^2 A_2}{\partial z^2}\right)\right]$$
$$+ \left[k\frac{\partial}{\partial z}\left\{\frac{\partial A_1}{\partial x} + \frac{\partial A_2}{\partial y} + \frac{\partial A_3}{\partial z}\right\} - k\left(\frac{\partial^2 A_3}{\partial x^2} + \frac{\partial^2 A_3}{\partial y^2} + \frac{\partial^2 A_3}{\partial z^2}\right)\right]$$
$$= \left(\bar{i}\frac{\partial}{\partial x} + \bar{j}\frac{\partial}{\partial y} + \bar{k}\frac{\partial}{\partial z}\right) \times \left(\frac{\partial A_1}{\partial x} + \frac{\partial A_2}{\partial y} + \frac{\partial A_3}{\partial z}\right) - \left(\frac{\partial^2}{\partial x^2} + \frac{\partial^2}{\partial y^2} + \frac{\partial^2}{\partial z^2}\right)(A_1\bar{i} + A_2\bar{j} + A_3\bar{k})$$

$$\nabla \times (\nabla \times \bar{A}) = \nabla(\nabla \cdot \bar{A}) - \nabla^2 \bar{A}.$$

Example 4: Prove that $\nabla \times (f\nabla g) = \nabla f \times \nabla g = -\nabla \times (g\nabla f)$ and deduce that $\nabla \times (f\nabla f) = 0$

Solution: $\nabla \times (f\nabla g) = \nabla f \times \nabla g + f\nabla \times \nabla g = \nabla f \times \nabla g$, also

$$-\nabla \times (g\nabla f) = -\nabla g \times \nabla f - g\nabla \times \nabla f$$

$$= \nabla f \times \nabla g - 0$$

since $\nabla \times \nabla g = 0$.

Taking $f = g$, $\nabla \times (f\nabla f) = \nabla f \times \nabla f = 0$.

Example 5: Prove that $\nabla \cdot (f\nabla g \times g\nabla f) = 0$.

Solution: Since $\nabla \cdot (\bar{A} \times \bar{B}) = \bar{B} \cdot (\nabla \times \bar{A}) - \bar{A} \cdot (\nabla \times \bar{B})$

$$\nabla \cdot (f\nabla g \times g\nabla f) = g\nabla f \cdot (\nabla \times (f\nabla g)) - f\nabla g \cdot (\nabla \times (g\nabla f)) = 0$$

Since $\nabla \times (f\nabla g) = \nabla f \times \nabla g$ and $\nabla \times (g\nabla f) = -\nabla f \times \nabla g$, from just above example.

Example 6: Prove that

$$\nabla \cdot (f\nabla \times \bar{A}) = \nabla f \cdot (\nabla \times \bar{A}).$$

Solution:

$$\nabla \cdot (f\nabla \times \bar{A}) = \nabla f \cdot (\nabla \times \bar{A}) + f\nabla \cdot (\nabla \times \bar{A}) = \nabla f \cdot (\nabla \times \bar{A})$$

Since $\nabla \cdot (\nabla \times \bar{A}) = 0$.

WORKED OUT EXAMPLES

Laplacian ∇^2

Example 1: Calculate $\nabla^2 f$ when $f = 3x^2z - y^2z^3 + 4x^3y + 2x - 3y - 5$ at the point (1, 1, 0).

Solution:

$$\nabla^2 f = \left(\frac{\partial^2}{\partial x^2} + \frac{\partial^2}{\partial y^2} + \frac{\partial^2}{\partial z^2}\right) \times \left(3x^2z - y^2z^3 + 4x^3y + 2x - 3y - 5\right)$$

Consider

$$\frac{\partial}{\partial x}(3x^2z - y^2z^3 + 4x^3y + 2x - 3y - 5) = 6xz + 12x^2y + 2$$

$$\frac{\partial^2}{\partial x^2}(3x^2z - y^2z^3 + 4x^3y + 2x - 3y - 5) = 6z + 24xy.$$

Similarly,

$$\frac{\partial^2}{\partial y^2} f = -2z^3$$

and

$$\frac{\partial^2}{\partial z^2} f = -6y^2z$$

Thus substituting these values, we have

$$\nabla^2 f = 6z + 24xy - 2z^3 - 6y^2z$$

$\nabla^2 f$ at the point (1, 1, 0) is $0+24 \cdot 1 \cdot 1+0+0 = 24$.

Example 2: Prove that

a. $\nabla^2 f(r) = \frac{d^2 f}{dr^2} + \frac{2}{r}\frac{df}{dr}$

b. Find $f(r)$ such that $\nabla^2 f(r) = 0$.

Solution:

a. $\nabla^2 f(r) = \nabla \cdot \nabla f(r)$

Since $\frac{\partial f}{\partial x} = \frac{\partial f}{\partial r}\frac{\partial r}{\partial x} = \frac{\partial f}{\partial r}\frac{x}{r}$

$$\nabla f(r) = \frac{\partial f}{\partial r}(xi + yj + zk)\frac{1}{r}$$

i.e., $$\nabla f(r) = \frac{\bar{r}}{r}\frac{df}{dr} = \bar{r}\left(\frac{f'(r)}{r}\right) \quad (1)$$

Using (1), we have

$$\nabla^2 f(r) = \nabla \cdot (\nabla f(r)) = \nabla \cdot \left(\bar{r}\frac{f'}{r}\right) = \nabla \cdot \left(\bar{r}\left(\frac{f'}{r}\right)\right)$$

Applying the result

$\nabla \cdot (\bar{A}f) = f(\nabla \cdot \bar{A}) + (\nabla f) \cdot \bar{A}$

$$\nabla^2 f = \nabla \cdot \left(\bar{r}\left(\frac{f'}{r}\right)\right) = \frac{f'}{r}\nabla \cdot \bar{r} + \left(\nabla \frac{f'}{r}\right) \cdot \bar{r} \quad (2)$$

Consider

$$\nabla\left(\frac{f'}{r}\right) = \nabla(f'r^{-1}) = r^{-1}\nabla f' + f'\nabla r^{-1}$$

Using (1) for f', we get

$$\nabla f' = \bar{r}\frac{f''}{r}$$

and $$\nabla r^{-1} = -1 \cdot r^{-1-2} \cdot \bar{r} = -\frac{\bar{r}}{r^3}$$

We have

$$\nabla\left(\frac{f'}{r}\right) = r^{-1}\frac{f''}{r}\bar{r} + f'\left(-\frac{\bar{r}}{r^3}\right) = \left(\frac{f''}{r^2} - \frac{f'}{r^3}\right)\bar{r} \quad (3)$$

Also $$\nabla \cdot \bar{r} = 3 \quad (4)$$

Substituting (3) and (4) in (2), we get

$$\nabla^2 f = 3\frac{f'}{r} + \left(\frac{f''}{r^2} - \frac{f'}{r^3}\right)\overline{r}\cdot\overline{r}$$

$$= 3\frac{f'}{r} + \left(\frac{f''}{r^2} - \frac{f'}{r^3}\right)r^2$$

$$\nabla^2 f(r) = f'' + \frac{2}{r}f' = \frac{d^2 f}{dr^2} + \frac{2}{r}\frac{df}{dr}$$

b. Since $\nabla^2 f(r) = \frac{d^2 f}{dr^2} + \frac{2}{r}\frac{df}{dr} = 0$

This is a 2nd order homogeneous equation which is separable in f'

$$\frac{df'}{dr} + \frac{2}{r}f' = 0$$

with solution $f' = \frac{c}{r^2}$.

Integrating w.r.t. r

$$f(r) = B + \frac{A}{r}$$

where A and B are arbitrary constants.

Example 3: Prove that

i. $\nabla^2 r^n = n(n+1)r^{n-2}$ where n is a constant

ii. $\nabla^2 r^2 = 6$

iii. $\nabla^2\left(\frac{1}{r}\right) = 0$

iv. $\nabla^2 \ln r = \frac{1}{r^2}$

vi. $\nabla^2(gh) = g\nabla^2 h + h\nabla^2 g$.

Solution:

i. With $f(r) = r^n$, from the result from above Example 3 (i), we get

$$\nabla^2 f = \nabla^2 r^n = \frac{d^2}{dr^2}(r^n) + \frac{2}{r}\frac{d}{dr}(r^n)$$

$$= \frac{d}{dr}(n\cdot r^{n-1}) + \frac{2}{r}\cdot nr^{n-1}$$

$$= n\cdot(n-1)r^{n-2} + 2nr^{n-2}$$

$$= nr^{n-2}[n-1+2] = n(n+1)r^{n-2}$$

ii. Put $n = 2$ in (i) $\nabla^2 r^2 = 2(2+1)r^{2-2} = 6$

iii. Put $n = -1$ in (i)

$$\nabla^2\left(\frac{1}{r}\right) = (-1)(-1+1)r^{-1-2} = 0$$

iv. With $f(r) = \ln r$

$$\nabla^2 f = \nabla^2 \ln r = \frac{d^2}{dr^2}\ln r + \frac{2}{r}\frac{d}{dr}\ln r$$

$$= -\frac{1}{r^2} + \frac{2}{r}\cdot\frac{1}{r} = \frac{1}{r^2}$$

v. With $f = gh$

$$\nabla^2 f = \nabla^2(gh) = \frac{d^2}{dr^2}(gh) + \frac{2}{r}\frac{d}{dr}(gh)$$

$$= 2\frac{dg}{dr}\frac{dh}{dr} + g\frac{d^2h}{dr^2} + h\frac{d^2g}{dr^2} + \frac{2}{r}\left\{g\frac{dh}{dr} + h\frac{dg}{dr}\right\}$$

$$= 2\frac{dg}{dr}\frac{dh}{dr} + g\left[\frac{d^2h}{dr^2} + \frac{2}{r}\frac{dh}{dr}\right] + h\left[\frac{d^2g}{dr^2} + \frac{2}{r}\frac{dg}{dr}\right]$$

$$= 2\nabla g\cdot\nabla h + g\nabla^2 h + h\nabla^2 g$$

where we have used result of above Example 2(a).

Aliter: The above examples can also be solved directly. For example consider

$$\nabla^2 \ln r = \left(\frac{\partial^2}{\partial x^2} + \frac{\partial^2}{\partial y^2} + \frac{\partial^2}{\partial z^2}\right)\ln r$$

$$= \left(\frac{\partial^2}{\partial x^2} + \frac{\partial^2}{\partial y^2} + \frac{\partial^2}{\partial z^2}\right)\ln\sqrt{(x^2+y^2+z^2)} \quad (1)$$

Consider

$$\frac{\partial}{\partial x}\ln\sqrt{x^2+y^2+z^2}$$

$$= \frac{1}{\sqrt{x^2+y^2+z^2}}\cdot\frac{1}{2}\cdot\frac{1}{\sqrt{x^2+y^2+z^2}}\cdot 2x$$

$$= \frac{x}{r^2}$$

$$\frac{\partial^2}{\partial x^2}\ln\sqrt{x^2+y^2+z^2}$$

$$= \frac{\partial}{\partial x}\left[\frac{x}{r^2}\right] = \frac{r^2\cdot 1 - x\cdot 2r\frac{\partial r}{\partial x}}{r^4}$$

$$= \frac{r^2 - 2xr\frac{x}{r}}{r^4} = \frac{r^2 - 2x^2}{r^4} \quad (2)$$

Similarly,

$$\frac{\partial^2}{\partial y^2}\ln\sqrt{x^2+y^2+z^2} = \frac{r^2 - 2y^2}{r^4} \quad (3)$$

and

$$\frac{\partial^2}{\partial z^2}\ln\sqrt{x^2+y^2+z^2} = \frac{r^2 - 2z^2}{r^4} \quad (4)$$

Substituting (2), (3), (4) in (1), we get

$$\nabla^2 \ln r = \frac{r^2-2x^2}{r^4} + \frac{r^2-2y^2}{r^4} + \frac{r^2-2z^2}{r^4}$$

$$= \frac{3r^2-2(x^2+y^2+z^2)}{r^4} = \frac{3r^2+2r^2}{r^4} = \frac{r^2}{r^4} = \frac{1}{r^2}.$$

Example 4: Prove that

$$\nabla^2(fg) = f\nabla^2 g + 2\nabla g \cdot \nabla f + g\nabla^2 f.$$

Solution:

$$\nabla^2(fg) = \nabla \cdot \nabla(fg)$$

$$= \nabla \cdot [f\nabla g + g\nabla f]$$

$$= \nabla \cdot (f\nabla g) + \nabla \cdot (g\nabla f)$$

$$= [f\nabla \cdot \nabla g + \nabla f \cdot \nabla g] + [g\nabla \cdot \nabla f + \nabla g \cdot \nabla f]$$

$$\nabla^2 fg = f\nabla^2 g + 2\nabla f \cdot \nabla g + g\nabla^2 f.$$

Example 5: Find the directional derivative of $\nabla \cdot (\nabla f)$ at the point $(1, -2, 1)$ in the direction of the normal to the surface $xy^2z = 3x + z^2$ where $f = 2x^3y^2z^4$.

Solution: Let

$$U(x, y, z) = \nabla \cdot \nabla f = \nabla^2 f = \nabla^2(2x^3y^2z^4)$$

$$= \left(\frac{\partial^2}{\partial x^2} + \frac{\partial^2}{\partial y^2} + \frac{\partial^2}{\partial z^2}\right)(2x^3y^2z^4)$$

$$= 2 \cdot [6xy^2z^4 + 2x^3z^4 + 12x^3y^2z^2]$$

Normal to the surface $g = xy^2z - 3x - z^2 = 0$ is

$$\nabla g = (y^2z-3)\bar{i} + (2xyz)\bar{j} + (xy^2-2z)\bar{k}$$

$$\nabla g\Big|_{\text{at}(1,-2,1)} = \bar{i} - 4\bar{j} + 2\bar{k}$$

Unit vector $\hat{a}$ in the direction of normal at the point P $(1, -2, 1)$ is

$$\hat{a} = \frac{\nabla g}{|\nabla g|} = \frac{i-4j+2k}{\sqrt{1+16+4}} = \frac{i-4j+2k}{\sqrt{21}}$$

Consider

$$\nabla U = \nabla(12xy^2z^4 + 4x^3z^4 + 24x^3y^2z^2)$$

$$= (12y^2z^4 + 12x^2z^4 + 72x^2y^2z^2)\bar{i}$$

$$+(24xyz^4 + 48x^3yz^2)j$$

$$+(48xy^2z^3 + 16x^3z^3 + 48x^3y^2z)\bar{k}$$

$$\nabla U\Big|_{\text{at}(1,-2,1)} = 348i - 144j + 400\bar{k}$$

Thus the required directional derivative is

$$\nabla U \cdot \hat{a} = (348\bar{i} - 144\bar{j} + 400\bar{k}) \cdot \frac{(i-4j+2k)}{\sqrt{21}} = \frac{1724}{\sqrt{21}}.$$

Example 6: Evaluate $\nabla^2\left[\nabla \cdot \left(\frac{\bar{r}}{r^2}\right)\right]$.

Solution: Consider

$$\nabla \cdot \left(\frac{\bar{r}}{r^2}\right) = \nabla \cdot (r^{-2}\bar{r})$$

$$= r^{-2}\nabla \cdot \bar{r} + \bar{r} \cdot \nabla r^{-2}$$

$$= 3r^{-2} + \bar{r} \cdot (-2r^{-4}\bar{r})$$

$$= 3r^{-2} - 2r^{-4}\bar{r} \cdot \bar{r}$$

$$\nabla \cdot \left(\frac{\bar{r}}{r^2}\right) = 3r^{-2} - 2r^{-4}r^2 = r^{-2}$$

Now

$$\nabla^2\left[\nabla \cdot \left(\frac{\bar{r}}{r^2}\right)\right] = \nabla^2(r^{-2})$$

Applying result (i) of Example 3 above with $n = -2$

$$= -2(-2+1)r^{-2-2} = 2r^{-4}.$$

EXERCISE

Laplacian ∇^2

1. Show that $\nabla \cdot (\nabla f) = \frac{\partial^2 f}{\partial x_2} + \frac{\partial^2 f}{\partial y^2} + \frac{\partial^2 f}{\partial z^2}$
2. Calculate $\nabla^2 f$ when $f = 4x^2 + 9y^2 + z^2$

Ans. 28

3. Find $\nabla^2 f$ at the point $(2, 3, 1)$ when $f = xy/z$

Ans. $2xy/z^3$; 12

4. If $\bar{F} = r^a\bar{r}$ prove that

$$\nabla^2\bar{F} = a(a+3)r^{n-2}\bar{F}$$

5. Show that ∇f is both solenoidal and irrotational if $\nabla^2 f = 0$.

15.7 CURVILINEAR COORDINATES: CYLINDRICAL AND SPHERICAL COORDINATES

It is more convenient in many problems to define the position of a point P in space by three numbers (q_1, q_2, q_3) instead of the three cartesian coordinates (x, y, z). Then q_1, q_2, q_3 are known as "curvilinear coordinates" of the point P. The three surfaces $q_1 = c_1, q_2 = c_2$ and $q_3 = c_3$, (refer Fig. 15.2)where c_1, c_2, c_3 are constants, are known as "coordinate surfaces" of the system of curvilinear coordinates. On these coordinates surfaces, say $q_1 = c_1$, one of the coordinates, here q_1, remains constant.

The "coordinate curves (lines) (axis)" are the curves (or lines) of intersection of any two coordinate surfaces. Thus on the coordinate curve say which is the intersection of $q_2 = c_2$ and $q_3 = c_3$, only q_1 varies, while q_2 and q_3 remain constant. Suppose the rectangular coordinates (x, y, z) of any point P in space be expressed as functions of (q_1, q_2, q_3) so that

$$x = x(q_1, q_2, q_3), y = y(q_1, q_2, q_3),$$
$$z = z(q_1, q_2, q_3) \quad (1)$$

Solving (1) for q_1, q_2, q_3 in terms of x, y, z, we get

$$q_1 = q_1(x, y, z), q_2 = q_2(x, y, z),$$
$$q_3 = q_3(x, y, z) \quad (2)$$

The set of Equations (1) and (2) are known as "transformation of coordinates".

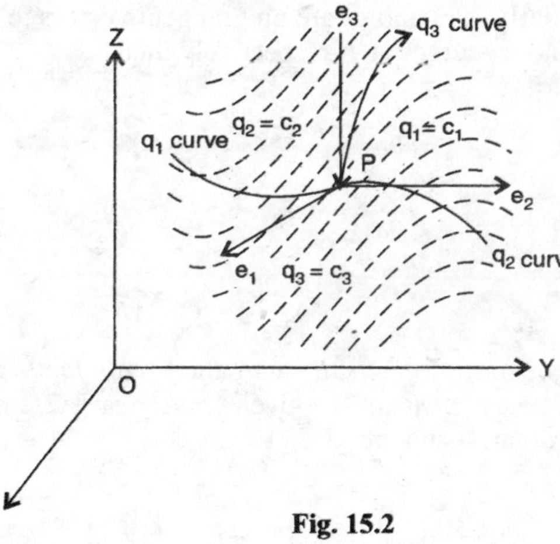

Fig. 15.2

If the coordinate surfaces intersect at right angles (and therefore the coordinate lines are at right angles), then the curvilinear coordinate system is known as "orthogonal curvilinear system of coordinates".

Let $\overline{e}_1, \overline{e}_2, \overline{e}_3$ be unit vectors directed along the tangents to the coordinates axes q_1, q_2, q_3 at the point P in the direction of increasing q_1, q_2, q_3 respectively, such that $\overline{e}_1, \overline{e}_2, \overline{e}_3$ form a right-handed trihedral (triad).

Example: Rectangular cartesian coordinate system x, y, z, where the three coordinate surfaces are planes $x = c_1, y = c_2, z = c_3$ which are mutually at right angles.

Note: The basic difference between curvilinear coordinates and cartesian coordinates is that the unit vectors $\overline{i}, \overline{j}, \overline{k}$ in the cartesian coordinate system remain constant and are same for all points of space, while in any other system the unit vectors $\overline{e}_1, \overline{e}_2, \overline{e}_3$, generally speaking, are not constant i.e., change their directions when passing from one point P to the other.

Example: Cylindrical coordinates (refer Fig. 15.3)

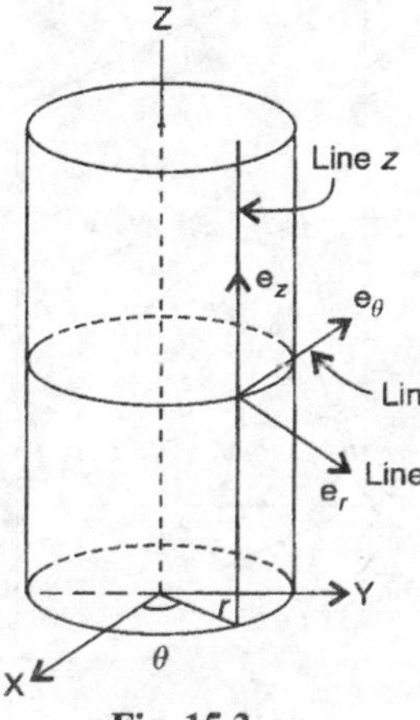

Fig. 15.3

$$\left.\begin{array}{ll} q_1 = r, & 0 \le r < \infty \\ q_2 = \theta, & 0 \le \theta < 2\pi \\ q_3 = z, & -\infty < z < +\infty \end{array}\right\} \quad (3)$$

coordinate surfaces are

$r =$ constant: circular cylinders coaxial with z-axis

$\theta =$ constant: half plane, adjoining z-axis, through z-axis

$z =$ constant: plane perpendicular to z-axis

coordinate lines (or axes) are

r: rays with origin on z-axis and perpendicular to z-axis

θ: circles with centre on z-axis and lying in planes perpendicular to z-axis

z: straight lines parallel to the z-axis

The transformation that relate cartesian coordinates to cylindrical coordinates are

$$\left.\begin{aligned} x &= r\cos\theta \\ y &= r\sin\theta \\ z &= z \end{aligned}\right\} \tag{4}$$

$r = \sqrt{x^2+y^2}$, $\theta = \text{arc}\,\tan\frac{y}{x}$, $z = z$

Example: Spherical coordinates (see Fig. 15.4)

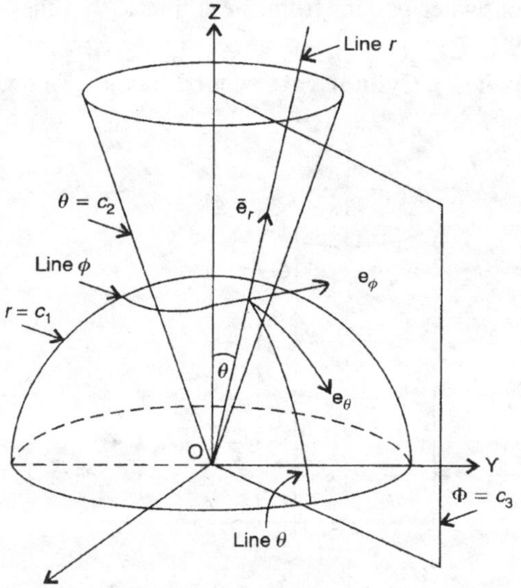

Fig. 15.4

$$\begin{aligned} q_1 &= r, & 0 \le r < +\infty \\ q_2 &= \theta, & 0 \le \theta \le \pi \\ q_3 &= \phi, & 0 \le \phi < 2\pi \end{aligned} \tag{5}$$

The coordinate surfaces are

$r = c_1$, spheres centred at origin 0

$\theta = c_2$, circular half-angle cones with z-axis with vertex at origin.

$\phi = c_3$, (half) planes adjoining the z-axis through z-axis.

coordinate lines are:

r: rays emanating from origin 0

θ: meridians on a sphere

ϕ: parallel on a sphere

cartesian coordinates are related to spherical coordinates as follows:

$$\left.\begin{aligned} x &= r\cos\phi\ \sin\theta \\ y &= r\sin\phi\ \sin\theta \\ z &= r\cos\theta. \end{aligned}\right\} \tag{6}$$

Unit Vectors in Curvilinear System

Suppose $\overline{r} = \overline{r}(q_1, q_2, q_3)$ be the position vector of a point P. A tangent vector of the q_1 curve at P (for which q_2 and q_3 are constants) is $\frac{\partial \overline{r}}{\partial q_1}$. Then a unit tangent vector $\overline{e}_1$ in this direction is

$$\overline{e_1} = \frac{\partial \overline{r}}{\partial q_1} \Big/ \left|\frac{\partial \overline{r}}{\partial q_1}\right|$$

so that

$$\frac{\partial \overline{r}}{\partial q_1} = h_1 \overline{e_1} \tag{7}$$

where

$$h_1 = \left|\frac{\partial \overline{r}}{\partial u_1}\right|.$$

Similarly if $\overline{e_2}$ and $\overline{e_3}$ are unit tangent vectors to the u_2 and u_3 curves at P respectively then

$$\frac{\partial \overline{r}}{\partial q_2} = h_2 \overline{e_2} \tag{8}$$

and

$$\frac{\partial \overline{r}}{\partial q_3} = h_3 \overline{e_3} \tag{9}$$

where

$$h_2 = \left|\frac{\partial \overline{r}}{\partial q_2}\right|, \quad h_3 = \left|\frac{\partial \overline{r}}{\partial q_3}\right|.$$

The quantities h_1, h_2, h_3 are called *scale factors or Lame coefficients* of the given curvilinear system of coordinates, and are given by

$$h_i = \sqrt{\left(\frac{\partial x}{\partial q_i}\right)^2 + \left(\frac{\partial y}{\partial q_i}\right)^2 + \left(\frac{\partial z}{\partial z_i}\right)^2};$$

$$i = 1, 2, 3 \tag{10}$$

Thus

$$d\overline{r} = \frac{\partial \overline{r}}{\partial q_1} dq_1 + \frac{\partial \overline{r}}{\partial q_2} dq_2 + \frac{\partial \overline{r}}{\partial q_3} dq_3$$

or

$$d\overline{r} = h_1 dq_1 \overline{e}_1 + h_2 dq_2 \overline{e}_2 + h_3 dq_3 \overline{e}_3 \quad (11)$$

Example: In rectangular coordinate system (q_1, q_2, q_3) is replaced by (x, y, z). Here

$$h_1 = h_2 = h_3 = 1, \quad \overline{e}_1 = \overline{i}, \quad \overline{e}_2 = \overline{j}, \quad \overline{e}_3 = \overline{k}$$

Example: Cylindrical coordinates

$$q_1 = r, \quad q_2 = \theta, \quad q_3 = z$$

By virtue of (10)

$$h_1 = H_r = \sqrt{\left(\frac{\partial x}{\partial r}\right)^2 + \left(\frac{\partial y}{\partial r}\right)^2 + \left(\frac{\partial z}{\partial r}\right)^2} = 1$$

$$h_2 = H_\theta = \sqrt{\left(\frac{\partial x}{\partial \theta}\right)^2 + \left(\frac{\partial y}{\partial \theta}\right)^2 + \left(\frac{\partial z}{\partial \theta}\right)^2} = r$$

$$h_3 = H_z = \sqrt{\left(\frac{\partial x}{\partial z}\right)^2 + \left(\frac{\partial y}{\partial z}\right)^2 + \left(\frac{\partial z}{\partial z}\right)^2} = 1$$

Example: Spherical coordinates

$$q_1 = r, q_2 = \theta, q_3 = \phi$$

Using (10), we get

$$h_1 = H_r = 1, h_2 = H_\theta = r, h_3 = H_\phi = r \sin\theta.$$

Expressions of Gradient, Divergence, Curl and Laplacian In Orthogonal Curvilinear, Spherical and Cylindrical Coordinates

WORKED OUT EXAMPLES

Example 1: Derive an expression for ∇f in orthogonal curvilinear coordinates. Hence deduce ∇ in rectangular cartesian coordinates.

Solution: Let

$$\nabla f = f_1 \overline{e}_1 + f_2 \overline{e}_2 + f_3 \overline{e}_3 \quad (1)$$

where f_1, f_2, f_3 are unknowns to be determined. Since

$$d\overline{r} = \frac{\partial \overline{r}}{\partial q_1} dq_1 + \frac{\partial \overline{r}}{\partial q_2} dq_2 + \frac{\partial \overline{r}}{\partial q_3} dq_3$$

$$d\overline{r} = h_1 \overline{e}_1 dq_1 + h_2 \overline{e}_2 dq_2 + h_3 \overline{e}_3 dq_3 \quad (2)$$

By taking dot product of (1) and (2), we have

$$df = \nabla f \cdot d\overline{r} = h_1 f_1 dq_1 + h_2 f_2 dq_2 + h_3 f_3 dq_3 \quad (3)$$

But by definition of differential,

$$df = \frac{\partial f}{\partial q_1} dq_1 + \frac{\partial f}{\partial q_2} dq_2 + \frac{\partial f}{\partial q_3} dq_3 \quad (4)$$

Equating (3) and (4), we get

$$f_1 = \frac{1}{h_1}\frac{\partial f}{\partial q}, \; f_2 = \frac{1}{h_2}\frac{\partial f}{\partial q_2}, \; f_3 = \frac{1}{h_3}\frac{\partial f}{\partial q_3} \quad (5)$$

Substituting (5) in (1), we get

$$\nabla f = \frac{\overline{e}_1}{h_1}\frac{\partial f}{\partial q_1} + \frac{\overline{e}_2}{h_2}\frac{\partial f}{\partial q_2} + \frac{\overline{e}_3}{h_3}\frac{\partial t}{\partial q_3}$$

Thus the nabla operator ∇ in orthogonal curvilinear coordinates is

$$\nabla = \frac{\overline{e}_1}{h_1}\frac{\partial}{\partial q_1} + \frac{\overline{e}_2}{h_2}\frac{\partial}{\partial q_2} + \frac{\overline{e}_3}{h_3}\frac{\partial}{\partial q_3} \quad (6)$$

Putting

$$h_1 = h_2 = h_3 = 1, \quad \overline{e}_1 = \overline{i}, \quad \overline{e}_2 = \overline{j}, \quad \overline{e}_3 = \overline{k}$$

and $q_1 = x, \quad q_2 = y, \quad q_3 = z$

(6) reduces to

$$\nabla = \overline{i}\frac{\partial}{\partial x} + \overline{j}\frac{\partial}{\partial y} + \overline{k}\frac{\partial}{\partial z}.$$

Example 2: Show that $\overline{e}_1 = h_2 h_3 \nabla q_2 \times \nabla q_3$.

Solution: From Example 1 with $f = q_1$, we have

$$\nabla f = \nabla q_1 = \frac{\overline{e}_1}{h_1}\frac{\partial q_1}{\partial q_1} + 0 + 0 = \frac{\overline{e}_1}{h_1}$$

Similarly, $\nabla q_2 = \dfrac{\overline{e}_2}{h_2}$

and $\nabla q_3 = \dfrac{\overline{e}_3}{h_3}$

Now

$$\nabla q_2 \times \nabla q_3 = \frac{\overline{e}_2}{h_2} \times \frac{\overline{e}_3}{h_3} = \frac{1}{h_2 h_3}\overline{e}_2 \times \overline{e}_3 = \frac{\overline{e}_1}{h_2 h_3}$$

So $\overline{e}_1 = h_2 h_3 \nabla q_2 \times \nabla q_3$.

In a similar way, we get

$$\overline{e}_2 = h_3 h_1 \nabla q_3 \times \nabla q_1$$

$$\overline{e}_3 = h_1 h_2 \nabla q_1 \times \nabla q_2.$$

Example 3: Derive an expression for $\nabla \cdot \overline{A}$ in orthogonal curvilinear coordinates. Deduce $\nabla \cdot \overline{A}$ in rectangular coordinates

Solution: Let

$$\bar{A} = A_1\bar{e}_1 + A_2\bar{e}_2 + A_3\bar{e}_3$$

So that

$$\nabla\cdot\bar{A} = \nabla\cdot(A_1\bar{e}_1 + A_2\bar{e}_2 + A_3\bar{e}_3)$$
$$= \nabla\cdot(A_1\bar{e}_1) + \nabla\cdot(A_2\bar{e}_2) + \nabla\cdot(A_3\bar{e}_3) \quad (1)$$

Consider,

$$\nabla\cdot(A_1\bar{e}_1) = \nabla\cdot(A_1h_2h_3\nabla q_2\times\nabla q_3)$$

Using result of above Example 2

$$= \nabla(A_1h_2h_3)\cdot\nabla q_2\times\nabla q_3 + A_1h_2h_3\nabla\cdot(\nabla q_2\times\nabla q_3)$$
$$= \nabla(A_1h_2h_3)\cdot\frac{\bar{e}_2}{h_2}\times\frac{\bar{e}_3}{h_3} + 0$$
$$= \nabla(A_1h_2h_3)\cdot\frac{\bar{e}_1}{h_2h_3}.$$

Using ∇f result from above Example 1.

$$\nabla\cdot(A_1\bar{e}_1) = \left[\frac{\bar{e}_1}{h_1}\frac{\partial}{\partial q_1}(A_1h_2h_3) + \frac{\bar{e}_2}{h_2}\frac{\partial}{\partial q_2}(A_1h_2h_3) + \frac{\bar{e}_3}{h_3}\frac{\partial}{\partial q_3}(A_1h_2h_3)\right]\cdot\frac{\bar{e}_1}{h_2h_3}$$

$$\nabla\cdot(A_1\bar{e}_1) = \frac{1}{h_1h_2h_3}\frac{\partial}{\partial q_1}(A_1h_2h_3) + 0 + 0 \quad (2)$$

Similarly, we get

$$\nabla\cdot(A_2\bar{e}_2) = \frac{1}{h_1h_2h_3}\frac{\partial}{\partial q_2}(A_2h_3h_1) \quad (3)$$

$$\nabla\cdot(A_3\bar{e}_3) = \frac{1}{h_1h_2h_3}\frac{\partial}{\partial q_3}(A_3h_1h_2) \quad (4)$$

Adding (2), (3), (4) and using (1), we get the required result as

$$\nabla\cdot\bar{A} = \frac{1}{h_1h_2h_3}\left[\frac{\partial}{\partial q_1}(A_1h_2h_3) + \frac{\partial}{\partial q_2}(A_2h_3h_1) + \frac{\partial}{\partial q_3}(A_3h_1h_2)\right].$$

Putting $h_1 = h_2 = h_3 = 1$

$$q_1 = x, \quad q_2 = y, \quad q_3 = z$$

$$\nabla\cdot\bar{A} = \frac{\partial A_1}{\partial x} + \frac{\partial A_2}{\partial y} + \frac{\partial A_3}{\partial z}$$

Example 4: Derive an expression for $\nabla\times\bar{A}$ in orthogonal curvilinear coordinates. Deduce $\nabla\times\bar{A}$ for cartesian coordinates.

Solution: If

$$\bar{A} = A_1\bar{e}_1 + A_2\bar{e}_2 + A_3\bar{e}_3$$

then

$$\nabla\times\bar{A} = \nabla\times(A_1\bar{e}_1 + A_2\bar{e}_2 + A_3\bar{e}_3)$$
$$\nabla\times\bar{A} = \nabla\times(A_1\bar{e}_1) + \nabla\times(A_2\bar{e}_2) + \nabla\times(A_3\bar{e}_3) \quad (1)$$

Consider

$$\nabla\times(A_1\bar{e}_1) = \nabla\times(A_1h_1\nabla q_1)$$

Since

$$\bar{e}_1 = h_1\nabla q_1$$
$$\nabla\times(A_1\bar{e}_1) = \nabla(A_1h_1)\times\nabla q_1 + A_1h_1\nabla\times\nabla q_1$$
$$= \nabla(A_1h_1)\times\frac{\bar{e}_1}{h_1} + 0$$

Substituting gradient value from above Example 1

$$\nabla\times(A_1\bar{e}_1) = \left[\frac{\bar{e}_1}{h_1}\frac{\partial}{\partial q_1}(A_1h_1) + \frac{\bar{e}_2}{h_2}\frac{\partial}{\partial q_2}(A_1h_1) + \frac{\bar{e}_3}{h_3}\frac{\partial}{q_3}(A_1h_1)\right]\times\frac{\bar{e}_1}{h_1}$$

$$= 0 - \frac{\bar{e}_3}{h_2h_1}\frac{\partial}{\partial q_2}(A_1h_1) + \frac{\bar{e}_2}{h_3h_1}\frac{\partial}{\partial q_3}(A_1h_1) \quad (2)$$

In a similar way, we get

$$\nabla\times(A_2\bar{e}_2) = \frac{\bar{e}_3}{h_1h_2}\frac{\partial}{\partial q_1}(A_2h_2) - \frac{\bar{e}_1}{h_2h_3}\frac{\partial}{\partial q_3}(A_2h_2) \quad (3)$$

and

$$\nabla\times(A_3\bar{e}_3) = \frac{\bar{e}_1}{h_2h_3}\frac{\partial}{\partial q_2}(A_3h_3) - \frac{\bar{e}_2}{h_3h_1}\frac{\partial}{\partial q_1}(A_3h_3) \quad (4)$$

Adding (2), (3), (4), we get the required expression for which can be written as

$$\nabla \times \bar{A} = \frac{1}{h_1h_2h_3}\begin{vmatrix} h_1\bar{e}_1 & h_2\bar{e}_2 & h_3\bar{e}_3 \\ \dfrac{\partial}{\partial q_1} & \dfrac{\partial}{\partial q_2} & \dfrac{\partial}{\partial q_3} \\ A_1h_1 & A_2h_2 & A_3h_3 \end{vmatrix}.$$

For cartesian system,

$$\nabla \times A = \begin{vmatrix} i & j & k \\ \dfrac{\partial}{\partial x} & \dfrac{\partial}{\partial y} & \dfrac{\partial}{\partial z} \\ A_1 & A_2 & A_3 \end{vmatrix}$$

Example 5: Express $\nabla^2 f$ in orthogonal curvilinear coordinates. Deduce for cartesian system

Solution: From Example 1,

$$\nabla f = \frac{\bar{e}_1}{h_1}\frac{\partial f}{\partial q_1} + \frac{\bar{e}_2}{h_2}\frac{\partial f}{\partial q_2} + \frac{\bar{e}_3}{h_3}\frac{\partial f}{\partial q_3}$$

If $\quad \bar{A} = A_1\bar{e}_1 + A_2\bar{e}_2 + A_3\bar{e}_3 = \nabla f$

Then equating coefficients of $\bar{e}_1, \bar{e}_2, \bar{e}_3$, we get

$$A_1 = \frac{1}{h_1}\frac{\partial f}{\partial q_1},\ A_2 = \frac{1}{h_2}\frac{\partial f}{\partial q_2},\ A_3 = \frac{1}{h_3}\frac{\partial f}{\partial q_3}$$

Thus

$$\nabla^2 f = \nabla \cdot \nabla f = \nabla \cdot \bar{A}$$

$$= \frac{1}{h_1h_2h_3}\left[\frac{\partial}{\partial q_1}(A_1h_2h_3) + \frac{\partial}{\partial q_2}(A_2h_3h_1) + \frac{\partial}{\partial q_3}(A_3h_1h_2)\right]$$

$$= \frac{1}{h_1h_2h_3}\left[\frac{\partial}{\partial q_1}\left(\frac{h_2h_3}{h_1}\frac{\partial f}{\partial q_1}\right) + \frac{\partial}{\partial q_2}\left(\frac{h_3h_1}{h_2}\frac{\partial f}{\partial q_2}\right) + \frac{\partial}{\partial q_3}\left(\frac{h_1h_2}{h_3}\frac{\partial f}{\partial q_3}\right)\right]$$

For cartesian system,

$$\nabla^2 f = \frac{\partial^2 f}{\partial x^2} + \frac{\partial^2 f}{\partial y^2} + \frac{\partial^2 f}{\partial z^2}.$$

Example 6: Express (a) ∇f, (b) $\nabla \cdot \bar{A}$, (c) $\nabla \times \bar{A}$, (d) $\nabla^2 f$ in cylindrical coordinates (r, θ, z).

Solution: For cylindrical coordinates (r, θ, z), we know that

$$q_1 = r,\ q_2 = \theta,\ q_3 = z, \bar{e}_1 = \bar{e}_r,\ \bar{e}_2 = \bar{e}_\theta,\ \bar{e}_3 = \bar{e}_z$$

and

$$h_1 = h_r = 1,\quad h_2 = h_\theta = r,\quad h_3 = h_z = 1$$

a. From Example 1

$$\nabla f = \frac{\bar{e}_1}{h_1}\frac{\partial f}{\partial q_1} + \frac{\bar{e}_2}{h_2}\frac{\partial f}{\partial q_2} + \frac{\bar{e}_3}{h_3}\frac{\partial f}{\partial q_3}$$

$$\nabla f = \frac{1}{1}\frac{\partial f}{\partial r}\bar{e}_r + \frac{1}{r}\frac{\partial f}{\partial \theta}\bar{e}_\theta + \frac{1}{1}\frac{\partial f}{\partial z}\bar{e}_z$$

b. From Example 3

$$\nabla \cdot \bar{A} = \frac{1}{h_1h_2h_3}\left[\frac{\partial}{\partial q_1}(h_2h_3A_1) + \frac{\partial}{\partial q_2}(h_3h_1A_2) + \frac{\partial}{\partial q_3}(h_1h_2A_3)\right]$$

$$= \frac{1}{1\cdot r\cdot 1}\left[\frac{\partial}{\partial r}(r\cdot 1\cdot A_r) + \frac{\partial}{\partial \theta}(1\cdot 1\cdot A_\theta) + \frac{\partial}{\partial z}(1\cdot r\cdot A_z)\right]$$

$$\nabla \cdot \bar{A} = \frac{1}{r}\left[\frac{\partial}{\partial r}(r\,A_r) + \frac{\partial}{\partial \theta}A_\theta + \frac{\partial}{\partial z}(r\,A_z)\right]$$

c.

$$\nabla \times \bar{A} = \frac{1}{h_1h_2h_3}\begin{vmatrix} h_1\bar{e}_1 & h_2\bar{e}_2 & h_3\bar{e}_3 \\ \frac{\partial}{\partial q_1} & \frac{\partial}{\partial q_2} & \frac{\partial}{\partial q_3} \\ h_1A_1 & h_2A_2 & h_3A_3 \end{vmatrix}$$

$$= \frac{1}{1\cdot r\cdot 1}\begin{vmatrix} \bar{e}_r & r\bar{e}_\theta & \bar{e}_z \\ \frac{\partial}{\partial r} & \frac{\partial}{\partial \theta} & \frac{\partial}{\partial z} \\ A_r & rA_\theta & A_z \end{vmatrix}.$$

d. From Example 5,

$$\nabla^2 f = \frac{1}{h_1h_2h_3}\left[\frac{\partial}{\partial q_1}\left(\frac{h_2h_3}{h_1}\frac{\partial f}{\partial q_1}\right) + \frac{\partial}{\partial q_2}\left(\frac{h_3h_1}{h_2}\frac{\partial f}{\partial q_2}\right) + \frac{\partial}{\partial q_3}\left(\frac{h_1h_2}{h_3}\frac{\partial f}{\partial q_3}\right)\right]$$

$$= \frac{1}{1\cdot r\cdot 1}\left[\frac{\partial}{\partial r}\left(\frac{r\cdot 1}{1}\frac{\partial f}{\partial r}\right) + \frac{\partial}{\partial \theta}\left(\frac{1\cdot 1}{r}\frac{\partial f}{\partial \theta}\right) + \frac{\partial}{\partial z}\left(\frac{1\cdot r}{1}\frac{\partial f}{\partial z}\right)\right]$$

$$\nabla^2 f = \frac{1}{r}\frac{\partial}{\partial r}\left(r\frac{\partial f}{\partial r}\right) + \frac{1}{r^2}\frac{\partial^2 f}{\partial \theta^2} + \frac{\partial^2 f}{\partial z^2}.$$

Example 7: Express (a) ∇f, (b) $\nabla\cdot\bar{A}$, (c) $\nabla\times\bar{A}$, (d) $\nabla^2 f$ in spherical curvilinear coordinates.

Solution: Here

$$q_1 = r,\quad q_2 = \theta,\quad q_3 = \phi,$$

$$\bar{e}_1 = \bar{e}_r,\quad \bar{e}_2 = \bar{e}_\theta,\quad \bar{e}_3 = \bar{e}_\phi$$

$$h_1 = h_r = 1,\quad h_2 = h_\theta = r,\quad h_3 = h_\phi = r\sin\theta.$$

a.
$$\nabla f = \frac{\bar{e}_1}{h_1}\frac{\partial f}{\partial q_1} + \frac{e_2}{h_2}\frac{\partial f}{\partial q_2} + \frac{\bar{e}_3}{h_3}\frac{\partial f}{\partial q_3}$$

$$\nabla f = \frac{\bar{e}_r}{1}\frac{\partial f}{\partial r} + \frac{\bar{e}_\theta}{r}\frac{\partial f}{\partial \theta} + \frac{\bar{e}_3}{r\sin\theta}\frac{\partial f}{\partial \phi}$$

b.
$$\nabla\cdot\bar{A} = \frac{1}{h_1h_2h_3}\left[\frac{\partial}{\partial q_1}(h_2h_3A_1) + \frac{\partial}{\partial q_2}(h_3h_1A_2) + \frac{\partial}{\partial q_3}(h_1h_2A_3)\right]$$

$$= \frac{1}{1\cdot r\cdot r\cdot\sin\theta}\left[\frac{\partial}{\partial r}(r\cdot r\sin\theta A_r) + \frac{\partial}{\partial\theta}(r\sin\theta\cdot 1\cdot A_\theta) + \frac{\partial}{\partial\phi}(1\cdot r\cdot A_\phi)\right]$$

$$\nabla\cdot\bar{A} = \frac{1}{r^2\sin\theta}\left[\frac{\partial}{\partial r}(r^2\sin\theta A_r) + \frac{\partial}{\partial\theta}(r\sin\theta A_\theta) + \frac{\partial}{\partial\phi}(rA_\phi)\right]$$

$$\nabla\cdot\bar{A} = \frac{1}{r^2}\frac{\partial}{\partial r}(r^2A_r) + \frac{1}{r\sin\theta}\frac{\partial}{\partial\theta}(\sin\theta A_\theta) + \frac{1}{r\sin\theta}\cdot\frac{\partial}{\partial\phi}(A_\phi).$$

c.
$$\nabla\times\bar{A} = \frac{1}{h_1h_2h_3}\begin{vmatrix} h_1\bar{e}_1 & h_2\bar{e}_2 & h_3\bar{e}_3 \\ \frac{\partial}{\partial q_1} & \frac{\partial}{\partial q_2} & \frac{\partial}{\partial q_3} \\ h_1A_1 & h_2A_2 & h_3A_3 \end{vmatrix}$$

$$\nabla\times\bar{A} = \frac{1}{1\cdot r\cdot r\sin\theta}\begin{vmatrix} e_r & re_\theta & r\sin\theta e_\phi \\ \frac{\partial}{\partial r} & \frac{\partial}{\partial\theta} & \frac{\partial}{\partial\phi} \\ A_r & rA_\theta & r\sin\theta A_\phi \end{vmatrix}$$

$$= \frac{1}{r^2\sin\theta}\left[\left\{\frac{\partial}{\partial\theta}(r\sin\theta A_\phi) - \frac{\partial}{\partial\phi}(rA_\theta)\right\}\bar{e}_r - \left\{\frac{\partial}{\partial r}(r\sin\theta A_\phi) - \frac{\partial}{\partial\phi}(A_r)\right\}r\bar{e}_\theta + \left\{\frac{\partial}{\partial r}(rA_\theta) - \frac{\partial}{\partial\theta}(A_r)\right\}r\sin\theta\bar{e}_\phi\right].$$

d.
$$\nabla^2 f = \frac{1}{h_1h_2h_3}\left[\frac{\partial}{\partial q_1}\left(\frac{h_2h_3}{h_1}\frac{\partial f}{\partial q_1}\right) + \frac{\partial}{\partial q_2}\left(\frac{h_3h_1}{h_2}\frac{\partial f}{\partial q_2}\right) + \frac{\partial}{\partial q_3}\left(\frac{h_1h_2}{h_3}\frac{\partial f}{\partial q_3}\right)\right]$$

$$\nabla^2 f = \frac{1}{1\cdot r\cdot r\sin\theta}\left[\frac{\partial}{\partial r}\left(\frac{r\cdot r\sin\theta}{1}\frac{\partial f}{\partial r}\right) + \frac{\partial}{\partial\theta}\left(\frac{r\sin\theta\cdot 1}{r}\frac{\partial f}{\partial\theta}\right) + \frac{\partial}{\partial\phi}\left(\frac{1\cdot r}{r\cdot\sin\theta}\frac{\partial f}{\partial\phi}\right)\right]$$

$$\nabla^2 f = \frac{1}{r^2\sin\theta}\left[\sin\theta\frac{\partial}{\partial r}\left(r^2\frac{\partial f}{\partial r}\right) + \frac{\partial}{\partial\theta}\left(\sin\theta\frac{\partial f}{\partial\theta}\right) + \frac{1}{\sin\theta}\frac{\partial^2 f}{\partial\phi^2}\right].$$

Example 8: Prove that a spherical coordinate system is orthogonal.

Solution: The position vector of any point in spherical coordinates is

$$\bar{r} = x\bar{i} + y\bar{j} + z\bar{k} = \rho\sin\theta\cos\phi\bar{i} + \rho\sin\theta\sin\phi\bar{j} + \rho\cos\theta\bar{k}$$

The tangent vectors to the ρ, θ, ϕ curves are given respectively by $\frac{\partial\bar{r}}{\partial\rho}, \frac{\partial\bar{r}}{\partial\theta}, \frac{\partial\bar{r}}{\partial\phi}$ where

$$\frac{\partial\bar{r}}{\partial\rho} = \sin\theta\cos\phi\bar{i} + \sin\theta\sin\phi\bar{j} + \cos\theta\bar{k}$$

$$\frac{\partial\bar{r}}{\partial\theta} = \rho\cos\theta\cos\phi\bar{i} + \rho\cos\theta\sin\phi\bar{j} - \rho\sin\theta\bar{k}$$

$$\frac{\partial\bar{r}}{\partial\phi} = -\rho\sin\theta\sin\phi\bar{i} + \rho\sin\theta\cdot\cos\phi\bar{j} + 0$$

The unit vectors in these directions are

$$\bar{e}_1 = \bar{e}_\rho = \frac{\partial\bar{r}/\partial\rho}{|\partial\bar{r}/\partial\rho|} = \frac{\sin\theta\cos\phi\bar{i} + \sin\theta\sin\phi\bar{j} + \cos\theta\bar{k}}{\sin^2\theta\cos^2\phi + \sin^2\theta\sin^2\phi + \cos^2\theta} = \frac{\sin\theta\cos\phi i + \sin\phi\sin\theta j + \cos\theta\bar{k}}{1}$$

$$\bar{e}_2 = \bar{e}_\theta = \frac{\partial\bar{r}/\partial\theta}{|\partial\bar{r}/\partial\theta|} = \frac{\rho\cos\theta\cos\phi\bar{i} + \rho\cos\theta\sin\phi\bar{j} - \rho\sin\theta\bar{k}}{\rho}$$

$$\bar{e}_3 = \bar{e}_\phi = \frac{\partial \bar{r}/\partial \phi}{|\partial \bar{r}/\partial \phi|} = \frac{-\rho \sin\theta \sin\phi \bar{i} + \rho \sin\theta \cos\phi \bar{j}}{\rho \sin\theta}$$

Then

$$\bar{e}_1 \cdot \bar{e}_2 = \sin\theta \cdot \cos\theta \cdot \cos^2\phi + \sin\theta \cdot \cos\theta \cdot \sin^2\phi - \cos\theta \sin\theta = 0$$

$$\bar{e}_1 \cdot \bar{e}_3 = -\sin\theta \cdot \cos\phi \sin\phi + \sin\theta \cdot \sin\phi \cdot \cos\phi = 0$$

$$\bar{e}_2 \cdot \bar{e}_3 = -\cos\theta \cdot \cos\phi \sin\phi + \cos\theta \sin\phi \cos\phi = 0.$$

So $\bar{e}_1, \bar{e}_2, \bar{e}_3$ are mutually perpendicular and the spherical coordinate system is orthogonal.

Example 9: Find the Jacobian of x, y, z with respect to the orthogonal curvilinear coordinates q_1, q_2, q_3.

Solution:

$$J\left(\frac{x, y, z}{q_1, q_2, q_3}\right) = \frac{\partial(x, y, z)}{\partial(q_1, q_2, q_3)} = \begin{vmatrix} \frac{\partial x}{\partial q_1} & \frac{\partial y}{\partial q_1} & \frac{\partial z}{\partial q_1} \\ \frac{\partial x}{\partial q_2} & \frac{\partial y}{\partial q_2} & \frac{\partial z}{\partial q_2} \\ \frac{\partial x}{\partial q_3} & \frac{\partial y}{\partial q_3} & \frac{\partial z}{\partial q_3} \end{vmatrix}$$

This determinant is triple scalar product given by

$$J = \left(\frac{\partial x}{\partial q_1}\bar{i} + \frac{\partial y}{\partial q_1}\bar{j} + \frac{\partial z}{\partial q_1}\bar{k}\right) \cdot \left(\frac{\partial x}{\partial q^2}\bar{i} + \frac{\partial y}{\partial q_2}\bar{j} + \frac{\partial z}{\partial q_2}\bar{k}\right) \times \left(\frac{\partial x}{\partial q_3}\bar{i} + \frac{\partial y}{\partial q_3}\bar{j} + \frac{\partial z}{\partial q_3}\bar{k}\right)$$

$$= \frac{\partial \bar{r}}{\partial q_1} \cdot \frac{\partial \bar{r}}{\partial q_2} \times \frac{\partial \bar{r}}{\partial q_3} = h_1\bar{e}_1 \cdot h_2\bar{e}_2 \times h_3\bar{e}_3$$

Jacobian $J = h_1 h_2 h_3 \bar{e}_1 \cdot \bar{e}_2 \times \bar{e}_3 = h_1 h_2 h_3.$

Note: If $J = 0$, $\bar{e}_1, \bar{e}_2, \bar{e}_3$ are coplanar and coordinate transformation breaks.

Example 10: Find the Jacobian $J\left(\frac{x,y,z}{q_1,q_2,q_3}\right)$ for

a. cylindrical,

b. spherical coordinates.

Solution:

a. $J = h_1 h_2 h_3 = 1 \cdot r \cdot 1 = r$

b. $J = h_1 h_2 h_3 = 1 \cdot r \cdot r \cdot \sin\theta = r^2 \sin\theta.$

Example 11: Find

$$\frac{\partial \bar{r}}{\partial q_1}, \frac{\partial \bar{r}}{\partial q_2}, \frac{\partial \bar{r}}{\partial q_3}, \nabla q_1, \nabla q_2, \nabla q_3$$

in cylindrical coordinates.

Solution:
$\bar{r} = xi + yj + zk = \rho \cos\theta i + \rho \sin\theta j + zk$

Then

$$\frac{\partial \bar{r}}{\partial q_1} = \frac{\partial \bar{r}}{\partial \rho} = \cos\theta \bar{i} + \sin\theta \bar{j}$$

$$\frac{\partial \bar{r}}{\partial q_2} = \frac{\partial \bar{r}}{\partial \theta} = -\rho \sin\theta \bar{i} + \rho \cos\theta \bar{j}$$

$$\frac{\partial \bar{r}}{\partial q_3} = \frac{\partial \bar{r}}{\partial z} = \bar{k}$$

$$\nabla q_1 = \nabla \rho = \frac{1}{h_1}\frac{\partial \rho}{\partial \rho}\bar{e}_\rho + \frac{1}{h_2}\frac{\partial \rho}{\partial \theta}\bar{e}_\theta + \frac{1}{h_3}\frac{\partial \rho}{\partial z}\bar{e}_z$$

$$= \frac{1}{1} \cdot 1 \cdot \bar{e}_\rho = \cos\theta \bar{i} + \sin\theta \bar{j}$$

Similarly,

$$\nabla q_2 = \nabla \theta = \frac{1}{h_2}\frac{\partial \theta}{\partial \theta}\bar{e}_\theta = \frac{1}{\rho}\bar{e}_\theta = \frac{-\sin\theta i + \cos\theta j}{\rho}$$

$$\nabla q_3 = \nabla z = \frac{1}{h_3}\frac{\partial z}{\partial z}\bar{e}_z = \bar{e}_z = \bar{k}.$$

Example 12: Prove that the surface area of a given region R of the surface $\bar{r} = \bar{r}(u, v)$ is $\int\int_R \sqrt{EG - F^2}\,du\,dv$. Use this to determine the surface area of a sphere.

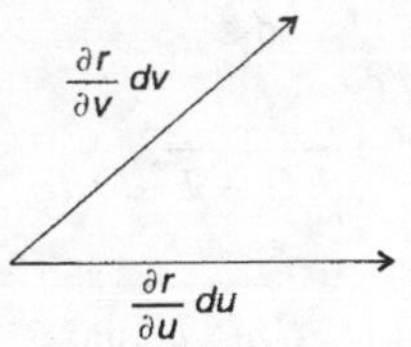

Fig. 15.5

Solution: Element of area is given by

$$dS = \left|\left(\frac{\partial \bar{r}}{\partial u}du\right) \times \frac{\partial \bar{r}}{\partial v}dv\right| = \left|\frac{\partial \bar{r}}{\partial u} \times \frac{\partial \bar{r}}{\partial v}\right| du\,dv$$

$$= \sqrt{\left(\frac{\partial \bar{r}}{\partial u} \times \frac{\partial \bar{r}}{\partial v}\right) \cdot \left(\frac{\partial \bar{r}}{\partial u} \times \frac{\partial \bar{r}}{\partial v}\right)}\, du\,dv$$

Using

$$(\bar{A} \times \bar{B}) \cdot (\bar{C} \times \bar{D}) = (\bar{A} \cdot \bar{C})(\bar{B} \cdot \bar{D}) - (\bar{A} \cdot \bar{D})(\bar{B} \cdot \bar{C}),$$

we have

$$dS = \sqrt{\left(\frac{\partial \bar{r}}{\partial u} \cdot \frac{\partial \bar{r}}{\partial u}\right)\left(\frac{\partial \bar{r}}{\partial v} \cdot \frac{\partial \bar{r}}{\partial v}\right) - \left(\frac{\partial \bar{r}}{\partial u} \cdot \frac{\partial \bar{r}}{\partial v}\right)\left(\frac{\partial \bar{r}}{\partial v} \cdot \frac{\partial \bar{r}}{\partial u}\right)} \times du\, dv$$

$$dS = \sqrt{EG - F^2}\, du\, dv$$

where

$$E = \frac{\partial \bar{r}}{\partial u} \cdot \frac{\partial \bar{r}}{\partial u}, \quad G = \frac{\partial \bar{r}}{\partial v} \cdot \frac{\partial \bar{r}}{\partial v}, \quad F = \frac{\partial \bar{r}}{\partial u} \cdot \frac{\partial \bar{r}}{\partial v},$$

Integrating over the region R, the surface area of the given region R is given by

$$S = \int\int_R \sqrt{EG - F^2}\, du\, dv$$

To find the surface area of a sphere of radius b the position vector $\bar{r}$ of any point on the surface is

$$\bar{r} = xi + yj + zk$$

In spherical coordinates

$$\bar{r} = b \sin\theta \cos\phi \bar{i} + b \sin\theta \sin\phi j + b \cos\theta k$$

$$\bar{r} = \bar{r}(\theta, \phi)$$

Differentiating partially w.r.t. θ and ϕ, we get

$$\frac{\partial \bar{r}}{\partial \theta} = b \cos\theta \cos\phi \bar{i} + b \cos\theta \sin\phi \bar{j} - b \sin\theta \bar{k}$$

$$\frac{\partial \bar{r}}{\partial \phi} = -b \sin\theta \sin\phi \bar{i} + b \sin\theta \cdot \cos\phi \bar{j}$$

$$E = \frac{\partial \bar{r}}{\partial \theta} \cdot \frac{\partial \bar{r}}{\partial \theta} = b^2, G = \frac{\partial \bar{r}}{\partial \phi} \cdot \frac{\partial \bar{r}}{\partial \phi} = b^2 \sin^2\theta,$$

$$F = \frac{\partial \bar{r}}{\partial \phi} \cdot \frac{\partial \bar{r}}{\partial \theta} = 0$$

$$S = \int\int_R \sqrt{EG - F^2} d\theta\, d\phi$$

$$= \int_0^{2\pi} \int_0^{\pi} \sqrt{b^2 \cdot b^2 \sin^2\theta - 0}\, d\theta\, d\phi$$

$$= 2\pi b^2(-1 - 1) = 4\pi b^2.$$

Example 13: Prove that curl of gradient $f = 0$ in any orthogonal curvilinear coordinate system.

Solution: In any orthogonal curvilinear coordinate

$$\nabla f = \frac{1}{h_1}\frac{\partial f}{\partial q_1}\bar{e}_1 + \frac{1}{h_2}\frac{\partial f}{\partial q_2}\bar{e}_2 + \frac{1}{h_3}\frac{\partial f}{\partial q_3}\bar{e}_3$$

Then

$$\nabla \times \nabla f = \frac{1}{h_1 h_2 h_3}\begin{vmatrix} h_1\bar{e}_1 & h_2\bar{e}_2 & h_3\bar{e}_3 \\ \frac{\partial}{\partial q_1} & \frac{\partial}{\partial q_2} & \frac{\partial}{\partial q_3} \\ h_1\frac{1}{h_1}\frac{\partial f}{\partial q_1} & h_2\frac{1}{h_2}\frac{\partial f}{\partial q_2} & h_3\frac{1}{h_3}\frac{\partial f}{\partial q_3} \end{vmatrix}$$

$$= \frac{1}{h_1 h_2 h_3}\left[h_1\bar{e}_1\left\{\frac{\partial^2 f}{\partial q_2 \partial q_3} - \frac{\partial^2 f}{\partial q_3 \partial q_2}\right\} - h_2\bar{e}_2\left\{\frac{\partial^2 f}{\partial q_1 \partial q_3} - \frac{\partial^2 f}{\partial q_3 \partial q_1}\right\} + h_3\bar{e}_3\left\{\frac{\partial^2 f}{\partial q_1 \partial q_2} - \frac{\partial^2 f}{\partial q_2 \partial q_1}\right\}\right] = 0.$$

Example 14: Find the square of the element of arc length in cylindrical coordinates and determine the corresponding scale factors.

Solution: The position vector is

$$\bar{r} = \rho \cos\theta \bar{i} + \rho \sin\theta \bar{j} + z\bar{k}$$

Then

$$d\bar{r} = \frac{\partial \bar{r}}{\partial \rho}d\rho + \frac{\partial \bar{r}}{\partial \theta}d\theta + \frac{\partial \bar{r}}{\partial z}dz$$

$$= (\cos\theta i + \sin\theta j)d\rho + (-\rho \sin\theta i + \rho \cos\theta j)d\theta + k\, dz$$

$$d\bar{r} = (\cos\theta d\rho - \rho \sin\theta d\theta)i + (\sin\theta d\rho + \rho \cos\theta d\theta)j + k\, dz$$

Thus

$$ds^2 = d\bar{r} \cdot d\bar{r} = (\cos\theta d\rho - \rho \sin\theta d\theta)^2 + (\sin\theta d\rho + \rho \cos\theta d\theta)^2 + (dz)^2$$

$$= (d\rho)^2 + \rho^2(d\theta)^2 + (dz)^2$$

Here

$$h_1 = h_\rho = 1, h_2 = h_\theta = \rho, h_3 = h_2 = 1$$

are the scale factors.

Example 15: A vector field is given in cylindrical coordinates as

$$\overline{A}(P) = \rho_\rho + \theta\overline{e}_\theta$$

Find the vector lines of the field.

Solution: It is given that $a_1 = 1, a_2 = \theta, a_3 = 0$. So that

$$\frac{d\rho}{1} = \frac{P\,d\theta}{\theta} = \frac{dz}{0}$$

whence

$$z = c_1$$

$$\rho = c_2\theta$$

which are Archimedean spirals lying in planes parallel to the xy-plane (i.e., $z =$ constant).

Example 16: Compute the gradient of the scalar field $f = \rho + z\cos\theta$ specified in cylindrical coordinates (ρ, θ, z).

Solution:

$$\nabla f = \frac{\partial f}{\partial \rho}\overline{e}_\rho + \frac{1}{\rho}\frac{\partial f}{\partial \theta}\overline{e}_\theta + \frac{\partial f}{\partial z}\overline{e}_z$$

$$= 1\cdot\overline{e}_\rho + \frac{1}{\rho}(-z\sin\theta)\overline{e}_\theta + \cos\theta\overline{e}_z.$$

Example 17: Compute the curl of $\overline{A}$ specified in cylindrical coordinates where

$$\overline{A} = \sin\theta\overline{e}_\rho + \frac{\cos\theta}{\rho}e_\theta - \rho z\overline{e}_z$$

Solution: Since

$$\nabla\times\overline{A} = \begin{vmatrix} \frac{1}{\rho}\overline{e}_\rho & \overline{e}_\theta & \frac{1}{\rho}\overline{e}_z \\ \frac{\partial}{\partial\rho} & \frac{\partial}{\partial\theta} & \frac{\partial}{\partial z} \\ a_1 & a_2 & a_3 \end{vmatrix} = \begin{vmatrix} \frac{1}{\rho}\overline{e}_\rho & \overline{e}_\theta & \frac{1}{\rho}\overline{e}_z \\ \frac{\partial}{\partial\rho} & \frac{\partial}{\partial\theta} & \frac{\partial}{\partial z} \\ \sin\theta & \cos\theta & -\rho z \end{vmatrix}$$

$$= \frac{1}{\rho}\overline{e}_\rho(0-0) - \overline{e}_\theta(-z-0) + \frac{1}{\rho}\overline{e}_z(0-\cos\theta)$$

$$\nabla\times\overline{A} = z\overline{e}_\theta - \frac{\cos\theta}{\rho}\overline{e}_z$$

Example 18: Show that the vector field $\overline{A}$ in spherical coordinates

$$\overline{A} = \frac{2\cos\theta}{r^3}\overline{e}_r + \frac{\sin\theta}{r^3}\overline{e}_\theta$$

is solenoidal.

Solution: We know that divergence in spherical coordinates

$$\nabla\cdot\overline{A} = \frac{1}{\rho^2}\frac{\partial(\rho^2 a_1)}{\partial\rho} + \frac{1}{\rho\sin\theta}\frac{\partial}{\partial\theta}(\sin\theta a_2) + \frac{1}{\rho\sin\theta}\frac{\partial a_3}{\partial\phi}$$

$$\nabla\cdot\overline{A} = \frac{1}{\rho^2}\frac{\partial}{\partial r}\left(\rho^2\cdot\frac{2\cos\theta}{\rho^3}\right)$$

$$+\frac{1}{\rho\sin\theta}\frac{\partial}{\partial\theta}\left(\sin\theta\cdot\frac{\sin\theta}{\rho^3}\right) + 0$$

$$= \frac{1}{\rho^2}\left(-\frac{2\cos\theta}{\rho^2}\right) + \frac{1}{\rho^4\sin\theta}\cdot 2\cdot\sin\theta\cos\theta = 0$$

wherever $r \neq 0$, which means that the vector field $\overline{A}$ is solenoidal at all points except at $r = 0$.

Example 19: Find the potential of

$$\overline{A} = \frac{1}{\rho}e^{\theta\phi}\overline{e}_r + \frac{\theta\ln\rho}{r\sin\theta}e^{\theta\phi}\overline{e}_\phi + \frac{\ln\rho}{\rho}\phi e^{\theta\phi}\overline{e}_\theta$$

given in spherical coordinates.

Solution: In spherical coordinates

$$\nabla\times\overline{A} = \frac{1}{\rho^2\sin\theta}\begin{vmatrix} \overline{e}_r & \rho\overline{e}_\theta & r\sin\theta\overline{e}_\phi \\ \frac{\partial}{\partial\rho} & \frac{\partial}{\partial\theta} & \frac{\partial}{\partial\phi} \\ \frac{1}{\rho}e^{\theta\phi} & \phi\ln r e^{\theta\phi} & \theta\ln r e^{\theta\phi} \end{vmatrix} = 0.$$

Thus $\overline{A}$ is a potential field in the region where $r > 0, \theta \neq n\pi(n = 0, \neq 1, \cdot s)$.

Let $\quad \overline{A} = \nabla f = \frac{\partial f}{\partial\rho}\overline{e}_\rho + \frac{\partial f}{\partial\theta}\overline{e}_\theta + \frac{\partial f}{\partial\phi}\overline{e}_\phi$

where $f = f(\rho, \theta, \phi)$ is the desired potential function, which is the solution of the following system of differential equations.

$$\frac{\partial f}{\partial\rho} = \frac{1}{\rho}e^{\theta\phi} \tag{1}$$

$$\frac{\partial f}{\partial\theta} = \phi e^{\theta\phi}\ln\rho \tag{2}$$

$$\frac{\partial f}{\partial\phi} = \theta e^{\theta\phi}\ln r \tag{3}$$

Integrating (1) w.r.t. ρ we get

$$f = e^{\theta\phi}\ln\rho + c_1(\phi, \theta) \tag{4}$$

Differentiating (4) w.r.t. 'θ' and equating it with (2)

$$\phi e^{\theta\phi}\ln r = \frac{\partial f}{\partial\theta} = \phi e^{\theta\phi}\ln r + \frac{\partial c_1}{\partial\theta}$$

i.e., $\frac{\partial c_1}{\partial \theta} = 0$ whence

$$c_1(\phi, \theta) = c_2(\phi) \quad (5)$$

Substituting (5) in (4)

$$f = e^{\theta\phi} \ln \rho + c_2(\phi) \quad (6)$$

Differentiating (6) w.r.t. ϕ and equating it with (3) we obtain

$$\theta e^{\theta\phi} \ln r = \frac{\partial f}{\partial \phi} = \theta e^{\theta\phi} \ln r + \frac{dc_2}{d\phi}$$

so $\frac{dc_2}{d\phi} = 0$ i.e., $c_2 = c =$ constant. The desired potential is

$$f(r, \theta, \phi) = e^{\theta\rho} \ln r + c$$

Example 20: Find all the solutions of the Laplace's equation $\nabla^2 f = 0$ that depend solely on the distance ρ.

Solution: Laplace's equation is spherical coordinates with spherical symmetry (f must not depend on θ or ϕ, i.e., $f = f(r)$). We have

$$\nabla^2 f = \frac{1}{\rho^2} \frac{\partial}{\partial \rho}\left(\rho^2 \frac{\partial f}{\partial \rho}\right) = 0$$

so that

$$\rho^2 \frac{\partial f}{\partial r} = c_1 = \text{const.}$$

whence

$$f = \frac{c_1}{\rho} + c_2$$

where c_1 and c_2 are constants.

Example 21: Find the Laplacian of

$$f(\rho, \theta, z) = \rho^2\theta + z^2\theta^3 - \rho\theta z$$

Solution:

$$\nabla^2 f = \frac{1}{\rho}\left[\frac{\partial}{\partial \rho}\left(\rho \frac{\partial f}{\partial \rho}\right) + \frac{\partial}{\partial \theta}\left(\frac{1}{\rho}\frac{\partial f}{\partial \theta}\right) + \frac{\partial}{\partial z}\left(\rho \frac{\partial f}{\partial z}\right)\right]$$

Here

$$\frac{\partial f}{\partial \rho} = 2\rho\theta + 0 - \theta z, \quad \frac{\partial^2 f}{\partial \rho^2} = 2\theta$$

$$\frac{\partial f}{\partial \theta} = \rho^2 + 3z^2\theta^2 - \rho z, \quad \frac{\partial^2 f}{\partial \theta^2} = 6z^2\theta$$

$$\frac{\partial f}{\partial z} = 2z\theta^3 - \rho\theta, \quad \frac{\partial^2 f}{\partial z^2} = 2\theta^3$$

$$\nabla^2 f = \frac{1}{\rho}\left[\frac{\partial f}{\partial \rho} + \rho\frac{\partial^2 f}{\partial \rho^2} + \frac{1}{\rho}\frac{\partial^2 f}{\partial \theta^2} + \rho\frac{\partial^2 f}{\partial z^2}\right]$$

Substituting the above partial derivatives

$$\nabla^2 f = \frac{1}{\rho}[2\overline{\rho\theta} - \theta\overline{z} + \rho\overline{2\theta} + \frac{1}{\rho}6\overline{z}^2\theta + \rho 2\overline{\theta}^3]$$

$$\nabla^2 f = 4\overline{\theta} - \frac{\theta z}{\rho} + \frac{6\theta z^2}{\rho^2} + 2\theta^3$$

Example 22: (a) Find the unit vectors $\overline{e}_\rho, \overline{e}_\theta, \overline{e}_\phi$ in spherical coordinate system in terms of i, j, k. (b) Solve for i, j, k in terms of e_r, e_θ, e_ϕ.

Solution: From Example 8

$$\overline{e_\rho} = \sin\theta\cos\phi\overline{i} + \sin\theta\sin\phi\overline{j} + \cos\theta\overline{k} \quad (1)$$

$$\overline{e_\theta} = \cos\theta\cos\phi\overline{i} + \cos\theta\sin\phi\overline{j} - \sin\theta\overline{k} \quad (2)$$

$$\overline{e_\phi} = -\sin\phi\overline{i} + \cos\phi\overline{j} \quad (3)$$

Solving (1), (2), (3) simultaneously

(1) × $\sin\theta$: $\sin\theta\overline{e}_\rho$

$$= \sin^2\theta(\cos\phi i + \sin\phi j) + \sin\theta\cos\theta\overline{k}$$

(2) × $\cos\theta$: $\cos\theta\overline{e_\theta}$

$$= \cos^2\theta(\cos\phi\overline{i} + \sin\phi\overline{j}) - \sin\theta\cos\theta\overline{k}$$

Adding

$$\sin\theta\overline{e}_\rho + \cos\theta\overline{e_\theta} = \cos\phi i + \sin\phi\overline{j} \quad (4) \times \sin\phi$$

$$\overline{e_\phi} = -\sin\phi\overline{i} + \cos\phi\overline{j} \quad (5) \times \cos\phi$$

Adding

$$\overline{j} = (\sin\theta\overline{e}_\rho + \cos\theta\overline{e_\theta})\sin\phi + \cos\phi\overline{e_\phi} \quad (6)$$

Substituting (6) in (3)

$$\overline{i} = \frac{1}{\sin\phi}(\cos\phi J - \overline{e_\phi})$$

$$\overline{i} = \frac{1}{\sin\phi}\Big[\{(\sin\theta\overline{e}_\rho + \cos\theta\overline{e_\theta})\sin\phi + \cos\phi\overline{e_\phi}\} \times \cos\phi - \overline{e_\phi}\Big]$$

$$\bar{i} = \{\sin\theta\bar{e}_\rho + \cos\theta\bar{e}_\theta + \cot\phi\bar{e}_\phi\} \times \cos\phi - \text{cosec}\phi\bar{e}_\phi \quad (7)$$

Substituting (6) and (7) in (2)

$$\bar{e}_\theta = \cos\theta \cdot \cos\phi\left[\{\sin\theta\bar{e}_\rho + \cos\theta\bar{e}_\theta + \cot\phi\bar{e}_\phi\} \times \cos\phi - \text{cosec}\phi\bar{e}_\phi\right] + \cos\theta\sin\phi \times \left[(\sin\theta\bar{e}_\rho + \cos\theta\bar{e}_\theta)\sin\phi + \cos\phi\bar{e}_\phi\right] - \sin\theta\bar{k}$$

Solving for $\bar{k}$,

$$\bar{k} = \cot\theta \cdot \cos\phi\left[\{\sin\theta\bar{e}_\rho + \cos\theta\bar{e}_\theta + \cot\phi\bar{e}_\phi\} \times \cos\phi - \text{cosec}\phi\bar{e}_\phi\right] + \cot\theta\sin\phi\left[(\sin\theta\bar{e}_\rho + \cos\theta\bar{e}_\theta)\sin\phi + \cos\phi\bar{e}_\phi\right] - \text{cosec}\theta \cdot \bar{e}_\theta \quad (8)$$

Example 23: Represent the vector $\bar{A} = 2yi - zj + 3xk$ in spherical coordinates and determine A_r, A_θ, A_ϕ.

Solution: Substituting (7), (6), (8) in $\bar{A} = 2y\,i - zj + 3x\bar{k}$ from the above Example 22, we get $\bar{A}$ in spherical coordinates as

$$\bar{A} = \left[\{\sin\theta\bar{e}_\rho + \cos\theta\bar{e}_\theta + \cot\phi\bar{e}_\phi\}\cos\phi - \text{cosec}\phi\bar{e}_\phi\right] \times 2r\sin\theta\sin\phi - \left[(\sin\theta\bar{e}_\rho + \cos\theta\bar{e}_\theta)\sin\phi + \cos\phi\bar{e}_\phi\right] \times r\cos\theta + 3r\sin\theta\cot\theta\cos\phi\left[\{\sin\theta\bar{e}_\rho + \cos\theta\bar{e}_\theta + \cot\phi\bar{e}_\phi\}\cos\phi - \text{cosec}\phi\bar{e}_\rho\right] + 3r\sin\theta \cdot \cos\phi \times \cot\theta \cdot \sin\phi\left[(\sin\theta\bar{e}_\rho + \cos\theta\bar{e}_\theta)\sin\phi + \cos\phi\bar{e}_\phi\right] - 3r\sin\theta \cdot \cos\phi\,\text{cosec}\,\theta\,\bar{e}_\theta.$$

Collecting the coefficients of $\bar{e}_\rho, \bar{e}_\theta$ and $\bar{e}_\phi$, we rewrite

$$\bar{A} = [2r\sin^2\theta \cdot \sin\phi\cos\phi - r\sin\theta\cos\theta\sin\phi + 3r\sin^2\theta\cot\theta\cos^2\phi\cos\phi + 3r\sin^2\theta\cot\theta \cdot \cos\phi\sin\phi \cdot \sin\phi]\bar{e}_\rho$$
$$+ [\cos\theta \cdot \cos\phi \cdot 2r \cdot \sin\theta \cdot \sin\phi - \cos\theta \cdot \sin\phi r\cos\theta + 3r\sin\theta \cdot \cos\phi \cdot \cot\theta \cdot \cos\phi\cos\theta \cdot \cos\phi + 3r\sin\theta \cdot \cos\phi \cdot \cot\theta \cdot \sin\phi\cos\theta\sin\phi - 3r\sin\theta \cdot \cos\phi \cdot \text{cosec}\theta]\bar{e}_\theta$$
$$+ \left[(\cot\phi\cos\phi - \text{cosec}\phi)2r\sin\theta\sin\phi - \cos\phi r\cos\theta + \{\cot\phi \cdot \cos\phi - \text{cosec}\phi\}3r\sin\theta \cdot \cot\theta \cdot \cos^2\phi + \cos\phi \cdot 3r\sin\theta \cdot \cot\theta \cdot \cos\phi \cdot \sin\phi\right]e_\phi$$

Simplifying the result, we get

$$A_r = 2r\sin^2\theta\sin\phi \cdot \cos\phi - r\sin\theta \cdot \cos\theta \cdot \sin\phi + 3r\sin\theta\cos\theta \cdot \cos\phi$$
$$A_\theta = 2r\sin\theta\cos\theta \cdot \sin\phi\cos\phi - r\cos^2\theta \cdot \sin\phi - 3r\sin^2\theta \cdot \cos\phi$$
$$A\phi = -2r\sin\theta\sin^2\phi - r\cos\theta\cos\phi$$

Example 24: Express the velocity $\bar{v}$ and acceleration $\bar{a}$ of a particle in spherical coordinates.

Solution: In rectangular coordinates the position vector, velocity and acceleration vectors are

$$\bar{r} = x\bar{i} + y\bar{j} + z\bar{k}$$
$$\bar{V} = \frac{d\bar{r}}{dt} = \dot{x}i + \dot{y}j + \dot{z}k$$
$$\bar{a} = \frac{d^2\bar{r}}{dt^2} = \ddot{x}i + \ddot{y}j + \ddot{z}k$$

In spherical coordinates

$$\bar{r} = x\bar{i} + y\bar{j} + z\bar{k} = r\sin\theta\cos\phi\bar{i} + r\sin\theta\sin\phi\bar{j} + r\cos\theta\bar{k}$$

Substituting $\bar{i}, \bar{j}, \bar{k}$ from (7), (6), (8) of the previous Example 22, we get

$$\bar{r} = r\sin\theta\cos\phi\left[(\sin\theta\bar{e}_\rho + \cos\theta\bar{e}_\theta)\cos\phi - \sin\phi\bar{e}_\phi\right] + r\sin\theta\sin\phi\left[(\sin\theta\bar{e}_\rho + \cos\theta\bar{e}_\theta)\sin\phi + \cos\phi\bar{e}_\phi\right] + r\cos\theta\left[\cot\theta \cdot \cos\phi\left\{(\sin\theta\bar{e}_\rho + \cos\theta\bar{e}_\theta)\cos\phi - \sin\phi\bar{e}_\phi\right\}\right] + r\cos\theta\left[\cot\theta\sin\phi\left\{(\sin\theta\bar{e}_\rho + \cos\theta\bar{e}_\theta)\sin\phi + \cos\phi\bar{e}_\phi\right\}\right] - r\cos\theta \cdot \text{cosec}\theta\bar{e}_\theta$$

Collecting the coefficients of $\bar{e}_\rho, \bar{e}_\theta, \bar{e}_\phi$

$$\bar{r} = [r\sin^2\theta \cdot \cos^2\phi + r\sin^2\theta \cdot \sin^2\phi + r\cos\theta \cdot \cot\theta \cdot \sin\theta \cdot \cos^2\phi + r\cos\theta \cdot \cot\theta \cdot \sin^2\phi \cdot \sin\theta]\bar{e}_\rho$$
$$+ \bar{e}_\theta[r\sin\theta \cdot \cos^2\phi \cdot \cos\theta + r\sin\theta\sin^2\phi \cdot \cos\theta + r\cos\theta \cdot \cot\theta \cdot \cos^2\phi\cos\theta + r\cos\theta \cdot \cot\theta \cdot \sin^2\phi\cos\theta - r\cos\theta\cos\theta]$$
$$+ [-r\sin\theta \cdot \cos\phi\sin\phi + r\sin\theta \cdot \sin\phi\cos\phi - r\cos\theta\cot\theta \cdot \cos\phi \cdot \sin\phi + r\cos\theta \cdot \cot\theta \cdot \sin\phi\cos\phi]\bar{e}_\phi$$

Simplifying, we get

$$\bar{r} = r\bar{e}_\rho + 0 \cdot \bar{e}_\theta + 0\bar{e}_\phi \qquad (1)$$

Differentiating (1) w.r.t. 't'

$$\text{Velocity} = \bar{V} = \frac{d\bar{r}}{dt} = \frac{dr}{dt}\bar{e}_\rho + r\frac{d}{dt}\bar{e}_\rho \qquad (2)$$

Here

$$\frac{d}{dt}\bar{e}_\rho = \frac{d}{dt}(\sin\theta\cos\phi\bar{i} + \sin\theta\sin\phi\bar{j} + \cos\theta\bar{k})$$
$$= \cos\theta \cdot \dot{\theta}\cos\phi i - \sin\theta\sin\phi \cdot \dot{\phi}\bar{i}$$
$$+ \cos\theta \cdot \dot{\theta} \cdot \sin\phi\bar{j} + \sin\theta \cdot \cos\phi\dot{\phi}\bar{j} - \sin\theta\dot{\theta}\bar{k}$$
$$= \dot{\theta}(\cos\theta \cdot \cos\phi\bar{i} + \cos\theta\sin\phi\bar{j} - \sin\theta\bar{k})$$
$$+ \dot{\phi}\sin\theta(-\sin\phi\bar{i} + \cos\phi\bar{j})$$

$$\frac{d}{dt}(\bar{e}_\rho) = \dot{\theta}\bar{e}_\theta + \dot{\phi}\sin\theta\bar{e}_\phi \qquad (3)$$

Substituting (3) in (1)

$$\bar{V} = \dot{\bar{r}}\bar{e}_\rho + r[\dot{\theta}\bar{e}_\theta + \dot{\phi}\sin\theta\bar{e}_\phi]$$
$$\bar{V} = v_\rho\bar{e}_\rho + v_\theta\bar{e}_\theta + v_\phi\bar{e}_\phi \qquad (4)$$

where

$$v_\rho = \dot{\bar{r}},\ v_\theta = \bar{r}\dot{\theta},\ v_\phi = \bar{r}\dot{\phi}\sin\theta$$

Differentiating (4) w.r.t. 't'

$$\text{Acceleration} = \bar{a} = \frac{d\bar{V}}{dt}$$
$$= \frac{d\bar{v}_\rho}{dt}\bar{e}_\rho + v_\rho\frac{d}{dt}\bar{e}_\rho + \frac{dv_\theta}{dt} \cdot \bar{e}_\theta$$
$$+ v_\theta \cdot \frac{d\bar{e}_\theta}{dt} + \frac{dv_\phi}{dt}\bar{e}_\phi + v_\phi \cdot \frac{d}{dt}\bar{e}_\phi \qquad (5)$$

Here

$$\frac{dv_\rho}{dt} = \frac{d}{dt}\dot{\bar{r}} = \ddot{r},$$
$$\frac{dv_\theta}{dt} = \frac{d}{dt}(\bar{r}\dot{\theta}) = \dot{r}\dot{\theta} + r\ddot{\theta}$$
$$\frac{dv_\phi}{dt} = \frac{d}{dt}(r\dot{\phi}\sin\theta) = \dot{r}\dot{\phi}\sin\theta + r\ddot{\phi}\sin\theta + r\dot{\phi}\dot{\theta}\cos\theta$$
$$\frac{d\bar{e}_\theta}{dt} = -\sin\theta \cdot \dot{\theta}\cos\phi\bar{i} - \cos\theta \cdot \sin\phi\dot{\phi}\bar{i}$$
$$- \sin\theta\dot{\theta}\sin\phi\bar{j} + \cos\theta \cdot \cos\phi\dot{\phi}j - \cos\theta\dot{\theta}\bar{k}$$
$$= -\dot{\theta}(\sin\theta \cdot \cos\phi\bar{i} + \sin\theta\sin\phi\bar{j} + \cos\theta\bar{k})$$
$$+ \dot{\phi}\cos\theta(-\sin\phi\bar{i} + \cos\phi\bar{j})$$

$$\frac{d\bar{e}_\theta}{dt} = -\dot{\theta}\bar{e}_\rho + \dot{\phi}\cos\theta\bar{e}_\phi$$
$$\frac{d\bar{e}_\phi}{dt} = \frac{d}{dt}(-\sin\phi i + \cos\phi j) = -\cos\phi\dot{\phi}i - \sin\phi\dot{\phi}j$$
$$= -\dot{\phi}(\cos\phi i + \sin\phi j)$$

Substituting these values in (5), we get and replacing $\bar{i}$ and $\bar{j}$ by (7) and (6) of previous Example 22

$$\bar{a} = \ddot{r}\bar{e}_\rho + \dot{r}(\dot{\theta}\bar{e}_\theta + \dot{\phi}\sin\theta\bar{e}_\phi) + (\dot{r}\dot{\theta} + r\ddot{\theta})\bar{e}_\theta$$
$$+ r\dot{\theta}(-\dot{\theta}\bar{e}_\rho + \dot{\phi}\cos\theta\bar{e}_\phi) + (\dot{r}\dot{\phi}\sin\theta + r\ddot{\phi}\sin\theta$$
$$+ r\dot{\phi}\dot{\theta}\cos\theta)\bar{e}_\phi - r\dot{\phi}^2\sin\theta \cdot \cos\phi\left[(\sin\theta\bar{e}_\rho\right.$$
$$\left.+ \cos\theta\bar{e}_\theta)\cos\phi - \sin\phi\bar{e}_\phi\right] - r\dot{\phi}^2\sin\theta\sin\phi$$
$$\times\left[(\sin\theta\bar{e}_\rho + \cos\theta\bar{e}_\theta)\sin\phi + \cos\rho\bar{e}_\phi\right]$$

Rearranging the terms

$$\bar{a} = \left[\ddot{r} - r\dot{\theta}^2 - r\dot{\phi}^2\sin^2\theta(\cos^2\phi + \sin^2\phi)\right]\bar{e}_\rho$$
$$+ \left[2\dot{r}\dot{\theta} + r\ddot{\theta} - r\dot{\phi}^2\sin\theta \cdot \cos\theta \cdot (\cos^2\phi\right.$$
$$\left.+ \sin^2\rho)\right]\bar{e}_\theta + [\dot{r}\dot{\phi}\sin\theta + 2r\dot{\theta}\dot{\phi}\cos\theta + \dot{r}\dot{\phi}\sin\theta$$
$$+ r\ddot{\phi}\sin\theta + r\dot{\phi}^2\sin\theta \cdot \cos\phi\sin\phi$$
$$- r\dot{\phi}^2\sin\theta \cdot \sin\phi\cos\phi]\bar{e}_\phi$$

Thus

$$\bar{a} = a_r\bar{e}_r + a_\theta\bar{e}_\theta + a_\phi\bar{e}_\phi$$

where
$$a_r = \ddot{r} - r\dot{\theta}^2 - r\sin^2\theta\dot{\phi}^2$$
$$a_\theta = \frac{1}{r}\frac{d}{dt}(r^2\dot{\theta}) - r\sin\theta\cos\theta\dot{\phi}^2$$
$$a_\phi = \frac{1}{r\sin\theta} \cdot \frac{d}{dt}(r^2\sin^2\theta\dot{\phi}).$$

Exercise

Find the equations of the vector fields $\bar{A}$ where:

1. $\bar{A} = \bar{e}_\rho + \frac{1}{\rho}\bar{e}_\theta + \bar{e}_z$ (cylindrical)

Ans. $\rho = \theta + c_1,\ \rho = z + c_2$

2. $\bar{A} = 2\alpha\frac{\cos\theta}{\rho^3}\bar{e}_r + \frac{\alpha\sin\theta}{\rho^3}\bar{e}_\theta,\ \alpha = \text{constant}$ (spherical)

Ans. $\phi = c_1,\ \rho = c_2\sin^2\theta$

Find the gradient of the scalar fields f:

3. $f = \rho z \cos\theta$ (cylindrical)

Ans. $(z\cos\theta)\bar{e}_\rho - (z\sin\theta)\bar{e}_\theta + (\rho\cos\theta)\bar{e}_z$

4. $f = \rho^2 \sin 2\theta \sin\phi$ (spherical)

Ans. $(2\rho\sin 2\theta\sin\phi)\bar{e}_\rho + (2\rho\cos 2\theta\sin\phi)\bar{e}_\theta + (2\rho\cos\theta\cos\phi)\bar{e}_\phi$

5. $f = xyz$ in (a) cylindrical (b) spherical coordinates

Ans. **a.** $(\rho z\sin 2\theta)\bar{e}_\rho + (\rho z\cos 2\theta)\bar{e}_\theta + \left(\frac{\rho^2}{2}\sin 2\theta\right)\bar{e}_z$

b. $(3\rho^2\sin^2\theta\cos\theta \cdot \sin\phi\cos\phi)\bar{e}_\rho + \frac{\rho^2}{2}\sin 2\phi\{-\sin^3\theta + 2\sin\theta\cos^2\theta\}\bar{e}_\theta + (\rho^2\sin\theta\cos\theta\cos 2\theta)\bar{e}_\phi$

6. $f = \rho^2 + 2\rho\cos\theta - e^z\sin\theta$ (in cylindrical)

Ans. $2(\rho + \cos\theta)\bar{e}_\rho - (2\sin\theta + \frac{1}{\rho}e^z\cos\theta)e_\theta - e^z\sin\theta e_z$

7. $f = 3\rho^2\sin\theta + e^\rho\cos\phi - r$ (in spherical)

Ans. $(6\rho \cdot \sin\theta + e^\rho\cos\phi - 1)\bar{e}_\rho + 3\rho\cos\theta\bar{e}_\theta - \frac{e^\rho\sin\phi}{\rho\sin\theta}e_\phi$

Compute the divergence of $\bar{A}$:

8. $\bar{A} = \theta\arctan\rho\bar{e}_\rho + 2\bar{e}_\theta - z^2e^z\bar{e}_z$ (in cylindrical)

Ans. $\frac{\theta}{\rho}\arctan\rho + \frac{\theta}{1+\rho^2} - (z^2+2z)e^z$

9. $\bar{A} = \rho^2\bar{e}_\rho - 2\cos^2\phi\bar{e}_\theta + \frac{\phi}{\rho^2+1}\bar{e}_\phi$ (in spherical)

Ans. $4\rho - \frac{2}{\rho}\cos^2\phi\cot\theta + \frac{1}{\rho(\rho^2+1)\sin\theta}$

Compute the curl of $\bar{A}$:

10. $\bar{A} = (2\rho + \alpha\cos\phi)\bar{e}_\rho - \alpha\sin\theta\bar{e}_\theta + \rho\cos\theta\bar{e}_\phi$, $\alpha =$ constant (in spherical)

Ans. $\frac{\cos 2\theta}{\sin\theta}\bar{e}_r - (2\cos\theta + \frac{\alpha\sin\phi}{\rho\sin\theta})\bar{e}_\theta - \frac{\alpha\sin\theta}{\rho}\bar{e}_\phi$

11. $\bar{A} = \cos\theta\bar{e}_\rho - \frac{\sin\theta}{\rho}\bar{e}_\theta + \rho^2\bar{e}_z$ (in cylindrical)

Ans. $-2\rho\bar{e}_\theta + \frac{\sin\theta}{\rho}\bar{e}_z$

12. **a.** Show that $\bar{A} = z\left\{(\sin\theta)\bar{e}_\rho + \cos\theta\bar{e}_\theta\right\} - \rho\cos\theta\bar{e}_z$ (in cylindrical) is solenoidal.

Hint: $\nabla \cdot \bar{A} = 0$

b. Show that $\bar{A} = (\rho z\sin 2\theta)\bar{e}_\rho + (\rho z\cos 2\theta)\bar{e}_\theta + \frac{\rho^2\sin^2\theta}{2}\bar{e}_z$ is irrotational.

Hint: $\nabla \times \bar{A} = 0$.

13. Show that $\bar{A}$ is a potential field where $\bar{A} = \frac{2\cos\theta}{\rho^3}\bar{e}_r + \frac{\sin\theta}{\rho^3}e_\theta$ (in spherical)

14. Show that $\bar{A} = f(\rho)\bar{e}_\rho$ is a potential field where f is any differentiable function.

Hint: $\bar{e}_1 = \cos\theta\bar{i} + \sin\theta\bar{j}$, $\bar{e}_2 = -\sin\theta\bar{i} + \cos\theta\bar{k}$, $\bar{e}_3 = \bar{k}$ prove that $\bar{e}_1 \cdot \bar{e}_2 = \bar{e}_2 \cdot \bar{e}_3 = \bar{e}_1 \cdot \bar{e}_3 = 0$.

15. In cylindrical coordinate ρ, θ, z, show that $\nabla(\log\rho)$ and $\nabla\theta$ are solenoidal vectors (if $\rho \neq 0, \theta \neq 0$).

16. Show that $\nabla^2 f = 2\rho^2\cos 2\theta$ when $f = \rho^2 z^2\cos 2\theta$ (in cylindrical)

17. Prove that $\nabla^2 f = 2\sin 2\theta + 2\cot\theta\cos 2\theta - \frac{2}{\rho^2}\operatorname{cosec}^2\theta\cos 2\phi$ if $f = \rho^2\sin 2\theta + \cos^2\phi$ (in spherical)

18. Represent the vector $\bar{A} = zi - 2xj + yk$ in cylindrical coordinates. Determine A_ρ, A_θ, A_z.

Ans. $A_\rho = z\cos\theta - 2\rho\cos\theta \cdot \sin\theta$

$A_\theta = -z\sin\theta - 2\rho\cos^2\theta$

$A_z = \rho\sin\theta$

19. Represent the vector $\bar{A} = xy\bar{i} - zj + xz\bar{k}$ in spherical coordinate system.

Ans. $A_\rho = \rho^2\sin^3\theta\sin\phi\cos^2\phi - \rho\sin\theta\cos\theta\sin\phi + \rho^2\sin\theta\cos^2\theta\cos\phi$

$A_\theta = \rho^2\sin^2\theta\cos\theta\sin\phi\cos^2\phi - \rho\cos^2\theta \cdot \sin\phi - \rho^2\sin^2\theta \cdot \cos\theta\cos\phi$

$A_\phi = -\rho^2\sin^2\theta\sin^2\phi\cos\phi - \rho\cos\theta\cos\phi$

20. Express $\bar{A} = 2y\bar{i} - z\bar{j} + 3x\bar{k}$ in spherical polar coordinate system.

Ans. $A_\rho = 2\rho\sin^2\theta \cdot \sin\phi\cos\phi - \rho\sin\theta\cos\theta \cdot \sin\phi + 3\rho\sin\theta\cos\theta\cos\rho$

$A_\theta = 2\rho\sin\theta \cdot \cos\theta\sin\phi\cos\phi$

Table 15.1 Table of Coordinates

	Cartesian Coordinates (x, y, z)	Curvilinear Coordinates	Spherical Coordinates (r, θ, ϕ)	Cylindrical Coordinates (ρ, θ, z)
1. Equations of transformation of coordinates	$x = x$ $y = y$ $z = z$	$q_1 = q_1(x, y, z)$ $q_2 = q_2(x, y, z)$ $q_3 = q_3(x, y, z)$	$x = r\sin\theta\cos\phi$ $y = r\sin\theta\sin\phi$ $z = r\cos\theta$	$x = \rho\cos\theta$ $y = \rho\sin\theta$ $z = z$
2. Scale factors	$h_1 = 1, h_2 = 1, h_3 = 1$	$h_1 h_2 h_3$	$h_1 = 1, h_2 = r, h_3 = r\sin\theta$	$h_1 = 1, h_2 = \rho, h_3 = 1$
3. Base vectors	$\bar{i}, \bar{j}, \bar{k}$	$\bar{e}_1, \bar{e}_2, \bar{e}_3$	$\bar{e}_r = (\sin\theta\cos\phi)i$ $+ (\sin\theta\sin\phi)j + (\cos\theta)k$ $e_\theta = (\cos\theta\cos\phi)i +$ $(\cos\theta\sin\phi)j - (\sin\theta)k$ $e_\phi = (-\sin\phi)i + (\cos\phi)j$	$e_\rho = (\cos\theta)i + (\sin\theta)j$ $e_\theta = (-\sin\theta)i + (\cos\theta)j$ $e_z = k$
4. Jacobian (J)	$\frac{\partial(x,y,z)}{\partial(x,y,z)} = 1$	$J\left(\frac{x,y,z}{q_1,q_2,q_3}\right) = h_1 h_2 h_3$	$\frac{\partial(x,y,z)}{\partial(r,\theta,\phi)} = r^2\sin\theta$	$\frac{\partial(x,y,z)}{\partial(\rho,\theta,z)} = \rho$
5. (Arc Length)^2	$(ds)^2 = (dx)^2 + (dy)^2 + (dz)^2$	$(ds)^2 = h_1^2 dq_1^2 + h_2^2 dq^2 + h_3^2 dq_3^2$	$(ds)^2 = (dr)^2 + r^2(d\theta)^2 +$ $r^2\sin^2\theta(d\phi)^2$	$ds^2 = (d\rho)^2 + \rho^2(d\theta)^2 +$ $(dz)^2$
6. Area elements on the coordinate surfaces	$ds_1 = dy\,dz$ $ds_2 = dz\,dx$ $ds_3 = dx\,dy$	$dA_1 = h_2 h_3 dq_2 dq_3$ $dA_2 = h_1 h_3 dq_1 dq_3$ $dA_3 = h_1 h_2 dq_1 dq_2$	$ds_r = r^2\sin\theta\,d\theta\,d\phi$ $ds_\theta = r\sin\theta\,d\phi\,dr$ $ds_\phi = r\,dr\,d\theta$	$ds_\rho = \rho d\theta\,dz$ $ds_\theta = dz\,d\rho$ $ds_z = \rho d\rho d\theta$
7. Volume element (dv)	$dv = dx\,dy\,dz$	$dv = h_1 h_2 h_3 dq_1 dq_2 dq_3$	$dv = r^2\sin\theta dr d\theta d\phi$	$dv = \rho d\rho d\theta dz$
8. Grad f	$\nabla f = \frac{\partial f}{\partial x}i + \frac{\partial f}{\partial y}j + \frac{\partial f}{\partial z}k$	$\frac{1}{h_1}\frac{\partial f}{\partial q_1}\bar{e}_1 + \frac{1}{h_2}\cdot\frac{\partial f}{\partial q_2}\bar{e}_2 + \frac{1}{h_3}\frac{\partial f}{\partial q_3}\bar{e}_3$	$\nabla f = \frac{\partial f}{\partial r}e_1 + \frac{1}{r}\frac{\partial f}{\partial \theta}e_\theta$ $+\frac{1}{r\sin\theta}\frac{\partial f}{\partial \phi}e_\phi$	$\nabla f = \frac{\partial f}{\partial \rho}e_\rho + \frac{1}{\rho}\frac{\partial f}{\partial \theta}e_\theta +$ $\frac{\partial f}{\partial z}e_z$
9. Div A	$\frac{\partial A_1}{\partial x} + \frac{\partial A_2}{\partial y} + \frac{\partial A_3}{\partial z}$	$\frac{1}{h_1 h_2 h_3}\left[\frac{\partial}{\partial q_1}(h_2 h_3 A_1) + \frac{\partial}{\partial q_2}(h_3 h_1 A_2)\right.$ $\left.+\frac{\partial}{\partial q_3}(h_1 h_2 A_3)\right]$	Div A $= \frac{1}{r^2\sin\theta}\left[\frac{\partial}{\partial r}(r^2\sin\theta A_1)\right.$ $\left.+\frac{\partial}{\partial \theta}(r\sin\theta A_2) + \frac{\partial}{\partial \phi}(rA_3)\right]$	$\frac{1}{\rho}\left[\frac{\partial}{\partial \rho}(\rho A_1) + \frac{\partial}{\partial \theta}(A_2)\right.$ $\left.+\frac{\partial}{\partial z}(\rho A_3)\right]$
10. Curl A	$\begin{vmatrix} i & j & k \\ \partial/\partial x & \partial\partial y & \partial/\partial z \\ A_1 & A_2 & A_3 \end{vmatrix}$	$\frac{1}{h_1 h_2 h_3}\begin{vmatrix} h_1\bar{e}_1 & h_2\bar{e}_2 & h_3\bar{e}_3 \\ \frac{\partial}{\partial q_1} & \frac{\partial}{\partial q_2} & \frac{\partial}{\partial q_3} \\ h_1 A_1 & h_2 A_2 & h_3 A_3 \end{vmatrix}$	$\frac{1}{r^2\sin\theta}\begin{vmatrix} e_r & re_\theta & (r\sin\theta)e_\phi \\ \partial/\partial r & \partial/\partial\theta & \partial/\partial\phi \\ A_1 & rA_2 & (r\sin\theta)A_3 \end{vmatrix}$	$\frac{1}{\rho}\begin{vmatrix} e_\rho & \rho e_\theta & e_z \\ \partial/\partial\rho & \partial\partial\theta & \partial/\partial_z \\ A_1 & \rho A_2 & A_3 \end{vmatrix}$
11. Laplacian $(\nabla^2 f)$	$\frac{\partial^2 f}{\partial x^2} + \frac{\partial^2 f}{\partial y^2} + \frac{\partial^2 f}{\partial z^2}$	$\frac{1}{h_1 h_2 h_3}\left[\frac{\partial}{\partial q}\left(\frac{h_2 h_3}{h_1}\frac{\partial f}{\partial q_1}\right) + \frac{\partial}{\partial q_2}\right.$ $\left.\left(\frac{h_3 h_1}{h_2}\frac{\partial f}{\partial q_2}\right) + \frac{\partial}{\partial q_3}\left(\frac{h_1 h_2}{h_3}\frac{\partial f}{\partial q_3}\right)\right]$	$\frac{\partial^2 f}{\partial r^2} + \frac{2}{r}\frac{\partial f}{\partial r} + \frac{1}{r^2}\frac{\partial^2 f}{\partial \theta^2} + \frac{\cot\theta}{r^2}\frac{\partial f}{\partial \theta}$ $+\frac{1}{r^2\sin^2\theta}\frac{\partial^2 f}{\partial \phi^2}$	$\frac{\partial^2 f}{\partial \rho^2} + \frac{1}{\rho}\frac{\partial f}{\partial \rho} + \frac{1}{\rho^2}\frac{\partial^2 f}{\partial \theta^2} + \frac{\partial^2 f}{\partial z^2}$

$$-\rho\cos^2\theta\sin\phi - 3\rho\sin^2\theta\cos\phi$$

$$A_\phi = -2\rho\sin\theta\sin^2\phi - \rho\cos\theta\cos\rho$$

21. Prove that a cylindrical coordinate system is orthogonal.

22. Find the square of the element of arc length in spherical coordinates and determine the corresponding scale factors.

Ans. $(ds)^2 = (dr)^2 + \rho^2(d\theta)^2 + \rho^2\sin^2\theta(d\phi)^2$

$h_1 = h_\rho = 1,\ h_2 = h_\theta = \rho,\ h_3 = h_\phi = \rho\sin\theta$

23. Prove that in any orthogonal curvilinear coordinate system, $\nabla\cdot\nabla\times\bar{A} = 0$

24. If q_1, q_2, q_3 are general coordinates, show that $\frac{\partial\bar{r}}{\partial q_1}, \frac{\partial\bar{r}}{\partial q_2}, \frac{\partial\bar{r}}{\partial q_3}$ and $\nabla q_1, \nabla q_2, \nabla q_3$ are reciprocal system of vectors.

Hint: Use $\nabla q_1\cdot d\bar{r} = dq_1 = \left(\nabla q_1\cdot\frac{\partial\bar{r}}{\partial q_1}\right)dq_1 + \left(\nabla q_1\cdot\frac{\partial\bar{r}}{\partial q_2}\right)dq_2 + \left(\nabla q_1\cdot\frac{\partial\bar{r}}{\partial q_3}\right)dq_3$

25. Prove that

$$\left\{\frac{\partial\bar{r}}{\partial q_1}\cdot\frac{\partial\bar{r}}{\partial q_2}\times\frac{\partial\bar{r}}{\partial q_3}\right\}\{\nabla q_1\cdot\nabla q_2\times\nabla q_3\} = 1.$$

Hint: $V\left(\frac{1}{V}\right) = 1$ or $J\left(\frac{x,y,z}{q_1,q_2,q_3}\right)\cdot J\left(\frac{q_1,q_2,q_3}{x,y,z_1}\right) = 1$

26. In cylindrical coordinate system ρ, θ, z prove that

$$\nabla\rho^n = n\rho^{n-1}\bar{e}_\rho$$

$$\nabla^2(\rho^n\cos n\theta) = 0.$$

27. In spherical coordinate system ρ, θ, ϕ, prove that

28. $\nabla\cdot[\bar{e}_\rho\cdot\cot\phi - 2\bar{e}_\phi] = 0$

29. $\nabla^2\left[\left(\rho + \frac{1}{\rho^2}\right)\cos\phi\right] = 0$

30. Find $\frac{\partial\bar{r}}{\partial q_1}, \frac{\partial\bar{r}}{\partial q_2}, \frac{\partial\bar{r}}{\partial q_3}, \nabla q_1, \nabla q_2, \nabla q_3$ in spherical coordinate system

Ans.

$$\frac{\partial\bar{r}}{\partial\rho} = \sin\theta\cos\phi i + \sin\theta\sin\phi j + \cos\theta k$$

$$\frac{\partial\bar{r}}{\partial\theta} = \rho\cos\theta\cos\rho i + \rho\cos\theta\sin\phi j - \rho\sin\theta k$$

$$\frac{\partial\bar{r}}{\partial\phi} = -\rho\sin\theta\sin\phi\bar{i} + \rho\sin\theta\cos\phi j$$

$$\nabla\rho = \sin\theta\cos\phi\bar{i} + \sin\theta\sin\phi\bar{j} + \cos\theta\bar{k}$$

$$\nabla\theta = \frac{\cos\theta\cdot\cos\phi\bar{i} + \cos\theta\cdot\sin\phi\bar{j} - \sin\theta\bar{k}}{\rho}$$

$$\nabla\phi = \frac{-\sin\phi\bar{i} + \cos\phi j}{\rho\sin\theta}$$

31. Express the velocity $\bar{V}$ and acceleration $\bar{a}$ of a particle in cylindrical coordinates.

Ans. $\bar{r} = \rho\bar{e}_\rho + z\bar{e}_z$

$\bar{V} = \dot{\rho}\bar{e}_\rho + \rho\dot{\theta}e_\theta + \dot{z}e_z$

$\bar{a} = (\ddot{\rho} - \rho\dot{\theta}^2)\bar{e}_\rho + (\rho\ddot{\theta} + 2\dot{\rho}\dot{\theta})\bar{e}_\theta + \ddot{z}\bar{e}_z$

Chapter 16

Vector Integral Calculus

INTRODUCTION

Vector integral calculus extends the concepts of (ordinary) integral calculus to vector functions. It has applications in fluid flow, design of underwater transmission cables, heat flow in stars, study of satellites. Line integrals are useful in the calculation of work done by variable forces along paths in space and the rates at which fluids flow along curves (circulation) and across boundaries (flux). In this chapter we consider three important integral theorems. Green's theorem, a great theorem of calculus, which converts line integrals to double integrals, evaluates flow and flux integrals across closed plane curves in non-conservative vector fields. Stokes theorem states that the circulation of a vector field around the boundary of a surface in space equals the integral of the normal component of the curl of the field over the surface. Gauss divergence theorem, which is important in electricity, magnetism and fluid flow, says that the outward flux of a vector field across a closed surface equals the triple integral of the divergence of the field over the region enclosed by the surface.

16.1 VECTOR INTEGRATION: INTEGRATION OF A VECTOR FUNCTION OF A SCALAR ARGUMENT

Definitions

Primitive

A vector function $\overline{F}(u)$ is the primitive of the vector function $\overline{f}(u)$ if

$$\frac{d\overline{F}(u)}{du} = \overline{f}(u)$$

Indefinite integral

Indefinite integral of the vector function $\overline{f} = \overline{f}(u)$ of a scalar argument u is the collection of all primitive functions of $\overline{f}(u)$ and is denoted by

$$\int \overline{f}(u)du = \int \frac{d\overline{F}(u)}{du}du = \overline{F}(u) + \overline{c}$$

where c is an arbitrary constant vector.

Properties

1. $\int \alpha \overline{f}(u)du = \alpha \int \overline{f}(u)du$,
   (α = numerical constant)
2. $\int [\overline{f}(u) \pm \overline{g}(u)]du = \int \overline{f}(u)du \pm \int \overline{g}(u)du$
3. If $\overline{f}(u) = f_1(u)i + f_2(u)j + f_3(u)k$ then
   $\int \overline{f}(u)du = i\int f_1(u)du + j\int f_2(u)du + k\int f_3(u)du$

Note: The integration of a vector function reduces to the evaluation of three ordinary real (scalar) integrals.

Definite integral between limits $u = a$ and $u = b$ is

$$\int_a^b \overline{f}(u)du = \overline{F}(u) + \overline{c}\Big|_a^b = \overline{F}(b) - \overline{F}(a)$$

WORKED OUT EXAMPLES

Example 1: Evaluate

a. $\int \overline{A}(u)\,du$ and

b. $\int_2^4 \bar{A}(u)\,du$

if $\bar{A}(u) = (3u^2 - u)i + (2 - 6u)j - 4uk.$

Solution:

a. $$\int \bar{A}(u)du = \int \left[(3u^2 - u)i + (2 - 6u)j - 4uk\right] du$$

$$= i\int (3u^2 - u)du + j\int (2 - 6u)du$$

$$+k\int -4udu$$

$$= \left(u^3 - \frac{u^2}{2}\right) i + (2u - 3u^2)j$$

$$-2u^2k + \bar{c}$$

b. $$\int_2^4 \bar{A}(u)du = \left(u^3 - \frac{u^2}{2}\right) i$$

$$+(2u - 3u^2)j - 2u^2k\Big|_2^4$$

$$= 50i - 32j - 24k.$$

Example 2: If $\bar{A}(u) = ui - u^2 j + (u - 1)k$ and $\bar{B}(u) = 2u^2 i + 6uk$ then evaluate $\int_0^2 \bar{A} \times \bar{B}\,du$.

Solution:

$$\bar{A} \times \bar{B} = \begin{vmatrix} i & j & k \\ u & -u^2 & (u-1) \\ 2u^2 & 0 & 6u \end{vmatrix}$$

$$\bar{A} \times \bar{B} = -6u^3 i + (2u^3 - 8u^2)j + 2u^4 k$$

$$\int_0^2 \bar{A} \times \bar{B}du = \int_0^2 \left[-6u^3 i + (2u^3 - 8u^2)j + 2u^4 k\right] du$$

$$= -\frac{6u^4}{4} i + \left(\frac{2u^4}{4} - \frac{8u^3}{3}\right) j + \frac{2u^5}{5} k \Big|_0^2$$

$$= -24i - \frac{40}{3} j + \frac{64}{5} k.$$

Example 3: Let $\bar{A} = ti - 3j + 2tk$, $\bar{B} = i - 2j + 2k$, $\bar{D} = 3i + tj - k$. Evaluate $\int_1^2 \bar{A} \cdot \bar{B} \times \bar{D}\,dt$.

Solution:

$$\bar{A} \cdot \bar{B} \times \bar{D} = \begin{vmatrix} t & -3 & 2t \\ 1 & -2 & 2 \\ 3 & t & -1 \end{vmatrix}$$

$$= t(2 - 2t) + 3(-1 - 6) + 2t(t + 6)$$

$$= +14t - 21$$

$$\int_1^2 (\bar{A} \cdot \bar{B} \times \bar{D})dt = \int_1^2 (14t - 21)dt = (7t^2 - 21t)\Big|_1^2$$

$$= 0.$$

Example 4: The acceleration $\bar{a}$ of a particle at any time $t \geq 0$ is given by

$$\bar{a}(t) = e^{-t} i - 6(t + 1)j + 3\sin t\, k$$

If the velocity $\bar{v}$ and displacement $\bar{r}$ are zero at $t = 0$, find $\bar{v}$ and $\bar{r}$ at any time t.

Solution: $\bar{a} = \frac{d\bar{v}}{dt} = e^{t} i - 6(t + 1)j + 3\sin tk.$
Integrating with respect to t

$$\bar{v}(t) = i\int e^{-t}dt - j\int 6(t + 1)dt + 3\bar{k}\int \sin t\,dt$$

$$= -e^{-t} i - j(3t^2 + 6t) - 3\bar{k}\cos t + \bar{c}$$

Given that $\bar{v} = 0$ when $t = 0$

so $$0 = -i - 3\bar{k} + \bar{c}$$

or $$\bar{c} = \bar{i} + 3\bar{k}$$

Thus

$$\bar{v}(t) = (1 - e^{-t})i - j(3t^2 + 6t) + 3k(1 - \cos t)$$

Integrating $\bar{v} = \frac{d\bar{r}}{dt}$ with respect to t

$$\bar{r}(t) = \int \bar{v}(t)dt = i\int (1 - e^{-t})dt - j\int (3t^2 + 6t)dt$$

$$+3k\int (1 - \cos t)dt$$

$$\bar{r} = i(t + e^{-t}) - j(t^3 + 3t^2) + 3k(t - \sin t) + \bar{c}_1$$

since $\bar{r} = 0$ at $t = 0$, we have

$$0 = i + \bar{c}_1 \quad \text{so} \quad \bar{c}_1 = -\bar{i}$$

Thus

$$\bar{r}(t) = i(-1 + t + e^{-t}) - j(t^3 + 3t^2) + (3t - 3\sin t)\bar{k}.$$

Example 5: If $\bar{A}(u) = ui + u^2 j + u^3 k$, $\bar{B}(u) = u^3 i + u^2 j + uk$

find $$\int_1^2 \left[\bar{A}(u) \times \frac{d\bar{B}}{du} + \frac{d\bar{A}}{du} \times \bar{B}(u)\right] du$$

Solution:

$$I = \int_1^2 \left[A \times \frac{dB}{du} + \frac{dA}{du} \times B\right] du$$

$$= \int_1^2 \frac{d}{du}(\overline{A} \times \overline{B})du = \overline{A} \times \overline{B} + \overline{c}\,\Big|_1^2$$

since $\frac{d}{du}(\overline{A} \times \overline{B}) = \overline{A} \times \frac{d\overline{B}}{du} + \frac{d\overline{A}}{du} \times \overline{B}$.

Here $\overline{A} \times \overline{B} = \begin{vmatrix} i & j & k \\ u & u^2 & u^3 \\ u^3 & u^2 & u \end{vmatrix}$

$$= (u^3 - u^5)i + (u^6 - u^2)j + (u^3 - u^5)k$$

$$I = -24i + 60j - 24k.$$

Example 6: Evaluate $\int_2^3 \overline{A} \cdot \frac{d\overline{A}}{dt}$ if $\overline{A}(2) = 2i - j + 2k$ and $\overline{A}(3) = 4i - 2j + 3k$.

Solution: Since $\overline{A} \cdot \frac{d\overline{A}}{dt} = \frac{A\,dA}{dt}$ where $A = |\overline{A}|$

$$\int_2^3 \overline{A} \cdot \frac{d\overline{A}}{dt}dt = \int_2^3 A\frac{dA}{dt}dt = \int_2^3 A\,dA$$

$$= \frac{A^2}{2}\Big|_2^3 = \frac{1}{2}[A^2(3) - A^2(2)]$$

$$= \frac{1}{2}[29 - 9] = 10$$

since

$$A(3) = |A|(3)| = \sqrt{16 + 4 + 9} = \sqrt{29},$$

$$A(2) = |\overline{A(2)}| = \sqrt{4 + 1 + 4} = \sqrt{9}.$$

EXERCISE

1. Find (a) $\int \overline{A}(u)du$ and (b) $\int_1^2 \overline{A}(u)du$ if $\overline{A}(u) = (u - u^2)i + 2u^3 j - 3k$.

Ans. **a.** $\left(\frac{u^2}{2} - \frac{u^3}{3}\right)i + \frac{u^4}{2}j - 3uk + \overline{c}$

b. $\frac{-5}{6}i + \frac{15}{2}j - 3k$

2. Evaluate $\int_0^{\pi/2}(3 \sin ui + 2 \cos uj)du$.

Ans. $3i + 2j$

3. Find $\int \overline{A}(u)du$ where $\overline{A}(u) = (3u^2 - \sin u)i + (e^u + \cos u)j + 4u^3 k$.

Ans. $(u^3 + \cos u)i + (e^u + \sin u)j + u^4 k + c$

4. Find $\int_0^1 t\overline{F}(t)dt$ when $\overline{F}(t) = 2ti - t^2 j + t^3 k$.

Ans. $\frac{2}{3}u^3 i - \frac{u^4}{4}j + \frac{u^5}{5}k\,\Big|_0^1 = \frac{2}{3}i - \frac{1}{4}j + \frac{1}{5}k$

5. Evaluate $\int_0^2 \overline{A} \cdot \overline{B}\,dt$ where $\overline{A} = ti - t^2 j + (t - 1)k$ and $B = 2t^2 i + 6tk$.

Ans. 12

6. Let $\overline{A} = ti - 3j + 2tk$, $\overline{B} = i - 2j + 2k$, $\overline{D} = 3i + tj - k$ then evaluate

$$\int_1^2 \overline{A} \times (\overline{B} \times \overline{D})\,dt.$$

Ans. $\frac{-87}{2}i - \frac{44}{3}j + \frac{15}{2}k$

7. Evaluate

$$\int \overline{A} \times \frac{d^2\overline{A}}{dt^2}dt$$

Hint:

$$\frac{d}{dt}\left(\overline{A} \times \frac{d\overline{A}}{dt}\right) = A \times \frac{d^2 A}{dt^2} + \frac{dA}{dt} \times \frac{dA}{dt}$$

$$= A \times \frac{d^2 A}{dt^2}.$$

Ans. $\overline{A} \times \frac{d\overline{A}}{dt} + \overline{c}$

8. Let $\overline{A}(t) = e^{-t} \sin t\ i + e^{-t} \cos tj + t^2 k$. Evaluate

$$\int_1^2 \overline{A} \cdot \frac{d\overline{A}}{dt}dt.$$

Hint: see Worked Out Example 6, on page 16.3, above.

Ans. $\frac{1}{2}[15 + e^{-4} - e^{-2}]$

9. The acceleration of a particle at any time $t \geq 0$ is given by $\overline{a} = 12 \cos 2ti - 8 \sin 2tj + 16tk$.

If the velocity $\overline{V}$ and displacement $\overline{r}$ are zero at $t = 0$, find $\overline{V}$ and $\overline{r}$ at any time t.

Ans. $\overline{V} = 6 \sin 2ti + (4 \cos 2t - 4)j + 8t^2 k$

$\overline{r} = (3 - 3 \cos 2t)i + (2 \sin 2t - 4t)j + 8\frac{t^3}{3}k$

10. Find the velocity and displacemant of a particle having acceleration $4t^3 i - 5t^4 j + 3t^2 k$ at any time t, given that acceleration and velocity are initially $i - j$ and $i + j + k$ respectively.

Ans. Velocity: $(t^4 + 1)i - (t^4 + 1)j + t^3 k$

displacement:

$$\left(\frac{t^5}{5} + t + 1\right)i - \left(\frac{t^6}{6} + t - 1\right)j + \left(\frac{t^4}{4} + 1\right)k$$

11. Find $\bar{A}$ if $\frac{d^2\bar{A}}{du^2} = 6ui - 12u^2 j + 4\cos uk$ and given that $\frac{d\bar{A}}{du} = -1 - 3k$ and $\bar{A} = 2i + j$ when $u = 0$.

Ans. $(u^3 - u + 2)i + (1 - u^4)j + (4 - 4\cos u - 3u)k$

12. Find the areal velocity of a particle which moves along the path $\bar{r} = a\cos wti + b\sin wt\bar{j}$ where a, b, w are constants and t is time.

Hint: Areal velocity $= \frac{1}{2}\bar{r} \times \bar{V}$, where $\bar{V} =$ velocity.

Ans. $\frac{1}{2}a\, bw\bar{k}$

16.2 LINE INTEGRALS: WORK DONE, POTENTIAL, CONSERVATIVE FIELD AND AREA

For the ordinary definite integral $\int_a^b f(x)dx$ the region of integration is an interval $a \le x \le b$ on the x-axis. i.e., we integrate along the x-axis from a to b.

This concept can be generalized to define a definite integral evaluated along a curve.

Line Integrals

Let c be curve defined from A to B with corresponding arc lengths $s = a$ and $s = b$ respectively. Divide c into n arbitrary portions (see Fig. 16.1).

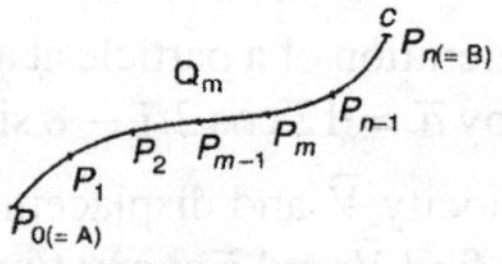

Fig. 16.1

Evaluate the given function f at a point in each of these portions and form the sum

$$J_n = \sum_{m=1}^{n} f(Q_m)\Delta s_m$$

where $\Delta s_m = s_m - s_{m-1}$. The limit of this sum as $n \to \infty$ is known as the line integral of f along c from A to B and is denoted by

$$\int_c f(p)ds = \int_a^b f(s)ds = \int_c f(x, y, z)ds \quad (1)$$

when P has coordinates $x(s)$, $y(s)$, $z(s)$. The line integral (1) is also known as curve integral or curvilinear integral.

Thus in a line integral, the integrand f is integrated (evaluated) along a curve (line). The curve c is known as path of integration. Its end points a and b are called the initial and terminal points.

The direction along the curve c from a to b is called the sense of integration.

Curve is said to be a closed curve (path) when the end points coincide. In such case the line integral is denoted as $\oint_c$.

Let the parametric equation of the curve c be

$$\bar{r}(t) = x(t)\bar{i} + y(t)\bar{j} + z(t)\bar{k}, \quad a \le t \le b \quad (2)$$

Properties of line integrals

Let $\bar{F} = F(\bar{r}) = F_1\bar{i} + F_2\bar{j} + F_3\bar{k}$ be a vector function. Then a line integral of $\bar{F}(\bar{r})$ along (taken over) the curve c is defined as

$$\int_c \bar{F}(\bar{r}) \cdot d\bar{r} = \int_c F_1 dx + F_2 dy + F_3 dz$$

$$= \int_c F_1\frac{dx}{dt} + F_2\frac{dy}{dt} + F_3\frac{dz}{dt}$$

$$= \int_a^b F_1(\bar{r}(t)) \cdot \frac{d\bar{r}}{dt}dt \quad (3)$$

Observation

Evaluation of a line integral reduces to evaluation of an ordinary integral along a coordinate axis (say x-axis).

For the line integral (3) the following properties follow from integral calculus:

1. $\int_c k\bar{F} \cdot d\bar{r} = k\int_c \bar{F} \cdot d\bar{r}$, $k =$ constant
2. $\int_c (\bar{F} \pm \bar{G}) \cdot d\bar{r} = \int_c \bar{F} \cdot d\bar{r} \pm \int_c \bar{G} \cdot d\bar{r}$
3. $\int_c \bar{F} \cdot d\bar{r} = \int_{c_1} \bar{F} \cdot d\bar{r} + \int_{c_2} \bar{F} \cdot d\bar{r}$

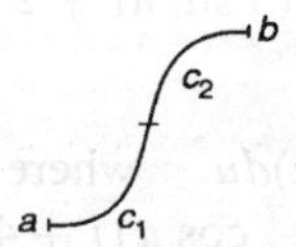

Fig. 16.2

where c is the sum of two curves c_1 and c_2 (see Fig. 16.2)

4. $\int_a^b \bar{F} \cdot d\bar{r} = -\int_b^a \bar{F} \cdot d\bar{r}$.

Applications of Line Integral

A. Work done by a force (work integral)

A natural application of the line integral is to define the work done by a force $\overline{F}$ in moving (displacing) a particle along a curve c from point P_1 to point P_2 as

$$\text{Work done} = \int_{P_1}^{P_2} \overline{F} \cdot d\overline{r} \quad (4)$$

When $\overline{F}$ denotes velocity of a fluid, then the *circulation* of $\overline{F}$ around a closed curve c is defined by

$$\text{Circulation} = \oint_c \overline{F} \cdot d\overline{r}$$

B. Independence of path; conservative field and scalar potential

If $\overline{F} = \nabla\phi$ then the line integral from P_1 and P_2 is independent of path joining P_1 to P_2 (Fig. 16.3)

$$\int_{P_1}^{P_2} \overline{F} \cdot d\overline{r} = \int_{P_1}^{P_2} \nabla\phi \cdot d\overline{r}$$

$$= \int_{P_1}^{P_2} \left(\frac{\partial\phi}{\partial x}dx + \frac{\partial\phi}{\partial y}dy + \frac{\partial\phi}{\partial z}dz \right)$$

$$= \int_{P_1}^{P_2} d\phi = \phi(P_2) - \phi(P_1)$$

Thus the line integral depends only on the end points P_1 and P_2 and not on the path joining them. Recall that when $\overline{F} = \nabla\phi$, then $\nabla \times \overline{F} = \nabla \times \nabla\phi = 0$. In such a case, $\overline{F}$ is called a conservative vector field and ϕ is called its scalar potential.

Note: That a conservative force field is also irrotational (since $\nabla \times \overline{F} = 0$).

Result 1: The work done in a conservative force field in moving a particle from P_1 to P_2 is independent of the path joining P_1 and P_2, but depends only on the end points P_1 and P_2. In such cases a scalar potential ϕ exists such that force field $\overline{F} = \nabla\phi$ and thus the work done from P_1 to $P_2 = \phi(P_2) - \phi(P_1)$ (without the need to evaluate the work integral).

Result 2: In a conservative field $\overline{F}$

$$\oint_c \overline{F} \cdot d\overline{r} = 0$$

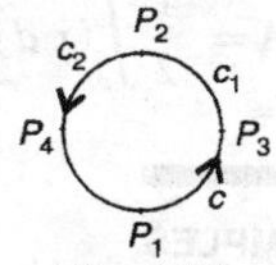

Fig. 16.3

along any closed curve c because

$$\oint_c \overline{F} \cdot d\overline{r} = \int_{P_1P_3P_2P_4} \overline{F} \cdot d\overline{r} = \int_{P_1P_3P_2} + \int_{P_2P_4P_1}$$

$$= \int_{P_1P_3P_2} \overline{F} \cdot d\overline{r} - \int_{P_1P_4P_2} \overline{F} \cdot d\overline{r} = 0.$$

which follows from the independence of path.

C. Test for exact differential

For $\overline{F} = F_1\overline{i} + F_2\overline{j} + F_3\overline{k}$, the necessary and sufficient condition that

$$F_1dx + F_2dy + F_3dz$$

be an exact differential is that $\overline{F}$ must be conservative i.e., $\nabla \times \overline{F} = 0$. When $\nabla \times \overline{F} = 0$, there exists a scalar ϕ such that $\overline{F} = \nabla\phi$. Then

$$F_1dx + F_2dy + F_3dz = \overline{F} \cdot d\overline{r} = \nabla\phi \cdot d\overline{r} = d\phi$$

$$= \text{Exact differential}$$

D. Area A of a regular region D

Bounded by a curve c (Fig. 16.4):

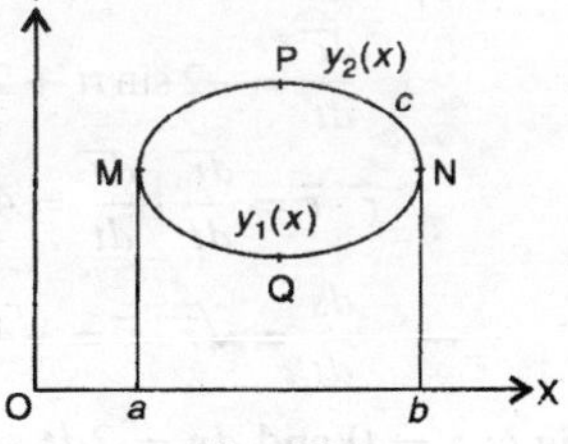

Fig. 16.4

$$A = \int_a^b y_2(x)d_x - \int_a^b y_1(x)dx$$

$$= -\int_{NPM} y\,dx - \int_{MQN} y\,dx = -\int_c ydx$$

Similarly, $A = \int_c x\, dy$

Thus $A = \frac{1}{2}\int_c (x\, dy - y\, dx)$

WORKED OUT EXAMPLES

Example 1: Evaluate $\int_c y^2 dx - 2x^2 dy$ along the parabola $y = x^2$ from (0, 0) to (2, 4).

Solution:

$$\int_c (y^2 dx - 2x^2 dy) = \int_0^2 (x^2)^2 dx - 2x^2 d(x^2)$$

$$= \int_0^2 x^4 dx - 4x^3 dx$$

$$= \frac{x^5}{5} - 4\frac{x^4}{4}\Big|_0^2 = -\frac{48}{5}.$$

Example 2: Evaluate the line integral

$$\int_c x^{-1}(y+z)ds$$

where c the arc of circle $x^2 + y^2 = 4$, $z = 0$ from (2, 0, 0) to $(\sqrt{2}, \sqrt{2}, 0)$ in the counterclockwise direction.

Solution: Equation of circle in parametric form is

$$x = 2\cos t,\ y = 2\sin t$$

when $x = 2$, then $t = 0$ and when $x = \sqrt{2}$, then $t = \frac{\pi}{4}$

$$\bar{r} = xi + yj + zk = 2\cos t i + 2\sin t j + 0$$

$$\frac{d\bar{r}}{dt} = -2\sin t i + 2\cos t j$$

$$\dot{\bar{r}} \cdot \dot{\bar{r}} = \frac{d\bar{r}}{dt} \cdot \frac{d\bar{r}}{dt} = 4\sin^2 t + 4\cos^2 t = 4$$

$$\frac{ds}{dt} = \sqrt{\dot{\bar{r}} \cdot \dot{\bar{r}}} = \sqrt{4} = 2$$

Along c: $z = 0$ and $ds = 2dt$ so that

$$\int_c x^{-1}(y+z)ds = \int \frac{y+0}{x} ds = \int_c \frac{2\sin t}{2\cos t} \cdot 2dt$$

$$= 2\int_0^{\pi/4} \tan t \cdot dt = 2 \ln \sec t \Big|_0^{\frac{\pi}{4}}$$

$$= 2 \ln \sqrt{2} = \ln 2.$$

Example 3: If $\bar{F} = (2x + y^2)i + (3y - 4x)\bar{j}$ evaluate $\oint_c \bar{F} \cdot d\bar{r}$ around a triangle ABC in the xy-plane with $A(0, 0)$, $B(2, 0)$, $C(2, 1)$ (refer Fig. 16.5). (a) in the counterclockwise direction (b) what is the value in the opposite direction?

Solution: **a.** In the counterclockwise direction:

$$I = \oint_c \bar{F} \cdot d\bar{r} = \int_{c_1} \bar{F} \cdot d\bar{r} + \int_{c_2} \bar{F} \cdot d\bar{r} + \int_{c_3} \bar{F} \cdot d\bar{r}$$

$$= I_1 + I_2 + I_3$$

Along c_1: The straight line from $A(0, 0)$ to $B(2, 0)$, $y = 0$, $z = 0$ and x varies from 0 to 2

Thus $\bar{r} = xi$, $d\bar{r} = \bar{i}dx$, $dy = 0$,

So with $y = 0$

$$I_1 = \int_{c1} \bar{F} \cdot d\bar{r} = \int_0^2 (2xi) \cdot idx = \int_0^2 2xdx = x^2\Big|_0^2 = 4$$

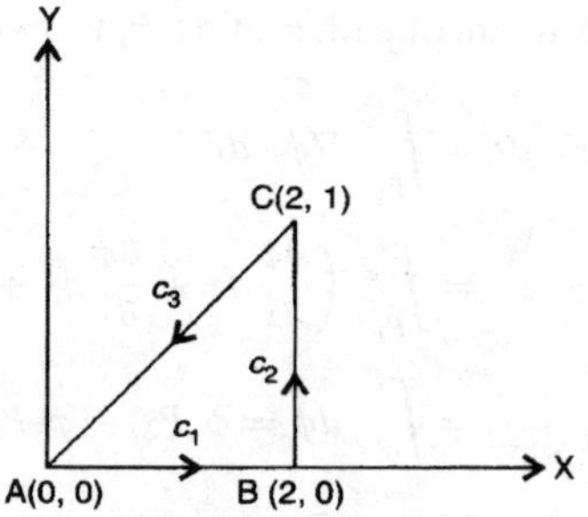

Fig. 16.5

Along c_2: The straight line from $B(2, 0)$ to $C(2, 1)$, $x = 2$, $z = 0$, y varies from 0 to 1

Thus $\bar{r} = 2i + yj$, $d\bar{r} = \bar{j}dy$

So with $x = 2$

$$I_2 = \int_{c_2} \bar{F} \cdot d\bar{r} = \int_0^1 \left[(4 + y^2)i + (3y - 8)j\right] \cdot [jdy]$$

$$= \int_0^1 (3y - 8)dy$$

$$= \frac{3y^2}{2} - 8y\Big|_0^1 = \frac{3}{2} - 8 = -\frac{13}{2}$$

Along c_3: Straight line from C(2,1) to A(0, 0) is $y = \frac{1}{2}x$, $dy = \frac{1}{2}dx$, x varies from 2 to 0

Thus $\bar{r} = xi + \frac{1}{2}x\bar{j}$, $\quad d\bar{r} = \left(i + \frac{1}{2}j\right)dx$

So with $y = \frac{x}{2}$

$$I_3 = \int_{c_3} \overline{F} \cdot d\overline{r} =$$

$$= \int_2^0 \left[\left(2x + \frac{x^2}{4}\right)i + \left(\frac{3x}{2} - 4x\right)j\right] \cdot \left[i + \frac{j}{2}\right] dx$$

$$= \int_2^0 \left(2x + \frac{x^2}{4} + \frac{3x}{4} - \frac{4x}{2}\right) dx = \frac{x^3}{12} + \frac{3x^2}{8} \Big|_2^0$$

$$= -\left(\frac{8}{12} + \frac{12}{8}\right) = -\frac{13}{6}$$

The required line integral in the counterclockwise direction

$$I = \oint_c \overline{F} \cdot d\overline{r} = I_1 + I_2 + I_3 = 4 - \frac{13}{2} - \frac{13}{6} = -\frac{14}{3}$$

b. Line integral value in the opposite direction is $\frac{14}{3}$.

Example 4: Evaluate $\int_c f d\overline{r}$ where $f = 2xy^2z + x^2y$ and c is the curve $x = t, y = t^2, z = t^3$ from $t = 0$ to 1.

Solution:

$$\overline{r} = xi + yj + zk = ti + t^2 j + t^3\overline{k}$$

$$d\overline{r} = (i + 2tj + 3t^2\overline{k})dt$$

Along c:

$$f = 2t \cdot (t^2)^2(t^3) + (t^2) \cdot t^2 = 2t^8 + t^4$$

$$\int_c f d\overline{r} = \int_0^1 (2t^8 + t^4)\left(\overline{i} + 2t\overline{j} + 3t^2\overline{k}\right) dt$$

$$= \overline{i}\int_0^1 (2t^8 + t^4)dt + \overline{j}\int_0^1 (4t^9 + 2t^5)dt$$

$$+\overline{k}\int_0^1 (6t^{10} + 3t^6)dt$$

$$= i\left(\frac{2t^9}{9} + \frac{t^5}{5}\right)\Big|_0^1 + \overline{j}\left(\frac{4t^{10}}{10} + \frac{2t^6}{6}\right)\Big|_0^1$$

$$+\overline{k}\left(\frac{6t^{11}}{11} + \frac{3t^7}{7}\right)\Big|_0^1$$

$$= \frac{19}{45}\overline{i} + \frac{11}{15}\overline{j} + \frac{75}{77}\overline{k}.$$

Example 5: Find the work done in moving a particle in the force field $\overline{F} = 3x^2\overline{i} + (2xz - y)j + z\overline{k}$ along

a. straight line from $A(0, 0, 0)$ to $B(2, 1, 3)$

b. space curve c: $x = 2t^2, y = t, z = 4t^2 - t$ from $t = 0$ to $t = 1$

c. curve c: defined by $x^2 = 4y, 3x^3 = 8z$ from $x = 0$ to $x = 2$.

Solution: **Work done along a curve c is $\int_c \overline{F} \cdot d\overline{r}$:**

a. $\overline{r} = 2ti + tj + 3tk$

$$d\overline{r} = (2i + j + 3k)dt$$

$$\overline{F} = 3x^2\overline{i} + (2xz - y)\overline{j} + z\overline{k}$$

$$= 12t^2\overline{i} + (12t^2 - t)\overline{j} + 3t\overline{k}$$

work done by $\overline{F}$ in moving along the straight line from $A(0, 0, 0)$ to $B(2, 1, 3)$

$$= \int_A^B \overline{F} \cdot d\overline{r}$$

$$= \int_0^1 \left[12t^2 i + (12t^2 - t)j + 3tk\right] \cdot [2i + j + 3k]dt$$

$$= \int_0^1 \left(24t^2 + 12t^2 - t + 9t\right) dt$$

$$= 12t^3 + 4t^2 \Big|_0^1 = 16$$

b. $\overline{r} = xi + yj + z\overline{k}$

$$\overline{r} = 2t^2 i + t\overline{j} + (4t^2 - t)\overline{k}$$

$$d\overline{r} = 4ti + j + (8t - 1)\overline{k}$$

$$\overline{F} = 3(2t^2)^2 i + \left[2 \cdot (2t^2)(4t^2 - t) - t\right]\overline{j} + (4t^2 - t)k$$

Work done

$$= \int_c \overline{F} \cdot d\overline{r} = \int_0^1 \left[12t^4 i + \left[4t^2(4t^2 - t) - t\right]j\right.$$

$$\left. + (4t^2 - t)k\right] \cdot [4t\,i + j + (8t - 1)k]$$

$$= \int_0^1 \left[48t^5 + (16t^4 - 4t^3 - t) + (8t - 1)(4t^2 - t)\right] dt$$

$$= 8t^6 + 16\frac{t^5}{5} + 7t^4 - 4t^3 \Big|_0^1 = 8 + \frac{16}{5} + 7 - 4$$

$$= 14.2$$

c. $\bar{r} = xi + y\bar{j} + z\bar{k} = xi + \frac{x^2}{4}j + \frac{3}{8}x^3\bar{k}$

$$= ti + \frac{t^2}{4}j + \frac{3}{8}t^3k$$

$$d\bar{r} = \left(\cdot i + \frac{t}{2}j + \frac{9}{8}t^2\bar{k}\right)$$

$$\bar{F} = 3x^2i + \left(2 \cdot x \cdot \frac{3}{8}x^3 - \frac{x^2}{4}\right)j + \frac{3}{8}x^3\bar{k}$$

Work done

$$= \int \bar{F} \cdot d\bar{r} = \int_0^2 \left[3t^2 + \frac{t}{2}\left(\frac{3}{4}t^4 - \frac{t^2}{4}\right) + \frac{27}{64}t^5\right] dt$$

$$= t^3 + \frac{t^6}{16} - \frac{t^4}{32} + \frac{27t^6}{384}\Bigg|_0^2 = 16$$

Example 6: If $\bar{A} = (y - 2x)i + (3x + 2y)j$, compute the circulation of $\bar{A}$ about a circle c in the xy plane with centre at the origin and radius 2, if c is traversed in the positive direction.

Solution: c : circle: $x^2 + y^2 = 4$

In parametric form $x = 2\cos t$, $y = 2\sin t$ with t varying 0 to 2π

$$\bar{A} = (2\sin t - 2(2\cos t))i + (3(2\cos t) + 2(2\sin t))\bar{j}$$

$$d\bar{r} = d(xi + yj) = dx\bar{i} + dy\bar{j}$$

$$d\bar{r} = (-2\sin t\, i + 2\cos t\, j)dt$$

By definition

circulation along curve $c = \int_c \bar{F} \cdot d\bar{r}$

$$= \int_0^{2\pi} [(2\sin t - 4\cos t)i$$

$$+(6\cos t + 4\sin t)j] \cdot [-2\sin t\, i + 2\cos t\, j]dt$$

$$= 4\int_0^{2\pi}\left[-\sin^2 t + 2\sin t\cos t + 3\cos^2 t + 2\sin t\cos t\right]dt$$

$$= 16\int_0^{2\pi} \sin t\, d(\sin t) - 4\int_0^{2\pi} \frac{1 - \cos 2t}{2}dt$$

$$+12\int_0^{2\pi} \frac{1 + \cos 2t}{2}dt$$

$$= 8\pi.$$

Example 7: Prove that $\bar{F} = (y^2\cos x + z^3)\bar{i} + (2y\sin x - 4)j + (3xz^2 + 2)\bar{k}$ is (a) conservative field (b) find scalar potential of $\bar{F}$ (c) find work done in moving an object in this field from $P_1(0, 1, -1)$ to $P_2\left(\frac{\pi}{2}, -1, 2\right)$.

Solution:

a. $\bar{F}$ is conservative if $\nabla \times \bar{F} = 0$

$$\nabla \times \bar{F} = \begin{vmatrix} i & j & k \\ \frac{\partial}{\partial x} & \frac{\partial}{\partial y} & \frac{\partial}{\partial z} \\ y^2\cos x + z^3 & 2y\sin x - 4 & 3xz^2 + 2 \end{vmatrix}$$

$$= i(0 - 0) - j(3z^2 - 3z^2)$$

$$+k(2y\cos x - 2y\cos x)$$

$$= 0$$

Hence $\bar{F}$ is conservative.

b. Let f be the scalar potential such that $\bar{F} = \nabla f$ then comparing the components of $\bar{i}, \bar{j}, \bar{k}$, we get

$$\frac{\partial f}{\partial x} = y^2\cos x + z^3 \quad (1)$$

$$\frac{\partial f}{\partial y} = 2y\sin x - 4 \quad (2)$$

$$\frac{\partial f}{\partial z} = 3xz^2 + 2 \quad (3)$$

Integrating (1) partially w.r.t. x,

$$f = y^2\sin x + xz^3 + g(y, z) \quad (4)$$

Differentiating (4) partially w.r.t. y and using (2)

$$2y\sin x + 0 + \frac{\partial g}{\partial y} = \frac{\partial f}{\partial y} = 2y\sin x - 4$$

Integrating w.r.t. y

$$g(y, z) = -4y + c_1(z) \quad (5)$$

Substituting (5) in (4)

$$f = y^2\sin x + xz^3 - 4y + c_1(z) \quad (6)$$

Differentiating (6) partially w.r.t. z and using (3)

$$0 + 3xz^2 - 0 + \frac{dc_1}{dz} = \frac{\partial f}{\partial z} = xz^2 + 2$$

Integrating w.r.t. z

$$c_1(z) = z^2 + c \quad (7)$$

Substituing (7) in (6)

$$f(x, y, z) = y^2 \sin x + xz^3 - 4y + z^2 + c$$

c. Work done

$$= f(P_2) - f(P_1) = f(\frac{\pi}{2}, -1, 2) - f(0, 1, -1)$$

$$= 12 + 4\pi.$$

Example 8: Show that $(z - e^{-x} \sin y)dx + (1 + e^{-x} \cos y)dy + (x - 8z)dz$ is an exact differential of a function f and find f.

Solution: Let

$$\bar{A} = Ai + A_2\bar{j} + A_3\bar{k} = (z - e^{-x} \sin y)\bar{i}$$
$$+(1 + e^{-x} \cos y)\bar{j} + (x - 8z)\bar{k}$$

Then

$$\nabla \times \bar{A} = \begin{vmatrix} i & j & k \\ \frac{\partial}{\partial x} & \frac{\partial}{\partial y} & \frac{\partial}{\partial z} \\ z - e^{-x} \sin y & 1 + e^{-x} \cos y & x - 8z \end{vmatrix}$$
$$= i(0 - 0) - j(1 - 1) + k(-e^{-x} \cos y$$
$$-(-e^{-x} \cos y)) = 0$$

Since $\nabla \times \bar{A} = 0$ (see Section 16.2 C)

$A_1dx + A_2dy + A_3dz$ will be an exact differential

$$(z - e^{-x} \sin y)dx + (1 + e^{-x} \cos y)dy$$
$$+ (x - 8z)dz = df$$

Regrouping

$$(zdx + xdz) - 8zdz + dy + (e^{-x} \cos ydy$$
$$- e^{-x} \sin y\, dx) = df$$

$$\because \quad df = d(xz) - d(4z^2) + dy + d(e^{-x} \sin y)$$

$$\therefore \quad f = xz - 4z^2 + y + e^{-x} \sin y.$$

Example 9: If $\bar{A} = (x - y)\bar{i} + (x + y)j$ evaluate $\oint_c \bar{A} \cdot d\bar{r}$ around the curve c consisting of $y = x^2$ and $y^2 = x$

Solution:

$$\because \quad I = \oint_c \bar{F} \cdot d\bar{r} = \int_{c_1} \bar{F} \cdot d\bar{r} + \int_{c_2} \bar{F} \cdot d\bar{r} = I_1 + I_2$$

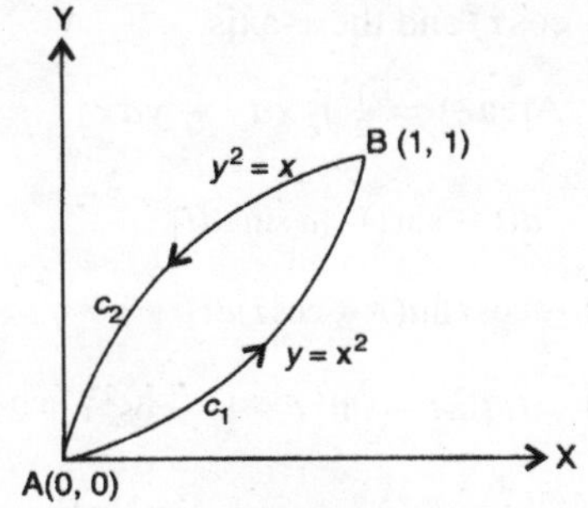

Fig. 16.6

where c_1: $y = x^2$ and $c_2 : y^2 = x$ as shown in Fig. 16.6 meet at points $A(0, 0)$ and $B(1, 1)$.

Along the curve $c_1 : y = x^2$, so that,

$$\bar{r} = xi + yj = xi + x^2 j = ti + t^2 j$$

$d\bar{r} = (i + 2tj)dt\cdot$ with t varying from 0 to 1

$$I_1 = \int_{c_1} (\bar{A} \cdot d\bar{r})$$
$$= \int_0^1 \left[(t - t^2)\bar{i} + (t + t^2)j\right] \cdot [i + 2tj]dt$$
$$= \int_0^1 \left[(t - t^2) + 2t(t + t^2)\right] dt = \int_0^1 (2t^3 + t^2 + t)dt$$

$$I_1 = \frac{2t^4}{4} + \frac{t^3}{3} + \frac{t^2}{2} \Big|_0^1 = \frac{1}{2} + \frac{1}{3} + \frac{1}{2} = 1 + \frac{1}{3} = \frac{4}{3}$$

Along the curve $c_2 : y^2 = x, y = \sqrt{x}$

$$\bar{r} = x\bar{i} + yj = xi + \sqrt{x}j = ti + \sqrt{t}j$$

$$d\bar{r} = (i + \frac{1}{2\sqrt{t}}j)dt$$

with t varying from 1 to 0

$$I_2 = \int_{c_2} \bar{A} \cdot d\bar{r}$$
$$= \int_1^0 \left[(t - \sqrt{t})\bar{i} + (t + \sqrt{t})j\right] \cdot \left[i + \frac{1}{2\sqrt{t}}j\right] dt$$
$$= \int_1^0 \left[(t - \sqrt{t}) + \frac{1}{2}(\sqrt{t} + 1)\right] dt$$
$$= \int_1^0 \left(t - \frac{\sqrt{t}}{2} + \frac{1}{2}\right) dt$$
$$= \frac{t^2}{2} - \frac{t^{\frac{3}{2}}}{3} + \frac{1}{2}t \Big|_1^0 = -\frac{1}{2} + \frac{1}{3} - \frac{1}{2} = -\frac{2}{3}$$

Line integral $I = I_1 + I_2 = \frac{4}{3} - \frac{2}{3} = \frac{2}{3}$.

Example 10: Compute the area of the region bounded by one arch of a cycloid $x = a(t - \sin t)$,

$y = a(1 - \cos t)$ and the x-axis.

Solution: Area $A = \frac{1}{2}\int_c xdy - ydx$

$$A = \frac{1}{2}\int_0^{2\pi} a(t - \sin t)\cdot[a \sin t dt]$$
$$-a(1-\cos t)[a(1-\cos t)dt]$$
$$= \frac{a^2}{2}\int_0^{2\pi}(t\sin t - \sin^2 t - 1 - \cos^2 t + 2\cos t)dt$$
$$= \frac{a^2}{2}\int_0^{2\pi}(-2 + t\sin t + 2\cos t)dt$$
$$= \frac{a^2}{2}[-4\pi - 2\pi + 0] = -\frac{-6\pi a^2}{2}.$$

Exercise

Evaluate the following line integrals:

1. $\int_c xy^3 ds$ where c is the segment of the line $y = 2x$ in the xy plane from $A(-1, -2, 0)$ to $B(1, 2, 0)$.

Ans. $16/\sqrt{5}$.

2. $\int_c (x^2 + xy)dx + (x^2 + y^2)dy$; c: square: $x \pm 1, y = \pm 1$

Ans. 0

3. $\int_c x^2 ydx + (x - z)dy + xyz\, dz$ where c is the arc of parabola $y = x^2$ in plane $z = 2$ from $A(0, 0, 2)$ to $B(1, 1, 2)$.

Ans. $-17/15$

4. $\int_c (x^2 + y^2 + z^2)^2 ds$ where c is the arc of circular helix $r(t) = \cos t\, i + \sin t\, j + 3t\, k$ from $A(1, 0, 0)$ to $B(1, 0, 6\pi)$.

Ans. $\sqrt{10}(2\pi + 6(2\pi)^3 + \frac{81}{5}(2\pi)^5)$

5. If $\bar{A}(t) = t\bar{i} - t^2 j + (t-1)k$, $\bar{B}(t) = 2t^2\bar{i} + 6t\bar{k}$ evaluate $\int_0^2 \bar{A}\cdot\bar{B}dt$.

Ans. 12

6. If $\bar{A}(t) = ti - 3j + 2t\bar{k}$, $\bar{B}(t) = i - 2j + 2k$, $\bar{c}(t) = 3i + tj - k$ then evaluate
   a. $\int_1^2 A\cdot(\bar{B}\times\bar{C})dt$
   b. $\int_1^2 (\bar{A}\times(\bar{B}\times\bar{C}))dt$

Ans. a. 0 b. $-\frac{87}{2}i - \frac{44}{3}j + \frac{15}{2}\bar{k}$

7. If $\bar{A}(2) = 2i - j + 2k$, $\bar{A}(3) = 4i - 2j + 3k$ then evaluate $\int_2^3 \bar{A}\cdot\frac{dA}{dt}dt$.

Ans. 10

8. Evaluate $\int_c \bar{A}\times d\bar{r}$ where $\bar{A} = 2y\bar{i} - z\bar{j} + x\bar{k}$ and c is the curve $x = \cos t$, $y = \sin t$, $z = 2\cos t$ from $t = 0$ to $\pi/2$.

Ans. $i(2 - \frac{\pi}{4}) + j(\pi - \frac{1}{2})$

9. Evaluate $\int_c \bar{A}\cdot d\bar{r}$ where
   a. $\bar{A} = 2xi + 4yj - 3zk$,
   c: curve: $\bar{r}(t) = \cos t\,\bar{i} + \sin t\, j + t\, k$ from $t = 0$ to π
   b. $\bar{A} = yi + zj + xk$
   c: circle $y^2 + z^2 = 1, x = 0$
   c. $\bar{A} = yzi + zxj + xy\bar{k}$
   c: curve from (0, 0, 0) to (1, 1, 0) along the curve $x = y^2, z = 0$ in xy-plane, followed by the straight line path from (1, 1, 0) to (1, 1, 1).

Ans. **a.** $-3\pi^2/2$ **b.** $-\pi$ **c.** 3/4

10. Determine whether the force field $\bar{F} = 2xz\bar{i} + (x^2 - y)\bar{j} + (2z - x^2)\bar{k}$ is conservative or not.

Ans. $\nabla\times\bar{F}\neq 0$ so non-conservative

11. **a.** Prove that $\bar{F} = (4xy - 3x^2z^2)\bar{i} + 2x^2 j - 2x^3 z\bar{k}$ is a conservative field.
    b. Find its scalar potential f.
    c. Also find the work done in moving an object in this field from (1, 1, 1) to (0, 0, 0).

Ans. **a.** $\nabla\times\bar{F} = 0$, so conservative
b. scalar potential $f = 2x^2y - x^3z^2 + c$.
c. work done $= f(1,1,1) - f(0,0,0) = 1$.

12. If $\bar{A} = (2xy + z^3)\bar{i} + x^2\bar{j} + 3xz^2\bar{k}$
    a. Prove that the line integral $\int_c \bar{A}\cdot d\bar{r}$ is independent of the curve c joining two given points $P_1(1, -2, 1)$ and $P_2(3, 1, 4)$.
    b. Show that there exists a scalar function f such that $\bar{A} = \nabla f$ and find f.
    c. Also find the work done in moving an object from P_1 to P_2.

Ans. **a.** $\nabla\times\bar{A} = 0$, $\bar{A}$ is conservative, so line integral is independent of path
b. $x^2y + xz^3 +$ constant
c. work done: 202

13. Find b such that the force field $\overline{A} = (e^x z - bxy)\overline{i} + (1 - bx^2)j + (e^x + bz)\overline{k}$ is conservative. Find the scalar potential f of $\overline{A}$ when $\overline{A}$ is conservative.

Ans. $b = 0,\ f = y + ze^x + c$

14. Find the scalar potential f of $\overline{F} = (z + \sin y)i + (-z + x\cos y)j + (x - y)\overline{k}$.

Ans. $f = xz + x\sin y - yz + c$

15. Find the total work done in moving a particle in a force field $\overline{A} = 3xy\overline{i} - 5zj + 10xk$ along the curve $x = t^2 + 1, y = 2t^2, z = t^3$ from $t = 1$ to $t = 2$.

Ans. 303

16. Calculate the work done in a force field given by $\overline{A} = (2y + 3)i + xzj + (yz - x)k$ when an object is moved from the point $P_1(0, 0, 0)$ to $P_2(2, 1, 1)$ along the curve $x = 2t^2$, $y = t,\ z = t^3$.

Ans. $\frac{288}{35}$

17. If $\overline{A} = (2x - y + 2z)\,\overline{i} + (x + y - z)j + (3x - 2y - 5z)\overline{k}$ calculate the circulation of $\overline{A}$ along the circle in the xy-plane of radius 2 and centre at origin.

Ans. Circulation $= \int \overline{A} \cdot d\overline{r} = 8\pi$.

18. Determine the circulation of $\overline{A} = yi + zj + xk$ around the curve $x^2 + y^2 = 1, z = 0$.

Ans. $-\pi$

19. If $\int_{P_1}^{P_2} \overline{A} \cdot d\overline{r}$ is independent of the path joining any two given points P_1 and P_2 in a given region then $\oint_c \overline{A} \cdot d\overline{r} = 0$ for all closed paths in the region passing through P_1 and P_2.

Hint: $P_1BP_2DP_1$ be any closed curve c

$$\oint_c = \int_{P_1BP_2DP_1} = \int_{P_1BP_2} + \int_{P_2DP_1}$$

$$= \int_{P_1BP_2} - \int_{P_1DP_2} = 0.$$

20. Prove that the work done in moving an object from P_1 to P_2 in a conservative force field $\overline{F}$ is independent of the path joining the two points P_1 and P_2.

Hint: Since $\overline{F}$ is conservative, $\overline{F} = \nabla f$

$$\int_{P_1}^{P_2} \overline{F} \cdot d\overline{r} = \int_{P_1}^{P_2} \nabla f \cdot d\overline{r}$$

$$= \int_{P_1}^{P_2} \frac{\partial f}{\partial x}dx + \frac{\partial f}{\partial y}dy + \frac{\partial f}{\partial z}dz$$

$$= \int_{P_1}^{P_2} df = f(P_2) - f(P_1)$$

Hint: Use $\frac{1}{2}\int_c xdy - ydx$ to compute area enclosed by c.

21. Compute the area of the ellipse $x = a\cos t$, $y = b\sin t$.

Ans. πab

22. Find the area under of one arch of the astroid $x = a\cos^3 t,\ y = a\sin^3 t$.

Ans. $3\pi a^2/8$

23. Find the area of the loop of the folium of descartes

$$x = \frac{3at}{1+t^3},\ y = \frac{3at^2}{1+t^3}$$

Ans. $3a^2/2$.

16.3 SURFACE INTEGRALS: SURFACE AREA AND FLUX

The concept of surface integral is a simple and natural generalization of a double integral

$$\iint_R f(x, y)dxdy$$

taken over a plane region R. In a surface integral $f(x, y)$ is integrated over a curved surface.

Let S be a two-sided surface with one side of S taken arbitrarily as the positive side (the outer side if S is closed) (refer Fig. 16.7). A unit normal $\overline{n}$ at any point of the positive side of S is known as positive outward drawn unit normal.

In the xyz-space, the equation of a surface S is

$$g(x, y, z) = 0$$

with unit normal $\overline{n} = \frac{\nabla g}{|\nabla g|}$

When S is represented in parametric form as

$$\overline{r}(u, v) = x(u, v)\overline{i} + y(u, v)j + z(u, v)\overline{k}$$

with the two parameters u and v varying in a region R of uv-plane, then the unit normal $\overline{n}$ to S at P is given by

$$\overline{n} = \frac{N}{|N|}$$

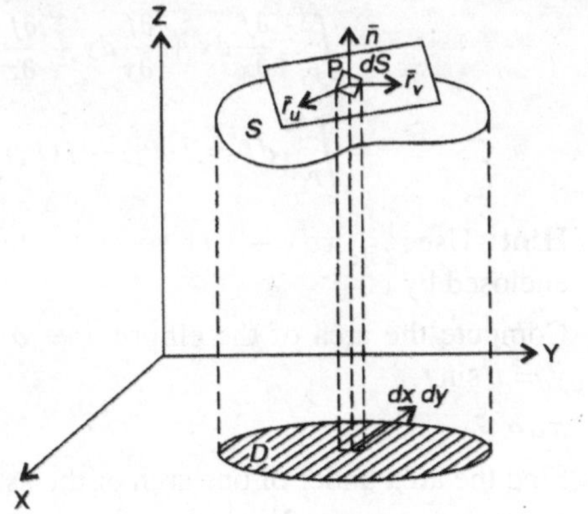

Fig. 16.7

where

$$\overline{N} = \overline{r}_u \times \overline{r}_v$$

The surface integral of a given vector function $\overline{F}$ taken over a surface S is defined as

$$\iint_S \overline{F} \cdot \overline{n}\, dS = \iint_S \overline{F} \cdot d\overline{S}$$

$$= \iint_R \overline{F}(\overline{r}(u, v)) \cdot \overline{N}(u, v) du dv$$

In the component form, where

$$\overline{F} = F_1\overline{i} + F_2\overline{j} + F_3\overline{k}$$

$$\overline{n} = \cos\alpha\overline{i} + \cos\beta\overline{j} + \cos\gamma\overline{k}$$

$$\overline{N} = N_1\overline{i} + N_2\overline{j} + N_3\overline{k}$$

the surface integral takes the form

$$\iint_S \overline{F} \cdot \overline{n}\, dS = \iint_S (F_1 \cos\alpha + F_2 \cos\beta + F_3 \cos\gamma) dS$$

$$= \iint_R (F_1N_1 + F_2N_2 + F_3N_3) du\, dv$$

Here α, β, γ are the angles between $\overline{n}$ and the positive directions of the coordinate axes (i.e, $\overline{n} \cdot \overline{i} = |\overline{n}||i| \cos\alpha = \cos\alpha$, etc.)

Apart from the normal surface integral

$$\iint_S \overline{F}(\overline{r}) \cdot \overline{n}\, dS$$

the other types of surface integrals are

$$\iint_S \phi\, dS, \iint_S \phi\, \overline{n}\, dS, \iint_S \overline{A} \times \overline{n}\, dS$$

where ϕ is a scalar function.

Evaluation of a Surface Integral

A surface integral is evaluated by reducing it to a double integral by projecting the given surface S onto one of the coordinate planes. Let D be the projection of S onto the xy-plane (see Fig. 16.7).

Then,

$$dS = \frac{dxdy}{|\overline{n} \cdot \overline{k}|}$$

Then,

$$\iint_S \overline{F} \cdot \overline{n}\, dS = \iint_D \overline{F}.\overline{n} \frac{dx\, dy}{|\overline{n} \cdot \overline{k}|}$$

where $\overline{n}$ is unit outward drawn normal to S. The R.H.S. double integral in x, y over the plane region D is evaluated as an two-fold iterated integral. In a similar way the surface integral can be evaluated by projecting S onto the Yz-plane as D_1 and Xz-plane as D_2 as follows

$$\iint_S \overline{F} \cdot \overline{n}\, dS = \iint_{D_1} \overline{F} \cdot \overline{n} \frac{dy\, dz}{|\overline{n} \cdot \overline{i}|}$$

$$\iint_S \overline{F} \cdot \overline{n}\, dS = \iint_{D_2} \overline{F} \cdot \overline{n} \frac{dx\, dz}{|\overline{n} \cdot \overline{j}|}$$

Surface Area of a Curved Surface

Let S be a surface represented by the equation

$$F(x, y, z) = 0 \tag{1}$$

Then the unit normal to the surface S is given by

$$\hat{n} = \frac{\nabla F}{|\nabla F|} = \frac{F_x\overline{i} + F_y\overline{j} + F_z\overline{k}}{\sqrt{F_x^2 + F_y^2 + F_z^2}}$$

where F_x, F_y, F_z are partial derivatives of F w.r.t. x, y, z respectively. Let D be the projection of S onto the xy-plane. Then

$$\text{Surface area of } S = \iint_S dS = \iint_D \frac{dx\, dy}{|\overline{n} \cdot \overline{k}|}$$

$$= \iint_D \frac{\sqrt{F_x^2 + F_y^2 + F_z^2}}{|F_z|} dxdy$$

since $$\overline{n} \cdot \overline{k} = \frac{F_z}{\sqrt{F_x^2 + F_y^2 + F_z^2}}$$

Corollary 1: If the equation of the surface S is $z = f(x, y)$ then

$$\text{Surface area} = \iint \sqrt{1 + z_x^2 + z_y^2}\, dxdy.$$

Flux

The normal component $\overline{F}\cdot\overline{n}$ is a scalar. Let ρ be the density, $\overline{V}$ be the velocity of a fluid and $\overline{F}=\rho\overline{V}$. Then flux of $\overline{F}$ represents the total quantity of fluid flowing in unit time through (across) the surface S in the positive direction. The flux of $\overline{F}$ across S is given by the flux integral

$$\text{Flux of } \overline{F} \text{ across } S=\iint_S \overline{F}\cdot\overline{n}\,dS.$$

Worked Out Examples

Example 1: Evaluate $\iint_S \overline{A}\cdot\overline{n}\,dS$ over the entire surface S of the region bounded by the cylinder $x^2+z^2=9$, $x=0$, $y=0$, $z=0$ and $y=8$ where $\overline{A}=6z\overline{i}+(2x+y)\overline{j}-x\overline{k}$ (see Fig. 16.8).

Solution: Here the entire surface S consists of 5 surfaces namely S_1 the curved (lateral) surface of the cylinder, $ABDCA$, $S_2: AOEC$, $S_3: OBDE$, $S_4: OAB$, $S_5: CDE$. Thus

$$\iint_S \overline{A}\cdot\overline{n}\,dS=\iint_{S_1+S_2+\cdots+S_5}$$
$$=\iint_{S_1}+\iint_{S_2}+\cdots+\iint_{S_5}$$
$$=SI_1+SI_2+SI_3+SI_4+SI_5$$

For the curved (lateral) surface S_1 of the cylinder

$$\text{unit normal } \overline{n}=\frac{\nabla(x^2+z^2)}{|\nabla(x^2+z^2)|}=\frac{2xi+2zk}{\sqrt{4x^2+4z^2}}=\frac{xi+zk}{3}$$

so

$$\overline{A}\cdot\overline{n}=(6zi+(2x+y)j-xk)\cdot\left(\frac{xi+zk}{3}\right)=\frac{5}{3}xz$$

$$\overline{n}\cdot\overline{k}=\frac{z}{3}$$

$$SI_1=\iint_{S_1}\overline{A}\cdot\overline{n}\,dS=\iint\frac{\overline{A}\cdot\overline{n}}{|\overline{n}\cdot\overline{k}|}dx\,dy$$
$$=\iint\frac{5}{3}\frac{xz}{(z/3)}dx\,dy$$
$$=5\int_0^8\int_0^3 x\,dx\,dy=180$$

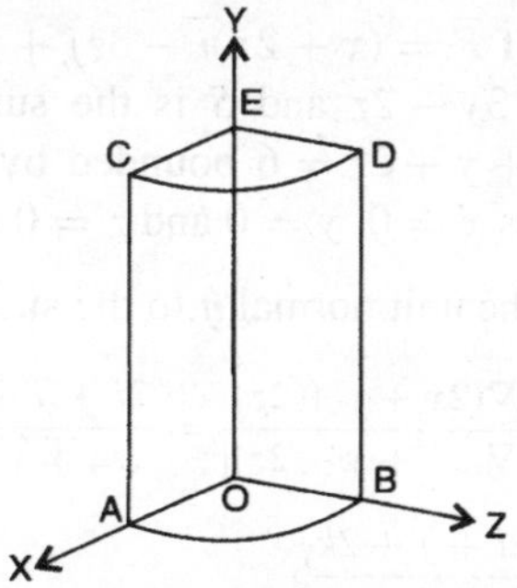

Fig. 16.8

On the plane $S_2: AOEC: z=0, \overline{n}=-k$, $\overline{A}\cdot\overline{n}=x$

$$SI_2=\iint_{S_2}\overline{A}\cdot\overline{n}\,dS=\int_0^8\int_0^3 x\,dx\,dy=36$$

On the plane S_3: $OBDE: x=0, \overline{n}=-\overline{i}$, $\overline{A}\cdot\overline{n}=-6z$

$$SI_3=\iint_{S_3}\overline{A}\cdot\overline{n}\,dS=\int_0^8\int_0^3 -6z\,dz\,dy=-216$$

On the sector $S_4: OAB: y=0, n=-j$, $\overline{A}\cdot\overline{n}=-(2x+y)=-2x$

$$SI_4=\iint_{S_4}\overline{A}\cdot\overline{n}\,dS=\iint_{OAB}-2x\,dxdz$$

In polar coordinates

$$SI_4=\int_0^{\frac{\pi}{2}}\int_0^3 -2\cdot r\cos t\cdot r\,dr\,dt=-18$$

On the sector $S_5: CDE: y=8, \overline{n}=j$, $\overline{A}\cdot\overline{n}=2x+y=2x+8$

$$SI_5=\iint_{S_5}\overline{A}\cdot\overline{n}\,dS=\iint_{CDE}(2x+8)dxdz$$

In polar coordinates

$$SI_5=\int_0^{\frac{\pi}{2}}\int_0^3(2r\cos t+8)r\,dr\,dt=18+18\pi$$

Thus the required surface integral is

$$SI=(180)+(36)+(-216)+(-18)+(18+18\pi)$$
$$=18\pi.$$

Example 2: Evaluate

a. $\iint_S(\nabla\times\overline{F})\cdot\overline{n}\,dS$ and

b. $\iint \phi \bar{n} dS$ if $\bar{F} = (x+2y)\bar{i} - 3zj + xk$, $\phi = 4x + 3y - 2z$ and S is the surface of the plane $2x + y + 2z = 6$ bounded by the coordinate planes $x = 0$, $y = 0$ and $z = 0$ (Fig. 16.9).

Solution: The unit normal $\hat{n}$ to the surface S is

$$\hat{n} = \frac{\nabla(2x+y+2z)}{|\nabla(2x+y+2z)|} = \frac{2i+j+2k}{\sqrt{4+1+4}}$$

$$\hat{n} = \frac{2i+j+2k}{3}$$

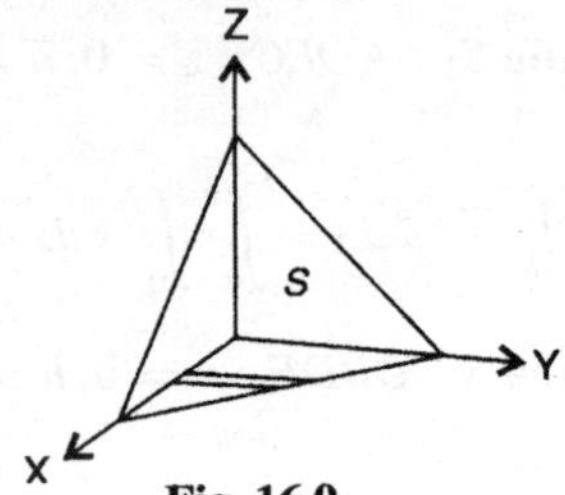

Fig. 16.9

a. $$\nabla \times \bar{F} = \begin{vmatrix} i & j & k \\ \frac{\partial}{\partial x} & \frac{\partial}{\partial y} & \frac{\partial}{\partial z} \\ x+2y & -3z & x \end{vmatrix} = 3i - j - 2k$$

so $(\nabla \times \bar{F}) \cdot \hat{n} = (3i - j - 2k) \cdot \left(\frac{2i+j+2k}{3}\right) = \frac{1}{3}$

$$\text{SI} = \iint_S (\nabla \times \bar{F}) \cdot \hat{n} dS = \frac{1}{3}\iint_S dS = \frac{1}{3}\iint \frac{dxdy}{|n \cdot k|}$$

$$= \frac{1}{3} \cdot \int_{x=0}^{3} \int_{y=0}^{6-2x} \frac{dydx}{2/3} = \frac{1}{2}\int_0^3 (6-2x)dx = \frac{9}{2}$$

b. $\text{SI} = \iint \phi \hat{n} dS = \iint (4x+3y-2z)\frac{(2i+j+2k)}{3} dS$

Eliminate z using, $z = \frac{6-2x-y}{2}$

$$\text{SI} = \frac{2i+j+2k}{3} \cdot \int_{x=0}^{3} \int_{y=0}^{6-2x} (6x+4y-6)\frac{dydx}{2/3}$$

$$= (2i+j+2k)\int_0^3 [3(x-1)(6-2x) + (6-2x)^2]dx$$

$$\text{SI} = 72i + 36j + 72k.$$

Example 3: Find the surface area of the plane $x + 2y + 2z = 12$ cut off by $x = 0$, $y = 0$, and $x^2 + y^2 = 16$ (refer Fig. 16.10).

Solution: Rewriting equation of plane

$$z = \frac{12 - x - 2y}{2}$$

we have $z_x = -\frac{1}{2}$, $z_y = -1$

$$\text{Surface area} = \iint_R \sqrt{1 + z_x^2 + z_y^2} dxdy$$

$$= \iint \sqrt{1 + 1 + \frac{1}{4}} dxdy$$

$$= \frac{3}{2}\iint dxdy$$

In polar coordinates $= \frac{3}{2}\int_0^{\frac{\pi}{2}} \int_0^4 r\, dr\, d\theta = 6\pi$.

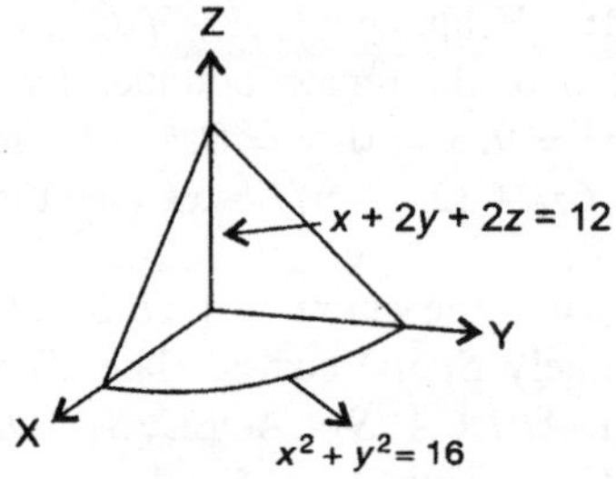

Fig. 16.10

Aliter: $F(x, y, z) = x + 2y + 2z - 12 = 0$, so $F_x = 1$, $F_y = 2$, $F_z = 2$

$$\sqrt{F_x^2 + F_y^2 + F_z^2} = \sqrt{1+4+4} = 3$$

$$\text{Surface area} = \iint \sqrt{\frac{F_x^2 + F_y^2 + F_z^2}{|F_z|}} dxdy$$

$$= \iint \frac{3}{2} dx\, dy = 6\pi.$$

Flux

Example 4: Find the flux of the vector field $\bar{A} = (x - 2z)\bar{i} + (x + 3y + z)\bar{j} + (5x + y)\bar{k}$ through the upper side of the triangle ABC with vertices at the points $A(1, 0, 0)$, $B(0, 1, 0)$, $C(0, 0, 1)$ (see Fig. 16.11).

Solution: Equation of the plane in which the triangle ABC lies is

$$x + y + z = 1$$

Unit normal $\hat{n}$ to ABC is

$$\frac{\nabla(x+y+z-1)}{|\nabla(x+y+z-1)|} = \frac{i+j+k}{\sqrt{3}}$$

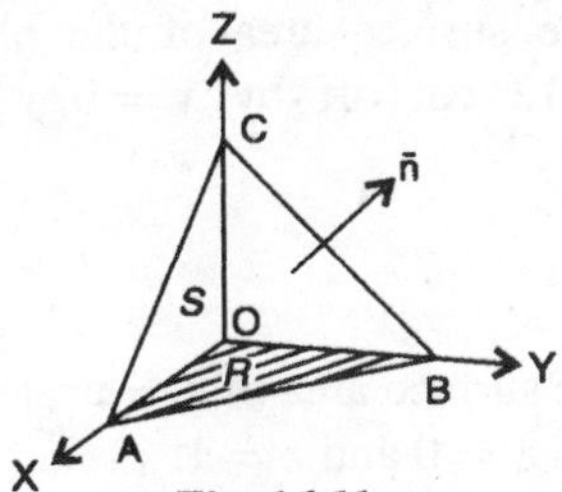

Fig. 16.11

$$\bar{A}\cdot\bar{n} = \frac{1}{\sqrt{3}}[(x-2z)+(x+3y+z)+(5x+y)]$$

$$= \frac{7x+4y-z}{\sqrt{3}}$$

Let *AOB* be the projection of *ABC* onto the *xy*-plane Then

$$dS = \frac{dx\,dy}{|n\cdot k|} = \sqrt{3}dxdy$$

Flux across the triangle $ABC = \iint_S \bar{A}\cdot\bar{n}\,dS$

$$= \iint_{AOB} \frac{7x+4y-z}{\sqrt{3}}\sqrt{3}dxdy$$

Replace z by $1-x-y$

$$= \iint [7x+4y-(1-x-y)]\,dxdy$$

$$= \int_{x=0}^{1}\int_{y=0}^{1-x}(8x+5y-1)dydx = \frac{5}{3}.$$

Exercise

1. If S is the surface $2x+y+2z=6$ bounded by $x=0, x=1, y=0, y=2$, evaluate (a) $\iint(\nabla\times\bar{F})\cdot\bar{n}dS$ and (b) $\iint_S \phi\bar{n}dS$.

Ans. **a.** 1 **b.** $2i+j+2k$

2. Evaluate $\iint_S \bar{A}\cdot\bar{n}\,dS$ where $\bar{A}=18zi-12j+3yk$ and S is that part of the plane $2x+3y+6z=12$ which is located in the first octant.

Ans. 24

3. Evaluate $\iint_S(\nabla\times\bar{F})\cdot\bar{n}dS$ where $\bar{F}=yi+(x-2xz)j-xyk$ and S is the surface of the sphere $x^2+y^2+z^2=a^2$ above the *xy*-plane.

Ans. 0

4. If S is the entire surface of the cube bounded by $x=0, x=b, y=0, y=b, z=0,$ and $z=b$ and $\bar{A}=4xz\bar{i}-y^2j+yz\bar{k}$ then evaluate $\iint_S \bar{F}\cdot\bar{n}dS$.

Ans. $3b^4/2$

5. Let S be the surface of the cylinder $x^2+y^2=16$ included in the first octant between $z=0$ and $z=5$. Evaluate $\iint_S \bar{A}\cdot\bar{n}dS$ where $\bar{A}=zi+xj-3y^2zk$.

Ans. 90

6. For the surface S defined in the previous problem 5, evaluate $\iint_S \phi\bar{n}dS$ where $\phi=\frac{3}{8}xyz$.

Ans. $100(i+j)$.

7. Find the surface integral over the parallelopiped $x=0, y=0, z=0, x=1, y=2, z=3$ when $\bar{A}=2xyi+yz^2j+xz\bar{k}$.

Ans. 33

8. If S is the surface of the sphere $x^2+y^2+z^2=d^2$ and $\bar{A}=axi+byj+czk$, evaluate $\iint_S \bar{F}\cdot\bar{n}dS$.

Hint: Project S onto *xoy*-plane and use symmetry.

Ans. $2\cdot\frac{2\pi d^3}{3}(a+b+c)$

9. Let S be the surface of the cylinder $x^2+y^2=a^2$ in the first octant between the planes $z=0$ and $z=h$. Evaluate $\iint \bar{A}\cdot\bar{n}dS$ where $\bar{A}=zi+xj-3zy^2k$.

Ans. $ah(a+h)/2$

Flux

10. Calculate the flux of water through the parabolic cylinder $y=x^2$, between the planes $x=0, z=0, x=3, z=2$ if the velocity vector is $\bar{A}=yi+2j+xzk$ m/sec.

Hint: Flux of $\bar{F}$ across S is $\iint_S \bar{F}\cdot\bar{n}dS$.

Ans. 69 m³/sec

11. Find the flux across the surface of the parabolic cylinder $y^2=8x$ in the first octant bounded by the planes $y=4$ and $z=6$ when the velocity vector $\bar{V}=2yi-zj+x^2k$.

Ans. 132

12. Find the flux of $\overline{A} = i - j + xyz\overline{k}$ through the circular region S obtained by cutting the sphere $x^2 + y^2 + z^2 = a^2$ with a plane $y = x$ (take the side of S facing the positive side of the x-axis).

Hint: S is bounded by the ellipse $2x^2 + z^2 = a^2$, $\hat{n} = (i - j)/\sqrt{2}$, $dS = \sqrt{2}dx\,dz$, area of the ellipse with semi axis $a/\sqrt{2}$ and a is $\pi a^2/\sqrt{2}$.

Ans. $\sqrt{2}\pi a^2$

13. Compute the flux of the vector field $\overline{A} = xi + yj + \sqrt{x^2 + y^2 - 1}\; k$ through the outer side of the hyperboloid of one sheet $z = \sqrt{x^2 + y^2 - 1}$ bounded by the planes $z = 0$ and $z = \sqrt{3}$.

Hint: $\hat{n} = \frac{xi+yj}{\sqrt{x^2+y^2-1}} - k$, $\overline{A}\cdot\overline{n} = \frac{1}{\sqrt{x^2+y^2-1}}$ with polar coordinates, flux $= \int_0^{2\pi}\int_1^2 \frac{r\,dr\,d\theta}{\sqrt{r^2-1}} = 2\sqrt{3}\pi$.

Ans. $2\sqrt{3}\pi$

14. Evaluate $\iint_S \overline{F}\cdot\overline{n}dS$ where $\overline{F} = \overline{r}/r^3$ and S is the sphere $x^2 + y^2 + z^2 = b^2$.

Ans. 4π

15. Calculate the surface integral of the vector function $\overline{A} = xi + yj$ over the portion of the surface of the unit sphere S: $x^2 + y^2 + z^2 = 1$ above the xy-plane $z \geq 0$.

Ans. $\frac{4\pi}{3}$

16. If S is the triangular surface with vertices $(2, 0, 0)$, $(0, 2, 0)$ and $(0, 0, 4)$ and $\overline{A} = xi + (z^2 - zx)j - xy\overline{k}$ then evaluate $\iint_S \overline{F}\cdot\overline{n}\,dS$.

Ans. $-\frac{22}{3}$

Surface area

17. What is the surface area of the surface S whose equation is $F(x, y, z) = 0$?

Ans. $$\iint_R \frac{\sqrt{F_x^2 + F_y^2 + F_z^2}}{|F_z|}dxdy$$

where R is the projection of S on xy-plane.

18. Find the surface area of the plane $x + 2y + 2z = 12$ cut off by $x = 0, y = 0, x = 1, y = 1$.

Ans. $\frac{3}{2}$

19. Find the surface area of $z = x^2 + y^2$ included between $z = 0$ and $z = 1$.

Ans. $\frac{\pi}{6}\left(\sqrt{125} - 1\right)$

20. Find the surface area of the region common to the intersecting cylinders $x^2 + y^2 = a^2$ and $x^2 + z^2 = a^2$.

Ans. $16\,a^2$.

16.4 VOLUME INTEGRALS

Let V be a region in space enclosed by a closed surface $\overline{r} = \overline{r}(u, v)$. Let $\overline{F}(\overline{r})$ be a vector point function. Then the triple integral

$$\iiint_V \overline{F}(\overline{r})dV \quad \text{or briefly} \quad \iiint_V \overline{F}dV$$

is known as volume integral or space integral. In the component form

$$\iiint_V \overline{F}dV = \overline{i}\iiint_V F_1 dxdydz + \overline{j}\iiint_V F_2 dxdydz + \overline{k}\iiint_V F_3 dxdydz$$

$\iiint_V \phi dV$ is another form of a volume integral. These integrals are evaluated as three-fold iterated integrals.

WORKED OUT EXAMPLES

Example 1: Evaluate $\iiint_V f\,dV$ where $f = 2x + y$, V is the closed region bounded by the cylinder $z = 4 - x^2$ and the planes $x = 0$, $y = 0$, $y = 2$ and $z = 0$ (see Fig. 16.12).

Solution: This closed region is covered if x and z

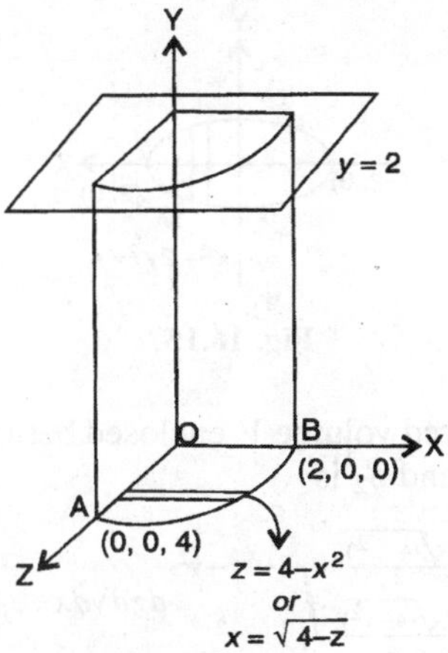

Fig. 16.12

varies covering the area OAB and y varies from 0 to 2. Thus

$$\iiint_V (2x+y)dV = \int_{y=0}^{2}\int_{z=0}^{4}\int_{x=0}^{\sqrt{4-z}} (2x+y)dxdzdy$$

$$= \int_0^2\int_0^4 (x^2+xy)\Big|_0^{\sqrt{4-z}} dzdy$$

$$= \int_0^2\int_0^4 \left[(4-z)+y\sqrt{4-z}\right]dzdy$$

$$= \int_0^2 4z - \frac{z^2}{2} - \frac{2}{3}y(4-z)^{\frac{3}{2}}\Big|_0^4$$

$$= \int_0^2 \left(8+\frac{16}{3}y\right)dy$$

$$= 8y + \frac{16}{6}y^2\Big|_0^2$$

$$= \frac{80}{3}.$$

Example 2: If V is the region in the first octant bounded by $y^2+z^2=9$ and the plane $x=2$ and $\overline{F}=2x^2yi - y^2j + 4xz^2k$. Then evaluate

$$\iiint_V (\nabla\cdot\overline{F})dV.$$

Solution: $\nabla\cdot\overline{F} = 4xy - 2y + 8xz$

The volume V of the solid region is covered by covering the plane region OAB while x varies

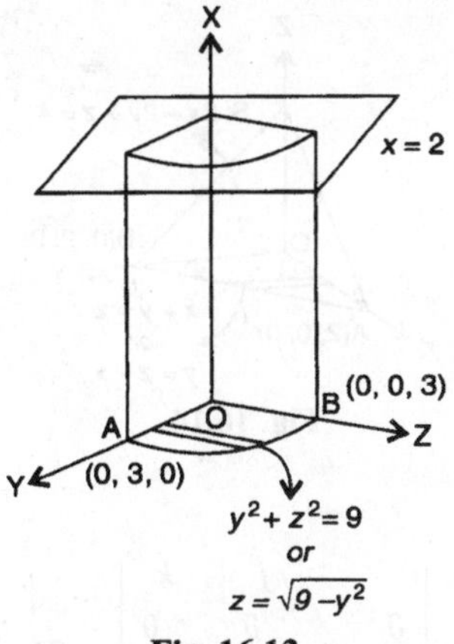

Fig. 16.13

from 0 to 2 (Fig. 16.13). Thus

$$\iiint_V (\nabla\cdot\overline{F})dV$$

$$= \int_{x=0}^{2}\int_{y=0}^{3}\int_{z=0}^{\sqrt{9-y^2}} (4xy-2y+8xz)dzdydx$$

$$= \int_0^2\int_0^3 4xyz - 2yz + 4xz^2\Big|_0^{\sqrt{9-y^2}} dydx$$

$$= \int_0^2\int_0^3 \left[(4xy-2y)\sqrt{9-y^2} + 4x(9-y^2)\right]dydx$$

$$= \int_0^2 (4x-2)(-\frac{1}{3}(9-y^2)^{\frac{3}{2}} + 4x\left(9y-\frac{y^3}{3}\right)\Big|_0^3 dx$$

$$= \int_0^2 [9(4x-2)+72x]dx$$

$$= 18x^2 - 18x + 36x^2\Big|_0^2 = 180.$$

Example 3: Evaluate $\iiint_V \nabla\times\overline{A}dV$ where $\overline{A} = (x+2y)i - 3zj + x\overline{k}$ and V is the closed region in the first octant bounded by the plane $2x+2y+z=4$

Solution: The solid region is covered by covering the plane region OAB in the xy-plane while z is varying from 0 to the plane $2x+2y+z=4$ (Fig. 16.14).

Thus z varies from 0 to $4-2x-2y$,

y varies from 0 to $2-x$

and x varies from 0 to 2.

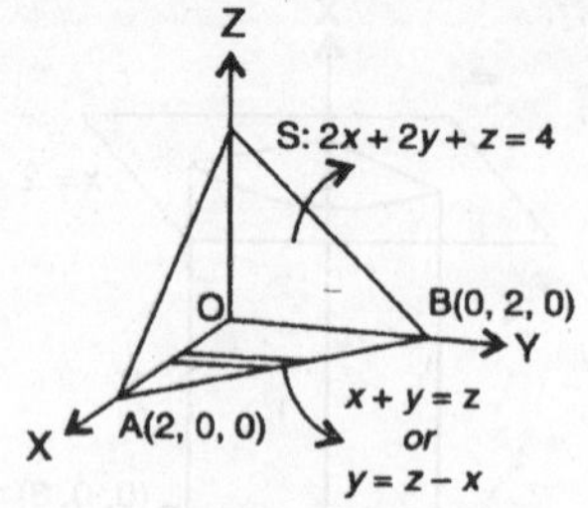

Fig. 16.14

Here

$$\nabla\times\bar{A}=\begin{vmatrix} i & j & k\\ \frac{\partial}{\partial x} & \frac{\partial}{\partial y} & \frac{\partial}{\partial z}\\ x+2y & -3z & x\end{vmatrix}=3i-j-2k$$

$$\iiint_V \nabla\times\bar{A}dV$$

$$=\int_{x=0}^{2}\int_{y=0}^{2-x}\int_{z=0}^{4-2x-2y}(3i-j+2k)dz\,dy\,dx$$

$$=(3i-j+2k)\int_0^2\int_0^{2-x}(4-2x-2y)dy\,dx$$

$$=2(3i-j+2k)\int_0^2\left((2-x)^2-\frac{(2-x)^2}{2}\right)dx$$

$$=(3i-j+2k)\left[4x+\frac{x^3}{3}-2x^2\right]_0^2$$

$$=\frac{8}{3}(3i-j+2k).$$

Example 4: Find the volume enclosed between the two surfaces $S_1: z=8-x^2-y^2$ and $S_2: z=x^2+3y^2$ (see Fig. 16.15).

Solution: Eliminating z from the given two surfaces S_1 and S_2, we get $8-x^2-y^2=z=x^2+3y^2$ i.e., $x^2+2y^2=4$. Thus the given two surfaces S_1 and S_2 intersect on the elliptic cylinder $x^2+2y^2=4$.

So the solid region between S_1 and S_2 is covered when

z varies from x^2+3y^2 to $8-x^2-y^2$,

y varies from $-\sqrt{\frac{4-x^2}{2}}$ to $\sqrt{\frac{4-x^2}{2}}$ and

x varies from -2 to 2.

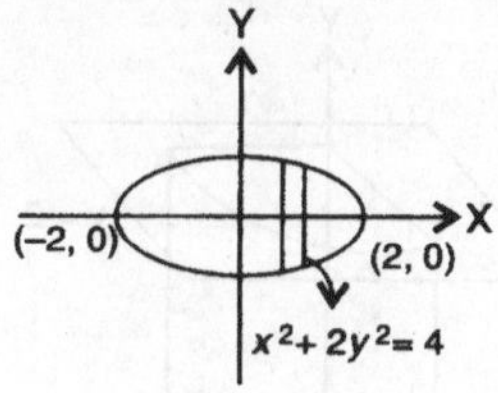

Fig. 16.15

So the required volume V enclosed between the two surfaces S_1 and S_2 is

$$V=\int_{-2}^{2}\int_{-\sqrt{\frac{(4-x^2)}{2}}}^{\sqrt{\frac{(4-x^2)}{2}}}\int_{x^2+3y^2}^{8-x^2-y^2}dzdydx$$

$$=\int_{-2}^{2}\int_{-\sqrt{\frac{(4-x^2)}{2}}}^{\sqrt{\frac{(4-x^2)}{2}}}(8-2x^2-4y^2)dydx$$

$$=\int_{-2}^{2}\left[2(8-2x^2)\sqrt{\frac{4-x^2}{2}}-\frac{8}{3}\left(\frac{4-x^2}{2}\right)^{\frac{3}{2}}\right]dx$$

$$V=\frac{4\sqrt{2}}{3}\int_{-2}^{2}(4-x^2)^{\frac{3}{2}}dx=8\pi\sqrt{2}.$$

EXERCISE

1. Evaluate $\iiint_V f\,dV$ where $f=45x^2y$ and V denotes the closed region bounded by the planes $4x+2y+z=8$, $x=0$, $y=0$, $z=0$.

Ans. 128

2. If $\bar{A}=(2x^2-3z)\bar{i}-2xyj-4x\bar{k}$ and V is the closed region bounded by the planes $x=0$, $y=0$, $z=0$ and $2x+2y+z=4$, evaluate $\iiint_V(\nabla\times\bar{A})dV$.

Ans. $\frac{8}{3}(j-k)$

3. Evaluate $\iiint_V \bar{A}dV$ where $\bar{A}=xi+yj+2zk$ and V is the volume enclosed by the planes $x=0$, $y=0$, $y=a$, $z=b^2$ and the surface $z=x^2$.

Ans. $\frac{ab^4}{4}i+\frac{a^2b^3}{3}j+\frac{4ab^5}{5}k$

4. Evaluate $\iiint_V \bar{B}\,dV$ where V is the region bounded by the surfaces $x=0$, $y=0$, $y=6$, $z=x^2$, $z=4$ and $\bar{B}=2xz\bar{i}-xj+y^2k$.

Ans. $128i - 24j + 384k$

5. If $\bar{A} = (x^3 - yz)\bar{i} - 2x^3yj + 2k$, evaluate $\iiint_V (\nabla \cdot \bar{A})dV$ over the volume of a cube of side b.

Ans. $\frac{1}{3}b^3$

6. Evaluate $\iiint_V (\nabla \cdot \bar{B})dV$ over the solid region of the sphere $x^2 + y^2 + z^2 = a^2$ when $\bar{B} = pxi + qyj + rzk$ where p, q, r are constants.

Ans. $\frac{4}{3}\pi a^2(p + q + r)$

Volume

7. Find the volume of the region common to the intersecting cylinders $x^2 + y^2 = a^2$ and $x^2 + z^2 = a^2$.

Ans. $16a^3/3$

8. Find the volume of the region bounded below by the paraboloid $z = x^2 + y^2$ and above by the plane $z = 2y$.

Ans. $\frac{\pi}{2}$

9. Find the volume cut from the sphere $x^2 + y^2 + z^2 = 4a^2$ by the cylinder $x^2 + y^2 = a^2$.

Ans. $4\pi a^3(8 - 3\sqrt{3})/3$

10. Find the volume bounded above by the sphere $x^2 + y^2 + z^2 = 2a^2$ and below by the paraboloid $az = x^2 + y^2$.

Ans. $(8\sqrt{2} - 7)\pi a^3/6$.

16.5 GREEN'S* THEOREM IN PLANE: TRANSFORMATION BETWEEN LINE INTEGRAL AND DOUBLE INTEGRAL AREA IN CARTESIAN AND POLAR COORDINATES

If R is a closed region in the xy-plane bounded by a simple closed curve c and if $M(x, y)$ and $N(x, y)$ are continuous functions of x and y having continuous derivatives in R, then

$$\oint_c M\,dx + N\,dy = \iint_R \left(\frac{\partial N}{\partial x} - \frac{\partial M}{\partial y}\right) dxdy$$

where c is traversed in the positive direction (refer Fig. 16.16).

Proof: Let the equations of the curves AEB and AFB by $y = Y_1(x)$ and $y = Y_2(x)$ respectively, consider

$$\iint_R \frac{\partial M}{\partial y} dx\,dy = \int_{x=a}^{b} \int_{y=Y_1(x)}^{Y_2(x)} \frac{\partial M}{\partial y} dy\,dx$$

$$= \int_a^b [M(x, Y_2) - M(x, Y_1)]dx$$

$$= -\int_b^a M(x, Y_2)dx - \int_a^b M(x, Y_1)dx$$

$$= -\int_{BFA} M(x, y)dx - \int_{AEB} M(x, y)dx$$

$$= -\int_{BFAEB} M(x, y)dx$$

$$= -\oint_c M(x, y)dx \qquad (1)$$

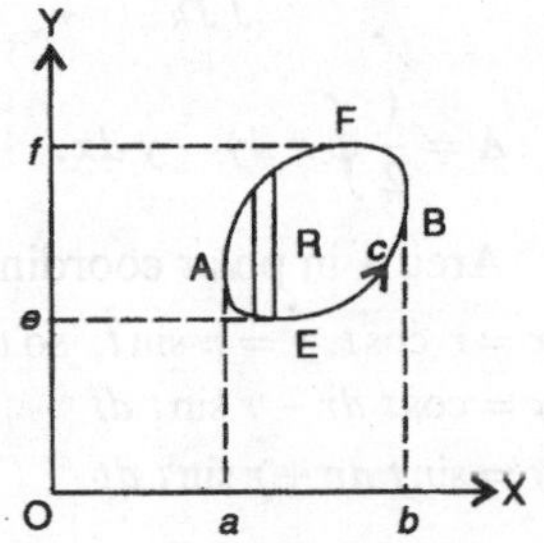

Fig. 16.16

Similarly let the equations of the curves EAF and EBF be $x = X_1(y)$ and $x = X_2(y)$ respectively. Then

$$\iint_R \frac{\partial N}{\partial x} dx\,dy = \int_{y=e}^{f} \int_{x=X_1(y)}^{X_2(y)} \frac{\partial N}{\partial x} dx\,dy$$

$$= \int_e^f [N(X_2, y) - N(X_1, y)]\,dy$$

$$= \int_e^f N(X_2, y)dy + \int_f^e N(X_1, y)dy$$

$$= \oint_c N(x, y)dy \qquad (2)$$

Adding (1) and (2), we get

$$\oint_c M\,dx + N\,dy = \iint_R \left(\frac{\partial N}{\partial x} - \frac{\partial M}{\partial y}\right) dxdy$$

*George Green (1793–1841) English mathematician.

Corollary 1: Vector notation of Green's theorem

Let $\bar{A} = Mi + Nj$ and $\bar{r} = x\bar{i} + y\bar{j}$ so that $\bar{A} \cdot d\bar{r} = M\,dx + N\,dy$

$$\nabla \times \bar{A} = \begin{vmatrix} i & j & k \\ \frac{\partial}{\partial x} & \frac{\partial}{\partial y} & \frac{\partial}{\partial z} \\ M & N & O \end{vmatrix} = \left(\frac{\partial N}{\partial x} - \frac{\partial M}{\partial y}\right)\bar{k}$$

Thus

$$\oint_c \bar{A} \cdot d\bar{r} = \iint_R (\nabla \times \bar{A}) \cdot \bar{k}\, dR$$

where $dR = dxdy$.

Corollary 2: Area A of the plane region R bounded by the simple closed curve c

Let $N = x$, $M = -y$ so that

$$\oint_c x\,dy - y\,dx = \iint_R (1+1)dxdy$$

$$= 2\iint_R dxdy = 2A$$

Thus

$$A = \frac{1}{2}\oint_c x\,dy - y\,dx.$$

Corollary 3: Area A in polar coordinates

Let $x = r\cos t$, $y = r\sin t$, so that

$dx = \cos t\, dr - r\sin t\, dt$

$dy = \sin t\, dr + r\sin t\, dt$

Thus

$$A = \frac{1}{2}\int_c r^2\, dt$$

Corollary 4: Green's theorem is valid for a doubly (multiply) connected domain R where c is the boundary of the region R consisting of c_1 and c_2 (several) curves all traversed in the positive direction.

Corollary 5: If $\frac{\partial M}{\partial y} = \frac{\partial N}{\partial x}$ then by Green's theorem

$$\oint_c M\,dx + N\,dy = 0.$$

Worked Out Examples

Green's theorem in plane

Example: Verify Green's theorem in plane for

$$\oint_c (x^2 - 2xy)dx + (x^2y + 3)dy$$

where c is the boundary of the region defined by $y^2 = 8x$ and $x = 2$ (refer Fig. 16.17).

Solution: Green's theorem states that

Line integral = Double integral.

a. The L.H.S. of the Green's theorem result is the line integral

$= LI = \oint_c (x^2 - 2xy)dx + (x^2y + 3)dy.$

Here c consists of the curves OA, ADB, BO, so

$$LI = \oint_c = \int_{OA+ADB+BO}$$

$$= \int_{OA} + \int_{ADB} + \int_{BO} = LI_1 + LI_2 + LI_3$$

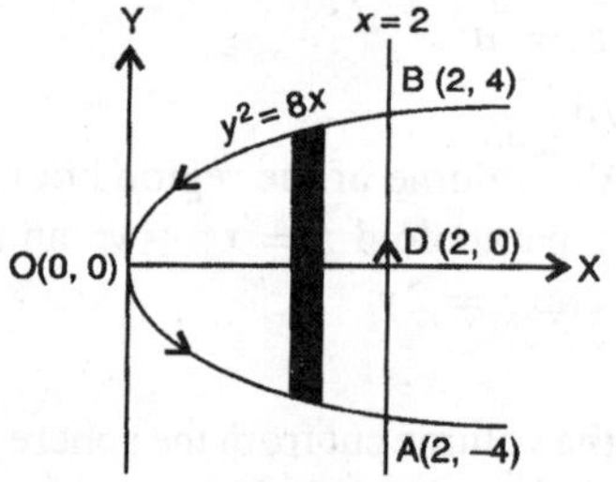

Fig. 16.17

Along OA: $y = -2\sqrt{2}\sqrt{x}$, so $dy = -\sqrt{\frac{2}{x}}dx$

$$LI_1 = \int_{OA} (x^2 - 2xy)dx + (x^2y + 3)dy$$

$$= \int_0^2 [x^2 - 2x(-2\sqrt{2}\sqrt{x})]dx$$

$$+[x^2(-2\sqrt{2}\sqrt{x}) + 3]\left(-\sqrt{\frac{2}{x}}\right)dx$$

$$= \int_0^2 (5x^2 + 4\sqrt{2}x^{\frac{3}{2}} - 3\sqrt{2}x^{-\frac{1}{2}})dx$$

$$= \frac{5x^3}{3} + 4\sqrt{3} \cdot \frac{2}{5} \cdot x^{\frac{5}{2}} - 3\sqrt{2} \cdot 2\sqrt{x}\Big|_0^2$$

$$= \frac{40}{3} + \frac{64}{5} - 12$$

Along ADB: $x = 2$, $dx = 0$

$$LI_2 = \int_{ADB} (x^2 - 2xy)dx + (x^2y + 3)dy$$

$$= \int_{-4}^{4} (4y + 3)dy = 24$$

Along BO: $y = 2\sqrt{2}\sqrt{x}$ with $x : 2$ to 0, $dy = \sqrt{\frac{2}{x}}dx$

$$LI_3 = \int_{BO} (x^2 - 2xy)dx + (x^2y + 3)dy$$

$$= \int_2^0 (5x^2 - 4\sqrt{2}x^{\frac{3}{2}} + 3\sqrt{2}x^{-\frac{1}{2}})dx$$

$$= -\frac{40}{3} + \frac{64}{5} - 12$$

$$LI = LI_1 + LI_2 + LI_3 = \left(\frac{40}{3} + \frac{64}{5} - 12\right) + (24)$$

$$+ \left(-\frac{40}{3} + \frac{64}{3} - 12\right) = \frac{128}{5}$$

b. Here

$$M = x^2 - 2xy, N = x^2y + 3,$$

$$\frac{\partial M}{\partial y} = -2x, \frac{\partial N}{\partial x} = 2xy.$$

So the R.H.S. of the Green's theorem is the double integral given by

$$\text{DI} = \iint_R \left(\frac{\partial N}{\partial x} - \frac{\partial M}{\partial y}\right) dxdy$$

$$= \iint_R [(2xy - (-2x))]dxdy$$

The region R is covered with y varying from $-2\sqrt{2}\sqrt{x}$ of the lower branch of the parabola to its upper branch $2\sqrt{2}\sqrt{x}$ while x varies from 0 to 2. Thus

$$\text{DI} = \int_{x=0}^{2} \int_{y=-\sqrt{8x}}^{\sqrt{8x}} (2xy + 2x)dydx$$

$$= \int_0^2 xy^2 + 2xy \Big|_{-\sqrt{8x}}^{\sqrt{8x}} dx$$

$$= 8\sqrt{2} \int_0^2 x^{\frac{3}{2}} dx = \frac{128}{5}$$

Since L.I. = D.I. the Green's theorem is thus verified.

Area of a plane region

Example 2: Using Green's theorem, find the area of the region in the first quadrant bounded by the curves $y = x$, $y = \frac{1}{x}$, $y = \frac{x}{4}$. (see Fig. 16.18)

Solution: By Green's theorem area A of the region bounded by a closed curve c is given by

$$A = \frac{1}{2} \oint_c xdy - ydx$$

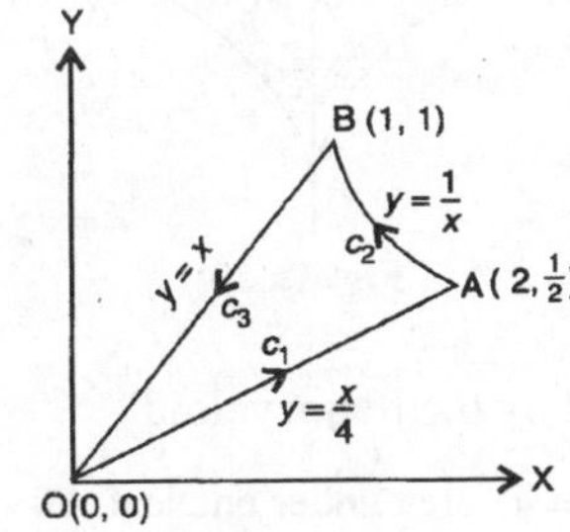

Fig. 16.18

Here c consists of the curves c_1: $y = \frac{x}{4}$, c_2: $y = \frac{1}{x}$ and c_3: $y = x$. So

$$A = \frac{1}{2}\oint_c = \frac{1}{2}\left[\int_{c_1} + \int_{c_2} + \int_{c_3}\right] = \frac{1}{2}\left[I_1 + I_2 + I_3\right]$$

Along c_1: $y = \frac{x}{4}$, $dy = \frac{1}{4}dx$, x: 0 to 2

$$I_1 = \int_{c_1} xdy - ydx = \int_{c_1} x\frac{1}{4}dx - \frac{x}{4}dx = 0$$

Along c_2: $y = \frac{1}{x}$, $dy = -\frac{1}{x^2}dx$, x: 2 to 1

$$I_2 = \int_{c_2} xdy - ydx = \int_2^1 x \cdot \left(-\frac{1}{x^2}\right) dx - \frac{1}{x}dx$$

$$= -2 \ln x \Big|_2^1 = 2 \ln 2$$

Along c_3: $y = x$, $dy = dx$; x: 1 to 0

$$I_3 = \int_{c_3} xdy - ydx = \int xdx - xdx = 0$$

$$A = \frac{1}{2}(I_1 + I_2 + I_3) = \frac{1}{2}(0 + 2 \ln 2 + 0) = \ln 2.$$

Example 3: Find the area bounded by the hypocycloid $x^{\frac{2}{3}} + y^{\frac{2}{3}} = a^{\frac{2}{3}}$ with $a > 0$ (see Fig. 16.19).

Solution: Parametric equations of the hypocycloid are

$$x = a \cos^3 t, y = a \sin^3 t$$

$$dx = -3a \cos^2 t \sin t dt,$$

$$dy = 3a \sin^2 t \cos t dt$$

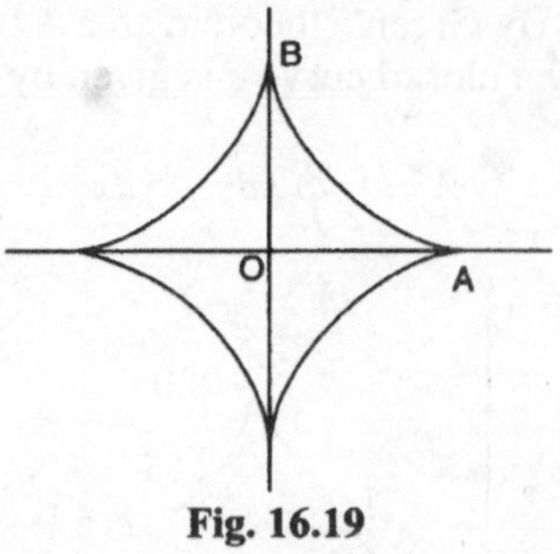

Fig. 16.19

Area bounded by the hypocycloid

$= 4 \cdot$ area under one leaf AB

$= 4$ area of the region AOB

Area of region $AOB = \frac{1}{2}\int_{ABOA} xdy - ydx$

$$= \frac{1}{2}\int_{AB} + \int_{BO} + \int_{OA}$$

$$= \frac{1}{2}\int_{AB} + 0 + 0$$

since $x = 0$ along BO and $y = 0$ along OA

$$= \frac{1}{2}\int_0^{\pi/2}\Big[a\cos^3 t \cdot (3a\sin^2 t\cos t dt)$$

$$-a\sin^3 t(-3a)\cos^2 t\sin t dt\Big]$$

$$= \frac{3a^2}{2}\int_0^{\pi/2} \sin^2 t\cos^4 t dt + \cos^2 t\sin^4 t dt$$

$$= \frac{3a^2}{2}\left[\frac{1\cdot 3\cdot 1}{6\cdot 4\cdot 2}\cdot\frac{\pi}{2}\right] + \frac{3a^2}{2}\left[\frac{3\cdot 1\cdot 1}{6\cdot 4\cdot 2}\frac{\pi}{2}\right] = 3\frac{\pi a^2}{32}$$

Area bounded by hypocycloid $= 4 \cdot \frac{3\pi a^2}{32} = \frac{3\pi a^2}{8}$.

Doubly connected region

Example 4: Verify Green's theorem in the plane for

$$\oint_c (2x - y^3)dx - xydy$$

where c is the boundary of the annulus (doubly connected) region enclosed by the circles $x^2 + y^2 = 1$ and $x^2 + y^2 = 9$ (refer Fig. 16.20).

Solution: Here $M = 2x - y^3, N = xy$ so that $\frac{\partial M}{\partial y} = -3y^2, \frac{\partial N}{\partial x} = y$

Thus R.H.S. of Green's theorem is

$$= \iint_R \left(\frac{\partial N}{\partial x} - \frac{\partial M}{\partial y}\right)dxdy$$

$$= \iint_R (y + 3y^2)dxdy$$

where R is the annulus region.

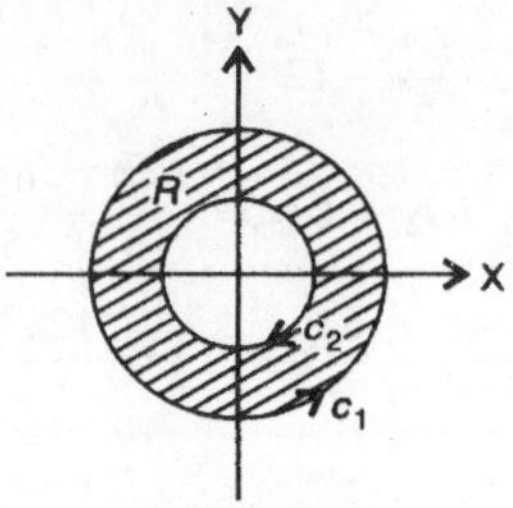

Fig. 16.20

Put $x = r\cos t$, $y = r\sin t$, so that t varies from 0 to 2π and r from 1 to 3

$$\text{R.H.S.} = \int_0^{2\pi}\int_1^3 (r\sin t + 3r^2\sin^2 t)r\,drdt$$

$$= \frac{26}{3}\int_0^{2\pi}\sin t dt + 60\int_0^{2\pi}\frac{1-\sin 2t}{2}dt = 60\pi$$

$$\text{L.H.S.} = \int_c Mdx + Ndy = \int_{c_1+c_2}(2x - y^3)dx - xydy$$

Changing to polar coordinate r, t

$$= \int(2r\cos t - r^3\sin^3 t)(-r\sin t dt) - \int r^3\cos^2 t\sin t dt$$

$$= r^4\frac{3\pi}{4}\Big|_1^3 = 60\pi.$$

EXERCISE

Use Green's theorem to evaluate the line integral $\oint_c Mdx + Ndy$ when $Mdx + Ndy$ equals to:

1. $-y^3dx + x^3dy$ where c: circle $x^2 + y^2 = 1$

Ans. $\frac{3\pi}{2}$

2. $x^{-1}e^y dx + (e^y\ln x + 2x)dy$ where c: the boundary of the region bounded by $y = 2$, $y = x^4 + 1$,

Ans. $\frac{16}{5}$

3. $(\cos x \sin y - xy)dx + \sin x \cos y \cdot dy$ where c: circle $x^2 + y^2 = 1$

Ans. 0

4. $(x^2 - \cosh y)dx + (y + \sin x)dy$ where c: the boundary of the rectangle $0 \le x \le \pi, 0 \le y \le 1$

Ans. $\pi(\cosh 1 - 1)$

5. $(3x^2 - 8y^2)dx + (4y - 6xy)dy$ where c: boundary of the region defined by $x = 0$, $y = 0, x + y = 1$.

Ans. $\frac{5}{3}$

6. $e^{-x}(\sin y dx + \cos y dy)$ where c: rectangle with vertices at $(0, 0)$, $(\pi, 0)$, $(\pi, \pi/2)$, $(0, \pi/2)$

Ans. $2(e^{-\pi} - 1)$

Verify Green's theorem or evaluate the line integral $\oint_c M dx + N dy$ (a) directly (b) using Green's theorem, where $M dx + N dy$ is:

7. $(xy + y^2)dx + x^2 dy$ with c: closed curve of the region bounded by $y = x$ and $y = x^2$

Ans. common value: $-\frac{1}{20}$

8. $(3x^2 - 8y^2)dx + (4y - 6xy)dy$ with c: boundary of the region defined by $y = \sqrt{x}$ and $y = x^2$

Ans. common value: $\frac{3}{2}$

9. $(2x - y^3)dx - xy dy$ with c: boundary of the region enclosed by the circles $x^2 + y^2 = 1$ and $x^2 + y^2 = 9$

Ans. common value: 60π

10. $(3x + 4y)dx + (2x - 3y)dy$ with c: $x^2 + y^2 = 4$

Ans. common value: -8π

Area using Green's theorem

11. Find the area of the region bounded by $y = x^2$ and $y = x + 2$.

Ans. $\frac{9}{2}$

12. Calculate the area bounded by the ellipse

$$\frac{x^2}{a^2} + \frac{y^2}{b^2} = 1$$

Deduce the area bounded by the circle $x^2 + y^2 = a^2$.

Hint: Put $x = a \cos t$, $y = b \sin t$.

Ans. Area of ellipse πab
put $a = b$, area of circle: πa^2

13. Find the area of the loop of the folium of descartes $x^3 + y^3 = 3axy$, $a > 0$.

Hint: Put $y = tx$, $t : 0$ to ∞.

Ans. $A = \frac{1}{2}\int x^2 dt = \frac{9}{2}\int_0^\infty \frac{a^2 t^2}{(1+t^3)^2} dt = \frac{3a^2}{2}$

14. Find the area of a loop of the four-leafed rose $\rho = 3 \sin 2\phi$.

Hint: $A = \frac{1}{2}\int_0^{\pi/2} \rho^2 d\phi = \frac{9\pi}{8}$.

15. Find the area of the cardioid $\rho = a(1 - \cos\theta)$, with $0 \le \theta \le 2\pi$.

Ans. $\frac{3\pi a^2}{2}$

16. Find the area bounded by one arch of the cycloid $x = a(\theta - \sin\theta)$, $y = a(1 - \cos\theta)$, $a > 0$ and the x-axis.

Ans. $3\pi a^2$

17. Evaluate $\int_c \bar{A} \cdot d\bar{r}$ where

$$\bar{A} = \alpha[-3a \sin^2 t \cos t i + a(2 \sin t - 3 \sin^3 t)j + b \sin 2t k]$$

and the curve c is given by

$$\bar{r} = a \cos t i + a \sin t j + bt k$$

and t varying from $\pi/4$ to $\pi/2$.

Ans. $\frac{\alpha}{2}(a^2 + b^2)$

18. Show that $\int_c f dg = \iint_R \left(\frac{\partial f}{\partial x}\frac{\partial g}{\partial y} - \frac{\partial f}{\partial y}\frac{\partial g}{\partial x}\right) dxdy$ where R is the region bounded by the simple closed curve c.

Hint: Use Green's theorem with $M = f\frac{\partial g}{\partial x}$ and $N = f\frac{\partial g}{\partial y}$.

19. Prove that $\int_c \frac{dF}{dn} dS = \iint_R \left(\frac{\partial^2 F}{\partial x^2} + \frac{\partial^2 F}{\partial y^2}\right) dxdy$ where $\frac{dF}{dn}$ is the directional derivative of F in the direction of the outer normal $\bar{n}$ to the curve c bounding the region R.

Hint: Choose $M = -\frac{\partial F}{\partial y}$, $N = \frac{\partial F}{\partial x}$ and note that $\bar{n} = \frac{dy}{ds} i - \frac{dx}{ds} j$.

20. If $\nabla^2 f = 0$ in R, show that

$$\iint_R \left[\left(\frac{\partial f}{\partial x}\right)^2 + \left(\frac{\partial f}{\partial y}\right)^2\right] dxdy = \int_c f \frac{\partial f}{\partial n} dS.$$

Hint: Take $M = -f\frac{\partial f}{\partial y}, N = f\frac{\partial f}{\partial x}$ and note that $\overline{n} = \frac{dy}{ds}i - \frac{dx}{ds}\overline{j}$.

21. Show that Green's theorem can be written in the form $\int_c \overline{F} \cdot \overline{n}ds = \iint_R \nabla \cdot \overline{F}dxdy$ where $\overline{F} = Mi - Nj$ and $\overline{n}$ is the outer unit normal to the curve c.

16.6 STOKES'* THEOREM

Transformation between line integral and surface integral. Let $\overline{A}$ be a vector function, having continuous first partial derivatives in a domain in space containing an open two sided surface S bounded by a simple closed curve c then

$$\iint_S (\nabla \times \overline{A}) \cdot \overline{n}dS = \oint_c \overline{A} \cdot d\overline{r} \qquad (1)$$

where $\overline{n}$ is a unit normal of S and c is traversed in the positive direction.

Proof: See Fig. 16.21. Assume that S can be represented as $z = f(x, y)$ or $x = g(y, z)$, or $y = h(x, z)$ where f, g, h are continuous, differentiable functions. Also assume that projections of S on the xy, yz, zx planes are regions bounded by simple closed curves. If $\overline{A} = A_1i + A_2j + A_3k$ then the result of Stokes' theorem (1) can be written as

$$\iint_S (\nabla \times \overline{A}) \cdot \overline{n}dS$$

$$= \iint_S (\nabla \times (A_1i + A_2j + A_3k)) \cdot \overline{n}dS = \oint_c \overline{A} \cdot d\overline{r}$$

$$= \oint_c (A_1i + A_2j + A_3k) \cdot (dx\overline{i} + dy\overline{j} + dz\overline{k})$$

$$= \oint_c A_1dx + A_2dy + A_3dz \qquad (2)$$

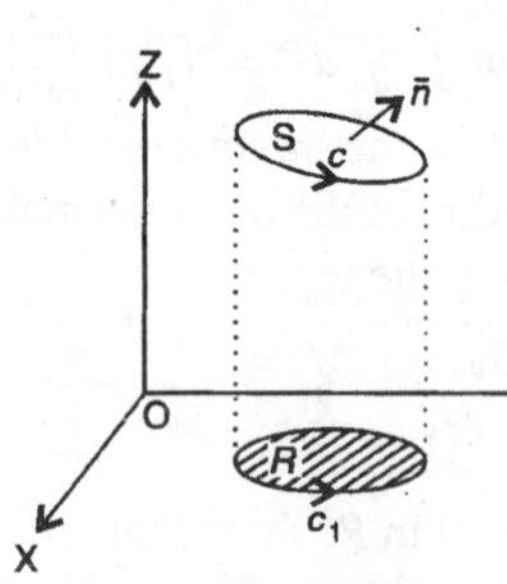

Fig. 16.21

*Sir George Gabriel Stokes (1819–1903) Irish mathematician.

To show that

$$\iint_S (\nabla \times A_1i) \cdot \overline{n}dS = \oint_c A_1dx \qquad (3)$$

Consider

$$\nabla \times A_1i = \begin{vmatrix} i & j & k \\ \frac{\partial}{\partial x} & \frac{\partial}{\partial y} & \frac{\partial}{\partial z} \\ A_1 & 0 & 0 \end{vmatrix} = \frac{\partial A_1}{\partial z}\overline{j} - \frac{\partial A_1}{\partial y}\overline{k}$$

so that

$$(\nabla \times A_1i) \cdot \overline{n}dS = \left(\frac{\partial A_1}{\partial z}\overline{n} \cdot j - \frac{\partial A_1}{\partial y}\overline{n} \cdot \overline{k}\right) dS \qquad (4)$$

Take the equation of S as $z = f(x, y)$. Then the position vector $\overline{r}$ to any point of S is

$$\overline{r} = x\overline{i} + y\overline{j} + z\overline{k} = x\overline{i} + y\overline{j} + f(x, y)\overline{k}$$

so $$\frac{\partial \overline{r}}{\partial y} = 0 + \overline{j} + \frac{\partial f}{\partial y}\overline{k} \qquad (5)$$

Now $$\overline{n} \cdot \frac{\partial \overline{r}}{\partial y} = 0$$

since normal $\overline{n}$ to S is perpendicular to the tangent $\frac{\partial \overline{r}}{\partial y}$ to S.

Thus taking dot product of (5) with $\overline{n}$, we have

$$0 = \overline{n} \cdot \frac{\partial \overline{r}}{\partial y} = \overline{n} \cdot \overline{j} + \frac{\partial f}{\partial y}\overline{n} \cdot \overline{k}$$

or $$\overline{n} \cdot j = -\frac{\partial f}{\partial y}\overline{n} \cdot \overline{k} = -\frac{\partial z}{\partial y}\overline{n} \cdot \overline{k} \qquad (6)$$

Substituting (6) in (4), we get

$$(\nabla \times A_1\overline{i}) \cdot \overline{n}dS = -\left(\frac{\partial A_1}{\partial z}\frac{\partial z}{\partial y} + \frac{\partial A_1}{\partial y}\right)\overline{n} \cdot \overline{k}dS \qquad (7)$$

Now on S,

$$A_1(x, y, z) = A_1(x, y, f(x, y)) = F(x, y)$$

so that $$\frac{\partial A_1}{\partial y} + \frac{\partial A_1}{\partial z}\frac{\partial z}{\partial y} = \frac{\partial F}{\partial y} \qquad (8)$$

Using (8) in (7), we get

$$\iint_S (\nabla \times A_1\overline{i}) \cdot \overline{n}dS = \iint_S -\frac{\partial F}{\partial y}\overline{n} \cdot \overline{k}dS$$

$$= -\iint_R \frac{\partial F}{\partial y}dxdy \qquad (9)$$

where R is the projection of S on xy-plane and

$$\overline{n} \cdot \overline{k}dS = dxdy$$

Applying Green's theorem in plane

$$-\iint_R \frac{\partial F}{\partial y} dx dy = \oint_{c_1} F dx = \oint_c A_1 dx \qquad (10)$$

since at each point (x, y) of c_1 the value of F is the same as the value of A_1 at each point (x, y, z) of c and since dx is same for both the curves c and c_1. Thus from (9) of (10), we arrive at

$$\iint_S (\nabla \times A_1 i) \cdot \overline{n} dS = \oint_c A_1 dx \qquad (3)$$

Similarly by projecting S on to other coordinate planes, we get

$$\iint_S (\nabla \times A_2 \overline{j}) \cdot \overline{n} dS = \oint_c A_2 dy \qquad (11)$$

and

$$\iint_S (\nabla \times A_3 \overline{k}) \cdot \overline{n} dS = \oint_c A_3 dz \qquad (12)$$

Adding (3), (11) and (12), we get (1) the result of Stokes' theorem.

Note 1: Stokes' theorem in rectangular form is

$$\iint_S \left[\left(\frac{\partial A_3}{\partial y} - \frac{\partial A_2}{\partial z} \right) \cos \alpha + \left(\frac{\partial A_1}{\partial z} - \frac{\partial A_3}{\partial x} \right) \cos \beta + \left(\frac{\partial A_2}{\partial x} - \frac{\partial A_1}{\partial y} \right) \cos \gamma \right] dS$$

$$= \oint_c A_1 dx + A_2 dy + A_3 dz$$

where $\overline{n} = \cos \alpha i + \cos \beta j + \cos \gamma k$, and α, β, γ are angles made by normal $\overline{n}$ with $\overline{i}, \overline{j}, \overline{k}$.

Note 2: Green's theorem in plane is a special case of Stoke's theorem.

Note 3: The *circulation* of $\overline{A}$ around a closed curve c is given by the line integral

$$\oint_c \overline{A} \cdot d\overline{r}$$

where $\overline{A}$ represents the velocity of a fluid circulation has applications in fluid mechanics and aerodynamics.

WORKED OUT EXAMPLES

Example 1: Prove that $\oint_c f d\overline{r} = \iint_S d\overline{S} \times \nabla f$.

Solution: Choose $\overline{A} = f\overline{c}$, where $\overline{c}$ is a constant vector, in the Stoke's theorem. Then

$$\text{L.H.S.} = \oint_c \overline{A} \cdot d\overline{r} = \oint_c f\overline{c} \cdot d\overline{r}$$

$$= \oint_c \overline{c} \cdot (f d\overline{r}) = \overline{c} \cdot \oint_c f d\overline{r}$$

Now $\nabla \times \overline{A} = \nabla \times (f\overline{c}) = (\nabla f) \times \overline{c} + f(\nabla \times \overline{c}) = \nabla f \times \overline{c}$ since $\nabla \times \overline{c} = 0$

So $\qquad (\nabla \times \overline{A}) \cdot \overline{n} = (\nabla f \times \overline{c}) \cdot \overline{n} = \overline{c} \cdot (\overline{n} \times \nabla f)$

$$\text{R.H.S.} = \iint_S (\nabla \times \overline{A}) \cdot \overline{n} ds = \iint_S \overline{c} \cdot (n \times \nabla f) dS$$

$$= \overline{c} \cdot \iint_S (\overline{n} \times \nabla f) dS$$

Thus

$$\overline{c} \cdot \oint_c f d\overline{r} = \overline{c} \cdot \iint_S (\overline{n} \times \nabla f) dS$$

Since this is true for any arbitrary constant $\overline{c}$, hence, we get the result.

$$\oint_c f d\overline{r} = \iint_S (\overline{n} \times \nabla f) dS = \iint_S \overline{n} dS \times \nabla f$$

$$= \iint_S d\overline{S} \times \nabla f.$$

Example 2: Evaluate $\iint_S (\nabla \times \overline{A}) \cdot \overline{n} dS$ over the surface of intersection of the cylinders $x^2 + y^2 = a^2$, $x^2 + z^2 = a^2$ which is included in the first octant, given that $\overline{A} = 2yzi - (x + 3y - 2)j + (x^2 + z)k$ (refer Fig. 16.22).

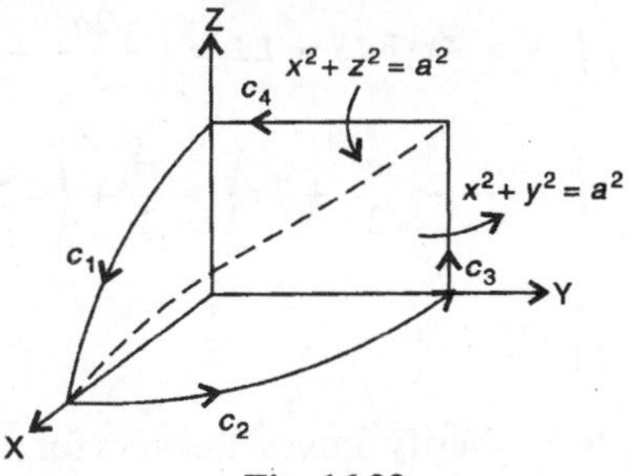

Fig. 16.22

Solution: By Stokes' theorem the given surface integral can be converted to a line integral i.e.,

$$SI = \iint_S (\nabla \times \bar{A}) \cdot \bar{n} dS = \oint_c \bar{A} \cdot d\bar{r} = LI$$

Here c is the curve consisting of the four curves c_1 : $x^2 + z^2 = a^2$, $y = 0$; $c_2 : x^2 + y^2 = a^2, z = 0$, $c_3 : x = 0, y = a, 0 \le z \le a$; $c_4 : x = 0, z = a, 0 \le y \le a$

$$LI = \oint_c \bar{A} \cdot d\bar{r} = \int_{c_1+c_2+c_3+c_4} = \int_{c_1} + \int_{c_2} + \int_{c_3} + \int_{c_4}$$

$$= LI_1 + LI_2 + LI_3 + LI_4$$

On the curve c_1: $y = 0$; $x^2 + z^2 = a^2$

$$LI_1 = \int_{c_1} \bar{A} \cdot d\bar{r} = \int_{c_1} (x^2 + z)dz$$

$$= \int_a^0 \left[(a^2 - z^2) + z\right] dz = -\frac{2}{3}a^3 - \frac{a^2}{2}$$

On the curve c_2: $z = 0, x^2 + y^2 = a^2$

$$LI_2 = \int_{c_2} \bar{A} \cdot d\bar{r} = \int_{c_2} -(x + 3y - 2)dy$$

$$= -\int_0^a \left(\sqrt{a^2 - y^2} + 3y - 2\right) dy$$

$$= -\frac{\pi a^2}{4} - \frac{3}{2}a^2 + 2a$$

On the curve c_3: $x = 0, y = a, 0 \le z \le a$

$$LI_3 = \int_{c_3} \bar{A} \cdot d\bar{r} = \int_0^a zdz = \frac{a^2}{2}$$

On c_4: $x = 0, z = a, 0 \le y \le a$

$$LI_4 = \int \bar{A} \cdot d\bar{r} = \int_a^0 (2 - 3y)dy = -2a + \frac{3a^2}{2}$$

$$SI = \iint_S (\nabla \times \bar{A}) \cdot \bar{n} dS = LI = \left(\frac{-2a^3}{3} - \frac{a^2}{2}\right)$$

$$+ \left(-\frac{\pi a^2}{4} - \frac{3a^2}{2} + 2a\right) + \frac{a^2}{2} + \left(-2a + \frac{3a^2}{2}\right)$$

$$SI = \frac{-a^2}{12}(3\pi + 8a).$$

Example 3: Verify Stokes' theorem for $\bar{A} = xzi - yj + x^2yk$ where S is the surface of the region bounded by $x = 0, y = 0, z = 0, 2x + y + 2z = 8$ which is not included in the xz-plane (Fig. 16.23).

Solution: Stokes' theorem states that

$$\oint_c \bar{A} \cdot d\bar{r} = \iint_S (\nabla \times \bar{A}) \cdot \bar{n} dS$$

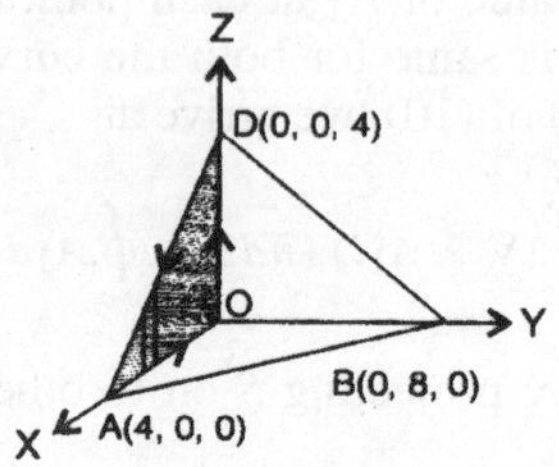

Fig. 16.23

Here c is curve consisting of the straight lines AO, OD and DA.

$$\text{L.H.S.} = \oint_c \bar{A} \cdot d\bar{r} = \int_{AO+OD+DA}$$

$$= \int_{AO} + \int_{OD} + \int_{DA} = LI_1 + LI_2 + LI_3$$

On the straight line AO: $y = 0, z = 0, \bar{A} = 0$ so

$$LI_1 = \int_{AO} \bar{A} \cdot d\bar{r} = 0$$

On the straight line OD: $x = 0, y = 0, \bar{A} = 0$ so

$$LI_2 = \int_{OD} \bar{A} \cdot d\bar{r} = 0$$

On the straight line DA: $x + z = 4$ and $y = 0$ so

$$\bar{A} = xzi = x(4 - x)i$$

$$LI_3 = \int_{DA} \bar{A} \cdot d\bar{r} = \int_0^4 x(4-x)i \cdot dxi = \int_0^4 x(4-x)dx = \frac{32}{3}$$

$$LI = 0 + 0 + \frac{32}{3} = \frac{32}{3}$$

Here the surface S consists of 3 surfaces (planes) $S_1 : OAB$, $S_2 : OBD$, $S_3 : ABD$, so that

$$\text{R.H.S.} = \iint_S (\nabla \times \bar{A}) \cdot \bar{n} dS = \iint_{S_1+S_2+S_3}$$

$$= \iint_{S_1} + \iint_{S_2} + \iint_{S_3} = SI_1 + SI_2 + SI_3$$

$$\nabla \times \bar{A} = \begin{vmatrix} i & j & k \\ \frac{\partial}{\partial x} & \frac{\partial}{\partial y} & \frac{\partial}{\partial z} \\ xz & -y & x^2y \end{vmatrix} = x^2i + x(1 - 2y)j$$

On the surface S_1: plane OAB: $z = 0, \bar{n} = -\bar{k}$, so

$$(\nabla \times \bar{A}) \cdot \bar{n} = \left[x^2 i + x(1-2y)j\right] \cdot (-k) = 0$$

$$SI_1 = \iint_{S_1} (\nabla \times \bar{A}) \cdot \bar{n} dS = 0$$

On surface S_2: plane OBD: plane $x = 0, \bar{n} = -i$ so

$$\nabla \times \bar{A} = 0$$

$$SI_2 = \iint_{S_2} (\nabla \times \bar{A}) \cdot \bar{n} dS = 0$$

On surface S_3: plane ABD: $2x + y + 2z = 8$. Unit normal $\hat{n}$ to the surface $S_3 = \frac{\nabla(2x+y+2z)}{|\nabla(2x+y+2z)|}$

$$\hat{n} = \frac{2i + j + 2k}{\sqrt{4+1+4}} = \frac{2i + j + 2k}{3}$$

$$(\nabla \times \bar{A}) \cdot \bar{n} = \frac{2}{3}x^2 + \frac{1}{3}x(1-2y)$$

To evaluate the surface integral on the surface S_3, project S_3 on to say xz-plane i.e., projection of ABD on xz-plane is AOD

$$dS = \frac{dxdz}{n \cdot j} = \frac{dxdz}{\frac{1}{3}} = 3dxdz$$

Thus

$$SI_3 = \iint_{S_3} (\nabla \times \bar{A}) \cdot \bar{n} dS$$

$$= \iint_{AOD} \left[\frac{2}{3}x^2 + \frac{x}{3}(1-2y)\right] 3dxdz$$

$$= \int_{x=0}^{4} \int_{z=0}^{4-x} \left[2x^2 + x(1-2y)\right] dzdx$$

since the region AOD is covered by varying z from 0 to $4 - x$, while x varies from 0 to 4. Using the equation of the surface S_3, $2x + y + 2z = 8$, eliminate y, then

$$SI_3 = \int_0^4 \int_0^{4-x} \left\{2x^2 + x[1 - 2(8 - 2x - 2z)]\right\} dzdx$$

$$= \int_0^4 \int_0^{4-x} (6x^2 - 15x + 4xz) dzdx$$

$$= \int_0^4 6x^2 z - 15xz + \left.\frac{4xz^2}{2}\right|_0^{4-x} dx$$

$$= \int_0^4 (23x^2 - 4x^3 - 28x) dx = \frac{32}{3}$$

Thus L.H.S = L.I. = R.H.S. = S.I.
Hence Stokes' theorem is verified.

Example 4: Verify Stokes' theorem for $\bar{A} = y^2 i + xyj - xzk$ where S is the hemisphere

$$x^2 + y^2 + z^2 = a^2, z \geq 0.$$

Solution: The curve c which is the boundary of the given hemisphere is the base circle (see Fig. 16.24)

$$x^2 + y^2 = a^2$$

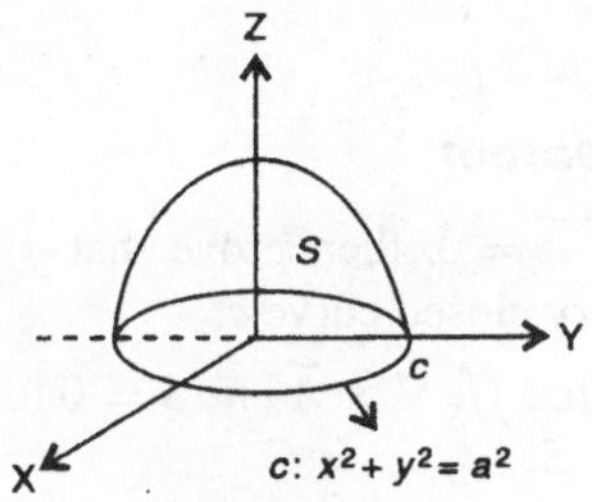

Fig. 16.24

On curve c: $z = 0, x^2 + y^2 = a^2$

$$\text{L.H.S.} = LI = \oint_c \bar{A} \cdot d\bar{r} = \int y^2 dx + xydy - xzdz$$

$$= \int y^2 dx + xydy$$

Introducing polar coordinates $x = a \cos t$, $y = a \sin t$, with t varying from 0 to 2π

$$LI = \int_0^{2\pi} a^2 \sin^2 t \, d(a \cos t) + a \cos t \cdot a \sin t \cdot d(a \sin t)$$

$$= a^3 \int_0^{2\pi} (-\sin^3 t + \cos^2 t \sin t) dt = 0$$

Now

$$\nabla \times \bar{A} = \begin{vmatrix} i & j & k \\ \frac{\partial}{\partial x} & \frac{\partial}{\partial y} & \frac{\partial}{\partial z} \\ y^2 & xy & -xz \end{vmatrix} = zj - yk$$

Unit normal $\hat{n}$ to the sphere is

$$\hat{n} = \frac{\nabla(x^2 + y^2 + z^2)}{|\nabla(x^2 + y^2 + z^2)|} = \frac{2xi + 2yj + 2zk}{\sqrt{4x^2 + 4y^2 + 4z^2}}$$

$$= \frac{xi + yj + zk}{a}$$

$$(\nabla \times \bar{A}) \cdot \bar{n} = (zj - yk) \cdot \left(\frac{xi + yj + zk}{a}\right)$$

$$= \frac{1}{a}(zy - zy) = 0$$

so

$$\text{R.H.S.} = SI = \iint_S (\nabla \times \bar{A}) \cdot \bar{n} dS = 0$$

Thus

$$\text{L.H.S.} = \text{L.I.} = 0 = \text{S.I.} = \text{R.H.S.}$$

Hence the Stokes' theorem is verified.

EXERCISE

Stokes' theorem

1. If $\nabla \times \bar{A} = 0$, then prove that $\oint_c \bar{A} \cdot d\bar{r} = 0$ for every closed curve c.
2. Prove that $\iint_S \nabla \times \bar{A} \cdot \bar{n} dS = 0$ for any closed surface S.
3. Prove that $\oint_c d\bar{r} \times \bar{B} = \iint_S (\bar{n} \times \nabla) \times \bar{B}\, ds$.

   Hint: Choose $\bar{A} = \bar{B} \times \bar{c}$, where $\bar{c}$ is a constant vector, and apply Stokes' theorem. Note that

   $$(\bar{n} \times \nabla) \times \bar{B} = \nabla(\bar{B} \cdot \bar{n}) - \bar{n}(\nabla \cdot \bar{B}).$$

4. Prove that $\oint_c f \nabla g \cdot d\bar{r} = \iint_S (\nabla f \times \nabla g) \cdot \bar{n}\, dS$ and deduce that $\oint_c f \nabla f \cdot d\bar{r} = 0$.

   Hint: Take $\bar{A} = f \nabla g$ in Stokes' theorem. Note that $\nabla \times \nabla g = 0$.

   For deduction, take $f = g$ and note that $\nabla f \times \nabla f = 0$.

5. Evaluate $\iint_S (\nabla \times \bar{A}) \cdot \bar{n}\, dS$ where S is the surface of the hemisphere $x^2 + y^2 + z^2 = 16$ above the xy-plane and $\bar{A} = (x^2 + y - 4)i + 3xyj + (2xz + z^2)k$.

Ans. -16π

6. If $\bar{A} = (y^2 + z^2 + x^2)i + (z^2 + x^2 - y^2)j + (x^2 + y^2 - z^2)k$ evaluate $\iint_S (\nabla \times \bar{A}) \cdot \bar{n}\, dS$ taken over the surface $S = x^2 + y^2 - 2ax + az = 0, z \geq 0$.

Ans. $2\pi a^3$

7. Evaluate $\iint_S \nabla \times (yi + zj + xk) \cdot \bar{n}\, dS$ over the surface of the paraboloid $z = 1 - x^2 - y^2, z \geq 0$.

Ans. π

8. Evaluate $\iint_S \nabla \times (yi + 2xj + zk) \cdot \bar{n}\, dS$ where S is the paraboloid $z = 1 - x^2 - y^2, z \geq 0$.

Ans. π

9. What is the surface integral of the normal component of the curl of the vector function $(x + y)i + (y - x)j + z^3k$ over the upper half of the sphere $x^2 + y^2 + z^2 = 1$.

Ans. -2π

10. Evaluate $\int_c y\,dx + z\,dy + x\,dz$ where c is the curve given by $x^2 + y^2 + z^2 - 2ax - 2ay = 0, x + y = 2a$, beginning at the point $(2a, 0, 0)$ and going at first below the z-plane.

Ans. $-2\sqrt{2}\pi a^2$

11. Evaluate $\oint_c \sin z\,dx - \cos x\,dy + \sin y\,dz$ where c: rectangle $0 \leq x \leq \pi, 0 \leq y \leq 1, z = 3$.

Ans. 2

12. Evaluate $\oint_c y\,dx + z\,dy + x\,dz$ where c is the curve of intersection of the sphere $x^2 + y^2 + z^2 = a^2$ and the plane $x + z = a$.

Ans. $-\pi a^2/\sqrt{2}$.

Verification of Stokes' theorem

13. Evaluate $\oint_c y\,dx + xz^3dy - zy^2dz$ (a) directly (b) using Stokes' theorem, given that c is the circle: $x^2 + y^2 = 4, z = -3$.

Ans. -28.4π

14. Evaluate (a) directly (b) using Stokes' theorem $\oint_c 4z\,dx - 2x\,dy + 2x\,dz$ where c is the ellipse $x^2 + y^2 = 1, z = y + 1$.

Ans. -4π

Verify Stokes' theorem in the following examples for:

15. $\bar{A} = (2x - y)i - yz^2j - y^2zk$ where S: upper half surface of the sphere $x^2 + y^2 + z^2 = 1$

    Hint: Here c: $x^2 + y^2 = 1, z = 0$.

Ans. π

16. $\bar{A} = x^2\bar{i} + xy\bar{j}$ where S is square $0 \leq x \leq a, 0 \leq y \leq a$ in the xy-plane

    Hint: c: square $0 \leq x \leq a, 0 \leq y \leq a, z = 0$

Ans. $a^3/2$

17. $\bar{A} = (x^2 + y^2)\bar{i} - 2xy\bar{j}$ taken around the rectangle bounded by $x = a$, $x = -a$, $y = 0$, $y = b$

Ans. $-4ab^2$

18. $\bar{A} = e^z(i + \sin yj + \cos yk)$ where S: $z = y^2$, $0 \le x \le 4, 0 \le y \le 2$

Ans. $\pm 4(1 - e^4)$ from $(0, 0, 0)$ to $(4, 0, 0)$
$\mp 4e^4$ from $(4, 2, 4)$ to $(0, 2, 4)$
The integrals over the parabolas cancel each other.

19. $\bar{A} = y^2i + z^2j + x^2k$ where S: portion of paraboloid $x^2 + y^2 = z, y \ge 0, z \le 1$

Ans. $\pm\frac{4}{3}$.

Work done around closed curve by Stokes' theorem

Find the work done by the force F in the displacement around the closed curve c where:

20. $\bar{F} = 2xy^3 \sin z\bar{i} + 3x^2y^2 \sin zj + x^2y^3 \cos z\bar{k}$
c: intersection of paraboloid $z = x^2 + y^2$ and cylinder $(x - 1)^2 + y^2 = 1$
Hint: $\nabla \times \bar{F} = 0$.

Ans. 0

21. $\bar{F} = x^3i + e^{3y}j + e^{-3z}k$, c: $x^2 + 9y^2 = 9$, $z = x^2$
Hint: $\nabla \times \bar{F} = 0$.

Ans. 0.

16.7 GAUSS* DIVERGENCE THEOREM

Transformation between surface integral and volume integral. Let $\bar{A}$ be a vector function of position, having continuous derivatives, in a volume V bounded by a closed surface S then

$$\oint\!\!\oint_S \bar{A} \cdot \bar{n} dS = \iiint_V \nabla \cdot \bar{A} dV \qquad (1)$$

where $\bar{n}$ is the outward drawn (positive) normal to S.

Proof: Assume that S is such that any line parallel to coordinate axes meets S in at most two points. Let S_1 and S_2 be the lower (below) and upper (top), portions of S having equations $z = f_1(x, y)$ and $z = f_2(x, y)$ and having $\bar{n}_1$ and n_2 as normals respectively (See Fig. 16.25). Let R be the projection of the surface S on the xy-plane. If $\bar{A}=A_1\bar{i}+A_2\bar{j}+A_3\bar{k}$, then the result of Gauss divergence theorem (1) in component form is

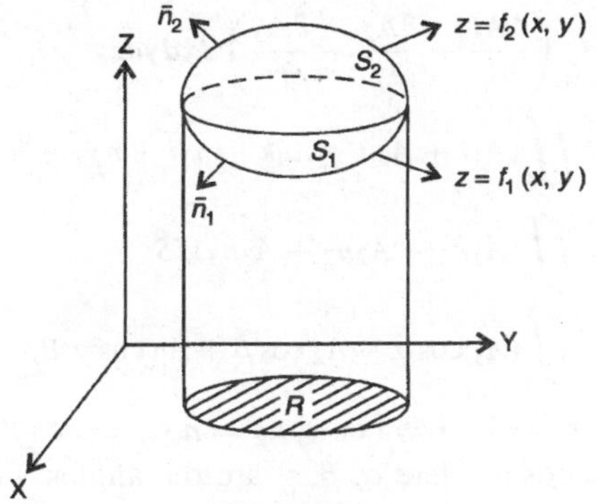

Fig. 16.25

$$\iint_S (A_1i + A_2j + A_3k) \cdot \bar{n} dS = \iiint_V \left(\frac{\partial A_1}{\partial x} + \frac{\partial A_2}{\partial y} + \frac{\partial A_3}{\partial z}\right) dV \qquad (2)$$

Consider

$$\iiint_V \frac{\partial A_3}{\partial z} dV = \iiint_V \frac{\partial A_3}{\partial z} dzdydx$$

$$= \iint_R \left[\int_{z=f_1}^{z=f_2} \frac{\partial A_3}{\partial z} dz\right] dydx$$

$$= \iint_R [A_3(x, y, f_2) - A_3(x, y, f_1)]\, dydx$$

$$= \iint_R A_3\bar{k} \cdot \bar{n}_2 dS_2 - \iint_R A_3\bar{k} \cdot (-\bar{n}_1) dS_1$$

since for upper surface S_2, $k \cdot \bar{n}_2 dS_2 = dydx$ while for lower surface S_1, $\bar{k} \cdot (-\bar{n}_1) dS_1 = dydx$. Thus

$$\iiint_V \frac{\partial A_3}{\partial z} dV = \iint_S A_3\bar{k} \cdot \bar{n} dS \qquad (3)$$

Similarly, projecting S on to yz-plane and xz-planes we have

$$\iiint_V \frac{\partial A_1}{\partial x} dV = \iint_S A_1\bar{i} \cdot \bar{n} dS \qquad (4)$$

$$\iiint_V \frac{\partial A_2}{\partial y} dV = \iint_S A_2\bar{j} \cdot \bar{n} dS \qquad (5)$$

*Karl Frierich Gauss (1777–1855), German mathematician.

Adding (3), (4), (5), we get the required result (2).

Note 1: Gauss divergence theorem (G.D.T.) transforms volume integrals to surface integrals and vice versa.

Note 2: G.D.T in rectangular form

$$\iiint_V \left(\frac{\partial A_1}{\partial x}+\frac{\partial A_2}{\partial y}+\frac{\partial A_3}{\partial z}\right)dxdydz$$

$$=\iint_S (A_1\bar{i}+A_2\bar{j}+A_3\bar{k})\cdot(n_1\bar{i}+n_2\bar{j}+n_3\bar{k})ds$$

$$=\iint_S (A_1n_1+A_2n_2+A_3n_3)dS$$

$$=\iint_S (A_1\cos\alpha+A_2\cos\beta+A_3\cos\gamma)dS$$

where $n_1 = n\cdot i = \cos\alpha$, $n_2 = n\cdot j = \cos\beta$, $n_3 = n\cdot k = \cos\gamma$. Here α, β, γ are the angles which $\bar{n}$ makes with the positive x, y, z axes.

Note 3: Apart from (1), G.D.T. can also be written in the following forms:

$$\iint_S \bar{n}\times\bar{A}\,dS=\iiint_V \nabla\times\bar{A}dV$$ (see Example 14 on Page 16.33)

$$\iint_S \bar{n}\phi dS=\iiint_V \nabla\phi dV$$ (see W.O.E. 7 on Page 16.31)

Note 4: G.D.T. is also known as "Green's theorem in space" because G.D.T. generalizes the "Green's theorem in plane" by replacing the (plane) region R and its closed boundary (curve) c by *a* (space) region V and its closed boundary (surface) S.

Note 5: When $\bar{A}=\bar{V}=$ velocity of a fluid then G.D.T. has the following physical interpretation:

Volume of fluid emerging (diverging) from a closed surfaces in unit time $=$ Volume of fluid supplied from within volume V in unit time

WORKED OUT EXAMPLES

Surface to volume integral using divergence theorem

Example 1: Find the volume V of a region bounded by a surface S.

Solution: By Gauss' divergence theorem

$$\iiint_V \nabla\cdot\bar{A}\,dV=\iint_S \bar{A}\cdot\bar{n}dS \quad (1)$$

Choose $\bar{A}=xi$, so that $\nabla\cdot\bar{A}=1$, with this (1) reduces to

$$V=\text{Volume}=\iiint 1\cdot dV$$

$$=\iint_S x(\bar{i}\cdot\bar{n})dS=\iint_S xdydz$$

Similarly by taking $\bar{A}=yj$ and $\bar{A}=z\bar{k}$, we get

$$V=\iint_S ydzdx \quad\text{and}\quad V=\iint_S zdxdy$$

or $$V=\frac{1}{3}\iint_S (xdydz+ydzdx+zdxdy).$$

Example 2: Evaluate $\iint_S e^x dydz - ye^x dzdx + 3zdxdy$ where S is the surface of the cylinder $x^2+y^2=c^2, 0\le z\le h$ (Fig. 16.26).

Solution: Here $A_1=e^x$, $A_2=-ye^x$, $A_3=3z$, so that $\nabla\cdot A=\frac{\partial A_1}{\partial x}+\frac{\partial A_2}{\partial y}+\frac{\partial A_3}{\partial z}=e^x-e^x+3=3$ using divergence theorem, given surface integral

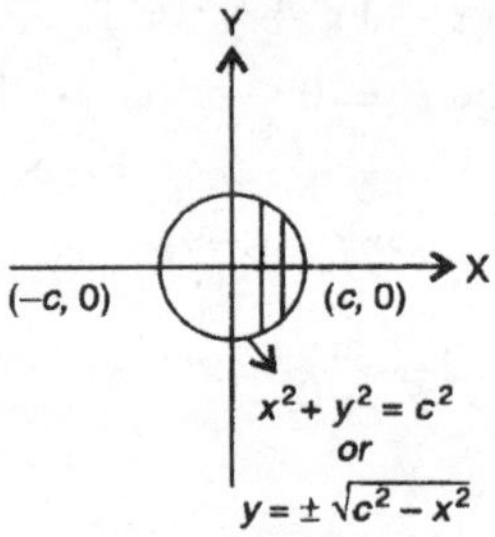

Fig. 16.26

$$=\iiint \nabla\cdot\bar{A}dV=3\iiint dxdydz$$

$$=3\int_{z=0}^{h}\int_{-c}^{c}\int_{-\sqrt{c^2-x^2}}^{+\sqrt{c^2-x^2}} dydxdz$$

$$=3\cdot 2\cdot 2\cdot \int_0^h\int_0^c\int_0^{\sqrt{c^2-x^2}} dydxdz$$

$$=12h\int_0^c \sqrt{c^2-x^2}dx \quad a=3\pi hc^2.$$

Example 3: Evaluate $\iint_S \bar{A}\cdot\bar{n}dS$ where $\bar{A}=2xyi+yz^2j+xzk$, and S is the surface of the region bounded by $x = 0$, $y = 0$, $z = 0$, $y = 3$ and $x + 2z = 6$ (refer Fig. 16.27).

Solution: $\nabla \cdot \bar{A} = 2y + z^2 + x$.
By Gauss' divergence theorem

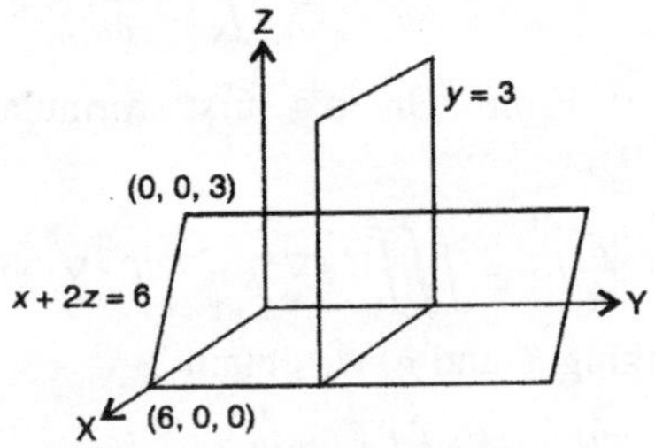

Fig. 16.27

$$SI = \iint_S \bar{A}\cdot\bar{n}\,dS = \iiint_V \nabla\cdot\bar{A}\,dV$$

$$= \iiint_V (2y + z^2 + x)\,dV$$

$$= \int_{y=0}^{3}\int_{x=0}^{6}\int_{z=0}^{\frac{6-x}{2}} (2y + z^2 + x)\,dz\,dx\,dy$$

$$= \int_0^3\int_0^6 (2y + x)z + \frac{z^3}{3}\Bigg|_0^{\frac{6-x}{2}}$$

$$= \int_0^3\int_0^6 \left[y(6-x) + \frac{6x - x^2}{2} + \frac{1}{24}(6-x)^3\right]dx$$

$$= \int_0^3 y\left(6x - \frac{x^2}{2}\right) + \frac{1}{2}\left(\frac{6x^2}{2} - \frac{x^3}{3}\right)$$

$$-\frac{1}{24}\left(\frac{6-x}{4}\right)^4\Bigg|_0^6$$

$$= \int_0^3 \left[18y + 216\left(\frac{1}{12} + \frac{1}{16}\right)\right]dy = \frac{351}{2}.$$

Example 4: Evaluate $\iint_S \frac{\bar{r}}{r^2}\cdot\bar{n}\,dS$.

Solution: Take $\bar{A} = \frac{\bar{r}}{r^2} = \frac{xi + yj + zk}{x^2 + y^2 + z^2}$ so that

$$\nabla\cdot\bar{A} = \nabla\cdot\left(\frac{\bar{r}}{r^2}\right)$$

$$= \frac{(r^2 - 2x^2)}{r^4} + \frac{(r^2 - 2y^2)}{r^4} + \frac{(r^2 - 2z^2)}{r^4}$$

$$= \frac{3r^2 - 2r^2}{r^4} = \frac{1}{r^2}$$

Applying Gauss divergence theorem, we get

$$\iint \frac{\bar{r}}{r^2}\cdot\bar{n}\,dS = \iint_S \bar{A}\cdot\bar{n}\,dS = \iiint_V \nabla\cdot\bar{A}\,dV$$

$$= \iiint_V \frac{1}{r^2}\,dV$$

Example 5: Evaluate $\iint_S r^5\bar{n}\,dS$

Solution: Put $f = r^5$ so that $\nabla f = \nabla r^5 = 5r^3\bar{r}$
Applying Gauss divergence theorem

$$\iint_S r^5\bar{n}\,dS = \iint f\bar{n}\,dS = \iiint_V \nabla f\,dV$$

$$= \iiint_V 5r^3\bar{r}\,dV$$

Example 6: Evaluate $\iint_S \bar{B}\cdot\bar{n}\,dS$ when $\bar{B} = \nabla\times\bar{A}$ and S is any closed surface.

Solution: By Gauss divergence theorem

$$\iint_S \bar{B}\cdot\bar{n}\,dS = \iiint_V \nabla\cdot\bar{B}\,dV$$

$$= \iiint_V \nabla\cdot(\nabla\times\bar{A})\,dV = 0$$

since $\nabla\cdot(\nabla\times\bar{A}) = 0$ for any $\bar{A}$.

Example 7: Prove that $\iiint \nabla f\,dV = \iint_S f\bar{n}\,dS$.

Solution: Choose $\bar{A} = f\bar{c}$ where $\bar{c}$ is a constant vector

so that $\nabla\cdot\bar{A} = \nabla\cdot(f\bar{c}) = \bar{c}\cdot\nabla f + f\nabla\cdot\bar{c} = \bar{c}\cdot\nabla f$

since $\nabla\cdot\bar{c} = 0$

Also $\bar{A}\cdot\bar{n} = (f\bar{c})\cdot\bar{n} = (f\bar{n})\cdot\bar{c} = \bar{c}\cdot(f\bar{n})$

Applying Gauss divergence theorem

$$\iiint_V \nabla\cdot\bar{A}\,dV = \iiint_V \bar{c}\cdot\nabla f\,dV$$

$$= \iint_S \bar{A}\cdot\bar{n}\,dS$$

$$= \iint_S \bar{c}\cdot(f\bar{n})\,dS$$

or $\quad \bar{c}\cdot\iiint_V \nabla f\,dV = \bar{c}\cdot\iint_S f\bar{n}\,dS$

Since $\bar{c}$ is arbitrary constant vector, the result follows.

Example 8: Prove that

$\iint_S \bar{r} \times d\bar{S} = \bar{0}$ for any closed surface S.

Solution: We know that

$$\iint_S \bar{n} \times \bar{B} dS = \iiint_V \nabla \times \bar{B} dV \qquad (1)$$

Consider

$$\iint_S \bar{r} \times d\bar{S} = \iint_S \bar{r} \times \bar{n} dS = -\iint \bar{n} \times \bar{r} dS$$

Choose $\bar{B} = -\bar{r}$ in the above result (1). Note that

$$\nabla \times \bar{B} = \nabla \times (-\bar{r}) = \begin{vmatrix} i & j & k \\ \frac{\partial}{\partial x} & \frac{\partial}{\partial y} & \frac{\partial}{\partial z} \\ -x & -y & -z \end{vmatrix} = 0$$

Thus

$$\iint \bar{r} \times d\bar{S} = -\iint \bar{n} \times \bar{r} dS$$
$$= -\iiint_V \nabla \times \bar{r} dV = 0$$

Green's formulas:
Green's first formula (identity)

Example 9: Prove that

$$\iint_S f \frac{\partial g}{\partial n} dS = \iiint_V (f\nabla^2 g + \nabla f \cdot \nabla g) dV$$

Solution: Choose $\bar{A} = f\nabla g$ in the divergence theorem then

$$\nabla \cdot \bar{A} = \nabla \cdot (f\nabla g) = f\nabla \cdot \nabla g + \nabla f \cdot \nabla g$$
$$= f\nabla^2 g + \nabla f \cdot \nabla g$$
$$\bar{A} \cdot \bar{n} = \bar{n} \cdot f\nabla g = f\bar{n} \cdot \nabla g = f\nabla g \cdot \bar{n} = f\frac{\partial g}{\partial n}$$

From divergence theorem

$$\iint_S \bar{A} \cdot \bar{n} dS = \iint_S f\frac{\partial g}{\partial n} dS = \iiint_V \nabla \cdot \bar{A} dV$$
$$= \iiint_V (f\nabla^2 g + \nabla f \cdot \nabla g) dV$$

since $\frac{\partial g}{\partial n} dS = \nabla g \cdot \bar{n} dS = \nabla g \cdot d\bar{S}$
Green's first identity can also be written as

$$\iint_S f\nabla g \cdot d\bar{S} = \iiint_V (f\nabla^2 g + \nabla f \cdot \nabla g) dV$$

Green's second formula (identity) or symmetrical theorem

Example 10: Show that

$$\iiint (f\nabla^2 g - g\nabla^2 f) dV = \iint_S \left(f\frac{\partial g}{\partial n} - g\frac{\partial f}{\partial n}\right) dS$$

Solution: From Green's first formula (above Example 9) we have

$$\iint_S f\frac{\partial g}{\partial n} dS = \iiint_V (f\nabla^2 g + \nabla f \cdot \nabla g) dV \qquad (1)$$

Interchanging f and g, we obtain

$$\iint_S g\frac{\partial f}{\partial n} dS = \iiint_V (g\nabla^2 f + \nabla g \cdot \nabla f) dV \qquad (2)$$

Subtracting (2) from (1), we get

$$\iint_S \left(f\frac{\partial g}{\partial n} - g\frac{\partial f}{\partial n}\right) dS = \iiint_V (f\nabla^2 g - g\nabla^2 f) dV$$

Note: Since

$$\iint_S [f(\nabla g \cdot \bar{n}) - g\nabla f \cdot \bar{n}] dS$$
$$= \iint_S (f\nabla g - g\nabla f) \cdot \bar{n} dS = \iint_S (f\nabla g - g\nabla f) \cdot d\bar{S}$$

Green's second identity can also be written as

$$\iint_S (f\nabla g - g\nabla f) \cdot d\bar{S} = \iiint_V (f\nabla^2 g - g\nabla^2 f) dV.$$

EXERCISE

Surface to volume integral, using divergence theorem

Using divergence theorem, evaluate the surface integral:

1. $\iint_S yz dy dz + zx dz dx + xy dx dy$ where S: $x^2 + y^2 + z^2 = 4$

Ans. 0

2. $\iint_S x^3 dy dz + x^2 y dz dx + x^2 z dx dy$ where S: closed surface consisting of the circular cylinder $x^2 + y^2 = a^2$, $(0 \le z \le b)$ and the circular disks $z = 0$ and $z = b$, $(x^2 + y^2 \le a^2)$.

Ans. $5\pi a^4 b/4$

3. $\iint_S \sin x\,dydz + (2-\cos x)y\,dzdx$ where S: parallelopiped $0 \le x \le 3, 0 \le y \le 2, 0 \le z \le 1$

Ans. 12

4. $\iint_S (ax^2 + by^2 + cz^2)dS$ where S: sphere of unit radius centered at origin.

Ans. $4\pi(a+b+c)/3$

5. $\iint_S (x^2 - yz)dzdy - 2x^2y\,dzdx + z\,dxdy$ where S: cube of side b and three of whose edges are along the axes.

Ans. $b^3(b^2+3)/3$

6. $\iint_S 9x\,dydz + y\cosh^2 x\,dzdx - z\sinh^2 x\,dxdy$ where S: ellipsoid $4x^2 + y^2 + 9z^2 = 36$

Ans. 480π

7. $\iint_S \sin x\,dydz + y\,dzdx + z\,dxdy$ where S: surface of $0 \le x \le \pi/2, x \le y \le z, 0 \le z \le 1$

Ans. $\frac{3}{2} - \frac{\pi^2}{4}$

8. $\iint_S \bar{r}\cdot\bar{n}\,dS$ where S: sphere of radius 2 with centre at origin

Ans. 32π

9. $\iint_S \bar{r}\cdot\bar{n}\,dS$ where S: surface of cube bounded by the planes $x=-1, y=-1, z=-1, x=1, y=1, z=1$

Hint: For examples 6 and 7 use result of worked example 4

Ans. 24

10. $\iint_S \bar{F}\cdot\bar{n}\,dS$ where $\bar{F} = 2xyi + yz^2j + xzk$ and S : surface of parallelopiped $0 \le x \le 2, 0 \le y \le 1, 0 \le z \le 3$

Ans. 30

11. If S is any closed surface enclosing a volume V and $\bar{A} = axi + byj + cz\bar{k}$, then evaluate $\iint_S \bar{A}\cdot\bar{n}\,dS$.

Ans. $(a+b+c)V$

12. If $\bar{n}$ is the unit outward drawn normal to any closed surface of area S, then evaluate $\iiint \nabla\cdot\bar{n}dV$.

Ans. S

13. Prove that $\iint_S \bar{n}dS = \bar{0}$ for any closed surface S.

Hint: Choose $f = 1$.

14. Prove that $\iint_S \bar{n}\times\bar{B}dS = \iiint_V \nabla\times\bar{B}dV$.

Hint: Take $\bar{A} = \bar{B}\times\bar{C}$ in divergence theorem, with $\bar{C}$ any arbitrary constant vector.

15. Evaluate $\iiint_V \nabla\times\bar{B}dV$ where V is the region bounded by a closed surface S and $\bar{B}$ is always normal to S.

Hint: Normal $\bar{n}$ to S and $\bar{B}$ are parallel, so $\bar{n}\times\bar{B} = 0$. Use result of Exercise Example 14

Ans. 0

16. Prove that $\iint_S \frac{\partial f}{\partial n}dS = \iiint_V \nabla^2 f dV$. Further if f is harmonic (solution of Laplace's equation) in a domain D, then evaluate $\iint \frac{\partial f}{\partial n}dS$.

Hint: Take $\bar{A} = \nabla f$ in divergence theorem and note that $\bar{A}\cdot\bar{n} = \nabla f\cdot\bar{n} = \frac{\partial f}{\partial n}$.

Ans. 0

17. If f and g are harmonic in V then evaluate $\iint_S \left(f\frac{\partial g}{\partial n} - g\frac{\partial f}{\partial n}\right)dS$.

Hint: Use Green's second identity (formula) and note that $\nabla^2 f = 0$ and $\nabla^2 g = 0$

Ans. 0

Gauss' theorem

18. Let S be a closed surface and let $\bar{r}$ be the position vector of any point (x, y, z) measured from an origin 0. Then prove that

$$\iint_S \frac{\bar{n}\cdot\bar{r}}{r^3}dS = 0 \quad \text{if o lies outside } S$$
$$= 4\pi \quad \text{if o lies inside } S$$

Hint: If 0 lies outside S, note that $r \ne 0$ and $\nabla\cdot\frac{\bar{r}}{r^3} = 0$. Now use divergence theorem
If 0 lies inside S, enclose 0 by *a* small sphere S^* of radius a then from the above result

$$\iint_{S+S^*} \frac{\bar{n}\cdot\bar{r}}{r^3}dS = 0$$

Note that $\iint_{S^*} \frac{\bar{n}\cdot\bar{r}}{r^3}dS = -4\pi$

19. Prove that $\iint_S \nabla(x^2+y^2+z^2)\cdot\bar{n}dS = 6V$ where S is any closed surface enclosing a volume V.

20. If $\bar{A} = (x^2+y-4)i + 3xyj + (2xz+z^2)k$ and S is the surface of the paraboloid $x^2+y^2+z = 4$ above the xy-plane, evaluate $\iint_S (\nabla\times\bar{A})\cdot\bar{n}dS$

Ans. -4π.

Verification of Gauss Divergence Theorem

WORKED OUT EXAMPLES

Example 1: Verify the divergence theorem for $\bar{A} = 2x^2y\bar{i} - y^2\bar{j} + 4xz^2\bar{k}$ taken over the region in the first octant bounded by the cylinder $y^2 + z^2 = 9$ and the plane $x = 2$ (refer Fig. 16.28).

Solution: Here $\nabla \cdot \bar{A} = 4xy - 2y + 8xz$

$$\text{R.H.S.} = \iiint_V \nabla \cdot \bar{A}\, dV$$

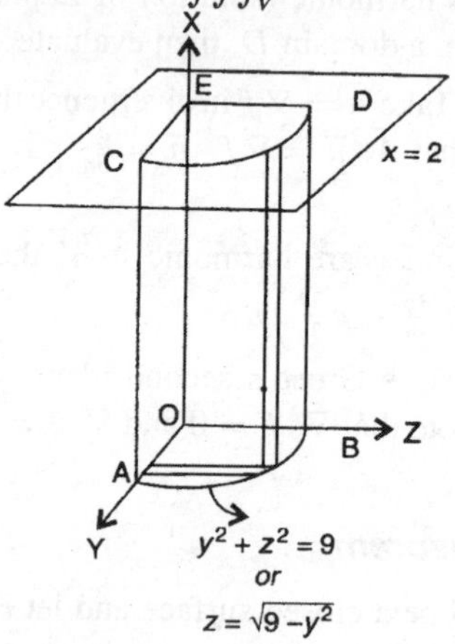

Fig. 16.28

The solid region is covered when z varies from 0 to $\sqrt{9-y^2}$, y varies from 0 to 3 and x varies from 0 to 2 (height of the cylinder) so

$$\text{R.H.S.} = \int_0^2 \int_0^3 \int_0^{\sqrt{9-y^2}} (4xy - 2y + 8xz)\,dz\,dy\,dx$$

$$= \int_0^2 \int_0^3 \left[(4xy - 2y)\left(\sqrt{9-y^2}\right) + 4x(9-y^2)\right] dy\,dx$$

$$= \int_0^2 \frac{(2-4x)}{2} \frac{(9-y^2)^{\frac{3}{2}}}{3/2}\Bigg|_0^3 dx$$

$$+ \int_0^2 36yx - 4x\frac{y^3}{3}\Bigg|_0^3 dx$$

$$= \int_0^2 -18(1-2x)dx + \int_0^2 (108x - 36x)dx$$

R.H.S. = 180

The entire surface S consists of five surfaces S_1, S_2, S_3, S_4, S_5. So

$$\text{L.H.S.} = \iint_S \bar{A} \cdot \bar{n}\, dS = \iint_{S_1} + \iint_{S_2} + \cdots$$

$$+ \iint_{S_5} = SI_1 + SI_2 + \cdots + SI_5$$

On S_1: OAB: $x = 0, \hat{n} = -\bar{i}, \bar{A} \cdot \bar{n} = 0$ so

$$SI_1 = \iint_{S_1} \bar{A} \cdot \bar{n}\, dS = 0$$

On S_2: CED: $x = 2, \hat{n} = \bar{i}, \bar{A} \cdot \bar{n} = 8y$ so

$$SI_2 = \iint_{S_2} \bar{A} \cdot \bar{n}\, dS = \iint_{S_2} 8y\,dy\,dz$$

$$= \int_0^3 \int_0^{\sqrt{9-z^2}} 8y\,dy\,dz = \int_0^3 4(9-z^2)dz = 72$$

On S_3: plane $OBDE$: $y = 0, \bar{n} = -\bar{j}, \bar{A} \cdot \bar{n} = 0$ so

$$SI_3 = \iint_{S_3} \bar{A} \cdot \bar{n}\, dS = 0$$

On S_4: plane: $OACE$: $z = 0, \bar{n} = -\bar{k}, \bar{A} \cdot \bar{n} = 0$, so

$$SI_4 = \iint_{S_4} \bar{A} \cdot \bar{n}\, dS = 0$$

On S_5: the curved surface $ABDC$ of the cylinder:

$$y^2 + z^2 = 9$$

unit normal $\hat{n}$ to S_5: $\frac{\nabla(y^2+z^2)}{|\nabla(y^2+z^2)|} = \frac{2y\bar{j}+2z\bar{k}}{\sqrt{4y^2+4z^2}}$

$$\hat{n} = \frac{y\bar{j} + z\bar{k}}{3}$$

so that $\quad \bar{A} \cdot \hat{n} = \dfrac{-y^3 + 4xz^3}{3}$

and $\quad \bar{n} \cdot \bar{k} = \dfrac{y\bar{j} + z\bar{k}}{3} \cdot \bar{k} = \dfrac{z}{3} = \dfrac{\sqrt{9-y^2}}{3}$

Projecting the surface S_5 on the yx-plane

$$SI_5 = \iint_{S_5} \bar{A} \cdot \bar{n}\, dS = \iint \frac{(4xz^3 - y^3)}{3} \cdot \frac{dx\,dy}{\bar{n} \cdot \bar{k}}$$

$$= \iint_R \frac{(4xz^3 - y^3)}{3\,\frac{\sqrt{9-y^2}}{3}} dx\,dy$$

$$= \int_{x=0}^2 \int_{y=0}^3 \left[4x(9-y^2) - y^3(9-y^2)^{-\frac{1}{2}}\right] dy\,dx$$

$$= \int_0^2 72x\,dx + 18\int_0^2 dx = 144 - 36 = 108$$

so

$$\text{L.H.S.} = 0 + 72 + 0 + 0 + 108 = 180$$

Hence the divergence theorem is verified.

Example 2: Compute the flux of the vector field

$$\bar{A} = \left(\frac{x^2 y}{1+y^2} + 6yz^2\right)\bar{i} + 2x \arctan y\, \bar{j} - \frac{2xz(1+y)+1+y^2}{1+y^2}\bar{k}$$

through the outer side of that part of the surface of the paraboloid of revolution $z = 1 - x^2 - y^2$ located above the xy-plane.

Solution: The flux of $\bar{A}$ through a surface S is given by the surface integral,

$$\text{flux} = \iint \bar{A} \cdot \bar{n}\, dS \quad (1)$$

Since the given surface of S_1 is the surface of the paraboloid of revolution $z = 1 - x^2 - y^2$, which is not a closed surface, so we close this surface from below with the circular portion S_2 of the xy-plane that is bounded by the circle $x^2 + y^2 = 1$, $z = 0$ (see Fig. 16.29).

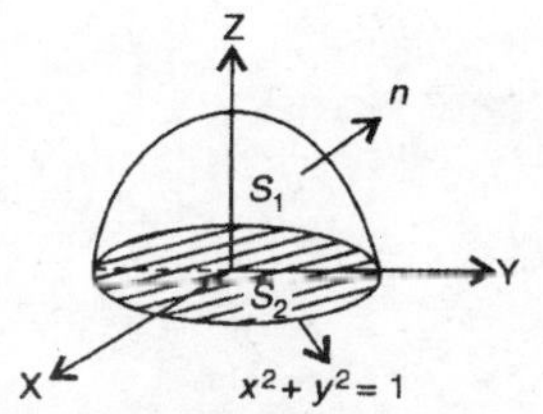

Fig. 16.29

Let V be the volume of the resulting solid bounded above by S_1 and below by S_2.
Now the flux (1) is calculated, using divergence theorem for the closed region V. Thus

$$\text{Flux} = \iint_S \bar{A} \cdot \bar{n}\, dS = \iiint_V (\nabla \cdot \bar{A})\, dV = 0$$

Since

$$\nabla \cdot \bar{A} = \frac{2xy}{1+y^2} + \frac{2x}{1+y^2} \frac{-2x(1+y)}{1+y^2} = 0$$

Flux across $S = S_1 + S_2$ is additive. So

$$\text{Flux} = \iint_S = \iint_{S_1+S_2} = \iint_{S_1} + \iint_{S_2} = 0$$

Thus

$$\iint_{S_1} \bar{A} \cdot \bar{n}\, dS = -\iint_{S_2} \bar{A} \cdot \bar{n}\, dS$$

i.e., flux across the required surface S_1 = −flux across the circular region S_2

On S_2: $z = 0$, $x^2 + y^2 \le 1$, $\bar{n} = -\bar{k}$ so that

$$\bar{A} \cdot \bar{n} = \left(\frac{x^2 y}{1+y^2} i + 2x \arctan y\, j - k\right) \cdot k = 1$$

$$\iint_{S_2} \bar{A} \cdot \bar{n}\, dS = \iint_{S_2} dS = S_2 = \text{area of the circular region}$$

$$= \pi r^2 = \pi \cdot 1^2 = \pi$$

Thus the required flux of $\bar{A}$ across the outer side of that part of the surface S_1 of the paraboloid of revolution $z = 1 - x^2 - y^2$ is $-\pi$

EXERCISE

Verify Gauss divergence theorem for:

1. $\bar{A} = 4xi - 2y^2 j + z^2 k$ taken over the region bounded by $x^2 + y^2 = 4$, $z = 0$ and $z = 3$.

Ans. common value: 84π

2. $\bar{A} = (x^3 - yz)i - 2x^2 yj + z\bar{k}$ taken over the entire surface of the cube $0 \le x \le a$, $0 \le y \le a$, $0 \le z \le a$.

Ans. common value: $\frac{a^5}{3} + a^3$.

3. $\bar{A} = axi + byj + czk$, theorem taken over the entire surface of the sphere of radius d and centered at origin.

Ans. common value: $\frac{4\pi}{3} d^3 (a + b + c)$

4. $\bar{A} = 2xyi + yz^2 j + xzk$ and S is the total surface of the rectangular parallelopiped bounded by the coordinate planes and $x = 1, y = 2, z = 3$.

Ans. common value: 33

5. $\bar{A} = 2xzi + yzj + z^2 k$ over the upper half of the sphere $x^2 + y^2 + z^2 = a^2$

Ans. common value: $5\pi a^4/4$

6. $\overline{A} = (x^2 - yz)\overline{i} + (y^2 - zx)j + (z^2 - xy)k$ taken over the rectangular parallelopiped bounded by the coordinate planes and $x = a$, $y = b$ and $z = c$

Ans. common value: $abc(a + b + c)$

7. $\overline{A} = x^2\overline{i} + y^2 j + z^2\overline{k}$ taken over the surface of the ellipsoid

$$\frac{x^2}{a^2} + \frac{y^2}{b^2} + \frac{z^2}{c^2} = 1$$

Ans. common value: 0

8. $\overline{A} = xi + yj$ taken over the upper half of the unit sphere $x^2 + y^2 + z^2 = 1$

Ans. common value: $4\pi/3$

9. $\overline{A} = x^3 i + x^2 yj + x^2 zk$ taken over the closed region of the cylinder $x^2 + y^2 = a^2$, bounded by the planes $z = 0$ and $z = b$

Ans. common value; $5\pi ba^4/4$

10. Compute the flux of the vector $\overline{A} = 4xi - yj + zk$ through the surface of a torus.

Hint: Volume of a torus with R_1 and R_2 as the inner and outer radii of the torus is $\frac{\pi^2}{4}(R_2 - R_1)^2(R_2 + R_1)$.

Ans. flux $= \pi^2(R_2 - R_1)^2(R_2 + R_1)$.

PART—V
FOURIER ANALYSIS AND PARTIAL DIFFERENTIAL EQUATIONS

Chapter 17

Fourier Series

INTRODUCTION

Fourier series introduced in 1807 by Fourier (after work by Euler and Daniel Bernoulli) was one of the most important developments in applied mathematics. It is very useful in the study of heat conduction, mechanics, concentrations of chemicals and pollutants, electrostatics, acoustics and in areas unheard of in Fourier's day such as computing and CAT scan (computer assisted tomography).

Fourier* series is an infinite series representation of periodic function in terms of the trigonometric sine and cosine functions. Fourier series is very powerful method to solve ordinary and partial differential equations particularly with periodic functions appearing as non-homogeneous terms. While Taylor's series expansion is valid only for functions which are continuous and differentiable, Fourier series is possible not only for continuous functions but for periodic functions, functions discontinuous in their values and derivatives. Further, because of the periodic nature, Fourier series constructed for one period is valid for all values. Harmonic analysis is the theory of expanding functions in Fourier series.

Periodic Function

A function $f(x)$ is said to be periodic if $f(x+T) = f(x)$ for all real x and for some positive number T. T is known as the period of $f(x)$.

* Jean-Baptiste Joseph Fourier (1768–1830), French physicist and mathematician.

Fundamental period

or primitive period or simply period of $f(x)$ is the smallest positive period of f.

Example: $\cos x$, $\sin x$, $\sec x$, $\operatorname{cosec} x$, are periodic functions with period 2π.

$\tan x$, $\cot x$ are periodic with period π.

Result 1: If T is the period of $f(x)$ then nT is also period of f for any integer n.

i.e. $\quad f(x+nT) = f(x) \qquad (n \neq 0)$

Example: $\cos(x+4\pi) = \cos(x+2\cdot 2\pi) = \cos((x+2\pi)+2\pi) = \cos(x+2\pi) = \cos 2\pi$.

Result 2: The function $h(x) = af(x) + bg(x)$ has period T if $f(x)$ and $g(x)$ have period T. Here a, b are constants.

Example: If $h(x) = a\cos x + b\sin x$, then

$$h(x+2\pi) = a\cos(x+2\pi) + b\sin(x+2\pi)$$
$$= a\cos x + b\sin x = h(x).$$

Result 3: If $f(x)$ is a periodic function of period T, then $f(ax)$ with $a \neq 0$, is a periodic function of period $\frac{T}{a}$.

Example: $\sin 2x$ has period $\frac{2\pi}{2} = \pi$

$\cos 3x$ has period $\frac{2\pi}{3}$ and so on.

Result 4: The period of a sum of a number of periodic functions is the least common multiple of the periods.

Example: $f(x) = \sin x + \frac{1}{2}\sin 2x + \frac{1}{3}\sin 3x + \frac{1}{4}\sin 4x$. Note that $\sin x$, $\sin 2x$, $\sin 3x$, $\sin 4x$ have periods $2\pi, \pi, \frac{2\pi}{3}$ and $\frac{\pi}{2}$ respectively. Then the period of $f(x)$ is 2π which is the L.C.M. of these periods.

Result 5: A constant function is periodic for any positive T.

Trigonometric Series is a functional series of the form

$$\frac{a_0}{2} + a_1 \cos x + b_1 \sin x + a_2 \cos 2x + b_2 \sin 2x + \ldots$$

or
$$\frac{a_0}{2} + \sum_{n=1}^{\infty} (a_n \cos nx + b_n \sin nx)$$

where the constants $a_0, a_n, b_n (n = 1, 2, 3, \ldots)$ are called the coefficients.

Result 1: Let n and m be integers, $n \neq 0, m \neq 0$.

For $m \neq n$

a. $\int_{\alpha}^{\alpha+2\pi} \cos mx \cdot \cos nx \, dx = 0$

b. $\int_{\alpha}^{\alpha+2\pi} \sin mx \cdot \sin nx \, dx = 0$

c. $\int_{\alpha}^{\alpha+2\pi} \sin mx \cdot \cos nx \, dx = 0$

d. $\int_{\alpha}^{\alpha+2\pi} \cos mx \, dx = 0$

e. $\int_{\alpha}^{\alpha+2\pi} \sin mx \, dx = 0$

For $m = n$

a. $\int_{\alpha}^{\alpha+2\pi} \cos mx \cdot \cos nx \, dx = \int_{\alpha}^{\alpha+2\pi} \cos^2 mx \, dx = \pi$

b. $\int_{\alpha}^{\alpha+2\pi} \sin^2 mx \, dx = \pi$

c. $\int_{\alpha}^{\alpha+2\pi} \cos mx \cdot \sin mx \, dx = 0.$

17.1 EULER'S (FOURIER–EULER) FORMULAE

Let $f(x)$, a periodic function with period 2π defined in the interval $(\alpha, \alpha + 2\pi)$, be the sum of a trigonometric series i.e.,

$$f(x) = \frac{a_0}{2} + \sum_{n=1}^{\infty} (a_n \cos nx + b_n \sin nx) \qquad (1)$$

To determine the coefficient a_0

Integrate both sides of (1) with respect to x in the interval α to $\alpha + 2\pi$. Then

$$\int_{\alpha}^{\alpha+2\pi} f(x) dx$$

$$= \int_{\alpha}^{\alpha+2\pi} \frac{a_0}{2} dx + \int_{\alpha}^{\alpha+2\pi} \left(\sum_{n=1}^{\infty} a_n \cos nx + b_n \sin nx \right) dx$$

$$= \frac{a_0}{2} \cdot x \Big|_{\alpha}^{\alpha+2\pi} + \sum_{n=1}^{\infty} \left[a_n \int_{\alpha}^{\alpha+2\pi} \cos nx \, dx + b_n \int_{\alpha}^{\alpha+2\pi} \sin nx \, dx \right]$$

From result 6, the last two integrals for all n will be zero. Thus

$$\int_{\alpha}^{\alpha+2\pi} f(x) dx = \frac{a_0}{2} \cdot 2\pi = \pi a_0$$

Hence

$$a_0 = \frac{1}{\pi} \int_{\alpha}^{\alpha+2\pi} f(x) dx \qquad (2)$$

To determine coefficient a_n for $n = 1, 2, 3 \ldots$: Multiplying both sides of (1) by $\cos mx$ and integrating w.r.t. x in $(\alpha, \alpha + 2\pi)$, we get

$$\int_{\alpha}^{\alpha+2\pi} f(x) \cos mx \, dx$$

$$= \int_{\alpha}^{\alpha+2\pi} \frac{a_0}{2} \cdot \cos mx \, dx + \int_{\alpha}^{\alpha+2\pi} \left(\sum_{n=1}^{\infty} a_n \cdot \cos nx \cdot \cos mx \right) dx + \int_{\alpha}^{\alpha+2\pi} \left(\sum b_n \sin nx \cdot \cos mx \right) dx$$

In R.H.S., except the integral of a_n (with $m = n$) all other integrals vanish (Result 6). For $m = n$, the integral of a_n is π. Thus

$$a_n = \frac{1}{\pi} \int_{\alpha}^{\alpha+2\pi} f(x) \cos nx \, dx \quad \text{for } n = 1, 2, 3 \cdots \quad (3)$$

Similarly,

$$b_n = \frac{1}{\pi} \int_{\alpha}^{\alpha+2\pi} f(x) \sin nx \, dx \quad \text{for } n = 1, 2, 3 \cdots \quad (4)$$

The formulae (2), (3), (4) are known as Euler (or Fourier–Euler) formulae which gives the coefficients a_0, a_n and b_n, which are known as **Fourier coefficients** of $f(x)$.

Fourier series

Fourier Series of a periodic function $f(x)$ with period 2π is the trigonometric series (1) with the Fourier coefficients a_0, a_n, b_n given by the Euler formulae (2), (3), (4). The individual terms in Fourier series are known as harmonics.

Dirichlet conditions

Dirichlet conditions for the expansion of $f(x)$ in Fourier series:

Let $f(x)$ be a periodic function with period 2π. Let $f(x)$ be piecewise continuous, and bounded in the interval $(\alpha, \alpha + 2\pi)$ with finite number of extrema. Then at the points of continuity, the Fourier series of $f(x)$ (R.H.S. of (1)) converges to $f(x)$ (L.H.S. of (1)). At a point of discontinuity x_0, the Fourier series of $f(x)$ converges to the arithmetic mean of the left and right hand limits of $f(x)$ at x_0 i.e., $f(x_0) = \frac{1}{2}[f(x_0 - 0) + f(x_0 + 0)]$.

Method of obtaining Fourier series of $f(x)$

I. $f(x) = \frac{a_0}{2} + \sum_{n=1}^{\infty} a_n \cos nx + \sum_{n=1}^{\infty} b_n \sin nx$

II. $a_0 = \frac{1}{\pi}\int_{\alpha}^{\alpha+2\pi} f(x)dx$

III. $a_n = \frac{1}{\pi}\int_{\alpha}^{\alpha+2\pi} f(x)\cos nx\, dx, \quad n = 1, 2, 3, \ldots$

IV. $b_n = \frac{1}{\pi}\int_{\alpha}^{\alpha+2\pi} f(x)\sin nx\, dx, \quad n = 1, 2, 3, \ldots$

Result 1: *Leibnitz's Rule:* $\int u\, v\, dx = uv_1 - u'v_2 + u''v_3 - u'''v_4 + \cdots$ where $'$ denotes differentiation and suffix integration w.r.t. x.

Result 2: $\cos n\pi = (-1)^n$; $\sin n\pi = 0$; $\cos(2n+1)\frac{\pi}{2} = 0$; $\sin(2n+1)\frac{\pi}{2} = (-1)^n$.

WORKED OUT EXAMPLES

Example 1: Find the Fourier series of

$$f(x) = \begin{cases} 0, & \text{when } -\pi \le x \le 0 \\ x^2, & \text{when } 0 \le x \le \pi \end{cases}$$

which is assumed to be periodic with period 2π.

Solution: The Fourier series is given by

$$f(x) = \frac{a_0}{2} + \sum_{n=1}^{\infty}(a_n \cos nx + b_n \sin nx)$$

Here

$$a_0 = \frac{1}{\pi}\int_{-\pi}^{\pi} f(x)dx = \frac{1}{\pi}\int_{-\pi}^{0} 0\, dx + \frac{1}{\pi}\int_{0}^{\pi} x^2 dx$$

$$a_0 = \frac{1}{\pi}\frac{x^3}{3}\Big|_0^{\pi} = \frac{\pi^3}{3\pi} = \frac{\pi^2}{3}$$

$$a_n = \frac{1}{\pi}\int_{-\pi}^{\pi} f(x)\cos nx\, dx$$

$$= \frac{1}{\pi}\int_{-\pi}^{0} 0 \cdot \cos nx\, dx + \frac{1}{\pi}\int_{0}^{\pi} x^2 \cos nx\, dx$$

Integrating by parts,

$$= \frac{1}{\pi}\left[x^2 \cdot \frac{\sin nx}{n} - (2x)\cdot\left(\frac{-\cos nx}{n^2}\right) + 2\cdot\left(\frac{-\sin nx}{n^3}\right)\right]\Bigg|_0^{\pi}$$

$$a_n = \frac{1}{\pi}\frac{2\pi}{n^2}\cos n\pi = \frac{2}{n^2}(-1)^n, \quad \text{for } n = 1, 2, 3, \ldots$$

$$b_n = \frac{1}{\pi}\int_{-\pi}^{\pi} f(x)\sin nx\, dx = \frac{1}{\pi}\int_{0}^{\pi} x^2 \sin nx\, dx$$

$$= \frac{1}{\pi}\left[x^2\left(\frac{-\cos nx}{n}\right) - 2x\cdot\left(\frac{-\sin nx}{n^2}\right) + 2\left(\frac{+\cos nx}{n^3}\right)\right]_0^{\pi}$$

$$b_n = \frac{1}{\pi}\left[\frac{-\pi^2}{n}\cos n\pi + \frac{2}{n^3}(\cos n\pi - 1)\right]$$

$$= \frac{-\pi}{n}(-1)^n + \frac{2}{\pi n^3}[(-1)^n - 1]$$

Substituting a_0, a_n and b_n we get the required Fourier series of $f(x)$ in the internal $(-\pi, \pi)$ as

$$f(x) = \frac{\pi^2}{6} + 2\sum_{n=1}^{\infty}\frac{(-1)^n}{n^2}\cos nx + \sum_{n=1}^{\infty}\left\{\frac{\pi}{n}(-1)^{n+1} + \frac{2}{\pi n^3}[(-1)^n - 1]\right\}\sin nx.$$

Example 2: Obtain the Fourier series expansion of $f(x) = e^{ax}$ in $(0, 2\pi)$.

Solution:

$$a_0 = \frac{1}{\pi}\int_0^{2\pi} f(x)dx = \frac{1}{\pi}\int_0^{2\pi} e^{ax}dx = \left.\frac{e^{ax}}{\pi}\right|_0^{2\pi}$$

$$a_0 = \frac{e^{2a\pi} - 1}{\pi}$$

$$a_n = \frac{1}{\pi}\int_0^{2\pi} f(x)\cos nx\, dx$$

$$= \frac{1}{\pi}\int_0^{2\pi} e^{ax}\cos nx\, dx$$

Using

$$\int e^{ax}\cos bx\, dx = e^{ax}[a\cos bx + b\sin bx]/(a^2 + b^2)$$

$$a_n = \frac{1}{\pi}\left[e^{ax}\left.\frac{\{a\cos nx + n\cdot\sin nx\}}{(a^2+n^2)}\right|_0^{2\pi}\right]$$

$$a_n = \frac{1}{\pi(a^2+n^2)}[ae^{2a\pi}\cos 2n\pi - e^0\cdot\cos 0]$$

$$= \frac{1}{\pi(a^2+n^2)}[ae^{2a\pi} - 1]$$

$$b_n = \frac{1}{\pi}\int_0^{2\pi} e^{ax}\sin nx\, dx$$

$$= \frac{1}{\pi}\left[e^{ax}\left.\frac{(a\sin nx - n\cos nx)}{a^2+n^2}\right]\right|_0^{2\pi}$$

Since

$$\int e^{ax}\sin bx\, dx = e^{ax}[a\sin bx - b\cos bx]/(a^2 + b^2)$$

$$b_n = \frac{n}{\pi(a^2+n^2)}\cdot[-e^{2a\pi}\cos 2n\pi + 1]$$

$$= \frac{n}{\pi(a^2+n^2)}(1 - e^{2a\pi})$$

Substituting these values the required Fourier series is

$$f(x) = \left(\frac{e^{2a\pi}-1}{\pi}\right)\left[\frac{1}{2} + \sum_{n=1}^{\infty}\left(\frac{1}{a^2+n^2}\right)(-n\sin nx)\right]$$

$$+\frac{ae^{2a\pi}-1}{\pi}\sum_{n=1}^{\infty}\frac{\cos nx}{(a^2+n^2)}.$$

Example 3: If $f(x) = x$ in $\left(\frac{-\pi}{2}, \frac{\pi}{2}\right)$ and $f(x) = 0$ in $\left(\frac{\pi}{2}, 3\pi/2\right)$ find the Fourier series of $f(x)$. Deduce that $\frac{\pi^2}{8} = \sum_{n=1}^{\infty}\frac{1}{(2n-1)^2}$ (refer Fig. 17.1).

Solution:

$$a_0 = \frac{1}{\pi}\int_{\frac{-\pi}{2}}^{\frac{3\pi}{2}} f(x)dx = \frac{1}{\pi}\int_{\frac{-\pi}{2}}^{\frac{\pi}{2}} x\, dx + \frac{1}{\pi}\int_{\frac{\pi}{2}}^{\frac{3\pi}{2}} 0\cdot dx$$

$$a_0 = \frac{1}{\pi}\left[\frac{x^2}{2}\right]_{\frac{-\pi}{2}}^{\frac{\pi}{2}} = \frac{1}{2\pi}\left[\frac{\pi^2}{4} - \frac{\pi^2}{4}\right] = 0$$

$$a_n = \frac{1}{\pi}\int_{\frac{-\pi}{2}}^{\frac{3\pi}{2}} f(x)\cos nx\, dx = \frac{1}{\pi}\int_{\frac{-\pi}{2}}^{\frac{\pi}{2}} x\cos nx\, dx$$

$$a_n = \frac{1}{\pi}\left[x\frac{\sin nx}{n} + \frac{1}{n}\cdot\frac{\cos nx}{n}\right]_{\frac{-\pi}{2}}^{\frac{\pi}{2}} = 0$$

$$b_n = \frac{1}{\pi}\int_{\frac{-\pi}{2}}^{\frac{3\pi}{2}} f(x)\sin nx\, dx = \frac{1}{\pi}\int_{\frac{-\pi}{2}}^{\frac{\pi}{2}} x\sin nx\, dx$$

$$= \frac{1}{\pi}\left[x\cdot\frac{(-\cos nx)}{n} + \frac{1}{n\pi}\frac{\sin nx}{n}\right]_{\frac{-\pi}{2}}^{\frac{\pi}{2}}$$

$$b_n = \frac{-1}{n}\left[\cos n\frac{\pi}{2}\right] + \frac{2}{n^2\pi}\sin\frac{n\pi}{2}$$

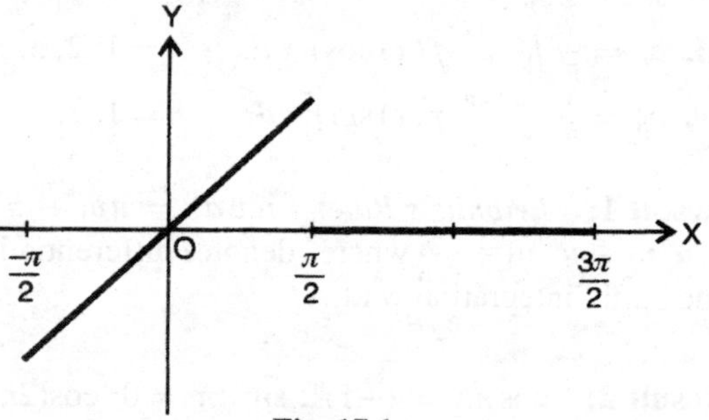

Fig. 17.1

For $n = 1, b_1 = \frac{2}{\pi}, b_2 = 0 + \frac{1}{2}, b_3 = -\frac{2}{9\pi}, b_4 = -\frac{1}{4}, b_5 = \frac{2}{25\pi}$ etc. Thus the Fourier series is

$$f(x) = 0 + 0 + \sum_{n=1}^{\infty}\left(\frac{-\cos n\frac{\pi}{2}}{n} + \frac{2}{n^2\pi}\sin n\frac{\pi}{2}\right)\sin nx$$

At $x = \frac{\pi}{2}$ which is a point of discontinuity,

$$f\left(\frac{\pi}{2}\right) = \frac{1}{2}\left[f\left(\frac{\pi}{2} - 0\right) + f\left(\frac{\pi}{2} + 0\right)\right]$$

$$= \frac{1}{2}\left[\frac{\pi}{2} + 0\right] = \frac{\pi}{4}.$$

Putting $x = \frac{\pi}{2}$ in the Fourier series expansion

$$\frac{\pi}{4} = f\left(\frac{\pi}{2}\right) = \sum_{n=1}^{\infty}\left(\frac{-\cos}{n}\left(\frac{n\pi}{2}\right) + \frac{2}{n^2\pi}\cdot\sin\frac{n\pi}{2}\right)\sin\frac{n\pi}{2}$$

$$= \sum_{n=1}^{\infty}\left(0 + \frac{2}{(2n-1)^2\pi}\right)(-1)^{2n}$$

or
$$\frac{\pi^2}{8} = \sum_{n=1}^{\infty}\frac{1}{(2n-1)^2}.$$

Example 4: Obtain the Fourier series of $f(x) = (\pi - x)/2$ in the interval $(0, 2\pi)$. Deduce

$$\frac{\pi}{4} = 1 - \frac{1}{3} + \frac{1}{5} - \frac{1}{7} + \cdots$$

Solution:

$$a_0 = \frac{1}{\pi}\int_0^{2\pi} f(x)dx = \frac{1}{\pi}\int_0^{2\pi}\left(\frac{\pi - x}{2}\right)dx$$

$$= \frac{1}{2\pi}\left[\pi x - \frac{x^2}{2}\right]_0^{2\pi} = 0$$

$$a_n = \frac{1}{\pi}\int_0^{2\pi} f(x)\cos nx\, dx$$

$$= \frac{1}{\pi}\int_0^{2\pi}\left(\frac{\pi - x}{2}\right)\cos nx\, dx$$

$$a_n = \frac{1}{2\pi}\left[(\pi - x)\cdot\frac{\sin nx}{n} - (-1)\frac{(-\cos nx)}{n^2}\right]_0^{2\pi} = 0$$

$$b_n = \frac{1}{\pi}\int_0^{2\pi}\left(\frac{\pi - x}{2}\right)\sin nx\, dx$$

$$= \frac{1}{2\pi}\left[(\pi - x)\left(\frac{-\cos nx}{n}\right) - (-1)\left(\frac{-1}{n}\right)\right.$$

$$\left.\times\frac{\sin nx}{n}\right]_0^{2\pi}$$

$$b_n = \frac{1}{2\pi}\left[\frac{\pi}{n} + \frac{\pi}{n}\right] = \frac{1}{n}$$

$$f(x) = \frac{\pi - x}{2} = 0 + 0 + \sum_{n=1}^{\infty}\frac{1}{n}\sin nx$$

$$= \sin x + \frac{\sin 2x}{2} + \frac{\sin 3x}{3} + \cdots$$

Put $x = \frac{\pi}{2}$ then $\frac{\pi - \frac{\pi}{2}}{2} = \frac{\pi}{4} = \sum_{n=1}^{\infty}\frac{1}{n}\sin n\frac{\pi}{2}$

$$= 1 - \tfrac{1}{3} + \tfrac{1}{5} - \tfrac{1}{7} + \cdots$$

Example 5: Sum of functions: suppose the Fourier series of $f(x) = \frac{a_0}{2} + \sum_{n=1}^{\infty}\left(a_n\cos\frac{n\pi x}{L} + b_n\sin\frac{nx\pi}{L}\right)$ and $g(x) = \frac{c_0}{2} + \sum_{n=1}^{\infty}\left(c_n\cos\frac{n\pi x}{L} + d_n\sin\frac{n\pi x}{L}\right)$ in the interval e to $e + 2L$ then find the Fourier series of $h(x) = \alpha f(x) + \beta g(x)$.

Solution:

$$h(x) = \alpha\left[\frac{a_0}{2} + \sum a_n\cos\frac{n\pi x}{L} + b_n\sin\frac{n\pi x}{L}\right]$$

$$+\beta\left[\frac{C_0}{2} + \sum c_n\cos\frac{n\pi x}{L} + d_n\sin\frac{n\pi x}{L}\right]$$

$$h(x) = \left(\alpha\frac{a_0}{2} + \beta\frac{c_0}{2}\right) + \sum(\alpha a_n + \beta c_n)\cos\frac{n\pi x}{L}$$

$$+\sum(\alpha b_n + \beta d_n)\sin\frac{n\pi x}{L}$$

$$h(x) = \frac{A_0}{2} + \sum_{n=1}^{\infty}\left[A_n\cdot\cos\left(\frac{n\pi x}{L}\right) + B_n\sin\left(\frac{n\pi x}{L}\right)\right]$$

where $A_0 = \alpha a_0 + \beta c_0, A_n = \alpha a_n + \beta c_n, B_n = \alpha b_n +$ $+\beta d_n$.

EXERCISE

$$f(x) = \frac{a_0}{2} + \sum_{n=1}^{\infty}(a_n\cos nx + b_n\sin nx)$$

Find the Fourier series of the following functions $f(x)$ which are periodic with period 2π and defined in the indicated interval.

1. Modified saw-toothed wave form

   $f(x) = 0$ for $-\pi < x \le 0$

   $= x$ for $0 < x \le \pi$.

   Deduce $\frac{\pi^2}{8} = \sum_{n=1}^{\infty} 1/(2n-1)^2$.

   Hint: $a_n = \frac{1}{\pi n^2}\cos nx\Big|_0^{\pi} = \frac{-2}{\pi n^2}$ for n odd, 0 when n is even.

   $b_n = -\frac{\cos n\pi}{n} = \frac{1}{n}$ for n odd, $\frac{-1}{n}$ for n even

   Put $x = 0$ in the equation below

   $0 = \frac{\pi}{4} - \frac{2}{\pi}\sum_{n=1}^{\infty}\frac{1}{(2n-1)^2}.$

Ans. $f(x) = \frac{\pi}{4} - \frac{2}{\pi}\left(\frac{\cos x}{1^2} + \frac{\cos 3x}{3^2} + \frac{\cos 5x}{5^2} + \cdots\right)$
$+ \left(\frac{\sin x}{1} - \frac{\sin 2x}{2} + \cdots\right).$

2. $f(x) = 0$ in $(-\pi, 0)$
$= \sin x$ in $(0, \pi)$
Deduce $\frac{\pi-2}{4} = \frac{1}{1.3} - \frac{1}{1.5} + \frac{1}{5.7} + \cdots$
Hint: $a_0 = \frac{2}{\pi}$, $a_1 = 0$, $a_n = \frac{1+\cos n\pi}{\pi(1-n^2)}$, $b_1 = \frac{1}{2}$, $b_n = 0$
Put $x = \frac{\pi}{2}$, $\sin\frac{\pi}{2} = 1 = \frac{1}{\pi} + \frac{1}{2}\sin\frac{\pi}{2} - \frac{2}{\pi}\sum_{n=1}^{\infty}\frac{1+\cos n\pi/2}{\pi(1-n^2)}$.
$\left(1 - \frac{1}{2} - \frac{1}{\pi}\right)\frac{\pi}{2} = \sum_{n=1}^{\infty}\frac{(-1)^{n+1}}{(2n+1)(2n-1)}$.

Ans. $f(x) = \frac{1}{\pi} + \frac{\sin x}{2} - \frac{2}{\pi}\left(\frac{\cos 2x}{3} + \frac{\cos 4x}{15} + \frac{\cos 6x}{35} + \frac{\cos 8x}{63} + \cdots\right).$

3. $f(x) = x$ in $(0, 2\pi)$.
Ans. $f(x) = x = \pi - 2\left(\sin x + \frac{1}{2}\sin 2x + \frac{1}{3}\sin 3x + \cdots\right)$
$a_0 = \pi$, $a_n = 0$, $b_n = \frac{1}{\pi}\left[-2\pi\frac{\cos 2n\pi}{n}\right] = -\frac{2}{n}$

4. $f(x) = x$ if $0 \le x \le \pi$
$= 2\pi - x$ if $\pi \le x \le 2\pi$.
Hint: $a_n = \frac{1}{\pi}\left\{\frac{\cos n\pi - 1}{n^2} - \frac{1-\cos n\pi}{n^2}\right\}$, $b_n = 0.$
Ans. $f(x) = \frac{\pi}{2} - \frac{4}{\pi}\left(\cos x + \frac{\cos 3x}{3^2} + \frac{\cos 5x}{5^2} + \cdots\right)$

5. $f(x) = e^{ax}$ in $(-\pi, \pi)$; Deduce when $x = 0$.
Ans. $a_0 = \frac{2\sinh a\pi}{a\pi}$, $a_n = \frac{2a(-1)^n\sinh a\pi}{\pi(n^2+a^2)}$
$b_n = -(2n)(-1)^n\sinh a\pi/[\pi(n^2+a^2)]$
when $x = 0$, $\frac{\pi}{2\sinh a\pi}$
$= \left\{\frac{1}{2a} - \frac{a}{a^2+1^2} + \frac{a}{a^2+2^2} - \frac{a}{a^2+3^2} + \cdots\right\}$

6. $f(x) = a\sin t$, if $0 \le t \le \pi$ (Half-wave
$= 0$, if $\pi \le t \le 2\pi$ rectifier)
Deduce $\frac{1}{2} = \frac{1}{1.3} + \frac{1}{3.5} + \frac{1}{5.7} + \cdots$
Ans. $f(x) = \frac{a}{\pi} + \frac{1}{2}a\sin x - \frac{2a}{\pi}\sum_{n=1}^{\infty}\frac{\cos 2nx}{4n^2-1}$
Put $t = \pi$, then $0 = a\sin\pi = \frac{a}{\pi} + \frac{a}{2}\sin\pi - \frac{2a}{\pi}\sum_{n=1}^{\infty}\frac{\cos 2n\pi}{4n^2-1}$

7. $f(x) = -\pi$, if $-\pi < x < 0$
$= x$, if $0 < x < \pi$
Deduce that $\frac{\pi^2}{8} = \frac{1}{1^2} + \frac{1}{3^2} + \frac{1}{5^2} + \cdots$
Ans. $a_0 = -\frac{\pi}{2}$, $a_n = \frac{1}{\pi n^2}(\cos n\pi - 1)$;
$b_n = \frac{1}{n}(1 - 2\cos n\pi)$

$$f(x) = -\frac{\pi}{4} - \frac{2}{\pi}\left(\cos x + \frac{\cos 3x}{3^2} + \frac{\cos 5x}{5^2} + \cdots\right)$$
$$+3\sin x - \frac{\sin 2x}{2} + \frac{3\sin 3x}{3} - \frac{\sin 4x}{4} + \cdots$$

$x = 0$ is a point of discontinuity. At $x = 0$,

$$-\frac{\pi}{2} = \frac{1}{2}[-\pi + 0] = \frac{1}{2}[f(0-0) + f(0+0)]$$
$$= -\frac{\pi}{4} - \frac{2}{\pi}\left[\frac{1}{1^2} + \frac{1}{3^2} + \frac{1}{5^2} + \cdots\right]$$

8. $f(x) = x^2$ when $0 < x < 2\pi$
Deduce $\frac{\pi^2}{6} = \frac{1}{1^2} + \frac{1}{2^2} + \frac{1}{3^2} + \cdots$
Ans. $f(x) = x^2 = \frac{4\pi^2}{3} + \sum_{n=1}^{\infty}\left(\frac{4}{n^2}\cos nx - \frac{4\pi}{n}\cdot\sin nx\right)$

$x = 0$ is a point of discontinuity. So

$$f(0) = \frac{1}{2}[f(0-0) + f(0+0)]$$
$$= \frac{1}{2}\left[(2\pi)^2 + 0^2\right] = 2\pi^2$$

Putting $x = 0$ in Fourier series expansion

$$2\pi^2 = f(0) = \frac{4\pi^2}{3} + \sum_{n=1}^{\infty}\left(\frac{4}{n^2}\cos 0 - 0\right)$$

9. $f(x) = e^{-x}$ in $0 < x < 2\pi$.
Ans. $e^{-x} = \frac{1-e^{-2\pi}}{\pi}\left[\frac{1}{2} + \left(\frac{1}{2}\cos x + \frac{1}{5}\cos 2x + \frac{1}{10}\cos 3x + \cdots\right) + \left(\frac{1}{2}\sin x + \frac{2}{5}\sin 2x + \frac{3}{10}\sin 3x + \cdots\right)\right]$

10. $f(x) = x + x^2$ in $(-\pi, \pi)$.
Hint: Treating $f(x) = x + x^2 = g(x) + h(x)$ as the sum of two functions $g(x) = x$ and $h(x) = x^2$, the Fourier series of $f(x)$ can be obtained as the sum of the Fourier series of $g(x)$ and $h(x)$ i.e. FS of $f(x) = FS$ of $g(x) + FS$ of $h(x)$ or

$$f(x)=\left[2\sum_{n=1}^{\infty}(-1)^{n+1}\frac{\sin x}{n}\right]+\left[\frac{\pi^2}{3}+4\sum_{n=1}^{\infty}(-1)^n\frac{\cos nx}{n^2}\right].$$

Ans. $x+x^2=\frac{\pi^2}{3}+2\sum_{n=1}^{\infty}\frac{(-1)^{n+1}}{n}\sin nx+4\sum_{n=1}^{\infty}\frac{(-1)^n}{n^2}\cos nx$

11. $f(x)=\left(\frac{\pi-x}{2}\right)^2$ in $(0, 2\pi)$.

Deduce $\frac{\pi^2}{6}=\sum_{n=1}^{\infty}\frac{1}{n^2}$.

Ans. $f(x)=\frac{\pi^2}{12}+\sum_{n=1}^{\infty}\frac{\cos nx}{n^2}$; put $x=0$ for deduction.

12. $f(x)=x\sin x$ in $0\le x\le 2\pi$.

Ans. $a_0=-2, a_1=-\frac{1}{2}, a_n=\frac{2}{n^2-1}, b_1=\pi, b_n=0$

$f(x)=-1-\frac{\cos x}{2}+\sum_{n=2}^{\infty}\left(\frac{2\cos nx}{(n^2-1)}\right)+\pi\sin x$

13. $f(x)=2x$ when $0\le x\le\pi$
$=x$ when $-\pi<x\le 0$.

Ans. $f(x)=\frac{\pi}{4}-\frac{2}{\pi}\left(\frac{\cos x}{1^2}+\frac{\cos 3x}{3^2}+\frac{\cos 5x}{5^2}+\cdots\right)+3\left(\frac{\sin x}{1}-\frac{\sin 2x}{2}+\frac{\sin 3x}{3}-\ldots\right)$

14. $f(x)=-x$ if $-\pi<x\le 0$
$=0$ if $0<x\le\pi$.

Ans. $f(x)=\frac{\pi}{4}-\frac{2}{\pi}\sum_{n=0}^{\infty}\frac{\cos(2n+1)x}{(2n+1)^2}-\sum_{n=1}^{\infty}(-1)^{n-1}\frac{\sin nx}{n}$

15. $f(x)=1$ if $-\pi<x\le 0$
$=-2$ if $0<x\le\pi$

Ans. $f(x)=-\frac{1}{2}-\frac{6}{\pi}\sum_{n=0}^{\infty}\frac{\sin(2n+1)x}{2n+1}$.

17.2 FOURIER SERIES FOR EVEN AND ODD FUNCTIONS

Definitions

Even function

A function $f(x)$ is said to be even if

$$f(-x)=f(x)\quad\text{for all } x.$$

Notes:

1. The graph of $f(x)$ is symmetric about y-axis.
2. $f(x)$ contains only even powers of x and may contain only $\cos x$, $\sec x$.
3. $\int_{-a}^{a} f(x)dx=2\int_0^a f(x)dx$ when $f(x)$ is even.
4. The sum of two even functions is even i.e., $h(x)=f(x)+g(x)$ is even when both $f(x)$ and $g(x)$ are even (or $e+e=e$).
5. Product of two even functions is even i.e., $h(x)=f(x)\cdot g(x)$ is even when both f and g are even (or $e\cdot e=e$).

Example: x^2, $\cos x$, $x^4+\cos 2x+2$

Odd function

A function $f(x)$ is said to be odd if

$$f(-x)=-f(x)\quad\text{for all } x.$$

Notes:

1. The graph of $f(x)$ is symmetric about the origin (lies in opposite quadrants Ist and IIIrd).
2. $f(x)$ contains only odd powers of x and may contain only $\sin x$, $\operatorname{cosec} x$.
3. $\int_{-a}^{a} f(x)dx=0$ when $f(x)$ is odd.
4. The sum of two odd functions is odd i.e., $h(x)=f(x)+g(x)$ is odd when both $f(x)$ and $g(x)$ are odd (or $0+0=0$).
5. Product of an odd function and even function is odd i.e., $h(x)=f(x)g(x)$ is odd when f is even and g is odd or vice versa (or) $0\cdot e=0$.
6. Product of two odd fucntions is even even i.e., $0\cdot 0=e$.

Example: x^3, $\sin x$.

Result: Most functions are neither even nor odd. But any function $f(x)$ can be written as the arithmetic mean of an even and odd function as

$$f(x)=\frac{1}{2}[f(x)+f(-x)]+\frac{1}{2}[f(x)-f(-x)].$$

Fourier Series for Even and Odd Functions

Let the Fourier series of $f(x)$ in $(-\pi,\pi)$ be

$$f(x)=\frac{a_0}{2}+\sum_{n=1}^{\infty}(a_n\cos nx+b_n\sin nx)\qquad(1)$$

where $$a_0=\frac{1}{\pi}\int_{-\pi}^{\pi}f(x)dx\qquad(2)$$

$$a_n = \frac{1}{\pi}\int_{-\pi}^{\pi} f(x)\cos nx\,dx \tag{3}$$

$$b_n = \frac{1}{\pi}\int_{-\pi}^{\pi} f(x)\sin nx\,dx \tag{4}$$

Case 1: Suppose $f(x)$ is even function in $(-\pi, \pi)$. Then all b_n's will be zero since the integrand in (4) is an odd function $(e \cdot o = o)$. Thus the Fourier series of an even function contains "only cosine terms" and is known as "**Fourier cosine series**" given by

$$f(x) = \frac{a_0}{2} + \sum_{n=1}^{\infty} a_n \cos nx \tag{5}$$

where

$$a_0 = \frac{1}{\pi}\int_{-\pi}^{\pi} f(x)dx = \frac{2}{\pi}\int_0^{\pi} f(x)dx \tag{6}$$

and

$$a_n = \frac{1}{\pi}\int_{-\pi}^{\pi} f(x)\cos nx\,dx = \frac{2}{\pi}\int_0^{\pi} f(x)\cos nx\,dx \tag{7}$$

Case 2: Suppose $f(x)$ is odd function in $(-\pi, \pi)$. Then all a_n's are zero because the integrand in (3) is an odd function $(o \cdot e = 0)$. Also a_0 is zero since f is odd. Thus the Fourier series of an odd function contains "only sine terms" and is known as "**Fourier sine series**" given by

$$f(x) = \sum_{n=1}^{\infty} b_n \sin nx \tag{8}$$

where

$$b_n = \frac{1}{\pi}\int_{-\pi}^{\pi} f(x)\sin nx\,dx = \frac{2}{\pi}\int_0^{\pi} f(x)\sin nx\,dx \tag{9}$$

(Note that the integrand in the above integral is even i.e. $0 \cdot 0 = e$).

Procedure

1. Identify whether the given function f is even or odd function in the given interval.
2. If f is even, calculate only a_0 and a_n's from (6) and (7) (no need to calculate b_n's). The Fourier cosine series is given by (5).
3. If f is odd, calculate only b_n's from (9). The Fourier sine series is given by (8).

Worked Out Examples

Example 1: Expand the function $f(x) = \frac{\pi^2}{12} - \frac{x^2}{4}$ in Fourier series in the interval $(-\pi, \pi)$ (Fig. 17.2).

Solution: $f(x)$ is even since $f(-x) = f(x)$. The Fourier series reduces to Fourier cosine series given by

$$f(x) = \frac{a_0}{2} + \sum_{n=1}^{\infty} a_n \cos nx$$

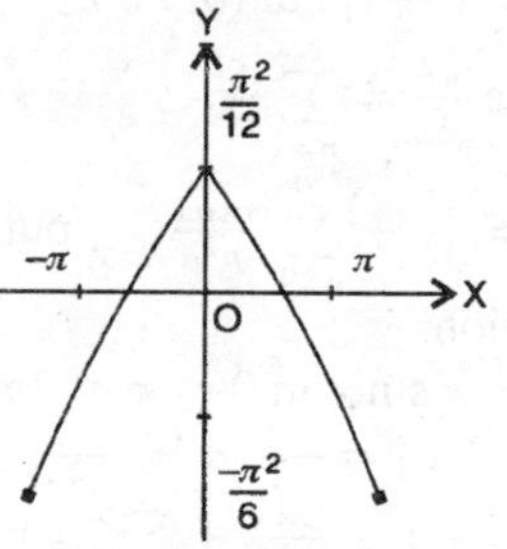

Fig. 17.2

where

$$a_0 = \frac{2}{\pi}\int_0^{\pi} f(x)dx = \frac{2}{\pi}\int_0^{\pi}\left(\frac{\pi^2}{12} - \frac{x^2}{4}\right)dx$$

$$= \frac{2}{\pi}\left[\frac{\pi^2}{12}x - \frac{x^3}{12}\right]_0^{\pi} = 0$$

$$a_n = \frac{2}{\pi}\int_0^{\pi}\left(\frac{\pi^2}{12} - \frac{x^2}{4}\right)\cos nx\,dx$$

$$= \frac{2}{\pi}\left[\left(\frac{\pi^2}{12} - \frac{x^2}{4}\right)\frac{\sin nx}{n} - \left(\frac{-2x}{4}\right) \times \left(\frac{-\cos nx}{n^2}\right) + \left(-\frac{1}{2}\right)\left(\frac{-\sin nx}{n^3}\right)\right]_0^{\pi}$$

$$a_n = \frac{2}{\pi}\left(\frac{1}{n^2}\right)\frac{\pi}{2}\cdot\cos n\pi = \frac{(-1)^{n+1}}{n^2}$$

$$f(x) = \sum_{n=1}^{\infty}\frac{(-1)^{n+1}}{n^2}\cdot\cos nx$$

$$= \cos x - \frac{\cos 2x}{2^2} + \frac{\cos 3x}{3^2} - \frac{\cos 4x}{4^2} + \cdots$$

Example 2: Find Fourier series of $f(x) = x^3$ in $(-\pi, \pi)$.

Solution: Since $f(x)$ is odd (i.e., $f(-x)=f(x)$) the series reduces to Fourier sine series given by

$$f(x)=\sum_{n=1}^{\infty} b_n \sin nx$$

where

$$b_n=\frac{2}{\pi}\int_0^{\pi} f(x)\sin nx\,dx=\frac{2}{\pi}\int_0^{\pi} x^3\sin nx\,dx$$

$$=\frac{2}{\pi}\left[x^3\left(\frac{-\cos nx}{n}\right)-3x^2\left(\frac{-\sin nx}{n^2}\right)\right.$$

$$\left.+6x(-1)(-)\frac{\cos nx}{n^3}\frac{-6}{n^4}\sin nx\right]_0^{\pi}$$

$$b_n=\frac{2}{\pi}\left[\frac{-\pi^3}{n}\cos n\pi+\frac{6\pi}{n^3}\cos n\pi\right]$$

$$=2(-1)^n\left[\frac{-\pi^2}{n}+\frac{6}{n^3}\right]$$

$$f(x)=2\sum_{n=1}^{\infty}\left(\frac{6}{n^3}-\frac{\pi^2}{n}\right)(-1)^n\sin nx$$

$$=2\left[\left(\frac{\pi^2}{1}-\frac{6}{1^3}\right)\sin x-\left(\frac{\pi^2}{2}-\frac{6}{2^3}\right)\sin 2x+\cdots\right].$$

EXERCISE

Find the Fourier series expansion of the following functions $f(x)$ in the indicated interval. $f(x)$ is periodic with period 2π:

1. *Saw-toothed wave form:* $f(x)=x$ in $(-\pi,\pi)$.

Ans. Odd function, $f(x)=\sum_{n=1}^{\infty}-\frac{2}{n}(-1)^n\sin nx$.

2. $f(x)=x^2$ in $(-\pi,\pi)$

Deduce that

a. $\frac{\pi^2}{12}=\frac{1}{1^2}-\frac{1}{2^2}+\frac{1}{3^2}-\frac{1}{4^2}+\cdots$

b. $\frac{\pi^2}{6}=\frac{1}{1^2}+\frac{1}{2^2}+\frac{1}{3^2}+\frac{1}{4^2}+\cdots$

c. $\frac{\pi^2}{8}=\frac{1}{1^2}+\frac{1}{3^2}+\frac{1}{5^2}+\frac{1}{7^2}+\cdots$

Ans. Even function, $f(x)=\frac{\pi^2}{3}+4\sum_{n=1}^{\infty}\frac{(-1)^n}{n^2}\cos nx$

(a) and (b) are obtained by putting $x=0$ and $x=\pi$ in Fourier series respectively. (c) is the arithmetic mean of results (a) and (b).

3. Triangular waveform

$$f(x)=1+\frac{2x}{\pi}\quad \text{if } -\pi\le x\le 0$$
$$=1-\frac{2x}{\pi}\quad \text{if } 0\le x\le \pi$$

Deduce that $\frac{\pi^2}{8}=\frac{1}{1^2}+\frac{1}{3^2}+\frac{1}{5^2}+\cdots$

Ans. $f(x)=0+\frac{4}{\pi^2}\sum_{n=1}^{\infty}\frac{[1-(-1)^n]}{n^2}\cos nx$

since f is even function.

Putting $x=0$, $1=f(0)=\frac{8}{\pi^2}\left(\frac{1}{1^2}+\frac{1}{3^2}+\frac{1}{5^2}+\cdots\right)$

4. $f(x)=|x|$ in $(-\pi,\pi)$ or

$$f(x)=-x \quad \text{in } (-\pi,0) \text{ and}$$
$$=x \quad \text{in } (0,\pi).$$

Ans. $f(x)=\frac{\pi}{2}+\frac{2}{\pi}\sum_{n=1}^{\infty}\frac{[(-1)^n-1]}{n^2}\cos nx$, even function.

5. *Square waveform*:

$$f(x)=-k,\quad \text{if } -\pi<x<0$$
$$=k,\quad \text{if } 0<x<\pi.$$

Ans. Odd function, $f(x)=\frac{2k}{\pi}\sum_{n=1}^{\infty}\frac{(1-\cos n\pi)}{n}\sin nx$

6. $f(x)=-(\pi+x)/2$ for $-\pi\le x<0$

$$=(\pi-x)/2 \quad \text{for } 0\le x<\pi.$$

Ans. Odd function, $f(x)=\sin x+\frac{1}{2}\sin 2x+\frac{1}{3}\sin 3x+\cdots$

7. $f(x)=|\cos x|$ in $(-\pi,\pi)$.

Ans. Even function, $f(x)=\frac{2}{\pi}+\frac{4}{\pi}\sum_{n=1}^{\infty}\frac{\cos n\frac{\pi}{2}}{1-n^2}\cos nx$.

8. $f(x)=|\sin x|$ in $(-\pi,\pi)$.

Ans. Even function, $f(x)=\frac{2}{\pi}-\frac{4}{\pi}\sum_{n=1}^{\infty}\frac{\cos 2nx}{(4n^2-1)}$.

9. $f(x)=\sin ax$ in $(-\pi,\pi)$, where a is not an integer.

Ans. Odd function, $f(x) = \frac{2\sin a\pi}{\pi}\sum_{n=1}^{\infty}\frac{(-1)^n n}{a^2-n^2}\sin nx$.

10. $f(x) = \cos ax$ in $(-\pi, \pi)$, where a is not an integer.

Ans. $f(x) = \frac{\sin a\pi}{\pi} + \frac{2a\sin a\pi}{\pi}\sum_{n=1}^{\infty}\frac{(-1)^n}{a^2-n^2}\cos nx$,

even function.

11. $f(x) = x\sin x$ in $(-\pi, \pi)$.
Deduce that $\frac{\pi-2}{4} = \frac{1}{1.3} - \frac{1}{3.5} + \frac{1}{5.7} - \frac{1}{7.9} + \cdots$

Ans. $f(x)=1-\frac{1}{2}\cos x+\sum_{n=2}^{\infty}\frac{2(-1)^{n+1}}{n^2-1}\cos nx$,

even function.

Put $x = \frac{\pi}{2}$,

$\frac{\pi}{2}\sin\frac{\pi}{2}=1-0-\frac{2}{1.3}(-1)+\frac{2}{2.4}\cdot(0)-\frac{2}{3.5}+\cdots$

12. $f(x) = x\cos x$ in $(-\pi, \pi)$.

Ans. $f(x) = -\frac{1}{2}\sin x + \sum_{n=2}^{\infty}\frac{(-1)^n}{(n^2-1)}\sin nx$,

odd function.

13. $f(x) = \sqrt{1-\cos x}$ in $(-\pi, \pi)$ or $-\sqrt{2}\sin\frac{x}{2}$ in $(-\pi, 0)$ and $\sqrt{2}\sin(x/2)$ in $(0, \pi)$.

Ans. Even, $f(x) = \frac{2\sqrt{2}}{\pi} - \frac{4\sqrt{2}}{\pi}\sum_{n=1}^{\infty}\cos nx/(4n^2-1)$

14. $f(x) = -x^2$ in $(-\pi, 0)$ and $= x^2$ in $(0, \pi)$.

Ans. $f(x) = \frac{2}{\pi}\sum_{n=1}^{\infty}\left\{\frac{\pi^2}{n}(-1)^{n+1} + \frac{2[1-(-1)^n]}{n^3}\right\}\sin nx$

odd function.

17.3 FOURIER SERIES FOR FUNCTIONS HAVING PERIOD 2L

Fourier series of periodic function with period 2π was considered so far. Now consider the Fourier series expansion of a function in an interval of length $2L$.

Let $f(x)$ be a periodic function with arbitrary period $2L$ defined in the interval $c < x < c + 2L$. Introduce a new variable z as

$$\frac{x}{2L} = \frac{z}{2\pi} \quad \text{or } x = \frac{Lz}{\pi} \quad \text{or } z = \frac{\pi x}{L} \tag{1}$$

At $x = c$, $z = \frac{\pi c}{L} = d$ say

At $x = c + 2L$, $z = \frac{\pi}{L}(c + 2L) = \frac{\pi c}{L} + 2\pi = d + 2\pi$

Thus as $c < x < c + 2L$, the new variable z lies in the interval
$d < z < d + 2\pi$

So z varies in the interval $(d, d + 2\pi)$ of length 2π. Substituting for x from (1)

$$f(x) = f\left(\frac{Lz}{\pi}\right) = F(z) \tag{2}$$

Let the Fourier series of $F(z)$ defined in the interval $(d, d + 2\pi)$ and with period 2π be

$$F(z) = \frac{a_0}{2} + \sum_{n=1}^{\infty}(a_n\cos nz + b_n\sin nz) \tag{3}$$

where $a_0 = \frac{1}{\pi}\int_d^{d+2\pi}F(z)\,dz$.
Changing the variable to x

$$a_0 = \frac{1}{\pi}\int_c^{c+2L} f(x)\cdot\left(\frac{\pi}{L}dx\right)$$

Since $$dz = \frac{\pi}{L}dx$$

Thus $$a_0 = \frac{1}{L}\int_c^{c+2L} f(x)\,dx \tag{4}$$

Similarly,

$$a_n = \frac{1}{\pi}\int_d^{d+2\pi} F(z)\cos(nz)\,dz$$
$$= \frac{1}{\pi}\int_c^{c+2L} f(x)\cdot\cos\left(\frac{n\pi x}{L}\right)\frac{\pi}{L}dx$$

So $$a_n = \frac{1}{L}\int_c^{c+2L} f(x)\cdot\cos\left(\frac{n\pi x}{L}\right)dx \tag{5}$$

In a similar way

$$b_n = \frac{1}{L}\int_c^{c+2L} f(x)\sin\left(\frac{n\pi x}{L}\right)dx \tag{6}$$

Hence the Fourier series expansion for a function $f(x)$ with period $2L$ is

$$f(x) = F(z)$$
$$= \frac{a_0}{2} + \sum_{n=1}^{\infty}\left[a_n\cdot\cos\left(\frac{n\pi x}{L}\right) + b_n\sin\left(\frac{n\pi x}{L}\right)\right]$$

with the Fourier coefficients a_0, a_n, b_n given by (4), (5), (6).

Result:

$$\int_{-L}^{L}\cos\left(\frac{m\pi x}{L}\right)\cdot\cos\left(\frac{n\pi x}{L}\right)dx=\begin{cases}2L & \text{if } m=n=0\\ L & \text{if } m=n\neq 0\\ 0 & \text{if } m\neq n\end{cases}$$

Result:

$$\int_{-L}^{L}\sin\left(\frac{m\pi x}{L}\right)\cdot\sin\left(\frac{n\pi x}{L}\right)dx=\begin{cases}0 & \text{if } m=n=0\\ L & \text{if } m=n\neq 0\\ 0 & \text{if } m\neq n\end{cases}$$

Result:

$$\int_{-L}^{L}\sin\left(\frac{m\pi x}{L}\right)\cdot\cos\left(\frac{n\pi x}{L}\right)dx=0 \quad \text{for all integers } m \text{ and } n.$$

Fourier Series for Even and Odd Functions

Defined in the interval $(-L, L)$ of length $2L$.

Case 1: $f(x)$ be even function in $(-L, L)$. Then the **Fourier cosine series** is

$$f(x)=\frac{a_0}{2}+\sum_{n=1}^{\infty}a_n\cos\frac{n\pi x}{L}$$

where

$$a_0=\frac{2}{L}\int_0^L f(x)dx$$

and

$$a_n=\frac{2}{L}\int_0^L f(x)\cos\frac{n\pi x}{L}dx.$$

Case 2: $f(x)$ be odd function in $(-L, L)$. Then the **Fourier sine series** is

$$f(x)=\sum_{n=1}^{\infty}b_n\sin\frac{n\pi x}{L}$$

where

$$b_n=\frac{2}{L}\int_0^L f(x)\sin\frac{n\pi x}{L}dx.$$

Worked Out Examples

Example 1: Find the Fourier series of $f(x)$ defined

$$f(x)=0 \quad \text{when } -c<x<0$$
$$=1 \quad \text{when } 0<x<c$$

Find the value of Fourier series at the point of discontinuity $x=0$.

Solution: The given interval is of length $2c$. The fourier series is

$$f(x)=\frac{a_0}{2}+\sum_{n=1}^{\infty}\left(a_n\cos\frac{n\pi x}{c}+b_n\sin\frac{n\pi x}{c}\right)$$

Here

$$a_0=\frac{1}{c}\int_{-c}^{c}f(x)\,dx=\frac{1}{c}\int_{-c}^{0}0\cdot dx+\frac{1}{c}\int_0^c 1\cdot dx$$

$$=\frac{1}{c}\cdot x\Big|_0^c=1$$

$$a_n=\frac{1}{c}\int_{-c}^{c}f(x)\cos\frac{n\pi x}{c}dx=\frac{1}{c}\int_0^c\cos\frac{n\pi x}{c}dx$$

$$a_n=\frac{1}{c}\cdot\left(\frac{c}{n\pi}\right)\sin\left(\frac{n\pi x}{c}\right)\Big|_0^c=\frac{1}{n\pi}[\sin n\pi-\sin 0]=0$$

$$b_n=\frac{1}{c}\int_{-c}^{c}f(x)\sin\frac{n\pi x}{c}dx=\frac{1}{c}\int_0^c\sin\frac{n\pi x}{c}dx$$

$$=\frac{1}{c}\left(\frac{c}{n\pi}\right)\left(\frac{-\cos n\pi x}{c}\right)\Big|_0^c=-\frac{1}{n\pi}[\cos n\pi-1]$$

$$b_n=\frac{1}{n\pi}[1-(-1)^n]$$

Thus the required Fourier series is

$$f(x)=\frac{1}{2}+\frac{1}{\pi}\sum_{n=1}^{\infty}\left[\frac{1-(-1)^n}{n}\right]\sin\left(\frac{n\pi x}{c}\right)$$

The sum of the series at $x_0=0$ is obtained by putting $x=0$ in the above series i.e., $f(0)=\frac{1}{2}+\frac{1}{\pi}\cdot 0=\frac{1}{2}$. At a point of discontinuity $x_0=0$, $f(0)=\frac{1}{2}[f(0+0)+f(0-0)]=\frac{1}{2}[1+0]=\frac{1}{2}$.

Example 2: Obtain the Fourier series expansion of $f(x)=(\pi-x)/2$ in $0<x<2$.

Solution: Here the length of interval is $2L=2$ (i.e., $L=1$)

$$a_0=\frac{1}{1}\int_0^2 f(x)\,dx=\int_0^2\frac{\pi-x}{2}dx$$

$$=\frac{1}{2}\left[\pi x-\frac{x^2}{2}\right]_0^2=(\pi-1)$$

$$a_n=\frac{1}{1}\int_0^2 f(x)\cos\frac{n\pi x}{1}dx$$

$$= \int_0^2 \left(\frac{\pi - x}{2}\right) \cos n\pi x\, dx$$

$$= \frac{1}{2}\left[(\pi - x)\left(\frac{1}{n\pi}\right) \sin n\pi x - (-1) \times \frac{1}{n\pi} \cdot (-1) \frac{\cos n\pi x}{n\pi}\right]_0^2$$

$$a_n = -\frac{1}{2n^2\pi^2}[\cos 2n\pi - \cos 0]$$

$$= -\frac{1}{2n^2\pi^2}[1 - 1] = 0$$

$$b_n = \frac{1}{1}\int_0^2 f(x) \sin \frac{n\pi x}{1} dx$$

$$= \int_0^2 \left(\frac{\pi - x}{2}\right) \sin n\pi x\, dx$$

$$= \frac{1}{2}\left[(\pi - x)\left(\frac{-\cos n\pi x}{n\pi}\right) - (-1) \cdot (-1) \frac{\sin n\pi x}{n^2\pi^2}\right]_0^2$$

$$b_n = \frac{-1}{2n\pi}[(\pi - 2)\cos 2n\pi - \pi \cdot \cos 0] = \frac{1}{n\pi}$$

$$f(x) = \frac{\pi - x}{2} = \frac{(\pi - 1)}{2} + 0 + \sum_{n=1}^{\infty} \frac{1}{n\pi} \cdot \sin n\pi x.$$

Example 3: Find the Fourier series of (Fig. 17.3)

$$f(t) = \begin{cases} 0 & \text{if } -2 \le t \le -1 \\ 1 + t & \text{if } -1 \le t \le 0 \\ 1 - t & \text{if } 0 \le t \le 1 \\ 0 & \text{if } 1 \le t \le 2. \end{cases}$$

Solution: $f(t)$ is defined in the interval $(-2, 2)$ of length $2L = 4$ (i.e., $L = 2$). Observe that $f(t)$ is an even function (symmetric about y-axis). Fourier series reduces to Fourier cosine series (with all $b_n = 0$).

$$f(t) = \frac{a_0}{2} + \sum_{n=1}^{\infty} a_n \cdot \cos \frac{n\pi t}{2}$$

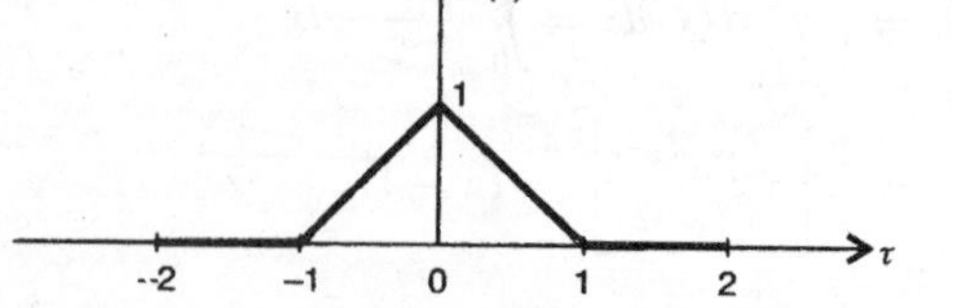

Fig. 17.3

Here

$$a_0 = \frac{2}{L}\int_0^L f(t)\,dt = \frac{2}{L}\int_0^2 f(t)\,dt$$

$$a_0 = \int_0^1 (1 - t)dt + \int_1^2 0 \cdot dt = \left(\frac{t - t^2}{2}\right)\Big|_0^1 = \frac{1}{2}$$

Now

$$a_n = \frac{2}{L}\int_0^L f(t) \cos\left(\frac{n\pi t}{2}\right) dt$$

$$= \frac{2}{2}\int_0^2 f(t) \cos\left(\frac{n\pi t}{2}\right) dt$$

$$= \int_0^1 (1 - t) \cos\left(\frac{n\pi t}{2}\right) dt + 0$$

$$= \left[(1 - t) \cdot \frac{2}{n\pi} \cdot \sin\left(\frac{n\pi t}{2}\right) - (-1)\frac{2}{n\pi} \cdot \frac{2}{n\pi} \cdot (-1) \cos\left(\frac{n\pi t}{2}\right)\right]\Big|_0^1$$

$$a_n = \frac{-4}{n^2\pi^2}\left[\cos\frac{n\pi}{2} - 1\right]$$

$$= 0 \qquad \text{when } n = 4, 8, 12, \ldots$$

$$= \frac{8}{n^2\pi^2} \qquad \text{when } n = 2, 6, 10, \ldots$$

$$= \frac{4}{n^2\pi^2} \qquad \text{when } n = 1, 3, 5, \ldots$$

Thus the Fourier series is

$$f(t) = \frac{1}{4} + \frac{4}{\pi^2}\sum_{n=1}^{\infty} \frac{1 - \cos(n\pi/2)}{n^2} \cdot \cos\left(\frac{n\pi t}{2}\right)$$

Example 4: Obtain Fourier series expansion of $f(x) = x \cdot \cos\left(\frac{\pi x}{L}\right)$ in the interval $-L \le x \le L$.

Solution: $f(x)$ is an odd function (since $0 \cdot e = 0$) in the interval $(-L, L)$ of length $2L$. So a_0 and a_n's are zero. The Fourier sine series is

$$f(x) = \sum_{n=1}^{\infty} b_n \cdot \sin\left(\frac{n\pi x}{L}\right)$$

For $n = 1$, $b_1 = \frac{2}{L}\int_0^L f(x) \sin\left(\frac{\pi x}{L}\right) dx$

$$b_1 = \frac{2}{L}\int_0^L x \cdot \cos\left(\frac{\pi x}{L}\right) \cdot \sin\left(\frac{\pi x}{L}\right) dx$$

$$= \frac{2}{L} \cdot \frac{L}{\pi} \cdot \int_0^L x \cdot \sin\left(\frac{\pi x}{L}\right) d\left(\frac{\sin \pi x}{L}\right)$$

$$= \frac{2}{\pi}\int_0^L \frac{x}{2}\cdot d\left(\frac{\sin^2 \pi x}{L}\right)$$

$$= \frac{1}{\pi}\left[x\cdot \sin^2\frac{\pi x}{2}\Big|_0^L - \int_0^L \sin^2\frac{\pi x}{L}dx\right]$$

$$= -\frac{1}{\pi}\int_0^L \sin^2\frac{\pi x}{L}dx$$

$$= -\frac{1}{\pi}\int_0^L \frac{1}{2}\left(1-\cos\frac{2\pi x}{L}\right)dx$$

$$= -\frac{1}{2\pi}\left[x - \frac{L}{2\pi}\cdot\sin\frac{2\pi x}{L}\right]_0^L$$

So $b_1 = -\frac{L}{2\pi}$

For $n \neq 1$

$$b_n = \frac{2}{L}\int_0^L \left(x\cdot\cos\frac{\pi x}{L}\right)\sin\left(\frac{n\pi x}{L}\right)dx$$

$$= \frac{2}{L}\int_0^L x\frac{1}{2}\left[\sin\left(\frac{(n-1)\pi x}{L}\right) + \sin\left(\frac{(n+1)\pi x}{L}\right)\right]dx$$

$$= I_1 + I_2$$

where

$$I_1 = \frac{1}{L}\int_0^L x\cdot\sin\left(\frac{(n-1)\pi x}{L}\right)dx$$

$$= \frac{1}{L}\left[x\cdot\frac{L}{(n-1)\pi}\left(-\cos\frac{(n-1)\pi x}{L}\right) - 1\cdot\frac{L^2}{(n-1)^2\pi^2}(-1)\cdot\sin\frac{(n-1)\pi x}{L}\right]_0^L$$

$$I_1 = \frac{L}{\pi(n-1)}(-1)(-1)^{n-1}$$

similarly,

$$I_2 = \frac{1}{L}\int_0^L x\cdot\sin\left(\frac{(n+1)}{L}\pi x\right)dx$$

$$I_2 = \frac{1}{L}\left[x\cdot\frac{L}{(n+1)\pi}\left(-\cos\frac{(n+1)\pi x}{L}\right) - \frac{L^2(-1)}{(n+1)^2\pi^2}\cdot\sin\left(\frac{(n+1)\pi x}{L}\right)\right]_0^L$$

$$I_2 = \frac{L}{\pi(n+1)}(-1)(-1)^{n+1}$$

Thus

$$b_n = I_1 + I_2 = \frac{L(-1)^n}{\pi}\left[\frac{1}{n-1} + \frac{1}{n+1}\right]$$

$$= \frac{2Ln}{\pi}\frac{(-1)^n}{(n^2-1)}$$

Hence

$$f(x) = -\frac{L}{2\pi}\sin\left(\frac{\pi x}{L}\right) + \frac{2L}{\pi}\sum_{n=2}^{\infty}\frac{(-1)^n n}{n^2-1}\sin\left(\frac{n\pi x}{L}\right).$$

EXERCISE

Find the Fourier series of $f(x)$ in the indicated interval:

1. $f(t) = \begin{cases} 0 & \text{if } -L < t < 0 \\ E\sin wt & \text{if } 0 < t < L \end{cases}$

Ans. $a_0 = \frac{2E}{\pi}$, $a_n = \frac{-2E}{(n^2-1)\pi}$ for $n = 2, 4, 6, \ldots$

$b_1 = E/2$, $b_n = 0$ for $n = 2, 3, 4, \ldots$

$$f(t) = \frac{E}{\pi} + \frac{E}{2}\sin wt - \frac{2E}{\pi}\left(\frac{1}{1.3}\cos 2wt + \frac{1}{3.5}\cos 4wt + \cdots\right)$$

2. $f(x) = e^x$ in $(-L, L)$

Ans. $f(x) = \sinh L\left[\frac{1}{L} + 2L\sum_{n=1}^{\infty}\frac{(-1)^n\cos(n\pi x/L)}{L^2+n^2\pi^2} + 2\pi\sum_{n=1}^{\infty}\frac{(-1)^{n-1}n\cdot\sin(n\pi x/L)}{L^2+n^2\pi^2}\right]$

3. $f(x) = x$ for $0 < x < 1$

$= 0$ for $1 < x < 2$

Ans. $f(x) = \frac{1}{4} + \frac{1}{\pi^2}\sum_{n=1}^{\infty}\left(\frac{\cos n\pi - 1}{n^2}\right)\cos n\pi x + \frac{1}{\pi}\sum_{n=1}^{\infty}\frac{(-1)^{n-1}}{n}\sin n\pi x$

4. $f(x) = x$ for $-1 < x \le 0$

$= x + 2$ for $0 < x \le 1$

Deduce that $\frac{\pi}{4} = 1 - \frac{1}{3} + \frac{1}{5} - \frac{1}{7} + \cdots$

Ans. $f(x) = 1 + \frac{2}{\pi}\sum_{n=1}^{\infty}\frac{1}{n}[1 - 2(-1)^n]\sin n\pi x$

Put $x = \frac{1}{2}$ on both sides to deduce the result.

5. $f(x) = 0$ when $-5 < x < 0$

$= 3$ when $0 < x < 5$

Ans. $f(x) = \frac{3}{2} + \sum_{n=1}^{\infty} \frac{3(1-\cos n\pi)}{n\pi} \sin\left(\frac{n\pi x}{5}\right)$

6. $f(x) = \pi x$ when $0 \le x \le 1$

$= \pi(2-x)$ when $1 \le x \le 2$

Ans. $f(x) = \frac{\pi}{2} + \frac{2}{\pi} \sum_{n=1}^{\infty} \frac{[(-1)^n - 1]}{n^2} \cos n\pi x$

7. $f(x) = \begin{cases} 1 & \text{if } 0 < x < 1 \\ 2 & \text{if } 1 < x < 2 \end{cases}$

Ans. $f(x) = 3 - \sum_{n=1}^{\infty} \frac{2}{(2n-1)\pi} \cdot \sin(2n-1)\pi x$

Even and odd functions

8. $f(x) = \begin{cases} 0 & \text{if } -2 < x < -1 \\ k & \text{if } -1 < x < 1 \\ 0 & \text{if } 1 < x < 2 \end{cases}$

Ans. even, $f(x) = \frac{k}{2} + \frac{2k}{\pi} \sum_{n=1}^{\infty} \frac{\sin(n\pi/2)}{n} \cdot \cos\left(\frac{n\pi x}{2}\right)$

9. $f(t) = 4 - t^2$ in $(-2, 2)$

Ans. even, $f(t) = \frac{8}{3} + \frac{16}{\pi^2} \sum \frac{(-1)^{n+1}}{n^2} \cos\frac{n\pi t}{2}$

10. $f(x) = \frac{1}{2} + x$ when $-1 \le x \le 0$

$= \frac{1}{2} - x$ when $0 \le x \le 1$

Ans. even, $f(x) = -\frac{4}{\pi^2} \sum_{n=1}^{\infty} \frac{\cos[(2n-1)\pi x]}{(2n-1)^2}$

11. $f(x) = |x|$ in $(-L, L)$

Ans. even, $f(x) = \frac{L}{2} - \frac{4L}{\pi^2} \sum_{n=\text{odd}} \frac{\cos(n\pi x/L)}{n^2}$

12. $f(x) = x^2$ in $(-L, L)$

Ans. even, $f(x) = \frac{L^3}{3} + \sum_{n=1}^{\infty} \frac{(-1)^n 4L^2}{n^2\pi^2} \cos\left(\frac{n\pi x}{L}\right)$

13. $f(x) = \pi x$ from $x = -c$ to $x = c$

Ans. $2c \sin\frac{\pi x}{c} - \frac{1}{2}\sin\frac{2\pi x}{c} + \frac{1}{3}\sin\frac{3\pi x}{c} - \frac{1}{4}\sin\left(\frac{4\pi x}{c}\right) + \cdots$

14. $f(x) = -a$ when $-c < x < 0$

$= a$ when $0 < x < c$

Ans. odd, $f(x) = \frac{2a}{\pi} \sum_{n=1}^{\infty} \left(\frac{1-\cos n\pi}{n}\right) \cdot \sin\left(\frac{n\pi x}{c}\right)$

15. $f(x) = \sin ax$ in $(-L, L)$

Ans. odd, $f(x) = \sin a\, L \sum_{n=1}^{\infty} \frac{(-1)^{n+1}}{n^2\pi^2 - a^2L^2} \sin\left(\frac{n\pi x}{L}\right)$

16. $f(x) = x + x^2$ in $(-1, 1)$

Hint: $f(x)$ may be treated as the sum of an odd function $g(x) = x$ and an even function $h(x) = x^2$. Fourier series of $f(x)$ is obtained by adding the Fourier series of $g(x)$ and $h(x)$.

Ans. $f(x) = \frac{1}{3} + \frac{4}{\pi^2} \sum_{n=1}^{\infty} \frac{(-1)^n}{n^2} \cos n\pi x + \frac{2}{\pi} \sum_{n=1}^{\infty} \frac{(-1)^{n+1}}{n} \sin n\pi x.$

17.4 HALF RANGE EXPANSIONS: FOURIER COSINE AND SINE SERIES

So far we have considered the Fourier series expansion of a function which is periodic, defined in an interval c to $c + 2L$ of length $2L$. Now we consider the procedure to expand a **non-periodic** function $f(x)$ defined in half of the above interval say $(0, L)$ of length L. Such expansions are known as **half range expansions or half range Fourier series.** In particular, a half range expansion containing only cosine terms is known as **half range Fourier cosine series** of $f(x)$ in the interval $(0, L)$. In a similar way **half range Fourier sine series** contains only sine terms.

Note that the given function $f(x)$ is neither periodic nor even nor odd. In order to obtain a Fourier cosine series for $f(x)$ in the interval $(0, L)$, we construct (define) a new function $g(x)$ such that

i. $g(x) \equiv f(x)$ in the interval $(0, L)$ (Fig. 17.4)

ii. $g(x)$ is even function in $(-L, L)$ and is periodic with period $2L$. (Fig. 17.5)

Such a function $g(x)$ is known as the "even periodic continuation (or extension) of $f(x)$". The Fourier cosine series for $g(x)$ valid in $(-L, L)$ (or in fact for all x) is readily obtained as

$$g(x) = \frac{a_0}{2} + \sum_{n=1}^{\infty} a_n \cos\left(\frac{n\pi x}{L}\right) \quad (1)$$

where

$$a_0 = \frac{2}{L}\int_0^L g(x)dx = \frac{2}{L}\int_0^L f(x)dx \quad (2)$$

$$a_n = \frac{2}{L}\int_0^L g(x)\cos\frac{n\pi x}{L}dx$$

or

$$a_n = \frac{2}{L}\int_0^L f(x)\cos\left(\frac{n\pi x}{L}\right)dx \qquad (3)$$

Since by construction $f(x)$ and $g(x)$ are equal in $(0, L)$, the required half-range Fourier cosine series (or half-range expansion in cosines) of $f(x)$ is given by (1) with a_0, a_n given by (2) and (3).

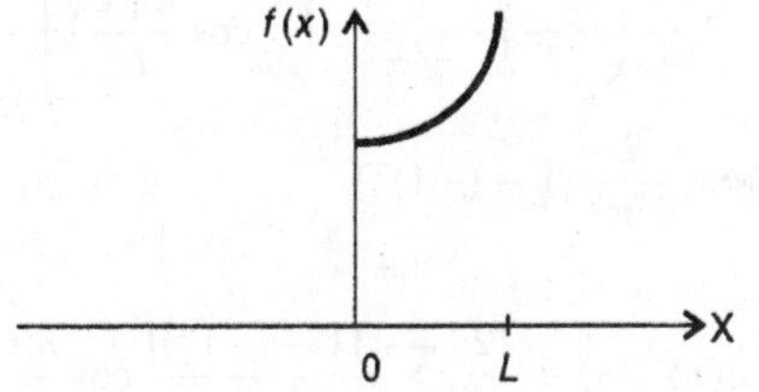

Fig. 17.4

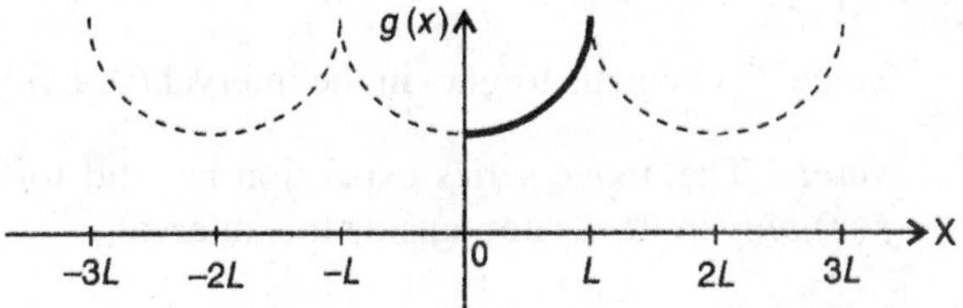

Even Periodic Continuation (Extension) of $f(x)$

Fig. 17.5

Here $\quad g(x) = f(x) \quad$ in $(0, L)$

$\qquad\qquad = f(-x) \quad$ in $(-L, 0)$

Important Note: The series expansion of $f(x)$ given by (1) is valid for $f(x)$ only in the interval $(0, L)$ but not outside this interval.

In a similar way, to obtain the half-range Fourier sine series (or half range expansion in sines) for $f(x)$ in $(0, L)$, define $g(x)$ such that

i. $g(x) = f(x)$ in $(0, L)$ (refer Fig. 17.6) and

ii. $g(x)$ is an odd function in $(-L, L)$, (refer Fig. 17.7) periodic with priod $2L$.

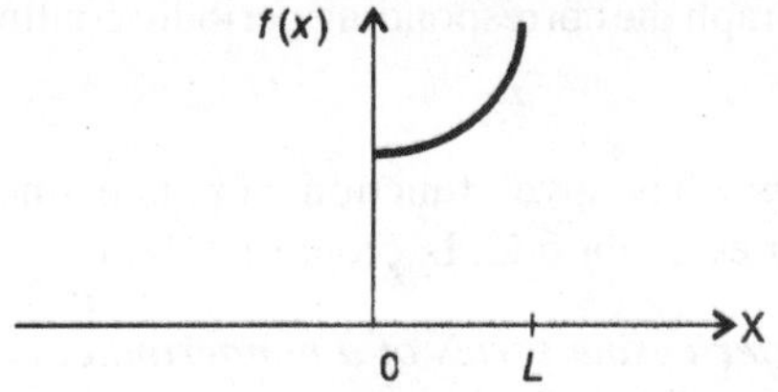

Fig. 17.6

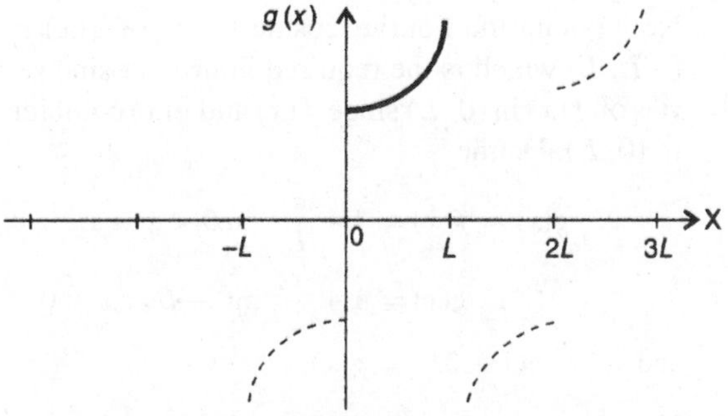

Odd Periodic Continuation (Extensions) of $f(x)$

Fig. 17.7

The new function $g(x)$ is called as on odd periodic continuation of $f(x)$. Now the Fourier sine series of $g(x)$ is given by

$$g(x) = \sum_{n=1}^{\infty} b_n \sin\frac{n\pi x}{L} \qquad (4)$$

with

$$b_n = \frac{2}{L}\int_0^L g(x)\sin\frac{n\pi x}{L}dx$$

or

$$b_n = \frac{2}{L}\int_0^L f(x)\sin\frac{n\pi x}{L}dx \qquad (5)$$

Since $f(x)$ and $g(x)$ are equal in $(0, L)$, the required half range Fourier sine series expansion of $f(x)$ in the interval $(0, L)$ is given by (4) with b_n's given by (5). This expansion (4) is valid for $f(x)$ only in the interval $(0, L)$. Thus a given non-periodic function $f(x)$ can be expanded in cosine series or sine series.

Worked Out Examples

Example 1: If $f(x) = 1 - \frac{x}{L}$ in $0 < x < L$ find (a) Fourier cosine series (b) Fourier sine series of

$f(x)$. Graph the corresponding periodic continuation of $f(x)$.

Solution: The given function $f(x)$ is neither periodic nor even nor odd. In order to obtain

a. *Fourier cosine series of a nonperiodic,* not even function $f(x)$. Define (or construct) a new function $g(x)$ such that (i) $g(x) = f(x)$ in $(0, L)$ (ii) $g(x)$ is even periodic function in $(-L, L)$. Now obtain the Fourier cosine series of $g(x)$ in $(-L, L)$ which is the required Fourier cosine series of $f(x)$ in $(0, L)$ since $f(x)$ and $g(x)$ coincide in $(0, L)$. Define

$$g(x) = f(x) = 1 - \frac{x}{L} \quad \text{in } 0 < x < L$$

$$g(x) = 1 + \frac{x}{L} \quad \text{in } -L < x < 0$$

and $\quad g(x + 2L) = g(x).$

Now $g(x)$ is even in $(-L, L)$ and is periodic with period $2L$. The graph in Fig. 17.9 is the even periodic continuation (or extension) of $f(x)$ shown in Fig. 17.8. The Fourier cosine series of $g(x)$ is

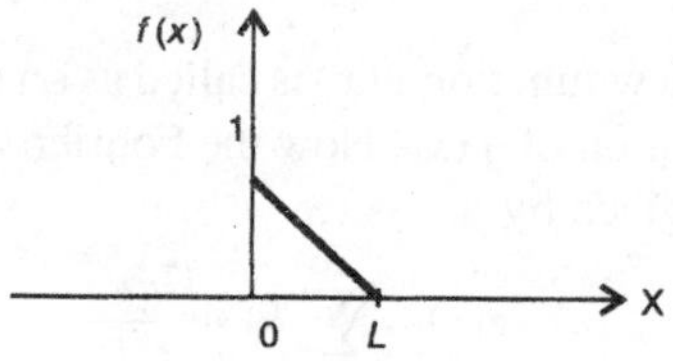

Fig. 17.8

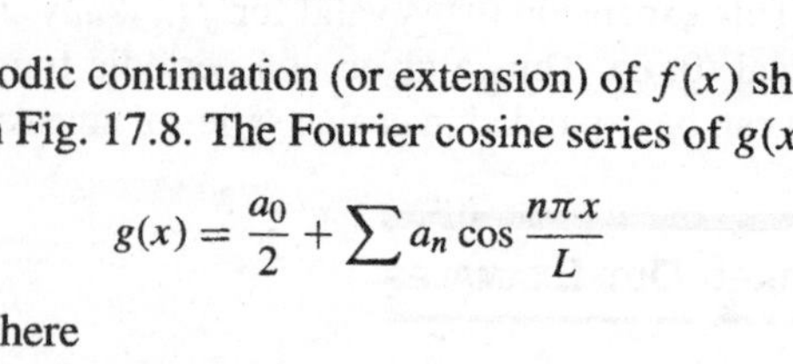

Fig. 17.9

$$g(x) = \frac{a_0}{2} + \sum a_n \cos\frac{n\pi x}{L}$$

where

$$a_0 = \frac{2}{L}\int_0^L f(x)dx = \frac{2}{L}\int_0^L \left(1 - \frac{x}{L}\right) dx$$

$$a_0 = \frac{2}{L}\left[\left(x - \frac{x^2}{2L}\right)\Big|_0^L\right] = \frac{2}{L}\left[L - \frac{L^2}{2L}\right] = 1$$

$$a_n = \frac{2}{L}\int_0^L f(x)\cos n\frac{\pi x}{L}dx$$

$$= \frac{2}{L}\int_0^L \left(1 - \frac{x}{L}\right)\cos n\frac{\pi x}{L}dx$$

$$= \frac{2}{L}\left[\left(1 - \frac{x}{L}\right)\cdot\frac{L}{n\pi}\sin\frac{n\pi x}{L} - \left(-\frac{1}{L}\right)\cdot\frac{L^2}{n^2\pi^2}\cdot\left(-\cos\frac{n\pi x}{L}\right)\right]_0^L$$

$$a_n = \frac{2}{n^2\pi^2}[1 - (-1)^n]$$

so

$$g(x) = \frac{1}{2} + \frac{2}{\pi^2}\sum \frac{[1 - (-1^n)]}{n^2}\cos\frac{n\pi x}{L}$$

Thus the required Fourier cosine series of $f(x)$ in $(0, L)$ is

$$f(x) = g(x) = \frac{1}{2} + \frac{2}{\pi^2}\sum_{n=1}^{\infty}\frac{[1 - (-1)^n]}{n^2}\cos\left(\frac{n\pi x}{L}\right)$$

(since $f(x)$ equals to $g(x)$ in the interval $(0, L)$).

Note: The above series expansion is valid for $f(x)$ *only* in $(0, L)$ not outside this interval.

b. *Fourier sine series of $f(x)$ in* $(0, L)$: On similar lines, define a new function $h(x)$ such that (i) $h(x) = f(x)$ in $(0, L)$ and (ii) $h(x)$ is odd periodic function. Define

$$h(x) = f(x) = 1 - \frac{x}{L} \quad \text{in } (0, L)$$

$$h(x) = -\left(1 + \frac{x}{L}\right) \quad \text{in } (-L, 0)$$

and $\quad h(x + 2L) = h(x) \qquad (6)$

Thus $h(x)$ is odd periodic function in $(-L, L)$ with period $2L$. The graph in Fig. 17.11 is the odd periodic continuation (or extension) of $f(x)$ shown in Fig. 17.10.

Now the Fourier sine series expansion of $g(x)$ in the interval $(-L, L)$ is

$$h(x) = \sum_{n=1}^{\infty} b_n \sin\frac{n\pi x}{L}$$

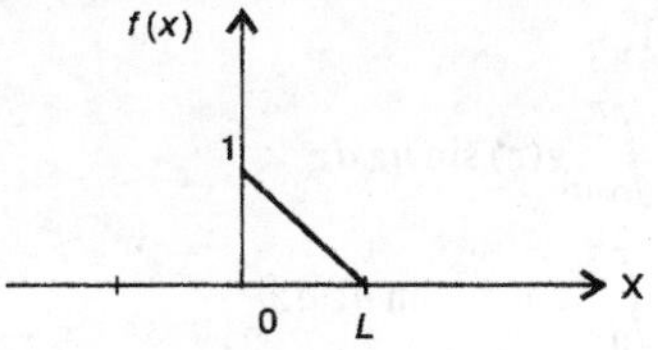

Fig. 17.10

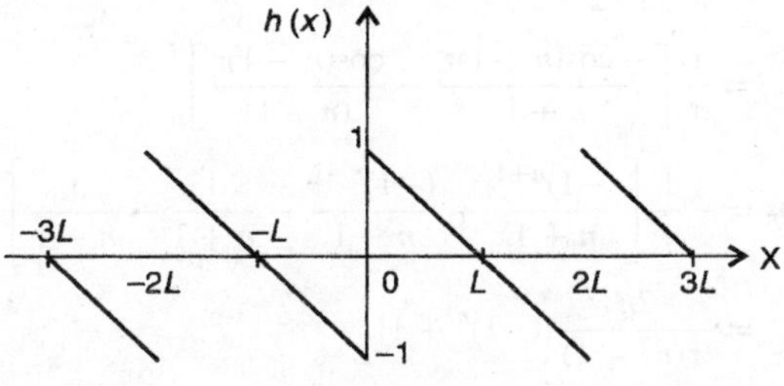

Fig. 17.11

where

$$b_n = \frac{2}{L}\int_0^L f(x)\sin\frac{n\pi x}{L}dx$$

$$b_n = \frac{2}{L}\int_0^L \left(1-\frac{x}{L}\right)\sin\frac{n\pi x}{L}dx$$

$$= \frac{2}{L}\left[\left(1-\frac{x}{L}\right)(-1)\left(\cos\frac{n\pi x}{L}\right)\cdot\frac{L}{n\pi}\right.$$

$$\left.-\left(\frac{-1}{L}\right)\frac{L^2}{n^2\pi^2}\sin\frac{n\pi x}{L}\right]_0^L$$

$$b_n = \frac{2}{n\pi}$$

so

$$h(x) = \frac{2}{\pi}\sum_{n=1}^{\infty}\frac{1}{n}\sin\left(\frac{n\pi x}{L}\right)$$

Thus the required Fourier sine series of $f(x)$ in the interval $(0, L)$ is

$$f(x) = h(x) = \frac{2}{\pi}\sum_{n=1}^{\infty}\frac{1}{n}\sin\left(\frac{n\pi x}{L}\right)$$

Note: Thus a non-periodic function $f(x)$ can be expanded in two different series in $(0, L)$.

Example 2: Represent $f(x) = \sin\frac{\pi x}{L}$ in $0 < x < L$ by a Fourier cosine series. Graph the corresponding periodic continuation of $f(x)$. (See Figs. 17.12 and 17.13)

Solution: In $0 < x < L$, $f(x)$ is neither periodic nor odd nor even. Construct

$$g(x) = f(x) = \sin\frac{\pi x}{L} \quad \text{in } 0 < x < L$$

$$g(x) = -\sin\frac{\pi x}{L} \quad \text{in } -L < x < 0$$

$g(x)$ is even, periodic with period $2L$.

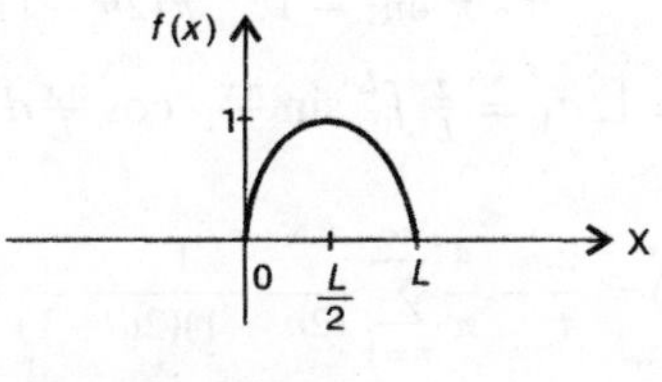

Fig. 17.12

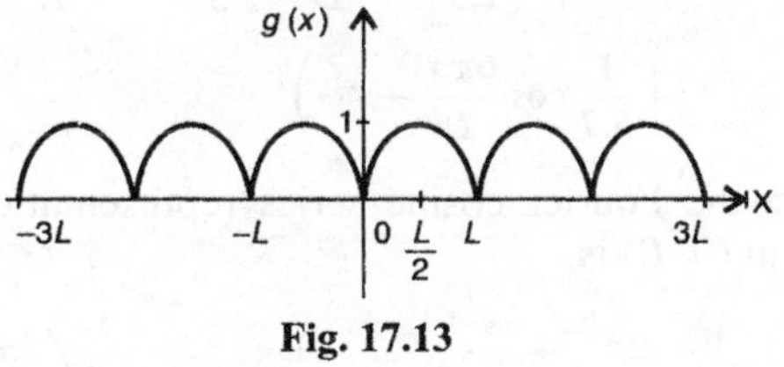

Fig. 17.13

Fourier cosine series of $g(x)$ is

$$g(x) = \frac{a_0}{2} + \sum_{n=1}^{\infty} a_n\cos\frac{n\pi x}{L} \quad \text{where}$$

$$a_0 = \frac{2}{L}\int_0^L f(x)dx = \frac{2}{L}\int_0^L \sin\left(\frac{\pi x}{L}\right)dx$$

$$= \frac{2}{L}\frac{L}{\pi}\left(-\cos\frac{\pi x}{L}\right)\Big|_0^L$$

$$a_0 = \frac{-2}{\pi}[-1-1] = \frac{4}{\pi}$$

For $n \neq 1$

$$a_n = \frac{2}{L}\int_0^L f(x)\cos\frac{n\pi x}{L}dx$$

$$= \frac{2}{L}\int_0^L \sin\left(\frac{\pi x}{L}\right)\cdot\cos\left(\frac{n\pi x}{L}\right)dx$$

$$= \frac{2}{L}\frac{1}{2}\int_0^L\left[\sin(1+n)\frac{\pi x}{L} + \sin(1-n)\frac{\pi x}{L}\right]dx$$

$$a_n = \frac{1}{L}\left[\frac{-L}{\pi(n+1)}\cdot\cos(1+n)\frac{\pi x}{L}\right.$$

$$\left.-\frac{L}{\pi(1-n)}\cos(1-n)\frac{\pi x}{L}\right]_0^L$$

$$= \frac{1}{\pi}\left[\frac{1}{n+1} - \frac{1}{n-1}\right]\{(-1)^n + 1\}$$

$$= -\frac{4}{\pi}\frac{1}{n^2-1} \quad \text{if } n \text{ is even}$$

$$= 0 \quad \text{if } n \text{ is odd}$$

i.e.,
$$a_{2n} = -\frac{4}{\pi}\frac{1}{4n^2-1} = \frac{-4}{\pi(2n-1)(2n+1)}$$

For $n = 1$, $a_1 = \frac{2}{L}\int_0^L \sin\frac{\pi x}{L}\cdot\cos\frac{\pi x}{L}dx = 0$.
Thus

$$g(x) = \frac{2}{\pi} - \frac{4}{\pi}\sum_{n=1}^{\infty}\frac{1}{(2n-1)(2n+1)}\cos 2n\frac{\pi x}{L}$$

$$= \frac{2}{\pi} - \frac{4}{\pi}\left(\frac{1}{1.3}\cos\frac{2\pi x}{L} + \frac{1}{3.5}\cos\frac{4\pi x}{L} + \frac{1}{5.7}\cos\frac{6\pi x}{L} + \cdots\right)$$

Hence the Fourier cosine series representation of $f(x)$ in $(0, L)$ is

$$f(x) = g(x) = \frac{2}{\pi} - \frac{4}{\pi}\sum_{n=1}^{\infty}\frac{1}{(2n-1)(2n+1)}\cos\left(\frac{2n\pi x}{L}\right)$$

Example 3: Show that in the interval (0, 1)

$$\cos\pi x = \frac{8}{\pi}\sum_{n=1}^{\infty}\frac{n}{4n^2-1}\sin 2n\pi x$$

(refer Figs. 17.14 and 17.15)

Solution: This is a Fourier sine series representation of $\cos\pi x$ in the interval $0 < x < 1$. Put $\pi x = z$, for $0 < x < 1$, then $0 < z < \pi$. Rewriting

$$\cos z = \frac{8}{\pi}\sum_{n=1}^{\infty}\frac{n}{4n^2-1}\sin 2nz$$

To expand $\cos z$ in Fourier sine series in $(0, \pi)$: Define

$$g(z) = f(z) = \cos z \quad \text{in } 0 < z < \pi$$

$$g(z) = -\cos z \quad \text{in } -\pi < z < 0$$

Now $g(z)$ is an odd function in $(-\pi, \pi)$ and is periodic of period 2π. Then

$$g(z) = \sum_{n=1}^{\infty} b_n \sin nz$$

For $n \neq 1$,

$$b_n = \frac{2}{\pi}\int_0^{\pi} g(z)\sin nz\,dz$$

$$= \frac{2}{\pi}\int_0^{\pi}\cos z\cdot\sin nz\,dz$$

$$= \frac{2}{\pi}\int_0^{\pi}\frac{1}{2}[\sin(n+1)z + \sin(n-1)z]dz$$

$$= \frac{1}{\pi}\left[\frac{-\cos(n+1)z}{n+1} - \frac{\cos(n-1)z}{(n-1)}\right]_0^{\pi}$$

$$b_n = -\frac{1}{\pi}\left[\frac{(-1)^{n+1}}{n+1} + \frac{(-1)^{n-1}}{n-1} - \frac{1}{n+1} - \frac{1}{n-1}\right]$$

$$= \frac{2n}{\pi(n^2-1)}\{(-1)^n + 1\}$$

$$b_n = 0 \quad \text{for } n \text{ odd}$$

$$b_1 = \frac{2}{\pi}\int_0^{\pi}\cos z\,\sin z\,dz = 0.$$

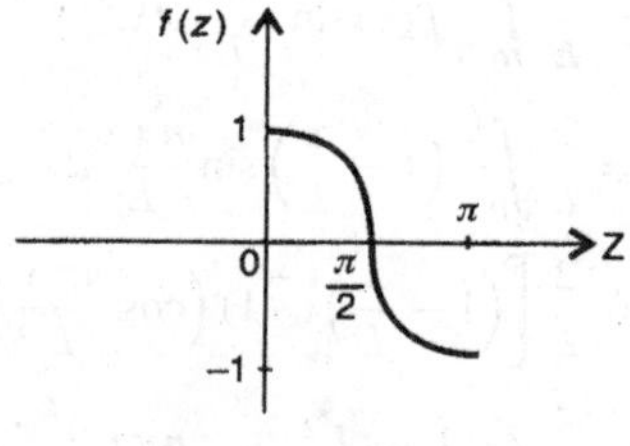

Fig. 17.14

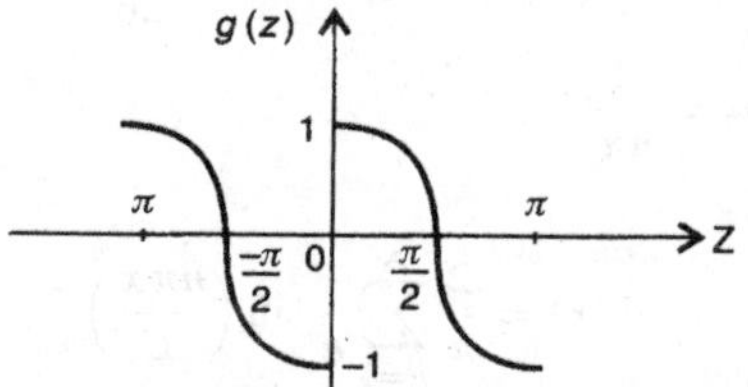

Fig. 17.15

The required Fourier sine series in $(0, \pi)$ is

$$f(z) = g(z) = \sum_{n=1}^{\infty}\frac{2n\{(-1)^n+1\}}{\pi(n^2-1)}\sin nz$$

$$= \sum_{n=1}^{\infty}\frac{2(2n)\cdot 2}{\pi[(2n)^2-1]}\sin 2nz$$

$$= \sum_{n=1}^{\infty}\frac{8n}{\pi(4n^2-1)}\sin 2nz$$

Replacing z by πx

$$\cos z = \cos \pi x = \sum_{n=1}^{\infty} \frac{8n}{\pi(4n^2-1)} \sin 2n\pi x \quad \text{in } 0 < x < 1.$$

Exercise

1. Expand $f(x) = x$ in $(0, \pi)$ by (a) Fourier sine series (Fig. 17.16) (b) Fourier cosine series (Fig. 17.17).

Ans. **a.** $g(x) = x$ in $(0, \pi)$
$= +x$ in $(-\pi, 0)$

$$f(x) = x = 2\left[\frac{\sin x}{1} - \frac{\sin 2x}{3} + \frac{\sin 3x}{3} \cdots\right]$$

$$= \sum_{n=1}^{\infty} \frac{2}{n}(-1)^{n+1} \sin nx$$

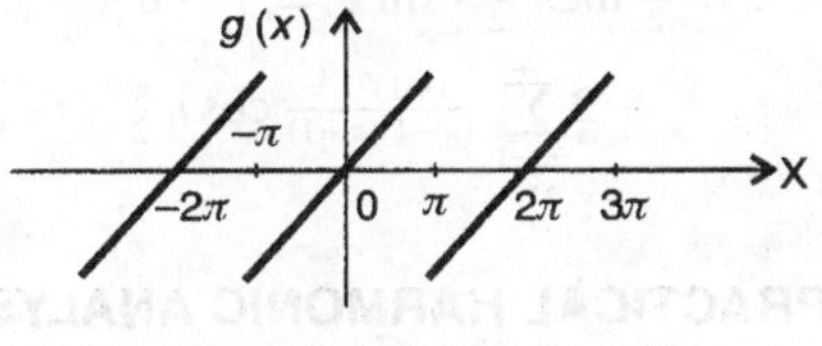

Fig. 17.16

b. $g(x) = |x|$ in $(-\pi, \pi)$

$$f(x) = x = \frac{\pi}{2} - \frac{4}{\pi}\left[\frac{\cos x}{1} + \frac{\cos 3x}{3^2} + \frac{\cos 5x}{5^2} + \cdots\right]$$

$$= \frac{\pi}{2} - \frac{4}{\pi}\sum_{n=1}^{\infty} \frac{\cos(2n-1)x}{(2n-1)^2}$$

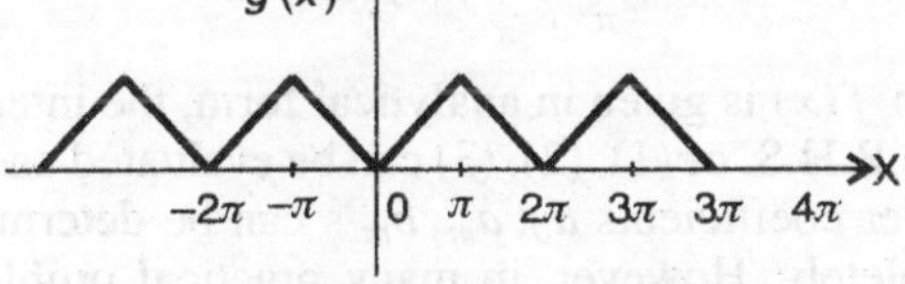

Fig. 17.17

2. Find the two half-range expansions of

$$fx = \begin{cases} 2kx/L & \text{if } 0 < x < (L/2) \\ 2k(L-x)/L & \text{if } \frac{L}{2} < x < L. \end{cases}$$

Ans. **a.** Even periodic extension: (Fourier cosine series) (refer Fig. 17.18).

$$f(x) = \frac{k}{2} + \frac{4k}{\pi^2}\sum_{n=1}^{\infty}\left(\frac{2\cos\frac{n\pi}{2} - \cos n\pi - 1}{n^2}\right) \times \cos\frac{n\pi x}{L}$$

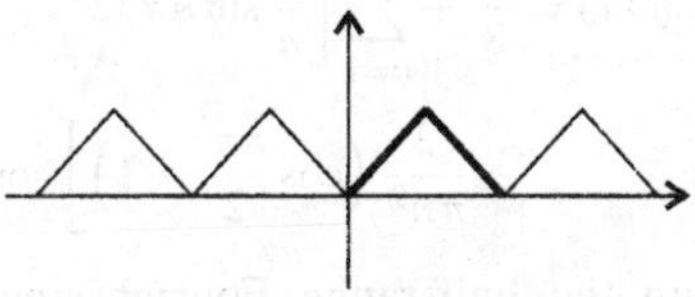

Fig. 17.18

b. Odd periodic extension: (Fourier sine series) (refer Fig. 17.19)

$$f(x) = \frac{8k}{\pi^2}\sum_{n=1}^{\infty}\frac{1}{n^2}\sin\frac{n\pi}{2}\cdot\sin\frac{n\pi x}{L}$$

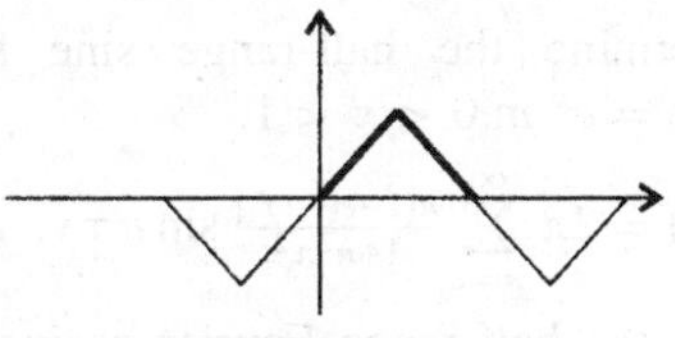

Fig. 17.19

3. Obtain the Fourier cosine series of $f(x) = \sin x$ in the interval $0 < x < \pi$.

Ans. $f(x) = \frac{2}{\pi} - \frac{4}{\pi}\sum_{n=1}^{\infty}\frac{\cos 2nx}{4n^2-1}$

4. Represent $f(x) = \begin{cases} 1 & \text{if } 0 < x < \frac{1}{2} \\ 0 & \text{if } \frac{1}{2} < x < 1 \end{cases}$

in (a) a Fourier sine series (b) Fourier cosine series (c) a Fourier series (with period 1).

Ans. **a.** $f(x) = \frac{2}{\pi}\sum_{n=1}^{\infty}(1 - \cos\frac{n\pi}{2})\frac{1}{n}\cdot\sin n\pi x$

b. $f(x) = \frac{1}{2} + \frac{2}{\pi}\sum_{n=1}^{\infty}\frac{1}{n}\cdot\sin\frac{n\pi}{2}\cdot\cos n\pi x$

c. $f(x) = \frac{1}{2} + \frac{1}{\pi}\sum[1-(-1)^n]\frac{1}{n}\cdot\sin n\pi x$

5. Find the Fourier sine and cosine series of

$$f(x) = \begin{cases} x & \text{when } 0 < x < \frac{\pi}{2} \\ 0 & \text{when } \frac{\pi}{2} < x < \pi. \end{cases}$$

Ans. Fourier sine series

$$f(x) = \frac{2}{\pi}\sum_{n=1}^{\infty}\left[\frac{1}{n^2}\sin\frac{n\pi}{2} - \frac{\pi}{2n}\cdot\cos\frac{n\pi}{2}\right]\sin nx$$

Fourier cosine series:

$$f(x)=\frac{\pi}{8}+\sum_{n=1}^{\infty}\left[\frac{1}{n}\sin n\pi/2\right.$$

$$\left.+\frac{2}{\pi n^2}\left(\cos\frac{n\pi}{2}-1\right)\right]\cos nx.$$

6. Obtain the half-range Fourier sine and cosine series for $f(x)=x$ in $0<x<\frac{\pi}{2}$ and $f(x)=\pi-x$ in $\frac{\pi}{2}<x<\pi$.

Ans. Sine series: $f(x)=\sum_{n=1}^{\infty}\frac{4}{\pi n^2}\sin\frac{n\pi}{2}\cdot\sin nx$

Cosine series:

$\frac{\pi}{4}+\frac{2}{\pi}\sum_{n=1}^{\infty}\frac{1}{n^2}\left[2\cos\frac{n\pi}{2}-1-(-1)^n\right]\cos nx$

7. Determine the half-range sine series for $f(x)=e^x$ in $0<x<1$.

Ans. $f(x)=2\pi\sum_{n=1}^{\infty}\frac{n[1-e(-1)^n]}{1+n^2\pi^2}\sin n\pi x.$

8. Find the half-range Fourier cosine series of $f(x)=x^3$ in $(0, L)$.

Ans. $f(x)=\frac{L^3}{4}+\frac{6L^3}{\pi^2}\sum_{n=1}^{\infty}\frac{1}{n^2}\left[(-1)^n+\frac{2}{n^2\pi^2}\{1-(-1)^n\}\right]\frac{\cos n\pi x}{L}.$

9. Represent $f(x)=x^2$ in $0<x<L$ by Fourier sine series.

Ans. $\frac{2L^2}{\pi}\sum_{n=1}^{\infty}\left[\frac{2}{n^3\pi^2}\{(-1)^n-1\}-\frac{(-1)^n}{n}\right]\sin\frac{n\pi x}{L}.$

10. Find the two half-range Fourier series of $f(x)=1$ in $0<x<L$.

Ans. Sine series:

$f(x)=1=\frac{2}{\pi}\sum_{n=1}^{\infty}\frac{1}{n}[1-(-1)^n]\sin\frac{n\pi x}{L}.$

cosine series $f(x)=1=\frac{2}{2}=1$

$(a_0=2, a_n=0 \quad \text{for } n\geq 1)$

11. Obtain two half-range Fourier series of $f(x)=x$ in $0<x<2$.

Ans. Sine series: $f(x)=\frac{4}{\pi}\sum_{n=1}^{\infty}\frac{(-1)^{n+1}}{n}\sin\frac{n\pi x}{2}$

$f(x)=1+\frac{4}{\pi^2}\sum_{n=1}^{\infty}\frac{1}{n^2}[(-1)^n-1]\cos\frac{n\pi x}{2}$ is cosine series.

12. Expand $f(x)=\frac{1}{4}-x$ if $0<x<0.5$

$=x-\frac{3}{4}$ if $0.5<x<1$

as Fourier series of sine terms

Ans. $\sum_{n=1}^{\infty}\left\{\frac{1}{2n\pi}[1-(-1)^n]-\frac{4\sin(n\pi/2)}{n^2\pi^2}\right\}\sin n\pi x$

13. Find the half-range Fourier sine series of $f(x)=\sin x$ when $0\leq x\leq\frac{\pi}{4}$ and $f(x)=\cos x$ in $\left(\frac{\pi}{4},\frac{\pi}{2}\right)$.

Ans. $f(x)=\frac{8}{\pi}\cos\frac{\pi}{4}\left[\frac{\sin 2x}{1.3}-\frac{\sin 6x}{5.7}+\frac{\sin 10x}{9.11}\cdots\right]$

14. Expand $x\sin x$ in $(0,\pi)$ as Fourier cosine series. Deduce that $1+\frac{2}{1\cdot 3}-\frac{2}{3\cdot 5}+\frac{2}{5\cdot 7}\cdots=\frac{\pi}{2}$

Ans. $f(x)=1-\frac{1}{2}\cos x+2\sum_{n=1}^{\infty}\frac{(-1)^{n-1}}{(n-1)(n+1)}\cos nx$

put $x=\frac{\pi}{2}$ then $\frac{\pi}{2}\cdot\sin\frac{\pi}{2}=1-0+$

$+2\sum_{n=1}^{\infty}\frac{(-1)^{n-1}}{(n-1)(n+1)}\cos n\frac{\pi}{2}.$

17.5 PRACTICAL HARMONIC ANALYSIS

Harmonic analysis is the theory of expanding a given function in Fourier series. We know that the Euler-Fourier coefficients of a function $f(x)$ with period 2π in the interval $(-\pi,\pi)$ are given by

$$a_0=\frac{1}{\pi}\int_{-\pi}^{\pi}f(x)dx \tag{1}$$

$$a_m=\frac{1}{\pi}\int_{-\pi}^{\pi}f(x)\cos mx dx \tag{2}$$

$$b_m=\frac{1}{\pi}\int_{-\pi}^{\pi}f(x)\sin mx dx. \tag{3}$$

When $f(x)$ is given in analytical form, the integrals in the R.H.S. of (1), (2), (3) can be evaluated and the Fourier coefficients a_o, a_m, b_m's can be determined completely. However, in many practical problems, the function $f(x)$ is in tabulated form. In such cases, practical harmonic analysis deals with the determination of the approximate values of the Fourier coefficients a_o, a_m, b_m. Divide the interval $[-\pi,\pi]$ into n equal parts with $(n+1)$ points

$$-\pi=x_0, x_1, x_2, \ldots, x_n=\pi$$

and subinterval size $h=\frac{\pi-(-\pi)}{n}=\frac{2\pi}{n}$. Let $y_i=f(x_i)$, for $i=0,1,2,\ldots n$.

Now the integrals on the R.H.S. of (1), (2), (3) are approximately evaluated using, say, rectangular formula (area = sum of n rectangles = $\sum_{i=1}^{n}$ width $h\times$ ordinate y_i). Then the Fourier coefficients (1), (2), (3) are determined approximately by the following:

$$a_0=\frac{1}{\pi}\int_{-\pi}^{\pi} f(x)dx=\frac{1}{\pi}\left[\sum_{i=1}^{n} h\cdot y_i\right]=\frac{1}{\pi}\cdot\frac{2\pi}{n}\sum_{i=1}^{n} y_i$$

or

$$a_0=\frac{2}{n}\sum_{i=1}^{n} y_i \quad (4)$$

Also

$$a_m=\frac{1}{\pi}\int_{-\pi}^{\pi} f(x)dx=\frac{1}{\pi}\left[\sum_{i=1}^{n} h\cdot y_i\cos mx_i\right]$$

$$=\frac{1}{\pi}\cdot\frac{2\pi}{n}\sum_{i=1}^{n} y_i\cos mx_i. \quad \text{Thus}$$

$$a_m=\frac{2}{n}\sum_{i=1}^{n} y_i\cos mx_i \quad (5)$$

Similarly,

$$b_m=\frac{2}{n}\sum_{i=1}^{n} y_i\sin mx_i \quad (6)$$

When $f(x)$ is given in the form of a graph, Fourier analyzer instruments determine the approximate values of the Fourier coefficients a_0, a_m's, b_m's.

Note: Choose n as a number divisibly by 4 since the values of sine and cosine are repeated in four quadrants. Usual choice $n=6, 12, 24$ (in which case (4),(5),(6) get simplified).

WORKED OUT EXAMPLES

Example 1: Compute approximately the Fourier coefficients a_0, a_1, a_2, a_3 and b_1, b_2, b_3 in the Fourier series expansion of function tabulated as follows. Find the amplitude of the first harmonic. Calculate $y(3)$

x:	0	1	2	3	4	5
y:	9	18	24	28	26	20

Solution: Here n = number of sub-intervals = 6. The interval $(0, 2\pi)$ is divided into 6 sub-intervals of size $\frac{2\pi}{6}=60°$.

Table 17.1

x	θ	$\cos\theta$	$\cos 2\theta$	$\cos 3\theta$	y	$y\cos\theta$	$y\cos 2\theta$	$y\cos 3\theta$
0	0°	1	1	1	9	9	9	9
1	60°	$+\frac{1}{2}$	$-\frac{1}{2}$	−1	18	−9	−9	−18
2	120°	$-\frac{1}{2}$	$-\frac{1}{2}$	1	24	−12	−12	24
3	180°	−1	+1	−1	28	−28	28	−28
4	240°	$-\frac{1}{2}$	$-\frac{1}{2}$	1	26	−13	−13	26
5	300°	$+\frac{1}{2}$	$-\frac{1}{2}$	−1	20	10	−10	−20
				Totals $\sum$	125	−25	−7	−7

Now

$$a_0=\frac{2}{n}\sum_{i=1}^{n} y_i=\frac{2}{6}(215)=41.666$$

$$a_1=\frac{2}{n}\sum_{i=1}^{n} y_i\cos x_i=\frac{2}{6}(-25)=-8.333$$

$$a_2=\frac{2}{n}\sum_{i=1}^{n} y_i\cos 2x_i=\frac{2}{6}(-7)=-2.333$$

$$a_3=\frac{2}{6}\sum_{i=1}^{n} y_i\cos 3x_i=\frac{2}{6}(-7)=-2.337$$

Similarly

Table 17.2

x	θ	$\sin\theta$	$\sin 2\theta$	$\sin 3\theta$	y	$y\sin\theta$	$y\sin 2\theta$	$y\sin 3\theta$
0	0	0	0	0	9	0	0	0
1	$\frac{\pi}{3}=60$	$\frac{\sqrt{3}}{2}$	$\frac{\sqrt{3}}{2}$	0	18	$9\sqrt{3}$	$9\sqrt{3}$	0
2	$\frac{2\pi}{3}=120$	$\frac{\sqrt{3}}{2}$	$-\frac{\sqrt{3}}{2}$	0	24	$12\sqrt{3}$	$-12\sqrt{3}$	0
3	$\pi=0$	0	0	0	28	0	0	0
4	$\frac{4\pi}{3}=240$	$-\frac{\sqrt{3}}{2}$	$\frac{\sqrt{3}}{2}$	0	26	$-13\sqrt{3}$	$13\sqrt{3}$	0
5	$\frac{5\pi}{3}=300$	$-\frac{\sqrt{3}}{2}$	$-\frac{\sqrt{3}}{2}$	0	20	$-10\sqrt{3}$	$-10\sqrt{3}$	0
				Totals	125	$-2\sqrt{3}$	0	0

$$b_1=\frac{2}{n}\sum_{i=1}^{n} y_i\sin x_i=\frac{2}{6}(-2\sqrt{3})=-1.1547$$

$$b_2=\frac{2}{n}\sum_{i=1}^{n} y_i\sin 2x_i=\frac{2}{6}(0)=0,\ b_3=0$$

Amplitude of the first harmonic $=\sqrt{a_1^2+b_1^2}=\sqrt{(-8.333)^2+(-1.1547)^2}=8.4126.$

The Fourier series of $y(x)$ containing the first 4 cos terms and 3 sine terms is

$$y=\frac{41.666}{2}+(-8.333)\cos x-2.333\cos 2x-$$
$$-2.337\cos 3x+(-1.1547)\sin x+0+0$$

At $x=3,\theta=\pi,\ y(3)=y(\theta=\pi)=\frac{41.666}{2}+(-8.33)(-1)-2.333(1)-(2.337)(-1)=29.166$ (Exact value: $y(3)=28$).

Exercise

1. Find the direct current part and amplitude of the first harmonic from the following table consisting of the variations periodic current

t sec :	0	$\frac{T}{6}$	$\frac{T}{3}$	$\frac{T}{2}$	$\frac{2T}{3}$	$\frac{5T}{6}$	T
A amp :	1.98	1.30	1.05	1.30	−.88	−.25	1.98

Ans. Direct current part $=\frac{a_0}{2}=0.75$, amplitude of the first harmonic $=\sqrt{a_1^2+b_1^2}=\sqrt{(0.373)^2+(1.005)^2}=1.072$

2. Compute $T(\theta=75^\circ)$ using the first four terms of the Fourier sine series representation of turning moment T given for a series of values of the crank angle $\theta^\circ=75^\circ$.

θ° :	0	30	60	90	120	150	180
T :	0	5224	8097	7850	5499	2626	0

Hint: $T(\theta)=785\sin\theta+150\sin 2\theta,\ b_3=0,\ b_4=0$

$$b_1=\frac{2}{6}\left[(5224+2626)\left(\frac{1}{2}\right)+(8097+5499)(.866)+7850\right]=785$$

$$b_2=\frac{2}{6}\left[(5224+8097)(.866)+(5499+2626)(-.866)\right]=150$$

Ans. 8332

3. Determine the first three coefficients of cosine and two coefficients of sine terms in the Fourier series expansion of the following tabulated function.

x :	0	$\frac{\pi}{6}$	$\frac{\pi}{3}$	$\frac{\pi}{2}$	$\frac{2\pi}{3}$	$\frac{5\pi}{6}$
y :	0	9.2	14.4	17.8	17.3	11.7

Hint: $y=11.733-7.733\cos 2x-2.833\cos 4x-1.566\sin 2x+0.116\sin 4x$, Length of interval is π; $L=\frac{\pi}{2}$.

Ans. $a_0 = 23.46, a_1 = -7.73, a_2 = -2.83, b_1 = -1.566, b_2 = +.116$

4. Find the harmonics $a_0, a_1, a_2, a_3, b_1, b_2, b_3$ of the Fourier series of the following data

x :	0	$\frac{\pi}{3}$	$\frac{2\pi}{3}$	π	$\frac{4\pi}{3}$	$\frac{5\pi}{3}$	2π
y :	1.0	1.4	1.9	1.7	1.5	1.2	1.0

Ans. $a_0 = 2.9, a_1 = -.37, a_2 = -0.1, a_3 = 0.03, b_1 = 0.17, b_2 = -0.06, b_3 = 0$

5. Compute a_0, a_1, a_2, a_3 in the Fourier cosine series for y which is tabulated below

x :	0	1	2	3	4	5
y :	4	8	15	7	6	2

Ans. $a_0 = 14, a_1 = -2.8, a_2 = -1.5, a_3 = 2.7$

6. Compute $a_0, a_1, a_2, a_3, b_1, b_2, b_3$ in the Fourier series expansion $f(\theta)$ tabulated below

θ :	0	30	60	90	120	150	180	210	240	270	300	330
$f(\theta)$:	298	356	373	337	254	155	80	51	60	93	147	221

Ans. $f(\theta) = 202 + 107\cos\theta - 13\cos 2\theta + 2\cos 3\theta + 121\sin\theta + 9\sin 2\theta - \sin 3\theta$.

Chapter 18

Partial Differential Equations

INTRODUCTION

Real world problems in general involve functions of several (independent) variables giving rise to partial differential equations more often than ordinary differential equations. Thus most problems in engineering and science abound with first and second order linear non homogeneous partial differential equations. In this chapter, we consider methods of obtaining solutions by Lagrange's and Charpits method for first order. The general solution of non homogeneous second order linear P.D.E. with constant coefficients is obtained as the sum of complementary function and particular integral. Monge's method is also considered for solving nonlinear second order P.D.E.

18.1 PARTIAL DIFFERENTIAL EQUATIONS

A partial differential equation is an equation involving two (or more) independent variables x, y and a dependent variable z and its partial derivatives such as $\frac{\partial z}{\partial x}, \frac{\partial z}{\partial y}, \frac{\partial^2 z}{\partial x^2}, \frac{\partial^2 z}{\partial x \partial y}$, etc.,

$$\text{i.e., } F\left(x, y, z, \frac{\partial z}{\partial x}, \frac{\partial z}{\partial y}, \frac{\partial^2 z}{\partial x^2}, \ldots\right) = 0$$

Standard notation

$$p = \frac{\partial z}{\partial x} = z_x, \quad q = \frac{\partial z}{\partial y} = z_y, \quad r = \frac{\partial^2 z}{\partial x^2} = z_{xx},$$

$$s = \frac{\partial^2 z}{\partial x \partial y} = z_{xy}, \quad t = \frac{\partial^2 z}{\partial y^2} = z_{yy}$$

Order of a partial differential equation (P.D.E.) is the order of the highest ordered derivative appearing in the P.D.E.

Formation of Partial Differential Equation

By elimination of arbitrary constants

Let

$$f(x, y, z, a, b) = 0 \tag{1}$$

be an equation involving two arbitrary constants a and b. Differentiating this equation partially with respect to x and y, we get

$$\frac{\partial f}{\partial x} + \frac{\partial f}{\partial z}\frac{\partial z}{\partial x} = 0 \tag{2}$$

$$\frac{\partial f}{\partial y} + \frac{\partial f}{\partial z}\frac{\partial z}{\partial y} = 0 \tag{3}$$

By eliminating a, b from (1), (2), (3), we get an equation of the form

$$F(x, y, z, p, q) = 0 \tag{4}$$

which is a partial differential equation of first order.

Note 1: If the number of arbitrary constants equals to the number of independent variables in (1), then the P.D.E. obtained by elimination is of first order.

Note 2: If the number of arbitrary constants is more than the number of independent variables then the P.D.E. obtained is of 2nd or higher orders.

By elimination of arbitrary functions of specific functions

a. *One arbitrary function* (resulting in first order P.D.E.):

Consider

$$z = f(u) \tag{5}$$

where $f(u)$ is an arbitrary function of u and u is a given (known) function of x, y, z i.e., $u = u(x, y, z)$.

Differentiating (5) partially w.r.t. x and y by chain rule

$$\frac{\partial z}{\partial x} = \frac{\partial f}{\partial u}\frac{\partial u}{\partial x} + \frac{\partial f}{\partial u}\frac{\partial u}{\partial z}\frac{\partial z}{\partial x} \tag{6}$$

$$\frac{\partial z}{\partial y} = \frac{\partial f}{\partial u}\frac{\partial u}{\partial y} + \frac{\partial f}{\partial u}\frac{\partial u}{\partial z}\frac{\partial z}{\partial y} \tag{7}$$

By eliminating the arbitrary function f from (5), (6), (7) we get a P.D.E. of first order.

b. Two arbitrary functions:

Differentiating twice or more, the elimination process results in a P.D.E. of 2nd or higher order.

Note: When n is the number of arbitrary functions, one may get several P.D. equations. But generally the one with the least order is chosen.

Example: For $z = x\,f(y) + y\,g(x)$ involving two arbitrary functions f and g, $\frac{\partial^4 z}{\partial x^2 \partial y^2} = 0$ is also a P.D.E. obtained by elimination. The other P.D.E. $xys = xp + yq - z$ of second order obtained by elimination may be chosen.

Elimination of Arbitrary Function *F*

from the equation

$$F(u, v) = 0 \tag{8}$$

where $u = u(x, y, z)$ and $v = v(x, y, z)$ are given functions of x, y, z.

Differentiating the Equation (8) partially w.r.t. x by chain rule, we get

$$\frac{\partial F}{\partial u}\left(\frac{\partial u}{\partial x} + \frac{\partial u}{\partial z}\frac{\partial z}{\partial x}\right) + \frac{\partial F}{\partial v}\left(\frac{\partial v}{\partial x} + \frac{\partial v}{\partial z}\frac{\partial z}{\partial x}\right) = 0 \tag{9}$$

Similarly, differentiating Equation (8) partially w.r.t. y, we get

$$\frac{\partial F}{\partial u}\left(\frac{\partial u}{\partial y} + \frac{\partial u}{\partial z}\frac{\partial z}{\partial y}\right) + \frac{\partial F}{\partial v}\left(\frac{\partial v}{\partial y} + \frac{\partial v}{\partial z}\frac{\partial z}{\partial y}\right) = 0 \tag{10}$$

Eliminating $\frac{\partial F}{\partial u}$ and $\frac{\partial F}{\partial v}$ from (9) and (10), we have

$$\begin{vmatrix} \frac{\partial u}{\partial x} + \frac{\partial u}{\partial z}\frac{\partial z}{\partial x} & \frac{\partial v}{\partial x} + \frac{\partial v}{\partial z}\frac{\partial z}{\partial x} \\ \frac{\partial u}{\partial y} + \frac{\partial u}{\partial z}\frac{\partial z}{\partial y} & \frac{\partial v}{\partial y} + \frac{\partial v}{\partial z}\frac{\partial z}{\partial y} \end{vmatrix} = 0$$

Rewriting

$$\begin{bmatrix} u_x + pu_z & v_x + pv_z \\ u_y + qu_z & v_y + qv_z \end{bmatrix} = 0$$

Expansion of this determinant results in a P.D.E. which is free of the arbitrary function F as

$$(u_x + pu_z)(v_y + qv_z) - (u_y + qu_z)(v_x + pv_z) = 0$$

or

$$Pq + Qq = R$$

which is a first order linear P.D.E. Here

$$P = u_z v_y - u_y v_z = \frac{\partial(u, v)}{\partial(y, z)}$$

$$Q = u_x v_z - u_z v_x = \frac{\partial(u, v)}{\partial(z, x)}$$

$$R = u_y v_x - u_x v_y = \frac{\partial(u, v)}{\partial(x, y)}.$$

WORKED OUT EXAMPLES

Elimination of arbitrary constants

Form (obtain) partial differential equation by eliminating the arbitrary constants/functions:

Example 1: $z = ax^2 + by^2$

Solution: Differentiating partially w.r.t. x and y, we get

$$z_x = 2ax, z_y = 2by \quad \text{or} \quad a = \frac{z_x}{2x}, b = \frac{z_y}{2y}.$$

Eliminating the two arbitrary constants a and b

$$z = \frac{z_x}{2x} \cdot x^2 + \frac{z_y}{2y} \cdot y^2 \quad \text{or } 2z = xz_x + yz_y = xp + yq$$

Example 2: $(x-a)^2 + (y-b)^2 = z^2 \cot^2 \alpha$ where α is a parameter.

Solution: Differentiating partially w.r.t. x and y, we get

$$2(x-a) + 0 = 2z \cdot z_x \cdot \cot^2 \alpha$$

$$2 \cdot 0 + 2(y-b) = 2z\, z_y \cot^2 \alpha$$

Substituting $(zz_x \cot^2 \alpha)^2 + (zz_y \cot^2 \alpha)^2 = z^2 \cot^2 \alpha$

$$p^2 + q^2 = \tan^2 \alpha$$

Example 3: Find the differential equation of all spheres whose centres lie on the z-axis.

Solution: Equation $x^2 + y^2 + (z-a)^2 = b^2$ where a and b are arbitrary constants.

Differentiating

$$2x + 0 + 2(z-a)z_x = 0 \tag{1}$$

$$2y + 2(z-a)z_y = 0 \tag{2}$$

From (2),

$$(z-a) = -\frac{y}{z_y} \tag{3}$$

Substituting (3) in (1)

$$x + \left(-\frac{y}{z_y}\right) \cdot z_x = 0$$

$$xz_y - yz_x = 0$$

or

$$xq - yp = 0.$$

Example 4: $ax + by + cz = 1$

Solution: Differentiating w.r.t. x, $a + 0 + cz_x = 0$. Differentiating again w.r.t. x, $0 + cz_{xx} = 0$, since $c \neq 0$, $z_{xx} = 0$. Similarly by differentiating w.r.t. y and z twice, we get $z_{yy} = 0$, $z_{xy} = 0$ so $r = 0$ or $s = 0$ or $t = 0$. Thus we get 3 PDE.

Elimination of one arbitrary functions

Example 5:

$$z = (x+y)\,\phi\,(x^2 - y^2) \tag{1}$$

Solution: Differentiating

$$z_x = 1 \cdot \phi + (x+y)2x \cdot \phi' \tag{2}$$

$$z_y = 1 \cdot \phi + (x+y)(-2y)\phi' \tag{3}$$

From (3),

$$\phi' = \frac{\phi - z_y}{2y(x+y)} \tag{4}$$

Substituting (4) in (2)

$$z_x = \phi + 2x(x+y) \cdot \left[\frac{\phi - z_y}{2y(x+y)}\right]$$

$$p = \phi + \frac{x}{y}(\phi - q)$$

$$p = \left(\frac{x+y}{y}\right)\phi - \frac{x}{y}q$$

From the given Equation (1), $\phi = \frac{z}{(x+y)}$

Substituting ϕ,

$$p = \frac{(x+y)}{y} \cdot \frac{z}{x+y} - \frac{x}{y}q = \frac{z}{y} - \frac{x}{y}q$$

or $\quad yp + xq = z$

Example 6: $z = x^n f\left(\frac{y}{x}\right)$

Solution: By differentiation,

$$z_x = nx^{n-1} \cdot f + x^n \cdot \left(-\frac{y}{x^2}\right) \cdot f'$$

$$z_y = x^n \cdot \frac{1}{x} \cdot f' \quad \text{or } f' = \frac{z_y}{x^{n-1}}$$

Eliminating f',

$$z_x = nx^{n-1} f - x^{n-2} \cdot y \cdot \frac{z_y}{x^{n-1}}$$

$$xp = nx^n f - yq$$

or

$$xp = nz - yq$$

Example 7: $xyz = f(x+y+z)$

Solution: Differentiating w.r.t. x and y

$$yz + xy\,z_x = 1 \cdot f' + f' \cdot z_x \tag{1}$$

$$xz + xy\,z_y = 1 \cdot f' + f' \cdot z_y \tag{2}$$

From (2),

$$f' = \frac{xz + xy\,z_y}{1 + z_y} = \frac{xz + xyq}{1+q} \tag{3}$$

Put (3) in (1)

$$yz + xyp = (1+p)f' = (1+p)\left(\frac{xz+xyq}{1+q}\right)$$

$$(1+q)(yz+xyp) = (1+p)(xz+xyq)$$

or $\quad x(y-z)p + y(z-x)q = z(x-y)$

Elimination of two arbitrary functions

Example 8: $z = f(x)\,g(y)$

Solution: Differentiating w.r.t. x and y, we get $z_x = f'g,\ z_y = fg'$

$$z_x \cdot z_y = f'g\,fg' = fg\,f'g' = z\,f'g'$$

But

$$z_{xy} = f'g' \quad \text{so}$$

$$z_x \cdot z_y = z \cdot z_{xy}$$

or $\quad pq = z \cdot s.$

Example 9: $z = f(x+y) \cdot g(x-y)$

Solution: Differentiating partially w.r.t. x and y, we get

$$p = z_x = f' \cdot 1 \cdot g + f \cdot 1 \cdot g' \tag{1}$$

$$q = z_y = f' \cdot 1 \cdot g + f \cdot (-1)g' \tag{2}$$

$$r = z_{xx} = f''g + f'g' + f'g' + fg''$$

$$= f''g + 2f'g' + fg'' \tag{3}$$

$$t = z_{yy} = f''g + f'g'(-1)$$

$$-f'g' - fg''(-1)$$

$$= f''g - 2f'g' + fg'' \tag{4}$$

$$s = z_{xy} = f''g + f'g'(-1)$$

$$+f'g' + fg''(-1)$$

$$= f''g - fg'' \tag{5}$$

Adding (1) and (2),

$$f' = \frac{p+q}{2g} \tag{6}$$

Subtracting (2) from (1),

$$g' = \frac{p-q}{2f} \tag{7}$$

Adding (3) and (4)

$$r + t = 2(f''g + fg'')$$

From (5), $\quad 2s = 2(f''g - fg'')$

Adding $\quad r + t + 2s = 4f''g \quad (8)$

Subtracting $\quad r + t - 2s = 4fg'' \quad (9)$

Substituting (6), (7), (8), (9) in (3)

$$r = \left(\frac{r+t+2s}{4}\right) + 2 \cdot \frac{(p+q)}{2g} \cdot \frac{(p-q)}{2f}$$

$$+\left(\frac{r+t-2s}{4}\right)$$

$$(r-t)z = (p+q)(p-q)$$

Example 10: $z = xf(ax+by) + g(ax+by)$

Solution: Differentiating w.r.t. x and y, we get

$$z_x = f + xaf' + ag'$$

$$z_{xx} = af' + af' + ax\,f''a + a^2g''$$

$$= a[2f' + axf'' + ag''] \tag{1}$$

$$z_y = bx\,f' + bg'$$

$$z_{yy} = b^2xf'' + b^2g'' = b^2[xf'' + g''] \tag{2}$$

$$z_{yx} = bf' + bx\,af'' + ba\,g''$$

$$= b[f' + a(xf'' + g'')] \tag{3}$$

Substituting (2) in (3)

$$\frac{z_{yx}}{b} = f' + a \cdot \frac{z_{yy}}{b^2}$$

Solving

$$f' = \frac{z_{yx}}{b} - \frac{a}{b^2}z_{yy} \tag{4}$$

Substituting (2) and (4) in (1)

$$z_{xx} = 2af' + a^2[xf'' + g'']$$

$$= 2a \cdot \left[\frac{z_{yx}}{b} - \frac{a}{b^2}z_{yy}\right] + a^2\left[\frac{z_{yy}}{b^2}\right]$$

$$b^2z_{xx} + a^2z_{yy} = 2ab\,z_{xy}$$

$$b^2r + a^2t = 2abs$$

Elimination of arbitrary function of specific functions

$F(u, v) = 0$ where u and v are given.

Example 11:

$$F(xy + z^2, x + y + z) = 0 \quad (1)$$

Solution:

$$\text{Let } u(x, y, z) = xy + z^2 \quad (2)$$

$$v(x, y, z) = x + y + z \quad (3)$$

Differentiating (1) partially w.r.t. x by chain rule

$$\frac{\partial F}{\partial u}\left[\frac{\partial u}{\partial x} + \frac{\partial u}{\partial z}\frac{\partial z}{\partial x}\right] + \frac{\partial F}{\partial v}\left[\frac{\partial v}{\partial x} + \frac{\partial v}{\partial z}\frac{\partial z}{\partial x}\right] = 0$$

i.e., $$F_u \cdot (y + 2z\, z_x) + F_v \cdot (1 + z_x) = 0$$

Differentiating w.r.t. 'y', we get

$$F_u(x + 2z\, z_y) + F_v(1 + z_y) = 0$$

Eliminating F_u and F_v (i.e., the coefficient matrix should be singular)

$$\begin{vmatrix} y + 2z\, z_x & 1 + z_x \\ x + 2z\, z_y & 1 + z_y \end{vmatrix} = 0$$

or $$(1 + q)[y + 2zp] - (1 + p)[x + 2zq] = 0$$

$$(2z - x)p + (y - 2z)q = x - y.$$

Example 12: $xyz = f(x + y + z)$

Solution: Put $u = x + y + z$, $v = xyz$ so that the given equation may be written as $F(u, v) = 0$

Differentiating w.r.t. x and y, we get

$$F_u \cdot (1 + z_x) + F_v(yz + xy\, z_x) = 0$$

$$F_u(1 + z_y) + F_v(xz + xy\, z_y) = 0$$

Eliminating F_u, F_v, we have

$$\begin{vmatrix} 1 + p & yz + xyp \\ 1 + q & xz + xyq \end{vmatrix} = 0$$

or $$(xz + xyq)(1 + p) - (1 + q)(yz + xyp) = 0$$

$$x(z - y)p + (x - z)yq + z(x - y) = 0.$$

Example 13: $F(x^2 + y^2 + z^2, z^2 - 2xy) = 0$

Solution: Let $u = x^2 + y^2 + z^2$, $v = z^2 - 2xy$

Differentiating F partially w.r.t. x, we get

$$\frac{\partial F}{\partial u}\frac{\partial u}{\partial x} + \frac{\partial F}{\partial u}\frac{\partial u}{\partial z}\frac{\partial z}{\partial x} + \frac{\partial F}{\partial v}\cdot\frac{\partial v}{\partial x} + \frac{\partial F}{\partial v}\frac{\partial v}{\partial z}\frac{\partial z}{\partial x} = 0$$

$$F_u \cdot 2x + F_u \cdot 2z \cdot p + F_v(-2y) + F_v\, 2z \cdot p = 0$$

Similarly w.r.t. y, we get

$$\frac{\partial F}{\partial u}\frac{\partial u}{\partial y} + \frac{\partial F}{\partial u}\frac{\partial u}{\partial z}\frac{\partial z}{\partial y} + \frac{\partial F}{\partial v}\frac{\partial v}{\partial y} + \frac{\partial F}{\partial v}\frac{\partial v}{\partial z}\frac{\partial z}{\partial y} = 0$$

$$F_u \cdot 2y + F_u\, 2z \cdot q + F_v(-2x) + F_v \cdot 2z \cdot q = 0$$

Solving

$$\begin{vmatrix} x + zp & -y + zp \\ y + zq & -x + zq \end{vmatrix} = 0$$

$$(x + zp)(zq - x) - (zp - y)(y + zq) = 0$$

$$xz(q - p) + yz(q - p) + (y^2 - x^2) = 0$$

$$(x + y)[z(q - p) + (y - x)] = 0.$$

EXERCISE

Form (obtain) partial differential equation by eliminating the arbitrary constants/functions:

Elimination of arbitrary constants

1. $2z = \frac{x^2}{a^2} + \frac{y^2}{b^2}$

Ans. $2z = xp + yq$

2. $z = ax + by + a^2 + b^2$

Ans. $z = px + qy + p^2 + q^2$

3. $z = (x^2 + a^2)(y^2 + b^2)$

Ans. $4xyz = pq$

4. $z = axy + b$

Ans. $px = qy$

5. $z = axe^y + \frac{1}{2}a^2e^{2y} + b$

Ans. $q = px + p^2$

6. $z = (x - a)^2 + (y - b)^2 + 1$

Ans. $4z = p^2 + q^2 + 4$

7. $z = a(x + y) + b(x - y) + abt + c$

Hint: Number of independent variables x, y, t are 3 = number of arbitrary constants a, b, c. So P.D.E. is of 1st order.

Ans. $z_x^2 - z_y^2 = 4z_t$

8. $\frac{x^2}{a^2} + \frac{y^2}{b^2} + \frac{z^2}{c^2} = 1$

Ans. $pz = xp^2 + xz\, r$

or $qz = yq^2 + zy\,t$

Note: Number of arbitrary constants a, b, c is three > number of independent variables x, y is two so P.D.E. is of 2nd order.

9. $z = ae^{-b^2t}\cos bx$

Ans. $\frac{\partial z}{\partial t} = \frac{\partial^2 z}{\partial x^2}$

10. $z = ae^{bx}\sin by$

Ans. $\frac{\partial^2 z}{\partial x^2} + \frac{\partial^2 z}{\partial y^2} = 0$

11. $z = xy + y\sqrt{x^2 + a^2} + b$

Ans. $pq = py + qx$

Find the differential equation of:

12. All planes which are at a constant distance b from the origin.

Hint: Equation $lx + my + nz = b$ with $l^2 + m^2 + n^2 = 1$.

Ans. $z = px + qy + b\sqrt{1 + p^2 + q^2}$

13. All planes having equal x and y intercepts.

Hint: Equation $\frac{x}{a} + \frac{y}{a} + \frac{z}{b} = 1$.

Ans. $p = q$

14. All spheres of given radius c having their centres in the xy-plane.

Hint: $(x - a)^2 + (y - b)^2 + z^2 = c^2$; a, b constants.

Ans. $z^2(p^2 + q^2 + 1) = c^2$

15. All cones with their vertices at the origin.

Ans. $px + qy = z$

Elimination of arbitrary functions

16. $z = f(x^2 - y^2)$

Ans. $yp + xq = 0$

17. $x + y + z = f(x^2 + y^2 + z^2)$

Ans. $(y - z)p + (z - x)q = x - y$

18. $z = y\,f\left(\frac{y}{x}\right)$

Ans. $z = px + qy$

19. $z = f(\sin x + \cos y)$

Ans. $p\sin y + q\cos x = 0$

20. $z = e^{ax+by}\cdot f(ax - by)$

Ans. $bp + aq = 2abz$

21. $z = y^2 + 2f\left(\frac{1}{x} + \ln y\right)$

Ans. $x^2p + yq = 2y^2$

22. $z = x + y + f(xy)$

Ans. $xp - yq = x - y$

23. $z = f\left(\frac{xy}{z}\right)$

Ans. $xp = yq$

24. $z = f(x + at) + g(x - at)$

Ans. $z_{tt} = a^2 z_{xx}$

25. $z = f(x) + e^y g(x)$

Ans. $t = q$

26. $z = f(x + iy) + g(x - iy)$

Ans. $z_{xx} + z_{yy} = 0$

27. $z = yf(x) + xg(y)$

Ans. $xys = px + qy - z$

28. $z = xf\left(\frac{y}{x}\right) + yg(x)$

Ans. $x\frac{\partial^3 z}{\partial x\partial y^2} + \frac{\partial^2 z}{\partial y^2} + y\frac{\partial^3 z}{\partial y^3} = 0$

29. $z = f(xy) + g(x + y)$

Ans. $rx(y - x) - s(y^2 - x^2) + t\,y(y - x) + (p - q)(x + y) = 0$

30. $z = [f(r - at) + g(r + at)]/r$

Ans. $z_{tt} = \frac{a^2}{r^2}\frac{\partial}{\partial r}\left(r^2\frac{\partial z}{\partial r}\right)$

Elimination of arbitrary function F(u, v) = 0

31. $F(x^2 + y^2, z - xy) = 0$

Ans. $xq - yp = x^2 - y^2$

32. $F(x + y + z, x^2 + y^2 - z^2) = 0$

Ans. $(y + z)p - (z + x)q = x - y$

33. $F(x^2 + y^2, x^2 - z^2) = 0$

Ans. $yp - xq = \frac{xy}{z}$

34. $F(ax + by + cz, x^2 + y^2 + z^2) = 0$

Ans. $(bz - cy)p + (cx - az)q = ay - bx$

35. $z = x^2\phi(x - y)$

Hint: Rewrite the given equation in the form

$F(\frac{z}{x^2}, x-y) = 0.$

Ans. $2z = xp + xq$.

18.2 PARTIAL DIFFERENTIAL EQUATIONS OF FIRST ORDER

The general form of a first order partial differential equation is

$$F(x, y, z, p, q) = 0 \quad (1)$$

where x, y are the two independent variables, z is the dependent variable and $p = z_x, q = z_y$.

Complete Solution

Any function

$$f(x, y, z, a, b) = 0 \quad (2)$$

involving two arbitrary constants a, b and satisfying the P.D.E. (1) is known as complete solution or complete integral or primitive.

General Solution

of P.D.E. (1) is any arbitrary function F of specific (given) functions u, v

$$F(u, v) = 0 \quad (3)$$

satisfying P.D.E. (1).

Here $u = u(x, y, z)$ and $v = v(x, y, z)$ are known functions of x, y, z.

Linear

A partial differential equation is said to be linear (after rationalization and cleared of fractions) if the dependent variable z and its derivatives are of degree (power) one and products of z and its derivatives do not appear in the equation.

Quasi-linear

P.D.E. is said to be quasi-linear if degree of highest ordered derivative is one and no products of partial derivatives of the highest order are present.

Example: $x^2p + y^2q = z$ is linear in z and of first order.

Example: $z\, z_{xx} + (z_y)^2 = 0$ is a quasi-linear of second order.

Non-linear

A P.D.E. which is not linear is known as non-linear P.D.E.

Example: $\left(\frac{\partial^2 u}{\partial x^2}\right)^2 + u^2\left(\frac{\partial u}{\partial y}\right) = f(x, y)$

is non-linear in u and of second order.

Homogeneous

if each term contains the dependent variable or its derivatives.

Otherwise non-homogeneous.

18.3 LINEAR PARTIAL DIFFERENTIAL EQUATIONS OF FIRST ORDER

The general form of a quasi-linear partial differential equation of the first order is

$$P(x, y, z)z_x + Q(x, y, z)z_y = R(x, y, z) \quad (1)$$

This Equation (1) is known as "**Lagrange's** linear equation".

If P and Q are independent of z and R is linear in z then (1) is a linear equation. The general solution of the Lagrange's linear P.D.E.

$$Pp + Qq = R \quad (1)$$

is given by the equation

$$F(u, v) = 0 \quad (2)$$

since the elimination of the arbitrary function F from (2) results in (1).

Here $u = u(x, y, z)$, $v = v(x, y, z)$ are specific (known) functions of x, y, z.

Method of Obtaining General Solution

1. Rewrite the equation in the standard form

$$Pp + Qq = R$$

2. Form the Lagrange's auxiliary equations (A.E.)

$$\frac{dx}{P} = \frac{dy}{Q} = \frac{dz}{R} \quad (3)$$

3. Nature of solution to the simultaneous equations of the form $\frac{dx}{P} = \frac{dy}{Q} = \frac{dz}{R}$:
$u(x, y, z) = c_1$ and $v(x, y, z) = c_2$ are said to be the complete solution of the system of simultaneous equations (provided u_1 and u_2 are linearly independent i.e., $u_1/u_2 \neq$ constant).

Case 1: One of the variables is either absent or cancells out from the set of auxiliary equations.

Case 2: If $u = c_1$ is known but $v = c_2$ is not possible by case I, then use $u = c_1$ to get $v = c_2$.

Case 3: Introducing Lagrange's multipliers P_1, Q_1, R_1, which are functions of x, y, z or constants, each fraction in (3) is equal to

$$\frac{P_1 dx + Q_1 dy + R_1 dz}{P_1 P + Q_1 Q + R_1 R} \quad (4)$$

If P_1, Q_1, R_1 are so chosen that $P_1 P + Q_1 Q + R_1 R = 0$ then $P_1 dx + Q_1 dy + R_1 dz = 0$ which can be integrated.

Case 4: Multipliers may be chosen (more than once) such that the numerator $P_1 dx + Q_1 dy + R_1 dz$ is an exact differential of the denominator $P_1 P + Q_1 Q + R_1 R$. Now combine (4) with a fraction of (3) to get an integral.

4. General solution of (1) is

$$F(u, v) = 0 \quad \text{or} \quad v = \phi(u).$$

Worked Out Examples

Solve the following:

Example 1: $xp + yq = 3z$

Solution: This is a linear P.D.E. of first order $Pp + Qq = R$ with $P = x$, $Q = y$ and $R = 3z$. The Lagrange's auxiliary equations are

$$\frac{dx}{x} = \frac{dy}{y} = \frac{dz}{3z}$$

Integrating the first two equations (or fractions) $\frac{dx}{x} = \frac{dy}{y}$, we get $\ln x = \ln y + c_1$ or $\frac{x}{y} = c$

Integrating first and the last equations $\frac{dx}{x} = \frac{dz}{3z}$, we have

$$3 \ln x = \ln z + c_z \quad \therefore x^3 = c_1 z$$

Thus the required solution is $x^3 = zf\left(\frac{x}{y}\right)$. The general solution can also be written as $F(x^3/z, x/y) = 0$.

Note: By integrating 2nd and 3rd equations $\frac{dy}{y} = \frac{dz}{3z}$, we also get $y^3 = c_2 z$ so the general solution is also given by $y^3 = zf\left(\frac{x}{y}\right)$.

Example 2: $yzp - xzq = xy$

Solution: Auxiliary equations are

$$\frac{dx}{yz} = \frac{dy}{-xz} = \frac{dz}{xy}$$

From first and second fractions, we get

$$\frac{dx}{yz} = \frac{dy}{-xz}$$

or

$$\frac{dx}{y} = \frac{dy}{-x}$$

or

$$xdx + ydy = 0$$

Integrating $x^2 + y^2 = c_1$
From first and third fractions

$$\frac{dx}{yz} = \frac{dz}{xy}$$

or

$$\frac{dx}{z} = \frac{dz}{x}$$

Integrating $x^2 - z^2 = c_2$
Thus the general solution is

$$F(x^2 + y^2, x^2 - z^2) = 0.$$

Example 3: $p - q = \ln(x + y)$

Solution: Auxiliary equations are

$$\frac{dx}{1} = \frac{dy}{-1} = \frac{dz}{\ln(x + y)}$$

Integrating the first two fractions $dx + dy = 0$ yields $x + y = c_1$
From first and last fractions $\ln(x + y)dx = dz$
Put $x + y = c_1$, then $\ln c_1 dx = dz$

Integrating $x \ln c_1 = z + c_2$

or $$x \cdot \ln(x+y) = z + c_2$$

The general solution is

$$F(x+y, x\ln(x+y) - z) = 0.$$

Example 4: $z(z^2 + xy)(px - qy) = x^4$

Solution: Auxiliary equations are

$$\frac{dx}{xz(z^2+xy)} = \frac{dy}{-yz(z^2+xy)} = \frac{dz}{x^4}$$

From first and second fractions, we get

$$\frac{dx}{x} = \frac{dy}{-y}$$

on integration $xy = c_1$

From first and third fractions

$$x^3 dx = (z^3 + xyz)dz$$

using $xy = c_1$, $x^3 dx = (z^3 + c_1 z)dz$

Integrating $\frac{x^4}{4} = \frac{z^4}{4} + c_1\frac{z^2}{2} + c_2$

or $$x^4 - z^4 - 2c_1 z^2 = c_2$$

Substituting for c_1, $x^4 - z^4 - 2(xy)z^2 = c_2$

The general solution is

$$F(xy, x^4 - z^4 - 2xyz^2) = 0$$

Example 5: $xzp + yzq = xy$

Solution: Auxiliary equations $\frac{dx}{xz} = \frac{dy}{yz} = \frac{dz}{xy}$

From one and two $\frac{dx}{xz} = \frac{dy}{yz}$ or $\frac{dx}{x} = \frac{dy}{y}$

Integrating $x = c_1 y$

Choosing the multipliers as $y, x, 2z$

$$\frac{ydx + xdy}{yxz + xyz} = \frac{2zdz}{2zxy}$$

or $$ydx + xdy - 2z\,dz = 0$$

$$d(xy) - d(z^2) = 0$$

Integrating $xy - z^2 = c_2$

The general solution is

$$F\left(\frac{x}{y}, xy - z^2\right) = 0.$$

Example 6: $(z-y)\,p + (x-z)\,q = y - x$

Solution: Auxiliary equations are

$$\frac{dx}{z-y} = \frac{dy}{x-z} = \frac{dz}{y-x}$$

Choosing multipliers as, 1, 1, 1

$$dx + dy + dz = (z-y) + (x-z) + (y-x) = 0$$

Integrating $x + y + z = c_1$

Choosing multipliers as x, y, z

$$xdx + ydy + zdz = x(z-y) + y(x-z) + z(y-x) = 0$$

Integrating $x^2 + y^2 + z^2 = c_2$

The general solution is

$$F(x+y+z, x^2+y^2+z^2) = 0$$

Example 7: $(y+zx)p - (x+yz)q = x^2 - y^2$

Solution: Auxiliary equations are

$$\frac{dx}{y+zx} = \frac{dy}{-(x+yz)} = \frac{dz}{x^2-y^2}$$

Choosing multipliers as $x, y, -z$

$$xdx + ydy - zdz$$
$$= x(y+zx) + y(-1)(x+yz) - z(x^2-y^2) = 0$$

Integrating $x^2 + y^2 - z^2 = c_1$

Choosing multipliers as $y, x, 1$, we get

$$ydx + xdy + dz$$
$$= y(y+zx) + x(-1)(x+yz) + (x^2-y^2) = 0$$

Integrating $xy + z = c_2$

The general solution is

$$F(x^2 + y^2 - z^2, xy + z) = 0.$$

Example 8: $(y^2 + z^2)p - xyq + zx = 0$

Solution: Auxiliary equations are

$$\frac{dx}{y^2+z^2} = \frac{dy}{-xy} = \frac{dz}{-zx}$$

From the 2nd and 3rd fractions

$$\frac{dy}{y} = \frac{dz}{z} \quad \text{or} \quad \frac{y}{z} = c_1$$

Choosing multipliers as x, y, z

$$xdx + ydy + zdz = x(y^2+z^2) + y(-xy) + z(-zx) = 0$$

Integrating $x^2+y^2+z^2=c_2$
The general solution is

$$F\left(\frac{y}{z}, x^2+y^2+z^2\right)=0.$$

Example 9: $px(x+y)=qy(x+y)-$
$-(2x+2y+z)(x-y)$

Solution: Auxiliary equations are

$$\frac{dx}{x(x+y)}=\frac{dy}{-y(x+y)}=\frac{dz}{-(x-y)(2x+2y+z)}$$

From first two fractions, cancelling $(x+y)$, we get

$$\frac{dx}{x}=-\frac{dy}{y} \quad \text{or} \quad d(\ln x)+d(\ln y)=c$$

which on integration gives $xy=c_1$

$$\frac{dx+dy}{x(x+y)-y(x+y)}=\frac{dx+dy}{(x+y)(x-y)}$$
$$=\frac{dz}{-(x-y)(2x+2y+z)}$$

Cancelling the $(x-y)$ term, we get

$$(2x+2y+z)(dx+dy)+(x+y)dz=0$$

or

$$(x+y+z)(dx+dy)+(x+y)(dx+dy)+(x+y)dz=0$$
$$(x+y+z)\,d(x+y)+(x+y)\,d(x+y+z)=0$$

i.e., $$d((x+y)(x+y+z))=0$$

Integrating $(x+y)(x+y+z)=c_2$
Thus the general solution is

$$F(xy, (x+y)(x+y+z))=0$$

Example 10: $(x^2-y^2-yz)p+$
$(x^2-y^2-zx)q=z(x-y)$

Solution: Auxiliary equations are

$$\frac{dx}{x^2-y^2-yz}=\frac{dy}{x^2-y^2-zx}=\frac{dz}{z(x-y)}$$

$$dx-dy-dz=x^2-y^2-yz-(x^2-y^2-zx)$$
$$-z(x-y)=0$$

Integrating $x-y-z=c_1$

From first and second

$$\frac{xdx-ydy}{x^3-xy^2-x^2y-y^3}=\frac{dz}{z(x-y)}$$

or $$\frac{xdx-ydy}{(x^2-y^2)(x-y)}=\frac{dz}{z(x-y)}$$

i.e., $$\frac{1}{2}d(\ln(x^2-y^2))=d(\ln z)$$

$\therefore$ $$(x^2-y^2)/z^2=c_2.$$

$\therefore$ The general solution is

$$f(x-y-z, (x^2-y^2)/z^2)=0$$

EXERCISE

Solve the following:

1. $y^2zp+x^2zq=xy^2$

Ans. $F(x^3-y^3, x^2-z^2)=0$

2. $p\tan x+q\tan y=\tan z$

Ans. $F\left(\frac{\sin x}{\sin y}, \frac{\sin y}{\sin z}\right)=0$

3. $xp+yq=z$

Ans. $F\left(\frac{x}{z}, \frac{y}{z}\right)=0$

4. $2p+3q=1$

Ans. $F(3x-2y, y-3z)=0$

5. $x^2p+y^2q=z^2$

Ans. $F\left(\frac{1}{x}-\frac{1}{y}, \frac{1}{y}-\frac{1}{z}\right)=0$

6. $pyz+qzx=xy$

Ans. $F(x^2-y^2, y^2-z^2)=0$

7. $x^2p+y^2q=(x+y)z$

Ans. $F\left(\frac{xy}{z}, \frac{x-y}{z}\right)=0$

8. $p+3q=5z+\tan(y-3x)$

Hint: Use the solution $y-3x=c$ obtained from $\frac{dx}{1}=\frac{dy}{3}$.

Ans. $F(y-3x, e^{-5x}\{5z+\tan(y-3x)\})=0$

9. $z(p-q)=z^2+(x+y)^2$

Hint: Use solution $x+y=c$ obtained from $\frac{dx}{1}=\frac{dy}{-1}$.

Ans. $F[x+y, e^{2y}\{z^2+(x+y)^2\}]=0$

10. $z(xp-yq)=y^2-x^2$

Hint: Use x, y, z as multipliers.

Ans. $F(xy, x^2+y^2+z^2)=0$

11. $(x^3+3xy^2)p+(y^3+3x^2y)q=2(x^2+y^2)z$

Hint: Choose multipliers $(1/x, 1/y, 0)$, $(1, 1, 0)$, $(1, -1, 0)$.

Ans. $F[(x-y)^{-2}-(x+y)^{-2}, xy/z^2]=0$

12. $x^2(y^3-z^3)p+y^2(z^3-x^3)q=z^2(x^3-y^3)$

Ans. $F(x^2+y^2+z^2, \frac{1}{x}+\frac{1}{y}+\frac{1}{z})=0$

13. $(2x^2+y^2+z^2-2yz-zx-xy)p+$ $(x^2+2y^2+z-yz-2zx-xy)q=$ $(x^2+y^2+2z^2-yz-2xy)$

Hint: $\frac{dx-dy}{x-y}=\frac{dy-dz}{y-z}=\frac{dz-dx}{z-x}$.

Ans. $F\left(\frac{x-y}{y-z}, \frac{y-z}{z-x}\right)=0$

14. $(mz-ny)p+(nx-lz)q=ly-mx$

Ans. $F(x^2+y^2+z^2, lx+my+nz)=0$

15. $(x^2-yz)p+(y^2-zx)q=z^2-xy$

Ans. $F\left(\frac{x-y}{y-z}, xy+yz+zx\right)=0$

16. $(x^2-y^2-z^2)p+2xyq=2xz$

Ans. $F(x^2+y^2+z^2, y/z)=0$

17. $(x+2z)p+(4zx-y)q=2x^2+y$

Hint: Multipliers $y, x, -2z$ and $2x, -1, -1$.

Ans. $F(xy-z^2, x^2-y-z)=0$

18. $x(y-z)p+y(z-x)q=z(x-y)$

Ans. $F(x+y+z, xyz)=0$

19. $(y-z)p+(x-y)q=(z-x)$

Ans. $F(x+y+z, \frac{x^2}{2}+yz)=0$

20. $(y+z)p+(z+x)q=x+y$

Ans. $F\left(\frac{x-y}{y-z}, \frac{y-z}{\sqrt{x+y+z}}\right)=0$

21. $x^2(y-z)p+y^2(z-x)q=z^2(x-y)$.

Hint: Multipliers: $1/x, 1/y, 1/z$ and $1/x^2, 1/y^2, 1/z^2$.

Ans. $F\left(xyz, \frac{1}{x}+\frac{1}{y}+\frac{1}{z}\right)=0$

22. $(z^2-2yz-y^2)p+(xy+zx)q=xy-zx$

Ans. $F(x^2+y^2+z^2, y^2-2yz-z^2)=0$

23. $px(z-2y^2)=(z-qy)(z-y^2-2x^3)$

Hint: Multiply A.E. by $(z-y^2-2x^3)$, use multipliers $1, 2xy, -x$, divide by x^2 throughout.

Ans. $F\left(y/z, \frac{y^2}{x}-\frac{z}{x}-x^2\right)=0$

24. $x(y^2-z^2)p+y(z^2-x^2)q=z(x^2-y^2)$

Hint: Multipliers x, y, z and $1/x, 1/y, 1/z$.

Ans. $F(x^2+y^2+z^2, xyz)=0$.

18.4 NON-LINEAR PARTIAL DIFFERENTIAL EQUATIONS OF FIRST ORDER

Non-linear P.D.E. of first order contains p and q of degree (power) other than one and/or product terms of p and q. Its complete solution is given by $f(x, y, z, a, b)=0$ where a and b are any two arbitrary constants. Some special types of non-linear first order P.D.E. are presented.

Form I: $f(p, q)=0$

i.e., equation contains only p and q (or x, y, z are absent)

Assume that $p=a$ then $f(a, q)=0$
Solving $q=\phi(a)$
Consider

$$dz=\frac{\partial z}{\partial x}dx+\frac{\partial z}{\partial y}dy=p\,dx+q\,dy$$

$$dz=a\,dx+\phi(a)dy$$

Integrating $z=ax+\phi(a)y+c$
where a and c are arbitrary constants.
Thus the complete solution is

$$z=ax+by+c$$

where a, b satisfy the equation $f(a, b)=0$

i.e., $$b=\phi(a)$$

Form II: $f(z, p, q) = 0$

i.e., equation does not involve the independent variables x and y.

Assume $q = ap$. Substituting q in the given equation $f(z, p, ap) = 0$ and solving for p, we get $p = \phi(z)$. Now

$$dz = pdx + qdy = pdx + ap\,dy$$

$$dz = p(dx + ady) = \phi(z)(dx + ady)$$

Integrating $x + ay = \int \frac{dz}{\phi(z)} + b$
where a and b are two arbitrary constants.

Form III: $f(x, p) = g(y, q)$

i.e., x, p and y, q are separable.

Assume $f(x, p) = g(y, q) = a =$ constant.
Solving each equation for p and q, we get

$$p = f_1(x, a) \quad \text{and} \quad q = g_1(y, a)$$

Now $dz = pdx + qdy = f_1(x, a)dx + g_1(y, a)dy$

Integrating $z = \int f_1(x, a)dx + \int g_1(y, a)dy + b$.

Form IV: Clairaut Equation: $z = px + qy + f(p, q)$

The complete solution of this equation is

$$z = ax + by + f(a, b)$$

which is obtained by replacing p by a and q by b in the given clairaut equation.

Note: All these four forms can be solved by Charpit's method.

WORKED OUT EXAMPLES

Form I: $f(p, q) = 0$

Solve the following:

Example 1: $p^3 - q^3 = 0$

Solution: The complete solution is

$$z = ax + by + c$$

where a, b are connected by $a^3 - b^3 = 0$ or $a = b$

Thus $z = ax + ay + c$ is the complete integral.

Example 2: $p^2 + q^2 = npq$

Solution: The complete solution is

$$z = ax + by + c$$

where a, b satisfy the equation $a^2 + b^2 = nab$.
Solving for b

$$b = \frac{-na \pm \sqrt{n^2a^2 - 4a^2}}{2} = +\frac{a}{2}\left[+n \pm \sqrt{n^2 - 4}\right]$$

so complete integral:

$$z = ax + \tfrac{a}{2}\left(n \mp \sqrt{n^2 - 4}\right) y + c$$

Example 3: $(x + y)(p + q)^2 + (x - y)(p - q)^2 = 1$

Solution: To reduce this equation to $f(P, Q) = 0$ form, put $x + y = X^2$ and $x - y = Y^2$. Then

$$X = \sqrt{x + y}, \quad Y = \sqrt{x - y}$$

$$p = z_x = \frac{\partial z}{\partial x} = \frac{\partial z}{\partial X}\frac{\partial X}{\partial x} + \frac{\partial z}{\partial Y} \cdot \frac{\partial Y}{\partial x}$$

$$= \frac{1}{2X}\frac{\partial z}{\partial X} + \frac{1}{2Y}\frac{\partial z}{\partial Y}$$

since

$$\frac{\partial X}{\partial x} = \frac{1}{2}\frac{1}{\sqrt{x + y}} = \frac{1}{2X} \quad \text{and} \quad \frac{\partial Y}{\partial x} = \frac{1}{2}\frac{1}{\sqrt{x - y}} = \frac{1}{2Y}$$

Similarly,

$$q = z_y = \frac{\partial z}{\partial y} = \frac{\partial z}{\partial X}\frac{\partial X}{\partial y} + \frac{\partial z}{\partial Y} \cdot \frac{\partial Y}{\partial y} = \frac{1}{2X}z_x - \frac{1}{2Y}z_y$$

Adding $p + q = z_x \cdot \frac{1}{X}$,
subtracting $p - q = z_y \frac{1}{Y}$.
Thus the given equation reduces to

$$X^2 \cdot \left(z_x \frac{1}{X}\right)^2 + Y^2\left(z_y \frac{1}{Y}\right)^2 = 1 \quad \text{or} \quad z_X^2 + z_Y^2 = 1$$

The complete solution is

$$z = aX + bY + c$$

where $a^2 + b^2 = 1$ or $b = \sqrt{1 - a^2}$

or $$z = a\sqrt{(x + y)} + \sqrt{1 - a^2}\sqrt{(x - y)} + c.$$

Example 4: $(x - y)(px - qy) = (p - q)^2$

Solution:

Put $$x + y = u \quad \text{and} \quad xy = v \qquad (1)$$

Then $$p = \frac{\partial z}{\partial x} = \frac{\partial z}{\partial u}\frac{\partial u}{\partial x} + \frac{\partial z}{\partial v}\frac{\partial v}{\partial x} = z_u + yz_v \qquad (2)$$

Since $u_x = 1, u_y = 1, v_x = y, v_y = x$

Similarly, $q = \frac{\partial z}{\partial y} = \frac{\partial z}{\partial u}\frac{\partial u}{\partial y} + \frac{\partial z}{\partial v}\frac{\partial v}{\partial y} = z_u + xz_v$ (3)

From (2) and (3)

$$p - q = (y - x)z_v \quad (4)$$

$$xp - qy = (x - y)z_u \quad (5)$$

Using (4), (5) the given equation transforms to

$$(x - y)(x - y)z_u = (y - x)^2 z_v^2$$

or

$$z_u = z_v^2$$

$$P = Q^2$$

Its complete solution is

$$z = au + bv + c$$

where $a = b^2$.

Using (1), replace u, v then

$$z = b^2(x + y) + bxy + c$$

where b, c are arbitrary constants.

Form II: $f(z, p, q) = 0$

Example 5: $zpq = p + q$

Solution: Assume $q = ap$.
Substituting in the given equation

$$zp \cdot ap = p + ap$$

Solving for p, we get

$$p = \frac{1 + a}{az}$$

We know that $dz = pdx + qdy = p(dx + ady)$

$$dz = \frac{1 + a}{az}(dx + ady) \quad \text{or } az\,dz = (1 + a)(dx + ady)$$

Integrating: $a\frac{z^2}{2} = (1 + a)[x + ay] + b$.

Example 6: $p^2z^2 + q^2 = p^2q$

Solution: Let $q = ap$ then the given equation reduces to

$$p^2z^2 + a^2p^2 = p^2 \cdot ap$$

Solving $p = (z^2 + a^2)/a$.

Then

$$dz = pdx + qdy = p(dx + ady)$$

$$dz = \frac{(z^2 + a^2)}{a}(dx + ady)$$

or

$$\frac{adz}{(z^2 + a^2)} = dx + ady$$

Integrating $\tan^{-1}\left(\frac{z}{a}\right) = x + ay + b$. Thus the complete solution is

$$z = a\tan(x + ay + b).$$

Example 7: $p^2x^2 = z(z - qy)$

Solution: To reduce this equation to $f(z, p, q) = 0$ form, put $X = \ln x$, $Y = \ln y$ so that

$$dx = \frac{dx}{x}, dy = \frac{dy}{y}$$

Rewriting

$$\left(x\frac{\partial z}{\partial x}\right)^2 = z\left(z - y\frac{\partial z}{\partial y}\right)$$

$$\left(\frac{\partial z}{\partial X}\right)^2 = z\left(z - \frac{\partial z}{\partial Y}\right)$$

Let $P = \frac{\partial z}{\partial X}$, $Q = \frac{\partial z}{\partial Y}$. Then the given equation reduces to

$$P^2 = z(z - Q)$$

which is of the form $f(z, P, Q) = 0$

Let $Q = aP$. Substituting this in the new equation

$$P^2 = z^2 - zQ = z^2 \quad zaP$$

or

$$P^2 + azP - z^2 = 0$$

Solving $P = \frac{-az \pm \sqrt{a^2z^2 + 4z^2}}{2} = z \cdot k$

where $k = \left(-a \pm \sqrt{a^2 + 4}\right)/2$

now

$$dz = P\,dX + Q\,dY = P(dX + adY)$$

or

$$dz = k \cdot z(dX + adY) \quad \text{since } P = kz$$

Replacing X and Y by $\ln x$ and $\ln y$

$$\frac{1}{k}\frac{dz}{z} = d\ln x + ad\ln y$$

Integrating $z^{\frac{1}{k}} = xy^a b$
where a, b are arbitrary constants and

$$k = \left(-a \pm \sqrt{a^2 + 4}\right)/2$$

Form III: $f(x, p) = g(y, q)$

Example 8: $yp + xq + pq = 0$

Solution: Rewriting $(x + p)q = -yp$

or $$\frac{x+p}{p} = -\frac{y}{q} = a, \quad \text{say}$$

Then solving for p and q

$$p = \frac{x}{a-1} \quad \text{and} \quad q = -\frac{y}{a}$$

Now $dz = pdx + qdy = \frac{x}{(a-1)}dx + \left(-\frac{y}{a}\right)dy$

Integrating $z = \frac{1}{(a-1)}\frac{x^2}{2} - \frac{y^2}{2a} + b.$

Example 9: $p^2q^2 + x^2y^2 = x^2q^2(x^2 + y^2)$

Solution: To transform this to the standard form, put $x^2 = X$ and $y^2 = Y$ so that

$$2x\,dx = dX, \quad 2y\,dy = dY$$

Then $$p = \frac{\partial z}{\partial x} = \frac{\partial z}{\partial X} \cdot \frac{\partial X}{\partial x} = 2x\,z_X, q = 2xP$$

Similarly, $$q = \frac{\partial z}{\partial y} = \frac{\partial z}{\partial Y} \cdot \frac{\partial Y}{\partial y} = 2y\,z_y = 2yQ$$

Substituting these values, the given equation reduces to

$$4x^2P^2 \cdot 4y^2Q^2 + x^2y^2 = x^24y^2Q^2(x^2 + y^2)$$

$$4XP^2 \cdot 4YQ^2 + XY = 4XYQ^2(X + Y)$$

or $$16P^2Q^2 + 1 = 4Q^2(X + Y)$$

Now rewrite this as

$$16P^2Q^2 - 4XQ^2 = 4YQ^2 - 1$$

$$(4P^2 - X) = \frac{4YQ^2 - 1}{4Q^2} = a^2 \quad \text{say}$$

This is in the standard form

$$f(P, X) = g(Q, Y).$$

Solving for P and Q, we get

$$P = \frac{1}{2}(X + a^2)^{\frac{1}{2}}$$

$$Q = \frac{1}{2}\frac{1}{(Y - a^2)^{\frac{1}{2}}}$$

Now $$dz = PdX + QdY$$

$$dz = \frac{1}{2}(X + a^2)^{\frac{1}{2}}dX + \frac{1}{2}(Y - a^2)^{-\frac{1}{2}}dY$$

Integrating $z = \frac{1}{3}(X + a^2)^{\frac{3}{2}} + (Y - a^2)^{\frac{1}{2}} + b$

or $$z = \frac{1}{3}(x^2 + a^2)^{\frac{3}{2}} + (y^2 - a^2)^{\frac{1}{2}} + b$$

where a, b are arbitrary constants.

Example 10: $zpy^2 = x(y^2 + z^2q^2)$.

Solution: To get rid of z in the equation, put $Z = \frac{z^2}{2}$ or $dZ = zdz$. Then

$$zp = \frac{z\partial z}{\partial x} = \frac{\partial Z}{\partial x} = P, \text{ similarly, } zq = z\frac{\partial z}{\partial y} = \frac{\partial Z}{\partial y} = Q$$

Now the equation gets transformed to

$$Py^2 = x(y^2 + Q^2)$$

or $$\frac{P}{x} = \frac{Q^2 + y^2}{y^2} = a \quad \text{say}$$

Solving $P = ax, \quad Q = \sqrt{a-1}y$

So $$dz = Pdx + Qdy = ax\,dx + \sqrt{a-1}y\,dy$$

or $$zdz = ax\,dx + \sqrt{a-1}y\,dy$$

Integrating $z^2 = ax^2 + \sqrt{a-1}y^2 + b$
where a, b are arbitrary constants.

Form IV: Clairaut's equation

Example 11: $z = px + qy + \ln pq$

Solution: The complete solution of this Clairaut's equation is

$$z = ax + by + \ln ab$$

Example 12: $(p - q)(z - px - qy) = 1$

Solution: Rewriting this in Clairaut's form

$$z = px + qy + \frac{1}{p-q}$$

The complete solution is

$$z = ax + by + \frac{1}{a-b}$$

EXERCISE

Find the complete solution of:

Form I: $f(p, q) = 0$

1. $pq = k$

Ans. $z = ax + k\frac{y}{a} + c$

2. $p^2 + q^2 = m^2$

Ans. $z = ax + \sqrt{m^2 - a^2}\, y + c$

3. $\sqrt{p} + \sqrt{q} = 1$

Ans. $z = ax + (1 - \sqrt{a})^2 y + c$

4. $p^2 - q^2 = 4$

Ans. $z = ax + \sqrt{a^2 - 4}\, y + c$

5. $p + q = pq$

Ans. $z = ax + ay/(a - 1) + c$

6. $p = e^q$

Ans. $z = ax + y \ln a + c$

7. $2p^2 + 6p + 2q + 4 = 0$

Ans. $z = ax - (2 + 3a + a^2/2)y + c$

8. $x^2p^2 + y^2q^2 = z^2$

Hint: Put $Z = \ln z$, $X = \ln x$, (where $a = \cos\alpha$, $b = \sqrt{1 - a^2} = \sin\alpha$,) $Y = \ln y$, which transforms the given equation to $P^2 + Q^2 = 1$.

Ans. $z = c^* x^{\cos\alpha} \cdot y^{\sin\alpha}$

9. $(x^2 + y^2)(p^2 + q^2) = 1$

Hint: Put $x = r\cos\theta$, $y = r\sin\theta$, (where $r = \sqrt{x^2 + y^2}$, $\tan\theta = y/x$, $R = \ln r$, then equation reduces to $\left(\frac{\partial z}{\partial R}\right)^2 + \left(\frac{\partial z}{\partial \theta}\right)^2 = 1$)

Ans. $z = a \ln r + \sqrt{1 - a^2}\,\theta + c$

10. $pq = x^m y^n z^l$

Ans. $z^p = p\left(\frac{ax^q}{q} + \frac{1}{a}\frac{y^r}{r} + c\right)$

where $p = 1 - \frac{l}{2}$, $q = m + 1$, $r = n + 1$

Hint: Put $Z = \frac{z^p}{p}$, $X = \frac{x^q}{q}$, $Y = \frac{y^r}{r}$

Then equation reduces to

$$\frac{\partial Z}{\partial X} \cdot \frac{\partial Z}{\partial Y} = 1.$$

Form II: $f(z, p, q) = 0$

1. $p^2z^2 + q^2 = 1$

Ans. $az(1 + a^2z)^{\frac{1}{2}} - \ln\left[az + (1 + a^2z^2)^{\frac{1}{2}}\right] = 2a(ax + y + b)$

2. $p(1 + q) = qz$

Ans. $\ln(az - 1) = x + ay + b$

3. $q^2 = z^2p^2(1 - p^2)$

Ans. $a^2z^2 = (y + ax + c)^2 + 1$

4. $p^3 + q^3 = 27z$

Ans. $(1 + a^3)z^2 = 8(x + ay + b)^3$

5. $z^2(p^2 + q^2 + 1) = \alpha^2$

Ans. $(1 + b^2)(\alpha^2 - z^2) = (x + by + c)^2$

6. $z^2 = 1 + p^2 + q^2$

Ans. $z = \cosh(x + ay + c/\sqrt{1 + a^2})$

7. $p(1 + q^2) = q(z - \alpha)$

Ans. $4a(z - \alpha) = 4 + (x + ay + c)^2$

8. $9(p^2z + q^2) = 4$

Ans. $(z + a^2)^3 = (x + ay + b)^2$

9. $z^2(p^2x^2 + q^2) = 1$

Hint: Put $X = \ln x$, equation reduces to $z^2\left[\left(\frac{\partial z}{\partial X}\right)^2 + \left(\frac{\partial z}{\partial y}\right)^2\right] = 1$.

Ans. $z^2\sqrt{1 + a^2} = \pm 2(\ln x + ay) + b$

10. $q^2y^2 = z(z - px)$

Hint: Put $X = \ln x$, $Y = \ln y$ then equation reduces to $Q^2 = z^2 - zP$

Ans. $xy^a \cdot b = z^{\frac{1}{k}}$

where $k = \left(-1 \pm \sqrt{1 + 4a^2}\right)/(2a^2)$

Form III: $f(x, p) = g(y, q)$

1. $p^2 \pm q^2 = x \pm y$

Ans. $z = \frac{2}{3}(a + x)^{\frac{3}{2}} + \frac{2}{3}(\mp a + y)^{\frac{3}{2}} + b$

2. $\sqrt{p} + \sqrt{q} = x + y$

Ans. $z = \frac{(a+x)^3}{3} + \frac{(y-a)^3}{3} + b$

3. $p + q = \sin x + \sin y$

Ans. $z = ax - \cos x - \cos y - ay + b$

4. $p^2 y(1 + x^2) = qx^2$

Ans. $z = a\sqrt{1+x^2} + \frac{1}{2}a^2y^2 + b$

5. $pe^y = qe^x$

Ans. $z = a(e^x + e^y) + b$

6. $p - q = x^2 + y^2$

Ans. $z = \frac{1}{3}(x^3 - y^3) + a(x + y) + b$

7. $y^2q^2 - xp + 1 = 0$

Ans. $z = (a^2 + 1)\ln x + a \ln y + b$

8. $z^2(p^2 + q^2) = x^2 + y^2$

Hint: Put $Z = \frac{1}{2}z^2$ equation reduces to $P^2 + Q^2 = x^2 + y^2$ where $P = zp$, $Q = zq$.

Ans. $z^2 = x\sqrt{x^2+a} + y\sqrt{y^2-a} + a \ln \frac{x+\sqrt{x^2+a}}{y+\sqrt{y^2-a}} + 2b$

9. $z(xp - yq) = y^2 - x^2$

Hint: Put $Z = \frac{z^2}{2}$ equation reduces to $xP - yQ = y^2 - x^2$ where $P = \frac{\partial Z}{\partial x}$, $Q = \frac{\partial Z}{\partial y}$.

Ans. $z^2 = 2a \ln xy - (x^2 + y^2) + 2b$

10. $z(p^2 - q^2) = x - y$

Hint: Put $Z = \frac{2}{3}z^{\frac{3}{2}}$, equation reduces to $P^2 - Q^2 = x - y$ where $P = \frac{\partial Z}{\partial x}$, $Q = \frac{\partial Z}{\partial y}$

Ans. $z^{\frac{3}{2}} = (a + x)^{\frac{3}{2}} + (a + y)^{\frac{3}{2}} + c.$

Form IV: $z = px + qy + f(p, q)$: Clairaut's equation

1. $2q(z - px - qy) = 1 + q^2$

Ans. $z = ax + by + \frac{b^2+1}{2b}$

2. $pqz = p^2(xq + p^2) + q^2(yp + q^2)$

Ans. $z = ax + by + \left(\frac{a^3}{b} + \frac{b^3}{a}\right)$

3. $z = px + qy \pm pq$

Ans. $z = ax + by \pm ab$

4. $(px + qy - z)^2 = d\,(1 + p^2 + q^2)$

Ans. $z = ax + by \pm \left(\sqrt{1 + a^2 + b^2}\right) d$

5. $(p + q)(z - xp - yq) = 1$

Ans. $z = ax + by + \frac{1}{a+b}$

6. $4xyz = pq + 2px^2y + 2q\,xy^2$

Hint: Put $X = x^2$, $Y = y^2$, equation reduces to $z = PX + QY + PQ$.

Ans. $z = ax^2 + by^2 + ab.$

18.5 CHARPIT'S METHOD

Charpit's method is a general method to find the complete solution of the first order non-linear P.D.E. of the form

$$f(x, y, z, p, q) = 0 \tag{1}$$

We know that

$$dz = \frac{\partial z}{\partial x}dx + \frac{\partial z}{\partial y}dy = p\,dx + q\,dy \tag{2}$$

Integrating (2), we get the complete solution of (1). In order to integrate (2), we must know p and q in terms of x, y, z. For this purpose, introduce another first order non-linear P.D.E. of the form

$$g(x, y, z, p, q, a) = 0 \tag{3}$$

involving an arbitrary constant "a" compatible with (1). Solving (1) and (3), we get

$$p = p(x, y, z, a), \quad q = q(x, y, z, a) \tag{4}$$

On substitution of (4) in (2), equation (2) becomes integrable, resulting in the complete solution of (1) in the form

$$F(x, y, z, a, b) = 0 \tag{5}$$

containing two arbitrary constants a and b.

Now differentiating (1) and (3) partially w.r.t. x and y and eliminating $\frac{\partial p}{\partial x}$ and $\frac{\partial q}{\partial x}$, we get after simplification, a Lagrange's linear equation of ϕ (as dependent variable) in terms of x, y, z, p, q (as independent variables) as

$$f_q\frac{\partial g}{\partial x} + f_q\frac{\partial g}{\partial y} + (pf_p + qf_q)\frac{\partial g}{\partial z} - (f_x + pf_z)\frac{\partial g}{\partial p} - (f_y + qf_z)\frac{\partial g}{\partial q} = 0 \tag{6}$$

The subsidiary equations of (6) are

$$\frac{dx}{f_p} = \frac{dy}{f_q} = \frac{dz}{pf_p + qf_q} = \frac{dp}{-(f_x + pf_z)} = \frac{dq}{-(f_y + qf_z)} \quad (7)$$

These Equations (7) are known as Charpit's equations. Solving (7), we get relations (4) of p and q, using which, the equation (2) is integrated resulting in the complete solution (5) of (1).

Note: *Not* all of the Charpit's Equations (7) need be used. Choose the simplest of (7) so that p and q are easily obtained.

WORKED OUT EXAMPLES

Example 1: Solve $z^2 = pq\,xy$.

Solution: Here $f(x, y, z, p, q) = z^2 - pq\,xy = 0$. Differentiating f partially w.r.t. x, y, z, p, q, form the auxiliary equations

$$\frac{dx}{f_p} = \frac{dy}{f_q} = \frac{dz}{pf_p + qf_q} = \frac{dp}{-(f_x + pf_z)} = \frac{dq}{-(f_y + qf_z)}$$

$$\frac{dx}{-qxy} = \frac{dy}{-pxy} = \frac{dz}{-2pqxy} = \frac{dp}{-(-pqy + p2z)} = \frac{dq}{-(-pqx + 2qz)}$$

Using the multipliers p, q, o, x, y, we have

$$\frac{p\,dx + x\,dp}{-pq\,xy + xpq\,y - 2pxz} = \frac{q\,dy + y\,dq}{-q\,pxy + ypq\,x - 2y\,qz} \quad \text{or}$$

$$\frac{p\,dx + x\,dp}{-2xpz} = \frac{q\,dy + y\,dq}{-2yqz}$$

$$\frac{d(xp)}{(xp)} = \frac{d(yq)}{(yq)}$$

Integrating $xp = a\,yq$.
Solving $q = \frac{xp}{ay}$. Substituting q in given P.D.E.

$$z^2 = p \cdot \left(\frac{xp}{ay}\right) xy = \frac{p^2x^2}{a} \quad \text{or } p = \sqrt{a} \cdot \frac{z}{x};$$

Then
$$q = \frac{xp}{ay} = \frac{x}{ay} \cdot \sqrt{a}\frac{z}{x} = \frac{z}{\sqrt{a}y}$$

Now
$$dz = p\,dx + q\,dy = \sqrt{a}\frac{z}{x}dx + \frac{1}{\sqrt{a}}\frac{z}{y}\,dy$$

$$\frac{dz}{z} = \sqrt{a}\,\frac{dx}{x} + \frac{1}{\sqrt{a}}\,\frac{dy}{y}.$$

Integrating, we get the complete solution as

$$z = b\,x^a\,y^{\frac{1}{a}}$$

where a and b are two arbitrary constants.

Example 2: Solve $2(z + xp + yq) = yp^2$.

Solution: Here $f = 2(z + xp + yq) - yp^2$
Forming the auxiliary equations

$$\frac{dx}{2x - 2yp} = \frac{dy}{2y} = \frac{dz}{2xp - 2yp^2 + 2qy} = \frac{dp}{-(2p + 2p)} = \frac{dq}{-(2q - p^2 + 2q)}$$

$$\frac{dx}{x - yp} = \frac{dy}{y} = \frac{dz}{xp - yP^2 + yq} = \frac{dp}{-2p} = \frac{dq}{-(2q - \frac{p^2}{2})}$$

Using second and fourth

$$\frac{dy}{y} = \frac{dp}{-2p} \quad \text{or} \quad p = ay^{-2} = \frac{a}{y^2}$$

Substituting p in the given P.D.E.

$$2\,yq = y\left(\frac{a}{y^2}\right)^2 - 2z - 2x\left(\frac{a}{y^2}\right) \quad \text{or}$$

$$q = \frac{a^2}{2y^4} - \frac{z}{y} - \frac{ax}{y^3}$$

$$\text{Now } dz = p\,dx + q\,dy = \frac{a}{y^2}dx + \left(\frac{a^2}{2y^4} - \frac{z}{y} - \frac{ax}{y^3}\right)dy$$

Regrouping the terms

$$\left(\frac{y\,dz + z\,dy}{y}\right) = \left(\frac{ay\,dx - ax\,dy}{y^3}\right) + \frac{a^2}{2y^4}\,dy$$

Multiplying throughout by y:

$$d(yz) = ad\left(\frac{x}{y}\right) + \frac{a^2}{2}\,\frac{dy}{y^3}.$$

Integrating
$$yz = a\frac{x}{y} + \frac{a^2}{2} \cdot \left(\frac{1}{-2y^2}\right) + b$$

$$z = \frac{ax}{y^2} - \frac{a^2}{4y^3} + \frac{b}{y}$$

is the required complete solution involving two arbitrary constants a and b.

EXERCISE

Solve (obtain the complete solution):

1. $16p^2z^2 + 9q^2z^2 + 4z^2 - 4 = 0$

   Hint: $\frac{dp}{32p^3z+18pq^2z+8pz} = \frac{dq}{32p^2qz+18q^3z+8qz} = \frac{-dx}{32pz^2} = \frac{-dy}{+18qz^2} = \frac{-dz}{32p^2z^2-18q^2z^2}$

   $4z\,dp + 0 \cdot dq + 1 \cdot dx + 0 \cdot dy + 4p\,dz = 0$, $x + 4pz = a$, $p = -\frac{x-a}{4z}$, $q = \frac{2}{3z}\sqrt{1 - z^2 - \frac{1}{4}(x-a)^2}$.

   Ans. $\frac{(x-a)^2}{4} + \frac{(y-b)^2}{9/4} + z^2 = 1$

2. $p(1+q^2) + (b-z)q = 0$.

   Hint: $\frac{dp}{pq} = \frac{dq}{q^2} = \frac{dz}{3pq^2+p+(b-z)q} = \frac{dx}{q^2+1} = \frac{dy}{-z+b+2pq}$ (i)(ii) $q = pc$, sub $q = \sqrt{c(z-b)-1}$.

   Ans. $2\sqrt{[c(z-b)-1]} = x + cy + a$; a, c are arbitrary constants.

3. $2xz - px^2 - 2qxy + pq = 0$.

   Hint: $q = a$, $p = \frac{2x(z-ay)}{x^2-a}$.

   Ans. $z - ay = b(x^2 - a)$

4. $q \mp px - p^2 = 0$.

   Hint: $q = a$, $p = \frac{1}{2}\left[\mp x \pm \sqrt{x^2+4a}\right]$.

   Ans. $z = \mp\frac{x^2}{4} \pm \frac{1}{2}\left[\frac{x}{2}\sqrt{x^2+4a} + 2a\ln\{x + \sqrt{x^2+4a}\}\right] + ay + b$

5. $px + qy \mp pq = 0$.

   Hint: $p = aq$, $q = -\frac{(y+ax)}{a}$, $p = -(y+ax)$.

   Ans. $az = -\frac{1}{2}(y+ax)^2 + b$

6. $qz - p^2y - q^2y = 0$.

   Hint: $p^2 + q^2 = a$, $q = \frac{ay}{z}$, $p^2 = \frac{az^2-a^2y^2}{z^2}$.

   Ans. $z^2 = a(y^2 + (x+b)^2)$

7. $\frac{p^2+qy}{2} = -(z+y^2)$.

   Hint: $p = -x+a$, $q = \frac{1}{y}[-2z-2y^2-(a-x)^2]$.

   Ans. $y^2[(x-a)^2 + 2z + y^2] = b$

8. $yz - p(xy+q) - qy = 0$.

   Hint: $p = a$, $q = \frac{y(z-ax)}{a+y}$.

   Ans. $(z-ax)(y+a)^a = be^y$

9. $2(xy - px - qy) + p^2 + q^2 = 0$.

   Hint: $dp+dq = dx+dy$, $(p-x)+(q-y) = a$

   $$p - x = 2a \pm \frac{\sqrt{4a^2 - 8\{a^2 - (x-y)^2\}}}{4}$$

   $$q = y + \frac{1}{2}\left[a \pm \sqrt{\{2(x-y)^2 - a^2\}}\right].$$

   Ans. $z = \frac{x^2}{2} + \frac{y^2}{2} - \frac{a}{2}(x+y) \pm \frac{1}{\sqrt{2}}\left(\frac{x-y}{2}\sqrt{\left\{(x-y)^2 - \frac{a^2}{2}\right\}} - \frac{a^2}{4}\ln\left[(x-y) + \sqrt{\left\{(x-y)^2 - \frac{a^2}{2}\right\}}\right]\right.$

10. $z - q^2y - p^2x = 0$.

    Hint: $2px\,dp + p^2dx = 2qy\,dq + q^2dy$, $p^2x = q^2ya$, $q^2 = \left[\frac{z}{1+a}\right]$, $p^2 = \frac{za}{(1+a)x}$.

    Ans. $\sqrt{(1+a)}\sqrt{z} = \sqrt{a}\sqrt{x} + \sqrt{y} + b$.

18.6 HOMOGENEOUS LINEAR PARTIAL DIFFERENTIAL EQUATIONS WITH CONSTANT COEFFICIENTS

Consider a partial differential equation of the form

$$\left(A_0\frac{\partial^n z}{\partial x^n} + A_1\frac{\partial^n z}{\partial x^{n-1}\partial y} + \ldots + A_n\frac{\partial^n z}{\partial y^n}\right) + \left(B_0\frac{\partial^{n-1} z}{\partial x^{n-1}} + B_1\frac{\partial^{n-1} z}{\partial x^{n-2}\partial y} + \ldots + B_{n-1}\frac{\partial^{n-1} z}{\partial y^{n-1}}\right) + \cdots + \left(M_0\frac{\partial z}{\partial x} + M_1\frac{\partial z}{\partial y}\right) + N_0 z = 0 \quad (1)$$

Here $A_0, A_1, \cdots A_n, B_0, B_1, \cdots, B_{n-1}, M_0, M_1, N_0$ are all constants. In this equation the dependent variable z and its derivatives are linear. Since each term (in the LHS) of (1) contains z or its derivatives, equation (1) is known as as homogeneous* linear partial differential equation of order n with constant coefficients. Introducing the notation $D_x = \frac{\partial}{\partial x}$ and $D_y = \frac{\partial}{\partial y}$ equation (1) takes the form $(A_0D_x^n + A_1D_x^{n-1}D_y + \cdots + A_nD_y^n)z + (B_0D_x^{n-1} + B_1D_x^{n-2}D_y + \cdots + B_{n-1}D_y^{n-1})z +$

*Some authors call an equation homogeneous of *each* term in the equation is of the *same order*.

$$\cdots + (M_0 D_x + M_1 D_y)z + N_0 z = 0$$

or
$$F(D_x, D_y)z = 0 \tag{2}$$

where $F(D_x, D_y)$ is a linear differential operator given by

$$F(D_x, D_y) = A_0 D_x^n + \cdots + A_n D_y^n + \cdots + M_0 D_x + M_1 D_y + N_0 \tag{3}$$

Reducible

The differential operator (3) is said to be reducible if (3) can be written as the product of linear factors of the form $(a_i D_x + b_i D_y + c_i)$ where a_i, b_i, c_i are constants. Otherwise (3) is said to be *irreducible*. Here we consider only reducible case. Then the reducible equation (2) can be written as $F(D_x, D_y)z = (a_1 D_x + b_1 D_y + c_1) \ldots (a_n D_x + b_n D_y + c_n) = 0$

$$= \prod_{i=1}^{n} (a_i D_x + b_i D_y + c_i)z = 0 \tag{4}$$

Here the order in which the linear factors in the LHS of (4) occurs is immaterial.

Book Work 1: Prove that the general solution (G.S.) of the first order linear homogeneous partial differential equation

$$(aD_x + bD_y + c)z = 0 \tag{5}$$

is given by

$$\boxed{z = e^{-cx/a}\phi(ay - bx), \ (a \neq 0)} \tag{6}$$

or

$$\boxed{z = e^{-cy/b}\psi(ay - bx), \ (b \neq 0)} \tag{7}$$

Proof: The given equation is of the form $P(x)p + Q(x)q = R(x)$ where $P(x) = a$, $Q(x) = b$, $R(x) = -cz$. Then the auxiliary system of (5) is

$$\frac{dx}{a} = \frac{dy}{b} = \frac{dz}{-cz}$$

From $\frac{dx}{a} = \frac{dy}{b}$, we get $ay - bx = A =$ const. From $\frac{dx}{a} = \frac{dz}{-cz}$, we get $\ln z = -\frac{c}{a}x + \ln B$ provided $a \neq 0$. Then $z = Be^{-cx/a}$. Thus the general solution of (5) involving one arbitrary function ϕ is

$$z = e^{-cx/a}\phi(ay - bx), \qquad a \neq 0$$

Instead from $\frac{dy}{b} = \frac{dz}{-cz}$, we get $z = ce^{-cx/b}$ provided $b \neq 0$. Thus the general solution of (5) involving one arbitrary function ψ is

$$z = e^{-cy/b}\psi(ay - bx), \qquad b \neq 0$$

Corollary 1: If $c = 0$, then the general solution

of
$$(aD_x + bD_y)z = 0 \tag{8}$$
is
$$z = \phi(ay - bx) \tag{9}$$

Corollary 2: If $c \neq 0, a = 0, b \neq 0$, then the general solution of

$$(bD_y + c)z = 0 \tag{10}$$

is
$$z = e^{-cy/b}\psi(x) \tag{11}$$

Corollary 3: If $c \neq 0, a \neq 0, b = 0$. Then G.S. of

$$(aD_x + c)z = 0 \tag{12}$$

is
$$z = e^{-cx/a}\phi(y) \tag{13}$$

Corollary 4: If $c = 0, a = 0, b \neq 0$ then G.S. of

$$bD_y z = 0 \tag{14}$$

is
$$z = \phi(x) \tag{15}$$

Corollary 5: If $c = 0, b = 0, a \neq 0$, then G.S. of

$$aD_x z = 0 \tag{16}$$

is
$$z = \psi(y) \tag{17}$$

Recall the *linearity principle* which states that if $z_1, z_2, \ldots, z_n$ are linearly independent solutions of the homogeneous linear partial differential equation $F(D_x, D_y)z = 0$, then $z = \sum_{i=1}^{n} c_i z_i$ is also a solution since

$$F(D_x, D_y)z = F(D_x, D_y)\left(\sum_{i=1}^{n} c_i z_i\right)$$

$$= \sum_{i=1}^{n} F(D_x, D_y)(c_i z_i)$$

$$= \sum_{i=1} c_i F(D_x, D_y) z_i = 0$$

Here $c_1, c_2, \cdots, c_n$ are arbitrary constants. Using the linearity principle and book work 1 which provides general solution to each linear factor of the form $(a_i D_x + b_i D_y + c_i)$, we now get the general solution of the linear homogeneous equation (4) as the sum of n arbitrary functions given by $z = e^{-\alpha_1 x}\phi_1(a_1 y - b_1 x) + e^{-\alpha_2 x}\phi_2(a_2 y - b_2 x) + \cdots + e^{-\alpha_n x}\phi_n(a_n y - b_n x)$

or $$z = \sum_{i=1}^{n} e^{-\alpha_i x}\phi_i(a_i y - b_i x) \tag{18}$$

where $\alpha_i = c_i/a_i$. We can also write the general solution of (4) as

or $$z = \sum_{i=1}^{n} e^{-\beta_i y}\psi_i(a_i y - b_i x) \tag{19}$$

where $\beta_i = c_i/b_i$.

Multiple Factors

The solution corresponding to a multiple factor say $(a_i D_x + b_i D_y + c_i)^2$ of multiplicity z, in the decomposition of $F(D_x, D_y)$ into linear factors is given by

$$z = e^{-c_i x/a_i}[\phi_1(a_i y - b_i x) + x\phi_2(a_i y - b_i x)].$$

Extending this to a factor $(a_i D_x + b_i D_y + c_i)^k$ of multiplicity k, the corresponding solution is given by

$$z = e^{-c_i x/a_i}[\phi_1(a_i y - b_i x) + x\phi_2(a_i y - b_i x) + \cdots + x^{k-1}\phi_k(a_i y - b_i x)] \tag{20}$$

or

$$z = e^{-c_i x/a_i}\sum_{j=1}^{k} x^{j-1}\phi_j(a_i y - b_i x).$$

Special case:

When the total degree of each term in $F(D_x, D_y)$ is same (i.e., the derivatives involved are *all* of the same order) then the above analysis gets simplified. Replacing D_x by m and D_y by 1 (or equivalently D_x/D_y by m) in the differential operator $F(D_x, D_y)$ we get an *auxiliary equation* $(A.E.)$ of the form

$$F(m, 1) = 0 \tag{21}$$

which is an nth degree polynomial in m having n roots. Then the general solution (G.S.) of $F(D_x, D_y)z = 0$ is given as follows depending on the nature of the roots of the auxiliary equation (21).

Case i : n distinct real roots: $m_1 \neq m_2 \neq \cdots \neq m_n$ then the general solution is

$$z = \phi_1(y + m_1 x) + \phi_2(y + m_2 x) + \cdots + \phi_n(y + m_n x).$$

Case ii : One root of multiplicity k, so that

$$m_1 = m_2 = \cdots = m_k \neq m_{k+1} \neq \cdots \neq m_n.$$

Then G.S. is

$$z = \phi_1(y + m_1 x) + x\phi_2(y + m_1 x) + x^2\phi_3(y + m_1 x) + \cdots + x^{k-1}\phi_k(y + m_1 x) + \phi_{k+1}(y + m_{k+1}x) + \cdots + \phi_n(y + m_n x).$$

Case iii : If m_1, m_2 are complex conjugate pair say $m_1 = a + bi$, $m_2 = a - bi$ then general solution is

$$z = \phi_1(y + ax + ibx) + \phi_1(y + ax - ibx) + i\{\phi_2(y + ax + ibx) - \phi_2(y + ax - ibx)\} + \phi_3(y + m_3 x) + \cdots + \phi_n(y + m_n x).$$

Worked Out Examples

Complementary functions

Example 1: Find the complementary function of the following P.D.E. (a to g)

(a) $(D_x^3 - 3D_x^2 D_y + 2D_y^2 D_x)z = 0$

Solution: Auxiliary equation (A.E.) is obtained by replacing D_x by m and D_y by 1. So the A.E. is $m^3 - 3m^2 \cdot 1 + 2 \cdot 1^2 \cdot m = 0$ or $m(m-1)(m-2) = 0$ i.e., roots are $m = 0, 1, 2$. Then the complementary function (C.F.) is

$$z = \phi_1(y + 0 \cdot x) + \phi_2(y + x) + \phi_3(y + 2x)$$

(b) $25r - 40s + 16t = 0$ or $(25D_x^2 - 40D_{xy} + 16D_y^2)z = 0$

Solution: A.E.: $25m^2 - 40m + 16 = 0$ or $(5m-4)^2 = 0$ so double roots are $m = \frac{4}{5}, \frac{4}{5}$. Then C.F. is

$$z = \phi_1\left(y + \frac{4}{5}x\right) + x\phi_2\left(y + \frac{4}{5}x\right)$$

or $z = f_1(5y+4x) + x f_2(5y+4x)$

(c) $(D_x^4 + D_x^3 D_y - 3D_x^2 D_y^2 - 5D_x D_y^3 - 2D_y^4)z = 0$

Solution: A.E. is $m^4 + m^3 - 5m - 2 = 0$, or $(m-2)(m+1)^3 = 0$ so roots are $m = 2, -1, -1, -1$ with -1 repeated three times. Then C.F. is

$$z = \phi_1(y+2x) + \phi_2(y-x) + x\phi_3(y-x) + x^2\phi_4(y-x)$$

(d) $r + b^2 t = 0$ or $(D_x^2 + b^2 D_y^2)z = 0$

Solution: A.E.: $m^2 + b^2 = 0$ with complex conjugate roots $m = \pm b_i$. The complementary function is

$$z = \phi_1(y+ax+ibx) + \phi_1(y+ax-ibx) + i[\phi_2(y+ax+ibx) - \phi_2(y+ax-ibx)]$$

Here $a = 0$. So the C.F. is

$$z = \phi_1(y+ibx) + \phi_1(y-ibx) + i[\phi_2(y+ibx) - \phi_2(y-ibx)]$$

(e) $(D_x + 2D_y - 3)(D_x + D_y - 1)z = 0$

Solution: If $aD_x + bD_y + c$ is a factor then the solution is $z = e^{-\frac{c}{a}x}\phi(bx - ay)$.
For the first factor $a = 1, b = 2, c = -3$ and for the second factor $a = 1, b = 1, c = -1$. So the C.F. is

$$z = e^{-\left(\frac{-3}{1}\right)x}\phi_1(2x-y) + e^{-\frac{(-1)x}{1}} \cdot \phi_2(x-y).$$

or

$$z = e^{3x}\phi_1(2x-y) + e^{x}\phi_2(x-y)$$

(f) $(D_x + 2D_y)(+D_x + 3D_y + 1)(D_x + 2D_y + 2)^2 z = 0$

Solution: Here the last factor is repeated twice. C.F. is

$$z = e^{0.x}\phi_1(2x-y) + e^{-1.x}\phi_2(3x-y) + e^{-2x}\{\phi_3(2x-y) + x\phi_4(2x-y)\}$$

(g) Irreducible:

$$(D_x^2 + D_x D_y - D_y^2 + D_x - D_y)z = 0$$

Solution: C.F. is $z = \sum_{i=1}^{\infty} c_i e^{a_i x + b_i y}$ where a_i, b_i satisfy the equation $a_i^2 + a_i b_i - b_i^2 + a_i - b_i = 0$ and c_i are arbitrary constants.

EXERCISE

Solve the following:

1. $(D_x^2 - D_x D_y - 6D_y^2)z = 0$

Ans. $z = f(y-2x) + g(y+3x)$

Hint: Roots of auxiliary equation are $-2, 3$

2. $(D_x^3 - D_x^2 D_y - 8D_x D_y^2 + 12D_y^3)z = 0$

Ans. $z = f(y+2x) + xg(y+2x) + h(y-3x)$

Hint: Roots are 2, 2 (double), -3

3. $(D_x^4 - D_x^3 D_y + 2D_x^2 D_y^2 - 5D_x D_y^3 + 3D_y^4)z = 0$

Ans. $z = f_1(y+x) + xf_2(y+x) + f_3\left[y - \frac{1}{2}(1+i\sqrt{11}x)\right] + f_3\left[y - \frac{1}{2}(1-i\sqrt{11}x)\right] + i\left[f_4\left\{y - \frac{1}{2}(1+i\sqrt{11}x)\right\} - f_4\left\{y - \frac{1}{2}(1-i\sqrt{11}x)\right\}\right]$

Hint: Roots 1, 1 (double), $\frac{-1}{2}(1 \pm i\sqrt{11})$ (complex conjugate).

4. $(2D_x^2 + 5D_x D_y + 2D_y^2)z = 0$

Ans. $z = f(y-2x) + g\left(y - \frac{1}{2}x\right)$
or $= f(y-2x) + h(2y-x)$

Hint: Roots $-2, -\frac{1}{2}$

5. $(D_x^2 + 6D_x D_y + 9D_y^2) = 0$

Ans. $z = f(y-3x) + xg(y-3x)$

Hint: $m = -3, -3$ (repeated root)

6. $(D_x^4 - D_y^4)z = 0$

Ans. $z = f_1(y+x) + f_2(y-x) + f_3(y+ix) + f_3(y-ix) + i[f_4(y+ix) - f_4(y-ix)]$

7. $(D_x^2 - a^2 D_y^2)z = 0$

Ans. $z = f(y+ax) + g(y-ax)$

Hint: Roots $\pm a$

8. $(9D_x^2 + 24D_x D_y + 16D_y^2)z = 0$

Ans. $z = f\left(y - \frac{4}{3}x\right) + xg\left(y - \frac{4}{3}x\right)$
or $z = f_1(3y-4x) + xg_1(3y-4x)$

Hint: Roots: $-\frac{4}{3}, -\frac{4}{3}$ (repeated)

9. $(2D_x + D_y + 1)(D_x^2 + 3D_xD_y - 3D_x)z = 0$

Ans. $z = f_1(y) + e^{-x/2}.f_2(2y - x) + e^{3x} f_3(y - 3x)$

or

$z = f_1(y) + e^{-y} f_2(2y - x) + f_3(y - 3x)$

10. $(2D_x + D_y + 5)(D_x - 2D_y + 1)^2 z = 0$

Ans. $z = e^{-5y} f_1(2y - x) + e^{-x}[f_2(y + 2x) + x f_3(y + 2x)]$

11. $(D_x^2 - D_y^2 + 3D_x - 3D_y)z = 0$

Ans. $z = f_1(y + x) + e^{-3x} f_2(y - x)$

12. $(2D_x + 3D_y - 1)^2(D_x - 3D_y + 3)^3 z = 0$

Ans. $z = e^{\frac{x}{2}}[f_1(2y - 3x) + x f_2(2y - 3x)] + e^y[f_3(y + 3x) + y f_4(y + 3x) + y^2 f_5(y + 3x)]$

13. $(D_x^4 + D_y^4 - 2D_x^2D_y^2)z = 0$ (biharmonic equation)

Ans. $z = x f_1(x - y) + f_2(x - y) + x\ f_3(x + y) + f_4(x + y)$

Hint: Root 1, −1 (both repeated twice)

14. $(D_x^3 - 3D_x^2D_y + 2D_xD_y^2)z = 0$

Ans. $z = f_1(y) + f_2(y + x) + f_3(y + 2x)$

Hint: Roots: 0, 1, 2

15. $(D_x^3 - 6D_x^2D_y + 11D_xD_y^2 - 6D_y^3)z = 0$

Ans. $z = f_1(y + x) + f_2(y + 2x) + f_3(y + 3x)$

18.7 NON-HOMOGENEOUS LINEAR PARTIAL DIFFERENTIAL EQUATIONS WITH CONSTANT COEFFICIENTS

A partial differential equation of the form

$$F(D_x, D_y) = f(x, y) \tag{1}$$

is known as a non-homogeneous linear equation with constant coefficients. The general solution of the corresponding homogeneous equation

$$F(D_x, D_y) = 0 \tag{2}$$

is known as the "complementary function (C.F.)" of (1). *Any* solution of (1) is known as the "particular integral" (P.I.) of (1). Then the general solution of the non-homogeneous equation (1) is the sum of the complementary function z_c and the particular integral z_p. Thus

$$\text{General solution} = \text{complementary function} + \text{particular integral}$$

$$(\text{G.S} = \text{C.F.} + \text{P.I.}) \tag{3}$$

Methods of obtaining complementary function were considered in detail in this section.

Methods of Obtaining Particular Integral

Suppose $F(D_x, D_y)$ is factored into m linear factors so that (1) takes the form

$$F(D_x, D_y) = (D_x - m_1D_y)(D_x - m_2D_y)\ldots$$
$$\ldots(D_x - m_nD_y) = f(x, y)$$

Then the particular integral is

$$z_p = \frac{1}{(D_x - m_1D_y)\ldots(D_x - m_nd_y)} f(x, y) \tag{4}$$

or $$z_p = \left(\frac{1}{(D_x - m_1D_y)}\right)\left(\frac{1}{D_x - m_2D_y)}\right)\cdots$$
$$\cdots\left(\frac{1}{D_x - m_{n-1}D_y}\right)\left(\frac{1}{D_x - m_nD_y}\right) f(x, y)$$

By introducing $u_1 = \frac{1}{D_x - m_nD_y} f(x, y)$, $u_2 = \frac{1}{D_x - m_{n-1}D_y} u_1, \ \ldots, \ z = u_n = \frac{1}{D_x - m_1D_y} u_{n-1}$ the solution of (4) reduces to solution of above n first order equations which can be solved by the following, book work.

Book work: Prove that the particular integral of $p - mq = g(x, y)$ is given by $z = \int g(x, a - mx)dx$ where a is replaced by $y + mx$ *after integration.* Also the constant of integration is omitted.

Proof: The auxiliary system of the first order linear equation

$$p - mq = g(x, y)$$

is $$\frac{dx}{1} = \frac{dy}{-m} = \frac{dz}{g(x, y)}$$

From $\frac{dx}{1} = \frac{dy}{-m}$, we get $y + mx = \text{constant} = a$.

From $\frac{dx}{1} = \frac{dz}{g(x,y)}$ we get

$$z = \int g(x, y)dx$$

Substituting $y = a - mx$,

$$z = \int g(x, a - mx)dx$$

Here the constant of integration is omitted and a is replaced by $y + mx$ *after* integration.

Special case: If $F(D_x, D_y)$ is a homogeneous function of D_x and D_y of degree n and RHS function $f(x, y)$ is of the form $\phi(ax + by)$, then a particular integral of equation

$$F(D_x, D_y) = f(x, y) \qquad (1)$$

is given by $z_p = \frac{1}{F(D_x,D_y)}\phi(ax + by)$

or $\boxed{z_p = \frac{1}{F(a,b)} \int\int \ldots \int \phi(v)dv^n}$, $F(a, b) \neq 0$

where $v = ax + by$, i.e., integrate $\phi(v)$ w.r.t v, n times and *after* integration replace v by $ax + by$, provided $F(a, b) \neq 0$.

If $F(a, b) = 0$ so that

$$F(D_x, D_y) = (bD_x - aD_y)^m G(D_x, D_y), G(a, b) \neq 0,$$

then the particular integral is

$$z_p = \frac{1}{(bD_x - aD_y)^m}\phi(ax + by)$$

$$\boxed{z_p = \frac{x^m}{b^m m!}\phi(ax + by)}$$

Short Methods of Obtaining Particular Integrals

The inverse operator short methods of obtaining particular integrals similar to those used in ordinary differential equations are listed here.

Case 1: $f(x, y) = e^{ax+by}$

$$\text{P.I.} : z = \frac{1}{F(D_x, D_y)}e^{ax+by} = \frac{1}{F(a, b)}e^{ax+by}$$

provided $F(a, b) \neq 0$.

Case 2: If $F(a, b) = 0$ then $F(D_x, D_y) = (D_x - \frac{a}{b}D_y)^r g(D_x, D_y)$ so that $g(a, b) \neq 0$. Thus

$$\text{P.I.} : z = \frac{1}{(D_x - \frac{a}{b}D_y)^r} \frac{1}{g(D_x, D_y)} e^{ax+by}$$
$$= \frac{1}{g(a, b)} \cdot \frac{x^r}{r!}e^{ax+by}$$

provided $g(a, b) \neq 0$.

Case 3: $f(x, y) = \sin(ax + by)$ or $\cos(ax + by)$. Replace D_x^2 by $-a^2$, D_y^2 by $-b^2$ and $D_x \cdot D_y$ by $-ab$. Thus

$$\frac{1}{F(D_x^2, D_xD_y, D_y^2)} \sin(ax + by)$$
$$= \frac{1}{F(-a^2, -ab, -b^2)} \sin(ax + by)$$

and

$$\frac{1}{F(D_x^2, D_xD_y, D_y^2)} \cos(ax + by)$$
$$= \frac{1}{F(-a^2, -ab, -b^2)} \cos(ax + by)$$

provided $F(-a^2, -ab, -b^2) \neq 0$.

Note: If $F(-a^2, -ab, -b^2) = 0$, then use the Book work stated above.

Case 4: $f(x, y) = x^m y^n$ where m and n are positive constants. Then particular integral is

$$z_p = \frac{1}{F(D_x, D_y)}x^m y^n = [F(D_x, D_y)]^{-1} x^m y^n$$

(a) If $n < m$, $\frac{1}{F(D_x,D_y)}$ is expanded in powers of $\frac{D_y}{D_x}$.

(b) If $m < n$, $\frac{1}{F(D_x,D_y)}$ is expanded in powers of $\frac{D_x}{D_y}$.

Note: Although the answers in (a) and (b) are different, the difference can be merged into the arbitrary functions of C.F.

Case 5: Exponential shift: $f(x, y) = e^{ax+by} \cdot V(x, y)$ where $V(x, y)$ is any function of x and y. Then

$$z_p = \frac{1}{F(D_x, D_y)} \left\{e^{ax+by} \cdot V(x, y)\right\}$$
$$= e^{ax+by} \frac{1}{F(D_x + a, D_y + b)} V(x, y)$$

WORKED OUT EXAMPLES

Example 1: Obtain the particular solution of $p - 2q = \sin(x + 2y)$

Solution: $\frac{dx}{1} = \frac{dy}{-2} = \frac{dz}{\sin(x+2y)}$

Integrating $y + 2x = c_1$. From (1) and (3), we have $z = \int \sin(x + 2y)dx = \int \sin(x + 2(c_1 - 2x))dx$ since $y = c_1 - 2x$. So $z = \int \sin(2c_1 - 3x)dx = \frac{\cos(3x-2c_1)}{3}$.
Replacing c_1 by $y + 2x$, $z = \frac{\cos(3x-2(y+2x))}{-3} = \frac{\cos(x+2y)}{3}$.
Thus the particular integral is $3z = \cos(x + 2y)$.

Example 2: Solve $(D_x^2 + 5D_xD_y + 6D_y^2)z = e^{x-y}$

Solution: A.E. is $m^2 + 5m + 6 = 0$ or $(m + 2)(m + 3) = 0$. So C.F. is $z_c = \phi_1(y - 2x) + \phi_2(y - 3x)$.

P.I. is $z_p = \frac{1}{f(D_x, D_y)} e^{x-y} = \frac{1}{(D_x^2 + 5D_xD_y + 6D_y^2)} e^{x-y}$. Here $a = 1$, $b = -1$, so replace D_x by $a = 1$, D_y by $b = -1$. Then

$$z_p = \frac{1}{1^2 + 5\cdot 1(-1) + 6(-1)^2} e^{x-y} = \frac{1}{2} e^{x-y}.$$

Then the general solution is

$$z = C.F. + P.I. = \phi_1(y - 2x) + \phi_2(y - 3x) + \frac{1}{2}e^{x-y}$$

Example 3: Solve $4r + 12s + 9t = e^{3x-2y}$

Solution: P.D.E. is $(4D_x^2 + 12D_xD_y + 9D_y^2)z = e^{3x-2y}$.
A.E. is $4m^2 + 12m + 9 = 0$ or $(2m + 3)^2 = 0$ i.e., $m = -\frac{3}{2}, -\frac{3}{2}$. (repeated roots)
So C.F. is $z_c = f_1\left(y - \frac{3}{2}x\right) + xf_2\left(y - \frac{3}{2}x\right) = \phi_1(2y - 3x) + x\phi_2(2y - 3x)$.

P.I.: $z_p = \frac{1}{4D_x^2 + 12D_xD_y + 9D_y^2} e^{3x-2y} = \frac{1}{4\left[D_x - \left(-\frac{3}{2}\right)D_y\right]^2} e^{3x-2y}$

Here $a = 3$, $b = -2$. Since $-\frac{3}{2}$ is a (double) root of the AE this is a failure case and D_x and D_y can *not* be replaced by 3 and -2 respectively because $f(+3, -2) = 0$. Applying the B.W. (on page 18.22) we get
P.I.: $z_p = \frac{1}{4} \cdot \frac{x^2}{2!} e^{3x-2y}$. Thus G.S $= z =$ C.F. + P.I. $= \phi_1(2y - 3x) + x\phi_2(2y - 3x) + \frac{1}{8}x^2e^{3x-2y}$

Aliter: Put $u = \left(D_x + \frac{3}{2}D_y\right) z$, then PDE reduces to

$$(4D_x^2 + 12D_xD_y + 9D_y^2)z = 4\left(D_x + \frac{3}{2}D_y\right)^2 z$$
$$= 4\left(D_x + \frac{3}{2}D_y\right) u = e^{3x-2y}.$$

Solution is $u = \int F(x, c - mx)dx$. Here $y - \frac{3}{2}x = c_1$ or $2y - 3x = c$. Then $u = \int e^{3x-2y}dx = \int e^{3x-(c+3x)}dx = e^{-c}\int dx = xe^{-c}$.
Thus $u = xe^{-2y+3x}$. Now $4\left(D_x + \frac{3}{2}D_y\right) z = u = xe^{3x-2y}$. Solution by BW (see page 18.22) $z = \frac{1}{4}\int F(x, c - mx)dx$. Here $2y - 3x = c$.

$$z = \frac{1}{4}\int xe^{3x-2y}dx = \frac{1}{4}\int xe^{-c}dx = \frac{e^{-c}}{4}\int x\,dx =$$
$$= e^{-c}\frac{x^2}{2} = \frac{1}{4}\frac{x^2}{2}.e^{-c} = \frac{x^2}{8}e^{3x-2y}$$

Example 4: Solve $\frac{\partial^2 z}{\partial x^2} - 2\frac{\partial^2 z}{\partial x \partial y} + \frac{\partial^2 z}{\partial y^2} = \sin x$.

Solution: A.E.: $m^2 - 2m + 1 = 0$ or $(m - 1)^2 = 0$ with repeated (double) roots 1, 1. So C.F. is

$$z_c = \phi_1(y + x) + x\phi_2(y + x).$$

$$\text{P.I. : } z_p = \frac{1}{D_x^2 - 2D_xD_y + D_y^2} \cdot \sin x.$$

Here $a = 1$, $b = 0$. Replace D_x^2 by $-m^2 = -1^2 = -1$, $D_y^2 = -n^2 = 0$ and $D_xD_y = -m \cdot n = -1.0 = 0$, then

$$P.I. : z_p = \frac{1}{-1 - 0 - 0} \sin x = -\sin x.$$

Thus

$$z = \text{G.S.} = \text{C.F.} + \text{P.I.} = \phi_1(y + x) + x\phi_2(y + x) - \sin x$$

Example 5: Solve $r - 2s = \sin x \cdot \cos 2y$

Solution: PDE is $\frac{\partial^2 z}{\partial x^2} - 2\frac{\partial^2 z}{\partial x \partial y} = \sin x \cdot \cos 2y$.

AE is $m^2 - 2m = m(m-2) = 0$ i.e., $m = 0, 2$

C.F. $:z_c = f_1(y+0x) + f_2(y+2x)$

P.I. $:z_p = \frac{1}{D_x^2 - 2D_x D_y} \sin x \cdot \cos 2y$

$$= \frac{1}{2} \cdot \frac{1}{D_x^2 - 2D_x D_y}[\sin(x+2y) + \sin(x-2y)]$$

Using the result (on page 18.23) (with $a = 1, b = 2$ for the first term and $a = 1, b = -2$ for the second term)

$$= \frac{1}{2}\left[\frac{1}{-1^2 - 2\cdot 0(-2)}\sin(x+2y) + \right.$$

$$\left. + \frac{1}{-1^2 - 2.0(+2)}\sin(x-2y)\right]$$

$$= +\frac{1}{2}\left[\frac{1}{3}(\sin x \cdot \cos 2y + \sin 2y \cdot \cos x)\right.$$

$$\left. -\frac{1}{5}(\sin x \cdot \cos 2y - \sin 2y \cdot \cos x)\right]$$

$$= \frac{1}{15}\sin x \cdot \cos 2y + \frac{4}{15}\sin 2y \cdot \cos x$$

$$z = \text{G.S.} = \text{C.F.} + \text{P.I.} = f_1(y) + f_2(y+2x)$$

$$+ \frac{1}{15}(\sin x \cos 2y + 4 \sin 2y . \cos x)$$

Example 6: Solve $(2D_x^2 - 5D_x D_y + 2D_y^2)z = 5\sin(2x+y)$

Solution: A.E. is $2m^2 - 5m + 2 = 0$ or $(2m-1)(m-2) = 0$ i.e., $m = 2, \frac{1}{2}$. So C.F. is $z_c = \phi_1(y + \frac{1}{2}x) + \phi_2(y+2x)$ or

$z_c = \phi_3(2y + x) + \phi_2(y+2x)$.

Here $a = 2, b = 1$. Note $a = 2$ is a root of the A.E.

$$\text{P.I.} : z_p = \frac{1}{2D_x^2 - 5D_x D_y + 2D_y^2} 5 \cdot \sin(2x+y)$$

$$= \frac{5}{2}\frac{1}{D_x^2 - \frac{5}{2}D_x D_y + D_y^2}\sin(2x+y)$$

$$= \frac{5}{2}\frac{1}{(D_x - \frac{1}{2}D_y)(D_x - 2D_y)}\sin(2x+y)$$

At $D_x^2 = -a^2 = -4$, $D_y^2 = -b^2 = -1$, $D_x D_y = -a \cdot b = -2$, $f(D_x, D_y) = 2D_x^2 - 5D_x D_y + 2D_y^2 = 0$. So rewriting

$$(2D_x - D_y)(D_x - 2D_y)z = 5\sin(2x+y)$$

Put $u = (D_x - 2D_y)z$ then $(2D_x - D_y)u = 5\sin(2x+y)$

$$u = \frac{5}{2D_x - D_y} \cdot \sin(2x+y).$$

Here $2y + x = x$, or $y = \frac{c-x}{2}$.

$$u = 5\int \sin\left(2x + \frac{c-x}{2}\right)dx$$

$$= 5\int \sin\left(\frac{3x+c}{2}\right)dx$$

Integrating, $u = 5 \cdot \frac{2}{3}\left(-\cos\left(\frac{3x+c}{2}\right)\right)$

$$= \frac{-10}{3} \cdot \cos\left(\frac{3x+2y+x}{2}\right) = \frac{-10}{3}\cos(2x+y)$$

Now P.I.:$= z_p = \frac{1}{D_x - 2D_y}[u] =$

$$\frac{1}{D_x - 2D_y}\left[\frac{-10}{3}\cos(2x+y)\right]$$

Using $y + 2x = c$,

$$z_p = \int \frac{-10}{3} \cdot \cos c dx = \frac{-10}{3}\cos c \int dx$$

$$= \frac{-10}{3} \cdot x \cdot \cos c = \frac{-10}{3} \cdot x \cdot \cos(y+2x)$$

Thus G.S. = C.F. + P.I.

$$z = f_1(y+2x) + f_2(2y x) - \frac{10}{3} \cdot x \cdot \cos(y+2x)$$

Example 7: Solve $(D_x^2 + D_y^2)z = x^2y^2$

Solution: A.E.: $m^2 + 1 = 0$ or $m = \pm i$. So

C.F.:$z_c = \phi_1(y+ix) + \phi_1(y-ix)$

$+i[\phi_2(y+ix) - \phi_2(y-ix)]$

P.I.: $z_p = \frac{1}{D_x^2 + D_y^2}x^2y^2 = \frac{1}{D_x^2}\frac{1}{\left[1 + \left(\frac{D_y}{D_x}\right)^2\right]}x^2y^2$

Expanding by binomial series

$$= \frac{1}{D_x^2}\left[1 - \left(\frac{D_y}{D_x}\right)^2 + \left(\frac{D_y}{D_x}\right)^4 + \ldots\right]x^2y^2$$

$$= \frac{1}{D_x^2}(x^2y^2) - \frac{1}{D_x^4}D_y^2(x^2y^2) + 0 + \ldots$$

Integrating and differentiating

$$= \frac{1}{D_x}\frac{x^3}{4}y^2 - \frac{1}{D_x^4}2x^2 = \frac{x^4}{12}y^2 - 2\cdot\frac{x^6}{3\cdot4\cdot5\cdot6}$$

$$z_p = \frac{1}{180}(15x^4y^2 - x^6).$$

Thus G.S. = C.F. + P.I. = $z_c + z_p$

$$z = \phi_1(y+ix) + \phi_1(y-ix)$$
$$+i[\phi_2(y+ix) - \phi_2(y-ix)]$$
$$+\frac{1}{180}(15x^4y^2 - x^6)$$

Exponential shift

Example 1: Solve $z_{xx} - z_{xy} - 2z_{yy} = (y-1)e^x$

Solution: $(D_x^2 - D_xD_y - 2D_y^2)z = (y-1)e^x$
A.E.: $m^2 - m - 2 = 0$ or $(m+1)(m-2) = 0$
$\therefore\ m = -1, 2$
C.F.: $z_c = f_1(y-x) + f_2(y+2x)$
P.I.: $z_p = \frac{1}{D_x^2 - D_xD_y - 2D_y^2}(y-1)e^x$
Applying exponential shift with $a = 1, b = 0$

$$z_p = \frac{e^x}{(D_x+1)^2 - (D_x+1)D_y - 2D_y^2}(y-1)$$

$$= e^x\frac{1}{1 + (D_x^2 + 2D_x - D_xD_y - D_y - 2D_y^2)}(y-1)$$

$$= e^x[1 - (D_x^2 + 2D_x - D_xD_y - D_y - 2D_y^2) + \ldots](y-1)$$

$$= e^x[(y-1) - (0+0-0-1-0) + \ldots] = ye^x$$

$\therefore$ G.S. : $z =$ C.F. + P.I. $= z_c + z_p = f_1(y-x) + f_2(y+2x) + ye^x$

Method of undetermined coefficients

Example 1: Solve $(D_x^2 + D_xD_y - 6D_y^2)z = x^2\sin(x+y)$

Solution: A.E.: $m^2 + m - 6 = 0$ or $(m-2)(m+3) = 0$ so $m = 2, -3$.
C.F.: $z_c = f_1(y+2x) + f_2(y-3x)$.
To find the particular integral, use the method of undetermined coefficients. So assume the P.I. as $z_p = x^2(A\sin(x+y) + B\cos(x+y)) + x(C\sin(x+y) + D\cos(x+y)) +(E\sin(x+y) + F\cos(x+y)$
or
$z_p = (Ax^2 + Cx + E)\sin(x+y) + (Bx^2 + D_x + F)\cos(x+y)$
Differentiating z partially w.r.t. x and y, we get
$z_x = (2Ax + C - Bx^2 - Dx - F)\quad \sin(x+y) + (Ax^2 + Cx + E + 2Bx + D)\cos(x+y)$
$z_{xx} = (2A - Ax^2 - Cx - E - 4Bx - 2D)\sin(x+y) +(4Ax + 2C + 2B - Bx^2 - Dx - F)\cos(x+y)$.
$z_{xy} = -(Ax^2 + Cx + E + 2Bx + D)\sin(x+y) + +(2Ax + C - Bx^2 - D\cdot x - F)\cos(x+y)$
$z_{yy} = -(Ax^2 + Cx + E)\sin(x+y) - (Bx^2 + D\cdot x +F)\cos(x+y)$.
Substituting these values in the given PDE, we obtain $[(2A - Ax^2 - Cx - E - 4Bx - 2D)\ +(-Ax^2 - Cx - E - 2Bx - D) + 6(Ax^2 + Cx + E)] \times \sin(x+y) + \cos(x+y)\ +[(4Ax + 2C + 2B - Bx^2 - D\cdot x - F)\ +(2Ax + C - Bx^2 - D\cdot x - F) + 6(Bx^2 + Dx + F)] = x^2\sin(x+y)$.
Now equating the coefficients on both sides, we have
$x^2\sin x : -A - A + 6A = 1\ \therefore\ A = \frac{1}{4}$
$x\sin x : -C - 4B - C - 2B + 6C = 0$
$\therefore 2C - 3B = 0$
$x^0\sin x : 2A - E - 2D - E - D + 6E = 0$
$\therefore 3D - 4E = \frac{1}{2}$
$x^2\cos x : -B - B + 6B = 0\ \therefore\ B = 0\ \therefore\ C = 0$
$x\cos x : 4A - D + 2A - D + 6D = 0\ \therefore\ D = -\frac{3}{8}$
$\therefore\ E = \frac{-13}{32}$
$x^0\cos x : 2C + 2B - F + C - F + 6F = 0$
$\therefore\ F = 0$
Thus $A = \frac{1}{4}$, $B = C = 0$, $D = \frac{-3}{8}$, $E = \frac{-13}{32}$, $F = 0$.
Hence the P.I. is
$z_p = \frac{1}{4}x^2\sin(x+y) - \frac{13}{32}\sin(x+y) - \frac{3}{8}x\cos(x+y)$
Thus G.S. = z = C.F. + P.I. = $z_c + z_p$.

Example 2: Solve $(D_x^2 - 2D_xD_y + D_y^2)z = \tan(y+x)$.

Solution: $(D_x - D_y)^2z = \tan(y+x)$. So the complementary function is $z_c = \phi_1(y+x) + x\phi_2(y+x)$. Now particular integral is $z_p = \frac{1}{(D_x - D_y)^2}\tan(y+x)$. Since $F(D_x, D_y) = (D_x - D_y)^2 = (1-1) = 0$. So applying the result, on page 18.23 $z_p = \frac{x^2}{2!}\tan(y+x)$. Thus the general solution is $z = \phi_1(y+x) + x\phi_2(y+x) + \frac{x^2}{2}\tan(y+x)$.

Example 3: Solve $(D_x^2+3D_xD_y+D_x+2D_y^2-2)z=e^{3x+4y}+y(1-2x)$.

Solution: Factorizing the L.H.S. we have $(D_x+D_y-1)(D_x+2D_y+2)z=e^{3x+4y}+y(1-2x)$. For the first factor (D_x+D_y-1), the associated solution is $e^x\phi_1(y-x)$ with $a=1, b=1, c=-1$ and for the second factor D_x+2D_y+2, the associated solution is $e^{-y}\phi_2(y-2x)$ with $a=1$, $b=2$, $c=2$. Then the complementary function z_p is $z_p=e^x\phi_1(y-x)+e^{-y}\phi_2(y-2x)$. Now the particular integral is

$z_p=$

$$\frac{1}{(D_x+D_y-1)(D_x+2D_y+2)}\cdot[e^{3x+4y}+y(1-2x)]$$

$$z_p=I_1+I_2$$

For $$I_1=\frac{1}{(D_x+D_y-1)(D_x+2D_y+2)}e^{3x+4y}$$

$$=\frac{1}{(3+4-1)(3+8+2)}e^{3x+4y}$$

with $a=3, b=4$

$$I_1=\frac{1}{78}e^{3x+4y}$$

Now

$$I_2=\frac{1}{(D_x+D_y-1)(D_x+2D_y+2)}y(1-2x)$$

Consider $\frac{1}{(D_x+D_y-1)}y(1-2x)=$

$\frac{-1}{1-(D_x+D_y)}y(1-2x) \quad =-[1+(D_x+D_y)+(D_x+D_y)^2+\ldots][y(1-2x)]$

$=-[(y-2xy)-2y+1-2x+0+0+2(-2)]$

$=(2xy+y+2x+3)$

Then $I_2=\frac{1}{D_x+2D_y+2}(2xy+y+2x+3)$

$$=\frac{1}{2}\frac{1}{1+(\frac{D_x}{2}+D_y)}(2xy+y+2x+3)$$

$$=\frac{1}{2}\left[1-\left(\frac{D_x}{2}+D_y\right)+\left(\frac{D_x}{2}+D_y\right)^2+\ldots\right]$$

$[2xy+y+2x+3]=\frac{1}{2}[(2xy+y+2x+3)-\frac{1}{2}(2y+2)-(2x+1)+\frac{1}{4}\cdot 0+0+2]$

$$=\frac{1}{2}[2xy+0+0+3]=xy+\frac{3}{2}.$$

Thus the particular integral is

$$z_p=\frac{1}{78}e^{3x+4y}+\left(xy+\frac{3}{2}\right)$$

Hence the general solution is

$z=z_c+z_p=e^x\phi_1(y-x)+e^{-y}\phi_2(y-2x)$

$+\frac{1}{78}e^{3x+4y}+xy+\frac{3}{2}$.

Example 4: Solve $(3D_xD_y-2D_y^2-D_y)z=\cos(3y+2x)$

Solution: Rewriting $D_y(3D_x-2D_y-1)z=\cos(3y+2x)$

For D_y the corresponding solution is $\phi_1(x)$ and for $(3D_x-2D_y-1)$ the corresponding solution is $e^{y/2}\phi_2(3y+2x)$ with $a=3, b=-2, c=-1$. Then the complementary function is

$$z_c=\phi_1(x)+e^{y/2}\phi_2(3y+2x).$$

The particular integral is

$$z_p=\frac{1}{3D_xD_y-2D_y^2-D_y}\cdot\cos(3y+2x)$$

Here $a=2, b=3$. So replace D_y^2 by $-b^2=-3^2=-9$ and replace D_xD_y by $-a\cdot b=-2\cdot 3=-6$. Then $z_p=\frac{1}{3(-6)-2(-9)-D_y}\cos(3y+2x)$

$$=-\frac{1}{D_y}\cos(3y+2x)$$

$$z_p=-\frac{\sin(3y+2x)}{3}$$

Thus the general solution is

$$z=z_c+z_p=\phi_1(x)+e^{y/2}\phi_2(3y+2x)-\frac{1}{3}\sin(3y+2x)$$

Example 5: Solve $(6D_x^2+5D_xD_y-6D_y^2)z=132\log(x+3y)$

Solution: Rewriting $(2D_x+3D_y)(3D_x-6D_y)z=132\log(x+3y)$. For the first factor $a=2$, $b=3$ so the corresponding solution is $\phi_1(2x+3y)$.

For the second factor $a=3$, $b=-2$ so the corresponding solution is $\phi_2(3x-2y)$. Thus the complementary function is

$$z_c=\phi_1(2x+3y)+\phi_2(3x-2y).$$

The particular integral is

$$z_p = \frac{1}{6D_x^2 + 5D_xD_y - 6D_y^2}[132\log(x+3y)]$$

Since the operator $F(D_x, D_y) = 6D_x^2 + 5D_xD_y - 6D_y^2$ is homogeneous of degree $n = 2$, the P.I. is obtained by integrating $132\log v$, twice w.r.t. v and dividing by $F(a, b)$. Here $v = x + 3y$ and $a = 1$, $b = 3$. Thus the particular integral is

$$z_p = \frac{132}{F(1,3)}\int\int \ln v\, dv\, dv$$

$$= \frac{132}{6\cdot 1^2 + 5\cdot 1\cdot 3 - 6\cdot 3^2 + \cdots}\int\int \ln v\, dv\, dv$$

$$= \frac{132}{-33}\int [v\ln v - v]\, dv = 4\left[\frac{v^2}{2} - \int v\ln v\, dv\right]$$

$$= 4\left[\frac{v^2}{2} - \int \ln v\, d\left(\frac{v^2}{2}\right)\right] = 4\left[\frac{v^2}{2} - \frac{v^2}{2}.\ln v + \frac{1}{2}\frac{v^2}{2}\right]$$

$$= v^2[3 - 2\ln v]$$

Replacing v by $x + 3y$, we get P.I. as

$$y_p = (x+3y)^2[3 - 2\ln(x+3y)]$$

Then the general solution is

$$z = z_c + z_p = \phi_1(2x+3y) + \phi_2(3x-2y) + (x+3y)^2[3 - 2\ln(x+3y)]$$

EXERCISE

Solve (obtain the general solution (G.S.))

1. $(D_x^2 - D_y^2)z = x - y$

Ans. $z = f_1(x+y) + f_2(x-y) + \frac{1}{4}x(x-y)^2$

2. $(D_x^2 + 3D_xD_y + 2D_y^2)z = x + y$

Ans. $z = f_1(y-x) + f_2(y-2x) + \frac{1}{36}(x+y)^2$

3. $(4D_x^2 - 4D_xD_y + D_y^2)z = 16\log(x+2y)$

Ans. $z = f_1(2y+x) + xf_2(2y+x) + 2x^2\log(x+2y)$

4. $(D_x^2 + D_y^2)z = \cos mx\cdot\cos ny$

Ans. $z = f_1(y+ix) + f_2(y-ix) - \frac{\cos mx\cdot\cos ny}{(m^2+n^2)}$

5. $(D_x^2 - 6D_xD_y + 9D_y^2)z = 12x^2 + 36xy$

Ans. $z = f_1(y+3x) + xf_2(y+3x) + 10x^4 + 6x^3y$

6. $(D_x^2 - D_y^2 + D_x + 3D_y - 2)z = e^{x-y} - x^2y$

Ans. $z = e^{-2x}f_1(y+x) + e^xf_2(y-x) - \frac{1}{4}e^{x-y} + \frac{1}{2}\left(x^2y + xy + \frac{3}{2}x^2 + \frac{3y}{2} + 3x + \frac{21}{4}\right)$

7. $(D_x - 3D_y - 2)^2z = 2e^{2x}\tan(y+3x)$

Ans. $z = e^{2x}[f_1(y+3x) + xf_2(y+3x)] + x^2e^{2x}\tan(y+3x)$

Hint: Use exponential shift.

8. $(D_x^2 + D_xD_y - 6D_y^2)z = y\cos x$

Ans. $z = f_1(y+2x) + f_2(y-3x) + \sin x - y\cos x$

9. $(D_x^2 + D_xD_y - 6D_y^2)z = \cos(2x+y)$

Ans. $z = \phi_1(y-3x) + \phi_2(y+2x) + \frac{x}{5}\sin(2x+y) + \frac{1}{25}\cos(2x+y)$

10. $(D_x^2 + 2D_xD_y + D_y^2 - 2D_x - 2D_y)z = \sin(x+2y)$

Ans. $z = f_1(y-x) + e^{2x}f_2(y-x) + \frac{1}{39}[2\cos(x+2y) - 3\sin(x+2y)]$

11. $D_x^2 + 5D_xD_y + 5D_y^2)z = x\cdot\sin(3x-2y)$

Ans. $z = f_1(y + (-5+\sqrt{5})\frac{x}{2}) + f_2(y + (-5-\sqrt{5})\frac{x}{2}) + x\sin(3x-2y) + 4\cos(3x-2y)$

Hint: Assume P.I. as $Ax\sin(3x-2y) + Bx\cos(3x-2y) + C\sin(3x-2y) + D\cos(3x-2y)$. By method of undetermined coefficients $A = 1$, $B = C = 0$, $D = 4$.

12. $(D_x^3 + D_x^2D_y - D_xD_y^2 - D_y^3)z = e^x\cos 2y$

Ans. $z = f_1(y-x) + xf_2(y-x) + f_3(y+x) + \frac{1}{25}e^x\cos 2y + \frac{2}{25}e^x\sin 2y$

Hint: a) Assume P.I. as $Ae^x\cos 2y + Be^x\sin 2y$ then $A = \frac{1}{25}$ and $B = \frac{2}{25}$

(b) Applying exponential shift

P.I. =

$$e^x\frac{1}{(D_x+1)^3 + (D_x+1)^2D_y - (D_x+1)D_y^2 - D_y^3}\cdot\cos 2y,$$

(replace $D_x^2 = 0$, $D_y^2 = -4$, $D_xD_y = 0$)

$$= e^x\frac{D_y}{D_y(5D_y + 3D_x + 5)}\cos 2y$$

$$= -\frac{2}{5}e^x\frac{1}{D_y - 4}\sin 2y = -\frac{2}{5}e^x\frac{D_y+4}{D_y^2 - 16}\sin 2y$$

$$= \frac{e^x}{50}(D_y + 4)\sin 2y$$

13. $(D_x^3 - 7D_xD_y^2 - 6D_y^3)z = \sin(x+2y) + e^{3x+y}$

Ans. $z = f_1(y-x) + f_2(y-2x) + f_3(y+3x) - \frac{1}{75}\cos(x+2y) + \frac{x}{20}e^{3x+y}$

14. $(D_x^2 + 2D_xD_y - 8D_y^2)z = \sqrt{2x+3y}$

Ans. $z = f_1(y+2x) + f_2(y-4x) - \frac{1}{210}(2x+3y)^{5/2}$

15. $(D_x^3 - 7D_xD_y^2 - 6D_y^3)z = \cos(x-y) + x^2 + xy^2 + y^3$

Ans. $z = f_1(y-x) + f_2(y-2x) + f_3(y+3x) + \frac{1}{4}x\cos(x-y) + \frac{5}{72}x^6 + \frac{1}{60}x^5(1+21y) + \frac{1}{24}x^4y^2 + \frac{1}{6}x^3y^3$

Find the particular integral (see pages 18.22 and 18.23)

16. $p + 3q = \cos(2x+y)$

Ans. $z = \frac{1}{5}\sin(2x+y)$

Hint: $y - 3x = c_1$,
$z = \int \cos(2x + c_1 + 3x)dx$

17. $p - 2q = (y+1)e^{3x}$

Ans. $z = \frac{1}{3}\left(y + \frac{5}{3}\right)e^{3x}$

Hint: $y + 2x = c_1$,
$z = \int [(c_1 - 2x) + 1]e^{3x}dx$

18. $(D_x^3 - 3D_x^2D_y - 4D_xD_y^2 + 12D_y^3)z = \sin(y+2x)$

Ans. $z = \frac{1}{4}x\sin(y+2x)$.

Hint: P.I.$= \frac{1}{(D_x+2D_y)(D_x-3D_y)(D_x-2D_y)} \cdot \sin(y+2x)$

$= \frac{1}{D_x^2 - D_xD_y - 6D_y^2} \cdot \frac{1}{(D_x - 2D_y)} \cdot \sin(y+2x)$

$= -\frac{1}{4}\left(\frac{1}{D_x - 2D_y}\right)\sin(y+2x)$.

Now $\frac{dx}{1} = \frac{dy}{-2} = \frac{dz}{\sin y + 2x}$, $y + 2x = c$,
$z = \int \sin(y+2x)dx$,
$z = \sin c \int dx = x \cdot \sin c = x \cdot (y+2x)$

19. $(D_x + D_y - 1)(D_x + D_y - 3)(D_x + D_y)z = e^{x+y+z}\cos(2x-y)$

Ans. $z = e^x\phi_1(y-x) + e^{3x}\phi_2(y-x) + \phi_3(y-x) - \frac{1}{10}e^{x+y+z}[\sin(2x-y) + 2\cos(2x-y)]$

Hint: Use exponential shift.

20. $[D_x^2 + D_xD_y - 2D_y^2]z = 8\ln(x+5y)$

Ans. $z = \phi_1(2x-y) + \phi_2(x+y) + \frac{1}{22}[3 - 2\ln(x+5y)](x+5y)^2$

Hint: Since $F(D_x, D_y)$ is homogeneous of degree 2, integrate $8\ln v$ twice w.r.t. v and divide divide by $F(1,5)$. Here $v = x + 5y$.

21. $(D_x^2 - D_xD_y)z = \cos x \cdot \cos 2y$

Ans. $z = \phi_1(y) + \phi_2(y+x) + \frac{1}{2}\cos(x+2y) - \frac{1}{6}\cos(x-2y)$

22. $(D_x^3 - 2D_x^2D_y)z = 2e^{2x} + 3x^2y$

Ans. $z = \phi_1(y) + x\phi_2(y) + \phi_3(y+2x) + \frac{1}{60}(15e^{2x} + 3x^5y + x^6)$

23. $(D_x^2 - 4D_xD_y + 4D_y^2)z = e^{2x+y}$

Ans. $z = \phi_1(y+2x) + x\phi_2(y+2x) + \frac{1}{2}x^2e^{2x+y}$

18.8 CAUCHY TYPE DIFFERENTIAL EQUATION

The partial differential equation of the form

$$F(xD_x, yD_y) = f(x,y)$$

with *variable* coefficients can be transformed to P.D.E. with constant coefficients by putting $x = e^u$ and $y = e^v$. Then $u = \ln x$, $v = \ln y$, $x\frac{\partial}{\partial x} = \frac{\partial}{\partial u}$, $x^2\frac{\partial^2}{\partial x^2} = \frac{\partial}{\partial u}\left(\frac{\partial}{\partial u} - 1\right)$ i.e, $xD_x = D_u$ and $x^2D_x^2 = D_u(D_u - 1)$. Similarly $yD_y = D_v$ and $y^2D_y^2 = D_v(D_v - 1)$. Here D_v is $\frac{\partial}{\partial v}$. The transformed equation in the new dependent variables u and v which is a D.E. with constant coefficients, is then solved using the above methods. In the solution $z(u,v)$, replace u by $\ln x$ and v by $\ln y$ to get the general solution $z(x,y)$.

WORKED OUT EXAMPLES

Euler-Cauchy equation

Example 1:

Solve $(x^2D_x^2 - y^2D_y^2)z = x^2y$.

Solution: Put $x = e^u$, $y = e^v$ then $x^2y = e^{2u+v}$. Also $xD_x = D_u$, $x^2D_x^2 = D_u(D_u - 1)$. Similarly $yD_y = D_v$, $y^2D_y^2 = D_v(D_v - 1)$. Then the given D.E. in the new variables u and v is

$$[D_u(D_u - 1) - D_v(D_v - 1)]z = e^{2u+v}$$
$$(D_u - D_v)(D_u + D_v - 1)z = e^{2u+v}$$

The complementary function is

$$z_c = \phi_1(v + u) + e^u\phi_2(v - u)$$

or $z_c = \phi_1(\ln y + \ln x) + x\phi_2(\ln y - \ln x)$

$$z_c = \phi_1(\ln xy) + x\phi_2(\ln(y/x))$$
$$z_c = \psi_1(xy) + x\psi_2\left(\frac{y}{x}\right)$$

Now the particular integral is

$$z_p = \frac{1}{D_u^2 - D_u - D_v^2 + D_v} \cdot e^{2u+v}$$

Here $a = 2$, $b = 1$. Replace D_u by $a = 2$, D_v by $b = 1$.

Then $z_p = \frac{1}{4-2-1+1} e^{2u+v} = \frac{1}{2}e^{2u+v}$

or $z_p = \frac{1}{2}x^2y$.

Hence the general solution is

$$z = z_c + z_p = \psi_1(xy) + x\psi_2\left(\frac{y}{x}\right) + \frac{1}{2}x^2y$$

EXERCISE

1. $(x^2D_x^2 - 4xyD_xD_y + 4y^2D_y^2 + 6yD_y)z = x^3y^4$

Ans. $z = \phi_1(x^2y) + xf_2(x^2y) + \frac{1}{30}x^3y^4$

Hint: Put $x = e^u$, $y = e^v$ then
$$(D_u - 2D_v)(D_u - 2D_v - 1)z = e^{3u+4v}$$
where $D_u = \frac{\partial}{\partial u}$ and $D_v = \frac{\partial}{\partial v}$.

2. $\left(\frac{1}{x^2}D_x^2 - \frac{1}{x^3}D_x - \frac{1}{y^2}D_y^2 + \frac{1}{y^2}D_y\right)z = 0$

Ans. $z = \phi_1(y^2 + x^2) + \phi_2(y^2 - x^2)$

Hint: Put $x^2 = 2u$, $y^2 = 2v$, then $(D_u^2 - D_v^2)z = 0$

3. $(x^2D_x^2 - y^2D_y^2 + xD_x - yD_y)z = \log x$

Ans. $z = \phi_1(xy) + \phi_2\left(\frac{y}{x}\right) + \frac{1}{6}(\log x)^3$

Hint: Put $x = e^u$, $y = e^v$ then $(D_u^2 - D_v^2)z = u$.

4. $(xD_x^3D_y^2 - yD_x^2D_y^3)z = 0$

Ans. $z = \psi_1(y) + \psi_2(x) + x\psi_3(y) + y\psi_4(x) + \psi_5(xy)$

Hint: Put $x = e^u$, $y = e^v$, $(x^3y^2D_x^3D_y^2 - x^2y^3D_x^2D_y^3)z = 0$
$$D_uD_v(D_u - 1)(D_v - 1)(D_u - D_v)z = 0$$

5. $(x^2D_x^2 + xyD_xD_y - 2y^2D_y^2 - xD_x - 6yD_y)z = 0$

Ans. $z = \phi_1(y/x^2) + x^2\phi_2(xy)$

6. $(x^2D_x^2 - 2xyD_xD_y - 3y^2D_y^2 + xD_x - 3yD_y)z = x^2y\sin(\ln x)^2$

Ans. $z = \phi_1(x^3y) + \phi_2(y/x) - \frac{1}{65}x^2y[4\cos(\ln x)^2 + 7\sin(\ln x)^2]$.

7. $(x^2D_x^2 + y^2D_y^2 + 2xyD_xD_y - nxD_x - nyD_y + n)z = x^2 + y^2$

Ans. $z = x\phi_1(y/x) + x^n\phi_2(y/x) + \frac{1}{(2-n)}(x^2 + y^2)$

Hint: Put $x = e^u$, $y = e^v$, then
$$(D_u + D_v - n)(D_u + D_v - 1)z = e^{2u} + e^{2v}$$

18.9 NON-LINEAR PARTIAL DIFFERENTIAL EQUATIONS OF SECOND ORDER: MONGE'S METHOD

The most general second order non-linear partial differential equation in two independent variables x and y, and z as the dependent variable has the form

$$F(x, y, z, p, q, r, s, t) = 0 \quad (1)$$

whose solution can be obtained by Monge's* method in special cases. The first step in Monge's method consists of finding one or two intermediate integrals (also known as first integrals) of the form

$$u_1 = \psi_1(v_1) \quad (2)$$

and/or $\quad u_2 = \psi_2(v_2) \quad (3)$

*Gaspard Monge (1746-1818), Professor at Paris.

Here u_1, u_2, v_1, v_2 are functions of x, y, z, p, q and the functions ψ_1 and ψ_2 are arbitrary. Note that an intermediate integral may not exist for PDE (1).

By differentiating

$$u = \psi(v) \tag{4}$$

w.r.t. x and y and eliminating ψ' we obtain the most general PDE of the form

$$Rr + Ss + Tt + U(rt - s^2) = V \tag{5}$$

which has (4) as an intermediate integral (I.I.). Here R, S, T, U, V are functions of x, y, z, p and q.

When $U = v_p u_q - v_q u_p = 0$ then (5) reduces to

$$Rr + Ss + Tt = V. \tag{6}$$

Since the coeffiecients R, S, T, V are functions of p and q as well as x, y, z, P.D.E. (6) is non-linear (also referred to as "*quasilinear*" or "*uniform non-linear*" equation unlike (5) which is known as "*non-uniform*" equation).

We know that

$$dz = \frac{\partial z}{\partial x}dx + \frac{\partial z}{\partial y}dy = pdx + qdy \tag{7}$$

$$dp = \frac{\partial p}{\partial x}dx + \frac{\partial p}{\partial y}dy = rdx + sdy \tag{8}$$

$$dq = \frac{\partial q}{\partial x}dx + \frac{\partial q}{\partial y}dy = s\,dx + t\,dy \tag{9}$$

Solving (8) and (9), we get

$$r = \frac{dp - sdy}{dx}, \quad t = \frac{dq - sdx}{dy} \tag{10}$$

Using (10) eliminate r and t from (6), we have

$$R\left(\frac{dp - sdy}{dx}\right) + Ss + T\left(\frac{dq - sdx}{dy}\right) = V$$

or

$$s\left[R(dy)^2 - S\,dxdy + T(dx)^2\right] = Rdydp + \\ + Tdxdq - Vdxdy$$

Thus we obtain

$$R(dy)^2 - Sdxdy + T(dx)^2 = 0 \tag{11}$$

$$Rdydp + Tdxdq - Vdxdy = 0 \tag{12}$$

The two simultaneous equations (11) and (12) are known as *Monge's (subsidiary) equations (M.E.)*. By solving the Monge's equations (11) and (12) one or two intermediate integrals of (6) are obtained.

In general, the quadratic (11) can be resolved into two (or one repeated) equations.

Case I: Suppose (11) can be resolved as

$$R(dy)^2 - Sdxdy + T(dx)^2 =$$
$$= (A_1dy + B_1dx)(A_2dy + B_2dx) = 0$$

where $A_1B_2 \neq A_2B_1$. Then we have two systems

$$\left.\begin{array}{l} A_1dy + B_1dx = 0 \\ \text{and} \\ R\,dydp + T\,dx\,dq - V dx\,dy = 0 \end{array}\right\} \tag{13}$$

and

$$\left.\begin{array}{l} A_2dy + B_2dx = 0 \\ Rdydp + Tdxdq - Vdxdy = 0 \end{array}\right\} \tag{14}$$

Integrating (13), we get two integrals $u_1 = a, v_1 = b$. Thus we get an intermediate integral of (6) as

$$u_1 = \psi_1(v_1) \tag{2}$$

Similarly from (14), we get another intermediate integral of (6) as

$$u_2 = \psi_2(v_2). \tag{3}$$

Solving (2) and (3), determine p and q as functions (in terms) of x, y and z. Substituting p and q in

$$dz = p\,dx + q\,dy \tag{7}$$

which on integration yields the required general solution of (6) involving two arbitrary functions.

Note 1: Some cases, the second intermediate integral can be obtained from the first one by inspection.

Note 2: When inspection fails, rearrange the first intermediate integral in the form $Pp + Qq = R^*$ and solve by Lagrange's method.

Case II: When (11) is a perfact square, i.e., $R(dy)^2 - S\,dx\,dy + T(dx)^2 = (Ady + Bdx)^2 = 0$, then we have only one system

$$\left.\begin{array}{l} Ady + Bdx = 0 \\ \text{and } R\,dy\,dp + T\,dx\,dq - V\,dx\,dy = 0 \end{array}\right\}. \tag{15}$$

Solving (15) we get only one intermediate integral $u = \psi(v)$ of the form $Pp + Qq = R^*$ which is integrated using Lagrange's method. (or Charpit's method).

Note: When equation (11) gives neither two factors nor a perfect square, then use the methods explained in Section 18.7 on page 18.22.

Worked Out Examples

Example 1: Solve $r + (a + b)s + abt = xy$

Solution: Comparing the given equation with

$$Rr + Ss + Tt = V$$

we get $R = 1$, $S = a + b$, $T = ab$, $V = xy$. Then the Monge's subsidiary equations

$$R(dy)^2 - S\,dx\,dy + T(dx)^2 = 0$$

and $\quad Rdp\,dy + T\,dq\,dx - V\,dx\,dy = 0$

reduces to

$$(dy)^2 - (a + b)dx\,dy + ab(dx)^2 = 0 \qquad (1)$$

and $\quad dpdy + ab\,dq\,dx - xydxdy = 0 \qquad (2)$

Factorising equation (1), we get

$$(dy - adx)(dy - bdx) = 0$$

which yields $\quad dy - adx = 0 \qquad (3)$

and $\quad dy - bdx = 0 \qquad (4)$

Integrating (3) and (4), we get

$$y - ax = c_1 \qquad (5)$$

$$y - bx = c_2 \qquad (6)$$

Substitute $dy = adx$ from (3) in (2) then

$$dp(a\,dx) + ab\,dq\,dx - x\,y\,dx(a\,dx) = 0$$

or $\quad dp + b\,dq - xy\,dx = 0$

Integrating we get

$$p + bq - y\frac{x^2}{2} = \text{constant} = \psi_1(y - ax) \qquad (7)$$

Similarly using $dy = bdx$ from (4) in (2), we get

$$dp(b\,dx) + ab\,dq\,dx - x\,y\,dx(b\,dx) = 0$$

or $\quad dp + a\,dq - xy\,dx = 0$

Integrating

$$p + aq - y\frac{x^2}{2} = \text{constant} = \psi_2(y - bx) \qquad (8)$$

Now solve (7) and (8) for p and q. Multiply (7) by 'a' and (8) by 'b' and subtract, then

$$(b - a)p - (b - a)\frac{x^2y}{2} = b\psi_2 - a\psi_1$$

or $p = \frac{b\psi_2 - a\psi_1}{b-a} + \frac{x^2y}{2} \qquad (9)$

Similarly subtracting (8) from (7), we get

$$q = \frac{\psi_2 - \psi_1}{a - b}. \qquad (10)$$

We know that

$$dz = p\,dx + q\,dy.$$

Substituting p and q from (9) and (10), we have

$$dz = \left(\frac{b\psi_2 - a\psi_1}{b - a} + \frac{x^2y}{2}\right)dx + \left(\frac{\psi_2 - \psi_1}{a - b}\right)dy$$

Rearranging

$$(a - b)dz = -\psi_1(dy - a\,dx) + \psi_2(dy - bdx) + (a - b)\frac{x^2y}{2}dx$$

Integrating

$$z = \phi_1(y - ax) + \phi_2(y - bx) + \frac{x^3y}{6}$$

which is the required general solution involving two arbitrary functions ϕ_1 and ϕ_2 which are functions of their arguments.

Example 2: Solve $q^2r - 2pqs + p^2t = pt - qs$.

Solution: Rewriting the equation in the standard form

$$q^2r - q(2p - 1)s + p(p - 1)t = 0$$

Here $R = q^2$, $S = q - 2pq$, $T = p^2 - p$, $V = 0$

The first Monge's equation (11) is

$$q^2(dy)^2 + (2pq - q)dx\,dy + (p^2 - p)(dx)^2 = 0$$

Rewriting

$$q^2(dy)^2 + q(p-1)dxdy + pq\,dx\,dy +$$
$$+p(p-1)(dx)^2 = 0$$

or $[p\,dx + q\,dy][q\,dy + (p-1)dx] = 0$

The two factors are

$$p\,dx + q\,dy = dz = 0 \text{ or } z = c_1$$

and

$$q\,dy + (p-1)dx = q\,dy + p\,dx - dx$$
$$= dz - dx = 0$$

or $z - x = c_2$

The second Monge's equation (12) is

$$q^2 dpdy + p(p-1)dq\,dx = 0.$$

Since $p\,dx + q\,dy = 0$, substitute $q\,dy = -p\,dx$, then

$$q\,dp(-p\,dx) + p(p-1)dq\,dx = 0$$

or $-q\,dp + p\,dq - dq = 0$

Rewriting, $\dfrac{q\,dp - p\,dq}{q^2} + \dfrac{dq}{q^2} = 0$

or $d\left(\dfrac{p}{q}\right) + d\left(-\dfrac{1}{q}\right) = 0.$

Integrating $\frac{p}{q} - \frac{1}{q} = \text{constant} = \psi_1(z)$

or $p - \psi_1(z)q = 1$

which is a Lagrange's equation, with auxiliary equations.

$\frac{dx}{1} = \frac{dy}{-\psi_1} = \frac{dz}{1}$

From the first and third, $x - z = \text{constant} = c_2$

From second and third

$$\psi_1(z)dz + dy = 0$$

Integrating

$$\psi_2(z) + y = \text{constant} = \psi_3(x - z).$$

Thus the general solution is

$$y = \psi_2(z) + \psi_3(x - z)$$

where ψ_2, ψ_3 are arbitrary functions.

Example 3: Solve $x^2 r - 2xs + t + q = 0$

Solution: Here $R = x^2, S = -2x, T = 1, V = -q$

The first Monge's equation is

$$x^2(dy)^2 + 2x\,dx\,dy + (dx)^2 = 0$$

or $(x\,dy + dx)^2 = 0$, repeated factors.

Then $x\,dy + dx = 0$ or $dy + \dfrac{dx}{x} = 0$

Integrating $y + \ln x = c_1$

Now the second Monge's equation is

$$x^2 dp\,dy + dq\,dx + q\,dx\,dy = 0$$

Substituting $xdy = -dx$

$$x\,dp(-dx) + dq\,dx + q\,dx\left(\frac{dx}{-x}\right) = 0$$

or $\quad -xdp + dq - q\dfrac{dx}{x} = 0$

i.e., $\quad -dp + \dfrac{dq}{x} - q\dfrac{dx}{x^2} = 0$

or $\quad -dp + d\left(\dfrac{q}{x}\right) = 0$

Integrating $p = \frac{q}{x} + c_1(y + \ln x)$

or $\quad p - \dfrac{q}{x} = c_1$

which is Lagrange's equation with auxiliary equations

$$\frac{dx}{1} = \frac{dy}{-\frac{1}{x}} = \frac{dz}{c_1}$$

From one and two, $\frac{1}{x}dx + dy = 0$ or $y + \ln x = c$

From one and three, $dz = c_1 dx$ or $z = c_1 x + c_2$.

Thus the general solution involving two arbitrary function is

$$z = x\psi_1(y + \ln x) + \psi_2(y + \ln x)$$

Example 4: Solve $(e^x - 1)(qr - ps) = pq\,e^x$

Solution: Rewriting in standard form

$$(e^x - 1)q \cdot r + p(1 - e^x)s = pqe^x$$

Here $R = q(e^x - 1)$, $S = p(1 - e^x)$, $T = 0$, $V = pqe^x$.

Monge's first equation is

$$q(e^x - 1)(dy)^2 - p(1 - e^x)dx\,dy = 0$$

$$(e^x - 1)dy[q\,dy + p dx] = 0$$

so $(e^x - 1)dy = 0$, $p\,dx + q\,dy = dz = 0$
Integrating $y = c_1$, $z = c_2$.
Now the Monge's second equation is

$$q(e^x - 1)dy\,dp - p\,q\,e^x dx\,dy = 0$$

Substituting $q\,dy = -p\,dx$, we get

$$(e^x - 1)(-p\,dx)\cdot dp - p\,q e^x dx\,dy = 0$$

or $$(e^x - 1)dp + qe^x dy = 0$$

i.e., $$(e^x - 1)dp - pe^x dx = 0$$

$$\frac{dp}{p} = \frac{e^x dx}{e^x - 1} = \frac{d(e^x - 1)}{(e^x - 1)}$$

Integrating

$$p = c_1(e^x - 1)$$

or $$\frac{\partial z}{\partial x} = p = \psi_1(z)(e^x - 1)$$

Integrating

$$\frac{dz}{\psi_1(z)} = (e^x - 1)dx$$

$$\psi_2(z) = e^x - x + \psi_3(y)$$

$$\therefore \quad x = \psi_3(y) - \psi_2(z) + e^x$$

Example 5: Solve $(1 - q)^2 r - 2(2 - p - 2q + pq)s + (2 - p)^2 t = 0$

Solution: Here $R = (1 - q)^2$, $S = -2(2 - p - 2q + pq)$, $T = (2 - p)^2$, $V = 0$.
Monge's first equation is

$$(1 - q)^2(dy)^2 + 2(2 - p - 2q + pq)dxdy +$$
$$+ (2 - p)^2(dx)^2 = 0$$

$$[(1 - q)dy + (2 - p)dx]^2 = 0$$

or $$(1 - q)dy + (2 - p)dx = 0$$

$$dy - q\,dy + 2dx - p\,dx = dy + 2dx -$$
$$- (p\,dx + q\,dy) = 0$$
$$\text{i.e. } dy + 2dx - dz = 0$$

Since $dz = p\,dx + q\,dy$.
Integrating $y + 2x - z = c_1$
Substituting $(1 - q)dy = -(2 - p)dx$ in the Monge's second equation

$$(1 - q)^2 dy\,dp + (2 - p)^2 dx\,dq = 0$$

reduces to

$$(1 - q)dp[-(2 - p)dx] + (2 - p)^2 dx\,dq = 0$$

or

$$(1 - q)dp + (2 - p)dq = 0$$

i.e., $$\frac{dp}{p - 2} = \frac{dq}{q - 1}$$
Integrating $(p - 2) = c_2(q - 1)$

$$q - 1 = c_3(p - 2) = \psi(y + 2x - z)(p - 2).$$

Then $c_3 p - q = 2c_3 - 1$
which is a Lagrange's equation with subsidiary equations.

$$\frac{dx}{c_3} = \frac{dy}{-1} = \frac{dz}{2c_3 - 1}$$

From first and second

$$dx + c_3 dy = 0$$

which on integration gives

$$x + y \cdot \psi(y + 2x - z) = \phi(y + 2x - z)$$

Example 6: Solve $r - 3s - 10t = -3$.

Solution: Here $R = 1$, $S = -3$, $T = -10$, $V = -3$. Monge's equations are

$$(dy)^2 + 3\,dx\,dy - 10(dx)^2 = 0 \tag{1}$$

and $$dy\,dp - 10dx\,dq + 3dxdy = 0 \tag{2}$$

Equation (1) can be factorized as

$$(dy + 5dx)(dy - 2dx) = 0$$

Thus $dy + 5dx = 0$ (3)

and $dy - 2dx = 0$ (4)

which on integration gives

$$y + 5x = c_1 \tag{5}$$

$$y - 2x = c_2 \tag{6}$$

Substituting $dy = -5dx$ from (3) in (2), we get

$$(-5\,dx)dp - 10dx\,dq + 3dx(-5\,dx) = 0$$

or $$dp + 2dq + 3dx = 0$$

Integrating we have

$$p + 2q + 3x = c_3 \tag{7}$$

Substituting $dy = 2\,dx$ from (4) in (2), we get

$$(2dx)dp - 10\,dx\,dq + 3dx(2dx) = 0$$

$$dp - 5dq + 3dx = 0$$

which on integration gives

$$p - 5q + 3x = c_4 \tag{8}$$

Thus $\quad p + 2q + 3x = f(y - 5x) \qquad (9)$

and $\quad p - 5q + 3x = g(y - 2x) \qquad (10)$

Solving (9) and (10), we get

$$7p = 5f + 2g - 21x$$
$$7q = f - g.$$

Substituting p and q in

$$dz = p\,dx + q\,dy$$
$$7dz = (5f + 2g - 21x)dx + (f - g)dy$$
$$7\,dz = (5dx + dy)f(5x + y) - (dy - 2dx)g(y - 2x)$$
$$-21x\,dx$$

Integrating $z = \phi(5x + y) + \phi(y - 2x) - \dfrac{3x^2}{2}$.

EXERCISE

Solve (obtain the general solution (G.S.)) of the following examples by Monge's method.

1. $qs - pt = q^3$

Ans. G.S.: $y + xz = f(z) + g(x)$

Hint: Monge's Equations (M.E.) are $q \quad dx \quad dy + p(dx)^2 = 0, \quad p\,dx\,dq + q^3 dx\,dy = 0; \quad dz = pdx + qdy = 0$ so $z = c_1$.
Intermediate integral (I.I.): $\frac{1}{q} + x = c_2 = \psi(z)$

2. $q^2 r - 2pqs + p^2 t = 0$

Ans. G.S.: $y + x\psi(z) = \phi(z)$

Hint: M.E.: $(pdx + qdy)^2 = 0$, $q^2 dp\,dy + p^2 dq\,dx = 0; dz = 0, z = c_1, q\,dp = p\,dq$ or $p = c_2 q$; Lagrange's $\frac{dx}{1} = \frac{dy}{-\psi(z)} = \frac{dz}{0}, z = c_1$, $y + xf(c_1) = c_2$

3. $r - t\cos^2 x + p\tan x = 0$

Ans. G.S.: $z = f(y - \sin x) + g(y + \sin x)$

Hint: M.E.: $(dy)^2 - \cos^2 x(dx)^2 = 0$, $dpdy - \cos^2 x\,dq\,dx + p\tan x \cdot dxdy = 0$; $y - \sin x = c_1$, I.I., $p\sec x - q = c_1 = f(y - \sin x)$, $y + \sin x = c_2$, I.I.: $p\sec x + q = c_2 = g(y + \sin x)$, solving $p = \frac{f+q}{2\sec x}, q = \frac{1}{2}(g - f)$.

4. $x(r + 2xs + x^2 t) = p + 2x^3$

Ans. G.S.: $z = \psi(x^2 - 2y) + \frac{x^2}{2}\phi(x^2 - 2y) + \frac{x^4}{4}$

Hint: M.E.: $(dy - xdx)^2 = 0$, $x\,dy\,dp + x^3 dx\,dq - (p + 2x^3)dx\,dy = 0$.
$x^2 - 2y = c_1$, I.F. $\frac{1}{x^2}$, I.I.: $p + xq = x^3 + xf(x^2 - 2y)$.
Lagrange's: $\dfrac{dx}{1} = \dfrac{dy}{x} = \dfrac{dz}{x^3 + xf(x^2 - 2y)}$

5. $(x - y)(xr - xs - ys + yt) = (x + y)(p - q)$

Ans. G.S.: $Z = f(xy) + g(x + y)$

Hint: M.E.: $(xdy + ydx)(dx + dy) = 0$, $x\,dp\,dy + ydq\,dx - \frac{x+y}{x-y} \cdot (p - q)\,dy\,dx = 0, xy = c_1$, $x + y = c_2$,
I.I.: $dp - dq - \frac{dx-dy}{x-y}(p - q) = 0$
i.e., $p - q = c_3(x - y)$; use Lagrange's $\frac{dx}{1} = \frac{dy}{-1} = \frac{dz}{(x-y)f(xy)}$

6. $y^2 r - 2ys + t - p - 6y = 0$

Ans. G.S.: $z = y^3 - yf(y^2 + 2x) + g(y^2 + 2x)$

Hint: M.E.: $(ydy + dx)^2 = 0$, $y^2 + 2x = c_1$, $ydp - dq + (p + 6y)dy = 0$
I.I.: $py - q + 3y^2 = c_2$, use Lagrange's auxiliary equations.

7. $2x^2 r - 5xys + 2y^2 t = -2(px + qy)$

Ans. G.S.: $z = f(yx^2) + g(xy^2)$

Hint: M.E.: $(x\,dy + 2y\,dx)(2x\,dy + y\,dx) = 0$, $x^2 y = c_1$, $xy^2 = c_2$, $2x^2 dpdy + 2y^2 dqdx + 2(px + qy)dxdy = 0$
I.I.: $2xp - yq = c_3$ use Lagrange's equations $\frac{dx}{2x} = \frac{dy}{-y} = \frac{dz}{f(yx^2)}$

8. $q(1 + q)r - (1 + 2q)(1 + p)s + (1 + p)^2 t = 0$

Ans. G.S.: $x = \psi(x + y + z) + \phi(x + z)$

Hint: M.E.: $q dy + (1+p)dx = 0$, $q(1+q)dydp + (1+p)^2 dxdq = 0$ $x + z = c_1$, I.I.: $\frac{1+p}{\phi} = f(x+z)\cdot(1+q)$; M.E.: $(1+q)dy + (1+p)dx = 0$, $q(1+q)dy\,dp + (1+p)^2 dx\,dq = 0$, $x+y+z = c_2$, I.I.: $1+p = g(x+y+z)\cdot q$. Solving $p = \frac{fg+f-g}{g-f}$, $q = \frac{f}{g-f}$, $dz = p\,dx + q\,dy$, $fgdx = -f(dx+dy+dz) + g(dx+dz)$,

$$dx = -\frac{dx+dy+dz}{g(x+y+z)} + \frac{dx+dz}{f(x+z)}$$

9. $xy(t-r) + (x^2-y^2)(s-2) - py + qx = 0$

Ans. G.S.: $z = xy + f(x^2+y^2) + g\left(\frac{y}{x}\right)$

Hint: M.E.: $(x\,dy - ydx)(ydy + x\,dx) = 0$, $x^2+y^2 = c_1$, $\frac{y}{x} = c_2$, $-xy\,dp\,dy + xydqdx$ $-[py - qx + 2(x^2-y^2)]dx\,dy = 0$ I.I.: $xp + yq = 2xy + \psi(x^2+y^2)$. Use Lagrange's A.E.

10. $y^2r + 2xys + x^2t + px + qy = 0$

Ans. G.S.: $z = f(x^2-y^2) + g(x^2-y^2)\log(x+y)$

Hint: M.E.: $(x\,dx - y\,dy)^2 = 0$, $x^2 - y^2 = c_1$, $y^2dp\,dy + x^2dq\,dx + (px+qy)dxdy = 0$, I.I.: $yp + xq = c_2 = \psi(x^2-y^2)$. Use Lagrange's A.E.: $\frac{dz}{\psi(x^2-y^2)} = \frac{dy}{x} = \frac{dy}{\sqrt{y^2+c_1}}$, integrating

$$z - \psi(c_1)\log\{y + \sqrt{y^2+c^2}\} = c_3$$

11. $qs - pt + s - t = 0$

Ans. G.S. $z = f(x) + g(x+y+z)$

Hint: M.E.: $-(p+1)dq\,dx = 0$, $-(q+1)dx\,dy - (p+1)(dx)^2 = 0$; $dq = 0$, $dx + dy + p\,dx + qdy = dx + dy + dz = 0$, $x+y+z = c_1$, *integrate* $q = c_2 = \psi(x+y+z)$ w.r.t. y.

12. $q^2r - 2pqs + p^2t - pq^2 = 0$

Ans. G.S.: $y = e^x f(z) + g(z)$

Hint: M.E.: $(p\,dx + q\,dy)^2 = 0$, $z = c_1$, $q^2dy\,dp + p^2dx\,dq - pq^2dx\,dy = 0$, $-\frac{dp}{p} + \frac{dq}{q} + dx = 0$, I.I.: $e^xq = p\psi(z)$. Use Lagrange's A.E.

13. $x^2r + 2xys + y^2t = 0$

Ans. $z = x\phi(y/x) + \psi(y/x)$

14. $qr - ps = p^3$

Ans. $x = yz - \phi(z) + \psi(y)$.

18.10 SOLUTION OF SECOND ORDER P.D.E.: MISCELLANEOUS

A large class of P.D.E. of first and second order can be integrated by methods similar to integration of O.D.E. . While integrating a P.D.E. partially w.r.t. say one independent variable x, the resulting constant of integration will be a function of the other independent variable(s) say y.

Type I

P.D.E. can be reduced to a linear equation in p and x, p and y, q and x or q and y which can be integrated as a linear first order ordinary D.E.

Type II

P.D.E. may be rewritten as partial derivatives w.r.t. a single independent variable say x which after integration can be solved by Lagrange's method.

Type III

P.D.E. can easily be integrated by inspection. Note that while integration partially w.r.t. one independent variable say x, the other independent variables(s) is (are) held constant.

WORKED OUT EXAMPLES

Reduciable to linear equation

Example 1: Solve $yt - q = xy$

Solution: Rewriting the given P.D.E. we get $y\frac{\partial q}{\partial y} - q = xy$ or $\frac{\partial q}{\partial y} - \frac{q}{y} = x$

which is linear in q and y. The integration factor for this D.E. is I.F. $= e^{\int -\frac{1}{y}dy} = \frac{1}{y}$. So the solution is $q \cdot \frac{1}{y} = \int x \cdot \frac{1}{y} dy + f(x)$ where $f(x)$ is an arbitrary function of x. Then

$$\frac{q}{y} = x \ln y + f(x)$$

$$\text{or} \quad q = xy \ln y + y f(x)$$

$$\text{i.e.,} \quad \frac{\partial z}{\partial y} = xy \ln y + y f(x)$$

Integrating partially w.r.t. 'y', we get

$$z = x \int y \ln y \, dy + \frac{y^2}{2} f(x) + g(x)$$

where $g(x)$ is an arbitrary function of x. Then

$$z = x\left[y^2 \ln y - \frac{y^2}{2}\right] + \frac{y^2}{2} f(x) + g(x)$$

Use of Lagrange's method

Example 1: Solve $p + r + s = 1$

Solution: Rewriting the given P.D.E.

$$\frac{\partial z}{\partial x} + \frac{\partial p}{\partial x} + \frac{\partial q}{\partial x} = 1$$

which is expressed as partial derivatives w.r.t. x. Integrating partially w.r.t. x, we have

$$z + p + q = x + f(y)$$

where $f(y)$ is an arbitrary function of y. Rewriting this as a linear first order P.D.E.

$$p + q = x + f(y) - z$$

We get the auxiliary equations as

$$\frac{dx}{1} = \frac{dy}{1} = \frac{dz}{x + f(y) - z}$$

Using first two, we get $x = y$ or $x - y = c_1$. From two and three, we get

$$\frac{dz}{dy} = x + f(y) - z$$

$$\text{or} \quad \frac{dz}{dy} + z = x + f(y)$$

which is a first order linear D.E. in z. Its I.F. is $e^{\int dy} = e^y$.

Then

$$e^y \cdot z = \int [x + f(y)] e^y dy + c_2$$

$$e^y \cdot z = x e^y + \int f(y) e^y dy + c_2.$$

Thus the general solution is

$$z = x + F(y) \cdot e^{-y} + c_2 e^{-y} = x + F(y) e^{-y} + e^{-y} \phi(x - y)$$

$$\text{where } F(y) = \int f(y) dy$$

Equations solvable by direct integration

Example 1: Solve

$$\frac{\partial^3 z}{\partial x^2 \partial y} = \cos(2x + 3y)$$

Solution: Integrating partially w.r.t. x, we get

$$\frac{\partial^2 z}{\partial x \partial y} = \int \cos(2x+3y) dx + f(y) = \frac{\sin(2x+3y)}{2} + f(y)$$

Integrating again partially w.r.t. x, we have

$$\frac{\partial z}{\partial y} = \int \frac{1}{2} \sin(2x + 3y) dx + \int x f(y) dx + g(y)$$

Here $f(y)$ and $g(y)$ are the arbitrary functions of y. Thus

$$\frac{\partial z}{\partial y} = -\frac{1}{4} \cos(2x + 3y) + \frac{x^2}{2} f(y) + g(y).$$

Now integrating partially w.r.t. 'y', we get

$$z = -\frac{1}{4} \int \cos(2x + 3y) dy + \int \frac{x^2}{2} f(y) dy + \int g(y) dy + h(x)$$

$$z = -\frac{1}{12} \sin(2x + 3y) + \frac{x^2}{2} \psi(y) + n(y) + h(x)$$

where

$$\psi(y) = \int f(y) dy, \; n(y) = \int g(y) dy \text{ and } h(x)$$

is an arbitrary function of x.

Example 2: Solve $z_x = 6x + 3y$ and

$$z_y = 3x - 4y.$$

Solution: Integrating $z_x = 6x + 3y$ partially w.r.t. x, we have $z = 3x^2 + 3xy + f(y)$ where $f(y)$ is an arbitrary function of y. To determine $f(y)$, differentiate the above equation partially w.r.t. y and equate it with the given second equation. Then

$$0+3x+\frac{df}{dy}=z_y=3x-4y$$

or $\frac{df}{dy}=-4y$. Integrating w.r.t. y, we get

$$f(y)=-2y^2+c$$

where c is a pure constant. Thus the solution is

$$z=3x^2+3xy-2y^2+c$$

Example 3: Solve $\frac{\partial^2 z}{\partial x^2}=a^2 z$ if $\frac{\partial z}{\partial x}=a\sin y$ and $\frac{\partial z}{\partial y}=0$ when $x=0$.

Solution: If z is a function of x alone, then

$$\frac{d^2 z}{dx^2}=a^2 z$$

whose auxiliary equation is $m^2-a^2=0$ and general solution is

$$z=c_1e^{ax}+c_2e^{-ax} \qquad (1)$$

But since z is actually a function of x and y, vary the constants c_1 and c_2 as functions of y. Thus

$$z=f(y)e^{ax}+g(y)e^{-ax}. \qquad (2)$$

Use the two conditions, to determine the two unknown functions $f(y)$ and $g(y)$.

$$a\sin y=\left.\frac{\partial z}{\partial x}\right|_{x=0}=af(y)e^{ax}+g(y)(-a)e^{-ax}\Big|_{x=0}$$

$$\sin y=f(y)-g(y) \qquad (3)$$

$$\text{Also } 0=\left.\frac{\partial z}{\partial y}\right|_{x=0}=f'(y)e^{ax}+g'(y)e^{-ax}\Big|_{x=0}$$

$$0=f'(y)-g'(y) \qquad (4)$$

Solving (3) and (4), we get
$f'(y)=\frac{1}{2}\cos y$ and $g'(y)=-\frac{1}{2}\cos y$.
Integrating w.r.t. y, we get
$f(y)=\frac{1}{2}\sin y+c_1,\ g(y)=-\frac{1}{2}\sin y+c_2 \qquad (5)$

Substituting (5) in (2), we have

$$z=\left(\frac{1}{2}\sin y+c_1\right)e^{ax}+\left(-\frac{1}{2}\sin y+c_2\right)e^{-ax}$$

$$=\sin y\cdot\sinh ax+c_1e^{ax}+c_2e^{-ax}$$

$$z=\sin y\cdot\sinh ax+c_3\cosh ax$$

Since $f-g=\left(-\frac{1}{2}\sin y+c_1\right)-\left(-\frac{1}{2}\sin y+c_2\right)=\sin y$, so $c_1=c_2$. Here $c_3=2c_1$.

EXERCISE

Solve the following P.D.E.

1. $\frac{\partial^2 z}{\partial x\partial y}=\sin x$

Ans. $z=-y\cos x+g(x)+F(y)$

2. $\frac{\partial^2 z}{\partial x\partial y}=\sin x\cdot\sin y$ given that $\frac{\partial t}{\partial y}=-2\sin y$ when $x=0$, and $z=0$ when y is an odd multiple of $\frac{\pi}{2}$.

Ans. $z=\cos x\cdot\cos y+\cos y$.

Hint: $\frac{\partial z}{\partial y}=-\cos x\cdot\sin y+f(y)$, use B.C. $f(y)=-\sin y$, $z=\cos x\cdot\cos y+\cos y+g(x)$, use B.C., $g(x)=0$.

3. $z_x=3x-y$ and $z_y=-x+\cos y$

Ans. $z=\frac{3x^2}{2}-xy+\sin y+c$

Hint: $z=\frac{3x^2}{2}-xy+f(y)$, use 2nd equation, $f'=\cos f(y)=\sin y+c$

4. $ys+p=\cos(x+y)-y\sin(x+y)$

Ans. $zy=y\sin(x+y)+F(y)+\phi(x)$

5. Show that the surface satisfying $t=6x^3y$ and containing the two lines $y=0, z=0, y=1, z=1$ is $z=x^3y^3+y(1-x^3)$

Hint: $z=x^3y^3+yf(x)+\phi(x)$, use conditions, $\phi(x)=0$, $1=x^3+f+\phi$

6. Prove that the surface $8axz=4ax-y^2$ passes through the parabolas $z=0, y^2=4ax$ and $z=1, y^2=-4ax$ and satisfies the equation $xr+2p=0$

7. Solve $\frac{\partial^3 z}{\partial x^2\partial y}+18xy^2+\sin(2x-y)=0$

Ans. $z=\frac{1}{4}\cos(2x-y)-x^3y^3+xf(y)+g(y)+h(x)$

8. Solve $\frac{\partial^2 z}{\partial x^2} + z = 0$ if $z = e^y$ and $\frac{\partial z}{\partial x} = 1$ when $x = 0$.

Ans. $z = \sin x + e^y \cos x$

9. Solve $t - xq = x^2$

Ans. $z = -xy + e^{xy}\frac{f(x)}{x} + g(x)$

Hint: Rewrite $\frac{\partial q}{\partial y} - xq = x^2$, I.F.: e^{-xy}

10. Solve $xs + q = 4x + 2y + 2$

Ans. $xyz = 2x^2y + y^2x + 2xy + \int f(y)dy + g(x)$

Hint: Rewrite $x\frac{\partial q}{\partial x} + q = 4x + 2y + z$, I.F.: x

11. Solve $t + s + q = 0$

Ans. $ze^x = F(x) + g(x - y)$ where $F(x) = \int f(x)e^x dx$

Hint: Rewrite $\frac{\partial q}{\partial x} + \frac{\partial p}{\partial y} + \frac{\partial z}{\partial y} = 0$. Integrate w.r.t. y, $p + q = f(x) - z$, $\frac{dx}{1} = \frac{dy}{1} = \frac{dz}{f(x)-z}$, $x - y = c_1$, $\frac{dz}{dx} + z = f(x)$.

12. Show that the equation of the surface satisfying $r + s = 0$ and touching the elliptic paraboloid $z = 4x^2 + y^2$ along its section by the plane $y = 2x + 1$ is $z + 4x^2 + y^2 - 8yx + 8x - 4y + 2 = 0$

Hint: Rewrite $\frac{\partial p}{\partial x} + \frac{\partial q}{\partial x} = 0$, integrate w.r.t. x, $p + q = f(y)$, $y - x = c_1$, $z = F(y) + \phi(y - x)$, $\phi(y - x) = 8(y - x) - 4(y - x)^2 + c_1$, $F(y) = 3y^2 - 4y + c_2$.

Chapter 19

Applications of Partial Differential Equations

INTRODUCTION

Several problems in fluid mechanics, solid mechanics, heat transfer, electromagnetic theory and other areas of physics are modeled as Initial Boundary Value Problems (IBVP) consisting of partial differential equations and initial conditions (I.C.s) specifying the state of the system at some initial time) and/or boundary conditions (B.Cs) (specifying the state of the system on the boundary say end points of an interval over which the solution is defined). Method of separation of variables is a powerful tool to solve such IBVP when PDE is linear and boundary conditions are homogeneous. In this chapter, we derive and solve by separation of variables technique some of the most important PDE's of one-dimensional heat equation, wave equation, Laplace's equation in two dimensions and in polar coordinates, two-dimensional heat equation, wave equation, vibrations of circular membrane, transmission lines. Unlike the ordinary differential equations, the general solution of PDE's involve arbitrary functions which require the knowledge of single and double Fourier series.

19.1 METHOD OF SEPARATION OF VARIABLES

Separation of variables is a powerful technique to solve P.D.E. For a P.D.E. in the function u of two independent variables x and y, assume that the required solution is separable, i.e.,

$$u(x, y) = X(x)Y(y) \tag{1}$$

where $X(x)$ is a function of x alone and $Y(y)$ is a function of y alone. Then substitution of u from (1) and its derivatives reduces the P.D.E. to the form

$$f(X, X', X'', \ldots) = g(Y, Y', Y'', \ldots) \tag{2}$$

which is separable in X and Y. Since the L.H.S. of (2) is a function of x alone and R.H.S. of (2) is a function of y alone, then (2) must be equal to a common constant say k. Thus (2) reduces to

$$f(X, X', X'', \ldots) = k \tag{3}$$

$$g(X, X', X'', \ldots) = k \tag{4}$$

Thus the determination of solution to P.D.E. reduces to the determination of solutions to two O.D.E. (with appropriate conditions).

WORKED OUT EXAMPLES

1. Solve $u_{xx} - u_y = 0$ by separation of variables.

Solution: Assume that $u(x, y) = X(x)Y(y)$. Differentiating w.r.t. x and y, we get

$$u_{xx} = X''Y, u_y = XY' \quad \text{so the P.D.E. reduces to}$$

$$2X''Y - XY' = 0$$

or $$\frac{2X''}{X} = \frac{Y'}{Y} = 2k = \text{constant}$$

Solving $X'' - kX = 0$, we have A.E.: $m^2 - k = 0$, solution is $X(x) = c_1 e^{\sqrt{k}x} + c_2 e^{-\sqrt{k}x}$

Solving $Y' - 2kY = 0$, we have solution as

$$Y(y) = c_3 e^{2ky}$$

Hence the required solution is

$$u(x, y) = X(x)Y(y) = \left(c_1 e^{\sqrt{k}x} + c_2 e^{-\sqrt{k}x}\right)(c_3 e^{2ky})$$

$$u(x, y) = \left(Ae^{\sqrt{k}x} + Be^{-\sqrt{k}x}\right)e^{2ky}$$

2. Use the separation of variables technique to solve $3u_x + 2u_y = 0$ with $u(x, 0) = 4e^{-x}$.

Solution: Assume $u(x, y) = X(x)Y(y)$. Then P.D.E. becomes $3X'Y + 2XY' = 0$

or $$\frac{X'}{X} = -\frac{2}{3}\frac{Y'}{Y} = k = \text{constant}$$

Solving $X' - kX = 0$, we get $X(x) = c_1 e^{kx}$

Similarly, $Y' + \frac{3}{2}kY = 0$, we get $Y(y) = c_2 e^{-\frac{3}{2}ky}$

So $u(x, y) = X(x)Y(y) = c_1 e^{kx} \cdot c_2 e^{-\frac{3}{2}ky} = ce^{\frac{k}{2}(2x-3y)}$

Given that $4e^{-x} = u(x, 0) = X(x)Y(0) = cee^{kx}$

Thus $c = 4$, $k = -1$. Hence the required solution is

$$u(x, y) = 4e^{-\frac{1}{2}(2x-3y)}.$$

Exercise

Solve the following P.D.E. by the method of separation of variables:

1. $4u_x + u_y = 3u$ and $u(0, y) = e^{-5y}$

Ans. $u(x, y) = e^{2x-5y}$

2. $2xz_x - 3yz_y = 0$

Ans. $z(x, y) = Ax^{3b}y^{2b}$ where $A = c_1c_2$, $k = 6b$

3. $u_x = 4u_y$, $u(0, y) = 8e^{-3y}$

Ans. $u(x, y) = 8e^{-12x-3y}$

4. $u_{xt} = e^{-t}\cos x$ with $u(x, 0) = 0$ and $\frac{\partial u(0,t)}{\partial t} = 0$

Ans. $u(x, t) = \sin x - e^{-t}\sin x$

5. $y^3 z_x + x^2 z_y = 0$

Ans. $z(x, y) = ce^{k\left(\frac{x^3}{3} - \frac{y^4}{4}\right)}$

6. $u_{xx} = u_y + 2u$, with $u(0, y) = 0$, $\frac{\partial u(0,y)}{\partial x} = 1 + e^{-3y}$

Ans. $u(x, y) = \frac{1}{\sqrt{2}}\sinh\sqrt{2}x + e^{-3y}\sin x$

7. $z_{xx} - 2z_x + z_y = 0$

Ans. $z(x, y) = [c_1 e^{ax} + c_2 e^{bx}]e^{-ky}$ where $a = (1 + \sqrt{1+k})$, $b = (1 - \sqrt{1+k})$

8. $u_x = 2u_t + u$ where $u(x, 0) = 6e^{-3x}$

Ans. $u(x, t) = 6e^{-(3x+2t)}$

19.2 CLASSIFICATION OF PARTIAL DIFFERENTIAL EQUATIONS OF SECOND ORDER

In the fields of wave propagation, heat conduction, vibrations, elasticity, boundary layer theory, etc., second order partial differential equations are of particular interest.

The general form of a second-order P.D.E. in the function u of the two independent variables x, y is given by

$$A(x, y)\frac{\partial^2 u}{\partial x^2} + B(x, y)\frac{\partial^2 u}{\partial x \partial y} + C(x, y)\frac{\partial^2 u}{\partial y^2} + f\left(x, y, u, \frac{\partial u}{\partial x}, \frac{\partial u}{\partial y}\right) = 0 \quad (1)$$

This equation is linear in second order terms. PDE (1) is said to be "linear or quasi-linear" according as f is linear or non-linear.

PDE (1) is classified as *elliptic, parabolic or hyperbolic* according as $B^2 - 4AC < 0, = 0$ or > 0.

Example:

Elliptic: $(B^2 - 4AC < 0)$

Laplace's equation in two dimensions

$$\frac{\partial^2 u}{\partial x^2} + \frac{\partial^2 u}{\partial y^2} = 0 \quad (2)$$

Poisson's equation

$$\frac{\partial^2 u}{\partial x^2} + \frac{\partial^2 u}{\partial y^2} = f(x, y) \quad (3)$$

Parabolic: $(B^2 - 4AC = 0)$

One dimensional heat-flow equation

$$a^2\frac{\partial^2 u}{\partial x^2} = \frac{\partial u}{\partial t} \quad (4)$$

Hyperbolic: $(B^2 - 4AC > 0)$

One-dimensional wave equation

$$\frac{\partial^2 u}{\partial t^2} = a^2\frac{\partial^2 u}{\partial x^2} \quad (5)$$

P.D.E. (2), (4), (5) are homogeneous while (3) is non-homogeneous.

Note: For an elliptic Equation (2) or (3) boundary conditions are prescribed in a closed region, whereas for parabolic or hyperbolic Equations (4) or (5) boundary conditions and initial conditions are prescribed in an open-ended region.

19.3 DERIVATION OF ONE-DIMENSIONAL HEAT EQUATION

Consider a heat conducting homogeneous rod of length L, placed along the x-axis with one end of the rod at $x = 0$ (origin) and the other end of the rod at $x = L$ (Fig. 19.1). Assume that the rod has constant density ρ and uniform cross section A. Also assume that the lateral surface of the rod is impenetrable to heat transfer i.e., rod is insulated laterally and therefore heat flows only in the x-direction. The rod is sufficiently thin so that the temperature is same at all points of any cross sectional area of the rod. Let $u(x, t)$ be the temperature of the cross section at the point x at any time t.

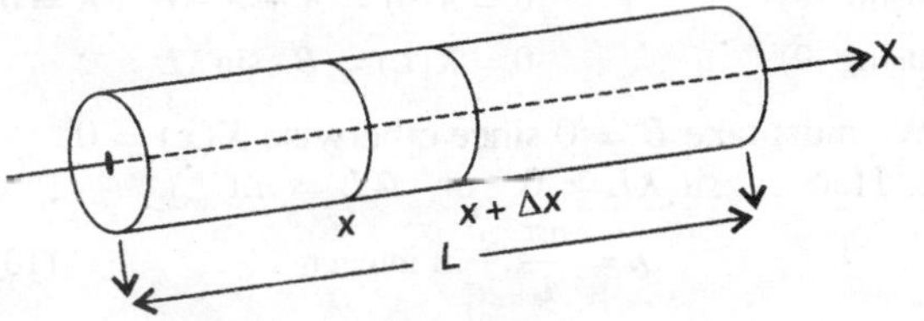

Fig. 19.1

The amount of heat $Q(t)$ in a small segment of the rod between the cross sections at x and $x + \Delta x$ is

$$Q(t) = \int_x^{x+\Delta x} (c)(\rho A)\, u(s, t)\, ds \qquad (1)$$

where c is specific heat of the rod. By Fourier law of heat conduction, rate of propagation of heat (i.e., the quantity of heat passing through a cross section at x in unit time) is

$$q = -k \frac{\partial u}{\partial x} \cdot A \qquad (2)$$

where k is the coefficient of thermal conductivity. Since heat flows in the direction of decreasing temperature, a negative sign appears in (2). The rate of heat flow at cross section $x + \Delta x$ is

$$-kA \frac{\partial u}{\partial x}(x + \Delta x, t)$$

The rate of change of heat content in the segment of the rod between x and $x + \Delta x$ must be equal to net heat flow into this segment of the rod. Thus

$$\frac{\partial Q}{\partial t} = kA \left[\left.\frac{\partial u}{\partial x}\right|_{x+\Delta x} - \left.\frac{\partial u}{\partial x}\right|_{x} \right] \qquad (3)$$

By mean value theorem,

$$\frac{\partial Q}{\partial t} = \frac{\partial}{\partial t} \int_x^{x+\Delta x} c\rho A u(s, t)\, dt = \int_x^{x+\Delta x} c\rho A \frac{\partial u}{\partial t}\, ds$$

$$= c\rho A \cdot u_t(\xi, t)\Delta x \qquad (4)$$

where ξ lies between x and $x + \Delta x$. Replacing L.H.S. of (3) by (4)

$$c\rho A \frac{\partial u}{\partial t}(\xi, t)\, \Delta x = kA \left[\left.\frac{\partial u}{\partial x}\right|_{x+\Delta x} - \left.\frac{\partial u}{\partial x}\right|_{x} \right]$$

Rewriting this and taking the limit as $\Delta x \to 0$,

$$\frac{\partial u(x, t)}{\partial t} = a^2 \frac{\partial^2 u}{\partial x^2} \qquad (5)$$

where $a^2 = \frac{k}{cp}$ known as diffusivity constant. Equation (5) is the one-dimensional heat equation which is second order, homogeneous and **parabolic** type.

19.4 SOLUTION OF ONE-DIMENSIONAL HEAT EQUATION

By separation of variables technique

Consider a long thin wire or rod or bar of constant cross section and homogeneous heat conducting material. Let the bar be of length L oriented along the x-axis with one end A coinciding with origin (Fig. 19.2). Suppose that the lateral surface of the bar is perfectly insulated. Then the heat flows in the bar along the x-direction only.

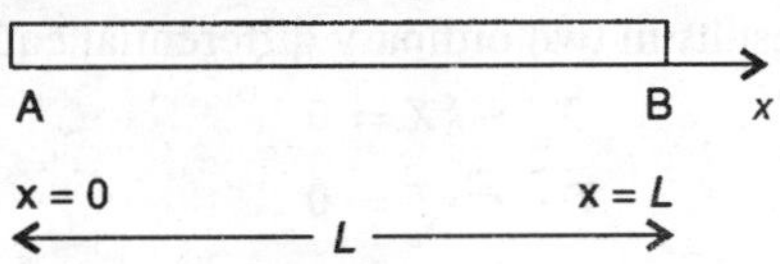

Fig. 19.2

Thus the temperature u of the bar depends only on x and t. This phenomenon is described by the initial boundary value problem consisting of one-dimensional heat equation

$$\frac{\partial u}{\partial t} = c^2 \frac{\partial^2 u}{\partial x^2} \tag{1}$$

where c^2 is the thermal diffusivity, the boundary conditions (B.C.'s) prescribed at the end points A and B.

$$x(0, t) = 0, \quad x(L, t) = 0 \tag{2}$$

i.e., the end points are assumed to be at zero temperature and the initial temperature distribution in the bar given by

$$u(x, 0) = f(x) \tag{3}$$

where $f(x)$ is a given (prescribed) function of x. Solution by the separation of variables technique or the method of separating variables or product method reduces the IBVP to the solution of two ordinary differential equations as follows:

Step I. Assume that the solution $u(x, t)$ is separable i.e.,

$$u(x, t) = X(x)T(t) \tag{4}$$

where $X(x)$ is a function of x alone and $T(t)$ is a function of t only.

Differentiating (4) w.r.t. t and x, we get

$$\frac{\partial u}{\partial t} = X\dot{T}, u_x = X'T, u_{xx} = X''T$$

where $'$ denotes differentiation w.r.t x and $\dot{}$ denotes differentiation w.r.t. t. Substituting these in the P.D.E. (1), we get

$$X\dot{T} = c^2 X''T$$

$$\frac{X''}{X} = \frac{\dot{T}}{c^2 T} \tag{5}$$

Since the L.H.S. of (5) is a function of x alone while R.H.S. of (5) is a function of t alone, both sides of (5) must be a constant. Thus

$$\frac{X''}{X} = \frac{\dot{T}}{c^2 T} = k = \text{constant}$$

which results in two ordinary differential equations

$$X'' - kX = 0 \tag{6}$$

and

$$\dot{T} - c^2 kT = 0 \tag{7}$$

Using (4) the boundary conditions (2) reduce to

$$0 = u(0, t) = X(0)T(t) \implies X(0) = 0 \tag{8}$$

$$0 = u(L, t) = X(L)T(t) \implies X(L) = 0 \tag{9}$$

since $T(t) \neq 0$ (otherwise if $T(t) = 0$, $u = 0$ for all t which is a trivial solution).

Step II.

Case 1: $k = 0$, $X'' = 0$

Solution is $\quad X(x) = ax + b$

using (8) and (9) $\quad 0 = X(0) = a \cdot 0 + b \quad \therefore\ b = 0,$

$0 = X(L) = a \cdot L \quad \therefore\ a = 0$

Thus $X(x) = 0$ for all x or $u = 0$ for all x which is a trivial solution.

Case 2: $k = \lambda^2 > 0, \quad X'' - \lambda^2 X = 0$

Solution is $\quad X(x) = Ae^{\lambda x} + Be^{-\lambda x}$

using (8) $\quad 0 = X(0) = A + B$ or $A = -B$

using (9) $\quad 0 = X(L) = A(e^{\lambda L} - e^{-\lambda L})$

$\implies A = 0 = B$

since $\quad e^{\lambda L} - e^{-\lambda L} = 2\sinh \lambda L = 0$ only when $\lambda = 0$.

Thus $X(x) = 0$ for all x or $u = 0$ for all x which is a trivial solution.

Case 3: $k = -\lambda^2$

Equation is $\quad X'' + \lambda^2 X = 0$

Solution is $\quad X(x) = A\cos\lambda x + B\sin\lambda X$

using (8) $\quad 0 = X(0) = A + B \cdot 0 \quad \therefore\ A = 0$

using (9) $\quad 0 = X(L) = B \cdot \sin\lambda L$

We must take $B \neq 0$ since otherwise $X(x) = 0$.

Hence $\sin\lambda L = 0$ or $\lambda L = n\pi$

$$\therefore \quad \lambda = \frac{n\pi}{L}, \quad n \text{ integer} \tag{10}$$

Assuming $B = 1$, we get infinitely many solutions

$$X_n(x) = \sin\frac{n\pi x}{L}, \quad n = 1, 2, 3, \ldots \tag{11}$$

Solving (7) $\quad \dot{T} + c^2\lambda^2 T = 0$, we get

$$T_n(t) = B_n e^{-\lambda_n^2 t} = B_n e^{-n^2\pi^2 t/L^2} \tag{12}$$

Hence the "eigen functions"

$$u_n(x, t) = X_n(x)T_n(t)$$

$$= B_n \frac{\sin n\pi x}{L} \cdot e^{-\lambda_n^2 t}, n = 1, 2, 3, \ldots \tag{13}$$

satisfy Equation (1) and boundary conditions (2).

Step III. In order to satisfy the initial condition (3) we invoke the *superposition* or *linearity principle* which states that

$$u(x,t)=\sum_{n=1}^{\infty} u_n(x,t)$$

$$=\sum_{n=1}^{\infty} B_n \sin\frac{n\pi x}{L}\cdot e^{-c^2n^2\pi^2t/L^2} \quad (14)$$

satisfies (1) and (2). To determine the unknown constants B_n, use initial condition (3) in (14)

$$f(x)=u(x,0)=\sum_{n=1}^{\infty} B_n\cdot\sin\frac{n\pi x}{L} \quad (15)$$

Since the R.H.S. of (15) is a half range Fourier sine series expansion of $f(x)$ in the interval $(0, L)$, the Fourier coefficients B_n's are given by

$$B_n=\frac{2}{L}\int_0^L f(x)\sin\frac{n\pi x}{L}dx, \quad n=1,2,\ldots \quad (16)$$

Thus the temperature distribution in the bar is given by the solution (14) with B_n's calculated from (16).

Note: Solution (14) is transient contains (negative exponential) i.e., u decreases as t increases which is consistent with physical nature of the heat conduction problem.

Steady-state condition

A condition (phenomena) is said to be steady-state if the (dependent) variables are independent (free) of time t.

Non-homogeneous boundary conditions are boundary conditions which are all not zero.

Non-homogeneous (Non-zero) Boundary Conditions

Suppose a bar AB of length L has its ends A and B maintained at A_0° and B_0° respectively until steady-state condition is reached. Then the temperatures at A and B are suddenly and simultaneously changed to A_1° and B_1° and maintained thereafter. To find the subsequent temperature distribution in the bar:

Step I. Initial temperature distribution in the bar is determined using the steady-state condition in which u will be independent of t $\left(\text{i.e., } \frac{\partial u}{\partial t}=0\right)$. Therefore the heat equation reduces to

$$\frac{d^2u}{dx^2}=0$$

Its solution is $u(x)=ax+b$

The boundary conditions in the steady-state are $u(0)=A_0$ and $u(L)=B_0$. Using these

$$A_0=u(0)=a\cdot 0+b \quad \therefore \quad b=A_0$$

$$B_0=u(L)=a\cdot L+A_0 \quad \therefore \quad a=(B_0-A_0)/L$$

Thus the initial temperature distribution in the bar is

$$u(x,0)=\frac{(B_0-A_0)}{L}x+A_0 \quad (17)$$

Now the boundary conditions (temperatures) at A and B have been changed to

$$u(0,t)=A_1 \quad \text{and} \quad u(L,t)=B_1 \quad (18)$$

Thus the heat flow problem is to solve the one-dimensional heat equation (1) with B.C.'s (18) and initial condition (17). Unlike the previous problem, here the boundary conditions are non-homogeneous (non-zero). Therefore assume that the required solution $u(x,t)$ as

$$u(x,t)=u_s(x)+u_{tr}(x,t) \quad (19)$$

where $u_s(x)$ is the steady-state solution and $u_{tr}(x,t)$ is the transient solution (containing negative exponential of time t).

Step II. To determine $u_s(x)$:

The equation is $\quad \frac{d^2u_s(x)}{dx^2}=0$

Its solution is $\quad u_s(x)=Ax+B$

Use (18) $\quad A_1=u_s(0)=A\cdot 0+B \;\therefore\; B=A_1$

$$B_1=u_s(L)=A\cdot L+B \;\therefore\; A=\frac{B_1-A_1}{L}$$

Thus $\quad u_s(x)=\frac{B_1-A_1}{L}x+A_1 \quad (20)$

Step III. To find $u_{tr}(x)t$:

From (19), $\quad u_{tr}(x,t)=u(x,t)-u_s(x)$

So $\quad u_{tr}(0,t)=u(0,t)-u_s(0)=A_1-A_1=0 \quad (21)$

$$u_{tr}(L,t)=u(L,t)-u_s(L)=B_1-B_1=0 \quad (22)$$

Thus the boundary conditions for $u_{tr}(x, t)$ are homogeneous (zero).

The initial condition is

$$u_{tr}(x, 0) = u(x, 0) - u_s(x)$$

$$= \left[\frac{B_0 - A_0}{L}x + A_0\right] - \left[\frac{B_1 - A_1}{L}x + A_1\right]$$

$$u_{tr}(x, 0) = [(A_1 - A_0) + (B_1 - B_0)] \cdot \frac{x}{L} + (A_0 - A_1)$$

$$= A_2\frac{x}{L} + A_3 \qquad (23)$$

where $A_2 = (A_1 - A_0) + B_1 - B_0,\ A_3 = A_0 - A_1$

Therefore $u_{tr}(x, t)$ satisfying one-dimensional heat equation, the zero B.C.'s (21), (22), and the initial condition (23) is given by

$$u_{tr}(x, t) = \sum A_n \sin\frac{n\pi x}{L} \cdot e^{-c^2n^2\pi^2 t/L^2} \qquad (24)$$

where $$A_n = \frac{2}{L}\int_0^L \cdot f(x) \cdot \sin\frac{n\pi x}{L}dx$$

$$A_n = \frac{2}{L}\int_0^L \left(\frac{A_2 x}{L} + A_3\right)\sin\frac{n\pi x}{L}dx$$

$$= \frac{2}{L}\left[\left(\frac{A_2 x}{L} + A_3\right)\left(\frac{-L}{n\pi}\right) \cdot \cos\frac{n\pi x}{L}\right.$$

$$\left. - \frac{A_2}{L} \cdot \left(\frac{-L^2}{n^2\pi^2}\right) \cdot \sin\frac{n\pi x}{L}\right]_0^L$$

$$A_n = \frac{-2}{n\pi}[(A_2 + A_3)\cos n\pi - A_3]$$

$$= \frac{2}{n\pi}[A_3 - (A_2 + A_3)(-1)^n] \qquad (25)$$

Hence the required temperature distribution in bar is given by (19), (20), (24), (25) i.e.,

$$u(x, t) = u_s(x) + u_{tr}(x, t)$$

$$= \left[(B_1 - A_1)\frac{x}{L} + A_1\right] +$$

$$+\frac{2}{\pi}\sum\frac{[A_3 - (A_2 + A_3)(-1)^n]}{n} \times$$

$$\times \sin\frac{n\pi x}{L} \cdot e^{-c^2n^2\pi^2 t/L^2}.$$

Bar with both ends insulated

When both ends of the bar are insulated, no heat flows through them and then the corresponding boundary conditions are

$$\left.\frac{\partial u}{\partial x}\right|_{x=0} = u_x(0, t) = 0,\quad \left.\frac{\partial u}{\partial x}\right|_{x=L} = u_x(L, t) = 0 \qquad (26)$$

Using (4), these boundary conditions reduce to

$$0 = u_x(0, t) = X'(0)T(t) \quad \therefore \quad X'(0) = 0 \qquad (27)$$

$$0 = u_x(L, t) = X'(L)T(t) \quad \therefore \quad X'(L) = 0 \qquad (28)$$

Differentiating $X(x) = A\cos\lambda x + B\sin\lambda x$ w.r.t. 'x'

$$X'(x) = -A\lambda\sin\lambda x + B\lambda\cos\lambda x$$

Using (27), $0 = X'(0) = B\lambda$ and then using (28)

$$0 = X'(L) = -A\lambda\sin\lambda L$$

Thus assuming $A = 1$ and $B = 0$, then $\lambda = \frac{n\pi}{L}$, n integer

Hence $$X_n(x) = \cos\frac{n\pi x}{L},\quad n = 0, 1, 2, 3, \ldots$$

No change in the solution for $T(t)$. The required eigen functions are $u_n(x, t) = X_n(x)T_n(t)$, with $n = 0, 1, 2, \ldots$

By the linearity principle, the required solution is

$$u(x, t) = \sum_{n=0}^{\infty} u_n(x, t) = \sum_{n=0}^{\infty} X_n(x) \cdot T_n(t)$$

$$u(x, t) = \sum_{n=0}^{\infty} A_n \cdot \cos\frac{n\pi x}{L} \cdot e^{-c^2n^2\pi^2 t/L^2}$$

where $$A_0 = \frac{2}{L}\int_0^L f(x)dx$$

$$A_n = \frac{2}{L}\int_0^L f(x)\cos\frac{n\pi x}{L}dx.$$

WORKED OUT EXAMPLES

Homogeneous (zero) boundary conditions

Example 1: Find the temperature in a bar of length 2 whose ends are kept at zero and lateral surface insulated if the initial temperature is $\sin\frac{\pi x}{2} + 3\sin\frac{5\pi x}{2}$.

Solution: Let $u(x, t)$ be the temperature in the bar. The boundary conditions are $u(0, t) = u(2, t) = 0$ for any t. The initial condition is $u(x, 0) = \sin\frac{\pi x}{2} + 3\sin 5\pi\frac{x}{2}$. Then the solution is

$$u(x,t)=\sum_{n=1}^{\infty} A_n \sin\frac{n\pi x}{l}\cdot e^{-n^2\pi^2c^2t/l^2}$$

Here l = length of bar = 2 and

$$A_n=\frac{2}{l}\int_0^l f(x)\sin\frac{n\pi x}{l}dx$$

$$A_n=\frac{2}{2}\int_0^2\left(\sin\frac{\pi x}{2}+3\sin\frac{5\pi x}{2}\right)\sin\frac{n\pi x}{2}dx$$

$A_1=1$, $A_{2,3,4}=0$, $A_5=3$, $A_n=0$ for $n\geq 6$

Thus

$$u(x,t)=\sin\left(\frac{\pi x}{2}\right)\cdot e^{-c^2\pi^2t/4}+3\sin\left(\frac{5\pi x}{2}\right)\cdot e^{-c^2\pi^2 25t/4}$$

Example 2: Find the temperature $u(x,t)$ in a homogeneous bar of heat conducting material of length L cm with its ends kept at zero temperature and initial temperature given by $dx(L-x)/L^2$.

Solution: The initial boundary value problem consists of

i. *P.D.E. heat equation*: $\frac{\partial u}{\partial t}=c^2\frac{\partial^2 u}{\partial x^2}$

ii. *Boundary conditions*: $u(x,0)=0$, $u(L,0)=0$, for any t

iii. *Initial condition*: $u(x,0)=dx(L-x)/L^2$, $0<x<L$

The solution is

$$u(x,t)=\sum_{n=1}^{\infty}A_n\cdot\sin\frac{n\pi x}{L}\cdot e^{-n^2\pi^2c^2t/L^2}$$

where

$$A_n=\frac{2}{L}\int_0^L f(x)\sin\frac{n\pi x}{L}dx \quad \text{for } n=1,2,3,\ldots$$

$$A_n=\frac{2}{L}\int_0^L\frac{dx(L-x)}{L^2}\sin n\frac{\pi x}{L}dx$$

$$=\frac{2d}{L^3}\left[x(L-x)\cdot\left(\frac{-L}{n\pi}\right)\cos\frac{n\pi x}{L}\right.$$

$$-(L-2x)\cdot\left(\frac{-L^2}{n^2\pi^2}\right)\cdot\sin\frac{n\pi x}{L}$$

$$\left.+(-2)\left(\frac{-L^3}{n^3\pi^3}\right)\left(-\cos\frac{n\pi x}{L}\right)\right]_0^L$$

$$=\frac{2d}{L^3}\left[0+0-\frac{2L^3}{n^3\pi^3}\{(-1)^n-1\}\right]$$

$$=\frac{8d}{n^3\pi^3}\quad\text{if } n \text{ is odd}$$

Therefore the temperature distribution in the bar is given by

$$u(x,t)=\frac{8d}{\pi^3}\sum_{n=1}^{\infty}\frac{1}{(2n-1)^3}\sin\frac{(2n-1)\pi x}{L}\times$$

$$\times\, e^{-(2n-1)^2\pi^2c^2t/L^2}.$$

Steady-state conditions and zero B.C.'s

Example 3: A bar of length L, laterally insulated, has its ends A and B kept at 0° and u_0° respectively until steady-state conditions prevail. If the temperature at B is then suddenly reduced to 0° and kept so while that of A is maintained at 0° find the temperature in the bar at any subsequent time.

Solution: Let $u(x,t)$ be the temperature in the bar AB. The initial temperature distribution in the bar is to be determined from the steady-state condition (in which u will be independent of time t). The equation governing this steady-state condition is

$$\frac{d^2u}{dx^2}=0$$

whose solution is

$$u(x)=ax+b$$

with a and b two arbitrary constants. Use the boundary conditions $u(x=0)=0$ and $u(x=L)=u_0$ to determine a and b.

$$0=u(0)=a\cdot 0+b \quad \therefore b=0$$

$$u_0=u(L)=a\cdot L+0 \quad \therefore a=\frac{u_0}{L}$$

Thus the initial temperature distribution in the bar is

$$u(x,0)=\frac{u_0}{L}x \tag{1}$$

and the boundary conditions are

$$u(0,t)=0,\quad u(L,t)=0 \tag{2}$$

The solution to the one-dimensional heat equation $u_t=c^2u_{xx}$ with the initial condition (1) and boundary conditions (2) is

$$u(x,t)=\sum_{n=1}^{\infty} A_n \cdot \sin\left(\frac{n\pi x}{L}\right)\cdot e^{-(n^2\pi^2c^2t/L^2)}$$

$$\text{where } A_n=\frac{2}{L}\int_0^L f(x)\sin\frac{n\pi x}{L}dx$$

$$=\frac{2}{L}\int_0^L \frac{u_0 x}{L}\cdot\sin\frac{n\pi x}{L}dx$$

$$=\frac{2u_0}{L^2}\left[-\frac{xL}{n\pi}\cos\frac{n\pi x}{L}+\frac{L^2}{n^2\pi^2}\sin\frac{n\pi x}{L}\right]_0^L$$

$$=-\frac{2u_0L^2}{L^2n\pi}(-1)^n$$

Thus the required solution (temperature distribution in the bar) is

$$u(x,t)=\frac{2u_0}{\pi}\sum_{n=1}^{\infty}\frac{(-1)^{n+1}}{n}\sin\frac{n\pi x}{L}\cdot e^{-(n^2\pi^2c^2t/L^2)}.$$

Non-homogeneous (non-zero) boundary conditions

Example 4: A bar AB of length 10 cm has its ends A and B kept at 30° and 100° temperatures respectively, until steady-state condition is reached. Then the temperature at A is lowered to 20° and that at B to 40° and these temperatures are maintained. Find the subsequent temperature distribution in the bar.

Solution: In order to find the initial temperature distribution in the bar, make use of the steady-state condition that the temperature $u(x,t)$ is independent of time t.
Then the one-dimensional heat equation

$$\frac{\partial u}{\partial t}=c^2\frac{\partial^2 u}{\partial x^2} \tag{1}$$

reduces to $\frac{d^2u}{dx^2}=0$ whose solution is given by $u(x)=ax+b$. The two arbitrary constants a and b are determined using the boundary conditions $u(0,t)=30°$ and $u(10,t)=100°$. Thus

$$30°=u(0,t)=0\cdot 0+b \quad \therefore b=30°$$

$$100°=u(10,t)=a\cdot 10+b=a\cdot 10+30° \quad \therefore a=7$$

So the initial temperature distribution in the bar is

$$u(x,0)=7x+30 \tag{2}$$

When steady-state condition is reached, the temperatures at the ends A and B has been changed to 20° and 40°. So the boundary conditions are

$$u(0,t)=20 \quad\text{and}\quad u(10,t)=40 \tag{3}$$

which are non-homogeneous. Therefore assume the solution as

$$u(x,t)=u_s(x)+u_{tr}(x,t) \tag{4}$$

To find $u_s(x)$: solve $\frac{d^2u_s}{dx_2}=0$ with $u_s(0)=20$, $u_s(10)=40$. The solution is $u_s(x)=a_1x+b_1$. Using boundary conditions, we get

$$20=u_s(0)=a_1\cdot 0+b_1 \quad \therefore b_1=20$$

$$40=u_s(10)=a_1\cdot 10+20 \quad \therefore a_1=2$$

Thus $$u_s(x)=2x+20 \tag{5}$$

To find $u_{tr}(x,t)$: From (4)

$$u_{tr}(x,t)=u(x,t)-u_s(x)$$

the boundary conditions are

$$u_{tr}(0,t)=u(0,t)-u_s(0)=20-20=0 \tag{6}$$

$$u_{tr}(10,t)=u(10,t)-u_s(10)=40-40=0 \tag{7}$$

i.e., boundary conditions are homogeneous. The initial condition is

$$u_{tr}(x,0)=u(x,0)-u_s(x)=(7x+30)-(2x+20)$$

$$u_{tr}(x,0)=5x+10 \tag{8}$$

Thus to determine $u_{tr}(x,t)$, solve the one-dimensional heat equation

$$\frac{\partial u_{tr}}{\partial t}=c^2\frac{\partial^2 u_{tr}}{\partial x^2}$$

with the homogeneous boundary conditions (6), (7) and initial condition (8). Hence the required solution $u_{tr}(x,t)$ is given by

$$u_{tr}(x,t)=\sum_{n=1}^{\infty}A_n\sin\frac{n\pi x}{L}\cdot e^{-c^2n^2\pi^2t/L^2}$$

where $$A_n=\frac{2}{L}\int_0^L f(x)\cdot\sin\frac{n\pi x}{L}dx$$

$$=\frac{2}{L}\int_0^L(5x+10)\sin\frac{n\pi x}{L}dx$$

$$=\frac{2}{L}\left[(5x+10)\cdot\left(\frac{-L}{n\pi}\right)\cos\frac{n\pi x}{L}\right.$$

$$-5\left(\frac{-L^2}{n^2\pi^2}\right)\cdot\sin\frac{n\pi x}{L}\Bigg]_0^L$$

$$A_n = \frac{20}{n\pi}[1 - 6(-1)^n] \quad \text{with } L = 10$$

Therefore the temperature distribution in bar is

$$u(x, t) = u_s(x) + u_{tr}(x, t)$$

$$= (2x + 20) + \frac{20}{\pi}\sum_{n=1}^{\infty}\left\{\frac{1 - 6(-1)^n}{n}\right\} \times$$

$$\times \sin\frac{n\pi x}{10}\cdot e^{-c^2\pi^2 n^2 t/100}$$

Both ends insulated

Example 5: Find the temperature in a thin metal rod of length L, with both the ends insulated (so that there is no passage of heat through the ends) and with initial temperature in the rod $\sin(\pi x/L)$.

Solution: Let $u(x, t)$ be the temperature in rod. Then the initial boundary value problem is

$$\frac{\partial u}{\partial t} = c^2\frac{\partial^2 u}{\partial x^2}, \quad \text{P.D.E.}$$

$$\left.\frac{\partial u}{\partial x}\right|_{x=0} = 0, \quad \left.\frac{\partial u}{\partial x}\right|_{x=L} = 0, \quad \text{B.C.}$$

$$u(x, 0) = \sin\left(\frac{\pi x}{L}\right), \quad \text{I.C.}$$

Its solution is

$$u(x, t) = \frac{c_0}{2} + \sum_{n=1}^{\infty} c_n \cos\frac{n\pi x}{L}\cdot e^{-n^2\pi^2c^2t/L^2}$$

where $$c_0 = \frac{2}{L}\int_0^L f(x)dx = \frac{2}{L}\int_0^L \sin\frac{n\pi x}{L}dx$$

$$c_0 = \frac{2}{L}\frac{L}{\pi}\left[-\cos\frac{\pi x}{L}\right]_0^L$$

$$= \frac{2}{\pi}[+1 + 1] = \frac{4}{\pi}$$

$$c_1 = \frac{2}{L}\int_0^L \sin\frac{\pi x}{L}\cdot\cos\frac{\pi x}{L}dx = 0$$

$$c_n = \frac{2}{L}\int_0^L \sin\frac{\pi x}{L}\cdot\cos\frac{n\pi x}{L}dx \quad \text{for } n \neq 1$$

$$= \frac{2}{L}\frac{1}{2}\int_0^L\left[\sin\left(1 + n\right)\frac{\pi x}{L}\right.$$

$$\left. + \sin\left(\frac{(1 - n)\pi x}{L}\right)\right]dx$$

$$= \frac{1}{L}\left[\frac{-L}{(n + 1)\pi}\cos(1 + n)\frac{\pi x}{L}\right.$$

$$\left. - \frac{L}{(1 - n)\pi}\cdot\cos(1 - n)\frac{\pi x}{L}\right]_0^L$$

$$= -\frac{[(-1)^{n+1} - 1]}{\pi}\left[\frac{1}{n + 1} - \frac{1}{n - 1}\right]$$

$$= \frac{-2[(-1)^{n+1} - 1]}{\pi(n^2 - 1)}$$

$$= \frac{4}{\pi(n^2 - 1)} \quad \text{where } n \text{ is even.}$$

Thus the temperature $u(x, t)$ in the rod

$$= \frac{2}{\pi} - \frac{4}{\pi}\sum_{n=1}^{\infty}\frac{1}{(4n^2 - 1)}\cdot\cos\left(\frac{2n\pi x}{L}\right)\cdot e^{-(4n^2c^2t\pi^2/L^2)}.$$

One end insulated

Example 6: Solve $u_t = c^2 u_{xx}$ when

i. $u \neq \infty$ as $t \to \infty$

ii. $u_x = 0$ when $x = 0$ for all t

iii. $u = 0$ when $x = L$ for all t

iv. $u = u_0 =$ constant

when $t = 0$ for $0 < x < L$.

Solution: The given equations represent heat flow in a rod of length L whose one end (at $x = 0$) is insulated and the initial temperature in the rod is constant. Assume that the temperature in the rod

$$u(x, t) = X(x)T(t)$$

The P.D.E. reduces to

$$\frac{X''}{X} = \frac{\dot{T}}{c^2T} = k = \text{constant}$$

with $$u_x(0, t) = x'(0)T = 0 \Rightarrow X'(0) = 0$$

$$u(L, t) = X(L)T = 0 \Rightarrow X(L) = 0$$

when $k = -\lambda^2 < 0$, non-trivial solutions exist. So

$$X'' + \lambda^2 X = 0, \quad X'(0) = 0, \quad X(L) = 0$$

$$X(x) = A\cos\lambda x + B\sin\lambda x$$

$$X'(x) = -A\lambda\sin\lambda x + B\lambda\cos\lambda x$$

$$0 = X'(0) = 0 + B\lambda \quad \therefore B = 0$$

$$0 = X(L) = A\cos\lambda L$$

$$\therefore \lambda = \frac{(2n-1)}{L}\frac{\pi}{2}, \quad n = 1, 2, \ldots$$

Thus $\quad X_n(x) = \cos\dfrac{(2n-1)}{L}\dfrac{\pi}{2}\cdot x$

$$T_n(t) = e^{-c^2\lambda_n^2 t}$$

Hence the general solution (by principle of superposition) is

$$u(x,t) = \sum_{n=1}^{\infty} A_n \cos(\lambda_n x)\cdot e^{-c^2\lambda_n^2 t}$$

To find A_n's use the initial condition

$$u_0 = u(x,0) = \sum_{n=1}^{\infty} A_n \cos(\lambda_n x)$$

So $\quad A_n = \dfrac{2}{L}\displaystyle\int_0^L f(x)\cdot\cos(\lambda_n x)dx$

$$A_n = \frac{2}{L}\int_0^L u_0\cdot\cos\left(\frac{(2n-1)}{L}\frac{\pi}{2}x\right)dx$$

$$= \frac{2u_0}{L}\cdot\frac{2L}{(2n-1)\pi}\cdot\sin\left(\frac{(2n-1)\pi x}{2L}\right)\Bigg|_0^L$$

$$= \frac{4u_0(-1)^{n-1}}{(2n-1)\pi}$$

$$\therefore u(x,t) = \frac{4u_0}{\pi}\sum_{n=1}^{\infty}\frac{(-1)^{n-1}}{(2n-1)}\cos\left(\frac{(2n-1)\pi x}{2L}\right)\times$$

$$\times e^{-(c^2(2n-1)^2\pi^2 t)/(4L^2)}.$$

EXERCISE

Homogeneous B.C.'(s)

Find the temperature $u(x,t)$ in a laterally insulated heat conducting bar of length L with its ends kept at 0° and with the initial temperature in the bar is $u(x,0)$:

1. $u(x,0) = 100\sin(\pi x/80)$; $L = 80$ cm

Ans. $u(x,t) = 100\sin(\pi x/80)\cdot e^{-c^2\pi^2 t/L^2}$

2. $u(x,0) = \sin\frac{\pi x}{2} - 3\sin 2\pi x$; $L = 2$

Ans. $u(x,t) = \sin\left(\frac{\pi x}{2}\right)\cdot e^{-\pi^2c^2t/4}$
$- 3\sin 2\pi x\cdot e^{-4^2\pi^2c^2t/4}$

3. "Triangular" temperature:

$$u(x,0) = \frac{2Tx}{L}, \quad \text{when } 0 \le x \le \frac{L}{2}$$
$$= \frac{2T}{L}(L-x), \quad \text{when } \frac{L}{2} \le x \le L$$

Ans. $u(x,t) = \frac{8T}{\pi^2}\sum_{n=1}^{\infty}\frac{1}{(2n-1)^2}\sin\frac{(2n-1)\pi}{2}\cdot\sin\frac{(2n-1)\pi x}{L}$
$\times e^{-(2n-1)^2\pi^2c^2t/L^2}$

4. $u(x,0) = x(L-x), 0 < x < L$

Ans. $u(x,t) = \sum_{n=1}^{\infty}\frac{8L^2}{(2n-1)^2\pi^2}\sin\left(\frac{(2n-1)\pi x}{L}\right)$
$\times e^{-(2n-1)^2\pi^2c^2t/L^2}$

5. $u(x,0) = u_0 =$ constant

Ans. $u(x,t) = \frac{4u_0}{\pi}\sum_{n=1}^{\infty}\frac{1}{(2n-1)}\cdot\sin\left(\frac{(2n-1)\pi x}{L}\right)$
$\times e^{-(2n-1)^2\pi^2c^2t/L^2}$

6. $u(x,0) = bx, 0 < x < L, b =$ constant

Ans. $u(x,t) = \sum_{n=1}^{\infty}\frac{2Lb}{\pi n}(-1)^{n+1}\sin\left(\frac{n\pi x}{L}\right)$
$\times e^{-\pi^2c^2n^2t/L^2}$

7. $u(x,0) = T_1$ when $0 < x < \frac{L}{2}$
$= T_2$ when $\frac{L}{2} < x < L$

where T_1, T_2 are constants.

Ans. $u(x,t) = \sum_{n=1}^{\infty}\left[\frac{2T_1}{n\pi}\left(1-\cos\frac{n\pi}{2}\right) + \frac{2T_2}{n\pi}\times\right.$
$\left.\times\left(\cos\frac{n\pi}{2} - \cos n\pi\right)\right]\sin\frac{n\pi x}{L}$
$\times e^{-n^2\pi^2c^2t/L^2}$

8. A bar of 30 cm length has its ends kept at 20° and 80° respectively until steady-state conditions prevail. The temperature at each end is then suddenly reduced to 0° and maintained thereafter. Find the temperature in bar.

Hint: $u(x,0) = 2x + 20$ obtained from steady-state.

Ans. $u(x,t) = \frac{40}{\pi}\sum_{n=1}^{\infty}\left[\frac{1-4(-1)^n}{n}\right]\sin\left(\frac{n\pi x}{30}\right)$
$\times e^{-c^2n^2\pi^2t/900}.$

Non-homogeneous B.C.'s

9. Solve $u_t = c^2 u_{xx}, 0 \le x \le 1$ with $u(0,t) = 2, u(1,t) = 3, u(x,0) = x(1-x)$.

Ans. $u(x,t) = x + 2 + 2\sum_{n=1}^{\infty}\left[\frac{3(-1)^n}{n\pi} - \frac{2(-1)^n}{n^3\pi^3} - \frac{2}{n\pi} + \frac{2}{n^3\pi^3}\right]\sin n\pi x \times e^{-n^2\pi^2c^2t}$

10. A bar of 10 cm long with its ends A and B kept at 20° and 40° respectively until steady-state conditions prevail. The temperature at A is then suddenly raised to 50° and at the same time at B is lowered to 10°. Find the subsequent temperature distribution. Show that the temperature at the middle point of the bar reamains unaltered for all time.

Ans. $u(x,t) = 50 - 4x - \frac{120}{\pi}\sum_{n=1}^{\infty}\frac{1}{2n}\sin\frac{2n\pi x}{10} \times e^{-4n^2\pi^2c^2t/100}$

At the mid-point $x = 5, u(5,t) = -20 + 50 + 0 = 30 =$ constant.

11. A bar of length L is laterally insulated with its ends A and B kept at 0° and 100° respectively until steady-state condition is reached. Then suddenly the temperature at A is raised to 20° and at B reduced to 80° simultaneously. Find the subsequent temperature distribution in the bar.

Hint: Initial condition $u(x,0) = 100x/L$. Assume $u(x,t) = u_s(x) + u_{tr}(x,t)$ where $u_s(x) = 20 + 60x/L$.

Ans. $u(x,t) = \frac{60x}{L} + 20 - \frac{40}{\pi}\sum_{n=1}^{\infty}\frac{1}{n}\sin\left(\frac{2n\pi x}{L}\right) \times e^{-4c^2n^2\pi^2t/L^2}$

12. A rod of length L has its ends A and B maintained at 0° and 100° respectively until steady-state conditions reached. Then suddenly the temperature at A and B have been changed to 25° and 75° respectively. Find the subsequent temperature in rod.

Hint: $u(x,0) = \frac{100x}{L}, u_s(x) = \frac{50x}{L} + 25$

Ans. $u(x,t) = \frac{50x}{L} + 25 - \frac{50}{\pi}\sum_{n=1}^{\infty}\frac{1}{n}\sin\frac{2n\pi x}{L} \times e^{-4c^2n^2\pi^2t/L^2}$.

Both edges insulated

Find the temperature in a laterally insulated bar of length L whose both ends are insulated and

13. with initial temperature $u(x,0) = x$ if $0 < x < \frac{L}{2}$ and $= L - x$ if $\frac{L}{2} < x < L$

Ans. $u(x,t) = \frac{L}{4} - \frac{2L}{\pi^2}\sum_{n=1}^{\infty}\frac{1}{n^2}\left(2\cos\frac{n\pi}{2} - \cos n\pi - 1\right) \times \cos\frac{n\pi x}{L} \cdot e^{-c^2n^2\pi^2t/L^2}$

14. with initial temperature $u(x,0) = 60$, $0 < x < 50, = 40, 50 < x < 100; L = 100$

Ans. $u(x,t) = 50 + \sum_{n=1}^{\infty}\left(\frac{40}{n\pi}\right)\sin\frac{n\pi}{2} \cdot \cos\frac{n\pi x}{100} \times e^{-c^2n^2\pi^2t/L^2}$

15. with initial temperature $u(x,0) = x(L-x)$

Ans. $u(x,t) = \frac{L^2}{6} - \frac{L^2}{\pi^2}\sum_{n=1}^{\infty}\frac{1}{n^2}\cos\left(\frac{2n\pi x}{L}\right) \times e^{-4n^2\pi^2c^2t/L^2}$

16. A rod of 100 cm length has its ends kept at 0° and 100° until steady-state conditions prevail. The two ends are then suddenly insulated and maintained so. Find the temperature in the rod. Show that the sum of the temperatures at any two points equidistant from the centre is always 100°.

Hint: $u(x,t) + u(l-x,t) = L = 100$ since $\cos\left(\frac{(2n-1)\pi(L-x)}{L}\right) = -\cos\frac{(2n-1)\pi x}{L}$

Ans. $u(x,t) = \frac{L}{2} - \frac{4L}{\pi^2}\sum_{n=1}^{\infty}\frac{1}{(2n-1)^2}\cos\left(\frac{(2n-1)\pi x}{L}\right) \times e^{-c^2(2n-1)^2\pi^2t/L^2}$

One edge insulated

Obtain the temperature in bar of length L and with one edge insulated:

17. $L = 1, u_x(1,t) = 0, u(0,t) = 10$, $u(x,0) = 1 - x, 0 < x < 1$

Ans. $u(x,t) = 10 + \sum_{n=1}^{\infty}\left[\frac{8(-1)^n}{(2n-1)^2\pi^2} - \frac{36}{(2n-1)\pi}\right]$

$\times \sin\frac{(2n-1)\pi x}{2} \cdot e^{-(2n-1)^2\pi^2c^2t/4}$

18. $u_x(0,t) = 0, u(L,t) = 0, u(x,0) = x$

Ans. $u(x,t) = \frac{8L}{\pi^2}\sum_{n=1}^{\infty}\frac{1}{(2n-1)^2}\left[\frac{(2n-1)\pi}{2}(-1)^{n+1} - 1\right]$

$\times \cos\frac{(2n-1)\pi x}{2L} \cdot e^{-(2n-1)^2\pi^2c^2t/(4L^2)}$.

19.5 DERIVATION OF ONE-DIMENSIONAL WAVE EQUATION

The classical one-dimensional wave equation, which is **hyperbolic**, arises in the study of transverse vibrations of an elastic string or torsional oscillations or longitudinal vibrations of a rod.

Vibrating string:

Consider an elastic string, stretched to its length L and alinged (placed) along the x-axis, with its two ends $x = 0$ and $x = L$ fixed. Let ρ be the constant density of the string (i.e., homogeneous string). Let the function $u(x,t)$ denote the displacement (or deflection) of string at any point x and at any time $t > 0$ from the equilibrium position (x-axis). When the string is distorted, then it vibrates. The small transverse vibrations of such a vibrating string are mathematically modelled by one-dimensional wave equation.

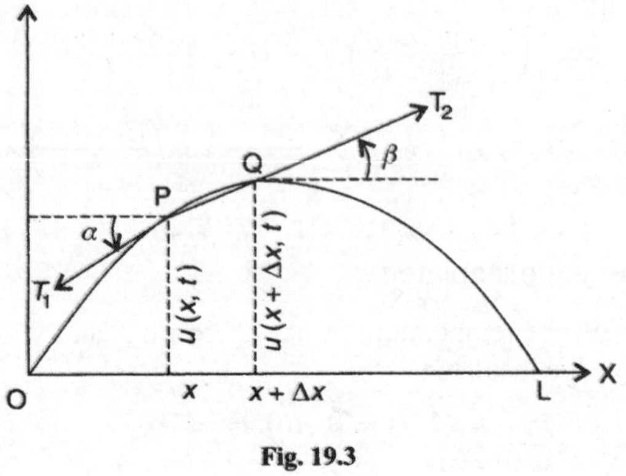

Fig. 19.3

Assume that the string is perfectly flexible and offers no resistance to bending. Applying Newton's second law of motion for a small portion of the string between x and $x + \Delta x$, the equation is derived. Let T_1 and T_2 be tension at the end points P and Q of this portion of the string.

Assuming that points on the string move only in the vertical direction, there is no motion in the horizontal direction. Thus the sum of the forces in the horizontal direction must be zero i.e.,

$$-T_1 \cos\alpha + T_2\cos\beta = 0$$

or $$T_1\cos\alpha = T_2\cos\beta = T = \text{constant} \tag{1}$$

Neglecting the gravitational force on the string, the only two forces acting on the string are the vertical components of tension $-T_1\sin\alpha$ at P and $T_2\sin\beta$ at Q with upward direction takes as positive. By Newton's second law

$$\text{resultant of forces} = \text{mass} \times \text{acceleration}$$

$$T_2\sin\beta - T_1\sin\alpha = (\rho\Delta x)\left(\frac{\partial^2 u}{\partial t^2}\right) \tag{2}$$

Dividing (2) by (1), we have

$$\frac{T_2\sin\beta}{T_2\cos\beta} - \frac{T_1\sin\alpha}{T_1\cos\alpha} = \frac{\rho\Delta x}{T}\frac{\partial^2 u}{\partial t^2}$$

$$\tan\beta - \tan\alpha = \frac{\rho\Delta x}{T}\frac{\partial^2 u}{\partial t^2} \tag{3}$$

Replace $\tan\alpha$ by $\frac{\partial u(x,t)}{\partial x}$ and $\tan\beta$ by $\frac{\partial u(x+\Delta x,t)}{\partial t}$ because they are slopes of the string at x and $x + \Delta x$. Then

$$\frac{\partial u(x+\Delta x,t)}{\partial t} - \frac{\partial u(x,t)}{\partial t} = \frac{\rho\Delta x}{T}\frac{\partial^2 u}{\partial t^2}$$

Rewriting and taking the limit as $\Delta x \to 0$

$$\lim_{\Delta x\to 0}\frac{1}{\Delta x}\left[\left.\frac{\partial u}{\partial x}\right|_{x+\Delta x} - \left.\frac{\partial u}{\partial x}\right|_{x}\right] = \frac{\rho}{T}\frac{\partial^2 u}{\partial t^2}$$

Thus

$$\frac{\partial^2 u}{\partial t^2} = c^2\frac{\partial^2 u}{\partial x^2} \tag{4}$$

where $c^2 = \frac{T}{\rho}$. Equation (4) is known as one-dimensional wave equation; which is second order, homogeneous, hyperbolic type.

19.6 SOLUTION OF ONE-DIMENSIONAL WAVE EQUATION BY SEPARATION OF VARIABLES:

Consider an elastic string, placed along the x-axis, stretched to length L between two fixed points $x = 0$ and $x = L$. Let $y(x, t)$ denote the deflection (displacement from equilibrium position). Then the small transverse vibrations of the string is governed by the one-dimensional wave equation

$$\frac{\partial^2 y}{\partial t^2} = a^2 \frac{\partial^2 y}{\partial x^2} \tag{1}$$

with boundary conditions

$$y(0, t) = 0, \ y(L, t) = 0 \tag{2}$$

The form of motion of the string will depend on the initial displacement (deflection at time $t = 0$)

$$y(x, 0) = f(x) \tag{3}$$

and the initial velocity

$$\left.\frac{\partial y}{\partial t}\right|_{t=0} = g(x) \tag{4}$$

The initial boundary value problem can be solved by the separation of variables techniques.
Step I. Assume that y is separable i.e.,

$$y(x, t) = X(x)T(t) \tag{5}$$

Substituting (5) in (1), we get

$$\frac{X''}{X} = \frac{\ddot{T}}{a^2 T} \tag{6}$$

Since L.H.S. of (6) is a function of x only and R.H.S. of (6) is a function of t only, both sides of (6) must be equal to common constant k. This results in two ordinary differential equations for X and T as

$$X'' - kX = 0 \tag{7}$$

$$\ddot{T} - ka^2 T = 0 \tag{8}$$

The boundary conditions (2) reduce to

$$0 = y(0, t) = X(0)T(t) \quad \text{i.e., } X(0) = 0 \tag{9}$$

$$0 = y(L, t) = X(L)T(t) \quad \text{i.e, } X(L) = 0 \tag{10}$$

It can be shown that for $k \geq 0$, only trivial solutions exist. So for non-trivial solutions consider $k = -\lambda^2 < 0$. Then the solution of $X'' + \lambda^2 X = 0$ is

$$X(x) = A \cos \lambda x + B \sin \lambda x$$

using (9), $0 = X(0) = A + B \cdot 0 \quad \therefore A = 0$
using (10), $0 = X(L) = B \sin \lambda L$
Since B should not be zero,

$$\sin \lambda L = 0$$

i.e., $\lambda L = n\pi$

or
$$\lambda_n = \frac{n\pi}{L} \quad \text{for } n = 1, 2, 3, \ldots \tag{11}$$

This results in infinitely many solutions to (7) given by

$$X_n(x) = \sin\left(\frac{n\pi x}{L}\right), n = 1, 2, 3, \ldots \tag{12}$$

With $k = -\lambda^2 = -\frac{n^2\pi^2}{L^2}$ solve (8). Its solution is

$$T_n(t) = A_n \cos \lambda_n at + B_n \sin \lambda_n at \tag{13}$$

The set of solutions satisfying (1) and (2) are

$$y_n(x, t) = X_n(x)T_n(t) \qquad n = 1, 2, 3, \ldots.$$

By the principle of superposition the general solution of (1) is

$$y(x, t) = \sum_{n=1}^{\infty} y_n(x, t)$$

$$= \sum_{n=1}^{\infty} (A_n \cos \lambda_n at + B_n \sin \lambda_n at) \sin \lambda_n x \tag{14}$$

The unknown constants A_n and B_n's are determined using the initial conditions

Case 1: *When initial displacement is given*: $y(x, 0) = f(x)$.
Put $t = 0$ in (14). Then

$$f(x) = y(x, 0) = \sum_{n=1}^{\infty} A_n \sin \frac{n\pi x}{L} \tag{15}$$

Thus A_n's are the Fourier coefficients in the half range Fourier sine series expansion of $f(x)$ in the interval $(0, L)$.
Hence

$$A_n = \frac{2}{L} \int_0^L f(x) \sin\left(\frac{n\pi x}{L}\right) dx \tag{16}$$

with $n = 1, 2, 3, \ldots$

Case 2: *When initial velocity is given:*

$$y_t(x, 0) = g(x).$$

Differentiate (14) w.r.t., t

$$\frac{\partial y}{\partial t} = \sum_{n=1}^{\infty} [-A_n\lambda_n a \sin \lambda_n at + B_n\lambda_n a \cos \lambda_n at] \sin \frac{n\pi x}{L}$$

Put $t = 0$

$$g(x) = \left.\frac{\partial y}{\partial t}\right|_{t=0} = \sum_{n=1}^{\infty} B_n \cdot \lambda_n a \sin\left(\frac{n\pi x}{L}\right)$$

$$\text{Then } B_n \cdot \lambda_n = \frac{2}{L}\int_0^L g(x) \cdot \sin\left(\frac{n\pi x}{L}\right) dx$$

or

$$B_n = \frac{2}{a \cdot n\pi}\int_0^L g(x) \sin\left(\frac{n\pi x}{L}\right) dx \quad (17)$$

Hence the general solution of the one-dimensional wave Equation (1) with boundary conditions (2) and initial conditions (3) and (4) is

$$y(x, t) = \sum_{n=1}^{\infty}\left[A_n \cdot \cos\left(\frac{n\pi at}{L}\right) + B_n \sin\left(\frac{n\pi at}{L}\right)\right] \times \sin\left(\frac{n\pi x}{L}\right)$$

where A_n and B_n's are given by (16) and (17).

Corollary 1: When only displacement is given, $f(x) \neq 0$ and $g(x) = 0$, in which case all B_n's are zero. Then the general solution is

$$y(x, t) = \sum_{n=1}^{\infty} A_n \cdot \cos\left(\frac{n\pi at}{L}\right) \cdot \sin\left(\frac{n\pi x}{L}\right)$$

with A_n's given by (16).

Corollary 2: When only initial velocity is prescribed i.e., $g(x) \neq 0$ and $f(x) = 0$ then the solution is

$$y(x, t) = \sum B_n \cdot \sin\left(\frac{n\pi at}{L}\right) \cdot \sin\left(\frac{n\pi x}{L}\right)$$

with B_n's given by (17).

WORKED OUT EXAMPLES

Initial displacement

Example 1: A string of length L is fastened at both ends A and C. At a distance 'b' from the end A, the string is transversely displaced to a distance 'd' and is released from rest when it is in this position. Find the equation of the subsequent motion (refer Fig. 19.4).

Solution: Let $y(x, t)$ denote the displacement of the string. The initial displacement is given by ABC.

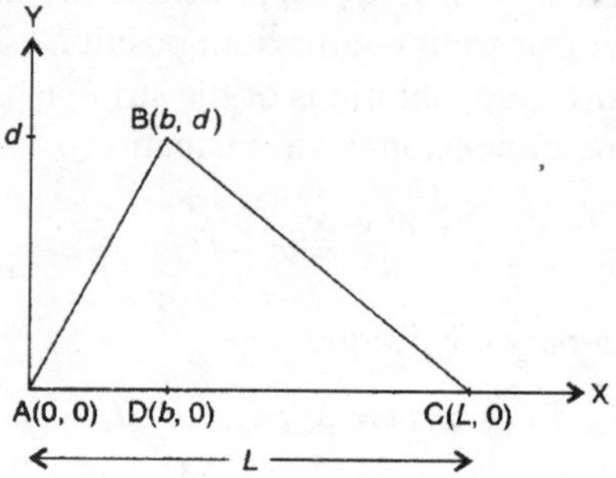

Fig. 19.4

Equation of AB is $y = \frac{dx}{b}$.
Equation of BC is $y = \frac{d(x-L)}{(b-L)}$. Thus the problem is to solve the one-dimensional wave equation

$$\frac{\partial^2 y}{\partial t^2} = a^2 \frac{\partial^2 y}{\partial x^2}$$

with boundary conditions

$$y(0, t) = 0, \quad y(L, t) = 0$$

and with initial displacement

$$y(x, 0) = f(x) = \begin{cases} \frac{dx}{b} & \text{if } 0 \le x \le b \\ \frac{d(x-L)}{(b-L)} & \text{if } b \le x \le L \end{cases}$$

The solution is given by

$$y(x, t) = \sum_{n=1}^{\infty} A_n \sin\frac{n\pi x}{L} \cdot \cos\frac{n\pi at}{L}$$

$$\text{where } A_n = \frac{2}{L}\int_0^L f(x) \cdot \sin\frac{n\pi x}{L} dx$$

$$A_n = \frac{2}{L}\int_0^b \frac{d}{b} x \cdot \sin\left(\frac{n\pi x}{L}\right) dx + $$

$$+ \frac{2}{L}\frac{d}{(b-L)}\int_b^L (x - L) \sin\frac{n\pi x}{L} dx$$

$$= \frac{2d}{bL}\left[x \cdot \left(\frac{-L}{n\pi}\right) \cdot \cos\frac{n\pi x}{L} - \right.$$

$$-\left(\frac{-L^2}{n^2\pi^2}\right)\cdot\sin\frac{n\pi x}{L}\Big]_0^b$$

$$+\frac{2d}{L(b-L)}\left[(x-L)\left(\frac{-L}{n\pi}\right)\cos\frac{n\pi x}{L}-\right.$$

$$\left.-\left(\frac{-L^2}{n^2\pi^2}\right)\cdot\sin\frac{n\pi x}{L}\right]_b^L$$

$$=-\frac{2dbL}{bL\,n\pi}\cdot\cos\frac{n\pi b}{L}+\frac{2dL^2}{bLn^2\pi^2}\cdot\sin\frac{n\pi b}{L}$$

$$+\frac{2d}{n\pi}\cdot\cos\frac{n\pi b}{L}-\frac{2dL^2}{L(b-L)n^2\pi^2}\cdot\sin\frac{n\pi b}{L}.$$

Thus $A_n=\dfrac{2dL^2}{b(L-b)n^2\pi^2}\cdot\sin\dfrac{n\pi b}{L}$

Hence the subsequent displacement of the string is given by the displacement function

$$y(x,t)=\frac{2dL^2}{b(L-b)\pi^2}\sum_{n=1}^{\infty}\frac{1}{n^2}\sin\frac{n\pi b}{L}$$

$$\times\sin\frac{n\pi x}{L}\cdot\cos\frac{n\pi at}{L}.$$

Initial velocity

Example 2: Find the displacement of a string stretched between two fixed points at a distance $2c$ apart when the string is initially at rest in equilibrium position and points of the string are given initial velocities v where

$$v=\begin{cases}\frac{x}{c}, & \text{when } 0<x<c\\ \frac{2c-x}{c}, & \text{when } c<x<2c\end{cases}$$

x being the distance measured from one end.

Solution: The initial displacement $f(x)=0$ while the initial velocity $g(x)=v(x)$. The solution is

$$y(x,t)=\frac{2c}{\pi a}\sum_{n=1}^{\infty}\frac{Bn}{n}\sin\left(\frac{n\pi x}{2c}\right)\sin\left(\frac{n\pi at}{2c}\right)$$

$$\text{where } B_n=\frac{2}{2c}\int_0^{2c}f(x)\cdot\sin\left(\left(\frac{n\pi x}{2c}\right)x\right)dx$$

$$B_n=\frac{1}{c}\int_0^c\frac{x}{c}\sin\left(\frac{n\pi x}{2c}\right)dx$$

$$+\frac{1}{c}\int_c^{2c}\left(\frac{2c-x}{c}\right)\cdot\sin\left(\frac{n\pi x}{2c}\right)dx$$

$$=\frac{1}{c}\left[\frac{x}{c}(-1)\left(\frac{2c}{n\pi}\right)\cos\left(\frac{n\pi x}{2c}\right)\right.$$

$$\left.-\frac{1}{c}(-1)\frac{4c^2}{n^2\pi^2}\cdot\sin\left(\frac{n\pi x}{2c}\right)\right]_0^c$$

$$+\frac{1}{c}\left[\left(\frac{2c-x}{c}\right)\cdot\left(\frac{-2c}{n\pi}\right)\cdot\cos\left(\frac{n\pi x}{2c}\right)\right.$$

$$\left.-\left(\frac{-1}{c}\right)\cdot\left(\frac{-4c^2}{n^2\pi^2}\right)\cdot\sin\left(\frac{n\pi x}{2c}\right)\right]_c^{2c}$$

$$=\frac{-2}{n\pi}\cos\frac{n\pi c}{2c}+\frac{4}{n^2\pi^2}\cdot\sin\frac{n\pi c}{2c}$$

$$+\frac{2}{n\pi}\cos\frac{n\pi}{2}+\frac{4}{n^2\pi^2}\frac{\sin n\pi}{2}$$

$$B_n=\frac{8}{n^2\pi^2}\cdot\sin\frac{n\pi}{2}$$

Hence the displacement function is given by

$$y(x,t)=\frac{2c}{\pi a}\cdot\frac{8}{\pi^2}\sum_{n=1}^{\infty}\frac{1}{n}\cdot\frac{1}{n^2}\sin\left(\frac{n\pi}{2}\right)$$

$$\times\sin\left(\frac{n\pi x}{2c}\right)\cdot\sin\left(\frac{n\pi at}{2c}\right)$$

$$y(x,t)=\frac{16c}{a\pi^3}\sum_{n=1}^{\infty}\frac{1}{n^3}\sin\left(\frac{n\pi}{2}\right)$$

$$\times\sin\left(\frac{n\pi x}{2c}\right)\cdot\sin\left(\frac{n\pi at}{2c}\right).$$

Note: Here a is the constant appearing in wave equations.

Both initial displacement and initial velocity

Example 3: Solve the vibrating string problem with:

i. $y(0,t)=0,\ y(L,t)=0$

ii. $y(x,0)=x$ when $0<x<\frac{L}{2}$

$=L-x$ when $\frac{L}{2}<x<L$

iii. $y_t(x,0)=x(x-L)$ when $0<x<L$

Solution: Here $f(x)=y(x,0)$ is the initial displacement and $g(x)=y_t(x,0)$ is the initial velocity. The solution is

$$y(x,t)=\sum_{n=1}^{\infty}\left[A_n\cos\left(\frac{n\pi}{L}at\right)\right.$$

$$+B_n \sin\left(\frac{n\pi}{L}at\right)\Big] \sin\frac{n\pi x}{L}$$

where $A_n = \dfrac{2}{L}\displaystyle\int_0^L f(x)\sin\left(\frac{n\pi x}{L}\right)dx$

$$A_n = \frac{2}{L}\left[\int_0^{\frac{L}{2}} x\sin\left(\frac{n\pi x}{L}\right)dx + \int_{\frac{L}{2}}^{L}(L-x)\sin\left(\frac{n\pi x}{L}\right)dx\right]$$

$$= \frac{2}{L}\left[x\cdot\left(\frac{-L}{n\pi}\right)\cdot\cos\left(\frac{n\pi x}{L}\right) - 1\cdot\left(\frac{-L^2}{n^2\pi^2}\right)\cdot\sin\left(\frac{n\pi x}{L}\right)\right]_0^{\frac{L}{2}}$$

$$+\frac{2}{L}\left[(L-x)\left(\frac{-L}{n\pi}\right)\cos\left(\frac{n\pi x}{L}\right) - (-1)\left(\frac{-L^2}{n^2\pi^2}\right)\cdot\sin\left(\frac{n\pi x}{L}\right)\right]_{\frac{L}{2}}^{L}$$

$$= \frac{2}{L}\left[\frac{-L^2}{2n\pi}\cdot\cos\frac{n\pi}{2} + \frac{L^2}{n^2\pi^2}\cdot\sin\frac{n\pi}{2} + \frac{L^2}{2n\pi}\cos\frac{n\pi}{2} + \frac{L^2}{n^2\pi^2}\sin\frac{n\pi}{2}\right]$$

$$A_n = \frac{4L}{n^2\pi^2}\sin\frac{n\pi}{2}$$

Now $B_n = \dfrac{L}{an\pi}\cdot\dfrac{2}{L}\displaystyle\int_0^L g(x)\cdot\sin\left(\frac{n\pi x}{L}\right)dx$

$$B_n = \frac{2}{n\pi a}\left[\int_0^L x(x-L)\sin\left(\frac{n\pi x}{L}\right)dx\right]$$

$$= \frac{2}{n\pi a}\left[x(x-L)\left(\frac{-L}{n\pi}\right)\cos\left(\frac{n\pi x}{L}\right) - (2x-L)\left(\frac{-L^2}{n^2\pi^2}\right)\cdot\sin\frac{n\pi x}{L} + (2)\left(\frac{-L^3}{n^3\pi^3}\right)\cos\frac{n\pi x}{L}\right]_0^L$$

$$B_n = \frac{2}{n\pi a}\left[-\frac{2L^3}{n^3\pi^3}\{(-1)^n - 1\}\right]$$

$$B_n = \frac{8L^3}{an^4\pi^4} \quad \text{when } n \text{ is odd.}$$

Thus the required solution is

$$y(x,t) = \frac{4L}{\pi^2}\sum_{n=1}^{\infty}\left[\frac{1}{n^2}\cdot\sin\frac{n\pi}{2}\cdot\cos\left(\frac{n\pi at}{L}\right)\cdot\sin\left(\frac{n\pi x}{L}\right)\right.$$

$$+\frac{8L^3}{a\pi^4}\sum_{n=1}^{\infty}\left[\frac{1}{(2n-1)^4}\cdot\sin\left(\frac{(2n-1)\pi at}{L}\right)\times\sin\frac{(2n-1)\pi x}{L}\right]$$

Example 4: Find the displacement of a string stretched between the fixed points (0, 0) and (1, 0) and released from rest from the position $A\sin\pi x + B\sin 2\pi x$.

Solution: Here initial deflection $y(x,0) = f(x) = A\sin\pi x + B\sin 2\pi x$

So $A_n = \dfrac{2}{1}\displaystyle\int_0^1 (A\sin\pi x + B\sin 2\pi x)\sin\frac{n\pi x}{1}\cdot dx$

On integration, $A_1 = A$, $A_2 = B$, $A_n = 0$ for $n \geq 3$. So the displacement is given by

$$y(x,t) = A\sin\pi x\cdot\cos\pi at + B\sin 2\pi x\cdot\cos 2\pi at.$$

EXERCISE

A string of length L is stretched and fastened to two fixed points. Find the solution of the wave equation $y_{tt} = a^2 y_{xx}$ when:

1. Initial displacement $y(x,0) = f(x) = b\sin\left(\frac{\pi x}{L}\right)$

Ans. $y(x,t) = b\sin\left(\frac{\pi x}{L}\right)\cdot\cos\left(\frac{\pi at}{L}\right)$

2. Initial displacement $y(x,0) = f(x)$ where

$$f(x) = \begin{cases}\frac{3b}{L}x & \text{if } 0 \leq x \leq \frac{L}{3}\\ \frac{3b}{L}(L-2x) & \text{if } \frac{L}{3} \leq x \leq \frac{2L}{3}\\ \frac{3b}{L}(x-L) & \text{if } \frac{2L}{3} \leq x \leq L\end{cases}$$

Ans. $y(x,t) = \frac{9b}{\pi^2}\displaystyle\sum_{n=1}^{\infty}\frac{1}{n^2}\sin\left(\frac{2n\pi}{3}\right)\cdot\sin\left(\frac{2n\pi x}{L}\right)\times\cos\left(\frac{2n\pi at}{L}\right)$

3. Triangular initial deflection

$$f(x) = \begin{cases}\frac{2kx}{L}, & \text{if } 0 < x < \frac{L}{2}\\ \frac{2k(L-x)}{L}, & \text{if } \frac{L}{2} < x < L\end{cases}$$

Ans. $y(x,t) = \frac{8k}{\pi^2}\sum_{n=1}^{\infty}\frac{1}{n^2}\sin\left(\frac{n\pi}{2}\right)\cdot\sin\frac{n\pi x}{L}$
$\times\cos\left(\frac{n\pi at}{L}\right)$

4. Initial displacement $y(x,0) = f(x) = b_0\sin^3\left(\frac{\pi x}{L}\right)$

Ans. $y(x,t) = \frac{3}{4}\sin\frac{\pi x}{L}\cdot\cos\frac{\pi at}{L} - \frac{1}{4}\sin\frac{3\pi x}{L}$
$\times\cos\frac{3\pi at}{L}$

5. Initial displacement $y(x,0) = f(x) = \frac{4bx(L-x)}{L^2}$

Ans. $y(x,t) = \frac{32b}{\pi^3}\sum_{n=1}^{\infty}\frac{1}{(2n-1)^3}\sin\left(\frac{(2n-1)\pi x}{L}\right)$
$\times\cos\left(\frac{(2n-1)\pi at}{L}\right)$

6. Initial displacement $y(x,0) = f(x) = k(Lx - x^2)$

Ans. $y(x,t) = \frac{8kL^2}{\pi^3}\sum_{n=1}^{\infty}\frac{1}{(2n-1)^3}\sin\left(\frac{(2n-1)\pi x}{L}\right)$
$\times\cos\left(\frac{(2n-1)\pi at}{L}\right)$

7. Initial displacement $y(x,0) = f(x)$

$$f(x) = \begin{cases} 0 & \text{if } 0 < x < \frac{1}{4} \\ 4h\left(\frac{x}{L} - \frac{1}{4}\right) & \text{if } \frac{L}{4} < x < \frac{L}{2} \\ 4h\left(\frac{3}{4} - \frac{x}{L}\right) & \text{if } \frac{L}{2} < x < \frac{3L}{4} \\ 0 & \text{if } \frac{3L}{4} < x < L \end{cases}$$

Ans. $y(x,t) = \frac{32h}{\pi^2}\sum_{n=1}^{\infty}\sin\left(\frac{n\pi}{2}\right)\sin^2\left(\frac{n\pi}{8}\right)\frac{1}{n^2}\sin\left(\frac{n\pi x}{L}\right)\times$
$\times\cos\left(\frac{n\pi at}{L}\right)$

8. Initial velocity $\left(\frac{\partial y}{\partial t}\right)_{t=0} = g(x) = b_0\sin^3\frac{\pi x}{L}$

Ans. $y(x,t) = \frac{Lb_o}{12a\pi}\left[9\sin\left(\frac{\pi x}{L}\right)\sin\left(\frac{a\pi t}{L}\right) - \sin\left(\frac{3\pi x}{L}\right)\times\right.$
$\left.\times\sin\left(\frac{3a\pi t}{L}\right)\right]$

9. Initial velocity $\left(\frac{\partial y}{\partial t}\right)_{t=0} = g(x) = bx(L-x)$

Ans. $y(x,t) = \frac{8bL^3}{a\pi^4}\sum_{n=1}^{\infty}\frac{1}{(2n-1)^4}\sin\left(\frac{(2n-1)\pi x}{L}\right)\times$
$\times\sin\left(\frac{(2n-1)\pi at}{L}\right)$

10. Initial velocity $\left(\frac{\partial y}{\partial t}\right)_{t=0} = g(x) = b\sin\left(\frac{3\pi x}{L}\right)\times$
$\times\cos\left(\frac{2\pi x}{L}\right)$

Ans. $y(x,t) = \frac{Lb}{2a\pi}\sin\left(\frac{\pi x}{L}\right)\cdot\sin\left(\frac{\pi at}{L}\right)$
$+\frac{Lb}{5a\pi}\sin\left(\frac{5\pi x}{L}\right)\cdot\sin\left(\frac{5\pi at}{L}\right)$

11. Initial velocity $\left(\frac{\partial y}{\partial t}\right)_{t=0} = g(x)$

$$= \begin{cases} cx, & 0 \le x \le \frac{L}{2} \\ c(L-x), & \frac{L}{2} \le x \le L \end{cases}$$

Ans. $y(x,t) = \frac{4cL^2}{a\pi^3}\sum_{n=1}^{\infty}\frac{1}{n^3}\sin\left(\frac{n\pi}{2}\right)\cdot\sin\left(\frac{n\pi x}{L}\right)$
$\times\sin\left(\frac{n\pi at}{L}\right)$

12. Initial velocity $y_t(x,0) = g(x) = 1,\ L = 1$

Ans. $y(x,t) = \frac{4}{a\pi^2}\sum_{n=1}^{\infty}\frac{1}{(2n-1)^2}\sin\left((2n-1)\pi x\right)$
$\times\sin\left((2n-1)\pi at\right).$

19.7 LAPLACE'S EQUATION or POTENTIAL EQUATION or TWO-DIMENSIONAL STEADY-STATE HEAT FLOW

The two-dimensional heat equation

$$\frac{\partial u}{\partial t} = a^2\left(\frac{\partial^2 u}{\partial x^2} + \frac{\partial^2 u}{\partial y^2}\right)$$

reduces to Laplace's equation or potential equation in two dimensions given by

$$\frac{\partial^2 u}{\partial x^2} + \frac{\partial^2 u}{\partial y^2} = 0 \tag{1}$$

when the heat-flow is in the steady-state (i.e., $\frac{\partial u}{\partial t} = 0$). The solution $u(x, y)$ of the Laplace's Equation (1) in a rectangular region can be obtained by the separation of variables technique both in the Dirichlet problem (where u is prescribed on the boundary) and in the Neumann problem (where derivative of u in the normal direction to the boundary is prescribed). A rectangular thin plate, with its two surfaces (faces) insulated, is considered so that the heat flow is purely two-dimensional. The boundary conditions are prescribed on the four edges of the plate. The steady-state heat flow in such a plate is obtained by solving Laplace's equation in two-dimensions.

WORKED OUT EXAMPLES

Example 1: Solve the Laplace's equation

$$\frac{\partial^2 u}{\partial x^2}+\frac{\partial^2 u}{\partial y^2}=0 \tag{1}$$

in a rectangle in the xy-plane, $0<x<a$ and $0<y<b$ (Fig. 19.5) satisfying the following boundary conditions

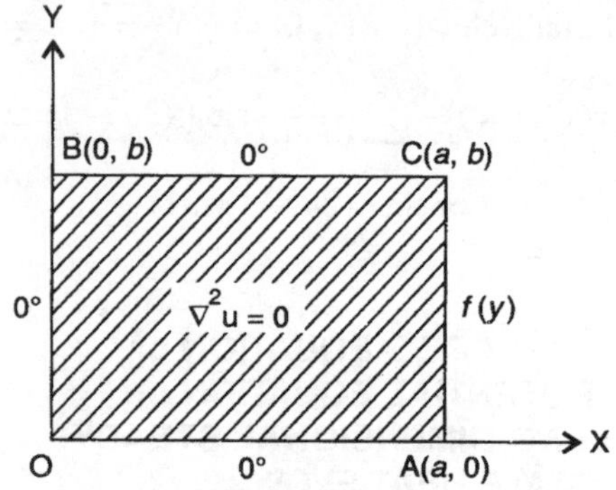

Fig. 19.5

$$u(x,0)=0 \quad (\text{on } OA) \tag{2}$$

$$u(x,b)=0 \quad (\text{on } BC) \tag{3}$$

$$u(0,y)=0 \quad (\text{on } OB) \tag{4}$$

$$u(a,y)=f(y) \quad (\text{on } AC) \tag{5}$$

i.e., u is zero on three sides OA, OB, BC and is prescribed by given function $f(y)$ on the fourth side AC of the rectangle $OACB$.

Solution: By separation of variables:
Step I. Assume that u is separable i.e.,

$$u(x,y)=X(x)\cdot Y(y) \tag{6}$$

Substituting (6) in (1), we get

$$X''X+X\ddot{Y}=0$$

where $'$ denotes differentiation w.r.t., x and $\cdot$ denotes differentiation w.r.t., y.

so
$$\frac{\ddot{Y}}{Y}=-\frac{X''}{X} \tag{7}$$

Both sides of (7) must be equal to a constant k since L.H.S. of (7) is a function of y only and the R.H.S. of (7) is a function of x only. Then

$$\frac{\ddot{Y}}{Y}=-\frac{X''}{X}=k$$

or the boundary value problem reduces to two ordinary differential equations in X and Y.

$$\ddot{Y}-kY=0 \tag{8}$$

$$X''+kX=0 \tag{9}$$

The boundary conditions (2) (3) (4) reduces to

$$0=u(x,0)=X(x)Y(0) \quad \text{i.e., } Y(0)=0 \tag{10}$$

$$0=u(x,b)=X(x)Y(b) \quad \text{i.e., } Y(b)=0 \tag{11}$$

$$0=u(0,y)=X(0)Y(y) \quad \text{i.e., } X(0)=0 \tag{12}$$

If $k\ge 0$, (8) will have only trivial solutions. So assume $k=-\lambda^2<0$. Then the Equation (8)

$$\ddot{Y}+\lambda^2 Y=0$$

has the general solution

$$Y(y)=A\cos\lambda y+B\sin\lambda y$$

Using (10), $0=Y(0)=A\cdot 1+B\cdot 0$ i.e., $A=0$

Using (11), $0=Y(b)=B\sin\lambda b$

Since $B\ne 0$, $\sin\lambda b=0$ i.e., $\lambda b=n\pi$

or
$$\lambda_n=\frac{n\pi}{b}, \quad \text{for } n=1,2,3,\ldots \tag{13}$$

Thus we get infinitely many solutions

$$Y_n(y)=\sin\frac{n\pi y}{b} \tag{14}$$

(assuming $B=1$). Now the solution of (9) is

$$X(x)=A^*e^{-\lambda x}+B^*e^{\lambda x}$$

Using (12), $0=X(0)=A^*+B^*$ i.e., $A^*=-B^*$

or
$$X(x)=2B^*\frac{(e^{+\lambda x}-e^{-\lambda x})}{2}=B\sinh\lambda x$$

Thus
$$X_n(x)=B_n\sinh\left(\frac{n\pi}{b}x\right) \tag{15}$$

Hence the solution to (1) with boundary conditions (2) (3) (4) is

$$u_n(x,y)=B_n\sinh\left(\frac{n\pi x}{b}\right)\cdot\sin\left(\frac{n\pi y}{b}\right)$$

By superposition principle:

$$u(x,y)=\sum u_n(x,y)$$
$$=\sum_{n=1}^{\infty} B_n \sinh\left(\frac{n\pi x}{b}\right)\sin\left(\frac{n\pi y}{b}\right) \qquad (16)$$

To find the unknown constants B_n, use the fourth boundary condition (5)

$$f(y)=u(a,y)=\sum\left[B_n\cdot\sinh\left(\frac{n\pi a}{b}\right)\right]\cdot\sin\left(\frac{n\pi y}{b}\right)$$

Thus B_n's are the Fourier coefficients of the Fourier half range sine series of $f(y)$ in the interval $(0, b)$ and are given by

$$B_n\cdot\sinh\left(\frac{n\pi a}{b}\right)=\frac{2}{b}\int_0^b f(y)\cdot\sin\left(\frac{n\pi y}{b}\right)dy$$

or $$B_n=\frac{2}{b\cdot\sinh\left(\frac{n\pi a}{b}\right)}\int_0^b f(y)\sin\left(\frac{n\pi y}{b}\right)dy,$$

for $n = 1, 2, 3, \ldots$ (17)

Hence the harmonic function $u(x, y)$ satisfying the Laplace's equation (1) and the four boundary conditions (2), (3), (4), (5) is given by (16) with B_n's determined by (17).

Example 2: Solve the above problem when

$$f(y)=ky(b-y),\ 0<y<b$$

and k is a constant.

Solution: Consider

$$I=\int_0^b f(y)\sin\left(\frac{n\pi y}{b}\right)dy$$
$$=\int_0^b ky(b-y)\sin\left(\frac{n\pi y}{b}\right)dy$$
$$=\Bigg[ky(b-y)\cdot\left(\frac{-b}{n\pi}\right)\cos\left(\frac{n\pi y}{b}\right)$$
$$-k(b-2y)\left(\frac{-b^2}{n^2\pi^2}\right)\cdot\sin\frac{n\pi y}{b}$$
$$+k(-2)\left(\frac{-b^3}{n^3\pi^3}\right)\cos\left(\frac{n\pi y}{b}\right)\Bigg]_0^b$$
$$=\frac{2b^3k}{n^3\pi^3}[(-1)^n-1]=\frac{-4kb^3}{n^3\pi^3}\text{ when } n \text{ is odd.}$$

Now substituting I in (17)

$$B_n=\frac{2}{b\sinh\left(\frac{n\pi a}{b}\right)}\int_0^b f(y)\sin\left(\frac{n\pi y}{b}\right)dy$$
$$=\frac{2}{b\cdot\sinh\left(\frac{n\pi a}{b}\right)}\left(\frac{-4kb^3}{n^3\pi^3}\right)\quad\text{for } n \text{ odd}$$

$$B_n=\frac{-8kb^2}{n^3\pi^3\sinh\left(\frac{n\pi a}{b}\right)}\quad\text{with } n \text{ odd}$$

Hence the required solution (16) becomes

$$u(x,y)=\sum_{n=\text{odd}}\frac{-8kb^2}{n^3\pi^3\sinh\left(\frac{n\pi a}{b}\right)}\times$$
$$\times\sinh\left(\frac{n\pi x}{b}\right)\cdot\sin\left(\frac{n\pi y}{b}\right)$$

$$u(x,y)=\frac{-8kb^2}{\pi^3}\sum_{n=1}^{\infty}\frac{1}{(2n-1)^3\sinh\left(\frac{(2n-1)\pi a}{b}\right)}\times$$
$$\times\sinh\left(\frac{(2n-1)\pi x}{b}\right)\cdot\sin\left(\frac{(2n-1)\pi y}{b}\right).$$

Example 3: Solve Laplace's equation in rectangle with $u(0, y) = 0$, $u(a, y) = 0$, $u(x, b) = 0$ and $u(x, 0) = f(x)$ (on OA) (refer Fig. 19.6).

Solution: Boundary conditions are

$$0=u(0,y)=X(0)\,Y(y)\quad\text{i.e., } X(0)=0$$
$$0=u(a,y)=X(a)\,Y(y)\quad\text{i.e., } X(a)=0$$
$$0=u(x,b)=X(x)\,Y(b)\quad\text{i.e., } Y(b)=0$$

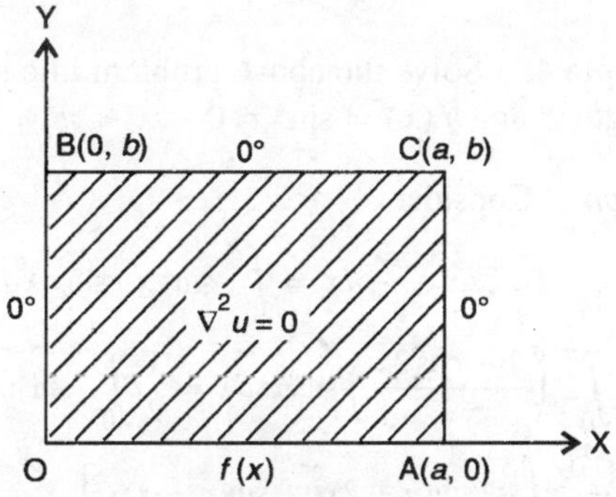

Fig. 19.6

The Equation (7) is taken as

$$\frac{X''}{X} = -\frac{\ddot{Y}}{Y} = k = -\lambda^2 < 0$$

Thus the problem reduces to solving

$$X'' + \lambda^2 X = 0 \quad \text{with } X(0) = 0, X(a) = 0$$

and $\quad \ddot{Y} - \lambda^2 Y = 0 \quad \text{with } Y(b) = 0$

Solving $\quad X_n(x) = \sin\frac{n\pi x}{a}, \quad \text{with } n = 1, 2, \ldots$

$$Y(y) = A^* e^{\lambda y} + B^* e^{-\lambda y} \quad \text{with } Y(b) = 0$$

$$0 = Y(b) = A^* e^{\lambda b} + B^* e^{-\lambda b}$$

or $\quad A^* = \frac{-B^* e^{-\lambda b}}{e^{\lambda b}}$

Rewriting

$$Y(y) = \frac{-B^* e^{-\lambda b}}{e^{\lambda b}} \cdot e^{\lambda y} + B^* e^{-\lambda y}$$

$$Y(y) = \frac{B^*}{e^{\lambda b}}\left[e^{\lambda b}e^{-\lambda y} - e^{-\lambda b}e^{\lambda y}\right]$$

$$= B \sinh\{\lambda(b - y)\}$$

Thus $\quad Y_n(y) = B_n \cdot \sinh\left\{\frac{n\pi}{a}(b - y)\right\}$

The required solution is

$$u(x, y) = \sum_{n=1}^{\infty} B_n \sin\left(\frac{n\pi x}{a}\right) \cdot \sinh\left(\frac{n\pi(b - y)}{a}\right)$$

Use the condition on side OA to find B_n's.

$$f(x) = u(x, 0) = \sum_{n=1}^{\infty} B_n \cdot \sin\left(\frac{n\pi x}{a}\right) \cdot \sinh n\pi\frac{(b)}{a}$$

Then $\quad B_n = \frac{2}{a \cdot \sinh\left(\frac{n\pi b}{a}\right)} \int_0^a f(x) \sin\left(\frac{n\pi x}{a}\right) dx.$

Example 4: Solve the above problem in a square of length π and $f(x) = \sin^2 x, 0 < x < \pi$.

Solution: Consider

$$I = \int_0^a f(x) \sin\frac{n\pi x}{a} dx = \int_0^{\pi} \sin^2 x \cdot \sin nx dx$$

$$= \int_0^{\pi} \left(\frac{1 - \cos 2x}{2}\right) \sin nx dx = \frac{1}{2}\int_0^{\pi} \sin nx dx$$

$$-\frac{1}{4}\int_0^{\pi}\left[\sin(n + 2)x + \sin(n - 2)x\right]$$

$$= \frac{1}{2}\left[\frac{-\cos nx}{n}\right]_0^{\pi} + \frac{1}{4}\left[\frac{\cos(n + 2)x}{n + 2}\right.$$

$$\left. + \frac{\cos(n - 2)x}{(n - 2)}\right]_0^{\pi} \quad \text{for } n \neq 2$$

$$= \frac{1}{2n}[1 - (-1)^n] + \frac{1}{4}\left\{(-1)^{n+2} - 1\right\} \times$$

$$\times \left[\frac{1}{n + 2} - \frac{1}{n - 2}\right]$$

$$I = -\frac{4}{n(n^2 - 4)} \quad \text{for } n \text{ odd.}$$

(Note for $n = 2$, even, $I = 0$)

Thus $\quad B_n = \frac{2}{\pi \sinh\left(n\pi \cdot \frac{\pi}{\pi}\right)} \int_0^{\pi} \sin^2 x \cdot \sin nx dx$

$$= \frac{2}{\pi \sinh n\pi}\left(\frac{-4}{n(n^2 - 4)}\right)$$

$$= \frac{-8}{n\pi(n^2 - 4)\sinh n\pi}$$

Hence the solution is

$$u(x, y) = -\frac{8}{\pi}\sum_{n=\text{odd}} \frac{\sin nx \cdot \sinh(n(\pi - y))}{(\sinh n\pi) \cdot n(n^2 - 4)}.$$

Example 5: Find the steady-state temperature distribution in a rectangular thin plate with its two surfaces insulated and with the conditions. (Fig. 19.7)

$$u(0, y) = 0, u(x, 0) = 0, u(a, y) = g(y), u(x, b) = f(x)$$

Solution: Superposition applied to boundary conditions dismantles the given problem to solution of two simpler problems each of which can readily be solved by separation of variables.

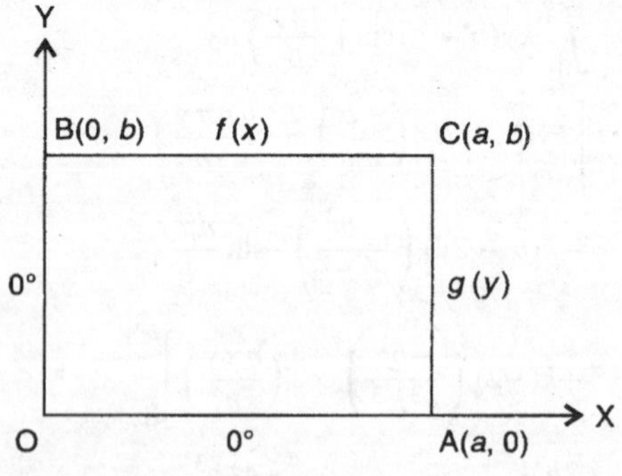

Fig. 19.7

Assume $u(x, y) = u_1(x, y) + u_2(x, y)$. Then

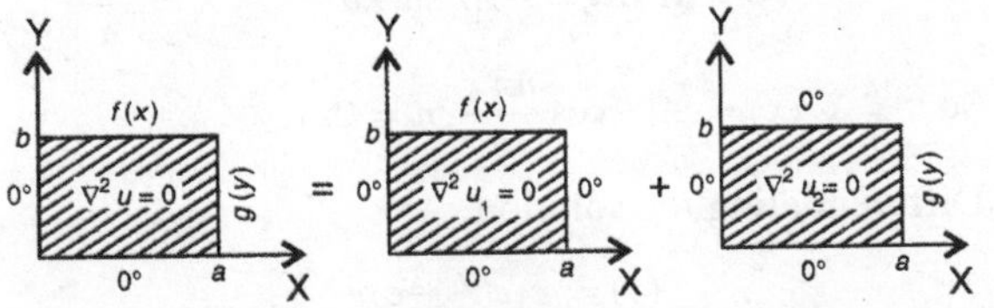

Fig. 19.8

Thus the two boundary value problems are to be solved. (Actually if one problem is solved, the other can be obtained by interchanging x by y, a by b and $f(x)$ by $g(y)$.)

I. $\dfrac{\partial^2 u_1}{\partial x^2} + \dfrac{\partial^2 u_1}{\partial y^2} = 0$

$u_1(0, y) = 0, u_1(a, y) = 0$

$u_1(x, 0) = 0, u_1(x, b) = f(x)$

II. $\dfrac{\partial^2 u_2}{\partial x^2} + \dfrac{\partial^2 u_2}{\partial y^2} = 0$

$u_2(0, y) = 0, u_2(x, 0) = 0$

$u_2(x, b) = 0, u_2(a, y) = g(y)$

Problem I: $X'' + \lambda X = 0, X(0) = X(a) = 0$

$$X_n(x) = \sin\frac{n\pi x}{a}$$

$$Y'' - \lambda Y = 0, Y(0) = 0,$$

$$Y(y) = c_1 e^{\lambda y} + c_2 e^{-\lambda y} \quad \therefore c_1 = -c_2$$

$$Y_n(y) = A_n \cdot \sinh\frac{n\pi y}{a}$$

Thus $u_1(x, y) = \sum A_n \sin\left(\frac{n\pi x}{a}\right) \cdot \sinh\left(\frac{n\pi y}{a}\right)$

Since $f(x) = u_1(x, b) = \sum\left[A_n \cdot \sinh\left(\frac{n\pi b}{a}\right)\right] \times \sin\left(\frac{n\pi x}{a}\right)$

Therefore $A_n = \dfrac{2}{a \cdot \sinh\left(\frac{n\pi b}{a}\right)} \displaystyle\int_0^a f(x) \cdot \sin\left(\frac{n\pi x}{a}\right) dx.$

In a similar way, the solution to problem II is obtained (by swapping x by y, a by b, $f(x)$ by $g(y)$)

$$u_2(x, y) = \sum_{n=1}^{\infty} B_n \cdot \sin\left(\frac{n\pi y}{b}\right) \cdot \sinh\left(\frac{n\pi x}{b}\right)$$

where $\quad B_n = \dfrac{2}{b \cdot \sinh\left(\frac{n\pi a}{b}\right)} \displaystyle\int_0^b g(y) \sin\left(\frac{n\pi y}{b}\right) dy.$

The required solution is

$$u(x, y) = u_1(x, y) + u_2(x, y).$$

Example 6: Solve the above problem when $u(a, y) = g(y) = 0$ and $u(x, b) = f(x) = \frac{2x}{a}$ when $0 < x < \frac{a}{2}$ and $f(x) = \frac{2(a-x)}{a}$ when $\frac{a}{2} < x < a$.

Solution: Since $g(y) = 0$, all B_n's are zero, $u_2(x, y) = 0$
Consider

$$I = \int_0^a f(x) \sin\left(\frac{n\pi x}{a}\right) dx$$

$$= \int_0^{\frac{a}{2}} \frac{2x}{a} \cdot \sin\left(\frac{n\pi x}{a}\right) dx + \int_{\frac{a}{2}}^a \frac{2(a-x)}{a} \sin\left(\frac{n\pi x}{a}\right) dx$$

$$I = \frac{2}{a}\left[x \cdot \left(\frac{-a}{n\pi}\right) \cos\left(\frac{n\pi x}{a}\right) - 1\left(\frac{-a^2}{n^2\pi^2}\right) \sin\left(\frac{n\pi x}{a}\right)\right]_0^{\frac{a}{2}}$$

$$+\frac{2}{a}\left[(a-x)\left(\frac{-a}{n\pi}\right) \cos\left(\frac{n\pi x}{a}\right) - (-1)\left(\frac{-a^2}{n^2\pi^2}\right) \sin\left(\frac{n\pi x}{a}\right)\right]_{\frac{a}{2}}^a$$

$$= \frac{2}{a}\left[\left\{\frac{a}{2} \cdot \left(\frac{-a}{n\pi}\right) \cos\frac{n\pi}{2} + \frac{a^2}{n^2\pi^2} \sin\frac{n\pi}{2}\right\} - \left\{\frac{a}{2}\left(\frac{-a}{n\pi}\right) \cos\frac{n\pi}{2} - \frac{a^2}{n^2\pi^2} \sin\left(\frac{n\pi}{2}\right)\right\}\right]$$

$$= \frac{4a}{n^2\pi^2} \sin\left(\frac{n\pi}{2}\right).$$

Substituting I in A_n, we get

$$A_n = \frac{2}{a \sinh\left(\frac{n\pi b}{a}\right)} \cdot \frac{4a}{n^2\pi^2} \sin\left(\frac{n\pi}{2}\right)$$

$$= \frac{8}{n^2\pi^2} \frac{(-1)^n}{\sinh\left(\frac{n\pi b}{a}\right)}$$

when n is odd.

Hence solution is

$$u(x, y) = \frac{8}{\pi^2} \sum_{n=1}^{\infty} \frac{(-1)^n}{(2n-1)^2} \times$$

$$\times \frac{\sinh\left(\frac{(2n-1)\pi y}{a}\right) \cdot \sin\left(\frac{(2n-1)\pi x}{a}\right)}{\sinh\left(\frac{(2n-1)\pi b}{a}\right)}.$$

Insulated edges

Example 7: Find the steady-state temperature in a rectangular plate $0<x<a, 0<y<b$ when the sides $x = 0, x = a, y = b$ are insulated while the edge $y = 0$ is kept at temperature $k\frac{\cos \pi x}{a}$ (Fig. 19.9).

Solution: The boundary conditions are

$$\frac{\partial u}{\partial x}|_{x=0} = 0; \quad \frac{\partial u}{\partial x}\bigg|_{x=a} = 0, \quad \frac{\partial u}{\partial y}\bigg|_{y=b} = 0$$

$$u(x, 0) = f(x) = k \cos\left(\frac{\pi x}{a}\right).$$

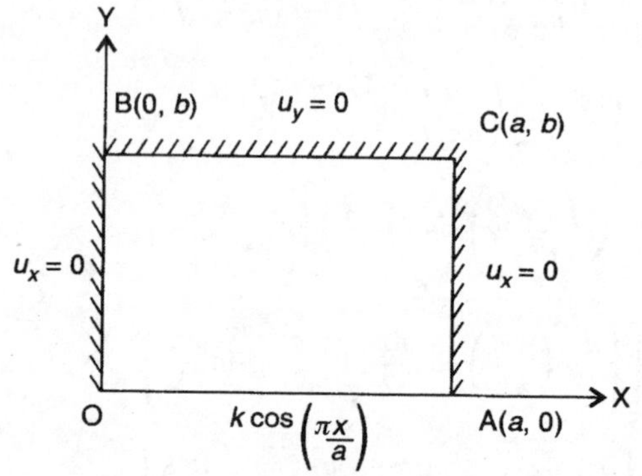

Fig. 19.9

The boundary conditions reduce to

$$X'(0) = 0, X'(a) = 0, Y'(b) = 0.$$

Differentiating the solution

$$X(x) = A \cos \lambda x + B \sin \lambda x$$

$$X'(x) = -A\lambda \sin \lambda x + B\lambda \cos \lambda x$$

Using, $\quad 0 = X'(0) = 0 + B\lambda 1 \quad \therefore B = 0$

$$0 = X'(a) = -A\lambda \sin \lambda a \quad \therefore \quad \lambda = \frac{n\pi}{a}$$

So $\quad X_n(x) = A_n \cdot \cos \frac{n\pi x}{a}, n = 0, 1, 2, \ldots$

Differentiating the solution

$$Y(y) = c_1 e^{\lambda y} + c_2 e^{-\lambda y}$$

$$Y'(y) = \lambda c_1 e^{\lambda y} - c_2 \lambda e^{-\lambda y}$$

Using, $\quad 0 = Y'(b) = \lambda(c_1 e^{\lambda b} - c_2 e^{-\lambda b})$

or $\quad c_1 = \frac{c_2 e^{-\lambda b}}{e^{\lambda b}}$

Thus $\quad Y(y) = c_2 \left(\frac{e^{\lambda(b-y)} + e^{-\lambda(b-y)}}{2}\right)$

$$= c_3 \cosh\left(\frac{n\pi(b-y)}{a}\right)$$

Hence the solution is

$$u(x, y) = \sum_{n=0}^{\infty} A_n \cos\left(\frac{n\pi x}{a}\right) \cdot \cosh\left(\frac{n\pi(b-y)}{a}\right)$$

where $\quad A_n = \frac{2}{a \cosh\left(\frac{n\pi b}{a}\right)} \int_0^a f(x) \cos\left(\frac{n\pi x}{a}\right) dx$

$$A_n = \frac{2}{a \cosh\left(\frac{n\pi b}{a}\right)} \int_0^a k \cdot \cos \frac{\pi x}{a}$$

$$\times \cos\left(\frac{n\pi x}{a}\right) dx$$

$$A_1 = \frac{2k}{a \cosh\left(\frac{\pi b}{a}\right)} \cdot \frac{\pi}{2} \cdot \frac{a}{\pi},$$

$$A_n = 0. \quad \text{for } n = 0, n \geq 2$$

$$u(x, y) = k \operatorname{sech}\left(\frac{\pi b}{a}\right) \cdot \cos\left(\frac{\pi x}{a}\right)$$

$$\times \cosh\left(\frac{\pi(b-y)}{a}\right).$$

Example 8: Find the steady-state temperature distribution in a thin sheet of metal plate which occupies the semi-infinite strip, $0 \le x \le L$ and $0 \le y < \infty$ when the edge $y = 0$ is kept at temperature $u(x, 0) = f(x) = kx(L - x), 0 < x < L$ while

i. The edges $x = 0$ and $x = L$ are kept at zero temperature.

ii. The edges $x = 0$ and $x = L$ are insulated. Assume that $u(x, \infty) = 0$.

Solution:

Case 1: $u(0, y) = 0, u(L, y) = 0$ for every y: (Fig. 19.10)
Solution of $X'' + \lambda^2 X = 0, X(0) = X(L) = 0$ is

$$X_n(x) = \sin\frac{n\pi x}{L}, n = 1, 2, 3, \ldots$$

Solution of $Y'' - \lambda^2 y = 0$ is $Y(y) = Ae^{\lambda y} + Be^{-\lambda y}$.
A must be zero since $y(\infty) = 0$. So

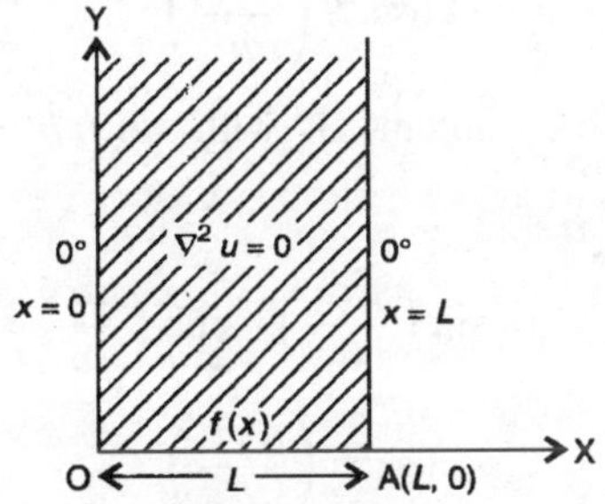

Fig. 19.10

$$Y(y) = B_n e^{-\lambda y}$$

Thus $$u(x, y) = \sum_{n=1}^{\infty} B_n e^{-n\pi y/L} \cdot \sin\frac{n\pi}{L}x$$

where $$B_n = \frac{2}{L}\int_0^L kx(L - x) \cdot \sin\frac{n\pi x}{L} dx$$

$$= \frac{2k}{L}\left[x(L - x) \cdot \left(\frac{-L}{n\pi}\right) \cos\frac{n\pi x}{L} - \right.$$

$$-(L - 2x)\left(\frac{-L^2}{n^2\pi^2}\right) \sin\frac{n\pi x}{L}$$

$$\left. + \frac{2L^3}{n^3\pi^3} \cos\frac{n\pi x}{L}\right]_0^L$$

$$B_n = \frac{2k}{L}\frac{2L^3}{n^3\pi^3}[(-1)^n - 1]$$

$$= \frac{-8kL^2}{n^3\pi^3} \text{ when } n \text{ is odd}$$

$$u(x, t) = \frac{-8kL^2}{\pi^3}\sum_{n=1}^{\infty}\frac{1}{(2n - 1)^3}e^{-(2n-1)\pi y/L} \times$$

$$\times \sin\left(\frac{(2n - 1)\pi x}{L}\right).$$

Case 2: Both edges insulated (Fig. 19.11):

$$\left.\frac{\partial u}{\partial x}\right|_{x=0} = 0, \quad \left.\frac{\partial u}{\partial x}\right|_{x=L} = 0$$

the solution of $X'' + \lambda^2 X = 0$ with the boundary conditions $X'(0) = 0$ and $X'(L) = 0$ is

$$X_n(x) = A\cos\frac{n\pi x}{L}, n = 0, 1, 2, \ldots$$

The solution of $Y'' - \lambda^2 Y = 0$ with $Y(\infty) = 0$ is

$$Y_n(y) = e^{-n\pi y/L}$$

Thus the solution to the problem is

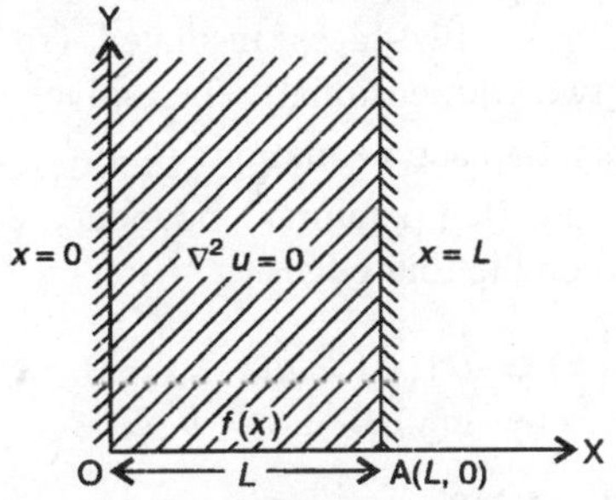

Fig. 19.11

$$u(x, y) = \sum_{n=0}^{\infty} A_n e^{-n\pi y/L} \cdot \cos\left(\frac{n\pi x}{L}\right)$$

where $$A_0 = \frac{1}{L}\int_0^L kx(L - x)dx = \frac{k}{L}\left[\frac{Lx^2}{2} - \frac{x^3}{3}\right]_0^L$$

$$A_0 = \frac{kL^2}{6}$$

$$A_n = \frac{2}{L}\int_0^L kx(L - x) \cdot \cos\left(\frac{n\pi x}{L}\right) dx$$

$$= \frac{2k}{L}\left[x(L - x) \cdot \left(\frac{-L}{n\pi}\right) \sin\frac{n\pi x}{L}\right.$$

$$-(L-2x)\cdot\left(\frac{-L^2}{n^2\pi^2}\right)\cdot\cos\frac{n\pi x}{L}$$

$$\left.+(-2)\left(\frac{-L^3}{n^3\pi^3}\right)\cdot\sin\left(\frac{n\pi x}{L}\right)\right]_0^L$$

$$A_n=\frac{2k}{L}\frac{L^3}{n^2\pi^2}[(-1)^n+1]$$

$$=\frac{4kL^2}{n^2\pi^2}\text{ where } n \text{ is even}$$

Hence the required solution is

$$u(x,y)=\frac{kL^2}{6}+\frac{4kL^2}{\pi^2}\sum\frac{1}{4n^2}e^{-(2n\pi y/L)}\cdot\cos\left(\frac{2n\pi x}{L}\right).$$

EXERCISE

Rectangular plate

Find the steady-state temperature distribution in a thin rectangular metal plate $0 < x < a$, $0 < y < b$ with its two-faces insulated (so that the flow is two dimensional) (or solve the two-dimensional Laplace's equation $\frac{\partial^2 u}{\partial x^2}+\frac{\partial^2 u}{\partial y^2}=0$) with the following (temperatures) boundary conditions prescribed on the four edges.

1. $u(0,y)=u(x,0)=u(x,b)=0$, $u(a,y)=g(y), 0<y<b$

Ans. $u(x,y)=\sum_{n=1}^{\infty}B_n\sinh\left(\frac{n\pi x}{b}\right)\cdot\sin\left(\frac{n\pi y}{b}\right)$

where

$$B_n=\frac{2}{b}\text{cosech}\left(\frac{n\pi a}{b}\right)\int_0^b g(y)\sin\left(\frac{n\pi y}{b}\right)dy$$

2. Solve problem 1, with $g(y)=100$.

Ans. $B_n=\frac{400}{(n\pi\sinh\left(\frac{n\pi a}{b}\right))}$ for n odd.

3. $u(0,y)=u(a,y)=u(x,0)=0, u(x,b)=f(x), 0<x<a$

Ans. $u(x,y)=\sum_{n=1}^{\infty}A_n\sin\left(\frac{n\pi x}{a}\right)\cdot\sinh\left(\frac{n\pi y}{a}\right)$ with

$$A_n=\frac{2}{a\sinh\left(\frac{n\pi b}{a}\right)}\int_0^a f(x)\sin\left(\frac{n\pi x}{a}\right)dx.$$

4. Solve problem 3 with $a=b=L, u(x,L)=f(x)=x(L-x)$ for $0\le x\le L$.

Ans. $u(x,y)=\frac{8L^2}{\pi^2}\sum_{n=\text{odd}}\frac{\sinh\left(\frac{n\pi y}{L}\right)\cdot\sin\left(\frac{n\pi x}{L}\right)}{n^3\sinh n\pi}$

5. Solve problem 3 with $u(x,b)=f(x)=100$

Ans. $u(x,y)=\frac{400}{\pi}\sum_{n=\text{odd}}\frac{1}{n}\sin\left(\frac{n\pi x}{a}\right)\cdot\sinh\left(\frac{n\pi y}{a}\right)$

$$\times\text{cosech}\left(\frac{n\pi b}{a}\right).$$

6. Solve problem 3 with $u(x,b)=f(x)=\sin\frac{n\pi x}{a}$.

Ans. $u(x,y)=\sin\left(\frac{n\pi x}{a}\right)\cdot\sinh\left(\frac{n\pi y}{a}\right)$

$$\times\text{cosech}\left(\frac{n\pi b}{a}\right)$$

7. $u(0,y)=u(a,y)=u(x,b)=0, u(x,0)=f(x)=x(a-x), 0<x<a$

Ans. $u(x,y)=\frac{8a^2}{\pi^3}\sum_{n=0}\frac{1}{m^3}\sin\left(\frac{m\pi x}{a}\right)\cdot\sinh\left(\frac{m\pi(b-y)}{a}\right)$

$$\times\text{cosech}\left(\frac{m\pi b}{a}\right)\text{ where } m=2n+1$$

8. $u(0,y)=u(a,y)=u(x,b)=0, u(x,0)=5\sin\frac{4\pi x}{a}+3\sin\frac{3\pi x}{a}$.

Ans. $u(x,y)=3\cdot\text{cosech}\left(\frac{3\pi b}{a}\right)\cdot\sin\left(\frac{3\pi x}{a}\right)\times$

$$\times\sinh\left(\frac{3\pi(b-y)}{a}\right)+5\cdot\text{cosech}\left(\frac{4\pi b}{a}\right)\times$$

$\times \sin\left(\frac{4\pi x}{a}\right) \cdot \sinh\left(\frac{4\pi(b-y)}{a}\right)$

9. $u(0, y) = u(a, y) = 0, \left(\frac{\partial u}{\partial y}\right)\Big|_{y=b} = 0$

(i.e., $y = b$ is insulated) $u(x, 0) = f(x)$

Ans. $u(x, y) = \sum B_n \cdot \sin\left(\frac{n\pi x}{a}\right) \cdot \cosh\left(\frac{n\pi(b-y)}{a}\right) \times$

$\times \operatorname{sech}\left(\frac{n\pi b}{a}\right)$ where

$$B_n = \frac{2}{a}\int_0^a f(x) \sin\left(\frac{n\pi x}{a}\right) dx$$

10. $u_x(0, y) = u_x(L, y) = 0 = u_y(x, 0), u(x, z_0)$
$= gz_0 + gcx$ (three faces insulated)

Ans. $u(x, y) = \left(gz_0 + \frac{gcL}{2}\right) -$

$-\frac{4gcL}{\pi^2}\sum_{n=0}^{\infty} \frac{\cos\left(\frac{m\pi x}{L}\right)\cdot\cosh\left(\frac{m\pi y}{L}\right)}{m^2\cosh\left(\frac{m\pi z_0}{L}\right)}$ where $m = 2n+1$.

Semi infinite strip

Find the steady-state temperature distribution in a thin sheet of metal plate which occupies the semi-infinite strip, $0 \le x \le L$ and $0 \le y < \infty$ with the boundary conditions

11. $u(0, y) = 0, u(\pi, y) = 0$ for all y, $L = \pi$

$u(x, \infty) = 0, u(x, 0) = u_0$ in $0 < x < \pi$

Ans. $u(x, y) = \frac{4u_0}{\pi}\sum_{n=1}^{\infty}\frac{1}{(2n-1)} \cdot \sin(2n-1)x$

$\times e^{-(2n-1)y}$

12. $u(0, y) = 0, u(a, y) = 100$ for all y, $L = a$

$u(x, \infty) = 0, u(x, 0) = f(x)$ for $0 < x < a$

Ans. $u(x, y) = 100\frac{x}{a} + \sum_{n=1}^{\infty} B_n e^{-n\pi y/a} \cdot \sin\left(\frac{n\pi x}{a}\right)$

where $B_n = \frac{2}{a}\int_0^a \left[f(x) - \frac{100x}{a}\right] \cdot \sin\frac{n\pi x}{a} dx$

13. $u(0, y) = 0, u(10, y) = 0$ for all y; $L - 10$

$u(x, \infty) = 0, u(x, 0) =$

$$f(x) = \begin{cases} 20x, & 0 \le x \le 5 \\ 20(10-x), & 5 \le x \le 10 \end{cases}$$

Ans. $u(x, y) = \frac{800}{\pi^2}\sum_{n=1}^{\infty}\frac{(-1)^{n+1}}{(2n-1)^2} \cdot \sin\left(\frac{(2n-1)\pi x}{10}\right)$

$\times e^{-(2n-1)\pi y/10}$

14. $u(x, 0) = 0, u(x, L) = 0$ for all x,

$u(\infty, y) = 0, u(0, y) = g(y)$ for $0 \le y \le L$

Ans. $u(x, y) = \sum_{n=1}^{\infty} B_n \sin\frac{n\pi y}{L} \cdot e^{\frac{-n\pi x}{L}}$ where

$$B_n = \frac{2}{L}\int_0^L g(y) \sin\left(\frac{n\pi y}{L}\right) dy$$

15. $u(0, y) = 0, u(8, y) = 0$ for all y, $L = 8$

$u(x, \infty) = 0, u(x, 0) = f(x) = 100 \sin\frac{\pi x}{8}$

Ans. $u(x, y) = 100 \sin\left(\frac{\pi x}{8}\right) \cdot e^{\frac{-\pi y}{8}}$

19.8 LAPLACE EQUATION IN POLAR COORDINATES

The two dimensional Laplace equation in Cartesian coordinates

$$\nabla^2 u = \frac{\partial^2 u}{\partial x^2} + \frac{\partial^2 u}{\partial y^2} = 0$$

gets transformed to the Laplace equation in polar coordinates r, θ as

$$\nabla^2 u = \frac{\partial^2 u}{\partial r^2} + \frac{1}{r}\frac{\partial u}{\partial r} + \frac{1}{r^2}\frac{\partial^2 u}{\partial \theta^2} = 0 \tag{1}$$

by the transformations

$$x = r\cos\theta, \; y = r\sin\theta$$

(see WE6 on page 15.25)
Consider the problem of solving Laplace equation (1), by the method of separation of variables in the plane region $a < r < b$, $0 < \theta < \alpha$ with the

following boundary conditions

$$\text{on } QR: \; u(r,0) = c, a < r < b \tag{2}$$

$$\text{on } TS: \; u(r,\alpha) = d, a < r < b \tag{3}$$

$$\text{on } TQ: \; u(a,\theta) = 0, 0 < \theta < \alpha \tag{4}$$

$$\text{on } SR: \; u(b,\theta) = f(\theta), 0 < \theta < \alpha \tag{5}$$

Assume that $u(r,\theta) = R(r)\phi(\theta)$ (6)

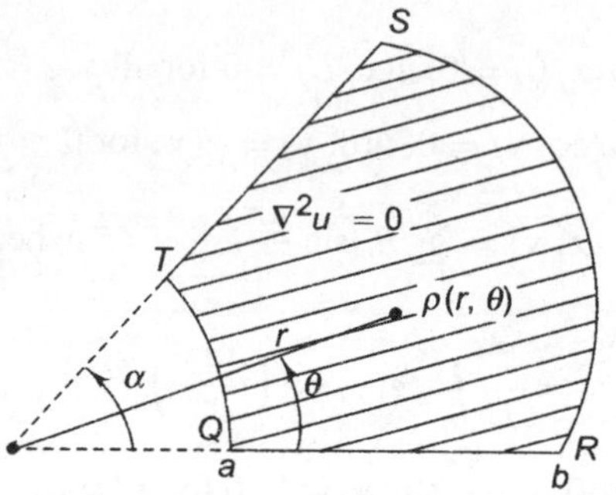

Fig. 19.12

Substituting (6) in (1), we get

$$\frac{r^2R''+rR'}{R}+\frac{\ddot{\phi}}{\phi}=0$$

Here $'$ denotes differentiation w.r.t. r and (.) denotes differentiation w.r.t. θ.

or

$$\frac{r^2R''+rR'}{R}=-\frac{\ddot{\phi}}{\phi}=\text{constant}=\lambda \tag{7}$$

since the R.H.S. is a function of ϕ alone and the L.H.S. is a function of 'r' alone, the equality holds good only when both of them are constant.

In order to have physically meaningful solutions satisfying the given boundary conditions choose the unknown constant $\lambda = +p^2$. This (7) given rise to two ordinary differential equations

$$r^2R''+rR'-p^2R=0 \tag{8}$$

and

$$\ddot{\phi}+p^2\phi=0 \tag{9}$$

Equation (8) is Euler-Cauchy equation which can be solved by substitution $x = e^t$. For $p = 0$, the general solutions of (8) and (9) are

$$R(r) = c_1 + c_2 \ln r$$

$$\phi(\theta) = c_3 + c_4\theta$$

For $p \neq 0$, the general solutions of (8) and (9) are

$$R(r) = c_5r^p + c_6r^{-p}$$

$$\phi(\theta) = c_7\cos p\theta + c_8 \sin p\theta$$

Using the principle of superposition, we get

$$u(r,\theta) = (c_1 + c_2\ln r)(c_3 + c_4\theta) + \\ +(c_5r^p + c_6r^{-p})(c_7\cos p\theta + c_8\sin p\theta) \tag{10}$$

The arbitrary constants $c_1, c_2.., c_8$ will be determined using the boundary conditions (2), (3), (4) and (5). Using (2) in (10), we have

$$c = u(r,0) = (c_1 + c_2\ln r)(c_3) + (c_5r^p + c_6r^{-p})c_7$$

Then $c_1c_3 = c$, $c_2 = 0$, $c_7 = 0$. Then (10) reduces to

$$u(r,\theta) = c + c_9\theta + (c_{10}r^p + c_{11}r^{-p})\sin p\theta \tag{11}$$

where $c = c_1c_3$, $c_9 = c_1c_4$, $c_{10} = c_5 \cdot c_8$, $c_{11} = c_6 \cdot c_8$. Using (3) in (11), we have

$$d = u(r,\alpha) = c + c_9\alpha + (c_{10}r^p + c_{11}r^{-p})\sin p\alpha$$

Then $d = c + c_9\alpha$ and $\sin p\alpha = 0$ or $c_9 = \frac{d-c}{\alpha}$ and $p\alpha = n\pi$ i.e., $p = \frac{n\pi}{\alpha}$, $n = 1, 2, \ldots$

Thus (11) reduces to

$$u(r,\theta) = c + \frac{(d-c)}{+}\alpha\theta + \\ + \sum_{n=1}^{\infty}(A_nr^{n\pi/\alpha} + B_nr^{-n\pi/\alpha})\sin\left(\frac{n\pi\theta}{\alpha}\right) \tag{12}$$

Using (4) in (12), we get

$$0 = u(a,\theta) = c + \left(\frac{d-c}{\alpha}\right)\theta + \\ + \sum_{n=1}^{\infty}(A_na^{n\pi/\alpha} + B_na^{-n\pi/\alpha})\sin\left(\frac{n\pi\theta}{\alpha}\right) \tag{13}$$

Therefore the third term in the R.H.S. of (13) is a half range Fourier since series expansion of the function $-c - \frac{(d-c)\theta}{\alpha}$ in the interval $0 < \theta < \alpha$. From Fourier series, we have

$$(A_na^p + B_na^{-p}) = \\ = \frac{2}{\alpha}\int_0^\alpha \left[-c - \frac{(d-c)}{\alpha}\theta\right] \cdot \sin p\theta d\theta \tag{14}$$

Finally using (5) in (12), we get

$$f(\theta)=u(b,\theta)=c+\left(\frac{d-c}{\alpha}\right)\theta+$$
$$+\sum(A_n b^p+B_n b^{-p})\sin p\theta$$

By similar analysis

$$(A_n b^p+B_n b^{-p})=\frac{2}{\alpha}\int_0^{\alpha}\left[f(\theta)-c-\frac{(d-c)}{\alpha}\theta\right]\sin p\theta d\theta \tag{15}$$

For given $c, d, \alpha, f(\theta)$, equations (14) and (15) constitute two equations for the two unknowns A_n and B_n. A non-trivial unique solution exists because the coefficient determinant is zero i.e.,

$$\begin{vmatrix} a^p & a^{-p} \\ b^p & b^{-p} \end{vmatrix}=\left(\frac{a}{b}\right)^p-\left(\frac{b}{a}\right)^p\neq 0 \qquad \because\ b\neq a$$

Thus the general solution to the Laplace equation (1) with boundary conditions (2), (3), (4), (5) is given by (12) with the coefficients A_n's and B_n's obtained by solving (14) and (15).

Dirichlet Problem for Circular Disk

Consider a thin circular disk of radius 'b' with the two faces insulated and the temperature on the circumference is prescribed as

$$u(b,\theta)=f(\theta),\quad (-\infty<\theta<\infty) \tag{16}$$

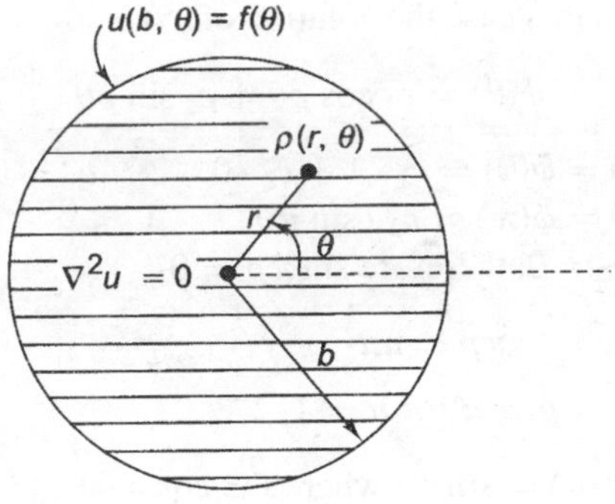

Fig. 19.13

By the method of separation of variables the solution to the Laplace equation (1) is given by (10) (of the previous section)

$$u(r,\theta)=(c_1+c_2\ln r)(c_3+c_4\theta)+$$
$$+(c_5r^p+c_6r^{-p})(c_7\cos p\theta+c_8\sin p\theta) \tag{10}$$

Apparently there is only one boundary condition (16) in the circular disk problem, unlike the previous problem where four boundary conditions (2), (3), (4), (5) are available. The circular disk (Fig. 19.13) may be considered as the limiting case of the region (Fig. 19.12) as $a\to 0$ and $\alpha\to 2\pi$. So assume that $u(r,\theta)$ is bounded as $r\to 0$. Using this in (10), as $r\to 0$, c_2 and c_6 must be zero (because $\ln r\to\infty$, $r^{-p}\to\infty$) as $r\to 0$. Thus

$$u(r,\theta)=c_9+c_{10}\theta+$$
$$+r^p(c_{11}\cos p\theta+c_{12}\sin p\theta) \tag{17}$$

where $c_9=c_1c_3$, $c_{10}=c_1c_4$, $c_{11}=c_5c_7$, $c_{12}=c_5c_8$. Note that the domain of θ is infinite in the present problem $(-\infty<\theta<\infty)$ unlike the previous problem where θ is finite $(0<\theta<\alpha)$. So we assume that $u(r,\theta)$ is periodic, i.e., $u(r,\theta+2\pi)=u(r,\theta)$ in order that $u(r,\theta)$ is single-valued function of θ. In equation (17), since $c_{10}\theta$ is not periodic, we take $c_{10}=0$. The constant c_9 is anyway periodic. Finally in order that the remaining two terms $\cos p\theta$ and $\sin p\theta$ are periodic, we must have

$$\cos p(\theta+2\pi)=\cos p\theta \tag{18}$$

and

$$\sin p(\theta+2\pi)=\sin p\theta \tag{19}$$

Expanding equation (18), we have $\cos p\theta\cdot\cos 2\pi p-\sin p\theta\cdot\sin 2\pi p=\cos p\theta$. Then $\cos 2\pi p=1$ for $k=1,2,3,\ldots$ and $\sin 2\pi p=0$ for $k=\frac{1}{2},1,\frac{3}{2},2\cdots$. Thus equation (18) is satisfied for the common values $k=1,2,3,\ldots$. Similar result holds good for equation (19) also. Now with $c_{10}=0$ and $p=n$, the equation (17) reduces to

$$u(r,\theta)=\frac{A_0}{2}+\sum_{n=1}^{\infty}r^p(A_n\cos n\theta+B_n\sin n\theta) \tag{20}$$

Here $c_9=\frac{A_0}{2}$

Now making use of the only boundary condition of the present problem (16) in (20), we have

$$f(\theta) = u(b, \theta) = \frac{A_0}{2} + \sum_{n=1}^{\infty} b^n (A_n \cos n\theta + B_n \sin n\theta) \tag{21}$$

Observe that the R.H.S. of equation (21) is the (full) Fourier series expansion of the periodic function $f(\theta)$ of period 2π. Then the Fourier coefficients are given by

$$A_0 = \frac{1}{\pi} \int_{-\pi}^{\pi} f(\theta) d\theta \tag{22}$$

$$A_n = \frac{1}{\pi b^n} \int_{-\pi}^{\pi} f(\theta) \cos n\theta d\theta \tag{23}$$

$$B_n = \frac{1}{\pi b^n} \int_{-\pi}^{\pi} f(\theta) \sin n\theta d\theta \tag{24}$$

Thus the solution to the circular disk problem is given by equation (20) where the coefficient A_0, A_n, B_n are determined by equation (22), (23) and (24).

Note: Putting $r = 0$ in equation (20), we get the value of u at the centre of the disk which equals to $\frac{A_0}{2}$.

WORKED OUT EXAMPLES

Determine the steady-state temperature distribution $u(r, \theta)$ in a semi-circular plate of radius 'b' cm with insulated faces. The bounding diameter is kept at 0°C and the temperature on the circumference is prescribed by $u(b, \theta) = k\theta(\pi - \theta)$ when $0 \le \theta \le \pi$.

Solution The boundary value problem consists of the two-dimensional Laplace equation in polar coordinates

$$\frac{\partial^2 u}{\partial r^2} + \frac{1}{r}\frac{\partial u}{\partial r} + \frac{1}{r^2}\frac{\partial^2 u}{\partial \theta^2} = 0, 0 < r < b \tag{1}$$

with boundary conditions

$$u(r, 0) = 0, \ u(r, \pi) = 0, \ 0 < r < b \tag{2}$$

$$u(b, \theta) = f(\theta) = k\theta(\pi - \theta), \ 0 \le \theta \le \pi \tag{3}$$

Assume $u(r, \theta) = R(r)\phi(\theta)$ (4)

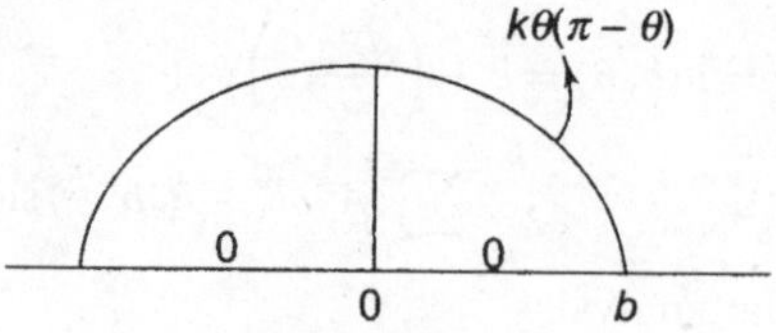

Fig. 19.14

Using (4) equation (1) takes the form

$$\frac{r^2 R'' + rR'}{R} = -\frac{\ddot{\phi}}{\phi} = \lambda = \text{constant}$$

consider $\frac{d^2\phi}{d\theta^2} + \lambda\phi = 0$, (5)

with boundary conditions $u(r, 0) = R(r)\phi(0) = 0 \therefore \phi(0)$ and

$$u(r, \pi) = R(r)\phi(\pi) = 0 \ \therefore \quad \phi(\pi) = 0$$

For $\lambda \le 0$, we get trivial solutions.

For $\lambda = 0$, the solution of (5) is $\phi(\theta) = c_1\theta + c_2$

Since $0 = \phi(0) = c_1 0 + c_2 \quad \therefore \quad c_2 = 0$

Since $0 = \phi(\pi) = c_1 \cdot \pi \quad \therefore \quad c_1 = 0$.

Thus $\phi(\theta) = 0$ for any θ.

Similarly for $\lambda = -p^2$, the solution of (5) is

$$\phi(\theta) = c_1 e^{p\theta} + c_2 e^{-p\theta}$$

since $0 = \phi(0) = c_1 + c_2$ or $c_1 = -c_2$

since $0 = \phi(\pi) = c_1 e^{p\pi} + c_2 e^{-p\pi} = c_1(e^{p\pi} - e^{-p\pi}) \therefore c_1 = 0$ and $c_2 = -c_1 = 0$.

Again $\phi(\theta) = 0$ for any θ.

So consider $\lambda = p^2 > 0$ to obtain non-trivial solutions. In this case the solution of (5) is

$$\phi(\theta) = c_1 \cos p\theta + c_2 \sin p\theta$$

using $0 = \phi(0) = c_1 \cdot 1 + c_2 \cdot 0 \quad \therefore \quad c_1 = 0$

using $0 = \phi(\pi) = c_2 \cdot \sin \pi\theta$

since $c_2 \ne 0$, therefore $\sin \pi p = 0$

or $\qquad \pi p = n\pi$

or $\qquad p = n$ for $n = 1, 2, 3 \ldots$ (6)

Thus $\phi_n(\theta) = \sin n\theta$ where the constant c_2 is taken as unity.

Now the D.E. in R is

$$r^2 \frac{d^2 R}{dr^2} + r\frac{dR}{dr} - p^2 R = 0$$

which is Euler-Cauchy equation.

Put $r = e^z$. Then $r^2\frac{d^2R}{dr^2} = \frac{dR^2}{dz^2} - \frac{dR}{dz}$ and $r\frac{dR}{dr} = \frac{dR}{dz}$

Then the D.E. reduces to

$$\frac{d^2R}{dz^2} - \frac{dR}{dz} + \frac{dR}{dz} - n^2R = 0 \because p = n.$$

The general solution of this equation is

$$R(z) = c_1e^{nz} + c_2e^{-nz}$$

$$\text{or} \quad R(r) = c_1r^n + c_2r^{-n}$$

since $u(r, \theta)$ is bounded, $\lim_{\text{as } r\to 0} u(r,\theta) =$ finite

Now since n is a positive integer $\lim_{\text{as } r\to 0} R(r)$ is finite provided $c_2 = 0$

Thus $R_n(r) = c_1r^n$

Then by superposition principle

$$u(r, \theta) = \sum_{n=1}^{\infty} B_nr^n \sin n\theta \tag{7}$$

To determine the unknown coefficients B_n's use the boundary condition $u(b, \theta)$, in (7). So

$$K\theta(\pi - \theta) = u(b, \theta) = \sum_{n=1}^{\infty} B_nb^n \sin n\theta$$

which is a half range Fourier sine series expansion of $K\theta(\pi - \theta)$ in the interval $(0, \pi)$. Then the Fourier coefficients are given by

$$b^nB_n = \frac{2}{\pi}\int_0^{\pi} K\theta(\pi - \theta)\sin n\theta d\theta$$

Integrating by parts

$$b^nB_n = \frac{2K}{\pi}\Bigg[\theta(\pi - \theta)\cdot\left(-\frac{1}{n}\cos n\theta\right) - (\pi - 2\theta)\left(\frac{-1}{n^2}\right)\sin n\theta + (-2)\cdot\left(\frac{1}{n^3}\right)\cos n\theta\Bigg]\Bigg|_{\theta=0}^{\pi}$$

$$b^nB_n = \frac{2K}{\pi}\left[\frac{2}{n^3}(1 - \cos n\pi)\right]$$

$$= \frac{4K}{\pi n^3}(1 - (-1)^n)$$

Thus the steady-state temperature distribution in the circular plate is

$$u(r, \theta) = \frac{4K}{\pi}\sum_{n=1}^{\infty}\frac{[1 - (-1)^n]}{b^nn^3}r^n \sin n\theta.$$

EXERCISE

1. A plate having the shape of a quadrant of a circle of radius 10 cm has insulated faces. The bounding radii $\theta = 0$ and $\theta = \frac{\pi}{2}$ are kept at 0°C while the temperature along the circular quadrant is kept at $100(\pi\theta - 2\theta^2)$ when $0 \le \theta \le \frac{\pi}{2}$ until steady-state conditions prevail. Find the temperature at the centroid of the plate having coordinates $\left(\frac{40\sqrt{2}}{3\pi}, \frac{\pi}{4}\right)$.

Ans. 45.8

Hint: $\phi(0) = \phi\left(\frac{\pi}{2}\right) = 0$, $p = 2n$, $u(r, \theta) = \sum_{n=1}^{\infty} c_nr^{2n}\sin 2n\theta$, $u(10, \theta) = 100(\pi\theta - 2\theta^2)$, $c_n = \frac{-400}{\pi n^3(10)^{2n}}$, for $n = 1$, at centroid, $u = \frac{400}{\pi}\left(\frac{4\sqrt{2}}{3\pi}\right)^2$

2. The bounding diameter of a semi-circular plate of radius 'b' cm, with insulated faces, is kept at 0°C. The semi-circumference is maintained at a temperature given by $\frac{K\theta}{\pi}$ in $0 \le \theta \le \frac{\pi}{2}$ and $\frac{K}{\pi}(\pi - \theta)$ in $\frac{\pi}{2} \le \theta \le \pi$. Determine the steady-state temperature distribution in the plate.

Ans: $u(r, \theta) = \frac{4K}{\pi^2}\sum_{n=1}^{\infty}\frac{(-1)^{n-1}}{(2n-1)^2}\left(\frac{r}{b}\right)^{2n-1}\cdot\sin(2n - 1)\theta.$

3. Find the steady-state temperature distribution in a semi-circular plate of redius 'b' cm with the bounding diameter kept at 0°C and with a constant temperature K on the semicircular boundary.

Ans. $u(r, \theta) = \frac{4K}{\pi}\sum_{n=1}^{\infty}\frac{1}{(2n-1)}\left(\frac{r}{b}\right)^{2n-1}\sin(2n - 1)\theta$

4. Solve the Laplace equation in two dimensions for $u(r,\theta)$ defined in the region $a \le r \le b$, $0 \le \theta \le \frac{\pi}{2}$ and with boundary conditions $u(r,0) = u\left(r, \frac{\pi}{2}\right) = 0$, $u(b,\theta) = 0$, $u(a,\theta) = \theta\left(\frac{\pi}{2} - \theta\right)$.

Ans. $$u(r,\theta) = \frac{2}{\pi}\sum_{m=1}^{\infty} \frac{\left(\frac{r}{b}\right)^n - \left(\frac{b}{r}\right)^n}{\left(\frac{a}{b}\right)^n - \left(\frac{b}{a}\right)^n} \frac{\sin n\theta}{(2m-1)}$$
where $n = 4m - 2$

5. Solve the Dirichlet problem:
$$u_{rr} + \frac{1}{r}u_r + \frac{1}{r^2}u_{\theta\theta} = 0$$
$$a < r < b, 0 < \theta < \alpha$$
$$u(r,0) = 0, \quad u(r,\alpha) = 0, \quad (a < r < b)$$
$$u(a,\theta) = 0, \quad u(b,\theta) = 100, \quad (0 < \theta < \alpha)$$

Ans. $$u(r,\theta) = \frac{400}{\pi}\sum_{n=1}^{\infty} \frac{1}{m} \frac{\left(\frac{r}{a}\right)^\lambda - \left(\frac{a}{r}\right)^\lambda}{\left(\frac{b}{a}\right)^\lambda - \left(\frac{a}{b}\right)^\lambda} \cdot \sin\lambda\theta$$
Here $m = 2n - 1$, $\lambda = (2n-1)\pi/\alpha$

6. Determine the steady-state temperature $u(r,\theta)$ in the annulus region $2 < r < 4$ with the temperatures along the boundaries given by $u(2,\theta) = 6\cos\theta + 10\sin\theta$, $u(4,\theta) = 15\cos\theta + 17\sin\theta$.

Ans. $u(r,\theta) = 4\left(r - \frac{1}{r}\right)\cos\theta + 4\left(r + \frac{1}{r}\right)\sin\theta$.

Hint: $$u(r,\theta) = (a_0 + b_0 \log r) + \sum_{n=1}^{\infty}(a_n r^n + b_n r^{-n})\cos n\theta + \sum_{n=1}^{\infty}(c_n r^n + d_n r^{-n})\sin n\theta$$

Dirichlet Problem for Disk

7. Find the steady-state temperature distribution in a circular plate of radius 'b', with insulated faces and with the temperature on circumference prescribed by
$$f(\theta) = u(a,\theta) = \begin{cases} 100, & 0 \le \theta \le \pi \\ 0, & \pi \le \theta \le 2\pi \end{cases}$$

Ans. $$u(r,\theta) = 50 + \frac{200}{\pi}\sum_{n=1}^{\infty} \frac{1}{(2n-1)}\left(\frac{r}{b}\right)^{2n-1}\sin(2n-1)\theta$$

Hint: $$u(r,\theta) = \frac{a_0}{2} + \sum_{n=1}^{\infty} r^n(a_n\cos n\theta + b_n \sin n\theta)$$
$$a_0 = \frac{1}{2\pi}\int_{-\pi}^{\pi} f(\theta)d\theta,\ a_n = \frac{1}{\pi b^n}\int_{-\pi}^{\pi} f(\theta)\cos n\theta d\theta$$
$$b_n = \frac{1}{\pi b^n}\int_{-\pi}^{\pi} f(\theta)\sin n\theta d\theta$$

Solve the Laplace equation $\nabla^2 u = 0$ for $u(r,\theta)$ with the following boundary conditions in the region specified.

8. $1 < r < 2$, $\quad 0 < \theta < \pi$, $\quad u(r,\pi) = 100$, $u(r,0) = u_r(1,\theta) = u(2,\theta) = 0$

Ans. $$u(r,\theta) = \frac{100\theta}{\pi} + \frac{200}{\pi}\sum_{n=1}^{\infty} \frac{1}{n}\frac{(-1)^n}{2^n + 2^{-n}}(r^n + r^{-n})\sin n\theta$$

9. $1 < r < 2$, $0 < \theta < \pi$, $u(1,\theta) = u_\theta(r,0) = u_\theta(r,\pi) = 0$, $u(2,\theta) = 100$

Ans. $u(r,\theta) = \frac{100\ln r}{\ln 2}$

10. $r < 1$, u bounded, $u(1,0) = 50 + 20\cos\theta$

Ans. $u(r,\theta) = 50 + 20r\cos\theta$

19.9 DERIVATION AND SOLUTION OF TWO-DIMENSIONAL HEAT EQUATION

Consider a thin flat metal rectangular plate of uniform thickness θ (cm) of a heat conducting material sandwiched between sheets of insulation. Choose the coordinate system such that one face of the plate is taken as the XOY– plane as shown in Fig. 19.15. Assume that the temperature u at any point of the plate depends on the position in the xy-plane and time t and is independent of the z-coordinate so that u is function of x, y, t i.e. $u = u(x,y,t)$. Thus, the flow is two-dimensional. Let r denote the rate of heat generation per unit volume, $\rho(\text{gr/cm}^3)$ be the density s(cal/gr deg) be the heat capacity and k(cal/cm sec deg) be the thermal conductivity of the plate. Let $q_x(x,y)$ $q_y(x,y)$ denote the heat flow rates in the x and y directions.

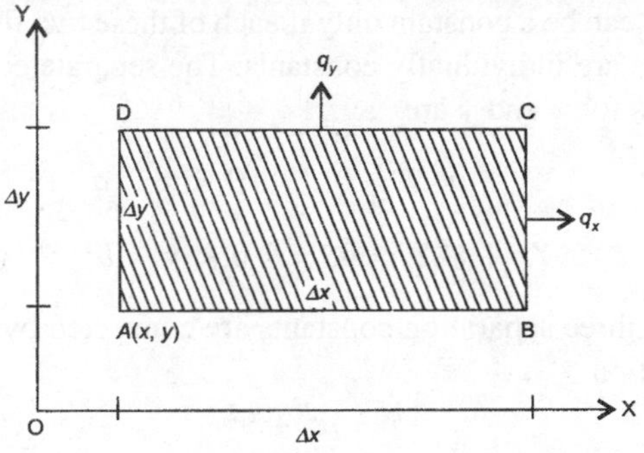

Fig. 19.15

Now the two-dimensional heat equation is derived by applying the law of conservation of energy in the rate form to a small rectangle $ABCD$ with sides Δx and Δy. The conservation of energy states that rate in + rate of generation = rate out + rate of storage or rate in − rate out = rate of storage − rate of generation

$$\left[q_x\left(x, y+\frac{1}{2}\Delta y\right) - q_x\left(x+\Delta x, y+\frac{1}{2}\Delta y\right)\right]\theta\,\Delta y$$

$$+\left[q_x\left(x+\frac{1}{2}\Delta x, y\right) - q_x\left(x+\frac{1}{2}\Delta x, y+\Delta y\right)\right]\theta\,\Delta x$$

$$= \left(-r+\rho s\frac{\partial u}{\partial t}\right)\Delta x\cdot\Delta y\cdot\theta.$$

Dividing throughout by $\theta\,\Delta x\,\Delta y$ and taking the limit as Δx and Δy tend to zero, we get

$$-\frac{\partial q_x}{\partial x} - \frac{\partial q_y}{\partial y} = -r+\rho s\frac{\partial u}{\partial t}$$

Using Fourier law,

$$\frac{\partial}{\partial x}\left(\kappa_x\frac{\partial u}{\partial x}\right)+\frac{\partial}{\partial y}\left(\kappa_y\frac{\partial u}{\partial y}\right) = \rho s\frac{\partial u}{\partial t} - r$$

Assuming that the material is isotropic, we have $\kappa_x = \kappa_y = \kappa$ we have the 2-dimensional heat equation

$$\kappa\left(\frac{\partial^2 u}{\partial x^2}+\frac{\partial^2 u}{\partial y^2}\right) = \rho s\frac{\partial u}{\partial t} - r$$

In the absence of source of heat generation, $r = 0$. The flow of thermal energy in a two-dimensional region is given by

$$\boxed{\frac{\partial u}{\partial t} = c^2\left(\frac{\partial^2 u}{\partial x^2}+\frac{\partial^2 u}{\partial y^2}\right)} \qquad (1)$$

Here $c^2 = \frac{\kappa}{\rho s}$, is a positive constant, is the diffusivity. The temperature distribution $u(x, y, t)$ in the metal plate at any time in the transient state is described by equation (1). In addition to (1), specification of temperature on the boundary (condition) of the plate and initial temperature distribution is required.

Steady-state

When u is independent of time t, temperature distribution is said to be in the steady-state and in this case the equation (1) reduces to

$$\frac{\partial^2 u}{\partial x^2}+\frac{\partial^2 u}{\partial y^2} = 0 \qquad (2)$$

Equation (2) is known as two-dimensional Laplace's equation or steady-state heat equation.

Solution of Two-dimensional Heat Equation by the Method of Separation of Variables

The diffusion of heat in a rectangular metal plate of uniform, isotropic material, with both faces insulated and with the four edges kept at zero temperature is given by the transient temperature $u(x, y, t)$ which satisfies the following initial boundary value problem (IBVP) consisting of

1. Partial Differential Equation (P.D.E.)

$$\frac{\partial u}{\partial t} = c^2\left(\frac{\partial^2 u}{\partial x^2}+\frac{\partial^2 u}{\partial y^2}\right) \qquad (1)$$

in the region of the plate

$$0 \le x \le a, \quad 0 \le y \le b, \quad 0 < t$$

2. With Boundary Conditions (B.C.)

$$u(x, 0, t) = 0, \quad u(x, b, t) = 0,$$
$$0 < x < a, \quad 0 < t \qquad (2)$$
$$u(0, y, t) = 0, \quad u(a, y, t),$$
$$0 < y < b, \quad 0 < t \qquad (3)$$

3. Initial Conditions (I.C.)

$$u(x, y, 0) = f(x, y), \quad 0 < x < a, \quad 0 < y < b \qquad (4)$$

B.C. (2) and (3) are homogeneous. Assume that

$$u(x, y, t) = \phi(x, y)T(t)$$

substituting u in (1), we have

$$\left(\frac{\partial^2\phi}{\phi x^2} + \frac{\partial^2\phi}{\partial y^2}\right) T = \frac{1}{c^2}\phi\dot{T}$$

or

$$\left(\frac{\partial^2\phi}{\phi x^2} + \frac{\partial^2\phi}{\partial y^2}\right)\frac{1}{\phi} = \frac{\dot{T}}{c^2 T}$$

The mutual value of the L.H.S. and R.H.S. of the above equation must be a constant. Since the time-dependent part of the product solution exponentially decays (if $\lambda > 0$), a separation constant in the form of $-\lambda^2$ may be introduced. Then the resulting equations are

$$\dot{T} + \lambda^2 c^2 T = 0, \qquad t > 0 \qquad (5)$$

$$\frac{\partial^2\phi}{\partial x^2} + \frac{\partial^2\phi}{\partial y^2} = -\lambda^2\phi, \quad 0 < x < a, \quad 0 < y < b \qquad (6)$$

The eigen value λ relates to the decay of the time-dependent part. The boundary conditions (2), (3) take the form

$$\phi(x, 0)T(t) = 0, \qquad \phi(x, b)T(t) = 0$$

$$\phi(0, y)T(t) = 0, \qquad \phi(a, y)T(t) = 0$$

If $T(t) = 0$, we get trivial solution that $u = 0$ for all t. Thus, the required boundary conditions are

$$\phi(x, 0) = 0, \quad \phi(x, b) = 0, \quad 0 < x < a \qquad (7)$$

$$\phi(0, y) = 0, \quad \phi(a, y) = 0, \quad 0 < y < b \qquad (8)$$

In the new two-dimensional eigen value problem consisting of equations (6), (7), (8) the P.D.E. and B.C. are linear and homogeneous. So separation of variable can be applied again. Assuming

$$\phi(x, y) = X(x)Y(y)$$

equation (6) becomes

$$\frac{X''}{X} + \frac{Y''}{Y} = -\lambda^2, \qquad 0 < x < a, \qquad 0 < y < b$$

The ratios $\frac{X''}{X}$ and $\frac{Y''}{Y}$ should be a negative constant, denoted by $-\mu^2$ and $-\nu^2$ respectively because the sum of a function of x and a function of y can be a constant only if each of these two functions are individually constants. The separate equations for x and y are

$$X'' + \mu^2 X = 0, \qquad 0 < x < a \qquad (9)$$

$$Y'' + \nu'' Y = 0, \qquad 0 < y < b \qquad (10)$$

The three separation constants are connected by the relation

$$\lambda^2 = \mu^2 + \nu^2 \qquad (11)$$

the boundary conditions (7) and (8) takes the form

$$X(x)Y(0) = 0, \quad X(x)Y(b) = 0, \quad 0 < x < a$$

$$X(0)Y(y) = 0, \quad X(a)Y(y) = 0, \quad 0 < y < b$$

Again $X(x) = 0$ for all x or $Y(y) = 0$ for all y, leads to trivial solution $u = 0$. Thus, the appropriate boundary conditions are

$$Y(0) = 0, \quad Y(b) = 0 \qquad (12)$$

$$X(0) = 0, \quad X(a) = 0 \qquad (13)$$

Equations (9) and (13) and Eqns (10) and (12) form two independent eigen value problems, with the following solutions:

$$X_m(x) = \sin\left(\frac{m\pi x}{a}\right), \quad \mu_m^2 = \left(\frac{m\pi}{a}\right)^2$$

and

$$Y_m(y) = \sin\left(\frac{n\pi y}{b}\right), \quad \nu_n^2 = \left(\frac{n\pi}{b}\right)^2$$

Here the indices n and m are independent. The solutions of the two-dimensional eigen value problems (6), (7), (8) are

$$\phi_{mn}(x, y) = X_m(x)Y_n(y)$$

with $\qquad \lambda_{mn}^2 = \mu_m^2 + \nu_n^2$

and the corresponding solution of (5) is

$$T_{mn} = \exp(-\lambda_{mn}^2 c^2 t)$$

The P.D.E. (1) and B.C. (2), (3) are satisfied by the function

$$u_{mn}(x, y, t) = \phi_{mn}(x, y) \cdot T_{mn}(t)$$

$$= \sin\left(\frac{m\pi x}{a}\right) \cdot \sin\left(\frac{n\pi y}{b}\right) \exp(-\lambda_{mn}^2 c^2 t)$$

for each pair of indices m, n (with $m = 1, 2, 3, \ldots$ and $n = 1, 2, 3, \ldots$).

Using the superposition principle, we obtain the double series

$$u(x, y, t) = \sum_{m=1}^{\infty} \sum_{n=1}^{\infty} A_{mn} \phi_{mn}(x, y) T_{mn}(t) \quad (14)$$

which satisfies equations (1), (2), (3). The unknown coefficients A_{mn} are determined using the initial condition (4) as follows:

The key idea here is the double Fourier series (also known as double trigonometric series).

In a problem involving a rectangle $0 < x < a$, $0 < y < b$, a double Fourier sine series is given by

$$\sum_{m=1}^{\infty} \sum_{n=1}^{\infty} A_{mn} \sin\left(\frac{m\pi x}{a}\right) \cdot \sin\left(\frac{n\pi y}{b}\right)$$

Similarly a double Fourier cosine series is given by

$$\sum_{m=1}^{\infty} \sum_{n=1}^{\infty} A_{mn} \cdot \cos\left(\frac{m\pi x}{a}\right) \cdot \cos\left(\frac{n\pi y}{b}\right)$$

clearly other combinations of sines and cosines could be considered.

All these double series are of the form

$$\sum_{m=1}^{\infty} \sum_{n=1}^{\infty} \phi_m(x) \psi_n(y)$$

where ϕ_m, ψ_n are the eigen functions of a Sturm-Liouville problem, satisfying the orthogonality relations. For example, in the case of double Fourier sine series, the orthogonality relation is

$$\int_0^a \int_0^b \sin\left(\frac{m\pi x}{a}\right) \sin\left(\frac{n\pi y}{b}\right) \sin\left(\frac{p\pi x}{a}\right) \sin\left(\frac{q\pi x}{b}\right) dx\, dy$$

$$= \begin{cases} \frac{ab}{4}, & \text{if both } m = p \text{ and } n = q \\ 0, & \text{otherwise} \end{cases}$$

Thus, many functions can be expressed as sums of multiple Fourier series. The expansion formula in the case of double Fourier sine series in the rectangle $0 < x < a, 0 < y < b$ is

$$f(x, y) = \sum_{m=1}^{\infty} \sum_{n=1}^{\infty} A_{mn} \sin\left(\frac{m\pi x}{a}\right) \sin\left(\frac{n\pi y}{b}\right)$$

Here the unknown coefficients A_{mn} are given by the *generalized Euler Formula.*

$$A_{mn} = \frac{4}{ab} \int_0^b \int_0^a f(x, y) \sin\left(\frac{m\pi x}{a}\right) \sin\left(\frac{n\pi y}{b}\right) dx\, dy \quad (15)$$

with $m = 1, 2, 3, \ldots$ and $n = 1, 2, 3, \ldots$.

Now consider the general solution given by equation (14) as

$$u(x, y, t) = \sum_{m=1}^{\infty} \sum_{n=1}^{\infty} A_{mn} \sin\left(\frac{m\pi x}{a}\right) \sin\left(\frac{n\pi y}{b}\right) \exp(-\lambda_{mn}^2 c^2 t). \quad (16)$$

Using the initial condition (4) in (16) we have

$$f(x, y) = u(x, y, 0) = \sum_{m=1}^{\infty} \sum_{n=1}^{\infty} A_{mn} \sin\left(\frac{m\pi x}{a}\right) \sin\left(\frac{n\pi y}{b}\right)$$

which is a double Fourier sine series. Here A_{mn} are determined by (15). Thus, the complete solution to the IBVP (1), (2), (3), (4) is given by (16) with coefficients A_{mn} determined by (15).

Worked Out Examples

Edges at zero temperature

Example 1: Determine the transient temperature in a rectangular metal plate of uniform isotropic material with both faces insulated, with the four edges maintained at zero temperature and with unitial temperature distributation given by xy.

Solution: Here the initial condition is $u(x, y, o) = f(x, y) = xy$

Then from (15),

$$A_{mn} = \frac{4}{ab} \int_0^b \int_0^a xy \sin\left(\frac{m\pi x}{a}\right) \sin\left(\frac{n\pi y}{b}\right) dx dy$$

$$= \frac{4}{ab} \left[\int_0^b \sin\left(\frac{m\pi y}{b}\right) dy\right] \left[\int_0^a x \sin\left(\frac{n\pi x}{z}\right) dx\right]$$

Integrating by parts

$$A_{mn} = \frac{4}{ab}\left[y\left(\frac{b}{n\pi}\right)(-1)\cos\frac{m\pi y}{b}\right.$$

$$\left.-1\frac{b^2}{n^2\pi^2}(-1)\sin\frac{n\pi y}{b}\right]\Bigg|_{y=0}^{b} \times$$

$$\times\left[x\left(\frac{a}{n\pi}\right)(-1)\cos\left(\frac{n\pi x}{a}\right)\right.$$

$$\left.-1\frac{a^2}{n^2\pi^2}(-1)\sin\left(\frac{n\pi y}{b}\right)\right]\Bigg|_{x=0}^{a}$$

$$= \frac{4}{ab}\left[\frac{-b^2}{m\pi}\cos m\pi\right]\left[\frac{-a^2}{n\pi}\cos n\pi\right]$$

$$= \frac{4ab}{mn\pi^2}\cos m\pi\cos n\pi$$

Thus, the required solution is

$$u(x, y, t) = \frac{4ab}{\pi^2}\sum_{m=1}^{\infty}\sum_{n=1}^{\infty}\frac{(-1)^{m+n}}{mn}\sin\frac{n\pi x}{a}$$

$$\sin\frac{m\pi y}{a}e^{-\lambda_{mn}^2 c^2 t}$$

where $\lambda_{mn}^2 = \mu_m^2 + \nu_n^2$.

Two edges insulated

Example 1: Find the temperature distribution in a rectangular plate $0 < x < 1$, $0 < y < 2$, with both its faces insulated, top and bottom edges insulated and right and left edges kept at zero temperature. The initial temperature distribution is given by

$$u(x, y, 0) = x(1-x)y^2(3-y),$$

$0 < x < 1$, $\quad 0 < y < 2$. Refer Fig. 19.16.

Solution: The initial boundary value problem consists of PDE

$$\frac{\partial u}{\partial t} = c^2\left(\frac{\partial^2 u}{\partial x^2} + \frac{\partial^2 u}{\partial y^2}\right) \tag{1}$$

B.C: Left edge OC : $u(0, y, t) = 0$, $\quad 0 < y < 2$, all t (2)

Right edge AB $u(1, y, t) = 0$, $\quad 0 < y < 2$, all t (3)

Bottom edge OA : $u_y(x, 0, t) = 0$, $\quad 0 < x < 1$,

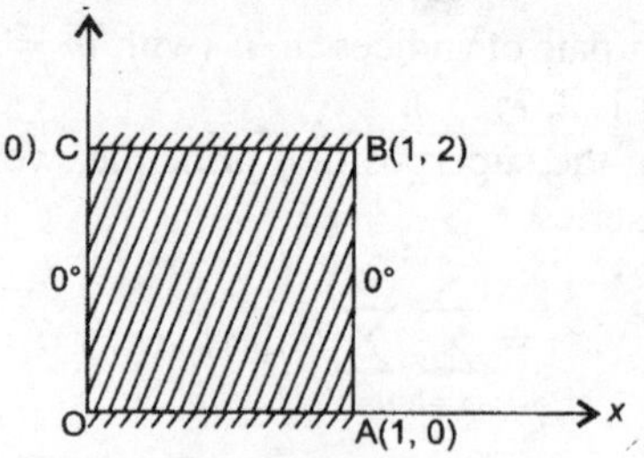

Fig. 19.16

all t (4)

Top edge CB : $u_y(x, 2, t) = 0$, $\quad 0 < x < 1$, all t (5)

I.C. : $u(x, y, 0) = x(1-x)y^2(3-y)$, $\quad 0 < x < 1$, $0 < y < 2$ (6)

substitution of

$$u(x, y, t) = X(x)Y(y)T(t) \tag{7}$$

into the P.D.E. (1) gives

$$XY\dot{T} = c^2(X''YT + XY''T)$$

or

$$\frac{X''}{X} = \frac{\dot{T}}{c^2T} - \frac{Y''}{Y}$$

Assuming the common value of the two sides of this equation as a constant $-\lambda^2$, we get

$$\frac{X''}{X} + \frac{Y''}{Y} = -\lambda^2 = \frac{\dot{T}}{c^2T}$$

Assuming the ratios $\frac{X''}{X}$ and $\frac{Y''}{Y}$ as negative constants $-\mu^2$ and $-\nu^2$ respectively, we get the following three ordinary differential equations.

$$X'' + \mu^2 X = 0 \tag{8}$$

$$Y'' + \nu^2 Y = 0 \tag{9}$$

$$\dot{T} + c^2\lambda^2 T = 0 \tag{10}$$

Using (7) in the boundary conditions (2), (3), (4), (5), we get

$$X(0) = X(1) = 0 \tag{11}$$

$$Y'(0) = Y'(2) = 0 \tag{12}$$

The non-trivial solutions to the eigenvalue problem (8), (11) are given by constant multiples of

$$X_m = \sin\frac{m\pi x}{1}$$

with $\mu_m = m^2\pi^2$, and m a positive integer. Similarly the corresponding non trivial solutions of (9), (12) are constant multiples of

$$Y_n = \cos\frac{n\pi y}{2}$$

with $\nu_n = \frac{n^2\pi^2}{4}$, and n a positive integer (here $y_0 = \cos 0 = 1$)
with $\lambda_{mn}^2 = \mu_m^2 + \nu_n^2 = m^2\frac{\pi^2}{1^2} + \frac{n^2\pi^2}{4}$, the equation (10) becomes

$$\dot{T} + \lambda_{mn}^2 c^2 T = 0$$

with solutions as constant multiples of

$$T_{mn}(t) = \exp(-\lambda_{mn}^2 c^2 t)$$

Then the most general solution satisfying (1), (2), (3), (4), (5) is given by

$$u(x, y, t) = \sum_{m=1}^{\infty}\sum_{n=0}^{\infty} A_{mn}\sin\frac{m\pi x}{1}\cdot\cos\frac{n\pi y}{2} \cdot e^{-\lambda_{mn}^2 c^2 t} \qquad (13)$$

The unknown coefficients A_{mn} are now determined by using the initial condition (6) in (13).

So, $u(x, y, 0) = \sum_{m=1}^{\infty}\sum_{n=0}^{\infty} A_{mn}\sin m\pi x\cdot\cos\frac{n\pi y}{2}$
$= x(1-x)y^2(3-y)$
which is a double trigonometric series. Now for each fixed value of y, express the initial temperature distribution as a sine series in x. Put

$$A_m(y) = \sum_{n=0}^{\infty} A_{mn}\cos\frac{n\pi y}{2}$$

Then

$$x(1-x)y^2(3-y) = \sum_{m=1}^{\infty} A_m(y)\cdot\sin m\pi x$$

with y treated as constant, we have

$$A_m(y) = \frac{2}{1}\int_{x=0}^{1} x(1-x)y^2(3-y)\cdot\sin m\pi x\,dx$$

$$= 2y^2(3-y)\int_0^1 x(1-x)\sin m\pi x\,dx$$

Integrating by parts

$$A_m(y) = 2y^2(3-y)\left[x(1-x)\left(\frac{-1}{m\pi}\right)\cos m\pi x\right.$$

$$-(1-2x)(-1)\frac{1}{m^2\pi^2}\sin m\pi x$$

$$\left.+(-2)(-1)(-1)\frac{1}{m^3\pi^3}\cdot\cos m\pi x\right]_{x=0}^{1}$$

$$= -\frac{2}{m^3\pi^3}[\cos m\pi - 1]\,2y^2(3-y)$$

$$A_m(y) = +\frac{8y^2(3-y)}{m^3\pi^3} \quad \text{when } m \text{ is odd}$$

$$= 0 \quad \text{when } m \text{ is even}$$

Now expand each function $A_m(y)$ in a cosine series of the form

$$A_m(y) = \sum_{n=0}^{\infty} A_{mn}\cos\frac{n\pi y}{2}$$

If m is even then $A_m(y) = 0$ so

$$a_{mn} = 0 \text{ for all } n.$$

If m is odd, then for $n = 0$, we have

$$A_{m0} = \frac{1}{2}\int_{y=0}^{2}\frac{8y^2(3-y)}{m^3\pi^3}dy$$

$$2A_{m0} = \frac{1}{m^3\pi^3}\left[24\frac{y^3}{3} - 8\frac{y^4}{y}\right]\Bigg|_{y=0}^{2} = \frac{32}{m^3\pi^3}.$$

For $n > 0$,

$$A_{mn} = \frac{2}{2}\int_{y=0}^{2}\frac{8y^2(3-y)}{m^3\pi^3}\cos\frac{n\pi y}{2}dy$$

$$A_{mn} = \frac{8}{m^3\pi^3}\left[y^2(3-y)\left(\frac{2}{n\pi}\right)\sin\frac{n\pi y}{2}\right.$$

$$-(6y-3y^2)\left(\frac{-2^2}{n^2\pi^2}\right)\cos\frac{n\pi y}{2}$$

$$+(6-6y)\left(\frac{-2^2}{n^2\pi^2}\right)\left(\frac{2}{n\pi}\right)\cdot\sin\frac{n\pi y}{2}$$

$$\left.-(6)(-1)\frac{2^4}{n^4\pi^4}(-1)\cdot\cos\frac{n\pi y}{2}\right]_{y=0}^{2}$$

$$= \frac{8}{m^3\pi^3}\left[(-1)\frac{16.6}{n^4\pi^4}(\cos n\pi - 1)\right]$$

$$= \begin{cases} \frac{1536}{m^3 n^4\pi^7} & \text{if } n \text{ is odd} \\ 0 & \text{if } n \text{ is even} \end{cases}$$

Thus, the required solution is

$$u(x, y, t) = \sum_{m=1}^{\infty}\sum_{n=0}^{\infty} A_{mn} \sin m\pi x \cdot \cos \frac{n\pi y}{2} e^{-c^2\lambda_{mn}^2 t}$$

where

$$A_{mm} = \begin{cases} 0 & \text{if } n \text{ is even or if } m > 0 \text{ is even} \\ \frac{16}{m^3\pi^3} & \text{if } n \text{ is odd and } m = 0 \\ \frac{1536}{m^3n^4\pi^7} & \text{if } n \text{ and } m \text{ are both odd} \end{cases}$$

EXERCISE

1. Find the temperature $u(x, y, t)$ in a rectangular plate $0 \le x \le 2, 0 \le y \le 3$ with both faces insulated and the four edges kept at zero temperature and with initial temperature given by

$$u(x, y, 0) = 4 \sin\left(\frac{3\pi x}{2}\right) \sin \pi y$$

$$-2 \sin \pi x \cdot \sin \frac{2\pi y}{3}$$

Ans. $u(x, y, t) = 4 \exp\left[-\pi^2\left\{\left(\frac{3}{2}\right)^2 + 1^2\right\} c^2 t\right] \sin \frac{3\pi x}{2} \times \sin \pi y - 2 \cdot \exp\left[-\pi^2\left\{1^2 + \left(\frac{2}{3}\right)^2\right\} c^2 t\right] \sin \pi x \times \sin \frac{2\pi y}{3}$.

2. Solve the problem (*two edges insulated*)

D.E. : $u_t = u_{xx} + u_{yy}, 0 \le x \le 1, 0 \le y \le 1, t \ge 0$

B.C. : $u(x, 0, t) = 0, u_y(x, 1, t) = 0, u_x(0, y, t) = 0, u(1, y, t) = 0$

I.C. : $u(x, y, 0) = \sin\left(\pi \frac{(3x + y)}{2}\right) - \sin\left(\frac{3\pi x}{2}\right) \cdot \cos\left(\frac{\pi y}{2}\right)$

Ans. $u(x, y, t) = \exp\left[-\pi^2\left\{\left(\frac{3}{2}\right)^2 + \left(\frac{1}{2}\right)^2 t\right\}\right] \cos\left(\frac{3\pi x}{2}\right) \times \sin\left(\frac{\pi y}{2}\right)$.

3. Solve the two-dimensional heat conduction problem in a rectangle $0 < x < a, 0 < y < b$ if all *the four edges are insulated* and with initial temperature distribution given as
(a) $u(x, y, 0) = 1$
(b) $u(x, y, 0) = x + y$
(c) $u(x, y, 0) = xy$

Ans. (a) $u(x, y, t) = 1$
The solution for (b) and (c) is of the form

$$u(x, y, t) = \sum_{m=0}^{\infty}\sum_{n=0}^{\infty} A_{mn} \cos \frac{m\pi x}{a} \times \cos\left(\frac{n\pi y}{b}\right) \exp\left(-\lambda_{mn}^2 c^2 t\right)$$

with $\lambda_{mn}^2 = \left(\frac{m\pi}{a}\right)^2 + \left(\frac{n\pi}{b}\right)^2$ and m and n varying from 0 to ∞.

(b) $A_{00} = \frac{a + b}{2}$, $A_{mo} = \frac{-2b(1 - \cos m\pi)}{m^2\pi^2}$,
$A_{0n} = \frac{-2a(1 - \cos n\pi)}{n^2\pi^2}$,
$A_{mn} = 0$, otherwise

(c) $A_{00} = \frac{ab}{4}$, $A_{mo} = \frac{-ab(1 - \cos m\pi)}{m^2\pi^2}$,
$A_{on} = \frac{-(ab)(1 - \cos n\pi)}{n^2\pi^2}$,
$A_{mn} = \frac{4ab(1 - \cos n\pi)(1 - \cos m\pi)}{m^2n^2\pi^2}$
when m and n are greater than zero.

4. Solve the two-dimensional heat conduction problem in $0 \le x \le L, 0 \le y \le M, t \ge 0$. With

B.C. : $u(x, 0, t) = 0, u_y(x, M, t) = 0$
$u_x(0, y, t) = 0, u_x(L, y, t) = 0$

(three edges insulated) and with

I.C. : $u(x, y, 0) = \cos\left(\frac{2\pi x}{L}\right) \cdot \sin\left(\frac{3\pi y}{2M}\right)$

Ans. $u(x, y, t) = \cos\left(\frac{2\pi x}{L}\right) \cdot \sin\left(\frac{3\pi y}{2M}\right) \times \exp\left[-\left\{\left(\frac{2}{L}\right)^2 + \left(\frac{3}{2M}\right)^2\right\}\pi^2 c^2 t\right]$

5. Solve the two-dimensional heat equation in a rectangular plate $0 < x < L_1, 0 < y < L_2$ with all the four edges maintained at zero temperature and with prescribed initial temperature distribution given by $u(x, y, 0) = 1$. Find the relaxation time.

Ans. $u(x, y, t) = \frac{4}{\pi^2} \sum_{m=1}^{\infty}\sum_{n=1}^{\infty} \left[\frac{1-(-1)^m}{m}\right]\left[\frac{1-(-1)^n}{n}\right] \times$

$\sin\frac{m\pi x}{L_1} \times \sin\frac{n\pi y}{L_2} \cdot e^{-\lambda_{mn}\cdot c^2 t}$

Relaxation time τ is $L_1^2 L_2^2 / \left[c^2\pi^2(L_1^2 + L_2^2)\right]$

6. Solve the initial value problem for the heat equation in $0 < x < a$, $0 < y < b$ with BC's $u_x(0, y, t) = 0$, $u_x(a, y, t) = 0$, $u(x, 0, t) = T_1$, $u(x, b, t) = T_2$ and IC $u(x, y, 0) = T_3$ where T_1, T_2, T_3 are constants.

Ans. $u(x, y, t) = \frac{T_2 y}{b} + \frac{(b-y)T_1}{a} + \sum_{n=1}^{\infty} A_n \sin\frac{n\pi y}{b} \cdot \exp\left[-\left(\frac{n\pi}{b}\right)^2 c^2 t\right]$

with $A_n = \frac{2(T_3-T_1)[1-(-1)^n]}{n\pi} + \frac{2(T_1-T_2)(-1)^n}{bn\pi}$

7. Solve the heat equation in a square plate of length a, with left, right, bottom edges at zero temperature and top edge at T_1 constant temperature and with zero initial temperature distribution.

Ans. $u(x, y, t) = \frac{T_1 y}{a} - \frac{4T_1}{\pi^2 a^2}\sum_{m=1}^{\infty}\sum_{n=1}^{\infty}\left[\frac{1-(-1)^m}{m}\right]\left[\frac{1-(-1)^{n+1}}{n}\right] \times \sin\left(\frac{m\pi x}{a}\right) \times \sin\left(\frac{n\pi y}{a}\right) e^{-\lambda_{mn} c^2 t}$

with $\lambda_{mn} = \frac{(m^2+n^2)\pi^2}{a^2}$

19.10 DERIVATION AND SOLUTION OF TWO-DIMENSIONAL WAVE EQUATION

The small transverse vibrations of a membrane tightly stretched over a flat frame (in the xy-plane) are governed by the two-dimensional wave equation. Let $u(x, y, t)$ denote the transverse displacement of the membrane from its equilibrium position at time t. Assume that (a) surface tension T is constant and is independent of position (b) membrane is homogeneous with ρ surface density (mass/unit area) constant (c) deflections are small so that all angles of inclination are small. The wave equation is derived by applying Newton's law of motion to a small rectangle of dimensions Δx and Δy of the membrane. On each edge of the rectangle a concentrated force of magnitude $T\Delta x$ or $T\Delta y$ is exerted. Refer Fig. 19.17.

Looking at the projections on the xu- and yu-planes (Figs. 19.18a and b), the resultant of horizontal forces in x-direction is $T\Delta y(\cos\beta - \cos\alpha)$

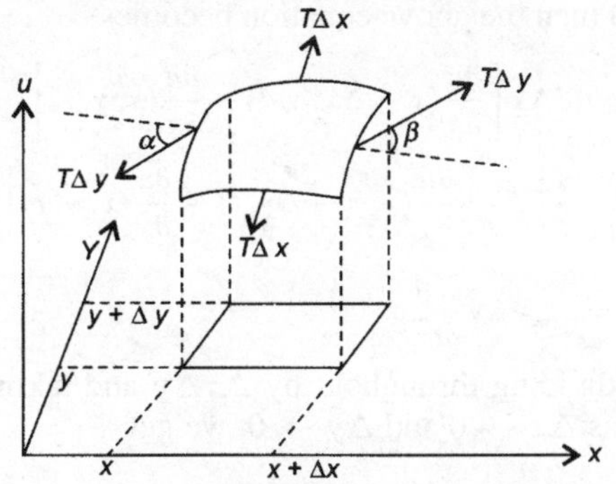

Fig.19.17 Vibrating membrane

and in the y-direction is $T\Delta x(\cos\delta - \cos\gamma)$. For small $\alpha, \beta, \delta, \gamma$, their cosines are close to 1 and hence, the horizontal components at opposite sides are equal.

Equating the sum of the forces in the vertical direction to mass times acceleration, we get

$$T\Delta y(\sin\beta - \sin\alpha) + T\Delta x(\sin\delta - \sin\gamma) = \rho\Delta x\Delta y \cdot \frac{\partial^2 u}{\partial t^2}$$

Assuming $\alpha, \beta, \delta, \gamma$ to be small, we have

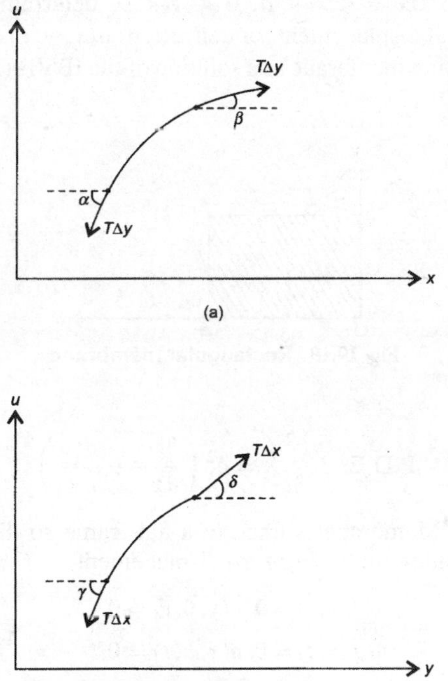

Fig. 19.18

etc. then the above equation becomes

$$T\Delta y\left[\frac{\partial u}{\partial x}(x+\Delta x, y, t)-\frac{\partial u}{\partial x}(x, y, t)\right]$$

$$+T\Delta x\left[\frac{\partial u}{\partial y}(x, y, +\Delta y, t)-\frac{\partial u}{\partial y}(x, y, t)\right]$$

$$=\rho\Delta x\Delta y\frac{\partial^2 u}{\partial t^2}$$

Now dividing throughout by $\Delta x\Delta y$ and taking the limit as $\Delta x\to 0$ and $\Delta y\to 0$, we get

$$T\left(\frac{\partial^2 u}{\partial x^2}+\frac{\partial^2 u}{\partial y^2}\right)=\rho\frac{\partial^2 u}{\partial t^2}$$

or

$$\frac{\partial^2 u}{\partial t^2}=c^2\left(\frac{\partial^2 u}{\partial x^2}+\frac{\partial^2 u}{\partial y^2}\right)$$

This equation is known as two-dimensional wave equation.
Here $c^2=\frac{T}{\rho}$ is a positive constant.

Vibrating Rectangular Membrane

To analyze the transverse vibrations of a rectangular membrane $0<x<a$, $0<y<b$, determine the vertical displacement (or deflection) $u(x, y, t)$ of the vibrating membrane as a solution of the IBVP. (Refer Fig. 19.19).

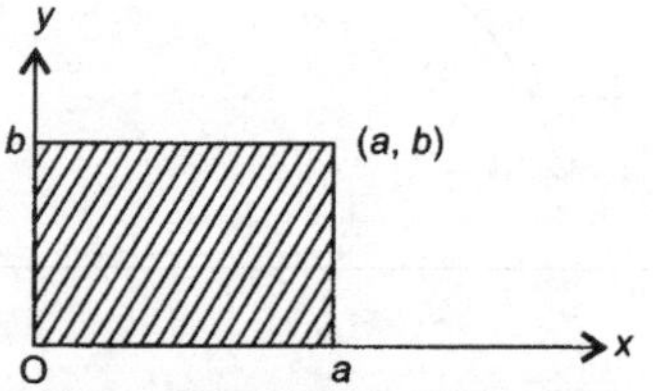

Fig. 19.19 Rectangular membrane

$$\text{P.D.E. : }\quad \frac{\partial^2 u}{\partial t^2}=c^2\left(\frac{\partial^2 u}{\partial x^2}+\frac{\partial^2 u}{\partial y^2}\right) \quad (1)$$

B.C.: Membrane is fixed to a flat frame so that all four sides are having zero displacement

$$u(0, y, t)=0,\ u(x, 0, t)=0 \quad (2)$$

$$u(a, y, t)=0,\ u(x, b, t)=0 \quad (3)$$

$$\text{I.C. : } u(x, y, 0)=\alpha(x, y) \quad (4)$$

where $\alpha(x, y)$ is the given initial displacement and

$$\frac{\partial u}{\partial t}(x, y, 0)=\beta(x, y) \quad (5)$$

where $\beta(x, y)$ is the given initial velocity. We apply the method of separation of variables since both the P.D.E. and B.C.'s are linear and homogeneous. First we separate only the time variable by seeking a product solution of the form

$$u(x, y, t)=h(t)\phi(x, y) \quad (6)$$

substituting (6) into wave equation (1), we get

$$\ddot{h}\phi=c^2(h\phi_{xx}+h\phi_{yy})$$

or

$$\frac{\ddot{h}}{c^2 h}=\frac{\phi_{xx}+\phi_{yy}}{\phi}=-\lambda^2$$

where we have introduced a separation constant $-\lambda^2$.
This results in an O.D.E.

$$\ddot{h}+c^2\lambda^2 h=0 \quad (7)$$

for the time furnction $h(t)$ and a P.D.E.

$$\phi_{xx}+\phi_{yy}+\lambda^2\phi=0 \quad (8)$$

for the amplitude function $\phi(x, y)$. Equation (8) is known as two-dimensional Helmholtz equation. Substitution of (6) in B.C.'s (2), (3) gives

$$\phi(0, y)=0, \qquad \phi(x, 0)=0 \quad (9)$$

$$\phi(a, y)=0, \qquad \phi(x, b)=0 \quad (10)$$

i.e., $\phi=0$ along the entire boundary. Equation (8) and B.C.'s (9) (10) constitute a two-dimensional eigen value problem. Since this eigen value problem has a linear homogeneous P.D.E. in two independent variables with homogeneous B.C.'s we can again apply the method of separation of variables. So we assume that

$$\phi(x, y)=f(x)g(y) \quad (11)$$

a product of functions of *each* independent variable. Using (11) in (8), we get

$$g(y)\frac{d^2 f}{dx^2}+f(x)\frac{d^2 g}{dy^2}=-\lambda^2 f(x)g(y)$$

a product of functions of *each* independent variable. Using (11) in (8), we get

$$g(y)\frac{d^2 f}{dx^2} + f(x)\frac{d^2 g}{dy^2} = -\lambda^2 f(x)g(y)$$

or

$$\frac{1}{f}\frac{d^2 f}{dx^2} = -\lambda^2 - \frac{1}{g}\frac{d^2 g}{dx^2} = -\mu^2$$

where we have introduced a *second* separation constant $-\mu^2$. If $\mu > 0$, we get oscillatory solutions. Thus, we get

$$\frac{d^2 f}{dx^2} + \mu^2 f = 0 \tag{12}$$

$$\frac{d^2 g}{dy^2} + (\lambda^2 - \mu^2)g = 0 \tag{13}$$

with the assumption of separability for the function $u(x, y, t) = f(x)g(y)h(t)$, for the P.D.E. in three variables, we obtain three O.D.E. one function for each independent variable, with only two separation constants. Using (11), the B.C.'s (9), (10) reduce to

$$f(0) = 0, \qquad f(a) = 0 \tag{14}$$

$$g(0) = 0, \qquad g(b) = 0 \tag{15}$$

Thus, we have one Sturm-Liouville eigen value problem in the x-variable namely (12), (14) with μ^2 as the eigen value and $f(x)$ as the eigen function. Similarly, we have another Sturm-Liouville eigen value problem in the y-variable namely (13), (15) with $\lambda^2 - \mu^2$ as the eigen value and $g(y)$ as the eigen function. In the familiar way, for the first eigen value problem, we obtain the eigen value as

$$\mu_n^2 = \left(\frac{n\pi}{L}\right)^2, n = 1, 2, 3, \ldots \tag{16}$$

with corresponding eigen functions as

$$f_n(x) = \sin\left(\frac{n\pi x}{a}\right) \tag{17}$$

For each value of μ_n in (16), There are an infinite number of eigen values λ. Thus, for the second eigen value problem we get the eigen values as

$$\lambda_{mn}^2 - \mu_n^2 = \left(\frac{m\pi}{b}\right)^2, m = 1, 2, 3, \ldots \tag{18}$$

with corresponding eigen functions as

$$g_{nm}(y) = \sin\left(\frac{m\pi y}{b}\right) \tag{19}$$

then

$$\lambda_{nm}^2 = \mu_n^2 + \left(\frac{m\pi}{b}\right)^2 = \left(\frac{n\pi}{a}\right)^2 + \left(\frac{m\pi}{b}\right)^2 \tag{20}$$

where $n = 1, 2, 3, \ldots$ and $m = 1, 2, 3, \ldots$ Thus, $\phi_{mn}(x, y) = \sin\left(\frac{n\pi x}{a}\right)$. $\sin\left(\frac{m\pi y}{b}\right)$ with $n = 1, 2, 3, \ldots$ and $m = 1, 2, 3, \ldots$ solving (7) we get the time-dependent part of the product solutions as $\sin c\lambda_{nm}t$ and $\cos c\lambda_{nm}(t)$ oscillations with natural frequencies $c\lambda_{nm} = c\sqrt{\left(\frac{n\pi}{a}\right)^2 + \left(\frac{m\pi}{b}\right)^2}$ with $n = 1, 2, 3 \ldots$ and $m = 1, 2, 3 \ldots$. The two doubly infinite families of product solutions

$$\sin\left(\frac{n\pi x}{a}\right)\sin\left(\frac{m\pi y}{b}\right)\sin(c\lambda_{mn}t)$$

and

$$\sin\left(\frac{n\pi x}{a}\right)\cdot\sin\left(\frac{m\pi y}{b}\right)\cdot\cos(c\lambda_{mn})t$$

are known as modes of vibration. As time varies, the shape remains the same but only the amplitude varies periodically.

Applying the principle of superposition, we get the displacement function as

$$\begin{aligned} u(x, y, t) &= \sum_{m=1}^{\infty}\sum_{n=1}^{\infty} u_{n,m}(x, y, t) \\ &= \sum_{m=1}^{\infty}\sum_{n=1}^{\infty} \sin\left(\frac{n\pi x}{a}\right)\cdot\sin\left(\frac{m\pi y}{b}\right)\times \\ &\quad \times [A_{mn}\cos(c\lambda_{mn}t) + B_{nm}\sin(c\lambda_{nm}t)] \end{aligned} \tag{21}$$

The function $u_{nm}(x, y, t)$ is known as the (n, m)-th harmonic of the rectangular drum.

Using the two initial conditions (4) and (5), the two families of coefficients A_{nm} and B_{nm} are determined. Putting $t = 0$ in (21) and using IC(4) we have a double Fourier series

$$\begin{aligned} \alpha(x, y) &= u(x, y, 0) \\ &= \sum_{m=1}^{\infty}\left(\sum_{n=1}^{\infty} A_{nm}\sin\left(\frac{n\pi x}{a}\right)\right)\sin\left(\frac{m\pi y}{b}\right) \end{aligned} \tag{22}$$

Put

$$A_m^*(x) = \sum_{n=1}^{\infty} A_{mn}\sin\left(\frac{n\pi x}{a}\right). \tag{23}$$

$< y < b$. Thus, the Fourier sine coefficients are given by

$$A_m^*(x) = \frac{2}{b}\int_0^b \alpha(x, y)\cdot \sin\left(\frac{m\pi y}{b}\right)dy \qquad (24)$$

for each m. R.H.S. of (24) is a function of x alone (since integration wrt y is performed) and is valid for all x. Then from (23)

$$\sum_{n=1}^{\infty} A_{nm}\sin\left(\frac{n\pi x}{a}\right) = A_m^*(x)$$

In other words A_{nm} are the coefficients in the Fourier sine series in x of $A_m^*(x)$ in $0 < x < a$. Then

$$A_{nm} = \frac{2}{a}\int_0^a A_m^*(x)\cdot \sin\left(\frac{n\pi x}{a}\right)dx \qquad (25)$$

substituting $A_m^*(x)$ from (24) in (25), we get

$$A_{nm} = \frac{2}{a}\int_0^a\left[\frac{2}{b}\int_0^b \alpha(x, y)\sin\left(\frac{m\pi y}{b}\right)dy\right]\times$$
$$\times \sin\left(\frac{n\pi x}{a}\right)dx$$

or as a double integral

$$\boxed{A_{nm} = \frac{4}{ab}\int_0^a\int_0^b \alpha(x, y)\sin\frac{m\pi y}{b}dy\cdot \sin\left(\frac{n\pi x}{a}\right)dx} \qquad (26)$$

To obtain the coefficients B_{nm}, differentiate (21) w.r.t. t and put $t = 0$ and use the IC (5). Then

$$\beta(x, y) = \left.\frac{\partial u}{\partial t}\right|_{(x,y,0)} = \sum_{m=1}^{\infty}\sum_{n=1}^{\infty}\sin\left(\frac{n\pi x}{a}\right)\cdot\sin\left(\frac{m\pi y}{b}\right)\times$$
$$\times[-c\lambda_{nm}A_{nm}\sin(c\lambda_{nm}\cdot 0)$$
$$+B_{nm}\cdot c\lambda_{nm}\cos(c\lambda_{nm}\cdot 0)]$$

or

$$\beta(x, y) = \sum_{n=1}^{\infty}\sum_{m=1}^{\infty} c\lambda_{nm}B_{nm}\sin\left(\frac{n\pi x}{a}\right)\cdot\sin\left(\frac{m\pi y}{b}\right)$$

The coefficients in this double Fourier series are given by

$$\boxed{c\cdot\lambda_{nm}\cdot B_{nm} = \frac{4}{ab}\int_0^a\int_0^b B(x, y)\cdot\sin\frac{m\pi y}{b}\cdot\sin\frac{n\pi x}{a}dydx} \qquad (27)$$

Thus, the solution to the IBVP (1), (2), (3), (4), (5) is the doubly infinite series given by (21) with the coefficients A_{nm} and B_{nm} determined by (26) and (27) respectively.

Corollary 1: If the membrane starts from rest, then $B(x, y) = 0$ so all B_{nm} are zero. The solution (21) reduces to

$$u(x, y, t) = \sum_{m=1}^{\infty}\sum_{n=1}^{\infty} A_{nm}\sin\left(\frac{n\pi x}{a}\right)\times$$
$$\times\sin\left(\frac{m\pi y}{b}\right)\times\cos(c\lambda_{nm}t) \qquad (28)$$

with A_{nm}'s determined by (26).

Corollary 2: If the initial displacement is zero, then $\alpha(x, y) = 0$ so all A_{nm}'s are zero. Then (21) reduced to

$$u(x, y, t) = \sum_{m=1}^{\infty}\sum_{n=1}^{\infty} B_{nm}\sin\left(\frac{n\pi x}{a}\right)\times$$
$$\times\sin\left(\frac{m\pi y}{b}\right)\cdot\sin(c\lambda_{nm}t) \qquad (29)$$

with B_{nm}'s determined by (27).

WORKED OUT EXAMPLES

Initial Displacement

Example 1: Find the deflection $u(x, y, t)$ of a rectangular membrane $0 < x < a$, $0 < y < b$ given that its entire boundary is fixed, initial velocity is zero (starts from rest) and initial deflection $\alpha(x, y) = \kappa xy(a - x)(b - y)$.

Solution: Here $\beta(x, y) = 0$ since membrane starts from rest. Therefore B_{nm} are all zero. The displacement (or deflection) function $u(x, y, t)$ is given by (28)

$$u(x, y, t) = \sum_{m=1}^{\infty}\sum_{n=1}^{\infty} A_{nm}\cdot\sin\left(\frac{n\pi x}{a}\right)\cdot\sin\left(\frac{m\pi y}{b}\right)\times$$
$$\times\cos(c\lambda_{nm}t)$$

where the coefficients A_{nm}'s are determined by

$$A_{nm} = \frac{4}{ab}\int_0^a\int_0^b \kappa xy(a - x)(b - y)\sin\left(\frac{n\pi x}{a}\right)\times$$
$$\times\sin\left(\frac{m\pi y}{b}\right)dy\cdot dx$$

where the coefficients A_{nm}'s are determined by

$$A_{nm} = \frac{4}{ab}\int_0^a\int_0^b \kappa xy(a-x)(b-y)\sin\left(\frac{n\pi x}{a}\right)\times$$
$$\times\sin\left(\frac{m\pi y}{b}\right)dy\cdot dx$$

$$\frac{ab}{4\kappa}A_{nm} = \left[\int_0^a x(a-x)\sin\left(\frac{n\pi x}{a}\right)dx\right]\times$$
$$\times\left[\int_0^b y(b-y)\sin\left(\frac{m\pi y}{b}\right)dy\right]$$
$$= I_1\times I_2$$

Integrating by parts,

$$I_1 = \left[x(a-x)\left(\frac{a}{n\pi}\right)\left(-\cos\left(\frac{n\pi x}{a}\right)\right)\right.$$
$$-(a-2x)\cdot\left(\frac{-a^2}{n^2\pi^2}\right)\cdot\sin\left(\frac{n\pi x}{a}\right)$$
$$\left.+(-2)(-1)\frac{a^3}{n^3\pi^3}(-1)\cdot\cos\frac{n\pi x}{a}\right]\Bigg|_{x=0}^{a}$$

$$I_1 = 0+0-\frac{2a^3}{n^3\pi^3}\cdot[\cos n\pi - 1]$$

similarly

$$I_2 = \frac{-2b^3}{m^3\pi^3}[\cos m\pi - 1]$$

Then

$$\frac{ab}{4\kappa}A_{nm} = I_1 I_2 =$$
$$\left\{\frac{-2a^3}{n^3\pi^3}[(-1)^n-1]\right\}\left\{\frac{-2b^3}{m^3\pi^3}[(-1)^m-1]\right\}$$

Thus,

$$A_{nm} = \begin{cases}\frac{64a^2b^2}{\pi^6 n^3 m^3}, & \text{when both } m \text{ and } n \text{ are odd}\\ 0, & \text{otherwise}\end{cases}$$

The required deflection is

$$u(x,y,t) = \frac{64a^2b^2}{\pi^6}\sum_{m=1}^{\infty}\sum_{n=1}^{\infty}\frac{1}{n^3}\sin\left(\frac{n\pi x}{a}\right)$$
$$\times\frac{1}{m^3}\cdot\sin\left(\frac{m\pi y}{b}\right)\cdot\cos(c\lambda_{nm}t)$$

where

$$\lambda_{nm}^2 = \left(\frac{\eta\pi}{a}\right)^2 + \left(\frac{m\pi}{b}\right)^2$$

Initial Velocity

Example 2: Determine the displacement function $u(x, y, t)$ of a rectangular membrane $0 < x < L_1, 0 < y < L_2$ with the entire boundary fixed and with initial conditions $u(x, y, 0) = \alpha(x, y) = 0$ and $u_t(x, y, 0) = \beta(x, y) = 1$.

Solution: In this problem the initial displacement is zero, so the solution is given by (29)

$$u(x,y,t) = \sum_{m=1}^{\infty}\sum_{n=1}^{\infty} B_{nm}\sin\left(\frac{n\pi x}{L_1}\right)\cdot\sin\left(\frac{m\pi y}{L_2}\right)$$
$$\times\sin(c\lambda_{nm}t)$$

with the coefficients B_{nm} given by (27) as

$$c\lambda_{nm}B_{nm} = \frac{4}{L_1\cdot L_2}\int_0^{L_1}\int_0^{L_2} 1\cdot\sin\left(\frac{m\pi y}{L_2}\right)\times$$
$$\times\sin\left(\frac{n\pi x}{L_1}\right)dydx$$

$$= \frac{4}{L_1L_2}\left[\left(\frac{-L_2}{m\pi}\right)\cos\left(\frac{m\pi y}{L_2}\right)\right]_{y=0}^{L_2}\times$$
$$\left[\left(\frac{-L_1}{n\pi}\right)\cos\left(\frac{n\pi x}{L_1}\right)\right]_{x=0}^{L_1}$$

$$= \frac{4}{mn\pi^2}[(-1)^m-1][(-1)^n-1]$$

So

$$B_{nm}\begin{cases}\frac{16}{mn\pi^2 c\lambda_{nm}} & \text{if both } m \text{ and } n \text{ are odd}\\ 0 & \text{otherwise}\end{cases}$$

the required solution is

$$u(x,y,t) = \frac{16}{\pi^2 c}\sum_{m=\text{odd}}^{\infty}\sum_{n=\text{odd}}^{\infty}\sin\left(\frac{n\pi x}{L_1}\right)\cdot\sin\left(\frac{m\pi y}{L_2}\right)$$
$$\times\frac{1}{mn\lambda_{nm}}\cdot\sin(c\lambda_{nm}t)$$

where

$$\lambda_{nm}^2 = \left(\frac{n\pi}{L_1}\right)^2 + \left(\frac{m\pi}{L_2}\right)^2$$

Solution: Here $\beta(x, y) = 0$ so all B_{nm} are zero. The solution is given by (28)

$$u(x, y, t) = \sum_{m=1}^{\infty}\sum_{n=1}^{\infty} A_{nm} \sin\left(\frac{n\pi x}{1}\right) \cdot \sin\left(\frac{m\pi y}{1}\right) \times \cos(c\lambda_{nm} t)$$

where

$$A_{nm} = \frac{4}{1.1}\int_0^1\int_0^1 \left(\sin\frac{m\pi y}{1} \cdot \sin\frac{n\pi x}{1}\right) \times k \cdot \sin 2\pi x \cdot \sin \pi y \, dy \, dx$$

$$A_{21} = 4k \cdot \frac{1}{2} \cdot \frac{1}{2} = k,$$

$$\lambda_{21}^2 = \left(\frac{2\pi}{1}\right)^2 + \left(\frac{1 \cdot \pi}{1}\right)^2 = 5\pi^2$$

All other coefficients A_{nm} are zero for $m \neq 2, n \neq 1$. Thus, the deflection function is

$$u(x, y, t) = k \cdot \sin 2\pi x \cdot \sin \pi y \cdot \cos(c\sqrt{5}\pi t)$$

EXERCISE

1. Find the deflection $u(x, y, t)$ of the square membrane with $a = b = 1$ and $c = 1$ if the initial velocity is zero and the initial deflection is $\alpha(x, y)$ where
   1. $k \sin \pi x \cdot \sin 2\pi y$
   2. $k \sin 3\pi x \cdot \sin 4\pi y$

Ans: 1. $k \cdot \cos \pi\sqrt{5}t \cdot \sin \pi x \cdot \sin 2\pi y$
2. $k \cdot \cos 5\pi t \cdot \sin 3\pi x \cdot \sin 4\pi y$

2. Solve the IVBP for the vibrating membrane in the square $0 < x < L, 0 < y < L$, with the initial condition

$$u(x, y, 0) = 3 \sin\frac{\pi x}{L} \cdot \sin\frac{2\pi y}{L} + 4 \sin\frac{3\pi x}{L} \cdot \sin\frac{5\pi y}{L}, \; u_t(x, y, 0) = 0$$

Ans: $u(x, y, t) = 3 \sin\frac{\pi x}{L} \cdot \sin\frac{2\pi y}{L} \times \cos(\pi ct\sqrt{5}\,/\,L) + 4 \sin\left(\frac{3\pi x}{L}\right) \cdot \sin\left(\frac{5\pi y}{L}\right) \times \cos(\pi ct\,\sqrt{34}\,/\,L)$

3. Find the separated solutions of the wave equation $u_{tt} = c^2(u_{xx} + u_{yy})$ in the square $0 < x < L,\ 0 < y < L$ with the B.C.'s $u_x(0, y, t) = 0, u_x(L, y, t) = 0, u_y(x, 0, t) = 0, u_y(x, L, t) = 0$ and I.C. $u(x, y, 0) = 0$.

Ans: $u_{mn}(x, y, t) = \cos\left(\frac{m\pi x}{L}\right)\cos\left(\frac{n\pi y}{L}\right) \cdot \sin\left(\lambda_{nm}\frac{\pi ct}{L}\right)$ with $\lambda_{nm}^2 = m^2 + n^2$ with $m, n = 0, 1, 2, \ldots$

4. Solve IBVP: $u_{tt} = c^2(u_{xx} + u_{yy}),\ 0 \le x \le 1, 0 \le y \le 1$.

$$\text{B.C.}: u(x, 0, t) = 0, u(x, 1, t) = 0, u(0, y, t) = 0, u(1, y, t) = 0$$

$$\text{I.C.}: u(x, y, 0) = 0, u_t(x, y, 0) = x(x-1)y(y-1)$$

Ans: $u(x, y, t) = \sum_{m=\text{odd}}\sum_{n=\text{odd}} \left[\frac{4}{mn\pi^2}\right]^3 \left[\pi c^2(n^2 + m^2)^{\frac{1}{2}}\right]^{-1} \times \sin[\pi c(n^2 + m^2)^{\frac{1}{2}} t] \sin(n\pi x) \cdot \sin(m\pi y)$

5. Solve the IBVP for a vibrating rectangular membrane $0 < x < a,\ 0 < y < b$ subject to $u(x, y, 0) = 0$ and $\frac{\partial u}{\partial t}(x, y, 0) = \beta(x, y)$ and $u_x(0, y, t) = 0, u_y(x, 0, t) = 0, u_x(a, y, t) = 0, u_y(x, b, t) = 0$.

Ans: $u(x, y, t) = \sum_{n=0}^{\infty}\sum_{m=0}^{\infty} A_{nm} \cos\frac{n\pi x}{a} \cdot \cos\frac{m\pi y}{b} \times \phi_{nm(t)}$

where

$$\phi_{nm}(t) = \begin{cases} t, & \text{for } n = 0, m = 0 \\ \sin \lambda_{nm} t & \text{otherwise} \end{cases}$$

Here $\lambda_{nm}^2 = c^2\left[\left(\frac{n\pi}{a}\right)^2 + \left(\frac{m\pi}{b}\right)^2\right]$

$$A_{nm}\phi'_{nm}(0) = \frac{\int\int \beta(x, y) \cos\frac{n\pi x}{a} \cdot \cos\frac{m\pi y}{b} dx dy}{\int\int \cos^2\left(\frac{n\pi x}{a}\right) \cdot \cos^2\left(\frac{m\pi y}{b}\right) dx dy}$$

19.11 VIBRATIONS OF CIRCULAR MEMBRANE

In engineering, circular membranes, which occur in drums, pumps, microphones, telephones and so on, are of great importance. The vibrations (or displacement) of a plane, thin elastic membrane, offering no resistance to bending, stretched tightly and fixed to a circular frame of radius R are modelled by the

19.11 VIBRATIONS OF CIRCULAR MEMBRANE

In engineering, circular membranes, which occur in drums, pumps, microphones, telephones and so on, are of great importance. The vibrations (or displacement) of a plane, thin elastic membrane, offering no resistance to bending, stretched tightly and fixed to a circular frame of radius R are modelled by the two-dimensional wave equation in polar coordinates given by

$$\frac{\partial^2 u}{\partial t^2} = c^2\left(\frac{\partial^2 u}{\partial r^2} + \frac{1}{r}\frac{\partial u}{\partial r} + \frac{1}{r^2}\frac{\partial^2 u}{\partial \theta^2}\right)$$

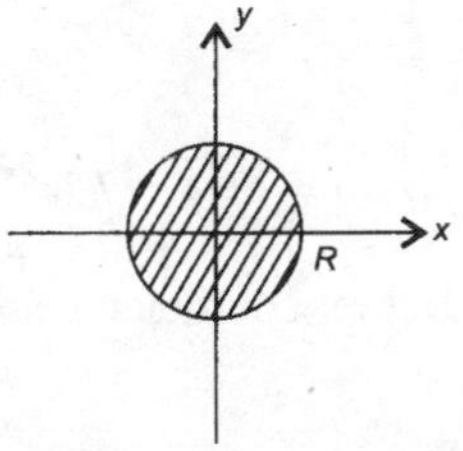

Fig. 19.20 Circular membrane

Assume that the initial conditions are circularly symmetric, (or radially or rotationally symmetric, Fig. 19.20). Then the displacement function $u = u(r, t)$ is independent of θ. Therefore, the above equation reduces to

$$\frac{\partial^2 u}{\partial t^2} = c^2\left(\frac{\partial^2 u}{\partial r^2} + \frac{1}{r}\frac{\partial u}{\partial r}\right), 0 < r < R \qquad (1)$$

with boundary condition (fixed along its entire boundary)

$$u(R, t) = 0, \quad \text{for} \quad t > 0 \qquad (2)$$

and with initial conditions

$$u(r, 0) = \alpha(r), \qquad 0 < r < R \qquad (3)$$

$$\frac{\partial u}{\partial t}(r, 0) = \beta(r), \qquad 0 < r < R \qquad (4)$$

Here $\alpha(r)$ is the initial deflection and $\beta(r)$ is the initial velocity.

Applying the method of separation of variables, we look for product solutions of the form

$$u(r, t) = \phi(r)h(t) \qquad (5)$$

substitution of (5) in (1) yields

$$\frac{1}{c^2}\frac{1}{h}\frac{d^2h}{dt^2} = \frac{1}{r\phi}\frac{d}{dr}\left(r\frac{d\phi}{dr}\right) = -\lambda^2$$

Here we have introduced a separation constant $-\lambda^2$ since displacement oscillates in time for $\lambda > 0$. The time dependent equation is

$$\frac{d^2h}{dt^2} + \lambda^2c^2h = 0 \qquad (6)$$

having solutions $\sin c\lambda t$ and $\cos c\lambda t$ when $\lambda > 0$. The eigen value problem for the separation constant λ is

$$\frac{d}{dr}\left(r + \frac{d\phi}{dr}\right) + \lambda^2 r\phi = 0$$

or

$$r\frac{d^2\phi}{dr^2} + \frac{d\phi}{dr} + \lambda^2 r\phi = 0 \qquad (7)$$

with B.C. $\qquad \phi(R) = 0 \qquad (8)$

since $r = 0$ is a singular point of D.E. (7) we add the requirement that $|\phi(r)|$ is bounded at $r = 0$

i.e., $\qquad |\phi(0)| < \infty \qquad (9)$

(7), (8), (9) form a Sturm-Liouville problem, with eigen functions orthogonal wrt weight r. Using the transformation

$$z = \lambda r$$

The O.D.E. (7) becomes

$$z^2\frac{d^2\phi}{dz^2} + z\frac{d\phi}{dz} + z^2\phi = 0$$

which is a Bessel's equation of order zero, with the general solution

$$\phi = A\,J_0(z) + BY_0(z) = A\,J_0(\lambda r) + BY_0(\lambda r)$$

a linear combination of the zeroth-order Bessel functions. Since $Y_0(\lambda r)$ has a logarithmic singularity at $r = 0$, the boundedness condition (9) demands that $B = 0$. Then

$$\phi(r) = A\,J_0(\lambda r)$$

Now using the B.C. (8) at $r = R$ we have

$$J_0(\lambda R) = 0$$

The I.C. (3) and (4) are satisfied if

$$u(r,0)=\sum_{n=1}^{\infty} A_n J_0(\lambda_n r)=\alpha(r) \quad (11)$$

for $\quad 0<r<R$

and

$$\frac{\partial u}{\partial t}(r,0)=\sum_{n=1}^{\infty} B_n \lambda_n C J_0(\lambda_n r)=\beta(r) \quad (12)$$

for $\quad 0<r<R$

Thus, (11) and (12) are the Fourier-Bessel series representation of $\alpha(r)$ and $\beta(r)$ in $0<r<R$ respectively. The Fourier-Bessel coefficients of order zero of these series are given by the integral formulas

$$A_n=\frac{1}{I_n}\int_0^R r\cdot\alpha(r)J_0(\lambda_n r)dr \quad (13)$$

$$B_n=\frac{1}{c\cdot\lambda_n\cdot I_n}\int_0^R r\cdot\beta(r)J_0(\lambda_n r)dr \quad (14)$$

where

$$I_n=\int_0^R r\cdot[J_0(\lambda_n r)]^2 dr \quad (15)$$

Using the orthogonality relation

$$\int_0^R r\cdot[J_m(\lambda r)]^2 dr=\frac{R^2}{2}[J_{m+1}(R\lambda)]^2 \quad (16)$$

with $m=0$, we have

$$I_n=\frac{R^2}{2}[J_1(R\lambda)]^2 \quad (17)$$

Thus, the displacement function $u(r,t)$ of the vibrating circular membrane is given by (10) with coefficients A_n's and B_n's determined by (13), (14), (17).

Worked Out Examples

Initial Displacement and Initial Velocity

Example 1: Determine the displacement $u(r,t)$ of circular membrane tightly stretched and fixed to a circular frame of radius R with initial displacement $\alpha(r)=3J_0(r\lambda_1)+J_0(r\lambda_3)$ and initial velocity $\beta(r)=J_0(r\lambda_2)$.

Solution: The displacement $u(r,t)$ is given by

$$u(r,t)=\sum_{n=1}^{\infty} A_n J_0(\lambda_n r)\cos(c\lambda_n t)+\sum_{n=1}^{\infty} B_n J_0(\lambda_n r)\sin(c\lambda_n t)$$

Here the initial conditions are

$$u(r,0)=\alpha(r)=3J_0(r\lambda_1)+J_0(r\lambda_3)$$

and

$$\frac{\partial u}{\partial t}(r,0)=\beta(r)=J_0(r\lambda_2).$$

So the Fourier-Bessel coefficients are given by (13) and (14) as

$$A_n=\frac{1}{I_n}\int_0^R r\cdot[3J_0(\lambda_1 r)+J_0(r\lambda_3)]J_0(\lambda_n r)dr$$

Using the orthoganality relation (16) we have $A_1=3$, $A_2=0$, $A_3=1$ and $A_n=0$ for $n>3$. From

$$B_n=\frac{1}{c\lambda_n I_n}\int_0^R rJ_0(r\lambda_2)J_0(\lambda_n r)dr$$

We have $B_1=0$, $B_2=1$, $B_n=0$ for $n>2$. Thus, the required displacement function is

$$u(r;t)=3(J_0\lambda_1)\cdot\cos(c\lambda_1 t)+J_0(r\lambda_3)\cdot\cos(c\lambda_3 t)+\frac{1}{c\lambda_2}J_0(r\lambda_2)\sin(c\lambda_2 t)$$

Exercise

1. Find the deflection $u(r,t)$ of a unit circular membrane (with $c=1$) and with initial velocity zero and initial displacement $\alpha(r)=k(1-r^2)$.

Ans: $u(r,t)=4k\sum_{n=1}^{\infty}\frac{J_2(\lambda_n)}{\lambda_n^2 J_1^2(\lambda_n)}\cos(\lambda_n t)\cdot J_0(\lambda_n r)$

Note: $J_2(\lambda_n)=[2J_1(\lambda_n)]/\lambda_n$

1. Find the deflection $u(r, t)$ of a unit circular membrane (with $c = 1$) and with initial velocity zero and initial displacement $\alpha(r) = k(1 - r^2)$.

Ans: $u(r, t) = 4k \sum_{n=1}^{\infty} \frac{J_2(\lambda_n)}{\lambda_n^2 J_1^2(\lambda_n)} \cos(\lambda_n t) \cdot J_0(\lambda_n r)$

Note: $J_2(\lambda_n) = [2\, J_1(\lambda_n)]/\lambda_n$

2. Determine the displacement $u(r, t)$ of a unit circular membrane (with $c = 1$) and with initial displacement zero and initial velocity $\beta(r) = k(1 - r^2)$.

Ans: $u(r, t) = 4k \sum_{n=1}^{\infty} \frac{J_2(\lambda_n)}{\lambda_n^3 J_1^2(\lambda_n)} \sin(\lambda_n t) \cdot J_0(\lambda_n r)$

19.12 TRANSMISSION LINE EQUATIONS

The purpose of transmisssion lines network is to transfer electric energy from generating units at various locations to the distribution system which ultimately supplies the load. All transmission lines in a power system exhibit the electrical properties of resistance R (ohms/km), inductance L(henries/km), capacitance to ground C (farads/km), and conductance to ground G (mhos/km) of the cable per unit length. Consider a long cable or telephone wire (see Fig. 19.21) that is imperfectly insulated so that leaks occur along the entire length l km of the cable (Fig. 19.21). The source S (the sending end) be at $x = 0$ and the terminal T (the receiving end) be at $x = l$. Let P be any point x km from the source S. The instantaneous current and voltage (or potential) at point P be $i(x, t)$ and $v(x, t)$, where t is time.

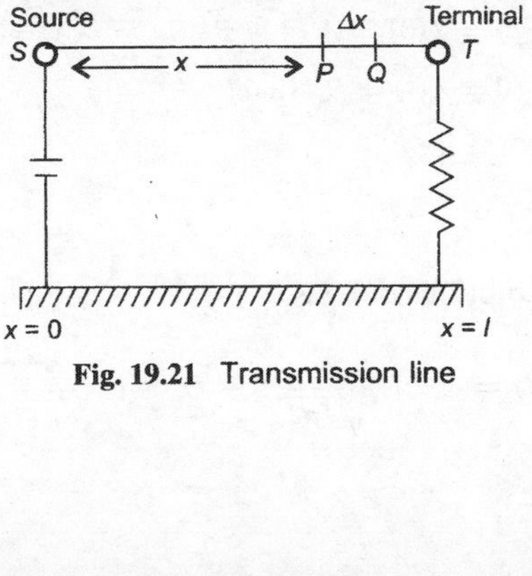

Fig. 19.21 Transmission line

Derivation of Transmission Line Equations

Let Q be at a distance Δx from P. Applying Kirchhoff's voltage law to a small portion PQ of the cable between x and $x + \Delta x$, we have difference of potentials (voltage) at x and $x + \Delta x$ = resistive drop + inductive drop, i.e.

$$-\Delta v = iR\Delta x + L\Delta x \cdot \frac{\partial i}{\partial t}$$

After dividing by Δx and taking limit as $\Delta x \to 0$ we have

$$\boxed{-\frac{\partial v}{\partial x} = Ri + L\frac{\partial i}{\partial t}} \qquad (1)$$

known as *first transmission line equation.*

Similarly applying Kirchhoff's current law: difference of the current at x and $x + \Delta x$ = Loss due to leakage to: ground + captive loss,

i.e., $$-\Delta i = Gv\Delta x + C\frac{\partial v}{\partial t}\Delta x$$

After dividing by Δx and taking limit as $\Delta x \to 0$ we have

$$\boxed{-\frac{\partial i}{\partial x} = Gv + C\frac{\partial v}{\partial t}} \qquad (2)$$

known as *second transmission line equation.*

Telephone Equations

Elimination of i or v from the two transmission line equations (1) and (2) leads to a second order P.D.E. in i or v known as telephone equations as follows: Rewrite (1) and (2) as

$$\left(R + L\frac{\partial}{\partial t}\right) i + \frac{\partial v}{\partial x} = 0 \qquad (3)$$

$$\left(G + C\frac{\partial}{\partial t}\right) v + \frac{\partial i}{\partial x} = 0 \qquad (4)$$

Operate (3) by $\frac{\partial}{\partial x}$ and (4) by $\left(R + L\frac{\partial}{\partial t}\right)$ and subtracting results in

$$\frac{\partial^2 v}{\partial x^2} - \left(R + L\frac{\partial}{\partial t}\right)\left(G + C\frac{\partial}{\partial t}\right) v = 0$$

or

$$v_{xx} = Lcv_{tt} + (RC + GL)v_t + RG\, v \qquad (5)$$

Similarly operate (3) by $\left(G + C\frac{\partial}{\partial t}\right)$ and (4) by $\frac{\partial}{\partial x}$ and subtracting leads to

Telegraph Equations

For a submarine cable, leakages are negligible so $G = 0$ and frequencies are low so $L = 0$. In this case with $G = 0$, $L = 0$, equations (5) and (6) reduces to

$$v_{xx} = RCv_t \tag{7}$$

and

$$i_{xx} = RCi_t \tag{8}$$

known as *submarine cable equations or telegraph equations.* Equations (7) and (8) are similar to one-dimensional heat equation.

High Frequency Line Equations

In the case of alternating currents of high frequencies we can neglect leakages and resistance. Thus, with $G = 0$ and $R = 0$, equations (5) and (6) reduces to

$$v_{xx} = LCv_{tt} \tag{9}$$

and $$i_{xx} = LC\, i_{tt} \tag{10}$$

known as *high-frequency line equations or radio equation.*

Equations (9) and (10) are similar to one dimensional wave equation $u_{tt} = a^2 u_{xx}$ where $a^2 = \frac{1}{LC}$. Thus, the general solution of (9) is

$$v(x, t) = f(x + at) + g(x - at)$$

$$v(x, t) = c_1 e^{\alpha x} \cos(\omega t + \beta x) + c_2 e^{-\alpha x} \cdot \cos(\omega t - \beta x)$$

So at any point along the lossless transmission line, the voltage $v(x, t)$ can be considered as the sum of an incident wave (progressive wave for which $e^{\alpha x}$ increases as x increases) and a reflected wave (receding wave for which $e^{-\alpha x}$ decreases as x increases) travelling with equal velocity a. Similarly current $i(x, t)$ can also be considered as the superposition of incident and reflected waves which behave like travelling waves similar to disturbance in water at some sending point.

WORKED OUT EXAMPLES

Radio equation

Example 1: Assuming R and G are negligible, find the voltage $v(x, t)$ and current $i(x, t)$ in a transmission line of length l, t seconds after the ends are suddenly grounded. The initial conditions are $v(x, 0) = v_0 \sin\left(\frac{\pi x}{l}\right)$ and $i(x, 0) = i_0$.

Solution: When $R = 0$ and $G = 0$, for v, we have P.D.E. the radio equation

$$v_{xx} = LC\, v_{tt} \tag{1}$$

the boundary conditions are

$$v(0, t) = 0 \quad \text{and} \quad v(l, t) = 0 \tag{2}$$

since the ends are suddenly grounded. The given initial condition is

$$v(x, 0) = v_0 \sin\left(\frac{\pi x}{l}\right) \tag{3}$$

we use the separation of variables technique. Assume that

$$v(x, t) = X(x)T(t) \tag{4}$$

Substituting (4) in (1), we get

$$\frac{X''}{X} = LC\frac{\ddot{T}}{T} = -\lambda^2$$

where we have taken the separation constant as $-\lambda^2$, results in the two ordinary differential equations

$$X'' + \lambda^2 X = 0 \tag{5}$$

and $$\ddot{T} + \frac{\lambda^2}{LC} T = 0 \tag{6}$$

General solution of (5) is

$$X(x) = A\cos\lambda x + B\sin\lambda x$$

Use B.C. (2) which take the form

$$X(0) = 0, \qquad X(l) = 0$$

So $$0 = X(0) = A \cdot 1 + B \cdot 0 \quad \therefore \quad A = 0$$

$$0 = X(l) = B \cdot \sin\lambda l$$

or $$\lambda l = n\pi,$$

i.e. $$\lambda = \frac{n\pi}{l}$$

General solution of (6) is

$$T(t) = C\cos\frac{\lambda}{\sqrt{LC}}t + D \cdot \sin\frac{\lambda}{\sqrt{LC}}t$$

i.e. $\lambda = \frac{n\pi}{l}$

General solution of (6) is

$$T(t) = C\cos\frac{\lambda}{\sqrt{LC}}t + D\cdot\sin\frac{\lambda}{\sqrt{LC}}t$$

Using superposition principle, the most general solution of IBVP is

$$v(x,t) = \sum_{n=1}^{\infty}\sin\frac{n\pi x}{l}\cdot(A_n\cdot\cos\mu t + B_n\sin\mu t) \qquad (7)$$

where $\mu = \frac{n\pi}{l\sqrt{LC}}$

From equation

$$-\frac{\partial i}{\partial x} = C\frac{\partial v}{\partial t} + Gv$$

with $G = 0$, we have

$$\frac{\partial v}{\partial t} = -\frac{1}{C}\frac{\partial i}{\partial x}$$

or

$$\left.\frac{\partial v}{\partial t}\right|_{(x,0)} = -\frac{1}{C}\frac{\partial i(x,0)}{\partial x} = -\frac{1}{C}\frac{\partial i_0}{\partial x} = 0 \qquad (8)$$

This is another initial condition. Differentiating (7) w.r.t. t, we get

$$\frac{\partial v}{\partial t} = \sum_{n=1}^{\infty}\sin\left(\frac{n\pi x}{l}\right)\cdot[-\mu A_n\cdot\sin\mu t + \mu B_n\cos\mu t]$$

Using (8)

$$0 = \left.\frac{\partial v}{\partial t}\right|_{(x,0)} = \sum_{n=1}^{\infty}\mu B_n\cdot\sin\frac{n\pi x}{l}$$

Thus, $B_n = 0$ for all n. Now (7) reduces to

$$v(x,t) = \sum_{n=1}^{\infty}A_n\sin\left(\frac{n\pi x}{l}\right)\cdot\cos\frac{n\pi t}{l\sqrt{LC}} \qquad (9)$$

The unknown coefficient A_n's are determined using the initial condition (3). Putting $t = 0$ in (9), we have

$$v_0\sin\left(\frac{\pi x}{l}\right) = v(x,0) = \sum_{n=1}^{\infty}A_n\sin\left(\frac{n\pi x}{l}\right)\cdot 1$$

so $A_1 = v_0$ and $A_n = 0$ for $n > 1$. Thus, the required voltage is

$$v(x,t) = v_0\cdot\sin\left(\frac{\pi x}{l}\right)\cdot\cos\left(\frac{\pi t}{l\sqrt{LC}}\right) \qquad (10)$$

To determine the current $i(x,t)$, use

$$-\frac{\partial i}{\partial x} = Gv + C\frac{\partial v}{\partial t}$$

with $G = 0$. So

$$\frac{\partial i}{\partial x} = -C\frac{\partial v}{\partial t} = -C\frac{\partial}{\partial t}\left[v_0\cdot\sin\left(\frac{\pi x}{l}\right)\cdot\cos\left(\frac{\pi t}{l\sqrt{LC}}\right)\right]$$

$$\frac{\partial i}{\partial x} = C\cdot v_0\cdot\sin\left(\frac{\pi x}{l}\right)\cdot\frac{\pi}{l\sqrt{LC}}\cdot\sin\left(\frac{\pi t}{l\sqrt{LC}}\right)$$

Integrating partially w.r.t. x we get

$$i(x,t) = Cv_0\cdot\frac{\pi}{l\sqrt{LC}}\cdot\sin\frac{\pi t}{l\sqrt{LC}}\cdot\left(\frac{-l}{\pi}\right)\cos\frac{\pi x}{l} + C_1$$

where C_1 is the constant of integration.
Using the initial condition

$$i_0 = i(x,0) = 0 + C_1 \quad \therefore \quad C_1 = i_0$$

then the current

$$i(x,t) = i_0 - v_0\sqrt{\frac{C}{L}}\cdot\cos\left(\frac{\pi x}{l}\right)\cdot\sin\left(\frac{\pi t}{l\sqrt{LC}}\right).$$

Telegraph Line

Example 1: In a telephone wire of length l, a steady voltage distribution of 20 volts at the source end and 12 volts at the terminal end is maintained. At time $t = 0$, the terminal end is grounded. Determine the voltage and current. Assume that $L = 0$ and $G = 0$.

Solution: The equation of telegraph line is

$$v_{xx} = RCv_t$$

or

$$v_t = \frac{1}{RC}\cdot v_{xx}$$

In steady-state voltage distribution, v is independent of time t, so that the above equation reduces to

$$\frac{\partial^2 v}{\partial x^2} = 0$$

Thus, initial steady-state voltage is given by

$$v_s(x) = \frac{-8}{l}x + 20 \qquad (1)$$

Now, at time $t = 0$ the terminal end is grounded. So the boundary conditions change to $v = 20$ at $x = 0$ and $v = 0$ at $x = l$. Using these new boundary conditions we have the steady-state voltage (*after* grounding the terminal end) as

$$v_s^*(x) = ax + b$$

Then $\quad 20 = a \cdot 0 + b \quad \therefore \quad b = 20$

$$0 = a \cdot l + b = a \cdot l + 20 \quad \therefore \quad a = \frac{-20}{l}$$

Thus, $v_s^*(x) = \frac{-20}{l}x + 20 = \frac{20}{l}(l - x)$. Assume that

$$v(x,t) = v_s^*(x) + v^{**}(x,t)$$

Here $v^{**}(x,t)$ is the transient solution of the IBVP consisting of telegraph equation (similar to one-dimensional heat equation with $\frac{1}{RC}$ as a^2) and zero boundary conditions i.e. both ends grounded and is given by

$$v^{**}(x,t) = \sum_{n=1}^{\infty} A_n \cdot \sin\left(\frac{n\pi x}{l}\right) e^{-\left[n^2\pi^2/(l^2 RC)\right]t}$$

Thus

$$v(x,t) = \frac{20}{l}(l - x) + \sum_{n=1}^{\infty} A_n \cdot \sin\left(\frac{n\pi x}{l}\right) \times$$
$$\times e^{-[n^2\pi^2/(l^2RC)]t} \qquad (2)$$

The unknown coefficients A_n's are determined using the initial condition (1). Thus, putting $t = 0$ in (2), we have

$$\frac{-8}{l}x + 20 = v_s(x) = v(x,0)$$
$$= \frac{20}{l}(l - x) + \sum_{n=1}^{\infty} A_n \cdot \sin\left(\frac{n\pi x}{l}\right)$$

or

$$\frac{12x}{l} = \sum_{n=1}^{\infty} A_n \cdot \sin\frac{n\pi x}{l}$$

so

$$A_n = \frac{2}{l}\int_0^l \left(\frac{12x}{l}\right) \sin\left(\frac{n\pi x}{l}\right) dx$$

$$A_n = \frac{24}{l^2}\left[x \cdot \left(\frac{-l}{n\pi}\right)\cos\frac{n\pi x}{l} - 1 \cdot \left(\frac{-l^2}{n^2\pi^2}\right)\sin\frac{n\pi x}{l}\right]\Bigg|_{x=0}^{l}$$
$$= \frac{24}{l^2}\left[l \cdot \left(\frac{-l}{n\pi}\right)\cos n\pi\right]$$
$$= \frac{24}{n\pi}(-1)^{n+1}$$

Thus, the required voltage distribution is

$$v(x,t) = \frac{20(l - x)}{l}$$
$$+ \frac{24}{\pi}\sum \frac{(-1)^{n+1}}{n} \sin\left(\frac{n\pi x}{l}\right) e^{-[n^2\pi^2/(l^2RC)]t}$$

To obtain current $i(x,t)$, use the equation

$$-\frac{\partial v}{\partial x} = Ri + L\frac{\partial i}{\partial t}$$

with $L = 0$, then

$$i(x,t) = -\frac{1}{R}\frac{\partial v}{\partial x}$$

$$i(x,t) = -\frac{1}{R}\frac{\partial}{\partial x}\left[\frac{20(l - x)}{l} + \frac{24}{\pi}\sum \frac{(-1)^{n+1}}{n} \sin\frac{n\pi x}{l} \times \right.$$
$$\left. \times e^{-[n^2\pi^2/l^2RC)t}\right]$$

$$= -\frac{1}{R}\left[-\frac{20}{l} + \frac{24}{\pi}\sum \frac{(-1)^{n+1}}{n}\frac{n\pi}{l}\cos\left(\frac{n\pi x}{l}\right) \times \right.$$
$$\left. \times e^{-[n^2\pi^2/l^2RC)t}\right]$$

Thus,

$$i(x,t) = \frac{20}{Rl} + \frac{24}{lR}\sum (-1)^n \cos\left(\frac{n\pi x}{l}\right) \times$$
$$\times e^{-[n^2\pi^2/(l^2RC)]t}$$

EXERCISE

1. Solve $v_{xx} = LCv_{tt}$ assuming that the initial voltage is $v_0 \cdot \sin\frac{\pi x}{l}$, $v_t(x,0) = 0$ and $v = 0$ at the ends, $x = 0$ and $x = l$ for all t.

Ans: $v(x,t) = v_0 \cos\left(\frac{\pi t}{l\sqrt{LC}}\right) \cdot \sin\left(\frac{\pi x}{l}\right)$

2. In the case of a submarine cable, assuming $L = C = 0$, find the voltage and current given that $v(0) = v_0$ and $i(0) = i_0$.

EXERCISE

1. Solve $v_{xx} = LCv_{tt}$ assuming that the initial voltage is $v_0 \cdot \sin \frac{\pi x}{l}$, $v_t(x, 0) = 0$ and $v = 0$ at the ends, $x = 0$ and $x = l$ for all t.

Ans: $v(x, t) = v_0 \cos\left(\frac{\pi t}{l\sqrt{LC}}\right) \cdot \sin\left(\frac{\pi x}{l}\right)$

2. In the case of a submarine cable, assuming $L = C = 0$, find the voltage and current given that $v(0) = v_0$ and $i(0) = i_0$.

Ans: $v(x) = v_0 \cosh \alpha x - i_0 z_0 \sinh \alpha x$,
$i(x) = i_0 \cosh \alpha x - \frac{v_0}{z_0} \cdot \sinh \alpha x$
where $\alpha = \sqrt{GR}$, $z_0 = \sqrt{\frac{R}{G}}$

Hint: Solve $v_{xx} = GRv$ and use $Ri = \frac{-\partial v}{\partial x}$ to find $i(x)$.

3. Determine the electromotive force $v(x, t)$ in a transmission line of length l, t seconds after the ends were suddenly grounded. Assume that R and G are negligible and initial conditions are $v(x, 0) = a_1 \sin \frac{\pi x}{l}$ $+ a_5 \sin \frac{5\pi x}{l}$ and $i(x, 0) = i_0$.

Ans: $v(x, t) = a_1 \sin \frac{\pi x}{l} \cdot \cos \frac{\pi t}{l\sqrt{LC}}$
$+ a_5 \sin \frac{5\pi}{l} \cdot \cos \frac{5\pi t}{l\sqrt{LC}}$

4. Consider a telephone line 3000 km long with resistance 4 ohms/km and capacitance of 5×10^{-7} farad/km. Assume that $L = 0$ and $G = 0$. Initially both the ends are grounded so that the transmission line is not charged. At time $t = 0$, a constant emf E_0 is applied at one end while the other end is left grounded. Find the steady state current at the grounded end after 1 second.

Ans: $i = (0.053)i_\infty = 5.3\%$ of i_∞

Hint: Solve $v_{xx} = RCv_t$, obtaining $v(x, t) = \frac{E_0 x}{l} + \frac{2E_0}{\pi} \sum_{n=1}^{\infty} \sin \frac{n\pi x}{l} \cdot e^{-n^2\pi^2 t/(RCl^2)} \cdot \frac{(-1)^n}{n}$
to compute, i, use $i = -\frac{1}{R}\frac{\partial v}{\partial x}$ obtaining
$i = -\frac{E_0}{lR} - \frac{2E_0}{lR} \sum_{n=1}^{\infty} (-1)^n e^{-n^2\pi^2 t/(RCl^2)}$

Put $t = 1$

Note: $e^{-\pi^2/(RCl^2)} = e^{-0.548} = 0.578$

5. Assuming $L = 0$, $G = 0$, find the potential $v(x, t)$ in a transmission line of 1000 km long which is initially under steady-state conditions with potential 1300 volts at the source end ($x = 0$) and 1200 volts at the terminal end ($x = 1000$). The terminal end of the line is suddenly grounded, while the potential at the source end is maintained at 1300 volts.

Ans: $v(x, t) = 1300 - 1.3x + \frac{2400}{\pi} \sum_{n=1}^{\infty} \frac{(-1)^{n+1}}{n} \sin \frac{n\pi x}{1000} e^{-[n^2\pi^2 t/(l^2 RC)]}$

Chapter 20

Fourier Integral, Fourier Transforms and Integral Transforms

INTRODUCTION

Just as the Fourier series decomposes a periodic function into a discrete set of contributions of various frequencies (all multiples of one fundamental frequency), the Fourier transform provides a continuous frequency resolution of a (possibly nonperiodic) function. Fourier transform is useful in the study of frequency response of a filter, solution of PDE, discrete Fourier transform and Fast Fourier transform in signal analysis.

Two more integral transforms Hankel transforms* useful in problems involving Bessel functions and Hilbert transform are considered.

20.1 FOURIER INTEGRAL THEOREM

A periodic function $f(x)$ defined in a finite interval $(-L, L)$ can be expressed in Fourier series. By extending this concept, non-periodic functions defined in $-\infty < x < \infty$ (for all x) can be expressed as a Fourier integral, since in practice periodic functions are fairly rare.

Theorem: *A function $f(x)$, which is piecewise continuous in every finite interval and is absolutely integrable on the x-axis, can be represented by a Fourier integral*

$$f(x) = \int_0^\infty [A(\alpha)\cos\alpha x + B(\alpha)\sin\alpha x]\,d\alpha \qquad (1)$$

which is valid at all points of continuity.

At a point of discontinuity x_0, the Fourier integral $= \frac{1}{2}\left[f(x_0-0)+f(x_0+0)\right]$ i.e., average of the left and right hand limits.

Proof: Consider the Fourier series expansion of $f(x)$ in any interval $[-L, L]$ given by

$$f(x) = \frac{a_0}{2} + \sum_{n=1}^{\infty}\left(a_n\cos\frac{n\pi x}{L} + b_n\sin\frac{n\pi x}{L}\right) \qquad (2)$$

where $$a_n = \frac{1}{L}\int_{-L}^{L} f(t)\cos\left(\frac{n\pi}{L}t\right)dt \qquad (3)$$

$$b_n = \frac{1}{L}\int_{-L}^{L} f(t)\sin\left(\frac{n\pi}{L}t\right)dt \qquad (4)$$

By substituting the coefficients a_n and b_n from (3) and (4), the Fourier series (1) takes the form

$$f(x) = \frac{1}{2L}\int_{-L}^{L} f(t)dt + $$

$$+\frac{1}{L}\sum_{n=1}^{\infty}\left[\int_{-L}^{L} f(t)\cos\left(\frac{n\pi}{L}t\right)dt\right]\cos\frac{n\pi x}{L}$$

$$+\frac{1}{L}\sum_{n=1}^{\infty}\left[\int_{-L}^{L} f(t)\sin\left(\frac{n\pi t}{L}\right)dt\right]\sin\left(\frac{n\pi x}{L}\right)$$

$$= \frac{1}{2L}\int_{-L}^{L} f(t)dt + \frac{1}{L}\sum_{n=1}^{\infty}\int_{-L}^{L} f(t)\left[\cos\left(\frac{n\pi t}{L}\right)\right.$$

$$\left.\times\cos\left(\frac{n\pi x}{L}\right) + \sin\left(\frac{n\pi t}{L}\right)\cdot\sin\left(\frac{n\pi x}{L}\right)\right]dt$$

$$f(x) = \frac{1}{2L}\int_{-L}^{L} f(t)dt$$

$$+\frac{1}{L}\sum_{n=1}^{\infty}\int_{-L}^{L} f(t)\cos\left(\frac{n\pi(t-x)}{L}\right)dt \qquad (5)$$

*Available on our Web site http://www.mhhe.com/ramanahem

Put $\alpha_n = \frac{n\pi}{L}$ and $\Delta\alpha_n = \alpha_{n+1} - \alpha_n$

$$= (n+1)\frac{\pi}{L} - \frac{n\pi}{L} = \frac{\pi}{L} \qquad (6)$$

Then

$$f(x) = \frac{1}{2L}\int_{-L}^{L} f(t)dt + \frac{1}{\pi}\sum_{n=1}^{\infty}\left[\int_{-L}^{L} f(t)\cos(\alpha_n(t-x))\,dt\right]\Delta\alpha_n \qquad (7)$$

As $L \to \infty$, $\frac{1}{L} \to 0$ and $\Delta\alpha_n = \frac{\pi}{L} \to 0$, the infinite series in (7) becomes an integral from 0 to ∞. Thus

$$f(x) = \frac{1}{\pi}\int_0^{\infty}\left[\int_{-\infty}^{\infty} f(t)\cdot\cos(\alpha(t-x))\,dt\right]d\alpha \qquad (8)$$

since the first term in the right hand side of (7) becomes zero because $f(x)$ is absolutely integrable. Thus as $L \to \infty$ the Fourier series becomes a Fourier Integral.

Expanding $\cos(\alpha(t-x))$, (8) is rewritten as

$$f(x) = \frac{1}{\pi}\int_0^{\infty}\left(\int_{-\infty}^{\infty} f(t)\cos\alpha t\,dt\right)\cos\alpha x\,d\alpha + \frac{1}{\pi}\int_{\alpha=0}^{\infty}\left(\int_{-\infty}^{\infty} f(t)\sin\alpha t\,dt\right)\sin\alpha x\,d\alpha \qquad (9)$$

The **Fourier integral** expansion of $f(x)$ is the right side (expression) of (8) or (9).

Introducing

$$A(\alpha) = \frac{1}{\pi}\int_{-\infty}^{\infty} f(t)\cos(\alpha t)dt \qquad (10)$$

$$B(\alpha) = \frac{1}{\pi}\int_{-\infty}^{\infty} f(t)\sin(\alpha t)dt \qquad (11)$$

(9) can be rewritten as

$$f(x) = \int_0^{\infty}[A(\alpha)\cos(\alpha x) + B(\alpha)\sin(\alpha x)]\,d\alpha \qquad (1)$$

Note: Fourier integral is very useful in solving differential equations and integral equations.

Particular cases of Fourier integral (10):

Fourier Cosine Integral

When $f(x)$ is an even function, then $B(\alpha) = 0$ and

$$A(\alpha) = \frac{2}{\pi}\int_0^{\infty} f(t)\cos(\alpha t)dt \qquad (12)$$

Then the Fourier integral (1) reduces to the Fourier cosine integral

$$f(x) = \int_0^{\infty} A(\alpha)\cos(\alpha x)d\alpha \qquad (13)$$

Fourier Sine Integral

When $f(x)$ is odd function, then $A(\alpha) = 0$ and

$$B(\alpha) = \frac{2}{\pi}\int_0^{\infty} f(t)\sin(\alpha t)dt \qquad (14)$$

Then the Fourier integral (1) reduces to the Fourier sine integral

$$f(x) = \int_0^{\infty} B(\alpha)\sin\alpha x\,d\alpha \qquad (15)$$

Suppose $f(x)$ is defined in the interval $(0, \infty)$. Then for $x > 0$, $f(x)$ can be represented by Fourier cosine integral (13) by redefining $f(x)$ in $(-\infty, 0)$ such that $f(x)$ is even function in $(-\infty, \infty)$.

Similarly, by redefining $f(x)$ in $(-\infty, 0)$ such that $f(x)$ is an odd function in $(-\infty, \infty)$, the given function $f(x)$ can be represented by the Fourier sine integral (15) valid for $x > 0$.

Fourier Integral in Complex Form

Since $\cos(\alpha(t-x))$ is an even function of α, then (8) can be written as

$$f(x) = \frac{1}{2}\cdot\frac{1}{\pi}\int_{-\infty}^{\infty}\left[\int_{-\infty}^{\infty} f(t)\cos\Big(\alpha(t-x)\Big)\,dt\right]d\alpha \qquad (16)$$

Since $\sin(\alpha(t-x))$ is an odd function of α, then

$$0 = \frac{1}{2}\cdot\frac{1}{\pi}\int_{-\infty}^{\infty}\left[\int_{-\infty}^{\infty} f(t)\sin\Big(\alpha(t-x)\Big)\,dt\right]d\alpha. \qquad (17)$$

Multiplying (17) by $-\frac{i}{2\pi}$ and adding it to (16), we get

$$f(x) = \frac{1}{2\pi}\int_{-\infty}^{\infty}\left[\int_{-\infty}^{\infty} f(t)\cdot\cos\Big(\alpha(t-x)\Big) - i\sin\Big(\alpha(t-x)\Big)\,dt\right]d\alpha$$

$$f(x) = \int_{-\infty}^{\infty}\left[\frac{1}{2\pi}\int_{-\infty}^{\infty} f(t)\cdot e^{-i\alpha t}dt\right]e^{i\alpha x}d\alpha \qquad (18)$$

or $f(x) = \frac{1}{\sqrt{2\pi}}\int_{-\infty}^{\infty}\left[\frac{1}{\sqrt{2\pi}}\int_{-\infty}^{\infty} f(t)e^{-i\alpha t}dt\right]e^{i\alpha x}d\alpha$.

Note: A list of standard results which are very often used are given below:

Standard Results

1. $\int e^{ax} \cdot \sin bx dx = \frac{e^{ax}}{a^2+b^2}(a \sin bx - b \cos bx)$
2. $\int_0^\infty e^{-ax} \cdot \sin bx dx = \frac{b}{a^2+b^2}$
3. $\int e^{ax} \cdot \cos bx dx = \frac{e^{ax}}{a^2+b^2}(a \cos bx + b \sin bx)$
4. $\int_0^\infty e^{-ax} \cos bx dx = \frac{a}{a^2+b^2}$
5. $\int_{-\infty}^\infty e^{-x^2} dx = \sqrt{\pi}$
6. $\int_0^\infty \frac{e^{ax}-e^{-ax}}{e^{\pi x}-e^{-\pi x}} dx = \frac{1}{2} \tan \frac{a}{2}$
7. $\int_0^\infty \frac{e^{ax}+e^{-ax}}{e^{\pi x}-e^{-\pi x}} dx = \frac{1}{2} \sec \frac{a}{2}$
8. $\int_0^\infty \frac{e^{-ax}}{x} \sin bx dx = \tan^{-1} \frac{b}{c}, c > 0, b > 0$
9. $\int_0^\infty \frac{\sin ax}{x} dx = \frac{\pi}{2}$ if $a > 0$.

20.2 FOURIER TRANSFORM

From the Fourier integral representations (18), (13) and (15), we get Fourier transform, (which is complex), Fourier cosine transform (which is real) and Fourier sine transform (which is real) of $f(x)$ as follows:

Fourier Transform of *f(x)*

The Fourier integral of $f(x)$ in the complex form given by (18)

$$f(x) = \int_{-\infty}^{\infty} \left[\frac{1}{2\pi} \int_{-\infty}^{\infty} f(t) e^{-i\alpha t} dt \right] e^{i\alpha x} d\alpha \qquad (18)$$

can be written as

$$f(x) = \int_{-\infty}^{\infty} F(\alpha) e^{i\alpha x} d\alpha \qquad (19)$$

where $$F(\alpha) = \frac{1}{2\pi} \int_{-\infty}^{\infty} f(x) e^{-i\alpha x} dx \qquad (20)$$

$F(\alpha)$ defined by (20) is known as the Fourier transform of $f(x)$.

$f(x)$ defined by (19) is known as the inverse Fourier transform of $F(\alpha)$. $F(\alpha)$ and $f(x)$ are known as Fourier transform pair which differ in form only in the sign of the exponent.

Note: The factor $\frac{1}{2\pi}$ can multiply the $f(x)$ integral (19) instead of the $F(\alpha)$ integral (20). Alternatively the factor $\frac{1}{\sqrt{2\pi}}$ can multiply each of the integrals in (19) and (20).

Fourier transform breaks up the function into a continuous spectrum of frequencies α.

Fourier transform method is the process of obtaining $F(\alpha)$ for a given function $f(x)$.

Fourier Cosine Transform of *f(x)*

The Fourier integral of an even function $f(x)$ reduces to Fourier cosine integral given by (13)

$$f(x) = \frac{2}{\pi} \int_0^\infty \left[\int_0^\infty f(t) \cos(\alpha t) dt \right] \cos(\alpha x) d\alpha$$

Put $$F_c(\alpha) = \int_0^\infty f(x) \cos \alpha x dx \qquad (21)$$

Then $$f(x) = \frac{2}{\pi} \int_0^\infty F_c(\alpha) \cos(\alpha x) d\alpha \qquad (22)$$

$F_c(\alpha)$ given by (21) is called the Fourier cosine transform of $f(x)$ in the interval $0 < x < \infty$ and $f(x)$ given by (22) as the inverse Fourier cosine transform of $F_c(\alpha)$.

Fourier Sine Transform of *f(x)*

The Fourier integral of an odd function $f(x)$ reduces to the Fourier sine integral given by (15)

$$f(x) = \frac{2}{\pi} \int_0^\infty \left[\int_0^\infty f(t) \sin(\alpha t) dt \right] \sin(\alpha x) d\alpha$$

Take $$F_s(\alpha) = \int_0^\infty f(x) \sin(\alpha x) dx \qquad (23)$$

Then $$f(x) = \frac{2}{\pi} \int_0^\infty F_s(\alpha) \sin(\alpha x) d\alpha \qquad (24)$$

The function $F_s(\alpha)$ defined by (23) is known as the Fourier sine transform of $f(x)$ in $0 < x < \infty$. The function $f(x)$ given by (24) is known as the **inverse Fourier sine transform** of $F_s(\alpha)$.

Linearity Property

Fourier transform, Fourier cosine transform and Fourier sine transform are all linear operations (since the integral operation is linear). For example for any two functions $f(x)$ and $g(x)$ and for any two constants a and b, the Fourier transform of $af(x) + bg(x)$ is given by

$$F(af(x) + bg(x)) = \frac{1}{2\pi} \int_{-\infty}^{\infty} \left[af(x) + bg(x) \right] e^{-i\alpha x} dx$$

$$= a \cdot \frac{1}{2\pi} \int_{-\infty}^{\infty} f(x) e^{-i\alpha x} dx$$

$$+ \frac{b}{2\pi} \int_{-\infty}^{\infty} g(x) e^{-i\alpha x} dx$$

$$= aF\left(f(x)\right) + bF\left(g(x)\right)$$

In a similar way

$$F_c(af(x)+bg(x))$$
$$=\int_0^\infty (af+bg)\cos\alpha x d\alpha$$
$$=a\int_0^\infty f\cos\alpha x d\alpha + b\int_0^\infty g\cos\alpha x\, d\alpha$$
$$=aF_c(f)+bF_c(g).$$

Fourier Transform of Derivatives

Fourier transform of a derivative of a function $f(x)$ corresponds to multiplication of the Fourier transform by $i\alpha$ i.e.,

$$F\left\{f'(x)\right\}=F\left\{\frac{df}{dx}\right\}=(i\alpha)F\left\{f(x)\right\}$$

Proof: By definition

$$F\left\{f'(x)\right\}=\frac{1}{2\pi}\int_{-\infty}^{\infty} f'(x)e^{-i\alpha x}dx$$

Integrating by parts

$$=\frac{1}{2\pi}\left[f\cdot e^{-i\alpha x}\Big|_{-\infty}^{\infty}-(-i\alpha)\int_{-\infty}^{\infty} f\cdot e^{-i\alpha x}dx\right]$$

Assuming that $f(x)\to 0$ as $|x|\to\infty$, we have

$$F\left\{f'(x)\right\}=(i\alpha)F\left\{f(x)\right\}$$

In general $F\left\{\frac{d^n f}{dx^n}\right\}=(i\alpha)^n F\{f\}$.

In particular

$$F\left\{\frac{d^2 f}{\partial x^2}\right\}=-\alpha^2F\{f\}. \tag{25}$$

Fourier Cosine and Sine Transforms of Derivatives

Prove that

a. $F_c\left\{f'(x)\right\}=\alpha F_s\left\{f(x)\right\}-f(0)$

b. $F_s\left\{f'(x)\right\}=-\alpha F_c\left\{f(x)\right\}$

Proof:

a. By definition and applying integration by parts

$$F_c\{f'\}=\int_0^\infty f'\cos(\alpha x)dx=f\cdot\cos\alpha x\Big|_0^\infty$$
$$+w\int_0^\infty f\cdot\sin\alpha x dx$$
$$F_c\{f'\}=-f(0)+wF_s\{f\}$$

b. Similarly,

$$F_s\{f'\}=\int_0^\infty f'\sin(\alpha x)d\alpha=f\sin\alpha x\Big|_0^\infty$$
$$-\alpha\int_0^\infty f\cos\alpha x dx$$
$$=0-\alpha F_c\{f\}.$$

Corollary 1:

$$F_c\{f''\}=\alpha F_s\{f'\}-f'(0)$$
$$=\alpha(-\alpha F_c\{f\})-f'(0)$$
$$F_c\{f''\}=-\alpha^2F_c\{f\}-f'(0) \tag{26}$$

Corollary 2:

$$F_s\{f''\}=-\alpha F_c\{f'\}=-\alpha\left[\alpha F_s\{f\}-f(0)\right]$$
$$F_s\{f''\}=-\alpha^2F_s\{f\}+\alpha f(0) \tag{27}$$

Application to Initial Boundary Value Problem (IBVP)

The solution of a IBVP consisting of a partial differential equation together with boundary and initial conditions can be solved by the Fourier transform method. If the boundary conditions are of the Dirichlet type where the function value is prescribed on the boundary, then the Fourier sine transform is used. If the boundary conditions are of the Neumann type where the derivative of function is prescribed on boundary, then Fourier cosine transform is applied. In either case, the P.D.E. reduces to an O.D.E. in Fourier transform which is solved. Then the inverse Fourier sine (or cosine) transform will give the solution to the problem.

20.3 CONVOLUTION

Convolution of two functions $f(x)$ and $g(x)$ denoted by $f*g$ is defined as

$$h(x)=(f*g)(x)=\int_{-\infty}^{\infty} f(s)g(x-s)ds$$
$$=\int_{-\infty}^{\infty} f(x-s)g(s)ds$$

Theorem: *The Fourier transform of the convolution of f and g is the product of their Fourier transforms.*

Proof: By definition

$$F\{f*g\}=\frac{1}{2\pi}\int_{-\infty}^{\infty}(f*g)e^{-i\alpha x}dx$$

$$=\frac{1}{2\pi}\int_{-\infty}^{\infty}\left[\int_{-\infty}^{\infty}f(s)g(x-s)ds\right]e^{-i\alpha x}dx$$

Interchange the order of integration

$$=\frac{1}{2\pi}\int_{-\infty}^{\infty}\left[\int_{-\infty}^{\infty}f(s)g(x-s)e^{-i\alpha x}dx\right]ds$$

Now put $x-s=q$, so $x=s+q$, with q as the new variable of integration instead of x.

$$=\frac{1}{2\pi}\int_{-\infty}^{\infty}\int_{-\infty}^{\infty}f(s)g(q)e^{-i\alpha(s+q)}dq\,ds$$

$$=\left[\frac{1}{2\pi}\int_{-\infty}^{\infty}f(s)e^{-i\alpha s}ds\right]\left[\frac{2\pi}{2\pi}\int_{-\infty}^{\infty}g(q)e^{-i\alpha q}dq\right]$$

$$=2\pi\cdot F\{f\}\cdot F\{g\}.$$

Note: Convolution is commutative $f*g=g*f$, associative $f*(g*h)=(f*g)*h$.

Worked Out Examples

Fourier integral

Example 1: Using Fourier integral representation show that

$$\int_0^{\infty}\frac{\cos x\alpha+\alpha\sin x\alpha}{1+\alpha^2}d\alpha=\begin{cases}0 & \text{if } x<0\\ \frac{\pi}{2} & \text{if } x=0\\ \pi e^{-x} & \text{if } x>0.\end{cases}$$

Solution: Consider the function defined by

$$f(x)=\begin{cases}0 & \text{if } x<0\\ \frac{1}{2} & \text{if } x=0\\ e^{-x} & \text{if } x>0.\end{cases}$$

Now find the Fourier integral representation of $f(x)$ in the exponential form: By definition

$$f(x)=\frac{1}{2\pi}\int_{-\infty}^{\infty}\left[\int_{-\infty}^{\infty}f(t)\cdot e^{-i\alpha t}dt\right]e^{i\alpha x}d\alpha$$

Consider $I=\int_{-\infty}^{\infty}f(t)e^{-i\alpha t}dt$

$$=\int_{-\infty}^{0}0+\int_0^{\infty}e^{-x}e^{-i\alpha t}dt$$

$$=\left.\frac{e^{-t(1+i\alpha)}}{-(1+i\alpha)}\right|_{t=0}^{\infty}=\frac{1}{1+i\alpha}=\frac{1-i\alpha}{1+\alpha^2}$$

So $\quad f(x)=\frac{1}{2\pi}\int_{-\infty}^{\infty}Ie^{i\alpha x}d\alpha$

$$=\frac{1}{2\pi}\int_{-\infty}^{\infty}\left(\frac{1-i\alpha}{1+\alpha^2}\right)e^{i\alpha x}d\alpha$$

$$=\int_{-\infty}^{\infty}\frac{1}{2\pi(1+\alpha^2)}(1-i\alpha)\times$$
$$\times(\cos\alpha x+i\sin\alpha x)d\alpha$$

$$=\int_{-\infty}^{\infty}\frac{1}{2\pi(1+\alpha^2)}\Big[(\cos\alpha x+\alpha\sin\alpha x)+$$
$$+i(\sin\alpha x-\alpha\cos\alpha x)\Big]d\alpha$$

The second integral on the right side is zero because the integrand is an odd function.

$$f(x)=\frac{2}{2\pi(1+\alpha^2)}\int_0^{\infty}(\cos\alpha x+\alpha\sin\alpha x)d\alpha$$

For $x>0$, $f(x)=e^{-x}$ so

$$e^{-x}=\frac{1}{\pi}\int_0^{\infty}\frac{\cos\alpha x+\alpha\sin\alpha x}{1+\alpha^2}d\alpha \qquad (1)$$

For $x<0$, $f(x)=0$ so

$$0=\frac{1}{\pi}\int_0^{\infty}\frac{\cos\alpha x+\alpha\sin\alpha x}{1+\alpha^2}d\alpha \qquad (2)$$

At $x=0$, $f(x)$ has a discontinuity. So

$$f(x)=\frac{1}{2}\Big[f(x+0)+f(x-0)\Big]=\frac{1}{2}[1+0]=\frac{1}{2}$$

For $x=0$

$$\frac{1}{2}=\frac{1}{\pi}\int_0^{\infty}\frac{\cos\alpha x+\alpha\sin\alpha x}{1+\alpha^2}d\alpha. \qquad (3)$$

From (1), (2), (3) the result follows.

Example 2: Find the (a) Fourier cosine integral and (b) Fourier sine integral (representation) of

$$f(x)=\sin x \quad \text{if } 0\le x\le\pi$$
$$=0 \quad \text{if } x>\pi$$

Solution: **a.** *The Fourier cosine integral of*

$$f(x) = \int_0^\infty A(\alpha)\cos\alpha x d\alpha \quad \text{where}$$

$$A(\alpha) = \frac{2}{\pi}\int_0^\infty f(t)\cos\alpha t\,dt$$

$$A(\alpha) = \frac{2}{\pi}\left[\int_0^\pi \sin t\cdot\cos\alpha t dt + \int_\pi^\infty 0\right]$$

$$= \frac{1}{2}\frac{2}{\pi}\int_0^\pi \left[\sin(1+\alpha)t + \sin(1-\alpha)t\right]dt$$

$$= -\frac{1}{\pi}\left[\frac{\cos(1+\alpha)t}{1+\alpha} + \frac{\cos(1-\alpha)t}{1-\alpha}\right]_{t=0}^{\pi}$$

$$= +\frac{1}{\pi}\left[\frac{1-\cos(1+\alpha)\pi}{1+\alpha} + \frac{1-\cos(1-\alpha)\pi}{1-\alpha}\right]$$

$$A(\alpha) = \frac{1}{\pi}\left[\frac{1+\cos\alpha\pi}{1+\alpha} + \frac{1+\cos\alpha\pi}{1-\alpha}\right]$$

$$= \frac{2(1+\cos\alpha\pi)}{\pi(1-\alpha^2)}$$

For $\alpha = 1$, $A(1) = 0$.
Substituting $A(\alpha)$, we get the Fourier cosine integral of $f(x)$ as

$$f(x) = \int_0^\infty \frac{2(1+\cos\alpha\pi)}{\pi(1-\alpha^2)}\cdot\cos\alpha x\, d\alpha$$

$$= \frac{2}{\pi}\int_0^\infty \frac{1+\cos\alpha\pi}{1-\alpha^2}\cos\alpha x\, d\alpha$$

b. Fourier sine integral of $f(x)$:

$$f(x) = \int_0^\infty B(\alpha)\sin\alpha x d\alpha, \quad \text{where}$$

$$B(\alpha) = \frac{2}{\pi}\int_0^\infty f(t)\cdot\sin\alpha t dt$$

$$= \frac{2}{\pi}\left[\int_0^\pi \sin t\cdot\sin\alpha t dt + \int_\pi^\infty 0\right]$$

$$= \frac{2}{\pi}\frac{1}{2}\int_0^\pi\left[\cos(1-\alpha)t - \cos(1+\alpha)t\right]dt$$

$$= \frac{1}{\pi}\left[\frac{\sin(1-\alpha)t}{1-\alpha} - \frac{\sin(1+\alpha)t}{1+\alpha}\right]_{t=0}^{\pi}$$

$$= \frac{1}{\pi}\left[\frac{\sin\alpha\pi}{1-\alpha} + \frac{\sin\alpha\pi}{1+\alpha}\right] = \frac{2\sin\alpha\pi}{\pi(1-\alpha^2)}$$

For $\alpha = 1$, $B(1) = 1$.

$$\text{Thus } f(x) = \int_0^\infty \frac{2\sin\alpha\pi}{\pi(1-\alpha^2)}\cdot\sin\alpha x d\alpha$$

$$= \frac{2}{\pi}\int_0^\infty \frac{\sin\alpha\pi}{(1-\alpha^2)}\sin\alpha x d\alpha.$$

Fourier transform

Example 3: Represent $f(x)$ as an exponential Fourier transform when

$$f(x) = \begin{cases}\sin x, & 0 < x < \pi\\ 0, & \text{otherwise}\end{cases}$$

show that the result can be written as

$$f(x) = \frac{1}{\pi}\int_0^\infty \frac{\cos\alpha x + \cos\alpha(x-\pi)}{1-\alpha^2}d\alpha$$

Solution: Fourier transform in the exponential form is given by

$$F\left\{f(x)\right\} = F(\alpha) = \frac{1}{2\pi}\int_{-\infty}^\infty f(x)\cdot e^{-i\alpha x}dx$$

$$= \frac{1}{2\pi}\left[\int_{-\infty}^0 0 + \int_0^\pi \sin x\cdot e^{-i\alpha x}dx + \int_\pi^\infty 0\right]$$

$$= \frac{1}{2\pi}\int_0^\pi\left(\frac{e^{ix}-e^{-ix}}{2i}\right)e^{-i\alpha x}dx$$

$$\text{since } \sin x = \frac{(e^{ix}-e^{-ix})}{2i}$$

$$= \frac{1}{4\pi i}\int_0^\pi\left[e^{i(1-\alpha)x} - e^{-i(\alpha+1)x}\right]dx$$

$$= \frac{1}{4\pi i}\left[\frac{e^{i(1-\alpha)x}}{i(1-\alpha)} - \frac{e^{-i(1+\alpha)x}}{-i(1+\alpha)}\right]_0^\pi$$

$$= \frac{1}{4\pi i}\left[\frac{e^{i(1-\alpha)\pi}}{i(1-\alpha)} + \frac{e^{-i(1+\alpha)\pi}}{i(1+\alpha)} - \frac{1}{i(1-\alpha)} - \frac{1}{i(1+\alpha)}\right]$$

$$= \frac{1}{2\pi}\frac{e^{-i\pi\alpha}+1}{1-\alpha^2} \quad \text{since } e^{\pm i\pi} = -1$$

$$\text{Then} \quad f(x) = \int_{-\infty}^\infty F(\alpha)e^{i\alpha x}dx$$

$$= \int_{-\infty}^\infty \frac{1}{2\pi}\left(\frac{e^{-i\pi\alpha}+1}{1-\alpha^2}\right)e^{i\alpha x}dx$$

is the required exponential Fourier transform representation. Expanding the integrand

$$(1+e^{-i\pi\alpha})e^{i\alpha x}$$
$$= [1+\cos\alpha\pi - i\sin\alpha\pi][\cos\alpha n + i\sin\alpha x]$$
$$= \cos\alpha x + \cos\alpha(\pi - x) + i[+\sin\alpha x \cdot \sin\alpha\pi]$$

$$f(x) = \frac{2}{2\pi}\int_0^\infty \frac{\cos\alpha x + \cos\alpha(\pi - x)}{(1-\alpha)^2}dx + \frac{1}{2\pi}\int_{-\infty}^\infty + \frac{\sin\alpha\pi \cdot \sin\alpha x}{1-\alpha^2}dx.$$

Since the 2nd integral is zero (odd function), result follows.

Example 4: Find the Fourier transform of

$$f(x) = \frac{1}{2a}, \quad \text{if } |x| \le a$$
$$= 0, \quad \text{if } |x| > a$$

Solution: By definition, the Fourier transform of $f(x)$ is

$$F\{f(x)\} = \frac{1}{2\pi}\int_{-\infty}^\infty f(x)\cdot e^{-i\alpha x}dx$$
$$= \frac{1}{2\pi}\left[\int_{-\infty}^{-a} + \int_{-a}^{a} + \int_a^\infty\right]$$
$$= \frac{1}{2\pi}\left[\int_{-\infty}^{-a} 0 + \int_{-a}^{a}\frac{1}{2a}e^{-i\alpha x}dx + \int_a^\infty 0\right]$$
$$= \frac{1}{4\pi a}\left.\frac{e^{-i\alpha x}}{-i\alpha}\right|_{x=-a}^{a}$$
$$= \frac{-1}{4\pi a\alpha i}\left[e^{-i\alpha a} - e^{+i\alpha a}\right]$$
$$= \frac{1}{2\pi a\alpha}\left[\frac{e^{i\alpha a} - e^{-i\alpha a}}{2i}\right] = \frac{\sin(\alpha a)}{2\pi a\alpha}.$$

Example 5: Find the (a) Fourier cosine and (b) sine transform of $f(x) = e^{-ax}$ for $x \ge 0$ and $a > 0$. Deduce the integrals known as "Laplace integrals" $\int_0^\infty \frac{\cos\alpha x}{a^2+\alpha^2}d\alpha$ and $\int_0^\infty \frac{\alpha\sin\alpha x}{a^2+\alpha^2}d\alpha$.

Solution: **a.** *By definition Fourier cosine transform of $f(x)$ is*

$$F_c\{f(x)\} = F_c(\alpha) = \int_0^\infty f(x)\cos\alpha x\,dx$$

$$F_c\{f\} = \int_0^\infty e^{-ax}\cos\alpha x dx = \frac{a}{a^2+\alpha^2}$$

The inverse Fourier cosine transform

$$f(x) = \frac{2}{\pi}\int_0^\infty F_c(\alpha)\cos\alpha x dx$$
$$= \frac{2}{\pi}\int_0^\infty \frac{a}{a^2+\alpha^2}\cos\alpha x d\alpha$$

Since $f(x) = e^{-ax}$, the above integral can be rewritten as

$$\int_0^\infty \frac{\cos\alpha x}{a^2+\alpha^2}d\alpha = \frac{\pi e^{-ax}}{2a}$$

b. *The Fourier sine transform of $f(x)$ is*

$$F_s\{f(x)\} = F_s(\alpha) = \int_0^\infty f(x)\sin\alpha x dx$$
$$= \int_0^\infty e^{-ax}\sin\alpha x dx = \frac{\alpha}{a^2+\alpha^2}$$

Now the inverse Fourier sine transform

$$f(x) = \frac{2}{\pi}\int_0^\infty F_s(\alpha)\sin\alpha x dx$$
$$= \frac{2}{\pi}\int_0^\infty \frac{\alpha}{a^2+\alpha^2}\sin\alpha x dx$$

with $f(x) = e^{-ax}$, this above integral takes the form

$$\int_0^\infty \frac{\alpha\sin\alpha x}{a^2+\alpha^2}d\alpha = \frac{\pi e^{-ax}}{2}$$

Note: For $a = 0$, $\int_0^\infty \frac{\alpha\sin\alpha x}{\alpha^2}d\alpha = \int_0^\infty \frac{\sin\alpha x}{\alpha}d\alpha = \frac{\pi}{2}$.

Example 6: Find the inverse Fourier sine transform of $\frac{1}{s}e^{-as}$.

Solution:

$$f(x) = F_s^{-1}\left\{\frac{e^{-as}}{s}\right\}$$

$$f = \frac{2}{\pi}\int_0^\infty \frac{e^{-as}}{s}\sin sx ds \quad (1)$$

Differentiating w.r.t., x

$$\frac{df}{dx} = \frac{2}{\pi}\int_0^\infty \frac{e^{-as}}{s}\frac{d}{dx}(\sin sx)ds$$
$$= \frac{2}{\pi}\int_0^\infty \frac{e^{-as}}{s}\cdot s\cdot\cos sx ds$$

$$\frac{df}{dx} = \frac{2}{\pi}\int_0^\infty e^{-as}\cos sx ds = \frac{2}{\pi}\frac{a}{x^2+a^2}$$

Integrating $f(x)=\frac{2}{\pi}\int a\frac{dx}{x^2+a^2}$

$$=\frac{2}{\pi}\tan^{-1}\frac{x}{a}+A \qquad (2)$$

From (1) at $x=0$, $f(0)=0$. Using this in (2)

$$0=f(0)=0+A \quad \therefore A=0$$

Thus $f(x)=\frac{2}{\pi}\tan^{-1}\frac{x}{a}$

is the required inverse Fourier sine transform.

Note: When $a=0$, $f(x)=F_s^{-1}\left\{\frac{1}{s}\right\}$ $=\frac{2}{\pi}\tan^{-1}\infty=1$

Example 7: Find $f(x)$ whose Fourier cosine transform is $\frac{\sin as}{s}$.

Solution: It is given $F_c\{f(x)\}=\frac{\sin sa}{s}$.

$$f(x)=F_c^{-1}\left\{\frac{\sin sa}{s}\right\}=\frac{2}{\pi}\int_0^\infty\frac{\sin sa}{s}\cos sx\,ds$$

$$=\frac{2}{\pi}\frac{1}{2}\int_0^\infty\frac{\sin(s(a+x))+\sin(s(a-x))}{s}ds$$

$$=\frac{1}{\pi}\int_0^\infty\frac{\sin s(a+x)}{s}ds+\frac{1}{\pi}\int_0^\infty\frac{\sin s(a-x)ds}{s}$$

$$=\frac{1}{\pi}\left(\frac{\pi}{2}+\frac{\pi}{2}\right) \quad \text{if } a-x>0 \text{ i.e., } x<a$$

$$=\frac{1}{\pi}\left(\frac{\pi}{2}-\frac{\pi}{2}\right) \quad \text{if } x-a>0 \text{ i.e., } x>a$$

since $\int_0^\infty\frac{\sin ax}{x}dx=\frac{\pi}{2}$ when $a>0$. (See note of WE 5 on page 20.7)

Thus

$$f(x)=\begin{cases}1 & \text{if } x<a\\ 0 & \text{if } x>a.\end{cases}$$

Example 8: Solve for $f(x)$ the integral equation

$$\int_0^\infty f(x)\sin xt\,dx=\begin{cases}1, & 0\le t<1\\ 2, & 1\le t<2\\ 0, & t\ge 2\end{cases}$$

Solution: By definition

$$F_s(t)=F_s\{f(x)\}=\int_0^\infty f(x)\sin xt\,dx=\begin{cases}1, & 0\le t<1\\ 2, & 1\le t<2\\ 0, & t\ge 2\end{cases}$$

$$f(x)=\frac{2}{\pi}\int_0^\infty F_s(t)\sin tx\,dt$$

$$=\frac{2}{\pi}\left[\int_0^1 1\cdot\sin tx\,dt+\int_1^2 2\sin tx\,dt+\int_2^\infty 0\right]$$

$$=\frac{2}{\pi}\left[\frac{-\cos tx}{x}\Big|_{t=0}^1-2\frac{\cos tx}{x}\Big|_{t=1}^2\right]$$

$$f(x)=\frac{2}{\pi x}[1+\cos x-2\cos 2x].$$

Example 9: Find the temperature distribution in semi-infinite bar with its end point and lateral surface insulated and with initial temperature distribution in the bar is prescribed by $f(x)$. Deduce the solution when $f(x)=e^{-ax}$.

Solution: This problem is represented by the one-dimensional heat equation

$$\frac{\partial u}{\partial t}=c^2\frac{\partial^2 u}{\partial x^2}$$

for $0<x<\infty$, $t>0$ with boundary condition

$$u_x(0,t)=0 \qquad \text{(insulated)}$$

and with initial condition

$$u(x,0)=f(x) \quad \text{(given)} \qquad \text{for } 0<x<\infty$$

Since the boundary condition is of the Neumann (derivative) type, apply Fourier cosine transformation to the equation

$$\int_0^\infty\frac{\partial u}{\partial t}\cos sx\,dx=c^2\int_0^\infty\frac{\partial^2 u}{\partial x^2}\cos sx\,dx$$

Integrating the R.H.S. by parts

$$\frac{\partial}{\partial t}\int_0^\infty u\cos sx\,dx=c^2\left[\frac{\partial u}{\partial x}\cos sx\Big|_0^\infty+\int_0^\infty s\sin sx\cdot\frac{\partial u}{\partial x}dx\right]$$

Assuming $\frac{\partial u}{\partial x}\to 0$ as $x\to\infty$ and using the boundary condition $\frac{\partial u}{\partial x}\Big|_{x=0}=0$, the first term in R.H.S. becomes zero. Integrating by parts again

$$\frac{\partial}{\partial t}\int_0^\infty u\cdot\cos sx\,dx$$

$$=c^2 s\left[u\cdot\sin sx\Big|_0^\infty-\int_0^\infty us\cdot\cos sx\,dx\right]$$

Assuming that u is bounded i.e., $u \to 0$ as $x \to \infty$ the first term is R.H.S. becomes zero.

Put $$U(t, s) = F_c\left\{u(x, t)\right\} = \int_0^\infty u(x, t) \cdot \cos sx dx$$

Then $$\frac{d}{dt}U = -s^2c^2U$$

Note that the result of Fourier transforming on x is to eliminate derivatives of x from the heat equation, thus leaving an ordinary differential equation w.r.t., 't'.

Integrating by separation of variables

$$U(s, t) = Ae^{-c^2s^2t}$$

At $t = 0, U(s, 0) = A$

Thus $$A = U(s, 0) = \int_0^\infty u(x, 0) \cdot \cos(sx)dx$$

From the initial condition $u(x, 0) = f(x)$,

then $$A = \int_0^\infty f(x)\cos(sx)dx$$

Taking inverse Fourier cosine transform

$$u(x, t) = F_c^{-1}\{U\} = F_c^{-1}\left\{Ae^{-c^2s^2t}\right\}$$

$$= \frac{2}{\pi}\int_0^\infty Ae^{-c^2s^2t} \cdot \cos(sx)ds$$

Special Case: when $f(x) = e^{-ax}$ then

$$A = \int_0^\infty e^{-ax}\cos(sx)dx = \frac{a}{a^2 + b^2}$$

Thus the solution in this case is

$$u(x, t) = F_c^{-1}\left\{\frac{a}{a^2 + b^2}e^{-c^2s^2t}\right\}$$

$$u(x, t) = \frac{2}{\pi}\int_0^\infty \frac{a}{a^2 + b^2} \cdot e^{-c^2s^2t} \cdot \cos(sx)ds$$

Example 10: Solve the Laplace's equation in the semi infinite strip shown in Fig. 20.1.

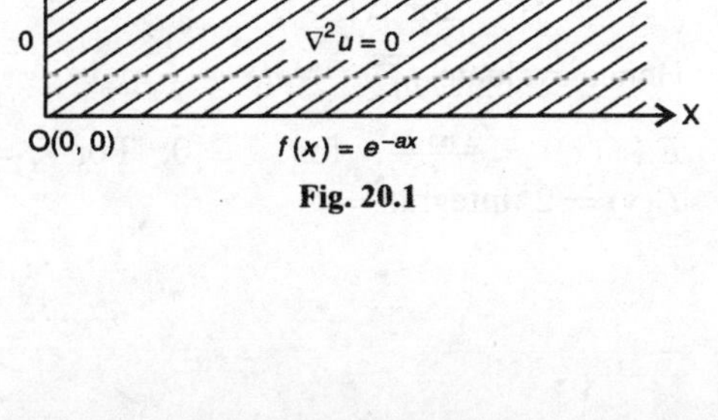

Fig. 20.1

Solution: To solve the Laplace's equation

$$\frac{\partial^2 u}{\partial x^2} + \frac{\partial^2 u}{\partial y^2} = 0 \tag{1}$$

with the boundary conditions

$$u(0, y) = 0, \quad \text{for } 0 < y < b \tag{2}$$

$$u(x, b) = 0 \quad \text{for } 0 < x < \infty \tag{3}$$

$$u(x, 0) = e^{-ax} \quad \text{with } a > 0. \tag{4}$$

Region is: $0 < x < \infty, 0 < y < b$.

Applying Fourier sine transform of (1) on both sides

$$\int_0^\infty \frac{\partial^2 u}{\partial x^2}\sin sx dx + \int_0^\infty \frac{\partial^2 u}{\partial y^2}\sin sx dx = 0 \tag{5}$$

The first integral is simplified by successive integration by parts and gives

$$\int_0^\infty \frac{\partial^2 u}{\partial x^2}\sin sx dx = \frac{\partial u}{\partial x} \cdot \sin sx\Big|_0^\infty$$

$$- \int_0^\infty \frac{\partial u}{\partial x} \cdot s \cdot \cos sx dx$$

Assuming that $\frac{\partial u}{\partial x} \to 0$ as $x \to \infty$

$$= -s\left[u \cdot \cos sx\Big|_0^\infty - \int_0^\infty u \cdot s \cdot (-\sin sx)dx\right]$$

using the boundary condition (2), $u(0, y) = 0$ and assuming u is bounded i.e., $u \to 0$ as $x \to \infty$

$$= -s^2\int_0^\infty u \sin sx dx$$

Put $U(s, y) = F_s\left\{u(x, y)\right\} = \int_0^\infty u(x, y)\sin sx dx$

Thus

$$\int_0^\infty \frac{\partial^2 u}{\partial x^2}\sin sx dx = -s^2U. \tag{6}$$

Substituting (6) in (5) and rewriting

$$-s^2U + \frac{d^2}{dy^2}\left[\int_0^\infty u(x, y)\sin sx\, dx\right] = 0$$

$$-s^2U + \frac{d^2U}{dy^2} = 0 \tag{7}$$

Thus Laplace Equation (1) has been reduced to an ordinary differential equation in y (with s as a parameter). Solution of (7) is

$$U(s, y) = A\cosh sy + B\sinh sy \tag{8}$$

when $y = b$, $U(s, b) = F_s\{u(x, b)\} = F_s\{0\} = 0$ since from (3), $u(x, b) = 0$. From (8), at $y = b$,

$$0 = U(s, b) = A \cosh sb + B \sinh sb \tag{9}$$

When $y = 0$, $U(s, 0) = \int_0^\infty u(x, 0) \sin sx dx$
From (4), $u(x, 0) = e^{-ax}$ so

$$U(s, 0) = \int_0^\infty e^{-ax} \sin sx dx = \frac{s}{a^2 + s^2} \tag{10}$$

Putting $y = 0$ in (8)

$$\frac{s}{a^2 + s^2} = U(s, 0) = A + B \cdot 0$$

or
$$A = \frac{s}{a^2 + s^2} \tag{11}$$

Substituting (11) in (9)

$$B = -\frac{s}{a^2 + s^2} \coth sb \tag{12}$$

Thus the solution (8) becomes

$$U(s, y) = \frac{s}{a^2 + s^2} \cosh sy - \frac{s}{a^2 + s^2} \coth sb \sinh sy \tag{13}$$

Taking inverse Fourier sine transform of (12)

$$u(x, y) = F_s^{-1}\{U(s, y)\}$$

$$= \frac{2}{\pi} \int_0^\infty \frac{s}{a^2 + s^2} (\cosh sy - \coth sb \cdot \sinh sy) \sin sx ds.$$

EXERCISE

Fourier integrals

Find the Fourier integral representation of $f(x)$:

1. $f(x) = \begin{cases} x, & |x| < 1 \\ 0, & |x| > 1 \end{cases}$

Ans. $f(x) = \int_{-\infty}^\infty \frac{\sin \alpha - \alpha \cos \alpha}{i\pi\alpha^2} e^{i\alpha x} d\alpha$

2. $f(x) = \begin{cases} \cos x, & \frac{-\pi}{2} < x < \frac{\pi}{2} \\ 0, & |x| > \frac{\pi}{2} \end{cases}$

Ans. $f(x) = \frac{1}{\pi} \int_{-\infty}^\infty \frac{\cos(\alpha\pi/2)}{1-\alpha^2} e^{i\alpha x} d\alpha$

3. $f(x) = \begin{cases} 1, & |x| \le 1 \\ 0, & |x| > 1 \end{cases}$. Hence evaluate

$$\int_0^\infty \frac{\sin \lambda \cos \lambda x}{\lambda} d\lambda = \begin{cases} \frac{\pi}{2}, & \text{if } |x| < 1 \\ \frac{\pi}{4}, & \text{if } |x| = 1 \\ 0, & \text{if } |x| > 1 \end{cases}$$

Ans. $f(x) = \frac{2}{\pi} \int_0^\infty \frac{\sin \lambda \cos \lambda x}{\lambda} d\lambda.$

4. Find the Fourier cosine and sine integrals of $f(x) = e^{-kx}$, for $x > 0, k > 0$.

Ans. Fourier cosine integral (F.C.I.)

$$f(x) = e^{-kx} = \frac{2k}{\pi} \int_0^\infty \frac{\cos sx}{k^2 + s^2} ds$$

Fourier sine integral (F.S.I.) $= f(x) = e^{-kx}$ $= \frac{2}{\pi} \int_0^\infty \frac{s \sin sx}{k^2+s^2} ds.$

5. Find the Fourier cosine integral of $f(x) = e^{-x} \cos x$.

Ans. $f(x) = e^{-x} \cos x = \frac{2}{\pi} \int_0^\infty \frac{(s^2+2) \cos sx ds}{s^4+4}$

6. Find the Fourier sine integral of $f(x) = e^{-ax} - e^{-bx}$.

Ans. $f(x) = e^{-ax} - e^{-bx} = \frac{2}{\pi} \int_0^\infty \frac{(b^2-a^2)s \sin sx ds}{(a^2+s^2)(b^2+s^2)}$, $a > 0, b > 0$

7. Using Fourier integral representation, show that

$$\int_0^\infty \frac{\sin s \cdot \cos xs}{s} ds = \begin{cases} \frac{\pi}{2}, & \text{if } 0 \le x < 1 \\ \frac{\pi}{4}, & \text{if } x = 1 \\ 0, & \text{If } x > 1. \end{cases}$$

Hint: Find Fourier integral of

$$f(x) = \begin{cases} 1, & \text{if } 0 \le x < 1 \\ 0, & \text{if } x \ge 1. \end{cases}$$

Fourier transforms

Find the Fourier transform of $f(x)$:

8. $f(x) = \begin{cases} 1, & \text{for } |x| < a \\ 0, & \text{for } |x| > a \end{cases}$.

Hence evaluate $\int_0^\infty \frac{\sin ax}{x} dx$

Ans. $F\{f(x)\} = \frac{2 \sin sa}{s}$, for $s \ne 0$, For $s = 0$, $F(s) = 2$, Integral $= \frac{\pi}{2}$

9. $f(x) = \begin{cases} 1 - x^2, & \text{if } |x| \le 1 \\ 0, & \text{if } |x| > 1 \end{cases}$. Hence evaluate $\int_0^\infty \frac{x \cos x - \sin x}{x^3} \cos \frac{x}{2} dx$

Ans. $F\{f(x)\} = -4(s \cos s - \sin s)/s^3$,
Integral $= \frac{-3\pi}{16}$

10. $f(x) = \begin{cases} x, & \text{for } |x| \le a \\ 0, & \text{for } |x| > a \end{cases}$

Ans. $\frac{2i}{s^2}(as \cos sa - \sin sa)$

11. $f(x) = e^{\frac{-x^2}{2}}, -\infty < x < \infty$

Ans. $\sqrt{2\pi} e^{-\frac{s^2}{2}}$

12. $f(x) = \begin{cases} 0, & -\infty < x < a \\ x, & a \le x \le b \\ 0, & x > b \end{cases}$

Ans. $\frac{1}{s}(ae^{isa} - be^{isb}) + \frac{1}{s^2}(e^{isb} - e^{isa})$

13. $f(x) = xe^{-x}, 0 \le x < \infty$

Ans. $\frac{1}{2\pi} \cdot \frac{(1+is)^2}{(1+s^2)^2}$

14. $f(x) = \begin{cases} \cos x, & \text{if } 0 < x < 1 \\ 0, & \text{otherwise} \end{cases}$

Ans. $\frac{1}{4\pi}\left[\frac{\sin(s+1)}{s+1} + \frac{\sin(s-1)}{s-1} + i\left\{\frac{1-\cos(s+1)}{s+1} + \frac{1-\cos(s-1)}{s-1}\right\}\right]$.

Fourier cosine and sine transforms

15. Find F.C.T. and F.S.T. of
$f(x) = \begin{cases} k, & \text{if } 0 < x < a \\ 0, & \text{if } x > a. \end{cases}$

Ans. $F_c\{f(x)\} = k \cdot \frac{\sin as}{s}$, $F_s\{f(x)\} = \frac{k(1-\cos as)}{s}$

16. Find F.S.T. of $e^{-|x|}$. Hence show that

$$\int_0^\infty \frac{x \sin mx}{1 + x^2} dx = \frac{\pi e^{-m}}{2}, m > 0.$$

Ans. F.S.T. $= \frac{s}{1+s^2}$

17. Find F.S.T. of $\frac{e^{-ax}}{x}$.

Ans. $\tan^{-1} \frac{s}{a} + c$

18. Find F.C.T. of $f(x) = \begin{cases} x, & 0 < x < 1 \\ 2 - x, & 1 < x < 2 \\ 0, & x > 2. \end{cases}$

Ans. $\frac{(2 \cos s - \cos 2s - 1)}{s^2}$

19. Find F.S.T. and F.C.T. of $2e^{-5x} + 5e^{-2x}$.

Ans. F.S.T.: $\frac{2s}{s^2+25} + \frac{5s}{s^2+4}$, F.C.T.: $\frac{10}{s^2+4} + \frac{10}{s^2+25}$

20. Find F.C.T. of $f(x) = e^{-ax} \cos ax$.

Ans. $\frac{a(s^2+2a^2)}{s^4+4a^4}$.

Inverse Fourier transform

21. Find $f(x)$ if its F.C.T. is $\frac{1}{1+s^2}$ and F.S.T. is $\frac{s}{1+s^2}$.

Ans. $f(x) = e^{-x}$, $f(x) = e^{-x}$

22. Find $f(x)$ if its F.C.T. is $\frac{1}{2\pi}\left(a - \frac{s}{2}\right)$ if $s < 2a$ and zero if $s \ge 2a$.

Ans. $(2 \sin^2 ax)/\pi^2 x^2$

23. Find the inverse F.S.T. of $s^n e^{-as}$

Ans. $\frac{2 \cdot n! \sin[(n+1)x]}{\pi \cdot (a^2+x^2)^{\frac{n+1}{2}}}$

24. Find the inverse Fourier transform of $e^{-|s|y}$

Ans. $\frac{y}{[\pi(y^2+x^2)]}$.

Integral equations

Solve the integral equations:

25. $\int_0^\infty \cdot f(x) \cdot \sin \alpha x \, dx = \begin{cases} 1 - \alpha, & 0 \le \alpha \le 1 \\ 0, & \alpha > 1 \end{cases}$

Ans. $f(x) = \frac{2(x - \sin x)}{(\pi x^2)}$

26. $\int_{-\infty}^\infty \frac{f(u) du}{(x-u)^2+a^2} = \frac{1}{x^2+b^2}, 0 < a < b$

Ans. $f(x) = \frac{(b-a)\alpha}{b\pi[x^2+(b-a)^2]}$

27. $\int_0^\infty f(x) \cos \alpha x dx = \begin{cases} 1 - \alpha, & 0 \le \alpha < 1 \\ 0, & \alpha > 1. \end{cases}$

Hence evaluate $\int_0^\infty \frac{\sin^2 t}{t^2} dt$

Ans. $f(x) = 2\frac{(1-\cos x)}{(\pi x^2)}$

28. $\int_0^\infty f(x) \cos \alpha x dx = e^{-\alpha}$

Ans. $\frac{2}{\pi(1+x^2)}$.

Solution to boundary value problems (B.V.P.): Laplace's equation

29. Find the steady-state temperature distribution $u(x, y)$ in an infinite metal plate covering the first quadrant with the edge along the y-axis held at 0° and the edge along the x-axis held at $u(x, 0) = \begin{cases} 100, & 0 < x < 1 \\ 0, & x > 1 \end{cases}$.

Hint: Use F.S.T. to solve Laplace's equation.

Ans. $u(x, y) = \frac{200}{\pi} \int_0^\infty \frac{1-\cos s}{s} e^{-sy} \sin sx ds$

or $u(x, y) = \frac{200}{\pi} \left[\text{arc tan}\left(\frac{x}{y}\right) - \frac{1}{2}\text{arc tan}\left(\frac{x+1}{y}\right) - \frac{1}{2}\text{arc tan}\left(\frac{x-1}{y}\right) \right]$

30. Solve $u_{xx} + u_{yy} = 0$ in the upper half-plane when the temperature along the x-axis $u(x, 0) = f(x)$. Deduce (a) when $u(x, 0) = 1$ if $|x| < 1$ (b) $u(x, 0) = \begin{cases} 2u_0, & x < -1 \\ u_0, & -1 < x < 1 \\ 0, & x > 1 \end{cases}$

Ans. $u(x, y) = \frac{1}{\pi} \int_{-\infty}^{\infty} \frac{yf(s)}{(x-s)^2+y^2} ds$ known as Poisson integral formula for the half-plane $y > 0$ or Schwarz integral formula.

a. $u(x, y) = \frac{1}{\pi} \left[\tan^{-1}\left(\frac{1-x}{y}\right) + \tan^{-1}\left(\frac{x+1}{y}\right) \right]$

b. $u(x, y) = \frac{u_0}{\pi} \left[\pi - \tan^{-1}\left(\frac{1+x}{y}\right) + \tan^{-1}\left(\frac{1-x}{y}\right) \right]$.

One-dimensional heat equation

31. Find the temperature distribution $u(x, t)$ in a semi-infinite metal bar with the end $x = 0$ at zero temperature and initial temperature distribution $f(x)$. Deduce when $f(x) = e^{-x}$.

Ans. $u(x, t) = \frac{2}{\pi} \int_0^\infty U(s, 0) e^{-c^2 s^2 t} \sin sx ds$

where $U(s, 0) = \int_0^\infty u(x, 0) \cdot \sin sx dx = \int_0^\infty f(x) \sin sx dx$

when $f(x) = e^{-x}$: $U(s, 0) = \frac{s}{1+s^2}$;

$u(x, t) = \frac{2}{\pi} \int_0^\infty \frac{se^{-c^2 p^2 t}}{1+s^2} \sin sx ds$

32. Find the temperature $u(x, t)$ in a semi-infinite bar with $\frac{\partial u}{\partial x} = \mu$ when $x = 0$ and with initial temperature 0. Assume that $\frac{\partial u}{\partial x} \to 0$ as $x \to \infty$ and is bounded.

Ans. $u(x, t) = \frac{2}{\pi} \int_0^\infty \frac{\mu}{s^2}(1 - e^{-s^2 c^2 t}) \cos sx ds$

33. Solve $\frac{\partial u}{\partial t} = c^2 \frac{\partial^2 u}{\partial x^2}$, $x > 0, t > 0$ with $\left.\frac{\partial u}{\partial x}\right|_{x=0} = 0$ and $u(x, 0) = \begin{cases} x, & 0 \le x \le 1 \\ 0, & x > 1 \end{cases}$.

Ans. $u(x, t) = \frac{2}{\pi} \int_0^\infty \left[\frac{\sin s}{s} + \frac{1}{s^2}(\cos s - 1) \right] e^{-s^2 c^2 t} \times \cos sx ds$

34. Solve $\frac{\partial u}{\partial t} = c^2 \frac{\partial^2 u}{\partial x^2}$, $x > 0, t > 0$ with $u(x, 0) = 0, x > 0$ and $u(0, t) = u_0$ for $t > 0$.

Ans. $u(x, t) = u_0 \left[1 - \frac{2}{\pi} \int_0^\infty \frac{e^{-s^2 c^2 t}}{s} \cdot \sin sx ds \right]$.

Vibrating string

35. An infinite string is initially at rest and has an initial transverse displacement $y(x, 0) = f(x), -\infty < x < \infty$. Show that the displacement $y(x, t)$ of the string is $y(x, t) = \frac{1}{2}[f(x + ct) + f(x - ct)]$.

20.4 FINITE FOURIER SINE AND COSINE TRANSFORMS

Let $f(x)$ be a function defined in a finite interval $0 < x < L$ i.e., when the range of one of the variables say x is finite. Suppose $f(x)$ is neither periodic nor even nor odd. Now by redefining $f(x)$ as an odd function in $-L < x < L$, the half range Fourier sine series expansion of $f(x)$ can be obtained as

$$f(x) = \sum_{n=1}^{\infty} b_n \sin\left(\frac{n\pi x}{L}\right)$$

where $$b_n = \frac{2}{L} \int_0^L f(x) \sin\left(\frac{n\pi x}{L}\right) dx.$$

Then the finite Fourier sine transform of $f(x)$ in $0 < x < L$ is defined as

$$F_s(n) = \int_0^L f(x) \sin\left(\frac{n\pi x}{L}\right) dx$$

which is a function of n, an integer.

The inverse finite Fourier sine transform of $F_s(n)$ is given by

$$f(x) = \frac{2}{L}\sum F_s(n)\cdot\sin\left(\frac{n\pi x}{L}\right)$$

In a similar way, the finite Fourier cosine transform of $f(x)$ in $0 < x < L$ is defined as

$$F_c(n) = \int_0^L f(x)\cos\left(\frac{n\pi x}{L}\right)dx$$

The inverse finite Fourier cosine transform of $F_c(n)$ is given by

$$f(x) = \frac{1}{L}F_c(0) + \frac{2}{L}\sum_{n=1}^{\infty} F_c(n)\cos\frac{n\pi x}{L}.$$

Result 1:

$$F_s\left\{\frac{\partial^2 f}{\partial x^2}\right\} = \frac{-n^2\pi^2}{L^2}F_s(f) - \frac{-n\pi}{L}[(-1)^n f(L,T) - f(0,t)]$$

By definition

$$F_s\left\{\frac{\partial^2 f}{\partial x^2}\right\} = \int_0^L\left(\frac{\partial^2 f}{\partial x^2}\right)\cdot\sin\frac{n\pi x}{L}dx,$$

integrating by parts

$$= \frac{\partial f}{\partial x}\cdot\sin\frac{n\pi x}{L}\bigg|_0^L - \frac{n\pi}{L}\int_0^L\frac{\partial f}{\partial x}\cos\frac{n\pi x}{L}dx$$

$$= 0 - \frac{n\pi}{L}\left\{\left[f(x,t)\cdot\cos\left(\frac{n\pi x}{L}\right)\right]_0^L + \frac{n\pi}{L}\int_0^L f(x,t)\cdot\sin\left(\frac{n\pi x}{L}\right)dx\right\}$$

$$F_s\left\{\frac{\partial^2 f}{\partial x^2}\right\} = -\frac{n\pi}{L}[(-1)^n f(L,t) - f(0,t)] - \frac{n^2\pi^2}{L^2}F_s(f).$$

Similarly,

Result 2:

$$F_c\left\{\frac{\partial^2 f}{\partial x^2}\right\} = -[f_x(0,t) - (-1)^n f_x(L,t)] - \frac{n^2\pi^2}{L^2}F_c(f).$$

By definition

$$F_c\left\{\frac{\partial^2 f}{\partial x^2}\right\} = \int_0^L\frac{\partial^2 f}{\partial x^2}\cdot\cos\left(\frac{n\pi x}{L}\right)dx,$$

integrating by parts

$$= \frac{\partial f}{\partial x}\cdot\cos\frac{n\pi x}{L}\bigg|_0^L + \frac{n\pi}{L}\int_0^L\frac{\partial f}{\partial x}\cdot\sin\frac{n\pi x}{L}dx$$

$$= [-f_x(0,t) + (-1)^n f_x(L,t)] + \frac{n\pi}{L}\left[f\cdot\sin\frac{n\pi x}{L}\bigg|_0^L - \frac{n\pi}{L}\int_0^L f\cos\frac{n\pi x}{L}dx\right]$$

$$F_c\left\{\frac{\partial^2 f}{\partial x^2}\right\} = [(-1)^n f_x(L,t) - f_x(0,t)] - \frac{n^2\pi^2}{L^2}F_c(f).$$

Note 3:

1. Apply finite Fourier sine transform when the boundary conditon f prescribed.
2. Apply finite Fourier cosine transform when the boundary condition f_x prescribed.

Worked Out Examples

Example 1: Find the finite Fourier sine and cosine transform of $f(x) = x(\pi - x)$ in $0 < x < \pi$.

Solution: Finite Fourier sine transform of $f(x)$ is

$$F_s(n) = \int_0^L f(x)\sin\frac{n\pi x}{L}dx$$

$$= \int_0^{\pi} x(\pi - x)\cdot\sin nx\,dx$$

$$= x(\pi - x)\cdot\left[\frac{-\cos nx}{n}\bigg|_0^{\pi} - -(\pi - 2x)\cdot\left(\frac{-\sin nx}{n^2}\right)\bigg|_0^{\pi} + (-2)\cdot\frac{\cos nx}{n^3}\right]_0^{\pi}$$

$$F_s(n) = 0 + 0 + \frac{2}{n^3}[1 - \cos n\pi]$$

$$= \frac{2}{n^3}[1 - (-1)^n]$$

Finite Fourier cosine transform of $f(x)$ is

$$F_c(n) = \int_0^L f(x)\cos\frac{n\pi x}{L}dx$$

$$= \int_0^\pi x(\pi - x)\cdot\cos nx\,dx$$

$$= \left[x(\pi - x)\cdot\frac{\sin nx}{n} - (\pi - 2x)\left(\frac{-\cos nx}{n^2}\right) + (-2)\frac{-\sin nx}{n^3}\right]_0^\pi$$

$$F_c(n) = 0 - \frac{(\pi\cos n\pi + \pi)}{n^2} = \frac{-\pi}{n^2}[1 + (-1)^n].$$

Example 2: Find the inverse finite Fourier cosine transform $f(x)$ if $F_c(n) = \frac{\sin(\frac{n\pi}{2})}{2n}$ for $n = 1, 2, 3, \ldots$ and $= \frac{\pi}{4}$ when $n = 0$ in $0 < x < \pi$

Solution:

$f(x)$ = Inverse finite Fourier cosine transform of $F_c(n)$

$$= \frac{1}{\pi}F_c(0) + \frac{2}{\pi}\sum_{n=1}^{\infty}F_c(n)\cdot\cos\left(\frac{n\pi x}{L}\right)$$

$$= \frac{1}{\pi}\cdot\frac{\pi}{4} + \frac{2}{\pi}\sum_{n=1}^{\infty}\sin\left(\frac{n\pi}{2}\right)\cdot\left(\frac{1}{2n}\right)\cdot\cos\left(\frac{n\pi x}{\pi}\right)$$

$$f(x) = \frac{1}{4} + \frac{1}{\pi}\sum\frac{1}{n}\cdot\sin\left(\frac{n\pi}{2}\right)\cdot\cos nx.$$

Example 3: Find the inverse finite Fourier sine transform $f(x)$ if $f_s(n) = \frac{2\pi(-1)^{n-1}}{n^2}$, $n = 1, 2, \ldots$ when $0 < x < \pi$.

Solution:

$$f(x) = f_s^{-1}\{F_s(n)\} = \frac{2}{L}\sum_{n=1}^{\infty}F_s(n)\cdot\sin\left(\frac{n\pi x}{L}\right)$$

$$= \frac{2}{\pi}\sum_{n=1}^{\infty}\left\{\frac{2\pi(-1)^{n-1}}{n^2}\right\}\cdot\sin nx$$

$$f(x) = 4\sum_{n=1}^{\infty}\frac{(-1)^{n-1}}{n^2}\sin nx.$$

Example 4: Find the temperature distribution in a bar of length L, with its both ends and lateral surface insulated when the initial temperature in the bar is $f(x)$. Deduce when $f(x) = x^2$ and $L = 10$.

Solution: This problem is to solve the one-dimensional heat equation $\frac{\partial u}{\partial t} = c^2\frac{\partial^2 u}{\partial x^2}$ for $0 < x < L$, with boundary conditions $u_x(0, t) = u_x(L, t) = 0$ and with initial condition $u(x, 0) = f(x)$. Since the boundary conditions are of the Neumann type (derivatives) apply finite Fourier cosine transform to both sides of the equation.

$$\int_0^L \frac{\partial u}{\partial t}\cdot\cos\left(\frac{n\pi x}{L}\right)dx = c^2\int_0^L\frac{\partial^2 u}{\partial x^2}\cos\left(\frac{n\pi x}{L}\right)dx \quad (1)$$

Put $U = F_c\{u(x, t)\}$ and use (2)

$$F_c\left\{\frac{\partial^2 u}{\partial x^2}\right\} = [(-1)^n u_x(L, t) - u_x(0, t)] - \frac{n^2\pi^2}{L^2}F_c(u)$$

the equation (1) reduces to

$$\frac{d}{dt}\int_0^L u\cdot\cos\left(\frac{n\pi x}{L}\right)dx$$

$$= c^2\left\{[(-1)^n u_x(L, t) - u_x(0, t)] - \frac{n^2\pi^2}{L^2}F_c(u)\right\}$$

or $$\frac{dU}{dt} = c^2\left(\frac{-n^2\pi^2}{L^2}\right)U \quad (3)$$

since the first term in the R.H.S. is zero because of the zero boundary conditions ($u_x(L, t) = u_x(0, t) = 0$) Here $U = U(n, t)$.
Solving (3), we get

$$U(n, t) = Ae^{-\left(\frac{n^2\pi^2c^2t}{L^2}\right)} \quad (4)$$

To determine the arbitrary constant A in (4) use the initial condition.
Put $t = 0$ in (4) this yields

$$A = U(n, 0) = F_c\{u(x, 0)\} = F_c\{f(x)\}$$

$$= \int_0^L f(x)\cdot\cos\left(\frac{n\pi x}{L}\right)dx \quad (5)$$

$$\therefore U(n, t) = \left[\int_0^L f(x)\cdot\cos\left(\frac{n\pi x}{L}\right)dx\right]e^{-n^2\pi^2c^2t/L^2} \quad (6)$$

Taking inverse finite Fourier cosine transform

$$u(x, t) = F_c^{-1}\{U(n, t)\}$$

$$= \frac{1}{L}F_c(0) + \frac{2}{L}\sum_{n=1}^{\infty}F_c(n)\cos\left(\frac{n\pi x}{L}\right)$$

where $F_c(0)=\int_0^L f(x)dx$

and $F_c(n)=F_c\{u(x,t)\}=U(n,t)$ given by (6)

Thus $u(x,t)=\frac{1}{L}\int_0^L f(x)dx+$

$$+\frac{2}{L}\sum_{n=1}^{\infty}\left[\int_0^L f(x)\cos\left(\frac{n\pi x}{L}\right)dx\right]\times$$

$$\times\cos\frac{n\pi x}{L}e^{-(n^2\pi^2c^2t/L^2)} \tag{7}$$

When $f(x)=x^2$, and $L=10$,

$$F_c(0)=\int_0^{10}x^2dx=\left.\frac{x^3}{3}\right|_0^{10}=\frac{1000}{3}$$

$$U(n,0)=F_c\{u(x,0)\}=\int_0^{10}x^2\cdot\cos\left(\frac{n\pi x}{10}\right)dx$$

$$=x^2\cdot\left(\frac{10}{n\pi}\right)\cdot\sin\left(\frac{n\pi x}{10}\right)-2x\cdot\left(\frac{-10^2}{n^2\pi^2}\right)$$

$$\times\cos\left(\frac{n\pi x}{10}\right)+2\left(\frac{-10^3}{n^3\pi^3}\right)\sin\left.\frac{n\pi x}{10}\right|_0^{10}$$

$$=\frac{2000}{n^2\pi^2}(-1)^n.$$

Thus $u(x,t)=\frac{100}{3}+\frac{2}{10}\cdot\frac{2000}{\pi^2}\sum_{n=1}^{\infty}\frac{(-1)^n}{n^2}$

$$\times e^{-(n^2\pi^2c^2t/100)}\cdot\cos\left(\frac{n\pi x}{10}\right).$$

Example 5: Solve the Laplace's equation $\frac{\partial^2u}{\partial x^2}+\frac{\partial^2u}{\partial y^2}=0$ in square plate of length L with the following conditions $u(0,y)=u(L,y)=0$, $u(x,0)=0$, $u(x,L)=f(x)$. Deduce when $f(x)=x^2$ and $L=\pi$.

Solution: Since the boundary conditions are Dirichlet type (u prescribed), take finite Fourier sine transform of the Laplace equation on both sides

$$\int_0^L\frac{\partial^2u}{\partial x^2}\sin\left(\frac{n\pi x}{L}\right)dx+\int_0^L\frac{\partial^2u}{\partial y^2}\sin\left(\frac{n\pi x}{L}\right)dx=0$$

Integrating by parts

$$\frac{\partial u}{\partial x}\cdot\sin\left.\left(\frac{n\pi x}{L}\right)\right|_0^{\pi}-\frac{n\pi}{L}\int_0^L\frac{\partial u}{\partial x}\cdot\cos\left(\frac{n\pi x}{L}\right)dx$$

$$+\frac{\partial^2}{\partial y^2}\int_0^L u(x,y)\sin\left(\frac{n\pi x}{L}\right)=0$$

Put $U=F_s\{u(x,y)\}=\int_0^L u(x,y)\sin\left(\frac{n\pi x}{L}\right)dx$

$$=U(n,y)$$

Then
$$\frac{d^2U}{dy^2}+0-\frac{n\pi}{L}\left[u(x,y)\cdot\cos\left.\frac{n\pi x}{L}\right|_0^L\right.$$

$$\left.+\frac{n\pi}{L}\int_0^L u(x,y)\cdot\sin\left(\frac{n\pi x}{L}\right)dx\right]=0$$

$$\frac{d^2U}{dy^2}-\frac{n\pi}{L}[u(L,y)(-1)^n-u(0,y)]$$

$$-\frac{n^2\pi^2}{L^2}U=0$$

Using the boundary condition $u(L,y)=u(0,y)=0$,

$$\frac{d^2U}{dy^2}-\lambda^2U=0\quad\text{where}\quad\lambda=\frac{n\pi}{L} \tag{1}$$

The solution of this 2nd order ordinary differential equation is

$$U(n,y)=A\cosh\cdot\lambda y+B\sinh\lambda y \tag{2}$$

Using the boundary condition $u(x,0)=0$

$$0=U(n,0)=A\cdot1+B\cdot0\quad\therefore A=0 \tag{3}$$

Using $u(x,L)=f(x)$

$$U(n,L)=F_s\{u(x,L)\}=F_s\{f(x)\}$$

$$=\int_0^L f(x)\sin\left(\frac{n\pi x}{L}\right)dx \tag{4}$$

Substituting (3), (4) in (2)

$$U(n,L)=B\cdot\sinh\lambda L$$

$$\therefore\quad B=\frac{U(n,L)}{\sinh\lambda L} \tag{5}$$

Thus (2) reduces to

$$U(n,y)=B\cdot\sinh\lambda y \tag{6}$$

where B and λ are given by (5) and (1).

Taking the inverse finite Fourier sine transform of (6), we get

$$u(x,y)=F_s^{-1}\{U\}=\frac{2}{L}\sum_{n=1}^{\infty}U(n,y)\cdot\sin\frac{n\pi x}{L}$$

$$u(x,y)=\frac{2}{L}\sum_{n=1}^{\infty}B\cdot\sinh\left(\frac{n\pi y}{L}\right)\cdot\sin\left(\frac{n\pi x}{L}\right) \tag{7}$$

When $f(x) = x^2$ and $L = \pi$, from (4), we get

$$U(n, L) = \int_0^L f(x)\sin\left(\frac{n\pi x}{L}\right)dx = \int_0^\pi x^2 \sin nx\, dx$$

$$= x^2 \cdot \left(\frac{-\cos nx}{n}\right) - 2x \cdot \left(-\frac{1}{n^2}\sin nx\right)$$

$$+2\left(+\frac{1}{n^3}\right)\cdot \cos nx\Big|_0^\pi$$

$$U(n, L) = \frac{-\pi^2(-1)^n}{n} + \frac{2}{n^3}(-1)^n - \frac{2}{n^3} \qquad (8)$$

Substituting (5), (8) in (7), we get

$$u(x, y) = \frac{2}{\pi}\sum_{n=1}^{\infty}\frac{1}{\sinh n\pi}\left[(-1)^n\left(\frac{2}{n^3} - \pi^2\right) - \frac{2}{n^3}\right]$$

$$\times \sinh ny \cdot \sin nx$$

EXERCISE

Find the finite Fourier sine transform (F.F.S.T.) and finite Fourier cosine transform (F.F.C.T.) of $f(x)$:

1. $f(x) = 2x$ in $0 < x < 4$

Ans. F.F.S.T.: $\frac{32}{s\pi}(-1)^{n+1}$, F.F.C.T.: $\frac{32}{s\pi}((-1)^n - 1)$

2. $f(x)\begin{cases} 1, & 0 < x < \pi/2 \\ -1, & \pi/2 < x < \pi \end{cases}$

Ans. F.F.S.T. $= (-2\cos(s\pi/2) + 1 + \cos s\pi/2)/s$

F.F.C.T. $= 2\sin(s\pi/2)/2, \quad s = 1, 2, 3, \ldots$

3. $f(x) = x^2$ in $0 < x < L$

Ans. F.F.S.T.: $\frac{-L^3\cos n\pi}{n\pi} + \frac{2L^3}{n^3\pi^3}(\cos n\pi - 1), \quad n = 1, 2, 3, \ldots$

F.F.C.T.: $2L^3(\cos n\pi - 1)/n^2\pi^2$

4. Find F.F.S.T. of $f(x) = e^{ax}$ in $(0, L)$

Ans. $s\pi L\left[(-1)^{s+1}\cdot e^{aL} + 1\right]/\left[a^2L^2 + s^2\pi^2\right]$

5. Find F.F.C.T. of $f(x) = \frac{x^2}{2\pi} - \frac{\pi}{6}$ in $(0, \pi)$

Ans. $(-1)^n/n^2$ for $n = 1, 2, 3$ and zero for $n = 0$

6. Determine the inverse F.F.S.T. of $\frac{16(-1)^{n-1}}{n^3}$, $n = 1, 2, 3, \cdots$ and $0 < x < 8$

Ans. $f(x) = \frac{2}{8}\sum_{n=1}^{\infty}\frac{16(-1)^{n-1}}{n^3}\sin\frac{n\pi x}{8}$

7. Determine the inverse F.F.C.T. of $\frac{6\sin\frac{n\pi}{2} - \cos n\pi}{(2n+1)\pi}$ for $n = 1, 2, 3, \ldots$ and equals to $\frac{2}{\pi}$ for $n = 0$ in $0 < x < 4$

Ans. $f(x) = \frac{1}{4}\cdot\frac{2}{\pi} + \frac{2}{4}\sum_{n=1}^{\infty}\frac{6(\sin(n\pi/2) - \cos n\pi)}{(2n+1)\pi}\cos\left(\frac{n\pi}{4}\right)$

8. Determine F.F.S.T. of $(1 - \cos n\pi)/n^2\pi^2$ where $0 < x < \pi$

Ans. $f(x) = \frac{2}{\pi^3}\sum_{n=1}^{\infty}\left(\frac{1-\cos n\pi}{n^2}\right)\sin nx$

Solve the one-dimensional heat equation $\frac{\partial u}{\partial t} = c^2\frac{\partial^2 u}{\partial x^2}$ with

9. $u(0, t) = u(4, t) = 0$, $u(x, 0) = 2x$

Ans. $u(x, t) = \frac{2}{4}\sum_{n=1}^{\infty}\frac{32(-1)^{n+1}}{n\pi}e^{-n^2\pi^2c^2t/16}\cdot\sin\frac{n\pi x}{4}$

10. $u_x(0, t) = u_x(6, t) = 0$, $u(x, 0) = 2x$

Ans. $u(x, t) = 6 + \frac{2\pi}{\pi^2}\sum_{n=1}^{\infty}\frac{(-1)^n - 1}{n^2}e^{-n^2\pi^2c^2t/36}\times$

$\times \cos\frac{n\pi x}{6}$

11. Solve the Laplace's equation $\frac{\partial^2 u}{\partial x^2} + \frac{\partial^2 u}{\partial y^2} = 0$ in square metal plate of side π with $u(x, \pi) = u_0$, $u(0, y) = 0, u(\pi, y) = 0, u(x, 0) = 0$. Assume $u(x, y)$ is bounded

Ans. $u(x, y) = \frac{4u_0}{\pi}\sum_{n=0}^{\infty}\frac{\sin h((2n+1)y)\sin(2n+1)x}{(2n+1)\cdot\sin h((2n+1)s\pi)}$

12. Find the displacement $y(x, t)$ of a string of length π governed by the wave equation $\frac{\partial^2 y}{\partial t^2} = \frac{c^2\partial^2 y}{\partial x^2}$ with one end fixed $y(\pi, t) = 0$, initially at rest $y(x, 0) = 0$, with initial displacement $u(0, t) = a\sin wt$.

Ans. $y(x, t) = a\sin wt\cdot\sin w\frac{(\pi - x)}{c}\operatorname{cosec}\left(\frac{\pi w}{c}\right) +$

$\left(\frac{2awc}{\pi}\right)\sum_{n=1}^{\infty}(w^2 - n^2c^2)^{-1}\sin nx\sin nct.$

20.5 PARSEVAL'S* IDENTITY FOR FOURIER TRANSFORMS

Let $F(\alpha)$ and $G(\alpha)$ be respectively the Fourier transforms of $f(x)$ and $g(x)$. Then

$$\int_{-\infty}^{\infty} F(\alpha)\overline{G(\alpha)}d\alpha = \frac{1}{2\pi}\int_{-\infty}^{\infty} f(x)\overline{g(x)}dx \qquad (1)$$

where bar implies the complex conjugate.

Proof: Using the inversion formula for Fourier transform for $\overline{g(x)}$ in the R.H.S. of (1),

$$\int_{-\infty}^{\infty} f(x)\overline{g(x)}dx = \int_{-\infty}^{\infty} f(x)\left\{\int_{-\infty}^{\infty} \overline{G(\alpha)e^{i\alpha x}}\,d\alpha\right\}dx$$

Interchanging the order of integrain in R.H.S.

$$= \int_{-\infty}^{\infty} \overline{G(\alpha)}\left\{\int_{-\infty}^{\infty} f(x)e^{i\alpha x}dx\right\}d\alpha$$

Observing that inner most integral is the Fourier transform of $f(x)$, we have

$$= \int_{-\infty}^{\infty} \overline{G(\alpha)}\,\{2\pi F(\alpha)\}\,d\alpha = 2\pi\int_{-\infty}^{\infty} F(\alpha)\overline{G(\alpha)}d\alpha$$

Thus $\frac{1}{2\pi}\int_{-\infty}^{\infty} f(x)\overline{g(x)}dx = \int_{-\infty}^{\infty} F(\alpha)\overline{G(\alpha)}dx$.

Corollary 1: Put $g(x) = f(x)$ and note that $z\overline{z} = |z|^2$. Then (1) reduces to

$$\int_{-\infty}^{\infty} F(\alpha)\overline{F(\alpha)}d\alpha = \int_{-\infty}^{\infty} |F(\alpha)|^2 d\alpha$$

$$= \frac{1}{2\pi}\int_{-\infty}^{\infty} f(x)\overline{f(x)}dx$$

$$= \frac{1}{2\pi}\int_{-\infty}^{\infty} |f(x)|^2 dx$$

i.e.,

$$\boxed{\int_{-\infty}^{\infty} |F(\alpha)|^2 d\alpha = \frac{1}{2\pi}\int_{-\infty}^{\infty} |f(x)|^2 dx} \qquad (2)$$

Results (2) and its generalization (1) are known as **Parseval's identity for Fourier transforms (integrals)**.

Corollary 2: The Parseval's identity for Fourier cosine sine transform are

(a) $\frac{2}{\pi}\int_0^{\infty} F_c(\alpha)G_c(\alpha)d\alpha = \int_0^{\infty} f(x)g(x)dx$

(b) $\frac{2}{\pi}\int_0^{\infty} |F_c(\alpha)|^2\alpha\alpha = \int_0^{\infty} |f(x)|^2 d\alpha$

Parseval's identity for Fourier sine transform are

(c) $\frac{2}{\pi}\int_0^{\infty} F_s(\alpha)G_s(\alpha)d\alpha = \int_0^{\infty} f(x)g(x)dx$

(d) $\frac{2}{\pi}\int_0^{\infty} |F_s(\alpha)|^2 d\alpha = \int_0^{\infty} |f(x)|^2 dx$

These results can be proved on similar lines as above.

WORKED OUT EXAMPLES

Example 1: Prove that $\int_0^{\infty} \frac{(x\cos x - \sin x)^2}{x^6}dx = \frac{\pi}{15}$.

Solution: Consider the function $f(x)$ defined as

$$f(x) = \begin{cases} a^2 - x^2 & \text{if } |x| \le a \\ 0 & \text{if } |x| > a \end{cases}$$

Fourier transform of $f(x) = F\{f(x)\} = F(\alpha) = \frac{1}{2\pi}\int_{-\infty}^{\infty} f(x)e^{-i\alpha x}dx = \frac{1}{2\pi}\left[\int_{-\infty}^{-a} + \int_{-a}^{a} + \int_{a}^{\infty}\right]$.
Since $f(x) = 0$ in $(-\infty, -a)$ and (a, ∞), we have

$$F(\alpha) = \frac{1}{2\pi}\int_{-a}^{a} (a^2 - x^2)e^{-i\alpha x}dx = \frac{1}{2\pi}[I_1 - I_2]$$

where

$$I_1 = \int_{-a}^{a} a^2 e^{-i\alpha x}dx - a^2\left(\frac{e^{-i\alpha x}}{-i\alpha}\right)\Bigg|_{-a}^{a} = \frac{a^2}{i\alpha}[e^{i\alpha a} - e^{-i\alpha a}]$$

and integrating by parts.

$$I_2 = \int_{-a}^{a} x^2 e^{-i\alpha x}dx$$

$$= \left(x^2\frac{e^{-i\alpha x}}{-i\alpha} - 2x\frac{e^{-i\alpha x}}{i^2\alpha^2} + 2\frac{e^{-i\alpha x}}{-i^3\alpha^3}\right)\Bigg|_{-a}^{a}$$

$$= \frac{a^2}{i\alpha}\left[e^{i\alpha a} - e^{-i\alpha a}\right] + \frac{2a}{\alpha^2}\left[e^{i\alpha a} + e^{-i\alpha a}\right] + $$

$$+ \frac{2}{i\alpha^3}\left[e^{-i\alpha a} - e^{i\alpha a}\right]$$

So

$$F(\alpha) = \frac{1}{2\pi}[I_1 - I_2] = \frac{1}{2\pi}\left[\frac{-2a}{\alpha^2}(2\cos\alpha a) + \frac{2}{\alpha^3}2\cdot\sin\alpha a\right]$$

$$F(\alpha) = +\frac{2}{\pi\alpha^3}[\sin\alpha a - a\alpha\cos\alpha a]$$

Marc Antoine Parseval (1755–1836) French mathematician.

Apply Parseval's identity for $f(x)$ and $F(\alpha)$.

$$\int_{-\infty}^{\infty} |F(\alpha)|^2 d\alpha = \int_{-\infty}^{\infty} \frac{4}{\pi^2}\left(\frac{\sin a\alpha - a\alpha\cos a\alpha}{\alpha^3}\right)^2 d\alpha$$

$$= \frac{1}{2\pi}\int_{-\infty}^{\infty} |f(x)|^2 dx$$

$$= \frac{1}{2\pi}\int_{-a}^{a} (|a^2 - x^2|)^2 dx$$

$$= \frac{1}{2\pi}\int_{-a}^{a} (a^4 + x^4 - 2a^2x^2) dx$$

$$= \frac{1}{2\pi}\left[a^4 x + \frac{x^5}{5} - 2a^2\frac{x^3}{3}\right]\Bigg|_{-a}^{a}$$

$$= \frac{1}{2\pi}\cdot 2a^2\left(1 + \frac{1}{5} - \frac{2}{3}\right)$$

$$= \frac{16a^5}{15}\cdot\frac{1}{2\pi} = \frac{8}{15}\frac{a^5}{\pi}$$

Thus

$$2\int_0^{\infty} \frac{4}{\pi^2}\left(\frac{ax\cos ax - \sin ax}{x^3}\right)^2 dx = \frac{8}{15}\frac{a^5}{\pi}$$

or

$$\int_0^{\infty} \frac{(ax\cos ax - \sin ax)^2}{x^6} dx = \frac{\pi a^5}{15}$$

For $a = 1$, we get the required result.

Example 2: Evaluate $\int_0^{\infty} \frac{x^2}{(a^2+x^2)(b^2+x^2)} dx$ and hence find $\int_0^{\infty}\left(\frac{x}{x^2+1}\right)^2 dx$.

Solution: We know that if $f(x)=e^{-ax}$ then Fourier sine transform of $f(x)=F_s\{f(x)\}=F_s(\alpha) = \frac{\alpha}{a^2+\alpha^2}$. Similarly for $g(x)=e^{-bx}$ then $G_s\{g(x)\} = \frac{\alpha}{b^2+\alpha^2}$. Recall Parseval's identity for sine transform

$$\frac{2}{\pi}\int_0^{\infty} F_s(\alpha)G_s(\alpha)d\alpha = \int_0^{\infty} f(x)g(x)dx$$

$$\int_0^{\infty}\left(\frac{\alpha}{a^2+\alpha^2}\right)\left(\frac{\alpha}{b^2+\alpha^2}\right)d\alpha$$

$$= \int_0^{\infty} e^{-ax}\cdot e^{-bx} dx = \frac{2}{\pi}\int_0^{\infty} e^{-(a+b)x} dx$$

$$= \frac{\pi}{2}\frac{e^{-(a+b)x}}{-(a+b)}\Bigg|_{x=0}^{\infty} = \frac{1}{(a+b)}\frac{\pi}{2}$$

Thus

$$\int_0^{\infty} \frac{x^2}{(a^2+x^2)(b^2+x^2)} dx = \frac{\pi}{2(a+b)}$$

For $a = b = 1$, $\int_0^{\infty}\left(\frac{x}{x^2+1}\right)^2 dx = \frac{\pi}{4}$.

EXERCISE

Solve the following problems using Parseval's identity for Fourier tansforms.

1. Show that $\int_0^{\infty} \frac{\sin^2 ax}{x^2} dx = \frac{\pi a}{2}$

   Hint: Consider $f(x) = 1$, $|x| < a$ and $f(x) = 0$ for $|x| > a$. Then $F\{f(x)\} = F(\alpha) = \frac{2\sin a\alpha}{\alpha}$, apply P.I. then $\int_{-a}^{a}(1)^2 dx = \frac{1}{2\pi}\int_{-\infty}^{\infty}\frac{4\sin^2 a\alpha}{\alpha^2}d\alpha$

2. Evaluate (a) $\int_0^{\infty}\frac{dx}{(a^2+x^2)(b^2+x^2)}$ and hence find (b) $\int_0^{\infty}\frac{dx}{(x^2+1)^2}$.

   Hint: $F_c\{e^{-ax}\} = \frac{a}{a^2+\alpha^2}$, $F_c\{e^{-bx}\} = \frac{b}{b^2+\alpha^2}$, use P.I.

   Ans. (a) $\frac{\pi}{2ab(a+b)}$ (b) For $a = b = 1$, $\frac{\pi}{4}$.

3. Prove that $\int_0^{\infty}\left(\frac{\sin x}{x}\right)^4 dx = \frac{\pi}{3}$

   Hint: Take $f(x) = 1 - |x|$, for $|x| < 1, 0$, otherwise, $F\{f(x)\} = \sqrt{\frac{2}{\pi}}\left(\frac{1-\cos\alpha}{\alpha^2}\right)$, use P.I.

4. Evaluate (a) $\int_0^{\infty}\left(\frac{1-\cos x}{x}\right)^2 dx$ (b) $\int_0^{\infty}\frac{\sin^4 x}{x^2}dx$

   Hint: Take $f(x) = 1$, for $0 \le x < 1, 0$ otherwise, then $F_c\{f(x)\} = \frac{1-\cos\alpha}{\alpha}$, $F_s\{f(x)\} = \frac{\sin\alpha}{\alpha}$, use P.I.

   Ans. (a) $\frac{\pi}{2}$ (b) $\frac{\pi}{2}$

5. Evaluate $\int_0^{\infty}\frac{\sin ax}{x(a^2+x^2)}dx$

   Hint: $F_c\{e^{-ax}\} = \frac{a}{a^2+\alpha^2}$, $F_c\{g(x)\} = \frac{\sin a\alpha}{\alpha}$ where $g(x) = 1, 0 < x < a$, zero for $x > a$. use (a) in cor. 2.

   Ans. $\frac{\pi}{2}\left(\frac{1-e^{-a^2}}{a^2}\right)$

Chapter 21

Linear Difference Equations and Z-Transforms

INTRODUCTION

Difference Equations

Difference equations model processes in which we know relationships between changes or differences rather than rates of changes which lead to differential equations. Thus a difference equation is an equation relating various terms of a sequence $a_0, a_1, a_2, \ldots$. A string loaded with a finite number of beads at equally spaced points leads to a difference equation. Recurrence relations obtained in the solution of DE by power series or Frobenius method are also difference equations. Numerical solution of DE also leads to difference equations.

Z-transforms

Solution of a discrete system, expressed as a difference equation is obtained using z-transform. Discrete analysis played important role in the development of communication engineering.

21.1 LINEAR DIFFERENCE EQUATIONS

Difference equations are functional equations that define sequences just as differential equations define functions. They arise in several situations as follows.

Finance: Compound Interest

Let P_0 be the amount of money deposited (invested) in a bank earning interest periodically say monthly or quarterly or annually. The conversion period r is the time period between interest payment. Let P_n denote the value of the deposit at the end of the nth conversion period, after n interest payments have accrued.

Then in case of simple interest,

$$P_{n+1} = P_n + r P_0$$

which is a first order difference equation

$$P_{n+1} - P_n = r P_0$$

with solution $P_n = P_0(1 + nr)$

In case of compound interest

$$P_{n+1} = P_n + r P_n$$

which is a homogeneous difference equation

$$P_{n+1} - (1 + r)P_n = 0$$

with solution $P_n = P_0(1 + r)^n$. In this discrete process P is a function of an integer n rather than a continuous variable.

Fibonacci Relation:

Suppose there is only one pair of rabbits male and female just born. Further suppose that every month each pair of rabbits that are one month old produce a new pair of offspring of opposite sexes. Then F_n the number of rabbits after n months is given by the recurrence relation.

$$F_n = F_{n-1} + F_{n-2} \quad \text{for } n \geq 2$$

with the initial conditions $F_0 = F_1 = 1$. This gives rise to a second order homogeneous difference equation

$$F_n - F_{n-1} + F_{n-2} = 0$$

By repeated application of the recurrence relation the equation can be solved recursively. Then we get $F_2 = 2, F_3 = 3, F_4 = 5, F_5 = 8, F_6 = 13, F_7 = 21, F_8 = 34$ etc. The disadvantage here is that F_n is calculated upto certain value of n and these values are also dependent on the initial conditions. Instead the general solution

$$F_n = c_1\left(\frac{1+\sqrt{5}}{2}\right)^n + c_2\left(\frac{1-\sqrt{5}}{2}\right)^n$$

is function of n and is independent of the initial conditions.

Differential Equations

In the numerical solution of ordinary differential equations, the derivatives are discretized by replacing them by the finite (forward) differences. This gives rise to difference equations of the higher order. Thus a continuous process described by a differential equation is approximated by a discrete process described by its counterpart a difference equation. (see Chapter on Numerical Analysis 33). For example, in a third order ordinary differential equation

$$a_3 y''' + a_2 y'' + a_1 y' + a_0 y = F(x)$$

the derivatives can be replaced by $y' = \dfrac{y_{n+1} - y_n}{h}$,

$$y'' = \frac{y_{n+2} - 2y_{n+1} + y_n}{h^2}$$

$y''' = \dfrac{y_{n+3} - 3y_{n+2} + 3y_{n+1} - y_n}{h^3}$, which results in a third order differences equation of the form

$$b_3 y_{n+3} + b_2 y_{n+2} + b_1 y_{n+1} + b_0 y_n = F(x)$$

Recall that a sequence is a numerical valued function whose domain of definition is the set of integers. It is denoted by $\{a_n\}$ or a_n or $a(n)$.

A kth order linear difference equation in the sequence y_n is of the form

$$a_k y_{n+k} + a_{k-1} y_{n+k-1} + \ldots + a_1 y_{n+1} + \\ + a_0 y_n = f(n) \qquad (1)$$

where $n = 0, 1, 2, 3, \ldots$. Thus (1) represents not just a single equation but an infinite system of equations one equation for every n. Here the coefficients $a_0, a_1, \ldots, a_k$ are all constants and do not depend on n. Here $f(n)$ depends only on n. When a_k is chosen as one, (1) is said to be in the standard form. If $f_n \neq 0$ for all n, then (1) is said to be non-homogeneous. Otherwise it is said to be homogeneous if $f_n = 0$ for all n. The *order* of the difference equation (1) is the positive integer k which is the greatest difference in the index of non-zero values of y. Equation (1) is linear because each term in (1) is of first degree (linear) in y_n. Thus (1) is a non-homogeneous kth order linear difference equation with constant coefficients.

Difference equation is also referred to as recurrence relation since it expresses y_{n+k} in terms of one or more of the previous terms (of the sequence) namely $y_{n+k-1}, \ldots, y_{n+1}, y_n$. In this case (1) can be written as $y_{n+k} = -a_{k-1} y_{n+k-1} \ldots - a_1 y_{n+1} - a_0 y_n + f(n)$. The difference equation (1) models a physical system. So f_n is known as system input (system excitation or forcing sequence or driving sequence) while y_n is referred to as system output (system response). The structure of the system is defined by the values of the coefficients and order of the equation. Thus any system output depends on the system input and the structure of the system. The general solution of (1) determines the output y_n which depends only on n (but no longer on the prior terms of the sequence) and describes the complete sequence y_n in the closed form. Thus any sequence y_n that satisfies the difference equation (1) is a solution of (1). The solution of (1) can be obtained by (a) classical approach similar to those used for solving linear non-homogeneous differential equations with constant coefficients (b) Laplace transform method (c) z-transform method (d) recursive method (which is a numerical solution yielding a finite number of terms of the sequence and is disadvantageous because it is influenced by the change of initial conditions). Here we consider only the classical approach since theory of difference equations is analogous to that of differential equations (which is considered in chapter 9).

Homogeneous Equations

First order homogeneous difference equation

Consider a first order linear homogeneous difference equation $y_{n+1} - by_n = 0$ for $n \geq 0$ and b is a constant. Assume the solution of the form $y_n = cr^n$ where $c \neq 0, r \neq 0$. Then $y_{n+1} = cr^{n+1}$. Substituting in the given difference equation, we have

$$cr^{n+1} - bcr^n = cr^n(r - b) = 0 \quad \Rightarrow r - b = 0$$

or $b = r$. Thus the general solution of the difference equation $y_{n+1} - by_n = 0$ is given by $y_n = cb^n$. In addition, if a boundary condition $y_0 = d$ then $d = y_0 = cb^0 \quad \therefore c = d$. Then the particular solution is $y_n = db^n$. The solution y_n defines a discrete function whose domain is the set N of all non-negative integers.

Second-order linear homogeneous difference equation with constant coefficients

Consider $a_2 y_{n+2} + a_1 y_{n+1} + a_0 y_n = 0$ (2)

Assume $y_n = cr^n$ (with $c \neq 0, r \neq 0$) (3)

as a solution of (2). Then substituting (3) in (2), we get

$$a_2 cr^{n+2} + a_1 cr^{n+1} + a_0 cr^n = 0$$

or

$$cr^n(a_2 r^2 + a_1 r + a_0) = 0$$

Thus (3) is solution of (2) if

$$a_2 r^2 + a_1 r + a_0 = 0 \tag{4}$$

since $c \neq 0$ and $r \neq 0$. The equation (4) which is a quadratic in r is known as the *characteristic or auxiliary equation* of (2). Three cases arise.

Case 1: When the roots of the characteristic equation are *real and distinct* given by r_1 and r_2 then r_1^n and r_2^n are two linearly independent solutions. Thus the general solution of (2) is

$$\boxed{y_n = c_1 r_1^n + c_2 r_2^n} \tag{5}$$

where c_1, c_2 are two arbitrary constants.

Case 2: If the roots are *real and equal*, say r, then the general solution of (2) is

$$\boxed{y_n = c_1 r^n + c_2 \cdot n \cdot r^n = (c_1 + n \cdot c_2) r^n} \tag{6}$$

Case 3: Suppose the roots of (4) are *complex conjugate* given by $a \pm bi$. Then the general solution of (2) is

$$\boxed{y_n = r^n(c_1 \cos n\theta + c_2 \sin n\theta)} \tag{7}$$

where $r = \sqrt{a^2 + b^2}$, $\tan\theta = \frac{b}{a}$.

This analysis can easily be extended to kth order difference equation by considering the nature of the k roots of the auxilary equation which is a kth degree polynomial. (see Chapter 9).

Note 1: The forward-difference or advancing-difference operator Δ is defined by $\Delta f_k = f_{k+1} - f_k$

Note 2: The shift operator E is defined as the operator that increases the argument of a function by one tabular interval. Thus

$$Ef_k = Ef(x_k) = f(x_k + h) = f(x_{k+1}) = f_{k+1}$$

Note 3: Δ and E are related by $E = 1 + \Delta$.

Note 4: The difference equation

$$a_k y_{n+k} + a_{k-1} y_{n+k-1} + \ldots + a_1 y_{n+1} + a_0 y_n = f(n) \tag{1}$$

can be written in terms of E as follows

$$(a_k E^k + a_{k-1} E^{k-1} + \ldots + a_1 E + a_0) y_n = f(n) \tag{8}$$

Non-homogeneous Equations

The general solution of a non-homogeneous linear difference equation with constant coefficients (1) is the sum of the complementary function and any particular solution. Here the complementary function (C.F.) of (1) is the general solution of the corresponding homogeneous equation (2). Particular solution, more often known as particular integral of (1) can be obtained by (a) method of undetermined coefficients (b) short cut inverse operator methods.

(a) In the method of undetermined coefficients

The particular integral is assumed in a particular form depending on the form of the RHS function f_n.

(b) Inverse operator methods

The non-homogeneous equation (8) can be written as

$$F(E) y_n = f(n) \tag{9}$$

where $F(E) = (a_k E^k + a_{k-1}E^{k-1} + \ldots + a_1 E + a_0)$ (10)

is a function of the operator E.

Then the particular integral is

$$\text{P.I.} = \frac{1}{F(E)} f(n)$$

Case 1: If $f(n) = a^n$ then

$\text{P.I.} = \frac{1}{F(E)} a^n = \frac{1}{F(a)} a^n$, provided $F(a) \neq 0$.

Case 2: Failure case: If $F(a) = 0$, then

$$\text{P.I.} = \frac{1}{(E-a)^3} a^n = \frac{n(n-1)(n-2)}{3!} a^{n-3}$$

Case 3: If $f(n) = \sin \alpha n$ then

$$\text{P.I.} = \frac{1}{F(E)} \sin \alpha n = \frac{1}{F(E)} \left(\frac{e^{i\alpha n} - e^{-i\alpha n}}{2i} \right)$$

$$= \frac{1}{2i} \left[\frac{1}{F(E)} a^n - \frac{1}{F(E)} b^n \right]$$

where $a = e^{i\alpha n}$ and $b = e^{-i\alpha n}$.

Similarly if $f(n) = \cos \alpha n$, replace

$$\cos \alpha_n = \frac{1}{2}(e^{i\alpha n} + e^{-i\alpha n}) = \frac{1}{2}(a^n + b^n)$$

Case 4: If $f(n) = n^m$ or polynomial in n. Replace E by $1 + \Delta$ and expand $1/F(1+\Delta)$ in binomial series in ascending powers of Δ upto Δ^m. Express $f(n)$ in factorials and use $\Delta[x]^n = n[x]^{n-1}$

Case 5: If $f(n) = a^n V(n)$ where $V(n)$ is a polynomial in n. Then

$$\text{P.I.} = \frac{1}{F(E)}\{a^n V(n)\} = a^n \frac{1}{F(aE)} V(n)$$

A. Homogeneous difference equations

Worked Out Examples

First-order difference equation

Example 1: Find the general solution of the first order difference equation $2a_n - 3a_{n-1} = 0$, $n \geq 1$, $a_4 = 81$.

Solution: The general solution of $a_n - \frac{3}{2}a_{n-1} = 0$ or $a_{n+1} - \frac{3}{2}a_n = 0$ for $n \geq 0$ is $a_n = Ad^n = A\left(\frac{3}{2}\right)^n$.

Since $81 = a_4 = A\left(\frac{3}{2}\right)^4$

$\therefore$ $A = 16$. Thus the unique solution is $a_n = 16\left(\frac{3}{2}\right)^n$ for $n \geq 0$.

Finance: Compound interest

Example 2: If Raju invests Rs 1000 at 6% interest compounded quarterly, how many month must he wait for his money to double (Note that Raju can *not* withdraw the money before the quarter is up). How many months it trebles.

Solution: Annual interest rate is 6% so the quarterly rate is $\frac{6\%}{4} = \frac{3}{2}\% = \frac{3}{200} = 0.015$. For $0 \leq n \leq 4$, P_n denotes the value of Raju's deposit at the end of n quarters.

Then $P_{n+1} = P_n + 0.015P_n$ where $0.015P_n$ is the interest earned on P_n during $(n+1)$th quarter. Here $P_0 = 1000$. The solution of the difference equation $P_{n+1} - 1.015P_n = 0$ is $P_n = P_0(1.015)^n = 1000(1.015)^n$. If the money doubles then $P_n = 2P_0 = 2000$. Then $2000 = 1000(1.015)^n$ or $\frac{\ln 2}{\ln(1.015)} = n$ or $n = 46.56 \approx 47$ quarters. So money doubles in $47 \times 3 = 141$ months.

If money trebles then $3000 = 1000(1.015)^n$ so

$$n = \frac{\ln 3}{\ln(1.015)} = 73.80 \approx 74,$$

i.e. in $74 \times 3 = 222$. Thus money trebles in 222 months.

Formation of difference equation

Example 3: Form the differences equation corresponding to the family of curves by eliminating the two arbitrary constants.

(a) $y_n = A3^n + B5^n$ (b) $y(x) = ax + b2^x$

Solution: (a) From $y_n = A3^n + B5^n$ with $n = n+1$ and $n+2$ we get $y_{n+1} = A3^{n+1} + B5^{n+1} = 3A3^n + 5B5^n$ and $y_{n+2} = A3^{n+2} + B5^{n+2} =$

$9A3^n + 25B5^n$. Eliminating A and B, we get

$$\begin{vmatrix} y_n & 1 & 1 \\ y_{n+1} & 3 & 5 \\ y_{n+2} & 9 & 25 \end{vmatrix} = 0.$$

Expanding the determinant

$$y_n(75-45) - 1(25y_{n+1} - 5y_{n+2}) + 1(9y_{n+1} - 3y_{n+2}) = 0$$

or the required differences equation is

$$30y_n - 16y_{n+1} - 2y_{n+2} = 0$$

(b) Rewriting $y_n = a_n + b2^n$, we get

$$\Delta y_n = y_{n+1} - y_n = [a(n+1) + b2^{n+1}] - [an + b2^n]$$

$$\Delta y_n = a + b2^n(2-1) = a + b2^n$$

Now $\Delta^2 y_n = \Delta y_{n+1} - \Delta y_n$

$$= (a + b2^{n+1}) - (a + b2^n)$$

$$= b2^n(2-1) = b2^n$$

Thus $b = \frac{\Delta^2 y_n}{2^n}$.

Substituting b, we get

$$\Delta y_n = a + b2^n = a + \Delta^2 y_n$$

or $\quad a = \Delta y_n - \Delta^2 y_n$

Now eliminating a and b

$$y_n = a_n + b2^n = (\Delta y_n - \Delta^2 y_n)n + \Delta^2 y_n$$

$$y_n = \Delta y_n + (1-n)\Delta^2 y_n$$

since $\Delta^2 y_n = \Delta y_{n+1} - \Delta y_n =$

$$(y_{n+2} - y_{n+1}) - (y_{n+1} + y_n)$$

$$= y_{n+2} - 2y_{n+1} + y_n$$

We get

$$y_n = (y_{n+1} - y_n) + (1-n)(y_{n+2} - 2y_{n+1} + y_n)$$

or $\quad (1-n)y_{n+2} + y_{n+1}(1 - 2 + 2n) + y_n(-2 + 1 - n)$

Thus the difference equation is

$$(1-x)y_{x+2} + (2x-1)y_{x+1} - (x+1)y_x = 0$$

Real, distinct roots

Example 4: Solve $y_{n+2} - 3y_{n+1} - 10y_n = 0$

Solution: The auxiliary equation (A.E.) is obtained by assuming $y_n = cr^n$ and substituting in the given difference equation. Thus $y_{n+1} = cr^{n+1}$, $y_{n+2} = cr^{n+2}$ so

$$cr^{n+2} - 3cr^{n+1} - 10cr^n = 0$$

$$\text{or} \quad r^2 - 3r - 10 = 0$$

is the characteristic equation with distinct real roots $r_1 = 5$ and $r_2 = -2$ (since $r^2 - 3r - 10 = (r-5)(r+2) = 0$). Symbolically the A.E. is obtained by replacing y_n by 1, y_{n+1} by r, y_{n+2} by r^2 (or equivalently E by r and E^2 by r^2).

Then $y_n = 5^n$ and $y_n = (-2)^n$ are two linearly independent solutions because one is *not* a multiple of the other. Thus the general solution of the given difference solution is

$$y_n = c_1 5^n + c_2(-2)^n$$

Equal roots

Example 5: Solve $a_n - 6a_{n-1} + 9a_{n-2} = 0$, $n \geq 2$, $a_0 = 5$, $a_1 = 12$.

Solution: The auxilary equation is $r^2 - 6r + 9 = 0$ or $(r-3)^2$. So the roots are $r = 3, 3$, real equal. The general solution is

$$a_n = c_1 3^n + c_2 \cdot n3^n$$

Since $5 = a_0 = c_1 \cdot 1 + c_2 \cdot 0 \cdot 1 \quad \therefore \quad c_1 = 5$

Since $12 = a_1 = 5 \cdot 3 + c_2 \cdot 1 \cdot 3 \quad \therefore \quad c_2 = -1$

The required (particular) solution satisfying the given two conditions is

$$a_n = 53^n - n3^n = (5-n)3^n$$

Complex conjugate roots

Example 6: Solve $(E^3 - E^2 + E - 1)y = 0$ with $y_0 = 1$, $y_1 = 0$ and $y_2 = 2$.

Solution: Here E is the *shift* operator. So the given equation is $y_{n+3} - y_{n+2} + y_{n-1} - y_n = 0$.

Its auxiliary equation is $m^3 - m^2 + m - 1 = 0$ or $(m-1)(m^2+1) = 0$. So $m = 1$ and $m = \pm i$ are the

three roots. For the complex conjugate roots $a = 0$, $b = 1$, so in the polar form $i = 1 \cdot e^{i\frac{\pi}{2}}$. Here $r = \sqrt{a^2 + b^2} = 1$, $\theta = \tan\frac{b}{a} = \infty \quad \therefore \quad \theta = \pi/2$.
Then the general solution is

$$y_n = c_1 1^n + r^n(c_2 \cos n\theta + c_3 \sin n\theta)$$

$$y_n = c_1 + 1^n\left(c_2 \cos n\frac{\pi}{2} + c_3 \sin n\frac{\pi}{2}\right)$$

Since $1 = y_0 = c_1 + c_2 + c_3 \cdot 0 \quad \therefore c_1 + c_2 = 1$
Since $0 = y_1 = c_1 + c_2 \cdot 0 + c_3 \quad \therefore c_1 + c_3 = 0$
Since $2 = y_2 = c_1 - c_2 + c_3 \cdot 0 \quad \therefore c_1 - c_2 = 2$
Solving $c_1 = \frac{3}{2}$, $c_2 = -\frac{1}{2}$, $c_3 = -\frac{3}{2}$.
Thus the particular solution is

$$y_n = \frac{3}{2} - \frac{1}{2}\cos\frac{n\pi}{2} - \frac{3}{2}\sin\frac{n\pi}{2}$$

Simultaneous equations

Example 7: Solve $x_{n+1} - 7x_n - 10y_n = 0$ and $y_{n+1} - x_n - 4y_n = 0$, with $x_0 = 3$ and $y_0 = 2$.

Solution: Introducing the shift operator E notation, the given two equations can be written as

$$(E - 7)x_n - 10y_n = 0 \tag{1}$$

$$-x_n + (E - 4)y_n = 0 \tag{2}$$

Multiplying (1) by (E-4) and (2) by 10 and adding, we get

$$(E - 7)(E - 4)x_n - 10x_n = 0$$

or $\quad (E^2 - 11E + 28 - 10)x_n$

$$= (E^2 - 11E + 18)x_n = 0.$$

Its auxiliary equation is $r^2 - 11r + 18 = 0$ with two real distinct roots 2 and 9. Then

$$x_n = c_1 2^n + c_2 9^n \tag{3}$$

Using $3 = x_0 = c_1 \cdot 1 + c_2 \cdot 1$, we get $c_1 + c_2 = 3$ (4)
Substituting (3) in (1) we have

$$(c_1 2^{n+1} + c_2 9^{n+1}) - 7(c_1 2^n + c_2 9^n) - 10y_n = 0$$

So $y_n = \frac{1}{10}[-5c_1 2^n + 2c_2 9^n]$

Using $2 = y_0 = \frac{1}{10}[-5 \cdot c_1 \cdot 1 + 2c_2 \cdot 1]$, we get

$$-5c_1 + 2c_2 = 20 \tag{5}$$

Solving (4) and (5) we have $c_1 = -2$, $c_2 = 5$. Thus the required solution is

$$x_n = -2 \cdot 2^n + 5 \cdot 9^n$$

and $y_n = \frac{1}{10}[102^n + 109^n] = 2^n + 9^n$

Determinant

Example 8: Express the following nth order determinant D_n as an explicit function of n. When $a > 2$, show that $D_n = \frac{\sinh(n+1)\mu}{\sinh \mu}$ where $\cos h\mu = \frac{a}{2}$. What is the value of D_n when (i) $a = 2$ (ii) $a = -2$ where

$$D_n = \begin{vmatrix} a & 1 & 0 & . & . & . & 0 & 0 \\ 1 & a & 1 & . & . & . & 0 & 0 \\ 0 & 1 & a & . & . & . & 0 & 0 \\ . & . & . & . & . & . & . & . \\ 0 & 0 & 0 & . & . & . & a & 1 \\ 0 & 0 & 0 & . & . & . & 1 & a \end{vmatrix}$$

Solution: Expanding D_n in terms of the elements in the first row (or column), we get

$$D_n = a \cdot \begin{vmatrix} a & 1 & 0 & \ldots & 0 & 0 \\ 1 & a & 1 & \ldots & 0 & 0 \\ 0 & 1 & a & \ldots & \ldots & \ldots \\ \ldots & \ldots & 0 & \ldots & \ldots & \ldots \\ 0 & 0 & 0 & \ldots & a & 1 \\ 0 & 0 & 0 & \ldots & 1 & a \end{vmatrix} -$$

$$-1 \cdot \begin{vmatrix} 1 & 1 & \ldots & \ldots & 0 & 0 \\ 0 & a & 1 & \ldots & 0 & 0 \\ 0 & 1 & a & 1\ldots & 0 & 0 \\ \ldots & \ldots & \ldots & \ldots & \ldots & \ldots \\ 0 & 0 & 0 & \ldots & a & 1 \\ 0 & 0 & 0 & \ldots & 1 & a \end{vmatrix}$$

The first determinant in RHS is identical in structure to D_n except that it contains only $n - 1$ rows and $n - 1$ columns. Thus the first determinant in RHS is D_{n-1}. The second determinant in RHS does not have the form of D_n. But expanding the second determinant by the first row, we get

$$D_n = a \cdot D_{n-1} - 1 \cdot 1 \begin{vmatrix} a & 1 & 0 & \dots & 0 & 0 \\ 1 & a & 1 & \dots & 0 & 0 \\ 0 & 1 & a & \dots & 0 & 0 \\ \dots & \dots & \dots & \dots & 0 & 0 \\ 0 & 0 & 0 & \dots & a & 1 \\ 0 & 0 & 0 & \dots & 1 & a \end{vmatrix}$$

Now the second determinant in RHS has the same structure as D_n but is of $(n-2)$ order. Thus

$$D_n = aD_{n-1} - D_{n-2}$$

$$\text{or } D_{n+2} - aD_{n+1} + D_n = 0$$

i.e., $(E^2 - aE + 1)D_n = 0$ (1)
The auxiliary equation is

$$m^2 - am + 1 = 0$$

Its roots are given by

$$m = \frac{a \pm \sqrt{a^2-4}}{2} = \frac{a \pm b}{2} \text{ where } b = \sqrt{a^2-4} \quad (2)$$

If $a > 2$ then $b > 0$ so the roots are real, distinct given by

$$m_1 = \frac{a+b}{2}, \; m_2 = \frac{a-b}{2} \quad (3)$$

Then the general solution of the difference equation (1) is

$$D_n = c_1 m_1^n + c_2 m_2^n$$

$$D_n = c_1\left(\frac{a+b}{2}\right)^n + c_2\left(\frac{a-b}{2}\right)^n \quad (4)$$

Thus (4) expresses the nth order determinant D_n as an explicit function of n (in terms of the parameter 'a').

Put $\cosh\mu = \frac{a}{2}$ then $b = a^2 - 4 = \sqrt{4\cosh^2\mu - 4}$ so 'b' $= 2\sinh\mu$. Then $m_1 = \frac{a+b}{2} = \frac{2\cdot\cosh\mu + 2\cdot\sinh\mu}{2} = e^{\mu}$

Similarly $m_2 = \frac{a-b}{2} = \cosh\mu - \sinh\mu = e^{-\mu}$, then the general solution (4) takes the form

$$D_n = c_1(e^{\mu})^n + c_2(e^{-\mu})^n$$

$$= c_1(\cosh\mu n + \sinh\mu n) + c_2(\cosh\mu n - \sinh\mu n)$$

$$D_n = A\cosh\mu n + B\sinh\mu n$$

where $A = c_1 + c_2$ and $B = c_1 - c_2$. Choosing $A = 1$ and $B = \coth\mu$, we get

$$D_n = \cosh\mu n + \coth\mu \cdot \sinh\mu n = \frac{\sinh(n+1)\mu}{\sinh\mu}$$

Now for $n = 1$, $D_1 = a$ and $n = 2$, $D_2 = \begin{vmatrix} a & 1 \\ 1 & a \end{vmatrix} = a^2 - 1$. Thus $D_1 = a = 2\cosh\mu$, $D_2 = 4\cosh^2\mu - 1 = 2\cosh 2\mu$. Using $2\cosh\mu = a = D_1 = A\cosh\mu + B\sinh\mu$. So $A = 2$ and $B = 0$. Using $2\cosh 2\mu = D_2 = 2\cdot\cosh 2\mu + 0\cdot\sinh\cdot 2\mu = 2\cosh 2\mu$ which is true.

Case 1: For $a = 2$, $b = 0$, the roots are real and equal given by $m_1 = m_2 = 1$. Then the general solution is

$$D_n = (c_1 + nc_2)\cdot 1^n = c_1 + nc_2$$

For $n = 1$, $D_1 = a = 2 = c_1 + 1\cdot c_2$.
For $n = 2$, $D_2 = a^2 - 1 = 4 - 1 = 3 = c_1 + 2c_2$
Solving $c_1 = c_2 = 1$. Thus when $a = 2$, $D_n = 1 + n$.

Case 2: When $a = -2$, the general solution is
$D_n = (c_1 + c_2\cdot n)(-1)^n$
For $n = 1$, $D_1 = a = -2 = (c_1 + c_2)(-1)$
For $n = 2$, $D_2 = a^2 - 1 = 3 = c_1 + 2c_2$
Solving $c_1 = c_2 = 1$. Thus when $a = -2$, we get

$$D_n = (1+n)(-1)^n$$

Example 9: A seed of a particular plant produces 8-fold when one year old and produces 18-fold when two or more years old. If a_n denotes the number of seeds produced at the end of the nth year, express a_n in term of n.

Solution: At the end of $(n+1)$th year, the a_n seeds which are one year old produces 8-fold. The a_{n-1} seeds which are two years old produces 18-fold. Similarly the $a_{n-2}, a_{n-3}, \dots a_2, a_1$ which are more than two years old also produces 18 fold. Thus the recurrence relation is

$$a_{n+1} = 8(a_n) + 18(a_{n-1} + a_{n-2} + \dots + a_2 + a_1)$$

Similarly

$$a_{n+2} = 8(a_{n+1}) + 18(a_n + a_{n-1} + \dots + a_3 + a_2 + a_1)$$

Subtracting

$$a_{n+2} - a_{n+1} = 8a_{n+1} + 18a_n - 8a_n$$

or the difference equation is

$$a_{n+2} - 9a_{n+1} - 10a_n = 0.$$

Its characteristic equation is

$$m^2 - 9m - 10 = (m - 10)(m + 1) = 0$$

with real distinct roots $m_1 = 10, m_2 = -1$. The general solution is

$$a_n = c_1(10)^n + c_2(-1)^n$$

Non-homogeneous difference equations

WORKED OUT EXAMPLES

Method of undetermined coefficients

Example 1: Determine a formula for the sum of the first n cubes of natural numbers.

Solution: Let S_n be the partial sum of the cubes of the first n natural numbers. Then

$$S_n = 1^3 + 2^3 + 3^3 + \ldots + n^3 = \sum_{i=1}^{n} i^3$$

Now $S_{n+1} = 1^3 + 2^3 + 3^3 + \ldots + n^3 + (n+1)^3$

$$= \sum_{i=1}^{n+1} i^3$$

Subtracting

$$S_{n+1} - S_n = (n+1)^3 \quad (1)$$

This is a first order non-homogeneous difference equation with auxiliary equation $m - 1 = 0$. So the complementary function is

$$S_c = c \cdot 1^n = c \quad (2)$$

where c is an arbitrary constant. To determine the particular integral, use the method of undetermined coefficients. Since the RHS is a function of the type

$$(n+1)^3 = n^3 + 3n^2 + 3n + 1$$

We assume the particular integral of the form

$$S_n = An^4 + Bn^3 + Cn^2 + Dn \quad (3)$$

(Even if we take an additional constant E, it will become redundant). Substituting (3) in (1), we get

$$S_{n+1} - S_n = [A(n+1)^4 + B(n+1)^3 + C(n+1)^2 + D(n+1)] - [An^4 + Bn^3 + Cn^2 + Dn] = (n+1)^3 \quad (5)$$

The unknown coefficients A, B, C, D will be determined by equating the coefficients of like powers of n on both sides of (5).

$$A(n^4+4n^3+6n^2+4n+1)+B(n^3+3n^2+3n+1)$$
$$+C(n^2+2n+1) - (An^4 + Bn^3 + Cn^2 + Dn)$$
$$= n^3 + 3n^2 + 3n + 1$$

or $(4A)n^3 + (6A + 3B)n^2 + (4A + 3B + 2C)n$

$$+(A + B + C + D) = n^3 + 3n^2 + 3n + 1$$

Then $4A = 1$, $6A + 3B = 3$, $4A + 3B + 2C = 3$, and $A + B + C + D = 1$

Solving $A = \frac{1}{4}, B = \frac{1}{2}, C = \frac{1}{4}, D = 0$. Thus the particular integral is $\frac{1}{4}n^4 + \frac{1}{2}n^3 + \frac{1}{4}n^2$. Then the general solution of the difference equation (1) is given by

$$S_n = \text{C.F.} + \text{P.I.} = c + \frac{1}{4}n^4 + \frac{1}{2}n^3 + \frac{1}{4}n^2 \quad (6)$$

For $n = 1$, we know that $S_1 = 1^3 = 1$. Using this in (6), we get

$$1 = S_1 = c + \frac{1}{4} + \frac{1}{2} + \frac{1}{4} \qquad \therefore\ c = 0$$

Thus the formula expressing the sum of cubes of the first n natural numbers is

$$S_n = \sum_{i=1}^{n} i^3 = 0 + \frac{1}{4}n^4 + \frac{1}{2}n^3 + \frac{1}{4}n^2$$

$$S_n = \frac{n^2(n+1)^2}{4}$$

Short cut inverse operator methods

Power function

Example 2: Solve $a_{n+2} - 6a_{n+1} + 5a_n = 2^n$ with $a_0 = 0, a_1 = 0$.

Solution: The auxiliary equation is $m^2 - 6m + 5 = 0$ with real distinct roots $m_1 = 1, m_2 = 5$. So the complementary function is $c_1 \cdot 1^n + c_2 5^n$. The particular integral is obtained by inverse operator method. The given difference equation is rewritten in terms of the shift operator E as $(E^2-6E+5)a_n = 2^n$. Then the particular integral is

$$\text{P.I.} = \frac{1}{E^2-6E+5}\cdot 2^n = \frac{1}{2^2-6\cdot 2+5}\cdot 2^n$$

$$= -\frac{1}{3}2^n$$

Here E is replaces by 2. Thus the general solution is

$$a_n = \text{C.F.} + \text{P.I.} = c_1 + c_2 5^n - \frac{1}{3}2^n$$

Since $0 = a_0 = c_1 + c_2 - \frac{1}{3}$ $\therefore$ $c_1 + c_2 = \frac{1}{3}$
Since $0 = a_1 = c_1 + 5c_2 - \frac{2}{3}$ $\therefore$ $c_1 + 5c_2 = \frac{2}{3}$
Solving $c_1 = \frac{1}{4}$, $c_2 = \frac{1}{12}$. Then

$$a_n = \frac{1}{4} + \frac{1}{12}5^n - \frac{1}{3}2^n$$

Failure case

Example 3: Solve $y_{n+3} - 12y_{n+2} + 48y_{n+1} - 64y_n = 5\cdot 4^n$.

Solution: A.E. is $m^3 - 12m^2 + 48m - 64 = 0$ or $(m-4)^3 = 0$. The roots are real, equal repeated thrice $m = 4, 4, 4$. So the

$$\text{C.F.} = c_1 4^n + c_2\cdot n\cdot 4^n + c_3 n^2 4^n$$

$$\text{Now P.I.} = \frac{1}{E^3 - 12E^2 + 48E - 64}5\cdot 4^n$$

$$= \frac{5}{(E-4)^3}\cdot 4^n$$

Here E can *not* be replaced by 4. Then by result (case 2 on page 21.4) we have

$$\text{P.I.} = 5\cdot\frac{n(n-1)(n-2)}{3!}4^{n-3}$$

Thus the general solution is

$$y_n = \text{C.F.} + \text{P.I.} = (c_1 + c_2\cdot n + c_3\cdot n^2)4^n +$$

$$+\frac{5n(n-1)(n-2)}{3!}4^{n-3}$$

Trigonometric function

Example 4: Find the general solution of $(E^2+4)y_n = \cos\alpha n$.

Solution: A.E. is $m^2+4=0$ with conjugate complex roots $\pm 2i$. So $a=0$, $b=2$, then $r=\sqrt{a^2+b^2}=2$, and $\tan\theta = \frac{2}{0} = \infty$. $\therefore$ $\theta = \frac{\pi}{2}$. Thus

$$\text{C.F.} = r^n(c_1\cos n\theta + c_2\sin n\theta)$$

$$\text{C.F.} = 2^n\left(c_1\cos\frac{n\pi}{2} + c_2\sin\frac{n\pi}{2}\right)$$

Now P.I. $= \frac{1}{E^2+4}\cos\alpha n = \frac{1}{(E+2i)(E-2i)}\cos\alpha n$.
But $\cos\alpha n = \frac{1}{2}(e^{in\alpha} + e^{-in\alpha})$. So

$$\text{P.I.} = \frac{1}{(E+2i)(E-2i)}\left\{\frac{1}{2}(e^{in\alpha} + e^{-in\alpha})\right\}$$

$$= \frac{1}{(E+2i)(E-2i)}\frac{1}{2}(a^n + b^n)$$

where $a = e^{i\alpha}$, $b = e^{-i\alpha}$

$$= \frac{1}{2}\frac{1}{(a+2i)(a-2i)}a^n + \frac{1}{2}\frac{1}{(b+2i)(b-2i)}b^n$$

$$= \frac{1}{2}\cdot\frac{1}{a^2+4}a^n + \frac{1}{2}\frac{1}{(b^2+4)}b^n$$

$$= \frac{1}{2}\left[\frac{1}{e^{2i\alpha}+4}e^{i\alpha n} + \frac{1}{e^{-2i\alpha}+4}e^{-i\alpha n}\right]$$

$$= \frac{1}{2}\left[\frac{(e^{-2i\alpha}+4)e^{i\alpha n} + (e^{2i\alpha}+4)e^{-i\alpha n}}{(e^{2i\alpha}+4)(e^{-2i\alpha}+4)}\right]$$

$$= \frac{4\cos\alpha n + \frac{1}{2}\{e^{i\alpha(n-2)} + e^{i\alpha(n-2)}\}}{1+4(e^{2i\alpha}+e^{-2i\alpha})+16}$$

$$= \frac{4\cos\alpha n + \cos\alpha(n-2)}{17+8\cos 2\alpha}$$

Thus the general solution is

$$y_n = 2^n\left(c_1\cos\frac{n\pi}{2} + c_2\sin\frac{\pi}{2}\right) +$$

$$+\left(\frac{4\cos\alpha n + \cos\alpha(n-2)}{17+8\cos 2\alpha}\right)$$

Polynomial

Example 5: Solve $a_{n+2} - 5a_{n+1} + 6a_n = 2n^2 - 6n - 1$.

Solution: A.E. is $m^2 - 5m + 6 = 0$ with real distinct roots 2 and 3. So C.F. $= c_1 2^n + c_2 3^n$. The given difference equation in shift oeprator E is

$$(E^2 - 5E + 6)a_n = 2n^2 - 6n - 1$$

So P.I. $= \frac{1}{E^2 - 5E + 6)}(2n^2 - 6n - 1)$

Replace E by $1 + \Delta$ then

$$E^2 - 5E + 6 = (1 + \Delta)^2 - 5(1 + \Delta) + 6$$

$$= 1 + \Delta^2 + 2\Delta - 5 - 5\Delta + 6 = \Delta^2 - 3\Delta + 2$$

Then

$$\text{P.I.} = \frac{1}{\Delta^2 - 3\Delta + 2}(2n^2 - 6n - 1)$$

$$= \frac{1}{2}\frac{1}{1 + \left(\frac{\Delta^2 - 3\Delta}{2}\right)}(2n^2 - 6n - 1)$$

Expanding in binomial series

$$= \frac{1}{2}\left[1 - \left(\frac{\Delta^2 - 3\Delta}{2}\right) + \left(\frac{\Delta^2 - 3\Delta}{2}\right)^2 + \ldots\right] \times$$

$$\times(2n^2 - 6n - 1)$$

Neglecting powers of Δ more than 2, we get

$$\text{P.I.} = \frac{1}{2}\left[1 + \frac{3}{2}\Delta + \frac{7}{4}\Delta^2\right][2n^2 - 2n - 4n - 1]$$

$$= \frac{1}{2}\left[1 + \frac{3}{2}\Delta + \frac{7}{4}\Delta^2\right]\{2[n]^2 - 4[n] - 1\}$$

where $[n]^2 = n(n-1)$. Then

$$\text{P.I.} = \frac{1}{2}\left[\{2[n]^2 - 4[n] - 1\} + \frac{3}{2}\{4[n] - 4\} + \frac{7}{4}\{4\}\right]$$

$$= \frac{1}{2}[\{2n^2 - 6n - 1\} + 6\{(n-1)\} + 7] = n^2$$

Therefore the general solution is

$$a_n = c_1 2^n + c_2 3^n + n^2$$

Exponential shift

Example 6: Solve $(E^2 + E - 56)a_n = 2^n(n^2 - 3)$

Solution: A.E. is $m^2 + m - 56 = (m+8)(m-7) = 0$ with distinct real roots -8, 7. So C.F. $= c_1(-8)^n + c_2(7)^n$.

Since the R.H.S. is of the form $a^n F(n)$, apply shift result for obtaining particular integral. So replace E by $2E$, we get

$$\text{P.I.} = \frac{1}{E^2 + E - 56}\{2^n(n^2 - 3)\}$$

$$= \frac{2^n}{(2E)^2 + 2E - 56}\{(n^2 - 3)\}.$$

$$= \frac{2^n}{2(2E^2 + E - 28)}(n^2 - 3)$$

$$= \frac{2^n}{2[2(1+\Delta)^2 + (1+\Delta) - 28)]}(n^2 - 3)$$

where E is replaced by $1 + \Delta$.

$$\text{P.I.} = \frac{2^n}{2[2\Delta^2 + 5\Delta - 25]}(n^2 - 3)$$

$$= -\frac{2^n}{50}\left[1 - \left(\frac{5\Delta + 2\Delta^2}{25}\right)\right]^{-1}(n^2 - 3)$$

Expanding in binomial series we get

$$\text{P.I.} = -\frac{2^n}{50}\left[1 + \frac{5\Delta + 2\Delta^2}{25} + \left(\frac{5\Delta + 2\Delta^2}{25}\right)^2 + \ldots\right] \times$$

$$\times(n^2 - 3).$$

Neglecting powers of Δ more than 2 we have

$$\text{P.I.} = -\frac{2^n}{50}\left[1 + \frac{5}{25}\Delta + \frac{3}{25}\Delta^2\right]\{[n]^2 + [n] - 3\}$$

where $n^2 - 3 = n^2 - n + n - 3 = n(n-1) + n - 3 = [n]^2 + [n] - 3$.

$$\text{P.I.} = -\frac{2^n}{50}[\{[n]^2 + [n] - 3\} + \frac{5}{25}\{2[n] + 1\} + \frac{3}{25}\{2\}]$$

$$\text{P.I.} = -\frac{2^n}{50}\left[n^2 + \frac{2}{5}n - \frac{64}{25}\right]$$

Therefore the general solution is

$$a_n = c_1(-8)^n + c_2(7)^n - \frac{2^{n-1}}{25}\left(n^2 + \frac{2}{5}n - \frac{64}{25}\right)$$

Homogeneous difference equations

EXERCISE

First order

Solve the difference equation

1. $a_n - 7a_{n-1} = 0$ when $n \geq 1$ and $a_2 = 98$

Ans. $a_n = 2(7)^n$ for $n \geq 0$

2. $a_{n+1}^2 - 5a_n^2 = 0$ when $a_n > 0$ for $n \geq 0$ and $a_0 = 2$. Also find a_{12}.

Ans. $a_n = 2(\sqrt{5})^n$, $a_{12} = 2(\sqrt{5})^{12} = 31,250$

Hint: $a_n^2 = b_n$ and solve for b_n.

3. How much will be the deposit of Rs 1000 a year later, if the bank pays 6% annual interest, compounding the interest monthly.

Ans. Rs 1061.68

Hint: $P_{n+1} = P_n + 0.005P_n$, $P_n = P_0(1.005)^n$, $P_0 = 1000$, $n = 12$

4. How much will be Rs 10,000 after 30 years in a bank yielding 11% per year interest compounded annually.

Ans. Rs 228,922.97

Hint: $P_{n+1} = P_n + 0.11P_n$, $P_n = P_0(1.11)^n$, $P_0 = 10{,}000$, $n = 30$

5. Solve the following difference equations
(a) $a_n - 3a_{n-1} = 0$, $a_0 = 2$

Ans. $a_n = 2 \cdot 3^n$

(b) $a_n - 2a_{n-1} + 1 = 0$, $a_0 = 1$

Ans. $a_n = 1$

(c) $a_n - 2na_{n-1} = 0$, $a_0 = 1$,

Ans. $a_n = 2^n \cdot n!$

Second and higher order

Solve the following difference equations

6. $y_{n+2} + y_{n+1} - 6y_n = 0$ with $a_0 = -1$, $a_1 = 8$

Ans. $y_n = 2^n - 2(-3)^n$

7. $F_{n+2} - F_{n+1} - F_n = 0$ for $n \geq 0$, with $F_0 = 0$, $F_1 = 1$.

Ans. $F_n = \frac{1}{a}\left[\left(\frac{1+a}{2}\right)^n - \left(\frac{1-a}{2}\right)^n\right], n \geq 0, a = \sqrt{5}$

8. $a_n - a_{n-1} + 2a_{n-2} = 0$, $a_0 = 2$, $a_1 = 7$

Ans. $a_n = 3 \cdot 2^n - (-1)^n$

9. $a_n - 6a_{n-1} + 9a_{n-2} = 0$, $a_0 = 1$, $a_1 = 6$

Ans. $a_n = (1+n)3^n$

10. $y_{n+3} - 6y_{n+2} + 11y_{n+1} - 6y_n = 0$ with initial conditions $y_0 = 2$, $y_1 = 5$, $y_2 = 15$.

Ans. $y_n = 1 - 2^n + 2 \cdot 3^n$

11. $y_{n+3} + 3y_{n+2} + 3y_{n+1} + y_n = 0$ with $y_0 = 1$, $y_1 = -2$ and $y_2 = -1$.

Ans. $y_n = (1 + 3n - 2n^2)(-1)^n$.

12. $y_{n+3} - 2y_{n+2} - y_{n+1} + 2y_n = 0$ with $y_0 = 0$, $y_1 = 1$, $y_2 = 1$. Also find y_{10}.

Ans. $y_n = \frac{1}{3}\{2^n - (-1)^n\}$, $y_{10} = 341$

13. $y_{n+2} - 2y_{n+1} + 2y_n = 0$, with $y_0 = 0$, $y_1 = 1$

Ans. $y_n = \frac{(1+i)^n - (1-i)^n}{2i}$

14. $y_{n+3} - 2y_{n+2} - 5y_{n+1} + 6y_n = 0$

Ans. $y_n = c_1 \cdot 1^n + c_2(-2)^n + c_3 3^n$

15. $y_{n+1} - 2y_n \cos\alpha + y_{n-1} = 0$

Ans. $y_n = c_1 \cos n\alpha + c_2 \sin n\alpha$

16. $a_{m+4} + 16a_m = 0$

Ans. $a_m = 2^m\left[c_1 \cos\frac{m\pi}{4} + c_2 \sin\frac{m\pi}{4} + c_3 \cos\frac{3m\pi}{4} + c_4 \cdot \sin\frac{3m\pi}{4}\right]$

17. $(E^2 + 2E + 4)y_n = 0$

Ans. $y_n = 2^n((c_1 \cos\frac{2n\pi}{3} + c_2 \sin\frac{2n\pi}{3})$

18. Determine D_n as a function of n where

$$D_n = \begin{vmatrix} 2 & 2 & 0 & 0 & \ldots & 0 & 0 \\ 1 & 2 & 2 & 0 & \ldots & 0 & 0 \\ 0 & 1 & 2 & 2 & \ldots & 0 & 0 \\ 0 & 0 & 1 & 2 & \ldots & 0 & 0 \\ \ldots & \ldots & \ldots & \ldots & \ldots & \ldots & \ldots \\ 0 & 0 & 0 & 0 & \ldots & 2 & 2 \\ 0 & 0 & 0 & 0 & \ldots & 1 & 2 \end{vmatrix}$$

Ans. $D_n = 2^{n/2}\left(\cos\frac{n\pi}{4} + \sin\frac{n\pi}{4}\right)$

Hint Solve $D_n - 2D_{n-1} + 2D_{n-2} = 0$.

Non-homogeneous difference equations

EXERCISE

Solve

1. Determine a formula for the sum of the squares of the first n natural numbers.

Ans. $S_n = n(n+1)(2n+1)/6$

Hint: Solve $S_{n+1} - S_n = (n+1)^2$, use $S_1 = 1$.

2. $y_n - y_{n-1} = 3n^2, n \geq 1, y_0 = 7$

Ans. $y_n = 7 + \frac{1}{2}n(n+1)(2n+1)$

3. $y_{n+1} - 3y_n = 357^n, y_0 = 2$

Ans. $y_n = \frac{5}{4}7^{n+1} - \frac{1}{4}3^{n+3}$

4. Tower of Hanoi: $a_{n+1} - 2a_n = 1, a_0 = 0$, find a_{64}.

Ans. $a_n = 2^n - 1$,
$a_{64} = 18, 446, 744, 073, 709, 551, 615$

5. A car loan of S rupees is to be paid back in T periods of time. Determine the equi (constant) payment P at the end of each period if r is the interest rate per period for the loan.

Ans. $P = (S \cdot r)/[1 - (1+r)^{-T}]$

Hint: Solve $a_{n+1} - a_n(1+r) = P$, $a_0 = S$, $a_T = 0$ and $0 \leq n \leq T - 1$

6. $(E^2 + 2E - 8)a_n = 5n + 3(2^n)$

Ans. $a_n = c_1 2^n + c_2(-4)^n - n - \frac{4}{5} + n2^{n-2}$

7. $a_{n+2} + 2a_{n+1} + a_n = n + 2, a_0 = 0, a_1 = 0$

Ans. $a_n = \frac{1}{4}(3n-1)(-1)^n + \frac{1}{4}(n+1)$

8. $a_{n+2} - 5a_{n+1} + 6a_n = 2n + 1$, $a_0 = 0$, $a_1 = 1$

Ans. $a_n = \frac{5}{2}3^n - 5 \cdot 2^n + n + \frac{5}{2}$

9. $a_{n+2} + 4a_{n+1} - 5a_n = 24n - 8, a_0 = 3, a_1 = -5$

Ans. $a_n = 2n^2 - 4n + 2 + (-5)^n$

10. $a_{n+2} - 4a_{n+1} + 3a_n = 5^n$

Ans. $a_n = c_1 + c_2 \cdot 3^n + \frac{1}{8}5^n$

11. $(E^2 - 4E + 4)a_n = 2^n$

Ans. $a_n = (c_1 + c_2 \cdot n)2^n + n(n-1)2^{n-3}$

12. $(E^2 - 2\cos\alpha E + 1)a_n = \cos\alpha n$

Ans. $a_n = c_1 \cos\alpha n + c_2 \sin\alpha n + n \cdot \frac{\sin(n-1)\alpha}{2\sin\alpha}$

13. $(E^2 - 4)a_n = n^2 + n - 1$

Ans. $a_n = c_1 2^n + c_2(-2)^n - \frac{n^2}{3} - \frac{7n}{9} - \frac{17}{27}$

14. $(E^2 - 2E + 1)a_n = 2^n \cdot n^2$

Ans. $a_n = c_1 + c_2 \cdot n + 2^n(n^2 - 8n + 20)$

15. $a_n - 7a_{n-1} + 10a_{n-2} = 7 \cdot 3^n$ for $n \geq 2$

Ans. $a_n = c_1 2^n + c_2 5^n + \left(\frac{-7}{2}\right)3^n$

16. $a_n - 6a_{n-1} + 8a_{n-2} = n \cdot 4^n$ with $a_0 = 8$, $a_1 = 22$

Ans. $a_n = 3 \cdot 4^n + 5 \cdot 2^n + n(n-1)4^n$

Simultaneous equations

Solve

17. $x_{n+1} - y_n = 1;\ y_{n+1} - x_n = 1;\ x_0 = 0,\ y_0 = -1$

Ans. $y_n = \frac{1}{2}\left[\sin\frac{n\pi}{2} - \cos\frac{n\pi}{2} - 1\right]$,
$x_n = \frac{1}{2}\left[1 - \cos\frac{n\pi}{2} - \sin\frac{n\pi}{2}\right]$.

18. $x_{n+1} + y_n - 3x_n = n$, $3x_n + y_{n+1} - 5y_n = 4^n$ with $x_1 = 2$, $y_1 = 0$

Ans. $x_n = (1.33)2^n - (0.0167)6^n - 0.8n - 0.76 + 4^{n-1}$,
$y_n = (1.33)2^n - (0.05)6^n - 0.6n - 1.36 - 4^{n-1}$

19. $x_{n+1} - 3x_n - 2y_n = -n$,
$-x_n + y_{n+1} - 2y_n = n$, $x_0 = 0$, $y_0 = 3$

Ans. $x_n = 2 \cdot 4^n - 2 - \frac{1}{2}n(n-1)$,
$y_n = 4^n + 2 + \frac{1}{2}n(n-1)$

21.2 Z-TRANSFORMS

Introduction

Z-transform is useful in solving difference equations which represent a discrete system. Z-transform operates on a sequence u_n of discrete integer-valued arguments $n = 0, \pm 1, \pm 2, \ldots$ unlike the Laplace transform which operates on continuous functions.

Thus Z-transform is the discrete analogue of Laplace transform. Therefore for every operational rule and application of Laplace transform, there corresponds an operational rule and application of Z-transform.

Definition

The Z-transform of a sequence u_n defined for discrete values $n = 0, 1, 2, \cdots$ (and $u_n = 0$ for $n < 0$) is denoted by $Z(u_n)$ and is defined as

$$Z(u_n) = \sum_{n=0}^{\infty} u_n z^{-n} = U(z) \qquad (1)$$

where U is a function of z.

Z-transforms exists only when the infinite series in (1) is convergent.

Inverse Z-transform

is denoted by $Z^{-1}(U(z)) = u_n$ determines the sequence u_n which generates the given Z-transform.

Results

Z-transform of some standard sequences

1. $U_n = \{a^n\} = \{1, a, a^2, a^3, \ldots, a^n, \ldots\}$

By definition,

$$Z(u_n) = Z(a^n) = \sum u_n z^{-n} \sum_{n=0}^{\infty} a^n z^{-n}$$

$$= 1 + \frac{a}{z} + \frac{a}{z^2} + \frac{a}{z^3} + \cdots + \frac{a^n}{z^n} + \cdots = \frac{1}{1 - \frac{a}{z}}$$

$$Z(a^n) = \frac{z}{z-a}.$$

Recurrence formula

2. $u_n = \{n^p\} = \{0, 1^p, 2^p, 3^p, \ldots, n^p, \ldots\}$ where p is a positive integer

By definition

$$Z(n^p) = \sum_{n=0}^{\infty} n^p z^{-n} \qquad \text{(i)}$$

Replacing p by $p-1$

$$Z(n^{p-1}) = \sum_{n=0}^{\infty} n^{p-1} z^{-n} \qquad \text{(ii)}$$

Differentiating (ii) w.r.t. z, we get

$$\frac{d}{dz}\left[Z(n^{p-1})\right] = \frac{d}{dz}\left[\sum_{n=0}^{\infty} n^{p-1} z^{-n}\right]$$

$$= \sum_{n=0}^{\infty} n^{p-1} \cdot (-n) z^{-n-1}$$

$$= -z^{-1} \sum_{n=0}^{\infty} n^p \cdot z^{-n}$$

$$= -z^{-1} Z(n^p), \text{ using (i)}$$

Thus $$Z(n^p) = -z \frac{d}{dz}\left[Z(n^{p-1})\right]$$

Special cases of result 1.

3. $Z(1) = \frac{z}{z-1}$ obtained by putting $a = 1$ in result 1.

4. $Z(k) = \frac{kz}{z-1}$ (where k is a constant) since $z(k) = \sum kz^{-n} = k\frac{1}{1-\frac{1}{z}} = k\frac{z}{z-1}$

5. $Z\left[(-1)^n\right] = \frac{z}{z+1}$ obtained by taking $a = -1$ in result 1.

Special cases of result 2.

6. $Z(n) = \frac{z}{(z-1)^2}$

since with $p = 1$, $Z(n) = -z \cdot \frac{d}{dz}[Z(1)] = -z\frac{d}{dz}\left(\frac{z}{z-1}\right) = -z \cdot \left[\frac{(z-1)\cdot 1 - z \cdot 1}{(z-1)^2}\right] = \frac{z}{(z-1)^2}$

7. $Z(n^2) = \frac{z^2+z}{(z-1)^3}$ by taking $p = 2$ in result 2.

8. $Z(n^3) = \frac{z^3+4z^2+z}{(z-1)^4}$

9. $Z(n^4) = \frac{z^4+11z^3+11z^2+z}{(z-1)^5}$

10. $Z\left(\frac{1}{n!}\right) = \sum_{n=0}^{\infty} \frac{1}{n!} z^{-n} = 1 + \frac{1}{z} + \frac{1}{2!}\frac{1}{z^2} + \frac{1}{3!}\frac{1}{z^3} + \cdots + \frac{1}{n!}\frac{1}{z^n} + \cdots = e^{\left(\frac{1}{z}\right)}$

11. Unit step sequence $u(n) = \begin{cases} 0, & \text{for } n < 0 \\ 1, & \text{for } n \geq 0 \end{cases}$

$$Z(u(n)) = \sum_{n=0}^{\infty} u(n) z^{-n} = \sum_{n=0}^{\infty} 1 \cdot z^{-n}$$

$$= 1 + \frac{1}{z} + \frac{1}{z^2} + \cdots + \frac{1}{z^n} + \cdots$$

$$Z[u(n)] = \frac{1}{1-\frac{1}{z}} = \frac{z}{z-1}$$

12. Unit impulse sequence $\delta(n) = \begin{cases} 1, & \text{for } n = 0 \\ 0, & \text{for } n \neq 0 \end{cases}$

$$Z[\delta(n)] = \sum_{n=0}^{\infty} \delta(n) z^{-n} = 1 + 0 + 0 + 0 \cdots = 1.$$

Properties

I. Linearity

$Z(au_n + bv_n) = aZ(u_n) + bZ(v_n)$

since $Z(au_n + bv_n) = \sum (au_n + bv_n) z^{-n}$

$$= a \sum u_n z^{-n} + b \sum v_n z^{-n}.$$

II. Change of scale (or damping rule)

If $Z(u_n) = U(z)$ then $Z(a^{-n}u_n) = U(az)$ since

$$Z(a^{-n}u_n) = \sum_{n=0}^{\infty} a^{-n} u_n z^{-n} = \sum_{n=0}^{\infty} u_n (az)^{-n} = U(az)$$

Similarly,

$$Z(a^n u_n) = U(z/a)$$

Results from application of damping rule:

13. $Z(na^n) = \frac{az}{(z-a)^2}$

because $Z(n) = \frac{z}{(z-1)^2}$ then $Z(na^n)$ is obtained by replacing z by z/a. So $Z(na^n) = \frac{z/a}{(z/a-1)^2} = \frac{az}{(z-a)^2}$

14. $Z(n^2 a^n) = \frac{az^2 + a^2 z}{(z-a)^3}$

since $Z(n^2) = \frac{z^2+z}{(z-1)^3}$ by damping rule. $Z(n^2a^n)$ is obtained by replacing z by z/a

i.e., $$Z(n^2 a^n) = \frac{(z/a)^2 + (z/a)}{[(z/a) - 1]^3} = \frac{a(z^2 + az)}{(z-a)^3}$$

15. $Z(\cos n\theta) = \frac{z(z - \cos\theta)}{z^2 - 2z\cos\theta + 1}$

Solution:

$$Z(e^{-in\theta}) = Z\left[(e^{-i\theta})^n\right] = Z\left[(e^{-i\theta})^n \cdot 1\right]$$

since $Z(1) = \frac{z}{z-1}$, apply damping rule and replace z by z/a where $a = e^{-i\theta}$ i.e., z by $z/e^{-i\theta}$ or $ze^{i\theta}$. Thus

$$Z\left[(e^{-i\theta})^n \cdot 1\right] = \left.\frac{z}{z-1}\right|_{ze^{i\theta}} = \frac{ze^{i\theta}}{ze^{i\theta} - 1}$$

$$= \frac{z}{z - e^{-i\theta}} = \frac{z(z - e^{i\theta})}{(z - e^{-i\theta})(z - e^{i\theta})}$$

$$= \frac{z[z - \cos\theta - i\sin\theta]}{z^2 - z(e^{i\theta} + e^{-i\theta}) + 1}$$

Thus

$$Z(e^{-in\theta}) = Z(\cos n\theta - i\sin n\theta)$$

$$= \frac{z(z - \cos\theta) - iz\sin\theta}{z^2 - 2z\cos\theta + 1}$$

Since $\quad \cos\theta = (e^{i\theta} + e^{-i\theta})/2$

Applying linearity property and comparing the real parts on both sides, we get

$$Z(\cos n\theta) = \frac{z(z - \cos\theta)}{z^2 - 2z\cos\theta + 1}$$

Similarly, comparing the imaginary parts

16. $Z(\sin n\theta) = \frac{z\sin\theta}{z^2 - 2z\cos\theta + 1}$

17. $Z(a^n \cos n\theta) = \frac{z(z - a\cos\theta)}{z^2 - 2az\cos\theta + a^2}$ obtained by replacing z by z/a in (15) by damping rule.

Similarly,

18. $Z(a^n \sin n\theta) = \frac{az\sin\theta}{z^2 - 2az\cos\theta + a^2}$

III. Shifting property

a. Shifting u_n to the right

If $Z(u_n) = U(z)$ then $Z(u_{n-k}) = z^{-k}U(z)$ for $n \geq k$, k positive integer.

This follows from the definition

$$Z(u_{n-k}) = \sum_{n=0}^{\infty} u_{n-k} z^{-n}$$

$$= u_0 z^{-k} + u_1 z^{-(k+1)} + u_2 z^{-(k+2)} + \cdots$$

since $u_m = 0$ for $m < 0$ (so $u_{-k}, u_{1-k}, u_{2-k}, \ldots, u_{k-1-k}$ are all 0)

put $n - k = m$, then

$$Z\{u_{n-k}\} = \sum_{m=0}^{\infty} u_m z^{-k-m} = z^{-k} \sum_{m=0}^{\infty} u_m z^{-m}$$

$$= z^{-k} \sum_{n=0}^{\infty} u_n z^{-n} = z^{-k} U(z)$$

b. Shifting to the left

If $Z(u_n) = U(z)$ then

$$Z(u_{n+k}) = z^k\left[U(z) - u_0 - \frac{u_1}{z} - \frac{u_2}{z^2}\cdots\frac{u_{k-1}}{z^{k-1}}\right]$$

Proof:

$$Z(u_{n+k}) = \sum_{n=0}^{\infty} u_{n+k}z^{-n} = z^k\sum_{n=0}^{\infty} u_{n+k}z^{-(n+k)}$$

$$= z^k\left[u_k z^{-k} + u_{1+k}z^{-(1+k)} + u_{2+k}z^{-(2+k)} + \cdots\right]$$

$$= z^k\left[(u_0 + u_1 z^{-1} + u_2 z^{-2} + \cdots + u_{k-1}z^{-(k-1)}) + \right.$$

$$\left. + u_k z^{-k} + u_{k+1}z^{-(k+1)+\cdots}\right] -$$

$$-z^k\left[u_0 + u_1 z^{-1} + u_2 z^{-2} + \cdots + u_{k-1}z^{-(k-1)}\right]$$

$$Z(u_{n+k}) = z^k\left[\sum_{n=0}^{\infty} u_n z^{-n} - \sum_{n=0}^{k-1} u_n z^{-n}\right]$$

$$Z(u_{n+k}) = z^k\left[U(z) - u_0 - u_1 z^{-1} - u_2 z^{-2}\cdots\right.$$

$$\left. - u_{k-1}z^{-(k-1)}\right]$$

In particular for $k = 1, 2, 3$

19. $Z(u_{n+1}) = z[U(z) - u_0]$

20. $Z(u_{n+2}) = z^2[U(z) - u_0 - u_1 z^{-1}]$

21. $Z(u_{n+3}) = z^3[U(z) - u_0 - u_1 z^{-1} - u_2 z^{-2}]$.

IV. Multiplication by n

If $Z(u_n) = U(z)$ then $Z(nu_n) = -z\frac{dU}{dz}(z)$.

Proof:

$$Z(nu_n) = \sum_{n=0}^{\infty} n\cdot u_n z^{-n} = -z\sum_{n=0}^{\infty} u_n(-n)z^{-n-1}$$

$$= -z\sum_{n=0}^{\infty} u_n\frac{d}{dz}(z^{-n}) = -z\sum\frac{d}{dz}(u_n z^{-n})$$

$$= -z\frac{d}{dz}\left(\sum u_n z^{-n}\right) = -z\cdot\frac{d}{dz}U(z)$$

In general (by mathematical induction)

$$Z(n^p u_n) = (-z)^p\frac{d^p U(z)}{dz^p}$$

Initial value and final value theorems determine the values of u_n for $n = 0$ and for $n \to \infty$ without the complete knowledge of u_n.

V. Initial value theorem

Theorem: *If $Z(u_n) = U(z)$ then $u_0 = \lim_{z\to\infty} U(z)$*

Proof: By definition

$$U(z) = Z(u_n) = u_0 + \frac{u_1}{z} + \frac{u_2}{z^2} + \cdots + \frac{u_n}{z^n} + \cdots$$

Taking the limit as $z \to \infty$

22. $\lim_{z\to\infty} U(z) = u_0 + 0 + 0\cdots = u_0$

23. $\lim_{z\to\infty} z[U(z) - u_0] =$

$\lim_{z\to\infty}\left[u_1 + \frac{u_2}{z} + \frac{u_3}{z^2} + \cdots\right] = u_1$

24. $\lim_{z\to\infty} z^2\left[u(z) - u_0 - \frac{u_1}{z}\right] =$

$\lim_{z\to\infty}\left[u_2 + \frac{u_3}{z} + \frac{u_4}{z^2} + \cdots\right] = u_2.$

VI. Final value theorem

Theorem: *If $Z(u_n) = U(z)$ then*

$$\lim_{n\to\infty} u_n = \lim_{z\to 1}\{(z-1)U(z).\}$$

Proof: Consider

$$Z(u_{n+1} - u_n) = Z(u_{n+1}) - Z(u_n)$$
$$= Z[U(z) - u_0] - U(z) \quad \text{using (19)}$$
$$= U(z)\cdot(z-1) - u_0 z$$

From definition and the above result, we have

$$(z-1)U(z) - u_0 z = Z(u_{n+1} - u_n) = \sum_{n=0}^{\infty}(u_{n+1} - u_n)z^{-n}$$

As $z \to 1$, we have

$$\lim_{z\to 1}[(z-1)U(z)] - u_0$$

$$= \lim_{z\to 1}\sum_{n=0}^{\infty}(u_{n+1} - u_n)z^{-n} = \sum_{n=0}^{\infty}(u_{n+1} - u_n)$$

$$= \lim_{n\to\infty}[(u_1 - u_0) + (u_2 - u_1) + \cdots + (u_{n+1} - u_n)]$$

$$= \lim_{n\to\infty}[u_{n+1} - u_0] = \lim_{n\to\infty} u_{n+1} - u_0$$

Thus

$$\lim_{n\to\infty} u_n = \lim_{z\to 1}[(z-1)U(z)].$$

VII. Convolution theorem

Theorem: *If $Z^{-1}[U(z)] = u_n$ and $Z^{-1}[V(z)] = v_n$ then*

$$Z^{-1}[U(z)\cdot V(z)] = u_n * v_n = \text{convolution of } u_n \text{ and } v_n = \sum_{m=0}^{n} u_m v_{n-m}$$

Proof:

$$U(z)\cdot V(z) = \left[\sum_{n=0}^{\infty} u_n z^{-n}\right]\left[\sum_{n=0}^{\infty} v_n z^{-n}\right]$$

$$= \left[u_0 + \frac{u_1}{z} + \frac{u_2}{z^2} + \cdots + \frac{u_n}{z^n} + \cdots\right] \times$$

$$\times \left[v_0 + \frac{v_1}{z} + \frac{v_2}{z^2} + \cdots + \frac{v_n}{z^n} \cdots\right]$$

Collecting the coefficients of z^{-n}

$$= \sum_{n=0}^{\infty} (u_0 v_n + u_1 v_{n-1} + u_2 v_{n-2} + \cdots + u_n v_0) z^{-n}$$

$$= Z(u_0 v_n + u_1 v_{n-1} + \cdots + u_n v_0) \quad \text{by definition}$$

$$= Z\left(\sum_{m=0}^{n} u_m v_{n-m}\right)$$

Taking inverse Z-transform the result follows.

VIII. Region of convergence (R.O.C.)

The region of convergence of Z-transform is the region in the z-plane where the infinite series convergence absolutely.

Thus the region of convergence of a one-sided Z-transform of a right-sided sequence i.e.,

$$U(z) = \sum_{n=0}^{\infty} u_n z^{-n}$$

is $|z| > a$ i.e., the exterior of the circle with centre at origin and of radius a.

Similarly the R.O.C. of

$$U(z) = \sum_{n=-\infty}^{0} u_n z^{-n}$$

is $|z| < a$. Finally the R.O.C. of two-sided Z-transform defined by

$$U(z) = \sum_{n=-\infty}^{\infty} u_n z^{-n}$$

is the annulus region $a < |z| < b$.

IX. Solution of difference equations

Step 1. Take Z-transform on both sides of the given difference equation.

Step 2. Use given conditions and solve for $U(z)$.

Step 3. Apply partial fractions method.

Step 4. Take inverse Z-transform on both sides which results in the given sequence.

Note: A list of standard Z-transforms is presented on page 21.30.

WORKED OUT EXAMPLES

Linearity property

Example 1: Find the Z-transform of $2n + 5\sin\frac{n\pi}{4} - 3a^4$.

Solution: By linearity property

$$Z\left(2n + 5\sin\frac{n\pi}{4} - 3a^4\right)$$

$$= 2Z(n) + 5Z\left(\sin\frac{n\pi}{4}\right) - 3a^4 Z(1)$$

$$= \frac{2\cdot z}{(z-1)^2} + \frac{5\cdot z\cdot \sin\frac{\pi}{4}}{z^2 - 2z\cdot\cos\frac{\pi}{4} + 1} - 3a^4\frac{z}{z-1}$$

$$= \frac{2z}{(z-1)^2} + \frac{5(z/\sqrt{2})}{z^2 - \sqrt{2}z + 1} - 3a^4\frac{z}{z-1}.$$

Example 2: Find $Z(\cos\theta + i\sin\theta)^n$.

Solution: We know that $(\cos\theta + i\sin\theta) = e^{i\theta}$

$$Z(\cos\theta + i\sin\theta)^n = Z\left[(e^{i\theta})^n\right] = \frac{z}{z - e^{i\theta}}$$

since $\quad Z(a^n) = \dfrac{z}{z-a}.$

Example 3: Find $Z\left[(n+1)^2\right]$.

Solution:

$$Z(n+1)^2 = Z(n^2 + 2n + 1) = Z(n^2) + 2Z(n) + Z(1)$$

$$= \frac{z^2 + z}{(z-1)^3} + \frac{2\cdot z}{(z-1)^2} + \frac{z}{z-1}$$

$$= \frac{(z^2+z) + 2z(z-1) + z(z-1)^2}{(z-1)^3} = \frac{z^3+z^2}{(z-1)^3}$$

Damping rule

Example 4: Find $Z(e^{-an} \cdot \sin n\theta)$.

Solution: We know that $Z(\sin n\theta) = \frac{z \sin\theta}{z^2 - 2z\cos\theta + 1}$

Now $Z(e^{-an} \cdot \sin n\theta) = Z\left[(e^a)^{-n} \sin n\theta\right]$

Applying damping rule $Z[a^{-n}u_n] = U(az)$

Replace z by az i.e., $e^a z$ then

$$z[e^{-an} \sin n\theta] = \left.\frac{z \sin\theta}{z^2 - 2z\cos\theta + 1}\right|_{z=e^a z}$$

$$= \frac{e^a z \cdot \sin\theta}{e^{2a}z^2 - 2e^a z\cos\theta + 1}.$$

Multiplication by n

Example 5: Find $Z(n\cos n\theta)$.

Solution: We know that $Z(nu_n) = -z\frac{d}{dz}(U(z))$

Here $u_n = \cos n\theta$

and $Z(u_n) = U(z) = \dfrac{z(z - \cos\theta)}{z^2 - 2z\cos\theta + 1}$

so $Z(n\cos n\theta) = -z \cdot \dfrac{d}{dz}\left\{\dfrac{z(z-\cos\theta)}{z^2 - 2z\cos\theta + 1}\right\}$

$$Z(n\cos n\theta) = -z \cdot \frac{[-z^2\cos\theta + 2z - \cos\theta]}{(z^2 - 2z\cos\theta + 1)^2}$$

$$= \frac{z^3\cos\theta - 2z^2 + z\cos\theta}{(z^2 - 2z\cos\theta + 1)^2}.$$

Shift and initial value theorems

Example 6: Find $Z(u_{n+2})$ if $Z(u_n) = \frac{z}{z-1} + \frac{z}{z^2+1}$.

Solution:

Here $Z(u_n) = U(z) = \frac{z}{z-1} + \frac{z}{z^2+1}$

To find u_0, u_1 by initial value theorem

$$u_0 = \lim_{z\to\infty} U(z) = \lim_{z\to\infty}\left[\frac{z}{z-1} + \frac{z}{z^2+1}\right] = 1$$

$$u_1 = \lim_{z\to\infty} \{z\,[U(z) - u_0]\}$$

$$= \lim_{z\to\infty} z \cdot \frac{2z^2 - z + 1}{(z-1)(z^2+1)} = 2$$

From shift to left we know that

$$Z(u_{n+2}) = z^2\left[U(z) - u_0 - u_1 z^{-1}\right]$$

$$Z(u_{n+2}) = z^2\left[\frac{z}{z-1} + \frac{z}{z^2+1} - 1 - \frac{2}{z}\right]$$

$$= z^3\left[\frac{1}{z-1} + \frac{1}{z^2+1} - \frac{(z+2)}{z^2}\right]$$

$$Z(u_{n+2}) = \frac{z(z^2 - z + 2)}{(z-1)(z^2+1)}.$$

Inverse transform by partial fractions

Example 7: Find the inverse Z-transform of $\frac{4z^2 - 2z}{z^3 - 5z^2 + 8z - 4}$.

Solution: Consider

$$\frac{4z^2 - 2z}{z^3 - 5z^2 + 8z - 4} = \frac{2z(2z-1)}{(z-1)(z-2)^2}$$

By partial fractions

$$\frac{2z - 1}{(z-1)(z-2)^2} = \frac{A}{z-1} + \frac{B}{z-2} + \frac{C}{(z-2)^2}$$

$$(2z - 1) = A(z-2)^2 + B(z-1)(z-2) + C(z-1)$$

when $z = 2$, $3 = C$; when $z = 1$, $A = 1$ and when $z = 0$

$$-1 = 4A + 2B - C \qquad \therefore \quad B = -1$$

Thus

$$Z^{-1}\left[\frac{4z^2 - 2z}{(z-1)(z-2)^2}\right]$$

$$= Z^{-1}\left[2z \cdot \frac{1}{z-1}\right] + Z^{-1}\left[2z\frac{-1}{z-2}\right]$$

$$+ Z^{-1}\left[2z \cdot \frac{3}{(z-2)^2}\right]$$

$$= 2 \cdot 1 - 2 \cdot 2^n + 3z^{-1}\left[\frac{2z}{(z-2)^2}\right]$$

$$= 2 - 2^{n+1} + 3n \cdot 2^n$$

since $Z(na^n) = \dfrac{az}{(z-a)^2}$

Region of convergence

Example 8: Find $Z^{-1}\{(z-5)^{-3}\}$ when $|z| > 5$. Determine the region of convergence.

Solution: Consider

$$(z-5)^{-3} = \frac{1}{(z-5)^3} = \frac{1}{z^3}\frac{1}{\left(1-\frac{5}{z}\right)^3}$$

Expanding by Binomial series which is valid when $|\frac{5}{z}| < 1$ or $|z| > 5$, we have

$$\frac{1}{(z-5)^3} = \frac{1}{z^3}\left[1 + 3\left(\frac{5}{z}\right) + \frac{3\cdot 4}{1\cdot 2}\cdot\left(\frac{5}{z}\right)^2 + \frac{3\cdot 4\cdot 5}{1\cdot 2\cdot 3}\cdot\left(\frac{5}{z}\right)^3 + \frac{3\cdot 4\cdot 5\cdot 6}{1\cdot 2\cdot 3\cdot 4}\left(\frac{5}{z}\right)^4 + \cdots\right]$$

$$\frac{1}{(z-5)^3} = \frac{1}{2}\cdot\frac{1}{z^3}\left[1\cdot 2\left(\frac{5}{z}\right)^0 + 2\cdot 3\left(\frac{5}{z}\right)^1 + 3\cdot 4\left(\frac{5}{z}\right)^2 + 4\cdot 5\left(\frac{5}{z}\right)^3 + 5\cdot 6\left(\frac{5}{z}\right)^4 + \cdots\right]$$

$$= \frac{1}{2}\sum_{m=0}^{\infty}(m+1)(m+2)5^m\cdot z^{-m-3}; \text{ put } m+3=n$$

$$U(z) = (z-5)^{-3} = \frac{1}{2}\sum_{n=3}^{\infty}(n-1)(n-2)5^{n-3}z^{-n}$$

$$= \sum_{n=0}^{\infty} u_n z^{-n}$$

Taking inverse Z-transform

$$Z^{-1}[U(z)] = Z^{-1}\left[(z-5)^{-3}\right] = u_n$$

$$= \frac{1}{2}(n-1)(n-2)5^{n-3} \quad \text{for } n \geq 3$$

$$= 0 \quad \text{for } n < 3$$

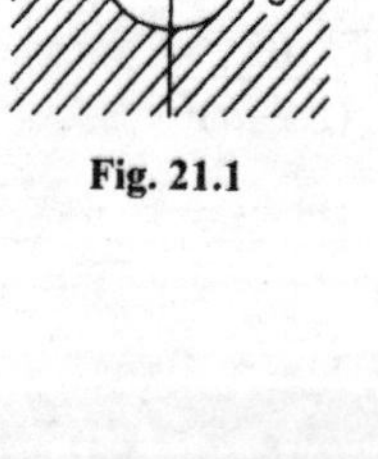

Fig. 21.1

The region of convergence is the exterior of the circle $|z| = 5$ i.e., with centre at origin and of radius 5 (Fig. 21.1).

Convolution

Example 9: Using convolution theorem, find the inverse Z-transform of $\left(\frac{z}{z-a}\right)^3$. Deduce for $\left(\frac{z}{z-1}\right)^3$.

Solution: We know that

$$Z^{-1}\left\{\frac{z}{z-a}\right\} = a^n,\ Z^{-1}\left\{\frac{z}{z-a}\right\} = a^n$$

So

$$Z^{-1}\left\{\frac{z^2}{(z-a)^2}\right\}$$

$$= Z^{-1}\left\{\frac{z}{z-a}\cdot\frac{z}{z-a}\right\} = a^n * a^n$$

by convolution theorem

$$= \sum_{m=0}^{n} a^m a^{n-m} = a^n\sum a^m\cdot a^{-m} = a^n\sum_{m=0}^{n} 1$$

$$= a^n\cdot[1+1+1+\cdots+1] = a^n(n+1).$$

Applying convolution theorem again

$$Z^{-1}\left\{\frac{z^3}{(z-a)^3}\right\} = Z^{-1}\left\{\frac{z^2}{(z-a)^2}\cdot\frac{z}{z-a}\right\}$$

$$= \left[a^n\cdot(n+1)\right] * a^n$$

$$= \sum_{m=0}^{n} a^m\cdot(m+1)\cdot a^{n-m}$$

$$= a^n\sum_{m=0}^{n}(m+1)$$

$$Z^{-1}\left[\left(\frac{z}{z-a}\right)^3\right] = a^n\cdot[1+2+3+\cdots+(n+1)]$$

$$= a^n\cdot\frac{1}{2}\cdot(n+1)(n+2)$$

Put $a = 1$, then

$$Z^{-1}\left\{\frac{z}{(z-1)^3}\right\} = \frac{1}{2}(n+1)(n+2).$$

Solution of difference equations

Example 10: Using Z-transform, solve the difference equation $u_{n+2} - 4u_{n+1} + 3u_n = 5^n$.

Solution: Take Z-transform on both sides of the difference equation

$$Z(u_{n+2}) - 4Z(u_{n+1}) + 3Z(u_n) = Z(5^n)$$

Denote $Z(u_n) = U(z)$, then

$$Z(u_{n+2}) = z^2\left[U(z) - u_0 - u_1 z^{-1}\right]$$

$$Z(u_{n+1}) = z[U(z) - u_0]$$

Substituting these values

$$z^2\left[U(z)-u_0-u_1z^{-1}\right]-4z\,[U(z)-u_0]+3U(z)=\frac{z}{z-5}$$

$$U(z)\left[z^2-4z+3\right]-u_0\left[z^2-4z\right]-u_1 z=\frac{z}{z-5}$$

Solving

$$\frac{U(z)}{z}=\frac{1}{(z-5)(z-1)(z-3)}+\frac{u_0(z-4)+u_1}{(z-1)(z-3)}$$

By partial fractions

$$\frac{U(z)}{z}=\left[\frac{A}{(z-1)}+\frac{B}{(z-3)}+\frac{C}{(z-5)}\right]+\left[\frac{D}{(z-1)}+\frac{E}{z-3}\right]$$

$$U(z)=\left[\frac{1}{8}\frac{z}{z-1}-\frac{1}{4}\frac{z}{z-3}+\frac{1}{8}\frac{z}{z-5}\right]+\left[\frac{(u_1-3u_0)z}{2(z-1)}+\frac{3z(u_0-u_1)}{2(z-3)}\right]$$

Taking inverse Z-transform on both sides

$$u_n=Z^{-1}[U(z)]=\frac{1}{8}\cdot 1-\frac{1}{4}3^n+\frac{1}{8}5^n+\left(\frac{u_1-3u_0}{2}\right)1+\frac{3(u_0-u_1)}{2}3^n$$

$$u_n=C_1+C_2 3^n+\frac{1}{8}5^n \quad \text{is the solution}$$

Here $\quad C_1=\frac{1}{8}+\frac{u_1-3u_0}{2},\ C_2=\frac{3(u_0-u_1)}{2}-\frac{1}{4}.$

Example 11: Solve $u_{n+2}+2u_{n+1}+u_n=n$ with $u_0=u_1=0$

Solution: Taking Z-transform on both sides

$$Z(u_{n+2})+2Z(u_{n+1})+Z(u_n)=Z(n)$$

$$z^2\left[U(z)-u_0-u_1z^{-1}\right]+2z[U(z)-u_0]+U(z)=\frac{z}{(z-1)^2}$$

Putting $u_0=u_1=0$ and solving

$$U(z)=\frac{z}{(z-1)^2(z+1)^2}$$

By partial fractions

$$\frac{1}{(z-1)^2(z+1)^2}=\frac{A}{(z-1)}+\frac{B}{(z-1)^2}+\frac{C}{(z+1)}+\frac{D}{(z+1)^2}$$

we get $A=-\frac{1}{4}$, $B=C=D=\frac{1}{4}$ so

$$U(z)=\frac{1}{4}\left[-\frac{z}{z-1}+\frac{z}{(z-1)^2}+\frac{z}{z+1}+\frac{z}{(z+1)^2}\right]$$

Taking inverse Z-transform on either side

$$u_n=\frac{1}{4}\left[-1^n+n+(-1)^n-n(-1)^n\right]$$

$$u_n=\left(\frac{n-1}{4}\right)\left[1-(-1)^n\right].$$

EXERCISE

Find the Z-transform:

1. $\frac{1}{(n+2)!}$ *Ans.* $z^2(e^{\frac{1}{z}}-1-z^{-1})$
2. $\cosh n\theta$ *Ans.* $\frac{z(z-\cosh\theta)}{z^2-2z\cosh\theta+1}$
3. $a^n\cdot\sinh n\theta$

   Hint: Use damping rule for $z(\sinh n\theta)=\frac{z\sinh\theta}{z^2-2z\cosh\theta+1}$

   Ans. $\frac{az\cdot\sinh\theta}{z^2-2az\cdot\cosh\theta+a^2}$
4. $a^n e^{-a}/n!$ *Ans.* $e^{a(z^{-1}-1)}$
5. $\sin(n+1)\theta$ *Ans.* $\frac{z^2\sin\theta}{z^2-2z\cos\theta+1}$
6. $\frac{1}{n+1}$ *Ans.* $z\ln\left(\frac{z}{z+1}\right)$
7. Find u_2, u_3 if $U(z)=(2z^2+5z+14)/(z-1)^4$.

   Ans. $u_0=0,\ u_1=0,\ u_2=2,\ u_3=13$
8. Find u_2, u_3 when $U(z)=(5z^2+3z+12)/(z-1)^4$.

   Ans. $u_0=0,\ u_1=0,\ u_2=5,\ u_3=23$
9. Determine u_2 where $U(z)=(2z^2+3z+4)/(z-3)^3$

Ans. $u_0 = 0, u_1 = 2, u_2 = 21$

10. Use convolution theorem to find the inverse Z-transform of $z^2/[(z-a)(z-b)]$.

Ans. $(a^{n+1} - b^{n+1})/(a-b)$

11. By convolution, evaluate $Z^{-1}[z^2/(z-a)^2]$

Ans. $(n+1)a^n$

Find the inverse Z-transform of

12. $\frac{2z^2+3z}{(z+2)(z-4)}$

Ans. $\frac{1}{6}(-2)^n + \frac{11}{6}4^n$

13. $\frac{z^3-20z}{(z-2)^3(z-4)}$

Ans. $\frac{1}{2}(2^n + 2 \cdot n^2 2^n) - 4^n$

14. $\frac{z}{z^2+11z+24}$

Ans. $\frac{1}{5}\,[(-3)^n - (-8)^n]$

15. $\frac{z}{z^3-z^2+z-1}$

Ans. $\frac{1}{2}\left[1 - \cos\left(\frac{n\pi}{2}\right) - \sin\left(\frac{n\pi}{2}\right)\right]$

16. $\frac{z}{(z+3)^2(z-2)}$

Ans. $-\frac{1}{25}(-3)^n - \frac{1}{5}n(-3)^n + \frac{1}{25}2^n$

17. $\frac{2(z^2-5z+6.5)}{(z-2)(z-3)^2}$ for $2 < |z| < 3$

Ans. $u_n = 2^{n-1}$, for $n \geq 1$

$u_n = -(n+2)3^{n-2}$, for $n \leq 0$

18. $\frac{1}{(z-2)(z-3)}$ for

i. $|z| < 2$

ii. $2 < |z| < 3$

iii. $|z| > 3$

Ans. **i.** $-\left(\frac{1}{3} + \frac{z}{3^2} + \frac{z^2}{3^3} + \frac{z^3}{3^4} + \cdots\right.$

$\left. + \left(\frac{1}{2} + \frac{z}{2^2} + \frac{z^2}{2^3} + \frac{z^3}{2^4} + \cdots\right)\right)$

ii. (-2^{n-1}) for $n > 0$

iii. $3^{n-1} - 2^{n-1}, n \geq 1, 0$ for $n < 0$

Using Z-transform solve the difference equation:

19. $u_{n+2} + 4u_{n+1} + 3u_n = 3^n$ with $u_0 = 0, u_1 = 1$

Ans. $u_n = \frac{3}{8}(-1)^n + \frac{1}{24}3^n - \frac{5}{12}(-3)^n$

20. $u_{n+2} + 6u_{n+1} + 9u_n = 2^n$ with $u_0 = u_1 = 0$

Ans. $u_n = \frac{1}{25}\left[2^n - (-3)^n + \frac{5}{3}n(-3)^n\right]$

21. $u_{n+2} - 5u_{n+1} + 6u_n = y_n$ with $u_0 = 0, u_1 = 1$ where $y(n) = 1$ for $n = 0, 1, 2, 3, \ldots$

Ans. $u_n = \frac{1}{2} - 2(2)^n + \frac{3}{2}(3)^n$

22. $u_n + \frac{1}{4}u_{n-1} = y_n + \frac{1}{3}y_{n-1}$ where y_n is a unit step sequence.

Ans. $\frac{1}{12}\left(-\frac{1}{4}\right)^{n-1}$

23. $u_{n+2} - 2u_{n+1} + u_n = 3n + 5$

Ans. $\frac{1}{2}n(n-1)(n+3) + C_0 + C_1 n$ where $C_0 = u_0, C_1 = u_1 - u_0$

24. $4u_n - u_{n+2} = 0$ with $u_0 = 0, u_1 = 2$

Ans. $u_n = (-2)^{n-1} + 2^{n-1}$

25. $u_{n+2} - 2u_{n+1} + u_n = 2^n$ with $u_0 = 2, u_1 = 1$

Ans. $u_n = 2^n + 1 - 2n$.

21.3 STANDARD Z-TRANSFORMS

	Sequence u_n (for $n \geq 0$)	$U(z)$: Z-transform
1.	n	$z/(z-1)^2$
2.	n^2	$(z^2+z)/(z-1)^3$
3.	n^p (p is positive integer)	$-z\frac{d}{dz}\left[Z(n^{p-1})\right]$
4.	a^n	$z/(z-a)$
5.	nu_n	$-z\frac{d}{dz}[U(z)]$
6.	$a^n u_n$	$U(z/a)$
7.	u_{n+1}	$z[U(z) - u_0]$
8.	u_{n+2}	$z^2\left[U(z) - u_0 - u_1 z^{-1}\right]$
9.	u_{n+3}	$z^2\left[U(z) - u_0 - u_1 z^{-1} - u_2 z^{-2}\right]$
10.	u_{n-k}	$z^{-k}U(z)$
11.	u_0	$\lim_{z\to\infty} U(z)$
12.	$\lim_{n\to\infty} u_n$	$\lim_{z\to 1}[(z-1)U(z)]$
13.	$\sin n\theta$	$\frac{z\sin\theta}{z^2-2z\cos\theta+1}$
14.	$\cos n\theta$	$\frac{z(z-\cos\theta)}{z^2-2z\cos\theta+1}$

HIGHER ENGINEERING MATHEMATICS

PART—VI

COMPLEX ANALYSIS

Chapter 22

Complex Function Theory

INTRODUCTION

Complex-valued functions or simply complex functions are functions which produce complex numbers from complex numbers. Complex analysis or complex function theory is the study of complex analytic functions. It is an elegant and powerful method useful in the study of heat flow, fluid dynamics and electrostatics. Two-dimensional potential problem can be solved using analytic functions since the real and imaginary part of an analytic function are solutions of two-dimensional Laplace's equation.

22.1 COMPLEX FUNCTION

Basic Definitions

Circle

$|z - z_0| = \rho$ represents a circle with centre at the point z_0 and of radius ρ.

Neighbourhood

Neighbourhood of a point z_0 is the set of all points z for which $|z - z_0| < \delta$ where δ is a positive constant (Fig. 22.1).

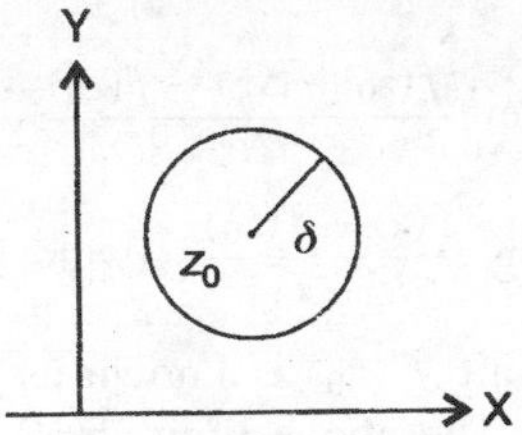

Fig. 22.1

Deleted neighbourhood of z_0 is $0 < |z - z_0| < \delta$.
Annulus of z_0 is : $\rho_1 < |z - z_0| < \rho_2$.

Interior point z_0 of a set S

If there exists some neighbourhood of z_0 which contains only points of S.

Boundary point z_0 of S

If every neighbourhood of z_0 contains both points in S and points not in S.

Connected set

If any two points of the set S are joined by a polygonal line, all the points of which lie in S.

Domain

A set S is said to be a domain if every point of S is an interior point and connected.

Boundary of a domain is the collection of all boundary points of S.

Region

is a domain together with some of its boundary points.

Closed region

is a region together with the boundary (all boundary points included).

Bounded region

is bounded if it can be enclosed in a circle of finite radius.

Examples:

1. $|z| \leq 1$ is closed bounded region.

2. Im $(z^2) > 0$ i.e., $xy > 0$ is open unbounded, unconnected region.

Complex variable is denoted by $z = x + iy$ where x and y are real variables.

Complex Function of a Complex Variable z

If for every z in a set S, a unique value w is associated, then w is said to be a function of z and is denoted by

$$w = f(z)$$

S is known as the domain of definition of f. Range of f is the totality (set of) all values of $f(z)$ corresponding to z in S.

Since w is complex, it is written as

$$w = f(z) = u(x, y) + iv(x, y)$$

Here $u(x, y)$ and $v(x, y)$, real valued functions of x and y, are known as the real and imaginary parts of the functions w(or $f(z)$).

Example:

$$\begin{aligned} f(z) &= 2z^2 - 3iz = 2(x + iy)^2 - 3i(x + iy) \\ &= 2(x^2 - y^2) + 4ixy - 3ix + 3y \end{aligned}$$

So $\quad u(x, y) = 2x^2 - 2y^2 + 3y, \quad v = 4xy - 3x$

Multi-valued function f is one in which for every z more than one value of w is associated. **We consider only single-valued functions.**

Polynomial in z of degree n (zero or positive integer) is the function

$$f(z) = P_n(z) = a_0 + a_1 z + \ldots + a_n z^n$$

where $a_0, a_1, \ldots, a_n$ are complex constants. Rational fractional function is the quotient of polynomials $P(z)/Q(z)$ defined for all z except those for which $Q(z) = 0$.

Limit: L is said to be the limit of $f(z)$ as z approaches z_0 and is denoted by

$$\lim_{z \to z_0} = f(z) = L$$

if for every $\epsilon > 0$, there exists $\delta > 0$ such that $|f(z) - L| < \epsilon$ whenever $|z - z_0| < \delta$.

Here z may approach z_0 from any direction.

22.2 CONTINUITY

A function $f(z)$ is said to be continuous at a point z_0 if $f(z_0)$ exists, $\lim_{z \to z_0} f(z)$ exists and

$$\lim_{z \to z_0} f(z) = f(z_0)$$

i.e., limiting value of $f(z)$ as z approaches z_0 coincides with the value $f(z_0)$.

A function is said to be continuous in a domain if it is continuous at every point of the domain.

A function which is not continuous at z_0($f(z_0)$ does not exist, or $\lim_{z \to z_0} f(z)$ does not exist or $\lim_{z \to z_0} f(z) \neq f(z_0)$) is known as discontinuous at z_0.

Result 1: If $f(z)$ and $g(z)$ are continuous functions in D, then their sum $f + g$, difference $f - g$, product fg, quotient f/g are all continuous in D. Continuous function of a continuous function is continuous.

Result 2: $f = u + iv$ is continuous if both u and v are continuous.

22.3 DIFFERENTIABILITY

A function $f(z)$ is said to be differentiable at a point z_0 if the limit

$$\begin{aligned} f'(z_0) &= \lim_{\Delta z \to 0} \frac{f(z_0 + \Delta z) - f(z_0)}{\Delta z} \\ &= \lim_{z \to z_0} \frac{f(z) - f(z_0)}{z - z_0} \quad (\text{with } z = z_0 + \Delta z) \end{aligned}$$

exists. The limit $f'(z_0)$ is known as the derivative of $f(z)$ at z_0. The above limit should be the same along any path from z to z_0. Thus differentiability of a complex function is a severe requirement.

Differentiation rules

of real calculus are valid in complex differentiation also.

1. $\frac{dc}{dz} = 0,\quad$ where c = complex constant
2. $\frac{d}{dz}[f \pm g] = \frac{df}{dz} \pm \frac{dg}{dz}$
3. $\frac{d}{dz}[cf(z)] = c\frac{df}{dz}$
4. $\frac{d}{dz}[f \cdot g] = f\frac{dg}{dz} + \frac{df}{dz} \cdot g$
5. $\frac{d}{dz}[\frac{f}{g}] = \frac{g\frac{df}{dz} - f\frac{dg}{dz}}{g^2}$
6. a. $\frac{d}{dz}[f(z)]^n = n[f(z)]^{n-1}\frac{df}{dz}$

   b. $\frac{d}{dz}z^n = nz^{n-1}$
7. Chain rule $\frac{dw}{dz} = \frac{dw}{d\zeta} \cdot \frac{d\zeta}{dz}$ if $w = f(\zeta)$, and $\zeta = g(z)$.

22.4 ANALYTICITY

A function $f(z)$ is said to be **analytic** at a point z_0 if f is differentiable not only at z_0 but at every point of some neighbourhood of z_0.

A function $f(z)$ is analytic in a domain if it is analytic at every point of the domain.

An analytic function is also known as "holomorphic", "regular", "monogenic".

Entire Function

A function which is analytic everywhere (for all z in the complex plane) is known as entire function (refer Fig. 22.2).

Example: Polynomials, rational functions are entire.

Example: $|z|^2$ is differentiable only at $z = 0$. So it is nowhere analytic.

Thus analyticity is a very stringent condition.

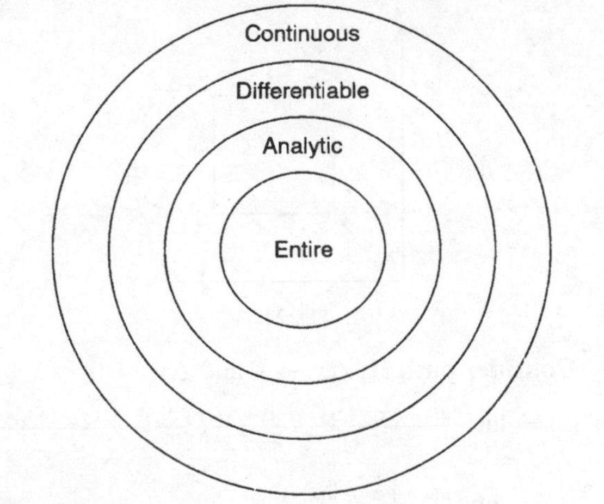

Fig. 22.2

22.5 CAUCHY-RIEMANN (C-R) EQUATIONS: IN CARTESIAN COORDINATES

Cauchy-Riemann equations (or conditions) are used to determine whether a complex function is analytic or not.

Theorem: *If $f(z) = u(x, y) + iv(x, y)$ is differentiable at z then at this point the first order partial derivatives of u and v exist and satisfy the Cauchy-Riemann equations*

$$\frac{\partial u}{\partial x} = \frac{\partial v}{\partial y} \text{ and } \frac{\partial u}{\partial y} = -\frac{\partial v}{\partial x} \qquad (1)$$

Proof: By hypothesis, f is differentiable, so f' exists i.e.,

$$f'(z) = \lim_{\Delta \to 0} \frac{f(z + \Delta z) - f(z)}{\Delta z} \qquad (2)$$

Since the limit in (2) exists, the limit value along any two paths must be equal, which results in the Cauchy-Riemann Equations (1) (see Fig. 22.3).

Consider path I: $\Delta y \to 0$ and $\Delta x \to 0$

$$f'(z) = \lim_{\substack{\Delta y \to 0 \\ \Delta x \to 0}} \frac{u(x+\Delta x, y+\Delta y) + iv(x+\Delta x, y+\Delta y) - u(x,y) - iv(x,y)}{\Delta x + i\Delta y}$$

$$= \lim_{\Delta x \to 0} \frac{u(x+\Delta x, y) - u(x,y)}{\Delta x} + $$

$$+ i \lim_{\Delta x \to 0} \frac{v(x+\Delta x, y) - v(x,y)}{\Delta x} = u_x + iv_x \qquad (3)$$

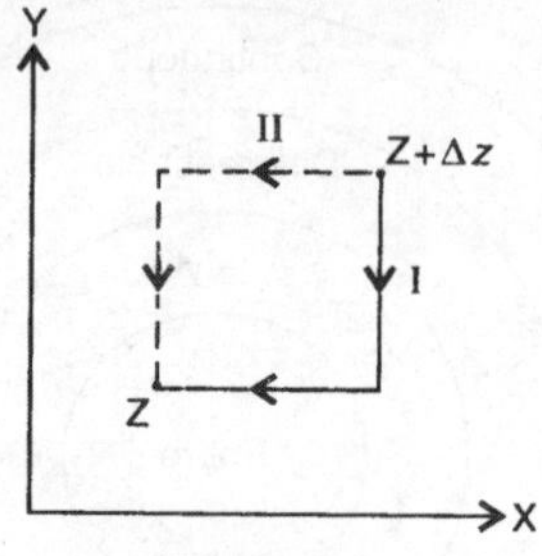

Fig. 22.3

Consider path II: $\Delta x \to 0$ and $\Delta y \to 0$

$$f'(z) = \lim_{\substack{\Delta x \to 0 \\ \Delta y \to 0}} \frac{u(x+\Delta x, y+\Delta y)+iv(x+\Delta x, y+\Delta y)-u(x,y)-iv(x,y)}{\Delta x + i\Delta y}$$

$$= \lim_{\Delta y \to 0} \frac{u(x,y+\Delta y)-u(x,y)}{i\Delta y} +$$

$$+ \lim_{\Delta y \to 0} \frac{v(x,y+\Delta y)-v(x,y)}{\Delta y}$$

$$f' = \tfrac{1}{i} u_y + v_y = -iu_y + v_y \qquad \because \tfrac{1}{i} = -i \qquad (4)$$

Equating the limit values (3) and (4) of f' along path I and II, we get

$$u_x + iv_x = f' = -iu_y + v_y$$

Thus $u_x = v_y$ and $u_y = -v_x$.

Corollary 1: If f is analytic in a domain D, then u, v satisfy C-R conditions at all points in D.

Corollary 2: Derivative f' can be calculated using (3) or (4).

Corollary 3: C-R conditions are necessary but not sufficient.

Corollary 4: C-R conditions are sufficient if the partial derivatives are continuous i.e., if $u(x, y)$, $v(x, y)$ have continuous first partial derivatives and satisfy C-R conditions then $f = u + iv$ is analytic.

$$f \text{ analytic} \Rightarrow \text{C-R conditions}$$
$$+$$
$$\text{analyticity} \Leftarrow \text{Continuous P.D.}$$

Properties of analytic functions

1. If $f(z)$ and $g(z)$ are analytic, then $f \pm g$, fg, f/g are analytic if $g(z) \neq 0$.
2. Analytic function of an analytic function is analytic.
3. An entire function of an entire function is entire.
4. If f is analytic, then it is continuous (analyticity $\Rightarrow$ differentiability $\Rightarrow$ continuity.
5. Derivative of an analytic function is itself analytic. ($f' = u_x + iv_x = U + iV$.
   f analytic, so $u_x = v_y$, $u_y = -v_x$, differentiating w.r.t x and y, $u_{xx} = v_{yx}$, $u_{yy} = -v_{xy}$ or $U_x = V_y$ and $U_y = -V_x$ i.e., U, V satisfy C-R conditions. Hence f' is analytic).
6. If $f = u + iv$ is analytic, then the family of curves $u(x, y) = c_1$ and $v(x, y) = c_2$ are mutually orthogonal i.e., $u = c_1$ are orthogonal trajectories of $v = c_2$ and vice versa.
   (By implicit differentiation of $u = c_1$, we get
   $$\frac{\partial u}{\partial x} + \frac{\partial u}{\partial y} \cdot \frac{dy}{dx} = 0 \quad \text{or} \quad \frac{dy}{dx} = \frac{-ux}{uy}$$
   Similarly $v_x + v_y \frac{dy}{dx} = 0$ or $\frac{dy}{dx} = \frac{-v_x}{v_y}$.
   Product of slopes $= \frac{-ux}{uy} \cdot \left(\frac{-v_x}{v_y}\right) = -1$ by C-R conditions.)

22.6 HARMONIC AND CONJUGATE HARMONIC FUNCTIONS

Harmonic Function

A function $A(x, y)$ is said to be a harmonic function if it satisfies the Laplace's equation i.e., $\nabla^2 A = 0$.

Theorem: *The real and imaginary parts of an analytic function are harmonic.*

Proof: Let $f(z) = u(x, y) + iv(x, y)$ be analytic. So C-R conditions $u_x = v_y$, $u_y = -v_x$ are satisfied. Differentiating partially w.r.t., x and y, we get

$$u_{xx} = v_{yx}$$
$$u_{yy} = -v_{xy}$$

Adding $\nabla^2 u = u_{xx} + u_{yy} = v_{yx} - v_{xy} = 0$

Thus u is a solution of Laplace's equation. Hence u is a harmonic function.

Similarly, differentiating w.r.t., y and x

$$u_{xy} = v_{yy}$$
$$-u_{yx} = v_{xx}$$

Adding $\nabla^2 v = v_{xx} + v_{yy} = u_{xy} - u_{yx} = 0$.
so v is harmonic.
Complex form of Laplace's equation $\frac{\partial^2 u}{\partial \bar{z} \partial z} = 0$
Laplacian operator $\frac{\partial^2}{\partial x^2} + \frac{\partial^2}{\partial y^2} = 4 \frac{\partial^2}{\partial \bar{z} \partial z}$.

Conjugate Harmonic Function

The real part u of an analytic function $f = u + iv$ is known as the conjugate harmonic function of v and vice versa (i.e., v is the conjugate harmonic of u). Conjugate of a given harmonic function is uniquely determined upto an arbitrarily real additive constant.

Note: Adjective conjugate here is not to be confused with conjugate $\bar{z} = x - iy$.

22.7 CAUCHY-RIEMANN EQUATIONS: IN POLAR COORDINATES

Let $x = r \cos\theta$, $y = r \sin\theta$. Then

$$z = x + iy = r\cos\theta + ir\sin\theta = re^{i\theta}.$$

So $u + iv = f(z) = f(r\,e^{i\theta})$

Differentiating partially w.r.t., r and θ,

$$\frac{\partial u}{\partial r} + i\frac{\partial u}{\partial r} = f'(r\,e^{i\theta}) \cdot e^{i\theta}$$

$$\frac{\partial u}{\partial \theta} + i\frac{\partial v}{\partial \theta} = f'(r\,e^{i\theta}) \cdot ire^{i\theta}$$

$$= ir\left(\frac{\partial u}{\partial r} + i\frac{\partial v}{\partial r}\right)$$

Equating the real and imaginary parts

$$\frac{\partial u}{\partial r} = \frac{1}{r}\frac{\partial v}{\partial \theta} \quad \text{and} \quad \frac{\partial v}{\partial r} = -\frac{1}{r}\frac{\partial u}{\partial \theta}$$

The derivative f' can be calculated using

$$f' = e^{-i\theta}(u_r + iv_r)$$

or

$$f' = \frac{-i}{re^{i\theta}}(u_\theta + iv_\theta)$$

Milne Thompson Method

I. If u is given, take $f' = u_x - iu_y$
If v is given, take $f' = v_y + iv_x$

II. Replace x by z and y by 0 in f'.

III. Integrate f' w.r.t., z.

WORKED OUT EXAMPLES

Limit: Stringent Condition

Example 1: Show that $\lim_{z\to 0} \frac{x^2 y}{x^4+y^2}$ does not exist even though this function approaches the same limit along every straight line through the origin.

Solution:
Path I.

$$\lim_{\substack{x\to 0\\ y\to 0}} \frac{x^2 y}{x^4 + y^2} = \lim_{y\to 0} 0 = 0$$

Path II.

$$\lim_{\substack{y\to 0\\ x\to 0}} \frac{x^2 y}{x^4 + y^2} = \lim_{x\to 0} 0 = 0$$

Path III. along any straight line through origin. Let $y = mx$.

$$\lim_{\substack{y=mx\\ x\to 0}} \frac{x^2 y}{x^4 + y^2} = \lim_{x\to 0} \frac{mx^3}{x^4 + m^2x^2} = \lim_{x\to 0} \frac{mx}{x^2 + m^2} = 0$$

choose **path IV** as $y = mx^2$, then

$$\lim_{\substack{y=mx^2\\ x\to 0}} \frac{x^2 y}{x^4 + y^2} = \lim_{x\to 0} \frac{m \cdot x^4}{x^4 + m^2x^4}$$

$$= \lim_{x\to 0} \frac{m}{1 + m^2} = \frac{m}{1 + m^2} \neq 0$$

and different for different values of m. Therefore the limit does not exist.

Continuity

Example 2: Determine where the given function is continuous (a) $\frac{1}{1+z^2}$ (b) $\frac{1}{z-1}$ inside a unit circle. How about in the complex plane.

Solution: $\frac{1}{1+z^2}$ is continuous everywhere except where $1 + z^2 = 0$ i.e., at $z = \pm i$. When unit circle is considered, $|z| < 1$, $z = \pm i$ are excluded. Thus $\frac{1}{1+z^2}$ is continuous inside $|z| = 1$.
Similarly, $\frac{1}{z-1}$ is also continuous inside $|z| = 1$. If the entire complex plane is considered, both $\frac{1}{1+z^2}$ and $\frac{1}{z-1}$ are discontinuous, at $z = \pm i$ and $z = 1$ respectively.

Example 3: Is $f(z) = z/|z|$ continuous at origin (defined for $z \neq 0$ and $f(0) = 0$).

Solution: $\lim_{z\to 0} f(z) = \lim_{z\to 0} \frac{x+iy}{\sqrt{x^2+y^2}}$

Path I.

$$\lim_{\substack{x\to 0\\ y\to 0}} \frac{x+iy}{\sqrt{x^2+y^2}} = i$$

Path II.

$$\lim_{\substack{y\to 0\\ x\to 0}} \frac{x+iy}{\sqrt{x^2+y^2}} = 1$$

since limit does not exist, f is discontinuous at $z_0 = 0$.

Example 4: Determine where the function

$$f(z) = \begin{cases} \frac{z^2+3iz-2}{z+i}, & \text{for } z \neq -i \\ 5, & \text{for } z = -i. \end{cases}$$

is continuous? Can the function be refined to make it continuous at $z = -i$?

Solution: $f(z) = \frac{g(z)}{h(z)}$ is continuous when $g(z)$ and $h(z)$ are continuous except at $h(z) = 0$. So $f(z)$ is continuous everywhere except at $z = -i$, since $g(z)$, $h(z)$ are continuous.

$$\lim_{z\to -i} f(z) = \lim_{\substack{x\to 0\\ y\to -1}} \frac{(x+iy)^2 + 3i(x+iy)-2}{x+iy+i}$$

$$= \lim_{y\to -1} \frac{-y^2-3y-2}{i(y+1)}$$

$$= \lim_{y\to -1} \frac{-2y-3}{i} = -\frac{1}{i} = i$$

Also $\lim_{z\to -i} f(z) = \lim_{\substack{y\to -1\\ x\to 0}} = \lim_{x\to 0} \frac{2(x-i)+3i}{1} = i$

Thus $\lim_{z\to -i} f(z) = i \neq 5 = f(-i)$

Hence f is not continuous at $z = -i$.
Suppose we redefine $f(z)$ as follows:

i.e., $f(-i) = i$ (instead of 5).

Then $f(z)$ is continuous at $z = i$ and is therefore continuous everywhere. $z = i$ is known as **removable discontinuity**.

Differentiability/Derivative

Example 5: Show that $f(z) = \text{Re}\, z = x$ is continuous but not differentiable.

Solution: Continuity: For any point z,

$$\lim_{z\to z_0} f(z) = \lim_{z\to z_0} x = \lim_{\substack{x\to x_0\\ y\to y_0}} x = x_0 = f(z_0)$$

Not differentiable: For any point z,

$$f'(z) = \lim_{\Delta z\to 0} \frac{f(z+\Delta z) - f(z)}{\Delta z}$$

$$= \lim_{\Delta z\to 0} \frac{x+\Delta x - x}{\Delta x + i\Delta y}$$

$$= \lim_{\substack{\Delta x\to 0\\ \Delta y\to 0}} \frac{\Delta x}{\Delta x + i\Delta y} = 0$$

while $$= \lim_{\substack{\Delta y\to 0\\ \Delta x\to 0}} \frac{\Delta x}{\Delta x + i\Delta y} = 1.$$

So limit does not exist i.e., f is not differentiable.

Example 6: Find the derivative from (a) definition (b) differentiation rules of $f(z) = 3z^2 + 4iz - 5 + i$ at $z = 2$.

Solution:

a. *From definition:*

$$f'(z) = \lim_{\Delta z\to 0} \frac{3(z+\Delta z)^2 + 4i(z+\Delta z) - 5 + i - 3z^2 - 4iz + 5 - i}{\Delta z}$$

$$= \lim_{\Delta z\to 0} \frac{3\Delta z^2 + 6z\Delta z + 4i\Delta z}{\Delta z}$$

$$= \lim_{\Delta z\to 0} (3\Delta z + 6z + 4i)$$

$$= 6z + 4i = 12 + 4i \text{ at } z = 2.$$

b. *From rules of differentiation:*

$$\frac{df}{dz} = \frac{d}{dz}(3z^2 + 4iz - 5 + i)$$

$$= 3\cdot 2z + 4i \cdot 1 + 0\Big|_{\text{at } z=2} = 12 + 4i.$$

C-R Equation/condition: Cartesian coordinates

Example 7: Prove that $\frac{d}{dz}(z^2\bar{z})$ does not exist anywhere.

Solution:

a. *From definition, for any z:*

$$\frac{d}{dz}(z^2\overline{z}) = \lim_{\Delta z\to 0}\frac{(z+\Delta z)^2(\overline{z+\Delta z}) - z^2\overline{z}}{\Delta z}$$

$$= \lim_{\Delta z\to 0}\frac{\overline{z}\Delta z^2+2z\overline{z}\Delta z+z^2\Delta\overline{z}+\Delta z^2\Delta\overline{z}+2z\Delta z\Delta\overline{z}}{\Delta z}$$

$$= \lim_{\Delta z\to 0}\left[0+2z\overline{z}+z^2\frac{\Delta\overline{z}}{\Delta z}+0+0\right]$$

$$= \lim_{\substack{\Delta x\to 0\\ \Delta y\to 0}}\left[2z\overline{z}+z^2\frac{\Delta\overline{z}}{\Delta z}\right] = 2z\overline{z}+iz^2$$

$$= \lim_{\substack{\Delta y\to 0\\ \Delta x\to 0}}\left[2z\overline{z}+z^2\frac{\Delta\overline{z}}{\Delta z}\right] = 2z\overline{z}+z^2$$

So limit does not exist. Nowhere differentiable.

b. *From Cauchy-Riemann conditions:*

$$z^2\overline{z} = (x+iy)^2(x-iy) = \left[x(x^2-y^2)+2xy^2\right]$$

$$+i\left[2x^2y+y(y^2-x^2)\right]$$

So $u = x(x^2-y^2)+2xy^2$, $v = 2x^2y+y(y^2-x^2)$

$u_x = 3x^2-y^2+2y^2$, $u_y = -2xy+4xy$

$v_x = 4xy-2xy$, $v_y = 2x^2+3y^2-x^2$

C-R conditions are not satisfied for any x, y so $f' = u_x+iv_x = v_y-iu_y$ does not exist for any x, y i.e., for any z.

Example 8: Show that every differentiable function is continuous (converse is not true i.e., a function may be continuous but not differentiable).

Solution: Let $f(z)$ be differentiable at z_0. Then

$$f'(z_0) = \lim_{z\to z_0}\frac{f(z)-f(z_0)}{z-z_0} \quad \text{exists.}$$

Therefore $f(z_0)$ is well defined.
Consider

$$\lim_{z\to z_0} f(z)-f(z_0) = \lim_{z\to z_0}\frac{f(z)-f(z_0)}{z-z_0}\cdot z-z_0$$

$$= \lim_{z\to z_0}\frac{f(z)-f(z_0)}{z-z_0}\lim_{z\to z_0}(z-z_0)$$

$$= f'(z_0)\cdot\lim_{z\to z_0}(z-z_0) = 0$$

Thus $\lim_{z\to z_0} f(z) = \lim_{z\to z_0} f(z_0) = f(z_0)$.

Therefore $f(z)$ is continuous at z_0.

Counter Example: (a) $f(z) = \overline{z}$, (b) $f(z) = |z|^2$, (c) $f(z) = \text{Im}\, z$ are continuous but not differentiable at (a) any point (b) at zero (c) any point.

Analyticity

Example 9: Determine where the Cauchy-Riemann equations are satisfied for the given functions. Determine the region of analyticity.

a. $f(z) = e^z = e^x(\cos y + i\sin y)$

Solution:

$$u(x,y) = e^x\cos y, \quad v(x,y) = e^x\sin y$$

$$u_x = e^x\cos y, \quad u_y = -e^x\sin y,$$

$$v_x = e^x\sin y, \quad v_y = e^x\cos y.$$

So $\quad u_x = v_y$ and $u_y = -v_x$

i.e., C-R conditions are satisfied for all x and y. Therefore given function e^z is analytic everywhere. Thus e^z is an entire function.

b. $f(z) = (x-y)^2 + 2i(x+y)$

Solution:

$$u = (x-y)^2, \quad v = 2(x+y)$$

$$u_x = 2(x-y),\ u_y = -2(x-y),\ v_x = 2,\ v_y = 2$$

So $u_x = 2(x-y) = v_y = 2$ if $(x-y) = 1$.

Also $u_y = -2(x-y) = -v_x = -2$ if $(x-y) = 1$.

Thus C-R equations are satisfied only along the straight line $x - y = 1$. So

$$f'(z) = u_x + iv_x = 2(x-y) + i2 = 2.1 + i2 = 2+2i$$

exists only along the line $x - y = 1$, not through any region (neighbourhood) R. Hence $f(z)$ is nowhere analytic.

c. $f(z) = e^{i\overline{z}} = e^y(\cos x + i\sin x)$.

Solution:

$$u = e^y \cos x, \quad v = e^y \sin x,$$

$$\text{So} \quad u_x = -e^y \sin x, \; u_y = e^y \cos x,$$
$$v_x = e^y \cos x, \quad v_y = e^y \sin x$$

Thus $u_x \neq v_y$ and $u_y \neq -v_x$ for any x and y. C-R conditions are not satisfied for any z. Hence $f(z) = e^{\bar{iz}}$ is nowhere analytic.

d. $f(z) = \cos x(\cosh y + a \sinh y) + i \sin x(\cosh y + b \sinh y)$ where a and b are constants.

Solution:

$$u = \cos x(\cosh y + a \sinh y)$$
$$v = \sin x(\cosh y + b \sinh y)$$
$$u_x = -\sin x(\cosh y + a \sinh y)$$
$$v_y = \sin x(\sinh y + b \cosh y)$$
$$u_y = \cos x(\sinh y + a \cosh y)$$
$$v_x = \cos x(\cosh y + b \sinh y)$$

So $u_x = v_y$ and $u_y = -v_x$ if $a = b = -1$.
f is analytic when $a = b = -1$.

e. $f(z) = \frac{(z+3i)^5}{(z^2-2z+5)^2}$.

Solution: $f(z) = \frac{g(z)}{h(z)}$ where $g(z) = (z + 3i)^5$ (polynomial) is analytic everywhere and $h(z) = (z^2 - 2z + 5)^2$ (polynomial) is analytic everywhere. The quotient $f(z) = \frac{g(z)}{h(z)}$ is analytic everywhere except when $h(z) = (z^2 - 2z + 5)^2 = 0$. Thus $f(z)$ is analytic everywhere except at $z^2 - 2z + 5 = 0$ i.e., at $z = 1 \pm 2i$.

Example 10: If $f(z)$ is analytic, show that

$$\left[\frac{\partial}{\partial x}|f|\right]^2 + \left[\frac{\partial}{\partial y}|f|\right]^2 = |f'|^2$$

Solution: $|f(z)| = |u(x, y) + iv(x, y)| = \sqrt{u^2 + v^2}$ partially differentiating w.r.t., x and y

$$\frac{\partial}{\partial x}|f| = \frac{1}{2}(u^2 + v^2)^{-\frac{1}{2}}[2uu_x + 2vv_x] = \frac{uu_x + vv_x}{|f|}$$

Similarly,

$$\frac{\partial}{\partial y}|f| = \frac{uu_y + vv_y}{|f|}.$$

Squaring and adding, we get

$$\left[\frac{\partial}{\partial x}|f|\right]^2 + \left[\frac{\partial}{\partial y}|f|\right]^2 = \frac{(uu_x + vv_x)^2 + (uu_y + vv_y)^2}{|f|^2}$$
$$= \Big[(u^2u_x^2 + v^2v_x^2 + 2uv\,u_x v_x)$$
$$+ (u^2u_y^2 + v^2v_y^2 + 2uv\,u_y v_y)\Big]/|f|^2$$

Since f is analytic, C-R conditions are satisfied. So $u_x = v_y, u_y = -v_x$. Then $2uv\,u_x\,v_x = -2uvu_yv_y$

$$= \frac{(u^2 + v^2)(u_x^2 + v_x^2)}{|f|^2} = u_x^2 + v_x^2 = |f'|^2$$

since $f' = u_x + ivx$ and $|f'| = \sqrt{u_x^2 + v_x^2}$.

Example 11: Is the function $u(x, y) = 2xy + 3xy^2 - 2y^3$ harmonic (i.e., solution of Laplace's equation)?

Solution: $u_x = 2y + 3y^2, u_{xx} = 0, u_y = 2x + 6xy - 6y^2, u_{yy} = 6x - 12y$. So $u_{xx} + u_{yy} \neq 0$. Therefore u is not harmonic.

Harmonic and conjugate harmonic functions

Example 12: Show that $v(x, y) = -\sin x \sinh y$ is harmonic. Find the conjugate harmonic of v (or find an analytic function $f = u + iv$).

Solution: Differentiating v partially w.r.t., x and y, we get

$$v_x = -\cos x \sinh y, \; v_{xx} = \sin x \sinh y$$
$$v_y = -\sin x \cdot \cosh y, \; v_{yy} = -\sin x \cdot \sinh y.$$

Then $v_{xx} + v_{yy} = \sin x \sinh y + (-\sin x \sinh y) = 0$.
Therefore v is harmonic.
To find conjugate harmonic u of v:
From C-R conditions $u_x = v_y$ and $u_y = -v_x$.

$$\text{So} \quad u_x = v_y = -\sin x \cdot \cosh y \qquad (1)$$

Integrating (1) partially w.r.t. x, we get

$$u(x, y) = \cos x \cdot \cosh y + c(y) \qquad (2)$$

Differentiating (2) partially w.r.t., y and using second C-R condition, $(u_y = -v_x)$, we have

$$\cos x \cdot \sinh y + \frac{dc}{dy} = \frac{\partial u}{\partial y} = -v_x = \cos x \cdot \sinh y$$

So $\frac{dc}{dy} = 0$ or $c =$ constant.
Hence the conjugate harmonic u of v is

$$u(x, y) = \cos x \cdot \cosh y + c$$

The required analytic function is

$$f(z) = \cos x \cosh y + c + i(-\sin x \cdot \sinh y).$$

Milne-Thompson method

Example 13: Find the analytic function $f(z) = u + iv$ where $u = e^x(x \cos y - y \sin y) + 2 \sin x \cdot \sinh y + x^3 - 3xy^2 + y$.

Solution: Differentiating partially w.r.t., x and y

$$u_x = e^x(x \cos y - y \sin y) + e^x(\cos y) + 2 \cos x \sinh y + 3x^2 - 3y^2$$

$$u_y = e^x(-x \sin y - \sin y - y \cos y) + 2 \sin x \cosh y - 6xy + 1.$$

We know that $f'(z) = u_x + iv_x = u_x - iu_y$
Replace x by z and $y = 0$, then

$$f'(z) = e^z(z \cdot 1 - 0) + e^z(1) + 0 + 3z^2 - 0 - 2i \sin z - i$$

Integrating w.r.t. 'z', we get

$$f(z) = \int (ze^z + e^z + 3z^2 - 2i \sin z - i)dz$$

$$f(z) = ze^z - e^z + e^z + z^3 + 2i \cos z - iz + c.$$

Example 14: Determine the analytic function $f(z)$ such that $\text{Re}\left(f'(z)\right) = 3x^2 - 4y - 3y^2$ and $f(1+i) = 0$.

Solution: $f'(z)$ is analytic since f is analytic. Let $f' = U + iV$.
Then $U = \text{Re}\left(f'(z)\right) = 3x^2 - 4y - 3y^2 . U_x = 6x$, $U_y = -4 - 6y$. Since U, V, satisfy C-R conditions, $U_x = 6x = V_y$.

Integrating w.r.t. y, we get

$$V = 6xy + c_1(x)$$

Differentiating V w.r.t. x and using second C-R condition $(V_x = -U_y)$, we have

$$6y + \frac{dc_1}{dx} = V_x = -U_y = 4 + 6y$$

$$c_1(x) = 4x + c_2$$

$$\text{Thus} \quad V(x, y) = 6xy + 4x + c_2$$

where c_2 is an arbitrary constant. Then

$$f'(z) = U + iV = (3x^2 - 4y - 3y^2) + i(6xy + 4x + c_2)$$

Applying Milne-Thompson method, replace x by z and y by 0, we have

$$f'(z) = 3z^2 + 4iz + c_2$$

Integrating w.r.t. z, we get

$$f(z) = z^3 + 2iz^2 + c_2 z + c_3$$

where c_3 is an arbitrary constant. Since $f(1+i) = 0$, we get

$$0 = f(1+i) = (1+i)^3 + 2i(1+i)^2 + c_2(1+i) + c_3$$

$$c_3 = -c_2(1+i) - 6 + 2i$$

Thus

$$f(z) = z^3 + 2iz^2 + c_2 z - c_2(1+i) - 6 + 2i.$$

C-R Equations not sufficient

Example 15: Show that for

$$f(z) = \frac{2xy(x+iy)}{x^2+y^2} \quad \text{if } z \neq 0$$
$$= 0 \quad \text{if } z = 0$$

The C-R are satisfied at origin but derivative of $f(z)$ at origin does not exist. (i.e., C-R conditions are not sufficient conditions for analyticity).

Solution: C-R conditions at origin:

$$f(z) = \frac{2xy(x+iy)}{x^2+y^2}, \text{ so } u = \frac{2x^2y}{x^2+y^2}, v = \frac{2xy^2}{x^2+y^2}$$

$$\frac{\partial u}{\partial x} = \lim_{x\to 0} \frac{u(x,0) - u(0,0)}{x} = \lim_{x\to 0} \frac{0-0}{x} = 0$$

$$\frac{\partial u}{\partial y} = \lim_{y\to 0} \frac{u(0,y) - u(0,0)}{y} = \lim_{y\to 0} \frac{0-0}{y} = 0$$

$$\frac{\partial v}{\partial x} = \lim_{x\to 0} \frac{v(x,0) - v(0,0)}{x} = \lim_{x\to 0} \frac{0-0}{x} = 0$$

$$\frac{\partial v}{\partial y} = \lim_{y\to 0} \frac{v(0,y) - v(0,0)}{y} = \lim_{y\to 0} \frac{0-0}{y} = 0.$$

So C-R conditions are satisfied at $z = 0$.
Derivative at $z = 0$:

$$f'(0) = \lim_{z\to 0} \frac{f(z) - f(0)}{z - 0} = \lim_{z\to 0} \frac{\frac{2xy(x+iy)}{x^2+y^2} - 0}{x + iy}$$

$$= \lim_{z\to 0} \frac{2xy}{x^2 + y^2} = 0 \quad \text{as} \quad \begin{matrix} x \to 0 & y \to 0 \\ y \to 0, & x \to 0 \end{matrix} \quad \text{and}$$

$$= \lim_{\substack{y\to mx \\ x\to 0}} \frac{2xy}{x^2+y^2} = \lim_{x\to 0} \frac{2mx^2}{x^2+m^2x^2} = \lim_{x\to 0} \frac{2m}{1+m^2}$$

$$= \frac{2m}{1 + m^2} \neq 0$$

Thus derivative of $f(z)$ does not exist at $z = 0$.

Orthogonal trajectories

Example 16: Find the orthogonal trajectories of the family of curves $x^3y - xy^3 = c =$ constant.

Solution: Take $u(x, y) = x^3y - xy^3$. Then the $v =$ constant family of curves will be the required orthogonal trajectories if $f(z) = u + iv$ is analytic. So

$$u_x = 3x^2y - y^3, \quad u_y = x^3 - 3xy^2.$$

Then $\quad v_y = u_x = 3x^2y - y^3,$

Integrating $\quad v = \dfrac{3x^2y^2}{2} - \dfrac{y^4}{4} + c(x)$

Differentiating

$$3xy^2 - 0 + \frac{dc}{dx} = v_x = -u_y = -x^3 + 3xy^2$$

$$c(x) = -\frac{x^4}{4} + c$$

where c is an arbitrary constant. Thus

$$v(x, y) = \frac{3x^2y^2}{2} - \frac{y^4}{4} - \frac{x^4}{4} + c$$

The required orthogonal trajectories

$$v = \text{constant}$$

or $x^4 + y^4 - 6x^2y^2 =$ constant.

C-R Equations: in polar coordinates

Example 17: Find the derivative of $f(z) = \frac{1}{z^n}$ for $n \neq -1$.

Solution: Introducing polar coordinates

$$f(z) = \frac{1}{(r\cos\theta + ir\sin\theta)^n} = \frac{1}{r^n}(\cos n\theta - i\sin n\theta)$$

so $u = r^{-n}\cos n\theta$, $v = -r^{-n}\sin n\theta$

$$f'(z) = e^{-i\theta}(u_r + iv_r)$$

$$= e^{-i\theta}\Big(-nr^{-n-1} \cdot \cos n\theta + i(-1)(-n)r^{-n-1} \cdot \sin n\theta \Big)$$

$$= \frac{-n}{r^{n+1}} \cdot e^{-i\theta}(\cos n\theta - i\sin n\theta)$$

$$= \frac{-n}{r^{n+1}} \cdot e^{-i\theta} \cdot e^{-in\theta} = \frac{n}{r^{n+1}} e^{-i(n+1)\theta}$$

$$f'(z) = \frac{-n}{r^{n+1} \cdot e^{i(n+1)\theta}} = \frac{-n}{z^{n+1}}, n \neq -1.$$

Example 18: If $v(r, \theta) = \left(r - \frac{1}{r}\right)\sin\theta, r \neq 0$, then find an analytic function $f(z) = u + iv$.

Solution: C-R conditions in polar coordinates are $u_r = \frac{1}{r}v_\theta$ and $u_\theta = -rv_r$.
Differentiating

$$u_\theta = -rv_r = -r \cdot \left(1 + \frac{1}{r^2}\right)\sin\theta = -\left(r + \frac{1}{r}\right)\sin\theta$$

Integrating w.r.t. θ, we get

$$u(r, \theta) = \left(r + \frac{1}{r}\right)\cos\theta + c_1(r)$$

Differentiating w.r.t. r and using $u_r = \frac{1}{r}v_\theta$, we get

$$\left(1 - \frac{1}{r^2}\right)\cos\theta + \frac{dc_1}{dr} = u_r = \frac{1}{r}v_\theta = \frac{1}{r}\left(r - \frac{1}{r}\right)\cos\theta$$

$$\therefore \quad \frac{dc_1}{dr} = 0 \quad \text{or} \quad c_1 = \text{constant}$$

Hence $\quad u(r, \theta) = \left(r + \dfrac{1}{r}\right)\cos\theta + c_1$

Thus $\quad f(z) = u + iv = \left(r + \dfrac{1}{r}\right)\cos\theta + i\left(r - \dfrac{1}{r}\right)\sin\theta + c$

Example 19: If f is analytic show that

$$f' = (\cos\theta - i\sin\theta)\frac{\partial f}{\partial r}.$$

Solution: $x = r\cos\theta$, $y = r\sin\theta$, $r = \sqrt{x^2+y^2}$, $\theta = \tan^{-1}\frac{y}{x}$.

$$r_x = \frac{1}{2}\frac{2x}{\sqrt{x^2+y^2}} = \frac{x}{r} = \cos\theta,\ r_y = \sin\theta,$$

$$\theta_x = \frac{1}{1+\left(\frac{y}{x}\right)^2}\cdot\left(\frac{-y}{x^2}\right) = \frac{-y}{x^2+y^2} = \frac{-y}{r^2}$$

$$\theta_x = \frac{-\sin\theta}{r},\quad \theta_y = \frac{\cos\theta}{r}.$$

We know that $f' = u_x + iv_x = u_x - iu_y$

$$f' = (u_r\cdot r_x + u_\theta\cdot\theta_x) - i(u_r r_y + u_\theta\theta_y).$$

since u is function of r, θ which are functions of x, y.

$$f' = \left(\cos\theta u_r - \frac{\sin\theta}{r}u_\theta\right) - i\left(u_r\cdot\sin\theta + \frac{\cos\theta}{r}\cdot u_\theta\right)$$

since f is analytic, C-R conditions are satisfied (in polar coordinates) $u_r = \frac{v_\theta}{r}$, $u_\theta = -rv_r$

$$f' = \left(\cos\theta\cdot u_r + \frac{\sin\theta}{r}\cdot rv_r\right)$$
$$-i(u_r\sin\theta - v_r\cos\theta)$$
$$= (\cos\theta - i\sin\theta)u_r + i(\cos\theta - i\sin\theta)v_r$$
$$= (\cos\theta - i\sin\theta)(u_r + iv_r)$$
$$f' = (\cos\theta - i\sin\theta)\frac{\partial f}{\partial r}$$

since $\quad \dfrac{\partial f}{\partial r} = \dfrac{\partial}{\partial r}(u+iv) = u_r + iv_r.$

Example 20: Find the orthogonal trajectories of the family of curves $r^2\cos 2\theta = c_1$.

Solution: Take $u(r,\theta) = r^2\cos 2\theta$,
so $u_r = 2r\cos 2\theta$, $u_\theta = -2r^2\sin 2\theta$.
From C-R conditions $v_\theta = ru_r$

$$v_\theta = 2r^2\cos 2\theta \quad\text{or}\quad v(r,\theta) = r^2\sin 2\theta + c(r).$$

Differentiating w.r.t. r,

$$2r\sin 2\theta + \frac{dc}{dr} = v_r = -\frac{1}{r}u_\theta$$
$$= -\frac{1}{r}(-2r^2)\sin 2\theta$$
$$\therefore\quad \frac{dc}{dr} = 0 \ \therefore c = \text{constant}.$$

Orthogonal trajectories : $v = r^2\sin 2\theta$.

EXERCISE

Complex function

1. Classify the following regions:
   a. $0 < |z| < 1$
   b. $0 < |z| \le 1$
   c. $1 < |z| < 2$
   d. $|z| < 1$ and $|z| > 2$
   e. $|\text{Re}\, z| < 2$
   f. $|z-4| > 3$
   g. $|z-1+3i| \le 1$

Ans. **a.** Open region.
b. Region.
c. Connected open region.
d. Unconnected.
e. Open unbounded region.
f. Open unbounded region.
g. Closed bounded region.

2. Determine the domains of definition of $f(z)$.
   a. $\frac{y}{x} + \frac{1}{1-y}i$
   b. $z^4 + 3z^2 + iz$
   c. $\frac{1}{(z^2+4)(z^2-9)}$
   d. $y\int_0^\infty e^{-xt}dt + i\sum_{n=0}^{\infty} y^n$

Ans. **a.** Entire complex plane except $x = 0$, $y = 1$.
b. For all z.
c. For all z except $z = \pm 2i, \pm 3$.
d. $x > 0$ and $-1 < y < 1$.

3. Find the real and imaginary parts u, v of $f = u + iv$ where $f(z)$ is
   a. $z + \frac{1}{z}$
   b. $\frac{(1-z)}{(1+z)}$
   c. $z^{\frac{1}{2}}$

Ans. **a.** $u = x + \frac{x}{(x^2+y^2)}$, $v = y - \frac{y}{(x^2+y^2)}$
b. $\frac{(1-x^2-y^2)}{[(1+x)^2+y^2]}, \frac{-2y}{[(1+x)^2+y^2]}$
c. $u = \sqrt{r}\cos\frac{\theta}{2}$, $v = \sqrt{2}\sin\frac{\theta}{2}$
where $x = r\cos\theta$, $y = r\sin\theta$.

Continuity

Determine whether the function $f(z)$ is continuous at origin. Give reason.

4. $f(z) = \frac{(x+y)^2}{(x^2+y^2)}, \quad z \neq 0$
   $= 0, \quad z = 0$

Ans. Not continuous. lim f along $y = mx$ has $\frac{(1+m)^2}{(1+m^2)}$ different values.

5. $f(z) = \frac{z \cdot \text{Re}\, z}{|z|}$, for $z \neq 0$
   $= 2$, for $z = 0$

Ans. Discontinuous at origin. But by redefining $f(0) = 0$, the function can be made continuous at origin.

6. Is the function $f(z) = (x + y^2) + ixy$ continuous?

Ans. Continuous everywhere.

7. Determine whether $f(z)$ is continuous. Redefine if necessary to make it continuous

$$f(z) = \begin{cases} z^2 + iz + 2, & z \pm i \\ i, & z = i \end{cases}$$

Ans. $f(z)$ is continuous everywhere except at $z = i$, since $\lim_{z \to i} f = 0 \neq f(i) = i$. By redefining $f(i) = 0$, f becomes continuous at $z = i$ also.

8. Show that $f(z) = P_n(z) = a_0 + a_1 z + a_2 z^2 + \cdots + a_n z^n$ is continuous everywhere.

9. Prove that

$$f(z) = \begin{cases} z^3, & z \neq z_0 \\ 2, & z = z_0 \end{cases} \quad \text{where } z_0 \neq 2^{\frac{1}{3}}$$

is discontinuous at z_0.

10. Is the function

$$f(z) = \frac{3z^4 - 2z^3 + 8z^2 - 2z + 5}{z - i}$$

continuous at $z = i$?

Ans. $f(i)$ is undefined. Discontinuous at $z = i$. Redefine $f(i) = \lim_{z \to i} f(z) = 4 + 4i$. Then f becomes continuous at $z = i$. It is known as removable discontinuity.

11. Determine for what values of z the given functions are continuous
   a. $f(z) = \frac{z}{(z^2+1)}$
   b. $csc z = \frac{1}{\sin z}$

Ans. **a.** Continuous everywhere except where the denominator $z^2 + 1 = 0$ i.e., at $z = \pm i$.
b. Continuous everywhere except where $\sin z = 0$ i.e., at $z = \pm n\pi, n = 0, 1, 2, 3, \ldots$

Differentiability

12. Find derivative of $f(z) = \frac{1+z}{1-z}$
   a. From definition.
   b. From differentiation rules at $z = 2$.

Ans. **a.** $f' = \lim_{\Delta z \to 0} \left[\frac{1 + (z + \Delta z)}{1 - (z + \Delta z)} - \frac{1+z}{1-z} \right] \frac{1}{\Delta z}$
$= \frac{2}{(1-z)^2}$
b. By Quotient rule at $z = 2$, $f' = 2$

13. Show that the following functions are continuous but not differentiable:
   (a) $\bar{z}$ (b) $|z|^2$ (c) Imz (d) $|z|$ (e) $\frac{1}{z}$

   Hint:
   a. Not differentiable anywhere.
   b. Differentiable only at origin.
   c. Nowhere differentiable.
   d. Nowhere differentiable.
   e. Not differentiable at 0.

14. Find the derivative at indicated points:
   a. $\frac{2z-i}{z+2i}$ at $z = -i$
   b. $3z^{-2}, z = 1 + i$

Ans. **a.** $-5i$ **b.** $\frac{3}{2}(1 + i)$.

Analyticity

Determine where C-R conditions are satisfied for the given function $f(z)$:

15. $f(z) = \frac{1}{2} \ln(x^2 + y^2) + i \tan^{-1} \frac{y}{x}$

Ans. For all z, everywhere (entire)

16. $f(z) = x + ay + i(bx + cy)$

Ans. $a = -b, c = 1$

17. $f(z) = xy + iy$

Ans. Nowhere (analytic).

18. $f(z) = |x^2 - y^2| + 2i|xy|$

Ans. **a.** For $f(z) = z^2, 0 < \theta < \frac{\pi}{4}, \pi < \theta < \frac{5\pi}{4}$

b. For $f(z) = -z^2, \frac{\pi}{2} < \theta < \frac{3\pi}{4}, \frac{3\pi}{2} < \theta < \frac{7\pi}{4}$

19. $f(z) = z\bar{z}$

Ans. Only at origin.

20. $f(z) = \sin x \cosh y + i \cos x \sinh y$

Ans. Everywhere.

21. Show that $f = x^2 + iy^3$ is nowhere analytic.

Hint: C-R conditions satisfied only at origin.

22. If $f(z) = xy^2 + ix^2y$, determine where

a. C-R conditions satisfied

b. f' exist

c. f is analytic

Ans. **a.** C-R satisfied only at origin

b. f' exist only at origin

c. f is nowhere analytic.

Show that the given function satisfies C-R conditions at origin but does not have a derivative at origin:

23. $f(z) = \begin{cases} \frac{(\bar{z})^2}{z}, & z \neq 0 \\ 0, & z = 0 \end{cases}$

24. $f(z) = \sqrt{|xy|}$

25. $f(z) = \begin{cases} \frac{(x^3-y^3)+i(x^3+y^3)}{x^2+y^2}, & \text{if } z \neq 0 \\ 0, & \text{if } z = 0. \end{cases}$

Harmonic and conjugate harmonic functions

Verify that the given function is harmonic and find its conjugate harmonic function. Express $u + iv$ as an analytic function $f(z)$:

26. $u = x^2 - y^2 - y$

Ans. $v = 2xy + x + c, f(z) = z^2 + iz + c$

27. $v = x^2 - y^2 + \frac{x}{x^2+y^2}$

Ans. $u = -2xy + \frac{y}{x^2+y^2} + c, w = i\left(z^2 + \frac{1}{z}\right) + c$

28. $u = \frac{\sin 2x}{(\cosh 2y - \cos 2x)}$

Ans. $v = \frac{-\sinh 2y}{(\cosh 2y - \cos 2x)}, f = \cot z + c$

29. $u = 3xy^2 - x^3$

Ans. $v = y^3 - 3x^2y + c, f = -z^3 + ic$

30. $v = y^2 - x^2$

Ans. $u = 2xy + c, f = -iz^2 + c$

31. $u = e^{-x}(x \sin y - y \cos y)$

Ans. $v = e^{-x}(y \sin y + x \cos y) + c$

$f(z) = ize^{-z}$

32. $u = \frac{x^2-y^2}{(x^2+y^2)^2}$

Ans. $v = \frac{-2xy}{(x^2+y^2)^2}, f(z) = \frac{1}{z^2} + c$

33. $u = \frac{1}{2}\ln(x^2 + y^2)$

Ans. $v = \arg z + c, f = \ln z + c$

34. $u = 3x^3y + 2x^2 - y^3 - 2y^2$

Ans. Not harmonic.

35. $u = e^{-2xy}\sin(x^2 - y^2)$

Ans. $v = -e^{-2xy}\cos(x^2 - y^2) + c$

$f(z) = -ie^{iz^2} + ci$

36. If $\text{Im}\left\{f'(z)\right\} = 6x(2y - 1)$ and $f(0) = 3 - 2i, f(1) = 6 - 5i$ find $(1 + i)$.

Hint: f' is analytic. Determine $\text{Re}\{f'\}$. Use Milne-Thompson method.

Ans. $6 + 3i$

37. If $u - v = \frac{\cos x + \sin x - e^{-y}}{2(\cos x - \cosh y)}$ and $f\left(\frac{\pi}{2}\right) = 0$ determine the analytic function $f(z) = u + iv$

Hint: Differentiating w.r.t., x, y, use C-R, find u_x, v_x, use Milne-Thompson method.

38. Determine constant 'b' such that $u = e^{bx}\cos 5y$ is harmonic. Find its conjugate harmonic.

Ans. $b = \pm 5, v = \pm e^{\pm 5x}\sin 5y + c$

39. If $f(z)$ is analytic in a domain D and $|f(z)| = k =$ constant in D, then show that $f(z) =$ constant in D.

40. If $f(z)$ is a regular function, show that

$$\left(\frac{\partial^2}{\partial x^2} + \frac{\partial^2}{\partial y^2}\right)|f(z)|^2 = 4|f'(z)|^2.$$

41. If $f'(z) = 0$ then show that $f(z)$ is constant.

42. If both $f(z)$ and $\overline{f(z)}$ are analytic show that $f(z)$ is constant.

43. If $f = u + iv$ is analytic show that $g = -v + iu$ is also analytic. Also show that u and $-v$ are conjugate harmonic.

Hint: f analytic, $u_x = v_y$, $u_y = -v_x$, so g satisfies C-R $-v_x = u_y$ and $-v_y = -u_x$. g analytic, $-v, u$, C.H.F.

44. If $f(z)$ is an analytic function, show that

$$\left(\frac{\partial^2}{\partial x^2} + \frac{\partial^2}{\partial y^2}\right) |\text{Res } f(z)|^2 = 2|f'(z)|^2$$

Hint: For 40, 44, $\nabla^2 = \frac{\partial^2}{\partial z \partial \bar{z}}$, write $|f(z)|^2 = f(z)\overline{f(z)}$ and Res $f(z) = \frac{\left(f(z)+\overline{f(z)}\right)}{2}$.

45. If $f(z)$ is analytic, prove that

$$\left(\frac{\partial^2}{\partial x^2} + \frac{\partial^2}{\partial y^2}\right) |f(z)|^n = n^2 |f(z)|^{n-2} \cdot |f'(z)|^2$$

46. Show that (a) $\nabla^2 \ln |f(z)| = 0$ (b) $\nabla^2 \arg f(z) = 0$ if f is analytic.

47. Show that $U(x, y) = e^u \cos v$, $V(x, y) = e^u \sin v$ are conjugate harmonic of each other if $f = u + iv$ is analytic.

Orthogonal trajectories, polar coordinates

Find the orthogonal trajectories of the family of curves:

48. $e^{-x}(x \sin y - y \cos y) = c_1$

Ans. $e^{-x}(y \sin y + x \cos y) = c_2$

49. $e^{-x} \cos y + xy = c_1$

Ans. $2e^{-x} \sin y + x^2 - y^2 = c_2$

50. $x^4 - 6x^2y^2 + y^4 = c_1$

Ans. $x^3y - xy^3 = c_2$

51. $x^3 - 3xy^2 = c_1$

Ans. $3x^2y - y^3 = c_2$

52. $(r^2 + 1) \cos\theta = c_1 r$

Ans. $(r^2 - 1) \sin\theta = c_2 r$

53. If $f(z)$ is analytic, show that

$$f' = \frac{-\sin\theta + i\cos\theta}{r} \cdot \frac{\partial f}{\partial \theta}$$

54. Show that $u(r, \theta) = e^{-\theta} \cos(\ln r)$ is harmonic. Find its conjugate harmonic function.

Hint: u is harmonic if it satisfies the Laplace's equation in polar coordinates.

$$r^2 u_{rr} + r u_r + u_{\theta\theta} = 0$$

Ans. $v(r, \theta) = e^{-\theta} \sin(\ln r) + c$

55. Find the conjugate harmonic function of $v(r, \theta) = r^2 \cos 2\theta - r \cos\theta + 2$. Show that v is harmonic.

Ans. $u(r, \theta) = -r^2 \sin 2\theta + r \sin\theta + c$.

56. Find the conjugate harmonic function of $u(r, \theta) = -r^3 \sin 3\theta$. Show that u is harmonic.

Ans. $v = r^3 \cos 3\theta + c$.

22.8 ELEMENTARY FUNCTIONS

Algebraic function of z is a function $w = f(z)$ which is a solution of the polynomial equation

$$P_0(z) \cdot w^n + P_1(z)w^{n-1} + \cdots + P_{n-1}(z)w + P_n(z) = 0$$

where $P_0 \neq 0$, $P_1(z), P_2(z), \ldots, P_n(z)$ are polynomials in z and n is a positive integer.

Polynomials and rational functions are special cases of algebraic functions. Transcendental functions are those which cannot be expressed as an algebraic function.

Logarithmic, trigonometric, hyperbolic and their corresponding inverses are transcendental (non-algebraic) functions. Elementary functions consist of algebraic functions and transcendental functions and functions derived from them by a finite number of algebraic operations of addition, subtraction, multiplication, division and root taking.

Exponential Function: e^z

For $z = x + iy$, exponential function e^z also written as exp z is defined as

$$e^z = e^{x+iy} = e^x(\cos y + i \sin y)$$

e^z may also be defined by a power series as

$$e^z = \sum_{n=0}^{\infty} \frac{z^n}{n!} = 1 + z + \frac{z^2}{2!} + \frac{z^3}{3!} + \cdots$$

which converges for all z.

Properties

1. e^z is an entire function

$$f(z) = u + iv = e^z = e^x(\cos y + i \sin y)$$

so $\quad u = e^x \cos y,\ v = e^x \sin y.$

Now $\quad u_x = e^x \cos y,\ u_y = -e^x \sin y,$

$$v_x = e^x \sin y,\ v_y = e^x \cos y.$$

Thus C-R conditions are satisfied for all z.

2. $\frac{d}{dz} e^z = u_x + i v_x = e^x \cos y + i e^x \sin y$

$$= e^x(\cos y + i \sin y) = e^{x+iy} = e^z.$$

3. e^z has no zeros i.e., $e^z \neq 0$ for any z.

$e^z = Re^{i\phi} \quad$ so $\quad R = e^x > 0,\ y = \phi,\ |e^{iy}| = 1$

so $\quad |e^z| = e^x \neq 0.$

4. $e^{2n\pi i} = \cos(2n\pi) + i \sin(2n\pi) = 1$

Thus $\quad e^{\pm 2n\pi i} = 1,\ e^{\pm \pi i} = -1$

$$e^{\frac{\pi i}{2}} = i,\ e^{-\frac{i\pi}{2}} = -i$$

5. Periodic function

$$e^{z \pm 2n\pi i} = e^z \cdot e^{\pm 2n\pi i} = e^z$$

e^z is a periodic function of imaginary period $2\pi i$.

6. From definition

a. $e^{z_1} \cdot e^{z_2} = e^{z_1 + z_2}$

b. $\frac{e^{z_1}}{e^{z_2}} = e^{z_1 - z_2}$

c. $(e^z)^n = e^{nz}$

7. $e^{\bar{z}} = \overline{e^z}$

$$e^{\bar{z}} = e^{x - iy} = e^x(\cos y - i \sin y)$$
$$= \overline{e^x(\cos y + i \sin y)} = \overline{e^z}.$$

8. Polar form of z in terms of exponential,

$$z = x + iy = r \cos\theta + ir \sin\theta = re^{i\theta}.$$

Worked Out Examples

Exponential function

Example 1: Separate into real and imaginary: (a) $e^{2 \pm 3\pi i}$ (b) $e^{5 + \frac{i\pi}{2}}$ (c) $e^{(5+3i)^2}$.

Solution:

a. $e^{2 \pm 3\pi i} = e^2 \cdot e^{\pm 3\pi i} = e^2 \cdot (-1) = -e^2$

b. $e^{5 + \frac{i\pi}{2}} = e^5 \cdot e^{\frac{i\pi}{2}} = e^5 \cdot i = ie^5$

c. $e^{(5+3i)^2} = e^{(25 - 9 + 30i)} = e^{16} \cdot e^{30i} = e^{16}$ cis 30

Example 2: Show that $\frac{d}{dz}(e^{iz}) = ie^{iz}$.

Solution:

$$e^{iz} = e^{i(x+iy)} = e^{-y+ix} = e^{-y}(\cos x + i \sin x)$$

$$\frac{d}{dz}(e^{iz}) = \frac{\partial}{\partial x}[e^{-y} \cos x] + i \frac{\partial}{\partial x}[e^{-y} \sin x]$$

since $\quad f' = u_x + i v_x$

$$= -e^{-y} \cdot \sin x + i e^{-y} \cos x$$
$$= i(e^{-y} \cos x + i e^{-y} \sin x) = ie^{iz}$$

Example 3: Find all values of z such that $e^z = -2$.

Solution: Express -2 in polar form: $x = -2$, $y = 0$

$$r = \sqrt{2^2 + 0^2} = 2,\ \cos\theta = -1,\ \sin\theta = 0,\ \therefore \theta = \pm\pi$$

$$e^z = -2 = 2 \cdot e^{\pm \pi i \pm 2n\pi i} = e^{\ln 2 \pm (2n+1)\pi i}$$

$\therefore z = \ln 2 \pm (2n + 1)\pi i,\ n = 0, 1, 2, \ldots$

Example 4: Show that $e^{\bar{z}}$ is nowhere analytic.

Solution: $f(z) = u + iv = e^{\bar{z}} = e^{x - iy}$
$= e^x(\cos y - i \sin y)$

so $\quad u = e^x \cos y,\ v = -e^x \sin y.$

Differentiating, $u_x = e^x \cos y,\ u_y = -e^x \sin y,$

$v_x = -e^x \sin y,\ v_y = -e^x \cos y.$

$e^{\bar{z}}$ is nowhere analytic since C-R conditions are not satisfied for any z.

Example 5: Show in two ways that e^{z^2} is entire. Find its derivative in two ways.

Solution:

a. z^2 is entire, e^w is entire, entire function of an entire function is entire. Therefore e^{z^2} is entire.
By chain rule differentiation

$$\frac{d}{dz}e^{z^2} = e^{z^2} \cdot \frac{d}{dz}(z^2) = 2ze^{z^2}$$

b. $f(z) = u + iv = e^{z^2} = e^{x^2-y^2+i2xy}$ so

$$u = e^{x^2-y^2} \cos 2xy,\ v = e^{x^2-y^2} \cdot \sin 2xy$$

$$u_x = 2xe^{x^2-y^2} \cdot \cos 2xy - 2ye^{x^2-y^2} \cdot \sin 2xy$$

$$u_x = 2xu - 2yv$$

Similarly, $u_y = -2yu - 2xv$

$$v_x = 2xv + 2yu,\ v_y = -2yv + 2xu$$

Thus C-R conditions are satisfied for all x, y. Hence e^{z^2} is an entire function.

$$f' = u_x + iv_x = (2xu - 2yv) + i(2xv + 2yu)$$

By Milne-Thompson method, replace x by z, y by 0

$$f'(z) = (2ze^{z^2} - 0) + i(0 + 0) = 2ze^{z^2}.$$

EXERCISE

Exponential function

1. Find e^z (in the form $u + iv$) and $|e^z|$ if z equals
(a) $3 - 5\pi i$ (b) $4\pi(2 + i)$ (c) $6 - \frac{\pi i}{2}$
(d) $1.4 - 0.6i$.

Ans. (a) $-e^3, e^3$ (b) $e^{8\pi}, e^{8\pi}$ (c) $-ie^6, e^6$
(d) $3.347 - 2.290i, 4.055$

2. Represent each of the following in the exponential form $re^{i\theta}$ (a) $3 + 4i$ (b) $-4i$ (c) -2.

Ans. (a) $5 \cdot \exp\left(i \text{ arc } \tan \frac{4}{3}\right)$ (b) $4 \cdot e^{-\frac{i\pi}{2}}$
(c) $2e^{\pm 2n\pi i}, n = 0, 1, 2, \ldots$

Find all solutions:

3. $\exp(2z - 1) = 1$

Ans. $z = \frac{1}{2} \pm n\pi i, n = 0, 1, 2, \ldots$

4. $e^z = 3 + 4i$

Hint: $\ln 5 = 1.609$, $\sin y = 0.8$, $y = 0.927$

Ans. $z = 1.609 + 0.927i \pm 2n\pi i, n = 0, 1, 2, \ldots$

5. $e^{3z} = 1$

Ans. $z = \frac{2k\pi i}{3}, k = 0, \pm 1, \pm 2$

6. $e^{4z} = i$

Ans. $z = \left(\frac{1}{8}\pi + \frac{k\pi}{2}\right) i, k = 0, \pm 1, \pm 2$

7. $e^{2z-1} = 1 + i$

Ans. $z = \frac{1}{2} + \frac{1}{4} \ln 2 + i\left(n\pi + \frac{\pi}{8}\right)$,
$n = 0, \pm 1, \pm 2, \pm 3, \ldots$

8. Show that e^z and $e^{\bar{z}}$ are conjugate.

9. Prove that $\frac{e^{z_1}}{e^{z_2}} = e^{z_1 - z_2}$.

Trigonometric Functions

Cos z and sin z are defined in terms of e^z as

$$\cos z = \frac{1}{2}(e^{iz} + e^{-iz})$$

$$\sin z = \frac{1}{2i}(e^{iz} - e^{-iz})$$

From these definitions, it follows that

$$e^{iz} = \cos z + i \sin z$$

which shows that Euler's formula is valid for complex z.

The other trigonometric functions are defined

$$\tan z = \frac{\sin z}{\cos z},\ \cot z = \frac{\cos z}{\sin z},\ \sec z = \frac{1}{\cos z},\ \text{cosec} z = \frac{1}{\sin z}.$$

Properties

1. $\sin z$, $\cos z$ are entire functions since e^z is entire.

2. $\tan z$, $\sec z$ and $\cot z$, cosec$z$ are analytic everywhere except where $\cos z$ is zero and $\sin z$ is zero respectively.

3. Since $\frac{d}{dz}e^z = e^z$ and $\frac{d}{dz}e^{iz} = ie^{iz}$, it follows that $\frac{d}{dz}(\sin z) = -\cos z$, $\frac{d}{dz}(\cos z) = \sin z$, $\frac{d}{dz}(\tan z)$ $= \sec^2 z$ etc.

4. Even and odd functions. $\sin z$, $\tan z$, cosec$z$ are odd functions, $\cos z$, $\sec z$ are even functions.

$$\sin(-z) = \frac{1}{2i}[e^{i(-z)} - e^{-i(-z)}] = -\frac{1}{2i}[-e^{-iz} + e^{iz}]$$

$$= -\sin z$$

5. Periodic functions $\cos z$, $\sin z$ are periodic functions with real period 2π while $\tan z$, $\cot z$ have period π

$$\cos(z+2\pi) = \frac{1}{2}[e^{i(z+2\pi)} + e^{-i(z+2\pi)}]$$
$$= \frac{1}{2}[e^{iz} + e^{-iz}] = \cos z,$$

since $e^{\pm 2\pi i} = 1$.

6. Real and imaginary parts

For $\cos z = \frac{1}{2}(e^{iz} + e^{-iz})$

$$= \frac{1}{2}[e^{i(x+iy)} + e^{-i(x+iy)}]$$
$$= \frac{1}{2}[e^{ix}e^{-y} + e^{-ix}\cdot e^{y}]$$
$$= \frac{1}{2}e^{-y}(\cos x + i\sin x)$$
$$+\frac{1}{2}e^{y}(\cos x - i\sin x)$$
$$= \frac{1}{2}(e^{-y} + e^{y})\cos x - \frac{1}{2}i(e^{y} - e^{-y})\cdot\sin x$$

$$\cos z = \cosh y\cdot\cos x - i\sinh y\cdot\sin x$$

Thus absolute value of $\cos z$ is

$$|\cos z| = \sqrt{\cosh^2 y\cos^2 x + \sinh^2 y\sin^2 x}$$
$$|\cos z|^2 = (1+\sinh^2 y)\cos^2 x + \sinh^2 y\sin^2 x$$
$$= \cos^2 x + \sinh^2 y.$$

Similarly,

$$\sin z = \sin x\cdot\cosh y + i\cos x\cdot\sinh y$$
$$|\sin z|^2 = \sin^2 x + \sinh^2 y.$$

Similarly,

$$\tan z = \frac{\sin 2x + i\sinh 2y}{\cos 2x + \cosh 2y}$$

(See WE 4,5 on pages 22.18, 22.19)

7. Zeros of $\sin z$ and $\cos z$ are given by $z = \pm 2n\pi$, and $z = \pm\frac{1}{2}(2n+1)\pi$, $n = 0, 1, 2, \ldots$ respectively.

8. All the general formulas for the real trigonometric functions are valid for complex trigonometric functions also.

$\cos^2 z + \sin^2 z$

$$= \left[\frac{1}{2}(e^{iz} + e^{-iz})\right]^2 + \left[\frac{1}{2}\frac{(e^{iz} - e^{-iz})}{i}\right]^2$$
$$= \left(\frac{e^{2iz} + 2 + e^{-2iz}}{4}\right) - \left(\frac{e^{2iz} - 2 + e^{-2iz}}{4}\right)$$
$$= 1.$$

Similarly,

$$\sin(z_1 \pm z_2) = \sin z_1\cdot\cos z_2 \pm \cos z_1\cdot\sin z_2$$
$$\cos(z_1 \pm z_2) = \cos z_1\cdot\cos z_2 \mp \sin z_1\cdot\sin z_2$$

9. Relations between trigonometric and hyperbolic:

$$\cos(iz) = \frac{1}{2}[e^{i(iz)} + e^{i(-iz)}] = \frac{1}{2}[e^{-z} + e^{-z}]$$

a. $\cos(iz) = \cosh z$

Also $\sin(iz) = \frac{1}{2i}[e^{i(iz)} - e^{-i(+iz)}]$

$$= \frac{1}{2i}(e^{-z} - e^{-z})$$

b. $\sin(iz) = i\sinh z$

c. $\tan(iz) = i\tanh z$.

Hyperbolic Functions

Complex hyperbolic sine and cosine are defined as

$$\sinh z = \frac{1}{2}(e^{z} - e^{-z})$$
$$\cosh z = \frac{1}{2}(e^{z} + e^{-z}).$$

Properties

1. Both $\sinh z$ and $\cosh z$ are entire functions since e^z is entire. Other functions are similarly defined

$$\tanh z = \frac{\sinh z}{\cosh z},\ \coth z = \frac{\cosh z}{\sinh z},\ \operatorname{sech} z = \frac{1}{\cosh z}$$

and $\operatorname{cosech} z = \frac{1}{\sinh z}$

2. *Derivatives*: $\frac{d}{dz}(\sinh z) = \frac{d}{dz}\left[\frac{1}{2}(e^{z} - e^{-z})\right]$

$$= \frac{1}{2}[e^{z} + e^{-z}] = \cosh z, \text{ etc.}$$

3. *Periodic*: $\sinh z$, $\cosh z$ are periodic functions of imaginary period $2\pi i$ since

$$\sinh(z+2\pi i) = \frac{1}{2}\left(e^{z+2\pi i} - e^{-(z+2\pi i)}\right)$$
$$= \frac{1}{2}(e^{z} - e^{-z}) = \sinh z.$$

5. *Even and odd*: $\cosh z$ is even, $\sinh z$ is odd function

6. Relation between hyperbolic and trigonometric functions:

   a. $\cosh iz = \frac{1}{2}(e^{iz} + e^{-iz}) = \cos z$

   $\sinh iz = \frac{1}{2}(e^{iz} - e^{-iz}) = \frac{i}{2i}(e^{iz} - e^{-iz})$

   b. $\sinh iz = i \sin z$

   c. $\tanh iz = i \tan z$ **e.** $\sec hiz = \sec z$

   d. $\coth iz = -i \cot z$ **f.** $\operatorname{cosec} iz = -i\operatorname{cosec} z$

7. Zeros of $\sinh z$ are $\pm n\pi i$ and zeros of $\cosh z$ are $z = \pm(n + \frac{1}{2})\pi i, n = 0, 1, 2 \ldots$
   (see Worked Out Example 10 on Page 22.19)

8. General formulas true for real are valid for complex functions also

   a. $\cosh^2 z - \sinh^2 z = \left(\frac{e^z+e^{-z}}{2}\right)^2 - \left(\frac{e^z-e^{-z}}{2}\right)^2 = 1$

   b. $1 - \tanh^2 z = \operatorname{sech}^2 z$

   c. $\cosh(z_1 + z_2) =$
   $\cosh z_1 \cdot \cosh z_2 + \sinh z_1 \cdot \sinh z_2$

   d. $\sinh(z_1 + z_2) =$
   $\sinh z_1 \cdot \cosh z_2 + \sinh z_2 \cdot \cosh z_1$.

9. Relations between trigonometric and hyperbolic:

$$\cos(iz) = \frac{1}{2}[e^{i(iz)} + e^{i(-iz)}] = \frac{1}{2}[e^{-z} + e^{-z}]$$

   a. $\cos(iz) = \cosh z$

Also $\quad \sin(iz) = \frac{1}{2i}[e^{i(iz)} - e^{-i(+iz)}]$

$$= \frac{1}{2i}(e^{-z} - e^{-z})$$

   b. $\sin(iz) = i \sinh z$ **e.** $\sec iz = \operatorname{sech} z$

   c. $\tan(iz) = i \tanh z$. **f.** $\operatorname{cosec} iz = -\operatorname{cosech} z$

   d. $\cot iz = -i \coth z$

WORKED OUT EXAMPLES

Trigonometric and hyperbolic functions

Example 1: Find all solutions of $\sin z = 3$

Solution: $\sin z = \frac{e^{iz}-e^{-iz}}{2i} = 3$

$(e^{iz})^2 - 6ie^{iz} - 1 = 0$

$$e^{iz} = \frac{6i \pm \sqrt{-36+4}}{2} = (3 \pm \sqrt{8})i = (3 \pm \sqrt{8})e^{i\frac{\pi}{2} \pm 2n\pi i}$$

$e^{iz} = e^{\ln(3\pm\sqrt{8})+i(\frac{\pi}{2}\pm 2n\pi)}, n = 0, 1, 2, 3 \ldots$

$$z = (-\frac{\pi}{2} \pm 2n\pi) + i\ln(3 \pm \sqrt{8}), n = 0, 1, 2, 3 \ldots$$

$(3 + \sqrt{8} = 5.828, 3 - \sqrt{8} = 0.1715)$.

Example 2: Find all solutions of $\cosh z = -2$.

Solution: $\cosh z = \cosh x \cdot \cos y + i \sinh x \cdot \sin y = -2$.

Since R.H.S. is real, $\sinh x \cdot \sin y = 0$, which is possible when $x = 0$ or $y = \pm n\pi$, $n = 0, 1, 2, 3, \ldots$.
Now $\cosh x \cdot \cos y = -2$.

Since $\cosh x \geq 1$ for any x, we must take $y = \pm n\pi$. So $\cosh x \cdot \cos(\pm n\pi) = -2$, n odd.
Or $\cosh x = 2$ or $x = \cosh^{-1} 2$.
So all solutions to the given equation are

$$z = x + iy = \cosh^{-1} 2 + (2m + 1)\pi i, m = 0, 1, 2, 3, \ldots.$$

Example 3: Find all roots of the equation: $\tanh z + 2 = 0$.

Solution: $\tanh z + 2 = \frac{\sinh z}{\cosh z} + 2 = 0$

$$\frac{e^z - e^{-z}}{e^z + e^{-z}} + 2 = 0 \quad \text{or} \quad e^z - e^{-z} + 2(e^z + e^{-z}) = 0$$

$$3e^z + e^{-z} = 0 \quad \text{or} \quad 3(e^z)^2 = -1$$

so $\quad e^z = \left(-\frac{1}{3}\right)^{\frac{1}{2}} = \frac{1}{\sqrt{3}}i = \frac{1}{\sqrt{3}}e^{i(\frac{\pi}{2}\pm 2n\pi)}$

$$e^z = e^{-\frac{1}{2}\ln 3 + i(\frac{\pi}{2}\pm 2n\pi)}, \quad n = 0, 1, 2, \ldots$$

$$\therefore \quad z = -\frac{1}{2}\ln 3 + i\pi\left(2n + \frac{1}{2}\right), n = 0, \pm 1, \pm 2, \ldots.$$

Example 4: Find the real and imaginary parts of $\cot z$.

Solution: $\cot z = \frac{\cos z}{\sin z} = \frac{\cos(x+iy)}{\sin(x+iy)}$

$= \frac{\cos x \cdot \cosh y - i \sin x \cdot \sinh y}{\sin x \cdot \cosh y + i \cos x \cdot \sinh y}$, rationalizing

$$= \frac{(\cos x \cosh y - i \sin x \sinh y)(\sin x \cdot \cosh y - i \cos x \sinh y)}{\sin^2 x \cdot \cosh^2 \cdot y + \cos^2 x \cdot \sinh^2 y}$$

$$= \frac{\cos x \cdot \sin x(\cosh^2 y - \sinh^2 y) - i \sinh y \cdot \cosh y(\sin^2 x + \cos^2 x)}{\left(\frac{1-\cos 2x}{2}\right)\left(\frac{1+\cosh 2y}{2}\right) + \left(\frac{1+\cos 2x}{2}\right)\left(\frac{\cosh 2y - 1}{2}\right)}$$

$$= \frac{\frac{1}{2}(\sin 2x - i \sinh 2y)}{\frac{1}{2}(\cosh 2y - \cos 2x)} = \frac{\sin 2x - i \sinh 2y}{\cosh 2y - \cos 2x}$$

Thus $\quad \operatorname{Re}(\cot z) = \dfrac{\sin 2x}{\cosh 2y - \cos 2x}$ and

$$\operatorname{Im}(\cot z) = \frac{-\sinh 2y}{\cosh 2y - \cos 2x}.$$

Example 5: Find the real and imaginary parts of sech z.

Solution: $\operatorname{sech} z = \frac{1}{\cosh z} = \frac{1}{\cosh x \cdot \cos y + i \sinh x \sin y}$
rationalizing, we get

$$\operatorname{sech} z = \frac{\cosh x \cdot \cos y - i \sinh x \cdot \sin y}{\cosh^2 x \cdot \cos^2 y + \sinh^2 x \cdot \sin^2 y}$$

$$= \frac{\cosh x \cdot \cos y - i \sinh x \cdot \sin y}{\left(\frac{1+\cosh 2x}{2}\right)\left(\frac{1+\cos 2y}{2}\right) + \left(\frac{\cosh 2x - 1}{2}\right)\left(\frac{1-\cos 2y}{2}\right)}$$

$$= \frac{\cosh x \cdot \cos y - i \sinh x \cdot \sin y}{\frac{1}{4}(2\cosh 2x + 2\cos 2y)}$$

$$\operatorname{Re}(\operatorname{sech} z) = \frac{2\cosh x \cdot \cos y}{\cosh 2x + \cos 2y}$$

$$\operatorname{Im}(\operatorname{sech} z) = \frac{-2\sinh x \cdot \sin y}{\cosh 2x + \cos 2y}.$$

Example 6: Prove that (a) $\overline{\sin z} = \sin \bar{z}$ (b) $\overline{\cos z} = \cos \bar{z}$ (c) $\overline{\tan z} = \tan \bar{z}$.

Solution: (a) $\sin \bar{z} = \sin(x - iy)$

$$= \sin x \cdot \cos(-iy) + \sin(-iy) \cos x$$

$$= \sin x \cdot \cos iy - \sin(iy) \cos x$$

$$\sin \bar{z} = \sin x \cdot \cosh y - i \sinh y \cdot \cos x \qquad (1)$$

since $\quad \cos iy = \cosh y, \sin iy = i \sinh y$

By definition

$$\sin z = \sin x \cosh y + i \cos x \cdot \sinh y$$

$$\text{so } \overline{\sin z} = \sin x \cosh y - i \sinh y \cos x \qquad (2)$$

Hence from (1) and (2), $\sin \bar{z} = \overline{\sin z}$
(b) $\cos \bar{z} =$
$\cos(x - iy) = \cos x \cos(-iy) - \sin x \cdot \sin(-iy)$

$$= \cos x \cdot \cos(iy) + \sin x \cdot \sin iy$$

$$= \cos x \cosh y + i \sinh y \cdot \sin x \qquad (3)$$

But $\quad \cos z = \cos x \cdot \cosh y - i \sin x \sinh y$

so $\quad \overline{\cos z} = \cos x \cosh y + i \sin x \sinh y \qquad (4)$

Hence from (3) and (4), $\cos \bar{z} = \overline{\cos z}$

(c) $\overline{\tan z} = \overline{\left\{\frac{\sin z}{\cos z}\right\}} = \frac{\overline{\sin z}}{\overline{\cos z}} = \frac{\sin \bar{z}}{\cos \bar{z}} = \tan \bar{z}$

Example 7: Find derivative of cos z.

Solution: $\frac{d}{dz}(\cos z) = \frac{d}{dz}\left(\frac{e^{iz}+e^{-iz}}{2}\right)$
But since $\frac{d}{dz}e^{iz} = ie^{iz}$, we have

$$\frac{d}{dz}(\cos z) = \frac{1}{2}i(e^{iz} - e^{-iz}) = \frac{i^2(e^{iz} - e^{-iz})}{2i}$$

$$= -\left(\frac{e^{iz} - e^{-iz}}{2i}\right) = -\sin z.$$

Example 8: Find the derivative of tanh z.

Solution: $\frac{d}{dz}(\tanh z) = \frac{d}{dz}\left(\frac{\sinh z}{\cosh z}\right)$

$$= \frac{\cosh z \cdot \cosh z - \sinh z \cdot \sinh z}{\cosh^2 z} = \frac{1}{\cosh^2 z} = \sec^2 hz.$$

Example 9: Find all zeros of (a) sin z (b) cos z.

Solution:

a. $\sin z = \sin x \cdot \cosh y + i \cos x \sinh y = 0$ so, $\sin x \cosh y = 0$ and $\cos x \cdot \sinh y = 0$.
Since $\cosh y \geq 1$, $\sin x = 0$ or $x = \pm n\pi, n = 0,$ $1, 2, \ldots$ Then $\cos x = \cos(n\pi) = -1 \neq 0$ so $\sinh y = 0$ or $y = 0$.
Thus the zeros of $\sin z$ are $z = x + iy = \pm n\pi, n = 0, 1, 2$

b. $\cos z = \cos x \cdot \cosh y - i \sin x \cdot \sinh y = 0$ so, $\cos x \cdot \cosh y = 0$ and $\sin x \cdot \sinh y = 0$.
Since $\cosh y \geq 1$, $\cos x = 0$ or $x = \pm(2n - 11)\frac{\pi}{2}, n = 1, 2, 3, \ldots$. Then $\sin x = \sin(2n - 1)\frac{\pi}{2} \neq 0$ so $\sinh y = 0$ or $y = 0$.
Thus the zeros of $\cos z$ are $z = x + iy = \pm(2n - 1)\frac{\pi}{2}, n = 1, 2, \ldots$

Example 10: Find all zeros of (a) sinh z (b) cosh z.

Solution:

a. $\sinh z = \frac{1}{2}(e^z - e^{-z}) = 0$ or $(e^z)^2 - 1 = 0$

so $\quad e^z = \pm 1 = e^{\pm 2n\pi i}, n = 0, 1, 2, \ldots.$

$$= e^{\pm n\pi i} \text{ for } n \text{ odd}$$

Thus zeros of sinh z are

$$z = \pm 2n\pi i, n = 0, 1, 2, 3 \ldots.$$

$$= \pm n\pi i, n \text{ odd}$$

b. $\cosh z = \frac{e^z + e^{-z}}{2} = 0$ or $(e^z)^2 + 1 = 0$
so $e^z = \pm i = 1 \cdot e^{\pm i\frac{\pi}{2} \pm 2n\pi i}, n = 0, 1, 2, \ldots$
zeros are $z = \pm(2n + \frac{1}{2})\pi i, n = 0, 1, 2, \ldots.$

Example 11: If $|z| = 1$ prove that $\frac{z^2-1}{z^2+1} = i\tan\theta$

Solution: Since $|z| = 1$, $z = e^{i\theta}$.
Substituting

$$\frac{z^2-1}{z^2+1} = \frac{e^{i2\theta}-1}{e^{i2\theta}+1} = \frac{e^{i\theta}-e^{-i\theta}}{e^{i\theta}+e^{-i\theta}} = i\frac{\left(\frac{e^{i\theta}-e^{-i\theta}}{2i}\right)}{\left(\frac{e^{i\theta}+e^{-i\theta}}{2}\right)}$$

$$= i\frac{\sin\theta}{\cos\theta} = i\tan\theta.$$

Example 12: If $\tan(x+iy) = A+iB$ show that $A^2+B^2+2A\cot 2x = 1$.

Solution: We know that

$$A = \text{Re}\,(\tan(x+iy)) = \frac{\sin 2x}{\cos 2x + \cosh 2y} \text{ and}$$

$$B = \text{Im}\,(\tan(x+iy)) = \frac{\sinh 2y}{\cos 2x + \cosh 2y}$$

Now $A^2+B^2+2A\cot 2x$

$$= \frac{\sin^2 2x}{(\cos 2x+\cosh 2y)^2} + \frac{\sinh^2 2y}{(\cos 2x+\cosh 2y)^2}$$

$$+ 2\cdot\frac{\sin 2x}{(\cos 2x+\cosh 2y)}\cdot\cot 2x$$

$$= \frac{[\sin^2 2x+\sinh^2 2y+2\cos 2x\cdot(\cos 2x+\cosh 2y)]}{(\cos 2x+\cosh 2y)^2}$$

$$= \frac{1+\sinh^2 2y+2\cos 2x\cdot\cosh y+\cos^2 2x}{(\cos 2x+\cosh 2y)^2}$$

$$= \frac{(\cos 2x+\cosh 2y)^2}{(\cos 2x+\cosh 2y)^2} = 1.$$

EXERCISE

Trigonometric and hyperbolic functions

1. Find the real and imaginary parts of (a) $\tan z$ (b) $\sec z$ (c) $\operatorname{cosec} z$ (d) $\tanh z$

Ans. **a.** $\frac{\sin 2x}{\cos 2x+\cosh 2y}, \frac{\sinh 2y}{\cos 2x+\cosh 2y}$

b. $\frac{2\cos x\cdot\cosh y}{\cos 2x+\cosh 2y}, \frac{2\sin x\cdot\sinh y}{\cos 2x+\cosh 2y}$

c. $\frac{2\sin x\cdot\cosh y}{\cosh 2y-\cos 2x}, \frac{-2\cos x\cdot\sinh y}{\cosh 2y-\cos 2x}$

d. $\frac{\sinh 2x}{\cosh 2x+\cos 2y}, \frac{\sin 2y}{\cosh 2x+\cos 2y}$

2. Find all zeros of (a) $\tan z$ (b) $\cot z$ (c) $\tanh z$ (d) $\coth z$.

Ans. Zeros are the zeros of the numerator
a. $z = \pm n\pi$,
b. $z = \pm(2n-1)\frac{\pi}{2}$
c. $\pm 2n\pi i, n = 0, 1, 2, 3$ and $\pm n\pi i$, n odd
d. $\pm\left(2n+\frac{1}{2}\right)\pi i, n = 0, 1, 2, \ldots$

3. If $\cosh(x+iy) = A+iB$, show that $\frac{A^2}{\cosh^2 x}+\frac{B^2}{\sinh^2 x} = 1$ and $\frac{A^2}{\cos^2 y}-\frac{B^2}{\sin^2 y} = 1$

4. If $\tan(x+iy) = A+iB$ show that
a. $A^2+B^2-2B\cot h2y+1 = 0$
b. $A\sinh 2y = B\sin 2x$

Find all the solutions of the equation

5. $\cos z = 5$

Ans. $\pm 2n\pi \pm 2.292i, n = 0, 1, 2, \ldots$

6. $\cos(2iz+13) = 0$

Ans. $[(2n-1)\pi+26]i/4, n = 0, \pm 1, \pm 2, \ldots$

7. $\cosh z = \frac{1}{2}$

Ans. $\left(\pm\frac{1}{3}\pm 2n\right)\pi i, n = 0, 1, 2, \ldots$

8. $\sin z = 2$

Ans. $2n\pi+\frac{\pi}{2}-i\ln(2+\sqrt{3}), n = 0, 1, 2, \ldots$

9. $\sinh z = i$

Ans. $i\left(2n+\frac{1}{2}\right)\pi, n = 0, 1, 2, \ldots$

10. $\sin z = \cosh 4$

Ans. $n\pi+(-1)^n\left(\frac{\pi}{2}-4i\right), n = 0, 1, 2, \ldots$

11. Show that $\tan z = z$ has only real roots.

Hint: From $\frac{\sin 2x}{\cos 2x+\cosh 2y} = x$, $\frac{\sinh 2y}{\cos 2x+\cosh 2y} = y$, we have $\frac{\sin 2x}{x} = \frac{\sinh 2y}{y}$, $\tan x = x$ has solutions but $\tanh y = y$ has no solutions.

12. If $\cos(\alpha+i\beta) = re^{i\theta}$, prove that

$$e^{2\beta} = \frac{\sin(\alpha-\theta)}{\sin(\alpha+\theta)}.$$

13. Show that $\cos z$, $\cosh z$ are even functions while $\sin z$, $\tan z$, $\sinh z$, $\tanh z$ are odd functions.

Hint: $\cos(-z) = \frac{e^{-z}+e^{-(-z)}}{2} = \frac{e^{-z}+e^{z}}{2} = \cos z$, etc.

Logarithm

For $z = x + iy \neq 0$, the exponential function $z = e^w \neq 0$ for any w. The natural logarithm of z, denoted by $\ln z$ is defined as the inverse of the exponential function $e^w = z$.

Let $w = u + iv$ and $z = x + iy = re^{i\theta}$.

Then $e^w = e^{u+iv} = e^u e^{iv} = z = re^{i\theta}$ so $e^u = r$ or $u = \ln r$, $v = \theta$.

Thus $\ln z = w = u + iv = \ln r + i(\theta + 2k\pi)$ where $k = 0, 1, 2, 3, \ldots$.

Here $\theta = \arg z$ can assume infinitely many values which differ by 2π. Hence the complex natural logarithm $\ln z$ is infinitely many valued (unlike the real ln which is single valued).

Ln z: principal value of ln z is denoted by Ln z is the value of $\ln z$ for the principal value of arg z, Θ which lies in $-\pi < \Theta \leq \pi$. Thus

$$\operatorname{Ln} z = \ln r + i\Theta, \quad (r > 0, -\pi < \Theta \leq \pi).$$

Ln z is single valued since Θ is unique. In general, $\ln z$ can be expressed in terms of Ln z as

$$\ln z = \operatorname{Ln} z \pm 2n\pi i, n = 1, 2, 3, \ldots.$$

Ln z becomes the real natural logarithm when z is positive real ($\because \quad \arg z = 0$).

Analytic

Ln z is analytic everywhere except on the negative real axis ($x = -\pi$).

Derivative

In cartesian form

$$w = \ln z = \ln(x + iy) = \ln r + i\theta$$

$$\ln z = \ln\sqrt{x^2 + y^2} + i\tan^{-1}\left(\frac{y}{x}\right)$$

so $$u = \ln\sqrt{x^2 + y^2}, v = \tan^{-1}\frac{y}{x}$$

Differentiating partially w.r.t. x

$$u_x = \frac{1}{\sqrt{x^2+y^2}} \cdot \frac{1}{2}\frac{1}{\sqrt{x^2+y^2}} \cdot 2x = \frac{x}{x^2+y^2}$$

$$v_x = \frac{1}{1 + \left(\frac{y}{x}\right)^2} \cdot \left(\frac{-y}{x^2}\right) = -\frac{y}{x^2+y^2}$$

Then

$$\frac{d}{dz}(\ln z) = u_x + iv_x$$

$$= \frac{x}{x^2+y^2} + i\frac{(-y)}{x^2+y^2} = \frac{x - iy}{x^2+y^2} = \frac{\bar{z}}{z\bar{z}} = \frac{1}{z}$$

Standard results

For any two complex numbers z_1 and z_2,

a. $\ln(z_1 z_2) = \ln z_1 + \ln z_2$ with $z_1 = r_1 e^{i\theta_1}$, $z_2 = r_2 e^{i\theta_2}$

$$\ln(z_1 z_2) = \ln(r_1 r_2 e^{i\theta_1} e^{i\theta_2}) = \ln(r_1 r_2 e^{i\theta_1 + \theta_2})$$
$$= \ln(r_1 r_2) + i(\theta_1 + \theta_2)$$
$$= (\ln r_1 + \ln r_2) + i\theta_1 + i\theta_2 = \ln z_1 + \ln z_2$$

b. $\ln(z_1/z_2) = \ln z_1 - \ln z_2$

c. $\ln z^{m/n} = \frac{m}{n}\ln z$.

WORKED OUT EXAMPLES

Example 1: Determine all values and the principal value of (a) $\ln(-4)$ (b) $\ln(3i)$ (c) $\ln(\sqrt{3} - i)$.

Solution:

a. $z = x + iy = -4$, $x = -4$, $y = 0$,
$r = x^2 + y^2 = \sqrt{(-4)^2 + 0} = 4$,
$-4 = x = r\cos\theta = 4\cos\theta$,
$\cos\theta = -1$, $\sin\theta = 0 \quad \therefore \theta = \pi$.
$\ln z = \ln(-4) = \ln r + i(\theta \pm 2k\pi) = \ln 4 + i(\pi \pm 2k\pi)$
principal value: $Lnz = \ln 4 + i\pi$

b. $\ln 3i = \ln 3 + i(\frac{\pi}{2} \pm 2k\pi)$, $k = 0, 1, 2, \ldots$
principal value: $\ln 3 + i\pi/2$

c. $\ln(\sqrt{3} - i)$; $z = \sqrt{3} - i$, $x = \sqrt{3}$, $y = -1$;
$r = \sqrt{(\sqrt{3})^2 + (-1)^2} = \sqrt{4} = 2$,
$\sqrt{3} = x = r\cos\theta = 2\cos\theta$.
$\cos\theta = \frac{\sqrt{3}}{2}$, $\sin\theta = -\frac{1}{2}$ so $\theta = 2\pi - \frac{\pi}{6} = \frac{11\pi}{6}$.
$\ln(\sqrt{3} - i) = \ln 2 + i(\frac{11\pi}{6} \pm 2k\pi)$,
$k = 0, 1, 2, \ldots$
principal value $\operatorname{Ln}(\sqrt{3} - i) = \ln 2 + i\frac{11\pi}{6}$.

Example 2: Find the modulus and argument of $(i)^{\sqrt{i}}$

Solution: $(i)^{\sqrt{i}} = e^{i \ln \sqrt{i}} = e^{\frac{1}{2} i \ln i}$

$$= e^{\frac{1}{2} i}\left\{\ln 1 \pm \left(i\frac{\pi}{2} \pm 2n\pi\right)\right\}$$

$$(i)^{\sqrt{i}} = e^{-\frac{1}{2}}\left(2n + \frac{1}{2}\right)\pi$$

modulus is $e^{-\left(2n+\frac{1}{2}\right)\pi}$, argument 0.

Example 3: Find the real and imaginary parts of $\ln \cos(x + iy)$.

Solution: Let $\ln(Re^{i\phi}) = \ln(R\cos\phi + Ri\sin\phi)$

$$= \ln \cos(x + iy)$$

$$= \ln(\cos x \cdot \cosh y - i \sin x \cdot \sinh y).$$

Thus $R\cos\phi = \cos x \cdot \cosh y$,
$R\sin\phi = -\sin x \sinh y$

Then squaring and adding

$$R^2 = \cos^2 x \cosh^2 y + \sin^2 x \cdot \sinh^2 y$$

$$= \left(\frac{1 + \cos 2x}{2}\right)\left(\frac{1 + \cosh 2y}{2}\right) + \left(\frac{1 - \cos 2x}{2}\right)\left(\frac{\cosh 2y - 1}{2}\right)$$

$$= \frac{1}{2}(\cos 2x + \cosh 2y)$$

Now

$$\frac{R\sin\phi}{R\cos\phi} = \tan\phi = \frac{-\sin x \sinh y}{\cos x \cosh y} = -\tan x \cdot \tanh y$$

Hence

$$\ln\cos(x + iy) = \ln(Re^{i\phi}) = \ln R + i\phi$$

$$= \ln\left\{\frac{1}{\sqrt{2}}\sqrt{\cos 2x + \cosh 2y}\right\} - i\tan^{-1}(\tan x \tan hy)$$

Example 4: If $\tan\ln(x + iy) = a + ib$ such that $a^2 + b^2 \neq 1$, show that

$$\tan\ln(x^2 + y^2) = \frac{2a}{1 - a^2 - b^2}.$$

Solution: Let $\tan\ln(x + iy) = \tan\ln(re^{i\theta})$

$$= \tan\{\ln r + i\theta\} = \tan(A + iB) = a + ib$$

We know that

$$a = \text{Re}\tan(A + iB) = \frac{\sin 2A}{\cos 2A + \cosh 2B} \text{ and}$$

$$b = \text{Im}\cdot\tan(A + iB) = \frac{\sinh 2B}{\cos 2A + \cosh 2B}$$

Consider $1 - a^2 - b^2$

$$= 1 - \frac{(\sin^2 2A + \sinh^2 2B)}{(\cos 2A + \cosh 2B)^2}$$

$$= \frac{\cos^2 2A + \cosh^2 2B + 2\cos 2A\cdot\cosh 2B - \sin^2 2A - \sinh^2 2B}{(\cos 2A + \cosh 2B)^2}$$

$$= \frac{\cos^2 2A + 1 - \sin^2 2A + 2\cos 2A\cdot\cosh 2B}{(\cos 2A + \cosh 2B)^2}$$

$$= \frac{2\cos^2 2A + 2\cos 2A\cdot\cosh 2B}{(\cos 2A + \cosh 2B)^2} = \frac{2\cos 2A}{(\cos 2A + \cosh 2B)}$$

So

$$\frac{2a}{1 - a^2 - b^2} = \frac{2\cdot\sin 2A}{\cos 2A + \cosh 2B}\cdot\frac{\cos 2A + \cosh 2B}{2\cos 2A} = \tan 2A$$

$$= \tan(2\ln r) = \tan(2\ln\sqrt{x^2 + y^2})$$

$$\frac{2a}{1 - a^2 - b^2} = \tan(\ln(x^2 + y^2)).$$

Exercise

Find all values and the principal value of:

1. $\ln 1$

Ans. $0, \pm 2n\pi i, n = 0, 1, 2, \ldots, \ln 1 = 0$

2. $\ln 4$

Ans. $1.386294 \pm 2n\pi i, \ln 4 = 1.386294$

3. $\ln(-1)$

Ans. $\pm n\pi i$, n odd, $\ln(-1) = \pi i$

4. $\ln(-4)$

Ans. $1.386294 \pm (2n + 1)\pi$, $\ln(-4) = 1.386 + \pi i$

5. $\ln i$

Ans. $\frac{\pi i}{2}, \frac{-3\pi i}{2}, \frac{5\pi i}{2}, \ldots, \ln i = \frac{\pi i}{2}$

6. $\ln(-4i)$

Ans. $1.386 - \frac{\pi}{2}i \pm 2n\pi i, \ln(-4i) = 1.386 - \frac{\pi i}{2}$

7. $\ln(3 - 4i)$

Ans. $\ln 5 + i \arg(3 - 4i)$

$= 1.609 - 0.927i \pm \cdot 2n\pi i$

$\ln(3-4i) = 1.609 - 0.927i$

8. $\ln i^i$

Ans. $e^{-(4n+1)\frac{\pi}{2}}$

9. $\ln(1-i)$

Ans. $\ln\sqrt{2} + \frac{7\pi i}{4} + 2k\pi i$

principal value: $\frac{1}{2}\ln 2 + \frac{7\pi i}{4}$

10. Find the modulus and argument of $i^{\ln(1+i)}$

Ans. $e^{-\pi^2/8}$, $\frac{\pi}{4}\ln e^2$

11. Determine the real and imaginary parts of $\ln\sin(x+iy)$.

Ans. $\frac{1}{2}\ln\left[\frac{1}{2}(\cosh 2y - \cos 2x)\right]$; $\tan^{-1}(\cot x \cdot \tanh y)$

12. Show that $\ln\left(\frac{a+ib}{a-ib}\right) = i2\tan^{-1}\left(\frac{b}{a}\right)$.

Hence prove that $\cos\left[i\ln\left(\frac{a+ib}{a-ib}\right)\right] = \frac{a^2-b^2}{a^2+b^2}$

General Powers: $w = z^c$

The general power of a complex number $z = x+iy$ with respect to a complex exponent c is defined by

$$z^c = e^{c\ln z} \quad \text{with} \quad z \neq 0$$

z^c is multivalued since $\ln z$ is infinitely many-valued. The principal value of z^c written as

$$\text{principal value : of } z^c = e^{c\,\text{Ln}\,z}$$

where Lnz is principal value of $\ln z$.

For any complex number c

$$c^z = e^{z\ln c}$$

Results

For c, d any complex numbers, $z \neq 0$,

1. $z^{-c} = \frac{1}{z^c}$
2. $z^c z^d = z^{c+d}$
3. $z^c/z^d = z^{c-d}$
4. $(z^c)^n = z^{cn}$ for any integer n.
5. $\frac{d}{dz}z^c = \frac{d}{dz}e^{c\ln z} = e^{c\ln z}\cdot c\cdot\frac{1}{z} = c\frac{e^{c\ln z}}{e^{\ln z}} = ce^{(c-1)\ln z} = cz^{c-1}$
6. Real and imaginary parts of $(\alpha+i\beta)^{x+iy}$: $(\alpha+i\beta)^{x+iy} = e^{(x+iy)\ln(\alpha+i\beta)} = e^{(x+iy)\{\ln r + i(\theta\pm 2n\pi)\}}$.

where $r = |\alpha+i\beta|$, $\theta = \arg(\alpha+i\beta)$.

$(\alpha+i\beta)^{x+iy} = e^{A+iB} = e^A(\cos B + i\sin B)$

where $B = y\ln r + x(2n\pi+\theta)$.

WORKED OUT EXAMPLES

Example 1: Determine all values of $(1+i)^i$.

Solution: $(1+i)^i = e^{i\ln(1+i)}$

Express $1+i$ in polar form: $x=1$, $y=1$, $r=\sqrt{2}$, $\cos\theta = \frac{1}{\sqrt{2}}$, $\sin\theta = \frac{1}{\sqrt{2}}$. Thus $\theta = \frac{\pi}{4} \pm 2n\pi$, $1+i = \sqrt{2}e^{i(\frac{\pi}{4}\pm 2n\pi)}$, $n = 0, 1, 2, 3, \ldots$.

Thus

$$(1+i)^i = e^{i\ln(1+i)} = e^{i\left[\ln\sqrt{2} + i\left(\frac{\pi}{4}\pm 2n\pi\right)\right]}$$

since

$$\ln(z) = \ln|z| + i\arg z.$$

$$= e^{i\ln\sqrt{2}}e^{-\left(\frac{\pi}{4}\pm 2n\pi\right)}$$

$$= e^{-\frac{\pi}{4}\pm 2k\pi}(\cos\ln\sqrt{2} + i\sin\ln\sqrt{2})$$

where $k = 0, 1, 2, 3, \ldots$.

Example 2: Find the principal value of $\sqrt{2i}$.

Solution: $\sqrt{2i} = (2i)^{\frac{1}{2}} = e^{\frac{1}{2}\ln 2i}$

Now $2i = 2\cdot i = 2\cdot e^{i\frac{\pi}{2}\pm 2n\pi i}$

$$\ln 2i = \ln 2 + i\left(\frac{\pi}{2} \pm 2n\pi\right), n = 0, 1, 2, 3, \ldots$$

$$\text{Ln } 2i = \ln 2 + i\frac{\pi}{2} \quad \text{since } -\pi < \frac{\pi}{2} \leq \pi$$

Principal value of $\sqrt{2i} = e^{\frac{1}{2}\text{Ln }2i}$

$$= e^{\frac{1}{2}\left[\ln 2 + i\frac{\pi}{2}\right]}$$

$$= e^{\frac{1}{2}\ln 2}\cdot e^{i\frac{\pi}{4}} = \sqrt{2}\cdot\left(\cos\frac{\pi}{4} + i\sin\frac{\pi}{4}\right)$$

$$= \sqrt{2}\left(\frac{1}{\sqrt{2}} + i\sqrt{2}\right) = 1 + i.$$

Example 3: Find the principal value of $\left\{\frac{1}{2}[e(-1-i\sqrt{3})]\right\}^{3\pi i}$.

Solution: $\left\{\frac{1}{2}[e(-1-i\sqrt{3})]\right\}^{3\pi i}$

$$= \exp\left[3\pi i\left\{\ln\left[\frac{1}{2}e(-1-i\sqrt{3})\right]\right\}\right]$$

$$= \exp\left[3\pi i\left\{\ln e + \ln\left(-\frac{1}{2} - i\frac{\sqrt{3}}{2}\right)\right\}\right]$$

Consider

$$-\frac{1}{2} - i\frac{\sqrt{3}}{2} = x + iy, \text{ so } x = -\frac{1}{2}, y = -\frac{\sqrt{3}}{2}$$

$$r = \sqrt{\frac{1}{4} + \frac{3}{4}} = 1, \cos\theta = -\frac{1}{2}, \sin\theta = -\frac{\sqrt{3}}{2},$$

$$\theta = \pi + \frac{\pi}{3},$$

Thus

$$-\frac{1}{2} - i\frac{\sqrt{3}}{2} = 1 \cdot e^{\frac{i4\pi}{3}}$$

$$\ln\left(-\frac{1}{2} - i\frac{\sqrt{3}}{2}\right) = \ln 1 + i\left(\frac{4\pi}{3} \pm 2n\pi\right), n = 0, 1, 2.$$

Principal argument which lies in $-\pi < \theta \leq \pi$ is $\frac{4\pi}{3} - 2\pi = -\frac{2\pi}{3}$. Then

$$\text{Ln}\left(-\frac{1}{2} - i\frac{\sqrt{3}}{2}\right) = \ln 1 - i\frac{2\pi}{3} = -\frac{i2\pi}{3}$$

$$\text{principal value of } \left\{\frac{1}{2}[e(-1 - i\sqrt{3})]\right\}^{3\pi i}$$

$$= \exp\left[3\pi i\left\{\ln e - \frac{i2\pi}{3}\right\}\right]$$

$$= \exp\left[3\pi i\left\{1 - \frac{2\pi i}{3}\right\}\right] = e^{3\pi i} \cdot e^{2\pi^2} = -e^{2\pi^2}.$$

EXERCISE

1. Determine all values of $1^{\sqrt{2}}$.

Ans. $\cos 2\sqrt{2}k\pi +$
$i \sin 2\sqrt{2}k\pi, k = 0, 1, 2, 3, \ldots.$

2. Find Re $\{(1 - i)^{1+i}\}$

Ans. $e^{\ln\sqrt{2} - 7\pi/4 - 2k\pi} \cdot \cos(7\pi/4 + \ln\sqrt{2})$.

3. Determine the modulus of $(-i)^{-i}$

Ans. $e^{\frac{3\pi}{2} + 2k\pi}, k = 0, 1, 2, 3, \ldots$

Find the prinicipal value of:

4. i^i

Ans. $e^{-\pi/2 \pm 2n\pi}$,
$n = 0$, principal value: $e^{-\frac{\pi}{2}}$

5. $(1 + i)^{2-i}$

Ans. $2e^{\pi/4 \pm 2n\pi}[\sin\ln\sqrt{2} + i\cos\ln\sqrt{2}]$
$n = 0$, principal value: $2e^{\pi/4}[\sin\ln\sqrt{2} + i\cos\ln\sqrt{2}]$
$= 1.490 + 4.126i$

6. 1^i

Ans. $e^{-2k\pi}, k = 0$, principalvalue : 1

Find all values of:

7. i^{-2i}

Ans. $\exp[(4n + 1)\pi], n = 0, \pm 1, \pm 2, \pm 3, \ldots.$.
principal value: e^{π}

8. $(1 - i)^{1+i}$

Ans. $\sqrt{2}(1 - i)e^{(2n + \frac{1}{4})\pi} e^{i\ln\sqrt{2}}$
$n = 0, \pm 1, \pm 2, \ldots,$
principal value: $(1 - i)e^{\pi/4}e^{i\ln\sqrt{2}}$

9. $\ln i^i$

Ans. $-(2n + \frac{1}{2}\pi), n = 0, \pm 1, \pm 2, \ldots.$

Find the real part of the principal value of:

10. $i^{\ln(1+i)}$

Ans. $e^{-\pi^2/8} \cdot \cos(\frac{\pi}{4}\ln 2)$

11. $1^{\sqrt{2}}$

Ans. $\cos 2k\sqrt{2\pi} + i\sin 2k\sqrt{2\pi}, k = 0, 1, 2, 3, \ldots$

Chapter 23

Complex Integration

INTRODUCTION

The advantage of complex integration is that certain complicated real integrals can be evaluated and properties of analytical functions can be established. In this chapter we consider Cauchy's integral theorem which is one of the fundamental theorems of complex function theory. It gives sufficient conditions for a line integral around a simple closed curve to be zero. An important consequence of this theorem is the Cauchy's integral formula in which the value $f(z_0)$ of an analytic function at z_0 is completely determined by an integral of $f(z)$ on any simple closed curve enclosing z_0. It is powerful tool in evaluating certain integrals. The complex Taylor series is a direct generalization of the Taylor series of real function. However Laurent series is different from any series in real calculus and is useful in evaluation of both real and complex integrals and in summation of series.

23.1 LINE INTEGRAL IN COMPLEX PLANE

Introduction

Continuous arc: The set of points (x, y) defined by $x = \phi(t)$, $y = \psi(t)$, with parameter t in the interval (a, b), define a continuous arc provided ϕ and ψ are continuous functions.

Smooth arc If ϕ and ψ are differentiable, arc is said to be smooth.

Simple curve is a curve having no self intersections i.e., no two distinct values of t correspond to the same point (x, y).

Closed curve is one in which end points coincide i.e., $\phi(a) = \phi(b)$ and $\psi(a) = \psi(b)$.

Simple closed curve is a curve having no self intersections and with coincident end points.

Contour is a continuous chain of a finite number of smooth arcs.

Closed contour is a piecewise smooth closed curve without points of self intersection.

Examples: Boundaries of circle, ellipse, rectangle, triangle.

Positive sense

of traversing a contour is the direction such that the interior domain bounded by the given closed contour remains on the left of the direction of motion.

Line Integral

Definite integral or complex line integral or simply line integral of a complex function $f(z)$ from z_1 to z_2 along a curve c is defined as

$$\int_c f(z)dz = \int_c (u + iv)(dx + idy)$$

$$= \int_c (udx - vdy) + i\int_c (vdx + udy)$$

$$= \int_{z_1}^{z_2} (u\dot{x} - v\dot{y})dt + i\int_{z_1}^{z_2} (v\dot{x} + u\dot{y})dt$$

where dot $(\cdot)$ denotes differentiation w.r.t. 't'.

Here c is known as the path of integration. If it is a closed curve, the line integral is denoted by $\oint_c$

(with zero or circle on the integral sign). When the direction is positive sense, it is indicated as $\int_{c+}$ or simply $\int_c$ while negative direction by $\int_{c-}$. Contour integral is an integral along a closed contour.

Basic Properties of Line Integrals

1. Linearity:

$$\int_c \Big(k_1 f(z) + k_2 g(z)\Big) dz = k_1 \int_c f(z)dz + k_2 \int_c g(z)dz$$

2. Sense reversal:

$$\int_A^B f(z)dz = -\int_B^A f(z)dz$$

3. Partitioning of path:

$$\int_c f(z)dz = \int_{c_1} f(z)dz + \int_{c_2} f(z)dz$$

where curve c consists of the curves c_1 and c_2

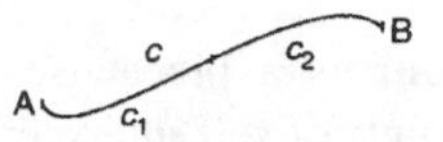

Fig. 23.1

4. *ML-inequality:*

$$\left| \int_c f(z)dz \right| \leq ML$$

where $|f(z)| \leq M$ everywhere on c and L is the length of the curve c.

Note: Although real definite integrals are interpreted as area, no such interpretation is possible for complex definite integrals.

Simply connected domain D is a domain such that every simple closed path in D contains only points of D.

Example 1: Interior of circle, rectangle, triangle, ellipse.

Multiply connected domain is one that is not simply connected (Fig. 23.2).

Example 2: Annulus region, regions with holes.

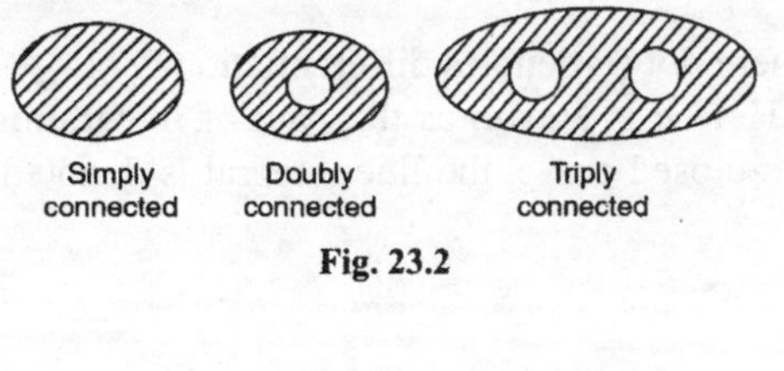

Fig. 23.2

Evaluation of Complex Line Integral

I. *By indefinite integration (of analytic functions):* If $f(z)$ is analytic in a simply connected domain D then

$$\int_{z_1}^{z_2} f(z)dz = F(z_2) - F(z_1)$$

where $F'(z) = \frac{dF}{dz} = f(z)$ in D. (See "Independence of path" on page 23.7.)

II. *By use of the path:*

If curve c is represented by $z = z(t), a \leq t \leq b$

Then

$$\int_c f(z)dz = \int_a^b f(z(t))\frac{dz}{dt}dt$$

i.e., the line integral is converted to an ordinary integral in t by making use of the property (nature) of the curve c.

WORKED OUT EXAMPLES

Example 1: Evaluate $\oint_c |z|^2 dz$ around the square with vertices at (0, 0), (1, 0), (1, 1), (0, 1) (refer Fig. 23.3).

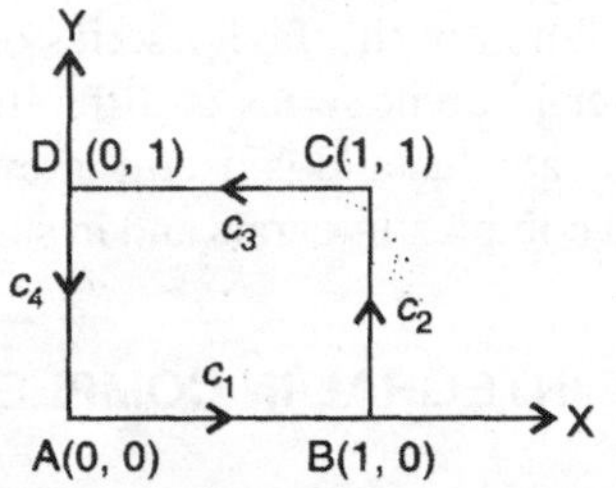

Fig. 23.3

Solution: Square c consists of four curves (lines) $c_1 : AB, c_2 : BC, c_3 : CD, c_4 : DA$.

$$I = \oint_c |z|^2 dz = \int_{c_1+c_2+c_3+c_4} (x^2 + y^2)(dx + idy)$$

$$= \int_{c_1} + \int_{c_2} + \int_{c_3} + \int_{c_4}$$

$$= I_1 + I_2 + I_3 + I_4$$

Along $c_1 : AB : y = 0, 0 \le x \le 1.$

$$I_1 = \int_{c_1} (x^2 + y^2)(dx + idy)$$

$$= \int_0^1 (x^2 + 0)dx = \frac{x^3}{3}\Big|_0^1 = \frac{1}{3}$$

Along $c_2 : BC : x = 1, 0 \le y \le 1$

$$I_2 = \int_{c_2} (x^2 + y^2)(dx + idy)$$

$$= \int_0^1 (1 + y^2)idy = i\left(y + \frac{y^3}{3}\right)\Big|_0^1 = \frac{4}{3}i$$

Along $c_3 : y = 1$ and x varies from 1 to 0

$$I_3 = \int_{c_3} (x^2 + y^2)(dx + idy)$$

$$= \int_1^0 (x^2 + 1)dx = \left(\frac{x^3}{3} + x\right)\Big|_1^0 = -\frac{4}{3}$$

Along $c_4 : x = 0$, y varies from 1 to 0

$$I_4 = \int_{c_4} (x^2 + y^2)(dx + idy)$$

$$= \int_1^0 y^2 idy = i\frac{y^3}{3}\Big|_1^0 = \frac{i}{3}$$

$$I = I_1 + I_2 + I_3 + I_4$$

$$= \frac{1}{3} + \frac{4}{3}i - \frac{4}{3} - \frac{i}{3} = -1 + i$$

Example 2: Evaluate $\int_c (z - z^2)dz$ where c is the upper half of the circle $|z - 2| = 3$ (Fig. 23.4). What is the value of the integral if c is the lower half of the above circle?

Solution: Equation of the circle is $z - 2 = 3e^{i\theta}$ or

$$z = 2 + 3e^{i\theta},\ z^2 = (2 + 3e^{i\theta})^2$$

$$z - z^2 = (2 + 3e^{i\theta}) - (2 + 3e^{i\theta})^2 = -2 - 9e^{i\theta} - 9e^{i2\theta}$$

$$dz = 3ie^{i\theta}d\theta.$$

$$\int_c (z - z^2)dz = -\int_0^\pi (2 + 9e^{i\theta} + 9e^{i2\theta})(i3e^{i\theta})d\theta$$

$$= -3i\left[\frac{2e^{i\theta}}{i} + \frac{9e^{i2\theta}}{2i} + \frac{9e^{i3\theta}}{3i}\right]_0^\pi$$

$$= -3[-4 + 0 - 6] = 30$$

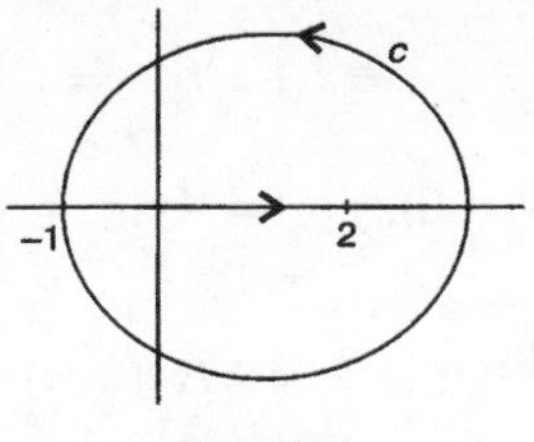

Fig. 23.4

For the lower semicircle

$$\int_c (z - z^2)dz = -\int_\pi^{2\pi} (2 + 9e^{i\theta} + 9e^{i2\theta})(3ie^{i\theta})d\theta$$

$$= -3\left[2e^{i\theta} + 9e^{i2\theta} + 3e^{i3\theta}\right]_\pi^{2\pi} = -30$$

Example 3: Evaluate $\int_c f(z)dz$, where

$$f(z) = \begin{cases} 4y & \text{when } y > 0 \\ 1 & \text{when } y < 0 \end{cases}$$ and c is the arc

from $z = -1 - i$ to $z = 1 + i$ of the cubical curve $y = x^3$ (see Fig. 23.5).

Solution:

$$\int_c f(z)dz = \int_{BOA} = \int_{BO} + \int_{OA} = \int_{c_1} + \int_{c_2}$$

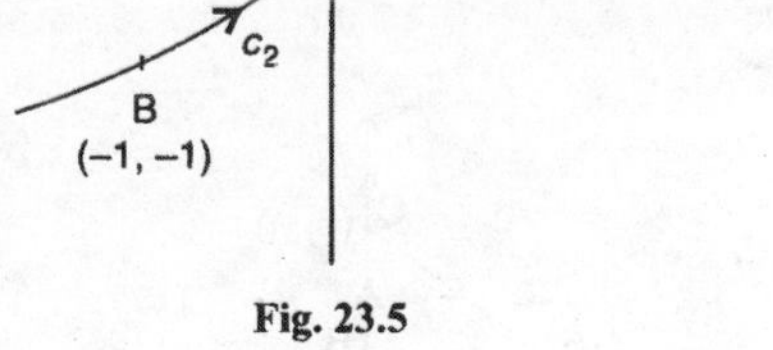

Fig. 23.5

Along $BO : c_2 : (y < 0),\ f(z) = 1,\ y = x^3$

$$dz = dx + idy = dx + i3x^2dx$$

x: varies from -1 to 0

$$\int_{c_2} f(z)dz = \int_{-1}^0 1 \cdot (1 + 3x^2 i)dx$$

$$= x + x^3 i\Big|_{-1}^{0} = 1 + i$$

Along $c_1 : (y > 0)$, $f(z) = 4y$, $y = x^3$, x : varies from 0 to 1

$$\int_{c_1} f(z)dz = \int_0^1 4y(dx + idy)$$

$$= \int_0^1 4y\left(\frac{1}{3}y^{-\frac{2}{3}} + i\right) dy$$

since $x = y^{\frac{1}{3}}$, $dx = \frac{1}{3}y^{-\frac{2}{3}}dy$.

$$= y^{\frac{4}{3}} + 2y^2 i\Big|_0^1 = 1 + 2i$$

Thus

$$\int_c f(z)dz = \int_{c_1} + \int_{c_2} = (1+i) + (1+2i) = 2 + 3i.$$

Example 4: Find an upper bound for the absolute value of the integral $\int_c (e^z - \bar{z})\,dz$ where c denotes the boundary of the triangle with vertices at the points $z = 0, -4, 3i$ (refer Fig. 23.6).

Solution: From the ML-inequality

$$\left|\oint_c f(z)dz\right| \le ML$$

$$\left|\oint_c (e^z - \bar{z})dz\right| = \left|\oint_c e^z dz - \oint_c \bar{z}dz\right|$$

$$\le \left|\oint_c e^z dz\right| + \left|\oint_c \bar{z}dz\right|$$

But $\oint_c e^z dz = 0$ since e^z is analytic everywhere, (see Page 23.6) so $\left|\oint_c \bar{z}dz\right| \le$ (Max $\cdot\,|\bar{z}|$ on c)$\cdot$ (Length of c).

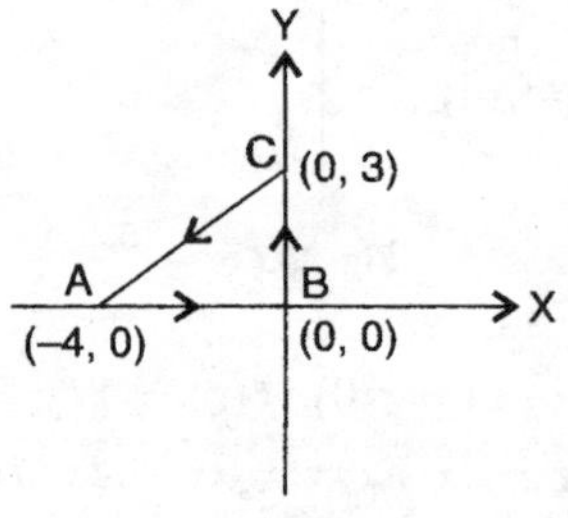

Fig. 23.6

Length of $c = AB + BC + CA = 4 + 3 + 5 = 12$

For straight line maximum occurs at the end points.

$$|\bar{z}| = \sqrt{x^2 + y^2}$$

Along $AB : y = 0$, $|\bar{z}| = 4$

Along $BC : x = 0$, $|\bar{z}| = 3$

Along CA : Max$|\bar{z}|$ = Max[C or A] = Max[4, 3] = 4

Thus maximum value of $\bar{z}$ on c is 4.

$$\text{Hence } \left|\oint_c (e^z - \bar{z})dz\right| \le \left|\oint_c \bar{z}dz\right| \le (4)(12) = 48$$

Example 5: Evaluate $\int_c Re\,z\,dz$ where c is (a) shortest path from $1+i$ to $3+2i$ (b) along the straight line from (1, 1) to (3, 1) and then from (3, 1) to (3, 2) (c) Are the two integrals in (a), (b) equal. If not give reason (see Fig. 23.7).

Solution:

a. The shortest path from A to B is the straight line $y = \frac{x+1}{2}$. So

$$\int_c Re\,z\,dz = \int_{AB} x(dx + i\,dy)$$

$$= \int_1^3 x\left(dx + i\frac{1}{2}dx\right)$$

$$= \frac{x^2}{2} + \frac{x^2}{4}i\Big|_1^3 = 4 + 2i$$

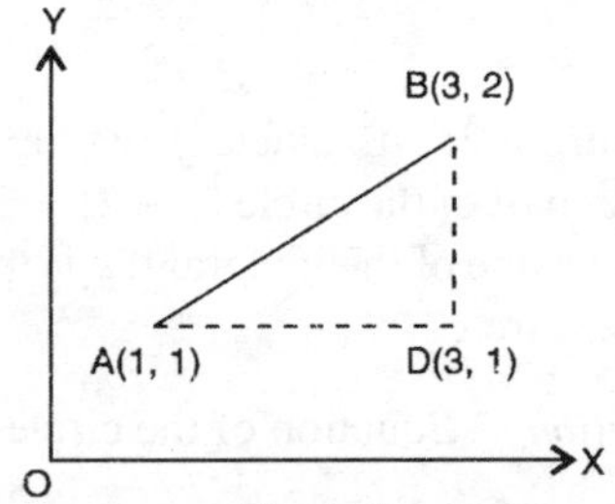

Fig. 23.7

b. $\int_c Re\,z\,dz = \int_{AD+DB} = \int_{AD} + \int_{DB} = I_1 + I_2$.

Along $AD : y = 1$, x : varies from 1 to 3, $dy = 0$

$$I_1 = \int_{AD} Re\,z\,dz = \int_1^3 x\,dx = \frac{x^2}{2}\Big|_1^3 = 4$$

Along $DB : x = 3$, y : varies from 1 to 2, $dx = 0$

$$I_2 = \int_{DB} Re\,z\,dz = \int_1^2 3\cdot i\,dy = 3iy\Big|_1^2 = 3i$$

$$\int_c Re\, z\, dz = I_1 + I_2 = 4 + 3i$$

c. Integral values along path AB is not equal to integral value along ADB since the integrand $f(z) = Re\, z = x$ is not analytic (see Page 609).

Example 6: Evaluate

a. $\int_i^1 (z+1)^2 dz = \int_i^1 (z+1)^2 d(z+1) = \left.\frac{(z+1)^3}{3}\right|_i^1$

$= \frac{8}{3} - \frac{(i+1)^3}{3} = \frac{10-2i}{3}$

b. $\int_0^i z e^{z^2} dz = \frac{1}{2}\int e^t\, dt$ where $t = z^2$

$= \frac{1}{2}e^t = \left.\frac{1}{2}e^{z^2}\right|_0^i = \frac{1}{2}[e^{-1} - 1]$

c. $\int_0^{\pi i} \cos z\, dz = \left.\sin z\right|_0^{\pi i} = \sin(i\pi) = i \sinh \pi$

d. $\int_{-\pi i}^{\pi i} \sin^2 z\, dz = \int_{-\pi i}^{\pi i} \frac{1-\cos 2z}{2} dz = \left.\frac{1}{2}z - \frac{\sin 2z}{4}\right|_{-\pi i}^{\pi i}$

$= \frac{1}{2}(\pi i + \pi i) - \frac{1}{4}[\sin(2\pi i) - \sin(-2\pi i)]$

$= \pi i - \frac{1}{2} i \sinh 2\pi$

e. $\int_0^{2i} \sinh z\, dz = \left.\cosh z\right|_0^{2i} = \cosh 2i - 1 = \cos 2 - 1.$

EXERCISE

Integrate the given function along the given curve c:

1. z^2,
   a. c_1: straight line segment $z = 0$ to $z = 2 + i$
   b. c_2: straight line from $(0, 0)$ to $(2, 0)$ and then from $(2, 0)$ to $(2, 1)$

Ans. (a) $\frac{2}{3} + \frac{11}{3}i$ (b) $\frac{2}{3} + \frac{11}{3}i$

2. $Re\, z$,
   a. c_1: straight line segment from $z = 0$ to $z = 1 + 2i$
   b. c_2: from $(0, 0)$ to $(1, 0)$ along real axis and then from $(1, 0)$ to $(1, 2)$ vertically.

Ans. (a) $\frac{1}{2} + i$ (b) $\frac{1}{2} + 2i$

3. $(\bar{z})^2$ from 0 to $2 + i$
   a. along the line $2y = x$
   b. from $(0, 0)$ to $(2, 0)$ along real axis and then from $(2, 0)$ to $(2, 1)$ vertically

Ans. (a) $\frac{5(2-i)}{3}$ (b) $\frac{(14+11i)}{3}$

4. $12z^2 - 4iz : c : y = x^3 - 3x^2 + 4x - 1$ joining $(1, 1)$ and $(2, 3)$
5. $z^2, c : x = 4 - y^2$ from $y = 2$ to -2,

Ans. $\frac{16i}{3}$

6. $|z|\bar{z}, c$: consisting of upper semicircle $|z| = 1$ and the line segment $-1 \le x \le 1$.

Ans. πi

7. $\bar{z}, c$: (a) circle $|z - 2| = 3$, (b) Square with vertices at $z = 0, 2, 2i, 2 + 2i$, (c) Ellipse $|z - 3| + |z + 3| = 10$.

Ans. (a) $18\pi i$ (b) $8i$ (c) $40\pi i$

8. $\frac{z+2}{z} c$:
   a. Circle of radius 2, centre origin.
   b. Upper semi-circle.
   c. Lower semi-circle.

   Hint: Take $z = 2e^{i\theta}$, θ varies from (a) $-\pi$ to π (b) 0 to π (c) 0 to $-\pi$.

Ans. (a) $4\pi i$ (b) $-4 + 2\pi i$ (c) $-4 - 2\pi i$

9. $8\bar{z} + 3z, c : x^{\frac{2}{3}} + y^{\frac{2}{3}} = a^{\frac{2}{3}}$

Ans. $6\pi i a^2$

10. $\pi e^{\pi \bar{z}}$, c : square with vertices $0, 1, 1 + i, i$

Ans. $4(e^{\pi} - 1)$.

Indefinite integration

Evaluate:

1. $\int_0^{1+i} z^2 dz$ *Ans.* $-\frac{2}{3} + \frac{2}{3}i$
2. $\int_{-\pi i}^{\pi i} \cos z dz$ *Ans.* $2i \sinh \pi$
3. $\int_{8+\pi i}^{8-\pi i} e^{\frac{z}{2}} dz$ *Ans.* 0
4. $\int_{-i}^{i} \frac{dz}{z}$ *Ans.* $i\pi$
5. $\int_{3i}^{1-i} 4z\, dz$ *Ans.* $18 - 4i$
6. $\int_0^1 z e^{2z} dz$ *Ans.* $\frac{1}{4}(e^2 + 1)$
7. $\int_0^{2\pi} z^2 \sin 4z dz$ *Ans.* $-\pi^2$
8. $\int_{2i}^{3} \sin z dz$ *Ans.* $\cosh 2 - \cos 3$
9. $\int_3^{-3} z^{\frac{1}{2}} dz$ *Ans.* $-2\sqrt{3}(1 + i)$

10. $\int_{-1}^{1} z^i dz$ *Ans.* $\frac{1+e^{-\pi}}{2}(1-i)$

11. $\int_{-i}^{i} z \cosh z^2 dz$ *Ans.* 0

12. $\int_{1}^{1+\pi i} e^z dz$ *Ans.* $-2e$

13. $\int_{0}^{1+i\pi}(z^2+\cosh 2z)dz$

Ans. $\frac{1}{3}-\pi^2+\frac{\sinh 2}{2}+\frac{i\pi}{3}(3-\pi^2)$.

ML-inequality

Find an upper bound for the absolute value of the integral, without actually evaluating the integral

1. $\int_c z^2 dz$, c: straight line from 0 to $1+i$

Ans. $2\sqrt{2}$, $M=2$, $L=\sqrt{2}$

2. $\int_c \sin z dz$, c: straight line from 3 to $2i$.

Ans. $\sqrt{13}\sqrt{\cosh 4}$, $M=\sqrt{\cosh^2 2+\sinh^2 2}$, $L=\sqrt{13}$

3. $\int_c \frac{dz}{z(z-2)^3}$, $c: |z|=5$.

Ans. $\frac{2\pi}{27}$, $L=2\pi 5$, $M=\frac{1}{(5)(27)}$

4. $\int_c \frac{(z^2+3)e^{iz}\ln z}{z^2-2}dz$, $c: z=2e^{i\theta}, 0\le\theta\le\frac{\pi}{3}$

Ans. $ML=\left[7(3\ln 2+\pi)\frac{\pi}{9}\right]\left[\frac{2\pi}{3}\right]$

5. $\int_c \frac{dz}{z^2+1}$, $c: |z|=2$ in 1st quadrant.

Ans. $ML=\left(\frac{1}{3}\right)\cdot(\pi)$

6. $\int_c \frac{dz}{z^4}$, c: Line segment from i to 1.

Hint: $y=1-x$, $|z^4|=\left[2\left(x-\frac{1}{2}\right)^2+\frac{1}{2}\right]^2\ge\frac{1}{4}$

Ans. $ML=4\cdot(\sqrt{2})$.

7. Show that $\left|\int_c z^2 dz\right|\le 10$, c: Line from i to $2+i$. Compare with exact value.

Ans. $ML=(5)(2)$, Exact value $=\sqrt{\frac{148}{9}}=4.055$.

23.2 CAUCHY'S INTEGRAL THEOREM

If $f(z)$ is analytic in a simply connected domain D then

$$\oint_c f(z)dz=0$$

for any simple closed curve c lying entirely within D (Fig. 23.8).

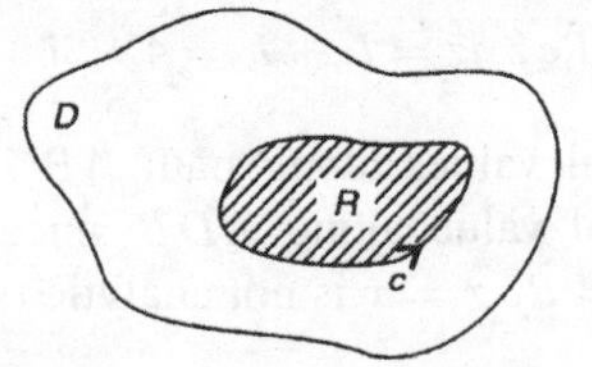

Fig. 23.8

Proof: Consider

$$\oint_c f(z)dz=\oint_c\Big(u(x,y)+iv(x,y)\Big)[dx+idy]$$

$$=\oint_c(u\,dx-v\,dy)+i\oint_c(u\,dy+v\,dx)$$

$$=I_1+I_2$$

u and v have continuous partial derivative in D because $f(z)$ is analytic and f' is assumed to be continuous. Applying Green's theorem in plane for I_1,

$$I_1=\oint_c u\,dx-v\,dy=\int\int_R\left(-\frac{\partial v}{\partial x}-\frac{\partial u}{\partial y}\right)dxdy=0$$

since from Cauchy-Riemann equations $u_y=-v_x$. Here R is the region bounded by c.
Similarly,

$$I_2=\oint_c udy-vdx=\int\int_R\left(\frac{\partial u}{\partial x}-\frac{\partial v}{\partial y}\right)dxdy=0$$

since $u_x=v_y$. Thus

$$\oint f(z)dz=I_1+I_2=0+0=0.$$

Thus Cauchy's theorem establishes one of the basic properties of an analytic function that the integral of an analytic function around any simple closed curve lying entirely in the simply connected domain of its analycity is zero.

Note 1: Cauchy's integral theorem is also known as Cauchy's theorem.

Note 2: Cauchy's theorem without the assumption that f' is continuous is known as Cauchy's-Goursat theorem.

Note 3: Simply connectedness is essential.

Example: $\int_c \frac{dz}{z-1}$, $c: 1<|z-1|<2$. Although $\frac{1}{(z-1)}$ is analytic in the annulus, integral is not zero, since annulus is not simply connected.

Independence of Path

Let $f(z)$ be analytic in a simply connected domain D. Let c_1 and c_2 be any two paths in D joining any two points Z_1 and Z_2 in D and having no further points in common (Fig. 23.9). Then

$$\int_{c_1} f(z)dz = \int_{c_2} f(z)dz$$

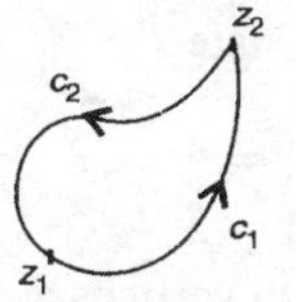

Fig. 23.9

both c_1 and c_2 traversed in the same direction i.e., the integral of $f(z)$ from z_1 to z_2 is independent of the path joining them (however it depends on the points z_1 and z_2).

Proof: The two curves c_1, c_2 together form a simple closed curve c in D. By Cauchy's theorem

$$\oint_c f(z)dz = \int_{c_1} f(z)dz + \int_{c_2} f(z)dz = 0$$

or $$\int_{c_1} f(z)dz = -\int_{c_2} f(z)dz = \int_{c_2} f(z)dz$$

where now c_1 and c_2 are both traversed in the same direction.

Principal of Deformation of Path

From independence of path, it appears as if the path c_2 is obtained by continuous deformation of path c_1.

Cauchy's Theorem for Multiply Connected Domains

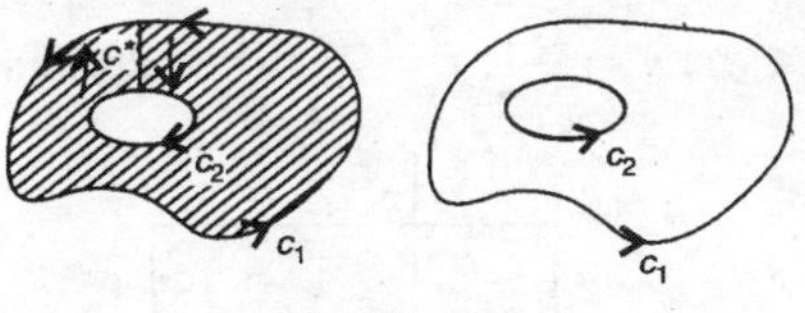

Fig. 23.10

A multiply connected domain can be cut such that the resulting domain is simply connected. For example by one cut c^* the doubly connected domain spreads into a simply connected domain (without any holes) (refer Fig. 23.10). Applying Cauchy's theorem along the oriented boundary B consisting of c_1, c^*, c_2, c^*, we have

$$\oint_B f(z)dz = \int_{c_1+c^*+c_2+c^*} f(z)dz = 0$$

or $$\oint_{c_1} + \int_{c^*} + \oint_{c_2} - \int_{c^*} = 0.$$

Thus $$\oint_{c_1} f(z)dz = \oint_{c_2} f(z)dz$$

where c_1 and c_2 are both traversed in the same direction. Similar result can be obtained for triply connected domain by introducing two cuts c_1^* and c_2^*, resulting

$$\int_{c_1} f(z)dz = \oint_{c_2} f(z)dz + \oint_{c_3} f(z)dz$$

where c_1, c_2, c_3 are all traversed in the same direction (see Fig. 23.11).

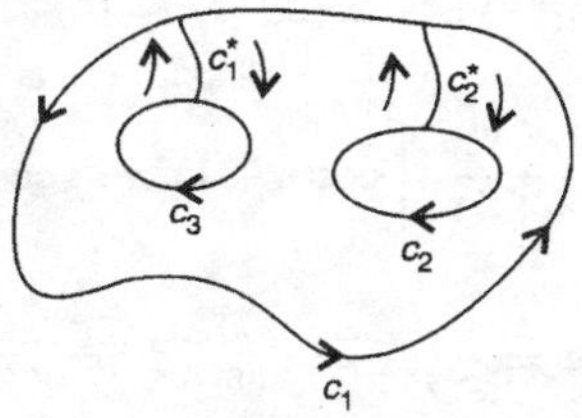

Fig. 23.11

Evaluation of line integrals by indefinite intergration: If $f(z)$ is analytic in simply connected domain D, there exists $F(z)$ such that $F'(z) = f(z)$ and

$$\int_{z_1}^{z_2} f(z)dz = f(z_1) - F(z_2)$$

along any path joining z_1 to z_2. (See WE 6 on page 23.5)

Integral of Integer Powers

Book Work: Prove that

$$I = \oint_c (z-a)^n dz = \begin{cases} 0 & \text{if } n \neq -1, \quad a \text{ inside } c \\ 2\pi i & \text{if } n = -1, \quad a \text{ inside } c \\ 0 & \text{if } n = -1, \quad a \text{ outside } c \end{cases}$$

Here n is any integer (positive, negative or zero) a is a complex number. Curve c is oriented counterclockwise and is

a. Circle with centre at a and of radius R.

b. Any arbitrary simple closed (piecewise smooth) curve (refer Fig. 23.12).

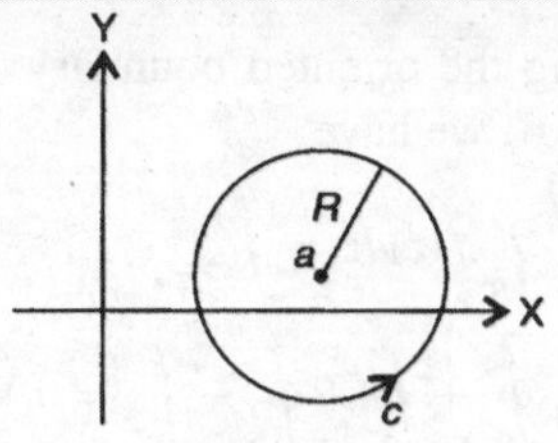

Fig. 23.12

Proof:

a. c: circle $z - a = Re^{i\theta}$, i.e., a is inside c.

$$I = \int_0^{2\pi} (R\,e^{i\theta})^n \cdot (i\,Re^{i\theta})d\theta$$

$$I = i\,R^{n+1} \int_0^{2\pi} e^{i(n+1)\theta}\,d\theta$$

Suppose $n \neq -1$,

$$I = i\,R^{n+1} \cdot \left. \frac{e^{i(n+1)\theta}}{i(n+1)} \right|_0^{2\pi}$$

$$I = \frac{i\,R^{n+1}}{(n+1)} \cdot \left[\cos(n+1)2\pi - \cos 0 \right] = 0$$

Suppose $n = -1$, $I = i\int_0^{2\pi} d\theta = 2\pi i$

Suppose $n = -1$ and a is outside circle c.

$I = \int_c \frac{dz}{z-a} = 0$ by Cauchy's theorem since $\frac{1}{z-a}$ is analytic inside c.

b. c: any simple closed curve enclosing a circle c^*: $z - a = Re^{i\theta}$ i.e., c^* is circle with centre at a and of radius R (Fig. 23.13).

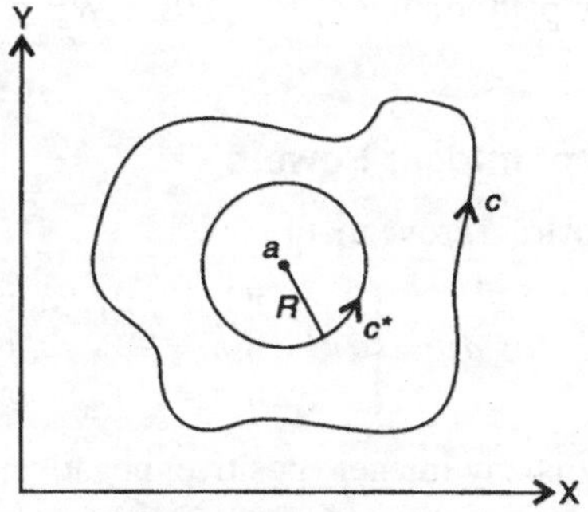

Fig. 23.13

Suppose a is inside c. Then $(z-a)^n$ for any n is analytic on and between c and c^*. By Cauchy's theorem for multiply connected domains.

$$\oint_c (z-a)^n dz = \int_{c^*} (z-a)^n dz = \begin{cases} 0, & n \neq -1 \\ 2\pi i, & n = -1. \end{cases}$$

Suppose a is outside c. Then by Cauchy's theorem

$$\int_c (z-a)^n dz = 0$$

Result: $\oint_c \frac{dz}{z} = 2\pi i$, where c: unit circle $|z| = 1$.

Example: Evaluate $I = \oint_c \frac{dz}{z-2}$ around

a. Circle $|z-2| = 4$
b. Circle $|z-1| = 5$
c. Rectangle with vertices at $3 \pm 2i$, $-2 \pm 2i$
d. Triangle with vertices at (0, 0), (1, 0), (0, 1).

Solution:

a. $2 \in c$, by above book work with $n = -1$, $I = 2\pi i$ (Fig. 23.14).

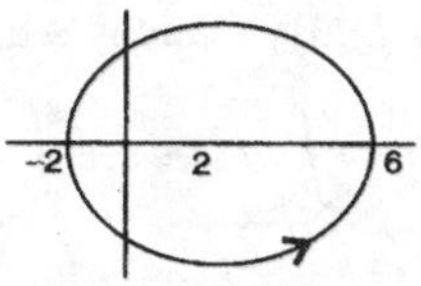

Fig. 23.14

b. $2 \in c : n = -1, I = 2\pi i$ (Fig. 23.15).

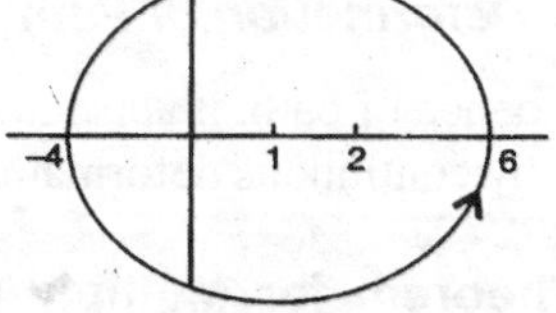

Fig. 23.15

c. $2 \in c, : n = -1, I = 2\pi i$ (Fig. 23.16)

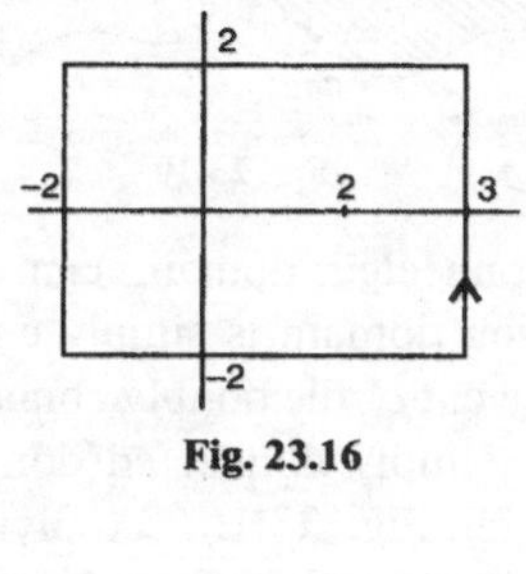

Fig. 23.16

d. $2 \notin$ triangle. $\frac{1}{z-2}$ is analytic in c. By Cauchy's theorem $I = 0$ (Fig. 23.17).

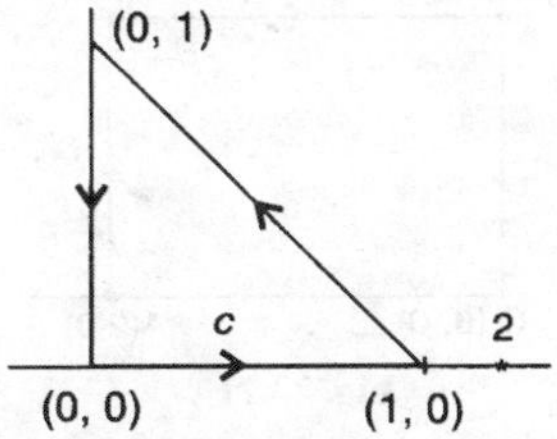

Fig. 23.17

Example: Evaluate $\int_c \frac{1}{z^3} dz, c : |z| = 1$
Integral is zero by above book work. Thus f is analytic is a sufficient condition but not necessary.

WORKED OUT EXAMPLES

Example 1: Evaluate $\oint_c (5z^4 - z^3 + 2)dz$ around

a. Unit circle $|z| = 1$
b. Square with vertices at $(0, 0), (1, 0), (1, 1), (1, 0)$
c. Curve consisting of the parabola $y = x^2$ from $(0, 0)$ to $(1, 1)$ and $y^2 = x$ from $(1, 1)$ to $(0, 0)$.

Solution: $f(z) = 5z^4 - z^3 + 2$ is analytic everywhere. So by Cauchy's integral theorem, $\oint_c f(z)dz$ equals to zero for any simple closed curve. The curves in (a), (b) and (c) are all simple closed (piecewise smooth) curves. Therefore

$$\oint_c (5z^4 - z^3 + 2)dz = 0$$

for c in (a), (b) and (c) i.e., in all cases.

Example 2: Verify Cauchy's integral theorem for $f(z) = z^2$ taken over the boundary of a square with vertices at $\pm 1 \pm i$ in counter-clockwise direction (Fig. 23.18).

Solution:

a. *By Cauchy's theorem,* $\oint_c z^2 dz = 0$ since z^2 is analytic everywhere and c the boundary of the square is a closed piecewise-smooth simple curve

b. *By direct evaluation,*

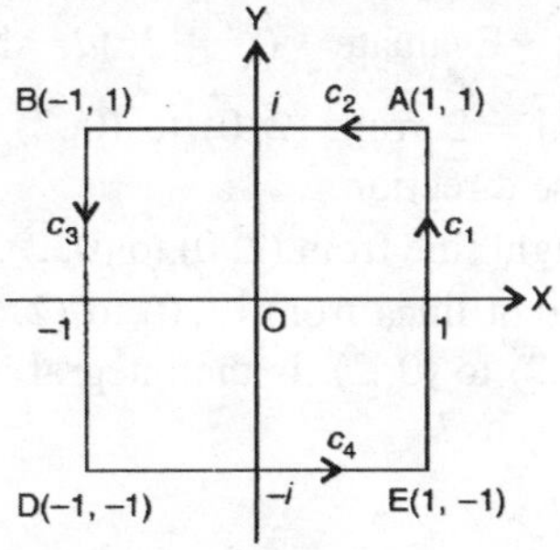

Fig. 23.18

The boundary of square c consists of four curves (lines) c_1, c_2, c_3, c_4. So

$$I = \oint_c z^2 dz = \int_{c_1} + \int_{c_2} + \int_{c_3} + \int_{c_4}$$
$$= I_1 + I_2 + I_3 + I_4.$$

Along c_1: $EA : x = 1, -1 \le y \le 1, dz = i dy$

$$I_1 = \int_{c_1} z^2 dz = \int_{-1}^{1} (1 + iy)^2 i dy$$
$$= \int_{-1}^{1} (1 - y^2 + 2iy) i dy = \frac{4}{3}$$

Along c_2: $AB, y = 1, -1 \le x \le 1, dz = x$

$$I_2 = \int_{c_2} z^2 dz = \int_{1}^{-1} (x + i)^2 dx$$
$$= \int_{1}^{-1} (x^2 - 1 + 2ix) dx = -\frac{4}{3}$$

Along c_3: $BD, x = -1, -1 \le y \le 1, dz = i dy$

$$I_3 = \int_{c_3} z^2 dz = \int_{1}^{-1} (-1 + iy)^2 i dy$$
$$= \int_{1}^{-1} (1 - y^2 - 2iy) i dy = \frac{4}{3}$$

Along c_4: $DE : y = -1, -1 \le x \le 1, dz = dx$

$$I_4 = \int_{c_4} z^2 dz = \int_{-1}^{1} (x - i)^2 dx$$
$$= \int_{1}^{-1} (x^2 - 1 - 2ix) dx = -\frac{4}{3}$$

So $$\oint_c z^2 dz = I_1 + I_2 + I_3 + I_4$$
$$= \frac{4}{3} - \frac{4}{3} + \frac{4}{3} - \frac{4}{3} = 0.$$

Example 3: Evaluate $\int_c (z^2+3z)dz$ along:

a. Circle $|z|=2$ from (2, 0) to (0, 2) in counter-clockwise direction.

b. The straight line from (2, 0) to (0, 2).

c. The straight lines from (2, 0) to (2, 2) and then from (2, 2) to (0, 2). Is the integral independent of path? Why?

Solution:

a. $c_1 : |z| = 2$ or $z = 2e^{i\theta}$ (Fig. 23.19).

$$\int_{c_1}(z^2+3z)dz$$

$$= \int_0^{\frac{\pi}{2}} [4e^{i2\theta}+6e^{i\theta}]2ie^{i\theta}d\theta = 4i\left[\frac{2e^{i3\theta}}{3i}+\frac{3e^{i2\theta}}{2i}\right]_0^{\frac{\pi}{2}}$$

$$= -\frac{8}{3}(i+1) - 12 = -\frac{44}{3} - \frac{8}{3}i$$

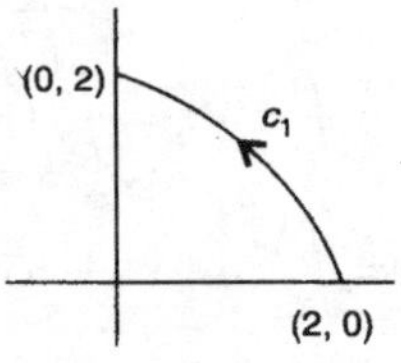

Fig. 23.19

b. $c_2 : x + y = 2; y = 2 - x, dz = (1 - i)dx$, (Fig. 23.19)

$$\int_2^0 \left\{\left[x+i(2-x)\right]^2 + 3\left[x+i(2-x)\right]\right\}(1-i)dx$$

$$= \int_2^0 \Big[x^2-(2-x)^2+2ix(2-x)+3x$$

$$+3i(2-x)\Big](1-i)dx = -\frac{44}{3}-\frac{8}{3}i$$

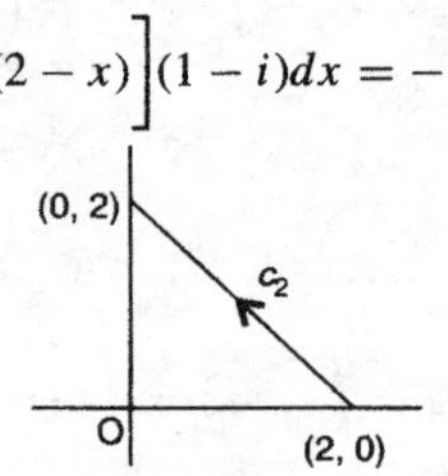

Fig. 23.20

c. $c : c_1 + c_2$ Along $c_1 : x = 2, 0 \le y \le 2$, $dz = idy$ (Fig. 23.21)

$$\int_{c_1} = \int_0^2 \left[(2+iy)^2 + 3(2+iy)\right]idy$$

$$= -14 + \frac{52}{3}i$$

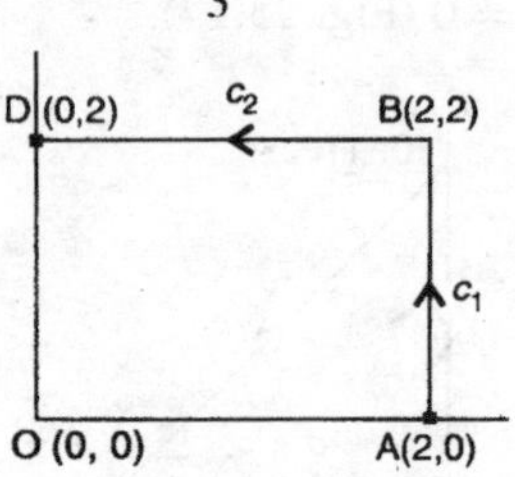

Fig. 23.21

Along c_2 : $y = 2, 0 \le x \le 2, dz = dx$

$$\int_{c_2} = \int_2^0 \left[(x+2i)^2+3(x+2i)\right]dx = -\left[\frac{2}{3}+20i\right]$$

$$\int_c = \int_{c_1} + \int_{c_2} = -14 + \frac{52}{3}i - \frac{2}{3} - 20i = -\frac{44}{3} - \frac{8}{3}i$$

Yes. The integral is independent of path because the integrand $f(z) = z^2 + 3z$ is analytic, everywhere.

Example 4: Evaluate $\oint_c \frac{2z^3+z^2+4}{z^4+4z^2}dz$ where c is the circle $|z-2| = 4$, clockwise.

Solution: Integrand has 3 singular points at $z = 0, \pm 2i$, all of which lie inside the circle c. By partial fractions (see Fig. 23.22).

$$\frac{2z^3+z^2+4}{z^4+4z^2} = \frac{Az+B}{z^2} + \frac{D}{z+2i} + \frac{E}{z-2i}$$

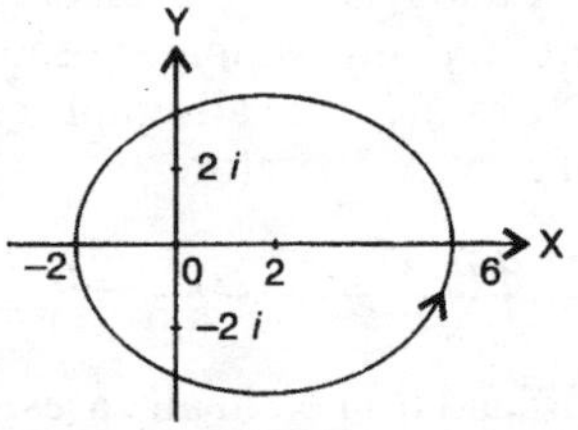

Fig. 23.22

$A = 0, B = D = E = 1.$

$$I = \oint_c \frac{2z^3+z^2+4}{z^4+4z^2}dz$$

$$= \oint_c \frac{dz}{z^2} + \oint_c \frac{dz}{z+2i} + \oint_c \frac{dz}{z-2i}$$

By principle of deformation of path

$$= \oint_{c_1} \frac{dz}{(z-0)^2} + \oint_{c_2} \frac{dz}{z-(-2i)} + \oint_{c_3} \frac{dz}{z-2i}$$

where c_1, c_2, c_3 are circles with centres at $0, -2i, 2i$ and such that they are non-overlapping inside c. By applying the book work

$$\int_c (z-a)^n dz = \begin{cases} 0, & n \neq -1 \\ 2\pi i, & n = -1 \end{cases}$$

$$I = 0 - 2\pi i - 2\pi i = -4\pi i$$

The minus sign is because curve is traced in the clockwise direction.

Example 5: If n is a positive integer, show that

$$\int_0^{2\pi} e^{\sin n\theta} \cos(\theta - \cos n\theta) d\theta$$

$$= \int_0^{2\pi} e^{\sin n\theta} \cdot \sin(\theta - \cos n\theta) d\theta = 0$$

Solution: Consider $\int_c e^{z^n} dz$ where $c : |z| = 1$. So

$z = x + iy = \sin\theta + i\cos\theta,\ z^n = \sin n\theta + i\cos n\theta,$
$dz = (\cos\theta - i\sin\theta)d\theta,\ \theta :$ varies from 0 to 2π.

Thus

$$\int_c e^{z^n} dz = \int_0^{2\pi} e^{(\sin\theta + i\cos\theta)^n} \cdot (\cos\theta - i\sin\theta) d\theta$$

$$= \int_0^{2\pi} e^{\sin n\theta + i\cos n\theta} \cdot (\cos\theta - i\sin\theta) d\theta$$

$$= \int_0^{2\pi} e^{\sin n\theta} \cdot \left[\cos(\cos n\theta) + i\sin(\cos n\theta)\right] \times [\cos\theta - i\sin\theta] d\theta$$

$$= \int_0^{2\pi} \left[e^{\sin n\theta} \cdot \cos(\theta - \cos n\theta) - i\sin(\theta - \cos n\theta)\right] d\theta$$

Since e^{z^n} is analytic, by Cauchy's theorem $\int_c = 0$. Hence the results follows.

Exercise

Using Cauchy's integral theorem evaluate the $\oint_c f(z)dz$ where $f(z)$ and c are:

1. **a.** e^z
   b. $\sin z$
   c. $\cos z$
   d. $z^n, n = 0, 1, 2, 3, \ldots$
   c is any simple clothed path.

Ans. (a), (b), (c) and (d) are all zero.

2. (a) $\sec z$, (b) $\frac{1}{z^2+4}; c : |z| = 1$

Ans. **a.** 0 since all singularities $z = \pm\frac{\pi}{2}, \pm\frac{3\pi}{2}, \ldots \notin c$
b. 0 since $z = \pm 2i \notin c$

3. $\frac{(z^2-z+1)}{(z-2)}, c : |z-1| = \frac{1}{2}$

Ans. 0, since $z = 2$ singularity $\notin c$.

4. $f(z) = \frac{1}{\left(z^2(z-2)(z-4)\right)}$

Hint: By partial fractions $\frac{3}{32}\frac{1}{z} + \frac{1}{8}\frac{1}{z^2} - \frac{1}{8}\frac{1}{z-2} + \frac{1}{32}\frac{1}{z-4}$ use book work and Cauchy's theorem.

Ans. $-\frac{\pi i}{16}$

5. $\frac{1}{\left(z^2(z^2+9)\right)}, c : 1 < |z| < 2$

Hint: Apply theorem for doubly connected domain.

Ans. 0

6. Verify Cauchy's theorem for
   a. c is the square with vertices at $\pm 1 \pm i$ and $f(z) = 3z^2 + iz - 4$, $5\sin 2x$, $3\cosh(z+2)$.

Ans. 0

   b. $f(z) = z^3 - iz^2 - 5z + 2i$.
   i. c is the circle $|z| = 1$
   ii. circle $|z-1| = 2$
   iii. ellipse $|z-3i| + |z+3i| = 20$

Ans. 0

   c. z^3 taken over the boundary of the rectangle with vertices at $-1, 1, 1+i, -1+i$.

Ans. 0

7. Show that $\int_0^{2\pi} e^{\cos\theta} \cdot \cos(\theta + \sin\theta) d\theta = \int_0^{2\pi} e^{\cos\theta} \sin(\theta + \sin\theta) d\theta = 0$.

Hint: $\int e^z dz = 0$ by theorem, $c : |z| = 1$, $z = \cos\theta + i\sin\theta$.

23.3 CAUCHY'S INTEGRAL FORMULA

Let $f(z)$ be analytic in a simply connected domain D. Let c be any simple closed curve in D enclosing any point z_0 in D. Then

$$\oint_c \frac{f(z)}{z-z_0}dz = 2\pi i\, f(z_0)$$

where c is traversed in counter clockwise direction.

Proof: Consider

$$\oint_c \frac{f(z)}{z-z_0}dz = \oint_c \frac{f(z_0)+\left[f(z)-f(z_0)\right]}{z-z_0}dz$$

$$= f(z_0)\oint_c \frac{dz}{z-z_0} + \oint_c \frac{f(z)-f(z_0)}{z-z_0}dz$$

$$= I_1 + I_2$$

$$= 2\pi i \cdot f(z_0) + I_2$$

since the integral value in I_1 is $2\pi i$ by book work. Now by showing I_2 is zero, the proof is complete. Construct a circle c_1 with z_0 as centre and of radius R_1 such that c_1 is non-overlapping with c and lies completely inside c (Fig. 23.23).

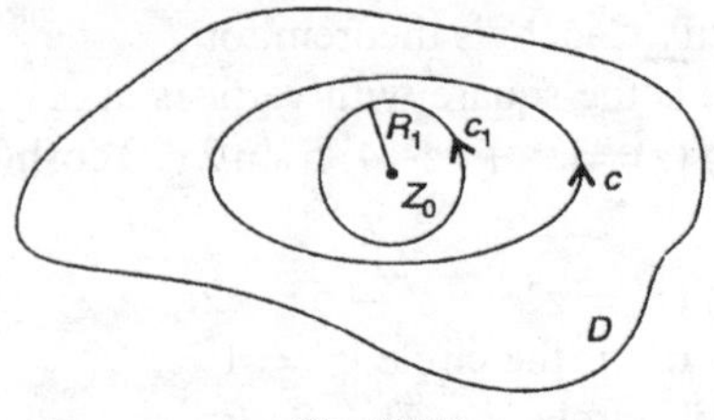

Fig. 23.23

Then by Cauchy's theorem for multiply connected domain

$$I_2 = \oint_c \frac{f(z)-f(z_0)}{z-z_0}dz = \oint_{c_1} \frac{f(z)-f(z_0)}{z-z_0}dz,$$

since $\frac{f(z)-f(z_0)}{z-z_0}$ is analytic in between c and c_1. Since f is analytic, it is continuous. For $\epsilon > 0$ there exists $\delta > 0$ such that

$$\left|f(z)-f(z_0)\right| < \epsilon \quad \text{whenever } |z-z_0| < \delta$$

choose $R_1 < \delta$. Then

$$\left|\frac{f(z)-f(z_0)}{z-z_0}\right| < \frac{\epsilon}{R_1}$$

Using ML-inequality

$$|I_2| = \left|\oint_{c_1} \frac{f(z)-f(z_0)}{z-z_0}dz\right| \le \frac{\epsilon}{R_1}\cdot 2\pi R_1 = 2\pi\epsilon$$

since length of the circle c_1 is $2\pi R_1$.
As $\epsilon \to 0$, $I_2 \to 0$. Hence the result.

Note 1: The Cauchy's integral formula is an important consequence of Cauchy's theorem which establishes relation between value of analytic function at interior point z_0 and boundary value of the function.

Note 2: Cauchy's integral formula also known as Cauchy's formula is useful in evaluating integrals, in obtaining Taylor series, etc.

WORKED OUT EXAMPLES

Example 1: Determine $F(2)$, $F(4)$, $F(-3i)$, $F'(i)$, $F''(-2i)$ if $F(\alpha) = \oint_c \frac{5z^2-4z+3}{z-\alpha}dz$ where c is the ellipse $16x^2+9y^2 = 144$ (Fig. 23.24).

Solution: Curve c is

$$\left(\frac{x}{3}\right)^2 + \left(\frac{y}{4}\right)^2 = 1$$

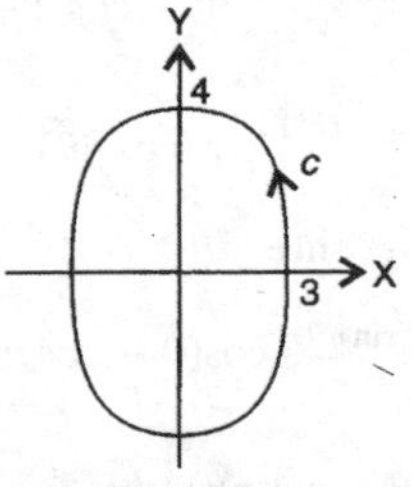

Fig. 23.24

use Cauchy's integral formula, with $f(z) = 5z^2 - 4z + 3$ and $z_0 = \alpha$.

a. $z_0 = 2$ lies inside c. So by Cauchy's integral formula

$$F(2) = \oint_c \frac{5z^2-4z+3}{z-2}dz = 2\pi i f(z)\Big|_{z=2}$$

$$F(2) = 2\pi i\cdot(5z^2-4z+3)\Big|_{\text{at } z=2} = 30\pi i$$

b. $z_0 = 4$ lies outside c. So by Cauchy's theorem $f(4) = 0$ since $\frac{(5z^2-4z+3)}{(z-4)}$ is analytic inside c.

c. $z_0 = -3i$, lies inside c, by Cauchy's integral formula

$$F(-3i) = 2\pi i \cdot (5z^2 - 4z + 3)\Big|_{\text{at } z=-3i}$$
$$= -\pi(6 + 90i)$$

d. For any α, by Cauchy's integral formula

$$F(\alpha) = \oint_c \frac{5z^2 - 4z + 3}{z - \alpha} dz$$
$$= 2\pi i(5z^2 - 4z + 3)\Big|_{\text{at } z=\alpha}$$
$$F(\alpha) = 2\pi i(5\alpha^2 - 4\alpha + 3)$$

So $\quad F'(\alpha) = 2\pi i(10\alpha - 4)$

and $\quad F''(\alpha) = 20\pi i.$

Thus $\quad F'(i) = 2\pi i(10i - 4) = -4\pi(5 + 2i)$

$$F''(-2i) = 20\pi i.$$

Example 2: Evaluate $\oint_c \frac{dz}{z^2+9}$ where c is (a) $|z - 3i| = 4$ (b) $|z + 3i| = 2$ (c) $|z| = 5$

Solution: The integrand $f(z) = \frac{1}{z^2+9}$ has two singular points at $z = \pm 3i$

a. $c_1 : |z - 3i| = 4$. The singular point $z_0 = 3i$ lies inside c_1 and the singular point $z = -3i$ lies outside c_1 (Fig. 23.25).

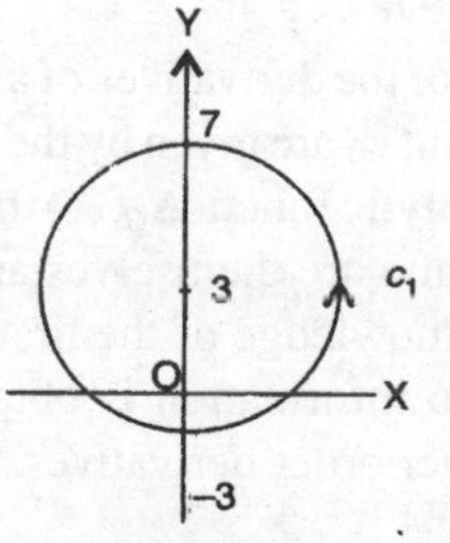

Fig. 23.25

By Integral formula

$$I_1 = \oint_{c_1} \frac{dz}{z^2 + 9} = \oint_{c_1} \frac{dz}{(z + 3i)(z - 3i)}$$
$$= \oint_{c_1} \left(\frac{1}{z + 3i}\right) \cdot \frac{dz}{z - 3i}$$

where $f(z) = \frac{1}{z+3i}$ and $z_0 = 3i$

$$I = 2\pi i \cdot f(3i) = 2\pi i \cdot \frac{1}{3i + 3i} = \frac{\pi}{3}.$$

b. $c_2 : |z + 3i| = 2$ contains $z = -3i$ but $z = 3i$ lies outside c_2 (Fig. 23.26).

So by integral formula

$$I_2 = \oint_{c_2} \left(\frac{1}{z - 3i}\right) \cdot \frac{dz}{z - (-3i)} = 2\pi i \frac{1}{z - 3i}\Big|_{z=-3i}$$
$$I_2 = 2\pi i \cdot \frac{1}{-3i - 3i} = -\frac{\pi}{3}$$

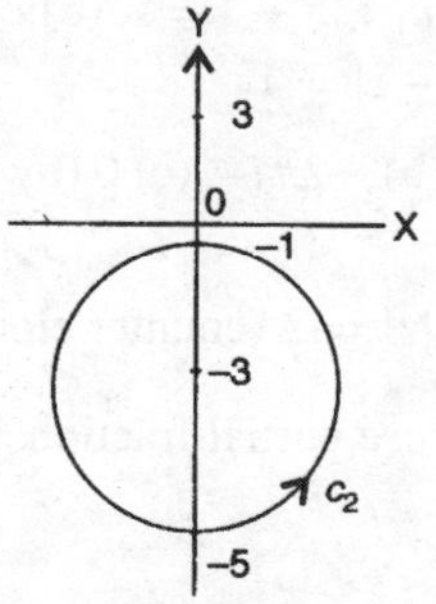

Fig. 23.26

c. $c : |z| = 5$, contains both the singular points $z = \pm 3i$. Construct two circles c_1 and c_2 with these points as centres such that c_1, c_2 do not overlap and lie completely inside c. Then $f(z) = \frac{1}{z^2+9}$ is analytic in the multiply connected domain in between c, c_1 and c_2 (Fig. 23.27).

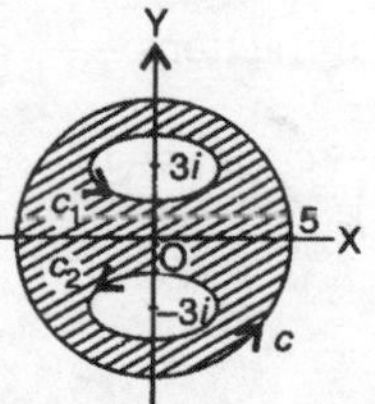

Fig. 23.27

So by Cauchy's theorem

$$\oint_c \frac{1}{z^2 + 9} dz = \oint_{c_1} \frac{dz}{z^2 + 9} + \oint_{c_2} \frac{dz}{z^2 + 9}$$
$$= \frac{\pi}{3} - \frac{\pi}{3} = 0$$

using (a) and (b) above.

EXERCISE

Integrate the given function around the given contour c:

1. $\frac{e^z}{(z^2+1)}$, (a) $c: |z-i|=1$ (b) $|z+i|=1$

Ans. (a) $\pi(\cos 1 + i\sin 1)$ (b) $-\pi(\cos 1 - i\sin 1)$

2. $\frac{(z^3-6)}{(2z-i)}$, $c: |z|=1$

Ans. $\frac{\pi}{8} - 6\pi i$

3. $\frac{(z^2+1)}{(z^2-1)}$, (a) $c_1: |z-1|=1$, (b) $c_2: |z+1|=1$, (c) $c_3: |z-i|=1$

Ans. (a) $2\pi i$ (b) $-2\pi i$ (c) 0 (by Cauchy's theorem)

4. $\frac{\tan z}{(z^2-1)}$, $c: |z| = \frac{3}{2}$ (counter clockwise)

Ans. $2\pi i \tan 1$, use partial fractions: $\frac{1}{z^2-1} = \frac{1}{z-1} - \frac{1}{z+1}$

5. $\frac{(z^2-z+1)}{(z-1)}$, (a) $c: |z|=1$, (b) $|z| = \frac{1}{2}$

Ans. (a) $2\pi i$ (b) 0 (by Cauchy's theorem)

6. $\frac{e^{2z}}{\left((z-1)(z-2)\right)}$, $c: |z|=3$

Ans. $2\pi i(e^4 - e^2)$

7. $\frac{\cos \pi z}{(z^2-1)}$, c : rectangle with vertices $\pm 2 \pm i$

Ans. 0 (use partial fractions $\frac{1}{z^2-1} = \frac{1}{z-1} - \frac{1}{z+1}$)

8. $\frac{\cos(2\pi z)}{\left((2z-1)(z-3)\right)}$, $c: |z|=1$

Ans. $\frac{2\pi i}{5}$

9. $\frac{\sin z}{(z^2-iz+2)}$ (a) $c_1: |z+2|=2$ (b) c_2: rectangle with vertices at $(1,0)$, $(1,3)$, $(-1,3)$, $(-1,0)$ (c) c_3: rectangle with vertices at $(2,0)$, $(2,3)(-2,3)$, $(-2,-3)$.

Ans. (a) 0 (by Cauchy's theorem)
(b) $i\frac{2\pi}{3} \cdot \sinh 2$ (c) $i\frac{2\pi}{3}(\sinh 2 + \sinh 1)$

10. $\frac{(z+4)}{(z^2+2z+5)}$, $c: |z+1-i|=2$

Ans. $\frac{\pi}{2}(3+2i)$

11. $\frac{(\sin \pi z^2 + \cos \pi z^2)}{(z-1)(z-2)}$, $c: |z|=3$

Ans. $4\pi i$

12. Determine $F(4)$, $F(+i)$, $F'(-1)$, $F''(-i)$ if

$$F(\alpha) = \int_c \frac{4z^2+z+5}{z-\alpha} dz,$$

$$c: \text{ellipse } \left(\frac{x}{2}\right)^2 + \left(\frac{y}{3}\right)^2 = 1$$

Ans. $0, 2\pi(i-1), -14\pi i, 16\pi i$.

23.4 DERIVATIVE OF ANALYTIC FUNCTIONS

Formally differentiating the Cauchy's formula under the integral sign w.r.t. z_0

$$\oint_c \frac{(-1)\cdot f(z)}{(z-z_0)^2} \cdot (-1) dz = \oint_c \frac{f(z)}{(z-z_0)^2} dz = 2\pi i f'(z_0).$$

Similarly differentiating once more w.r.t., z_0

$$2\pi i f''(z_0) = \oint_c \frac{f(z)}{(z-z_0)^3} \cdot 2!\, dz$$

In general, after n differentiations w.r.t., z_0, we get **the generalized Cauchy's integral formula**

$$\oint_c \frac{f(z)}{(z-z_0)^{n+1}} dz = \frac{2\pi i}{n!} f^{(n)}(z_0)$$

or

$$f^{(n)}(z_0) = \frac{n!}{2\pi i} \oint_c \frac{f(z)}{(z-z_0)^{n+1}} dz, \quad n = 1, 2, \ldots$$

Thus the values of the derivatives of an analytic function $f(z)$ at a point z_0 are given by the above formula. Hence for an analytic function $f(z)$ the derivatives of all orders exist and are themselves analytic. In contract, from the knowledge of the first derivative of a real function, no information is obtained about the second and higher order derivatives.

Note: In Cauchy's integral formula for multiply connected domains, c is replaced by the oriented boundary B of the multiply connected domain.

WORKED OUT EXAMPLES

Example 1: Evaluate $\oint_c \frac{e^{3z}dz}{(z-\ln 2)^4}$ where c is the square with vertices at $\pm 1 \pm i$ (Fig. 23.28).

Solution: Here $z_0 = \ln 2 = 0.69315$ lies inside the square c. Using generalized Cauchy's integral formula

$$I = \oint_c e^{3z} \cdot \frac{dz}{(z - \ln 2)^{3+1}} = \frac{2\pi i}{3!} \cdot \frac{d^3}{dz^3}(e^{3z})\bigg|_{\text{at } z=\ln 2}$$

$$= \frac{\pi i}{3} 27e^{3z}\bigg|_{z=\ln 2} = 9\pi i e^{3\ln 2} = 72\pi i$$

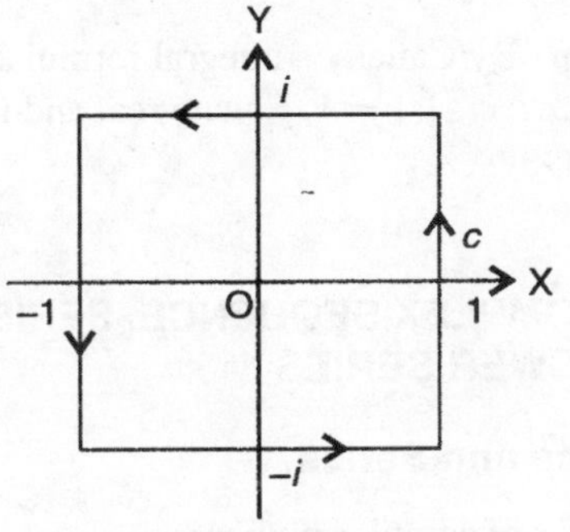

Fig. 23.28

Example 2: Evaluate $\oint_c \frac{e^z dz}{z(1-z)^3}$ where c is (a) $|z| = \frac{1}{2}$ (b) $|z - 1| = \frac{1}{2}$ (c) $|z| = 2$ (Fig. 23.29).

Solution: $z = 0, 1$ are the singular points

a. $c_1 : |z| = \frac{1}{2}$ contains 0 but does not contain $z = 1$.

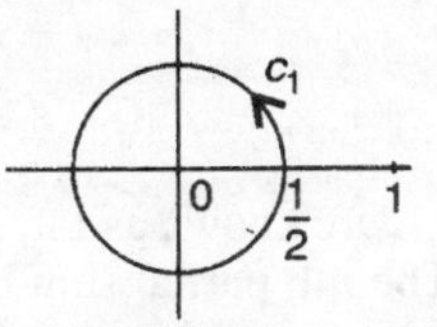

Fig. 23.29

Using integral formula

$$I_1 = \oint_{c_1} \frac{e^z}{(1-z)^3} \cdot \frac{dz}{z - 0} = 2\pi i \cdot \left[\frac{e^z}{(1-z)^3}\right]_{\text{at } z=0}$$

$$= 2\pi i$$

since $\frac{e^z}{(1-z)^3}$ is analytic inside c_1.

b. $c_2 : |z - 1| = \frac{1}{2}$ contains 1 but not 0.

Rewriting

$$I_2 = \oint_{c_2} \frac{e^z}{z} \cdot \frac{dz}{(1-z)^{2+1}}.$$

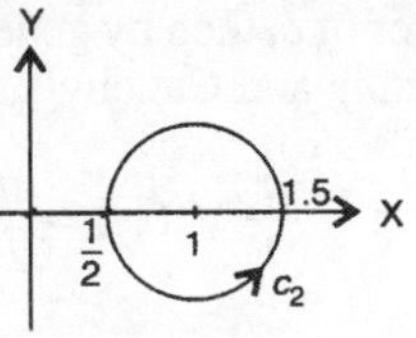

Fig. 23.30

Since $\frac{e^z}{z}$ is analytic inside c_2 (Fig. 23.30), applying generalized formula with $n = 2$,

$$I_2 = \oint_{c_2} -\frac{e^z}{z} \frac{dz}{(z-1)^{2+1}} = -\frac{2\pi i}{2!} \frac{d^2}{dz^2}\left(\frac{e^z}{z}\right)\bigg|_{\text{at } z=1}$$

$$I_2 = -\pi i \left(\frac{ze^z - e^z}{z^2} - \frac{z^2e^z - 2ze^z}{z^4}\right)_{\text{at } z=1} = -\pi e i$$

c. $c : |z| = 2$ contains both the singular points $z = 0, 1$. Construct two circles c_1, and c_2 with centres at 0 and 1 such that they do not overlap and lie completely inside c (Fig. 23.31). Then $f(z) = \frac{e^z}{(z(1-z)^3)}$ is analytic in the region between c, c_1, and c_2. By Cauchy's theorem for multiply connected domain

$$\oint_c \frac{e^z}{z(1-z)^3} dz = \oint_{c_1} \frac{e^z}{z(1-z)^3} dz + \oint_{c_2} \frac{e^z}{z(1-z)^3} dz$$

$$= 2\pi i - \pi e i = \pi i(2 - e)$$

where results (a) and (b) above are used.

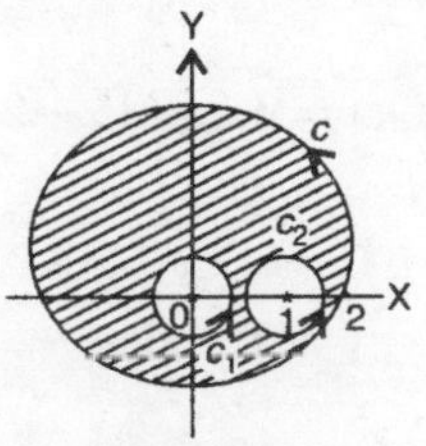

Fig. 23.31

Example 3: Show that $\int_c \frac{f'(z)dz}{z-z_0} = \int_c \frac{f(z)dz}{(z-z_0)^2}$ if f is analytic within and on a simple closed curve c and z_0 is not on c.

Solution:

i. If z_0 is outside c then $\frac{f'}{(z-z_0)}$ and $\frac{f}{(z-z_0)^2}$ are both analytic in c and by Cauchy's theorem both the integrals are zero.

ii. If z_0 is interior of c, then by generalized Cauchy's integral formula and Cauchy's integral formula

$$\int_c \frac{f(z)}{(z-z_0)^2}dz = \frac{2\pi i}{1!}f'(z_0) = \int_c \frac{f'(z)}{z-z_0}dz$$

EXERCISE

Integrate the given function around the given curve traversed in counter-clockwise direction:

1. $\frac{\cos z}{(z-\pi i)^2}$, $c : |z| = 5$

Ans. $2\pi \sinh \pi$

2. $\frac{(z^4-3z^2+6)}{(z+i)^3}$, $c : |z| = 2$

Ans. $-18\pi i$

3. $\frac{e^z}{(z-1)^2(z^2+4)}$, $c : |z-1| = \frac{1}{2}$

Ans. $\frac{6e\pi}{25}$

4. $\frac{\sin^2 z}{\left(z-\frac{\pi}{6}\right)^3}$, $c : |z| = 1$

Ans. πi

5. $\frac{e^{2z}}{(z+1)^4}$, $c : |z| = 3$

Ans. $\frac{8\pi i e^{-2}}{3}$

6. $\frac{e^z}{(z^2+\pi^2)^2}$, $c : |z| = 4$

Ans. $\frac{i}{\pi}$

7. $\frac{(z^3-z)}{(z-2)^3}$, $c_1 : |z| = 3$, $c_2 : |z-2| = 1$, $c_3 : |z| = 1$

Ans. $12\pi i, 12\pi i, 0$

8. $\frac{e^z}{z^3}$, $c : |z| = 1$

Ans. πi

9. $\frac{\sin 2z}{(z+3)(z+1)^2}$, c : rectangle with vertices at $3+i, -2+i, -2-i, 3-i$

Ans. $\frac{\pi i(4\cos 2+\sin 2)}{2}$

10. $z^{-2n-1}\cos z$, $c : |z| = 1$

Hint: 2nth derivative of $\cos z$ is $(-1)^n \cos z$

Ans. $(-1)^n \frac{2\pi i}{(2n)!}$

11. $\frac{e^{zt}}{(z^2+1)^2}$, $c : |z| = 3, t > 0$

Ans. $\pi i(\sin t - t\cos t)$

12. $\frac{\tan\left(\frac{z}{2}\right)}{(z-1-i)^2}$, c : rectangle with vertices at $\pm 2 \pm 2i$

Ans. $\pi i \sec^2\left(\frac{(1+i)}{2}\right)$

13. $\frac{\cosh z}{z^4}$, $c : |z| = \frac{1}{2}$

Ans. 0

14. Show that $\int_0^\pi e^{k\cos\theta}\cdot\cos(k\sin\theta)d\theta = \pi$ and $\int_0^\pi e^{k\cos\theta}\cdot\sin(k\sin\theta)d\theta = 0$.

Hint: By Cauchy's integral formulae $\int_c \frac{e^{kz}}{z}dz = 2\pi i$, $c : |z| = 1$. Equate real and imaginary parts.

23.5 COMPLEX SEQUENCE, SERIES AND POWER SERIES

Sequence and Series

A **sequence** of complex numbers $z_1, z_2, z_3, \ldots z_n, \ldots$ is an assignment to each positive integer n a complex number z_n called the term of the sequence. A sequence denoted by $\{z_n\}$ is said to be convergent if $\lim_{n\to\infty} z_n = c =$ (finite quantity) otherwise it is said to be divergent.

An **infinite complex series or infinite series or simply series** is the sum of the terms of a given sequence of complex numbers $z_1, z_2, \ldots z_n, \ldots$ denoted by

$$\sum_{n=1}^{\infty} z_n = z_1 + z_2 + z_3 + \cdots + z_n + \cdots.$$

$z_1, z_2, \ldots z_n \ldots$ are known as 1st, 2nd, ..., nth ... term of series. The nth partial sum of a series is the sum of the first n terms of the series, denoted by

$$S_n = z_1 + z_2 + \cdots + z_n = \sum_{m=1}^{n} z_m$$

A series is said to be convergent to a sum S if $\lim_{n\to\infty} S_n = S$ i.e., if sequence of partial sums converges. Otherwise the series is said to be divergent.

Power Series

A **power series** in powers of $(z-z_0)$ is a series of the form

$$\sum_{n=0}^{\infty} a_n(z-z_0)^n = a_0 + a_1(z-z_0) + a_2(z-z_0)^2 + \cdots \quad (1)$$

Here $a_0, a_1, a_2 \cdots$ are complex (or real) constants known as coefficients of the series, (1) z is a complex variable and z_0 is called the centre of the series. (1) is also known as power series about the point z_0.

Power series in powers of z is

$$\sum_{n=0}^{\infty} a_n z^n = a_0 + a_1 z + a_2 z^2 + \cdots \quad (2)$$

obtained as a particular case with $z_0 = 0$ in (1).

Region of convergence

of a series is the set of all points z for which the series converges.

Power series converges in a disk

Three distinct possibilities exist regarding the region of convergence of a power series (1).

i. The series converges only at the point $z = z_0$.

ii. The series converges for all z (i.e., in the whole plane).

iii. The series converges everywhere inside a circular disk $|z - z_0| < R$ and diverges everywhere outside the disk $|z - z_0| > R$. Here R is known as the radius of convergence and the circle $|z - z_0| = R$ as the circle of convergence.

Note: Series may converge or diverge at the points on the circle of convergence.

Examples:

1. Geometric series: $\sum_{m=0}^{\infty} z^m = 1 + z + z^2 + \cdots$ converges absolutely when $|z| < 1$ and diverges when $|z| > 1$ i.e., $R = 1$.

2. Power series: $\sum_{n=0}^{\infty} \frac{z^n}{n!}$ converges for all z i.e., $R = \infty$

3. $\sum_{n=0}^{\infty} n^n z^n$ diverges for all z except $z_0 = 0$ i.e., $R = 0$.

Termwise integration of power series:

$$\int_C \sum_{n=0}^{\infty} c_n (z - z_0)^n dz = \sum_{n=0}^{\infty} c_n \int_C (z - a)^n dz$$

Power series represent analytic functions. Conversely every analytic function can be represented as a power series known as the Taylor series.

A function $f(z)$ can be expanded about a singular point z_0 as a Laurent series containing both positive and negative integer powers of $(z - z_0)$.

23.6 TAYLOR'S SERIES (THEOREM)

A function $f(z)$ which is analytic at all points within a circle c_2 with centre at z_0 and of radius R_2 can be represented uniquely as a convergent power series in c_2 known as the **Taylor's series** given by

$$f(z) = \sum_{n=0}^{\infty} a_n (z - z_0)^n \quad (1)$$

where $a_n = \frac{f^{(n)}(z_0)}{n!}$.

Proof: Let z be any arbitrary point inside a circle c_1 with centre at z_0 and of radius $R_1 < R_2$. (see Fig. 23.32).

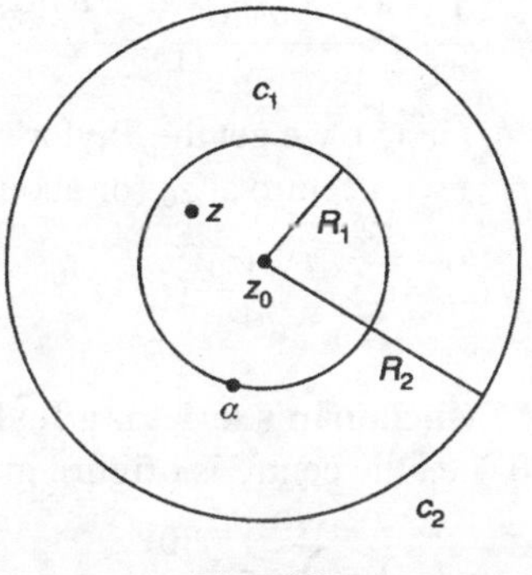

Fig. 23.32

By Cauchy's integral formula, we have

$$f(z) = \frac{1}{2\pi i} \int_{c_1} \frac{f(\alpha)}{\alpha - z} d\alpha \quad (2)$$

The integrand of (2) can be written using binomial series as

$$\frac{1}{\alpha - z} = \frac{1}{\alpha - z_0 + z_0 - z} = \frac{1}{\alpha - z_0} \cdot \frac{1}{1 - \frac{z - z_0}{\alpha - z_0}}$$

$$= \frac{1}{\alpha - z_0} \sum_{n=0}^{\infty} \left(\frac{z - z_0}{\alpha - z_0} \right)^n$$

$$= \sum_{n=0}^{\infty} \frac{(z-z_0)^n}{(\alpha - z_0)^{n+1}} \tag{3}$$

because $\left|\frac{z-z_0}{\alpha-z_0}\right| < 1$ (i.e., $|z-z_0| < |\alpha - z_0|$)

Putting (3) in (2) and integrating termwise

$$f(z) = \frac{1}{2\pi i}\int_{c_1} f(\alpha)\left(\sum_{n=0}^{\infty}\frac{(z-z_0)^n}{(\alpha-z_0)^{n+1}}\right) d\alpha$$

$$= \sum_{n=0}^{\infty} \frac{1}{2\pi i}\left[\int_{c_1}\frac{f(\alpha)\cdot d\alpha}{(\alpha-z_0)^{n+1}}\right](z-z_0)^n \tag{4}$$

$$f(z) = \sum_{n=0}^{\infty} a_n (z-z_0)^n \tag{5}$$

where $a_n = \frac{1}{2\pi i}\int_{c_1}\frac{f(\alpha)d\alpha}{(\alpha-z_0)^{n+1}}$

$$= \frac{1}{2\pi i}\int_{c}\frac{f(\alpha)d\alpha}{(\alpha-z_0)^{n+1}} \tag{6}$$

Since $f(z)$ is analytic within c_2, by Cauchy's theorem the circle c_1 in (4) can be replaced by any closed contour c completely lying inside c_2. By Cauchy's integral formula for derivatives

$$a_n = \frac{1}{2\pi i}\int_{c}\frac{f(\alpha)d\alpha}{(\alpha-z_0)^{n+1}} = \frac{f^{(n)}(z_0)}{n!} \tag{7}$$

Substituting (7) in (5), we get the Taylor's series expansion of $f(z)$ with centre at z_0 (or about z_0) as

$$f(z) = \sum_{n=0}^{\infty} \frac{f^{(n)}(z_0)}{n!}(z-z_0)^n \tag{8}$$

Corollary: **Maclaurin's series** is a Taylor's series about $z_0 = 0$ (i.e., the centre is origin), given by

$$f(z) = \sum_{n=0}^{\infty} \frac{f^{(n)}(0)}{n!} z^n \tag{9}$$

Standard Maclaurin's Series

1. $e^z = \sum_{n=0}^{\infty} \frac{z^n}{n!}$ for $|z| < \infty$

   since $f^{(n)}(z) = \frac{d^n}{dz^n}e^z = e^z$, $f^{(n)}(0) = e^0 = 1$, for any n.

2. $\sin z = \sum_{n=0}^{\infty} (-1)^n \frac{z^{2n+1}}{(2n+1)!}$ for $|z| < \infty$.

3. $\cos z = \sum_{n=0}^{\infty} (-1)^n \frac{z^{2n}}{(2n)!}$ for $|z| < \infty$.

4. $\sinh z = \sum_{n=0}^{\infty} \frac{z^{2n+1}}{(2n+1)!}$ for $|z| < \infty$.

5. $\cosh z = \sum_{n=0}^{\infty} \frac{z^{2n}}{(2n)!}$ for $|z| < \infty$.

6. $\frac{1}{1-z} = \sum_{n=0}^{\infty} z^n$ for $|z| < 1$

   (known as **geometric series**).

7. $\text{Ln}(1+z) = z - \frac{z^2}{2} + \frac{z^3}{3!} + \cdots$ for $|z| < 1$

   Hint: nth derivative $= (-1)^{n-1}\cdot(n-1)!(1+z)^{-n}$.

8. $\frac{1}{(1-z)^m} = \sum_{n=0}^{\infty} \frac{(m+n-1)!}{(m-1)!n!} z^n$ for $|z| < 1$

   $= 1 + mz + \frac{m(m+1)}{2!}z^2 + \frac{m(m+1)(m+2)}{3!}z^3 + \cdots$

   (**known as binomial series** for any positive integer m).

9. $\frac{1}{(1+z)^m} = \sum_{n=0}^{\infty}\binom{-m}{n} z^n = 1 - mz + \frac{m(m+1)}{2!}z^2 - \frac{m(m+1)(m+2)}{3!}z^3 + \cdots$

WORKED OUT EXAMPLES

Find the Taylor's series expansion of $f(z)$ about the indicated point. Determine the region of convergence:

Example 1: e^z about $z = a$

Solution: *Direct method:*

Taylor's series of $f(z)$ is

$$f(z) = \sum \frac{f^{(n)}(a)}{n!}(z-a)^n$$

Here $f(z) = e^z$, $f^{(n)}(z) = e^z$, $f^{(n)}(a) = e^a$

So

$$e^z = \sum \frac{e^a}{n!}(z-a)^n = e^a \sum_{n=0}^{\infty} \frac{(z-a)^n}{n!}$$

Indirect method:

$$e^z = e^{z-a+a} = e^a \cdot e^{z-a}$$

$$= e^a \cdot \sum_{n=0}^{\infty} \frac{(z-a)^n}{n!}, \quad |z-a| < \infty.$$

Example 2: $f(z) = \frac{a}{bz+c}$ about z_0

Solution:

$$f(z) = \frac{a}{bz+c} = \frac{a}{bz - bz_0 + bz_0 + c}$$

$$= \frac{1}{(bz_0 + c)}\left[\frac{a}{1 + \frac{b(z-z_0)}{bz_0+c}}\right]$$

put $bz_0 + c = d,\ \frac{b}{d} = e$

$$= \frac{a}{d}\left[\frac{1}{1 + e(z - z_0)}\right]$$

Expanding by binomial series

$$= \frac{a}{d}\left[\sum_{n=0}^{\infty}(-1)^n e^n (z - z_0)^n\right] \quad \text{if } |e(z - z_0)| < 1$$

$$f(z) = \frac{a}{bz_0 + c}\sum_{n=0}^{\infty}(-1)^n\left(\frac{b}{bz_0 + c}\right)^n (z - z_0)^n$$

if $|z - z_0| < \frac{1}{e} = \frac{d}{b}$.

Example 3: $f(z) = \frac{1}{4-3z}$, $z_0 = 1 + i$

Solution: In the previous problem

$a = 1, b = -3, c = 4, z_0 = 1 + i, d = b\,z_0 + c$

$= -3(1 + i) + 4 = 1 - 3i$

$$f(z) = \frac{1}{1 - 3i}\sum_{n=0}^{\infty}(-1)^n\left(\frac{-3}{1 - 3i}\right)^n\left(z - (1 + i)\right)^n.$$

Region of convergence

$$|z - (1 + i)| < \frac{bz_0 + c}{b} = \left|\frac{1 - 3i}{-3}\right|$$

or
$$\left|z - (1 + i)\right| < \frac{\sqrt{10}}{3}.$$

Example 4: $f(z) = \frac{1}{z^2 - z - 6}$ about (a) $z = -1$, (b) $z = 1$.

Solution: By partial fractions

$$f(z) = \frac{1}{z^2 - z - 6} = \frac{1}{(z-3)(z+2)}$$

$$= \frac{1}{5}\left[\frac{1}{z-3} - \frac{1}{z+2}\right]$$

a. About $z = -1$ i.e., in powers of $(z + 1)$

$$f(z) = \frac{1}{5}\left[\frac{1}{z-3} - \frac{1}{z+2}\right]$$

$$= \frac{1}{5}\left[\frac{1}{z + 1 - 4} - \frac{1}{1 + (z + 1)}\right]$$

$$= \frac{1}{5}\frac{1}{-4\left[1 - \left(\frac{z+1}{4}\right)\right]} - \frac{1}{5}\frac{1}{1 + (z + 1)}$$

Expanding by binomial series

$$= -\frac{1}{20}\sum_{n=0}^{\infty}\left(\frac{z+1}{4}\right)^n - \frac{1}{5}\sum(-1)^n(z + 1)^n$$

$$= \frac{1}{20}\sum_{n=0}^{\infty}\frac{(-4)^{n+1} - 1}{4^n}(z + 1)^n$$

Valid for $|z + 1| < 1$ i.e., region of convergence is the interior of circle with centre at $z = -1$ and radius 1.

b. About $z = 1$ i.e., in powers of $(z - 1)$

$$f(z) = \frac{1}{5}\left[\frac{1}{z-3} - \frac{1}{z+2}\right]$$

$$= \frac{1}{5}\left[\frac{1}{z - 1 - 2} - \frac{1}{z - 1 + 3}\right]$$

$$= \frac{1}{5}\left[\frac{1}{-2\left(1 - \left(\frac{z-1}{2}\right)\right)}\right] - \frac{1}{5}\cdot\frac{1}{3}\frac{1}{1 + \left(\frac{z-1}{3}\right)}$$

Expanding by binomial series

$$f(z) = -\frac{1}{10}\sum_{n=0}^{\infty}\left(\frac{z-1}{2}\right)^n - \frac{1}{15}\sum_{n=0}^{\infty}(-1)^n\left(\frac{z-1}{3}\right)^n$$

valid when $\left|\frac{z-1}{2}\right| < 1$ and $\left|\frac{z-1}{3}\right| < 1$
Region of convergence is $|z - 1| < 2$ (common region).

Example 5: $f(z) = \frac{1}{(2z+1)^3}$ about (a) $z = 0$ (b) about $z = 2$.

Solution:

a. About $z = 0$

From binomial series

$$\frac{1}{(1-z)^m}=\sum_{n=0}^{\infty}\frac{(m+n-1)!}{(m-1)!n!}, \qquad |z|<1$$

with $m=3$, we get

$$\frac{1}{(1+2z)^3}=\sum_{n=0}^{\infty}\frac{(-1)^n(3+n-1)!}{2!n!}(2z)^n \quad \text{if } |2z|<1$$

$$=\sum\frac{(-1)^n}{2}(n+2)(n+1)\cdot 2^n\cdot z^n$$

$$\text{if } |z|<\frac{1}{2}$$

b. About $z=2$

$$\frac{1}{(1+2z)^3}=\frac{1}{(1+2z-4+4)^3}$$

$$=\frac{1}{5^3}\frac{1}{\left[1+\frac{2}{5}(z-2)\right]^3}$$

$$=\frac{1}{125}\sum_{n=0}^{\infty}(-1)^n\cdot\frac{(n+2)(n+1)}{2}\times$$

$$\times\left(\frac{2}{5}\right)^n(z-2)^n$$

with region of convergence $\left|\frac{2}{5}(z-2)\right|<1$

$$\text{i.e.,}\quad |z-2|<\frac{5}{2}.$$

Example 6: $f(z)=\cosh z$ (a) z_0 (b) $z_0=\pi i$

Solution: **a.**

$$f(z)=\cosh z=\frac{e^z+e^{-z}}{2}$$

$$=\frac{e^{z-z_0+z_0}+e^{-z+z_0-z_0}}{2}$$

$$=\frac{1}{2}e^{z_0}\cdot e^{z-z_0}+\frac{e^{-z_0}}{2}\cdot e^{-(z-z_0)}$$

$$=\frac{1}{2}e^{z_0}\cdot\sum\frac{(z-z_0)^n}{n!}+\frac{e^{-z_0}}{2}\cdot\sum\frac{(-1)^n(z-z_0)^n}{n!}$$

b. put $z_0=\pi i$

$$f(z)=-\frac{1}{2}\sum\frac{(z-\pi i)^n}{n!}-\frac{1}{2}\sum\frac{(z-\pi i)^n\cdot(-1)^n}{n!}$$

$$f(z)=-\sum_{n=0}^{\infty}\frac{(z-\pi i)^{2n}}{(2n)!} \quad \text{if } |z-\pi i|<\infty.$$

Example 7: $f(z)=e^z\sin z$, about $z=0$

Solution:

$$f(z)=e^z\cdot\sin z=\left[\sum_{n=0}^{\infty}\frac{z^n}{n!}\right]\left[\sum_{n=0}^{\infty}\frac{(-1)^n z^{2n+1}}{(2n+1)!}\right]$$

$$=\left[1+z+\frac{z^2}{2!}+\frac{z^3}{3!}+\cdots\right]\left[z-\frac{z^3}{3!}+\frac{z^5}{5!}+\cdots\right]$$

$$=z+z^2+\frac{z^3}{3}-\frac{1}{30}z^5+\cdots$$

Example 8: Find Maclaurin's series by termwise integrating the integrand: $f(z)=e^{z^2}\int_0^z e^{t^2}dt$.

Solution:

$$e^{t^2}=\sum_{n=0}^{\infty}\frac{(t^2)^n}{n!}=\sum_{n=0}^{\infty}\frac{t^{2n}}{n!}$$

$$\int_0^z e^{t^2}dt=\int_0^z\left[1+t^2+\frac{t^4}{2!}+\frac{t^6}{3!}+\frac{t^8}{4!}+\cdots\right]dt$$

$$=z+\frac{z^3}{3}+\frac{1}{2!}\frac{z^5}{5}+\frac{1}{3!}\frac{z^7}{7}+\frac{1}{4!}\frac{z^9}{9}+\cdots$$

Then

$$e^{z^2}\int_0^z e^{t^2}dt=\left[1+z^2+\frac{z^4}{2!}+\frac{z^6}{3!}+\frac{z^8}{4!}+\cdots\right]$$

$$\left[z+\frac{z^3}{3!}+\frac{1}{2!}\frac{z^5}{5}+\frac{1}{3!}\frac{z^7}{7}+\cdots\right]$$

$$=z+\frac{7}{6}z^3+\frac{23}{30}z^5+\cdots$$

EXERCISE

Find the Taylor's series expansion of $f(z)$ about the indicated point z_0. Determine the region of convergence:

1. $\frac{1}{1+z^2}$, $z_0=0$

Ans. $\sum_0^{\infty}(-1)^n z^2 n$, $\quad |z|<1$

2. $\tan^{-1}$, $z_0=0$

Hint: $(\tan^{-1}z)'=\frac{1}{1+z^2}$, integrate series in the above Example 1.

Ans. $z-\frac{z^3}{3}+\frac{z^5}{5}-+\cdots$, $\quad |z|<1$

3. $\frac{1}{\alpha - z}, z = z_0$

Ans. $\frac{1}{\alpha - z_0}\sum_{n=0}^{\infty}\left(\frac{z-z_0}{\alpha-z_0}\right)^n, |z - z_0| < |\alpha - z_0|$

4. $\frac{2z^2+9z+5}{z^3+z^2-8z-12}, z_0 = 1$

Hint: By partial fractions $= \frac{1}{(z+2)^2} + \frac{2}{z-3}$, use binomial series $= \frac{1}{[3+(z-1)]^2} - \frac{2}{2-(z-1)}$.

Ans. $\sum_{n=0}^{\infty}\left[\frac{(-1)^n(n+1)}{3^{n+2}} - \frac{1}{2^n}\right](z-1)^n, \quad |z-1| < 2$

5. $\frac{3}{3z-z^2}, z_0 = 1$

Ans. $\frac{3}{2} - \frac{3}{4}(z-1) + \frac{9}{8}(z-1)^2 + \cdots\left[\frac{1}{2^{n+1}} + (-1)^n\right](z-1)^n + \cdots$

6. $\sinh z, z_0 = -i\pi$

Ans. $i\sum_{n=0}^{\infty}\frac{(z+i\pi)^{2n+1}}{(2n+1)!}$

7. $\frac{1}{(3-z)^2}, z_0 = i$

Ans. $\frac{1}{(3-i)^2}\sum_{n=0}^{\infty}(n+1)\left(\frac{z-i}{3-i}\right)^n, |z-i| < |3-i|$

8. $\ln\left(\frac{1+z}{1-z}\right), z_0 = 0$

Ans. $\sum_{n=0}^{\infty}\frac{2z^{2n+1}}{2n+1}, |z| < 1$

9. $\sin z, z_0 = \frac{\pi}{4}$

Ans. $\frac{1}{\sqrt{2}}\left[1 + \left(z - \frac{\pi}{4}\right) - \frac{\left(z-\frac{\pi}{4}\right)^2}{2!} - \frac{\left(z-\frac{\pi}{4}\right)^3}{3!} + \cdots, |z| < \infty\right]$

10. $\frac{1}{(z+1)^2}, z_0 = -i$

Ans. $\frac{i}{2}\left[1 + \sum_{n=1}^{\infty}\frac{(-1)^n(n+1)(z+i)^n}{(1-i)^n}\right]$

11. $\frac{2z^3+1}{z^2+z}, z_0 = i$

Hint: Partial fractions: $(2i-2) + 2(z-i) + \frac{1}{z} + \frac{1}{z+1}$.

Ans. $\left(\frac{i}{2} - \frac{3}{2}\right) + \left(3 + \frac{i}{2}\right)(z-i) + \sum_{n=2}^{\infty}(-1)^n\left\{\frac{1}{i^{n+1}} + \frac{1}{(1+i)^{n+1}}\right\}(z-i)^n$

12. $\frac{1}{(z-1)(z-2)}, z_0 = 0$

Ans. $\frac{1}{2} + \frac{3}{4}z + \frac{7}{8}z^2 + \frac{15}{16}z^3 + \cdots, |z| < 1$

13. $\frac{e^z}{\cos z}, z_0 = 0$

Hint: Method of undetermined coefficients: Assume $\frac{e^z}{\cos z} = a_0 + a_1 z + a_2 z^2 + a_3 z^3 + \cdots$

$$\sum_{n=0}^{\infty}\frac{z^n}{n!} = \left(\sum_{n=0}^{\infty}a_n z^n\right)\left(\sum_{n=0}^{\infty}\frac{(-1)^n z^{2n}}{(2n)!}\right)$$

Equate coefficients of like powers of z.

Ans. $1 + z + z^2 + \frac{2}{3}z^3 + \cdots$

14. $\frac{z-1}{z^2}, z_0 = 1$

Ans. $\sum_{n=1}^{\infty}(-1)^{n+1}\cdot n\cdot(z-1)^n$

15. $e^z\cdot\cosh z, z_0 = 0$

Hint: Multiply the series $\left(\sum\frac{z^n}{n!}\right)\cdot\left(\sum\frac{z^{2n}}{2n!}\right)$.

Ans. $1 + z + z^2 + \frac{5}{6}z^3 + \cdots$

16. $\frac{e^z}{z(z+1)}, z_0 = 2$

Ans. $\frac{1}{6} - \frac{5}{9}(z-2) + \frac{19}{27}(z-2)^2 + \cdots, |z-2| < 1$

17. Evaluate the integral of the series $\sum_{n=0}^{\infty}z^n$ along a straight-line path from $z = 0$ to $z = \frac{1+i}{2}$.

Hint: $\int_c\sum_{0}^{\infty}z^n dz = \int_0^{\frac{1+i}{2}}\frac{dz}{1-z}$

$$= -\ln(1-z)\Big|_{z=0}^{z=\frac{(1+i)}{2}}$$

Ans. $\left(\frac{1+i}{2}\right) + \frac{1}{2}\left(\frac{1+i}{2}\right)^2 + \frac{1}{3}\left(\frac{1+i}{2}\right)^3 + \cdots$

23.7 LAURENT SERIES

Let $f(z)$ be analytic on two concentric circles c_1 and c_2 with centre z_0 and radii R_1 and R_2 and in the annulus region $R_1 < |z - z_0| < R_2$ (Fig. 23.33). Then $f(z)$ is uniquely represented by a convergent **Laurent series** given by

$$f(z) = \sum_{n=0}^{\infty}a_n(z-z_0)^n + \sum_{n=1}^{\infty}\frac{b_n}{(z-z_0)^n} \tag{1}$$

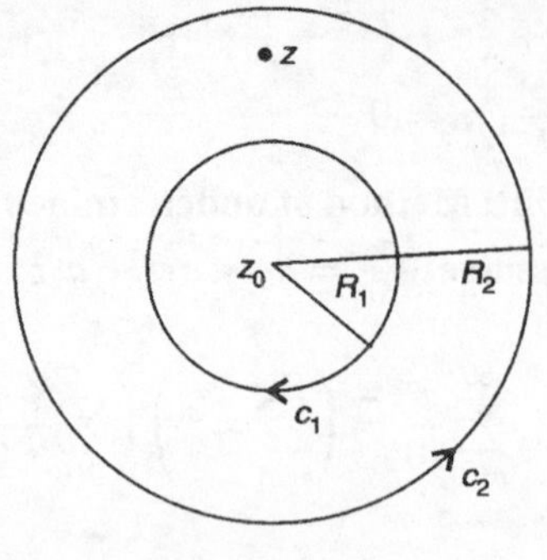

Fig. 23.33

Proof: For any arbitrary point z_0 in the annulus region, applying Cauchy's integral formula for multiply connected domain, we have

$$f(z) = \frac{1}{2\pi i}\int_{c_2}\frac{f(\alpha)}{\alpha - z}d\alpha + $$

$$+\frac{1}{2\pi i}\int_{c_1}\frac{f(\alpha)}{\alpha - z}d\alpha = I_1 + I_2 \qquad (2)$$

Since $\left|\frac{z-z_0}{\alpha-z_0}\right| < 1$ for any α on c_2, the integrand $\frac{1}{\alpha-z}$ of the first integral I_1 can be represented as

$$\frac{1}{\alpha - z} = \frac{1}{\alpha - z_0 - (z - z_0)} = \frac{1}{\alpha - z_0}\frac{1}{1 - \frac{z-z_0}{\alpha-z_0}}$$

$$= \frac{1}{\alpha - z_0}\cdot\sum_{n=0}^{\infty}\left(\frac{z - z_0}{\alpha - z_0}\right)^n \qquad (3)$$

Using (3) in I_1 and performing term by term integration, we get

$$I_1 = \frac{1}{2\pi i}\int_{c_2}\frac{f(\alpha)}{\alpha - z}d\alpha$$

$$= \sum_{n=0}^{\infty}\left[\frac{1}{2\pi i}\int_{c_2}\frac{f(\alpha)}{(\alpha - z_0)^{n+1}}d\alpha\right](z - z_0)^n$$

$$I_1 = \sum_{n=0}^{\infty}a_n(z - z_0)^n \qquad (4)$$

where $$a_n = \frac{1}{2\pi i}\int_c\frac{f(\alpha)}{(\alpha - z_0)^{n+1}}d\alpha \qquad (5)$$

Here c is any simple closed curve completely lying in the annulus region and traversed in counterclockwise direction.

For any α on c_1, $\left|\frac{\alpha-z_0}{z-z_0}\right| < 1$. So the integrand of I_2 can be expressed as

$$\frac{1}{\alpha - z} = \frac{1}{\alpha - z_0 - (z - z_0)} = \frac{-1}{(z - z_0)}\frac{1}{1 - \left(\frac{\alpha-z_0}{z-z_0}\right)}$$

$$= \frac{-1}{(z - z_0)}\sum_{n=0}^{\infty}\left(\frac{\alpha - z_0}{z - z_0}\right)^n = -\sum_{n=0}^{\infty}\frac{(\alpha - z_0)^n}{(z - z_0)^{n+1}}$$

$$= -\sum_{m=1}^{\infty}\frac{(\alpha - z_0)}{(z - z_0)^m} \qquad (6)$$

Using (6) in I_2 and performing integration

$$I_2 = \frac{-1}{2\pi i}\int_{c_1}\frac{f(\alpha)}{\alpha - z}d\alpha$$

$$= +\sum_{m=1}^{\infty}\left[\frac{1}{2\pi i}\int_{c_1}f(\alpha)\cdot(\alpha - z_0)^{m-1}d\alpha\right]\frac{1}{(z - z_0)^m}$$

Here c_1 is traversed in counterclockwise direction, so minus sign is absorbed.

$$I_2 = \sum_{n=1}^{\infty}b_n\cdot\frac{1}{(z - z_0)^n} \qquad (7)$$

where

$$b_n = \frac{1}{2\pi i}\int_c\frac{f(\alpha)}{(\alpha - z_0)^{-n+1}}d\alpha \qquad (8)$$

with c traversed in the counterclockwise direction.

Thus using (4) and (7) in (2), we get the required Laurent series (1) where a_n, b_n are given by (5) and (8).

Note: Laurent series (1) can also be written as

$$f(z) = \sum_{n=-\infty}^{\infty}c_n(z - z_0)^n \qquad (9)$$

where $$c_n = \frac{1}{2\pi i}\int_c\frac{f(\alpha)}{(\alpha - z_0)^{n+1}}d\alpha \qquad (10)$$

Note 1: If $f(z)$ is analytic at all points inside c_1 (i.e., no singular points inside c_2) then by Cauchy's theorem $b_n = 0$ for all $n - 1 \geq 0$.

Hence the Laurent series reduces to Taylor series. Thus Laurent series expansion about an analytic point z_0 is Taylor series about z_0.

Note 2: The region of convergence (validity) of Laurent series is the annulus region,

$$R_1 < |z - z_0| < R_2$$

Note 3: If z_0 is the only singular point inside c_1, then series is convergent in the deleted neighbourhood $0 < |z - z_0| < R_1$.

Note 4: In practice, the Laurent series is obtained by rearrangement, manipulation and using the standard series expansions (both Taylor's and Maclaurin's) and not by formulae (5) and (8).

Note 5: If $f(z)$ has more than one singular point, then several (more than one) Laurent series expansions can be obtained about the same singular point by appropriately considering analytic regions about (centered) at z_0.

WORKED OUT EXAMPLES

Example 1: Find Laurent series of $f(z) = \frac{\sinh 3z}{z^3}$ for $0 < |z| < \infty$.

Solution:

$$\frac{\sinh 3z}{z^3} = \frac{1}{z^3} \cdot \sum_{n=0}^{\infty} \frac{(3z)^{2n+1}}{(2n+1)!}$$

$$= \sum_{n=0}^{\infty} \frac{3^{2n+1} \cdot z^{2n-2}}{(2n+1)!} = \frac{3}{z^2} + \frac{9}{2} + \frac{81}{40} z^2 + \cdots$$

Example 2: Find Laurent series of $f(z) = \frac{1}{z^2+1}$ about its singular points. Determine the region of convergence.

Solution: $f(z)$ has two singular points at $z = \pm i$

a. Laurent series about $z = i$

$$f(z) = \frac{1}{z^2+1} = \frac{1}{(z-i)(z+i)} = \frac{1}{(z-i)} \cdot \frac{1}{z-i+2i}$$

$$= \frac{1}{(z-i)} \frac{1}{2i} \frac{1}{1 + \left(\frac{z-i}{2i}\right)} \quad \text{provided } \left|\frac{z-i}{2i}\right| < 1$$

$$= \frac{1}{2i} \frac{1}{z-i} \cdot \sum_{n=0}^{\infty} (-1)^n \left(\frac{z-i}{2i}\right)^n$$

if $|z - i| < |2i| = 2$

$$f(z) = \sum_{n=0}^{\infty} (-1)^n \frac{(z-i)^{n-1}}{(2i)^{n+1}} \quad \text{if } |z - i| < 2$$

Region of convergence is $|z - i| < 2$

b. Laurent series about $z = -i$

$$f(z) = \frac{1}{z^2+1} = \frac{1}{(z+i)} \frac{1}{(z-i)}$$

$$= \frac{1}{(z+i)} \cdot \frac{1}{z+i-2i}$$

$$= \frac{1}{z+i} \frac{1}{(-2i)\left(1 - \frac{z+i}{2i}\right)}$$

$$= \frac{1}{z+i} \left(\frac{-1}{2i}\right) \cdot \sum_{n=0}^{\infty} \left(\frac{z+i}{2i}\right)^n \quad \text{if } \left|\frac{z+i}{2i}\right| < 1$$

$$= -\sum_{n=0}^{\infty} \frac{(z+i)^{n-1}}{(2i)^{n+1}}$$

Region of convergence is $|z + i| < 2$.

Example 3: Find Laurent series of $f(z) = \frac{e^z}{z(1-z)}$ about $z = 1$. Find region of convergence.

Solution:

$$f(z) = \frac{e^z}{z(1-z)} = \frac{1}{e} \cdot e^{z-1} \cdot \frac{1}{(z-1+1)} \cdot \frac{1}{(1-z)}$$

$$= \frac{1}{e} \frac{1}{1-z} \cdot e^{z-1} \cdot \frac{1}{1+(z-1)}$$

$$= \frac{-1}{e(z-1)} \left[\sum_{n=0}^{\infty} \frac{(z-1)^n}{n!}\right] \left[\sum_{n=0}^{\infty} (-1)^n (z-1)^n\right]$$

if $|z - 1| < 1$

$$= -\frac{1}{e} \frac{1}{(z-1)} \left[1 + (z-1) + \frac{(z-1)^2}{2!} + \frac{(z-3)^3}{3!} + \cdots\right] \times$$

$$\times \left[1 - (z-1) + (z-1)^2 - (z-1)^3 + \cdots\right]$$

$$= -\frac{1}{e} \left[\frac{1}{z-1}\right] \left[1 + \frac{3}{2}(z-1)^2 - \frac{1}{3}(z-1)^3 + \cdots\right]$$

$$f(z) = \frac{1}{e} \left[-\frac{1}{z-1} - \frac{3}{2}(z-1) + \frac{1}{3}(z-1)^2 + \cdots\right]$$

Region of convergence is $|z-1|<1$

Example 4: Find all possible Laurent series of $f(z)=\frac{7z^2+9z-18}{z^3-9z}$ about its singular points.

Solution: $f(z)$ has three singular points $z=0,-3$ and 3. By partial fractions

$$f(z)=\frac{7z^2+9z-18}{z(z+3)(z-3)}=\frac{A}{z}+\frac{B}{z+3}+\frac{C}{z-3}$$

$$=\frac{2}{z}+\frac{1}{z+3}+\frac{4}{z-3}$$

Case 1: The two analytic regions about $z=0$ are (a) $0<|z-0|<3$ and (b) $|z-0|>3$. So we get two Laurent series of $f(z)$ about the singular point $z=0$ (Fig. 23.34).

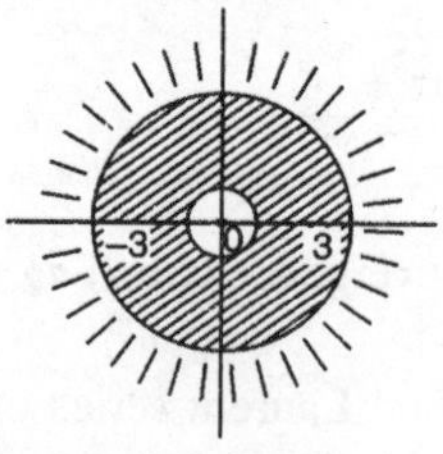

Fig. 23.34

a. For $0<|z|<3$

$$f(z)=\frac{2}{z}+\frac{1}{z+3}+\frac{4}{z-3}$$

$$=\frac{2}{z}+\frac{1}{3}\frac{1}{1+\frac{z}{3}}-\frac{4}{3}\frac{1}{1-\frac{z}{3}}$$

$$=\frac{2}{z}+\frac{1}{3}\sum_{n=0}^{\infty}(-1)^n\left(\frac{z}{3}\right)^n-\frac{4}{3}\sum\left(\frac{z}{3}\right)^n$$

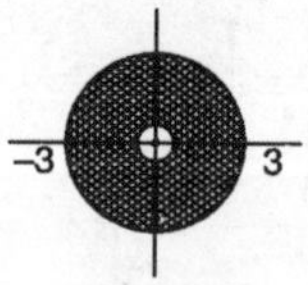

Fig. 23.35

provided $\left|\frac{z}{3}\right|<1$ or $0<|z|<3$

$$f(z)=\frac{2}{z}+\sum_{n=0}^{\infty}\left[(-1)^n-4\right]\frac{z^n}{3^{n+1}},$$

b. For $|z|>3$

$$f(z)=\frac{2}{z}+\frac{1}{z+3}+\frac{4}{z-3}$$

$$=\frac{2}{z}+\frac{1}{z}\frac{1}{1+\frac{3}{z}}+\frac{4}{z}\frac{1}{1-\frac{3}{z}}$$

$$=\frac{2}{z}+\frac{1}{z}\sum_{n=0}^{\infty}(-1)^n\left(\frac{3}{z}\right)^n+\frac{4}{z}\sum_{n=0}^{\infty}\left(\frac{3}{z}\right)^n$$

$$f(z)=\frac{2}{z}+\sum_{n=0}^{\infty}\left[(-1)^n+4\right]\frac{3^n}{z^{n+1}}$$

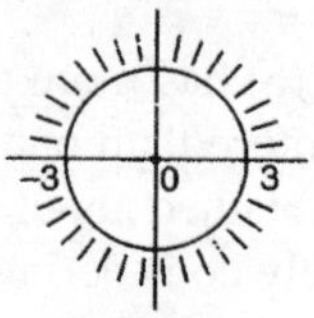

Fig. 23.36

Case 2: (refer Fig. 23.37). The three analytic regions of $f(z)$ about $z=3$ are

a. $0<|z-3|<3$, I

b. $3<|z-3|<6$, II

c. $|z-3|>6$, III

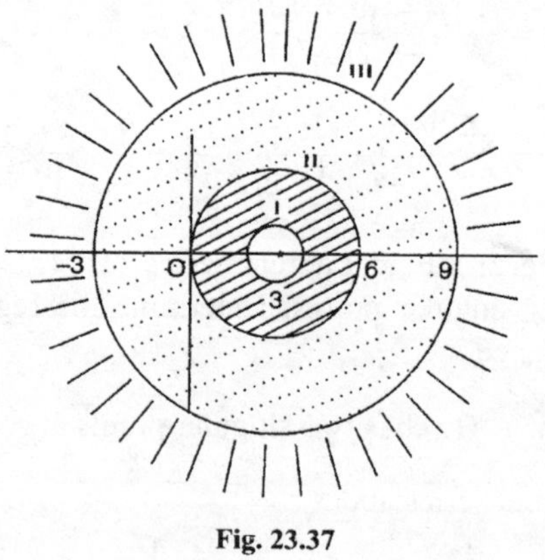

Fig. 23.37

So three Laurent series of $f(z)$ about $z=3$ can be obtained as follows:

a. For $0<|z-3|<3$. (Region I in Fig. 23.37)

$$f(z)=\frac{2}{z}+\frac{1}{z+3}+\frac{4}{z-3}$$

$$= \frac{2}{z-3+3} + \frac{1}{z-3+6} + \frac{4}{z-3}$$

$$= \frac{2}{3}\frac{1}{1+\left(\frac{z-3}{3}\right)} + \frac{1}{6}\frac{1}{1+\left(\frac{z-3}{6}\right)} + \frac{4}{z-3}$$

$$f(z) = \frac{2}{3}\sum(-1)^n\left(\frac{z-3}{3}\right)^n + \frac{1}{6}\sum(-1)^n\left(\frac{z-3}{6}\right)^n + \frac{4}{z-3}$$

if $\left|\frac{z-3}{3}\right| < 1$ and $\left|\frac{z-3}{6}\right| < 1$
i.e., $|z-3| < 3$ and $|z-3| < 6$,
the common region is $0 < |z-3| < 3$.

b. For $3 < |z-3| < 6$ (Region II in Fig. 23.37)

$$f(z) = \frac{2}{z-3+3} + \frac{1}{z-3+6} + \frac{4}{z-3}$$

$$= \frac{1}{z-3}\cdot\frac{2}{1+\left(\frac{3}{z-3}\right)} + \frac{1}{6}\frac{1}{1+\left(\frac{z-3}{6}\right)} + \frac{4}{z-3}$$

$$= \frac{1}{z-3}\cdot 2\sum_{n=0}^{\infty}(-1)^n\left(\frac{3}{z-3}\right)^n$$

$$+\frac{1}{6}\sum_{n=0}^{\infty}(-1)^n\left(\frac{z-3}{6}\right)^n + \frac{4}{z-3}$$

if $\left|\frac{3}{z-3}\right| < 1$ and $\left|\frac{z-3}{6}\right| < 1$ i.e., $3 < |z-3| < 6$

c. For $|z-3| > 6$ (Region III in Fig. 23.37)

$$f(z) = \frac{2}{z-3+3} + \frac{1}{z-3+6} + \frac{4}{z-3}$$

$$= \frac{1}{z-3}\frac{2}{1+\left(\frac{3}{z-3}\right)} + \frac{1}{z-3}\frac{1}{1+\left(\frac{6}{z-3}\right)} + \frac{4}{z-3}$$

$$= \frac{2}{z-3}\cdot\sum_{n=0}^{\infty}(-1)^n\left(\frac{3}{z-3}\right)^n +$$

$$+\frac{1}{z-3}\sum(-1)^n\left(\frac{6}{z-3}\right)^n + \frac{4}{z-3}$$

if $\left|\frac{3}{z-3}\right| < 1$ and $\left|\frac{6}{z-3}\right| <$ i.e., $|z-3| > 3$ and $|z-3| > 6$.

The common region is $|z-3| > 6$.

Case 3: Three Laurent series of $f(z)$ can be obtained about $z = -3$ in the regions, I, II, III shown in Fig. 23.38.

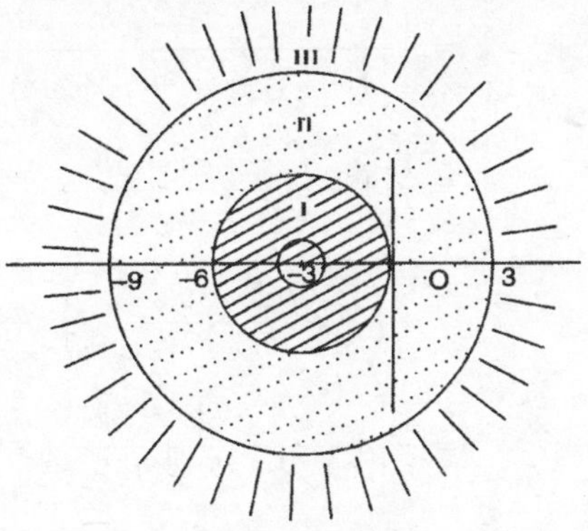

Fig. 23.38

a. For $0 < |z+3| < 3$

$$f(z) = \frac{2}{z} + \frac{1}{z+3} + \frac{4}{z-3}$$

$$= \frac{2}{z+3-3} + \frac{1}{z+3} + \frac{4}{z+3-6}$$

$$= -\frac{2}{3}\sum_{n=0}^{\infty}\left(\frac{z+3}{3}\right)^n + \frac{1}{z+3} - \frac{4}{6}\sum_{n=0}^{\infty}\left(\frac{z+3}{6}\right)^n$$

b. For $3 < |z+3| < 6$

$$f(z) = \frac{2}{z+3}\sum_{n=0}^{\infty}\left(\frac{3}{z+3}\right)^n + \frac{1}{z+3}$$

$$-\frac{4}{6}\sum_{n=0}^{\infty}\left(\frac{z+3}{6}\right)^n$$

c. For $|z+3| > 6$

$$f(z) = \frac{2}{z+3}\sum_{n=0}^{\infty}\left(\frac{3}{z+3}\right)^n + \frac{1}{z+3}$$

$$-\frac{4}{(z+3)}\sum_{n=0}^{\infty}\left(\frac{6}{z+3}\right)^n$$

Example 5: Show that $\text{cosec} z = \frac{1}{z} + \frac{1}{3!}z + \frac{7}{360}z^3 + \cdots$

Solution: $\csc z = \frac{1}{\sin z}$ has singular points at $z = 0, \pm n\pi$, n integer. Expand in series in $0 < |z| < \pi$.

$$\text{cosec} z = \frac{1}{\sin z} = \frac{1}{z - \frac{z^3}{3!} + \frac{z^5}{5!} - \frac{z^7}{7!} + \cdots}$$

$$= \frac{1}{z}\frac{1}{\left[1-\left(\frac{z^2}{3!}-\frac{z^4}{5!}+\frac{z^6}{7!}+\cdots\right)\right]}$$

$$= \frac{1}{z}\left[1+\left(\frac{z^2}{3!}-\frac{z^4}{5!}+\frac{z^6}{7!}+\cdots\right)\right.$$

$$\left.+\left(\frac{z^2}{3!}-\frac{z^4}{5!}+\frac{z^6}{7!}+\cdots\right)^2+\cdots\right]$$

$$= \frac{1}{z}\left[1+\frac{z^2}{3!}-\frac{z^4}{5!}+\frac{z^4}{(3!)^2}+\cdots\right]$$

$$= \frac{1}{z}+\frac{1}{3!}z^2+\left(\frac{1}{3!^2}-\frac{1}{5!}\right)z^3+\cdots$$

Example 6: Prove that for k real, $k^2 < 1$

$$\sum_{n=0}^{\infty} k^n \sin(n+1)\theta = \frac{\sin\theta}{(1-2k\cos\theta+k^2)}$$

$$\sum_{n=0}^{\infty} k^n \cos(n+1)\theta = \frac{\cos\theta-k}{(1-2k\cos\theta+k^2)}$$

Solution: $\frac{1}{z-k} = \frac{1}{z}\frac{1}{1-\frac{k}{z}} = \frac{1}{z}\sum_0^{\infty}\left(\frac{k}{z}\right)^n = \sum_0^{\infty}\frac{k^n}{z^{n+1}}; |z|>k$

Put $z = e^{i\theta} = \cos\theta + i\sin\theta$

$$\frac{1}{z-k} = \frac{1}{e^{i\theta}-k} = \sum k^n \cdot e^{-(n+1)\theta i}$$

$$= \sum k^n \cos(n+1)\theta - i\sin(n+1)\theta \quad (*)$$

But $\frac{1}{e^{i\theta}-k} = \frac{1}{\cos\theta+i\sin\theta-k} = \frac{(\cos\theta-k)-i\sin\theta}{1-2k\cos\theta+k^2}$

Comparing real and imaginary parts of (*), the result is obtained.

Exercise

Find the Laurent series of $f(z)$ about the indicated point. State the region of convergence:

1. $z^{-5}\cdot\sin z$, $z_0 = 0$

Ans. $\sum_{n=0}^{\infty}\frac{(-1)^n}{(2n+1)!}z^{2n-4}$, $|z| > 0$

2. $z^2e^{1/z}$, $z_0 = 0$

Ans. $z^2+z+\frac{1}{2}+\frac{1}{3!}\frac{1}{z}+\frac{1}{4!}\frac{1}{z^2}+,\ldots, |z| > 0$

3. $\frac{1}{1-z}$, $z_0 = 0, 1$

Ans. $\sum_{n=0}^{\infty} z^n$(Taylor's) $|z|<1$, $-\sum_{n=0}^{\infty}\frac{1}{z^{n+1}}$, $|z| > 1$(Laurent), $\frac{-1}{z-1}$, L.S., for all z

4. $\frac{-2z+3}{z^2-3z+2}$, $z_0 = 0$

Ans. Three Laurent series

a. $|z| < 1$, $\sum_0^{\infty}\left(1+\frac{1}{2^{n+1}}\right)z^n$

b. For $1 < |z| < 2$, $\sum_0^{\infty}\frac{1}{2^{n+1}}z^n - \frac{1}{z}\sum_0^{\infty}\frac{1}{z^n}$

c. For $|z| > 2$, $-\sum_0^{\infty}(2^n+1)\frac{1}{z^{n+1}}$

5. $\frac{7z-2}{(z+1)(z)(z-2)}$, $z_0 = -1$

Ans. Three Laurent series expansions

a. For $0 < |z+1| < 1$, $\frac{-3}{z+1}-\frac{5}{3}-\frac{11}{9}(z+1)$ $-\frac{29}{27}(z+1)^2-\frac{83}{81}(z+1)^3+\ldots.$

b. For $1 < |z+1| < 3$, $\ldots+\frac{1}{(z+1)^3}+\frac{1}{(z+1)^2}$ $-\frac{2}{z+1}-\frac{2}{3}-\frac{2}{9}(z+1)-\frac{2}{27}(z+1)^2-$ $-\frac{2}{81}(z+1)^3+\cdots.$

c. For $|z+1| > 3$, $+\cdots+19\frac{1}{(z+1)^3}+7\frac{1}{(z+1)^2}$

6. $\frac{1}{(1-z)(z-2)}$, $z_0 = 0$, two Laurent series

a. For $1 < |z| < 2$, $\sum_0^{\infty}\frac{1}{z^{n+1}}+\sum_0^{\infty}\frac{z^n}{2^{n+1}}$

b. For $|z| > 2$, $\sum_0^{\infty}\frac{1-2^n}{z^{n+1}}$

7. Expand $\frac{1}{(z+1)(z+3)}$ in Laurent series valid for

a. $1 < |z| < 3$

b. $|z| > 3$

c. $0 < |z+1| < 2$

d. $|z| < 1$

Ans. **a.** $-\frac{1}{2z^4}+\frac{1}{2z^3}-\frac{1}{2z^2}+\frac{1}{2z}-\frac{1}{6}+\frac{z}{18}-\frac{z^2}{54}+$ $\frac{z^3}{162}+\cdots$

b. $\frac{1}{z^2}-\frac{4}{z^3}+\frac{13}{z^4}-\frac{40}{z^5}+\cdots.$

c. $\frac{1}{2(z+1)}-\frac{1}{4}+\frac{1}{8}(z+1)-\frac{1}{16}(z+1)^2+\cdots.$

d. $\frac{1}{3}-\frac{4}{9}z+\frac{13}{27}z^2-\frac{40}{81}z^3+\cdots$ (Taylor's series)

8. Find Laurent series about the indicated singu-

larity:

a. $\frac{e^{2z}}{(z-1)^3}, z = 1$

b. $(z-3)\sin\left(\frac{1}{z+2}\right), z = -2$

c. $\frac{z-\sin z}{z^3}, z = 0$

d. $\frac{z}{(z+1)(z+2)}, z = -2$

e. $\frac{1}{z^2(z-3)^2}, z = 3$

Ans. **a.** $\frac{e^2}{(z-1)^3} + \frac{2e^2}{(z-1)^2} + \frac{2e^2}{(z-1)} + \frac{4e^2}{3} + \frac{2e^2}{3}(z-1) + \cdots$.

b. $1 - \frac{5}{z+2} - \frac{1}{6(z+2)^2} + \frac{5}{6(z+2)^3} + \frac{1}{120(z+2)^4} + \cdots$.

c. $\frac{1}{3!} - \frac{z^2}{5!} + \frac{z^4}{7!} + \cdots$.

d. $\frac{2}{2+z} + 1 + (z+2) + (z+2)^2 + \cdots$.

e. $\frac{1}{9(z-3)^2} - \frac{2}{27(z-3)} + \frac{1}{27} - \frac{4(z-3)}{243} + \cdots$.

9. $\frac{1}{[(z-a)(z-b)]}, 0 < |a| < |b|$, around $z = 0, a, \infty$ and annulus $|a| < |z| < |b|$.

a. $\frac{1}{b-a} \cdot \sum_{n=0}^{\infty} \frac{b^{n+1} - a^{n+1}}{a^{n+1}b^{n+1}} \cdot z^n$ for $|z| < a$

b. $\frac{1}{a-b}\left[\frac{1}{z-a} + \sum_{n=0}^{\infty} \frac{(z-a)^n}{(b-a)^{n+1}}\right]$ for $0 < |z-a| < |b-a|$

c. $\frac{1}{b-a}\sum_{2}^{\infty} \frac{b^{n-1} - a^{n-1}}{z^n}$ for $|z| > b$

d. $\frac{1}{a-b}\sum_{0}^{\infty}\left(\frac{z^n}{b^{n+1}} + \frac{a^n}{z^{n+1}}\right), |a| < |z| < |b|$

23.8 ZEROS AND POLES

Zeros

A point z_0 is called a zero of an analytic function $f(z)$ if $f(z_0) = 0$.

Zero of kth order (multiplicity k)

z_0 is called a zero of kth order if not only f but also the derivatives $f', f'', \ldots, f^{(k-1)}$ are all zero at z_0 and $f^{(k)}(z_0) \neq 0$.

Simple zero is a zero of first order $k = 1$ (i.e., $f(z_0) = 0$, and $f'(z_0) \neq 0$).

By expanding $f(z)$ about z_0 in power series

$$f(z) = \sum_{n=0}^{\infty} a_n(z-z_0)^n$$

it follows that when z_0 is a simple zero, $a_0 = 0$. For a kth order zero, $a_0, a_1, a_2, \ldots, a_{k-1}$ are all zero while $a_k \neq 0$.

Poles

A regular point of $f(z)$ is a point where $f(z)$ is analytic.

A singular point or singularity of $f(z)$ is a point z_0 at which $f(z)$ is not analytic. However, every neighbourhood of z_0 contains points at which $f(z)$ is analytic.

Isolated singular point (ISP)

A singular point z_0 is said to be an isolated singular point if there exists a neighbourhood of z_0 which does not contain any other singular point of $f(z)$.

Classification of Isolated Singular Point z_0

Expand $f(z)$ about the singular point z_0 in Laurent series in the annulus region $0 < |z-z_0| < R$, we get

$$f(z) = \sum_{n=0}^{\infty} a_n(z-z_0)^n + \sum_{n=1}^{\infty} \frac{b_n}{(z-z_0)^n}$$

Then, the series of negative integer powers of $(z-z_0)$ namely $\sum_{n=1}^{\infty} \frac{b_n}{(z-z_0)^n}$ is known as the **"principal part"** of the Laurent series expansion of $f(z)$ about z_0.

Removable singularity(ISP)

If all the coefficients b_n's are zero, i.e., Laurent series does not contain negative integer powers of $(z-z_0)$ then z_0 is called a removable singularity i.e., $f(z)$ can be made analytic by redefining $f(z_0)$ suitably i.e., if $\lim_{z\to z_0} f(z)$ exists.

Pole of order m

If the principal part contains a finite number of terms $\leq m$ involving negative power of $(z-z_0)$, then z_0 is called a pole of order m, of $f(z)$ i.e., $b_m \neq 0$, but

$b_n = 0$ for $n > m$. Then $|f| \to \infty$ as $z \to z_0$.
Simple pole is a pole of order one.

Essential singularity

If the principal part contains infinite number of terms containing negative powers of $(z - z_0)$ i.e., all b_n's are non-zero. i.e., Lim $f(z)$ as $z \to z_0$ does not exist.

Analytic (or singular) at infinity: The behaviour of $f(z)$ at ∞ can be investigated by introducing $z = 1/w$. Then $f(z)$ is said to be analytic (*or singular*) at infinity if $g(w)$ is analytic (*or singular*) at $w = 0$.

Worked Out Examples

Zeros

Example 1: Find the orders of all zeros of the given functions:

a. $z^2 + 9$

Ans. $z = \pm 3i$ are simple zeros

b. $\frac{z^2+8}{z^4}$

Ans. $z = \pm\sqrt{8}i$ are simple zeros. $z = \infty$ is a zero of second order, put $w = \frac{1}{z}$.

$$\frac{\frac{1}{w^2} + 8}{\frac{1}{w^4}} = \frac{1 + 8w^2}{w^2} \cdot w^4 = w^2(1 + 8w^2)$$

$w = 0$ is a zero of second order, so $z = \infty$ is a zero of second order

c. $f = z \sin z$.

Ans. $z = 0$ is a zero of second order, since $f'(0) = \sin z + z\cos z = 0$, while $f''(0) = 2\cos z - \sin z = 2 \neq 0$.

Also $z = \pm k\pi$, $k = 1, 2, 3, \ldots$ are zeros of first order.

d. $f = (1 - e^z)(z^2 - 4)^3$

Ans. $z = \pm 2$ are zeros of 3rd order. $z = 2k\pi$, $k = 0, \pm 1, \pm 2, \pm \cdots$ are zeros of first order

e. $e^{\tan z}$

Ans. no zeros since $e^z \neq 0$ for any z

f. $z^4 \sin^2(1/z)$

Ans. $z = \pm\frac{1}{n\pi}$, double zeros for $n = 1, 2, 3, \ldots$

g. $\frac{(z^2-\pi^2)\sin z}{z^7}$

Ans. $z = \pm\pi$ are 3rd order zeros

$z = k\pi$, $k = 0, \pm 2, \pm 3, \ldots$ are first order zeros.

Poles

Example 2: Determine and classify all singularities of the given functions. (See more examples on page 24.2)

1. $\frac{1}{z-z^3}$

Ans. $z = 0, \pm 1$ are poles of first order

$z = \infty$ is a regular point (zero of 3rd order)

2. $\frac{z^4}{1+z^4}$

Ans. $z = \frac{\pm 1 \pm i}{\sqrt{2}}$ are simple poles

$z = \infty$ is a regular point

3. $\frac{z^5}{(z-2)^2(z^2+4)^3(z-5)^4}$

Ans. $z = 2$ double pole,

$z = \pm 2i$ triple pole

$z = 5$ is pole of 4th order

$z = \infty$ is a regular point (zero of 7th order)

4. $\frac{e^z}{1+z^2}$

Ans. $z = \pm i$ are simple poles

$z = \infty$ is an essential singularity

(put $z = \frac{1}{w}$, $\frac{e^z}{1+z^2} = \frac{w^2 e^{\frac{1}{w}}}{w^2+1}$, $w = 0$ is an essential singularity)

5. $\cot\frac{1}{z} - \frac{1}{z}$

Ans. $z = \frac{1}{k\pi}$, $k = \pm 1, \pm 2, \ldots$ are poles of first order, $z = 0$ is a limit point for poles: $z = 0$ is a non-isolated singular point, $z = \infty$ is a pole of first order.

6. $\frac{1-e^{2z}}{z^4}$

Ans. $z = 0$ a pole of order 3

$$\left(\text{since Laurent series } -\sum_{n=1}^{\infty}\frac{2^n}{n!} \cdot z^{n-4}\right)$$

7. $\frac{1-\cos z}{z}$

Ans. $z = 0$, removable singularity

8. $e^{z/(z-2)}$

Ans. $z = 2$ is essential singularity

$z = \infty$ is removable (put $z = \frac{1}{w}$, $e^{\frac{1}{1-2w}}$).

Exercise

Zeros

Determine the location and order of the zeros of the given functions:

1. $z^2 + 1$

Ans. $z = \pm i$ simple zeros

2. $(1 - z^4)^2$

Ans. $z = \pm 1, \pm i$ are double zeros

3. $1 - \cos z$

Ans. $z = 2k\pi$, $(k = 0, \pm 1, \pm 2, \ldots)$ are zeros of 2nd order

4. $(1 - \cos z)^2$

Ans. 4th order zeros at $z = 2k\pi$, where $k = 0, \pm 1, \pm 2, \ldots$

5. $e^z - e^{2z}$

Ans. $\pm 2n\pi i, n = 0, 1, 2, \ldots$ simple zeros

6. $\frac{1-\cot z}{z}$

Ans. $z = \frac{\pi}{4} + k\pi, k = 0, \pm 1, \pm 2, \ldots$ simple

7. $\cos^3 z$

Ans. $z = (2k + 1)\frac{\pi}{2}, k = 0, \pm 1, \pm 2, \ldots$, 3rd order

8. $\cos z^3$

Ans. $z = ((2k + 1)\frac{\pi}{2})^{\frac{1}{3}}$ and

$z = \frac{1}{2}\sqrt{(2k+1)\frac{\pi}{2}}\,(1 \pm i\sqrt{3})$ with

$k = 0, \pm 1, \pm 2, \ldots$ simple zeros.

9. e^z

Ans. no zeros

10. $\frac{\sin^3 z}{z}$

Ans. $z = 0$, 2nd order

$z = k\pi, k = \pm 1, \pm 2, \ldots$, 3rd order.

Poles

Determine and classify all the singularities of the given functions:

1. $\frac{z^8+z^4+2}{(z-1)^3(3z+2)^2}$

Ans. $z = 1$ pole of order 3

$z = -2/3$ pole of order 2

$z = \infty$ is pole of order 3

2. $\frac{z+1}{z^3(z^2+1)}$

Ans. $z = 0$, pole of order 3,

$z = \pm i$ simple poles

3. $\frac{1}{z(z-2)^5} + \frac{3}{(z-2)^2}$

Ans. $z = 0$ simple pole,

$z = 2$ pole of order 5

4. $\frac{1}{\sin(\frac{\pi}{z})}$

Ans. $z = \pm\frac{1}{n}, n = 1, 2, 3 \ldots$ are simple poles

$z = 0$ is a non-isolated singular point

5. $\frac{\sinh z}{z^4}$

Ans. $z = 0$ pole of order 3

6. $\cosh \frac{1}{z}$

Ans. $z = 0$ is essential singularity

7. $\cot z$

Ans. $z = 0, \pm n\pi\,(n = 1, 2, 3, \ldots)$ simple

8. $\frac{1}{z(e^z-1)}$

Ans. $z = 0$ second order pole

9. $\frac{z-\sin z}{z^2}$

Ans. $z = 0$ is removable singularity

10. $(z + 1)\sin\frac{1}{z-2}$

Ans. $z = 2$ is an essential singularity

11. $\frac{1}{\cos z-\sin z}$

Ans. $z = \frac{\pi}{4}$ is simple pole

12. $ze^{\frac{1}{z^2}}$

Ans. $z = 0$ essential singularity

13. $\frac{z^2-1}{(z-1)^2}$

Ans. $z = 1$ is simple pole

14. $\frac{1}{e^z}$

Ans. no singular points

15. $\frac{1}{1+1/(1+z)}$

Ans. $z = -2$ simple pole

16. $\frac{1}{z^2}$

Ans. analytic at $z = \infty$

2nd order zero at ∞

17. e^z, $\sin z$, $\cos z$

Ans. essential singularity at $z = \infty$.

Chapter 24

Theory of Residues

INTRODUCTION

Residue theorem is a very powerful and elegant theorem in complex integration. Using the residue theorem many complicated real integrals can be evaluated. It is also used to sum a real convergent series and to find the inverse Laplace transform.

24.1 RESIDUE

Residue of an analytic function $f(z)$ at an isolated singular point $z = z_0$ is the coefficient say b_1 of $(z-a)^{-1}$ in the Laurent series expansion of $f(z)$ about z_0. Residue of $f(z)$ at z_0 is denoted by $\underset{\text{at } z=z_0}{\text{Res}}\ f(z)$ or $\underset{z=z_0}{\text{Res}} f(z)$. From Laurent series, we know that the coefficient b_1 is given by

$$b_1 = \frac{1}{2\pi i}\oint_c f(z)dz$$

Thus

$$\text{Residue of } f(z) \text{ at } z=z_0 = \underset{z=z_0}{\text{Res}}\ f(z) = b_1 = \frac{1}{2\pi i}\oint_c f(z)dz$$

where c is any closed contour enclosing z_0 (and such that f is analytic on and within c).

Calculation of Residue at Simple Pole

Formula I. $\underset{z=z_0}{\text{Res}} f(z) = b_1 = \lim_{z\to z_0}\Big[(z-z_0)f(z)\Big]$
The Laurent series of $f(z)$ about z_0 is

$$f(z) = \frac{b_1}{z-z_0} + \sum_{n=0}^{\infty} a_n(z-z_0)^n$$

is multiplied by $(z-z_0)$ which gives

$$(z-z_0)f(z) = b_1 + (z-z_0)\sum_{n=0}^{\infty} a_n(z-z_0)^n$$

As $z \to z_0$, $\lim_{z\to z_0}(z-z_0)f(z) = b_1 + 0 = b_1$.

Formula II. Suppose $f(z) = \frac{P(z)}{Q(z)}$ has a simple pole at z_0 such that $P(z_0) \neq 0$. Then

$$\underset{z=z_0}{\text{Res}}\ f(z) = \underset{z=z_0}{\text{Res}}\ \frac{P(z)}{Q'(z)} = \frac{P(z_0)}{Q'(z_0)}.$$

Calculation of Residue at a Multiple Pole

The Laurent series expansion of $f(z)$ about a pole z_0 of order $m > 1$ is

$$f(z) = \frac{b_m}{(z-z_0)^m} + \frac{b_{m-1}}{(z-z_0)^{m-1}} + \cdots + \frac{b_1}{(z-z_0)} + \sum_{n=0}^{\infty} a_n(z-z_0)^n$$

Multiply both sides by $(z-z_0)^m$

$$(z-z_0)^m f(z) = b_m + b_{m-1}(z-z_0)\cdots + b_1(z-z_0)^{m-1} + \sum_{n=0}^{\infty} a_n(z-z_0)^{n+m}$$

Differentiating both sides $(n-1)$ times and let $z \to z_0$

$$\lim_{z\to z_0}\frac{d^{m-1}}{dz^{m-1}}(z-z_0)^m f(z) = (m-1)!\, b_1 + 0 + \cdots$$

Thus residue of $f(z)$ at z_0 is

$$b_1 = \frac{1}{(m-1)!} \lim_{z\to z_0}\left[\frac{d^{m-1}}{dz^{m-1}}\{(z-z_0)^m\cdot f(z)\}\right]$$

For simple pole $(m = 1)$

$$b_1 = \operatorname*{Res}_{z=z_0} f(z) = \lim_{z\to z_0}[(z-z_0)f(z)]$$

Note: When $z = z_0$ is an essential singularity of $f(z)$ then the above formulae fail. In such cases the residue at an essential singularity is obtained by expanding $f(z)$ about z_0 in Laurent series (and identify b_1, the coefficient of $(z-z_0)^{-1}$).

24.2 RESIDUE THEOREM

Residue theorem, which is very powerful (but whose proof is very simple) is useful in evaluating contour integrals where the closed contour contains several singularities inside.

Theorem 1: *Let $f(z)$ be analytic within and on a simple closed path c except at a finite number of singular points $z_1, z_2, \ldots, z_n$ inside c. Then*

$$\oint_c f(z)\,dz = 2\pi i \sum_{i=1}^{n} \operatorname*{Res}_{z=z_i} f(z)$$

where the integral is taken counterclockwise around c. (Fig. 24.1)

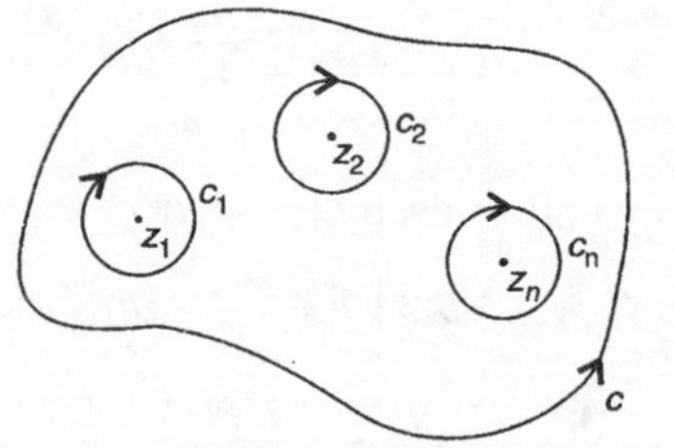

Fig. 24.1

Proof: Enclose each of the singular points z_i in a circle c_i of small radius such that all these n circles and curve c are all separated. Then $f(z)$ is analytic in the multiply connected domain D bounded by B consisting of c and $c_1, c_2, \ldots, c_n$ and on the entire boundary B of D. Now by Cauchy's integral theorem (for multiply connected domain), we have

$$\oint_B = \oint_{c+c_1+c_2+\cdots+c_n} = \oint_c f(z)\,dz + \oint_{c_1} f(z)\,dz + \cdots + \oint_{c_n} f(z)\,dz = 0$$

where the integral along c taken in the counterclockwise direction and along $c_1, c_2, \ldots, c_n$ in the clockwise direction. Thus

$$\oint_c f(z)\,dz = \oint_{c_1} f(z)\,dz + \cdots + \oint_{c_n} f(z)\,dz$$

Here $c_1, c_2, \ldots, c_n$ are also taken along counterclockwise direction (the minus sign reverses the sense of integration from clockwise to counterclockwise).

By definition of residues

$$\oint_{c_i} f(z)\,dz = 2\pi i \cdot \operatorname*{Res}_{z=z_i} f(z),$$

for $i = 1, 2, 3, \ldots, n$

Thus

$$\oint_c f(z)\,dz = 2\pi i\left[\operatorname*{Res}_{z=z_1} f(z) + \cdots + \operatorname*{Res}_{z=z_n} f(z)\right]$$

$$\oint_c f(z)\,dz = 2\pi i \sum_{i=1}^{n} \operatorname*{Res}_{z=z_i} f(z).$$

WORKED OUT EXAMPLES

Residues

Example 1: Expand each of the following function in a Laurent series about $z = 0$, name the type of singularity in each case. Find the residues at $z = 0$: (a) $z^2e^{-z^4}$ (b) $\frac{(1-\cos z)}{z}$ (c) $\frac{e^{z^2}}{z^3}$ (d) $z^{-1}\cosh z^{-1}$.

Solution:

a. $$z^2e^{-z^4} = z^2\sum_{n=0}^{\infty}\frac{(-z^4)^n}{n!} = \sum_{n=0}^{\infty}\frac{(-1)^n z^{4n+2}}{n!}$$

$$= z^2 - z^6 + \frac{z^{10}}{2!} - \frac{z^{14}}{3!} + \cdots$$

Laurent series contains only positive powers of z. Thus $z = 0$ is an ordinary point. Residue at $z = 0$ is 0.

b. $\frac{1-\cos z}{z} = \frac{1}{z}\left[1 - \left(1 - \frac{z^2}{2!} + \frac{z^4}{4!} - \cdots \frac{(-1)^{n-1}z^{2n-2}}{(2n-2)!} + \cdots\right)\right]$

$$= \frac{1}{z}\left[\frac{z^2}{2!} - \frac{z^4}{4!} + \cdots + \frac{(-1)^n z^{2n-2}}{(2n-2)!} + \cdots\right]$$

$$= \frac{z}{2!} - \frac{z^3}{4!} + \frac{z^5}{6!} + \cdots$$

Although $\frac{1-\cos z}{z}$ is not defined at $z = 0$,

$$\lim_{z\to 0}\frac{1-\cos z}{z} = 0$$

So $z = 0$ is a removable singularity. Residue is 0.

c. $\frac{e^{z^2}}{z^3} = \frac{1}{z^3}\sum_{n=0}^{\infty}\frac{(z^2)^n}{n!} = \frac{1}{z^3}\left[1 + \frac{z^2}{1!} + \frac{z^4}{2!} + \frac{z^6}{3!} + \cdots\right]$

$$= \frac{1}{z^3} + \frac{1}{z} + \frac{z}{2!} + \frac{z^3}{3!} + \frac{z^5}{4!} + \frac{z^7}{5!} + \cdots$$

So the principal part contains finite number of terms with the highest power $\frac{1}{z^3}$. Thus $z = 0$ is a pole of order 3. Residue is 1.

d. $z^{-1}\cosh z^{-1} = \frac{1}{z}\sum_{n=1}^{\infty}\left(\frac{1}{z}\right)^{2n-2}\cdot\frac{1}{(2n-2)!}$

$$= \sum_{n=1}^{\infty}\left(\frac{1}{z}\right)^{2n-1}\cdot\frac{1}{(2n-2)!}$$

$$= \frac{1}{z} + \frac{1}{2!}\frac{1}{z^3} + \frac{1}{4!}\frac{1}{z^5} + \frac{1}{6!}\frac{1}{z^7} + \cdots$$

Since the principal part contains infinite number of terms, so $z = 0$ is an essential singularity. Residue is 1.

Example 2: Determine and classify the singularities (a) $\frac{z}{(e^z-1)}$ (b) $\frac{1}{(2\sin z-1)^2}$.

Solution:

a. Singularities are the zeros of the denominator i.e., $e^z - 1 = 0$ or $e^z = 1$ or $e^{\pm 2n\pi i} = 1$.

Therefore the singular points are $z = 2m\pi i, m = \pm 1, \pm 2, \ldots$ which are simple poles. But $z = 0$ is removable singularity since

$$\lim_{z\to 0}\frac{z}{e^z-1} = \lim_{z\to 0}\frac{z}{1+z+\frac{z^2}{2!}+\frac{z^3}{3!}+\cdots-1}$$

$$= \lim_{z\to 0}\frac{1}{1+\frac{z}{2!}+\frac{z^2}{3!}+\cdots} = 1$$

$z = \infty$ is an essential singularity: put $w = \frac{1}{z}$,

$$\frac{z}{e^z-1} = \frac{1}{w}\cdot\frac{1}{e^{\frac{1}{w}}-1}$$

$$= \frac{1}{w}\cdot\frac{1}{\left[1+\frac{1}{w}+\frac{1}{w^2}\frac{1}{2!}+\frac{1}{w^3}\frac{1}{3!}+\cdots-1\right]}$$

$$= \frac{1}{w}\cdot\frac{1}{\frac{1}{w}+\frac{1}{w^2}\frac{1}{2!}+\frac{1}{w^2}\frac{1}{3!}+\cdots}$$

$$= \frac{1}{w^2}\left[1+\frac{1}{w}\frac{1}{2!}+\frac{1}{w^2}\frac{1}{w!}+\cdots\right]^{-1}$$

$$= \frac{1}{w^2}\left[1-\left(\frac{1}{w}\frac{1}{2!}+\frac{1}{w^2}\frac{1}{3!}+\cdots\right) + \left(\frac{1}{w}\frac{1}{2!}+\frac{1}{w^2}\frac{1}{3!}\cdots\right)^2+\cdots\right]$$

So $w = 0$ is an essential singularity. Therefore $z = \infty$ is an essential singularity.

b. Zeros of $2\sin z - 1$ are solutions of $\sin z = \frac{1}{2}$

$\frac{e^{iz}-e^{-iz}}{2i} = \frac{1}{2}$, $(e^{iz})^2 - ie^{iz} - 1 = 0$,

$$e^{iz} = \frac{i\pm\sqrt{3}}{2} = e^{i}\left(2m\pi + \frac{\pi}{6}\right)$$

$$e^{iz} = e^{\frac{(i+\sqrt{3})}{2}} = e^{i\left(2m+\frac{\pi}{6}\right)},$$

$$e^{iz} = e^{\frac{i-\sqrt{3}}{2}} = e^{i\left((2m+1)\pi-\frac{\pi}{6}\right)}$$

Thus $\frac{\pi}{6} + 2m\pi, (2m+1)\pi - \frac{\pi}{6}, m = 0, \pm 1, \pm 2, \pm 3, \ldots$ are poles of order 2.

Example 3: Determine the residues at the poles:

(a) $\frac{2z+1}{z^2-z-2}$ (b) $\left(\frac{z+1}{z-1}\right)^3$ (c) $\frac{z+1}{(z^2-16)(z+2)}$ (d) $\frac{\sin z}{z^2}$

(e) $\frac{1-e^{2z}}{z^4}$ (f) $\frac{(\operatorname{csc}z\cdot\operatorname{csch}z)}{z^3}$ (g) $\frac{z^2}{(z^2+3z+2)^2}$.

Solution:

a. $\frac{2z+1}{z^2-z-2} = \frac{2z+1}{(z+1)(z-2)}$ so $z = -1, 2$ are poles of order 1 (simple poles)

$$\operatorname*{Res}_{z=a} f(z) = \lim_{z\to a}\left[(z-a)f(z)\right]$$

$$\operatorname*{Res}_{\text{at } z=-1}\frac{2z+1}{(z+1)(z-2)} = \lim_{z\to -1}\left[(z+1)\cdot\frac{2z+1}{(z+1)(z-2)}\right]$$

$$= \frac{1}{3}$$

$$\underset{\text{at } z=2}{\text{Res}} \frac{2z+1}{(z+1)(z-2)} = \lim_{z\to 2}\left[(z-2)\cdot\frac{2z+1}{(z+1)(z-2)}\right]$$

$$= \frac{5}{3}.$$

b. $\left(\frac{z+1}{z-1}\right)^3$: $z = 1$ is pole of order 3

Rewriting $\left(\frac{z+1}{z-1}\right)^3 = \left(\frac{z-1+2}{z-1}\right)^3 = \left(1+\frac{2}{z-1}\right)^3$

$$= 1 + 3\cdot 1\cdot\left(\frac{2}{z-1}\right)^2 + 3\cdot 1\cdot\left(\frac{2}{z-1}\right) + \left(\frac{2}{z-1}\right)^3$$

Residue of pole $z = 1$ is the coefficient of $\frac{1}{z-1}$ i.e., 6.

c. $z = \pm 4, -2$ are simple poles:

$$\underset{z=z_0}{\text{Res}}\, f(z) = \frac{p(z_0)}{q'(z_0)}$$

Here

$$p(z) = z+1,\ q(z) = (z^2-16)(z+2)$$

$$q'(z) = 2z(z+2) + (z^2-16)$$

$$= 3z^2 + 4z - 16$$

$$\underset{\text{at } z=4}{\text{Res}}\, f(z) = \frac{p(4)}{q'(4)} = \frac{4+1}{3\cdot 4^2 + 4\cdot 4 - 16} = \frac{5}{48}$$

$$\underset{\text{at } z=-4}{\text{Res}} = \frac{p(-4)}{q'(-4)} = \frac{-4+1}{3(-4)^2+4(-4)-16} = \frac{-3}{16}$$

$$\underset{\text{at } z=-2}{\text{Res}}\, f(z) = \frac{p(-2)}{q'(-2)} = \frac{-2+1}{3(-2)^2+4(-2)-16} = \frac{1}{12}.$$

d. $\frac{\sin z}{z^2}$ has a pole of order two at $z = 0$:

Expanding

$$\frac{\sin z}{z^2} = \frac{1}{z^2}\sum_{n=0}^{\infty}\frac{(-1)^n z^{2n+1}}{(2n+1)!}$$

$$= \frac{1}{z^2}\left[z - \frac{z^3}{3!} + \frac{z^5}{5!} + \cdots\right]$$

$$= \frac{1}{z} - \frac{z}{3!} + \frac{z^3}{5!} + \cdots$$

Residue is the coefficient of $\frac{1}{z}$ i.e., 1.

e. $\frac{1-e^{2z}}{z^4}$ has a pole of order 3 (not 4) at $z = 0$:

Expanding

$$\frac{1-e^{2z}}{z^4} = \frac{1-\left[1+2z+\frac{(2z)^2}{2!}+\frac{(2z)^3}{3!}+\cdots\right]}{z^4}$$

$$= -\left[\frac{2}{z^3} + \frac{2}{z^2} + \frac{8}{6}\frac{1}{z} + \frac{16}{24} + \cdots\right]$$

Residue is $-\frac{8}{6} = -\frac{4}{3}$ which is the coefficient of $\frac{1}{z}$.

f. Expanding

$$\frac{\csc z\cdot \text{csch} z}{z^3} = \frac{1}{z^3}\cdot\frac{1}{\sin z\cdot\sinh z}$$

$$= \frac{1}{z^3}\frac{1}{\left[z-\frac{z^3}{3!}+\frac{z^5}{5!}+\cdots\right]\left[z+\frac{z^3}{3!}+\frac{z^5}{5!}+\cdots\right]}$$

$$= \frac{1}{z^3}\frac{1}{\left[z^2+0+\left(\frac{2}{5!}-\frac{1}{(3!)^2}\right)z^6+\cdots\right]}$$

$$= \frac{1}{z^5}\left[1-\frac{1}{90}z^4+\cdots\right]^{-1}$$

$$= \frac{1}{z^5}\left[1-\frac{1}{90}z^4+\cdots\right] = \frac{1}{z^5} - \frac{1}{90}\frac{1}{z} + \cdots$$

Residue is $\frac{-1}{90}$.

g. $\frac{z^2}{(z^2+3z+2)^2}$ has poles of order 2 at $z = -1, -2$. (since $z^2+3z+2 = (z+1)(z+2)$.

$$\underset{\text{at } z=a}{\text{Res}}\, f(z) = \frac{1}{(m-1)!}\lim_{z\to a}\frac{d^{m-1}}{dz^{m-1}}\left\{(z-a)^m\cdot f(z)\right\}$$

$$\underset{z=-1}{\text{Res}}\, f(z) = \lim_{z\to -1}\frac{d}{dz}\left\{(z+1)^2\cdot\frac{z^2}{(z+1)^2(z+2)^2}\right\} = -4$$

$$\underset{z=-2}{\text{Res}}\, f(z) = \lim_{z\to -2}\frac{d}{dz}\left\{(z+2)^2\cdot\frac{z^2}{(z+1)^2(z+2)^2}\right\} = 4.$$

Residue theorem

Example 4: Evaluate $I = \oint_c \frac{z^2+4}{z^3+2z^2+2z}$ where c is

(a) $|z| = 1$ (b) $|z+1-i| = 1$

(c) $|z+1+i| = 1$ (d) $|z-1| = 5$

(e) rectangle with vertices at $2+i, 6+i, 2+4i$, and $6+4i$.

Solution: $f(z) = \frac{z^2+4}{z^3+2z^2+2z}$ has simple poles at $z = 0, -1+i, -1-i$ which are the zeros of

$$z^3+2z^2+2z = z(z^2+2z+2) = z(z-z_1)(z-z_2)$$

where $z_2 = -1+i$, $z_3 = -1-i$

$$k_1 = \underset{\text{at } z=0}{\text{Res}}\, f(z) = \frac{p(0)}{q'(0)} = \frac{0+4}{0+0+2} = 2$$

Here $p(z) = z^2 + 4, q(z) = z^3 + 2z^2 + 2z, q'(z) = 3z^2 + 4z + 2$

$$k_2 = \underset{\text{at } z=z_2=-1+i}{\text{Res}} f(z) = \frac{(-1+i)^2+4}{3(-1+i)^2+4(-1+i)+2}$$

$$k_2 = \frac{4-2i}{-2(i+1)} = \frac{(i-2)}{(i+1)} = \frac{3i-1}{2} = -\frac{1}{2}(1-3i)$$

$$k_3 = \underset{\text{at } z=z_3=-1-i}{\text{Res}} f(z) = \frac{(-1-i)^2+4}{3(-1-i)^2+4(-1-i)+2}$$

$$= -\frac{1}{2}(1+3i).$$

a. $c: |z| = 1$ is unit circle enclosing only one singular point $z = 0$. By residue theorem

$$I = 2\pi i \cdot \underset{\text{at } z=0}{\text{Res}} f(z) = 2\pi i \cdot k_1 = 2\pi i \cdot 2 = 4\pi i.$$

b. $c: |z+1-i| = 1$ is a circle with centre at $-1+i$ and radius 1. So c encloses only one pole $-1+i$

$$I = 2\pi i \cdot \underset{z=z_2}{\text{Res}} f(z) = 2\pi i \cdot k_2 = 2\pi i\left(\frac{3i-1}{2}\right)$$

$$I = -\pi(3+i).$$

c. $c: |z+1+i| = 1$ is circle centered at $-1-i$ and of radius 1. So c encloses only one pole $-1-i$

$$I = 2\pi i \cdot \underset{z=z_3}{\text{Res}} f(z) = 2\pi i \cdot k_3 = 2\pi i \frac{(1+3i)}{-2}$$

$$= \pi(3-i).$$

d. $c: |z-1| = 5$ is circle with centre at 1 and radius 5. So c enclosed all the three poles $z = 0, -1+i, -1-i$. By residue theorem

$$I = 2\pi i[k_1 + k_2 + k_3]$$

$$= 2\pi i\left[2 + \left(\frac{3i-1}{2}\right) - \frac{(3i+1)}{2}\right]$$

$$I = 2\pi i.$$

e. c: Rectangle does not include any of the three poles. $f(z)$ is analytic within c. By Cauchy's theorem, $I = 0$.

Example 5: $I = \oint_c \frac{dz}{\sinh 2z}$ where $c: |z| = 2$.

Solution: Zeros of $\sinh 2z$ are $z = \pm\frac{n\pi i}{2}$. Thus out of the infinite number of poles of $f(z) = \frac{1}{\sinh 2z}$ only three poles $z = 0, \pm\frac{\pi i}{2}$ lies inside the circle $|z| = 2$. Now

$$k_1 = \underset{z=0}{\text{Res}} f(z) = \lim_{z\to 0} \frac{z}{\sinh 2z} = \frac{1}{2}$$

$$k_2 = \underset{z=\frac{\pi i}{2}}{\text{Res}} f(z) = \lim_{z\to\frac{\pi i}{2}} \frac{\left(z-\frac{\pi i}{2}\right)}{\sinh 2z}$$

$$= \lim \frac{1}{2\cdot\cosh 2z} \text{ is } \left(\frac{0}{0}\right) \text{ form}$$

$$= -\frac{1}{2} \qquad \text{(using L' Hospital's Rule)}$$

$$k_3 = \underset{z=-\frac{\pi i}{2}}{\text{Res}} f(z) = \lim_{z\to\frac{-\pi i}{2}} \frac{\left(z+\frac{\pi i}{2}\right)}{\sinh 2z}$$

$$= \lim \frac{1}{2\cdot\cosh 2z} = -\frac{1}{2}$$

By residue theorem,

$$I = 2\pi i\left[\frac{1}{2} - \frac{1}{2} - \frac{1}{2}\right] = -\pi i.$$

Example 6: Evaluate $I = \oint_c e^{-\frac{1}{z}} \cdot \sin\left(\frac{1}{z}\right) dz$ where c is the circle $|z| = 1$.

Solution: $z = 0$ is an essential singularity, which is enclosed in the circle $|z| = 1$. Expanding

$$e^{-\frac{1}{z}} \cdot \sin\left(\frac{1}{z}\right) = \left[\sum_{n=0}^{\infty}\left(-\frac{1}{z}\right)^n \frac{1}{n!}\right]$$

$$\times \left[\sum_{n=1}^{\infty}(-1)^{n-1}\cdot\left(\frac{1}{z}\right)^{2n-1} \frac{1}{(2n-1)!}\right]$$

$$= \left[1 - \frac{1}{z} + \frac{1}{z^2}\cdot\frac{1}{2!} - \frac{1}{z^3}\cdot\frac{1}{3!}\cdots\right]$$

$$\times \left[\frac{1}{z} - \frac{1}{z^3}\cdot\frac{1}{3!} + \frac{1}{5!}\frac{1}{z^5} + \cdots\right]$$

$$= \left[\frac{1}{z} - \frac{1}{z^3}\frac{1}{3!} + \frac{1}{5!}\frac{1}{z^5} + \cdots \right.$$

$$\left. -\frac{1}{z^2} + \frac{1}{z^4}\frac{1}{3!} + \cdots\right]$$

Residue at $z = 0$ is 1.
By residue theorem,

$$I = 2\pi i \cdot 1 = 2\pi i.$$

EXERCISE

Residue theorem

1. Find the nature of singularity and find the residue:
   a. $\frac{e^{2z}}{(z-1)^3}$
   b. $(z-3)\sin\left(\frac{1}{z+2}\right)$
   c. $\frac{(z-\sin z)}{z^3}$
   d. $\frac{1}{(z^2)(z-3)^2}$
   e. $e^{\frac{z}{(z-2)}}$

Ans. **a.** $z = 1$, pole of order 3, Res $2e^2$
b. $z = -2$ is essential singularity, Res -5
c. $z = 0$ is removable singularity, Res 0
d. $z = 0, 3$ are poles of order 2
Res at $z = 0$ is $\frac{2}{27}$, at $z = 3$ is $-\frac{2}{27}$
e. $z = 2$ is essential singularity, Res at $z = 2$ is $2e$
$z = \infty$ is removable singularity.

2. Determine the nature of singularities of $f(z)$:
   a. $\frac{(z-\sin z)}{z^2}$
   b. $(z+1)\sin\left(\frac{1}{z-2}\right)$
   c. $\frac{1}{(\cos z-\sin z)}$
   d. $\frac{z}{(e^{\frac{1}{z}}-1)}$
   e. $\frac{z^4}{(1+z^4)}$
   f. $\tanh z$

Ans. **a.** $z = 0$ is removable singularity
b. $z = 2$ is essential singularity
c. $z = \frac{\pi}{4}$ is simple pole
d. $z = \frac{i}{2m\pi}$, $m = \pm 1, \pm 2, \ldots$ simple poles, $z = 0$ is essential, $z = \infty$ is pole of order 2
e. $z = \frac{\pm 1 \pm i}{\sqrt{2}}$ are simple poles
f. $z = n\pi i$, $n = 0, \pm 1, \pm 2, \ldots$ simple poles.

3. Find the residues at all its poles in finite plane:
   a. $\frac{z^2-2z}{(z+1)^2(z^2+4)}$
   b. $e^z \cdot \csc^2 z$
   c. $\frac{\cot z \cdot \coth z}{z^3}$

   Hint. Expand, $\cot z$, $\coth z$ in powers of z.

   d. $\frac{z^2}{\left((z-1)^2(z+2)\right)}$
   e. $\frac{(9z+i)}{(z(z^2+1))}$
   f. $\frac{50z}{\left[(z+4)(z-1)^2\right]}$
   g. $\frac{(1+z)}{(1-\cos z)}$

Ans. **a.** $z = -1$, double pole, Res $-\frac{14}{25}$
$z = \pm 2i$, simple, $\frac{7+i}{25}, \frac{7-i}{25}$
b. $z = m\pi$, $m = 0, \pm 1, \pm 2, \ldots$ double poles
Res $e^{m\pi}$
c. $z = 0$, Res $-\frac{7}{45}$
d. $z = -2$, Res $\frac{4}{9}$
$z = 1$, double pole, Res $\frac{5}{9}$
e. $z = 0$, Res i
$z = \pm i$, Res $-5i, 4i$
f. $z = -4$, Res -8
$z = 1$, double, Res 8
g. $z = 0$, Res 2

Using residue theorem, evaluate:

4. $\oint_c \frac{\tan z}{(z^2-1)} dz$, $c : |z| = \frac{3}{2}$

Ans. $I = 2\pi i \tan 1$, simple pole $z = \pm 1$, Res $\frac{\tan 1}{2}, \frac{\tan 1}{2}$

5. $\oint_c \frac{4-3z}{z^2-z} dz$, c : any simple closed path such that
(a) $0, 1 \in c$ (b) $0 \in c, 1 \notin c$
(c) $1 \in c, 0 \notin c$ (d) $0, 1 \notin c$

Ans. $z = 0, 1$ are simple poles, Res at $z = 0$, is -4, Res at $z = 1$ is 1
(a) $2\pi i(-4+1) = -6\pi i$ (b) $-8\pi i$
(c) $2\pi i$ (d) 0

6. $I = \oint_c \left(\frac{ze^{\pi z}}{z^4-16} + ze^{\frac{\pi}{z}}\right) dz$,
c : ellipse $9x^2 + y^2 = 9$

Ans. $I = I_1, +I_2$, For $I_1, z = \pm 2i, \pm 2$ are simple poles $z = \pm 2i \in c$, $z = \pm 2 \notin c$, Res at $z = 2i, -\frac{1}{16}$, at $z = -2i, -\frac{1}{16}$
For I_2, $z = 0$, essential singularity,
Res at $z = 0$ is $\frac{\pi^2}{2}$

Ans. $I = 2\pi i\left(-\frac{1}{16} - \frac{1}{16}\right) + 2\pi i\left(\frac{\pi^2}{2}\right)$
$= \pi\left(\pi^2 - \frac{1}{4}\right) i$

7. $\int_c \frac{z-3}{z^2+2z+5} dz$
   a. $c : |z| = 1$

b. $|z+1-i|=2$
c. $|z+1+i|=2$

Ans. $z_{1,2}=-1\pm 2i$ simple poles
a. $z_{1,2}\notin c$, $I=0$
b. $z_1=-1+2i\in c$, $z_2=-1-2i\notin c$,
$I=2\pi i\left(i+\frac{1}{2}\right)=\pi(i-2)$
c. $z_2=-1-2i\in c$, $z_1=-1+2i\notin c$,
$I=2\pi i\left(\frac{1}{2}-i\right)=\pi(2+i)$

8. $\int_c \tan z\,dz$, $c:|z|=2$

Ans. $z=(2n+1)\frac{\pi}{2}$, $n=0,\ \pm 1,\ \pm 2,\ldots$ simple poles
$z=\pm\frac{\pi}{2}$ only $\in c$, Res $z=\frac{\pi}{2}$ is -1, Res at $z=-\frac{\pi}{2}$ is -1
$I=2\pi i(-1-1)=-4\pi i$

9. $\int_c \frac{\sin\pi z^2+\cos\pi z^2}{(z-1)^2(z-2)}dz,\quad c:|z|=3$

Ans. $z=1$ double pole, Res $2\pi+1$, $z=2$ is simple pole, Res 1
$I=2\pi i\,[(2\pi+1)+1]=4\pi(\pi+1)i$

10. $\int_c \frac{4-3z}{z(z-1)(z-2)}dz$, $c:|z|=\frac{3}{2}$

Ans. $z=0,1,2$ simple poles, Res $2,-1$, $(z=2\notin c)$
$I=2\pi i\big(2+(-1)\big)=2\pi i$

11. $\int_c \frac{dz}{(z^2+4)^2}dz$, $c:|z-i|=2$

Ans. $z=\pm 2i$ double poles, $z=2i\in c$, Res is $-\frac{2}{64i^3}$, $I=\frac{\pi}{16}$

12. $\int_c \frac{5z-2}{z(z-1)}dz$, $c:|z|=2$

Ans. $z=0,1$ simple poles, $\in c$; Res 2, 3, $I=10\pi i$

13. $\int \frac{(3z^2+2)dz}{(z-1)(z^2+9)}$, c : (a) $|z-2|=2$
(b) $c:|z|=4$

Ans. $z=1,\pm 3i$ simple poles,
a. $z=1\in c$, Res $\frac{1}{2}$, $I=\pi i$
b. all $\in c$, Res $\frac{1}{2}$, $\frac{2\mp 81i}{6i(3i-1)}$, $I=6\pi i$

14. $\int_c \frac{e^{zt}}{z^2(z^2+2z+2)}dz$, $c:|z|=3$

Ans. $z=0$, double pole, Res $\frac{t-1}{2}$, $z=-1\pm i$ simple poles, Res $\frac{e^{(-1\pm i)t}}{4}$,
$I=2\pi i\left[\frac{t-1}{2}+\frac{1}{2}e^{-t}\cos t\right]$.

24.3 EVALUATION OF REAL INTEGRALS

Real integrals of the following types can be evaluated using the residue theorem.

Type I: Evaluation of Real Definite Integral of Rational Function of $\cos\theta$ and $\sin\theta$

By integration around a unit circle

Consider

$$I=\int_0^{2\pi} F(\sin\theta,\cos\theta)\,d\theta \tag{1}$$

where F is real rational function of $\cos\theta$ and $\sin\theta$ and is finite in the interval $(0,2\pi)$. Put $z=e^{i\theta}$ so $dz=ie^{i\theta}d\theta=izd\theta$. Also

$$\cos\theta=\frac{e^{i\theta}+e^{-i\theta}}{2}=\frac{z+\frac{1}{z}}{2}\quad\text{and}$$

$$\sin\theta=\frac{e^{i\theta}-e^{-i\theta}}{2i}=\frac{z-\frac{1}{z}}{2i}$$

Note that $|z|=|e^{i\theta}|=1$ which represents a unit circle c (with centre at origin). Substituting these values (1) takes the form

$$I=\oint_c f(z)\frac{dz}{iz}=\oint_c G(z)dz \tag{2}$$

Here $f(z)=F(\sin\theta,\cos\theta)$. The complex integral (2) is evaluated around the unit circle c (as θ varies from 0 to 2π).

Here $G(z)$ is a rational function of z. Now by residue theorem

$$I=2\pi i\sum \text{Res } G(z) \tag{3}$$

where the summation is taken for all poles of $G(z)$ which are within the unit circle $c:|z|=1$.

Type II: Evaluation of Improper Real Integral (of first kind) of Rational Function

By integration around a semi-circle

Consider

$$I=\int_{-\infty}^{\infty} f(x)\,dx \tag{4}$$

where $f(x) = \frac{p(x)}{q(x)}$. Here $p(x)$ and $q(x)$ are polynomials. Integral (4) converges if

i. $q(x)$ has no real zeros (i.e., $q(x) \neq 0$, for any x) and

ii. the degree of q is at least two greater than the degree of p.
Assuming these conditions, (4) can be evaluated by integrating around a semi-circle as follows:
Consider a simple closed curve c consisting of the straight line L along the real axis from $-R$ to R and the semi-circle S of radius R and with centre at origin (Fig. 24.2). By residue theorem

$$\oint_c f(z)\,dz = 2\pi i \sum \text{Res } f(z) \tag{5}$$

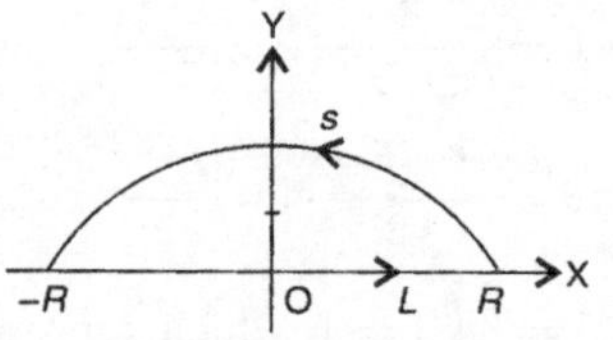

Fig. 24.2

where the summation is for the residues of $f(z)$ at all the singular points (poles) within c. Since $c = L + R$, (5) is rewritten as

$$\int_c = \int_{L+S} = \int_L + \int_S = \int_{-R}^{R} f(x)\,dx + \int_S f(z)\,dz$$
$$= 2\pi i \sum \text{Res} f(z) \tag{6}$$

As $R \to \infty$, the semi-circle S engulfs and becomes the entire upper half plane ($y > 0$). Then (6) takes the form

$$\lim_{R\to\infty} \int_{-R}^{R} f(x)\,dx = \lim_{R\to\infty} 2\pi i \sum \text{Res } f(z) - \lim_{R\to\infty} \int_S f(z)\,dz$$

$$\int_{-\infty}^{\infty} f(x)dx = 2\pi i \sum \text{Res } f(z) - \lim_{R\to\infty} \int_S f(z)\,dz \tag{7}$$

In RHS of (7) the summation of residues is for all poles of $f(z)$ in the upper half plane.

By assumption $|f(z)| < \frac{k}{|z|^2}$. So

$$\left|\int_S f(z)\,dz\right| \le M \cdot L = \frac{k}{|z|^2} \cdot 2\pi R = \frac{k}{R^2} 2\pi R = \frac{2\pi k}{R}$$

which tends to zero as $R \to \infty$. Thus (7) reduces to

$$\int_{-\infty}^{\infty} f(x)\,dx = 2\pi i \times [\text{sum of the residues of } f(z) \text{ at all the poles of } f(z) \text{ in the upper half plane}]$$
$$= 2\pi i \sum \text{Res } f(z) \tag{8}$$

Type III: Fourier Integral: Improper Integral Involving Trigonometric Functions

The integrals

$$\int_{-\infty}^{\infty} f(x) \cos mx dx \tag{9}$$

$$\int_{-\infty}^{\infty} f(x) \sin mx dx \tag{10}$$

can be evaluated by integrating

$$f(z)e^{imz}$$

around the contour c discussed in II. By residue theorem and similar analysis as in II

$$\int_{-\infty}^{\infty} f(x)e^{imx}dx = 2\pi i \sum \text{Res}\left(f(z)e^{imz}\right) \tag{11}$$

where the summation extends to all poles of $f(z)e^{imz}$ in the upper half=plane (since $|e^{imz}| = |e^{imx}||e^{-my}| = e^{-my} \le 1$ for $m > 0, y \ge 0$ and $|f(z)e^{imz}| = |f(z)||e^{imz}| \le |f(z)|$ it follows that $\int_S \to 0$ as $R \to \infty$). Equating the real and imaginary parts of (11), we get

$$\int_{-\infty}^{\infty} f(x) \cos mx\,dx = -2\pi \sum \text{Im.Res}\left[f(z)e^{imz}\right] \tag{12}$$

$$\int_{-\infty}^{\infty} f(x) \sin mx\,dx = 2\pi \sum \text{Re.Res}\left[f(z)e^{imz}\right] \tag{13}$$

Type IV: Evaluation of Improper Integral (of the second kind) whose Integrand Becomes Infinite

By indenting the contours having poles on the real axis

Consider

$$\int_A^B f(x)\,dx$$

where the integrand $f(x)$ become infinite at a point b between A and B. i.e.,

$$\lim_{x\to b} f(x) = \infty$$

Then the Cauchy's principal value of the integral is denoted by pr. v. and is defined as

$$\text{pr.v.} \int_A^B f(x)\,dx = \lim_{\epsilon\to 0}\left[\int_A^{b-\epsilon} f(x)\,dx + \int_{b+\epsilon}^B f(x)\,dx\right]$$

Simple poles on real axis

Delete the pole $z = b$ on the real axis by indenting the contour by drawing a semi-circle of radius r and with centre at b.

Result 1: If $f(z)$ has a simple pole at $z = b$ on the real axis, then

$$\lim_{r\to 0}\int_{c_1} f(z)\,dz = \pi i \operatorname*{Res}_{z=b} f(z) \qquad (14)$$

where c_1 is circle of radius r and with centre at $z = b$.
Poles in the upper half plane and simple poles on real axis:

Result 2: If $f(z)$ has several simple poles on the real axis, then

$$\text{pr. v.} \int_{-\infty}^{\infty} f(x)\,dx = 2\pi i \sum \text{Res } f(z) + \pi i \sum \text{Res } f(z) \qquad (15)$$

In the R.H.S. of (15), the first summation extends overall poles of $f(z)$ in the upper half plane and the second summation overall simple poles on the real axis.

Theorems on limiting contours

These theorems are useful in evaluating $\int_{C_R} f(z)\,dz$ where C_R is an arc of a circle whose radius $R \to \infty$ or $\to 0$.

Theorem 1: *If* $z \cdot f(z) \to 0$ *uniformly as* $R \to \infty$ *then*

$$\lim_{R\to\infty}\int_{C_R} f(z)\,dz \to 0$$

where C_R *is a circular arc of radius* R *and with centre at the origin.*

Theorem 2: ***(Jordan's Lemma)***
If $f(z) \to 0$ *uniformly as* $R \to \infty$ *then*

$$\lim_{R\to\infty}\int_{C_R} e^{imz} f(z)\,dz = 0, \quad (m > 0)$$

where C_R *is a circular arc (with radius* R *and centre at origin) in the first and/or second quadrants.*

Theorem 3: *If* $(z-a)f(z) \to 0$ *uniformly as* $r \to 0$ *then*

$$\lim_{r\to 0}\int_{C_r} f(z)\,dz = 0$$

where C_r *is a circular arc with radius* r *and with centre at* $z = a$.

Type V: Result

Let $f(z)$ be analytic everywhere in the z-plane except at a finite number of poles. Assume that $f(z)$ has no poles on the positive half of the real axis. If $z^a f(z) \to 0$ as $z \to 0$, then as $z \to \infty$

$$\int_0^{\infty} x^{a-1} f(x)\,dx = \frac{\pi}{\sin a\pi} \times$$

$$\times \sum \text{Residues of}\{(-z)^{a-1} \cdot f(z)\}$$

at all its poles.

Worked Out Examples

Type I

Example 1: Evaluate $I = \int_0^{2\pi} \frac{d\theta}{1-2p\sin\theta+p^2}$, $|p|<1$.

Solution: Put $z = e^{i\theta}$, $dz = ie^{i\theta}d\theta = iz\,d\theta$

$$\sin\theta = \frac{e^{i\theta} - e^{-i\theta}}{2i} = \frac{z - \frac{1}{z}}{2i} = \frac{z^2 - 1}{2iz}$$

$|z| = \left|e^{i\theta}\right| = 1$. Thus the given integral reduces to

$$I = \oint_c \frac{1}{1 - 2p\frac{(z^2-1)}{2iz} + p^2}\frac{dz}{iz} = -\frac{1}{p}\int_c \frac{dz}{z^2 - \frac{i(1+p^2)}{p}z - 1}$$

where $c : |z| = 1$.

The integrand has simple poles at

$$z = \frac{i(1+p^2) \pm i(1-p^2)}{2p} = \frac{i}{p}, ip$$

$z_1 = \frac{i}{p}$ lies outside while $z_1 = ip$ lies inside the unit circle c since $|z_1| = |ip| = |i||p| = |p| < 1$.

$$\operatorname*{Res}_{\text{at } z=z_2} f(z) = \lim_{z\to z_2}(z - z_2)\cdot\frac{1}{(z - z_1)(z - z_2)} = \frac{1}{z_2 - z_1}$$

$$= \frac{1}{ip - \frac{i}{p}} = \frac{p}{i(p^2 - 1)} = \frac{ip}{1 - p^2}$$

By residue theorem

$$I = -\frac{1}{p}\int_c \frac{dz}{(z - z_1)(z - z_2)} = -\frac{1}{p}\cdot 2\pi i \operatorname*{Res}_{\text{at } z=z_2} f(z)$$

$$= -\frac{1}{p}\cdot 2\pi i\cdot\left(\frac{ip}{i - p^2}\right) = \frac{2\pi}{1 - p^2}.$$

Example 2: Evaluate $I = \int_0^{\pi}\frac{\cos^2 3\theta\, d\theta}{5 - 4\cos 2\theta}$.

Solution: $I = \frac{1}{2}\int_{-\pi}^{\pi}\frac{\cos^2 3\theta\, d\theta}{5 - 4\cos 2\theta}$

Put $z = e^{i\theta}$, $\cos\theta = \frac{e^{i\theta} + e^{-i\theta}}{2} = \frac{z + \frac{1}{z}}{2} = \frac{z^2+1}{2z}$

Using $\cos^2 3\theta = \frac{1 + \cos 6\theta}{2}$, $\cos 6\theta = \frac{z^6 + \frac{1}{z^6}}{2} = \frac{z^{12}+1}{2z^6}$

and $\cos 2\theta = \frac{z^2 + \frac{1}{z^2}}{2} = \frac{z^4+1}{2z^2}$.

The integral reduces to

$$I = \frac{1}{2}\oint_c \frac{1}{2}\left(1 + \frac{z^{12}+1}{2z^6}\right)\frac{1}{\left[5 - 4\cdot\left(\frac{z^4+1}{2z^2}\right)\right]}\cdot\frac{dz}{iz}$$

$$= -\frac{1}{16i}\oint_c \frac{z^{12} + 2z^6 + 1}{z^5\left(z^4 - \frac{5}{2}z^2 + 1\right)}dz = -\frac{1}{16i}\oint_c f(z)\,dz$$

The singular points of the integrand are the zeros of the denominator i.e., $z^5 = 0$ and $z^4 - \frac{5}{2}z^2 + 1 = 0$, or $z^2 = 2, \frac{1}{2}$. Thus $z = 0$ is pole of order 5 and $z = \pm\frac{1}{\sqrt{2}}$ are two simple poles, which lie inside c, while $z = \pm\sqrt{2}$ lies outside c.

Residue by Laurent series:

Expanding $\left(1 - \left(\frac{5}{2}z^2 - z^4\right)\right)$ in series, we get

$$f(z) = \frac{z^{12} + 2z^6 + 1}{z^5}\cdot\left[1 - \left(\frac{5}{2}z^2 - z^4\right)\right]^{-1}$$

$$f(z) = \frac{z^{12} + 2z^6 + 1}{z^5}\left[1 + \frac{5}{2}z^2 - z^4 + \frac{25}{4}z^4 + z^8 - 5z^6 + \cdots\right]$$

Residue of $f(z)$ at $z = 0$ is the coefficient of z^{-1} in this Laurent series expansion, i.e.,

$$k_1 = \operatorname*{Res}_{\text{at } z=0} f(z) = -1 + \frac{25}{4} = \frac{21}{4}$$

$$k_2 = \operatorname*{Res}_{\text{at } z=z_1=+\frac{1}{\sqrt{2}}} f(z) = \lim_{z\to z_1}\frac{(z^6+1)^2(z - z_1)}{z^5(z^2 - 2)(z - z_1)(z - z_2)}$$

$$= \frac{(z_1^6+1)^2}{z_1^5(z_1^2 - 2)(z_1 - z_2)}$$

$$= \frac{81}{64}\cdot\frac{1}{\frac{1}{4\sqrt{2}}\left(-\frac{3}{2}\right)\left(\frac{2}{\sqrt{2}}\right)} = -\frac{27}{8}$$

$$k_3 = \operatorname*{Res}_{\text{at } z=z_2=-\frac{1}{\sqrt{2}}} f(z)$$

$$= \lim_{z\to z_2}\frac{(z^6+1)^2}{z^5(z^2 - 2)(z - z_1)(z - z_2)}\cdot(z - z_2) = -\frac{27}{8}$$

Thus by residue theorem,

$$I = -\frac{1}{16i}\cdot\{2\pi i(\text{Sum of the residues})]$$

$$= -\frac{1}{16i}\cdot 2\pi i\left(\frac{21}{4} - \frac{27}{8} - \frac{27}{8}\right) = -\frac{1}{16i}2\pi i\left(-\frac{6}{4}\right)$$

$$I = \frac{3\pi}{16}.$$

Type II

Example 3: Evaluate $\int_0^{\infty}\frac{dx}{(a^2+x^2)^2}$.

Solution: Consider a curve c consisting of semicircle S and the line L from $-R$ to R along the real axis (refer Fig. 24.3). Consider

$\int_c \frac{dz}{(a^2+z^2)^2} = \oint_c f(z)\,dz$ where $f(z) = (a^2 + z^2)^{-2}$

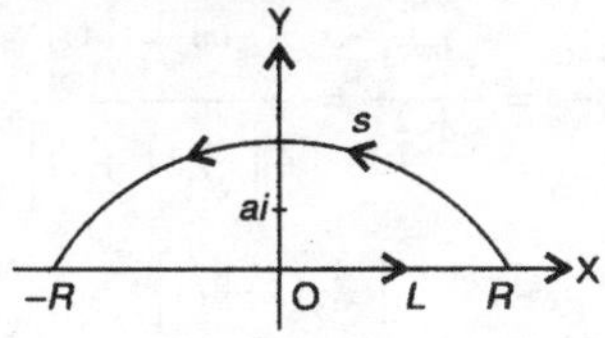

Fig. 24.3

Rewriting

$$\oint_c f(z)\,dz = \int_L + \int_S = \int_{-R}^{R} f(x)\,dx + \int_S f(z)\,dz \quad (1)$$

Since $|a^2 + z^2| > |z|^2 - |a|^2$

on S: $\quad |a^2 + z^2| > R^2 - a^2$

or $\quad \left|\frac{1}{(a^2+z^2)^2}\right| < \frac{1}{(R^2-a^2)^2}$

So $\quad \left|\int_S f(z)\,dz\right| \le \frac{1}{(R^2-a^2)^2} \cdot \pi R \quad (2)$

As $R \to \infty$ $\int_S f(z)\,dz \to 0$
As $R \to \infty$ (1) becomes

$$\lim_{R\to\infty} \oint_c f(z)\,dz = \lim_{R\to\infty} \int_{-R}^{R} f(x)\,dx + \lim_{R\to\infty} \int_S f(z)\,dz$$

$$\therefore \quad \int_{-\infty}^{\infty} \frac{dx}{(a^2+x^2)^2} dx = \lim_{R\to\infty} \oint_c f(z)\,dz, \text{ using (2)}$$

As $R \to \infty$, c engulfs the entire upper half plane. Now evaluating the R.H.S. integral by residue theorem, $= 2\pi i \cdot$ Res $f(z)$ at singular points which are in the upper half plane.

For $f(z) = \frac{1}{(z^2+a^2)^2}$ has singular points at $z = \pm ai$ which are poles of order 2. But only $z = ai$ lies in the upper half-plane.

$$\operatorname*{Res}_{z=ai} f(z) = \lim_{z\to ai} \cdot \frac{d}{dz}(z-ai)^2 \cdot \frac{1}{(z-ai)^2(z+ai)^2}$$

$$= \lim_{z\to ai} -\frac{-2}{(z+ai)^3} = -\frac{-2}{8a^3 i^3} = -\frac{-i}{4a^3}$$

Hence $\quad \int_0^{\infty} \frac{dx}{(a^2+x^2)^2} = \frac{1}{2}\int_{-\infty}^{\infty} \frac{dx}{(a^2+x^2)^2}$

$$= \frac{1}{2} \cdot 2\pi i \cdot \left(-\frac{i}{4a^3}\right) = \frac{\pi}{4a^3}.$$

Example 4: Evaluate $I = \int_{-\infty}^{\infty} \frac{x\,dx}{(x^2+1)(x^2+2x+2)}$

Solution: Let c be a simple closed curve consisting of semicircle S and line L from $-R$ to R (see Fig. 24.4).

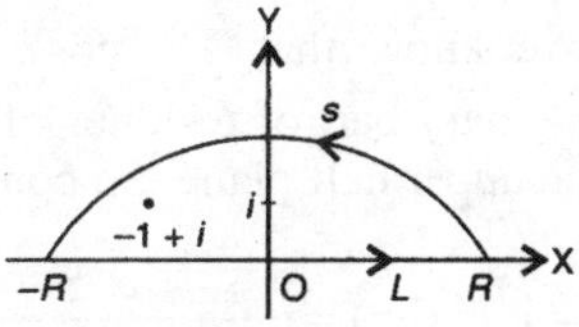

Fig. 24.4

Let $f(z) = \frac{z}{((z^2+1)(z^2+2z+2))}$

Consider

$$\oint_c f(z)\,dz = \oint_c \frac{z\,dz}{(z^2+1)(z^2+2z+2)}$$

This integral can be evaluated by residue theorem.

Rewriting

$$\oint_c f(z)\,dz = \int_L + \int_S$$

$$= \int_{-R}^{R} \frac{x\,dx}{(x^2+1)(x^2+2x+2)} + \int_S f(z)\,dz$$

As $R \to \infty$, the 2nd integral on the right becomes 0. So

$$\lim_{R\to\infty} \oint_c f(z)\,dz = \int_{-\infty}^{\infty} \frac{x\,dx}{(x^2+1)(x^2+2x+2)}$$

As $R \to \infty$, c encloses the entire upper half plane. Thus by applying residue theorem

$$I = \int_{-\infty}^{\infty} \frac{x\,dx}{(x^2+1)(x^2+2x+2)} = 2\pi i \cdot \sum \text{Res } f(z)$$

The singular points of $f(z) = \frac{z}{(z^2+1)(z^2+2z+2)}$ are $z = \pm i$ and $z = -1 \pm i$. Out of these only $z = i$ and $z = -1 + i$ lies in the upper half plane.

$$\operatorname*{Res}_{z=i} f(z) = \lim_{z\to i}(z-i) \cdot \frac{z}{(z-i)(z+i)(z^2+2z+2)}$$

$$= \frac{i}{2i(-1+2i+2)} = \frac{1}{2(2i+1)}$$

$$\operatorname*{Res}_{z=-1+i} f(z) = \lim_{z\to -1+i} (z+1-i) \cdot \frac{z}{(z^2+1)(z+1-i)(z+1+i)}$$

$$= \frac{(-1+i)}{[(-1+i)^2+1][-1+i+1+i]} = \frac{-1+i}{2i(1-2i)}$$

Thus

$$I = 2\pi i \left[\frac{1}{2(2i+1)} + \frac{(-1+i)}{2i(1-2i)}\right] = -\frac{\pi}{5}.$$

Type III

Example 5: Evaluate $\int_{-\infty}^{\infty} \frac{\cos mx}{(x^2+a^2)(x^2+b^2)} dx$.

Solution: We know that $\int_{-\infty}^{\infty} \cos mx\, Q(x) dx = -2\pi \cdot \sum$ imaginary part of residues of $e^{imz}Q(z)$ at its poles in the upper half plane. So consider

$$f(z) = e^{imz}Q(z) = \frac{e^{imz}}{(z^2+a^2)(z^2+b^2)}$$

$f(z)$ has simple poles at $z = \pm ai, z = \pm bi$ out of which $z = ai, bi$ lies in the upper half plane.

$$k_1 = \underset{\text{at } z=ai}{\text{Res}}\ f(z) = \lim_{z\to ai} \frac{(z-ai)\cdot e^{imz}}{(z-ai)(z+ai)(z^2+b^2)}$$

$$= \frac{e^{-am}}{2ai(b^2-a^2)} = \frac{-ie^{-am}}{2a(b^2-a^2)}$$

Similarly,

$$k_2 = \underset{\text{at } z=bi}{\text{Res}}\ f(z) = \frac{e^{-bm}}{2bi(a^2-b^2)} = \frac{-ie^{-bm}}{2b(a^2-b^2)},$$

$$\int_{-\infty}^{\infty} \frac{\cos mx}{(x^2+a^2)(x^2+b^2)} dx$$

$$= -2\pi \left[-\frac{e^{-am}}{2a(b^2-a^2)} - \frac{e^{-bm}}{2b(a^2-b^2)} \right]$$

$$= \frac{\pi}{a^2-b^2} \left[\frac{e^{-bm}}{b} - \frac{e^{-am}}{a} \right].$$

Example 6: $\int_{-\infty}^{\infty} \frac{x \sin mx}{1+x^4} dx$.

Solution: We know that $\int_{-\infty}^{\infty} \sin mx\, Q(x) dx = 2\pi \sum$ real part of residues of $e^{imz}Q(z)$ at its poles in the upper half plane. So consider $f(z) = e^{imz} \cdot \frac{z}{1+z^4}$ which has singular points at $z = (-1)^4 = e^{i\left(\frac{\pi+2k\pi}{4}\right)}$ for $k = 0, 1, 2, 3$. Out of these four, only $z_1 = e^{\frac{i\pi}{4}} = \frac{1}{\sqrt{2}}(1+i)$ and $z_2 = e^{\frac{i3\pi}{4}} = \frac{1}{\sqrt{2}}(-1+i)$ lies in the upper half plane.

$$k_1 = \underset{\text{at } z=z_1}{\text{Res}}\ f(z) = \lim_{z\to z_1} (z-z_1) \cdot \frac{z \cdot e^{imz}}{(1+z^4)},$$

which is $\frac{0}{0}$ form. Applying L' Hostpital's rule

$$= \lim_{z\to z_1} \frac{[(z-z_1)\{e^{imz} + z\cdot im \cdot e^{imz}\} + ze^{imz}]}{4z^3}$$

$$= \frac{z_1 e^{imz_1}}{4z_1^3} = \frac{e^{imz_1}}{4z_1^2} = \frac{e^{im\frac{1}{\sqrt{2}}(1+i)}}{4\left[\frac{1}{\sqrt{2}}(1+i)\right]^2}$$

$$= \frac{e^{im\frac{1}{\sqrt{2}}(1+i)} \left[\frac{1}{\sqrt{2}}(1-i)\right]^2}{4\left[\frac{1}{\sqrt{2}}(1+i)\right]^2 \left[\frac{1}{\sqrt{2}}(1-i)\right]^2}$$

$$k_1 = \frac{e^{-\frac{m}{\sqrt{2}}} \cdot \left(\cos\frac{m}{\sqrt{2}} + i\sin\frac{m}{\sqrt{2}}\right) \cdot \frac{1}{2}(-2i)}{4\cdot\frac{1}{2}(1+1)}$$

Real part of $k_1 = \frac{1}{4} e^{-\frac{m}{\sqrt{2}}} \cdot \sin\frac{m}{\sqrt{2}}$

Similarly,

$$k_2 = \underset{\text{at } z=z_2}{\text{Res}}\ f(z) = \frac{e^{im\frac{1}{\sqrt{2}}(-1+i)} \left[\frac{1}{\sqrt{2}}(-1-i)\right]^2}{4\left[\frac{1}{\sqrt{2}}(-1+i)\right]^2 \left[\frac{1}{\sqrt{2}}(-1-i)\right]^2}$$

$$k_2 = \frac{e^{-\frac{m}{\sqrt{2}}} \cdot \left(\cos\frac{m}{\sqrt{2}} - i\sin\frac{m}{\sqrt{2}}\right) \cdot \frac{1}{2}(2i)}{4\cdot\frac{1}{2}(1+1)}$$

Real part of $k_2 = \frac{1}{4} e^{-\frac{m}{\sqrt{2}}} \cdot \sin\frac{m}{\sqrt{2}}$.

Then

$$\int_{-\infty}^{\infty} \frac{x \sin mx}{1+x^4} dx = 2\pi[Re(k_1) + Re(k_2)]$$

$$= 2\pi \left[e^{-\frac{m}{\sqrt{2}}} \cdot \sin\frac{m}{\sqrt{2}} \left(\frac{1}{4} + \frac{1}{4}\right) \right]$$

$$= \pi e^{-\frac{m}{\sqrt{2}}} \cdot \sin\frac{m}{\sqrt{2}}.$$

Integration around a rectangular contour

Example 7: Evaluate $\int_{-\infty}^{\infty} \frac{\cos mx}{e^x+e^{-x}} dx$.

Solution: Consider $\oint_c \frac{e^{imz}}{e^z+e^{-z}} dz$ where c is the rectangle having vertices at $-R, R, R+\pi i, -R+\pi i$ and consisting of the lines c_1, c_2, c_3, c_4 as shown in Fig. 24.5.

The poles of $\frac{e^{imz}}{e^z+e^{-z}}$ are simple and occur when $e^z + e^{-z} = 0$ i.e., $e^{2z} = -1 = e^{\pm(2n+1)\pi i}$ i.e., $z = \left(n+\frac{1}{2}\right)\pi i$ with $n = 0, \pm1, \pm2, \ldots$. Out of these only $z = \left(\frac{\pi i}{2}\right)$ lies inside the rectangle c.

By residue theorem

$$I = \oint_c \frac{e^{imz}}{e^z+e^{-z}} dz = \frac{1}{2} \oint_c \frac{e^{imz}}{\left(\frac{e^z+e^{-z}}{2}\right)} = \frac{1}{2} \oint_c \frac{e^{imz}}{\cosh z}$$

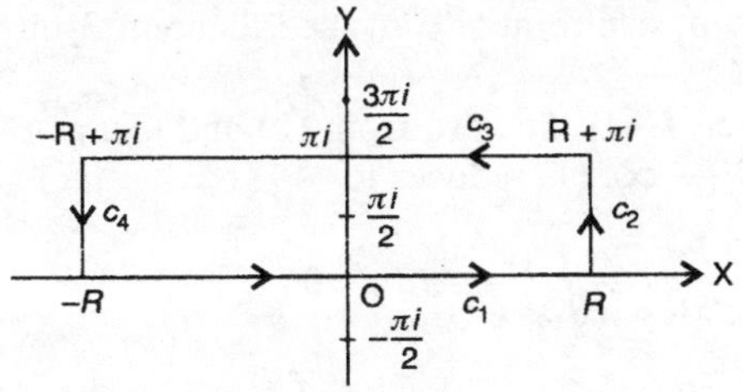

Fig. 24.5

$$= \frac{1}{2}2\pi i\left(\text{Res}\frac{e^{imz}}{\cosh z} \text{ at } z = \frac{\pi i}{2}\right) \tag{1}$$

Now

$$\underset{\text{at } z=\frac{\pi i}{2}}{\text{Res}} \frac{e^{imz}}{\cosh z} = \lim_{z\to\frac{\pi i}{2}} \left(z-\frac{\pi i}{2}\right)\cdot\frac{e^{imz}}{\cosh z}, \left(\frac{0}{0}\text{ form}\right)$$

$$= \lim_{z\to\frac{\pi i}{2}} \frac{\left(z-\frac{\pi i}{2}\right)\cdot ime^{imz}+1\cdot e^{imz}}{\sin hz}$$

$$= \frac{e^{im\cdot\frac{\pi i}{2}}}{\sinh\left(\frac{\pi i}{2}\right)} = \frac{e^{-\frac{m\pi}{2}}}{i\sin(\pi/2)} = -ie^{-\frac{m\pi}{2}} \tag{2}$$

Therefore substituting (2) in (1)

$$I = \oint_c \frac{e^{imz}}{e^z+e^{-z}}dz = \frac{1}{2}\cdot 2\pi i\cdot(-ie - \frac{m\pi}{2}) = \frac{\pi}{e^{\frac{m\pi}{2}}}$$

$$I = \oint_c \frac{e^{imz}dz}{2\cdot\cosh z} = \frac{\pi}{e^{\frac{m\pi}{2}}}$$

$$\therefore \oint_c \frac{e^{imz}}{\cosh z} = \frac{2\pi}{e^{\frac{m\pi}{2}}} \tag{3}$$

Since $c = c_1 + c_2 + c_3 + c_4$, we have

$$\oint_c f(z)\,dz = \int_{c_1} + \int_{c_2} + \int_{c_3} + \int_{c_4} \text{ where } f(z) = \frac{e^{imz}}{\cosh z}$$

$$= \int_{-R}^{R} f(x)\,dx + \int_0^{\pi} f(R+iy)i\,dy$$

$$+ \int_R^{-R} f(x+\pi i)dx + \int_{\pi}^{0} f(-R+iy)i\,dy$$

$$= \int_{-R}^{R} \frac{e^{imx}}{\cosh x}dx + \int_0^{\pi} \frac{e^{im(R+iy)}}{\cosh(R+iy)}i\,dy$$

$$+ \int_R^{-R} \frac{e^{im(x+\pi i)}}{\cosh(x+\pi i)}dx$$

$$+ \int_{\pi}^{0} \frac{e^{im(-R+iy)}}{\cosh(-R+iy)}i\,dy \tag{4}$$

$$= I_1 + I_2 + I_3 + I_4.$$

To show that I_2 and $I_4 \to 0$ as $R\to\infty$: Note $|e^{im(R+iy)}| = |e^{imR}|\cdot|e^{-my}| = |e^{-my}| < 1$ for $y > 0$

Since $|\cosh(R+iy)|$

$$= \frac{e^{R+iy}+e^{-R-iy}}{2} \ge \frac{1}{2}\left\{\left|e^{R+iy}\right| - \left|e^{-R-iy}\right|\right\}$$

$$= \frac{1}{2}(e^R - e^{-R}) \ge \frac{1}{4}e^R$$

from triangle inequality $|z_1+z_2| \ge |z_1| - |z_2|$.

$$|I_2| = \left|\int_0^{\pi} \frac{e^{im(R+iy)}}{\cosh(R+iy)}i\,dy\right| \le \int_0^{\pi} \frac{1}{\frac{e^R}{4}}dy = 4e^{-R}\cdot\pi$$

As $R\to\infty$, $I_2 \to 0$.

Similarly as $R\to\infty$, $I_4\to 0$.

Now (4) reduces to

$$\lim_{R\to\infty}\left[\int_{-R}^{R}\frac{e^{imx}}{\cosh x} + e^{-m\pi}\int_{-R}^{R}\frac{e^{imx}}{\cosh x}dx\right] = \oint_c \frac{e^{imz}}{\cosh z}dz$$

since $\cosh(x+\pi i) = \cosh x\cdot\cos\pi - 0 = -\cosh x$

$$(1+e^{-m\pi})\cdot\int_{-\infty}^{\infty}\frac{e^{imx}}{\cosh x}dx = \oint_c \frac{e^{imz}}{\cosh z}dz = \frac{2\pi}{e^{\frac{m\pi}{2}}}$$

from (3)

$$\int_{-\infty}^{\infty}\frac{e^{imx}}{\cosh x}dx = \frac{2\cdot\pi}{e^{\frac{m\pi}{2}}(1+e^{-m\pi})}$$

$$\int_{-\infty}^{\infty}\frac{e^{imx}dx}{(e^x+e^{-x})} = \frac{\pi}{(e^{\frac{m\pi}{2}}+e^{\frac{-m\pi}{2}})}$$

Type IV By Indentation

Example 8: Show that

$$\int_0^{\infty}\frac{\cos x}{\sqrt{x}}dx = \int_0^{\infty}\frac{\sin x}{\sqrt{x}}dx = \sqrt{\frac{\pi}{2}}.$$

Solution: Consider the contour c consisting of c_1, c_2, c_3, c_4 where c_1 : AB: straight-line from r to R along x-axis, c_2 : BD: arc of the circle of radius R in the first quadrant, c_3 : DE: straight line from R to r along y-axis, c_4 : EA: arc of circle of radius r lying in the first quadrant (Fig. 24.6).

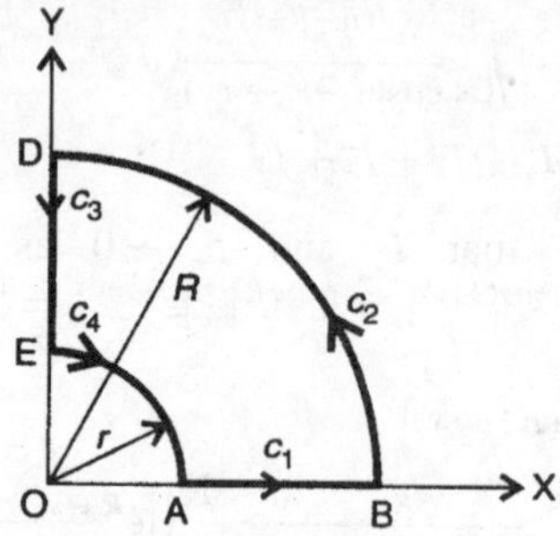

Fig. 24.6

The function $f(z)=\frac{e^{iz}}{\sqrt{z}}$ has a singular point at $z=0$ which lies on both the x-axis and y-axis. To delete this singular point from the region, indent the contour by drawing a quadrant circle with this singular point $z=0$ as the centre.

Since the only singular point $z=0$ of $f(z)$ is deleted $f(z)$ has no singular point within c, therefore by Cauchy's theorem,

$$\oint_c f(z)\,dz=0$$

Since $c=c_1+c_2+c_3+c_4$, we have

$$\int_c=\int_{c_1}+\int_{c_2}+\int_{c_3}+\int_{c_4}=I_1+I_2+I_3+I_4=0$$

or
$$\int_r^R f(x)\,dx+\int_{c_2} f(z)\,dz+\int_R^r f(iy)i\,dy+$$

$$+\int_{c_4} f(z)\,dz=0 \tag{1}$$

As $R\to\infty$, $z\to\infty$ then $f(z)\to 0$.
By Jordan's Lemma (see page 24.9)

$$I_2=\int_{c_2} f(z)\,dz\to 0 \quad \text{as} \quad R\to\infty \tag{2}$$

(* $\lim_{R\to\infty}\int_{c_R} e^{imz} f(z)\,dz=0$ when $f(z)\to 0$ as $R\to\infty$)

Also as $r\to 0$, $(z-0)\cdot\frac{e^{iz}}{\sqrt{z}}=\sqrt{z}e^{iz}\to 0$ then

$$I_4=\int_{c_4} f(z)\,dz\to 0 \quad \text{as} \quad r\to 0 \tag{3}$$

(If $(z-a)f(z)\to 0$ as $\rho\to 0$ then $\lim_{\rho\to 0}\int_{c_\rho} f(z)\,dz=0$ where c_ρ is circular arc of radius ρ, centre at $z=a$. (see theorem 3 on page 24.9)

Since I_2, I_4 are zero from (2) and (3), as $r\to 0$ and $R\to\infty$, (1) reduces to

$$\int_0^\infty \frac{e^{ix}}{\sqrt{x}}dx+\int_\infty^0 \frac{e^{i(iy)}}{\sqrt{iy}}i\,dy=0$$

$$\int_0^\infty \frac{e^{ix}}{\sqrt{x}}dx=\int_0^\infty e^{-y}\cdot y^{-\frac{1}{2}}\cdot i^{+\frac{1}{2}}dy$$

$$=(e^{\frac{i\pi}{2}})^{\frac{1}{2}}\int_0^\infty e^{-y}y^{\frac{1}{2}-1}dy$$

since $i=e^{\frac{i\pi}{2}}$.

$$\int_0^\infty \frac{e^{ix}}{\sqrt{x}}dx=e^{\frac{i\pi}{4}}\Gamma\left(\frac{1}{2}\right)=\left(\cos\frac{\pi}{4}+i\sin\frac{\pi}{4}\right)\sqrt{\pi}$$

$$=\sqrt{\frac{\pi}{2}}(1+i)$$

since $\int_0^\infty e^{-y}y^{\frac{1}{2}-1}dy=\Gamma\left(\frac{1}{2}\right)=\sqrt{\pi}$

Equating the real and imaginary parts on both sides, we get

$$\int_0^\infty \frac{\cos x}{\sqrt{x}}dx=\int_0^\infty \frac{\sin x}{\sqrt{x}}dx=\sqrt{\frac{\pi}{2}}.$$

Example 9: Find the Cauchy principal value of

$$I=\int_{-\infty}^\infty \frac{dx}{x^2-ix}.$$

Solution: The function $f(z)=\frac{1}{z^2-iz}=\frac{1}{z(z-i)}$ has simple poles at $z=0$ and $z=i$. While $z=i$ lies in the upper half=plane, $z=0$ lies on the real axis. Then

$$\text{pr. v.}\int_{-\infty}^\infty f(x)\,dx=2\pi i\sum \text{Res}\,f(z)+\pi i\sum\text{Res}\,f(z)$$

where the first sum extends to poles in upper half=plane, second sum to poles on real axis.

$$\operatorname*{Res}_{z=0} f(z)=\lim_{z\to 0} z\cdot\frac{1}{z(z-i)}=\frac{1}{-i}=i$$

$$\operatorname*{Res}_{z\to i} f(z)=\lim_{z\to i}(z-i)\cdot\frac{1}{z(z-i)}=\frac{1}{i}=-i$$

$$\text{pr. v.}\int_{-\infty}^\infty \frac{dx}{x^2-ix}=2\pi i(-i)+\pi i(i)=2\pi-\pi=\pi.$$

Type V

Example 10: Evaluate

$$I = \int_0^\infty \frac{x^{a-1}}{1+x^3}dx, \quad 0 < a < 3.$$

Solution: We know that

$\int_0^\infty x^{a-1} f(x)\,dx = \frac{\pi}{\sin a\pi} \sum$ Res of $(-z)^{a-1} f(z)$ at all its poles.

$f(z) = \frac{1}{1+z^3}$ has simple poles at $z_1 = -1$ and $z_2 = \frac{1+\sqrt{3}i}{2}$, $z_3 = \frac{1-\sqrt{3}i}{2}$ (where z_2, z_3 are roots of $z^2 - z + 1 = 0$).

$$k_1 = \text{Res at } z = z_1 = \lim_{z \to -1} \frac{(-z)^{a-1}}{1+z^3} \cdot (z+1)$$

$$= \frac{1^{a-1}}{1+1+1} = \frac{1}{3}$$

$$k_2 = \text{Res at } z = z_2 = \lim_{z \to z_2} \frac{(-z)^{a-1}}{(z+1)(z-z_2)(z-z_3)}(z - z_2)$$

$$= \frac{(-z_2)^{a-1}}{(z_2+1)(z_2-z_3)}$$

$$= -\frac{1}{z_2} \cdot \frac{1}{(z_2+1)(z_2-z_3)} \cdot \left(-\frac{1}{2} - \frac{\sqrt{3}}{2}i\right)^a$$

$$k_2 = \frac{1}{3}\left(e^{\frac{i4\pi}{3}}\right)^a = \frac{1}{3}e^{\frac{i4\pi a}{3}}$$

since $z_2 \cdot (z_2+1)(z_2-z_3)$

$$= \left(\frac{1+\sqrt{3}i}{2}\right)\left(\frac{3+\sqrt{3}i}{2}\right)(\sqrt{3}i) = 3,$$

Similarly,

$$k_3 = \text{Res at } z = z_3 = \lim_{z \to z_3} \frac{(-z)^{a-1}}{(z+1)(z-z_2)(z-z_3)}(z - z_3)$$

$$= \frac{(-z_3)^{a-1}}{(z_3+1)(z_3-z_2)}$$

$$= -\frac{1}{z_3(z_3+1)(z_3-z_2)} \left(-\frac{1}{2} + \frac{\sqrt{3}}{2}i\right)^a$$

$$= \frac{1}{3}\left(e^{\frac{i2\pi}{3}}\right)a = \frac{1}{3}e^{\frac{i2\pi a}{3}}$$

Then $\quad I = \frac{\pi}{\sin a\pi}\left[\frac{1}{3} + \frac{1}{3}e^{\frac{i4\pi a}{3}} + \frac{1}{3}e^{\frac{i2\pi a}{3}}\right]$

$$= \frac{\pi}{3\sin a\pi}\left[1 + 2\cos\frac{2\pi a}{3}\right].$$

Exercise

Type I

Evaluate

1. $\int_0^{2\pi} \frac{d\theta}{\sqrt{2}-\cos\theta}$

Ans. 2π, $z_1 = \sqrt{2} - 1$, Res $\frac{1}{2}$

2. $\int_0^{2\pi} \frac{\cos 2\theta\, d\theta}{1-2p\cos\theta+p^2}$

Ans. $\frac{2\pi p^2}{1-p^2}$, $z = 0, p$, Res $\frac{1+p^2}{2ip^2}, \frac{1+p^4}{2ip^2(1-p^2)}$

3. $\int_0^{2\pi} \frac{\cos 3\theta\, d\theta}{5-4\cos\theta}$

Ans. $\frac{\pi}{12}$, $z = 0, \frac{1}{2}$, Res $\frac{21}{8}, -\frac{65}{24}$

4. $\int_0^{2\pi} \frac{\sin^2\theta\, d\theta}{a+b\cos\theta}$, $a > b > 0$

Ans. $\frac{2\pi}{b^2}[a - \sqrt{a^2-b^2}]$, $z = 0, -\frac{a}{b} + \frac{\sqrt{a^2-b^2}}{b}$,

Res $-\frac{2a}{b^2}, \frac{2\sqrt{a^2-b^2}}{b^2}$

5. $\int_0^{2\pi} \frac{d\theta}{\frac{5}{4}+\sin\theta}$

Ans. $\frac{8\pi}{3}$, $z = -\frac{1}{2}i$, Res $\frac{4}{3i}$

6. $\int_0^{2\pi} \frac{d\theta}{3-2\cos\theta+\sin\theta}$

Ans. π, $z = (2-i)/5$, Res $\frac{1}{2i}$

7. $\int_0^{2\pi} \frac{d\theta}{(5-3\sin\theta)^2}$

Ans. $\frac{5\pi}{32}$, $z = \frac{i}{3}$, Res $-\frac{5}{256}$

8. $\int_0^{2\pi} \frac{d\theta}{a+b\sin\theta}$

Ans. $\frac{2\pi}{\sqrt{a^2-b^2}}$, if $|a| > |b|$; $\frac{-a+\sqrt{a^2-b^2}}{b}i$,

Res $\frac{1}{\sqrt{a^2-b^2}}i$

9. $\int_0^{\pi} \frac{d\theta}{(a+\cos\theta)^2}$

Ans. $\pi a(a^2-1)^{-\frac{3}{2}}$,

$a > 1, -a + \sqrt{a^2-1}$, Res $\frac{2a}{2^3(a^2-1)^{\frac{3}{2}}}$

10. $\int_0^{\pi} \sin^{2n}\theta\, d\theta$

Ans. $\frac{(2n)!\pi}{2^{2n}(n!)^2}$, $z = 0$ of order $2n+1$, Res $\frac{(-1)^n \cdot 2n!}{(n!)^2}$

11. $\int_{-\pi}^{\pi} \frac{d\theta}{1+\sin^2\theta}$

Ans. $\pi\sqrt{2}, z = 3 - 2\sqrt{2}$, Res $-1/(4\sqrt{2})$

Type II

Evaluate

12. $\int_0^\infty \frac{dx}{(x^2+1)^n}$, n = natural number

Ans. For $n = 1$

$\frac{\pi}{2}, z = i$, Res $\frac{1}{2i}$

Ans. For $n > 1$, $I = \frac{1 \cdot 3 \cdot 5 \cdots (2n-3)}{2 \cdot 4 \cdot 6 \cdots (2n-2)} \cdot \frac{\pi}{2}$

13. $\int_0^\infty \frac{dx}{(x^2+9)(x^2+4)^2}$

Ans. $\frac{\pi}{200}, z = 2i, 3i$, Res $-\frac{13i}{200}, -\frac{3}{50i}$

14. $\int_0^\infty \frac{dx}{a^4+x^4}$

Ans. $\frac{\pi}{a^3 2\sqrt{2}}, z = ae^{\frac{i\pi}{4}}, ae^{\frac{i3\pi}{4}}$, Res $-\frac{e^{\frac{i\pi}{4}}}{4a^3}, \frac{e^{\frac{-i\pi}{4}}}{4a^3}$

15. $\int_{-\infty}^\infty \frac{x^2 dx}{(x^2+1)^2(x^2+2x+2)}$

Ans. $\frac{7\pi}{50}, z = i, -1+i$, Res $\frac{9i-12}{100}, \frac{3-4i}{25}$

16. $\int_{-\infty}^\infty \frac{x^2 dx}{(x^2+a^2)(x^2+b^2)}$

Ans. $\frac{\pi}{a+b}, z = ai, bi$, Res $\frac{a}{2i(a^2-b^2)}, \frac{b}{2i(b^2-a^2)}$

17. $\int_0^\infty \frac{dx}{x^6+1}$

Ans. $\frac{\pi}{3}, z = e^{\frac{\pi i}{6}}, e^{\frac{3\pi i}{6}}, e^{\frac{5\pi i}{6}}$ Res $\frac{1}{6}e^{\frac{-5\pi i}{6}}, \frac{1}{6}e^{\frac{-5\pi i}{2}}$, $\frac{1}{6}e^{-\frac{25\pi i}{6}}$

18. $\int_{-\infty}^\infty \frac{x^2 dx}{(x^2+1)(x^2+4)}$

Ans. $\frac{\pi}{3}, z = i, 2i$, Res $\frac{i}{6}, -\frac{i}{3}$

19. $\int_{-\infty}^\infty \frac{dx}{(1+x^2)^3}$

Ans. $\frac{3\pi}{8}, z = i$, order 3, Res $\frac{3}{16i}$

20. $\int_{-\infty}^\infty \frac{x dx}{(x^2+4x+13)^2}$

Ans. $-\frac{\pi}{27}$

21. $\int_0^\infty \frac{x^2 dx}{(x^2+a^2)^2}$

Ans. $\frac{\pi}{4a}$

Type III

22. $\int_{-\infty}^\infty \frac{\cos mx}{k^2+x^2}$ $(m > 0, k > 0)$

Ans. $\frac{\pi}{k}e^{-km}, z = ik$, Res $\frac{e^{-km}}{2ik}$

23. $\int_{-\infty}^\infty \frac{\sin mx}{k^2+x^2}$

Ans. 0

24. $\int_{-\infty}^\infty \frac{x \cdot \sin \pi x}{x^2+2x+5}$

Ans. $-\pi e^{-2\pi}, z = -1+2i$, Res $\frac{\pi}{2}\frac{(1-2i)}{e^{2\pi}}$

25. $\int_0^\infty \frac{x \sin x\, dx}{(x^2+1)(x^2+4)}$

Ans. $\frac{\pi(e-1)}{6e^2}, z = i, 2i$, Res $+\frac{1}{(6e)}, \frac{-1}{(6e^2)}$

26. $\int_{-\infty}^\infty \frac{\cos x\, dx}{(x+a)^2+b^2}$

Ans. $\frac{\cos a}{be^b}, z = -a+bi, \frac{e^{-ia-b}}{2bi}$

27. $\int_0^\infty \frac{\cos mx\, dx}{(a^2+x^2)^2}$

Ans. $\pi e^{-am}\frac{(am+1)}{(4a^3)}$

28. $\int_{-\infty}^\infty \frac{\sin mx\, dx}{(x-a)^2+b^2}$

Ans. $\frac{(e^{-mb} \cdot \sin ma)}{b}$

29. $\int_{-\infty}^\infty \frac{x \cos x\, dx}{x^2-2x+10}$

Ans. $\pi\frac{(\cos 1 - 3\sin 1)}{(3e^3)}$

Rectangle contour

30. $\int_{-\infty}^\infty \frac{e^{ax} dx}{e^x+1}$

Hint: c: rectangle with vertices at A(R, 0), B(R, 2π), C($-$R, 2π), D($-$R, 0) and $f(z) = e^{az}/(e^z+1)$, $z = \pi i$ only pole in C. Res $-e^{a\pi i}$. Prove integrals from A to B and C to D as zero. Take $R \to \infty$.

Ans. $\frac{\pi}{\sin a\pi}$

31. $\int_0^\infty \frac{\cosh ax}{\cosh x} dx$

Hint: Integrate $\frac{e^{az}}{\cosh z}$ along a rectangle with vertices at $-R, R, R+\pi i, -R+\pi i$, only pole $\frac{\pi i}{2}$, Res $-ie^{\frac{a\pi i}{2}}$. Integrals from R to $R+\pi i$, $-R+\pi i$ to $-R$ are zero as $R \to \infty$.

Ans. $\frac{\pi}{2\cos(\pi\frac{a}{2})}$ where $|a| < 1$

32. Prove that $\int_0^\infty \frac{\sin ax}{e^{2\pi x}-1}dx = \frac{1}{4}\coth\frac{a}{2} - \frac{1}{2a}$.

Hint: Integrate $\frac{e^{aiz}}{(e^{2\pi z}-1)}$ around a rectangle with vertices at $0, R, R+i, i$ and take $R \to \infty$.

33. Prove that $\int_0^\infty \sin x^2 dx = \int_0^\infty \cos x^2 dx = \frac{1}{2}\sqrt{\frac{\pi}{2}}$.

Hint: Integrate e^{iz^2} around $c: \theta = 0$, arc of circle radius R, centre origin, $\theta = \frac{\pi}{4}$.

Type IV

34. Show that $\int_0^\infty \frac{\sin mx}{x} dx = \frac{\pi}{2}$ with $m > 0$.

Hint: Integrate $\frac{e^{iz}}{z}$ around $c: c_1 + c_2 + c_3 + c_4$ (Fig. 24.7).

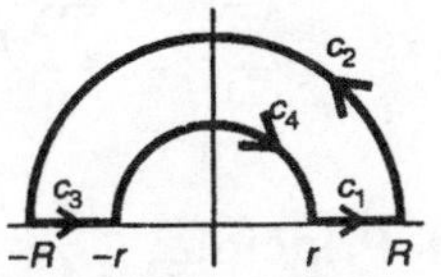

Fig. 24.7

35. Find principal value of $\int_{-\infty}^{\infty} \frac{dx}{(x^2-3x+2)(x^2+1)}$

Ans. $\frac{\pi}{10}$, $z = 1, 2, i$, Res $-\frac{1}{2}, \frac{1}{5}, \frac{3-i}{20}$

36. Show that pr. v. $\int_{-\infty}^{\infty} \frac{\sin x \, dx}{(x^2+4)(x-1)} = \frac{\pi}{5}\left(\cos 1 - \frac{1}{e^2}\right)$.

37. Prove that pr. v. $\int_{-\infty}^{\infty} \frac{x}{8-x^3} dx = -\frac{\sqrt{3}\pi}{6}$.

38. Prove that $\int_0^\infty \frac{\ln(x^2+1)}{x^2+1} dx = \pi \ln 2$

Hint: Integrate $\frac{\ln(z+i)}{z^2+1}$ around c: real axis $-R$ to R and semicircle of radius R.

39. Prove that $\int_{-\infty}^{\infty} \frac{e^{kx} dx}{1+e^x} = \frac{\pi}{\sin k\pi}$, $0 < k < 1$.

Hint: Integrate $\frac{e^{kz}}{(1+e^z)}$ around the rectangle $y = 0$, $y = 2\pi$, $x = \pm a$ as $a \to \infty$.

Type V

40. Show that $\int_0^\infty \frac{x^{a-1} dx}{1+x^2} = \frac{\pi}{2\sin\left(\frac{a\pi}{2}\right)}$ for $0 < a < 2$.

41. Prove that $\int_0^\infty \frac{x^{a-1}}{1+x} dx = \frac{\pi}{\sin a\pi}$, $0 < a < 1$.

24.4 ARGUMENT PRINCIPLE

Argument principle determines the number (how many) of zeros or poles of a function in a given region which is useful in the stability criteria of linear systems.

Argument Theorem

Theorem: *Let $f(z)$ be analytic on and within a simple closed curve c except for a pole $z = a$ of order (multiplicity) p inside c. Further suppose that $f(z)$ has only one zero $z = b$ of order (multiplicity) n and no other zeros on c.*

Then

$$\frac{1}{2\pi i} \oint_c \frac{f'(z)}{f(z)} dz = n - p \tag{1}$$

Proof: Enclose $z = a$ and $z = b$ by non-overlapping circles c_1 and c_2 respectively (Fig. 24.8). Then

$$\frac{1}{2\pi i} \oint_c \frac{f'(z)}{f(z)} dz = \frac{1}{2\pi i} \oint_{c_1} \frac{f'}{f} dz + \frac{1}{2\pi i} \oint_{c_2} \frac{f'}{f} dz \tag{2}$$

Since $f(z)$ has a pole of order p at $z = a$, we have

$$f(z) = \frac{F(z)}{(z-a)^p} \tag{3}$$

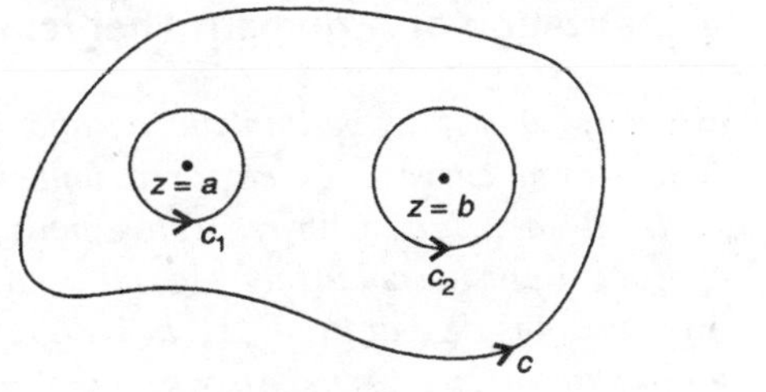

Fig. 24.8

where $F(z)$ is analytic and non-vanishing on and within c_1. Taking log of (3) and differentiating w.r.t., z, we get

$$\frac{f'(z)}{f(z)} = \frac{F'(z)}{F(z)} - \frac{p}{z-a} \tag{4}$$

Thus $$\frac{1}{2\pi i} \oint_{c_1} \frac{f'}{f} dz = \frac{1}{2\pi i} \oint_{c_1} \frac{F'}{F} dz - \frac{p}{2\pi i} \oint_{c_1} \frac{dz}{z-a} \tag{5}$$

Since F is analytic, so is F' and therefore $\frac{F'}{F}$ is analytic in c_1. Therefore by Cauchy's theorem $\oint_{c_1} \frac{F'}{F} dz = 0$. For a circle c_1 with centre at $z = a$, $\oint_{c_1} \frac{dz}{z-a} = 2\pi i$. With these values, (5) reduces to

$$\frac{1}{2\pi i} \oint_{c_1} \frac{f'(z)}{f(z)} dz = 0 - \frac{p}{2\pi i} \cdot 2\pi i = -p \tag{6}$$

Since $f(z)$ has a zero of order n at $z = b$, we have

$$f(z) = (z-b)^n G(z) \tag{7}$$

where $G(z)$ is analytic and non-vanishing on and within c_2. Taking log of (7) and differentiating, we get

$$\frac{f'(z)}{f(z)} = \frac{G'(z)}{G(z)} + \frac{n}{z-b} \quad (8)$$

Thus

$$\frac{1}{2\pi i}\oint_{c_2} \frac{f'}{f} dz = \frac{1}{2\pi i}\oint_{c_2} \frac{G'}{G} dz + \frac{n}{2\pi i}\oint_{c_2} \frac{dz}{z-b}$$

$$= 0 + \frac{n}{2\pi i} \cdot 2\pi i = n \quad (9)$$

Substituting (6) and (9) in (2), we get

$$\frac{1}{2\pi i}\oint_c \frac{f'(z)}{f(z)} dz = n - p$$

Generalization of Argument Theorem

Theorem: *Let $f(z)$ be analytic on and within a simple closed curve c except at a finite number of poles $a_1, a_2, \ldots a_j$ with respective multiplicities $p_1, p_2, \ldots p_j$ inside c. Further $f(z)$ has a finite number of zeros inside c at $b_1, b_2, \ldots b_k$ with respective multiplicities $n_1, n_2, \ldots n_k$. Also $f(z) \neq 0$ on c.*

Then

$$\frac{1}{2\pi i}\oint_c \frac{f'(z)}{f(z)} dz = N - P$$

where $N = \sum_{r=1}^{k} n_r$ = total *number of zeros of* $f(z)$ *inside* c, *counting their multiplicities and* $P = \sum_{r=1}^{j} p_r$ = total *number of poles of* $f(z)$ *inside* c, *counting multiplicities.*

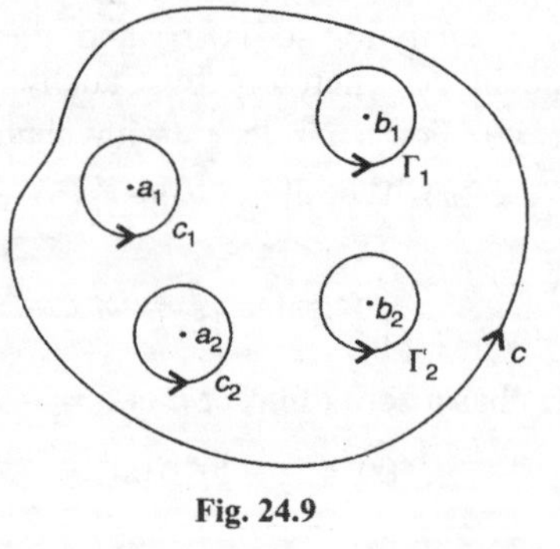

Fig. 24.9

Proof: Enclose $a_1, a_2, \ldots a_j$ and $b_1, b_2, \ldots b_k$ by non-overlapping circles $c_1, c_2, \ldots c_j$ and $\Gamma_1, \Gamma_2, \ldots \Gamma_k$ respectively and use the argument theorem above (Fig. 24.9).

$$\frac{1}{2\pi i}\oint_c \frac{f'}{f} dz = \sum_{r=1}^{k} \frac{1}{2\pi i}\oint_{\Gamma_r} \frac{f'}{f} dz + \sum_{r=1}^{j} \frac{1}{2\pi i}\oint_{c_r} \frac{f'}{f} dz$$

$$\frac{1}{2\pi i}\oint_c \frac{f'(z)}{f(z)} dz = \sum_{r=1}^{k} n_r - \sum_{r=1}^{j} \cdot p_r = N - P.$$

Argument Principle: (or Principle of Argument (Angle))

From Argument theorem

$$N - P = \frac{1}{2\pi i}\oint_c \frac{f'(z)}{f(z)} dz = \frac{1}{2\pi i}\oint_c d[\ln f(z)]$$

$$= \frac{1}{2\pi i} \text{[Variation of } \ln f(z) \text{ in going completely (once) around (along) } c]$$

$$= \frac{1}{2\pi i} \text{ var } [\ln |f(z)| + i \arg f(z)]$$

Since $\ln |f|$ is same at the beginning and at the end of one full circuit around c, therefore

$$N - P = \frac{i}{2\pi i} \text{ var of arg } f = \frac{1}{2\pi} \text{ var of arg } f(z)$$

$$= \frac{1}{2\pi} \text{ [variation of argument of } f(z) \text{ (i.e., change in the angle } f(z)) \text{ as curve } c \text{ is traversed completely once]}$$

Corollary: If $f(z)$ is analytic everywhere (so $P = 0$) then

$$N = \text{Number of zeros of } f(z) \text{ inside } c$$

$$= \frac{1}{2\pi} \cdot \text{ net variation of argument of } f(z) \text{ as } z \text{ traverses the closed curve } c.$$

Geometrically the number of zeros of $f(z)$ is the number of times the locus c^* of $w = f(z)$ encircles the origin.

Note: If $w = 0$, origin $\notin c^*$, then N = net variation of argument $f(z)$ is zero.

24.5 ROUCHE'S THEOREM

Introduction

The counting of the total number of zeros of an analytic function in a given domain can be appreciably simplified by Rouche's theorem.

Theorem: *Let the functions $f(z)$ and $g(z)$ be analytic within and on a simple closed curve c bounding a region R. Further $|f(z)| > |g(z)|$ on c and f and g are non-vanishing on c. Then the total number of zeros inside c (in region R) of $f(z) + g(z)$ is equal to the total number of zeros of $f(z)$.*

Proof: Let $F(z) = f(z) + g(z)$. Then on c $|F(z)| = |f(z) + g(z)| \geq |f(z)| - |g(z)| > 0$. Thus both $f(z)$ and $F(z)$ are analytic in c and non-vanishing on c. Therefore by argument principle, number of zeros $F(z) = \frac{1}{2\pi}$ variation of the argument of $F(z)$ as z traverses along the closed curve c completely once i.e.,

$$N[f(z) + g(z)] = \frac{1}{2\pi} \text{ var } [\arg (f + g)]_c$$

Similarly,

$$N[f(z)] = \frac{1}{2\pi} \text{ var } [\arg f]_c$$

Taking the difference

$$N[f + g] - N[f]$$

$$= \frac{1}{2\pi} \text{ var } [\arg (f + g) - \arg f]_c$$

$$= \frac{1}{2\pi} \text{ var } \left[\arg \left(\frac{f+g}{f}\right)\right]_c$$

$$= \frac{1}{2\pi} \text{ var } \left[\arg \left(1 + \frac{g}{f}\right)\right]_c = \frac{1}{2\pi} \text{ var } [\arg w]_c$$

where

$$w = 1 + \frac{g(z)}{f(z)} = w(z). \text{ Then } |w - 1| = \left|\frac{g}{f}\right| < 1.$$

As the point z traverses c completely in the z-plane the corresponding point w describes a closed curve c^* in the w-plane. Since $|w - 1| < 1$, c^* lies entirely inside some circle $|w - 1| = \rho_0$. So the point $w = 0$ lies outside c^*. Consequently the var $[\arg w]_{c^*} = 0$.

Hence the total number of zeros inside c of $f + g$ and f are same.

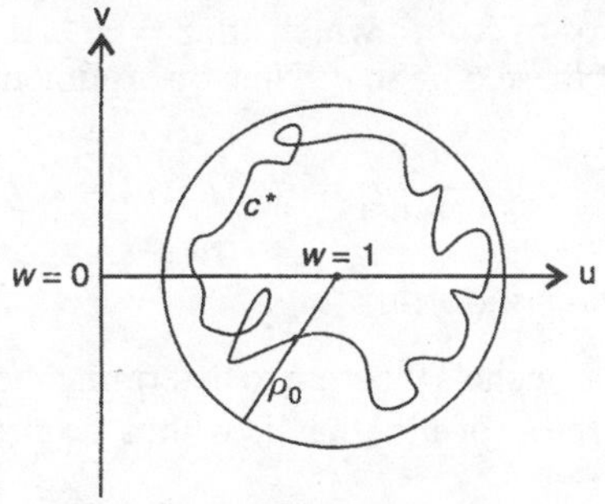

Fig. 24.10

Note: If $w = 0$ origin lies inside c^*, then the variation of argument of w is not zero, but is determined by the number of total circuits about the point $w = 0$ performed by w in its complete motion along the closed contour c^* (Fig. 24.10).

WORKED OUT EXAMPLES

Argument principle

Evaluate $\int_c \frac{f'(z)}{f(z)} dz$ where c is a simple closed curve:

Example 1: $f(z) = z^5 - 3iz^2 + 2z - 1 + i$

Solution: By fundamental theorem of algebra, $f(z)$ has 5 zeros. But $f(z)$ has no poles.
By argument principle

$$\int_c \frac{f'(z)}{f(z)} dz = 2\pi i[N - P] = 2\pi i[5 - 0] = 10\pi i.$$

Example 2: $f(z) = (z^2 + 1)^2/(z^2 + 2z + 2)^3$, $c : |z| = 4$

Solution: $z = \pm i$ are zeros of $f(z)$ each of order 2 (i.e., multiplicity 2). Poles of $f(z)$ are zeros of $z^2 + 2z + 2$ i.e., $z = -1 \pm i$. Thus $-1 \pm i$ are poles of $f(z)$ each of order 3. By argument principle $\oint_c \frac{f'(z)}{f(z)} dz = 2\pi i[2 \times 2 - 2 \times 3] = -4\pi i$.

Example 3: $f(z) = \tan \pi z$, $c : |z| = \pi$.

Solution: Since $\tan \pi z = \frac{\sin \pi z}{\cos \pi z}$, the zeros of $\tan \pi z$ are zeros of $\sin \pi z$ which are $z = \pm n$, $n = 0, 1, 2, 3, \cdots$ of these only $z = 0, \pm 1, \pm 2, \pm 3$ (seven of them) lies in c. Similarly, poles of $\tan \pi z$

are zeros of $\cos \pi z$ which are $z = \pm\frac{1}{2}, \pm\frac{3}{2}, \pm\frac{5}{2}, \ldots$, etc. Of these 6 lie in c. Thus by argument principle

$$\oint_c \frac{f'(z)}{f(z)} dz = 2\pi i[7 - 6] = 2\pi i.$$

Rouche's theorem

Apply Rouche's theorem to determine the number of roots (zeros) of $p(z)$ that lie within the circle/annulus region:

Example 4: $p(z) = z^9 - 2z^6 + z^2 - 8z - 2$, $c : |z| = 1$

Solution: Choose $f(z) = -8z$, $g(z) = z^9 - 2z^6 + z^2 - 2$

$$\text{on } c : |z| = 1, |f| = |-8z| = |8||z| = 8$$

$$\begin{aligned}\text{on } c : |z| = 1, |g| &= |z^9 - 2z^6 + z^2 - 2| \\ &\le |z^9| + |2||z^6| + |z^2| + |2| \\ &= 1 + 2 + 1 + 2 = 6\end{aligned}$$

Thus $|f| = 8 > 6 = |g|$ on c, f, g are analytic on and within c. So by Rouche's theorem f and $p = f + g$ has same number of roots within c. But f has only one zero in c. Hence $p(z)$ has one zero inside $c : |z| = 1$.

Example 5: $p(z) = z^4 - 5z + 1$, annulus region $1 < |z| < 2$.

Solution: $c_1 : |z| = 1$. Take $f(z) = -5z$, $g(z) = z^4 + 1$.
Then on c, $|f| = 5 > 2 = |g|$.

So by Rouche's theorem $p = f + g$ has one zero inside c_1 since f has one zero in c_1.
Take $f = z^4, g = -5z + 1$. Then on $c_2, |z| = 2$, $|f| = 2^4 = 16 > 11 = |g|$.

So by Rouche's theorem, $p = f + g$ has 4 zeros within c_2 since f has 4 zeros within c_2. Hence there are $4 - 1 = 3$ zeros of p in the annulus region $1 < |z| < 2$.

Example 6: $p(z) = e^z - 4z^n + 1, c : |z| = 1$

Solution: Take $f = -4z^n, g = e^z + 1$

$$\text{on } c : |z| = 1, |f| = 4|z^n| = 4 > |g| = |e^z| + 1 = e + 1$$

Therefore $p = f + g$ has n zeros within c since f has n zeros within c.

EXERCISE

Argument principle

Evaluate $\oint_c \frac{f'(z)}{f(z)} dz$ where c is a simple closed curve:

1. $f(z) = (z^2 - 1)/(z^2 + z)^2, c : |z| = 2$

Ans. $-4\pi i$

2. $f(z) = (z^2 + 2)^3/(z^3 + 2z^2 + 2z)^4$, $c : |z| = 10$

Ans. $-12\pi i$

3. $f(z) = \sin \pi z, c : |z| = \pi$

Ans. $14\pi i$

4. $f(z) = \cos \pi z, c : |z| = \pi$

Ans. $12\pi i$.

Rouche's theorem

Use Rouche's theorem to determine the number roots (zeros) of $f(z)$ within the circle c/annulus region indicated:

5. $f(z) = 2z^5 - z^3 + 3z^2 - z + 8, c : |z| = 1$

Ans. No zeros

6. $f(z) = z^7 - 5z^4 + z^2 - 2, c : |z| = 1$

Ans. 4 zeros

7. $f(z) = z^6 - 5z^4 + z^3 - 2z, c : |z| = 1$

Ans. 4

8. $f(z) = 2z^4 - 2z^3 + 2z^2 - 2z + 9, c : |z| = 1$

Ans. 0

9. $f(z) = z^7 - 4z^3 + z - 1, c : |z| = 1$

Ans. 3

10. $f(z) = z^7 - z^3 + 12$, annulus region $1 < |z| < 2$

Ans. 7

11. $f(z) = z^4 - 8z + 10, 1 < |z| < 3$

Ans. 4

12. $f(z) = 2z^5 - 6z^2 + z + 1, 1 < |z| < 2$

Ans. 3

13. $f(z) = az^n - e^z, c : |z| = 1, |a| > e$

Ans. n

14. $f(z) = 2z^5 + 8z - 1, c : |z| = 2$

Ans. 5

15. $f(z) = z^5 + 15z + 1, \frac{3}{2} < |z| < 2$

Ans. 4.

24.6 FUNDAMENTAL THEOREM OF ALGEBRA

Theorem 1: *Every polynomial of degree n in the complex plane has at least one zero.*

Proof: (By Liouville's theorem)
Let $p(z)$ be a polynomial in z of degree $n > 1$ given by

$$p(z) = a_0 + a_1 z + a_2 z^2 + \cdots + a_n z^n$$

where $a_n \neq 0$

Suppose that $p(z)$ has no zeros, then $p(z) \neq 0$ for any value of z. Now $f(z) = \frac{1}{p(z)}$ is analytic everywhere. Also $|f(z)| \to 0$ as $|z| \to \infty$ so that $|f(z)|$ is bounded for all z. Liouville's theorem states that a function $f(z)$ which is analytic everywhere and bounded is constant. This is a contradiction since $p(z)$ is not constant. Hence $p(z)$ is zero for at least one value of z.

Aliter

Every polynomial of degree n in the complex plane has n zeros, counting the multiplicities.

Proof: (By Rouche's theorem)
Let $p(z) = a_0 + a_1 z + a_2 z^2 + \cdots + a_n z^n$ be the polynomial with $a_n \neq 0$. Choose $f(z) = a_n z^n$ and $g(z) = a_0 + a_1 z + \cdots + a_{n-1} z^{n-1}$. Let c be a circle with centre at origin of radius $R (> 1)$. Then on c

$$\left|\frac{g(z)}{f(z)}\right| = \frac{|a_0 + a_1 z + \cdots + a_{n-1} z^{n-1}|}{|a_n z^n|}$$

$$\leq \frac{|a_0| + |a_1|R + |a_2|R^2 \cdots + |a_{n-1}|R^{n-1}}{|a_n|R^n}$$

$$\leq \frac{|a_0|R^{n-1} + |a_1|R^{n-1} + \cdots + |a_{n-1}|R^{n-1}}{|a_n|R^n}$$

$$= \frac{|a_0| + |a_1| + \cdots + |a_{n-1}|}{|a_n|R}$$

For any specified values of the coefficients $a_0, a_1, \cdots a_n$ we can choose R such that

$$0 < \left|\frac{g(z)}{f(z)}\right|_{\text{on } c} = \left|\frac{g(z)}{f(z)}\right|_{|z|=R} < 1 \text{ i.e., } |g(z)| < |f(z)|$$

By Rouche's theorem, the total number of zeros of the polynomial $p(z) = f(z) + g(z)$ in the circle c is same as the number of zeros of $f(z) = a_n z^n$ in c. But the function $f(z) = a_n z^n$ has (n-fold zero) n zeros all located at $z = 0$. Hence $p(z) = f(z) + g(z)$ has n zeros in c. Since R is arbitrary, this is true for the entire complex plane.

24.7 LIOUVILLE* THEOREM

Theorem: *If $f(z)$ is entire and $|f(z)|$ is bounded for all z, then $f(z)$ is constant.*

Proof: Consider the generalized Cauchy's integral formula

$$f^{(n)}(z_0) = \frac{n!}{2\pi i} \oint_c \frac{f(z)}{(z - z_0)^{n+1}} dz \tag{1}$$

with C chosen as a circle with centre at z_0 and of radius r. Since $|f(z)|$ is bounded, $|f(z)| \leq M$.

Applying M-L inequality to (1)

$$|f^{(n)}(z_0)| = \frac{n!}{2\pi} \left| \oint_C \frac{f(z)}{(z - z_0)^{n+1}} dz \right| \leq \frac{n!}{2\pi} \frac{M}{r^{n+1}} 2\pi r,$$

since $C : |z - z_0| = r$ and $L =$ Length of $C = 2\pi r$

or
$$|f^{(n)}(z_0)| \leq \frac{n! M}{r^n} \tag{2}$$

which is known as **Cauchy's inequality**. From (2) for $n = 1$, we have

$$|f'(z_0)| \leq \frac{M}{r} \tag{3}$$

Since f is entire (analytic everywhere) and $|f|$ is bounded for all z, (3) is true for any r. As $r \to \infty$

$$|f'(z_0)| \to 0$$

Thus since z_0 is arbitrary, $f'(z) = 0$ for all z and therefore $f(z)$ is a constant.

---

*Joseph Liouville (1809–1882) French mathematician.

24.8 DETERMINATION OF ZEROS OF A COMPLEX POLYNOMIAL

WORKED OUT EXAMPLES

Example 1: Using argument principle prove that the complex polynomial equation

$$2z^4 - 3z^3 + 3z^2 - z + 1 = 0$$

has no roots on the real and imaginary axes and has one complex root in each quadrant.

Solution: Consider the complex polynomial function

$$f(z) = u + i\,v = 2z^4 - 3z^3 + 3z^2 - z + 1$$

(i) To prove that $f(z)$ has no real roots. For $z = x, (x > 0)$ we have

$$f(x) = 2x^4 - 3x^3 + 3x^2 - x + 1.$$

Rewriting $f(x) = x(x-1)^3 + (x^4+1) > 0$ for $x > 1$. Rewriting $f(x) = 2x^4 + 3x^2 + 2x + 1 + 3x(x+1)(1-x) > 0$ when $0 < x < 1$. Thus $f(x) > 0$ for all positive x. Also $f(-x) = 2x^4 + 3x^3 + 3x^2 + x + 1 > 0$ for all negative x. Thus $f(x) > 0$ for all real x. Hence $f(z) = 0$ has no root on the real axis.

(ii) To prove that $f(z)$ does not have purely imaginary roots, put $z = iy$. Then

$$f(z) = f(iy) = 2(iy)^4 - 3(iy)^3 + 3(iy)^2 - (iy) + 1$$
$$= u + iv = (2y^4 - 3y^2 + 1) + i(3y^3 - y)$$

So $f(z) = 0$ when $u = 2y^4 - 3y^2 + 1 = 0$ and $v = y(3y^2 - 1) = 0$ simultaneously. These two equations have no common solution in y. Thus $f(z) = 0$ has no purely imaginary roots.

(iii) To find the number of roots in the first quadrant. Let $z = re^{i\theta}, 0 \le \theta \le \frac{\pi}{2}$. Let C be the closed curve $OABO$. As $r \to \infty$, $OABO$ defines the first quadrant. $f(z)$ is analytic for every z (except $z = \infty$) because $f(z)$ is a polynomial. So $f(z)$ is analytic on and within the curve C.

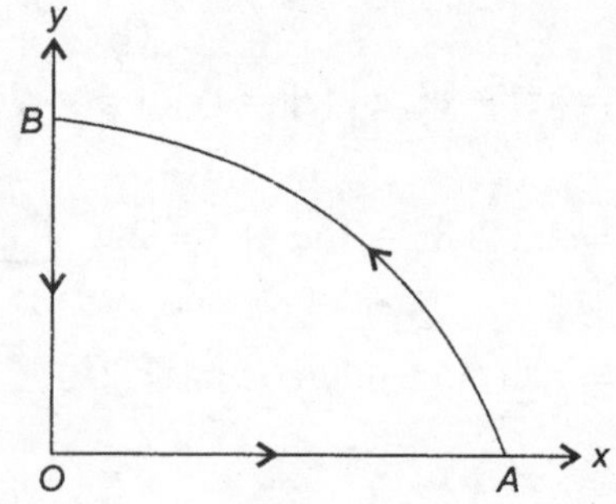

(a) Along $OA: z = x$ and x varies from 0 to ∞. Then

$$f(z) = f(x) = u + iv = 2x^4 - 3x^3 + 3x^2 - x + 1$$

so $u = 2x^4 - 3x^3 + 3x^2 - x + 1$, and $v = 0$. Hence argument of $f = \tan^{-1}\frac{v}{u} = \tan^{-1}\frac{0}{u} = 0$. Thus the net variation of argument of f along the straight line OA is $\Delta_{OA} \arg f = \left[\tan^{-1}\frac{0}{u}\right]_{OA} = \tan^{-1}\frac{0}{u}\Big|_{x=0}^{\infty} = 0$

(b) Along arc $AB: z = re^{i\theta}, 0 < \theta < \frac{\pi}{2}$, so

$$f(z) = u + iv = 2r^4e^{i4\theta} - 3r^3e^{i3\theta} + 3r^2e^{i\theta} - re^{i\theta} + 1$$
$$= r^4e^{i4\theta}\left[2 - \frac{3}{re^{i\theta}} + \frac{3}{r^2e^{i2\theta}} - \frac{1}{r^3e^{i3\theta}} + \frac{1}{r^4e^{i4\theta}}\right]$$

As $r \to \infty$, $f(z) \to r^4e^{i4\theta}$. Then $\arg f = \tan^{-1}\frac{v}{u} = \tan^{-1}\left(\frac{r^4 \sin 4\theta}{r^4 \cos 4\theta}\right)$

$$= \tan^{-1}(\tan 4\theta) = 4\theta$$

$$\Delta \arg f \text{ along } AB = 4\theta\Big|_{\theta=0}^{\frac{\pi}{2}} = 4\left(\frac{\pi}{2} - 0\right) = 2\pi$$

(c) Along $BO: z = iy$ and y varies from ∞ to 0. For $z = iy$

$$f(iy) = u + iv = (2y^4 - 3y^2 + 1) + iy(3y^2 - 1).$$

Then $\arg f = \tan^{-1}\frac{v}{u} = \tan^{-1}\left\{\frac{y(3y^2-1)}{2y^4-3y^2+1}\right\}_{y=\infty}^{0}$

This is zero at both the limits. So

$$\Delta \arg \text{ along } BO = 0$$

Hence the net variation of argument of $f(z)$ as we traverse along the curve $C: OABO$ is

$$\Delta_c \arg f(z) = \Delta_{OA} \arg f + \Delta_{AB} \arg f +$$
$$+\Delta_{BO} \arg f$$
$$= 0 + 2\pi + 0 = 2\pi.$$

Using argument principle, the number of zeros of $f(z)$ within C is given by

$$N = \frac{1}{2\pi}\Delta_C \arg f = \frac{1}{2\pi}\cdot 2\pi = 1.$$

Thus $f(z)$ has one zero in the first quadrant. The second zero lies in the fourth quadrant because the complex roots of a polynomial equation with real coefficients occur in conjugate pairs.

From fundamental theorem of algebra the equation, being of fourth degree, has four roots and the remaining two roots will lie in the second and third quadrants. Thus there is one zero of $f(z)$ in each quadrant.

EXERCISE

1. Using argument principle, show that the polynomial equation

$$z^4 + 2z^3 + 3z^2 + 4z + 5 = 0$$

has no zeros on the real and imaginary axes and has one zero in each quadrant.

Hint: $f(x) = (x^2+2)(x-1)^2 + 3 > 0$, no real root, $f(iy) = (y^4 - 3y^2 + 5) + 2i(2y - y^3) = u + iv$. $u = 0, v = 0$ have no common solution, so no imaginary roots. $N = \frac{1}{2\pi}\Delta_C \arg f = \frac{1}{2\pi}\cdot[\Delta_{OA} \arg f + \Delta_{AB} \arg f + \Delta_{BO} \arg f] = \frac{1}{2\pi}[0 + 2\pi + 0] = 1$.

2. Show that $z^4 + 2z^3 + 3z^2 + z + 2$ has exactly two zeros in the right half plane.

Hint: $f(iy) = (y^2-2)(y^2-1) + iy(1 - 2y^2)$, $\lim_{R\to\infty} \arg f(iy)|_{-R}^{R} = 0$.

3. In what quadrants are the zeros of the equation $z^4 + z^3 + 4z^2 + 2z + 3 = 0$?

Ans. There are two zeros each in the 2nd and 3rd quadrants.

4. Show that the equation $z^4 + 4(1+i)z + 1 = 0$ has one root in each quadrant.

5. Prove that one root of the equation $z^4 + z^3 + 1 = 0$ lies in the first quadrant and the equation has no zeros on the real and imaginary axes.

6. How many roots has the equation

$$z^6 + z^5 + 6z^4 + 5z^3 + 8z^2 + 4z + 1 = 0$$

in the right half-plane

Ans: 0

7. Show that $z^5 + z^3 + 2z + 3$ has just one zero in the first quadrant.

24.9 SUMMARY

Argument principle states that if N and P are the number of zeros and poles of an analytic function $f(z)$ within simple closed curve c then $N - P = \frac{1}{2\pi}\cdot$ variation of argument of $f(z)$ as z traverses c completely once.

Rouche's theorem states that the number of zeros of $f(z) + g(z)$ and $f(z)$ are same inside c if f, g are analytic and $|f| > |g|$ on c.

Fundamental theorem of algebra assures that every polynomial of degree n has n zeros.

Liouville's theorem states that an entire and bounded function $f(z)$ is constant.

Chapter 25

Conformal Mapping

INTRODUCTION

Suppose we are able to solve some problem for a simple domain such as a disk or half plane D. Further suppose we map this domain D conformally to another domain D^* in which the solution is sought. Then using such a mapping, from solution of D, we get a solution for D^*. Conformal mapping, which preserves angles in magnitude and sense is useful in solving boundary value problems in two-dimensional potential theory by transforming a complicated region to a simpler region. i.e., conformal mapping preserves solutions of two-dimensional Laplace equation. Bilinear transformation, mappings by z^n, e^z, $\sin z$, $\cos z$ are often used. Schwarz-Christoffel transformation maps polygons to upper half-plane and consequently to a unit disk.

25.1 MAPPING (or TRANSFORMATION or OPERATOR)

A real function $y = f(x)$ involving two variables x and y can be plotted as a plane graph in the xy-plane.

Since a complex function

$$w = f(z) = u(x, y) + iv(x, y)$$

of a complex variable $z = x + iy$ involves four real variables, x, y, $u(x, y)$, $v(x, y)$, two planes z-plane and w-plane are needed for the geometrical representation. The values of z are plotted in z-plane and the corresponding function values w, known as images of z, are plotted in the w-plane. In general the points on a curve c in the z-plane get mapped or transformed to points on an image curve c^* in the w-plane. Thus the function $w = f(z)$ is said to be a mapping or transformation from z-plane into w-plane.

Critical point of a function $w = f(z)$ is a point z_0 where $f'(z_0) \neq 0$.

25.2 CONFORMAL MAPPING

A mapping $w = f(z)$ is said to be conformal if the angle between any two smooth curves c_1, c_2 in the z-plane intersecting at the point z_0 is equal in magnitude and sense to the angle between their images c_1^*, c_2^* in the w-plane at the point $w_0 = f(z_0)$ (see Fig. 25.1).

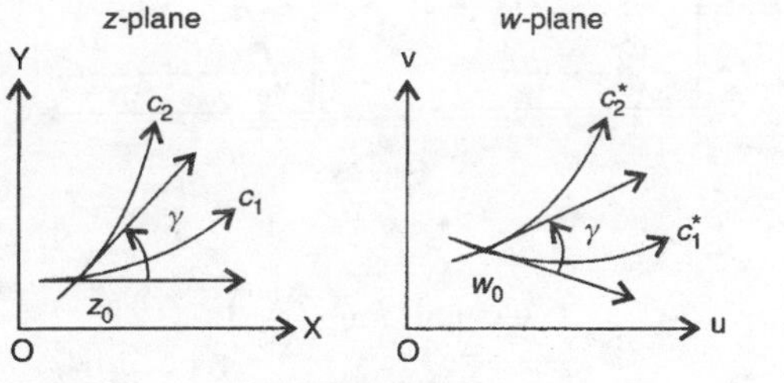

Fig. 25.1

Thus conformal mapping preserves angles both in magnitude and sense [also known as conformal mapping of the first kind. Conformal mapping of the second kind (isogonal mappings) preserve angles only in magnitude but not in sense, which is reversed, like $w = \bar{z}$, where $\arg \bar{z} = -\arg z$].

Conformal mapping is used to map complicated regions conformally onto simpler, standard regions such as circular disks, half planes and strips for which the boundary value problems are easier.

Given two mutually orthogonal one-parameter families of curves say $\phi(x, y) = c_1$ and $\psi(x, y) = c_2$, their image curves in the w- plane $\phi(u, v) = c_3$ and $\psi(u, v) = c_4$ under a conformal mapping are also mutually orthogonal. Thus conformal mapping preserves the property of mutual orthogonality of system of curves in the plane.

Condition for Conformality

A mapping $w = f(z)$ is conformal at each point z_0 where $f(z)$ is analytic and $f'(z_0) \neq 0$ (Fig. 25.2).

Proof: Since f is analytic, f' exists and since $f' \neq 0$, we have at a point z_0

$$R_0 e^{i\theta_0} = f'(z_0) = \lim_{\Delta z \to 0} \frac{f(z_0 + \Delta z) - f(z)}{\Delta z}$$

$$= \lim_{\Delta z \to 0} \frac{\Delta w}{\Delta z}$$

$$= \lim_{\Delta z \to 0} \left(\left| \frac{\Delta w}{\Delta z} \right| + i \arg \frac{\Delta w}{\Delta z} \right)$$

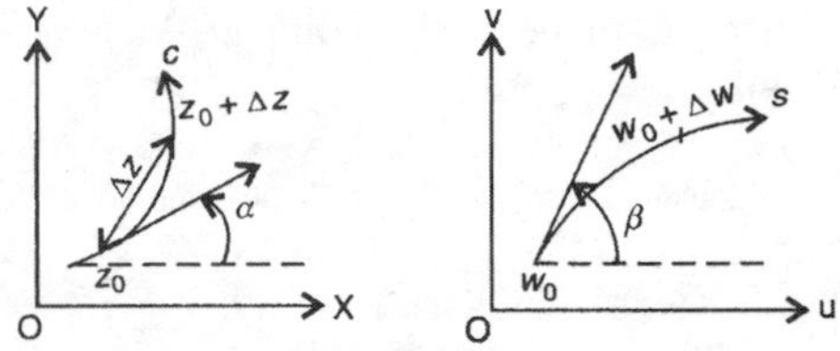

Fig. 25.2

So $$\theta_0 = \lim_{\Delta z \to 0} \left(\arg \frac{\Delta w}{\Delta z} \right)$$

Since $$\Delta w = \frac{\Delta w}{\Delta z} \cdot \Delta z,$$

$$\arg \Delta w = \arg \frac{\Delta w}{\Delta z} + \arg \Delta z$$

As $\Delta z \to 0$

$$\beta = \theta_0 + \alpha$$

Thus the directed tangent to curve c at z_0 is rotated through an angle $\theta_0 = \arg f'(z_0)$, which is same for all curves through z_0. Let α_1, α_2 be angles of inclination of two curves c_1 and c_2 and β_1 and β_2 be the corresponding angles for their images S_1 and S_2,

Then $$\beta_1 = \alpha_1 + \theta_0 \quad \text{and} \quad \beta_2 = \alpha_2 + \theta_0$$

Thus $$\beta_2 - \beta_1 = \alpha_2 - \alpha_1 = \gamma.$$

Hence the angle γ between the curves c_1 and c_2 and their images S_1 and S_2 is same both in magnitude and sense.

Result: An analytic function $f(z)$ is conformal everywhere except at its critical points where $f'(z) \neq 0$.

Note: Solutions of Laplace's equation are invariant under conformal transformation.

25.3 CONFORMAL MAPPING BY ELEMENTARY FUNCTIONS

General Linear Transformation or simply linear transformation defined by the function

$$w = f(z) = az + b \tag{1}$$

($a \neq 0$, and b are arbitrary complex constants) maps conformally the extended complex z-plane onto the extended w-plane, since this function is analytic and $f'(z) = a \neq 0$ for any z. If $a = 0$, (1) reduces to a constant function.

Special cases of linear transformation are

i. Identity transformation

$$w = z \tag{2}$$

for $a = 1, b = 0$, which maps a point z onto itself.

ii. Translation

$$w = z + b \tag{3}$$

for $a = 1$, which translates (shifts) z through a distance $|b|$ in the direction of b.

iii. Rotation

$$w = e^{i\theta_0} \cdot z \tag{4}$$

for $a = e^{i\theta_0}, b = 0$ which rotates (the radius vector of point) z through a scalar angle θ_0 (counterclockwise if $\theta_0 > 0$, while clockwise if $\theta_0 < 0$).

iv. Stretching (scaling)

$$w = az \tag{5}$$

for 'a' real stretches if $a > 1$ (contracts if $0 < a < 1$) the radius vector by a factor 'a'.

Thus the linear transformation (1) consists of rotation through angle arg a, scaling by factor $|a|$, followed by translation through vector b. This transformation is used for constructing conformal mappings of "similar" figures.

Result: Linear transformation preserves circles i.e., a circle in the z-plane under linear transformation maps to a circle in the w-plane.

Consider any circle in the z-plane

$$A(x^2 + y^2) + Bx + Cy + D = 0 \tag{6}$$

From (1)

$$u + iv = w = az + b = a(x + iy) + (b_1 + ib_2)$$

or $\quad u = ax + b_1, \; v = ay + b_2$

or $\quad x = \dfrac{u - b_1}{a}, \; y = \dfrac{v - b_2}{a}, \; a \neq 0 \qquad (7)$

Substituting (7) in (6), we get

$$A^*(u^2 + v^2) + B^*u + c^*v + D^* = 0 \tag{8}$$

which is a circle in the w-plane.

Here

$$A^* = \frac{A}{a^2}, \; B^* = \frac{B - 2Ab_1}{a}, \; C^* = \frac{C - 2Ab_2}{a},$$

$$D^* = D + A\left(\frac{b_1^2 + b_2^2}{a^2}\right) - \frac{Bb_1}{a} - \frac{Cb_2}{a}.$$

Thus circles are invariant under translation, rotation and stretching.

Inversion and Reflection Transformation

$$w = \frac{1}{z} \quad \text{for } z \neq 0 \tag{9}$$

In polar coordinates

$$Re^{i\phi} = \frac{1}{re^{i\theta}} = \frac{1}{r}e^{-i\theta}$$

so $R = \frac{1}{r}, \phi = -\theta$. Thus this transformation consists of an inversion in the unit circle ($Rr = 1$) followed by a mirror reflection about the real axis. Also $|w| = \frac{1}{|z|}$. So the unit circle $|z| = 1$ maps onto the unit circle $|w| = \frac{1}{1} = 1$. Further the interior of the unit circle $|z| = 1$ (points lying within $|z| = 1$) are transformed to the exterior of the unit circle $|w| = 1$ (points lying outside $|w| = 1$) or vice versa (Fig. 25.3).

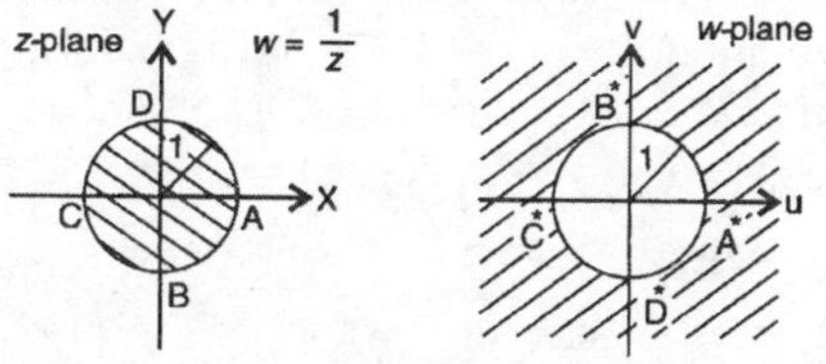

Fig. 25.3

By associating $z = 0$ to $w = \infty$ (also $z = \infty$ to $w = 0$) (9) is valid for the extended complex plane.

Result: Circles are invariant under $w = \frac{1}{z}$.

In terms of cartesian coordinates

$$u(x, y) + iv(x, y) = w = \frac{1}{z} = \frac{1}{x + iy}$$

$$= \frac{x - iy}{(x + iy)(x - iy)} = \frac{x - iy}{x^2 + y^2}$$

Thus $\quad u = \dfrac{x}{x^2 + y^2}, \; v = -\dfrac{y}{x^2 + y^2}.$

Similarly, $\quad x = \dfrac{u}{u^2 + v^2}, \; y = -\dfrac{v}{u^2 + v^2} \qquad (10)$

Substituting (10) in the equation of any circle in z-plane given by (6), we get

$$D(u^2 + v^2) + Bu - Cv + A = 0 \tag{11}$$

which is a circle in w-plane.

Observations: From (6) and (11), note that

i. $A \neq 0, D \neq 0$, circles not passing through

origin in z-plane maps to circles not passing through origin in w-plane.

ii. $A = 0, D \neq 0$, straight lines (considered as the limiting case of circles) in z-plane maps to circles through origin in w-plane.

iii. $A = 0, D = 0$, straight lines in z-plane maps to straight lines in w-plane and so on. Thus circles under $w = \frac{1}{z}$ are preserved.

WORKED OUT EXAMPLES

$w = az + b$ and $w = \frac{1}{z}$

Example 1: Find and plot the image of triangular region with vertices at (0, 0), (1, 0), (0, 1) under the transformation $w = (1 - i)z + 3$ (Fig. 25.4).

Solution: $u + iv = w = (1 - i)(x + iy) + 3$

So $u(x, y) = x + y + 3$, $v(x, y) = y - x$

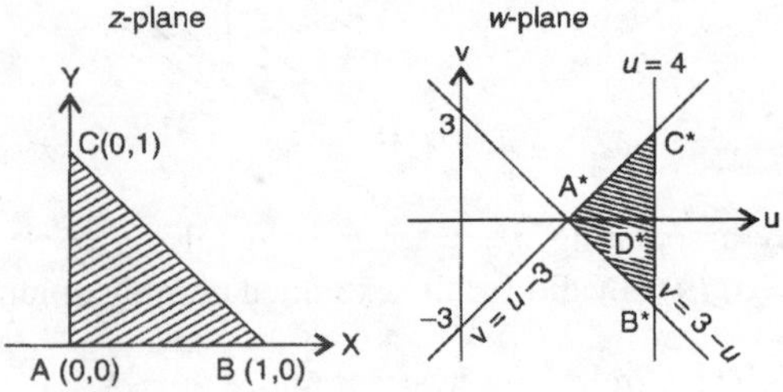

Fig. 25.4

AB: $y = 0, u = x + 3, v = -x$

or $u = -v + 3 \quad \therefore v = 3 - u : A^*B^*$

AC: $x = 0, u = y + 3, v = y,$

or $u = v + 3 \quad \therefore v = u - 3 : A^*C^*$

BC: $x + y = 1$, or substituting $u = (x + y) + 3$

$= 1 + 3 = 4,$

i.e., $u = 4 : B^*C^*$

So the image is the triangular region with vertices at $A^*(3, 0)$, $B^*(4, -1)$, $C^*(4, 1)$. Let $D\left(\frac{1}{4}, \frac{1}{4}\right)$ be any interior point of ABC. Its image is $D^*(3.5, 0)$ which is also an interior point of $A^*B^*C^*$.

Note 1: The angles $\frac{\pi}{2}, \frac{\pi}{4}, \frac{\pi}{4}$ at A, B, C are preserved as $\frac{\pi}{2}, \frac{\pi}{4}, \frac{\pi}{4}$ at vertices A^*, B^*, C^* since given function is conformal (everywhere).

Note 2: Since $z_1 = 1 - i$, $r = \sqrt{2}$, $\theta = \frac{-\pi}{4}$ or $\frac{7\pi}{4}$.

Thus $1 - i = \sqrt{2}e^{i7\pi/4}$.

Rewriting

$$w = (1 - i)z + 3 = \sqrt{2}e^{\frac{i7\pi}{4}} \cdot z + 3$$

the above transformation first rotates the triangle ABC in the z-plane clockwise by $\frac{\pi}{4}$ (or anticlockwise $\frac{7\pi}{4}$) and stretches the triangle by a scaling factor $\sqrt{2}$ and then finally translates the triangle to distance 3 units to the right, resulting in the triangle $A^*B^*C^*$ in the w-plane. ($AB = 1$, $AC = 1$, $BC = \sqrt{2}$ while $A^*B^* = \sqrt{2}$, $A^*C^* = \sqrt{2}$, $BC = 2$).

Example 2: Find the graph the strip $1 < x < 2$ under the mapping $w = \frac{1}{z}$ (Fig. 25.5).

Solution: $u + iv = w = \frac{1}{z} = \frac{x - iy}{x^2 + y^2}$

$$\text{so} \quad x = \frac{u}{u^2 + v^2}, \quad y = \frac{-v}{u^2 + v^2}$$

Since $1 < x < 2$ so $1 < \frac{u}{u^2 + v^2} < 2$

or $u^2 + v^2 - u < 0$ and $2(u^2 + v^2) - u > 0$

Rewriting

$$\left(u - \frac{1}{2}\right)^2 + v^2 < \frac{1}{4} \quad \text{and} \quad \left(u - \frac{1}{4}\right)^2 + v^2 > \frac{1}{16}$$

$$\text{or} \quad \left|w - \frac{1}{2}\right| < \frac{1}{2} \quad \text{and} \quad \left|w - \frac{1}{4}\right| > \frac{1}{4}$$

i.e., interior of the circle with centre at $\left(\frac{1}{2}, 0\right)$ and radius $\frac{1}{2}$ and exterior of the circle with centre at $\left(\frac{1}{4}, 0\right)$ and radius $\frac{1}{4}$.

Thus the infinite strip maps to the region shaded in the w-plane.

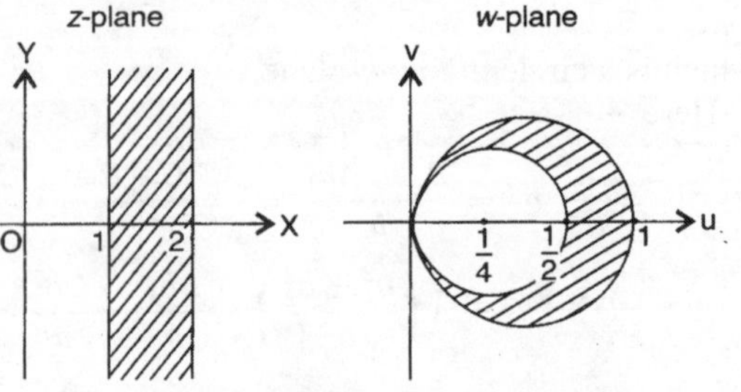

Fig. 25.5

EXERCISE

$w = az + b$ and $w = \frac{1}{z}$

1. Show that $w = iz$ is a rotation of the z-plane through an angle $\frac{\pi}{2}$ in the counterclockwise direction. Find and plot the image of the regions:

   (a) $0 < x < 1$ (b) $x > 2$ (c) $2 < x < 3$ (d) $1 < x < 2$ and $2 < y < 3$

Ans. $w = iz = e^{\frac{i\pi}{2}} \cdot re^{i\theta} = re^{i(\theta+\frac{\pi}{2})}$,

so $\phi = \theta + \frac{\pi}{2}$, $u = -y$, $v = x$,

(a) $0 < v < 1$ (b) $v > 2$ (c) $2 < v < 3$ (d) $-2 > u > -3$, $1 < v < 2$.

2. Find and plot the rectangular region $0 \le x \le 2$ $0 \le y \le 1$ under the transformations:

   a. $w = z + (1 - 2i)$

   b. $w = \sqrt{2}e^{\frac{\pi i}{4}} z$

   c. $w = \sqrt{2}e^{\frac{\pi i}{4}} z + (1 - 2i)$.

Ans. **a.** $u = x + 1$, $v = y - 2$, $1 \le u \le 3$, $-1 \ge v \ge -2$, translation

b. $u = x - y$, $v = x + y$, $u = -v$, $u = v$, $v + u = 4$, $v - u = 2$ rotation through $\frac{\pi}{4}$ and stretching by $\sqrt{2}$

c. $u = x - y + 1$, $v = x + y - 2$, $u + v = -1$, $u + v = 3$, $u - v = 1$, $u - v = 3$, rotation, stretching followed by translation.

3. Find and plot the image of the circle $|z| = c_1$ under the transformation $w = (1+i)z+3+2i$.

Ans. $|w - (3 + 2i)| = \sqrt{2}c_1$, circle with centre at $3 + 2i$ and radius $\sqrt{2}c_1$

4. Determine the image of the regions under $w=\frac{1}{z}$

   a. $x > 1$, $y > 0$ b. $0 < y < \frac{1}{2c}$.

Ans. (a) $\left|w - \frac{1}{2}\right| < \frac{1}{2}$ (b) $u^2 + (v + c)^2 > c^2$

5. Determine and sketch the image of $|z - 3| = 5$ under $w = \frac{1}{z}$.

Ans. $\left|w + \frac{3}{16}\right| = \frac{5}{16}$.

6. Prove that the image of the hyperbola $x^2 - y^2 = 1$ under $w = \frac{1}{z}$ is the lemniscate $r^2 = \cos 2\theta$.

   Hint: $R = \frac{1}{r}$, $\phi = -\theta$, $x = r\cos\theta$, $y = r\sin\theta$. Substitute in $x^2 - y^2 = 1$.

7. Find and draw the image of the infinite horizontal strip $2 < y < 4$ under $w = \frac{1}{z}$.

Ans. $\left|w + \frac{1}{4}\right| < \frac{1}{4}$ and $\left|w + \frac{1}{8}\right| > \frac{1}{8}$ region between the two circles with centre at $-\frac{1}{4}$, radius $\frac{1}{4}$ and with centre at $-\frac{1}{8}$ and radius $\frac{1}{8}$

8. Find the critical points of the mappings:

   (a) $w = z^4$ (b) $w = e^{z^2}$ (c) $w = e^z$ (d) $w = \sin z$ (e) $w = z^2 + az + b$ (f) $w = z + \frac{1}{z}$ (g) $w = z^4 - z^2$ (h) $w = z^2 + \frac{1}{z^2}$.

Ans. (a) $z = 0$ (b) $z = 0$ (c) none (d) $z = \frac{n\pi}{2}$, n odd (e) $z = -\frac{a}{2}$ (f) $z = \pm 1$ (g) $z = 0, \pm\frac{1}{\sqrt{2}}$ (h) $w = \pm 1, \pm i$

9. Find an analytic function $w = u + iv = f(z)$ which maps the half plane $x \ge 0$ onto the region $u \ge 2$ such that $z = 0$ corresponds to $w = 2 + i$.

   Hint: $w_1 = z$, $w_2 = w_1 + 2$, $w = w_2 + i$.

Ans. $w = z + 2 + i$.

25.4 TRANSFORMATION: $w = z^n$

where n is integer greater than 1

Rewriting, $Re^{i\phi} = w = z^n = (re^{i\theta})^n$

$$\text{or} \quad R = r^n \quad \text{and} \quad \phi = n\theta \tag{12}$$

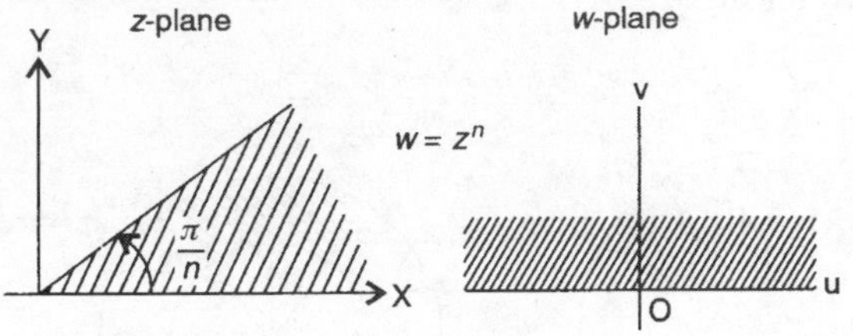

Fig. 25.6

Thus angular region sector with central angle $\alpha = \frac{\pi}{n}$, $r > 0$, $0 \le \theta \le \frac{\pi}{n}$ in the z-plane under $w = z^n$ maps to the upper half plane ($v \ge 0$) in the w-plane (since $0 \leq n\theta = \phi \leq \pi$, $R > 0$ i.e., $v \geq 0$) (see Fig. 25.6).

The transformation $w = z^n$ is conformal every except at $z = 0$ and ∞ since z^n is entire and since $f'(z) = nz^{n-1}$ is non-zero and bounded everywhere except at $z = 0$ and ∞. This transformation maps the sector $\theta_0 < \arg z = \theta < \theta_0 + \frac{2\pi}{n}$ onto the w-plane (cut along the ray $\arg w = \phi = n\theta_0$) since both the boundaries I and II of the sector say $0 \leq \theta \leq \frac{2\pi}{n}$ of the z-plane maps to the positive real axis of the w-plane) (see Fig. 25.7).

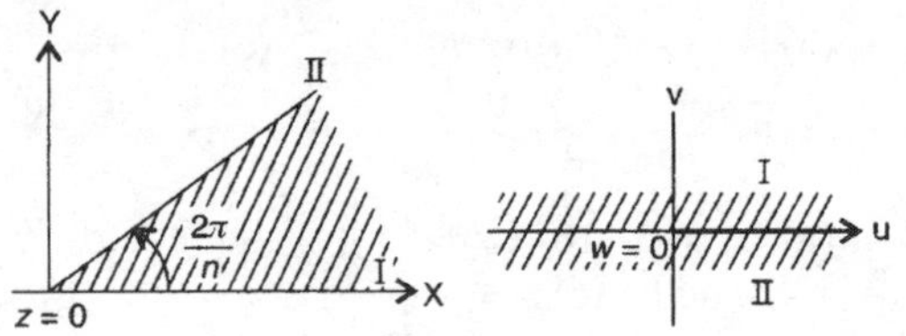

Fig. 25.7

Note: The angles at the origin are multiplied by a factor n in this mapping and the angular region is spread onto a half plane.

25.5 MAPPING $w = z^2$

In polar coordinates

$$Re^{i\phi} = w = (re^{i\theta})^2 = r^2 e^{i2\theta} \tag{13}$$

i.e., $R = r^2$, $\phi = 2\theta$

thus the angles at the origin are doubled. For example, the first quadrant in z-plane $\left(0 \leq \theta \leq \frac{\pi}{2}\right)$ maps to the upper half plane in the w-plane $(0 \leq \phi \leq \pi)$ (refer Fig. 25.8).

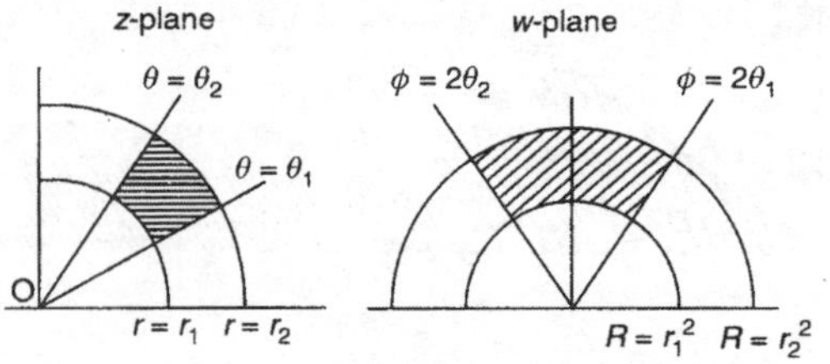

Fig. 25.8

The circle $r = r_0$ maps to circle $R = r_0^2$

The ray $\theta = \theta_0$ maps to ray $\phi = 2\theta_0$

In cartesian coordinates

$$u(x, y) + iv(x, y) = w = z^2 = (x + iy)^2$$
$$= (x^2 - y^2) + i2xy$$

So $$u(x, y) = x^2 - y^2, \; v(x, y) = 2xy \tag{14}$$

Case 1: If $u = u_0 =$ constant and $v = v_0 =$ constant then $x^2 - y^2 = u_0$ and $2xy = v_0$ represent equilateral hyperbolas (with the lines $y = \pm x$ and the coordinate axes $x = 0$, $y = 0$ as asymptotes respectively) which are orthogonal trajectories of each other (refer Fig. 25.9).

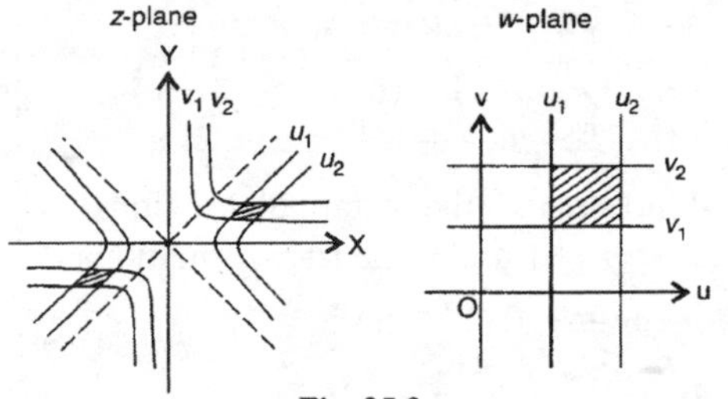

Fig. 25.9

Case 2: If $x = c_1 =$ constant and $y = c_2 =$ constant then eliminating x and y from (14) for $x = c_1$,

$$u = c_1^2 - y^2, \; v = 2c_1 y \quad \text{so} \quad u = c_1^2 - \frac{v^2}{4c_1^2}$$

which is a parabola with focus at origin, $v = 0$ as axis and open to the left. Similarly, $y = c_2$, $u = x^2 - c_2^2$, $v = 2c_2 x$, so $u = \frac{v^2}{4c_2^2} - c_2^2$ parabola open to the right (Fig. 25.10).

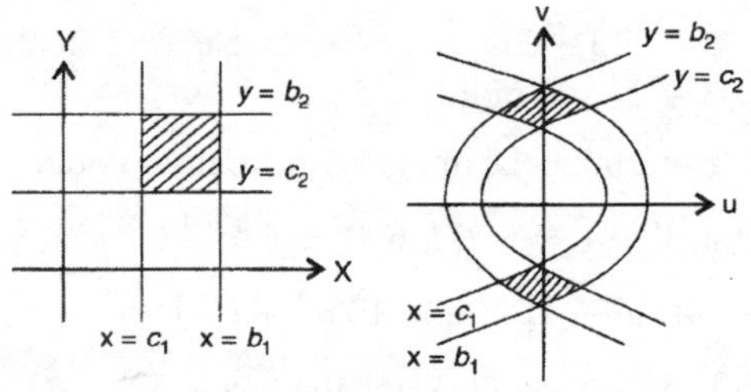

Fig. 25.10

These parabolas are orthogonal to each other. $w = z^2$ is conformal everywhere except at $z = 0$ where $w' = 2z = 0$.

Worked Out Examples

$w = z^n$ and $w = z^2$

Example 1: Describe the region onto which the sector $r < a, 0 \le \theta \le \frac{\pi}{4}$ is mapped by (a) $w = z^2$ (b) $w = z^3$ (c) $w = z^4$ (d) $w = iz^2$ (e) $w = \frac{i}{z^2}$

Solution: $Re^{i\phi} = w = z^2 = r^2 e^{i2\theta}$ so $R = r^2$, $\phi = 2\theta$

a. $R = r^2 < a^2, \quad 0 \le \phi \le \frac{\pi}{2}$ (Fig. 25.11)

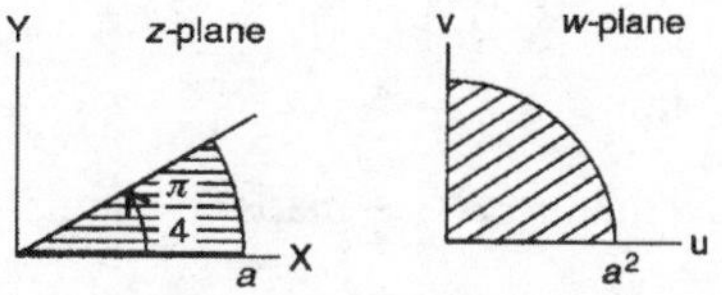

Fig. 25.11

b. $w = z^3, \quad R = r^3, \quad \phi = 3\theta, \quad R = r^3 < a^3,$ $0 \le \phi \le \frac{3\pi}{4}$ (Fig. 25.12)

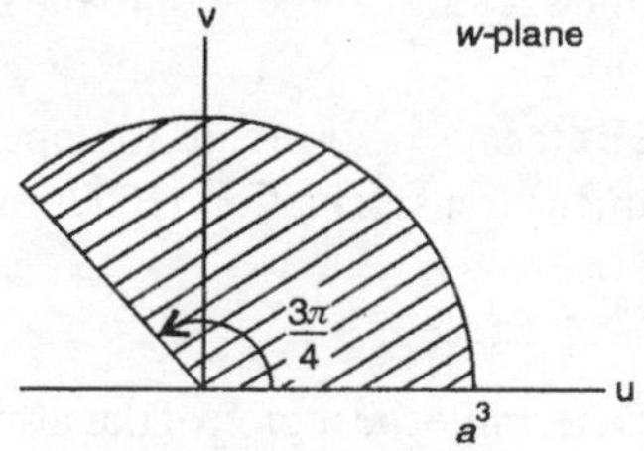

Fig. 25.12

c. $w = z^4, \quad R = r^4 < a^4, \quad 0 \le \phi \le \pi$ (refer Fig. 25.13)

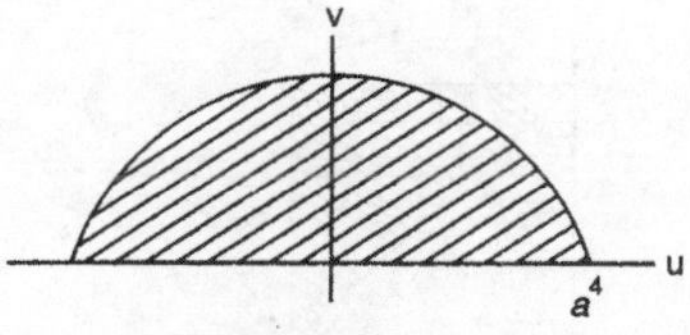

Fig. 25.13

d. $w = iz^2$, rotation of (Fig of a) by $\frac{\pi}{2}$ in counter-clockwise (Fig. 25.14).

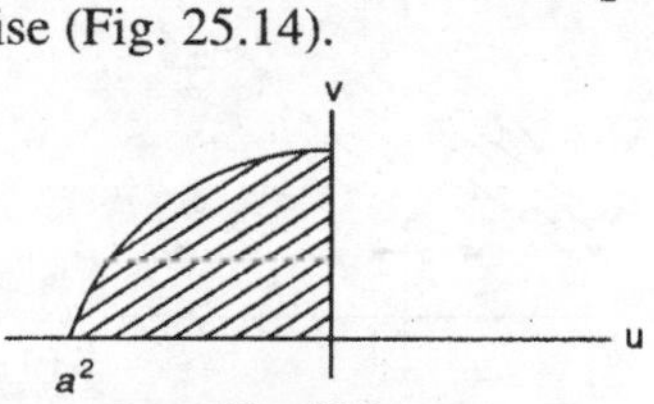

Fig. 25.14

e. $Re^{i\phi} = w_1 = \frac{1}{z^2} = \frac{1}{r^2 e^{i2\theta}} = \frac{1}{r^2} e^{-i2\theta}$

so $R = \frac{1}{r^2} > \frac{1}{a^2}$ and $\phi = -2\theta$ i.e., $0 \ge \theta \ge -\frac{\pi}{2}$ (refer Fig. 25.15).

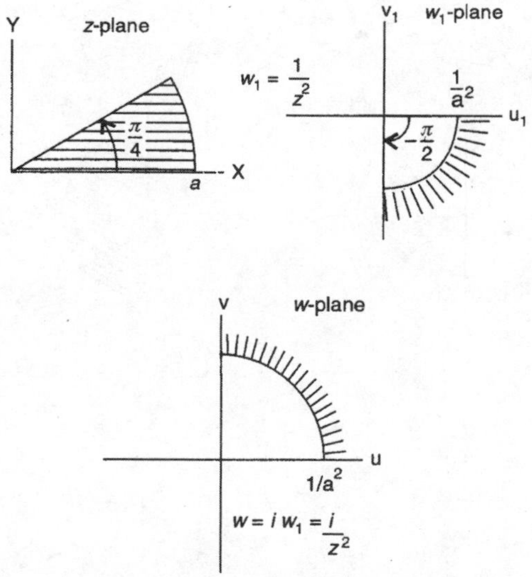

Fig. 25.15

Sector in z-plane is inverted in w_1-plane and then rotated through $\frac{\pi}{2}$ in counterclockwise direction.

Example 2: Find an analytic function $w = u + iv = f(z)$ such that the angular region $0 < \arg z < \frac{\pi}{3}$ maps onto the region $u \le 1$ (refer Fig. 25.16).

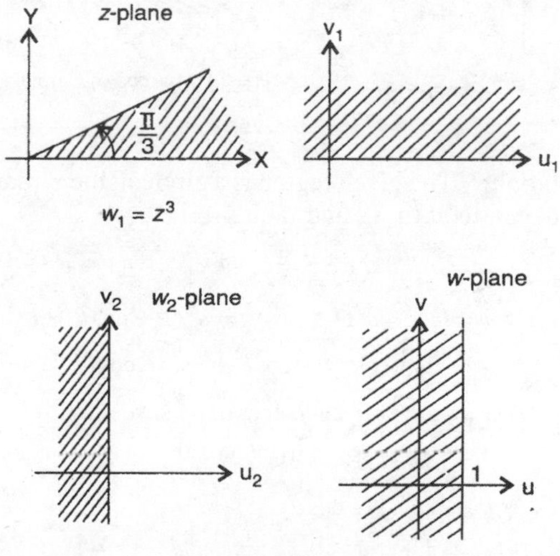

Fig. 25.16

Solution: z^3 maps the given region onto upper half plane in w_1-plane, which rotated through $\frac{\pi}{2}$ in w_2-plane and translated to the right by 1 in the w-plane.

Example 3: Plot the image of the region $2 < |z| < 3$ and $|\arg z| < \frac{\pi}{4}$ under $w = z^2$ (see Fig. 25.17).

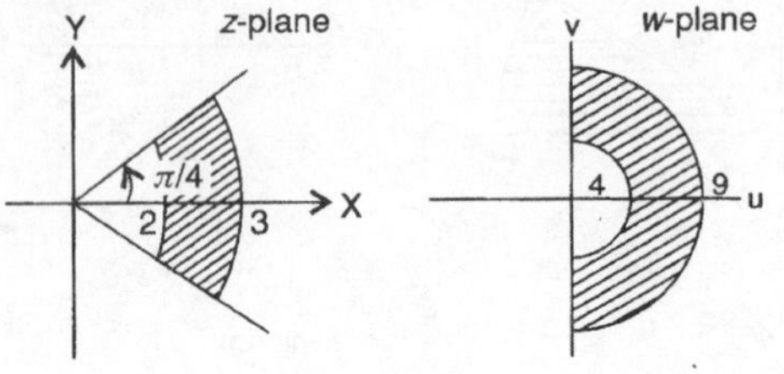

Fig. 25.17

Solution: $R = r^2$ so for $2 < r < 3, 4 < R < 9$
Since $-\frac{\pi}{4} < \theta < \frac{\pi}{4}$
therefore

$$-\frac{\pi}{2} < \phi < \frac{\pi}{2}$$

Example 4: Determine and graph the image of $|z - a| = a$ under $w = z^2$ (Fig. 25.18).

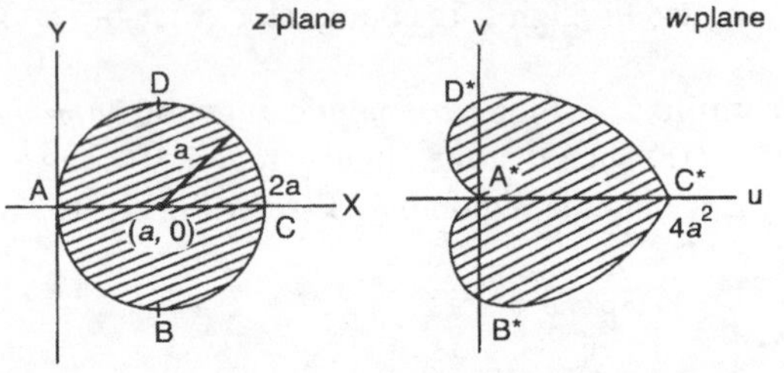

$|z - a| = a$ $\qquad$ $R = 2a^2\,(1 + \cos\phi)$ Cardioid

Fig. 25.18

Solution: The given region is a circle in the z-plane with centre at $(a, 0)$ and radius a. i.e.,

$$z - a = ae^{i\theta} \quad \text{or} \quad z = a + ae^{i\theta} = a(1 + e^{i\theta})$$

So $\quad w = z^2 = a^2(1 + e^{i\theta})^2 = a^2(1 + \cos\theta + i\sin\theta)^2$

$$= 2a^2(\cos^2\theta + \cos\theta + i\sin\theta\cos\theta + i\sin\theta)$$

$$Re^{i\phi} = w = 2a^2(1 + \cos\theta)(\cos\theta + i\sin\theta)$$

$$= 2a^2(1 + \cos\theta)e^{i\theta}$$

Thus $\quad R = 2a^2(1 + \cos\theta)$

$$= 2a^2(1 + \cos\phi) \quad \text{since } \phi = \theta$$

EXERCISE

$w = z^n$ and $w = z^2$

Determine and plot the images of the regions under the transformation $w = z^2$:

1. $|z| > 2$
   Ans. $|w| > 4$
2. $|\arg z| \le \frac{\pi}{2}$
   Ans. $|\arg w| \le \pi$
3. $\frac{1}{2} < |z| < 2$, $Re\, z \ge 0$
   Ans. $\frac{1}{4} < |w| < 4$, $-\pi \le \phi \le \pi$
4. Show that $w = \frac{z^4 - i}{z^4 + i}$ maps $0 < \arg z < \frac{\pi}{4}$ onto $|w| = 1$.
   Hint: $w_1 = z^4$ onto upper half plane, $w = \frac{w_1 - i}{w_1 + i}$ onto unit circle $|w| = 1$.
5. Find a transformation which will map an infinite sector of angle $\frac{\pi}{3}$ onto the interior of a unit circle.
   Hint: $w_1 = z^3$ spreads to upper half plane, bilinear transformation (B.T.) maps to unit circle.
   Ans. $w = \frac{(z^3 - i)}{(z^3 + i)}$
6. Determine the region of the w-plane into which the region bounded by $x = 1$, $y = 1$, $x + y = 1$ is mapped by $w = z^2$. Show that angles are preserved (refer Fig. 25.19).

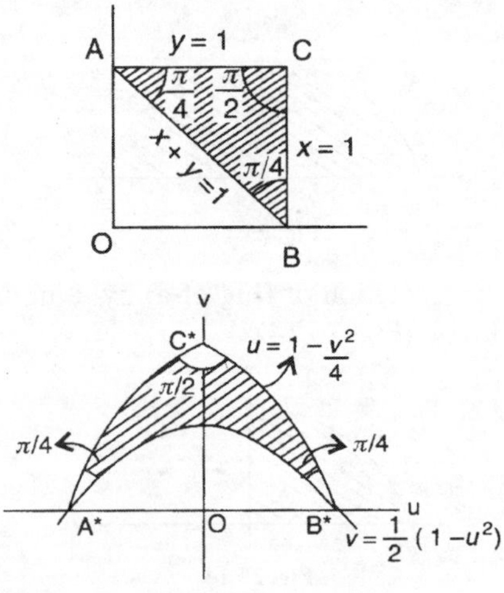

Fig. 25.19

Ans. $u = \frac{v^2}{4} - 1, u = 1 - \frac{v^2}{4}, v = \frac{1}{2}(1 - u^2)$

7. Find the region in z-plane whose image under $w = z^2$ is the rectangular domain in w-plane bounded by the lines $u = 1, u = 2, v = 1, v = 2$.

Ans. $1 < x^2 - y^2 < 2, \frac{1}{2} < xy < 1$ rectangular hyperbolas

8. Determine the image of the rectangle $a \le x \le b, c \le y \le d$ under $w = \sqrt{z}$.

   Hint: Consider $w^2 = z$ and use above problem 7.

Ans. $a \le u^2 - v^2 \le b$ and $c \le 2uv \le d$ rectangular hyperbolas

9. Show that the image of the unit circle $|z| = 1$ under $w = 2z + z^2$ is a cardioid

   $R = 2(1 + \cos\phi)$.

25.6 TRANSFORMATION $w = e^z$

Rewriting

$$Re^{i\phi} = w = e^z = e^{x+iy} = e^x \cdot e^{iy}$$

Therefore $R = e^x$ and $\phi = y$ (1)

i.e., modulus of w is e^x and argument of w is y. The line $x = c =$ constant maps onto the circle $R = e^c$ (refer Fig. 25.20).

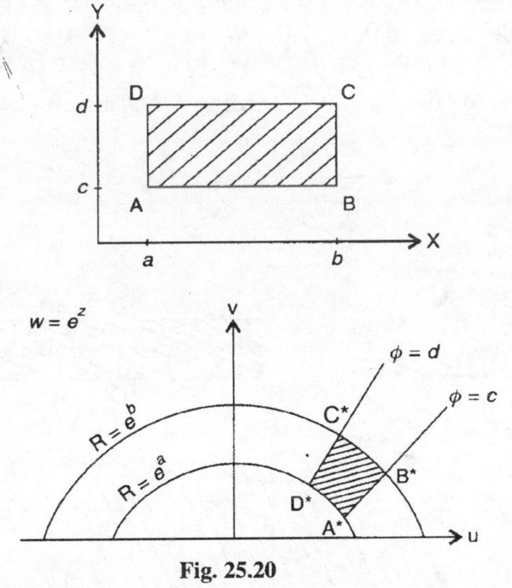

Fig. 25.20

The line $y = c$ maps onto the ray $\phi = c$. Thus the rectangular region $a \le x \le b, c \le y \le d$ in the z-plane is mapped to the region $A^*B^*C^*D^*$ in the w-plane bounded by the concentric circles $R = e^b$ and $R = e^a$ and by the rays $\phi = c$ and $\phi = d$.

Note 1: Since $e^z \neq 0$, $w = 0$ is not mapped. Thus the origin in w-plane is excluded.

Note 2: This mapping is one to one if $d - c < 2\pi$.

Particular case: $c = 0, d = \pi$

Consider the rectangular region in z-plane $a \le x \le b, 0 \le y \le \pi$ (see Fig. 25.21). By (1),

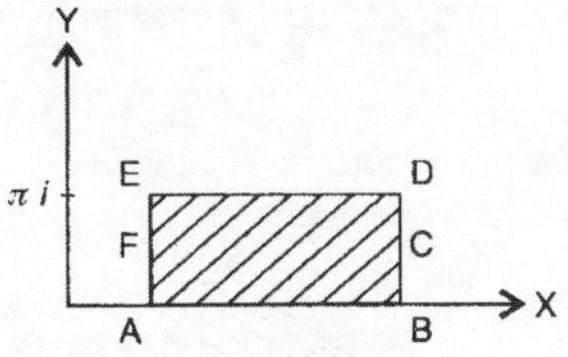

Fig. 25.21

$$e^a \le R = e^x \le e^b \quad \text{and} \quad 0 \le \phi = y \le \pi.$$

Thus the rectangular region maps onto the upper half of the annulus ring $e^a < R < e^b, 0 \le \phi \le \pi$ (refer Fig. 25.22).

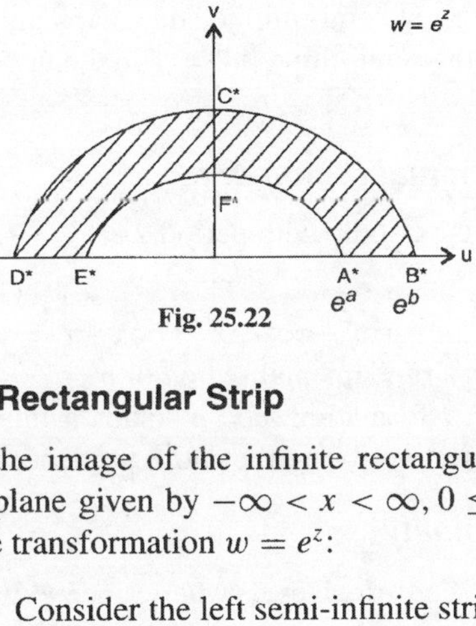

Fig. 25.22

Infinite Rectangular Strip

To find the image of the infinite rectangular strip in the z-plane given by $-\infty < x < \infty, 0 \le y \le \pi$ under the transformation $w = e^z$:

Case 1: Consider the left semi-infinite strip

$-\infty < x \le 0, 0 \le y \le \pi$. By (1), $x = 0, R = e^0 = 1$

and as $x \to -\infty$, $R = e^x \to 0$. Also $0 \le \phi \le \pi$

AB: $-\infty < x < 0, \quad y = 0 \quad$ then $0 < R < 1$ and $\phi = 0$

BCD: $x = 0, \quad 0 \le y \le \pi \quad$ then $R = 1$

and $0 \leq \phi \leq \pi$

DE: $-\infty < x < 0, y = \phi$ then $0 < R < 1$

and $\phi = \pi$.

Thus the left semi-infinite strip $ABCDE$ maps onto the semi circle $0 < R \leq 1, 0 \leq \phi \leq \pi$ given by $A^*B^*C^*D^*E^*$ in the w-plane (Fig. 25.23).

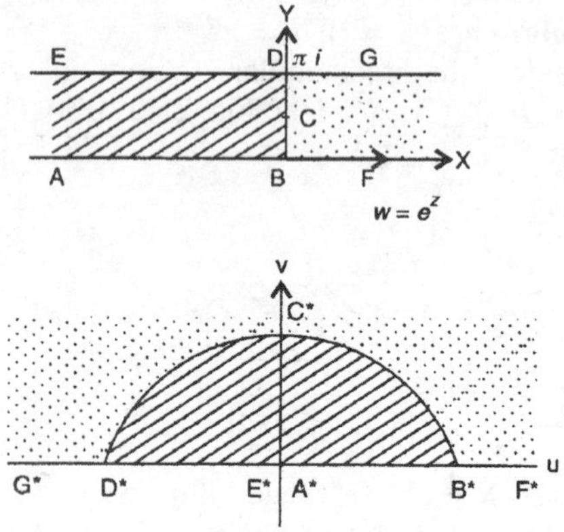

Fig. 25.23

Case 2: Right semi-infinite strip $0 \leq x < \infty$, $0 \leq y \leq \pi$. By (1), $1 \leq R < \infty$ and $0 \leq \phi \leq \pi$. Thus the right semi-infinite strip maps onto the exterior of the semi-circle $|w| = 1$ in the upper half of w-plane.

Fundamental region of e^z

Since e^z is periodic with period $2\pi i$

$$e^{z+2\pi i} = e^z$$

The infinite strip $-\infty < x < \infty, -\pi < y \leq \pi$, is known as a fundamental region of e^z.

The transformation $w = e^z$ maps a fundamental region bijectively onto the entire w-plane.

Conformality

$w = e^z$ is conformal everywhere since e^z is analytic everywhere and has no critical points ($w' = e^z \neq 0$ for any z).

Mapping of Logarithmic Function

Since logarithm is the inverse of exponential function, the mapping logarithm can be easily obtained from those of the exponential function by interchanging the roles of z (z-plane) and w (w-plane).

Worked Out Examples

Example 1: Find and draw the image of the rectangular region $-1 \leq x \leq 3, -\pi \leq y \leq \pi$ in the z-plane under the transformation $w = e^z$.

Solution: By transformation $R = e^x, y = \phi$, we get for $-1 \leq x \leq 3, e^{-1} \leq R \leq e^3$ and for $-\pi \leq y \leq \pi, -\pi \leq \phi \leq \pi$. Thus the image is the annulus region bounded by the circles of radii e^{-1} and e^3 (Fig. 25.24).

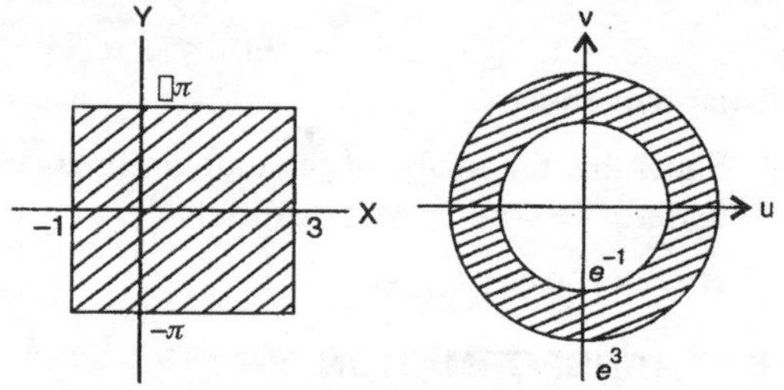

Fig. 25.24

Example 2: Find an analytic function which maps the region R bounded by the positive x and y-axes and the hyperbola $xy = \frac{\pi}{2}$ in the first quadrant onto the upper half plane.

Solution: The transformation $w_1 = z^2$ maps the given region onto the semi-infinite strip $-\infty < u_1 < \infty, 0 \leq v_1 \leq \pi$ in the w_1-plane. [$u_1 = x^2 - y^2, v_1 = 2xy$ so OA: $x = 0, v_1 = 0$, OB: $y = 0, v_1 = 0$ so BOA maps $v_1 = 0$ i.e., to $B^*O^*A^*$, CDE: $xy = \frac{\pi}{2}$ so $v_1 = 2xy = 2 \cdot \frac{\pi}{2} = \pi$, so CDE onto $C^*D^*E^*$: $v_1 = \pi$] (Fig. 25.25).

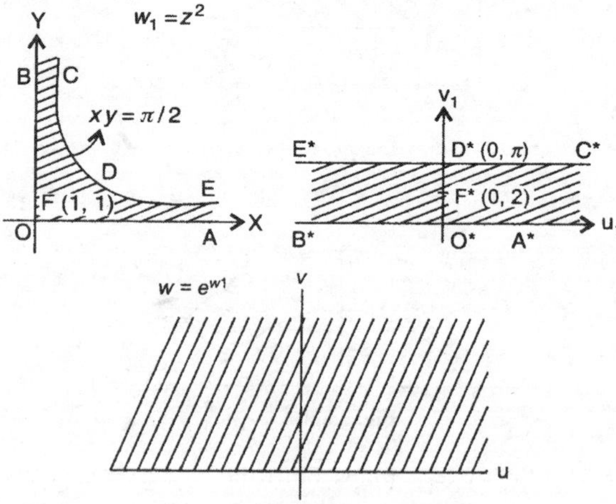

Fig. 25.25

Now the transformation $w = e^{w_1} = e^{z^2}$ transforms the semi-infinite strip onto the upper half plane in the w-plane.

EXERCISE

Find and graph the images of the regions under the mapping $w = e^z$:

1. $-1 < x < 1, \frac{-\pi}{2} < y < \frac{\pi}{2}$

Ans. $e^{-1} < R < e, \frac{-\pi}{2} < \phi < \frac{\pi}{2}$

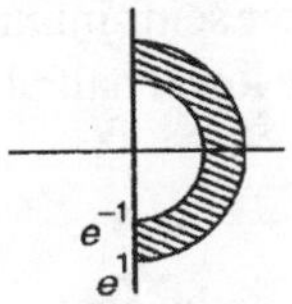

Fig. 25.26

2. $-2 \le x \le 2, -\pi \le y \le \frac{-\pi}{2}$

Ans. $e^{-2} \le R \le e^2, -\pi \le \phi \le \frac{-\pi}{2}$

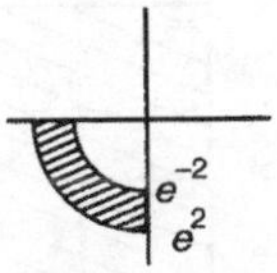

Fig. 25.27

3. Find the transformation which conformally maps the horizontal strip $0 < y < \pi$ onto the disk $|w| < 1$.

Ans. $w = f(z) = \frac{(e^z - i)}{(e^z + i)}$. Put $Z = e^z$ which transforms horizontal strip onto upper half plane, then B.T. $w = \frac{(z-i)}{(z+i)}$ maps upper half plane onto disk $|w| < 1$.

4. Find the transformation which maps the infinite strip $0 < y < a$ in the z-plane into the upper half plane of w-plane.

Ans. $w = e^{\frac{\pi z}{a}}$

5. Find the transformation which maps the annulus region $a < R < b$ in z-plane onto a rectangle in the w-plane.

Hint: $e^w = z$ transforms rectangles in w-plane onto annulus region in z-plane.

Ans. $w = \ln z$

6. Show that the transformation $w = \tan z$ transforms $|x| < \frac{\pi}{4}$ onto unit disk $|w| < 1$.

Hint: $w = \tan z = \frac{1}{i}\frac{e^{iz} - e^{-iz}}{e^{iz} + e^{-iz}} = \frac{-ie^{2iz} + i}{e^{2iz} + 1}$

Put $Z = e^{i2z}$, $w = \frac{-iZ + i}{Z + 1}$.

Vertical strip to right half plane to unit circle.

25.7 TRANSFORMATION $w = \sin z$

We know that

$$u + iv = w = f(z) = \sin z$$

$$= \sin x \cdot \cosh y + i \cos x \sinh y \qquad (1)$$

so

$$u(x, y) = \sin x \cdot \cosh y \qquad (2)$$

$$v(x, y) = \cos x \sinh y \qquad (3)$$

If $-\frac{\pi}{2} \le x \le \frac{\pi}{2}$, then the mapping in one-to-one. If $x = c =$ constant then from (2) and (3)

$$u = \sin c \cosh y, \; v = \cos c \sinh y,$$

so

$$1 = \cosh^2 y - \sinh^2 y = \left(\frac{u}{\sin c}\right)^2 - \left(\frac{v}{\sin c}\right)^2$$

Thus the images of the vertical lines $x =$ constant are hyperbolas given by

$$\frac{u^2}{\sin^2 x} - \frac{v^2}{\cos^2 x} = 1 \qquad (4)$$

If $y = c$ then from (2) and (3)

$$u = \sin x \cdot \cosh c, \; v = \cos x \cdot \sinh c,$$

$$\text{so } 1 = \sin^2 x + \cos^2 x = \left(\frac{u}{\cosh c}\right)^2 + \left(\frac{v}{\sinh c}\right)^2$$

Thus the images of the horizontal lines $y = c$ are ellipses given by

$$\frac{u^2}{\cosh^2 y} + \frac{v^2}{\sinh^2 y} = 1 \qquad (5)$$

The focii for the ellipses (5) and hyperbolas (4) are same given by $w = \pm 1$ (independent of the constant c) (Fig. 25.28).

Hence $w = \sin z$ transforms $x = c$ and $y = c$ lines into confocal hyperbolas (4) and confocal ellipses (5) respectively. The families of hyperbolas and ellipses are orthogonal to each other.

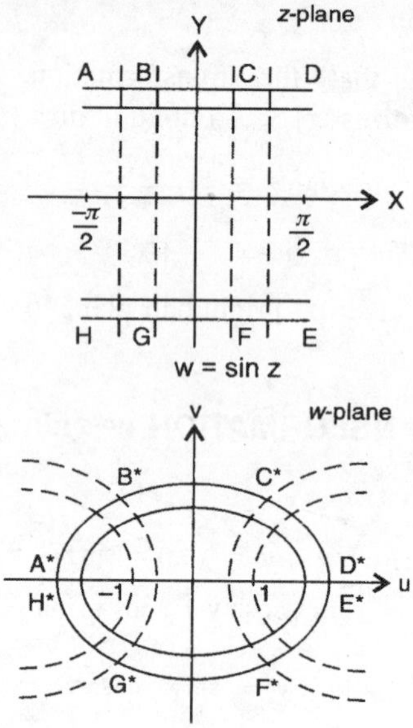

Fig. 25.28

Semi-infinite strip: $-\frac{\pi}{2} \le x \le \frac{\pi}{2}, y \ge 0.$
(see Fig. 25.29)

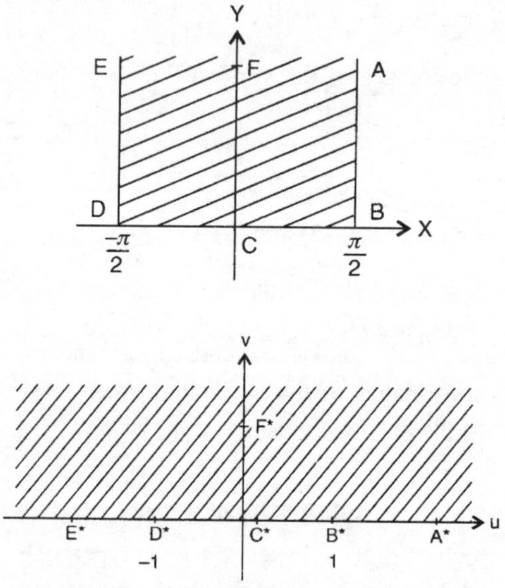

Fig. 25.29

From (2) and (3), $u = \sin x \cdot \cosh y$,
$v = \cos x \sinh y$

AB: $x = \frac{\pi}{2}, y \ge 0$ so $v = 0, u = \cosh y \ge 1$

BC: $y = 0, 0 \le x < \frac{\pi}{2}$ so $v = 0, u = \sin x,$

thus $0 \le u \le 1$

CD: $y = 0, \frac{-\pi}{2} \le x \le 0,$ so $v = 0, u = \sin x,$

thus $-1 \le u = 0$

DE: $x = \frac{-\pi}{2}, y \ge 0,$ so $v = 0, u = -\cosh y \le -1$

CF: $x = 0, y \ge 0,$ so $u = 0, v = \sinh y \ge 0.$

Thus the upper semi-infinite strip in the z-plane under $w = \sin z$ transforms to the upper half plane in the w-plane.

Similarly, the lower semi-infinite strip $-\frac{\pi}{2} \le x \le \frac{\pi}{2}, y \le 0$ maps to the lower half plane of the w-plane.

Cut

To find the image of the rectangle in the z-plane $-\frac{\pi}{2} \le x \le \frac{\pi}{2}, -1 \le y \le 1$; under $w = \sin z$ (see Fig. 25.30)

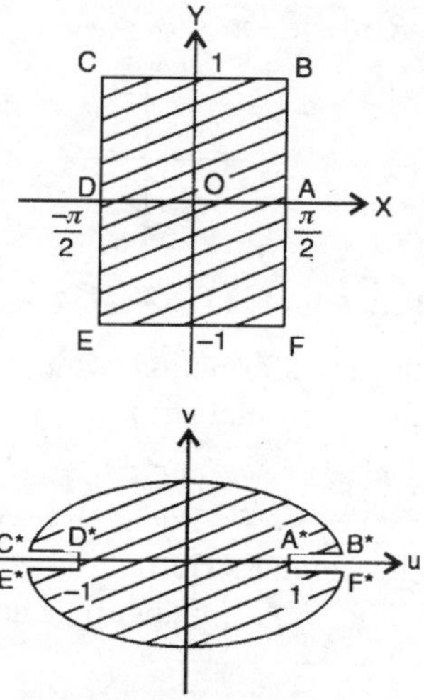

Fig. 25.30

Line by line correspondence:

CB: $y = 1, \frac{-\pi}{2} \le x \le \frac{\pi}{2}$, upper portion of the ellipse

$$\frac{u^2}{\cosh^2 1} + \frac{v^2}{\sinh^2 1} = 1$$

since $v = \cos x \cdot \sinh 1 \ge 0$ for $\frac{-\pi}{2} \le x \le \frac{\pi}{2}$

EF: $y = -1, \frac{-\pi}{2} \le x \le \frac{\pi}{2}$, lower portion of the above ellipse since $v = -\cos x \cdot \sinh 1 \le 0$ for $\left(\frac{-\pi}{2}, \frac{\pi}{2}\right)$

$$BA: x = \frac{\pi}{2}, y > 0, \quad \text{so } v = 0, u = \cosh y \geq 1$$

$$AF: x = \frac{\pi}{2}, y < 0, \quad \text{so } v = 0, u = \cosh(-y)$$

$$= \cosh y \geq 1.$$

Thus BA and AF both get mapped onto the same line segment $v = 0, u \geq 1$ in the w-plane. $B^*A^*F^*$ is known as cut along the real axis.

$$CD: x = -\frac{\pi}{2}, y > 0, \quad \text{so } v = 0, u = -\cosh y \leq -1$$

$$DE: x = -\frac{\pi}{2}, y < 0, \quad \text{so } v = 0, u = -\cosh \leq -1.$$

Thus CD and DE both map to the cut $C^*D^*E^*$. Hence the upper and lower sides of the rectangle are mapped onto semi-ellipses while the vertical sides onto $-\cosh 1 \leq u \leq -1$ and $1 \leq u \leq \cosh 1$ $(v = 0)$.

Mapping $w = \sin z$ is conformal everywhere except at $z = \pm 1$ where it is not one-to-one. In general, $w = \sin z$ is conformal everywhere since $w = \sin z$ is analytic everywhere except at the critical points $z = (2n-1)\frac{\pi}{2}$ where

$$w' = \cos z = 0 \cdot (n = 0, \pm 1, \pm 2, \ldots).$$

Successive Transformations

from one plane to another are equivalent to a single transformation.

1. $w = \cos z = \sin\left(z + \frac{\pi}{2}\right)$
   with $z^* = z + \frac{\pi}{2}$, $w = \cos z = \sin z^*$.
   Thus the cosine transformation is the same mapping as sine preceded by a translation to the right through $\frac{\pi}{2}$ units.
2. $w = \sinh z = -i \sin(iz)$
   with $w_1 = iz$, $w_2 = \sin w_1$, $w = -iw_2$
   Hyperbolic sine consists of counterclockwise rotation through $\frac{\pi}{2}$, followed by sine transformation followed by closewise rotation through $\frac{\pi}{2}$.
3. $w = \cosh z = \cos(iz)$
   with $w_1 = iz$, $w = \cos(w_1)$
   Hyperbolic cosine consists of counterclockwise rotation through $\frac{\pi}{2}$ followed by cosine transformation.

Worked Out Examples

Example 1: Find and graph the image of the region $0 < x < 2\pi$, $1 < y < 2$ in the z-plane under the mapping $w = \sin z$.

Solution: $u = \sin x \cosh y$; $v = \cos x \cdot \sinh y$.

The line $AGEKB$ in the z-plane given by $y = 1$ and $0 < x < 2\pi$ maps to the (inner) ellipse (Fig. 25.31).

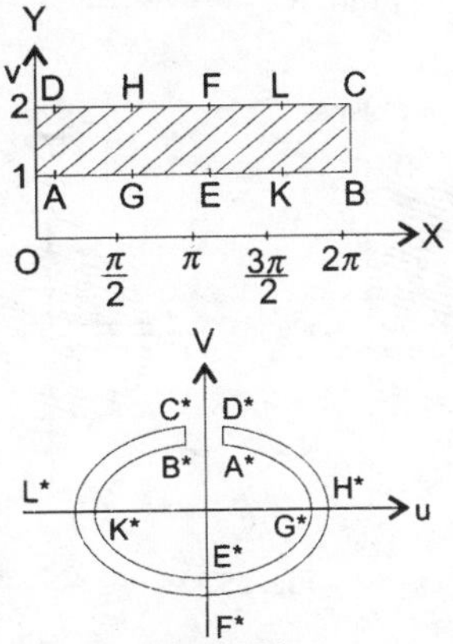

Fig. 25.31

$$\frac{u^2}{\cosh^2 1} + \frac{v^2}{\sinh^2 1} = 1$$

Similarly, the line $CLFHD$ in the z-plane given by $y = 2$ and $0 < x < 2\pi$ maps to the (outer ellipse)

$$\frac{u^2}{\cosh^2 2} + \frac{v^2}{\sinh^2 2} = 1$$

The shaded rectangular strip in the z-plane maps to the elliptical annulus bounded the above two ellipses with a cut along the positive imaginary axis.

Line $AD: x = 0, 1 < y < 2$ so $u = 0$,
$v = \sinh y > 0$
For $1 < y < 2$, $\sinh 1 < v < \sinh 2$
Line $BC: x = 2\pi, 1 < y < 2$ so $u = 0$,
$v = \sinh y > 0$ for $1 < y < 2$, $\sinh 1 < v < \sinh 2$.
Thus both the line segments AD and BC of z-plane gets mapped onto the same line segment $u = 0, \sinh 1 < v < \sinh 2$. Thus there exists a cut along the positive imaginary axis.

$$GH: x = \frac{\pi}{2}, 1 < y < 2, \quad \text{so } v = 0, u = \cosh y$$

$$\text{thus } \cosh 1 < u < \cosh 2$$

$EF: x = \pi, 1 < y < 2,\quad$ so $u = 0, v = -\sinh y$

thus $-\sinh 1 < v < \sinh 2$

$KL: x = \frac{3\pi}{2}, 1 < y < 2\quad$ so $v = 0, u = -\cosh y$

thus $-\cosh 1 < u < -\cosh 2.$

Example 2: Find and graph the image of the region $0 \le x \le \frac{\pi}{2}$ in the z-plane under the mapping $w = \tan^2 \frac{z}{2}$ (Fig. 25.32).

Solution: $w = \tan^2 \frac{z}{2} = \frac{\sin^2 \frac{z}{2}}{\cos^2 \frac{z}{2}} = \frac{1-\cos z}{1+\cos z}$

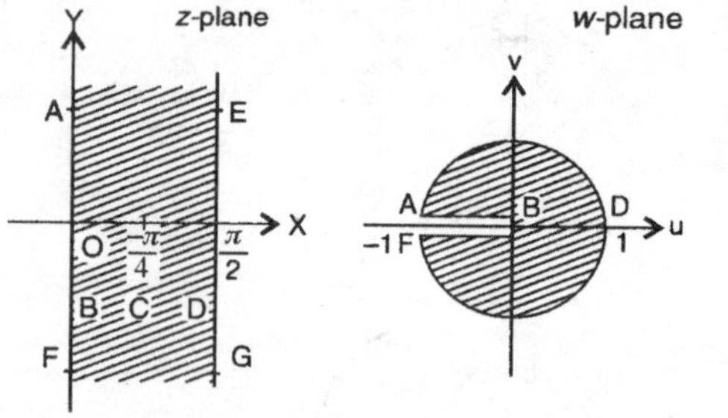

Fig. 25.32

We know that $\cos z = \cos x \cdot \cosh y - i \sin x \sinh y$
$ED: x = \frac{\pi}{2}, y \ge 0, \cos z = -i \sinh y$

$$w = \frac{1-\cos z}{1+\cos z} = \frac{1+i\sinh y}{1-i\sinh y}$$

$$|w| = \left|\frac{1+i\sinh y}{1-i\sinh y}\right| = \frac{\sqrt{1+\sinh^2 y}}{\sqrt{1+\sinh^2 y}} = 1 \quad \text{for any } y.$$

Thus the line $x = \frac{\pi}{2}$ is mapped onto the unit circle $|w| = 1$ in the w-plane.

The y-axis: $x = 0$ is mapped onto

$$w = \frac{1-\cosh y}{1+\cosh y}$$

w is purely real.

At $y = 0$, $w = u = 0$, $y > 0$ as $y \to \infty$ $u \to -1$. Similarly as $y \to -\infty$ also $u \to -1$.

Thus both AB and BF maps onto the same interval (a cut) $-1 \le u \le 0$.

Any line $x = \alpha$, where $0 < \alpha < \frac{\pi}{2}$

$$|w| = \left|\frac{1-\cos(\alpha+iy)}{1+\cos(\alpha+iy)}\right|$$

$$= \sqrt{\frac{(1-\cos\alpha\cosh y)^2 + (\sin\alpha\sinh y)^2}{(1+\cos\alpha\cosh y)^2 + (\sin\alpha\sinh y)^2}} < 1$$

Since $\quad -2\cos\alpha\cosh y \le 2\cos\alpha\cosh y$

or $\quad 4\cos\alpha\cosh y \ge 1$

which is true for $\cos\lambda$ with $0 < \alpha < \frac{\pi}{2}$ and $y \ge 0$. Hence any line $x = \alpha$ gets mapped to interior of $|w| = 1$.

EXERCISE

Find and graph the images of the following regions under $w = \sin z$: (Fig. 25.33)

1. $-\frac{\pi}{2} < x < \frac{\pi}{2}, 1 < y < 2$

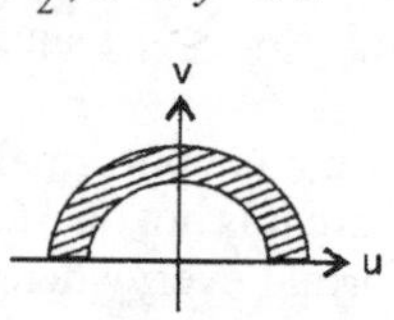

Fig. 25.33

Ans. $y = c$, $\frac{u^2}{\cosh^2 c} + \frac{v^2}{\sinh^2 c} = 1$ where $c = 1, 2$ bounded by inner and outer ellipses in the upper half plane.

2. Rectangular region $-\frac{\pi}{2} \le x \le \frac{\pi}{2}, 0 \le y \le \theta_0$ (refer Fig. 25.34).

Ans.

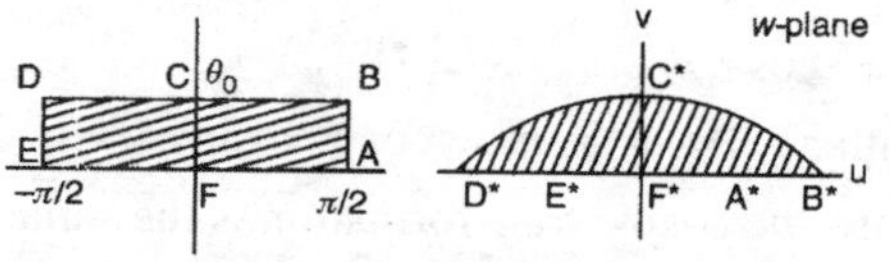

Fig. 25.34

semi elliptic region in the upper half plane.

3. $-\pi \le x \le \pi, \theta_1 \le y \le \theta_2$ (refer Fig. 25.35).

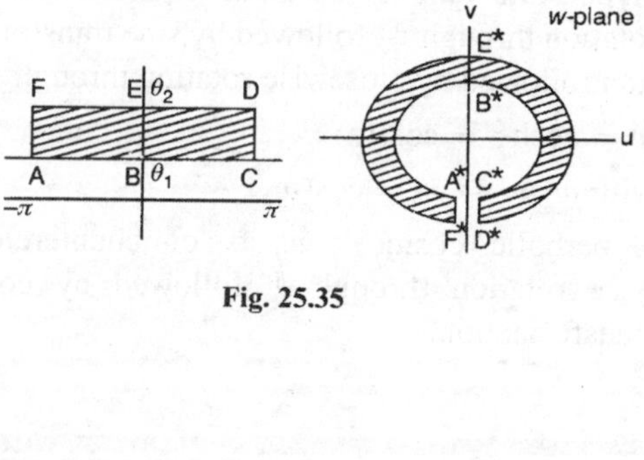

Fig. 25.35

Region bounded by confocal ellipses with a cut along the negative imaginary axis.

25.8 JOUKVOWSKI'S (ZHUKOVSKY'S) TRANSFORMATION

$$w = z + \frac{a^2}{z} \qquad (1)$$

is used in solutions of problems in hydro- and aerodynamics. (1) is analytic everywhere except at a simple pole $z = 0$. Since

$$\frac{dw}{dz} = \frac{z^2 - a^2}{z^2},$$

w' is non-zero everywhere except at $z = \pm a$.

Thus the Joukvowski's transformation (function) is conformal everywhere except at the points $z = \pm a$ which correspond to $w = \pm 2a$ in the w-plane. Solving (1) for z, we get

$$z = \frac{w \pm \sqrt{w^2 - 4a^2}}{2}$$

so z is a double-valued function of w.

Rewriting (1) in polar coordinates

$$u + iv = w = re^{i\theta} + \frac{a^2}{re^{i\theta}} = re^{i\theta} + \frac{a^2}{r}e^{-i\theta}$$

$$= r(\cos\theta + i\sin\theta) + \frac{a^2}{r}(\cos\theta - i\sin\theta)$$

$$= \left(r + \frac{a^2}{r}\right)\cos\theta + i\left(r - \frac{a^2}{r}\right)\sin\theta$$

Thus $u(r,\theta) = \left(r + \frac{a^2}{r}\right)\cos\theta, \quad v = \left(r - \frac{a^2}{r}\right)\sin\theta \qquad (2)$

To find the image of circle $|z| = r_0$, eliminate the parameter θ from (2), then we have

$$\left(\frac{u}{r + \frac{a^2}{r}}\right)^2 + \left(\frac{v}{r - \frac{a^2}{r}}\right)^2 = \cos^2\theta + \sin^2\theta = 1 \qquad (3)$$

For $r = r_0 =$ constant, (3) represents an ellipse with focii at

$$\sqrt{\left(r + \frac{a^2}{r}\right)^2 - \left(r - \frac{a^2}{r}\right)^2} = \sqrt{4a^2} = \pm 2a$$

which are independent of r.

Thus the Joukvowski's function (1) maps the family of concentric circles $|z| = r_0$ of the z-plane onto the family of confocal ellipses of the w-plane with focii at $w = \pm 2a$.

Special Case 1: $|z| = a$, circle.

As $r \to a$, $v=0$, $u = 2a\cos\theta$. Since $|\cos\theta| \le 1$, $-2a \le u \le 2a$.

Thus the circle $|z| = a$ in the z-plane maps to the degenerated ellipse (3) which flattens to the line segment $v = 0$, $-2a \le u \le 2a$ on the real u-axis traversed twice.

Case 2: As $r \to 0$, the ellipse (3) is transformed into a circle of infinitely large radius (Fig. 25.36).

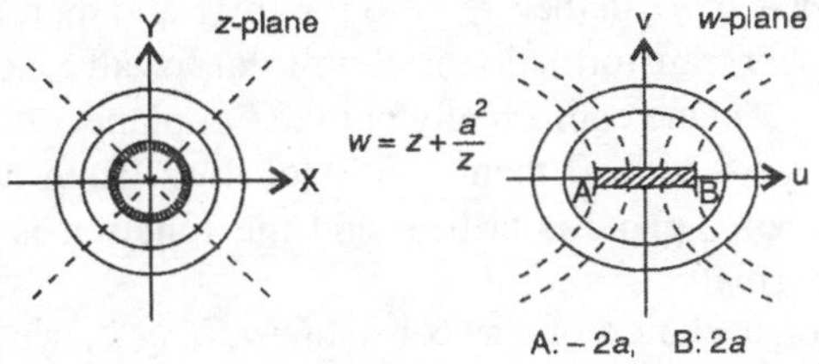

Fig. 25.36

To find the image of the ray $\arg z = \theta_0$, eliminate the parameter r from (2), and $\arg z = \theta = \theta_0 =$ const, then we get

$$\left(\frac{u}{\cos\theta}\right)^2 - \left(\frac{v}{\sin\theta}\right)^2 = \left(r + \frac{a^2}{r}\right)^2 - \left(r - \frac{a^2}{r}\right)^2 = 4a^2$$

or

$$\frac{u^2}{4a^2\cos^2\theta_0} - \frac{v^2}{4a^2\sin^2\theta_0} = 1 \qquad (4)$$

which represent hyperbola with focii at $w = \pm 2a$. Thus Joukvowski's function defines a transformation of the orthogonal system of polar coordinates in z-plane, into an orthogonal curvilinear system of coordinates whose coordinate lines are the confocal families of ellipses and hyperbolas in the w-plane.

25.9 BILINEAR TRANSFORMATION

Bilinear transformation is the function w of a complex variable z of the form

$$w = f(z) = \frac{az + b}{cz + d} \qquad (1)$$

where a, b, c, d are complex or real constants subject to $ad - bc \neq 0$.

If $ad - bc = 0$, $f(z)$ would be identically constant. When (1) is cleared of fractions, it takes the form

$$Azw + Bz + Cw + D = 0$$

which is linear in z, linear in w or bilinear in z and w. Bilinear transformation (B.T.) (1) is also known as linear fractional transformation or Möbius transformation.

Differentiating (1) w.r.t. z, we get

$$\frac{dw}{dz} = \frac{ad - bc}{(cz + d)^2} \tag{2}$$

If $ad - bc \neq 0$, then $\frac{dw}{dz} \neq 0$ for any z and therefore Bilinear transformation is conformal for all z i.e., it maps z-plane conformally onto the w-plane.

If $ad - bc = 0$, then $\frac{dw}{dz} = 0$ for any z. Then every point of z-plane is critical and the function is not conformal.

For a choice of the constants a, b, c, d, we get special cases of Bilinear transformation as

$w = z + b$ Translation

$w = az$ Rotation

$w = az + b$ Linear transformation

$w = \frac{1}{z}$ Inversion in the unit circle

Thus B.T. can be considered as combination of these transformations.

Solving (1) for z, we find that inverse of the Bilinear transformation is

$$z = \frac{dw - b}{-cw + a} \tag{3}$$

which is again a Bilinear transformation.

From (1), observe that the point $z = -\frac{d}{c}$ corresponds to $w = \infty$, point at infinity in the w-plane. Similarly from (3), the point $w = \frac{a}{c}$ corresponds to $z = \infty$, point at infinity in the z-plane.

Fixed (or Invariant) Points

Fixed (or invariant) points of function $w = f(z)$ are points which are mapped onto themselves. i.e., $w = f(z) = z$.

Example:

$w = z$ has every point a fixed point,

$w = \bar{z}$, infinitely many

$w = \frac{1}{z}$ has two

$w = z + b$ has no fixed point.

To obtain the fixed points of (1), solve

$$z = \frac{az + b}{cz + d}$$

which is a quadratic in z given by

$$cz^2 - (a - d)z - b = 0 \tag{4}$$

Thus the roots say α, β of (4) are the fixed points of (1). If the two roots of (4) are equal then the Bilinear transformation is said to be parabolic.

The quadratic with α, β as roots is

$$z^2 - (\alpha + \beta)z + \alpha\beta = 0$$

For any complex constant γ,

$$z^2 - (\alpha + \beta)z + \gamma z - \gamma z + \alpha\beta = 0$$

$$z\Big(z - (\alpha + \beta - \gamma)\Big) = \gamma z - \alpha\beta$$

$$z = \frac{\gamma z - \alpha\beta}{z - (\alpha + \beta) + \gamma}$$

Thus the bilinear transformations whose fixed points α, β are given by

$$w = \frac{\gamma z - \alpha\beta}{z - (\alpha + \beta) + \gamma} \tag{5}$$

For various values γ, (5) gives B.T. with fixed points α, β.

Theorem 1: *Circles are transformed into circles under Bilinear transformation.*

Proof: By division,

$$w = \frac{az + b}{cz + d} = \frac{a\left(z + \frac{d}{c}\right) + b - \left(\frac{ad}{c}\right)}{c\left(z + \frac{d}{c}\right)}$$

or $$w = \frac{a}{c} + \frac{bc - ad}{c^2} \cdot \frac{1}{z + \frac{d}{c}}.$$

Put $w_1 = z + \frac{d}{c}$, $w_2 = \frac{1}{w_1}$, $w_3 = \frac{bc - ad}{c^2} w_2$

Then $w = w_3 + \frac{a}{c}$.

Here w_1, w_2, w_3, w are translation, inversion, rotation and translation respectively. All these transformations preserve circles. Thus the bilinear transformation can be considered as combination of translation, rotation, stretching and inversion passing from z-plane to w_1-plane to w_2-plane to w_3-plane to w-plane. Hence under B.T. circles transform into circles.

The *cross-ratio* or *anharmonic ratio* of four numbers z_1, z_2, z_3, z_4 is the linear fraction given by

$$\frac{(z_1 - z_2)(z_3 - z_4)}{(z_1 - z_4)(z_3 - z_2)}.$$

Theorem 2: *The cross-ratio of four points is invariant under a bilinear transformation.*

Proof: Suppose w_1, w_2, w_3, w_4 are respectively the images of z_1, z_2, z_3, z_4 under the bilinear transformation

$$w = \frac{az + b}{cz + d}$$

Then
$$w_i - w_j = \frac{az_i + b}{cz_i + d} - \frac{az_j + b}{cz_j + d}$$

$$w_i - w_j = \frac{(ad - bc)(z_i - z_j)}{(cz_i + d)(cz_j + d)} \tag{6}$$

Now consider the cross-ratio of w_1, w_2, w_3, w_4, and use (6). Then

$$\frac{(w_1 - w_2)(w_3 - w_4)}{(w_1 - w_4)(w_3 - w_2)} = \frac{\frac{(ad-bc)(z_1-z_2)}{(cz_1+d)(cz_2+d)} \frac{(ad-bc)(z_3-z_4)}{(cz_3+d)(cz_4+d)}}{\frac{(ad-bc)(z_1-z_4)}{(cz_1+d)(cz_4+d)} \frac{(ad-bc)(z_3-z_2)}{(cz_3+d)(cz_2+d)}}$$

$$= \frac{(z_1 - z_2)(z_3 - z_4)}{(z_1 - z_4)(z_3 - z_2)}$$

Hence the cross-ratio is preserved under bilinear transformation.

Determination of Bilinear Transformation

A bilinear transformation can be uniquely determined by three given conditions. Although four constants a, b, c, d appear in (1), essentially they are three ratios of three of these constants to the fourth one.

To find the unique bilinear transformation which maps three given distinct points z_1, z_2, z_3 onto three distinct images w_1, w_2, w_3, consider w which is the image of a general point z under this transformation. Now by Theorem 2, the cross-ratio of the four points w_1, w_2, w_3, w must be equal to the cross-ratio of z_1, z_2, z_3, z. Hence the unique bilinear transformation that maps three given points z_1, z_2, z_3 onto three given images w_1, w_2, w_3 is given by

$$\frac{(w_1 - w_2)(w_3 - w)}{(w_1 - w)(w_3 - w_2)} = \frac{(z_1 - z_2)(z_3 - z)}{(z_1 - z)(z_3 - z_2)}.$$

Note 1: If one of these points is ∞ then the quotient of the two differences containing this point must be replaced by 1.

Note 2: The B.T. $w = e^{i\alpha}\left(\frac{z-z_0}{z-\bar{z}_0}\right)$ maps the half plane $y \geq 0$ onto the unit disk $|w| \leq 1$ for any arbitrary real α and $\text{Im}(z_0) > 0$.

WORKED OUT EXAMPLES

Example 1: Find the bilinear transformation that maps the points $0, 1, i$ in z-plane onto the points $1 + i, -i, 2 - i$ in the w-plane.

Solution: The required bilinear transformation is

$$\frac{(w_1 - w_2)(w_3 - w)}{(w_1 - w)(w_3 - w_2)} = \frac{(z_1 - z_2)(z_3 - z)}{(z_1 - z)(z_3 - z_2)}$$

$$\frac{(1 + i + i)(2 - i - w)}{(1 + i - w)(2 - i + i)} = \frac{(0 - 1)(i - z)}{(0 - z)(i - 1)}$$

$$\frac{(1 + 2i)}{2}\frac{(2 - i - w)}{(1 + i - w)} = (i - 1)\left(\frac{i - z}{z}\right)$$

$$\frac{2 - i - w}{1 + i - w} = \frac{2(3i + 1)}{5}\left(\frac{i - z}{z}\right)$$

Solving for w,

$$5z(2 - i - w) = 2(3i + 1)(1 + i - w)(i - z)$$

or
$$w = \frac{(6i + 2)(1 + i)(i - z) - (2 - i)5z}{-5z + (6i + 2)(i - z)}$$

$$w = \frac{z(6 + 3i) + (8 + 4i)}{z(7 + 6i) + (6 - 2i)}.$$

Example 2:

a. Determine the linear fractional transformation that sends the points $z = 0, -i, 2i$ into the points $w = 5i, \infty, -\frac{i}{3}$ respectively.

b. What are the invariant points of this transformation. Find the image of $|z| < 1$ (interior of a unit circle) under this transformation.

Solution:

a. $\frac{(0+i)(2i-z)}{(0-z)(2i+i)} = \frac{(5i-w_2)\left(-\frac{i}{3}-w\right)}{(5i-w)\left(-\frac{i}{3}-w_2\right)}$

where $w_2 = \infty$. So

$$\frac{z-2i}{3z} = \lim_{w_2 \to \infty} \frac{\left(\frac{5i}{w_2}-1\right)\left(+\frac{i}{3}+w\right)}{(5i-w)\left(-\frac{i}{3w_2}+1\right)}$$

$$\frac{z-2i}{3z} = \frac{\frac{i}{3}+w}{w-5i}.$$

Note: If one of points, in this case $w_2 = \infty$, then the quotient of the two differences which contain w_2 i.e., $\left(\frac{5i-w_2}{-\frac{i}{3}-w_2}\right)$ is replaced by 1 (which gives the above result).

Solving for w, we get

$$\frac{2i}{z} = \frac{2w+6i}{5i-w}$$

or $$w = \frac{-3iz-5}{z+i} = \frac{-3z+5i}{-iz+1}$$

b. Invariant points are given by

$$w = z = \frac{-3z+5i}{-iz+1}$$

$z^2 + 4iz + 5 = 0$ which has two roots.

$z = \frac{-4i \pm 6i}{2} = i, -5i$ are the invariant points.

c. Rewriting the bilinear transformation (Fig. 25.37)

$$z = \frac{w-5i}{iw-3}$$

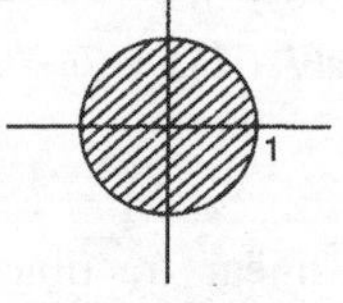

Fig. 25.37

The image of $|z| < 1$ is given by

$$|z| = \left|\frac{w-5i}{iw-3}\right| < 1$$

$$|w-5i| < |iw-3|$$

$$|u + i(v-5)| < |-(3+v) + iu|$$

or $$u^2 + (v-5)^2 < (3+v)^2 + u^2$$

$$1 < v$$

Thus the interior of the unit circle $|z| = 1$ in the z-plane is mapped to the upper half plane above the line $v = 1$ (Fig. 25.38).

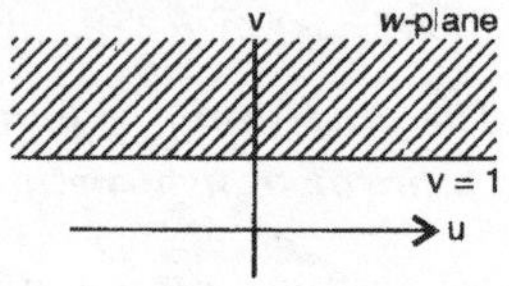

Fig. 25.38

Example 3: Determine the Möbius transformation having 1 and i as fixed (invariant) points and maps 0 to -1.

Solution: The Möbius transformation having α and β as fixed points is given by

$$w = \frac{\gamma z - \alpha\beta}{z - \alpha - \beta + \gamma}$$

for various values of γ. For $\alpha = 1, \beta = i$, we have

$$w = \frac{\gamma z - i}{z - 1 - i + \gamma}$$

Since $z = 0$ is mapped to $w = -1$,

$$-1 = \frac{0-i}{0-1-i+\gamma}$$

or $\gamma = 2i + 1$

Thus the required transformation is

$$w = \frac{(2i+1)z - i}{z+i}$$

Example 4: Find a Bilinear transformation which maps the upper half of the z-plane into the interior of a unit circle in the w-plane. Verify the transformation (Fig. 25.39).

Solution: Suppose any three points in the upper half of z-plane say $A: -1$, $B: 0$, $C = 1$ gets mapped to any three points in the interior of the circle $|w| = 1$

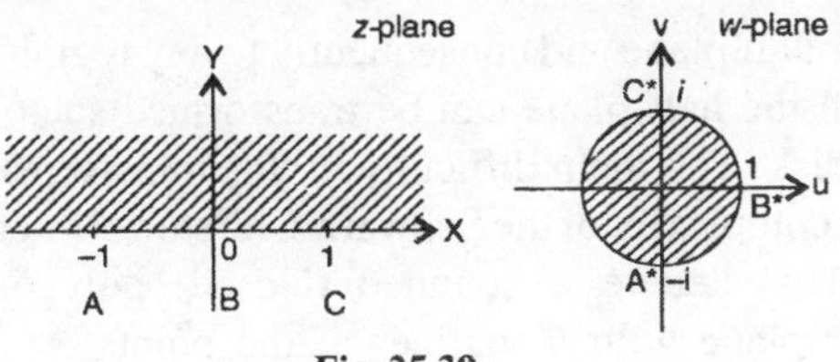

Fig. 25.39

in the w-plane, say $A^* : -i$, $B^* : 1$, $C^* : i$. Thus the required bilinear transformation is the one which maps $-1, 0, 1$ from z-plane to $-i, 1, i$ in the w-plane. This is

$$\frac{(-1-0)(1-z)}{(-1-z)(1-0)} = \frac{(-i-1)(i-w)}{(-i-w)(i-1)}$$

or
$$\frac{1-z}{1+z} = \frac{1+iw}{i+w}$$

Solving
$$w = \frac{i-z}{i+z}$$

Verification :
$$|w| = \left|\frac{i-z}{i+z}\right| \le 1$$

or
$$|i-z| \le |i+z|$$

$$\sqrt{x^2+(1-y)^2} \le \sqrt{x^2+(1+y)^2}$$

$$4y \ge 0$$

Thus the bilinear transformation $w = \frac{i-z}{i+z}$ transforms interior of unit circle in w-plane onto the upper half plane in z-plane.

Also
$$|w| = \left|\frac{i-z}{i+z}\right| = \sqrt{\frac{x^2+(1-y)^2}{x^2+(1+y)^2}}$$

For $y=0$, $|w| = \frac{x^2+1}{x^2+1} = 1$. Thus the real axis ($y=0$) gets mapped to the unit circle $|w| = 1$.

EXERCISE

1. Represent $w = \frac{z+i}{iz+4}$ as a composite of mappings.

Ans. $w = w_4 - i$, $w_4 = 5i\,w_3$, $w_3 = \frac{1}{w_2}$, $w_2 = w_1 + 4$, $w_1 = iz$, w_4, w, are rotations, w_2 is translation, w_3 is inversion.

2. Determine the cross-ratio (C.R.) of
   a. the fourth roots of -1.

Ans. $z_1 = \frac{1+i}{\sqrt{2}}$, $z_2 = \frac{1-i}{\sqrt{2}}$, $z_3 = \frac{-1-i}{\sqrt{2}}$, $z_4 = \frac{-1+i}{\sqrt{2}}$, $CR = -1$.

   b. Four complex sixth roots of 1.

Ans. $z_{1,2,3,4} = \pm\frac{1}{2} \pm \frac{\sqrt{3}}{2}i$, $CR = -\frac{1}{3}$.

3. Find the invariant (fixed) points of the transformation:

   a. $w = \frac{z-1}{z+1}$.
   b. $w = \frac{6z-9}{z}$.
   c. $w = (z-i)^2$.
   d. $w = z^2$.
   e. $w = \frac{(2z-5)}{(z+4)}$.

Ans. **a.** $z = \pm i$.
b. $z = 3$.
c. $z = \frac{(1+2i)\pm\sqrt{1+4i}}{2}$.
d. $z = 0, 1$.
e. $z = -1 \pm 2i$.

4. Determine the bilinear transformations whose fixed points are

   a. $1, -1$
   b. $i, -i$.
   c. $1, 1$.

Ans. **a.** $w = \frac{\gamma z+1}{z+\gamma}$, for various γ.
i.e., $\gamma = 0$, $w = \frac{1}{z}$,
$\gamma = 1$, $w = \frac{z+1}{z+1}$, $\gamma = 2$, $w = \frac{2z-1}{z+2}$.
b. $w = \frac{\gamma z-1}{z+\gamma}$.
c. $w = \frac{\gamma z-1}{z-2+\gamma}$.

Find the bilinear transformation that maps z_1, z_2, z_3 onto w_1, w_2, w_3 respectively:

5. $z = -1, 0, 1$ onto $w = 0, i, 3i$

Ans. $w = \frac{-3i(z+1)}{(z-3)}$

6. $z = 0, -i, -1$ onto $w = i, 1, 0$

Ans. $w = -i\left(\frac{z+1}{z-1}\right)$

7. $z = 1, i, -1$ onto $w = 2, i, -2$

Ans. $w = \frac{-6z+2i}{iz-3}$

8. $z = \infty, i, 0$ onto $w = 0, i, \infty$

Ans. $w = -\frac{1}{z}$

9. $z = 1, 0, -1$ onto $w = i, 1, \infty$

Ans. $w = \frac{(-1+2i)z+1}{z+1}$

10. $z = 0, 1, \infty$ onto $w = -1, -i, 1$

Ans. $w = \frac{z-i}{z+i}$

11. $z = 0, i, \infty$ onto $w = 0, \frac{1}{2}, \infty$

Ans. $w = -\frac{iz}{2}$

12. $z = -1, i, 1+i$ onto $w = 0, 2i, 1-i$

Ans. $w = \frac{-2i(z+1)}{4z-1-5i}$

13. $z = -1, \infty, i$ onto $w = \infty, i, 1$

Ans. $w = \frac{iz+2+i}{z+1}$

Find the bilinear transformation

14. whose fixed points are $\frac{1}{2}$ and 2 and maps $\frac{(5+3i)}{4}$ into ∞.

Ans. $w = \frac{z(1-4i)-2(1-i)}{2z(1-i)-(4-i)}$

15. having i as double fixed point and 1 goes to ∞

Ans. $w = \frac{\left[z(3-i)-(1+i)\right]}{\left[(1+i)(1-z)\right]}$

16. which maps $-1, 0, 1$ into $1, i, -1$.

Determine the image of the upper half plane.

Ans. $w = \frac{(z-i)}{(iz-1)}$, unit circle

17. which maps $z = 1, i, -1$ onto $w = i, 0, -i$. Find the image of $|z| < 1$. Determine fixed points.

Ans. $w = \frac{(1+iz)}{(1-iz)}$, $u > 0$, entire right half plane, fixed points are $-\frac{\left[1+i\pm\sqrt{6i}\right]}{2}$.

25.10 SCHWARZ*-CHRISTOFFEL** TRANSFORMATION

The determination of the specific function, which transforms conformally one given region to another, is a very difficult task. However, Schwarz-Christoffel transformation determines functions which conformally maps bounded (or unbounded) polygons to upper half plane and consequently to any region into which the half-plane can be transformed such as a unit disk. The main difficulty in this transformation is the complexity of the Schwarz-Christoffel integral.

Let P denote a bounded (closed) polygon in the w-plane with n vertices at the points A_1, A_2, ..., A_n. Let $\alpha_1\pi$, $\alpha_2\pi$, ..., $\alpha_n\pi$ be the interior angles at these vertices respectively. Here α_1, α_2, ..., α_n are positive constants. For a closed polygon, the sum of the interior angles is $(n-2)\pi$. Thus $(\alpha_1\pi + \alpha_2\pi + \ldots + \alpha_n\pi) = (\alpha_1 + \alpha_2 + \ldots + \alpha_n)\pi = (n-2)\pi$. So $\sum_{i=1}^{n} \alpha_i = n-2$ with $n > 2$ and $0 < \alpha_i < 2$. Then the Schwarz-Christoffel transformation is defined by the function

$$w = f(z) = c\int_0^z \prod_{k=1}^{n} (z-a_k)^{\alpha_k - 1} dz + c_1 \quad (1)$$

transforms conformally the upper half plane in the z-plane onto the interior of the polygon P (see Fig. 25.40). Here $a_1, a_2, \ldots, a_n$ are points on the x-axis (in z-plane), corresponding to the vertices A_1, A_2, ..., A_n of the polygon P. The real numbers a_1, a_2, ..., a_n are arranged in increasing such that $-\infty < a_1 < a_2 < \cdots < a_n < \infty$.

Also c and c_1 are complex constants.

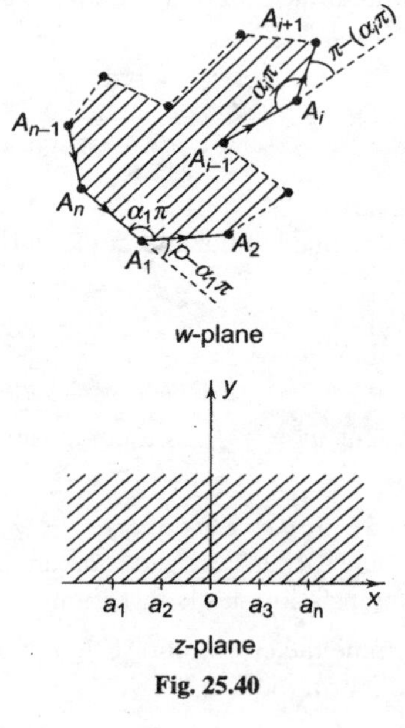

Fig. 25.40

*H.A. Schwarz (1843-1921), German mathematician
**E.B. Christoffel (1892-1900), Swiss mathematician

The points a_i lying on the real axis in the z-plane are singularities of the function (1). Points a_i's are transformed by the function (1) into points A_i's of the w-plane. In order to show that the line segments $a_k < x < a_{k+1}$ of the real axis in the z-plane are mapped by (1) onto straight line segments $A_k A_{k+1}$ of the polygon P of the w-plane, consider

$$\frac{dw}{dz} = f'(z) = c(z-a_1)^{\alpha_1-1}(z-a_2)^{\alpha_2-1}\times \cdots \times (z-a_n)^{\alpha_n-1} \quad (2)$$

we need the following two propositions.

Proposition I: Recall that a transformation $w - A_1 = (z-a_1)^{\alpha_1}$ with $\frac{dw}{dz} = \alpha_1(z-a_1)^{\alpha_1-1}$ maps the straight line Ba_1D on the real axis to $B^x A_1 D^*$ with $(\alpha_1)\pi$ angle at the vertex A_1.

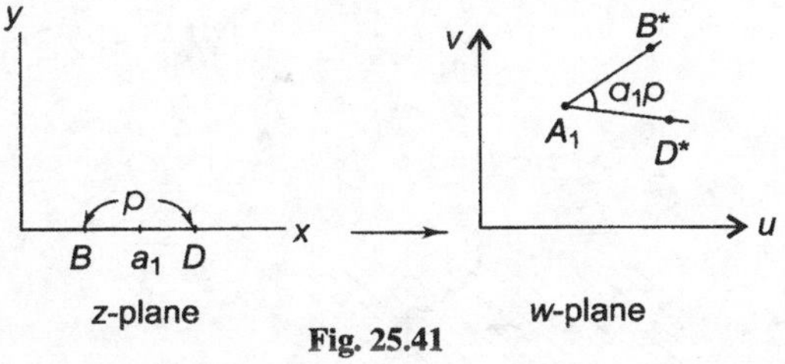

Fig. 25.41

Proposition II: Recall that if S at w_0 is the image of the curve c at z_0 under the transformation $w = f(z)$ which is analytic and $f'(z_0) \neq 0$, then the tangent to S at w_0 is rotated through an angle arg $f'(z_0)$. Thus the size of the angle through which tangent to S at w_0 to be rotated is determined by arg $f'(z_0)$.

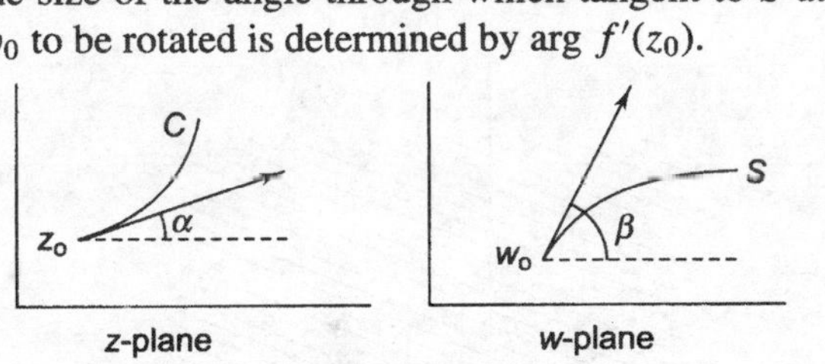

Fig. 25.42

(i.e.,) $\beta = \alpha + \arg f'(z_0)$. If c is straight line along x-axis, then $\alpha = 0$ and if $f'(z_0) =$ constant, then S is also a straight line in w-plane.

Now from (2) $\arg dw = \arg c + (\alpha_1 - 1)\arg(z - a_1) + (\alpha_2 - 1)\arg(z - a_2) + \ldots + (\alpha_n - 1)\arg(z - a_n) + \arg dz$ (3)

To determine the size of the angles between adjacent segments say $A_{i-1}A_i$ and A_iA_{i+1} of the polygonal line, consider the variation of the argument of $f'(z)$.

For $z = x < a_1$, the numbers $(z-a_1)$, $(z-a_2)$, ..., $(z-a_n)$ are all negative real numbers and hence have π as their arguments. Generalizing this, we have

$$\arg(z-a_i) = \begin{cases} \pi \text{ if } z < a_i \\ 0 \text{ if } z > a_i \end{cases}$$

Note that dz is positive and its argument is zero. Thus when $z < a_1$, all the terms in (3) are constants and therefore arg dw remains constant. Thus the image point w traces a straight line in the w-plane.

Now as soon as z passes through a_1, a_1 becomes less than z, so $(z-a_1)$ changes abruptly from negative to positive and therefore arg $(z-a_1)$ changes abruptly from π to 0. Thus straight line A_nA_1 bends through an angle $\pi - \alpha_1\pi$ since argument of dw changes by $\pi - \alpha_1\pi$. Thus the interior angle at A_1 of the polygon P is $\alpha_1\pi$. Now for $a_1 < z < a_2$, arg dw remains constant, but when z passes through a_2, then $a_2 < z$ so $(z-a_2)$ is positive with argument zero. Then arg dw changes by $\pi - \alpha_2\pi$ and so on. Thus when the point z traverses the entire real axis in the positive direction, its image w makes a complete circuit of the boundary of a closed polygon with interior angles $\alpha_1\pi, \alpha_2\pi, \ldots \alpha_n\pi$. Hence (1) maps the real axis of the z-plane onto some closed polygon line $A_1A_2\ldots A_n$ whose sides are the straight line segments A_kA_{k+1}. Since w is analytic, (1) is conformal. Thus the Schwarz-Christoffel transformation which maps the upper half plane in z-axis onto the interior of a closed polygon in the w-plane is

$$\boxed{w = f(z) = c\int_0^z (t-a_1)^{\alpha_1-1}(t-a_2)^{\alpha_2-1}\times \ldots \times (t-a_n)^{\alpha_n-1}dt + c} \quad (5)$$

Here t is a dummy variable. The transformation (5) may be written as

$$w = f(z) = c\, g(z) + c_1$$

where $g(z)$ maps upper half plane in z-plane to the interior of some polygon in the w-plane. The second transformation $c\,g(z) + c_1$ then translates and rotates (contract $c < 0$ or magnify for $c > 0$) the polygon.

Note: To determine (5), three points of $a_1, a_2, \ldots, a_n$ may be chosen arbitrarily.

In the case of an unbounded (open) polygon, the vertex A_n may be considered as a point at ∞. Then (5) gets modified to

$$w = f(z) = c\int_0^z (t-a_1)^{\alpha_1-1}(t-a_2)^{\alpha_2-1}\times \ldots \times (t-a_{n-1})^{\alpha_{n-1}-1}dt + c_1 \quad (6)$$

The integrand in the Schwarz-Christoffel integral (6) does *not* contain the factor corresponding to the vertex A_n at infinity.

Proposition III: Recall that the bilinear transformation $w = e^{i\theta}\frac{z-z_0}{z-\bar{z}_0}$, $\text{Im} z_0 > 0$, $\theta \in R$ maps the upper half plane onto the interior of a unit disc $c : |w| = 1$.

To show that an interior point of the polygon in w-plane is mapped onto a interior point in the upper half plane, because of the above Proposition III, it is sufficient to prove that the interior point of the polygon is mapped onto an interior point of the unit disc $c : |z| = 1$. By Cauchy's integral formula, for any interior point b of the polygon P, we have

$$\frac{1}{2\pi i}\int_P \frac{1}{w-b}dw = 1.$$

Since $w - b = f(z) - b$ and $dw = f'(z)dz$, we have

$$1 = \frac{1}{2\pi i}\int_P \frac{dw}{w-b} = \frac{1}{2\pi i}\int_c \frac{f'(z)dz}{f(z)-b}$$

From the argument principle, the number of zeros of $f(z) - b$ inside c is one. Thus there is an interior point z_0 of c such that $f(z_0) = b$. Thus the interior point b of the polygon is mapped onto an interior point z_0 of c and hence the interior point of the upper half plane.

Worked Out Examples

Example 1: Determine the transformation that will map the region in the w-plane shown in the Fig. 25.43 onto the upper half plane of the z-plane. Obtain the transformation for (a) $\alpha = 0$ (b) $\alpha = \frac{\pi}{2}$.

Solution: From the Fig. 25.43, the interior angles of the polygon at the points $A(0, 0)$ is $\alpha_1\pi = \frac{\pi}{2} = \frac{1}{2}$ and at the point $B(0, b)$ is $\alpha_2\pi = (\pi + \alpha)$. The point A maps to $A^*(0, 0)$ and B maps to $B^*(1, 0)$ in the z-plane. Therefore the Schwarz-Christoffel transformation takes the form

$$w(z) = c\int (z-0)^{k_1}(z-1)^{k_2}dz + c_1$$

Here $k_1 = \alpha_1 - 1 = \frac{1}{2} - 1 = -\frac{1}{2}$, $k_2 = \alpha_2 - 1 = \frac{\pi+\alpha}{\pi} - 1 = \frac{\alpha}{\pi}$

Thus $w(z) = c\int_0^z t^{-\frac{1}{2}}(t-1)^{\frac{\alpha}{\pi}}dt$ where t is the new dummy variable. Since B maps to B^*, $w = ib$ when $z = 1$. So

$$ib = c\int_0^1 t^{-\frac{1}{2}}(t-1)^{\frac{\alpha}{\pi}}dt$$

or $c = ib/I$ where $I = \int_0^1 t^{-\frac{1}{2}}(t-1)^{\frac{\alpha}{\pi}}dt$

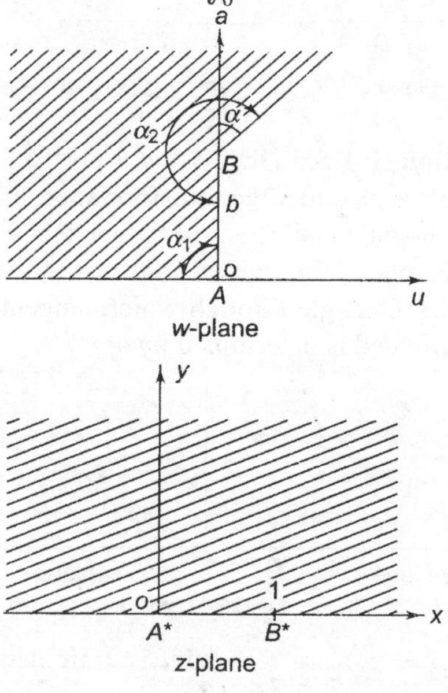

Fig. 25.43

Thus the required Schwarz-Christoffel transformation is

$$w = c\int_0^z t^{-\frac{1}{2}}(t-1)^{\frac{\alpha}{\pi}}dt$$

where $c = \frac{ib}{I}$ with $I = \int_0^1 t^{-\frac{1}{2}}(t-1)^{\frac{\alpha}{\pi}}\,dt.$

Case i For $\alpha = 0$, $I = \int_0^1 t^{-\frac{1}{2}}dt = 2\sqrt{t}\big|_0^1 = 2$ so $c = \frac{ib}{2}$. Then

$$w = \frac{ib}{2}\int_0^z t^{-\frac{1}{2}}dt = \frac{ib}{2}\,2\sqrt{t}\Big|_0^z = ib\sqrt{z}$$

Case ii For $\alpha = \frac{\pi}{2}$, $I = \int_0^1 t^{-\frac{1}{2}}(t-1)^{\frac{1}{2}}dt$

or $I = i\int_0^1 t^{\frac{1}{2}-1}(1-t)^{\frac{3}{2}-1}dt = i\beta\left(\frac{1}{2},\frac{3}{2}\right) = \frac{\Gamma\left(\frac{1}{2}\right)\Gamma\left(\frac{3}{2}\right)}{\Gamma\left(\frac{4}{2}\right)}$.

$I = i\frac{\sqrt{\pi}\cdot\frac{1}{2}\sqrt{\pi}}{1} = i\frac{\pi}{2}$. Then $c = \frac{ib}{i\frac{\pi}{2}} = \frac{2b}{\pi}$

Now $\int_0^z t^{-\frac{1}{2}}(t-1)^{\frac{1}{2}}dt = i\int_0^z t^{-\frac{1}{2}}(1-t)^{\frac{1}{2}}dt$

put $\sqrt{t} = x$, $\frac{1}{2}\frac{1}{\sqrt{t}}dt = dx$ so

$$i\int_0^z (1-t)^{\frac{1}{2}}\frac{dt}{\sqrt{t}} = 2i\int_0^z \sqrt{1^2 - x^2}\,dx$$

$$= 2i\left[\frac{1}{2}\sin^{-1}x + \frac{x\sqrt{1-x^2}}{2}\right]\Bigg|_{x=0}^{\sqrt{z}}$$

$$= 2i\left[\frac{1}{2}\sin^{-1}\sqrt{z} + \frac{\sqrt{z}\sqrt{1-z}}{2}\right]$$

Thus the transformation is

$$w(z) = \frac{2b}{\pi}\left[i\left\{\sin^{-1}\sqrt{z} + \sqrt{z(1-z)}\right\}\right]$$

Example 2: Determine the integral which maps the rectangle in the w-plane shown in Fig. 25.44 onto the upper half of the z-plane.

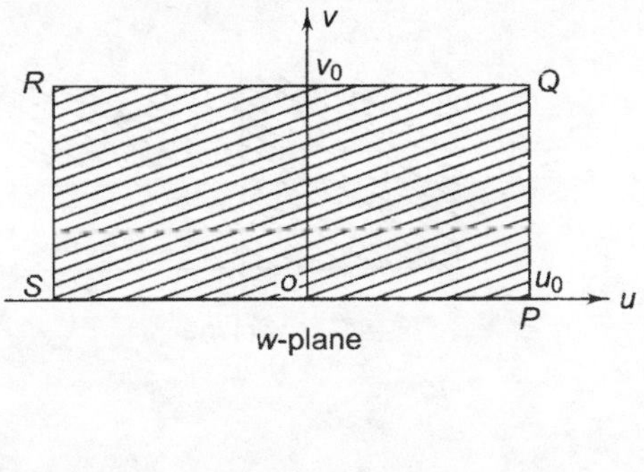

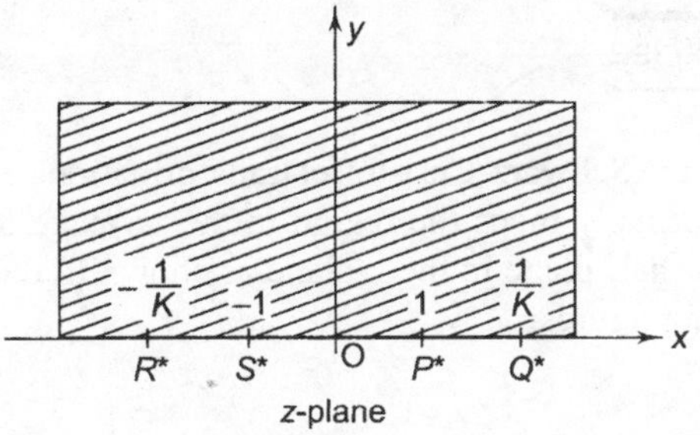

Fig. 25.44

Solution: Let the vertices of the rectangle $PQRS$ be $P(u_0, 0)$, $Q(u_0, v_0)$, $R(-u_0, v_0)$, $S(-u_0, 0)$. The corresponding interior angles in the rectangle $PQRS$ are each $\frac{\pi}{2}$ i.e., $\alpha_1\pi = \alpha_2\pi = \alpha_3\pi = \alpha_4\pi = \frac{\pi}{2}$. Suppose the vertices P, Q, R, S maps to $P^*(1, 0)$, $Q^*\left(\frac{1}{k}, 0\right)$. $S^*(-1, 0)$, $R^*\left(-\frac{1}{k}, 0\right)$ respectively. Then $k_i = \alpha_i - 1 = \frac{1}{2} - 1 = -\frac{1}{2}$, for i, 1, 2, 3, 4.
Then the required Schwarz-Christoffel transformation is

$$w(z) = c\int_0^z (z-1)^{-\frac{1}{2}}\left(z-\frac{1}{k}\right)^{-\frac{1}{2}}(z+1)^{-\frac{1}{2}}\left(z+\frac{1}{k}\right)^{-\frac{1}{2}}dz$$

$$= c\int_0^z \frac{dt}{\sqrt{(t^2-1)\left(t^2-\frac{1}{k^2}\right)}}$$

where t is the dummy variable.

$$\text{or}\quad w(z) = kc\int_0^z \frac{dt}{\sqrt{(1-t^2)(1-k^2t^2)}}.$$

Since $P(u_0, 0)$ maps to $P^*(1, 0)$, we have $w = u_0$ when $z = 1$. Then $u_0 = kc\int_0^1 \frac{dt}{\sqrt{1-t^2}\sqrt{1-k^2t^2}}$.

Then $w(z) = \frac{u_0}{I}\int_0^z \frac{dt}{\sqrt{(1-t^2)(1-k^2t^2)}}$ where

$$I = \int_0^1 \frac{dt}{\sqrt{(1-t^2)(1-k^2t^2)}}$$

This integral $w(z)$ is known as *elliptic integral* of the first kind, which can *not* be expressed in terms of elementary functions.

EXERCISE

Find the Schwarz-Christoffel transformations which conformally maps the region in the w-plane to the upper half plane in the z-plane as shown in the following figures:

1.

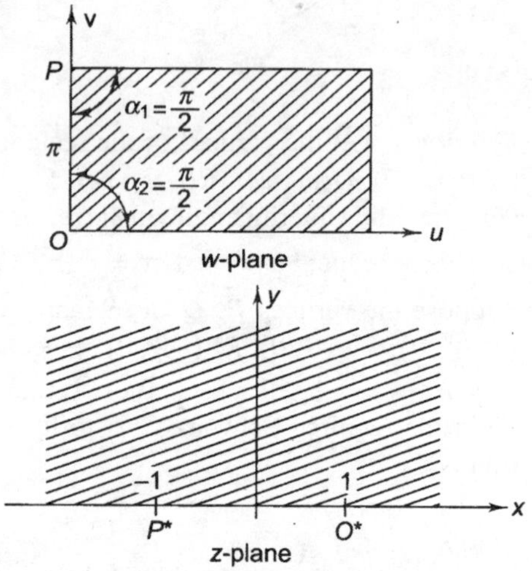

Fig. 25.45

Semi-unifinte (right-open) strip $0 \le v \le \pi$.

Ans. $w = \cosh^{-1} z$ (or $z = \cosh w$).

Hint: $k_1 = k_2 = -\frac{1}{2}$, $w = k\int_0^z \frac{dt}{\sqrt{(t-1)(t+1)}}$, $k = 1$ since $P(0, i\pi)$ maps to $P^*(-1, 0)$ i.e., $w = i\pi$ when $z = -1$.

2.

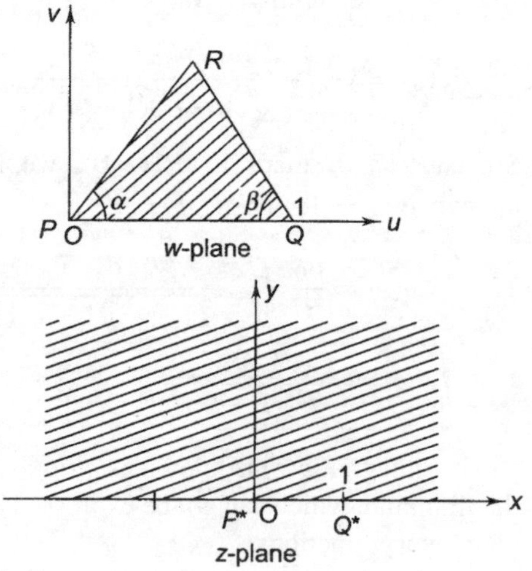

Fig. 25.46

Ans. $w(z) = \frac{1}{c_1}\int_0^z t^{k_1}(1-t)^{k_2}dt$ where $k_1 = \frac{\alpha}{\pi} - 1$, $k_2 = \frac{\beta}{\pi} - 1$, $c_1 = \beta\left(\frac{\alpha}{\pi}, \frac{\beta}{\pi}\right)$.

Hint: For interior angles α, β; $k_1 = \frac{\alpha}{\pi} - 1$, $k_2 = \frac{\beta}{\pi} - 1$, $w(z) = c\int_0^z (t-0)^{k_1}(t-1)^{k_2}dt$.
Note: For an isosceles triangle with $\alpha = \beta$, the transformation reduces to $w(z) = \frac{1}{c_1}\int_0^z \{t(1-t)\}^{k_1}\,dt$ where $k_1 = \frac{\alpha}{\pi} - 1$, $c_1 = \beta\left(\frac{\alpha}{\pi}, \frac{\alpha}{\pi}\right)$

3.

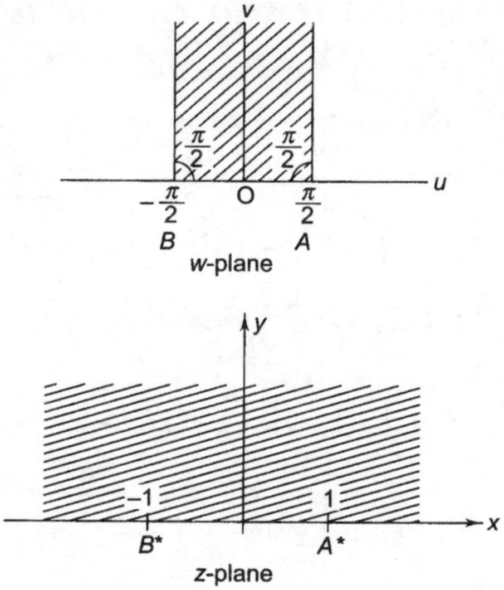

Fig. 25.47

Ans. $w = \sin^{-1} z$ or $z = \sin w$

Hint: Interior angles at A and B are $\frac{\pi}{2}$ each. So $k_1 = k_2 = \frac{\alpha_i}{\pi} - 1 = \frac{\pi}{2}\cdot\frac{1}{\pi} - 1 = -\frac{1}{2}$. Then $w(z) = c\int_0^z (z-1)^{-\frac{1}{2}}(z+1)^{-\frac{1}{2}}dz$ $= c\int_0^z \frac{dt}{\sqrt{(t^2-1)}}$, $w(z) = ci\sin^{-1} z$, $c = 0$ since B maps to B^*, $u = -\frac{\pi}{2}i$ when $x = -1$.

4.

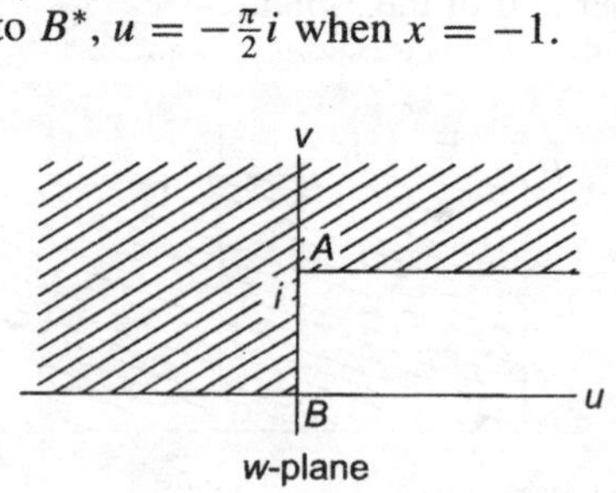

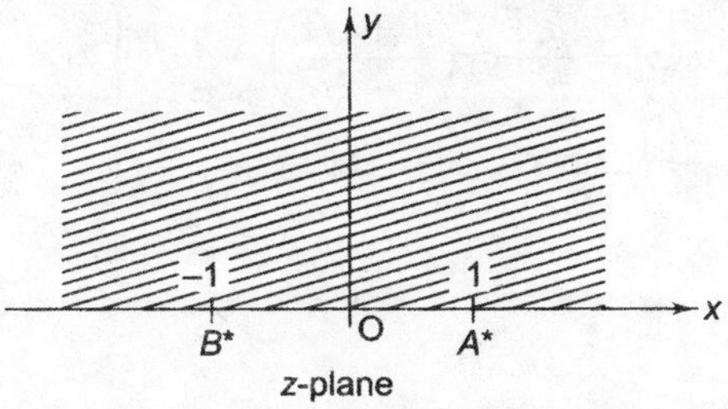

Fig. 25.48

Open region above (and to the left the) polygonal boundary in the w-plane.

Ans. $w(z) = c \int_0^z \sqrt{\dfrac{t-1}{t+1}}\,dt$

Hint: Here interior angles at B, $\alpha_1 = \frac{\pi}{2}$, at $A, \alpha_2 = \frac{3\pi}{2}$ so $k_1 = \frac{\alpha_1}{\pi} - 1 = \frac{\pi}{2}\frac{1}{\pi} - 1 = -\frac{1}{2}$, $k_2 = \frac{\alpha_2}{\pi} - 1 = \frac{3\pi}{2}\frac{1}{\pi} - 1 = \frac{1}{2}$.

5.

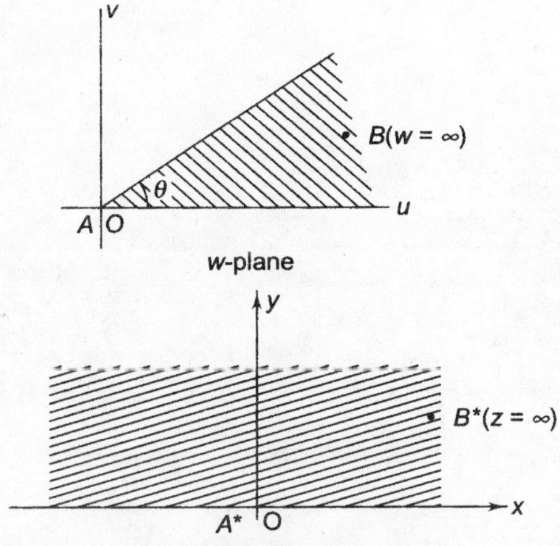

Fig. 25.49

Sector: $0 < \arg w = \theta < \alpha\pi, 0 < \alpha < 2$.

Ans. $w(z) = \frac{c}{\alpha} z^\alpha$

Hint: Sector is polygon with vertices at $A(w = 0)$ and $B(w = \infty)$, mapped to $A^*(z = 0)$ and $B^*(z = \infty)$. Then $w(z) = c \int_0^z (t-0)^{k_1} dt$ where $k_1 = \dfrac{\alpha\pi}{\pi} - 1 = \alpha - 1$

so $w(z) = c \displaystyle\int_0^z t^{\alpha - 1} dt = c \frac{t^\alpha}{\alpha}\Big|_0^z = \frac{c}{\alpha} z^\alpha.$

Note: In the integrand of the Schwarz-Christoffel integral, the factor corresponding to the point B is omitted.

6.

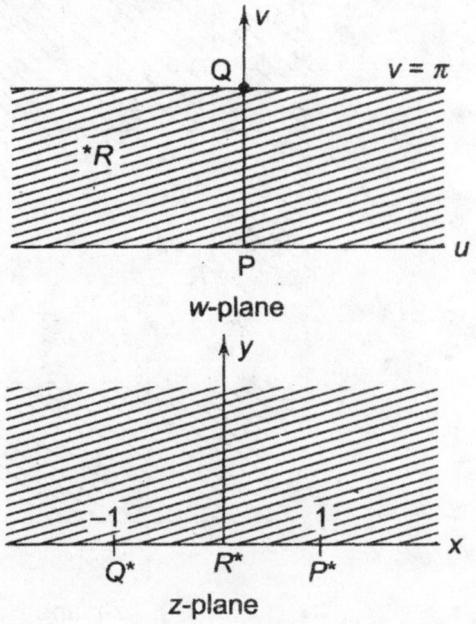

Fig. 25.50

Ans. $w = \ln z$

Hint: Interior angles at $P(0, 0)$ and $Q(0, \pi)$ are π each so $k_1 = \frac{\alpha_1}{\pi} - 1 = \frac{\pi}{\pi} - 1 = 0, k_2 = 0$. The interior angle for point $R(w = \infty)$ is 0 so $k_3 = \frac{\alpha_3}{\pi} - 1 = \frac{0}{\pi} - 1 = -1$. Then $w(z) = \int_0^z (t-1)^0 (t-0)^{-1} (t+1)^0 dt$, $w = \int_0^z \frac{dt}{t} = \ln z$.

Note: The infinite strip may be considered as the limiting form of a rhombus with two of its opposite vertices moved to infinity.

7.

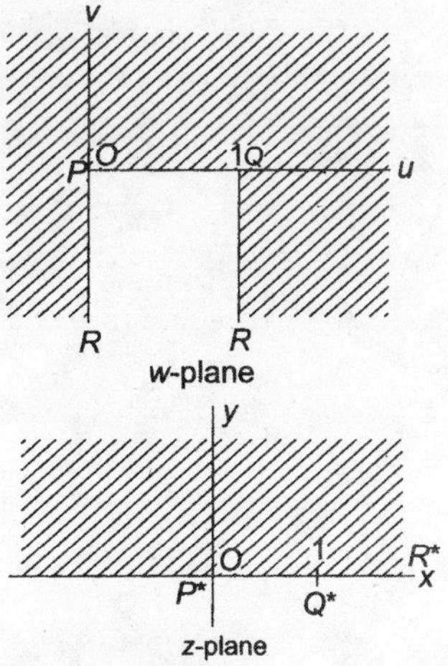

Fig. 25.51

Ans. $w(z) = \frac{2}{\pi}[\sin^{-1}\sqrt{z} - (1-2z)\sqrt{z-z^2}]$

8.

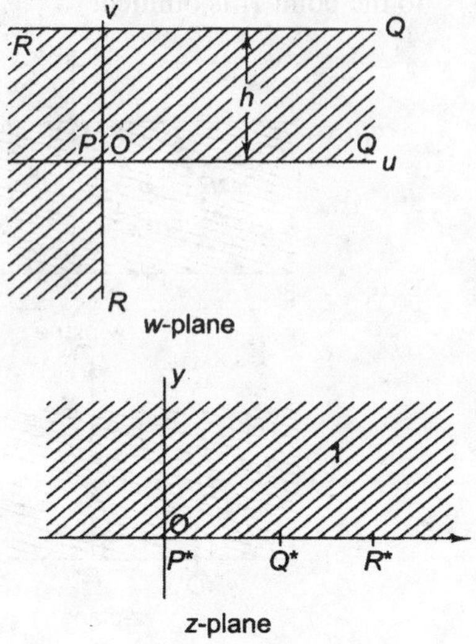

Fig. 25.52

Ans. $w(z) = \frac{h}{\pi}\left[\ln\left(\frac{1+\sqrt{z}}{1-\sqrt{z}}\right) - 2\sqrt{z}\right] = \frac{2h}{\pi}\left[\tanh^{-1}\sqrt{z} - \sqrt{z}\right]$

PART—VII

PROBABILITY AND STATISTICS

Chapter 26

Probability

INTRODUCTION

In random phenomena, past information no matter how voluminous, will not allow to formulate a rule to determine precisely (uniquely) what will happen in future. The theory of probability is the study of such random phenomena which are not deterministic. In analyzing and interpreting data that involves an element of "chance" or uncertainty, probability theory plays a vital role in the theory and application of statistics. Blaise Pascal in the middle of 17th century was the first to use probability in problems of gambling. Laplace, De Moivre, Gauss, Poisson and Kolmogorov greatly contributed to the development of probability theory which finds application in engineering, biology, economics, computer science, politics, traffic control, medicine, meteorology, psychology, agriculture, geography and management of natural resources. We consider the Baye's theorem (also known as theorem of inverse probability) which determines the probability of "causes."

26.1 REVIEW OF SET THEORY

Set

A *Set* is well-defined collection of objects. Sets are denoted by capital letters $A, B, C, \ldots$ and the objects also known as *elements* or *members* of the set by small letters $x, y, \ldots$. If x is a member or element of A, it is denoted by $x \in A$. If y is *not* an element of B, then $y \notin B$.

Example 1: V = set of all vowels in English alphabet = $\{a, e, i, o, u\}$
2. N = set of natural numbers = $\{0, 1, 2, 3, \ldots\}$
3. E = set of even numbers = $\{0, 2, 4, 6, 8, \ldots\} = \{2n | n \in N\}$

Null set

A ***Null set*** or ***empty set*** is the unique set containing *no* elements. It is denoted by ϕ or $\{\}$.

Example 1: The set of women presidents of India is an empty set.
2. $\{n \in N : 3 < n < 4\}$ is an empty set
3. $\{x \in \text{Real} : x^2 + 3 = 0\}$ is an empty set.

Equality

Two sets are said to be equal if they contain the same elements.

Note: The order of listing is irrelevant. There is no advantage (or harm) in listing the elements more than once.

Example: Sets $A = \{1, 3, 5, 6\}$, $B = \{6, 5, 3, 1\}$, and $C = \{3, 5, 3, 6, 1, 6, 1, 5, 5\}$ are all equal. Neither order nor repetition is relevant.

Finite and Infinite Sets

If the number of elements in a set A is finite, then the set is said to be a finite set. The number of distinct elements in a finite set A is known as the *cardinality* or *size* of A and is denoted by $|A|$.

Example: |Set of vowels| = 5, Ex : $|\phi| = 0$, Ex : $|A| = |B| = |C| = 4$, Ex : $D = \{x^2 | x \in N, x^2 < 30\} = \{1, 4, 9, 16, 25\}$, then $|D| = 5$ when the number of elements is infinite, the set is said to be an infinite set.

Example: Set of natural numbers, set of real numbers are infinite sets.

Universal Set

Universal Set denoted by U is set containing all objects under study (or consideration).

Subset

A is a subset of B and is written as $A \subseteq B$ if every element of A as an element of B.

Example: ϕ is subset of any set A, i.e., $\phi \subseteq A$.

Example: Set of even numbers is a subset of set of integers.

Example: A is subset of itself: $A \subseteq A$.

Proper Subset

If B contains elements not in A, then the subset A is said to be proper subset of B, denoted by $A \subset B$.

Example: $A = \{4, 5, 6, \}$, $B = \{6, 7\}$, $C = \{4, 5, 6, 7\}$ then (a) $A \subseteq C$ (b) $A \subset C$ (c) $B \subseteq C$ (d) $B \subset C$ (e) $B \nsubseteq A$ (f) $A \subseteq A$ (g) $A \not\subset A$ (i.e., A is *not* a proper subset of A itself) (h) $\phi \subset A$ (i) $\phi \subset B$ (k) $\phi \subset C$

Note: Only a set can be a subset of another set, while only elements can be members of a set.

Power Set

Power Set of A, denoted by $P(A)$, is the set (collection) of *all* subsets of A.

Example: $A = \{4, 5, 6, 7\}$ then the power set of A is $P(A) = \{\phi, \{4\}, \{5\}, \{6\}, \{7\}, \{4, 5\}, \{4, 6\}, \{4, 7\}, \{5, 6\}, \{5, 7\}, \{6, 7\}, \{4, 5, 6\}, \{4, 5, 7\}, \{4, 6, 7\}, \{5, 6, 7\}, \{4, 5, 6, 7\}\}$.

Result 1: If $|A| = n$, then the power set of a set A with n elements has 2^n elements i.e., $|P(A)| = 2^{|A|} = 2^n$

Example: $|A| = 4$, $P(A) = 2^4 = 16$
$P(\phi) = \{\phi\}$
$P(\{\phi\}) = \{\phi, \{\phi\}\}$

Result 2: For any $0 \le k \le n$, there are n_{c_k} subsets of size k.

Example: $A = \{4, 5, 6, 7\}$, $|A| = n = 4$, let $k = 2$. Then there are $n_{c_k} = 4_{c_2} = 6$ subsets of each size 2 namely $\{4, 5\}, \{4, 6\}, \{4, 7\}, \{5, 6\}, \{5, 7\}, \{6, 7\}$.
Ex: If a set A has 63 proper subsets then $|A| = 6$ since A is not a proper subset of A, there are 64 subsets = 2^6 so $|A| = 6$.
Ex: If set B has 64 subsets of odd cardinality, i.e., B has $2^6 (= 64)$ subsets of odd cardinality, so $|B| = 6 + 1 = 7$.

In general if B has 2^n subsets of odd cardinality then $|B| = n + 1$.

Set Operations

Union of A and B

$A \cup B = \{x | x \in A \text{ or } x \in B \text{ or both}\}$

Intersection of A and B

$A \cap B = \{x | x \in A \text{ and } x \in B\}$
Symmetric difference of A and B:
$A \Delta B = \{x | x \in A \cup B \text{ and } x \notin A \cap B\}$

Disjoint

A and B are disjoint or mutually disjoint when $A \cap B = \phi$.

Example: $A = \{4, 5, 6, 7, 8\}$, $B = \{3, 4, 6, 7\}$, $C = \{7, 8, 9\}$, $D = \{9, 10, 11\}$
Then (a) $A \cup B = \{3, 4, 5, 6, 7, 8\}$ (b) $A \cap B = \{4, 6, 7\}$ (c) $A \cup C = \{4, 5, 6, 7, 8, 9\}$, (d) $A \cap C = \{7, 8\}$, (e) $B \cap C = \{7\}$ (f) $A \cap D = \phi$ (g) $B \cap D = \phi$ (h) $C \cap D = \{9\}$ (i) $A \Delta B = \{3, 5, 8\}$ (j) $A \Delta C = \{4, 5, 6, 9\}$.

***Relative Complement of** A **in** B* is
$B - A = \{x \mid x \in B \text{ and } x \notin A\}$

Example: (a) $B - A = \{3\}$ (b) $A - B = \{5, 8\}$
(c) $C - D = \{7, 8\}$ (d) $D - C = \{10, 11\}$

Complement

Complement of A (w.r.t. the universal set U) is denoted by $\overline{A}$ of A^C or $(U - A)$ is given by
$\{x | x \in U \text{ and } x \notin A\}$.

Inverse Laws

(a) $A \cup A^C = U$ (b) $A \cap A^C = \phi$

Example: $U = \{1, 3, 7, 9, 11\}$, $A = \{1, 3, 9\}$, $B = \{9, 11\}$, $C = \{1\}$ then
(a) $A^C = \{7, 11\}$ (b) $B^C = \{1, 3, 7\}$
(c) $C^C = \{3, 7, 9, 11\}$ (d) $A \cup A^C = \{1, 3, 9\} \cup \{7, 11\} = \{1, 3, 7, 9, 11\} = U$ (e) $A \cap A^C = \{1, 3, 9\} \cap \{7, 11\} = \phi$.

Result: De Morgan's Laws
(a) $(A \cup B)^C = A^C \cap B^C$
(b) $(A \cap B)^C = A^C \cup B^C$

Counting

Result: (a) $|A \cup B| = |A| + |B| - |A \cap B|$
(b) $|A \cup B \cup C| = |A| + |B| + |C| - |A \cap B| - |B \cap C| - |C \cap A| + |A \cap B \cap C|$.

Example: In a class, if 30 are studying mathematics, 25 are studying computer science and 15 are studying both, how many students are in the class?
$|A| = 30$, $B = 25$
$|A \cap B| = 15$
$|A \cup B| = |A| + |B| - |A \cap B|$
$|A \cup B| = 30 + 25 - 15 = 40$ = No. of students in the class.

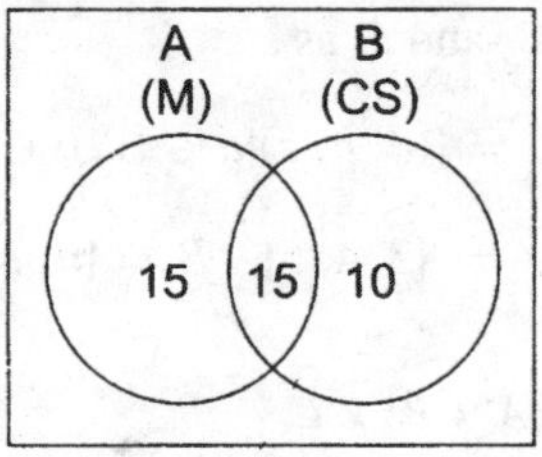

Fig. 26.1

Example: In a class of 57, if 23 students are studying Mathematics, 26 Physics, 30 Chemistry, 7 studying both Maths and Physics, 8 both Maths and Chemistry, 10 both Physics and Chemistry, find how many students are studying all three—Maths, Physics and Chemistry.
$|M \cup P \cup C| = 57$

$|M| = 23$, $|P| = 26$,
$|C| = 30$, $|M \cap P| = 7$
$|M \cap C| = 8$, $|P \cap C| = 10$
So

$$|M \cup P \cup C| = |M| + |P| + |C| - |M \cap P| - |M \cap C| - |P \cap C| + |M \cap P \cap C|$$

Then

$$57 = 23 + 26 + 30 - 7 - 8 - 10 + |M \cap P \cap C|$$

$|M \cap P \cap C| = 3 =$ no. students studying all three—Maths, Physics and Chemistry.

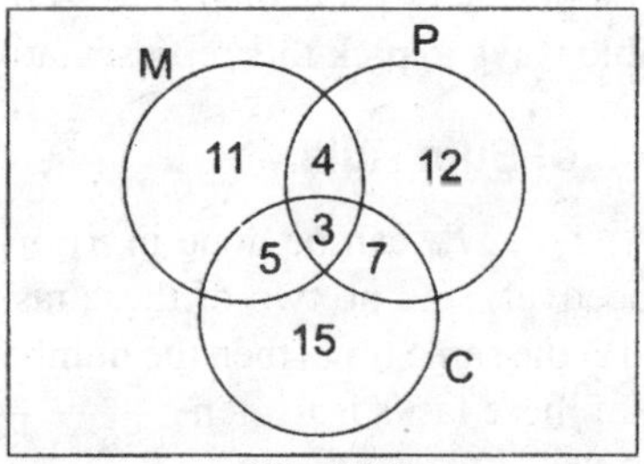

Fig. 26.2

Cartesian Product

Cartesian Product of two sets A and B, denoted by $A \times B$ is the set of all ordered pairs (a, b) where $a \in A$ and $b \in B$.

Example: $A = \{1, 2, 3\}$, $B = \{a, b\}$ then $A \times B = \{(1, a), (1, b), (2, a), (2, b), (3, a), (3, b)\}$

Note: It $|A| = m$, $|B| = n$, then $A \times B$ has $m \cdot n$ ordered pairs. Similarly we have cartesian product of three sets A, B, and C as

$$A \times B \times C = \{(a, b, c) | a \in A, b \in B, c \in C\}$$

Example: $A = \{3, 4, 5\}$, $B = \{d, e\}$, $c = \{0, 1\}$ then

$$\begin{aligned} A \times B \times C &= (A \times B) \times C \\ &= \{(3, d), (3, e), (4, d), (4, e), (5, d), (5, e)\} \times C \\ &= \{(3, d, 0), (3, e, 0), (4, d, 0), (4, e, 0), (5, d, 0), \\ &\quad (5, e, 0), (3, d, 1), (3, e, 1), (4, d, 1), (4, e, 1), \\ &\quad (5, d, 1), (5, e, 1)\} \end{aligned}$$

26.2 REVIEW OF COUNTING

The Sum Rule

If a first task can be done in n_1 ways and a second task in n_2 ways and if these two tasks cannot be performed simultaneously, then there are $n_1 + n_2$ ways of performing either task.

Example: Suppose a university representative is to be chosen either from 200 teaching or 300 non-teaching employees. Then there are $200 + 300 = 500$ possible ways to pick this representative.

Extension of Sum Rule

If tasks $T_1, T_2 \ldots T_m$ can be done in $n_1, n_2, \ldots, n_m$ ways respectively and no two of these tasks can be performed at the same time, then the number of ways to do *one* of these tasks is $n_1 + n_2 + \cdots + n_m$.

Example: If a student can choose a project from either 20 from Mathematics or 35 from computer science or 15 from Engineering then the student can choose *a* project in $20 + 35 + 15 = 70$ ways.

Fundamental principle of counting:

The Product Rule

Suppose a procedure can be broken down into two tasks T_1 and T_2. If the first task T_1 can be performed in n_1 ways and the second task T_2 can be performed in n_2 ways *after* the first task T_1 has been done, then the total procedure can be carried out, in the designated order, in $n_1 \cdot n_2$ ways.

Example: A tourist can travel from Hyderabad to Tirupati in 4 ways (by plane, train, bus or taxi). He can travel from Tirupati to Tirumala hills in 5 ways (by bus, taxi, walk, rope way or motor cycle). Then the tourist can travel from Hyderabad to Tirumala hills in $4 \times 5 = 20$ ways.

Extension of Product Rule

Suppose a procedure consists of performing tasks $T_1, T_2, \ldots, T_m$ in that order. Suppose task T_i can be performed in n_i ways, *after* the tasks $T_1, T_2, \ldots, T_{i-1}$ are perform then the number of ways the procedure can be executed in the designated order is

$$n_1 \cdot n_2 \cdot n_3 \ldots n_m.$$

Example 1: 'Charmas' brand shirt comes in 12 colors, has a male and female version, comes in 4 sizes for each sex comes in three makes economy, standard and luxury. Then the number of different types of shirts produced are $12 \times 2 \times 4 \times 3 = 288$ types.

Example 2: The number of elements (n-tuples) in the cartesian product of finite sets $A_1, A_2, \ldots, A_m$ with $n_1, n_2, \ldots, n_m$ elements respectively is $n_1, n_2 \ldots n_m$ i.e., $|A_1 \times A_2 \times \ldots \times A_m| = |A_1| \cdot |A_2| \cdots |A_m| = n_1 \cdot n_2 \ldots n_m$.

Example 3: A hotel offers 12 kinds of sweets, 10 kinds of hot tiffins and 5 kinds of beverages (hot tea, hot coffee, juice, coke, icecream). The breakfast consists of a sweet and a hot beverage or a hot tiffin and cold beverage. The number of ways in which the above breakfast can be ordered is $12 \times 2 + 10 \times 3 = 24 + 30 = 54$. Here we have applied both product rule and sum rule.

Permutation

A *Permutation* of a set of n distinct objects is an ordered arrangement of these n objects.

An *r-permutation* is an ordered arrangement of r elements taken from the n objects.

Example: $A = \{a, b, c, d\}$. Arrangements $dcba$, $cdba$ are permutation of A. Arrangements abc, abd, bcd, dbc, etc. are 3-permutations of A. Arrangements ab, ba, cd, dc, etc. are 2-permutations of A.

Result 1: The *number* of r-permutations of a set with n distinct elements is denoted by $P(n, r)$ and is given by

$$P(n, r) = n(n-1)(n-2)\cdots(n-r+1)$$

$$\boxed{P(n, r) = \frac{n!}{(n-r)!}}, 0 \leq r \leq n$$

Result 2: When $r = n$, $P(n, n) = n!$

Result 3: When $r = 0$, $P(n, 0) = 1$

Note that $P(n, r)$ counts the (linear) arrangements in which the objects *cannot* be repeated.

Example: The number of "words" of three distinct letters can be formed from the letters of the word JNTU is $P(4, 3) = 4P_3 = \frac{4!}{(4-3)!} = 24$.

Example: In how many ways can 8 men and 8 women be seated in a row if

(a) any person may sit next to any other

(b) men and women must occupy alternate seats

(c) Generalize this result for n men and n women.

Solution:

(a) Here men and women are indistinguishable, so these are 16 objects. The number of permutations from 16 objects with 16 chosen is $P(16, 16) = 16! = 20922789890000$

(b) Here men and women are distinct (different)

M	W	M	W	M	W	M	W	M	W	M	W	M	W	M	W
8.	8.	7.	7.	6.	6.	5	5	4	4	3	3	2	2	1	1

The number of ways = $8!8!$

Alternatively women sits first followed by man, which gives another $8!8!$ ways.

Thus the number of ways men and women occupy alternatively is $8!8! + 8!8! = 2(8!)^2 = 3251404800$.

(c) Any person may sit: $(2n)!$

Men and women sit alternatively $2(n!)^2$.

Combination

An r-combination is an unordered selection or combination of r elements from a set with n distinct elements.

The number of combinations of size r from a set of size n is denoted by $C(n, r)$ and is given by

$$\boxed{C(n, r) = \frac{P(n, r)}{r!} = \frac{n!}{r!(n-r)!}}, 0 \leq r \leq n$$

Example: $A = \{a, b, c, d, e\}$. The number of 3-combinations are $C(5, 3) = \frac{5!}{3!2!} = 10$. They are $\{a, b, c\}$, $\{a, b, d\}$, $\{a, b, e\}$, $\{b, c, d\}$, $\{b, c, e\}$, $\{c, d, e\}$, $\{a, c, e\}$, $\{a, c, d\}$ $\{b, d, e\}$, $\{d, e, a\}$. Observe that the order is irrelevant in combinations. Thus $\{a, b, c\}$, $\{a, c, b\}$, $\{b, a, c\}$ $\{b, c, a\}$, $\{c, a, b\}$, $\{c, b, a\}$ are *all* one and the same 3-combination of a, b, c.

Example: A committee of 12 is to be selected from 10 men and 10 women. In how many ways can the selection be carried out if

(a) There are no restrictions

(b) There must be 6 men and 6 women

(c) There must be an even number of women

(d) There must be more women than men

(e) There must be at least 8 men.

Solution:

(a) No distinction between men and women. Problem is to choose 12 out of a set of 20 objects. So the number of ways 12 chosen out of 20 is $C(20, 12) = \frac{20!}{12!8!} = 125970$.

(b) First stage to choose 6 men out of 10, given by $C(10, 6)$. Second stage to choose 6 women out of 10 again $C(10, 6)$. Using product rule, the number of ways in which the committee will consist of 6 men and 6 women is $C(10, 6) \cdot C(10, 6) = (210)(210) = 44100$.

(c) If $2i$ even number of women are chosen, then the remaining $12 - 2i$ members of the committee should be men. By product rule $C(10, 2i)C(10, 12 - 2i)$ then the total number of ways is

$$\sum_{i=1}^{5} \binom{10}{12-2i}\binom{10}{2i}$$

(d) Since the strength of the committee is 12, there must be 7 or more women in the committee so that there are more women than men in the committee. Using product rule, the number of ways is

$$\sum_{i=7}^{10} \binom{10}{i}\binom{10}{12-i}$$

(e) By similar argument

$$\sum_{i=8}^{10} \binom{10}{i}\binom{10}{12-i}$$

Permutation with Repetition

The number of arrangements of r objects from n objects with repetition is n^r.

Example: *String* is an arrangement made up of prescribed *alphabet* symbols.

By the rule of product there are 4^n strings of length n for the alphabet 0, 1, 2, 3. Thus the collection of all strings of length 10 made up from the alphabet 0,1, 2, 3 is 4^{10}.

0	0	0	1	2	3	1	2	3	3

is one such string.

Combination with Repetition

The number of combinations of n objects taken r at a time, with repetition is $C(n+r-1, r) = \frac{(n+r-1)!}{r!(n-1)!}$

1. Permutation: $P(n, r) = \frac{n!}{(n-r)!}$, $0 \le r \le n$ (order important, no repetition)
2. Arrangement: n^r, $n, r \ge 0$ (order important, repetition allowed)
3. Combination: $C(n, r) = \frac{n!}{r!(n-r)!}$ for $0 \le r \le n$. (order irrelevant, repetition *not* allowed)
4. Combination with repetition

$$\binom{n+r-1}{r} \quad \text{with} \quad n, r \ge 0$$

(order irrelevant)

26.3 INTRODUCTION TO PROBABILITY

A deterministic experiment is an experiment whose outcome or result is known with certainty or predictable, i.e., result is unique.

Example: Ohm's law $I = \frac{E}{R}$ determines the current uniquely (with certainty).

Trial is a single performance of an experiment.

A probablistic or non-deterministic or random experiment is an experiment whose outcome or result is not unique and therefore cannot be predicted with certainty.

Examples:

i. Tossing of a coin, head or tail may occur.
ii. Throwing a die, 1, 2, 3, 4, 5, or 6 may appear.
iii. Tensile strength of beam.
iv. Life-time of a computer system.

Probability is a measure of certainty.

Sample space S of a random experiment is the set of all possible outcomes of the experiment.

Examples:

i. Tossing of coin: $S = \{H, T\}$
ii. Throwing a die: $S = \{1, 2, 3, 4, 5, 6\}$
iii. Tensile strength of beam: $S = \{r \ge 0, r \text{ real}\}$.

Sample or sample point is a particular (outcome) element of S.

Event

Even is a subset of a sample space.

Example: Tossing of a die,
$E_1 = \{\text{odd number}\} = \{1, 3, 5\}$
$E_2 = \{\text{even number}\} = \{2, 4, 6\}$
$E_3 = \{\text{prime number}\} = \{2, 3, 5\}$
$E_4 = \{\text{number greater than } 2\} = \{3, 4, 5, 6\}$

Mutually exclusive events

Two events A and B are mutually exclusive if A and B can not happen (occur) simultaneously, i.e., $A \cap B = \phi$, i.e., A and B are disjoint.

Collectively exhaustive events

A list of events $A_1, A_2, \ldots, A_n$ are said to be collectively exhaustive if $\bigcup_{i=1}^{n} A_i = S$.

Universal event

The entire sample space S is called a universal (or certain or sure) event.

The null set ϕ is called the null or impossible event.

Mathematical or classical or 'a priori' probability

If an event E can happen m ways out of possible n mutually exclusive, collectively exhaustive and equally likely ways then probability of event E, denoted by, $P(E)$ is defined as

$$P(E) = p = \frac{m}{n} = \frac{\text{Favourable cases for } E}{\text{Total cases}}$$

The probability of non-occurrence of event E (called its failure), denoted by $P(\text{not } E)$ or

$$q = p(\text{not} E) = P(\overline{E}) = P(E^c) = P(\sim E) = \frac{n-m}{n}$$
$$= 1 - \frac{m}{n} = 1 - p = 1 - P(E)$$

Thus, $p + q = 1$ and $0 \le p \le 1$ and $0 \le q \le 1$.
Probability of a certain (sure) event is $\frac{n}{n} = 1$.
Probability of an impossible (null) event is $\frac{0}{n} = 0$.

Note: This definition fails when (i) The outcomes are not equally likely and (ii) Number of outcomes is infinite (not exhaustive).

Statistical or Empirical or Estimated (von mises) Probability

$$p = P(E) = \lim_{n \to \infty} \frac{m}{n}$$

where m is the number of times event E happens (occurs) in n trials assuming that the trials are performed under essentially homogeneous and identical conditions.

Note: This definition fails when (i) limit does not exist (ii) or is not unique.

A.N. Kolmogorov (in 1933) developed the axiomatic probability theory which includes the above two approaches as special cases (the classical theory corresponding to equiprobable spaces).

Axioms of Probability

1. For any event A of S

$$0 \le P(A) \le 1$$

i.e., probability is a numerical value lying between 0 and 1.

2. $P(S) = 1$ (sure event).

3. For any two mutually exclusive events A and B in S

$$P(A \cup B) = P(A) + P(B)$$

i.e., probability of sum is the sum of the probabilities (special addition rule).

Based on the above axioms (rules) the remaining probability theory is developed. For example, for any sequence of mutually exclusive events $A_1, A_2, A_3, \ldots$ of S, we have

3a. $P(A_1 \cup A_2 \cup A_3 \cdots) = P(A_1) + P(A_2) + P(A_3) + \cdots$.

Elementary Theorems

Theorem 1: $P(\phi) = 0$.

Proof:

$$S = S \cup \phi$$
$$P(S) = P(S \cup \phi) = P(S) + P(\phi)$$

by axiom 3 since S and ϕ are mutually exclusive

or $\qquad P(\phi) = 0$.

Theorem 2: $P(A^c) = 1 - P(A) \le 1$.

Proof: A and A^c are mutually exclusive and

$$S = A \cup A^c$$
$$1 = P(S) = P(A \cup A^c) = P(A) + P(A^c)$$

where we have used axiom (2) and (3). Hence

$$P(A^c) = 1 - P(A) \le 1$$

since $\qquad P(A) \ge 0$.

Additive Theorem or Rule or General Addition Rule of Probabilities

Theorem 3: *If A and B are any two arbitrary events of S then*

$$P(A \cup B) = P(A) + P(B) - P(A \cap B)$$

Proof:

$$A = (A \cap B^c) \cup (A \cap B)$$
$$B = (A \cap B) \cup (A^c \cap B)$$

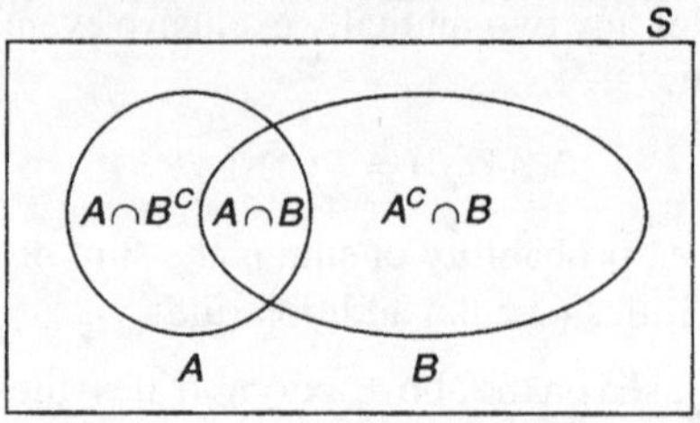

Fig. 26.3

Note that $A \cap B^c$, $A \cap B$ and $A^c \cap B$ are mutually exclusive (i.e., they are mutually disjoint) (refer Fig. 26.3).

Applying axiom (3)

$$P(A) = P(A \cap B^c) + P(A \cap B)$$

$$P(B) = P(A \cap B) + P(A^c \cup B)$$

Adding

$$P(A) + P(B) = \left[P(A \cap B^c) + P(A \cap B) + P(A^c \cap B)\right] + P(A \cap B)$$

Observe that

$$A \cup B = (A \cap B^c) \cup (A \cap B) \cup (A^c \cap B)$$

so $\quad P(A) + P(B) = P(A \cup B) + P(A \cap B)$

or $\quad P(A \cup B) = P(A) + P(B) - P(A \cap B).$

Corollary 1: Since $P(A \cap B) \geq 0$ it follows that

$$P(A \cup B) \leq P(A) + P(B).$$

Corollary 2: General additive rule

Theorem 4: *For any three arbitrary events A, B, C, (Fig. 26.4)*

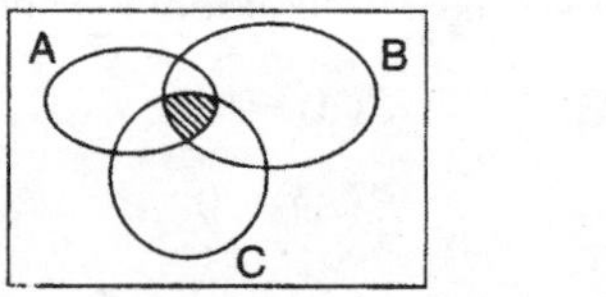

Fig. 26.4

$$P(A \cup B \cup C) = P(A) + P(B) + P(C) - P(A \cap B) - P(B \cap C) - P(C \cap A) + P(A \cap B \cap C)$$

Proof: Applying additive theorem for A and $B \cup C$, we have

$$P(A \cup (B \cup C)) = P(A) + P(B \cup C) - P\left\{A \cap (B \cup C)\right\}$$

Again applying additive theorem for $B \cup C$, we get

$$P(A \cup B \cup C) = P(A) + P(B) + P(C) - P(B \cap C) - P\left\{(A \cap B) \cup (A \cap C)\right\}$$

Now using additive theorem for $(A \cap B) \cup (A \cap C)$, we have

$$P(A \cup B \cup C) = P(A) + P(B) + P(C) - P(B \cap C) - \left\{P(A \cap B) + P(A \cap C) - P\left[(A \cap B) \cap (A \cap C)\right]\right\}$$

$$= P(A) + P(B) + P(C) - P(A \cap B) - P(B \cap C) - P(C \cap A) + P(A \cap B \cap C)$$

since $(A \cap B) \cap (A \cap C) = A \cap B \cap C$

Theorem 5: *If $B \subset A$ then $P(B) \leq P(A)$.*

Proof: $A = B \cup (A \cap B^c)$ and $B, A \cap B^c$ are mutually exclusive

so $\quad P(A) = P(B \cup (A \cap B^c)) = P(B) + P(A \cap B^c)$

i.e., $\quad P(B) \leq P(A).$

Conditional Probability

Although probability of an event E is with reference to the sample space S, on many occasions one is interested in finding the probability of E with respect to a *reduced* sample space. For example, consider the following table:

	Employed	Unemployed	Total
Male	160	140	300
Female	40	80	120
Total	200	220	420

Total sample space is 420 persons. If a person is male, what is the probability that he is unemployed? In this case we should consider only the reduced sample space of males (only) (since it is given or known that the person is male). Thus the required probability is

$$\frac{140}{300} = \frac{\text{unemployed and male}}{\text{male}}$$

Conditional Probability

Conditional probability of an event A given that B has happened, denoted by, $P(A/B)$ is defined as

$$P(A/B) = \frac{P(A \cap B)}{P(B)} \quad \text{if } P(B) > 0.$$

Also read as conditional probability of A given B. Thus for the above example,

$$P(U/M)=\frac{140}{300},\ P(E/M)=\frac{160}{300},$$

$$P(M/E)=\frac{160}{200},\ P(F/E)=\frac{40}{200},\ P(E/F)=\frac{40}{120}\ \text{etc.}$$

General multiplicative rule

$$P(A\cup B)=P(B)\,P(A/B)$$

Similarly, $P(B/A)=\frac{P(A\cap B)}{P(A)}$ if $P(A)>0$

So $P(A\cap B)=P(A)\,P(B/A)$.

Corollary For any events $A_1, A_2, \ldots A_n$ $P(A_1\cap A_2\cap\cdots\cap A_n)=P(A_1)P(A_2/A_1)\times P(A_3/A_1\cap A_2)\cdots P(A_n/A_1\cap A_2\cdots\cap A_{n-1})$

Independent Events

Two events A and B are said to be independent if

$$P(B/A)=P(B)$$

or

$$P(A/B)=P(A)$$

i.e., the occurrence (or non-occurrence) of event A has no influence (or impact) on the occurrence (or non-occurrence) of B. Otherwise they are said to be *dependent*.

Special multiplication rule

If A and B are independent events then

$$P(A\cap B)=P(A)P(B)$$

i.e., in a sense probability of the product is the product of the probabilities.

Corollary For independent events $A_1, A_2, A_3, \ldots$ $P(A_1\cap A_2\cap A_3\cap\cdots)=P(A_1)\,P(A_2)\,P(A_3)\cdots$

Partition

The family of sets $A_1, A_2, \ldots A_n$ is said to be a partition of S if

i. $\bigcup_{i=1}^{n} A_i=S$ (collectively exhaustive)

ii. $A_i\cap A_j=\phi$ for any i, j (mutually disjoint).

Worked Out Examples

Permutations

Example 1: (a) How many car number-plates can be made if each plate contains two different letters followed by three different digits? (b) Solve the problem if the first digit cannot be 0. Solve (a) and (b) with repetitions and without repetitions.

Solution: With repetitions:

a. $26\times 26\times 10\times 10\times 10=6,76,000$

b. $26\times 26\times 9\times 10\times 10=6,08,400$

without repetitions:

a. $26\times 25\times 10\times 9\times 8=4,68,000$

b. $26\times 25\times 9\times 9\times 8=4,21,200.$

Example 2: Determine the number of permutations that can be formed from all the letters of each word (i) queue (ii) committee (iii) proposition (iv) baseball.

Solution:

i. $n=5, n_1=2, n_2=2, \frac{5!}{2!2!1!}=30$

ii. $n=9, n_1(m,m)=2, n_2(t,t)=2, n_3(ee)=2,$ $\frac{9!}{2!2!2!1!1!1!}=45360$

iii. $n=11, n_1=(p,p)=2, n_2=(0,0,0)=3,$ $n_2=(i,i)=2, \frac{11!}{2!3!2!}=1,66,3200$

iv. $\frac{8!}{2!2!2!}=5040$

Combinations

Example 3: From five statisticians and six economists a committee consisting of three statisticians and two economists is to be formed. How many different committees can be formed if

a. No restrictions are imposed?

b. Two particular statisticians must be on the committee?

c. One particular economist can not be on the committee?

Solution:

a. $5c_3\cdot 6c_2=\frac{5!}{2!3!}\frac{6!}{2!4!}=150.$

b. Two particular statisticians are chosen. To choose one statistician out of the remaining three statisticians $3_{C_1} \cdot 6_{C_2} = 45$.

c. One particular economist is barred. So to choose two economists from the remaining five economists $5_{C_3} 5_{C_2} = 100$.

Probability

Example 4: Determine the probability for each of the following events:

a. A non-defective bolt will be found if out of 600 bolts already examined, 12 were defective.

b. At least one head appears in a four tosses of a fair coin.

c. The sum 8 appears in a single toss of pair of fair dice.

d. The sum 7 or 8 or 12 appears in a single toss of a pair of fair dice.

e. A king, ace, jack of clubs or queen of diamonds appear in drawing in a single card from a well shuffled ordinary deck of cards (i.e., without joker).

Solution:

a. $P(\text{defective}) = \frac{12}{600} = \frac{1}{50}$

$P(\text{non-defective}) = 1 - P(\text{defective})$

$= 1 - \frac{1}{50} = \frac{49}{50} = 0.98.$

b. $P(TTTT) = P(T)P(T)P(T)P(T) = \frac{1}{2}\frac{1}{2}\frac{1}{2}\frac{1}{2} = \frac{1}{16}$

$P(\text{at least one head}) = 1 - P\ (\text{all tails})$

$= 1 - P(TTTT)$

$= 1 - \frac{1}{16} = \frac{15}{16}.$

Note: Sample space $= \{H, T\} \times \{H, T\} \times \{H, T\} \times \{H, T\} = 2^4 = 16$.

{*HHHH, HHHT, HHTH, HTHH, HHTT, HTHT, HTTH, HTTT*
THHH, THHT, THTH, TTHH, THTT, TTHT, TTTH, TTTT}

c. The sum 8 can appear in the following cases

(6, 2), (5, 3), (4, 4), (2, 6), (3, 5) i.e., 5 cases

Total number of cases $= 6 \times 6 = 36$ ways.

$P(\text{sum } 8) = \frac{\text{Favourable cases}}{\text{Total cases}} = \frac{5}{36}$

d. Sum 7 appears 6 ways as

(6, 1), (1, 6), (5, 2), (2, 5), (4, 3), (3, 4)

Sum 12 appears in one way as (6, 6)

$P(7 \text{ or } 8 \text{ or } 12) = P(7) + P(8) + P(12) =$

$\frac{6}{36} + \frac{5}{36} + \frac{1}{36} = \frac{12}{36} = \frac{1}{3}.$

e. $P(\text{king}) = \frac{4}{52}$, $P(\text{ace}) = \frac{4}{52}$,

$P(\text{jack of clubs}) = \frac{1}{52}$,

$P(\text{queen of diamonds}) = \frac{1}{52}$,

P(king or ace or jack of clubs or queen of diamonds)

$= P(K) + P(A) + P(\text{jack of clubs})$

$+ P(\text{Queen of diamond})$

$= \frac{4}{52} + \frac{4}{52} + \frac{1}{52} + \frac{1}{52} = \frac{10}{52} = \frac{5}{26}.$

Example 5: Find the probability that at least two 9's appear (as a sum) in four tosses of a pair of fair dice.

Solution: The sum 9 appears in the following four cases: (3, 6), (6, 3), (4, 5), (5, 4).

$p = P$ (sum 9 occurring in a single throw of a pair of fair dice) $= \frac{4}{36} = \frac{1}{9}.$

$q = P(\text{sum 9 not occurring}) = 1 - p = 1 - \frac{1}{9} = \frac{8}{9}.$

Sum 9 appears in all four tosses with probability

$= P(9 \text{ and } 9 \text{ and } 9 \text{ and } 9) = P(9)P(9)P(9)P(9)$

$= \frac{1}{9} \cdot \frac{1}{9} \cdot \frac{1}{9} \cdot \frac{1}{9}.$

Sum 9 appears in three tosses and fails in one toss.

Prob $(9 \text{ and } 9 \text{ and } 9 \text{ and } \tilde{9}) = \frac{1}{9} \cdot \frac{1}{9} \cdot \frac{1}{9} \cdot \frac{8}{9}.$

This can happen in 4_{C_1}, ways.

Sum 9 appears in two tosses and fails in two tosses

Prob $(9 \text{ and } 9 \text{ and } \tilde{9} \text{ and } \tilde{9}) = \frac{1}{9}\frac{1}{9}\frac{8}{9}\frac{8}{9}.$

This can happen 4_{C_2} ways.

Thus the probability of at least two 9's in four tosses $= P$ (all 9's)$+P$ (Three 9's and one $\tilde{9}$) $+P$(Two 9's and two $\tilde{9}$)

$= \frac{1}{9}\frac{1}{9}\frac{1}{9}\frac{1}{9} + 4_{C_1}\left(\frac{1}{9}\frac{1}{9}\frac{1}{9}\frac{8}{9}\right) + 4_{C_2}\left(\frac{1}{9}\frac{1}{9}\frac{8}{9}\frac{8}{9}\right)$

$= \frac{1}{6561}[1 + 4 \cdot 8 + 6 \cdot 64] = \frac{417}{6561}.$

Example 6: If 4 tickets are drawn from tickets numbered 1 to 30 inclusive, determine the probability that the tickets marked 1 and 2 are among the four of them.

Solution: Total cases: 30_{C_4}

Since 1 and 2 must appear among the four, only two other tickets are to be chosen from the remaining 28 tickets in 28_{C_2} ways. Thus probability $= \frac{28_{C_2}}{30_{C_4}}$.

Example 7: Two marbles are drawn in succession from a box containing 10 red, 30 white, 20 blue and 15 orange marbles, with replacement being made after each drawing. Find the probability that (a) both are white (b) first is red and second is white (c) neither is orange. Solve this problem with no replacement after each drawing.

Solution: Total number of marbles $= 10 + 30 + 20 + 15 = 75$. Let $R =$ red, $W =$ white, $B =$ blue, $O =$ orange.

a. $P(\text{first marble drawn is white}) = \frac{30}{75}$.

After replacement, $P(\text{second is white}) = \frac{30}{75}$.

$P(\text{first is white and second is white})$

$P(W) \cdot P(W) = \frac{30}{75}\frac{30}{75} = \frac{4}{25}$

since the first and second drawings of marbles are independent.

b. $P(\text{first is red}) = \frac{10}{75}$.

After replacement, $P(\text{second white}) = \frac{30}{75}$.

$P(\text{first is red and second is white}) = \frac{10}{75} \cdot \frac{30}{75} = \frac{4}{75}$.

c. $P(\text{first is orange}) = \frac{15}{78}$, $P(\text{first is not orange}) = 1 - \frac{15}{75} = \frac{60}{75}$.

$P(\text{second is orange}) = \frac{15}{75}$.

$P(\text{second is not orange}) = 1 - \frac{15}{75} = \frac{60}{75}$.

$P(\text{neither is orange}) = \frac{60}{75} \cdot \frac{60}{75} = \frac{16}{25}$.

When no replacement is made:

a. $P(\text{first is white}) = \frac{30}{75}$.

Since no replacement is made, the second drawing (event) depends (or influenced by) on the first drawing. Thus

$P(\text{second is white}) = \frac{29}{74}$.

$P(\text{both white}) = \frac{30}{75} \cdot \frac{29}{74} = \frac{29}{185}$.

b. $P(\text{first is red}) = \frac{10}{75}$.

$P(\text{second is white}) = \frac{30}{74}$
(similar argument as above).

$P(\text{first red and second white}) = \frac{10}{75} \cdot \frac{30}{74} = \frac{2}{37}$.

c. $P(\text{first is orange}) = \frac{15}{75}$.

$P(\text{first is not orange}) = 1 - \frac{15}{75} = \frac{60}{75}$.

Since no replacement is made, there are 74 marbles of which 15 are orange marbles because no orange marble was drawn in the first drawing.

$P(\text{second is not orange}) = 1 - \frac{15}{74} = \frac{59}{74}$.

$P(\text{neither is orange}) = \frac{60}{75}\frac{59}{74} = \frac{118}{165}$.

Finite probability spaces

Example 8: Three students A, B and C are in a swimming race. A and B have the same probability of winning and each is twice as likely to win as C. Find the probability that B or C wins.

Solution: Given that $P(A) = P(B) = 2P(C)$. Since only three students A, B, C are in race, the probability space is finite,

i.e., $P(A) + P(B) + P(C) = 1$

or $(2 + 2 + 1)P(C) = 1$ so $P(C) = \frac{1}{5}$

and $P(A) = \frac{2}{5} = P(B)$

Now $P(B \cup C) = P(B) + P(C) = \frac{2}{5} + \frac{1}{5} = \frac{3}{5}$

Finite equiprobable spaces

Example 9: Of 10 girls in a class, 3 have blue eyes. If 2 of the girls are chosen at random, what is the probability that (i) both have blue eyes? (ii) neither has blue eyes? (iii) at least 1 has blue eyes?

Solution: All the girls are equally likely or equiprobable (indistinguishable). The number of ways 2 girls can be chosen from 10 girls is 10_{C_2}.

i. Favourable cases for both girls to have blue eyes: 3_{C_2}

Prob (both girls with blue eyes) $= \frac{\text{Favourable cases}}{\text{Total cases}}$

$= \frac{3_{C_2}}{10_{C_2}} = \frac{1}{15}$.

ii. Favourable cases for non-blue eyed girls are 7_{C_2}.

$P(\text{neither has blue eyes}) = \frac{7_{C_2}}{10_{C_2}} = \frac{7}{15}$.

iii. $P(\text{at least one blue eye})$

$= 1 - P(\text{neither}) = 1 - \frac{7}{15} = \frac{8}{15}$.

Example 10: A class consists of 6 girls and 10 boys. If a committee of 3 is chosen at random from the class, find the probability that (i) 3 boys are selected (ii) exactly two boys are selected (iii) at least one boy is selected (iv) exactly two girls are selected.

Solution:

i. $P(3B) = \frac{10_{C_3}}{16_{C_3}} = \frac{3}{14}$

since favourable cases are 10_{C_3} (3 out of 10 boys) and total cases are 16_{C_3} (3 out of total 16 students).

ii. $P(2B, 1G) = \frac{10_{C_2} \cdot 6_{C_1}}{16_{C_3}} = \frac{27}{56}$.

Boys selection is in 10_{C_2} ways, while girls selection is in 6_{C_1} ways. By counting principle, the number of ways these two can happen is $10_{C_2} \cdot 6_{C_1}$.

iii. $P(3G) = \frac{6_{C_3}}{16_{C_3}} = \frac{1}{28}$.

$P(\text{at least one boy}) = 1 - P(3G) = 1 - \frac{1}{28} = \frac{27}{28}$.

iv. $P(2G) = \frac{6_{C_2} \cdot 10_{C_1}}{16_{C_3}} = \frac{15}{56}$.

Independence

Example 11: If A and B are independent, prove that (a) A and B^c are independent (b) A^c and B are independent.

Solution: Given that A and B are independent i.e., $P(A \cap B) = P(A)P(B)$ (see Fig. 26.5)

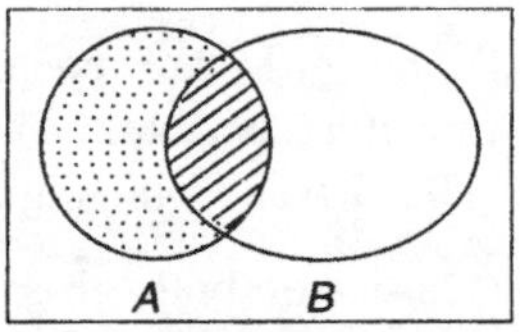

Fig. 26.5

a. $P(A \cap B^c) = P(A) - P(A \cap B)$

$= P(A) - P(A)P(B)$

$= P(A)[1 - P(B)]$

$= P(A)P(B^c)$

so A and B^c are independent.

b. $P(B \cap A^c) = P(B) - P(A \cap B) = P(B) - P(A)P(B)$

$= P(B)[1 - P(A)] = P(B)P(A^c)$

so A and B^c are independent.

Example 12: Two aeroplanes bomb a target in succession. The probability of each correctly scoring a hit is 0.3 and 0.2 respectively. The score will bomb only if the first misses the target. Find the probability that

i. the target is hit
ii. the target is hit by the second plane
iii. both fail to score hits.

Solution: A_1: first aeroplane, A_2: second aeroplane.

i. $P(\text{target is hit}) = P(A_1 \text{ hits})$ or (A_1 fails and A_2 hit) $= P(A_1 \text{ hits}) + P(A_1 \text{ fails and } A_2 \text{ hits})$, by addition theorem.

Since A_1 fails and A_2 hits are independent events, applying multiplication rule, we have

$P(\text{target is hit})$
$= P(A_1 \text{ hits}) + P(A_1 \text{ fails}) \cdot P(A_2 \text{ hits})$
$= 0.3 + (1 - 0.3)(0.2) = 0.44.$

ii. $P(\text{target hit by } A_2) = P(A_1 \text{ fails and } A_2 \text{ hits})$
$= P(A_1 \text{ fails}) \cdot P(A_2 \text{ hits}) = (0.7)(0.2) = 0.14$

iii. $P(\text{both fails}) = P(A_1 \text{ fails and } A_2 \text{ fails})$
$= P(A_1 \text{ fails}) \cdot P(A_2 \text{ fails})$
$= (1 - 0.3)(1 - 0.2) = (0.7)(0.8) = 0.56.$

Example 13: A man alternatively tosses a coin and throws a die beginning with coin. What is the probability that he will get a head before he gets a 5 or 6 on the die?

Solution: $P(\text{Head}) = P(H) = \frac{1}{2} = P\,(\text{Tail}) = P(T)$
$P(5 \text{ or } 6 \text{ on die}) = P(5) + P(6) = \frac{1}{6} + \frac{1}{6} = \frac{1}{3}$

so $P(\text{not getting 5 or 6 on die}) = P(\widetilde{5 \text{ or } 6}) = \frac{2}{3}$

Now he will succeed if

H or (T and $\widetilde{5 \text{ or } 6}$) or [($T$ and $\widetilde{5 \text{ or } 6}$) and ($T$ and $\widetilde{5 \text{ or } 6}$)$H$]

or $\cdots$

These events H, (T and $\widetilde{5 \text{ or } 6}$), etc. are mutually exclusive events.

By addition theorem,

$$P(\text{success}) = P(H \text{ or } (T\ \widetilde{5 \text{ or } 6}\ H) \text{ or } (T\widetilde{5 \text{ or } 6}\ T\ \widetilde{5 \text{ or } 6}\ H)$$
$$\text{or } \cdots$$
$$= P(H) + P(T \text{and } \widetilde{5 \text{ or } 6} H) +$$

$$+P(T \text{ and } \widetilde{5 \text{ or } 6} \text{ and } T \text{ and } \widetilde{5 \text{ or } 6} \text{ and } H)$$
$$+\cdots$$

Since T, $\widetilde{5 \text{ or } 6}$ are independent events, apply multiplication rule,

$$P(\text{success}) = P(H) + P(T)P(\widetilde{5 \text{ or } 6})\,P(H)$$
$$+P(T)P(\widetilde{5 \text{ or } 6})P(T)P(\widetilde{5 \text{ or } 6})P(H) + \cdots$$
$$= \frac{1}{2} + \frac{1\,2\,1}{2\,3\,2} + \frac{1\,2\,1\,2\,1}{2\,3\,2\,3\,2} + \cdots$$
$$= \frac{1}{2}\left[1 + \frac{1}{3} + \left(\frac{1}{3}\right)^2 + \left(\frac{1}{3}\right)^3 + \cdots\right]$$
$$= \frac{1}{2}\left[\frac{1}{1-\frac{1}{3}}\right] = \frac{3}{4}$$

since sum of geometric series is $\frac{a}{1-r}$, $a = 1$ first term, $r = \frac{1}{3}$ common ratio.

Example 14: Box A contains 5 red and 3 white marbles and box B contains 2 red and 6 white marbles (a) If a marble is drawn from each box, what is the probability that they are both of the same colour? (b) If 2 marbles are drawn from each box, what is the probability that all four marbles are of the same colour?

Solution: R_A: red marble from box A.
R_B: red marble from box B, W_A, W_B denote white marble from box A and B respectively.

a. Probability of both marbles same colour=
$P((R_A \text{ and } R_B) \text{ or } (W_A \text{ and } W_B))=$
$P(R_A \cap R_B \text{ or } W_A \cap W_B)$
$= P(R_A \cap R_B) + P(W_A \cap W_B)$
Since R_A, R_B and W_A, W_B are independent events,
$= P(R_A)P(R_B) + P(W_A)P(W_B)$
$= \frac{5}{8}\cdot\frac{2}{8} + \frac{3}{8}\cdot\frac{6}{8} = \frac{7}{16}.$

b. Probability of each of the two marbles same colour
= [(first and second from A are red) and (first and second from B are red)]
or [(first and second from A are white) and (first and second from B are red)]

$$= P((R_{1_A} \cap R_{2_A} \cap R_{1_B} \cap R_{2_B}) \text{ or } (W_{1_A} \cap W_{2_A} \cap W_{1_B} \cap W_{2_B}))$$
$$= P(R_{1_A} \cap R_{2_A} \cap R_{1_B} \cap R_{2_B}) + P(W_{1_A} \cap W_{2_A} \cap W_{1_B} \cap W_{2_B})$$
$$= P(R_{1_A})P(R_{2_A}/R_{1_A}) \cdot P(R_{1_B}) \cdot P(R_{2_B}/R_{1_B})$$
$$+P(W_{1_A})P(W_{2_A}/W_{1_A}) \cdot P(W_{1_B})P(W_{2_B}/W_{1_B}).$$

Note that R_{1_A}, R_{2_A} are dependent events R_{1_B}, R_{2_B} are dependent events but $R_{1_A} \cap R_{2_A}$ and $R_{1_B} \cap R_{2_B}$ are independent events etc.

$$= \left(\frac{5}{8}\cdot\frac{4}{7}\right)\left(\frac{2}{8}\cdot\frac{1}{7}\right) + \left(\frac{3}{8}\cdot\frac{2}{7}\right)\left(\frac{6}{8}\cdot\frac{5}{7}\right) = \frac{55}{784}.$$

Conditional probability

Example 15: A die is tossed. If the number is odd, what is the probability that it is prime?

Solution: 2, 4, 6, are even, 1, 3, 5 are three odd numbers, of which two are prime numbers (namely 3 and 5) (refer Fig. 26.6).

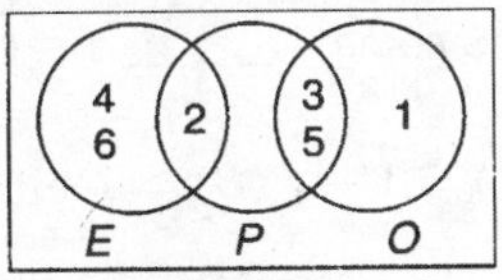

Fig. 26.6

Conditional probability is w.r.t. the reduced sample space of odd numbers only. Thus

$$P(\text{Prime given odd}) = P(P/O) = \frac{\text{Favourable cases}}{\text{Total cases}} = \frac{2}{3}.$$

Example 16: Two digits are selected at random from the digits 1 through 9 (a) If the sum is odd, what is the probability that 2 is one of the numbers selected (b) If 2 is one of the digits selected, what is the probability that the sum is odd?

Solution: We know that even + even = even, odd + odd = even and odd + even = odd.

a. odd = $\{1, 3, 5, 7, 9\}$, even = $\{2, 4, 6, 8\}$
Since the sum is odd it must be the sum of an odd number and an even number i.e., odd= odd+ even

$$P(2 \text{ selected}) = \frac{\text{Favourable cases for 2}}{\text{Total (even) cases}} = \frac{1}{4}.$$

b. 2 is selected, so the other number must be odd number

$P(\text{sum is odd/2 is selected})$

$$= \frac{\text{Favourable case for odd}}{\text{Total cases}} = \frac{5}{8}.$$

Sample space is reduced to 8 since 2 is already selected (removed from 9 digits).

Example 17: If A and B be events with $P(A) = \frac{1}{3}$, $P(B) = \frac{1}{4}$ and $P(A \cup B) = \frac{1}{2}$. Find (a) $P(A/B)$ (b) $P(B/A)$ (c) $P(A/B^c)$ (d) $P(A/B^c)$.

Solution:

a. $P(A/B) = \frac{P(A \cap B)}{P(B)} = \frac{\frac{1}{12}}{\frac{1}{4}} = \frac{1}{3}$

Since $P(A \cup B) = P(A) + P(B) - P(A \cap B)$

$\frac{1}{2} = \frac{1}{3} - \frac{1}{4} - P(A \cap B)$ or $P(A \cap B) = \frac{1}{12}$.

b. $P(B/A) = \frac{P(A \cap B)}{P(A)} = \frac{\frac{1}{12}}{\frac{1}{3}} = \frac{1}{4}$.

c. $P(A \cap B^c) = P(A \backslash B) = P(A) - P(A \cap B) = \frac{1}{3} - \frac{1}{12} = \frac{1}{4}$.

d. $P(A/B^c) = \frac{P(A \cap B^c)}{P(B^c)} = \frac{\frac{1}{4}}{\frac{3}{4}} = \frac{1}{3}$

since $P(B^c) = 1 - P(B) = 1 - \frac{1}{4} = \frac{3}{4}$.

Example 18: In a certain town 40% have brown hair, 25% have brown eyes and 15% have both brown hair and brown eyes. A person is selected at random from the town (Fig. 26.7).

a. If he has brown hair, what is the probability that he has brown eyes also?

b. If he has brown eyes, determine the probability that he does not have brown hair.

c. Determine the probability that he has neither brown hair nor brown eyes.

Solution: Let BH: Brown hair, BE: Brown eyes.

a. Probability that person has brown eyes given that he has brown hair is

$P(BE/BH) = \frac{P(BE \cap BH)}{P(BH)} = \frac{15}{40} = \frac{3}{8}$.

b. $P(BH^c/BE) = \frac{P(BE \cap BH^c)}{P(BE)} = \frac{10}{25} = \frac{2}{5}$.

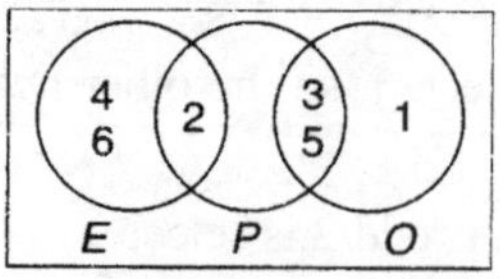

Fig. 26.7

c. $P(\text{neither } BH \text{ nor } BE)$

$= P(BH^c \cap BE^c)$

Using DeMorgan's law $A^c \cap B^c = (A \cup B)^c$, we have

$= P((BH \cup BE)^c) = 1 - P(BH \cup BE)$

by complementation.

$= 1 - \frac{50}{100} = 1 - \frac{1}{2} = \frac{1}{2}$.

Example 19: A box contains 9 tickets numbered 1 to 9 inclusive. If 3 tickets are drawn from the box one at a time, find the probability they are alternatively either odd, even, odd or even, odd, even.

Solution: 4 even numbers $= \{2, 4, 6, 8\}$,

5 odd numbers $= \{1, 3, 5, 7, 9\}$.

Let e denote even and o denote odd number

$$P(oeo \text{ or } eoe) = P(oeo) + P(eoe)$$

since oeo and eoe are mutually exclusive. Now Prob(odd and even and odd) $= P$ (odd) P(even/given odd) P(odd/given odd and even)

i.e., $P(oeo) = P(o)P(e/o)P(o/oe)$

$$= \frac{5}{9} \cdot \frac{4}{8} \cdot \frac{4}{7} = \frac{10}{63}$$

Similarly,

$$P(eoe) = P(e)P(o/e)P(e/eo)$$

$$P(eoe) = \frac{4}{9} \cdot \frac{5}{8} \cdot \frac{3}{7} = \frac{5}{42}$$

So $P(oeo \text{ or } eoe) = \frac{10}{63} + \frac{5}{42} = \frac{5}{18}$.

Example 20: A class has 10 boys and 5 girls. Three students are selected at random, one after the other (see Fig. 23.8). Find probability that

a. first two are boys and third is girl.

b. first and third boys and second is girl.

c. first and third of same sex and the second is of opposite sex.

Solution: $B =$ Boy, $G =$ Girl,
$B_1 =$ first is boy, $B_2 =$ second is boy, etc.

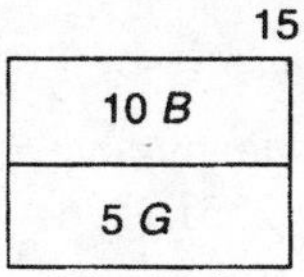

Fig. 26.8

a. $P(B_1 \text{ and } B_2 \text{ and } G_3) = P(B_1B_2G_3)$

$= P(B_1)P(B_2/B_1)P(G/B_1B_2)$

$= \frac{10}{15}\cdot\frac{9}{14}\cdot\frac{5}{13} = \frac{15}{91}.$

b. $P(B_1G_2B_3) = P(B_1)P(G_2/B_1)P(B_3/B_1G_2)$

$= \frac{10}{15}\cdot\frac{5}{14}\cdot\frac{9}{13} = \frac{15}{91}.$

c. $P(B_1G_2B_3 \text{ or } G_1B_2G_3) =$
$P(B_1G_2B_3) + P(G_1B_2G_3)$

since $B_1G_2B_3$ and $G_1B_2G_3$ are mutually exclusive

$= P(B_1)P(G_2/B_1)P(B_3/B_1G_2)$
$+P(G_1)P(B_2/G_1)P(G_3/G_1B_2)$

$= \frac{10}{15}\cdot\frac{5}{14}\cdot\frac{9}{13} + \frac{5}{15}\cdot\frac{10}{14}\cdot\frac{4}{13} = \frac{5}{21}.$

Example 21: Determine whether sex and blood group are independent from the following table:

Blood Group	Male	Female	Total
O	113	113	226
A	103	103	206
B	25	25	50
AB	10	10	20
Total	251	251	502

Solution:

$$P(O) = \frac{226}{502} = \frac{113}{251} = P(O/M)$$

$$P(M) = \frac{251}{502} = \frac{1}{2} = P(M/O) = \frac{113}{226}$$

So the events of blood group O and Male are independent events.
Similarly,

$$P(A) = \frac{206}{502} = P(A/M) = \frac{103}{251}$$

$$P(M) = \frac{251}{502} = \frac{1}{2} = P(M/A) = \frac{103}{206}$$

So group *A* and male sex are independent.
Similarly,

$$P(F) = \frac{251}{502} = \frac{1}{2} = P(F/O) = \frac{113}{226}$$

$$P(O) = \frac{226}{502} = \frac{113}{251} = P(O/F) = \frac{113}{251} \text{ etc.}$$

So *O* group and female sex are independent.

EXERCISE

Permutation

1. Let $A = \{a, b\}$, $B = \{4, 3, 5\}$, $C = \{0, 1\}$. Construct the "tree diagram" of $A \times B \times C$ and then find $A \times B \times C$ (refer Fig. 26.9).

Ans. "tree" is constructed from left to right. $A \times B \times C$ consists of the ordered triples listed to the right of the "tree".

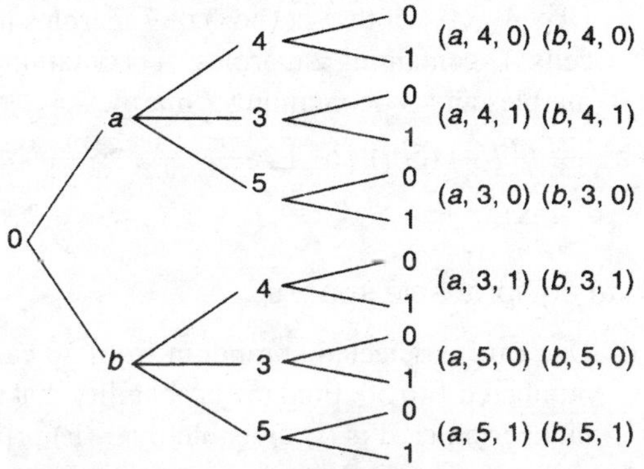

Fig. 26.9

2. If repetitions are not permitted (i) How many 3-digit numbers can be formed from the six digits 2, 3, 5, 6, 7 and 9? (ii) How many of these are less than 400? (iii) How many are even? (iv) How many are odd? (v) How many are multiples of 5?

Ans. (i) $6\cdot5\cdot4 = 120$ (ii) $2\cdot5\cdot4 = 40$ (iii) $5\cdot4\cdot2 = 40$ (iv) $5\cdot4\cdot4 = 80$ (v) $5\cdot4\cdot1 = 20$

3. (a) In how many ways can 3 boys and 2 girls sit in a row? (b) In how many ways can they sit in a row if the boys and girls are each to sit

together? In how many ways can they sit in a row if just girls sit together?

Ans. **a.** $5 \cdot 4 \cdot 3 \cdot 2 \cdot 1 = 5! = 120$
b. *BBBGG* or *GGBBB*: $3! \cdot 2! \cdot 2 = 24$
c. *GGBBB, BGGBB, BBGGB, BBBGG* $\cdot$ $3! \cdot 2! \cdot 4 = 48$.

Combinations

4. A student is to answer 8 out of 10 questions in an exam. (i) How many choices has he? (ii) How many if he must answer the first three questions? (iii) How many if he must answer at least four of the first five questions?

Ans. (i) $10_{C_8} = 10_{C_2} = 45$ ways (ii) $7_{C_5} = 7_{C_2} = 21$ (iii) All the first five, $5_{C_3} = 10$ ways, 4 out of first five, $5_{C_4} =$ five ways and choose the other four out of last five, $5_{C_1} =$ five ways. Thus $5 \cdot 5 = 25$ ways. Hence

$10 + 25 = 35$ ways

5. Compute the number of ordered partitions from $\{A_1, A_2, A_3\}$ of the set (box) of 7 marbles into cells A_1 containing 2 marbles, A_3 containing 3 marbles and A_3 containing 2 marbles.

Ans. $\frac{7!}{2!5!} \cdot (7_{C_2})(5_{C_3})(2_{C_2}) = \frac{7!}{2!5!} \cdot \frac{5!}{3!2!} \cdot \frac{2!}{2!10!} = \frac{7!}{2!3!2!}$.

Finite equiprobable spaces

6. One card is selected at random from 50 cards numbered 1 to 50. Find the probability that the number on card is (i) divisible by 5 (ii) prime (iii) ends in digit 2.

Ans. **i.** P (divisible by 5) $= \frac{10}{50} = \frac{1}{5}$
ii. P (prime) $= \frac{15}{50}$
iii. P (ends in digit 2) $= \frac{5}{50}$

Hint:
i. 5, 10, 15, 20, 25, 30, 35, 40, 45, 50: 10
ii. 2, 3, 5, 7, 11, 13, 17, 19, 23, 29, 31, 37, 41, 43, 47: 15
iii. 2, 12, 22, 32, 42:5.

7. A pair of fair dice is tossed. Find probability that maximum of the two numbers is greater than 4.

Ans. $\frac{20_{C_1}}{36_{C_1}} = \frac{20}{36} = \frac{5}{9}$

Hint: 20 favourable cases such as $(1, 5), (1, 6), (2, 5), (2, 6) \cdots (6, 5), (6, 6)$.

8. Let A, B be events with $P(A \cup B) = \frac{7}{8}$, $P(A \cap B) = \frac{1}{4}$ and $P(A^c) = \frac{5}{8}$. Determine (a) $P(A)$ (b) $P(B)$ (c) $P(A \cap B^c)$, (d) $P(A^c \cap B)$ (e) $P(A^c \cup B^c)$ (f) $P(A^c \cap B^c)$.

Ans. **a.** $P(A \cup B) = \frac{7}{8} = P(A) + P(B) - P(A \cap B)$,
b. $P(A) + P(B) = \frac{9}{8}$, $P(A) = 1 - P(A^c) = 1 - \frac{5}{8} = \frac{3}{8}$, $P(B) = \frac{6}{8}$
c. $P(A \cup B^c) = P(A \backslash B) = P(A) - P(A \cap B) = \frac{3}{8} - \frac{1}{4} = \frac{1}{8}$
d. $P(A^c \cap B) = \frac{1}{2}$
e. $P(A^c \cup B^c) = \frac{3}{4}$
f. $\frac{1}{8}$.

9. **a.** Two digits are drawn in succession from tickets numbered 1 to 5. Determine the probability that an odd digit will be selected (i) first time (ii) second time (iii) both times.
b. From 25 tickets marked 1 to 25 inclusive, one is drawn at random. Find the probability that
i. it is a multiple of 5 or 7
ii. it is a multiple of 3 or 7

Ans. **a.** (i) $\frac{12}{20} = \frac{3}{5}$ (ii) $\frac{12}{20} = \frac{3}{5}$ (iii) $\frac{6}{20} = \frac{3}{10}$
b. (i) $\frac{8}{25}$ (ii) $\frac{10}{25}$.

Probability

10. A lot contains 10 good articles, 4 with minor defects and 2 with major defects. 2 articles are chosen from the lot at random (without replacement). Find the probability that (i) both are good (ii) both have major defects (iii) at least one is good (iv) at most one is good (v) exactly one is good (vi) neither has major defects (vii) neither is good.

Ans. **i.** P (both are good) $= \frac{10_{C_2}}{16_{C_2}} = \frac{3}{8}$
ii. P (both have major defects) $= \frac{2_{C_2}}{16_{C_2}} = \frac{1}{120}$
iii. P (at least one is good) $= \frac{10_{C_1} \times 6_{C_1} + 10_{C_2}}{16_{C_2}} = \frac{7}{8}$

iv. P (at most one is good)$= \frac{10_{C_0}\times 6_{C_2}+10_{C_1}\times 6_{C_1}}{16_{C_2}}$
$= \frac{5}{8}$

v. P (exactly one is good) $= \frac{10_{C_1}\times 6_{C_1}}{16_{C_2}} = \frac{1}{2}$

vi. P (neither has major defects) $= \frac{14_{C_2}}{16_{C_2}} = \frac{91}{120}$

vii. P (neither is good) $= P$ (both are defected)
$\frac{6_{C_2}}{16_{C_2}} = \frac{1}{8}$

11. What is the probability that (a) non leap year (b) leap year should have 53 sundays?

Ans. (a) $\frac{1}{7}$ (b)$\frac{2}{7}$

Hint:

a. $\frac{365}{7} = 52$ weeks $+1$ day:

b. $\frac{366}{7} = 52$ weeks $+2$ days:

$\overset{*}{S}M, MT, TW, WTh, ThF, FS, S\overset{*}{S}$

12. A bag contains 40 tickets numbered 1, 2, 3,... 40 of which 4 are drawn at random and arranged in ascending order $t_1 < t_2 < t_3 < t_4$. Find the probability of t_3 being 25.

Ans. $\frac{24_{C_2}\times 15_{C_1}}{40_{C_4}} = \frac{414}{9139}$

Hint: t_1, t_2 can come in 24_{C_2} ways t_4 in 15_{C_1} ways.

13. A bag contains eight white and six red marbles. Find the probability of drawing two marbles of the same colour.

Ans. $\frac{8_{C_2}}{14_{C_2}} + \frac{6_{C_2}}{14_{C_2}} = \frac{28}{91} + \frac{15}{91} = \frac{43}{91}$

14. A box I contains four tickets numbered 1, 2, 3, 4 and another box II contains six tickets numbered 2, 4, 6, 7, 8, 9. If one of the two boxes is chosen at random and a ticket is drawn at random from the chosen box, find the probabilities that the ticket drawn is numbered (i) 2 or 4 (ii) 3 (iii) 1 or 9.

Ans. **i.** $\frac{1}{2}\cdot\frac{2}{4} + \frac{1}{2}\cdot\frac{2}{6} = \frac{5}{12}$

ii. $\frac{1}{2}\cdot\frac{1}{4} + \frac{1}{2}\cdot 0 = \frac{1}{8}$

iii. $\frac{1}{2}\cdot\frac{1}{4} + \frac{1}{2}\cdot\frac{1}{6} = \frac{5}{24}$.

Independent trials

15. A missile hits its target with probability 0.3. How many missiles should be fired so that there is at least an 80% probability of hitting a target?

Ans. At least five missiles should be fired.

Hint: $1-(\text{no hit}) = 1-(1-0.3)^n$
$= 1-(0.7)^n > 0.8$

16. A cricket team wins (W) with probability 0.6, loses (L) with probability 0.3 and draws (D) with probability 0.1. The team plays 3 games.
(i) Determine the elements of the event A that the team wins at least twice and does not lose, and find $P(A)$. (ii) Determine the elements of the event B that the team wins, loses and draws and find $P(B)$.

Ans. **i.** $P(A) = P(WWW) + P(WWD) + P(WDW) + P(DWW) = (0.6)(0.6)(0.6) + (0.6)(0.6)(0.1) + (0.6)(0.1)(0.6) + (0.1)(0.6)(0.6) = 0.324$

ii. $P(B) = P(WLD) + P(WDL) + P(LDW) + P(DWL) + P(DLW)$
$= 6(0.6)(0.3)(0.1) = 0.108$

17. A, B, C hit a target with probabilities $\frac{1}{2}, \frac{2}{3}, \frac{3}{4}$ respectively. If all of them fire at the target, find the probability P that (i) none of them hits the target (ii) at least one of them hits the target.

Ans. **i.** $P(\bar{A}\cap\bar{B}\cap\bar{C}) = P(\bar{A})P(\bar{B})P(\bar{C}) = \frac{1}{2}\frac{1}{3}\frac{1}{4} = \frac{1}{24}$

ii. P (at least one hits) $= 1 - P(\text{no hit}) = 1 - \frac{1}{24} = \frac{23}{24}$.

18. A and B alternately throw pair of dice. A wins if he throws six before B throws seven and B wins if he throws seven before A throws six. If A begins, show that his chance of winning is $\frac{30}{61}$.

Ans. $P(6) = \frac{5}{36}, P(7) = \frac{1}{6},$

$P(A) = P(A \text{ or } \bar{A}\bar{B}A \text{ or } \bar{A}\bar{B}\bar{A}\bar{B}A \text{ or } \ldots)$

$P(A) = P(A)+P(\bar{A}\bar{B}A)+P(\bar{A}\bar{B}\bar{A}\bar{B}A)+\cdots$

$= P(A) + P(\bar{A})\cdot P(\bar{B})\cdot P(A) + \cdots$

$= \frac{5}{36} + \frac{31}{36}\cdot\frac{5}{6}\cdot\frac{5}{36} + \cdots$

$= \frac{5/36}{1-(155)/(216)} = \frac{30}{61}$.

19. A, B, C can hit a target with probability $\frac{3}{5}, \frac{2}{5}, \frac{3}{4}$ respectively. Determine the probability that (i) two shots hit (ii) at least two shots hit.

Ans. **i.** $\frac{3}{5}\frac{2}{5}(1-\frac{3}{4})+\frac{2}{5}\frac{3}{4}(1-\frac{3}{5})+\frac{3}{4}\frac{3}{5}(1-\frac{2}{5})=0.45$
ii. $0.45 + P$ (all hit) $= 0.45+\frac{3}{5}\frac{2}{5}\frac{3}{4}=0.63$

Conditional probability

20. A pair of fair dice is thrown. Find the probability p that the sum is 10 or greater if (i) a 5 appears on the first die (ii) a 5 appears on at least one of the dice.

Ans. **i.** $p=\frac{2}{6}$

Hint: 5 appears on first die:
$(5,1)(5,2)(5,3)(5,4)(5,5)(5,6)$: 6
Favourable cases for sum ≥ 10 is $(5,5)$, $(5,6)$: 2

ii. $p=\frac{3}{11}$.

Hint: 5 appears on at least one of the dice (reduced sample space)
$(5,1), (5,2), (5,3), (5,4), (5,5), (5,6)(1,5)$, $(2,5), (3,5), (4,5), (6,5): 11$

Favourable cases for sum ≥ 10 is $(5, 5)$, $(5,6), (6,5)$: 3.

21. 2 digits are selected at random from the digits 1 through 9. If the sum is even, find the probability p that both the numbers are odd.

Ans. $p=\frac{10}{16}$

Hint: $e+e=e, o+d=e, 4_{C_2}=6$ ways of two e, $5_{C_2}=10$ ways of two o. 16 ways to choose two for even sum.

22. A box contains 4 bad and 6 good tubes. 2 tubes are drawn out from the box at a time. One of them is found to be good. Determine the probability that the other one is also good.

Ans. $P(B/A)=\frac{P(A\cap B)}{P(A)}=\frac{1/3}{6/10}=\frac{5}{9}$ where

A : first is good

B : second is good.

23. A bag contains 10 gold and 8 silver coins. Two successive drawings of 4 coins are made such that (i) coins are replaced before the second trial (ii) coins are not replaced before the second trial. Determine the probability that the first drawing will give 4 gold and the second four silver coins.

Ans. **i.** $P(A\cap B)=P(A)P(B)=\frac{{}^{10}C_4}{{}^{18}C_4}\cdot\frac{{}^{8}C_4}{{}^{18}C_4}$

ii. $P(A\cap B)=P(A)P(B/A)=\frac{{}^{10}C_4}{{}^{18}C_4}\cdot\frac{{}^{8}C_4}{{}^{14}C_4}$

24. Bag I contains 4 white, 3 black marbles and bag II contains 3 white, 5 black marbles. One marble is drawn from the bag I and placed unseen in the bag II. Determine the probability that a marble *now* drawn from bag II is black.

Ans. $P[(B_1\cap B_2)$ or $(W_1\cap B_2)]$
$=P(B_1\cap B_2)+P(W_1\cap B_2)$
$=P(B_1)P(B_2/B_1)+P(W_1)P(B_2/W_1)$
$=\frac{3}{7}\cdot\frac{6}{9}+\frac{4}{7}\cdot\frac{5}{9}=\frac{38}{63}$.

26.4 THEOREM OF TOTAL PROBABILITY (or THE RULE OF ELIMINATION)

Theorem: *Let $B_1, B_2, \ldots, B_k$ constitute a partition of the sample space S with $P(B_i)\neq 0$ for $i=1,2,\ldots,k$. Then for any event A of S*

$$P(A)=\sum_{i=1}^{k}P(B_i\cap A)=\sum_{i=1}^{k}P(B_i)P(A/B_i)$$

Proof: Since $B_1, B_2, \ldots, B_k$ constitute a partition

$$S=\bigcup_{i=1}^{k}B_i$$

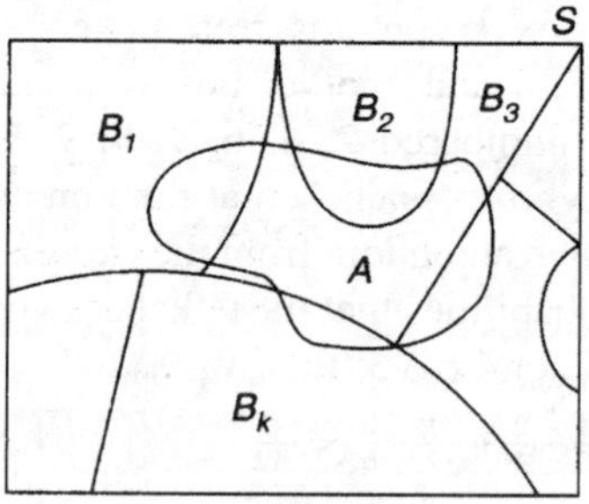

Fig. 26.10

and $B_i\cap B_j=\phi$ for any i and j, i.e., their union is S and B_i's are mutually disjoint sets. Now (refer Fig. 26.10)

$$A=A\cap S=A\cap\left(\bigcup_{i=1}^{k}B_i\right)$$

$$= A \cap (B_1 \cup B_2 \cup \cdots \cup B_k)$$
$$= (A \cap B_1) \cup (A \cap B_2) \cup \cdots \cup (A \cap B_k).$$

The sets $A \cap B_1, A \cap B_2, \ldots, A \cap B_k$ are all mutually disjoint sets. Applying the additive rule for mutually exclusive events, we have

$$P(A) = P[(A \cap B_1) \cup (A \cap B_2) \cup \cdots \cup (A \cap B_k)]$$
$$= P(B_1 \cap A) + P(B_2 \cap A) + \cdots + P(B_k \cap A)$$

Now applying multiplicative rule

$$P(A) = \sum_{i=1}^{k} P(B_i \cap A) = \sum_{i=1}^{k} P(B_i)P(A/B_i).$$

Worked Out Examples

Theorem on total probability

Example 1: Police plan to enforce speed limits by using radar traps at 4 different locations within the city limits. The radar traps at each of these locations L_1, L_2, L_3, L_4 are operated for 40%, 30%, 20% and 30% of the time. If a person who is speeding on his way to work has probabilities of 0.2, 0.1, 0.5 and 0.2 respectively of passing through these locations, what is the probability that he will be fined (for over speed)?

Solution: A: event of passing through locations and caught by radar traps.
By theorem on total probability

$$P(A) = P(L_1)P(A/L_1) + P(L_2)P(A/L_2)$$
$$+ P(L_3)P(A/L_3) + P(L_4)P(A/L_4).$$
$$= (0.4)(0.2) + (0.3)(0.1) + (0.2)(0.5) + (0.3)(0.2)$$
$$= 0.27.$$

Example 2: Suppose colored balls are distributed in three indistinguishable boxes as follows:

	Box 1	Box 2	Box 3
Red	2	4	3
White	3	1	4
Blue	5	3	3
Total	10	8	10

A box is selected at random from which a ball is selected at random. What is the probability that the ball is colored (i) red (ii) white (iii) black?

Solution: A : colour of the ball (red, white or black)

$$\text{Box 1} = B_1, \text{Box 2} = B_2, \text{Box 3} = B_3$$

By theorem on total probability

$$P(\text{red colour}) = P(B_1)P(A/B_1) + P(B_2)P(A/B_2)$$
$$+ P(B_3)P(A/B_3)$$

Since the three boxes are indistinguishable, probability of choosing them is $\frac{1}{3}$ i.e., $P(B_1) = P(B_2) = P(B_3) = \frac{1}{3}$. Thus

$$P(R) = \frac{1}{3} \cdot \frac{2}{10} + \frac{1}{3} \cdot \frac{4}{8} + \frac{1}{3} \cdot \frac{3}{10} = \frac{1}{3}$$

since P (Red given box 1) $= \frac{2}{10}$, $P(R/B_2) = 4/8$, $P(R/B_3) = 3/10$ etc.
Similarly,

$$P(W) = \frac{1}{3} \cdot \frac{3}{10} + \frac{1}{3} \cdot \frac{1}{8} + \frac{1}{3} \cdot \frac{4}{10} = \frac{198}{720}$$
$$P(B) = \frac{1}{3} \cdot \frac{5}{10} + \frac{1}{3} \cdot \frac{3}{8} + \frac{1}{3} \cdot \frac{3}{10} = \frac{94}{240}.$$

Exercise

Theorem on total probability

1. Three machines A, B and C produce respectively 50%, 30% and 20% of the total number of items of a factory. The percentage of defective output of these machines are 3%, 4% and 5%. If an item is selected at random, find the probability that the item is defective.

Ans. $P(\text{defective})$

$$= P(A)P(D/A) + P(B)P(D/B) + P(C)P(D/C)$$
$$= (0.5)(0.03) + (0.3)(.04) + (0.2)(0.05)$$
$$= 0.037$$

2. The probability that X, Y, Z will be elected as president of a club are 0.3, 0.5 and 0.2 respectively. The probability that membership fees of club are increased is 0.8 if X is elected president, is 0.1 if Y is elected and is 0.4 if Z is elected. What is the probability that there will be an increase in membership fee?

Ans. $P(\text{fee increased}) = (0.3)(0.8) + (0.5)(0.1) + (0.2)(0.4) = 0.37$

3. Box I contains 10 white and 3 black balls while Box II contains 3 white and 5 black balls. Two balls are drawn at random from Box I and placed in Box II. Then 1 ball is drawn at random from box II. What is the probability that it is a white ball?

Ans. 59/130

Hint: B_1: event of drawing 2 white balls, from Box I

B_2: $2B$ from Box I, B_3: one W and one B from Box I.

A: drawing W from box II (after transfer)

$$P(A) = P(B_1)P(A/B_1) + P(B_2)P(A/B_2) + P(B_3)P(A/B_3)$$

$$= \frac{10_{C_2}}{13_{C_2}} \cdot \frac{5}{10} + \frac{3_{C_2}}{13_{C_2}} \cdot \frac{3}{10}\ \frac{10_{C_1} 3_{C_1}}{13_{C_2}} \cdot \frac{4}{10}.$$

26.5 BAYES' THEOREM (or BAYES' RULE)

Theorem 1: *Let the sample space S be partitioned into k subsets $B_1, B_2, \ldots, B_k$ with $P(B_i) \neq 0$ for $i = 1, 2, \ldots, k$. For any arbitrary event A in S with $P(A) \neq 0$,*

$$P(B_r/A) = \frac{P(B_r \cap A)}{\sum_{i=1}^{k} P(B_i \cap A)} = \frac{P(B_r)P(A/B_r)}{\sum_{i=1}^{k} P(B_i)P(A/B_i)} \quad (1)$$

for $r = 1, 2 \cdots, k$

Proof: From the definition of conditional probability

$$P(B_r/A) = \frac{P(B_r \cap A)}{P(A)} \quad (2)$$

From the theorem of total probability or the rule of elimination

$$P(A) = \sum_{i=1}^{k} P(B_i \cap A) = \sum_{i=1}^{k} P(B_i)P(A/B_i) \quad (3)$$

Also from the multiplication rule

$$P(B_r \cap A) = P(B_r)P(A/B_r) \quad (4)$$

Substituting (3), (4) in (2), we get (1).

Note: Bayes' theorem is also known as formula for the probability of "causes", i.e., probability of a particular (cause) B_r given that event A has happened (already).

$P(B_i)$ is 'a priori probability' known even before the experiment, $P(A/B_i)$ "likelihoods" and $P(B_i/A)$ 'posteriori probabilites' determined after the result of the experiment.

WORKED OUT EXAMPLES

Bayes' theorem

Example 1: In a certain college 25% of boys and 10% of girls are studying mathematics. The girls constitute 60% of the student body. (a) What is the probability that mathematics is being studied? (b) If a student is selected at random and is found to be studying mathematics, find the probability that the student is a girl? (c) a boy?

Solution: Given that $P(\text{Boy}) = P(B) = \frac{40}{100} = \frac{2}{5}$; $P(\text{Girl}) = P(G) = \frac{60}{100} = \frac{3}{5}$, probability that maths is studied given that the student is a boy $= P(M/B) = \frac{25}{100} = \frac{1}{4}$. Similarly, $P(M/G) = \frac{10}{100} = \frac{1}{10}$

a. Probability that maths is studied

$$= P(M) = P(G)P(M/G) + P(B)P(M/B)$$

by theorem on total probability

$$P(M) = \frac{3}{5} \cdot \frac{1}{10} + \frac{2}{5} \cdot \frac{1}{4} = \frac{4}{25}$$

b. Bayes' thorem:
Probability that a maths student is a girl

$$= P(G/M) = \frac{P(G)P(M/G)}{P(M)} = \frac{\frac{3}{5}\cdot\frac{1}{10}}{\frac{4}{25}} = \frac{3}{8}$$

c. Probability that a maths student is a boy

$$= P(B/M) = \frac{P(B)P(M/B)}{P(M)} = \frac{\frac{2}{5}\cdot\frac{1}{4}}{\frac{4}{25}} = \frac{5}{8}.$$

Example 2: A businessman goes to hotels X, Y, Z 20%, 50%, 30% of the time, respectively. It is known that 5%, 4%, 8% of the rooms in X, Y, Z hotels have faulty plumbing. (a) Determine the probability that the businessman goes to hotel with faulty plumbing (b) What is the probability that businessman's room having faulty plumbing is assigned to hotel Z?

Solution: A : event of faulty plumbing

$$B_1 = X, B_2 = Y, B_3 = Z$$

a. By theorem on total probability

$$P(\text{Faulty plumbing}) = P(A) = \sum_{i=1}^{3} P(B_i)P(A/B_i)$$

$$= P(X)P(A/X) + P(Y)P(A/Y) + P(Z)P(A/Z)$$

It is known (given) that

$$P(B_1) = P(X) = \frac{20}{100} = 0.2,$$

$$P(B_2) = P(Y) = \frac{50}{100} = 0.5,$$

$$P(B_3) = P(Z) = 0.3$$

$$P(A/X) = \frac{5}{100} = 0.05,$$

$$P(A/Y) = \frac{4}{100} = 0.04,$$

$$P(A/Z) = \frac{8}{100} = 0.08.$$

Thus

$$P(\text{Faulty plumbing}) = P(A) = (0.2)(0.05)$$
$$+(0.5)(0.04) + (0.3)(0.08)$$
$$= 0.054.$$

b. P(Assigned to hotel Z given that room has faulty plumbing) $= P(Z/A)$.

By Bayes' theorem

$$P(Z/A) = \frac{P(Z)P(A/Z)}{P(A)} = \frac{(0.3)(0.08)}{0.054} = \frac{4}{9}.$$

EXERCISE

Bayes' theorem

1. Companies B_1, B_2, B_3 produce 30%, 45% and 25% of the cars respectively. It is known that 2%, 3% and 2% of the cars produced from B_1, B_2 and B_3 are defective. (a) What is the probability that a car purchased is defective? (b) If a car purchased is found to be defective what is the probability that this car is produced by company B_3?

Ans: **a.** $P(\text{Defective}) = (0.3)(0.02)+ (0.45)(0.03) + (0.25)(0.02) = 0.0245$

b. $P(B_3/D) = \frac{(0.25)(0.02)}{P(D)} = 10/49.$

2. Of the three men, the chances that a politician, a businessman and an academician will be appointed as a vice-chancellor of a university are 0.50, 0.30 and 0.20 respectively. Probability that research is promoted by these people if they are appointed as V.C. are 0.3, 0.7 and 0.8 respectively. (a) Determine the probability that research is promoted in the university. (b) If research is promoted in the university, what is the probability that the V.C. is an academician? (c) A businessman?

Ans: **a.** $P(\text{Research}) = (0.5)(0.3) + (0.3)(0.7)+ (0.2)(0.8) = 0.52$

b. $P(\text{Academician/Research}) = \frac{(0.2)(0.8)}{0.52} = 0.30769$

c. $P(\text{Bussinessman/Research}) = \frac{(0.3)(0.7)}{0.52} = 0.4038.$

3. Suppose three companies X, Y, Z produce T.V's. X produce twice as many as Y while Y and Z produce the same number. It is known that 2% of X, 2% of Y and 4% of Z are defective. All the TV's produced are put into one shop and then one TV is chosen at random.

a. What is the probability that the TV is defective?

b. Suppose a TV chosen is defective, what is the probability that this TV is produced by company X?

Ans: **a.** $P(D) = P(X)P(D/X) + P(Y)P(D/Y) + P(Z)P(D/Z)$
$= \frac{1}{2}(0.02) + \frac{1}{4}(0.02) + \frac{1}{4}(0.04) = 0.025$

b. $P(X/D) = \frac{(0.02)\left(\frac{1}{2}\right)}{0.025} = 0.40$

4. Box I contains 1 white, 2 red, 3 green balls, Box II contains 2 white, 3 red, 1 green balls, Box III contains 3 white, 1 red, 2 green balls. Two balls are drawn from a box chosen at random. These are found to be one white and one red. Determine the probability that the balls so drawn came from box II.

Ans: $\frac{6}{11}$

Hint: $P(I) = P(II) = P(III) = \frac{1}{3}$,
A: event: 2 balls white and red drawn

$$P(A/I) = (1c_1)(2c_1)/6c_2 = \frac{2}{15},$$

$$P(A/II) = (2c_1)(3c_1)/6c_2 = \frac{2}{5},$$

$$P(A/III) = (3c_1)(1c_1)/6c_2 = \frac{1}{5},$$

$$P(II/A) = \frac{\frac{1}{3}\cdot\frac{2}{5}}{\frac{1}{3}\cdot\frac{2}{15} + \frac{1}{3}\cdot\frac{2}{5} + \frac{1}{3}\cdot\frac{1}{5}} = \frac{6}{11}.$$

5. For a certain binary communication channel, the probability that a transmitted '0' is received as a'0' is 0.95 and the probability that a transmitted '1' is received as '1' is 0.90. If the probability that a '0' is transmitted is 0.4, find the probability that (i) a'1' is received (ii) a '1' was transmitted given that a '1' was received.

Ans: **i.** 0.56 **ii.** 27/28

Hint: A = event of transmitting '1', $\bar{A}$ = event of transmitting '0'

B = event of receiving '1', $\bar{B}$ = event of receiving '0'.

i. $P(B) = P(A)P(B/A) + P(\bar{A})(B/\bar{A}) = (0.6)(0.9) + (0.4)(.05) = 0.56$

ii. $P(A/B) = \frac{P(A)P(B/A)}{P(B)} = \frac{(0.6)(0.9)}{0.56} = \frac{27}{28}$.

6. A student has to answer a multiple-choice question with 5 alternatives. What is the probability that the student knew the answer given that he answered it correctly?

Ans: $5p/(4p+1)$ where p = probability that he knew the correct answer.

Hint: B_1: knew right answer, B_2: guesses right answer, A: gets the right answer

$$P(B_1/A) = \frac{(p)(1)}{(p)(1) + (1-p)\left(\frac{1}{5}\right)}$$

since $P(A/B_2) = \frac{1}{5}$.

Chapter 27

Probability Distributions

INTRODUCTION

Probability distribution is the theoretical counterpart of frequency distribution, and plays an important role in the theoretical study of populations. A probability model can be developed, for a given idealized conditions in a game of chance by incorporating all the factors that have a bearing on this game. In building such model, the empirical data of frequency distribution, A.M., variance etc. are to be taken into account. In the discrete case we consider discrete uniform distribution, Binomial, Hypergeometric, Poisson distributions. The continuous probability distributions we study are uniform distribution, normal distribution, exponential, gamma, Weibull distributions which are of great practical importance.

Recall that in a random experiment, the outcomes (or results) are governed by chance mechanism and the sample space S of such a random experiment consists of all outcomes of the experiment. When the elements (outcomes/events) of the sample space are non-numeric, they can be quantified by assigning a real number to every event of the sample space. This assignment rule, known as the random variable (R.V.), provides the power of abstraction and thus discards unimportant finest-grain description of the sample space.

A **random variable** X on a sample space S is a function $X: S \to R$ from S to the set of real numbers R, which assigns a real number $X(s)$ to each sample point s of S (refer Fig. 27.1).

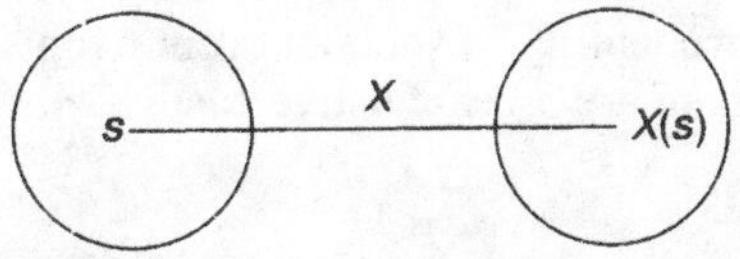

S: Sample Space R_x: Possible Values of x

Fig. 27.1

Range space R_X: is the set of all possible values of X is a subset of real numbers R.

Although X is called a random "variable" note that it is infact a "single-valued function".

Notation: If R.V. is denoted by X, then x (corresponding small letter) denotes one of its values.

Discrete

A R.V. X is said to be discrete R.V. if its set of possible outcomes, the sample space S, is countable (finite or an unending sequence with as many elements as there are whole numbers).

Continuous

A R.V. X is said to be continuous R.V. if S contains infinite numbers equal to the number of points on a line segment.

27.1 PROBABILITY DISTRIBUTIONS

Discrete Probability Distributions

Each event in a sample space has certain probability (or chance) of occurrence (or happening). A formula representing all these probabilities which a discrete R.V. assumes, is known as the discrete probability distribution.

Example: Let X denote the discrete R.V. which denotes the minimum of the two numbers that appear in a single throw of a pair of fair dice. Then X is a function from the sample space S consisting of 36 ordered pairs $\{(1, 1), (1, 2), \ldots, (6, 6)\}$ to a subset of real numbers $\{1, 2, 3, 4, 5, 6\}$.

The event minimum 5 can appear in the following cases (occurrences) (5, 5), (5, 6), (6, 5). Thus R.V. X assigns to this event of the sample space a real number 3. The probability of such an event happening is $\frac{3}{36}$ since there are 36 exhaustive cases. This is represented as

$$P(X = x_i) = p_i = f(x_i) = P(X = 5) = f(5) = \frac{3}{36}.$$

Calculating in a similar way the other probabilities, the distribution of probabilities of this discrete R.V. is denoted by the discrete probability distribution as follows:

$X = x_i$	1	2	3	4	5	6
$P(X = x_i) = f(x_i) = p_i$	$\frac{11}{36}$	$\frac{9}{36}$	$\frac{7}{36}$	$\frac{5}{36}$	$\frac{3}{36}$	$\frac{1}{36}$

Discrete probability distribution, probability function or probability mass function of a discrete R.V. X is the function $f(x)$ satisfying the following conditions:

i. $f(x) \geq 0$

ii. $\sum_x f(x) = 1$

iii. $P(X = x) = f(x)$.

Thus probability distribution is the set of ordered pairs $(x, f(x))$, i.e., outcome x and its probability (chance) $f(x)$.
Cumulative distribution or simply *distribution* of a discrete R.V. X is $F(x)$ defined by

$$F(x) = P(X \leq x) = \sum_{t \leq x} f(t) \quad \text{for } -\infty < x < \infty.$$

It follows that

$$F(-\infty) = 0, \quad F(+\infty) = 1,$$

$$p(x_j) = P(X = x_j) = F(x_j) - F(x_{j-1}).$$

Continuous Probability Distributions

For a continuous R.V. X, the function $f(x)$ satisfying the following, is known as the probability density function (P.D.F.) or simply density function (Fig. 27.2).

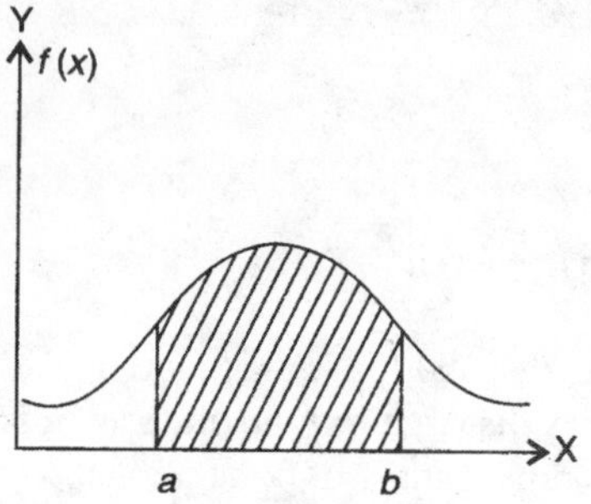

Fig. 27.2

i. $f(x) \geq 0$

ii. $\int_{-\infty}^{\infty} f(x)dx = 1$

iii. $P(a < X < b) = \int_a^b f(x)dx$ = area under $f(x)$ between ordinates $x = a$ and $x = b$.

Note 1: $P(a < X < b) = P(a \leq X < b)$ $= P(a < X \leq b) = P(a \leq X \leq b)$

i.e., inclusion or non-inclusion of end points, does not change the probability, which is *not* the case in the discrete distributions.

Note 2: Probability at a point, $P(X = a) = \int_{a-\Delta x}^{a+\Delta x} f(x)dx$.

Cumulative Distribution

For a continuous R.V. X, with P.D.F. $f(x)$, the cumulative distribution $F(x)$ is defined as

$$F(x) = P(X \leq x) = \int_{-\infty}^{x} f(t)dt, \quad -\infty < x < \infty$$

It follows that

$$F(-\infty) = 0, \quad F(+\infty) = 1, \quad 0 \leq F(x) \leq 1$$

for $-\infty < x < \infty$.

$$f(x) = \frac{dF(x)}{dx} = F'(x) \geq 0 \text{ and}$$

$$P(a < X < b) = F(b) - F(a).$$

Expectation

The behaviour of a R.V. (either discrete or continuous) is completely characterized by the distribution function $F(x)$ or density $f(x)$ [$P(x_i)$ in discrete case]. Instead of a function, a more compact description can be made by a single numbers such as mean (expectation), median and mode known as measures of central tendency of the R.V. X.

Expectation or mean or expected value

Expectation or mean or expected value of a random variable X, denoted by $E(X)$ or μ, is defined as

$$E(X) = \begin{cases} \sum_i x_i f(x_i), & \text{if } X \text{ is discrete} \\ \int_{-\infty}^{\infty} x f(x) dx, & \text{if } X \text{ is continuous} \end{cases}$$

Note 1: x is median if $P(X < x) \leq \frac{1}{2}$ and $P(X > x) \leq \frac{1}{2}$.

Note 2: x is mode for which $f(x)$ or $P(x_i)$ attains its maximum.

Variance

Variance characterizes the variability in the distributions, since two distributions with *same* mean can still have different dispersion of data about their means.

Variance of R.V. X is

$$\sigma^2 = E[(X-\mu)^2] = \sum_x (X-\mu)^2 f(x), \text{ for } X \text{ discrete}$$

$$\sigma^2 = E[(X-\mu)^2] = \int_{-\infty}^{\infty} (x-\mu)^2 f(x) dx,$$

for X continuous.

Standard deviation (S.D.) denoted by σ, is the positive square root of variance.

Result: $\sigma^2 = E(X^2) - \mu^2$

$$\text{Since } \sigma^2 = \sum_x (x-\mu)^2 f(x) = \sum_x (x^2 - 2\mu x + \mu^2) f(x)$$

$$= \sum x^2 f(x) - 2\mu \sum x f(x) + \mu^2 \sum f(x)$$

$$= E(X^2) - 2\mu \cdot \mu + \mu^2 \cdot 1 = E(X^2) - \mu^2$$

Since $\mu = \sum x f(x), \sum f(x) = 1$.

Similar result follows for continuous R.V. X, with $\sum$ replaced by integration from $-\infty$ to ∞.

Note 1: In a gambling game, expected value E of the game is considered to be the value of the game to the player. Game is favourable to the player if $E > 0$, unfavourable if $E < 0$, fair if $E = 0$.

Note 2: Mathematical expectation $E = a_1 p_1 + a_2 p_2 + \cdots + a_k p_k$ where the probabilities of obtaining the amounts $a_1, a_2, \ldots$ or a_k are $p_1, p_2, \ldots p_k$ respectively.

27.2 CHEBYSHEV'S THEOREM

Theorem: *Let μ and σ be the mean and standard deviation of a random variable X with probability density $f(X)$. Then the probability that X will assume a value within k standard deviations of the mean is at least $1 - \frac{1}{k^2}$, for any positive constant k. Symbolically*

$$P(\mu - k\sigma < X < \mu + k\sigma) = P(|X-\mu| < k\sigma) \geq 1 - \frac{1}{k^2}$$

Proof: By definition

$$\sigma^2 = \text{variance} = E\left[(X-\mu)^2\right] = \int_{-\infty}^{\infty} (x-\mu)^2 f(x) dx$$

$$= \int_{-\infty}^{\mu-k\sigma} (x-\mu)^2 f(x) dx + \int_{\mu-k\sigma}^{\mu+k\sigma} (x-\mu)^2 f(x) dx$$

$$+ \int_{\mu+k\sigma}^{\infty} (x-\mu)^2 f(x) dx$$

Since the second integral on the R.H.S. is non-negative, we get an inequality of the form

$$\sigma^2 \geq \int_{-\infty}^{\mu-k\sigma} (x-\mu)^2 f(x) dx + \int_{\mu+k\sigma}^{\infty} (x-\mu)^2 f(x) dx$$

For the first integral $x \leq \mu - k\sigma$ and for the second integral $x \geq \mu + k\sigma$. In either case, we have

$$x - \mu \leq -k\sigma \quad \text{or} \quad x - \mu \geq k\sigma \quad \text{i.e.,} \quad |x-\mu| \geq k\sigma$$

$$\text{or} \quad (x-\mu)^2 \geq k^2\sigma^2.$$

Replacing $(x-\mu)^2$ by $k^2\sigma^2$ in the two integrals,

$$\sigma^2 \geq \int_{-\infty}^{\mu-k\sigma} k^2\sigma^2 f(x) dx + \int_{\mu+k\sigma}^{\infty} k^2\sigma^2 f(x) dx$$

Rewriting

$$P(|X-\mu| \geq k^2\sigma^2) = \int_{-\infty}^{\mu-k\sigma} f(x)dx + \int_{\mu+k\sigma}^{\infty} f(x)dx \leq \frac{1}{k^2}$$

By complimentation rule,

$$\int_{\mu-k\sigma}^{\mu+k\sigma} f(x)dx = 1 - \left[\int_{-\infty}^{\mu-k\sigma} f(x)dx + \int_{\mu+k\sigma}^{\infty} f(x)dx\right]$$

Hence $P(|X-\mu| < k\sigma) = P(\mu - k\sigma < X < \mu + k\sigma)$

$$= \int_{\mu-k\sigma}^{\mu+k\sigma} f(x)dx \geq 1 - \frac{1}{k^2}$$

Note 1: Put $k\sigma = C > 0$ then

$$P\left\{|X-\mu| \geq C\right\} \leq \frac{\sigma^2}{C^2}$$

i.e., $$P\left\{|X - E(X)| \geq C\right\} \leq \frac{\text{var}(X)}{C^2}$$

or $$P\left\{|X-\mu| < C\right\} \geq 1 - \frac{\sigma^2}{C^2}$$

i.e., $$P\left\{|X - E(X)| < C\right\} \geq \frac{1-\text{var}(X)}{C^2}$$

Note 2: Chebyshev's theorem (1853) is "distribution-free" since it is applicable to any unknown distribution and gives only the lower bound for the probability.

Worked Out Examples

Chebyshev's theorem

Example 1: Determine the smallest value of k in the Chebyshev's theorem for which the probability is (a) at least 0.95 (b) at least 0.99.

Solution: From Chebyshev's theorem, we have the probability as

$$P\left(|X-\mu| < k\sigma\right) \geq 1 - \frac{1}{k^2}$$

a. Probability $= 0.95 \geq 1 - \frac{1}{k^2}$ or $k^2 \leq \frac{1}{0.05}$
$\therefore k = \sqrt{20} = 4.472$

b. Probability $= 0.99 \geq 1 - \frac{1}{k^2}$ or $k^2 \leq \frac{1}{0.01}$
$\therefore k = 10$

Example 2: Find the probability, using Chebyshev's theorem, that the number of driving licences X issued by Road Transport Authority (R.T.A.) in a specific month is between 64 and 184 if the number of driving licences issued X is a random variable with $\mu = 124$ and $\sigma = 7.5$.

Solution: $X = \mu \pm k\sigma$,

For 64, $64 = 124 - k(7.5) \quad \therefore \quad k = 8$

For 184, $184 = 124 - k(7.5) \quad \therefore \quad k = 8$

$$P(|X-\mu| < \sigma) \geq 1 - \frac{1}{k^2} = 1 - \frac{1}{8^2} = 0.984375$$

Example 3: Suppose the amount of thiamine in a slice of 'modern' bread is a random variable X with $\mu = 0.260$ mg and $\sigma = 0.005$ mg. Using Chebyshev's theorem, between what values must be the thiamine content of (a) at least $\frac{35}{36}$ of all slices of 'modern' bread (b) at least $\frac{143}{144}$ of all slices of the 'modern' bread, lies.

Solution: By Chebyshev's theorem

$$P\left(|X-\mu| < k\sigma\right) \geq 1 - \frac{1}{k^2}$$

a. Given $\frac{35}{36} = 0.972 \geq 1 - \frac{1}{k^2} \quad \therefore \quad k \leq 6$,
$\mu = 0.26, \sigma = 0.005$

For $\mu - k\sigma = 0.260 - 6(0.005) = 0.230$

For $\mu + k\sigma = 0.260 + 6(0.005) = 0.290$

Thus at least $\frac{35}{36}$ of all slices of bread will contain thiamine between 0.230 and 0.290.

b. Given $\frac{143}{144} = 0.993 \geq 1 - \frac{1}{k^2} \quad \therefore k \leq 12$

For $\mu - k\sigma = 0.260 - 12(0.005) = 0.200$

For $\mu + k\sigma = 0.260 + 12(0.005) = 0.320$

Thus at least $\frac{143}{144}$ of all slices of bread will contain thiamine between 0.200 and 0.320.

Exercise

Chebyshev's theorem

1. Suppose X is a random variable such that $E(X) = 3$ and $E(X^2) = 13$. Calculate a lower bound for the probability that X lies between -2 and 8 using Chebyshev's theorem.

   Hint: $E(X) = \mu = 3, \sigma^2 = E(X^2) - \left\{E(X)\right\}^2 = 13 - 9 = 4, \sigma = 2$.

$P\{3-2k < X < 3+2k\} \geq 1 - \frac{1}{k^2}$, choose $k = \frac{5}{2}$.

Ans. $P(-2 < X < 8) \geq 1 - \frac{1}{\left(\frac{5}{2}\right)^2} = \frac{21}{25} = 0.84$

2. The number of patients requiring I.C.U. in a hospital in a random variable with $\mu = 18$ and $\sigma = 2.5$. Determine the probability that there will be between 8 and 28 patients.

Ans. $k = \frac{28-18}{2.5} = \frac{18-8}{2.5} = 4$, the probability is at least $1 - \frac{1}{4^2} = \frac{15}{16}$

3. Let X be a random variable with an unknown probability distribution, with mean $\mu = 8$ and variance $\sigma^2 = 9$. Determine
(i) $P(|X-8| \geq 6)$ (ii) $P(-4 < X < 20)$

Ans. **i.** $P\left(|X-8| \geq 6\right) = 1 - P\left(|X-8| < 6\right)$
$= 1 - P(-6 < (X-8) < 6)$
$= 1 - P\left[8-(2)(3) < X < 8+(2)(3)\right] \leq \frac{1}{4}$

ii. $P(-4 < X < 20) =$
$P\left[8-(4)(3) < X < 8+(4)(3)\right] \geq \frac{15}{16}$.

4. Let X be the discrete random variable denoting the number appearing in a single throw of a fair die. Let $E(X) = \mu$
Using Chebyshev's theorem prove that

$$P\left[|X-\mu| > 2.5\right] < 0.47$$

while the actual probability is zero.

Ans. $E(X) = \frac{1}{6}(1+2+3+4+5+6) = \frac{7}{2}$, $E(X^2) = \frac{91}{6}$

$$\text{var}(X) = E(X^2) - \left[E(X)\right]^2 = \frac{91}{6} - \frac{49}{4} = 2.9167$$

Choose $k = 2.5$

$$P\left\{|X-\mu| > 2.5\right\} < \frac{2.9167}{6.25} = 0.47$$

Actual probability $= p = P\{|X-3.5| > 2.5\}$

$= P\{X$ lies outside the limits $(3.5-2.5, 3.5+2.5)\}$

$= P\left\{X \text{ lies outside } (1, 6)\right\} = 0$

since impossible event.

5. Apply Chebyshev's theorem to calculate
(i) $P(5 < X < 15)$ (ii) $P(|X-10| \geq 3)$
(iii) $P(|X-10| < 3)$
for a random variable X with mean $\mu = 10$ and variance $\sigma^2 = 4$.

Hint:

i. $\mu - k\sigma = 10 - k2 = 5$
$\therefore k = \frac{5}{2}$, $\mu + k\sigma = 10 + k2 = 15$
$P(5 < X < 15) = P(10-2k < X < 10+2k) \geq 1 - \frac{1}{k^2} = 1 - \frac{4}{25} = \frac{21}{25}$

ii. $|X-10| \geq 3$ or $-3 < X-10 < 3$ or $10-3 < x < 10+3$
$7 < x < 13 : \mu + k\sigma = 10 + 2k = 13$ and $10 - 2k = 7$
Choose $k = \frac{3}{2}$
$P(|X-10| \geq 3) \leq \frac{1}{k^2} = \frac{4}{9}$

iii. $P(|X-10| < 3) \leq 1 - \frac{4}{9} = \frac{5}{9}$

Ans. (i) at least $\frac{21}{25}$ (ii) at most $\frac{4}{9}$ (iii) at least $\frac{5}{9}$

Theoretical probability distributions

Generally, frequency distributions are formed from the observed or experimental data. However, frequency distributions of certain populations can be deduced mathematically by fitting a theoretical probability distributions under certain assumptions.

Example: The shoes-industry should know the 'sizes' of foot of the population, the food industry the 'tastes' (menu) of the population, etc.

Three such important theoretical probability distributions in order of their discovery are:

i. Binomial (due to James Bernoulli, 1700).
ii. Normal (due to De-Moivre 1733) also credited to Laplace (1774), Gauss (1809).
iii. Poisson (due to S.D. Poisson 1837).

Discrete probability distributions: Binomial, Poisson, geometric, negative binomial, hypergeometric, multinomial, multivariate hypergeometric distributions.

Continuous probability distributions: Uniform (rectangular), normal, Gamma, exponential, χ^2, Beta, bivariate normal, 't', 'F', distributions.

WORKED OUT EXAMPLES

Discrete probability distributions

Example 1: Prove that (a) $E(kX) = kE(X)$ (b) $E(X+k) = E(X)+k$ (c) $E(X+Y) = E(X)+E(Y)$.

Solution:

a. $E(kX) = \frac{\sum kf_i x_i}{\sum f_i} = k\frac{\sum f_i x_i}{\sum f_i} = kE(X)$

b. $E(X+k) = \frac{\sum f_i(x_i+k)}{\sum f_i} = \frac{\sum f_i x_i}{\sum f_i} + k\frac{\sum f_i}{\sum f_i}$

$= E(X) + k$

c. $E(X+Y) = \frac{\sum f_i(x_i+y_i)}{\sum f_i} = \frac{\sum f_i x_i}{\sum f_i} + \frac{\sum f_i y_i}{\sum f_i}$

$= E(X) + E(Y)$

Note 1: Above results can be proved for continuous case by 'replacing' $\sum$ by $\int_{-\infty}^{\infty}$.

Note 2: Above results are rewritten in 'μ' notation as (a)$\mu_{kX} = k\mu_X$ (b) $\mu_{X+k} = \mu_X + k$ (c)$\mu_{X+y} = \mu_X + \mu_Y$.

Example 2: Prove that (a) Var $(X+k) =$ Var (X) (b) Var $(kX) = k^2$ Var (X). Hence $\sigma_{X+k} = \sigma_X$ and $\sigma_{kX} = |k|\sigma_X$.

Solution:

a. Var $(X+k) = \sum(x_i+k)^2 f(x_i) - \mu_{X+k}^2$ by using the result Var $(X) = E(X^2) - \mu_X^2$

$= \sum(x_i^2 + k^2 + 2kx_i)f(x_i) - (\mu_X + k)^2$

$= \sum x_i^2 f_i + k^2 \sum f_i + 2k \sum x_i f_i - (\mu_X^2 + k^2 + 2\mu_X k)$

$= \sum x_i^2 f_i + k^2 + 2k\mu_X - \mu_X^2 - 2k\mu_X - k^2$

$= (\text{Var}\,(X) + \mu_X^2) - \mu_X^2 = \text{Var}\,(X)$

b. Var $(kX) = \sum(kx_i)^2 f_i - \mu_{kx}^2$

$= k^2 \sum x_i^2 f_i - (k\mu_X)^2 = k^2\left(\sum x_i^2 f_i - \mu_X^2\right)$

$= k^2 \text{Var}(X).$

Example 3: Determine the discrete probability distribution, expectation, variance, S.D. of a discrete random variable (D.R.V.) X which denotes the minimum of the two numbers that appear when a pair of fair dice is thrown once.

Solution: The total number of cases are $6 \times 6 = 36$. The minimum number could be 1, 2, 3, 4, 5, 6, i.e., $X(s) = X(a, b) = \min\{a, b\}$. The number 6 will appear only in one case (6, 6), so

$$f(6) = P(X = 6) = P(\{(6, 6)\}) = \frac{1}{36}.$$

For minimum 5, favourable cases are (5, 5), (5, 6), (6, 5) so

$$f(5) = P(X = 5) = \frac{3}{36}.$$

For minimum 4, favourable cases are (4, 4), (4, 5), (4, 6), (5, 4) so

$$f(4) = P(X = 4) = \frac{5}{36}.$$

For minimum 3: (3, 3), (3, 4), (3, 5), (3, 6), (6, 3), (5, 3), (4, 3) so

$$f(3) = P(X = 3) = \frac{7}{36}.$$

For minimum 2: (2, 2), (3, 3), (2, 4), (2, 5), (2, 6), (6, 2), (5, 2), (4, 2), (3, 2) so

$$f(2) = P(X = 2) = \frac{9}{36}.$$

Similarly,

$$f(1) = P(X = 1) = \frac{11}{36}.$$

Thus the required discrete probability distribution

$X = x_i$	1	2	3	4	5	6
$P(X = x_i) = f(x_i) = f_i$	$\frac{11}{36}$	$\frac{9}{36}$	$\frac{7}{36}$	$\frac{5}{36}$	$\frac{3}{36}$	$\frac{1}{36}$

Mean = Expectation = $E(X) = \sum x_i f_i$

$$E(X) = 1 \cdot \frac{11}{36} + 2 \cdot \frac{9}{36} + 3.\frac{7}{36} + 4 \cdot \frac{5}{36} + 5 \cdot \frac{3}{36} + 6 \cdot \frac{1}{36} = 2.5$$

$$\text{Var}\,(X) = \sum x^2 f_i - \mu^2$$

$$= 1 \cdot \frac{11}{36} + 4 \cdot \frac{9}{36} + 9 \cdot \frac{7}{36} + 16 \cdot \frac{5}{36} + 25 \cdot \frac{3}{36} + 36 \cdot \frac{1}{36} - (2.5)^2$$

$\sigma^2 = 1.9745$, so $\sigma =$ S.D. $= 1.4$.

Example 4: A player tosses 3 fair coins. He wins Rs. 500 if 3 heads occur, Rs. 300 if 2 heads occur, Rs. 100 if one head occurs. On the other hand, he loses Rs. 1500 if 3 tails occur. Find the value of the game to the player. Is it favourable?

Solution: Let X = D.R.V. = number of heads occurring in 3 tosses of a fair coin. The sample space S is

$$S = \{H, T\} \times \{H, T\} \times \{H, T\}$$
$$= \{HHH, HHT, HTH, HTT, THH, THT, TTH, TTT\}.$$

Probability of all 3 heads $= P(X = 3) = \frac{1}{8}$
Probability of all 3 tails $= P(X = 0) = \frac{1}{8}$
Probability of 2 heads $= P(X = 2) = \frac{3}{8}$,
$P(X = 1) = \frac{3}{8}$
Discrete probability distribution is

$X = x_i$	0	1	2	3
$P(X = x_i) = f(x_i)$	$\frac{1}{8}$	$\frac{3}{8}$	$\frac{3}{8}$	$\frac{1}{8}$

Expected value of the game

$$= 500 \cdot \frac{1}{8} + 300 \cdot \frac{3}{8} + 100 \cdot \frac{3}{8} - 1500 \cdot \frac{1}{8}$$

$$= \frac{200}{8} = 25 \text{ rupees.}$$

Game is favourable to the player since $E > 0$.

Continuous probability distributions

Example 5: Suppose a continuous R.V. x has the probability density

$$f(x) = \begin{cases} k(1 - x^2) & \text{for } 0 < x < 1 \\ 0 & \text{elsewhere} \end{cases}$$

(a) Find k (b) Find $P(0.1 < x < 0.2)$ (c) $P(x > 0.5)$ Using distribution function, determine the probabilities that (d) x is less than 0.3 (e) between 0.4 and 0.6 (f) Calculate mean and variance for the probability density function.

Solution:

a. Since $\int_{-\infty}^{\infty} f(x)dx = 1$ so

$$\int_{-\infty}^{\infty} f(x) = \int_0^1 k(1 - x^2)dx = k\left(x - \frac{x^3}{3}\right)\Big|_0^1$$
$$= \frac{2}{3}k = 1 \quad \therefore k = \frac{3}{2}.$$

b. $P(0.1 < x < 0.2) = \int_{0.1}^{0.2} k(1 - x^2)dx$

$$= \frac{3}{2}\left(x - \frac{x^3}{3}\right)\Big|_{0.1}^{0.2} = 0.1465.$$

c. $P(x > 0.5) = \int_{0.5}^{\infty} f(x)dx = \int_{0.5}^{1} f(x)dx$

$$= \frac{3}{2}\left(x - \frac{x^3}{3}\right)\Big|_{0.5}^{1} = 0.3125.$$

d. Distribution function:

$F(x) = \int_{-\infty}^{x} f(t)dt$ so

$$F(x) = \int_0^x \frac{3}{2}(1 - x^2)dx = \frac{3}{2}\left(x - \frac{x^3}{3}\right)$$

$$F(x < 0.3) = \int_{-\infty}^{0.3} f(t)dt = \frac{3}{2}\left(x - \frac{x^3}{3}\right)\Big|_0^{0.3}$$
$$= 0.4365$$

e. $F(0.4 < x < 0.6) = F(b) - F(a)$
$= F(0.6) - F(0.4)$

$$= \frac{3}{2}\left(x - \frac{x^3}{3}\right)\Big|_{0.4}^{0.6} = 0.224.$$

f. Mean $= \mu = \int_{-\infty}^{\infty} xf(x)dx = \int_0^1 x\{\frac{3}{2}(1 - x^2)\}dx$

$$= \frac{3}{2}\left(\frac{x^2}{2} - \frac{x^4}{4}\right)\Big|_0^1 = \frac{3}{8}$$

Variance $= \int_{-\infty}^{\infty}(x - \mu)^2 f(x)dx$

$$= \int_0^1 \left(x - \frac{3}{8}\right)^2 \frac{3}{2}(1 - x)^2 dx = \frac{19}{320}$$

or variance $= \int_0^1 x^2\{k(1 - x^2)\}dx - \mu^2$

$$= k\left(\frac{x^3}{3} - \frac{x^5}{5}\right)\Big|_0^1 - \mu^2 = \frac{19}{320}.$$

Example 6: The daily consumption of electric power (in millions of kW-hours) is a R.V. having the

$$\text{P.D.F.} \qquad f(x) = \begin{cases} \frac{1}{9}xe^{-x/3}, & x > 0 \\ 0, & x \le 0 \end{cases}$$

If the total production is 12 million kW-hours, determine the probability that there is power cut (shortage) on any given day.

Solution: Probability that the power consumed is between 0 to 12 is

$$P(0 \le x \le 12) = \int_0^{12} f(x)dx = \int_0^{12} \frac{1}{9} x e^{-x/3} dx$$

$$= -\frac{x}{3} e^{-x/3} - e^{-x/3}\Big|_0^{12} = 1 - 5e^{-4}$$

Power supply is inadequate if daily consumption exceeds 12 million kW, i.e.,

$$P(x > 12) = 1 - P(0 \le x \le 12) = 1 - [1 - 5e^{-4}] = 5e^{-4}$$

$$= 0.0915781$$

Example 7: (a) Find the mean and variance of a uniform (rectangular) distribution (b) Determine its cumulative distribution function (Fig. 27.3).

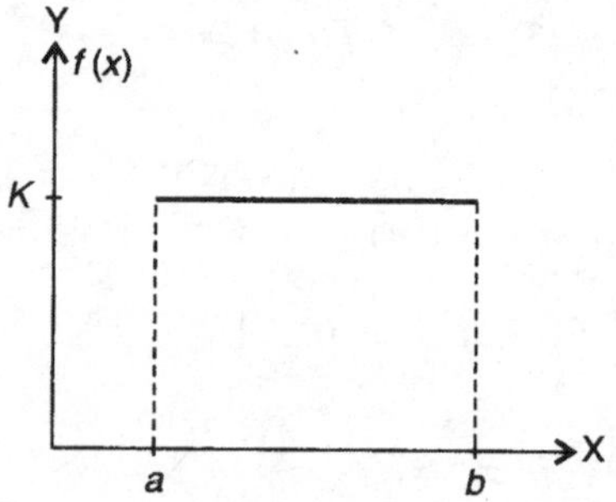

Fig. 27.3

Solution: The uniform distribution is defined by

$f(x) = k =$ constant in (a, b)
$= 0$ elsewhere

Since $\quad 1 = \int_{-\infty}^{\infty} f(x)dx = \int_a^b k dx = k(b-a),$

so $\quad k = \frac{1}{b-a}$

Thus $\quad f(x) = \begin{cases} \frac{1}{b-a}, & \text{for } a < x < b \\ 0, & \text{elsewhere} \end{cases}$

$$\text{Mean} = \mu = \int_a^b x f(x)dx = \int_a^b x \cdot \frac{dx}{b-a}$$

$$= \frac{1}{b-a} \cdot \frac{x^2}{2}\Big|_a^b = \frac{b^2 - a^2}{2(b-a)}$$

$$\mu = \frac{b+a}{2}$$

$$\text{Variance} = \int_a^b x^2 f(x)dx - \mu^2 = \frac{1}{b-a} \cdot \int_a^b x^2 dx - \mu^2$$

$$\sigma^2 = \frac{1}{b-a} \frac{x^3}{3}\Big|_a^b - \mu^2 = \frac{b^3 - a^3}{3(b-a)} - \left(\frac{b+a}{2}\right)^2$$

$$= \frac{(b-a)^2}{12}$$

Cumulative distribution $F(x)$:

i. When $x \le a$, $\quad F(x) = \int_{-\infty}^{x} = \int 0 = 0$

so $F(x) = 0$

ii. When $a < x < b$,

$F(x) = \int_{-\infty}^{x} = \int_{-\infty}^{a} 0 \cdot dx + \int_a^x \frac{1}{b-a} dx$

$F(x) = \frac{x-a}{b-a}$

iii. When $x \ge b$, $\quad F(x) = \int_{-\infty}^{x} = \int_{-\infty}^{a} 0$

$+ \int_a^b \frac{1}{b-a} dx + \int_b^x 0 = \frac{b-a}{b-a} = 1.$

Thus

$$f(x) = \begin{cases} 0, & \text{when } x \le a \\ \frac{x-a}{a-b}, & \text{when } a < x < b \\ 1, & \text{when } x \ge b. \end{cases}$$

Exercise

Probability distributions

1. Calculate μ, σ^2, σ for

a.

x_i	2	3	8
f_i	$\frac{1}{4}$	$\frac{1}{2}$	$\frac{1}{4}$

b.

x_i	−1	0	1	2	3
f_i	.3	.1	.1	.3	.2

Ans. **a.** $\mu = 2 \cdot \frac{1}{4} + 3 \cdot \frac{1}{2} + 8 \cdot \frac{1}{4} = 4,$
$\sigma^2 = (2-4)^2 \frac{1}{4} + (3-4)^2 \frac{1}{2} + (8-4)^2 \frac{1}{4}$
$= 5.5$
$\sigma = 2.3$ [or $\sigma^2 = 4 \cdot \frac{1}{4} + 9 \cdot \frac{1}{2} + 64 \cdot \frac{1}{4} - 4^2$
$= 5.5$]

b. $\mu = -1 \cdot (.3) + 0 \cdot (.1) + 1 \cdot (0.1)$
$+ 2 \cdot (0.3) + 3 \cdot (0.2) = 1.0$
$\sigma^2 = 4(.3) + 0(.1) + 1(.1) + 4(.3) + 9(.2)$
$= 2.4, \sigma = 1.5$

2. Determine the expected number of families to have (a) 2 boys and 2 girls (b) at least one boy (c) no girls (d) at most two girls, out of 800 families with 4 children each. Assume equal probabilities for boys and girls.

Ans. (a) 37.5% (b) 93.75% (c) 6.25% (d) 68.75%

Hint: X: No boys in a family. P.D.F.

X	0	1	2	3	4
$p(x_i)$	$\frac{1}{16}$	$\frac{4}{16}$	$\frac{6}{16}$	$\frac{4}{16}$	$\frac{1}{16}$

Percentage of families $= \frac{6}{16} \times 100 = 37.5\%$ for (a) etc.

3. A box contains 8 items of which 2 are defective. A person draws 3 items from the box. Determine the expected number of defective items he has drawn.

Ans. $0 \cdot \frac{20}{56} + 1 \cdot \frac{30}{56} + 2 \cdot \frac{6}{56} = \frac{3}{4}$

Hint:

X	0	1	2
$f(x_i)$	$\frac{6c_3 2c_0}{8c_3}$	$\frac{6c_2 2c_1}{8c_3}$	$\frac{6c_1 2c_2}{8c_3}$

Here X is the number of defectives.

4. A stake of Rs. 44 is to be won between 2 players A and B, whoever gets 6 in a throw of die alternatively. Determine their respective expectations if A starts the game.

Ans. $E(A) = \frac{6}{11} \times 44 = \text{Rs. } 24$,

$E(B) = \frac{5}{11} \times 44 = \text{Rs. } 20$

Hint: $P(A \text{ wins}) = \frac{1}{6} + \frac{5}{6} \cdot \frac{5}{6} \cdot \frac{1}{6} + \frac{5}{6} \cdot \frac{5}{6} \cdot \frac{5}{6} \cdot \frac{5}{6} \cdot \frac{1}{6} + \cdots = \frac{1}{6} \frac{1}{1-(\frac{5}{6})^2} = \frac{6}{11}$

$p =$ probability of $6 = \frac{1}{6}$, prob no $6 = \frac{5}{6}$.

5. A person wins Rs. 80 if 3 heads occur, Rs. 30 if 2 heads occur, Rs. 10 if only one head occurs in a single toss of 3 fair coins. If the game is to be fair, how much should he lose if no heads occur?

Ans. Rs. 200

Hint: $0 =$ Expectation $= 80 \cdot \frac{1}{8} + 30 \cdot \frac{3}{8} + 10 \cdot \frac{3}{8} - x \cdot \frac{1}{8}$

6. Find the mean and variance of P.D.F.

$$f(x) = \begin{cases} \frac{1}{4} e^{-x/4} & \text{for } x > 0 \\ 0 & \text{elsewhere.} \end{cases}$$

Ans. $\mu = \int_0^\infty \frac{1}{4} e^{-x/4} dx = \frac{e^{-x/4}}{-1/4} = 4 \cdot 1 = 4$

$\sigma^2 = \int_0^\infty (x-4)^2 \frac{1}{4} e^{-x/4} dx = \int_0^\infty x^2 \frac{1}{4} e^{-x/4} dx - \mu^2 = 96 - 16 = 80$

7. If P.D.F. $f(x) = k(x+3)$ in $(2, 8)$, determine (a) $P(3 < x < 5)$ (b) $P(x \geq 4)$ (c) $P(|x-5| < 0.5)$

Ans. $k = \frac{1}{48}$, (a) $\frac{7}{24}$ (b) $\frac{3}{4}$

(c) $P(4.5 < x < 5.5) = \frac{1}{6}$.

8. Find the mean and variance of the "exponential" distribution $f(x) = \frac{1}{b} e^{-x/b}$ for $x > 0, b > 0$.

Ans. $\mu = b, \sigma^2 = b^2$

Hint: $\mu = \int_0^\infty x \cdot \frac{1}{b} e^{-x/b} dx = b\Gamma(2) = b$,

$\sigma^2 = \int_n^\infty x^2 \frac{1}{b} e^{-x/b} dx - \mu^2 = b^2\Gamma(3) - b^2$

$= 2b^2 - b^2 = b^2$

27.3 DISCRETE UNIFORM DISTRIBUTION

Discrete uniform distribution is the simplest of all discrete probability distribution. The discrete random variable assumes each of its values with the same (equal) probability. In this equiprobable or uniform space each sample point is assigned equal probabilities.

Example 1: In the tossing of a fair die, each sample point in the sample space $\{1, 2, 3, 4, 5, 6\}$ is assigned with the same (uniform) probability $\frac{1}{6}$ i.e. $p(x) = \frac{1}{6}$ for $x = 1, 2, 3, 4, 5, 6$. In particular, if the sample space S contains k points, then probability of each point is $\frac{1}{k}$. Thus in the discrete uniform distribution, the discrete random variable X assigns equal probabilities to all possible values of X. Therefore the probability mass function $f(X)$ has the form

$$f(X) = \frac{1}{k} \text{ for } X = x_1, x_2, \ldots, x_k \tag{1}$$

or equivalently

X	x_1	x_2	x_3	...	x_k
$f(X)$	$\frac{1}{k}$	$\frac{1}{k}$	$\frac{1}{k}$	...	$\frac{1}{k}$

Here the constant K which is the *parameter* completely determines the discrete uniform distribution (1). The mean and variance of (1) are given by

$$\mu = E(X) = \sum_{i=1}^{k} x_i f(x_i) = \sum x_i \cdot \frac{1}{k} = \frac{\sum_{i=1}^{k} x_i}{k}$$

and

$$\sigma^2 = E((X-\mu)^2) = \sum_{i=1}^{k} (x_i - \mu)^2 f(x_i)$$

$$= \sum_{i=1}^{k} \frac{(x_i-\mu)^2}{k}$$

$$\sigma^2 = \frac{1}{k}\sum_{i=1}^{k} ()x_i - \mu)^2$$

The discrete uniform distribution is of particular importance in lotteries.

Worked Out Examples

Example 2: If a ticket is drawn from a box containing 10 tickets numbered 1 to 10 inclusive. Find the probability that the number x drawn is (a) less that 4 (b) even number (c) prime number (d) find the mean and variance of the random variable X.

Solution: (a) since each ticket has the same probability for being drawn, the probability distribution is discrete uniform distribution given by $f(x) = \frac{1}{10}$ for $x = 1, 2, 3, \ldots, 10$. Now

$$P(x<4) = \sum_{x=0}^{3} P(x) = \sum_{x=0}^{3} \frac{1}{10} = \frac{1}{10}(1+1+1) = \frac{3}{10}$$

(b) 2, 4, 6, 8, 10 are even numbers each with probability $\frac{1}{10}$ probability of even number $= \frac{5}{10} = \frac{1}{2}$ (c) 2, 3, 5, 7 are prime prob. of prime $= \frac{4}{10} = \frac{2}{5}$. (d) Mean $= E(x) = \sum_{x=1}^{10} xP(x) = \sum_{x=1}^{10} x \cdot \frac{1}{10} = \frac{1}{10}(1+2+3+\cdots+10)$

Mean $= \mu = \frac{1}{10} \cdot \frac{10(10+1)}{10\ 2} = \frac{11}{2} = 5\cdot 5$

Variance $= \sigma^2 = \sum_{x=1} (x-\mu)^2 p(x)$

$$= \frac{1}{10}[(1-5\cdot5)^2 + (2-5\cdot5)^2 + (3-5\cdot5)^2$$

$$+ \ldots + (10-5\cdot5)^2]$$

$$\sigma^2 = 8.25$$

(or $\sigma^2 = E(x^2) - \{E(x)\}^2 = \sum x^2 \cdot P(x) - (5\cdot5)^2$

$$= \frac{1}{10}\cdot\frac{n(n+1)(2n+1)}{6} = \frac{1}{10}\frac{10(10+1)(20+1)}{6}$$

$$-(5\cdot5)^2$$

$38.5 - 30.25 = 8.25$)

Exercise

1. Deteremine the probability that an odd number appears in the toss of a fair die.

Ans. $\frac{3}{6}$

2. Find the probability that at least one head appears in the throw of three fair coins.

Ans. $\frac{7}{8}$

3. If a card is selected at random from an ordinary pack of 52 cards, find the probability that (a) card is a spade (b) card is a face card i.e., Jack, queen or king (c) card is a spade face card.

Ans. $\frac{13}{52}$ (b) $\frac{12}{52}$ (c) $\frac{3}{52}$

Hint: (a) 13 spade cards available. So $\binom{13}{1}/\binom{52}{1}$

(b) 12 face cards so $\binom{12}{1}/\binom{52}{1}$

(c) 3 spade face cards so $\binom{3}{1}/\binom{52}{1}$.

Note: Each card has the same probability $\frac{1}{52}$.

4. Find the probability that a card drawn at random from 50 cards numbered 1 to 50 is (a) prime, (b) ends in the digit 2, (c) divisible by 5.

Ans. (a) $\frac{3}{10}$ (b) $\frac{1}{10}$ (c) $\frac{1}{5}$

5. Two marbles are drawn from a box containing 4 red and 8 black marbles. Find the probability that (a) both are red (b) both are black (c) at least one is red.

Ans. (a) $\binom{4}{2}/\binom{12}{2} = \frac{1}{11}$,

(b) $\binom{8}{2}/\binom{12}{2} = \frac{14}{33}$ (c) $1 - \frac{14}{33} = \frac{19}{33}$

6. If two cards are drawn from an ordinary pack of 52 cards, determine the probability that (a) both are spades (b) one is a spade and one is a heart.

Ans. (a) $\binom{13}{2}/\binom{15}{2} = \frac{78}{1326} = \frac{1}{17}$

(b) $(13.13)\Big/\binom{15}{2} = \frac{13}{102}$

27.4 BINOMIAL DISTRIBUTION

Binomial distribution (B.D.) due to James Bernoulli (1700) is a discrete probability distribution. The Bernoulli process has the following properties:

i. An experiment is repeated n number of times, called n trials where n is a fixed integer.

ii. The outcome of each trial is classified into two mutually exclusive (dichotomus) categories arbitrarily called a "success" and a "failure".

iii. Probability of success, denoted by p, remains constant for all trials.

iv. The outcomes are independent (of the outcomes of the previous trials).

Each trial in the Bernoulli process is known as Bernoulli trial.

The binomial random variable X is the *number* of successes in n Bernoulli trials. X is discrete since X takes only integer values (we 'count' the number of successes).

Binomial distribution is thus the probability distribution of this discrete random variable X, and is given by

$$b(x; n, p) = \binom{n}{x} p^x q^{n-x}, x = 0, 1, 2, \ldots, n \quad (1)$$

where n is the number of trials and p is the probability of success in any trial. The probability of x successes is p^x and remaining failures is q^{n-x}. This can happen in n_{C_x} ways. By multiplication rule, the probability of x successes in n trials is $\binom{n}{x} p^x q^{n-x}$.

Note that the $(n+1)$ terms of the binomial expansion

$$(q+p)^n = \binom{n}{0} q^n + \binom{n}{1} pq^{n-1} + \binom{n}{2} p^2 q^{n-2} + \cdots + \binom{n}{n} p^n$$

$$= b(0; n, p) + b(1; n, p) + b(2; n, p) + \cdots + b(n; n, p)$$

$$= \sum_{x=0}^{n} b(x; n, p)$$

correspond to various values of $b(x; n, p)$ for $x = 0, 1, 2, \ldots, n$.

Since $p + q = 1$, it follows that

$$\sum_{x=0}^{n} b(x; n, p) = 1.$$

B.D. is characterized by the parameter p and the number of trials n.

The mean μ of B.D. is np and the variance σ^2 of B.D. is npq. (see Worked Out Example 1 on page 27.11)

The binomial sums

$$B(r; n, p) = \sum_{x=0}^{r} b(x; n, p)$$

are tabulated (see A2 to A7) since

$$\frac{n_{C_{x+1}}}{n_{C_x}} = \frac{n!}{(x+1)!(n-x-1)!} \cdot \frac{x!(n-x)!}{n!} = \left(\frac{n-x}{x+1}\right)$$

The recurrence relation for B.D. is

$$b(x+1; n, p) = \left(\frac{n-x}{x+1}\right)\left(\frac{p}{q}\right) b(x; n, p).$$

WORKED OUT EXAMPLES

Binomial distribution

Example 1: Find the (a) mean and (b) variance of B.D.

Solution:

a. mean $= \mu =$ expectation $= \sum_{x=0}^{n} x P(x)$

$$= \sum_{x=0}^{n} x b(x; n, p) = \sum_{x=0}^{n} x \frac{x!}{x!(n-x)!} p^x q^{n-x}$$

$$= np \sum_{x=1}^{n} \frac{(n-1)!}{(n-x)!(x-1)!} p^{x-1} q^{n-x}, \text{ put } x - 1 = x^*$$

$$= np \sum_{x^*=0}^{n-1} \frac{(n-1)!}{(n-x^*-1)!x^*!} p^{x^*} q^{n-x^*-1}$$

$$= np(p+q)^{n-1} = np \quad \text{since } p + q = 1.$$

b. Variance $\sigma^2 = \sum_{x=0}^{n}(x-\mu)^2 p(x)$

$$= \sum_{x=0}^{n}(x^2 - 2\mu x + \mu^2)p(x)$$

$$= \sum_{x=0}^{n} x^2 p(x) - 2\mu \sum_{x=0}^{n} xp(x) + \mu^2 \sum_{x=0}^{n} p(x)$$

$$= \sum_{x=0}^{n} x^2 p(x) - 2\mu \cdot np + \mu^2 \cdot 1$$

$$\because \sum xp(x) = \mu, \quad \sum p(x) = 1$$

$$= \sum_{x=0}^{n} x^2 p(x) - n^2 p^2 \quad \text{since } \mu = np$$

Consider

$$\sum_{x=0}^{n} x^2 p(x) = \sum_{x=1}^{n} x^2 \frac{n!}{(n-x)!x!} p^x q^{n-x}$$

$$= \sum_{x=1}^{n} [x(x-1) + x] \frac{n!}{(n-x)!x!} p^x q^{n-x}$$

$$= \sum_{x=2}^{n} \frac{x(x-1)n!}{(n-x)!x!} p^x q^{n-x} + \sum_{x=1}^{N} x \frac{n!}{(n-x)!x!} p^x q^{n-x}$$

$$= \sum_{x=2}^{n} \frac{n!}{(n-x)!(x-2)!} p^x q^{n-x} + np$$

$$= n(n-1)p^2 \sum_{x=2}^{n} \frac{(n-2)!}{(n-x)!(x-2)!} p^{x-2} q^{n-x} + np$$

$$= n(n-1)p^2 \cdot (q+p)^{n-2} + np$$

$$= n(n-1)p^2 + np \quad \text{since } p + q = 1$$

Thus

$$\sigma^2 = n(n-1)p^2 + np - n^2p^2 = np - np^2$$

$$\sigma^2 = np(1-p) = npq.$$

Example 2: Determine the probability of getting 9 exactly twice in 3 throws with a pair of fair dice.

Solution: In a single throw of a pair of fair dice, 9 can occur in 4 ways: (6, 3), (3, 6), (5, 4), (4, 5) out of $6 \times 6 = 36$ ways. Thus

p = probability of occurrence of 9 in one throw= $\frac{4}{36} = \frac{1}{9}$.

n = number of trials = 3.

Probability of getting 9 exactly twice in 3 throws

$$= b\left(2; 3, \frac{1}{9}\right) = 3_{C_2}\left(\frac{1}{9}\right)^2\left(\frac{8}{9}\right)^{3-2} = 3 \cdot \frac{1}{9} \cdot \frac{1}{9} \cdot \frac{8}{9} = \frac{8}{243}.$$

Example 3: Out of 800 families with 5 children each, how many would you expect to have (a) 3 boys (b) 5 girls (c) either 2 or 3 boys. Assume equal probabilities for boys and girls.

Solution: Probability of boy $= P(B) = p = \frac{1}{2}$, and probability of girl $= P(G) = q = \frac{1}{2}$.

n = number of trials = 5, X = no. of boys in a family

a. Probability of a family having 3 boys

$$= P(X=3) = 5_{C_3}\left(\frac{1}{2}\right)^3\left(\frac{1}{2}\right)^{5-2}$$

$$= \frac{5!}{2!3!}\left(\frac{1}{2}\right)^5 = \frac{10}{32} = \frac{5}{16}$$

Expected number of families having 3 boys out of 5 children $= 800\left(\frac{5}{16}\right) = 250$, i.e., 250 families have 3 boys out of 5 children.

b. $P(X=0) = P$ (all girls) $= \left(\frac{1}{2}\right)^5 = \frac{1}{32}$

Expectation $= 800 \times \frac{1}{32} = 25$.

c. $P(X=2) = P(2 \text{ boys}) = 5_{C_2}\left(\frac{1}{2}\right)^5 = \frac{5}{8}$

Expectation $= 800 \times \frac{5}{8} = 500$.

Example 4: Determine the probability distribution of the number of bad eggs in a box of 6 chosen at random if 10% of eggs are bad, in a large consignment.

Solution: Probability of a bad egg $= p = \frac{10}{100} = 0.1$. Let X = number of bad eggs, $n = 6$. The required B.D. $= b(x; 6, 0.1) = 6_{C_x}(.1)^x(.9)^{6-x}$, for $x = 0, 1, 2, 3, 4, 5, 6$.

X:	0	1	2	3	4	5	6
$P(X)$:	.5311	.35429	0.098	0.015	0.001215	0.000054	0

Example 5: Assume that 50% of all engineering students are good in mathematics. Determine the probabilities that among 18 engineering students (a)

exactly 10 (b) at least 10 (c) at most 8 (d) at least 2 and at most 9, are good in maths.

Solution: Let X = number of engineering students who are good in maths:
p = prob of good in maths $= \frac{50}{100} = \frac{1}{2}$, $n = 18$
$b(x; n, p) = b(x; 18, \frac{1}{2}) = 18_{C_x}\left(\frac{1}{2}\right)^x\left(\frac{1}{2}\right)^{18-x}$

a. Exactly 10 students out of 18 are good in maths

$$P(X = 10) = 18_{C_{10}}\left(\frac{1}{2}\right)^{10}\left(\frac{1}{2}\right)^{8} = .1670$$

From tables (A2 to A7)

$$P(X = 10) = \sum_{x=0}^{10} b\left(x; 18, \frac{1}{2}\right) - \sum_{x=0}^{9} b\left(x; 18, \frac{1}{2}\right)$$

$$= .7597 - .5927 = .1670$$

b. $$P(X \geq 10) = \sum_{x=10}^{18} 18_{C_x}\left(\frac{1}{2}\right)^x\left(\frac{1}{2}\right)^{18-x}$$

$$= \sum_{x=0}^{18} b\left(x; 18, \frac{1}{2}\right) - \sum_{x=0}^{9} b\left(x; 18, \frac{1}{2}\right)$$

$$= 1 - .5927 = .4073$$

c. $$P(X \leq 8) = \sum_{x=0}^{8} 18_{C_x}\left(\frac{1}{2}\right)^x\left(\frac{1}{2}\right)^{18-x} = .4073$$

from table with $n = 18$, $x = 8$, $p = \frac{1}{2}$.

d. $$P(2 \leq x \leq 9) = \sum_{x=2}^{9} 18_{C_x}\left(\frac{1}{2}\right)^x\left(\frac{1}{2}\right)^{18-x}$$

$$= \sum_{x=0}^{9} b\left(x; 18, \frac{1}{2}\right) - \sum_{x=0}^{1} b\left(x; 18, \frac{1}{2}\right).$$

$$= .5927 - .0007 = .5920.$$

Example 6: The probability of a man hitting a target is $\frac{1}{3}$. (a) If he fires 5 times, what is the probability of his hitting the target at least twice? (b) How many times must he fire so that the probability of his hitting the target at least once is more than 90%?

Solution: Probability of hitting $= p = \frac{1}{3}$

probability of no hit (or failure) $= q = \frac{2}{3}$

a. X = number of hits (successes), $n = 5$

$$P(X \geq 2) = \sum_{x=2}^{5} 5_{C_x}\left(\frac{1}{3}\right)^x\left(\frac{2}{3}\right)^{5-x}$$

$$= 1 - \sum_{x=0}^{1} 5_{C_x}\left(\frac{1}{3}\right)^x\left(\frac{2}{3}\right)^{5-x}$$

$$= 1 - \left(\frac{2}{3}\right)^5 - 5_{C_1}\left(\frac{1}{3}\right)\left(\frac{2}{3}\right)^4 = \frac{131}{243}.$$

b. The probability of not hitting the target is q^n in n trials (fires). Thus to find the smallest n for which the probability of hitting at least once $1 - q^n$ is more than 90%.

i.e., $$1 - q^n > 0.9$$

or $$1 - \left(\frac{2}{3}\right)^n > 0.9 \quad \text{i.e.,} \quad \left(\frac{2}{3}\right)^n < 0.1$$

For $n = 6$, $2^6 = 64 < (0.1)3^6 = 72.9$ this is true. In other words, he must fire 6 times.

Example 7: If X be a binomially distributed random variable with $E(X) = 2$ and Var $(X) = \frac{4}{3}$, find the distribution of X.

Solution: We know that $E(X)$ = mean $= np = 2$ and Var (X) = Variance $= npq = \frac{4}{3}$. or $\frac{npq}{np} = \frac{4}{3}\frac{1}{2} = \frac{2}{3}$ or $q = \frac{2}{3}$ so $p = \frac{1}{3}$. Thus $n\frac{1}{3} = 2$ or $n = 6$.
Hence the B.D. is $b(x; 6, \frac{1}{3})$

x_i	0	1	2	3	4	5	6
$f(x_i)$	$\frac{64}{729}$	$\frac{192}{729}$	$\frac{240}{729}$	$\frac{160}{729}$	$\frac{60}{729}$	$\frac{12}{729}$	$\frac{1}{729}$

Example 8: Fit a binomial distribution to the following data:

X	0	1	2	3	4
f	30	62	46	10	2

Solution: Here n = no. of trials = 4 and N = total frequency $= \sum_{i=0}^{4} f_i = 30 + 62 + 46 + 10 + 2 = 150$.

Mean of the binomial distribution is

$$\mu = np = \frac{\sum f_i x_i}{\sum f_i} = \frac{0(30)+1(62)+2(46)+3(10)+4(12)}{150}$$

$$np = 4p = \frac{192}{150} \quad \therefore p = \frac{192}{600} = 0.32$$

Thus the binomial distribution that fits the given data is $b(x; 4, 0.32) = 4_{C_x}(.32)^x(.68)^{4-x} = p(x)$

x:	0	1	2	3	4
$P(x)$:	.2138	0.4	0.2866	0.0866	0.0133
Expected frequency = $N \times P(x)$ = $(150)P(x)$	32	60	43	13	2

Hint: Use the recurrence relation

$$b(x+1; n, p) = \left(\frac{n-x}{x+1}\right)\frac{p}{q} b(x; n, p)$$

EXERCISE

Binomial distribution

1. A fair coin is tossed 6 times. Find the probability of getting (a) exactly 2 heads (b) at least four heads (c) no heads (d) at least one head.

Ans. (a) $\frac{15}{64}$ (b) $\frac{11}{32}$ (c) $\frac{1}{64}$ (d) $\frac{63}{64}$

Hint:

a. $b(2; 6, \frac{1}{2}) = 6_{C_2}\left(\frac{1}{2}\right)^2\left(\frac{1}{2}\right)^4$

b. $\sum_{x=4}^{6} b\left(x; 6, \frac{1}{2}\right) = \frac{15}{64} + \frac{6}{64} + \frac{1}{64}$

c. $1 - q^6 = 1 - \left(\frac{1}{2}\right)^6$

2. A fair die is tossed 7 times. Determine the probability that a 5 or a 6 appears (a) exactly 3 times (b) never occurs.

Ans. (a) $\frac{560}{2187}$ (b) $\frac{2059}{2187}$

Hint: (a) $b\left(3; 7, \frac{1}{3}\right)$ (b) $1 - q^7 = 1 - \left(\frac{2}{3}\right)^7$

3. Team A has probability $\frac{2}{3}$ of winning whenever it plays. If A plays 4 games, find the probability that A wins (i) exactly 2 games (ii) at least 1 game (iii) more than half of the games.

Ans. (i) $P(2) = b\left(2; 4, \frac{2}{3}\right) = \frac{8}{27}$ (ii) $1 - q^4 = 1 - \left(\frac{1}{3}\right)^4 = \frac{80}{81}$ (iii) $P(3)+P(4) = \frac{32}{81} + \frac{16}{81} = \frac{16}{27}$

4. How many dice must be thrown so that there is a better than even chance of obtaining a six?

Ans. 4 dice

Hint: Find n such that $\left(\frac{5}{6}\right)^n < \frac{1}{2}$.

5. A man hits a target with probability $\frac{1}{4}$. (i) Determine the probability of hitting at least twice when he fires 7 times (ii) How many times must he fire so that the probability of his hitting the target at least once is greater than $\frac{2}{3}$?

Ans. **i.** $1 - P(0) - P(1) = 1 - \frac{2187}{16384} - \frac{5103}{16384} = \frac{4547}{8192}$

ii. $n = 4$

Hint: Find n such that $1 - q^n > \frac{2}{3}$.

6. The probability that a pen manufactured by a company will be defective is 0.1. If 12 such pens are examined, find the probability that (a) exactly two (b) at least two (c) none, will be defective.

Ans. (a) 0.2301 (b) 0.3412 (d) 0.2833

7. In sampling a large number of parts manufactured by a machine, the mean number of defectives in a sample of 20 is 2. Out of 1000 such samples, how many would be expected to contain at least 3 defective parts?

Ans. 323

Hint: Mean $= 2 = np = 20p, \ p = 0.1$,

$$P(X \geq 3) = 1 - \sum_{x=0}^{2} b(x; 20, 0.1) = 0.323$$

Expected number $= 1000 \times 0.323 = 323$.

8. The probability that a patient recovers from a disease is 0.4. If 15 persons have such a disease, determine the probability that (a) exactly 5 survive (b) at least 10 survive (c) from 3 to 8 survive.

Ans. **a.** $P(X=5) = b(5; 15, 0.4) = \sum_{x=0}^{5} b(x, 15, 0.4)$
$- \sum_{x=0}^{4} b(x; 15, 0.4) = 0.4032 - 0.2173$

$= 0.1859.$

b. $P(X \geq 10) = 1 - P(X < 10)$

$= 1 - \sum_{x=0}^{9} b(x; 15, 0.4) = 1 - 0.9662$

$= 0.0338.$

c. $P(3 \leq X \leq 8) = \sum_{x \geq 3}^{\infty} = \sum_{x=0}^{\infty} - \sum_{x=0}^{2}$

$= 0.9050 - 0.0271 = 0.8779.$

9. A manufacturer of fax machine claims that only 10% of his machines require repairs within one year. If 5 of 20 of his machines required repaires within 1 year, does this tend to support or refute the claim?

Ans. Reject (refute) the claim since probability is very small.

Hint: $\sum_{x=5}^{20} b(x; 20, 0.10) = 1 - \sum_{x=0}^{4} b(x; 20, 0.10)$

$= 1 - 0.9568 = 0.0432.$

10. Two dice are thrown 120 times. Find the average number of times in which the number on first dice exceeds the number on the second dice.

Ans. $E(X) = np = 120\left(\frac{5}{12}\right) = 50$

Hint: Successful are (2, 1), (3, 1), (3, 2), (4, 1), (4, 2), (4, 3), (5, 1), (5, 2), (5, 3), (5, 4), (6, 1), (6, 2), (6, 3), (6, 4), (6, 5), $p = \frac{15}{6.6} = \frac{5}{12}$.

11. A communication system consists of n components, each of which will independently function with probability p. The total system will be able to operate effectively if at least one half of its components function. For what values of p is a 5-component system more likely to operate effectively than a 3-component system?

Ans. $p \geq \frac{1}{2}$

Hint: $P(X \geq 3) = \sum_{x=3}^{5} b(x; 5, p) \geq P(X \geq 2)$

$= \sum_{x=2}^{3} b(x; 3, p).$

Fitting of binomial distribution

Fit a B.D. to the following data:

12.

x:	0	1	2	3	4	5	6
f:	5	18	28	12	7	6	4

Ans. 4 15 25 22 11 3 0

Hint: $n = 6$, $p = 0.4$, $N = 80$

13.

x:	0	1	2	3	4	5	6	7	8	9	10
f:	6	20	28	12	8	6	0	0	0	0	0

Ans. 6.9 19.1 24 17.8 8.6 2.9 .7 .1 0 0 0

Hint: $n = 10$, $N = 80$, $p = 0.2175$

14.

x:	0	1	2	3	4	5
f:	38	144	342	287	164	25

Ans. 33.2 161.9 316.2 308.7 150.7 29.4

Hint: $n = 5$, $p = 0.494$

15. Seven coins are tossed and number of heads noted. The experiment is repeated 128 times with the following data:

No. of heads	0	1	2	3	4	5	6	7
Frequencies	7	6	19	35	30	23	7	1

Fit a binomial distribution assuming
i. coin is unbiased
ii. nature of coin is not known.

Ans. **i.** 1, 7, 21, 35, 35, 21, 7, 1

Hint: $p = \frac{1}{2}$, $N = 128$.

ii. 1, 8, 23, 36, 34, 19, 6, 1

Hint: $n = 7$, $N = 128$,.
$np = \frac{433}{128} = 3.3828$, $p = 0.48326$

27.5 HYPERGEOMETRIC DISTRIBUTION

The binomial distribution quite frequently arises from a random experiment in which sampling is done *with* replacement. In contrast, the hypergeometric distribution arises from random experiments in which sampling is done *without* replacement. It is very useful in quality control and analysis of the opinion surveys. Consider a population of N units in which each unit is classified into two dischotomus classes (arbitrarily known as "success" and "failure") according to whether the unit does or does not posses

a certain property under consideration. Let k be the number of successes and $N-k$ be the failures in the population.
Draw a random sample of size n *without* replacement from the population. Let X be the discrete random variable which denotes the number of successes in the sample. Then the probability distribution of x is known as hypergeometric distribution and is given by $h(x; N, n, k) = \dfrac{\binom{k}{x}\binom{N-k}{n-x}}{\binom{N}{n}}$,

$x = 0, 1, 2, \ldots, n$ (1)

The integer x should lie in the interval $\max(0, n-N+k) \le x \le \min(n, k)$.

Here $\binom{N}{n}$ is the number of ways of choosing a sample of size n from a population N, $\binom{k}{x}$ is the number of ways in which x successes is chosen from a total of k successes and finally $\binom{N-k}{n-x}$ gives the number of ways of getting $(n-x)$ failures out of the (remaining) $N-k$ failures. The hypergeometric distribution (1) has N, n, k as the three parameters.

Book work I Prove that mean of hypergeometric distribution is $\dfrac{nk}{N}$.

Proof: Mean $= E(x) = \sum_{x=0}^{n} x \dfrac{\binom{k}{x}\binom{N-k}{n-x}}{\binom{N}{n}} =$

$$k\sum_{x=1}^{n} \frac{\frac{(k-1)!}{(x-1)!(k-x)!}\binom{N-k}{n-x}}{\binom{N}{n}} = k\sum_{x=1}^{n}\frac{\binom{k-1}{x-1}\binom{N-k}{n-x}}{\binom{N}{n}}$$

Now $\binom{N}{n} = \frac{N!}{n!(N-n)!} = \frac{N}{n}\binom{N-1}{n-1}$ so

$$E(x) = \frac{nk}{N}\sum_{y=0}^{n-1}\frac{\binom{k-1}{y}\binom{N-1-(k-1)}{n-1-y}}{\binom{N-1}{n-1}}$$

$$= \frac{nk}{N}\cdot 1 = \frac{nk}{N}$$

Here $y = x-1$ and the summation represents the sum of all probabilities in hypergeometric experiment $(n-1)$ sample is drawn from a population of $(N-1)$ containing $(k-1)$ successes.

Book work II Prove that the variance of hypergeometric distribution is $\dfrac{nk(N-k)(N-n)}{N^2(N-1)}$.

Proof: $\text{Var}(X) = E(X^2) - \{E(X)\}^2$

Now $E(X^2) = E\{X(X-1)+X\} = E\{X(X-1)\} + E(X)$

Consider

$$E(X(X-1)\} = \sum_{x=0}^{n} x(x-1)\frac{\binom{k}{x}\binom{N-k}{n-x}}{\binom{N}{n}}$$

$$= \sum\nolimits_{x=0}^{n} x(x-1)\cdot\frac{k!}{x!(k-x)!}\frac{\binom{N-k}{n-x}}{\binom{N}{n}}$$

$$= \frac{k(k-1)}{\binom{N}{n}}\sum_{x=2}^{n}\frac{(k-2)!}{(x-2)!(k-2-(x-2))!}\times$$

$$\times\binom{N-2-(k-2)}{n-2-(x-2)}$$

Now

$$\binom{N}{n} = \frac{N!}{n!(N-n)!} = \frac{N(N-1)\cdot(N-2)!}{n(n-1)\cdot(n-2)!(N-n)!}$$

So,

$$E\{X(X-1)\} = \frac{k(k-1)n(n-1)}{N(N-1)}\times$$

$$\times\sum_{y=0}^{n-2}\frac{\binom{k-2}{y}\binom{N-2-(k-2)}{n-2-y}}{\binom{N-2}{n-2}}$$

$$= \frac{k(k-1)n(n-1)}{N(N-1)}\cdot 1$$

Here $y = x - 2$ and the summation is one because it is the sum of the probabilities for $y = 0$ to $n - 2$. Thus

$$\text{var}(X) = E(X^2) - \{E(X)\}^2 = E\{X(X-1)\} + E(X) - \{E(X)\}^2$$

$$= \frac{k(k-1)n(n-1)}{N(N-1)} + \frac{nk}{N} - \frac{n^2k^2}{N^2} =$$

$$= \frac{nk}{N^2(N-1)}\{N(N-1)+N(k-1)(n-1)-(N-1)nk\}$$

$$= \frac{nk}{N^2(n-1)}\{N^2 - Nn - Nk + nk\}.$$

$$\sigma^2 = \text{var}(X) = \frac{nk(N-k)(N-n)}{N^2(N-1)}$$

$$= \frac{N-n}{N-1} \cdot n \cdot \frac{k}{N}\left(1 - \frac{k}{N}\right)$$

Introducing $p = \frac{k}{N}$ which is the proportion successes in the population, the mean and variance of the hypergeometric distribution are written as

$$\mu = n \cdot \frac{k}{N} = np \quad \text{and}$$

$$\sigma^2 = \left(\frac{N-n}{N-1}\right) n \cdot p(1-p) = \left(\frac{N-n}{N-1}\right) npq.$$

Observe that the mean of the hypergeometric distribution and the mean of binomial distribution are same, while the variances differ by the factor $\left(\frac{N-n}{N-1}\right)$, known as "*finite population correction factor*", which tends 1 as $N \to \infty$. Thus binomial distribution may be viewed as a large population edition of the hypergeometric distribution, since sampling from the finite population with replacement amounts to sampling from the infinite population (without replacement).

Approximation of the Hypergeometric Distribution by the Binomial Distribution

Book work III Show that hypergeometric distribution tends to binomial distribution as $N \to \infty$ and $\frac{k}{N} \to p$.

Proof: Consider

$$h(x; N, n, k) = \frac{\binom{k}{x}\binom{N-k}{n-x}}{\binom{N}{n}}$$

$$= \frac{k!}{x!(k-x)!} \frac{(N-k)!}{(n-x)!(N-k-n+x)!} \frac{n!(N-n)!}{N!}$$

$$= \frac{k(k-1)(k-2)\ldots(k-(x-1))}{x!} \times$$

$$\times \cdot \frac{(N-k)(N-k-1)\ldots(N-k-(n-k-1))}{(n-k-1)!} \times$$

$$\times \frac{n!}{N(N-1)(N-2)\ldots(N-(n-1))}$$

$$= \frac{n!}{x!(n-x)!} \times$$

$$\left[\left(\frac{k}{N}\right)\left(\frac{k}{N} - \frac{1}{N}\right)\left(\frac{k}{N} - \frac{2}{N}\right)\cdots\left(\frac{k}{N} - \frac{(x-1)}{N}\right)\right]$$

$$\times \frac{\left(1 - \frac{k}{N}\right)\left(1 - \frac{k}{N} - \frac{1}{N}\right)\cdots\left(1 - \frac{(n-k-1)}{N}\right)}{\left(1 - \frac{1}{N}\right)\left(1 - \frac{2}{N}\right)\cdots\left(1 - \frac{n-1}{N}\right)}$$

Letting $N \to \infty$ and putting $\frac{k}{N} = p$ we get

$$\lim_{N\to\infty} h(x; N, n, k) = \binom{n}{x} p \cdot (p-0)(p-0)\ldots(p-0) \times (1-p)(1-p-0)\ldots(1-p)$$

$$= \binom{n}{x} p^k (1-p)^k = b(x; p, 1-p)$$

Worked Out Examples

Example 1: Out of 60 applicants to a university 40 are from the south. If 20 applicants are selected at random, find the probability that (a) 10 (b) not more than 2, are from south.

Solution: The total number of ways in which 20 applicants are selected from 60 is $\binom{60}{20}$.

(a) The number of ways in which 10 applicants from south are selected from 40 south applicants is $\binom{40}{10}$.

Now out of the total 20 selected the remaining 10 non-south applicants will be selected from 20(60–40 south) non-south applicants, which is $\binom{20}{10}$.

Thus the probability that 10 out of 20 applicants are from south is $\frac{\binom{40}{10}\binom{20}{10}}{\binom{60}{20}} = \frac{\left(\frac{40!}{10!30!}\right)\left(\frac{20!}{10!10!}\right)}{\left(\frac{60!}{20!40!}\right)} =$ 0.0373613

or population size $N = 60$,

sample size $= n = 20$

number of south applicants $k = 40$. Let x denote the number of applicants from the south. Then the hypergeometric distribution is $h(x; N, n, k) =$

$$\frac{\binom{k}{x}\binom{N-k}{n-x}}{\binom{N}{n}} \quad \text{for } x = 0, 1, 2, \ldots n.$$

Here $\quad h(10; 60, 20, 40) = \frac{\binom{40}{10}\binom{60-40}{20-10}}{\binom{60}{20}} =$

0.03736

(b) Probability that $x \leq 2$ is

$$P(x \leq 2) = \sum_{x=0}^{2} h(x; N, n, k) = \sum_{x=0}^{2} h(x; 60, 20, 40)$$

$$= h(0; 60, 20, 40) + h(1; 60, 20, 40) + h(2; 60, 20, 40)$$

$$= \frac{\binom{40}{0}\binom{60-40}{20-0}}{\binom{60}{20}} + \frac{\binom{40}{1}\binom{60-40}{20-1}}{\binom{60}{20}} + \frac{\binom{40}{2}\binom{60-40}{20-2}}{\binom{60}{20}}$$

$$= \frac{(40c_0)(20c20) + (40c1)(20c19) + (40c2)(20c18)}{60c20}$$

Example 2: Solve (a) of the above problem using Binomial approximation. Explain the accuracy obtained. What is the finite population correlation factor.

Solution: As $N \to \infty$, Binomial approximation is $P(X = x) = \binom{n}{x} p^x q^{n-x}$.

Here the probability of (south) success is $p = \frac{40}{60} = \frac{2}{3}$ and probability of (non-south) failure is $q = \frac{20}{60} = \frac{1}{3}$ and $n = 20$ and $x = 10$.

Now using binomial approximation we have $P(X = x) =$ probability of 10 south applicants out of 20 applicants $= \binom{20}{10}\left(\frac{2}{3}\right)^{10}\left(\frac{1}{3}\right)^{10}$

$= 0.0542591$

From hypergeometric distribution the probability is 0.03736. The accuracy is less since $N = 60$ is not very large more so compared with $n = 20$. Finite population correction factor $= \frac{N-n}{N-1} = \frac{60-20}{60-1} =$ 0.6779

Example 3: Find the mean and variance and standard deviation of the random variable X in (a) of above example.

Solution: Mean $= \frac{nk}{N} = \frac{20(40)}{(60)} = 1.3333$

$$\text{Variance} = \frac{N-n}{N-1} \cdot n \cdot \frac{k}{N} \cdot \left(1 - \frac{k}{N}\right)$$

$$= \left(\frac{60-20}{60-1}\right) \cdot 20 \frac{(40)}{(60)} \cdot \left(1 - \frac{40}{60}\right) = 0.30131$$

Standard deviation = 0.5489

Exercise

1. Find the probability of selecting 5 cards of which 3 are red and 2 are black from an ordinary deck of 52 playing cards.

Ans. $\binom{26}{3}\binom{26}{2}\Big/\binom{52}{5} = 0.3251$

2. Determine the probability that exactly one defective is found in a sample of 5 from a lot of 40 components containing 3 defectives (in the entire lot).

Ans. $\binom{3}{1}\binom{37}{4}\Big/\binom{40}{5} = 0.3011$

3. Find the mean and variance of the random variable X in the above example 2.

Ans. Mean $= \frac{5\cdot 3}{40} = 0.375$, $\sigma^2 = 0.3113$

4. (a) Determine the probability that exactly 3 computers are defective out of 10 computers purchased from a lot of 5000 computers containing 1000 defective computers. (b) Use binomial approximation and explain the accuracy. (c) what is the finite population correction factor.

Ans. (a) $h(3; 5000, 10, 1000) = 0.2015$
(b) $h(3; 5000, 10, 1000) \simeq b(3; 10, 0.2)$
$= \sum_{x=0}^{3} b(x; 10, 0.2) - \sum_{x=0}^{2} b(x; 10, 0.2) =$
$0.8791 - 0.6778 = 0.2015$
since $N = 5000$ large compared to sample $n = 10$, the accuracy achieved is high.
(c) $\frac{N-n}{N-1} = \frac{5000-10}{5000-1} = 0.9982$

5. TV's are shipped in lots of 50. A shipment is accepted if a sample of 5 TV's inspected from this lot, does not contain any defective TV. If one or more are found defective, the entire shipment is rejected . Suppose the lot of 50 contains 3 defective TV's. Determine the probability that 100% inspection is required?

Ans. $P(X \geq 1) = 1 - P(X = 0)$
$= 1 - \left\{\binom{3}{0}\binom{47}{5}\Big/\binom{50}{5}\right\} = 0.28$ where X is the number of defective TV's.

6. Suppose a shipment of 100 cars contain 25 defectives. Determine the probability that 2 cars out of a sample of 10 cars drawn from the lot are defective. Use binomial approximation also.

Ans. $h(2; 10, 25, 100) = 0.292$
$b(2; 10, 0.25) = 0.282$

7. Find the probability of getting 3 blue marbles if 5 marbles are drawn (one after the other without replacement) from a box containing 6 blue and 4 red marbles.

Ans. $\binom{6}{3}\binom{4}{2}\Big/\binom{10}{5} = \frac{10}{21} = 0.4762$

8. A committee of 2 is chosen from five faculty members out of which 3 are doctorate and 2 are post-graduates. If X denotes the number of doctorates in the committee, obtain the probability distribution of X.

Ans.

X	0	1	2
$P(x)$	$\frac{2}{20}$	$\frac{12}{20}$	$\frac{6}{20}$

Hint: $h(x; 5, 2, 3)$ for $x = 0, 1, 2$.

9. Let X be the number of defective motors in a sample of 6 drawn from a lot of 12 motors containing four defectives. Compute (a) $P(X = 1)$ (b) $P(X \geq 4)$ (c) $P(1 \leq X \leq 3)$.

Ans. (a) 0.242 (b) 0.030 (c) 0.939

10. A box contains 12 red and 8 black marbles. If 5 marbles are drawn successively (without replacement) from the box, find the probability that (a) 3 are red and 2 are black (b) at least 3 are red (c) all the 5 are of the same colour.

Ans. (a) $\frac{385}{969}$ (b) $\frac{682}{969}$ (c) $\frac{53}{969}$

Hint: (a) $\binom{12}{3}\binom{8}{2}\Big/\binom{20}{35}$

(b) $\sum_{x=3}^{5}\binom{12}{x}\binom{8}{5-x}\Big/\binom{20}{5}$

(c) $\left\{\binom{12}{5}\binom{8}{0} + \binom{12}{0}\binom{8}{5}\right\}\Big/\binom{20}{5}$

27.6 POISSON DISTRIBUTION

Poisson* distribution is the discrete probability distribution of a discrete random variable x, which has no upper bound. It is defined for non negative values of x as follows:

$$f(x, \lambda) = P(X = x) = \frac{\lambda^x e^{-\lambda}}{x!} \quad \text{for } x = 0, 1, 2, \ldots \quad (1)$$

Here $\lambda > 0$ is called the parameter of the distribution. Note that in binomial distribution the number of successes (occurrence of an event) out of a total definite number of n trials is determined, whereas in Poisson distribution the number of successes at a random point of time and space is determined.

* Simeon Denis Poisson (1781–1840) French mathematician.

Poisson distribution (P.D.) is suitable for 'rare' events for which the probability of occurrence p is very small and the number of trials n is very large. Also binomial distribution can be approximated by poisson distribution when $n \to \infty$ and $p \to 0$ such that $\lambda = np =$ constant.

Examples of rare events:

i. Number of printing mistakes per page.
ii. Number of accidents on a highway.
iii. Number of defectives in a production centre.
iv. Number of telephone calls during a particular (odd) time.
v. Number of bad (dishonoured) cheques at a bank.

Result 1: Since $\sum_{x=0}^{\infty} f(x,\lambda) = \sum_{x=0}^{\infty} p(X=x) = \sum_{x=0}^{\infty} \frac{e^{-\lambda}\lambda^x}{x!} = e^{-\lambda}\sum_{x=0}^{\infty}\frac{\lambda^x}{x!} = e^{-\lambda}e^{\lambda} = 1$, Therefore (1) is a probability function.

Result 2: Arithmetic mean of Poisson distribution

$$\overline{X} = E(X) = \sum_{x=0}^{\infty} xP(X=x) = \sum_{x=0}^{\infty} x\cdot\frac{e^{-\lambda}\lambda^x}{x!}$$

$$= \lambda e^{-\lambda}\sum_{x=1}^{\infty}\frac{\lambda^{x-1}}{(x-1)!} = \lambda e^{-\lambda}\sum_{m=0}^{\infty}\frac{\lambda^m}{m!}$$

$$= \lambda e^{-\lambda}\cdot e^{\lambda} = \lambda$$

Thus the parameter λ is the A.M. of P.D.

Result 3: Variance of Poisson distribution

$$= E\left[(x-\overline{X})^2\right] = \sum_{x=0}^{\infty}(x-\overline{X})^2 P(X=x)$$

$$= \sum\left(x^2+\overline{X}^2-2\overline{X}x\right)P = \sum x^2 P + \overline{X}^2\sum P$$

$$-2\overline{X}\sum xP$$

$$= \sum x^2P + \overline{X}^2 - 2\overline{X}\overline{X} = \sum x^2 P + \lambda^2 - 2\lambda^2$$

$$= \sum x^2 P - \lambda^2.$$

But

$$\sum x^2 P = \sum_{x=0}^{\infty}[x(x-1)+x]e^{-\lambda}\frac{\lambda^x}{x!}$$

$$= e^{-\lambda}\sum_{x=0}^{\infty}\frac{\lambda^x\cdot x(x-1)}{x!} + e^{-\lambda}\sum_{x=0}^{\infty}\frac{\lambda^x\cdot x}{x!}$$

$$= \lambda^2 e^{-\lambda}\sum_{x=2}^{\infty}\frac{\lambda^{x-2}}{(x-2)!} + \lambda e^{-\lambda}\sum_{x=1}^{\infty}\frac{\lambda^{x-1}}{(x-1)!}$$

$$= \lambda^2 e^{-\lambda}e^{\lambda} + \lambda e^{-\lambda}e^{\lambda} = \lambda^2 + \lambda$$

Thus

$$\text{Variance} = \sum x^2 p - \lambda^2 = (\lambda^2+\lambda) - \lambda^2 = \lambda$$

Hence the variance of P.D. = mean of P.D.

Result 4: Recurrence formula

$$\frac{P(x+1)}{P(x)} = \frac{e^{-\lambda}\lambda^{x+1}}{(x+1)!}\cdot\frac{x!}{e^{-\lambda}\lambda^x} = \frac{\lambda}{x+1}$$

Thus

$$P(x+1) = \left(\frac{\lambda}{x+1}\right)P(x).$$

Result 5: Poisson distribution function

$$F(x;\lambda) = \sum_{k=0}^{x}\frac{e^{-\lambda}\lambda^k}{k!}$$

has been tabulated (see A8 to A11)

Then $\quad f(x;\lambda) = F(x;\lambda) - F(x-1;\lambda).$

Theorem: *Prove that Poisson distribution is the limiting case of binomial distribution for very large trials with very small probability, i.e.,* $f(x;\lambda) = \lim_{\substack{n\to\infty\\ p\to 0}} b(x;n,p)$ *such that* $\lambda = np =$ *constant.*

Proof: Put $p = \frac{\lambda}{n}$ in binomial distribution

$$b(x;n,p) = \frac{n!}{x!(n-x)!}\cdot\left(\frac{\lambda}{n}\right)^x\left(1-\frac{\lambda}{n}\right)^{n-x}$$

$$= \frac{n(n-1)(n-2)\cdots(n-(x-1))}{x!}\cdot\frac{\lambda^x}{n^x}\times$$

$$\times\left(1-\frac{\lambda}{n}\right)^{n-x}$$

$$= \frac{n^x\cdot 1\cdot\left(1-\frac{1}{n}\right)\left(1-\frac{2}{n}\right)\cdots\left(1-\frac{(x-1)}{n}\right)}{x!}\frac{\lambda^x}{n^x}\times$$

$$\times\left(1-\frac{\lambda}{n}\right)^{n-x}$$

Taking the limit as $n \to \infty$, we have

$$\lim_{n\to\infty} b(x; n, p) = \frac{1}{x!}\lambda^x \cdot e^{-\lambda} = \frac{\lambda^x e^{-\lambda}}{x!}$$

since

$$\lim_{n\to\infty}\left(1-\frac{1}{n}\right)\left(1-\frac{2}{n}\right)\cdots\left(1-\frac{x-1}{n}\right) = 1\cdot 1\cdots 1 = 1$$

and

$$\lim_{n\to\infty}\left(1-\frac{\lambda}{n}\right)^{n-x} = \left[\left(1-\frac{\lambda}{n}\right)^{n/\lambda}\right]^{\lambda} \times \left[1-\frac{\lambda}{n}\right]^{-x} = e^{-\lambda}.$$

Note: $\lim_{n\to\infty}\left(1-\frac{x}{n}\right)^n = e^{-x}$

Thus binomial probabilities for large n and small p are often approximated by means of poisson distribution with mean $\lambda = np$.

Example: For $n = 3000$, $p = 0.005$, the probability of 18 successes by binomial distribution is given by $b(18; 3000, .005) = 3000_{C_{18}}(.005)^{18}(.995)^{2982}$ which involves prohibitive amount of work. Instead using Poisson distribution as an approximation, we get $\lambda = 3000 \times .005 = 15$. Probability of 18 success $= f(18, 15) = 0.8195$ from table (A8 to A11).

General rule: Poisson approximation to B.D. is used whenever $n \geq 20$ and $p \leq 0.05$. For $n \geq 100$, approximation is excellent provided $\lambda = np \leq 10$.

WORKED OUT EXAMPLES

Poisson distribution

Example 1: A distributor of bean seeds determines from extensive tests that 5% of large batch of seeds will not germinate. He sells the seeds in packets of 200 and guarantees 90% germination. Determine the probability that a particular packet will violate the guarantee.

Solution: The probability of a seed not germinating $= p = \frac{5}{100} = 0.05$

$\lambda =$ mean number of seeds, in a sample of 200, which do not germinate

$= np = 200 \times 0.05 = 10$

Let X = R.V. = number of seeds that do not germinate

A packet will violate guarantee if it contains more than 20 non-germinating seeds.
Probability that the guarantee is violated

$$= P(X > 20) = 1 - P(X \leq 20) = 1 - \sum_{x=0}^{20}\frac{e^{-10}10^x}{x!}$$

$$= 1 - F(20, 10) = 1 - .9984 = 0.0016$$

where cummulative distribution function F is read for x=20 and λ=10 from the tables (A8 to A11).

Example 2: The average number of phone calls/minute coming into a switch board between 2 and 4 PM is 2.5. Determine the probability that during one particular minute there will be (a) 0 (b) 1 (c) 2 (d) 3 (e) 4 or fewer (f) more than 6 (g) at most 5 (h) at least 20 calls.

Solution: $\lambda = 2.5$, $f(x;\lambda) = f(x; 2.5) = \frac{(2.5)^x(e^{-2.5})}{x!}$
Let X = R.V. = number of phone calls/minute during that (odd) 2 and 4 PM.

a. $f(0; 2.5) = e^{-2.5} = .08208$

b. $f(1; 2.5) = .2052$

c. $f(2; 2.5) = .2565$

d. $f(3; 2.5) = .2138$

e. $P(X \leq 4) = \sum_{x=0}^{4} f(x; 2.5) = F(4; 2.5) = .8912$

(read from tables A8 to A11)

f. $P(X > 6) = 1 - P(X \leq 6) = 1 - \sum_{x=0}^{6} f(x; 2.5)$

$= 1 - F(6; 2.5) = 1 - .9858 = 0.0142$

g. $P(X \leq 5) = \sum_{x=0}^{5} f(x; 2.5) = F(5; 2.5) = .9580$

h. $P(X \geq 2.0) = 1 - P(X \leq 19) = 1 - \sum_{x=0}^{19} f(x; 2.5)$

$= 1 - F(19; 2.5) = 1 - 1 = 0.$

Example 3: Suppose that on the average one person in 1000 makes a numerical error in preparing income tax return (ITR). If 10000 forms are selected at random and examined, find the probability that 6, 7 or 8 of the forms will be in error.

Solution: Let X = R.V. = number ITR forms containing a numerical error. Essentially this is a binomial experiment with 10000 trials and probability (of success) $p = \frac{1}{1000} = 0.001$. So by B.D. probability of 6, 7 or 8 error forms = $P(X = 6, 7 \text{ or } 8)$

$$= P(6) + P(7) + P(8) = \sum_{x=6}^{8} b(x; 10000, 0.001)$$

$$= \sum_{x=6}^{8} {}^{10000}C_x (.001)^x (.999)^{10000-x}$$

which involves cumbersome lengthy calculations. Since n is large and p is small, approximate the binomial probabilities by Poisson distribution with $\lambda = np = 10000 \times \frac{1}{1000} = 10$.

Probability of 6, 7 or 8 error ITR forms = $P(X = 6, 7 \text{ or } 8)$

$$= \sum_{x=6}^{8} f(x; 10) = \sum_{x=6}^{\infty} \frac{e^{-10}(10)^x}{x!}$$

$$= e^{-10}\left[\frac{10^6}{6!} + \frac{10^7}{7!} + \frac{10^8}{8!}\right] = .2657$$

Instead, using tables A8 to A11, we get the result in a simpler way

$$= F(8; 10) - F(5; 10) = .3328 - .0671 = .2657.$$

Example 4: Fit a Poisson distribution to the following data:

X_i:	0	1	2	3	4
Observed frequencies f_i	30	62	46	10	2

Solution: To fit a Poisson distribution, determine the only parameter λ of the distribution from the given data. Since λ is the arithmetic mean,

$$\lambda = \frac{\sum_{i=0}^{4} f_i X_i}{\sum f_i X_i} = \frac{0 \times 30 + 1 \times 62 + 2 \times 46 + 3 \times 10 + 4 \times 2}{150}$$

$$= \frac{192}{150} = 1.28$$

Thus the Poisson distribution that "fits" to the given data is $P(X) = \frac{e^{-1.28}(1.28)^X}{X!}$.

Here total frequency $N = \sum_{i=0}^{4} f_i = 150$

Expected frequency = (Total frequency) × Probability

X_i:	0	1	2	3	4
$P(X_i)$:	0.27803	.35588	.22776	.09718	.031097
$(N)(P(X_i))$ =Expected frequency	41.7045 ≈ 42	53.382 ≈ 53	34.164 ≈ 34	14.577 ≈ 15	4.6646 ≈ 5

EXERCISE

Poisson distribution

1. Determine the probability that 2 of 100 books bound will be defective if it is known that 5% of books bound at this bindery are defective. (a) use B.D. (b) use Poisson approximation to B.D.

Ans. **a.** $b(2; 100, 0.5) = \binom{100}{2}(0.05)^2(.95)^{98} = 0.081$

b. $f(2; 5) = \frac{5^2 e^{-5}}{2!} = 0.084$ with $\lambda = np = 100(0.05) = 5$

2. Find the probabilities that 0, 1, 2, 3, 4, ... of 3840 generators fail if the probability of failure is $\frac{1}{1200}$.

Ans.

x:	0	1	2	3	4	5	6	7	8	9	10
$f(x, 3.2)$:	0.041	.130	.209	.223	.178	.114	.06	.028	.011	.004	.002

Hint: $\lambda = 3840 \times \frac{1}{1200} = 3.2$. Use tables (A8 to A11) and the identity $f(x; \lambda) = F(x; \lambda) - F(x-1; \lambda)$.

3. On an average, 1.3 gamma particles/millisecond come out of a radioactive substance, determine (a) mean (b) variance

(c) probability of more than one gamma particles emanate from the substance.

Ans. (a)(b): $\lambda = \sigma^2 = 1.3$ (c) $1 - P(X = 0) = 1 - e^{-1.3} = 0.727$

4. Determine the probability p that there are 3 defective items in a sample of 100 items if 2% of items made in this factory are defective.

Ans. $p = f(3; 2) = \frac{2^3 e^{-2}}{3!} = 0.180$ with $\lambda = np = 100(0.02) = 2$

5. Suppose 300 misprints are distributed randomly throughout a book of 500 pages. Find the probability P that a given page contains (i) exactly two misprints (ii) two or more misprints.

Ans. **i.** $f(2; 0.6) = \frac{(0.6)^2 e^{-0.6}}{2!} = 0.0988 \approx 0.1$

ii. $P = 1 - P(0 \text{ or } 1 \text{ misprint})$
$= 1 - (0.549 + 0.329) = 0.122$

6. In a factory producing blades, the probability of any blade being defective is 0.002. If blades are supplied in packets of 10, determine the number of packets containing (a) no defective (b) one defective and (c) two defective blades respectively in a consignment of 10000 packets.

Ans. **a.** $10000 \times P(0) = 10000 \times e^{-0.02}$

$= 10000 \times .9802 = 9802$

i.e., 9802 packets do not have any defective blades.

b. $10000 \times (0.02)(.9802) = 196$

c. $10000 \times \frac{(.02)^2}{2!} \cdot 9802 = 2$

Hint: $\lambda = np = 10 \times 0.002 = 0.02$.

7. A manufacturer of cotter pins knows that 5% of his product is defective. Pins are sold in boxes of 100. He guarantees that not more than 10 pins will be defective. Determine the probability that a box will fail to meet the guarantee.

Ans. $P(X > 10) = 1 - P(X \le 10)$

$= 1 - \sum_{x=0}^{10} \frac{e^{-5} 5^x}{x!} = 1 - F(10, 5)$

$= 1 - .9863 = 0.0137$

Hint: $\lambda = np = 100 \times 0.05 = 5$

8. On an average 20 red blood cells are found in a fixed volume of blood for a normal person. Determine the probability that the blood sample of a normal person will contain less than 15 red cells.

Ans. $P(X < 15) = \sum_{x=0}^{14} \frac{e^{-20}(20)^x}{x!} = F(14, 20)$

$= 0.105$

Hint: $\lambda = 20$.

9. Two shipments of computers are received. The first shipment contains 1000 computers with 10% defectives and the second shipment contains 2000 computers with 5% defectives. One shipment is selected at random. Two computers are found good. Find the probability that the two computers are drawn from the first shipment.

Ans. 0.183

Hint: $q_1 = 0.1$, $p_1 = 0.9$, $q_2 = 0.05$, $p_2 = 0.95$

$\lambda_1 = n_1 p_1 = (1000)(0.9) = 900$,

$\lambda_2 = n_2\, p_2 = (2000)(.95) = 1900$

C: two computers good, A: first shipment, B: second shipment.

$$P(A/C) = \frac{P(A)P(C/A)}{P(A)P(C/A) + P(B) \cdot P(C/B)}$$

where

$$P(A) = P(B) = \frac{1}{2},\ P(C/A) = \frac{e^{-900}(900)^2}{2!},$$

$$P(C/B) = \frac{e^{-1900}(1900)^2}{2!}.$$

10. Given that the probability of an accident in an industry is 0.005 and assuming the accidents are independent (a) determine the probability that in any given period of 400 days, there will be an accident one day? (b) What is the probability that there are at most three days with an accident?

Ans. (a) $P(X = 1) = e^{-2} 2^1 = 0.271$

(b) $P(X \leq 3) = \sum_{x=0}^{3} \frac{e^{-2}2^x}{x!} = 0.857$

Hint: $\lambda = np = 400(0.005) = 2$.

11. If one in every 1000 of computers produced is defective, determine the probability that a random sample of 8000 will yield fewer than 7 defective computers?

Ans. $P(X < 7) = \sum_{x=0}^{6} b(x; 8000, 0.001)$

$\simeq \sum_{x=0}^{6} f(x; 8) = 0.3134$

Hint: B.D. calculation is very hard, approximate it by P.D. with $\lambda = np = (8000)(0.001) = 8$.

12. Suppose the average number of telephone calls coming into a telephone exchange between 10 AM to 11 AM is 2, while between 11 AM to 12 noon is 6, determine the probability that more than five calls come in between 10 AM to 12 noon, assuming that calls are independent.

Ans. $P(x > 5) = 1 - P(x \leq 5) = 1 - \sum_{x=0}^{5} \frac{e^{-8}8^x}{x!} = 1 - 0.1912 = 0.8088$

Hint: P.D. is additive: $X = X_1 + X_2$, $\lambda = \lambda_1 + \lambda_2 = 2 + 6 = 8$.

Fitting of Poisson distribution

Fit a Poisson distribution to the following data:

1.

x:	0	1	2	3	4	5	6	7	8
Observed frequency f_i	56	156	132	92	37	22	4	0	1

Ans. 69.6 137.25 135.33 88.95 43.85 17.29 5.68 1.60 0.3942

Hint: $\lambda = \frac{\sum f_i x_i}{N} = \frac{986}{500} = 1.972$.

2.

x:	0	1	2	3	4
f_i:	122	60	15	2	1

Ans. 121 61 15 2 0

Hint:

$\lambda = \sum \frac{f_i x_i}{N} = \frac{60+36+6+4}{200} = 0.5; e^{-.5} = 0.61$

3.

x:	0	1	2	3	4	5
f_i:	142	156	69	27	5	1

Ans. 147.15 147.15 73.58 24.53 6.13 1.23

Hint: $\lambda = \frac{\sum f_i x_i}{\sum f_i} = \frac{400}{400} = 1$.

4. Determine the number of pages expected with 0, 1, 2, 3, and 4 errors in 1000 pages of a book if on the average two errors are found in five pages.

Ans.

x:	0	1	2	3	4
$P(x)$:	.6703	.26812	.053624	.0071	.00071
Expected number of pages	670	268	54	7	1

Hint: $\lambda = 2/5 = 0.4$, $e^{-0.4} = .6703$,

Expected number of pages $= 1000 \times P(x)$.

5.

x:	0	1	2	3	4
f:	109	65	22	3	1

Ans. 108.7 66.3 20.2 4.1 0.7

Hint: $\lambda = \frac{65+44+9+4}{200} = \frac{122}{200} = 0.61$.

27.7 POISSON PROCESS

Poisson process is a random process in which the number of events (or successes) x occurring in a time interval of length say T is counted. It is a continuous parameter, discrete state process. By dividing T into n equal parts of length Δt, we have $T = n \cdot \Delta t$.

Assume that

1. The probability of success (or occurrence of an event) in a given time interval is proportional to the length of the interval, i.e., $p \propto \Delta t$ or $p = \alpha \Delta t$ where α is the proportionality constant.
2. The occurrences of events are independent, i.e., probability of success in an interval of time (or space) does not depend on the what happened prior to that time or any other interval.
3. The probability of more than one success during a small time interval Δt is negligible.

As $n \to \infty$, the probability of x successes (or occurrence of an event) during a time interval T is governed by the Poisson distribution with the parameter

$$\lambda = n \cdot p = \left(\frac{T}{\Delta t}\right)(\alpha \Delta t) = \alpha T$$

Thus α is the average (mean) number of successes (occurrences) per unit time.

WORKED OUT EXAMPLES

Poisson process

Example: Average rate of arrival of persons in a queue is 1.5 per minute. Determine the probability that (a) at most four persons will arrive in any given minute (b) at least five will arrive during an interval of 2 minutes (c) at most 20 will arrive during an interval of 6 minutes.

Solution: α = arrival rate = 1.5.
Let X be number of persons arriving in queue

a. T = time interval = 1 minute

$$\lambda = \alpha T = (1.5)(1) = 1.5$$

$$P(X \leq 4) = \sum_{x=0}^{4} f(x; 1.5) = F(4; 1.51) = .981.$$

b. T = time interval = 2 minutes

$$\lambda = \alpha T = (1.5)(2) = 3.0$$

$$P(X \geq 5) = 1 - P(X < 5) = 1 - \sum_{x=0}^{4} f(x, 3)$$

$$= 1 - F(4, 3) = 1 - .815 = .185.$$

c. T = time interval = 6 minutes

$$\lambda = \alpha T = (1.5)(6) = 9$$

$$P(X \leq 20) = \sum_{x=0}^{20} f(x; 9) = F(20, 9) = 1.0.$$

EXERCISE

Poisson process

1. The average rate of phone calls received is 0.6 calls per minute at an office. Determine the probability that (a) there will be one or more calls in a minute (b) there will be at least tnree calls during 4 minutes.

Ans. **a.** $f(x \geq 1; 0.6) = 1 - F(0, 0.6)$
$= 1 - .549 = .451$

b. $f(x \geq 3; 2.4) = 1 - F(2, 2.4)$
$= 1 - .570 = .430$

2. On an average six bad cheques per day are received by a bank. Find the probability that the bank will receive (a) on any given day four bad cheques (b) 10 bad cheques on any two consecutive days.

Ans. **a.** $f(4; 6) = e^{-6}6^4/4! = 0.135$

b. $f(10; 12) = F(10, 12) - F(9, 12)$
$= .347 - .242 = 0.105$

3. At an airport, the average number of aeroplanes arriving is 10. There are only 15 runways in the airport. Determine the probability that an aeroplane will be refused landing on any given day.

Ans. $P(X \geq 15) = 1 - \sum_{x=0}^{15} f(x, 10) = 1 - F(15, 10)$
$= 1 - .09513 = .0487$

4. The number of e-mails received by a computer is at the rate of two per 3 minutes. Determine the probability that five or more e-mails are received in a duration of 9 minutes.

Ans. $\sum_{x=5}^{\infty} f(x, 6) = 1 - F(4; 6) = 1 - 0.285$
$= 0.7149$

Hint: By reproductive property of Poisson process
$\lambda = \lambda_1 + \lambda_2 + \lambda_3 = 2 + 2 + 2 = 6$ for R.V. X_1, X_2, X_3 in 3 minutes duration, so $X = X_1 + X_2 + X_3$.

5. On an average two emergency cases are received in a week (7 days) period at a hospital. Determine the probability that there are
   a. three or less emergency cases in 2 weeks period
   b. exactly eight emergency cases in 3 weeks period.

Ans. **a.** $F(3; 4) = \sum_{x=0}^{3} f(x; 4) = 0.4335$

b. $f(8;6) = F(8;6) - F(7;6)$
$= .8472 - .7440 = 0.1032.$

27.8 CONTINUOUS UNIFORM DISTRIBUTION

The probability density function $f(X)$ of a continous random variable X having *uniform* distribution over the interval $[a, b]$ is given by

$$U[a, b] = f(X) = \begin{cases} K, \text{ constant } a \le x \le b \\ 0, \text{ otherwise} \end{cases}$$

Here X is uniformly distributed over the interval $[a, b]$. Since $\int_{-\infty}^{\infty} f(x)dx = 1$ we have $\int_a^b K dx = 1$ or $(b-a)K = 1$ or $K = \frac{1}{b-a}$. So the constant k is the reciprocal of the length of the interval. Thus the continuous uniform distribution takes the form

$$U[a, b] = f(X) = \begin{cases} \frac{1}{b-a} & \text{if } a \le x \le b \\ 0 & \text{otherwise} \end{cases} \quad (1)$$

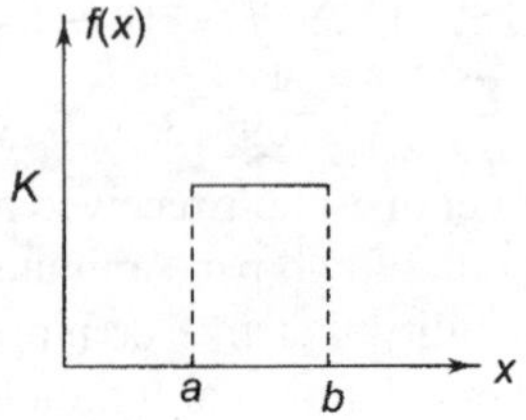

Fig. 27.4

The density (1) is uniform in value over the interval $[a, b]$. The uniform distribution (1) is also known as *rectangular* distribution since the graph of the distribution is rectangular. The constants a and b, which are known as the parameters, completely determines the distribution (1). Since

$$E(X^K) = \int_a^b x^K f(x)dx = \int_a^b x^K \left(\frac{1}{b-a}\right) dx$$

$$= \frac{1}{b-a}\left(\frac{b^{K+1} - a^{K+1}}{K+1}\right)$$

we get the mean and variance of the uniform distribution as

$$\mu = E(X) = \frac{b^2 - a^2}{(b-a)^2} = \frac{b+a}{2}$$

and

$$\sigma^2 = E(X^2) - \mu^2 = \frac{b^3 - a^3}{(b-a) \cdot 3} - \left(\frac{b+a}{2}\right)^2$$

$$= \frac{(b-a)^2}{12}$$

Solving $a = \mu - \sqrt{3}\sigma$, $b = \mu + \sqrt{3}\sigma$

The cumulative distribution function $F(x)$:

(i) when $x \le a$, $F(x) = \int_{-\infty}^{x} 0\,dx = 0$

(ii) when $a \le x \le b$, $F(x) = \int_{-\infty}^{x} = \int_{-\infty}^{a} 0 + \int_a^x \frac{1}{b-a} \cdot dx = 0 + \frac{1}{b-a}(x-a) = \frac{x-a}{b-a}$

(iii) when $x > b$, $F(x) = \int_{-\infty}^{x} = \int_{-\infty}^{a} 0 + \int_a^b \frac{1}{b-a} dx + \int_b^x 0 = \frac{1}{b-a} \cdot (b-a) = 1$

Thus

$$F(x) = \begin{cases} 0 & \text{when } x < a \\ \frac{x-a}{b-a} & \text{when } a \le x \le b \\ 1 & \text{when } x > b \end{cases}$$

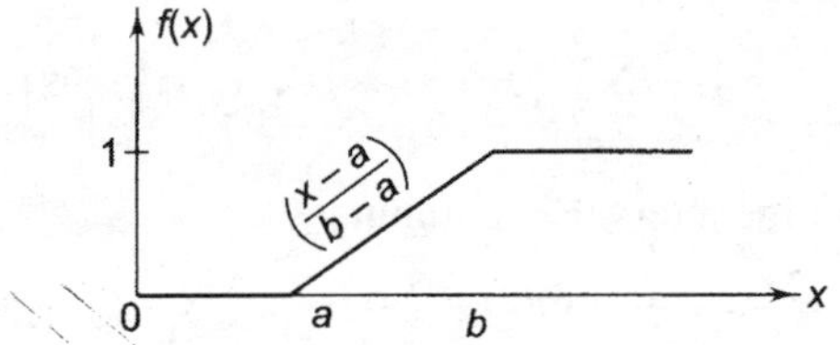

Fig. 27.5

Now for any subinterval $[c, d]$ where $a \le c < d \le b$. The probability that x lies in the interval $[c, d]$ is given by

$$P(c \le X \le d) = \int_c^d f(x)dx$$

$$= \int_c^d \frac{1}{b-a} dx = \frac{d-c}{b-a}.$$

Thus the probability depends only on the length of the interval $(d-c)$ but not on the location of that interval in $[a, b]$. Therefore in continuous uniform distribution, the probability is *same* (uniform) for *all* subintervals having the *same length*.

WORKED OUT EXAMPLES

Example 1: If X is uniformly distributed in $-2 \leq x \leq 2$, find (a) $P(X < 1)$ (b) $P|X - 1| \geq 1/2)$.

Solution: (a) Since $X < 1$, it lies in the interval $[-2, 1]$, of length 3. Then

$$P(X < 1) = \frac{1-(-2)}{2-(-2)} = \frac{3}{4}$$

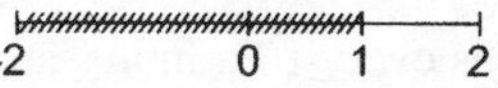

(b) If $|X - 1| \geq \frac{1}{2}$ then $X \geq \frac{3}{2}$ and $x \leq \frac{1}{2}$ i.e. x lies in the two intervals $[\frac{3}{2}, 2]$ and $[-2, \frac{1}{2}]$. So

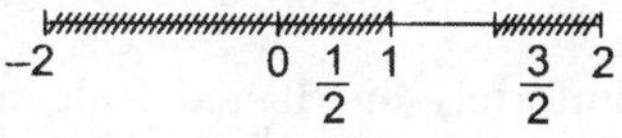

$$P\left(|X-1| \geq \frac{1}{2}\right) = P\left(-2 \leq X \leq \frac{1}{2}\right) + P\left(\frac{3}{2} \leq X \leq 2\right)$$

$$= \frac{\frac{1}{2}-(-2)}{2-(-2)} + \frac{2-\frac{3}{2}}{2-(-2)}$$

$$= \frac{\frac{5}{2}+\frac{1}{2}}{4} = \frac{3}{4}$$

Example 2: If X is uniformly distributed in $[-\alpha, \alpha]$ with $\alpha > 0$ then determine α such that $P(X > 1) = \frac{1}{3}$.

Solution: If $\alpha < 1$, then $P(X > 1)$ should be zero since X lies outside the given interval $[-\alpha, \alpha]$. Therefore α must be greater than 1. Now $P(X > 1) = \frac{\alpha-1}{\alpha-(-\alpha)} = \frac{\alpha-1}{2\alpha} = \frac{1}{3}$ (given). So $1 - \frac{1}{\alpha} = \frac{2}{3}$ or $\alpha = 3$.

Example 3: A bus travels between two cities A and B which are 100 miles apart. If the bus has a breakdown, the distance X of the point of breakdown from city A has a uniform distribution $U[0, 100]$.

(a) There are service garages in the city A, city B and midway between cities A and B. If a breakdown occurs, a tow truck is sent from the garage closest to the point of breakdown. What is the probability that the tow truck has to travel more than 10 miles to reach the bus.

(b) Would it be more "efficient" if the three service garages were placed at 25, 50 and 75 miles from city A? Explain.

Solution:

(a) If the bus breaksdown in the intervals [10, 40] miles or [60, 90] miles, then the bus have to be towed for more than 10 miles.

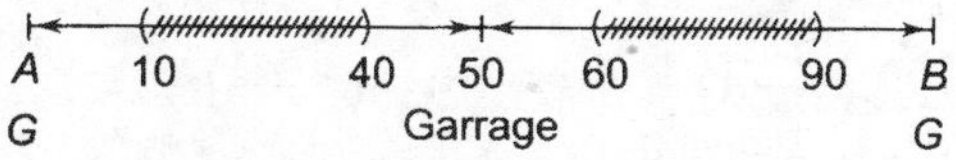

So probability that the bus has to be towed for more than 10 miles = probability that X lies in the intervals [10, 40] or [60, 90]. Thus the required probability is given by

$$P(10 < X < 40 \text{ or } 60 < X < 90)$$

$$- P(10 < X < 40) + P(60 < X < 90)$$

$$= \frac{40-10}{100-0} + \frac{90-60}{100-0}$$

$$= \frac{3}{10} + \frac{3}{10} = \frac{3}{5}$$

(b) Suppose three garages are placed at 25, 50 75 miles from city A.

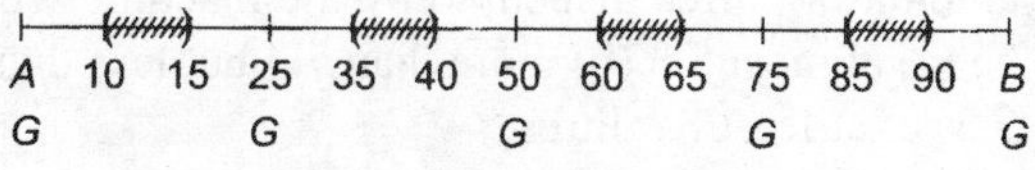

In this case, the bus is to be towed for more than 10 miles if the bus breaksdown in any one of the four intervals (10, 15), (35, 40), (60, 65) or (85, 90) miles. Probability is given by

$$P(10 < X < 15 \text{ or } 35 < X < 40$$

$$\text{or} \quad 60 < X < 65 \text{ or } 85 < X < 90)$$

$= P(10 < X < 15) + P(35 < X < 40)$

$+P(60 < X < 65) + P(85 < x < 90)$

$$= \frac{15-10}{100-0} + \frac{40-35}{100-0} + \frac{65-60}{100-0} + \frac{90-85}{100-0}$$

$$= \frac{20}{100} = \frac{1}{5}.$$

Since the probability is small $\left(\frac{1}{5}\right)$ compared to $\left(\frac{3}{5}\right)$ in the case a, b is more "effective".

EXERCISE

1. A point is chosen at random from the line segment [0, 2]. What is the probability that the chosen point lies (a) $1 \le X \le \frac{3}{2}$ (b) $X \ge \frac{3}{2}$ (c) $X \le 1$ (d) $x \ge 3$

Ans. (a) $\left(\frac{3}{2} - 1\right)\frac{1}{2} = \frac{1}{4}$ (b) $\left(2 - \frac{3}{2}\right)\frac{1}{2} = \frac{1}{4}$ (c) $(1-0)\frac{1}{2} = \frac{1}{2}$ d) 0.

Hint: $f(x) = \frac{1}{2-0} = \frac{1}{2}$

2. If X is uniformly distributed in $[-\alpha, \alpha]$ with $\alpha > 0$. Then find α such that $P(X < \frac{1}{2}) = 0.7$.

Ans. $\alpha = \frac{5}{4}$

3. If a conference room cannot be reserved for more than 4 hours, find the probability that a given conference lasts more than three hours.

Ans. $\frac{1}{4}$

Hint:: $f(x) = \frac{1}{4}$, $P(X \ge 3) = (4-3)\frac{1}{4} = \frac{1}{4}$.

4. The daily amount X of coffee, in liters dispensed by a machine is uniformly distributed with $a = 7, b = 10$. Determine the probability that the amount of coffee dispensed by the machine will be (a) at most 8.8 (b) more than 7.4 but less than 9.5 (c) at least 8.5 litres.

Ans. a) 0.6 b) 0.7 (c) 0.5

Hint: $f(x) = \frac{1}{10-7} = \frac{1}{3}$

5. The driving time X from house to bus station is uniformly distributed $U[10, 50]$. If it takes 2 minutes to board the bus, determine the probability that person catches the 6.00 pm bus if he starts at 5.43 pm at his house.

Ans. $P(X \le 15) = \frac{15-10}{50-10} = \frac{1}{8} = 0.125$

Hint: Maximum time to catch bus is 6.0 – 5.43 – 0.2 = 15 minutes.

6. Find the third and fourth moment about the mean of a uniform distribution.

Ans. 0, $(b-a)^4/80$

Hint: $\mu_r = E\{(X-\mu)^r\} = \int_a^b \left\{x - \frac{(b+a)}{2}\right\}^r \frac{1}{b-a}dx = \begin{cases} 0 & \text{if } r \text{ is odd} \\ \frac{1}{r+1}\left(\frac{b-a}{2}\right)^r & \text{if } r \text{ is even} \end{cases}$

7. A bus arrives every 10 minutes at a bus stop. Assuming waiting time X for bus is uniformly distributed find the probability that a person has to wait for the bus (a) for more than 7 minutes (b) between 2 and 7 minutes

Ans. (a) $\frac{10-7}{10-0} = \frac{3}{10} = 0.3$ (b) $\frac{7-2}{10-0} = \frac{5}{10} = \frac{1}{2} = 0.5$

8. If X is uniformly distributed with mean 1 and variance $\frac{4}{3}$ find $P(X < 0)$.

Ans. $\frac{1}{4}$

Hint: Mean $= \frac{b+a}{2} = 1$, variance $= \frac{(b-a)^2}{12} = \frac{4}{3}$, $a = -1, b = 3$ $f(x) = \frac{1}{3-(-1)} = \frac{1}{4}$, $P(X < 0) = \frac{0-(-1)}{1}\frac{1}{4} = \frac{1}{4}$.

27.9 NORMAL DISTRIBUTION

Normal probability distribution or simply normal distribution is the probability distribution of a continuous random variable X, known as normal random variable or normal variate. It is given by

$$N(\overline{X}, \sigma) = f(X) = Y(X) = \frac{1}{\sigma\sqrt{2\pi}} e^{-\frac{1}{2}(X-\overline{X})^2/\sigma^2} \quad (1)$$

Here $\overline{X}$ = Arithmetic mean, σ = standard deviation, are the two parameters of the continuous distribution (1). Normal distribution (N.D.) is also known as Gaussian distribution (due to Karl Friedrich Gauss and also credited to de Moivre and Laplace). This theoretical distribution (1) is most important, simple, useful and is the corner stone of modern statistics because (a) discrete probability distributions such as Binomial, Poisson,

Hypergeometric can be approximated by N.D. (b) sampling distributions 't', F, χ^2 tend to be normal for large samples and (c) it is applicable in statistical quality control in industry.

Properties of Normal Distribution (N.D.)

1. The graph of the N.D. $y = f(X)$ in the XY-plane is known as normal curve (N.C.). N.C. is (a) symmetric about y-axis (b) it is bell shaped (c) the mean, median and mode coincide and therefore N.C. is unimodal (has only one maximum point). (d) N.C. has inflection points at $\overline{x} \pm \sigma$. (e) N.C. is asymptotic to both positive x-axis and negative x-axis (see Fig. 27.6).

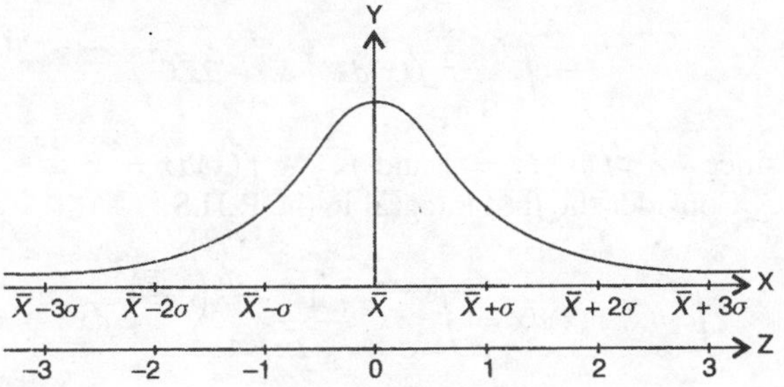

Fig. 27.6

2. Area under the normal curve is unity.
3. Probability that the continuous random variable X lies between X_1 and X_2 is denoted by probability $(X_1 \leq X \leq X_2)$ and is given by

$$P(X_1 \leq X \leq X_2) = \int_{X_1}^{X_2} \frac{1}{\sigma\sqrt{2\pi}} e^{-\frac{1}{2}\left(\frac{x-\overline{x}}{\sigma}\right)^2} dx \tag{2}$$

Since (2) depends on the two parameters $\overline{x}$ and σ, we get different normal curves for different values of $\overline{x}$ and σ and it is an impractiable task to plot all such normal curves. Instead, by introducing

$$Z = \frac{x - \overline{x}}{\sigma}$$

the R.H.S. integral in (2) becomes independent (dimensionless) of the two parameters $\overline{x}$ and σ. Here Z is known standard (or standardized) variable (variate).

4. Change of scale from x-axis to z-axis.

$$P(X_1 \leq X \leq X_2) = \int_{X_1}^{X_2} \frac{1}{\sigma\sqrt{2\pi}} e^{-\frac{1}{2}\left(\frac{x-\overline{x}}{\sigma}\right)^2} dx$$

$$P(Z_1 \leq Z \leq Z_2) = \int_{Z_1}^{Z_2} \frac{1}{\sigma\sqrt{2\pi}} e^{-Z^2/2} \sigma\, dZ$$

$$= \int_{Z_1}^{Z_2} \frac{1}{\sqrt{2\pi}} e^{-Z^2/2} dZ \tag{3}$$

where

$$Z_1 = \frac{X_1 - \overline{X}}{\sigma}, \quad Z_2 = \frac{X_2 - \overline{X}}{\sigma}.$$

5. Error function or probability integral is defined as

$$P(Z) = \frac{1}{\sqrt{2\pi}} \int_0^Z e^{-Z^2/2} dZ \tag{4}$$

Now (3) can be written using (4) as

$$P(Z_1 \leq Z \leq Z_2) = \int_{Z_1}^{Z_2} \frac{1}{\sqrt{2\pi}} e^{-Z^2/2} dZ$$

$$= P(Z_2) - P(Z_1) \tag{5}$$

Normal distribution $N(\overline{x}, \sigma)$ transformed by the standard variable Z is given by

$$N(0, 1) = Y(Z) = \frac{1}{\sqrt{2\pi}} e^{-Z^2/2}$$

with mean 0 and standard deviation 1. $N(0, 1)$ is known as "Standard Normal Distribution" and its normal curve as standard normal curve (Fig. 27.7). The probability integral (4) is tabulated for various values of Z varying from 0 to 3.9 and is known as normal table (see A12). Thus the entries in the normal table gives (represents) the area under the normal curve between the ordinates $Z = 0$ to Z (shaded in the figure). Since normal curve is symmetric about y-axis, the area from 0 to $-Z$ is same as the area from 0 to Z. For this reason, normal table is tabulated only for positive values of Z. Hence the determination of normal probabilities (3) reduce to the determination of areas under

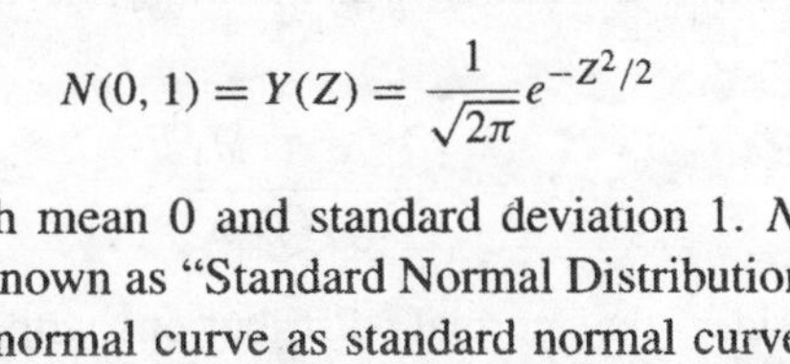

Fig. 27.7

Fig. 27.7). Therefore

$$P(X_1 \le X \le X_2) = P(Z_1 \le Z \le Z_2) = P(Z_2) - P(Z_1)$$

= (Area under the N.C. from 0 to Z_2)

− (Area under the N.C. from 0 to Z_1)

6. Area under the N.C. is distributed as follows:
68.27% area lies between $\overline{X} - \sigma$ to $\overline{X} + \sigma$
i.e., between $-1 \le Z \le 1$
94.45% area lies between $\overline{X} - 2\sigma$ to $\overline{X} + 2\sigma$
i.e., between $-2 \le Z \le 2$
99.73% area lies between $\overline{X} - 3\sigma$ to $\overline{X} + 3\sigma$
i.e., between $-3 \le Z \le 3$

Note: 50% area in the Z-interval (−.745, +.745)
99% area in the Z-interval (−2.58, +2.58)

Arithmetic Mean of Normal Distribution

By definition
the A.M. of a continuous distribution $f(x)$ is given by

$$\text{A.M.} = \frac{\int_{-\infty}^{\infty} x\, f(x)dx}{\int_{-\infty}^{\infty} f(x)dx}.$$

Consider the normal distribution with B, C as the parameters, i.e., $N(B, C) = f(x) = \frac{1}{c\sqrt{2\pi}} e^{-\frac{1}{2}\left(\frac{x-B}{c}\right)^2}$. Then

$$\text{A.M.} = \overline{X} = \int_{-\infty}^{\infty} x \cdot \frac{1}{c\sqrt{2\pi}} e^{-\frac{1}{2}\left(\frac{x-B}{c}\right)^2} dx$$

since $\int_{-\infty}^{\infty} f(x)dx$ = area under the normal curve = 1

Put $\frac{x-B}{c} = z$ so $x = B + cz$, $dx = cdz$

$$\text{So } \overline{X} = \int_{-\infty}^{\infty} (B + cz)\frac{1}{c\sqrt{2\pi}} e^{-\frac{1}{2}z^2} cdz$$

$$= B\int_{-\infty}^{\infty} \frac{1}{\sqrt{2\pi}} e^{-\frac{z^2}{2}} dz + \frac{c}{\sqrt{2\pi}} \int_{-\infty}^{\infty} ze^{-\frac{z^2}{2}} dz$$

$$= B + \frac{c}{\sqrt{2\pi}} \int_{-\infty}^{\infty} e^{-\frac{z^2}{2}} d\left(\frac{z^2}{2}\right)$$

since $\int_{-\infty}^{\infty} f(x)dx = \int_{-\infty}^{\infty} \frac{1}{\sqrt{2\pi}} e^{-\frac{z^2}{2}} dz = 1.$

$$= B + \frac{c}{\sqrt{2\pi}} \left.\frac{e^{-\frac{z^2}{2}}}{-1}\right|_{-\infty}^{\infty} = B + 0$$

So $\overline{X} = B$.

Variance for Normal Distribution

By definition

$$\text{Variance} = \int_{-\infty}^{\infty} (x - \overline{x})^2 f(x)dx$$

$$= \int_{-\infty}^{\infty} x^2 f(x)dx + \overline{x}^2 \int_{-\infty}^{\infty} f(x)dx$$

$$-2\overline{x}\int_{-\infty}^{\infty} x\, f(x)dx$$

$$= \int_{-\infty}^{\infty} x^2 f(x)dx + \overline{x}^2 - 2\overline{x}\overline{x}$$

since $\int_{-\infty}^{\infty} f(x)dx = 1$ and $\int_{-\infty}^{\infty} x\, f(x)dx = \overline{x}$.
Consider the first integral in the R.H.S.

$$\int_{-\infty}^{\infty} x^2 f(x)dx = \int_{-\infty}^{\infty} x^2 \frac{1}{c\sqrt{2\pi}} e^{-\frac{1}{2}\left(\frac{x-\overline{x}}{c}\right)^2} dx$$

Put $\frac{x-\overline{x}}{c} = z$ so $x = \overline{x} + cz$, $dx = cdz$

$$\int_{-\infty}^{\infty} x^2 f(x)dx = \int_{-\infty}^{\infty} (\overline{x} + cz)^2 \frac{1}{c\sqrt{2\pi}} e^{-\frac{z^2}{2}} cdz$$

$$= \frac{1}{\sqrt{2\pi}}\left[c^2 \int_{-\infty}^{\infty} z^2 e^{-\frac{1}{2}z^2} dz\right.$$

$$\left. + \overline{x}^2 \int_{-\infty}^{\infty} e^{-\frac{z^2}{2}} dz + 2c\overline{x}\int_{-\infty}^{\infty} ze^{-\frac{z^2}{2}} dz\right]$$

$$= \frac{-c^2}{\sqrt{2\pi}} \int_{-\infty}^{\infty} zd\left(e^{-\frac{z^2}{2}}\right) + \overline{x}^2 \cdot 1 + 2c\overline{x} \cdot 0$$

$$= \frac{-c^2}{\sqrt{2\pi}} \left. ze^{-\frac{z^2}{2}} \right|_{-\infty}^{\infty} + \frac{c^2}{\sqrt{2\pi}} \int_{-\infty}^{\infty} e^{-\frac{z^2}{2}} dz + \overline{x}^2$$

$$= 0 + c^2 \cdot 1 + \overline{x}^2$$

Substituting this value

$$\text{Variance} = \int_{-\infty}^{\infty} x^2 f(x)dx - \overline{x}^2 = [c^2 + \overline{x}^2] - \overline{x}^2 = c^2$$

Thus the standard deviation (s.d.), of N.D. is c.

Book Work: Show that the area under the normal curve is unity.

Proof: Normal probability distribution is given by

$$y = f(x) = \frac{1}{\sigma\sqrt{2\pi}} e^{-(x-\bar{x})^2/2\sigma^2}$$

Then the area A under the normal curve is

$$A = \int_{-\infty}^{\infty} \frac{1}{\sigma\sqrt{2\pi}} e^{-(x-\bar{x})^2/2\sigma^2} dx$$

Put $\frac{x-\bar{x}}{\sigma} = z$, so $dx = \sigma dz$

$$\text{or} \quad A = \int_{-\infty}^{\infty} \frac{1}{\sigma\sqrt{2\pi}} e^{-z^2/2} \sigma dz = \sqrt{\frac{2}{\pi}} \int_0^{\infty} e^{-z^2/2} dz$$

$$\text{or} \quad A \cdot A = \left[\sqrt{\frac{2}{\pi}} \int_0^{\infty} e^{-x^2/2} dx\right]\left[\sqrt{\frac{2}{\pi}} \int_0^{\infty} e^{-y^2/2} dy\right]$$

Here x, y are dummy variables (Fig. 27.8).

$$A^2 = \frac{2}{\pi} \int_0^{\infty}\int_0^{\infty} e^{-(x^2+y^2)/2} dx\, dy$$

Put $x = r\cos\theta$, $y = r\sin\theta$, J = Jacobian $= r$
Limits for r : 0 to ∞, $\theta = 0$ to 2π (to cover the first quadrant $0 < x < \infty$, $0 < y < \infty$)

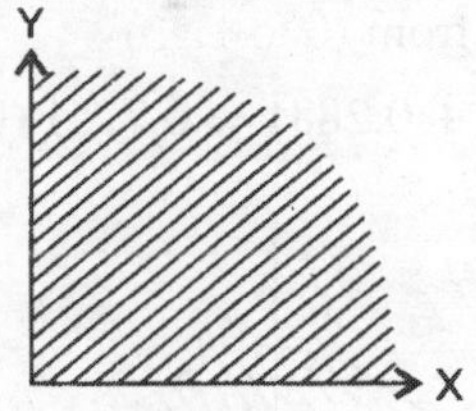

Fig. 27.8

$$\text{So} \quad A^2 = \frac{2}{\pi} \int_0^{\frac{\pi}{2}} \int_0^{\infty} e^{-\frac{r^2}{2}} r\, dr\, d\theta$$

$$= \int_0^{\infty} e^{-\frac{r^2}{2}} d\left(\frac{r^2}{2}\right) = \left.\frac{e^{-\frac{r^2}{2}}}{-1}\right|_0^{\infty} = 1$$

Thus A = area under the normal curve = 1.

Book Work: Prove that for normal distribution the mean deviation from the mean equals to $\frac{4}{5}$ of standard deviation approximately.

Proof: Let $\bar{x}$ and σ be the mean and standard deviation of the normal distribution. Then by definite mean deviations from the mean $= \int_{\infty}^{\infty} |x-\bar{x}| f(x) dx$

$$= \frac{1}{\sigma\sqrt{2\pi}} \int_{-\infty}^{\infty} |x-\bar{x}| e^{-\frac{(x-\bar{x})^2}{(2\sigma^2)}} dx$$

$$= \frac{1}{\sigma\sqrt{2\pi}} \int_{-\infty}^{\infty} \sigma |z| e^{-\frac{1}{2}z^2} \sigma dz$$

where $z = \frac{x-\bar{x}}{\sigma}$ and $dx = \sigma dz$.

$$= \sigma\sqrt{\frac{2}{\pi}} \int_0^{\infty} z e^{-\frac{z^2}{2}} dz = \sigma\sqrt{\frac{2}{\pi}} \left[-e^{-\frac{z^2}{2}}\right]_0^{\infty} = \sqrt{\frac{2}{\pi}}\sigma$$

$$= 0.7979\sigma \approx 0.8\sigma = \frac{8}{10}\sigma = \frac{4}{5}\sigma$$

Fitting of Normal Distribution

Given any frequency distribution, a normal distribution (i.e., a normal curve) can be fitted to it using

$$N(\bar{X}, \sigma) = \frac{1}{\sigma\sqrt{2\pi}} e^{-\frac{1}{2}\left(\frac{X-\bar{X}}{\sigma}\right)^2}.$$

Here $\bar{X}$ = A.M. and σ = s.d. are calculated from the given frequency distribution.

Procedure

Consider a frequency distribution (F.D.)

$L_1 - U_1$	f_1
$L_2 - U_2$	f_2
$\vdots$	
$n - U_n$	f_n

and N = total frequency $= \sum_{i=1}^{n} f_i$

Let $\bar{X}$ and σ be the A.M. and S.D. of the F.D.

Here L_i, U_i are the true lower and upper limits of the ith class.

I. Compute standard variable $z_i = \frac{X_i - \bar{X}}{\sigma}$ for each of the true lower limit X_i of the n classes (there will be $n+1$ such quantities).

II. Compute area under N.C. (from normal table A12) from 0 to z_i.

III. Normal probability of a class is obtained by taking the difference between the successive areas calculated in step II. (when z_i's are of opposite sign, add the successive areas).

IV. Expected or theoretical frequencies are obtained by multiplying probabilities in III by N, the total frequency of the F.D.

Worked Out Examples

Normal distribution

Example 1: Find the area A under the normal curve:

a. to the left of $z = -1.78$

b. to the left of $z = 0.56$

c. to the right of $z = -1.45$

d. corresponding to $z \geq 2.16$

e. corresponding to $-0.80 \leq z \leq 1.53$

f. to the left of $z = -2.52$ and to right of $z = 1.83$

Solution: Refer to normal table (A12)

a. $A = 0.5 -$ Area (0 to $-$ 1.78) (Fig. 27.9)

$= 0.5 -$ Area (0 to 1.78) due to symmetry

$= 0.5 - 0.4625 = 0.0375$ (from table)

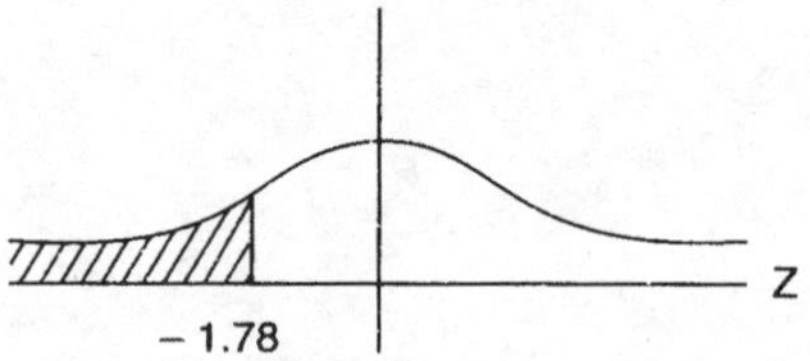

Fig. 27.9

b. $A = 0.5 +$ Area from 0 to 0.56 (Fig. 27.10)

$= 0.5 + 0.2123$ (from table)

$= 0.7123$

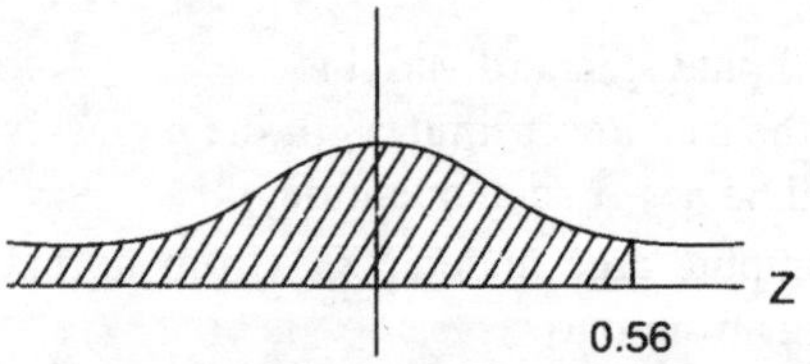

Fig. 27.10

c. $A = 0.5 +$ Area from 0 to $-$ 1.45 (Fig. 27.11)

$= 0.5 +$ Area from 0 to 1.45

$= 0.5 + 0.4265 = 0.9265$

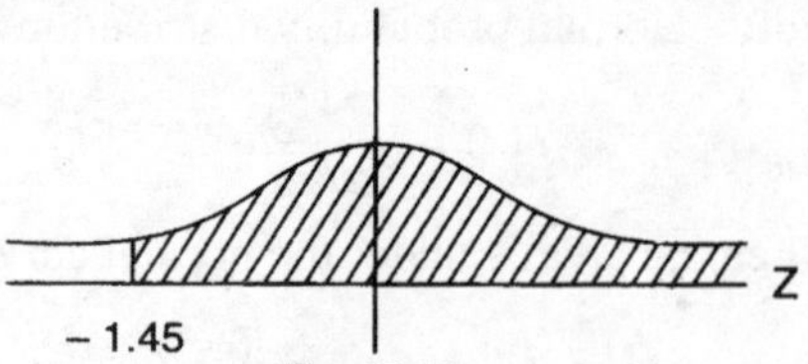

Fig. 27.11

d. $A = 0.5 - A$ (0 to 2.16)

$= 0.5 - 0.4846 = 0.0154$ (Fig. 27.12)

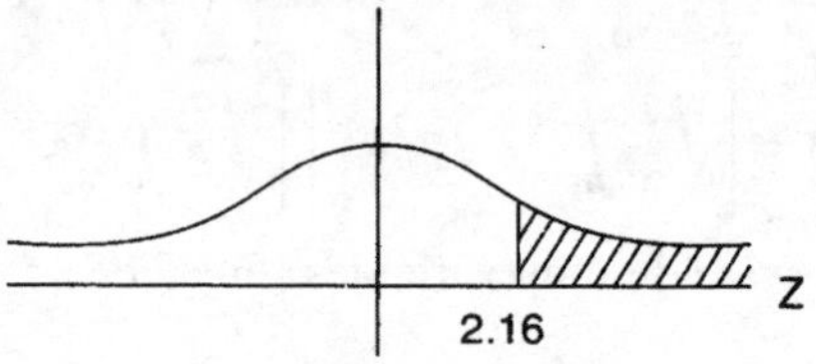

Fig. 27.12

e. $A =$ Area from (0 to $-$ 0.8)

+Area from (0 to 1.53)

$=$ Area from (0 to 0.8)

+Area from (0 to 1.53)

$= 0.4370 + 0.2881 = 0.7251$ (Fig. 27.13)

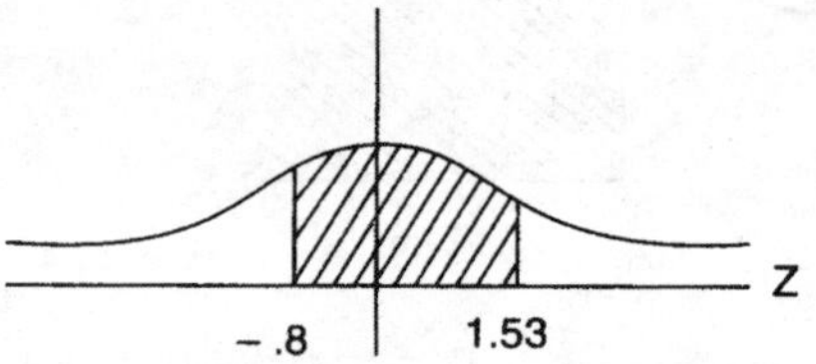

Fig. 27.13

f. $A = [0.5 - A(0 \text{ to } 2.52)] + [0.5 - A(0, 1.83)]$

$= (0.5 - 0.4941) + (0.5 - 0.4664)$

$= 0.0059 + 0.0336$

$= 0.0395$ (Fig. 27.14)

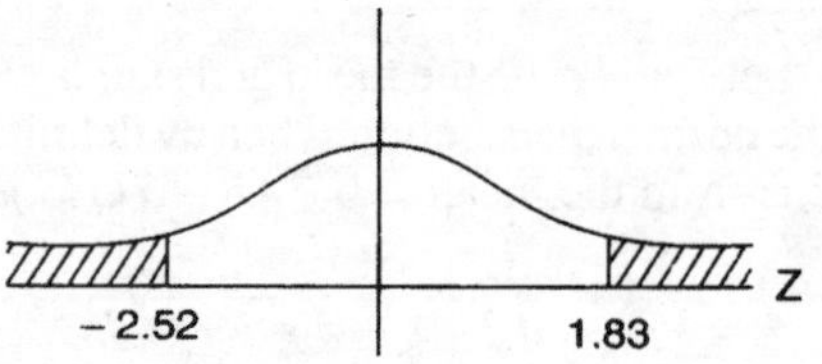

Fig. 27.14

Example 2: If z is normally distributed with mean 0 and variance 1, find

a. $P(z \geq -1.64)$

b. $P(-1.96 \leq z \leq 1.96)$

c. $P(z \leq 1)$

d. $P(z \geq 1)$

Solution:

a. $P(z \geq -1.64) = 0.5 + A(0 \text{ to } -1.64)$ (Fig. 27.15)

$= 0.5 + A(0 \text{ to } 1.64)$

$= 0.5 + 0.4495 = 0.9495$

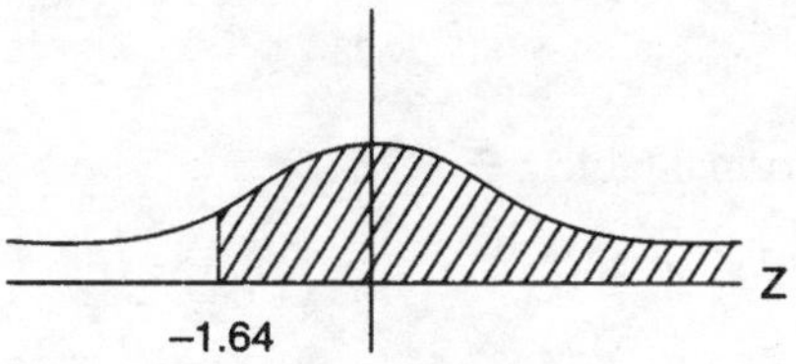

Fig. 27.15

b. $P(-1.96 \leq z \leq 1.96)$

$= 2A(0 \text{ to } 1.96)$ by symmetry

$= 2(0.4750) = 0.9500$ (Fig. 27.16)

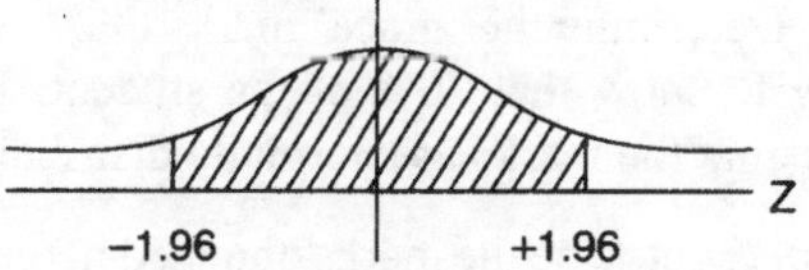

Fig. 27.16

c. $P(z \leq 1) = 0.5 + A(0 \text{ to } 1)$

$= 0.5 + 0.3413$

$= 0.8413$ (Fig. 27.17)

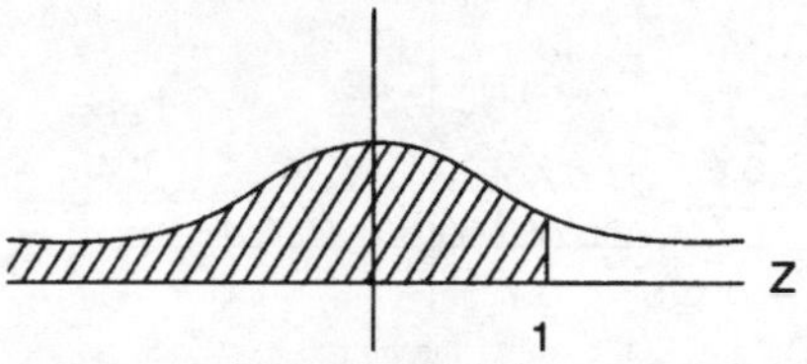

Fig. 27.17

d. $P(z \geq 1) = 0.5 - A(0 \text{ to } 1)$

$= 0.5 - 0.3413 = 0.1587$ (Fig. 27.18).

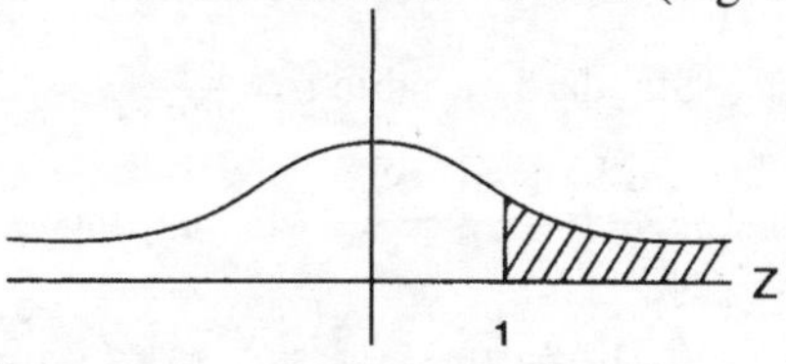

Fig. 27.18

Example 3: Determine the value of z such that (a) area to the right of z is 0.2266 (b) area to the left of z is 0.0314.

Solution: Here the areas (entries of the normal table) are given, the values of z (1st column) are determined.

a. Since area $0.2266 < \frac{1}{2}$ is to the right of z, z must be positive such that area from 0 to z is $0.5 - 0.2266 = 0.2734$. From normal table for area 0.2734, the value of z is 0.75 (Fig. 27.19).

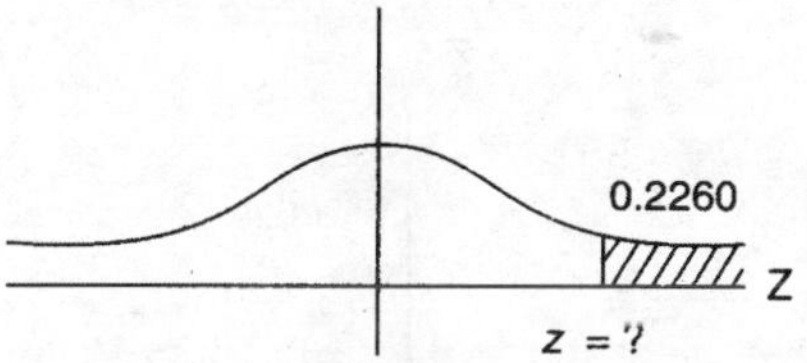

Fig. 27.19

b. Since area $0.134 < \frac{1}{2}$ is to the left of z, z must be negative. So determine z such that area from 0 to z is $0.5 - 0.134 = 0.4686$. From table A12, $z = -1.86$ (Fig. 27.20).

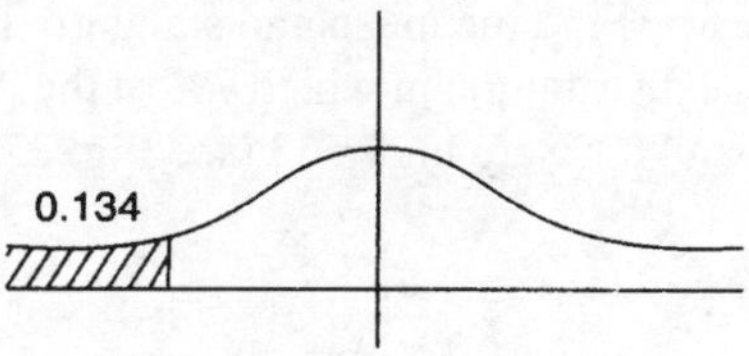

Fig. 27.20

Example 4: Find the (a) mean and (b) standard deviation of an examination in which grades 70 and

88 correspond to standard scores of −0.6 and 1.4 respectively.

Solution: Standard variable $z = \frac{X-\overline{X}}{\sigma}$.

Here $$-0.6 = \frac{70-\overline{X}}{\sigma} \quad \text{so} \quad \overline{X} - 0.6\sigma = 70$$

$$1.4 = \frac{88-\overline{X}}{\sigma} \quad \text{so} \quad \overline{X} + 1.4\sigma = 88$$

Solving $\overline{X} = 75.4$, $\sigma = 9$ are the mean and standard deviation.

Example 5: Determine the minimum mark a student must get in order to receive an A grade if the top 10% of the students are awarded A grades in an examination where the mean mark is 72 and standard deviation is 9.

Solution: The 0.1 area to the right of z corresponds to the top 10% of the students (see Fig. 27.21). From table if area from 0 to z is 0.4, then $z = 1.28$. Given $\overline{X} = 72$, $\sigma = 9$, we have $1.28 = z = \frac{X-\overline{X}}{\sigma} = \frac{X-72}{9}$,

$$X = 72 + 11.52 = 83.52 \simeq 84$$

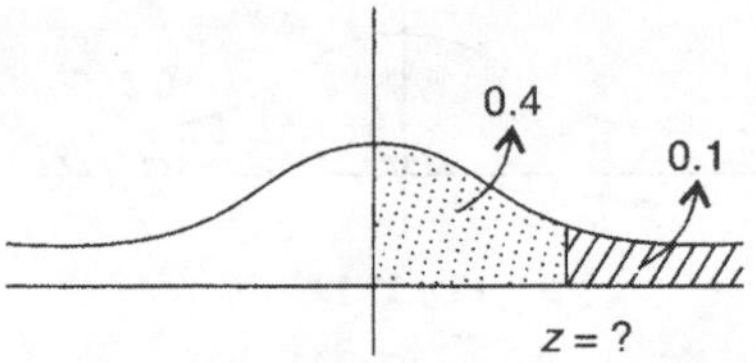

Fig. 27.21

So a student must get a minimum (or more) of 84 marks to get an A grade.

Example 6: Find the mean and standard deviation of a normal distribution in which 7% of the items are under 35 and 89% are under 63 (see Fig. 27.22).

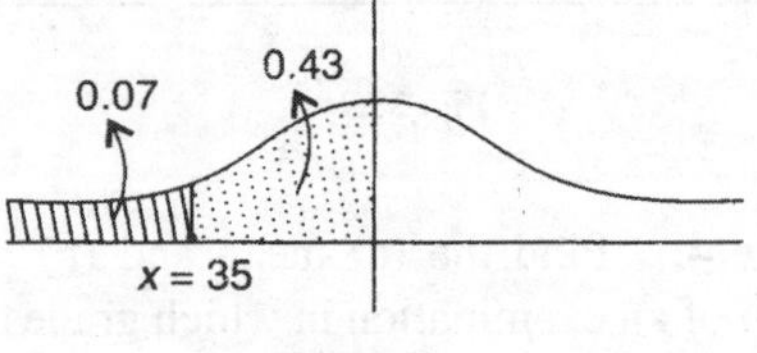

Fig. 27.22

Solution: Let X be the continuous random variable. Given that $P(X < 35) = 0.07 < \frac{1}{2}$. So z must be negative such that area from 0 to z is $0.5 - 0.07 = 0.43$. From normal table $z = -1.48$.
Given that $P(X < 63) = 0.89 > \frac{1}{2}$. So z must be positive such that area from 0 to z is $0.89 - 0.5 = 0.39$ (Fig. 27.23).

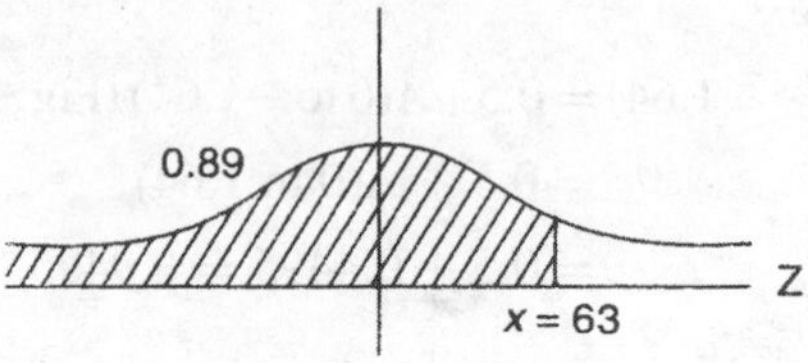

Fig. 27.23

From normal table $z = 1.23$.
Since $z = \frac{X-\overline{X}}{\sigma}$, we have

$$-1.48 = \frac{35-\overline{X}}{\sigma} \quad \text{or} \quad \overline{X} - 1.48\sigma = 35$$

$$1.23 = \frac{63-\overline{X}}{\sigma} \quad \text{or} \quad \overline{X} + 1.23\sigma = 63$$

Solving the arithmetic mean $\overline{X} = 50.3$ and standard deviation $\sigma = 10.33$.

Example 7: When the mean of marks was 50% and S.D. 5% then 60% of the students failed in an examination. Determine the 'grace' marks to be awarded in order to show that 70% of the students passed. Assume that the marks are normally distributed.

Solution: Let X be the marks obtained in the exam. Given $\overline{X} = 0.5$, $\sigma = \text{s.d.} = 0.05$.

Before grace marks were awarded, 60% failed. Since 60% failure corresponds 0.6 area, z_1 must be positive (Fig. 27.24). Determine z_1 such that the area to its left is 0.6. The value of z_1 for which the area is 0.1 is 0.25.

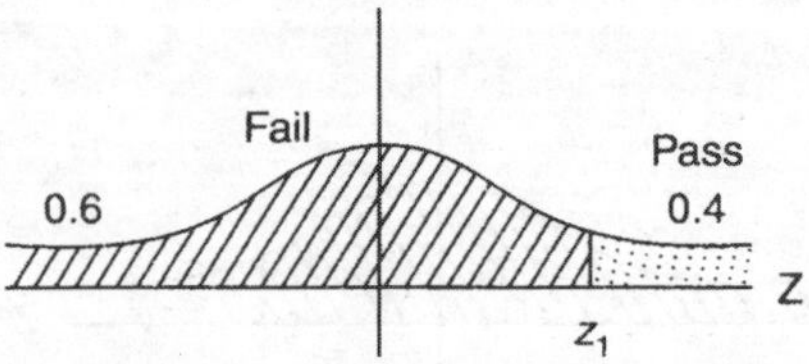

Fig. 27.24

$$0.25 = z_1 = \frac{X_1 - 0.5}{0.05} \quad \text{so} \quad X_1 = 0.5125$$

After grace marks were awarded, 70% passed examination. The area 0.70 $\left(> \frac{1}{2}\right)$ corresponds pass students (Fig. 27.25). Determine z_2 such that the area to its right is 0.7. So z_2 must be negative and from table, $z_2 = -0.52$. Then

$$z_2 = -0.52 = \frac{X - 0.5}{0.05} \quad \text{or} \quad X_2 = 0.4740$$

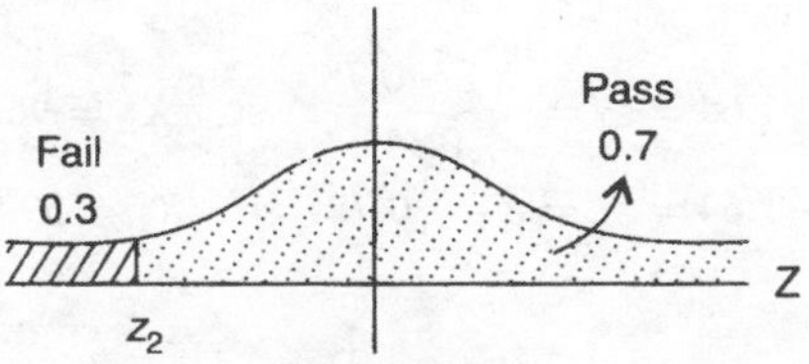

Fig. 27.25

Thus the minimum pass mark for a student is 51.25 before grace while the minimum pass mark is 47.40 after grace. So grace marks awarded is 51.25 – 47.40 = 3.85.

Example 8: Assume that the 'reduction' of a person's oxygen consumption during a period of Transcendenta Meditation (T.M.) is a continuous random variable X normally distributed with mean 37.6 cc/mt and s.d. 4.6 cc/mt. Determine the probability that during a period of T.M. a person's oxygen consumption will be reduced by (a) at least 44.5 cc/mt (b) at most 35.0 cc/mt (c) anywhere from 30.0 to 40.0 cc/mt.

Solution: $z = \frac{X - \overline{X}}{\sigma} = \frac{X - 37.6}{4.6}$

a. For $X = 44.5$, $z = \frac{44.5 - 37.6}{4.6} = 1.5$ (Fig. 27.26)

$P(X \geq 44.5) = P(z \geq 1.5) = 0.5 - 0.4332 = 0.068$

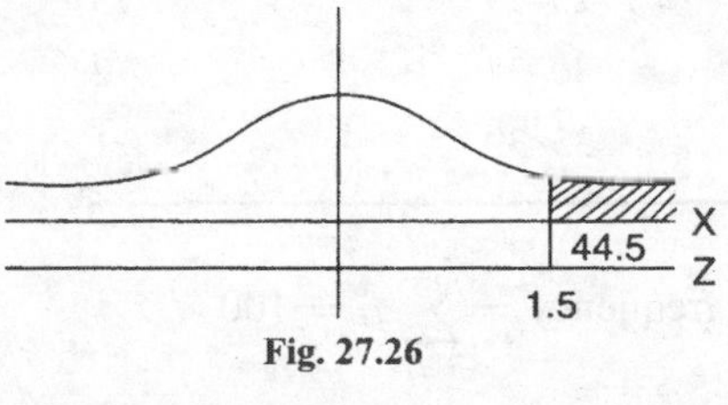

Fig. 27.26

b. For $X = 35.0$, $z = \frac{35.0 - 37.6}{4.6} = -0.5652$ (Fig. 27.26)

$$P(X \leq 35) = P(z \leq -0.5652)$$
$$= 0.5 - 0.2157$$
$$= 0.2843.$$

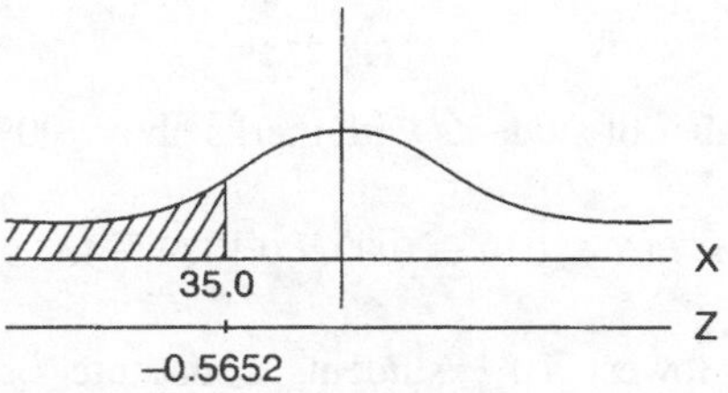

Fig. 27.27

c. For $X_1 = 30$, $z_1 = \frac{30 - 37.6}{4.6} = -1.6521$

For $X_2 = 40$, $z_2 = \frac{40 - 37.6}{4.6} = 0.52173$ (Fig. 27.26)

$$P(30 \leq X \leq 40) = P(-1.6521 \leq z \leq 0.52173)$$
$$= 0.4505 + 0.1985$$
$$= 0.6490.$$

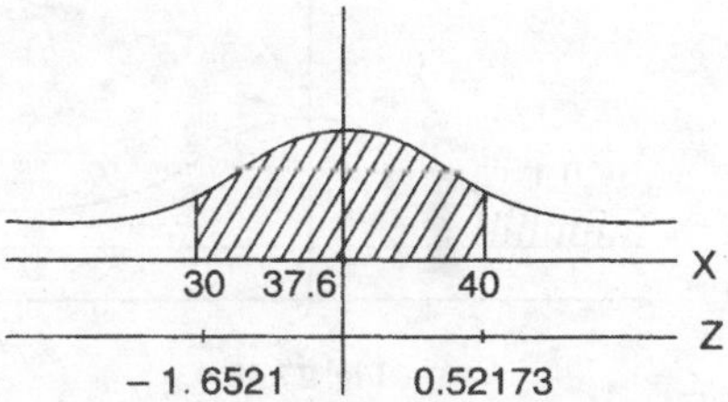

Fig. 27.28

Example 9: The marks X obtained in mathematics by 1000 students in normally distributed with mean 78% and s.d. 11% (Fig. 27.29). Determine (a) how many students got marks above 90%? (b) what was the highest mark obtained by the lowest 10% of students? (c) semi-inter quartile range (d) within what limits did the middle 90% of students lie?

Solution: Here $z = \frac{X - \overline{X}}{\sigma} = \frac{X - 0.78}{0.11}$

a. For $X = 0.9$, $z = \frac{0.9 - 0.78}{0.11} = 1.09.$

$P(X > 0.9) = P(z > 1.09) = 0.5 - 0.3621 = 0.1379$

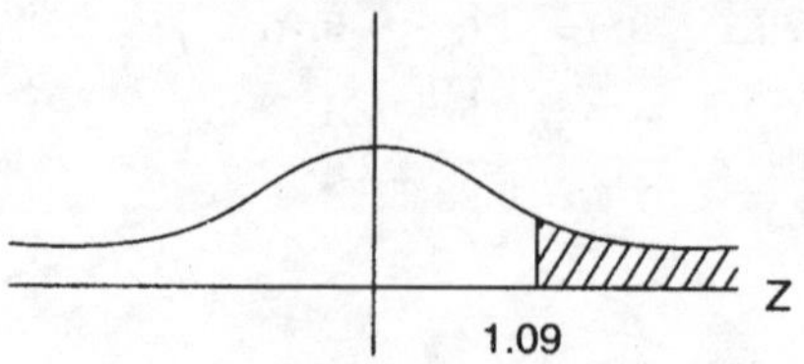

Fig. 27.29

Number of students with marks above 90%

$= 1000 \times P(X > 0.9) = 1000 \times 0.1379 = 137.9 \approx 138.$

b. The lowest 10% students constitute 0.1 area $\left(< \frac{1}{2}\right)$ of extreme left tail. So z_1 must be negative. From table $0.4 = 0.5 - 0.1 =$ 0.5− Area 0.1 from 0 to z_1 so $z_1 = -1.28$.

Thus $\quad -1.28 = z_1 = \dfrac{X - 0.78}{0.11} \quad$ or $X = 0.6392$

(see Fig. 27.30)

Thus the highest mark obtained by the lowest 10% of students is $63.92 \approx 64\%$.

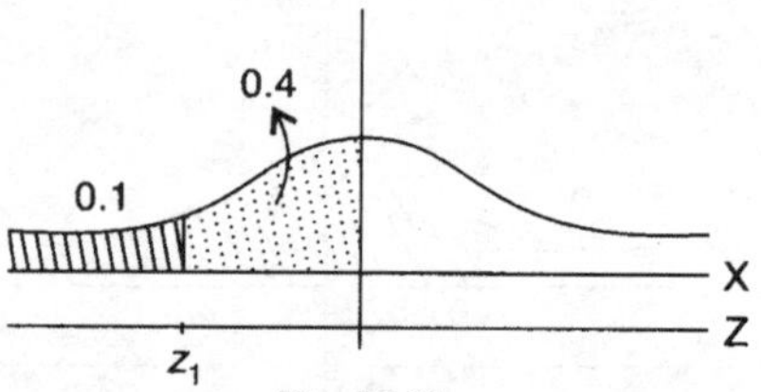

Fig. 27.30

c. Quartiles Q_1, Q_2, Q_3 divide the area into four equal parts. The value of z_1 corresponding to the first quartile Q_1 is such that the area to its left is 0.25. From table $z_1 = -0.67$. Similarly, $z_3 = 0.67$ corresponding to Q_3. Now $-0.67 = z_1 = \frac{X_1 - 0.78}{0.11}$. So the quartile mark is $X_1 = 0.7063 = 70.63\%$. Similarly, $X_3 = 85.37\%$. Thus the semi-inter quartile range $= \frac{Q_3 - Q_1}{2} = \frac{85.37 - 70.63}{2} = 7.37$ (Fig. 27.31).

d. Middle 90% correspond to 0.9 area, leaving 0.05 area on both sides. Then the corresponding z's are ± 1.64 (refer Fig. 27.32).

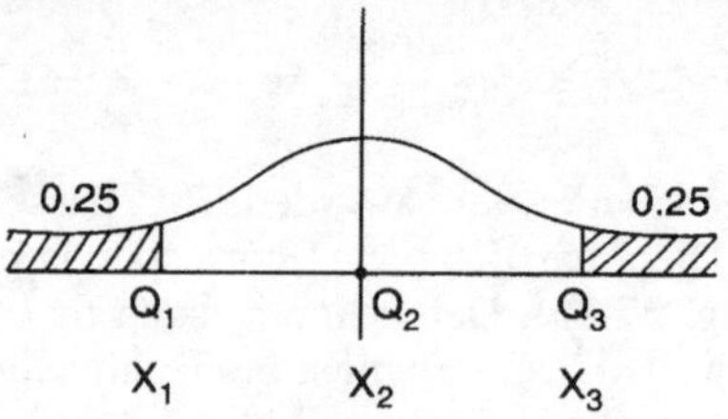

Fig. 27.31

$$1.64 = z_2 = \frac{X_2 - 0.78}{0.11} \quad \text{so} \quad X_2 = 96.04$$

$$-1.64 = z_1 = X_1 - 0.78 \quad \text{so} \quad X_1 = 59.96$$

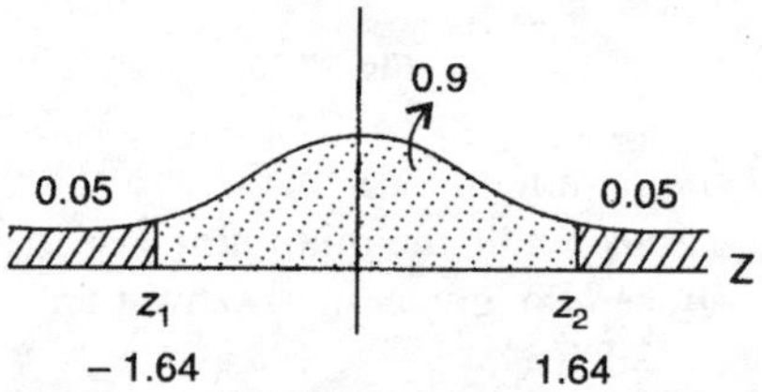

Fig. 27.32

Thus the middle 90% have marks in between 60 to 96.

Example 10: Fit a normal distribution to the following data (frequency distribution):

S. No.	Class	Observed frequency f_i
1	5–9	1
2	10–14	10
3	15–19	37
4	20–24	36
5	25–29	13
6	30–34	2
7	35–39	1

$$\text{Total frequency} = \sum_{i=1}^{7} f_i = 100$$

Solution:

1	2	3	4	5	6	7	8
S. No.	Class	Frequency f_i	True lower class limit X_i	Standard variate $z_i = \frac{X_i-20}{5}$	Area from 0 to z_i	Area for class (Probability P)	Expected or theoretical frequency $= NP = 100P$
1	5–9	1	4.5	−3.1	0.4990	0.0169	$1.69 \approx 2$
2	10–14	10	9.5	−2.1	0.4821	0.1178	$11.78 \approx 12$
3	15–19	37	14.5	−1.1	0.3643	0.3245	$32.45 \approx 32$
			19.5	−0.1	0.0398		
4	20–24	36	24.5	0.9	0.3159	0.3557	$35.57 \approx 36$
5	25–29	13	29.5	1.9	0.4713	0.1554	$15.54 \approx 16$
6	30–34	2	34.5	2.9	0.4981	0.0268	$2.68 \approx 3$
7	35–39	1	39.5	3.9	0.5000	0.0019	$0.19 \approx 0$
		Total Frequency $N = 100$					

Note: Entries in column 7 are obtained by subtracting successive values in column 6 whenever they (in 6) are of the same sign. Add the values in column 6 when they are of opposite sign.

EXERCISE

Normal distribution

1. Determine the area under the normal curve
   a. between $z = -1.2$ and $z = 2.4$
   b. between $z = 1.23$ and $z = 1.87$
   c. between $z = -2.35$ and $z = -0.5$
   d. to the left of $z = -1.90$
   e. to the left of $z = 1.0$
   f. to the right of $z = -2.40$
   g. to the left of $z = -3.0$ and to the right of $z = 2.0$.

Ans. (a) 0.8767 (b) 0.0786 (c) 0.2991 (d) 0.0287 (e) 0.8413 (f) 0.9918 (g) 0.0241

2. Find the value of z such that
   a. area between -0.23 and z is 0.5722
   b. area between 1.15 and z is 0.0730
   c. area between $-z$ and z is 0.9.

Ans. (a) $z = 2.08$ (b) $z = 0.1625$ (c) $z = -1.65$ to $+1.65$

3. **a.** Calculate the standard marks of two students whose marks are 93 and 62 in an examination given that the mean mark is 78 and s.d. is 10.
   b. If the standard marks of two students are -0.6 and 1.2, determine their respective marks.

Ans. (a) $z = 1.5, -1.6$ (b) $X = 72, 90$

4. Determine the probability that the amount of cosmic radiation X a pilot of jet plane will be exposed is more than 5.20 m rem if X is normally distributed with mean 4.35 m rem and s.d. 0.59 m rem.

Ans. $P(X > 5.20) = P(z > 1.44)$ $= 0.5 - 0.4251 = 0.0749$.

5. Suppose the life span X of certain motors is normally distributed with mean 10 years and s.d. 2 years. If the manufacturer is ready to replace only 3% of motors that fail, how many years of guarantee can he offer (Fig. 27.33).

Ans. $-1.88 = z = \frac{X-\overline{X}}{\sigma} = \frac{X-10}{2}$, $X = 6.24$ years

6. Determine the expected number of boys whose weight is
   a. between 65 and 70 kg
   b. greater than or equal to 72 kg

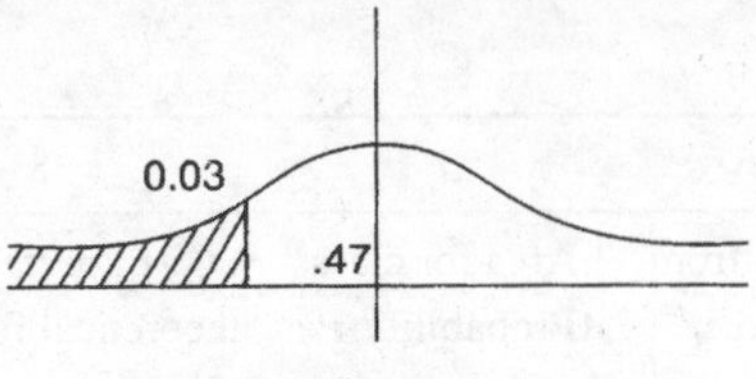

Fig. 27.33

if the weight X of 800 boys follows normal distribution with $\overline{X} = 66$, $\sigma = 5$.

Ans. **a.** $P(65 \le X \le 70) = P(-0.20 < z < 0.80)$
$= 0.0793 + 0.2881 = 0.3674$

Number $= 800(0.3674) = 294$

b. $P(X \ge 72) = P(z \ge 1.2) = 0.5 - 0.3849$
$= 0.1151$

Number of boys $= 800(0.1151) = 92$

7. Calculate the mean and s.d. of a normal distribution in which 31% are under 45 and 8% are over 64.

Ans. $\overline{X} = 50$, $\sigma = 10$

8. Assume that the average life span of computers produced by a company is 2040 hours with s.d. of 60 hours. Find the expected number of computers whose life span is
   a. more than 2150 hours
   b. less than 1950 hours
   c. more than 1920 hours and less than 2160 hours

   from a pool of 2000 computers assuming that the life span X is normally distributed.

Ans. **a.** $P(X > 2150) = P(z > 1.833) =$
$0.5 - 0.4664 = 0.0336$.

Expected number of computers whose life span is more than 2150 hours $=$ $2000(0.0336) = 67$.

b. $P(X < 1950) = P(z < -1.33)$
$= 0.5 - 0.40821 = 0.0918$

Expected number $= 2000(0.0918) = 184$

c. $P(1920 \le X \le 2160) = P(-2 \le z \le 2)$
$= 2(0.4772) = 0.9544$.

Expected number $= 2000 \times 0.9544 = 1909$.

9. If the top 15% of the students receives A grade and bottom 10% receives F grades in a mathematics examination, determine the
   a. minimum mark to get an A grade
   b. minimum mark to pass (not to get F grade). Assume that the marks are normally distributed with mean 76 and s.d. 15.

Ans. (a) 92 (b) 57

10. A university awards distinction, first class, second class, third class or pass class according as the student gets 80% or more; 60% or more; between 45% and 60%; between 30% and 45%; or 30% or more marks respectively. If 5% obtained distinction and 10% failed, determine the percentage of students getting second class. Assume that marks X are normally distributed.

Ans. 34% second class.

Hint: $P(X < 30) = 0.10$ failed; $P(X \ge 80) = 0.05$ distinction, $\frac{30-\overline{X}}{\sigma} = -1.28$, $\frac{80-\overline{X}}{\sigma} = 1.64$, $\overline{X} = 52$, $\sigma = 17.12$

$P(45 < X < 60) = P(-0.41 < z < 0.47) \le 0.1591 + 0.1808 = 0.3399$.

11. The amount of pollutant X released by an industry should lie between 30 and 35. Assume that X is normally distributed with mean $\overline{X} = 33$ and s.d. $\sigma = 3$. The industry gets a profit of Rs. 100 when $30 < X < 35$; Rs. 50 when $25 < X \le 30$ or $35 \le X < 40$ and incurs a fine of Rs. 60 otherwise. Determine the expected profit for the industry.

Ans. $100(0.5890) + 50(0.396) - 60(0.0137)$
$=$ Rs. 79.

EXERCISE

Fitting of normal distribution

1. Fit a normal curve to the following data:

Class	60–62	63–65	66–68	69–71	72–74
Frequency	5	18	42	27	8

Ans. $4.13 \approx 4$ $20.68 \approx 21$ $38.92 \approx 39$

$27.71 \approx 28 \quad 7.43 \approx 7$

Hint: $\overline{X} = 67.45, \sigma = 2.92, N = 100.$

2. Fit a normal distribution to the following frequency distribution

x:	2	4	6	8	10
f:	1	4	6	4	1

Ans. $0.97 \approx 1 \quad 3.9 \approx 4 \quad 6.1 \approx 6 \quad 3.9 \approx 4$
$0.97 \approx 1.0$

Hint: $\overline{X} = 6, \sigma = 2, N = 16$, x is taken as the mid value of the class, i.e., 2 is mid value of the class (1, 3), etc.

3. Fit a normal curve to the following observed data:

Class	9.3–9.7	9.8–10.2	10.3–10.7	10.8–11.2
f	2	5	12	17

Class	11.3–11.7	11.8–12.2	12.3–12.7	12.8–13.2
f	14	6	3	1

Ans. $1.704 \approx 2 \quad 5.562 \approx 6 \quad 11.7420 \approx 12$
$15.624 \approx 16 \quad 13.942 \approx 14 \quad 7.62 \approx 8$
$2.712 \approx 3 \quad 0.168 \approx 1$

Hint: $\overline{X} = 11.09$, s.d. $= 0.733, N = 60.$

4. Fit a normal distribution to the following data:

Class	150–158	159–167	168–176	177–185
f	9	24	51	66

Class	186–194	195–203	204–212	213–221	222–230
f	72	48	21	6	3

Ans. 9.0 25.4 51.5 71.2 67.8 44.6 20.2
6.3 1.4

Hint: $\overline{X} = 184.3, \sigma = 14.54, N = 300.$

27.10 NORMAL APPROXIMATION TO BINOMIAL DISTRIBUTION

For large n, the calculation of binomial probabilities is very cumbersome. In such cases they are computed by approximation procedures. For $n \to \infty$ and $p \to 0$ B.D. can be approximated by Poisson distribution with $\lambda = np$.
For $n \to \infty$ and $p \not\to 0$, i.e., p *not* close to 0 or 1, B.D. can be approximated by normal distribution.

Theorem: *Let X be a binomial random variable with mean* $\overline{X} = np$ *and s.d.* $= \sqrt{npq}$ *then the limiting form of the distribution of*

$$z = \frac{X - np}{\sqrt{npq}}$$

as $n \to \infty$ *is standard normal distribution* $N(z; 0, 1)$.

Normal approximation to B.D. will be fairly good even when

a. *n is small and p is close to* $\frac{1}{2}$

b. *both np and nq are* ≥ 5.

WORKED OUT EXAMPLES

Normal approximation to B.D.

Example 1: If 10% of the truck drivers on road are drunk determine the probability that out of 400 drivers randomly checked

a. at most 32

b. more than 49

c. at least 35 but less than 47 drivers are drunk on the road.

Solution: Here p = probability of a driver drunk $= \frac{10}{100} = 0.1$ and $n = 400$ = no. of trials.
Let X = number of truck drivers drunk.
Here X is a binomial random variable with B.D. $= b(x; 400, 0.1)$. This B.D. can be approximated by normal distribution with A.M. $= \overline{X} = np = 400 \times \frac{10}{100} = 40$ and s.d. $= \sigma = \sqrt{npq} = \sqrt{400 \times \frac{10}{100} \times \frac{90}{100}} = 6.$

a. For $X = 32, z = \frac{X - \overline{X}}{\sigma} = \frac{31.5 - 40}{6} = -\frac{8.5}{6} = -1.416$ since X is treated as continuous variable, values upto and more that 31.5 will be rounded up to 32 (Fig. 27.34).

$$P(X < 32) = P(z \leq -1.416) = 0.5 - 0.4222$$
$$= 0.0778$$

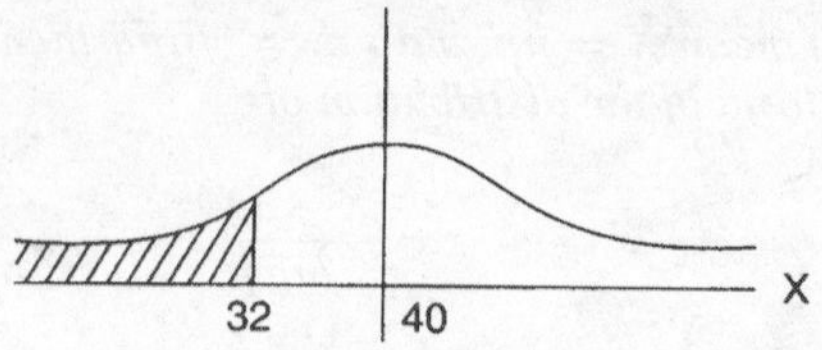

Fig. 27.34

b. Since 49.5 and above are rounded to 50, for $X > 49$, $z = \frac{49.5-40}{6} = 1.58$ (Fig. 27.35)

$$P(X > 49) = P(z \geq 1.58) = 0.5 - 0.4429$$
$$= 0.0571$$

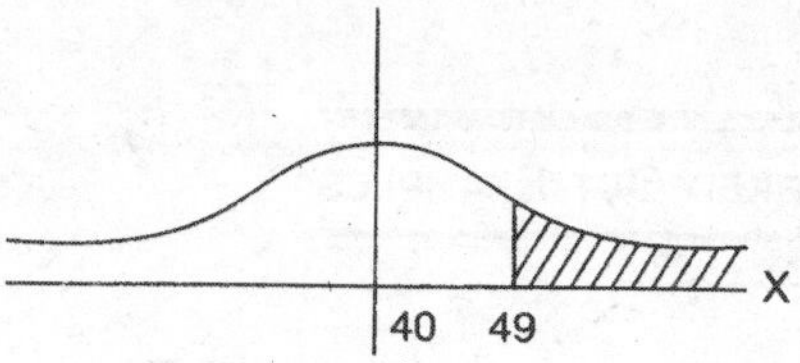

Fig. 27.35

c. $X \geq 35$ includes values of X upto 34.5 and $x < 47$ includes values of X upto 46.5 (Fig. 27.36).

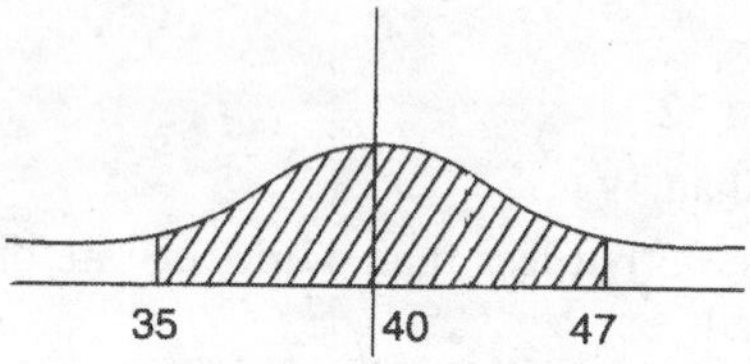

Fig. 27.36

$$z_1 = \frac{34.5 - 40}{6} = -0.916,$$

$$z_2 = \frac{46.5 - 40}{6} = 1.083$$

$$P(35 \leq X < 47) = P(-0.916 \leq z \leq 1.083)$$
$$= 0.3212 + 0.3599 = 0.681.$$

Example 2: A pair of dice is rolled 180 times. Determine the probability that a total of 7 occurs

a. at least 25 times

b. between 33 and 41 times inclusive

c. exactly 30 times.

Solution: Sum 7 occurs in a single throw of a pair of dice as follows: (1, 6), (6, 1), (2, 5), (5, 2), (3, 4), (4, 3).

p = probability of 7 occurring $= \frac{6}{36} = \frac{1}{6}$, $n = 180$; X = number of occurrences of a sum of 7 in a pair of dice = a binomial random variable.

Treating X as a continuous R.V., B.D. $= b\left(x; 180, \frac{1}{6}\right)$ can be approximated by normal distribution with A.M. $= \overline{X} = np = 180 \times \frac{1}{6} = 30$, σ = s.d. $= \sqrt{npq} = \sqrt{180 \cdot \frac{1}{6} \cdot \frac{5}{6}} = 5$

a. For $X \geq 25$, X includes up to 24.5. Thus

$$z = \frac{24.5 - 30}{5} = -1.1$$

$$P(X \geq 25) = P(z \geq -1.1) = 0.5 + 0.3643$$
$$= 0.8643 \text{ (Fig. 27.37)}$$

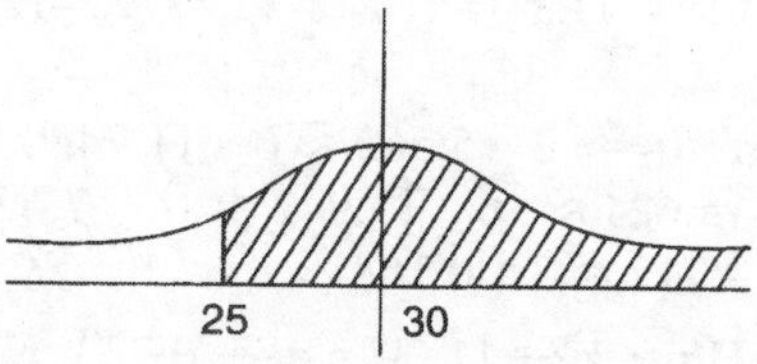

Fig. 27.37

b. $P(33 \leq X \leq 41) = P(0.5 \leq z \leq 2.3)$
$= 0.4893 - 0.1915 = 0.2978$ (Fig. 27.38)

since $z_1 = \frac{32.5-30}{5} = 0.5$, $z_2 = \frac{41.5-30}{5} = 2.3$

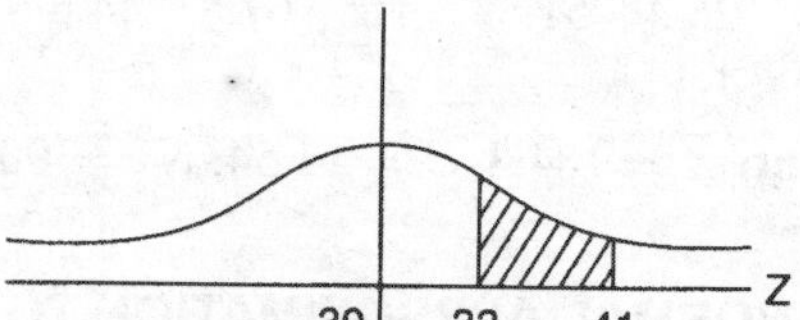

Fig. 27.38

c. $P(X = 30) = P(29.5 \leq X \leq 30.5)$
$= P(-0.1 \leq z \leq 0.1)$
$= 2(0.0398) = 0.0796$ (Fig. 27.39)

since $z_1 = \frac{29.5-30}{5} = -0.1$, $z_2 = \frac{30.5-30}{5} = 0.1$.

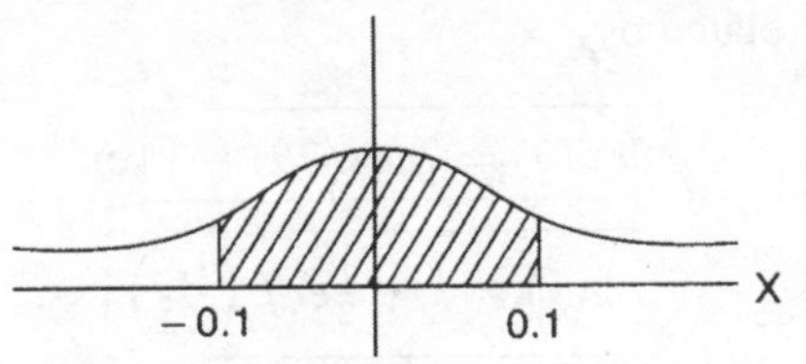

Fig. 27.39

EXERCISE

Normal approximation of B.D.

1. Determine the probability that by guess-work a student can correctly answer 25 to 30 questions in a multiple-choice quiz consisting of 80 questions. Assume that in each question with four choices, only one choice is correct and student has no knowledge.

Ans. $P(25 \le X \le 30) = P(1.16 \le z \le 2.71) = 0.9960 - 0.8770 = 0.1196.$

Hint: $\overline{X} = np = (80)\left(\frac{1}{4}\right) = 20,$

$\sigma = \sqrt{npq} = \sqrt{80 \cdot \frac{1}{4} \cdot \frac{3}{4}} = 3.873$

$z_1 = \frac{24.5-20}{3.873} = 1.16, z_2 = \frac{30.5-20}{3.573} = 2.71.$

2. Find the probability that out of 100 patients
   a. between 84 and 95 inclusive
   b. fewer than 86,
   will survive a heart-operation given that the chances of survival is 0.9.

Ans. **a.** $P(84 \le X \le 95) = P(-2.166 \le z \le 1.833) = 0.4850 + 0.4664 = 0.9514$

Hint: $\overline{X} = np = (100)(0.9) = 90,$
$\sigma = \sqrt{npq} = \sqrt{(100)(0.9)(0.1)} = 3$
$z_1 = \frac{83.5-90}{3} = -2.1666, z_2 = \frac{95.5-90}{3} = 1.8333$

b. $P(X < 86) = P(z_1 \le -1.5) = 0.5 - 0.4332 = 0.0668$

Hint: $z_1 = \frac{85.5-90}{3} = -1.5.$

3. Find the probability P that the number of heads occurring, when a fair coin is tossed 12 times, is between 4 and 7 inclusive by (a) B.D. (b) normal approximation to B.D.

Ans. **a.** $P = \sum_{x=4}^{7} b\left(x; 12, \frac{1}{2}\right) = \sum_{x=4}^{7} {}^{x}C_4 \left(\frac{1}{2}\right)^x \left(\frac{1}{2}\right)^{12-x}$
$= \frac{495+4096+924+792}{4096} = 0.7332$

b. $P = P(3.5 \le X \le 7.5) = P(-1.45 \le z \le -0.87) =$
$= 0.4265 + 0.3078 = 0.7343$

Hint: $\overline{X} = np = 12 \cdot \frac{1}{2} = 6,$

$\sigma = \sqrt{npq} = \sqrt{12\frac{1}{2} \cdot \frac{1}{2}} = 1.73.$

4. The probability that a patient needs an ICU is 0.05 in a hospital with 600 patients. How many ICU's should be available so that the probability of none of the patients of the hospital are turned away due to lack of ICU's is more than 0.90.

Ans. $P(x < x_1) = P(0 < z < z_1) > 0.90,$

$z_1 = \frac{x_1-30}{5.3}$

so $z_1 = 1.28$ or $\frac{x_1-30}{5.3} = z_1 > 1.28$ so $x_1 > 36.784 \approx 37$

Hint: $\overline{X} = np = (600)(0.05) = 30,$
$\sigma = \sqrt{npq} = \sqrt{(600)(0.5)(0.95)} = 5.3.$

27.11 ERROR FUNCTION

Error function of x (also known as error integral), denoted by $erf\,x$, is defined as

$$erf(x) = \frac{2}{\sqrt{\pi}} \int_0^x e^{-t^2} dt \qquad (1)$$

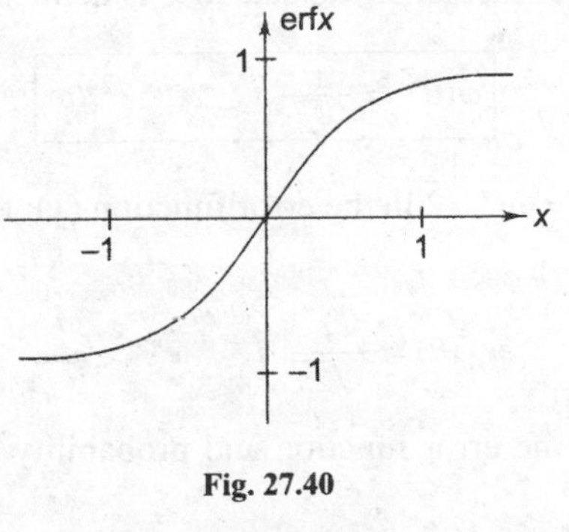

Fig. 27.40

It occurs in probability theory, thermodynamics, heat conduction problems. *Erf x* is known as a 'special function', since (1) can not be evaluated in terms of 'elementary functions' by the usual methods of calculus.

Properties

1. $erf(0) = 0$
2. $erf(\infty) = \frac{2}{\sqrt{\pi}} \int_0^\infty e^{-t^2} dt = \frac{2}{\sqrt{\pi}} \frac{\sqrt{\pi}}{2} = 1$
3. It is defined for all x, $-\infty < x < \infty$, monotonically increasing in the interval $(0, \infty)$; passes through origin. Asymptotic to $y = \pm 1$.
4. It is an odd function since

$$erf(-x) = \frac{2}{\sqrt{\pi}} \int_0^{-x} e^{-t^2} dt = \frac{2}{\sqrt{\pi}} \int_0^x e^{-v^2} (-dv)$$

where $v = -t$

$$= -\frac{2}{\sqrt{\pi}} \int_0^x e^{-v^2} dv = -erf(x)$$

5. $erf(-\infty) = -erf(\infty) = -1$
6. $erf(x) + erf(-x) = erf(x) - erf(x) = 0$
7. Complementary error function of x, denoted by

$$erfc(x) = 1 - erf(x) = erf(\infty) - erf(x)$$

$$= \frac{2}{\sqrt{\pi}} \int_0^\infty e^{-t^2} dt - \frac{2}{\sqrt{\pi}} \int_0^x e^{-t^2} dt$$

$$erfc(x) = \frac{2}{\sqrt{\pi}} \int_x^\infty e^{-t^2} dt$$

8. $erfc(x) + erfc(-x) = [1 - erf(x)] + [1 - erf(-x)] = 2 - erf(x) + erf(x) = 2$
9. Probability integral (Normal distribution function) of mathematical statistics is defined as

$$\boxed{\phi(\theta) = \frac{1}{\sqrt{2\pi}} \int_{-\infty}^{\theta} e^{-\omega^2/2} d\omega}$$

put $t = \omega/\sqrt{2}$ in the error function (1); then

$$erf(\theta) = \frac{2}{\sqrt{2\pi}} \int_0^{\sqrt{2}\theta} e^{-\omega^2/2} d\omega$$

Thus the error function and probability integral are related by

$$\boxed{erf(\theta) = 2\phi(\sqrt{2}\theta) - 1} \text{ or}$$

$$\boxed{\phi(x) = \tfrac{1}{2} + \tfrac{1}{2} erf\left(\tfrac{x}{\sqrt{2}}\right)}$$

10. Approximate formula (due to C. Hastings Jr.)

$$\boxed{erf(x) \approx 1 - (a_1 p + a_2 p^2 + a_3 p^3) e^{-x^2}}$$

where $p = \frac{1}{1+0.47047x}$, $a_1 = 0.3480242$, $a_2 = -0.0958798$, $a_3 = 0.7478556$, accurate upto ± 0.000025.

Note The factor $\frac{2}{\sqrt{\pi}}$ is included in the definition of error function to normalize it so that $erf(\infty) = 1$ since $\int_0^\infty e^{-t^2} dt = \frac{\sqrt{\pi}}{2}$.

11. $P(\mu - K\sigma \le X \le \mu + K\sigma) = erf\left\{\frac{K}{\sqrt{2}}\right\}$.

Let X denote the measured quantity in a certain experiment. Then the measurement error is indicated by the probability of an event such as $\mu - K\sigma \le X \le \mu + K\sigma$. Thus

$$P(\mu - K\sigma \le X \le \mu + K\sigma)$$
$$= N(\mu + K\sigma) - N(\mu - K\sigma)$$
$$= \Phi(K) - \Phi(-K)$$
$$= 2\Phi(K) - 1$$
$$= erf\left\{\frac{K}{\sqrt{2}}\right\} \quad \text{using above result 9.}$$

For example, for $K = 3$, we get

$$P(\mu - 3\sigma \le X \le \mu + 3\sigma) = 0.977.$$

Thus, on the average in only 0.3 per cent of the trials, the Gaussian random variable deviates from its mean by more than ± 3 standard deviations.

WORKED OUT EXAMPLES

Example 1: Expand $erf(x)$ in ascending powers of x.

Solution: By definition

$$erf(x) = \frac{2}{\sqrt{\pi}} \int_0^x e^{-t^2} dt = \frac{2}{\sqrt{\pi}} \int_0^x \sum_{n=0}^{\infty} \frac{(-t^2)^n}{n!} dt$$

since $e^t = \sum_{n=0}^{\infty} \frac{t^n}{n!}$. Now carrying term by term integration, we have

$$erf(x) = \frac{2}{\sqrt{\pi}}\left[\int_0^x \left(1 - \frac{t^2}{1!} + \frac{t^4}{2!} - \frac{t^6}{3!} + \frac{t^8}{4!} + \cdots\right) dt\right]$$

$$= \frac{2}{\sqrt{\pi}}\left[t - \frac{t^3}{3} + \frac{t^5}{5\cdot 2!} - \frac{t^7}{7\cdot 3!} + \frac{t^9}{9\cdot 4!} + \cdots\right]\Bigg|_{t=0}^{x}$$

$$erf(x) = \frac{2}{\sqrt{\pi}}\left[x - \frac{x^3}{3} + \frac{x^5}{10} - \frac{x^7}{42} + \frac{x^9}{216} - \cdots\right]$$

Example 2: Find $\frac{d}{dx}[erf(ax)]$.

Solution: From the above result replacing x by ax we have

$$erf(ax) = \frac{2}{\sqrt{\pi}}\left[ax - \frac{a^3x^3}{3} + \frac{a^5x^5}{10} - \frac{a^7x^7}{42} + \frac{a^9x^9}{216} - \cdots\right]$$

Differentiating both sides w.r.t. 'x' we get

$$\frac{d}{dx}[erf(ax)] = \frac{2}{\sqrt{\pi}}\left[a - a^3\cdot x^2 + a^5\frac{x^4}{2!} - a^7\frac{x^6}{3!} + a^9\frac{x^8}{4!} - \cdots\right]$$

$$= \frac{2a}{\sqrt{\pi}}\left[1 - a^2x^2 + \frac{(a^2x^2)^2}{2!} - \frac{(a^2x^2)^3}{3!} + \frac{(a^2x^2)^4}{4!} - \cdots\right]$$

$$= \frac{2a}{\sqrt{\pi}}e^{-a^2x^2}$$

Example 3: Compute $erf(0.5)$ correct to three decimal places.

Solution: Putting $x = 0.5$ in example 1, above, we have

$$erf(0.5) = \frac{2}{\sqrt{\pi}}\left[\frac{1}{2} - \left(\frac{1}{2}\right)^3\frac{1}{3} + \left(\frac{1}{2}\right)^5\frac{1}{10} - \left(\frac{1}{2}\right)^7\frac{1}{42} + \left(\frac{1}{2}\right)^9\frac{1}{216} - \cdots\right]$$

$$= \frac{0.922544642}{1.77245384} = 0.52049$$

Example 4: Show that

$$\int_0^\infty e^{-x^2-2ax}dx = \frac{\sqrt{\pi}}{2}e^{a^2}[1 - erf(a)]$$

Solution: $\int_0^\infty e^{-x^2-2ax}dx =$ $\int_0^\infty e^{-(x^2+2ax+a^2)}\cdot e^{a^2}\,dx = e^{a^2}\int_0^\infty e^{-(x+a)^2}dx$. Put $x + a = t$, so t varies from a to ∞ and $dx = dt$

$$= e^{a^2}\int_a^\infty e^{-t^2}dt = e^{a^2}\cdot\frac{\sqrt{\pi}}{2}erfc(a)$$

$$= e^{a^2}\cdot\frac{\sqrt{\pi}}{2}[1 - erf(a)]$$

EXERCISE

1. Prove that $\frac{d}{dx}[erfc(ax)] = -\frac{2a}{\sqrt{\pi}}e^{-a^2x^2}$.

   Hint: $\frac{d}{dx}[erfc(ax)] = \frac{d}{dx}[1 - erf(ax)] = -\frac{d}{dx}erf(ax)$ use W.E. 2., above

2. Compute $erf(0.3)$ correct to three decimal places.

Ans. 0.3248

   Hint: Put $x = 0.3248$ in W.E. 1.

3. Prove that
   (a) $\int_a^b e^{-t^2}dt = \frac{\sqrt{\pi}}{2}[erf(b) - erf(a)]$
   (b) $\int_{-b}^b e^{-t^2}dt = \sqrt{\pi}erf(b)$
   Hint: (a) $\int_a^b = \int_a^0 + \int_0^b = \int_0^b - \int_0^a$
   (b) $\int_{-b}^b = \frac{\sqrt{\pi}}{2}[erf(b) - erf(-b)] = \frac{\sqrt{\pi}}{2}[erf(b) + erf(b)]$

4. Show that

$$\int_0^t erfc(ax)dx = t\cdot erfc(at) - \frac{e^{-a^2t^2}}{a\sqrt{\pi}} + \frac{1}{a\sqrt{\pi}}$$

   Hint: Integrating by parts

$$x\cdot erfc(ax)|_0^t - \int_0^t x\cdot d(erfc(ax))$$

   Use result in exercise example 1, then

$$= t\cdot erf(at) + \frac{a}{\sqrt{\pi}}\cdot\frac{e^{-a^2x^2}}{a^2}\Bigg|_t^0$$

27.12 THE EXPONENTIAL DISTRIBUTION

Many scientific experiments involve the measurement of the duration of time X between an initial point of time and the occurrence of some phenomenon of interest. For example X is the life time of a light bulb which is turned on and left until it burns out. The continuous random variable X having the probability density function

$$f(x) = \begin{cases} \lambda e^{-\lambda x} & \text{if } x \geqslant 0 \\ 0 & \text{if } x < 0 \end{cases}$$

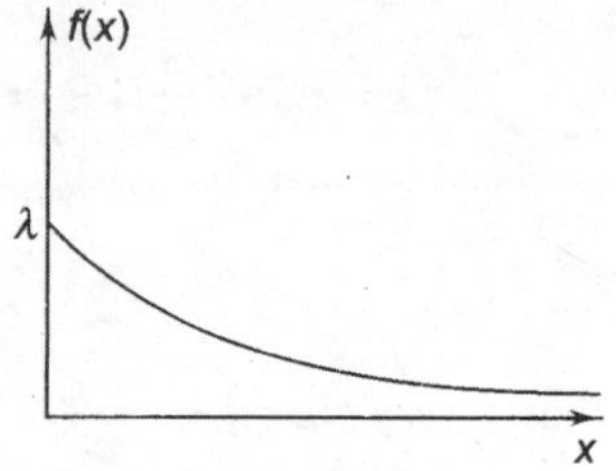

Fig. 27.41

is said to have an *exponential distribution.* Here the only parameter of the distribution is λ which is greater than zero. This distribution, also known as the *negative exponential distribution*, is a special case of the gamma distribution (with $r = 1$). Examples of random variables modeled as exponential are

a. (inter-arrival) time between two successive job arrivals

b. duration of telephone calls

c. life time (or time to failure) of a component or a product

d. service time at a server in a queue

e. time required for repair of a component

The exponential distribution occurs most often in applications of **Reliability Theory** and **Queuing Theory** because of the memoryless property and relation to the (discrete) **Poisson Disribution.** Exponential distribution can be obtained from the Poisson distribution by considering the inter-arrival times rather than the number of arrivals.

Mean and Variance

For any $r \geq 0$,

$$E(X^r) = \int_0^\infty x^r f(x)dx = \int_0^\infty x^r \lambda e^{-\lambda x} dx$$

put $\lambda x = t$, $x = \frac{t}{\lambda}$, $dx = \frac{1}{\lambda}dt$. Then

$$E(X^r) = \int_0^\infty \left(\frac{t}{\lambda}\right)^r \cdot \lambda \cdot e^{-t} \cdot \frac{1}{\lambda} dt = \frac{1}{\lambda^r} \int_0^\infty e^{-t} t^r dt$$

$$E(X^r) = \frac{\Gamma(r+1)}{\lambda^r}$$

In particular with $r = 0$,

$$\int_0^\infty f(x)dx = \Gamma(1) = 1$$

(i.e., $f(x)$ is a probability density function).
With $r = 1$, mean $= \mu = E(X) = \frac{\Gamma(2)}{\lambda} = \frac{1}{\lambda}$

with $r = 2$, variance $= \sigma^2 = E(X^2) - \{E(X)\}^2 = \frac{\Gamma(3)}{\lambda^2} - \frac{1}{\lambda^2}$

$$\sigma^2 = \frac{2}{\lambda^2} - \frac{1}{\lambda^2} = \frac{1}{\lambda^2}$$

Note: Both the mean and standard deviation of the exponential distribution are equal to $\frac{1}{\lambda}$.

Cumulative Distribution Function

$$F(x) = \int_0^x f(t)dt = \int_0^x \lambda e^{-\lambda t} dt = \left. \frac{\lambda e^{-\lambda t}}{-\lambda} \right|_{t=0}^{x}$$

$$F(x) = 1 - e^{-\lambda x} \quad \text{for } x \geqslant 0,$$

and $F(x) = 0$ when $x < 0$

$F(x)$ gives the probability that the "system" will "die" before x units of time have passed.

Probability Calculations

For any $a \geqslant 0$,

$$P(X \geq a) = P(X > a) = 1 - F(a) = e^{-\lambda a}$$

$$P(a \leq X \leq b) = P(a \leq X < b) = P(a < X < b)$$

$$= P(a < X \leq b) = F(b) - F(a)$$

$$= e^{-\lambda a} - e^{-\lambda b}$$

In table (A22) in appendix, the values of e^{-t} are tabulated for $t = 0.00(0.01)7.99$.

Corollary 1: $P\left(X > \frac{1}{\lambda}\right) = e^{-\lambda\frac{1}{\lambda}} = e^{-1} = 0.368 < \frac{1}{2}$

Survival Function

It gives the probability that the "system" survives more than x units of time and is given by

$$P(X > x) = 1 - F(x) = \begin{cases} 1 & \text{if } x < 0 \\ e^{-\lambda x} & \text{if } x \geq 0 \end{cases}$$

Memoryless or Markov Property

Among all distributions of non-negative continuous variables, only the exponential distributions have "no memory" (like the discrete geometric distribution) which results in analytical tractibility.

For any $s > 0,\ t > 0$

$$P(X > s + t | X > s) = \frac{P(\{X > s + t\} \cap \{X > s\})}{P(X > s)}$$

$$= \frac{P(X > s + t)}{P(X > s)} \quad (1)$$

When $X > s + t$ then X is also greater than s i.e., $X > s$. Since $\{X > s + t\} \cap \{X > s\} = \{X > s + t\}$

Thus the event $X > s$ in the numerator is redundant because both events can occur iff $X > s + t$.

Now

$$P(X > s + t | X > s) = \frac{e^{-\lambda(s+t)}}{e^{-\lambda s}} = e^{-\lambda t}$$

$$P(X > s + t | X > s) = P(X > t) \quad (2)$$

This memoryless property asserts that the conditional probability of additional waiting time is the *same* as the unconditional probability of the original waiting time. Thus the distribution of additional lifetime is exactly the same as the original distribution of lifetime, so at each point of time the component shows no effect of wear. In other words the distribution of "remaining" lifetime is independent of current age. In this sense, the exponential distribution has "*no memory*" of the past.

Combining (1) and (2) we have

$$P(X > s + t) = P(X > s) \cdot P(X > s + t | X > s)$$

$$= P(X > s) \cdot P(X > t)$$

which yields the famous functional equations known as *Cauchy equation.*

$$h(s + t) = h(s)h(t), \quad s > 0, t > 0$$

Here $h(s) = P\{X > s\},\ s > 0$.

Example: Suppose when a person arrives, one telephone booth has just been occupied (engaged) while another telephone booth has been occupied since (say 110 minutes) long. Then the probability distribution of the length of waiting time (to use the phone) will be the same for either phone booths. Therefore it does not matter which phone booth the person descides to wait!

WORKED OUT EXAMPLES

Example 1: Let the mileage (in thousands of miles) of a particular tyre be a random variable X having the probability density

$$f(x) = \begin{cases} \frac{1}{20} e^{-x/20} & \text{for } x > 0 \\ 0 & \text{for } x \leq 0 \end{cases}$$

Find the probability that one of these tyres will last (1) at most 10,000 miles (b) anywhere from 16,000 to 24,000 miles (c) at least 30,000 miles. (d) Find the mean (e) Find the variance of the given probability density function.

Solution: (a) Probability that a tyre will last almost 10,000 miles

$$= P(X \leq 10) = \int_0^{10} f(x)dx$$

$$= \int_0^{10} \frac{1}{20} e^{-x/20} dx$$

$$= \frac{1}{20} \cdot e^{-x/20} \cdot \left(\frac{-20}{1}\right)\Bigg|_0^{10}$$

$$= 1 - e^{-\frac{1}{2}} = 0.3934$$

(b) $$P(16 \leq X \leq 24) = \int_{16}^{24} f(x)dx$$

$$= \int_{16}^{24} \frac{1}{20} e^{-x/20} dx$$

$$= -e^{-\frac{x}{20}}\Big|_{16}^{24} = e^{-\frac{4}{5}} - e^{-\frac{6}{5}}$$

$$= 0.148$$

(c) $$P(X \geq 30) = \int_{30}^{\infty} f(x)dx$$

$$= \int_{30}^{\infty} \frac{1}{20} e^{-x/20} dx = -e^{x/20}\Big|_{30}^{\infty} = e^{-\frac{3}{2}}$$

$$= 0.2231$$

(d) $$\text{Mean} = \mu = \int_{-\infty}^{\infty} x \cdot f(x)dx$$

$$= \int_{0}^{\infty} x \cdot \frac{1}{20} e^{-\frac{x}{20}} dx$$

$$= -\int_{0}^{\infty} x \cdot d\left(e^{-\frac{x}{20}}\right)$$

$$= -xe^{-\frac{x}{20}} - 20e^{-\frac{x}{20}}\Big|_{0}^{\infty} = 0 - (-20)$$

$$\mu = 20 = \frac{1}{\lambda}$$

(e) $$\text{Variance} = \sigma^2 = \int_{-\infty}^{\infty} (x-\mu)^2 f(x)dx$$

$$= \int_{-\infty}^{\infty} x^2 f(x)dx - \mu^2$$

Consider

$$\int_{-\infty}^{\infty} x^2 f(x)dx = \int_{0}^{\infty} x^2 \frac{1}{20} e^{-\frac{x}{20}} dx$$

$$= -x^2 e^{-\frac{x}{20}}\Big|_{0}^{\infty} + 2 \cdot 20 \cdot \int_{0}^{\infty} \frac{1}{20} \cdot xe^{-x/20} dx$$

$$= 0 + 2 \cdot 20 \cdot \mu = 2.20.20 = 2.20^2$$

Then $$\sigma^2 = \int_{0}^{\infty} x^2 f(x)dx - \mu^2 = 2.20^2 - 20^2$$

$$= 20^2 = \frac{1}{\lambda^2}$$

Example 2: The length of time for one person to be served at a cafeteria is a random variable X having an exponential distribution with a mean of 4 minutes. Find the probability that a person is served in less than 3 minutes on at least 4 of the next 6 days.

Solution: The probability that a person is served at a cafeteria in less than 3 minutes is

$$P(T < 3) = 1 - P(T \geq 3)$$

Since the mean $\mu = \frac{1}{\lambda} = 4$ or $\lambda = \frac{1}{4}$, the exponential distribution is $\frac{1}{4}e^{-\frac{x}{4}}$. Now

$$P(T < 3) = 1 - P(T \geq 3) = 1 - \int_{3}^{\infty} \frac{1}{4} e^{-\frac{t}{4}} dt$$

$$P(T < 3) = 1 - \frac{1}{4} e^{-\frac{t}{4}} \cdot \left(\frac{-4}{1}\right)\Big|_{3}^{\infty} = 1 - e^{-\frac{3}{4}}$$

Let X represent the number of days on which a person is served in less than 3 minutes. Then using the binomial distribution, the probability that a person is served in less than 3 minutes on at least 4 of the next 6 days is

$$P(X \geq 4) = \sum_{x=4}^{6} {}^6C_x (1 - e^{-3/4})^x (e^{-3/4})^{6-x} = 0.3968$$

EXERCISE

1. Let T be the time (in years) to failure of certain components of a system. The random variable T has exponential distribution with mean time to failure $\beta = 5$. If 5 of these components are in different systems, find the probability that at least 2 are still functioning at the end of 8 years.

Ans. 0.2627

Hint: $P(T > 8) = \frac{1}{5}\int_8^{\infty} e^{-t/5} dt$ $= e^{-8/5} \simeq 0.2$, $P(X \geq 2) = \sum_{x=2}^{\infty} b(x; 5, 0.2) =$ $1 - \sum_{x=0}^{1} b(x, 5, 0.2) = 1 - 0.7373$

2. If a random variable X has the exponential distribution with mean $\mu = \frac{1}{\lambda} = \frac{1}{2}$ calculate the probabilities that (a) X will lie between 1 and 3 (b) X is greater than 0.5 (c) X is at most 4.

Ans. (a) 0.133 (b) 0.368 (c) 0.98168

Hint: PDF $f(x) = 2e^{-2x}$ (a) $\int_1^3 2e^{-2x} dx = e^{-2} - e^{-6}$
(b) $\int_{0.5}^{\infty} 2e^{-2x} dx = e^{-1}$ (c) $\int_0^4 2e^{-2x} dx = 1 - e^{-4}$

3. The life (in years) of a certain electrical switch has an exponential distribution with an average life of $\frac{1}{\lambda} = 2$. If 100 of these switches are installed in

different systems, find the probability that at most 30 fail during the first year.

Hint: $P(T > 1) = \int_1^\infty \frac{1}{2}e^{-\frac{t}{2}}\, dt = +e^{-\frac{1}{2}} = 0.606$

Ans. $P(X \le 30) = \sum_{x=0}^{30} b(x, 2, 0.606) = \sum_{x=0}^{30} {}^{100}C_x (0.606)^x\, (0.39346)^{100-x}$

4. Suppose the life length X (in hours) of a fuse has exponential distribution with mean $\frac{1}{\lambda}$. Fuses are manufactured by two different processes. Process I yields an expected life length of 100 hours and process II yields an expected life length of 150 hours. Cost of production of a fuse by process I is Rs. C while by the Process II it is Rs 2C. A fine of Rs K is levied if a fuse lasts less than 200 hours. Determine which process should be preferred?

Ans. Prefer Process I if $C > 0.13K$

Hint: $c_1 = c$ if $X > 200$
$= c + k$ if $X \le 200$

$$E(c_1) = c \cdot P(X > 200) + (c+k)P(X \le 200)$$

$$= c \cdot e^{-\frac{1}{100}\cdot 200} + (c+k)(1 - e^{-\frac{1}{100}\cdot 200})$$

$$= k(1 - e^{-2}) + c$$

$E(c_2) = k(1 - e^{-4/3}) + 2c, \quad E(c_2) - E(c_1) = c - 0.13k$

5. Suppose N_t be a discrete random variable denoting the number of arrivals in time interval $(0, t]$. Let X be the time of the next arrival, so X is the elapsed time between the occurrences of two successive events. Assuming that N_t is Poisson distributed with parameter λt, show that X is exponentially distributed.

   Here λ is the expected numbers of events occurring in one unit of time.

Ans. $P(X > t) = P(N_t = 0) = \dfrac{e^{-\lambda t}(\lambda t)^0}{0} = e^{-\lambda t}$

6. If the average rate of job submission is $\lambda = 0.1$ jobs/second, find the probability that an interval of 10 seconds elapses without job submission.

Ans. $P(X \ge 10) = \int_{10}^\infty 0.1e^{0.1t}dt = e^{-1} = 0.368$

Hint: Assume that the number of arrivals/unit time is poisson distributed and the inter arrival time X is exponentially distributed with parameter λ.

7. Let the mileage (in thousands of miles) of a certain radial tyre is a random variable with exponential distribution with mean 40,000 miles. Determine the probability that the tyre will last (a) at least 20,000 km (b) at most 30,000 km.

Ans. (a) $P(X \ge 20,000) = e^{-0.5} = 0.6065$
(b) $P(X \le 30,0001 = 1 - e^{-0.75} = 0.5270$

8. The amount of time (in hours) required to repair a T.V. is exponentially distributed with mean $\frac{1}{2}$. Find the (a) probability that the repair time exceeds 2 hours (b) the conditional probability that repair takes at least 10 hours given that already 9 hours have been spent repairing the TV.

Ans. (a) $P(X > 2) = e^{-1} = 0.3679$
(b) $P(X \ge 10 | X > 9) = P(X > 1) = e^{-0.5} = 0.6065$
(because of the memoryless property).

9. The duration of time X in seconds between presses of the white rat on a bar, which are periodically conditioned, has an exponential distribution with parameter $\lambda = 0.20$. Find the probability that the duration is more than one second but less than 3 seconds (b) more than 3 seconds.

Ans. (a) $P(1 \le X \le 3) = e^{-0.2(1)} - e^{-(0.2)3} = 0.819 - 0.549 = 0.270$
(b) $P(X > 3) = e^{-0.2(3)} = 0.549$

10. The time X (seconds) that it takes a certain on-line computer terminal (the elapsed time between the end of user's inquiry and the beginning of the system's response to that inquiry) has an exponential distribution with expected time 20 seconds. Compute the probabilities (a) $P(X \le 30)$ (b) $P(X \ge 20)$ (c) $P(20 \le X \le 30)$ (d) For what value of t is $P(X \le t) = 0.5$ (i.e., t is the fiftieth percentile of the distribution)

Ans. (a) 0.777 (b) 0.368 (c) 0.145 (d) 13.863

27.13 THE GAMMA DISTRIBUTION

Consider a system consisting of one original component and $(r-1)$ spare components such that when the original component fails, one of the $(r-1)$ spare components is used. If this component fails, one of the $(r-2)$ spare components is used. System fails only when the original component and all the $(r-1)$ spare components fail. Assume that the lifetimes $X_1, X_2, \ldots, X_r$ of the r duplicates of the essential components have infinite lifetimes (except for the original component). Suppose each of the random variables $X_1, X_2, \ldots, X_r$ have the same exponential distribution with parameter λ and are probabilistically independent. Then the lifetime (time until failure) of the entire system is the sum $Y = \sum_{i=1}^{r} X_i$ having the *gamma distribution* with density function

$$f(y) = \begin{cases} \frac{\lambda^r y^{r-1} e^{-\lambda y}}{\Gamma(r)}, & \text{if } y \geq 0 \\ 0, & \text{if } y < 0 \end{cases} \quad (1)$$

(1) is a skewed distribution.

The two parameters of (1) are the positive numbers λ and r (although r need *not* be an integer). If r is a positive integer, then gamma distribution (1) is known as *Erlang distribution*. Introducing $V = \lambda y$, (1) reduces to

$$f(v) = \frac{1}{\lambda} f\left(\frac{v}{\lambda}\right) = \frac{1}{\lambda}\left\{\lambda^r \left(\frac{v}{\lambda}\right)^{r-1} \cdot \frac{e^{-v}}{\Gamma(r)}\right\}$$

$$= \begin{cases} \frac{v^{r-1} e^{-v}}{\Gamma(r)} & \text{if } v \geq 0 \\ 0 & \text{if } v < 0 \end{cases} \quad (2)$$

The probability density function of the random variable V given by (2) is known as the "*standard gamma function*" with parameter r (and is independent of λ). When $r = 1$, the density function (2) reduces to the density function of exponential distribution with the parameter $\lambda = 1$. For large r (say $r \geq 50$) (2) resembles a normal distribution with mean and variance approximately equal to r. The gamma distribution with parameter $\lambda = \frac{1}{2}$ and $r = \frac{\nu}{2}$ (where ν is a positive integer) reduces to the chi-squared distribution with ν degrees of freedom credited to Karl Pearson (1857–1936) and F.R. Helmert (1843–1917).

The *incomplete gamma function* defined by

$$F_V(t) = \int_0^t \frac{v^{r-1}e^{-v}}{\Gamma(r)} dv = I_r(t), \quad t \geq 0 \quad (3)$$

is tabulated in the tables of the appendix A23 to A28 for $r = 1(1)5$, $t = 0.2(0.2)8.0(0.5)15.0$ and for $r = 6(1)10$, $t = 1.0(0.2)8.0(0.5)17.0$.

Now

$$P(Y \geq a) = P(Y > a) = 1 - F(a) = 1 - I_r(\lambda a)$$

and

$$P(a \leq Y \leq b) = F(b) - F(a) = I_r(\lambda b) - I_r(\lambda a)$$

Moments of the Gamma Distribution

For any $k \geq 0$,

$$E(Y^k) = E\left(\frac{V^k}{\lambda^k}\right) = \frac{1}{\lambda^k} E(V^k)$$

$$= \frac{1}{\lambda^k} \int_0^\infty v^k \cdot \left(\frac{v^{r-1}e^{-v}}{\Gamma(r)}\right) dv$$

$$= \frac{1}{\lambda^k \Gamma(r)} \int_0^\infty e^{-v} v^{k+r-1} dv$$

Then $$E(Y^k) = \frac{\Gamma(r+k)}{\lambda^k \Gamma(r)} \quad (4)$$

For $k = 0$ in (4) we have $\int_0^\infty f(y)dy = 1$ so $f(y)$ is a probability density function.

For $k = 1$, in (4) we get the mean $= \mu = E(Y) = \frac{\Gamma(r+1)}{\lambda \Gamma(r)} = \frac{r\Gamma(r)}{\lambda \Gamma(r)}$.

So $\mu = \frac{r}{\lambda}$. (5)

With $k = 2$ in (4), we get

variance $= \sigma^2 = E(Y^2) - \{E(Y)\}^2$

$$= \frac{\Gamma(r+2)}{\lambda^2 \Gamma(r)} - \frac{r^2}{\lambda^2}$$

$$\sigma^2 = \frac{r(r+1)\Gamma(r)}{\lambda^2 \Gamma(r)} - \frac{r^2}{\lambda^2} = \frac{r}{\lambda^2} \quad (6)$$

Thus the parameter r and λ are determined from (5) and (6) as

$$\lambda = \frac{\mu}{\sigma^2}, \quad r = \frac{\mu^2}{\sigma^2} \quad (7)$$

Relation Between Exponential, Gamma and Poisson Distributions

Suppose the lifetimes of a batch of components each have exponential distribution with parameter λ. Stating at time $t = 0$, the first component is used until its extinction (until it "dies" or "fails"). Replace the component by another instantaneously and wait unit this new component also fails. Continuing this Process, stop at a given time t. Then the number of failed components L is a random variable having the Poisson distribution with $\lambda^* = \lambda t$. Also the lifetime of the entire process $Y = \sum_{i=1}^{k} X_i$ follows gamma distribution with parameters λ and k.

Let $X_1, X_2, \ldots, X_k$ be the lifetimes of the first k components that have failed. Assume that each lifetime X_i has an exponential distribution with parameter λ and are probabilistically independent. Recall that, then $Y = \sum_{i=1}^{k} X_i$ has a gamma distribution with parameters λ and k. If $X_1 + X_2 + \ldots + X_k$ (total lifetime of the process) $\leq t$ then in this case at least k components have all failed, (one after the other) before the time is up.

Now probability of the event that at least k components have failed at the time of termination of the experiment is given by

$$P(L \geq k) = P\{Y \leq t\} = I_k(\lambda t)$$

$$= \int_0^{\lambda t} \frac{v^{k-1} e^{-v}}{\Gamma(k)} dv$$

Integrating by parts, we have

$$= \frac{v^{k-1}}{\Gamma(k)} \cdot \left(\frac{e^{-v}}{-1}\right)\Bigg|_{v=0}^{\lambda t} + \int_0^{\lambda t} \frac{(k-1)v^{k-2} \cdot e^{-v}}{\Gamma(k)} dv$$

$$= -\frac{(\lambda t)^{k-1} e^{-\lambda t}}{(k-1)!} + \int_0^{\lambda t} \frac{v^{k-2} \cdot e^{-v}}{\Gamma(k-1)} dv.$$

Integrating by parts $(k-1)$ more times,

$$P(L \geq k) = -\sum_{i=1}^{k-1} \frac{(\lambda t)^i e^{-\lambda t}}{i!} + \int_0^{\lambda t} \frac{e^{-v}}{\Gamma(1)} dv$$

But $\Gamma(1) = 1$ and $\int_0^{\lambda t} e^{-v} dv = \frac{e^{-v}}{-1}\Big|_0^{\lambda t} = 1 - e^{-\lambda t}$.

$$\text{Thus } P(L \geq k) = 1 - \sum_{i=0}^{k-1} \frac{(\lambda t)^i e^{-\lambda t}}{i!}$$

But

$$P(L = k) = P(L \geq k) - P(L \geq k+1)$$

$$= \left[1 - \sum_{i=0}^{k-1} \frac{(\lambda t)^i e^{-\lambda t}}{i!}\right] - \left[1 - \sum_{i=0}^{k} \frac{(\lambda t)^i e^{-\lambda t}}{i!}\right]$$

$$P(L = k) = \frac{(\lambda t)^k e^{-\lambda t}}{k!} = \frac{\lambda^{*k} \cdot e^{-\lambda^*}}{k!}, \lambda^* = \lambda t$$

Thus the probability distribution of the discrete random variable L is a Poisson (discrete) distribution with parameter $\lambda^* = \lambda t$.

The cumulative distribution of the gamma distribution of Y can be calculated in terms of tabulated cumulative distribution of the Poisson distribution from

$$F(t) = P(L \geq k) = P(Y \leq t) = 1 - \sum_{i=0}^{k-1} \frac{(\lambda t)^i e^{-\lambda t}}{i!}$$

Worked Out Examples

Example 1: The daily consumption of electric power (in millions if kW-hours) in a certain city is a random variable X having the probability density

$$f(x) - \begin{cases} \frac{1}{9} x e^{-x/3} & \text{if } x > 0 \\ 0 & \text{if } x \leq 0 \end{cases}$$

Find the probability that the power supply is inadequate on any given day if the daily capacity of the power plant is 12 million kW hours.

Solution: Observe that this is gamma distribution with $r = 2$ and $\lambda = \frac{1}{3}$. The power supply is inadequate when $X > 12$. Now $P(X > 12) = 1 - F(12) = 1 - I_2\left(\frac{1}{3} 12\right) = 1 - I_2(4)$. Using table A23 to A28, we get

$$P(X > 12) = 1 - 0.90892 = 0.09108$$

Example 2: The lifetime X (in months) of a computer has a gamma distribution with mean 24 months and standard deviation 12 months. Find the probability that the computer will

(a) last between 12 and 24 months.

(b) last at most 24 months.

(c) Determine the median lifetime of X.

(d) suppose that the test will actually be terminated after t months. Determine the value of t such that only one-half of 1% of all computers would still be functioning at termination.

Solution: Here $\mu = 24, \sigma = 12$. Then from (7)

$\lambda = \frac{24}{12^2} = \frac{1}{6}, r = \left(\frac{\mu}{\sigma}\right)^2 = \left(\frac{24}{12}\right)^2 = 4$

(a) $P(12 < X < 24) = F(24) - F(12)$

$= I_4\left(\frac{1}{6}\cdot 24\right) - I_4\left(\frac{1}{6}\cdot 12\right)$

$= I_4(4) - I_4(2)$

Using table A23 to A28 we get

$P(12 < X < 24) = 0.56653 - 0.14288 = 0.42365$.

(b) $P(X \le 24) = I_4\left(\frac{1}{6}\cdot 24\right) = I_4(4) = 0.56653$

(c) median $\hat{x}$ is such that $P(X \le \hat{x}) = \frac{1}{2}$

Then $I_4\left(\frac{\hat{x}}{6}\right) = 0.5$. From table (see entry in the table A23 to A28) we get $\frac{\hat{x}}{6} = 4$) or $\hat{x} = 24$ months

(d) At the termination time t, only one-half of 1% computers are still functioning. So

$$P(X \le t) = 1 - \frac{1}{2}(0.01) = 0.995$$

But $P(X \le t) = I_4\left(\frac{1}{6}\hat{t}\right) = 0.995$ (given). From table A23 to A28, we have $I_4(11) = 0.995$. Therefore $\frac{\hat{t}}{6} = 11$ or $\hat{t} = 66$ months.

EXERCISE

1. Suppose the reaction time X has a standard gamma distribution with $r = 2$. Find (a) $P(3 \le X \le 5)$ (b) $P(X > 4)$.

Ans. (a) $I_2(5) - I_2(3) = 0.95957 - 0.80085 = 0.15872$ (b) $P(X > 4) = 1 - P(X \le 4) = 1 - I_2(4) = 1 - 0.90842 = 0.09158$

2. Suppose that the time (in hours) taken by a home-owner to mow his lawns is a random variable X having a gamma distribution with parameters $r = 2$ and $\lambda = 2$. Find the probability that it takes (a) at most 1 hour (b) at least 2 hours (c) between 0.5 and 1.5 hours to mow the lawn.

Ans. (a) 0.594 (b) 0.092 (c) 0.537

3. The survival time X (in weeks) of a male mouse exposed to radiation has a gamma distribution with $r = 8$ and $\lambda = \frac{1}{15}$. Find the probability that the mouse survives (a) between 60 and 120 weeks (b) at least 30 week. Find (c) mean (d) variance of X.

Ans. (a) $P(60 \le X \le 120) = F\left(\frac{120}{15}, 8\right) - F\left(\frac{60}{15}, 8\right) = I_8(8) - I_8(4) = 0.547 - 0.051 = 0.496$

(b) $P(X \ge 30) = 1 - P(X < 30) = 1 - F\left(\frac{30}{15}, 8\right) = 1 - 0.00110 = 0.9989$

4. If Y has gamma distribution with $\lambda = 0.40$ and $r = 5$, find (a) $P(Y > 30)$ (b) $P(15 \le Y \le 20)$

Ans. (a) $1 - I_5(12) = 1 - 0.9924 = 0.0076$

(b) $I_5(8) - I_5(6) = 0.90037 - 0.71494 = 0.18543$

5. If a random variable X has the gamma distribution with $r = 2$ and $\lambda = \frac{1}{2}$ find (a) the mean (b) standard deviation (c) the probability that X will take a value less than 4.

Ans. (a) 4 (b) 2.828 (c) 0.5940

6. The gross sales X in thousands of rupees is a random variable having gamma distribution with $\lambda = \frac{1}{2}$ and $r = 80000\sqrt{n}$ where n is the number of employees in the company. If the sales cost is Rs. 8000 per employee, how many employees should the company employ to maximise the expected profit.

Ans. $n = 100$,

Hint: $\mu = \frac{r}{\lambda} = 160000\sqrt{n}$, y = profit = total expected sales − total cost = $160000\sqrt{n} - 8000n$, $\frac{dy}{dn} = 0$ for $n = 100$.

$$\frac{d^2y}{dn^2} = -\frac{40000}{n^{3/2}} < 0$$

27.14 THE WEIBULL DISTRIBUTION

Lifetimes, waiting times, learning times, travelling times, duration of epidemics are some of the important examples of non-negative random variables whose variability can be explained in many cases by exponential and gamma distributions. However in certain cases Weibull distribution provides good probability model for describing "*length of life*" of objects having the '*weakest link*' property. An object, composed of a large number of separate parts, put under stress is said to have the *weakest link property* if the lifetime of the object is equal to the minimum lifetime of any of its parts.

Example: A chain is as strong as its weakest link.

The Weibull distribution was introduced in 1939 by the Swedish physicist *Waloddi Weibull* and is given by probability density function

$$f(x) = \begin{cases} \frac{\beta}{\alpha}\left(\frac{x-\nu}{\alpha}\right)^{\beta-1} \exp\left[-\left(\frac{x-\nu}{\alpha}\right)^{\beta}\right] & \text{if } x \geq \nu \\ 0 & \text{if } x < \nu \end{cases} \quad (3)$$

The three constants $\beta > 0$, $\alpha > 0$ and $\nu \geq 0$ are the parameters of the distribution. The smallest possible value of X is given by ν. The constant β determines the shape of the density function (1). When $\beta = 1$ and $\nu = 0$, the Weibull distribution (1) reduces to the exponential distribution with the parameter $\lambda = \frac{1}{\alpha}$. If X has a Weibull distribution with parameters α, β, ν then $Y = \left[\frac{(X-\nu)}{\alpha}\right]^{\beta}$ has an exponential distribution with the parameter $\lambda = 1$.

Since $X = \alpha Y^{1/\beta} + \nu$ so $\frac{dX}{dY} = \frac{\alpha}{\beta} Y^{\frac{1}{\beta}-1}$ and $f(X) = \frac{\beta}{\alpha} Y^{(\beta-1)/\beta} e^{-Y}$. Then the probability density function of Y is

$$f(y) = f(x)\frac{dx}{dy} = \frac{\beta}{\alpha} y^{\frac{\beta-1}{\beta}} \cdot e^{-y} \cdot \frac{\alpha}{\beta} y^{\frac{1}{\beta}-1}$$

$$= \begin{cases} e^{-y} & \text{if } y \geq 0 \\ 0 & \text{if } y < 0 \end{cases}$$

which is an exponential distribution with parameter $\lambda = 1$.

The cumulative distribution of X is

$$F(x) = P(X \leq x) = \begin{cases} 1 - \exp\left[-\left(\frac{x-\nu}{\alpha}\right)^{\beta}\right] & \text{if } x \geq \nu \\ 0 \text{ if } x < \nu \end{cases} \quad (2)$$

Probabilites are calculated using $F(x)$.

Mean and Variance

$$\text{Mean} = \mu = E(X) = E(\alpha Y^{\frac{1}{\beta}} + \nu) = \alpha E(Y^{\frac{1}{\beta}}) + \nu$$

Since $E(X^r) = \frac{\Gamma(r+1)}{\lambda^r}$, with $r = \frac{1}{\beta}$ and $\lambda = 1$,

$$\text{Mean} = \mu = \frac{\alpha\Gamma\left(\frac{1}{\beta}+1\right)+\nu}{1} \quad (3)$$

Consider

$$E(X^2) = E\left(\left(\alpha Y^{\frac{1}{\beta}} + \nu\right)^2\right) = E(\alpha^2 Y^{2/\beta} + \nu^2 + 2\alpha\nu Y^{\frac{1}{\beta}})$$

$$= \alpha^2 E(Y^{2/\beta}) + 2\nu\alpha E(Y^{1/\beta}) + \nu^2$$

Now with $r = \frac{2}{\beta}$ and $\lambda = 1$, we get
Variance $= \sigma^2 = E(X^2) - \{E(X)\}^2$

$$= \left[\alpha^2 \frac{\Gamma\left(\frac{2}{\beta}+1\right)}{1^2} + 2\nu\alpha\frac{\Gamma\left(\frac{1}{\beta}+1\right)}{1} + \nu^2\right] -$$

$$-\left[\alpha\Gamma\left(\frac{1}{\beta}\cdot 1\right)+\nu\right]^2$$

$$\sigma^2 = \alpha^2\left\{\Gamma\left(1+\frac{2}{\beta}\right) - \left[\Gamma\left(1+\frac{1}{\beta}\right)\right]^2\right\} \quad (4)$$

Median: $P(X \leq x) = F(x) = \frac{1}{2}$ or $1 - \exp\left[-\left(\frac{x-\nu}{\alpha}\right)^{\beta}\right] = \frac{1}{2}$. Solving $x = \alpha(.693)^{\frac{1}{\beta}} + \nu$

Observe that variance depends upon α and β but independent of ν.

The *survival function* is given by

$$1 - F(x) = \begin{cases} 1 & \text{if } x < \nu \\ \exp\left[-\left(\frac{x-\nu}{\alpha}\right)^{\beta}\right] & \text{if } x \geq \nu \end{cases} \quad (5)$$

Exponential distribution is a special case of both the gamma and Weibull distributions.

Note that gamma distribution with λ and $r = 1$ is an exponential distribution with parameter λ. Similarly the Weibull distribution with $\alpha = \frac{1}{\lambda}$, $\beta = 1$ and $\nu = 0$ is an exponential distribution with parameter λ. Thus the gamma and Weibull distributions are generalization of the exponential distribution. However gamma distribution with $r \neq 1$ is *not* a Weibull distribution. Also Weibull distributions with $\beta \neq 1$ or $\nu \neq 0$ is not gamma distribution.

With $\nu = 0$ the above results takes the following form:

The Weibull distribution is

$$f(x) = \begin{cases} \frac{\beta}{\alpha^\beta} x^{\beta-1} e^{-\frac{x^\beta}{\alpha^\beta}}, & \text{if } x \geq 0 \\ 0, & \text{if } x < 0 \end{cases}$$

put $\alpha^* = \alpha^{-\beta}$($or\ \alpha = (\alpha^*)^{-\frac{1}{\beta}}$) then

$$f(x) = \begin{cases} \alpha^* \beta e^{-\alpha^* x^\beta} \cdot x^{\beta-1}, & \text{if } x \geq 0 \\ 0, & \text{if } x < 0 \end{cases} \quad (6)$$

The cumulative distribution is

$$F(x) = \begin{cases} 1 - e^{-\alpha^* x^\beta}, & \text{if } x \geq 0 \\ 0, & \text{if } x < 0 \end{cases} \quad (7)$$

$$\text{Mean}: \mu = (\alpha^*)^{-\frac{1}{\beta}} \Gamma\left(\frac{1}{\beta} + 1\right) \quad (8)$$

$$\text{Variance}: \sigma^2 = (\alpha^*)^{\frac{-2}{\beta}} \left\{ \Gamma\left(1 + \frac{2}{\beta}\right) - \left[\Gamma\left(1 + \frac{1}{\beta}\right)\right]^2 \right\}$$

$$\text{Median}: \hat{x} = \alpha^{*^{-\frac{1}{\beta}}} (0.693)^{\frac{1}{\beta}} = \left(\frac{0.693}{\alpha^*}\right)^{\frac{1}{\beta}} \quad (10)$$

The Weibull Failure Law

The (instantaneous) failure rate Z (also known as "*hazard function*") associated with the random variable T is given by

$$Z(t) = \frac{f(t)}{1 - F(t)}$$

where $f(t)$ and $F(t)$ are the probability density function and cumulative distribution. $Z(t)$ is also known as mortality curve, life characteristic or lambda characteristic.

Suppose the life length T (or time to failure) of a component has the Weibull distribution

$$f(t) = \alpha^* \beta e^{-\alpha^* t^\beta}$$

with α^* and β as parameters and consequently has the cumulative distribution $F(t) = 1 - e^{-\alpha^* t^\beta}$, then the Weibull failure law is given by

$$Z(t) = \frac{f(t)}{1 - F(t)} = \frac{\alpha^* \beta e^{-\alpha^* t^\beta} \cdot t^{\beta-1}}{e^{-\alpha^* t^\beta}}$$

$$Z(t) = \alpha^* \beta t^{\beta-1} \quad (11)$$

Here the failure rate is proportional to t (unlike the only exponential distribution whose exponential failure law where $Z(t)$, the failure rate is constant).

WORKED OUT EXAMPLES

Example: Suppose the life time X (in hours) of a semiconductor is a random variable having the Weibull distribution with parameters. $\alpha^* = (200)^{-2.5}$, $\beta = 2.5$, $\nu = 0$. Determine the probability that the lifetime of the semiconductor is (a) at most 200, (b) less than 200, (c) more than 300, (d) between 100 and 200 hours. Find (e) the mean (f) variance (g) median of life time X.

Solution: Recall that the cumulative distribution with $\alpha^* = (200)^{-2.5}$, $\beta = 2.5$ is

$$F(x) = P(X \leq x) = 1 - e^{-\alpha^* x^\beta} = 1 - e^{-\left(\frac{x}{200}\right)^{2.5}}$$

(a) Probability that life time X of the semiconductor is at most 200 hours

$$= P(X \leq 200) = F(200)$$

$$= 1 - e^{-\left(\frac{200}{200}\right)^{2.5}} = 1 - e^{-1} = 1 - 0.36787 = 0.632$$

(b) $P(X < 200) = P(X \leq 200) = 0.632$

(c) $P(X > 300) = 1 - F(300) = e^{-\left(\frac{200}{200}\right)^{2.5}} = e^{-\left(\frac{3}{2}\right)^{2.5}} = 0.0635$

(d) $P(100 < X < 200) = F(200) - F(100)$

$$= \left[1 - e^{-\left(\frac{200}{200}\right)^{2.5}}\right] - \left[1 - e^{-\left(\frac{100}{200}\right)^{2.5}}\right]$$

$$= e^{-\left(\frac{1}{2}\right)^{2.5}} - e^{-1} = 0.83796 - 0.36789$$

$$= 0.4700$$

(e) Mean: $\mu = (\alpha^*)^{-\frac{1}{\beta}} \Gamma\left(\frac{1}{\beta} + 1\right) = \frac{\Gamma\left(\frac{1}{2.5}+1\right)}{200} = 200 \cdot \Gamma\left(\frac{2}{5} + 1\right) = \frac{400}{5} \Gamma\left(\frac{2}{5}\right) = 80 \cdot \Gamma(2/5)$

(f) Variance = $(\alpha^*)^{-2/\beta} \left[\Gamma\left(1 + \frac{2}{\beta}\right) - \left\{\Gamma\left(1 + \frac{1}{\beta}\right)\right\}^2\right] =$

$$= \left(\frac{1}{200}\right)^{-2} \left[\Gamma\left(1 + \frac{4}{5}\right) - \left\{\Gamma\left(1 + \frac{2}{5}\right)\right\}^2\right]$$

$= 40000\left[\frac{4}{5}\cdot\Gamma\left(\frac{4}{5}\right) - \left\{\frac{2}{5}\Gamma\left(\frac{2}{5}\right)\right\}^2\right]$

(g) Median: $\hat{x}$ middlemost value such that $P(X \le \hat{x}) = \frac{1}{2}$. Then ,

$$F(\hat{x}) = 1 - e^{-\alpha^* \hat{x}^\beta} = \frac{1}{2}$$

So

$$\hat{x} = \text{median} = (\alpha^*)^{-\frac{1}{\beta}}(.693)^{\frac{1}{\beta}}$$

$$= 200(.693)^{\frac{1}{2.5}} = 200(.693)^{\frac{2}{5}}$$

$$= 200(.86356) = 172.7123$$

EXERCISE

1. Suppose the service life, in years, of a system is a random variable having a Weibull distribution with $\alpha^* = \frac{1}{2}, \beta = 2$. Find the probability that such a system will still be functioning after 2 years.

Ans. $P(X \ge 2) = 1 - F(2) = e^{-\alpha^* x^\beta} = e^{-\frac{1}{2}(2)^2} = e^{-2}$

2. If the lifetime X of a hearing aid battery has Weibull distribution with $\alpha^* = 0.1$ and $\beta = 0.5$ determine the probability that such a battery (a) will function for more than 300 hours (b) will not last 100 hours. Find (c) mean (d) variance (e) median of X.

Ans. (a) $P(X \ge 300) = e^{-0.1(300)^{0.5}} = 0.177$

(b) $P(X < 100) = 1 - e^{-0.1(100)^{0.5}} = 0.6321$

(c) $\mu = \text{mean} = (0.1)^{-2}\Gamma\left(1 + \frac{1}{0.5}\right) = 200$ hours

(d) Variance $= (0.1)^{-4}[\Gamma(1+4) - \Gamma^2(3)] = 20(.1)^{(-4)} = 200000$ hours

(e) median $= (0.1)^{-2}(.693)^2 = 100(0.480) = 48$ hours

3. Suppose the tensile strength X has Weibull distribution with $\alpha^* = (100)^{-20}$, $\beta = 20$. Find (a) $P(X \le 105)$ (b) $P(98 \le X \le 102)$

Ans. (a) $1 - e^{-(105/100)^{20}} = 1 - 0.070 = 0.930$

(b) $F(102) - F(98) = e^{-(.98)^{20}} - e^{(1.02)^{20}} = .513 - .226 = 0.287$

4. Let X, the corrosion weight loss of an alloy follows Weibull distribution with $\alpha = 4$, $\beta = 2$, $\nu = 3$. Find (a) $P(X > 3.5)$ (b) $P(7 \le X \le 9)$.

Ans. (a) $e^{-0.156} = 0.985$ (b) $e^{-1} - e^{-2.25} = 0.89 - 0.632 = 0.263$

Hint: Use (5): on page 27.51 $F(x) = 1 - \exp\left(\frac{x-3}{4}\right)^2$

5. Suppose each of the 36 transistors in a system has life length, in years, having the Weibull distribution with $\alpha = 25$, $\beta = 2$. Find the probability that no transistor will have to be replaced during the first 2 months of use assuming that the transistors are functionally independent.

Hint:

$$P\left(X > \frac{2}{12}\right) = e^{-25\left(\frac{1}{6}\right)^2} = e^{-25/36}$$

Ans. $P(\text{no transistor replaced}) = \left(e^{-25/36}\right)^{36} = e^{-25} = (1.38879) \times 10^{-11}$

6. If the probability that the life length X (in years) of a computer exceeds 5 years is $e^{-0.25}$, determine α of the Weibull distribution of X with $\beta = 2$. Find the mean and variance of X.

Ans. $\alpha = \frac{1}{100}$, $\mu = 5\sqrt{\pi}$, $\sigma^2 = 100\left(1 - \frac{\pi}{4}\right)$

Hint: $P(X > 5) = e^{-25\alpha} = e^{-0.25}$ (given)

7. Suppose the time to failure X in hours of a component is modeled by a Weibull distribution with parameter $\beta = 2$. If 15% of the components that have lasted 90 hours fail before 100 hours, find the parameter α.

Ans. $\alpha = 0.00008554 = (-\ln 0.85/1900)$

Hint:

$$F(x) = 1 - e^{-\alpha x^\beta} = 1 - e^{-\alpha x^2}$$

$$P(X < 100 | X > 90) = \frac{P(90 < X < 100)}{P(X > 90)}$$

$$= \frac{F(100) - F(90)}{1 - F(90)}$$

$$= \frac{e^{-\alpha(90)^2} - e^{-\alpha(100)^2}}{e^{-\alpha(90)^2}}$$

$$= 0.15 \text{ (given)}$$

Chapter 28

Sampling Distribution

INTRODUCTION

Sampling distribution of a statistic is the theoretical probability distribution of the statistic which is easy to understand and is used in inferential or inductive statistics. A statistic is a random variable since its value depends on observed sample values which will differ from sample to sample. Whereas its particular value depends on a given set of sample values. Thus determination of sampling distribution of a statistic is essentially a mathematical problem.

Suppose we wish to compare the mean I.Q of students of one university with another university or to compare proportion of alcoholics among men with that among women or compare the variances of lifetimes of T.V. produced by one company with the variance of another company. In such statistical investigation of the study of two (or more) populations we compare their respective parameters.

The nearness of two quantities can be measured as follows. If the difference of the two quantities is close to zero, then the two quantities are very close to each other. Alternatively when the ratio of the two quantities is nearly equal to zero, then the two quantities are near to each other, otherwise they are far apart. By taking two random samples one from each population, we compare the corresponding sample analogues.

Statistical methods are used to study a process by analyzing the data, discrete or continuous, recorded as either numerical value or a descriptive representation to improve the "quality" of the process. Thus statistician is mainly concerned with the analysis of data about the characteristics of persons or objects or observations.

28.1 POPULATION AND SAMPLE

Population is the set or collection or totality of objects, animate or inanimate, actual or hypothetical, under study. Thus mainly population consists of sets of numbers, measurements or observations which are of interest.

Size

Size of the population N is the number of objects or observations in the population.

Population is said to be finite or infinite depending on the size N being finite or infinite.

Since it is impracticable or uneconomical or time consuming to study the entire population, a finite subset of the population known as *sample* is studied. Size of the sample is denoted by n. *Sampling* is the process of drawing samples from a given population.

Examples: (i) Population of India, Population of A.P. Estate (sample) (ii) Engineering colleges recognised by AICTE, Engineering colleges affiliated to JNTU (sample) (iii) Cars produced in India, Matiz cars (sample) (iv) Healthcare expenditure by Central Govt., Expenditure of A.P. Govt. (sample), Expenditure by district (sub sample), Expenditure in a district hospital (sub-sample) etc. (v) Number of software specialists by 2020.

Large sampling

If $n \geq 30$, the sampling is said to be large sampling.

Small sampling

If $n < 30$, the sampling is said to be small sampling or **exact sampling**.

Statistical Inference

Statistical inference or inductive statistics deals with the methods of drawing (arriving at) valid or logical generalizations and predictions about the population using the information contained in the sample alone, with an indication of the accuracy of such inferences.

Parameters

Statistical measures or constants obtained from the population are known as population parameters or simply parameters.

Example: Population mean, population variance. Similarly, statistical quantities computed from sample observations are known as sample statistics or briefly "**statistics**".

Thus parameters refer to population while statistics refer to sample.

Examples: sample mean, sample variance etc.

Notation

μ, σ, p represent the population mean, population standard deviation, population proportion. Similarly, $\overline{X}, s, P$ denote sample mean, sample s.d., sample proportion.

Population *f*(*x*)

Population $f(x)$ is a population whose probability distribution is $f(x)$.

Example: If $f(x)$ is binomial, Poisson or normal, then the corresponding population is known as binomial population, Poisson population or normal population.

Since the samples must be representative of the population, sampling should be random.

Random sampling is one in which each member of the population has equal chances or probability of being included in the sample.

In **sampling with replacement**, each member of the population may be chosen more than once, since the member is replaced in the population.

Thus sampling from finite population with replacement can be considered theoretically as sampling from infinite population. Whereas, in **sampling without replacement**, an element of the population can not be chosen more than once, as it is not replaced.

Statistic

Statistic is a real valued function of the random sample. So statistic is a function of one or more random variables not involving any unknown parameter. Thus statistic is a function of samples observations only and is itself a random variable. Therefore a statistic must have a probability distribution.

Sample mean and sample variance are two important statistics which are measures of a random sample $X_1, X_2, \ldots X_n$ of size n.

$$\textbf{Sample mean} = \overline{X} = \frac{\sum_{i=1}^{n} X_i}{n}$$

(measure of central tendency)

Sample Variance

$$= S^2 = \frac{\sum_{i=1}^{n}(X_i - \overline{X}^2)}{n-1} = \frac{n\sum X_i^2 - (\sum X_i)^2}{n(n-1)}$$

(measure of variability of data about the mean).

Sample standard deviation is the positive square root of the sample variance.

Degrees of Freedom (dof) of a statistic is a positive integer, denoted by ν, equals to $n - k$ where n is the number of independent observations of the random sample and k is the number of population parameters which are calculated using the sample data. Thus dof $\nu = n - k$ is the difference between n the sample size and k the number of independent constraints imposed on the observations in the sample.

28.2 SAMPLING DISTRIBUTION

Draw all possible samples of size n, from a given finite population of size N. Then the total number of *all* possible samples each of the same size n, which can be drawn from the population is given by $N_{C_n} = \frac{N!}{n!(N-n)!} = k$. Compute a statistic S (such as the mean, s.d., median, mode etc.) for each of these sample using the sample data $x_1, x_2, \ldots, x_n$ by $S = S(x_1, x_2, x_3, \ldots, x_n)$

Sample number	1	2	3	...	k
Statistic S	S_1	S_2	S_3	...	S_k

Sampling distribution (S.D.) of the statistic is the set of values $\{S_1, S_2, \ldots, S_k\}$ of the statistic S obtained one for each sample. Thus sampling distribution describes how a statistic S will vary from one sample to the other of the same size. Although all the k samples are drawn from the given population, the members included in different samples are different. The difference in the value of the statistics, attributed to chance, is known as **sampling fluctuations**.

When the number of samples (each of the same size n) is infinitely large (i.e., sampling without replacement) then the probability distribution of the statistic is the sampling distribution of the statistic.

If the statistic S is mean, then the corresponding distribution of the statistics is known as sampling distribution of means. Thus if S is variance, proportion or medians etc., the associated distribution is known as sampling distribution of variances, sampling distribution of proportions etc. Now for each of these sampling distributions, the statistics mean, variance etc., can be computed as follows:

Mean of the sampling distribution of $S = \overline{S} = \frac{1}{k}\sum_{i=1}^{k} S_i$ variance of $S = \frac{1}{k}\sum_{i=1}^{k}(S_i - \overline{S})^2$

Thus we can have mean of the sampling distribution of means, variance of the sampling distribution of means, variance of the sampling distribution of variances etc.

Standard Error (S.E.)

Standard Error (S.E.) is the standard deviation of the sampling distribution of a statistic S. It gives an index of the precision of the estimate of the parameters. As the sample size n increases, S.E. decreases. Standard error plays an important role in large sample theory and forms the basis in tests of hypothesis.

Sampling distribution of statistics helps to learn information about the corresponding population parameters.

28.3 SAMPLING DISTRIBUTION OF MEANS: (σ KNOWN)

Sampling distribution of means (S.D.M.) $\overline{X}$ is the probability distribution of $\overline{X}$.

Finite Population

Consider a finite population of size N with mean μ and standard deviation σ. Draw *all* possible samples of size *without replacement* from this population. Let $\mu_{\overline{X}}$ and $\sigma_{\overline{X}}$ denote the mean and standard deviation of the sampling distribution of means. Suppose $N > n$. Then

$$\mu_{\overline{X}} = \mu \quad \text{and}$$

$$\sigma_{\overline{X}} = \frac{\sigma}{\sqrt{N}}\sqrt{\frac{N-n}{N-1}}$$

Here $\frac{N-n}{N-1}$ is known as finite population correction factor.

Infinite Population

Suppose the samples are drawn from an infinite population or sampling is done *with replacement* then

$$\mu_{\overline{X}} = \mu \quad \text{and}$$

$$\sigma_{\overline{X}} = \frac{\sigma}{\sqrt{n}}.$$

Standard error of mean, $\sigma_{\overline{X}} = \frac{\sigma}{\sqrt{n}}$, measure the reliability of the mean as an estimate of the population mean μ.

$$\text{Standardized sample mean,} \quad Z = \frac{\overline{X} - \mu}{\sigma/\sqrt{n}}$$

Non-normal Population (Large Sample)

Consider a population with unknown (non-normal) distribution. Let the population mean μ and population variance σ be both finite. Let the population be finite or infinite. In case the population is finite assume that the population size N is at least twice the sample size n. Draw all possible samples of size n. Then the sampling distribution of $\overline{X}$ is approximately normally distributed with mean $\mu_{\overline{X}} = \mu$ and variance $\sigma^2_{\overline{X}} = \sigma^2/n$ provided the sample size is large (i.e., $n \geq 30$).

Central Limit Theorem

Whenever n is large, the sampling distribution of $\overline{X}$ is approximately (nearly) normal with mean μ and variance σ^2/n regardless of the form of the population distribution. This is established by central limit theorem stated below (without proof).

Theorem: *If $\overline{X}$ is the mean of a sample of size n drawn from a population with mean μ and finite* *variance σ^2 then the standardized sample mean*

$$Z = \frac{\overline{X} - \mu}{\sigma/\sqrt{n}}$$

is a random variable whose distribution function approaches that of the standard normal distribution $N(Z; 0, 1)$ as $n \to \infty$.

Normal Population (Small Sample)

Sampling distribution of $\overline{X}$ is normally distributed even for small samples of size $n < 30$ provided sampling is from normal population.

Worked Out Examples

Sampling distribution of means

Example 1: A population consists of four numbers 2, 3, 4, 5. Consider all possible distinct samples of size two with replacment. Find (a) the population mean (b) the population standard deviation (s.d.) (c) the sampling distribution of means (d) the mean of the S.D. of means (e) s.d. of S.D. of means. Verify (c) and (e) directly from (a) and (b) by use of suitable formulae.

Solution:

a. Mean of population

$$\mu = \frac{2+3+4+5}{4} = \frac{14}{4} = 3.5$$

b. s.d. of population

$$\sigma = \sqrt{\frac{(2-3.5)^2+(3-3.5)^2+(4-3.5)^2+(5-3.5)^2}{4}}$$

$$\sigma = \sqrt{\frac{(1.5)^2+(0.5)^2+(0.5)^2+(1.5)^2}{4}}$$

$$= \sqrt{\frac{5}{4}} = \sqrt{1.25} = 1.118033$$

c. Sampling with replacement (infinite population): The total number of samples with replacement is $N^n = 4^2 = 16$. Here N = population size and n = sample size. Listing all possible samples of size 2 from population 2, 3, 4, 5 with replacement, we get 16 samples:

(2, 2)	(2, 3)	(2, 4)	(2, 5)
(3, 2)	(3, 3)	(3, 4)	(3, 5)
(4, 2)	(4, 3)	(4, 4)	(4, 5)
(5, 2)	(5, 3)	(5, 4)	(5, 5)

Now compute the statistic the arithmetic mean for each of these 16 samples.

The set of 16 means $\overline{X}$ of these 16 samples, gives rise to the distribution of means of the samples known as sampling distribution of means (S.D.M.):

2	2.5	3	3.5
2.5	3	3.5	4
3	3.5	4	4.5
3.5	4	4.5	5

This S.D.M. can also be arranged in the form of frequency distribution

Sample mean: $\overline{X}_i$	2	2.5	3	3.5	4	4.5	5
Frequency: f_i	1	2	3	4	3	2	1

d. The mean of these 16 means is known as mean of the sampling distribution of means

$$\mu_{\overline{X}} = \frac{2+2(2.5)+3(3)+4(3.5)+3(4)+2(4.5)+5}{16}$$

$$\mu_{\overline{X}} = \frac{56}{16} = 3.5$$

e. The variance of S.D.M. is

$$\sigma^2_{\overline{X}} = \frac{\sum f_i(\overline{X}_i - \mu_{\overline{X}})^2}{n}$$

$$= \frac{1}{16}\left[1(2-3.5)^2 + 2(2.5-3.5)^2 + \cdots + 1(5-3.5)^2\right]$$

$$= \frac{(1.5)^2+2(1)^2+3(.5)^2+0+3(.5)^2+2(1)^2+(1.5)^2}{16}$$

$$= \frac{10}{16} = 0.625$$

s.d. of S.D.M. $\sigma_{\overline{X}} = 0.7905694$

Verification:

i. $\mu = 3.5 = \mu_{\overline{X}}$

ii. $\sigma_{\overline{X}} = 0.79056 = \frac{\sigma}{\sqrt{n}} = \frac{1.118033}{\sqrt{2}} = \frac{1.118033}{1.4142} = 0.79057$

Example 2: Solve the above Example 1 without replacement:

Solution:

a. $\mu = 3.5$

b. $\sigma = 1.118083$

c. Sampling without replacement (finite population). The total number of samples without replacement is $NC_n = \frac{4!}{2!2!} = 6$. The six samples are

(2, 3) (2, 4) (2, 5) (3, 4) (3, 5) (4, 5)

compute the statistic A.M. for each of these samples:

$\overline{X}_i$: 2.5 3 3.5 3.5 4 4.5

The S.D.M. is

$\overline{X}_i$:	2.5	3	3.5	4	4.5
f_i:	1	1	2	1	1

d. $\mu_{\overline{X}}$ = mean of S.D.M.

$$= \frac{2.5 + 3.0 + 2(3.5) + 4 + 4.5}{6} = \frac{21}{6}$$

$$\mu_{\overline{X}} = 3.5$$

e. $\sigma^2_{\overline{X}} = \frac{(2.5-3.5)^2+(3-3.5)^2+2(3.5-3.5)^2+(4-3.5)^2+(4.5-3.5)^2}{6} =$

$$= \frac{2.5}{6} = 0.4166.$$

so $\sigma_{\overline{X}} = 0.645497$

Verification:

i. $\mu_{\overline{X}} = 3.5 = \mu$.

ii. $\sigma_{\overline{X}} = 0.4166 = \frac{\sigma}{\sqrt{n}}\sqrt{\frac{N-n}{N-1}} = \frac{1.11803}{\sqrt{2}}\sqrt{\frac{4-2}{4-1}}$
$= 0.4166$

Sampling distribution of variances

Example 3: Find the mean and s.d. of sampling distribution of variances (S.D.V.) for the population 2, 3, 4, 5 by drawing samples of size two (a) with replacement (b) without replacement.

Solution:

a. with replacement: 16 samples with their corresponding means are

(2, 2)	(2, 3)	(2, 4)	(2, 5)
2	2.5	3	3.5
(3, 2)	(3, 3)	(3, 4)	(3, 5)
2.5	3	3.5	4
(4, 2)	(4, 3)	(4, 4)	(4, 5)
3	3.5	4	4.5
(5, 2)	(5, 3)	(5, 4)	(5, 5)
3.5	4	4.5	5

compute the statistic variance for each of these 16 samples:

variance for sample (2, 2) with mean 2 is

$$= \frac{(2-2)^2 + (2-2)^2}{2} = 0$$

Similarly, variance for sample (2, 3) with mean

$$= \frac{(2-2.5)^2 + (3-2.5)^2}{2} = 0.25$$

Thus the variance for each of the 16 samples

0	0.25	1	2.25
0.25	0	0.25	1
1	0.25	0	0.25
2.25	1	0.25	0

Thus the S.D. of variances (with replacement) is

S^2:	0	0.25	1	2.25
Frequency:	4	6	4	2

Mean of S.D. of variance $= \frac{4(0)+6(0.25)+4(1)+2(2.25)}{16}$

$$\mu_{S^2} = \frac{10}{16} = 0.625$$

variance of S.D. of variance

$$= \frac{4(0-0.625)^2+6(0.25-0.625)^2+4(1-0.625)^2+2(2.25-0.625)^2}{16}$$

$$= \frac{8.25}{16} = 0.515625$$

s.d. of S.D. of variance $= \sqrt{0.515625} = 0.7180$

b. without replacement:

Samples:	(2, 3)	(2, 4)	(2, 5)	(3, 4)	(3, 5)	(4, 5)
Means:	2.5	3	3.5	3.5	4	4.5
Variances:	0.25	1	2.25	0.25	1	0.25

Thus the S.D. of variances (without replacement)

σ_{S^2}:	0.25	1	2.25
Frequency:	3	2	1

mean of S.D. of variances $= \frac{3(0.25)+2(1)+1(2.25)}{6}$

$$\mu_{S^2} = \frac{5}{6} = 0.8333$$

variance of S.D. of variances is

$$\sigma^2_{S^2} = \frac{3(0.25-0.8333)^2+2(1-0.8333)^2+1(2.25-0.8333)^2}{6}$$

$$= \frac{3.08333}{6} = 0.51388$$

s.d. of S.D. of variance $= \sqrt{0.51388} = 0.71686.$

EXERCISE

Sampling distribution

1. Construct S.D. of means for the population 3, 7, 11, 15 by drawing samples of size two with replacement.

   Determine (a) μ (b) σ (c) S.D.M. (d) $\mu_{\overline{X}}$ (e) $\sigma_{\overline{X}}$. Verify the results.

   Hint: $4^2 = 16$ samples (3, 3), (3, 7), (3, 11), ······, (15, 11), (15, 15)

Ans. **a.** $\mu = \frac{36}{4} = 9$

b. $\sigma^2 = \frac{80}{4} = 20, \sigma = 4.4721$

c.

Means:	3	5	7	9	11	13	15
Frequency:	1	2	3	4	3	2	1

d. $\mu_{\overline{X}} = \frac{144}{16} = 9$

e. $\sigma^2_{\overline{X}} = \frac{160}{16} = 10, \sigma_{\overline{X}} = 3.16227.$

Verification:

i. $\mu_{\overline{X}} = 9 = \mu$

ii. $\sigma_{\overline{X}} = 3.162 = \frac{\sigma}{\sqrt{n}} = \frac{4.4721}{\sqrt{2}} = 3.162$

2. Solve the above problem if sampling is without replacement.

   Hint: $N_{C_n} = 4_{C_2} = 6$ samples are (3, 7), (3, 11), (3, 15), (7, 11), (7, 15), (11, 15).

Ans. S.D.M.: 5, 7, 9, 9, 11, 13

$$\mu_{\overline{X}} = \frac{54}{6} = 9, \sigma_{\overline{X}} = \sqrt{\frac{40}{6}} = 2.58198$$

3. Find standard error of sample means of size 2 drawn from a population 2, 3, 6, 8, 11 with replacement.

   Hint: $N^n = 5^2 = 25$ samples (2, 2), (2, 3), (2, 6), ······, (11, 8), (11, 11).

With means:	2	2.5	3	4	4.5	5	5.5	6	6.5	7	8	8.5	9.5	11
Frequency:	1	2	1	2	2	2	2	1	2	4	1	3	1	1

$$\mu_{\overline{X}} = \frac{150}{25} = 6.0, \mu = \frac{30}{5} = 6,$$

$$\sigma = \sqrt{10.8} = 3.29, \sigma^2_{\overline{X}} = \sqrt{\frac{135}{25}} = 2.32$$

Ans. S.E. $= \sigma_{\overline{X}} = 3.29$

Verification:

$$\mu_{\overline{X}} = 6 = \mu, \sigma^2_{\overline{X}} = 5.40 = \frac{\sigma^2}{n} = \frac{10.8}{2} = 5.40$$

4. Solve the above problem without replacement.

Ans. $N_{C_n} = 5_{C_2} = 10$ samples (2,3), (2, 6), (2,8), ······, (6, 11), (8, 11).

means: 2.5, 4, 5, 6.5, 4.5, 5.5, 7, 7, 8.5, 9.5

$$\mu_{\overline{X}} = 6 = \mu, \sigma^2_{\overline{X}} = 4.05$$

$$= \frac{\sigma^2}{n}\left(\frac{N-n}{N-1}\right) = \frac{10.8}{2}\left(\frac{5-2}{5-1}\right) = 4.05$$

5. Construct S.D.M. for the infinite population with distribution given by

x:	1	2	3	4
$f(x)$:	0.25	0.25	0.25	0.25

   with $n = 2$, verify

   Hint: $N^n = 4^2 = 16$ samples (1, 1), (1, 2), (1, 3) ······ (4, 3), (4, 4).

Ans. $\mu = \frac{10}{4} = 2.5, \sigma^2 = \frac{5}{4} = 1.25$

Means:	1	1.5	2	2.5	3	3.5	4
Frequency:	1	2	3	4	3	2	1

$$\mu_{\overline{X}} = \frac{40}{16} = 2.5, \sigma^2_{\overline{X}} = \frac{10}{16} = 0.625 = \frac{\sigma^2}{n} = \frac{1.25}{2}$$

6. Determine the mean and s.d. of S.D. of variances for the population 3, 7, 11, 15 with $n = 2$ and with sampling (a) with replacement (b) without replacement

Hint: $N^n = 4^2 = 16$ samples (3, 3), (3, 7), ..., (15, 11), (15, 15)

With means:	3	5	7	9	11	13	15
Frequency:	1	2	3	4	3	2	1
Variances:	0	4	16	36			

Ans. (a) $\mu_{S^2} = 10$, (b) $\sigma^2_{S^2} = \sqrt{\frac{2112}{16}} = 11.489$

7. Calculate the s.d. of S.D. of means for the population 16, 4, 12, 8, 24, 20 by drawing samples of size 2 without replacement. Verify the results.

Hint: S.D.M.:

Means:	6	8	10	12	14	16	18	20	22
Frequency:	1	1	2	2	3	2	2	1	1

$N_{C_n} = 6_{C_2} = 15$ samples.

Ans. $\mu = \frac{84}{6} = 14, \mu_{\overline{X}} = \frac{210}{15} = 14;$
$\sigma = 6.8313, \sigma_{\overline{X}} = 4.32.$

8. Find mean and s.d. of S.D.M. for population 1, 2, 3 with $n = 2$ and sampling (a) with replacement (b) without replacement. Verify results.

Ans. $\mu = 2, \sigma^2 = \frac{2}{3}$

a. $\mu_{\overline{X}} = \frac{18}{9} = 2, \sigma^2_{\overline{X}} = \frac{1}{3} = \frac{\sigma^2}{2}$

b. $\mu_{\overline{X}} = 2, \sigma^2_{\overline{X}} = \frac{1}{6} = \frac{\sigma^2}{n}\left(\frac{N-n}{N-1}\right) =$
$= \frac{2}{3} \cdot \frac{1}{2} \cdot \left(\frac{3-2}{3-1}\right) = \frac{1}{6}$ with $N = 3, n = 2$

WORKED OUT EXAMPLES

Sampling distribution of means (σ known)

Example 1: Determine the mean and s.d. of the sampling distribution of means of 300 random samples each of size $n = 36$ are drawn from a population of $N = 1500$ which is normally distributed with mean μ=22.4 and s.d. σ of 0.048, if sampling is done (a) with replacement and (b) without replacement.

Solution:

a. with replacement

$$\mu_{\overline{X}} = \mu = 22.40$$

$$\sigma_{\overline{X}} = \frac{\sigma}{\sqrt{n}} = \frac{0.048}{\sqrt{36}} = 0.008$$

b. without replacement

$$\mu_{\overline{X}} = \mu = 22.40$$

$$\sigma_{\overline{X}} = \frac{\sigma}{\sqrt{n}}\sqrt{\frac{N-n}{N-1}} = \frac{0.048}{\sqrt{36}}\sqrt{\frac{1500-36}{1500-1}} = 0.00790605$$

$$\sigma_{\overline{X}} \approx 0.008$$

Example 2: Determine the expected number of random samples having their means (a) between 22.39 and 22.41 (b) greater than 22.42 (c) less than 22.37 (d) less than 22.38 or more than 22.41, for the above Example 1.

Solution: N = size of population = 1500
n = sample size = 36
N_S = number of samples = 300
μ = population mean = 22.4,
σ = population s.d. = 0.48

use the standardized variable

$$Z = \frac{\overline{X} - \mu}{\sigma/\sqrt{n}} = \frac{\overline{X} - 22.4}{0.008}$$

a. For $\overline{X} = 22.39$, $Z = \frac{22.39-22.4}{0.0079} = -1.26$ etc.

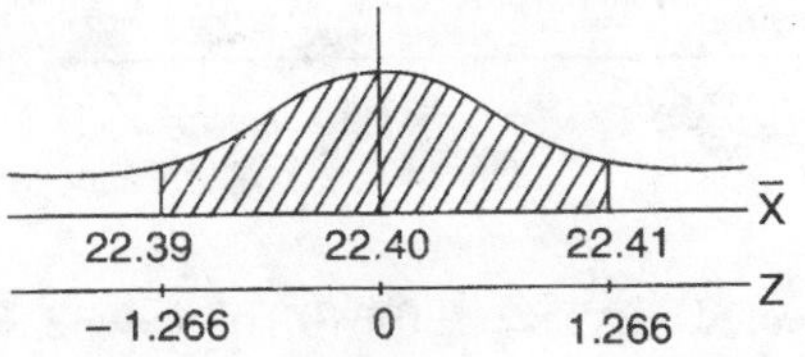

Fig. 28.1

$$P(22.39 < \overline{X} < 22.41) = P(-1.26 < Z < 1.26)$$
$$= 2(0.3962) = 0.7888$$

Expected number of samples = (Total number of samples) × (probability) = $(N_S)P(\overline{X})$

Expected number of samples who have mean lying between 22.39 to 22.41 is (300)(0.7921) = 236.6 ≈ 237

b. $P(\overline{X} > 22.42) = P(Z > 2.50) =$
$= 0.5 - 0.4933 = 0.0062$

Expected number of samples = (0.00057) (300) = 1.86 ≈ 2

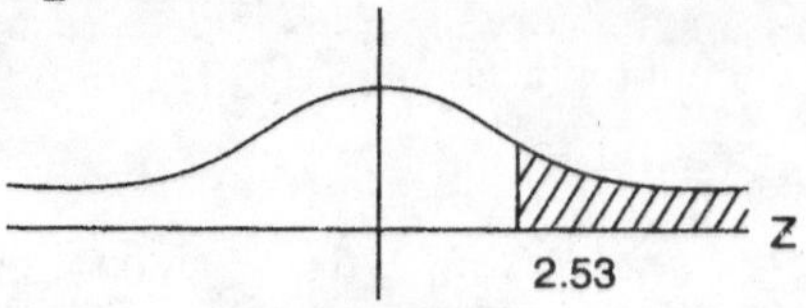

Fig. 28.2

c. $P(\overline{X} < 22.37) = P(Z < -3.8) =$
$= 0.5 - 0.4999 = 0.0001$

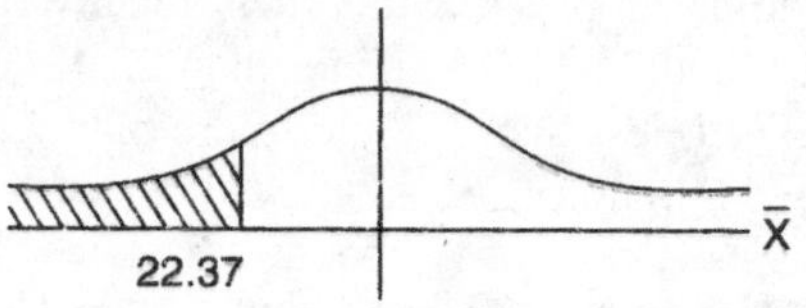

Fig. 28.3

Expected number of samples = (300) (0.0001) = 0.03 ≈ 0

d. $P(\overline{X} < 22.38)$ and $\overline{X} > 22.41)$

$= P(Z < -2.53$ and $Z > 1.26)$
$= 0.0057 + 0.1038 = 0.1095$

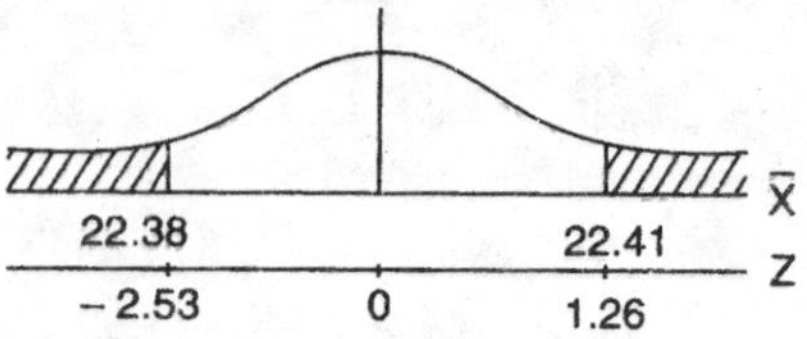

Fig. 28.4

Expected number = (300) (0.1095) = 32.85 ≈ 33.

EXERCISE

1. Determine the probability that the sample mean area covered by a sample of 40 of 1 litre paint boxes will be between 510 to 520 square feet given that a 1 litre of such paint box covers on the average 513.3 square feet with s.d. of 31.5 s.ft.

Hint: $Z_1 = \frac{510-513.3}{31.5/\sqrt{40}} = -0.66$, $Z_2 = \frac{520-513.3}{31.5/\sqrt{40}} = 1.34$

Ans. $P(510 < \overline{X} < 520) = P(-0.66 < Z < 1.34) = 0.6553$

2. Calculate the probability that a random sample of 16 computers will have an average life of less than 775 hours assuming that length of life of computers is approximately normally distributed with mean 800 hours and s.d. 40 hours.

Ans. $P(\overline{X} < 775) = P\left(Z < \frac{775-800}{10} = -2.5\right) = 0.0062$

3. Determine the probability that $\overline{X}$ will be between 75 and 78 if a random sample of size 100 is taken from an infinite population having the mean $\mu = 76$ and the variance $\sigma^2 = 256$.

Hint: $Z_1 = \frac{75-76}{16/\sqrt{100}} = -0.625$, $Z_2 = \frac{78-76}{16/\sqrt{100}} = 1.25$

Ans. $P(75 < \overline{X} < 78) = P(-0.625 < Z < 1.25) = 0.3944 + 0.2324 = 0.6268$

4. Find $P(\overline{X} > 66.75)$ if a random sample of size 36 is drawn from an infinite population with mean $\mu = 63$ and s.d. $\sigma = 9$.

Hint: $Z = \frac{66.75-63}{9/\sqrt{36}} = 2.5$

Ans. $P(\overline{X} > 66.75) = P(Z > 2.5) = 0.5 - 0.4938 = 0.0062$

5. Calculate the mean and s.d. of the sampling distribution of means of 80 samples each of size 25 by sampling (a) with replacement (b) without replacement from a normal population of 3000 with mean 68 and s.d. 3.

Ans. **a.** $\overline{\mu}_X = 68 = \mu, \sigma_{\overline{X}} = \frac{\sigma}{\sqrt{n}} = \frac{3}{\sqrt{25}} = 0.6$

b. $\mu_{\overline{X}} = 68 = \mu, \sigma_{\overline{X}} = \frac{\sigma}{\sqrt{n}}\sqrt{\frac{N-n}{N-1}}$
$= \frac{3}{\sqrt{25}}\sqrt{\frac{3000-25}{3000-1}} = 0.59759 \approx 0.6$

6. Determine the expected number of samples whose mean (a) lies between 66.8 and 68.3 (b) is less than 66.4 for the above Example 5

Hint: $z_1 = \frac{(66.8-68.0)}{0.6} = -2.0$, $z_2 = \frac{68.3-68}{0.6} = 0.5$

Ans. **a.** $P(66.8 < \overline{X} < 68.3) = P(-2.0 < Z < 0.5) = 0.4772 + 0.1915 = 0.6687$

Expected number of samples with mean between 66.8 and 68.3 is = (number of samples) (probability) = (80)(0.6680) = 53.504 ≈ 54

Hint: $z = \frac{66.4-68}{0.6} = -2.67$.

b. $P(\overline{X} < 66.4) = P(z < -2.67)$
$= 0.5 - 0.4962 = 0.0038$

Expected number of samples = (80)(0.0038) = 0.304 ≈ 0.

28.4 SAMPLING DISTRIBUTION OF PROPORTIONS

Let p be the probability of occurrence of an event (called its success) and $q = 1 - p$ is the probability of non-occurrence (failure). Draw all possible samples of size n from an infinite population. Compute the proportion P of successes for each of these samples. Then the mean μ_P and variance σ_P^2 of the sampling distribution of proportions are given by

$$\mu_P = P \quad \text{and}$$

$$\sigma_P^2 = \frac{pq}{n} = \frac{p(1-p)}{n}$$

while population is binomially distributed, the sampling distribution of proportion is normally distributed whenever n is large. For finite population (with replaement) of size N,

$$\mu_P = p \quad \text{and}$$

$$\sigma_P^2 = \frac{pq}{n}\left(\frac{N-n}{N-1}\right).$$

28.5 SAMPLING DISTRIBUTION OF DIFFERENCES AND SUMS

Let μ_{S_1} and σ_{S_1} be the mean and standard deviation of a sampling distribution of statistic S_1 obtained by computing S_1 for all possible samples of size n_1 drawn from population A. Similarly μ_{S_2} and σ_{S_2} be the mean and standard deviation of sampling distribution of statistic S_2 obtained by computing S_2 for all possible samples of size n_2 drawn from another different population B. Now compute the statistic $S_1 - S_2$, the difference of the statistic from all possible combinations of these samples from the two populations A and B. Then the mean $\mu_{S_1-S_2}$ and the standard deviation $\sigma_{S_1-S_2}$ of the sampling distribution of differences are given by

$$\mu_{S_1-S_2} = \mu_{S_1} - \mu_{S_2} \quad \text{and}$$

$$\sigma_{S_1-S_2} = \sqrt{\sigma_{S_1}^2 + \sigma_{S_2}^2}$$

assuming that the samples are independent. *Sampling distribution of sum of statistics* has mean $\mu_{S_1+S_2}$ and standard deviation $\sigma_{S_1+S_2}$ given by

$$\mu_{S_1+S_2} = \mu_{S_1} + \mu_{S_2} \quad \text{and}$$

$$\sigma_{S_1+S_2} = \sqrt{\sigma_{S_1}^2 + \sigma_{S_2}^2}$$

For example, for infinite population the sampling distribution of sums of means has mean $\mu_{\overline{X}_1+\overline{X}_2}$ and $\sigma_{\overline{X}_1+\overline{X}_2}$ given by

$$\mu_{\overline{X}_1+\overline{X}_2} = \mu_{\overline{X}_1} + \mu_{\overline{X}_2} = \mu_1 + \mu_2 \quad \text{and}$$

$$\sigma_{\overline{X}_1+\overline{X}_2} = \sqrt{\sigma_{\overline{X}_1}^2 + \sigma_{\overline{X}_2}^2} = \sqrt{\sigma_{\frac{1}{n_1}}^2 + \sigma_{\frac{2}{n_2}}^2}$$

For sampling distribution of differences of proportions we have

$$\mu_{P_1-P_2} = \mu_{P_1} - \mu_{P_2} = p_1 - p_2 \quad \text{and}$$

$$\sigma_{P_1-P_2} = \sqrt{\sigma_{P_1}^2 + \sigma_{P_2}^2} = \sqrt{\frac{p_1q_1}{n_1} + \frac{p_2q_2}{n_2}}$$

WORKED OUT EXAMPLES

Sampling distribution of differences and sums

Example 1: Let $U_1 = \{2, 7, 9\}$, $U_2 = \{3, 8\}$. Find (a) μ_{U_1} (b) μ_{U_2} (c) $\mu_{U_1+U_2}$ (d) $\mu_{U_1-U_2}$ (e) σ_{U_1} (f) σ_{U_2} (g) $\sigma_{U_1+U_2}$ (h) $\sigma_{U_1-U_2}$. Verify that (i) $\mu_{U_1+U_2} = \mu_{U_1} + \mu_{U_2}$ (j) $\mu_{U_1-U_2} = \mu_{U_1} - \mu_{U_2}$ (k) $\sigma_{U_1 \pm U_2} = \sqrt{\sigma_{U_1}^2 + \sigma_{U_2}^2}$.

Solution:

a. $\mu_{U_1} = \frac{2+7+9}{3} = \frac{18}{3} = 6$

b. $\mu_{U_2} = \frac{3+8}{2} = \frac{11}{2} = 5.5$

c. Population consisting of the sums of any member of U_1 and any member of U_2 is

$$2 + 3 = 5, 7 + 3 = 10, \quad 9 + 3 = 12$$

$$2 + 8 = 10, 7 + 8 = 15, \quad 9 + 8 = 17$$

i.e., $= U_1 + U_2 = \{5, 10, 12, 10, 15, 17\}$

$$\mu_{U_1+U_2} = \frac{5+10+12+10+15+17}{6} = \frac{69}{6} = 11.5$$

$$= 11.5 = 6 + 5.5 = \mu_{U_1} + \mu_{U_2}$$

d. Population consisting of differences of any member of U_1 and any member of U_2 is

$$2-3=-1, 7-3=4, 9-3=6$$

$$2-8=-6, 7-8=-1, 9-8=1$$

i.e., $U_1 - U_2 = \{-1, 4, 6, -6. -1, 1\}$

$$\mu_{U_1-U_2} = \frac{-1+4+6-6-1+1}{6} = \frac{3}{6}$$

$$= 0.5 = 6 - 5.5 = \mu_{U_1} - \mu_{U_2}$$

e. $\sigma^2_{U_1}$ = Variance of population U_1 (with mean 6)

$$= \frac{(2-6)^2+(7-6)^2+(9-6)^2}{3} = \frac{26}{3} = 8.66$$

so $\sigma_{U_1} = \sqrt{8.66} = 2.9439$

f. $\sigma^2_{U_2}$ = Variance of population U_2 (with mean 5.5)

$$= \frac{(3-5.5)^2+(8-5.5)^2}{2} = 6.25$$

so $\sigma_{U_2} = \sqrt{6.25} = 2.5$

g. $\sigma^2_{U_1+U_2}$ = Variance of population $U_1 + U_2$ (with mean 11.5)

$$= \frac{(5-11.5)^2+2(10-11.5)^2+(12-11.5)^2+(15-11.5)^2+(17-11.5)^2}{6}$$

$$= 14.9166$$

so $\sigma_{U_1+U_2} = \sqrt{14.9166} = 3.86220$

$$= \sqrt{\sigma_{U_1^2} + \sigma_{U_2^2}} = \sqrt{8.66+6.25} = \sqrt{14.91}$$

$$= 3.86220$$

h. $\sigma^2_{U_1-U_2}$ = Variance of population $U_1 - U_2$ (with mean 0.5)

$$= \frac{89.50}{6} = 14.9166$$

so $\sigma_{U_1-U_2} = \sqrt{14.9166} = 3.8622$

$$= \sqrt{\sigma_{U_1^2} + \sigma_{U_2^2}} = \sqrt{8.66+6.25} = \sqrt{14.91}$$

$$= 3.86220.$$

Example 2: The mean voltage of a battery is 15 and s.d. is 0.2. Find the probability that four such batteries connected in series will have a combined voltage of 60.8 or more volts.

Solution: Let mean voltage of batteries A, B, C, D be $\overline{X}_A, \overline{X}_B, \overline{X}_C, \overline{X}_D$. Then mean of the series of the four batteries connected is

$$\mu_{\overline{X_A+X_B+X_C+X_D}} = \mu_{\overline{X}_A} + \mu_{\overline{X}_B} + \mu_{\overline{X}_C} + \mu_{\overline{X}_D}$$

$$= 15 + 15 + 15 + 15 = 60$$

$$\sigma_{A+B+C+D} = \sqrt{\sigma_A^2 + \sigma_B^2 + \sigma_C^2 + \sigma_D^2}$$

$$= \sqrt{4(0.2)^2} = 0.4.$$

Let X be the combined voltage of the series.

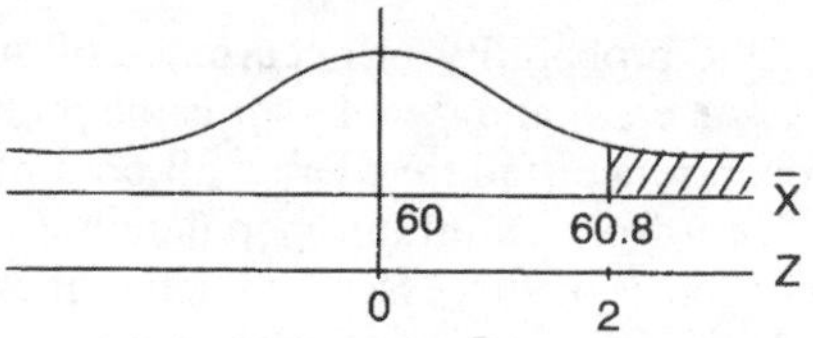

Fig. 28.5

60.8 in standard units is

$$\frac{X-\mu}{\sigma} = \frac{60.8-60}{0.4} = 2$$

so probability that combined voltage is more than 60.8

$$= P(X \geq 60.8) = P(Z > 2) = 0.5 - 0.4772 = 0.0228.$$

Example 3: Suppose the diameter of motor shafts in a lot have a mean of 0.249 inches and s.d. if 0.003 inches. The inner dia of bearings in another lot have a mean of 0.255 inches and s.d. of 0.002 inches.

a. Find the mean and s.d. of clearances between shafts and bearings selected from these lots.

b. If a shaft and bearing are selected at random, find the probability that the shaft will not fit inside the bearing. Assume that both dimensions are normally distributed.

Solution: Let $\overline{X}_B$ = mean diameter of bearing
$\overline{X}_S$ = mean diameter of shaft.
It is given $\overline{X}_B = 0.255$, $\overline{X}_S = 0.249$, $\sigma_B = 0.002$, $\sigma_S = 0.003$.

a. Then mean diameter of the difference in diameters of the bearing and the shaft is $\overline{X}_d$ given by

$$\overline{X}_d = \overline{X}_{B-S} = \overline{X}_B - \overline{X}_S = \mu_B - \mu_S$$

$= 0.255 - 0.249 = 0.006$

$$\sigma_d = \sigma_{\overline{X}_B - \overline{X}_S} = \sqrt{\sigma^2_{\overline{X}_B} + \sigma^2_{\overline{X}_S}}$$

$$= \sqrt{(0.003)^2 + (0.002)^2} = 0.00360$$

b. Shaft will not fit inside bearing if $d < 0$.
0 in standard units is $z = \frac{0-0.006}{0.0036055} = -1.664$.
Probability that shaft will not fit inside bearing

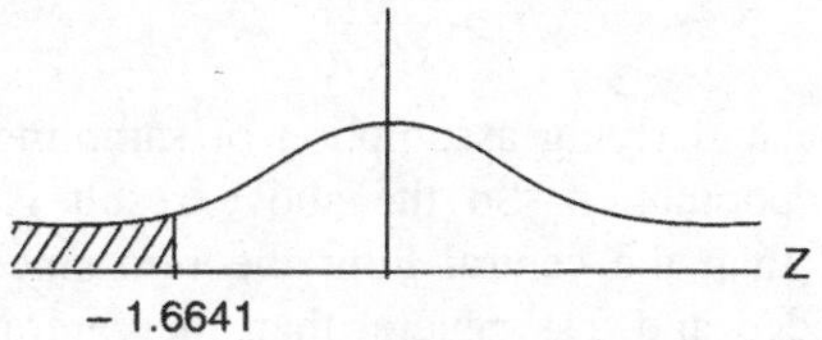

Fig. 28.6

$= P(d < 0) = P(Z < -1.6641)$

$= 0.5 - 0.4515 = 0.0485.$

EXERCISE

Sampling distribution of differences and sums

1. Let $U_1 = \{3, 7, 8\}$ and $U_2 = \{2, 4\}$. Calculate (a) μ_{U_1} (b) μ_{U_2} (c) $\mu_{U_1-U_2}$ (d) σ_{U_1} (e) σ_{U_2} (f) $\sigma_{U_1-U_2}$

   Hint: $U_1 - U_2 : \{1, 5, 6, -1, 3, 4\}$

Ans. **a.** $\mu_{U_1} = \frac{18}{3} = 6$,
b. $\mu_{U_2} = \frac{6}{2} = 3$,
c. $\mu_{U_1-U_2} = 3 = \mu_{U_1} - \mu_{U_2} = 6 - 3$, ·
d. $\sigma_{U_1} = \sqrt{\frac{14}{3}}$,
e. $\sigma_{U_2} = .1$,
f. $\sigma_{U_1-U_2} = \sqrt{\frac{17}{3}} = \sqrt{\sigma^2_{U_1} + \sigma^2_{U_2}}$

2. Three masses are measured as 62.34, 20.48, 35.97 kgs with s.d. 0.54, 0.21, 0.46 kgs. Find the mean and s.d. of the sum of the masses.

Ans. $\mu_{A+B+C} = \mu_A + \mu_B + \mu_C = 62.34 + 20.48 + 35.97 = 118.79$

$$\sigma_{A+B+C} = \sqrt{\sigma^2_A + \sigma^2_B + \sigma^2_C}$$

$$= \sqrt{(0.54)^2 + (0.21)^2 + (0.46)^2} = 0.74$$

3. The mean life time of light bulbs produced by a company is 1500 hours and s.d. of 150 hours. Find the probability that lighting will take place for (a) at least 5000 h (b) at most 4200 h if three bulbs are connected such that when one bulb burns out, another bulb will go on. Assume that life times are normally distributed.

Ans. $\mu_{L_1+L_2+L_3} = \mu_{L_1} + \mu_{L_2} + \mu_{L_3} = 1500 + 1500 + 1500 = 4500$

$\sigma_{L_1+L_2+L_3} = \sqrt{\sigma^2_{L_1} + \sigma^2_{L_2} + \sigma^2_{L_3}} = \sqrt{3(150)^2}$
$= 260$

a. $P(X > 5000) = P\left(Z > \frac{5000-4500}{260}\right)$
$= P(Z > 1.92) = 0.5 - 0.4726 = 0.0274$

b. $P(X < 4200) = P\left(Z < \frac{4200-4500}{260}\right) =$
$P(Z < -1.15) = 0.5 - 0.3749 = 0.1251$

4. Determine the probability that the mean breaking strength of cables produced by company *B* will be (a) at least 600 N more than (b) at least 450 N more than the cables produced by company *A*, if 100 cables of brand *A* and 50 cables of brand *B* are tested.

Company	Mean breaking strength	s.d.	Sample size
A	4000 N	300 N	100
B	4500 N	200 N	50

Ans. $\mu_{\overline{X}_B - \overline{X}_A} = \mu_{\overline{X}_B} - \mu_{\overline{X}_A} = 4500 - 4000$
$= 500$ N

$$\sigma_{\overline{X}_B - \overline{X}_A} = \sqrt{\frac{\sigma^2_B}{N_B} + \frac{\sigma^2_A}{N_A}} = \sqrt{\frac{(200)^2}{50} + \frac{(300)^2}{100}}$$

$$= \sqrt{1700} = 41.23$$

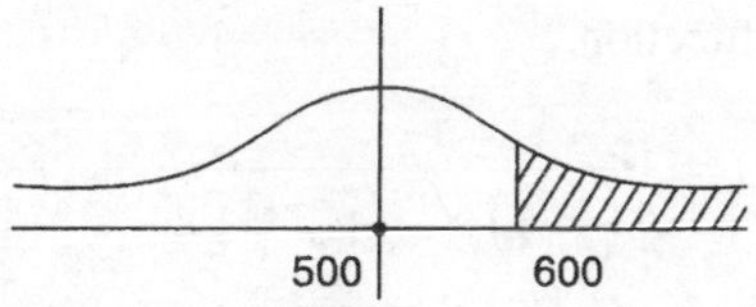

Fig. 28.7

a. $P(\overline{X}_B - \overline{X}_A > 600)$
$= P\left(Z > \frac{600-500}{41.23}\right)$
$= P(Z > 2.4254) = 0.5 = 0.4922 = 0.0078$

b. $P(\overline{X}_B - \overline{X}_A > 450) = P\left(Z > \frac{450-500}{41.23}\right)$
$= P(Z > -1.2127) = 0.5 + 0.3869$
$= 0.8869$

5. Let $\overline{X}_A$ and $\overline{X}_B$ be the average drying times of two types of paints A and B, for samples of size

$n_A = n_B = 18$. Suppose $\sigma_A = \sigma_B = 1$. Find $P(\overline{X}_A - \overline{X}_B > 1.0)$ assuming that the mean drying time is equal for the two types of paints.

Hint: $\sigma^2_{\overline{X}_A\overline{X}_B} = \frac{\sigma_A^2}{n_A} + \frac{\sigma_B^2}{n_B} = \frac{1}{18} + \frac{1}{18} = \frac{1}{9}$

Ans. $P(\overline{X}_A - \overline{X}_B > 1) = P\left(Z > \frac{1-(\mu_A-\mu_B)}{\sigma_{X_A - X_B}/\sqrt{n}}\right) = P\left(\frac{1-0}{\sqrt{1/9}}\right) = P(Z > 3) = 1 - 0.9987 = 0.0013$

28.6 SAMPLING DISTRIBUTION OF MEAN (σ UNKNOWN): *t*-DISTRIBUTION

Earlier in problems of inference on a population mean or the difference between two population means it was assumed that the population standard deviation σ is *known*. When σ is *unknown*, for large $n(\geq 30)$, σ can be replaced by the sample standard deviation s, calculated using the sample mean $\overline{x}$ by the formula $s^2 = \sum_{i=1}^{n} \frac{(x_i - \overline{x})^2}{n-1}$. For small sample of size $n(< 30)$, the *unknown* σ can be substituted by s, provided we make an assumption that the sample is drawn from a normal population.

Result: Let $\overline{x}$ be the mean of a random sample of size n drawn from a normal population with mean μ and variance σ^2 then

$$t = \frac{\overline{x} - \mu}{s/\sqrt{n}}$$

is a random variable having the t-distribution with $\nu = n - 1$ degrees of freedom with probability density function.

$$f(t) = \frac{1}{\beta\left(\frac{1}{2}, \frac{\nu}{2}\right)} \frac{1}{\sqrt{\nu}} \frac{1}{\left(1 + \frac{t^2}{\nu}\right)^{(\nu+1)/2}}, \quad -\infty < t < \infty$$

W.S. Gosset* first published in 1908, the probability distribution of t under the pseudonym "student". So the t-distribution is also known as "student t-distribution". 1925, R.A. Fisher used t-distribution to test the regression coefficient.

*William Sealy Gosset (1876–1937) English statistician.

Here sample variance s is given by

$$s^2 = \frac{1}{n-1} \sum_{i=1}^{n} (x_i - \overline{x})^2.$$

Thus for *small* samples ($n < 30$) and with σ *unknown*, a natural statistic for infernce on population mean μ is

$$t = \frac{\overline{x} - \mu}{s/\sqrt{n}}$$

with the underlying assumption of sampling from normal population. So the above result is more general than the central limit theorem since σ is not needed and less general than the central limit theorem since population is assumed to be normal.

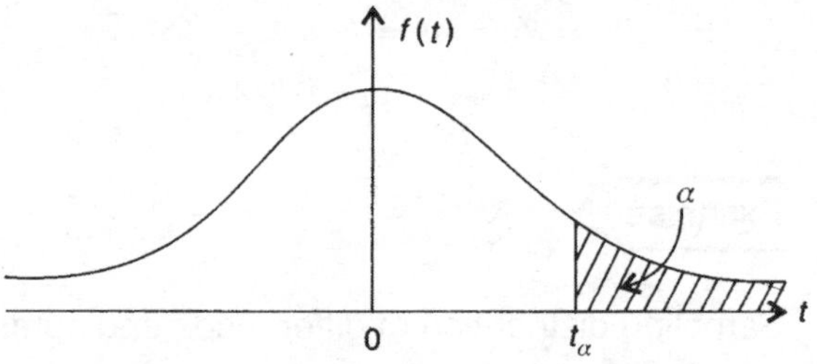

t-distribution

Fig. 28.8

The t-distribution curve is symmetric about the mean 0, unimodal, bell shaped and asymptotic on both sides of t-axis. Thus the t-distribution curve is similar to normal curve. While the variance for normal distribution is 1, the variance for the t-distribution is more than one since it depends on the parameter ν. So the t-distribution is more variable. As $n \to \infty$, variance of t-distribution approaches 1.

Thus as $\nu = (n - 1) \to \infty$, t-distribution approaches the standard normal distribution. Infact for $n \geq 30$, standard normal distribution provides a good approximation to the t-distribution.

Critical values of t-distribution (see A13 to A14) is denoted by t_α which is such that the area under the curve to the **right** of t_α equals to α. Since the t-distribution is symmetric, it follows that

$$t_{1-\alpha} = -t_\alpha$$

i.e., the t-value leaving an area of $1 - \alpha$ to the right and therefore an area α to its left, is equal to the negative t-value which leaves an area α in the right tail of the distribution.

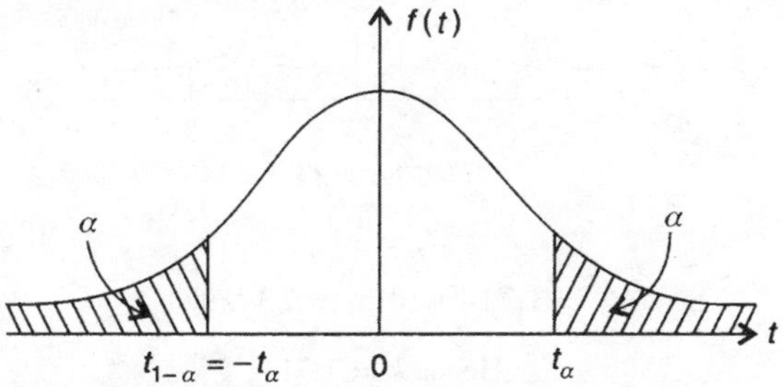

Fig. 28.9

Symmetry property of t-distribution.

The critical values t_α are tabulated in tables (A13 to A14) for various values of the parameter ν. In these tables, the left-hand column contains values of ν the column headings are areas α in the right-hand tail of the t-distribution, the entries are values of t_α.

Note 1: In the tables (A13 to A14), the areas are now the column heading and the entries are the t-values (which is the opposite in the normal tables where the entries are areas and column headings are z-values).

Note 2: Exactly 95% of the values of a t-distribution with $\nu = n - 1$ dof lies between $t_{-0.02}$ and $t_{0.025}$.

The t-distribution is extensively used in tests of hypothesis about one mean, or about equality of two means when σ is unknown.

Worked Out Examples

Example 1: Find (a) $t_{0.025}$ when $\nu = 14$ (b) $-t_{0.01}$ when $\nu = 10$ (c) $t_{0.995}$ when $\nu = 7$.

Solution: From tables (A13 to A14)

a. $t_{0.025} = 2.145$

b. $-t_{0.01} = -2.764$

c. $t_{0.995} = t_{1-0.005} = -t_{0.005} = -3.499$

Example 2: Find (a) $P(t < 2.365)$ when $\nu = 7$ (b) $P(t > 1.318)$ when $\nu = 24$ (c) $P(-1.356 < t < 2.179)$ with $\nu = 12$ (d) $P(t > -2.567)$ when $\nu = 17$.

Solution:

a. When $t < 2.365$, $P(t < 2.365)$ is given by the area to the left of $t = 2.365$.

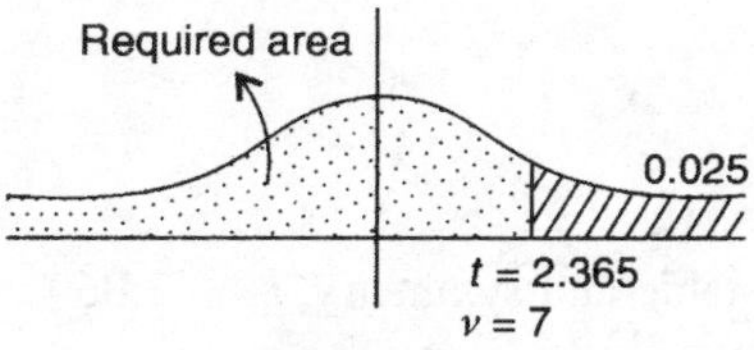

Fig. 28.10

From table, $t_\alpha = 2.365$ for $\nu = 7$ dof then $\alpha = 0.025$

$$\therefore P(t < 2.365) = 1 - 0.025 = 0.975$$

b. When $t_\alpha = 1.318$ with $\nu = 24$ then $\alpha = 0.10$

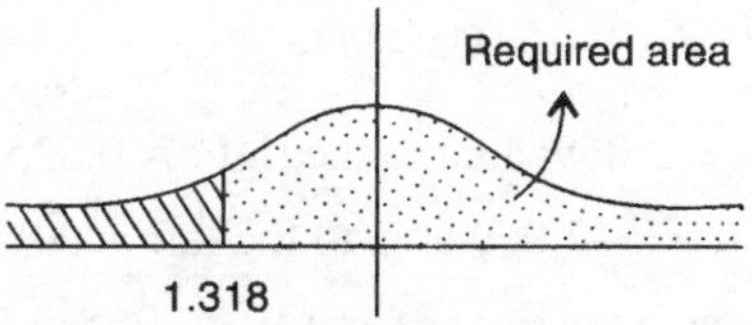

Fig. 28.11

$\therefore P(t > 1.318)$ is area to the right of 1.318 is 0.1

c. When $t < 2.179$ with $\nu = 12$

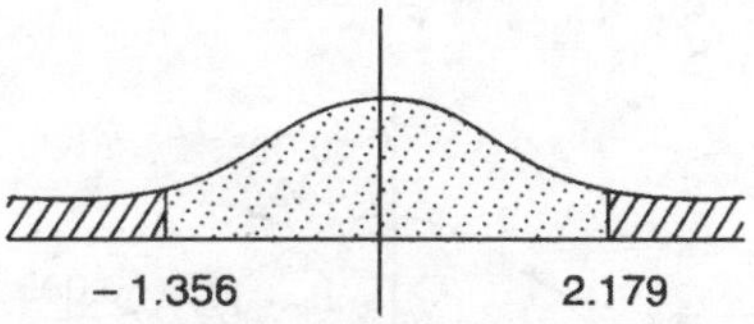

Fig. 28.12

Area to the right of 2.179 is 0.025 when $t > -13.36$ with $\nu = 12$, area to the left is 0.10

$\therefore$ When $-1.356 < t < 2.179$ the area is

$$1 - 0.10 - 0.025 = 0.875$$

$$\therefore P(-1.356 < t < 2.179) = 0.875$$

d. $P(t > -2.567) = 1 - 0.01 = 0.99$

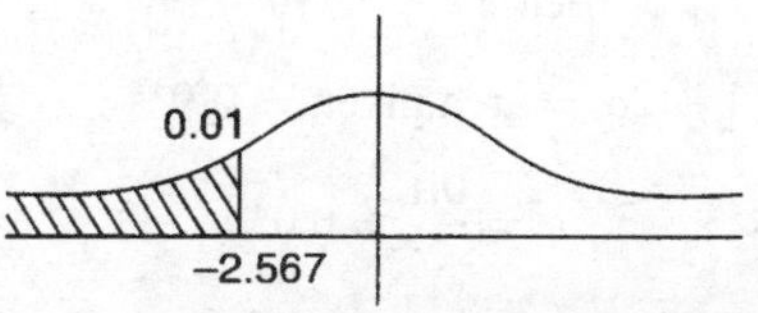

Fig. 28.13

Example 3: Find k for a random sample of size 24 from a normal distribution such that

a. $P(-2.069 < t < k) = 0.965$

b. $P(k < t < 2.807) = 0.095$

c. $P(-k < t < k) = 0.90$

Solution:

a. From table and symmetry, $t_\alpha = 2.069$ with $\nu = 24 - 1 = 23$,

$$\alpha = 0.025$$

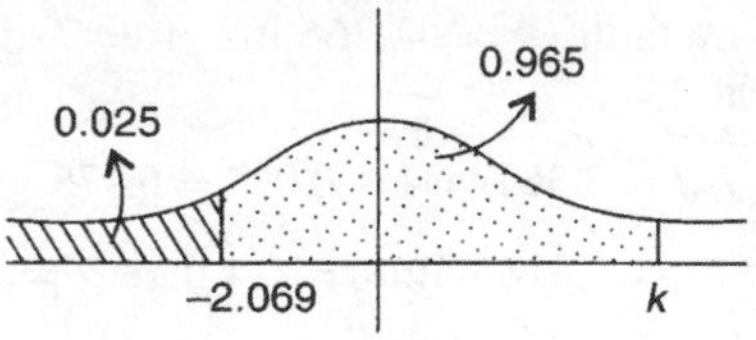

Fig. 28.14

For $t_\alpha = 2.069$, area to the right is 0.025

For $-t_\alpha = -2.069$ area to the left is 0.025

The area to the right of k is $1 - 0.965 - 0.025 = 0.01$

$\therefore t_{0.01} = 2.50$ with $\nu = 23$ dof

$$\therefore k = 2.500$$

b. $t_\alpha = 2.807$ with $\nu = 23$ dof

$\alpha = 0.005$

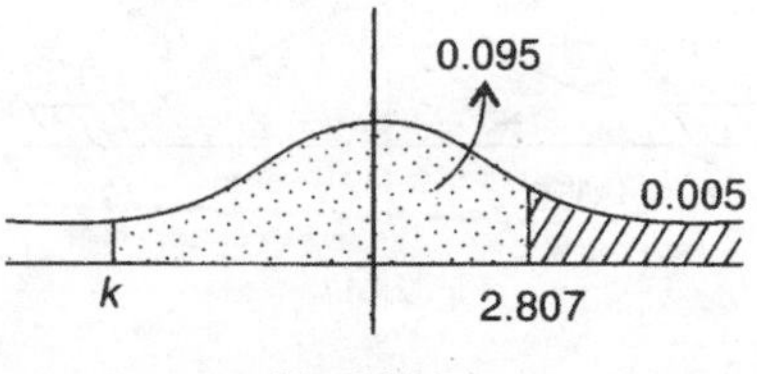

Fig. 28.15

Area to the left of k is

$$1 - 0.005 - 0.095 = 0.9$$

i.e., area to the right of k is 0.1.

$\therefore t_{0.1} = 1.319$ with $\nu = 23$

$$\text{Hence } k = 1.319$$

c. Since $1 - 2\alpha$ = area given = 0.9

$$\therefore \alpha = \frac{0.1}{2} = 0.05$$

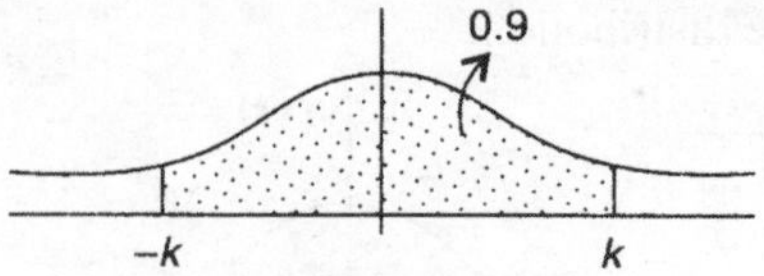

Fig. 28.16

$$\text{So } t_{0.05} = 1.714 \text{ with } \nu = 23 \text{ dof}$$

$$\text{Hence } k = 1.714$$

Example 4: A process for making certain ball bearings is under control if the diameters of the bearings have a mean of 0.5000 cm. If a random sample of 10 of these bearings has a mean diameter of 0.5060 cm and s.d. of 0.0040 cm, is the process under control?

Solution:

$\overline{x} = 0.5060$ = sample mean,

$\mu = 0.5000$ = population mean,

n = sample size = 10

S = sample s.d. = 0.0040

Then $t = \frac{\overline{x}-\mu}{s/\sqrt{n}} = \frac{0.5060-0.5000}{0.0040/\sqrt{10}} = 4.7434.$

Here $\nu = n - 1 = 10 - 1 = 9$ dof

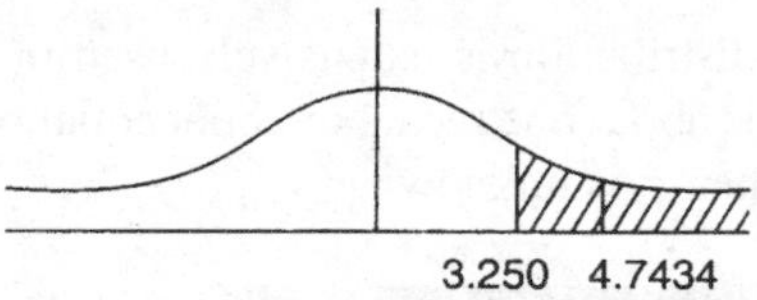

Fig. 28.17

Since $t_{cal} = 4.7334 > 3.250 = t_\alpha$ with $\alpha = 0.005$ and $\nu = 9$ dof, the process is not under control.

EXERCISE

1. Find the t-value with $\nu = 14$ dof that leaves an area 0.025 to the left.

Ans. $t_{0.975} = -t_{0.025} = -2.145$

2. Find $P(-t_{0.025} < t < t_{0.05})$

Hint: $t_{0.05}$ leaves an area 0.05 to the right, $-t_{0.025}$ leaves an area of 0.025 to the left. Total area $= 1 - 0.05 - 0.025 = 0.925$.

Ans. 0.925

3. Find k such that $P(k < t < -1.761) = 0.045$ for a random sample of size 15 selected from a normal distribution.

   Hint: 1.761 corresponds to $t_{0.05}$ with $\nu = 14$, so $-t_{0.05} = -1.761$ since k is to left, $-t_{0.05} = -1.761$ so $0.045 = 0.05 - \alpha$ or $\alpha = 0.005$, hence $k = -t_{0.005} = -2.977$.

Ans. $k = -t_{0.005} = -2.977$

4. Determine (a) $t_{0.01}$ with $\nu = 18$ (b) $t_{0.05}$ with $\nu = 12$ (c) $t_{-0.10}$ with $\nu = 15$

Ans. (a) 2.878 (b) 2.179 (c) 1.753

5. Find the 90th percentile of t-distribution with 10 dof.

Ans. 1.372

6. Find $P(t \geq 2.086)$ with 20 dof

Ans. 0.05

7. Find $P(-2.583 \leq t \leq 2.583)$ with 16 dof

Ans. $1 - 2(0.01) = 0.98$

8. Fuses produced by a company will blow in 12.40 minutes on the average when overloaded. Suppose the mean blow time of 20 fuses subjected to overload is 10.63 minutes and s.d. 2.48 mts. Does this information tend to support or refute the claim that the population mean blow time is 12.40 mts?

Ans. $t = \frac{10.63-12.4}{2.48/\sqrt{20}} = -3.19$, $\nu = 20 - 1 = 19$ dof

Data refutes the producer's claim since

$$t = -3.19 < -2.861$$

with probability $\alpha = 0.005$

9. A company claims that the mean life time of tube lights is 500 hours. Is the claim of the company tenable if a random sample of 25 tube lights produced by the company has mean 518 hours and s.d. 40 hours. Company is satisfied if t falls between $-t_{0.01}$ and $t_{0.01}$.

Ans. $t = \frac{\bar{x}-\mu}{s/\sqrt{n}} = \frac{518-500}{40/\sqrt{25}} = 2.25 < t_\alpha = 2.492$

Accept the claim of the company, $\nu = 24$ dof.

10. A random sample of size 25 from a normal population has the mean $\bar{x} = 47.5$ and s.d. $= 8.4$. Does this information tend to support or refute the claim that the mean of the population is 42.1.

Ans. Does not support the claim.

Ans. $t = \frac{\bar{x}-\mu}{s/\sqrt{n}} = \frac{47.5-42.1}{8.4/\sqrt{25}}$, $\nu = 24$ dof.

28.7 CHI-SQUARED DISTRIBUTION

Chi-squared distribution plays a very important role in estimation and hypothesis testing. It is a continuous probability distribution of a continuous random variable X, with density function given by

$$f(x) = \frac{1}{2^{\nu/2}\Gamma(\nu/2)} x^{\frac{\nu}{2}-1} e^{-x/2} \quad \text{for} \quad x > 0$$

which is a special case of gamma distribution with $\alpha = \nu/2$ and $\beta = 2$. Here ν is a positive integer is the only single parameter of the distribution. ν is known as "degrees of freedom" (dof). Chi-squared distribution is extensively used in sampling distributions, analysis of variance (ANOVA) and non-parametric tests. It was first discovered by Helmert in 1876 and later independently by Karl Pearson in 1900.

It was mainly used as a measure of goodness of fit and to test the independence of attributes. Chi-squared distribution is denoted as χ^2-distribution.

Properties of χ^2-distribution

1. χ^2-distribution curve is not a normal curve and lies completely in the first quadrant since χ^2 varies from 0 to ∞. i.e., χ^2-distribution is not symmetrical.
2. It depends only on ν, the dof.
3. It is unimodal curve with mode at $\chi^2 = (k - 1)$.
4. It is additive i.e., if χ_1^2 and χ_2^2 are two independent distributions with ν_1 and ν_2 dof then $\chi_1^2 + \chi_2^2$ will be chi-squared distribution with $(\nu_1 + \nu_2)$ dof.

Here α denotes the area under the chi-square distribution to the **right** of χ_α^2. Thus α denotes the prob-

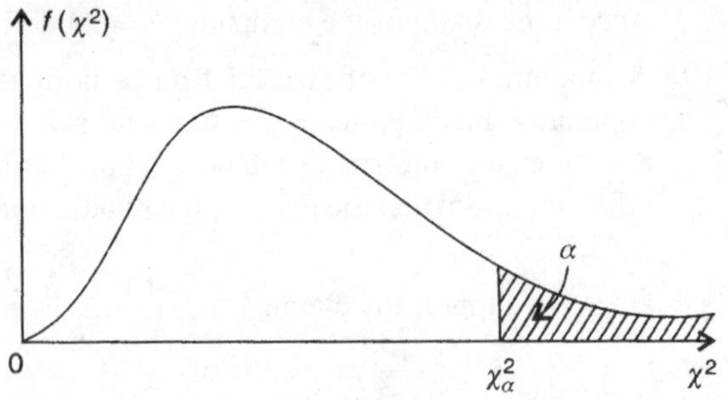

χ^2 - distribution

Fig. 28.18

ability that a random sample produces a χ^2-value $> \chi^2_\alpha$. So χ^2_α represents the χ^2-value such that the area under the chi-square curve to its (χ^2_α's) right is equal to α.

For various values of α and ν, the values of χ^2_α are presented in the tables (A15 to A16)

In χ^2 table, the left-hand column contains values of ν, dof, the column headings are areas α in the right hand tail of χ^2-distribution curve, the table entries are values of χ^2.

28.8 SAMPLING DISTRIBUTION OF VARIANCE s^2

The theoretical sampling distribution of the sample variance for random samples from normal population is related to the chi-squared distribution as follows:

Let s^2 be the variance of a random sample of size n, taken from a normal population having the variance σ^2. Then

$$\chi^2 = \frac{(n-1)s^2}{\sigma^2} = \sum_{i=1}^{n} \frac{(X_i - \overline{X})^2}{\sigma^2}$$

is a value of a random variable having the χ^2-distribution with $\nu = n - 1$ dof.

Exactly 95% of χ^2-distribution lies between $\chi^2_{0.975}$ and $\chi^2_{0.025}$. When σ^2 is too small, χ^2-value falls to the right of $\chi^2_{0.025}$ and when σ^2 is too large, χ^2 falls to the left of $\chi^2_{0.975}$. Thus when σ^2 is correct, χ^2-values falls to the left of $\chi^2_{0.975}$ or to the right of $\chi^2_{0.025}$.

28.9 F-DISTRIBUTION

Let s_1^2 be the sample variance of an independent sample of size n_1 drawn from a normal population $N(\mu_1, \sigma_1^2)$. Similarly, let s_2^2 be the sample variance in an independent sample of size n_2 drawn from another normal population $N(\mu_2, \sigma_2^2)$. Thus s_1^2 and s_2^2 are two variances of two random samples of sizes n_1 and n_2 respectively drawn from two normal population. In order to determine whether the two samples come from two populations having equal variances, consider the sampling distribution of the ratio of the variances of the two independent random sample defined by

$$F = \frac{s_1^2/\sigma_1}{s_2^2/\sigma_2^2} = \frac{\sigma_2^2 s_1^2}{\sigma_1^2 s_2^2}$$

which is an F-distribution with $\nu_1 = n_1 - 1$ and $\nu_2 = n_2 - 1$ degrees of freedom. Here F is used as mark of respect for Sir R.A. Fisher. F-distribution was worked out by G.W. Snedecor. Under the hypothesis that two normal populations have the same variance i.e., $\sigma_1^2 = \sigma_2^2$, we have

$$F_{\nu_1,\nu_2} = \frac{s_1^2}{s_2^2}$$

F determines whether the ratio of two sample variances s_1 and s_2 is too small or too large.

When F is close to 1, the two sample variances s_1 and s_2 are nearly same. It is customary, to take the larger sample variance as the numerator. F-distribution is related to the Beta distribution and the two parameters ν_1 and ν_2 are known as numerator and denominator degrees of freedom. The curve of the F-distribution depends not only on the two parameters ν_1 ande ν_2 but also on the order in which they are stated. F-distribution is also known as *variance ratio distribution*. F is always a positive number. The F-distribution curve lies entirely in first quadrant and is unimodal.

$F_\alpha(\nu_1, \nu_2)$ is the value of F with ν_1 and ν_2 dof such that the area under the F-distribution curve to the **right** of F_α is α. In tables (A17 to A20) F_α is tabulated (prepared by Snedecor) for $\alpha = 0.05$ and $\alpha = 0.01$ for various combinations of the dof ν_1 and ν_2. The values of $F_{0.95}$ and $F_{0.99}$ can be calculated from the above tables (A17 to A20) by using the following result.

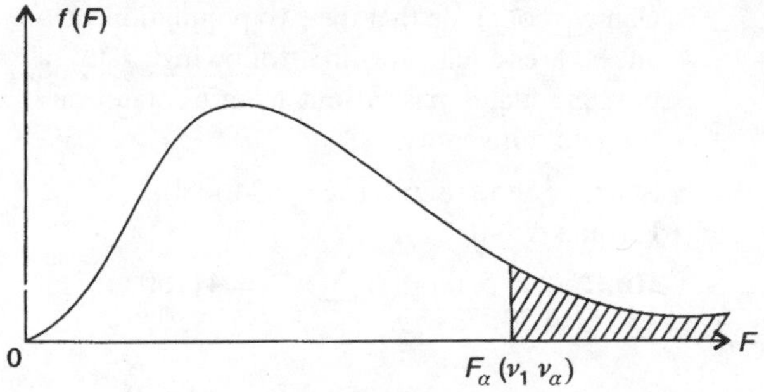

F - distribution

Fig. 28.19

$$F_{1-\alpha}(\nu_1, \nu_2) = \frac{1}{F_\alpha(\nu_2, \nu_1)}$$

F-distribution is extremely useful in testing the equality of several population means, comparing sample variances, and forms the backbone of analysis of variance (ANOVA).

WORKED OUT EXAMPLES

F-distribution

Example 1: For an F-distribution find

a. $F_{0.05}$ with $\nu_1 = 7$ and $\nu_2 = 15$

b. $F_{0.01}$ with $\nu_1 = 24$ and $\nu_2 = 19$

c. $F_{0.95}$ with $\nu_1 = 19$ and $\nu_2 = 24$

d. $F_{0.99}$ with $\nu_1 = 28$ and $\nu_2 = 12$

Solution:

a. From table $F_{0.05}$ with $\nu_1 = 7$ and $\nu_2 = 15$ is 2.71

b. $F_{0.01}$ with $\nu_1 = 24$, $\nu_2 = 19$ is 2.92

c. $F_{0.95}(19, 24) = \frac{1}{F_{0.05}(24,19)} = \frac{1}{2.11} = 0.473933$

d. $F_{0.99}(28, 12) = \frac{1}{F_{0.01}(12,28)} = \frac{1}{2.90} = 0.34482$

Example 2: Determine the probability that the variance of the first sample of size $n_1 = 9$ will be at least 4 times as large as the variance of the second sample of size $n_2 = 16$ if the two samples are independent random samples from a normal population.

Solution: From table (A17 to A20) $F_{0.01} = 4.0$ for $\nu_1 = n_1 - 1 = 9 - 1$, $\nu_2 = n_2 - 1 = 16 - 1 = 15$, the desired probability is 0.01.

Example 3: The household net expenditure on health care in south and north India, in two samples of households, expressed as percentage of total income is shown the following table

South	15.0, 8.0, 3.8, 6.4, 27.4, 19.0, 35.3, 13.6
North	18.8, 23.1, 10.3, 8.0, 18.0, 10.2, 15.2, 19.0, 20.2

Test the equality of variances of households net expenditure on health care in south and north India.

Solution: Let the net expenditure of the south and north be considered as $N(\mu_1, \sigma_1^2)$ and $N(\mu_2, \sigma_2^2)$ respectively.

1. Null Hypothesis: $H_0 : \sigma_1^2 = \sigma_2^2$, i.e., equality of variances

2. Alternate Hypothesis: $H_1 : \sigma_1^2 \neq \sigma_2^2$, i.e., difference in variances

3. Computations: $n_1 = 8$, $n_2 = 9$,

$$\sum_{i=1}^{8} x_{1i} = 15 + 8 + 3.8 + 6.4 + 27.4 + 19$$

$$+35.3 + 13.6 = 128.5$$

$$\sum_{i=1}^{8} x_{1i}^2 = (15)^2 + 8^2 + (3.8)^2 + (6.4)^2 + \cdots$$

$$+(13.6)^2 = 2887.21$$

$$\sum_{i=1}^{9} x_{2i} = 142.8, \sum_{i=1}^{9} x_{2i}^2 = 2485.26.$$

$$\text{so} \quad s_1^2 = \frac{1}{n_1 - 1}\left\{\sum_{i=1}^{8} x_{1i}^2 - \frac{(\sum x_{1i})^2}{n_1}\right\}$$

$$= \frac{1}{7}\left\{2887.21 - \frac{(128.5)^2}{8}\right\} = 117.59696$$

$$\text{also} \quad s_2^2 = \frac{1}{8}\left\{2485.26 - \frac{(142.8)^2}{9}\right\} = 27.4375$$

4. Test statistic: $F = \frac{s_1^2}{s_2^2} = \frac{117.59696}{27.4375} = 4.28599$. Variances are significantly different.

EXERCISE

F-distribution

1. Find the value of
   a. $F_{0.05}$ for $\nu_1 = 15$ and $\nu_2 = 7$
   b. $F_{0.95}$ for 12 and 15 dof
   c. $F_{0.99}$ for 6 and 20 dof
   d. $F_{0.95}$ for $\nu_1 = 10$, $\nu_2 = 20$

Ans. **a.** 3.51
b. $F_{0.95}(12, 15) = \frac{1}{F_{0.05}}(15, 12) = \frac{1}{2.62} = 0.38$
c. $F_{.99}(6, 20) = \frac{1}{F_{0.01}}(20, 6) = \frac{1}{7.4} = 0.135135$
d. $F_{0.95}(10, 20) = \frac{1}{F_{0.05}}(20, 10) = \frac{1}{2.77} = 0.36$

2. Find the probability that the variance of the first sample will be at least 3 times as large as that of the second sample if two independent random samples of size $n_1 = 7$ and $n_1 = 13$ are taken from a normal population.

   Hint: From table $F_{0.05} = 3.00$ for $\nu_1 = 7 - 1 = 6$ and $\nu_2 = 13 - 1 = 12$

Ans. 0.05

3. If independent random samples of size $n_1 = n_2 = 8$ come from normal populations having the same variance, what is the probability that either sample variance will be at least seven times as large as the other.

   Hint: From table $F_{0.01} = 6.99 \approx 7$ for $\nu_1 = 8 - 1 = 7$, $\nu_2 = 8 - 1 = 7$

Ans. 0.01

4. Can we conclude that the two population variances are equal for the following data of post graduates passed out from a 'state' and 'private' university.

   State: 8350 8260 8130 8340 8070
   Private: 7890 8140 7900 7950 7840 7920

   Hint: $n_1 = 5$, $n_2 = 6$, $\sum x_{1i} = 41150$, $\sum x_{1i}^2 = 338727500$

   $$s_1^2 = \frac{63000}{4} = 15750, \sum x_{2i} = 47640,$$

   $$\sum x_{2i}^2 = 378316200$$

   $$s_2^2 = \frac{54600}{5} = 10920, F = \frac{15750}{10920} = 1.442$$

Ans. F-ratio is 1.44. The variances are not significantly different

5. Is there reason to believe that the life expected in south and north India is same or not from the following data.

   South: 34.0, 39.2, 46.1, 48.7, 49.4, 45.9, 55.3, 42.7, 43.7
   North: 49.7, 55.4, 57.0, 54.2, 50.4, 44.2, 53.4, 57.5, 61.9, 56.6, 58.2

   Hint: $s_1^2 = \frac{1}{8}\left\{18527.78 - \frac{(405)^2}{9}\right\} = 37.848$

   $$s_2^2 = \frac{1}{10}\left\{32799.91 - \frac{(598.5)^2}{11}\right\} = 23.607$$

Ans. Yes, variances of life expectancy is same for the south and north since $F = \frac{s_1^2}{s_2^2} = \frac{37.843}{23.607} = 1.603$

Chapter 29

Estimation and Test of Hypothesis

INTRODUCTION

By "estimate" we mean "judgement or opinion of the approximate size or amount". For a trip from kakinada to Hyderabad by car we estimate the distance as 550 km, mileage/litre as 12 km, price/litre petrol as Rs 50 from which eventually estimate the entire cost of the trip. Also we might estimate the distance being between 500 to 600 km, mileage/litre as between 10 to 15 km, cost of petrol between Rs 45 to 55. In the first case, we are estimating distance, mileage, price as specific values or points. So this method of estimate is known as "point estimation" where as the second method is known as "interval estimation" since we are estimating the parameters enclosed in an interval.

Hypothesis testing main aim is to provide rules that lead to decision resulting in acceptance or rejection of statements about the population parameters. A man going to office on a cloudy day is in a dilemma whether to carry his umbrella or not. When the day begins he has to take the decision although he is not aware whether it rains or not. In any case, one of the following is going to happen.

(i) He takes his umbrella and it rains (wise decision)

(ii) He does not take his umbrella and it does not rain (wise decision again)

(iii) He takes umbrella and does not rain (wrong decision)

(iv) He does not take his umbrella and it rains (wrong decision).

The situations (iii) and (iv) are undesirable.

Test of goodness of fit is a statistical test which determines whether the sample data are in conformity with the hypothesized distribution. In fact this test literally tests how good the fit is. It is based on how close are the observed numbers and the numbers that we expect from the hypothesized distribution.

29.1 POINT ESTIMATION

Statistical Estimation

It is a part of statistical inference where a population parameter is estimated from the corresponding sample statistics. An estimate of the unknown true or exact value of the parameter or an interval in which the parameter is to be determined on the basis of sample data from the population.

Unbiased estimator

A statistic $\hat{\theta}$ is known as an unbiased estimator of the corresponding parameter θ if

$$E(\hat{\theta}) = E(\text{statistic}) = \text{parameter} = \theta$$

i.e., the mean of the sampling distribution of estimator equals to θ. Unbiasedness property is desirable, although not essential.

Point estimation

Point estimation of a parameter is a statistical estimation where the parameter is estimated by a single number (or value) from sample data.

Maximum Error of Estimate E

Since the sample mean estimate very rarely equals to the mean of population μ, a point estimate is generally accompanied with a statement of error which gives difference between estimate and the quantity to be estimated, the estimator. Thus error $= \bar{x} - \mu$. For large n, the random variable $\frac{\bar{x}-\mu}{\sigma/\sqrt{n}}$ is normal variate approximately. Then the inequality

$$-Z_{\alpha/2} \leq \frac{\bar{x}-\mu}{\sigma/\sqrt{n}} \leq Z_{\alpha/2}$$

is satisfied with probability $(1-\alpha)$

i.e., $|\bar{x}-\mu| \leq Z_{\alpha/2} \cdot \sigma/\sqrt{n}$

Confidence interval for μ

A $(1-\alpha)$ 100% confidence interval for μ is given by

$$\bar{x} - Z_{\alpha/2}\frac{\sigma}{\sqrt{n}} < \mu < \bar{x} + Z_{\alpha/2}\frac{\sigma}{\sqrt{n}}$$

So the maximum error of estimate E with $(1-\alpha)$ probability is given by

$$E = Z_{\alpha/2} \cdot \frac{\sigma}{\sqrt{n}}$$

Thus in the point estimation of population mean μ with sample mean $\bar{x}$ for a large random sample $(n \geq 30)$, one can assert with probability $(1-\alpha)$ that the error $|\bar{x}-\mu|$ will not exceed $Z_{\alpha/2}\sigma/\sqrt{n}$.

Sample Size

When α, E, σ are known, the sample size n is given by

$$n = \left(\frac{Z_{\alpha/2}\,\sigma}{E}\right)^2$$

when σ is unknown (or $n < 30$ small sample):
In this case, σ is replaced by s, the standard deviation of sample to determine E.
Thus the maximum error estimate

$$E = t_{\alpha/2}\frac{s}{\sqrt{n}}$$

with $(1-\alpha)$ probability.
Here t-distribution is with $n-1$ degrees of freedom.

Worked Out Examples

Example 1: The efficiency expert of a computer company tested 40 engineers to estimate the average time it takes to assemble a certain computer component, getting a mean of 12.73 minutes and s.d. of 2.06 minutes. (a) If $\bar{x} = 12.73$ is used as a point estimate of the actual average time required to perform the task, determine the maximum error with 99% confidence (b) construct 98% confidence intervals for the true average time it takes to do the job (c) with what confidence can we assert that the sample mean does not differ from the true mean by more than 30 seconds.

Solution: Here $\bar{x} = 12.73$, $s = 2.06$, For 99%, $Z_{\alpha/2} = 2.575$

a. Maximum error of estimate $E = Z_{\alpha/2}\frac{\sigma}{\sqrt{n}}$

$$= (2.575)\frac{(2.06)}{(\sqrt{40})} = 0.8387$$

b. For 98% confidence, $E = (2.33)\frac{(2.06)}{\sqrt{40}} = 0.758915$, 98% confidence interval limits are $\bar{x} \pm Z_{\alpha/2}\frac{\sigma}{\sqrt{n}}$

$$= \bar{x} \pm E = 12.73 \pm 0.7589$$

i.e., confidence interval is (11.97, 13.4889)

c. $\frac{30}{60}$ mts $= \frac{1}{2}$ minute $= E = Z_{\alpha/2}\frac{\sigma}{\sqrt{n}} = Z_{\alpha/2}\frac{2.06}{\sqrt{40}}$

$$\therefore \quad Z_{\alpha/2} = 1.5350$$

From normal table (A850), the area corresponding to $Z_{\alpha/2} = 1.5350$ is 0.4370. Then the area between $Z_{-\alpha/2}$ to $Z_{\alpha/2}$ is $2(0.4370) = 0.8740$.
Thus we can ascertain with 87.4% confidence.

Example 2: To estimate the average amount of time visitors take to move from one building to another in an office complex, the mean of a random sample of size n is used. Given $\sigma = 1.40$ minutes, determine how large should be the sample size if it is ascertained with 99% confidence that the error E is at most 0.25.

Solution:

$$n = \left(\frac{Z_{\alpha/2}\,\sigma}{E}\right)^2 = \left[\frac{(2.575)(1.40)}{0.25}\right]^2 = 207.98 \approx 208.$$

Example 3: Find the degree of confidence to assert that the average salary of school teachers is between Rs. 272 and Rs. 302 if a random sample of 100 such teachers revealed a mean salary of Rs. 287 with s.d. of Rs. 48.

Solution: Standard variable corresponding to Rs. 272 is

$$Z_1 = \frac{272 - 289}{48/\sqrt{100}} = -3.125$$

For Rs. 302 is $Z_2 = \frac{302-287}{48/\sqrt{100}} = 3.125$

Let X be the mean salary of teacher, then

$$P(272 < X < 302) = P(-3.125 < Z < 3.125)$$

$$= 2(.499) = 0.9982$$

Thus we can ascertain with 99.82% confidence.

EXERCISE

1. Using the mean of a random sample of size 150 to estimate the mean mechanical aptitude of mechanics of a large workshop and assuming $\sigma = 6.2$, what can we assert with 0.99 probability about the maximum size of the error.

   Hint: $n = 150$, $\sigma = 6.2$, $Z_{0.005} = 2.575$, $E = 2.575\left(\frac{6.2}{\sqrt{150}}\right) = 1.30$.

Ans: Can assert with 0.99 probability that the error will be at most 1.30.

2. Assuming that the population standard deviation is 0.3, calculate the (a) 95% and (b) 99% confidence intervals for the mean lead concentration in a river if the mean lead concentration recovered from a sample of lead measurements in 36 different locations is 2.6 gms/ml.

   Hint: $\overline{x} \pm Z_{\alpha/2}\frac{\sigma}{\sqrt{n}}$, $\overline{x} = 2.6$, $n = 36$, $\sigma = 0.3$, $Z_{0.025} = 1.96$, $Z_{0.005=2.575}$ from normal tables A11.

Ans: (a) $2.50 < \mu < 2.70$ (b) $2.47 < \mu < 2.73$

3. For the above Example 2, if we want to be 99% confident that our estimate of μ is off by less than 0.01, how large a sample should be chosen?

Ans: $n = \left[\frac{(2.575)(2.3)}{0.01}\right]^2 = 5967.5625 \approx 5968$

i.e., we can be 99% confident that a random sample of size 5968 will provide an estimate of $\overline{x}$ differing from μ by an amount less than 0.01.

4. (**σ unknown**) Determine 99% confidence interval for the mean of contents of soft drink bottles if contents of 7 such soft drink bottles are 10.2, 10.4, 9.8, 10.0, 9.8, 10.2, 9.6 ml.

   Hint: $\overline{x} = 10.0$, $s = 0.283$ are the mean and s.d. for given data. $t_{0.005} = 3.707$ with 6 dof

Ans: $10 \pm (3.707)\left(\frac{0.283}{\sqrt{7}}\right)$ i.e., 10 ± 2.64575 or $(7.354, 12.6458)$

5. The pulse rate of 50 yoga practitioners decreased on the average by 20.2 beats/minute with s.d. of 3.5. (a) If $\overline{x} = 20.2$ is used as a point estimate of the true average decrease in the pulse rate, what can we assert with 95% confidence about the maximum error E. (b) Construct 99% confidence intervals for the true average decrease in pulse rate.

Ans: **a.** $E = Z_{\alpha/2}\frac{\sigma}{\sqrt{n}} = Z_{0.025}\frac{3.5}{\sqrt{50}} = 1.96\frac{(3.5)}{\sqrt{50}} = 0.97015$

b. C.I.: $\overline{x} \pm Z_{\alpha/2}\frac{\sigma}{\sqrt{n}} = 20.2 \pm 2.575\frac{3.5}{\sqrt{50}}$; $19.229 < \mu < 21.17015)$

6. For the above Example 5, how large a sample should we take in order to assert with 95% confidence that the mean of the sample is off by at most 0.50?

Ans: $n = \left[\frac{(1.96)(3.5)}{0.50}\right]^2 = (13.72)^2 = 188.23 \approx 188.$

29.2 INTERVAL ESTIMATION

Point estimates rarely coincide with quantities they are intended to estimate. So instead of point estimation where the quantity to be estimated is replaced by a sigle value a better way of estimation is interval estimation, which determines an interval in which the parameter lies. For a given sample of values of a population, interval estimation consists of determining an interval whose two end points are computed from the sample data. The interval estimate thus constructed is such that the probability of the parameter lying in the interval can be determined. Accuracy of the estimate is indicated by the length of the interval. Thus **interval estimates** are intervals for which one can be $(1-\alpha)$ 100% confident that the parameter under investigation lies in this interval. Such an interval is known as **confidence interval** for the parameter with (having) $1-\alpha$ or $(1-\alpha)$ 100% degree of confidence. The two end points of the confidence interval are known as **confidence limits** or **fiducial limits** or **critical values** or confidence coefficients. **Confidence level** denoted by α is the percentage of confidence.

Consider a large random sample of size $n(\geq 30)$ from a population with *unknown* mean μ and *known* variance σ^2. Then the **large-sample confidence interval for μ**:

$$\bar{x} - Z_{\alpha/2}\frac{\sigma}{\sqrt{n}} < \mu < \bar{x} + Z_{\alpha/2}\frac{\sigma}{\sqrt{n}}$$

Note 1: Confidence interval is exact for random samples from normal populations.

Note 2: Confidence interval provides good approximation for large samples ($n \geq 30$) from non-normal populations also.

Small-sample confidence interval for μ: (when $n < 30$, assuming sampling from normal population)

$$\bar{x} - t_{\alpha/2}\frac{s}{\sqrt{n}} < \mu < \bar{x} + t_{\alpha/2}\frac{s}{n}.$$

Confidence level α	99.73%	99%	98%	96%	95.5%	95%	90%	80%	68.27%	50%
$Z_{\alpha/2}$	3.00	2.58	2.33	2.05	2.00	1.96	1.645	1.28	1.00	0.675

Worked Out Examples

Example: A random sample of 10 ball bearings produced by a company have a mean diameter of 0.5060 cm with s.d 0.004 cm. Find the maximum error estimate E and 95% confidence interval for the actual mean dia of ball bearings produced by this company assuming sampling from normal population.

Solution: Sample size $= n = 10 < 30$, so use t-distribution (small sampling).
Maximum error estimate at 95% confidence is

$$E = t_{\alpha/2}\frac{\sigma}{\sqrt{n}} = (2.262)\frac{(0.004)}{\sqrt{10}} = 0.00286$$

since $t_{0.025}$ with $n-1 = 10-1 = 9$ dof is 2.262.
95% confidence interval limits are $\bar{x} \pm t_{\alpha/2}\frac{\sigma}{\sqrt{n}}$
$= 0.5060 \pm (2.262)\frac{(0.004)}{\sqrt{10}}$.
95% confidence interval is (0.5031, 0.5089).

Exercise

1. If on the average, the test strips painted across heavily travelled roads in 15 different locations, disappeared after they had been crossed by 146692 cars with s.d. 14380 cars, calculate 99% confidence intervals for the true average number of cars it takes to wear off the paint, assuming normal population.

   Hint: $n = 15 < 30$, $t_{0.005}$ with 14 dof is 2.977, $\bar{x} = 146692$, $\sigma = 14380$,
   C.I. $(146692 \pm \frac{(2.977)(14380)}{\sqrt{15}})$.

Ans: $135639 < \mu < 157745$

2. A random sample of 20 fuses subjected to overload has mean time for blow of 10.63 minutes with s.d. of 2.48 mt. What can we assert with 95% confidence about the maximum error if we use $\bar{x} = 10.63$ mts as a point estimate of true average it takes such fuses for blow when subjected to overload.

Ans: $E = t_{\alpha/2}\frac{\sigma}{\sqrt{n}} = 2.093\frac{(2.48)}{\sqrt{20}} = 1.16$ mt

3. Construct a 99% confidence interval for the true mean weight loss if 16 persons on diet control after one month had a mean weight loss of 3.42 kgs with s.d. of 0.68 kgs.

   Hint: $n = 16$, $\bar{x} = 3.42$, $s = 0.68$,
   $t_{0.005} = 2.947$ for 15 dof
   $3.42 \pm 2.947\frac{(0.68)}{\sqrt{16}}$.

Ans: $2.92 < \mu < 3.92$

29.3 BAYESIAN ESTIMATION

Personal or subjective probability is the new concept introduced in Bayesian methods. Also, parameters are viewed as random variables in Bayesian methods.

To estimate the mean of a population, μ is treated as a random variable whose distribution is indicative of the "strong feelings" or assumption of a person about the possible value of μ. Let μ_0 and σ_0 be the mean and standard deviation of such a subjective "prior distribution".

Bayesian Estimation

Combining the prior feelings about the possible values of μ with direct sample evidence, the "posterior" distribution of μ in Bayesian estimation is approximated by normal distribution with

$$\mu_1 = \frac{n\bar{x}\sigma_0^2 + \mu_0\sigma^2}{n\sigma_0^2 + \sigma^2} \quad \text{and}$$

$$\sigma_1 = \sqrt{\frac{\sigma^2\sigma_0^2}{n\sigma_0^2 + \sigma^2}}$$

Here μ_1 and σ_1 are known as the mean and standard deviation of the posterior distribution. In the com putation of μ_1 and σ_1, σ^2 is assumed to be *known*. When σ^2 is unknown, which is generally the case, σ^2 is replaced by sample variance s^2 provided $n \geq 30$ (large sample).

Bayesian interval for μ:

A $(1 - \alpha)100\%$ Bayesian interval for μ is given by

$$\mu_1 - Z_{\alpha/2}\sigma_1 < \mu < \mu_1 + Z_{\alpha/2}\sigma_1.$$

WORKED OUT EXAMPLES

Example: A professor's feelings about the mean mark in the final examination in "probability" of a large group of students is expressed subjectively by normal distribution with $\mu_0 = 67.2$ and $\sigma_0 = 1.5$. (a) If the mean mark lies in the interval (65.0, 70.0), determine the prior probability the professor should assign to the mean mark. (b) Find the posterior mean μ_1 and posterior s.d. σ_1 if the examination is conducted on a random sample of 40 students yielding mean 74.9 and s.d. 7.4. Use $s = 7.4$ as an estimate of σ. (c) Determine the posterior probability which he will thus assign to the mean mark being in the interval (65.0, 70.0), using results obtained in (b)· (d) construct a 95% Bayesian interval for μ.

Solution:

a. Here $\mu_0 = 67.2$, $\sigma_0 = 1.5$, $n = 40$

standard variable corresponding to 65.0 is

$$Z_1 = \frac{65.0 - 67.2}{1.5} = -1.466,$$

Similarly, $$Z_2 = \frac{70 - 67.2}{1.5} = 1.866$$

Let X be the mean mark obtained in the final examination.

$$\text{Prior probability} = P(65 < X < 70)$$
$$= P(-1.47 < Z < 1.87)$$
$$= 0.4292 + 0.4693 = 0.8985$$

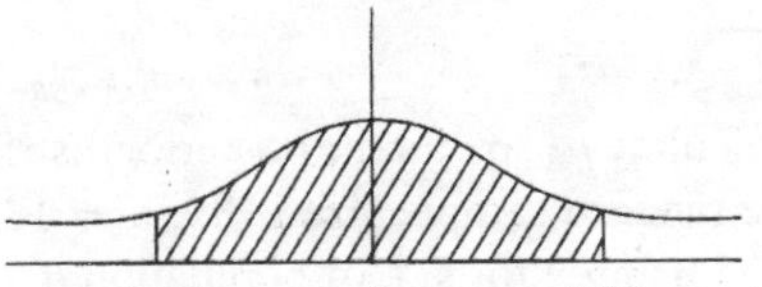

Fig. 29.1

b. Here $\bar{x} = 74.9$, $\sigma = s = 7.4$

$$\text{posterior mean } \mu_1 = \frac{n\bar{x}\sigma_0^2 + \mu_0\sigma^2}{n\sigma_0^2 + \sigma^2}$$

$$= \frac{40(74.9)(1.5)^2 + (67.2)(7.4)^2}{40(1.5)^2 + (7.4)^2} = 71.987 \approx 72$$

posterior s.d.

$$\sigma_1 = \sqrt{\frac{\sigma^2\sigma_0^2}{n\sigma_0^2 + \sigma^2}} = \sqrt{\frac{(7.4)^2(1.5)^2}{40(1.5)^2}} = 0.922568 \approx 0.923$$

c. Here $\mu_1 = 72$, $\sigma_1 = 0.923$.

Standard variable corresponding to 650 is

$$Z_1 = \frac{65 - 72}{0.923} = -7.5839,$$

similarly, corresponding to 70.0 is

$$Z_2 = \frac{70 - 72}{0.923} = -2.16684$$

Posterior probability $= P(65 < X < 70)$

$= P(-7.584 < Z < -2.167)$

$= 0.5 - 0.4850 = 0.0150$

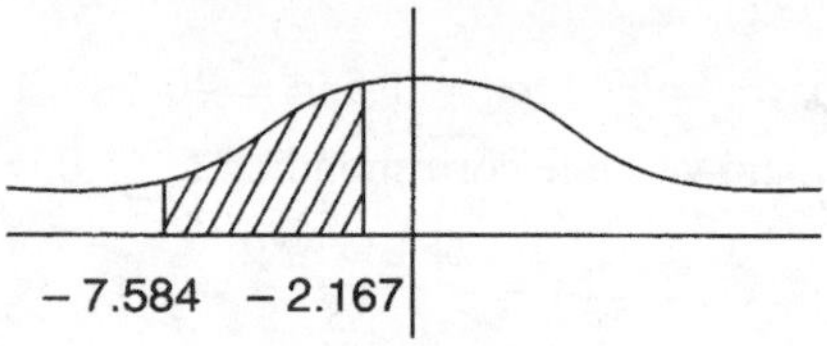

Fig. 29.2

d. 95% Bayesian interval limits are

$$\mu_1 \pm Z_{\alpha/2}\sigma_1 = 71.987 \pm (1.96)(0.922568)$$

Thus the Bayesian interval is

$$(70.17876, 72.909568).$$

EXERCISE

1. Calculate μ_1, σ_1 for the posterior distribution if the random sample size is 80, $\overline{x} = 18.85$, $s = 5.55$ using s for s.d. of population σ.

Ans: $\mu_1 = 18.77, \sigma_1 = 0.60$

2. An insurence agent feelings about the average monthly commission of insurance policies may be described by means of normal distribution with $\mu_0 =$ Rs. 3800 and $\sigma_0 =$ Rs. 260. (a) What probability is the agent thus assigning to the true average monthly commission being in the interval of Rs. 3,500 to Rs. 4000. (b) How does the probability in part (a) is affected if the mean commission is Rs. 3702 with s.d. Rs. 390 for 9 months? Use $s = 390$ as an estimate of σ.

Hint: $P(3500 < X < 4000) = P(-1.154 < Z < 0.77) = 0.3749 + 0.2794.$

Ans: **a.** 0.6543

Hint: $\overline{x} = 3702$, $s = \sigma = 390$, $n = 9$, $\mu_0 = 3800$, $\sigma_0 = 260$

$$\mu_1 = \frac{n\overline{x}\sigma_0^2 + \mu_0\sigma^2}{n\sigma_0^2 + \sigma^2} = 3721.6 \approx 3722,$$

$$\sigma_1 = \sqrt{\frac{\sigma^2\sigma_0^2}{n\sigma_0^2 + \sigma^2}} = 116.3 \approx 116$$

$P(3500 < X < 4000) = P(-1.91 < Z < 2.40) = 0.4719 + 0.4918 = 0.9637$

Ans: **b.** 0.9637

3. The mean mark in mathematics in common entrance test will vary from year to year. If this variation of the mean mark is expressed subjectively by a normal distribution with mean $\mu_0 = 72$ and variance $\sigma_0^2 = 5.76$. (a) What probability can we assign to the actual mean mark being somewhere between 71.8 and 73.4 for the next year's test. (b) construct a 95% Bayesian interval for μ if the test is conducted for a random sample of 100 students from the next incoming class yielding a mean mark of 70 with s.d. of 8. (c) What posterior probability should we assign to the event of part (a).

Hint: $P(71.8 < X < 73.4) = P(-0.083 < Z < 0.583)$

$= 0.0319 + 0.2190 = 0.2509$

Ans: **a.** 0.2509

Hint: $n = 100$, $\overline{x} = 70$, $\sigma_0 = 2.4$, $\mu_0 = 72$, $\sigma = 8$, so

$\mu_1 = 70.2, \sigma_1 = 0.7589$

C.I.: $\mu_1 \pm Z_{\alpha/2}(\sigma_1) = 70.2 \pm 1.96(0.7589)$

Ans: **b.** $68.71 < \mu < 71.69$

Hint: $P(71.8 < X < 73.4) = P(2.105 < Z < 4.21) = 0.5 - 0.4821 = 0.0179$

since $Z_1 = \frac{71.8-70.2}{0.76} = 2.105,$

$Z_2 = \frac{73.4-70.2}{0.76} = 4.2105$

Ans: **c.** 0.0179

4. A producer of TV's believes from past experience that the mean length of life of TV's μ is a normal random variable with mean $\mu_0 = 800$ hours and standard deviation $\sigma_0 = 10$ hours. It is known that TV's have mean length of life that is approximately normally distributed with a standard deviation of 100 hours. Construct a 95% Bayesian interval for μ if a random sample of 25 TV's has an average life of 780 hours.

Hint: $\mu_1 = \frac{(25)(780)(10)^2+(800)(100)^2}{25(10)^2+(100)^2} = 796$

$$\sigma_1 = \sqrt{\frac{(10)^2(100)^2}{25(10)^2+(100)^2}} = \sqrt{80},$$

B.I.: $780 \pm 1.96\left(\frac{100}{\sqrt{25}}\right)$

Here $Z_{\alpha/2} = 1.96, n = 25, \sigma = 100,$

$\sigma_0 = 10, \bar{x} = 780.$

Ans: $740.8 < \mu < 819.2$

29.4 TEST OF HYPOTHESIS

The principal objective of statistical inference is to draw inferences (or generalize) about the population on the basis of data collected by sampling from the population. Statistical inference consists of two major areas, estimation and tests of hypothesis. Estimation was discussed in Sections 29.1, 29.2 and 29.3. In tests of hypothesis, a postulate or conjecture or statement about a parameter of the population is tested for its validity or truthfulness.

Statistical decisions

Statistical decisions are decisions or conclusions about the population parameters on the basis of a random sample from the population.

Statistical Hypothesis

It is an assumption or conjecture or guess about the parameter(s) of population distribution(s). The statistical hypothesis is established before hand and may or may not be true. When more than one population is considered, statistical hypothesis consists of relationship between the parameters of the populations.

Null Hypothesis

It is (N.H.) denoted by H_0 is the statistical hypothesis which is to be actually tested for acceptance or rejection. N.H. is the hypothesis which is tested for possible rejection under the assumption that it is true (R.A. Fisher).

Alternative Hypothesis

(A.H.) denoted by H_1, is any hypothesis other than the null hypothesis. Neyman originated the concept of alternative hypothesis.

Test of Hypothesis

Test of hypothesis or test of significance or rules of decision is a procedure to decide whether to accept or reject the (null) hypothesis. This test determines whether observed samples differ significantly from expected results. Acceptance of hypothesis merely indicates that the data do not give sufficient evidence to refute the hypothesis. Whereas, rejection is a firm conclusion where the sample evidence refutes it.

When N.H. is accepted, result is said to be non-significant i.e., observed differences are due to 'chance' caused by the process of sampling. When N.H. is rejected (i.e., A.H. is accepted) the result is said to be significant. Thus test of hypothesis decides whether a statement concerning a parameter is true or false instead of estimating the value of the parameter. Since the test is based on sample observations, the decision of acceptance or rejection of the null hypothesis is always subjected to some error i.e., some amount of risk.

Types of errors in test of hypothesis:

	Accept H_0	Reject H_0
H_0 is true	correct decision	Type I error
H_0 is false	Type II error	correct decision

Type I error involves rejection of null hypothesis when it should be accepted (as true).

Type II error involves acceptance of the null hypothesis when it is false and should be rejected.

Level of significance

(L.O.S.) of a test denoted by α is the probability of committing type I error. Thus L.O.S. measures the amount of risk or error associated in taking decisions. It is costomary to fix α before sample information is collected and to choose (take) generally α as 0.05 or 0.01. L.O.S. $\alpha = 0.01$ is used for high precision and $\alpha = 0.05$ for moderate precision. L.O.S. is also expressed as percentage. Thus L.O.S. $\alpha = 5\%$ means there are 5 chances in 100 that N.H. is rejected when it is true or one is 95% confident that a right decision is made. L.O.S. is also known as the **size of the test.** Thus $\alpha =$ probability of committing type I error $= P$ (reject H_0/H_0) $= \alpha$ and $\beta =$ prob (type II error)$= P$ (accept H_0/H_1) $= \beta$.

Power of the test is computed as $1 - \beta$.

Note 1: When the size of the sample is increased, the probability of committing both types of errors I and II i.e., α and β can be reduced simultaneously.

Note 2: α and β are known as producer's risk and consumer's risk respectively.

Note 3: When both α and β are small, the test procedure is good one giving good chance of making the correct decision.

Simple Hypothesis

It is a statistical hypothesis which completely specifies an exact parameter. Null hypothesis is always a simple hypothesis stated as an equality specifying an exact value of the parameter (includes any value not stated by A.H.).

Examples:

1. N.H. $= H_0 : \mu = \mu_0$

   i.e., population mean equals to a specified constant μ_0.

2. N.H. $= H_0 : \mu_1 - \mu_2 = \delta$

   i.e., the difference between the sample means equals to a constant δ.

Composite Hypothesis

It is stated in terms of several possible values i.e., by an inequality.

Alternative Hypothesis

It is a composite hypothesis involving statements expressed as inequalities such as $<, >$ or $\neq$.

Examples:

1. A.H.: $H_1; \mu > \mu_0$
2. A.H.: $H_1 : \mu < \mu_0$
3. A.H.: $H_1 : \mu \neq \mu_0$

Critical Region (C.R.)

In any test of hypothesis, a test statistic S^*, calculated from the sample data, is used to accept or reject the null hypothesis of the test. Consider the area under the probability curve of the sampling distribution of the test statistic S^* which follows some known (given) distribution. This area under the probability curve is divided into to dichotomous regions, namely the region of rejection (significant region or critical region) where N.H. is rejected, and the region of acceptance (non-significant region or non-critical region) where N.H. is accepted. Thus **critical region** is the region of rejection of N.H. The area of the critical region equals to the level of significance α. Note that C.R. always lies on the tail (s) of the distribution. Depending on the nature of A.H., C.R. may lie on one side or both sides of the tails (s).

Critical value(s) or significant value(s)

It is (area) the value of the test statistic S_α^* (for given level of significance α) which divides (or separates) the area under the probability curve into critical (or rejection) region and non-critical (or acceptance) region.

One tailed test (O.T.T.) and two tailed test (T.T.T.)

Right one tailed test (R.O.T.T.): When the alternative hypothesis (A.H.): H_1 is of the greater than type i.e., $H_1 : \mu > \mu_0$ or $H_1 : \sigma_1^2 > \sigma_2^2$ etc., then the entire criticial region of area α lies on the right side tail of the probability density curve as shown shaded in the Fig. 29.3. In such case, the test of hypothesis (T.O.H.) is known as **right one tailed test.**

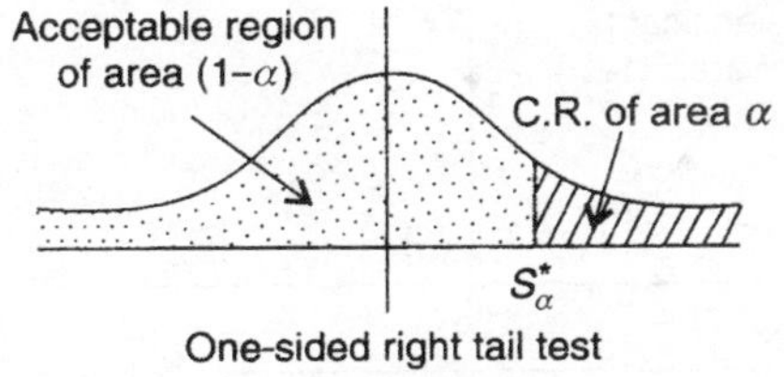

Fig. 29.3

Left one tailed test (L.O.T.T.)

When the A.H.: H_1 is of the less than type i.e., H_1 $\mu_1 < \mu_0$ or $H_1 : \sigma_1^2 < \sigma_2^2$ etc. then the entire C.R. of area α lies on the left side tail of the curve as shown in Fig. 29.4.

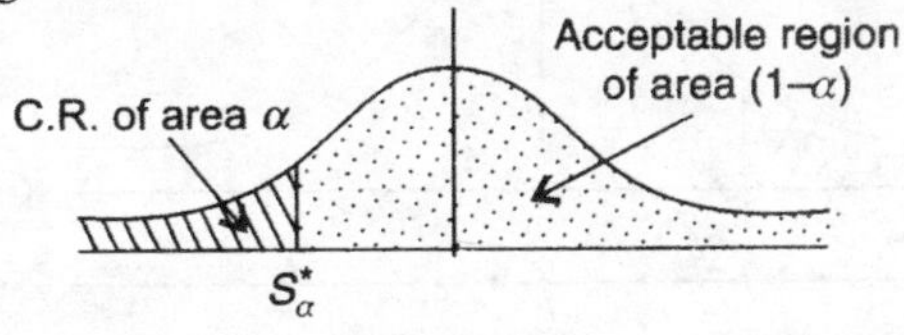

Fig. 29.4

Two tailed test (T.T.T.)

If A.H. is of the not equals type i.e., $H_1 : \mu_1 \neq \mu_2$ or $H_1 : \sigma_1 \neq \sigma_2$ etc. then the C.R. lies on both sides of the right and left tails of the curve such that the C.R. of area $\frac{\alpha}{2}$ lies on the right tail and C.R. of area $\frac{\alpha}{2}$ lies on the left tail, as shown in Fig. 29.5.

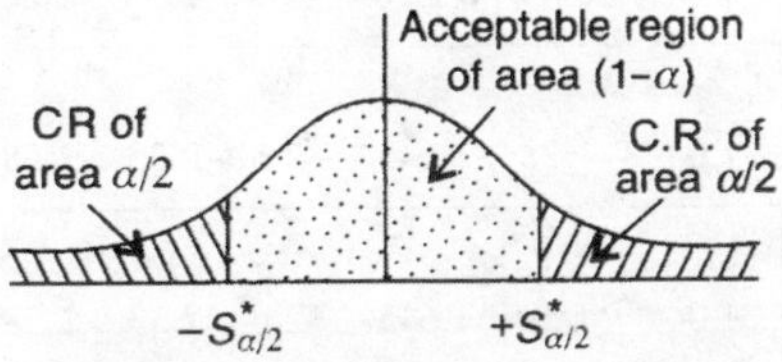

Fig. 29.5

Thus the test of hypothesis or test of significance or rule of decision consists of the following six steps.

1. Formulate N.H.: H_0.
2. Formulate A.H.: H_1.
3. Choose L.O.S.: α.
4. C.R.: is determined by the critical value S_α^* and the kind of A.H. (based on which the test is R.O.T.T. or L.O.T.T. or T.T.T.).
5. Compute the test statistic S^* using the sample data.
6. Decision: Accept or reject N.H. depending on the relation between S^* and S_α^*.

P-Value

In tests of hypothesis, preselection of a significance level α does not account for values of test statistics that are "close" to the critical region. Thus a test statistic value that is non-significant say for $\alpha = 0.05$ may become significant for $\alpha = 0.01$. In applied statistics, *P*-value approach is designed to give the user an alternative (in terms of probability) to a mere "reject" or "do not reject" conclusion.

***P*-Value** is the lowest level (of significance) at which the observed value of the test statistic is significant.

In the significance testing by *P*-Value approch, α is not pre-determined but the conclusion is based on the size of the *P*-Value which is computed using the value of test statistic.

29.5 TEST OF HYPOTHESIS CONCERNING SINGLE POPULATION MEAN μ: (WITH KNOWN VARIANCE σ^2: LARGE SAMPLE)

Let μ and σ^2 be the mean and variance of a population from which a random sample of size n is drawn. Let $\overline{x}$ be the mean of the sample. Then for large samples ($n \geq 30$), from central theorem it follows that the sampling distribution of $\overline{x}$ is approximately normally distributed with mean $\mu_{\overline{x}} = \mu$ and $\sigma_{\overline{x}}^2 = \frac{\sigma^2}{n}$.

The test statistic for single mean with known variance is $Z = \frac{\overline{x}-\mu_0}{\sigma/\sqrt{n}}$.

To test whether the population mean μ equals to a specified constant μ_0 or not, formulate the test of hypothesis as follows:

1. N.H.: $\mu = \mu_0$.
2. A.H.: $\mu \neq \mu_0$.
3. L.O.S.: α.
4. C.R.: Since the A.H. is a not equal to type, a T.T.T. is considered. For given α, critical values

$-Z_{\alpha/2}$ and $+Z_{\alpha/2}$ are determined from normal table since normal distribution is assumed. For example, for $\alpha = 5\%$ or 0.05 from normal table $-Z_{0.025} = -1.96$ and $Z_{0.025} = 1.96$. Thus the critical region consists of the two shaded regions in Fig. 29.6 i.e., reject N.H.: H_0 if $Z < -Z_{\alpha/2}$ or $Z > Z_{\alpha/2}$.

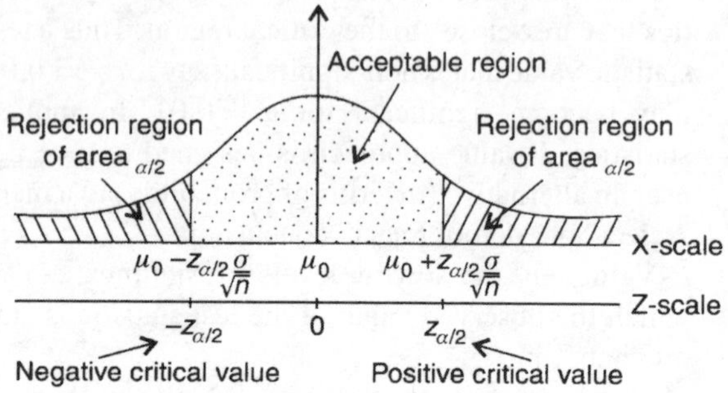

Fig. 29.6

5. Compute the test statistic Z, denoted by Z_{cal} or simply Z by

$$Z = \frac{\overline{x} - \mu_0}{\sigma/\sqrt{n}}$$

Here $\overline{x}$, the mean of the sample of size n, is calculated from the sample data.

6. Conclusion: Reject H_0 if Z_{cal} or Z falls in the critical region i.e., observed sample statistic is probably significant or highly significant at α level. Otherwise accept H_0 (if $-Z_{\alpha/2} < Z < Z_{\alpha/2}$).

Note 1: Suppose the A.H. is $H_1 : \mu > \mu_0$. Then the critical region is given by $Z > Z_\alpha$ since we consider a right one tail test is this case, i.e., reject H_0 if $Z > Z_\alpha$ otherwise accept H_0 (if $Z < Z_\alpha$) (see Fig. 29.7).

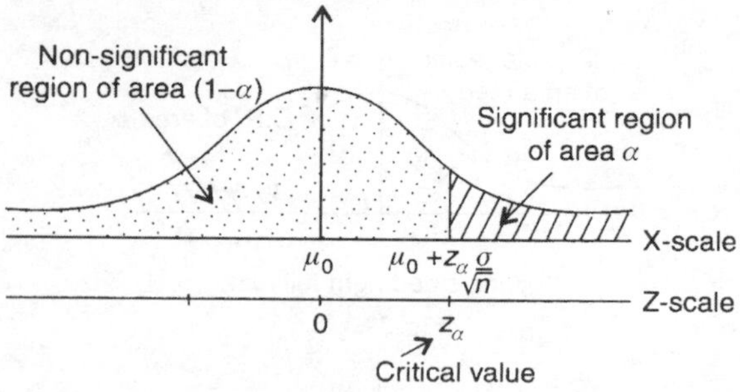

Fig. 29.7

Note 2: If A.H. is $H_1 : \mu < \mu_0$ then consider a left one tail test with C.R. given by $Z < -Z_\alpha$ as shown in Fig. 29.8, i.e., reject H_0 is $Z < -Z_\alpha$ otherwise accept H_0 (if $Z > -Z_\alpha$).

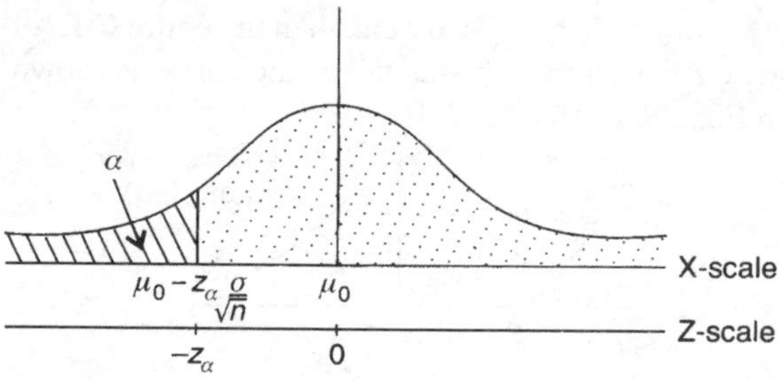

Fig. 29.8

Note 3: Reference table of critical values for a given L.O.S. α for T.T.T., R.O.T.T. and L.O.T.T.

α% α	15% 0.15	10% 0.1	5% 0.05	4% 0.04	1% 0.01	.5% 0.005	.2% 0.002
$-Z_{\alpha/2}$ and $+Z_{\alpha/2}$ for T.T.T.	−1.44 and 1.44	−1.645 and 1.645	−1.96 and 1.96	−2.06 and 2.06	−2.58 and 2.58	−2.81 and 2.81	−3.08 and 3.08
$-Z_\alpha$ for L.O.T.T.	−1.04	−1.28	−1.645	−2.6	−2.33	−2.58	−2.88
Z_α for R.O.T.T.	1.04	1.28	1.645	2.6	2.33	2.58	2.88

Note 4: For large sample $n \geq 30$, even if σ is *unknown*, σ can be replaced by sample variances (which can be computed from sample information).

Worked Out Examples

Test of hypothesis: For one mean

Example 1: The length of life X of certain computers is approximately normally distributed with mean

800 hours and standard deviation 40 hours. If a random sample of 30 computers has an average life of 788 hourse, test the null hypothesis that $\mu = 800$ hours against the alternative that $\mu \neq 800$ hours at (a) 0.5% (b) 1% (c) 4% (d) 5% (e) 10% (f) 15% level of significance.

Solution:

Case a:

1. Null hypothesis: $\mu = 800$ hours
2. Alternate hypothesis : $\mu \neq 800$ hours
3. α level of significance = .5% = .005 (case (a))
4. Critical region: since alternate hypothesis is $\neq$ type, the test is two tailed and the critical region is

$$-2.81 < Z < 2.81$$

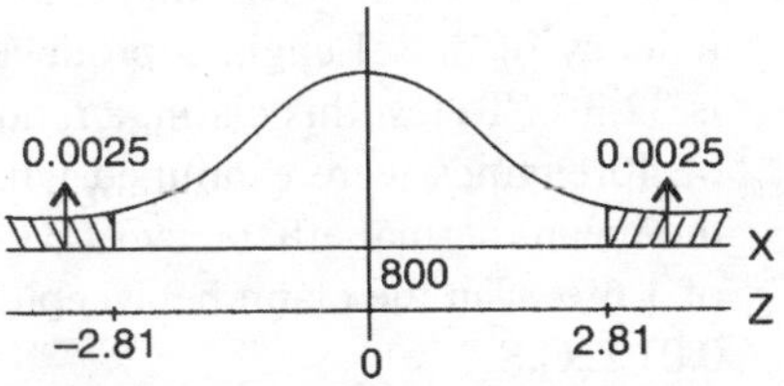

Fig. 29.9

5. Calculation of statistic:

Here $\bar{x}$ = mean of the sample = 788

n = sample size = 30

standard deviation $\sigma = 40$,

so $Z = \frac{\bar{x}-\mu_0}{\sigma/\sqrt{n}} = \frac{788-800}{40/\sqrt{30}} = -1.643$

6. Decision: Accept the null hypothesis H_0 since

$$Z = -1.643 > -2.81 = Z_{\alpha/2} = Z_{0.0025}$$

Case b: α = level of significance = 1% = 0.01 critical region

$$-2.58 < Z < 2.58$$

Desision: Accept H_0 since $Z = -1.643 > -2.58 = Z_{\alpha/2} = Z_{0.005}$ (Fig. 29.10)

Case c: $\alpha = 4\% = 0.04$, C.R.: $-2.06 < Z < 2.06$. Accept H_0 since $Z = -1.643 > -2.06$ (Fig. 29.11)

Case d: $\alpha = 5\% = 0.05$, C.R.: $-1.96 < Z < 1.96$. Accept H_0 since $Z = -1.643 > -1.96$ (Fig. 29.12)

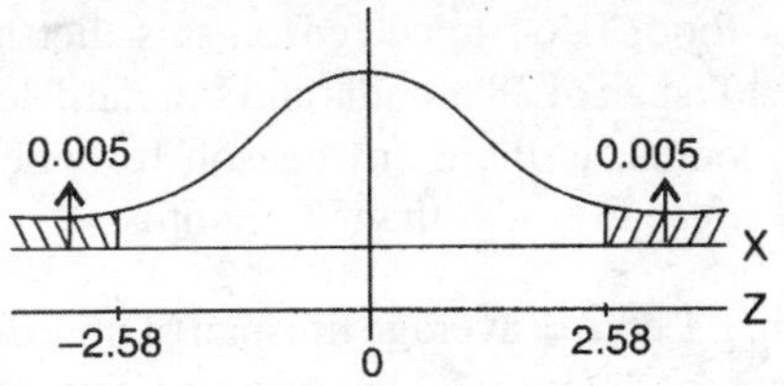

Fig. 29.10

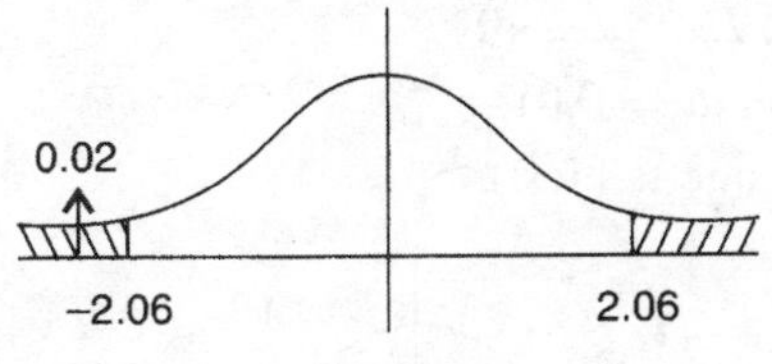

Fig. 29.11

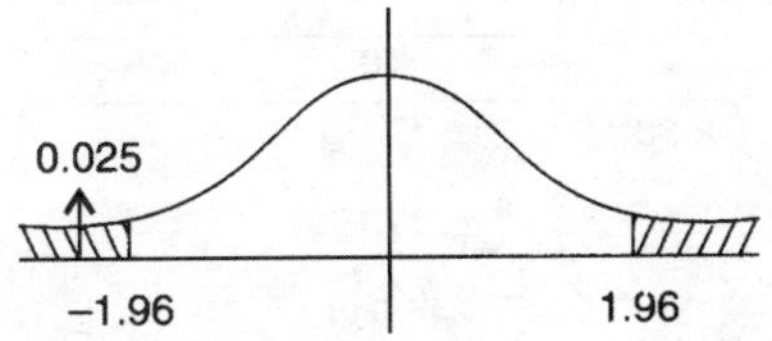

Fig. 29.12

Case e: $\alpha = 10\% = 0.10$, C.R.: $-1.645 < Z < 1.645$. Accept H_0 since $Z = -1.643 > -1.645$

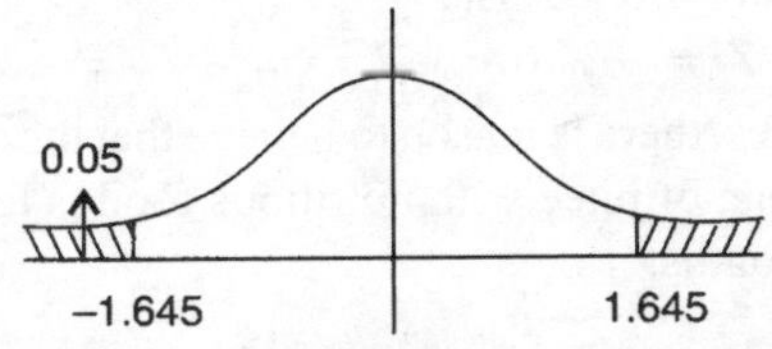

Fig. 29.13

Case f: $\alpha = 15\% = 0.15$, C.R.: $-1.44 < Z < 1.44$. Reject H_0 since $Z = -1.643 < -1.44$

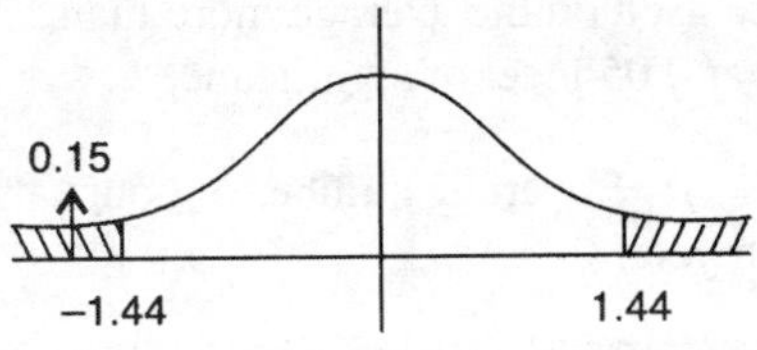

Fig. 29.14

Test is significant.

Example 2: Mice with an average lifespan of 32 months will live upto 40 months when fed by a certain nutrious food. If 64 mice fed on this diet have an average lifespan of 38 months and standard deviation of 5.8 months, is there any reason to believe that average lifespan is less than 40 months.

Solution: Let μ = average lifespan of mice fed with nutritious food. Use 0.01 level of significance

1. N.H.: $H_0 : \mu = 40$ months
2. A.H.: $H_1 : \mu < 40$
3. L.O.S.: $\alpha = 0.01$
(Left one-tail test)

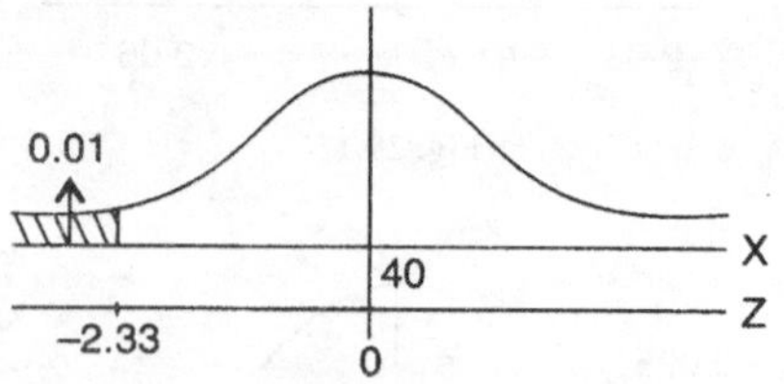

Fig. 29.15

4. C.R.: $Z < -Z_\alpha = -Z_{0.01} = -2.33$
5. Computation: Here $\overline{x} = 38, \sigma = 5.8, n = 64$

$$Z = \frac{38 - 40}{5.8/\sqrt{64}} = -2.76$$

6. Decision: Reject H_0,
since $Z = -2.76 < -2.33 = -Z_\alpha = -Z_{0.01}$
i.e., yes, there is reason to believe that the average lifespan of mice with nutritious food is less than 40 months.

Example 3: A machine runs on an average of 125 hours/year. A random sample of 49 machines has an annual average use of 126.9 hours with standard deviation 8.4 hours. Does this suggest to believe that machines are used on the average more than 125 hours annually at 0.05 level of significance?

Solution: μ = average number of hours a machine runs in an year.

1. $H_0 \cdot \mu = 125$ hours/year
2. $H_1 \cdot \mu > 125$
3. L.O.S.: $\alpha = 0.05$
4. C.R.: $Z > Z_\alpha = Z_{0.05} = 1.64$
5. Calculation: $Z = \frac{\overline{x}-\mu}{\sigma/\sqrt{n}} = \frac{126.9-125}{8.4/\sqrt{49}} = 1.58$

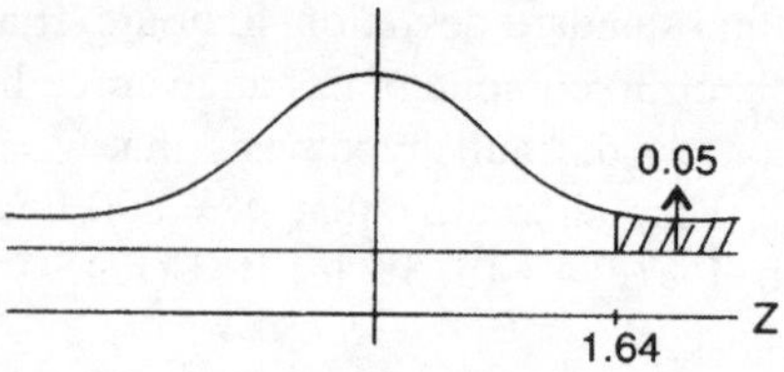

Fig. 29.16

6. Decision: Accept H_0 since $Z = 1.58 < 1.64 = Z_{0.05}$ i.e., can not believe that machine works more than 125 hours in an year.

EXERCISE

Test of hypothesis: For one mean

1. A company claims that the mean thermal efficiency of diesel engines produced by them is 32.3%. To test this claim, a random sample of 40 engines were examined which showed the mean thermal efficiency of 31.4% and s.d of 1.6%. Can the claim be accepted or not, at 0.01 L.O.S.?

Hint: $Z = \frac{\overline{x}-\mu_0}{\sigma/\sqrt{n}} = \frac{31.4-32.3}{1.6/\sqrt{40}} = -3.56 < -2.81 = Z_\alpha = Z_{0.01}$

Ans: Reject N.H.: $H_0 \cdot \mu_0 = 32.3$ from a T.T.T. against A.H.: $H_1 : \mu_0 \neq 32.3$

2. It has previously been recorded that the average depth of ocean at a particular region is 67.4 fathoms. Is there reason to believe this at 0.01 L.O.S. if the readings at 40 random locations in that particular region showed a mean of 69.3 with s.d. of 5.4 fathoms?

Hint: $Z = \frac{69.3-67.4}{5.4/\sqrt{40}} = 2.23 < 2.58 = Z_{\alpha/2} = Z_{0.005}$

Ans: Accept $H_0 : \mu_0 = 67.4$ against A.H.: $H_1; \mu_0 \neq 67.4$ by a T.T.T.

3. To determine whether the mean breaking strength of synthetic fibre produced by a certain company is 8 kg or not, a random sample of 50 fibres were tested yielding a mean breaking strength of 7.8 kg. If s.d. is 0.5 kg, test at 0.01 L.O.S.

Hint: $Z = \frac{78-8}{0.5/\sqrt{50}} = -2.83 > -2.575 = Z_{\alpha/2} = Z_{0.005}$

Ans: Reject $H_0 : \mu = 8$, accept A.H.: $H_1 : \mu \neq 8$ by a T.T.T.

4. Can it be concluded that the average lifespan of Indian is more than 70 years if a random sample of 100 Indians has an average lifespan of 71.8 years with a s.d. of 8.9 years.

Hint: $Z = \frac{71.8-70}{8.9/\sqrt{100}} = 2.02 > 1.645 = Z_\alpha = Z_{0.05}$ by right O.T.T.

Ans: Yes, average lifespan is more than 70 years.

5. A company producing computers states that the mean lifetime of its computers is 1600 hours. Test this claim at 0.01 L.O.S. against the A.H.: $\mu < 1600$ hours if 100 computers produced by this company has mean lifetime of 1570 hours with s.d. of 120 h.

Hint: $Z = \frac{(1570-1600)}{120/\sqrt{100}} = -2.50 < -2.33 = Z_{0.01}$ by left O.T.T.

Ans: Reject H_0 i.e., claim is not tenable

6. A manufacturer of tyres guarantees that the average lifetime of its tyres is more than 28000 miles. If 40 tyres of this company tested, yields a mean lifetime of 27463 miles with s.d. of 1348 miles, can the guarantee be accepted at 0.01 L.O.S.?

Hint: $Z = \frac{27463-28000}{1348/\sqrt{40}} = -2.52 < -2.33 = Z_{0.01}$ by left O.T.T.

Ans: No, tyres run for < 28000 miles.

29.6 TEST OF HYPOTHESIS CONCERNING TWO MEANS

When variances σ_1 and σ_2 are known or large samples

Let $\overline{x}_1$ be the mean of a random sample of size n_1 drawn from a population with mean μ_1 and variance σ_1^2. Let $\overline{x}_2$ be the mean of an independent random sample of size n_2 drawn from another population with mean μ_2 and variance σ_2^2. To test the hypothesis for difference of means, consider the null hypothesis $\mu_1 - \mu_2 = \delta =$ given constant. So when $\delta = 0$, there is no difference between the means i.e., the two populations have the same means. If $\delta \neq 0$, the means of the two populations are different. In these cases, the test statistic will depend on the difference between the sample means $\overline{x}_1 - \overline{x}_2$ and is given by

$$Z = \frac{(\overline{x}_1 - \overline{x}_2) - \delta}{\sigma_{\overline{x}_1 - \overline{x}_2}}$$

Here it is assumed that both samples are drawn from normal populations with known variances.

Here Z follows standard normal distribution. If the two populations are infinite then the variance of the sampling distribution of the difference between the sample means is

$$\sigma^2_{\overline{x}_1 - \overline{x}_2} = \frac{\sigma_1^2}{n_1} + \frac{\sigma_2^2}{n_2}$$

Substituting we have the statistic for test concerning difference between two means as

$$Z = \frac{(\overline{x}_1 - \overline{x}_2) - \delta}{\sqrt{\frac{\sigma_1^2}{n_1} + \frac{\sigma_2^2}{n_2}}}$$

Note: When the two variances σ_1^2 and σ_2^2 are unknown, they can be replaced by sample variances s_1^2 and s_2^2 provided both the samples are large ($n_1, n_2 \geq 30$). In this case the test statistic is

$$Z = \frac{(\overline{x}_1 - \overline{x}_2) - \delta}{\sqrt{\frac{s_1^2}{n_1} + \frac{s_2^2}{n_2}}}$$

The critical regions for testing $\mu_1 - \mu_2 = \delta$ are:

1. A.H.: $\mu_1 - \mu_2 \neq \delta$, Reject H_0 if $Z < -Z_{\alpha/2}$ or $Z > Z_{\alpha/2}$
2. A.H.: $\mu_1 - \mu_2 > \delta$, Reject H_0 if $Z > Z_\alpha$
3. A.H.: $\mu_1 - \mu_2 < \delta$, Reject H_0 if $Z < -Z_\alpha$.

The A.H. 2 and 3 are used to determine whether one product (population) is better than (superior to) the other product.

WORKED OUT EXAMPLES

Example 1: In a random sample of 100 tube lights produced by company A, the mean lifetime (mlt) of tube light is 1190 hours with standard deviation of 90 hours. Also in a random sample of 75 tube lights from company B the mean lifetime is 1230 hours with standard deviation of 120 hours. Is there a difference between the mean lifetimes of the two brands of tube lights at a significance level of (a) 0.05 (b) 0.01?

Solution: Let X_A, X_B denote the lifetime (in hours) of tube lights produced by company A and B respectively. It is given that the mean lifetime of tube lights of company A is $\overline{X}_A = 1190$, standard deviation for tube lights of A is $s_A = 90$. Similarly $\overline{X}_B = 1230$, $s_B = 120$, $n_A =$ sample size of tube lights from $A = 100$, $n_B =$ sample size from $B = 75$

1. Null hypothesis: $H_0 : \mu_1 - \mu_2 = \delta = 0$ i.e., no difference.
2. Alternate hypothesis: $H_1 : \mu_1 - \mu_2 \neq 0$ i.e., there is difference.
3. L.O.S.: α : (a) 0.05 (b) 0.01.
4. Critical region: two tailed test.

If

a. $-1.96 < Z < 1.96$ Accept N.H.

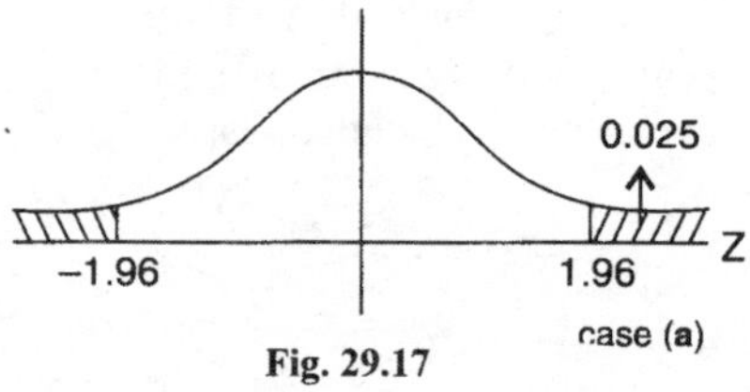

Fig. 29.17

b. $-2.57 < Z < 2.57$ Accept N.H.

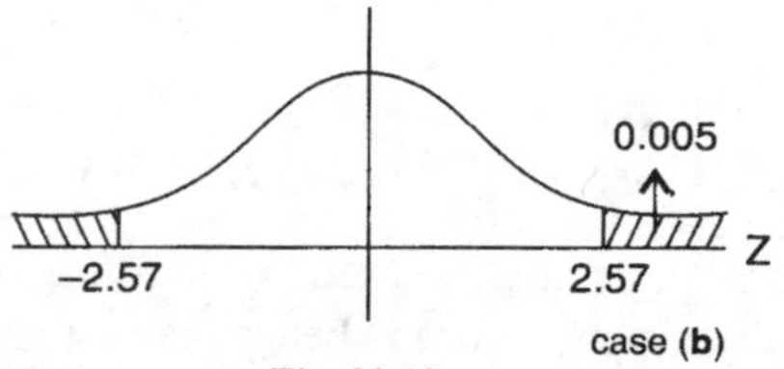

Fig. 29.18

5. Computation: $\mu_{\overline{X}_A - \overline{X}_B} = \mu_{\overline{X}_A} - \mu_{\overline{X}_B} = \mu_A - \mu_B = 0$

$$\sigma_{\overline{X}_A - \overline{X}_B} = \sqrt{\sigma^2_{\overline{X}_A} + \sigma^2_{\overline{X}_B}} = \sqrt{\frac{s_A^2}{n_A} + \frac{s_B^2}{n_B}}$$

$$= \sqrt{\frac{(90)^2}{100} + \frac{(120)^2}{75}} = 16.5227$$

Test statistic:

$$Z = \frac{(\overline{X}_A - \overline{X}_B) - (\mu_A - \mu_B)}{\sigma_{\overline{X}_A - \overline{X}_B}} = \frac{1190 - 1230}{16.5227} = -2.421$$

6. Decision:
   a. For $\alpha = 0.05$
   Reject N.H. since $Z = -2.421 < -1.96 = Z_{\alpha/2} = Z_{0.025}$ i.e., yes, there is difference between the mean lifetimes of the tube lights produced by A and B.
   b. For $\alpha = 0.01$
   Accept N.H. since $Z = -2.421$ lies in the acceptable region $-2.57 < Z < 2.57$ i.e., no, there is no difference between $\overline{X}_A$ and $\overline{X}_B$.

Example 2: To test the effects a new pesticide on rice production, a farm land was divided into 60 units of equal areas, all portions having identical qualities as to soil, exposure to sunlight etc. The new pesticide is applied to 30 units while old pesticide to the remaining 30. Is there reason to believe that the new pesticide is better than the old pesticide if the mean number of kgs of rice harvested / unit using new pesticide (N.P.) is 496.31 with s.d. of 17.18 kgs while for old pesticide (O.P.) is 485.41 kgs and 14.73 kgs. Test at a level of significance (a) $\alpha = 0.05$ (b) 0.01.

Solution: Let the subscripts N and O denote respectively the quantities related the new pesticide and old pesticide.

1. N.H.: $H_0 : \mu_N - \mu_O = \delta = 0$ i.e., no difference.
2. A.H.: $H_1 : \mu_N - \mu_O = \delta > 0$ i.e., new pesticide is superior to (better than) old pesticide.
3. L.O.S.: (a) $\alpha = 0.05$ (b) 0.01.
4. Critical region: Right one tailed test.

Case a: $\alpha = 0.05$

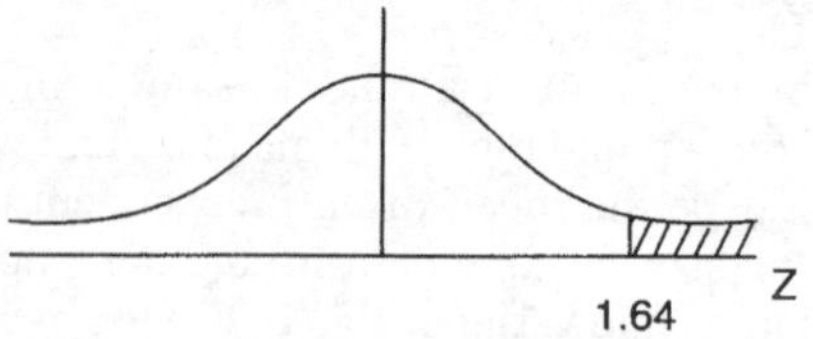

Fig. 29.19

Accept N.H. if $Z < Z_\alpha = Z_{0.05} = 1.64$

Case b: $\alpha = 0.01$
Accept N.H. if $Z < Z_\alpha = Z_{0.01} = 2.33$

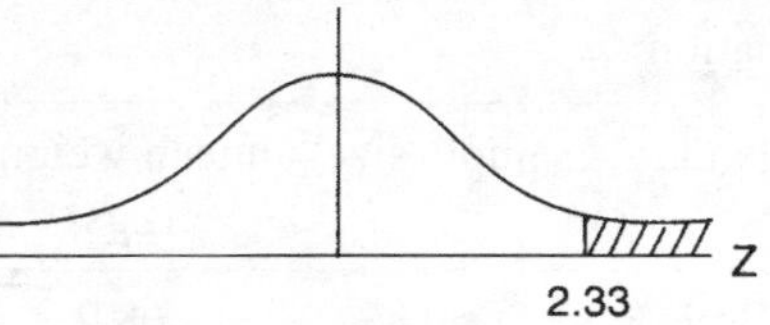

Fig. 29.20

5. Computation: Given data is
$\overline{X}_N = 496.31, \overline{X}_0 = 485.41, s_N = 17.18, s_0 = 14.73, n_N = 30, n_0 = 30$
Test statistic is

$$Z = \frac{(\overline{X}_N - \overline{X}_0) - (\mu_N - \mu_0)}{\sigma_{\overline{X}_N - \overline{X}_0}} = \frac{(\overline{X}_N - \overline{X}_0) - 0}{\sqrt{\frac{(s_N)^2}{n_N} + \frac{(s_0)^2}{n_0}}}$$

$$= \frac{(496.31 - 485.41) - 0}{\sqrt{\frac{(17.18)^2}{30} + \frac{(14.73)^2}{30}}} = 2.63814$$

6. Decision
Case a: $\alpha = 0.05$
Reject N.H. since $Z = 2.638 > Z_\alpha = Z_{0.05} = 1.64$ i.e., accept A.H. or new pesticide is superior to old pesticide.
Case b: $\alpha = 0.01$
Reject N.H. since $Z = 2.638 > Z_\alpha = Z_{0.01} = 2.33$ i.e., Accept A.H. or new pesticide is better than the old pesticide.

EXERCISE

1. A random sample of 40 'geyers' produced by company A have a mean lifetime (mlt) of 647 hours of continuous use with a s.d. of 27 hours, while a sample 40 produced by another company B have mlt of 638 hours with s.d. 31 hours. Does this substantiate the claim of company A that their 'geyers' are superior to those produced by company B at (a) 0.05 (b) 0.01 L.O.S.

Hint: N.H.: $\mu_A = \mu_B = 0$, A.H.: $\mu_A - \mu_B > 0$, reject N.H. if $Z > 1.645$.

$$Z_{\text{cal}} = \frac{(647 - 638) - 0}{\sqrt{\frac{(27)^2}{40} + \frac{(31)^2}{40}}} = 1.38, \text{ accept N.H.}$$

Ans: **a.** No, there is no difference between 'geyers' produced by the two companies A and B.
b. Accept N.H. since $1.38 < Z_\alpha = 2.33$

2. Test at 0.05 L.O.S. a manufacturer's claim that the mean tensile strength (mts) of a thread A exceeds the mts of thread B by at least 12 kgs. if 50 pieces of each type of thread are tested under similar conditions yielding the following data:

	sample size	mts (kgs)	s.d. (kgs)
Type A	50	86.7	6.28
Type B	50	77.8	5.61

Hint: $H_0 \cdot \mu_A - \mu_B \geq 12$, $H_1 \cdot \mu_A - \mu_B < 12$, reject H_0 if $Z < Z_\alpha = -1.64$

$$Z = \frac{(86.7 - 77.8) - 12}{\sqrt{\frac{(6.28)^2}{50} + \frac{(5.61)^2}{50}}} = -2.60, \text{ reject } H_0,$$

Accept $\mu_1 - \mu_2 < 12$.

Ans: Claim not tenable.

3. Test the N.H.: $\mu_A - \mu_B = 0$ against the A.H.: $\mu_A - \mu_B \neq 0$ at 0.01 L.O.S. for the following data:

	sample size	mts (kgs)	s.d. (kgs)
Type A	40	247.3	15.2
Type B	30	254.1	18.7

Hint: $Z = \frac{(247.3 - 254.1) - 0}{\sqrt{\frac{(15.2)^2}{40} + \frac{(18.7)^2}{30}}} = -1.62866$,

Acceptable region: $-2.58 < Z < 2.58$, Accept N.H.

Ans: Accept N.H. i.e., no difference between type A and B

4. If random sample data show that 42 men earn on the average $\overline{x}_1 = 744.85$ with s.d. $s_1 = 397.7$ while 32 women earn on the average $\overline{x}_2 = 516.78$ with s.d. $s_2 = 162.523$, test at 0.05 level of significance whether the average

income for men and women is same or not.

Hint: $H_0 : \mu_1 = \mu_2,\ H_1 : \mu_1 \neq \mu_2,$

$Z = \frac{(744.85-516.78)-0}{\sqrt{\frac{158165.43}{42}+\frac{26413.61}{32}}} = 3.36$

Since $Z = 3.36 > 1.96 = Z_\alpha = Z_{0.05}$, Reject N.H. H_0.

Ans: Not same.

5. A company claims that alloying reduces resistance of electric wire by more than 0.050 ohm. To test this claim samples of standard wire and alloyed wire are tested yielding the following results:

Type of wire	sample size	mean resistance (ohms)	s.d. (ohms)
Standard	32	0.136	0.004
Alloyed	32	0.083	0.005

Can the claim be substantiated at 0.05 L.O.S.

Hint: $H_0 : \mu_1 - \mu_2 = 0.05,\ H_1 : \mu_1 - \mu_2 > 0.05,\ Z_\alpha = Z_{0.05} = 1.645$

$$Z = \frac{(0.136 - 0.083) - (0.05)}{\sqrt{\frac{(0.004)^2}{32} + \frac{(0.005)^2}{32}}} = 2.65,$$

Reject N.H. since $Z = 2.65 > 1.645 = Z_{0.05}$

Ans: Data substantiate the claim.

6. To test the claim that men are taller than women, a survey was conducted resulting in the following data:

Gender	sample size	mean height (cm)	s.d. (cm)
Men	1600	172	6.3
Women	6400	170	6.4

Is the claim tenable at 0.01 L.O.S.

Hint: $H_0 : \mu_1 = \mu_2,\ H_1 : \mu_1 > \mu_2,$

$Z = \frac{172-170}{\sqrt{\frac{(6.3)^2}{1600}+\frac{(6.4)^2}{6400}}} = 11.32$

Reject H_0 since $Z = 11.32 > 2.33 = Z_\alpha = Z_{0.01}$.

Ans: Yes, men are taller than women.

7. Test the claim that teen-age boys are heavier than teen-age girls given the following information:

Gender	sample size	mean weight (kgs)	s.d. (kgs)
Boys	50	68.2	2.5
Girls	50	67.5	2.8

Use L.O.S. (a) 0.05 (b) 0.01

Hint: $Z = \frac{\overline{X}_1-\overline{X}_2}{\sigma_{\overline{X}_1-\overline{X}_2}} = \frac{68.2-67.5}{0.53} = 1.32,$

$$\sigma_{\overline{X}_1-\overline{X}_2} = \sqrt{\frac{(2.5)^2}{50} + \frac{(2.8)^2}{50}} = 0.53$$

Ans: **a.** Accept N.H. i.e., no difference between mean weights.

b. Reject N.H. i.e., boys are heavier than girls.

29.7 TEST FOR ONE MEAN (SMALL SAMPLE: *t*-DISTRIBUTION)

For "expensive" populations such as satelites, aeroplanes, nuclear reactors, super computers, etc. the investigation of characteristics of large samples ($n \geq 30$) is uneconomical, impractiable and time consuming. In all such cases, the size of the sample, drawn is small (i.e., $n < 30$). For σ unknown and for small sample size, the test statistic cannot be used. Then the decision criterion is based on the t-distribution with $\nu = n - 1$ degrees of freedom. Thus the test statistic for small sample test (with σ unknown) concerning one mean is

$$t = \frac{\overline{x} - \mu_0}{s/\sqrt{n}}$$

This is also known as "one-sample t-test". So the test procedure for small samples is similar to the procedure for large samples except that 't' values are used in place of Z values and σ is replaced by s. For example, when A.H. is $\mu \neq \mu_0$ then the C.R. is $t < -t_{\alpha/2}$ or $t > t_{\alpha/2}$ etc.

WORKED OUT EXAMPLES

Examples: An ambulance service company claims that on an average it takes 20 minutes between a call for an ambulance and the patient's arrival at the hospital. If in 6 calls the time taken (between a call and arrival at hospital) are 27, 18, 26, 15, 20, 32. Can the company's claim be accepted?

Solution: Here $n = 6$. Let X be the time taken between a call and patient's arrival at hospital. From given data $\overline{X}$ = average time taken

$$\overline{X} = \frac{27 + 18 + 26 + 15 + 20 + 32}{6} = \frac{138}{6} = 23$$

standard deviation: $s = \sqrt{\frac{\sum(X_i - \overline{X})^2}{n-1}}$

$$s^2 = \frac{(27-23)^2+(18-23)^2+(26-23)^2+(15-23)^2+(20-23)^2+(32-23)^2}{6-1}$$

$s^2 = 40.8, s = 6.38748$

1. N.H.: $X = 20$ minutes
2. A.H.: $X > 20$
3. L.O.S.: $\alpha = 0.10$
4. Critical region:

   Reject N.H. if $t > t_\alpha = 1.476$ where $t_{0.10}$ with $\nu = n - 1 = 6 - 1 = 5$ degrees of freedom.
5. Calculation: $t = \frac{\overline{X}-\mu}{s/\sqrt{n}} = \frac{23-20}{6.39/\sqrt{6}} = 1.15$

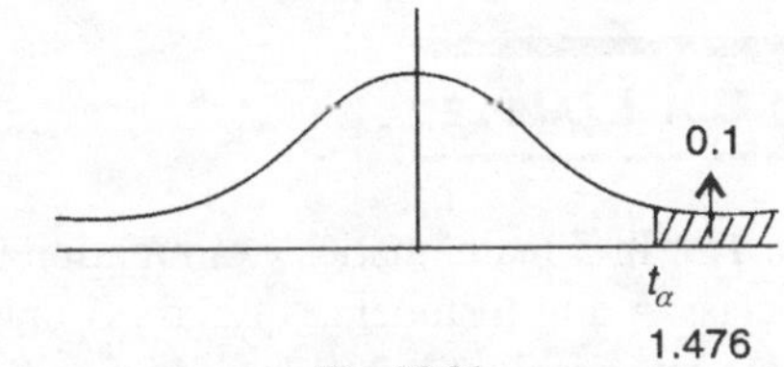

Fig. 29.21

6. Decision: Accept H_0 since $t = 1.15 < 1.476 = t_{0.1}$ with 5 dof i.e., accept the claim of the company.

EXERCISE

1. Mean lifetime (mlt) of computers manufactured by a company is 1120 hours with standard deviation of 125 hours. (a) Test the hypothesis that mean lifetime of computers has not changed if a sample of 8 computers has a mlt of 1070 hours (b) Is there decrease in mlt? Use (i) 0.05 (ii) 0.01 L.O.S.

   Hint: N.H.: $\mu = 1120$, A.H.: $\mu \neq 1120$, $t = \frac{1070-1120}{125/\sqrt{8}} = -1.1313$, $t_{0.005}$ with 7 dof is ± 3.499, $t_{0.025}$ with 7 dof is ± 2.365: Accept H_0 in both cases.

   Ans: **a.** Two-tailed test indicates that there is no reason at either level to belive that mlt has changed.

   Hint: A.H.: $\mu < 1120$ (i) $t_{0.05}$ with 7 dof is -1.895 (ii) $t_{0.01}$ with 7 dof is -2.998.

   b. One-tail test indicates no decrease in mlt at either of the L.O.S.

2. Producer of 'gutkha', claims that the nicotine content in his 'gutkha' on the average is 1.83 mg. Can this claim be accepted if a random sample of 8 'gutkhas' of this type have the nicotine contents of 2.0, 1.7, 2.1, 1.9, 2.2, 2.1, 2.0, 1.6 mg?

   Hint: $\overline{x} = \frac{15.6}{8} = 1.95, s = \frac{\sqrt{0.3}}{7} = 0.20702$, $t = \frac{1.95-1.83}{.207/\sqrt{8}} = 1.6395$, $t_\alpha = t_{0.05}$ with 7 dof is 1.895.

   N.H.: $\mu = 1.83$, A.H.: $\mu > 1.83$

   Ans: Yes, the producer's claim can be accepted with 95% confidence.

3. In 1950 in India the mean life expectancy was 50 years. If the life expectancies from a random sample of 11 persons are 58.2, 56.6, 54.2, 50.4, 44.2, 61.9, 57.5, 53.4, 49.7, 55.4, 57.0, does it confirm the expected view.

   Hint: H_0: $\mu = 50$, $H_1 : \mu \neq 50$, $\overline{x} = \frac{598.5}{11} = 54.41$, $s = 4.859$, $t = 3.01$, reject H_0 since $t = 3.01 > 2.228 = t_{0.0025}$ with 10 dof.

   Ans: No, the life expectancy is more than 50 years.

4. An auditor claims that he takes on an average 10.5 days to file income tax returns (I.T. returns). Can this claim be accepted if a random sample shows that he took 13, 19, 15, 10, 12, 11, 14, 18 days to file I.T. returns? Use (a) 0.01 (b) 0.05 L.O.S.

   Hint: N.H.: $\mu = 10.5$, A.H.: $\mu > 10.5$,

   $\overline{x} = \frac{112}{8} = 14, s = \sqrt{\frac{72}{7}} = 3.207$,

$t = \frac{14-10.5}{3.207/\sqrt{8}} = 3.0869$, (a) $t_{0.01}$ with 7 dof is 2.998 (b) $t_{0.05}$ with 7 dof is 1.895.

Ans: Reject the claim, i.e., it takes more than 10.5 days to file I.T. returns.

5. If 5 pieces of certain ribbon selected at random have mean breaking strength of 169.5 pounds with s.d. of 5.7, do they confirm to the specification mean breaking strength of 180 pounds.

Hint: H_0: $\mu = 180$, $H_1 : \mu < 180$, L.O.S.: $\alpha = 0.01$, $t = -4.12$,

$t_\alpha = t_{0.01}$ with 4 dof is -3747, so reject N.H.

Ans: They do not confirm to specification i.e., mbs is below.

6. In a random sample of 10 bolts produced by a machine the mean length of bolt is 0.53 mm and standard deviation 0.03 mm. Can we claim from this that the machine is in proper working order if in the past it produced bolts of length 0.50 mm? Use (a) 0.05 (b) 0.01 L.O.S.

Hint: $H_0 : \mu = 0.50$, $H_1 : \mu \neq 0.50$,

$t = \frac{\overline{x}-\mu}{s}\sqrt{n-1} = \frac{0.53-0.50}{0.03}\sqrt{10-1} = 3.0$

Acceptable region $-2.26 < t < 2.26$

Ans: **a.** At 0.05 L.O.S., by a T.T.T., reject H_0.

b. At 0.01 L.O.S., by T.T.T., accept H_0

Acceptable region $-3.25 < t < 3.25$.

29.8 SMALL-SAMPLE TEST CONCERNING DIFFERENCE BETWEEN TWO MEANS

Suppose the two sample sizes n_1, n_2 or both are small ($n < 30$) and two samples are drawn from two normal populations with population variances σ_1^2 and σ_2^2 *unknown* but equal (i.e., $\sigma_1 = \sigma_2 = \sigma$). Then the pooling varience σ^2 is given by

$$\sigma^2 = \frac{\sum(x_{i1} - \overline{x}_1)^2 + \sum(x_{2_i} - \overline{x}_2)^2}{n_1 + n_2 - 2}$$

$$= \frac{(n_1 - 1)s_1^2 + (n_2 - 1)s_2^2}{n_1 + n_2 - 2}$$

where $\overline{x}, s_1^2$ and $\overline{x}_2, s_2^2$ are the mean and variance of two samples of size n_1 and n_2 respectively. In a test concerning the difference between the means for small samples, the t-test statistic is

$$t = \frac{(\overline{x}_1 - \overline{x}_2) - \delta}{\sigma_{\overline{x}_1 - \overline{x}_2}}$$

with $n_1 + n_2 - 2$ degrees of freedom. This test is also known as two-sample pooled t-test. Rewriting

$$t = \frac{(\overline{x}_1 - \overline{x}_2) - \delta}{\sqrt{(n_1 - 1)s_1^2 + (n_2 - 1)s_2^2}}\sqrt{\frac{n_1 n_2(n_1 + n_2 - 2)}{n_1 + n_2}}$$

with $n_1 + n_2 - 2$ dof.

The critical regions with this t-distribution can be obtained in a similar way. For example when A.H. is $\mu_1 - \mu_2 \neq \delta$, then the critical region (Reject H_0) is

$$t < -t_{\alpha/2}, n_1 + n_2 - 2 \quad \text{or} \quad t > t_{\alpha/2}, n_1 + n_2 - 2.$$

Note 1: The critical values are given when $n_1 + n_2 - 2 \geq 30$ (although n_1 and n_2 are small).

Note 2: The two-sample t-test can not be used if $\sigma_1 \neq \sigma_2$.

Note 3: The two-sample t-test can not be used for "before and after" kind of data, where the data is naturally paired. In other words the samples must be "independent" for two sample t-test.

WORKED OUT EXAMPLES

Example 1: In a mathematics examination 9 students of class A and 6 students of class B obtained the following marks. Test at 0.01 level of significance whether the performance in mathematics is same or not for the two classes A and B. Assume that the samples are drawn from normal populations having same variance.

A	44	71	63	59	68	46	69	54	48
B	52	70	41	62	36	50			

Solution: Let X_A and X_B be the marks obtained in mathematics of class A and class B. Then from the given data $\overline{X}_A = \frac{\sum x_i}{n_1} = \frac{522}{9} = 58$, $\overline{X}_B = \frac{311}{6} = 51.83$

$$s_A^2 = \frac{\Sigma(X_i - \overline{X})^2}{n_1 - 1} = \frac{872}{8} = 109, s_A = 10.44,$$

$$s_B^2 = \frac{804.8334}{5}, \qquad s_B = 12.687$$

Here n_A = sample size from (population) class A = 9

n_B = sample size from (population) class B = 6

1. N.H.: $\mu_1 - \mu_2 = 0$ i.e., no difference in performance.
2. A.H.: $\mu_1 - \mu_2 \neq 0$ i.e., there is difference.
3. L.O.S.: $\alpha = 0.01$.
4. Critical region: Two-tailed test. Reject N.H. if $t < -t_{\alpha/2}$ or $t > t_{\alpha/2}$ where $t_{\alpha/2} = t_{0.005}$ with $n_1 + n_2 - 2 = 9 + 6 - 2 = 13$ degrees of freedom. From table, $t_{0.005}$ is 3.012.
5. Computation: Test statistic

$$t = \frac{(\overline{X}_A - \overline{X}_B) - (\mu_1 - \mu_2)}{\sqrt{(n_A - 1)s_A^2 + (n_B - 1)s_B^2}}\sqrt{\frac{n_A n_B(n_A + n_B - 2)}{n_A + n_B}}$$

$$t = \frac{(58-51.83)-0}{\sqrt{(9-1)(109)+(6-1)(160.96)}}\sqrt{\frac{(9)(6)(13)}{9+6}} = 1.030.$$

6. Decision: Accept N.H. since

$$t = 1.03 < 3.012 = t_{\alpha/2} = t_{0.005}$$

i.e., there is no difference between the two classes A and B in performance in mathematics examination.

Example 2: Under quality improvement programme some teachers are trained by instruction methodology A and some by methodology B. In a random sample of size 10, taken from a large group of teachers exposed to each of these two methods, the following marks are obtained in an appropriate achievement test

Method A 65 69 73 71 75 66 71 68 68 74

Method B 78 69 72 77 84 70 73 77 75 65

Assuming that populations sampled are approximately normally distributed having same variance, test the claim that method B is more effective at 0.05 level of significance.

Solution: Let subscripts A and B denote data pertaining to methodology A and B respectively. Then from the given data, $n_A = n_B = 10$,

$\overline{X}_A$ = average marks obtained in appropriate achievement test by teachers trained under methodology A is $\frac{700}{10} = 70$. Similarly, $\overline{X}_B = \frac{740}{10} = 74$

$$s_A^2 = \frac{102}{9} = 11.33, \quad s_A = 3.366,$$

$$s_B^2 = \frac{262}{9} = 29.11, \quad s_B = 5.3954$$

1. N.H.: $H_0 : \mu_1 - \mu_2 = 0$ i.e., no difference in teaching methodologies
2. A.H.: $H_1 : \mu_1 - \mu_2 < 0$ i.e., method B is more effective (superior) than method A
3. L.O.S.: $\alpha = 0.05$
4. Critical region (left one tailed test)

   Reject H_0 if $t < -t_\alpha = -t_{0.05}$ with $n_A + n_B - 2 = 10 + 10 - 2 = 18$ degrees of freedom. From table $t_{0.05} = -1.734$
5. Computation

$$t = \frac{(\overline{X}_A - \overline{X}_B) - (\mu_1 - \mu_2)}{\sqrt{(n_A - 1)s_A^2 + (n_{B-1})s_B^2}}\sqrt{\frac{n_A n_B(n_A + n_B - 2)}{n_A + n_B}}$$

$$= \frac{(70-74)-0}{\sqrt{9(11.33)+9(29.11)}}\sqrt{\frac{(10)(10)(18)}{10+10}} = -1.989$$

6. Decision: Reject N.H. since $t = -1.989 < -1.734 = t_{0.05}$ i.e., accept the claim that method B is more effective (better) than the method A.

Example 3: Out of a random sample of 9 mice, suffering with a disease, 5 mice were treated with a new serum while the remaining were not treated. From the time of commencement of experiment, the following are the survival times:

Treatment	2.1	5.3	1.4	4.6	0.9
No treatment	1.9	0.5	2.8	3.1	

Test whether the serum treatment is effective in curing the disease at 0.05 L.O.S., assuming that the two distributions are normally distributed with equal variances.

Solution: Let μ_T and μ_{NT} be the mean survival times of the mice treated and not treated with serum respectively.

1. N.H.: $H_0 : \mu_T - \mu_{NT} = 0$ i.e., not effective

2. A.H.: $H_1 : \mu_T - \mu_{NT} > 0$ i.e., serum is effective

3. L.O.S.: $\alpha = 0.05$

4. Critical region: Reject N.H. if $z > t_{0.95,7} = 1.90$ since the dof is $\nu = n_T + n_{NT} - 2 = 5 + 4 - 2 = 7$.

5. Computation: $n_T = 5$, $\overline{X}_T = \frac{14.3}{5} = 2.86$

$$s_T^2 = \frac{15.532}{4} = 3.883, s_T = 1.9705,$$

$$n_{NT} = 4, \overline{X}_{NT} = \frac{8.3}{4} = 2.075,$$

$$s_{NT}^2 = \frac{4.0875}{3} = 1.3625,$$

$$s_{NT} = 1.16726,$$

$$S_p^2 = \frac{(n_T - 1)s_T^2 + (n_{NT} - 1)s_{NT}^2}{n_T + n_{NT} - 2} =$$

$$S_p^2 = \frac{(5-1)(1.9705)^2 + (4-1)(1.16726)^2}{5+4-2}$$

$$= 2.802, S_p = 1.674$$

$$t = \frac{(2.86 - 2.075) - 0}{1.674\left(\frac{1}{5} + \frac{1}{4}\right)} = 0.6990 \approx 0.7$$

6. Decision: Accept N.H. since $t = 0.7 < 1.9 = t_{0.95,7}$ i.e., serum treatment is not effective.

EXERCISE

1. Random samples of specimens of coal from two mines A and B are drawn and their heat-producing capacity (in millions of calories/ton) were measured yielding the following results:

Mine A: 8350, 8070, 8340, 8130, 8260

Mine B: 7900, 8140, 7920, 7840, 7890, 7950

Is there significant difference between the means of these two samples at 0.01 L.O.S.

Hint: N.H.: $\mu_1 - \mu_2 = 0$, A.H.: $\mu_1 - \mu_2 \neq 0$, $t_{0.005}$ with $5 + 6 - 2 = 9$ dof is 3.250

Accept if $-3.250 < t < 3.250$, $\overline{x}_1 = \frac{41150}{5} = 8230$, $\overline{x}_2 = \frac{47640}{6} = 7940$

$s_1^2 = \frac{63000}{4} = 15750$, $s_2^2 = \frac{54600}{5} = 10920$,

$$t = \frac{8230-7940}{\sqrt{63000+54600}}\sqrt{\frac{5.6.9}{11}} = 4.19$$

Reject N.H. since $t = 4.19 > t_{0.005} = 3.250$.

Ans. Yes, there is significant difference.

2. To test the claim that substrate concentration (S.C.) causes an increase in the mean velocity (M.V.) of a chemical reaction by more than 0.5 m/l/30 minutes a study is conducted resulting in the following data:

Reaction with S.C. of	No. of runs	Mean velocity	Sample s.d.
1.5 moles/litre	15	7.5	1.5
2.0 moles/litre	12	8.8	1.2

Is the claim tenable at 0.01 L.O.S. assuming that the populations are normally distributed with equal variances.

Hint: $H_0 : \mu_1 - \mu_2 = \delta = 0.5$, $H_1 : \mu_1 - \mu_2 > \delta = 0.5$, $t_{0.01,25} = 2.485$

$$S_p^2 = \frac{14(1.5)^2 + 11(1.2)^2}{15 + 2 - 2} = 1.8936,$$

$$t = \frac{(8.8 - 7.5) - (0.5)}{1.376\sqrt{\frac{1}{15} + \frac{1}{12}}} = 1.50$$

Reject N.H. since $t = 1.50 < 2.48$ (Right O.T.T.)

Ans. No, claim not tenable.

3. A study is conducted to determine whether the wear of material A exceeds that of B by more than 2 units. If test of 12 pieces of material A yielded a mean wear of 85 units and s.d. of 4 while test of 10 pieces of material B yielded a mean of 81 and s.d. 5, what conclusion can be drawn at 0.05 L.O.S. Assume that populations are approximately normally distributed with equal variances.

Hint: $H_0 : \mu_1 - \mu_2 = 2$, $H_1 : \mu_1 - \mu_2 > 2$, $t_{0.05}$ with 20 dof is 1.725

$$S_p^2 = \frac{11(16) + 9(25)}{12 + 10 - 2} = 20.052,$$

$$t = \frac{(85 - 81) - 2}{4.478\sqrt{\frac{1}{12} + \frac{1}{10}}} = 1.04$$

Accept H_0 since $t = 1.04 < 1.725 = t_{0.05,20}$.

Ans. Wear of A does not exceed that of B by 2 units.

4. To determine whether vegetarian and non-vegetarian diets effects significantly on increase in weight a study was conducted yielding the following data of gain in weight.

Vegetarian: 34, 24, 14, 32, 25, 32, 30, 24, 30, 31, 35, 25

Non-vegetarian: 22, 10, 47, 31, 44, 34, 22, 40, 30, 32, 35, 18, 21, 35, 29

Can we claim that the two diets differ pertaining to weight gain, assuming that samples are drawn from normal populations with same variance.

Hint: $\bar{X}_V = \frac{336}{12} = 28, n_V = 12, \bar{X}_{NV} = \frac{450}{15} = 30, n_{NV} = 15$

$$S^2 = 71.6, t = \frac{\bar{X}_V - \bar{X}_{NV}}{\sqrt{S^2\left(\frac{1}{n_V} + \frac{1}{n_{NV}}\right)}} = \frac{28-30}{\sqrt{71.6\left(\frac{1}{12}+\frac{1}{15}\right)}}$$

$= -0.609$

Accept N.H.: $\mu_V = \mu_{NV}$ against A.H.: $\mu_V \neq \mu_{NV}$ at 0.05 L.O.S.

Since $-0.609 > -2.06$ and $t_{0.05}$ at $12 + 15 - 2 = 25$ dof is 2.06

Ans. Vegetarian and non-vegetarian diets do not differ significantly as far as their effect on increase in weight.

5. In a study on the influence of habitation, the intelligent quotients (IQs) of 16 students from urban area was found to have a mean of 107 and s.d. of 10, while the IQs of 14 students from a rural area showed a mean of 112 and s.d. 8. Determine whether the IQs differ significantly at (a) 0.01 (b) 0.05 levels.

Hint:

$$S_p^2 = \frac{n_1 s_1^2 + n_2 s_2^2}{n_1 + n_2 - 2} = \frac{16(10)^2 + 14(8)^2}{16 + 14 - 2} = 89.1136$$

$$t = \frac{112 - 107}{9.44\sqrt{\frac{1}{16} + \frac{1}{14}}} = 1.45$$

a. Accept $H_0 : \mu_U = \mu_R$, Acceptable region $(-2.76, 2.76)$

b. Accept H_0, since 1.45 lies in the acceptable region $(-2.05, 2.05)$

Ans. No, habitation has influence on IQs.

6. To test the claim that application of pesticide increases production of rice, a study was conducted as follows. Out of 24 plots of equal areas, equal soil conditions, same exposure to sunlight, 12 plots were treated with pesticides while the remaining 12 plots were left untreated. The mean increase in rice production was 4.8 kgs and s.d. of 0.40 kgs for treated plots and 5.1 kgs mean and 0.36 s.d. for untreated plots. Is there significant increase in rice production due to pesticide application at (a) 0.01 (b) 0.05 L.O.S.

Hint:

$$S^2 = \frac{12(0.40)^2 + 12(0.36)^2}{12 + 12 - 2} = 0.157609$$

$$t = \frac{5.1 - 4.8}{0.397\sqrt{\frac{1}{12} + \frac{1}{12}}} = 1.85$$

$t_{0.99}$ at 22 dof, right O.T.T. is 2.51

$t_{0.95}$ at 22 dof, right O.T.T. is 1.72

Ans. (a) Accept $H_0 : \mu_1 = \mu_2$ (b) Reject H_0, significant.

29.9 PAIRED-SAMPLE *t*-TEST

Paired observations arise in a very special experimental situation where each homogeneous experimental unit receives both population conditions. As a result, each experimental unit has a pair of observations, one for each population. Thus the paired observations are on the same unit or matching units.

Examples: To test the effectiveness of "insulin" some 10 diabetic patients sugar level in blood is measured "before" and "after" the insulin is injected. Here the individual diabetic patient is the experimental unit and the two populations are blood sugar level "before" and "after" the insulin is injected.

So for each observation is one sample, there is a corresponding observation in the other sample pertaining to the same character. Thus the two samples are *not* independent. Paired t-test is applied for n paired observations (which are dependent) by taking the (signed) differences $d_1, d_2, \ldots, d_n$ of the paired data. To test whether the differences d form a random sample from a population with $\mu_D = d_O$ use large sample test (on Page 766) or one-sample t-test (on Page 773) when sample is small (the one sample t-test in this case is known as the paired-sample t-test). The test statistic is

$$\frac{\overline{d} - \mu_d}{S_d/\sqrt{n}}$$

with $\nu = n - 1$ dof and $\overline{d}$ and S_d^2 are the mean and variance of the differences $d_1, d_2, \ldots, d_n$.

WORKED OUT EXAMPLES

Examples: In a study of usefulness of yoga in weight reduction, a random sample of 16 persons undergoing yoga were examined of their weight before (without) and after (with) yoga with the following results:

Weight before	209	178	169	212	180	192	158	180	170	153	183	165	201	179	243	144
Weight after	196	171	170	207	177	190	159	180	164	152	179	162	199	173	231	140

Test whether yoga is useful in weight reduction at 0.01 level of significance.

Solution: Let μ be the mean of population of differences,

1. N.H.: $\mu = 0$ i.e., not useful.

2. A.H.: $\mu > 0$ i.e., yoga is useful in weight reduction.

3. L.O.S.: $\alpha = 0.01$.

4. Critical region: Right one tailed test.
Reject N.H. if $t > t_{0.01}$ with $16 - 1 = 15$ degrees of freedom. From table $t_{0.01} = 2.602$.

5. Calculation: differences d_i's are

13, 7, −1, 5, 3, 2, −1, 0, 6, 1, 4, 3, 2, 6, 12, 4

$$\overline{x} = \text{mean of differences of sampled data} = \frac{66}{16} = 4.125$$

$$s^2 = \frac{247.73}{15} = 16.516, s = 4.064$$

$$t = \frac{\overline{x} - \mu_0}{s/\sqrt{n}} = \frac{4.125 - 0}{4.064/\sqrt{16}} = 4.06$$

6. Decision: Reject N.H. since $t = 4.06 > 2.602 = t_{0.01}$ i.e., yoga is useful in weight reduction.

EXERCISE

1. Use paired sample test at 0.05 level of significance to test from the following data whether

Weight	Scale I	11.23	14.36	8.33	10.50	23.42	9.15	13.47	6.47	12.40	19.38
in gms	Scale II	11.27	14.41	8.35	10.52	23.41	9.17	13.52	6.46	12.45	19.35

the differences of the means of the weights obtained by two different scales (weighting machines) is significant.

Hint: $\overline{x} = -\frac{0.2}{10} = -0.02$, $s = 0.028674$, $n = 10$, $t = \frac{-0.02-0}{0.028/\sqrt{10}} = -2.21$, $t_\alpha = t_{0.05}$ with 9 dof is 1.833.

Ans. No significant difference in the two scales.

2. The average weekly losses of man-hours due to strikes in an institute before and after a disciplinary program was implemented are as follows:

Before	45	73	46	124	33	57	83	34	26	17
After	36	60	44	119	35	51	77	29	24	11

Is there reason to believe that the disciplinary program is effective at 0.05 level of significance?

Hint: $\overline{x} = 5.2$, $s = 4.08$, $n = 10$, $t = 4.03$, $t_{0.05}$ with 9 dof is 1.833.

Ans. Yes, program is effective.

3. The pulsality index (P.I.) of 11 patients before and after contracting a disease are given below. Test at 0.05 level of significance whether there is a significant increase of the mean of P.I. values.

Before	0.4	0.45	0.44	0.54	0.48	0.62	0.48	0.60	0.45	0.46	0.35
After	0.5	0.60	0.57	0.65	0.63	0.78	0.63	0.80	0.69	0.62	0.68

Hint: $\overline{x} = \frac{188}{11} = 0.171, s = 0.065, n = 11,$ $t = 8.72, t_{0.05}$ with 10 dof is 1.812.

Ans. Yes, there is significant increase in P.I. values.

4. The following data gives the amount of androgen present in blood of 15 deers before and 30 minutes after a certain drug in injected to them.

Before	2.76	5.18	2.68	3.05	4.10	7.05	6.60	4.79	7.39	7.30	11.78	3.9	26	67.48	17.04
After	7.02	3.1	5.44	3.99	5.21	10.26	13.91	18.53	7.91	4.85	11.1	3.74	94.03	94.03	41.7

Test at 0.05 L.O.S. whether there is significant change in the concentration levels of androgen in blood.

Hint: $\overline{x} = 9.848, s = 18.474, t = 2.06$, critical region: $t < -2.145$ and $t > 2.145$ (with 14 dof).

Ans. Yes, there is difference in mean circulating levels of androgen in the blood of deer.

5. The blood pressure (B.P.) of 5 women before and after intake of a certain drug are given below:

Before	110	120	125	132	125
After	120	118	125	136	121

Test at 0.01 L.O.S. whether there is significant change in B.P.

Hint: $\overline{x} = \frac{10}{5} = 2, s = 5.477, t = 0.817,$ $t_{0.01}$ with 4 dof is 3.747.

Ans. No significant change in B.P.

6. Marks obtained in mathematics by 11 students before and after intensive coaching are given below:

Before	24	17	18	20	19	23	16	18	21	20	19
After	24	20	22	20	17	24	20	20	18	19	22

Test at 0.05 L.O.S. whether the intensive coaching is useful?

Hint: $\overline{x} = -1, s = 2.296, t = -1.38,$ $t_{0.05}$ with 10 dof is 1.812.

Ans. Not useful.

29.10 TEST OF HYPOTHESIS: ONE PROPORTION: SMALL SAMPLES

The quality control engineer wants to know the proportion of defective products (items) in his industry, a university the percentage of first classes and a electronic component manufacturer the probability that a component works for a certain period and so on. In these cases, the observations on various items or objects are classified into two mutually exclusive (dichotomus) classes (forming a binomial population).

Let X, the number of successes be a binomial random variable. Let p be the parameter of the binomial distribution. Then the test of hypothesis concerning one proportion for *small samples* is as follows:

1. N.H.: $H_0 : p = p_0$ i.e., a proportion (percentage or probability) equals some given constant p_0.
2. A.H.: $H_1 : p \neq p_0$ (or $p < p_0$ or $p > p_0$).
3. L.O.S.: α
4. Test statistic: Binomial variable X with $p = p_0$.
5. Computation: Let x be the number of successes in a sample of size n.

   Compute p-value:

   a. A.H.: $p \neq p_0$:
   $P = 2P(X \leq x \text{ when } p = p_0)$ if $x < np_0$.
   $P = 2P(X \geq x \text{ when } p = p_0)$ if $x > np_0$

   b. A.H.: $p < p_0 : P = P(X \leq x \text{ when } p = p_0)$

   c. A.H.: $p > p_0 : P = P(X \geq x \text{ when } p = p_0)$
6. Decision: Reject N.H.: H_0 if $P \leq \alpha$.

WORKED OUT EXAMPLES

Example 1: If 6 out of 20 cigarette smokers randomly chosen preferred 'charminar' cigarettes, test the claim at 0.05 L.O.S., that 20% of the smokers prefer 'charminar'.

Solution:

1. N.H.: H_0: Proportion of smokers preferring 'charminar' brand $= p = 0.2$.
2. A.H.: $H_1 : p \neq 0.2$.
3. L.O.S.: $\alpha = 0.05$.
4. Test statistic: Let X be the discrete binomial random variable which is the number of 'charminar' smokers with $p = 0.2$ and $n = 20$.
5. Computations: $X = 6$, $np_0 = (20)(0.2) = 4$. Since $X = 6 > np_0 = 4$

$$P = 2\,[\text{probability that } X \geq 6 \text{ when } p = 0.2]$$
$$= 2P(x \geq 6 \text{ with } p = 0.2)$$
$$= 2\left[1 - \sum_{X=0}^{5} b(X; 20, 0.2)\right]$$
$$= 2[1 - 0.8042] = 2(0.1958) = 0.3916.$$

Since $P = 0.3916 > 0.05$, accept H_0 i.e., $p = 0.2$.

Example 2: Past experience shows that 40% of Indian youth favored 'cricket'. If in a random sample of 15 Indian youth, 8 favoured cricket, is there reason to believe that the proportion of Indian youth favoring cricket today has increased. Use 0.05 L.O.S.

Solution: Let X be *discrete random variable*: number of Indian youth favouring cricket.

1. $H_0 : p = 0.4$
2. $H_1 : p > 0.4$
3. $\alpha : 0.05$
4. Binomial variable X with $p = 0.4$, $n = 15$.
5. Computation: $X = 8$, $np_0 = 15(0.4) = 16$ $\therefore 8 = X > np_0 = 6$.

$$P = P(X \geq 8 \text{ when } p = 0.4) = 1 - P(X \leq 7)$$
$$= 1 - \sum_{X=0}^{7} b(X; 15, 0.4)$$
$$P = 1 - 0.7869 = 0.2131 > 0.05.$$

6. Conclusion: Accept N.H.: $p = 0.4$ i.e., no, there is no increase in the proportion of Indian youth favouring cricket.

EXERCISE

1. To test the claim of a flat builder that mosquito nets were installed in 70% of the flats, a random survay was conducted and found that 8 out of 15 flats had mosquito nets. Is the claim valid at 0.10 L.O.S.

   Hint:
   $$P = 2P(X \leq 8 \text{ when } p = 0.7) = 2\sum_{x=0}^{8} b(x; 15, 0.7)$$
   $$= 2(0.1311) = 0.2622,\ x = 8 < 10.5 = np_0$$
   $$= (15)(0.7)$$

   Ans. Accept H_0, claim of the builder is valid since $P = 0.2622 > 0.10$ by a T.T.T.

2. Test the N.H.: that a coin is fair at 0.03 L.O.S. against an A.H. that heads occur less than 50% of the time if 5 heads occur when it is tossed 20 times.

   Ans. Reject N.H., i.e., coin is unpair (unbalanced) since $P = P(X \leq 5, p = 0.5) = \sum_{x=0}^{5} b(x; 20, 0.5) = 0.0207 < 0.03.$

3. If 9 out 20 'pizza' eaters like the native variety over the Italian, can the claim that 40% of 'pizza' eaters like native variety is tenable?

   Hint: $P = 0.4044$ (use O.T.T.)

   Ans. Accept the claim (i.e., claim is not refuted).

29.11 TEST OF HYPOTHESIS: ONE PROPORTION: LARGE SAMPLE

Assume that sample size n is large. Then the binomial distribution can be approximated by normal distribution with the parameters mean $\mu = np_0$ and variance

$\sigma^2 = np_0q_0$. The test statistic for testing $p = p_0$ is given by

$$z = \frac{x - np_0}{\sqrt{np_0q_0}}$$

where x is the number of successes in a sample of size n and $q_0 = 1 - p_0$. This statistic can also be written as

$$z = \frac{P - p}{\sqrt{pq/n}}$$

where $P = \frac{x}{n}$ = proportion of successes in the sample, p = actual population proportion of successes. As usual, the critical regions are

a. $z < -z_{\alpha/2}$ or $z > z_{\alpha/2}$ for A.H.: $p \neq p_0$
b. $z > z_\alpha$ for A.H.: $p > p_0$
c. $z < -z_\alpha$ for A.H.: $p < p_0$

Worked Out Examples

Example 1: If in a random sample of 600 cars making a right turn at a certain trafic junction 157 drove into the wrong lane, test whether actually 30% of all drivers make this mistake or not at this given junction. Use (a) 0.05 (b) 0.01 L.O.S.

Solution: Let X be *discrete random variable* denoting the number of cars driving into the wrong lane at a junction.

1. N.H.: $H_0 : p = 0.3$
2. A.H.: $H_1 : p \neq 0.3$
3. L.O.S.: (a) $\alpha = 0.05$
(b) $\alpha = 0.01$
4. Acceptable region: $\begin{cases} \text{a) } -1.96 < Z < 1.96 \\ \text{b) } -2.57 < Z < 2.57 \end{cases}$
5. Computation: $\mu = np = (600)(0.3) = 180$,

$$\sigma = \sqrt{npq} = \sqrt{600(0.3)(0.7)} = \sqrt{126} = 11.225$$

157 in standard variable is $\frac{157-180}{11.225} = -2.0489$

6. Conclusion:
a. since $-2.0489 < -1.96$, reject N.H.
b. since $-2.0489 > -2.57$, accept N.H.

Example 2: Test the claim of a manufacturer that 95% of his 'stabilizers' confirm to ISI specifications if out of a random sample of 200 stabilizers produced by this manufacturer 18 were faulty. Use (a) 0.01 (b) 0.05 L.O.S.

Solution: Let p = probability that the stabilizers are of ISI standard i.e., good

1. N.H.: $H_0 : p = 0.95$
2. A.H.: $H_1 : p < 0.95$
3. L.O.S. (a) $\alpha = 0.01$ (b) 0.05
critical region (a) $z < -2.33$ (b) $z < -1.645$
4. If H_0 is true, $\mu = np = (200)\left(\frac{95}{100}\right) = 190$,

$$\sigma = \sqrt{npq} = \sqrt{200 \cdot \frac{95}{100} \cdot \frac{5}{100}} = 3.082$$

5. Computation: $200 - 18 = 182$ in standard unit is $\frac{182-190}{3.082} = -2.5957$
6. Conclusion:
a. since $z = -2.5957 < -2.33$, reject N.H.
b. since $z = -2.5957 < -1.645$, reject N.H.

i.e., reject the claim of the manufacturer at both levels using left O.T.T.

Exercise

1. If in a random sample of 200 persons suffering with 'headache' 160 persons got cured by a drug, can we accept the claim of the manufacturer that his drug cures 90% of the sufferers. Use 0.01 L.O.S.
Hint: $\mu = np = (200)(0.9) = 180$,
$\sigma = \sqrt{npq} = \sqrt{200(.9)(.1)} = 4.23$

Ans. Reject N.H., claim is not tenable since $z = \frac{160-180}{4.23} = -4.73 < -2.33 = z_\alpha = z_{0.01}$ by a left O.T.T.

2. A student answers by guess 32 questions correctly in an examination with 50 true or false questions. Are the results significant at (a) 0.05 L.O.S. (b) 0.01 L.O.S.
Hint: $\mu = np = (50)(0.5) = 25$,
$\sigma = \sqrt{50(0.5)10.5} = \sqrt{12.5} = 3.54$

N.H.: $p = 0.5$, student is guessing i.e., results are due to chance, A.H.: $p > 0.5$

Ans. **a.** Reject N.H. since $z = \frac{32-25}{3.54} = 1.98 > 1.645 = z_{0.05}$ by a right O.T.T.

b. Accept N.H. since $z = \frac{32-25}{3.54} = 1.98 < 2.33 = z_\alpha = z_{0.01}$, by a right O.T.T.

3. If a random sample of 120 tractors produced by a company 47 are defective, is the claim, by the company that at most 30% of the tractors are defective, tenable. Use 0.05 L.O.S.

Hint: N.H.: $p = 0.3$, A.H.: $p > 0.3$

Ans. Reject N.H., claim not tenable (valid) since $z = \frac{47-120(0.3)}{\sqrt{12(0.3)(0.7)}} = 2.191 > 1.645 = z_{0.05}$ by a right O.T.T.

4. In a sample of 90 university professors 28 own a computer. Can we conclude at 0.05 L.O.S. that at most $\frac{1}{4}$ of the professors own a computer?

Ans. Accept N.H.: $p = \frac{1}{4}$ against $H_1 : p > \frac{1}{4}$ since $z = \frac{28-(90)(0.25)}{\sqrt{90(0.25)(0.75)}} = 1.3388 < 1.645 = z_{0.05}$ by right O.T.T.

5. It is observed that 174 out of a random sample of 200 truck drivers on highway during night are drunk. Is it valid to state that at least 90% of the truck drivers are drunk. Use 0.05 L.O.S.

Ans. Accept N.H.: $p = 0.9$ against A.H.: $p < 0.9$ since $z = \frac{174-(200)(0.9)}{\sqrt{200(.9)(.1)}} = -1.41 > -1.645 = z_{0.05}$ by a left O.T.T.

6. A hospital claims that at least 40% of the patients admitted are for 'emergency' ward. Is there reason to believe this claim if the records shows that only 49 of 150 patients are for 'emergency' ward. Use 0.01 L.O.S.

Ans. Accept N.H.: $p = 0.4$ against $p < 0.4$ since $z = \frac{49-(150)(0.4)}{\sqrt{150(.4)(.6)}} = -1.833 > -2.33 = z_{0.01}$ by a left O.T.T.

29.12 TEST OF HYPOTHESIS: TWO PROPORTIONS

Suppose there are two distinct populations A and B. Let each item (member) of these two populations belongs to two mutually exclusive classes depending on whether the item has (possess) an attribute c (success) or not (failure).

	Classes having		
	attribute c (success)	without c (failures)	Total
Sample from population A	x_1	$n_1 - x_1$	n_1
Sample from population B	x_2	$n_2 - x_2$	n_2
Total	$x_1 + x_2$	$n_1 + n_2 - x_1 - x_2$	

Let x_1 and x_2 be the number of items having (possessing) attribute c (successes), in random samples of sizes n_1 and n_2 drawn from the two populations A and B respectively. Then $p_1 = \frac{x_1}{n_1}$ and $p_2 = \frac{x_2}{n_2}$ are the sample proportions. Let P_1 and P_2 be population proportions of populations A and B respectively. To determine whether the proportion of items having attribute c (success) is same in the both the populations, test the null hypothesis.

$$H_0 : P_1 = P_2$$

or

$$H_0 : P_1 - P_2 = 0$$

i.e., there is "no difference" between the two population proportions, against the A.H.: $P_1 \neq P_2$ or A.H.: $P_1 > P_2$ or A.H.: $P_1 < P_2$. For large samples (when both $n_1, n_2 \geq 30$), p_1 and p_2 are asymptotically normally distributed and therefore the sampling distribution of differences in proportions $(p_1 - p_2)$ will be approximately normally distributed with mean $\mu_{p_1-p_2} = 0$ and $\sigma_{p_1-p_2} = \sqrt{\hat{p}(1-\hat{p})\left(\frac{1}{n_1}+\frac{1}{n_2}\right)}$.

Here an unbiased pooled estimate of the population proportion $\hat{p}$ is

$$\hat{p} = \frac{n_1 p_1 + n_2 p_2}{n_1 + n_2} = \frac{x_1 + x_2}{n_1 + n_2}$$

obtained by pooling the data from both the samples. Thus the z-value for testing $p_1 = p_2$ is

$$z = \frac{p_1 - p_2}{\sigma_{p_1-p_2}} = \frac{\frac{x_1}{n_1} - \frac{x_2}{n_2}}{\sqrt{\hat{p}(1-\hat{p})\left(\frac{1}{n_1}+\frac{1}{n_2}\right)}}$$

Using the critical points of the standard normal curve, the critical regions are determined as before depending on the appropriate alternative hypothesis.

Test of Hypothesis: Difference Between Proportions

Suppose the difference between two population proportions equals to some constant δ. Then the test of hypothesis consists of
N.H.: H_0: $P_1 - P_2 = \delta$
A.H.: H_1: $P_1 - P_2 \neq \delta$ or $P_1 - P_2 > \delta$ or $P_1 - P_2 < \delta$ and the test statistic

$$z = \frac{\left(\frac{x_1}{n_1} - \frac{x_2}{n_2}\right) - \delta}{\sqrt{\frac{\frac{x_1}{n_1}\left(1 - \frac{x_1}{n_1}\right)}{n_1} + \frac{\frac{x_2}{n_2}\left(1 - \frac{x_2}{n_2}\right)}{n_2}}}$$

which for large samples is a random variable having the standard normal distribution.

Worked Out Examples

Test of hypothesis: Two proportions

Example 1: Out of two vending machines at a 'super bazar', the first machine fails to work 13 times in 250 trials and second machine fails to work 7 times in 250 trials. Test at 0.05 L.O.S. whether the difference between the corresponding sample proportions is significant.

Solution:

1. N.H.: $H_0 : p_1 = p_2$ i.e., no difference
2. A.H.: $H_1 : p_1 \neq p_2$ i.e., there is difference
3. L.O.S.: $\alpha = 0.05$
4. Critical region: Reject H_0 if

   $z < -z_{\alpha/2} = -1.96$ or if $z > z_{\alpha/2} = 1.96$
5. Computations: $x_1 = 237$, $x_2 = 243$, $n_1 = 250$, $n_2 = 250$

$$\hat{p} = \frac{x_1 + x_2}{n_1 + n_2} = \frac{237 + 243}{250 + 250} = \frac{24}{25}$$

$$z = \frac{\frac{x_1}{n_1} - \frac{x_2}{n_2}}{\sqrt{\hat{p}(1 - \hat{p})\left(\frac{1}{n_1} + \frac{1}{n_2}\right)}}$$

$$= \frac{\frac{237}{250} - \frac{243}{250}}{\sqrt{\frac{24}{25}\left(1 - \frac{24}{25}\right)\left(\frac{1}{250} + \frac{1}{250}\right)}} = -1.3698$$

6. Decision: Since $z = -1.3698$ lies between -1.96 and 1.96 accept H_0 i.e., there is no significant difference between the two vending machines.

Example 2: If 57 out of 150 patients suffering with certain disease are cured by allopathy and 33 out of 100 patients with same disease are cured by homeopathy, is there reason to believe that allopathy is better than homeopathy at 0.05 L.O.S.

Solution: Let p_1 and p_2 be proportion of patients cured by allopathy and homeopathy respectively.

1. N.H.: $H_0 : p_1 = p_2$ i.e., no difference
2. A.H.: $H_1 : p_1 > p_2$ i.e., allopathy is superior to homeopathy
3. L.O.S.: $\alpha = 0.05$
4. Criterion: Reject N.H. if $z > 1.645$ by a right O.T.T.
5. Calculations: $n_1 = 150$, $x_1 = 57$, $n_2 = 100$, $x_2 = 33$

$$\hat{p} = \frac{x_1 + x_2}{n_1 + n_2} = \frac{57 + 33}{150 + 100} = \frac{9}{25} = 0.36$$

$$z = \frac{\frac{x_1}{n_1} - \frac{x_2}{n_2}}{\sqrt{\hat{p}(1 - \hat{p})\left(\frac{1}{n_1} + \frac{1}{n_2}\right)}}$$

$$= \frac{\frac{57}{150} - \frac{33}{100}}{\sqrt{(0.36)(0.64)\left(\frac{1}{150} + \frac{1}{100}\right)}} = 0.807 \simeq 0.81$$

6. Decision: N.H. can not be rejected or accept N.H. since

$$z = 0.81 < 1.645 = z_\alpha = Z_{0.05}$$

i.e., there do not appear to be any significant difference at 0.05 level between the two treatments of allopathy and homeopathy.

Test of hypothesis: Difference between proportions

Example 3: A question in a true-false quiz is considered to be smart if it discriminates between intelligent person (IP) and average person (AP). Suppose 205 of 250 IP's and 137 of 250 AP's answer a quiz question correctly. Test at 0.01 L.O.S. whether for the given question, the proportion of correct answers can be expected to be at least 15% higher among IP's than among the AP's.

Solution: Let p_1 and p_2 be the proportion of correct answers by IP's and AP's respectively. Then

1. N.H.: $p_1 - p_2 = \delta$
2. A.H.: $p_1 - p_2 > \delta = \frac{15}{100} = 0.15$
3. L.O.S.: $\alpha = 0.01$
4. CR: Reject H_0 if $z > z_\alpha = 2.33$
5. Calculation:

$$z = \frac{\left(\frac{x_1}{n_1} - \frac{x_2}{n_2}\right) - \delta}{\sqrt{\frac{\frac{x_1}{n_1}\left(1 - \frac{x_1}{n_1}\right)}{n_1} + \frac{\frac{x_2}{n_2}\left(1 - \frac{x_2}{n_2}\right)}{n_2}}}$$

$$= \frac{\left(\frac{205}{250} - \frac{137}{250}\right) - \frac{15}{100}}{\sqrt{\frac{(.82)(.18)}{250} + \frac{(.548)(.452)}{250}}} = 3.068$$

6. Decision: Reject H_0 since $z = 3.068 > 2.33 = z_\alpha$ i.e., yes, the proportion of correct answers by IP is 15% more than those by AP's.

EXERCISE

Test of hypothesis: Two proportions

1. A study of TV viewers was conducted to find the opinion about the mega serial 'Ramayana'. If 56% of a sample of 300 viewers from south and 48% of 200 viewers from north preferred the serial, test the claim at 0.05 L.O.S. that (a) there is a difference of opinion between south and north (b) 'Ramayana' is preferred in the south.

Hint: $z = \frac{0.560-0.480}{0.0456} = 1.75$,
$\hat{p} = \frac{n_1p_1+n_2p_2}{n_1+n_2} = \frac{(300)(0.56)+(200)(.48)}{300+200}$,
$\hat{p} = 0.528$, $q = 1 - \hat{p} = 1 - .528 = .472$,
$\sigma_{p_1-p_2} = \sqrt{pq\left(\frac{1}{n_1} + \frac{1}{n_2}\right)} = 0.0456$

Ans. **a.** Accept N.H. i.e., no significant difference between north and south viewers since by a T.T.T., $z = 1.75$ lies in $(-1.96, 1.96)$.

b. Reject N.H. i.e., 'Ramayana' is preferred in the south since $z = 1.75 > 1.645$ by right O.T.T.

2. In a survey of A.C. machines produced by company A it was found that 19 machines were defective in a random sample of 200 while for company B 5 were defective out of 100. At 0.05 L.O.S. is there reason to believe that (a) there is significant difference in performance of A.C. machines between the two companies A and B (b) products of B are superior to products of A.

Hint: $\hat{p} = \frac{n_1p_1+n_2p_2}{n_1+n_2} = \frac{200(.905)+100(.95)}{200+100} = 0.92$, $q = 1 - \hat{p} = 1 - .92 = 0.08$,
$\sigma_{p_1-p_2} = \sqrt{pq\left(\frac{1}{n_1} + \frac{1}{n_2}\right)} = 0.033$,
$z = \frac{p_1-p_2}{\sigma_{p_1-p_2}} = \frac{.905-.95}{0.033} = -1.363636$.

Ans. **a.** Accept N.H. i.e., no difference between performance of A.C. machines since $z = -1.363636$ lies in acceptable region $(-1.90, 1.90)$.

b. Accept N.H. i.e., B is not superior (better) in an A, by a left O.T.T. $z_\alpha = z_{0.05} = -1.645$.

3. In a random sample of 200 parents from urban areas 120, while 240 of 500 parents from rural areas preferred 'private' professional colleges, can we conclude that parents from urban areas prefer 'private' colleges at 0.025 L.O.S.

Hint: $\hat{p}_1 = \frac{120}{200} = 0.6$, $\hat{p}_2 = \frac{240}{500} = 0.48$,
$\hat{p} = \frac{x_1+x_2}{n_1+n_2} = \frac{120+240}{200+500} = 0.51$,
$z = \frac{0.6-0.48}{\sqrt{(0.51)(0.49)\left(\frac{1}{200}+\frac{1}{500}\right)}} = 2.9$.

Ans. Reject N.H. i.e., urban parents prefer 'private' colleges since $z = 2.97 > z_\alpha = z_{0.025} = 1.96$.

4. If 48 out of 400 persons in rural area possessed 'cell' phones while 120 out 500 in urban area can it be accepted that the proportion of 'cell' phones in the rural and urban area is same or not. Use 5% L.O.S.

Hint: $p_1 = \frac{48}{300} = \frac{3}{25}, p_2 = \frac{120}{500} = \frac{6}{25}, \hat{p} = \frac{168}{900} = \frac{14}{75}, z = \frac{\left|\frac{3}{25} - \frac{6}{25}\right|}{\sqrt{\frac{14}{75} \cdot \frac{61}{75} \cdot \left(\frac{1}{400} + \frac{1}{500}\right)}} = 4.8$

Ans. Reject H_0 since $z = 4.8 > 1.96 = z_\alpha$ by a T.T.T. with $\alpha = 0.05$.

Test of hypothesis: Difference between proportions

5. In a study of the effect of drugs on 'cancer', two groups of 80 such patients were considered. One group was treated with allopathic drug while the other group with homeopathic drug. It was observed that 23 in the first group and 41 in the second group were cured. At 0.05 L.O.S. test whether the true percentage of patients cured is at least 8% less for those who were treated by homeopathic drug?

Hint: p_1 = Allopathy, p_2 = Homeopathy, N.H.: $p_1 - p_2 = \delta = 0.08$ A.H.: $p_1 - p_2 > 0.08$, $z = \frac{\left(\frac{57}{80} - \frac{39}{80}\right) - \frac{8}{100}}{\sqrt{\frac{(.57)(.43)}{80} + \frac{(.39)(.61)}{80}}} = 1.92358$.

Reject H_0 since $z = 1.92358 > 1.645 = z_\alpha = z_{0.05}$ by right O.T.T.

Ans. Reject H_0 i.e., allopathy drug is more effective or proportion of patients cured by homeopathy is at least 8% less than those treated by allopathy drug.

29.13 TEST OF HYPOTHESIS FOR SEVERAL PROPORTIONS

Consider k binomial populations with parameters $p_1, p_2, \ldots p_k$. To test whether the population proportions of these k populations are all equal, consider the N.H.: $p_1 = p_2 = \ldots = p_k$, against A.H.: these proportions are not all equal. Now draw k independent random samples of sizes $n_1, n_2, \ldots, n_k$ one from each of the k populations. Let $x_1, x_2, \ldots x_k$ denote the number of items possessing the attribute (i.e., success)

	Sample 1	Sample 2	...	Sample k	Total
Success	x_1	x_2	...	x_k	x
Failure	$n_1 - x_1$	$n_2 - x_2$	...	$n_k - x_k$	$n - x$
Total	n_1	n_2	...	n_k	n

Here x and n denote the total number of successes and total number of trials for all samples combined. The expected cell frequencies e_{ij} are calculated by

$$e_{1j} = \frac{n_j \cdot x}{n}$$

$$e_{2j} = \frac{n_j(n - x)}{n}$$

Test statistic concerning difference among proportions is given by $\chi^2 = \sum_{i=1}^{2}\sum_{j=1}^{k} \frac{(o_{ij} - e_{ij})^2}{e_{ij}}$.

Reject N.H. if $\chi^2 > \chi^2_\alpha$ with $(k - 1)$ dof.

WORKED OUT EXAMPLES

Example: Test whether there is significant difference at 0.05 level in the quality of teaching among four engineering colleges A, B, C, D of a technological university if the number of failures are 26, 23, 15, 32 respectively. Assume that each college has a strength of 200 students.

Solution: Let p_1, p_2, p_3, p_4 be the proportion of students who passed (successful) from the A, B, C, D engineering colleges respectively.

1. N.H.: $H_0 : p_1 = p_2 = p_3 = p_4$, i.e., no difference
2. A.H.: $H_1 : p_1, p_2, p_3, p_4$ are not all equal i.e., there is difference in quality of teaching.
3. L.O.S.: $\alpha = 0.05$
4. Criterion: Reject N.H. if $\chi^2 > \chi^2_{0.05}$ with $k - 1 = 4 - 1 = 3$ dof. From table, $\chi^2_{0.05}$ with 3 dof is 7.815.

5. Computation:

	Engineering Colleges					
	A	B	C	D	Total	
Success	174	177	185	168	704	
Failure	26	23	15	12	76	
Total	200	200	200	200	800	Grand Total

$e_{ij} = n_j \frac{x}{n}, e_{2j} = n_j \frac{(n-x)}{n}$ so

$e_{11} = 704\left(\frac{200}{800}\right) = 176 = e_{12} = e_{13} = e_{14}$

$e_{21} = 200 - 176 = 24 = e_{22} = e_{23} = e_{24}$

$$\chi^2 = \sum_{i=1}^{2}\sum_{j=1}^{4}(o_{ij} - e_{ij})^2/e_{ij}$$

$$= \frac{(176-174)^2}{176} + \frac{(177-176)^2}{176} + \cdots + \frac{(32-24)^2}{24}$$

$$= \frac{4}{176} + \frac{1}{176} + \frac{81}{176} + \frac{64}{176} + \frac{4}{24} + \frac{1}{24} + \frac{81}{24} + \frac{64}{24}$$

$$= 7.10$$

6. Decision:

Accept H_0 since $\chi^2 = 7.10 < \chi^2_{0.05} = 7.815$

Exercise

1. A study was conducted to estimate the proportion of wives who regularly watch TV 'serials' yielding the following data:

	PG wives	Graduate wives	Illiterate wives	Total
Watch	52	31	37	120
Do not watch	148	119	113	380
Total	200	150	150	500

Is there reason to believe at 0.05 L.O.S. that there is no difference among the true proportions of wives with different educational background who watch TV 'serials'.

Hint: $e_{11} = 48$, $e_{12} = 36$, $e_{13} = 36$, $e_{21} = 152$, $e_{22} = 114$, $e_{23} = 114$, $\chi^2 = 1.3888 < 5.991 = \chi^2_{0.05}$ with $3 - 1 = 2$ dof.

Ans. Accept N.H. i.e., no difference or wives watch TV serials irrespective of their level of literacy.

2. Three cough syrups A, B, C were used on patients with cough with the following results:

	Cough syrup A	B	C	Total
Cured	41	27	22	90
Not cured	79	53	78	210
Totals	120	80	100	300

Can we conclude whether there is significant (at 0.05 level) difference among the proportion of patients cured by the three brands A, B, C?

Hint: $e_{11} = 36$, $e_{12} = 24$, $e_{13} = 30$, $e_{21} = 84$, $e_{22} = 56$, $e_{23} = 70$

Ans. There is no difference among the three brand A, B, C in curing cough i.e., their effect is same, or accept N.H. since $\chi^2 = 4.575 < 5.991 = \chi^2_{0.05}$ with $3 - 1 = 2$ dof.

3. A survey was conducted to determine whether three categories of employees prefer pension scheme or not resulting the table given below

	Teaching	Non-teaching	Administrative	Totals
For pension	67	84	109	260
Against pension	33	66	41	140
Totals	100	150	150	400

At 0.01 L.O.S. test whether the proportions of employees favouring pension scheme are same.

Hint: $e_{11} = 65$, $e_{12} = 67.5$, $e_{13} = 97.5$, $e_{21} = 35$, $e_{22} = 52.5$, $e_{23} = 52.5$.

Ans. Reject N.H., i.e., not same since $\chi^2 = 9.392 > 9.21 = \chi^2_{0.01}$ with $3 - 1 = 2$ dof.

4. If a can containing 500 dry fruits is selected at random from each of three different companies A, B, C of mixed dry fruits and there are 345, 313 and 359 cashew nuts respectively in each of the cans, can we conclude at 0.01 L.O.S. that the mixed dry fruits of three companies contain equal proportions of cashew nuts.

Hint: $e_{11} = 339$, $e_{12} = 339$, $e_{13} = 339$, $e_{21} = 161 = e_{22} = e_{23}$.

Ans. Reject N.H. i.e., proportions of cashew nuts are different since $\chi^2 = 10.187 > \chi^2_{0.01}$ with $3 - 1 = 2$ dof $= 9.392$.

29.14 ANALYSIS OF $r \times c$ TABLES (CONTIGENCY TABLES)

A **manifold classification** is a classification in which attributes are divided into more than two classes (categories). Suppose attribute A is divided into r classes $A_1, A_2, \ldots, A_r$ and another attribute B is divided into c classes $B_1, B_2, \ldots, B_c$. Then the various cell frequencies can be expressed in the form of a table known as $r \times c$ manifold contigency table where A_i is the number of items possessing the attribute A_i with $i = 1, 2, \ldots, r$ and B_j is the number of items having attribute B_j with $j = 1, 2, \ldots, c$ and o_{ij} known as observed frequencies denotes the number of items possessing both the attributes A_i and B_j (with $i = 1, 2, \ldots, r$, and $j = 1, 2, \ldots, c$). Here the total frequency $N = \sum_{i=1}^{r} A_i = \sum_{j=1}^{c} B_j$.

	B						
A	B_1	B_2	...	B_j	...	B_c	Row Total
A_1	O_{11}	O_{12}		O_{1j}		O_{1c}	$RT1$
A_2	O_{21}	O_{22}		O_{2j}		O_{2c}	$RT2$
⋮							
A_i	O_{i1}	O_{i2}		O_{ij}		O_{ic}	RTi
⋮							
A_r	O_{r1}	O_{r2}		O_{rj}		O_{rc}	RTr
Column total	CT_1	CT_2		CT_j		CT_c	$N = \sum_{i=1}^{r} A_i = \sum_{j=1}^{c} B_j$

Here RT and CT denotes row totals and column totals respectively also known as marginal frequencies.

Thus $r \times c$ table is expressed in matrix form with r rows and and c columns containing mn cells with cell frequencies O_{ij}. These tables arise in essentially two kinds of problems.

Test for independence (TFI)

In this problem 'c' samples from one population with each item are classified with respect to two (usually qualitative) attributes. The row totals and column totals are *not fixed, but random.* Only the grand total N is fixed. The null hypothesis consist of testing whether the two attributes are independent. Then

$$p_{ij} = (\text{prob of getting value belonging to } i\text{th row}) \times (\text{prob of getting a value belonging to } j\text{th column})$$

The alternative hypothesis is that the two attributes are not independent (i.e., dependent).

Test for homogenity (TFH)

In this problem samples from several (c) populations are considered with each trial permitting more than two possible outcomes. Here both the marginal frequencies i.e., row totals and column totals are *fixed beforehand.* To test whether an attribute is common to all the populations i.e., to determine whether the c populations are "homogeneous" with respect to an attribute, consider the null hypothesis

$$\text{N.H.: } p_{i1} = p_{i2} = \cdots = p_{ic}$$

for $i = 1, 2, \ldots r$ i.e., probability of obtaining an observation in the rth row is the *same* for each column i.e., $\sum_{i=1}^{r} p_{ij} = 1$ for each column.

The alternate hypothesis is p's are not all equal for at least one row (i.e., non homogeneous).

In either of the problems the **expected cell frequencies** denoted by e_{ij} are calculated by

$$e_{ij} = \begin{cases} (\text{total observed frequencies in the } j\text{th column}) \times \\ (\text{total observed frequencies in the } r\text{th row}) \div \\ (\text{total of all cell frequencies}) \end{cases}$$

i.e.,
$$e_{ij} = \frac{(\text{column total}) \times (\text{row total})}{\text{grand total}}$$

Statistic for analysis of $r \times c$ tables is

$$\chi^2 = \sum_{i=1}^{r} \sum_{j=1}^{c} \frac{(O_{ij} - e_{ij})^2}{e_{ij}}$$

with $(r - 1)(c - 1)$ dof.

Reject N.H. if the calculated value of test statistic χ^2_{cal} or simply χ^2 exceeds χ^2_α with $(r - 1)(c - 1)$ dof.

i.e.,

Reject N.H. if $\chi^2 > \chi^2_\alpha$

WORKED OUT EXAMPLES

Test for independence (categorical data)

Examples: Test the hypothesis at 0.05 L.O.S. that the presence or absence of hypertension (HT) is independent of smoking habits from the following experimental data on 180 persons

	Non smokers	Moderate smokers	Heavy smokers
HT	21	36	30
No HT	48	26	19

Solution:

1. N.H.: H_0: Hypertension and smoking habits are independent.
2. A.H.: H_1: Hypertension and smoking habits are not independent.
3. L.O.S.: $\alpha = 0.05$.
4. Criterion: Reject N.H. if $\chi^2 > \chi^2_{0.05}$ with $\nu = (r-1)(c-1) = (2-1)(3-1) = 2$ dof.

   From table $\chi^2_{0.05} = 5.991$
5. Computations:

	NS	MS	HS	Total	
HT	21 [33.35]	36 [29.97]	30 [23.68]	87	
No HT	48 [35.65]	26 [32.03]	19 [25.32]	93	
Total	69	62	49	180	Grand total

$e_{11} = 69\left(\frac{87}{180}\right) = 33.35,\quad e_{12} = 62\left(\frac{87}{180}\right) = 29.97,\ e_{21} = 35.65,\ e_{22} = 32.03$ etc.

$\chi^2 = \frac{(21-33.35)^2}{33.35} + \frac{(36-29.97)^2}{29.97} + \cdots + \frac{(19-25.32)^2}{25.32} = 14.4644$

6. Conclusion: Reject N.H. since $\chi^2 = 14.46 > 5.99 = \chi^2_{0.05}$ i.e., Hypertension and smoking habits are dependent (i.e., not independent).

EXERCISE

Test for independence

1. A study was conducted to determine whether physical handicapness (P.H.) affects the performance of worker's in an industry with the following results:

Performance

	Good	Satisfactory	Not satisfactory	Total
Blind	21	64	17	102
Deaf	16	49	14	79
No handicap	29	93	28	150
Total	66	206	59	331

Test the claim that handicaps have no effect on performance at 0.05 L.O.S.

Hint: $e_{11} = 20.34,\ e_{12} = 63.5,\ e_{13} = 18.18,\ e_{21} = 15.75,\ e_{22} = 49.17,\ e_{23} = 14.08,\ e_{31} = 29.90,\ e_{32} = 93.35,\ e_{33} = 26.74$

$\chi^2 = .19472 \approx .195 < 9.488 = \chi^2_{0.05}$ with $(3-1)(3-1) = 4$ dof.

Ans. Accept N.H. i.e., Handicaps have no effect on performance or performance and handicaps are independent.

2. The following 'police' records shows the type of crime in four regions of a country.

Type of crime

Region	Physical assault	Murder	Rape	Homicide	Total
East	162	118	451	18	749
West	310	196	996	25	1527
North	258	193	458	10	919
South	280	175	390	19	864
Total	1010	682	2295	72	4059

Determine whether the incidence of crime depended on the region. Use 0.01 L.O.S.

Hint: $e_{11} = 186.37,\ e_{12} = 125.85,\ e_{13} = 423.49,\ e_{14} = 13.29,\ e_{21} = 379.96,\ e_{22} = 256.57,\ e_{23} = 863.38,\ e_{24} = 27.09,$

$e_{31} = 228.68,\quad e_{32} = 154.41,\quad e_{33} = 519.6,$
$e_{34} = 16.30,\quad e_{41} = 215,\quad e_{42} = 145.17,$
$e_{43} = 488.51,\quad e_{44} = 15.33$

Ans. Reject N.H. i.e., incidence of crime depends on the region since $\chi^2 = 124.5 > 21.666 = \chi^2_{0.00}$ with $(4-1)(4-1) = 9$ dof

3. In order to determine whether 'efficiency' in job depends on the 'academic performance', 400 persons were examined yielding the following data:

Efficiency \ Academic performance	Excellent	Good	Satisfactory	Total
Excellent	23	60	29	112
Good	28	79	60	167
Satisfactory	9	49	63	121
Total	60	188	152	400

Hint: $e_{11} = 16.8,\quad e_{12} = 52.6,\quad e_{13} = 42.6,$ $e_{21} = 25.0,\quad e_{22} = 78.5,\quad e_{23} = 63.5,\quad e_{31} = 18.2,\ e_{32} = 56.9,\ e_{33} = 45.9$

Ans. Yes, there is dependence between 'efficiency' in job and academic performance since $\chi^2 = 20.34 > 13.277 = \chi^2_{0.01}$ with $(3-1)(3-1) = 4$ dof (Reject N.H.)

Worked Out Examples

Test for homogeneity

Examples: A study was conducted with parents 200 from north, 150 from south, 100 from east and 100 from west regions of India to determine the current attitudes about prayers in public schools. Test at 0.01 L.O.S. for homogenity of attitudes of parents among the four regions concerning prayers in the public schools.

Solution:

1. N.H.: $p_1 = p_2 = p_3 = p_4$
2. A.H.: Not all P_i's are equal
3. L.O.S.: 0.01.
4. Critical region: Reject N.H. H_0 if $\chi^2 > \chi^2_{0.01}$ with $\nu = (3-1)(4-1) = 6$ dof i.e., if $\chi^2 > 16.812$
5. Computation:

Attitude of parents \ Region	North	South	East	West	Total
Favour	65 (74.55)	66 (55.90)	40 (37.27)	34 (37.27)	205
Oppose	42 (53.45)	30 (40.09)	33 (26.72)	42 (26.72)	147
No opinion	93 (72)	54 (54)	27 (36)	24 (36)	198
Total	200	150	100	100	550

$$\chi^2 = \sum_{i=1}^{3}\sum_{j=1}^{4} \frac{(O_{ij} - e_{ij})^2}{e_{ij}}$$

$$= \frac{(65 - 74.55)^2}{74.55} + \cdots + \frac{(24 - 36)^2}{36}$$

$$= 31.11636 \approx 31.17$$

6. Decision: Reject N.H. H_0 since $\chi^2 = 31.17 > 16.812 = \chi^2_{0.01}$ with 6 dof i.e., attitudes of parents about prayer are *not* homogeneous (not same).

Exercise

Test for homogenity

1. In order to find out the opinion about "ragging" in professional colleges, a campus study was conducted with the following results.

Ragging	Faculty	Students	Administration	Total
For	82	70	62	214
Against	93	62	67	222
Undecided	25	18	21	64
Total	200	150	150	500

Determine whether the three categories of persons are *homogeneous* w.r.t. their opinions pertaining to "ragging".

Hint: $e_{11} = 85.6,\quad e_{12} = 64.2,\quad e_{13} = 64.2,$ $e_{21} = 88.8,\quad e_{22} = 66.6,\quad e_{23} = 66.6,\quad e_{31} = 25.6,\ e_{32} = 19.2,\ e_{33} = 19.2$

Ans. Accept N.H.: homogeneous opinion since

$\chi^2 = 1.53 < 9.488$ where $\chi^2_{0.05}$ with $\nu = (3-1)(3-1) = 4$ dof is 9.488.

2. To determine the effectiveness of drugs against "aids", three types of medicines, allopathic, homeopathic and ayurvedic were tested on 50 persons with the following results.

Effectiveness	Drug: Allopathy	Homeopathy	Ayurved	Total
No relief	11	13	9	33
Some relief	32	28	27	87
Total relief	7	9	14	30
Total	50	50	50	150

Hint: $e_{11} = 11, e_{12} = 11, e_{13} = 11, e_{21} = 29, e_{22} = 29, e_{23} = 29, e_{31} = 10, e_{32} = 10, e_{33} = 10$.

Ans. Accept N.H. since $\chi^2 = 3.8100313 < 9.488 = \chi^2_{0.05}$ with $\nu = (3-1)(3-1) = 4$ dof i.e., three drugs are equally effective or the drugs are homogeneous.

3. The following table shows the opinions of voters before and after a presidential election.

	Before	After	Total
For ruling party	79	91	170
For opposition	84	66	150
Undecided	37	43	80
Total	200	200	400

Test the claim at 0.05 L.O.S. whether there has been a change of opinion of voters.

Hint: $e_{11} = e_{12} = 85, e_{21} = e_{22} = 75, e_{31} = e_{32} = 40$

Ans. No, there is no change in opinion of voters since $\chi^2 = 3.46 < 5.991 = \chi^2_{0.05}$ at $(3-1)(2-1) = 2$ dof (i.e., accept N.H.)

29.15 GOODNESS OF FIT TEST

To determine if a population follows a specified known theoretical distribution such as normal distribution, binomial distribution or Poisson distribution, the χ^2 (chi-square) test is used to assertion how closely the actual distribution approximate the assumed theoretical distributions. This test, which is based on how good a fit is there between the observed frequencies (o_i from the sample) and the expected frequencies (e_i from the theoretical distribution) is known as **"goodness-of-fit-test"**. This test judges whether the sample is drawn from a certain hypothetical distribution i.e., whether the observed frequencies follow a postulated distribution or not.

The statistic χ^2 is a measure of the discrepancy existing between the observed and expected (or theoretical) frequencies.

$$\left.\begin{array}{l}\text{Statistic for test of}\\ \text{"goodness of fit"}\end{array}\right\} \chi^2 = \sum_{i=1}^{k} \frac{(o_i - e_i)^2}{e_i} \qquad (1)$$

Here O_i and e_i are the observed and expected frequencies of the ith cell (or class interval), such that $\sum O_i = \sum e_i = N =$ Total frequency.

k is the number of cells or class intervals, in the given frequency distribution.

Here χ^2 is a random variable which is very closely approximated with ν degrees of freedom.

Degrees of Freedom (dof) for χ^2-Distribution

Let k be the number of terms in the formula (1) for χ^2. Then the dof for χ^2 is:

a. $\nu = k - 1$ if e_i can be calculated *without* having to estimate population parameters from sample statistics.

b. $\nu = k - 1 - m$ if e_i can be calculated *only by* estimating m number of population parameters from sample statistics.

Examples:

i. B.D.: p is the parameter, $m = 1$, $\nu = k - 1 - m = k - 1 - 1 = k - 2$

ii. P.D.: λ is the parameter, $m = 1$, $\nu = k - 2$

iii. N.D.: μ, σ are two parameters, $m = 2$, $\nu = k - 1 - 2 = k - 3$

Test for Goodness-of-Fit

1. N.H.: good-fit exists between the theoretical distribution and given data (of observed frequencies).
2. N.H.: no good fit.
3. L.O.S.: α (prescribed).
4. Critical region: Reject N.H. if $\chi^2 > \chi^2_\alpha$ with ν dof, i.e., theoretical distribution is a poor fit.
5. Compute χ^2 from (1).
6. Decision: Accept N.H. if $\chi^2 < \chi^2_\alpha$. i.e., the theoretical distribution is a good fit to the data.

Note 1: If $\chi^2 = 0$ then O_i and e_i agree exactly.

Note 2: when $\chi^2 > 0$.

a. χ^2 small: O_i are close to e_i, indicating "good" fit.

b. χ^2 large: O_i differ considerably from e_i indicating "poor" fit.

Conditions for Validity of χ^2-Test

1. Sample size n should be large (i.e., $n \geq 50$).
2. If individual frequencies O_i (or e_i) are small say $O_i < 10$ then combine neighbouring frequencies so that combined frequency O_i (or e_i) is ≥ 10.
3. The number of classes k should be neither too small nor too large. In general $4 \leq k \leq 16$.

WORKED OUT EXAMPLES

Example 1: Test for goodness of fit of a poisson distribution at 0.05 L.O.S. to the following frequency distribution:

Number of patients arriving/hour: (x)	0	1	2	3	4	5	6	7	8
Frequency	52	151	130	102	45	12	5	1	2

Solution: Mean of the given frequency distribution is

$$\lambda = \frac{0\times52+1\times151+2\times130+3\times102+4\times45+5\times12+6\times5+7\times1+8\times2}{52+151+130+102+45+12+5+1+2}$$

$$\lambda = \frac{1010}{500} = 2.02$$

The Poisson distribution that fits to the data with this parameter $\lambda = 2.02$ is

$$P(x, \lambda) = \frac{e^{-\lambda}\lambda^x}{x!} = \frac{e^{-2.02}(2.02)^x}{x!} x = 0, 1, 2, \ldots, 8$$

x:	0	1	2	3	4	5	6	7	8
$P(x)$	0.1326	0.26796	0.2706	0.1822	0.092	0.037	0.0125	0.0036	0.00091
Expected frequency $=500\times P(x)$	66.32 $\approx$ 66	133.94 $\approx$ 134	135.32 $\approx$ 135	91.116 $\approx$ 91	46.01 $\approx$ 46	18.5896 $\approx$ 19	6.25 $\approx$ 6	1.806 $\approx$ 2	0.456 $\approx$ 0

1. N.H.: H_0: R.V. x has Poisson distribution with $\lambda = 2.02$
2. A.H.: H_1: R.V. does *not* have P.D.
3. L.O.S. $\alpha = 0.05$
4. Critical region: Reject N.H. if $\chi^2 > 14.067$ where $\chi^2_{0.05}$ with $k - 1 - m = 9 - 1 - 1 = 9 - 2 = 7$ dof is 14.067 (since only one parameter λ is needed to calculate expected frequencies)
5. Calculation:

$$\chi^2 = \sum_i \frac{(O_i - e_i)^2}{e_i}$$

$$= \frac{(52-66)^2}{66} + \frac{(151-134)^2}{134} + \frac{(130-135)^2}{135} + \frac{(102-91)^2}{91} + \frac{(45-46)^2}{46} + \frac{(12-9)^2}{19} + \frac{[(5+1+2)-(6+2+0)]^2}{8}$$

$$\chi^2 = 9.2419$$

6. Decision: Accept N.H. i.e., Poisson distribution with $\lambda = 2.02$ is a good fit to the given frequency distribution since $\chi^2 = 9.2419 < 14.067 = \chi^2_{0.05}$ with 7 dof.

Example 2: Use 0.05 L.O.S. to test that the following given data may be treated as a random sample from a normal population

Class	Frequency
5.0–8.9	3
9.0–12.9	10
13.0–16.9	14
17.0–20.9	25
21.0–24.9	17
25.0–28.9	9
29.0–32.9	2
Total	80

Solution: A.M. $= \bar{x} = 18.85$, $\sigma =$ s.d. $= 5.5$.
The normal distribution with these two parameters is given below:

X Class Mark	$z = \frac{X-\bar{X}}{\sigma} = \frac{X-18.85}{5.5}$	Area under N.C.	Probability of a class	Expected frequency = probx 80
4.95	−2.52	.4941	0.03	2.4
8.95	−1.8	.4641	.1064	8.512
12.95	−1.072	.3577	.221	17.67
16.95	−0.345	.1368	.285	22.8
20.95	0.3818	.1480	.216	17.304
24.95	1.109	.3643	.105	8.4
28.95	1.836	.4693		
32.95	2.563	.4948	.0255	2.04

$$\chi^2 = \frac{(3-2.4)^2}{2.4} + \frac{(10-8.5)^2}{8.5} + \frac{(14-17.7)^2}{17.7} + \frac{(25-22.8)^2}{22.8} + \frac{(17-17.3)^2}{17.3} + \frac{(9-8.4)^2}{8.4} + \frac{(2-2.04)^2}{2.04} = 1.4524$$

Cannot reject H_0 i.e., accept H_0 since $\chi^2 = 1.4524 < 9.488 = \chi^2_{0.05}$ with $k - 1 - 2 = 7 - 3 = 4$ dof (since 2 parameters $\bar{X}, \sigma$ are needed).

EXERCISE

1. Test for goodness of fit of a Poisson distribution at 0.01 L.O.S. to the following observed data of e-mails received:

No. of e-mails	0	1	2	3	4	5	6	7	8	9	10	11	12	13
Observed frequency	3	15	47	76	68	74	46	39	15	9	5	2	0	1

Hint: $\lambda = \frac{1814}{400} = 4.535 \approx 4.6$

$e_0 = 4.0$, $e_1 = 18.4$, $e_2 = 42.8$, $e_3 = 65.2$, $e_4 = 74.8$, $e_5 = 69.2$, $e_6 = 52.8$, $e_7 = 34.8$, $e_8 = 20$, $e_9 = 10$, $e_{10} = 4.8$, $e_{11} = 2.0$, $e_{12} = 0.8$, $e_{13} = 0.4$

$$\chi^2 = \frac{(18-22.4)^2}{22.4} + \frac{(47-42.8)^2}{42.8} + \cdots + \frac{(8-8.0)^2}{8.0} = 6.749.$$

Ans. P.D. with $\lambda = 4.6$ provides a good fit since $\chi^2 = 6.749 < 16.919 = \chi^2_{0.01}$ with 9 dof.

2. Test for goodness of fit of a uniform distribution to the following data obtained when a die is tossed 120 times.

Face	1	2	3	4	5	6
Observed	20	22	17	18	19	24
Expected	20	20	20	20	20	20

Use 0.05 L.O.S.

Hint: $\chi^2 = \frac{(20-20)^2}{20} + \frac{(22-20)^2}{20} + \frac{(17-20)^2}{20} + \frac{(18-20)^2}{20} + \frac{(19-20)^2}{20} + \frac{(24-20)^2}{20} = 1.7.$

Ans. Uniform distribution is a good fit to the data. Accept N.H. that die is balanced since $\chi^2 = 1.7 < 11.070 = \chi^2_{0.05}$ with $6 - 1 = 5$ dof.

3. Test for goodness of fit of normal distribution to the following frequency table:

Class	1.45–1.95	1.95–2.45	2.45–2.95	2.95–3.45	3.45–3.95	3.95–4.45	4.45–4.95
Frequency O_i	2	1	4	15	10	5	3

Hint: $e_1 = 0.5 + 2.1 + 5.9 = 8.5$, $e_2 = 10.3$, $e_4 = 7.0 + 3.5 = 10.5$, $e_3 = 10.7$

$$\chi^2 = \frac{(7-8.5)^2}{8.5} + \frac{(15-10.3)^2}{10.3} + \frac{(10-10.7)^2}{10.7} + \frac{(8-10.5)^2}{10.5} = 3.05.$$

Ans. Normal distribution with $\overline{x} = 3.5$ and $\sigma = 0.7$ is a good fit since $\chi^2 = 3.05 < 7.815 = \chi^2_{0.05}$ for 3 dof (Here first three classes are clubbed to have 7 frequency and last two classes to have 8 frequency, so $k = 4$ classes $- 1 = 3$.

4. Test for goodness of fit of a binomial distribution to the data given below:

X_i	0	1	2	3	4	5	6
O_i	5	18	28	12	7	6	4

Hint: $\mu = 2.4 = 6p$, $p = 2.4$

e_i: 4 15 25 22 11 3 0

Clubbing: e_i 19 25 22 14

Ans. Reject N.H. i.e., B.D. is not a good fit since $\chi^2 = 6.39 > 5.99 = \chi^2_{0.05}$ with $4 - 2 = 2$ dof.

5. Test for goodness of fit of a Poisson distribution to the following data:

X	0	1	2	3	4	5
O_i	275	138	75	7	4	1

Hint: $\lambda = \frac{330}{500} = 0.66$, $e_1 = 258$, $e_2 = 171$, $e_3 = 56$, $e_4 = 12.4$, $e_5 = 2.05$, $e_6 = 0.25$,

$$\chi^2 = \frac{(275-258)^2}{258} + \frac{(138-171)^2}{171} + \frac{(75-56)^2}{56} + \frac{(12-15)^2}{15} = 14.534.$$

Ans. P.D. is not a good fit since $\chi^2 = 14.534 > 5.991 = \chi^2_{0.05}$ with 2 dof.

6. Test for goodness of fit of normal distribution to the following data:

Class	0–100	100–250	250–500	500–750	750–1000	1000–1250	1250–1500
Frequency	7	9	19	12	8	5	4

Hint: $e_1 = 7.62$, $e_2 = 6.32$, $e_3 = 15.26$, $e_4 = 16.17$, $e_5 = 11.52$, $e_6 = 5.24$, $e_7 = 1.88$, club the last two classes.

Ans. Normal distribution with $\overline{X} = 541.8$ and $\sigma = 375.46$ is a good fit to the given data since $\chi^2 = 4.751 < 7.81 = \chi^2_{0.05}$ with $6 - 2 - 1 = 3$ dof.

29.16 ESTIMATION OF PROPORTIONS

Engineering problems dealing with proportions, percentages or probabilities are one and the same because a proportion multiplied by 100 is percentage and proportion with number of trials very large is interpreted as probability.

Examples: Percentage of engineering students getting first class, proportion of students gaining useful employment or the probability that a first class graduate is employed.

Sample proportion $= \frac{X}{n}$ where X is the number of times an event occurs in n trials.

Sample proportion is an unbiased estimator of the of true proportion, the binomial parameter p.

Large Sample Confidence Interval for p

When n is large, normal approximation is used for binomial distribution to construct confidence interval for p from the inequality

$$-z_{\alpha/2} < \frac{X - np}{\sqrt{np(1-p)}} < z_{\alpha/2}$$

by replacing $\frac{x}{n}$ by p. Thus the confidence interval for p, when n is large, is

$$\frac{x}{n} - z_{\alpha/2}\sqrt{\frac{\frac{x}{n}\left(1-\frac{x}{n}\right)}{n}} < p < \frac{x}{n} + z_{\alpha/2}\sqrt{\frac{\frac{x}{n}\left(1-\frac{x}{n}\right)}{n}}.$$

Maximum Error of Estimate

The magnitude of error committed in using sample proportion $\frac{X}{n}$ for true proportion p is given by the maximum error of estimate E, where

$$E = z_{\alpha/2}\sqrt{\frac{p(1-p)}{n}}$$

Sample size:

a. when p is known, the sample size n is given by

$$n = p(1-p)\left[\frac{z_{\alpha/2}}{E}\right]^2$$

b. when p is unknown, then

$$n = \frac{1}{4}\left[\frac{z_{\alpha/2}}{E}\right]^2.$$

One-sided Confidence Interval

For $p \to 0$ and $n \to \infty$, binomial distribution is approximated with Poisson distribution with $\lambda = np$ and instead of the usual confidence interval, one-sided confidence interval of the form

$$p < \frac{1}{2n}\chi_\alpha^2$$

is used. Here χ_α^2 is with $2(x+1)$ degrees of freedom.

Worked Out Examples

Example 1: For an ABC insurance company, 84 insurance claims for "accident" cars out of a random sample of 200 such claims, claim amount exceeded Rs. 100000. Construct 95% confidence interval for the true proportion of insurance claims filed against the ABC company which exceeded Rs. 100000.

Solution: Here $n = 200, x = 84, z_{\alpha/2} = z_{0.025} = 1.96$ $(1-\alpha)100\%$ confidence interval for the true proportion p of insurance claims for accident cars exceeding Rs. 100000 amount is

$$\frac{x}{n} - z_{\alpha/2}\sqrt{\frac{\frac{x}{n}\left(1-\frac{x}{n}\right)}{n}} < p < \frac{x}{n} + z_{\alpha/2}\sqrt{\frac{\frac{x}{n}\left(1-\frac{x}{n}\right)}{n}}$$

i.e.,

$$\frac{84}{200} \pm 1.96\sqrt{\frac{\frac{84}{200}\left(1-\frac{84}{200}\right)}{200}}$$

or Confidence interval: $0.352 < p < 0.488$.

Example 2: Using the sample proportion as an estimate of the true proportion of insurance claims in the above example, find the maximum error of estimate with 99% confidence.

Solution:

$$\text{Maximum error} = E = z_{\alpha/2}\sqrt{\frac{\frac{x}{n}\left(1-\frac{x}{n}\right)}{n}}$$

$$= 2.575\sqrt{\frac{\frac{84}{200}\left(1-\frac{84}{200}\right)}{200}} = 0.0898671.$$

Example 3: To estimate the percentage of all "lorry" drivers exceeding 60 kmph speed on NH 5, determine the size of the smallest sample required to be at least 99% confidence that the error of estimate (sample percentage) is at most 3.5%.

Solution: Here $E = 0.035$, $z_{\alpha/2} = z_{0.005} = 2.575$ when p is unknown, sample size $n = \frac{1}{4}\left[\frac{z_{\alpha/2}}{E}\right]^2$

$$n = \frac{1}{4}\left[\frac{2.575}{0.035}\right]^2 = 1353.1887 \approx 1353.$$

Example 4: If the percentage (of all drivers) p to be estimated in the above example is known and is at most 40%, how large a sample is required?

Solution: when p is known,

$$\text{sample size} \quad n = p(1-p)\left[\frac{z_{\alpha/2}}{E}\right]^2$$

Here $p = 0.4$, $E = 0.035$, $z_{\alpha/2} = 2.575$ so

$$n = (0.4)(1-0.4)\left[\frac{2.575}{0.035}\right]^2 = 1299.06 \approx 1299.$$

Example 5: Construct an upper 95% confidence limit for the probability that a rocket will explode upon ignition, if it is found that in a random sample of 4000 such rocket firings, 10 exploded upon ignition.

Solution: Here: $n = 4000$, $\chi_\alpha^2 = \chi_{0.05}$ with $2(10+1) = 22$ dof is 33.294. So

$$p < \frac{1}{2n}\chi_\alpha^2 = \frac{1}{2(4000)}\cdot(33.294) = 0.00416175.$$

Exercise

1. Construct 95% confidence interval for the true proportion of computer literates if 36 out of 100 persons from rural areas are computer literates.

Ans. $0.266 < p < 0.454$

2. Find maximum error estimate with 99% confidences using the sample proportion as an estimate of the true proportion of parents if 136 of 400 parents for "privatization" of education.

Ans. $E = 2.575\sqrt{\frac{(.34)(.66)}{400}} = 0.061$

3. Construct 99% confidence interval for the true proportion of road accidents, if in a random sample of 400 road accidents, 231 were due to lack of traffic 'sense'.

Ans. C.I.: $\frac{231}{400} \pm 2.575\sqrt{\frac{(.5775)(.4225)}{400}} =$ (.5139, .64109)

4. Find maximum error with 95% confidence for the above Example 3 using the sample proportion to estimate the corresponding true proportion

Ans. M.E. $= 1.96\sqrt{\frac{(.5775)(.4225)}{400}} = 0.04840$

5. Construct 98% confidence interval for the probability that buyers are lured by 'discount sales' if it is found that 204 buyers out of random sample of 300 were lured by such 'discount sales'

Ans. C.I.$=\frac{204}{300} \pm 2.33\sqrt{\frac{\frac{204}{300}\frac{96}{300}}{300}} = (.61725, .74275)$

6. Find the maximum error with sample proportion as an estimate of true proportion of ragging incidents if 69 out of a random sample of 120 professional colleges reported cases of ragging.

Ans. $E = 1.96\sqrt{\frac{(.575)(.425)}{120}} = 0.0884491$

7. Determine the size of smallest sample required to estimate an unknown proportion of blind students to within a maximum error of 0.06 with at least 95% confidence.

Ans. $n = \frac{1}{4}\left[\frac{z_{\alpha/2}}{E}\right]^2 = \frac{1}{4}\left[\frac{1.96}{.06}\right]^2 = 266.77 \approx 267$

8. Given that the proportion of blind students to be estimated in the above Example 7 is at least 0.75 what should be the required sample size.

Ans. $n = p(1-p)\left[\frac{z_{\alpha/2}}{E}\right]^2 =$ $(0.75)(0.25)\left[\frac{1.96}{.06}\right]^2 = 200.08 \approx 200$

9. To estimate the true proportion of 'substandard' computers from a large consignment and to be at least 95% confident that the error is at most 0.04, what should be the size of the smallest sample required if (a) true proportion does not exceed 0.12 (b) if true proportion is not known

a. $p = 0.12$, $n = (0.12)(0.88)\left[\frac{1.96}{.04}\right]^2 = 253.55 \approx 254$

b. $n = \frac{1}{4}\left[\frac{1.96}{0.04}\right]^2 = 600.25 \approx 601$

10. In a study conducted for 500 days, only on 4 days it was recorded that 'lead' content in a famous river exceeded 200 mg/cm. Construct an upper 99% confidence limit for the probability that the 'lead' content in the river will exceed 200 mg/cm on any one day.

Ans. $p < \frac{1}{2n}\chi^2_\alpha = \frac{1}{2(500)}\chi^2_{0.01}$ with $2(4+1)$ dof

$= \frac{1}{1000}34.805 = 0.034805.$

Chapter 30

Curve Fitting, Regression and Correlation Analysis

INTRODUCTION

Approximating curve is the graph of data obtained through measurement or observation. Curve fitting is the process of finding the "best fit" curve since different approximating curves can be obtained for the same data. Least squares method is the best curve fitting method and is easily implemented on computers than the other methods like method of moments, method of group averages, graphical method.We also consider curve fitting by a sum of exponentials, linear weighted and non-linear weighted least squares approximation.

In the unvariate case, a single variable say the height of an Indian, is analyzed. Whereas in the bivariate case, two "numerical" variables are measured resulting in a pair of measurements for each member, say the height and weight of an Indian; the age and the blood pressure of a person; amount spent on advertising and volume of sales; intake of nutrious food and I.Q of a student etc. In the correlation analysis, one wish to find whether a (mathematical) relationship exists and measure the strength of such relationship. In the regression analysis, the exact nature and form of mathematical equation (of the relation) is obtaned. While the correlation coefficient measures the "closeness", the "regression equation" is used for prediction (or estimation).

30.1 CURVE FITTING

It is the method of finding equation of a curve that approximates a given set of data. On the basis of this mathematical equation, predictions can be made in many statistical investigations.

Relationship

Relationship (or association) between two (or more) variables may exist.

Examples: Blood pressure and age, rainfall and crop yield, volume of a cube and length of its side, consumption of food and weight gain, intake of drug and heart rate, height and weight, income and medical care, nutrition and I.Q.

Scatter Diagram

To find a mathematical relationship (equation) between say two variables X and Y, plot the set of given N paired observations of X and Y i.e., $(X_1, Y_1), (X_2, Y_2), \ldots, (X_N, Y_N)$ in the XY-plane. The resulting set of points is known as a scatter diagram.

Approximating Curve

It is a smooth curve that approximates the given set of N data points plotted in the scatter diagram.

Collocational polynomial For unequally spaced X_i's N coefficients $a_0, a_1, a_2, \ldots, a_{N-1}$ in the collocational polynomial

$$Y(X) = a_0 + a_1 X + a_2 X^2 + \cdots + a_{N-1} X^{N-1}$$

can be determined so that the given set of N data points $(X_1, Y_1), \ldots (X_N, Y_N)$ lies (collocates) on the curve (satisfies the above equation)

Lagrange's interpolation formula is used when X_i's are equally spaced. Collocation becomes laborious when the number of data points N is large and the empirical or observed data contains errors (i.e., when X or Y or both are random variables). In such cases data may be approximated by some function $Y = f(x)$ containing few unknown parameters.

Best fitting curve by method of least squares Let $(X_1, Y_1), (X_2, Y_2), \ldots (X_N, Y_N)$ be a given set of N data points. Let $d_i = Y_i - \hat{Y}_i$ denotes the difference between Y_i and the corresponding value

$$Y_{i\ \text{estimate}} = Y_{i\ \text{est}} = \hat{Y}_i = f(X_i)$$

determined from the curve c: $y = f(X)$ (Fig. 30.1). The d_i's known as deviations, errors or residuals which may be positive or negative or zero. Then the Legendre's principle of least squares (L.S.) or least squares criteria states that, of all the curves approximating a given set of data points, the curve having the least or minimum sum of the squares of the deviations is the "**best fitting**" curve, i.e., $\sum_{i=1}^{N} d_i^2$ the residuals or error sum of squares, is minimum.

Such a curve is known as a least squares (L.S.) curve. Thus the least squares criteria is the measure "goodness of fit". If the curve is a straight line fitted according to least squares sense then it is known as least squares straight line; if it is a parabola, it is known as least squares parabola, etc.

Some standard approximating curves

1. $Y = a_0 + a_1 X$ straight line
2. $Y = a_0 + a_1 X + a_2 X^2$

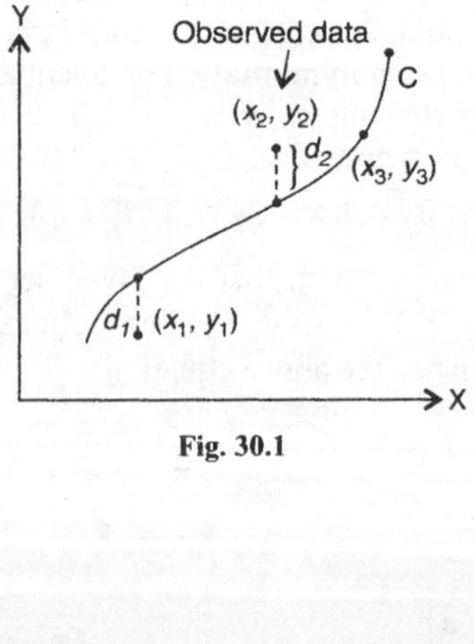

Fig. 30.1

parabola or quadratic curve

3. $Y = a_0 + a_1 X + \cdots + a_N X^n$ nth degree (polynomial) curve
4. $Y = AB^X$ exponential curve
5. $Y = AX^B$ geometric curve
6. $Y = \frac{1}{a_0 + a_1 X}$ hyperbola.

Curve Fitting by Least Squares

(i) Least squares straight line

For a given set of N data points (X_1, Y_1), $(X_2, Y_2), \ldots, (X_N, Y_N)$, assume that the straight line

$$Y = a_0 + a_1 X = f(X) \tag{1}$$

fits to the data in the least squares sense.

To determine the two unknowns a_0 and a_1 in (1), use the L.S. criteria that $\sum d_i^2$ is minimum, i.e.,

$$\sum_{i=1}^{N} d_i^2 = \sum_{i=1}^{N} (Y_i - f(X_i))^2 = \sum_{i=1}^{N} (Y_i - a_0 - a_1 X_i)^2 \tag{2}$$

is minimum. Differentiating (2) partially w.r.t. a_0 and a_1 and equating to zero, we get

$$\frac{\partial \sum d_i^2}{\partial a_0} = \frac{\partial}{\partial a_0}\left\{\sum (Y_i - a_0 - a_1 X_i)^2\right\}$$

$$= 2\sum (Y_i - a_0 - a_1 X_i)^{2-1} \cdot (-1) = 0$$

or $\quad \sum (Y_i - a_0 - a_1 X_i) = 0$

i.e.,
$$\sum_{i=1}^{N} Y_i = N a_0 + a_1 \sum_{i=1}^{N} X_i \tag{3}$$

Similarly,

$$\frac{\partial}{\partial a_1}\left(\sum d_i^2\right) = \frac{\partial}{\partial a_1}\left(\sum (Y_i - a_0 - a_1 X_i)^2\right)$$

$$= 2\sum (Y_i - a_0 - a_1 X_i)^{2-1} \cdot (-X_i) = 0$$

or
$$\sum Y_i X_i = a_0 \sum X_i + a_1 \sum X_i^2 \tag{4}$$

Thus the two unknown parameters a_0, a_1 of (1) are determined from the two equations

$$\sum Y_i = N a_0 + a_1 \sum X_i \tag{3}$$

$$\sum X_i Y_i = a_0 \sum X_i + a_1 \sum X_i^2 \tag{4}$$

known as "**normal equations**". In such a case Equation (1) represents a least squares straight line.

Note: Here and in the following normal equations, the summation is for $i = 1$ to n.

(ii) Least squares quadratic curve (parabola)

Assume that

$$Y = a_0 + a_1 X + a_2 X^2 \tag{5}$$

approximates the data according to L.S. principle. Then the unknown three parameters a_0, a_1, a_2 are determined from the following three normal equations obtained in a similar way as above.

$$\sum Y_i = Na_0 + a_1 \sum X_i + a_2 \sum X_i^2$$
$$\sum X_i Y_i = a_0 \sum X_i + a_1 \sum X_i^2 + a_2 \sum X_i^3$$
$$\sum X_i^2 Y_i = a_0 \sum X_i^2 + a_1 \sum X_i^3 + a_2 \sum X_i^4.$$

(iii) Nonlinear curves

Nonlinear curves (4), (5), (6) can be transformed to a linear curve straight line. For the exponential curve (4)

$$Y = AB^X$$

taking logarithm on both sides, we get

$$\ln Y = \ln A + X \ln B$$

or put $Y^* = \ln Y$, $\ln A = A^*$, $\ln B = B^*$, then

$$Y^* = A^* + BX$$

which is a straight line.

In a similar way by putting $Y = \frac{1}{Y^*}$ in (6), we get $Y^* = a_0 + a_1 X$ which is linear and can be solved as in *i*.

Note: Normal equations of the *n*th degree Equation (3) can be obtained formally by multiplying (3) on both sides by $1, X, X^2, \ldots, X^N$ and summing upto N terms. This results in $(N + 1)$ normal equations to determine the $(N + 1)$ unknown parameters $a_0, a_1, a_2, \ldots, a_N$ of (3).

Linear interpolation (Extrapolation) is to find Y corresponding to a value of X included between two (or outside or exterior to the) given values of X.

Result 1: Show that least squares line always passes through the point $(\overline{X}, \overline{Y})$.

Solution: Let

$$Y = a_0 + a_1 X \tag{1}$$

be the least square line (L.S.L.) of Y on X. Its normal equation is

$$\sum Y = Na_0 + a_1 \sum X$$

dividing by N,
$$\frac{\sum Y}{N} = a_0 + a_1 \frac{\sum X}{N}$$

or
$$\overline{Y} = a_0 + a_1 \overline{X} \tag{2}$$

i.e., L.S.L. passes through $(\overline{X}, \overline{Y})$.

Similarly, for

$$X = b_0 + b_1 Y \tag{3}$$

dividing by N its normal equation, we get

$$\frac{\sum X}{N} = b_0 + b_1 \frac{\sum Y}{N} \quad \text{or} \quad \overline{X} = b_0 + b_1 \overline{Y} \tag{4}$$

So L.S.L passes through $(\overline{X}, \overline{Y})$.

Result 2: Prove that L.S.L. Y on X can be expressed as

$$y = \left(\frac{\sum xy}{\sum x^2}\right) x \tag{5}$$

where $x = X - \overline{X}$, $y = Y - \overline{Y}$.

Solution: Let L.S.L. be $Y = a_0 + a_1 X$. Since it passes through $(\overline{X}, \overline{Y})$, $\overline{Y} = a_0 + a_1 \overline{X}$. Subtracting

$$Y - \overline{Y} = a_1 (X - \overline{X})$$

or
$$y = a_1 x$$

So it is enough to show that $a_1 = \frac{\sum xy}{\sum x^2}$

Solving the normal equations

$$\sum Y = Na_0 + a_1 \sum X$$
$$\sum XY = a_0 \sum X + a_1 \sum X^2,$$

we get
$$a_1 = \frac{N \sum XY - \sum X \sum Y}{N \sum X^2 - (\sum X)^2}$$

Substituting $X = x + \overline{X}, \quad Y = y + \overline{Y}$

$$a_1 = \frac{N\sum(x+\overline{X})(y+\overline{Y}) - \sum(x+\overline{X})\sum(y+\overline{Y})}{N\sum(x+\overline{X})^2 - \left(\sum(x+\overline{X})\right)^2}$$

$$= \frac{N\sum(xy+x\overline{Y}+y\overline{X}+\overline{X}\overline{Y}) - \sum(xy+x\overline{Y}+y\overline{X}+\overline{X}\overline{Y})}{N\sum(x^2+\overline{X}^2+2x\overline{X}) - \left(\sum x+N\overline{X}\right)^2}$$

$$= \frac{N\sum xy + N\overline{Y}\sum x + N\overline{X}\sum y + N^2\overline{XY} - \left(\sum x+N\overline{X}\right)\left(\sum y+N\overline{Y}\right)}{N\sum x^2 + 2N\overline{X}\sum x + N^2\overline{X}^2 - \left(\sum x + \sum\overline{X}\right)^2}$$

Since $\quad \sum x = \sum(X - \overline{X}) = 0,$

$$\sum y = \sum(Y - \overline{Y}) = 0$$

$$a_1 = \frac{N\sum xy + N^2\overline{X}\overline{Y} - N^2\overline{X}\overline{Y}}{N\sum x^2 + N^2\overline{X}^2 - N^2\overline{X}^2} = \frac{\sum xy}{\sum x^2} \qquad (6)$$

Result 3: Similarly, L.S.L. of X on Y can be expressed as

$$x = \left(\frac{\sum xy}{\sum y^2}\right) y$$

Result 4: Solving the normal equations

$$a_0 = \frac{\sum X^2 \sum Y - \sum X \sum YX}{N\sum X^2 - (\sum X)^2} \qquad (7)$$

Introducing

$$S_{XX} = N\sum X^2 - \left(\sum X\right)^2$$

$$S_{YY} = N\sum Y^2 - \left(\sum Y\right)^2$$

$$S_{XY} = N\sum XY - \left(\sum X\right)\left(\sum Y\right)$$

we have

$$a_1 = \frac{N\sum XY - \sum X \sum Y}{N\sum X^2 - (\sum X)^2} = \frac{S_{XY}}{S_{XX}}$$

$$a_0 = \frac{\sum X^2 \sum Y - \sum X \sum XY}{S_{XX}}$$

or from $\overline{Y} = a_0 + a_1\overline{X}$, we get

$$a_0 = \overline{Y} - a_1\overline{X} \quad \text{where} \quad a_1 = \frac{S_{XY}}{S_{XX}}.$$

30.2 REGRESSION ANALYSIS

In regression analysis, the nature (or form) of actual relationship if it exists, between two (or more variables) is studied by determining the mathematical equation between the variables. It is mainly used to predict or estimate one (the dependent) variable (response) in terms of the other (independent) variable(s), (regressor(s)). It is also used in optimization, to determine the values of independent variable(s) for which the dependent variable attains maximum or minimum. Sir Francis Galton (1822–1911) used regression analysis to study, whether the offspring having either short or tall parents, revert (regress) back to average height of the general population.

Simple Regression

It establishes the relationship between two variables (one dependent variable and one independent variable). In **multiple regression** the number of variables is more than two (with one dependent variable and two or more independent variables).

Linear Regression

In linear regression, the relationship between the variables, is linear and is represented by a straight line, known as a regression line or the line of average relationship or prediction equation.

Regression Line of *Y* on *X*

Suppose in the study of relationship between two variables X and Y if Y is dependent on X then the simple linear relation

$$Y = a_0 + a_1 X$$

is known as regression line of Y on X. Similarly, if X depends on Y, then

$$X = b_0 + b_1 Y$$

is known as regression line of X on Y.
In multiple regression the equation is

$$Y = f(X_1, X_2, X_3, \ldots, X_k)$$

In multiple linear regression, f is linear

i.e., $\quad Y = b_0 + b_1X_1 + b_2X_2 + \cdots + b_kX_k$

In multiple nonlinear regression, f is nonlinear, for example,

i.e., $\; Y = b_0 + b_1X_1 + b_2X_2 + b_3X_1X_2 + b_4X_1^2 + b_5X_2^2.$

30.3 INFERENCES BASED ON THE LEAST SQUARES ESTIMATION

Simple linear regression model consists of

$$Y = \alpha + \beta x + \epsilon$$

where α and β are unknown intercept and slope parameters respectively. Here ϵ, known as, random error or random disturbance is assumed to be normally distributed with mean $E(\epsilon) = 0$ and variance σ^2. The quantity σ^2 is known as residual variance or error variance. To estimate the regression coefficients α and β, a regression line $\hat{Y} = a_0 + a_1 x$ is fitted according to the principle of least squares. Here $\hat{Y}$ is the predicted or fitted or estimated value. The least square estimates of α and β, are a_0 and a_1 given by (6) and (7) in 27.1. The slope of regression line β is the change in mean of Y's corresponding to a unit increase in x.

Result: Gauss-Markov theorem: The least squares estimates a_0 and a_1 for the actual regression coefficients α and β have the smallest variance and hence most reliable among all unbiased estimators.

Confidence Intervals

A $(1-\alpha)100\%$ confidence interval for the parameter β is

$$a_1 - \frac{t_{\alpha/2} s_\epsilon \sqrt{n}}{\sqrt{S_{xx}}} < \beta < a_1 + \frac{t_{\alpha/2} s_\epsilon \sqrt{n}}{\sqrt{S_{xx}}}$$

where $t_{\alpha/2}$ is the value of the t-distribution with $n-2$ degrees of freedom.

A $(1-\alpha)100\%$ confidence interval for α is

$$a_0 - t_{\alpha/2} s_\epsilon \sqrt{\frac{S_{xx} + (n\overline{x})^2}{n S_{xx}}} < \alpha$$

$$< a_0 + t_{\alpha/2} s_\epsilon \sqrt{\frac{S_{xx} + (n\overline{x})^2}{n S_{xx}}}$$

Here $s_\epsilon^2 =$ (unbiased) estimate of σ^2

$$= \frac{S_{xx} \cdot S_{yy} - (S_{xy})^2}{n(n-2)S_{xx}}$$

Testing of hypothesis Statistics for inferences about α and β.

For slope β:

$$t = \frac{a_1 - \beta}{s_\epsilon}\sqrt{\frac{S_{xx}}{n}}$$

For intercept α:

$$t = \frac{a_0 - \alpha}{s_\epsilon}\sqrt{\frac{n S_{xx}}{S_{xx} + (n\overline{x})^2}}$$

Here t distribution is of $n-2$ degrees of freedom.

WORKED OUT EXAMPLES

Curve fitting: Least square straight line

Example 1: Find a least squares straight line for the following data:

X:	1	2	3	4	5	6
Y:	6	4	3	5	4	2

and estimate (predict) Y at $X = 4$ and X at $Y = 4$.

	X	Y	X^2	Y^2	XY
	1	6	1	36	6
	2	4	4	16	8
	3	3	9	9	9
	4	5	16	25	20
	5	4	25	16	20
	6	2	36	4	12
Total	21	24	91	106	75

so $\sum X = 21$, $\sum Y = 24$, $\sum X^2 = 91$, $\sum Y^2 = 106$, $\sum XY = 75$, $N = 6$

Assume that the least squares straight line of Y on X is $Y = a_0 + a_1 X$.
Its normal equations are

$$\sum Y = N a_0 + a_1 \sum X$$

$$\sum XY = a_0 \sum X + a_1 \sum X^2$$

Substituting the values

$$24 = 6a_0 + 21a_1$$

$$75 = 21a_0 + 91a_1$$

Solving $a_0 = 5.7999$, $a_1 = -0.51428571$
Thus the least square straight line Y on X is

$$Y = 5.7999 - 0.514X$$

$$Y_{\text{estimate}} = Y(\text{at } X = 4) = 5.7999 - 0.514(4) = 3.743$$

Similarly, L.S.S.L. of X on Y is assumed to be

$$X = b_0 + b_1 Y$$

where b_0, b_1 are obtained as solutions of the normal equations

$$\sum X = Nb_0 + b_1 \sum Y$$
$$\sum XY = b_0 \sum Y + b_1 \sum Y^2$$

or
$$21 = 6b_0 + 24b_1$$
$$75 = 24b_0 + 106b_1$$

Solving $b_0 = 7.1$, $b_1 = -0.9$ so

$$X = 7.1 - 0.94\ Y$$

$X_{\text{estimate}} = X$ (at $Y = 4$) $= 7.1 - 0.9(4) = 3.5$.

Least squares parabola

Example 2: Fit a least squares quadratic curve to the following data

X	1	2	3	4
Y	1.7	1.8	2.3	3.2

Estimate $Y(2.4)$

Solution: Assume the L.S. quadratic curve (parabola) as

$$Y = a_0 + a_1 X + a_2 X^2$$

The normal equations are

$$\sum Y = Na_0 + a_1 \sum X + a_2 \sum X^2$$
$$\sum XY = a_0 \sum X + a_1 \sum X^2 + a_2 \sum X^3$$
$$\sum X^2 Y = a_0 \sum X^2 + a_1 \sum X^3 + a_2 \sum X^4$$

Here $N = 4$

	X	Y	X^2	XY	X^3	X^4	X^2Y
	1	1.7	1	1.7	1	1	1.7
	2	1.8	4	3.6	8	16	7.2
	3	2.3	9	6.9	27	81	20.7
	4	3.2	16	12.8	64	256	51.2
Total	10	9.0	30	25.0	100	354	80.8

Substituting these sums into normal equations, we have

$$9.0 = 4a_0 + 10a_1 + 30a_2$$
$$25 = 10a_0 + 30a_1 + 100a_2$$
$$80.8 = 30a_0 + 100a_1 + 354a_2$$

Solving $\quad a_0 = 2,\ a_1 = -0.5,\ a_2 = 0.2$

Thus the required L.S. quadratic curve (parabola) is

$$Y(X) = 2 - 0.5X + 0.2X^2$$

Estimate: $Y(2.4) = 2 - 0.5(2.4) + 0.2(2.4)^2 = 1.952$

Inferences based on least square estimates

Example 3: Fit a least squares straight line (L.S.S.L.) to the following data:

X:	2	7	9	1	5	12
Y:	13	21	23	14	15	21

Solution: Here $n = 6$, $\sum X = 36$, $\sum Y = 107$, $\sum X^2 = 304$, $\sum Y^2 = 2001$, $\sum XY = 721$, $\bar{X} = \frac{36}{6} = 6$, $\bar{Y} = \frac{107}{6} = 17.833$

$$S_{xx} = n\sum_{i=1}^{n} x_i^2 - \left(\sum x_i\right)^2 = 6(304) - (36)^2 = 528$$

$$S_{yy} = n\sum y_i^2 - \left(\sum y_i\right)^2 = 6(2001) - (107)^2 = 557$$

$$S_{xy} = n\sum x_i y_i - \left(\sum x_i\right)\left(\sum y_i\right)$$
$$= 6(721) - (36)(107) = 474$$

Regression coefficient $b = \frac{S_{xy}}{S_{xx}} = \frac{474}{528} = 0.8977$

Intercept $\quad a = \bar{Y} - b\bar{X} = (17.8333) - (0.8977)6 = 12.447$

L.S.S.L: $Y = a + bX = 12.45 + 0.8977X$.

Example 4: (a) For the above Example 3, find the standard error of estimate s_ϵ^2. (b) Test for null hypothesis $\beta = 1.2$ against $\beta < 1.2$ at 0.05 level of significance.

Solution: **a.** The standard error of estimate

$$s_\epsilon^2 = \frac{S_{xx} \cdot S_{yy} - (S_{xy})^2}{n(n-2)S_{xx}} = \frac{(528)(557) - (474)^2}{6(6-4)(528)} = 5.47822$$

so $\quad s_\epsilon = 2.3405596 \approx 2.341$

b. 1. Null hypothesis: $\beta = 1.2$

2. Alternate hypothesis: $\beta < 1.2$
3. Level of significance: $\alpha = 0.05$
4. Critical region: Left One-tailed test.

   Reject N.H. if $t < -t_\alpha = -t_{0.05}$ with $n - 2$ degrees of freedom. From t-table $t_{0.05}$ with 4 D.O.F. is 2.132.

   Thus reject N.H. if $t < -2.132$

   where $t = \frac{b-\beta}{s_\epsilon}\sqrt{\frac{S_{XX}}{n}}$.
5. Computation: $n = 6,\ b = 0.8977,\ \beta = 1.2,$

   $s_\epsilon = 2.341,\ S_{XX} = 528$ so

   $$t = \frac{0.8977 - 1.2}{2.341}\left(\sqrt{\frac{528}{6}}\right) = -1.21137.$$
6. Decision: Accept N.H. (i.e., can not reject N.H.) since $t = -1.21137 > t_\alpha$ with 4 D.O.F. $= -2.132$.

Example 5: Construct a 95% confidence interval for (a) α and (b) β, for the above Example 3.

Solution:

a. 95% confidence limits for α are

$$a \pm t_{\alpha/2} s_\epsilon \sqrt{\frac{S_{XX} + (n\bar{x})^2}{nS_{XX}}}$$

$$= 12.45 \pm (2.776)(2.34)\sqrt{\frac{528 + ((6)(6))^2}{6(528)}}$$

$$= 12.45 \pm 4.93$$

Thus the 95% confidence interval for α is

$$(7.52,\ 17.379)$$

Here for $\alpha = 0.05$, $t_{\alpha/2} = t_{0.025} = 2.776$ from table.

b. 95% confidence limits for β are

$$b \pm t_{\alpha/2} s_\epsilon \sqrt{\frac{n}{S_{XX}}}$$

$$= 0.8977 \pm (2.776)(2.34)\sqrt{\frac{6}{528}}$$

$$= 0.8977 \pm 0.6925$$

Thus the 95% confidence interval for β is

$$(0.205,\ 1.59).$$

EXERCISE

Inference based on the least squares estimators

1. **a.** Predict Y at $X = 5$ by fitting a least squares straight line to the following data:

X	2	4	6	8	10	12
Y	1.8	1.5	1.4	1.1	1.1	0.9

   b. Construct a 95% confidence interval for α.

   c. Test null hypothesis $\beta = -0.12$ against $\beta > -0.12$ at 0.01 level of significance.

   Hint: $N = 6, \sum X = 42, \sum Y = 7.8, \sum X^2 = 364, \sum XY = 48.6, \sum Y^2 = 10.68$ $S_{XX} = 420, S_{YY} = 3.24, S_{XY} = -36, s_\epsilon = 0.08017$.

   Ans. **a.** $Y(X) = 1.9 - 0.086X,\ Y(5) = 1.47$

   b. confidence interval for α (1.6933, 2.1067)

   c. $t = 3.58$, reject null hypothesis

2. **a.** Estimate Y at $X = 25$ given that $N = 33$, $\sum X_i = 1104$, $\sum Y_i = 1124$, $\sum X_iY_i = 41355$, $\sum X_i^2 = 41086$.

   b. Determine a 95% confidence interval for α, β.

   c. Test the hypothesis $\beta = 1.0$ against $\beta < 1.0$.

   d. Test the hypothesis that $\alpha = 0$ against $\alpha \neq 0$ at 0.05 level of significance.

   Ans. **a.** $Y = 3.8296 + 0.9036X,\ Y(25) = 26.4196$

   b. $0.8012 < \beta < 1.0061$, $0.2132 < \alpha < 7.4461$

   c. reject N.H. $\beta = 1.0$; $t = -1.92$.

   d. reject N.H. $\alpha = 0$; $t = 2.17$.

   Hint: $S_{XX} = 4152.18,\ S_{XY} = 3752.09,$ $s_\epsilon = 3.2295$.

3. **a.** Predict Y when $X = 210$ by fitting a L.S.S.L. to the given data:

X	20	60	100	140	180	220	260	300	340	380
Y	0.18	.37	.35	.78	.56	.75	1.18	1.36	1.17	1.65

   b. Determine 95% confidence interval for α and β.

c. Test N.H. $\beta = 0$ against $\beta \neq 0$ at 0.05 level of significance.

Hint: $N = 10, \sum X = 2000, \sum X^2 = 532000, \sum Y = 8.35, \sum XY = 21754$

$S_{XX} = 1320000, S_{YY} = 21.3745,$
$S_{XY} = 5054, S_\epsilon = 0.0253.$

Ans. **a.** $Y = 0.069 + 0.0038X, Y(210) = 0.867$
b. $-0.164 < \alpha < 0.302,$
$0.00348 < \beta < 0.004119$
c. reject N.H.: $\beta = 0, (t = 8.36 > 2.306)$

30.4 CURVILINEAR (OR NONLINEAR) REGRESSION

In simple curvilinear (or nonlinear) regression, the regression equation $y = f(x)$ is non linear. Polynomial, exponential, power, reciprocal functions are some examples of nonlinear functions.

Polynomial Regression

Let $\quad Y = a_0 + a_1X + a_2X^2 + \cdots + a_NX^N$

represent a polynomial in X of degree N.

For a given set of N pair of observations (X_i, Y_i) the unknowns $a_0, a_1, a_2, \ldots, a_N$ are estimated by least square method by minimizing

$$d_i^2 = \sum_{i=1}^{N}\left[Y_i - (a_0 + a_1X_i + \cdots + a_NX_i^N)\right]^2$$

This results in the following $(N+1)$ normal equations for the determination of $(N+1)$ unknowns $a_0, a_1, a_2, \ldots, a_N$.

Normal equations

$$\sum Y_i = Na_0 + a_1\sum X_i + a_2\sum X_i^2 + \cdots + a_N\sum X_i^N$$

$$\sum X_iY_i = a_0\sum X_i + a_1\sum X_i^2 + a_2\sum X_i^3 + \cdots + a_N\sum X_i^{N+1}$$

....

....

$$\sum X_i^NY_i = a_0\sum X_i^N + a_1\sum X_i^{N+1} + a_2\sum X_i^{N+2} + \cdots + a_N\sum X_i^{2N}$$

Some special cases of polynomial curve are:

a. $N = 2$, parabola or quadratic curve

$$Y = a_0 + a_1X + a_2X^2$$

(generally used for relationship between the production of a crop and the quantity of fertilizer applied/unit area).

b. $N = 3$, cubic curve

$$Y = a_0 + a_1X + a_2X^2 + a_3X^3$$

c. $N = 4$, quartic curve

$$Y = a_0 + a_1X + a_2X^2 + a_3X^3.$$

Note: When the exact functional form $f(x)$ of the regression equation is not known, polynomial curve fitting is used.

The following simple curvilinear regression equations such as

a. Exponential growth curve: $Y = AB^X$

b. Exponential decay curve: $Y = AB^{-X}$

c. Power (geometric) curve: $Y = AX^B$

d. Reciprocal curve: $Y = \frac{1}{A+BX}$

can be transformed to simple linear regression equations by taking logarithms (or by substitution $Y = \frac{1}{Y^*}$ in case of (d)).

For example, taking logarithm $Y = AX^B$, we get

$$\ln Y = \ln A + B\ln X$$

putting $Y^* = \ln Y, X^* = \ln X, \ln A = A^*$, we have

$$Y^* = A^* + BX^*$$

which is linear X^* and Y^*.

WORKED OUT EXAMPLES

Curvilinear regression

Exponential curve

Example 1: Estimate the chlorine residual in a swimming pool 5 hours after it has been treated with chemicals by fitting an exponential curve of the form $Y = AB^X$ to the following data:

No. of hours X	2	4	6	8	10	12
Chlorine residual parts/million Y	1.8	1.5	1.4	1.1	1.1	0.9

Solution: Taking logarithm of the non-linear curve

$$Y = AB^X,$$

we get $\ln Y = \ln A + X \ln B$
Put $Y^* = \ln Y$, $\ln A = A^*$, $\ln B = B^*$
Then $Y^* = A^* + B^* X$
which is a linear equation in X. Its normal equations are

$$\sum Y^* = NA^* + B^* \sum X$$

$$\sum XY^* = A^* \sum X + B^* \sum X^2$$

	X	Y	$Y^* = \ln Y$	X^2	XY^*
	2	1.8	0.5878	4	0.1756
	4	1.5	0.4055	16	1.622
	6	1.4	0.3365	36	2.019
	8	1.1	0.0953	64	0.7264
	10	1.1	0.0953	100	0.953
	12	0.9	-0.10536	144	−1.26432
Total	42		1.415	364	5.26752

From the above table, $N = 6$, $\sum X = 42$, $\sum Y^* = 1.415$, $\sum X^2 = 364$, $\sum XY^* = 5.2675$. Thus the normal equation are

$$1.415 = 6A^* + 42B^*$$

$$5.268 = 42A^* + 364B^*$$

Solving $A^* = 0.3038$, $B^* = -0.02877$
or $A = 2.013$, $B = 0.936$.
The required least squares exponential curve is

$$Y = 2.013(0.936)^X$$

Prediction: chlorine content after 5 hours:

$$Y(X = 5) = 2.013(0.936)^5 = 1.4462 \text{ parts/million}$$

Geometric curve (power function)

Example 2: Fit a power function (geometric curve) of the form $Y = aX^b$ to the following data and estimate Y at $X = 12$:

Price X	20	16	10	11	14
Demand Y	22	41	120	89	56

Solution: Taking logarithm of the equation $Y = aX^b$, we get

$$\ln Y = \ln a + b \ln X$$

put $Y^* = \ln Y$, $A^* = \ln a$, $X^* = \ln X$, then

$$Y^* = A^* + bX^*$$

	X	Y	$X^* = \ln X$	$Y^* = \ln Y$	X^{*2}	X^*Y^*
	20	22	2.996	3.091	8.9760	9.2606
	16	41	2.77	3.7135	7.6729	10.2864
	10	120	2.30	4.7875	5.29	11.011
	11	89	2.398	4.4886	5.7504	10.763
	14	56	2.64	4.02535	6.9696	10.627
Total			13.107	20.106	34.66	51.95

So
$$20.106 = 5A^* + 13.107\,b$$

$$51.95 = 13.107A^* + 34.66\,b$$

Solving $A^* = 10.254$, $a = 28491.416$, $b = -2.37948$
Thus the least squares geometric curve is

$$Y = 28491X^{-2.38}$$

Estimate: $Y(X=12) = 28491(12)^{-2.38} = 76.956 \approx 77$.

Reciprocal function

Example 3: Estimate Y at $X = 5$ by fitting a least squares curve of the form $Y = \frac{b}{X(X-a)}$ to the following data:

X:	3.6	4.8	6.0	7.2	8.4	9.6	10.8
Y:	0.83	0.31	0.17	0.10	0.07	0.05	0.04

Solution: Rewriting the given equation

$$\frac{1}{Y} = -\frac{a}{b}X + \frac{1}{b}X^2$$

Put $Y^* = \frac{1}{Y}, \quad B^* = \frac{1}{b}, \quad A^* = -\frac{a}{b}$

so $Y^* = A^*X + B^*X^2 = a_2X + a_3X^2$

where $a_2 = A^*, \quad a_3 = B^*.$

The corresponding normal equations are

$$\sum Y^* = a_2\sum X + a_3\sum X^2$$

$$\sum XY^* = a_2\sum X^2 + a_3\sum X^3$$

$$\sum X^2Y^* = a_2\sum X^3 + a_3\sum X^4$$

Here $N = 7, \quad \sum X = 50.4, \quad \sum Y^* = 80$

$$\sum X^2 = 403, \quad \sum X^3 = 3484, \quad \sum X^4 = 31758$$

$$\sum XY^* = 709, \quad \sum X^2Y^* = 7893$$

So $709 = 403a_2 + 3484a_3$

$7893 = 3484a_2 + 31758a_3$

Solving $a_2 = -7.5472, a_3 = 1.0765$ or $a = 2.00562, b = 3.77396$
Required equation is

$$Y = \frac{3.774}{X(X-2)}$$

Now $Y(5) = 0.2516.$

EXERCISE

Parabola (quadratic curve)

1. Fit a least squares parabola to the following data:

X:	0.0	0.2	0.4	0.7	0.9	1.0
Y:	1.016	0.768	0.648	0.401	0.272	0.193

Hint: $n = 6, \sum X = 3.2, \sum X^2 = 2.5, \sum X^3 = 2.144, \sum X^4 = 1.9234, \sum Y = 3.298, \sum XY = 1.1313, \sum X^2Y = 0.74421.$

Ans. $Y = 0.999 - 1.006X + 0.210X^2$

2. Find the quadratic equation that fits the follwing data by least squares method:

X:	1	2	3	4	5	6
Y:	13235	11528	11600	12747	14940	18400

Hint: $n = 6, \sum X = 21, \sum X^2 = 91, \sum Y = 82450, \sum XY = 307179, \sum X^3 = 441, \sum X^4 = 2275, \sum X^2Y = 1403599.$

Ans. $Y = 11953 + 531.5X + 153.3X^2$

3. Fit a least squares curve of the form $Y = a_0 + a_2X^2$ for the following data:

X:	1	2.5	3.5	4.0
Y:	3.8	15.0	26.0	33.0

Hint: $n = 4, \sum X = 11, \sum Y = 77.8, \sum X^2 = 35.5, \sum X^4 = 446.125, \sum X^2Y = 944.05.$

Ans. $Y = 2.27 + 1.93X^2$

4. Using least squares method, fit a second degree polynomial. Estimate Y at $X = 6.5$

X:	0	1	2	3	4	5	6	7	8
Y:	12.0	10.5	10.0	8.0	7.0	8.0	7.5	8.5	9.0

Hint: $n = 9, \sum X = 36, \sum X^2 = 204, \sum X^3 = 1296, \sum X^4 = 8772, \sum Y = 80.5, \sum XY = 299, \sum X^2Y = 1697.$

Ans. $y = 12.2 - 1.85X + 0.183X^3, Y(6.5) = 7.9$

Exponential curve

5. Fit an exponential curve of the form $Y = Ae^{BX}$ for the following data:

X:	1	2	3	4
Y:	7	11	17	27

Ans. $Y = 4.48e^{0.45X}$

6. Predict the mean radiation dose at an altitude of 3000 feet by fitting an exponential curve to the given data:

Altitude x	50	450	780	1200	4400	4800	5300
Dose of radiation y	28	30	32	36	51	58	69

Ans. $Y = 4.26737e^{1.000067X}$

(or $\ln Y = 1.4512 + 0.000067X$)

$Y(\text{at } X = 3000) = 44.9$

7. Estimate Y at $X = 7$ by fitting $Y = AB^X$ to the data below:

Number of hours X	0	1	2	3	4	5	6
No. of bacteria/ unit volume Y	32	47	65	92	132	190	275

Ans. $Y = 32.14(1.427)^X$, $Y(7) = 387.3$

8. Fit an exponential curve by least squares

X:	1	2	5	10	20	30	40	50
Y:	98.2	91.7	81.3	64.0	36.4	32.6	17.1	11.3

Estimate Y when $X = 25$.

Ans. $Y = 100(0.96)^X$, $Y(25) = 33.9$

Geometric curve: $Y = AX^B$

9. Estimate γ by fitting the ideal gas law $PV^\gamma = c$ to the following data:

Pressure P (lb/in^2) (Y)	16.6	39.7	78.5	115.5	195.3	546.1
Volume V (in^3) (X)	50	30	20	15	10	5

Hint: $n = 6$, $\sum X = 7.352$, $\sum Y = 11.805$, $\sum X^2 = 9.63$, $\sum XY = 13.53$.

Ans. $\gamma = 1.504$, $c = 6476.33$

10. Fit a power function $Y = aX^b$ to the following data pertaining to demand for a product and its price charged at five different cities. Predict the demand when the price of the product is Rs. 12.

Price (Rs.) X	20	16	10	11	14
Demand (1000 units) Y	22	41	120	89	56

Ans. $a = 28491$, $b = -2.38$, $Y(12) = 76.9560$

11. Fit a geometric curve to the following data:

X:	1	2	4	6
Y:	6	4	2	2

Estimate $Y(2.5)$.

Hint: $n = 4$, $\sum X^* = \ln X = 3.87$, $\sum X^{*2} = 5.6$, $\sum Y^* = \ln Y = 4.56$, $\sum X^*Y^* = 3.16$.

Ans. $Y = 5.965X^{-0.672}$, $Y(2.5) = 3.2225$

12. Predict Y at $X = 3.75$ by fitting a power curve to the given data:

X:	1	2	3	4	5	6
Y:	2.98	4.26	5.21	6.10	6.80	7.50

Hint: $n = 6$, $\sum X^* = \sum \ln X = 2.8574$, $\sum Y^* = \sum \ln Y = 4.3133$, $\sum X^{*2} = 1.7749$, $\sum X^*Y^* = 2.2671$.

Ans. $Y = 2.978X^{0.5143}$, $Y(3.75) = 5.8769$

Reciprocal function

13. Estimate Y at $X = 2.25$ by fitting an indifference curve of the form $XY = AX + B$ to the following data:

X:	1	2	3	4
Y:	3	1.5	6	7.5

Hint: $Y = A + \frac{B}{X}$, put $X^* = \frac{1}{X}$, then $Y = A + BX^*$.

Ans. $XY = 1.3X + 1.7$, $Y(2.25) = 4.625$.

30.5 CURVE FITTING BY A SUM OF EXPONENTIALS

For a given set of data points

$$(x_1, y_1), (x_2, y_2), \ldots (x_n, y_n),$$

consider

$$f(x) = \sum_{i=1}^{n} A_i e^{\lambda_i x}$$

$$= A_1 e^{\lambda_1 x} + A_2 e^{\lambda_2 x} + \cdots + A_n e^{\lambda_n x} \quad (1)$$

be a sum of exponentials. The unknowns A_i and λ_i are determined since f in (1) satisfies the nth order differential equation

$$\frac{d^n y}{dx^n} + a_1 \frac{d^{n-1} y}{dx^{n-1}} + a_2 \frac{d^{n-2} y}{dx^{n-2}} + \cdots + a_n y = 0 \quad (2)$$

Here the coefficients $a_1, a_2 \ldots a_n$ are unknown constants. Froberg (1965) suggested the derivatives $y^{(n)}$, $y^{(n-1)} \ldots$ at n data points are evaluated numerically and then substituted in (2) thus obtaining a system of n linear equation for the n unknowns $a_1, a_2, \ldots a_n$. The unknown constants $\lambda_1, \lambda_2, \ldots \lambda_n$ are obtained as the roots of the algebraic equation

$$\lambda^n + a_1\lambda^{n-1} + a_2\lambda^{n-2} + \cdots + a_n = 0$$

Using the averaging technique, the unknown constants $A_1, A_2 \ldots A_n$ are calculated from (1) or using least squares method. However the disadvantage in this method is the calculation of higher order derivatives, involving round off errors, resulting in unreliable values.

The following procedure is a modification of an earlier method due to prony [see Clenshaw (1970)]. Let us assume that $f(x)$ is the sum of two exponentials i.e.,

$$y = f(x) = A_1 e^{-\lambda_1 x} + A_2 e^{-\lambda_2 x} \qquad (3)$$

The data points y_i are given at $x = 0, h, 2h, 3h$ i.e., $x = ih$, for $i = 0, 1, 2, 3$ where $h =$ spacing constant.

Substituting the set of four data points (x_0, y_0), (x_1, y_1), (x_2, y_2), (x_3, y_3) in (3), we get

$$y_i = f(x_i) = A_1 e^{-\lambda_1(ih)} + A_2 e^{-\lambda_2(ih)} \qquad (4)$$

Introducing $\eta_j = e^{-\lambda_i h}$ for $j = 1, 2,$ (5)
we rewrite (4) as

$$y_i = A_1(e^{-\lambda_1 h})^i + A_2(e^{-\lambda_2 h})^i$$

$$y_i = A_1 n_1^i + A_2 n_2^i \quad \text{for} \quad i = 0, 1, 2, 3 \qquad (6)$$

Here n_1 and n_2 are real positive roots of the quadratic

$$1 + p_1\eta + p_2\eta^2 = 0 \qquad (7)$$

From (6)

$$y_{i+1} = A_1\eta_1^{i+1} + A_2\eta_2^{i+1}$$

or

$$y_{i+1} = A_1\eta_1\eta_1^i + A_2\eta_2\eta_2^i \qquad (8)$$

From (6)

$$y_{i+2} = A_1\eta_1^{i+2} + A_2\eta_2^{i+2}$$

or

$$y_{i+2} = A_1\eta_1^2\eta_1^i + A_2\eta_2^2\eta_2^i \qquad (9)$$

Adding (6), (8), (9) we get

$$\begin{aligned} y_i + y_{i+1} \cdot p_1 + p_2 y_{i+2} &= (A_1 n_1^i + A_2 n_2^i) \\ &\quad + p_1(A_1 . n_1 n_1^i + A_2 n_2 n_2^i) \\ &\quad + p_2(A_1 n_1^2 n_1^i + A_2 n_2^2 n_2^i) \\ &= A_1 n_1^i(1 + p_1 n_1 + p_2 n_1^2) \\ &\quad + A_2(n_2^i)(1 + n_2 p_1 + n_2^2 p_2) \end{aligned}$$

since n_1 and n_2 are roots of (7), the R.H.S. above reduces to zero yielding

$$\boxed{y_i + y_{i+1} \cdot p_1 + p_2 y_{i+2} = 0} \qquad (10)$$

For $i = 0, 1$, from (10), we have

$$y_0 + p_1 y_1 + p_2 y_2 = 0 \qquad (11)$$

$$y_1 + p_2 y_2 + p_2 y_3 = 0 \qquad (12)$$

Substituting the given data y_0, y_1, y_2, y_3 in (11) and (12) and solving we get p_1 and p_2. With these values of p_1 and p_2, the quadratic equation (7) yields the two roots n_1 and n_2. From (5) we determine λ_j as

$$\lambda_j = -\frac{1}{h} \ln(n_j) \quad \text{for} \quad j = 1, 2 \qquad (13)$$

Now to determine A_1 amd A_2 we substitute the values of y_i and n_1 and y_2 in (6).

For $i = 0, 1, 2, 3$, we have

$$A_1 + A_2 = y_0$$

$$A_1 n_1 + A_2 n_2 = y_1$$

$$A_1 n_1^2 + A_2 n_2^2 = y_2$$

$$A_1 n_1^3 + A_2 n_2^3 = y_3$$

Introducing

$$P = \begin{bmatrix} 1 & 1 \\ n_1 & n_2 \\ n_1^2 & n_2^2 \\ n_1^3 & n_2^3 \end{bmatrix}_{4\times 2}, \quad B = \begin{bmatrix} y_0 \\ y_1 \\ y_2 \\ y_3 \end{bmatrix}_{4\times 1}$$

and

$$A = \begin{bmatrix} A_1 \\ A_2 \end{bmatrix}_{2\times 1}$$

the above system of 4 linear equations in 2 unknowns can be written as

$$PA = B. \qquad (14)$$

Now premultiplying (14) by P^T, this overdetermined system can be solved.

$$P^T P A = P^T B$$

or $\quad CA = D$

where $\quad C_{2\times 2} = P^T_{2\times 4} P_{4\times 2}$

and

$$D_{2\times 1} = P^T_{2\times 4} B_{4\times 1}$$

Now

$$A_{2\times 1} = \begin{bmatrix} A_1 \\ A_2 \end{bmatrix} = C^{-1}_{2\times 2} D_{2\times 1} \tag{15}$$

Putting the values of A_1, A_2 from (15) and λ_1, λ_2 from (13) in (3), we get the required fit by sum of exponentials.

Another computational technique, due to Moore (1974, *Int. J. Num. meth. in Engg.*, Vol. 8, p. 271) is presented here. It gives reliable results but involves more (numerical integration) computation. Consider a function

$$y(x) = A_1 e^{\lambda_1 x} + A_2 e^{\lambda_2 x} \tag{16}$$

which is the sum of two exponentials. Function (16) satisfies the second order differential equation

$$\frac{d^2 y}{dx^2} = a_1 \frac{dy}{dx} + a_2 y \tag{17}$$

Here a_1, a_2 are unknown constants to be determined. Assuming 'a' as the initial value of x, integrate (17) from 'a' to x. Then

$$\frac{dy}{dx} - \frac{dy(a)}{dx} = a_1 y(x) - a_1 y(a) + a_2 \int_a^x y(x)dx$$

Integrating the above equation again from 'a' to x we get

$$\begin{aligned} y(x) - y(a) - y'(a)\cdot(x-a) = a_1 &\int_a^x y(x)dx \\ &-a_1(x-a)y(a) \\ &+a_2 \int_a^x \int_a^x y(x)dx\,dx \end{aligned} \tag{18}$$

Since

$$\int_a^x \int_a^x \cdots \int_a^x f(x)dx \cdots dx = \frac{1}{(n-1)!} \int_a^x (x-t)^{n-1} f(t)dt$$

equation (18) reduces to

$$\begin{aligned} y(x) - y(a) - (x-a)y'(a) = a_1 &\int_a^x y(x)dx \\ &-a_1(x-a)y(a) \\ &+a_2 \int_a^x (x-t)y(t)dt \end{aligned} \tag{19}$$

Choosing two data points x_1 and x_2 such that $a - x_1 = x_2 - a$, we obtain from (19)

$$\begin{aligned} y(x_1) - y(a) - (x_1-a)y'(a) = a_1 &\int_a^{x_1} y(x)dx \\ &-a_1(x_1-a)y(a) \\ &+a_2 \int_a^{x_1} (x_1-t)y(t)dt \end{aligned}$$

and

$$\begin{aligned} y(x_2) - y(a) - (x_2-a)y'(a) = a_1 &\int_a^{x_2} y(x)dx \\ &-a_1(x_2-a)y(a) \\ &+a_2 \int_a^{x_2} (x_2-t)y(t)dt \end{aligned}$$

Eliminating $y'(a)$ from the above two equations we get

$$\begin{aligned} y(x_1) + y(x_2) - 2y(a) - a_1 &\left[\int_a^{x_1} y(x)dx + \int_a^{x_2} y(x)dx\right] \\ &+a_2 \left[\int_a^{x_1} (x_1-t)y(t)dt \right. \\ &\left. + \int_a^{x_2} (x_2-t)y(t)dt\right] \end{aligned} \tag{20}$$

For a choice of two pairs of (x_1, x_2), equation (20) yields two linear equations in a_1, and a_2, which are solved for a_1, and a_2. The values of λ_1 and λ_2 are obtained from $\lambda^2 = a_1\lambda + a_2$. Finally the values of A_1 and A_2 are obtained by the method of least squares.

WORKED OUT EXAMPLES

Example 1: Fit a curve by a sum of exponentials to the following data

x:	0	0.1	0.2	0.3
y:	1.175	1.336	1.510	1.693

Solution: Assume that

$$y = f(x) = A_1 e^{-\lambda_1 x} + A_2 e^{-\lambda_2 x} \qquad (1)$$

be a sum of two exponentials. We should determine the four unknown constants A_1, A_2, λ_1, λ_2 by fitting the given data to (1). Here the data x_i is evenly spaced. $x_i = x_0 + ih = 0 + ih = 0 + i(0.1)$ for $i = 0, 1, 2, 3$.
Here $h = 0.1$
Substituting the set of data points

$$(x_0 = 0, y_0 = 1.175), (x_1 = 0.1, y_1 = 1.336),$$

$$(x_2 = 0.2, y_2 = 1.510), (x_3 = 0.3, y_3 = 1.698),$$

in the equation (1), we get

$$y_i = f(x_i) = A_1 e^{-\lambda_1 x_i} + A_2 e^{-\lambda_2 x_i}$$

$$y_i = A_1 e^{-\lambda_1 ih} + A_2 e^{-\lambda_2 ih}$$

for $i = 0, 1, 2, 3$.
Put $n_s = e^{-\lambda_s h}$ for $s = 1, 2$
Then

$$y_i = A_1 n_1^i + A_2 n_2^i \quad \text{for} \quad i = 0, 1, 2, 3.$$

Using

$$y_i + p_1 y_{i+1} + p_2 y_{i+2} = 0 \quad \text{for} \quad i = 0, 1$$

we get

$$y_0 + p_1 y_1 + p_2 y_2 = 0$$

and

$$y_1 + p_1 y_2 + p_2 y_3 = 0$$

For the given data

$$1.175 + 1.336 p_1 + 1.510 p_2 = 0$$

and

$$1.336 + 1.510 p_1 + 1.698 p_2 = 0$$

solving $p_1 = -1.9193$, $p_2 = 0.91998$
substituting p_1, p_2 in

$$1 + np_1 + n^2 p_2 = 0$$

we get

$$1 - 1.9193n + 0.92n^2 = 0$$

with roots $n_1 = 1.0762$, $n_2 = 1.01$. To determine A_1 and A_2, use

$$y_i = A_1 n_1^i + A_2 n_2^i, i = 0, 1, 2, 3$$

thus

$$y_0 = A_1 + A_2$$
$$y_1 = A_1 n_1 + A_2 n_2$$
$$y_2 = A_1 n_1^2 + A_2 n_2^2$$
$$y_3 = A_1 n_1^3 + A_2 n_2^3$$

yielding

$$A_1 + A_2 = 1.75$$
$$1.076A_1 + 1.01A_2 = 1.336$$
$$1.1578A_1 + 1.02A_2 = 1.510$$
$$1.246A_1 + 1.03A_2 = 1.698$$

or $PA = B$

where

$$\begin{bmatrix} 1 & 1 \\ 1.076 & 1.01 \\ 1.1578 & 1.02 \\ 1.246 & 1.03 \end{bmatrix} \begin{bmatrix} A_1 \\ A_2 \end{bmatrix} = \begin{bmatrix} 1.75 \\ 1.336 \\ 1.510 \\ 1.698 \end{bmatrix}$$

$$PA = B$$

This overdetermined system can be solved by premultiplying both sides by P^T.

$$P^T PA = P^T B$$

Here

$$C = P^T P = \begin{bmatrix} 1 & 1.076 & 1.578 & 1.246 \\ 1 & 1.01 & 1.02 & 1.03 \end{bmatrix} \begin{bmatrix} 1 & 1 \\ 1.076 & 1.01 \\ 1.1578 & 1.02 \\ 1.246 & 1.03 \end{bmatrix}$$

$$C = \begin{bmatrix} 6.2 & 4.98 \\ 4.98 & 4.12 \end{bmatrix}$$

$$P^T B = \begin{bmatrix} 1 & 1.076 & 1.578 & 1.246 \\ 1 & 1.01 & 1.02 & 1.03 \end{bmatrix} \begin{bmatrix} 1.75 \\ 1.336 \\ 1.510 \\ 1.698 \end{bmatrix}$$

$$= \begin{bmatrix} 7.111 \\ 5.8135 \end{bmatrix}$$

From

$$CA = P^T B$$

$$A = C^{-1} P^T B$$

Now

$$C^{-1} = (1.345) \begin{bmatrix} 4.12 & -4.98 \\ -4.98 & 6.2 \end{bmatrix}$$

Then

$$A = \begin{bmatrix} A_1 \\ A_2 \end{bmatrix} = C^{-1}(P^T B)$$

$$= (1.345)\begin{bmatrix} 4.12 & -4.98 \\ -4.98 & 6.2 \end{bmatrix}\begin{bmatrix} 7.111 \\ 5.8135 \end{bmatrix}$$

$$\begin{bmatrix} A_1 \\ A_2 \end{bmatrix} = \begin{bmatrix} 0.4655 \\ 0.8486 \end{bmatrix}$$

Thus, $A_1 = 0.4655$, $A_2 = 0.8486$. Now

$$1.0762 = n_1 = e^{-\lambda_1(0.1)} \quad \text{or} \quad \lambda_1 = -0.3189$$

$$1.01 = n_2 = e^{-\lambda_2(0.1)} \quad \text{or} \quad \lambda_2 = -0.0995$$

Thus the required curve, which is a sum of exponentials, is

$$y = 0.4655e^{0.3189x} + 0.8486e^{0.0995}$$

Example 2: Fit a function $y = f(x) = A_1e^{\lambda_1 x} + A_2e^{\lambda_2 x}$, a sum of exponentials to the following data using the Moore technique.

x:	2	2.2	2.4	2.6	2.8
y:	3.63	4.46	5.47	6.70	8.19
x:	3.0	3.2	3.4	3.6	
y:	10.02	12.25	14.97	18.29	

Solution: Choose $x_1 = 2.0$, $x_2 = 2.8$ such that $a - x_1 = x_2 - a$ i.e., $a - 2.0 = 2.8 - a$ $\therefore$ $a = 2.4$.
Using equation (20), we get

$$y(2)+y(2.8)-2y(2.4) = a_1\left[\int_{2.4}^{2} y(x)dx + \int_{2.4}^{2.8} y(x)dx\right]$$
$$+a_2\left[\int_{2.4}^{2}(2-t)y(t)dt\right]$$
$$+\left[\int_{2.4}^{2.8}(2.8-t)y(t)dt\right]$$

The integrals on the RHS are evaluated using Simpson's $\frac{1}{3}$ rule, on two intervals with three points. For example:

$$I_1 = -\int_2^{2.4} y(x)dx = -\frac{0.2}{3}[3.63 + 4(4.46) + 5.47]$$

$$I_2 = \int_{2.4}^{2.8} y(x)dx = \frac{0.2}{3}[5.47 + 4(6.70) + 8.19]$$

$$I_3 = -\int_2^{2.4}(2-t)y(t)dt$$

$$I_3 = -\frac{0.2}{3}[(2-2)3.63 + 4(2-2.2)4.46 + (2-2.4)5.47]$$

$$I_4 = \frac{0.2}{3}[(2.8-2.4)5.47 + 4(2.8-2.6)6.70$$
$$+(2.8-2.8)8.19]$$

etc.
After simplification this yields one equation for a_1 and a_2 as

$$0.88 = 0.90a_1 + 0.887a_2 \qquad (1)$$

Similarly choosing $x_1 = 2.8$, $x_2 = 3.6$, we get from $a - x_1 = x_2 - a$, $a - 2.8 = 3.6 - a$ as $a = 3.2$. Substituting this data in (20), we get

$$y(2.8)+y(3.6)-2y(3.2) = a_1\left[\int_{3.2}^{2.8} y(x)dx + \int_{3.2}^{3.6} y(x)dx\right]$$
$$+a_2\left[\int_{3.2}^{2.8}(2.8-t)y(t)dt\right.$$
$$\left.+\int_{3.2}^{3.6}(3.6-t)y(t)dt\right]$$

After evaluation of R.H.S. integrals using Simpson's $\frac{1}{3}$ rule we get second equation for a_1 and a_2 as

$$1.98 = 0.6733a_1 + 1.986133a_2 \qquad (2)$$

solving (1) and (2), we get

$$a_1 - 0.0071115, a_2 = 0.999324$$

substituting a_1, a_2 in $\lambda^2 - a_1\lambda - a_2 = 0$, we get

$$\lambda^2 + 0.007\lambda - 0.9993 = 0$$

whose roots are $\lambda_1 = -1.0031$ and $\lambda_2 = 0.99615$. The equation takes the form

$$y = A_1e^{-1.0031x} + A_2e^{0.99615x}$$
$$= A_1e^{-x} + A_2e^{0.99x}$$

The values of A_1 and A_2 are determined by method of least squares by solving the two normal equations.

$$A_1 \sum_{k=1}^{9} e^{-2x_k} + A_2 \sum_k e^{-0.01x_k} = \sum_k y_k \cdot e^{-x_k}$$

and

$$A_1 \sum_{k=1}^{9} e^{-0.01x_k} + A_2 \sum_k e^{1.98x_k} = \sum_k y_k \cdot e^{0.99x_k}$$

The values of $A_1 = -0.51$, $A_2 = 0.4999$, thus the required function is

$$y = f(x) = -0.51e^{-x} + 0.499e^x$$

Note that the given data is obtained by taking the exact function $y = \sin hx$ whose values are rounded off to two decimal places.

EXERCISE

1. Fit a curve by a sum of exponentials, $y = A_1 e^{-\lambda_1 x} + A_2 e^{-\lambda_2 x}$

x:	0	0.2	0.4	0.6
y:	2.513	1.123	0.534	0.272

Ans. $p_1 = -4.1307$, $p_2 = 3.9809$,

$n_1 = 0.6529$, $n_2 = 0.3848$,

$A_1 = 0.5810$, $A_2 = 1.9322$, $\lambda_1 = 2.1316$, $\lambda_2 = -4.7712$

$y = 0.5810e^{-2.1316x} + 1.9322e^{4.7712x}$

Exact function:

$$f(x) = 0.0951e^{-x} - 0.8607e^{-3x} + 1.5576e^{-5x}$$

$$P = \begin{bmatrix} 1 & 1 \\ 0.6529 & 0.3848 \\ 0.4263 & 0.1481 \\ 0.2783 & 0.0570 \end{bmatrix}, \; P^T P = \begin{bmatrix} 1.6855 & 1.3302 \\ 1.3302 & 1.1732 \end{bmatrix}$$

2. Fit $y = f(x) = A_1 e^{\lambda_1 x} + A_2 e^{\lambda_2 x} + A_3 e^{\lambda_3 x}$ to the following data.

x:	1.0	1.1	1.2	1.3	1.4	1.5
y:	−1.543	−1.668	−1.811	−1.971	−2.151	−2.352
x:	1.6	1.7	1.8	1.9	2.0	
y:	−2.578	−2.828	−3.108	−3.418	−3.762	

Ans. $A_1 = \frac{1}{2}$, $A_2 = -\frac{1}{2}$, $A_3 = -\frac{1}{2}$, $\lambda_1 = 1$, $\lambda_2 = -1$, $\lambda_3 = -1$

3. Fit a function of the form $y = A_1 e^{\lambda_1 x} + A_2 e^{\lambda_2 x}$ to the following data:

x:	1.0	1.1	1.2	1.3	1.4
y:	1.54	1.67	1.81	1.97	2.15
x:	1.5	1.6	1.7	1.8	
y:	2.35	2.58	2.83	3.11	

Ans. $A_1 = 0.499$, $A_2 = 0.491$, $\lambda_1 = 0.988$, $\lambda_2 = -0.96$ $y = 0.499e^{0.988x} + 0.491e^{-0.96x}$

Exact function: $y = \cos hx$; with $A_1 = A_2 = \frac{1}{2}$ and $\lambda_1 = 1.0$ and $\lambda_2 = -1.0$

Hint. Solve $1.81a_1 + 2.180a_2 = 2.10$ and $2.88a_1 + 3.104a_2 = 3.00$ to get $a_1 = 0.03204$, $a_2 = 0.9364$. Solving $\lambda^2 = a_1\lambda + a_2$ gives λ_1 and λ_2.

30.6 LINEAR WEIGHTED LEAST SQUARES APPROXIMATION

Data are generally not exact. They are subject to measurement erros (known as noise in signal processing). Modeling of data aims at condensing and summarizing a given set of observations (data) by fitting it to a model, a "merit function" that depends on adjustable parameters. The parameters of the model are then adjusted to achieve a minimum in the merit function, yielding "best-fit" parameters. Least squares fitting is a maximum likelihood estimation of the fitted parameters if the measurement errors are independent and normally distributed with constant standard deviation. The least squares principle is to minimize the sum of the squares of the errors. For a given set of data, it gives a unique solution.

For discrete data $(x_0, y_0), (x_1 y_1) \ldots, (x_N y_N)$, *weights* w_i are positive numbers prescribed according to the relative accuracy of the data points. When a discrete data point (x^*, y^*) is more reliable (or accurate) than the other data, a larger weight is assigned to (x^*, y^*). If all the data points have the same accuracy then equal weights are assigned i.e., $w_i = 1$ for $i = 1$ to N.

For continuous (data) function, an integrable function $w(x)$ is called a *weight function* on $[a, b]$ of $w(x) \geq 0$ for $x \in [a, b]$. The purpose of a weight function is to assign varying degrees of importance to approximations on certain portions of the interval.

Example 1: Weight function $w(x) = (1 - x^2)^{-\frac{1}{2}}$ assigns more emphasis when $|x|$ is near one and less emphasis near the center of the interval $(-1, 1)$.

General weighted least squares approximation: Suppose the function $y = f(x)$ is known only at $(N+1)$ tabulated points (x_0, y_0), $(x_1, y_1) \ldots, (x_N, y_N)$ in the form a discrete data, with weights $w_1, w_2, \ldots w_N$; respectively.

x:	x_0	x_1	x_2	$\ldots$	x_N
y	y_0	y_1	y_2	$\ldots$	y_N
w:	w_0	w_1	w_2	$\ldots$	w_N

Then the function $f(x)$ can be approximated by a function of the form

$$p(x) = a_0\phi_0(x) + a_1\phi_1(x) + \ldots + a_m\phi_m(x) \quad (1)$$

where the set of functions $\{\phi_0, \phi_1 \ldots \phi_m\}$ are lineraly independent. These functions $\phi_i(x)$ are known as "basis" or "coordinate" functions, can be wildly nonlinear functions of x, usually chosen as $\phi_i(x) = x^i$ for $i = 0, 2, \ldots m$. The function $p(x)$ is said to be the weighted least squares approximation of $f(x)$ if the $(m+1)$ unknown coefficients $a_0, a_1, \ldots, a_m$ in (1) are determined such that the error of approximation

$$E(a_0, a_1, \cdots a_m) = \sum_{k=0}^{N} w_k \left[f(x_k) - \sum_{i=0}^{m} a_i\phi_i(x_k) \right]^2 \quad (2)$$

is minimum. The necessary conditions for the numbers $a_0, a_1, a_2, \ldots, a_m$ to minimize E are

$$\frac{\partial E}{\partial a_j} = 0 \qquad \text{for } j = 0, 1, 2, \ldots m$$

Differentiating (2) partially wrt a_j, we get $(m+1)$ linear equations in $(m+1)$ unknown $a_0, a_1, \ldots, a_m$, known as *normal equations* given by

$$\sum_{k=0}^{N} w_k \left[f(x_k) - \sum_{i=0}^{m} a_i\phi_i(x_k) \right]^2 \phi_j(x_k) \quad (3)$$

for $j = 0, 1, 2, \ldots m$. When the function $f(x)$ is known continuous function on $[a, b]$ then the normal equations take the form

$$\int_a^b w(x) \left[f(x) - \sum_{i=0}^{m} a_i\phi_i(x) \right] \phi_j(x)dx = 0$$

$$\text{for } j = 0, 1, \ldots m \quad (4)$$

Discrete (Data) Case

Suppose

$$P_1(x) = a_0 + a_1x \quad (5)$$

be the linear weighted least squares straighted line fitted to the following discrete data

$x: x_0 \; x_1 \; x_2 \ldots x_N$
$y: y_0 \; y_1 \; y_2 \ldots y_N$
$w: w_0 \; w_1 \; w_2 \ldots w_N$

In (1) we have taken $m = 1$ and $\phi_1(x) = x$. Then normal equations (3) reduce to

$$a_0 \sum_{i=0}^{N} w_i + a_1 \sum_{i=0}^{N} w_ix_i = \sum_{i=0}^{N} w_iy_i \quad (6)$$
$$a_0 \sum w_ix_i + a_1 \sum w_ix_i^2 = \sum w_ix_iy_i \quad (7)$$

Solving (6) and (7) we get a_0 and a_1 which when substituted in (5) gives the required linear weighted least squares approximation.

Continuous Function (Case)

Suppose $f(x)$ is a known continuous function defined on the interval $[a, b]$, then the normal equations (4) reduce to

$$a_0 \int_a^b w(x)dx + a_1 \int_a^b x \cdot w(x)dx = \int_a^b w(x)y(x)dx \quad (8)$$
$$a_0 \int_a^b xw(x)dx + a_1 \int_a^b x^2w(x)dx = \int_a^b xw(x)y(x)dx \quad (9)$$

Solving (8) and (9), we get a_0 and a_1. Substituting these values in $y = a_0 + a_1x$ we get the linear weighted least squares approximation in the continuous case.

WORKED OUT EXAMPLES

Linear Weighted Least Squares Approximation

Example 1: (Discrete Data) Fit a linear weighted least squares straight line to the following data

x:	−2	0	2	4	6
y:	1	3	6	8	13
w:	2	5	10	1	4

Solution: Let $y = a_0 + a_1 x$ be the LS line. Then the normal equations are

$$a_0 \sum_{i=1}^{5} w_i + a_i \Sigma w_i x_i = \Sigma w_i y_i$$

and

$$a_0 \Sigma w_i x_i + a_1 \sum w_i x_i^2 = \sum w_i x_i y_i$$

	x	y	w	wx	wx^2	wy	wxy
	−2	1	2	−4	8	2	−4
	0	3	5	0	0	15	0
	2	6	10	20	40	60	120
	4	8	1	4	16	8	32
	6	13	4	24	144	52	312
Σ			22	44	208	137	460

Thus $N = 5$, $\sum_{i=1}^{5} w_i = 22$, $\sum w_i x_i = 44$, $\sum w_i x_i^2 = 208$, $\sum w_i y_i = 137$, $\sum w_i x_i y_i = 460$.
The two normal equations are

$$22a_0 + 44a_1 = 137$$
$$44a_0 + 208a_1 = 460$$

Solving $a_1 = 1.55$, $a_0 = 3.127$. Thus the linear weighted least squares straight line fit to the given data is

$$y = 3.127 + 1.55x$$

At $x = 1$, $y(1) = 4.677$

Note: Linear (non-weighted) least squares line for the above data (with weights $w_1 = w_2 = w_3 = w_4 = w_5 = 1$) is

$$y = 3.30 + 1.45x$$

and $y(1) = 4.75$.

Example 2: (Continous function) Fit a linear weighted least squares straight line to the function: $f(x) = \frac{1}{x}$ on $[1, 3]$ with $w(x) = 1$.

Solution: Let $y = a_0 + a_1 x$ be the L.S. straight line. The normal equations are

$$a_0 \int_1^3 dx + a_1 \int_1^3 x\,dx = \int_1^3 \frac{1}{x} dx$$

$$a_0 \int_1^3 x\,dx + a_1 \int_1^3 x^2\,dx = \int_1^3 x \cdot \frac{1}{x} dx$$

or

$$2a_0 + 4a_1 = \ln 3$$
$$4a_0 + \frac{26}{3} a_1 = 2$$

Solving $a_1 = -0.2959$, $a_0 = 1.140$. The required linear LS line is

$$y = 1.140 - 0.2958x$$

At $x = 2$, $y(2) = 0.5484$. At $x = 2$, $f(x) = \frac{1}{x}$, $f(\frac{1}{2}) = 0.50$.

EXERCISE

Linear Weighted Least Squares Approximation (Discrete Data)

Fit a linear weighted least squares straight line $y = a + bx$ for the following data with the appropriate given weights.

1.

x:	−2	−1	0	1	2
y:	1	2	3	3	4
w:	1	2	3	4	5

Ans. $y = 2.63 + 0.657x$, $y\left(\frac{1}{2}\right) = 2.9585$. Without weights i.e., $w_1 = w_2 = w_3 = w_4 = w_5 = 1$, $y = y\frac{1}{2}$

Hint: $\sum w_i = 15$, $\sum w_i x_i = 10$, $\sum w_i x_i^2 = 30$, $\sum w_i y_i = 46$, $\sum w_i x_i y_i = 46$

2.

x:	−4	−2	0	2	4
y:	1.2	2.8	6.2	7.8	13.2
w:	2	3	5	3	2

Ans. $y = 6.1 + 1.43x$, $y(1) = 7.53$. Without weights i.e., $w_1 = w_2 = w_3 = w_4 = w_5 = 1$, $y = y(1)$

Hint: $\sum w_i = 15$, $\sum w_i x_i = 0$, $\sum w_i y_i = 91.6$, $\sum w_i x_i^2 = 88$, $\sum w_i x_i y_i = 126$

3.

x :	0	1	3	6	8
y :	1	3	2	5	4
w :	1	2	10	2	1

Ans. $y = 1.2 + 0.42x$

Hint: $\sum w_i = 16$, $\sum w_i x_i = 52$, $\sum w_i y_i = 41$, $\sum w_i x_i^2 = 228$, $\sum w_i x_i y_i = 158$

4.

x :	0	2	5	7
y :	−1	5	12	20
w :	1	1	10	1

Ans. $y = -1.349345 + 2.73799x$, $y(5) = 12.34061$

Hint: $\sum w_i = 13$, $\sum w_i x_i = 59$, $\sum w_i x_i^2 = 303$, $\sum w_i y_i = 144$, $\sum w_i x_i y_i = 750$. Without weights i.e., $w_1 = w_2 = w_3 = w_4 = 1$,

$$y = -1.0334 + 2.6222x$$

$$y(5) = 13.3449$$

5.

x :	0	2	5	7
y :	−1	5	12	20
w :	1	1	100	1

Ans. $y = -1.41258 + 2.6905x$, $y(5) = 12.0402$

Hint: $\sum w_i = 103$, $\sum w_i x_i = 509$, $\sum w_i x_i y_i = 6150$, $\sum w_i x_i^2 = 2553$, $\sum w_i y_i = 1224$. Without weights $y = -1.0334 + 2.622x$, $y(5) = 13.3449$.

Linear Weighted Least Squares Approximation (Continuous Function)

Fit a least squares approximation of degree one to $f(x)$ in $[a, b]$

1. $f(x) = x^2 - 2x + 3$, $[0, 1]$

Ans. $-x + 2.8333$

2. $f(x) = x^3 - 1$, $[0, 2]$

Ans. $3.6x - 2.6$

3. $f(x) = e^{-x}$, $[0, 1]$

Ans. $e^{-1}[6(e - 3)x + 2(4 - e)]$

4. $f(x) = \cos \pi x$, $[0, 1]$

Ans. $-2.4317x + 1.2159$

5. $f(x) = \ln x$, $[1, 2]$

Ans. $0.68223x - 0.63706$

30.7 NON-LINEAR WEIGHTED LEAST SQUARES APPROXIMATION

Given a set of $(N + 1)$ data points, we can fit a non linear mth degree polynomial of the form.

$$y = a_0 + a_1 x + a_2 x^2 + \ldots + a_m x^m \tag{10}$$

by minimizing the error function.

$$E(a_0, a_1, \ldots a_m) = \sum w_i [y_i - (a_0 + a_1 x_i + \ldots + a_m x_i^m)]^2 \tag{11}$$

The necessary conditions for minimization of (11) gives the following $(m + 1)$ normal equations.

$$a_0 \sum_{i=0}^{N} w_i + a_1 \sum_i x_i w_i + \cdots + a_m \sum_i x_i^m w_i = \sum_i y_i w_i .$$

$$a_0 \sum x_i w_i + a_1 \sum x_i^2 w_i + \cdots + a_m \sum x_i^{m+1} w_i = \sum x_i y_i w_i \tag{12}$$

— — — — — — — — — — — — — — — —

$$a_0 \sum x_i^m w_i + a_1 \sum x_i^{m+1} w_i + \cdots + a_m \sum x_i^{2m} w_i = \sum x_i^m y_i w_i .$$

If x_i are distinct data points and when $m < (N + 1)$, then the above set of $(m + 1)$ equations for the $(m + 1)$ unknowns $a_0, a_1, \ldots, a_m$ has *unique* solution.

Discret case

Suppose $y = a_0 + a_1 x + a_2 x^2$ is the non linear weighted least squares approximation to a given discrete set of $(N + 1)$ data points. Then normal equations (12) take the form.

$$a_0 \sum_i w_i + a_1 \sum x_i w_i + a_2 \sum x_i^2 w_i = \sum y_i w_i \tag{13}$$

$$a_0 \sum x_i w_i + a_1 \sum x_i^2 w_i + a_2 \sum x_i^3 w_i = \sum x_i y_i w_i \tag{14}$$

$$a_0 \sum x_i^2 w_i + a_1 \sum x_i^3 w_i + a_2 \sum x_i^4 w_i = \sum x_i^2 y_i w_i \tag{15}$$

The best fit parameters a_0, a_1, a_2 are obtained by solving the above three equations (13), (14) and (15).

Continuous function (case)

When $f(x)$ is a continuous function defined on the interval $[a, b]$ with a weight function $w(x)$, then the non-linear weighted least squares approximation is to minimize

$$E(a_0, a_1, \cdots a_m) = \int_a^b w(x)[y(x) - (a_0 + a_1 x + \cdots + a_m x^m)]^2 dx \quad (16)$$

Differentiating (16) partially w.r.t. the parameters $a_0, a_1, \ldots, a_m$ we get the following $(m+1)$ normal equations

$$a_0 \int_a^b w(x)dx + a_1 \int_a^b xw(x)dx + \cdots + a_m \int_a^b x^m w(x)dx = \int_a^b y(x)w(x)dx.$$

$$a_0 \int_a^b xw(x)dx + a_1 \int_a^b x^2 w(x)dx + \cdots + a_m \int_a^b x^{m+1} w(x)dx = \int_a^b xy(x)w(x)dx.$$

$$a_0 \int_a^b x^m w(x)dx + a_1 \int_a^b x^{m+1} w(x)dx + \cdots + \int_a^b x^{2m} w(x)dx = \int_a^b x^m y(x)w(x)dx.$$

When we consider a second degree polynomial of the form

$$y = a_0 + a_1 x + a_2 x^2$$

then the above normal equations take the form:

$$a_0 \int_a^b w(x)dx + a_1 \int_a^b xw(x)dx + a_2 \int_a^b x^2 w(x)dx = \int_a^b w(x) \cdot y(x)dx \quad (17)$$

$$a_0 \int_a^b xw(x)dx + a_1 \int_a^b x^2 w(x)dx + a_2 \int_a^b x^3 w(x)dx = \int_a^b xy(x)w(x)dx. \quad (18)$$

$$a_0 \int_a^b x^2 w(x)dx + a_1 \int_a^b x^3 w(x)dx + a_2 \int_a^b x^4 w(x)dx = \int_a^b x^2 y(x)w(x)dx. \quad (19)$$

Solving (17), (18) and (19) for a_0, a_1, a_2 gives the required non-linear weighted least squares approximation.

WORKED OUT EXAMPLES

Discrete (function) data

Example 1: Fit a non-linear weighted least squares (parabola) second degree polynomial $y = a_0 + a_1 x + a_2 x^2$ to the following data.

x:	−3	−1	1	3
y:	15	5	1	5
w:	2	5	10	20

Solution: The normal equations are

$$a_0 \sum_{i=1}^{4} w_i + a_1 \sum w_i x_i + a_2 \sum w_i x_i^2 = \sum w_i y_i$$

$$a_0 \sum w_i x_i + a_1 \sum w_i x_i^2 + a_2 \sum w_i x_i^3 = \sum w_i x_i y_i$$

$$a_0 \sum w_i x_i^2 + a_1 \sum w_i x_i^3 + a_2 \sum w_i x_i^4 = \sum w_i x_i^2 y_i$$

The data is shown in the following table

$N = 4, \sum w_i = 37, \sum w_i x_i = 59,$

$\sum x_i^2 w_i = 78, \sum w_i x_i^3 = 491, \sum w_i x_i^4 = 1797,$

$\sum w_i y_i = 165, \sum w_i x_i y_i = 195, \sum w_i x_i^2 y_i = 1205$

x	y	w	wx	wx^2	wx^3	wx^4	wy	wxy	wx^2y
−3	15	2	−6	18	−54	162	30	−90	270
−1	5	5	−5	5	−5	5	25	−25	25
1	1	10	10	10	10	10	10	10	10
3	5	20	60	45	540	1620	100	300	900
0	26	37	59	78	491	1797	165	195	1205

Thus the three normal equations are

$$37a_0 + 59a_1 + 78a_2 = 165$$

$$59a_0 + 78a_1 + 491a_2 = 195$$

$$78a_0 + 491a_1 + 1797a_2 = 1205$$

solving $a_0 = 0.38$, $a_1 = 2.65$, $a_2 = -0.07$
Thus the non-linear weighted least squares quadratic fit is

$$y = 0.38 + 2.65x - 0.07x^2$$

with $y(1) = 2.96$.
The corresponding least squares parapola fit (without weights i.e., $w_1 = w_2 = w_3 = w_4 = 1$) is

$$y = 2.125 - 1.70x + 0.875x^2$$

with $y(1) = 1.3$.

Note: See example II in exercise with different weights.

EXERCISE

Discrete (function) data

1. Fit a non-linear weighted least squares (parabola) second degree polynomial $y = a_0 + a_1x + a_2x^2$ to the following data.

I.

x:	−2	−1	0	1	2
y:	15	1	1	3	19
w:	1	3	10	3	1

Ans. $y = 0.1166 + x + 3.85x^2$, $y(0.5) = 1.5791$

Hint: $N = 5$, $\sum w_i = 18$, $\sum w_ix_i = 0$, $\sum w_ix_i^2 = 14$, $\sum w_ix_i^3 = 0$, $\sum w_ix_i^4 = 38$, $\sum w_iy_i = 56$, $\sum w_ix_iy_i = 14$, $\sum w_ix_i^2y_i = 148$

Note: (Non-weighted) corresponding LS parabola for the above data (with weights $w_1 = w_2 = w_3 = w_4 = w_5 = 1$) is

$$y = -1.057 + x + 4.43x^2$$

$$y(0.5) = 1.0505$$

II.

x:	−3	−1	1	3
y:	15	5	1	5
w:	1	2	10	5

Ans. $y = 1.91 - 1.72x + 0.91x^2$, $y(1) = 1.1$

Hint: $N = 4$, $\sum w_i = 18$, $\sum w_ix_i = 20$, $\sum w_ix_i^2 = 66$, $\sum w_ix_i^3 = 116$, $\sum w_ix_i^4 = 498$, $\sum w_iy_i = 60$, $\sum w_ix_iy_i = 30$, $\sum w_ix_i^2y_i = 380$

WORKED OUT EXAMPLES

Continuous function

Examples: Fit a non-linear weighted least squares (parabola) second degree polynomial $y = a_0 + a_1x + a_2x^2$ to the function $y(x) = e^x$ on the interval [0, 1] with respect to the weight function $w(x) = x$.

Solution: The three normal equations are

$$a_0\int_a^b w(x)dx + a_1\int_a^b xw(x)dx + a_2\int_a^b x^2w(x)dx = \int_a^b w(x)y(x)dx. \quad (1)$$

$$a_0\int_a^b xw(x)dx + a_1\int_a^b x^2w(x)dx + a_2\int_a^b x^3w(x)dx = \int_a^b xw(x)y(x)dx. \quad (2)$$

$$a_0\int_a^b x^2w(x)dx + a_1\int_a^b x^3w(x)dx + a_2\int_a^b x^4w(x)dx = \int_a^b x^2w(x)y(x)dx. \quad (3)$$

Here $a = 0$, $b = 1$, $w(x) = x$, $y(x) = e^x$.
So the normal equations are

$$a_0\int_0^1 xdx + a_1\int_0^1 x^2dx + a_2\int_0^1 x^3dx = \int_0^1 xe^xdx$$

$$a_0\int_0^1 x^2dx + a_1\int_0^1 x^3dx + a_2\int_0^1 x^4dx = \int_0^1 x^2e^xdx$$

$$a_0\int_0^1 x^3dx + a_1\int_0^1 x^4dx + a_2\int_0^1 x^5dx = \int_0^1 x^3e^xdx$$

After integration, we get

$$\frac{a_0}{2} + \frac{a_1}{3} + \frac{a_2}{4} = 1$$

$$\frac{a_0}{3} + \frac{a_1}{4} + \frac{a_2}{5} = e - 2$$

$$\frac{a_0}{4}+\frac{a_1}{5}+\frac{a_2}{6}=6-2e$$

or

$$6a_0+4a_1+3a_2=12$$

$$20a_0+15a_1+12a_2=60(e-2)$$

$$15a_0+12a_1+10a_2=60(6-2e).$$

Solving $a_0=(1632-600e)$, $a_1=2340e-6360$, $a_2=5220-1920e$.

Thus the required non-linear weighted least squares parabola fit to e^x with wieght function x on [0, 1] is

$$y(x)=(1632-600e)+(2340e-6360)x+(5220-1920e)x^2$$

At $x=\frac{1}{2}$, $xy(x)=\frac{1}{2}y\left(\frac{1}{2}\right)=\frac{1.6452}{2}=0.8226$

At $x=\frac{1}{2}$, $xe^x=\frac{1}{2}e^{\frac{1}{2}}=0.82436.$

Exercise

Fit a non-linear weighted least squares (parabola) second degree polynomial $y=a_0+a_1x+a_2x^2$ to the function $f(x)$ on the interval $[a, b]$ w.r.t. the weight function $w(x)$.

1. $f(x)$, e^x, $[0, 1]$, $w(x)=1$.

Ans. $y=(39e-105)+(588-216e)+$
$+(210e-570)x^2$

$y\left(\frac{1}{2}\right)=1.64838$, $e^{\frac{1}{2}}=1.64872$

Hint: Normal equations are

$$6a_0+3a_1+2a_2=6(e-1)$$

$$6a_0+4a_1+3a_2=12$$

$$20a_0+15a_1+12a_2=60(e-2)$$

2. $f(x)=\cos x$, $\left[0, \frac{\pi}{2}\right]$, $w(x)=1$

Ans. $$\cos x=\left(\frac{6}{\pi}+\frac{144}{\pi^2}-\frac{480}{\pi^3}\right)$$
$$+\left(-\frac{96}{\pi^2}+\frac{5760}{\pi^4}-\frac{1536}{\pi^3}\right)x$$
$$+\left(-\frac{11520}{\pi^5}+\frac{2880}{\pi^4}+\frac{240}{\pi^3}\right)x^2$$

At $x=\frac{\pi}{4}$, $\cos\left(\frac{\pi}{4}\right)=-\frac{3}{\pi}-\frac{60}{\pi^2}+\frac{240}{\pi^3}=$ 0.70616

Hint: Normal equations are

$$12a_0+3a_1\pi+a_2\pi^2=\frac{24}{\pi}$$

$$8a_0+\frac{8}{3}a_1\pi+a_2\pi^2=\frac{32(\pi-2)}{\pi^2}$$

$$\frac{20}{3}a_0+\frac{5}{2}a_1\pi+a_2\pi=\frac{40(\pi^2-8)}{\pi^3}$$

3. $f(x)=\sin(x)$, $\left[0, \frac{\pi}{2}\right]$, $w(x)=1$,

Ans. $$\sin x=\left(\frac{18}{\pi}+\frac{96}{\pi^2}-\frac{480}{\pi^3}\right)$$
$$+\left(-\frac{144}{\pi^2}-\frac{1344}{\pi^3}+\frac{5760}{\pi^4}\right)x$$
$$+\left(\frac{240}{\pi^3}+\frac{2880}{\pi^4}-\frac{11520}{\pi^5}\right)x^2$$

At $x=\frac{\pi}{4}$; $\sin\left(\frac{\pi}{4}\right)=-\frac{3}{\pi}-\frac{60}{\pi^2}+\frac{240}{\pi^3}=$ 0.70616

Hint: Normal equations are

$$12a_0\pi+3a_1\pi^2+a_2\pi^3=24$$

$$8a_0\pi^2+\frac{8}{3}a_1\pi^3+a_2\pi^4=64$$

$$\frac{20}{3}a_0\pi^3+\frac{5}{2}a_1\pi^4+a_2\pi^5=320\left(\frac{\pi}{2}-1\right)$$

4. $f(x)=\sin\pi x$, $[0, 1]$, $w(x)=1$

Ans. $P_2(x)=-0.050465+4.12251x-4.12251x^2$

Hint: Normal equations

$$a_0+\frac{a_1}{2}+\frac{a_2}{3}=\frac{2}{\pi},\ \frac{a_0}{2}+\frac{a_1}{3}+\frac{a_2}{4}=\frac{1}{\pi}$$

$$\frac{a_0}{3}+\frac{a_1}{4}+\frac{a_2}{5}=\frac{(\pi^2-4)}{\pi^3}$$

$$a_0=(12\pi^2-120)/\pi^3,\ a_1=-a_2$$
$$=(720-60\pi^2)/\pi^3.$$

30.8 MULTIPLE REGRESSION

It is known in agriculture that, the crop yield (Y) not only depends on the amount of rainfall (X_1) but also on the amount of fertilizer (X_2) applied, pesticides

(X_3) used, quality of seeds (X_4), quality of soil (X_5), etc. Thus in multiple regression, the dependent variable Y is a function of more than one independent variables, i.e.,

$$Y = f(X_1, X_2, X_3, \ldots, X_k).$$

In **multiple nonlinear regression**, f is nonlinear. In **multiple linear regression** f is linear i.e., $Y = \beta_0 + \beta_1 X_1 + \beta_2 X_2 + \cdots + \beta_k X_k$. Response surface analysis deals with statistical methods of prediction and optimization.

Linear Multiple Regression

Suppose Y depends on two independent variables X_1 and X_2, i.e.,

$$Y = \beta_0 + \beta_1 X_1 + \beta_2 X_2 \qquad (1)$$

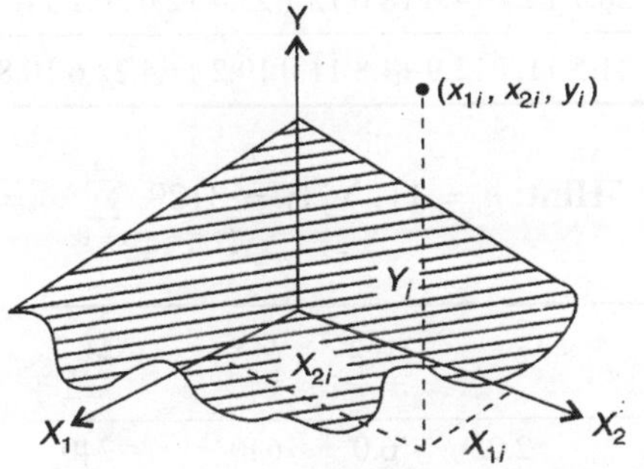

Fig. 30.2

Then the linear multiple regression problem is to fit the regression plane given by Equation (1) to a given set of N triples (X_{1i}, X_{2i}, Y_i) (Fig. 30.2). To estimate the coefficient $\beta_0, \beta_1, \beta_2$, apply the least squares method to minimize

$$\sum_{i=1}^{N} \left\{ Y_i - (b_0 + b_1 X_{1i} + b_2 X_{2i}) \right\}^2.$$

This results in three normal equations given by

$$\sum_{i=1}^{N} Y_i = N b_0 + b_1 \sum_{i=1}^{N} X_{1i} + b_2 \sum_{i=1}^{N} X_{2i}$$

$$\sum_{i=1}^{N} X_{1i} Y_i = b_0 \sum X_{1i} + b_1 \sum X_{1i}^2 + b_2 \sum X_{1i} X_{2i}$$

$$\sum X_{2i} Y_i = b_0 \sum X_{2i} + b_1 \sum X_{1i} X_{2i} + b_2 \sum X_{2i}^2$$

Here b_0, b_1, b_2 are the least squares estimates of $\beta_0, \beta_1, \beta_2$.

Note: By introducing $\overline{X}_1 = \sum \frac{X_{1i}}{N}$, $\overline{X}_2 = \frac{\sum X_{2i}}{N}$, $\overline{Y} = \frac{\sum Y_i}{N}$, $\overline{Y}_i = Y_i - \overline{Y}$, $\overline{X}_{1i} = X_{1i} - \overline{X}_1$, $\overline{X}_{2i} = X_{2i} - \overline{X}_2$, the above three normal equations reduces to 2 normal equations

$$\sum \overline{X}_{1i} \overline{Y}_i = b_1 \sum (\overline{X}_{1i})^2 + b_2 \sum \overline{X}_{1i} \overline{X}_{2i}$$

$$\sum \overline{X}_{2i} \overline{Y}_i = b_1 \sum \overline{X}_{1i} \overline{X}_{2i} + b_2 \sum (\overline{X}_{2i})^2.$$

Linear multiple regression in k-independent variables: The above analysis can be generalized to fit $N(k+1)$ tuples $(X_{1i}, X_{2i}, X_{3i}, \ldots, X_{ki})$ (with i varying from 1 to N) to the equation

$$Y = \beta_0 + \beta_1 X_1 + \beta_2 X_2 + \cdots + \beta_k X_k.$$

The $(k+1)$ normal equations are

$$\sum_{i=1}^{N} Y_i = N b_0 + b_1 \sum_{i=1}^{N} X_{1i} + b_2 \sum X_{2i} + \cdots + b_k \sum_{i=1}^{N} X_{ki}$$

$$\sum X_{1i} Y_i = b_0 \sum X_{1i} + b_1 \sum X_{1i}^2 + b_2 \sum X_{1i} X_{2i} + \cdots + b_k \sum X_{1i} X_{ki}$$

....

$$\sum X_{ki} Y_i = b_0 \sum X_{ki} X_{1i} + b_1 \sum X_{ki} X_{2i} + \cdots + b_k \sum X_{ki}^2.$$

Worked Out Examples

Examples: Fit a regression plane to estimate $\beta_0, \beta_1, \beta_2$ to the following data of a transport company on the weights of 6 shipments, the distances they were moved and the damage of the goods that was incurred. Estimate the damage when a shipment of 3700 kg is moved to a distance of 260 km.

Weight x_1 (1000 kg)	4.0	3.0	1.6	1.2	3.4	4.8
Distance x_2 (100 km)	1.5	2.2	1.0	2.0	0.8	1.6
Damage y (Rs.)	160	112	69	90	123	186

Solution: Let the dependent variable damage be y, the two independent variables be weight x_1 and distance x_2. Thus assume the equation of the regression plane as

$$y = b_0 + b_1 x_1 + b_2 x_2$$

where b_0, b_1, b_2 are the estimates of $\beta_0, \beta_1, \beta_2$. The three normal equations are

$$\sum_{i=1}^{6} y_i = nb_0 + b_1 \sum_{i=1}^{6} x_{1i} + b_2 \sum_{i=1}^{6} x_{2i}$$

$$\sum x_{1i} y_i = b_0 \sum x_{1i} + b_1 \sum x_{1i}^2 + b_2 \sum x_{1i} x_{2i}$$

$$\sum x_{2i} y_i = b_0 \sum x_{2i} + b_1 \sum x_{1i} x_{2i} + b_2 \sum x_{2i}^2.$$

Here $n = 6$. Substitute the data from the following table, in the normal equations:

	x_1 weight (1,000 kg)	x_2 distance (100 km)	y damage (Rs.)	x_1^2	x_2^2	x_1x_2	x_1y	x_2y
	4.0	1.5	160	16	2.25	6.0	640	240
	3.0	2.2	112	9	4.84	6.6	336	246.4
	1.6	1.0	69	2.56	1.0	1.6	110.4	69
	1.2	2.0	90	1.44	4.0	2.4	108	180
	3.4	0.8	123	11.56	0.64	2.72	418.2	98.4
	4.8	1.6	186	23.04	2.56	7.68	892.8	297.6
Total	18	9.1	740	63.6	15.29	27	250.54	1131.4

$n = 6, \sum x_{1i} = 18, \sum x_{2i} = 9.1, \sum y_i = 740,$
$\sum x_{1i}^2 = 63.6, \sum x_{2i}^2 = 15.29, \sum x_{1i} x_{2i} = 27,$
$\sum x_{1i} y_i = 250.54, \sum x_{2i} y_i = 1131.4.$

So $740 = 6b_0 + 18b_1 + 9.1b_2$

$250.54 = 18b_0 + 63.6b_1 + 27b_2$

$1131.4 = 9.1b_0 + 27b_1 + 15.29b_2$

Solving we get $b_0 = 14.56$, $b_1 = 30.109$, $b_2 = 12.16$. Thus the required regression plane is

$$y = 14.56 + 30.109x_1 + 12.16x_2$$

Estimate: For a weight of 3700 kg ($x_1 = 3.7$) and for a distance of 260 km ($x_2 = 2.6$), the damage incurred in rupees is

$$y(x_1 = 3.7, x_2 = 2.6) = 14.56 + 30.109(3.7) + 12.16(2.6)$$
$$= \text{Rs. } 714.5798 \approx \text{Rs. } 715.$$

EXERCISE

1. Estimate Y for given $X_1 = 12$ and $X_2 = 10$ by fitting a regression plane to the following data:

Y	412	226	292	323	233	368	239	382	218	222	214
X_1	28.7	13.4	14.6	18.0	12.1	23.4	12.6	30.2	11.6	12.0	12.4
X_2	21.5	11.7	12.9	14.8	11.0	19.2	11.4	22.6	10.8	10.2	10.1

Hint: $n = 11, \sum Y_i = 3129, \sum_{i=1}^{11} X_{1i} = 189.0,$ $\sum X_{2i} = 156.2, \sum X_{1i}^2 = 3737.50, \sum X_{2i}^2 = 2437.64, \sum X_{1i} X_{2i} = 3010.03,$ $\sum X_{1i} Y_i = 58754.7, \sum X_{2i} Y_i = 47816.$

Ans. $Y = 40.96 - 6.30X_1 + 24.77X_2,$
$Y(12, 10) = 213.06$

2. Find Y when $X_1 = 10$ and $X_2 = 6$ from the least squares regression equation of Y on X_1 and X_2 for the following data:

Y	90	72	54	42	30	12
X_1	3	5	6	8	12	14
X_2	16	10	7	4	3	2

Hint: $n = 6, \sum Y_i = 300, \sum X_{1i} = 48,$
$\sum X_{2i} = 42, \sum X_{1i}X_{2i} = 236, \sum X_{1i}^2 = 474,$
$\sum X_{2i}^2 = 434, \sum X_{1i}Y_i = 1818,$
$\sum X_{2i}Y_i = 2820.$

Ans. $Y = 61.40 - 3.65X_1 + 2.54X_2,$
$Y(10, 6) = 40.14 \approx 40$

3. Determine the equation of the regression plane connecting x_1, x_2 and y. Estimate y at $x_1 = 1.8, x_2 = 112$.

Diffusion time (hours) x_1	1.5	2.5	0.5	1.2	2.6	0.3	2.4	2.0	0.7	1.6
Sheet-resistance Ohms-cm x_2	66	87	69	141	93	105	111	78	66	123
Current gain y	5.3	7.8	7.4	9.8	10.8	9.1	8.1	7.2	6.5	12.6

Hint: $n = 10, \sum x_{1i} = 15.3, \sum x_{2i} = 939,$
$\sum y_i = 84.6, \sum x_{1i}^2 = 29.85, \sum x_{2i}^2 = 94131,$
$\sum x_{1i}x_{2i} = 1458.9, \sum x_{1i}y_i = 132.27,$
$\sum x_{2i}y_i = 8320.2.$

Ans. $y = 2.27 + 0.22x_1 + 0.062x_2,$
$y(1.8, 112) = 9.61$

4. Fit a regression plane of y on x_1 and x_2 given $\sum(\overline{x}_{1i})^2 = 38.4$, $\sum(\overline{x}_{2i})^2 = 3.4$, $\sum \overline{x}_{1i}\overline{y}_i = 29.76$, $\sum \overline{x}_{2i}\overline{y}_i = 8.94$, $\sum \overline{x}_{1i}\overline{x}_{2i} = 9.6$ where $\overline{y}_i = y_i - \overline{y}$, $\overline{x}_{1i} = x_{1i} - \overline{x}_1$, $\overline{x}_{2i} = x_{2i} - \overline{x}_2$ are the deviation of the data from their respective means, $\overline{y} = 55.2, \overline{x}_1 = 20.1, \overline{x}_2 = 6.4$

Hint: Solve only two normal equations (see Page 808).

Ans. $y = 37.56 + 0.4x_1 + 1.5x_2$

5. Fit a multiple linear regression equation to the following data and predict wear (y) when oil viscosity (x_1) is 30 and load (x_2) is 1400.

Wear y	193	230	172	91	113	125
Oil viscosity x_1	1.6	15.5	22.0	43.0	33.0	40.0
Load x_2	851	816	1058	1201	1357	1115

Hint: $n = 6, \sum y_i = 924, \sum x_{1i} = 155.1,$
$\sum x_{2i} = 6398, \sum x_{1i}x_{2i} = 178309.6,$
$\sum x_{1i}^2 = 5264.81, \sum x_{2i}^2 = 7036496,$
$\sum x_{1i}y_i = 20299.8, \sum x_{2i}y_i = 935906.$

Ans. $y = 350.9943 - 1.2702x_1 - 0.1539x_2,$
$y(x_1 = 30, x_2 = 1400) = 97.4283.$

30.9 CORRELATION ANALYSIS

In correlation analysis, the degree (or strength) of relationship between two variables, say X and Y, is measured by a single number r called a correlation coefficient formed by Karl Pearson in 1896. Here both X and Y are assumed to be random variables unlike the regression analysis where the dependent variable Y is assumed to be a random variable and the regressor (independent) variable x to be a physical or scientific or mathematical variable but not a random variable. Besides Pearsonian correlation, the other types of correlation include rank correlation, biserial correlation, intraclass correlation.

Examples:

a. Volume of a cube $V = L^3$, perfectly correlated.

b. Rainfall and crop yield, correlated.

c. Two coins being tossed simultaneously, uncorrelated.

Types of Correlation

By ploting a given set of n pairs of random variables (X_i, Y_i), for $i = 1, 2, 3, \ldots, n$, as a scatter diagram, the correlation is said to be

Positve or direct if Y increases as X increases.

Negative or inverse if Y decreases as X increases.

Linear if all the n points lie near a straight line.

Non-linear if the points lie on some non-linear curve.

Examples:

a. Income and expenditure: positively correlated.

b. Age and IQ: negatively correlated.

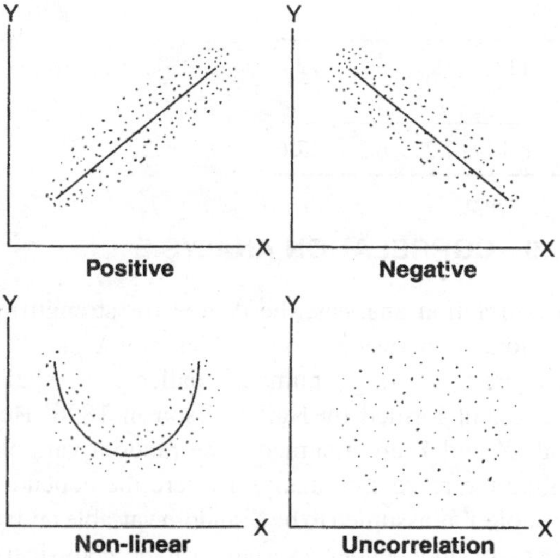

Fig. 30.3

Simple

The correlation between two variables is said to be simple correlation.

Multiple

The correlation between more than two variables is known as multiple correlation. If $r = \pm 1$, there is a perfect positive (or negative) correlation. If $r = 0$ there is no (linear) correlation; but a nonlinear correlation may exist. Similarly, a high correlation due to a third (lurking) variables is known as a spurious correlation (refer Fig. 30.3).

Example: Poverty and crime are highly correlated but the spurious correlation is due to the lurking variable illiteracy.

Standard error of estimate

Y on X is denoted by $S_{Y,X}$ is defined as

$$S_{Y,X} = \sqrt{\frac{\sum (Y - Y_{est})^2}{N}}$$

where Y_{est} is the estimated or predicted value of Y from the least squares regression line $Y = a_0 + a_1 X$. Similarly, the standard error of estimate X on Y is

$$S_{X,Y} = \sqrt{\frac{\sum (X - X_{est})^2}{N}}$$

In general, $S_{Y,X} \neq S_{X,Y}$.

Result: $\sum (Y - \overline{Y})^2 = \sum (Y - Y_{est})^2 + \sum (Y_{est} - \overline{Y})^2$, i.e., Total variation = unexplained variation + explained variation.

Coefficient of determination

$$= r^2 = \frac{\text{Explained variation}}{\text{Unexplained variation}}$$

Coefficient of correlation

$$= r = \pm \sqrt{\frac{\text{Explained variation}}{\text{Unexplained variation}}}$$

i.e., $$r = \pm \sqrt{\frac{\sum (Y_{est} - \overline{Y})^2}{\sum (Y - \overline{Y})^2}}.$$

The +ve and −ve signs correspond to positive and negative correlation respectively.

Properties of *r*

i. r lies in the interval $[-1, 1]$, i.e., $-1 \leq r \leq 1$.

ii. r is independent of origin.

iii. r is independent of (scale of measurements) unit.

Karl Pearson product-moment formula or simply sample correlation coefficient for the linear correlation coefficient r:

From the least squares regression line of Y on X

$$\hat{Y} = Y_{\text{est}} = a_0 + a_1 X$$

Put $y_{\text{est}} = Y_{\text{est}} - \overline{Y}$, $x = X - \overline{X}$, $y = Y - \overline{Y}$ then

$$y_{\text{est}} = a_1 x = \frac{\sum xy}{\sum x^2} x.$$

We know that

$$r^2 = \frac{\sum y_{\text{est}}^2}{\sum y^2} = \frac{\sum a_1^2 x^2}{\sum y^2} = a_1^2 \frac{\sum x^2}{\sum y^2}$$

$$= \left(\frac{\sum xy}{\sum x^2}\right)^2 \frac{\sum x^2}{\sum y^2} = \frac{(\sum xy)^2}{\sum x^2 \sum y^2}.$$

Therefore

$$r = \pm \frac{\sum xy}{\sqrt{\sum x^2 \sum y^2}}$$

The $\pm$ sign can be omitted without any loss of generality since y_{est} increases (decreases) as x increases (decreases).

Thus the coefficient of linear correlation is

$$r = \frac{\sum xy}{\sqrt{\sum x^2 \sum y^2}}$$

Introducing covariance S_{XY} of X and Y by

$$S_{XY} = \frac{\sum xy}{N}$$

and s.d. of X and Y by

$$S_X = \sqrt{\frac{\sum x^2}{N}}, \quad S_Y = \frac{\sum y^2}{N}.$$

we can write

$$r = \frac{S_{XY}}{S_X S_Y}.$$

Computational Formula

$$r = \frac{N\sum XY - \sum X \sum Y}{\sqrt{\left[N\sum X^2 - \left(\sum X\right)^2\right]\left[N\sum Y^2 - \left(\sum Y\right)^2\right]}}$$

Regression lines and the linear correlation coefficient:

The least squares regression line Y on X

$$Y = a_0 + a_1 X$$

can be written as

$$y = \left(\frac{\sum xy}{\sum x^2}\right) x$$

where $y = Y - \overline{Y}$, $\quad x = X - \overline{X}$.

Since $\quad r = \dfrac{\sum xy}{\sqrt{\sum x^2 \sum y^2}}$

so $\quad \dfrac{\sum xy}{\sum x^2} = \dfrac{r\sqrt{\sum x^2 \sum y^2}}{\sum x^2} = r\sqrt{\dfrac{\sum y^2}{\sum x^2}} = r\dfrac{S_Y}{S_X}.$

Thus $\quad y = a_1 x = \left(\dfrac{\sum xy}{\sum x^2}\right) x = r\dfrac{S_Y}{S_X} x.$

In a similar way, we get

$$x = r\frac{S_X}{S_Y} y.$$

Result: Show that the coefficient of correlation r is the geometric mean between the regression coefficients.

Solution: Let $Y = a_0 + a_1 X$ and $X = b_0 + b_1 Y$ be the least squares regression lines, with a_1 and b_1 as regression coefficients. We know that $a_1 = \frac{rS_Y}{S_X}$ and $b_1 = r\frac{S_X}{S_Y}$.

So $\quad a_1 b_1 = \left(r\dfrac{S_Y}{S_X}\right)\left(r\dfrac{S_X}{S_Y}\right) = r^2$

or $\quad r = \sqrt{a_1 b_1}.$

Test of Hypothesis for Correlation Coefficient

Let $\rho = \pm\sqrt{1 - \frac{\sigma^2}{\sigma_z^2}}$ denote the population correlation coefficient where $\sigma_z^2 = \frac{1}{n-3}$, n is number of pairs of observations. The Fisher Z transformation is defined by

$$Z = \frac{1}{2} \ln \frac{1+r}{1-r} = 1.1513 \log_{10}\left(\frac{1+r}{1-r}\right).$$

Here the statistic Z is a value of random variable approximately normally distributed with mean $\mu_Z = \frac{1}{2} \ln \frac{1+\rho}{1-\rho}$ and variance $\sigma_Z^2 = \frac{1}{n-3}$.

Statistic for inference about $\rho(\neq 0)$

$$z = \frac{Z - \mu_Z}{\left(\frac{1}{\sqrt{n-3}}\right)} = \frac{\sqrt{n-3}}{2} \ln\left\{\frac{(1+r)(1-\rho)}{(1-r)(1+\rho)}\right\}$$

Test of hypothesis for no correlation: $\rho = 0$

$$z = \sqrt{n-3} \quad Z = \frac{\sqrt{n-3}}{2} \ln\left(\frac{1+r}{1-r}\right)$$

values of Z are tabulated (A21) for various values of $r = 0$ to 0.99.

Note: When r is negative, read Z corresponding to $-r$ and then take $-Z$.

WORKED OUT EXAMPLES

Correlation:

Example 1: (a) Estimate (predict) the blood pressure (B.P.) of a woman of age 45 years from the following data which shows the ages X and systolic B.P. Y of 12 women. (b) Are the two variables ages X and B.P. Y correlated?

Age (X)	56	42	72	36	63	47	55	49	38	42	68	60
B.P. (Y)	147	125	160	118	149	128	150	145	115	140	152	155

Solution: **a.** To estimate B.P., determine the prediction equation which is the least squares regression equation of Y on X. Assume it to be $Y = a_0 + a_1 X$. Its normal equations are

$$\sum Y = N a_0 + a_1 \sum X$$

$$\sum XY = a_0 \sum X + a_1 \sum X^2$$

From the given data, $\sum X = 628$; $\sum Y = 1684$; $\sum X^2 = 34416$; $\sum Y^2 = 238822$; $\sum XY = 89894$; $N = 12$.
Substituting

$$1684 = 12a_0 + 628a_1$$

$$89894 = 628a_0 + 34416a_1$$

Solving $a_0 = 80.77738$, $a_1 = 1.138005$.

so the prediction equation is $Y = 80.777 + 1.138X$. The B.P. of a woman with age $X = 45$ is obtained as

$$Y(45) = 80.777 + 1.138(45) = 131.987225 \approx 132.$$

b. To find the association between age and B.P., determine the correlation coefficient r by

$$r = \frac{N\sum XY - \sum X \sum Y}{\sqrt{\left[N\sum X^2 - (\sum X)^2\right]\left[N\sum Y^2 - (\sum Y)^2\right]}}$$

$$= \frac{12(89894) - (628)(1684)}{\sqrt{\left[(12)(34416) - (628)^2\right]\left[(12)(238822) - (1684)^2\right]}}$$

$$= 0.8961$$

Age X and B.P. Y are strongly positively correlated.

Example 2: Test the hypothesis that there is no linear association among two variables x, air velocity cm/sec, and y, evaporation coefficient mm^2/sec of a burning fuel droplets in an impulse engine, with $n = 10$ and $r = 0.9515$.

Solution:

1. Null hypothesis $\rho = 0$
2. Alternate hypothesis $\rho \neq 0$
3. Level of significance: $\alpha = 0.05$
4. Critical region: (two tailed test)
   Reject null hypothesis if
   $$Z > Z_{\alpha/2} = 1.96 \quad \text{or} \quad Z < -Z_{\alpha/2} = -1.96$$
5. Calculation: For $r = 0.95$ from table
   $$Z^* = \frac{1}{2} \ln\left(\frac{1+r}{1-r}\right) = 1.832$$
   so $\quad Z = \sqrt{n-3}\, Z^* = \sqrt{10-3}(1.832) = 4.847$
6. Decision: Reject N.H. since $Z = 4.847 > Z_{\alpha/2} = 1.96$, i.e., there is linear relation between x and y.

Example 3: In a certain paired data $n = 18$ and $r = 0.44$, test the N.H. $\rho = 0.30$ against $\rho > 0.30$ at 0.01 level of significance.

Solution:

1. Null hypothesis: $\rho = 0.30$
2. Alternate hypothesis: $\rho > 0.30$
3. Level of significance: $\alpha = 0.01$
4. Critical region: (Right one tailed test)

   Reject null hypothesis of $Z > 2.33$
5. Calculation: For $r = 0.44$, from table A21

$$Z^* = \frac{1}{2}\ln\left(\frac{r+1}{r-1}\right) = 0.472$$

$$\text{so} \quad Z = \sqrt{n-3}\, Z^* = \sqrt{18-3}(0.472) = 1.828$$

6. Decision: Accept the null hypothesis since

$$Z = 1.828 < Z_\alpha = 2.33$$

Example 4: Determine 95% confidence limits and the confidence intervals for a correlation coefficient which is computed to be 0.60 from a sample of 28.

Solution: $r = 0.60, n = 28, \alpha = 0.05$
Confidence limits (C.L.) for μ_Z are $Z \pm 1.96\sigma_Z$ where $Z = 0.693$ from table A21 (for $r = 0.60$) and

$$\sigma_Z = \frac{1}{\sqrt{n-3}} = \frac{1}{\sqrt{28-3}} = \frac{1}{5} = 0.2. \text{ Thus}$$

μ_Z has class limits as

$$0.693 - 1.96\,(0.2), 0.693 + 1.96\,(0.2)$$

i.e., confidence interval for μ_Z is:
(0.3011516, 1.0851516).
If $0.30115 = \mu_Z = 1.1513 \ln\left(\frac{1+\rho}{1-\rho}\right)$ then
$\rho = \frac{.826}{2.826} = 0.29236$
If $1.0851 = \mu_Z = 1.1513 \ln\left(\frac{1+\rho}{1-\rho}\right)$ then
$\rho = \frac{7.7608}{9.7608} = 0.7951.$
Thus the confidence limits for ρ are 0.29236, 0.7951 and the confidence interval is (0.29236, 0.7951).

Least squares regression lines, standard error estimate

Example 5: For the following data determine (a) least squares regression line of y on x (b) y (3) (c) least squares regression line of x on y (d) $x(4)$ (e) S_{yx} (f) S_{xy} (g) total variation in y (h) unexplained variation in y (i) explained variation in y.

x	6	5	8	8	7	6	10	4	9	7
y	8	7	7	10	5	8	10	6	8	6

Solution:

x	y	x^2	y^2	xy
6	8	36	64	48
5	7	25	49	35
8	7	64	49	56
8	10	64	100	80
7	5	49	25	35
6	8	36	64	48
10	10	100	100	100
4	6	16	36	24
9	8	81	64	72
7	6	49	36	42
Total 70	75	520	587	540

so $\sum x = 70$, $\sum y = 75$, $\sum x^2 = 520$, $\sum y^2 = 587$, $\sum xy = 540$, $N = 10$

a. Assume that the required L.S.R.L. of y on x is

$$y = a_0 + a_1 x$$

whose normal equations are

$$\sum y = Na_0 + a_1 \sum x$$

$$\sum xy = a_0 \sum x + a_2 \sum x^2$$

Substituting

$$75 = 10\,a_0 + 70\,a_1$$

$$540 = 70\,a_1 + 520\,a_1$$

Solving $a_1 = 0.5, a_0 = 4.$
So L.S.R.L. of y on x is

$$y = 4 + 0.5x \tag{1}$$

b. $y(3) = 4 + 0.5(13) = 4 + 1.5 = 5.5$
c. Assume that L.S.R.L. of x on y is

$$x = b_0 + b_1 y$$

with normal equations

$$\sum x = Nb_0 + b_1 \sum y$$

$$\sum xy = b_0 \sum y + b_1 \sum y^2$$

or $70 = 10b_0 + 75b_1$

$540 = 75b_0 + 587b_1$

Solving

$$b_1 = 0.612,\ b_0 = 2.41$$

So L.S.R.L. of x on y is

$$x = 2.41 + 0.612y \quad (2)$$

$(y_{est} - \bar{y})$	0.5	−1	0.5	0.5	0	−0.5	1.5	−1.5	1	0	Total
$(y_{est} - \bar{y})^2$	0.25	1	0.25	0.25	0	0.25	2.25	2.25	1	0	7.50
$(y - \bar{y})$	0.5	−0.5	−0.5	2.5	−2.5	0.5	2.5	−1.5	0.5	−1.5	
$(y - \bar{y})^2$	0.25	0.25	0.25	6.25	6.25	0.25	6.25	2.25	0.25	2.25	24.50

d. $x(4) = 2.41 + 0.612(4) = 4.858$

e. From (1), estimated value of $y = y_{est} = 4 + 0.5x$

x	y	y_{est}	$y - y_{est}$	$(y - y_{est})^2$
6	8	7	1	1
5	7	6.5	0.5	0.25
8	7	8	−1	1
8	10	8	2	4
7	5	7.5	−2.5	6.25
6	8	7	1	1
10	10	9	1	1
4	6	6	0	0
9	8	8.5	−0.5	0.25
7	6	7.5	−1.5	2.25
			Total	17.0

Now $S_{yx} = \sqrt{\frac{\sum(y-y_{est})^2}{N}} = \sqrt{\frac{17.0}{10}} = 1.30384$

f. From (2) $x_{est} = 2.41 + 0.612y$

y	x	x_{est}	$x - x_{est}$	$(x - x_{est})^2$
8	6	7.306	−1.306	1.705636
7	5	6.694	−1.694	2.869636
7	8	6.694	1.306	1.705636
10	8	8.53	−0.53	0.2809
5	7	5.47	1.53	2.3409
8	6	7.306	−1.306	1.705636
10	10	8.53	1.47	2.1609
6	4	6.082	−2.082	4.334724
8	9	7.306	1.694	2.869636
6	7	6.082	0.918	0.842724
			Total	20.816328

Now

$$S_{xy} = \sqrt{\frac{\sum(x - x_{est})^3}{N}} = 1.44278647 \approx 1.443$$

g, h, i. $\bar{y} = \frac{\sum y}{N} = \frac{75}{10} = 7.5.$

Total variation = unexplained variation + explained variation.

$$\sum(y - \bar{y})^2 = \sum(y - y_{est})^2 + \sum(y_{est} - \bar{y})^2 \quad (3)$$

Total variation $= \sum(y - \bar{y})^2 = 24.50.$
Explained variation $= \sum(y_{est} - \bar{y})^2 = 7.50.$
From (3) unexplained variation $= \sum(y - y_{est})^2 = 24.50 - 7.50 = 17.$

Example 6: Calculate the coefficient of correlation r for the above data in 4 ways.

Solution:

a.
$$r = \pm\sqrt{\frac{\text{explained variation}}{\text{total variation}}}$$

$$= \pm\sqrt{\frac{7.50}{24.50}} = 0.55328335.$$

b. Correlation coefficient is the geometric mean between the regression coefficients, i.e.,

$$r = \sqrt{b_{xy} \cdot b_{yx}}$$

From Equation (1) the regression coefficient of y on x is $b_{xy} = 0.5$. Similarly, regression coefficient of x on y is $b_{xy} = 0.612$

$$\text{Now } r = \sqrt{(0.5)(0.612)} = 0.553172667.$$

c. By product-moment formula

$$r = \frac{N\sum xy = \sum x \sum y}{\sqrt{[N\sum x^2 - (\sum x)^2][N\sum y^2 - (\sum y)^2]}}$$

$$r = \frac{10(540)-(70)(75)}{\sqrt{[10(520)-(70)^2][10(587)-(75)^2]}} = \frac{150}{\sqrt{73500}}$$

$= 0.55328$

d. By formula

$$r=\frac{\sigma_x^2+\sigma_y^2-\sigma_{x-y}^2}{2\sigma_x\sigma_y}$$

where $\sigma_x^2, \sigma_y^2, \sigma_{x-y}^2$ are variances of $x, y, x-y$ respectively.

We know that

$$\text{variance} = \frac{\sum d^2}{N}-\left(\frac{\sum d}{N}\right)^2$$

Where d is the deviation of data from an assumed (class mark) origin.

	x	y	$X = x-10$	$Y = y-10$	X^2	Y^2	$y-x$	$(x-y)^2$
	6	8	−4	−2	16	4	2	4
	5	7	−5	−3	25	9	2	4
	8	7	−2	−3	4	9	−1	1
	8	10	−2	0	4	0	2	4
	7	5	−3	−5	9	25	−2	4
	6	8	−4	−2	16	4	2	4
	10	10	0	0	0	0	0	0
	4	6	−6	−4	36	16	2	4
	9	8	−1	−2	1	4	−1	1
	7	6	−3	−4	9	16	−1	1
Total	70	75	−30	−25	120	87	5	27

Now

$$\sigma_x^2=\frac{\sum X^2}{N}-\left(\frac{\sum X}{N}\right)^2=\frac{120}{10}-\left(\frac{-30}{10}\right)^2$$

$$=12-9=3$$

$$\sigma_y^2=\frac{\sum Y^2}{N}-\left(\frac{\sum Y}{N}\right)^2=\frac{87}{10}-\left(\frac{-25}{10}\right)^2=2.45$$

$$\sigma_{x-y}^2=\frac{\sum(x-y)^2}{N}-\left(\frac{\sum(x-y)}{N}\right)^2=\frac{27}{10}-\left(\frac{5}{10}\right)^2$$

$$=2.45$$

By formula

$$r=\frac{\sigma_x^2+\sigma_y^2-\sigma_{x-y}^2}{2\sigma_x\sigma_y}=\frac{3+2.45-2.45}{2\sqrt{3}\sqrt{2.45}}$$

$$=0.55328335.$$

Example 7: Calculate (a) the standard error estimate of y (b) the standard error estimate of x for the above data in Example 6.

Solution: **a.** Standard error estimate of y

$$=S_y=\sigma_y\sqrt{(1-r^2)}=\sqrt{2.45}\sqrt{1-(0.55328335)^2}$$

$$S_y=1.086089.$$

Similarly, standard error estimate of x

$$S_x=\sigma_x\sqrt{(1-r^2)}=3\sqrt{1-(0.55328335)^2}$$

$$=2.081630.$$

Example 8: From 10 pairs of observations for x and y the following data is obtained: $n=10, \sum x = 66, \sum y = 69, \sum x^2=476, \sum y^2=521, \sum xy = 485$. It was later found that two pairs of (correct) values

x	y		x	y
4	6	were (erroneously) copied down as	2	3.
9	8		7	5

Calculate the correct value of the coefficient of correlation.

Solution: In order to get the correct data, subtract the incorrect values and add the corresponding correct values.

Thus $\sum x = 66-2-7+4+9=70,$

$$\sum y = 69-3-5+6+8=75,$$

$$\sum x^2 = 476-4-49+16+81=520,$$

$$\sum y^2 = 521-9-25+36+64=587,$$

$$\sum xy = 485-6-35+24+72=540.$$

Hence

$$r=\frac{N\sum xy-(\sum x)(\sum y)}{\sqrt{\left[N\sum x^2-(\sum x)^2\right]\left[N\sum y^2-(\sum y)^2\right]}}$$

$$=\frac{10(540)-(70)(75)}{\sqrt{[10(520)-(70)^2][10(587)-(75)^2]}}=0.55328.$$

EXERCISE

1. Calculate correlation coefficient r for the following data:

   X: 63, 50, 55, 65, 55, 70, 64, 70, 58, 68, 52, 60

Y: 87, 74, 76, 90, 85, 87, 92, 98, 82, 91, 77, 78

Hint: $N = 12, \sum X = 730, \sum Y = 1017,$
$\sum X^2 = 44932, \sum Y^2 = 86801,$
$\sum XY = 62352.$

Ans. $r = 0.86$

2. Compute r for the data given below:

X:	1	2	3	4	5	6
Y:	6	4	3	5	4	2

Hint: $N = 6, \sum X = 21, \sum Y = 24,$
$\sum X^2 = 91, \sum Y^2 = 106, \sum XY = 75.$

Ans. $r = -0.68$

3. Determine r for the following data:

X:	50	60	70	90	100
Y:	65	51	40	26	8

Ans. $r = -0.99$

4. Determine the least squares regression line of (a) y on x and (b) x on y (c) Find r using the regression coefficients. (d) Find $y(8)$ (e) Find $x(16)$.

x:	12	10	14	11	12	9
y:	18	17	23	19	20	15

Ans. **a.** $y = 1.913 + 1.478x$

b. $x = 0.60714y$

c. $r = \sqrt{(1.478)(0.60714)} = 0.947287$

≈ 0.95

d. $y(8) = 13.737$

e. $x(16) = 9.71424$

5. Use $r = \frac{\sigma_x^2+\sigma_y^2-\sigma_{x-y}^2}{2\sigma_x\sigma_y}$ to find r for the following data:

x:	21	23	30	54	57	58	72	78	87	90
y:	60	71	72	83	110	84	100	92	113	135

Hint: $\sigma_x^2 = 584.6, \sigma_y^2 = 468.8,$
$\sigma_{x-y}^2 = 134.6.$

Ans. $r = 0.876$

6. In a paired data for x, y with $N = 25, \sum x = 127, \sum y = 100, \sum x^2 = 760, \sum y^2 = 449, \sum xy = 500$, it was found later that two pairs of correct values

x	y
8	12
6	8

were (erroneously) copied down as

x	y
8	14.
8	6

Determine the correlation coefficient for the correct data.

Hint: $\sum x = 127 + 8 + 6 - 8 - 8 = 125,$
$\sum y = 100 + 12 + 8 - 14 - 6 = 100$
$\sum x^2 = 760 + 64 + 36 - 64 - 64 = 732,$
$\sum y^2 = 449 + 144 + 64 - 39 - 196 = 425$
$\sum xy = 500 + 96 + 48 - 48 - 112 = 484.$

Ans. -0.30984

7. For $n = 10$ and $r = 0.732$ test the null hypothesis $\rho = 0$ against alternate hypothesis $\rho \neq 0$ at the 0.05 level of significance.

Ans. $Z = \sqrt{10 - 3}(0.933) = 2.47 > 1.96$, Reject N.H.

8. Construct a 95% confidence interval for the population correlation coefficient ρ given $r = 0.70$ and $n = 30$.

Ans. $0.867 \pm \frac{1.96}{\sqrt{30-3}}$ i.e., $(0.490 < \mu_Z < 1.244)$ so $(0.45 < \rho < 0.85)$

9. At 0.05 level of significance, test the null hypothesis $\rho = 0.9$ against the alternative $\rho > 0.9$, for $n = 29, r = 0.9435$.

Ans. $Z = \frac{\sqrt{26}}{2} \ln\left[\frac{(1+0.9435)(0.1)}{(1-0.9435)(1.9)}\right] = 1.51 < 1.645$, accept N.H.

30.10 RANK CORRELATION or SPEARMAN'S CORRELATION

Although data is measured numeric (quantitative) in several cases the data turns out to be non-numeric (qualitative).

Examples:

a. Appearance: beautiful, ugly.

b. Efficiency: excellent, good, average, bad.

c. Temperament: wild, composed, dosile.

In such cases, the data is ranked according to that particular character instead of taking numeric measurements on them, and therefore the usual Pearsonian correlation coefficient can not be calculated. Instead Charles Edward Spearman (1906) a psychologist developed a nonparametric counterpart of the conventional correlation coefficient as follows:

For a given set of n paired observations (X_i, Y_i), for $i = 1$ to n; ranks $1, 2, \dots, n$ are assigned to the X observations in order of magnitude and similarly to the Y observations. Then these ranks are substituted for the actual numerical values. The correlation coefficient calculated in this manner is called the "**rank correlation coefficient or Spearman's correlation coefficient**" and is given by

$$r_{\text{rank}} = r_s = 1 - \frac{6\sum_{i=1}^{n} d_i^2}{n(n^2-1)}$$

Here d_i = difference between ranks assigned to X_i and Y_i; n = number of pairs of data.

Note 1: If there are ties among either X or Y observations, substitute for each of the tied observations, the mean of the ranks that they jointly occupy.

Note 2: r_s lies between -1 and 1.

Worked Out Examples

Example: Determine rank correlation for the following data which shows the marks obtained in two quizes in mathematics:

Marks in 1st quiz (X)	6	5	8	8	7	6	10	4	9	7
Marks in 2nd quiz (Y)	8	7	7	10	5	8	10	6	8	6

Solution: Assigning ranks to the data of X, we get

X:	4,	5,	6,	6,	7,	7,	8,	8,	9,	10
Rank:	1,	2,	3,	4,	5,	6,	7,	8,	9,	10
or:	1,	2,	3.5,	3.5,	5.5,	5.5,	7.5,	7.5,	9,	10
Similarly, Y:	5,	6,	6,	7,	7,	8,	8,	8,	10,	10
Rank:	1,	2,	3,	4,	5,	6,	7,	8,	9,	10
or:	1,	2.5,	2.5,	4.5,	4.5,	7,	7,	7,	9.5,	9.5

Data assigned with ranks is

X	3.5	2	7.5	7.5	5.5	3.5	10	1	9	5.5
Y	7	4.5	4.5	9.5	1	7	9.5	2.5	7	2.5
D	−3.5	−2.5	3	−2	4.5	−3.5	0.5	−1.5	2	3
D^2	12.25	6.25	9	4	20.25	12.25	0.25	2.25	4	9

$$\text{Rank correlation} = 1 - \frac{6\sum D^2}{N(N^2-1)} = 1 - \frac{6(79.5)}{10(99)}$$

$$= 1 - .4818181 = 0.5181818.$$

Exercise

Rank correlation

Find the rank correlation for the following data:

1.

X:	56	42	72	36	63	47	55	49	38	42	68	60
Y:	147	125	160	118	149	128	150	145	115	140	152	155

Ans. $r_{\text{rank}} = \frac{1-6(19.5)}{12(143)} = 0.931818$

2.

X:	2	4	5	6	8	11
Y:	18	12	10	8	7	5

Ans. $r_{\text{rank}} = 1 - \frac{6(70)}{6(35)} = -1$

3.

X:	14	17	28	17	16	13	24	25	18	31
Y:	0.9	1.1	1.6	1.3	1.0	0.8	1.5	1.4	1.2	2.0

Ans. $r_{\text{rank}} = 1 - \frac{6(5.50)}{(10)(99)} = 0.967$

4.

X:	11.1	10.3	12.0	15.1	13.7	18.5	17.3	14.2	14.8	15.3
Y:	10.9	14.2	13.8	21.5	13.2	21.1	16.4	19.3	17.4	19.0

Ans. $r_s = 0.697$

5. Two judges gave the following ranks to 11 girls in a beauty contest:

Girl	1	2	3	4	5	6	7	8	9	10	11
Judge A	3	4	1	2	5	10	11	7	9	8	6
Judge B	2	4	3	1	7	9	6	11	10	5	8

Ans. $r_s = 1 - \frac{6(66)}{11(120)} = 0.70$

6. Rank by Judge

A:	1	6	5	10	3	2	4	9	7	8
B:	3	5	8	4	7	10	2	1	6	9
C:	6	4	9	8	1	2	3	10	5	7

Hint: $\sum d_1^2 = 200$, $\sum d_2^2 = 60$, $\sum d_3^2 = 214$, $n = 10$.

Ans. $r(A, B) = -\frac{7}{33}, r(A, C) = \frac{7}{11}, r(B, C) = -\frac{49}{65}$, Judges A and C have nearest common approach (judgement) since $r(A, C)$ is maximum.

30.8 CORRELATION FOR BIVARIATE FREQUENCY DISTRIBUTION

When the data is very large, it is arranged into a bivariate frequency table (or bivariate frequency distribution) as follows:

Class intervals for X

Class intervals for Y: Y \ X	$X_{L_1} - X_{U_1}$	$\cdots$	$X_{L_k} - X_{U_k}$	Totals
$Y_{L_1} - Y_{U_1}$	f_{11}	$\cdots$	f_{1k}	
$\vdots$	$\vdots$	f_{ij}	$\vdots$	
$Y_{L_m} - Y_{U_m}$	f_{m1}	$\cdots$	f_{mk}	
Totals				N

CORRELATION TABLE

Mid value Y	Mid value X / U_X \ U_Y	X_1 $\cdots$ X_k	f_Y	$f_Y U_Y$	$f_Y U_Y^2$	Sum of corner (boxed) numbers in each row
Y_1 $\vdots$ Y_m	$\vdots$					
f_X			$N = \Sigma f_X = \Sigma f_Y$	$\Sigma f_Y U_Y$	$\Sigma f_Y U_Y^2$	$\Sigma f U_X U_Y$
$f_X U_X$			$\Sigma f_X U_X$			
$f_X U_X^2$			$\Sigma f_X U_X^2$			
Sum of corner (boxed) numbers in each column			$\Sigma f U_X U_Y$			

CHECK

The bivariate frequency table is rewritten as (see Page 30.36)

Assume that

X is grouped into k classes

Y is grouped into m classes

f_{ij} or simply f is the cell frequency of the ith X-class interval and jth Y-class interval.

X_{L_1}, X_{U_1} denotes the lower and upper limits of the 1st class, Y_{L_m}, Y_{U_m} denotes the lower and upper limits of the mth class etc. Blank cell denotes zero cell frequency.

$$N = \text{Total frequency} = \sum_{j=1}^{k}\sum_{i=1}^{m} f_{ij}$$

Let

X_i be the mid value (class mark) of the ith X class

Y_j be the mid value of the jth Y class.

$$\text{Put } U_X = \frac{X - A}{C_1}, \quad U_Y = \frac{Y - B}{C_2}$$

where C_1 = class size of X intervals

C_2 = class size of Y intervals

A = Assumed class mark for X-classes

B = Assumed class mark for Y-classes.

f_X: marginal frequencies of X (column sums of f_{ij}'s)

f_Y: marginal frequencies of Y (row sums of f_{ij}'s).

Note that $\sum f_X = \sum f_Y = N$.

$f U_X U_Y$ is denoted by a number in corner of each cell as □.

Now the correlation coefficient takes the form as

$$r = \frac{N\sum fU_XU_Y - (\sum f_XU_X)(\sum f_YU_Y)}{\sqrt{\left[N\sum f_XU_X^2 - (\sum f_XU_X)^2\right]\left[N\sum f_YU_Y^2 - (\sum f_YU_Y)^2\right]}}$$

Note:

1. U_X, U_Y turns out to be $0, \pm 1, \pm 2, \pm 3, \ldots$
2. In general, $C_1 = C_2$.
3. Check: total frequency $N = \sum f_X = \sum f_Y$.

4. Check: $\sum fU_XU_Y = \sum fU_YU_X$ obtained from row sums and column sums.

Worked Out Examples

Correlation for bivariate frequency distribution

Example 1: The following table shows the bivariate frequency distribution of marks obtained by 25 students in mathematics X and computer science Y. Determine the coefficient of correlation r. Test the null hypothesis $\rho = 0$ against the alternative hypothesis $\rho \neq 0$ at 0.05 level of significance. Determine whether there is a relationship between marks in the two subjects.

Marks in mathematics: X

Marks in C. S: Y	21–25	26–30	31–35	36–40	41–45
21–25	1				
26–30		3	1		
31–35		2	5	2	
36–40			1	4	1
41–45			1	3	
46–50					1

Solution: Here N = total frequency = 25.
From the bivariate correlation table (overleaf on Page 821), we get $\sum f_XU_X = 6$, $\sum f_YU_Y = 11$, $\sum f_XU_X^2 = 26$, $\sum f_YU_Y^2 = 39$, $\sum fU_XU_Y = 25$.
We know that the correlation coefficient for a bivariate frequency distribution is given by

$$r = \frac{N\sum fU_XU_Y - (\sum f_XU_X)(\sum f_YU_Y)}{\sqrt{\left[N\sum f_XU_X^2 - (\sum f_XU_X)^2\right]\left[N\sum f_YU_Y^2 - (\sum f_YU_Y)^2\right]}}$$

$$= \frac{(25)(25) - (6)(11)}{\sqrt{\left[(25)(26) - (6)^2\right]\left[(25)(39) - (11)^2\right]}} = 0.7719669.$$

Test of hypothesis:

1. H_0: $\rho = 0$ Null hypothesis
2. H_1: $\rho \neq 0$ Alternative hypothesis
3. $\alpha = 0.05$ Level of significance
4. Critical region: Reject the null hypothesis if $Z < -1.96$ or $Z > 1.96$, where

$$Z = \sqrt{n-3}Z^*.$$

5. Computation: The value of Z^* corresponding to $r = 0.772$ is $Z^* = \frac{1}{2}\ln\frac{1+r}{1-r} = \frac{1}{2}\ln\frac{1+0.772}{1-0.772}$ $= 1.020$
so $Z = \sqrt{n-3}Z^* = \sqrt{25-3}(1.020)$
$= 4.80889 \approx 4.81$
6. Decision: Reject the null hypothesis of "no linear association" because $Z = 4.81 > 1.96$. So conclude that there is a relationship between the marks obtained in the subjects mathematics and computer science.

Example 2: Test the null hypothesis $\rho = 0.9$ against the alternative that $\rho > 0.9$ at 0.05 level of signification for the above data with $r = 0.772$.

Solution:

1. H_0: $\rho = 0.9$
2. H_1: $\rho > 0.9$
3. α: 0.05
4. Critical region: $Z > 1.645$
5. Calculation:

$$Z = \frac{\sqrt{n-3}}{2}\ln\left\{\frac{(1+r)(1-\rho_0)}{(1-r)(1+\rho_0)}\right\}$$

$$Z = \frac{\sqrt{25-3}}{2}\ln\left[\frac{(1+0.772)(0.1)}{(1-0.772)(1.9)}\right] = -2.14$$

6. Decision: Since $Z = -2.14 < Z_\alpha = 1.645$ accept null hypothesis i.e., there is some evidence that correlation coefficient does not exceed 0.9.

Exercise

Find the correlation coefficient for the following bivariate frequency distribution:

1.

Y \ X	59–62	63–66	67–70	71–74	75–78
90–109	2	1			
110–129	7	8	4	2	
130–149	5	15	22	7	1
150–169	2	12	63	19	5
170–189		7	28	32	12
190–209		2	10	20	7
210–229			1	4	2

BIVARIATE CORRELATION TABLE

Y \ X		23	28	33	38	43	f_Y	$f_Y U_Y$	$f_Y U_Y^2$	Sum of corner numbers in each row
	U_Y \ U_X	−2	−1	0	1	2				
23	−2	1 [4]					1	−2	4	4
28	−1		3 [3]	1 [0]			4	−4	4	3
33	0		2 [0]	5 [0]	2 [0]		9	0	0	0
38	1			1 [0]	4 [4]	1 [2]	6	6	6	6
43	2			1 [0]	3 [6]		4	8	16	6
48	3					1 [6]	1	3	9	6
f_X		1	5	8	9	2	$\sum f_X = \sum f_Y = 25$	$\sum f_Y U_Y = 11$	$\sum f_Y U_Y^2 = 39$	$\sum f U_X U_Y = 25$
$f_X U_X$		−2	−5	0	9	4	$\sum f_X U_X = 6$			
$f_X U_X^2$		4	5	0	9	8	$\sum f_X U_X^2 = 26$			
Sum of corner numbers in each column		4	3	0	10	8	$\sum f U_X U_Y = 25$			

CHECK

Hint: $N = 300$, $\sum fU_XU_Y = 208$,
$\sum f_X\, U_X = 61$, $\sum f_YU_Y = 77$,
$\sum f_XU_X^2 = 301$, $\sum f_YU_Y^2 = 459$.

Ans. 0.54075

2.

Y \ X	150–154	155–159	160–164	165–169	170–174	175–179
51–53	1	1				
54–56	1	2	1			
57–59		2	2	5		
60–62		1	15	23	1	
63–65			6	18	1	
66–68			1	3	7	1
69–71				1	3	4

Hint: $N = 100$, $\sum fU_XU_Y = 131$,
$\sum f_X\, U_X = 79$, $\sum f_YU_Y = 50$,
$\sum f_XU_X^2 = 157$, $\sum f_YU_Y^2 = 188$.

Ans. 0.7367

3.

Y \ X	20–24	25–29	30–34	35–39
20–24	20	10	3	2
25–29	4	28	6	4
30–34		5	11	
35–39			2	
40–44				5

Hint: $N = 100$, $\sum fU_XU_Y = 138$,
$\sum f_X\, U_X = -80$, $\sum f_YU_Y = -100$,
$\sum f_XU_X^2 = 150$, $\sum f_YU_Y^2 = 204$.

Ans. $r = 0.613$

4.

Y \ X	18	19	20	21
10–20	4	2	2	
20–30	5	4	6	4
30–40	6	8	10	11
40–50	4	4	6	8
50–60		2	4	4
60–70		2	3	1

Hint: $N = 100$, $\sum f_XU_X = 68$,
$\sum f_YU_Y = 25$, $\sum f_XU_X^2 = 162$,
$\sum f_YU_Y^2 = 167$, $\sum fU_XU_Y = 52$.

Ans. $r = 0.25$

5.

Y \ X	15–25	25–35	35–45	45–55	55–65	65–75
15–25	1	1				
25–35	2	12	1			
35–45		4	10	1		
45–55			3	6	1	
55–65				2	4	2
65–75					1	2

Hint: $\sum fU_XU_Y = 86$, $\sum f_XU_X = 10$,
$\sum f_YU_Y = 16$, $\sum f_XU_X^2 = 98$,
$\sum f_YU_Y^2 = 92$, $n = 53$.

Ans. $r = 0.91$

6.

Y \ X	1–3	3–5	5–7	7–9	9–11	11–13
20–30	2	8	14		1	
30–40	5	9	6	3		
40–50	6	7	5			1
50–60	2	6	3	1		
60–70	1					

Hint: $N = 80$, $\sum fU_XU_Y = 23$,
$\sum f_XU_X = -59$, $\sum f_YU_Y = -53$,
$\sum f_XU_X^2 = 139$, $\sum f_YU_Y^2 = 111$.

Ans. $r = -0.19$

SUMMARY

Curve fitting is the method of finding equation of a curve that approximates a given set of data.

For a given set of N data points (X_1, Y_1), $(X_2, Y_2) \cdots (X_N, Y_N)$ the best fitting curve by the method of least squares is obtained by minimizing the sum of the squares of the errors (deviations). The normal equations for a least squares (L.S.) straight line are given by

$$\sum Y_i = Na_0 + a_1 \sum X_i$$

$$\sum X_iY_i = a_0 \sum X_i + a_1 \sum X_i^2$$

Nonlinear curves such as exponential curve $Y = AB^X$, geometric curve $Y = AX^B$, hyperbola $Y = \frac{1}{a_0 + a_1 X}$ can be transformed to linear curve (straight line). L.S. straight line of Y on X can be expressed as

$$y = \left(\frac{\sum xy}{\sum x^2}\right) x$$

where $x = X - \bar{X}$, $y = Y - \bar{Y}$

Simple linear regression of Y on X establishes the relation between Y and X by the equation

$$Y = a_0 + a_1 X$$

which is used for prediction (estimation).

$A(1-\alpha)100\%$ confidence intervals for the parameters regression coefficients α and β are given respectively by

$$a_0 - t_{\frac{\alpha}{2}} s_\epsilon \sqrt{\frac{S_{xx} + (n\bar{x})^2}{nS_{xx}}} < \alpha < a_0 + t_{\frac{\alpha}{2}} s_\epsilon \sqrt{\frac{S_{xx} + (n\bar{x})^2}{nS_{xx}}}$$

$$\text{and} \quad a_1 - t_{\frac{\alpha}{2}} \frac{s_\epsilon \sqrt{n}}{\sqrt{S_{xx}}} < \beta < a_1 + t_{\frac{\alpha}{2}} \frac{s_\epsilon \sqrt{n}}{\sqrt{S_{xx}}}$$

In curvilinear regression, for a polynomial curve of degree N

$$Y = a_0 + a_1 X + a_2 X^2 + \cdots + a_N X^N \qquad (1)$$

the $(N+1)$ normal equations to solve for the $(N+1)$ unknowns $a_0, a_1, a_2, \ldots, a_N$ are obtained by formally multiplying the regression polynomial (1) respectively by $1, X_i, X_i^2, \ldots, X_i^N$ and summing up all the terms from $i = 1$ to N.

The normal equations in linear multiple regression for the equation

$$Y = b_0 + b_1 X + b_2 X_2$$

are given by

$$\sum Y_i = N b_0 + b_1 \sum X_{1i} + b_2 \sum X_{2i}$$

$$\sum X_{1i} y_i = b_0 \sum X_{1i} + b_1 \sum X_{1i}^2 + b_2 \sum X_{1i} X_{2i}$$

$$\sum X_{2i} Y_i = b_0 \sum X_{2i} + b_1 \sum X_{1i} X_{2i} + b_2 \sum X_{2i}^2$$

Correlation coefficient r measures the strength of relationship between two or more variables. It is given by the Karl Pearson product-moment formula

$$r = \frac{\sum xy}{\sqrt{\sum x^2 \sum y^2}}$$

$$= \frac{N \sum XY - \sum X \sum Y}{\sqrt{\left[N \sum X^2 - (\sum X)^2\right]\left[N \sum Y^2 - (\sum Y)^2\right]}}$$

r is the geometric mean of the regression coefficients. Zero correlation can be tested by the statistic

$$Z = \sqrt{n-3}\, \frac{1}{2} \ln \left(\frac{1+r}{1-r}\right)$$

Statistic for inference about $\rho (\neq 0)$ is

$$Z = \sqrt{n-3}\, \frac{1}{2} \ln \left\{\frac{(1+r)(1-\rho)}{(1-r)(1+\rho)}\right\}.$$

Spearman's rank correlation is a non-parametric counterpart of the conventional correlation coefficient. It is given by

$$r_{\text{rank}} = r_S = 1 - \frac{6 \sum_{i=1}^{n} d_i^2}{n(n^2 - 1)}$$

where d_i = difference between ranks assigned to data and n is the size of the data.

For bivariate frequency distribution, the correlation coefficient r takes the form

$$r = \frac{N \sum f U_X U_Y - \left(\sum f_X U_X\right)\left(\sum f_Y U_Y\right)}{\sqrt{\left[N \sum f_X U_X^2 - \left(\sum f_X U_X\right)^2\right]\left[N \sum f_Y U_Y^2 - \left(\sum f_Y U_Y\right)^2\right]}}$$

(see correlation table on Page 819 for notation).

Chapter 31

Joint Probability Distribution and Markov Chains

INTRODUCTION

So far in the univariate case, we restricted our attention to probability distribution of a single random variable. However, in problems in economics, biology or social sciences we will be interested in the study of statistical methods analyzing two or more (bivariate or multivariate) variables. In such cases the concept of joint probability distribution is required. Markov* chains, involving calculation of high power of matrices, are powerful tool for forecasting future events.

31.1 JOINT PROBABILITY DISTRIBUTION

Let S be a sample space associated with a random experiment ε. Let $X = X(s)$ and $Y = Y(s)$ be two random variables on the same sample space S. The two real valued functions $X(s)$ and $Y(s)$ each assign a real number to each outcome s of the sample space S, with respective image sets given by

$$X(s) = \{X(s_1), X(s_2), \ldots, X(s_n)\} = \{X_1, X_2, \ldots, X_n\}$$

and $Y(s) = \{Y(s_1), Y(s_2), \ldots, Y(s_m)\} = \{Y_1, Y_2, \ldots, Y_m\}$

Here the two random variables X and Y are discrete since the possible values of X and Y are finite (or countably infinite).

Example:

a. X: Age, Y: Blood pressure of a person.

b. X: Crop yield, Y: Rain fall in an area.

c. X: I.Q., Y: Nutrition of an individual.

Consider the product set

$$X(S) \times Y(S) = \{(x_1, y_1), (x_1, y_2), \ldots, (x_n, y_m)\}.$$

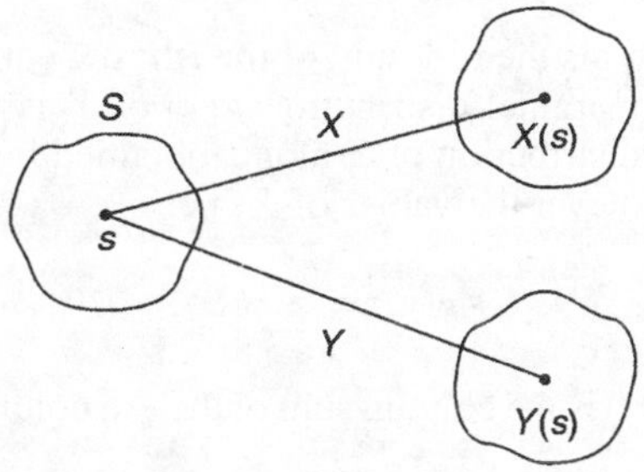

Fig. 31.1

The joint probability distribution or joint distribution or probability mass function or joint probability function of two discrete random variables X and Y is a function $h(x, y)$ difined on the product set $X(s) \times Y(s)$ assigning probability to each of the ordered pairs (x_i, y_i).

Thus $h(x_i, y_i) = P(X = x_i, Y = y_j)$ gives the probability for the simultaneous occurrences of the outcomes x_i and y_i. Further $h(x, y)$ satisfies the following:

1. $h(x_i, y_j) \geq 0$
2. $\sum_{j=1}^{m} \sum_{i=1}^{n} h(x_i, y_j) = 1$

The set of triplets $(x_i, y_j, h(x_i, y_j))$ for $i = 1$ to n, $j = 1$ to m, known as the probability distribution of

* Andrei Andrejevtich Markov (1856-1922), Russian mathematician

the two random variables X, Y is in general represented in the form of a rectangular table as follows:

Joint Probability distribution

$X \backslash Y$	y_1	y_2	$\cdots$	y_m	Row sums
x_1	$h(x_1, y_1)$	$h(x_1, y_2)$	$\cdots$	$h(x_1, y_m)$	$f(x_1)$
x_2	$h(x_2, y_1)$	$h(x_2, y_2)$	$\cdots$	$h(x_2, y_m)$	$f(x_2)$
$\vdots$	$\cdots$	$\cdots$	$\cdots$	$\cdots$	$\vdots$
x_n	$h(x_n, y_1)$	$h(x_n, y_2)$	$\cdots$	$h(x_n, y_m)$	$f(x_n)$
Column sums	$g(y_1)$	$g(y_2)$	$\cdots$	$g(y_m)$	

Marginal Distributions

Marginal distribution $f(x)$ of X is the probability distribution of X alone; obtained by summing $h(x, y)$ over the values of Y. i.e.,

$$f(x_i) = \sum_{j=1}^{m} h(x_i, y_j)$$

or $f(x_i)$ is the row sum of the ith row entries. Similarly, marginal distribution $g(y)$ of Y is the probability distribution of Y alone, obtained by summing $h(x, y)$ over the values of X. i.e.,

$$g(y_j) = \sum_{i=1}^{n} h(x_i, y_j)$$

or $g(y_j)$ is the column sum of the jth column entries.

Conditional Probability Distributions

Conditional distribution of random variable Y given that $X = x$ is

$$h(y|x) = \frac{h(x, y)}{f(x)}, \qquad \text{provided } f(x) > 0.$$

Similarly, the conditional distribution of X given that $Y = y$ is

$$h(x|y) = \frac{h(x, y)}{g(y)}, \qquad \text{provided } g(y) > 0.$$

Thus

$$\text{P}(a < X < b|Y = y) = \sum_{x} h(x|y)$$

where the summation is for all values of X between a and b.

Statistical Independence

The random variables X and Y are said to be statistically independent or simply independent if

$$h(x, y) = f(x) \cdot g(y)$$

for all (x, y) i.e.,

$$h(x_i, y_j) = \text{P}(X = x_i, Y = y_j) = \text{P}(X = x_i) \cdot \text{P}(Y = y_j) = f(x_i) \cdot g(y_j)$$

for all $i = 1$ to n and $j = 1$ to m. In other words, when X and Y are independent, each entry $h(x_i, y_j)$ is obtained as the product of its marginal entries. The joint distribution of X and Y can be determined from their marginal distribution functions. However, in the case of dependent variables, the joint distribution can <u>**not**</u> be determined in this simple fashion.

Covariance

The covariance of X and Y, denoted by Cov (x, y) is given by

$$\text{Cov}\,(X, Y) = \sum_j \sum_i (x_i - \mu_X)(y_j - \mu_Y) h(x_i, y_j)$$
$$= \text{E}\,((X - \mu_X)(Y - \mu_Y))$$
$$= \sum_j \sum_i x_i \cdot y_j \cdot h(x_i, y_j) - \mu_X \cdot \mu_Y$$
$$\text{Cov}\,(X, Y) = \text{E}(XY) - \mu_X \mu_Y$$

Correlation

The correlation of X and Y is

$$\rho(X, Y) = \frac{\text{Cov}\,(X, Y)}{\sigma_X \sigma_Y}$$

The dimensionless number ρ satisfy the following properties:

i. $\rho(X, Y) = \rho(Y, X)$ symmetry

ii. $\rho(X, X) = 1$ perfectly correlated

$\rho(X, -X) = -1$ negatively correlated

iii. $-1 \le \rho \le 1$ $\begin{cases} -1 \text{ if } X = -aY, (a > 0) \\ 0 \text{ if } X \text{ and } Y \text{ are uncorrelated} \\ 1 \text{ if } X = aY, (a > 0) \end{cases}$

iv. $\rho(aX + b, cY + d) = \rho(X, Y)$ if $a, c \neq 0$.

Result: If X and Y are independent, then

a. $\text{E}(XY) = \text{E}(X)\text{E}(Y)$

b. $\text{Var}\,(X + Y) = \text{Var}\,(X) + \text{Var}\,(Y)$

c. $\text{Cov}\,(X, Y) = 0$

Proof:

a. $\mathrm{E}(XY) = \sum_{i,j} x_i y_j h(x_i, y_j) = \sum x_i y_j f(x_i) g(y_j)$

$= \sum_i x_i f(x_i) \sum_j y_j g(x_j) = \mathrm{E}(X) \cdot \mathrm{E}(Y)$

b. $\mathrm{Var}\,(X + Y)$

$= \sum_{i,j} (x_i + y_j)^2 h(x_i, y_j) - \mu^2_{X+Y}$

$= \sum x_i^2 h(x_i, y_j) + 2 \sum x_i y_j h(x_i, x_j)$

$+ \sum_{i,j} y_j^2 h(x_i, y_j) - (\mu_X + \mu_Y)^2$

$= \sum_i x_i^2 f(x_i) + 2 \sum_i x_i f(x_i) \cdot \sum_j y_j g(x_j) +$

$+ \sum_j y_j^2 g(y_j) - \mu_X^2 - 2\mu_X\mu_Y - \mu_Y^2$

$= \left[\sum_i x_i^2 f(x_i) - \mu_X^2\right] + \left[\sum_j y_j^2 g(y_j) - \mu_Y^2\right]$

$= \mathrm{Var}\,(X) + \mathrm{Var}(Y)$

c. $\mathrm{Cov}\,(X, Y) = \mathrm{E}(XY) = \mu_X \mu_Y$

$= \mathrm{E}(X)\mathrm{E}(Y) - \mu_X \mu_Y$

since $\mathrm{E}(XY) = \mathrm{E}(X)\mathrm{E}(Y)$

$\mathrm{Cov}\,(X, Y) = \mu_X\mu_Y - \mu_X\mu_Y = 0.$

Worked Out Examples

Example 1: Find (a) marginal distributions $f(x)$ and $g(y)$, (b) $\mathrm{E}(X)$ and $\mathrm{E}(Y)$, (c) $\mathrm{Cov}\,(X, Y)$, (d) σ_X, σ_Y and (e) $\rho(X, Y)$ for the following joint distribution, (f) Are X and Y independent random variables?

X \ Y	−4	2	7
1	$\frac{1}{8}$	$\frac{1}{4}$	$\frac{1}{8}$
5	$\frac{1}{4}$	$\frac{1}{8}$	$\frac{1}{8}$

Solution: **(a)** The marginal distribution $f(x)$ is obtained by row sums and $g(y)$ by column sums.

X \ Y	−4	2	7	Row sums
1	$\frac{1}{8}$	$\frac{1}{4}$	$\frac{1}{8}$	$\frac{1}{2}$
5	$\frac{1}{4}$	$\frac{1}{8}$	$\frac{1}{8}$	$\frac{1}{2}$
Column sums	$\frac{3}{8}$	$\frac{3}{8}$	$\frac{1}{4}$	CHECK 1

The marginal distribution $f(x)$:

x_i:	1	5
$f(x_i)$:	$\frac{1}{2}$	$\frac{1}{2}$

The marginal distribution $g(y)$:

y_i:	−4	2	7
$g(y_i)$:	$\frac{3}{8}$	$\frac{3}{8}$	$\frac{1}{4}$

(b) $\mathrm{E}(X) = \mu_X = \sum_{i=1}^{2} x_i f(x_i) = 1 \cdot \frac{1}{2} + 5 \cdot \frac{1}{2} = \frac{6}{2} = 3$

$\mathrm{E}(Y) = \mu_Y = \sum_{j=1}^{3} y_j\, g(y_j)$

$\mathrm{E}(Y) = -4 \cdot \frac{3}{8} + 2 \cdot \frac{3}{8} + 7 \cdot \frac{1}{4} = 1.$

(c) We know that $\mathrm{Cov}\,(X, Y) = \mathrm{E}(XY) - \mu_X \cdot \mu_Y$.

Now

$\mathrm{E}(XY) = \sum_{j=1}^{3} \cdot \sum_{i=1}^{2} x_i\, y_j\, h\,(x_i, y_j)$

$= 1 \cdot (-4) \cdot \frac{1}{8} + 1 \cdot 2 \cdot \frac{1}{4} + 1 \cdot 7 \cdot \frac{1}{8} +$

$+5 \cdot (-4) \cdot \frac{1}{4} + 5 \cdot 2 \cdot \frac{1}{8} + 5 \cdot 7 \cdot \frac{1}{8}$

$\mathrm{E}(XY) = \frac{3}{2} = 1.5.$

So $\mathrm{Cov}\,(X, Y) = \mathrm{E}(XY) - \mu_X \cdot \mu_Y = 1.5 - 3 \cdot 1 = -1.5.$

(d) $\sigma_X^2 = \mathrm{Var}\,(X) = \mathrm{E}(X^2) - \mu_X^2 = 13 - 3^2 = 4, \sigma_X = 2$

$\left(\because\ \mathrm{E}(X^2) = 1 \cdot \frac{1}{2} + 25 \cdot \frac{1}{2} = 13\right)$

$\sigma_Y^2 = \mathrm{Var}\,(Y) = \mathrm{E}(Y^2) - \mu_Y^2 = 19.75 - 1 = 18.75, \sigma_y = 4330$

$\left(\because\ \mathrm{E}(Y^2) = 16 \cdot \frac{3}{8} + 4 \cdot \frac{3}{8} + 49 \cdot \frac{1}{4} = \frac{158}{8} = 19.75\right)$

$$\rho(X, Y) = \frac{\text{Cov}(X, Y)}{\sigma_X \cdot \sigma_Y} = \frac{-1.5}{2(4.33)} = -0.1732.$$

(e) Note that $P(X = 1, Y = -4) = \frac{1}{8}$ from table.

$P(X = 1) = \frac{1}{2}$, $P(Y = -4) = \frac{3}{8}$ from marginal distributions. Thus

$$P(X = 1, Y = -4) = \frac{1}{8} \neq P(X = 1) \cdot P(Y = -4)$$
$$= \frac{1}{2} \cdot \frac{3}{8} = \frac{3}{16}$$

$\therefore$ X and Y are **not** independent.

Example 2: Find the joint distribution of X and Y, which are independent random variables with the following respective distributions:

x_i:	1	2
$f(x_i)$:	.7	.3

and

y_j:	−2	5	8
$g(y_j)$:	.3	.5	.2

Show that Cov $(X, Y) = 0$.

Solution: Since X and Y are independent random variables,

$$h(x_i, y_j) = f(x_i)\, g(y_j).$$

Thus the entries of the joint distribution are the products of the marginal entries

X \ Y	−2	5	8	Sum
1				.7
2				.3
Sum	.3	.5	.2	

X \ Y	−2	5	8	Sum
1	.21	.35	.14	.7
2	.09	.15	.06	.3
Sum	.3	.5	.2	CHECK1

Now $E(XY) = 1 \cdot (-2)(.21) + 1 \cdot 5(.35) + 1 \cdot 8(.14) + 2 \cdot (-2)(.09) + 2 \cdot 5(.15) + 2 \cdot 8(.06) = 4.55$

$$\mu_X = 1(.7) + 2(.3) = 1.3,$$
$$\mu_Y = -2(.31) + 5(.5) + 8(.2) = 3.5$$

So $\text{Cov}(X, Y) = E(XY) - \mu_X \cdot \mu_Y$

$$= 4.55 - (1.3)(3.5) = 0.$$

Example 3: Two cards are selected at a random from a box which contains five cards numbered 1, 1, 2, 2 and 3. Find the joint distribution of X and Y where X denotes the sum and Y, the maximum of the two numbers drawn. Also determine Cov (X, Y) and $\rho(X, Y)$.

Solution: The possible pair of numbers are (1,1), (1,1), (1,2), (2,1), (1,2), (2,1), (1,3), (3,1), (2,2), (2,3), (3,2). Sum of two numbers are 2, 3, 4, 5 while maximum numbers are 1, 2, 3. Thus

Distribution of X is

x_i:	2	3	4	5
$f(x_i)$:	$\frac{1}{10} = 0.1$	$\frac{4}{10} = 0.4$	$\frac{3}{10} = 0.3$	$\frac{2}{10} = 0.2$

Distribution of Y is

y_j:	1	2	3
$g(y_j)$:	$\frac{1}{10} = 0.1$	$\frac{5}{10} = 0.5$	$\frac{4}{10} = 0.4$

Joint distribution

X \ Y	1	2	3	Sum
2	.1	0	0	.1
3	0	.4	0	.4
4	0	.1	.2	.3
5	0	0	.2	.2
Sum	.1	.5	.4	CHECK 1

$$\text{Cov}(X, Y) = E(XY) - \mu_X \cdot \mu_Y = 8.8 - (3.6)(2.3)$$
$$= 0.52 \neq 0.$$

Therefore X, Y are **not** independent.

$\sigma_X^2 = \text{Var}(X) = E(X^2) - \mu_X = 13.8 - (3.6)^2 = 0.84$

$(\because E(X^2) = .4 + 3.6 + 4.8 + 5 = 13.8)$

$\sigma_Y^2 = \text{Var}(Y) = E(Y^2) - \mu_Y = 5.7 - (2.3)^2 = 0.41$

$(\because E(Y^2) = .1 + 2 + 3.6 = 5.7)$

Thus $\sigma_X = 0.9165$, $\sigma_Y = 0.6403$

$$\therefore \rho(X, Y) = \frac{\text{Cov}(XY)}{\sigma_X \sigma_Y} = \frac{0.52}{(0.9165)(0.6403)} = 0.886 \simeq 0.9$$

Example 4: Evaluate the conditional distributions $h(x|1)$ for the following joint distribution. Show that X and Y are not independent

X \ Y	1	2	3
1	$\frac{1}{12}$	$\frac{1}{6}$	0
2	0	$\frac{1}{9}$	$\frac{1}{5}$
3	$\frac{1}{18}$	$\frac{1}{4}$	$\frac{2}{15}$

Solution: Adding rowwise and columnwise rewrite the joint distribution with the marginal distributions. Note that the entries of the above joint distribution are *not* obtained as the products of the marginal entries because it is not known whether X and Y are independent (In fact, it will be proved below that X and Y are dependent).

Explanation for the rowwise entries The number of pairs having the sum 2 and maximum number 1 is (1, 1) only. Thus $h(2, 1) = \frac{1}{10} = 0.1$. Also, there is no pair where the sum is 2 and maximum number is also 2. So this is an impossible event, therefore $h(2, 2) = 0$. The number of pairs whose sum is 3 and maximum number is 2 are (1, 2), (2, 1), (1, 2), (2, 1): four: $h(3, 2) = \frac{4}{10} = 0.4$, similarly, $h(4, 2) = \frac{1}{10}$ (only (2, 2)), $h(4, 3) = \frac{2}{10}$ (only {(1, 3) and (3,1)}). Similarly, $h(5, 1) = 0$, $h(5, 2) = 0$, $h(5, 3) = \frac{2}{10} = 0.2$ {(2, 3) and (3, 2) are the two pairs}.

$$\mu_X = 2(.1) + 3(.4) + 4(.3) + 5(.2) = 3.6$$

$$\mu_Y = 1(.1) + 2(.5) + 3(.4) = 2.3$$

$$E(XY) = 2\cdot 1\cdot(.1) + 3\cdot 2\cdot(.4) + 4\cdot 2\cdot(.1) + 4\cdot 3\cdot(.2) + 5\cdot 3\cdot(.2) = 8.8$$

Y \ X	1	2	3	Row sum
1	$\frac{1}{12}$	$\frac{1}{6}$	0	$\frac{1}{4}$
2	0	$\frac{1}{9}$	$\frac{1}{5}$	$\frac{14}{45}$
3	$\frac{1}{18}$	$\frac{1}{4}$	$\frac{2}{15}$	$\frac{201}{540}$
Column sum	$\frac{10}{72}$	$\frac{19}{36}$	$\frac{5}{15}$	CHECK 1

Conditional distribution $h(x|y) = \frac{h(x,y)}{g(y)}$

Now $h(x|1) = \frac{h(x,1)}{g(1)} = 4h(x, 1)$

since from the table $g(1) = \frac{1}{4}$. So

$$h(1|1) = \frac{h(1, 1)}{g(1)} = 4\,h(1, 1) = 4\left(\frac{1}{12}\right) = \frac{1}{3}$$

$$h(2|1) = 4\,h(2, 1) = 4\cdot\frac{1}{6} = \frac{2}{3}$$

$$h(3|1) = 4\,h(3, 1) = 4\cdot 0 = 0$$

Conditional distribution $h(x|1)$:

x_i:	1	2	3
$h(x\|1)$:	$\frac{1}{3}$	$\frac{2}{3}$	0

Dependence: From table

$$h(2, 3) = P(X = 2, Y = 3) = \frac{1}{4}.$$

From marginal distributions

$P(X = 2) = f(2) = \frac{19}{36}$ and

$P(Y = 3) = g(3) = \frac{201}{540}$. But

$$h(2, 3) = P(X = 2, Y = 3) = \frac{1}{4} \neq \frac{19}{36}\cdot\frac{201}{540} = f(2)g(3)$$

$$= P(X = 2)\cdot P(Y = 3)$$

$\therefore$ X and Y are not independent.

EXERCISE

1. Determine (a) marginal distributions of X and Y (b) Cov (X, Y) (c) $\rho(X, Y)$, for the following joint distribution. (d) Determine whether X and Y are independent.

X \ Y	−3	2	4
1	.1	.2	.2
3	.3	.1	.1

Hint: $\mu_X = 2$, $\mu_Y = 0.6$, $E(XY) = 0$, $E(X^2) = 5$, $\sigma_X^2 = 1$, $E(Y^2) = 9.6$, $\sigma_Y^2 = 9.24$, $\sigma_Y = 3.0$ (d) $h(1, -3) = .1 \neq .2 = f(1), g(-3) = (.5)(.4)$.

Ans. (a)

x_i	1	3
$f(x_i)$	.5	.5

,

y_j	−3	2	4
$g(y_j)$	.4	.3	.3

;

(b) Cov $(X, Y) = -1.2$, (c) $\rho(X, Y) = -4$, (d) not independent

2. If X and Y are independent random variables, find the joint distribution of X and Y with the

following marginal distribution of X and Y.

x_i	1	2
$f(x_i)$	.6	.4

,

y_j	5	10	15
$g(y_j)$	.2	.5	.3

Hint: $h(x_i, y_j) = f(x_i)g(y_j)$.

Ans.

X \ Y	5	10	15	Row sums
1	.12	.30	.18	.6
2	.08	.20	.12	.4
Column sums	.2	.5	.3	

3. A fair coin is tossed three times. Let X denote 0 or 1 according as a head or a tail occurs on the first toss. Let Y denote the number of heads which occur. (a) Find the marginal distributions of X and Y, (b) Determine the joint distribution of X and Y and (c) Cov (X, Y).

Hint:

$S = \{H, T\} \times \{H, T\} \times \{H, T\} : 2^3 = 8$ points

$= \{HHH, HHT, HTH, HTT, THH, THT,$

$TTH, TTT\}, X(HHH) = 0, X(THH)$

$= 1$ etc., $Y(HHH) = 3, Y(TTT) = 0$ etc.

(c) $\mu_X = \frac{1}{2}, \mu = \frac{3}{2}, \text{E}(XY) = \frac{1}{2}$

Ans. **(a)**

x_i	0	1
$f(x_i)$	$\frac{1}{2}$	$\frac{1}{2}$

y_j	0	1	2	3
$g(y_j)$	$\frac{1}{8}$	$\frac{3}{8}$	$\frac{3}{8}$	$\frac{1}{8}$

(b)

X \ Y	0	1	2	3	Sums
0	0	$\frac{1}{8}$	$\frac{2}{8}$	$\frac{1}{8}$	$\frac{1}{2}$
1	$\frac{1}{8}$	$\frac{2}{8}$	$\frac{1}{8}$	0	$\frac{1}{2}$
Sums	$\frac{1}{8}$	$\frac{3}{8}$	$\frac{3}{8}$	$\frac{1}{8}$	

(c) Cov $(X, Y) = -\frac{1}{4}$

4. The joint distributions of two pairs of random variables X, Y and X^*, Y^* are given below. Find the marginal distributions of X, Y and X^*, Y^*, Cov (X, Y), Cov (X^*, Y^*), $\rho(X, Y)$, $\rho(X^*, Y^*)$. Comment.

Ans. Marginal distributions of X and X^* are same. Also, marginal distributions of Y and Y^* are same.

x_i	1	3
$f(x_i)$	$\frac{1}{2}$	$\frac{1}{2}$

,

y_j	4	10
$g(x_j)$	$\frac{1}{2}$	$\frac{1}{2}$

$\mu_X = \mu_{X^*} = 2$, $\mu_Y = \mu_{Y^*} = 7$, Cov $(X, Y) = 0$, Cov $(X^*, Y^*) = -3$, $\rho(X, Y) = 0$, $\rho(X^*, Y^*) = -\frac{3}{2 \cdot 7} = -\frac{3}{14} = -0.21428$.

Comment: Althouth the marginal distributions are identical, covariances and correlations coefficients are different.

5. Given the joint distribution

Y \ X	0	1	2
0	.1	.4	.1
1	.2	.2	0

(a) Determine the marginal distributions of X and Y, (b) Find the conditional probability distribution $h(x|y = 1)$, (c) Are X and Y independent?

Ans. **(a)**

x_i	0	1	2
$f(x_i)$	.3	.6	.1

,

y_j	0	1
$g(y_j)$	.6	.4

;

(b) $h(0|1) = \frac{h(0,1)}{g(1)} = \frac{.2}{.4} = 0.5$, $h(1|1) = .5$, $h(2|1) = 0$;

(c) Dependent, since $f(0, 1) = 0.2 \neq (.3)(.4) = f(0)g(1)$.

6. Find the marginal distributions of X and Y and find $\text{P}(Y = 3|X = 2)$ if the joint distribution is

Y \ X	1	2	3
1	0.05	0.05	0.1
2	0.05	0.1	0.35
3	0	0.2	0.1

Ans.

x_i:	1	2	3
$f(x_i)$:	0.1	0.35	0.55

,

y_j:	1	2	3
$g(y_j)$:	0.2	0.5	0.3

$$\text{P}(Y = 3|X = 2) = \frac{0.2}{0.35} = 0.5714.$$

7. Two marbles are selected at random from a box containing 3 blue, 2 red and 3 green marbles. If X is the number of blue marbles and Y is the number of red marbles selected, find (a) joint

probability function $h(x, y)$, (b) $P[(X, Y) \in A]$ where A is the region $\{(x, y)|x + y \leq 1\}$, (c) the marginal distributions of X and Y.

Hint: (a) $h(x, y) = \left({}^3C_x\right)\left({}^2C_y\right)\left({}^{3}_{2-x-y}\right)/\left({}^8C_2\right)$

(b) $x = 0, 1, 2;\ y = 0, 1, 2;\ 0 \leq x + y \leq 2$

$$P[(x, y) \in A] = P(X + Y \leq 1)$$
$$= h(0, 0) + h(0, 1) + h(1, 0)$$
$$= \frac{3}{28} + \frac{3}{14} + \frac{9}{28} = \frac{9}{14}.$$

(c)

x_i:	0	1	2
$f(x_i)$:	$\frac{5}{14}$	$\frac{15}{28}$	$\frac{3}{28}$

,

y_j:	0	1	2
$g(y_j)$:	$\frac{15}{28}$	$\frac{3}{7}$	$\frac{1}{28}$

Ans.

Y \ X	0	1	2	Sums
0	$\frac{3}{28}$	$\frac{9}{28}$	$\frac{3}{28}$	$\frac{15}{28}$
1	$\frac{3}{14}$	$\frac{3}{14}$	0	$\frac{3}{7}$
2	$\frac{1}{28}$	0	0	$\frac{1}{28}$
Sums	$\frac{5}{14}$	$\frac{15}{28}$	$\frac{3}{28}$	CHECK 1

8. For the joint distribution in problem 7 (above). (a) Find the conditional distribution of X, given that $Y = 1$, (b) Determine $P(X = 0|Y = 1)$, (c) Show that the random variables X and Y are *not* statistically independent.

Hint:

$$g(1) = \sum_{x=0}^{2} h(x, 1) = \frac{3}{14} + \frac{3}{14} + 0 = \frac{3}{7}$$

$$h(x|1) = \frac{h(x, 1)}{g(1)} = \frac{7}{3} \cdot h(x, 1) \quad \text{for} \quad x = 0, 1, 2$$

Ans.

x_i	0	1	2
(a) $h(x_i|1)$	$\frac{1}{2}$	$\frac{1}{2}$	0

(b) $p(X = 0|Y = 1) = h(0, 1) = \frac{1}{2}$

(c) $h(0, 1) = \frac{3}{14} \neq \frac{5}{14} \cdot \frac{3}{7} = f(0)g(1)$.

31.2 MARKOV CHAINS

Suppose a box A contains 5 red, 3 white and 8 black marbles while box B contains 3 red and 5 white marbles. A fair die is tossed and if 2 or 5 occurs a marble is chosen from B otherwise from A. Further in box A, two red, one white and 4 black marbles are defective while in box B one red and 2 white marbles are defective. To determine the probability that a marble drawn at random is say a defective red marble, we have to conduct a sequence of experiments in which each experiment has a finite number of outcomes with given probabilities as shown in the tree diagram below (Fig. 31.2).

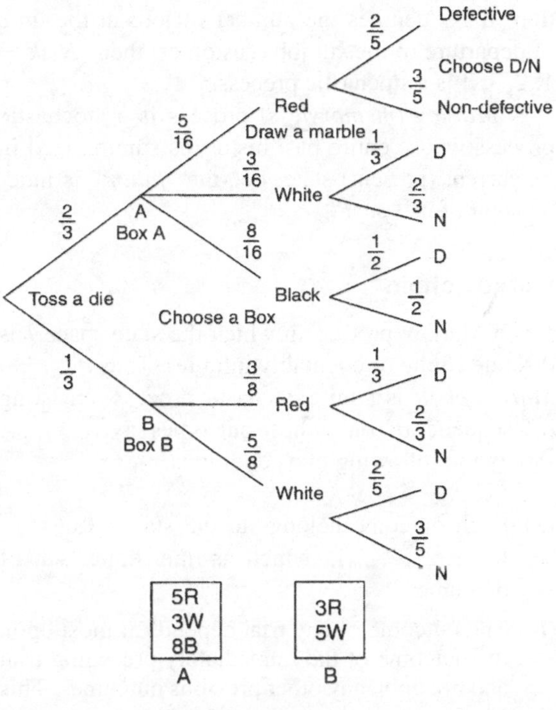

Fig. 31.2

The four experiments are: toss a die, choose a box, draw a marble and decide whether it is defective. Here probability of a defective red marble is $\frac{2}{3} \cdot \frac{5}{16} \cdot \frac{2}{5} + \frac{1}{3} \cdot \frac{3}{8} \cdot \frac{1}{3}$.

Stochastic Process (or Chance or Random Process)

It is a family of random variables $\{X(t)|t \in T\}$ defined on a common sample space S and indexed by the parameter t, which varies on an index set T.

The values assumed by the random variables $X(t)$ are called *states*, and the set of all possible values from the *state space* of the process is denoted by I. If the state space is discrete, the stochastic process is known as a *chain*. In this case the state space is assumed to be $I = \{0, 1, 2, \ldots\}$. Thus a (finite) stochastic process consists of a sequence of experiments in which each experiment has a finite number of outcomes with given probabilities.

Example: Jobs arrive at random points in time, queue for service and depart after service completion. If N_k denotes the number of jobs at the time of departure of the kth job (customer) then $\{N_k | k = 1, 2, \ldots\}$ is a stochastic process.

A *Markov (memoryless)* process is a stochastic process whose entire past history is summarized in its current (present) state. i.e., the "*future*" is independent of its "*past*".

Markov chain

It is a Markov process in which the state space I is discrete (finite or countably infinite). Thus *a (finite) Markov chain* is a finite stochastic process consisting of a sequence of trials whose outcomes say $x_1, x_2, \ldots$ satisfy the following two conditions:

(a) Each outcome belongs to the state space $I = \{a_1, a_2, \ldots, a_m\}$, which is the finite set of outcomes.

(b) The outcome of any trial depends at most upon the outcome of the immediately preceding trial and not upon any other previous outcomes. This Markov property can be stated as

$$P\left(X_n = i_n \middle| X_0 = i_o, X_1 = i_1, \ldots, X_{n-1} = i_{n-1}\right)$$
$$= P\left(X_n = i_n \middle| X_{n-1} = i_{n-1}\right).$$

Now the system is said to be in state 'a_i' at time n or at the nth step if a_i is the outcome on the nth trial. Associated with each ordered pair of states (a_i, a_j), the number p_{ij} gives the probability that system changes from ith state to jth state. In other words, p_{ij} is the probability that a_j occurs immediately after a_i occurs. The numbers p_{ij} are known as transition probabilities.

Transition matrix

P is square matrix of the transition probabilities p_{ij}:

$$P = \begin{array}{c} \\ a_1 \\ a_2 \\ \vdots \\ a_m \end{array} \begin{array}{c} \begin{array}{ccccc} a_1 & a_2 & a_3 & \ldots & a_m \end{array} \\ \begin{pmatrix} p_{11} & p_{12} & p_{13} & & p_{1m} \\ p_{21} & p_{22} & p_{23} & & p_{2m} \\ \ldots & \ldots & \ldots & \ldots & \ldots \\ p_{m1} & p_{m2} & p_{m3} & \ldots & p_{mm} \end{pmatrix} \end{array}.$$

The ith row of P namely $(p_{i1}, p_{i2}, \ldots, p_{im})$ represents the probabilities of that system will change from a_i to $a_1, a_2, a_3, \ldots, a_m$.

Probability vector

It is a vector $v = (v_1, v_2, \ldots, v_n)$, if $v_i \geq 0$ for every i and $\sum_{i=1}^{n} v_i = 1$.

Note: A vector whose components are non-negative, but their sum is not one, can be converted into a probability vector by dividing each component by the sum of the components.

Stochastic matrix

P is a square matrix with each row being a probability vector. In other words, all the entries of P are non-negative and the sum of the entries of any row is one.

A vector v is said to be a **fixed vector** or a fixed point of a matrix A if $vA = v$ and $v \neq 0$.

Obviously if v is a fixed vector of A, so is kv since $(kv)A = k(vA) = k(v) = kv$.

Theorem 1: *If $v = (v_1 v_2 v_3)$ is a probability vector of a stochastic matrix $P = \begin{pmatrix} a_1 & b_1 & c_1 \\ a_2 & b_2 & c_2 \\ a_3 & b_3 & c_3 \end{pmatrix}$ then vP is also a probability vector.*

Proof:

$$vP = (v_1 v_2 v_3)_{1\times 3} \begin{pmatrix} a_1 & b_1 & c_1 \\ a_2 & b_2 & c_2 \\ a_3 & b_3 & c_3 \end{pmatrix}_{3\times 3}$$
$$= (v_1 a_1 + v_2 a_2 + v_3 a_3, v_1 b_1 + v_2 b_2 + v_3 b_3, v_1 c_1 + v_2 c_2 + v_3 c_3).$$

Since a_i, b_i, c_i, v_i are all non-negative for any i, the components of vP are all non-negative. Now the sum of the components of vP is

$$(v_1a_1 + v_2a_2 + v_3a_3) + (v_1b_1 + v_2b_2 + v_3b_3) + (v_1c_1 + v_2c_2 + v_3c_3)$$
$$= v_1(a_1+b_1+c_1) + v_2(a_2+b_2+c_2) + v_3(a_3+b_3+c_3)$$
$$= v_1 \cdot 1 + v_2 \cdot 1 + v_3 \cdot 1 = v_1 + v_2 + v_3 = 1, \text{ since}$$

P is a stochastic matrix and v is given probability vector.

General result: If $v = (v_1 v_2 v_3 \ldots v_n)$ is a probability vector of a n square stochastic matrix P then vP is also a probability vector.

Theorem 2: *If P and Q are stochastic matrices then their product PQ is also stochastic matrix. Thus P^n is stochastic matrix for all positive integer values of n.*

Proof: The ith row of PQ is the product of ith row of P with matrix Q. Since P and Q are stochastic matrices, ith row of P is a probability vector and by previous Theorem 1, ith row of P with matrix Q is also a probability vector and hence PQ is a stochastic matrix. If $P = Q$, then $PQ = P^2$ is stochastic and in general P^n is stochastic for n positive integer.

Theorem 3: *Let $t = (t_1 t_2 \ldots t_m)$ be a vector and T be a square matrix whose rows are each the same vector t. Then $pT = t$. if $p = (p_1 p_2 \ldots p_m)$ is a probability vector.*

Proof:

$$pT = (p_1 p_2 \ldots p_m)_{1\times m} \begin{pmatrix} t_1 & t_2 & t_3 & \ldots & t_m \\ t_1 & t_2 & t_3 & \ldots & t_m \\ -- & -- & -- & -- & -- \\ t_1 & t_2 & t_3 & \ldots & t_m \end{pmatrix}_{m\times m}$$

$$= (t_1(p_1 + p_2 + \cdots + p_m), t_2(p_1 + p_2 + \cdots + p_m), \ldots, t_m(p_1 + p_2 + \cdots + p_m))$$
$$= (t_1 \cdot 1, t_2 \cdot 1, \ldots, t_m \cdot 1) = (t_1 t_2 \ldots t_m) = t$$

since p is a probability vector (then $\sum_{i=1}^{m} p_i = 1$).

Theorem 4: *The transition matrix P of a Markov chain is a stochastic matrix.*

Proof: All the entries of a transition matrix P are non negative because p_{ij} are probabilities. The sum of the elements $(p_{i1}, p_{i2,\ldots,p_{im}})$ of any ith row is one, because they represent the probabilities of all the possible outcomes of transition from state a_i to the states $a_1, a_2, \ldots, a_i, \ldots, a_m$. Thus each row of P is a probability vector. Therefore P is a stochastic matrix.

A Stochastic matrix P is said to be *regular* if all the entries of some power P^m are positive.

Theorem 5: *Let P be a regular stochastic matrix. Then*

(a) *P has a unique fixed probability vector t and the components of t are all positive.*

(b) *The sequence $P, P^2, P^3 \ldots$ of powers of P approaches the matrix T whose rows are each the fixed point t.*

(c) *If p is any probability vector, then the sequence of vectors $pP, pP^2, pP^3, \ldots$ approaches the fixed point t.*

Note: Here matrix A approaches matrix B, means every entry of A approaches the corresponding entry of B.

Higher Transition Probabilities

One-step transition probabilities

The entry p_{ij} in the transition probability matrix P is the probability that the system moves from the state a_i to the state a_j in one step i.e., $a_i \to a_j$. The one-step transition probabilities in P can also be described by a directed graph known as state-transition diagram or simply transition diagram of the Markov chain. A node labelled i of the transition diagram represents state i of the Markov chain. A branch labeled p_{ij} from node i to j represents the conditional probability (or the one-step transition probabilities) defined by

$$p_{ij} = P\left(X_n = j \middle| X_{n-1} = i\right)$$

n-step Transition Probabilities

The probability that a Markov chain will move from state i to state j in exactly n steps, is denoted by $p_{ij}(n)$ or $p_{ij}^{(n)}$ and is given by

$$p_{ij}^{(n)} = p_{ij}(n) = P\left(X_{m+n} = j \middle| X_m = i\right)$$

i.e., $a_i \to a_{k_1} \to a_{k_2} \to \cdots \to a_{k_{n-1}} \to a_j$.

Evaluation of n-step Transition Probability Matrix $P^{(n)}$ or $P(n)$

using Chapman-Kolmogorov equation

$$p_{ij}(m+n) = \sum_k p_{ik}(m) p_{kj}(n).$$

Let $P^{(n)}$ or $P(n)$ represent a matrix whose (i, j)th entry is $p_{ij}^{(n)}$ or $p_{ij}(n)$. Putting $m = 1$ and $n = n-1$ in the above C-K equation, $P^{(n)}$ or $P(n)$ the n-step transition probabilities matrix can be written as

$$P^{(n)} = P(n) = P \cdot P(n-1) = P P P(n-2) = P^n.$$

Thus the matrix of n-step transition probabilities $P^{(n)}$ is obtained by multiplying the matrix of one step transition probabilities P by itself $n-1$ times.

Theorem 6: *If P is the transition matrix of a Markov chain, then the n-step transition matrix $P^{(n)}$ is equal to the nth power of P, i.e., $P^{(n)} = P^n$.*

In other words, the problem of finding the n-step transition probabilities is reduced to one of forming powers of a given matrix.

Probability distribution of the system at some arbitrary time is denoted by the probability vector.

$$p = (p_1, p_2, p_i, \ldots, p_m)$$

where p_i denotes the probability that the system is in state a_i. At time $t = 0$, when the process begins, the corresponding probability vector

$$p^{(0)} = (p_1^{(0)}, p_2^{(0)}, \ldots, p_i^{(0)}, \ldots, p_m^{(0)})$$

denotes the initial probability distribution. Similarly, the nth step probability distribution i.e., the distribution after the first n-steps is denoted by

$$p^{(n)} = (p_1^{(n)}, p_2^{(n)}, \ldots, p_m^{(n)}).$$

Now the (marginal) pmf of the random variable X_n can be obtained from the n-step transition probabilities and the initial distribution as follows

$$\boxed{p^{(n)} = p^{(0)} P^{(n)} = p^{(0)} P^n}$$

Thus the probability distributions of a homogeneous Markov chain are completely determined from the one-step transition probability matrix P and the initial probability distribution $p^{(0)}$.

Theorem 7: *The probability distribution of the system n-steps later is given by*

$$p^{(n)} = p^{(0)} P^n$$

i.e., $p^{(1)} = p^{(0)} P$; $p^{(2)} = p^{(1)} P = p^{(0)} P P = p^{(0)} P^2$

$p^{(3)} = p^{(2)} P = p^{(0)} P^2 P = p^{(0)} P^3$ etc.

Stationary Distribution of Regular Markov Chains

Theorem 7: *Let P be a regular transition matrix of a Markov chain. Then in the long run, the probability that any state a_j occurs is approximately equal to the component t_j of the unique fixed probability vector t of P.*

Proof: Suppose the Markov chain is regular, i.e., P is regular, then by Theorem 5, the sequence of n-step transition matrices P^n approaches the matrix T, whose rows are each the unique fixed probability vector t of P. Hence the probability $p_{ij}(n)$ that a_j occurs for sufficiently large n is independent of the original state a_i and it approaches the component t_j of t.

Stationary distribution

Stationary distribution of a Markov chain is the unique fixed probability vector t of the regular transition matrix P of the Markov chain because every sequence of probability distributions approaches t.

Absorbing States

A state a_i of a Markov chain is said to be an absorbing state if the system remains in the state a_i once it enters there, i.e., a state a_i is absorbing if $p_{ii} = 1$. Thus once a Markov chain enters such an absorbing state, it is destined there to remain forever. In other words the ith row in P has 1 at the main diagonal (i, i) position and zeros everywhere else.

Theorem 8: *A stochastic matrix P is not regular if a 1 occurs in the principal main diagonal.*

Proof: Suppose a_i is the absorbing state of the given Markov chain whose transition matrix is P. Then 1 occurs in the (i, i) position and the ith row of P is of the form $(0, 0, \ldots, 0, 1, 0, \ldots, 0)$. When powers of P are calculated the ith row of P^n persists to contain $(0\,0\,0 \ldots, 1, 0 \ldots 0)$. Thus for $i \neq j$ (non diagonal elements), the n-step transition probability $p_{ij}^{(n)} = 0$ for any n. Thus every power of P contains some zero elements. Therefore P is not regular.

WORKED OUT EXAMPLES

Probability vector and stochastic matrix

Example 1: Which vectors are probability vectors

(i) $\left(\frac{1}{4}, \frac{3}{2}, -\frac{1}{4}, \frac{1}{2}\right)$ (ii) $\left(\frac{5}{2}, 0, \frac{8}{3}, \frac{1}{6}, \frac{1}{6}\right)$

(iii) $\left(\frac{1}{12}, \frac{1}{2}, \frac{1}{6}, 0, \frac{1}{4}\right)$ (iv) $(3, 0, 2, 5, 3)$

Solution:

(i) is not a probability vector because negative entry $\left(-\frac{1}{4}\right)$

(ii) is not because the sum of the components do not add up to 1

(iii) is a probability vector because all the entries are non-negative and sum $\frac{1}{12}+\frac{1}{2}+\frac{1}{6}+\frac{1}{4}=\frac{12}{12}=1$.

(iv) Dividing by $3+0+2+5+3 = 13$, we get the probability vector $\left(\frac{3}{13}, 0, \frac{2}{13}, \frac{5}{13}, \frac{3}{13}\right)$.

Example 2: Which matrices are stochastic

(i) $\begin{pmatrix} 0 & 1 & 0 \\ \frac{1}{2} & \frac{1}{4} & \frac{1}{4} \end{pmatrix}$ **(ii)** $\begin{pmatrix} 1 & 0 \\ 0 & 1 \end{pmatrix}$ **(iii)** $\begin{pmatrix} 0 & 1 \\ \frac{1}{2} & \frac{1}{4} \end{pmatrix}$

(iv) $\begin{pmatrix} \frac{1}{2} & \frac{1}{2} \\ \frac{1}{2} & \frac{1}{2} \end{pmatrix}$ **(v)** $\begin{pmatrix} 0 & 1 \\ -\frac{1}{2} & \frac{3}{2} \end{pmatrix}$.

Solution: (i), (iii), (v) are not stochastic because (i) is not square (iii) last row sum is not 1 (v) negative entry (ii) & (iv) are stochastic matrices: each row sum is one, entries non-negative.

Example 3: Which of the stochastic matrices are regular

(i) $A = \begin{pmatrix} \frac{1}{2} & \frac{1}{4} & \frac{1}{4} \\ 0 & 1 & 0 \\ \frac{1}{2} & 0 & \frac{1}{2} \end{pmatrix}$ **(ii)** $B = \begin{pmatrix} \frac{1}{2} & \frac{1}{2} & 0 \\ \frac{1}{2} & \frac{1}{2} & 0 \\ \frac{1}{4} & \frac{1}{4} & \frac{1}{2} \end{pmatrix}$

(iii) $C = \begin{pmatrix} 0 & 0 & 1 \\ \frac{1}{2} & 0 & \frac{1}{2} \\ 0 & 1 & 0 \end{pmatrix}$

Solution:

(i) not regular since 1 lies on the main diagonal.

(ii) $B^2 = B \cdot B = \begin{pmatrix} \frac{1}{2} & \frac{1}{2} & 0 \\ \frac{1}{2} & \frac{1}{2} & 0 \\ \frac{3}{8} & \frac{3}{8} & \frac{1}{4} \end{pmatrix}$,

$$B^3 = B^2 B = \begin{pmatrix} \frac{1}{2} & \frac{1}{2} & 0 \\ \frac{1}{2} & \frac{1}{2} & 0 \\ \frac{7}{16} & \frac{7}{16} & \frac{1}{8} \end{pmatrix}$$

since entries b_{13}, b_{23} are zero, B is not regular

(iii) $C^2 = C \cdot C = \begin{pmatrix} 0 & 1 & 0 \\ 0 & \frac{1}{2} & \frac{1}{2} \\ \frac{1}{2} & 0 & \frac{1}{2} \end{pmatrix}$,

$$C^3 = C \cdot C^2 = \begin{pmatrix} \frac{1}{2} & 0 & \frac{1}{2} \\ \frac{1}{4} & \frac{1}{2} & \frac{1}{4} \\ 0 & \frac{1}{2} & \frac{1}{2} \end{pmatrix}$$

$$C^4 = C^3 \cdot C = \begin{pmatrix} 0 & \frac{1}{2} & \frac{1}{2} \\ \frac{1}{4} & \frac{1}{4} & \frac{1}{2} \\ \frac{1}{4} & \frac{1}{2} & \frac{1}{4} \end{pmatrix},$$

$$C_5 = \begin{pmatrix} \frac{1}{4} & \frac{1}{2} & \frac{1}{4} \\ \frac{1}{8} & \frac{1}{2} & \frac{3}{8} \\ \frac{1}{4} & \frac{1}{4} & \frac{1}{2} \end{pmatrix}.$$

Since all the entries of some power of C are positive, C is regular stochastic matrix.

Fixed probability vectors

Example 4: Show that $v = (b\ a)$ is fixed point of the stochastic matrix $P = \begin{pmatrix} 1-a & a \\ b & 1-b \end{pmatrix}$.

Solution: $vP = (b\ a)\begin{pmatrix} 1-a & a \\ b & 1-b \end{pmatrix} =$
$(b - ab + ab \quad ba + a - ab) = (b\ a) = v.$

Example 5:

(a) Find the unique fixed probability vector t of

$$P = \begin{pmatrix} 0 & \frac{3}{4} & \frac{1}{4} \\ \frac{1}{2} & \frac{1}{2} & 0 \\ 0 & 1 & 0 \end{pmatrix}$$

(b) What matrix does P^n approach?

(c) What vector does $\left(\frac{1}{4}, \frac{1}{4}, \frac{1}{2}\right) P^n$ approach?

Solution:

(a) Let $t = (x, y, z)$ be the fixed probability vector. By definition $x + y + z = 1$. So $t = (x, y, 1 - x - y)$, t is said to be fixed vector, if $tP = t$

$$(x\ \ y\ \ 1-x-y)\begin{pmatrix} 0 & \frac{3}{4} & \frac{1}{4} \\ \frac{1}{2} & \frac{1}{2} & 0 \\ 0 & 1 & 0 \end{pmatrix} = (x\ \ y\ \ 1-x-y)$$

Solving

$$\frac{1}{2}y = x$$

$$\frac{3}{4}x + \frac{1}{2}y + 1 - x - y = y$$

$$\frac{1}{4}x = 1 - x - y$$

Solving $\quad y = 2x,\ x = \frac{4}{13},\ y = \frac{8}{13},\ z = \frac{1}{13}$

Required fixed probability vector is

$$t = (x, y, z) = \left(\frac{4}{13}, \frac{8}{13}, \frac{1}{13}\right)$$
$$= (0.3077, 0.6154, 0.077)$$

(b)

$$P^2 = P \cdot P = \begin{pmatrix} \frac{3}{8} & \frac{5}{8} & 0 \\ \frac{1}{4} & \frac{5}{8} & \frac{1}{8} \\ \frac{1}{2} & \frac{1}{2} & 0 \end{pmatrix}$$

$$P^3 = P^2 P = \begin{pmatrix} \frac{5}{16} & \frac{19}{32} & \frac{3}{32} \\ \frac{5}{16} & \frac{5}{8} & \frac{1}{16} \\ \frac{1}{4} & \frac{5}{8} & \frac{1}{8} \end{pmatrix}$$

$$P^4 = P^3 P = \begin{pmatrix} \frac{19}{64} & \frac{40}{64} & \frac{5}{64} \\ \frac{20}{64} & \frac{39}{64} & \frac{5}{64} \\ \frac{20}{64} & \frac{40}{64} & \frac{4}{64} \end{pmatrix}$$

$$P^5 = P^4 \cdot P = \frac{1}{64}\begin{pmatrix} 20 & \frac{196}{4} & \frac{19}{4} \\ \frac{39}{2} & \frac{79}{2} & 5 \\ 20 & 39 & 5 \end{pmatrix}$$

$$= \frac{1}{256}\begin{pmatrix} 80 & 196 & 19 \\ 78 & 158 & 20 \\ 80 & 156 & 20 \end{pmatrix} = \begin{pmatrix} 0.3125 & 0.7656 & 0.0742 \\ 0.304 & 0.61718 & 0.078 \\ 0.3125 & 0.61718 & 0.078 \end{pmatrix}$$

Thus $P^n \to T = \begin{pmatrix} t \\ t \\ t \end{pmatrix}$, where $t = (0.3077, 0.6154, 0.077)$

(c) $\left(\frac{1}{4}\frac{1}{4}\frac{1}{2}\right) P^n = \left(\frac{1}{4}\frac{1}{4}\frac{1}{2}\right)\begin{pmatrix} 0.3125 & 0.7656 & 0.0742 \\ 0.304 & 0.61718 & 0.078 \\ 0.3125 & 0.61718 & 0.078 \end{pmatrix}$

$$= (0.310375, 0.654285, 0.07705) \approx t$$

Finite stochastic process

Example 6: An urn A contains 5 red, 3 white and 8 green marbles while urn B contains 3 red and 5 white marbles (Fig. 31.3). A fair die is tossed; if 3 or 6 appears a marble is chosen from B otherwise from A. Find the probability that (a) a red marble is chosen (b) a white marble is chosen (c) a green marble is chosen.

Solution: p = probability that 3 or 6 appears in the toss of a die

$$p = \frac{1}{6} + \frac{1}{6} = \frac{2}{6} = \frac{1}{3}$$

$$q = 1 - p = \frac{2}{3}$$

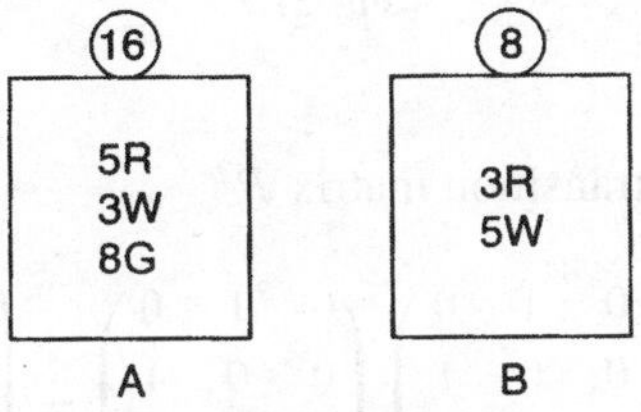

Fig. 31.3

probability of a red marble chosen from urn $A = P(R/A) = \frac{5}{16}$, $P(\text{white from } A) = P(W/A) = \frac{3}{16}$, $P(G/A) = \frac{8}{16}$, probability of a red marble chosen from urn $B = P(R/B) = \frac{3}{8}$, $P(W/B) = \frac{5}{8}$.

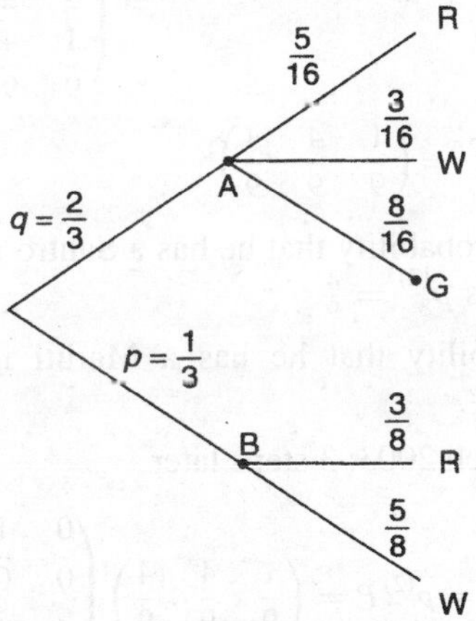

Fig. 31.4

Here we perform a sequence of two experiments. First, toss a die and choose the box. Second, choose a marble from the (chosen) box.

(a) Probability that a red marble is chosen

$$= P(R) = P(A) \cdot P(R/A) + P(B) \cdot P(R/B)$$

$$= \frac{2}{3} \cdot \frac{5}{16} + \frac{1}{3} \cdot \frac{3}{8} = \frac{16}{48} = \frac{1}{3}$$

(b) $$P(W) = P(A) \cdot P(W/A) + P(B) \cdot P(W/B)$$

$$= \frac{2}{3} \cdot \frac{3}{16} + \frac{1}{3} \cdot \frac{5}{8} = \frac{16}{48} = \frac{1}{3}$$

(c) $$P(G) = P(A)P(G/A) = \frac{2}{3} \cdot \frac{8}{16} = \frac{1}{3}.$$

Transition matrix and transition diagram

Example 7: Figure 31.5 shows four compartments with door leading from one to another. A mouse in any compartment is equally likely to pass through each of the doors of the compartment. Find the transition matrix of the Markov chain. Draw the transition diagram.

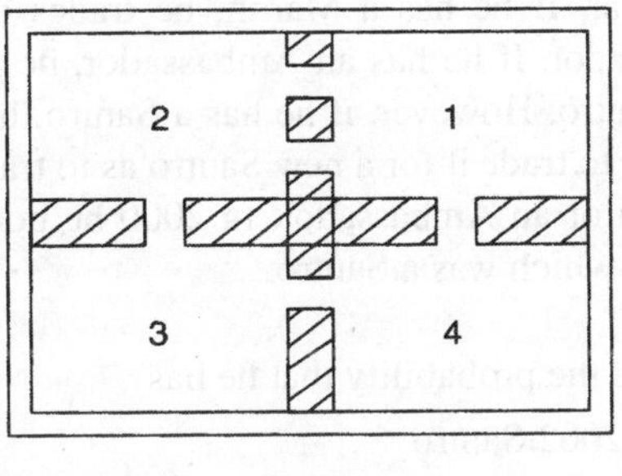

Fig. 31.5

Solution: The 4 rooms are considered as four states say 1, 2, 3, 4. Since mouse is moving, it does not stay in the same room. From room 1 it can go to 4 or 2 with probability $\frac{1}{3}$ or $\frac{2}{3}$. It can not go from 1 to 3. Then the first row consists of 0, $\frac{2}{3}$, 0, $\frac{1}{3}$. Thus the transition matrix is

$$\begin{array}{c} \\ 1 \\ 2 \\ 3 \\ 4 \end{array} \begin{array}{c} \begin{array}{cccc} 1 & 2 & 3 & 4 \end{array} \\ \begin{pmatrix} 0 & \frac{2}{3} & 0 & \frac{1}{3} \\ \frac{2}{3} & 0 & \frac{1}{3} & 0 \\ 0 & \frac{1}{2} & 0 & \frac{1}{2} \\ \frac{1}{2} & 0 & \frac{1}{2} & 0 \end{pmatrix} \end{array}$$

Figure 31.6 gives the transition diagram:

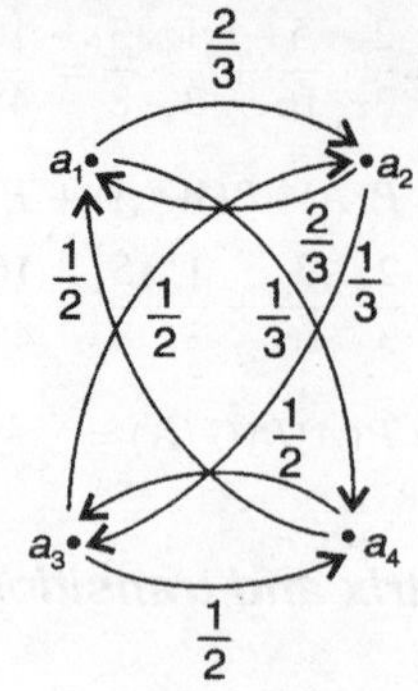

Fig. 31.6

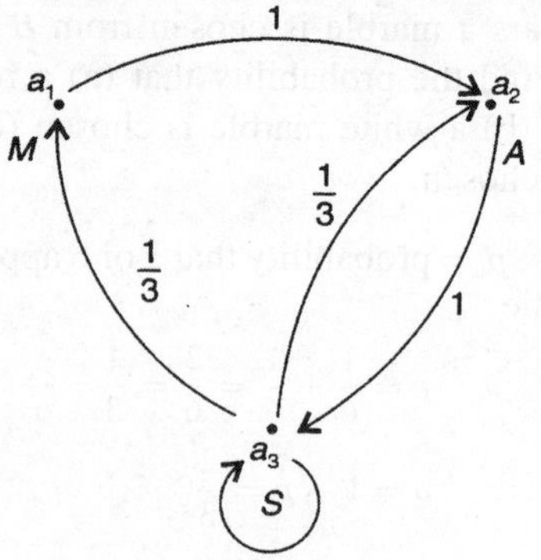

Fig. 31.7

Markov chain

Example 8: Every year, a man trades his car for a new car. If he has a Maruti, he trades it for an Ambassador. If he has an Ambassador, he trades it for a Santro. However, if he has a Santro, he is just as likely to trade it for a new Santro as to trade it for a Maruti or an Ambassador. In 2000 he bought his first car, which was a Santro.

(i) Find the probability that he has

(a) 2002 Santro

(b) 2002 Maruti

(c) 2003 Ambassador

(d) 2003 Santro

(ii) In the long run, how often will he have a Santro.

Solution: **(i)** Define 3 states a_1, a_2, a_3 as follows a_1: state of having Maruti car, a_2: having Ambassador, a_3: having Santro. Then the transition matrix is

$$P = \begin{matrix} & a_1 & a_2 & a_3 \\ a_1 & 0 & 1 & 0 \\ a_2 & 0 & 0 & 1 \\ a_3 & \frac{1}{3} & \frac{1}{3} & \frac{1}{3} \end{matrix}$$

2000: Initial state $= P^{(0)} = (0, 0, 1)$ since he has Santro car in 2000 (his first purchase).

(a) To reach 2002 year, (2-steps later) compute the 2-step transition matrix P^2

$$P^2 = \begin{pmatrix} 0 & 1 & 0 \\ 0 & 0 & 1 \\ \frac{1}{3} & \frac{1}{3} & \frac{1}{3} \end{pmatrix} \begin{pmatrix} 0 & 1 & 0 \\ 0 & 0 & 1 \\ \frac{1}{3} & \frac{1}{3} & \frac{1}{3} \end{pmatrix} = \begin{pmatrix} 0 & 0 & 1 \\ \frac{1}{3} & \frac{1}{3} & \frac{1}{3} \\ \frac{1}{9} & \frac{4}{9} & \frac{4}{9} \end{pmatrix}$$

Then

$$p^{(2)} = p^{(0)} P^2 = (0 \quad 0 \quad 1) \begin{pmatrix} 0 & 0 & 1 \\ \frac{1}{3} & \frac{1}{3} & \frac{1}{3} \\ \frac{1}{9} & \frac{4}{9} & \frac{4}{9} \end{pmatrix}$$

$$p^{(2)} = \begin{pmatrix} \frac{1}{9} & \frac{4}{9} & \frac{4}{9} \end{pmatrix}$$

The probability that he has a Santro in the year 2002 is $p_3^{(2)} = \frac{4}{9}$.

(b) Probability that he has a Maruti in 2002 is $p_1^{(2)} = \frac{1}{9}$.

(c) To reach 2003: 3 steps later

$$p^{(3)} = p^{(2)} P = \begin{pmatrix} \frac{1}{9} & \frac{4}{9} & \frac{4}{9} \end{pmatrix} \begin{pmatrix} 0 & 1 & 0 \\ 0 & 0 & 1 \\ \frac{1}{3} & \frac{1}{3} & \frac{1}{3} \end{pmatrix}$$

(or equivalently $p^{(3)} = p^{(0)} P^3$).

$$p^{(3)} = \begin{pmatrix} \frac{4}{27} & \frac{7}{27} & \frac{16}{27} \end{pmatrix}$$

Probability that he has an Ambassador in 2003 is $p_2^{(3)} = \frac{7}{27}$.

(d) Probability that has a Santro in 2003 is

$$p_3^{(3)} = \frac{16}{27}.$$

(ii) To discover what happens in the long run, we must find a fixed probability vector t of P. Let $t = (x, y, 1 - x - y)$.

Then $tP = t$

$$(x \quad y \quad 1-x-y)\begin{pmatrix} 0 & 1 & 0 \\ 0 & 0 & 1 \\ \frac{1}{3} & \frac{1}{3} & \frac{1}{3} \end{pmatrix} = (x \quad y \quad 1-x-y)$$

or $\quad 1 - x - y = 3x$

$$3x + (1 - x - y) = 3y$$

$$3y + (1 - x - y) = 3(1 - x - y)$$

Solving $\; y = \frac{1}{3}, x = \frac{1}{6}, z = \frac{3}{6} = \frac{1}{2}$

Thus $\quad t = \left(\frac{1}{6}, \frac{1}{3}, \frac{1}{2}\right)$

In the long run, he has a Santro 50% $\left(\frac{1}{2}\right)$ of the time.

Example 9: Suppose an urn A contains 2 white marbles and urn B contains 4 red marbles. At each step of the process, a marble is selected at random from each urn and the two marbles selected are interchanged. Let X_n denote the number of red marbles in urn A after n interchanges.

(i) Find the transition matrix P.

(ii) What is the probability that there are 2 red marbles in urn A after 3 steps.

(iii) In the long run, what is the probability that there are 2 red marbles in urn A.

(iv) What is the stationary distribution of the system.

Solution: There are three states a_0, a_1 and a_2 as shown in Fig. 31.8 below:

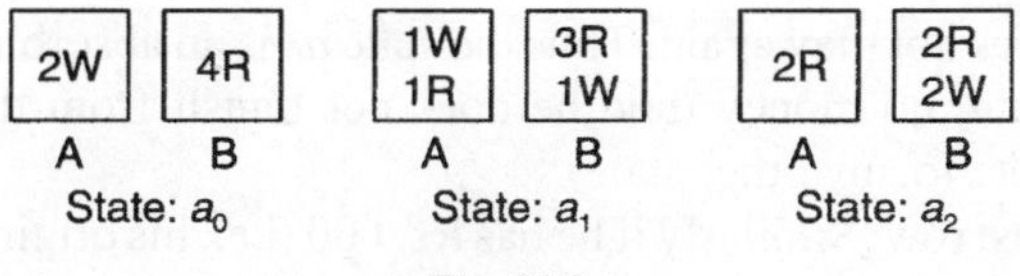

Fig. 31.8

(i) Transition matrix

If the system is in the state a_0, then a white marble from A and a red from B must be selected, so that the system will now move to state a_1. Accordingly the first row of the transition matrix (T.M.) is (0, 1, 0). Now suppose the system is in a_1. It can move to state a_0, iff red from A and white from B with probability $\frac{1}{2} \cdot \frac{1}{4} = \frac{1}{8}$. Thus $p_{10} = \frac{1}{8}$. The system can move from a_1 to a_2, iff white from A and red from B with probability $\frac{1}{2} \cdot \frac{3}{4} = \frac{3}{8}$ i.e., $p_{12} = \frac{3}{8}$, probability that system will remain in a_1 itself is $1 - \frac{1}{8} - \frac{3}{8} = \frac{1}{2}$. (Note: white from A and white from B with probability $\frac{1}{2} \cdot \frac{1}{4} = \frac{1}{8}$ or red from A and red from B with probability $\frac{1}{2} \cdot \frac{3}{4} = \frac{3}{8}$. Thus the probability that system will remain in state a_1 itself is $\frac{1}{8} + \frac{3}{8} = \frac{1}{2}$). Thus the 2nd row of T.M. is $\left(\frac{1}{8}, \frac{1}{2}, \frac{3}{8}\right)$.

Finally, suppose the system is in state a_2. Note that the system can never move from state a_2 to a_0. However, it may remain in a_2 itself, if a red from A and red from B is chosen. In this case the probability is $\frac{1}{1} \cdot \frac{2}{4} = \frac{1}{2}$. Lastly, if a red from A and white from B is chosen, then system moves from a_2 to a_1 with probability $\frac{2}{4} = \frac{1}{2}$. Thus third row of the T.M. is $\left(0, \frac{1}{2}, \frac{1}{2}\right)$. The Transition Matrix and transition diagram are shown in Fig. 31.9

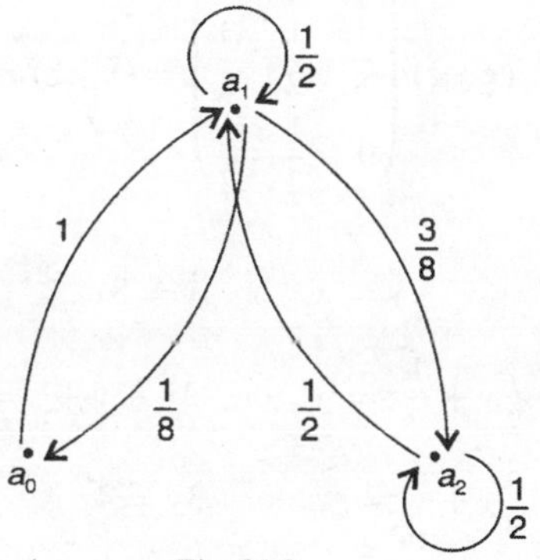

Fig. 31.9

$$\begin{array}{c} \\ a_0 \\ a_1 \\ a_2 \end{array} \begin{array}{c} \begin{array}{ccc} a_0 & a_1 & a_2 \end{array} \\ \begin{pmatrix} 0 & 1 & 0 \\ \frac{1}{8} & \frac{1}{2} & \frac{3}{8} \\ 0 & \frac{1}{2} & \frac{1}{2} \end{pmatrix} \end{array}$$

(ii) The system starts in state a_0, so that $p^{(0)} = (1, 0, 0)$ is the initial state. Now

$$p^{(1)} = p^{(0)}P = (1\,0\,0)\begin{pmatrix} 0 & 1 & 0 \\ \frac{1}{8} & \frac{1}{2} & \frac{3}{8} \\ 0 & \frac{1}{2} & \frac{1}{2} \end{pmatrix} = (0\,1\,0)$$

$$p^{(2)} = p^{(1)}P = (0\,1\,0)\begin{pmatrix} 0 & 1 & 0 \\ \frac{1}{8} & \frac{1}{2} & \frac{3}{8} \\ 0 & \frac{1}{2} & \frac{1}{2} \end{pmatrix} = \left(\frac{1}{8}\;\frac{1}{2}\;\frac{3}{8}\right)$$

$$p^{(3)} = p^{(2)}P = \left(\frac{1}{8}\;\frac{1}{2}\;\frac{3}{8}\right)\begin{pmatrix} 0 & 1 & 0 \\ \frac{1}{8} & \frac{1}{2} & \frac{3}{8} \\ 0 & \frac{1}{2} & \frac{1}{2} \end{pmatrix} = \left(\frac{1}{16}\;\frac{9}{16}\;\frac{6}{16}\right)$$

Probability that there are two red in A i.e., in state a_2 *after* three steps is $\frac{6}{16} = \frac{3}{8}$.

(iii) To study the system in the long run, we should find a unique fixed probability vector t of the transition matrix P. Let t be (x, y, z) or $(x, y, 1 - x - y)$.

Then $$tT = t$$

$$(x\;y\;z)\begin{pmatrix} 0 & 1 & 0 \\ \frac{1}{8} & \frac{1}{2} & \frac{3}{8} \\ 0 & \frac{1}{2} & \frac{1}{2} \end{pmatrix} = (x\;y\;z)$$

Solving $$\frac{1}{8}y = x \quad \text{or} \quad y = 8x$$

$$x + \frac{1}{2}y + \frac{1}{2}z = y \quad \text{or} \quad 2x - y + z = 0$$

$$\frac{3}{8}y + \frac{1}{2}z = z \quad \text{or} \quad 3y = 4z$$

Now $3y = 4z = 4(1 - x - y)$

$$7y = 4 - 4x \quad \text{or} \quad 56x + 4x = 4$$

$$\therefore \quad x = \frac{4}{60}, y = \frac{8}{15}, z = \frac{6}{15}$$

Therefore, the fixed vector

$$t = \left(\frac{1}{15}\;\;\frac{8}{15}\;\;\frac{6}{15}\right)$$

Hence the system in the long run stays in the state a_2, 40% of the time $\left(\frac{6}{15} = \frac{2}{5}\right)$ (i.e., there will be 2 red in A, 40% of the time).

(iv) The fixed unique probability vector $t = \left(\frac{1}{15}, \frac{8}{15}, \frac{6}{15}\right)$ is the stationary distribution, since P^n approaches t, in the long run.

Markov chain with absorbing states:

Example 10: A player has Rs. 300. At each play of a game, he losses Rs. 100 with probability $\frac{3}{4}$ but wins Rs. 200 with probability $\frac{1}{4}$. He stops playing if he has lost his Rs. 300 or he has won at least Rs. 300.

(a) Determine the transition probability matrix of the Markov chain.

(b) Find the probability that there are at least 4 plays to the game.

Solution: (a) This is random walk with absorbing barriers at states 0 and 6. The transition probability matrix P is

$$P = \begin{array}{c} \\ a_0 \\ a_1 \\ a_2 \\ a_3 \\ a_4 \\ a_5 \\ a_6 \end{array}\begin{array}{c} \begin{array}{ccccccc} a_0 & a_1 & a_2 & a_3 & a_4 & a_5 & a_6 \end{array} \\ \begin{pmatrix} 1 & 0 & 0 & 0 & 0 & 0 & 0 \\ \frac{3}{4} & 0 & 0 & \frac{1}{4} & 0 & 0 & 0 \\ 0 & \frac{3}{4} & 0 & 0 & \frac{1}{4} & 0 & 0 \\ 0 & 0 & \frac{3}{4} & 0 & 0 & \frac{1}{4} & 0 \\ 0 & 0 & 0 & \frac{3}{4} & 0 & 0 & \frac{1}{4} \\ 0 & 0 & 0 & 0 & \frac{3}{4} & 0 & \frac{1}{4} \\ 0 & 0 & 0 & 0 & 0 & 0 & 1 \end{pmatrix} \end{array}$$

Explanation for the probability matrix P: There are seven states $a_0, a_1, a_2, a_3, a_4, a_5, a_6$ where a_i indicates the state that he has Rs. i hundreds (i.e., a_3 indicate he has Rs. 300, a_5 he has 500 etc.).

First row: If he is in state a_0 he has zero money (lost his initial amount of Rs. 300) and therefore he stops the game. Then he remains in that state forever. He does not play again. Thus the state a_0 is an absorbing state, no money (and he does not transit from this state to any other state).

Last row: Similarly if he has Rs. 600 (i.e., his original Rs. 300 and winning amount of Rs. 300) he stops the

game and remains in the state (he does not play). Thus the state a_6 is also an absorbing state, all money.
Second row: Suppose he has Rs. 100 i.e., he is in state a_1. Then the probability that he will lose Rs. 100 is $\frac{3}{4}$ and therefore transfers to a_0 state (no money). But he can not transfer to states a_2, a_3, a_5, a_6. But by winning Rs. 200 with probability $\frac{1}{4}$ he can have Rs. 300 (Rs. 100 original + Rs. 200 winning). Thus the probability of going from state a_1 to a_3 is $\frac{1}{4}$, and from a_1 to a_0 is $\frac{3}{4}$. Thus the probability vector (second row) is $\left(\frac{3}{4}\ 0\ 0\ \frac{1}{4}\ 0\ 0\ 0\right)$.

Third row: Starting with Rs. 200 (a_2 state) he can lose Rs. 100 with probability $\frac{3}{4}$ thereby go to state a_1 or win Rs. 200 with probability $\frac{1}{4}$, thereby go to state a_4. Thus third row $\left(0\ \frac{3}{4}\ 0\ 0\ \frac{1}{4}\ 0\ 0\right)$.
Similarly, other rows of P are obtained.
(b) The initial probability distribution is

$$p^{(0)} = (0\ 0\ 0\ 1\ 0\ 0\ 0)$$

because he has started the game with an initial amount of Rs. 300 and is therefore in state a_3. To find the probability that the game has at least 4 plays, we compute $p^{(4)}$, which gives the probability distribution of the system after 4 steps (i.e., 4 games). Now

$$p^{(1)} = p^{(0)}P =$$

$$= (0\ 0\ 0\ 1\ 0\ 0\ 0)\begin{pmatrix} 1 & 0 & 0 & 0 & 0 & 0 & 0 \\ \frac{3}{4} & 0 & 0 & \frac{1}{4} & 0 & 0 & 0 \\ 0 & \frac{3}{4} & 0 & 0 & \frac{1}{4} & 0 & 0 \\ 0 & 0 & \frac{3}{4} & 0 & 0 & \frac{1}{4} & 0 \\ 0 & 0 & 0 & \frac{3}{4} & 0 & 0 & \frac{1}{4} \\ 0 & 0 & 0 & 0 & \frac{3}{4} & 0 & \frac{1}{4} \\ 0 & 0 & 0 & 0 & 0 & 0 & 1 \end{pmatrix}$$

$$p^{(1)} = \left(0, 0, \frac{3}{4}, 0, 0, \frac{1}{4}, 0\right)$$

$$p^{(2)} = p^{(1)}P = \left(0, \frac{9}{16}, 0, 0, \frac{6}{16}, 0, \frac{1}{16}\right)$$

$$p^{(3)} = p^{(2)}P = \left(\frac{27}{64}, 0, 0, \frac{27}{64}, 0, 0, \frac{10}{64}\right)$$

$$p^{(4)} = p^{(3)}P = \left(\frac{27}{64}, 0, \frac{81}{256}, 0, 0, \frac{27}{256}, \frac{10}{64}\right)$$

He plays 4 or more games, if after 4 steps he is not in any one of the absorbing states a_0 or a_6. Thus the probability that there at least 4 plays in the game is

$$0 + \frac{81}{256} + 0 + 0 + \frac{27}{256} = \frac{108}{256} = \frac{27}{64}.$$

Exercise

1. Which vectors are probability vectors (i) $\left(\frac{1}{2}, \frac{1}{3}, 0, -\frac{1}{5}\right)$; (ii) (3 4 5 0); (ii) $\left(\frac{1}{4}, \frac{1}{2}, 0, \frac{1}{4}\right)$.

Ans. (i) not (since negative component); (ii) not (since do not add upto 1); (iii) yes

2. Find a scalar multiple of each vector, which is a probability vector.
(i) $\left(2, \frac{1}{2}, 0, \frac{1}{4}, \frac{3}{4}, 0, 1\right)$; (ii) $\left(\frac{1}{3}, 2, \frac{1}{2}, 0, \frac{1}{4}, \frac{2}{3}\right)$;
(iii) (1 2 3 4 5 6); (iv) $\left(\frac{1}{2}, \frac{2}{3}, 0, 2, \frac{5}{6}\right)$.

Ans. (i) $\frac{4}{18}$; (ii) $\frac{12}{45}$; (iii) $\frac{1}{21}$. Then required probability vectors are $\frac{4}{18}\left(2, \frac{1}{2}, 0, \frac{1}{4}, \frac{3}{4}, 0, 1\right) = \left(\frac{8}{18}, \frac{2}{18}, 0, \frac{1}{18}, \frac{3}{18}, 0, \frac{4}{18}\right)$;
(iv) multiply 6 : (3, 4, 0, 12, 5). Divide by $3 + 4 + 0 + 12 + 5 = 24$. Then probability vector $\left(\frac{3}{24} = \frac{1}{8}, \frac{1}{6}, 0, \frac{1}{2}, \frac{5}{54}\right)$.

3. Which matrices are stochastic

(a) $\begin{pmatrix} \frac{1}{3} & \frac{2}{3} & \frac{4}{3} \\ \frac{1}{2} & 1 & \frac{1}{2} \end{pmatrix}$ (b) $\begin{pmatrix} \frac{15}{16} & \frac{1}{16} \\ \frac{2}{3} & \frac{4}{3} \end{pmatrix}$

(c) $\begin{pmatrix} 1 & 0 \\ \frac{1}{2} & \frac{1}{2} \end{pmatrix}$ (d) $\begin{pmatrix} \frac{1}{2} & -\frac{1}{2} \\ \frac{1}{4} & \frac{3}{4} \end{pmatrix}$

Ans. (a) not (square); (b) not adding to 1; (c) yes; (d) No (negative entry)

4. Which of the following matrices are regular

(a) $A = \begin{pmatrix} \frac{1}{2} & \frac{1}{2} \\ 0 & 1 \end{pmatrix}$ (b) $B = \begin{pmatrix} 0 & 1 \\ 1 & 0 \end{pmatrix}$;

(c) $C = \begin{pmatrix} \frac{1}{2} & \frac{1}{4} & \frac{1}{4} \\ 0 & 1 & 0 \\ \frac{1}{2} & \frac{1}{2} & 0 \end{pmatrix}$ (d) $D = \begin{pmatrix} 0 & 0 & 1 \\ \frac{1}{2} & \frac{1}{4} & \frac{1}{4} \\ 0 & 1 & 0 \end{pmatrix}$

Ans. **(a)** Not regular, since 1 appears in the main diagonal.

(b) $B^2 = I$, $B^3 = B$, B is not regular, since 1 appears in the main diagonal

(c) c not regular, 1 appears on diagonal

(d) $D^2=\begin{pmatrix} 0 & 1 & 0 \\ \frac{1}{8} & \frac{5}{16} & \frac{9}{16} \\ \frac{1}{2} & \frac{1}{4} & \frac{1}{4} \end{pmatrix}$, $D^3=\begin{pmatrix} \frac{1}{2} & \frac{1}{4} & \frac{1}{4} \\ \frac{5}{32} & \frac{41}{64} & \frac{13}{64} \\ \frac{1}{8} & \frac{5}{16} & \frac{9}{16} \end{pmatrix}$

D is regular since all the entries of D^3 are positive.

5. Find the unique fixed probability vector of each matrix

(a) $A=\begin{pmatrix} \frac{2}{3} & \frac{1}{3} \\ \frac{2}{5} & \frac{3}{5} \end{pmatrix}$ (b) $B=\begin{pmatrix} \frac{1}{4} & \frac{3}{4} \\ \frac{5}{6} & \frac{1}{6} \end{pmatrix}$

(c) $C=\begin{pmatrix} 0.2 & 0.8 \\ 0.5 & 0.5 \end{pmatrix}$

Ans. (a) $\left(\frac{6}{11}, \frac{5}{11}\right)$; (b) $\left(\frac{10}{19}, \frac{9}{19}\right)$; (c) $\left(\frac{5}{13}, \frac{8}{13}\right)$.

6. Find the unique fixed probability vector of

(a) $A=\begin{pmatrix} 0 & \frac{1}{2} & \frac{1}{2} \\ \frac{1}{3} & \frac{2}{3} & 0 \\ 0 & 1 & 0 \end{pmatrix}$, (b) $B=\begin{pmatrix} 0 & 1 & 0 \\ \frac{1}{2} & 0 & \frac{1}{2} \\ \frac{1}{2} & \frac{1}{4} & \frac{1}{4} \end{pmatrix}$

Ans. (a) $\left(\frac{2}{9}, \frac{6}{9}, \frac{1}{9}\right)$ (b) $\left(\frac{5}{15}, \frac{6}{15}, \frac{4}{15}\right)$

7. Given $P=\begin{pmatrix} 0 & 1 \\ \frac{1}{2} & \frac{1}{2} \end{pmatrix}$ (a) find a unique fixed probability vector; (b) what matrix does P^n approach; (c) what vector does $\left(\frac{1}{4} \;\; \frac{3}{4}\right) P^n$ approach.

Ans. (a) $\left(\frac{1}{3}, \frac{2}{3}\right)$; (b) $P^5=\begin{pmatrix} 0.31 & 0.69 \\ 0.34 & 0.66 \end{pmatrix} \to$ (0.33, 0.66); (c) $\left(\frac{43}{128} \;\; \frac{85}{128}\right) \approx \left(\frac{1}{3} \;\; \frac{2}{3}\right)$

8. **(a)** Find the unique fixed probability vector t of

$$P=\begin{pmatrix} 0 & \frac{1}{2} & \frac{1}{2} & 0 \\ \frac{1}{2} & \frac{1}{4} & 0 & \frac{1}{4} \\ 0 & 0 & 0 & 1 \\ 0 & \frac{1}{2} & 0 & \frac{1}{2} \end{pmatrix}$$

(b) What matrix does P^n approach

(c) What vector does $\left(\frac{1}{4}\; 0\; \frac{1}{2}\; \frac{1}{4}\right) P^n$ approach

(d) What vector does $\left(\frac{1}{2}\; 0\; 0\; \frac{1}{2}\right) P^n$ approach.

Hint: $t=(x, y, z, 1-x-y-z)$, $y=2x$, $x=2z$, $y=4z$

Ans. **(a)** $t=\left(\frac{2}{11}, \frac{4}{11}, \frac{1}{11}, \frac{4}{11}\right)=(0.1818, 0.3636, 0.0909, 0.3636)$

(b) $P^5=\frac{1}{1024}\begin{pmatrix} 196 & 546 & 136 & 540 \\ 306 & 569 & 100 & 766 \\ 240 & 440 & 96 & 592 \\ 200 & 436 & 80 & 552 \end{pmatrix} \sim t$

(c) $\sim t$

(d) $\sim t$

9. Find the transition matrix

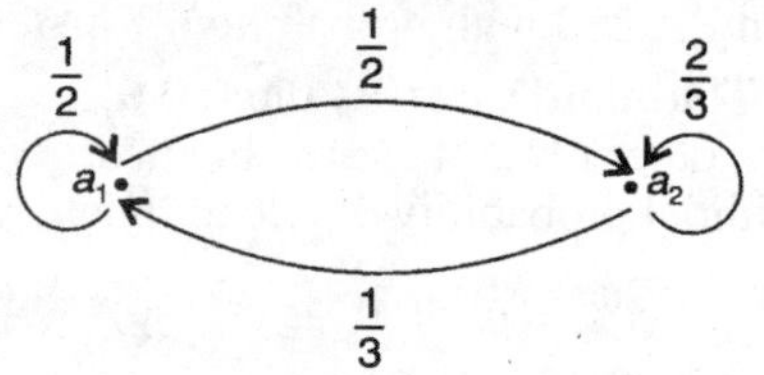

Fig. 31.10

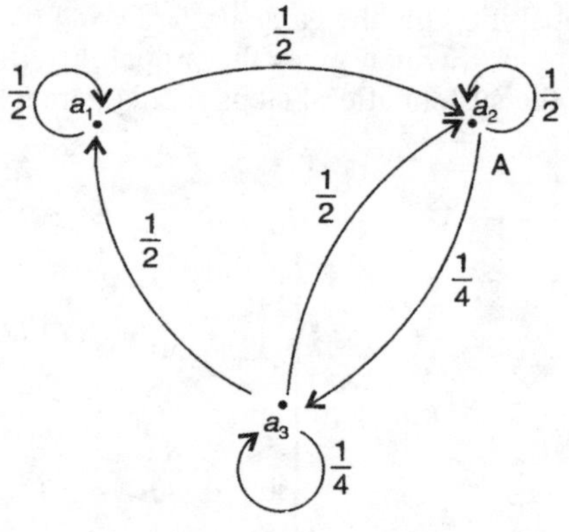

Fig. 31.11

Ans. **(i)**
$$\begin{array}{c c} & a_1 \quad a_2 \\ \begin{array}{c} a_1 \\ a_2 \end{array} & \begin{pmatrix} \frac{1}{2} & \frac{1}{2} \\ \frac{1}{3} & \frac{2}{3} \end{pmatrix} \end{array}$$

(ii)
$$\begin{array}{c c} & a_1 \quad a_2 \quad a_3 \\ \begin{array}{c} a_1 \\ a_2 \\ a_3 \end{array} & \begin{pmatrix} \frac{1}{2} & \frac{1}{2} & 0 \\ 0 & \frac{1}{2} & \frac{1}{2} \\ \frac{1}{2} & \frac{1}{4} & \frac{1}{4} \end{pmatrix} \end{array}$$

10. For a Markov chain, the transition matrix $P = \begin{pmatrix} \frac{1}{2} & \frac{1}{2} \\ \frac{3}{4} & \frac{1}{4} \end{pmatrix}$ with initial distribution $p^{(0)} = \left(\frac{1}{4} \;\; \frac{3}{4}\right)$. Find

(a) $p_{21}^{(2)}$ (b) $p_{12}^{(2)}$

(c) $p^{(2)}$ (d) $p_1^{(2)}$

(e) the vector $p^{(0)}P^n$ approaches

(f) P^n approaches

Ans. $P^2 = \begin{pmatrix} \frac{5}{8} & \frac{3}{8} \\ \frac{9}{16} & \frac{7}{16} \end{pmatrix}$ (a) $p_{21}^{(2)} = \frac{9}{16}$ (b) $p_{12}^{(2)} = \frac{3}{8}$

(c) $p^{(2)} = p^{(0)}P^2 = \left(\frac{37}{64} \;\; \frac{27}{64}\right)$ (d) $p_1^{(2)} = \frac{37}{64}$

(e) $p^{(0)}P^n$ approaches fixed vector $t = \left(\frac{3}{5} \;\; \frac{2}{5}\right)$

(f) $P^n \to T = \begin{pmatrix} t \\ t \end{pmatrix} = \begin{pmatrix} \frac{3}{5} & \frac{2}{5} \\ \frac{3}{5} & \frac{2}{5} \end{pmatrix}$

11. A man's smoking habits are as follows. If he smokes filter cigarettes one week, he switches to nonfilter cigarettes the next week with probability 0.2. On the other hand if he smokes nonfilter cigarettes one week, there is a probability of 0.7 that he will smoke nonfilter cigarettes the next week as well. In the long run how often does he smoke filter cigarettes.

Ans. P :

$$\begin{array}{c} \\ F \\ NF \end{array}\begin{array}{c} \begin{array}{cc} F & NF \end{array} \\ \begin{pmatrix} 0.8 & 0.2 \\ 0.3 & 0.7 \end{pmatrix} \end{array}, t = \text{fixed} = (x, 1 - x)$$

$$x = \frac{3}{5}, \quad y = \frac{2}{5}, \quad t = \left(\frac{3}{5}, \frac{2}{5}\right)$$

Man smokes filter cigarettes 60% $\left(\frac{3}{5}\right)$ time in the long run.

12. A saleman's territory consists of 3 cities A, B and C. He never sells in the same city on successive days. If he sells in city A, then the next day he sells in city B. However if he sells in either B or C, then the next day he is twice as likely to sell in city A as in other city. In the long run, how often does he sell in each of the cities.

Ans. $$P = \begin{array}{c} \\ A \\ B \\ C \end{array}\begin{array}{c} \begin{array}{ccc} A & B & C \end{array} \\ \begin{pmatrix} 0 & 1 & 0 \\ \frac{2}{3} & 0 & \frac{1}{3} \\ \frac{2}{3} & \frac{1}{3} & 0 \end{pmatrix} \end{array}, \quad \begin{array}{l} \frac{2}{3}y + \frac{2}{3}z = z \\ x + \frac{1}{3}z = y \\ \frac{1}{3}y = z \end{array}$$

$t = \left(\frac{2}{5}, \frac{9}{20}, \frac{3}{20}\right)$. In the long run he sells 40% of time in city A, 45% in B, 15% of time in C.

13. There are 2 white marbles in box A and 3 red marbles in box B. At each step of the process a marble is selected from each box and the two marbles selected are interchanged. Let the state a_i of the system be the number i of red marbles in box A.

(a) Find the transition matrix P.

(b) What is the probability that there are 2 red marbles in box A after 3 steps

(c) In the long run, what is the probability that there are 2 red marbles in box A.

Hint:

A	B	A	B	A	B
2W	3R	1W 1R	1W 2R	2R	2W 1R
a_0		a_1		a_2	

(a) $$P = \begin{array}{c} \\ a_0 \\ a_1 \\ a_2 \end{array}\begin{array}{c} \begin{array}{ccc} a_0 & a_1 & a_2 \end{array} \\ \begin{pmatrix} 0 & 1 & 0 \\ \frac{1}{6} & \frac{1}{2} & \frac{1}{3} \\ 0 & \frac{2}{3} & \frac{1}{3} \end{pmatrix} \end{array}$$

Initial distribution: $p^{(0)} = (1, 0, 0)$

$$p^{(1)} = p^{(0)}P = (0, 1, 0), \; p^{(2)} = p^{(1)}P = \left(\frac{1}{6} \; \frac{1}{2} \; \frac{1}{3}\right)$$

$$p^{(3)} = p^{(2)}P = \left(\frac{1}{12} \; \frac{23}{36} \; \frac{5}{18}\right).$$

(b) Probability that there are 2 red marbles in box A after 3 steps is $\frac{5}{18}$.

(c) Fixed probability vector: $t = (0.1, 0.6, 0.3)$.

Ans. (c) In the long run, 30% of the time, there will be 2 red marbles in box A.

Finite stochastic process

14. Box A contains 3 red and 5 white marbles, Box B contains 2 red and 1 white marbles, Box C contains 2 red and 3 white marbles. One box is selected at random and a marble is drawn from the box. If the marble is red, what is the probability that it came from box A.

Hint: $P(A/R) = \frac{P(A \cap R)}{P(R)} = \frac{\frac{1}{8}}{\frac{173}{360}}$

Ans. $\frac{45}{173}$

15. Box A contains cards numbered 1 to 9. Box B contains cards numbered 1 to 5. One box is chosen at random and a card is drawn. If the card is even, another card from the same box is drawn, if odd the card is drawn from the other box. Find

(a) the probability that both cards are even.

(b) if both cards are even, find the probability that they came from box A.

(c) what is the probability that both cards are odd?

Ans. **(a)** $\frac{1}{2} \cdot \frac{4}{9} \cdot \frac{3}{8} + \frac{1}{2} \cdot \frac{2}{5} \cdot \frac{1}{4} = \frac{1}{12} + \frac{1}{20} = \frac{2}{15}$

(b) $\frac{\frac{1}{12}}{\frac{2}{15}} = \frac{5}{8}$

(c) $\frac{1}{2} \cdot \frac{5}{9} \cdot \frac{3}{5} + \frac{1}{2} \cdot \frac{3}{5} \cdot \frac{5}{9} = \frac{1}{6} + \frac{1}{6} = \frac{1}{3}$.

16. A box contains 3 coins, two of them fair and one two-headed. A coin is selected at random and tossed twice. If head appears both times, what is the probability that the coin is two headed.

Ans. $\frac{\frac{1}{3} \cdot 1 \cdot 1}{\frac{1}{3} \cdot \frac{1}{2} \cdot \frac{1}{2} + \frac{1}{3} \cdot \frac{1}{2} \cdot \frac{1}{2} + \frac{1}{3} \cdot 1 \cdot 1} = \frac{2}{3}$

Absorbing states

17. A player has Rs. 200. He bets Rs. 100 at a time and wins Rs. 100 with probability $\frac{1}{2}$. He stops playing if he loses the Rs. 200 or wins Rs. 400.

(a) Find the probability that he has lost his money at the end of at most 5 days.

(b) Determine the probability that the game lasts more than 7 plays.

Hint:

$$\begin{array}{c} \\ a_0 \\ a_1 \\ a_2 \\ a_3 \\ a_4 \\ a_5 \\ a_6 \end{array} \begin{array}{c} \begin{array}{ccccccc} a_0 & a_1 & a_2 & a_3 & a_4 & a_5 & a_6 \end{array} \\ \begin{pmatrix} 1 & 0 & 0 & 0 & 0 & 0 & 0 \\ \frac{1}{2} & 0 & \frac{1}{2} & 0 & 0 & 0 & 0 \\ 0 & \frac{1}{2} & 0 & \frac{1}{2} & 0 & 0 & 0 \\ 0 & 0 & \frac{1}{2} & 0 & \frac{1}{2} & 0 & 0 \\ 0 & 0 & 0 & \frac{1}{2} & 0 & \frac{1}{2} & 0 \\ 0 & 0 & 0 & 0 & \frac{1}{2} & 0 & \frac{1}{8} \\ 0 & 0 & 0 & 0 & 0 & 0 & 1 \end{pmatrix} \end{array}$$

$p^{(0)} = (0, 0, 1, 0, 0, 0, 0)$,

$$p^{(1)} = p^{(0)} P = \left(0, \frac{1}{2}, 0, \frac{1}{2}, 0, 0, 0\right)$$

$$p^{(2)} = p^{(1)} P = \left(\frac{1}{4}\ 0\ \frac{1}{2}\ 0\ \frac{1}{4}\ 0\ 0\right),\ p^{(5)} = p^{(4)} P$$

$$p^{(5)} = \left(\frac{3}{8}, \frac{5}{32}, 0, \frac{9}{32}, 0, \frac{1}{8}, \frac{1}{16}\right).$$

$p^{(5)}$: probability that he has no money after 5 plays is $\frac{3}{8}$, $p^{(7)} = \left(\frac{29}{64}, \frac{7}{64}, 0, \frac{27}{128}, 0, \frac{13}{128}, \frac{1}{8}\right)$

Ans. (a) $\frac{3}{8}$ (b) $\frac{7}{64} + \frac{27}{128} + \frac{13}{128} = \frac{27}{64}$.

HIGHER ENGINEERING MATHEMATICS

PART—VIII

NUMERICAL ANALYSIS

Root

Any value ξ for which $f(\xi) = 0$ is known as the root or solution of the equation (2) or ξ is called the zero of the function $f(x)$. In the case of the algebraic Equation (1), the roots can be determined in analytical (literal) form when $n = 1, 2, 3, 4$ (i.e., for linear, quadratic, cubic and biquadratic equations). For $n \geq 5$ no such results exists for the roots of (1). Also from the fundamental theorem of algebra, (1) has exactly n roots, real or complex whereas the number and analytical form of roots of a transcendental equation are not known at all.

Example: $\sin x = 2$ has no real roots, $\sin x = \frac{1}{2}$ has infinite number of roots, while $\sin x = \frac{x}{2}$ has three real roots (see Fig. 32.9 on page 32.6).

Therefore only approximate solutions (roots) of an algebraic or transcendental Equation (2) are to be found by numerical methods consisting of (a) isolating the roots (b) and then improving the value of the approximate roots. Here it is assumed that the roots of (2) are isolated i.e., for any root ξ of (2) there exists an interval containing no other root except ξ.

Geometrically, the root of Equation (2) is the point where the graph (or curve) of $y = f(x)$ crosses the x-axis (i.e., $y = f(x) = 0$). Although the roots can be isolated by drawing the graphs of the curve, these graphical methods are cumbersome. The following theorem is very useful in isolating the roots of (2).

Theorem 1: *If a continuous function $f(x)$ assumes values of opposite sign at the end points of an interval $[\alpha, \beta]$ i.e., $f(\alpha)f(\beta) < 0$, then the interval will contain at least one root of the equation $f(x) = 0$ i.e., there exists $\xi \in (\alpha, \beta)$ such that $f(\xi) = 0$.*

Note 1: Root ξ will be unique (only one) in the interval (α, β) if $f'(x)$ has the same sign in the interval (α, β) (i.e., $f' > 0$ or $f' < 0$ in $\alpha < x < \beta$).

Note 2: For an nth degree algebraic Equation (1), we get $(n + 1)$ sign changes.

Note 3: Descarte's rule of sign: The number of positive roots of $f(x) = 0$ can not exceed the number of

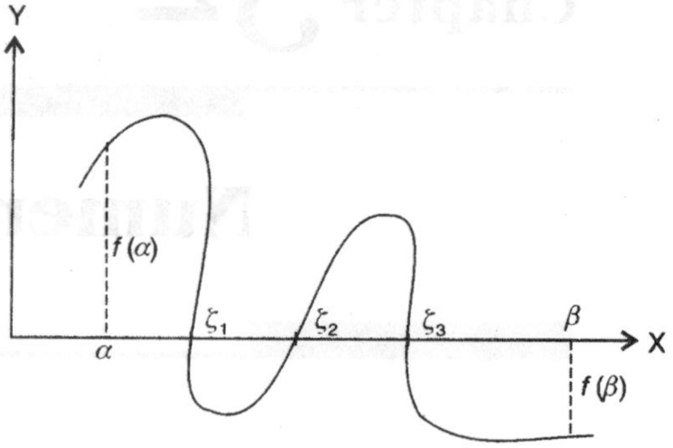

Fig. 32.1

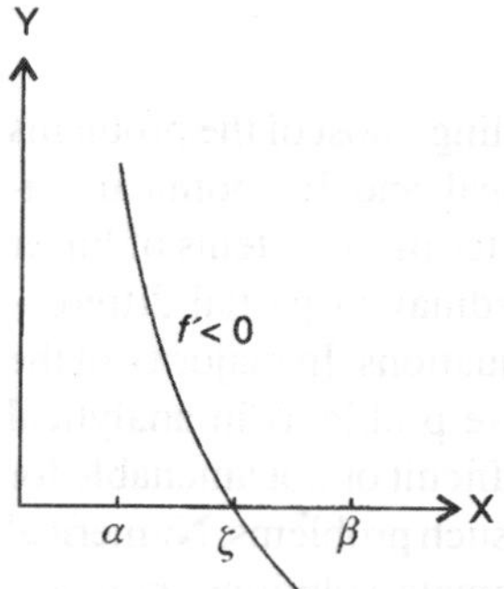

Fig. 32.2

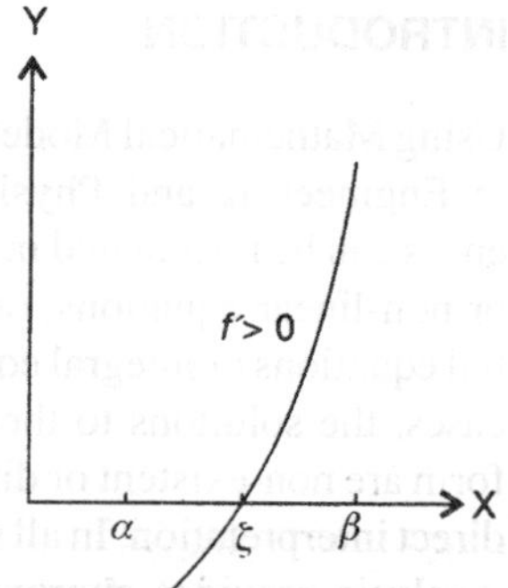

Fig. 32.3

changes of sign in $f(x)$. Also the number of negative roots of $f(x)$ can not exceed the number of changes of sign in $f(-x)$.

Example: $f(x) = x^5 - 6x^2 - 4x + 5 = 0$.
$f(x) : + - - + : 2$ changes of sign; no more than 2 positive roots.
$f(-x) : - - ++$: one change of sign; no more than one negative root.

Bisection Method (or Halving Method or Bolzano Method)

Consider the equation

$$f(x) = 0 \tag{2}$$

in $[a, b]$. Assume that $f(a)f(b) < 0$. The bisection method isolates the root in $[a, b]$ by halving process, approximately dividing the given interval $[a, b]$ into two, four, eight etc. equal parts. Thus in order to find a root of (2) lying in the interval $[a, b]$, divide the interval in half. If $f\left(\frac{a+b}{2}\right)=0$, then $\xi=\frac{a+b}{2}$ is the required root. If $f\left(\frac{a+b}{2}\right)\neq 0$, then choose that half $\left[a, \frac{a+b}{2}\right]$ or $\left[\frac{a+b}{2}, b\right]$, at the end points of which $f(x)$ has opposite signs. The newly reduced interval $[a, b]$ is again bisected and the above process is repeated. Thus we get a sequence of nested intervals $[a_1, b_1], [a_2, b_2], \ldots, [a_n, b_n]$ such that

$$f(a_n)f(b_n)<0 \quad \text{for} \quad n=1,2,\ldots$$

and

$$b_n-a_n=\frac{1}{2^n}(b-a).$$

The required root $\xi=\frac{1}{2}(a_n+b_n)$.

Rugula-Falsi Method (or Method of False Position or Linear Interpolation or Method of Chords or Method of Proportional Parts)

This method is probably the oldest, faster method more generally applicable.

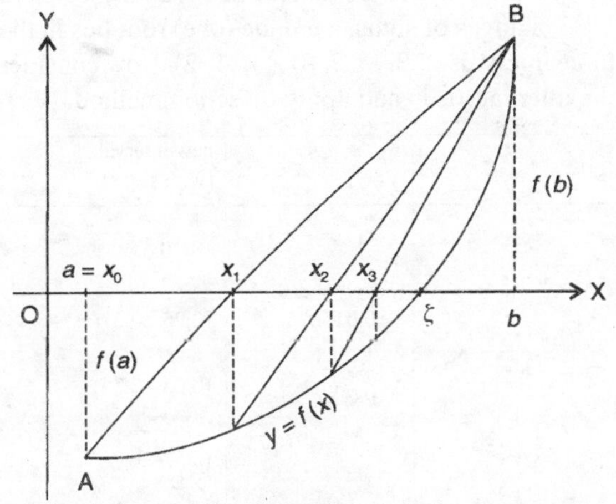

Fig. 32.4

To find root ξ of the equation $f(x)=0$ in the interval $[a, b]$ assume that $f(a)<0$ and $f(b)>0$ so that $f(a)\cdot f(b)<0$.

Geometrically, this method is equivalent to replacing the curve $y=f(x)$ by a chord that passes through the points $A(a, f(a))$ and $B(b, f(b))$. The equation of the chord AB is

$$\frac{x-a}{b-a}=\frac{y-f(a)}{f(b)-f(a)}$$

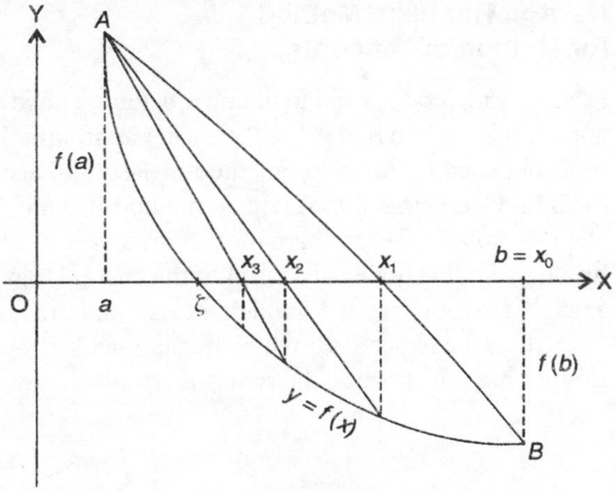

Fig. 32.5

If chord AB meets x-axis at $x=x_1$, then $y=0$.

$$\frac{x_1-a}{b-a}=\frac{0-f(a)}{f(b)-f(a)} \quad \text{or}$$

$$x_1=a-\frac{f(a)}{f(b)-f(a)}(b-a)$$

If $f(a)<0$, then the end point b is fixed and the successive approximations

$$x_0=a$$

$$x_{n+1}=x_n-\frac{f(x_n)}{f(b)-f(x_n)}(b-x_n),$$

(for $n=0,1,2,3,\ldots$) form a bounded increasing monotonic sequence and

$$x_0<x_1<x_2\cdots<x_n<x_{n+1}<\cdots<\xi<b.$$

If $f(a)>0$, end point a is fixed and successive approximations are

$$x_0=b$$

$$x_{n+1}=x_n-\frac{f(x_n)}{f(x_n)-f(a)}(x_n-a)$$

and

$$a<\xi<\cdots<x_{n+1}<x_n<\cdots<x_1<x_0.$$

Note 1: Fix the end point for which sign of f and f'' is same.

Note 2: Successive approximations x_n lie on the side of root ξ where sign of f is opposite to the sign of f''.

Newton-Raphson Method (or Method of Tangents)

Newton's method is used to obtain a better (refined) approximation of a root using the earlier approximations obtained by bisection method or Regula-Falsi method. Geometrically, Newton's method is equivalent to replacing a small arc of the curve $y = f(x)$ by a tangent line drawn at a point of the curve. Draw a tangent to the curve at B_0 which meets x-axis at x_1. Then draw a tangent at B_1 which meets x-axis at x_2. Continuing this process, the root ξ is obtained.

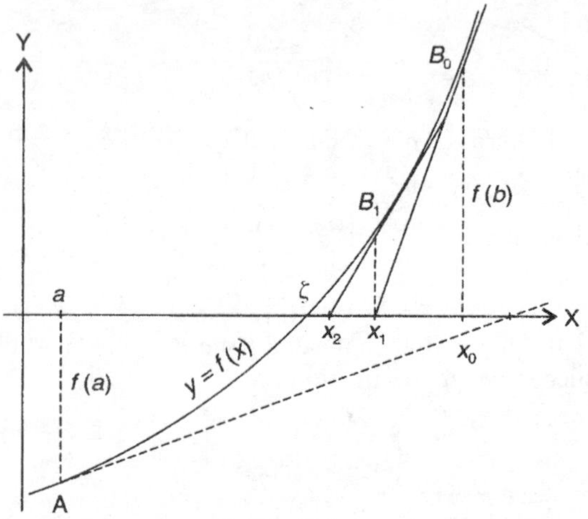

Fig. 32.6

Suppose $\xi = x + h$ where h is a small quantity. Then applying Taylor's formula

$$0 = f(x+h) \approx f(x) + hf'(x)$$

or
$$h = -\frac{f(x)}{f'(x)}$$

Thus
$$\xi = x + h = x - \frac{f(x)}{f'(x)}.$$

In general,

$$\boxed{x_{n+1} = x_n - \frac{f(x_n)}{f'(x_n)}}, \quad n = 0, 1, 2, \ldots.$$

Note 1: A root ξ of equation $f(x) = 0$ can be computed to any degree of accuracy if a 'good' initial approximation x_0 is chosen for which

$$f(x_0) f''(x_0) > 0$$

i.e., choose the end point of the interval at which f and f'' have the same sign.

Note 2: Newton's method converge slow if f' is small (fails when $f' = 0$ because tangent in this case is parallel to x-axis and will never meet it).

Worked Out Examples

Example 1: Isolate the roots of the equation

$$x^3 - 4x + 1 = 0$$

Find all the roots using bisection method.

Solution: Here $f(x) = x^3 - 4x + 1$.

x:	$-\infty$	-3	-2	-1	0	1	2	3	4	∞
sign of $f(x)$:	$-$	-14	1	4	1	-2	1	16	49	∞

Since equation is cubic (degree three) and since there are 3 changes of signs, a unique (one) root lies in the three intervals $(-3, -2)$, $(0, 1)$, $(1, 2)$. Now consider the interval $(0, 1)$ and apply bisection method.

x_i	$f(x_i)$	new interval (with signs)
$x_1 = \frac{1+0}{2} = \frac{1}{2}$	$f(x_1) = -8.75$	$+ \quad -$ $(0,\ 0.5)$
$x_2 = \frac{0+0.5}{2} = \frac{1}{4}$	$f(x_2) = f(0.25)$ $= 0.015625$	$+ \quad -$ $(0.25,\ 0.5)$
$x_3 = \frac{.25+.5}{2} = .35$	$f(.35) =$ -0.357	$+ \quad -$ $(0.25,\ 0.35)$
$x_4 = \frac{.25+.35}{2} = .3$	$f(.3) = -0.173$	$+ \quad -$ $(.25,\ .3)$
$x_5 = \frac{.25+.3}{2} = \frac{.55}{2}$	$f(.275) =$ -0.0792	$+ \quad -$ $(.25,\ .275)$
$x_6 = \frac{.25+.275}{2}$ $= .2625$	$f(.2625) =$ -0.0319	$+ \quad -$ $(.25,\ .2625)$
$x_7 = \frac{.25+.2625}{2}$ $= 0.25625$	$f(.25625) =$ -0.00817	$+ \quad -$ $(0.25,\ .25625)$
$x_8 = \frac{.25+.25625}{2}$ $= 0.253125$	$f(.253125) =$ 0.0100	$+ \quad -$ $(0.253125,\ .25625)$
$x_9 =$ $\frac{.253125+.25625}{2}$ $= 0.2546875$	$f(.2546875) =$ -0.00222	$+ \quad -$ $(.253125,\ .2546875)$
$x_{10} =$ $\frac{0.253125+0.2546875}{2}$ $= 0.25390625$	$f(0.25390625)$ $= 0.000743$	$+ \quad -$ $(.25390625,\ .2546875)$

The approximate root is

$$\xi = \frac{0.25390625 + 0.2546875}{2} = 0.254296875 \approx 0.2540$$

$$f(0.254296875) = -0.00074290925$$

By synthetic division

0.254	1	0	−4	1
		0.254	0.064516	−0.99612
	1	0.254	−3.93546	0.000387 remainder

Roots of the quadratic equation

$$x^2 + 0.254x - 3.93546 = 0$$

are $x = -2.11475 \approx -2.115$, $x = 1.86075 \approx 1.8608$.

Example 2: Using Regula-Falsi method, compute the real root of the equation $x^3 - 4x - 9 = 0$.

Solution: Here $f(x) = x^3 - 4x - 9$.

x:	0	1	2	3
$f(x)$:	−9	−12	−9	6

By bisection method, $f\left(\frac{2+3}{2}\right) = f(2.5) = -3.375$ so a root of $f(x)$ lies in the interval (2.5, 3). Here $f' = 3x^2 - 4$, $f'' = 6x$, $f''(3) = 18 > 0$, $f''(2.5) = 15 > 0$. Since sign of f and f'' is same at $x = 3$, fix the point $b = 3$ and since $f(a) = f(2.5) = -3.375 < 0$, use

$$x_{n+1} = x_n - \frac{f(x_n)}{f(b) - f(x_n)}(b - x_n)$$

$$x_{n+1} = x_n - \frac{f(x_n)}{6 - f(x_n)}(3 - x_n)$$

Then $x_0 = 2.5$, $f(x_0) = -3.375$, $n = 0$, then

$$x_1 = x_0 - \frac{f(x_0)(3 - x_0)}{6 - f(x_0)} = 2.5 + \frac{3.375}{6 + 3.375}(3 - 2.5)$$

$$x_1 = 2.5 + 0.18 = 2.68$$

At $x_1 = 2.68$, $f(2.68) = -0.471168$.

$$x_2 = 2.68 + \frac{0.471168}{6 + 0.471168}(3 - 2.68)$$

$$= 2.68 + 0.0232993$$

$$x_2 = 2.7033$$

At x_2, $f(x_2) = f(2.7033) = -0.05794$

$$x_3 = 2.7033 + \frac{0.05794}{6 + 0.05794}(3 - 2.7033)$$

$$x_3 = 2.7033 + 0.0028377 = 2.706$$

$$f(2.706) = -0.009488$$

$$x_4 = 2.706 + \frac{0.009488}{6.009488}(3 - 2.706) = 2.70646$$

$$x_4 = 2.70650.$$

Example 3: Determine the root of $xe^x - 2 = 0$ by method of false position.

Solution: Here $f(x) = xe^x - 2$.

x:	0	0.5	0.8	0.9	1.0
$f(x)$:	−2	−1.1756	−0.2196	0.2136	0.718

A root lies between 0.8 and 0.9. Now

$$f'(x) = e^x + xe^x, \qquad f''(x) = 2e^x + xe^x$$

$$f''(0.8) = 6.2315, \qquad f''(0.9) = 7.1328.$$

Since f and f'' have the same (positive) sign at $x = 0.9$, fix this point $b = 0.9$. Note that $f(0.8) < 0$.

$$x_{n+1} = x_n - \frac{f(x_n)}{f(0.9) - f(x_n)}(0.9 - x_n)$$

Take $n = 0$, $x_0 = 0.8$, $f(x_0) = f(0.8) = -0.2196$

$$x_1 = 0.8 + \frac{0.2196}{0.2136 + 0.2196}(0.9 - 0.8)$$

$$x_1 = 0.8 + 0.05069252 = 0.851$$

$$f(x_1) = f(0.851) = -0.00697$$

$$x_2 = 0.851 + \frac{0.00697}{0.2136 + 0.00697}(0.9 - 0.851)$$

$$x_2 = 0.851 + 0.0015484 = 0.85256$$

$$f(x_2) = -0.0001977$$

$$x_3 = 0.85256 + \frac{0.0001977}{0.2136 + 0.0001977}(0.9 - 0.85256)$$

$$x_3 = 0.85256 + 0.000043868 = 0.8526$$

$$f(x_3) = -0.0000239$$

$$x_4 = 0.8526 + \frac{0.0000239}{0.2136 + 0.0000239}(0.9 - 0.8526)$$

$$x_4 = 0.8526 + 0.0000053 = 0.8526$$

∴ Approximate root is 0.8526.

Example 4: Find the approximate value of the real root of the equation $2x - \log_{10}^{x} - 7 = 0$ by using Newton-Raphson method.

Solution: Here $f(x) = 2x - \log_{10}^{x} - 7$. Since log function is involved, x must be positive

x:	1	2	3	4
$f(x)$:	−5	−3.301	−1.4771	0.3979

Roots lies in between 3 and 4. Now

x:	3.5	3.7	3.8
$f(x)$:	−0.5441	−0.1682	0.0202

Here $f'(x) = 2 - \frac{\log_{10}^{e}}{x}$ and $f''(x) = \frac{\log_{10}^{e}}{x^2} = \frac{0.4343}{x^2}$. Since f and f'' have the same sign at $x = 3.8$, choose $x_0 = 3.8$. Then $f(x_0) = 0.0202$

$$x_{n+1} = x_n - \frac{f(x_n)}{f'x_n} = x_n - \frac{(2x_n - \log_{10} x_n - 7)}{\left(2 - \frac{\log_{10}^{e}}{x_n}\right)}$$

Take $n = 0$, $x_1 = 3.8 - \frac{0.0202}{1.88571} = 3.8 - 0.010712$

$x_1 = 3.7893$, $f(x_1) = 0.000041$, then

$$x_2 = 3.7893 - \frac{0.000041}{1.88538} = 3.7893 - 0.00002175$$

$x_2 = 3.789278 \approx 3.7893$ is the required root.

Example 5: Find a positive root of $x^4 - x = 10$ using Newton-Raphson's method.

Solution: Here $f(x) = x^4 - x - 10$.

x :	0	1	2
$f(x)$:	−10	−10	4

A root lies between 1 and 2

x :	1	1.5	1.75	1.8	1.9	2
$f(x)$:	−10	−6.4375	−2.3711	−1.3024	1.1321	4

Root lies between 1.8 and 1.9. Now

$$f'(x) = 4x^3 - 1. \text{ So}$$

$$x_{n+1} = x_n - \frac{x_n^4 - x_n - 10}{4x_n^3 - 1}.$$

Since both f and f'' have the same sign at $x = 1.9$, choose $x_0 = 1.9$ as the starting point. Now

$$x_1 = x_0 - \frac{x_0^4 - x_0 - 10}{4x_0^3 - 1}$$

$$= 1.9 - \frac{f(1.9)}{f'(1.9)} = 1.9 - \frac{1.1321}{26.436}$$

$$x_1 = 1.9 - 0.042824 = 1.8572$$

$f(x_1) = 0.03972$

$$x_2 = 1.8572 - \frac{0.03972}{24.623} = 1.8572 - 0.0016131$$

$$x_2 = 1.855586897$$

$f(x_2) = +0.000058169$.

So $x_2 = 1.855587$ is the required root.

EXERCISE

1. Using graphical method find an approximate roots of the transcendental equations:

   (a) $3x - \cos x - 1 = 0$; (b) $e^x - 3x = 0$; (c) $\sin x = 2$; (d) $\sin x = \frac{1}{2}$; (e) $\sin x = \frac{1}{2}x$.

   Hint:

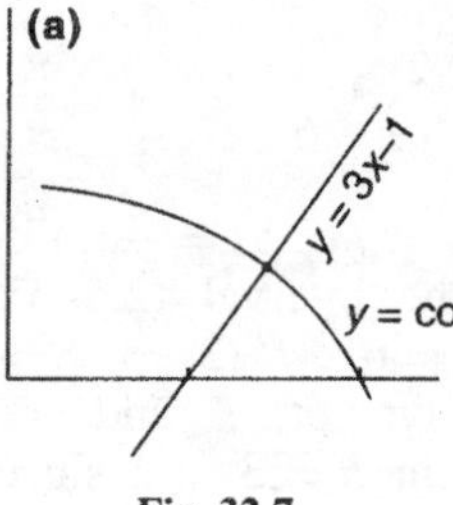

Fig. 32.7

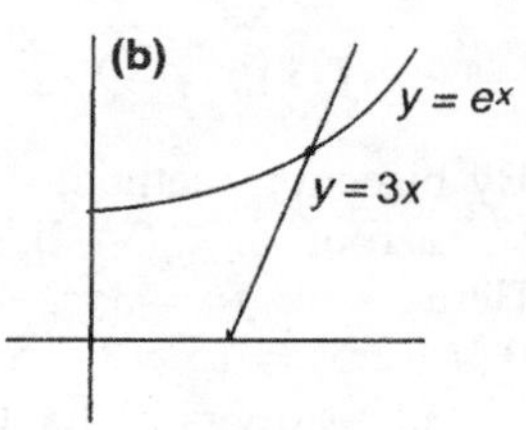

Fig. 32.8

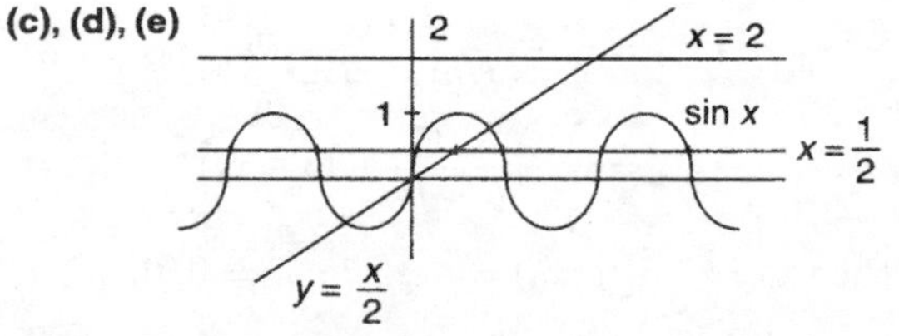

Fig. 32.9

Ans. (a) 0.6; (b) 0.619; (c) no root; (d) infinite number of roots; (e) $x = 30°$

2. Obtain an approximate root using bisection method, for the following equations:

   (a) $x^4 - x - 10 = 0$; (b) $x^3 - x - 1 = 0$; (c) $x^4 - 4x - 9 = 0$; (d) $x^4 + 2x^3 - x - 1 = 0$.

Ans. (a) (1.85546875, 1.857421855) at 9th iteration; (b) $x_5 = 1.38125$; (c) $x_4 = 2.6875$; (d) x_6: 0.867

3. Isolate the roots of the equations:

   (a) $x^3 - 6x + 2 = 0$; (b) $x + e^x = 0$.

Ans. **a.** $(-3, -1), (0, 1), (1, 3)$
b. $(-\infty, 1), (1, +\infty)$

4. Using Regula-Falsi method find an approximate root of the following equations:

   (a) $x^3 - 4x + 1 = 0$; (b) $x^3 - 0.2x^2 - 0.2x - 1.2 = 0$; (c) $x^3 - 2x - 5 = 0$; (d) $xe^x - 3 = 0$; (e) $x \log_{10}^x - 2 = 0$; (f) $xe^x - \cos x = 0$.

Ans. (a) 0.2541; (b) 1.198; (c) 2.094548; (d) 1.050; (e) 2.74065; (f) 0.5177

5. Using Newton-Raphson method, find an approximate root of the following equations:

   (a) $(x - 1)\sin x - x = 1$; (b) $e^{-x} = \sin x$; (c) $x^3 - 25 = 0$; (d) $x^4 - x - 1 = 0$; (e) $x + e^x = 0$; (f) $3x^3 + 5x - 40 = 0$; (g) $x - e^{-x} = 0$; (h) $2\sin x = x$; (i) $x^2 + 4\sin x = 0$; (j) $x^3 - 3x - 5 = 0$; (k) $x^4 - 3x^2 + 75x - 10,000 = 0$.

Ans. (a) −0.42036; (b) 0.5885; (c) 2.924; (d) 1.22138; (e) −0.567; (f) 0.5635; (g) 0.5671; (h) 1.895494; (i) −1.9338; (j) 2.7984; (k) −10.261

6. Use Newton's method to find the smallest positive root of the equation $\tan x = x$.

   Hint: Roots lies in $\left(\pi, \frac{3\pi}{2}\right)$.

Ans. 4.49343

7. Find the real root of $x \log_{10}^x = 1.2$ using Newton's iterative method.

   Hint: Root lies in (2, 3).

Ans. 2.74065

8. Apply Newton-Raphson method to evaluate approximately $\sqrt{12}$.

   Hint: Solve $x^2 - 12 = 0$, root lies in (3, 4).

Ans. 3.4641.

32.2 FINITE DIFFERENCES

Let $y = f(x)$ be a function and $\Delta x = h$ denote the increment in the independent variable x. Assume that Δx, increment in the argument x (also known as *interval or spacing*) is fixed. i.e., h = constant. Then the first finite difference of the function y is defined

$$\Delta y \equiv \Delta f(x) = f(x + \Delta x) - f(x).$$

Similarly, finite differences of higher orders are defined as follows

$$\Delta^2 y = \Delta(\Delta y) = \Delta(f(x + \Delta x) - f(x))$$

$$= \Delta(f(x + \Delta x)) - \Delta f(x)$$

$$= \left[f(x + 2\Delta x) - f(x + \Delta x)\right] - \left[f(x + \Delta x) - f(x)\right]$$

$$\Delta^2 y = f(x + 2\Delta x) - 2f(x + \Delta x) + f(x).$$

In general,

$$\boxed{\Delta^n y = \Delta(\Delta^{n-1} y)}, \quad \text{for } n = 2, 3, 4, \ldots$$

Now consider the function $y = f(x)$ specified by tabular values $y_i = f(x_i)$ for a set of ***equidistant*** points x_i where $i = 0, 1, 2, \ldots$ and $\Delta x_i = x_{i+1} - x_i = h$ = constant. Thus the tabulated function consists of ordered pairs $(x_0, y_0), (x_1, y_1), (x_2, y_2), \ldots, (x_k, y_k), \ldots$. Here y_k's are known as entries.

Forward differences

The first forward difference is denoted by Δy_k and defined as

$$\Delta y_k = y_{k+1} - y_k \quad (1)$$

Here the symbol Δ is the forward differences operator, having the following properties:

(i) $\Delta c = 0$ (differences of a constant function are zero)

(ii) $\Delta(cy_k) = c(\Delta y_k)$ where c is a constant

(iii) $\Delta(u_k + v_k) = \Delta u_k + \Delta v_k$

(iv) $\Delta(u_k v_k) = v_{k+1}\Delta u_k + u_k \Delta v_k$

(v) $\Delta^m(\Delta^n y_k) = \Delta^{m+n} y_k$

where m and n are non-negative integers and $\Delta^0 y_k = y_k$ (by definition). Note that because of (ii) and (iii) Δ is a linear operator.

The higher order forward differences are similarly defined: The second order forward difference of y_k is

$$\Delta^2 y_k = \Delta(\Delta y_k) = \Delta y_{k+1} - \Delta y_k$$

$$-\ -\ -\ -\ -\ -\ -\ -\ -\ -\ -\ -\ -\ -$$

In general,

$$\Delta^n y_k = \Delta(\Delta^{n-1} y_k) = \Delta^{n-1} y_{k+1} - \Delta^{n-1} y_k \quad (2)$$

defines the nth order forward differences. For example, $\Delta^3 y_k = \Delta^2 y_{k+1} - \Delta^2 y_k = (\Delta y_{k+2} - \Delta y_{k+1}) - (\Delta y_{k+1} - \Delta y_k) = y_{k+3} - 3y_{k+2} + 3y_{k+1} - y_k$.

From (1) $\quad y_{k+1} = y_k + \Delta y_k = (1+\Delta) y_k$

Similarly, $\quad y_{k+2} = y_{k+1} + \Delta y_{k+1} = (1+\Delta) y_{k+1}$

or $\quad y_{k+2} = (1+\Delta)(1+\Delta) y_k = (1+\Delta)^2 y_k$

Thus $\quad y_{k+3} = (1+\Delta)^3 y_k$

$$-\ -\ -\ -\ -\ -\ -\ -\ -\ -\ -\ -\ -\ -$$

$$y_{k+n} = (1+\Delta)^n y_k.$$

Expanding $(1+\Delta)^n$ by binomial theorem

$$y_{k+n} = y_k + n_{c_1} \Delta y_k + n_{c_2} \Delta^2 y_k + \cdots + \Delta^n y_k.$$

Conversely, we have

$$\Delta^n y_k = \left[(1+\Delta) - 1\right]^n y_k$$

$$= (1+\Delta)^n y_k - n_{c_1}(1+\Delta)^{n-1} y_k +$$

$$+ n_{c_2}(1+\Delta)^{n-2} y_k - \cdots + \cdots + (-1)^n y_i$$

or $\quad \Delta^n y_k = y_{n+k} - n_{c_1} y_{n+k-1}$

$$+ n_{c_2} y_{n+k-2} + \cdots + (-1)^n y_k \quad (3)$$

Thus any higher order forward differences can be expressed in terms of the successive values y_k's of the function. For example,

$$\Delta^2 y_k = y_{k+2} - 2y_{k+1} + y_k$$

$$\Delta^3 y_k = y_{k+3} - 3y_{k+2} + 3y_{k+1} - y_k \quad \text{etc.}$$

Finite differences of various orders are conveniently arranged in the form of a forward (diagonal) differences table (Table 32.1).

In general, y_0, the first entry is known as the leading terms, and Δy_0, $\Delta^2 y_0$, $\Delta^3 y_0$, $\Delta^4 y_0$, $\Delta^5 y_0$, $\Delta^6 y_0$, $\Delta^7 y_0$ are known as the leading differences.

Table 32.1 Forward differences table

x	y	Δy	$\Delta^2 y$	$\Delta^3 y$	$\Delta^4 y$	$\Delta^5 y$	$\Delta^6 y$	$\Delta^7 y$
x_0	y_0							
		Δy_0						
x_1	y_1		$\Delta^2 y_0$					
		Δy_1		$\Delta^3 y_0$				
x_2	y_2		$\Delta^2 y_1$		$\Delta^4 y_0$			
		Δy_2		$\Delta^3 y_1$		$\Delta^5 y_0$		
x_3	y_3		$\Delta^2 y_2$		$\Delta^4 y_1$		$\Delta^6 y_0$	
		Δy_3		$\Delta^3 y_2$		$\Delta^5 y_1$		$\Delta^7 y_0$
x_4	y_4		$\Delta^2 y_3$		$\Delta^4 y_2$		$\Delta^6 y_2$	
		Δy_4		$\Delta^3 y_3$		$\Delta^5 y_2$		
x_5	y_5		$\Delta^2 y_4$		$\Delta^4 y_3$			
		Δy_5		$\Delta^3 y_4$				
x_6	y_6		$\Delta^2 y_5$					
		Δy_6						
x_7	y_7							

Differences of a Polynomial

Book Work: If $P_n(x) = a_0 x^n + a_1 x^{n-1} + \cdots + a_n$ is an nth degree polynomial, then prove that

$$\Delta^n P_n(x) = n! a_0 h^n = \text{constant} \quad (4)$$

where $\Delta x = h$.

Proof: Consider $\Delta P_n(x) = P_n(x+h) - P_n(x)$

$$= [a_0(x+h)^n + a_1(x+h)^{n-1} + \cdots + a_n] -$$

$$-[a_0 x^n + a_1 x^{n-1} + \cdots + a_n]$$

$$= a_0[(x+h)^n - x^n] + a_1[(x+h)^{n-1} - x^{n-1}] +$$

$$+ \cdots + a_{n-1}[x+h-x].$$

Expanding by binomial theorem

$$(x+h)^n - x^n = x^n + nhx^{n-1} + n_{c_2} h^2 x^{n-2} + \cdots - x^n$$

Similar terms in the other brackets. Thus

$$Q_{n-1}(x) = \Delta P_n(x) = b_0 x^{n-1} + b_1 x^{n-2} + \cdots + b_{n-1}$$

which is a polynomial of $(n-1)$th degree. Here $b_0 = nha_0$. Now

$$\Delta^2 P_n(x) = \Delta(\Delta P_n(x)) = Q_{n-1}(x+h) - Q_{n-1}(x).$$

Expanding by binomial theorem, observe that $\Delta^2 P_n(x)$ is a polynomial of degree $n-2$:

$$\Delta^2 P_n(x) = c_0 x^{n-2} + c_1 x^{n-3} + \cdots + c_{n-2} \quad \text{where}$$
$$c_0 = (n-1)hb_0 = n(n-1)h^2 a_0.$$

Continuing in this manner successively, we get

$$\Delta^n P_n(x) = n! a_0 h^n = \text{constant}$$

Thus the nth order differences of a polynomial of nth degree are constant.

Corollary:

$$\Delta^S P_n(x) = 0 \qquad \text{for } s > n \tag{5}$$

i.e., $(n+1)$th order differences of a polynomial of nth degree are zero.

Converse: If the nth differences of a tabulated function are constant when the values of the independent variable (argument) are taken in arithmetic progression (i.e., at equal intervals apart), then the function is a polynomial of degree n.

Backward Differences

The first (order) backward difference is denoted by ∇ and defined as

$$\nabla y_k = y_k - y_{k-1} \tag{6}$$

Second (order) backward difference

$$\nabla^2 y_k = \nabla(\nabla y_k) = \nabla y_k - \nabla y_{k-1}$$

In general,

$$\nabla^n y_k = \nabla(\nabla^{n-1} y_k) = \nabla^{n-1} y_k - \nabla^{n-1} y_{k-1}.$$

Now $\nabla^3 y_k = \nabla^2 y_k - \nabla^2 y_{k-1}$

$$= \nabla y_k - \nabla y_{k-1} - \nabla y_{k-1} + \nabla y_{k-2}$$
$$= y_k - y_{k-1} - 2y_{k-1} + 2y_{k-2} + y_{k-2} - y_{k-3}$$
$$= y_k - 3y_{k-1} + 3y_{k-2} - y_{k-3}$$

In general,

$$\nabla^n y_k = \sum_{i=0}^{n} (-1)^i n_{c_i} y_{k-i} \tag{7}$$

Here y_4, the last entry, is the leading term and $\nabla y_4, \nabla^2 y_4, \nabla^3 y_4, \nabla^4 y_4$ are known as leading backward differences.

Table 32.2 Backward differences table

x	y	∇y	$\nabla^2 y$	$\nabla^3 y$	$\nabla^4 y$
x_0	y_0				
		∇y_1			
x_1	y_1		$\nabla^2 y_2$		
		∇y_2		$\nabla^3 y_3$	
x_2	y_2		$\nabla^2 y_3$		$\nabla^4 y_4$
		∇y_3		$\nabla^3 y_4$ ↗	
x_3	y_3		$\nabla^2 y_4$ ↗		
		∇y_4 ↗			
x_4	y_4 ↗				

Generalized Power or Factorial

The generalized nth power of a number x, denoted by $x^{[n]}$ or $[x]^n$, is defined as the product of n consecutive factors, the first of which is equal to x and each subsequent factor is h less than the preceding:

$$x^{[n]} = [x]^n = x \cdot (x-h) \cdot (x-2h) \cdots (x-(n-1)h) \tag{8}$$

Here h is some fixed constant $[x]^n$ is also known as a "factorial function". Here $[x]^0 = x^{[0]} = 1$.

Corollary 1: For $h = 0$, the generalized power coincides with the ordinary power i.e., $x^{[n]} = [x]^n = x^n$.

Corollary 2: For $h = 1$, $[x]^n = x(x-1)(x-2)\cdots(x-n+1)$.

Differences of a Generalized Power

The first difference of the factorial function is

$$\Delta[x]^n = [x+h]^n - [x]^n$$
$$= \Big\{(x+h)(x)(x-h)\cdots(x-(n-2)h)\Big\} - \Big\{x(x-h)\cdots(x-(n-1)h)\Big\}$$
$$= \Big\{x(x-h)\cdots(x-(n-2)h)\Big\} \times \Big\{x+h-(x-(n-1)h)\Big\}$$
$$= \Big\{x(x-h)\cdots(x-(n-2)h)\Big\} nh = nh[x]^{n-1}.$$

Thus $\Delta[x]^n = nh[x]^{n-1}$.

Now the second difference

$$\Delta^2[x]^n = \Delta(\Delta[x]^n) = \Delta(nh[x]^{n-1})$$
$$= nh\Delta[x]^{n-1} = nh\cdot(n-1)\cdot h\cdot[x]^{n-2}$$
$$\Delta^2[x]^n = n(n-1)h^2[x]^{n-2}.$$

By mathematical induction

$$\Delta^k[x]^n = n(n-1)\cdots(n-(k-1))h^k[x]^{n-k} \quad (9)$$

where $k = 1, 2, 3, \ldots, n$.

Corollary 1: For $k = n$, from (9), we have

$$\Delta^n[x]^n = n(n-1)(n-2)\cdots 2\cdot 1\cdot h^n[x]^0$$
$$\Delta^n[x]^n = n!h^n = \text{constant} \quad (10)$$

since $[x]^0 = 1$.

Corollary 2: For $k = n+1$, from (9),

$$\Delta^{n+1}[x]^n = \Delta(\Delta^n[x]^n) = n!h^n - n!h^n = 0.$$

Thus

$$\Delta^s[x]^n = 0 \quad \text{when } s > n \quad (11)$$

Corollary 3: When $h = 1$, from (10), we have

$$\Delta^n[x]^n = n! \quad (12)$$

i.e., the result of *differencing* $[x]^n$ is analogous to that of *differentiating* x^n. In other words, the operator Δ is equivalent the differential operator $D = \frac{d}{dx}$ i.e., $\Delta \equiv D$.

Corollary 4: Every polynomial of degree n can be expressed as a factorial polynomial of the same degree and vice versa.

Worked Out Examples

Example 1: Show that

$$\Delta(u_k v_k) = v_{k+1}\Delta u_k + u_k\Delta v_k.$$

Solution: $\Delta(u_k v_k) = u_{k+1}v_{k+1} - u_k v_k$. We know that $\Delta u_k = u_{k+1} - u_k$, $\Delta v_k = v_{k+1} - v_k$

$$\therefore\ u_{k+1}\cdot v_{k+1} = (u_k + \Delta u_k)(v_k + \Delta v_k)$$
$$= u_k v_k + u_k\Delta v_k + v_k\Delta u_k + \Delta u_k\Delta v_k.$$

Thus

$$\Delta(u_k v_k) = u_{k+1}\cdot v_{k+1} - u_k v_k$$
$$= u_k\Delta v_k + \Delta u_k\{v_k + v_{k+1} - v_k\}$$
$$= u_k\Delta v_k + v_{k+1}\Delta u_k.$$

Note: $\Delta(u_k v_k) = v_k\Delta u_k + u_{k+1}\Delta v_k$ is true.

Example 2: Evaluate $\Delta^n(e^{3x+5})$.

Solution: $\Delta(e^{3x+5}) = e^{3(x+1)+5} - e^{3x+5}$
$= e^{3x+5}(e^3 - 1)$. Now

$$\Delta^2(e^{3x+5}) = \Delta(\Delta e^{3x+5}) = \Delta\left\{(e^3-1)(e^{3x+5})\right\}$$
$$= (e^3-1)\Delta(e^{3x+5}), \text{ using the first result}$$
$$= (e^3-1)(e^3-1)e^{3x+5} = (e^3-1)^2e^{3x+5}.$$

By induction $\Delta^n(e^{3x+5}) = (e^3-1)^n e^{3x+5}$.

Example 3: Evaluate $\Delta^2\left\{\frac{4x^2-25x+31}{(x-1)(x-2)(x-3)}\right\}$.

Solution: Resolving into partial fractions

$$\Delta^2\left(\frac{4x^2-25x+31}{(x-1)(x-2)(x-3)}\right)$$
$$= \Delta^2\left\{\frac{5}{x-1} + \frac{3}{x-2} - \frac{4}{x-3}\right\}$$
$$= \Delta\left[\Delta\left(\frac{5}{x-1}\right) + \Delta\left(\frac{3}{x-2}\right) + \Delta\left(\frac{-4}{x-3}\right)\right]$$
$$= \Delta\left[5\left(\frac{1}{x} - \frac{1}{x-1}\right) + 3\left(\frac{1}{x-1} - \frac{1}{x-2}\right)\right.$$
$$\left. -4\left(\frac{1}{x-2} - \frac{1}{x-3}\right)\right]$$
$$= 5\left\{\left(\frac{1}{x+1} - \frac{1}{x}\right) - \left(\frac{1}{x} - \frac{1}{x-1}\right)\right\}$$
$$+3\left\{\left(\frac{1}{x} - \frac{1}{x-1}\right) - \left(\frac{1}{x-1} - \frac{1}{x-2}\right)\right\}$$
$$-4\left\{\left(\frac{1}{x-1} - \frac{1}{x-2}\right) - \left(\frac{1}{x-2} - \frac{1}{x-3}\right)\right\}$$
$$= \frac{5}{x+1} - \frac{7}{x} - \frac{5}{x-1} + \frac{11}{x-2} - \frac{4}{x-3}.$$

Example 4: If $f(x) = x^3 + 5x - 7$, then form the table of backward differences for $x =$

$-1, 0, 1, 2, 3, 4, 5$. Continue the table to obtain $f(6)$.

Solution: For $x = -1, 0, 1, 2, 3, 4, 5$, the values of $f(x)$ are respectively $-13, -7, -1, 11, 35, 77, 143$. The backward differences table is

Table 32.3

i	x	$y = f(x)$	∇y	$\nabla^2 y$	$\nabla^3 y$	$\nabla^4 y$	$\nabla^5 y$	$\nabla^6 y$
0	-1	-13						
			6					
1	0	-7		0				
			6		6			
2	1	-1		6		0		
			12		6		0	
3	2	11		12		0		0
			24		6		0	
4	3	35		18		0		$\boxed{0}$
			42		6		$\boxed{0}$	
5	4	77		24		$\boxed{0}$		
			66		$\boxed{6}$			
6	5	143		$\boxed{30}$				
			$\boxed{96}$					
7	$\boxed{6}$	$\boxed{239}$						

Values obtained by continuation at $f(6)$ are squared boxes.

Example 5: Construct the missing values in the following table

$x:$	0	5	10	15	20	25
$y:$	6	10	—	17	—	31

Solution: Name the missing values as y_2 and y_4. Then the differences table is Table 32.4

Since only four entries y_0, y_1, y_3, y_5 are given, y can be represented by third degree polynomial. Consequently its fourth order differences are zero. Thus

$$\Delta^4 y_0 = y_4 + 6y_2 - 102 = 0 \quad \text{and}$$

$$\Delta^4 y_1 = 143 - 4y_4 - 4y_2 = 0$$

Solving $\quad y_2 = 13.25,\ y_4 = 22.5.$

Generalized power (factorial function)

Example 6: Express $x^3 - 2x^2 + x - 1$ in generalized powers (or into factorial polynomial). Hence

Table 32.4

x	y	Δy	$\Delta^2 y$	$\Delta^3 y$	$\Delta^4 y$
0	6				
		4			
5	10		y_2-14		
		y_2-10		$41-3y_2$	y_4+6y_2-102
10	y_2		$27-2y_2$		
		$17-y_2$		y_4+3y_2-61	
15	17		y_4-34+y_2		$143-4y_4-4y_2$
		y_4-17		$82-3y_4-y_2$	
20	y_4		$48-2y_4$		
		$31-y_4$			
25	31				

show that $\Delta^4 f(x) = 0$.

Solution: Assume that

$$f(x) = x^3 - 2x^2 + x - 1 = [x]^3 + B[x]^2 + C[x] + D$$
$$= x(x-1)(x-2) + Bx(x-1) + cx + D.$$

Put $x = 0$, then $-1 = D$

Put $x = 1$, then $C + D = -1 \quad \therefore C = 0$

Put $x = 2$, then $2B + C + D = 1 \quad \therefore B = 1$

Thus $f(x) = [x]^3 + [x]^2 - 1$

Since differencing of power function $[x]^r$ amounts differentiation of x^r, we have

$$\Delta f(x) = \Delta[x]^3 + \Delta[x]^2 - \Delta[x]^0$$
$$= \frac{d}{dx}x^3 + \frac{d}{dx}x^2 - \frac{d}{dx}1 = 3[x]^2 + 2[x]$$
$$\Delta^2 f = 6[x] + 2$$
$$\Delta^3 f = 6 \quad \text{and} \quad \Delta^4 f(x) = 0.$$

Example 7: Obtain the function whose first difference is $2x^3 + 3x^2 - 5x + 4$.

Solution: Let $\Delta f(x) = 2x^3 + 3x^2 - 5x + 4$. Now like the above problem

$$2x^3 + 3x^2 - 5x + 4 = 2[x]^3 + B[x]^2 + C[x] + D$$

Solving $D = 4$, $C = 0$, $B = 9$. Thus

$$\Delta f(x) = 2[x]^3 + 9[x]^2 + 0 + 4$$

Integrating, $f(x) = \frac{2[x]^4}{4} + 9\frac{[x]^3}{3} + 4[x] + C$

$$f(x) = \frac{1}{2}x(x-1)(x-2)(x-3) +$$
$$+3(x)(x-1)(x-2) + 4x + C$$

where C is constant of integration.

Example 8: Evaluate $\Delta^3[(1-x)(1-2x)(1-3x)]$.

Solution:

$$\Delta^3[(1-x)(1-2x)(1-3x)]$$
$$= \Delta^3[(-1)(-2)(-3)x^3 + (11)x^2 + (-6)x + 1]$$
$$= -6\Delta^3 x^3 + 0 + 0 + 0 = -6 \cdot 3! = -6 \cdot 6 = -36$$
$$\because \Delta^n f(x) = a \cdot n! h^n.$$

EXERCISE

1. Evaluate (a) $\Delta \tan^{-1}\left(\frac{n-1}{n}\right)$; (b) $\Delta^2 \cos 2x$; (c) $\Delta^n(e^{2x+3})$.

Ans. (a) $\tan^{-1}\frac{1}{2n^2}$; (b) $-4\sin^2 h \cos(2x+2h)$; (c) $(e^2-1)e^{2x+3}$

2. Evaluate

   a. $\Delta^2\left(\frac{5x+12}{x^2+5x+16}\right)$

   b. $\Delta^2\left(\frac{1}{(x^2)+5x+6}\right)$

   with internal spacing $h = 1$.

Ans. (a) $\frac{2(5x+16)}{(x+2)(x+3)(x+4)(x+5)}$; (b) $\frac{-2}{(x+2)(x+3)(x+4)}$

3. Evaluate (a) $\Delta^{10}[(1-x)(1-2x^2)(1-3x^3)\times \times(1-4x^4)]$ if the interval of spacing is 2.

   Hint: L.H.S.$=\Delta^{10}[(-1)(-2)(-3)(-4)x^{10} +$ terms containing powers of x less than $10] = 24\Delta^{10}[x^{10}] + 0 + \cdots + 0 = 24 \cdot 10!2^{10}$ since $\Delta^n f(x) = n!h^n$.

Ans. $24 \times 2^{10} \times 10!$

4. Show that $\Delta\left(\frac{u_k}{v_k}\right) = \frac{v_k \Delta u_k - u_k \Delta v_k}{v_k v_{k+1}}$.

   Hint: $\Delta\left(\frac{u_k}{v_k}\right) = \frac{u_{k+1}}{v_{k+1}} - \frac{u_k}{v_k} = \frac{u_{k+1}v_k - u_k v_{k+1}}{v_k v_{k+1}}$

   $= \frac{(u_{k+1}v_k - u_k v_k + u_k v_k - u_k v_{k+1})}{(u_k v_{k+1})}$

5. Express $f(x) = 2x^3 - 3x^2 + 3x - 10$ in generalized power (in factorial notation) and hence find $\Delta^3 y$.

Ans. $f(x) = 2[x]^3 + 3[x]^2 + 2[x] - 10$, $\Delta^3 y = 12$

6. Compute the missing values in the following table:

x :	45	50	55	60	65
y :	3.0	—	2.0	—	−2.4

   Hint: Solve $\Delta^3 y_0 = 3y_1 + y_3 - 9 = 0$, $\Delta^3 y_1 = y_1 + 3y_2 - 3.6 = 0$

Ans. $y_1 = 2.925$, $y_3 = 0.225$

7. Represent $f(x) = x^4 - 12x^3 + 42x^2 - 30x + 9$ and its successive differences in generalized power.

Ans. $f(x) = [x]^4 - 6[x]^3 + 13[x]^2 + [x]^1 + 9$, $\Delta y = 4[x]^3 - 18[x]^2 + 26[x] + 1$, $\Delta^4 y = 24$.

8. Obtain the function whose first difference is $x^3 + 3x^2 + 5x + 12$.

Ans. $\frac{1}{4}[x]^4 + 2[x]^3 + \frac{9}{2}[x]^2 + 125[x]^1 + C$.

32.3 INTERPOLATION

To fix the height of dam across a river, an engineer utilizes a set of data of the form (x_i, y_i) where x_i denotes the year and y_i, the peak flood level across the river; to estimate the highest possible flood level in future. From the recorded data of (t_i, A_i), time and altitude of a rocket, a physicist would like to estimate what was the altitude at a particular time $t_0(\neq t_i)$. Thus most of the experimental or observed data is in the form a set of say $(n+1)$ ordered pairs $(x_0, y_0), (x_1, y_1), \ldots, (x_n, y_n)$ which is the tabular form of an unknown function $y = f(x)$. The process of determining the value of y for an $x \in [x_0, x_n]$ is known as interpolation. Here $x_0, x_1, \ldots, x_n$ are called interpolation (or mesh) points. Note that x differs from the interpolation points. Thus interpolation is the "art of reading *between* the lines of a table." In extrapolation value of y is determined for an $x \notin [x_0, x_n]$ i.e., for x outside the interval $[x_0, x_n]$. But generally interpoaltion includes extrapolation also.

The problem of interpolation is to construct a new (interpolating) function $F(x)$ which collocates (coincides) with the unknown function $f(x)$ at the tabulated $(n+1)$ interpolation points

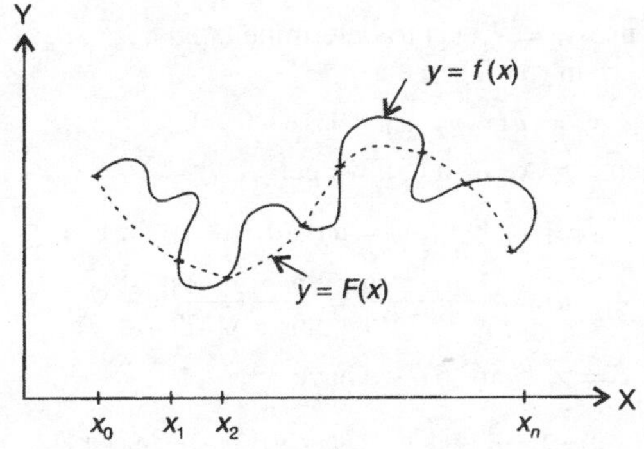

Fig. 32.10

i.e., $y_0 \equiv f(x_0) = F(x_0), \quad y_1 \equiv f(x_1) = F(x_1), \ldots$

$y_n \equiv f(x_n) = F(x_n).$

Geometrically, this means that graphs of $y = f(x)$ and $y = F(x)$ coincide at the $n+1$ points. Since a unique straight line passes through two given points, a uniquc parabola through three given points, a unique polynomial $F(x)$ of degree n can be determined (passing) satisfying the given set of $(n+1)$ points. In this case it is called a polynomial interpolation. Now from $y = F(x)$, y can be computed for a given x. This interpolating polynomial will be expressed in terms of finite differences in several forms leading to the Newton-Gregory forward, backward and central differences formulae of Gauss, Stirling, Bessel, Everett etc. Lagrange's interpolation and Newton's divided differences formulae are applicable for unequally spaced interpolation points x_i.

32.4 NEWTON-GREGORY FORWARD INTERPOLATION FORMULA

Suppose the values of $y_i = f(x_i)$ are given for equally spaced values of the independent variable (argument) $x_i = x_0 + ih$ for $i = 0, 1, 2, \ldots, n$. Here h, known as the size of the interval or spacing, is constant. Assume that the nth degree interpolating polynomial is given by

$$F(x) = a_0 + a_1(x-x_0) + a_2(x-x_0)(x-x_1) + \cdots + a_n(x-x_0)(x-x_1)\cdots(x-x_{n-1}) \quad (1)$$

Using the $n+1$ conditions, $y_i = F(x_i)$ for $i = 0, 1, 2, \ldots, n$, we determine the $(n+1)$ unknown coefficients $a_0, a_1, a_2, \ldots, a_n$ in (1).

Putting $x = x_0$ in (1), we get

$$y_0 = F(x_0) = a_0 + 0 + \cdots + 0 \quad \therefore \quad a_0 = y_0$$

Putting $x = x_1$ in (1), we have

$$y_1 = F(x_1) = a_0 + a_1(x_1 - x_0)$$

But $\quad a_0 = y_0 \quad$ and $\quad x_1 - x_0 = h$

$$\therefore \quad a_1 = \frac{y_1 - a_0}{x_1 - x_0} = \frac{y_1 - y_0}{h} = \frac{1}{h}\Delta y_0$$

Now with $x = x_2$ in (1), we get

$$y_2 = F(x_2) = a_0 + a_1(x_2 - x_0) + a_2(x_2 - x_0)(x_2 - x_1)$$
$$= a_0 + a_1 \cdot 2h + a_2 \cdot 2h \cdot h$$

$$\text{so} \quad a_2 = \frac{y_2 - a_0 - 2h \cdot a_1}{2h^2} = \frac{y_2 - y_0 - 2h\frac{1}{h}\Delta y_0}{2h^2}$$

$$a_2 = \frac{y_2 - y_0 - 2(y_1 - y_0)}{2h}$$
$$= \frac{y_2 - 2y_1 + y_0}{2h} = \frac{1}{2h}\Delta^2 y_0.$$

Similarly, at $x = x_3$, we have

$$y_8 = F(x_3) = a_0 + a_1(x_3 - x_0) + a_2(x_3 - x_0)(x_3 - x_1) + a_3(x_3 - x_0)(x_3 - x_1)(x_3 - x_2)$$
$$= a_0 + a_1 \cdot 3h + a_2 \cdot 3h \cdot 2h + a_3 \cdot 3h \cdot 2h \cdot h$$

$$\text{Solving} \quad a_3 = \frac{y_3 - 3y_2 + 3y_1 - y_0}{3!h^2} = \frac{1}{3!h^2}\Delta^3 y_0.$$

This way, we get

$$a_4 = \frac{1}{4!h^4}\Delta^4 y_0, \quad a_5 = \frac{1}{5!h^5}\Delta^5 y_0 \quad \text{etc. and}$$

$$\boxed{a_n = \frac{1}{n!h^n}\Delta^n y_0} \tag{2}$$

Substituting these values of $a_0, a_1, \ldots, a_n$ in (1), we get the Newton-Gregory forward interpolation formula (also known as *Newton's first interpolation formula*) as

$$y = F(x) = y_0 + \frac{\Delta y_0}{h}(x - x_0) + \frac{\Delta^2 y_0}{2!h^2}(x - x_0)(x - x_1) +$$
$$+ \frac{\Delta^3 y_0}{3!h^3}(x - x_0)(x - x_1)(x - x_2) + \cdots$$
$$+ \frac{\Delta^n y_0}{n!h^n}(x - x_0)(x - x_1)(x - x_2)\cdots(x - x_{n-1}) \tag{3}$$

Introducing $q = \frac{x - x_0}{h}$, the above formula (3) can be written in more convenient way.

Now $\dfrac{x - x_1}{h} = \dfrac{x - (x_0 + h)}{h} = \dfrac{x - x_0}{h} - 1 = q - 1,$

$$\frac{x - x_2}{h} = \frac{x - (x_0 + 2h)}{h} = \frac{x - x_0}{h} - 2 = q - 2 \text{ etc.}$$

$$\frac{x - x_{n-1}}{h} = \frac{x - (x_0 + (n-1)h)}{h}$$
$$= q - (n - 1) = q - n + 1.$$

Substituting these values, we have

$$F(x) = F(x_0 + hq) = g(q) = y_0 + \Delta y_0 \cdot q +$$
$$+ \frac{\Delta^2 y_0}{2!}q(q - 1) + \frac{\Delta^3 y_0}{3!}q(q - 1)(q - 2) + \cdots$$
$$+ \frac{q(q - 1)\cdots(q - n + 1)}{n!}\Delta^n y_0 \tag{4}$$

Note that the coefficients of Δ's are binomial coefficients. Since (4) involves only the "forward differences" $\Delta y_0, \Delta^2 y_0, \ldots, \Delta^n y_0$, Newton-Gregory forward interpolation formula given by (4) is most often used to interpolate (and extrapolate) for values of y at the ***begining*** of a set of tabular data. For $n = 1$ in (4), we get linear interpolation

$$P_1(x) = y_0 + q\Delta y_0.$$

For $n = 2$ in (4), we have parabolic interpolation

$$P_2(x) = y_0 + q\Delta y_0 + \frac{q(q - 1)}{2}\Delta^2 y_0.$$

Newton-Gregory Backward Interpolation Gormula

It is mainly useful to interpolate near the ***end*** of the table. Assume the polynomial as

$$y = F(x) = a_0 + a_1(x - x_n) + a_2(x - x_n)(x - x_{n-1}) +$$
$$+ a_3(x - x_n)(x - x_{n-1})(x - x_{n-2}) + \cdots +$$
$$+ a_n(x - x_n)(x - x_{n-1})\cdots(x - x_1) \tag{5}$$

We use $y_i = F(x_1)$ to determine $a_0, a_1, \ldots, a_n$. Put $x = x_n$ in (5). Then

$$y_n = F(x_n) = a_0 + 0 \cdots + 0 \quad \therefore a_0 = y_n.$$

When $x = x_{n-1}$ in (5), we get

$$y_{n-1} = F(x_{n-1}) = a_0 + a_1(x_{n-1} - x_n)$$

or $\quad a_1 = \dfrac{y_{n-1} - a_0}{x_{n-1} - x_n} = \dfrac{y_n - y_{n-1}}{x_n - x_{n-1}} = \dfrac{1}{h}\nabla y_n$

For $x = x_{n-2}$ in (5), we have

$$y_{n-2} = F(x_{n-2}) = a_0 + a_1(x_{n-2} - x_n) +$$
$$+ a_2(x_{n-2} - x_n)(x_{n-2} - x_{n-1})$$
$$a_2 = \frac{y_{n-2} - 2y_{n-1} + y_n}{2h^2} = \frac{1}{2h^2}\nabla^2 y_n.$$

Similarly, we get

$$\boxed{a_n = \frac{1}{n!h^n}\nabla^n y_n}$$

Substituting these values of $a_0, a_1, \ldots, a_n$ in (5), we get the Newton-Gregory backward interpolation formula (also known as *Newton's second interpolation formula*) as

$$y = F(x) = y_n + \frac{(x - x_n)}{h}\nabla y_n +$$
$$+ \frac{(x - x_n)(x - x_{n-1})}{2!h^2}\nabla^2 y_n + \cdots$$
$$+ \frac{(x - x_n)(x - x_{n-1})\cdots(x - x_1)}{n!h^n}\nabla^n y_n \tag{6}$$

Introducing $q = \frac{x - x_n}{h}$ and noting that

$$\frac{x - x_{n-1}}{h} = q + 1, \quad \frac{x - x_{n-2}}{h} = q + 2, \ldots$$

$$\frac{x - x_n}{h} = \frac{x - (x_n - (n - 1)h)}{h} = q + n - 1.$$

$$y = F(x) = F(x_n + hq) = y_n + q\nabla y_n + \frac{q(q+1)}{2!}\nabla^2 y_n + \frac{q(q+1)(q+2)}{3!}\nabla^3 y_n + \cdots + q(q+1)(q+2)\cdots(q+n-1)\cdot\frac{1}{n!}\nabla^n y_n \quad (7)$$

Generally (4) is used for forward interpolation and backward extrapolation and (7) is used for backward interpolation and forward extrapolation.

WORKED OUT EXAMPLES

Example 1: Compute (a) $y(9)$; (b) $y(7)$; (c) $y(17)$ and (d) $y(19)$ from the following data:

x :	8	10	12	14	16	18
y :	10	19	32.5	54	89.5	15.4

Solution: Since $x = 9$ and $x = 7$ are at the beginning of the table (data), use Newton-Gregory Forward Interpolation Formula (NGFIF) for interpolation at $x = 9$ and extrapolation at $x = 7$. Similarly, since $x = 17$ and $x = 19$ occur at the end of the data, use Newton-Gregory Backward Interpolation Formula (NGBIF) for interpolation at $x = 17$ and extrapolation at $x = 19$. Here interval size is $h = 2$ (points x_i are equidistant).
The finte differences table (from which both forward and backward differences can be read) is obtained below:

Table 32.5

i	x	y	Δy	$\Delta^2 y$	$\Delta^3 y$	$\Delta^4 y$	$\Delta^5 y$
0	8	10					
			9				
1	10	19		4.5			
					3.5		
			13.5			2.5	
2	12	32.5		8			6.5
			21.5		6		
3	14	54		14		9	
			35.5		15		
4	16	89.5		29			
			64.5				
5	18	154					

Newton-Gregory forward interpolation formula is

$$y = y_0 + q\Delta y_0 + \frac{q(q-1)}{2!}\Delta^2 y_0 + \frac{q(q-1)(q-2)}{3!}\Delta^3 y_0 + \frac{q(q-1)(q-2)(q-3)}{4!}\Delta^4 y_0 + \frac{q(q-1)(q-2)(q-3)(q-4)}{5!}\Delta^5 y_0.$$

From the difference table, $y_0 = 10$, $\Delta y_0 = 9$, $\Delta^2 y_0 = 4.5$, $\Delta^3 y_0 = 3.5$, $\Delta^4 y_0 = 2.5$ and $\Delta^5 y_0 = 6.5$.

a. To interpolate at $x = 9$: $q = \frac{x - x_0}{h}$ so $q = \frac{9-8}{2} = \frac{1}{2}$.
Substituting the above data in the formula

$$y(9) = 10 + \frac{1}{2}(9) + \frac{1}{2}\left(\frac{1}{2} - 1\right)\frac{1}{2!}(4.5) + \frac{\frac{1}{2}\left(\frac{1}{2}-1\right)\left(\frac{1}{2}-2\right)}{3!}(3.5) + \frac{\frac{1}{2}\left(\frac{1}{2}-1\right)\left(\frac{1}{2}-2\right)\left(\frac{1}{2}-3\right)}{4!}(2.5) + \frac{\frac{1}{2}\left(\frac{1}{2}-1\right)\left(\frac{1}{2}-2\right)\left(\frac{1}{2}-3\right)\left(\frac{1}{2}-4\right)}{5!}(6.5)$$

$y(9) = 14.455.$

b. To extrapolate at $x = 7$: $q = \frac{7-8}{2} = -\frac{1}{2}$.

$$y(7) = 10 - \frac{9}{2} + \left(\frac{1}{2}\right)\left(\frac{3}{2}\right)\frac{1}{2!}(4.5) - \frac{1}{2}\left(\frac{3}{2}\right)\left(\frac{5}{2}\right)\frac{1}{3!}(3.5) + \frac{1}{2}\left(\frac{3}{2}\right)\left(\frac{5}{2}\right)\left(\frac{7}{2}\right)\frac{1}{4!}(2.5) - \frac{1}{2}\left(\frac{3}{2}\right)\left(\frac{5}{2}\right)\left(\frac{7}{2}\right)\left(\frac{9}{2}\right)\frac{1}{5!}(6.5)$$

$y(7) = 5.1777.$

Newton-Gregory backward interpolation formula is

$$y = y_n + q\nabla y_n + \frac{q(q+1)}{2!}\nabla^2 y_n + \frac{q(q+1)(q+2)}{3!}\nabla^3 y_n +$$

$$+\frac{q(q+1)(q+2)(q+3)}{4!}\nabla^4 y_n + \frac{q(q+1)(q+2)(q+3)(q+4)}{5!}\nabla^5 y_n.$$

From the difference table, $x_n = 18$, $y_n = 154$, $\nabla y_n = 64.5$, $\nabla^2 y_n = 29$, $\nabla^3 y_n = 15$, $\nabla^4 y_n = 9$, $\nabla^5 y_n = 6.5$.

c. To interpolate at $x = 17$, we have

$$q = \frac{x - x_n}{h} = \frac{17-18}{2} = -\frac{1}{2}$$

Substituting these values in the formula

$$y(17) = 154 + \left(-\frac{1}{2}\right)(64.5) + \left(-\frac{1}{2}\right)\left(-\frac{1}{2}+1\right)\cdot\frac{1}{2!}(29) + \left(-\frac{1}{2}\right)\left(-\frac{1}{2}+1\right)\left(-\frac{1}{2}+2\right)\frac{1}{3!}15$$

$$+\frac{\left(-\frac{1}{2}\right)\left(-\frac{1}{2}+1\right)\left(-\frac{1}{2}+2\right)\left(-\frac{1}{2}+3\right)(9)}{4!} + \frac{\left(-\frac{1}{2}\right)\left(-\frac{1}{2}+1\right)\left(-\frac{1}{2}+2\right)\left(-\frac{1}{2}+3\right)\left(-\frac{1}{2}+4\right)(6.5)}{5!}$$

$$y(17) = 126.841.$$

d. To extrapolate at $x = 19$, take $q = \frac{19-18}{2} = \frac{1}{2}$. Then

$$y(19) = 154 + \frac{1}{2}(64.5) + \frac{1}{2}\left(\frac{3}{2}\right)\frac{1}{2!}(29) + \frac{1}{2}\left(\frac{3}{2}\right)\left(\frac{5}{2}\right)\frac{1}{3!}(15) + \frac{1}{2}\left(\frac{3}{2}\right)\left(\frac{5}{2}\right)\left(\frac{7}{2}\right)\frac{1}{4!}(9) + \frac{1}{2}\left(\frac{3}{2}\right)\left(\frac{5}{2}\right)\left(\frac{7}{2}\right)\left(\frac{9}{2}\right)\frac{1}{5!}(6.5)$$

$$y(19) = 219.208.$$

Example 2: Fit a polynomial of degree three which takes the following values:

x :	3	4	5	6
y :	6	24	60	120

Solution: The Newton-Gregory forward interpolating polynomial collocates (takes or geometrically passes through) the given set of points (x_0, y_0), (x_1, y_1), (x_2, y_2), (x_3, y_3) and is given by

$$y = y_0 + q\Delta y_0 + \frac{q(q-1)}{2!}\Delta^2 y_0 + \frac{q(q-1)(q-2)}{3!}\Delta^3 y_0$$

Here $q = \frac{x-x_0}{h}$ and $h = 1$ (given).
The finite differences table is

Table 32.6

x	y	Δy	$\Delta^2 y$	$\Delta^3 y$
3	6			
		18		
4	24		18	
		36		6
5	60		24	
		60		
6	120			

From the table, $x_0 = 3$, $y_0 = 6$, $\Delta y_0 = 18$, $\Delta^2 y_0 = 18$, $\Delta^3 y_0 = 6$, $q = \frac{x-x_0}{h} = \frac{x-3}{1} = x - 3$
Substituting these values

$$y = 6 + (x-3)(18) + (x-3)\frac{(x-3-1)}{2!}(18) + (x-3)(x-3-1)\frac{(x-3-2)}{3!}(6)$$

$$= 6 + 18(x-3) + 9(x-3)(x-4) + (x-3)(x-4)(x-5)$$

$$y(x) = x^3 - 3x^2 + 2x$$

is the required cubic polynomial (which takes the given data for example, $y(3) = 3^3 - 3(3^2) + 2(3) = 6$ etc.).

Note: Using Newton's backward interpolating polynomial (using backward differences), we get the same cubic polynomial

$$y = y_n + q\nabla y_n + \frac{q(q+1)}{2!}\nabla^2 y_n + \frac{q(q+1)(q+2)}{3!}\nabla^3 y_n$$

$$= 120 + (x-6)60 + \frac{(x-6)(x-5)}{2!}(24) + \frac{(x-6)(x-5)(x-4)}{3!}6$$

$$y(x) = x^3 - 3x^2 + 2x.$$

EXERCISE

1. Using Newton's forward formula compute the pressure of the steam at temperature 142° from the following steam table:

Temperature:	140	150	160	170	180
Pressure:	3.685	4.854	6.302	8.076	10.225

Hint: Forward differences 1.169, 0.279, 0.047, 0.002.

Ans. 3.899

2. Using Newton's backward formula compute $f(43)$, $f(84)$ from the following table. Fit a polynomial

x :	40	50	60	70	80	90
$f(x)$:	184	204	226	250	276	304

Hint: Backward differences: 28, 2, 0, 0, 0.

Ans. $f(43) = 190$, $f(84) = 287$; $f(x) = 84 + 2.9x - 0.01x^2$

3. Fit a cubic polynomial which takes the following values. Hence find $y(4)$.

x :	0	1	2	3
y :	1	0	1	10

Ans. $x^3 - 2x^2 + 1$, $y(4) = 33$, forward differences: $-1, 2, 6$

4. Estimate the population in 1895 and 1925 from the following statistics:

Year x :	1891	1901	1911	1921	1931
Population y :	46	66	81	93	101

Hint: Forward differences: 20, −5, 2, −3;

Backward differences: 8, −4, −1, −3.

Ans. 54.85, 96.84

5. Compute the first and tenth term of the following series assuming that values of y are consecutive.

x :	3	4	5	6	7	8	9
y :	2.7	6.4	12.5	21.6	34.3	51.2	72.9

Hint: Forward differences: 3.7, 2.4, 0.6, 0;

Backward differences: 21.7, 4.8, 0.6, 0.

Ans. $y(1) = 0.1$, $y(10) = 100$

6. Construct an empirical formula for the function y specified in the following table. Hence find $y(-1)$, $y(0.5)$,

x :	0	1	2	3	4	5
y :	5.2	8.0	10.4	12.4	14.0	15.2

Ans. $y = 5.2 + 3x - 0.2x^2$, $y(-1) = 2$, $y(0.5) = 6.65$

7. Determine $\log_{10} 1044$ using backward formula from the data below:

x:	1000	1010	1020
$\log_{10} x$:	3	3.0043214	3.0086002
x:	1030	1040	1050
$\log_{10} x$:	3.0128372	3.017033	3.601193

Hint: Backward differences: 0.004156, 0.0000401, 0.0000008.

Ans. 3.0187005

Extrapolation

8. Compute sin 14° and sin 56° from the data below:

x:	15°	20°	25°	30°	35°
$\sin x$:	0.2588	0.342	0.4226	0.5	0.5736
x:	40°	45°	50°	55°	
$\sin x$:	0.6428	0.7071	0.706	0.8192	

Hint: Forward differences: 0.0832, −0.0026, −0.0006

Backward differences: 0.0532, −0.0057, −0.0003

Ans. sin 14° = 0.24192, sin 56° = 0.82904

9. Find the polynomial of degree four which takes the following values:

x :	2	4	6	8	10
y :	0	0	1	0	0

Hint: Differences: 0, 1, −3, 6.

Ans. $\frac{1}{64}[x^4 - 24x^3 + 196x^2 - 624x + 640]$

10. Estimate the number of students who secured marks between 40 and 45 from the following table:

Marks x :	30 to 40	40 to 50	50 to 60
No. Students y :	31	42	51
Marks x :	60 to 70	70 to 80	
No. Students y :	35	31	

Ans. $y(45) = 47.87 \approx 48$, number of students between 40 and 45 is $48 - 31 = 17$

11. Compute $y(1.5)$ and $y(8.5)$ from the following data:

x :	1	2	3	4	5	6	7	8
y :	1	8	27	64	125	216	343	512

Hint: F.D.: 7, 12, 6, 0, 0, B.D.: 169, 42, 6, 0, 0.

Ans. $y(1.5) = 3.375$, $y(8.5) = 614.125$

12. If $F(\phi) = \int_0^{\phi} \frac{dt}{\sqrt{1-\frac{1}{2}\sin^2 t}}$ (elliptic integral), compute $F(23.5)$ by both N-G forward and backward formulae.

ϕ :	21	22	23	24	25	26
$F(\phi)$:	0.3706	0.3887	0.4068	0.425	0.4433	0.4616

Hint: F.D.: $\phi = 22$, $F(22) = 0.3887, 0.0181, 0.0001$

B.D.: $\phi = 25$, $F(25) = 0.4433$, 0.0183, 0, -0.0001.

Ans. Forward: $F(23.5) = 0.4159$, Backward: 0.41588.

32.5 CENTRAL DIFFERENCES

Observe that the Newton-Gregory forward difference formula involves only the leading differences $\Delta y_0, \Delta^2 y_0, \Delta^3 y_0$ etc. corresponding to the leading term (initial value) x_0. Thus only those values of function were used which lie on one side of the chosen initial value. Similarly, the Newton-Gregory backward difference formula used one-sided values $\nabla y_n, \nabla^2 y_n, \nabla^3 y_n$ which lie on one side of y_n. Interpolation formulae that contain both preceding and succeeding values of the funtion w.r.t., the initial value, give very useful results. For this purpose, the differences located in a horizontal row (line) corresponding to the initial values x_0 and y_0 in a diagonal difference table are used. These are known as central differences and are used in the Gauss, Stirling and Bessel interpolation formulae, **for interpolation near the middle of the table**, taken as the initial value. Central differences are defined by the operator δ as follows:

Table 32.7 Central differences table

x	y	δy	$\delta^2 y$	$\delta^3 y$	$\delta^4 y$	$\delta^5 y$	$\delta^6 y$
x_0	y_0						
		$\delta y_{\frac{1}{2}}$					
x_1	y_1		$\delta^2 y_1$				
		$\delta y_{\frac{3}{2}}$		$\delta^3 y_{\frac{3}{2}}$			
x_2	y_2		$\delta^2 y_2$		$\delta^4 y_2$		
		$\delta y_{\frac{5}{2}}$		$\delta^3 y_{\frac{5}{2}}$		$\delta^5 y_{\frac{5}{2}}$	
x_3	y_3		$\delta^2 y_3$		$\delta^4 y_3$		$\delta^6 y_3$
		$\delta y_{\frac{7}{2}}$		$\delta^3 y_{\frac{7}{2}}$		$\delta^5 y_{\frac{7}{2}}$	
x_4	y_4		$\delta^2 y_4$		$\delta^4 y_4$		
		$\delta y_{\frac{9}{2}}$		$\delta^3 y_{\frac{9}{2}}$			
x_5	y_5		$\delta^2 y_5$				
		$\delta y_{\frac{11}{2}}$					
x_6	y_6						

$$\delta y_{\frac{1}{2}} = y_1 - y_0, \delta y_{\frac{3}{2}} = y_2 - y_1, \ldots, \delta y_{n-\frac{1}{2}} = y_n - y_{n-1}$$

Higher order central differences are

$$\delta^2 y_1 = \delta y_{\frac{3}{2}} - \delta y_{\frac{1}{2}}, \delta^2 y_2 = \delta y_{\frac{5}{2}} - \delta y_{\frac{3}{2}}, \quad \text{etc.}$$

Note 1: The subscript of δy for any difference is the average of the subscripts of the two members of the difference.

Note 2: Differences on the same horizontal line have the same suffix.

Note 3: Odd order differences are known only for half values of the suffix (having fractional suffix) while even order differences for only integral values of the suffix.

Note 4: Observe that $\Delta y_0 = y_1 - y_0 = \nabla y_1 = \delta y_{\frac{1}{2}}$. Also $\Delta^3 y_2 = \nabla^3 y_5 = \delta^3 y_{\frac{7}{2}}$ etc. i.e., same numbers occur in the same positions but identified in different ways.

In order to derive the central-differences interpolation formulas of Gauss, Stirling and Bessel, we rearrange the central differences table with the *initial values x_0 and y_0 placed at the middle of the table* as shown below. Here

$$x_i = x_0 + ih \quad \text{for } i = 0, \pm 1, \pm 2, \ldots$$

$$y_i = f(x_i) \quad \text{and}$$

$$\Delta y_i = y_{i+1} - y_i,\ \Delta^2 y_i = \Delta y_{i+1} - \Delta y_i \quad \text{etc.}$$

32.6 STIRLING AND BESSEL'S INTERPOLATION FORMULAE

Gaussian Interpolation Formulae

Consider $(2n+1)$ equally spaced points

$$x_{-n}, x_{-(n-1)}, \ldots, x_{-1}, x_0, x_1, \ldots, x_{n-1}, x_n$$

where

$$\Delta x_i = x_{i+1} - x_i = h = \text{constant}$$

with $i = -n, -(n-1), \ldots, -1, 0, 1, \ldots, n-1, n$.

Let $P(x)$ be a polynomial of degree $2n$, given by

$$P(x) = a_0 + a_1(x-x_0) + \cdots + a_{2n}(x-x_{-(n-1)})\cdots(x-x_{-1}) \times (x-x_0)(x-x_1)\cdots(x-x_{n-1})(x-x_n).$$

If $P(x_i) = y_i = f(x_i)$ for $i = 0, \pm 1, \pm 2, \ldots, \pm n$ then $a_0 = y_0$, $a_1 = \frac{\Delta y_0}{h}$, $a_2 = \frac{\Delta^2 y_{-1}}{2!h^2}, \ldots, a_{2n} = \frac{\Delta^{2n} y_{-n}}{(2n)!h^{2n}}$.

Introducing $q = \frac{x-x_0}{h}$, we get Gauss first (GI) interpolation formula

$$P(x) = y_0 + q\Delta y_0 + \frac{q(q-1)}{2!}\Delta^2 y_{-1} + \frac{(q+1)q(q-1)}{3!}\Delta^2 y_{-1} + \frac{(q+1)q(q-1)(q-2)}{4!}\Delta^2 y_{-2} + \cdots + \frac{(q+n-1)\cdots(q-n)}{(2n)!}\Delta^{2n} y_{-n} \quad (1)$$

Similarly, Gaussian second (GII) interpolation formula is

$$P(x) = y_0 + q\Delta y_{-1} + \frac{(q+1)q}{2!}\Delta^2 y_{-1} + \frac{(q+1)(q-1)}{3!}\Delta^3 y_{-2} + \frac{(q+2)(q+1)q(q-1)}{4!}\Delta^4 y_{-2} + \cdots + \frac{(q+n)(q+n-1)\cdots(q-n+1)}{(2n)!}\Delta^{2n} y_{-n} \quad (2)$$

To shift the initial value from (x_0, y_0) to (x_1, y_1) replace q by $q-1$ and increase the indices in R.H.S. of (2) by (1). This yields another Gaussian

Table 32.8 Central differences table

x	y	Δy	$\Delta^2 y$	$\Delta^3 y$	$\Delta^4 y$	$\Delta^5 y$	$\Delta^6 y$
x_{-4}	y_{-4}						
		Δy_{-4}					
x_{-3}	y_{-3}		$\Delta^2 y_{-4}$				
		Δy_{-3}		$\Delta^3 y_{-4}$			
x_{-2}	y_{-2}		$\Delta^2 y_{-3}$		$\Delta^4 y_{-4}$		
		Δy_{-2}		$\Delta^3 y_{-3}$		$\Delta^5 y_{-4}$	
x_{-1}	y_{-1}		$\Delta^2 y_{-2}$		$\Delta^4 y_{-3}$		$\Delta^6 y_{-4}$
		Δy_{-1}		$\Delta^3 y_{-2}$		$\Delta^5 y_{-3}$	
x_0	y_0		$\Delta^2 y_{-1}$		$\Delta^4 y_{-2}$		$\Delta^6 y_{-3}$
		Δy_0		$\Delta^3 y_{-1}$		$\Delta^5 y_{-2}$	
x_1	y_1		$\Delta^2 y_0$		$\Delta^4 y_{-1}$		$\Delta^6 y_{-2}$
		Δy_1		$\Delta^3 y_0$		$\Delta^5 y_{-1}$	
x_2	y_2		$\Delta^2 y_1$		$\Delta^4 y_0$		
		Δy_2		$\Delta^3 y_1$			
x_3	y_3		$\Delta^2 y_2$				
		Δy_3					
x_4	y_4						

------ G_2
—— G_1
.... G_3

interpolation (GIII) formula

$$P(x)=y_1+(q-1)\Delta y_0+\frac{q(q-1)}{2!}\Delta^2 y_0 +\frac{q(q-1)(q-2)}{3!}\Delta^3 y_{-1} +\frac{(q+1)q(q-1)(q-2)}{4!}\Delta^4 y_{-1}+\cdots +\frac{(q+n-1)\cdots(q-n)}{(2n)!}\Delta^{2n} y_{-(n-1)} \quad (3)$$

Stirling's Interpolation Formula

is the arithmetic mean of GI and GII i.e., of (1) and (2)

$$P(x)= y_0+q\frac{\Delta y_{-1}+\Delta y_0}{2}+\frac{q^2}{2}\Delta^2 y_{-1} +\frac{q(q^2-1)}{3!}\frac{\Delta^3 y_{-2}+\Delta^3 y_{-1}}{2} +\frac{q^2(q^2-1)}{4!}\Delta^4 y_{-2} +\frac{q(q^2-1)(q^2-2^2)}{5!}\frac{\Delta^5 y_{-3}+\Delta^5 y_{-2}}{2}+\cdots \quad (4)$$

Bessel's Interpolation Formula

is the arithmetic mean of GI and GIII i.e., of (1) and (3).

$$P(x)=y_0+q\Delta y_0+\frac{q(q-1)}{2}\frac{\Delta^2 y_{-1}+\Delta^2 y_0}{2} +\frac{(q-\frac{1}{2})q(q-1)}{3!}\Delta^3 y_{-1} +\frac{q(q-1)(q+1)(q-2)}{4!}\frac{\Delta^4 y_{-2}+\Delta^4 y_{-1}}{2}+\cdots.$$

Note 1: It is advised to use Stirling's formula for $|q|\le 0.25$ and Bessel's formula for $0.25\le q\le 0.75$.

Note 2: Stirling's formula consists of the even differences along the central line and mean of the odd differences just above and below the central line

$$\begin{array}{ccccccc} & \Delta y_{-1} & & \Delta^3 y_{-2} & & \Delta^5 y_{-3} & \\ -y_0 --- & & \Delta^2 y_{-1} --- & & \Delta^4 y_{-2} --- & & \Delta^6 y_{-3} - \text{central line.} \\ & \Delta y_0 & & \Delta^5 y_{-1} & & \Delta^5 y_{-2} & \end{array}$$

Note 3: Bessel's formula consists of means of even differences along and below the central line and odd differences below the central line

$$\begin{array}{ccccccc} y_0 -- & \Delta^2 y_{-1} --- & & \Delta^4 y_{-2} --- & & \Delta^6 y_{-3} -- \text{central line.} \\ \Delta y_0 & & \Delta^3 y_{-1} & & \Delta^5 y_{-2} & \\ & \Delta^2 y_0 & & \Delta^4 y_{-1} & & \Delta^6 y_{-2}. \end{array}$$

WORKED OUT EXAMPLES

Example 1: Using *Stirling's* formula, compute $f(1.22)$ from the following data:

x :	1.0	1.1	1.2	1.3	1.4
$f(x)$:	0.841	0.891	0.932	0.963	0.985

Solution: Choose origin as $x_0=1.2$ given $h=$ interval size $=0.1$ then $q=\frac{x-x_0}{h}=\frac{1.22-1.2}{0.1}=0.2$ The central differences table is 32.9 below. Since $q=0:2$ lies in between $-\frac{1}{4}$ and $\frac{1}{4}$, Stirling's formula is applicable.

Table 32.9

x	q	y	Δy	$\Delta^2 y$	$\Delta^3 y$	$\Delta^4 y$
1.0	−2	$y_{-2}=0.841$				
			$\Delta y_{-2}=0.05$			
1.1	−1	$y_{-1}=0.891$		$\Delta^2 y_{-2}=-0.009$		
			$\Delta y_{-1}=0.041$		$\Delta^3 y_{-2}=-0.001$	
1.2	0	$y_0=0.932$		$\Delta^2 y_{-1}=-0.01$		$\Delta^4 y_{-2}=0.002$
			$\Delta y_0=0.031$		$\Delta^3 y_{-1}=+0.001$	
1.3	1	$y_1=0.963$		$\Delta^2 y_0=-0.009$		
			$\Delta y_1=0.022$			
1.4	2	$y_2=0.985$				

Stirling's formula is

$$y = y_0 + q\left(\frac{\Delta y_{-1} + \Delta y_0}{2}\right)$$
$$+\frac{q^2}{2!}\Delta^2 y_{-1} + \frac{q(q^2-1)}{3!}\left(\frac{\Delta^3 y_{-2} + \Delta^3 y_{-1}}{2}\right)$$
$$+\frac{q^2(q^2-1)}{4!}\Delta^4 y_{-2} + \cdots$$

Substituting the values from the table (underlined)

$$y(1.22) = 0.932 + 0.2\frac{(.041 + .031)}{2} + \frac{(0.2)^2}{2!}(-0.01)$$
$$+\frac{(0.2)[(0.2)^2-1]}{6}\frac{(-0.001+0.001)}{2}$$
$$+\frac{(0.2)^2[(0.2)^2-1]}{24}(0.002)$$
$$y(1.22) = 0.9389968.$$

Example 2: Use *Bessel's* formula to compute $f(1.95)$ from the following data:

x:	1.7	1.8	1.9	2.0	2.1	2.2	2.3
$f(x)$:	2.979	3.144	3.283	3.391	3.463	3.997	4.491

Solution: Choose the origin at $x_0 = 2.0$, given $h = 0.1$ so $q = \frac{x-x_0}{h} = \frac{1.95-2.0}{0.1} = -0.5$. Since $q = -0.5$ lies between $-\frac{1}{4}$ and $\frac{3}{4}$, Bessel's formula is applicable. The central difference table is:

Table 32.10

x	q	y	Δy	$\Delta^2 y$	$\Delta^3 y$	$\Delta^4 y$	$\Delta^5 y$
1.7	−3	2.979					
			0.165				
1.8	−2	3.144		−0.026			
			0.139		−0.005		
1.9	−1	3.283		−0.031		0	
			0.108		−0.005		0.503
2.0	0	3.391		−0.036		0.503	
			0.072		0.498		−1.503
2.1	1	3.463		0.462		−1.0	
			0.53		−0.502		
2.2	2	3.997		−0.04			
			0.494				
2.3	3	4.491					

The Bessel's formula is

$$y = y_0 + q\Delta y_0 + \frac{q(q-1)}{2!}\left(\frac{\Delta^2 y_{-1} + \Delta^2 y_0}{2}\right)$$
$$+\frac{\left(q-\frac{1}{2}\right)q(q-1)}{3!}\Delta^3 y_{-1}$$
$$+\frac{(q+1)(q)(q-1)(q-2)}{4!}\left(\frac{\Delta^4 y_{-2} + \Delta^4 y_{-1}}{2}\right)$$

Substituting the values from the table (underlined)

$$y(1.95) = 3.391 + (-0.5)(0.072)$$
$$+\frac{(-0.5)(-0.5-1)}{2}\frac{(-0.036+0.462)}{2}$$
$$+\frac{\left(-0.5-\frac{1}{2}\right)(-0.5)(-0.5-1)}{6}(0.498)$$
$$+\frac{(-0.5+1)(0.5)(-0.5-1)(-0.5-2)}{24}\frac{(0.503-1.0)}{2}$$
$$y(1.95) = 3.362917.$$

Exercise

1. Compute sinh 1.41710 from the following data:

x:	1.0	1.1	1.2	1.3	1.4
$\sinh x$:	1.1752	1.33565	1.50946	1.69838	1.90430
x:	1.5	1.6	1.7	1.8	
$\sinh x$:	2.12928	2.37557	2.64563	2.94217	

Hint: Use Stirlings with $x_0 = 1.4$, $y_0 = 1.90430$, differences 0.20592, 0.22498, 0.01906, 0.00206, 0.00225.

Ans. 1.94136

2. Calculate $f(0.5437)$ for the probability integral $f(x) = \frac{2}{\sqrt{\pi}}\int_0^x e^{-t^2}dt$ with the following data:

x:	0.51	0.52	0.53	0.54
$f(x)$:	0.5292437	0.5378987	0.5464641	0.5549392
x:	0.55	0.56	0.57	
$f(x)$:	0.5633233	0.5716157	0.5798158	

Hint: Use Stirlings with $x_0 = 0.54$, $y_0 = 0.5549392$, differences 0.0084751, 0.0083841, −0.0000910, −0.0000007, 0.

Ans. 0.5580520

3. Compute e^x when $x = 0.644$ by (a) Stirling; (b) Bessel's formula from the following data. Also find $e^{0.638}$ by (a) Stirling; (b) Bessel's formulae.

x :	0.61	0.62	0.63	0.64
e^x :	1.840431	1.858928	1.877610	1.896481
x :	0.65	0.66	0.67	
e^x :	1.9515541	1.934792	1.954237	

Hint:
a, b. Differences: $x_0 = 0.64$, $y_0 = 1.896481$, 0.018871, 0.019060, 0.000189, 0, 0.000002, 0.000002.
c. $x_0 = 0.64$, $x_n = 0.638$, $q = -0.2$
d. $x_0 = 0.63$, $x_n = 0.638$, $q = 0.8$

Ans. **a.** 1.904082
b. 1.904082
c. 1.892692
d. 1.89262

4. Find $f(16)$ by Stirling's formula from the following data:

x :	0	5	10	15	20	25	30
$f(x)$:	0	0.0875	0.1763	0.2679	0.364	0.4663	0.5774

Hint: $x_0 = 15$, $y_0 = 0.2679$, Differences: 0.0916, 0.0961, 0.0045, 0.0017, 0.0017, 0, −0.0002, 0.0009.

Ans. 0.2867

5. Compute $y(x = 5)$ using Bessel's formula

x :	0	4	8	12
y :	143	158	177	199

Ans. 162.41

6. Find $y(12.2)$ using Stirling's formula

x :	10	11	12	13	14
y :	0.23967	0.28060	0.31788	0.35209	0.38368

Hint: $x_0 = 12$, $h = 1$, $p = 12.2 - 12 = 0.2$, $y_0 = 0.31788$, Differences: 0.03728, 0.03421, −0.00365, −0.00307, 0.00058, −0.00045, −0.00013.

Ans. 0.32497

7. Compute $y(25)$ using Bessel's formula

x :	20	24	28	32
y :	2854	3162	3544	3992

Hint: $x_0 = 24$, $h = 4$, $p = \frac{25-24}{4} = \frac{1}{4}$; $y_0 = 3162$, Differences: 382, 74, 66, −8.

Ans. 3250.875

32.7 LAGRANGE'S INTERPOLATION

The interpolation formulae derived so far, are applicable only when argument x is equally spaced (equidistant $x_{i+1} - x_i = h =$ constant). Lagrange's interpolation formula is a more general one and can be applied for arbitrarily specified points i.e., for unequally spaced argument (i.e., $x_{i+1} - x_i = h =$ variable). For example tables of empirical data contian variable intervals.

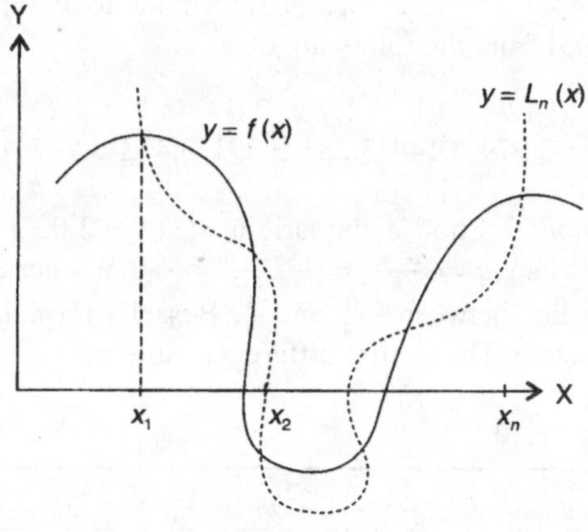

Fig. 32.11

In order to construct a polynomial $L_n(x)$ having (i) degree not exceeding n and (ii) satisfying the $(n+1)$ set of points i.e., $L_n(x_i) = y_i$ for $i = 0, 1, 2, \ldots, n$ choose

$$L_n(x) = \sum_{i=0} P_i(x) \cdot y_i \tag{1}$$

Geometrically, the second condition implies that the curve $y = L_n(x)$ meets the curve $y = f(x)$ at these $(n+1)$ points $x_0, x_1, x_2, \ldots, x_n$. Since the desired polynomial $L_i(x)$ vanishes at n points

$x_0, x_1, \ldots, x_{i-1}, x_{i+1,\ldots,x_n}$ and $L_i(x_i) = y_i$ choose

$$P_i(x) = c_i(x - x_0)(x - x_1)\cdots(x - x_{i-1}) \times \times (x - x_{i+1})\cdots(x - x_n) \quad (2)$$

where c_i is a constant coefficient. To satisfy the first condition, at $x = x_j$

$$L_n(x_j) = \sum_{i=0}^{n} P_i(x_j)y_j = y_j$$

we must have

$$P_i(x_j) = \delta_{ij} = \begin{Bmatrix} 1, \text{ if } & j = i \\ 0, \text{ if } & j \neq i \end{Bmatrix}$$

This determines c_i, put $x = x_i$ in (2), then

$$1 = P_i(x_i) = c_i(x_i - x_0)(x_i - x_1)\cdots(x_i - x_{i-1}) \cdot (x_i - x_{i+1})\cdots(x_i - x_n)$$

or

$$c_i = \frac{1}{(x_i - x_0)(x_i - x_1)\cdots(x_i - x_{i-1})(x_i - x_{i+1})\cdots(x_i - x_n)}$$

Substituting c_i in (2), we get the Lagrangian coefficient functions

$$P_i(x) = \frac{(x-x_0)(x-x_1)\cdots(x-x_{i-1})(x-x_{i+1})\cdots(x-x_n)}{(x_i-x_0)(x_i-x_1)\cdots(x_i-x_{i-1})(x_i-x_{i+1})\cdots(x_i-x_n)} \quad (3)$$

The degree of the polynomial $P_i(x)$ is not higher than n (i.e., $\leq n$). Substituting (3) in (1), we get the required Lagrange's interpolation formula

$$y = L_n(x) = \sum_{i=0}^{n} y_i \cdot \frac{(x-x_0)(x-x_1)\cdots(x-x_{i-1})(x-x_{i+1})\cdots(x-x_n)}{(x_i-x_0)(x_i-x_1)\cdots(x_i-x_{i-1})(x_i-x_{i+1})\cdots(x_i-x_n)}$$

or in the expanded form

$$y = L_n(x) = \frac{(x - x_1)(x - x_2)\cdots(x - x_n)}{(x_0 - x_1)(x_0 - x_2)\cdots(x_0 - x_n)} y_0$$

$$+ \frac{(x - x_0)(x - x_2)\cdots(x - x_n)}{(x_1 - x_0)(x_1 - x_2)\cdots(x_1 - x_n)} y_1 +$$

— — — — — — — — — — — — — — — — — — — —

$$+ \frac{(x-x_0)(x-x_1)\cdots(x-x_{i-1})(x-x_{i+1})\cdots(x-x_n)}{(x_i-x_0)(x_i-x_1)\cdots(x_i-x_{i-1})(x_i-x_{i+1})\cdots(x_i-x_n)} y_i$$

— — — — — — — — — — — — — — — — — — — —

$$+ \frac{(x-x_0)(x-x_1)(x-x_2)\cdots(x-x_{n-1})}{(x_n-x_0)(x_n-x_1)(x_n-x_2)\cdots(x_n-x_{n-1})} y_n \quad (4)$$

Note: If the points are equally spaced, Lagrange's interpolation polynomial coincids with the corresponding Newton's interpolation polynomial.

To determine y at a point x^*, put $x = x^*$ in (4). The unique polynomial (4) containing y_i explicitly is applicable for both unequally spaced and equally spaced points, abscissa $x_0, x_1, \ldots, x_n$ need not be in order, but inconvenient to move from one interpolation polynomial to another of degree one greater.

Note: The given function $y = f(x)$ can be split into partial fractions by dividing (4) throughout by $(x - x_0)(x - x_1)\cdots(x - x_i)\cdots(x - x_n)$ having $n + 1$ factors. Then

$$\frac{f(x)}{(x - x_0)\cdots(x - x_i)\cdots(x - x_n)}$$

$$= \frac{y_0}{(x_0 - x_1)(x_0 - x_2)\cdots(x_0 - x_n)} \cdot \frac{1}{(x - x_0)} +$$

$$+ \cdots \frac{y_1}{(x_1 - x_0)(x_1 - x_2)\cdots(x_1 - x_n)} \cdot \frac{1}{x - x_1} +$$

— — — — — — — — — — — — — — — — — — — —

$$+ \frac{y_n}{(x_n - x_0)(x_n - x_1)\cdots(x_n - x_{n-1})} \cdot \frac{1}{(x - x_n)}$$

$$= \frac{A_0}{x - x_0} + \frac{A_1}{x - x_1} + \cdots + \frac{A_n}{x - x_n}$$

where $A_0, A_1, \ldots, A_n$ coefficients are completely determined by the points $(x_0, y_0), (x_1, y_1) \cdots (x_n, y_n)$.

32.8 INVERSE INTERPOLATION USING LAGRANGE'S INTERPOLATION FORMULA

In the inverse interpolation, for a given value of y, the corresponding value of x is to be determined. For example to find value of the root x for which the function $y = f(x)$ becomes zero is an inverse interpolation problem. Inverse interpolation problem is similar to direct interpolation since the roles of x and y are interchanged.

Now interchanging the roles of x and y, the Lagrange's interpolation formula for inverse inter-

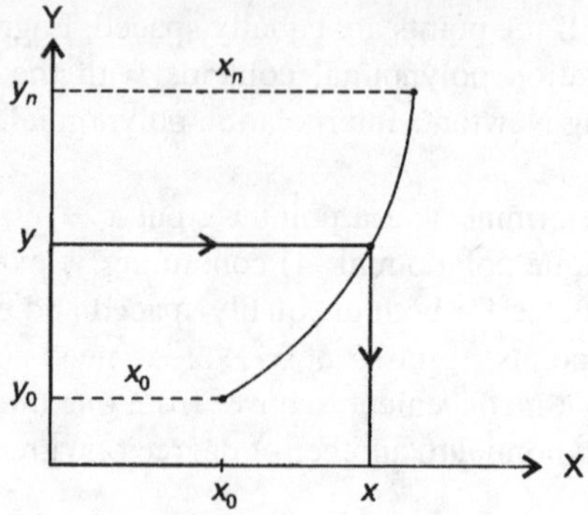

Fig. 32.12

polation takes the form given by

$$x=\sum_{i=0}^{n}\frac{(y-y_1)(y-y_2)\cdots(y-y_{i-1})(y-y_{i+1})\cdots(y-y_n)}{(y_i-y_1)(y_i-y_2)\cdots(y_i-y_{i-1})(y_i-y_{i+1})\cdots(y_i-y_n)}\cdot x_i$$

or in the expanded form

$$x=\frac{(y-y_1)(y-y_2)\cdots(y-y_n)}{(y_0-y_1)(y_0-y_2)\cdots(y_0-y_n)}x_0+$$
$$+\frac{(y-y_0)(y-y_2)\cdots(y-y_n)}{(y_1-y_0)(y_1-y_2)\cdots(y_1-y_n)}\cdot x_1+\cdots+$$
$$+\frac{(y-y_1)(y-y_2)\cdots(y-y_{i-1})(y-y_{i+1})\cdots(y-y_n)}{(y_i-y_1)(y_i-y_2)\cdots(y_i-y_{i-1})(y_i-y_{i+1})\cdots(y_i-y_n)}\cdot x_i+$$
$$+\cdots+\frac{(y-y_0)(y-y_1)\cdots(y-y_{n-1})}{(y_n-y_0)(y_n-y_1)\cdots(y_n-y_{n-1})}\cdot x_n \qquad (5)$$

Putting $y=y^*$ in (5), we get the value of x^*.

Worked Out Examples

Example 1: Use Lagrange's interpolation formula to fit a polynomial to the following data. Hence find $y(-2)$, $y(1)$ and $y(4)$.

x :	-1	0	2	3
y :	-8	3	1	2

Solution: Here $x_0=-1$, $x_1=0$, $x_2=2$, $x_3=3$ and $y_0=-8$, $y_1=3$, $y_2=1$, $y_3=2$. Using Lagrange's interpolation formula to this

$$y=f(x)=\frac{(x-0)(x-2)(x-3)}{(-1-0)(-1-2)(-1-3)}(-8)+$$
$$+\frac{(x+1)(x-2)(x-3)}{(0+1)(0-2)(0-3)}\cdot 3+\frac{(x+1)(x-0)(x-3)}{(2+1)(2-0)(2-3)}\cdot 1+$$
$$+\frac{(x+1)(x-0)(x-2)}{(3+1)(3-0)(3-2)}\cdot 2$$
$$=\frac{2}{3}x(x-2)(x-3)+\frac{1}{2}(x+1)(x-2)(x-3)-$$
$$-\frac{1}{6}x(x+1)(x-3)+\frac{1}{6}(x+1)x(x-2)$$
$$=(x-2)(x-3)\left(\frac{7x+3}{6}\right)+\frac{1}{6}x(x+1)$$
$$y=\frac{1}{6}[7x^3-31x^2+28x+18]$$

is the required 3rd degree polynomial.

$$y(-2)=\frac{1}{6}[-56-124-56+18]=-\frac{218}{6}=-36.33$$
$$y(1)=\frac{1}{6}[7-31+28+18]=\frac{22}{6}=3.666$$
$$y(4)=\frac{1}{6}[448-496+112+18]=\frac{82}{6}=13.666.$$

Partial fractions

Example 2: Express the function

$$\frac{x^2+6x-1}{(x^2-1)(x-4)(x-6)}$$

as a sum of partial fractions, using Lagrange's formula.

Solution: Consider $f(x)=x^2+6x-1$. Tabulate $f(x)$ at the roots of the denominator of the given expression i.e., at $x=-1,+1,4,6$.

x :	-1	1	4	6
$f(x)$:	-6	6	39	71

We fit a polynomial using Lagranges formula

$$f(x)=\frac{(x-1)(x-4)(x-6)}{(-1-1)(-1-4)(-1-6)}\cdot(-6)+$$
$$+\frac{(x+1)(x-4)(x-6)}{(1+1)(1-4)(1-6)}\cdot 6+\frac{(x+1)(x-1)(x-6)}{(4+1)(4-1)(4-6)}\cdot 39+$$
$$+\frac{(x+1)(x-1)(x-4)}{(6+1)(6-1)(6-4)}\cdot 71$$

$$f(x) = \frac{3}{35}(x-1)(x-4)(x-6) + \frac{1}{5}(x+1)(x-4)(x-6) - \frac{13}{10}(x+1)(x-1)(x-6) + \frac{71}{70}(x+1)(x-1)(x-4)$$

Dividing both sides by $(x^2 - 1)(x - 4)(x - 6)$,

$$\frac{f(x)}{(x^2-1)(x-4)(x-6)} = \frac{x^2+6x-1}{(x^2-1)(x-4)(x-6)} = \frac{3}{35}\frac{1}{x+1} + \frac{1}{5}\frac{1}{x-1} - \frac{13}{10}\frac{1}{x-4} + \frac{71}{70}\frac{1}{x-6}.$$

Inverse interpolation

Example 3: Compute the value of x, when $y = 8$ by inverse interpolation using Lagrange's formula

x :	-2	-1	1	2
y :	-7	2	0	11

Solution: Here $x_0 = -2, x_1 = 1, x_2 = 1, x_3 = 2$ and $y_0 = -7, y_1 = 2, y_2 = 0, y_3 = 11$. Now by Lagrange's formula, we have

$$x = \frac{(y-2)(y-0)(y-11)}{(-7-2)(-7-0)(-7-11)}(-2) + \frac{(y+7)(y-0)(y-11)}{(2+7)(2-0)(2-11)}(-1) + \frac{(y+7)(y-2)(y-11)}{(0+7)(0-2)(0-11)}(1) + \frac{(y+7)(y-2)(y-0)}{(11+7)(11-2)(11-0)} \cdot (2)$$

Put $y = 8$, then

$$x = \frac{6(8)(-3)}{(-9)(-7)(-18)}(-2) + \frac{(15)(8)(-3)}{(9)(2)(-9)}(-1) + \frac{(15)(6)(-3)}{7(-2)(-11)}(1) + \frac{(15)(6)8}{(18)(9)(11)}2$$

$$x = -\frac{8}{21} - \frac{20}{9} - \frac{135}{77} + \frac{80}{99} = -3.5483.$$

EXERCISE

Lagrange's interpolation

Use Lagrange's interpolation formula to solve the following problems:

1. Given (1, 2), (3, 5), (7, 12), (13, 20) find (4, ?).

Hint: $-\frac{27}{144}(2) + \frac{81}{80}(5) + \frac{27}{144}(12) - \frac{9}{720}(20)$.

Ans. 6.6875

2. Fit a polynomial of third degree and find $y(0.2)$ from the following data:

x_i :	0	0.1	0.3	0.5
y_i :	-0.5	0	0.2	1

Ans. -0.15004, $\frac{125}{3}x^3 - 30x^2 + \frac{73}{12}x - 0.5$

3. Fit a polynomial of 3rd degree

x :	0	1	3	4
y :	-12	0	6	12

Ans. $x^3 - 7x^2 + 18x - 12$

4. Determine $p(v = 21)$ given the following:

v :	10	15	22.5	33.75	50.625	75.937
p :	0.3	0.675	1.519	3.417	7.689	17.3

Ans. 1.323

5. Compute $f(9)$ from the following data:

x :	5	7	11	13	17
$f(x)$:	150	392	1452	2366	5202

Ans. $-\frac{50}{3} + \frac{3136}{15} + \frac{3872}{3} - \frac{2366}{3} + \frac{578}{5} = 810$

6. Determine $f(323.5)$ from the data given below:

x :	321.0	322.8	324.2	325.0
$f(x)$:	2.50651	2.50893	2.51081	2.51188

Ans. 2.50987

7. Given (300, 2.4771), (304, 2.4829), (305, 2.4843), (307, 2.4871), find (301, ?).

Ans. 2.4786

8. Compute $f(27)$ from the data below:

x :	14	17	31	35
$f(x)$:	68.7	64.0	44.0	39.1

Ans. 49.3.

Inverse interpolation

9. Determine $x(7)$ from the tabulated data:

x :	1	3	4
y :	4	12	19

Ans. ?

10. Find $x(12)$ using Lagrange's technique

x :	1.2	2.1	2.8	4.1	4.9	6.2
y :	4.2	6.8	9.8	13.4	15.5	19.6

Ans. 3.55

11. Let $y(x) = \frac{2}{\sqrt{11}} \int_0^x e^{-t^2} dt$ be the probability integral. Given:

x :	0.46	0.47	0.48	0.49
$y(x)$:	0.4846555	0.4937452	0.5027498	0.5116683

Determine for what value of x, $y(x) = \frac{1}{2}$.

Ans. 0.476937

12. Find x corresponding to $y = 10$

x :	10	15	17	20
y :	3	7	11	17

Ans. $x = 16.641$

13. Compute the value of x when $y = 15$

x :	5	6	9	11
y :	12	13	14	16

Ans. $\frac{5}{4} - 6 + \frac{27}{2} + \frac{11}{4} = \frac{46}{4} = 11.5$

14. Determine $x(0)$ by inverse interpolation

x :	1	2	2.5	3
y :	−6	−1	5.625	16

Ans. 2.122.

32.9 DIVIDED DIFFERENCES

In constructing the finite (forward and backward) differences tables, it is assumed that the independent variable x is equally spaced, i.e., $x_{i+1} - x_i = h =$ constant for $i = 0, 1, 2, \ldots$. When x is unequally spaced, with variable interval, the concept of finite differences is generalized to "divided differences", which takes into consideration (account) of the changes in the values of the argument x.

Let the tabulated function $y = f(x)$ consists of $(x_0, y_0), (x_1, y_1), \ldots, (x_n, y_n) \ldots$ with unequal intervals, $x_{i+1} - x_i \neq 0$ for $i = 0, 1, 2, \ldots$. Then the first order divided difference, denoted by $[x_i, x_{i+1}]$, is the ratio given by

$$[x_i, x_{i+1}] = \frac{y_{i+1} - y_i}{x_{i+1} - x_i}, \quad i = 0, 1, 2, \ldots$$

For example, $[x_0, x_1] = \frac{y_1 - y_0}{x_1 - x_0}$, $[x_1, x_2] = \frac{y_2 - y_1}{x_2 - x_1}$ etc

Similarly, the second order divided differences

$$[x_i, x_{i+1}, x_{i+2}] = \frac{[x_{i+1}, x_{i+2}] - [x_i, x_{i+1}]}{x_{i+2} - x_i}$$

for $i = 0, 1, 2, \ldots$

For example, $$[x_0, x_1, x_2] = \frac{[x_1, x_2] - [x_0, x_1]}{x_2 - x_0}.$$

In general, the nth order divided differences:

$$[x_i, x_{i+1}, \ldots, x_{i+n}] = \frac{[x_{i+1}, \ldots, x_{i+n}] - [x_i, \ldots, x_{i+n-1}]}{x_{i+n} - x_i}$$

for $n = 1, 2, 3, \ldots,$; $\quad i = 0, 1, 2, \ldots$.

Special case

If $x_{i+1} - x_i = h =$ constant i.e., when x is equally spaced, then divided differences reduce to forward (or backward) differences. Consider

$$[x_0, x_1] = \frac{y_1 - y_0}{x_1 - x_0} = \frac{1}{h}(y_1 - y_0) = \frac{1}{h}\Delta y_0,$$

$$[x_0, x_1, x_2] = \frac{[x_1, x_2] - [x_0, x_1]}{x_2 - x_0} = \frac{1}{2h}\left[\frac{\Delta y_1}{h} - \frac{\Delta y_0}{h}\right]$$

$$= \frac{1}{h^2 \cdot 2!}\Delta^2 y_0.$$

In general.

$$[x_0, x_1, x_2, \ldots, x_n] = \frac{1}{h^n n!}\Delta^n y_0 \quad \text{or}$$

$$[x_k, x_{k+i}, \ldots, x_{k+n}] = \frac{1}{h^n \cdot n!} \cdot \Delta^n y_k.$$

Properties

1. The divided differences operator denoted by $\Delta$is linear because $\Delta cf(x) = c\Delta f(x)$ and $\Delta(f(x) \pm g(x)) = \Delta f(x) \pm \Delta g(x)$, c being a constant.
2. The divided differences are symmetrical i.e.,

$$[x_0, x_1] = \frac{y_1 - y_0}{x_1 - x_0} = \frac{y_0}{x_0 - x_1} + \frac{y_1}{x_1 - x_0}$$

$$= \frac{y_0 - y_1}{x_0 - x_1} = [x_1, x_0]$$

Similarly,

$$[x_0, x_1, x_2] = \frac{[x_1, x_2] - [x_0, x_1]}{x_2 - x_0}$$

$$= \frac{1}{x_2 - x_0}\left\{\left[\frac{y_1}{x_1 - x_2} + \frac{y_2}{x_2 - x_1}\right] - \left[\frac{y_0}{x_0 - x_1} + \frac{y_1}{x_1 - x_0}\right]\right\}$$

$$= \frac{y_0}{(x_0 - x_1)(x_0 - x_2)} + \frac{y_1}{(x_1 - x_0)(x_1 - x_2)} + \frac{y_2}{(x_2 - x_0)(x_2 - x_1)}$$

$$= [x_1, x_2, x_0] = [x_2, x_0, x_1].$$

3. **Lemma:** If $y = P(x)$ is an nth degree polynomial, then its divided difference of nth order is constant. Consequently, the $(n + 1)$th order divided differences is zero i.e.,

$$[x, x_0, \ldots, x_{n-1}] = c = \text{constant and}$$

$$[x, x_0, \ldots, x_{n-1}, x_n] = \frac{c - c}{x - x_n} = 0$$

which follows from the fact that

$$[x, x_0] = \frac{P(x) - P(x_0)}{x - x_0}$$

is a polynomial of degree $(n - 1)$ in x, and

$$[x, x_0, x_1] = \frac{p(x, x_0) \quad p(x_0, x_1)}{(x - x_1)}$$

is a polynomial of degree $(n - 2)$ in x, and so on. Thus $[x, x_0, x_1, \ldots, x_n, x_{n-1}]$ is a polynomial of degree $(n - n)$ i.e., zero degree in x.

Note: Order of any divided difference is one less than the number of values of the argument in it i.e., 3rd order differences contain four values of x.

32.10 NEWTON'S DIVIDED DIFFERENCES FORMULA

When another interpolation point is added to the tabulated data, the Lagrangian coefficients are to be recalculated, resulting a different Lagrange's polynomial of higher degree. This difficulty is removed in the Newton's divided differences formula (or *Newton's general interpolation formula or Newton's interpolation formula for unequally spaced values of the argument*) in which polynomial of higher degree is obtained simply by addition of new terms (to the already existing formula).

Consider

$$[x, x_0] = \frac{y - y_0}{x - x_0} \quad \text{or}$$

$$y = y_0 + (x - x_0)[x, x_0] \tag{1}$$

Now $$[x, x_0, x_1] = \frac{[x, x_0] - [x_0, x_1]}{x - x_1} \quad \text{or}$$

$$[x, x_0] = [x_0, x_1] + (x - x_1)[x, x_0, x_1] \tag{2}$$

Substituting (2) in (1), we have

$$y = y_0 + (x - x_0)[x_0, x_1] + (x - x_0)(x - x_1)[x, x_0, x_1] \tag{3}$$

Again

$$[x, x_0, x_1, x_2] = \frac{[x, x_0, x_1] - [x_0, x_1, x_2]}{x - x_2} \quad \text{or}$$

$$[x, x_0, x_1] = [x_0, x_1, x_2] + (x - x_2)[x, x_0, x_1, x_2] \tag{4}$$

Substituting (4) in (3), we get

$$y = y_0 + (x-x_0)[x_0, x_1] + (x-x_0)(x-x_1)[x_0, x_1, x_2] + (x - x_0)(x - x_1)(x - x_2)[x, x_0, x_1, x_2].$$

Continuing this way, we arrive at the Newton's divided differences formula as

$$y = f(x) = y_0 + (x - x_0)[x_0, x_1] + (x - x_0)(x - x_1)[x_0, x_1, x_2] + (x - x_0)(x - x_1)(x - x_2)[x, x_0, x_1, x_2] + \cdots + (x-x_0)(x-x_1)\cdots(x-x_{n-1})[x_0, x_1, \ldots, x_n] \tag{5}$$

Note that the approximate relation (5) is obtained by suppressing the error term $E(x)$ where

$$E(x) = (x - x_0)\cdots(x - x_n)[x, x_0, x_1, \ldots, x_n].$$

Note: Various interpolation formulae obtained so far are 'different' forms of the same (unique) polynomial.

WORKED OUT EXAMPLES

Difference formula

Example 1: Find Newton's divided differences polynomial for the data in the table below. Also find $f(2.5)$

x :	−3	−1	0	3	5
$f(x)$:	−30	−22	−12	330	3458

Solution: The divided differences table is:

Table 32.11

x	$y=f(x)$	Divided differences of order 1	2	3	4
−3	−30				
		$\frac{-22-(-30)}{-1-(-3)}=\frac{8}{2}=4$			
−1	−22		$\frac{10-4}{0-(-3)}=\frac{6}{3}=2$		
		$\frac{-12-(-22)}{0-(-1)}=\frac{10}{1}=10$		$\frac{26-2}{3-(-3)}=4$	
0	−12		$\frac{114-10}{3-(-1)}=26$		$\frac{44-4}{5-(-3)}=5$
		$\frac{330-(-12)}{3-0}=114$		$\frac{290-26}{3-(-1)}=44$	
3	330		$\frac{1564-114}{5-0}=290$		
		$\frac{3458-330}{5-3}=1564$			
5	3458				

Here $x_0=-3,\ x_1=-1,\ x_2=0,\ x_3=3,\ x_4=5$, and $y_0=-30,\ y_1=-22,\ y_2=-12,\ y_3=330,\ y_4=3458$. The divided differences are

$$[x_0,x_1]=4,\ [x_1,x_2]=10,$$

$$[x_2,x_3]=114,\ [x_3,x_4]=1564,$$

$$[x_0,x_1,x_2]=\frac{[x_1,x_2]-[x_0,x_1]}{x_2-x_0}=2,\ [x_1,x_2,x_3]=26,$$

$$[x_2,x_3,x_4]=290.\quad \text{Now}$$

$$[x_0,x_1,x_2,x_3]=\frac{[x_1,x_2,x_3]-[x_0,x_1,x_2]}{x_3-x_0}=4$$

$$[x_1,x_2,x_3,x_4]=\frac{[x_2,x_3,x_4]-[x_1,x_2,x_3]}{x_4-x_1}=44$$

and finally

$$[x_0,x_1,x_2,x_3,x_4]=\frac{[x_1,x_2,x_3,x_4]-[x_0,x_1,x_2,x_3]}{x_4-x_0}=5$$

The Newton's divided difference polynomial is

$$y=f(x)=y_0+(x-x_0)[x_0,x_1]+(x-x_0)(x-x_1)[x_0,x_1,x_2]+$$
$$+(x-x_0)(x-x_1)(x-x_2)[x_0,x_1,x_2,x_3]+$$
$$+(x-x_0)(x-x_1)(x-x_2)(x-x_3)[x_0,x_1,x_2,x_3,x_4]$$

Substituting the above data

$$y=f(x)=-30+(x+3)(4)+(x+3)(x+1)(2)+$$
$$+(x+3)(x+1)(x-0)(4)+(x+3)(x+1)x(x-3)(5)$$
$$y=f(x)=5x^4+9x^3-27x^2-21x-12$$

is the required 4th degree polynomial. Now when $x=2.5$,

$$y=f(2.5)=5(2.5)^4+9(2.5)^3-27(2.5)^2-21(2.5)-12$$
$$=102.6785.$$

EXERCISE

Newton's divided differences formula

Using Newton's divided differences formula (Newton's general interpolation formula) solve the following problems:

1. Find $f(8)$.

x:	4	5	7	10	11	13
$f(x)$:	48	100	294	900	1210	2028

Hint: Divided differences are 52, 15, 1, 0.

Ans. 448

2. Fit a polynomial and find $f(1)$ and $f(8)$.

x:	−1	0	3	6	7
$f(x)$:	3	−6	39	822	1611

Hint: Divided differences are −9, 6, 5, 1.

Ans. $x^4-3x^3+5x^2-6$, $f(1)=-3$, $f(8)=2874$.

3. Find value of y for $x = 5.6075$.

x:	5.6	5.602	.5605	.5607	.5608
y:	.77556588	.77682686	.77871250	.779965	.78059114

Hint: Divided differences are 0.6305, −0.38668, −0.9485714, −20.595237.

Ans. 0.77729893

4. Find log 323.5.

x:	321.0	322.8	324.2	325.0
$\log x$:	2.50651	2.50893	2.51081	2.51188

Hint: Divided differences are 0.00134444, −0.00000158, −0.00000022.

Ans. 2.50987

5. Fit a cubic polynomial.

x:	0	1	2	5
$f(x)$:	2	3	12	147

Hint: Divided differences are 1, 4, 1.

Ans. $x^3 + x^2 - x + 2$

6. Obtain the Newton's divided difference interpolating polynomial and hence find $f(3)$.

x:	0	1	2	4	5	6
$f(x)$:	1	14	15	5	6	19

Hint: Divided differences are 13, −6, 1, 0, 0.

Ans. $x^3 - 9x^2 + 21x + 1$, $f(3) = 10$

7. Fit a cubic polynomial and find $f(6)$.

x:	3	7	9	10
$f(x)$:	168	120	72	63

Hint: Divided differences are −12, −2, 1.

Ans. $x^3 - 21x^2 + 119x - 27$, $f(6) = 147$

8. Form the table of divided differences:

x:	4	5	7	10	11	13
$f(x)$:	48	100	294	900	1210	2028

Extend the table to include values $x = 2$ and $x = 15$.

Hint: Divided differences: 52, 15, 1, 0, 0.

Ans. $x^3 - x^2$, $y(2) = 4$, $y(15) = 3150$.

9. Fit a polynomial to the data (−4, 1245), (−1, 33), (0, 5), (2, 9), (5, 1335). Hence find $f(1)$ and $f(7)$.

Ans. $3x^4 - 5x^3 + 6x^2 - 14x + 5$, $f(1) = -5$, $f(7) = 5689$

10. Calculate $f(1.5)$.

x:	1	2	3	4	5
$f(x)$:	0	7	26	63	124

Hint: Divided differences are 7, 12, 6, 0.

Ans. $f(1.5) = 2.25$

11. Calculate $f(9)$.

x:	5	7	11	13	17
$f(x)$:	150	392	1452	2366	5202

Hint: Divided differences are 121, 24, 1.

Ans. 810

12. Determine the Newton's general interpolation polynomial and hence find $f(-3)$ and $f(9)$.

x:	−4	−1	0	2	5
$f(x)$:	1245	33	5	9	1335

Hint: Divided differences are −404, 94, −14, 3.

Ans. $3x^4 - 5x^3 + 6x^2 - 14x + 5$, $f(-3) = 479$, $f(9) = 16403$.

32.11 ERRORS IN POLYNOMIAL INTERPOLATION

Computing of values for a tabulated function at points *not* in the table is known as interpolation. Thus, interpolation means to estimate a missing function value by taking a weighted average of known (given) function values at neighbouring points.

Suppose the function $y = f(x)$ is known (given) at $(n+1)$ points $(x_0, y_0), (x_1, y_1) \dots (x_n, y_n)$. A polynomial of degree n, $P_n(x)$ is known as the interpolating polynomial when $P_n(x)$ passes through (satisfies) these $(n+1)$ points. $P_n(x)$ can be constructed using only the numerical values x_k and y_k and does not need any order derivatives of $f(x)$. The approximation $P_n(x)$ is known as interpolated value when $x_0 < x < x_n$ and extrapolated value when $x < x_0$ or $x_n < x$.

The error is zero at tabulated points $x_0, x_1, \ldots, x_n$; while it may be nonzero at points *not* in the table. Error may increase when number of interpolation is increased (classical example: $f(x) = (1+25x^2)^{-1}$ in $[-1, 1]$).

The error term associated with the nth degree polynomial $P_n(x)$ is obtained by using the generalized Rolle's theorem and is given by

$$E(x) = f(x) - P_n(x)$$

$$= (x-x_0)(x-x_1)\ldots(x-x_n)\cdot\frac{f^{(n-1)}(\xi)}{(n+1)!} \quad (1)$$

Here ξ is any value in the smallest interval that contains $\{x, x_0, x_1, \ldots, x_n\}$. The disadvantage with this error expression is that the derivative of f is not known (since the function $f(x)$ itself is not known except at the tabulated points). Note that the error term (1) is same for Lagrange polynomial and interpolating polynomial obtained from divided-difference table.

Error estimation when $f(x)$ is unknown: When $f(x)$ is unknown, $\frac{f^{(n)}(x)}{n!}$ can be approximated by the nth order divided difference

i.e., $\frac{f^{(n)}(x)}{n!} = f[x_0, x_1, \ldots, x_n] = f_0^{[n]} = n$th order divided difference.

Thus the error of the interpolation is approximately given by the next term that would be added.

The *next-term* rule is a most valuable rule for estimating the error of interpolation as follows:

$E_n(x) \approx$ value of the next term that would be added to $P_n(x)$.

$$= f_0^{[n]}(x-x_0)(x-x_1)(x-x_2)\ldots(x-x_{n-1}).$$

The interpolating polynomial that fits a divided difference table at $x = x_0, x_1, x_2, \ldots, x_n$ is given by

$$P_n(x) = f_0^{[0]} + (x-x_0)f_0^{[1]} + (x-x_0)(x-x_1)f_0^{[2]}$$
$$+(x-x_0)(x-x_1)(x-x_3)f_0^{[3]} + \ldots$$
$$+(x-x_0)(x-x_1)\ldots(x-x_{n-1})f_0^{[n]}$$

Thus for $P_1(x)$ the next term to be added is $f_0^{[2]}\cdot(x-x_0)(x-x_1)$.

Worked Out Examples

Example 1: Find the error of the interpolates for $f(1.25)$ using polynomial of degree 1, 2, 3, 4, (*b*). Estimate the error using the next-term rule. Here $f(x) = e^{x^2-1}$ and $x_0 = 1$, $x_1 = 1.1$, $x_2 = 1.2$, $x_3 = 1.3$, $x_4 = 1.4$

Solution: The divided difference table is

x_i	$f(x_i)$	$f_i^{[1]}$	$f_i^{[2]}$	$f_i^{[3]}$	$f_i^{[4]}$
1	1	2.3368	4.2675	6.105	7.65
1.1	1.23368	3.1903	6.099	9.165	
1.2	1.55271	4.4101	8.8485		
1.3	1.99372	6.1798			
1.4	2.61170				

Here

$$f_0^{[1]} = f[x_0, x_1] = \frac{f_1 - f_0}{x_1 - x_0},$$

$$f_0^{[2]} = f[x_0, x_1, x_2] = \frac{f(x_1, x_2) - f(x_0, x_1)}{x_2 - x_0} \text{etc.}$$

The interpolating polynomials of degree 1, 2, 3 are

$$P_1(x) = f_0 + f_0^{[1]}(x-x_0) = 1 + 2.3368(x-1)$$

$$P_2(x) = f_0 + f_0^{[1]}(x-x_0) + f_0^{[2]}(x-x_0)(x-x_1)$$

$$P_2(x) = 1 + 2.3368(x-1) + 4.2675(x-1)(x-1.1)$$

$$P_3(x) = f_0 + f_0^{[1]}(x-x_0) + f_0^{[2]}(x-x_0)(x-x_1)$$
$$+ f_0^{[3]}(x-x_0)(x-x_1)(x-x_2)$$
$$= 1 + 2.3368(x-1) + 4.2675(x-1)(x-1.1)$$
$$+ 6.105(x-1)(x-1.1)(x-1.2)$$

$$P_4(x) = 1 + 2.3368(x-1) + 4.2675(x-1)(x-1.1)$$
$$+6.105(x-1)(x-1.1)(x-1.2)$$
$$+7.65(x-1)(x-1.1)(x-1.2)(x-1.3)$$

Interpolated values at $x = 1.25$ are

$P_1(1.25) = 1.5842$, $P_2(1.25) = 1.7442$,

$P_2(1.25) = 1.7556468$, $P_4(1.25) = 1.75496$;

The exact value $f(1.25) = e^{(1.25)^2-1}$
$= 1.7550546571$

Here Ist, 2nd, 3rd, 4th, 5th derivatives are

$f'(x) = 2xe^{x^2-1}$, $f'' = 2(1+2x)e^{x^2-1}$,

$f''' = 4(x + 2x^2 + 1)e^{x^2-1}$,

$f^{IV} = 4(4x^3 + 2x^2 + 6x + 1)e^{x^2-1}$;

$f^{V} = 8e^{x^2-1}(4x^4 + 2x^3 + 12x^2 + 3x + 3)$.

Error using

$$E_n(x) = \frac{f^{(n+1)}(\xi)}{(n+1)!}(x - x_0)(x - x_1)\cdots(x - x_n)$$

with $\xi = 1.3$.

$E_1(x) = 0.89717$

$E_2(x) = 0.0035388$, $E_3(x) = -0.000130682$

$E_4(x) = -0.0000133947$

Errors of interpolation for $f(1.25)$

Degree	Interpolated value	Actual error	Error	
			using $E_n(x)$	next-term rule
1	1.5842	0.17085467	0.89717	(4.2675)(1.25–1) × (1.25–1.1) = 0.1600
2	1.7442	0.01085467	– 0.0035388	0.0114
3	1.7556468	– 0.00059213	– 0.00013068	– 0.000717
4	1.75496	0.00009467	– 0.000013394	

Exact value: $f(1.25) = 1.755054657$

EXERCISE

1. Find the error of the interpolates for $f(1.75)$ using polynomials of degree 1, 2, 3. (b) Estimate the error using the next-term rule. Here $f(x) = x^2e^{-x/2}$.

Ans. Divided difference table

x_i	$f(x_i)$	$f_i^{[1]}$	$f_i^{[2]}$	$f_i^{[3]}$	$f_i^{[4]}$
1.10	0.6981	0.8593	−0.1755	0.0032	0.0027
2.00	1.4715	0.4381	−0.1631	0.0191	
3.50	2.1287	−0.0511	−0.0657		
5.00	2.0521	−0.2877			
7.10	1.4480				

Errors of interpolation for $f(1.75)$

Degree	Interpolated value	Actual error	Estimate from next-term rule	Max. f^{i+1}	Min. f^{i+1}	Upper bound	Lower bound
1.	1.25668	0.01996	0.02852	−.3679	.0594	.0299	−.00483
2.	1.28520	−0.00856	0.00091	−0.8661	0.1249	.0059	−0.0408
3.	1.28611	−0.00947	−0.00249	1.1398	−0.0359	.0014	−0.0439

2. Find an approximation to $\sin 0.34$ and find a bound for the error using a polynomial or degree 3 and given that $\sin 0.30 = 0.29552$, $\sin 0.32 = 0.31457$, $\sin 0.33 = 0.32404$, $\sin 0.35 = 0.34290$.

Ans. $\sin 0.34 = 0.33348$, Error bound: 1.2×10^{-9}

3. Find a bound for the error in approximation $f(1.09)$ using 3rd degree polynomial and the data $f(1.00) = 0.1924$, $f(1.05) = 0.2414$ $f(1.10) = 0.2933$, $f(1.15) = 0.3492$ for the function $f(x) = \log_{10} \tan x$.

Ans. $f(1.09) = 0.2826$, Error bound: 7.4×10^{-6}

4. Approximate $f(1.03)$ using interpolating polynomial of degree 1, 2, 3 using $x_0 = 1$, $x_1 = 1.05$, $x_2 = 1.07$, $x_3 = 1.1$, $x_4 = 1.15$ compare the actual error to the error bound and to error obtained by next-term rule of $f(x) = 3xe^x - e^{2x}$.

5. Write the error term $E_3(x)$ for cubic Lagrange interpolation to $f(x) = x^5 - 5x^4$ where interpolation is to be exact at the four nodes $x_0 = -1$, $x_1 = 0$ $x_2 = 3$, $x_3 = 4$.

Ans. $f^{(4)}(c) = 120(c - 1)$ for all c, so that $E_3(x) = 5(x + 1)(x - 3)(x - 4)(c - 1)$

6. For $f(x) = \sin x$ on $[0, 1]$, determine the step size h so that linear Lagrange interpolation has an accuracy of 10^{-6}.

Ans. $|E_1(x)| = 5 \times 10^{-7}$, $|f^{(2)}(c)| \leq |-\sin(1)| = 0.84147098 = M_2$, $h^2 \frac{M_2}{8} = \frac{h^2}{8}(0.84147098) < 5 \times 10^{-7}$ or $h < 0.00218027$

7. Let $f(x) = \log(1 + x^2)$. Consider the following divided difference table. Estimate the error for interepolate $f(2)$ using a quadric interpolate with the points 1.91, 1.97, 2.02.

x_i	$f(x_i)$	$f_i^{[1]}$	$f_i^{[2]}$	$f_i^{[3]}$
1.88	1.511693			
1.91	1.536459	0.825333		
1.97	1.585330	0.8145167	−0.1224067	
2.02	1.625390	0.8012000	−0.1210609	0.96128×10^{-2}
2.11	1.696001	0.7845555	−0.1188893	1.0858×10^{-2}
2.18	1.749617	0.7659571	−0.1162400	1.26157×10^{-2}

Ans. $P_2(x) = 1.536459 + 0.8145167(x - 1.91)$
$-0.1210609(x - 1.91)(x - 1.97)$
Error $\sim 10^{-2}(0.09)(0.03)(0.02) = 0.54 \times 10^{-6}$
Interpolated value: $P^2(2) = 1.609439$
Exact value: $f(2) = \log 5 = 1.6094379$

Note: Third order divided differences are all approximately equal to 10^{-2}.

8. Develop a divided differences table for the following data.

x:	0.3	0.42	0.5	0.58	0.66	0.72
y:	1.04403	1.08462	1.11803	1.15603	1.19817	1.23223

Estimate error at $x = 0.55$ using linear interpolation, at $x = 0.60$ using quadratic interpolation, at $x = 0.46$ using cubic interpolation. Exact values are $f(0.46) = 1.10073$, $f(0.55) = 1.14127$, $f(10.6) = 1.16619$.

Ans. $P_1(0.55) = 1.14178$ based on 0.5, 0.58
$P_2(0.6) = 1.16618$ based on 0.5, 0.58, 0.66
$P_3(0.46) = 1.04403$ based on 0.3, 0.42, 0.5, 0.58

32.12 SYMBOLIC RELATIONS AND SEPARATION OF SYMBOLS

Let $x_0, x_1, \ldots, x_n$ be a set of tabular points which are equally spaced i.e.,

$$x_i = x_0 + ih, i = 0, 1, 2, \ldots n.$$

Then

$$\nabla f(x_i) = f(x_i) - f(x_i - h)$$

is the backward-difference operator,

$$\Delta f(x_i) = f(x_i + h) - f(x_i)$$

is the forward-difference operator

$$\delta f(x_i) = f\left(x_i + \frac{h}{2}\right) - f\left(x_i - \frac{h}{2}\right)$$

is the central-difference operator. The shift operators are defined as

$$E(f(x_i)) = f(x_i + h)$$
$$E^{-1}(f(x_i)) = f(x_i - h)$$

	Δ	∇	E	δ
Δ	Δ	$(1-\nabla)^{-1} - 1$	E^{-1}	$\frac{1}{2}\delta^2 + \delta\sqrt{1 + \frac{1}{4}\delta^2}$
∇	$1 - (1+\Delta)^{-1}$	∇	$1 - E^{-1}$	$-\frac{1}{2}\delta^2 + \delta\sqrt{1 + \frac{1}{4}\delta^2}$
E	$\Delta + 1$	$(1-\nabla)^{-1}$	E	$\frac{1}{2}\delta^2 + \delta\sqrt{1 + \frac{1}{4}\delta^2} + 1$
δ	$\Delta(1+\Delta)^{-\frac{1}{2}}$	$\nabla(1-\nabla)^{-\frac{1}{2}}$	$E^{\frac{1}{2}} - E^{-\frac{1}{2}}$	δ
μ	$(1+\frac{\Delta}{2})(1+\Delta)^{\frac{1}{2}}$	$(1-\frac{\nabla}{2})(1-\nabla)^{-\frac{1}{2}}$	$\frac{E^{\frac{1}{2}}+E^{-\frac{1}{2}}}{2}$	$\sqrt{1 + \frac{1}{4}\delta^2}$

The averaging operator is defined as

$$\mu(f(x_i)) = \frac{1}{2}\left[f\left(x_i + \frac{h}{2}\right) + f\left(x_i - \frac{h}{2}\right)\right]$$

Higher order differences are obtained by repeated applications of the difference operators. Let $f_i = f(x_i)$. Then

$$\nabla^n f(x_i) = \nabla^{n-1} f_i - \nabla^{n-1} f_{i-1}$$
$$\Delta^n f(x_i) = \Delta^{n-1} f_{i+1} - \Delta^{n-1} f_i$$
$$E^n f(x_i) = f(x_i + nh)$$
$$E^{-n} f(x_i) = f(x_i - nh)$$
$$\delta^n f(x_i) = \delta^{n-1} f_{i+\frac{1}{2}} - \delta^{n-1} f_{i-\frac{1}{2}}$$
$$= \sum_{j=0}^{n} (-1)^j \frac{n!}{j!(n-j)!} f_{i+\frac{n}{2}-j}$$

Note that the differences $\Delta^k f_0$ lie on a straight line sloping downward to the right. The differences $\nabla^k f_3$ lie on a straight line sloping upward to the right. The differences $\delta^{2k} f_2$ lie on a horizontal line. Thus the entries in the difference table can be interpreted as either forward or backward differences.

Symbolic relations are established using these symbolic operators. Standing alone, all these operators are without any significance and are meaningless unless operated on a function, like $\int$ or $\frac{d}{dx}$ or $\sqrt{\ }$. These operators are linear operators i.e., say

$$\Delta(c_1 f_1 + c_2 f_2) = c_1 \Delta f_1 + c_2 \Delta f_2$$

and obey laws of algebra.

Relationship between the Operators

Some of the above relations can be established as follows:

(a) $\Delta(f(x_0)) - f(x_0 + h) - f(x_0) = E(f(x_0)) - f(x_0)) = (E - 1)(f(x_0))$
Thus $\Delta = (E - 1)$

(b) $\nabla(f(x)) = f(x) - f(x + h) = f(x) - E^{-1}(f(x)) = (1 - E^{-1})(f(x))$
Thus $\nabla = (1 - E^{-1})$

(c) $\delta(f(x)) = f\left(x + \frac{h}{2}\right) - f\left(x - \frac{h}{2}\right) =$
$= E^{\frac{1}{2}} f(x) - E^{-\frac{1}{2}} f(x) = \left(E^{\frac{1}{2}} - E^{-\frac{1}{2}}\right)(f(x))$
Thus $\delta = \left(E^{\frac{1}{2}} - E^{-\frac{1}{2}}\right)$

(d) $\mu(f(x)) = \frac{1}{2}\left[f\left(x + \frac{h}{2}\right) + f\left(x - \frac{h}{2}\right)\right]$

$$= \frac{1}{2}\left[E^{\frac{1}{2}} f(x) + E^{-\frac{1}{2}} f(x)\right]$$

$$= \frac{1}{2}\left[E^{\frac{1}{2}} + E^{-\frac{1}{2}}\right](f(x))$$

Thus $\delta = E^{\frac{1}{2}} - E^{-\frac{1}{2}}$

Similarly other results in the above table can be proved.

Now using laws of algebra, we can prove the following.

(i) $E\nabla = E(1 - E^{-1}) = E - 1 = \Delta$
$\nabla E = (1 - E^{-1})E = E - 1 = \Delta$
Thus $E^n\nabla^n = \nabla^n E^n = \Delta^n$

(ii) Since $\Delta = E - 1$

$$\Delta^n = (E-1)^n = E^n - nE^{n-1} + \binom{n}{2}E^{n-2} - \binom{n}{3}E^{n-3} + \ldots$$

$$\Delta^n f(x) = E^n f - nE^{n-1} f + \binom{n}{2}E^{n-2} f - \binom{n}{3}E^{n-3} f + \ldots$$
$$= f(x+nh) - nf[x+(n-1)h] + \binom{n}{2} f[x+(n-2)h] + \ldots$$

Thus

$$\Delta^n f = f_n - nf_{n-1} + \binom{n}{2} f_{n-2} - \binom{n}{3} f_{n-3} + \ldots$$

(iii) *Newton-Gregory forward formula*: We know that $E = 1 + \Delta$, so $E^n = (1+\Delta)^n$
Then

$$f_n = E^n f_0 = (1+\Delta)^n f_0$$
$$= [1 + n\Delta + \binom{n}{2}\Delta^2 + \binom{n}{3}\Delta^3 + \ldots] f_0$$
$$f_n = f_0 + n\Delta f_0 + \binom{n}{2}\Delta^2 f_0 + \binom{n}{3}\Delta^3 f_0 + \ldots$$

(iv) *Newton-Gregory backward formula*
Since $E = (1-\nabla)^{-1}$, $E^n = (1-\nabla)^{-n}$. Then

$$f_n = E^n f_0 = (1-\nabla)^{-n} f_0$$
$$= \left[1 + n\nabla + \binom{n+1}{2}\nabla^2 + \binom{n+2}{3}\nabla^3 + \ldots\right] f_0$$
$$f_n = f_0 + n\nabla f_0 + \binom{n+1}{2}\nabla^2 f_0 + \binom{n+2}{3}\nabla^3 f_0 + \ldots$$
$$= f_0 + n\nabla f_{-1} + \binom{n+1}{2}\Delta^2 f_{-2} + \binom{n+2}{3}\Delta^3 f_{-3} + \ldots$$

Worked Out Examples

Example 1: Prove that $\nabla^6 y_8 = \Delta^6 y_2$.

Solution: $\nabla^6 = (1 - E^{-1})^6$
$= (1 - 6E^{-1} + 15E^{-2} - 20E^{-3} + 15E^{-4} - 6E^{-5} + E^{-6})$

So

$$\nabla^6 y_8 = (1 - 6E^{-1} + 15E^{-2} - 20E^{-3} + 15E^{-4} - 6E^{-5} + E^{-6})y_8$$
$$\nabla^2 y_8 = y_8 - 6y_7 + 15y_6 - 20y_5 + 15y_4 - 6y_3 + y_2$$

Also

$$\nabla^6 y_2 = (E-1)^6 y_2 = (E^6 - 6E^5 + 15E^4 - 20E^3 + 15E^2 - 6E + 1)y_2$$
$$= y_8 - 6y_7 + 15y_6 - 20y_5 + 15y_4 - 6y_3 + y_2$$

Thus $\nabla^6 y_8 = \Delta^6 y_2$

Example 2: Show that $\nabla^r f_k = \Delta^r f_{k-r}$

Solution:
RHS $= \Delta^r f_{k-r} = (E-1)^r f_{k-r}$ $(\because \Delta = E - 1)$
$= E^r(1 - E^{-1})^r f_{k-r}$ $(\because \nabla = 1 - E^{-1})$
$= E^r\nabla^r f_{k-r} = \nabla^r E^r f_{k-r} = \nabla^2 f_k$
$\because$ $E^r f_{k-r} = f_{k-r+r}$

Example 3: Show that $\delta = 2\sin h\left(\frac{hD}{2}\right)$.

Solution: $\delta = E^{\frac{1}{2}} - E^{-\frac{1}{2}} = e^{\frac{hD}{2}} - e^{-\frac{hD}{2}}$
$(\because E = e^{hD})$ $= 2\sin h\left(\frac{hD}{2}\right)$.

Example 4: Prove that $(\Delta + 1)(1 - \nabla) = 1$

Solution: Since $\Delta = E - 1$, $\Delta + 1 = E$ and $\nabla = 1 - E^{-1}$ so $1 - \nabla = E^{-1}$
Then $(\Delta + 1)(1 - \nabla) = E \cdot E^{-1} = 1$

Example 5: Show that $\Delta\nabla = \delta^2$

Solution:
$$\begin{aligned}\Delta\nabla &= (E-1)(1-E^{-1})\\ &= E-1-1+E^{-1}\\ &= (E-E^{-1})^2\end{aligned}$$

Now $\delta = (E^{\frac{1}{2}} - E^{-\frac{1}{2}})$

$$\begin{aligned}\delta^2 &= (E^{\frac{1}{2}} - E^{-\frac{1}{2}})^2 = (E-2+E^{-1})\\ &= (E-E^{-1})^2\end{aligned}$$

Hence $\Delta\nabla = \delta^2$

Example 6: Show that $\frac{\Delta^2}{E}x^3 = 6xh^2$

Solution:

LHS $= \frac{\Delta^2}{E}x^3 = \frac{(E-1)^2}{E}x^3$

$$\begin{aligned}&= \left(\frac{E^2+1-2E}{E}\right)x^3 = (E-2+E^{-1})x^3\\ &= (x+h)^3 - 2x^3 + (x-h)^3\\ &= (x^3+3x^2h+3xh^2+h^3) - 2x^3\\ &\quad +(x^3-3x^2h+3xh^2-h^3)\\ &= 6xh^2\end{aligned}$$

Example 7: Show that $(\Delta+\nabla)^2(x^2+x) = 8$ with $h = 1$

Solution: Since $\Delta = E - 1$ and $\nabla = 1 - E^{-1}$. We have

$$\begin{aligned}(\Delta+\nabla)^2(x^2+x) &= (E-1+1-E^{-1})^2 = (E-E^{-1})^2\\ &= E^2+E^{-2}-2\end{aligned}$$

So

$$\begin{aligned}(\Delta+\nabla)^2 &= (E^2+E^{-2}-2)(x^2+x)\\ &= E^2(x^2+x)+E^{-2}(x^2+x)-2(x^2+x)\\ &\quad [(x+2)^2+(x+2)]+[(x-2)^2+(x-2)]\\ &\quad -2(x^2+x) = 8\end{aligned}$$

Example 8: Show that $u_0 + Eu_0 \cdot x + E^2u_0 \cdot x^2 + E^3u_0 \cdot x^3 + \ldots$

$$= \left[\frac{1}{1-x} + \frac{x}{(1-x)^2}\Delta + \frac{x^2}{(1-x)^3}\Delta^2 + \ldots\right]u_0$$

Solution: Rewriting LHS as

$$\text{LHS} = (1 + xE + x^2E^2 + x^3E^3 + \ldots)u_0$$

$$= \left(\frac{1}{1-xE}\right)u_0 = \left[\frac{1}{1-x(1+\Delta)}\right]u_0$$

since $E = 1 + \Delta$

$$\begin{aligned}&= \left[\frac{1}{1-x-x\Delta}\right]u_0 = \frac{1}{1-x}\left[\frac{1}{1-\frac{x\Delta}{1-x}}\right]u_0\\ &= \frac{1}{1-x}[1+\frac{x\Delta}{1-x}+\frac{x^2\Delta^2}{(1-x)^2}+\ldots]u_0\\ &= \left[\frac{1}{1-x}+\frac{x}{(1-x)^2}\Delta+\frac{x^2}{(1-x)^3}\Delta^2+\ldots\right]u_0\end{aligned}$$

Example 9: Show that $\delta E^{\frac{1}{2}} = \Delta$

Solution: $E^{\frac{1}{2}}f(x) = f(x+\frac{h}{2})$

$$\begin{aligned}\delta E^{\frac{1}{2}}(f(x)) &= \delta[f(x+\tfrac{h}{2})]\\ &= f(x+\tfrac{h}{2}+\tfrac{h}{2}) - f(x+\tfrac{h}{2}-\tfrac{h}{2})\\ &= f(x+h)-f(x)\\ &= \Delta f(x)\end{aligned}$$

Example 10: Prove that $\nabla - \Delta = -\Delta\nabla$

Solution:

$$\begin{aligned}\text{LHS}: \nabla - \Delta &= (1-E^{-1})-(E-1)\\ &= 2-E-E^{-1}\\ \text{RHS}: -\Delta\nabla &= -(E-1)(1-E^{-1})\\ &= -[E-1-1+E^{-1}]\\ &= 2-E-E^{-1}\end{aligned}$$

Example 11: Show that $\nabla + \Delta = \frac{\Delta}{\nabla} - \frac{\nabla}{\Delta}$

Solution:
$$\begin{aligned}\text{LHS} = \nabla+\Delta &= (1-E^{-1})+(E-1)\\ &= E-E^{-1}\end{aligned}$$

Now

$$\frac{\Delta}{\nabla} = \frac{E-1}{1-E^{-1}} = \frac{E(E-1)}{(E-1)} = E$$

$$\frac{\nabla}{\Delta} = \frac{1-E^{-1}}{E-1} = \frac{1}{E}\left(\frac{1-E^{-1}}{1-E^{-1}}\right) = E^{-1}$$

$$\text{RHS}: \frac{\Delta}{\nabla} - \frac{\nabla}{\Delta} = E - E^{-1}$$

EXERCISE

1. Show that $e^x = \left(\frac{\Delta^2}{E}\right)e^x \cdot \frac{Ee^x}{\Delta^2 e^x}$ taking h as the interval of difference.

Hint: $\left(\frac{\Delta^2}{E}\right)e^x = \Delta^2 E^{-1}e^x = \Delta^2 e^{x-h} = e^{-h}$
$\Delta^2\, e^x$

RHS: $e^{-h}\Delta^2 e^x \cdot \frac{Ee^x}{\Delta^2 e^x} = e^{-h}e^{x+h} = e^x$

2. Prove that
   (a) $\Delta^3 y_2 = \nabla^3 y_5$
   (b) $\Delta = \frac{1}{2}\delta^2 + \delta\sqrt{1+\frac{\delta^2}{4}}$
   (c) $2+\Delta = (E^{\frac{1}{2}}+E^{-\frac{1}{2}})(1+\Delta)^{\frac{1}{2}}$
   (d) $hD = \log(1+\Delta) = -\log(1-\Delta) = \sinh^{-1}(\mu\delta)$
3. Prove that
   (a) $\delta = \Delta(1+\Delta)^{-\frac{1}{2}} = \nabla(1-\nabla)^{-\frac{1}{2}}$
   (b) $\mu^2 = 1+\frac{\delta^2}{4}$
   (c) $\delta(E^{\frac{1}{2}}+E^{-\frac{1}{2}}) = \Delta E^{-1}+\Delta$
4. Show that $\nabla y_{n+1} = h[1+\frac{1}{2}\nabla+\frac{5}{12}\nabla^2+\ldots]y_n'$
5. Show that
   (a) $u_1x+u_2x^2+u_3x^3+\ldots = \left(\frac{x}{1-x}\right)^2\Delta u_1 + \left(\frac{x}{1-x}\right)^3\Delta^2 u_1+\ldots$
   (b) $u_0+\frac{u_1x}{1!}+\frac{u_2x^2}{2!}+\frac{u_3x^3}{3!}+\ldots = e^x(u_0+x\Delta u_0+\frac{x^2}{2!}\Delta^2u_0+\frac{x^3}{3!}\Delta^3u_0+\ldots)$
6. Prove that $E^{\frac{1}{2}} = \left(\frac{1}{2}+\frac{\delta^2}{4}\right)^{\frac{1}{2}}+\frac{\delta}{2}$
7. Show that $\nabla^2 = h^2D^2 - h^3D^3 + \frac{7}{12}h^4D^4+\ldots$
8. Show that $E = e^{hD}$

   Hint: $E(f(x)) = f(x+h)$, expand by Taylor's series, $E(f(x)) = f(x)+hf'+\frac{h^2}{2!}f''+\frac{h^3}{3!}f''+\ldots$

   $= f(x)+hDf+\frac{h^2}{2!}D^2f+\frac{h^3}{3!}D^3f+\ldots$

   $= \left[1+hD+\frac{h^2D^2}{2!}+\frac{h^3D^3}{3!}+\ldots\right]f$

   $= e^{hD}(f(x))$

   Prove the following.
9. $\Delta(f_ig_i) = f_i\Delta g_i + g_{i+1}\Delta f_i$
10. $\Delta f_i^2 = (f_i+f_{i+1})\Delta f_i$
11. $\Delta\left(\frac{1}{f_i}\right) = -\frac{\Delta f_i}{f_if_{i+1}}$
12. $\Delta\left(\frac{f_i}{g_i}\right) = \frac{(g_i\Delta f_i - f_i\Delta g_i)}{g_ig_{i+1}}$

32.13 NUMERICAL DIFFERENTIATION

Numerical differentiation or approximate differentiation is used when the function $y = f(x)$ is (i) given in tabular form (ii) function is highly complex. The basic idea in numerical differentiation is to replace the given function $y = f(x)$ on the interval $[a, b]$ by an interpolating polynomial $P(x)$ and set $f'(x) = P'(x)$; $f''(x) = P''(x)$ etc. Numerical differentiation is less exact than interpolation. Although $f(x_1) = P(x_1)$, $f'(x_1)$ need not be equal to $p'(x_1)$. (see figure).

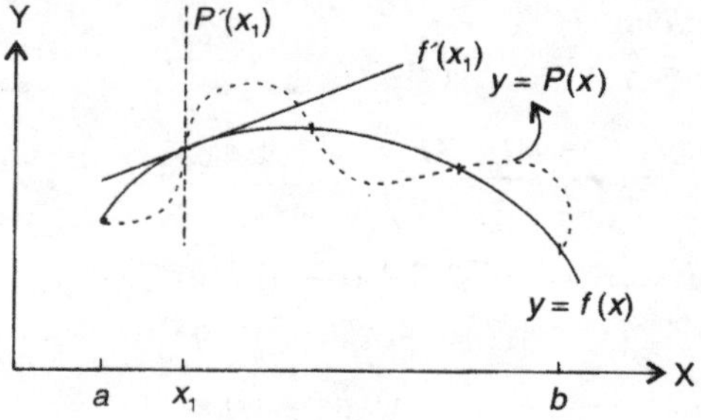

Fig. 32.13

Numerical Differentiation Using Newton's Forward Formula

Suppose $y = f(x)$ is specified in an interval $[a, b]$ at equally spaced points $x_i = x_0 + ih (i = 1, 2, \ldots n)$ by means of the values $y_i = f(x_i)$. Replace the tabulated function $y = f(x)$ by Newton's forward interpolation polynomial passing through the $(n+1)$ points $x_0, x_1, \ldots, x_n$. Thus

$$y(x) = y_0 + q\Delta y_0 + \frac{q(q-1)}{2!}\Delta^2 y_0 + \frac{q(q-1)(q-2)}{3!}\cdot\Delta^3 y_0 + \frac{q(q-1)(q-2)(q-3)}{4!}\Delta^4 y_0+\cdots \quad (1)$$

where $q = \frac{x-x_0}{h}$ and $h = x_{i+1} - x_i$, for $i = 0, 1, \ldots$ Here q is a function of x and $\frac{dq}{dx} = \frac{1}{h}$. Rewriting (1)

$$y(x) = y_0 + q\Delta y_0 + \frac{q^2-q}{2}\Delta^2 y_0 + \frac{q^3-3q^2+2q}{6}\cdot\Delta^3 y_0 + \frac{q^4-6q^3+11q^2-6q}{24}\Delta^4 y_0+\cdots \quad (1^*)$$

Differentiating (1*) w.r.t. x, we have

$$\frac{dy}{dx} = \frac{dy}{dq}\cdot\frac{dq}{dx} = \frac{1}{h}\frac{dy}{dq} =$$

$$= \frac{1}{h}\left[\Delta y_0 + \frac{2q-1}{2}\Delta^2 y_0 + \frac{3q^2-6q+2}{6}\Delta^3 y_0 + \frac{4q^3-18q^2+22q-6}{24}\Delta^4 y_0 + \cdots\right] \quad (2)$$

Similarly, differentiating (2) once again w.r.t. x, we get

$$\frac{d^2y}{dx^2} = \frac{d}{dx}\left(\frac{dy}{dx}\right) = \frac{1}{h}\frac{d}{dq}\left(\frac{dy}{dx}\right)$$

$$= \frac{1}{h^2}\left[\Delta^2 y_0 + (q-1)\Delta^3 y_0 + \frac{6q^2-18q+11}{12}\Delta^4 y_0 + \cdots\right] \quad (3)$$

Higher order derivatives can be computed similarly.

Special case

When derivative is required at a basic tabulated point x_i, then choose $x = x_0$, so $q = 0$ (since each tabular value may be taken as the initial value x_0). Thus

$$\left.\frac{dy}{dx}\right|_{x=x_0} = y'(x_0) = \frac{1}{h}\left[\Delta y_0 - \frac{1}{2}\Delta^2 y_0 + \frac{\Delta^3 y_0}{3} - \frac{\Delta^4 y_0}{4} + \frac{\Delta^5 y_0}{5} - \cdots\right] \quad (4)$$

and

$$\left.\frac{d^2y}{dx^2}\right|_{x=x_0} = y''(x_0) = \frac{1}{h^2}\left[\Delta^2 y_0 - \Delta^3 y_0 + \frac{11}{12}\Delta^4 y_0 - \frac{5}{6}\Delta^5 y_0 + \cdots\right] \quad (5)$$

Similarly,

$$y'''(x_0) = \frac{1}{h^3}\left[\Delta^3 y_0 - \frac{3}{2}\Delta^4 y_0 + \frac{7}{4}\Delta^5 y_0 + \cdots\right] \quad (6)$$

$$y''''(x_0) = \frac{1}{h^4}\left[\Delta^4 y_0 - 2\Delta^5 y_0 + \cdots\right] \quad (7)$$

Thus formulas (2), (3) (and (4), (5) in special case) given the derivatives of $y(x)$.

Numerical Differentiation Using Newton's Backward Formula

In this case, we replace $y(x)$ by Newton backward interpolation formula:

$$y(x) = y_n + q\nabla y_n + \frac{q(q+1)}{2!}\nabla^2 y_n + \frac{q(q+1)(q+2)}{3!}\nabla^3 y_n + \cdots$$

Differentiating w.r.t. x,

$$y'(x) = \frac{1}{h}\left[\nabla y_n + \frac{1}{2}(2q+1)\nabla^2 y_n + \frac{3q^3+6q+2}{6}\nabla^3 y_n + \cdots\right] \quad (8)$$

$$y''(x) = \frac{1}{h^2}\left[\nabla^2 y_n + (q+1)\nabla^3 y_n + \frac{6q^2+18q+11}{12}\nabla^4 y_n + \cdots\right] \quad (9)$$

Special case: When $x = x_n$, then $q = 0$.

$$y'(x_n) = \frac{1}{h}\left[\nabla y_n + \frac{1}{2}\nabla^2 y_n + \frac{1}{3}\nabla^3 y_n + \frac{1}{4}\nabla^4 y_n + \cdots\right] \quad (10)$$

$$y''(x_n) = \frac{1}{h^2}\left[\nabla^2 y_n + \nabla^3 y_n + \frac{11}{12}\nabla^4 y_n + \cdots\right] \quad (11)$$

WORKED OUT EXAMPLES

Example: Compute $f'(x)$ and $f''(x)$ at (a) $x = 16$; (b) $x = 15$; (c) $x = 24$; (d) $x = 25$ from the following table:

x:	15	17	19	21	23	25
$f(x) = \sqrt{x}$:	3.873	4.123	4.359	4.583	4.796	5.8

Compare with the exact values.

Solution:

a. To find f', f'' at $x = 16, 15$, use the forward differences results.

$$f'(x) = \frac{1}{h}\left[\Delta y_0 + \frac{2q-1}{2}\Delta^2 y_0 + \frac{3q^2-6q+2}{6}\cdot\Delta^3 y_0 +\right.$$

Table 32.12: The finite difference table

x	$y=\sqrt{x}$	Δy	$\Delta^2 y$	$\Delta^3 y$	$\Delta^4 y$	$\Delta^5 y$
15	3.873					
		0.250				
17	4.123		−0.014			
		0.236		0.002		
19	4.359		−0.012		−0.001	
		0.224		0.001		0.002
21	4.583		−0.011		0.001	
		0.213		0.002		
23	4.796		−0.009			
		0.204				
25	5.000					

$$+\frac{4q^3-18q^2+22q-6}{24}\Delta^4 y_0\Bigg]$$

Here $q=\frac{x-x_0}{h}$, $x_0=15$, $h=2$ and $x=16$, then $q=\frac{16-15}{2}=\frac{1}{2}=0.5$,
Also $\Delta y_0=0.250$, $\Delta^2 y_0=-0.014$, $\Delta^3 y_0=0.002$, $\Delta^4 y_0=-0.001$.

$$\therefore\ f'(16)=\frac{1}{2}\left[0.250+0+\frac{3(.5)^2-6(.5)+2}{6}(.002)\right.$$
$$\left.+\frac{4(.5)^3-18(.5)^2+22(.5)-6}{24}(-0.001)\right]$$
$$f'(16)=0.1249375 \quad \text{(Exact value: 0.125)}$$

Similarly,

$$f''(16)=\frac{1}{2^2}[-0.014+(0.5-1)(.002)$$
$$+\frac{6(.5)^2-18(.5)+11}{12}(-.001)$$
$$f''(16)=-0.0038229 \quad \text{(Exact value} = -0.00390625)$$

b. When $x=15$ (which is a tabulated point), then $q=0$

$$f'(15)=\frac{1}{2}\left[0.250-\frac{1}{2}(-0.014)+\frac{1}{3}(0.002)-\frac{1}{4}(-0.001)\right.$$
$$f'(15)=0.128958 \quad \text{(Exact value: 0.12909)}$$
$$f''(15)=\frac{1}{4}\left[-0.014-0.002+\frac{11}{12}(-0.001)\right]$$
$$f''(15)=-0.004229 \quad \text{(Exact value: } -0.0043033).$$

For (c) and (d) use Newton's backward formulae since $x=24$ and $x=25$ are at the end of the table.

c. Here $q=\frac{x-x_n}{h}$, $x=24$, $x_n=25$, $h=2$, $q=\frac{24-25}{2}=-0.5$, $q=-0.5$. Here $\nabla y_n=.204$, $\nabla^2 y_n=-0.009$, $\nabla^3 y_n=0.002$, $\nabla^4 y_n=0.001$.

$$f'(24)=\frac{1}{2}\left[.204+\frac{1}{2}(2(.5)+1)(-0.009)\right.$$
$$\left.+\frac{3(-.5)^3+6(-.5)+2}{6}(0.002)\right]$$
$$=0.09727 \quad \text{(Exact value: 0.10206)}$$
$$f''(24)=\frac{1}{2^2}\left[-0.009+(-.5+1)(0.002)\right.$$
$$\left.+\frac{6(-.5)^2+18(-.5)+11}{12}(0.001)\right]$$
$$=-0.00242708 \text{ (Exact value : } -0.002126293)$$

d. When $x=25$ which is a tabulated value, $q=0$. Then

$$f'(25)=\frac{1}{2}\left[.204+\frac{1}{2}(-0.009)+\frac{1}{3}(.002)+\frac{1}{4}(.001)\right]$$
$$=0.100208 \quad \text{(Exact value: 0.100)}$$
$$f''(25)=\frac{1}{2^2}\left[-0.009+0.002+\frac{11}{12}(.001)\right]$$
$$=-0.00225 \text{ (Exact value: } -0.002)$$

Note: $y=\sqrt{x}$, $y'(x)=\frac{1}{2\sqrt{x}}$, $y''=\frac{-1}{4x^{3/2}}$.

EXERCISE

1. Using forward differences find the first and second derivatives of y at $x=2$ for the data given below:

x:	2	4	6	8	10
y:	0	0	1	0	0

Hint: Differences are 0, 1, −3, 6.

Ans. −1.5, 2.375

2. Compute $y'(1)$, $y''(1)$ and $y'(3)$ from the following data:

x:	1	2	3	4	5	6	7	8
y:	2.105	2.808	3.614	4.604	5.857	7.451	9.467	11.985

Hint: Differences: .703, .103, .081, −0.002.

Ans. $y'(1)=0.6925$, $y''(1)=0.0201$; $y'(3)=.883$

3. Determine $y'(2)$, $y''(2)$, $y'''(2)$, $y''''(2)$, $y'(1.5)$, $y''(1.5)$ from the data below, compare with exact values:

x:	0	1	2	3	4	5
y:	0	1	8	27	64	125

Hint: Differences: 1, 6, 6, 0, 0.

Ans. $y'(2) = 12$, $y''(2) = 12$, $y'''(2) = 6$, $y''''(2) = 0$, $y'(1.5) = 6.75$, $y''(1.5) = 9$. Exact value: $y = x^3$, $y' = 3x^2$, $y'' = 6x$, $y''' = 6$, $y'''' = 0$.

4. Compute $y'(1.05)$, $y''(1.05)$, $y'(1.25)$, $y''(1.25)$ from the following data:

x:	1.00	1.05	1.10	1.15	1.20	1.25	1.30
y:	1.00	1.0247	1.04881	1.07238	1.09544	1.11803	1.14017

Ans. $y'(1.05) = 0.48763$, $y''(1.05) = -0.2144$, $y'(1.25) = 0.44733$, $y''(1.25) = -0.158332$

5. Find $y'(50)$ of the tabulated function $y = \log_{10} x$, given below:

x:	50	55	60	65
y:	1.6990	1.7404	1.7782	1.8129

Hint: Forward differences: 414, −36, 5, $h = 5$. Exact value: 0.0087.

Ans. $0.0087 \left(= \frac{1}{5}(0.0414 + 0.008 + 0.0002)\right)$

6. The following table contains the path $y = f(t)$ traversed in time t by a point moving in a straight line. Using finite forward differences upto order five inclusive, find the velocity $v = \frac{dy}{dt}$ and acceleration $A = \frac{d^2y}{dt^2}$ at the point of times: $t = 0, 0.01, 0.02, 0.03, 0.04$.

time t in sec:	0	0.01	0.02	0.03	0.04	0.05
Path $y(t_i)$ in cm:	0	1.519	6.031	13.397	23.396	35.721

t:	0.06	0.07	0.08	0.09
$y(t_i)$	50	65.798	82.635	100

Hint: Forward differences are 2.993, −0.139, −0.082, −0.004. Exact: $y = 100(1 - \cos\frac{50\pi t}{9})$, $V = \frac{dy}{dt} = \frac{5000\pi}{9}\sin\left(\frac{50\pi t}{9}\right)$, $W = \frac{d^2y}{dt^2} = \frac{250000\pi^2}{81}\cos\left(\frac{50\pi t}{9}\right)$.

Ans.

Table 32.13

t	V	W	V_{exact}	W_{exact}
0.00	0.4	30600	0.0	30462
0.01	303.6	29780	303.08	30001
0.02	596.3	28780	596.98	28625
0.03	873.2	26250	872.66	26381
0.04	1121.7	23360	1121.9	23340

7. Compute $y'(1.1)$, $y''(1.1)$, $y'(1.6)$, $y''(1.6)$ given:

x:	1.0	1.1	1.2	1.3	1.4	1.5	1.6
y:	7.989	8.403	8.781	9.129	9.451	9.750	10.031

Hint: Forward differences: .414, −0.36, 0.006, −0.002, 0.002, −0.003, backward differences .281, −0.018, 0.005, −0.001, −.001, −0.003.

Ans. $y'(1.1) = 3.946$, $y''(1.1) = -3.545$, $y'(1.6) = 2.727$, $y''(1.6) = -1.703$

8. Determine $y'(0)$, $y''(0)$ from following data:

x:	0	1	2	3	4	5
y:	4	8	15	7	6	2

Ans. $y'(0) = -27.9$, $y''(0) = 117.67$.

32.14 NUMERICAL INTEGRATION

Let $f(x)$ be continuous on the interval $[a, b]$ and its antiderivative $F(x)$ is known. Then the definite integral of $f(x)$ from a to b may be evaluated using Newton-Leibnitz formula

$$\int_a^b f(x)dx = F(b) - F(a) \tag{1}$$

where $f'(x) = f(x)$.

However, computation of the definite integral by (1) becomes difficult or practically impossible when (i) the antiderivative $F(x)$ can not be found by elementary means or is too involved (ii) when the integrand $f(x)$ is specified in tabular form.

The numerical integration of a single integral is known as **mechanical quadrature** and uses the geometrical interpretation of the definite integral $\int_a^b f(x)dx$ as the area under the curve $y = f(x)$ between the ordinates $x = a$ and $x = b$. The basic integration rule is to replace $f(x)$ by a simple polynomial $\phi(x)$, say Lagrange's interpolation polynomial in $[a, b]$. Thus

$$\int_a^b f(x)dx \sim \int_a^b \phi(x)dx \tag{2}$$

Choosing interval size $h = \frac{b-a}{n}$, divide the interval $[a, b]$ into n intervals by means of $(n+1)$ equally spaced points $x_0 = a$, $x_i = x_0 + ih$, $i = 1, 2, 3, \ldots, n-1$, $x_n = b$. Let $y_i = f(x_i)$ for $i = 0, 1, 2, \ldots, n$.

Integrating the integral in the R.H.S. of (2), we get Newton-Cotes formula of the form

$$\int_a^b f(x)dx \sim c \cdot h \cdot \sum_{i=0}^{n} w_i \cdot f(x_i) \tag{3}$$

[Since the end points x_0, x_n are used in (3), these formula are known as 'closed' type.]

Here m is the degree of the polynomial $\phi(x)$, c is the coefficient and w_i are weights, given in the following table:

Table 32.14

Case	m	c	w_0	w_1	w_2	w_3	Name of the rule
I.	0	1	1				Rectangular and mid point rules
II.	1	$\frac{1}{2}$	1	1			Trapezoidal rule
III.	2	$\frac{1}{3}$	1	4	1		Simpson's $\frac{1}{3}$ rule
IV.	3	$\frac{3}{8}$	1	3	3	1	Simpson's $\frac{3}{8}$ rule

The degree m of the polynomial $\phi(x)$ used in the derivation of the above formula are 0, 1, 2, 3 respectively.

Derivation of Trapezoidal, Simpson's $\frac{1}{3}$ and Simpson's $\frac{3}{8}$ Rule Formula:

The basic idea in numerical integration is to replace the unknown tabulated function $y = f(x)$ by an nth degree polynomial $P_n(x)$ say Newton-Gregory forward interpolation formula and carry on the integration. Thus

$$I = \int_{x_0}^{x_n} f(x)dx \simeq \int_{x_0}^{x_n} P_n(x)dx$$

$$\simeq \int_{q=0}^{n} \left[y_0 + q\Delta y_0 + \frac{q(q-1)}{2!}\Delta^2 y_0 + \frac{q(q-1)(q-2)}{3!}\Delta^3 y_0 + \cdots + \frac{q(q-1)(q-2)\cdots(q-n+1)}{n!}\Delta^n y_0 \right] dq$$

Here the new variable is $q = \frac{x-x_0}{h}$, so $dq = \frac{dx}{h}$.

Trapezoidal Rule

Take $n = 1$ (two points, one internal)

$$\int_0^1 y\,dx = h\int_0^1 (y_0 + q\Delta y_0)dq = h\left[y_0 q + \Delta y_0 \cdot \frac{q^2}{2} \right]\Bigg|_{q=0}^{1}$$

$$= h\left[y_0 + \frac{\Delta y_0}{2} \right] = h\left[y_0 + \frac{y_1 - y_0}{2} \right]$$

$$= \frac{h(y_1 + y_0)}{2} = \frac{h}{2}(y_1 + y_0).$$

Applying this to successive intervals, we get

$$\int_{x_0}^{x_n} y\,dx = \int_{x_0}^{x_1} y\,dx + \int_{x_1}^{x_2} y\,dx + \cdots + \int_{n_{n-1}}^{x_n} y\,dx$$

$$= \frac{h}{2}(y_0+y_1) + \frac{h}{2}(y_1+y_2) + \cdots + \frac{h}{2}(y_{n-1}+y_n)$$

$$\boxed{I = \frac{h}{2}\left[y_0 + 2(y_1 + y_2 + \cdots + y_{n-1}) + y_n \right]}$$

Simpson's $\frac{1}{3}$ Rule

Take $n = 2$ (three points, two intervals, curve (polynomial) parabola).

$$\int_{x_0}^{x_2} y\,dx = h\int_0^2 \left[y_0 + q\Delta y_0 + \frac{q(q-1)}{2}\Delta^2 y_0 \right] dq$$

$$= h\left[y_0 \cdot q + \frac{q^2}{2}\Delta y_0 + \frac{\Delta^2 y_0}{2}\left(\frac{q^3}{3} - \frac{q^2}{2} \right) \right]\Bigg|_{q=0}^{2}$$

$$= h\left[2y_0 + 2\Delta y_0 + \frac{1}{3}\Delta^2 y_0 \right]$$

$$= h\left[2y_0 + 2(y_1 - y_0) + \frac{1}{3}(y_2 - 2y_1 + y_0)\right]$$

$$= \frac{h}{3}[y_0 + 4y_1 + y_2].$$

Applying for successive intervals (*the number of intervals must be even or the number of points is odd*).

$$\int_{x_0}^{x_n} y\,dx = \int_{x_0}^{x_2} y\,dx + \int_{x_2}^{x_4} y\,dx + \cdots + \int_{x_{n-2}}^{x_n} y\,dx$$

$$= \frac{h}{3}[y_0 + 4y_1 + y_2] + \frac{h}{3}[y_2 + 4y_3 + y_4] + \cdots$$

$$+ \frac{h}{3}[y_{n-2} + 4y_{n-1} + y_n]$$

$$\boxed{I = \frac{h}{3}\left[y_0 + 2(y_2 + y_4 + \cdots + y_{n-2}) + 4(y_1 + y_3 + \cdots + y_{n-1}) + y_n\right]}$$

Simpson's $\frac{3}{8}$ Rule

Take $m = 3$, number of intervals 3 and number of points 4: (number of intervals should be multiples of 3 i.e., $3N$).

$$\int_{x_0}^{x_3} y\,dx = h\int_0^3 \left[y_0 + q\Delta y_0 + \frac{q(q-1)\Delta^2 y_0}{2} + \frac{q(q-1)(q-2)}{3!}\Delta^3 y_0\right] dq$$

$$= h\left[3y_0 + \frac{9}{2}\Delta y_0 + \frac{9}{4}\Delta^2 y_0 + \frac{3}{8}\Delta^3 y_0\right]$$

$$= h\left[3y_0 + \frac{9}{2}(y_1 - y_0) + \frac{9}{4}(y_2 - 2y_1 + y_0) + \frac{3}{8}(y_3 - 3y_2 + 3y_1 - y_0)\right]$$

$$= \frac{3}{8}h[y_0 + 3y_1 + 3y_2 + y_3].$$

Applying repeatedly

$$\int_{x_0}^{x_n} y\,dx = \frac{3}{8}h\left[y_0 + 2(y_3 + y_6 + \cdots + y_{n-3}) + 3(y_1 + y_2 + y_4 + y_5 + \cdots + y_{n-2} + y_{n-1}) + y_n\right].$$

Thus

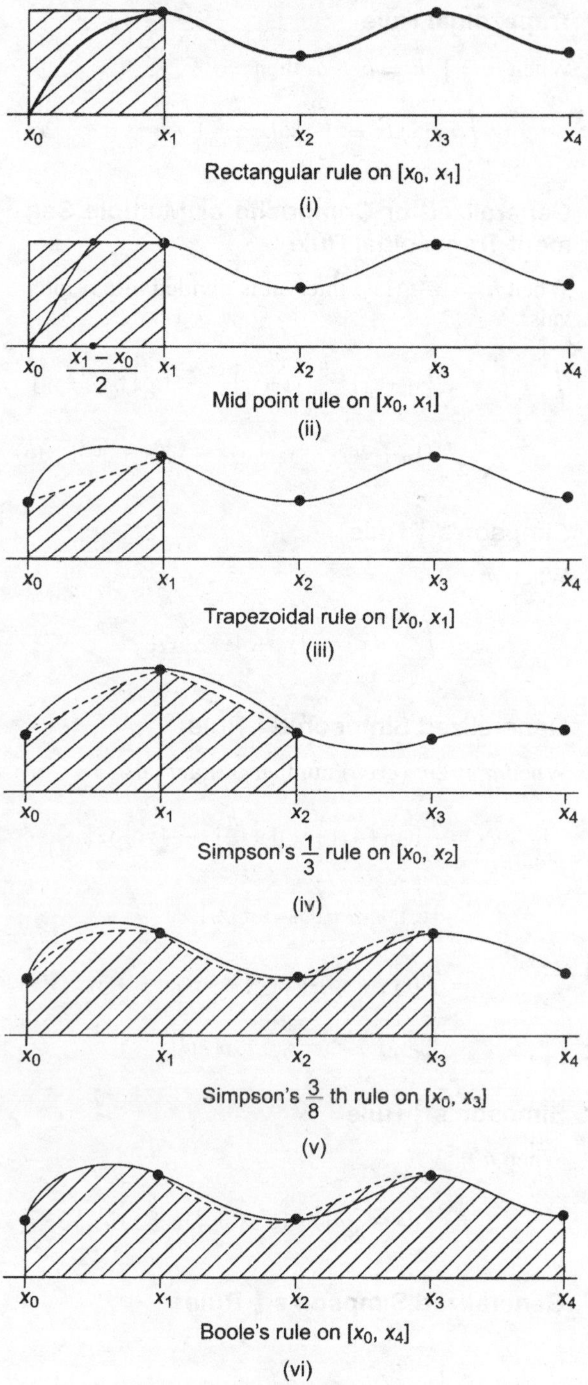

Fig. 32.14

Trapezoidal Rule

When $n = 1$, $h = b - a$, then

$$\int_a^b f(x)dx = \int_a^b y\,dx \simeq \frac{h}{2}[y_0 + y_1]. \quad (4)$$

Generalized or Composite or Multiple Segment Trapezoidal Rule

When $h = \frac{b-a}{n}$ (i.e., interval is divided into n intervals).

$$\int_a^b y\,dx = \frac{h}{2}[y_0+y_1]+\frac{h}{2}[y_1+y_2]+\cdots+\frac{h}{2}[y_{n-1}+y_n]$$

$$= \frac{h}{2}[y_0 + 2(y_1 + y_2 + \cdots + y_{n-1} + y_n)]. \quad (5)$$

Simpson's $\frac{1}{3}$ Rule

When $n = 2$,

$$\int_a^b y\,dx = \frac{h}{3}[y_0 + 4y_1 + y_2].$$

Generalized Simpson's $\frac{1}{3}$ Rule

When $n = 2m$ (*even*) number of intervals

$$\int_a^b y\,dx = \frac{h}{3}[(y_0+4y_1+y_2)] + \frac{h}{3}[y_2+4y_3+y_4] + \cdots$$

$$+\frac{h}{3}[y_{n-2} + 4y_{n-1} + y_n]$$

$$= \frac{h}{3}\left[(y_0 + y_n) + 2(y_2 + y_4 + \cdots + y_{n-2}) + 4(y_1 + y_3 + \cdots + y_{n-1})\right].$$

Simpson's $\frac{3}{8}$ Rule

When $n = 3$

$$\int_a^b y\,dx = \frac{3}{8}h[y_0 + 3y_1 + 3y_2 + y_3].$$

Generalized Simpson's $\frac{3}{8}$ Rule

$$\int_a^b y\,dx = \frac{3}{8}h[y_0 + 3y_1 + 3y_2 + y_3]$$

$$+\frac{3}{8}h[y_3 + 3y_4 + 3y_5 + y_6] + \cdots$$

$$+\frac{3}{8}h[y_{n-3} + 3y_{n-2} + 3y_{n-1} + y_n]$$

$$= \frac{3}{8}h\left[(y_0 + y_n) + 2(y_3 + y_6 + \cdots + y_{n-3}) + 3(y_1 + y_2 + y_4 + y_5 + \cdots + y_{n-2} + y_{n-1})\right].$$

Weddle Rule

When ($n = 6$),

$$\int_a^b y\,dx = \frac{3h}{10}[y_0 + 5y_1 + y_2 + 6y_3 + y_4 + 5y_5 + y_6].$$

Generalized Weddle Rule

$$\int_{a=x_0}^{b=x_0+nh} y\,dx = \frac{3h}{10}[y_0+5y_1+y_2+6y_3+y_4+5y_5+y_6]$$

$$+\frac{3h}{10}[y_6+5y_7+y_8+6y_9+y_{10}+5y_{11}$$

$$+y_{12}]+\cdots+\frac{3h}{10}[y_{n-6}+5y_{n-5}+y_{n-4}$$

$$+6y_{n-3}+y_{n-2}+5y_{n-1}+y_n].$$

Note: While there is no restriction for the number of intervals in trapezoidal rule, number of sub intervals n in the case of Simpson's $\frac{1}{3}$ rule must be even, for Simpson's $\frac{3}{8}$ rule must be multiple of 3, for Weddle's rule must be multiple of 6.

WORKED OUT EXAMPLES

Example 1: Given that $y = \log x$, and

x:	4.0	4.2	4.4	4.6	4.8	5.0	5.2
y:	1.3863	1.4351	1.4816	1.5261	1.5686	1.6094	1.6487

evaluate $I = \int_4^{5.2} \log x\,dx$ by (a) Trapezoidal rule; (b) Simpson's $\frac{1}{3}$ rule; (c) Simpson's $\frac{3}{8}$ rule; (d) Weddle's rule; (e) Compare it with exact value.

Solution:

a. By trapezoidal rule, ($h = 0.2$)

$$I = \frac{0.2}{2}[1.3863 + 1.6487 + 2(1.4351 + 1.4816 + 1.5261 + 1.5686 + 1.6094)]$$

$I = 1.8276551$

b. By Simpson's $\frac{1}{3}$ rule

$$I = \frac{0.2}{3}[1.3863 + 1.6487 + 2(1.4816 + 1.5686) + 4(1.4351 + 1.5261 + 1.6094)]$$

$I = 1.8278472$

c. By Simpson's $\frac{3}{8}$ rule

$$I = \frac{3}{8}(0.2)[1.3863 + 1.6487 + 2(1.5261) + 3(1.4351 + 1.4816) + (1.5686 + 1.6094)]$$

$I = 1.8278470$

d. By Weddle's rule

$$I = \frac{3}{10}(0.2)[1.3863 + 5(1.4351 + 1.6094) + 6(1.5261) + 1.4816 + 1.5686 + 1.6487]$$

$I = 1.8278474$

e. By integrating by parts

$$I = \int_5^{4.2} \ln x\,dx = x\cdot\ln x - \int x\cdot\frac{1}{x}dx = x\cdot\ln x - x\Big|_4^{5.2}$$

$$= (5.2\ln 5.2 - 5.2) - (4\ln 4 - 4) = 3.373 - 1.54517$$

$I = 1.827822556.$

Example 2: The half ordinates in fect of the mid ship section of a vessel are 12.5, 12.8, 12.9, 13, 13, 12.8, 12.4, 11.8, 10.4, 6.8, 0.5 and the ordinates are 2 feet apart. Find the centre of gravity of the section.

Solution: Centre of gravity $(\overline{x}, 0)$.

x:	0	2	4	6	8	10
half ordinates:	12.5	12.8	12.9	13	13	12.8
y:	25	25.6	25.8	26	26	25.6
xy:	0	51.2	103.2	156	208	256

x:	12	14	16	18	20
half ordinates:	12.4	11.8	10.4	6.8	0.5
y:	24.8	23.6	20.8	13.6	1
xy:	297.6	330.4	332.8	244.8	20

Here $\overline{x} = \dfrac{\int_0^{20} xy\,dx}{\int_0^{20} y\,dx} = \dfrac{I_1}{I_2}$

where by Simpson's $\frac{1}{3}$ rule

$$I_1 = \frac{1}{3}(2)[20 + 332.8 + 4(51.2 + 156 + 256 + 330.4 + 244.8) + 2(103.2 + 208 + 297.6 + 332.8)]$$

$I_1 = 4037.8666$

$$I_2 = \frac{1}{3}(2)[25 + 1 + 4(25.6 + 26 + 25.6 + 23.6 + 13.6) + 2(25.8 + 26 + 24.8 + 20.8)]$$

$I_2 = 452.2666$

$$\therefore \overline{x} = \frac{4037.8666}{452.2666} = 8.928066154 \approx 8.93.$$

Example 3: The speeds of an electric train at various times after leaving one station until it stops at the next station are given in the following table (Table 11.15). Find the distance between the two stations.

Table 32.15

Speed in mph	0	13	33	$39\frac{1}{2}$	40	40	36	15	0
Time in minutes	0	$\frac{1}{2}$	1	$1\frac{1}{2}$	2	$2\frac{1}{2}$	3	$3\frac{1}{4}$	$3\frac{1}{2}$

Solution: Let $v = \frac{ds}{dt}$ be the velocity of the train at any time t. Then s, the distance between the two stations (with a lapse of $3\frac{1}{2}$ minutes) is

$$s = \int_{t=0}^{3\frac{1}{2}} \frac{ds}{dt}\,dt = \int_{t=0}^{3\frac{1}{2}} v\,dt = \int_0^3 v\,dt + \int_3^{3\frac{1}{2}} v\,dt$$

since the time interval are not the same. By applying Simpson's $\frac{1}{3}$ rule with $h = \frac{1}{2}$ and $h = \frac{1}{4}$, we have

Table 32.16

Speed in mph	0	13	33	$39\frac{1}{2}$	40	40	36	15	0
Time in hours	0	$\frac{1}{120}$	$\frac{1}{60}$	$\frac{1}{40}$	$\frac{1}{30}$	$\frac{1}{24}$	$\frac{1}{20}$	$\frac{13}{240}$	$\frac{7}{480}$

$$s = s_1 + s_2 = \frac{1}{3}\cdot\frac{1}{120}\left[0 + 36 + 4(13 + 39\frac{1}{2} + 40) + 2(33 + 40)\right] + \frac{1}{3}\cdot\frac{1}{240}[36 + 0 + 4(15)]$$

$$= \frac{23}{15} + \frac{2}{15} = \frac{25}{15} = \frac{5}{3} = 1\frac{2}{3} = 1.666 \text{ miles}$$

Thus the distance between the two stations is 1.666 miles.

EXERCISE

Evaluate approximately the following integrals:

Trapezoidal rule

1. Use trapezoidal rule to evaluate (a) $\int_1^{10} \frac{dx}{x}, n =$ 10; (b) $\int_0^1 \frac{dx}{1+x^2}, n = 4$; (c) $\int_0^1 \frac{\sin x}{x} dx, n = 4$.

Ans. **a.** $\frac{1}{2}\left[1 + 0.1 + 2\left(\frac{1}{2} + \frac{1}{3} + \frac{1}{4} + \frac{1}{5} + \frac{1}{6} + \frac{1}{7} + \frac{1}{8} + \frac{1}{9}\right)\right] = 2.3788$

b. $\frac{1}{4} \cdot \left[\frac{1}{2}\left(1 + \frac{1}{2}\right) + \frac{16}{17} + \frac{4}{5} + \frac{16}{25}\right] = 0.7828$

c. $\frac{1}{4} \cdot \frac{1}{2}[1 + 0.84 + 2(.9898 + 0.95885 + .90885)] = 0.9445$

2. Using the following data evaluate by T.R.

a.

x:	1	2	3	4	5	6	7
y:	2.105	2.808	3.614	4.604	5.857	7.451	9.467

b.

x:	1.0	1.1	1.2	1.3	1.4	1.5	1.6	1.7	1.8
$f(x)$:	1.543	1.668	1.811	1.971	2.151	2.352	2.577	2.828	3.107

with (i) $h = 0.1$; (ii) $h = 0.2$; (iii) $h = 0.4$.

Ans. **a.** $\int_1^7 y\,dx = \frac{1}{2}[2.105+2(2.808+3.614+4.604+5.851+.7451)+9.467]=30.120$

b. $\int_1^{1.8} f(x)dx = 1.7683 (h=0.1), 1.7728 (h = 0.2), 1.7904 (h = 0.4)$.

3. Integrate $\int_0^{\pi} \sin x\,dx, n = 10$.

Hint:

x:	0	$\frac{\pi}{10}$	$\frac{2\pi}{10}$	$\frac{3\pi}{10}$	$\frac{4\pi}{10}$	$\frac{5\pi}{10}$	$\frac{6\pi}{10}$	$\frac{7\pi}{10}$	$\frac{8\pi}{10}$	$\frac{9\pi}{10}$	π
$\sin x$:	0	.309	.5878	.809	.951	1	.951	.809	.5878	.309	0

Ans. 6.3138

4. Evaluate $\int_0^1 \frac{dx}{1+x^2}, n = 10$.

Hint:

x:	0	.1	.2	.3	.4	.5	.6	.7	.8	.9	1
$f(x)$:	1	.99	.96	.917	.862	.8	.735	.67	.61	.55	.5

Ans. 0.6547

Simpson's $\frac{1}{3}$ rule

5. Integrate approximately $\int_1^{1.04} f(x)dx$ from the following data:

x:	1	1.01	1.02	1.03	1.04
$f(x)$:	3.953	4.066	4.182	4.300	4.421

Hint: $h = 0.1$.

Ans. 0.16734, $I = \frac{0.1}{3}[(3.953) + (4.421) + 4(4.066 + 4.182 + 4.3)]$

6. Evaluate $\log e^7$ by Simpson's $\frac{1}{3}$ rule.

Hint: $I = \int_0^6 \frac{dx}{1+x}$.

Ans. 1.9588

7. Compute $\int_0^{\frac{\pi}{2}} \sqrt{\sin x}\,dx$.

Hint: $n = 6, h = \frac{\pi}{12}$.

Ans. 1.18728, $I = \frac{\pi}{3} \cdot \frac{1}{12}[0 + 1 + 4(.50874 + 0.8409 + 0.98282) + 2(0.70711 + 0.9306)]$

8. Compute $I = \int_0^1 \frac{dx}{1+x}, n = 10, h = 0.1$.

Ans. 0.69315

9. Integrate approximately $\int_0^{\pi} \sin x\,dx, n = 10$.

Ans. 3.0778

10. Compute $\int_0^1 \frac{dx}{1+x^2}, n = 10$.

Ans. 0.785425

11. A river is 80 feet wide. The depth d in feet at a distance x feet from one bank is given by:

x:	0	10	20	30	40	50	60	70	80
d:	0	4	7	9	12	15	14	8	3

Find approximately the area of cross section of the river.

Hint: $I = \int_0^{80} d\,dx = \frac{1}{3}(10)[0 + 3 + 4(4 + 9 + 15 + 8) + 2(7 + 12 + 14)]$

Ans. 710 sq. feet

12. A body of weight w tons is acted upon by a variable force F tons weight. It acquires a speed of v mph after travelling y yards where $v^2 = \frac{89.8}{w} \int F(y)dy$. If $w = 600$ and F is given by the table below, estimate the velocity at the end of 400 yards from the starting point.

y:	0	50	100	150	200	300	400
$f(y)$:	90	62	45	34	26	15	8

Hint: $\int_0^{400} F\,dy = \int_0^{200} + \int_{200}^{400}$ with $h_1 = 50$, $h_2 = 100$.

Ans. $v = 44.16$ mph

13. Find the approximate mileage travelled by a train between 11.50 AM to 12.30 PM from the following data:

time t:	11.50 AM	12	12.10	12.20	12.30
speed (mph):	24.2	35	41.3	42.8	39.2

Hint: $I = s\int_{t=11.50}^{12.30} v\,dt = \frac{1}{6}\left(\frac{1}{3}\right)[24.2 + 39.2 + 4(35.0 + 42.8) + 2(41.3)]$.

Ans. $s =$ distance $= 25.4$ miles

Simpson's $\frac{3}{8}$ rule

14. Evaluate $\log_e 7$ by Simpson's $\frac{3}{8}$th rule.

Hint: $I = \int_0^6 \frac{dx}{1+x} = \log_e(1+x)\Big|_0^6 = \log_e 7$

x:	0	1	2	3	4	5	6
$\frac{1}{1+x}$:	1	$\frac{1}{2}$	$\frac{1}{3}$	$\frac{1}{4}$	$\frac{1}{5}$	$\frac{1}{6}$	$\frac{1}{7}$

Ans. 1.966

15. Integrate $\int_0^1 \frac{dx}{1+x^2}$, $n = 12$.

Hint:

x:	0	$\frac{1}{12}$	$\frac{1}{6}$	$\frac{1}{4}$	$\frac{1}{3}$	$\frac{5}{12}$	$\frac{1}{2}$
$f(x) = \frac{1}{1+x^2}$:	1	.993	.973	.94	.9	.85	.8
x:	$\frac{7}{12}$	$\frac{2}{3}$	$\frac{3}{4}$	$\frac{5}{6}$	$\frac{11}{12}$	1	
$f(x) = \frac{1}{1+x^2}$:	.8	.746	.692	.64	.59	.543	0.5

Ans. 0.78539

16. Evaluate $\int_0^1 e^x dx$ approximately.

Hint: $n = 9$, $h = \frac{1}{9} = 0.1111$

x:	0	$\frac{1}{9}$	$\frac{2}{9}$	$\frac{1}{3}$	$\frac{4}{9}$	$\frac{5}{9}$
e^x:	1	1.1175	1.2488	1.3956	1.5596	1.743
x:	$\frac{6}{9}$	$\frac{7}{9}$	$\frac{8}{9}$	1		
e^x:	1.9477	2.176	2.432	2.718		

Ans. 1.7182888

17. Integrate $\int_0^{40} a(t)dt$ from the data given below:

t:	0	5	10	15	20	25	30	35	40	45
$a(t)$:	40	45.25	48.5	51.25	54.35	59.48	61.5	64.3	68.7	70

Hint: $5\left(\frac{3}{8}\right)$ [40 + 70 + 2(51.25 + 61.5) + 3(45.25 + 48.5 + 54.35 + 59.48 + 64.3 + 68.7)].

Ans. $5\left(\frac{3}{8}\right)(1357.24) = 2544.825$

Weddle's rule

18. Integrate approximately $\int_0^{1.1} e^x dx$, $n = 12$.

Hint: $I = \frac{3}{10}\left(\frac{1.1}{12}\right)[1 + 5(1.1051) + 1.2214 + 6(1.3498) + 1.4918 + 5(1.6487) + 2(1.822) + 5(2.0137) + 2.2225 + 6(2.4596) + 5(2.71828) + 3.004166]$

Exact value: $e^x\Big|_0^{1.1} = 2.004166$

Ans. 2.0040219

19. Evaluate approximately $\int_1^7 y\,dx$, $n = 6$.

x:	1	2	3	4	5	6	7
y:	2.157	3.519	4.198	4.539	4.708	4.792	4.835

Ans. 25.4061

20. Integrate $\int_{0.2}^{1.4}(\sin x - \ln x + e^x)dx$, $n = 12$.

Hint:

x:	0.2	0.3	0.4	0.5	0.6	0.7	
y:	3.0295	2.8493	2.797	2.821	2.8976	3.01465	
x:	0.8	0.9	1.0	1.1	1.2	1.3	1.4
y:	3.166	3.348	3.559	3.8	4.0698	4.3705	4.3704

$I = \frac{3}{10}(0.1)$ [21.0584 + 5(13.5828) + 6(6.62137) + 2(3.16605)]

Exact value: $-\cos x - x(\ln x - 1) + e^x\Big|_{0.2}^{1.4} =$ 4.05095

Ans. 4.05098

21. Evaluate $\int_0^1 \frac{dx}{1+x^2}$, $n = 12$.

Hint:

x:	0	$\frac{1}{12}$	$\frac{1}{6}$	$\frac{1}{4}$	$\frac{1}{3}$	$\frac{5}{12}$	$\frac{1}{2}$
$f(x)$:	1	0.993	0.9729	0.94	0.9	0.852	0.8
x:	$\frac{7}{12}$	$\frac{2}{3}$	$\frac{3}{4}$	$\frac{5}{6}$	$\frac{11}{12}$	1	
$f(x)$:	.746	.6923	.64	.5906	.5434	0.5	

Ans. 0.7603725

Boole's Rule of Integration

The goal in numerical integration is to approximate $\int_a^b f(x)dx$ using $(N+1)$ sample points (x_0, f_0), $(x_1, f_1)\ldots, (x_N, f_N)$ where f_k denotes $f(x_k)$.

Boole's rule is obtained by using a 4th degree Lagrange interpolating polynomial based on the nodes x_0, x_1, x_2, x_3 and x_4. Consider the approximation formula

$$\int_0^4 g(t)dt = w_0 g(0) + w_1 g(1) + w_2 g(2) + w_3 g(3) + w_4 g(4) \quad (1)$$

we determine the weights w_0, w_1, w_2, w_3, w_4 assuming that (1) is exact for the functions $g(t) = 1, t, t^2, t^3, t^4$. Thus when $g(t) = 1$ then $g(0) = g(1) = g(2) = g(3) = g(4) = 1$. From (1)

$$\int_1^4 1 \cdot dt = 4 = w_0 + w_1 + w_2 + w_3 + w_4 \quad (2)$$

For $g(t) = t$, $g(0) = 0$, $g(1) = 1$, $g(2) = 2$, $g(3) = 3$, $g(4) = 4$ so from (1)

$$\int_0^4 t\, dt = \left.\frac{t^2}{2}\right|_0^4 = 8 = 0 + w_1 + 2w_2 + 3w_3 + 4w_4 \quad (3)$$

Similarly for $g(t) = t^2, t^3, t^4$ from (1) we get

$$\frac{64}{3} = w_1 + 4w_2 + 9w_3 + 16w_4 \quad (4)$$

$$64 = w_1 + 8w_2 + 27w_3 + 64w_4 \quad (5)$$

$$\frac{1024}{5} = w_1 + 16w_2 + 81w_3 + 256w_4 \quad (6)$$

Solving equations (1) to (6) we get $w_0 = \frac{14}{45}$, $w_1 = \frac{64}{45}$, $w_2 = \frac{24}{45}$, $w_3 = \frac{64}{45}$, $w_4 = \frac{14}{45}$.

Thus the Boole's rule for the interval $[x_0, x_4]$

$$\int_{x_0}^{x_4} f(x)dx \approx \frac{2h}{45}[7f_0 + 32f_1 + 12f_2 + 32f_3 + 7f_4] \quad (7)$$

Composite Boole's Rule

When Boole's rule is repeated over 4 M subintervals of $[a, b]$, we get the composite Boole's rule as

$$\int_a^b f(x)dx \approx \frac{2h}{45}\sum_{k=1}^{M}[7f_{4k-4} + 32f_{4k-3} + 12f_{4k-2} + 32f_{4k-1} + 7f_{4k}] \quad (8)$$

Note: To apply Boole's rule, the number of subintervals should be multiples of 4 (i.e., 4, 8, 12...) and the numbers of nodes must be odd.

Worked Out Examples

Example 1: Evaluate $\int_1^5 \frac{dx}{x}$ by Boole's rule.

Solution: Take $h = \frac{5-1}{4} = 1$ so that there 4 subintervals.

x	1	2	3	4	5
$f(x) = \frac{1}{x}$	1	$\frac{1}{2}$	$\frac{1}{3}$	$\frac{1}{4}$	$\frac{1}{5}$

Using Boole's rule

$$\int_1^5 \frac{dx}{x} \approx \frac{2.1}{45}\left[7.1 + 32\left(\tfrac{1}{2}\right) + 12\left(\tfrac{1}{3}\right) + 32\left(\tfrac{1}{4}\right) + 7\cdot\tfrac{1}{5}\right]$$

$$= \frac{364}{225} = 1.61777$$

while the exact value is $\ln 5 - \ln 1 = 1.60943$

Example 2: Find the distance travelled by a submarine under polar ice cap from the following table showing the velocity at variance times

t time (hr)	0.00	0.25	0.50	0.75	1.00	1.25	1.50	1.75	2.00
Velocity $v(t)$ km/hr	6.0	7.5	8.0	9.0	8.5	10.5	9.5	7.0	6.0

Solution: We know that

$$d = \text{distance travelled} = \int_{t=0}^{2} v(t)dt$$

We evaluate this definite integral by Boole's rule with $4 \cdot 2 = 8$ subintervals. Here $h = \frac{2-0}{8} = \frac{1}{4} = 0.25$ (and 9 node points). Applying the composite Boole's rule we get

$$d = \frac{2h}{45}[7f_0 + 32f_1 + 12f_2 + 32f_3 + 7f_4] + \frac{2h}{45}[7f_4 + 32f_5 + 12f_6 + 32f_7 + 7f_8]$$

$$= \frac{2(0.25)}{45}[7(6) + 32(7.5) + 12(8) + 32(9) + 7(8.5)] + \frac{2(0.25)}{45}[7(8.5) + 32(10.5) + 12(9.5) + 32(7.0) + 7(6.0)]$$

$$= 16.6777$$

EXERCISE

1. Evaluate $\int_0^1 \frac{dx}{1+x^2}$ by Boole's rule with 12 subintervals.

Ans. Exact value: $\frac{\pi}{4} = 0.7854$, by Boole's rule: 0.8236

Hint: $\frac{2}{45} \cdot \frac{1}{12}[7\{1 + 2(0.9 + 0.746 + 0.692) + 0.5\} \; +32\{0.993 + 0.94 + 0.85 + 0.746 + 0.64 +0.543\} \; +12\{0.973 + 0.8 + 0.59\}] = 0.8236$

2. Evaluate $\int_{0.2}^{1.4} (\sin x - \ln x + e^x)dx$

Ans. Exact value: 4.05025, $n = 12$, $h = 0.1$, by Boole's: 4.04235

Hint: $\frac{2(0.1)}{45}[7\{3.0295 + 2(2.8976 + 3.559) +4.3704\} + 32\{2.8493 + 2.821 + 3.01465 +3.348 + 3.8 + 4.3705\} \; +12(2.797 + 3.166 + 4.0698)] = 4.04235$

3. A river is 80 feet wide. The depth d in feet at a distance x from one bank (side) of the river is given below.

x :	0	10	20	30	40	50	60	70	80
d :	0	4	7	9	12	15	14	8	3

Find approximately the area of cross section of the river.

Ans. 708

Hint: $n = 8$, $h = 10$

Area $= \frac{2(10)}{45}[7\{0 + 2(12) + 3\} \; +32 \; \{4 + 9 + 15 + 8\} +12(7 + 14)] = 708$

4. Find the distance between two stations from the following data consisting of the speeds of an electric train at various times after leaving one station until it stops at the next station.

Speed of train $v(t)$ in mph	0	13	33	$39\frac{1}{2}$	40	40	36	15	0
t in mt	0	$\frac{1}{2}$	1	$1\frac{1}{2}$	2	$2\frac{1}{2}$	3	$3\frac{1}{2}$	4

Ans. 1.788148 miles

Hint: Distance $= \int_0^4 v(t)dt$, $n = 8$, $h = \frac{1}{2}$mt $= \frac{1}{120}$hr, $= \frac{2}{45} \cdot \frac{1}{120}[7(0 + 40 + 40 + 0) + 32(13 + 39\frac{1}{2} +40 + 15) + 12(33 + 36)]$

32.15 SPLINE INTERPOLATION

In analyzing experimental data, in ascertaining the relations among variables and in design work, the common problem encountered is fitting a curve through specified points in a plane. *Interpolating curve* is a curve that passes through a set of points in a plane and the curve is said to *interpolate* at these points.

It is observed that approximation of arbitrary functions on a closed interval becomes oscillatory for higher-degree polynomials. By dividing the given interval into a collection of subintervals, *piecewise polynomial approximation* consists of constructing a (generally) different approximating polynomial on *each* subinterval. *Cubic spline interplation* is the most common piecewise approximation using cubic polynomial, known as *spline function*, connecting each pair of data points (or nodes or knots). Cubic splines are third-order curves employed to connect each pair of data points.

Consider a set of $(n + 1)$ data points, which need not be evenly spaced, $(x_0, f(x_0))$, $(x_1, f(x_1))$, $(x_2, f(x_2)) \ldots (x_n, f(x_n))$

or $\quad (x_i, f_i)$ for $i = 0, 1, 2, \ldots n$. $\quad$ (1)

Between each pair of adjacent points from x_i to x_{i+1}, we fit an nth degree polynomial $s_i(x)$. Thus the spline (function) curve can be of any degree. By far, cubic splines are the most popular and are useful version in engineering practice since the cubic spline interpolation gives an interpolation formula that is smooth in the first derivative and continuous, both within an interval and its boundaries (end points x_0 and x_n). The name 'spline' is originated from the draftman's spline which is a drafting aid consisting of a thin flexible rod or strip of wood that is bent to draw a smooth curve through the points to be interpolated.

Linear Splines

Linear splines are first-order splines defined by linear functions. Geometrically they are straight lines connecting two adjacent data points. For a given set of $(n + 1)$ data points (1), the set of linear functions defining linear splines in the 'n' intervals are:

$S_1(x) = f(x_0) + m_0(x - x_0)$; $x_0 \le x \le x_1$.

$S_2(x) = f(x_1) + m_1(x - x_1)$; $x_1 \le x \le x_2$.

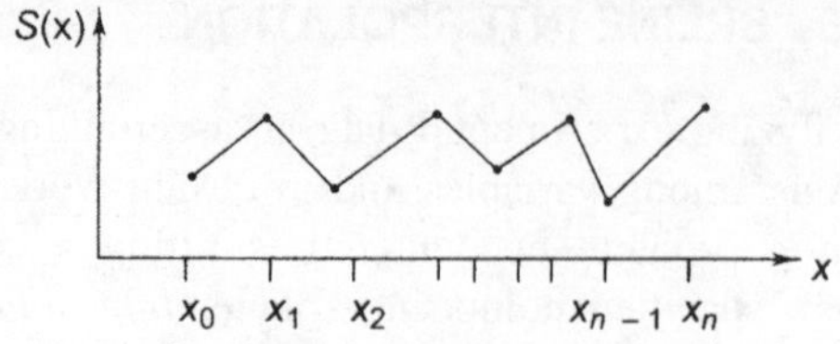

Fig. 32.15 Piecewise linear interpolation

....

$S_n(x) = f(x_n - 1) + m_{n-1}(x - x_{n-1})$;

$x_{n-1} \le x \le x_n$

Here m_i is the slope of the straight line connecting the data points (or nodes) x_i and x_{i+1} and is given by

$$m_i = \frac{f(x_{i+1}) - f(x_i)}{x_{i+1} - x_i} \qquad (2)$$

Fitting of linear splines is piecewise linear interpolation and the interpolating curve is broken curve joining 'n' straight line segments. Disadvantage of linear splines is that at the point (known as *knot*) where two splines meet, the slope changes abruptly i.e., first derivative is discontinuous.

Quadratic Splines

Interpolation of second-order polynomials leads to quadratic splines having continuous first derivatives at the knots.

For each interval we fit a parabola given by

$$s_i(x) = a_i x^2 + b_i x + c_i$$

For $n + 1$ data points we have n intervals. So we should determine $3 \times n$ unknown constants (a's, b's and c's). The $3n$ condition to find the $3n$ unknowns are given by the following equations.

1. At the interior knots, the function values must be equal.

$$a_{i-1}x_{i-1}^2 + b_{i-1}x_{i-1} + c_{i-1} = f(x_{i-1})$$

$$a_i x_{i-1}^2 + b_i x_{i-1} + c_i = f(x_{i-1})$$

for $i = 2, n$ giving raise to $2(n-1) = 2n - 2$ conditions.

2. The first and last functions must pass through the two end points x_0 and x_n.

$$a_1 x_0^2 + b_1 x_0 + c_1 = f(x_0)$$

$$a_n x_n^2 + b_n x_n + c_n = f(x_n)$$

giving 2 conditions.

3. At the interior knots, the first derivatives $f'(x) = 2ax + b$ must be equal.

$$2a_{i-1}x_i + b_{i-1} = 2a_i x_i + b_i$$

for $i = 2$ to n giving $n - 1$ conditions

4. Assume that the second derivative is zero at the first point x_0. Then the first two points are connected by a straight line, so $a_1 = 0$.

Thus solving $(2n - 2) + 2 + (n - 1) + 1 = 3n$ conditions above we get the $3n$ unknown constants a, b, c's. The drawback is the straight line connecting the first two points.

Cubic Splines

A spline of at least $(m + 1)$ order should be used in order to make sure that the mth derivatives are continuous at knots. Cubic splines employing third order polynomials are very popular in practice because they ensure that the first and second derivatives are continuous.

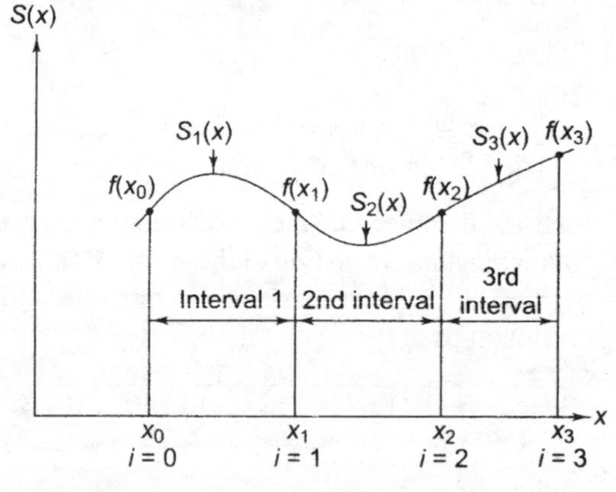

Fig. 32.16 Piecewise cubic spline approximation

Let us fit a (different) third order polynomial

$$S_i(x) = a_i x^3 + b_i x^2 + c_i x + d_i \qquad (3)$$

for **each** of the n intervals between the $(n + 1)$ data points (knots) $(i = 0, 1, 2, \ldots n)$. The problem is to find the $4 \times n = 4n$ unknown constants

($a_i's$, $b_i's$, $c_i's$ and $d_i's$). We have the following conditions

a. $2n-2$ conditions since the function values must be equal at the $(n-1)$ interior knots.
b. 2 conditions since the first and last functions must pass through the two end points x_0 and x_n.
c. $(n-1)$ conditions since the first derivatives must be equal at the $(n-1)$ interior points.
d. $(n-1)$ conditions since the second derivatives must be equal at the $(n-1)$ interior knots.

The above conditions are summed to $(2n-2)+(2)+(n-1)+(n-1)=4n-2$, thus short of 2 more conditions to solve the $4n$ unknown constants. Two end point constraint are assumed as follows:

Natural spline for which the end point constraints are assumed as "second derivatives at the end knots are zero" i.e., $S''(x_0)=S''(x_n)=0$. Natural splines are more frequently used in which the end cubic (in the first and last intervals) approach linearity (straight line) at their extremities. This matches precisely with the drafting spline.

The cubic spline for *each* interval (x_{i-1}, x_i) is given by

$$S_i(x)=\frac{f''(x_{i-1})}{6(x_i-x_{i-1})}(x_i-x)^3+\frac{f''(x_i)}{6(x_i-x_{i-1})}(x-x_{i-1})^3+\left[\frac{f(x_{i-1})}{(x_i-x_{i-1})}-\frac{f''(x_{i-1})(x_i-x_{i-1})}{6}\right](x_i-x)+\left[\frac{f(x_i)}{(x_i-x_{i-1})}-\frac{f''(x_i)\cdot(x_i-x_{i-1})}{6}\right](x-x_{i-1}). \quad (4)$$

Equation (4) contains only two unknowns $f''(x_{i-1})$ and $f''(x_i)$, the second derivatives at the end points of the interval (x_{i-1}, x_i), which are determined from the following equation.

$$(x_i-x_{i-1})f''(x_{i-1})+2(x_{i+1}-x_{i-1})f''(x_i)+(x_{i+1}-x_i)f''(x_{i+1})=\frac{6}{(x_{i+1}-x_i)}[f(x_{i+1})-f(x_i)]+\frac{6}{(x_i-x_{i-1})}[f(x_{i-1})-f(x_i)] \quad (5)$$

Equation (5) written for the $(n-1)$ interior points results in $(n-1)$ simultaneous equations for the second derivatives. Note that these $(n-1)$ equations form a tridiagonal system which can be solved in an extremely efficient manner using computer algorithms.

WORKED OUT EXAMPLES

Example 1: Fit a linear spline to the following data:

x:	1	3	6	8
$y=f(x)$:	4	5.5	7	9.5

Estimate the value at $x=2, 4, 7$.

Solution: There are $n=4$ points and 3 intervals. Assume that

$x_0=1,\quad x_1=3,\quad x_2=6,\quad x_3=8,$
$f(x_0)=4,\quad f(x_1)=5.5,\quad f(x_2)=7,\quad f(x_3)=9.5$

The three first order linear splines in the 3 intervals are:

$f(x)=f(x_0)+m_0(x-x_0),\ x_0\le x\le x_1$
$f(x)=f(x_1)+m_1(x-x_1),\ x_1\le x\le x_2$
$f(x)=f(x_2)+m_2(x-x_2),\ x_2\le x\le x_3$

Here $m_i=\frac{f(x_{i+1})-f(x_i)}{x_{i+1}-x_i}$ is the slope of straight line connecting the two points. For the given data

$$m_0=\frac{f(x_1)-f(x_0)}{x_1-x_0}=\frac{5.5-4}{3-1}=\frac{1.5}{2}=0.75$$

$$m_1=\frac{f(x_2)-f(x_1)}{x_2-x_1}=\frac{7-5.5}{6-3}=\frac{1.5}{3}=0.5$$

$$m_2=\frac{f(x_3)-f(x_2)}{x_3-x_2}=\frac{9.5-7}{8-6}=\frac{2.5}{2}=1.25$$

Thus the three linear splines are

$f(x)=4+0.75(x-1)$, when $1\le x\le 3$
$f(x)=5.5+0.5(x-3)$, when $3\le x\le 6$
$f(x)=7+1.25(x-6)$, when $6\le x\le 8$

Since $x=2$ lies in the first interval $1\le x\le 3$, we

get $f(2)$ from the first equation.

$$f(2) = 4 + 0.75(2 - 1) = 4.75$$

Similarly $f(7) = 7 + 1.25(7 - 6) = 8.25$.
Since $x = 4$ lies in the second interval $3 \le x \le 6$, we get $f(4)$ from the second equation:

$$f(4) = 5.5 + 0.5(4 - 3) = 6$$

Example 2: Fit quadratic splines to the following data. Use the results to estimate the value at $x = 2, 4, 7$.

x:	1	3	6	8
$f(x)$:	4	5.5	7	9.5

Solution: For the 4 data points in 3 intervals we fix three quadratic splines of the form

$$S_i(x) = a_i x^2 + b_i x + c_i$$

we have to determine $3 \times 3 = 9$ unknown $a_1, b_1, c_1, a_2, b_2, c_2, a_3, b_3, c_3$. We need $3 \cdot 3 = 9$ equations or conditions to determine these 9 unknowns. There are two interior points x_1 and x_2.

1. Function values must be equal at the two interior points, giving rise to 4 equations.

$$a_{i-1}x_{i-1}^2 + b_{i-1}x_i + c_{i-1} = f(x_{i-1})$$

$$a_i x_{i-1}^2 + b_i x_{i-1} + c_i = f(x_{i-1})$$

for $i = 2$ to 3.

For $i = 2$, $a_1 x_1^2 + b_1 x_1 + c_1 = f(x_1)$

$$a_2 x_1^2 + b_2 x_1 + c_2 = f(x_1)$$

i.e., $$9a_1 + 3b_1 + c_1 = 5.5 \quad (1)$$

$$9a_2 + 3b_2 + c_2 = 5.5 \quad (2)$$

For $i = 3$, $a_2 x_2^2 + b_2 x_2 + c_2 = f(x_2)$

$$a_3 x_2^2 + b_3 x_2 + c_3 = f(x_2)$$

i.e., $$36a_2 + 6b_2 + c_2 = 7 \quad (3)$$

$$36a_3 + 6b_3 + c_3 = 7 \quad (4)$$

2. The first and last functions must pass through the end points x_0 and x_0 giving rise to 2 equations.

$$a_1 x_0^2 + b_1 x_0 + c_1 = f(x_0)$$

$$a_n x_n^2 + b_n x_n + c_n = f(x_n)$$

i.e.,

$$a_1 + b_1 + c_1 = 4 \quad (5)$$

$$64a_3 + 8b_3 + c_3 = 9.5 \quad (6)$$

3. The first derivatives at the interior knots must be equal.

$$f'(x) = 2ax + b$$

$$2a_{i-1}x_i + b_{i-1} = 2a_i x_i + b_i$$

for $i = 2, 3$

$$2a_1 x_2 + b_1 = 2a_2 x_2 + b_2$$

$$2a_2 x_3 + b_2 = 2a_3 x_3 + b_3$$

$$\text{i.e., } 6a_1 + b_1 = 6a_2 + b_2 \quad (7)$$

$$12a_2 + b_2 = 12a_3 + b_3 \quad (8)$$

4. Assume that the second derivative is zero at the first point i.e.,

$$a_1 = 0 \quad (9)$$

Using (9) $a_1 = 0$ in (1) and (5) we get

$$3b_1 + c_1 = 5.5$$

$$b_1 + c_1 = 4$$

solving $b_1 = 0.75$, $c_1 = 3.25$

From (2) and (3) we get

$$9a_2 + 3b_2 + c_2 = 5.5$$
$$36a_2 + 6b_2 + c_2 = 7$$
$$\overline{27a_2 + 3b_2 \qquad = 1.5}$$

Form (7) $6a_2 + b_2 = 0 + b_1 = 0.75$
Solving $a_2 = \frac{-0.75}{9}$, $b_2 = 1$, $c_2 = 4$
From (4) and (6)

$$36a_3 + 6b_3 + c_3 = 7$$
$$64a_3 + 8b_3 + c_3 = 9.5$$

Subtracting $28a_3 + 2b_3 = 2.5$
From (8): $12a_3 + b_3 = 12a_2 + b_2 = 0$
Solving $a_3 = \frac{2.5}{4}$, $b_3 = -7.5$, $c_3 = 29.5$
Thus the values of the 9 coefficients are
$a_1 = 0$, $b_1 = 0.75$, $c_1 = 3.25$,
$a_2 = \frac{-0.75}{9}$, $b_2 = 1$, $c_2 = 4$,
$a_3 = \frac{2.5}{4}$, $b_3 = -7.5$, $c_3 = 29.5$,
Then the three required quadratic splines are
$S_1(x) = 0 + 0.75x + 3.25$, in $1 \le x \le 3$
$S_2(x) = \frac{-0.75}{9}x^2 + x + 4$, in $3 \le x \le 6$

$S_3(x) = \frac{2.5}{4}x^2 - 7.5x + 29.5$, in $6 \le x \le 8$
At $x = 2$, $S_1(2) = 4.75$
At $x = 4$, $S_2(4) = 6.6666$
At $x = 7$, $S_3(7) = 7.625$

Example 3: Fit cubic splines to the following data and estimate the value at $x = 2, 4, 7$ and $y'(4)$.

x:	1	3	6	8
$y = f(x)$:	4	5.5	7	9.5

Solution: Assume a cubic spline a third order polynomial

$$S_i(x) = a_i x^3 + b_i x^2 + c_i x + d_i$$

For the 4 data points and 3 intervals, we have to determine $4 \times 3 = 12$ unknown constants $a_1, b_1, c_1, d_1, a_2, b_2, c_2, d_2; a_3, b_3, c_3, d_3$.

1. The second derivatives at the end knots are zero; i.e., $f''(1) = f''(8) = 0$.
2. The second derivatives at the end of each interval (at the two interior points) are given by

$$(x_i - x_{i-1})f''(x_{i-1}) + 2(x_{i+1} - x_{i-1})f''(x_i) + (x_{i+1} - x_i)f''(x_{i+1})$$
$$= \frac{6}{(x_{i+1} - x_i)}[f(x_{i+1}) - f(x_i)] + \frac{6}{(x_i - x_{i-1})}[f(x_{i-1}) - f(x_i)] \quad (1)$$

For $i = 1$, from (i) we have

$$(x_i - x_0)f''(x_0) + 2(x_2 - x_0)f''(x_1) + (x_2 - x_1)f''(x_2)$$
$$= \frac{6}{(x_2 - x_0)}[f(x_2) - f(x_1)] + \frac{6}{(x_1 - x_0)}[f(x_0) - f(x_1)]$$

with $x_0 = 1, x_1 = 3, x_2 = 6, f_0 = 4, f_1 = 5.5, f_2 = 7$, we have

$$(3-1)f''(1) + 2(6-1)f''(3) + (6-3)f''(6)$$
$$= \frac{6}{(6-1)}[7 - 5.5] + \frac{6}{(3-1)}[4 - 5.5].$$

For natural spline, $f''(1) = 0$. So above equation reduces to

$$10f''(3) + 3f''(6) = -2.7 \quad (2)$$

Similarly for $i = 2$, from (1) we have

$$(x_2 - x_1)f''(x_1) + 2(x_3 - x_1)f''(x_2) + (x_3 - x_2)f''(x_3)$$
$$= \frac{6}{(x_3 - x_2)}[f(x_3) - f(x_2)] + \frac{6}{(x_2 - x_1)}[f(x_1) - f(x_2)]$$

with $x_1 = 3, x_2 = 6, x_3 = 8, f_1 = 5.5, f_2 = 7, f_3 = 9.5$, we get $(6-3)f''(3) + 2(8-3)f''(6) + (8-6)f''(8)$

$$= \frac{6}{(8-6)}[9.5 - 7] + \frac{6}{(6-3)}[5.5 - 7].$$

For natural spline $f''(8) = 0$. So above equation reduces to

$$3f''(3) + 10f''(6) = 4.5 \quad (3)$$

solving (2) and (3) we get
$f''(3) = -0.4451$, $f''(6) = 0.58352$

The equation of the cubic spline is

$$S_i = \frac{f''(x_{i-1})}{6(x_i - x_{i-1})}(x_i - x)^3 + \frac{f''(x_i)}{6(x_i - x_{i-1})}(x - x_{i-1})^3$$
$$+ \left[\frac{f(x_{i-1})}{(x_i - x_{i-1})} - \frac{f''(x_{i-1})(x_i - x_{i-1})}{6}\right](x_i - x)$$
$$+ \left[\frac{f(x_i)}{(x_i - x_{i-1})} - \frac{f''(x_i)(x_i - x_{i-1})}{6}\right](x - x_{i-1}) \quad (4)$$

For $i = 1$,

$$S_1(x) = \frac{f''(x_0)}{6(x_1 - x_0)}(x_1 - x)^3 + \frac{f''(x_1)}{6(x_1 - x_0)}(x - x_0)^3$$
$$+ \left[\frac{f(x_0)}{x_1 - x_0} - \frac{f''(x_0)(x_1 - x_0)}{6}\right](x_1 - x)$$
$$+ \left[\frac{f(x_1)}{x_1 - x_0} - \frac{f''(x_1)(x_1 - x_0)}{6}\right](x - x_0)$$

with $x_0 = 1, x_1 = 3, x_2 = 6, f_0 = 4, f_1 = 5.5, f_2 = 7$, $f''(1) = 0$, $f''(3) = -0.4451$, $f''(6) = 0.58352$ the above equation reduces to

$$S_1(x) = 0 + \frac{(-0.4451)}{6(3-1)}(x-1)^3 +$$

$$+\left[\frac{4}{(3-1)}-0\right](3-x)$$

$$+\left[\frac{5.5}{(3-1)}-(-0.4451)\frac{(3-1)}{6}\right](x-1) \qquad (5)$$

Thus the cubic spline in the first interval [1, 3] is

$$S_1(x) = -0.0371(x-1)^3 + 2(3-x) + 2.898(x-1)$$

For $i = 2$, form (4) we get

$$S_2(x) = \frac{f''(x_1)}{6(x_2-x_1)}(x_2-x)^3 + \frac{f''(x_2)}{6(x_2-x_1)}(x-x_1)^3$$
$$+\left[\frac{f(x_1)}{(x_2-x_1)}-\frac{f''(x_1)(x_2-x_1)}{6}\right](x_2-x)$$
$$+\left[\frac{f(x_2)}{x_2-x_1}-\frac{f''(x_2)(x_2-x_1)}{6}\right](x-x_1)$$

with the substitution of data, we get

$$S_2(x) = \frac{-0.4451}{6(6-3)}(6-x)^3 + \frac{0.58352}{6(6-3)}(x-3)^3$$
$$+\left[\frac{5.5}{(6-3)}-\frac{(-0.4451)(6-3)}{6}\right](6-x)$$
$$+\left[\frac{7}{6-3}-\frac{(0.58352)(6-3)}{6}\right](x-3) \qquad (6)$$

Thus the second cubic spline in the second interval [3, 6] is

$$S_2(x) = -0.02473(6-x)^3 + 0.03242(x-3)^3$$
$$+(2.05588)(6-x) + (2.04154)(x-3)$$

Similarly for $i = 3$, from (4) we get

$$S_3(x) = \frac{f''(x_2)}{6(x_3-x_2)}(x_3-x)^3 + \frac{f''(x_3)}{6(x_3-x_2)}$$
$$+\left[\frac{f(x_2)}{x_3-x_2}-\frac{f''(x_2)}{6}(x_3-x_2)\right](x_3-x)$$
$$+\left[\frac{f(x_3)}{x_3-x_2}-\frac{f''(x_3)(x_3-x_2)}{6}\right](x-x_2)$$

with the given data

$$f_3(x) = \frac{0.58352}{6(8-6)}(8-x)^3 + 0$$
$$+\left[\frac{7}{8-6}-\frac{0.58352}{6}(8-6)\right](8-x)$$
$$+\left(\frac{8}{8-6}-0\right)(x-6)$$

Thus the third cubic spline in the 3rd interval [6, 8] is

$$S_3(x) = 0.04863(8-x)^3 + 3.3055(8-x)$$
$$+4(x-6) \qquad (7)$$

Now using (5), (6), (7), we get

$$f_1(2) = 4.8609$$
$$f_2(4) = 5.98788$$
$$f_3(7) = 7.35413$$

Since 4 is in the interval [3, 6] differentiate $f_2(x)$ to find $f'(4)$. So

$$f_1'(x) = +0.07419(6-x)^2 + 0.09726(x-3)^2$$
$$-2.05588 + 2.04154$$

Thus $f'(4) = 0.37968$

EXERCISE

1. Obtain the natural cubic spline approximation for the following data

x:	0	1	2	3
$f(x)$:	1	2	33	244

Ans. $f_1(x) = -4x^3 + 5x + 1$ in [0, 1]
$f_2(x) = 50x^3 - 162x^2 + 167x - 53$ in [1, 2]
$f_3(x) = -46x^3 + 414x^2 - 985x + 715$ in [2, 3]

2. Fit a natural cubic spline to the following data

x:	1	2	3	4
y:	1	5	11	8

Hence compute $y(1.5)$, $y'(2)$

Ans. $y(1.5) = 103/40$, $y'(2.0) = 94/15$
$f_1(x) = \frac{1}{15}[17x^3 - 51x^2 + 94x - 45]$ in $1 \le x \le 2$
$f_2(x) = \frac{1}{15}[-55x^3 + 381x^2 - 770x + 531]$ in $2 \le x \le 3$
$f_3(x) = \frac{1}{15}[38x^3 - 456x^2 + 1741x - 1980]$ in $3 \le x \le 4$

3. Fit a natural cubic spline and evaluate the spline at 0.66, 1.75.

x:	0.0	1.0	1.5	2.25
$f(x)$:	2.000	4.4366	6.7134	13.9130

Ans. $f_1(x) = 0.3820(x-0)^3 + 0(x-0)^2$
$+2.0546(x-0) + 2.000$ in $[0, 1.0]$
$f_2(x) = 3.1199(x-1)^3 + 1.146(x-1)^2$
$+3.2005(x-1) + 4.4366$ in $[1, 1.5]$
$f_3(x) = -2.5893(x-1.5)^3 + 5.8259(x-1.5)^2$
$+6.6866(x-1.5) + 6.7134$ in $[1.5, 2.25]$
$f_1(0.66) = 3.4659$ Exact: 3.4340
$f_3(1.75) = 8.7087$ Exact: 8.4467
Exact $f(x) = 2e^x - x^2$

4. Fit a natural cubic spline to the following data

x:	0	1	2	3
y:	0	0.5	2	1.5

Ans. $S_1(x) = 0.4x^3 + 0.1x$ in $0 \le x \le 1$
$S_2(x) = -(x-1)^3 + 1.2(x-1)^2 +$
$+1.3(x-1) + 0.5$ in $1 \le x \le 2$
$S_3(x) = 0.6(x-2)^3 - 1.8(x-2)^2 +$
$+0.7(x-2) + 2.0$ in $2 \le x \le 3$.

5. Fit a cubic natural spline

x:	0	1	2
y:	4	1	2

Ans. $s_1(x) = 4 - 4x + x^3$ in $[0, 1]$
$s_2(x) = 1 - (x-1) + 3(x-1)^2 - (x-1)^3$ in $[1, 2]$

6. Fit (a) linear spline (b) quadratic spline (c) cubic natural spline to the following data.

x:	3.0	4.5	7.0	9.0
$y = f(x)$:	2.5	1.0	2.5	0.5

Evaluate the function at $x = 5$.

Ans. (a) Linear splines
$f_1(x) = 2.5 - (x-3)$ in $[3, 4.5]$
$f_2(x) = 1.0 + 0.6(x-4.5)$ in $[4.5, 7]$
$f_3(x) = 2.5 - 1(x-7)$ in $[7, 9]$
$f_2(5) = 1.3$
(b) Quadratic splines
$f_1(x) = -x + 5.5$ in $[3, 4.5]$
$f_2(x) = 0.64x^2 - 6.76x + 18.46$ in $[4.5, 7.0]$
$f_3(x) = -1.6x^2 + 24.6x - 91.3$ in $[7.0, 9.0]$
$f_2(5) = 0.66$
(c) Cubic natural splines
In $[3, 4.5]$
$f_1(x) = 0.193939(x-3)^3 + 1.66667(4.5-x)$
$+0.23030(x-3)$
$f_2(x) = 0.1163647(7-x)^3 - 0.116364(x-4.5)^3 \quad -0.327273(7-x) + 1.727273(x-4.5)$
in $[4.5, 7]$
$f_3(x) = -0.145455(9-x)^3 + 1.831818(9-x)$
$+0.25(x-7)$ in $[7, 9]$
$f_2(5) = 1.1255$

7. Fit natural cubic splines for the following data.

x:	-10	0	10	20	30
y:	.99815	.99987	.99973	.99823	.99567

Ans. In $-10 \le x \le 0$.
$f_1(x) = -0.00000042(x+10)^3 +$
$0.000214(x+10) + .99815$
In $0 \le x \le 10$
$f_2(x) = 0.00000024x^3 - 0.0000126x^2 +$
$.000088x + .99987$
In $10 \le x \le 20$
$f_3(x) = -0.00000004(x-10)^3 -$
$0.000054(x-10)^2 - 0.000092(x-10) + .9973$
In $20 \le x \le 30$
$f_4(x) = .00000022(x-20)^3 - 0.0000066(x-20)^2 - 0.000212(x-20) + .99823$

32.16 NUMERICAL METHODS IN LINEAR ALGEBRA: GAUSS-SEIDEL METHOD

Solutions to system of non-homogeneous linear equations in n unknowns can be obtained by direct (or reduction or elimination) methods such as Cramer's rule, matrix inversion method, Gaussian elimination, Gauss-Jordan method, Cholosky's (Crout's) method. The solutions can also be obtained by indirect (or iterative methods) which include the Gauss-Seidel method, (South Well's) relaxation method, Sweep method.

Diagonal:

A system of simultaneous linear equations is called a diagonal system if in each equation the coefficient of a different unknown is greater in absolute value than the sum of the absolute values of the other coefficients.

Usually, the large coefficient appears in the main diagonal position a_{ii}. Gauss-Seidel method is an iterative method which gives approximate solution by successive approximations. This method will always converge rapidly if the system is diagonal.

Gauss-Seidel method

Step I: Solve each of the n equations of the system for the unknown with the largest coefficient. Suppose a_{11} is the largest coefficient of x_1 in the first equation, then divide Equation (1) throughout by a_{11}, resulting in a new equation with x_1 expressed in terms of the remaining variables $x_2, x_3, \ldots, x_n$. Similarly, dividing the second equation by a_{22}, 3rd equation by $a_{33}\ldots$ and nth equation by a_{nn}, Then

$$\left.\begin{aligned} x_1 &= a_{12}^*x_2 + a_{13}^*x_3 + \cdots + a_{1n}^*x_n + k_1 \\ x_2 &= a_{21}^*x_1 + a_{23}^*x_3 + \cdots + a_{2n}^*x_n + k_2 \\ &\cdots\cdots\cdots\cdots\cdots\cdots \\ x_n &= a_{n1}^*x_1 + a_{n2}^*x_2 + \cdots + a_{n,n-1}^*x_{n-1} + k_n \end{aligned}\right\} \quad (1)$$

Step II Assume an initial solutions (guess) $(x_1^{(0)}, x_2^{(0)}, x_3^{(0)}, \ldots, x_n^{(0)})$. Here 0 indicates the starting solution. It may be taken as $(0, 0, 0, \ldots, 0)$.

Step III Substitute $x_i^{(0)}$ in the right hand side members of (1) which results in the new values $x_i^{(1)}$, which is first approximation. Now substitute $x_i^{(1)}$ in the R.H.S. of (1) which yields the second approximation $x_i^{(2)}$. Repeating this procedure, we arrive at $(x_1^{(m)}, x_2^{(m)}, \ldots, x_n^{(m)})$ at the mth iteration. Step III is repeated until required accuracy is obtained.

Note 1: Always use the last calculated (i.e., latest) values for the other unknown, say

$$x_2^{(k+1)} = a_{21}^*x_1^{(k+1)} + a_{23}^*x_3^{(k)} + \cdots + a_{2n}^*x_n^{(k)} + k_1$$

Note 2: Choice of initial value will not affect the solution (we get same solution) but will affect the number of iterations for convergence.

WORKED OUT EXAMPLES

Example: Using Gauss-Seidel method, solve the following system of equations starting with initial solution as (a) $(0, 0, 0)$ (b) $\left(\frac{9}{5}, \frac{4}{5}, \frac{6}{5}\right)$.

$$\begin{bmatrix} 5 & -1 & 0 \\ -1 & 5 & -1 \\ 0 & -1 & 5 \end{bmatrix}\begin{bmatrix} x_1 \\ x_2 \\ x_3 \end{bmatrix} = \begin{bmatrix} 9 \\ 4 \\ -6 \end{bmatrix}$$

Solution: **(a)** System is

$$5x_1 - x_2 = 9$$
$$-x_1 + 5x_2 - x_3 = 4$$
$$-x_2 + 5x_3 = -6$$

Step I: Rewriting the system

$$\left.\begin{aligned} x_1 &= \frac{9}{5} + \frac{1}{5}x_2 \\ x_2 &= \frac{4}{5} + \frac{1}{5}x_1 + \frac{1}{3}x_3 \\ x_3 &= -\frac{6}{5} + \frac{1}{5}x_2 \end{aligned}\right\} (*)$$

Step II: Assume initial solution as

$$(x_1 = 0, x_2 = 0, x_3 = 0)$$

Step III: Using the initial solution in R.H.S. of (*)

$$x_1^{(1)} = \frac{9}{5} + \frac{1}{5}\cdot 0 = \frac{9}{5} = 2.032$$
$$x_2^{(1)} = \frac{4}{5} + \frac{1}{5}\cdot\frac{9}{5} = \frac{29}{25} = 1.0128$$
$$x_3^{(1)} = \frac{-6}{5} + \frac{1}{5}\cdot\frac{29}{25} = -\frac{121}{125} = -0.99744$$

Note that in the calculation of $x_2^{(1)}$, the latest (available) value $x_1^{(1)} = \frac{9}{5}$ is used (but not $x_1^{(0)} = 0$). Similarly, in the calculation of $x_3^{(1)}$ the latest value $x_2^{(1)} = \frac{29}{25}$ is used (but not $x_2^{(0)} = 0$). Thus the first approximation is $\left(\frac{9}{5}, \frac{29}{25}, \frac{-121}{125}\right)$.
Second iteration:

$$x_1^{(2)} = \frac{9}{5} + \frac{1}{5}\left(\frac{29}{25}\right) = \frac{254}{125} = 2.032$$
$$x_2^{(2)} = \frac{4}{5} + \frac{1}{5}\left(\frac{254}{125}\right) - \frac{1}{5}\left(\frac{121}{125}\right) = \frac{633}{625} = 1.0128$$
$$x_3^{(2)} = \frac{-6}{5} + \frac{1}{5}\cdot\frac{633}{625} = \frac{-3117}{3125} = -0.99744$$

Third iteration:

$$\left\{\begin{aligned} x_1^{(3)} &= \frac{9}{5} + \frac{1}{5}(1.0128) = 2.00256 \\ x_2^{(3)} &= \frac{4}{5} + \frac{1}{5}(2.00256) - \frac{1}{3}(-0.99744) = 0.868032 \\ x_3^{(3)} &= \frac{-6}{5} + \frac{1}{5}(.868032) = -1.0263936. \end{aligned}\right.$$

Similarly, we get at the 4th iteration $x_1^{(4)} = 2.002048$, $x_2^{(4)} = 1.0000829$, $x_3^{(4)} = -0.99998$

At 5th iteration: $x_1^{(5)} = 2.0000165$, $x_2^{(5)} = 1.0000073$, $x_3^{(5)} = -0.99999$

(b) Initial approximation: $x_1^{(0)} = \frac{9}{5}$, $x_2^{(0)} = \frac{4}{5}$, $x_3^{(0)} = \frac{-6}{5}$

1st iteration : 1.96, 0.952, −1.0096

2st iteration : 1.9904, 0.99616, −1.000768

3st iteration : 1.999232, 0.99969, −1.000064

4st iteration : 1.999938, 0.99997, 1.0000049

5st iteration : 1.999994, 0.99999, −1.0000004

Note 1: Exact solution is 2, 1, −1

EXERCISE

Solve the following system of equations by Gauss-Seidel method:

1. $10x_1 + 8x_2 + 6x_3 = 16.4$, $10x_2 + 8x_3 + 4x_4 = -3.8$, $2x_1 + 10x_3 + 2x_4 = 36.9$, $x_1 + 6x_3 + 10x_4 = 30.9$

Ans. $x_1 = 2.4$, $x_2 = -3.2$, $x_3 = 3$, $x_4 = 1.05$. At 6th iteration: $x_1 = 2.3721$, $x_2 = -3.19096$, $x_3 = 3.00525$, $x_4 = 1.049632$

2. $8x_1 + x_2 - x_3 = 8$, $2x_1 + x_2 + 9x_3 = 12$, $x_1 - 7x_2 + 2x_3 = -4$.

Ans. $x_1 = 1, x_2 = 1, x_3 = 1$

3. Start with (2, 2, −1) and solve $5x_1 - x_2 + x_3 = 10$, $2x_1 + 4x_2 = 12$, $x_1 + x_2 + 5x_3 = -1$.

Ans. $x_1 = 2.5555$, $x_2 = 1.7222$, $x_3 = -1.0555$

4. $10x_1 + x_2 + x_3 = 12$, $2x_1 + 10x_2 + x_3 = 13$, $2x_1 + 2x_2 + 10x_3 = 14$.

Ans. $x_1 = 1, x_2 = 1, x_3 = 1$

5. $10x_1 - 2x_2 - x_3 - x_4 = 3$, $-2x_1 + 10x_2 - x_3 - x_4 = 15$, $-x_1 - x_2 + 10x_3 - 2x_4 = 27$, $-x_1 - x_2 - 2x_3 + 10x_4 = -9$.

Ans. $x_1 = 1, x_2 = 2, x_3 = 3, x_4 = 0$

6. $4x_1 + 2x_2 + x_3 = 11$, $-x_1 + 2x_2 = 3$, $2x_1 + x_2 + 4x_3 = 16$.

Ans. $x_1 = 1, x_2 = 2, x_3 = 3$

7. $20x_1 + x_2 - 2x_3 = 17$, $3x_1 + 20x_2 - x_3 = -18$, $2x_1 - 3x_2 + 20x_3 = 25$.

Ans. $x_1 = 1, x_2 = -1, x_3 = 1$.

8. $3x_1 - 0.1x_2 - 0.2x_3 = 7.85$, $0.1x_1 + 7x_2 - 0.3x_3 = -19.3$, $0.3x_1 - 0.2x_2 + 10x_3 = 71.4$.

Ans. $x_1 = 3$, $x_2 = -2.5$, $x_3 = 7$

9. $27x_1 + 6x_2 - x_3 = 85$, $6x_1 + 15x_2 + 2x_3 = 72$, $x_1 + x_2 + 54x_3 = 110$.

Ans. $x_1 = 2.4255$, $x_2 = 3.5730$, $x_3 = 1.9260$

10. $x_1 - 8x_2 + 3x_3 = -4$, $2x_1 + x_2 + 9x_3 = 12$, $8x_1 + 2x_2 - 2x_3 = 8$.

Ans. $x_1 = 1, x_2 = 1, x_3 = 1$.

32.17 LARGEST EIGEN VALUE AND THE CORRESPONDING EIGEN VECTOR: BY POWER METHOD

Bounds for eigen values λ of an arbitrary n rowed square matrix $A = (a_{jk})$ are given by circular disks from **Gershgorin theorem**

$$|a_{kk}-\lambda| \le |a_{k1}|+|a_{k2}|+\cdots+|a_{k,k-1}|+|a_{k,k+1}|+\cdots+|a_{kn}|$$

for some k where $1 \le k \le n$. Thus all the eigen values lie within these n circular disks (some may be identical) with centres at the diagonal elements a_{kk} and radii equal to the sum of the absolute values of the elements of the kth row baring a_{kk} (except a_{kk}).

Ex: $A = \begin{pmatrix} 5 & 1 \\ 1 & 5 \end{pmatrix}$, eigen values 4, 6. Gershgorin circular disks : 2 : centres 5, 5, radii 1, 1.

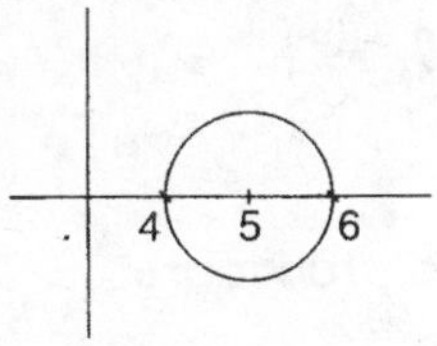

Fig. 32.18

However, to find the approximate values of all the eigen values and eigen vactors, iteration mathod or power mathod is used. Power method is particularly used when only the largest and/or the smallest eigen values of a matrix are desired. The advantage of this method is that the eigen vector associated with the largest eigen value is also obtained simultaneously. Also the intermediate (remaining) eigen values can also be found, after "sweeping" the already found largest and/or smallest eigen values.

Power Method

Let $\lambda_1, \lambda_2, \ldots, \lambda_n$ be the eigen values and $V_1, V_2, \ldots V_n$ be the corresponding eigen vactors of a n-rowed square matrix A. Assume that λ_1 is the absolutely largest eigen value of A i.e.,

$$|\lambda_1| > |\lambda_2| \geq |\lambda_3| \geq \ldots, \geq |\lambda_n|.$$

Consider $V = \alpha_1 V_1 + \alpha_2 V_2 + \cdots + \alpha_n V_n$ where $\alpha_1, \alpha_2, \cdots \alpha_n$ are scalars. Then

$$AV = \alpha_1 AV_1 + \alpha_2 AV_2 + \cdots + \alpha_n AV_n$$

$$= \alpha_1 \lambda_1 V_1 + \alpha_2 \lambda_2 V_2 + \cdots + \alpha_n \lambda_n V_n$$

Pre multiplying by A

$$A^2 V = \alpha_1 \lambda_1^2 V + \alpha_2 \lambda_2^2 V_2 + \cdots + \alpha_n \lambda_n^2 V_n.$$

Thus for any positive integer p

$$A^p V = \alpha_1 \lambda_1^p V_1 + \alpha_2 \lambda_2^p V_2 + \cdots + \alpha_n \lambda_n^p V_n$$

$$A^p V = \lambda_1^p \left[\alpha_1 V_1 + \alpha_2 \left(\frac{\lambda_2}{\lambda_1} \right)^p V_2 + \cdots + \alpha_n \left(\frac{\lambda_n}{\lambda_1} \right)^p V_n \right]$$

provided $\alpha_1 \neq 0$. All terms in R.H.S. except the first term have limit zero since λ_1 is the dominant value and p is large.

Now

$$\frac{A^{p+1} V}{A^p V} = \frac{\lambda_1^{p+1} \cdot \alpha_1 V_1}{\lambda_1^p \alpha_1 V_1} = \lambda_1$$

Also the required eigen vector corresponding to this largest eigen value is

$$\lambda_1^{-p} A^p V = \alpha_1 V_1.$$

Power method procedure

Step I: Choose an arbitrary real vector $x_0 \neq 0$. Generally, X_0 is chosen as

$$X_0 = \begin{bmatrix} 1 \\ 1 \\ 1 \\ 1 \end{bmatrix}$$

or any other value like $X_0 = [10000]^T$ etc.

Step II: Compute $X_1 = AX_0$, $X_2 = AX_1$, $X_3 = AX_2, \ldots, X_s = AX_{s-1}$. Put $X = X_{s-1}$, $Y = X_s$

Step III: Compute $m_0 = x^T x$, $m_1 = x^T y$, $m_2 = Y^T Y$

Step IV: Largest eigen value $= \lambda_1 = \frac{m_1}{m_0}$,

ϵ, the error in $\lambda_1 : |\epsilon| \leq \sqrt{\frac{m_2}{m_0} - \lambda_1^2}$,

Eigen vector corresponding to λ_1 is $Y = X_s$.

[This eigen vector can be normalized either by dividing the eigen vector by the magnitude of its first component or by its largest component or by normalizing to a unit length.]

Determination of Smallest Eigen Value

$$AX = \lambda X$$

$$A^{-1} AX = A^{-1} \lambda X$$

$$\frac{1}{\lambda} X = A^{-1} X$$

Thus if λ is eigen value of A, then the reciprocal $\frac{1}{\lambda}$ is the eigen value of A^{-1}. Then the reciprocal of the largest eigen value of A^{-1} will be the smallest eigen value of A.

Procedure:

Step I: Determine A^{-1}

Step II: Calculate λ^* dominant eigen value of A^{-1}

Step III: $\frac{1}{\lambda^*}$ is the smallest eigen value of A.

Determination of Intermediate (remaining) Eigen Values

$$Ax = \lambda x$$

$$(A - \lambda_1 I)x = (\lambda - \lambda_1)x$$

Then $(\lambda - \lambda_1)$ is an eigen value of $(A - \lambda_1 I)$. Thus 'sweeping' the dominant eigen value, $(A - \lambda_1 I)$ will have remaining eigen values dominant, say λ_2 which can be found by power method.

WORKED OUT EXAMPLES

Example 1: Find the largest eigen value and the corresponding eigen vector of the matrix

$$A = \begin{bmatrix} -2 & 0 & -1 \\ 1 & -1 & 1 \\ 2 & 2 & 0 \end{bmatrix}$$

Find the error in the value of the largest eigen value.

Solution: Choose the initial vector $X_0 = \begin{bmatrix} 1 \\ 1 \\ 1 \end{bmatrix}$,

Then $X_1 = AX_0 = \begin{bmatrix} -2 & 0 & -1 \\ 1 & -1 & 1 \\ 2 & 2 & 0 \end{bmatrix}\begin{bmatrix} 1 \\ 1 \\ 1 \end{bmatrix} = \begin{bmatrix} -3 \\ 1 \\ 4 \end{bmatrix}$.

Now $X_2 = AX_1 = \begin{bmatrix} -2 & 0 & -1 \\ 1 & -1 & 1 \\ 2 & 2 & 0 \end{bmatrix}\begin{bmatrix} -3 \\ 1 \\ 4 \end{bmatrix} = \begin{bmatrix} 2 \\ 0 \\ -4 \end{bmatrix}$.

Then $X_3 = Ax_2 = \begin{bmatrix} -2 & 0 & -1 \\ 1 & -1 & 1 \\ 2 & 2 & 0 \end{bmatrix}\begin{bmatrix} 2 \\ 0 \\ -4 \end{bmatrix} = \begin{bmatrix} 0 \\ -2 \\ 4 \end{bmatrix}$.

continuing these iterations, we get

$$X_4 = AX_3 = \begin{bmatrix} -4 \\ 6 \\ -4 \end{bmatrix}, \quad X_5 = AX_4 = \begin{bmatrix} 12 \\ -14 \\ 4 \end{bmatrix},$$

$$X_6 = AX_5 = \begin{bmatrix} -28 \\ 30 \\ -4 \end{bmatrix}, \quad X_7 = AX_6 = \begin{bmatrix} 60 \\ -62 \\ 4 \end{bmatrix},$$

$$X_8 = AX_7 = \begin{bmatrix} -124 \\ 126 \\ -4 \end{bmatrix} \text{ and } X_9 = AX_8 = \begin{bmatrix} 252 \\ -254 \\ 4 \end{bmatrix}.$$

Put $X = X_8$ and $Y = X_9$. Then

$$m_0 = X^T X = [-124 \;\; 126 \;\; -4]\begin{bmatrix} -124 \\ 126 \\ -4 \end{bmatrix} = 31268$$

$$m_1 = X^T y = [-124 \;\; 126 \;\; -4]\begin{bmatrix} 252 \\ -254 \\ 4 \end{bmatrix} = -63268$$

$$m_2 = Y^T Y = [252 \;\; -254 \;\; 4]\begin{bmatrix} 252 \\ -254 \\ 4 \end{bmatrix} = 128036.$$

The largest eigen value $= \lambda_1 = \frac{m_1}{m_0} = \frac{-63268}{31268}$

$$\lambda_1 = -2.0234105$$

(at the 9th iteration).
The corresponding eigen vector is

$$X_9 = \begin{bmatrix} 252 \\ -254 \\ 4 \end{bmatrix} \quad \text{or} \quad X_9 = \begin{bmatrix} 1 \\ -1.007936 \\ 0.0015 \end{bmatrix}.$$

The error in the largest eigen value is ϵ

$$|\epsilon| \le \sqrt{\frac{m_2}{m_0} - \lambda_1^2} = \sqrt{\frac{128036}{31268} - (2.023415)^2}$$

$$= \sqrt{4.094793 - 4.094208} = \sqrt{0.00058473}$$

$$= 0.02418.$$

Note that the exact three eigen values of A are $-2, -1, 0$.

Example 2: Find the absolutely smallest eigen value of the matrix

$$A = \begin{bmatrix} 2 & -1 & 0 \\ -1 & 2 & -1 \\ 0 & -1 & 2 \end{bmatrix}$$

Solution: We know that if λ is an eigen value of A then the reciprocal $\frac{1}{\lambda}$ is the eigen value of A^{-1}. Thus the largest eigen value of A^{-1} is the smallest eigen value of A.

Choose $X_0 = \begin{bmatrix} 1 \\ 1 \\ 1 \end{bmatrix}$ and apply power method, for A^{-1}. Now

$$B = A^{-1} = \frac{1}{4}\begin{bmatrix} 3 & 2 & 1 \\ 2 & 4 & 2 \\ 1 & 2 & 3 \end{bmatrix}$$

$$X_1 = BX_0 = \frac{1}{4}\begin{bmatrix} 3 & 2 & 1 \\ 2 & 4 & 2 \\ 1 & 2 & 3 \end{bmatrix}\begin{bmatrix} 1 \\ 1 \\ 1 \end{bmatrix} = \frac{1}{4}\begin{bmatrix} 6 \\ 8 \\ 6 \end{bmatrix} = \frac{1}{2}\begin{bmatrix} 3 \\ 4 \\ 3 \end{bmatrix}.$$

$$X_2 = B^2 X_0 = \frac{1}{4}\begin{bmatrix} 3 & 2 & 1 \\ 2 & 4 & 2 \\ 1 & 2 & 3 \end{bmatrix}\frac{1}{2}\begin{bmatrix} 3 \\ 4 \\ 3 \end{bmatrix} = \frac{1}{8}\begin{bmatrix} 20 \\ 28 \\ 20 \end{bmatrix} = \frac{1}{2}\begin{bmatrix} 5 \\ 7 \\ 5 \end{bmatrix}$$

$$X_3 = BX_2 = B^3 X_0 = \frac{1}{4}\begin{bmatrix} 3 & 2 & 1 \\ 2 & 4 & 2 \\ 1 & 2 & 3 \end{bmatrix}\frac{1}{2}\begin{bmatrix} 5 \\ 7 \\ 5 \end{bmatrix}$$

$$= \frac{1}{8}\begin{bmatrix}34\\48\\34\end{bmatrix} = \frac{1}{4}\begin{bmatrix}17\\24\\17\end{bmatrix}$$

$$X_4 = B^4X_0 = \frac{1}{4}\begin{bmatrix}3&2&1\\2&4&2\\1&2&3\end{bmatrix}\frac{1}{4}\begin{bmatrix}17\\24\\17\end{bmatrix}$$

$$= \frac{1}{16}\begin{bmatrix}116\\260\\116\end{bmatrix} = \frac{1}{16}\begin{bmatrix}116\\164\\116\end{bmatrix}$$

$$X_4 = \frac{1}{4}\begin{bmatrix}29\\41\\29\end{bmatrix},$$

$$X_5 = bX_4 = \frac{1}{4}\begin{bmatrix}3&2&1\\2&4&2\\1&2&3\end{bmatrix}\frac{1}{4}\begin{bmatrix}29\\41\\29\end{bmatrix} = B^5X_0$$

$$X_5 = \frac{1}{16}\begin{bmatrix}198\\280\\198\end{bmatrix}.$$

$$X_6 = BX_5 = B^6X_0 = \frac{1}{64}\begin{bmatrix}1352\\1912\\1352\end{bmatrix} = \frac{1}{8}\begin{bmatrix}169\\239\\169\end{bmatrix}$$

Put $X = X_5, Y = X_6, m_0 = X^TX = \frac{1}{64}\cdot(39202)$
$m_1 = x^Ty = \frac{1}{64}(66922)$.
Then the largest eigen value of B is $\frac{m_1}{m_0} = 1.70710$
The reciprocal $\frac{1}{1.70710} = 0.58578$ is the smallest eigen value of A.

Example 3: Find the bounds for eigen values of A.

where $A = \begin{bmatrix}1&0&-1\\1&2&1\\2&2&3\end{bmatrix}$.

Solution: The bounds for eigen values are given by three Gershgorin circular disks c_1, c_2, c_3
c_1: centre 1 and radius $|0| + |-1| = 1$,
c_2: centre 2 and radius $1 + 1 = 2$
c_3: centre 3 and radius $2 + 2 = 4$
The actual eigen values 1, 2, 3 of A lies within these three circular disks c_1, c_2 and c_3.

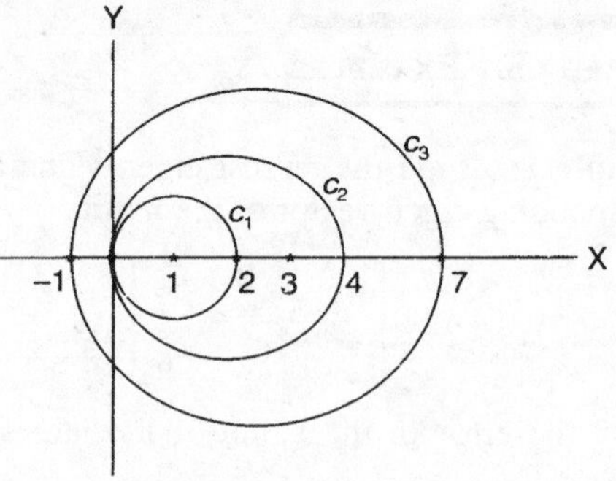

Fig. 32.19

EXERCISE

Find the largest eigen value of the matrix A and the error in the largest eigen value $|\epsilon| \le \sqrt{\frac{m_2}{m_0} - \lambda_1^2}$.

(Ex. 1 to 5)

1. $A = \begin{bmatrix}2&2&1\\1&3&1\\1&2&2\end{bmatrix}$

Hint: $X_0 = \begin{bmatrix}1\\1\\1\end{bmatrix}, X_1 = AX_0 = \begin{bmatrix}5\\5\\5\end{bmatrix} = 5\begin{bmatrix}1\\1\\1\end{bmatrix},$

$X_2 = 25\begin{bmatrix}1\\1\\1\end{bmatrix} X_3 = 125\begin{bmatrix}1\\1\\1\end{bmatrix}, X_4 = 625\begin{bmatrix}1\\1\\1\end{bmatrix},$

$X_5 = 3125X_0, X_6 = 15625X_0X_7 = 78125X_0$.
Put $m_0 = X_6,\ m_1 = X_7,\ \lambda_1 = \frac{m_1}{m_0} = \frac{(12625)(78125)(3)}{(12625^2)(3)} = 5,\ \frac{m_2}{m_0} = 25,\ |\epsilon| = 0$

Ans. 5 (Actual eigen values are 1, 1, 5)

2. $\begin{bmatrix}-3&2&1\\1&-2&1\\1&2&-3\end{bmatrix}$

Hint: $X_0 = \begin{bmatrix}1\\1\\1\end{bmatrix}, X_1 = \begin{bmatrix}-3\\1\\1\end{bmatrix}, X_2 = \begin{bmatrix}12\\-4\\-4\end{bmatrix},$

$X_3 = \begin{bmatrix}-48\\16\\16\end{bmatrix}, X_4 = \begin{bmatrix}192\\-64\\-64\end{bmatrix}, m_2 = 45056,$

$m_1 = -11264, m_0 = 2816, \frac{m_2}{m_0} = 16, \frac{m_1}{m_0} = \lambda_1 = -4, |\epsilon| = 0$

Ans. -4

3. $A = \begin{bmatrix}1&0&-1\\1&2&1\\2&2&3\end{bmatrix}$

Hint: $X_1 = \begin{bmatrix} 0 \\ 4 \\ 7 \end{bmatrix}, X_2 = \begin{bmatrix} -7 \\ 15 \\ 29 \end{bmatrix}, X_3 = \begin{bmatrix} -36 \\ 52 \\ 103 \end{bmatrix},$

$X_4 = \begin{bmatrix} -139 \\ 171 \\ 341 \end{bmatrix}, X_5 = \begin{bmatrix} -480 \\ 544 \\ 1087 \end{bmatrix}, X_6 = \begin{bmatrix} -1567 \\ 1695 \\ 3389 \end{bmatrix}.$

$m_2 = 16813835$, $m_1 = 5358083$, $m_0 = 1707905$, $q = \frac{m_1}{m_0}$. Exact eigen values are 1, 1, 3.

Ans. 3.131722.

4. $A = \begin{bmatrix} 2 & 0 & 1 & 0 \\ 0 & 0 & 3 & 1 \\ 1 & 3 & 4 & -2 \\ 0 & 1 & -2 & 0 \end{bmatrix}$

Hint: $X_1 = [3\ 4\ 6\ -1]^T$, $X_2 = [12\ 17\ 41\ -8]^T$, $X_3 = [65\quad 115\quad 243\quad -65]$, $m_2 = 80724$, $m_1 = 13218$, $m_0 = 2178$, $|\epsilon| = 0.48184$.

Ans. 6.0688705

5. $A = \begin{bmatrix} 9 & 10 & 8 \\ 10 & 5 & -1 \\ 8 & -1 & 3 \end{bmatrix}$

Hint:

$X_1 = [27\ 14\ 10]^T$, $X_2 = [463\ 330\ 232]^T$,

$X_3 = [9323\ 6048\ 4070]^T$,

$X_4 = [176947\ 119400\ 80746]^T$

$X_5 = [3432491\ 2285724\ 1538414]^T$

$X_6 = [66056971\ 44215116\ 29789446]$

$X_7 = [1.2749 \times 10^9,\ 8.5185 \times 10^8,\ 5.7360 \times 10^8]$

$m_0 = 7.2059 \times 10^{15}$, $m_1 = 1.3896 \times 10^{17}$.

Ans. 19.284197

6. Determine the largest eigen value and the corresponding eigen vector of the matrix

$$A = \begin{bmatrix} 2 & -1 & 0 \\ -1 & 2 & -1 \\ 0 & -1 & 2 \end{bmatrix}$$

Hint: $x_0 = [1\,0\,0]^T$, $X_1 = [1, -0.5\,0]^T$, $X_2 = 2.8[1\ -1\ .43]^T$ $X_3 = 3.43[.87\ -1\ 0.54]^T$, $X_4 = 3.41[.8\ -1\ .61]^T$, $X_5 = 3.41[.76\ -1\ .65]^T$, $X_6 = 3.41[.74\ -1\ .67]$

Ans. 3.41, $[0.74, -1, 0.67]^T$

7. Find the largest eigen value and the corresponding eigen vector of the matrix A where

$$A = \begin{bmatrix} 4 & 1 & 0 \\ 1 & 2 & 1 \\ 0 & 1 & 1 \end{bmatrix}$$

Hint: [1 1 1], [5 4 2], [24 15 6], [111 60 21], [504 252 81], [2268 1089 333], [10161 4779 1492], [45433, 21141, 6201], [202833, 93906, 27342], [905238, 417987, 121248], [4038939, 1862460, 539235].

Ans. 4.46, $[0.9\ 0.42\ .12]^T$

8. $A = \begin{bmatrix} 1 & 2 & 0 \\ 2 & 1 & 0 \\ 0 & 0 & -1 \end{bmatrix}$

Ans. 3.0, [1 1 0]

9. $A = \begin{bmatrix} 1 & 2 \\ 3 & 2 \end{bmatrix}$

Hint: Exact value: 4, $\left[\frac{2}{3}\ 1\right]^T$

Ans. 3.987, $[.668\ 1]^T$

10. $A = \begin{bmatrix} 1 & 6 & 1 \\ 1 & 2 & 0 \\ 0 & 0 & 3 \end{bmatrix}$

Hint: $X_0 = [1\ 0\ 0]$, $X_1 = [1\ 1\ 0]$, $X_2 = 3[2.3\ 1\ 0]$, $X_3 = 4[2.1\ 1.1\ 0]$, $X_4 = 4[2.2\ 1.1\ 0]$, $X_5 = 4.4[2\ 1\ 0]$, $X_6 = 4[2\ 1\ 0]$, $X_7 = 4[2\ 1\ 0]$.

Ans. 4, $[2\ 1\ 0]^T$

11. Determine the bounds for the eigen values of the matrix A where

(a) $A = \begin{pmatrix} 0 & 1 \\ 0 & 0 \end{pmatrix}$

Ans. $\lambda = 0$; Circles c_1, c_2; centres 0, 0 and radii 1, 0 respectively.

(b) $A = \begin{bmatrix} 3 & 4 \\ 4 & -3 \end{bmatrix}$

Ans. $\lambda = \pm 5$; centres: 3, −3; radii 4, 4 respectively

(c) $A = \begin{bmatrix} 6 & 0 & -3 \\ 0 & 6 & 3 \\ -3 & 3 & 2 \end{bmatrix}$

Ans. $\lambda = 6, 4 \pm \sqrt{22} = 6, 8.6904, -0.6904$ circular disks with centres at 6, 6, 2 and radii $|-3|, 3, |-3| + 3$ respectively.

(d) $A = \begin{bmatrix} 26 & -2 & 2 \\ 2 & 21 & 4 \\ 4 & 2 & 28 \end{bmatrix}$

Ans. $\lambda = 30, 25, 20$; circles: centers 26, 21, 28 radii: $|-2| + 2, 2 + 4, 4 + 2$.

Chapter 33

Numerical Solutions of ODE and PDE

INTRODUCTION

Analytical solutions can be obtained only for selected class of ODE & PDE, using series solutions, Laplace transform, Fourier transform, separation of variables technique etc. For certain problems, analytical solutions can not be obtained. However numerical solutions can be obtained to the desired degree of accuracy using computers, in all the above cases. The advantage is that numerical solutions can be obtained for problems involving irregularly shaped boundaries and is easy to program on computer. The only serious disadvantage with these numerical solutions is that they lack the generality of the analytical solutions. When the initial conditions are changed, the numerical solution must be calculated again. Taylor's series, Picard's, Modified Euler's, Runge-Kutta 4th order, Milne's predictor-corrector and Adame-Bashforth-Milne's methods for solution of first order ODE are considered in this chapter. Numerical solutions of PDE: one dimensional heat equation, wave equation and two-dimensional Laplace's equation are studied.

33.1 NUMERICAL SOLUTIONS OF FIRST ORDER ORDINARY DIFFERENTIAL EQUATIONS

Differential equation is the most important mathematical model for physical phenomena such as the motion of objects, fluids, heat flow, bending and cracking of materials, vibrations, chemical reactions. Since an nth order D.E. can in general, be reduced to a system of n first order D.E., in essence, it would be sufficient to examine a first order O.D.E.

$$\frac{dy}{dx} = f(x, y) \tag{1}$$

together with an initial condition (I.C.)

$$y(x_0) = y_0 \tag{2}$$

Numerical solutions of first order O.D.E. (1) plays a vital role, since analytical solutions of (1) are rare (except in standard forms).

Taylor's Series Method

Consists of expanding the function $y(x)$ in powers of $(x - x_0)$:

$$y(x) = y(x_0) + y'(x_0) \cdot (x - x_0) + \frac{y''(x_0)}{2!}(x - x_0)^2 + \frac{y'''(x_0)}{3!}(x-x_0)^3 + \cdots + \frac{y^{(n)}(x_0)}{n!}(x-x_0)^n + \cdots \tag{3}$$

In (3), $y(x_0)$ is known from I.C. (2). The remain ing coefficiently $y'(x_0)$, $y''(x_0)$, …, $y^{(n)}(x_0)$ etc. are obtained by successively differentiating (1) and evaluating at x_0. Substituting these values in (3), $y(x)$ at any point can be calculated from (3), provided $h = x - x_0$ is small.

When $x_0 = 0$, then Taylor's series (3) takes simpler form

$$y(x)=y(0)+y'(0) \cdot x+y''(0) \cdot \frac{x^2}{2!}+\cdots+y_{(0)}^{(n)} \cdot \frac{x^n}{n!}+\cdots \tag{4}$$

Modified Euler's Method

For small changes Δx in x, the change in y is $\Delta y \approx \frac{dy}{dx}\Delta x$. Thus

$$f(y+\Delta y) - f(y) = \Delta y = \frac{dy}{dx}\Delta x.$$

Using $\frac{dy}{dx} = f(x, y)$ at (x_i, y_i), we have

$$y_i' = f(x_i, y_i) \tag{5}$$

$$\boxed{\tilde{y}_{i+1} = y_i + h\, f(x_i, y_i)} \quad \text{Predictor} \tag{6}$$

Equation (6) is known as *Euler's method.* Using (5) as a predictor equation find

$$\tilde{y}_{i+1}' = f(x_i + 1, \tilde{y}_{i+1}). \tag{7}$$

Using the average of the slopes (5) and (7), we get

$$y_{i+1} = y_i + \frac{y_i' + \tilde{y}_{i+1}'}{2}.$$

$$\boxed{y_{i+1} = y_i + h\frac{f(x_i, y_i) + f(x_{i+1}, \tilde{y}_{i+1})}{2}} \quad \text{Corrector} \tag{8}$$

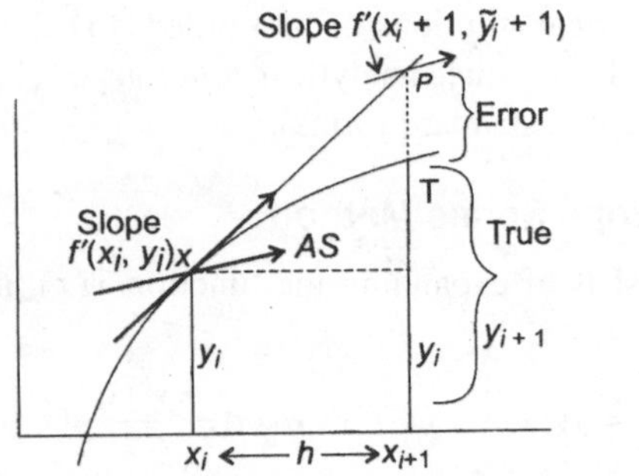

P: Predicted value, T: True value, AS: Average slope

Fig. 33.1

The Equation (8) is known as the corrector equation. Thus the method using the predictor (6) and corrector (8) is known as the *modified Euler's method* or Heun's method or Euler-Cauchy method or Euler-Trapezoidal method or improved Euler method or Euler's predictor-corrector method.

Runge-Kutta 4th Order Method

The one-step methods like Taylor series method, Euler's method or modified Euler's method evaluate y at a pivota point x_i, thus giving the starting values for the multi-step methods which use recurrence equations, at these preceding pivotal points. Runge-Kutta method is another one such one-step (or single step) method, which is also self-starting like the Taylor's Euler's, methods. Runge-Kutta methods achieves the accuracy of a Taylor series method without requiring the calculation of higher derivatives. A two parameter h, k family of formulas of fourth order accuracy are known as Runge-Kutta fourth order method and are given by the following formulae:

$$y_{i+1} = y_i + \frac{1}{6}(k_1 + 2k_2 + 2k_3 + k_4) \tag{9}$$

where

$$\left.\begin{aligned} k_1 &= h\, f(x_i, y_i) \\ k_2 &= h\, f\left(x_i + \tfrac{h}{2}, y_i + \tfrac{k_1}{2}\right) \\ k_3 &= h\, f\left(x_i + \tfrac{h}{2}, y_i + \tfrac{k_2}{2}\right) \\ k_4 &= h\, f(x_i + h, y_i + k_3) \end{aligned}\right\} \tag{10}$$

Note: When f is independent of y, (9) reduce's to Simpson's $\frac{1}{3}$ rule,

$$\left(\frac{h}{2}\right)\frac{1}{3}\left[f(x_i) + 4f\left(x_i + \frac{h}{2}\right) + f(x_i + h)\right].$$

The only disadvantage with R-K method is the repeated evaluation of function $f(x, y)$ at several points.

Milne's Predictor-Corrector Method

Is a multi-step method which uses the previously calculated four values of the dependent variable y and its derivative y' in the subsequent steps. Integrating

$$\int_{x_{n-3}}^{x_{n+1}} \frac{dy}{dx}dx = \int_{x_{n-3}}^{x_{n+1}} f(x, y)dx$$

where R.H.S. is evaluated by Simpson's $\frac{1}{3}$ rule for the extended interval x_{n-2} to x_n to x_{n-3} to x_{n+1}.

$$y_{n+1} - y_{n-3} = \frac{4h}{3}\left(2y_{n-2}' - y_{n-1}' + 2y_n'\right) \quad \text{or}$$

$$y_{n+1} = y_{n-3} + \frac{4h}{3}(2y_{n-2}^{\text{i}} - y_{n-1}^{\text{i}} + 2y_n^{\text{i}}) \tag{11}$$

(11) is known as the predictor formula. Using (11), calculate $y_{n+1}^{\text{i}} = f(x_{n+1}, y_{n+1})$. Integrating between x_{n-1} to x_{n+1} using Simpson's $\frac{1}{3}$ rule.

$$y_{n+1} = y_{n-1} + \frac{h}{3}(y_{n-1}^{\text{i}} + 4y_n^{\text{i}} + y_{n+1}^{\text{i}}) \tag{12}$$

(12) is known as the corrector formula. The next pair of values y_{n+2}, y_{n+2}^{i} are obtained in a similar way first using (11) to find y_{n+2}, then y_{n+2}^{i} by $f(x_{n+2}, y_{n+2})$.

With this value the corrector value of y_{n+2} is obtained from (12).

WORKED OUT EXAMPLES

Taylor's series method

Example 1: Using Taylor series expansion evaluate the integral of $y' - 2y = 3e^x$, $y(0) = 0$ at (a) $x = 0.1(0.1)0.3$; (b) $x = 1.0; 1.1$.

Solution: Rewriting $y' = 2y + 3e^x$, $y(0) = 0$. Differentiating and evaluating at $x = 0$,

$$y'(0) = 2y(0) + 3e^0 = 2\cdot 0 + 3\cdot 1 = 3$$
$$y'' = 2y' + 3e^x, \quad y''(0) = 2y'(0) + 3 = 2\cdot 3 + 3 = 9$$
$$y''' = 2y'' + 3e^x, \quad y'''(0) = 2y''(0) + 3 = 2\cdot 9 + 3 = 21$$
$$y'''' = 2y''' + 3e^x, \quad y''''(0) = 2y'''(0) + 3 = 2\cdot 21 + 3 = 45$$
$$y''''' = 2y'''' + 3e^x, \quad y'''''(0) = 2y'''' + 3 = 2\cdot 45 + 3 = 93$$

In general,

$y^{(n+1)}(x) = 2y^{(n)}(x) + 3e^x$ or $y^{(n+1)}(0) = 2y^{(n)}(0) + 3$.

The Taylor's series expansion of $y(x)$ about 0 is

$$y(x) = y(0) + xy'(0) + x^2\frac{y''(0)}{2!} + x^3\frac{y'''(0)}{3!} + x^4\frac{y''''(0)}{4!} + x^5\frac{y'''''(0)}{5!} + \cdots$$

Substituting the values of $y'(0)$, $y''(0)$, ...

$$y(x) = 0 + 3x + \frac{9}{2}x^2 + \frac{21}{6}x^3 + \frac{45}{24}x^4 + \frac{93}{120}x^5 + \cdots$$
$$y(x) = 3x + 4.5x^2 + 3.5x^3 + \frac{15}{8}x^4 + \frac{31}{40}x^5 + \cdots \quad (1)$$

Now put $x = 0.1$ in (1)

$$y(0.1) = 3(0.1) + 4.5(0.1)^2 + 3.5(0.1)^3 + \frac{15}{8}(0.1)^4 + \frac{31}{40}(.1)^5$$
$$y(0.1) = 0.34869$$

Similarly, put $x = 0.2$ in (1)

$$y(0.2) = 0.6 + 0.18 + 0.028 + 0.003 + 0.000248 = 0.811244.$$

$$y(0.3) = 0.9 + 0.405 + .0945 + .015187 + .0018832 = 1.4165702$$

Put $x = 1$ in (1)

$$y(1) = 3 + \frac{9}{2} + \frac{7}{2} + \frac{15}{8} + \frac{31}{40} = 13.65$$

Put $x = 1.1$ in (1)

$$y(1.1) = 3.3 + 5.445 + 4.6585 + 2.745 + 1.248 = 17.39.$$

Modified Euler's method

Example 2: Intensity of radiation is directly proportional to the amount of remaining radioactive substance. The DE is $y' = -ky$, where $k = 0.01$, $t_0 = 0$, $y_0 = 100$ g. Determine how much substance will remain at the moment $t = 100$ sec. Find the solution by (a) Euler's method, (b) Modified Euler's method with $h = 25$.

Solution: Here $h = 25$, $f(t, y) = -ky$, $f(t, y) = -0.01\,y$, since $\frac{dy}{dt} = f(t, y) = -ky$. (a) Euler's method (see Table 33.1)

Table 33.1 Euler's method

t	y_i	$f_i = f(t_i, y_i)$	$\Delta y_i = hf_i$	$y_{i+1} = y_i + \Delta y_i$
0	100	−1	−25	$100 + (-25) = 75$
25	75	−0.75	−18.75	$75 - 18.75 = 56.25$
50	56.25	−0.5625	−14.0625	$56.25 - 14.0625 = 42.1875$
75	42.1875	−0.421875	−10.546875	$42.1875 - 10.5468 = 31.640625$
100	31.640625			

(b) Modified Euler's method: $h = 25$, $f(t, y) = -0.01y$ (see Table 33.2)

Runge-Kutta 4th order method

Example 3: Using Runge-Kutta 4th order method find the solution of $\frac{dy}{dx} = y - x$ with initial condition

Table 33.2 Modified Euler's method

t	y_i	$f_i = f(t_i, y_i)$	$\bar{y}_{i+1} = y_i + hf_i$	$\bar{f}_{i+1} = f(x_{i+1}, \bar{y}_{i+1})$	$\tilde{\Delta y}_i = \frac{h}{2}(f_i + \bar{f}_{i+1})$	$y_{i+1} = y_i + \tilde{\Delta y}_i$
0	100	-1	$100 + 25(-1)$ $= 75$	$(-0.01)(75)$ $= -0.75$	$\frac{25}{2}(-1 - 0.75)$ $= -21.875$	$100 - 21.875$ $= 78.125$
25	78.125	-0.78125	$78.125 + 25(-0.78)$ $= 58.59$	$(-0.01)(58.59)$ $= -0.5859$	$\frac{25}{2}(-0.78 - 0.59)$ $= -17.09$	$78.125 - 17.09$ $= 61.034$
50	61.034	-0.61	45.7755	-0.4578	-13.35	47.68
75	47.68	-0.4768	35.76	-0.3576	-10.43	37.25
100	37.25					

Note: Exact solution: $y = 100e^{-kt}$, At $t = 100$, $y = 36.7879$.

$y(0) = 1.5$ on $[0, 1]$.

Solution: Here choose $h = 0.2$ given $f(x, y) = y - x$, $x_0 = 0$, $y_0 = 1.5$.
See Table 33.3 on p. 33.5.

Milne's predictor-corrector method

Example 4: Using Milne's predictor-corrector method evaluate the integral of $y' - 4y = 0$ at $x = 0.4, 0.5$ given that $y(0) = y_0 = 1$, $y_1 = y(0.1) = 1.492$, $y_2 = y(0.2) = 2.226$, $y_3 = y(0.3) = 3.320$.

Solution: Here $\frac{dy}{dx} = 4y$, so $f(x, y) = y' = 4y$, $h = 0.1, x_0 = 0, x_1 = 0.1, x_2 = 0.2, x_3 = 0.3, x_4 = 0.4, x_5 = 0.5$.

Predictor: $y_{n+1} = y_{n-3} + \frac{4h}{3}(2y^{\text{i}}_{n-2} - y^{\text{i}}_{n-1} + 2y^{\text{i}}_n)$

with $n = 3$,

$$y_4 = y_0 + \frac{4h}{3}(2y^{\text{i}}_1 - y^{\text{i}}_2 + 2y^{\text{i}}_3).$$

Here $h = 0.1$, $y_0 = 1$, $y^{\text{i}}_1 = 4y_1 = 4(1.492) = 5.968$.

$$y^{\text{i}}_2 = 4y_2 = 4(2.226) = 8.904, \quad y^{\text{i}}_3 = 4y_3 = 4(3.32) = 13.28$$

$$\therefore\ y_4 = y(x_4) = y(0.4) = 1 + \frac{4(0.1)}{3}[2(5.968) - 8.904 + 2(13.28)]$$

$$y_4 = 1 + \frac{4(0.1)}{3}(29.592) = 1 + 3.9456 = 4.9456$$

So $y^{\text{i}}_4 = 4y_4 = 4(4.9456) = 19.7824$

Corrector

$$y_{n+1} = y_{n-1} + \frac{h}{3}(y^{\text{i}}_{n-1} + 4y^{\text{i}}_n + y^{\text{i}}_{n+1})$$

with $n = 3$,

$$y_4 = y_2 + \frac{h}{3}(y^{\text{i}}_2 + 4y^{\text{i}}_3 + y^{\text{i}}_4)$$

$$\therefore\quad y_4 = y(x = 0.4) = 2.226 + \frac{0.1}{3}(2.226 + 4(13.28) + 19.7824)$$

$$y_4 = 2.226 + 2.72688 = 4.95288$$

with $n = 4$, predictor

$$y_5 = y_1 + \frac{4h}{3}[2y^{\text{i}}_2 - y^{\text{i}}_3 + 2y^{\text{i}}_4]$$

Since $y' = 4y$, $y^{\text{i}}_i = 4y_i$

$$y_5 = 1.492 + \frac{4(0.1)}{3}4[2y_2 - y_3 + 2y_4]$$

$$= 1.492 + \frac{16(0.1)}{3}[2(2.226) - 3.320 + 2(4.95288)]$$

$$y_5 = 7.3788$$

Correct with $n = 4$,

$$y_5 = y_3 + \frac{h}{3}[y^{\text{i}}_3 + 4y^{\text{i}}_4 + y^{\text{i}}_5]$$

$$y_5 = y_3 + \frac{4h}{3}[y_3 + 4y_2 + y_5] \quad (\text{since } y'_i = 4y_i)$$

$$= 3.320 + \frac{4(0.1)}{3}[3.320 + 4(4.95288) + 7.3788]$$

$$y_5 = y(x = 0.5) = 3.320 + 4.06804 = 7.3880426.$$

Note: The exact solution is $y = e^{4x} + c$, with $y(0) = 1$, $c = 0$ so $y = e^{4x}$. Then $y(0.4) = e^{4(0.4)} = 4.953032425$ and $y(0.5) = e^{4(0.5)} = 7.389056099$.

Table 33.3 Runge-Kutta 4th order method

	x	y	$k_v = h\,f(x, y)$	Correction
	x_0	y_0	$k_1 = h\,f(x_0, y_0)$	$k = \frac{1}{6}(k_1 + 2k_2 + 2k_3 + k_4)$
	$x_0 + \frac{h}{2}$	$y_0 + \frac{k_1}{2}$	$k_2 = h\,f\left(x_0 + \frac{h}{2}, y_0 + \frac{k_1}{2}\right)$	
	$x_0 + \frac{h}{2}$	$y_0 + \frac{k_2}{2}$	$k_3 = h\,f\left(x_0 + \frac{h}{2}, y_0 + \frac{k_2}{2}\right)$	
	$x_0 + h$	$y_0 + k_3$	$k_4 = h\,f(x_0 + h, y_0 + k_3)$	
1st iteration	x	y	$k_v = 0.2(x + y)$	$\Delta y = \frac{1}{6}(k_1 + 2k_2 + 2k_3 + k_4)$
	0.0	0.0	$k_1 = 0$	$k_1 = 0$
	0.1	0.0	$k_2 = 0.02$	$2k_2 = 0.04$
	0.1	0.01	$k_3 = 0.022$	$2k_3 = 0.044$
	0.2	0.022	$k_4 = 0.0444$	$k_4 = 0.0444$
				$\Delta y = 0.0214$
IInd iteration		$\therefore y = 0 + 0.0214 = 0.0214$		
	0.2	0.0214	0.04428	
	0.3	0.04354	0.068708	
	0.3	0.055754	0.071151	
	0.4	0.092551	0.098510	
				$\Delta y = 0.070418$
		$\therefore y = 0.0214 + 0.070418 = 0.091818$		
IIIrd iteration	0.4	0.0918	0.0983	
	0.5	0.141	0.1282	
	0.5	0.1559	0.1311	
	0.6	0.223	0.1646	
				$\Delta y = 0.13028$
		$\therefore y = 0.0918 + 0.13028 = 0.222107$		
IV iteration	0.6	0.222	0.1644	
	0.7	0.3043	0.2008	
	0.7	0.3225	0.2045	
	0.8	0.4266	0.245	
				$\Delta y = 0.20341$
Vth iteration		$\therefore y = 0.222 + 0.20341 = 0.425522$		
	0.8	0.42552	0.2451	
	0.9	0.548	0.2896	
	0.9	0.5703	0.2940	
	1.0	0.71958	0.3439	
				$\Delta y = 0.29273$

$\therefore y(1) = 0.425522 + 0.29273 = 0.718253.$

Note: Exact solution: $y(x) = e^x - 1 - x$, $y(1) = e^1 - 1 - 1 = 0.718281828$.

EXERCISE

Taylor series

1. Evaluate the integral of $y'' + yy' - x^2 = 0$, $y(0) = 1$, $y'(0) = 1$ at $x = 0.1(0.1)0.3$ by Taylor series.
   Hint: $y(x) = 1 + x - \frac{x^2}{2!} + \frac{5x^4}{4!} - \frac{8x^5}{5!} + \cdots$.

Ans. $y(0.1) = 1.095$, $y(0.2) = 1.180$, $y(0.3) = 1.257$

2. Find the first five terms of the expansion in a power series of the solution $y = y(x)$, $z = z(x)$ of the system $y'(x) = y \cos x - z \sin x$, $z^{\text{i}}(x) = y \sin x + 2 \cos x$, with initial conditions $y(0) = 1$, $z(0) = 0$.

Ans. $y(x) = 1 + x + \frac{x^2}{2} - 5\frac{x^4}{4!} + \cdots$, $z(x) = \frac{x^2}{2} + \frac{x^3}{2} + \frac{5}{4!}x^4 + \cdots$

3. Find $y(0.1)$ by Taylor's series expansion when $y' = x - y^2$, $y(0) = 1$.
   Hint: $y(x) = 1 - x + \frac{3}{2}x^2 - \frac{4}{3}x^3 - \frac{17}{12}x^4 - \frac{31}{20}x^5 + \cdots$.

Ans. $y(0.1) = 0.9138$

4. Given $y' = x^2 + y^2$, $y(0) = 0$, determine the first three non-zero terms in Taylor series and hence obtain $y(1)$.

Ans. $y(x) = \frac{1}{3}x^3 + \frac{1}{63}x^7 + \frac{2}{2079}x^{11}$, $y(1) = 0.3502$.

5. Determine the three terms in the Taylor's series solution to the Blasius equation $y''' + yy'' = 0$, $y(0) = 0$, $y'(0) = 0$, $y''(0) = 1$.

Ans. ?

6. Compute $y(0.1)$ and $y(0.2)$ by Taylor's series method if $y' = x^2y - 1$, $y(0) = 1$.
   Hint: $y(x) = 1 - x + \frac{x^3}{3} - \frac{x^4}{4} \cdots$.

Ans. $y(0.1) = 0.9033$, $y(0.2) = 0.80227$.

Euler's method

7. Evaluate the integral of $y' - y^2 = 0$ by Euler's predictor-corrector method at $x = 0.1$, 0.2 with initial condition $y(0) = 1$.

Ans. $y(0.1) = 1.1105$, $y(0.2) = 1.24827$.

8. Evaluate by Euler's method
   a. $y' - y^2 = 0$, $y(0) = 1$, at $x = 0.1, 0.2, 0.3$
   b. $y' = y - x$, $y(0) = 2$, at $x = 0.2, 0.4, 0.6$.

Ans. **a.** $y(0.1)$=1.1, $y(0.2)$=1.221, $y(0.3)$=1.37
b. $y(0.2) = 2.4$, $y(0.4) = 2.84$, $y(0.6) = 3.328$

9. Use Heun's method to integrate $y' = 4e^{0.8x} - 0.5y$ from $x = 0$ to 4 with a step size of 1 and with $y(0) = 2$.
   Hint: $y = \frac{4}{1.3}(e^{0.8x} - e^{-0.5x}) + 2e^{-0.5x}$ is exact solution.

Ans. $y(1) = 6.70108$, $y(2) = 16.319$, $y(3) = 37.199$, $y(4) = 83.377$

10. Solve the following initial value problems by modified Euler's method.
    a. $y' = x + y$, $y(0) = 1$, $h = 0.1$, $x = 0(0.1)0.3$
    b. $y' = \ln(x + y)$, $y(0) = 2$, $h = 0.2$, $x = 0(0.2)0.8$
    c. $y' = x + |\sqrt{y}|$, $y(0) = 1$, $h = 0.2$, $x = 0(0.2)0.6$

Ans. **a.** 1.1105, 1.2432, 1.4004
b. 2.0656, 2.1416, 2.2272, 2.3217
c. 1.2309, 1.5253, 1.8861

Runge-Kutta 4th order method

11. Integrate the D.E. $y' = y - \frac{2x}{y}$, $y(0) = 1$ from 0 to 2 by R-K method with $h = 0.2$.

Ans. $y(0.2) = 1.1832$, $y(0.4) = 1.3416$, $y(0.6) = 1.4832$, $y(0.8) = 1.61251$, $y(1.0) = 1.7321$, $y(1.2) = 1.844$, $y(1.4) = 1.9495$, $y(1.6) = 2.049$, $y(1.8) = 2.145$, $y(2.0) = 2.2366$

12. Integrate D.E. $y' = x + y$, $y(0) = 0$, $h = 0.2$ by R-K method, from 0 to 1.

Ans. $y(0.2) = 0.0214$, $y(0.4) = 0.0918$, $y(0.6) = 0.222$, $y(0.8) = 0.4255$, $y(1.0) = 0.718$

13. Using R-K method, find the solution of
$$y' = 0.25\,y^2 + x^2$$
with initial condition $y(0) = -1$ on [0, 0.5] with $h = 0.1$.

Ans. −0.97528, −.94978, −.92154, −.8887, −.84945

14. Integrate D.E. by R-K method given $y' = -2xy^2$, $y(0) = 1$, $h = 0.2$ on [0, 1].

Ans. $y(0.2) = .9615$, $y(0.4) = .862$, $y(.6) = .73527$, $y(.8) = .6097$, $y(1.0) = .500073$

15. Apply R-K method to solve

a. $y' = x + y$, $y(0) = 1$, $h = 0.2$, and find $y(0.2)$

b. $y' = \frac{y^2-x^2}{y^2+x^2}$, $y(0) = 1$, $h = 0.2$, $x = 0(0.2)0.4$

Ans. **a.** $y(0.2) = 0.2428$

b. $y(0.2) = 1.196$, $y(0.4) = 1.3752$.

Milne's predictor-corrector method

16. Use Milne's predictor-corrector method to integrate $y' = 4e^{0.8x} - 0.5y$ from $x = 0$ to $x = 4$ with a step size of 1 and with $y(0) = 2$.

Ans. 6.20485, 14.86, 33.7242, 75.4329

17. Compute $y(0.8)$ and $y(1.0)$ by Milne's method if $y' = 1 + y^2$, $y(0) = 0$, $y(0.2) = 0.2027$, $y(0.41) = 0.4228$, $y(0.6) = 0.6841$.

Hint: $y'(0) = 1$, $y'(0.2) = 1.0411$, $y'(0.4) = 1.1787$, $y'(0.6) = 1.4681$, $y_{\text{predictor}}(0.8) = 1.0239$, $y'(0.8) = 2.048$

Ans. $y(0.8) = 1.0294$, $y(1) = 1.5549$

18. Find $y(0.4)$ and $y(0.5)$ by Milne's method if $y' + y = 2e^x$, $y_0 = 2$, $y_1 = 2.010$, $y_2 = 2.04$, $y_3 = 2.09$, $h = 0.1$.

Ans. $\tilde{y}_4 = 2.162$, $y_4 = 2.162$, $\tilde{y}_5 = 2.255$, $y_5 = 2.2546$

19. Calculate y at $x = 0.4$, and $x = 0.5$, if $y' = x + y$ and $y_0 = 1$, $y_1 = 1.1103$, $y_2 = 1.2428$, $y_3 = 1.3997$, $h = 0.1$.

Ans. $\tilde{y}_4 = 1.5836$, $y_4 = 1.5836$, $\tilde{y}_5 = 1.7974$, $y_5 = 1.7974$

20. Determine $y(0.8)$, $y(1.0)$ by Milne's P-C method when $y' = x - y^2$, $y(0) = 0$.

Hint: $y_1 = 0.02$, $y_1^i = 0.1996$, $y_2 = 0.0795$, $y_2^i = 0.3937$, $y_3 = 0.1762$, $y_3^i = 0.5689$
$\left(y = \frac{x^2}{2} - \frac{x^5}{20} + \frac{x^8}{160} - \frac{x^{11}}{4400}\right)$.

Ans. $y(0.8) = 0.3046$, $y(1.0) = 0.4555$.

33.2 PICARD'S METHOD OF SUCCESSIVE APPROXIMATION*

A problem involving ODEs is not completely specified by its equations. Boundary conditions (B.C.s), which are algebaric conditions on the values of the function $y(x)$, play crucial role in determining how to attack the problem numerically. Initial value problems involve condition(s) specified at some starting value x_0, while in (two-point) boundary value problems boundary conditions are specified at more than one point x.

Picard's method of successive approximation and Taylor's series are single step methods in which the solutions are obtained in analytical form as power serier in x.

Picard's Method of Successive Approximation

The particular solution of the initial value problem consisting of an ordinary first order differential equation

$$\frac{dy}{dx} = f(x, y) \tag{1}$$

with initial condition $y(x_0) = y_0$ is obtained by integrating (1). Then

$$\int_{y_0}^{y} dy = \int_{x_0}^{x} f(x, y)dx$$

or

$$y(x) = y_0 + \int_{x_0}^{x} f(x, y)dx \tag{2}$$

In the Picard's method, the first approximation y_1 is obtained by replacing y in $f(x, y)$ in RHS of (2) by y_0 and then evaluating the integral (which is now a function of x) wrt x. The second approximation y_2

*Emile Picard (1856–1941), French mathematician.

is then obtained by replacing y in $f(x, y)$ of RHS of (2) by y_1 and integrating wrt x. Successive approximations are obtained similarly. In practice, Picard's method is restricted to a limited class of problems in which the integral in RHS of (2) can be evaluated easily. Thus

$$y_1 = y_0 + \int_{x_0}^{x} f(x, y_0)dx$$

$$y_2 = y_0 + \int_{x_0}^{x} f(x, y_1)dx$$

$$y_3 = y_0 + \int_{x_0}^{x} f(x, y_2)dx$$

and so on yielding a sequence of functions $y_1, y_2, y_3, \ldots$ giving rise to a better approximation.

Worked Out Examples

Example 1: Solve the differential equation

$$\frac{dy}{dx} = x^2 - y, \; y(0) = 1$$

by Picard's method of successive approximations to get the value of y at $x = 1$. Use terms through x^5, compare it with the exact analytical solution.

Solution: Integrating the D.E., we obtain

$$\int_{y_0}^{y} dy = \int_{x_0}^{x} (x^2 - y)dx$$

$y(x) = y(x_0) + \int_{x_0}^{x} (x^2 - y)dx = 1 + \int_0^x (x^2 - y)dx$
The first approximation y_1 is obtained by replacing y by $y_0 = 1$ in the integrand in RHS integral. Thus

$$y_1 = 1 + \int_0^x (x^2 - 1)dx = 1 + \frac{x^3}{3} - x$$

To obtain the second approximation y_2, replace y by $1 - x + \frac{x^3}{3}$ in the integrand of the RHS integral. Thus

$$y_2 = 1 + \int_0^x \left[x^2 - (1 - x + \frac{x^3}{3})\right]dx$$

$$y_2 = 1 + \frac{x^3}{3} - x + \frac{x^2}{x} - \frac{x^4}{12}$$

Now the third approximation y_3 is obtained by replacing y by $y_2 = 1 - x + \frac{x^2}{2} + \frac{x^3}{3} - \frac{x^4}{12}$. Thus

$$y_3 = 1 + \int_0^x \left[x^2 - \left\{1 - x + \frac{x^2}{2} + \frac{x^3}{3} - \frac{x^4}{12}\right\}\right]dx$$

Integrating we get

$$y_3 = 1 + \frac{x^3}{3} - x - \frac{x^4}{12} + \frac{x^2}{2} - \frac{x^3}{6} + \frac{x^5}{60}$$

$$y_3 = 1 - x + \frac{x^2}{2} + \frac{x^3}{6} - \frac{x^4}{12} + \frac{x^5}{60}$$

Similarly replacing y by y_3 and integrating wrt x we get the fourth approximation y_4 as

$$y_4 = 1 + \int_0^x \left[x^2 - \left\{1 - x + \frac{x^2}{2} + \frac{x^3}{6} - \frac{x^4}{12} + \frac{x^5}{60}\right\}\right]dx$$

$$y_4 = 1 - x + \frac{x^2}{2} + \frac{x^3}{6} - \frac{x^4}{24} + \frac{x^5}{60} - \frac{x^6}{360} \quad (*)$$

The exact solution of given DE is

$$y = x^2 - 2x + 2 - e^{-x}$$

Expanding e^{-x} in x, the exact solution is

$$y(x) = x^2 - 2x + 2 - \left[1 - x + \frac{x^2}{2!} - \frac{x^3}{3!} - \frac{x^4}{4!} + \frac{x^5}{5!} - \frac{x^6}{6!} + \ldots\right]$$

$$y(x) = 1 - x + \frac{x^2}{2} + \frac{x^3}{6} - \frac{x^4}{24} + \frac{x^5}{120} - \frac{x^6}{720} + \ldots \quad (**)$$

Thus the fourth approximation $y_4(*)$ concides with the exact analytical solution $y(**)$ up to the 4th power of x. Now

$$y_4(x = 1) = 1 - 1 + \frac{1}{2} + \frac{1}{6} - \frac{1}{24} + \frac{1}{60} - \frac{1}{360} = 0.638888$$

$$y(x = 1) = 1 - 1 + \frac{1}{2} + \frac{1}{6} - \frac{1}{24} + \frac{1}{120} - \frac{1}{720} = 0.6319444$$

Example 2: Obtain the Picard's second approximation for the initial value problem

$$\frac{dy}{dx} = \frac{x^2}{y^2+1}, \ y(0) = 0.$$

Find $y(1)$.

Solution: The first approximation is

$$y_1 = y_0 + \int_{x_0}^{x} \frac{x^2}{y^2+1} dx = 0 + \int_0^x \frac{x^2}{y^2+1} dx$$

Relpacing y by $y_0 = 0$, we get

$$y_1 = \int_0^x \frac{x^2}{0^2+1} dx = \frac{x^3}{3}$$

Replacing y by $y_1 = \frac{x^3}{3}$ we get the second approximation y_2 as

$$y_2 = \int_0^x \frac{x^2}{\left(\frac{x^3}{3}\right)^2 + 1} dx = \tan^{-1}\left(\frac{x^3}{3}\right) - 0$$

In the third approximation

$$y_3 = \int_0^x \frac{x^2}{\left(\tan^{-1}\frac{x^3}{3}\right)^2 + 1}$$

The integration is difficult. This is the drawback of the method.
Expanding $\tan^{-1}$, we get from second approximation

$$y_2 = \tan^{-1}\left(\frac{x^3}{3}\right) = \left(\frac{x^3}{3}\right) - \left(\frac{x^3}{3}\right)^3 \frac{1}{3} + \left(\frac{x^3}{3}\right)^5 \frac{1}{5} \ldots$$

$$y_2 = \frac{x^3}{3} - \frac{x^9}{81} + \frac{x^{15}}{1215} \ldots$$

At $x = 1$, $y_2(1) = \frac{1}{3} - \frac{1}{81} + \frac{1}{1215} = 0.321810699.$

EXERCISE

Using Picard's successive approximation, solve the following initial value problems.

1. $y' = x - y^2$, $y(0) = 1$. Find $y(0.1)$.

Ans. $y_2 = 1 - x + \frac{3}{2}x^2 - \frac{2}{3}x^3 + \frac{x^4}{4} - \frac{x^5}{20}$, $y(0.1) = 0.914357$

2. $y' = x^2 + y^2$, $y(0) = 0$, find $y(0.4)$

Ans. $y_3 = \frac{x^3}{3} + \frac{1}{63}x^7 + \frac{2}{2079}x'' + \frac{1}{59535}x^{15}$, $y(0.4) = 0.02135938$

3. $y' = x + y$, $y(0) = 1$. Find $y(1)$. Compare with exact solution.

Ans. $y_5 = 1 + x + x^2 + \frac{x^3}{3} + \frac{x^4}{12} + \frac{x^5}{60} + \frac{x^6}{720}$
Exact solution: $y = 1 + x + x^2 + \frac{x^3}{3} + \frac{x^4}{12} + \frac{x^5}{60} + \frac{x^6}{360} + \ldots$
$y_5(1) = 3.434$, $y(1) = 3.44$

4. $y' = \frac{y-x}{y+x}$, $y(0) = 1$. Find $y(0.1)$.

Ans. $y_1 = 1 - x + 2\log(1+x)$, y_2 is difficult for evaluation. $y_1(0.1) = 0.9828$

5. $y' = x + y + xy$, $y(0) = 1$, find $y(0.1)$

Ans. $y_3 = 1 + x + x^2 + \frac{x^3}{3} + \frac{5}{12}x^4 + \frac{11}{60}x^5 + \frac{x^6}{24}$
$y_3(0.1) = 1.1103768$

33.3 ADAMS-BASHFORTH-MOULTION METHOD (ABM METHOD)

Predictor-corrector is often loosely used to denote the multistep or multivalue integration technique for ODEs. The most popular predictor-corrector methods are probabily the Adams-Bashforth-Moulton schemes, which have good stability properties unlike the Milne's method which is unstable due to the corrector. The predictor-corrector method consists of three separate processes. The predictor step P, evaluation of derivative y'_{n+1} from the latest value of y, call as E and the corrector step, call as C. Iterating m times with the corrector in this notation, may be written as $P(EC)^m$. Since E is superior, the strategy is PECE i.e., predict, evaluate, correct and evaluate.

Adams-Bashforth-Moulton (ABM) Method is a non self-starting four step method which uses four initial points (x_0, y_0), (x_1, y_1), (x_2, y_2) and (x_3, y_3), obtained already by any one of the single-step methods to generate the points (x_n, y_n) for $n \geq 4$. It requires only two function evaluations of $f(x, y)$ per step.

A Lagrange cubic polynomial based on the four points from x_{n-3} to x_n, is integrated over one step

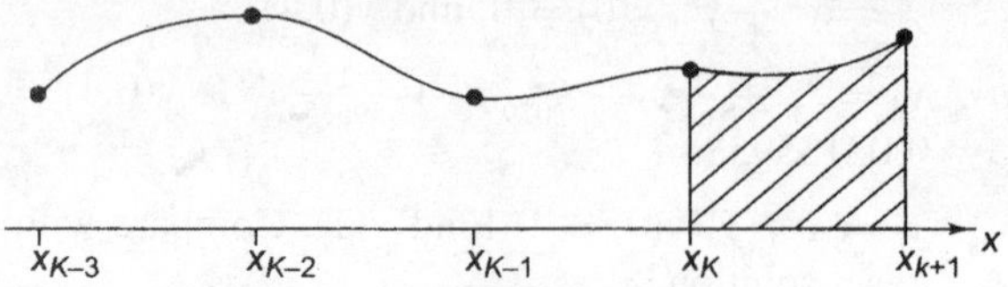

Extrapolation is used in the Adams-Bashforth predictor

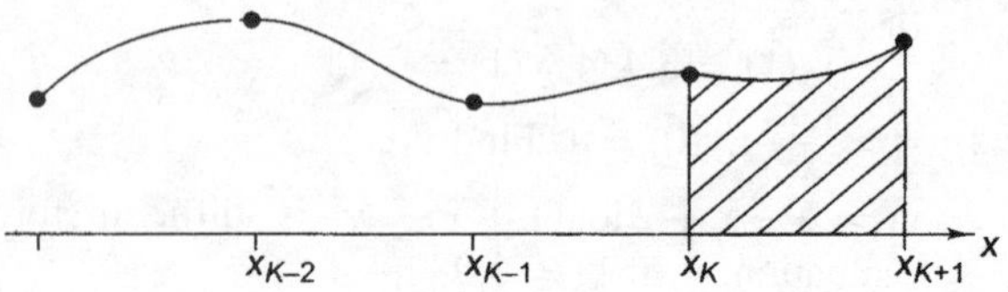

Interpolation is used in the Adams-Moulton corrector

Fig. 33.3 Integration on the interval x_k, x_{k+1}

from x_n to x_{n+1} yielding the *Adams-Bashforth predictor*.

$$y_{n+1} = y_n + \frac{h}{24}\left[55 f_n - 59 f_{n-1} + 37 f_{n-2} - 9 f_{n-3}\right] \quad (1)$$

The tentative computed value of y at x_{n+1} obtained from the predictor formula (1), together with values of y at x_n, x_{n-1}, x_{n-2} are used to construct a second cubic polynomial, which is then integrated over the interval x_n to x_{n+1}, yielding the *Adams-Moulton corrector:*

$$y_{n+1} = y_n + \frac{h}{24}[9 f_{n+1} + 19 f_n - 5 f_{n-1} + f_{n-2}] \quad (2)$$

The step size h, should be decreased, when the difference between predicted and corrected values reaches or exceeds the accuracy criterion.

WORKED OUT EXAMPLES

Example 1: Solve $y' = x^2 - y$, $y(0) = 1$ on the interval [0, 0.5] with step size $h = 0.05$ and with initial values $y(0.05) = 0.95127058$, $y(0.10) = 0.90516258$, $y(0.15) = 0.86179202$. Compare with true (exact) solution.

Solution: Here $f(x, y) = x^2 - y$. $h = 0.05$ $x_0 = 0$, $y_0 = 1$, $f(x_0, y_0) = f_0 = 0^2 - 1 = -1$
$x_1 = 0.05$, $y_1 = 0.95127058$, $f(x, y) = f_1 = (0.05)^2 - 0.95127058$ so $f_1 = -0.94877058$.
$x_2 = 0.10$, $y_1 = 0.90516258$, so $f(x_2, y_2) = f_2 = (0.10)^2 - 0.90516258 = -0.89516258$
$x_3 = 0.15$, $y_3 = 0.86179202$ so $f(x_3, y_3) = f_3 = (0.15)^2 - 0.86179202 = -0.83929202$
Use Adams-Bashforth predictor ($n = 3$)

$$y_4 = y_3 + \frac{h}{24}[55 f_3 - 59 f_2 + 37 f_1 - 9 f_0]$$

substituting the above starting values

$$y_4 = 0.86179202 + \frac{0.05}{24}[55(-0.8392920) - 59(-0.89516258) + 37(-0.94877058) - 9(-1)]$$

$$= 0.86179202 + \frac{0.05}{24}[-19.45097924]$$

$$y_4 = 0.86179202 - 0.040522873 = 0.821269146$$

Using this tentative predictor value, compute $f_4 = f(x_4, y_4) = f(0.20, 0.821269146)$

$$f_4 = (0.20)^2 - 0.821269146 = -0.781269146$$

Now we utilize this value of f_4 in the Adams-Moulton corrector ($n = 3$)

$$y_4 = y_3 + \frac{h}{24}[9 f_4 + 19 f_3 - 5 f_2 + f_1]$$

substituting the above data, we get

$$y_4 = 0.86179202 + \frac{0.05}{24}[9(-0.781269146) + 19(-0.83929202) - 5(-0.89516258) + (-0.94877058)]$$

$$= 0.86179202 + \frac{0.05}{24}[-19.45092837]$$

$$y_4 = 0.86179202 - 0.040522767 = 0.821269252$$

The exact (true) solution is

$$y(x) = x^2 - 2x + 2 - e^{-x}$$

so $y(0.20) = 0.821269246$. Calculating further values with ABM method we get the follwing data:

n	x	Adams-Bashforth Moulton method $y(x)$	Exact solution $y(x)$	Error
4	0.20	0.821269252	0.821269246	6×10^{-9}
5	0.25	0.7836992	0.783699216	1.69×10^{-8}
6	0.30	0.7491818	0.749181779	$+2.068 \times 10^{-8}$
7	0.40	0.6896800	0.689679954	4.6036×10^{-8}
8	0.50	0.6434694	0.64346934	5.9716×10^{-8}

Note: For $n = 4$, we have the following predictor corrector formulas.

Predictor: $y_5 = y_4 + \frac{h}{24}[55 f_4 - 55 f_3 + 37 f_2 - 9 f_1]$

Corrector: $y_5 = y_4 + \frac{h}{24}[9 f_5 + 19 f_4 - 5 f_3 + f_2]$.

Here

$f_4 = f(0.2, 0.821269252) =$
$(0.2)^2 - 0.821269252 = -0.781269252$

EXERCISE

ABM Method

Solve by ABM method using the given initial values. Compare with exact solution.

1. $y' = -2x - y$, $y(0) = -1$, find $y(0.4)$, $y(0.5)$
   $y(0) = -1$, $y(0.1) = -0.9145122$, $y(0.2) = -0.8561923$, $y(0.3) = -0.8224547$

Ans. At 0.4, Predictor value $= -0.8109687$
Corrector value : -0.8109652
Exact solution: $y(x) = -3e^{-x} - 2x + 2$
$y(0.4) =$ Exact value $= -0.8109601$
At 0.5 $= P: -0.8195978$, $C: -0.819505$, $E: -0.8195920$

2. $y' = (x - y)/2$, $y(0) = 1$, $h = 0.125$, find y on $[0, 1]$. Compare with exact solution. Initial values are : $y(0.125) = 0.94323919$, $y(0.250) = 0.89749071$, $y(0.375) = 0.86208736$.
   Exact solution: $y(x) = 3e^{-x/2} - 2 + x$

Ans. $y(0.5) = 0.83640227$, $y(0.625) = 0.81984673$, $y(0.75) = 0.81186762$, $y(0.875) = 0.81194530$, $y(1.0) = 0.81959166$

3. $y' = 2xy^2$, $y(0) = 1$, $h = 0.05$, find y on $[0, 0.5]$ Compare with exact solution.
   Initial values are $y(0.05) = 1.0025063$, $y(0.10) = 1.0101010$, $y(0.15) = 1.0230179$.
   Exact solution: $y(x) = 1/(1 - x^2)$

Ans. $y(0.2) = 1.0416675$ $y(0.25) = 1.0666688$
$y(0.3) = 1.0989052$, $y(0.4) = 1.1904878$,
$y(0.5) = 1.3333631$

4. $y' = y^2 \sin x$, $y(0) = 1$, find y on $[0.05]$, $h = 0.05$. Initial values: $y(0.05) = 1.0012513$, $y(0.10) = 1.0050209$, $y(0.15) = 1.0113564$.
   Exact solution: $y(x) = \sec x$

Ans. $y(0.2) = 1.0203389$, $y(0.25) = 1.0320852$,
$y(0.3) = 1.0467519$, $y(0.4) = 1.0857051$
$y(0.5) = 1.1394953$.

5. $y' = -\frac{x}{y}$, $y(1) = 1$, find y on $[1, 1.4]$. $h = 0.05$ Initial values $y(1.05) = 0.94736477$, $y(1.10) = 0.88881944$, $y(1.15) = 0.82310388$
   Exact solution: $y(x) = (2 - x^2)^{1/2}$

Ans. $y(1.2) = 0.7483205$, $y(1.25) = 0.6613998$,
$y(1.3) = 0.5566583$, $y(1.35) = 0.4208572$,
$y(1.4) = 0.1974740$.

6. $y' = x^3 + y^2$, $y(0) = 0$, $h = 0.1$. Find y on $[0, 1.2]$. Initial values are $y(0.2) = 0.0004$, $y(0.4) = 0.0064$, $y(0.6) = 0.0325$

Ans. $y(0.8) = 0.1035$, $y(0.9) = 0.1669$, $y(1.0) = 0.2574$, $y(1.1) = 0.3836$, $y(1.2) = 0.5581$

33.4 NUMERICAL SOLUTIONS TO PARTIAL DIFFERENTIAL EQUATIONS

Many physical phenomena can be modelled mathematically by partial differential equations, whose theory is quite difficult and numerical methods most difficult and needs extensive computation. Although analytical solutions can be obtained by separation of variable technique and integral transforms (Fourier transforms), in many complicated cases one has to resort to numerical methods. Initial value problems or time dependent problems consists of two types: hyperbolic problems (containing both f_{xx} and f_{tt} term as in wave equation) and parabolic problems (which contain only f_{xx} and f_t but no f_{tt} term as in the heat equation). For these problems, the solution is known at an initial time and then propagates through space in time. The domain is open (goes to infinity) somewhere for example, in time. The other type of problems is the boundary value problem where the domain is closed and is entirely surrounded by boundary values. These problems are called elliptic (and are time independent as in the case of Laplace equation).

The dimension of a partial differential equation is the number of space variables x, y, z appearing in the equation as independent variables (apart from the independent variable time t). Thus a one-dimensional heat equation or one-dimensional wave equation involves only one space variable say x (and of course time t). Similarly two-dimensional Laplace equation contains two space variables say x and y.

The central idea in numerical methods to partial differential equations is to discretize the partial differential equation by replacing it, approximately, by a finite system of algebraic equations. One such important discretization method is finite differences, in which the derivatives are replaced by finite differences. The entire given domain is discretized by a discrete set of points, as follows:

Introducing $x_i = x_0 + ih$, $i = 0, \pm 1, \pm 2, \pm 3, \ldots$ and $t_j = t_0 + jk$, $j = 0, \pm 1, \pm 2, \pm 3, \ldots$, the (numerical) solution is obtained at the discrete points (x_i, t_j) known as grid or nodal or lattice or mesh or pivotal points of the computational grid. Here $h =$ constant is spacing (or size of interval) in x-direction while $k =$ constant is spacing in t-direction.

33.5 NUMERICAL SOLUTION TO ONE-DIMENSIONAL HEAT EQUATION

Consider the initial boundary value problem of the parabolic one-dimensional heat equation.

P.D.E: $$\frac{\partial u}{\partial t} = c^2 \frac{\partial^2 u}{\partial x^2}, 0 \le x \le L, t \ge 0 \qquad (1)$$

Two boundary conditions (B.C.'s):

$$u(0, t) = 0 \quad u(L, t) = 0 \qquad (2)$$

One initial condition (I.C.)

$$u(x, 0) = f(x) \qquad (3)$$

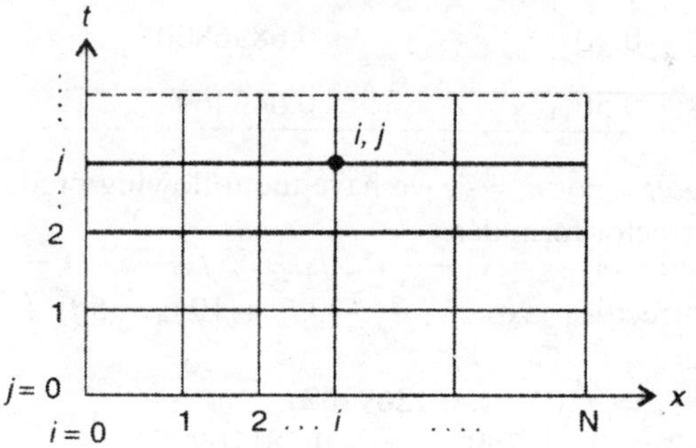

Fig. 33.3

Replacing u_t and u_{xx} by finite differences, equation (1) is approximated by a finite difference equation.

$$\frac{1}{k}\left(u_{i,j+1} - u_{ij}\right) = \frac{c^2}{h^2}\left(u_{i+1,j} - 2u_{ij} + u_{i-1,j}\right) \qquad (4)$$

Here h is mesh size in x-direction i.e., $h = \frac{L}{N}$ and k is mesh size in t-direction. Rewriting (4),

$$u_{i,j+1} = (1 - 2r)u_{ij} + r(u_{i+1,j} + u_{i-1,j}) \qquad (5)$$

where $r = \frac{c^2 k}{h^2}$. This method is known as *Schmidt explicit method* because the unknown $u_{i,j+1}$ is explicitly expressed in terms of the known (given) values $u_{i,j}; u_{i+1,j}; u_{i-1,j}$; whereas in the *implicit method* such as the Crank-Nicolson method the equations are implicit i.e., coupled and one has to solve a system of linear equations. Formula (5) provides a marching scheme to compute u at the mesh points. This method is convergent and stable if $r = \frac{c^2 k}{h^2} \le \frac{1}{2}$. In particular for $r = \frac{1}{2}$, (5) reduces to

$$u_{i,j+1} = \frac{1}{2}(u_{i+1,j} + u_{i-1,j}) \qquad (6)$$

which is known as *Bender-Schmidt recurrence relation*. So the values of h, k are so chosen (c is already given) that $r \le \frac{1}{2}$. For $r > \frac{1}{2}$ the method is divergent and unstable.

Stencil (pattern, molecule or star)

Finite difference approximations are often presented in terms of stencils or coefficient schemes which shows the pattern of connection in the difference equations. The entries in the stencil are coefficients in the difference equation.

For example the explicit method given by (5) is represented as a stencil shown below:

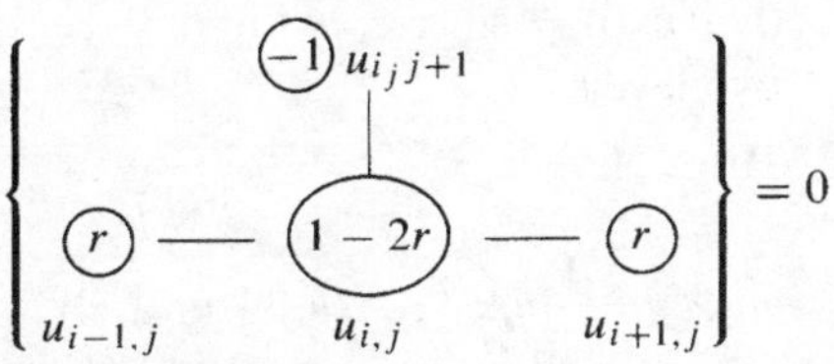

i.e., $(-1)u_{i,j+1} + r(u_{i-1,j}) + (1-2r)u_{ij} + ru_{i+1,j} = 0$

WORKED OUT EXAMPLES

Example 1: Solve the initial boundary value problem:

$$\frac{\partial f}{\partial t} = 2\frac{\partial^2 f}{\partial x^2}$$

$$f(0, t) = 10 \quad (1)$$

$$f(6, t) = 18 \quad (2)$$

$$f(x, 0) = \frac{x^2}{2} \quad (3)$$

with $h = 1$ and $\tau = \frac{1}{8}$ by *explicit* method.

Solution: $m = \frac{a\tau}{h^2}$. Here $a = 2, h = 1, \tau = \frac{1}{8}$ so $m = \frac{1}{4}$. The explicit formula $f_{i,j+1} = (1-2m)$ $f_{ij} + m(f_{i+1,j} + f_{i-1,j})$ with $m = \frac{1}{4}$ becomes

$$f_{i,j+1} = \frac{1}{2}f_{ij} + \frac{1}{4}(f_{i+1,j} + f_{i-1,j}) \quad (4)$$

Here first row $t = 0$, consists of the initial values given by $IC(3): f(x, 0) = \frac{x^2}{2}$ at $x = 1, 2, 3, 4, 5$. The first column $x = 0$, gives the first boundary condition $f(0, t) = 10$. The last column $x = 6$, gives the second boundary condition $f(6, t) = 18$. The remaining values of f are calculated using (4). As $t \to \infty$, the process reaches steady state, so $f_{xx} = 0, f(x) = c_1 x + c_2, 10 = f(0) = 0 + c_2 \therefore c_2 = 10, 18 = f(6) = 6.c_1 + 10$ $\therefore c_1 = \frac{4}{3}$, analytical exact solution as $t \to \infty$ is $f(x) = \frac{4}{3}x + 10$. These are plotted for $t \to \infty$ in the last row of the Table 33.7.

Table 33.7

		B.C. (1)						B.C. (2)
	$t_j \backslash x_i$	0	1	2	3	4	5	6
I.C.(3)	$t = 0$	10	$\frac{1}{2}$	2	4.5	8	12.5	18
	$t = \frac{1}{8}$	10	3.25	2.25	4.75	8.25	12.75	18
	$t = \frac{1}{4}$	10	4.69	2.88	5.0	8.5	12.94	18
	$t = \frac{3}{8}$	10	5.56	3.86	5.34	8.73	13.09	18
	$t = \frac{1}{2}$	10	6.25	4.66	5.82	8.98	13.23	18
	$t = \frac{5}{8}$	10	6.79	5.35	6.32	9.25	13.36	18
	$t = \frac{3}{4}$	10	7.23	5.95	6.81	9.55	13.49	18
	$t = \frac{7}{8}$	10	7.60	6.49	7.28	9.85	13.63	18
	$t = 1$	10	7.92	6.97	7.73	10.15	13.78	18
	$t = 1\frac{1}{8}$	10	8.20	7.40	8.15	10.45	13.93	18
	$t = 1\frac{1}{4}$	10	8.45	7.79	8.54	10.75	14.08	18
	$t = 1\frac{3}{8}$	10	8.67	8.14	8.91	11.03	14.23	18
	$t = 1\frac{1}{2}$	10	8.87	8.47	9.25	11.30	14.37	18
	$t = \infty$	10	$11\frac{1}{3}$	$12\frac{2}{3}$	14	$15\frac{1}{2}$	$16\frac{2}{3}$	18

Example 2: Solve the above problem by Bender-Schmidt recurrence equation; with $h = 1$ and τ determined accordingly.

Solution: $m = \frac{a\tau}{h^2} = \frac{2\tau}{1}$. For Bender-Schmidt $m = \frac{1}{2} \therefore \quad \tau =$ time interval size $= \frac{1}{4}$. (See Table 33.8.)

Non-dimensional form

Example 3: Transform the heat equation

$$U_T = c^2 U_{XX}, \text{ where } 0 \le X \le L$$

to non-dimensional form.

Solution: Introducing

$$u = \frac{U}{u_0} \quad \text{and} \quad x = \frac{X}{L}, \; t = \frac{c^2 T}{L^2}, \quad \text{we have}$$

$$\frac{\partial u}{\partial x} = \frac{\partial U}{\partial X} \cdot \frac{\partial X}{\partial x} = L\frac{\partial U}{\partial X}, \frac{\partial^2 u}{\partial x^2} = \frac{\partial}{\partial X}\left(L\frac{\partial U}{\partial X}\right)\cdot$$

Table 33.8

	$t_j \backslash x_i$	0 (B.C.(1))	1	2	3	4	5	6 (B.C.(2))
I.C.(3)	$t=0$	10	$\frac{1}{2}$	2	4.5	8	12.5	18
	$t=\frac{1}{4}$	10	6	2.5	5.0	8.5	13	18
	$t=\frac{1}{2}$	10	6.25	5.5	5.5	9	13.25	18
	$t=\frac{3}{4}$	10	7.75	5.88	7.25	9.38	13.5	18
	$t=1$	10	7.94	7.50	7.625	10.38	13.69	18
	$t=\frac{5}{4}$	10	8.75	7.78	8.94	10.66	14.19	18
	$t=\frac{3}{2}$	10	8.89	8.84	9.22	11.56	14.33	18
	$t=\frac{7}{4}$	10	9.42	9.05	10.20	11.77	14.78	18
	$t=2$	10	9.53	9.81	10.41	12.49	14.89	18
	$t=\frac{9}{4}$	10	9.90	9.97	10.15	12.65	15.25	18
	$t=\frac{5}{2}$	10	9.99	10.53	11.31	13.20	15.33	18

$$\frac{\partial^2 u}{\partial x^2} = L\frac{\partial^2 U}{\partial X^2}\cdot L = L^2\frac{\partial^2 U}{\partial X^2}.$$ Similarly,

$$\frac{\partial u}{\partial t} = \frac{\partial U}{\partial T}\frac{\partial T}{\partial t} = \frac{L^2}{c^2}\frac{\partial U}{\partial T}.$$ Then

$$U_T = \frac{\partial U}{\partial T} = \frac{c^2}{L^2}\frac{\partial u}{\partial t} = c^2 U_{XX} = c^2\frac{\partial^2 U}{\partial X^2} = c^2\cdot\frac{1}{L^2}\frac{\partial^2 u}{\partial x^2}$$

Thus $\frac{\partial u}{\partial t} = \frac{\partial^2 u}{\partial x^2}$ i.e., $u_t = u_{xx}$

for $0 \le \frac{X}{L} \le \frac{L}{L}$ i.e., $0 \le x \le 1$. This form is known as non-dimensional standard form. Similarly, the wave equation

$$U_{TT} = a^2 U_{XX} \quad \text{where} \quad 0 \le X \le L$$

can be transformed to non-dimensional standard form

$$u_{tt} = u_{xx} \quad \text{where} \quad 0 \le x \le 1.$$

Exercise

Numerical solution to one-dimensional heat equation

1. Using Schmidt explicit method, solve $u_t = u_{xx}$, $0 \le x \le 1$, $u(0,t) = 0$, $u(1,t) = 0$, $u(x,0) = 2x$ when $0 \le x \le 0.5$ and $u(x,0) = 2(1-x)$ when $0.5 \le x \le 1$ for $h = 1$, $k = 0.001$ at $t = 0.001, 0.002$.

Ans.

x:	0	.1	.2	.3	.4	.5	.6	.7	.8	.9	1.0
$u(x, 0.001)$:	0	.2	.4	.6	.8	.96	.8	.6	.4	.2	0
$u(x, 0.002)$:	0	.2	.4	.6	.796	.928	.796	.6	.4	.2	0

2. Use Schmidt explicit method to obtain numerical solution of $u_t = u_{xx}$, $u(0,t) = u(1,t) = 0$, $u(x,0) = \sin\pi x$, with $h = \frac{1}{3}$, $k = \frac{1}{36}$ for two levels.

Hint: Here $r = \frac{kc^2}{h^2} = \frac{1}{36}\cdot\frac{1}{\left(\frac{1}{3}\right)^2} = \frac{1}{4}$ so

$$u_{i,j+1} = \frac{1}{4}[u_{i-1,j} + 2u_{ij} + u_{i+1,j}]$$

Ans.

$x =$	0	$\sqrt{3}/2$	$\sqrt{3}/2$	0
$t = \frac{1}{36}$:	0	.65	.65	0
$t = \frac{2}{36}$:	0	.49	.49	0

3. Use Schmidt explicit method to solve $u_t = u_{xx}$, $u(0,t) = u(1,t) = 0$, $u(x,0) = \sin\pi x$; for $0 \le t \le 0.2$, with $r = 0.25$.

Hint: $r = k/h^2 = 0.25$, so $k = 0.01$, $u_{i,j+1} = 0.25\left(u_{i-1,j} + 2u_{ij} + u_{i+1,j}\right)$. Value at $x = 0.6$ is same as at $x = 0.4$; Value at $x = 0.8$ is same as at $x = 0.2$ due to symmetry.

Ans.

Table 33.9

	$x = 0.2$	$x = 0.4$
$t = 0$:	0.588	0.951
$t = 0.04$	0.393	0.637
$t = 0.08$	0.263	0.426
$t = 0.12$	0.176	0.285
$t = 0.16$	0.118	0.191
$t = 0.20$	0.079	0.128

4. Using Bender-Schmidt (explicit method with $r = \frac{1}{2}$) solve $u_{xx} - 2u_t = 0$, $u(0,t) = u(4,t) = 0$, $u(x,0) = x(4-x)$, with $h = 1$, $k = 1$.

Ans.

5. Solve $2u_{xx} - u_t = 0$, $u(x,0) = 50(4-x)$, $u(0,t) = 0$, $u(4,t) = 0$, with $h = 1$, $\tau = 0.25$.

Hint: use Bender-Schmidt scheme.

Ans. At $t = 1.50$; $u = 0, 6.25, 12.5, 6.25, 0$ (after six iterations)

Table 33.10

$t\backslash x$	0	1	2	3	4
0	0	3	4	3	0
1	0	2	3	2	0
2	0	1.5	2	1.5	0
3	0	1	1.5	1	0
4	0	.75	1	.75	0
5	0	.5	.75	.5	0

6. Compute $u(x,t)$ at $x = 0, 1, 2, \ldots 7, t = \frac{1}{8}j$, $j = 0, 1, 2, 3, 4, 5$ by solving $u_t = 4u_{xx}$, $u(0,t) = u(8,t) = 0, u(x,0) = 4x - \frac{x^2}{2}$.

Hint: $r = \frac{1}{2}$, use Bender-Schmidt recurrence relation.

Ans.

Table 33.11

$j\backslash x$	0	1	2	3	4	5	6	7	8
0	0	3.5	6	7.5	8	7.5	6	3.5	0
1	0	3	5.5	7	7.5	7	5.5	3	0
2	0	2.75	5	6.5	7	6.5	5		0
3	0	2.5	4.625	6	6.5	6	4.625		0
4	0	2.3	4.25	5.56	6	5.56	4.25		0
5	0	2.125	3.94	5.125	5.56	5.125	3.94	2.125	0

33.6 NUMERICAL SOLUTION TO ONE-DIMENSIONAL WAVE EQUATION

Consider the initial boundary value problem of hyperbolic one-dimensional wave equation

P.D.E :
$$\frac{\partial^2 u}{\partial t^2} = a^2 \frac{\partial^2 u}{\partial x^2}, 0 \le x \le L, t \ge 0 \qquad (1)$$

Two boundary conditions:

$$u(0,t) = u(L,t) = 0 \qquad (2)$$

Two initial conditions: Prescribed initial displacement:

$$u(x,0) = f(x) \qquad (3)$$

Prescribed initial velocity:

$$u_t(x,0) = g(x) \qquad (4)$$

The finite difference approximation of (1) is

$$\frac{1}{k^2}(u_{i,j+1} - 2u_{ij} + u_{i,j-1}) = \frac{a^2}{h^2}(u_{i+1,j} - 2u_{ij} + u_{i-1,j}) \qquad (5)$$

This explicit method converges and stable for $0 < r \le 1$. Here h is mesh size in x-direction and k is the mesh size in t-direction. For a given a^2, choose h and k such that

$$r = \frac{a^2k^2}{h^2} = 1.$$

Then Equation (5) simplifies to

$$u_{i,j+1} = u_{i-1,j} + u_{i+1,j} - u_{ij-1} \qquad (6)$$

The numerical solution is generally represented in the form of an open ended table, consisting of n columns, headed by $x_i = x_0 + ih$, where the mesh size $h = \frac{L-0}{n}$. Here $x_0 = 0, x_n = L$.

Step I All the entries in the first column are zeros, given by the first boundary condition $u(0,t) = 0$.

Step II Similarly, all the entries in last column are zeros, given by the second boundary condition.

$$u(L,t) = 0.$$

Step III Each row represents a time step. Entries in first row, are calculated using (3) the first initial condition $u(x,0) = f(x)$ for $x_i = x_0 + ih, i = 1, 2, 3, \cdots, n$.

Step IV Observe that formula (6) involves 3 time steps namely the future $(j+1)$ in terms of the present (j) and the past $j-1$. At this stage, from step III, the first row gives only on time step $(j = 0)$. So use (4) the second initial condition which gives rise to another (second) row which will be the second time step $(j = 1)$.

Now from I.C. (4) in difference form is

$$\frac{1}{2k}(u_{i1} - u_{i,-1}) = g_i$$

or
$$u_{i,-1} = u_{i1} - 2\,k\,g_i \qquad (7)$$

where $g_i = g(ih)$. For $t = 0$ (i.e., $j = 0$), we have from (6)

$$u_{i1} = u_{i-1,0} + u_{i+1,0} - u_{i,-1} \qquad (8)$$

Substitute $u_{i,-1}$ from (7) into (8), we get

$$u_{i1} = u_{i-1,0} + u_{i+1,0} - (u_{i1} - 2k\,g_i) \quad \text{or}$$

$$u_{i1} = \frac{1}{2}(u_{i-1,0} + u_{i+1,0}) + kg_i \qquad (9)$$

Equation (9) expresses u_{i1} in terms of the initial data. For $i = 0, 1, 2, \ldots, n$, Equation (9) gives the entries in the second row.

Step V Now that we have two rows, one from I.C.

(3) and one from I.C. (4), we can calculate third row entries by formula (6) using the first ($j = 0$) and second row ($j = 1$) entries. Continuing this process the 4th row is calculated by (6) using 2nd and 3rd row entries. Subsequent rows are obtained similarly. Stencil for formula (6)

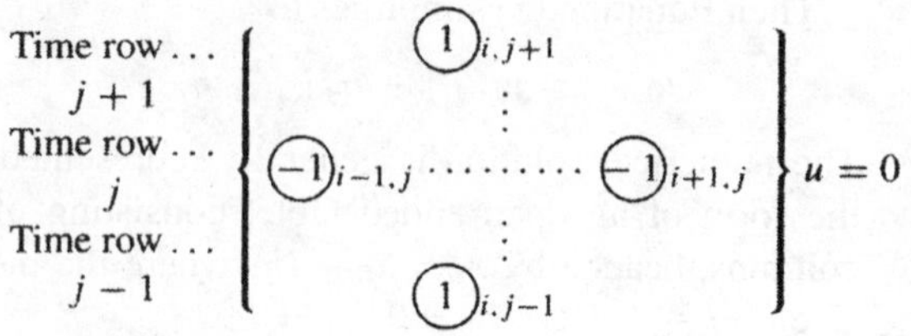

i.e., $u_{i,j+1} + u_{i,j-1} = u_{i-1,j} + u_{i+1,j}$

Worked Out Examples

Numerical solution to one-dimensional wave equation

Examples: Compute numerically the solution of the one-dimensional wave equation $25 f_{xx} - f_{tt} = 0$ with boundary conditions $f(0, t) = f(5, t) = 0$ and with initial conditions $f(x, 0) = 20x$, when $0 \le x \le 1$ and $f(x, 0) = 25\left(1 - \frac{x}{5}\right)$ when $1 \le x \le 5$; $\frac{\partial f}{\partial t}\Big|_{t=0} = 0$.

Solution: Here $a^2 = 25$, choose $h = 1$ mesh size in x. Since

$$\frac{a^2 t^2}{h^2} = 1 \text{ so } \frac{25\tau^2}{1} = 1 \text{ or } \tau = \frac{1}{5}, \text{ mesh size in } t.$$

Inltial values: At $t = 0$, (initial displacement).

$$f(x, 0) = \begin{cases} 20x, & 0 \le x \le 1 \\ 25\left(1 - \frac{x}{5}\right), & 1 \le x \le 5 \end{cases} \qquad \text{IC (1)}$$

For $x = 0$, $f_{00} = 0$, for $x = 1$, $f_{10} = 20$, for $x = 2$, $f_{20} = 15$, $f_{30} = 10$, $f_{40} = 5$, $f_{50} = 0$. These six values form the first row in the table.

Since at $\frac{\partial f}{\partial t}(x, 0) = 0$, I.C. (2)
the initial velocity is zero so

$$f_{i1} = \frac{1}{2}\left(f_{i-1,0} + f_{i+1,0}\right). \qquad \text{IC (2)}$$

For $i = 1$, $f_{11} = \frac{1}{2}(f_{00} + f_{20}) = \frac{1}{2}(0 + 15) = 7.5$,
For $i = 2$, $f_{21} = \frac{1}{2}(f_{10} + f_{30}) = \frac{1}{2}(20 + 10) = 15$
For $i = 3$, $f_{31} = \frac{1}{2}(f_{20} + f_{40}) = \frac{1}{2}(15 + 5) = 10$
For $i = 4$, $f_{41} = \frac{1}{2}(f_{30} + f_{50}) = \frac{1}{2}(10 + 0) = 5$
These six values form the second row in the table. The remaining displacements are calculated from

$$f_{i,j+1} = f_{i-1,j} + f_{i+1,j} - f_{i,j-1} \qquad (*)$$

which involves 2 time steps (at J and $J - 1$). Then the resulting (Table 33.12) is given below:

Table 33.12

	$t\backslash x$	0 ($f(0,t=0)$ B.C.(1)↓)	1	2	3	4	5 ($f(5,t)=0$ ↓B.C.(2))
I.C.(1)	0	$f_{00} = 0$	$f_{10} = 20$	$f_{20} = 15$	$f_{30} = 10$	$f_{40} = 5$	$f_{50} = 0$
I.C.(2)	1	0	$f_{11} = 7.5$	$f_{21} = 15$	$f_{31} = 10$	$f_{41} = 5$	0
	2	0	−5	2.5	10	5	0
	3	0	−5	−10	−2.5	5	0
	4	0	−5	−10	−15	−7.5	0
	5	0	−5	−10	−15	−20	0
	6	0	−5	−10	−15	−7.5	0
	7	0	−5	−10	−2.5	5	0
	8	0	−5	2.5	10	5	0
	9	0	7.5	15	10	5	0
	10	0	20	15	10	5	0

The values in the 2nd, 3rd, 4th, . . ., 10th row are calculated using the formula (*). The displacement function $f(x, t)$ goes through a complete oscillation (cycle) in 10 steps, indicating a periodic behaviour with period $T = 10\tau = 10\left(\frac{1}{5}\right) = 2$.

Exercise

Numerical solution to one-dimensional wave equation

1. Evaluate the pivotal values of $16u_{xx} = u_{tt}$, $u(0, t) = u(5, t) = 0$, $u(x, 0) = x^2(5 - x)$, $u_t(x, 0) = 0$, with $h = 1$ and up to one half of the period of vibration.

   Hint: Period of vibration $= \frac{2L}{a} = \frac{2.5}{4}$ since $a^2 = 16$, $L = 5$. Period: $\frac{5}{2}$ secs or compute up to $t = \frac{5}{4}$ secs. From $u_t(x, 0) = 0$, $u_{i1} = u_{i0}$ so 1st and 2nd row are same.

Ans.

Table 33.13

$t\backslash x$	0	1	2	3	4	5
0	0	4	12	18	16	0
1	0	4	12	18	16	0
2	0	8	10	10	2	0
3	0	6	6	−6	−6	0
4	0	−2	−10	−10	−8	0
5	0	−16	−18	−12	−4	0

2. Determine the displacement $u(x, t)$ of a string if $u_{tt} = u_{xx}$, $u(x, 0) = \frac{x(10-x)}{100}$, $u(0, t) = 0$, $u(10, t) = 0$, $u_t(x, 0) = 0$ for $x = 0(1)10$, $t = 0(1)5$.

Ans.

x:	0	1	2	3	4	5
$t = 0$:	0	0.09	0.16	0.21	0.24	0.25
$t = 1$:	0	.08	.15	.20	.23	.24
$t = 4$:	0	.02	.04	.06	.08	.09
$t = 5$:	0	0	0	0	0	0

x:	6	7	8	9	10
$t = 0$:	.24	.21	.16	.9	0
$t = 1$:	.23	.20	.15	.08	0
$t = 4$:	.08	.06	.04	.02	0
$t = 5$:	0	0	0	0	0

3. Compute numerically the solution of $4u_{xx} - u_{tt} = 0$, $u(x, 0) = x(4 - x)$; $u(0, t) = 0$, $u(4, t) = 0$, with $h = 1$, $\tau = \frac{1}{2}$.

Hint: Complete one oscillation in 8 steps; Period $= 8\frac{1}{2} = 4$.

Ans.

Table 33.14

$t\backslash x$	0	1	2	3	4
0	0	3	4	3	0
1	0	2	3	2	0
2	0	0	0	0	0
3	0	−2	−3	−2	0
4	0	−3	−4	−3	0
5	0	−2	−3	−2	0
6	0	0	0	0	0
7	0	2	3	2	0
8	0	3	4	3	0

4. Calculate the numerical solution with $h = k = 0.2$ by solving $u_{tt} = u_{xx}$, $u(x, 0) = \sin \pi x$, $u(0, t) = u(1, t) = 0$, $u_t(x, 0) = 0$.
Hint: Symmetric $u(x = 0.2, t) = u(x = 0.8, t)$ and $u(x = 0.4, t) = u(x = 0.6, t)$.

Ans.

Table 33.15

	$x = 0$	0.2	0.4
$t = 0$	0	.588	.951
$t = 0.2$	0	0.476	0.769
$t = 0.4$	0	0.182	0.294
$t = 0.6$	0	−0.182	−0.294
$t = 0.8$	0	−0.476	−0.769
$t = 1.0$	0	−0.588	−0.951

5. Evaluate the pivotal values for $\frac{1}{2}$ period of vibration by solving $25u_{xx} - u_{tt} = 0$, $u(0, t) = u(5, t) = 0$, $u(x, 0) = 2x$, $0 \le x \le 2.5$, $u(x, 0) = 10 - 2x$, $2.5 \le x \le 5$.

Hint: $h = 1$, $\tau = \frac{1}{5}$.

Ans.

Table 33.16

$j\backslash i$	0	1	2	3	4	5
0	0	2	4	4	2	0
1	0	2	3	3	2	0
2	0	1	1	1	1	0
3	0	−1	−1	−1	−1	0
4	0	−2	−3	−3	−2	0
5	0	−2	−4	−4	−2	0

33.7 NUMERICAL SOLUTION TO TWO-DIMENSIONAL LAPLACE EQUATION

Consider the boundary value problem of elliptic Laplace equation in two-dimensions defined in a rectangular domain with prescribed boundary conditions as follows:

$$\text{P.D.E}: \quad \frac{\partial^2 u}{\partial x^2} = \frac{\partial^2 u}{\partial y^2} = 0, \quad \begin{array}{l} 0 < x < a, \\ 0 < y < b \end{array} \tag{1}$$

Four B.C's

$$u(0, y) = p(y), u(x, 0) = q(x),$$

$$u(a, y) = r(y), u(x, b) = s(x) \tag{2}$$

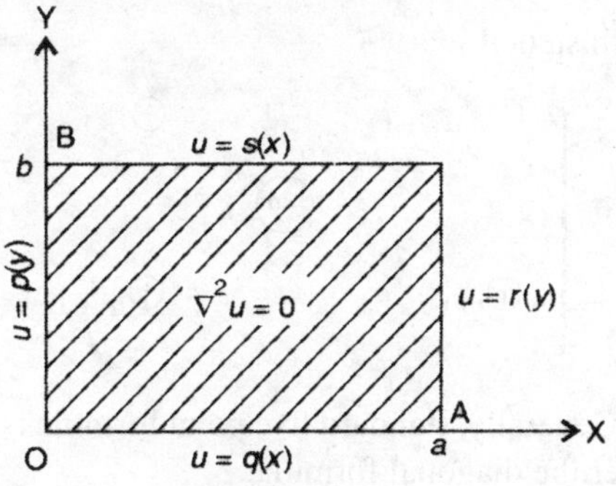

Fig. 33.4

The rectangular domain is discretized by dividing OA into M qual parts of size $\Delta x = \frac{a}{M} = h$, and dividing OB into N equal parts of size $\Delta y = \frac{b}{N} = k$. Thus the function $u(x, y)$ is to be determined at the nodal points $p_{ij} = (x_i, y_j) = (i\Delta x, j\Delta y) = (ih, jk)$ for $i = 0, 1, 2, \ldots M$ and $j = 0, 1, 2, \ldots N$. Replacing (1) by finite difference, we have

$$\frac{u_{i-1,j} - 2u_{ij} + u_{i+1,j}}{h^2} + \frac{u_{i,j-1} - 2u_{i,j} + u_{i,j+1}}{k^2} = 0$$

For a square domain, $h = k$. The above equation reduces to

$$u_{ij} = \frac{1}{4}\left[u_{i-1,j} + u_{i+1,j} + u_{i,j-1} + u_{i,j+1}\right] \quad (3)$$

(3) is known as the **standard five points formula** and represented by the stencil.

$$\left\{\begin{array}{ccc} & (-1)_{i,j+1} & \\ (-1)_{i,j-1} & (-4)_{i,j} & (-1)_{i,j+1} \\ & (-1)_{i,j-1} & \end{array}\right\} u = 0$$

Since Laplace's equation is invariant under rotation of axes through 45°, we get a **five point diagonal formula**

$$u_{ij} = \frac{1}{4}\left[u_{i-1,j-1} + u_{i+1,j-1} + u_{i+1,j+1} + u_{i-1,j+1}\right] \quad (4)$$

with stencil

$$\left\{\begin{array}{ccc} (1)_{i-1,j-1} & & (1)_{i-1,j+1} \\ & (-4)_{i,j} & \\ (1)_{i+1,j-1} & & (1)_{i+1,j+1} \end{array}\right\} u = 0$$

Generally, standard five point formula is preferred over the diagonal formula.

Procedure: Suppose the square domain is divided into a grid as follows:

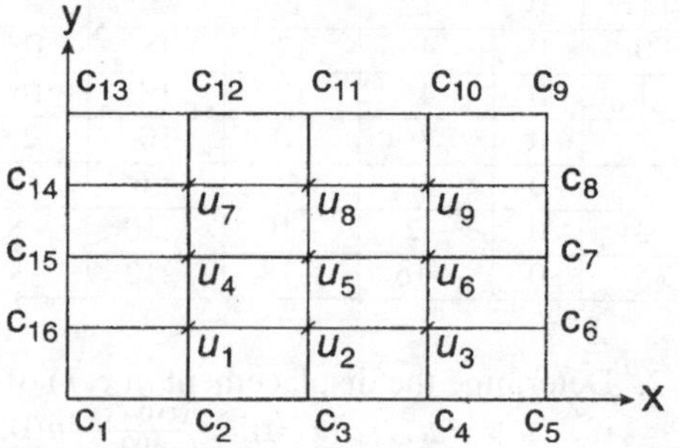

Here the boundary values $c_1, c_2, \cdots, c_{16}$ are given. To find the unknown function u at the 9 grid or mesh points $1, 2, 3, \ldots, 9$, first use the diagonal formula to compute first u_5 and then u_7, u_9, u_1, u_3 in that order. They are given by

$$u_5 = \frac{1}{4}[c_1 + c_5 + c_9 + c_3]$$

$$u_7 = \frac{1}{4}[c_{15} + u_5 + c_{11} + c_{13}]$$

$$u_9 = \frac{1}{4}[u_5 + c_7 + c_9 + c_{11}]$$

$$u_1 = \frac{1}{4}[c_1 + c_3 + u_5 + c_{15}]$$

$$u_3 = \frac{1}{4}[c_3 + c_5 + c_7 + u_5]$$

Now making use of these values u_5, u_7, u_9, u_1, u_3 compute the remaining values u_8, u_4, u_6, u_2 (in any order) by standard five points formula, as follows:

$$u_8 = \frac{1}{4}[u_5 + u_9 + c_{11} + c_7]$$

$$u_4 = \frac{1}{4}[u_1 + u_5 + u_7 + c_{15}]$$

$$u_6 = \frac{1}{4}[u_3 + c_7 + u_9 + u_5]$$

$$u_2 = \frac{1}{4}[c_3 + u_3 + u_5 + u_1]$$

The accuracy of these 9 values u_1 to u_9 can be improved by using iterative methods such as (i) **Jacobi's method** (ii) **Gauss-Seidel method** (iii) **Successive over relaxation method**. In the **Gauss-Seidel method (also known as Leibman's method)** a standard five point formula is used wherein the latest (last available) values of u are utilized. If n denotes the iteration number then the value of u at the $(n+1)$th iteration is given by

$$u_{i,j}^{n+1} = \frac{1}{4}\left[u_{i-1,j}^{(n+1)} + u_{i+1,j}^{(n)} + u_{i,j-1}^{(n+1)} + u_{i,j+1}^{(n)}\right] \qquad (5)$$

Observe that in R.H.S. of (5), the latest values

$$u_{i-1,j}^{n+1} \text{ and } u_{i,j-1}^{n+1} \quad \text{are utilized.}$$

Note: Solution of Laplace equation is known as **harmonic** function.

WORKED OUT EXAMPLES

Example 1: Compute a solution to the Laplace's equation $\nabla^2 f = \frac{\partial^2 f}{\partial x^2} + \frac{\partial^2 f}{\partial y^2} = 0$ at all mesh (grid) points of the following square in which the boundary values are indicated, using five point formula.

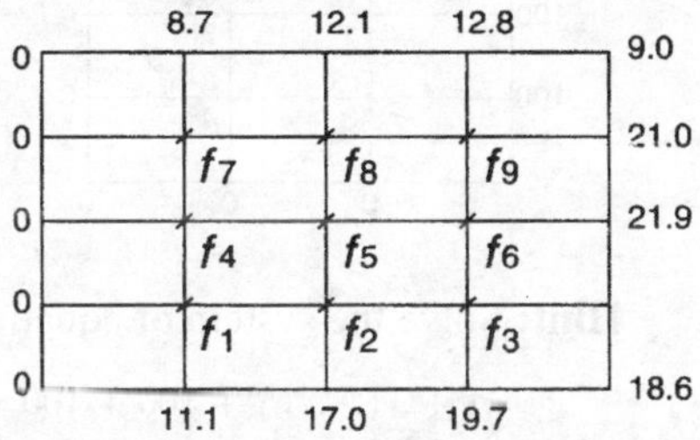

Solution: Let us index the mesh points as $f_1, f_2, f_3, \ldots, f_9$. First calculate f_5 using standard five point formula and then calculate f_1, f_3, f_7, f_9 using diagonal formula. Now the remaining values f_2, f_4, f_6, f_8 are again calculated by standard five point formula. Thus by standard formula

$$f_5 = \frac{1}{4}[0 + 17.0 + 21.9 + 12.1] = 12.525.$$

Now use diagonal formula

$$f_1 = \frac{1}{4}[0 + 0 + 17.0 + 12.525] = 7.38$$

$$f_3 = \frac{1}{4}[17.0 + 18.6 + 21.9 + 12.525] = 17.28$$

$$f_7 = \frac{1}{4}[12.1 + 0 + 0 + 12.525] = 6.16$$

$$f_9 = \frac{1}{4}[12.1 + 12.525 + 21.9 + 9] = 13.66$$

Resorting to standard formula

$$f_2 = \frac{1}{4}[7.38 + 17 + 17.28 + 12.525] = 13.55$$

$$f_4 = \frac{1}{4}[0 + 7.38 + 12.525 + 6.16] = 6.52$$

$$f_6 = \frac{1}{4}[12.525 + 17.28 + 21 + 13.66] = 16.12$$

$$f_8 = \frac{1}{4}[6.16 + 12.525 + 13.66 + 12.1] = 11.11.$$

Example 2: Solve the above problem, using the values of $f_1, f_2, \ldots, f_9$ as the initial approximation, by Gauss-Seidel (or Leibman's) method.

Solution:

$$f_{i,j}^{(n+1)} = \frac{1}{4}\left[f_{i-1,j}^{(n+1)} + f_{i+1,j}^{(n)} + f_{i+1,j}^{(n+1)} + f_{i,j+1}^{(n)}\right]$$

1st iteration:

$$f_1 = \frac{1}{4}[0 + 11.1 + 13.55 + 6.52] = 7.7925 \simeq 7.79$$

$$f_2 = \frac{1}{4}[7.79 + 17.0 + 17.28 + 12.525] = 13.65$$

$$f_3 = \frac{1}{4}[13.65 + 19.7 + 21.9 + 16.12] = 17.84$$

$$f_4 = \frac{1}{4}[0 + 7.79 + 12.525 + 6.16] = 6.62$$

$$f_5 = \frac{1}{4}[16.12 + 11.11 + 13.65 + 6.62] = 11.875$$

$$f_6 = \frac{1}{4}[21.0 + 13.66 + 17.84 + 11.875] = 16.09$$

$$f_7 = \frac{1}{4}[11.11 + 8.7 + 0 + 6.62] = 6.607$$

$$f_8 = \frac{1}{4}[13.66 + 12.1 + 11.875 + 6.607] = 11.06$$

$$f_9 = \frac{1}{4}[17.0 + 12.8 + 16.09 + 11.06] = 14.238$$

2nd iteration

$$f_1 = \frac{1}{4}[0 + 13.65 + 6.62 + 11.1] = 7.84$$

$$f_2 = \frac{1}{4}[7.84 + 17.84 + 11.87 + 17] = 13.639 \simeq 13.64$$

$$f_3 = \frac{1}{4}[13.64 + 16.09 + 19.7 + 21.9] = 17.83$$

$$f_4 = \frac{1}{4}[0 + 7.84 + 6.607 + 11.873] = 6.58$$

$$f_5 = \frac{1}{4}[6.58 + 13.64 + 11.01 + 16.09] = 11.843$$

$$f_6 = \frac{1}{4}[11.843 + 17.833 + 14.238 + 21] = 16.2285$$

$$f_7 = \frac{1}{4}[6.58 + 11.059 + 8.7 + 0] = 6.58$$

$$f_8 = \frac{1}{4}[12.11 + 6.587 + 14.238 + 11.843] = 11.19$$

$$f_9 = \frac{1}{4}[11.19 + 16.25 + 17 + 12.8] = 14.31.$$

Note that in every calculation, the latest (last available) values of f has been taken into the right hand side of the formula.

Ans: 7.8, 13.6, 17.8, 6.6, 11.9, 16.2, 6.6, 11.2, 14.3.

EXERCISE

1 (a) Determine the values of the interior lattice points of a square region of the harmonic function u with boundary conditions prescribed as shown in the figure.

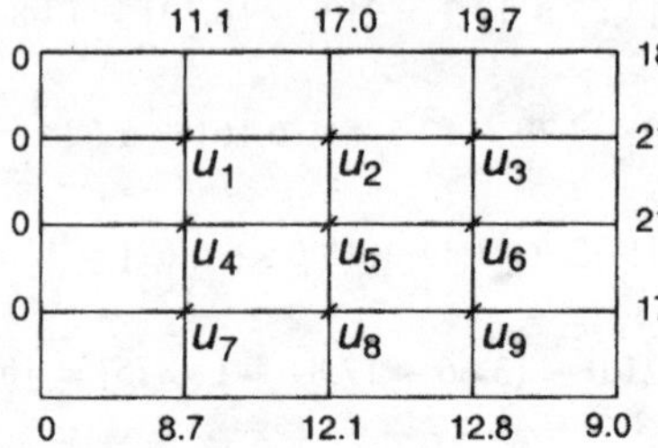

(b) Iterate by Gauss-Seidel (Leibman's) method.

Ans. (a) $u_1 = 7.4, u_2 = 13.6, u_3 = 17.3, u_4 = 6.5, u_5 = 12.5, u_6 = 16.1, u_7 = 6.2, u_8 = 11.1, u_9 = 13.7$

(b) $u_1 = 7.9, u_2 = 13.7, u_3 = 17.9, u_4 = 6.6, u_5 = 11.9, u_6 = 16.3, u_7 = 6.6, u_8 = 11.2, u_9 = 14.3$ (after 3 iterations)

2 (a) Compute solution of Laplace's equation in the following square grid with prescribed boundary conditions.

(b) Iterate the values by Gauss-Seidel method.

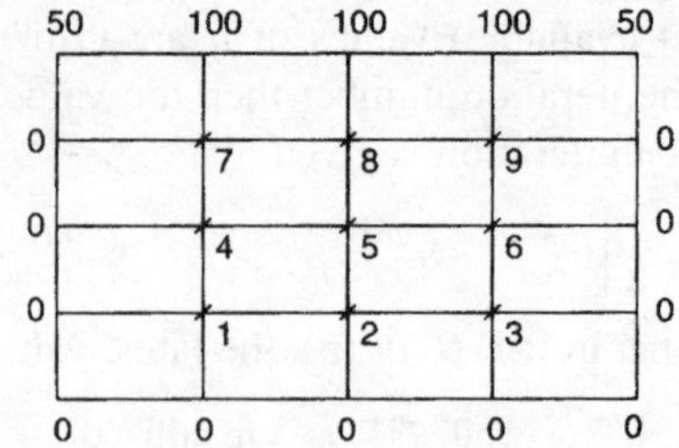

Ans. (a) $u_1 = 6.25, u_2 = 9.38, u_3 = 6.25, u_4 = 18.75, u_5 = 25, u_6 = 18.75, u_7 = 43.75, u_8 = 53.12, u_9 = 43.75$

(b) 7.17, 9.86, 7.16, 18.78, 25.04, 18.77, 42.88, 52.70, 42.87 (after 3 iterations)

3. Compute solution of Laplace's equation in the square grid.

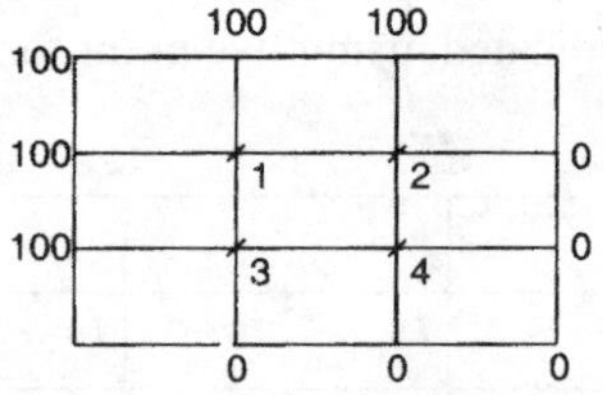

Hint: Solve the system of equations

$$u_2 + u_3 - 4u_1 + 100 + 100 = 0$$

$$u_1 + u_4 - 4u_2 + 100 + 0 = 0$$

$$u_1 + u_4 - 4u_3 + 100 + 0 = 0$$

$$u_2 + u_3 - 4u_4 + 0 + 0 = 0$$

Ans. $u_1 = 75, u_2 = 50, u_3 = 50, u_4 = 25$

4 (a) Determine $u_1, u_2, u_3, \ldots, u_9$ if u satisfies Laplace's equation in the grid with given boundary condition.

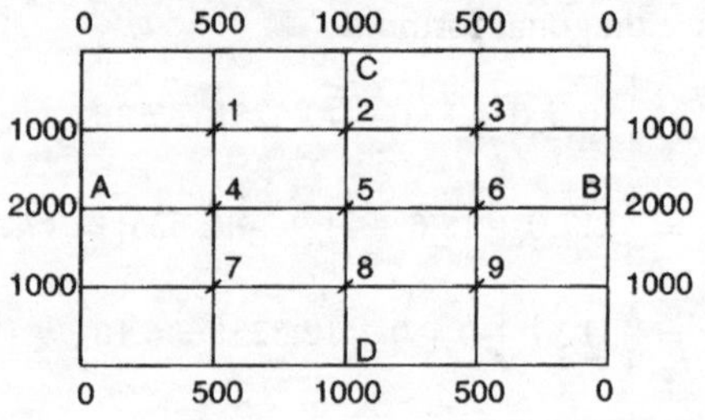

(b) Iterate by Gauss-Seidel method.

Hint: Use symmetry about AB and CD.

Ans. (a) $u_1 = u_3 = u_7 = u_9 = 1125, u_2 = u_8 = 1188, u_5 = 1500, u_6 = u_4 = 1438$

(b) $u_1 = 939, u_2 = 1001, u_4 = 1251, u_5 = 1126$ (after 12 iterations)

5. Solve Laplace's equation.

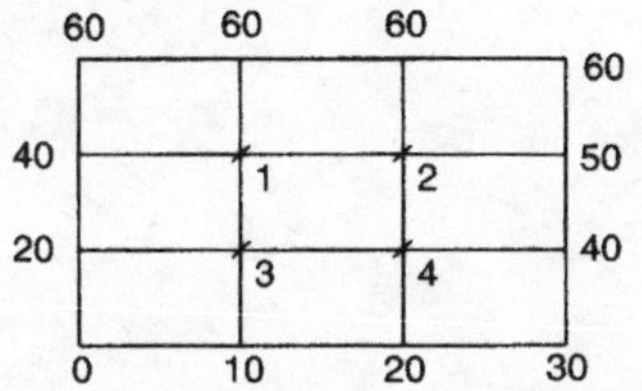

Hint: Solve the four equations

$$40 + 60 + u_2 + u_3 - 4u_1 = 0$$

$$60 + 50 + u_1 + u_4 - 4u_2 = 0$$

$$20 + 10 + u_1 + u_4 - 4u_3 = 0$$

$$20 + 40 + u_3 + u_2 - 4u_4 = 0$$

Ans. $u_1 = 43.33, u_2 = 46.66, u_3 = 26.66, u_4 = 33.33$

6. If $\nabla^2 u = 0$, find u at the mesh points.

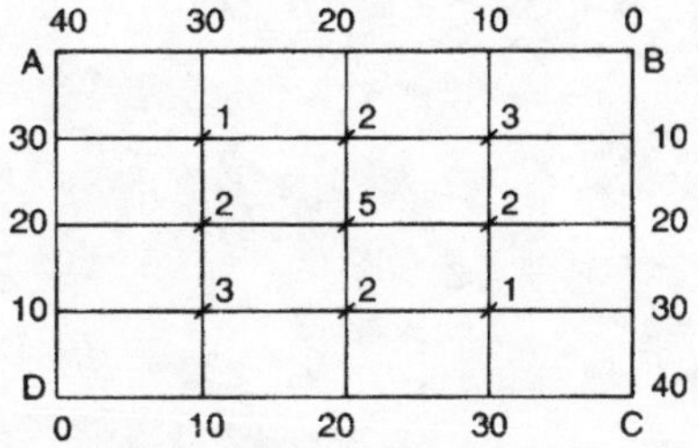

Hint: Square is symmetric about both the diagonaly AC and BD.

Ans. $u_1 = 25, u_3 = 15, u_2 = 20, u_5 = 20$

APPENDICES

Appendix A

Statistical Tables

A.2 — STATISTICAL TABLES

1. Binomial Distribution Function

Binomial Probability Sums: $\sum_{x=0}^{r} b(x;n,p) = \sum_{x=0}^{r} \binom{n}{k} p^x (1-p)^{n-x}$

		p									
n	*r*	0.10	0.20	0.25	0.30	0.40	0.50	0.60	0.70	0.80	0.90
1	0	0.9000	0.8000	0.7500	0.7000	0.6000	0.5000	0.4000	0.3000	0.2000	0.1000
	1	1.0000	1.0000	1.0000	1.0000	1.0000	1.0000	1.0000	1.0000	1.0000	1.0000
2	0	0.8100	0.6400	0.5625	0.4900	0.3600	0.2500	0.1600	0.0900	0.0400	0.0100
	1	0.9900	0.9600	0.9375	0.9100	0.8400	0.7500	0.6400	0.5100	0.3600	0.1900
	2	1.0000	1.0000	1.0000	1.0000	1.0000	1.0000	1.0000	1.0000	1.0000	1.0000
3	0	0.7290	0.5120	0.4219	0.3430	0.2160	0.1250	0.0640	0.0270	0.0080	0.0010
	1	0.9720	0.8960	0.8438	0.7840	0.6480	0.5000	0.3520	0.2160	0.1040	0.0280
	2	0.9990	0.9920	0.9844	0.9730	0.9360	0.8750	0.7840	0.6570	0.4880	0.2710
	3	1.0000	1.0000	1.0000	1.0000	1.0000	1.0000	1.0000	1.0000	1.0000	1.0000
4	0	0.6561	0.4096	0.3164	0.2401	0.1296	0.0625	0.0256	0.0081	0.0016	0.0001
	1	0.9477	0.8192	0.7383	0.6517	0.4752	0.3125	0.1792	0.0837	0.0272	0.0037
	2	0.9963	0.9728	0.9492	0.9163	0.8208	0.6875	0.5248	0.3483	0.1808	0.0523
	3	0.9999	0.9984	0.9961	0.9919	0.9744	0.9375	0.8704	0.7599	0.5904	0.3439
	4	1.0000	1.0000	1.0000	1.0000	1.0000	1.0000	1.0000	1.0000	1.0000	1.0000
5	0	0.5905	0.3277	0.2373	0.1681	0.0778	0.0312	0.0102	0.0024	0.0003	0.0000
	1	0.9185	0.7373	0.6328	0.5282	0.3370	0.1875	0.0870	0.0308	0.0067	0.0005
	2	0.9914	0.9421	0.8965	0.8369	0.6826	0.5000	0.3174	0.1631	0.0579	0.0086
	3	0.9995	0.9933	0.9844	0.9692	0.9130	0.8125	0.6630	0.4718	0.2627	0.0815
	4	1.0000	0.9997	0.9990	0.9976	0.9898	0.9688	0.9222	0.8319	0.6723	0.4095
	5		1.0000	1.0000	1.0000	1.0000	1.0000	1.0000	1.0000	1.0000	1.0000
6	0	0.5314	0.2621	0.1780	0.1176	0.0467	0.0156	0.0041	0.0007	0.0001	0.0000
	1	0.8857	0.6554	0.5339	0.4202	0.2333	0.1094	0.0410	0.0109	0.0016	0.0001
	2	0.9841	0.9011	0.8306	0.7443	0.5443	0.3438	0.1792	0.0705	0.0170	0.0013
	3	0.9987	0.9830	0.9624	0.9295	0.8208	0.6563	0.4557	0.2557	0.0989	0.0158
	4	0.9999	0.9984	0.9954	0.9891	0.9590	0.8906	0.7667	0.5798	0.3447	0.1143
	5	1.0000	0.9999	0.9998	0.9993	0.9959	0.9844	0.9533	0.8824	0.7379	0.4686
	6		1.0000	1.0000	1.0000	1.0000	1.0000	1.0000	1.0000	1.0000	1.0000
7	0	0.4783	0.2097	0.1335	0.0824	0.0280	0.0078	0.0016	0.0002	0.0000	
	1	0.8503	0.5767	0.4449	0.3294	0.1586	0.0625	0.0188	0.0038	0.0004	0.0000
	2	0.9743	0.8520	0.7564	0.6471	0.4199	0.2266	0.0963	0.0288	0.0047	0.0002
	3	0.9973	0.9667	0.9294	0.8740	0.7102	0.5000	0.2898	0.1260	0.0333	0.0027
	4	0.9998	0.9953	0.9871	0.9712	0.9037	0.7734	0.5801	0.3529	0.1480	0.0257
	5	1.0000	0.9996	0.9987	0.9962	0.9812	0.9375	0.8414	0.6706	0.4233	0.1497
	6		1.0000	0.9999	0.9998	0.9984	0.9922	0.9720	0.9176	0.7903	0.5217
	7			1.0000	1.0000	1.0000	1.0000	1.0000	1.0000	1.0000	1.0000

Binomial Probability Sums: $\sum_{x=0}^{r} b(x; n, p)$

		p									
n	*r*	0.10	0.20	0.25	0.30	0.40	0.50	0.60	0.70	0.80	0.90
8	0	0.4305	0.1678	0.1001	0.0576	0.0168	0.0039	0.0007	0.0001	0.0000	
	1	0.8131	0.5033	0.3671	0.2553	0.1064	0.0352	0.0085	0.0013	0.0001	
	2	0.9619	0.7969	0.6785	0.5518	0.3154	0.1445	0.0498	0.0113	0.0012	0.0000
	3	0.9950	0.9437	0.8862	0.8059	0.5941	0.3633	0.1737	0.0580	0.0104	0.0004
	4	0.9996	0.9896	0.9727	0.9420	0.8263	0.6367	0.4059	0.1941	0.0563	0.0050
	5	1.0000	0.9988	0.9958	0.9887	0.9502	0.8555	0.6846	0.4482	0.2031	0.0381
	6		0.9991	0.9996	0.9987	0.9915	0.9648	0.8936	0.7447	0.4967	0.1869
	7		1.0000	1.0000	0.9999	0.9993	0.9961	0.9832	0.9424	0.8322	0.5695
	8				1.0000	1.0000	1.0000	1.0000	1.0000	1.0000	1.0000
9	0	0.3874	0.1342	0.0751	0.0404	0.0101	0.0020	0.0003	0.0000		
	1	0.7748	0.4362	0.3003	0.1960	0.0705	0.0195	0.0038	0.0004	0.0000	
	2	0.9470	0.7382	0.6007	0.4628	0.2318	0.0898	0.0250	0.0043	0.0003	0.0000
	3	0.9917	0.9144	0.8343	0.7297	0.4826	0.2539	0.0994	0.0253	0.0031	0.0001
	4	0.9991	0.9804	0.9511	0.9012	0.7334	0.5000	0.2666	0.0988	0.0196	0.0009
	5	0.9999	0.9969	0.9900	0.9747	0.9006	0.7461	0.5174	0.2703	0.0856	0.0083
	6	1.0000	0.9997	0.9987	0.9957	0.9750	0.9102	0.7682	0.5372	0.2618	0.0530
	7		1.0000	0.9999	0.9996	0.9962	0.9805	0.9295	0.8040	0.5638	0.2252
	8			1.0000	1.0000	0.9997	0.9980	0.9899	0.9596	0.8658	0.6126
	9					1.0000	1.0000	1.0000	1.0000	1.0000	1.0000
10	0	0.3487	0.1074	0.0563	0.0282	0.0060	0.0010	0.0001	0.0000		
	1	0.7361	0.3758	0.2440	0.1493	0.0464	0.0107	0.0017	0.0001	0.0000	
	2	0.9298	0.6778	0.5256	0.3828	0.1673	0.0547	0.0123	0.0016	0.0001	
	3	0.9872	0.8791	0.7759	0.6496	0.3823	0.1719	0.0548	0.0106	0.0009	0.0000
	4	0.9984	0.9672	0.9219	0.8497	0.6331	0.3770	0.1662	0.0474	0.0064	0.0002
	5	0.9999	0.9936	0.9803	0.9527	0.8338	0.6230	0.3669	0.1503	0.0328	0.0016
	6	1.0000	0.9991	0.9965	0.9894	0.9452	0.8281	0.6177	0.3504	0.1209	0.0128
	7		0.9999	0.9996	0.9984	0.9877	0.9453	0.8327	0.6172	0.3222	0.0702
	8		1.0000	1.0000	0.9999	0.9983	0.9893	0.9536	0.8507	0.6242	0.2639
	9				1.0000	0.9999	0.9990	0.9940	0.9718	0.8926	0.6513
	10					1.0000	1.0000	1.0000	1.0000	1.0000	1.0000
11	0	0.3138	0.0859	0.0422	0.0198	0.0036	0.0005	0.0000			
	1	0.6974	0.3221	0,1971	0.1130	0.0302	0.0059	0.0007	0.0000		
	2	0.9104	0.6174	0.4552	0.3127	0.1189	0.0327	0.0059	0.0006	0.0000	
	3	0.9815	0.8369	0.7133	0.5696	0.2963	0.1133	0.0293	0.0043	0.0002	
	4	0.9972	0.9496	0.8854	0.7897	0.5328	0.2744	0.0994	0.0216	0.0020	0.0000
	5	0.9997	0.9883	0.9657	0.9218	0.7535	0.5000	0.2465	0.0782	0.0117	0.0003
	6	1.0000	0.9980	0.9924	0.9784	0.9006	0.7256	0.4672	0.2103	0.0504	0.0028
	7		0.9998	0.9988	0.9957	0.9707	0.8867	0.7037	0.4304	0.1611	0.0185
	8		1.0000	0.9999	0.9994	0.9941	0.9673	0.8811	0.6873	0.3826	0.0896
	9			1.0000	1.0000	0.9993	0.9941	0.9698	0.8870	0.6779	0.3026
	10					1.0000	0.9995	0.9964	0.9802	0.9141	0.6862
	11						1.0000	1.0000	1.0000	1.0000	1.0000

(Contd.)

Binomial Probability Sums: $\sum_{x=0}^{r} b(x; n, p) = \sum_{x=0}^{r} \binom{n}{x} p^x (1-p)^{n-x}$

		p									
n	r	0.10	0.20	0.25	0.30	0.40	0.50	0.60	0.70	0.80	0.90
12	0	0.2824	0.0687	0.0317	0.0138	0.0022	0.0002	0.0000			
	1	0.6590	0.2749	0.1584	0.0850	0.0196	0.0032	0.0003	0.0000		
	2	0.8891	0.5583	0.3907	0.2528	0.0834	0.0193	0.0028	0.0002	0.0000	
	3	0.9744	0.7946	0.6488	0.4925	0.2253	0.0730	0.0153	0.0017	0.0001	
	4	0.9957	0.9274	0.8424	0.7237	0.4382	0.1938	0.0573	0.0095	0.0006	0.0000
	5	0.9995	0.9806	0.9456	0.8821	0.6652	0.3872	0.1582	0.0386	0.0039	0.0001
	6	0.9999	0.9961	0.9857	0.9614	0.8418	0.6128	0.3348	0.1178	0.0194	0.0005
	7	1.0000	0.9994	0.9972	0.9905	0.9427	0.8062	0.5618	0.2763	0.0726	0.0043
	8		0.9999	0.9996	0.9983	0.9847	0.9270	0.7747	0.5075	0.2054	0.0256
	9		1.0000	1.0000	0.9998	0.9972	0.9807	0.9166	0.7472	0.4417	0.1109
	10				1.0000	0.9997	0.9968	0.9804	0.9150	0.7251	0.3410
	11					1.0000	0.9998	0.9978	0.9862	0.9313	0.7176
	12						1.0000	1.0000	1.0000	1.0000	1.0000
13	0	0.2542	0.0550	0.0238	0.0097	0.0013	0.0001	0.0000			
	1	0.6213	0.2336	0.1267	0.0637	0.0126	0.0017	0.0001	0.0000		
	2	0.8661	0.5017	0.3326	0.2025	0.0579	0.0112	0.0013	0.0001		
	3	0.9658	0.7473	0.5843	0.4206	0.1686	0.0461	0.0078	0.0007	0.0000	
	4	0.9935	0.9009	0.7940	0.6543	0.3530	0.1334	0.0321	0.0040	0.0002	
	5	0.9991	0.9700	0.9198	0.8346	0.5744	0.2905	0.0977	0.0182	0.0012	0.0000
	6	0.9999	0.9930	0.9757	0.9376	0.7712	0.5000	0.2288	0.0624	0.0070	0.0001
	7	1.0000	0.9980	0.9944	0.9818	0.9023	0.7095	0.4256	0.1654	0.0300	0.0009
	8		0.9998	0.9990	0.9960	0.9679	0.8666	0.6470	0.3457	0.0991	0.0065
	9		1.0000	0.9999	0.9993	0.9922	0.9539	0.8314	0.5794	0.2527	0.0342
	10			1.0000	0.9999	0.9987	0.9888	0.9421	0.7975	0.4983	0.1339
	11				1.0000	0.9999	0.9983	0.9874	0.9363	0.7664	0.3787
	12					1.0000	0.9999	0.9987	0.9903	0.9450	0.7458
	13						1.0000	1.0000	1.0000	1.0000	1.0000
14	0	0.2288	0.0440	0.0178	0.0068	0.0008	0.0001	0.0000			
	1	0.5846	0.1979	0.1010	0.0475	0.0081	0.0009	0.0001			
	2	0.8416	0.4481	0.2811	0.1608	0.0398	0.0065	0.0006	0.0000		
	3	0.9559	0.6982	0.5213	0.3552	0.1243	0.0287	0.0039	0.0002		
	4	0.9908	0.8702	0.7415	0.5842	0.2793	0.0898	0.0175	0.0017	0.0000	
	5	0.9985	0.9561	0.8883	0.7805	0.4859	0.2120	0.0583	0.0083	0.0004	
	6	0.9998	0.9884	0.9617	0.9067	0.6925	0.3953	0.1501	0.0315	0.0024	0.0000
	7	1.0000	0.9976	0.9897	0.9685	0.8499	0.6047	0.3075	0.0933	0.0116	0.0002
	8		0.9996	0.9978	0.9917	0.9417	0.7880	0.5141	0.2195	0.0439	0.0015
	9		1.0000	0.9997	0.9983	0.9825	0.9102	0.7207	0.4158	0.1298	0.0092
	10			1.0000	0.9998	0.9961	0.9713	0.8757	0.6448	0.3018	0.0441
	11				1.0000	0.9994	0.9935	0.9602	0.8392	0.5519	0.1584
	12					0.9999	0.9991	0.9919	0.9525	0.8021	0.4154
	13					1.0000	0.9999	0.9992	0.9932	0.9560	0.7712
	14						1.0000	1.0000	1.0000	1.0000	1.0000

Binomial Probability Sums: $\sum_{x=0}^{r} b(x; n, p)$

		p									
n	*r*	0.10	0.20	0.25	0.30	0.40	0.50	0.60	0.70	0.80	0.90
15	0	0.2059	0.0352	0.0134	0.0047	0.0005	0.0000				
	1	0.5490	0.1671	0.0802	0.0353	0.0052	0.0005	0.0000			
	2	0.8159	0.3980	0.2361	0.1268	0.0271	0.0037	0.0003	0.0000		
	3	0.9444	0.6482	0.4613	0.2969	0.0905	0.0176	0.0019	0.0001		
	4	0.9873	0.8358	0.6865	0.5155	0.2173	0.0592	0.0094	0.0007	0.0000	
	5	0.9978	0.9389	0.8516	0.7216	0.4032	0.1509	0.0338	0.0037	0.0001	
	6	0.9997	0.9819	0.9434	0.8689	0.6098	0.3036	0.0951	0.0152	0.0008	
	7	1.0000	0.9958	0.9827	0.9500	0.7869	0.5000	0.2131	0.0500	0.0042	0.0000
	8		0.9992	0.9958	0.9848	0.9050	0.6964	0.3902	0.1311	0.0181	0.0003
	9		0.9999	0.9992	0.9963	0.9662	0.8491	0.5968	0.2784	0.0611	0.0023
	10		1.0000	0.9999	0.9993	0.9907	0.9408	0.7827	0.4845	0.1642	0.0127
	11			1.0000	0.9999	0.9981	0.9824	0.9095	0.7031	0.3518	0.0556
	12				1.0000	0.9997	0.9963	0.9729	0.8732	0.6020	0.1841
	13					1.0000	0.9995	0.9948	0.9647	0.8329	0.4510
	14						1.0000	0.9995	0.9953	0.9648	0.7941
	15							1.0000	1.0000	1.0000	1.0000
16	0	0.1853	0.0281	0.0100	0.0033	0.0003	0.0000				
	1	0.5147	0.1407	0.0635	0.0261	0.0033	0.0003	0.0000			
	2	0.7892	0.3518	0.1971	0.0994	0.0183	0.0021	0.0001			
	3	0.9316	0.5981	0.4050	0.2459	0.0651	0.0106	0.0009	0.0000		
	4	0.9830	0,7982	0.6302	0.4499	0.1666	0.0384	0.0049	0.0003		
	5	0.9967	0.9183	0.8103	0.6598	0.3288	0.1051	0.0191	0.0016	0.0000	
	6	0.9995	0.9733	0.9204	0.8247	0.5272	0.2272	0.0583	0.0071	0.0002	
	7	0.9999	0.9930	0.9729	0.9256	0.7161	0.4018	0.1423	0.0257	0.0015	0.0000
	8	1.0000	0.9985	0.9925	0.9743	0.8577	0.5982	0.2839	0.0744	0.0070	0.0001
	9		0.9998	0.9984	0.9929	0.9417	0.7728	0.4728	0.1753	0.0267	0.0005
	10		1.0000	0.9997	0.9984	0.9809	0.8949	0.6712	0.3402	0.0817	0.0033
	11			1.0000	0.9997	0.9951	0.9616	0.8334	0.5501	0.2018	0.0170
	12				1.0000	0.9991	0.9894	0.9349	0.7541	0.4019	0.0684
	13					0.9999	0.9979	0.9817	0.9006	0.6482	0.2108
	14					1.0000	0.9997	0.9967	0.9739	0.8593	0.4853
	15						1.0000	0.9997	0.9967	0.9719	0.8147
	16							1.0000	1.0000	1.0000	1.0000

(Contd.)

Binomial Probability Sums: $\sum_{x=0}^{r} b(x; n, p) = \sum_{x=0}^{r} \binom{n}{x} p^x (1-p)^{n-x}$

		p									
n	*r*	0.10	0.20	0.25	0.30	0.40	0.50	0.60	0.70	0.80	0.90
17	0	0.1668	0.0225	0.0075	0.0023	0.0002	0.0000				
	1	0.4818	0.1182	0.0501	0.0193	0.0021	0.0001	0.0000			
	2	0.7618	0.3096	0.1637	0.0774	0.0123	0.0012	0.0001			
	3	0.9174	0.5489	0.3530	0.2019	0.0464	0.0064	0.0005	0.0000		
	4	0.9779	0.7582	0.5739	0.3887	0.1260	0.0245	0.0025	0.0001		
	5	0.9953	0.8943	0.7653	0.5968	0.2639	0.0717	0.0106	0.0007	0.0000	
	6	0.9992	0.9623	0.8929	0.7752	0.4478	0.1662	0.0348	0.0032	0.0001	
	7	0.9999	0.9891	0.9598	0.8954	0.6405	0.3145	0.0919	0.0127	0.0005	
	8	1.0000	0.9974	0.9876	0.9597	0.8011	0.5000	0.1989	0.0403	0.0026	0.0000
	9		0.9995	0.9969	0.9873	0.9081	0.6855	0.3595	0.1046	0.0109	0.0001
	10		0.9999	0.9994	0.9968	0.9652	0.8338	0.5522	0.2248	0.0377	0.0008
	11		1.0000	0.9999	0.9993	0.9894	0.9283	0.7361	0.4032	0.1057	0.0047
	12			1.0000	0.9999	0.9975	0.9755	0.8740	0.6113	0.2418	0.0221
	13				1.0000	0.9995	0.9936	0.9536	0.7981	0.4511	0.0826
	14					0.9999	0.9988	0.9877	0.9226	0.6904	0.2382
	15					1.0000	0.9999	0.9979	0.9807	0.8818	0.5182
	16						1.0000	0.9998	0.9977	0.9775	0.8332
	17						1.0000	1.0000	1.0000	1.0000	1.0000
18	0	0.1501	0.0180	0.0056	0.0016	0.0001	0.0000				
	1	0.4503	0.0991	0.0395	0.0142	0.0013	0.0001				
	2	0.7338	0.2713	0.1353	0.0600	0.0082	0.0007	0.0000			
	3	0.9018	0.5010	0.3057	0.1646	0.0328	0,0038	0.0002			
	4	0.9718	0.7164	0.5787	0.3327	0.0942	0.0154	0.0013	0.0000		
	5	0.9936	0.8671	0.7175	0.5344	0.2088	0.0481	0.0058	0.0003		
	6	0.9988	0.9487	0.8610	0.7217	0.3743	0.1189	0.0203	0.0014	0.0000	
	7	0.9998	0.9837	0.9431	0.8593	0.5634	0.2403	0.0576	0.0061	0.0002	
	8	1.0000	0.9957	0.9807	0.9404	0.7368	0.4073	0.1347	0.0210	0.0009	
	9		0.9991	0.9946	0.9790	0.8653	0.5927	0.2632	0.0596	0.0043	0.0000
	10		0.9998	0.9988	0.9939	0.9424	0.7597	0.4366	0.1407	0.0163	0.0002
	11		1.0000	0.9998	0.9986	0.9797	0.8811	0.6257	0.2783	0.0513	0.0012
	12			1.0000	0.9997	0.9942	0.9519	0.7912	0.4656	0.1329	0.0064
	13				1.0000	0.9987	0.9846	0.9058	0.6673	0.2836	0.0282
	14					0.9998	0.9962	0.9672	0.8354	0.4990	0.0982
	15					1.0000	0.9993	0.9918	0.9400	0.7287	0.2662
	16						0.9999	0.9987	0.9858	0.9009	0.5497
	17						1.0000	0.9999	0.9984	0.9820	0.8499
	18							1.0000	1.0000	1.0000	1.0000

Binomial Probability Sums: $\sum_{x=0}^{r} b(x; n, p)$

		p									
n	r	0.10	0.20	0.25	0.30	0.40	0.50	0.60	0.70	0.80	0.90
19	0	0.1351	0.0144	0.0042	0.0011	0.0001					
	1	0.4203	0.0829	0.0310	0.0104	0.0008	0.0000				
	2	0.7054	0.2369	0.1113	0.0462	0.0055	0.0004	0.0000			
	3	0.8850	0.4551	0.2631	0.1332	0.0230	0.0022	0.0001			
	4	0.9648	0.6733	0.4654	0.2822	0.0696	0.0096	0.0006	0.0000		
	5	0.9914	0.8369	0.6678	0.4739	0.1629	0.0318	0.0031	0.0001		
	6	0.9983	0.9324	0.8251	0.6655	0.3081	0.0835	0.0116	0.0006		
	7	0.9997	0.9767	0.9225	0.8180	0.4878	0.1796	0.0352	0.0028	0.0000	
	8	1.0000	0.9933	0.9713	0.9161	0.6675	0.3238	0.0885	0.0105	0.0003	
	9		0.9984	0.9911	0.9674	0.8139	0.5000	0.1861	0.0326	0.0016	
	10		0.9997	0.9977	0.9895	0.9115	0.6762	0.3325	0.0839	0.0067	0.0000
	11		0.9999	0.9995	0.9972	0.9648	0.8204	0.5122	0.1820	0.0233	0.0003
	12		1.0000	0.9999	0.9994	0.9884	0.9165	0.6919	0.3345	0.0676	0.0017
	13			1.0000	0.9999	0.9969	0.9682	0.8371	0.5261	0.1631	0.0086
	14				1.0000	0.9994	0.9904	0.9304	0.7178	0.3267	0.0352
	15					0.9999	0.9978	0.9770	0.8668	0.5449	0.1150
	16					1.0000	0.9996	0.9945	0.9538	0.7631	0.2946
	17						1.0000	0.9992	0.9896	0.9171	0.5797
	18							0.9999	0.9989	0.9856	0.8649
	19							1.0000	1.0000	1.0000	1.0000
20	0	0.1216	0.0115	0.0032	0.0008	0.0000					
	1	0.3917	0.0692	0.0243	0.0076	0.0005	0.0000				
	2	0.6769	0.2061	0.0913	0.0355	0.0036	0.0002	0.0000			
	3	0.8670	0.4114	0.2252	0.1071	0.0160	0.0013	0.0001			
	4	0.9568	0.6296	0.4148	0.2375	0.0510	0.0059	0.0003			
	5	0.9887	0.8042	0.6172	0.4164	0.1256	0.0207	0.0016	0.0000		
	6	0.9976	0.9133	0.7858	0.6080	0.2500	0.0577	0.0065	0.0003		
	7	0.9996	0.9679	0.8982	0.7723	0.4159	0.1316	0.0210	0.0013	0.0000	
	8	0.9999	0.9900	0.9591	0.8867	0.5956	0.2517	0.0565	0.0051	0.0001	
	9	1.0000	0.9974	0.9861	0.9520	0.7553	0.4119	0.1275	0.0171	0.0006	
	10		0.9994	0.9961	0.9829	0.8725	0.5881	0.2447	0.0480	0.0026	0.0000
	11		0.9999	0.9991	0.9949	0.9435	0.7483	0.4044	0.1133	0.0100	0.0001
	12		1.0000	0.9998	0.9987	0.9790	0.8684	0.5841	0.2277	0.0321	0.0004
	13			1.0000	0.9997	0.9935	0.9423	0.7500	0.3920	0.0867	0.0024
	14				1.0000	0.9984	0.9793	0.8744	0.5836	0.1958	0.0113
	15					0.9997	0.9941	0.9490	0.7625	0.3704	0.0432
	16					1.0000	0.9987	0.9840	0.8929	0.5886	0.1330
	17						0.9998	0.9964	0.9645	0.7939	0.3231
	18						1.0000	0.9995	0.9924	0.9308	0.6083
	19							1.0000	0.9992	0.9885	0.8784
	20								1.0000	1.0000	1.0000

2. Poisson Distribution Function

Poisson Probability Sums: $F(x;\lambda) = \sum_{k=0}^{x} e^{-\lambda} \frac{\lambda^k}{k!}$

λ \ x	0	1	2	3	4	5	6	7	8	9
0.02	0.980	1.000								
0.04	0.961	0.999	1.000							
0.06	0.942	0.998	1.000							
0.08	0.923	0.997	1.000							
0.10	0.905	0.995	1.000							
0.15	0.861	0.990	0.999	1.000						
0.20	0.819	0.982	0.999	1.000						
0.25	0.779	0.974	0.998	1.000						
0.30	0.741	0.963	0.996	1.000						
0.35	0.705	0.951	0.994	1.000						
0.40	0.670	0.938	0.992	0.999	1.000					
0.45	0.638	0.925	0.989	0.999	1.000					
0.50	0.607	0.910	0.986	0.998	1.000					
0.55	0.577	0.894	0.982	0.998	1.000					
0.60	0.549	0.878	0.977	0.997	1.000					
0.65	0.522	0.861	0.972	0.996	0.999	1.000				
0.70	0.497	0.844	0.966	0.994	0.999	1.000				
0.75	0.472	0.827	0.959	0.993	0.999	1.000				
0.80	0.449	0.809	0.953	0.991	0.999	1.000				
0.85	0.427	0.791	0.945	0.989	0.998	1.000				
0.90	0.407	0.772	0.937	0.987	0.998	1.000				
0.95	0.387	0.754	0.929	0.984	0.997	1.000				
1.00	0.368	0.736	0.920	0.981	0.996	0.999	1.000			
1.1	0.333	0.699	0.900	0.974	0.995	0.999	1.000			
1.2	0.301	0.663	0.879	0.966	0.992	0.998	1.000			
1.3	0.273	0.627	0.857	0.957	0.989	0.998	1.000			
1.4	0.247	0.592	0.833	0.946	0.986	0.997	0.999	1.000		
1.5	0.223	0.558	0.809	0.934	0.981	0.996	0.999	1.000		
1.6	0.202	0.525	0.783	0.921	0.976	0.994	0.999	1.000		
1.7	0.183	0.493	0.757	0.907	0.970	0.992	0.998	1.000		
1.8	0.165	0.463	0.731	0.891	0.964	0.990	0.997	0.999	1.000	
1.9	0.150	0.434	0.704	0.875	0.956	0.987	0.997	0.999	1.000	
2.0	0.135	0.406	0.677	0.857	0.947	0.983	0.995	0.999	1.000	

Poisson Distribution Function: $F(x, \lambda)$

λ \ x	0	1	2	3	4	5	6	7	8	9
2.2	0.111	0.355	0.623	0.819	0.928	0.975	0.993	0.998	1.000	
2.4	0.091	0.308	0.570	0.779	0.904	0.964	0.988	0.997	0.999	1.000
2.6	0.074	0.267	0.518	0.736	0.877	0.951	0.983	0.995	0.999	1.000
2.8	0.061	0.231	0.469	0.692	0.848	0.935	0.976	0.992	0.998	0.999
3.0	0.050	0.199	0.423	0.647	0.815	0.916	0.966	0.988	0.996	0.999
3.2	0.041	0.171	0.380	0.603	0.781	0.895	0.955	0.983	0.994	0.998
3.4	0.033	0.147	0.340	0.558	0.744	0.871	0.942	0.977	0.992	0.997
3.6	0.027	0.126	0.303	0.515	0.706	0.844	0.927	0.969	0.988	0.996
3.8	0.022	0.107	0.269	0.473	0.668	0.816	0.909	0.960	0.984	0.994
4.0	0.018	0.092	0.238	0.433	0.629	0.785	0.889	0.949	0.979	0.992
4.2	0.015	0.078	0.210	0.395	0.590	0.753	0.867	0.936	0.972	0.989
4.4	0.012	0.066	0.185	0.359	0.551	0.720	0.844	0.921	0.964	0.985
4.6	0.010	0.056	0.163	0.326	0.513	0.686	0.818	0.905	0.955	0.980
4.8	0.008	0.048	0.143	0.294	0.476	0.651	0.791	0.887	0.944	0.975
5.0	0.007	0.040	0.125	0.265	0.440	0.616	0.762	0.867	0.932	0.968
5.2	0.006	0.034	0.109	0.238	0.406	0.581	0.732	0.845	0.918	0.960
5.4	0.005	0.029	0.095	0.213	0.373	0.546	0.702	0.822	0.903	0.951
5.6	0.004	0.024	0.082	0.191	0.342	0.512	0.670	0.797	0.886	0.941
5.8	0.003	0.021	0.072	0.170	0.313	0.478	0.638	0.771	0.867	0.929
6.0	0.002	0.017	0.062	0.151	0.285	0.446	0.606	0.744	0.847	0.916
	10	11	12	13	14	15	16			
2.8	1.000									
3.0	1.000									
3.2	1.000									
3.4	0.999	1.000								
3.6	0.999	1.000								
3.8	0.998	0.999	1.000							
4.0	0.997	0.999	1.000							
4.2	0.996	0.999	1.000							
4.4	0.994	0.998	0.999	1.000						
4.6	0.992	0.997	0.999	1.000						
4.8	0.990	0.996	0.999	1.000						
5.0	0.986	0.995	0.998	0.999	1.000					
5.2	0.982	0.993	0.997	0.999	1.000					
5.4	0.977	0.990	0.996	0.999	1.000					
5.6	0.972	0.988	0.995	0.998	0.999	1.000				
5.8	0.965	0.984	0.993	0.997	0.999	1.000				
6.0	0.957	0.980	0.991	0.996	0.999	0.999	1.000			

(Contd.)

Poisson Distribution Function: $F(x, \lambda) = \sum_{k=0}^{x} e^{-\lambda} \frac{\lambda^k}{k!}$

λ \ x	0	1	2	3	4	5	6	7	8	9
6.2	0.002	0.015	0.054	0.134	0.259	0.414	0.574	0.716	0.826	0.902
6.4	0.002	0.012	0.046	0.119	0.235	0.384	0.542	0.687	0.803	0.886
6.6	0.001	0.010	0.040	0.105	0.213	0.355	0.511	0.658	0.780	0.869
6.8	0.001	0.009	0.034	0.093	0.192	0.327	0.480	0.628	0.755	0.850
7.0	0.001	0.007	0.030	0.082	0.173	0.301	0.450	0.599	0.729	0.830
7.2	0.001	0.006	0.025	0.072	0.156	0.276	0.420	0.569	0.703	0.810
7.4	0.001	0.005	0.022	0.063	0.140	0.253	0.392	0.539	0.676	0.788
7.6	0.001	0.004	0.019	0.055	0.125	0.231	0.365	0.510	0.648	0.765
7.8	0.000	0.004	0.016	0.048	0.112	0.210	0.338	0.481	0.620	0.741
8.0	0.000	0.003	0.014	0.042	0.100	0.191	0.313	0.453	0.593	0.717
8.5	0.000	0.002	0.009	0.030	0.074	0.150	0.256	0.386	0.523	0.653
9.0	0.000	0.001	0.006	0.021	0.055	0.116	0.207	0.324	0.456	0.587
9.5	0.000	0.001	0.004	0.015	0.040	0.089	0.165	0.269	0.392	0.522
10.0	0.000	0.000	0.003	0.010	0.029	0.067	0.130	0.220	0.333	0.458

	10	11	12	13	14	15	16	17	18	19
6.2	0.949	0.975	0.989	0.995	0.998	0.999	1.000			
6.4	0.939	0.969	0.986	0.994	0.997	0.999	1.000			
6.6	0.927	0.963	0.982	0.992	0.997	0.999	0.999	1.000		
6.8	0.915	0.955	0.978	0.990	0.996	0.998	0.999	1.000		
7.0	0.901	0.947	0.973	0.987	0.994	0.998	0.999	1.000		
7.2	0.887	0.937	0.967	0.984	0.993	0.997	0.999	1.000	1.000	
7.4	0.871	0.926	0.961	0.980	0.991	0.996	0.998	0.999	1.000	
7.6	0.854	0.915	0.954	0.976	0.989	0.995	0.998	0.999	1.000	
7.8	0.835	0.902	0.945	0.971	0.986	0.993	0.997	0.999	1.000	
8.0	0.816	0.888	0.936	0.966	0.983	0.992	0.996	0.998	0.999	1.000
8.5	0.763	0.849	0.909	0.949	0.973	0.986	0.993	0.997	0.999	0.999
9.0	0.706	0.803	0.876	0.926	0.959	0.978	0.989	0.995	0.998	0.999
9.5	0.645	0.752	0.836	0.898	0.940	0.967	0.982	0.991	0.996	0.998
10.0	0.583	0.697	0.792	0.864	0.917	0.951	0.973	0.986	0.993	0.997

	20	21	22
8.5	1.000		
9.0	1.000		
9.5	0.999	1.000	
10.0	0.998	0.999	1.000

Poisson Distribution Function: $F(x, \lambda)$

λ \ x	0	1	2	3	4	5	6	7	8	9
10.5	0.000	0.000	0.002	0.007	0.021	0.050	0.102	0.179	0.279	0.397
11.0	0.000	0.000	0.001	0.005	0.015	0.038	0.079	0.143	0.232	0.341
11.5	0.000	0.000	0.001	0.003	0.011	0.028	0.060	0.114	0.191	0.289
12.0	0.000	0.000	0.001	0.002	0.008	0.020	0.046	0.090	0.155	0.242
12.5	0.000	0.000	0.000	0.002	0.005	0.015	0.035	0.070	0.125	0.201
13.0	0.000	0.000	0.000	0.001	0.004	0.011	0.026	0.054	0.100	0.166
13.5	0.000	0.000	0.000	0.001	0.003	0.008	0.019	0.041	0.079	0.135
14.0	0.000	0.000	0.000	0.000	0.002	0.006	0.014	0.032	0.062	0.109
14.5	0.000	0.000	0.000	0.000	0.001	0.004	0.010	0.024	0.048	0.088
15.0	0.000	0.000	0.000	0.000	0.001	0.003	0.008	0.018	0.037	0.070

	10	11	12	13	14	15	16	17	18	19
10.5	0.521	0.639	0.742	0.825	0.888	0.932	0.960	0.978	0.988	0.994
11.0	0.460	0.579	0.689	0.781	0.854	0.907	0.944	0.968	0.982	0.991
11.5	0.402	0.520	0.633	0.733	0.815	0.878	0.924	0.954	0.974	0.986
12.0	0.347	0.462	0.576	0.682	0.772	0.844	0.899	0.937	0.963	0.979
12.5	0.297	0.406	0.519	0.628	0.725	0.806	0.869	0.916	0.948	0.969
13.0	0.252	0.353	0.463	0.573	0.675	0.764	0.835	0.890	0.930	0.957
13.5	0.211	0.304	0.409	0.518	0.623	0.718	0.798	0.861	0.908	0.942
14.0	0.176	0.260	0.358	0.464	0.570	0.669	0.756	0.827	0.883	0.923
14.5	0.145	0.220	0.311	0.413	0.518	0.619	0.711	0.790	0.853	0.901
15.0	0.118	0.185	0.268	0.363	0.466	0.568	0.664	0.749	0.819	0.875

	20	21	22	23	24	25	26	27	28	29
10.5	0.997	0.999	0.999	1.000						
11.0	0.995	0.998	0.999	1.000						
11.5	0.992	0.996	0.998	0.999	1.000					
12.0	0.988	0.994	0.997	0.999	0.999	1.000				
12.5	0.983	0.991	0.995	0.998	0.999	0.999	1.000			
13.0	0.975	0.986	0.992	0.996	0.998	0.999	1.000			
13.5	0.965	0.980	0.989	0.994	0.997	0.998	0.999	1.000		
14.0	0.952	0.971	0.983	0.991	0.995	0.997	0.999	0.999	1.000	
14.5	0.936	0.960	0.976	0.986	0.992	0.996	0.998	0.999	0.999	1.000
15.0	0.917	0.947	0.967	0.981	0.989	0.994	0.997	0.998	0.999	1.000

3. Areas under the Standard Normal Curve from 0 to z
(Normal Tables)

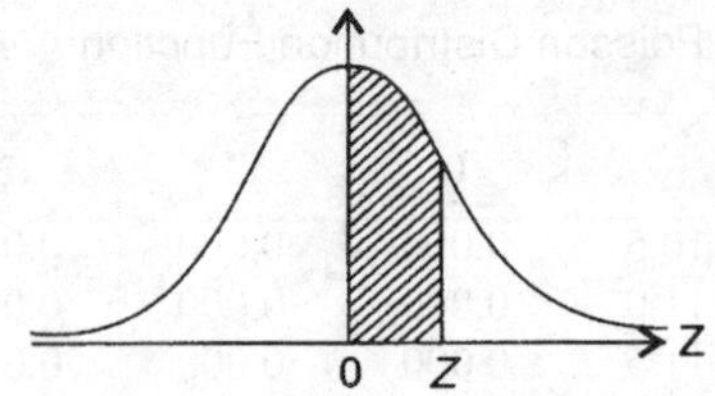

z	0	1	2	3	4	5	6	7	8	9
0.0	.0000	.0040	.0080	.0120	.0160	.0199	.0239	.0279	.0319	.0359
0.1	.0398	.0438	.0478	.0517	.0557	.0596	.0636	.0675	.0714	.0754
0.2	.0793	.0832	.0871	.0910	.0948	.0987	.1026	.1064	.1103	.1141
0.3	.1179	.1217	.1256	.1293	.1331	.1368	.1406	.1443	.1480	.1517
0.4	.1554	.1591	.1628	.1664	.1700	.1736	.1772	.1808	.1844	.1879
0.5	.1916	.1950	.1985	.2019	.2054	.2088	.2123	.2157	.2190	.2224
0.6	.2258	.2291	.2324	.2357	.2389	.2422	.2454	.2486	.2518	.2649
0.7	.2580	.2612	.2642	.2673	.2704	.2734	.2764	.2794	.2823	.2852
0.8	.2881	.2910	.2939	.2967	.2996	.3023	.3051	.3078	.3106	.3133
0.9	.3159	.3186	.3212	.3238	.3264	.3289	.3315	.3340	.3365	.3389
1.0	.3413	.3438	.3461	.3485	.3508	.3531	.3554	.3577	.3599	.3621
1.1	.3643	.3665	.3686	.3708	.3729	.3749	.3770	.3790	.3810	.3830
1.2	.3849	.3869	.3888	.3907	.3925	.3944	.3962	.3980	.3997	.4015
1.3	.4032	.4049	.4066	.4082	.4099	.4115	.4131	.4147	.4162	.4177
1.4	.4192	.4207	.4222	.4236	.4251	.4265	.4279	.4292	.4306	.4319
1.5	.4332	.4345	.4357	.4370	.4382	.4394	.4406	.4418	.4429	.4441
1.6	.4452	.4463	.4474	.4484	.4495	.4505	.4515	.4525	.4535	.4545
1.7	.4654	.4564	.4573	.4582	.4591	.4599	.4608	.4616	.4625	.4633
1.8	.4641	.4649	.4656	.4664	.4671	.4678	.4686	.4693	.4699	.4706
1.9	.4713	.4719	.4726	.4732	.4738	.4744	.4750	.4756	.4761	.4767
2.0	.4772	.4778	.4783	.4788	.4793	.4798	.4803	.4808	.4812	.4817
2.1	.4821	.4826	.4830	.4834	.4838	.4842	.4846	.4850	.4854	.4857
2.2	.4861	.4864	.4868	.4871	.4875	.4878	.4881	.4884	.4887	.4890
2.3	.4893	.4896	.4898	.4901	.4904	.4906	.4909	.4911	.4913	.4916
2.4	.4918	.4920	.4922	.4925	.4927	.4929	.4931	.4932	.4934	.4936
2.5	.4938	.4940	.4941	.4943	.4945	.4946	.4948	.4949	.4951	.4952
2.6	.4953	.4955	.4956	.4957	.4959	.4960	.4961	.4962	.4963	.4964
2.7	.4965	.4966	.4967	.4968	.4969	.4970	.4971	.4972	.4973	.4974
2.8	.4974	.4975	.4976	.4977	.4979	.4978	.4979	.4979	.4980	.4981
2.9	.4981	.4982	.4982	.4983	.4984	.4984	.4985	.4985	.4986	.4986
3.0	.4987	.4987	.4987	.4988	.4988	.4989	.4989	.4989	.4990	.4990
3.1	.4990	.4991	.4991	.4991	.4992	.4992	.4992	.4992	.4993	.4993
3.2	.4993	.4993	.4994	.4994	.4994	.4994	.4994	.4995	.4995	.4995
3.3	.4995	.4995	.4995	.4996	.4996	.4996	.4996	.4996	.4996	.4997
3.4	.4997	.4997	.4997	.4997	.4997	.4997	.4997	.4997	.4997	.4998
3.5	.4998	.4998	.4998	.4998	.4998	.4998	.4098	.4998	.4998	.4998
3.6	.4998	.4998	.4999	.4999	.4999	.4999	.4999	.4999	.4999	.4999
3.7	.4999	.4999	.4999	.4999	.4999	.4999	.4999	.4999	.4999	.4999
3.8	.4999	.4999	.4999	.4999	.4999	.4999	.4999	.4999	.4999	.4999
3.9	.5000	.5000	.5000	.5000	.5000	.5000	.5000	.5000	.5000	.5000

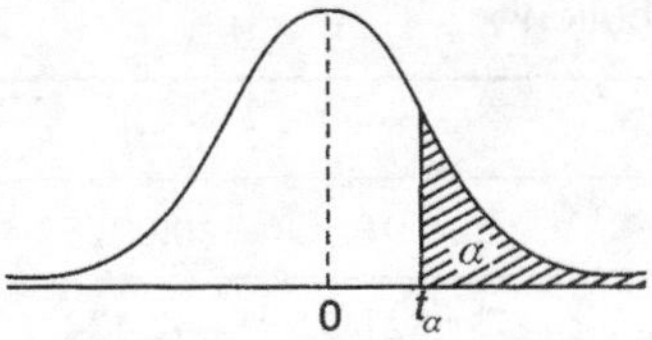

4. t_α-Critical Values of the t-Distribution

	α						
ν	0.40	0.30	0.20	0.15	0.10	0.05	0.025
1	0.325	0.727	1.376	1.963	3.078	6.314	12.706
2	0.289	0.617	1.061	1.386	1.886	2.920	4.303
3	0.277	0.584	0.978	1.250	1.638	2.353	3.182
4	0.271	0.569	0.941	1.190	1.533	2.132	2.776
5	0.267	0.559	0.920	1.156	1.476	2.015	2.571
6	0.265	0.553	0.906	1.134	1.440	1.943	2.447
7	0.263	0.549	0.896	1.119	1.415	1.895	2.365
8	0.262	0.546	0.889	1.108	1.397	1.860	2.306
9	0.261	0.543	0.883	1.100	1.383	1.833	2.262
10	0.260	0.542	0.879	1.093	1.372	1.812	2.228
11	0.260	0.540	0.876	1.088	1.363	1.796	2.201
12	0.259	0.539	0.873	1.083	1.356	1.782	2.179
13	0.259	0.537	0.870	1.079	1.350	1.771	2.160
14	0.258	0.537	0.868	1.076	1.345	1.761	2.145
15	0.258	0.536	0.866	1.074	1.341	1.753	2.131
16	0.258	0.535	0.865	1.071	1.337	1.746	2.120
17	0.257	0.534	0.863	1.069	1.333	1.740	2.110
18	0.257	0.534	0.862	1.067	1.330	1.734	2.101
19	0.257	0.533	0.861	1.066	1.328	1.729	2.093
20	0.257	0.533	0.860	1.064	1.325	1.725	2.086
21	0.257	0.532	0.859	1.063	1.323	1.721	2.080
22	0.256	0.532	0.858	1.061	1.321	1.717	2.074
23	0.256	0.532	0.858	1.060	1.319	1.714	2.069
24	0.256	0.531	0.857	1.059	1.318	1.711	2.064
25	0.256	0.531	0.856	1.058	1.316	1.708	2.060
26	0.256	0.531	0.856	1.058	1.315	1.706	2.056
27	0.256	0.531	0.855	1.057	1.314	1.703	2.052
28	0.256	0.530	0.855	1.056	1.313	1.701	2.048
29	0.256	0.530	0.854	1.055	1.311	1.699	2.045
30	0.256	0.530	0.854	1.055	1.310	1.697	2.042
40	0.255	0.529	0.851	1.050	1.303	1.684	2.021
60	0.254	0.527	0.848	1.045	1.296	1.671	2.000
120	0.254	0.526	0.845	1.041	1.289	1.658	1.980
∞	0.253	0.524	0.842	1.036	1.282	1.645	1.960

t_α-Critical Values of the t-Distribution

	α						
ν	0.02	0.015	0.01	0.0075	0.005	0.0025	0.0005
1	15.895	21.205	31.821	42.434	63.657	127.322	636.590
2	4.849	5.643	6.965	8.073	9.925	14.089	31.598
3	3.482	3.896	4.541	5.047	5.841	7.453	12.924
4	2.999	3.298	3.747	4.088	4.604	5.598	8.610
5	2.757	3.003	3.365	3.634	4.032	4.773	6.869
6	2.612	2.829	3.143	3.372	3.707	4.317	5.959
7	2.517	2.715	2.998	3.203	3.499	4.029	5.408
8	2.449	2.634	2.896	3.085	3.355	3.833	5.041
9	2.398	2.574	2.821	2.998	3.250	3.690	4.781
10	2.359	2.527	2.764	2.932	3.169	3.581	4.587
11	2.328	2.491	2.718	2.879	3.106	3.497	4.437
12	2.303	2.461	2.681	2.836	3.055	3.428	4.318
13	2.282	2.436	2.650	2.801	3.012	3.372	4.221
14	2.264	2.415	2.624	2.771	2.977	3.326	4.140
15	2.249	2.397	2.602	2.746	2.947	3.286	4.073
16	2.235	2.382	2.583	2.724	2.921	3.252	4.015
17	2.224	2.368	2.567	2.706	2.898	3.222	3.965
18	2.214	2.356	2.552	2.689	2.878	3.197	3.922
19	2.205	2.346	2.539	2.674	2.861	3.174	3.883
20	2.197	2.336	2.528	2.661	2.845	3.153	3.849
21	2.189	2.328	2.518	2.649	2.831	3.135	3.819
22	2.183	2.320	2.508	2.639	2.819	3.119	3.792
23	2.177	2.313	2.500	2.629	2.807	3.104	3.768
24	2.172	2.307	2.492	2.620	2.797	3.091	3.745
25	2.167	2.301	2.485	2.612	2.787	3.078	3.725
26	2.162	2.296	2.479	2.605	2.779	3.067	3.707
27	2.158	2.291	2.473	2.598	2.771	3.057	3.690
28	2.154	2.286	2.467	2.592	2.763	3.047	3.674
29	2.150	2.282	2.462	2.586	2.756	3.038	3.659
30	2.147	2.278	2.457	2.581	2.750	3.030	3.646
40	2.125	2.250	2.423	2.542	2.704	2.971	3.551
60	2.099	2.223	2.390	2.504	2.660	2.915	3.460
120	2.076	2.196	2.358	2.468	2.617	2.860	3.373
∞	2.054	2.170	2.326	2.432	2.576	2.807	3.291

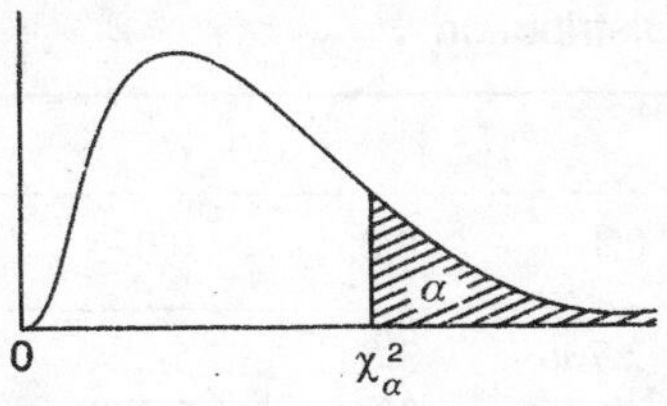

5. χ^2_α-Critical Values of the Chi-squared Distribution

	α									
ν	0.995	0.99	0.98	0.975	0.95	0.90	0.80	0.75	0.75	0.50
1	0.04393	0.03157	0.03628	0.03982	0.00393	0.0158	0.0642	0.102	0.148	0.455
2	0.0100	0.0201	0.0404	0.0506	0.103	0.211	0.446	0.575	0.713	1.386
3	0.0717	0.115	0.185	0.216	0.352	0.584	1.005	1.213	1.424	2.366
4	0.207	0.297	0.429	0.484	0.711	1.064	1.649	1.923	2.195	3.357
5	0.412	0.554	0.752	0.831	1.145	1.610	2.343	2.675	3.000	4.351
6	0.676	0.872	1.134	1.237	1.635	2.204	3.070	3.455	3.828	5.348
7	0.989	1.239	1.564	1.690	2.167	2.833	3.822	4.255	4.671	6.346
8	1.344	1.646	2.032	2.180	2.733	3.490	4.594	5.071	5.527	7.344
9	1.735	2.088	2.532	2.700	3.325	4.168	5.380	5.899	6.393	8.343
10	2.156	2.558	3.059	3.247	3.940	4.865	6.179	6.737	7.267	9.342
11	2.603	3.053	3.609	3.816	4.575	5.578	6.989	7.584	8.148	10.341
12	3.074	3.571	4.178	4.404	5.226	6.304	7.807	8.438	9.034	11.340
13	3.565	4.107	4.765	5.009	5.892	7.042	8.634	9.299	9.926	12.340
14	4.075	4.660	5.368	5.629	6.571	7.790	9.467	10.165	10.821	13.339
15	4.601	5.229	5.985	6.262	7.261	8.547	10.307	11.036	11.721	14.339
16	5.142	5.812	6.614	6.908	7.962	9.312	11.152	11.912	12.624	15.338
17	5.697	6.408	7.255	7.564	8.672	10.085	12.002	12.792	13.531	16.338
18	6.265	7.015	7.906	8.231	9.390	10.865	12.857	13.675	14.440	17.338
19	6.844	7.633	8.567	8.907	10.117	11.651	13.716	14.562	15.352	18.338
20	7.434	8.260	9.237	9.591	10.851	12.443	14.578	15.452	16.266	19.337
21	8.034	8.897	9.915	10.283	11.591	13.240	15.445	16.344	17.182	20.337
22	8.643	9.542	10.600	10.982	12.338	14.041	16.314	17.240	18.101	21.337
23	9.260	10.196	11.293	11.688	13.091	14.848	17.187	18.137	19.021	22.337
24	9.886	10.856	11.992	12.401	13.848	15.659	18.062	19.037	19.943	23.337
25	10.520	11.524	12.697	13.120	14.611	16.473	18.940	19.939	20.867	24.337
26	11.160	12.198	13.409	13.844	15.379	17.292	19.820	20.843	21.792	25.336
27	11.808	12.879	14.125	14.573	16.151	18.114	20.703	21.749	22.719	26.336
28	12.461	13.565	14.847	15.308	16.928	18.939	21.588	22.657	23.647	27.336
29	13.121	14.256	15.574	16.047	17.708	19.768	22.475	23.567	24.577	28.336
30	13.787	14.953	16.306	16.791	18.493	20.599	23.364	24.478	25.508	29.336

χ^2_α-Critical Values of the Chi-squared Distribution

	α									
ν	0.30	0.25	0.20	0.10	0.05	0.025	0.02	0.01	0.005	0.001
1	1.074	1.323	1.642	2.706	3.841	5.024	5.412	6.635	7.879	10.827
2	2.408	2.773	3.219	4.605	5.991	7.378	7.824	9.210	10.597	13.815
3	3.665	4.108	4.642	6.251	7.815	9.348	9.837	11.345	12.838	16.268
4	4.878	5.385	5.989	7.779	9.488	11.143	11.668	13.277	14.860	18.465
5	6.064	6.626	7.289	9.236	11.070	12.832	13.388	15.086	16.750	20.517
6	7.231	7.841	8.558	10.645	12.592	14.449	15.033	16.812	18.548	22.457
7	8.383	9.037	9.803	12.017	14.067	16.013	16.622	18.475	20.278	24.322
8	9.524	10.219	11.030	13.362	15.507	17.535	18.168	20.090	21.955	26.125
9	10.656	11.389	12.242	14.684	16.919	19.023	19.679	21.666	23.589	27.877
10	11.781	12.549	13.442	15.987	18.307	20.483	21.161	23.209	25.188	29.588
11	12.899	13.701	14.631	17.275	19.675	21.920	22.618	24.725	26.757	31.264
12	14.011	14.845	15.812	18.549	21.026	23.337	24.054	26.217	28.300	32.909
13	15.119	15.984	16.985	19.812	22.362	24.736	25.472	27.688	29.819	34.528
14	16.222	17.117	18.151	21.064	23.685	26.119	26.873	29.141	31.319	36.123
15	17.322	18.245	19.311	22.307	24.996	27.488	28.259	30.578	32.801	37.697
16	18.418	19.369	20.465	23.542	26.296	28.845	29.633	32.000	34.267	39.252
17	19.511	20.489	21.615	24.769	27.587	30.191	30.995	33.409	35.718	40.790
18	20.601	21.605	22.760	25.989	28.869	31.526	32.346	34.805	37.156	42.312
19	21.689	22.718	23.900	27.204	30.144	32.852	33.687	36.191	38.582	43.820
20	22.775	23.828	25.038	28.412	31.410	34.170	35.020	37.566	39.997	45.315
21	23.858	24.935	26.171	29.615	32.671	35.479	36.343	38.932	41.401	46.797
22	24.939	26.039	27.301	30.813	33.924	36.781	37.659	40.289	42.796	48.268
23	26.018	27.141	28.429	32.007	35.172	38.076	38.968	41.638	44.181	49.728
24	27.096	28.241	29.553	33.196	36.415	39.364	40.270	42.980	45.558	51.179
25	28.172	29.339	30.675	34.382	37.652	40.646	41.566	44.314	46.928	52.620
26	29.246	30.434	31.795	35.563	38.885	41.923	42.856	45.642	48.290	54.052
27	30.319	31.528	32.912	36.741	40,113	43.194	44.140	46.963	49.645	55.476
28	31.391	32.620	34.027	37.916	41,337	44.461	45.419	48.278	50.993	56.893
29	32.461	33.711	35.139	39.087	42.557	45.722	46.693	49.588	52.336	58.302
30	33.530	34.800	36.250	40.256	43.773	46.979	47.962	50.892	53.672	59.703

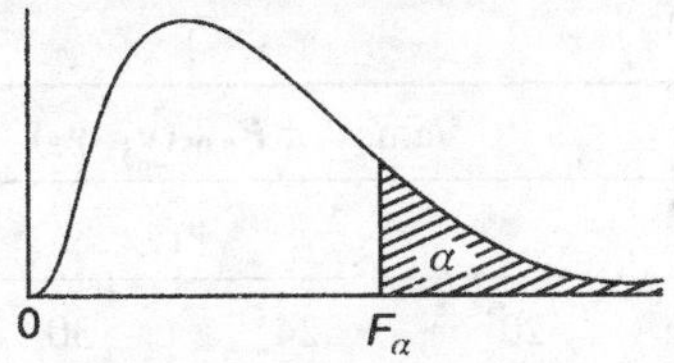

6. Critical Values of the *F*-Distribution

Values of $F_{0.05}(\nu_1, \nu_2)$

ν_2 \ ν_1	1	2	3	4	5	6	7	8	9
1	161.4	199.5	215.7	224.6	230.2	234.0	236.8	238.9	240.5
2	18.51	19.00	19.16	19.25	19.30	19.33	19.35	19.37	19.38
3	10.13	9.55	9.28	9.12	9.01	8.94	8.89	8.85	8.81
4	7.71	6.94	6.59	6.39	6.26	6.16	6.09	6.04	6.00
5	6.61	5.79	5.41	5.19	5.05	4.95	4.88	4.82	4.77
6	5.99	5.14	4.76	4.53	4.39	4.28	4.21	4.15	4.10
7	5.59	4.74	4.35	4.12	3.97	3.87	3.79	3.73	3.68
8	5.32	4.46	4.07	3.84	3.69	3.58	3.50	3.44	3.39
9	5.12	4.26	3.86	3.63	3.48	3.37	3.29	3.23	3.18
10	4.96	4.10	3.71	3.48	3.33	3.22	3.14	3.07	3.02
11	4.84	3.98	3.59	3.36	3.20	3.09	3.01	2.95	2.90
12	4.75	3.89	3.49	3.26	3.11	3.00	2.91	2.85	2.80
13	4.67	3.81	3.41	3.18	3.03	2.92	2.83	2.77	2.71
14	4.60	3.74	3.34	3.11	2.96	2.85	2.76	2.70	2.65
15	4.54	3.68	3.29	3.06	2.90	2.79	2.71	2.64	2.59
16	4.49	3.63	3.24	3.01	2.85	2.74	2.66	2.59	2.54
17	4.45	3.59	3.20	2.96	2.81	2.70	2.61	2.55	2.49
18	4.41	3.55	3.16	2.93	2.77	2.66	2.58	2.51	2.46
19	4.38	3.52	3.13	2.90	2.74	2.63	2.54	2.48	2.42
20	4.35	3.49	3.10	2.87	2.71	2.60	2.51	2.45	2.39
21	4.32	3.47	3.07	2.84	2.68	2.57	2.49	2.42	2.37
22	4.30	3.44	3.05	2.82	2.66	2.55	2.46	2.40	2.34
23	4.28	3.42	3.03	2.80	2.64	2.53	2.44	2.37	2.32
24	4.26	3.40	3.01	2.78	2.62	2.51	2.42	2.36	2.30
25	4.24	3.39	2.99	2.76	2.60	2.49	2.40	2.34	2.28
26	4.23	3.37	2.98	2.74	2.59	2.47	2.39	2.32	2.27
27	4.21	3.35	2.96	2.73	2.57	2.46	2.37	2.31	2.25
28	4.20	3.34	2.95	2.71	2.56	2.45	2.36	2.29	2.24
29	4.18	3.33	2.93	2.70	2.55	2.43	2.35	2.28	2.22
30	4.17	3.32	2.92	2.69	2.53	2.42	2.33	2.27	2.21
40	4.08	3.23	2.84	2.61	2.45	2.34	2.25	2.18	2.12
60	4.00	3.15	2.76	2.53	2.37	2.25	2.17	2.10	2.04
120	3.92	3.07	2.68	2.45	2.29	2.17	2.09	2.02	1.96
∞	3.84	3.00	2.60	2.37	2.21	2.10	2.01	1.94	1.88

Critical Values of the *F*-Distribution

Values of $F_{0.05}(\nu_1, \nu_2)$

ν_2	ν_1 10	12	15	20	24	30	40	60	120	∞
1	241.9	243.9	245.9	248.0	249.1	250.1	251.1	252.2	253.3	254.3
2	19.40	19.41	19.43	19.45	19.45	19.46	19.47	19.48	19.49	19.50
3	8.79	8.74	8.70	8.66	8.64	8.62	8.59	8.57	8.55	8.53
4	5.96	5.91	5.86	5.80	5.77	5.75	5.72	5.69	5.66	5.63
5	4.74	4.68	4.62	4.56	4.53	4.50	4.46	4.43	4.40	4.36
6	4.06	4.00	3.94	3.87	3.84	3.81	3.77	3.74	3.70	3.67
7	3.64	3.57	3.51	3.44	3.41	3.38	3.34	3.30	3.27	3.23
8	3.35	3.28	3.22	3.15	3.12	3.08	3.04	3.01	2.97	2.93
9	3.14	3.07	3.01	2.94	2.90	2.86	2.83	2.79	2.75	2.71
10	2.98	2.91	2.85	2.77	2.74	2.70	2.66	2.62	2.58	2.54
11	2.85	2.79	2.72	2.65	2.61	2.57	2.53	2.49	2.45	2.40
12	2.75	2.69	2.62	2.54	2.51	2.47	2.43	2.38	2.34	2.30
13	2.67	2.60	2.53	2.46	2.42	2.38	2.34	2.30	2.25	2.21
14	2.60	2.53	2.46	2.39	2.35	2.31	2.27	2.22	2.18	2.13
15	2.54	2.48	2.40	2.33	2.29	2.25	2.20	2.16	2.11	2.07
16	2.49	2.42	2.35	2.28	2.24	2.19	2.15	2.11	2.06	2.01
17	2.45	2.38	2.31	2.23	2.19	2.15	2.10	2.06	2.01	1.96
18	2.41	2.34	2.27	2.19	2.15	2.11	2.06	2.02	1.97	1.92
19	2.38	2.31	2.23	2.16	2.11	2.07	2.03	1.98	1.93	1.88
20	2.35	2.28	2.20	2.12	2.08	2.04	1.99	1.95	1.90	1.84
21	2.32	2.25	2.18	2.10	2.05	2.01	1.96	1.92	1.87	1.81
22	2.30	2.23	2.15	2.07	2.03	1.98	1.94	1.89	1.84	1.78
23	2.27	2.20	2.13	2.05	2.01	1.96	1.91	1.86	1.81	1.76
24	2.25	2.18	2.11	2.03	1.98	1.94	1.89	1.84	1.79	1.73
25	2.24	2.16	2.09	2.01	1.96	1.92	1.87	1.82	1.77	1.71
26	2.22	2.15	2.07	1.99	1.95	1.90	1.85	1.80	1.75	1.69
27	2.20	2.13	2.06	1.97	1.93	1.88	1.84	1.79	1.73	1.67
28	2.19	2.12	2.04	1.96	1.91	1.87	1.82	1.77	1.71	1.65
29	2.18	2.10	2.03	1.94	1.90	1.85	1.81	1.75	1.70	1.64
30	2.16	2.09	2.01	1.93	1.89	1.84	1.79	1.75	1.68	1.62
40	2.08	2.00	1.92	1.84	1.79	1.74	1.69	1.64	1.58	1.51
60	1.99	1.92	1.84	1.75	1.70	1.65	1.59	1.53	1.47	1.39
120	1.91	1.83	1.75	1.66	1.61	1.55	1.50	1.43	1.35	1.25
∞	1.83	1.75	1.67	1.57	1.52	1.46	1.39	1.32	1.22	1.00

Critical Values of the *F*-Distribution

Values of $F_{0.01}(\nu_1, \nu_2)$

ν_2 \ ν_1	1	2	3	4	5	6	7	8	9
1	4052	4999.5	5403	5625	5764	5859	5928	5981	6022
2	98.50	99.00	99.17	99.25	99.30	99.33	99.36	99.37	99.39
3	34.12	30.82	29.46	28.71	28.24	27.91	27.67	27.49	27.35
4	21.20	18.00	16.69	15.98	15.52	15.21	14.98	14.80	14.66
5	16.26	13.27	12.06	11.39	10.97	10.67	10.46	10.29	10.16
6	13.75	10.92	9.78	9.15	8.75	8.47	8.26	8.10	7.98
7	12.25	9.55	8.45	7.85	7.46	7.19	6.99	6.84	6.72
8	11.26	8.65	7.59	7.01	6.63	6.37	6.18	6.03	5.91
9	10.56	8.02	6.99	6.42	6.06	5.80	5.61	5.47	5.35
10	10.04	7.56	6.55	5.99	5.64	5.39	5.20	5.06	4.94
11	9.65	7.21	6.22	5.67	5.32	5.07	4.89	4.74	4.63
12	9.33	6.93	5.95	5.41	5.06	4.82	4.64	4.50	4.39
13	9.07	6.70	5.74	5.21	4.86	4.62	4.44	4.30	4.19
14	8.86	6.51	5.56	5.04	4.69	4.46	4.28	4.14	4.03
15	8.68	6.36	5.42	4.89	4.56	4.32	4.14	4.00	3.89
16	8.53	6.23	5.29	4.77	4.44	4.20	4.03	3.89	3.78
17	8.40	6.11	5.18	4.67	4.34	4.10	3.93	3.79	3.68
18	8.29	6.01	5.09	4.58	4.25	4.01	3.84	3.71	3.60
19	8.18	5.93	5.01	4.50	4.17	3.94	3.77	3.63	3.52
20	8.10	5.85	4.94	4.43	4.10	3.87	3.70	3.56	3.46
21	8.02	5.78	4.87	4.37	4.04	3.81	3.64	3.51	3.40
22	7.95	5.72	4.82	4.31	3.99	3.76	3.59	3.45	3.35
23	7.88	5.66	4.76	4.26	3.94	3.71	3.54	3.41	3.30
24	7.82	5.61	4.72	4.22	3.90	3.67	3.50	3.36	3.26
25	7.77	5.57	4.68	4.18	3.85	3.63	3.46	3.32	3.22
26	7.72	5.53	4.64	4.14	3.82	3.59	3.42	3.29	3.18
27	7.68	5.49	4.60	4.11	3.78	3.56	3.39	3.26	3.15
28	7.64	5.45	4.57	4.07	3.75	3.53	3.36	3.23	3.12
29	7.60	5.42	4.54	4.04	3.73	3.50	3.33	3.20	3.09
30	7.56	5.39	4.51	4.02	3.70	3.47	3.30	3.17	3.07
40	7.31	5.18	4.31	3.83	3.51	3.29	3.12	2.99	2.89
60	7.08	4.98	4.13	3.65	3.34	3.12	2.95	2.82	2.72
120	6.85	4.79	3.95	3.48	3.17	2.96	3.79	2.66	2.56
∞	6.63	4.61	3.78	3.32	3.02	2.80	2.64	2.51	2.41

Critical Values of the F-Distribution

Values of $F_{0.01}(\nu_1, \nu_2)$

ν_2	ν_1 = 10	12	15	20	24	30	40	60	120	∞
1	6056	6106	6157	6209	6235	6261	6287	6313	6339	6366
2	99.40	99.42	99.43	99.45	99.46	99.47	99.47	99.48	99.49	99.50
3	27.23	27.05	26.87	26.69	26.60	26.50	26.41	26.32	26.22	26.13
4	14.55	14.37	14.20	14.02	13.93	13.84	13.75	13.65	13.56	13.46
5	10.05	9.89	9.72	9.55	9.47	9.38	9.29	9.20	9.11	9.02
6	7.87	7.72	7.56	7.40	7.31	7.23	7.14	7.06	6.97	6.88
7	6.62	6.47	6.31	6.16	6.07	5.99	5.91	5.82	5.74	5.65
8	5.81	5.67	5.52	5.36	5.28	5.20	5.12	5.03	4.95	4.86
9	5.26	5.11	4.96	4.81	4.73	4.65	4.57	4.48	4.40	4.31
10	4.85	4.71	4.56	4.41	4.33	4.25	4.17	4.08	4.00	3.91
11	4.54	4.40	4.25	4.10	4.02	3.94	3.86	3.78	3.69	3.60
12	4.30	4.16	4.01	3.86	3.78	3.70	3.62	3.54	3.45	3.36
13	4.10	3.96	3.82	3.66	3.59	3.51	3.43	3.34	3.25	3.17
14	3.94	3.80	3.66	3.51	3.43	3.35	3.27	3.18	3.09	3.00
15	3.80	3.67	3.52	3.37	3.29	3.21	3.13	3.05	2.96	2.87
16	3.69	3.55	3.41	3.26	3.18	3.10	3.02	2.93	2.84	2.75
17	3.59	3.46	3.31	3.16	3.08	3.00	2.92	2.83	2.75	2.65
18	3.51	3.37	3.23	3.08	3.00	2.92	2.84	2.75	2.66	2.57
19	3.43	3.30	3.15	3.00	2.92	2.S4	2.76	2.67	2.58	2.49
20	3.37	3.23	3.09	2.94	2.86	2.78	2.69	2.61	2.52	2.42
21	3.31	3.17	3.03	2.88	2.80	2.72	2.64	2.55	2.46	2.36
22	3.26	3.12	2.98	2.83	2.75	2.67	2.58	2.50	2.40	2.31
23	3.21	3.07	2.93	2.78	2.70	2.62	2.54	2.45	2.35	2.26
24	3.17	3.03	2.89	2.74	2.66	2.58	2.49	2.40	2.31	2.21
25	3.13	2.99	2.85	2.70	2.62	2.54	2.45	2.36	2.27	2.17
26	3.09	2.96	2.81	2.66	2.58	2.50	2.42	2.33	2.23	2.13
27	3.06	2.93	2.78	2.63	2.55	2.47	2.38	2.29	2.20	2.10
28	3.03	2.90	2.75	2.60	2.52	2.44	2.35	2.26	2.17	2.06
29	3.00	2.87	2.73	2.57	2.49	2.41	2.33	2.23	2.14	2.03
30	2.98	2.84	2.70	2.55	2.47	2.39	2.30	2.21	2.11	2.01
40	2.80	2.66	2.52	2.37	2.29	2.20	2.11	2.02	1.92	1.80
60	2.63	2.50	2.35	2.20	2.12	2.03	1.94	1.84	1.73	1.60
120	2.47	2.34	2.19	2.03	1.95	1.86	1.76	1.66	1.53	1.38
∞	2.32	2.18	2.04	1.88	1.79	1.70	1.59	1.47	1.32	1.00

7. Fisher's *Z*-Transformation

Values of $Z = \frac{1}{2} \ln \frac{1+r}{1-r}$

r	0.00	0.01	0.02	0.03	0.04	0.05	0.06	0.07	0.08	0.09
0.0	0.000	0.010	0.020	0.030	0.040	0.050	0.060	0.070	0.080	0.090
0.1	0.100	0.110	0.121	0.131	0.141	0.151	0.161	0.172	0.182	0.192
0.2	0.203	0.213	0.224	0.234	0.245	0.255	0.266	0.277	0.288	0.299
0.3	0.310	0.321	0.332	0.343	0.354	0.365	0.377	0.388	0.400	0.412
0.4	0.424	0.436	0.448	0.460	0.472	0.485	0.497	0.510	0.523	0.536
0.5	0.549	0.563	0.576	0.590	0.604	0.618	0.633	0.648	0.662	0.678
0.6	0.693	0.709	0.725	0.741	0.758	0.775	0.793	0.811	0.829	0.848
0.7	0.867	0.887	0.908	0.929	0.950	0.973	0.996	1.020	1.045	1.071
0.8	1.099	1.127	1.157	1.188	1.221	1.256	1.293	1.333	1.376	1.422
0.9	1.472	1.528	1.589	1.658	1.738	1.832	1.946	2.092	2.298	2.647

* For negative values of *r* put a minus sign in front of the corresponding *Z*'s, and vice versa.

Table 8: Table of the Exponential function, e^{-x}, for $x > 0$

	0.00	0.01	0.02	0.03	0.04	0.05	0.06	0.07	0.08	0.09
0.0	1.00000	0.99005	0.98020	0.97045	0.96079	0.95123	0.94176	0.93239	0.92312	0.91393
0.1	0.90484	0.89583	0.88692	0.87810	0.86936	0.86071	0.85214	0.84366	0.83527	0.82696
0.2	0.81873	0.81058	0.80252	0.79453	0.78663	0.77880	0.77105	0.76338	0.75578	0.74826
0.3	0.74082	0.73345	0.72615	0.71892	0.71177	0.70469	0.69768	0.69073	0.68386	0.67706
0.4	0.67032	0.66365	0.65705	0.65051	0.64404	0.63763	0.63128	0.62500	0.61878	0.61263
0.5	0.60653	0.66050	0.59452	0.55860	0.58275	0.57695	0.57121	0.56553	0.55990	0.55433
0.6	0.54881	0.54335	0.53794	0.53259	0.52729	0.52204	0.51685	0.51171	0.50662	0.50158
0.7	0.49659	0.49164	0.48675	0.48191	0.47711	0.47237	0.46767	0.46301	0.45841	0.45384
0.8	0.44933	0.44486	0.44043	0.43605	0.43171	0.42741	0.42316	0.41895	0.41478	0.41066
0.9	0.40657	0.40252	0.39852	0.39455	0.39063	0.38674	0.38289	0.37908	0.37531	0.37158

	0.0	0.1	0.2	0.3	0.4	0.5	0.6	0.7	0.8	0.8
1.0	0.36788	0.33287	0.30119	0.27253	0.24660	0.22313	0.20190	0.18268	0.16530	0.14957
2.0	0.13534	0.12246	0.11080	0.10026	0.09072	0.08208	0.07427	0.06721	0.06081	0.05502
3.0	0.04979	0.04505	0.04076	0.03688	0.03337	0.03020	0.02732	0.02472	0.02237	0.02024
4.0	0.01832	0.01657	0.01500	0.01357	0.01228	0.01111	0.01005	0.00910	0.00823	0.00745
5.0	0.00674	0.00610	0.00552	0.00499	0.00452	0.00409	0.00370	0.00335	0.00303	0.00274
6.0	0.00248	0.00224	0.00203	0.00184	0.00166	0.00150	0.00136	0.00123	0.00111	0.00101

Note: Values of e^{-t} for larger values of t are smaller than 0.0004.

Table 9 Values of the Incomplete Gamma Function $I_r(\tau)$ for use in the Computation of the Cumulative Gamma Distribution Function

τ	r = 1	2	3	4	5
0.2	0.18127	0.01752	0.00115	0.00006	0.00000
0.4	0.32968	0.06155	0.00793	0.00078	0.00006
0.6	0.45119	0.12190	0.02312	0.00336	0.00039
0.8	0.55067	0.19121	0.04742	0.00908	0.00141
1.0	0.63212	0.26424	0.08030	0.01899	0.00366
1.2	0.69881	0.33737	0.12051	0.03377	0.00775
1.4	0.75340	0.40817	0.16650	0.05372	0.01425
1.6	0.79810	0.47507	0.21664	0.07881	0.02368
1.8	0.83470	0.53716	0.26938	0.10871	0.03641
2.0	0.86466	0.59399	0.32332	0.14288	0.05265
2.2	0.88920	0.64543	0.37729	0.18065	0.07250
2.4	0.90928	0.69156	0.43029	0.22128	0.09587
2.6	0.92573	0.73262	0.48157	0.26400	0.12258
2.8	0.93919	0.76892	0.53055	0.30806	0.15232
3.0	0.95021	0.80085	0.57681	0.35277	0.18474
3.2	0.95924	0.82880	0.62010	0.39748	0.21939
3.4	0.96663	0.85316	0.66026	0.44164	0.25582
3.6	0.97268	0.87431	0.69725	0.48478	0.29356
3.8	0.97763	0.89262	0.73110	0.52652	0.33216
4.0	0.98168	0.90842	0.76190	0.56653	0.37116
4.2	0.98500	0.92202	0.78976	0.60460	0.41017
4.4	0.98772	0.93370	0.81486	0.64055	0.44882
4.6	0.98995	0.94371	0.83736	0.67429	0.48677
4.8	0.99177	0.95227	0.85746	0.70577	0.52374
5.0	0.99326	0.95957	0.87535	0.73497	0.55951
5.2	0.99448	0.96580	0.89121	0.76193	0.59387
5.4	0.99548	0.97109	0.90524	0.78671	0.62669
5.6	0.99630	0.97559	0.91761	0.80938	0.65785
5.8	0.99697	0.97941	0.92849	0.83004	0.68728
6.0	0.99752	0.98265	0.93803	0.84880	0.71494

$$I_r(\tau) = F_V(\tau) = \int_0^{\tau} \frac{v^{r-1}e^{-v}}{\Gamma(r)} dv, \tau \geq 0$$

Table 9 *(continued)*

τ	r				
	1	2	3	4	5
6.2	0.99797	0.98539	0.94638	0.86577	0.74082
6.4	0.99834	0.98770	0.95368	0.88108	0.76493
6.6	0.99864	0.98966	0.96003	0.89485	0.78730
6.8	0.99889	0.99131	0.96556	0.90719	0.80797
7.0	0.99909	0.99270	0.97036	0.91823	0.82701
7.2	0.99925	0.99388	0.97453	0.92808	0.84448
7.4	0.99939	0.99487	0.97813	0.93685	0.86047
7.6	0.99950	0.99570	0.98124	0.94463	0.87506
7.8	0.99959	0.99639	0.98393	0.95152	0.88833
8.0	9.99966	0.99698	0.98625	0.95762	0.90037
8.5	0.99980	0.99807	0.99072	0.96989	0.92564
9.0	0.99988	0.99877	0.99377	0.97877	0.94504
9.5	0.99993	0.99921	0.99584	0.98514	0.95974
10.0	0.99995	0.99950	0.99723	0.98966	0.97075
10.5	0.99997	0.99968	0.99817	0.99285	0.97891
11.0	0.99998	0.99980	0.99879	0.99508	0.98490
11.5	0.99999	0.99987	0.99920	0.99664	0.98925
12.0	0.99999	0.99992	0.99948	0.99771	0.99240
12.5	1.00000	0.99995	0.99966	0.99845	0.99465
13.0	1.00000	0.99997	0.99978	0.99895	0.99626
13.5	1.00000	0.99998	0.99986	0.99929	0.99740
14.0	1.00000	0.99999	0.99991	0.99953	0.99819
14.5	1.00000	0.99999	0.99994	0.99968	0.99875
15.0	1.00000	1.00000	0.99996	0.99979	0.99914

Table 9 *(continued)*

	r				
τ	6	7	8	9	10
1.0	0.00059	0.00008	0.00001		
1.2	0.00150	0.00025	0.00004		
1.4	0.00320	0.00062	0.00011	0.00002	
1.6	0.00604	0.00134	0.00026	0.00005	0.00001
1.8	0.01038	0.00257	0.00056	0.00011	0.00002
2.0	0.01656	0.00453	0.00110	0.00024	0.00005
2.2	0.02491	0.00746	0.00198	0.00047	0.00010
2.4	0.03567	0.01159	0.00334	0.00086	0.00020
2.6	0.04904	0.01717	0.00533	0.00149	0.00038
2.8	0.06511	0.02441	0.00813	0.00243	0.00066
3.0	0.08392	0.03351	0.01190	0.00380	0.00110
3.2	0.10541	0.04462	0.01683	0.00571	0.00176
3.4	0.12946	0.05785	0.02307	0.00829	0.00271
3.6	0.15588	0.07327	0.03079	0.01167	0.00402
3.8	0.18444	0.09089	0.04011	0.01598	0.00580
4.0	0.21487	0.11067	0.05113	0.02136	0.00813
4.2	0.24686	0.13254	0.06394	0.02793	0.01113
4.4	0.28009	0.15635	0.07858	0.03580	0.01489
4.6	0.31424	0.18197	0.09505	0.04507	0.01953
4.8	0.34899	0.20920	0.11333	0.05582	0.02514
5.0	0.38404	0.23782	0.13337	0.06809	0.03183
5.2	0.41909	0.26761	0.15508	0.08193	0.03967
5.4	0.45387	0.29833	0.17834	0.09735	0.04875
5.6	0.48814	0.32974	0.20302	0.11432	0.05913
5.8	0.52169	0.36161	0.22897	0.13281	0.07084
6.0	0.55432	0.39370	0.25602	0.15276	0.08392
6.2	0.58589	0.42579	0.28398	0.17409	0.09838
6.4	0.61626	0.45767	0.31268	0.19669	0.11420
6.6	0.64533	0.48916	0.34192	0.22044	0.13136
6.8	0.67302	0.52008	0.37151	0.24523	0.14982

Table 9 *(continued)*

	r				
τ	6	7	8	9	10
7.0	0.69929	0.55029	0.40129	0.27091	0.16950
7.2	0.72410	0.57964	0.43106	0.29733	0.19035
7.4	0.74744	0.60804	0.46067	0.32435	0.21226
7.6	0.76932	0.63538	0.48996	0.35181	0.23515
7.8	0.78975	0.66159	0.51879	0.37956	0.25889
8.0	0.80876	0.68663	0.54704	0.40745	0.28338
8.5	0.85040	0.74382	0.61440	0.47689	0.34703
9.0	0.88431	0.79322	0.67610	0.54435	0.41259
9.5	0.91147	0.83505	0.73134	0.60818	0.47817
10.0	0.93291	0.86986	0.77978	0.66718	0.54207
10.5	0.94962	0.89837	0.82149	0.72059	0.60287
11.0	0.96248	0.92139	0.85681	0.76801	0.65949
11.5	0.97227	0.93973	0.88627	0.80941	0.71121
12.0	0.97966	0.95418	0.91050	0.84497	0.75761
12.5	0.98510	0.96543	0.93017	0.87508	0.79857
13.0	0.98927	0.97411	0.94597	0.90024	0.83419
13.5	0.99227	0.98075	0.95852	0.92100	0.86474
14.0	0.99447	0.98577	0.96838	0.93794	0.89060
14.5	0.99606	0.98955	0.97606	0.95162	0.91224
15.0	0.99721	0.99237	0.98200	0.96255	0.93015
15.5	0.99803	0.99446	0.98654	0.97121	0.94481
16.0	0.99862	0.99599	0.99000	0.97801	0.95670
16.5	0.99903	0.99712	0.99261	0.98331	0.96626
17.0	0.99933	0.99794	0.99457	0.98741	0.97388

Table 9 *(continued)*

	r				
τ	11	12	13	14	15
4.0	0.00284	0.00091	0.00027	0.00008	0.00002
4.5	0.00667	0.00240	0.00081	0.00025	0.00007
5.0	0.01370	0.00545	0.00202	0.00070	0.00023
5.5	0.02525	0.01099	0.00445	0.00169	0.00060
6.0	0.04262	0.02009	0.00883	0.00363	0.00140
6.5	0.06684	0.03388	0.01603	0.00710	0.00296
7.0	0.09852	0.05335	0.02700	0.01281	0.00572
7.5	0.13776	0.07924	0.04267	0.02156	0.01026
8.0	0.18411	0.11192	0.06380	0.03418	0.01726
8.5	0.23664	0.15134	0.09092	0.05141	0.02743
9.0	0.29401	0.19699	0.12423	0.07385	0.04147
9.2	0.31797	0.21682	0.13926	0.08438	0.05999
9.4	0.34236	0.23743	0.15524	0.09581	0.05590
9.6	0.36705	0.25876	0.17212	0.10815	0.06428
9.8	0.39195	0.28072	0.18988	0.12139	0.07346
10.0	0.41696	0.30322	0.20844	0.13554	0.08346
10.2	0.44197	0.32618	0.22777	0.15055	0.09429
10.4	0.46687	0.34951	0.24779	0.16641	0.10596
10.6	0.49159	0.37310	0.26843	0.18309	0.11847
10.8	0.51603	0.39687	0.28963	0.20054	0.11318
11.0	0.54011	0.42073	0.31130	0.21871	0.14596
11.2	0.56376	0.44459	0.33337	0.23756	0.16090
11.4	0.58690	0.46837	0.35576	0.25702	0.17661
11.6	0.60949	0.49198	0.37839	0.27703	0.19305
11.8	0.63146	0.51535	0.40117	0.29754	0.21019
12.0	0.65277	0.53840	0.42403	0.31846	0.22798
12.2	0.67338	0.56108	0.44690	0.33974	0.24637
12.4	0.69327	0.58331	0.46968	0.36130	0.26531
12.6	0.71239	0.60504	0.49232	0.38307	0.28474
12.8	0.73075	0.62623	0.51475	0.40498	0.30462

Table 9 *(continued)*

	r				
τ	11	12	13	14	15
13.0	0.74832	0.64684	0.53690	0.42696	0.32487
13.2	0.76510	0.66681	0.55870	0.44893	0.34543
13.4	0.78108	0.68614	0.58012	0.47084	0.36625
13.6	0.79628	0.70478	0.60110	0.49262	0.38725
13.8	0.81068	0.72273	0.62158	0.51421	0.40838
14.0	0.82432	0.73996	0.64154	0.53555	0.42956
14.2	0.83720	0.75647	0.66094	0.55659	0.45075
14.4	0.84934	0.77225	0.67975	0.57728	0.47188
14.6	0.86076	0.78731	0.69793	0.59756	0.49289
14.8	0.87149	0.80164	0.71549	0.61741	0.51373
15.0	0.88154	0.81525	0.73239	0.63678	0.53435
15.5	0.90388	0.84622	0.77173	0.68292	0.58459
16.0	0.92260	0.87301	0.80688	0.72549	0.63247
16.5	0.93813	0.89593	0.83790	0.76426	0.67746
17.0	0.95088	0.91533	0.86498	0.79913	0.71917
17.5	0.96126	0.93160	0.88835	0.83013	0.75736
18.0	0.96963	0.94511	0.90833	0.85740	0.79192
18.5	0.97635	0.95624	0.92525	0.88114	0.82286
19.0	0.98168	0.96533	0.93944	0.90160	0.85025
19.5	0.98589	0.97269	0.95125	0.91908	0.87427
20.0	0.98919	0.97861	0.96099	0.93387	0.89514
20.5	0.99176	0.98335	0.96897	0.94630	0.91310
21.0	0.99375	0.98710	0.97545	0.95664	0.92843
21.5	0.99528	0.99005	0.98069	0.96520	0.94141
22.0	0.99645	0.99237	0.98488	0.97222	0.95231
22.5	0.99735	0.99418	0.98823	0.97794	0.96140
23.0	0.99802	0.99557	0.99088	0.98275	0.96893
23.5	0.99853	0.99665	0.99297	0.98630	0.97512
24.0	0.99892	0.99748	0.99460	0.98928	0.98018
24.5	0.99920	0.99811	0.99587	0.99166	0.98428

Appendix B

Basic Results

1. Exponential function e^x:

$$e = 2.71828\ 18284$$

$$e^x y^y = e^{x+y}, \quad e^x/e^y = e^{x-y}, \quad (e^x)^y = e^{xy}$$

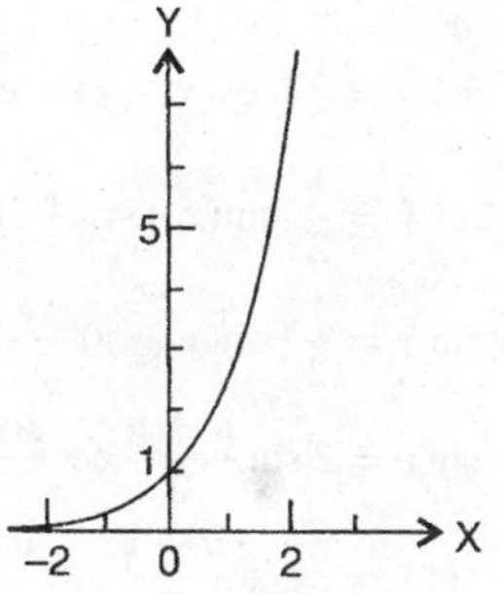

Exponential function e^x

2(a) Natural logarithm:

ln x is the inverse of e^x and has base e and $e^{\ln x} = x$, $e^{-\ln x} = e^{\ln(1/x)} = 1/x$.

$$\ln(xy) = \ln x + \ln y,$$

$$\ln(x/y) = \ln x - \ln y,$$

$$\ln(x^a) = a \ln x$$

(b) Logarithm of base ten $\log_{10} x$ or simply $\log x$ (known as common loarithm)

$\log x$ is the inverse of 10^x, and $10^{\log x} = x$, $\quad 10^{-\log x} = 1/x$.

$$\log x = M \ln x, \quad M = \log e = 0.43429$$

$$\ln x = \frac{1}{M} \log x, \quad \frac{1}{M} = 2.30258$$

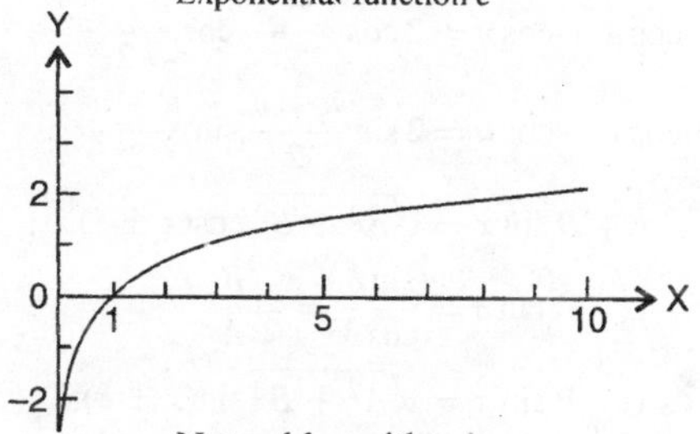

Natural logarithm ln x

3(a) Sine and cosine functions:

$\sin x$ is odd, $\quad \sin(-x) = -\sin x$

$\cos x$ is even, $\cos(-x) = \cos x$

Note: Angles are measured in radians in calculus, so that $\sin x$ and $\cos x$ have period 2π.

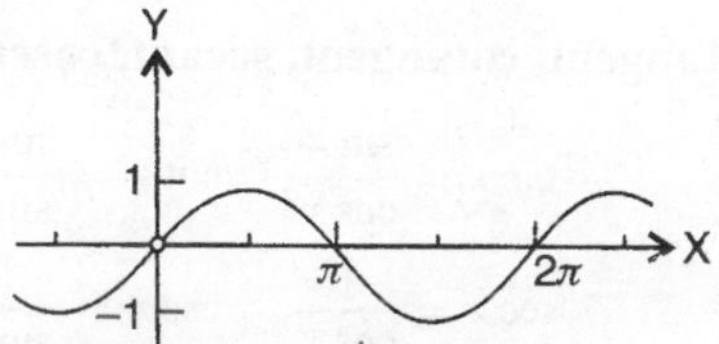

sin x

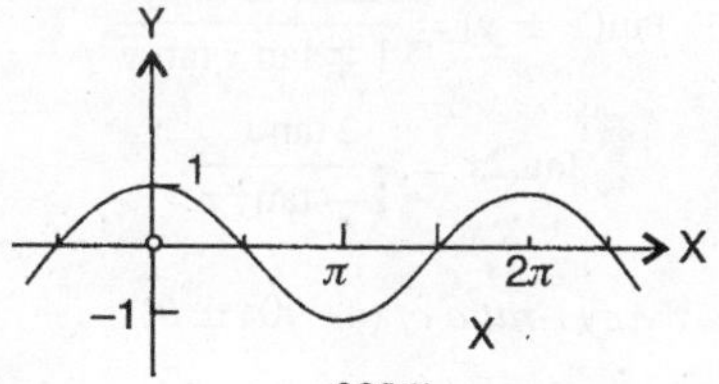

cos x

$$\sin^2 x + \cos^2 x = 1$$

$$\sin(x \pm y) = \sin x \cos y \pm \cos x \sin y$$

$$\cos(x \pm y) = \cos x \cos y \mp \sin x \sin y$$

$$\sin 2x = 2 \sin x \cos x = 2 \tan x/(1 + \tan^2 x),$$

$$\cos 2x = \cos^2 x - \sin^2 x = 1 - 2\sin^2 x =$$

$$= 2\cos^2 x - 1 = \frac{1 - \tan^2 x}{1 + \tan^2 x}$$

$$\sin x = \cos\left(x - \frac{\pi}{2}\right) = \cos\left(\frac{\pi}{2} - x\right)$$

$$\cos x = \sin\left(x + \frac{\pi}{2}\right) = \sin\left(\frac{\pi}{2} - x\right)$$

$$\sin(\pi - x) = \sin x, \quad \cos(\pi - x) = -\cos x$$

$$\cos^2 x = \frac{1}{2}(1 + \cos 2x), \ \sin^2 x = \frac{1}{2}(1 - \cos 2x)$$

$$\sin x \sin y = \frac{1}{2}[-\cos(x + y) + \cos(x - y)]$$

$$\cos x \cos y = \frac{1}{2}[\cos(x + y) + \cos(x - y)]$$

$$\sin x \cos y = \frac{1}{2}[\sin(x + y) + \sin(x - y)]$$

$$\cos x \sin y = \frac{1}{2}[\sin(x + y) - \sin(x - y)]$$

$$\sin u + \sin v = 2 \sin\frac{u + v}{2} \cos\frac{u - v}{2}$$

$$\sin u - \sin v = 2 \cos\frac{u + v}{2} \sin\frac{u - v}{2}$$

$$\cos u + \cos v = 2 \cos\frac{u + v}{2} \cos\frac{u - v}{2}$$

$$\cos v - \cos u = 2 \sin\frac{u + v}{2} \sin\frac{u - v}{2}$$

$$A \cos x + B \sin x = \sqrt{A^2 + B^2} \cos(x \pm \delta),$$

$$\tan\delta = \frac{\sin\delta}{\cos\delta} = \mp\frac{B}{A}$$

$$A \cos x + B \sin x = \sqrt{A^2 + B^2} \sin(x \pm \delta),$$

$$\tan\delta = \frac{\sin\delta}{\cos\delta} = \mp\frac{A}{B}$$

(b) Tangent, cotangent, secant, cosecant:

$$\tan x = \frac{\sin x}{\cos x}, \quad \cot x = \frac{\cos x}{\sin x},$$

$$\sec x = \frac{1}{\cos x}, \quad \csc x = \frac{1}{\sin x}$$

$$\tan(x \pm y) = \frac{\tan x \pm \tan y}{1 \mp \tan x \tan y},$$

$$\tan 2x = \frac{2 \tan x}{1 - \tan^2 x}$$

Any t-ratio of $(n \cdot 90° \pm \theta)$

$= \pm$ *same ratio of* θ *when n is even.*

$= \pm$ *co-ratio of* θ, *when n is odd.*

The sign + or − is to be decided from the quadrant in which $(n \cdot 90° \pm \theta)$ lies.

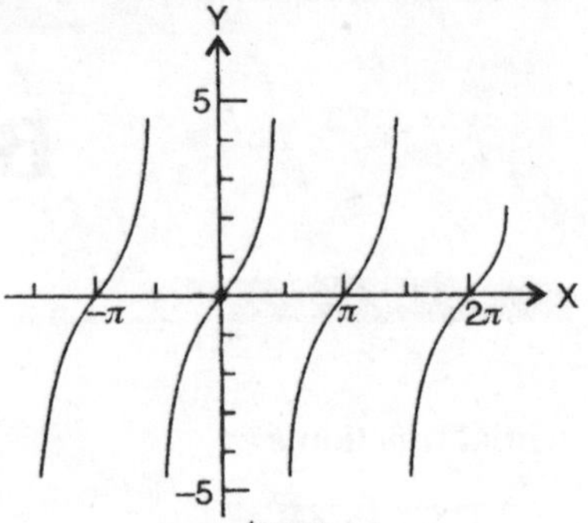

tan x

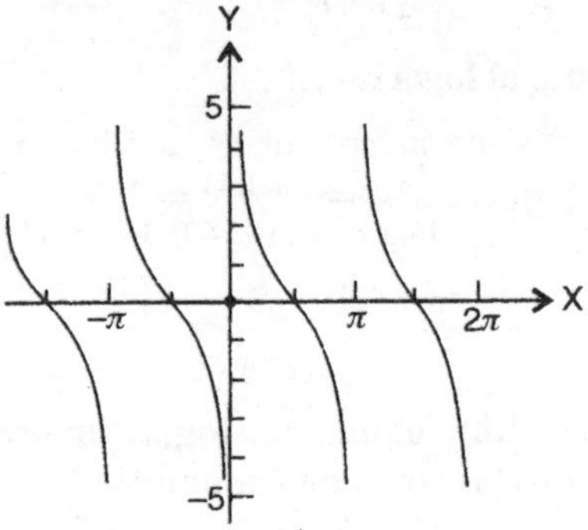

cot x

Examples:

$$\sin 570° = \sin(6 \times 90° + 30°) = -\sin 30° = -\frac{1}{2};$$

$$\tan 315° = \tan(3 \times 90° + 45°) = -\cot 45° = -1$$

In any ΔABC, $a/\sin A = b/\sin B = c/\sin C$ and $\cos C = \frac{a^2+b^2-c^2}{2ab}$.

4. Hyperbolic functions:

$$\sinh x = \tfrac{1}{2}(e^x - e^{-x}), \quad \cosh x = \tfrac{1}{2}(e^x + e^{-x})$$

$$\tanh x = \frac{\sinh x}{\cosh x}, \quad \coth x = \frac{\cosh x}{\sinh x}$$

$$\cosh x + \sinh x = e^x, \quad \cosh x - \sinh x = e^{-x}$$

$$\cosh^2 x - \sinh^2 x = 1$$

$$\sinh^2 x = \tfrac{1}{2}(\cosh 2x - 1), \quad \cosh^2 x = \tfrac{1}{2}(\cosh 2x + 1)$$

$$\begin{cases} \sinh(x \pm y) = \sinh x \cosh y \pm \cosh x \sinh y \\ \cosh(x \pm y) = \cosh x \cosh y \pm \sinh x \sinh y \end{cases}$$

$$\tanh(x \pm y) = \frac{\tanh x \pm \tanh y}{1 \pm \tanh x \tanh y}$$

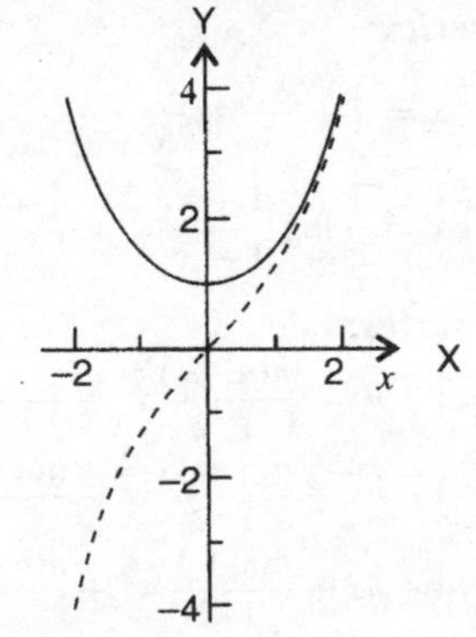

sinh x (dashed) and cosh x (solid)

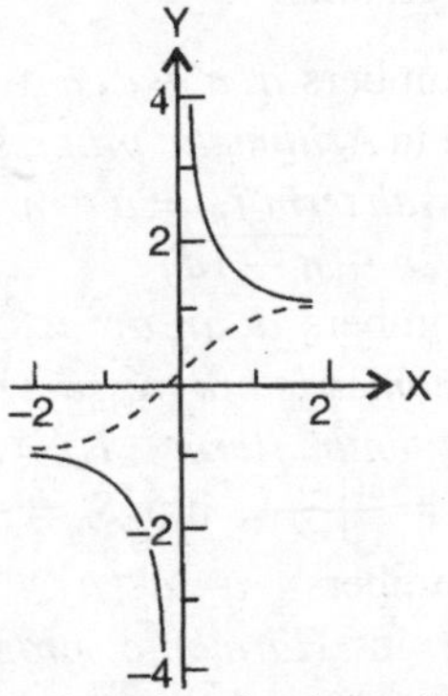

tanh x (dashed) and coth x (solid)

5. Differentiation:

$\frac{d}{dx}(uv) = u\frac{dv}{dx} + v\frac{du}{dx}$ $\qquad$ $\frac{d}{dx}\left(\frac{u}{v}\right) = \frac{v\,du/dx - u\,dv/dx}{v^2}$

$\frac{du}{dx} = \frac{du}{dy}\cdot\frac{dy}{dx}$ *(Chain Rule)* $\qquad$ $\frac{d}{dx}(ax+b)^n = n(ax+b)^{n-1}\cdot a$

$\frac{d}{dx}(e^x) = e^x$ $\qquad$ $\frac{d}{dx}(a^x) = a^x \log_e a$

$\frac{d}{dx}(\log_e x) = 1/x$ $\qquad$ $\frac{d}{dx}(\log_a x) = \frac{1}{x \log a}$

$\frac{d}{dx}(\sin x) = \cos x$ $\qquad$ $\frac{d}{dx}(\cos x) = -\sin x$

$\frac{d}{dx}(\tan x) = \sec^2 x$ $\qquad$ $\frac{d}{dx}(\cot x) = -\text{cosec}^2 x$

$\frac{d}{dx}(\sec x) = \sec x \tan x$ $\qquad$ $\frac{d}{dx}(\text{cosec}\, x) = -\text{cosec}\, x \cot x$

$\frac{d}{dx}(\sin^{-1} x) = \frac{1}{\sqrt{(1-x^2)}}$ $\qquad$ $\frac{d}{dx}(\cos^{-1} x) = -\frac{1}{\sqrt{(1-x^2)}}$

$\frac{d}{dx}(\tan^{-1} x) = \frac{1}{1+x^2}$ $\qquad$ $\frac{d}{dx}(\cot^{-1} x) = \frac{-1}{1+x^2}$

$\frac{d}{dx}(\sec^{-1} x) = \frac{1}{x\sqrt{(x^2-1)}}$ $\qquad$ $\frac{d}{dx}(\text{cosec}^{-1} x) = \frac{-1}{x\sqrt{(x^2-1)}}$

$\frac{d}{dx}(\sinh x) = \cosh x$ $\qquad$ $\frac{d}{dx}(\cosh x) = \sinh x$

$\frac{d}{dx}(\text{sech} x) = -\tanh x\, \text{sech} x$ $\qquad$ $\frac{d}{dx}(\tanh x) = \text{sec}^2 \text{h}x$

$\frac{d}{dx}(\text{cosech} x) = -\cot \text{h}x \cdot \text{cosech} x$ $\qquad$ $\frac{d}{dx}(\cot \text{h}x) = -\text{cosech}^2 x$

$D^n(e^{mx}) = m^n e^{mx}$ $\qquad$ $D^n(a^{mx}) = m^n(\log a)^n \cdot a^{mx}$

$$D^n(ax+b)^n = m(m-1)\quad(m-2)\ldots(m-n+1)(ax+b)^{m-n}$$

$$D^n \log(ax+b) = (-1)^{n-1}(n-1)!a^n/(ax+b)^n$$

$$D^n \sin(ax+b) = a^n \sin(ax+b+n\pi/2)$$

$$D^n \cos(ax+b) = a^n \cos(ax+b+n\pi/2)$$

$$D^n\left\{e^{ax}\sin(bx+c)\right\} = (a^2+b^2)^{n/2} e^{ax} \sin(bx+c+n\tan^{-1} b/a)$$

$$D^n\left\{e^{ax}\cos(bx+c)\right\} = (a^2+b^2)^{n/2} e^{ax} \cos(bx+c+n\tan^{-1} b/a)$$

$$D^n(uv) = u_n + {}^nC_1 u_{n-1}v + {}^nC_2 u_{n-2}v_2 + \cdots + {}^nC_r u_{n-r}v_r + \cdots + {}^nC_n v_n$$

6. Integration:

$\int x^n\,dx = \frac{x^{n+1}}{n+1}(n \neq -1)$ $\qquad$ $\int \frac{1}{x}dx = \log_e x = \ln x$

$\int e^x\,dx = e^x, \int e^{ax+b}dx = \frac{e^{ax+b}}{a}$ $\qquad$ $\int a^x dx = a^x/\log_e a, a > 0$

$\int \sin x\,dx = -\cos x$ $\qquad$ $\int \cos x\,dx = \sin x$

$\int \tan x\,dx = -\log \cos x$ $\qquad$ $\int \cot x\,dx = \log \sin x$

$\int \sec x\,dx = \log(\sec x + \tan x) = \log\tan(\frac{\pi}{4}+\frac{x}{2})$ $\qquad$ $\int \text{cosec}\, x\,dx = \log(\text{cosec}\, x - \cot x) = \log|\tan\frac{x}{2}|$

$\int \sec^2 x\,dx = \tan x$ $\qquad$ $\int \text{cosec}^2 x\,dx = -\cot x$

$\int \sinh x\,dx = \cosh x$ $\qquad$ $\int \cosh x\,dx = \sinh x$

$\int \frac{dx}{a^2+x^2} = \frac{1}{a}\tan^{-1}\frac{x}{a}$ $\qquad$ $\int \frac{dx}{\sqrt{(a^2-x^2)}} = \sin^{-1}\frac{x}{a}$

$\int \frac{dx}{a^2-x^2} = \frac{1}{2a}\log\frac{a+x}{a-x}$ $\qquad$ $\int \frac{dx}{\sqrt{(a^2+x^2)}} = \sinh^{-1}\frac{x}{a} = \ln\{x+\sqrt{a^2+x^2}\}$

$\int \frac{dx}{x^2-a^2} = \frac{1}{2a}\log\frac{x-a}{x+a}$ $\qquad$ $\int \frac{dx}{\sqrt{(x^2-a^2)}} = \cosh^{-1}\frac{x}{a} = \ln\{x+\sqrt{x^2-a^2}\}$

$\int \frac{dx}{ax+b} = \frac{1}{a}\ln|ax+b|$

$$\int\sqrt{(a^2-x^2)}\,dx = \frac{x\sqrt{(a^2-x^2)}}{2} + \frac{a^2}{2}\sin^{-1}\frac{x}{a}$$

$$\int\sqrt{(a^2+x^2)}\,dx = \frac{x\sqrt{(a^2+x^2)}}{2} + \frac{a^2}{2}\sinh^{-1}\frac{x}{a}$$

$$\int\sqrt{(x^2-a^2)}\,dx = \frac{x\sqrt{(x^2-a^2)}}{2} - \frac{a^2}{2}\cosh^{-1}\frac{x}{a}$$

$$\int e^{ax}\sin bx\,dx = \frac{e^{ax}}{a^2+b^2}(a\sin bx - b\cos bx)$$

$$\int_0^\infty e^{-ax}\sin bx dx = \frac{b}{a^2+b^2}$$

$$\int e^{ax}\cos bx\,dx = \frac{e^{ax}}{a^2+b^2}(a\cos bx + b\sin bx)$$

$$\int_0^\infty e^{-ax}\cos bx dx = \frac{a}{a^2+b^2}$$

$$\int_{-\infty}^{\infty} e^{-x^2}dx = \sqrt{\pi}$$

$$\int_0^\infty \frac{e^{-ax}}{x}\sin bx dx = \tan^{-1}\frac{b}{c},\quad c > 0,\quad b > 0$$

$$\int_0^\infty \frac{\sin ax}{x}dx = \frac{\pi}{2}\quad \text{if } a > 0$$

$$\int \frac{f'(x)}{f(x)}\,dx = \ln|f(x)|$$

$$\int (f(x))^n f'(x)dx = \frac{[f(x)]^{n+1}}{n+1}, n \neq -1$$

$$\int_0^{\infty} \frac{e^{ax} - e^{-ax}}{e^{-\pi x} - e^{-\pi x}}dx = \frac{1}{2}\tan\frac{a}{2}$$

$$\int_0^{\infty} \frac{e^{ax} + e^{-ax}}{e^{-\pi x} - e^{-\pi x}}dx = \frac{1}{2}\sec\frac{a}{2}$$

$$\int_0^{\pi/2} \sin^n x\,dx = \int_0^{\pi/2} \cos^n x\,dx$$

$$= \frac{(n-1)(n-3)(n-5)\ldots}{n(n-2)(n-4)\ldots}$$

Note: R.H.S. is multiplied by $\pi/2$ when n is even

$$\int_0^{\pi/2} \sin^m x \cos^n x\,dx$$

$$= \frac{(m-1)(m-3)\ldots \times (n-1)(n-3)\ldots}{(m+n)(m+n-2)(m+n-4)\ldots}$$

Note: R.H.S. is multiplied by $\pi/2$ when both m and n are even

$$\int_{-a}^{a} f(x)\,dx = \begin{cases} 2\int_0^a f(x)\,dx, & \text{if } f(x) \text{ is an } \textit{even} \text{ function.} \\ 0 & \text{if } f(x) \text{ is an } \textit{odd} \text{ function.} \end{cases}$$

$$\int_0^{2a} f(x)\,dx = \begin{cases} 2\int_0^a f(x)dx, & \text{if } f(2a-x) = f(x) \\ 0 & \text{if } f(2a-x) = -f(x). \end{cases}$$

$$\int u\,dv = uv - \int v\,du \qquad \textbf{\textit{(Integration by parts)}}$$

$$\int u(x)v(x)\,dx = uv_1 - u'v_2 + u''v_3 - u'''v_4 + \cdots$$

(Leibnitz General rule of integration by parts)

Note: Superscript $'$ denotes differentiation, i.e., u'' denotes differentiation of u twice. Subscript number denotes number of times integration of v, i.e., v_3 denotes integration of v thrice.

7. Series:

Exponential series: $e^x = 1 + \frac{x}{1!} + \frac{x^2}{2!} + \frac{x^3}{3!} + \cdots$

sin, cos, sinh, cosh series:

$\sin x = x - \frac{x^3}{3!} + \frac{x^5}{5!} - \cdots,\quad \cos x = 1 - \frac{x^2}{2!} + \frac{x^4}{x!} - \cdots$

$\sinh x = x + \frac{x^3}{3!} + \frac{x^5}{5!} + \cdots,\quad \cosh x = 1 + \frac{x^2}{2!} + \frac{x^4}{4!} + \cdots$

Log series:

$$\log(1+x) = x - \frac{x^2}{2} + \frac{x^3}{3} - \cdots,$$

$$\log(1-x) = -\left(x + \frac{x^2}{2} + \frac{x^3}{3} + \cdots\right)$$

Gregory series:

$$\tan^{-1} x = x - \frac{x^3}{3} + \frac{x^5}{5} - \cdots,$$

$$\tanh^{-1} x = \frac{1}{2}\log\frac{1+x}{1-x} = x + \frac{x^3}{3} + \frac{x^5}{5} + \cdots$$

Binomial series:

$$(1+x)^n = 1 + nx + \frac{n(n-1)}{1\cdot 2}x^2 + \frac{n(n-1)(n-2)}{1\cdot 2\cdot 3}x^3 + \cdots$$

$$(1+x)^{-1} = 1 - nx + \frac{n(n+1)}{1\cdot 2}x^2 - \frac{n(n+1)(n+2)}{1\cdot 2\cdot 3}x^2 + \cdots$$

$$(1-x)^{-n} = 1 + nx + \frac{n(n+1)}{1\cdot 2}x^2 + \frac{n(n+1)(n+2)}{1\cdot 2\cdot 3}x^3 + \cdots$$

8(a) Progressions:

i. Numbers $a, a+d, a+2d, \ldots$ are said to be in *Arithmetic progression (A.P.)* Its *nth term* $T_n = a + \overline{n-1}d$ and *sum* $S_n = \frac{n}{2}(2a + \overline{n-1}d)$

ii. Numbers $a, ar, ar^2, \ldots$ are said to be in *Geometric progression (G.P.)* Its *nth term* $T_n = ar^{n-1}$ and *sum* $S_n = \frac{a(1-r^n)}{1-r}$, $\lim\limits_{n\to\infty} S_n = \frac{a}{1-r}$ when $|r| < 1$.

iii. Numbers $a_1, a_2, a_3, \ldots$ are said to be in *Harmonic progression (H.P.)* if $1/a_1, 1/a_2, 1/a_3, \ldots$ are in *A.P.*

iv. For any two numbers a and b, their *Arithmetic mean* $= \frac{1}{2}(a+b)$, *Geometric mean* $= \sqrt{ab}$, *Harmonic mean* $= \frac{2ab}{(a+b)}$.

v. For the first n *natural numbers* $1, 2, 3, \ldots, n$,

$$\sum n = \frac{n(n+1)}{2},$$

$$\sum n^2 = \frac{n(n+1)(2n+1)}{6},$$

$$\sum n^3 = \left\{\frac{n(n+1)}{2}\right\}^2$$

vi. ***Stirling's approximation***. When n is large $n! \sim \sqrt{2\pi n}\cdot n^n e^{-n}$.

(b) Permutations and combinations:

$${}^nP_r = \frac{n!}{(n-r)!};$$

$${}^nC_r = \frac{n!}{r!(n-r)!} = \frac{{}^nP_r}{r!};$$

$${}^nC_{n-r} = {}^nC_r, {}^nC_0 = 1 = {}^nC_n$$

9. Matrices:

$$A^{-1} = \frac{1}{|A|}\text{adj}A, \quad (AB)^{-1} = B^{-1}A^{-1},$$

$$(AB)^T = B^T A^T, \quad (A^T)^{-1} = (A^{-1})^T$$

10. Ordinary differential equations:

First order linear:

$y'+p(x)y=q(x)$, I.F$=e^{\int p(x)dx}$

G.S: $y \cdot$ (I.F) $= \int$(I.F)$(Q|x)dx$

Bessel equation: $x^2y''+xy'+(x^2 - v^2)y=0$

Legendre equation: $(1-x^2)y''-2xy'+\lambda y=0$

θ (degrees)	0	30	45	60	90	180	270	360
θ (radians)	0	$\frac{\pi}{6}$	$\frac{\pi}{4}$	$\frac{\pi}{3}$	$\frac{\pi}{2}$	π	$\frac{3\pi}{2}$	3π
$\sin\theta$	0	$\frac{1}{2}$	$\frac{1}{\sqrt{2}}$	$\frac{\sqrt{3}}{2}$	1	0	-1	0
$\cos\theta$	1	$\frac{\sqrt{3}}{2}$	$\frac{1}{\sqrt{2}}$	$\frac{1}{2}$	0	-1	0	1
$\tan\theta$	0	$\frac{1}{\sqrt{3}}$	1	$\sqrt{3}$	∞	0	$-\infty$	0

11. (a) Law of exponents

$a^m \cdot a^n = a^{m+n}$; $(ab)^m = a^m b^m$, $(a^m)^n = a^{mn}$, $a^{\frac{m}{n}} = \sqrt[n]{a^m}$

If $a \neq 0$, $\frac{a^m}{a^n} = a^{m-n}$, $a^0 = 1$, $a^{-m} = \frac{1}{a^m}$

(b) Difference of like integer power, $n > 1$:

$a^n - b^n = (a-b)(a^{n-1} + a^{n-2}b + a^{n-3}b^2 + \cdots + ab^{n-2} + b^{n-1})$

For example

$$a^3 - b^3 = (a-b)(a^2 + ab + b^2)$$

$$a^4 - b^4 = (a-b)(a^3 + a^2b + ab^2 + b^3)$$

12. (a) Hyperbolic functions:

$\sinh x = \frac{e^x - e^{-x}}{2}$, $\cosh x = \frac{e^x + e^{-x}}{2}$, $\tanh x = \frac{\sinh x}{\cosh x}$

(b) Complex number: $z = x + iy = r(\cos\theta + i\sin\theta)$

Euler's Theorem: $\cos\theta + i\sin\theta = \text{cis}\,\theta = e^{i\theta}$

So $z = r\text{cis}\,\theta = re^{i\theta}$

De Moivre's Theorem:

$(\cos\theta + i\sin\theta)^n = \cos n\theta + i\sin n\theta$

$\sin x = \frac{e^{ix} - e^{-ix}}{2i}$, $\cos x = \frac{e^{ix} + e^{-ix}}{2}$

(c) Relations between hyperbolic and trigonometric functions:

$\sin ix = i\sinh x$, $\cos ix = \cosh x$,

$\tan ix = i\tanh x$,

$\sinh iz = i\sin z$, $\cosh iz = \cos z$, $\tanh iz = i\tan z$

13. Frequently used limits

1. $\lim_{x\to 0} \frac{\sin x}{x} = 1$, 2. $\lim_{x\to 0} \frac{a^x - 1}{x} = \log_e a$
3. $\lim_{x\to a} \frac{x^n - a^n}{x - a} = na^{n-1}$, n any rational number
4. (a) $\lim_{n\to\infty} \frac{1}{n} = 0$ (b) $\lim_{n\to\infty} \frac{1}{n^2} = 0$, (c) $\lim_{n\to\infty} \frac{1}{\sqrt{n}} = 0$
5. $\lim_{n\to\infty} n^{\frac{1}{n}} = 1$, 6. $\lim_{n\to\infty} \frac{\log n}{n} = 0$
7. $\lim_{n\to\infty} \left(1 + \frac{x}{n}\right)^n = e^x$ for any x
8. $\lim_{n\to\infty} x^{1/n} = 1$ for $x > 0$ } x remains fixed as $n \to \infty$
9. (a) $\lim_{n\to\infty} \frac{x^n}{n!} = 0$ for any $x > 0$ } x remains fixed as $n \to \infty$

   (b) $\lim_{n\to\infty} x^n = 0$ for $|x| < 1$ i.e., $-1 < x < 1$.

Appendix C

Bibliography

1. Ahlberg, *The Theory of Splines and Their Applications*, A.P. 1967.
2. Atkinson, K.E., *Introduction to Numerical Analysis*, Wiley, 1978.
3. Boyce et. al, *Elements of Differential Equations and Boundary Value Problems*, Wiley, 1973.
4. Burghes, *Modelling with Differential Equations*, Horwood, 1982.
5. Carrier, G.F., *Partial Differential Equations: Theory and Practice*, A.P., 1976.
6. Carslaw, H.S., *Introduction to Theory of Fourier Series and Integral*, Dover, 1960.
7. Chorlton, F., *Vector and Tensor Methods*, Horwood, 1976.
8. Clenshaw, C.W., *Numerical Approximations to Functions and Data*, (Ed.) Hayes, University of London, 1970.
9. Coddington, E.A., *An Introduction to Ordinary Differential Equations*, Prentice-Hall, 1968.
10. Elsgole, L.E. *Calculus of Variations*, Pergamon Press, 1961.
11. For, L. and I.B. Parkes, *Chebyshev Polynomials in Numerical Analysis*, Oxford Press, 1968.
12. Fraleigh, J.B., *Calculus with Analytical Geometry*, Addison-Wesley, 1980.
13. Froberg, C.E., *Introduction to Numerical Analysis*, Addison-Wesley, 1965.
14. Gerald, F.C. and Wheatley, D.P., *Applied Numerical Analysis*, 5th edition, Addison-Wesley, 1996.
15. Greenspan, II.P., *Calculus*, McGraw-Hill, 1973.
16. Godunov, S.K. and V.S. Ryabenski, *Theory of Differences Schemes*, North Holland, 1964.
17. Greenberg, D. *Advanced Engineering Mathematics*, Pearson, 2002.
18. Hildebrand, F.B., *Introduction to Numerical Analysis*, McGraw-Hill, 1956.
19. Hyslop, J.H., *Infinite Series*, Oliver, 1965.
20. Isaacson, E. and Keller, H.B., *Analysing Numerical Method*, Wiley, 1966.
21. Kreyzig, E. *Advanced Engineering Mathematics*, 8th edition, Wiley, 2000.
22. Lang, S., *Linear Algebra*, Springer, 1987.
23. Lass, H., *Vector and Tensor Analysis*, McGraw-Hill, 1950.
24. Levy, H. and Baggott, *Numerical Solutions of Differential Equations*, Dover, 1950.
25. Noble, B., *Applied Linear Algebra*, Prentice-Hall, 1977.
26. Pipes, L.A., *Applied Mathematics for Engineering and Physics*, McGraw-Hill, 1958.
27. Prenter, P.M., *Splines and Variational Methods*, Wiley, 1975.

28. Churchill, R.V., *Complex Variables and Applications*, McGraw-Hill, 1990.

29. Churchill, R.V., *Fourier Series and Boundary Value Problems*, McGraw-Hill, 1987.

30. Smith, G.D., *Numerical Solutions of Partial Differential Equations*, Oxford University Press, 1969.

31. Spigel, M.R., *Vector Analysis*, Schaum Series, 1959.

32. Spigel, M.R., *Laplace Transform*, Schaum Series, 1965.

33. Spigel M.R., *Advanced Calculus*, Schaum Series, 1963.

34. Spain, D. *Vector Analysis*, D. Van Nostrand, 1967.

35. Sokolnikoff and Redheffer, *Mathematical Physics and Modern Engineering*, McGraw-Hill, 1966.

36. Meyer, P.L., *Introduction to Probability and Statistics*, Oxford, 1970.

37. Miller and Freund, *Probability and Statistics for Engineers*, Pearson, 2001.

38. Piskunov, N., *Differential and Integral Calculus*, Vol I and II, Mir, 1974.

39. Ross, S.L., *Introduction to Ordinary Differential Equations*, 4th edition, Wiley, 1989.

40. Walpole, R.E. and Meyer, R.H., *Probability and Statistics for Engineers and Scientists*, Pearson, 2002.

41. Taha, H., *Operations Research*.

42. Thomas, G.B., *Calculus with Analytical Geometry*. Addison Wesley, 10th edition, 2003.

43. Wylie, C.R., *Advanced Engineering Mathematics*, McGraw-Hill, 1992, 5th edition.

44. Wider, D.V. "Advanced Calculus" Prentice-Hall, 1976.

45. Wilkinson, J.H. "Algebraic Eigenvalue Problems" Oxford, 1965.

Index

S